MW00423236

Scramjet Propulsion

Scramjet Propulsion

Edited by
E.T. Curran
Department of the Air Force
Dayton, OH
S.N.B. Murthy
Purdue University
West Lafayette, IN

Volume 189
PROGRESS IN
ASTRONAUTICS AND AERONAUTICS

Paul Zarchan, Editor-in-Chief
Charles Stark Draper Laboratory, Inc.
Cambridge, Massachusetts

Published by the
American Institute of Aeronautics and Astronautics, Inc.
1801 Alexander Bell Drive, Reston, Virginia 20191-4344

ISBN 1-56347-322-4

Progress in Astronautics and Aeronautics

Contents

Preface

Sometime in 1990, when various new approaches started being explored for realizing cost-effective space launch and hypersonic cruise and weapon systems, it seemed appropriate to provide a status review on propulsion systems, which certainly hold the key to success in such flight. Dr. E.T. Curran, then director of the Air Force Aero-Propulsion and Power Directorate, and I thought that a couple of volumes could be planned and realized to bring out the major aspects of world developments in this area. Finally a mini-series of three volumes evolved, and the well-known series Progress in Aeronautics and Astronautics of the American Institute of Aeronautics and Astronautics was chosen, and accepted, for the production of the volumes.

The first volume, entitled *High Speed Flight Propulsion Systems*, No. 137 in the Series, appeared in 1991. The second volume, entitled *Developments in High Speed Flight Propulsion Systems* was published as No. 165 in 1996. Neither of those dealt with scramjets in any detail, although it was stated in both of them that scramjets eventually held the greatest opportunity for realizing the full potential of airbreathing propulsion for high Mach flight. The current volume, the third in the miniseries, is devoted exclusively to scramjet propulsion.

Even as plans were being made for this volume, it became strikingly clear that substantial developments were being made in many parts of the world in supersonic combustion and associated engines. Thus one major part of the volume is devoted to a survey of developments in six countries, including the USA. This part on international developments is particularly noteworthy as it describes several initiatives in flight testing in Australia, France, Germany, Russia, and the USA, a much-needed effort in gaining and displaying confidence in this technology.

The other three parts of the volume deal with specific components of the engine, some possibilities for the future in detonation engines and electromagnetic field interactions, and integration methodology of the overall scramjet engine system.

It is a matter of extraordinary pleasure to record how readily the authors agreed to participate in the volume, and the effort they put in for preparing the various chapters. We must also acknowledge the timeliness with which most of the writing was done with great personal sacrifice of their limited free time.

Among the contributors to the volume, J. Vandenkerkhove, the doyen of modern developments in airbreathing propulsion for high speed flight in the European community, passed away in April 1995. Paul Czysz collaborated closely with J.V. in his endeavors in the airbreathing engines area, and the chapter started jointly by them appears in this volume as finished by Paul Czysz in his own style.

We are much indebted to each of the authors, and the reviewers who read the different drafts of the chapters.

With accomplishment of this final volume in the three-volume mini-series, it is my pleasant duty to acknowledge the invaluable role of Tom Curran as an editor. It can be truly said that he has read every word in the three volumes at least once. He has brought his extensive knowledge, clear thinking, and simple direct style of expressing and explaining technical matters to each article as he sought to encourage the authors to ensure they were satisfied in their presentation. For me, it has been a matter of much joy and enlightenment to work with him. In particular in the case of this volume his own contributions in scramjet engines (as dual-mode engines, supersonic flow engines, and part of combined cycle engines) have spread over nearly all of the decades of their development history. He has brought that wealth of experience to bear on the chapters in his readings.

This volume started in 1993 has been rather delayed in preparation. It is very much a reflection of the times for aerospace industry. No serious contributor in the field is a master of his own time and effort in any week.

Finally, at the AIAA, over the period of these three volumes there have been three changes in the Series Editor and countless other changes in the editorial staff. Nevertheless, whatever their own exasperations, they have encouraged us in the preparation of the volumes and given their enthusiastic help in the editorial work and preparation of the volumes. We again apologize to them for the extraordinary delays in the case of some parts of this volume, and wish to place on record our sincere appreciation for their personal understanding of our, and the writers', problems and for their devoted and outstanding service to the technical community at large.

S.N.B. Murthy
Co-Editor
January 2001

Introduction

This volume is the third in a short series, within the larger AIAA "Progress in Astronautics and Aeronautics" series, devoted to high-speed propulsion. In particular, this volume is almost exclusively concerned with supersonic combustion ramjet (scramjet) engine technologies, and international progress in developing such technologies.

I. International Efforts

A general review of the emergence and maturing of scramjet engine technologies during the period 1957 to 1997 was recently given by Curran[1]. Efforts completed or proceeding in the USA, Russia, France, Germany, Japan and other countries were briefly covered in that review. Furthermore, updated descriptions of such programs were given in Ref. 2. In this present volume, many more details are given of various national scramjet programs.

In the former Soviet Union, a significant scramjet program has been in progress for many years. Knowledge of the history of this program has been restricted in the past by the limited availability of documentation in the English language. Fortunately, for this book, a most valuable review on scramjet research and development in Russia has been contributed by leading Russian researchers. This chapter is an invaluable reference concerning the historical and technical evolution of the foundational research on scramjet engines performed in the former Soviet Union, and of the contributions of individual scientists.

Similarly, the evolution of scramjet work in France is covered by Falempin in the chapter entitled, "Scramjet Developments in France." This chapter traces the evolution of scramjet work from the initial combustion studies of Mestre and Viaud[3] to the comprehensive research program (PREPHA) addressing advanced hypersonic propulsion. In particular, the many achievements of the research teams at ONERA and Aerospatiale are outlined.

Regarding work in progress in the United States, the chapter on scramjet performance by Anderson, McClinton, and Weidner gives in-depth coverage of the extensive propulsion research conducted at the NASA Langley Research Center. This research evolved from the earlier NASA Hypersonic Ramjet Engine and airframe-integrated scramjet engine programs, and has more recently addressed very high speed engines operating with hypersonic combustion flows.

Activities and accomplishments in Australia, Germany, and Japan are also reported in this initial section of the book. The reader is also referred to another recent publication, edited and compiled by Townend[4], which includes articles on hypersonic studies by both British and other authors.

II. Inlets, Combustors, and Fuels

The scramjet engine literature is largely dominated by combustor studies and the related area of combustor-inlet interactions. In this present book, however, an article entitled "Scramjet Inlets" authored by Van Wie was solicited. This definitive treatment covers inlet design issues and, in particular, experimental methods of determining inlet performance at hypersonic speeds.

Many current scramjet combustor articles are concerned with computational modeling of the basic physical processes occurring in such combustors. The review article entitled "Supersonic Flow Combustors" by Kutschenreuter is based on his long experience of integrating such combustors into practical engine configurations, and gives a fresh perspective on combustor design tradeoffs. In regard to such tradeoffs, a key question is determining what criteria should be used to assess the merit of combustor performance within the context of overall vehicle design. Two current approaches are the available work (or exergy) method, expounded by Murthy[5], and the thrust-work-potential method treated by Riggins[6]. A further contribution, authored by Murthy, is included in this volume and is entitled, "Basic Performance Assessment of Scram Combustors."

A pervasive problem in dual-mode scramjet engine development is the occurrence of combustor-inlet interactions; usually an isolator section is placed between the inlet exit and the combustor entry to control such effects. A general treatment of combustor-inlet interactions, authored by Heiser and Pratt, is included in this book and is entitled, "Aerothermodynamics of the Dual-Mode Combustion System." This chapter is an updated version of the analytical approach given in their well-known textbook[7]

The continuing evolution of the Aerojet Strutjet concept is described in detail in the chapter by Siebenhaar, Bulman, and Bonnar, and in particular its integration as a rocket based combined cycle system.

This section concludes with a chapter covering the increasingly important area of hydrocarbon fuels for scramjet engines. Endothermic fuel technology is enabling for the active cooling of hypersonic engines, but achieving an effective combustion process over a wide speed range is challenging, as discussed in the chapter by Maurice, Edwards, and Griffiths. Further details of chemical kinetic models for aviation fuels are given in Ref 8.

III. Overall Systems

The first chapter in this section on Engine-Airframe Integration is by Hunt and Martin of NASA Langley, who have a long history of evaluating hypersonic vehicle designs; the chapter is entitled, "Rudiments and Methodology of Design /Analysis for Hypersonic Airbreathing Vehicles."

A major study of the problem of sizing reusable launch vehicles, authored by Czysz and Vandenkerckhove, is also included in this book and is entitled, "Transatmospheric Launcher Sizing." This chapter covers the sizing of both single-stage-to-orbit and two-stage-to-orbit vehicles, and discusses their relative payload capabilities and performance sensitivities. The potential for LOX collection for both vehicle classes is discussed.

An original and fundamental approach to vehicle integration is found in the chapter by Ortwerth, entitled "Scramjet-Flowpath Integration." The starting point for this chapter is the well-known Brayton cycle analysis developed by Builder[9]. This one-dimensional propulsion analysis is then coupled to the vehicle aerodynamic characteristics, yielding a broad prospective on overall vehicle design.

IV. Future Developments

The detonation wave ramjet engine was one of the earliest concepts proposed for supersonic combustion systems. The detonation wave provides significant compression and potentially intense combustion with much shorter combustion lengths; its anticipated performance appears to exceed that of the scramjet at higher hypersonic Mach numbers. However, the practical engineering of detonation wave engines appears very formidable. Such challenges are discussed by Sislian in this book.

Another evolving technology that has appeared in the open literature is the application of electromagnetic fields to partially ionized propulsion working fluids to enhance vehicle performance. In regard to external flows, several studies have appeared in the Russian literature, which indicate that appreciable reductions in vehicle drag and aerodynamic heating may be obtained by such techniques. Relatively little has appeared concerning the utilization of such ionized gases in internal engine flows. One application of such techniques is discussed by two Russian authors, Vatazhin, and Kopechenov, in the chapter entitled, "Problem of Hypersonic Flow Deceleration by Magnetic Field." This chapter, together with two early United States articles[10,11], give insight into an emerging research area, which may significantly impact future vehicle design.

More general studies of the engine cycle proposed for the AJAX class of vehicle have recently been published. In the Russian literature this engine concept is termed a Magneto Plasma Chemical Engine (MPCE) and can be considered a variant of the scramjet engine. A basic analysis of the internal engine flow path has been given by Fraishtadt, Kuranov, and Sheikin[12], which shows some potential for improved scramjet performance. In the United States the MPCE is more commonly referred to as an MHD energy-bypass engine, and preliminary thermodynamic and performance analyses have been published by both United States and Russian authors[13,14,15]. The emerging set of AJAX technologies holds promise of radical change in air vehicle characteristics as this new degree of freedom in electromagnetic flow interactions is explored.

V. Closing Comments

This volume has been largely concerned with scramjet engine technology. The discerning reader will note that there is much to be done to mature scramjet technology. It is to be hoped that this review of work in progress in many countries will stimulate increasing international collaboration to solve the technical problems of scramjet development, and, in particular, to develop and apply both refined computational tools and sophisticated nonintrusive instrumentation diagnostics to uncover the physics of the flow. The unexpected complexities of the flow in scramjet combustors and, specifically, the adverse combustor-inlet interactions have frustrated progress. As noted in Ref. 1, it is important to select scramjet combustor geometries that lend themselves to structured engineering approaches that will more precisely anchor, and spatially control, sensitive flow processes such as heat release, flow separation and reattachment, and other similar flow phenomena, resulting in "healthy" flows. Concentrating effort on such flows, which lend themselves to more precise modeling and simulation, should permit more rapid progress in achieving efficient systems.

E.T. Curran
August 2000

References

[1]Curran, E. T., "Scramjet Engines: The First Forty Years" ISABE Paper 97-7005, XIII International Symposium on Air Breathing Engines, 7–12 Sep 1997, Chattanooga, Tennessee.

[2]Curran, E. T. "Emerging Propulsion Technologies for the Next Century" AIAA Wright Brothers Lecture, 1999.

[3]Mestre, A. and Viaud, L., "Combustion Supersonique dans un Canal Cylindrique," *Supersonic Flow, Chemical Processes and Radiative Transfer*, MacMillan Co., 1964, pp. 93–111.

[4]Townend, L. H.,"Hypersonic aircraft: lifting re-entry and launch." Phil. Trans. R. Soc. Lond. A 357, No. 1759, 2317–2334, 15 Aug 1999.

[5]Murthy, S. N. B., "Effectiveness of a Scram Engine," AIAA Paper 94-3087, June 1994.

[6]Riggins, D. W., "Evaluation of Performance Loss Methods for High-Speed Engines and Engine Components" *Journal of Propulsion and Power,* Vol. 13, No. 2, March-April 1997.

[7]Heiser, W. H., and Pratt, D. T., *Hypersonic Airbreathing Propulsion*, AIAA, Washington, D.C., 1994.

[8]Maurice,L.Q., "Detailed Chemical Kinetic Models for Aviation Fuels " Doctoral Thesis, Imperial College of Science, Technology and Medicine, October 1996.

[9]Builder, C. H., "On the Thermodynamic Spectrum of Airbreathing Propulsion" AIAA Paper 64-243, 1st AIAA Annual Meeting, Washington, D.C., July 1964.

[10]Harsha, P., and Gurijanov, E. P., "AJAX: New Directions in Hypersonic Technology" AIAA Paper 96-4609, 7th Aerospace Planes and Hypersonic Technologies Meeting, Norfolk, Virginia, April 1996.

[11]Bruno, C., Czysz, P. A., and Murthy, S. N. B., "Electro-Magnetic Interactions in Hypersonic Propulsion Systems" AIAA Paper 97-3389. 33rd AIAA/SAE/ASEE Joint Propulsion Conference & Exhibit, 6–9 July 1997, Seattle, Washington.

[12]Fraishtadt, V. L., Kuranov, A. L.,and Sheikin, "Use of MHD Systems in Hypersonic Aircraft," *Technical Physics* Vol. 43, No. 11, Nov. 1998.

[13]Bityurin, V .A., Lineberry, J. T., Litchford, R J., and Cole, J.W., "Thermodynamic

Analysis of the AJAX Propulsion Concept" AIAA Paper 2000-0445, 38th AIAA Aerospace Sciences Meeting, January 2000, Reno, Nevada.

[14]Burakhanov, B., et. al. "Advancement of Scramjet MHD Concept," AIAA Paper 2000-0614, 38th AIAA Aerospace Sciences Meeting, January 2000, Reno, Nevada.

[15]Chase, R.L., et. al. "Comments on an MHD Energy Bypass Engine Powered Spaceliner," AIAA Paper 99-4975, 3rd Weakly Ionized Gases Workshop, Norfolk, VA, November 1999.

Chapter 1

Scramjet Testing in the T3 and T4 Hypersonic Impulse Facilities

A. Paull* and R. J. Stalker†
The University of Queensland, 4072 Australia

Nomenclature

A = area
a = local speed of sound
E = energy per kilogram
E_c = energy fluxes added by combustion
E_f = energy fluxes added by the addition of fuel
F = force
F_m = measured force
f_c = force due to skin friction on the combustion chamber walls
f_e = sum of all frictional forces
I = incremental specific impulse
I_m = measured incremental specific impulse
M = Mach number
$\dot{m}$ = mass flux of air into the combustion chamber
$\dot{m}f$ = fuel mass flux
P = static pressure
T = temperature in Kelvin
T_f = the total fuel momentum flux at the exit of the injector
t_i = initiation time
t_r = reaction time
u = velocity
γ = specific heat ratio
ρ = density
ϕ = equivalence ratio

---

*Senior Research Fellow, Department of Mechanical Engineering.
†Emeritus Professor, Department of Mechanical Engineering.

I. History, Aims, and Developments

THE T3 and T4 facilities are free piston-driven shock tunnels that are capable of simulating flight conditions up to orbital velocity. Although Mach-8 flight conditions can be reached with continuous facilities, shock tunnels are one of the few ground-based facilities that can reach higher flight Mach numbers. Hence, as ground-based testing will eventually extend past Mach 8, these types of facilities are expected to become more extensively used in the future. Thus, the experiments summarized in this chapter are not only unique and important in their own right, but will also form a basis for expanding ground testing to these higher Mach numbers.

One of the unique features of the groups surrounding these facilities is that they have been established within a university environment where research of a fundamental, rather than developmental, nature is encouraged, although the line between the two is somewhat hazy at times. Nevertheless, the result of this approach has been to investigate simple configurations in an attempt to understand or, at the very least, give more information about a particular topic. Consequently, the reader will observe in this chapter that simple two-dimensional or axisymmetrical models are used in general. Furthermore, simple analytical tools have been developed to give a better understanding of the role of the different parameters. In writing this chapter, we do not claim that this is the first time such an approach has been adopted: however, this chapter endeavors to show that such an approach is an important one that complements existing developmental projects and the use of computational fluid dynamics.

The T3 (at The Australian National University) and T4 (at The University of Queensland) facilities were commissioned in 1968 and 1987, respectively. Primarily, financial support for these facilities has been provided by the institutions in which they reside and the Australian Research Council. One of the past and present strengths of the group attached to the T3 facility is the development and deployment of optical techniques for understanding many of the fundamentals of high enthalpy flows. In contrast, whereas optics has played a role in the T4 facility, traditionally, mechanical devices, such as pressure sensors and force balances, are the primary tools for investigation. These different, but complementary approaches, have developed as a result of the different histories of the two groups; the T3 group is based in a physics department whereas the T4 group grew up in a mechanical engineering department.

Early scramjet experiments in a shock tunnel were done by Osgerby *et al.* (1970) at Arnold Air Force Station, Tennessee. The scramjet work undertaken in Australia started under the leadership of R. J. Stalker. The first experiments were made in 1981 by Stalker and Morgan (1985) who were residents at The University of Queensland. However, at the time T4 had not been commissioned, and the existing facilities in Queensland were inadequate for the experimental program. Hence, these experiments were done in the T3 facility, which Stalker had developed before moving to Queensland. On a regular basis and for approximately three months at a time, trips would be made to the T3 facility to do experiments. All the instrumentation, models, and some fuels would be packed into various cars and vans and transported over the 1300-km trip between Brisbane and Canberra. In retrospect, these experiments were very important because the foundations for a basic understanding by this group of supersonic combustion were formed from them. In addition, this was a period where basic skills were learned with many of the techniques and some of the equipment developed at this time, still in use today. This arrangement continued until approximately 1987 when the T4 facility was commissioned. The T4 facility's primary mission was to study scramjets; thus, subsequent to 1987 the majority of experiments have been made at The University of Queensland.

Initially, simple models consisting of a combustion chamber, an injector, and a thrust surface were used to evaluate various aspects of combustion at supersonic speeds. The initial experiments were commissioned by NASA Langley Research Center, which requested quick answers be provided to help give direction to certain aspects of the National Aerospace Plane (NASP) program. Unfortunately, quick results usually result in less than perfect ones, and in the scramble for unclassified experimental data, associated with the following years of NASP, the original intent of the experiments was forgotten by some users of the data. To make the problem worse, although some measurements on scramjets had been previously made in shock tunnels at other facilities, at the time the majority of measurements in use came from continuous facilities. Thus, for many (including ourselves), these measurements on scramjet performance were the first that had been encountered from an impulse facility. So replacing the mostly understood traditional disparities of continuous facilities with flight conditions was a whole new set of disparities that produced new uncertainties.

The experiments were often criticized because atomic oxygen existed in the freestream. However, after considerable effort, Bakos (1994) showed, through a joint program with GASL (formerly General Applied Science Laboratories), that this affected only the time for the energy to be released from the fuel and not the amount of energy released.

Another criticism of the original experiments was the perceived inability of the T3 facility to produce a steady flow. The facility was run in an undertailored mode, which resulted in a large initial pressure, but one which fell during the test time. This was done to reduce the possibility of test-gas contamination by the driver gas and to increase the freestream pressure. Although some criticism was made about using this less-than-perfect test environment, the falling pressure during the test time made little difference to the outcome of the experiments that were being performed. When the T4 facility came on line, this complication was removed, as steady conditions could be maintained at the appropriate pressure level, and any uncertainties in the test time because of contamination have been removed by Skinner (1994), Paull (1996) and Boyce (1997) who measured the contamination level time history.

Of course, the primary criticism of using impulse facilities for scramjet research is that there is insufficient time for the flow to reach a steady state. However, typically, the test slug is greater than 10 model lengths, which should allow the boundary layer to settle, and the kinetic reaction rates are so great that, under most circumstances, a steady state (if it exists) is reached. Furthermore, recently completed experiments in collaboration with the National Aeronautical Laboratory, (NAL) Kakuda, and independent experiments indicate that boundary-layer separation occurs in T4 under similar conditions to those observed in continuous facilities. There may, perhaps, be some phenomenon that will not come to a steady state during the test time, but for the T4 studies of scramjets to date, this has not been the case; moreover, the theoretical predictions based on the steady state are generally in agreement with the measurements. For many years the question has been whether or not the test time in a shock tunnel is long enough. As a result of the work done over the past two decades, the question that should now be asked is "Why do you need a long test time?"

Over the past 17 years that the group has been active, the scramjet models designed for testing in T3 and T4 have progressed from simple generic models of a combustor with an injector to sophisticated models that involve the complete flow path. The experiments with combustors have mainly been confined to systems that are fueled with hydrogen, for flight Mach numbers between 7 and 14; however, some experiments have also been performed with hydrocarbons and silane. Experiments have also been extended to measure the total thrust developed by a fully integrated scramjet model. This was a major achievement for a number of reasons. First, a force balance had to be developed for an impulse facility, and, second, this balance had to be made sufficiently robust to be used on an engine (where there is the added complication of fuel and its valving); both of which were not particularly easy to achieve. Furthermore, when these experiments were started (early 1990s), the T4 group had virtually no experience with integrating the intake, combustion chamber, and thrust surface on a scramjet. Hence, a conservative model was designed in the hope that many of the complications of this integration could be avoided. As it turned out, this was the case, and the final result was that a model scramjet was made to develop sufficient thrust to overcome its drag. The scramjet had a net specific impulse of approximately 150 s. Although this impulse is very small, the importance of this achievement was two-fold; first, it established that such a measurement could be made, which opened the way for future research (and development) of this type on scramjets in impulse facilities, and second, a milestone had been reached as the model scramjet could indeed accelerate forward under its own power [Paull (1993) and Paull *et al.* (1995[a,b])]. This occasion was the first time in the open literature that a scramjet was reported to do so.

Although this result is very encouraging, there is still a long way to go before a scramjet becomes economically viable. Hence, the main aim of the current research has not changed and is still to develop a greater understanding of the physics and principles that govern the performance of the engine, with the aim of increasing its performance.

II. Facility and Instrumentation

In this section a brief overview of the operation of a free-piston facility is given. In addition, methods used to determine the freestream properties and the instrumentation used in the facility are reviewed.

The T4 facility is simply a shock tunnel that consists of two tubes (compression and shock tubes) and a nozzle which is attached to the shock tube and vents into a test section. The compression and shock tubes are separated by the primary diaphragm and contain the driver and test gases, respectively. The driver gas is generally some mixture of helium and argon, the ratio of which depends on the conditions being targeted, and the test gas is usually air, although nitrogen is used when the effects of combustion are to be suppressed.

To increase the temperature and pressure of the driver gas, it is compressed in the compression tube using a piston that is driven by compressed air. Once the primary diaphragm ruptures, a shock travels down the shock tube to its end where the nozzle is located. At the end it reflects and consequently stagnates the test gas in this region. Typically total temperatures in this region can range between 2500 and 8000 K, and the total pressure can be as high as 50 MPa (although modifications are currently being made to double the total pressure). This stagnation region then acts as a reservoir, and the test gas is exhausted through a nozzle and over the model that is mounted in the test section. Typical test times range from 0.5 to 5 ms, depending on the total enthalpy. The Mach number of the flow is dependent on the nozzle, and at T4, Mach numbers from 4 to 10 can be accommodated by using four different nozzles. Typically one run of the facility can be made every 90 min, and the facility averages 500 runs every year.

Measurements that are routinely made include the shock speed in the shock tube, the reservoir or stagnation pressure in the reflected shock region, and the Pitot pressure in the test section. The freestream conditions are determined from these measurements, using them as inputs to the codes ESTC [McIntosh (1968)] and NENZF [Lordi *et al.* (1965)].

Models are generally instrumented with pressure transducers or heat-flux gauges. The pressure transducers have been bought off the shelf; however, the mounting techniques have been modified because of the large amounts of vibration experienced by the models. To avoid vibration problems, models should be made from steel. Aluminum should only be used in force-balance experiments (see Sec. IX) where it is important that stress waves (or vibrations) are not suppressed.

Specialized instrumentation has also been developed for the facility and, in particular, for scramjet experiments. This instrumentation includes a mass spectrometer, skin friction gauges, and force balances.

The mass spectrometer has primarily been used to measure the level of contamination by the driver gas in the test gas and also to measure the extent of the different constituents of the test gas, in particular the concentrations of atomic oxygen. These results are still under discussion. However, the extent of driver-gas contamination has been determined by Boyce

(1997) and Skinner (1994) and is shown to be in agreement with the results obtained by Paull (1996) who used a gas-dynamic technique to determine the time at which contamination exceeded a certain level (7.5% by volume). Although this has been the main use of the mass spectrometer, one of the important applications for which this instrument was designed was measurement of species concentrations in the combustion chamber of the scramjet. Some initial results have been obtained by Skinner (1994), and the extension of these results is the subject of current research.

A point skin-friction gauge for use in T4 was first developed by Kelly *et al.* (1992) and has been subsequently improved by Goyne *et al.* (1997) so that the local skin-friction levels in a combustor could be measured. At this stage of development, the gauge has been shown to give a good measure of the local skin-friction load, but there are still some difficulties with mounting, which makes the gauge less robust than desired. This is also an area of current research within the T4 group.

One of the major contributions to the measurement techniques has been the development of a force balance. Discussions of the history of its development will be given in Sec. IX; however, to summarize, the balance has now been developed so that three components of force (lift, drag, and pitching moment) can be measured on an inclined cone. [Mee *et al.* (1996)]. To measure the drag on an active and complex quasisymmetric body such as a scramjet is considered as routine. The technique has also been applied to measure the skin-friction load on an axisymmetric combustion chamber [Tanno and Paull (1997)]: The results will be discussed in Sec. X.

Finally, no system is complete unless a suitable data-recording system is available, and data can be reduced quickly and effectively. The data acquisition system used at T4 is 12 bit with a 1-Mhz clocking rate on each of the 21 channels. Each channel is capable of being multiplexed with four channels. The system was made in house (Daniels and Allsop). The data-reduction program used at T4 runs under the title of MONC [Paull (1998)] and has been developed to give an experimenter information quickly.

Having given a brief overview of the T4 facility, particular experiments that have been important to the groups development of an understanding of the scramjet will now be discussed. No attempt has been made to include all of the experiments that have been undertaken in the past 17 years; however, most of the main references have been provided, and the reader is encouraged to pursue these references (in particular, the NASA contractor reports) to gain a greater appreciation of the extent of hypersonic research undertaken at The University of Queensland.

III. Fuel-Injection Systems

To develop an efficient combustion chamber, a suitable means of delivering the fuel had to be obtained. In this section, three different injection systems are discussed in terms of their advantages and disadvantages and their effectiveness in distributing (or mixing) the fuel.

In all of the experiments except those described as using hot fuel, the fuel was stored prior to injection in a ludwig tube at room temperature. When hot fuel was used, the fuel was heated immediately before the test flow using a gun tunnel.

The normal procedure in the T4 facility is to measure the pressure distributions in a combustion chamber (and thrust surface, if used) when fuel is injected into either nitrogen or air and when fuel is not injected at all (fuel off). The change in pressure distribution from the fuel-off case to that when fuel is injected into nitrogen results from the additional flux of mass (which is usually small), momentum, and energy from the fuel. If the combustion chamber is sufficiently long so that the fuel and air become fully mixed, then the average pressure may either rise or fall over the average pressure of the fuel-off case, depending on the balance of the fuel momentum and energy (see Sec. V). In general, there will be a change in the average pressure of the fuel-air mixture, which will depend not only on the properties of the fuel at injection but also on the extent of mixing. In this way, the change in the pressure distribution can be used to indicate the extent of mixing; however, in general, this is inaccurate.

However, the main use of the tests with nitrogen is to separate out the effects of combustion. By comparing the pressure distributions for fuel injection into nitrogen and air, the increase in pressure produced when the fuel is injected into air can be obtained, and this increase can be attributed to the release of energy from the fuel. From these measurements an estimate of the performance of the scramjet is obtained (see Sec. V and VI).

When injecting fuel, the first aim clearly is to mix the fuel and air, and this can be achieved by either a diffusion process (e.g., a step/slot in the combustor wall) or by using a penetration mechanism, either by a physical strut or by injecting with suitable momentum at an angle to the chamber wall. If a diffusion process is used, the combustion chamber is generally long, and therefore combustion chamber skin-friction losses may become significant. The use of a penetration mechanism usually results in a shorter combustion chamber; however, shocks are produced by the penetration mechanism, and the total pressure losses generated by these shocks are important. Hence, both systems have losses associated with them.

To develop an understanding of the effectiveness of different injection systems, the following three means of delivering fuel have been used: 1) wall injection—injection from a downstream-facing step in the combustion chamber wall (Fig. 1); 2) central injection—injection from a downstream-facing strut centrally located in the combustion chamber (Fig. 2); and 3) port injection—injection from a hole or holes in the combustion chamber wall (Fig. 3).

In all of the preceding systems, fuel enters the freestream either sonically or supersonically. Wall injection will be classed as one that uses a purely diffusion process to obtain mixing, whereas port injection uses penetration mechanisms. Central injection may be seen as using a combination of the two.

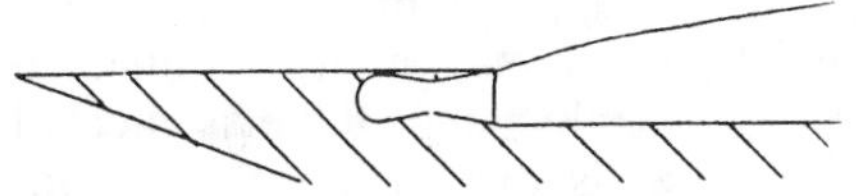

Fig. 1 Wall-injection configuration.

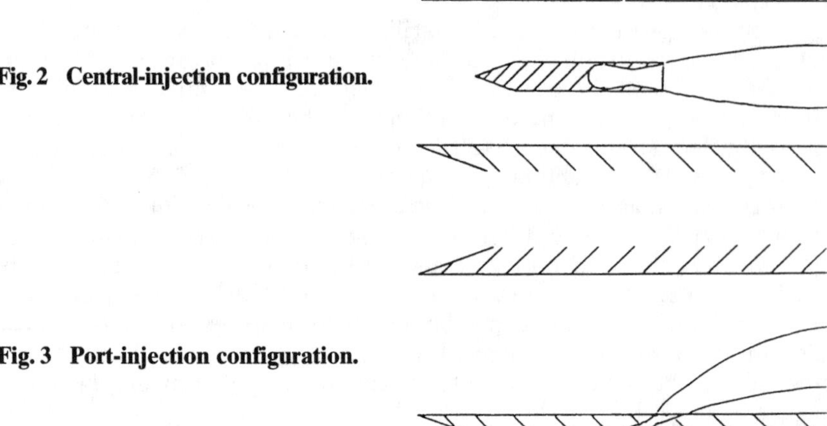

Fig. 2 Central-injection configuration.

Fig. 3 Port-injection configuration.

A. Wall-Injection Combustion Results

Some of the advantages of using wall injection over central injection are as follows:

1) It provides film cooling to the combustion chamber walls.

2) Freestream pressure losses are reduced.

3) There is no central strut, which eliminates any problems associated with its cooling.

Clearly, in an impulse facility, cooling of the engine is not an issue as the test times are short, and so advantages 1) and 3) apply to more practical engines. However, the pressure loss produced by the injector is important, for, as already discussed, the thrust developed by injecting the fuel is a significant part of the net thrust. If this comes at the price of a large total pressure loss in the freestream, then the benefits may be lost.

The ultimate disadvantage of wall-injection systems that have been used is that fuels, such as hydrogen, simply do not burn in a chamber of reasonable length. This disadvantage is highlighted in Fig. 4 [Brescianini (1993)] which displays the pressures measured along the centreline of a constant area rectangular combustion chamber when fuel is injected from either a wall injector or a central injector. The pressure rise from fuel injected from behind a step is substantially less than that from the central injector. Moreover, when the pressure rise in air produced using a step injector was compared with that in nitrogen, it was concluded that it had not burned [Morgan *et al.* (1996)].

The reasons for this lack of burning are not clear. Obviously, the main difference between wall-and central-injection systems is that wall injected fuel is injected close to a cold wall, whereas centrally injected fuel is injected into the middle of the freestream. Clearly, one disadvantage of the wall-in-

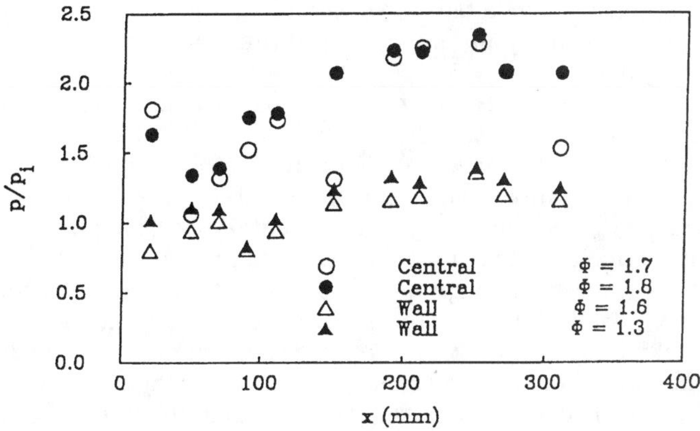

Fig. 4 Static pressures in a constant-area combustor for different injection systems. P_i is the intake pressure.

jection systems tested here is that the fuel injected from the wall has only one diffusion boundary whereas centrally injected fuel has two, giving it the advantage that it can mix with the same quantity of air more rapidly. This would account for a possible slowing down of combustion but would not account for its complete suppression. The presence of the cold wall adversely affects the diffusion and burning of the fuel.

In light of this, Morgan *et al.* (1987) postulated that the cold wall reduces the temperature of the fuel-air mixture close to the wall, which is not a problem with central injection. In the case of wall injection, the majority of the fuel lies in this region. Therefore, the fuel simply may not reach its self-ignition temperature (i.e., it is quenched by the presence of the cold wall). This hypothesis was consistent with the results of Morris (1989) who used a mixture of silane and hydrogen as a fuel. Silane has a very low ignition temperature (see Sec. VIII. B.) so that this quenching should not limit combustion. Indeed, it was found that a significant amount of burning occurs when a silane-hydrogen mixture was injected from the wall, and, furthermore, burning could be observed even at low equivalence ratios. This added weight to the hypothesis that mixing of the freestream and the fuel has occurred and that it is the quenching effect of the wall that hinders the burning. However, a numerical study made by Brescianini (1993), in which the combustion chamber wall temperature was varied, indicates that this is not the case. This study indicates that the reduction in temperature produced by the wall is not significant, and there is sufficient temperature for hydrogen to burn close to the wall. This study concluded that the failure to burn when fuel is injected from behind a step results primarily from a lack of mixing.

In an attempt to enhance the mixing process when using wall injection, Brescianini (1990) placed a number of mixing wedges downstream of the injector. However, it would appear that these devices did little to increase combustion. Only a slight, if any, increase in pressure rise was observed. The fuel temperature immediately downstream of the injector was approximately 175°K, and the wall temperature throughout the tests remained at room temperature.

This apparent lack of mixing occurred at an equivalence ratio close to one. However, Morgan (1985) and Brescianini *et al.* (1993) have both observed independently that wall injection from behind a step can produce combustion at equivalence ratios greater than approximately two; this is displayed in Fig. 5 [Brescianini (1993)] where the incremental specific impulse (see Sec. V and VI) is plotted as a function of equivalence ratio. Investigations into the cause for this have not been pursued; however, if the combustion mixing is limited, one possible reason for the increased burning is that the presence of more fuel along the wall will increase the dispersion rate of the fuel through the air. Therefore more fuel will come in contact with the air and presumably burn. A degree of penetration possibly may occur as the result of greater fuel pressure at the injector exit. If the pressure at the exit is greater than in the freestream, then fuel will be directed away from the wall and into the freestream, thus providing a penetrating mechanism. If on the other hand it is quenching produced by the cold wall that limits the combustion, it may be that the addition of more fuel simply displaces more fuel away from the effects of the wall. At this stage the reasons for the lack of combustion when hydrogen is injected from a step in the wall are inconclusive.

B. Wall-Injection Film-Cooling Results

Even though not all of the fuel would burn when a wall injector was used, to assume this type of injection system is inefficient may be incorrect because when the overall performance of the engine is assessed, the advantages provided by the layer of fuel next to the wall, which include reduced heating and possibly skin-friction loads on the combustor, may well outweigh the inefficiency associated with the wasted fuel.

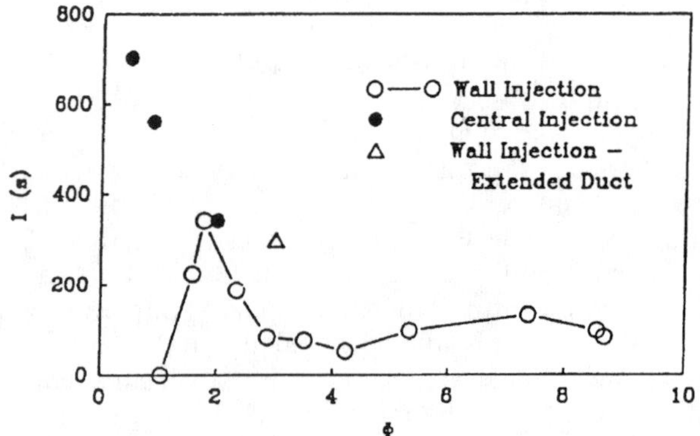

Fig. 5 Incremental specific impulse as a function of equivalence ratio for different injector configurations.

To gain an insight into the heat protection provided by a layer of hydrogen along the wall, heat-transfer measurements have been obtained by Stalker *et al.* (1990) in a scramjet combustion chamber, with a wall-injection system, as in the preceding section. Figure 6 displays the heat-transfer rate to the wall downstream of the wall injector in the presence and absence of fuel. The stagnation enthalpy and Mach number of the freestream are 4.2 MJ/kg and 3.5, respectively. The fuel is injected at a mass flow rate that produces an equivalence ratio of 1.4. The combustion chamber had a cross-sectional area of 27 × 54 mm, and the fuel was injected along the 54-mm surface as indicated in Fig. 6.

In the absence of combustion, hydrogen is a very effective coolant. The heat transfer rate was effectively reduced to zero for a downstream length of approximately two duct heights, and a further distance of six duct heights was required before the heat-transfer rate reached the unfueled value. Unfortunately, at this fuel mass flow rate, the fuel did not burn, so that the result may be considerably different if combustion had occurred. These results will also be different in a engine in flight because the temperature of the wall will be much higher than the cold wall used here.

The effectiveness of a film of fuel in reducing the skin-friction load on a combustion chamber is part of the current research.

C. Port-Injection Results

For a working scramjet port injection would appear to be a reasonable option for the delivery of fuel to the air. If the dynamic pressure is sufficiently high compared with that of the freestream, then penetration of the fuel into the freestream will be achieved from a port in the side of the combustion chamber if the size of the injection hole is not too small. The penetration can be shown to be relatively insensitive to the injection angle provided the injection angle is greater than 15 deg. Hence, fuel need not be injected perpendicular to the wall but can be injected at an angle to the flow,

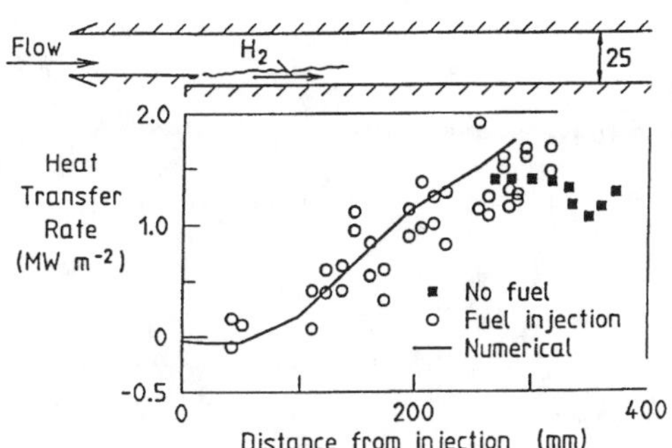

Film Cooling with Hydrogen (Stagnation enthalpy = 4.2 MJ/kg, M = 3.5, H_2 equivalence ratio = 1.4).

Fig. 6 Film cooling with Hydrogen (H_S = 4.2 MJ/kg, M = 3.5, φ = 1.4, F = 1.4).

so that a component of the fuel momentum is directed downstream, which will add to the thrust.

Experiments have been done by Morgan (1990) to determine the efficiencies of different configuration of port injectors. In these experiments the injection configurations shown in Figs. 7a and b were examined. The combustion chamber had a cross-sectional area of 25 × 50 mm. In both injection systems hydrogen fuel was injected downstream and at an angle of 30 deg. to the freestream. In the first configuration (Fig. 7a) fuel was injected sonically from one hole that had a diameter of 7 mm. In the second configuration (Fig. 7b) fuel was injected sonically from 4 holes each 3.45 mm in diameter. The size of these injectors was chosen so that the mass flow rate in both cases would be approximately the same for the same total fuel pressure. As might be expected, when fuel was injected through multiple injectors, a greater release of energy from the fuel was obtained. This occurred because there was greater mixing. When burning occurred, it was far superior to wall injection and comparable to that generated from central injection.

Additional experiments have been done by Wendt and Stalker (1996) who have compared the rate at which the pressure rises in a rectangular duct when hydrogen fuel is injected either centrally or through a single hole in the side wall. From theoretical and empirical estimates we see that the mixing efficiencies of cold hydrogen injected from a central injection system are between 20 and 35% better than that from a single hole. When hot fuel (total temperature between 600 and 700K) is used, the mixing efficiencies are approximately the same. The accuracy of the experimental results is limited, but the theoretical and experimental results are at least consistent with one another. These results are also consistent with the observations made by Morgan (1990), already described.

In some recent unpublished experiments a large number of small holes were used as an injection system, and a distinct lack of mixing was observed. (Pressure rises in the combustion chamber were not observed at conditions

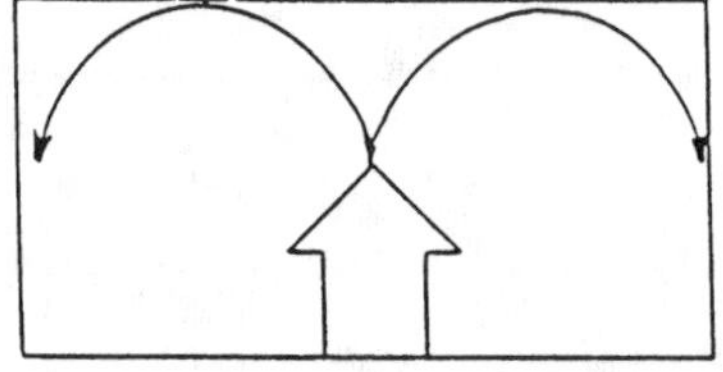

Fig. 7a Port injection through a single orifice.

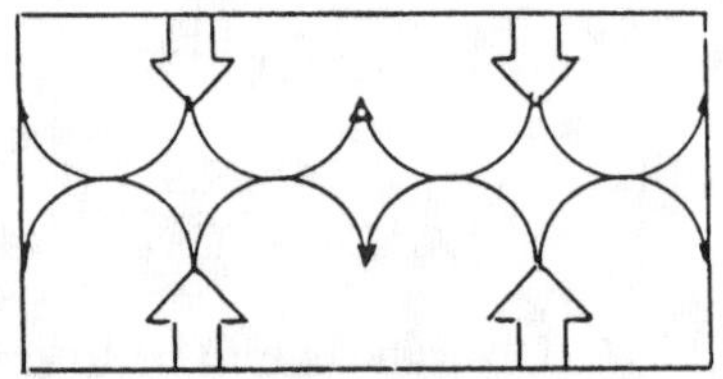

Fig. 7b Port injection through multiple orifices.

where burning should have occurred. This was attributed to a lack of mixing.) As the number of holes were decreased and the size of the holes was increased, better mixing was observed. The rule of thumb (proposed by Anderson (1998)) which states that the ports should be separated by twice the required penetration distance is a good means of determining the spread of ports.

D. Central Injection

In a central-injection system the fuel is delivered from a strut that is mounted across the combustion chamber. A large portion of the experiments undertaken in T3 and T4 have used this type of injector as it has proved to be an effective and relatively simple means of delivering the fuel [e.g. see Stalker and Morgan (1984), Paull (1992), Casey *et al.* (1992), Brescianini (1993), Wendt (1996), Pulsonetti (1996)].

In the next two sections various results that contribute to the understanding of this type of fuel delivery system and the development of a suitable thrust nozzle for such an injector are presented.

Combustion-Wake Profile with Central Injection

The processes by which fuel mixes and burns after being injected from a central injector are not completely understood. There are empirical fits and rules of thumb that prescribe mixing rates [(e.g., Northam and Anderson (1986) and Anderson *et al.* (1990))], which have been used in numerical analysis with varying degrees of success [e.g., Wendt and Stalker and Brescianini *et al.* (1993)]. However, a complete theoretical understanding of this type of injector is not currently available.

Meanwhile, Casey (1995) has obtained a simple model that approximates the Mach-number profile of the combustion region downstream of a central injector when hydrogen is used as the fuel. Casey uses the incompressible velocity profile developed by Weinstein *et al.* (1956) for the mixing of two coflowing streams of air. This profile is then extended to compressible flow using the Howarth transformation, and to include the effects of combustion, the Schwarb–Zeldovich transformation is used. The application of these transformations to the incompressible profile resulted in a first-order approximation for the Mach-number profile downstream of the injector.

In this approximation there is one parameter, the mixing parameter, that is not determined. Once this parameter has been obtained, it is a simple matter to draw the Mach-number profile. To determine this parameter, a number of different approaches can be pursued. One method is to use different mixing and combustion models to obtain numerical solutions for the mixing parameter, and another is to determine it experimentally. Casey chose the latter method.

Pitot-pressure profiles were obtained by Casey (1995) at different locations downstream of the injector. These profiles were obtained for the following three cases: 1) fuel injected into air (Figs. 8g–i), 2) fuel injected

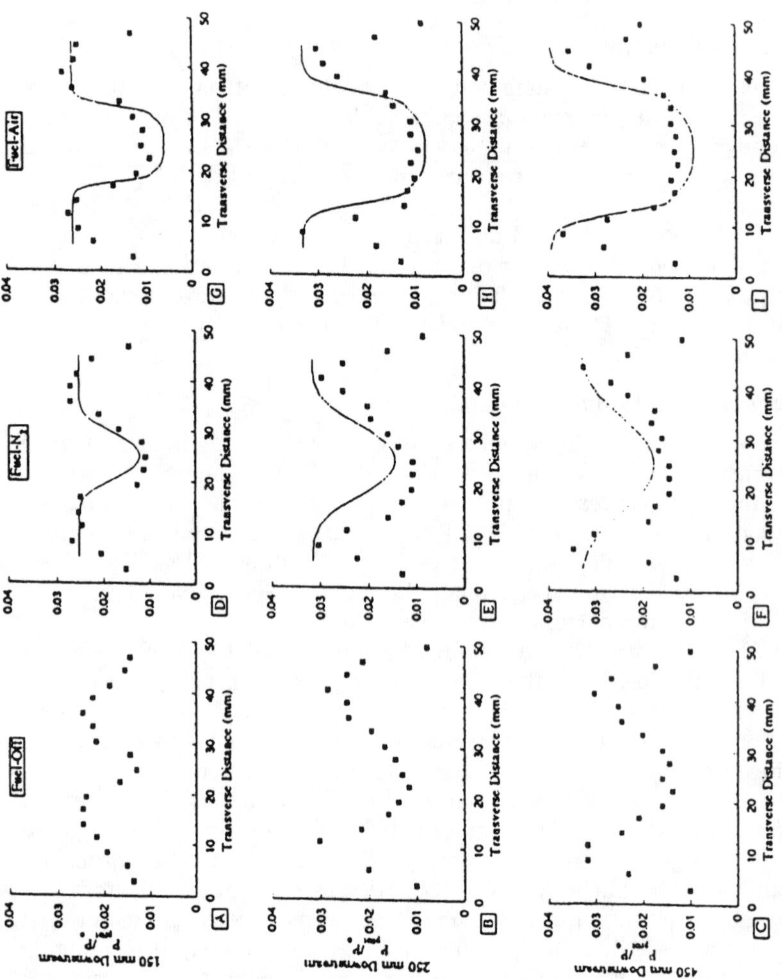

Fig. 8 Pitot-pressure measurements downstream of a central injector. Line is best fit using analysis of Casey (1995): a–c—fuel off; d–f—Nitrogen test gas, fuel on; and g–i—air test gas, fuel on.

into nitrogen (Figs. 8d–f), and 3) fuel not injected, with an air freestream (Figs. 8a–c).

The theoretical Mach number profile using Weinstein's profile and the aforementioned transformations were then fitted to these results by adjusting the value of the mixing parameter. The theoretical profiles for the best fit are displayed in Figs. 8d–i.

By a suitable choice of the mixing parameter, a reasonable agreement is obtained between the theoretical and measured profiles. That is to say, the change in pitot pressure (or Mach number) across the combustion chamber is reasonably well correlated with the width of the mixing region. This result is significant, as it shows that a simple relationship exists between the width of the combustion region and the change in Mach number between the freestream and the combustion region. Therefore, if the centerline Mach number is known, the extent of mixing can be determined, or, alternatively, if the extent of mixing is known, the centerline Mach number can be determined.

As can be seen, this result is not developed from a fundamental level and therefore gives little insight into the mechanics of mixing and burning. However, it is still useful as it can be employed as an engineering design tool, in a similar way to the empirical fits already described. In addition, apart from this engineering application, this result is interesting in its own right, as it shows how the Howarth transformation can be implemented to obtain useful compressible results from an equivalent problem in incompressible flow.

Obviously this result is redundant if the fuel is fully mixed and burned, as in this case the Mach number is uniform across the duct and can be determined from the freestream and fuel properties. However, it is not clear at this stage, if the maximum net thrust is produced by a scramjet that has a combustion chamber of sufficient length to ensure complete burning of the fuel. It may be that the skin-friction losses in such a long combustion chamber offset any benefits obtained by burning all of the fuel. A better understanding of the optimum length of the combustion chamber will be obtained when measurements of the skin-friction coefficient inside a combustion chamber are made. (Recent measurements of the skin friction have been made (see Sec. X). Additional measurements are currently being undertaken, and in the near future this calculation should be possible.) However, until then, understanding the effect that short combustion chambers have on the design of the nozzle is important and has been investigated and is the subject of the next section.

Two-Dimensional Effects of Short Combustion Chambers with Central Injection

If the combustion chamber is sufficiently long so that the Mach-number profile across the exit of the combustion chamber is uniform, then the development of a suitable exhaust nozzle is straightforward. However, if the combustion chambers are short, then the development of the contour for the exhaust nozzle is more complex, as a two-dimensional (or possibly three-dimensional) nonhomentropic flow exists at the entrance to the nozzle. Clearly, once this concept is realized, it is not a major problem when using computational fluid dynamics (CFD) to develop a suitable contour. However, from a historical point of view, it was a significant step to understand the mechanisms that govern the thrust production from a scramjet with a short combustion chamber. The understanding of these mechanisms paves the way to the correct analysis when applying CFD. (The results of Casey could be used as an input for the conditions at the entrance to the nozzle.)

An understanding of the mechanisms by which the pressure distribution on the thrust surface is generated was obtained by Stalker and Morgan (1984) by measuring the pressure distributions on an inclined flat surface (Fig. 9a) that was located downstream of a short, two-dimensional, centrally injected, constant-area combustion chamber. A schlieren photograph of the flow was also obtained (Fig. 9b). In these experiments hydrogen was used as the fuel. From the photograph one can see that there are two regions of interest: the region in the wake downstream of the injector where the fuel and air mix and burn and the freestream region. From the results of Casey (1995), discussed in the preceding section, the Mach number in the wake region is seen to be lower than the surrounding freestream. This is partially because the fuel is injected at a lower Mach number and partially because the heat released through combustion increases the local speed of sound.

The shape of the pressure profile on the thrust surface is understood by realizing that the Prandtl–Meyer expansion fan, which is centered at the corner of the combustion chamber and the thrust surface, is reflected at the interface between the wake and the freestream. This can be seen in the

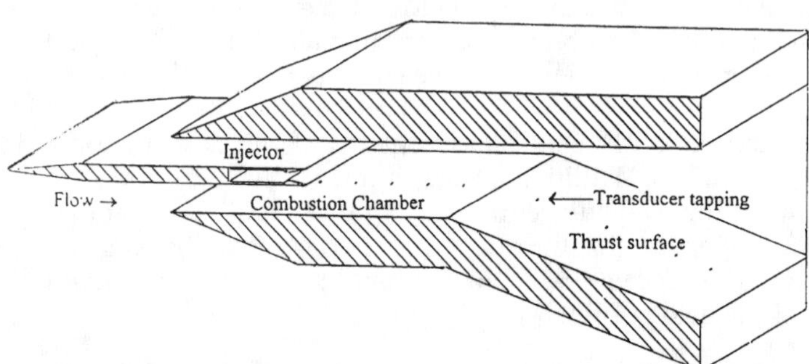

Fig. 9a Schematic (not to scale) of the generic scramjet used in T3 and T4. The combustion chamber, thrust surface lengths, and the thrust surface angle were adjustable. The injector cross-sectional width was 5 mm, and it spanned a 54-mm-wide combustion chamber. It is located centrally on the 27-mm sidewall of the combustion chamber. The leading edge of the injector was located sufficiently upstream to produce uniform flow at the entrance to the combustor. The trailing edge of the injector was 84 mm downstream of the entrance to the combustion chamber.

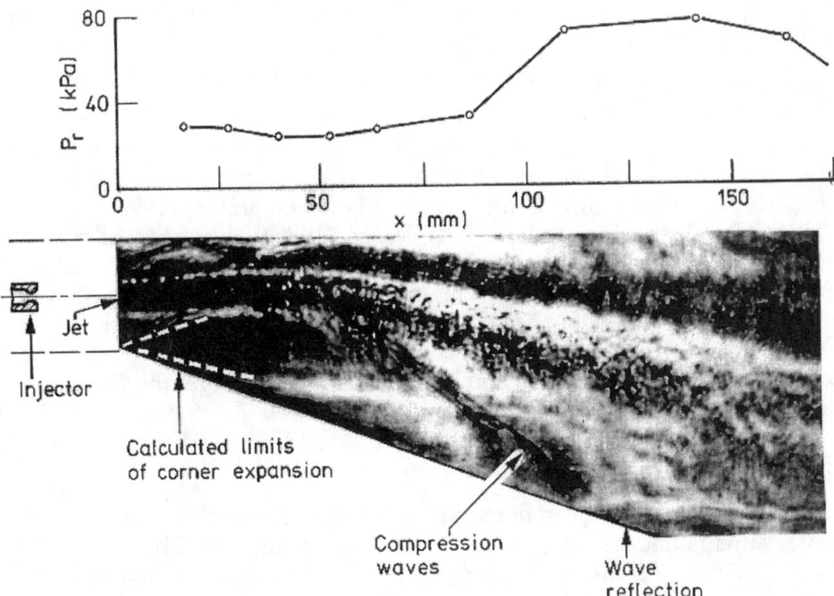

Fig. 9b Schlieren photograph and static pressure on the thrust surface downstream of a short combustor.

schlieren photograph. At this interface the Mach number decreases; thus, the expansion is reflected as a compression that travels back down toward the thrust surface. The arrival of this compression wave gives the initial and most substantial rise in the pressure on the thrust surface (at approximately 100 mm in Fig. 9b).

Stalker *et al.* (1988) have extended this work by developing a numerical scheme to simulate this wave process. They show that by properly considering the two-dimensional effects, already discussed, agreement can be reached between experiment and theory. This reference also gives some good physical explanations of the process by which the thrust is developed. An alternative understanding of this mechanism can be obtained from the numerical work of Allen (1993). This work was oriented toward discovering the optimum angle of the thrust surface.

IV. Combustion/Mixing Processes

Although empirical fits and the results of Casey (1995) that globally describe the effects produced by mixing and combustion are available, the coupling between the burning process and the mixing process is not understood. Mixing is required before combustion can occur. However, it is not clear to what extent combustion enhances the mixing process. There are two rates that control the burning of the fuel, namely, the mixing rate and the chemical kinetic rate (the rate at which a mixed fuel will oxidize). These will be considered in turn.

A. Mixing Controlled Combustion

The most commonly observed combustion/mixing process with hydrogen fuel is one that results in an almost linear pressure rise (in the mean) with distance from the ignition point (Fig. 10, Wendt and Stalker (1996)) (see also Bakos (1994) and Northam and Anderson (1986)). This type of combustion/mixing is consistent with the conceptual ideas associated with diffusion flames. The mixing process is slower than the time to complete oxidization, and, thus, the release of energy from the fuel is limited by the rate at which fuel can mix. This mixing limited combustion has been observed at free-stream pressures between 20 and 120 kPa and over a large range of equivalence ratios.

B. Kinetically Controlled Combustion

In contrast to this almost linear pressure rise, Pulsonetti (1992), Fig. 11, has observed an almost step-like rise in the pressure produced by the combustion process at a combustion chamber intake pressure of nominally 17 kPa. A similar result has also been observed by Casey *et al.* (1992) at an intake pressure of 40 kPa. Both experiments used hydrogen as fuel. The rise observed by Pulsonetti occurred at 700 mm downstream from the injector, whereas Casey observed a less dramatic rise at approximately 250 mm. These results were obtained at an equivalence ratio close to one using a central injector and with intake temperatures close to 1100 K. Further investigations into these results are being pursued, but preliminary analysis would suggest that some mixing occurs without immediate combustion so that combustion occurs at a much later time. However, when it does occur, the rate at which energy is released, and therefore the rate at which the pressure will rise, is now limited only by the kinetic rate and not by the much

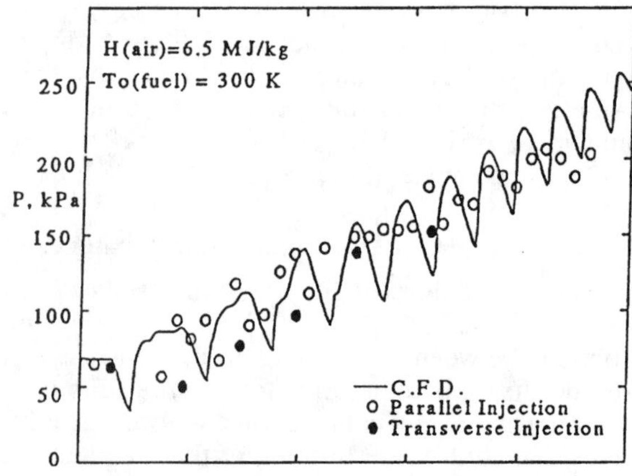

Fig. 10 Typical pressure measurements in a constant-area combustion chamber when burning is mixing limited.

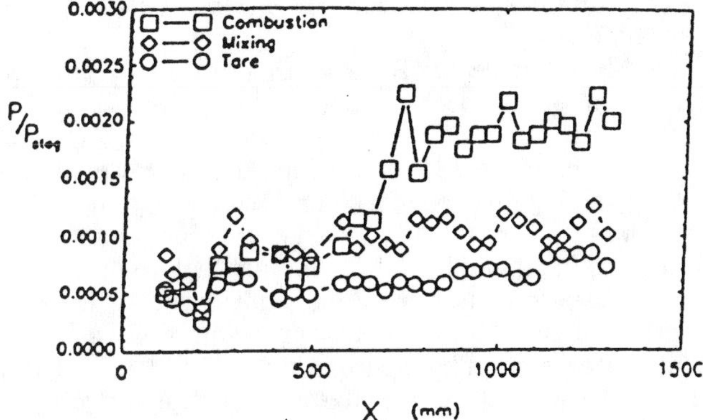

Fig. 11 Typical pressure measurements in a constant area combustion chamber when burning is kinetically limited.

slower mixing rate. Hence, the rise in pressure is very sharp and contrasts dramatically with that for mixing limited combustion.

The mechanisms that trigger the sudden release of the energy in these cases have not been identified; however, this issue is just one of the many problems related to the mechanisms that control mixing and burning in a messy environment such as a combustion chamber. These issues have been further examined by Buttsworth (1994), as discussed in the following.

C. Shock-Induced Ignition

If fuel is injected in the combustion chamber, then it is required to both mix and burn in the length of the chamber. In the experiments in T4 where the fuel was centrally injected or injected through a port in the wall with sufficient momentum, mixing does not appear to be difficult to produce in a reasonable distance. However, as noted in Sec. III, these types of injection systems suffer a pressure loss produced by the bow shock generated by either the injector or the fuel itself. To minimize and possibly eliminate this shock, wall injection behind a step was considered; however, as also noted in Sec. III.A, this type of injection suffered from severe burning limitations, and long combustion chambers are required.

One way of eliminating these long combustion chambers is to inject the fuel in the intake where the temperature and pressure are sufficiently low so that combustion does not occur. By injecting in the intake, the fuel has time to mix before its pressure and temperature are finally raised as it enters the combustion chamber. This increase in pressure and temperature is generated by passing a shock through the mixture. The shock would normally be present as it would be created by the intake as it compresses the air. This type of ignition process where the fuel is first allowed to mix and then combustion is induced by the presence of a shock will be referred to as shock-induced ignition and is effectively another example of kinetically controlled burning.

Studies have been made by Buttsworth (1994) into the feasibility of shock-induced ignition. In these experiments hydrogen was injected into a duct where the temperature and pressure of the fuel-air mixture were insufficient for combustion to occur. Pressure measurements were taken to verify this. The flow was then subjected to a compression generated by a long ramp, which turned the flow through 10°. Pressure measurements were taken on the ramp to determine if and how long the fuel took to burn.

A number of experiments were done, and in one typical example the freestream pressure and temperature prior to injection were 10.4 kPa and 682 K and after compression by the ramp were 37.5 kPa and 1020K, respectively. Hence, upstream of the corner to the compression ramp, the temperature is insufficient for combustion. However, downstream, the pressure and temperature are high enough so that combustion can occur. At approximately 600 mm downstream of the compression corner, the pressure started to rise on the ramp above that observed when fuel was injected into nitrogen, and furthermore, with injection into air, at approximately 800 mm combustion is essentially complete. The corresponding elapse time was consistent with the time predicted using an approximation given by Huber *et al.* (1979) who proposed that the time for 95% of the energy to be released in a hydrogen-air reaction at constant pressure can be satisfactorily determined from the sum of the hydrogen-air ignition time t_i and the reaction time t_r where

$$t_i = \frac{8 x 10^{-9} e^{9600/T}}{P} \qquad t_r = \frac{0.000105 e^{-1.12T/1000}}{P^{1.7}} \tag{1}$$

Here, t_i and t_r are in seconds, P is the mixture pressure in atmospheres, and T is the temperature in Kelvin.

In many ways these results are not surprising as they simply confirm the approximation determined by Huber *et al.* (1979). However, the significance of these results is that they show that the technique just described, which allows mixing in the intake, is one that is conceptually simple and one that can be implemented relatively easily.

D. Shock-Induced Mixing

Another important aspect of the work completed by Buttsworth (1996) shows that shocks cannot only be used to promote ignition, but they can also be used to generate vorticity as they cross the fuel-air mixing region, and, thus, can be used to enhance mixing essentially through a Richtmyer–Meshkov instability, Lebo *et al.* (1995). It is shown analytically that to ensure that an increase in vorticity is produced by a shock the gradients of the velocity and density within the mixing region must be in the same direction. Experimental results were obtained to verify that greater mixing had occurred when a shock was allowed to impinge on a suitable mixing region.

Additional results that highlight the effects disturbances have on mixing were obtained by McIntyre *et al.* (1995). A schlieren photograph (Fig. 12) of the flow in a combustion chamber was obtained when hydrogen was injected from a central injector into a rectangular duct into a freestream of nitrogen. The flow is from left to right. Two separate disturbances that are almost symmetric about the centerline can be seen propagating through the flow. The more interesting is the second disturbance that crosses the fuel stream about halfway along the photograph. Upstream of this point, a well-structured flow pattern can be seen; however, downstream the flow in the fuel jet is much less structured, and further downstream, after the first disturbance again crosses what remains of the fuel jet, there is very little structure left in the fuel jet. The nonexistence of such a structure indicates that a significant amount of mixing has occurred.

These are important results when considering the problem of mixing. For in a combustion chamber there are usually more shocks and disturbances generated by boundary layers, injectors, or intakes, etc., than one cares to deal with! Thus, there is ample opportunity for vorticity to be generated, and therefore there is ample opportunity for mixing to be enhanced in a working combustor.

V. Simple Theoretical Combustor and Thrust Model

The performance (as measured by the thrust and incremental specific impulse) obtained from a combustion chamber and thrust surface depends on the combustor intake conditions, geometry, and, to a lesser extent, on the fuel stagnation temperature and freestream conditions at injection. To discuss trends and effects produced by variation of these parameters, a simple theoretical model is given. These results are similar to those derived by Bakos (1994) and are an approximation to the exact results of Kerrebrock (1992)

We will assume the following: 1) the combustion chamber has a constant area cross-section; 2) all compressions and expansions are isentropic, and all

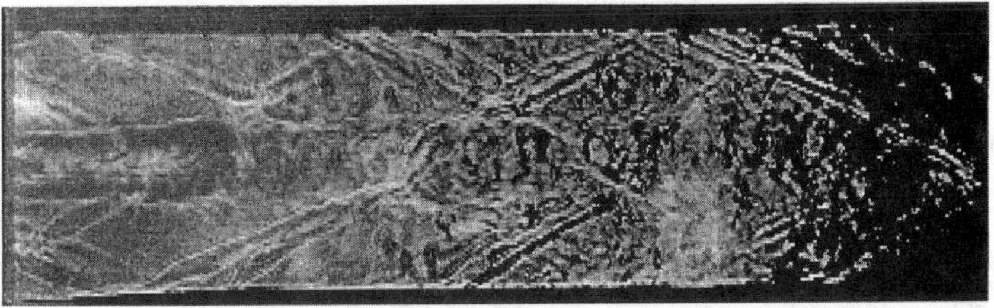

Fig. 12 Schlieren/shadow graph of flow downstream of a central injector. Nitrogen test gas.

gases are perfect; 3) the specific heat ratio remains constant throughout the combustor; 4) the fuel mass flow rate is negligible compared with the air mass flow rate; 5) the Mach number M throughout the combustor remains high; 6) the flow is one-dimensional; 7) the flow is steady; 8) the walls of the scramjet are insulated; 9) the combustion process adds only energy and not momentum to the flow; and 10) the intake and exhaust area ratios are the same.

The limitations of these approximations are clear; however, the results obtained in the following equations using these approximations help to provide an insight into the way in which a scramjet works by highlighting the importance of certain parameters. An exact solution to the one-dimensional calculations that does not contain all of the approximations just listed can be obtained very simply by using numerical methods. However, unless a large number of cases are examined, these results will not display the underlying physics to the results, whereas the analytical results derived next do.

The aim of this section will be to determine the thrust and incremental specific impulse of a scramjet for different freestream conditions. Specific relationships for skin-friction losses will not be given; however, the effects of skin friction will be included.

Consider the idealized scramjet depicted in Fig. 13. Subscripts 0, 1, 2, and 3 will refer to the freestream properties, properties at the combustor entrance and exit, and properties at the nozzle exit, respectively. If the fuel mass flow rate is small and therefore neglected, then by the conservation of mass

$$\rho_0 u_0 A_0 = \rho_1 u_1 A_1 = \rho_2 u_2 A_2 = \rho_3 u_3 A_3 \tag{2}$$

where ρ, u, and A are the density, velocity, and area, respectively, at the different locations. From the momentum equation and recalling that the assumption that the areas of the intake and exhaust are the same, the net forward force applied to the scramjet F, which includes the reaction force produced by injecting fuel (Depending on the situation, it can account for 200–300 s of a net specific impulse, which is usually significant.), is

$$F = \left(\rho_3 u_3^2 + P_3\right)A_0 - \left(\rho_0 u_0^2 + P_0\right)A_0 - f_e \tag{3}$$

where P is the static pressure and f_e is the sum of all frictional forces on all surfaces except the combustion chamber, intake, and nozzle.

If the combustion chamber is aligned parallel to the flow and its cross-sectional area does not change along its length, then the total flux of momentum leaving the chamber must be equal to the total air momentum flux

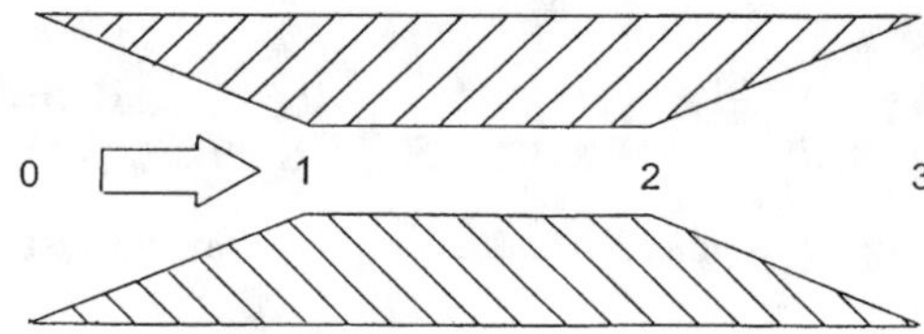

Fig. 13 Idealized scramjet.

entering the chamber, plus the total fuel momentum flux at the exit of the injector T_f (which is equal to the force on the injector produced by the fuel), and less the force due to skin friction on the combustion chamber walls f_c. That is to say,

$$\left(\rho_2 u_2^2 + P_2\right)A_1 = \left(\rho_1 u_1^2 + P_1\right)A_1 - f_c + T_f \tag{4}$$

If the flow in the intake is now assumed to be isentropic and perfect with constant specific heat, then, in the usual manner, relationships can be found between the flow properties at the entrance of the combustor and the entrance of the intake. Similar relationships exist between the flow properties at the exit of the combustor and the exit of the nozzle. If approximations are then sought by expanding in terms of M^{-2} and if terms of order $M_1{}^{-4}$ and higher are ignored, using Eqs. (2) and (4) show that Eq. (3) can be rewritten as

$$F = \frac{1}{\gamma - 1} A_1 (P_2 - P_1)\left(1 - \left(\frac{A_1}{A_0}\right)^{\gamma - 1}\right) - (f_e + f_c) + T_f \tag{5}$$

where γ is the specific heat ratio.

The difference between the pressure at the exit and entrance of the combustion chamber is obtained from the conservation of mass (Eq. (2)), momentum [Eq. (4)], and energy along the combustion chamber. If there is no energy loss to the walls of the combustion chamber (insulated walls), then the conservation of energy states that

$$\dot{m}_2\left[u_2^2 + 2a_2^2 / (\gamma - 1)\right] = \dot{m}_1\left[u_1^2 + 2a_1^2 / (\gamma - 1)\right] + 2(E_c + E_f) \tag{6}$$

where $\dot{m}$ is the mass flux of air into the combustion chamber, a is the local speed of sound, and E_c and E_f are the energy fluxes added by combustion and from the addition of fuel, respectively. If there is a loss of energy to the walls, then this would appear as a negative term in the last bracket of the right-hand side in Eq. (6).

If the mass flux of fuel is small compared with that of the air, then if terms of order $(M_1)^{-4}$ are ignored, one can see that

$$P_2 - P_1 = \frac{(\gamma - 1)}{A_1}\left\{\frac{E_c + E_f}{u_1} + f_c - T_f\right\} \tag{7}$$

Hence, from Eqs. (5) and (7),

$$F = \frac{(E_c + E_f)}{u_1}\left[1 - \left(\frac{A_1}{A_0}\right)^{\gamma - 1}\right] - f_e + (T_f - f_c)\left(\frac{A_1}{A_0}\right)^{\gamma - 1} \tag{8}$$

From Eq. (5) the net force on a scramjet is driven by the difference in pressure between the exit and entrance of the combustion chamber. As this quantity increases, then so does the forward thrust. A frictional force applied in the combustion chamber will increase the pressure at the end of the combustion chamber; thus, to some degree, the drag on the combustion chamber walls will be offset by greater thrust on the thrust surface. In fact from Eq. (8), one can see that to first order the momentum loss due to the skin friction with the combustion chamber wall is completely offset as the nozzle area ratio approaches infinity. Obviously, this cannot be achieved in reality, as f_e and other sources of drag would also increase. In addition, one may recall that the energy loss due to the flux of heat to the wall has not been included here.

The force applied by the injection of fuel will decrease the pressure at the combustion chamber exit, and this is offset by an increase in pressure produced by the addition of energy from the fuel, and, in the absence of combustion, the exit pressure will either rise or fall depending on the freestream pressure and fuel temperature. As the area ratio approaches infinity, the effect due to the force applied by the injector is reduced to zero. The net forward force on the scramjet is then purely a function of the energy put into the air by the fuel, either through its combustion or its initial internal energy. Conversely, as the area ratio approaches one, the propulsive force is only that from the force acting on the injector, and none of the energy released from the fuel is converted into mechanical work as should be expected in this limit.

Of equal importance as the thrust is the efficiency of the engine. One measure of the efficiency is the incremental specific impulse, which is defined as the ratio of the thrust developed to the mass flux of fuel (and divided by $g = 9.8\ \mathrm{m/s^2}$).

$$I = F(\dot{m}_f g)^{-1} \tag{9}$$

where $\dot{m}_f$ is the fuel mass flux. Hence,

$$I = \frac{\mathrm{E}}{gu_1}\left(1 - \left(\frac{A_1}{A_0}\right)^{\gamma-1}\right) - \left\{f_e + (T_f - f_c)\left(\frac{A_1}{A_0}\right)^{\gamma-1}\right\} \Big/ g\dot{m}_f \tag{10}$$

where $E = (E_c + E_f)/\dot{m}$ is the energy per kilogram of fuel added to the freestream by the fuel (the fuel energy density). If the fuel mixes completely with the freestream and the equivalence ratio is less than one, then this energy density is proportional to the equivalence ratio. (In actuality there will be a slight change in the energy released from combustion because the pressure and temperature of the fully mixed flow will change as the equivalence ratio is increased.) If the equivalence ratio is greater than one, this is not true because all of the available oxygen has been consumed at an equivalence ratio of one, so that additional fuel will not produce any additional release of energy from the combustion process.

By substituting Eq. (8) into Eq. (9), one can see that the two quantities which govern the efficiency of the engine are the entrance velocity of the freestream to the combustion chamber and the fuel energy density. If the combustion chamber entrance velocity is increased, then the efficiency decreases. Thus, a scramjet would be expected to be less efficient when it operates at higher velocities.

The second quantity governing the efficiency, the fuel energy density, is dependent upon both the temperature and pressure of the fuel, air, and combustion products mixture. For hydrogen if ignition occurs as the combustion chamber entrance temperature increases, so does the equilibrium temperature of the reaction products. Hence, less water is formed, and therefore less energy is released. Conversely, if the pressure is increased, more energy is released.

Thus, the efficiency of a scramjet will be optimized for a fixed flight speed by adjusting the combustor intake temperature, pressure, and velocity. If in the simple geometry of Fig. 13 the combustion chamber cross-sectional area is increased and the intake cross-sectional area remains unchanged, then the combustion chamber entrance velocity will increase, and the pressure will decrease. Both of these will reduce the efficiency. However, the temperature will decrease, and provided that the temperature is still greater than the ignition temperature this will produce an increase in efficiency. Thus, there are two competing effects that will determine the size of the most efficient combustion chamber.

Unfortunately, such an optimization is only applicable to one flight speed. If a scramjet is to operate efficiently over a range of flight conditions, the combustion chamber intake pressure temperature and velocity will have to be adjusted in some way. Hence, this will require an intake or a combustion chamber that can offer variable geometry. This is a substantial problem for future research.

VI. Experimental Results of Specific Impulse

Experiments to measure the thrust generated by burning various fuels have been made using the configuration shown in Fig. 9a. One wall of the combustion chamber is inclined, downstream of a uniform cross-sectional area. This inclined surface is called the thrust surface. The thrust is determined by first measuring the pressure at various locations along the thrust surface. The surface pressure immediately surrounding the sensor is assumed to be equal to that at the sensor. Thus, a local net force can be obtained from multiplying the local pressure with the surrounding area. The thrust is then determined by summing all of these local forces on the thrust surface and taking the component directed upstream. In the experiments reported in the following paragraphs, unless otherwise stated, the combustion chamber was 27 mm high and 54 mm wide. The injector was 5 mm high and located centrally along the side that is 27 mm high. It spanned the full 54 mm width.

Morgan *et al.* (1985) noted that the net thrust was insensitive to the angle of the thrust surface when the thrust surface was varied between 5 and 15 deg. In these experiments the area ratio to which the gases expanded remained constant. If the engine was meant for practical purposes, then the shape and length of the thrust surface will be constrained by the final area ratio, so it is important to maintain this ratio, rather than the length. Numerical results by Allen Jr. (1993) were consistent with this observation for a variation of thrust surface angle between 7 and 13 deg. Although the distribution and the magnitude of pressure changed as the thrust surface angle changed, provided the area ratio was sufficiently large, the net thrust was approximately constant, regardless of the thrust surface angle.

Figure 14 (Paull (1993)) displays a typical thrust time history when 1) fuel (hydrogen) is injected at an equivalence ratio of one and 2) in the absence of fuel. The model configuration is the same as that in Fig. 9a [see Bakos and Morgan (1992) for axisymmetric configurations]. The length of the combustion chamber was 575 mm, and the injector was 84 mm downstream of the entrance to the combustor. The combustion chamber was 27 mm high and 54 mm wide, and the thrust surface diverged at 11 deg to the freestream. The wall-pressure measurements (the lines joining the measurements are not meant to indicate the actual pressure but have been included to help distinguish the different results), from which this thrust was derived, are shown in Fig. 15. The wall-pressure measurements are a snapshot of the pressure distribution at a time equal to 1.8 ms in Fig. 14. At this time, the freestream pressure and temperature are 80 kPa and 1000 K, respectively, and the Mach number and freestream velocity are 4.5 and 2.78 km/s, respectively.

At this point, a slight digression to look in some detail at the results presented in Fig. 15 is worthwhile. The rise in pressure because of the reflection of the expansion fan off the lower Mach-number region in the center of the combustor (see sec. III.D) can be seen at approximately 650 mm. At first sight this rise appears to be not as great as in Fig. 9b. However, this observation is misleading for only the gradient of the rise has been reduced, and by comparing with the fuel-off result, the actual rise is seen to be still substantial. This occurs because the combustion chamber is longer and therefore the combustion region has had a chance to diffuse further across the duct. Consequently, the expansion from the corner of the thrust

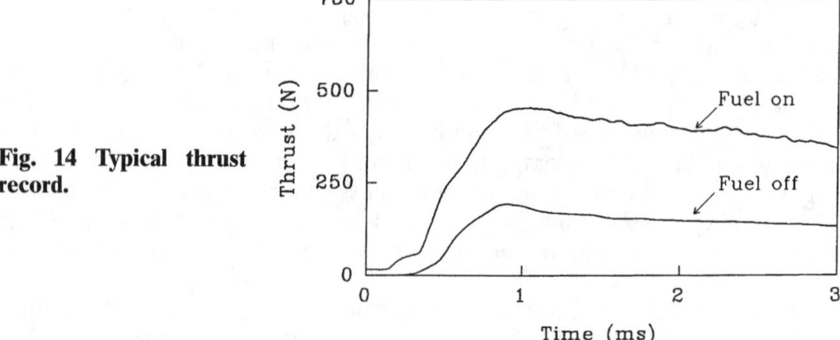

Fig. 14 Typical thrust record.

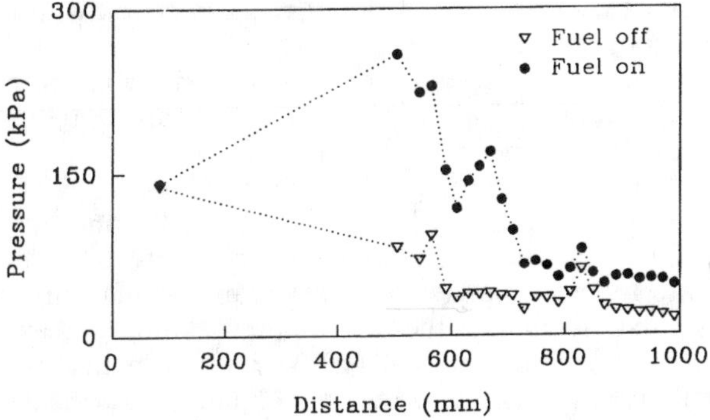

Fig. 15 Typical pressure measurements in a combustion chamber and along the thrust surface for both fuel on and fuel off.

surface and combustion chamber will move through less sharply defined regions of Mach number and will be reflected back as a compression wave but with a more gentle slope.

Although the rise in pressure is smeared out, the fact that the pressure does not fall monotonically from the corner is still important, which indicates that the Mach number at the exit of the combustor is not uniform, even though the combustion chamber is quite long. Hence, it is clear that for combustion chambers of reasonable length the results of Casey (1995) for predicting the Mach-number distribution at the combustion chamber exit will be required for developing efficient thrust surfaces.

From the thrust measurements one can see that after an initial starting process the thrust settles down to a reasonably steady level. There is a slight fall in thrust with time because of an equivalent fall in the total pressure of the freestream. The level of the thrust that would be used here to determine the incremental specific impulse is that at approximately 1.8 ms.

An estimate of the incremental specific impulse I_m is made by subtracting the thrust generated over the thrust surface when fuel is not burned from that when fuel is burned and dividing by the fuel mass flow rate and g. This measurement does not include the thrust developed by the injector and does not account for the reduction in thrust that results from the skin-friction losses. Hence, if the load on the intake is equal to the force on the exhaust nozzle when the fuel is off, then in terms of the preceding notation, this measured difference F_m is equal to the total force F minus both the loss due to the friction $(f_c + f_e)$ and the increase due to the stream thrust of the fuel. Hence,

$$F_m = F + f_c + f_e - T_f \tag{11}$$

Hence, from eqs. (8–10)

$$I_m = \left(1 - \left(A_1 / A_0\right)^{\gamma - 1}\right)\left(\frac{\mathsf{E}}{u_1} - \frac{T_f}{\dot{m}}\right) \Big/ g \tag{12}$$

This estimate does not account for the losses associated with the skin friction, so in this sense it is an upper limit to the incremental specific impulse. Moreover, it also does not account for the force on the injector. Hence, this estimate of the incremental specific impulse is only an approximation to the upper bound and should be treated as such.

Experiments using the configuration shown in Fig. 9a have been done by Stalker and Morgan (1984) and Paull (1993) to determine the incremental specific impulse as a function of the total enthalpy. Changing the total enthalpy of the test gas changes the freestream temperature and velocity and to a lesser extent changes the freestream pressure. The extent of these changes for the experiments of Paull (1993) can be seen in Figs. 16a–c [Paull (1995)], which display nonequilibrium, one-dimensional calculations [Lordi *et al.* (1965)] of the change in the freestream properties with total enthalpy. The combustor intake Mach number and static pressure used by Stalker and Morgan (1984) were nominally 3.7 and 110 kPa, respectively, whereas Paull (1993) used a nominal Mach number of 4.5 and recorded the incremental specific impulse at the nominal pressures of 33 and 100 kPa. Hydrogen was used as the fuel, and the equivalence ratios were approximately one-half or one. The thrust surface used by Paull was inclined at 11 deg to the freestream whereas Stalker and Morgan used a 15-deg thrust surface.

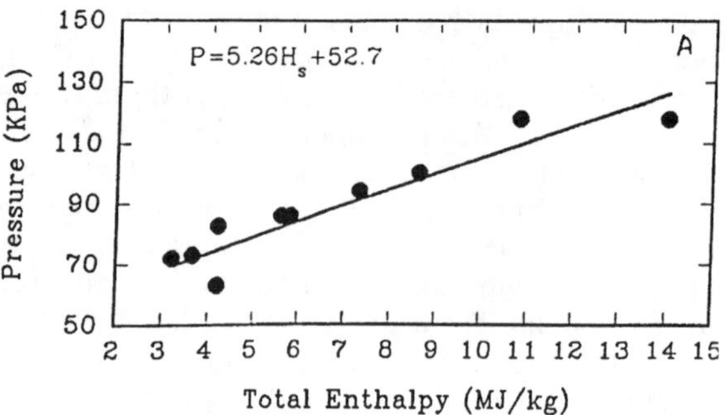

Fig. 16a Variation of combustion chamber entrance pressure with freestream total enthalpy in T4. Entrance Mach number nominally 4.5.

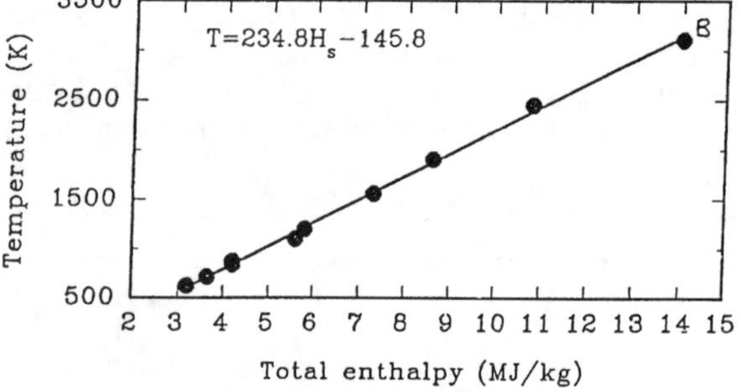

Fig. 16b Variation of combustion chamber entrance temperature with freestream total enthalpy in T4. Entrance Mach number nominally 4.5.

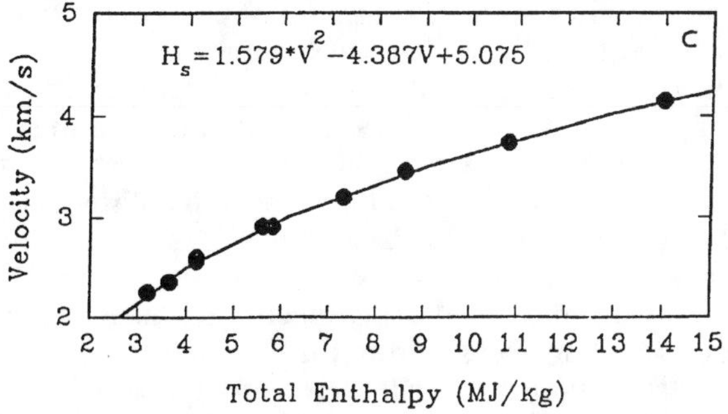

Fig. 16c Variation of combustion chamber entrance velocity with freestream total enthalpy in T4. Entrance Mach number nominally 4.5.

Figure 17 displays the dependence of incremental specific impulse on flight speed (where total enthalpy is half the square of the flight speed) for different combustor intake Mach numbers, pressures, and equivalence ratios. Recall from the preceding section that the incremental specific impulse is dependent on both the freestream velocity and temperature. Hence, both parameters must be known before any comparison can be made between results.

In Stalker and Morgan's results, as the velocity and temperature decrease, the incremental specific impulse increases and reaches a maximum of 1750s. A similar trend is observed by Paull, however, at the higher Mach number the peak has dropped to approximately 1500 s. This maximum occurs at the same temperature in both cases. Hence, at the higher Mach number the velocity at the entrance to the combustor has increased, which from Eq. (10)

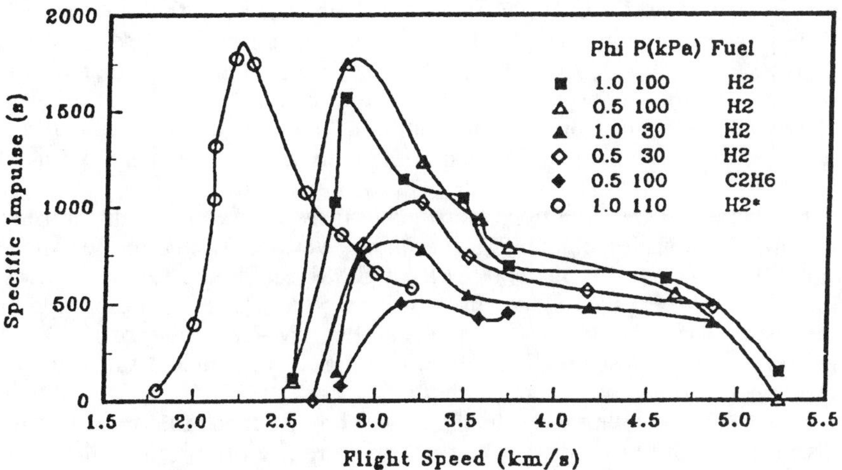

Fig. 17 Incremental specific impulse as a function of equivalent flight speed at combustion intake Mach numbers of 3.7 and 4.5.

should lower the incremental specific impulse. Thus, it is not unexpected that the observed peak should be lower at the higher Mach number.

The sharp drop in incremental specific impulse at the lower temperatures results because the temperature drops below the self-ignition temperature. The self-ignition temperature is approximately 850 K at the higher pressures and a little higher at the lower pressures. As the flight speed increases, one can see that the incremental specific impulse decreases and is eventually zero at approximately 5 km/s. This occurs for two reasons. As just described, a decrease in the incremental specific impulse is expected because of the increase in velocity; however, one can also see from Eq. (10) that a decrease in incremental specific impulse will also occur if the energy released from the fuel is also decreased. A decrease in the energy released from the fuel can occur in two ways. If the intake temperature is sufficiently high, then when the chemical process reaches equilibrium no or little water is formed, as it remains in its dissociated state. Therefore, no chemical energy is released from the fuel to the flow. As the flight speed increases, one can see from Fig. 16b that the temperature increases, and so at the higher flight speeds less energy is released from the fuel because of this kinetic limitation. It is also possible that less energy will be released because the residence time in the combustor is less as the flow velocity increases. As the flow velocity increases, there is less time for mixing to occur, and therefore less energy will be released from the available fuel.

VII. Effects of Atomic Oxygen and Nitric Oxide in the Freestream

Reflected shock tunnels operating with air test gas produce flows with concentrations of atomic oxygen and nitric oxide above that of equilibrium air. At enthalpies less than 7 MJ/kg, the calculated molar fraction of atomic oxygen and nitric oxide is less than 2 and 10% of the available oxygen, respectively (Lordi *et al.* (1965)). (Some doubt has been thrown on these values by preliminary results from a mass spectrometer (Skinner and Stalker (1994) and Boyce (1997.)) However, this change in chemical composition becomes more pronounced as the total enthalpy is increased. The effects of this change on the combustion characteristics of hydrogen in a scramjet were investigated both experimentally and theoretically by Bakos (1994).

Experimental investigations were made in two different facilities using the same combustion chamber. Experiments were performed in the expansion tube HYPULSE (Bakos (1996)), at GASL and in the reflected shock tunnel T4. The HYPULSE facility produces 52% of the atomic oxygen in the test flow that T4 does at an equivalent flight Mach number of 13. Other properties of the test flow in the two facilities were matched as closely as possible. Hydrogen was used as the fuel.

From the experimental results the conclusion was made that there was no discernible difference between the net pressure rise produced by the combustor when run in the different facilities. However, there was evidence to show that the combustion was completed earlier when more atomic oxygen was present.

The early completion of burning in T4 is consistent with theoretical expectations. However, the similar rise in the overall pressure may at first sight appear to be intuitively incorrect. When hydrogen reacts with oxygen molecules in a simplified reaction scheme such as given below,

$$\begin{aligned} H_2 &\leftrightarrow H \\ 1/2O_2 &\leftrightarrow O \\ H + O &\leftrightarrow OH \\ H + OH &\leftrightarrow H_2O \end{aligned}$$

oxygen molecules first have to dissociate into two oxygen atoms. This is an endothermic reaction. The exothermic reaction of importance is the final reaction that combines hydrogen atoms with hydroxide radicals. One could argue that if the oxygen is already partially dissociated before it enters the combustion chamber then less endothermic reactions need to occur before combustion is complete, and therefore, if the same exothermic reactions occur, there is more energy apparently released into the combustion chamber. Hence, the pressure rise should be greater when there is atomic oxygen in the freestream.

As this observation was contrary to the observations made by Bakos, a numerical investigation of the combustion in a combustion chamber of constant cross-sectional area was undertaken to determine the reasons for this observation. The hydrogen-air reactions were modeled using a 24-reaction mechanism described by Oldenborg *et al.* (1990). Bakos showed from his numerical investigations that there was in fact good agreement between theory and experiment and that the preceding intuitive argument was incorrect, and he showed that the intuitive argument fails to address the effect that the greater apparent heat release has on the final composition of the combustion products.

From the numerical results the observation was made that, as the combustion products reach equilibrium, when atomic oxygen is initially in greater concentrations in the freestream there is less water formed and therefore less energy released. So the energy saved by having atomic oxygen in the freestream is effectively lost because less energy is released because of less water being formed.

The physical reason for the reduced concentration of water products is as follows. As the intuitive argument suggests, the temperature of the combustion products tends to be greater because of less endothermic reactions, but as the temperature tends to be higher the equilibrium concentration of water tends to be lower.

One can also see that there is a large percentage of nitrous oxide in the freestream of T4, which is also present in HYPULSE at a similar level; hence, a comparison of the effects of nitric oxide on the performance of the combustor could not be made. However, Bakos[47] showed numerically that nitric oxide is essentially inert because the reaction rates associated with nitric oxide are very slow. However, it does act as a third body in some reactions, which will alter their reaction rates. If the reader wishes to pursue this aspect of the chemistry in greater depth, refer to the work by Jachimoswski. Rate constants supplied by Jachimowski can be found in Anderson (1989).

It is important to realize that nitric oxide is basically inert because this means that equivalence ratios should be (and are) based on the oxygen that is available in atomic and molecular form only. Oxygen tied up in the nitric oxide is unavailable for reaction.

VIII. Different Fuels

The majority of experiments made at T3 and T4 have used hydrogen as fuel. However, experiments have also been done with a number of hydrocarbon fuels as well as silane-hydrogen mixture. In this section the results from these experiments have been collated. There are additional results obtained by Morgan and McGregor (1995) for liquid hydrocarbon fuels, which are not included here. Results for heated fuels can also be obtained from Wendt (1995).

A. Hydrocarbon Fuels

Although hydrogen is generally a more energetic fuel than hydrocarbon fuels (i.e., the hydrogen has a greater energy density), some hydrocarbons offer the advantage that they can be liquefied without the use of cryogenic cooling and can also be contained within a smaller volume. When considering the overall costs of a launcher, both of these properties make the use of hydrocarbons attractive (Lewis and Guptà (1997) give a good review of this problem). However, to ascertain whether or not such reductions in cost would be sufficient to offset the losses associated with a less energetic fuel, performance data on hydrocarbon fuels are required. A preliminary set of data has been obtained by Paull (1995).

Tests have been done to determine the performance on two hydrocarbon fuels: ethylene, C_2H_4 and ethane, C_2H_6. One important aspect of these fuels is their ability to be liquefied at room temperature. The critical temperatures and pressures for ethylene and ethane are 282K and 5MPa and 305K and 4.9 Mpa, respectively.

Experiments were performed using the generic scramjet described in Sec. VI. As with the hydrogen experiments described in that section, the incremental specific impulse was determined over a range of total enthalpies at a nominal Mach number of 4.5 and a nominal static pressure of 90 kPa. The equivalence ratio was nominally 0.75.

The incremental specific impulses of ethylene and ethane as functions of equivalent flight speed are plotted in Fig. 18. An example of the fuel-off and fuel-on pressures for ethylene are displayed in Fig. 19. The fuel-on pressures for ethane were similar; however, there was a greater delay in the ignition time. Figures 17 and 18 show that the incremental specific impulse for hydrogen is approximately twice that for ethylene. (There is a slight change in equivalence ratio between the hydrogen and hydrocarbon tests that

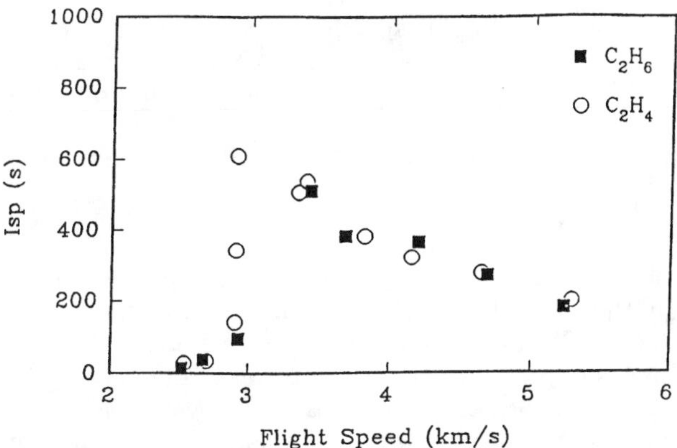

Fig. 18 Incremental specific impulse as a function of flight speed for ethane and ethylene. Nominal intake pressure of 100 kPa.

should be noted here.) Paull (1995) also noted that although the incremental specific impulse for ethylene is considerably less than that of hydrogen the thrust generated by these hydrocarbons was approximately the same as that generated by hydrogen. This is reasonably consistent with theory as it is expected that a stoichiometric mixture of ethane would release approximately the same energy as that for hydrogen (less energy is released from ethylene), and therefore to first order the same thrust should be developed.

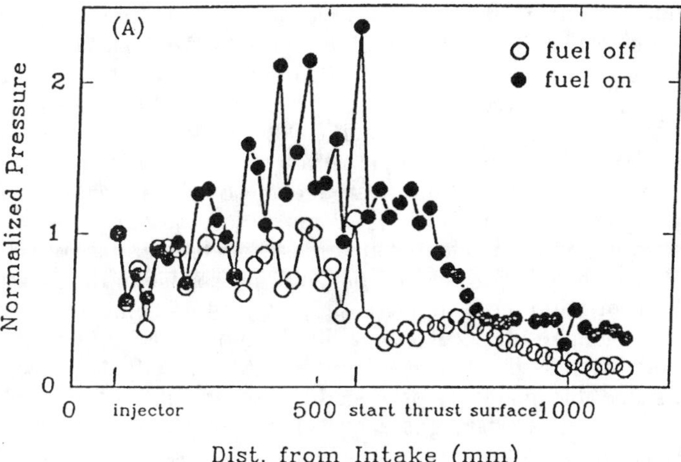

Fig. 19 Combustion chamber and thrust surface pressure measurements normalized by intake pressure (90kPa) for ethylene fuel and in the absence of fuel. Freestream velocity and temperature 2.9 km/s and 1100 K, respectively.

Thus, the drop in incremental specific impulse is not unexpected and is caused primarily because the hydrocarbons are much heavier than hydrogen. Figures 16b, 17, and 18 show that the self-ignition temperature of ethylene is close to that of hydrogen, at approximately 900 K, whereas the ignition temperature of ethane is higher, at approximately 1100K.

B. Silane-Enriched Fuels

Silane (SiH_4) is a pyrophoric gas that can been added to hydrogen to decrease ignition delay times of the fuel. Typically, silane concentrations between 5% and 20% by volume have been examined. This type and quantity of additive is useful when the combustion chambers are either short or the combustion chamber temperature is low. However, it has the disadvantage that, with a molecular weight of 32, the specific impulse developed by this fuel and additive will be low.

Morris (1989) (see also Morris *et al.* (1987) made an experimental investigation into the ignition limits of different mixtures of silane and hydrogen. A centrally injected constant-area duct combustion chamber (Fig. 2) was used with pressure measurements made downstream of the injector. The combustion chamber was 25 $\times$ 50 mm in cross-section and 600 mm long. Injection was made 85 mm downstream of the intake of the combustor.

If the fuel contained 20% by volume of silane, then the lowest temperature at which the fuel could be made to burn in this duct was approximately 375 K (the freestream intake pressure was nominally 25 kPa). At temperatures just above this limit, the flow in the combustion chamber became unsteady as a result of thermal choking. The lowest temperature at which burning was observed was very pressure sensitive when combustion chamber intake pressures were below 25 kPa; at an intake pressure of 16 kPa, the fuel would not burn in the combustion chamber for temperatures less than 700 K, which is over 300 higher than that measured at 25 kPa.

The ignition properties for different concentrations of silane at different freestream temperatures were also investigated by Morris.[21,53] Freestream temperatures of 350, 410, 540, 650, and 1000K and fuels constituted with 2.5, 5, 10, and 20% by volume of silane with hydrogen making up the balance were studied for both ignition delay and the overall increase in pressure produced by combustion. The freestream pressure was 20 kPa.

The observation was made that at a freestream temperature of 1000 K the addition of silane did not produce a significant change in the pressure, but as the concentration of silane increased, the ignition delay decreased marginally. A similar result occurred at 650 K although the ignition delay was significantly decreased between 2.5 and 5%. At 540 K the lower concentrations of silane unexplainably produced higher pressures after combustion, and the ignition delay was significantly greater than at the higher temperatures. At 410 K the flows were unsteady and at 350 K there was no ignition.

In Sec. VI hydrogen could not be made to burn when the temperature was less than 850 K. However, the results from Morris show that by adding silane to hydrogen burning was observed at as low as 410 K. The conclusion can be made from these results that silane is a useful additive to produce ignition in hydrogen when the intake temperatures of the combustion chamber are below that where spontaneous combustion of hydrogen would normally occur. Furthermore, the addition of silane will also decrease the ignition delay time. This is an important feature of silane, and it will be seen in the next section that this feature was used to advantage when small completely integrated scramjets were tested.

IX. Integrated Scramjet Measurements

The experiments already discussed in this chapter were fundamental in nature. They were designed to explore different aspects affecting scramjet performance. There was no attempt to measure the overall performance of an integrated scramjet, involving an intake, combustion chamber, and thrust nozzle. This is clearly the ultimate objective of a scramjet experiment.

The difficulty associated with measurement of thrust developed by an integrated scramjet in an impulse facility is that conventional force balances cannot be used, as they will not reach static equilibrium during the test time. To overcome this difficulty, Sanderson and Simmons (1991) have developed a force balance that uses the time history of stress waves transmitted into a long sting (or stress bar) from a model which is subjected to a load. This allows the force on a model to be measured without requiring static equilibrium to be achieved in the model or the sting. Initially, the drag produced by a small cone was measured. In this case the cone was sufficiently short so that static equilibrium was achieved in the model, but it was not achieved in the stress bar. This work was then extended by Tuttle (1990), and the drag on a long cone was measured. In this case static equilibrium was not achieved in either the model or the stress bar, but the load distribution on the model was almost uniform. The stress wave force balance was subsequently used to measure the drag on a long blunted cone (Porter *et al.* (1994)) where the load distribution was no longer uniform and neither the model nor the stress bar were in static equilibrium. It has now been developed to the point where the load on a complex and relatively large model such as a scramjet can be measured. The drag on an unfueled scramjet was first measured by Porter (1995). Shortly afterwards, the same model was connected to a fuel tank, and the thrust developed by a fueled scramjet was measured by Paull *et al.* (1995a,b). Figure 20 is a picture of the first (in the open literature) recorded simulated flight of a scramjet model in which a net positive thrust was measured (Paull *et al.* (1995a)).

A schematic of the scramjet, fuel tank, and force balance is given in Figs. 21 and 22. The scramjet was conceptually complete, in the sense that it had intakes, combustion chambers, and thrust surfaces. It was approximately 300 mm long and had a diameter of 67 mm. The forebody was a cone with a 9-deg semivertex angle. Six identical combustion chambers were evenly distributed around the outside. The combustion chambers were constant in area and had lengths of 59 mm. Fuel was injected from a 2-mm orifice located at the upstream end of each combustion chamber. The conical afterbody of the scramjet had a semivertex angle of 10 deg and was terminated when the radius was 10 mm. A cylindrical section then connected the

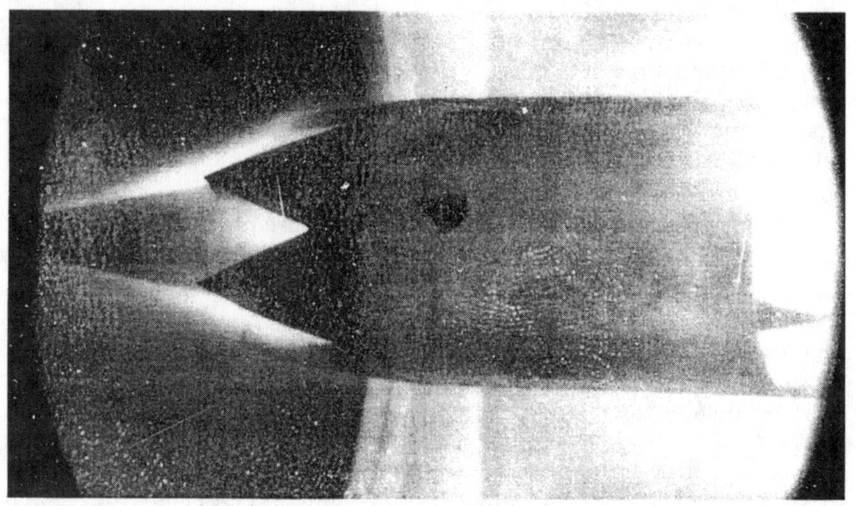

Fig. 20 The first recorded flight of a scramjet model in a ground-based facility.

scramjet's afterbody with the fuel tank and valve. The fuel tank and valve were shielded from the flow by a shroud that sealed on the cylindrical connecting pipe, so that the force which was measured was that only on the scramjet and the pipe. The drag on this pipe was negligible.

Originally, hydrogen was used as the fuel; however, the combustion chambers were found to be too short for combustion to occur. To promote faster ignition, a fuel comprising 13% silane (SiH_4) (by volume) with hydrogen making the balance was used. The results of these experiments are displayed in Figs. 23 and 24. In Fig. 23 three different traces can be seen. These traces correspond to the net force on the scramjet when 1) air was the test gas and fuel was not injected, 2) nitrogen was the test gas and fuel was injected, and 3) air was the test gas and fuel was injected (equivalence ratio of 0.8).

Trace 1) gives the total drag on the model. When the fuel is injected, there is a net increase in thrust because the fuel is injected downstream. This

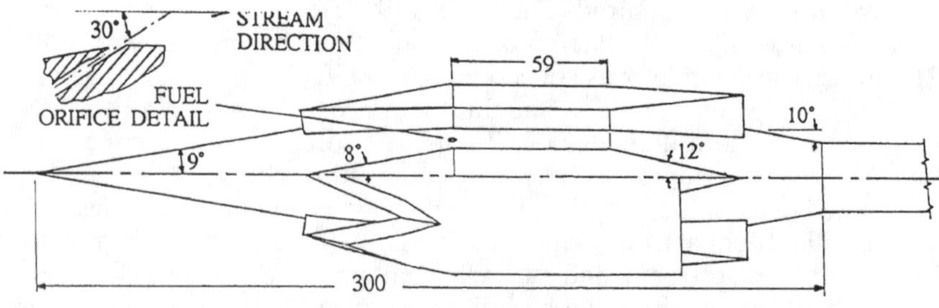

Fig. 21 The integrated scramjet model.

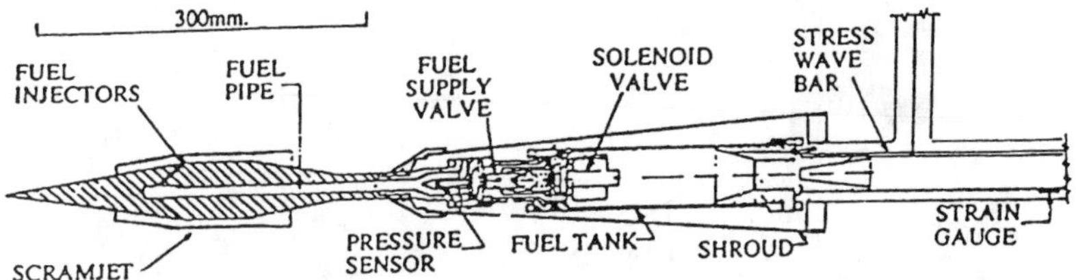

Fig. 22 Schematic of scramjet model, fuel system, and force balance.

increase in thrust can be estimated from trace 2. As the test gas is nitrogen, combustion will not occur; hence, the decrease in drag results primarily from the interaction of the fuel with the test gas. (There was some variation in the freestream conditions between the air and nitrogen test gases that also contributes slightly to the difference.)

Finally, trace 3 displays the net thrust produced by the scramjet when combustion occurs. As the flow first establishes itself over the scramjet, the drag on the model is similar to that experienced when nitrogen was used. However, after approximately 300 ms the effect of combustion can be seen. The thrust produced by the combustion process is sufficient to offset the drag, and a net thrust of approximately 50 N is obtained. The mass flow rate of the fuel was monitored, from which the incremental specific impulse was determined to be 150 s.

Figure 24 displays the decrease in the net thrust produced as the total enthalpy of the test gas is increased. The thrust decreases as the total enthalpy increases because of the following:

1) The temperature of the air at the entrance of the combustion chamber increases, and therefore the amount of energy released from the combustion decreases (see Sec. VI).
2) The amount of energy released from the fuel becomes a smaller percentage of the total energy of the air flow, so that the left-hand coefficient in Eq.

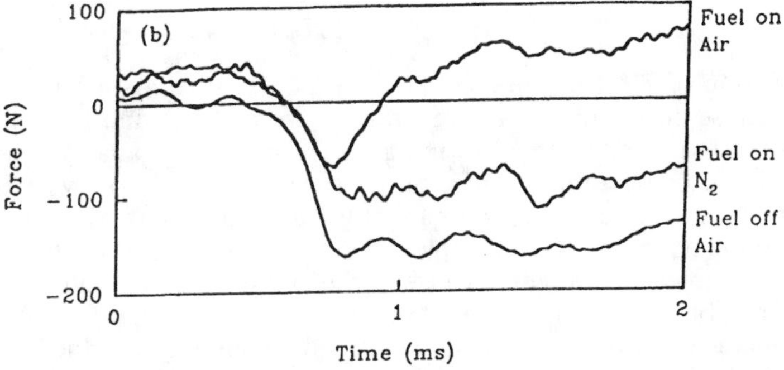

Fig. 23 Force balance measurements for the integrated scramjet.

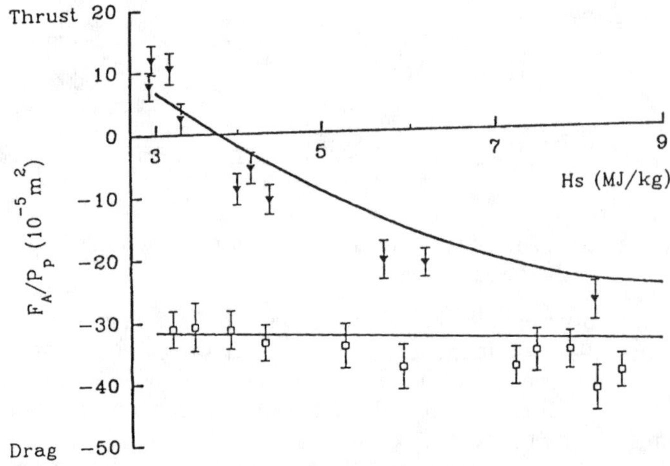

Fig. 24 Measured and predicted thrust/drag as a function of total enthalpy for the integrated scramjet.

(8) is reduced; therefore, the thrust (or force) produced by the engine is also reduced.

These results simply reflect what was observed in the generic models that were used in Sec. VI.

Theoretical predictions of the thrust/drag have also been made by the Paull, *et al.* (1995b), and are displayed in Fig. 24. The predictions of the inviscid thrust were made assuming that the fuel was completely mixed and the fuel-air mixture had reacted and reached equilibrium. The assumption was made that the silane burned according to the reaction

$$SiH_4 + 2O_2 \rightarrow SiO_2 + 2H_2O$$

and the hydrogen according to the reaction

$$2H_2 + O_2 \rightarrow 2H_2O$$

Dissociation of H_2O in the combustion products was taken into account, and, because postcombustion temperatures were above that for condensation of SiO_2, it was assumed to remain as a vapor. The analysis was conducted in the conventional manner, using the energy equation, with the heat of formation included in the enthalpy together with the momentum equation and the equation of state applied across a shock like discontinuity. The change in molecular weight because of the addition of fuel and the formation of combustion products were also taken into account.

Estimates of skin-friction drag were made assuming that the boundary layer on the intakes and the cowl's outer surface were laminar, but on the walls of the combustion chamber and thrust surfaces the boundary layer was assumed to be turbulent. The skin-friction coefficient for the turbulent boundary layer was determined using the theory of Spalding and Chi (1964); and the laminar skin-friction coefficient was determined using results in Hayes and Probstein (1959).

One can see that a reasonable correlation exists between the experiments and these predictions. These predictions were made using techniques that would now be considered almost analytical.

Although these experiments have shown that a net positive thrust can be obtained from a fully integrated scramjet, which in itself is exciting and encouraging, an important aspect of these experiments is that it has now been demonstrated that the net force on an intricate model can be measured in an impulse facility. In the past the inability to use a force balance in an impulse facility has been a major limitation of these facilities. Without a force balance only limited ground testing for hypersonic flight could be accomplished. These experiments have paved the way for measuring overall forces using impulse ground testing facilities.

Research that is currently being pursued in this area involves the development of a model which is fueled by hydrogen. Silane-hydrogen fuel was used in the model just discussed primarily because the combustion chambers were too short to complete combustion. To decrease the ignition time, the pressure in the combustion chambers was increased (Pressure-scaling studies have been made by Pulsonetti and Stalker (1996)), by substantially increasing the compression ratio of the intakes. A number of failures resulted from this approach because of the boundary layer separating in the intakes and the combustion chambers, which resulted in choked flow in the combustion chambers. However, a model that had less pressure rise on the intakes and that had a more moderate contraction ratio did not choke and produced sufficient thrust to balance its drag when fueled with hydrogen (Stalker and Paull (1998)). The configuration of this cruise model is given in Fig. 25. This model compromised complete burning for the sake of maintaining supersonic com-

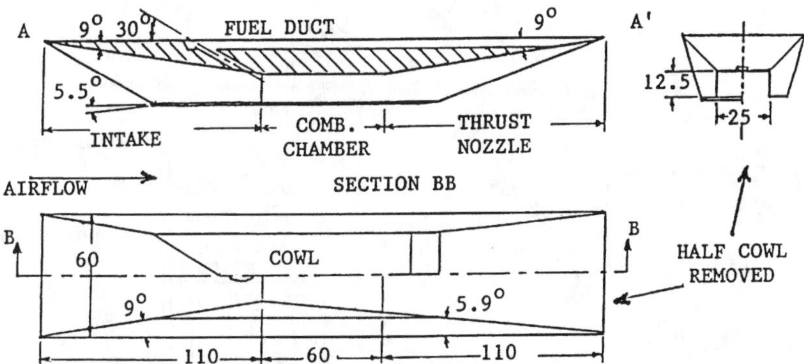

Fig. 25 Cruise model configuration. Dimensions in mm. The model tested was symmetric about AA'.

bustion. The contraction ratio was such that not all of the fuel and air had reacted in the length of the combustion chamber; however sufficient had done so to develop enough thrust to balance the total drag of the model.

The current research in this area has concentrated on reducing the combustion chamber length and its net contribution to the drag of the scramjet. Additional research is also being made to control boundary-layer separation in the combustion chamber. A review of this work can be found in Paull and Stalker (1998).

X. Skin-Friction Measurements

From the theoretical analysis of Paull *et al.* (1995b), one can clearly see that the skin-friction drag in the combustor is a major component of the total drag. It is therefore important to obtain an accurate measurement of this component.

Two approaches have been adopted to measure the skin-friction drag. The first approach was to develop a skin-friction gauge that had a response time of the order of 10–100 μs, which would measure the local skin-friction values. This type of gauge was first developed by Kelly *et al.* (1992) to measure the skin friction on a flat plate. Subsequently, Goyne *et al.* (1997) modified this gauge to one similar in design to that shown in Fig. 26. It consists of an invar cap mounted on a piezoceramic sensing element that in turn is mounted to a rigid base. Below the base is a mirror configuration of the sensing element and cap. This part of the gauge is not subjected to the flow, and the signal from this sensor is used for vibration compensation.

The gauges have been mounted at different locations in a combustor of a scramjet where the flow was either laminar, fully turbulent, or in transition.

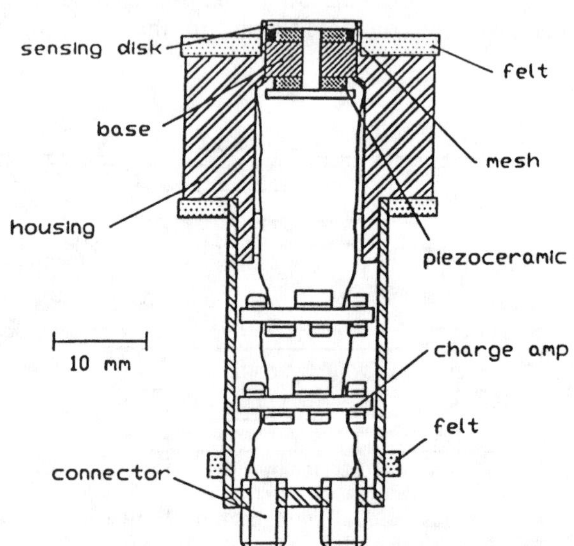

Fig. 26 Schematic of skin friction developed by Goyne *et al.* (1997).

One of the results obtained by Goyne is shown in Fig. 27 where the coefficient of skin friction is plotted against the Reynolds number based on the distance from the leading edge of the combustor. The total enthalpy of the flow is nominally 3.5 MJ/kg. In the laminar boundary layer the theory of van Driest (1952) fits the data reasonably well, whereas in the turbulent region, the theory of van Driest (1956) with a virtual origin recommended by Bertram and Cary (1974) best fits the data.

The second approach adopted to measure the skin-friction drag of a combustor was to couple the force balance used for the integrated scramjet measurements to a combustor that was decoupled from the rest of the model. In this way the total drag on the combustor alone could be measured. As the only force on the combustion chamber in this configuration is the skin-friction drag, an integrated measurement that avoids any inaccuracies which may be associated with local variations can be obtained. The technique was implemented by Tanno and Paull (1997). Figure 28 shows the time history of the skin-friction load on a circular combustor (ID = 33 mm, 500 mm long) for different equivalence ratios. The fuel was injected perpendicularly from holes upstream of the combustor. The results have been compared with the theory of van Driest (1952). The interesting point to note in these results is that there is very little variation in the skin-friction load with the amount of fuel injected. This is consistent with the theoretical analysis given by Paull *et al.* (1995b) who also used the theories of van Driest.

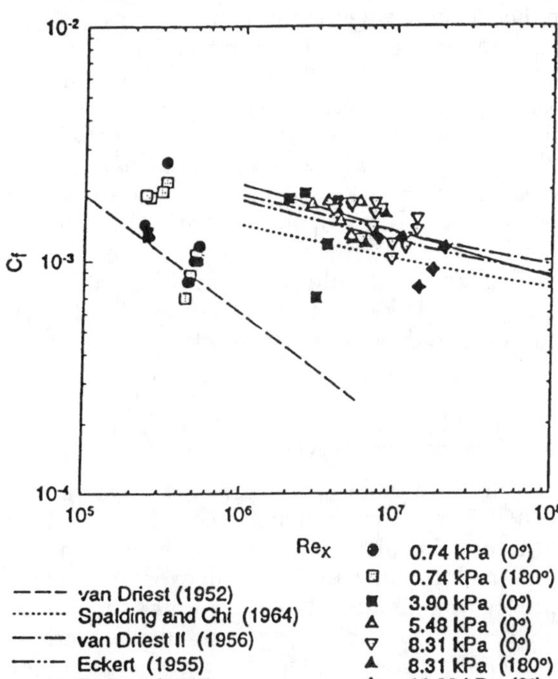

Fig. 27 Measured skin-friction coefficients along an instrument flat plate in T4. Nominal freestream total enthalpy of 3.5 MJ/kg. Goyne *et al.* (1997)

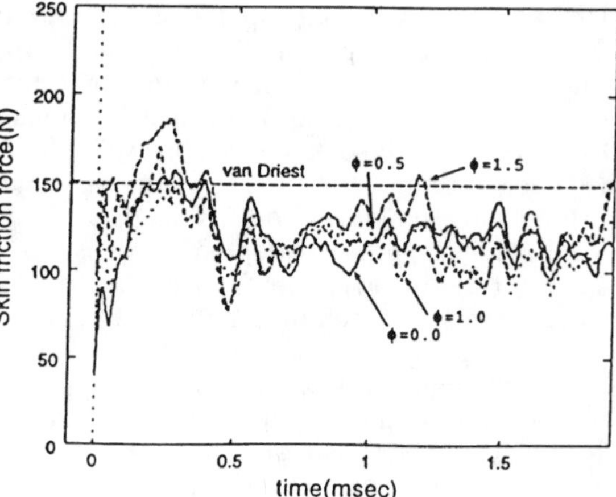

Fig. 28 Time history skin-friction measures in a scramjet combustor for different equivalence ratios made using a force balance. Entrance static pressure 85kPa, Mach number 4.3, velocity 3.3km/s, and temperature 1600 K.

The lack of variation in the skin-friction drag occurs because both the coefficient of skin friction for a turbulent boundary layer and the dynamic pressure are reasonably constant throughout the combustor. The lack of variation of the skin-friction coefficient with changing freestream conditions along the combustion chamber occurs because the skin-friction coefficient is inversely proportional to the Reynolds number raised to the power of 0.2. So significant variations in the freestream conditions can be tolerated before any significant change in the skin-friction coefficient is observed. Furthermore, as the flow velocity is high, the dominant term in the total enthalpy is the velocity, and as expected with high enthalpy flows the velocity will not change significantly, even though large variations in pressure and temperature occur along the combustion chamber. Hence, because the duct is of constant area, the conservation of mass [Eq. (2)] implies that there is also no significant change in the density. Thus, the combustor runs at a nearly constant density and velocity throughout its length, and therefore the dynamic pressure is also essentially constant. The net result is that only small variations in the skin friction are observed when fuel is injected into the combustion chamber.

XI. Discussion and Review

As the T3 and T4 facilities are located at universities, the research into scramjets has been approached from a fundamental point of view, rather than being constrained to the development of a particular configuration, which has lead to the development of a variety of different experiments that either provide empirical information about well-defined topics or provide an insight into important issues and possibly the physics that underpins them. Many of the experiments often generated more questions than they answered. However, some answers have been obtained regarding the combustion of hydrogen, and techniques have been developed for measuring many of the important parameters required for scramjet research.

Experiments have matured to the point where fully integrated scramjets are now being tested on a routine basis. This is a significant advance, as it has demonstrated that flight simulation can now be undertaken using impulse facilities rather than developing flight-test programs to resolve all difficulties. Thus, a relatively affordable means is available that can be used to test prototypes before they are subjected to the rigors and cost of a flight-test program. This should significantly reduce the cost of developing scramjets.

Apart from these tests, considerable experimentation has been completed from which an appreciation of some of the difficulties associated with burning hydrogen in a constant-area combustion chamber has been obtained. The effectiveness of different injection systems has been evaluated, and analytical results have been developed that describe their behavior. The effect on the performance of a scramjet as the flight speed is increased is also understood, and the experiments show trends that are consistent with what is expected theoretically.

There is still a lot of work that is required to be done to develop an efficient scramjet because the fully integrated experiments developed an incremental specific impulse of 150 s, which, although this was disproportionally low because silane was added, the generic tests still indicate that a incremental specific impulse in excess of 1500 s should be obtained with hydrogen. Thus, an order of magnitude increase in performance may be possible. The goal is therefore clear.

Acknowledgments

This work was completed under grants from the Australian Research Council, The Australian Space Office, and NASA Langley Research Center. The work has also be on possible through numerous scholarships from The University of Queensland and the Australian Postgraduate Awards.

Bibliography

Allen, Jr., G. A., "Parametric Study on Thrust Production in the Two-Dimensional Scramjet, Shock Tunnel Studies for Scramjet Phenomena," Supplement 6, NASA CR 191428, 1993, pp. 129–136; also AIAA Paper 91-0227, 1991.

Anderson, Jr., J. D., *Hypersonic and High Temperature Gas Dynamics,* McGraw–Hill, New York, 1989.

Anderson, G. Y., private communication, NASA Langdey Research Center, 1988.

Anderson, G. Y., Kumar, A., and Erdos, J., "Progress in Hypersonic Combustion Technology with Computation and Experiment," AIAA Paper 90-5254, 1990.

Bakos, R. J., "An Investigation of Test Flow Nonequilibrium Effects on Scramjet Combustion," Ph.D. Thesis, Univ. of Queensland, Australia, 1994.

Bakos, R. J., Castrogiovanni, A., Calleja, J. F., Nucci, L., and Erdos, J. I., "Expansion of the Scramjet Ground Test Envelope of the HYPULSE Facility," AIAA Paper 96-4506, 1996.

Bakos, R. J., and Morgan, R.G., "Axisymmetric Scramjet Thrust Production," *Proceedings of the 11th Aust. Fluid Mech. Conf.,* 1992, pp. 295–298.

Bertram, A. M. and Cary, M. H. "Engineering Prediction of Turbulent Skin Friction and Heat Transfer in High Speed Flow," NASA TN D-7507, 1974.

Boyce, R. R., "Mass Spectrometric Measurements of the Freestream Flow in the T4 Free-Piston Shock-Tunnel," *Proc. 21st Int. Symp. on Shock Waves,* 1997, pp. 429–434. (Paper 2899).

Brescianini, C., "Wall Injected Scramjet Experiments, Shock Tunnel Studies for Scramjet Phenomena," Supplement 5, NASA CR 182096, 1990, pp. 15–19.

Brescianini, C. P., "An Investigation of the Wall-Injected Scramjet," Ph.D. Thesis, Univ. of Queensland, Australia, 1993.

Brescianini, C., Stalker, R. J., and Morgan, R. G., "Shock Tunnel Studies for Scramjet Phenomena," Supplement 6, NASA CR 191428, 1993, pp. 75–112.

Buttsworth, D. R., "Shock Induced Mixing and Combustion in Scramjets," Ph.D. Thesis, Univ. of Queensland, Australia, 1994.

Buttsworth, D. R., "Interaction of Oblique Shock Waves and Planar Mixing Regions," *Journal of Fluid Mechanics,* 61, 306, 1996, pp. 43–57.

Casey, R. T., Stalker, R. J., and Brescianini, C., "Hydrogen Combustion in a Hypersonic Airstream," *J. Aero. Soc.,* May, 1992, pp. 200–202.

Casey, R. T., and Stalker, R. J., "Hydrogen Mixing and Combustion in a High Enthalpy Hypersonic Stream," *Proceedings of the 19th International Symposium on Shock Waves,* Vol. 7, Springer–Verlag, Berlin, 1995, pp. 151–156.

Daniels, B. and Alsop, B. The Dept. of Mechanical Engineering, Univ. of Queensland, Australia.

Eckert, E. R. G., "Engineering Relations for Friction and Heat Transfer to Surfaces in High Velocity Flow, *J. Aero Sc.,* 22 1955, pp. 585–587.

Goyne, C. P., Stalker, R. J., and Paull, A., "Skin Friction Measurements in the T4 Shock Tunnel," *Proceedings of the 21st International Symposium on Shock Waves* 1997, pp. 1125–1130 (Paper 2480).

Hayes, W. D., and Probstein, R. F., *Hypersonic Flow Theory,* Academic, 1959, p. 296.

Huber, P. W., Schexnayder, C. J., Jr., and McClinton, C. R., "Criteria for Self-Ignition of Supersonic Hydrogen-Air Mixtures," NASA TP 1457, 1979.

Kelly, G. M, Simmons, J. M., and Paull, A., "Skin Friction Gauge for Use in Hypervelocity Impulse Facilities," *AIAA Journal,* Vol. 30, No. 3, 1992, pp. 844, 845.

Kerrebrock, J. L., "Some Readily Quantifiable Aspects of Scramjet Engine Performance," *Journal of Propulsion and Power,* Vol. 8, No. 5, 1992, pp. 1116–1122.

Lebo, I. G., Rozanov, V. B., Tishkin, V. F., and Favorsky, A. P., "Numerical Simulations of Richtmyer–Meshkov Instability" *Proceedings of the 20th International Symposium on Shock Waves,* Vol. 1, 1995 pp. 605–610.

Lewis, M. J., and Gupta, A., "Impact of Fuel Selection on Hypersonic Vehicle Optimization," *Proceedings 13th ISABE,* edited by F. Billig, Vol. 2, 1997, pp. 1456–1463.

Lordi, J. A., Mates, R. E., and Moselle, J. R., "Computer Program for the Numerical Solution of Non-Equilibrium Expansions of Reacting Gas Mixtures," Cornell Aeronautical Lab., CAL Report AD-1689-A-6, 1965.

McIntosh, M. K., "Computer Program for the Numerical Calculation of Frozen and Equilibrium Conditions in Shock Tunnels," Dept. of Physics, ANU, Australia, 1968.

McIntyre, T. J., Rabbath, P. A. B., and Houwing, A. F. P., "Imaging of Combustion Processes in a Supersonic Combustion Ramjet," *Proceedings of the 12th ISABE,* Vol. 2, 1995, pp. 1163–1173.

Mee, D. J., Daniel, J. T., and Simmons, J. M., "Three-Component Force Balance for Flows of Millisecond Duration," *AIAA Journal,* Vol. 34, No. 3, 1996, pp. 590–595.

Morgan, R. G., Paull, A., Morris, N., and Stalker, R. J., "Scramjet Sidewall Burning—Preliminary Shock Tunnel Results," Dept. of Mechanical Engineering, Univ. of Queensland, Australia, Dec. 1985.

Morgan, R. G., Paull, A., Stalker, R. J., Jacobs, P., Morris, N., Stringer, I., and Brescianini, C., "Shock Tunnel Studies of Scramjet Phenomena," Supplement 3, Dept. of Mechanical Engineering, Univ. of Queensland, Australia, 1987.

Morgan, R. G., "Supersonic Combustion with Transverse, Circular Wall Jets, Shock Tunnel Studies for Scramjet Phenomena," Supplement 5, NASA CR 182096, 1990, pp. 20–39.

Morgan, R. G., and McGregor, W., "Liquid Fuelled Scramjet, Shock Tunnel Experiments," presented at 12th ISABE, 1995.

Morris, N., Morgan, R. G., Paull, A., and Stalker, R. J., "Silane as an Ignition Aid in Scramjets," AIAA Paper 87-1636, June 1987.

Morris, N. A., "Silane as an Ignition Aid in Scramjets," Ph.D. Thesis, Univ. of Queensland, Australia, 1989.

Northam, G. B., and Anderson, G. Y., "Supersonic Combustion Ramjet Research at Langley," AIAA Paper 86-0159, 1986.

Oldenberg, R., Chintz, W., Friedman, Jaffe, R., Jachimowski, C., Rabinowitz, M., and Schott, G., "Hypersonic Combustion Kinetics," Status rept. of the Rate Constant Committee, NASP High Speed Propulsion Technology Team, NASP TM-1107, 1990.

Osgerby, I. T., Smithson, H. K., and Wagner, D. A., "Supersonic Combustion Test with a Double-Oblique-Shock Scramjet in a Shock Tunnel," *AIAA* Journal, Vol. 8, No. 9, 1970, pp. 1703–1705.

Paull, A., "Hypersonic Ignition in a Scramjet," *11th Aust. Fluid Mech. Conf.,* Vol. 1, 1992, pp. 423–426.

Paull, A., "Hypersonic Ignition and Thrust Production in a Scramjet," AIAA Paper 93-2444, June 1993.

Paull, A., Stalker, R. J., and Mee, D. J., "Scramjet Thrust Measurement in a Shock Tunnel," *Aero. J. R. Aero. Soc.,* April 1995, pp. 161–163.

Paull, A., Stalker, R. J., and Mee, D. J., "Experiments on Supersonic Combustion Ramjet Propulsion in a Shock Tunnel," *Journal of Fluid Mechanics,* Vol. 296, 1995, pp. 159–183.

Paull, A., "Report on Hydrocarbon Scramjet Combustion Tests," Dept. of Mechanical Engineering, Univ. of Queensland, Australia, April 1995.

Paull, A., "A Simple Shock Tunnel Driver Gas Detector Shock Waves," *Shock Waves Journal,* Vol. 6, No. 5, 1996, pp. 309–312.

Paull, A., and Stalker, R. J., "Scramjet Testing in the T4 Impulse Facility," AIAA 98-1533, 1998.

Porter, L. M., Paull, A., Mee, D. J., and Simmons, J. M., "Shock Tunnel Measurements of Hypervelocity Blunted Cone Drag," *AIAA Journal,* Vol. 32 No. 12, 1994, p. 2476.

Porter, L. M., "High Enthalpy, Hypersonic Drag Measurements on Blunt Cones in an Impulse Facility," Ph.D. Thesis, Univ. of Queensland, Australia, 1995.

Pulsonetti, M., "Scaling and Ignition Effects in Scramjets," *Proc. 11th Aust. F. Mech. Conf.* Vol. 1, 1993. pp. 431–434.

Pulsonetti, M. V., and Stalker, R. J., "A Study of Scramjet Scaling," AIAA Paper 96-4533, 1996.

Sanderson, S. R., and Simmons, J. M., "Drag Balance for Hypervelocity Impulse Facilities," AIAA Journal, Vol. 29, No. 12, 1991, p. 2185.

Skinner, K. A., and Stalker, R. J., "A Time of Flight Mass Spectrometer for Impulse Facilities," *AIAA Journal,* Vol. 32, No. 11, 1994, pp. 2325–2328.

Spalding, D. B., and Chi, S. W., "The Drag on a Compressible Turbulent Boundary Layer on a Smooth Flat Plate with and Without Heat Transfer," *Journal of Fluid Mechanics,* Vol. 18, 1964, pp. 117–143.

Stalker, R. J., Morgan, R. G., Paull, A., and Brescianini, C. P., "Scramjet Experiments in Free Piston Shock Tunnels," NASP CR 1100, 1990.

Stalker, R. J., and Morgan, R. G., "Supersonic Hydrogen Combustion with a Short Thrust Nozzle," *Combustion and Flame,* Vol. 57, 1984, pp. 55–70.

Stalker, R. J., Morgan, R. G., and Netterfield, M. P., "Wave Processes in Scramjet Thrust Generation," *Combustion and Flame,* Vol. 71, 1988, pp. 63–77.

Stalker, R. J., and Paull, A., "Experiments on Cruise Propulsion with a Hydrogen Scramjet," *Aero. J. R. Aero. Soc.,* Vol. 102, Jan. 1998, pp. 37–43.

Tanno, H., and Paull, A., "Skin Friction Measurement of a Supersonic Combustor with Stress Wave Force Balance," *Proceedings of the 21st International Symposium on Shock Waves,* Vol. 2, 1997, pp. 1131–1136 (Paper 5640).

Tuttle, S. L., "A Drag Measurement Technique for Hypervelocity Impulse Facilities," M. Sc. Thesis, Dept. of Mechanical Engineering, Univ. of Queensland, Australia, 1990.

van Driest, E. R., "Investigation of Laminar Boundary-Layer in Compressible Fluids Using the Crocco Method," NACA TN-2597, 1952.

van Driest, E. R., "The Problem of Aerodynamic Heating," *Aero Eng Rev,* Vol. 15, 1956, pp. 26–41.

Weinstein, A. S., Osterle, J. F., and Forstall, W., "Momentum Diffusion from a Slot Jet into a Moving Secondary," *Journal of Applied Mechanics,* Sept. 1956, pp. 437–443.

Wendt, M. N., and Stalker, R. J., "Effect of Fuel Stagnation Temperature on Supersonic Combustion with Transverse Injection, *11th Aust. F. Mech. Conf,* 1992, pp. 439–441.

Wendt, M. N., and Stalker, R. J., "Transverse and Parallel Injection of Hydrogen with Supersonic Combustion in a Shock Tunnel," *Shock Waves,* Vol. 6, No. 1, June 1996, pp. 53–59.

Chapter 2

Scramjet Developments in France

F. H. Falempin*

Aerospatiale Matra Missiles, Chatillon, France

I. Historical Overview†

IN FRANCE, as in the world, the 1950s were a very rich period for hypersonic vehicle projects. Missiles, interceptors, and bombers that used airbreathing propulsion were among the most important envisaged applications.

At the beginning of the 1960s, American studies, conducted principally by A. Ferri[2] and G.L. Dugger[3], showed the interest of the supersonic combustion for the propulsion of the hypersonic vehicles flying beyond Mach 6–7: 1) lower combustor temperature than subsonic combustion resulting in a decrease of chemical dissociation problems and 2) lower air compression corresponding to better pressure recovery and lighter mechanical loads than in a conventional ramjet.

Studies are then undertaken in France to verify that combustion is feasible in a supersonic air flow.[4] At Ecole Nationale Supérieure de Mécanique et d'Aérothermique (ENSMA - Poitiers) and at Ecole Centrale des Arts et Manufacture (ECAM - Paris) studies are focused on shock-induced combustion. This combustion mode, proposed by M. Roy[5] in 1946, had been successfully tested by ONERA between 1948 and 1953. After 1960 ONERA focused its studies on diffusion flame combustion which appeared to be more favorable than the shock-induced combustion. During the period 1962–67, combustion tests were performed at ONERA Palaiseau Center with two fuels (kerosene and gaseous hydrogen), different conditions (combustion Mach number: 2.5–3; flight Mach number: 6.4 – 11; altitude: 36 – 56 km), in two typical geometries (constant area and divergent ducts). This

*Head, Propulsion and Pyrotechnics Department.

†The historical part of this chapter was inspired by the synthesis work[1] realized by Ph. Novelli from ONERA, whom I thank for his help.

basic research demonstrated the feasibility of stable supersonic combustion and established experience in the scramjet combustor design.

At this time, Ferri's studies[6] showed supersonic combustion that defined a fixed geometry ramjet operating on a large Mach number range (the corresponding performances decrease being, a priori, balanced by the simplification of the structural design). Theoretical studies realized by Mestre and Moreau[7] prove the possibility of using supersonic combustion from Mach 4 (with some ignition problems in the meantime). Marguet and Huet[8] arrived at the same conclusion: supersonic combustion, feasible from Mach 4.5, is less efficient than continuously adapted subsonic combustion before Mach 7 but its performances decrease more slowly than the subsonic combustion performances in a fixed geometry.

Since it appears possible to successively use the two combustion modes in the same combustor, the specialists developed a ramjet in which supersonic combustion follows subsonic combustion (transition occurs at Mach 5.5) in order to obtain the maximum performances with a fixed geometry. The feasibility of this kind of ramjet depends on specific parameters (efficiency, combustion completion, thermodynamic equilibrium) hardly bounded with the engine design, which were not well known with the basic research of this time. That's why the ESOPE program was decided in December 1966.

ESOPE program is similar to the Hypersonic Research Engine (HRE) program launched by NASA. The purpose of this program is to develop an experimental significant-scale engine using the previously described principle.

Two steps are initially foreseen: 1) validation of the thrust-drag balance in supersonic combustion mode at flight Mach 7 with a first engine ESOPE A (ground testing and flight demonstration as passenger of a ballistic missile), and 2) development of the engine ESOPE C operating from Mach 3 to Mach 7 (demonstration of the transition phase and autonomous flight test).

In fact, a little part of the program is realized. Basic tests are performed at the ONERA Palaiseau Center up to Mach 7 with vitiated air and the combustor of the engine, defined with the help of theoretical and experimental results obtained by O. Leuchter, is tested at Mach 6 with non-vitiated air in connected pipe in the high enthalpy wind tunnel S4MA at ONERA Modane-Avrieux Center.

During the two-test series, performed in S4MA (1970 and 1972), ignition difficulties were encountered and only "transonic combustion" (combustion starting in subsonic and continuing in supersonic mode after a thermal throat), was initially foreseen prior to Mach 5 exploration.

After this first experience, studies on hypersonic airbreathing propulsion were adjourned for the benefit of rocket propulsion for space launchers and liquid fueled subsonic ramjets for missiles[9] for two main reasons:

1) Hypersonic flight requires the use of very high temperature materials, which were not available.
2) An important missile development program — ASMP (middle range air-to-ground missile) — summoned up a large part of the French potential.

So the test at Mach 7, which would have allowed obtaining a supersonic combustion, was never realized.

Between 1987 and 1991 the revival of international research activity in the field of hypersonic airbreathing propulsion lead to the National Space Agency (CNES) and the Ministry of Defense (DGA) to undertake, with national industry and research centers, system studies on space launchers powered by a partially airbreathing propulsion system and high-speed airbreathing military systems. At the end of these studies, the French authorities decided to launch a joint research and technology program (PREPHA),[10] mainly focused on the scramjet techniques but which also tried to reinforce the national mastery of critical technologies needed by hypersonic vehicles.

Thanks to PREPHA, a new generation of engineers can acquire a first glance at hypersonic airbreathing propulsion. PREPHA also provides the opportunity to develop a large part of the needed test facilities and numerical means. PREPHA ended in 1999 and few activities have been continued by Industry and Research Centers to preserve human and material investments and to improve the mastery of high-speed airbreathing propulsion technology for future long-term applications.

II. Basic Research on Diffusion Flame Combustion (1962–1967)

This experimental work was realized essentially by A. Mestre and P. Moreau at ONERA Palaiseau Center. The primary aim of this work was to demonstrate the feasibility of the supersonic combustion which was discussed due to the very short staying time in the combustor. The experiments allowed a first analysis of the phenomena present in a supersonic combustor, which were not accessible by the numerical codes available at the time: 1) air/fuel mixing process, 2) time of combustion ignition, and 3) fuel jets penetration.

They also permitted verification of the validity of the assumption of a globally one-dimensional flow used for the theoretical performance calculations and to evaluate the effect of parameters neglected in these calculations such as skin friction, shocks losses, and heat transfers.

Three kinds of test were performed: 1) cylindrical duct test, 2) free jet test, and 3) divergent duct test.

A. Combustion in a Cylindrical Duct

This kind of test is very interesting because the walls pressure integral is zero. Then a 0D calculation is possible without hypothesis on the pressure evolution along the combustor walls, which depends directly on air/fuel mixing and chemical kinetics.

1. General Test Description

Initially these experiments used kerosene because this fuel seemed more interesting for a potential application, particularly for missiles. Nevertheless, its self-inflammation characteristics and chemical kinetics led quickly to the

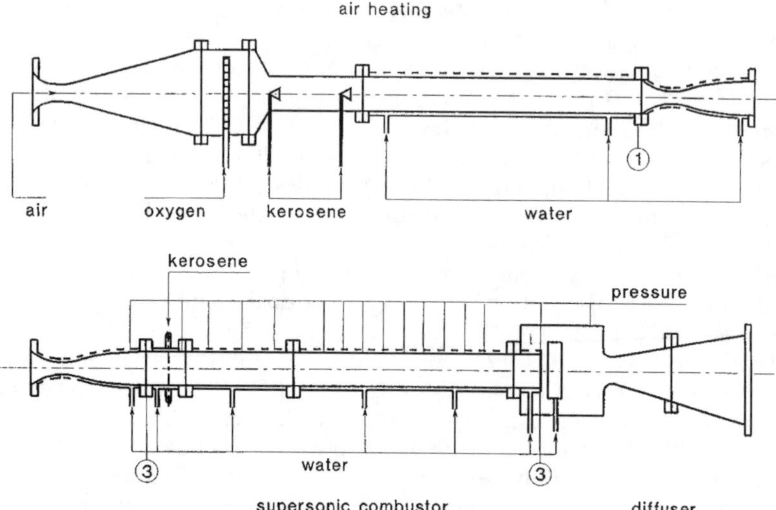

Fig. 1 Cylindrical duct—test device.

use of hydrogen, which offered, in addition, an important potential for structures cooling.

For both kerosene and hydrogen it was necessary to pre-heat the air in order to ensure spontaneous inflammation (1100 K for kerosene and 1000 K for hydrogen). The preheating was provided either by combustion of kerosene or hydrogen before reoxygenation or by an electric arc.

The test device was constituted by a cylindrical water-cooled duct following a nozzle and three test series were performed (see Table 1).

Three kinds of injection systems were studied: 1) wall injection by eight equidistant holes in the same normal section, 2) injection in the air flow by struts, and 3) wall injection by slots.

2. *Results*

The test results were compared with theoretical evaluations based on a 1D equivalent flow model, which took into account chemical dissociation

Table 1 Test Series of Cylindrical Water-cooled Duct

	Airflow				Combustor	
Fuel	Mach	Temp., K	Pressure, Pa	Preheating	Diameter, mm	Length, m
Kerosene	2.48	1,170	27,700	Kerosene	90	0.5 to 2
	2.8	2,200	25,000	Electric arc	53	0.6
Hydrogen	2.8	2,300	25,000	Hydrogen	95	0.6

and isentropic expansion with chemical equilibrium. During the first experiment (kerosene with Mach number of 2.48 at the entrance of combustor), the supersonic combustion was possible on a limited range of fuel equivalence ratio: 1) the low equivalence ratio by extinction and 2) the high equivalence ratio by the chocking of the duct created by a thermal throat following a subsonic combustion (see Table 2).

These results show the important effect of skin friction, which reduces the maximum equivalence ratio before chocking (test realized at Mach 3 without combustion indicated a loss of 11% of stream thrust at entrance of combustor which is probably higher with combustion).

The increase of the combustion efficiency contributes also to the chocking as the previous table shows: when the combustor is extended (L = 0.48 → 1.00) or the mixing accelerated (wall injection → struts injection).

With this geometry the Mach number at the end of combustor was always near 1. A divergent duct placed after the combustor allowed verifying that the flow was actually supersonic. Therefore, the heterogeneous distribution of total pressures (see Fig. 2) in the exit section makes the results analysis uncertain.

In order to realize supersonic combustion with an equivalence ratio of 1, the electric arc heater was used to reach a Mach number of 2.8 at the entrance of combustor. Supersonic combustion at ER=1 was effectively obtained with a Mach number between 1.5 and 2 at the end of combustor. Figures 3, 4, and 5 show typical pressure evolution along the duct. In Figure 5, for example, the highest curve corresponds with chocking. Pressure presents a negative gradient characteristic of a subsonic combustion. The other curves present three parts: 1) a part with a weak pressure increase, which indicates a supersonic flow with skin friction; 2) a part with high positive gradient of pressure, which corresponds with supersonic combustion; and 3) a final part where pressure increases again slowly, in which the combustion is achieved and the flow is supersonic with skin friction.

By analysis of these curves it is possible to define experimental lengths of ignition and mixing/combustion. These lengths are of course dependant on the test conditions, but they illustrate the influence of some parameters on combustion process.

So the ignition length for kerosene depends dramatically on the equiva-

Table 2 First Kerosene Experiment

	Wall injection		Struts injection	
L, m	ER[a] extinction	ER chocking	ER extinction	ER chocking
0.48	0.44	0.69	0.44	0.56
0.77	0.33	0.53	0.33	0.46
1.00	0.30	0.47	0.25	0.44

[a]ER = equivalence ratio.

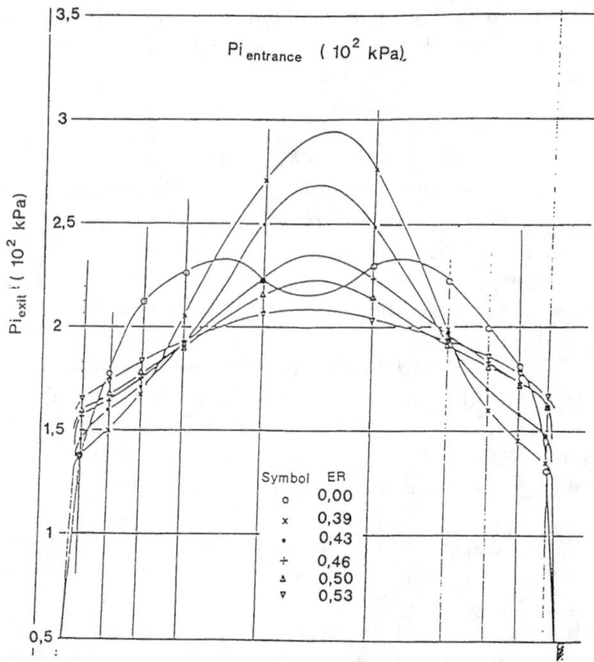

Fig. 2 Cylindrical duct - total pressure distribution along diameter of combustor exit (kerosene – L = 0.77 m – wall injection – $\mathbf{Pi_{entrance}}$ = 5.10² kPa).

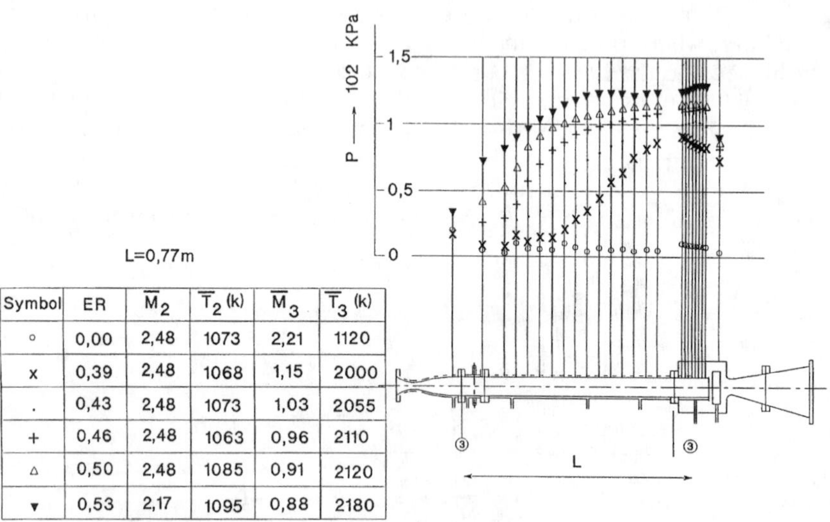

Symbol	ER	$\overline{M}_2$	$\overline{T}_2$ (k)	$\overline{M}_3$	$\overline{T}_3$ (k)
o	0,00	2,48	1073	2,21	1120
x	0,39	2,48	1068	1,15	2000
.	0,43	2,48	1073	1,03	2055
+	0,46	2,48	1063	0,96	2110
▵	0,50	2,48	1085	0,91	2120
▾	0,53	2,17	1095	0,88	2180

Fig. 3 Cylindrical duct - pressure distribution along the combustor (kerosene – L = 0.77 m – wall injection – $\mathbf{Pi_{entrance}}$ = 5.10² kPa).

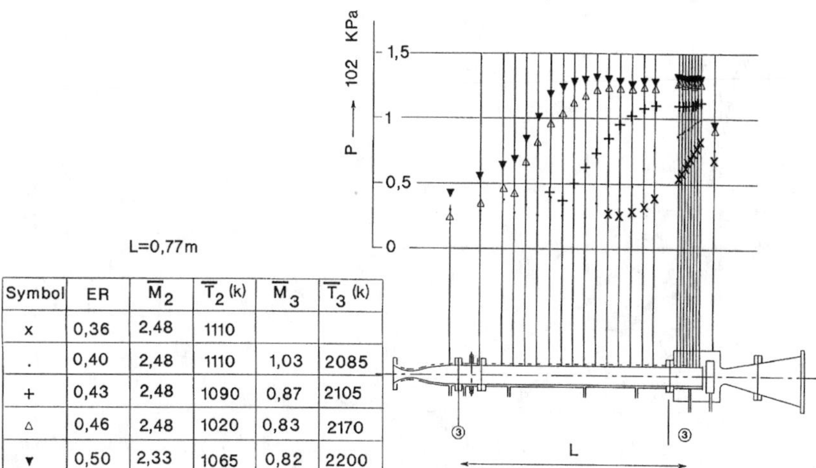

Symbol	ER	$\overline{M}_2$	$\overline{T}_2$ (k)	$\overline{M}_3$	$\overline{T}_3$ (k)
x	0,36	2,48	1110		
.	0,40	2,48	1110	1,03	2085
+	0,43	2,48	1090	0,87	2105
▵	0,46	2,48	1020	0,83	2170
▾	0,50	2,33	1065	0,82	2200

Fig. 4 Cylindrical duct - pressure distribution along the combustor (kerosene – L = 0.77 m – wall injection – $Pi_{entrance}$ = 5.10² kPa).

lence ratio (see Fig. 6). This ignition length decreases quickly with the temperature. For the test realized with electric arc heater, the combustion starts immediately after the injection. Moreover the mixing/combustion length is reduced from 500 mm (200 mm for ignition and 300 mm for combustion) to 300 mm in spite of the increase of flow speed (1600 $ms^{-1} \rightarrow 2450\ ms^{-1}$).

The effect of injection device appears on Figure 6, which compares mix-

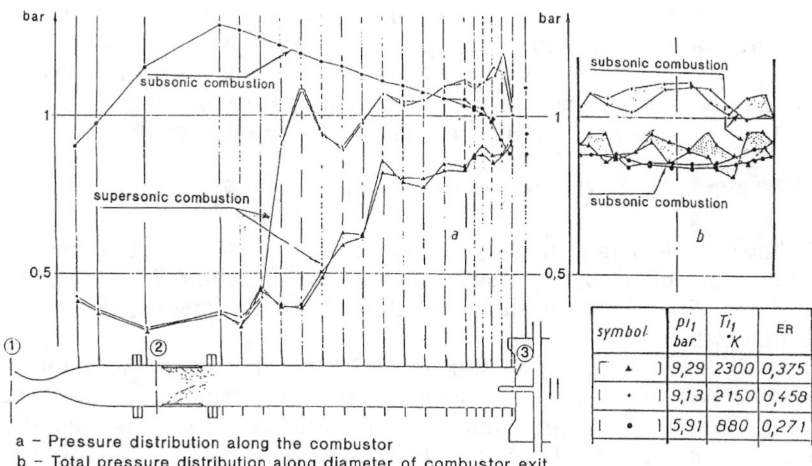

symbol	pi_1 bar	Ti_1 °K	ER
▲	9,29	2300	0,375
·	9,13	2150	0,458
•	5,91	880	0,271

Fig. 5 Cylindrical duct—pressure distribution (hydrogen-wall injection).

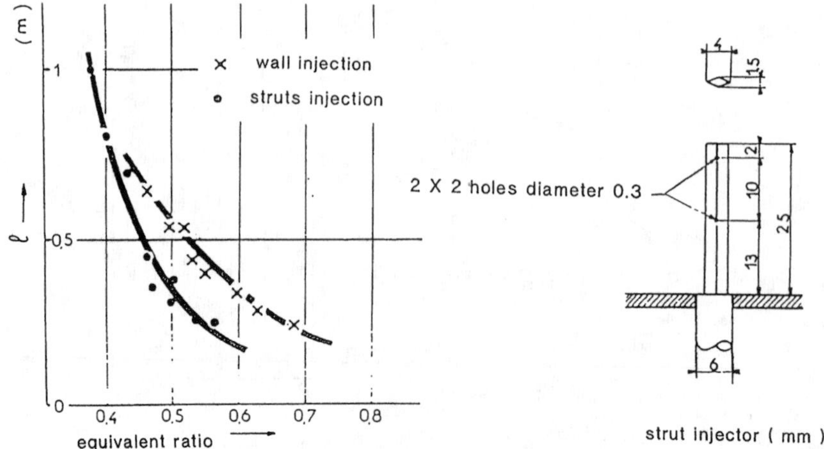

Fig. 6 Mixing/combustion length as function of equivalence ratio.

ing/combustion lengths for wall injection and struts injection (test at Mach 2.48).

For hydrogen, the pressure evolution is not so regular because of the shocks created by the gaseous hydrogen injection which corresponds with a much higher speed and, consequently, important momentum. The total mixing/combustion length is near 300 mm. When the equivalence ratio reaches 0.5, a very high gradient of pressure just after the injection shows a very early ignition.

The test results have been analyzed with the help of experimental data published by B.P. Mullins[11] and M.W. Patch and N. Montchiloff[12] concerning the ignition times of kerosene and hydrogen as a function of temperature, pressure, equivalence ratio, etc.... (these data give only tendencies because the ignition time is directly dependant on experimental conditions and particularly injection geometry). In the first approximation these experimental results showed that the ignition time follows the law:

$$D_i = Ke^{-mT}/P^n \qquad \text{with } 1 < n < 2$$

The available data were used to determine the ignition length in the case of kerosene combustion at Mach 2.48 with wall injection. With the conditions $T = 1200$ K, the calculated length is 12 m even when test indicates a length of 0.5 m.

Considering the flow conditions in the boundary layer (higher temperature and slower speed than in unviscous flow) the same calculation gives 0.42 m. This result confirms that the ignition takes place in the boundary layer: the interaction between the shocks system, issued from the injection, and the boundary layer induces temperature rising which reduces the ignition time.

With regards to jet penetration, a slot is more efficient than a circular hole or perpendicular slot but the created shock-boundary layer is not so heavy.

Concerning struts injection, the boundary layer is not as thick as on the combustor walls. Therefore, the shocks system, generated by the streamlined shape of the strut, is essential if the flow conditions before struts do not allow spontaneous ignition. Experiments realized with very thin lenticular struts by A. Mestre led to ignition difficulties with low equivalence ratio.

B. Freejet Test

These experiments were performed to compare different kinds of injection systems and to study the fuel penetration in the main airflow. Because of the limited analysis means at this time, their results are essentially qualitative, but allowed a better understanding of the ignition phenomena.

1. General Test Description

The test facility was roughly the same as in the cylindrical duct test. The nozzle emerged in a box fitted with windows in which the pressure is adjusted to modify the divergence of the jet. The fuel used was gaseous hydrogen.

2. Results Examples

Figure 7 shows the penetration of wall jets coming from slots. When the box pressure is low, the flow is very stratified. The penetration increases with the length of slots and/or the injection pressure. In some cases, it is possible to obtain the junction of the mixed jets on the center line.

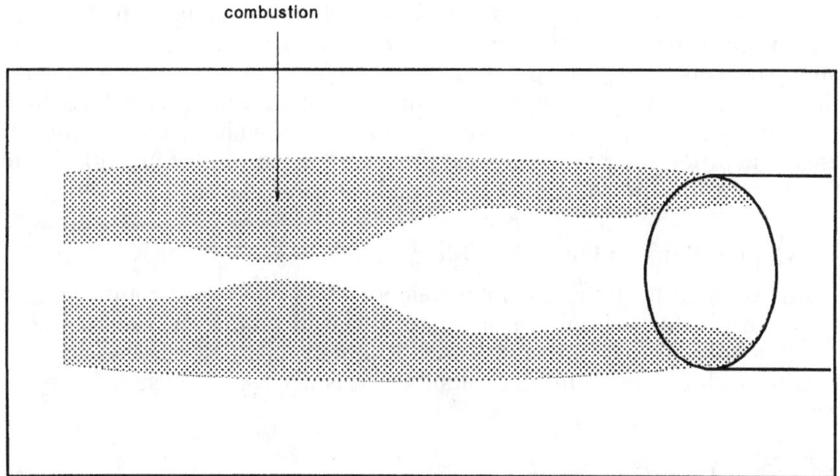

Fig. 7 Free jet test—wall injection (schematic view derived from original photography).

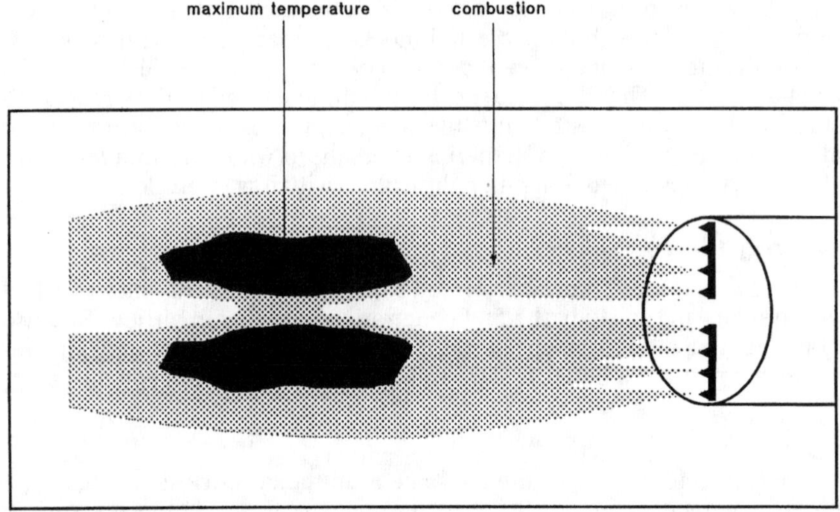

Fig. 8 Free jet test—struts injection (schematic view derived from original photography).

Struts injection allows a deeper penetration because the fuel is directly laid on the airflow. Nevertheless, as Fig. 8 shows, the jet's divergence is weak and it is necessary to inject the fuel in the whole section (it is worth noting that during this test, the hydrogen mass flow was sufficient to ensure the struts cooling).

These experiments were also used to analyze the effect of divergence on combustion ignition. Figure 9 shows a case where the expansion is so high that it stops the combustion process by cooling. The combustion restarts after Mach disc. The jet divergence angles measured by this test characterize the turbulent diffusion of hydrogen during the combustion. They were able to define the best spacing of two consecutive injection holes for the ESOPE engine in order to guarantee a good fuel distribution and the junction of the flame fronts in a given length of the combustor.

C. Combustion in a Divergent Duct

It is difficult to obtain high equivalence ratio supersonic combustion in a cylindrical duct without an increase in the incoming air Mach number which involves heavy pressure losses. To obtain high ER supersonic combustion with an optimum Mach number, it is necessary to use a divergent combustor.

1. General Test Description

The combustor was axisymmetric and burned kerosene (see Fig. 10). The exit to the entrance area ratio was 2.72. The injector was initially placed on

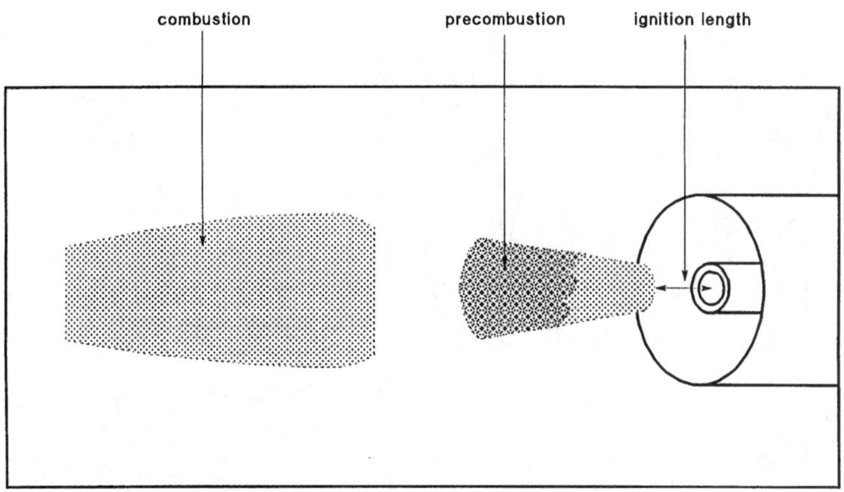

Fig. 9 Free jet test—combustion in over-expanded flow (schematic view derived from original photography).

the centerline of the combustor at the exit of the nozzle (Mach 2.5), but it was shifted upstream to make the ignition easier (Mach 2.0).

Another combustor, characterized by a first part with constant area, was tested with hydrogen with incoming air Mach number of 2.2. The exit to entrance area ratio was 2.50. The injection was realized by 8 equidistant wall slots.

2. *Results*

Figure 11 shows wall pressure evolution along the kerosene combustor for different equivalence ratio between 0.0 and 0.91.

Whatever the equivalence ratio may be, an expansion occurs in the divergent combustor because of the slow heat release which can not ensure an isobaric combustion.

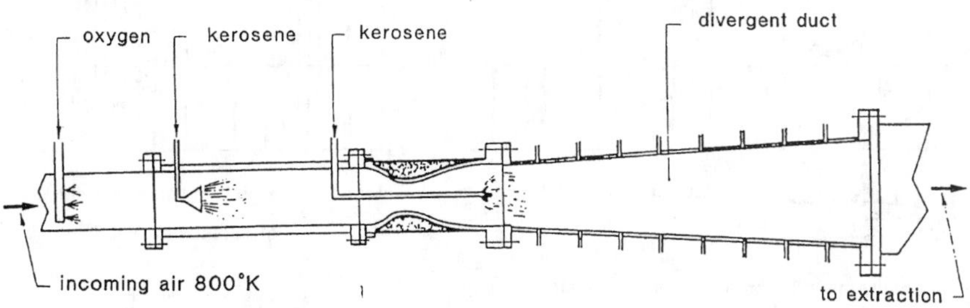

Fig. 10 Divergent duct—test device.

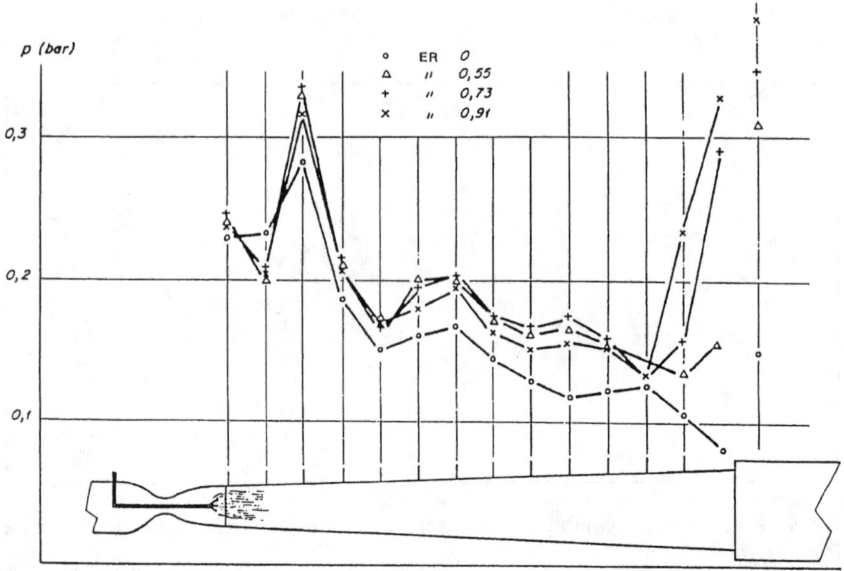

Fig. 11 Divergent duct—pressure distribution along the combustor (kerosene—central injection).

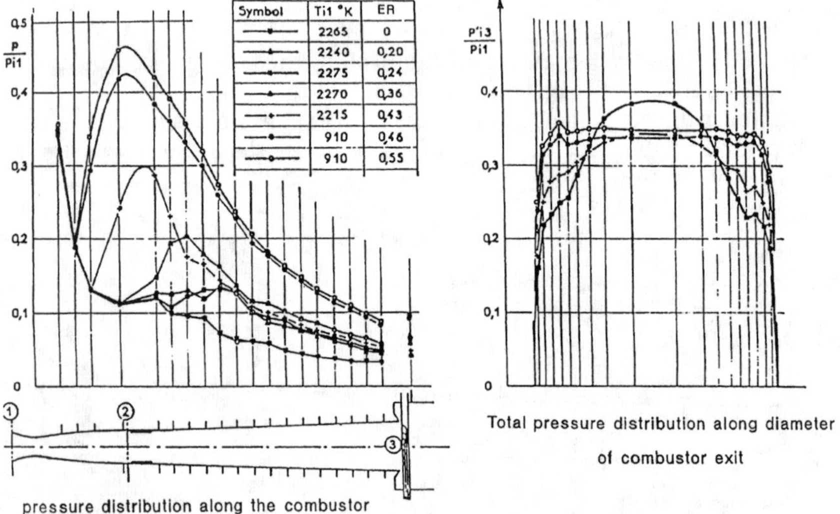

Fig. 12 Divergent duct—pressure distribution (hydrogen - wall injection).

In the case of a hydrogen combustor (see Fig. 12), the chocking was obtained since the equivalence ratio passed beyond 0.43. The first constant area of the combustor accelerates the ignition and then the heat release is able to balance the divergence effect. The isobaric combustion was reached with an equivalence ratio of 0.25.

D. Synthesis

These works demonstrated the feasibility of supersonic combustion in reasonable length. Indeed, the ignition is helped by the presence in the flow of boundary layers and shock waves due to the injection which reduces the time ignition by the local increase in temperature and pressure they produce. The combustion length decreases rapidly when the temperature increases.

But these experiments also showed the important stratification of the flow and the difficulty to obtain combustion with the whole air mass flow. From this point of view, the supersonic combustion problem is more the mixing than the chemical kinetic, and the struts injection seems to be more advisable than the wall injection.

The area evolution along the combustor must be optimized to allow 1) in its upstream part, the ignition of the combustion (~constant area duct), and 2) in its main rear part, the balancing of heat release to avoid chocking problem up to high equivalence ratio.

Finally, it is possible to realize subsonic and supersonic combustion in the same combustor, subsonic combustion being reached by thermal chocking with a high equivalence ratio.

III. ESOPE Program (1966–1973)

A. Origin and Principal Aims

ESOPE program was started in 1966. At that time, the feasibility of the supersonic combustion was demonstrated, but its interest for an operational vehicle was only evaluated with theoretical estimations going on unverified assumptions. Indeed, scramjet performances depend strongly on three parameters not well known at that time: 1) pressure losses in the combustor, 2) chemical burned gas recombination during their expansion in the nozzle, and 3) fuel/air mixing and combustion achievement, which were estimated (pressure losses) or badly approximated (burned gas in chemical equilibrium or frozen) or neglected (complete combustion).

These parameters are strongly dependant on the design of the engine. Therefore, definition, realization, and testing of a significant scaled engine appeared necessary to obtain a realistic aero-propulsive balance of the supersonic combustion. Studies were then started to define an experimental vehicle able to perform a flight point at Mach 7. The STATALTEX vehicle, which had flown up to Mach 5 with a ramjet engine (Fig. 13), was to be accelerated by a rocketbooster up to Mach 7 and had to allow comparison between ground and flight testing results with different fuels.

At this time, the ramjet (subsonic combustion) development was limited by the highly variable geometries that this kind of engine needs to preserve

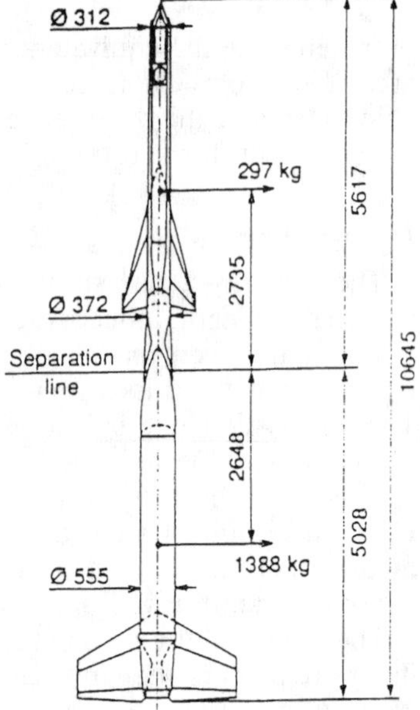

Fig. 13 STATALTEX—Geometric characteristic.

maximum performances on an extended range of flight Mach numbers and which carries a strong increase in the complexity and cost of the vehicle.

Ferri had already suggested that it was possible to define a fixed geometry scramjet operating on a large Mach number range.[6] Then, Marguet and Huet compared the theoretical performances of fixed geometry ramjet using subsonic or supersonic combustion in the flight Mach numbers range 3 to 7 (Fig. 14).[8] This comparison was realized with some arbitrary data: inlet performances, trajectory, engine sizing, equivalence ratio (= 1), nozzle efficiency, internal pressure losses (2%), and drag as function of the flight speed (STATALTEX results). They concluded that, for the high Mach numbers, the performances decreased slower with supersonic combustion and, consequently, that supersonic combustion is more efficient than subsonic combustion beyond Mach 5.5 if the possibility to adapt the section where the combustion starts is preserved and beyond Mach 6 if this section is fixed. Then, they envisaged to combine the two combustion modes in the same engine to obtain interesting performances with a limited variation of geometry (Fig. 15). Nevertheless, this option implies the passing of the supersonic combustion gas through the throat necessary for the subsonic combustion. The Mach number up to this gas has sufficient momentum to cross over the throat to determine the Mach number of the transition between the two combustion modes.

By another way, the best performances with subsonic combustion are obtained with constant area combustor. But, as we saw previously (§ 2), it is necessary for the supersonic combustion to place the injection just before a

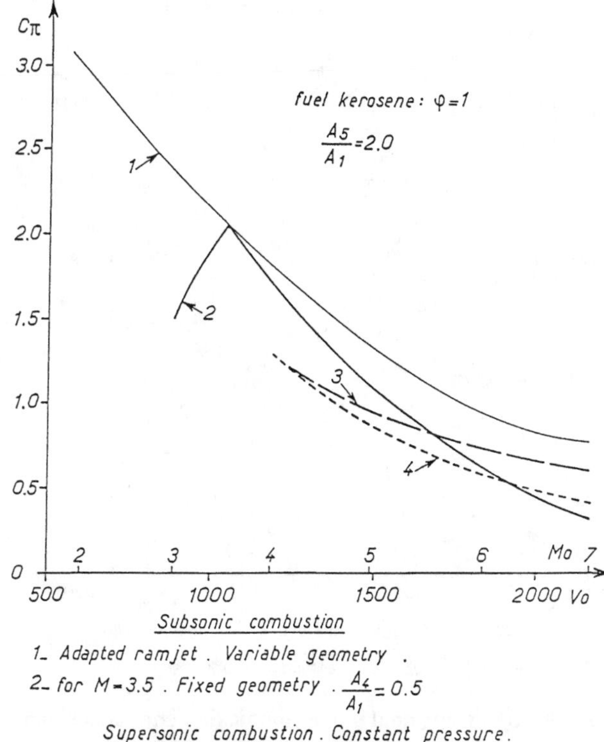

Fig. 14 Comparison of different ramjets.

divergent. Different zones of fuel injection must then be provided to adapt the injection with the combustion mode.

The solution proposed by Marguet and Huet is shown in Fig. 16.

During subsonic combustion, fuel is injected at the end of the inlet diffuser and combustion is realized in a constant area combustor followed by the throat without flameholder to allow the supersonic combustion. From this point of view the annular geometry is very favorable because it permits flame stabilization by the wall injection jets.

During supersonic combustion, the fuel is injected just after the minimum section of the inlet in a supersonic flow and the combustion is supersonic if the throat section is larger than the critical section of the burned gas.

This kind of engine was chosen in 1968 for the ESOPE program, which aimed at testing in flight a dual mode ramjet between Mach 3 and Mach 7 with a transition between the two modes at Mach 5.5. Two engines were to be designed: ESOPE A, demonstrating in flight the supersonic combustion at Mach 7, and ESOPE C, covering the Mach number range.

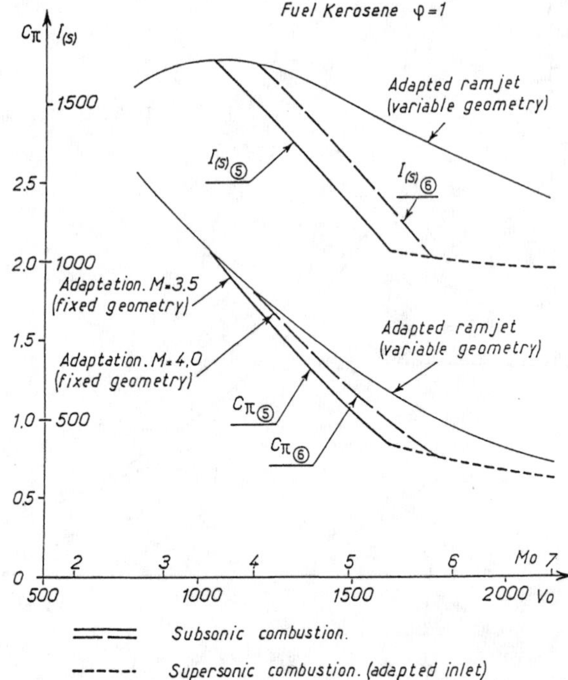

Fig. 15 Ramjet using dual mode combustion and fixed geometry.

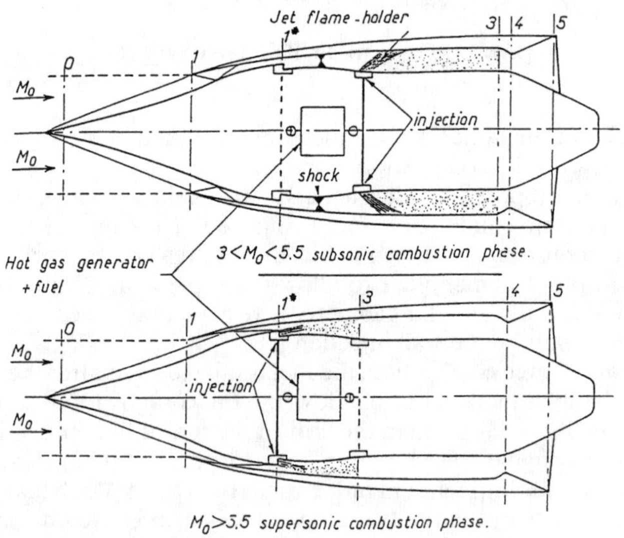

Fig. 16 Dual mode ramjet.

This program was very ambitious for a country like France and, in fact, only a part of it was realized: budget and timing restrictions led people to verify the feasibility of the dual mode ramjet on ground. This renunciation of flight test limited the explored Mach number range. Indeed, the available test facilities did not allow them to simulate Mach 7 flight conditions with pure air: Mach 7 simulation was possible at Palaiseau ONERA Center but with vitiation given by hydrogen-air preburning needed to reach the right enthalpy, test with pure air were possible at Modane ONERA Center (S4MA facility), but the Mach number was limited by the maximum temperature feasible with the alumina pebble bed heater (<1800 K) which corresponds to a flight Mach number of 6. The Mach 6 conditions were unfavorable to the ignition and the stabilization of the flame, but the test with pure air seemed to be essential to determine the vitiation effect particularly near the transition between the two modes which study was preserved.

Budget considerations also led people to renounce testing a complete engine with its inlet. No extractor was available for the S4MA facility and its development in the scope of ESOPE program was not possible. Inlet and combustor were then separately tested in spite of the additional problems that provided 1) flow speed and enthalpy at the combustor entrance, which were respected but not their gradients, and 2) the actual flow structure was not well simulated in particular for the shock waves.

Thereby, the model technology was simplified and the dimensional constraints reduced. It was also easier to vary simulated Mach number and altitude by modifying the pressure and total temperature.

So, the test program defined in July 1969 focused the studies on Mach 6. The aims of these studies were to determine the performances of a total engine (inlet, combustor, nozzle) at Mach 6 and to realize a transition between the two combustion modes around Mach 5.5. Three steps were planned: 1) a first test series at Palaiseau ONERA Center aiming to tune the engine (general design, injection system, thermal behavior). Considering the capacity of the test facilities, this study had to allow an exploration of the flight Mach number range 5 – 7 and a transition test; 2) test of the combustor in S4MA facility at Modane ONERA Center with the evaluation of the thrust; 3) study of the full-scale inlet in S4MA.

This last step was cancelled because of budget restrictions and only small-scale experiments were performed.

The results obtained during these studies had to constitute a database for the numerical codes developed in order to extrapolate the scramjet performances between Mach 5 and Mach 7.

B. Studies Results

For the reasons we have previously seen, the engine was designed to operate at Mach 6. At this point, the foreseen conditions at the entrance of the combustor were: Mach 2.5; total pressure ~ 1.5 MPa, total temperature ~1500/1700 K, and the combustor design had to ensure a completely supersonic combustion. The annular configuration was chosen for the combustor because it simplified the injection and the mixing of the fuel with the air.

Sized to be compatible with S4MA facility, the combustor had a diameter of 420 mm and a height at the entrance of the duct, defined by the inlet studies, of 10 mm.

A comparison between different fuels was realized by Huet in 1967 (Ref. 13). He concluded that for a missile operating between Mach 3 and Mach 7 hydrogen was not gainful when compared with liquid hydrocarbon fuel because its energetic advantage was compensated by the complexity and the overseeing of the vehicle which reduced the launcher specific impulse.

Nevertheless, hydrogen has very good characteristics in terms of ignition, combustion kinetics and cooling capacity, which led to its selection for the ESOPE program in spite of Huet's conclusions.

1. Concept and Design of the Combustor

The first possible concept for the combustor was the one proposed by Marguet and Huet (Fig. 16), which was previously discussed. But this solution presents some disadvantages: 1) uncertainties existed on the self-ignition of the H_2/air mixture at Mach 5.5, which were able to compromise the feasibility of the transition, and 2) pressure losses due to the physical throat risk reduced the performance of the supersonic combustion at Mach 6.

Another possibility was to use thermal chocking due to combustion to avoid the necessity of a physical throat. In such a solution, the combustor was constituted of a cylindrical duct, followed by a divergent duct and the two modes of combustion are described as follows:

1) Mach inferior to the transition Mach number: fuel is injected in the cylindrical duct and the heat release causes a thermal chocking. The combustion starts in a subsonic flow and eventually continues in a supersonic flow in the divergent duct beyond the thermal throat if the combustion is not completely achieved in the cylindrical part of the duct (transonic combustion).

2) Mach superior to the transition Mach number: a part of the fuel is injected in the cylindrical part of the duct and the heat release causes a slowing up of the supersonic flow up to a Mach number just superior to 1 at the end of the cylindrical part. The remaining fuel mass flow is then injected in the divergent duct where the combustion continues in supersonic regime.

This solution was chosen because it presented the advantage of an unbroken transition between the two modes of combustion, obtained by reducing the fuel mass flow in the cylindrical part of the duct, and ensured an average combustion Mach number around 1.5, which limited performance losses.

Unfortunately, the preliminary experiments at Palaiseau Center showed that ignition was possible only with an equivalence ratio in the first injection level (cylindrical duct) superior to 0.3. But this value also corresponded with the thermal chocking of the duct and, therefore, it was not possible to obtain pure supersonic combustion at Mach 6. The studies were then focused on the transonic combustion which appeared to be an interesting solution for the design of fixed geometry ramjets.

The combustor defined is described by Fig. 17. It is constituted by a first constant area part beginning after a rearward-facing step, used as an isolator for the inlet and a stabilizer for the final shock. A first level of injection, placed in this duct, induces thermal chocking as soon as fuel equivalence ratio exceeds 0.3 with a total temperature of the incoming air of 1650 K. It is followed by a slowly diverging part with a second injection level which provides the left fuel (the divergence is fixed to 3.5° to avoid a new thermal chocking).

Mestre's database was completed by experiments concerning the wall injection of hydrogen in a supersonic flow: Mach 2.5; total temperature 1650 K; total pressure 0.9 MPa. These experiments allowed defining the spacing of the injectors and the combustor length to obtain the junction of the individual flame fronts.

The nozzle made a trade-off between the drag and the expansion rate in the flight Mach number range 3 – 7. Its exit area was two times as big as the let capture area. It was constituted by two truncated cones with divergence angles of 20° on central body and 10° on external body.

2. *S4MA Test Principle*

The capacity of S4MA, and particularly of its alumina pebbles bed heater, led to realize only direct connected pipe test. An annular nozzle provided the right flight conditions for the incoming airflow: Mach 2.5; total temperature between 1500 and 1670 K; total pressure between 0.9 and 1.5 MPa (Fig. 18).

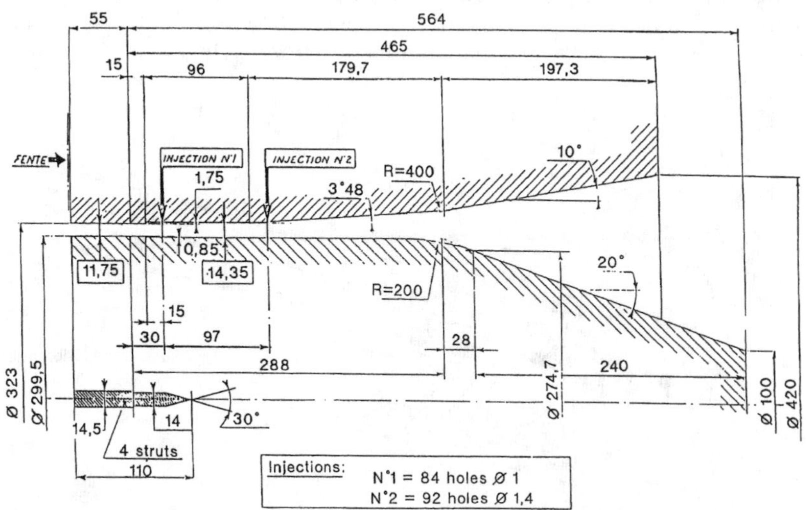

Fig. 17 ESOPE combustor—internal geometry.

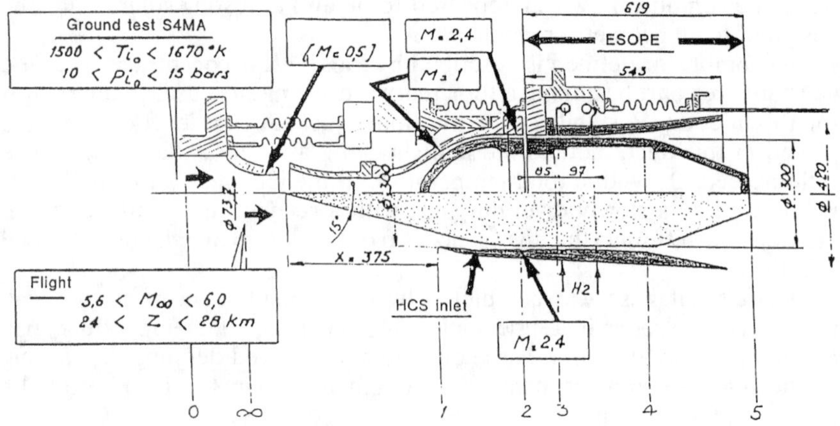

Fig. 18 ESOPE—flight simulation.

To determine the total thrust of the engine, a dynamometer measured the thrust balance of the combustor. Two bellows systems separated the combustor from the fixed parts of the test device. The dynamometer had a maximum capacity of 4,000 daN and a rigidity of 22,000 daN/mm. The precision of the whole mechanical system (dynamometer and separating bellows) was 0.1%.

Walls of the combustor were heated to minimize losses and to favor the combustion by a hot boundary layer. This solution limited the test time to 10 s of combustion (20 s of total run time).

The complete start of the combustor diffuser needed a pressure ratio of about 45. The possible presence of hydrogen, excluding the use of vacuum spheres, was tested with an exhaust induction nozzle (Fig. 19).

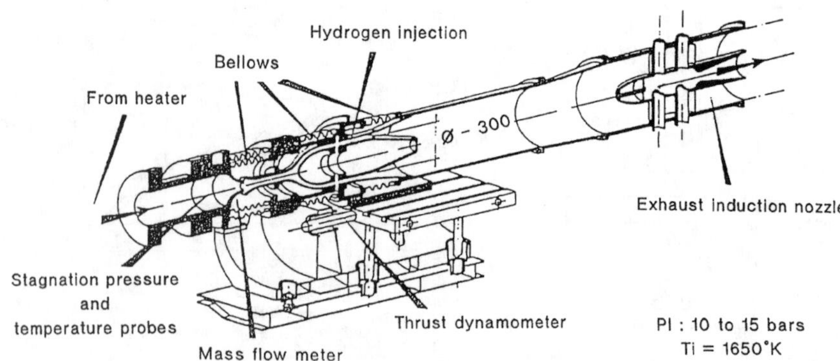

Fig. 19 Engine test in S4MA facility.

3. *Instrumentation and Measures Analysis*

In addition to the thrust dynamometer, pressure and temperature measures were installed on the combustor and the air-supplying system (Fig. 20). The total temperature was measured just after the heater with a precision of +/− 15 K. Air mass flow was determined by a false throat fitted out with four pressure and two temperature measures after a calibration, taking into account the throat area variation due to the temperature.

The measure analysis aimed to determine the performances of the total engine (inlet-combustor-nozzle). Available data were wall static pressure and temperature along the combustor, initial total temperature, air mass flow, hydrogen mass flow, and combustor thrust balance.

Two parallel ways were used to determine the thrust:

1) Thrust dynamometer: The measured thrust, corrected with bases and bellows forces, gave the internal thrust of the combustor. It was then added with stream thrust at the exit of the inlet corrected with the friction losses in the short cylindrical duct placed between the inlet and the combustor.

2) Pressure and temperature measures: The flow was assumed to be monodimensional and constituted by three phases: air, hydrogen, and combustion products. The combustion process was simulated by the development of this third phase to the detriment of the others. Thrust was determined by the integration, step-by-step, of pressure and skin friction forces along the combustor from the exit of the air-supplying nozzle to the end of the engine nozzle. Indeed, studies had shown that if the integration was started in a

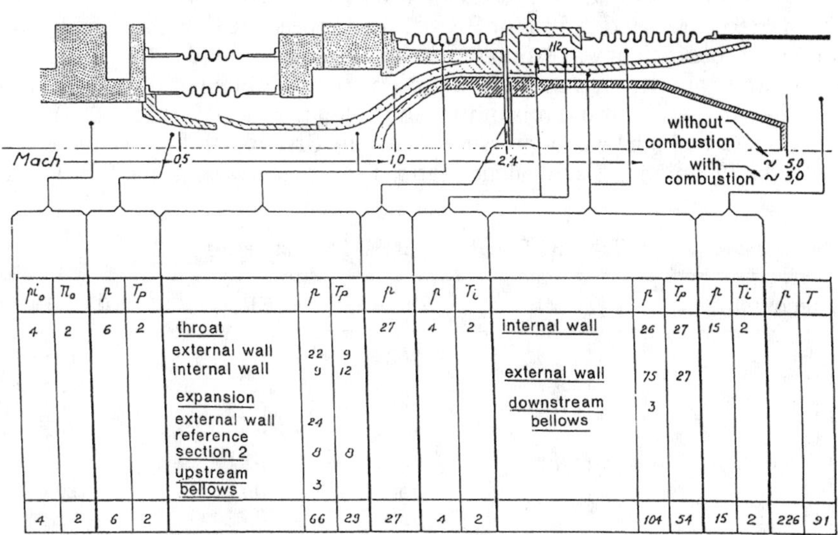

Fig. 20 ESOPE—Measurement device.

section where the flow was uniform (as at the end of the air-supplying nozzle), then the results of the calculation were acceptable in every section where the flow is uniform in spite of the very heterogeneous zones which separated these sections. The skin friction coefficient was not accessible in the combustion zone and an average value, issued from previous experiments, was chosen (0.004). Heat transfer coefficient was issued of Mestre's studies (0.002). These values were validated by the test results analysis.

4. *First Test Series Results*

After a preliminary test series at Palaiseau Center, the first test series was realized in S4MA facility in December 1970 (Ref. 14). The conditions for each test are given in Table 3.

During the tests, two problems occurred:

1) Air circulation appeared in the upstream cut between the combustor and the air-supplying nozzle, which made the estimation of the corresponding base force difficult;

2) A low performance of the exhaust induction nozzle led to a partial chocking of the engine nozzle, which complicated the results analysis (a complete expansion was substituted for the detached flow by assuming the combustion was finished as that was showed by the low temperature level: ~1200 K).

These problems strongly reduced the precision of the conventional thrust estimation obtained by dynamometric measures (± 26 % for ER = 0.25 and ± 10 % for ER = 1.0) or by pressure and skin friction forces integration (± 15 % for ER = 0.25 and ± 5 % for ER = 1.0). At the same time the combustion efficiency coefficient was estimated with a precision around ± 10%.

As an example, Figs. 21 and 22 present results obtained during the run number 141 with three levels of fuel equivalence ratio: 0.0; 0.25; 1.09. These curves show a high level of pressure above the first injection which increases with the injection n°2 equivalence ratio. That indicates subsonic flow in the

Table 3 Conditions of S4MA Test Series

Run number	Pt_0, bar	Tt_0, K	ER_1[a]	$ER_1 + ER_2$[b]
139	14.8	1623	0.25	0.25
141 – 1	14.9	1594	0.25	0.25
141 – 2	14.9	1599	0.25	1.09
142	9.3	1595	0.23	0.95
143	14.9	1671	0.21	0.93
144 – 1	14.9	1500	0.21	0.93
144 – 2	14.9	1506	0.22	0.22

[a]ER_1 = first injection fuel equivalence ratio. [b]ER_2 = second injection fuel equivalence ratio.

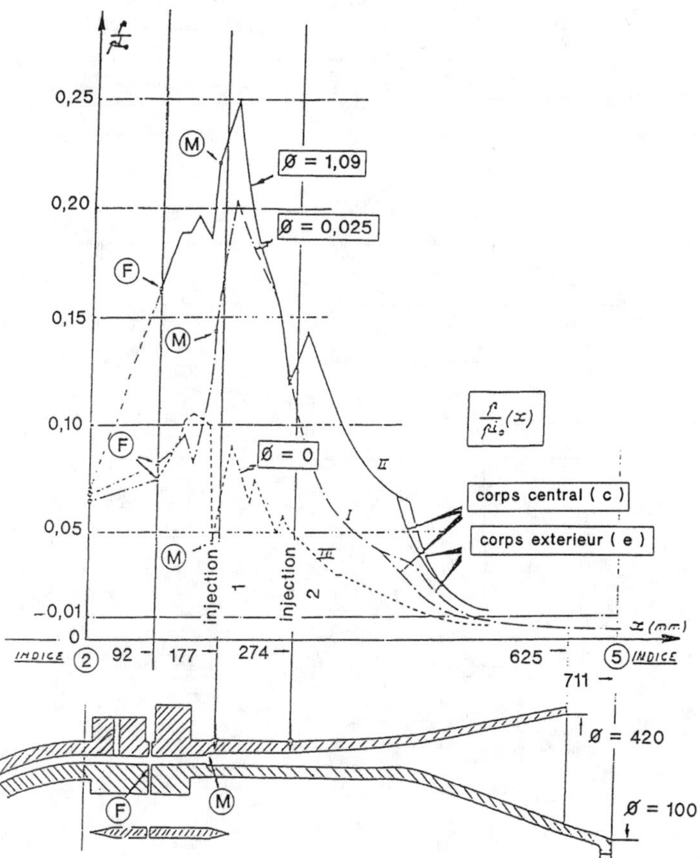

Fig. 21 Pressure evolution along the combustor run 141.

cylindrical duct, which is confirmed by the negative pressure gradient going with the combustion in this zone. The first injection causes thermal chocking of the combustor duct but not of the air-supplying nozzle. The abrupt increase of the pressure just after the second injection indicates a supersonic flow just before this injection. So, a transonic combustion had been obtained: the heat release caused by the first injection created a thermal throat at the end of the cylindrical part of the combustor, which accelerated the flow beyond Mach 1.

The combustion of the second injection took place in a complete supersonic flow. There was combustion as proved by the higher pressure and lower Mach number which were obtained with this second injection. Nevertheless, this second combustion started slowly (the temperature increased just before the nozzle) and it was rapidly stopped by the nozzle expansion. The heat release and, consequently, the combustion efficiency were then

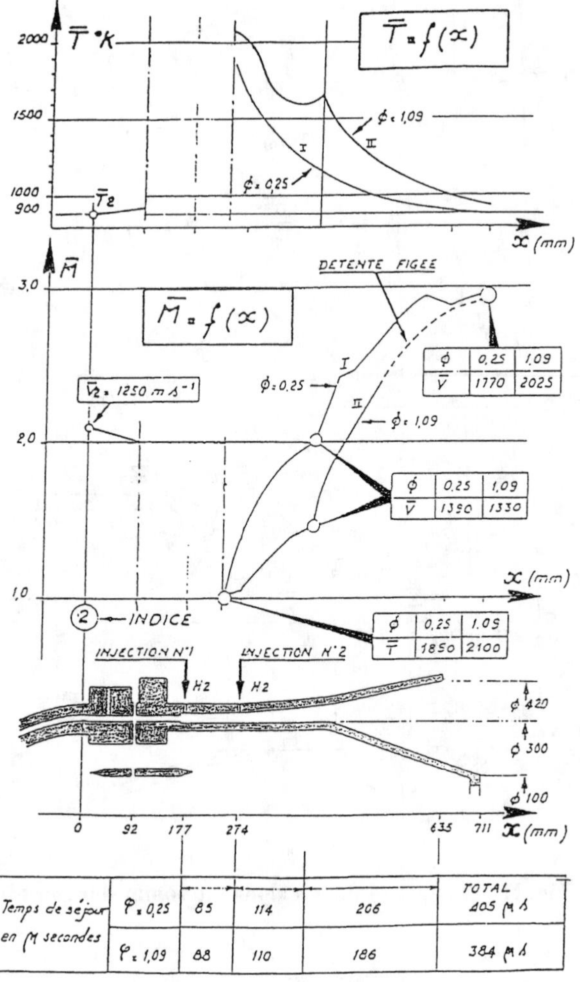

Fig. 22 ESOPE—first test series—temperature and Mach number along the combustor.

limited. The comparison between external and internal wall temperatures shows clearly that the fuel jets did not penetrate to the center body, particularly for the second injection.

Engine performances are summarized by Figs. 23 and 24. For low equivalence ratio, the conventional thrust coefficient was 0.2 (95 daN) with a specific impulse around 2600 s. The combustion efficiency was very good (subsonic combustion) but friction and inlet collecting drags were high (respectively, 0.5 and 1.8 conventional thrust). For high equivalence ratio, the thrust coefficient was constant (= 0.61) up to Mach 5.9 before rapidly

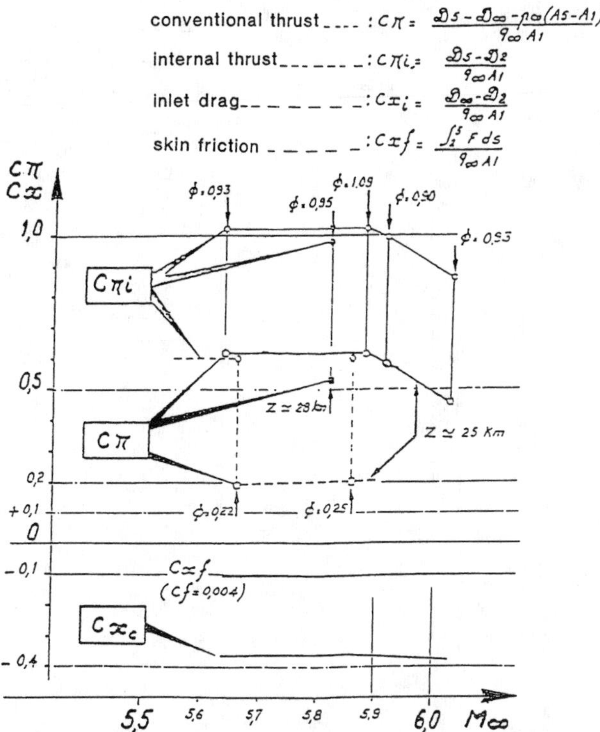

Fig. 23 ESOPE—first test series—thrust coeffecients.

decreasing to 0.45 at Mach 6.03. The combustion efficiency decreased strongly in the same time as shown by Table 4.

The low efficiency of the second injection can be caused by 1) the weak fuel penetration showed by the external and internal wall's temperature evolution; 2) the place of the second injection on the same wall as the first, which implies an injection into preburned air; 3) a decrease of the temperature caused by the injection of an important hydrogen mass flow; and 4) a too-short divergent duct.

Lastly, the experiments showed an important influence of flight altitude on the combustion efficiency, which was 0.6 at z = 25 km and only 0.5 at z = 28 km.

5. *Combustor Optimization*

The low efficiency of the second injection appeared to result from the uncomplete mixing and the presence of first injection combustion products. From test results, and by using the developed monodimensional numerical simulation, some modifications of combustor and injection system were envisaged. Indeed, simulations showed that, subject to ensure a good com-

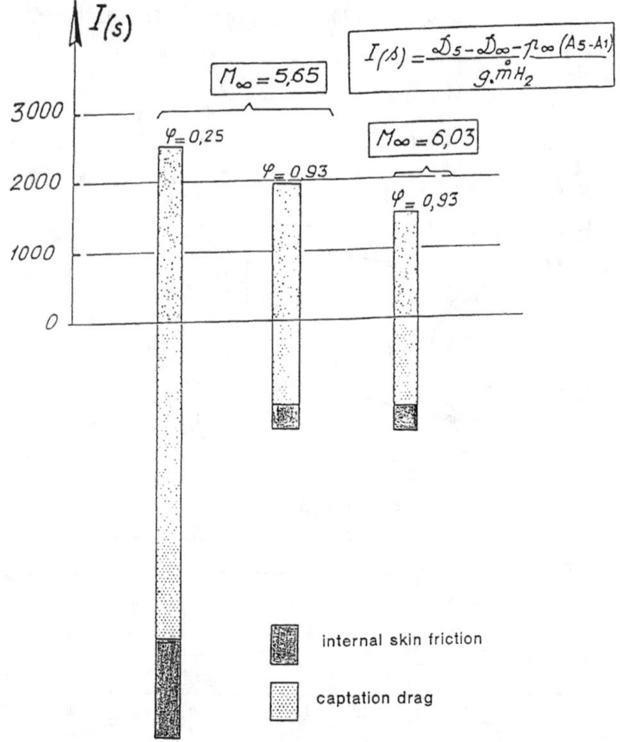

Fig. 24 ESOPE—specific impulse obtained during the first test series.

bustion efficiency (0.94), it was possible to obtain interesting performances in the Mach number range 5 – 7 (Fig. 25).

A second test series was realized in June and October 1972. The new combustor was extended in length: cylindrical part + 56% and divergent part + 30% (Fig. 26). At the same time, the cut separating the combustor and the air-supplying nozzle was put forward 50 mm in order to avoid interaction between combustion and thrust measure.

The combustor was fitted with two injection systems (Fig. 27).

Table 4 Decrease of Combustion Efficiency

Mach	$ER_1{}^+{}_2$	Thrust	Specific impulse	Thrust coeff.	Combust. efficiency	Friction / thrust	Collecting / thrust
5.64	0.93	277	1920	0.61	0.60	0.17	0.60
6.03	0.93	200	1500	0.45	0.46	0.23	0.46

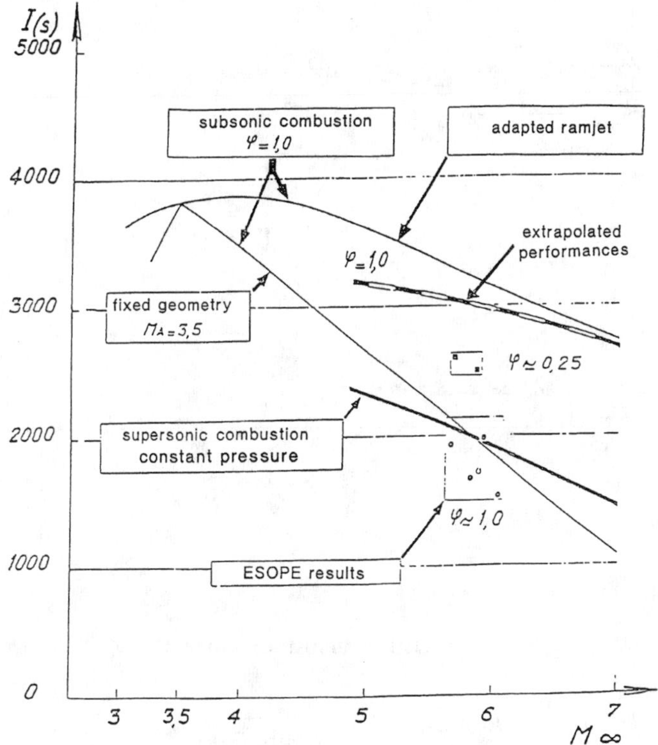

Fig. 25 ESOPE—performances extrapolation.

The first system, named E or "low penetration system," added two new injections points (3 and 4) on the center-body. The distances between 1 and 3, then 3 and 4, were determined to ensure the end of the combustion due to the upstream injection. The distance between 2 and 4 was chosen to avoid the interaction of the shock issued from 2 to 4. All the injection holes were 1 mm in diameter and separated about 10 mm on the circumference. The combustion spread out in the wall boundary layer detached by the injection jets and the flow might be stratified (subsonic and supersonic) or completely subsonic because of shock upstream of the combustion.

The second injection system was issued from the basic research undertaken by Leuchter during the ESOPE program. For this system, named A or "high penetration system," the holes of the four injections (5, 6, 7, and 8) were larger in diameter and more spaced on the circumference. Because of technological constraints, these injections were placed on the external body just after the first injection system and, therefore, did not take advantage of the combustor extension. The aim was to ensure that the combustion spread out in the whole duct height to limit the chocking risk.

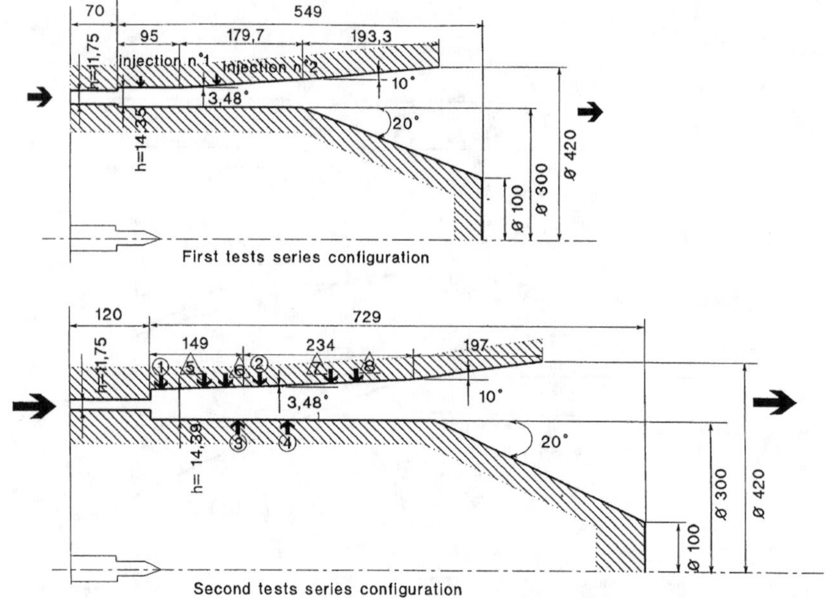

Fig. 26 ESOPE combustor modifications.

6. *Second Test Series Results*

Test device and analysis means were the same as for the first test series. Nevertheless, a gas-sampling probe was added for direct estimation of combustion efficiency.

Fifteen runs were realized, which allowed tests on about fifty injection schemes (Ref. 15 and 16).

For the first test series, the cylindrical duct was always thermally chocked. Table 5 summarizes the combustion efficiencies obtained with different injection schemes of the "low penetration system" in the Mach 6 test conditions.

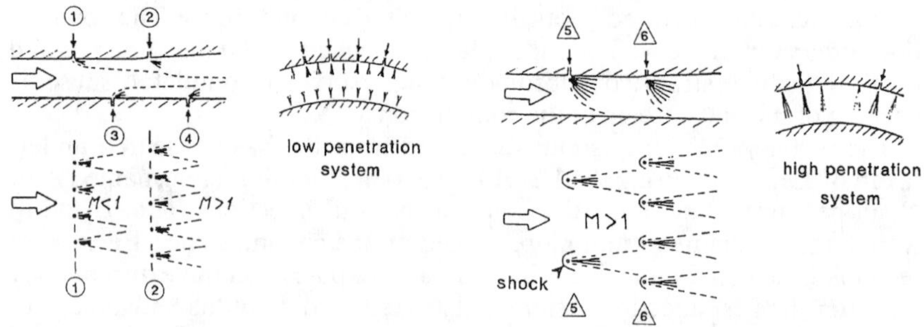

Fig. 27 ESOPE—second test series—injection systems.

Table 5 Combustion Efficiencies in the Mach 6 Test Conditions

	Equivalence ratio / ramp				
Run	1	3	2	4	Combustion efficiency
183	0.26	0	0.62	0	0.60
187[a]	0	0.27	0.79	0	0.70
188	0.26	0	0	0.84	0.60
195[a]	0.10	0.25	0.42	0.24	0.78

[a]Important pressure increase upstream the first injection.

These results confirmed the improvement of combustion efficiency given by the combustor extension: in the case of an injection realized only from the external body (run 183) combustion efficiency increased from 0.45 to 0.60.

The best results were obtained with the scheme of run 195; its performances were constant between Mach 5.6 and 6.0 (Fig. 28).

With the "high penetration" injection system *(A)*, the combustion efficiency obtained at Mach 6 was about 0.69 with ER = 0.8 and 0.64 with ER = 1.03 (Fig. 28). Thrust coefficient and specific impulse are given by Fig. 29. Figure 30 compares these performances with the theoretical estimations. The performances obtained during the second test series are close to the estimated performances of the optimized scramjet (flow speed reduced to Mach just superior to 1 at the end of the cylindrical duct, combustion continuing in constant supersonic Mach flow).

7. *Basic Research Supporting ESOPE Program*

Basic studies on injection and ignition were realized parallel to ESOPE program by Leuchter in Onera. These works, theoretical and experimental, aimed at giving usable information for the ESOPE combustor design and led in particular to the "high penetration injection system" definition. They concerned 1) numerical simulation of the mixing of two supersonic air and

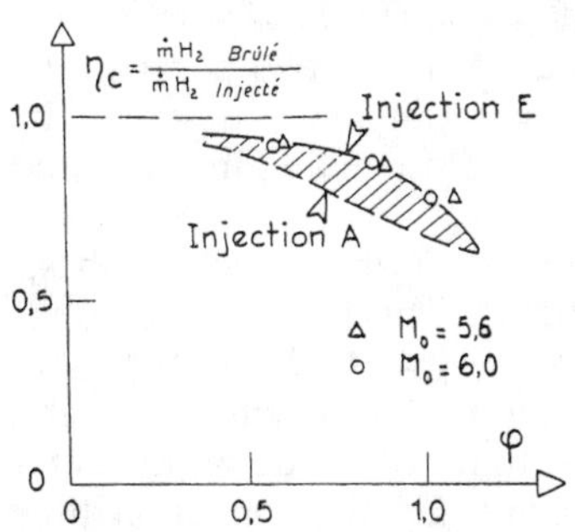

Fig. 28 "Low penetration injection system"—combustion efficiency.

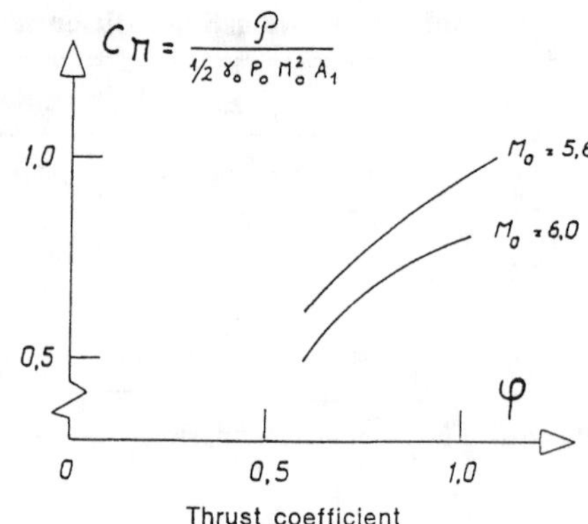

Fig. 29 "High penetration injection system"—performances.

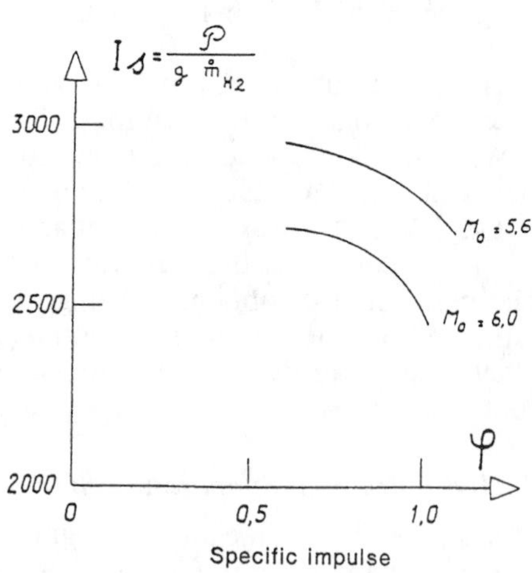

hydrogen flows and 2) experimental study of wall injection of hydrogen in a supersonic flow.

Numerical simulation of the mixing of two supersonic flows. This study aimed at determining the conditions of the self-ignition in the air-fuel mixing layer and was limited to the ideal case of the mixing of two parallel and uniform flows without initial boundary layer. The problem was described by the equations of an isobaric stationary turbulent flow taking into account the chemical process between air and hydrogen.[17]

The air-hydrogen combustion is characterized by a first induction phase, in the course of which temperature evolves slightly. This phase corresponds to dissociation and diffusion involving principally H, O, OH, and H_2O species. Temperature increase is due to recombination, which occurs at the end of induction phase. Induction and combustion times obtained by Leuchter for an air-hydrogen mixing at 1000 K are given Fig. 31. Induction time

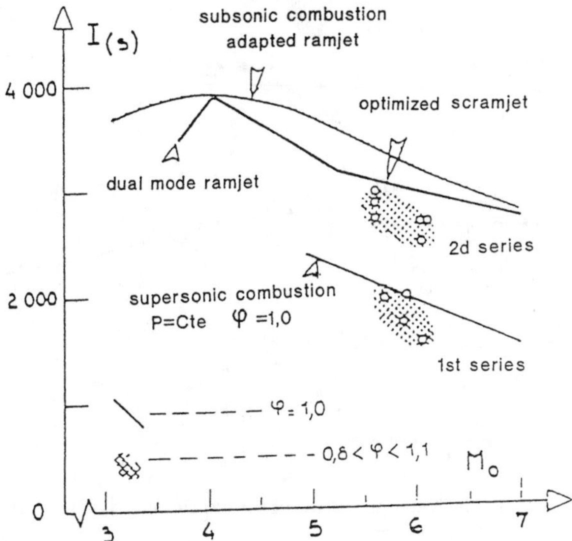

Fig. 30 ESOPE—Results of the two test series.

corresponds to a temperature increase of 10% and combustion time to an increase of 90%. Induction time increases dramatically around 1000 K (Fig. 32). The combustion time increases very slowly at the same time because recombination reactions are weak, depending on temperature.

Wall injection study. This study aimed at the definition and the optimization of the injection system for the cylindrical duct of ESOPE combustor. A completely supersonic combustion regime was envisaged at Mach 7 conditions. The first injection had to slow down the flow just above Mach 1 at the end of the cylindrical duct. According to Leuchter,[18] an important discretisation of the injection was necessary to ensure a good penetration using the shocks created upstream injection holes which gave favorable conditions, approaching those of a subsonic combustor, for ignition.

An adiabatic compression test facility allowed runs of a few tenths of a second with pure air up to 1650 K (Ref. 19). Hydrogen, heated up to 1000 K, was injected in a Mach 1.5 flow. Experiments concerning penetration and mixing were realized in cold flow to allow long duration test and continuous flow exploration. Leuchter proposed a law characterizing the penetration (Fig. 33):

$$\frac{Z_{\max}}{d} = 0.78\left(\frac{\rho_{\mathrm{H_2}}\cdot U_{\mathrm{H_2}}^2}{\rho_{\mathrm{air}}\cdot U_{\mathrm{air}}^2}\right)^{0.5}\cdot\left(\frac{x}{d}\right)^{0.35}$$

$$\frac{Z_a}{d} = 1.45\left(\frac{\rho_{\mathrm{H_2}}\cdot U_{\mathrm{H_2}}^2}{\rho_{\mathrm{air}}\cdot U_{\mathrm{air}}^2}\right)^{0.5}\cdot\left(\frac{x}{d}+0.5\right)^{0.35}$$

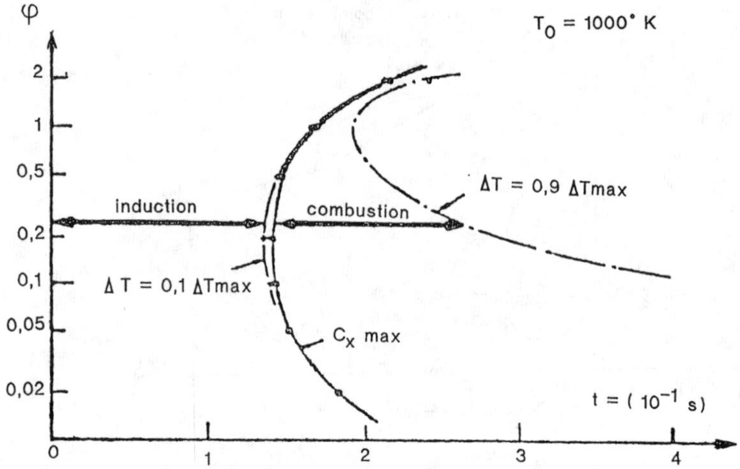

Fig. 31 Numerical simulation of a mixing layer ignition delay.

where Z_{max} is the line of maximum H_2 concentration and Z_a is the external jet boundary (H_2 concentration = 1%).

Leuchter also proposed an optimum injection hole spacing on the circumference:

$$\frac{\Delta}{h} = 0.063. \frac{\dot{m}_{air}}{\dot{m}_{H_2}} \cdot \frac{U_{air}}{U_{H_2}} \cdot \tau^2$$

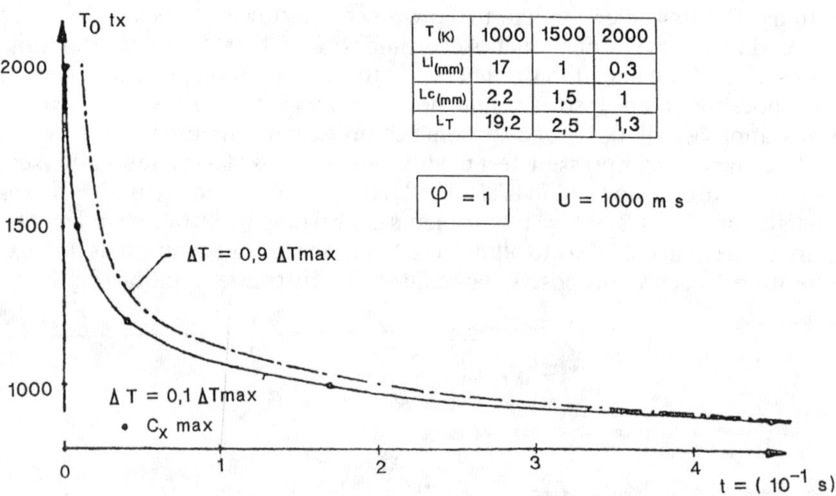

Fig. 32 Numerical simulation of a mixing layer chemical delay—H_2-air reaction.

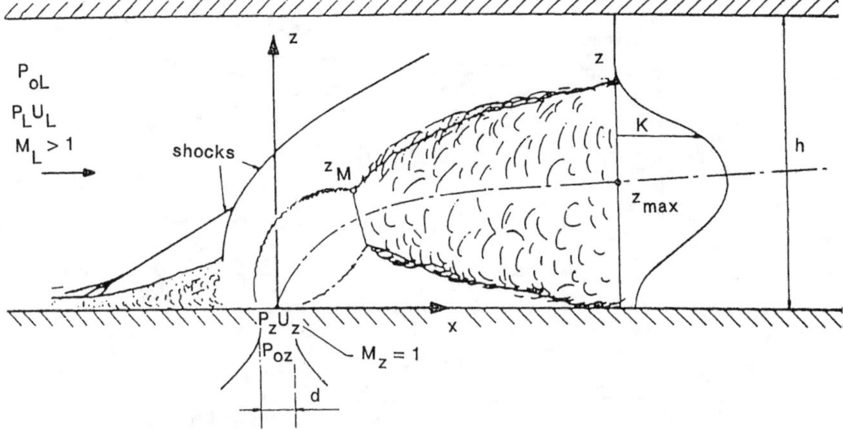

Fig. 33 Airflow-Hydrogen jet interaction.

In these optimum conditions, mixing analysis was performed by measurement of the fuel concentration. Maximum concentration decreases rapidly and moves to the opposed wall (Fig. 34). For hydrogen to air pressure ratio of 2, the ignition length depends only on fuel and air total temperatures (Fig. 35). Theoretical estimations gave an ignition length of 20 cm for the upstream flow conditions. Test results showed the favorable effect of flow perturbations. Reducing fuel temperature led to a low ignition length extension because ignition occurs in low hydrogen concentration zones.

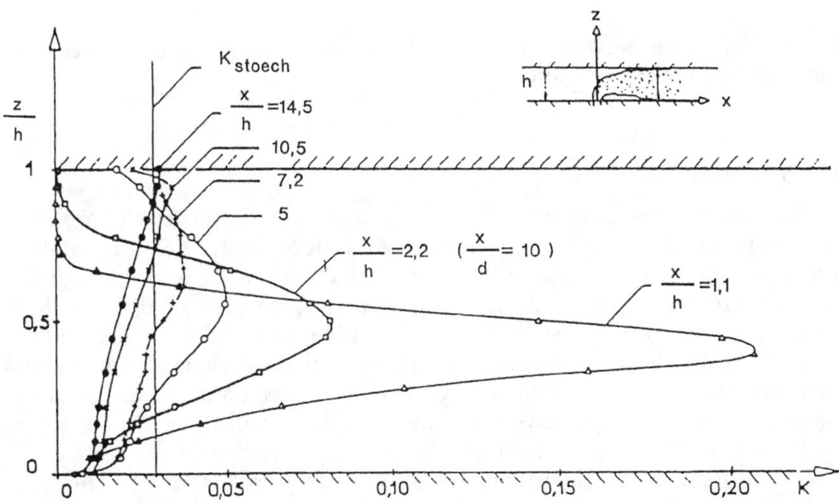

Fig. 34 Experimental study of the wall injection hydrogen concentration.

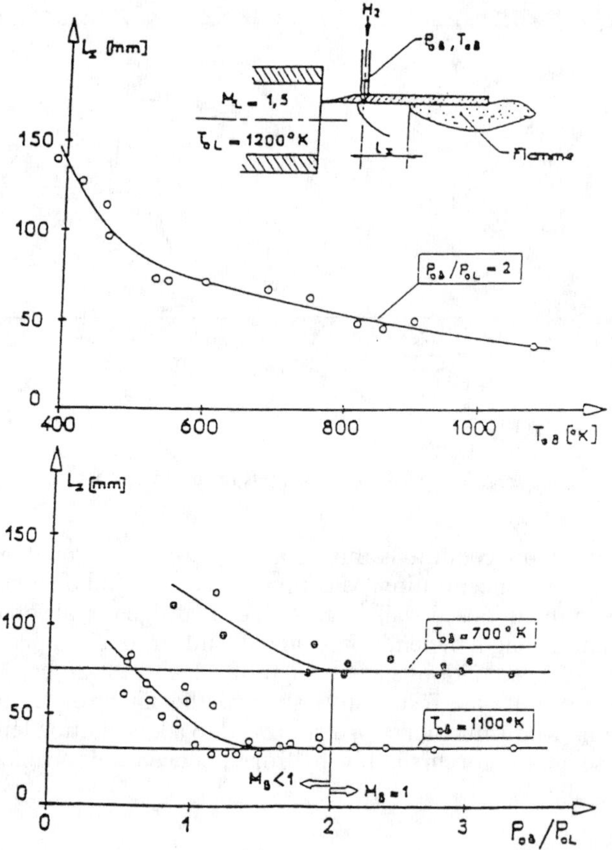

Fig. 35 Experimental study of the wall injection ignition delay as a function of injection pressure and temperature.

C. Synthesis

The ESOPE program would seem unsuccessful insofar as a completely supersonic combustion was not performed. Indeed, in the ESOPE combustor, the combustion due to the second injection was supersonic but the ignition process was realized in subsonic regime because of the low flight Mach number, which test facilities were able to simulate. Nevertheless, because of the shocks, it is very likely that it will always exist in some zones in a scramjet combustor where speed will be reduced and the temperature increased and, consequently, where the conditions will be favorable to the ignition, like in a subsonic combustion ramjet.

By another way, it is clear that the ESOPE results are not directly transposable to a large scramjet because of the low penetration of fuel jets in the supersonic flow, which leads to the use of strut injectors to ensure the fuel-air mixing in the whole air-flow. Moreover, the annular geometry is certainly not the best way to limit the heat and skin of friction losses considering the important ratio of each wall's area on the combustor section it corresponds to.

However, the ESOPE program allowed testing the transonic combustion mode, which appears to be a natural extension of the supersonic combustion to the low Mach numbers, particularly suitable for fixed geometry. Indeed, in a more or less divergent combustor, the regulation of the equivalence ratios injected upstream and downstream of the thermal throat allows to obtain, for every flight Mach number, the right thermal chocking and then to reach, without combustor variable geometry, the same performances as with a geometrically variable throat ramjet.

IV. Studies on Shock-Induced Combustion

A. Principle

As we have seen, studies led by ONERA between 1960 and 1975 were focused on diffusion flame combustion. The ignition of a pre-mixed fuel-air flow by a shock wave can be also envisaged. This combustion mode, proposed by M. Roy in 1946, was successfully tested by ONERA between 1948 and 1953. At the beginning of the 1960s, research was undertaken by Ecole Nationale Supérieure de Mécanique et d'Aérothermique (ENSMA) in Poitiers and by Laboratoire d'Aérothermodynamique de l'Ecole Centrale des Arts et Manufactures (LATECAM) in Paris to apply this technology to the hypersonic airbreathing propulsion.

In scramjets envisaged by ONERA or by Ferri *et al.* in the USA, combustion is controlled by the fuel-air mixing process in the combustor. Indeed, for high Mach numbers, fuel injected downstream of the inlet, ignites spontaneously and the combustion depends only on the diffusion of the fuel in the air. For lower Mach numbers (4.5 to 7), a pilot combustor could be necessary to ensure ignition. Combustion is then controlled by the convective heat transfer from pilot flame to main "cold" supersonic flow.

Shock-induced supersonic combustion can take the place of pilot combustor to generate the flame. Fuel must be injected upstream of the combustor to obtain an air-fuel mixture at the entrance of it. Ignition occurs just after the shock in a short time because of the very favorable local conditions. This kind of combustion needs 1) right pressure and temperature conditions for the injection to avoid preburning before the shock and 2) right shock intensity to obtain enough pressure and temperature for ensuring ignition.

But the shock is strongly dependant on the downstream combustion process. Thereby, this kind of scramjet is very sensitive to perturbations and its design is very difficult.

The previous conditions limit the use of this kind of combustion to the low flight Mach numbers (4 to 6) for which the ignition is not sure with diffusion flame combustion. Beyond Mach 7, ignition is not as critical and the shock-induced combustion is not advantageous because it does not work out the problem of the mixing which becomes the major parameter for scramjet performances.

B. ENSMA and LATECAM Studies

Studies were mainly led by Bellet and Soustre in Poitiers[20–24] and by Reingold and Serruys in Paris[25] with similar aims and experiments.

1. *Test Description*

The different experiments consisted of generating a normal shock in a supersonic flow (Mach between 2.5 and 3.5) by the intersection of two oblique shock waves. Fuel injection upstream of the shock allowed a flame to stabilize just after it and to analyze this flame with probes or optical measures. The aim of the test was to obtain knowledge about the structure and the main aerothermodynamic characteristics of the combustion process. Experiments were principally realized with kerosene and hydrogen but also with ethylene by ENSMA and propane by LATECAM. They determined ignition time, air vitiation effects, species concentrations, and flame temperatures.

The facilities provided air heated by precombustion. Therefore, experiments were realized with vitiated air with reoxygenation. The ENSMA facility allowed a connected pipe test (Fig. 36) while LATECAM realized free jet test (Fig. 37). The flow structures obtained were very different and modified the minimum total temperature giving combustion: in the ENSMA

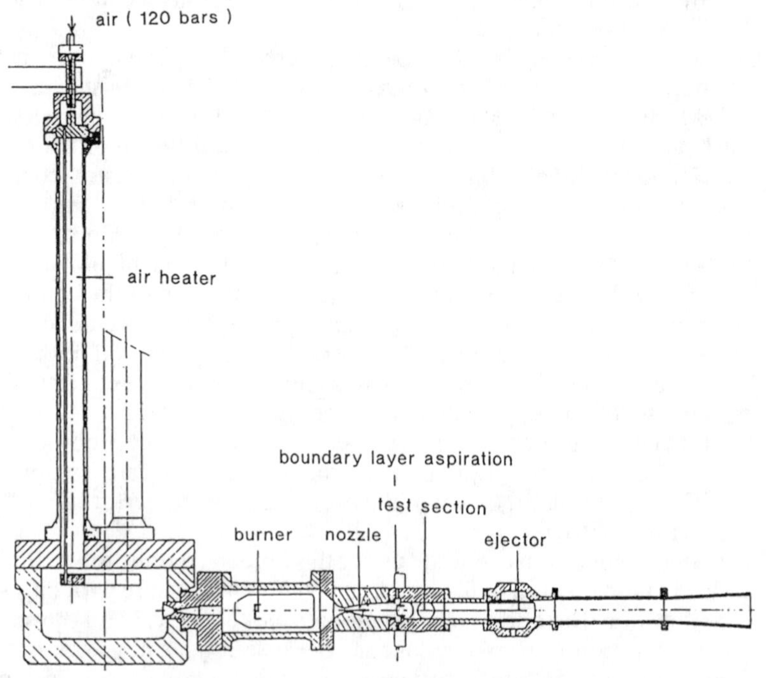

Fig. 36 ENSMA's facility for studies on shock-induced combustion.

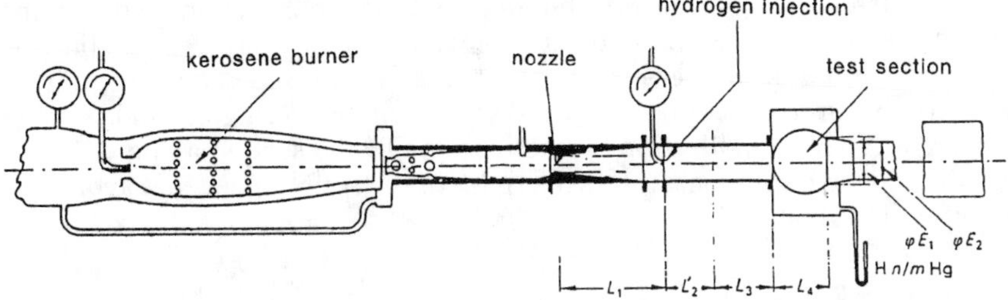

Fig. 37 LATECAM's facility for studies on shock-induced combustion.

facility, an expansion, occurring just after the shock, stopped the exothermic reactions for the low initial total temperatures.

2. Shock-Induced Combustion Structure

After the shock, subsonic flow is accelerated by the heat release until a sonic line, beyond which the aerodynamic expansion of the flow is more important than the heat release, allows a supersonic acceleration. The obtained combustion is then a transonic combustion. Figure 38 gives the evolutions of Mach number and hydrogen concentration for a vitiated air flow at Mach 2.7 and total temperature 1100 K. The combustion efficiency curve defines an ignition length (a few millimeters) and a combustion length

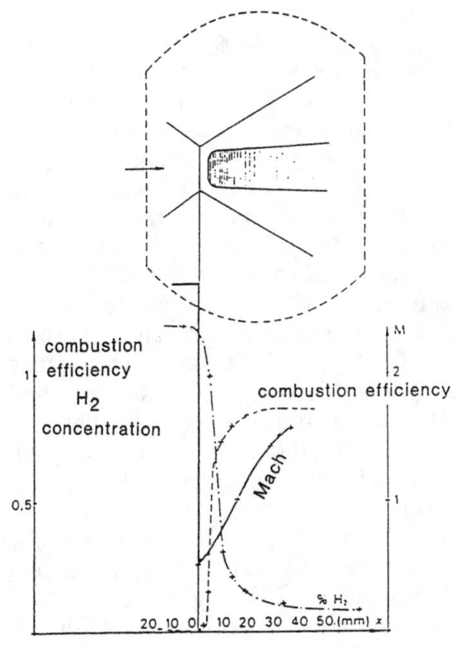

Fig. 38 Shock-induced combustion hydrogen-air.

around 10 mm (the combustion length cannot be directly compared with the results in chapter 3 because it does not include the mixing). The combustion efficiency is important because fuel and air are premixed and a large part of the combustion is realized in a subsonic regime. Reingold and Serruys explained flame stabilization by the increase of the section along the reactive zone which compensates the increase of the sound speed and avoids the climb of the flame.

3. *Ignition Time*

Bellet and Soustre obtained, with hydrogen ignition time, results which varied rapidly with the total temperature: ~1.10^{-5} s at 1 265 K; ~$1.2.10^{-6}$ s at 1 590 K. These results were in accordance with the experimental and theoretical values obtained by Nicholls and confirmed that the air vitiation reduces the ignition time probably because of the presence of OH radicals issued from the air pre-heating.

V. Prepha Program (1992–1997)

A. Origin and Principal Aims

In the spring of 1989, the General Delegation for Space (DGE) instituted a committee whose mission was to analyze the development of the hypersonic flight technology in the most advanced aeronautical nation. In June 1990, this committee's report recommended starting a large hypersonic technology program in order to maintain France's position in aeronautics and space launch technology.

At the same time, Ministry for Research and Technology (MRT) entrusted ONERA with the coordination of a working group, named G5, including Aerospatiale, Dassault Aviation, SEP, and SNECMA. This group had to review knowledge on hypersonic flight, to analyze the feasible applications and to define the necessary studies to complete the programs already in progress. In summer 1990, the five companies suggested a five-year program covering all the hypersonic flight and high-speed propulsion problems. After internal discussions, the French authorities meant to start a Research and Technology Program for Advanced Hypersonic Propulsion (PREPHA).

This five-year program was to be financed with 525 millions of French francs (~ 95 millions of USD): 400 MF coming from French authorities (General Delegation for Armament (DGA), which is the Executive Agency of the program; National Space Agency (CNES); Ministry for Research and Technology (MENESR) and 125 MF coming from the industrial partners and ONERA. Considering the available means, the priority of the program was to study and to test on ground the scramjet, because the mastery of scramjet is necessary for hypersonic flight.

In particular, people were convinced that for space launchers, the airbreathing phase must be widely extended beyond Mach 6 to have a potential chance to ensure the feasibility of reusable airbreathing vehicles. It was reasonable to limit the scramjet experimentation to ground testing during this short program, but it was also necessary to prepare for the future and define what could be an experimental vehicle able to validate the scramjet and its real propulsion capacities.[26]

Five axes of work were identified: 1) design and ground testing of an experimental scramjet; 2) development of the associated test facilities; 3) numerical codes and physical models; 4) materials; and 5) system studies.

Main outlines of the technical contents of these axes of work were specified by French authorities with the contribution of ONERA as state adviser. G5 partners concluded an agreement about data exchanges, financial contribution, working organization, and defined a detailed program proposition which was globally approved by the Executive Agency.[27]

PREPHA started in 1992. Budgetary restriction (525 MF $\rightarrow$ 380 MF) led to limited system studies and to drastically reduced light and cooled structures technology development. Nevertheless, the program continued until late 1999 and its main objective, defined as design and ground test of scramjet components with associated experimental and numerical means development, was preserved.[28]

B. System Studies

PREPHA's concerns are large scramjets used for operational vehicles like space launchers. Such scramjets cannot be tested full-scale on ground because of the complexity and cost of the needed test facilities. Thereby, a particular development methodology, based on very close cooperation of numerical and experimental approaches, must be defined for these engines. The program aims at acquiring numerical and experimental means, which should be necessary for such methodology, and to apply these means to a generic case.

A SSTO vehicle has been chosen as a generic case because it covers the largest range of Mach numbers and, consequently, the largest range of problems. The system studies realized on this vehicle allow to define some specifications for scramjet components studies and numerical and experimental development but also to synthesize the acquired results.

During the two first years of the program, four concepts of combined propulsion systems, using slush hydrogen as fuel, have been considered: two twin-duct concepts which are turborocket-scramjet-rocket and turbojet-dual mode ramjet (subsonic/transonic combustion until Mach 5/6 then supersonic combustion)-rocket; two one-duct concepts which are rocket-dual mode ramjet-rocket and ejector dual mode ramjet-rocket.[29]

After integration of these propulsion systems on the generic vehicle, first performance evaluation has been realized by trajectories simulation. It concluded that no concept was able to place into orbit a positive payload. Nevertheless, little advantage appeared for the one-duct concepts which compensate their bad specific impulse at low Mach number by an important thrust at the same time and by a lighter dry mass due to the own mass of the propulsion system but also to the high corresponding density of the propellant.[30]

Taking into account the first obtained results, a new design of the propulsion system has been optimized while the fuselage was enlarged (10 m to 12 m for a constant total length of 65 m corresponding with a maximum inlet entrance area of 24 m^2). At the same time, some evolution has been introduced in the trajectory requirements and propulsion modes transition: 1) initial acceleration phase at higher dynamic pressure (maximum dynamic pressure: 0.08 MPa to 0.14 MPa) to allow rapid extinction of the rocket engines because of the sufficient thrust of the ramjet; 2) optimization of the altitude of the elliptic transfer orbit perigee (110 km to 80 km for a constant 500 km circular final orbit); 3) optimization of scramjet/rocket engine transition (parallel use of scramjet and rocket engines for the Mach number range 10 to 12, optimization of the scramjet fuel equivalence ratio law).

After this evolution, a new comparison between turbojet-dual mode ramjet-rocket, ejector dual mode ramjet-rocket and rocket-dual mode ramjet-rocket concepts was performed.[31] For each concept and for different values of take-off mass, the height of the fuselage is adapted to obtain the volume needed by propellants. Comparison between the mass placed into orbit, resulting from the trajectory simulation, and the mass to be placed into orbit, resulting from the mass budget, gives the payload mass or the design margin for the vehicle structures.

The obtained results showed that the rocket-dual mode ramjet-rocket concept is the more efficient: for a given general sizing of the vehicle, it is able to place into orbit the heavier maximum payload mass—more than 10 metric tons—and it needs the lower variation of take-off mass to increase the payload mass) (Fig.39).[32]

Nevertheless, previous studies did not take into account the stability and trim constraints. Thereby, the design study has been refined and focused on the one-duct rocket-dual mode ramjet-rocket concept. A new design of the propulsion system has been performed for a dual mode ramjet using a

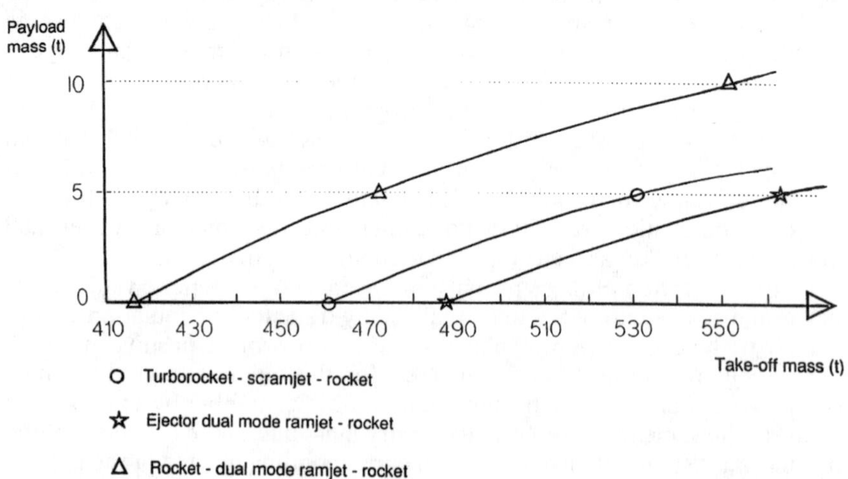

Fig. 39 Global performance comparison.

double thermal throat.[31] Then, under the given hypothesis, a new iteration on the global vehicle design led to a vehicle able to fulfill the mission with a total take-off mass of 487 metric tons without payload or 540 t with a payload of 5 metric tons (Fig. 40)[34]

These results correspond with an assisted horizontal take-off vehicle. Considering difficulty and cost to develop an assisting trolley, vertical take-off appears as an attractive alternative. Performance studies showed that a VTO vehicle having a total take-off mass of 503 t and using a 700 t thrust rocket system would be equivalent with the previous 487 t HTO vehicle.[34]

It is clear that previous results integrate intrinsic performance of combined propulsion system but also the effect of the provided technology level. A short study was made for a vertical take-off/horizontal landing vehicle, exclusively powered by rockets, with the given technology level. This study showed that a 600 metric tons take-off mass vehicle without payload results in a deficit of mass of 14.7 t (difference between mass placed into orbit,

Fig. 40 Generic vehicle.

determined by trajectory simulation, and mass which should be placed into orbit according to the mass budget).[34]

In another respect, a few studies have been performed to evaluate potential interest of in-flight oxygen collection. These studies showed that a high-speed cruise (Mach 5) would drastically reduce the needed initial on-board oxygen and then would make it easier to design a heavy payload vehicle with a reasonable take-off mass.[34]

All these system studies confirmed the potential interest of combined propulsion, using dual mode ramjet, for space launchers. Use of such a propulsion system could improve performance of reusable vehicles whatever the feasible technology level may be:

1) If this level does not allow realization of a fully reusable vehicle exclusively powered by rocket propulsion, combined propulsion could help to break the technology barrier.

2) If it is possible to design a fully reusable rocket powered vehicle, combined propulsion could increase design margins permitting safety, reliability and cost-effectiveness improvement.

C. Development of New Test Facilities

A large part of the effort has been dedicated to the development of new test facilities in order to allow[35] 1) improvement of knowledge on physical phenomena for high-speed air and/or reactive mixtures flows, 2) performing tests of representative scramjet combustor, and 3) evaluation of high-temperature materials in representative environmental conditions.

1. Combustion Test Facilities at ONERA

ATD cells. Several test benches allow to perform connected pipe test for ducted rocket or liquid fueled ramjet up to Mach 4.5 flight conditions (1150 K).

In the scope of PREPHA, a new test bench, named ATD 5, was added to perform connected pipe tests up to Mach 7.5 flight conditions (2400 K) for liquid fueled or hydrogen fueled scramjet combustor. The air temperature level is obtained by a heat exchanger (1000 K) and hydrogen burner with re-oxygenation. Main characteristics are 1) air mass flow: 5 kg/s, 2) air stagnation pressure: 4 MPa, 3) air stagnation temperature: 2400 K, 4) hydrogen mass flow: 300 g/s, 5) hydrogen stagnation pressure: 8 MPa, 6) hydrogen stagnation temperature: 300 K, and 7) test section 100×100 mm^2.

This test bench can be used for combustor testing but also for basic studies on injection system, testing of injection strut concepts or evaluation of materials.

LAERTE. The LAERTE laboratory is devoted to the detailed combustion study with extended use of optical measurement methods (CARS, LIF, Rayleigh scattering, etc.).

A specific test bench has been developed under the PREPHA program to enlarge database on helium/air supersonic mixing and hydrogen/air supersonic mixing and combustion process. The maximum values are 1) air stagnation temperature: 1800 K, 2) air stagnation pressure; 0.8 MPa, 3) He stagnation temperature: 650 K, 4) He stagnation pressure: 0.85 MPa, 5) H2 stagnation temperature: 500 K, 6) H2 stagnation pressure: 0.8 MPa, and 7) test section: 45×45 mm^2.

S4MA wind tunnel. S4MA is a hypersonic wind tunnel generally used for aerodynamic test on reentry vehicles.

Thanks to an alumina pebble bed heater, S4MA wind tunnel can provide 30 kg/s of air up to Mach 6.5 flight conditions (1800 K) and stagnation pressure of 15 MPa without water vitiation.

Studies have been done to define a few modifications of the wind tunnel to allow the test of a large scale scramjet engine up to Mach 7.5 conditions: H2/O2 burner increasing incoming air temperature from 1800 to 2400 K, test device permitting to measure the thrust of an engine equipped with a part of inlet and a part of nozzle.

But, due to budget restrictions, this modification was cancelled. Nevertheless, today, the S4MA wind tunnel is still operational and its upgrade is always feasible, particularly the test of the liquid fueled scramjet, which needs a limited amount of hydrogen (only used for the vitiator) reducing the safety difficulties and the cost of the modification.

2. *Combustion Test Facilities at Aerospatiale Matra*

Hypersonic line. Ramjet test facilities are concentrated in Aerospatiale Matra Missiles Bourges Subdray Center. Two test benches (owned by Celerg—joint venture created by Aerospatiale Matra and SNPE) are dedicated to subsonic combustion ramjet development. They allow connected pipe and semi-freejet tests for full scale military ramjet up to 410 mm in diameter.

Within the frame of Prepha Program, a new test bench (directly owned by Aerospatiale Matra Missiles) has been developed for testing large scale scramjet combustors. Main characteristics are 1) air mass flow: 100 kg/s, 2) air stagnation pressure: 8 MPa, 3) air stagnation temperature: 1800 K, 4) hydrogen mass flow: 1.5 kg/s, 5) hydrogen stagnation pressure: 6 MPa, and 6) hydrogen stagnation temperature: 300 K.

These characteristics give the possibility to test large scale scramjet combustors with hydrogen or hydrocarbon fuels. Generally used for combustion tests, this facility can also be used for technology validation: thermomechanical validation of variable geometry inlet or nozzle, injection strut, strutjets, etc.

Adiabatic compression tube. To be able to perform combustion tests with large scale scramjet combustors up to Mach 8 conditions with nonvitiated air, Aerospatiale Matra proposed and presented an original concept of adiabatic compression tube without piston.

Full scale design and «low» temperature test of a subscale model demonstrated the feasibility and the efficiency of the concept, with a very reasonable budget, it could be possible to run combustion test at Mach 8 flight conditions more than 30 s with a large scale combustor (0.05 m^2 at the combustor entrance section).

Materials test facilities. Previously presented ONERA ATD 5 and Aerospatiale Matra Hypersonic test bench allow material tests. That is the way ATD 5 has been used to realize test of cooling systems made with thermostructural composite materials.

Another way to perform more basic material evaluation, a specific laser oxydation test bench was developed during Prepha Program. This test facility, called BLOx 4, provides representative and accurate environment conditions to analyze oxydation process into materials which could constitute engine and vehicle structures.

Measurement techniques. Prepha Program gave the opportunity to develop measurement techniques, usable in industrial environment, for high-temperature reacting flows: thrust sensor, thin film heat flux gauges, skin-friction measurements, temperature measurements by spectrography, cooled Pitot rake, and spontaneous OH emission.[36]

D. Numerical Means

Scramjet components studies described hereafter have been performed with different existing numerical codes which have been adapted for the particular applications studied.

Nevertheless, a specific effort has been performed to improve and validate the physical models and to evaluate the updated numerical codes. For aero-thermo-chemistry, as for aerodynamics, this effort has been focused on two codes, initially developed by ONERA and usable by industrial partners.

1. Aerodynamics

For external aerodynamics, numerical developments have been mainly focused on the FLU3M code. These developments aim at bringing together all potentialities in a single aerodynamics code (Euler, PNS, FNS, turbulence models, real gases effects, and multi-grid.)[37,38]

Basic experimental studies have been performed by ONERA to improve the physical models related to 1) the boundary layer transition and turbulent boundary layer development along the forebody, 2) the shock-shock and shock-boundary layer interactions in the inlet region, 3) the boundary layer evolution along the nozzle expansion ramp, and 4) the interaction between the propulsive jet and the external boundary layer at the exit of the nozzle.

For this last subject, an experiment was realized at relatively low Mach number with a simple axi-symmetric model in which the external boundary layer thickness and the base between the external afterbody and the jet are representative of a real case (Fig. 41).

2. Aero-Thermochemistry

The effort has been focused on the MSD code, initially developed by ONERA, which has been upgraded to perform supersonic reactive flows

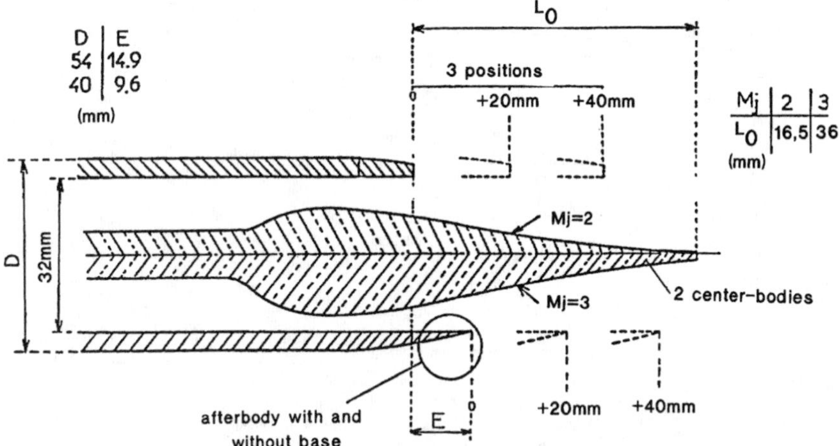

Fig. 41 Experimental study of jet/external boundary layer interaction.

simulation. It solves the unsteady, 3D, averaged Navier-stokes equations by a finite volume algorithm on multi-domain structured curvilinear grids.

The spatial discretisation is derived from the Roe scheme with a MUSCL approach and some applications to improve the robustness in presence of strong shocks or expansion waves and large differences between species density. Different turbulence models (two transport equations for kinetic energy and dissipation model type) and different combustion models (finite rate chemistry and turbulent combustion) can be used.[39,40]

Mixing and combustion process understanding and modeling have been improved thanks to basic research, led by different laboratories of the National Scientific Research Center (CNRS) and ONERA on its dedicated test bench in Laerte which gave, in a first step, detailed analysis of the hydrogen/air mixing and combustion process in the simple case of a parallel cylindrical fuel jet by using optical diagnostic techniques (LIF, CARS) (Figs. 42 and 43). It is clear that further effort has to be continued during several years to get all the benefit of the Laerte test facility potential.

3. *State of the Art*

Of course a lot of improvements are still necessary to have on disposal the very accurate numerical codes which would be indispensable for developing a high-speed airbreathing vehicle.

Nevertheless, for aero-thermochemistry as for aerodynamics, existing codes will soon be used daily for subsonic combustion ramjet missiles and strongly contribute to their development.

In addition the experimentation/computation comparisons made during the PREPHA Program for each scramjet component showed that the available codes are already able to provide an efficient support to the design studies.

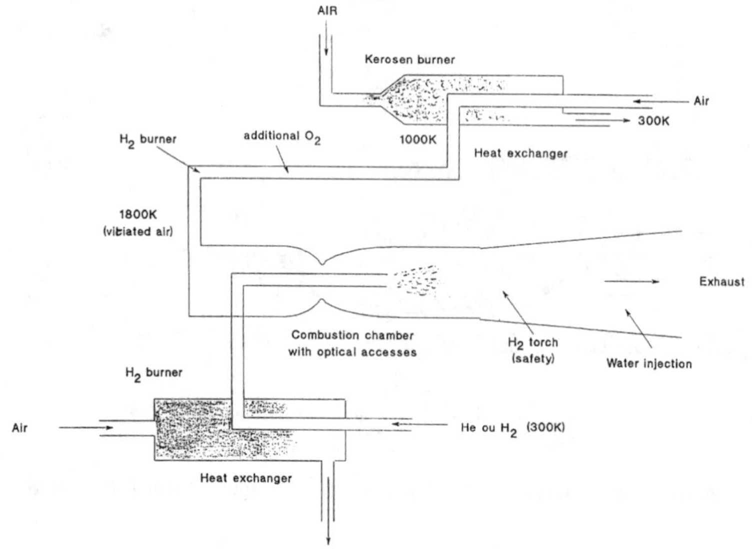

Fig. 42 LAERTE test facility—set-up design.

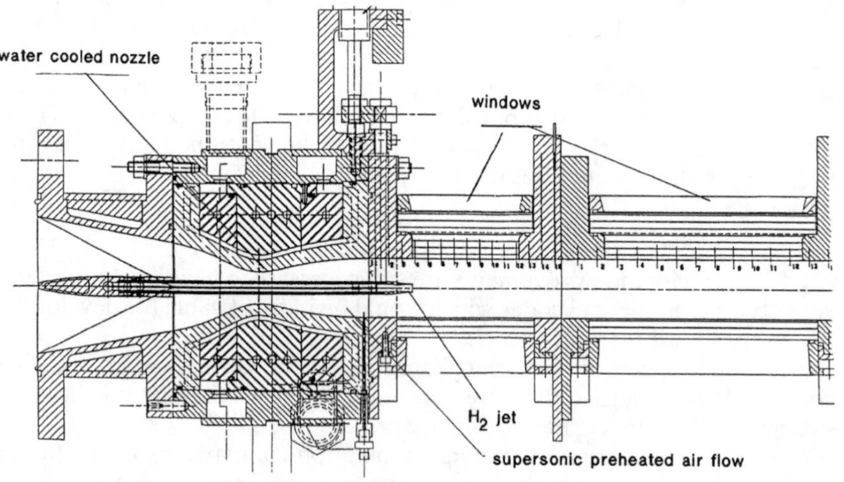

Fig. 43 LAERTE—detailed view of the test section.

E. Development of Scramjet Components

1. *Forebody*

Numerical studies have been performed to analyze the effect of different design parameters: bluntness radius of the nose, windward surface curvature, spatula effect, etc.Even if no definitive conclusion has been obtained, it can be noted that the forebody of the generic vehicle was designed with a low compression windward surface (effect increased by angle of attack) and a limited, but real, effect of spatula.

In addition, a generic forebody has been tested in ONERA S4MA wind tunnel with wall pressure and thermal flux measurements, total and static pressure rake in the plan of inlet entrance and infra-red thermography. Obtained results gave a good evaluation of the capability of numerical codes to predict the flow at the entrance of the inlet and the thermal conditions along the forebody.

2. *Inlet*

Studies are essentially focused on the range of Mach numbers from 6 to 12 (specially for the experimental approach) but the constraints induced by operating at low Mach numbers are taken into account. Particularly at ONERA, studies have been oriented to the definition of one-duct engine inlets able to provide air needed by the engine between Mach 2 and Mach 12. For a very large part of work, it has been chosen to consider fixed combustor geometry in order to limit the technological difficulties. Then, the adaptation of the propulsive duct has been realized with the use of variable capture area inlet (geometry is fixed beyond Mach 6) which ensures the right geometrical contraction ratio evolution with the Mach number. This kind of inlet has been studied by ONERA for several years,[41–43] and the basic concepts have been adapted to the needs and the constraints of the generic vehicle (Fig. 44). Several internal boundary layer bleeds are used below Mach 5/6.

Inlet concepts studies have been supported by a first experiment on a modular model which was tested at Mach 5 and 7 in the R2CH ONERA windtunnel (Fig. 45).[44] This model, which represented only the compression ramps and the injection system, was fitted with a starting device, which allowed to start the inlet with a small geometrical contraction ratio and to increase it after and whose interest and efficiency have been confirmed during the first test series. Experiments concerned parametric studies about contraction ratio, effects of forebody boundary layer thickness, maximum deviation angle on the compression ramps, etc.Results showed that it is possible to obtain a geometrical contraction ratio superior to 4 in spite of high deviation angles needed to limit the length, and then the mass, of an operational inlet.

After this first experimental approach, specifically focused on scramjet mode operation of the inlet, a new model has been defined and manufactured for a second experiment from Mach 2 to 5.5 (ONERA S3MA wind

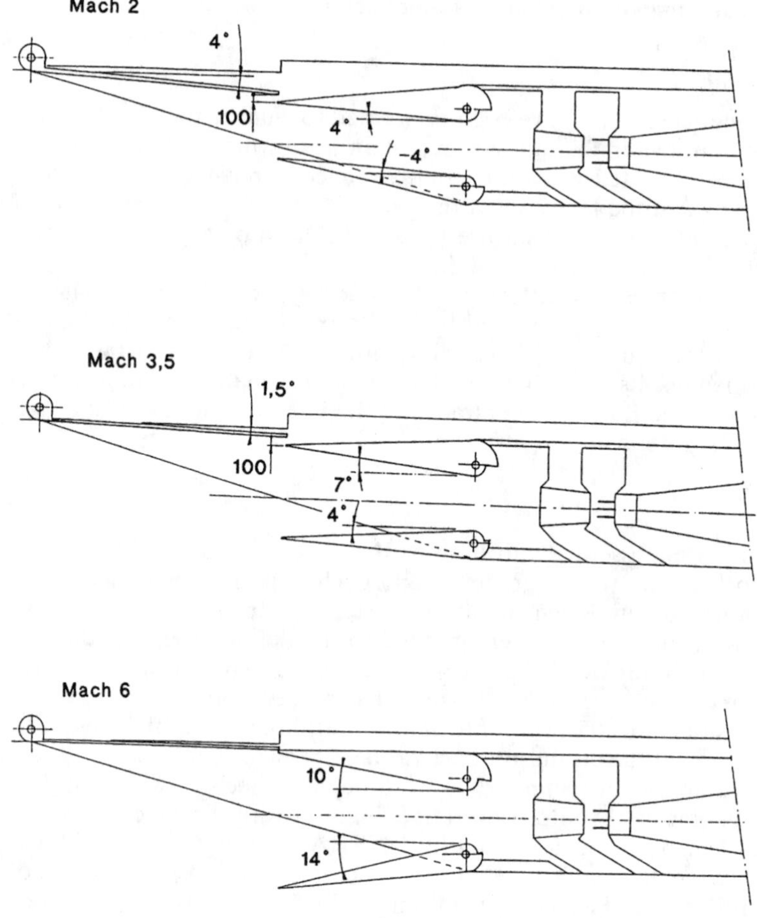

Fig. 44 Variable capture area inlets concepts.

tunnel). This new model is taking into account the last evolution concerning injection struts arrangement (injection struts are placed downstream the minimum section of the inlet). Instrumentation and test device are defined to explore the complete inlet characteristic curve from supercritical to subcritical regime with flowfield analysis in minimum section and after injection struts.

Again, the experiment aimed at defining the maximum contraction ratio, the effects of forebody boundary layer thickness, the maximum deviation angle on the compression ramps, etc. It resulted in a relatively good agreement between measured performances and assumed performances for the generic SSTO vehicle in spite of a lack of time to complete a real optimization. Results confirmed that it would not be reasonable to assume the feasibility of a very high performance level in the whole Mach number range

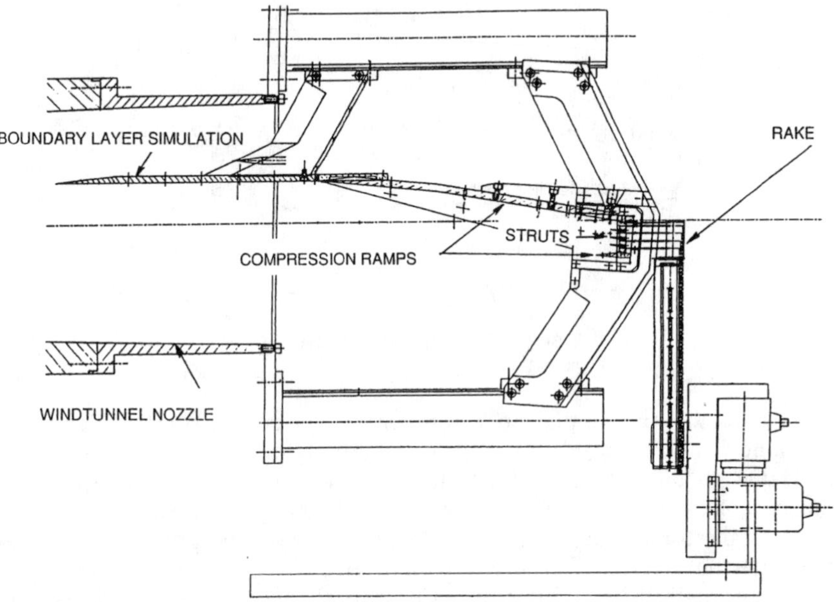

Fig. 45 Inlet test device.

from Mach 1.5 to 12 if the mass has to be drastically limited as in the case of the SSTO vehicle.

Further tests should be performed to take advantage of the large modularity of the model to explore in detail the effect of different design parameters. Moreover, the performed test series showed that the boundary layer bleeds placed on upper and lower compression ramps are able to absorb reflecting oblic shock and, then, could be a key point for reaching the needed performances during the subsonic combustion phase. Again, further tests would permit understanding of the conditions in which these boundary layer bleeds can be efficient.

3. *Nozzle and Afterbody*

Studies were focused on 2D Single Expansion Ramp Nozzle (SERN) concept. A numerical approach has been used to determine the influence of different design parameters: length of the movable or fixed cowl flap, expansion ramp evolution, etc. The global trade-off on the vehicle led us to choose a short fixed inferior flap and a limited height of the expansion ramp.[45]

In order to confirm these results and to obtain some global data allowing a general evaluation of the FLU3M code in the case of nozzle and afterbody, a generic model has been designed and manufactured by Dassault Aviation and ONERA for a test series performed at Mach 6.4 in the S4MA wind tunnel in Modane ONERA Center (Fig. 44).[46] In this model, an H_2 burner

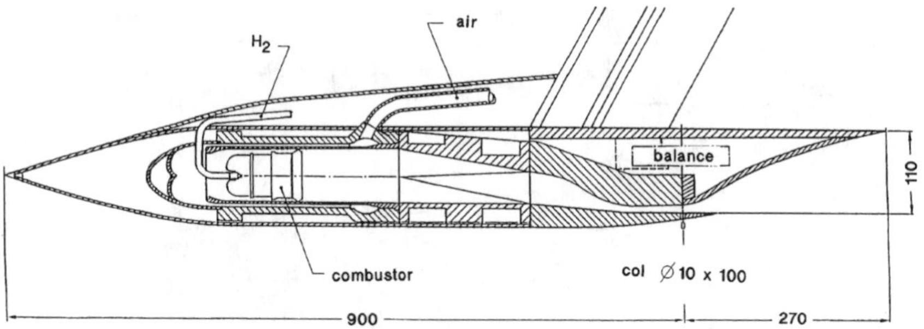

Fig. 46 Afterbody/nozzle test device.

makes possible the simulation of temperature effect up to 1500 K. Initially, a balance was envisaged to weigh the afterbody or the expansion ramp. In fact, due to the technical difficulties attached to this solution (pressure measurement in the interface between weighed and fixed parts), the evaluation of the forces has been ensured only by a detailed pressure measurement on the external and internal walls. This test allowed to evaluate the codes, particularly FLU3M, in the case of nozzle and afterbody and to check the effect of different parameters on the propulsive jet interaction with external airframe boundary layer.[47]

The generic model was designed to be compatible with ONERA S3MA wind-tunnel capacities in order to perform test at low Mach number to understand over-expanded operation of a SERN system. Unfortunately, budget restriction led to cancel this extension of the experimental investigation. This point needs to be completed because a variation of the thrust direction at low Mach number can dramatically affect the global performance of an SSTO vehicle.

4. *Combustor*

A large part of the program was focused on the study of the scramjet combustor with the development of two experimental combustors: the Chamois combustor and the MONOMAT combustor.

On the base of a common numerical conceptual design effort, the Chamois combustor has been developed by Aerospatiale. In spite of its limited dimensions (entrance area of 212×212 mm^2), the experimental combustor Chamois presents as far as possible the same difficulties as a large operational combustor (Fig. 47): 1) fuel injection by struts, 2) wall/injection strut interaction, 3) injection strut/injection strut interaction, and 4) upstream flow nonuniformity (boundary layer and shock waves); but the wall losses (heat transfer and skin friction) are oversized.

In a first step, test series were performed in ONERA ATD 5 test facility to obtain data on fuel jet penetration, air/hydrogen ignition, fuel/air mixing and combustion stability improvement techniques.

Numerical studies have been used to compare different concepts of injec-

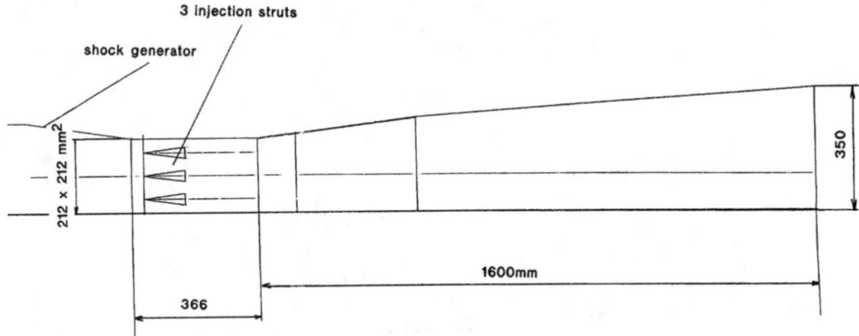

Fig. 47 Experimental combustor geometry.

tion struts (Figs. 48 and 49).[48] Two concepts have been selected and tested in isolated configuration with water cooled models in the ATD test facility in ONERA Palaiseau (Figs. 50 and 51).[49] This study has been completed by experimental basic research concerning fuel/air mixing improvement techniques (upstream secondary fuel jet and shape of the strut) and by numerical studies on the three injection struts configuration of the Chamois combustor aiming at optimizing the struts definition and at increasing the global mixing efficiency.

A specific study has been led by SEP to define and realize a metallic injection strut, completely cooled by the injected hydrogen. This injection strut was tested in ONERA ATD test facility in 1996 (Ref. 50).

The second step concerned the integration of the injection struts in the

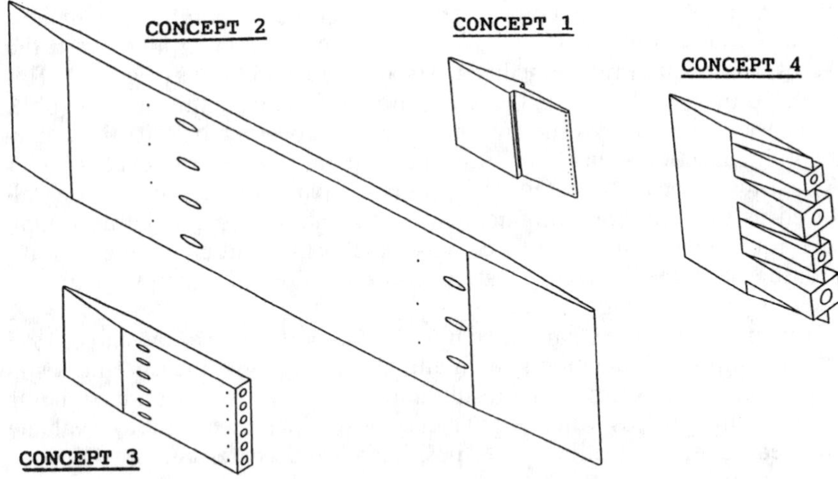

Fig. 48 Injection struts concepts.

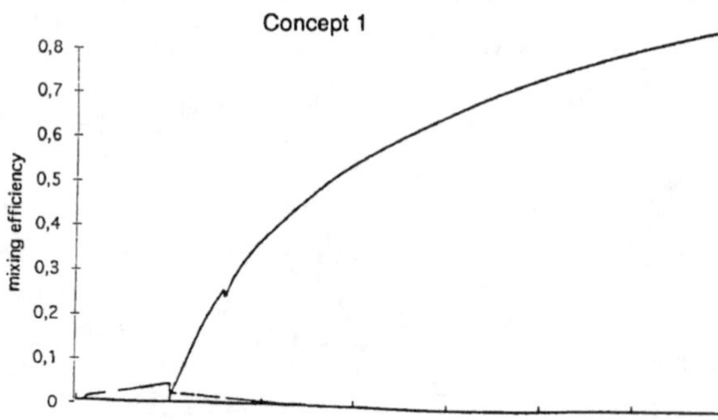

Fig. 49 Mixing efficiency.

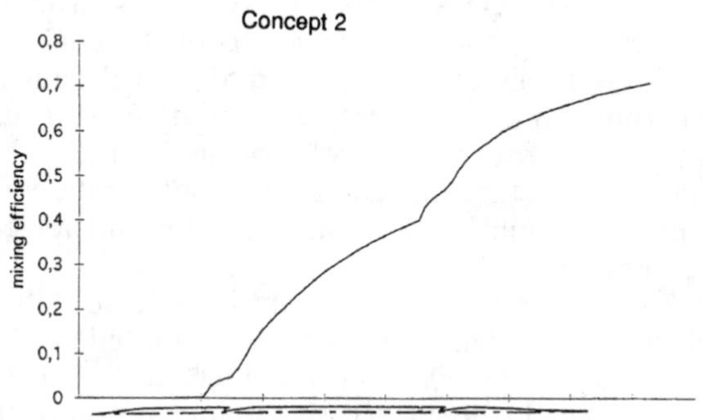

combustor. After numerical studies allowed to define the combustor geometry, a first direct connected pipe test of the experimental combustor Chamois has been successfully performed with a uniform incoming air flow in the Aerospatiale Subdray test facility at Mach 6 flight conditions (Fig. 52).[51] This work has been completed in the case of nonuniform incoming air flow (oblic shock wave generator at the entrance of the combustor) to verify the capacity of the injection struts to operate in a large range of upstream conditions. A first test series, associated with numerical parametric investigations, allowed us to study how the injection configuration changes the maximum fuel equivalence ratio before combustor chocking (number and relative position of injection struts, location, and mass flow of injection points on each strut).[52]

The final test of the Chamois combustor was performed at the end of 1997 with an improved injection system and all the available measurement systems.[53] All these experimental results allowed to improve the understanding of chocking process when equivalence ratio increases and to evaluate adapted relative injection struts positioning and corresponding injection distribution.

Initially, the S4MA test facility in ONERA Modane was to be upgraded

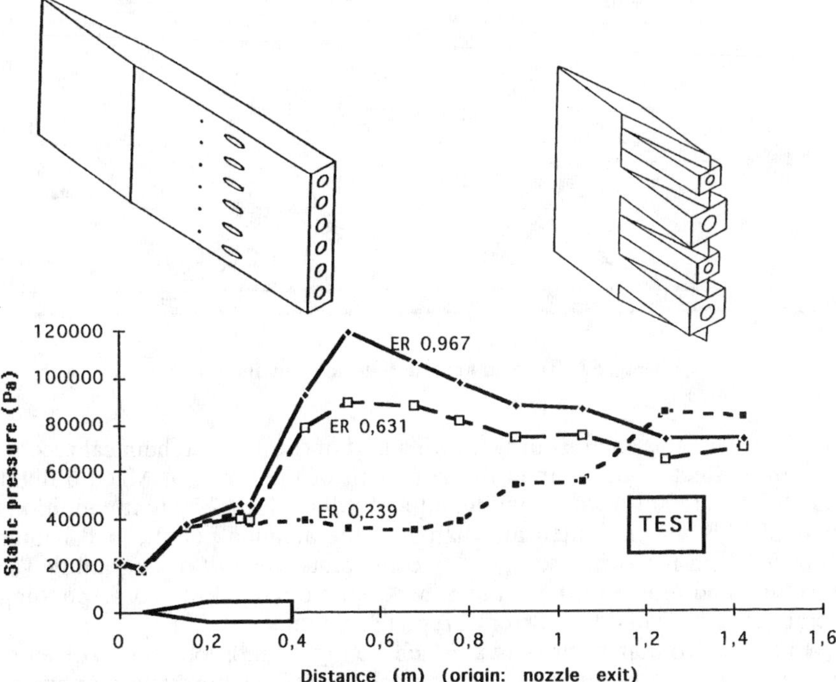

Fig. 50 Selected injection struts.

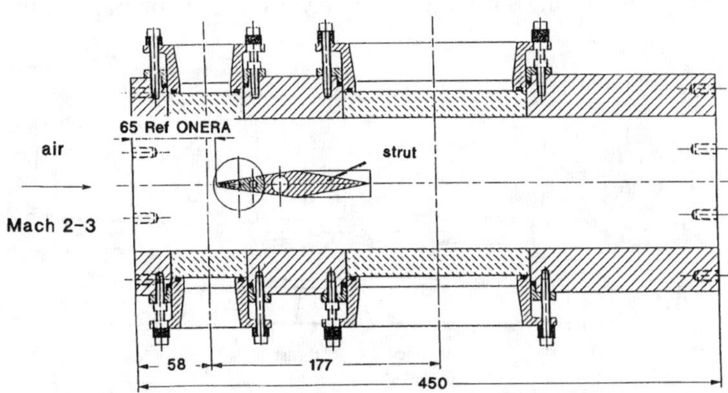

Fig. 51 ATD test facility—Top view of the test section.

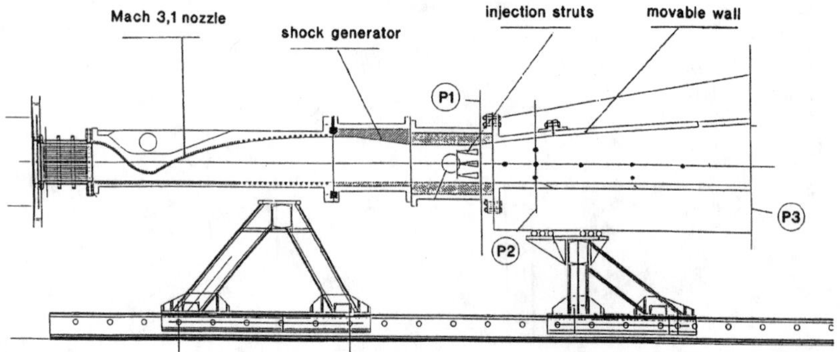

Fig. 52 Tests in aerospatiale subdray facility.

to allow Chamois combustor test with a part of the nozzle (chemical recombination effects) and a direct measurement of the thrust at Mach 6 flight conditions (1800 K) with nonvitiated air and at Mach 7.5 flight conditions with slightly water-vitiated air, thanks to the alumina pebbles bed heater (Fig. 53).[35] In fact, because of safety constraints, the cost of upgrading was too high and led us to defer the S4MA upgrading and the corresponding scramjet test beyond the current Prepha program.

In order to obtain some data at Mach 7.5 flight conditions and to observe the effects of water vitiation, a new small single injection strut combustor (100×100 mm^2 at the entrance), called Monomat, was defined by ONERA to perform complementary experiments in ATD 5 facility at Mach 7.5 conditions with vitiated air and at Mach 6 conditions with more or less vitiated air thanks the heat exchanger supplying the test facility (1000 K non-vitiated air). In other respects, Monomat combustor was used to analyze the transonic combustion mode in order to confirm the feasibility of the thermal chocked dual mode ramjet.[54]

Initially, the Chamois combustor thrust was to be directly measured by a weighing system during S4MA test. A study was led to compare accuracy of

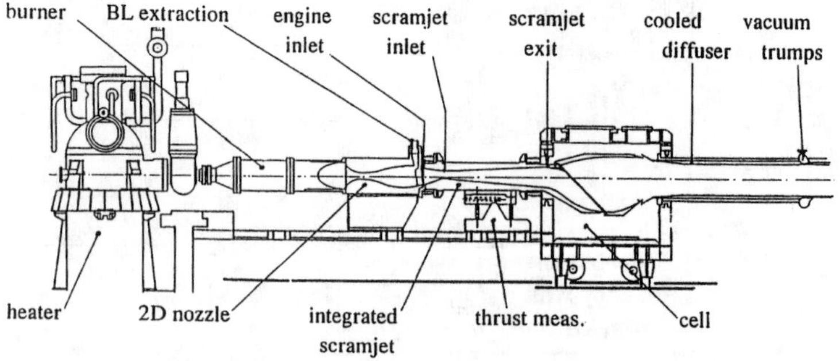

Fig. 53 Tests in S4MA facility.

Fig. 7b (Chapter 4) The engine model installed in the test cell. Heat-sink model.

Fig. 29a (Chapter 5) Uncooled heat-sink hydrogen- and kerosene-fueled scramjet combustors made of niobium alloy with a protective antioxidant coating (from V. Avrashkov et al., 1995): a) hydrogen-fueled combustor during direct-connect tests in MAI.

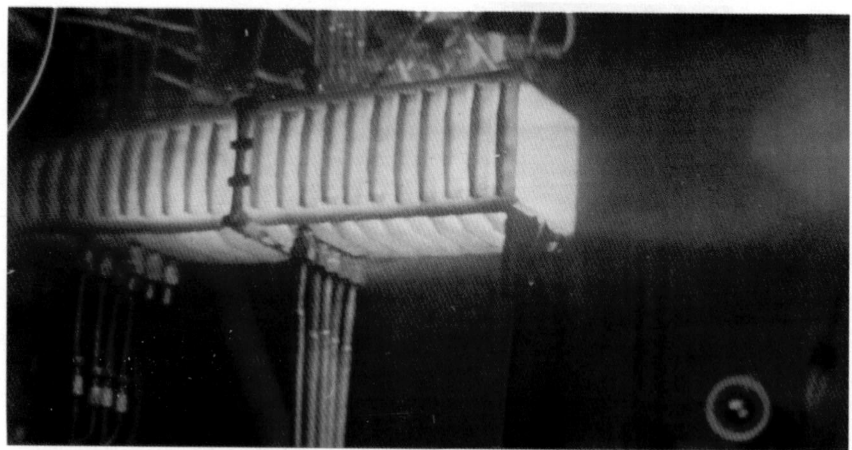

Fig. 29a (Chapter 5) Uncooled heat-sink hydrogen- and kerosene-fueled scramjet combustors made of niobium alloy with a protective antioxidant coating (from V. Avrashkov et al., 1995); b) kerosene-fueled combustor installed at direct-connect test facility.

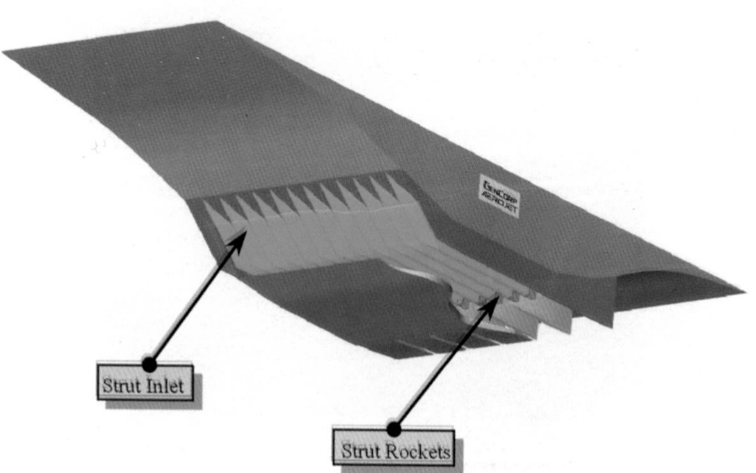

Fig. 2 (Chapter 11) Strutjet engine concept.

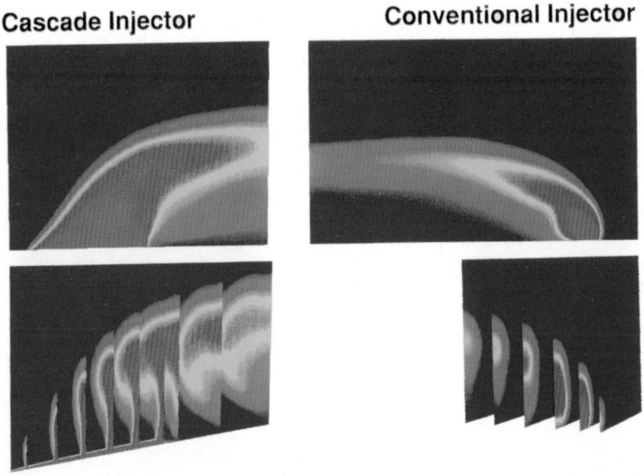

Cascade Injector

Conventional Injector

Fig. 13 (Chapter 11) Cascade injector penetrates deeper than conventional injector.

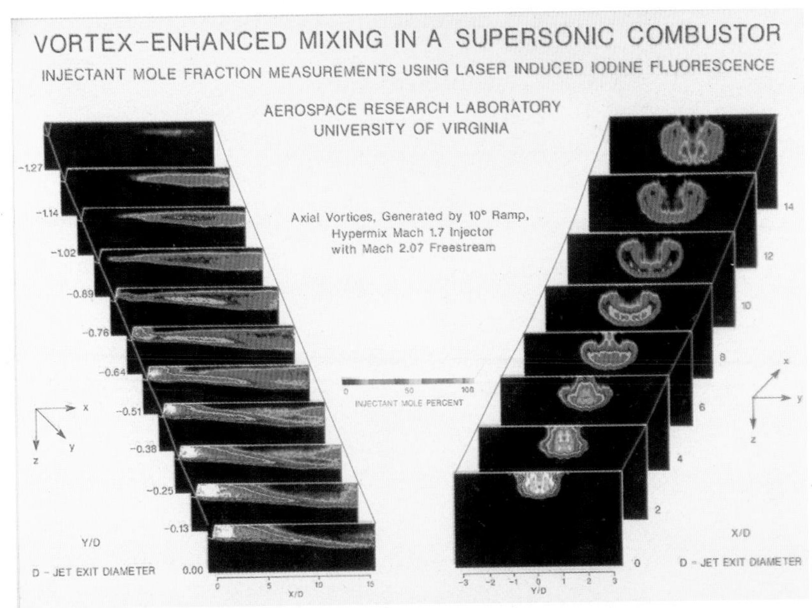

Fig. 48a (Chapter 17) Ramp mixing PLIF data at M = 2.2, stream wize laser sheet (left), normal to stream laser sheet (right).

two thrust measurement methods: 1) global combustor weighing with treatment of connection between fixed and weighed parts (bellows) and drag of resulting bases; and 2) wall pressure integration taking into account heat and skin friction losses (respectively measured by thin film transducers, developed by ONERA, and skin friction balances, developed by University of Virginia[55]).

Because of the impossibility of fitting injection struts with extensive instrumentation, the second solution needs to separately weigh them. Under this condition, the second solution appeared slightly less accurate than the first but less expensive and giving more usable data for test/calculation comparison. That is why an injection strut weighing system was developed by ONERA and has been tested in the Monomat combustor in 2000.

In addition, a simple thrust balance, measuring the variation of the exit momentum, was developed by Aerospatiale Matra and currently used for Chamois combustor tests. It permits a rapid comparison between different combustor or injection configuration.[52]

5. *Propulsion Integration*

For the prediction of the thrust-drag balance, Dassault Aviation has undertaken specific validation studies in collaboration with the Russian Academy of Science (ITAM in Novosibirsk). A ram/scramjet generic configuration has been tested in a free-jet hypersonic tunnel at Mach 8 with hydrogen combustion and the axial force has been measured.[56]

F. Materials and Cooled Structures

Materials studies have been undertaken at two levels. The first one aimed at the adaptation of existing composite materials (fibers: carbon and silicon carbide; matrix: silicon carbide, carbon or ceramic glass) to the specific environment of the high-speed airbreathing propulsion using hydrogen. Numerical means were developed to predict oxydation process[57] and protections were tested in the new BLOx4 test facility.

In addition, ATD 5 test facility was used to perform test series to verify the feasibility of composite parts, cooled by hydrogen: convective or transpiration (with protective effect against oxydation) systems.[58]

A second level of work concerned more basic studies for not very well known materials: intermetallic compounds and metallic or intermetallic matrix composites. The potential of intermetallic materials was evaluated at the laboratory level while metallic matrix composites were essentially studied from the point of view of the manufacturing of a small part typical of fuselage.[59]

G. Flight Testing

Considering the difficulties of simulating the flight conditions on ground, the cost of facilities and tests allowing the study of full-scale engine and the extreme sensitivity of the aeropropulsive balance for an operational vehicle, it is clear that the development of the scramjet technology needs a large phase of flight experimentation. A demonstrator of an operational vehicle being very expensive and the associated technical risk being very high, such flight experimentation should begin with the development of one or several small experimental vehicles.

A first step was made thanks to the basic flight experiments undertaken with the CIAM from Moscow on a boosted axisymmetric hydrogen fueled engine (Figs. 54 and 55).[60] After a first test realized in November 1991, CIAM performed a new test in November 1992 with a limited technical participation of the French Prepha partners. During this test, it has been possible to obtain a supersonic combustion in the flight Mach number range 5 – 5.3. But, the design of the tested engine is very close to the ESOPE combustor design and the limited height of the combustion chamber (~ 15 mm at the entrance) is not representative of a large operational scramjet. Moreover, the engine being boosted, is a good evaluation of the aeropropulsive balance is not possible. Indeed, these tests are to be considered a preliminary experience allowing 1) to acquire a first know-how on simple technologies usable for experimental scramjet, 2) to implement a methodology of flight data analysis, 3) to compare results of ground and flight tests, and 4) to evaluate numerical codes with a real case.

Beyond this first step, an analysis of the needs of flight experiments has been undertaken in Prepha.[61] For each component of an high-speed airbreathing vehicle and for each concerned scientific subject, a list of problems has been defined. For each of them, the capacity of the ground tests to

Fig. 54 Axisymmetric Scramjet from CIAM—general view of the experimental system.

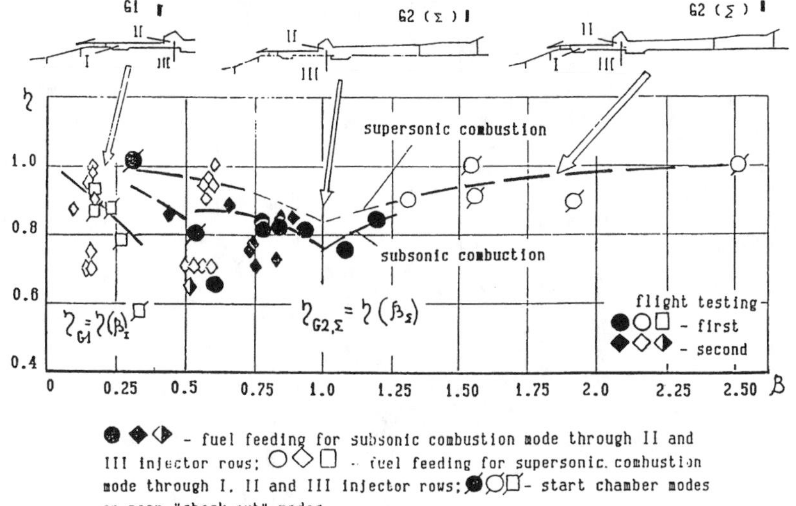

Fig. 55 CIAM's Scramjet—combustion efficiency = *f* (ER).

provide necessary data and their limitations have been identified. Then, the potential interest of flight tests and the corresponding requirements in terms of representativity (design and flight path) and instrumentation have been established. The capacity of a large range of typical experimental vehicles to fill these requirements has been evaluated and compared. Finally, a "critical" level of each problem for each possible application of the hypersonic airbreathing propulsion (space launcher, military aircraft, missile) has been determined to balance the previous results in order to establish a global comparison of the different possible types of flight experimentation as shown in Table 6.

Assuming the results of this analysis, projects of self-powered experimental vehicles have been sketched by Aerospatiale Matra (type 4, Fig. 56) and ONERA (type 5, Figs. 57 and 58).[62] Budgetary limits led Prepha partners to stop this kind of study in the scope of Prepha and then, it has not been possible to refine the comparison on the base of detailed projects especially integrating a complete reflection about on-board instrumentation.

VI. Perspectives

Possible middle and long-term applications of the high-speed airbreathing propulsion are clearly identified: missiles and reusable space launchers.

In both these cases, the better engine will probably be the dual-mode ramjet whose study has been started during the Prepha Program.

By taking advantage of the lessons learned thanks to this Program (with resulting numerical and experimental means) and of their large several decades long experience in the field of subsonic combustion ramjet technol-

Table 6 Types of Flight Experimentation

Type	Part of needs covered by different experimental vehicles, %								
	Forebody	Inlet	Combustor	Afterbody nozzle	Fuel system	Materials structures	NGC system	Vehicle	Total
Type 1[a]	45	5	4	0	0	22	10	11	12
Type 2[b]	21	50	52	33	45	31	20	11	30
Type 3[c]	92	5	4	0	0	22	70	37	32
Type 4[d]	92	95	54	64	50	31	90	76	74
Type 5[e]	92	95	80	64	100	63	100	100	88

[a]Type 1: flying test bench without airbreathing engine on board of an existing servicing vehicle or booster.
[b]Type 2: flying test bench with airbreathing engine on board of an existing servicing vehicle or booster.
[c]Type 3: autonomous experimental vehicle without airbreathing engine boosted before the test phase.
[d]Type 4: autonomous experimental vehicle with airbreathing engine boosted before the test phase.
[e]Type 5: autonomous experimental vehicle with airbreathing engine self-accelerated during the test phase.

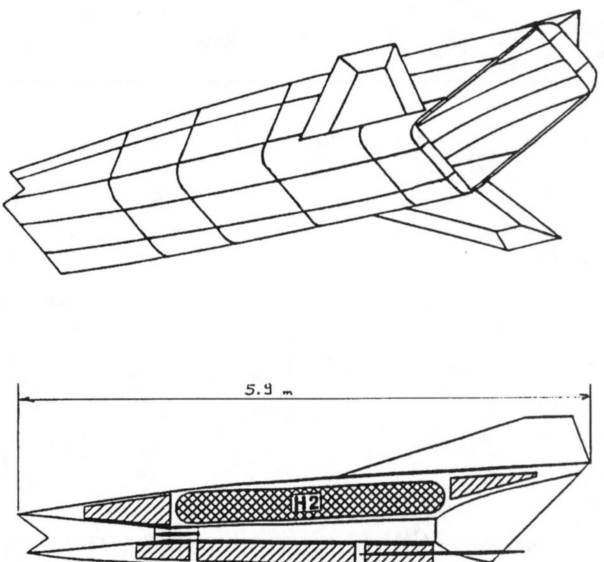

Fig. 56 Aerospatiale's experimental vehicle project.

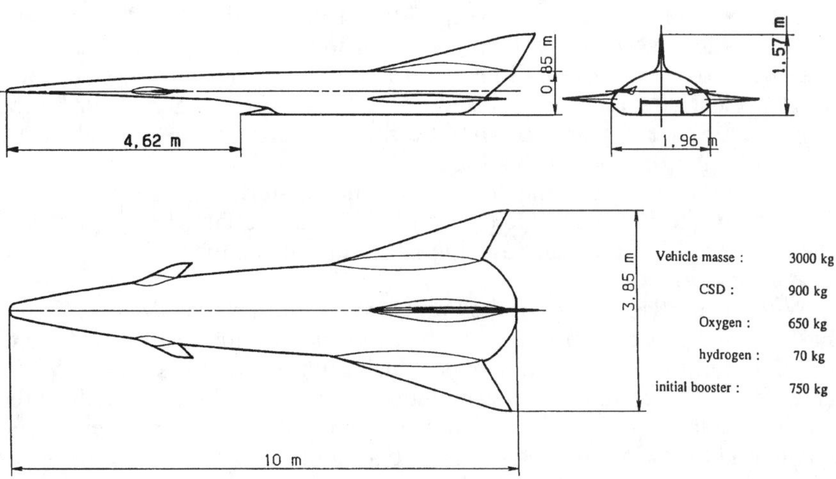

Fig. 57 Experimental vehicle—ARCHYTAS project from ONERA.

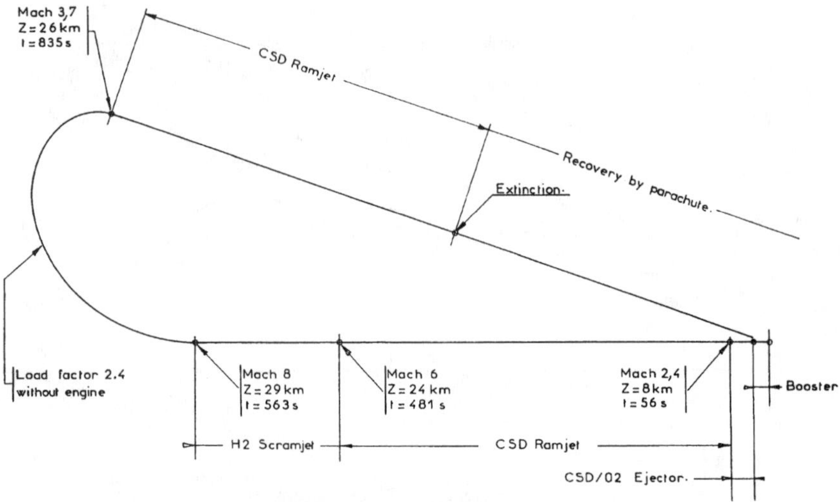

Fig. 58 ARCHYTAS project—typical trajectory.

ogy, Aerospatiale Matra Missiles and ONERA are today carrying on the development of the dual-mode ramjet technology through different research and technology efforts in order to improve the mastery of high-speed airbreathing propulsion in the view of its future civilian and military applications.

A. Space Application

The Prepha Program definitively ended in 1999. Today, July 2000, in spite of the possible interest of combined propulsion for fully reusable launchers, Europe is only considering long-term applications and no follow-up program is clearly foreseen by the French National Space Agency nor by the European Space Agency to continue the development of the high-speed airbreathing propulsion technology.

In order to preserve the intellectual and material investment and to improve mastery of combined propulsion for space launcher application, Aerospatiale Matra Missiles and ONERA have taken the initiative in starting further works.

With the partial support of the French MoD, Aerospatiale Matra Missiles has undertaken a cooperation with the Moscow Aviation Institute (MAI) to develop and test on ground a Wide Mach number Range Ramjet (WRR) (Fig. 59).[63,64] Preliminary experimental and numerical studies led to the design of a large scale WRR Prototype (0.05m^2 at the combustor entrance) having the following characteristics: 1) operation from at least Mach 3 up to Mach 12, 2) use of movable panels during operation along the trajectory, 3) optimization of propulsive performance (checked with theoretical studies, CFD work, and thrust measurement during tests), 4) modification of the

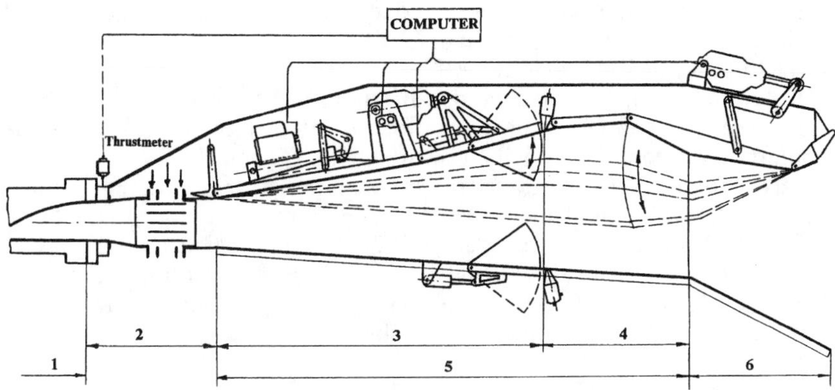

Fig. 59 Wide range ramjet - principle.

internal geometry by a control-command computer connected with sensors on the engine in order to maximize the performance in real time, 5) use of subsonic and then supersonic combustion, 6) use of kerosene and then hydrogen as fuels, 7) structural ability to reach at least Mach 12 flight conditions, 8) realization of a fuel-cooled structure and sealing of movable elements, 9) large scale engine (entrance area of 0.05m^2, several meters length, and injection system in the duct), and 10) possible extension to the higher speeds by the use of ODWE mode.

The cooperation aims at developing the prototype and at testing it to obtain the expected performances by optimizing the geometry at each flight condition up to Mach 6.5.

The possibility of testing an extended set of geometry will be used to understand in detail the working process of a two-mode ramjet and how to manage the transition between subsonic and supersonic combustion modes. Indeed, having at our disposal a fully variable geometry combustion chamber, it is possible to analyze the way to obtain the maximum thrust under any set of inlet conditions and any equivalence ratio. This should provide a large amount of information concerning the choice of the combustion chamber geometry for any engine project, which includes a two-mode ramjet.

Beyond this first goal, many more general results for scientific and technological respects are expected and will be usable for other types of two-mode ramjets which could be developed in the future (dual-mode ramjet tuning, using hydrogen bubbled kerosene, fuel-cooled structures).

Preliminary tests have been performed in order to check combustion at different regimes and to validate the structures which will equip the Prototype. Cooled panels, leading edges, injection struts have been designed, manufactured and systematically tested under hard conditions.[65] The WRR Prototype is assembled and functional test are in progress in MAI.[66] It was delivered at Aerospatiale Matra Missiles Bourges center in October 2000 for future combustion test up to flight Mach number 6.5 conditions. Considering the cost of this kind of test and the need to explore higher flight conditions, a additional subscale engine will be developed in 2000 for testing at MAI up to Mach 7.5 conditions.

In parallel with these technological studies, system studies have been performed for demonstrating the potential interest of the fully variable geometry for a combined engine powering a SSTO vehicle.[67]

Aerospatiale Matra is developing other national or international cooperations: 1) development and use of measurement techniques with the MAI, the French National Research Center (CNRS) and specialized company, 2) basic research on supersonic combustion with French National Research Center (CNRS), DLR, CIAM and Kazakhstan State National University, 3) inlet studies with ITAM in Novosibirsk,[68] and 4) experimental vehicles design with CIAM and LII.

Under internal funding, Aerospatiale Matra has also developed and tested a carbon/carbon hydrogen cooled injection strut, called "St LEME" (Fig. 60).[69] This strut has been successfully tested at (stagnation pressure 2.7 MPa and stagnation temperature 1400 K).

In addition, ONERA and German DLR decided to join their efforts in a common research program (1998–2001).[70,71] This program aims at studying an airbreathing propulsion system working in a Mach number range from 4 to 8, using subsonic then supersonic combustion, and at developing a methodology for ground and flight performance demonstration.

In this view, the common work is focused on 1) hydrogen fueled dual mode ramjet development: a) basic research and modeling for aero-thermo-

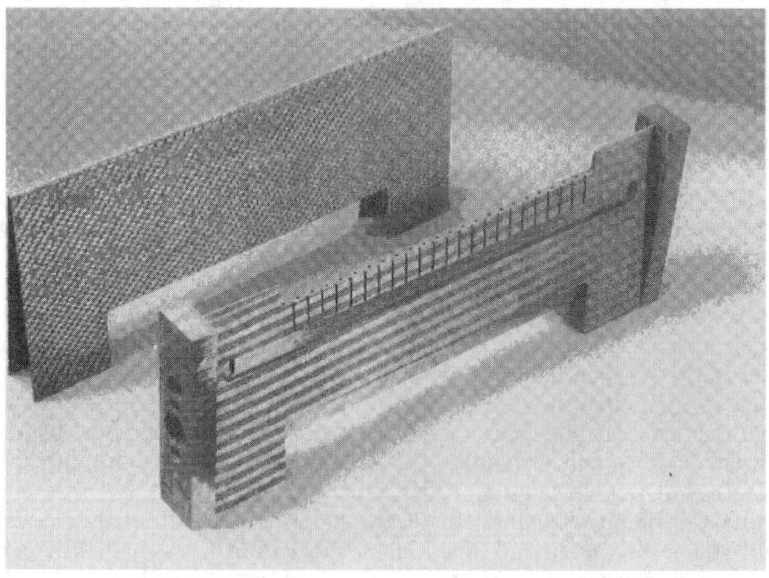

Fig. 60 SAINT-ELME carbon/carbon injection strut.

chemistry and internal aerodynamics, b) codes development and validation, c) conceptual design and test of dual mode ramjet components (inlet, combustor, injectors, nozzle), d) development and adaptation of advanced measurement techniques; and 2) defining and validating a methodology to assess, on ground and in flight, performances of an airbreathing engine (particularly vehicle aero-propulsive balance): a) development of the methodology, b) application to a generic experimental vehicle (Mach 4 to Mach 8) used as lead configuration (Fig. 61), c) assessment of in-flight instrumentation and measurement techniques.

The considered engine is a dual mode ramjet using two consecutive combustion chambers with two stages of strut injectors (Fig. 62). The downstream chamber is used for subsonic combustion regime with a thermal throttling while the upstream chamber is used for supersonic combustion regime.

A combustion chamber model has been developed and has been tested in 2000 in the ATD 5 test facility at ONERA Palaiseau center. In 1999, some studies were used to define the vehicle shape and to optimize its aerodynamic performances.[72] Particularly, the concept of the inlet, directly derived from Prepha program, has been tested in S3MA windtunnel at ONERA Modane test center.

B. Missile Application

At this time, it is not clearly established that cruising faster than Mach 6 gives a decisive advantage for the vulnerability which compensates the complexity and cost. Nevertheless, before the evaluation of operational applications, it is necessary to develop specific technologies for liquid fueled dual mode ramjet in order to define the right input for future system studies.

Aerospatiale Matra Missiles has led a few system studies to define what could be a missile powered by a dual-mode ramjet.[73]

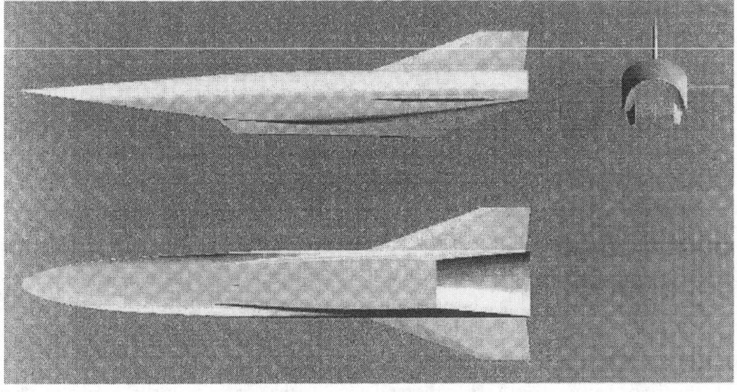

Fig. 61 General configuration of the JAPHAR vehicle.

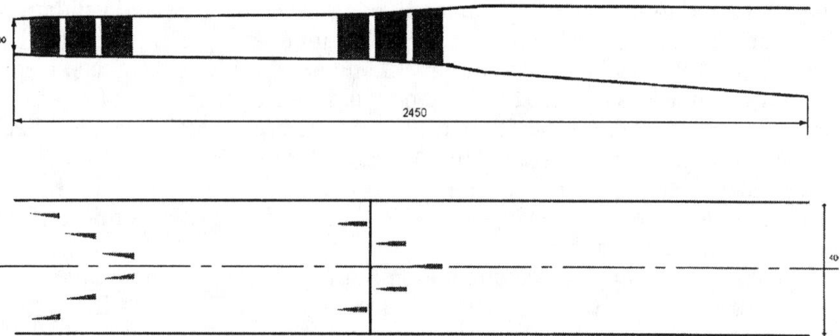

Fig. 62 Scheme of the JAPHAR engine.

Limited activities have already been performed by Aerospatiale Matra Missiles and ONERA to acquire a first know-how in the field of supersonic kerosene combustion. Particularly, late in 1997, Aerospatiale Matra Missiles performed a first combustion test with Chamois scramjet combustor, fueled by kerosene,[73] while ONERA tested an injection strut in ATD 5.

The "St ELME" injection strut development and test demonstrated the mastery of carbon/carbon technology for hydrogen cooled panels.[69] Under internal funding, Aerospatiale Matra is continuing to develop the technology of fuel cooled structure. The results of the "St ELME" program and of the experiments performed during PREPHA Program led to the expectation that thermostructural composite structures, cooled by endothermic fuels are feasible from the mechanical and thermal point of view with reasonable cost and mass. That is why Aerospatiale Matra is now developing and testing a specific technology for endothermic fuel cooled composite structures, taking advantage of its mastery of specific process for manufacturing the fibers pre-form of composite materials. This technology will give the capacity to realize a monolithic fuel cooled combustion chamber with the following advantages, by comparison with the "classical" technology of assembled cooled panels: 1) two-dimensional chamber without particular treatment for the corners, 2) limited connection problem, 3) limited weight, and 4) large design margin.[74]

By another way, ONERA is co-operating with Pratt&Whitney and SNECMA in the A3CP program to develop a composite materials technology for fuel-cooled structures under the aegis of DGA and USAF.[75]

Beyond these first steps, a research and technology program, called Promethee, started in 1999, under the aegis of French MoD.[76] This program aims at improving the mastery of the hydrocarbon fueled dual mode ramjet technology. It is oriented by the conceptual design of a generic air-to-ground missile (Fig. 63). A global comparison has been performed between different dual-mode ramjet concepts (impact on missile performances, technological risks, operational use, cost). It led to the selection of a variable geometry (one axis of rotation) two-mode ramjet operating from Mach 1.8 to Mach 8 (Fig.

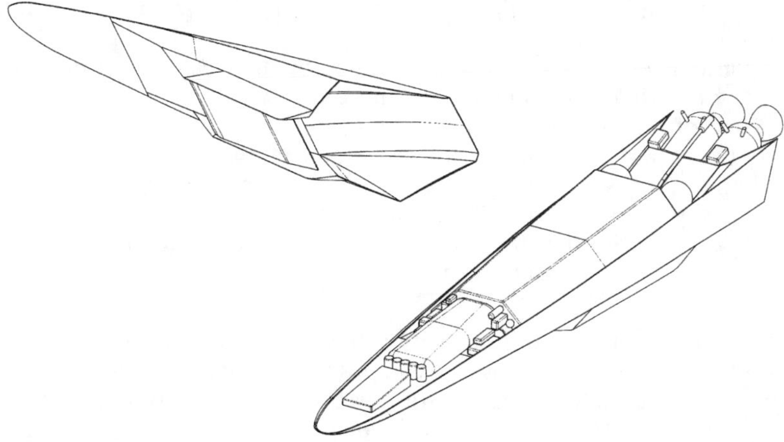

Fig. 63 Generic Promethee missile.

64). After a preliminary design provided a first estimate of performances and mass budget, numerical flight simulations have been performed to evaluate the achievable global performances of the missile.

In parallel, a full-scale model (with a reduced width) has been defined and will be manufactured in the view of experimental evaluation and optimization of the selected airbreathing propulsion system concept. 1) by separately testing the inlet in a low temperature windtunnel and the combustion chamber in a connected pipe configuration in the ATD5 test facility at ONERA Palaiseau Center (step 1) and 2) by globally testing of the inlet and the combustion chamber in a semi-free jet configuration at Aerospatiale Matra Missiles Bourges center, or ONERA S4MA, or CEPr R5 test facility (step 2).

The step 1 will be performed during the present phase of PROMETHEE

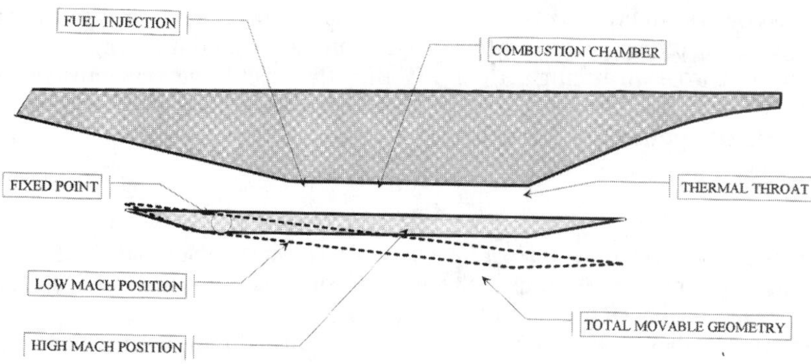

Fig. 64 Scheme of the Promethee engine.

(1999–2001) while step 2 will have to be performed in a second phase of the program (2002–2004).

Beyond the essential problem constituted by the feasibility of the supersonic hydrocarbon fuel combustion and the resulting aeropropulsive balance, one of the key technology problems to be addressed is the thermal ability to the structures, and particularly the combustion chamber walls, to sustain the rarely so severe long duration environment conditions.

The best solution (even if it weren't the only one) is to use the endothermic capacity of the fuel to cool, at least, the combustion chamber walls. Indeed, this solution presents some advantages:

1) It ensures a long duration capacity of the combustion chamber with some conceptual margin in the case of composite materials structure (e.g., this margin could be useful in the case of reusable engine for strategic drone),
2) It provides reformed gaseous fuel which is more capable of mixing and burning in the very short residence time in the combustion chamber,
3) It increases the global performance by limiting the heat losses through the combustion chamber walls.

Considering that ONERA and Aerospatiale Matra Missiles are respectively working on the adapted composite materials structures in A3Cprogram and in-house studies, the effort in Promethee is focused on the acquiring of knowledge for catalytic or thermal reforming process. The fuels reforming being a very large scientific and technical domain, the studies aim at selecting as soon as possible a limited set of solutions, well adapted to the specific Promethee requirements, then at developing one fuel solution (fuel + eventual catalyst + additives) and at starting to bring it into operation for combustion test.

As complement of the relatively short-term studies, related to the main problems of the high-speed airbreathing propulsion described above, a small part of the activity is dedicated to the definition of the on-ground and in-flight demonstration program which should be performed in the future before any operational development. This approach takes advantage of all the complementary works in France in the field of high-speed airbreathing propulsion to merge all the present efforts into a single future and mandatory flight experimentation program aiming at demonstrating the positive aeropropulsive balance of an airbreathing high-speed engine and to precisely determine the different components of this balance.

References

[1]Novelli, Ph., "Synthèse Bibliographique sur la Combustion Subsonique et Supersonique en France" (Bibliographic Synthesis on Subsonic and Supersonic Combustion in France), ONERA RT 1/6137 EY, France, 1991.

[2]Ferri, A., Libby, P. A., and Zakkay, V. "Theoretical and Experimental Investigation of Supersonic Combustion," *Proceedings of the International Council of the Aeronautical Sciences,* 1962.

[3]Dugger, G. L., "Comparison of Hypersonic Ramjet Engines with Subsonic and Supersonic Combustion," AGARD, 1960.

[4]Mestre, A., and Viaud, L., "Combustion Supersonique dans un Canal Cylindrique" (Supersonic Combustion in a Constant Section Duct), *Supersonic flow Chemical Processes and Radiative Transfer,* edited by D. B. Olfe and V. Zakkay, Pergamon, 1964.

[5]Roy, M., "Thermodynamique des Systèmes Propulsifs à Réaction et de la Turbine à Gaz" (Thermodynamic of Jet Engines and Gas Turbine), Librairie Dunod, Paris.

[6]Ferri, A., "Review of Problems in Application of Supersonic Combustion," *Journal of the Royal Aeronautical Society,* 1964

[7]Mestre, A., "Calcul des Performances de Statoréacteurs Hypersoniques à Foyer de Combustion Supersonique" (Calculation of Hypersonic Ramjet with Supersonic Combustor Performances), TN 6/7583 EY, 1964.

[8]Marguet, R., and Huet, Ch., "Recherche d'une Solution Optimale de Statoréacteur à Géométrie Fixe, de Mach 3 à Mach 7, Avec Combustion Subsonique Puis Supersonique" (Optimal Design of a Fixed Geometry Ramjet Using Subsonic then Supersonic Combustion from Mach 3 to Mach 7), Onera, TP No. 656, France, 1968.

[9]Marguet, R., "Ramjet Research and Applications in France," International Symposium on Airbreathing Engines, 1989.

[10]Debout, B., "French Research and Technology Program on Advanced Hypersonic Propulsion," AIAA Paper 91-5003, 1991.

[11]Mullins, B. P., "Studies on the Spontaneous Ignition of Fuels Injected into a Hot Air Stream - Part II: the Effect of Physical Factors upon the Ignition Delay of Kerosene-Air Mixtures," NGTER 90.

[12]Montchiloff, N., "Calculation of Ignition Delay for Hypersonic Ramjets," *9th Symposium on Combustion,* 1962.

[13]Huet, Ch., "Etude Comparative des Combustibles pour un Premier Étage de Lanceur Atmosphérique" (Comparison Between Different Fuels for the First Stage of an Atmospheric Launcher), ONERA, TP No. 497, France, 1967.

[14]Soulier, Ch., and Laverre, J., "Utilisation de la Soufflerie Hypersonique S4MA pour les Essais de Combustion Supersonique dans les Statoréacteurs" (Scramjet Combustion Tests in S4MA Hypersonic Facility), ONERA, TP No. 924, France, 1971.

[15]Contensou, P., Marguet, R., and Huet, Ch., "Etude théorique et Expérimentale d'un Statoréacteur à Combustion Mixte (Domaine de vol Mach 3,5/7)" (Theoretical and Experimental Study of a Dual Mode Ramjet—Mach 3.5 to 7), *La Recherche Aérospatiale* - No. 5, 1973

[16]Hirsinger, F., "Optimisation des Performances d'un Statoréacteur Supersonique—Étude Théorique et Expérimentale" (Optimization of Scramjet Performances—theoretical and Experimental Study), ONERA, TP No. 1106, France, 1972.

[17]Leuchter, O., "Etude des Évolutions Chimiques dans une Couche de Mélange Hydrogène-Air" (Study of Chemical Phenomena in a Hydrogen-Air Mixing Layer), ONERA, TP No. 981, France, 1971.

[18]Leuchter, O., "Problèmes de Mélange et de Combustion Supersonique D'Hydrogène dans un Statoréacteur Hypersonique" (Mixing and Supersonic Combustion of Hydrogen in a Scramjet), ONERA, TP No. 973, France, 1971.

[19]Leuchter, O., "An Experimental Method for Studying Air-Hydrogen Supersonic Combustion," ONERA TP, No. 682, France 1969.

[20]Bellet, J. C., and Deshayes, G., "Structure and Propagation of Detonations in Gaseous Mixtures in Supersonic Flow," *Astronautica Acta,* Vol. 15, 1970, pp. 465–469.

[21]Bellet, J. C., Soustre, J., Kageyama, T., and Manson, N., "Deux Souffleries pour L'Étude de la Combustion en Écoulement Supersonique" (Two Tests Facilities for the Study of Combustion in Supersonic Flow), *Entropie,* No. 32, 1970, p. 42.

[22]Bellet, J. C., Soustre, J., and Manson, N., "Développement des Réactions de Combustion en Présence d'un Gradient de Vitesse en Aval D'une onde de Choc Stationnaire" (Combustion Process Development Downstream a Stationary Shock Wave with a Gradient of Speed), *Compte Rendu de l'Académie des Sciences,* t. 272, 1972.

[23]Bellet, J. C., Donzier, H. P., Soustre, J., and Manson, N., "Influence of Aerodynamic Field on Shock-Induced Combustion of Hydrogen and Ethylene in Supersonic Flow," 14th Symposium on Combustion, 1972.

[24]Laurent, F., Bellet, J. C., Soustre, J., and Manson, N., "Shock-Induced Combustion of Kerosene with the Use of Isopropyl Nitrate Additive," Combustion Inst. European Symposium, - *University of Sheffield* - 1973.

[25]Reingold, L., and Serruys, M., "La Combustion Stabilisée sur une Onde de Choc Dans un Écoulement Supersonique Permanent" (Shock-Induced Combustion in a Supersonic Flow), *Entropie,* No. 22, 1968, p. 21.

[26]Debout, B., Mathieu, C., "French PREPHA Program: Status Report," AIAA Paper 92 - 5107, 1992.

[27]Sancho, M., "The French Hypersonic Research Program PREPHA: Progress Review," AIAA Paper 93 - 5160, 1993.

[28]Sancho, M., Colin, Y., and Johnson, C., "The French Hypersonic Research Program PREPHA," AIAA, 1996.

[29]Zendron, R., Bellande, P., Forrat, B., and Scherrer, D., "Comparison of Different Propulsive Systems for Air-Breathing Launcher," AIAA, Paper 95 - 6077, 1995.

[30]Falempin, F., Lacaze, H., and Viala, P., "Reference and Generic Vehicle for the French Hypersonic Technology Program," AIAA Paper 95 - 6008, 1995.

[31]Rothmund, C., Scherrer, D., Levy, F., and Bouchez, M., "Propulsion System for Airbreathing Launcher in the French PREPHA Program," AIAA Paper 96 - 4498, 1997.

[32]Falempin, F., Laruelle, G., Ramette, Ph., Lepelletier, M., and Hauvette, J., "Les Défis Techniques du Superstatoréacteur" (The SCRAMJET Technical Challenges), Association Aeronautique et Astronautique de France AAAF, 1996.

[33]Bonnefond, T., Falempin, F., and Viala, P., "Study of a Generic SSTO Vehicle Using Airbreathing Propulsion," AIAA Paper 96 - 4490, 1996.

[34]Falempin, F., "French Hypersonic Program PREPHA — System Studies Synthesis," International Symposium on Air Breathing Engines, 1997.

[35]Chevalier, A., and Falempin, F., "Review of New French Facilities for PREPHA Program," AIAA Paper 95 - 6128, 1995.

[36]Bouchez, M., "Status of Measurement Techniques for Supersonic and Hypersonic Ramjets in Industrial Test Facilities," International Symposium on Air Breathing Engines, 99 - 7168, 1999.

[37]Garnero, P., Auneau, I., and Duveau, Ph., "Design and Optimization Methods for Scramjet Inlets," AIAA Paper 95 - 6017, 1995.

[38]Penanhoat, O., and Darracq, D., "Simplified Model and Navier Stokes Calcula-

tions for Optimization and Prediction of an Hypersonic Air Intake," AIAA Paper 95 - 6016, 1995.

[39]Novelli, Ph., Scherrer, D., and Gaffie, D., "Scramjet Flowfields Investigation by Numerical Simulation," International Symposium on Air Breathing Engines, 1995.

[40]Jourdren, Ch., and Dessornes, O., "One Injection Strut Scramjet Combustion Chamber Study in the Frame of PREPHA Program," AIAA Paper 98 - 1560, 1998.

[41]Falempin, F., "Prises d'Air Pour Lanceurs Aérobies" (Airbreathing Space Launchers Inlets), Association Aeronautique et Astronautique de France, 1988.

[42]Falempin, F., Duveau, Ph., "Prises d'Air à Section de Captation Variable—Application aux Launceurs Aérobies" (Variable Capture Area Inlets—Application to Airbreathing Space Launchers), AGARD CP 498, 1991.

[43]Falempin, F., "Overview of French Research Center ONERA Activities on Hypersonic Airbreathing Propulsion," *Australian Mechanical Engineering Transactions,* Vol. ME20, No 4, 1995.

[44]Laruelle, G., Auneau, I., Duveau, Ph., and Penanhoat, O., "Défis de Réalisation des Entrées d'Air Pour Lanceurs Spatiaux" (Air Intakes for Space Launchers), Association Aeronautique et Astronautique de France, 1997.

[45]Dufour, A., "Some single Expansion Ramp Nozzles Studies," AIAA Paper 93 - 5061, 1993.

[46]Perrier, P., Rapuc, M., Rostand, P., Hallard, R., Regard, D., Dufour, A., and Penanhoat, O., "Nozzle and Afterbody Design for Hypersonic Airbreathing Vehicles," AIAA Paper 96 - 4548, 1996.

[47]Le Bozec, A., Rostand, Ph., Rouy, F., Dessornes, O. Fontaine, J., Dufour, A., and Penanhoat, O. "Afterbody Testing and Comparison to CFD Simulation," AIAA Paper 98 - 1596, 1998.

[48]Scherrer, D., Montmayeur, N., Ferrandon, O., and Tonon, D., "Scramjet Injectors Calculation and Design," AIAA Paper 93 - 5171, 1993.

[49]Scherrer, D., Dessornes, O., Montmayeur, N., and Ferrandon, O., "Injection Studies in the French Hypersonic Technology Program," AIAA Paper 95 - 6096, 1995.

[50]Rene-Corail, M., Rothmund, Ch., Ferrandon, O., and Dessornes, O., "A Hydrogen Cooled Injection Strut Design for Scramjet," AIAA Paper 96 - 4511, 1996.

[51] Bouchez, M., Montmayeur, N., and Leboucher, C., "Scramjet Combustor Design in France" AIAA Paper 95 - 6094, 1995.

[52]Bouchez, M., Hachemin, J-V., Leboucher, C., Scherrer, D., and Saucereau, D., "Scramjet Combustor Design in French PREPHA Program—Status in 96," AIAA Paper 96 - 4582, 1996.

[53] Bouchez, M., Kergaravat, Y., Scherrer, D., Souchet, M., and Saucereau, D., "Scramjet Combustor Design in French PREPHA Program—Final status in 1988," AIAA Paper 98 - 1534, 1998.

[54] Dessornes, O., Jourdren, C., and Scherrer, D., "One Strut Scramjet Chamber Tests in the Frame of the PREPHA Program," International Symposium on Air Breathing Engines, 99 - 7137, 1999.

[55] Chadwick, K., and Schetz J., "Direct Measurements of Skin Friction in High-enthalpy High-Speed Flows," AIAA Paper 92 - 5036, 1992.

[56] Perrier, P., Stoufflet, B., Rostand, Ph., Baev, V. K., Latypov, A. F., Shumaky, V. V., and Iaroslavtaev, M. I., "Integration of an Hypersonic Airbreathing Vehicle: Assess-

ment of Overall Aerodynamic Performances and of Uncertainties," AIAA Paper 95 - 6100, 1995.

[57] Darwin, S., Bacos, M., Dorvaux, J., and Lavigne, O., "Oxidation Model for Carbon-Carbon Composites," AIAA Paper 92 - 5016, 1992.

[58] Valazza, J., Protat, V., Ferrier, C., and Nugeyre, J. C., "Combustion Chamber, Thermal Insulation, Active Cooling by Air for Ramjet," AGARD PEP, 1997.

[59] Berton, B., Caron, P., Thomas, M., Renard, P., Franchet J. M., and Lecornu, J. P., "Intermetallic Compounds for Hypersonic Space Planes," AIAA Paper 96 - 4564, 1996.

[60] Roudakov, A., Schickhman, Y., Semenov, V., Novelli, Ph., and Fourt, O., "Flight Testing an Axisymmetric Scramjet—Russian Recent Advances," International Astronautical Federation, and 93 - S.4.485, 1993.

[61] Falempin, F., Vancamberg, Ph., Thevenot, R., Girard, Ph., and Joubert, H., "Hypersonic Airbreathing Propulsion: Flight Tests Needs," AIAA Paper 95 - 6013, 1995.

[62] Falempin, F., Forrat, M., Baldeck, J., and Hermant, E., "Flight Test Vehicles: A Mandatory Step in Scramjet Development," AIAA Paper 92 - 5052, 1992.

[63] Chevalier, A., and Levin, V., "French-Russian Partnership on Hypersonic Wide Range Ramjets," AIAA Paper 96 - 4554, 1996.

[64] Falempin, F., Levine, V., Avrashkov, V., Davidenko, D., and Bouchez, M., "French-Russian Partnership on Hypersonic Wide Range Ramjets—Status in 1998," AIAA Paper 98 - 1545, 1998.

[65] Falempin, F., Bouchez, M., Levine, V., Avrashkov, V., and Davidenko, D., "MAI/AEROSPATIALE Co-Operation on a Hypersonic Wide Range Ramjet—Evaluation of Thermal Protection Systems," International Symposium on Air Breathing Engine, 99 - 7140, 1999.

[66] Bouchez, M., Levine, V., Falempin, F., Avrashkov, V., and Davidenko, D., "French-Russian Partnership on Hypersonic Wide Range Ramjets—Status in 1999," AIAA Paper 99 - 4845, 1999.

[67] Bouchez, M., Levine, V., Falempin F., Avrashkov, V., and Davidenko, D., "Airbreathing Space Launcher Interest of a Fully Variable Geometry Propulsion System," AIAA Paper 99 - 2376, 1999.

[68] Falempin, F., Montazel, X., Goldfeld, M., Nestoulia, R., and Starov, A., "Investigation of Two-Mode Ramjet/Scramjet Inlet," International Symposium on Air Breathing Engine, 99 - 7040, 1999.

[69] Bouchez, M., Saunier, E., Peres, P., and Lansalot, J., "Advanced Carbon/Carbon Injection Strut for Actual Scramjet," AIAA Paper 96 - 4567, 1996.

[70] Falempin, F., and Koschel, W., "Combined Rocket and Air-Breathing Propulsion (European Perspectives)," Third International Symposium on Space Propulsion, 1997.

[71] Novelli, Ph., and Koschel, W., "JAPHAR—a Joint ONERA-DLR Research Project for High-Speed Airbreathing Propulsion," International Symposium on Air Breathing Engines, 99 - 7091, 1999.

[72] Duveau, Ph., Hallard, R., Novelli, Ph., and Eggers, T., "Aerodynamic Performances Analysis on the Hypersonic Airbreathing Vehicle JAPHAR," International Symposium on Air Breathing Engines, 99 - 7286, 1999.

[73] Bouchez, M., Montazel, X., and Dufour, E., "Hydrocarbon Fueled Scramjets for Hypersonic Vehicles," AIAA Paper 98 - 1589, 1998.

[74] Falempin, F., Salmon, T., Avrashkov, V., "Fuel-cooled composite materials structures—Status at AEROSPATIALE MATRA," AIAA Paper 2000-3343, 2000.

[75] Medwick, D. G., Castro, J. H., Sobel, D. R., Boyet, G., and Vidal, J. P., "Direct Fuel Cooled Composite Structure," International Symposium on Air Breathing Engines, 99 - 7284, 1999.

[76] Falempin, F., and Serre, L., "The French PROMETHEE Program Status in 2000," AIAA Paper 2000-3341, 2000.

Chapter 3

SCRAM-Jet Investigations Within the German Hypersonics Technology Program (1993–1996)

Norbert C. Bissinger*
Dasa, Ottobrunn, Germany
Wolfgang Koschel†
DLR, German Aerospace Research Center, Lampoldshausen, Germany
Peter W. Sacher‡
Dasa, Ottobrunn, Germany
and
Rainer Walther§
MTU Motoren-und Turbinen-Union, München, Germany

I. German Hypersonics Technology Program and SCRAM-Jet-Related Activities

A. German Hypersonics Technology Program

IN 1987 the German Ministry for Research and Technology (BMFT) initiated a Hypersonics Technology Program (HTP) oriented toward the SÄNGER concept, a two-stage fully reusable, future space transportation system. This concept was the result of a concerted effort by BMFT, industry and research institutes.[1]

As shown in Fig. 1, a system concept study and a propulsion concept study

*Team Manager Inlets/Afterbodies, Military A/C Division.

†Director, Institute for Propulsion.

‡Senior Manager Hypersonics Technologies, Military A/C Division.

§Senior Manager Hypersonics Technologies.

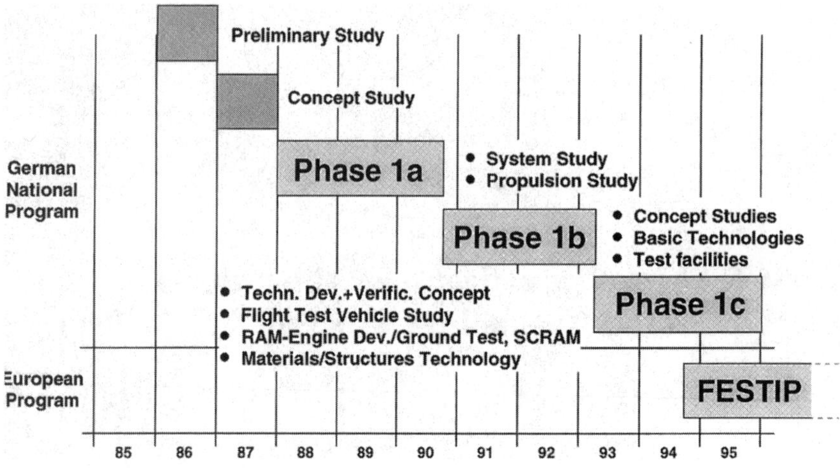

Fig. 1 Schedule of the German Hypersonics Technology Program.

were first performed in Phase 1a to define the overall SÄNGER space transportation system and its subsystems to be investigated for the lower and upper stages. In Phase 1b, the system concept studies were continued and, in parallel, basic technology work was performed, existing test facilities were again made operational and new test facilities were built up. As a most important result of all system related activities in the first two phases of the program, the most critical "Key Technologies" for the realization of a SÄNGER transportation system were identified in the technological areas of propulsion, aerothermodynamics, materials and structures and subsystems. Consequently, in the last Phase 1c of the HTP, a Technology Development and Verification Plan (TDVP) was established, comprising all efforts needed for ground and inflight experimentation to proof the technological readiness level of these most critical "Key Technologies."[2]

B. SCRAM-Jet Related Activities Within the HTP

As one of the most important technological challenges of the SÄNGER concept the airbreathing propulsion system of the first stage was identified. Although the SÄNGER baseline propulsion system of the first stage was a combined cycle TURBO-RAM concept, the trade-offs with regard to the optimum stage separation Mach-number led also to the consideration of a dual-mode RAM-SCRAM-Jet propulsion system. Consequently, the question of an optimum engine-airframe integration to obtain maximum thrust minus drag for acceleration also at the high end Mach-number, became more and more important. So in Phase 1c the major part of the available budget was focused on propulsion related work. Consequently in 1993 theoretical and experimental activities started to investigate SCRAM-Jet propulsion.

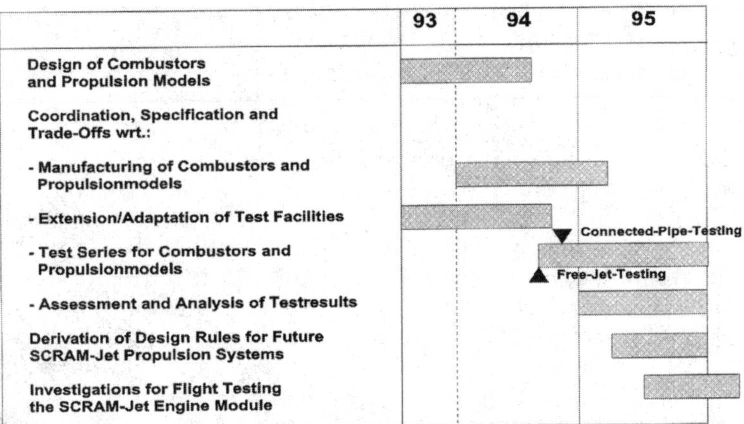

Fig. 2 SCRAM-Jet related activites within the German National HTP.

A joint effort (see Fig. 2) for "SCRAM-Jet Technology" was initiated under the lead of Dasa-MTU,[3] with the following institutions participating:

	Involved in
Dasa-BCE	Performance Synthesis & Combustor Specification
Dasa-MTU	Program Coordination Combustor and Engine Model Design
Dasa-M/RWTH Aachen	Intake-Design
DLR-Cologne	Laser Measurement Techniques
UNI Stuttgart/DLR Lampoldshausen	Fuel Injection Devices
TsAGI	Model Manufacturing Laser-Measurement Techniques, Connected Pipe Tests Free-Jet Testing

In parallel concept study work on flight testing, the modules tested on ground at TsAGI were performed by Dasa-M/OHB/TSAGI/RADUGA and RWTH Aachen/TsAGI. The Tupolev 22M-launched RADUGA-D2 was chosen as an already available low cost "Flying Testbed" to carry the engine as "Passenger."

II. Theoretical Investigations for SCRAM-Jet Intake Designs

A. Activities at Dasa-MT633

1. Numerical Approach and Calculation Method

The numerical investigations of the flow in the SCRAM-Jet model(s) has been concentrated onto the flow of the intake although the flow inside most or the complete SCRAM-Jet model has been calculated. Combustion within the SCRAM-Jet has not been considered. Due to the time and monetary constraints only a very limited number of calculations could be performed. Therefore a scientific approach (although highly desirable) has not been possible.

Calculations have been performed with air as flow medium. In most of the calculations a constant value for the ratio of specific heats γ has been assumed. Using measured data or combustion calculations the value for that parameter was estimated by TsAGl and the RWTH Aachen. Calculations with different values for γ showed noticeable effects on the flow. In one calculation the air has been treated as an equilibrium gas with variable γ. It has not been tried to simulate the real test conditions of the windtunnel in which kerosene has been burned to reach the high total temperatures needed.

The code applied by Dasa-M was the Navier-Stokes code NSFLEX in its two-dimensional version.[4] This code is based on the Euler code EUFLEX,[5] which is a 2nd to 3rd order finite volume upwind scheme. The flow equations are solved by an implicit procedure using a local time-stepping approach. Turbulent viscosity was calculated using the Baldwin-Lomax turbulence model.

The number of grid-points inside the intake and the SCRAM-Jet duct characterizes the density of the mesh-grid used for the Navier-Stokes calculations. Two different grid densities have been applied. Inside the duct the coarsest grid consisted of 331×54 points and the finest grid of 658×106 points, which amounts to a nearly doubling of points in each direction.

2. *Investigations and Results*

The first proposal for a SCRAM-Jet intake design presented by TsAGl can be seen in Fig. 3. The intake possesses three external ramps which produce a total flow deflection of 24°40′. At the intake lip station there exists a 6° expansion on the ramp side. Downstream of the leading edge of the

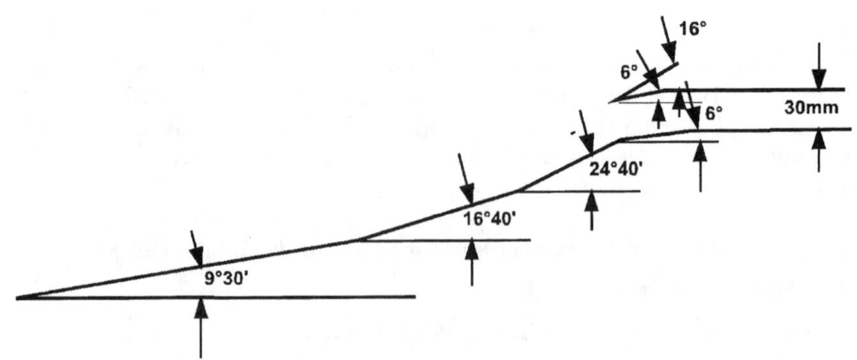

Fig. 3 SCRAM-Jet geometry (TsAGI).

intake-lip, the wall on the lip-side is parallel to that of the ramp-side for a length of 59 mm. Therefore, the flow from the third ramp experiences a further compression at the lip due to the local turning angle of 18°40′. The height of this constant area duct piece is 30 mm. It is followed by a diffuser, which serves as an isolator for the internal shock system. Downstream of this diffuser various geometries have been investigated in order to analyze supersonic and subsonic combustion.

The two-dimensional Navier-Stokes calculation for this geometry at $M = 6.3$, and $Re = 6.608{*}e^{+6}$ produced the lines of supersonic Mach numbers as given in Fig. 4 for the close-up of the intake area around the intake entrance. Whereas the first and second shock just touch the tip of the intake cowl lip the third ramp shock enters the intake.

The expansion originating at the end of the third external ramp interacts with the lip shock, which produces a curvature of this shock. The lip shock is strong enough to produce a large area of separated flow at the point where it hits the ramp side. The area of subsonic flow within this separation bubble is characterized in this figure by the large white area on the ramp side. The reflections of this shock induce a further, but much smaller separation zone, just visible at the right hand end of the part of the intake duct shown. Inside the intake duct the subsonic region of the boundary layer on the ramp side is much thicker than on the cowl lip side.

The thickness of the separation region inside the intake can be appreciated in Fig. 5. In this figure the velocity vectors around the separation are plotted. The separation spans half the intake duct height. An indication for the total pressure losses associated with this separation is given in Fig. 6. In this figure the "isentropic" total pressure at constant x-stations along the intake duct are plotted. Most of the losses occur in the first third of the duct from $\bar{x} = 0.42$ to $\bar{x} = 0.61$. It is interesting to note, that, downstream of this

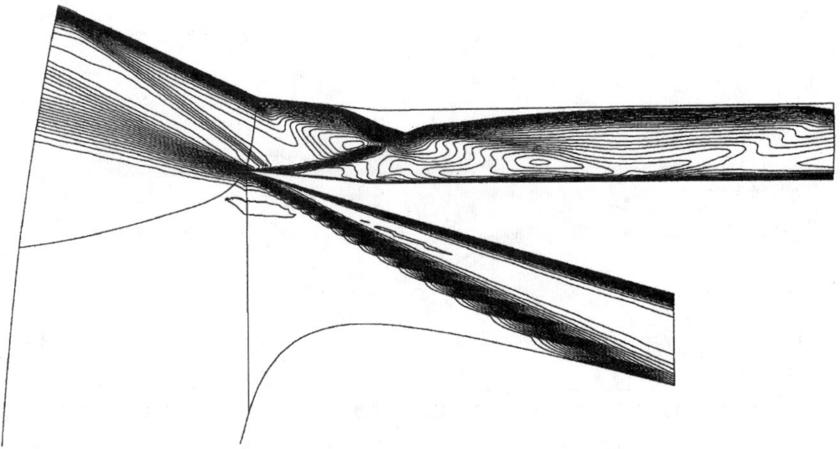

Fig. 4 Isolines of supersonic Mach numbers at M = 6.3.

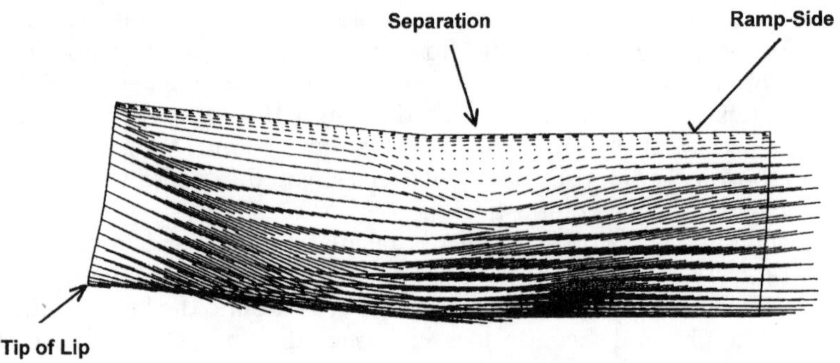

Fig. 5 Velocity vectors at intake entrance, M = 6.3.

area, the losses in the cowl-lip-side boundary layer are larger than in the ramp-side boundary layer. Downstream of the separation all the shocks and flow disturbances are slowly damped-out. At the end of the calculation domain even the static pressure on both the cowl and ramp wall are nearly identical (see Fig. 7). The Mach number at the exit is around $M = 3.0$.

The results of this calculation prompted a re-design of this SCRAM-Jet intake. For the continuation of the SCRAM-Jet tests two intakes with a design Mach number of M = 6.0 and 6.5, respectively, were selected. Whereas all the angles of the ramps and at the duct entrance were the same for both intakes, their lengths of the ramps differed as shown in Fig. 8. The

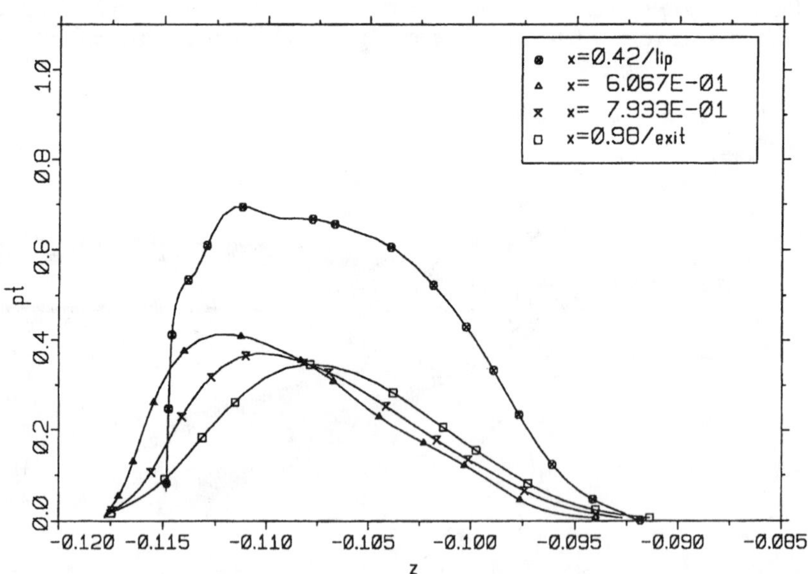

Fig. 6 Isentropic total pressure profiles, M = 6.3.

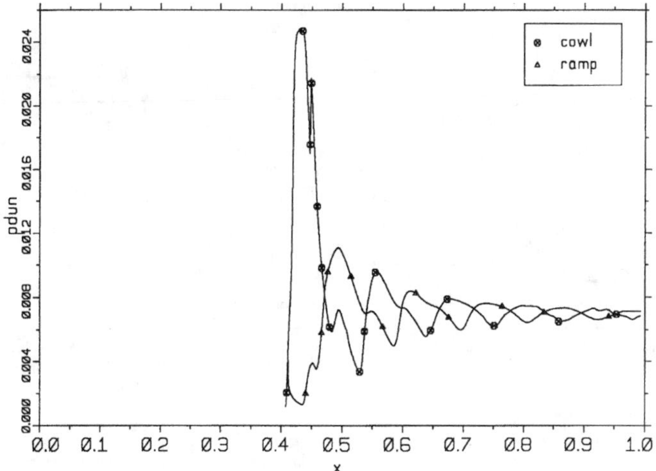

Fig. 7 Normalized static pressure on ramp and lip, M= 6.3.

total flow deflection by the external ramps has been reduced by about 4° to 20°. Due to this geometry change the flow deflection by the cowl lip reduced from 15° to 18°. With this lip angle the Mach number downstream of the lip shock is still supersonic ($Ma \approx 1.83$) even at free stream Mach numbers as low as $M = 3.5$.

The Navier-Stokes calculation for the MD60-intake (design Mach number $MD = 6.0$) at $M = 6.25$ and Re = $4.844e^{+6}$ produced the lines of supersonic Mach numbers presented in Fig. 9. The shocks from the second and third ramp coalesce on the tip of the cowl lip. The shock of the first ramp passes the intake lip by some distance thus producing some small amount of mass spillage. The interaction region between the cowl lip shock and the ramp expansion is located further downstream than for the original design. The separation region induced by the lip shock has been drastically reduced (see also the velocity vectors in Fig. 10). The losses in "isentropic" total pressure of the external shocks also has been reduced (Fig. 11). The maximum total pressure at the duct exit increased from about $pt/pt_\infty \approx 0.43$ to $\approx$ 0.49. However, at the exit the flow seems to be much more non-uniform than before which is reflected in the distribution of the static pressure along the cowl and ramp side, respectively, in Fig. 12. Compared with the original design the mean of the static pressure at the intake exit is somewhat lower.

The flow calculation for the MD65-intake (design Mach number $MD = 6.5$) at $M = 6.25$ and $Re = 4.844e^{+6}$ produced very similar results inside the intake duct. The lines of supersonic Mach numbers in Fig. 13 reveal an increased distance between all external shocks and the cowl lip which is indicative of an even further increased mass flow spillage.

The flow around these two intakes has been calculated for several off-design Mach numbers. The highest Mach number considered has been M = 8.0.

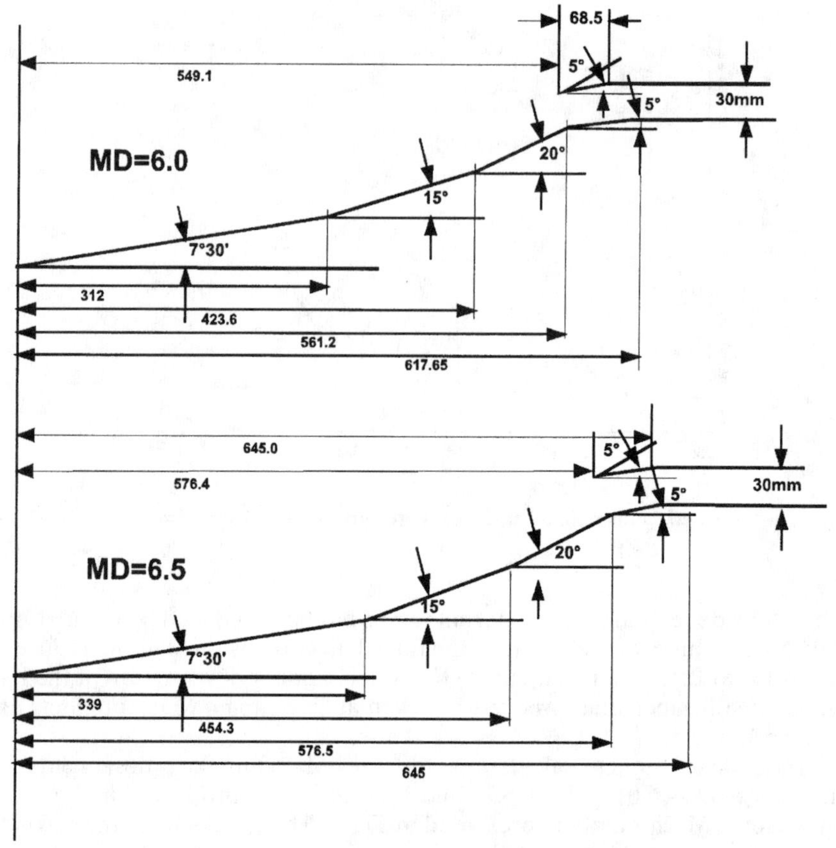

Fig. 8 Redesigned intake geometries.

For this Mach number Fig. 14 depicts lines of constant supersonic Mach numbers just in front and inside the intake duct for the MD60 (Re = 1.789e^{+6}). All external shocks enter the intake. In this calculation along the first 4 cm of the intake lip laminar flow has been assumed in order to investigate a possibly enlarged danger of separation due to the shock interaction with this laminar flow. A separation at that point was produced. However this separation was extremely small. The flow Mach number of M = 8.0 was reduced to a mean Mach number at the x-station X729 (which is one of the hydrogen injection positions) of M = 3.61. The lip shock was highly curved and produced a separation at the cowl side quite similar in size as for the other flow cases presented sofar.

At the lower off-design Mach number of M = 5.0 all the shocks become steeper. Because of this the reflection point of the cowl shock moves upstream along the ramp side of the intake duct. This can be recognized in Fig. 15 which presents lines of constant supersonic Mach numbers for the flow

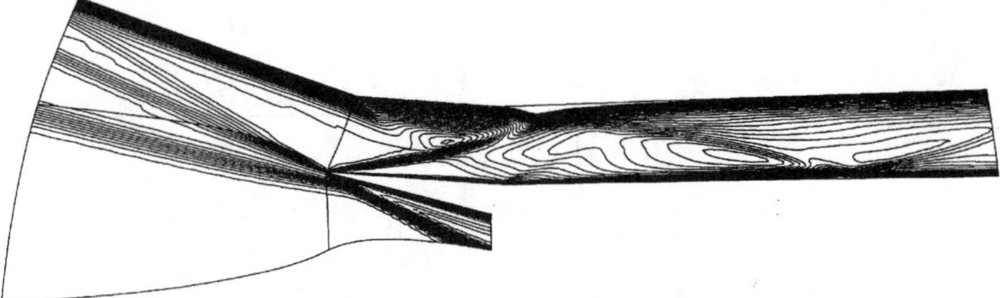

Fig. 9 Isolines of supersonic Mach numbers M= 6.25 (MD = 6.0).

of the MD60 intake at $M = 5.0$ and $Re = 9.8632e^{+6}$. The ramp separation moves even upstream of the hydrogen injection position X729. The mean Mach number at that position is predicted to be $M = 3.0$. The losses in total pressure, referenced to the free stream total pressure, is reduced, compared with the flow at $M = 6.25$ (compare Fig. 16 with Fig. 11). The static pressure referenced to the total pressure of the free stream is somewhat higher at the intake exit at the $M = 5.0$ flow conditions (compare Fig. 12 with Fig. 15).

For the $M = 5.0$ (MD60 intake) case a grid sensitivity study has been conducted by calculating its flow with a grid for which the points of each coordinate has been doubled. Results from this calculation are presented in Fig. 18 and Fig. 19. The lines of constant supersonic Mach numbers in Fig. 18 from the fine grid calculations reveal much more details than those of the medium grid calculation (Fig. 15). However, the additional flow features (shocks) appear to be weak. Clearly a thickening of the ramp boundary layer at each shock reflection point can be observed. The total pressure losses in both the ramp and cowl boundary layer are larger in the medium grid result (compare Fig. 16 with Fig. 19).

For the investigation of subsonic combustion the SCRAM-Jet model has been supplemented by a convergent-divergent nozzle in order to choke the flow. To be able to have some variability in the amount of choking the nozzle throat has been designed larger than necessary for choking and, in addition, high-pressure air was injected at the nozzle to accomplish complete choking. The lines of constant supersonic Mach numbers in Fig. 20 (MD60, M =

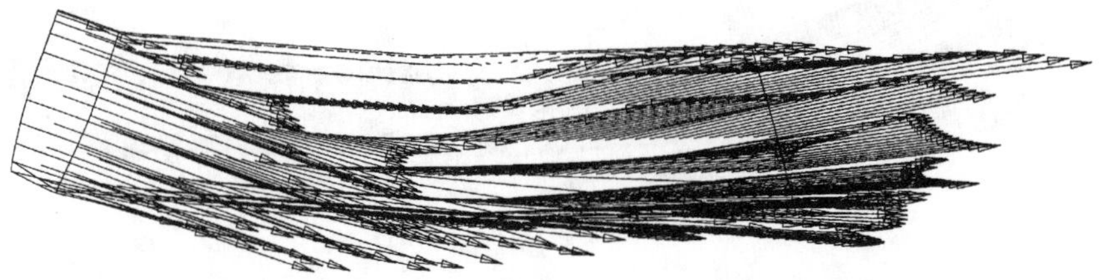

Fig. 10 Velocity vectors at intake entrance M = 6.25 (MD = 6.0).

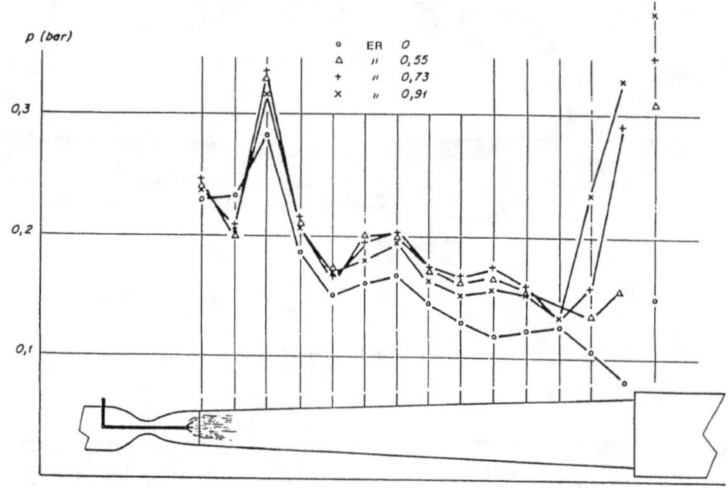

Fig. 11 Isentropic total pressure at the intake duct M = 6.25 (MD = 6.0).

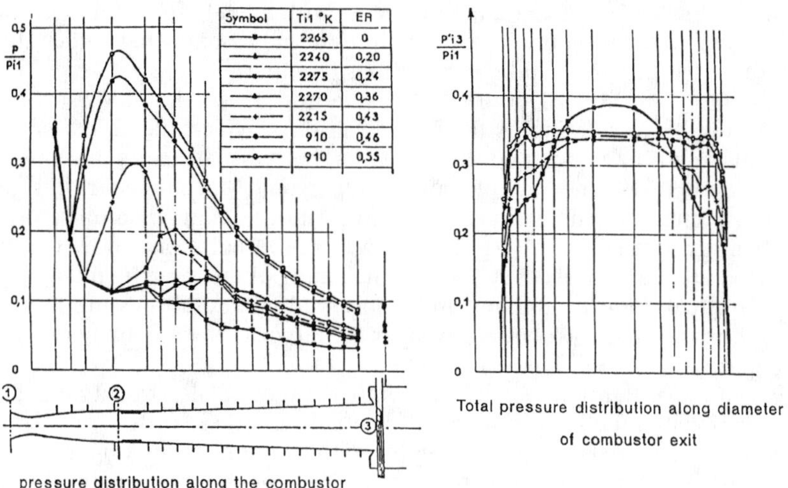

Fig. 12 Normalized static pressure on ramp and lip M = 6.25 (MD = 6.0).

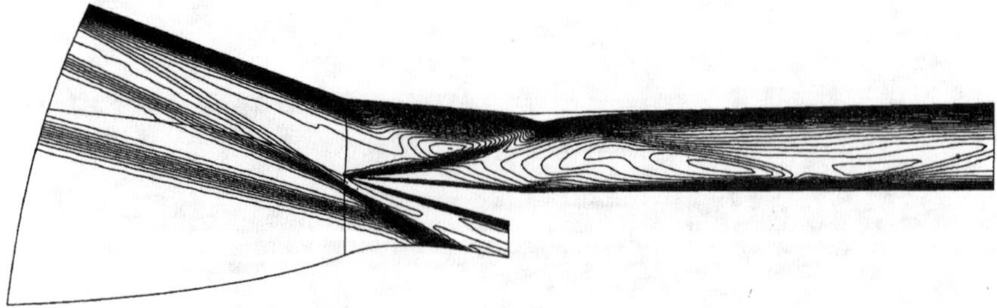

Fig. 13 Isolines of supersonic Mach numbers M= 6.25 (MD = 6.5).

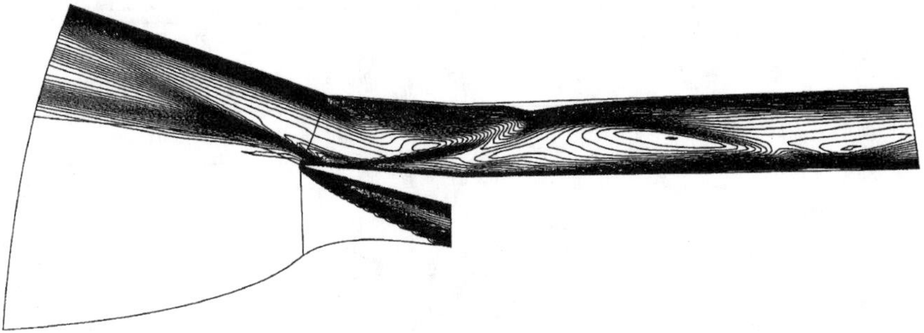

Fig. 14 Isolines of supersonic Mach numbers M= 8.0 (MD = 6.0).

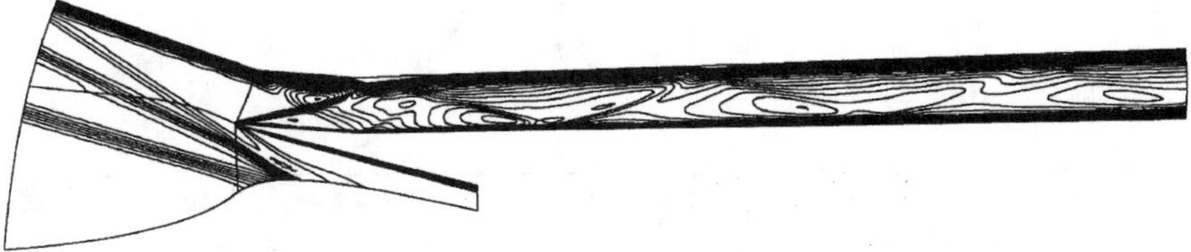

Fig. 15 Isolines of supersonic Mach numbers M= 5.0 (MD = 6.0).

5.139, $Re = 7.238e^{+6}$) represent the flow inside the SCRAM-Jet model without air injection. It is interesting to see the relatively large separations at both the ramp and cowl wall in the convergent part of the nozzle. The Mach number in the nozzle throat has been reduced to about $M = 2.0$.

The most obvious effect of using the ratio of specific heats of equilibrium air compared to a constant ratio can be seen in Figs. 21 and 22. In the equilibrium result all shocks are steeper. For example, the external ramp shocks no longer enter the intake. This effect has consequences especially inside the SCRAM-Jet duct because of its reflected shock system. For fixed injections positions, this means changed flow conditions as can be seen in the following figures. At the two injection positions of the MD60-SCRAM-Jet equilibrium air produces produces fuller Mach number profiles with a lower peak Mach number (compare Fig. 23 with 24). The profile of the normalized static pressure at these stations is completely changed (see Figs. 25 and 26). The peak temperatures at the walls are reduced but the minimum temperature is slightly larger (compare Fig. 27 with 28).

3. Concluding Remarks

For the support of the design and testing of SCRAM-Jet models two-dimensional Navier-Stokes calculations have been performed. Especially during the design these results have been very valuable. There exist some limitations of these results for the analysis of the test results because the actual test conditions (vitiated air by burning kerosene) could not be simulated. Three-dimensional flow calculations, including this effect are neces-

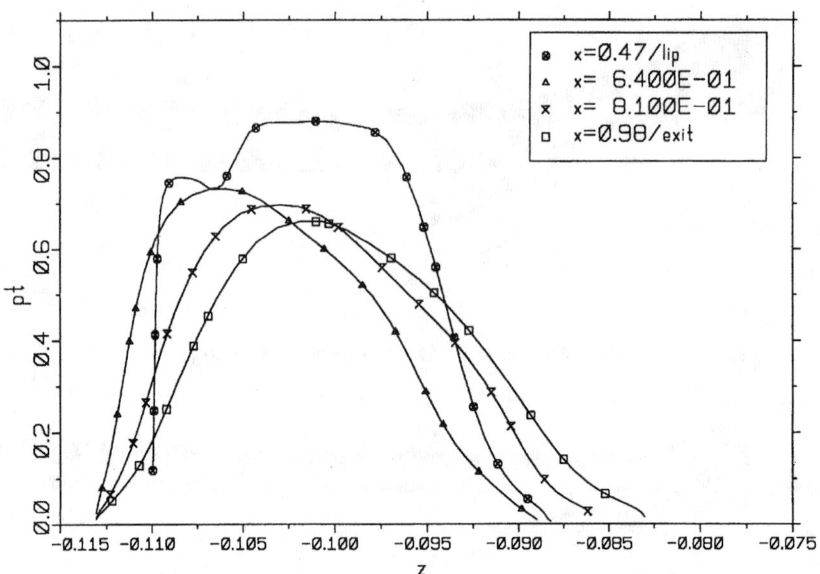

Fig. 16 Isentropic total pressure profiles at M= 5.0 (MD = 6.0).

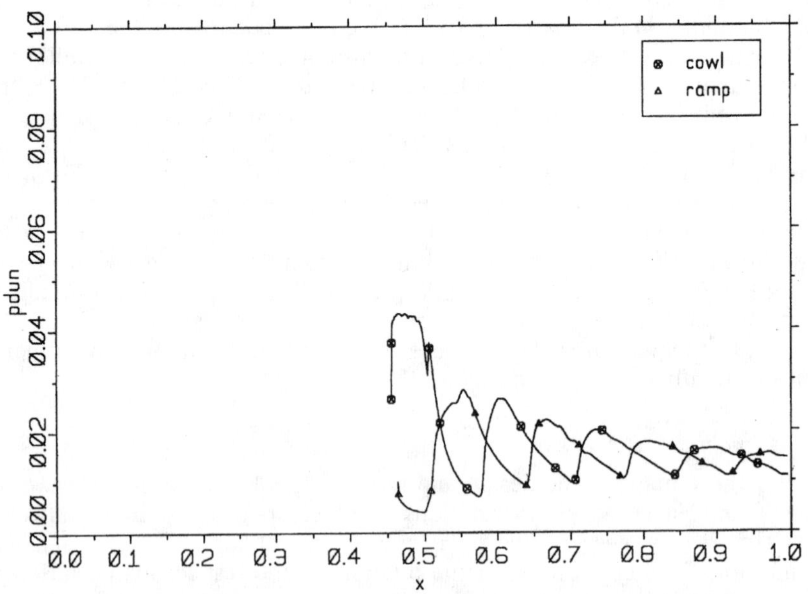

Fig. 17 Normalized static pressure on ramp and lip at M= 5.0 (MD = 6.0).

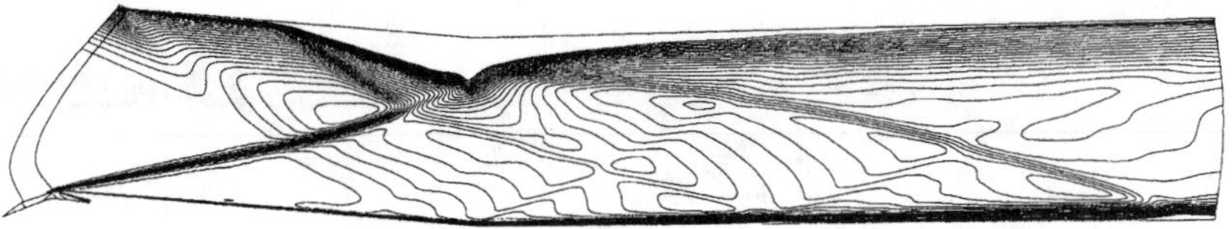

Fig. 18 Fine grid-isolines of supersonic Mach numbers at M= 5.0 (MD = 6.0).

sary to really understand the data measured during testing and to further optimize the SCRAM-Jet design.

B. Activities at RWTH Aachen

During the German National Hypersonics Technology Program the RWTH-Aachen played an important role as partner of the industry to define, design, manufacture and testing both at DLR in Germany and at TsAGI in Russia several alternative components for a SCRAM-Jet engine module which is described later in subsequent chapters. According to,[6] three main tasks had to be performed at the university in Aachen:

- Selection of adequate intake concepts (together with TsAGI and Dasa-M). It was decided that TsAGI concentrates on intakes with mixed internal-external compression, and RWTH-Aachen focussed on intake designs with internal compression only. The part of Dasa-M was already described in the preceding chapter.

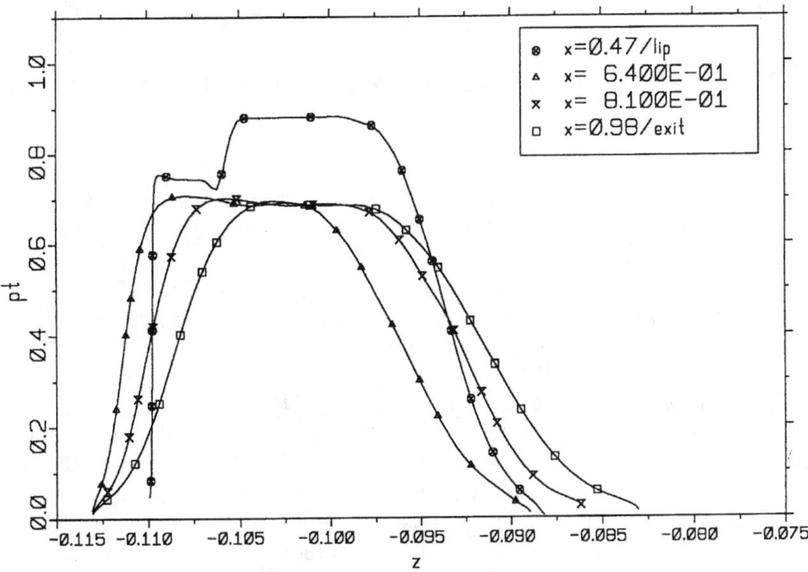

Fig. 19 Fine grid-isolines total pressure profiles at M= 5.0 (MD = 6.0).

Fig. 20 Condi-nozzle-isolines of supersonic Mach numbers at M= 5.139 (MD = 6.0).

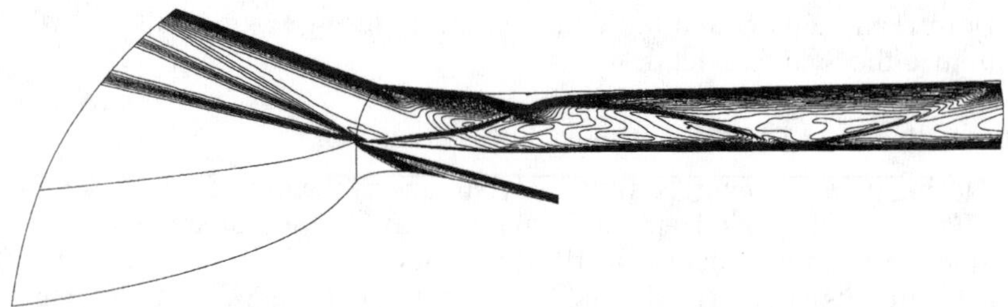

Fig. 21 Isolines of supersonic Mach numbers at M= 6.37 (MD = 6.0) and constant γ.

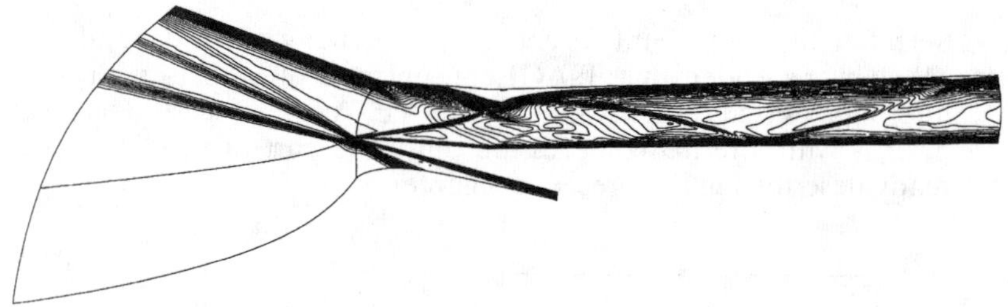

Fig. 22 Isolines of supersonic Mach numbers at M= 6.37 (MD = 6.0) and equilibrium air.

- Investigations of an intake model with internal compression at M= 3.0.
- Numerical analysis of internal flows for the propulsion system model designed by TsAGI.

At the University a test facility was modified and equipped with a new Laval-Nozzle and Ejector to extent the test range up to Mach=3.5. The test facility is shown in Fig. 29.

The rationale for the selection of the inlet types to be investigated is shown in Fig. 30. The performance of the inlet diffuser is usually characterised by the total pressure loss recovery coefficient (the ratio of the total pressure at diffuser front and diffuser end cross section). As the figure shows, the internal compression inlet promises for higher Mach numbers the best values for an achievable pressure loss coefficient.

Since in the frame of a preliminary design study the inlet diffuser design might not yet be selected, it is very difficult to assume a reasonable value for

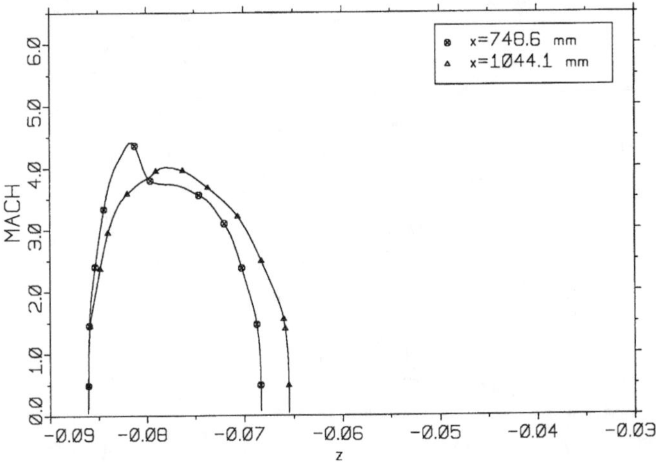

Fig. 23 Mach number profiles at M= 6.37 (MD = 6.0) and constant γ.

the pressure loss coefficients. A first guess was made to adjust the assumptions for achievable pressure loss coefficients of diffusers plotted in Fig. 30. Design constraints of mixed compression diffusers for a wide Mach-number range will lead to 3 up to 5 diffuser configurations in order to fulfill the requirements of an optimized turning of the flow.

Finally, viscous flow effects will govern the performance of mixed compression inlet diffusers, which is illustrated in Fig. 31.

To investigate the potential of an intake with complete internal compression, a symmetric "Side-Wall Compression Intake" was chosen, shown in Fig. 32. This was based on the results of a numerical flow simulation using an

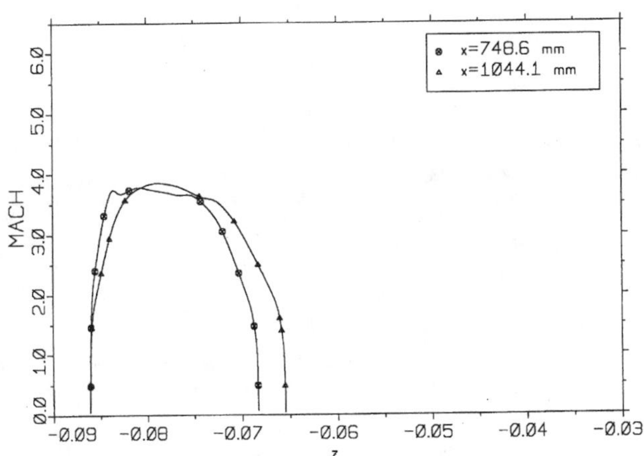

Fig. 24 Mach numbers profiels at M= 6.37 (MD = 6.0) and equilibrium air.

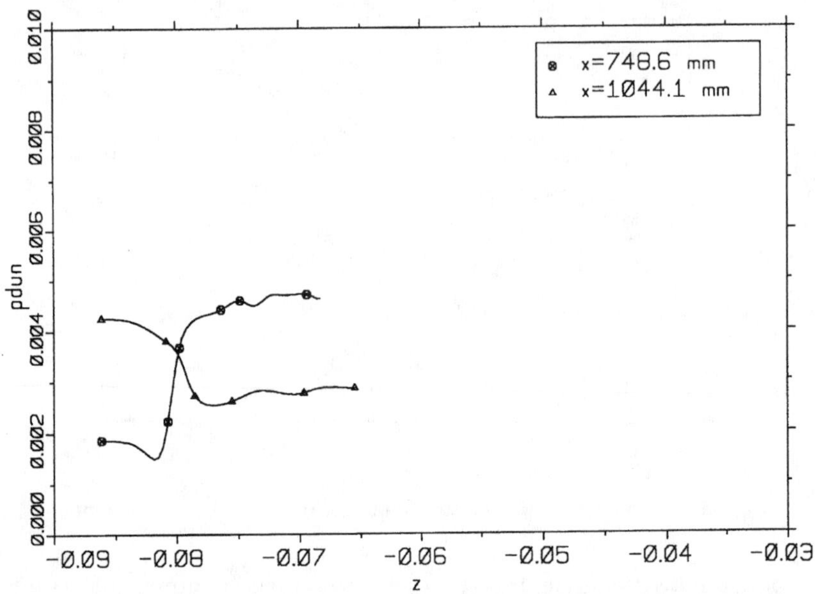

Fig. 25 Normalized static pressure profiles at M= 6.37 (MD = 6.0) and constant γ.

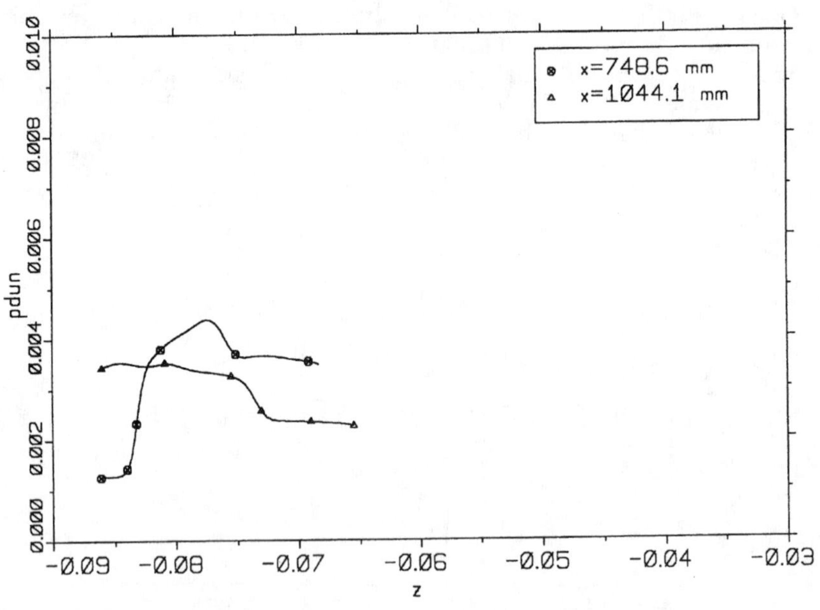

Fig. 26 Normalized static pressure profiles at M= 6.37 (MD = 6.0) and equilibrium air.

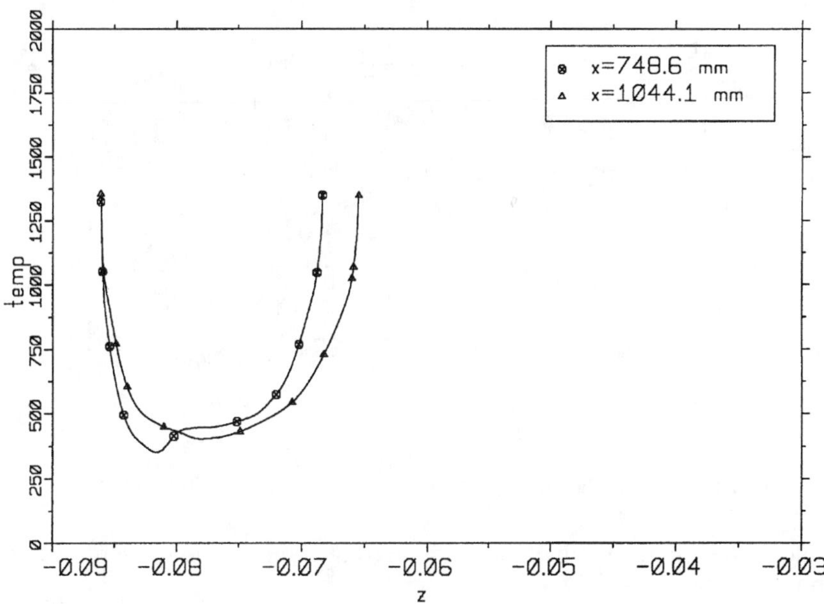

Fig. 27 Temperature profiles at M = 6.37 (MD = 6.0) and constant γ.

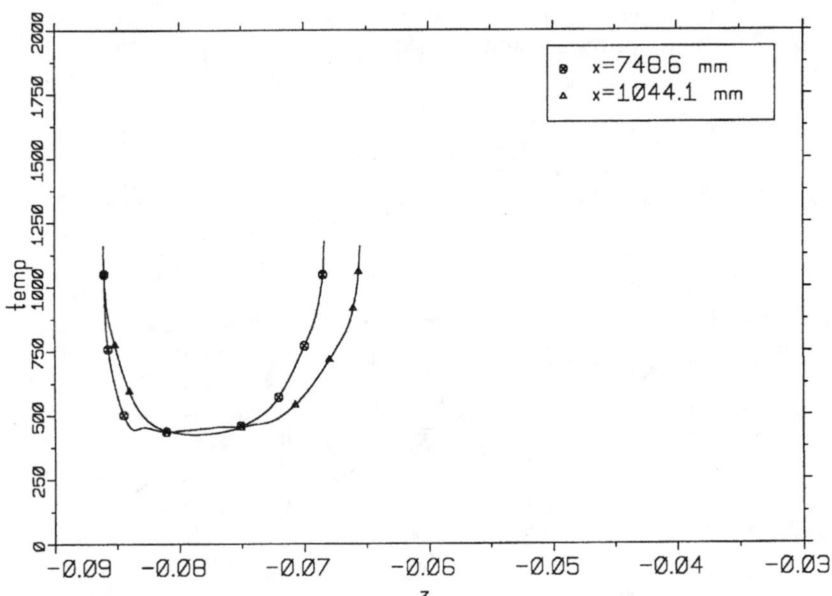

Fig. 28 Temperature profiles at M = 6.37 (MD = 6.0) and equilibrium air.

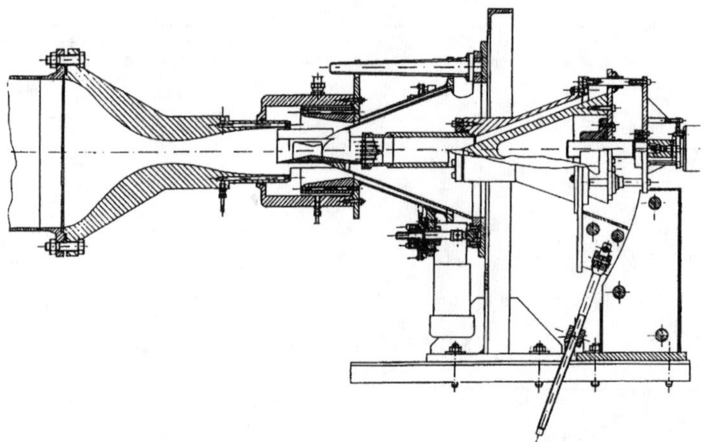

Fig. 29 The test facility at the RWTH Aachen used for intake testing.

Euler code applied to an intake with internal compression.[7] The results are shown in Figs. 33 and 34.

This model was build and tested and the experiment let to significant unstart problems due to the above mentioned viscous boundary layer/shock interactions during the transition of the Mach number from 2 < Mach < 3.5 which requires different optimum ratios of captured and throat area.

In principle if unstart occurs, there are two possibilities to overcome the problem. One solution is to use slots in the parallel walls, which will allow spillage for parts of the air and therefore "simulate" variable ratios of the

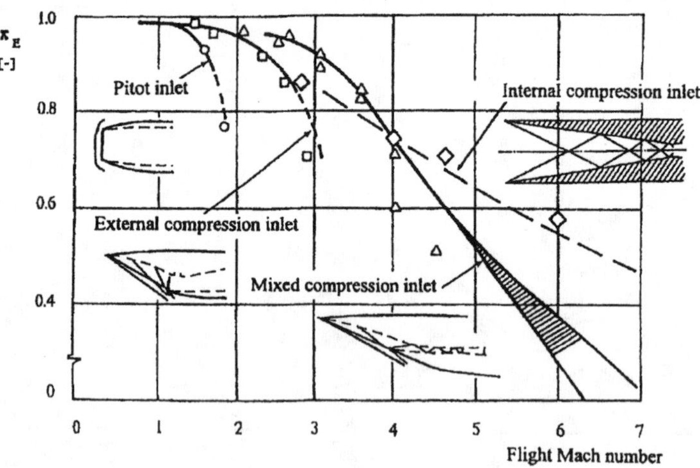

Fig. 30 Achievable total pressure loss coefficients for different types of inlet diffusers.

captured area and the throat area of the inlet diffuser. This will lead of course to increased losses due to spillage.

Therefore, the second way to avoid the unstart problem was chosen, the design of an internal compression intake with variable geometry, shown in Fig. 35.

At that time the national Hypersonics Technology Program was terminated due to the budget constraints. Further experimental investigations to prove the successful design of the variable internal compression intake by the experiment were not conducted.

III. Theoretical and Experimental Investigations of Scramjet Combustion at TsAGI and DLR Lampoldshausen

A large number of systematic connected-pipe test series using a geometrically variable combustor model was performed. For this purpose, the T-131V connected-pipe test facility at TsAGI was used. The test bed is capable of providing flow stagnation temperatures up to 2350 K, stagnation pressures up to 110 bar and a maximum air flow of 10 kg/sec. A description of the test facility and the test conducted at TsAGI is found in[8], the results also reported in[9].

A. Combustor Model Design

The geometry of the as-designed subscale combustor model is shown in Fig. 36. The model consists of four segments:

1) A slightly divergent rectangular isolator segment, 300 mm long, with a 0,5 degree expansion to account for boundary layer growth.
2) A rectangular channel module 500 mm long, replaceable with four dif-

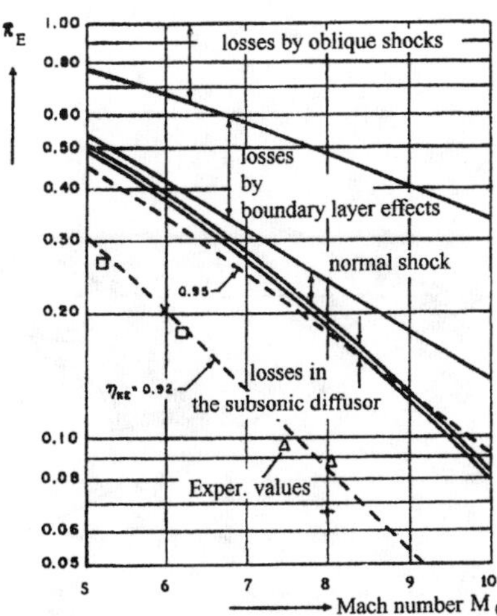

Fig. 31 Analysis of total pressure losses in mixed compression inlet diffusor types.

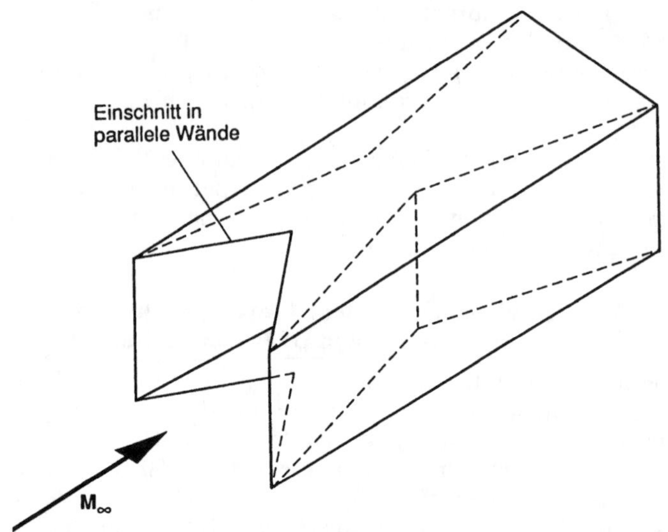

Fig. 32 Proposal for an intake model with internal compression and fixed geometry.

ferent variants, with the upper and lower walls each featuring 1-, 2-, 3- and 4-degree expansions.

3) A rectangular, variable geometry segment, 500 mm long, with the upper and lower walls variable from 0- to 4-degree expansions in 1-degree increments. In addition, unsteady cross-section steps can be installed on the top and bottom walls up to an area ratio of 2.

4) A diffuser segment, 500 mm long, with vertically adjustable upper and lower walls.

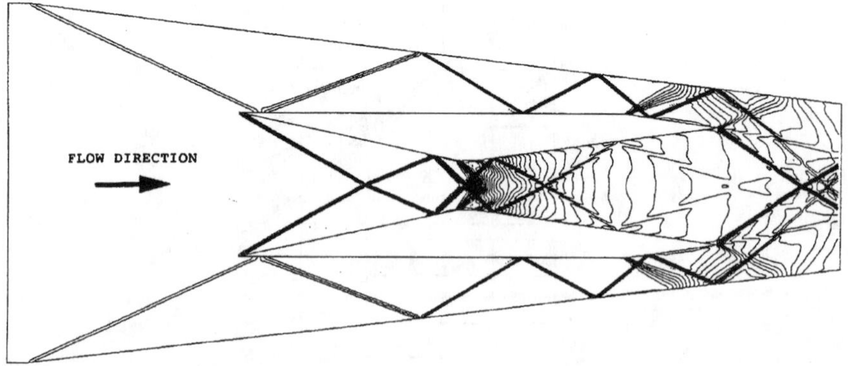

Fig. 33 Internal compression intake - lines of constant density at Mach = 3.0.

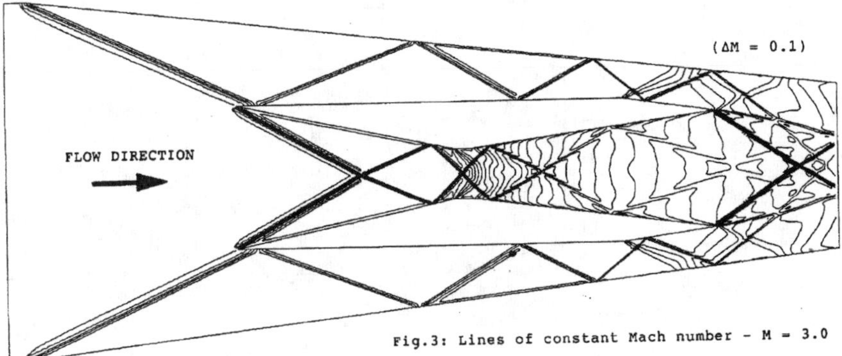

Fig. 34 Internal compression intake - lines of constant local Mach numbers at M = 3.0.

The isolator entrance cross-section is sized 30 × 100 mm². The combustor parts are manufactured from heat-resistant 8-mm steel plate, with no active cooling devices provided. To boost the structural stiffness, additional stiffeners were welded on to the exterior walls of the variable-geometry segments. To measure the static wall pressure distributions, the combustor walls were fitted with pressure tapping points spaced about 30-50 mm apart. At the rear end of the third combustor section, normal air injection can be provided to throttle the main flow at the instant of combustor ignition for the duration

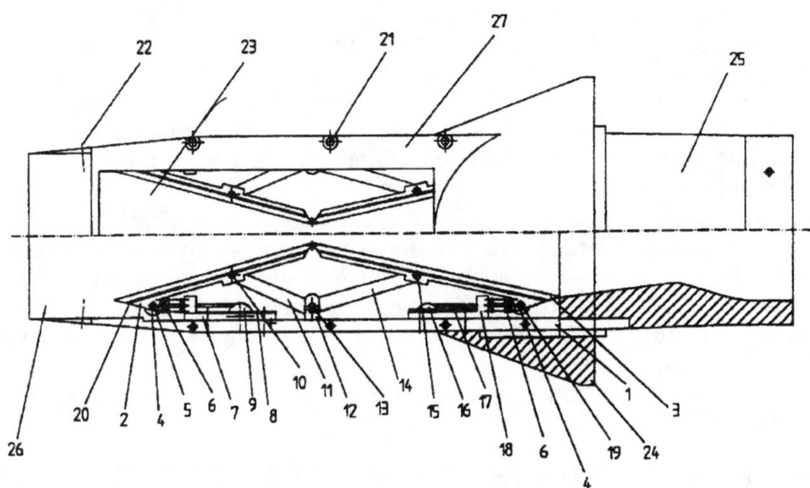

Fig. 35 Proposal for an intake model with internal compression and variable geometry.

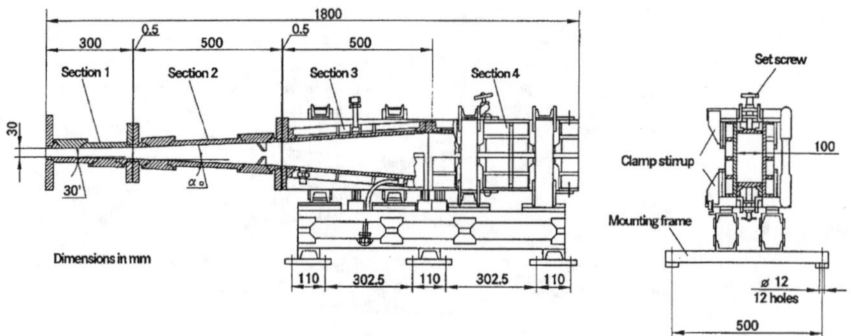

Fig. 36 Subscale combustor model.

of 1 sec maximum in case combustor entrance temperatures are too low for spontaneous ignition.

Fuel can be injected at two axially staged positions within the second combustor segment, using several injection system variants.

B. Fuel-Injection Modules

Three different fuel-injection variants were designed and manufactured (Fig. 37):

1) A tube injection system supplying the near-wall flow through wall orifices and the core flow through inclined tubes. Two injection modules are mounted in oppositely arranged pairs at the upper and lower walls, featuring a staggered holes and tubes configuration in cross-steam direction.
2) A wedge-shaped fin injection system with fuel supply normal to the flow through six orifices located on each lateral surface.
3) A system of swept wall-mounted ramps of the type proposed by NASA Langley Research Center. With this injector type, increased mixing is anticipated due to the vorticity and fuel entrainment generated by the swept ramp trailing edges.

The injection devices were manufactured from copper alloy and are cooled by the hydrogen fuel flow supplied to the combustor at ambient temperature. At the two staged stations in streamwise arrangement, the injection modules can be mounted in oppositely arranged pairs at the upper and lower wall.

C. Test Results

The primary objective of these tests was to provide data on the impact of 1) fuel supply, and fuel staging, 2) inlet Mach number and inlet stagnation temperature, and 3) combustion chamber geometry on gas dynamic flow conditions in the supersonic combustor. For the purpose, some 150 con-

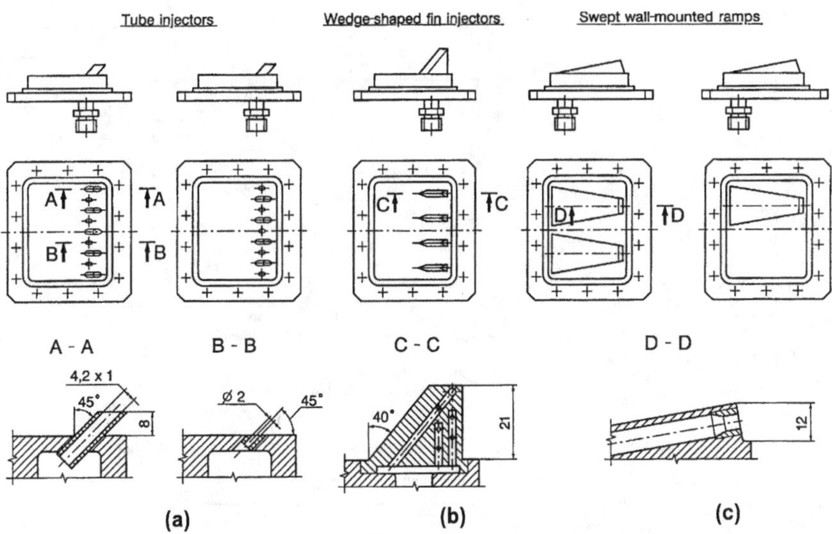

Fig. 37 (a) Tube injection system; (b) wedge-shaped fin injection system; (c) swept wall-mounted ramp injection system.

nected-pipe test runs were made with various combustion chamber geometries and fuel supply rates at diverse inlet conditions. On account of the uncooled combustor structure, typical test times were about 6 seconds.

The example below reflects some typical test data obtained using a divergent combustor geometry, with 1-degree divergence angle for both the upper and the lower walls.

Figure 38 illustrates static pressure distributions along the upper and lower channel walls. The pressure refers to the static pressure in the preheater. Hydrogen was injected at the first injection point using the tube injection system. At the fuel supply point, an upstream distortion of the pressure field that reaches to the isolator exit. The pressure peak and the upstream pressure distortion grow together with the equivalence ratio. Downstream of the fuel supply point, within the divergent channel section, the pressure distribution is characterized by expanding supersonic flow attend by combustion.

Figure 39 shows the results obtained with a fuel staging setup. At both supply stations the tube injection system was mounted. For the case $F_{\text{II}}>F_{\text{I}}$ (F_{I}, F_{II} are cross sections at the injection location according to Figs. 38 and 39) the absolute pressure peak is clearly reduced indicating a smoother combustion and heat release which in turn reduces thermal and mechanical loads on the combustor walls.

For the dimensioning and performance assessment of a SCRAM-Jet propulsion system, knowledge of a realistic combustion efficiency is of particular relevance. For some of the connected-pipe test runs, therefore, the achieved combustion efficiencies were determined.

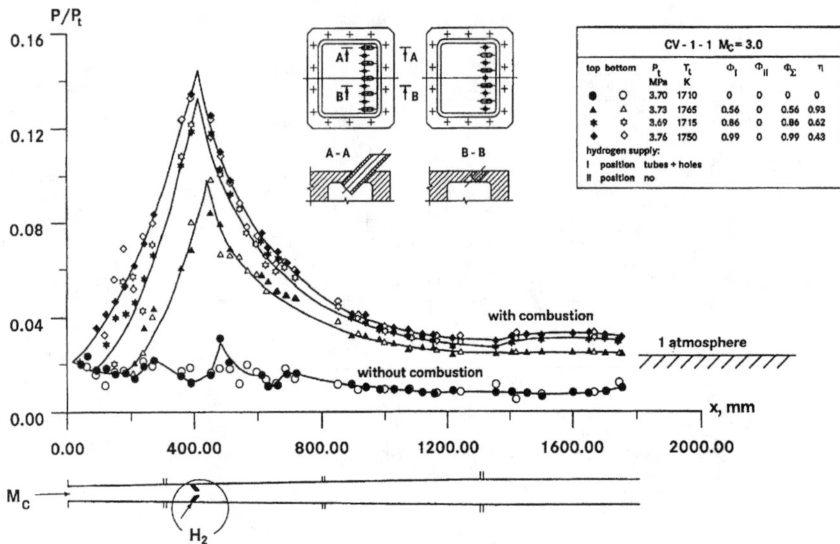

Fig. 38 Pressure distributions using tube injectors at position I.

Figure 40 shows the efficiencies achieved with the various injector systems. The efficiencies were determined with the aid of a 1-D analytical model using the observed static pressure distributions as input quantities. As it will become apparent the combustion efficiency diminishes as the equivalence ratio rises, for all injectors. This tendency had surfaced in earlier investigations and is attributed to the mixing problems associated with increasing fuel fractions. The comparison shows that the wall-mounted ramps give the best combustion efficiencies, while the fin injectors do the worst. If this is related to stagnation pressure losses, it will become apparent that the wall-mounted ramps with the best combustion efficiencies also are associated with the largest stagnation pressure losses, while the fin injectors, resulting in the worst efficiencies, caused the least stagnation pressure losses. For the injector types here investigated, then, it is apparent that good mixing of the fuel with the core flow, a prerequsite to a favorable efficiency, carries the penalty of the high stagnation pressure losses.

The combustor model design featuring a highly variable geometry demonstrated its durability in more than 150 test runs and proved a useful tool to study supersonic combustion processes at various operation conditions. The results obtained allow quantification of such essential aerothermodynamic characteristics as combustion-induced stagnation pressure losses, pressure peak generation at fuel injection points, as well as upstream propagation of pressure distortions and mitigation of the latter two. In addition, the specific impacts of combustor inlet stagnation temperature, inlet Mach number and combustor geometry on the aerothermodynamic flow characteristics were evaluated.

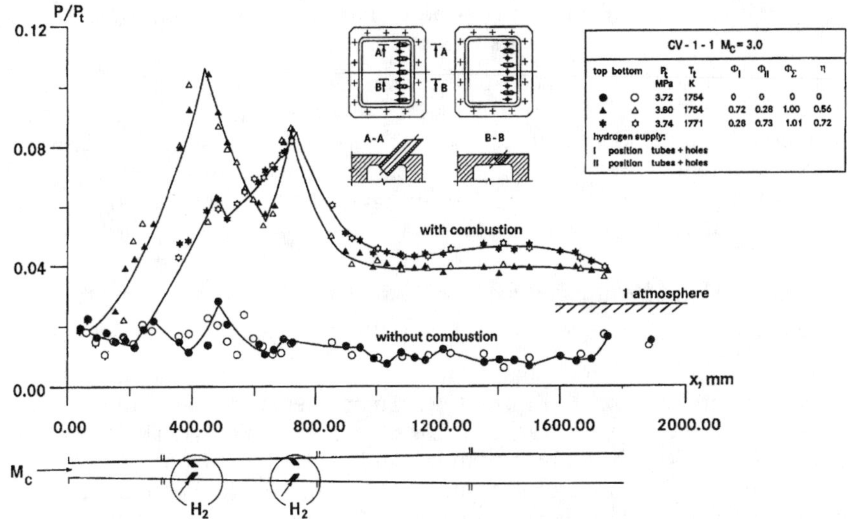

Fig. 39 Pressure distribution using a fuel staging setup.

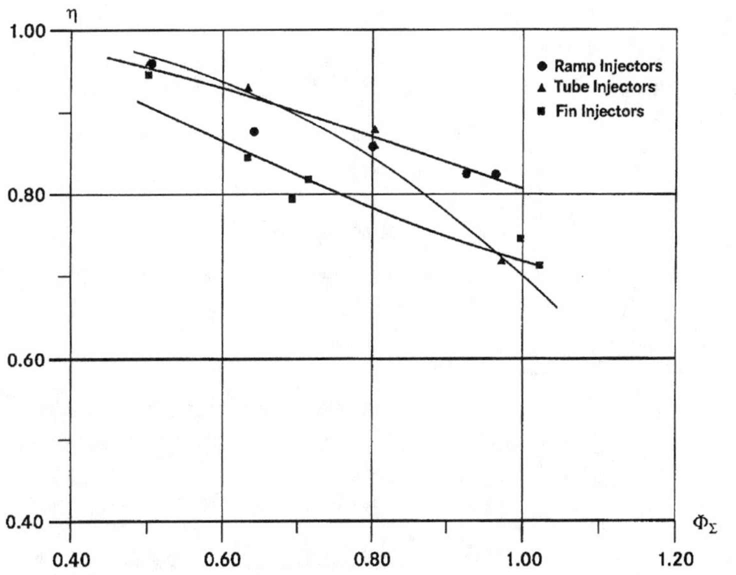

Fig. 40 Combustion efficiencies obtained by different injector types at position II.

IV. Freejet Wind-tunnel Testing of SCRAM-Jet Propulsion Systems at TsAGI

In addition to the connected-pipe tests, a series of freejet tests simulating a Mach 5 to 7 envelope for a complete subscale engine model was performed. The T-131 B freejet test facility at TsAGI was used. The kerosene fueled pre-heater provides stagnation temperatures up to 2350K at stagnation pressures up to 110 bar. A description of the test facility and the tests conducted at TsAGI is found in Ref. 8, the results are reported in Ref. 9.

A. SCRAM-Jet Propulsion System Model Concept

1. Engine Model Design

The 2-D SCRAM-Jet engine model design consists of a three-ramp, mixed-compression inlet, an isolator segment between inlet exit and first fuel injection point, a divergent nozzle segment, see Fig. 41. The cross-section is 100×150 mm^2 and the total model length is about 1700 mm.

The engine is manufactured from a chrome-nickel alloy containing 60% nickel and is heat resistant up to 1400 K. To measure the static wall pressure distribution, the model is fitted with 70 pressure tapping points spaced along the upper and lower channel walls. Static pressure fluctuations, wall temperatures and wall heat fluxes are measured by three pressure pulsation sensors, four thermo-couples and four heat flux sensors, respectively (Fig. 42).

The model features two fuel supply stations, the first 200 mm downstream of the inlet cowl, and the second 40 mm upstream of the lower wall step. To inject the hydrogen fuel, various injection modules were used.

2. Fuel-Injection Modules

As for the combustor model, three fuel-injection variants were designed and manufactured:

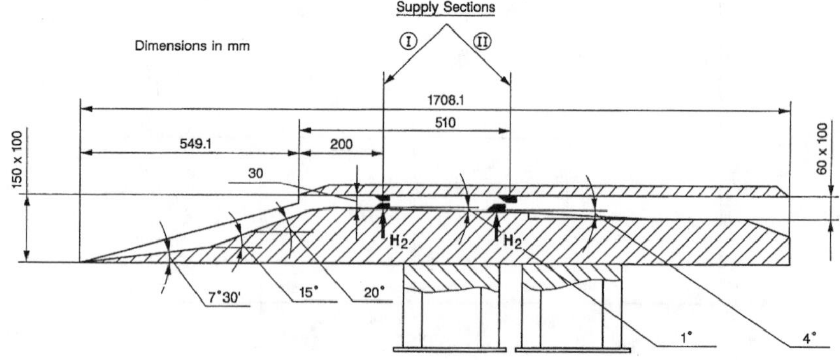

Fig. 41 Longitudinal section of subscale engine model.

1) A tube injection system similar to the type used for the combustor model, but with no wall holes drilled between tubes.
2) An injection module with wall holes inclined 60 degrees relative to the main stream. This injector variant is mainly used in combination with variant (1).
3) A wedge-shaped fin injection system with hydrogen supply normal to the main flow through two orifices on each lateral surface.

As for the combustor model, the injector modules are manufactured from copper alloy and are cooled by the hydrogen fuel flow supplied at ambient temperature. At the two staged stations the injection modules can be mounted in oppositely arranged pairs at the upper and lower walls.

B. Testing Focus

A total of 50 freejet test runs were made. The test program focused on the 1) impact of the total fuel supply and fuel staging on the aerothermodynamic interaction between combustor and inlet flows, 2) impact of various injector systems on the mixture formation and combustion efficiency, and 3) investigation of the dual-mode operation at Mach 5.

C. Test Results

1. Mach 6 Test Results

Figure 43 shows the results obtained with a fuel staging setup. Whereas at the first supply station, tube injectors are mounted, the fuel is injected through wall holes at the second station. Downstream of the first fuel injection station, the rising pressure curve indicates the presence of super-

Fig. 42 Subscale engine model.

sonic flow that remains supersonic downstream of the second injection point, due to the 4-degree channel expansion, which amply cancels the heat release effect. With the total temperature, total pressure and overall equivalence ratio being close to the parameters of the non-staged example, as was found with the fuel-staged connected-pipe combustor tests, the absolute pressure peak is clearly reduced by staged fuel injection (*P*max=1.8 down from 2.6 bar).

Figure 44 shows another test run with fuel staging applied. Whereas the test parameters Tt, pt are close to the values of the run shown in 3.3-Fig.3, the fuel flows injected at the first and second positions were increased by 25 and 32%, respectively. As a consequence, thermal choking occurred, witnessed by a pronounced pressure peak. Under these conditions, the pressure disturbance evidently reached the inlet and a strong shock wave occurred in front of the cowl edge.

2. *Mach 7 Test Results*

Figure 45 shows the pressure distribution for a test run at a free-stream Mach number of 7.2. For high-fidelity simulation of flight conditions, the inlet stagnation temperature was made 2053 K. The fuel was staged through tube injectors at the first fuel injection point and through wall ports at the second injection point. The pressure profile is normalized to the total pressure in the pre-heater. Downstream of the first injection point the rising pressure profile is indicative of supersonic flow, which is maintained also downstream of the second fuel injection point as a result of the 4-degree

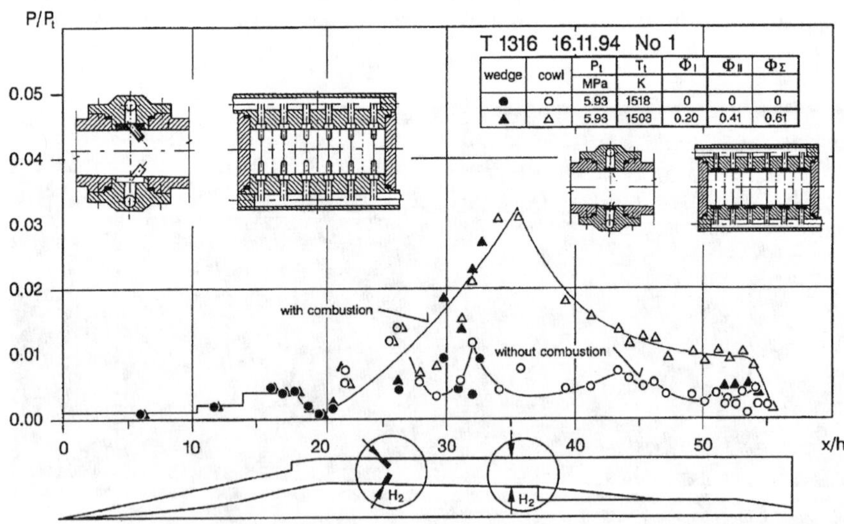

Fig. 43 Pressure distribution using a fuel staging setup Tt = 1503K, Pt = 0.20, FII = 0.41.

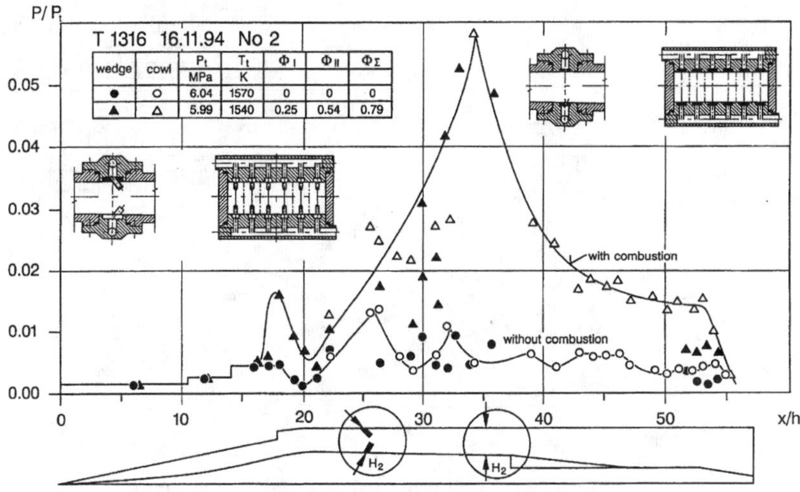

Fig. 44 Thermal choking causing disturbance of the inlet flow.

divergent channel portion, which overcompensates the deceleration through heat input. In terms of pressure distribution, therefore, we qualitatively have the same pattern as for the test runs conducted at free-stream Mach 6.

3. Thrust Measurements

For the subscale engine model, thrust measurements were taken in free-jet testing at free-stream Mach number 6.4, stagnation temperature 1450 K and a total fuel supply according to an equivalence ratio of 0.5 to 1.

One of the greatest challenges to be resolved in the thrust measuring effort was the conception, manufacture and calibration of a six-component thrust balance. This work was performed by TsAGI.

Thrust measurements taken on an engine model in the test section of the free-jet test bed might be compromised by test bed-induced distortions. This would involve, e.g., shock waves emanating from the forward portion of the engine model and being reflected from the test section walls to re-impact on the model in its aft portion. To minimize such test data corruptions, the model was enveloped in a shroud structure that closely followed, but never contacted with the model's outer contour.

Figure 46 exemplifies some of the thrust measurement data. To eliminate the drag of the engine model, the axial force acting along the engine centerline as measured at the same inlet flow conditions, first with fuel supply and combustion on, and second with no fuel supplied. The resultant difference represents the axial thrust shown versus fuel supply in Fig. 46. As expected, the generated thrust grows evenly with the fuel flow.

The subscale SCRAM-Jet engine model with is adaptable combustor geometry its operability in more than 50 test runs simulating a free-stream

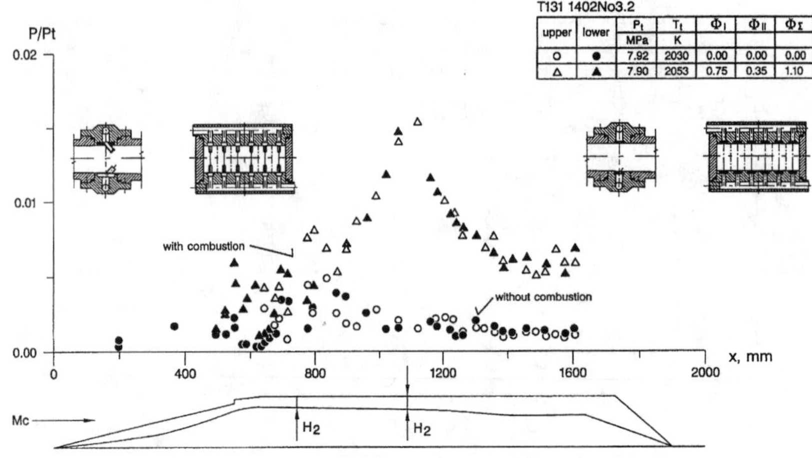

Fig. 45 Pressure distribution using a fuel staging setup.

Mach number range between 5.0 and 7.2. The test runs demonstrated the presence of intensive aerothermodynamic coupling between combustion and inlet performance. Uncontrolled heat release can cause excessive combustion induced pressure peaks with an extreme upstream propagation of pressure distortions resulting in unacceptable disturbances of the inlet flow. Thrust measurements were obtained at a free-stream Mach number of 6.4 using a six-component thrust balance.

The German-Russian program's objective, to expand the available

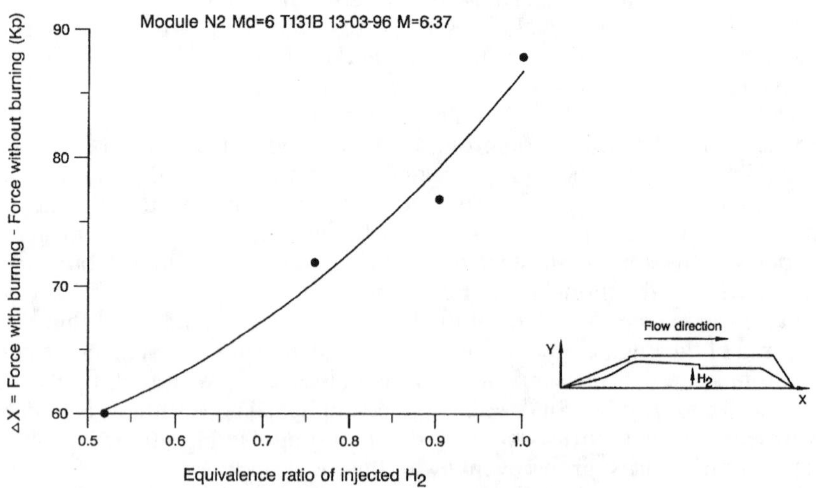

Fig. 46 Axial thrust versus fuel supply.

SCRAM-Jet technology base by application-oriented investigations of generic combustor and engine models, was fully achieved. The good progress of the program was in large part made possible by the Russian partner, by TsAGI's cooperative assistance in the design and manufacture of the combustor and engine models and the performance of the methodical connected-pipe and freejet test series. The SCRAM-Jet program was stopped at the end of 1995, when the German Hypersonics Technology Program was terminated. Further basis research on the subject continued at DLR and the Universities of Aachen, Munich and Stuttgart ("Centers of Excellence" for Hypersonics Technologies) but without the German Industry being involved.

V. Considerations for Flight Testing Small-Scale SCRAM-Jet Modules Using the RADUGA-D2 Flying Testbed

A. Objectives for Flight Testing

As already mentioned a general RAM/SCRAM-Jet propulsion testing Plan for performance validation and demonstration on ground and in flight has been established at the end of Phase 1b. The following picture, Fig. 47, outlines schematically the planning for an international cooperation. The important role of flight testing can be clearly identified in the figure. Until the end of 1995, the tests were performed on schedule, the analysis of experimental results were documented in mid-1996.

As shown in the previous chapters, both connected pipe and free-jet testing of the SCRAM-Jet engine module has provided the database for successful operation at static conditions by simulation using vitiated air. In addition thrust measurements have been made leading to a first extrapolation to estimate the required size of a SCRAM-Jet propulsion engine for accelerating a flying test vehicle.

The overall objective for testing the same SCRAM-Jet modules under realistic flight conditions can be summarized as follows:

1) performance of stable SCRAM-Jet operation at non-uniform inflow conditions
2) assessment of in-flight measured pressures and temperatures
3) completion of the ground-tested engine module(s) by adding a nozzle
4) assessment of in-flight measured Net-Thrust

B. RADUGA-D2 Flying Testbed

Compare Flight Test With Ground Test Data

The RADUGA-D2, shown in Fig. 48, is a winged vehicle, launched by a Tupolev TU 22M at supersonic speed and equipped with a liquid propellant rocket engine. The vehicle was designed and developed by the Russian company RADUGA and has a record of more than 500 successful flights. It has already been used as a "Flying Laboratory" for hypersonics technologies. The vehicle is not recoverable in its present configuration. In a common study between the Russian companies RADUGA and TsAGI, Dasa-M, OHB,[10] and

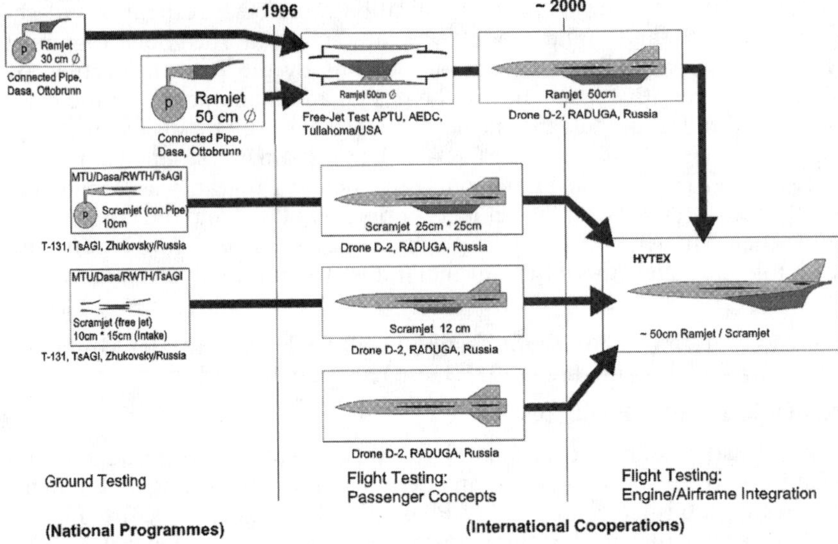

Fig. 47 General planning of propulsion testing with the German HTP.

RWTH Aachen, RADUGA and TsAGI,[11] the feasibility of using the D2 for flight testing the SCRAM-Jet engine was investigated in detail.

The flight regime of the RADUGA-D2 is shown in Fig. 49. The flight altitude corresponds up to Mach-number of 6 to the typical ascent trajectory of airbreathing space transportation systems. The major objective for flight testing the SCRAM-Jet engine module is the comparison with the results from ground testing. Therefore, during flight it is necessary that the inlet upstream flow parameters are similar to those conditions which have been found during the windtunnel tests at Mach = 5.25 and 6.25. These parameters are local Mach-number, dynamic pressure, angle of attack and the flow field nonuniformity level wrt to the intake entry cross section. Flight testing time should allow for thermal steady equilibrium operation and therefore exceed at least 10 seconds. According to these requirements, two segments of the flight domain of the flying testbed have been chosen appropriately as will be specified and discussed in the next chapter.

C. Flight Test Trajectory and Integration of SCRAM-Jet in the RADUGA-D2

Two options of flight testing profiles were chosen from the flight domain of the RADUGA-D2 shown in the previous figure. Both correspond to a large extent to the wind-tunnel flow conditions.

In the first option (see Fig. 50) the SCRAM-Jet operation will be performed at the descent part of the trajectory at segments corresponding to Mach-numbers in front of the intake cross section of 6.25 and 5.25 which correspond to flight Mach-numbers 6.5 and 5.5.

Max. Mach Number	**6.3**
Max. Altitude	**90 km**
Max. Range	**570 km**
Max Thrust	**70 kN**
Total Length	**11,67 m**
Wing Span	**3.00 m**
Fuselage Diameter	**0.92 m**
Maximum Mass	**5800 kg**
Propellant Mass	**3045 (Fuel&Oxidizer)**
Max. Payload Mass	**800 kg**

Fig. 48 RADUGA-D2—main characteristics.

1) During the first part of the angle-of-attack control was chosen to achieve at the final point the following flight parameters: Mach number M_∞=6.5, altitude H=24 km and the flight profile slope angle Θ=0. The duration of the first part of flight trajectory is 103 sec.. The maximum thrust of liquid fuel rocket engine is used during this part.

2) The second part of the flight profile is appropriate for SCRAM-Jet testing with M_∞=6.5. The throtling regime of the rocket engine is used. The duration of the horizontal flight at the altitude H=24 km is 14 sec, with angle-of-attack a=2.15 deg. As follows from the calculated local flow upstream of the inlet, this flight regime closely corresponds to ground tests of the SCRAM-Jet module at M_∞=6.25, with a dynamic pressure at the ramjet entry of 70000-83000 Pa.

3) During the third part the test-bed with rocket engine switched off glides to the altitude H=20 km., and to Mach number M_∞=5.5. Angle-of-attack control leads to flight profile slope angle Θ=0 at the final point of this part of the flight profile.

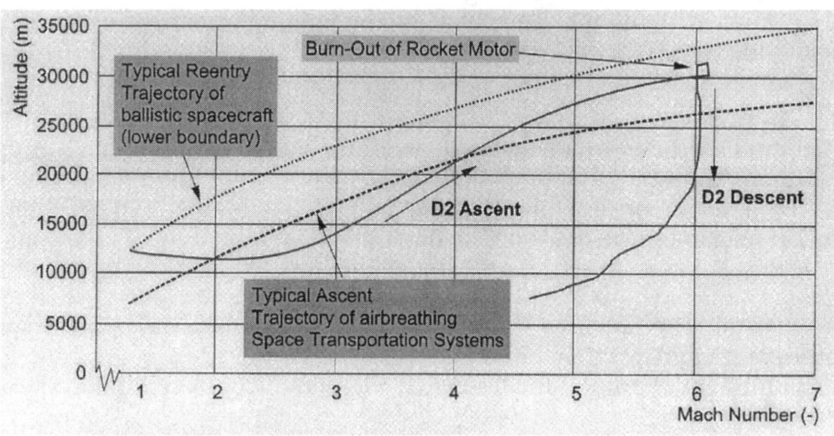

Fig. 49 Flight domain of the Russian RADUGA-D2 flying testbed.

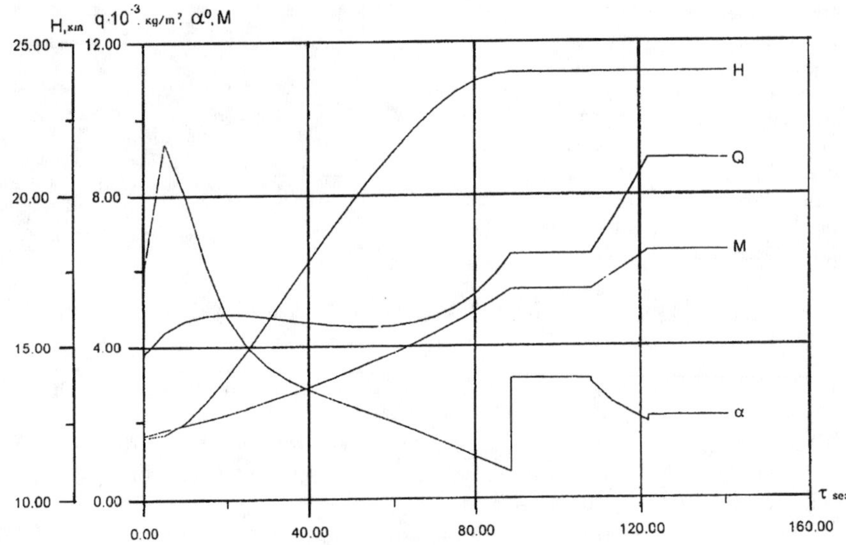

Fig. 50 Flight test trajectory (Option 1).

4) The fourth part satisfies conditions of SCRAM-Jet testing with M_∞=5.5. The throttling regime of the testbed engine is used. The horizontal flight duration at the altitude H=19.0 km is 11 sec., with α=1.7 deg. This regime corresponds to ground tests with M_∞=5.25 and with dynamic pressure at the ramjet entry of q=85000-114000 Pa. At the final point of the fourth part the rocket engine fuel runs out.

The second option for the flight profile allows to provide SCRAM-Jet testing first at M_∞=5.5 and later at M_∞=6.5. Flight trajectory also consists of four parts (see Fig. 51).

1) The first part leads to the final point with M_∞=5.5, H=24 km and Θ=0. The duration of this part is 89 sec, with the rocket engine running with maximum thrust.
2) The second part is appropriate for SCRAM-Jet testing at M_∞=5.5 the rocket engine throttled down. The duration of horizontal flight at the altitude H=24 km is 19 sec. at a=3.1 deg. This flight regime corresponds to ground tests of the SCRAM-Jet engine wrt Mach number only. The local Mach number at the SCRAM-Jet inlet location is near 5.25, and the dynamic pressure is approximately 57000 Pa.
3) During the third part the testbed accelerates at constant altitude up to M_∞=6.5.
4) The fourth part of the flight profile is appropriate for SCRAM-Jet testing at M_∞=6.5 and a=2.1 deg.. The duration of this part is 18 sec., and flight

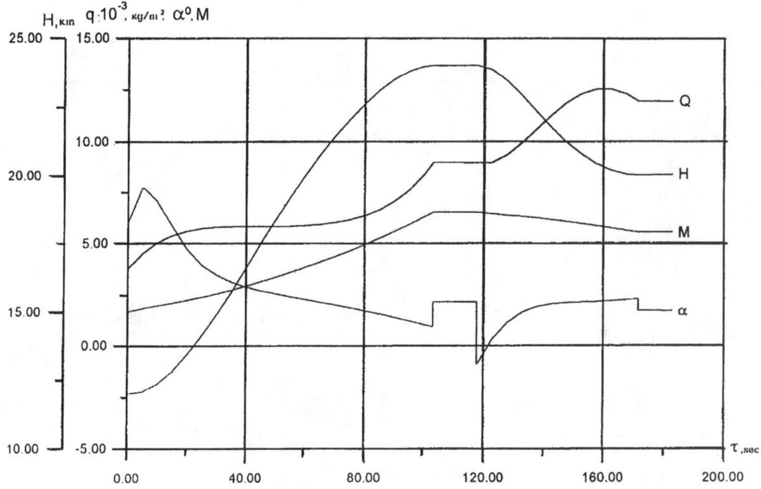

Fig. 51 Flight test trajectory (Option 2).

regime completely corresponds to ground testing with M_∞=6.25 and 85000-11400 Pa.

The range of the powered flight of the test-bed for the first option of the flight profile is 250 km with a net duration of the four parts 183 sec. (approx. 3 min.) and, for the second option, 175 km with a net duration of the four parts of 140 sec (2.3 min.). After the completion of the flight-tests the testbed glides with the rocket engine switched off. For recovery of the SCRAM-Jet engine test module, the Russian partners recommended a highly sophisticated three-stage parachute system which turns the RADUGA-D2 upside down for landing, as Fig. 52 demonstrates.

A detailed investigation was necessary to assess the changes needed for integrating the SCRAM-Jet engine module on the lower surface of the flying testbed RADUGA-D2. Due to the external loads during flight, some modifications are necessary on the SCRAM-Jet ground test module:

1) The load carrying bar of the inlet, combustor and nozzle has to be strengthened and therefore manufactured from alloy XH60BT,
2) The manufacturing of the foreword and backward suspension installed on the spar for the connection of the module with the airframe,
3) Extension of the internal cavity of the module for placement of start and cut-off equipment and of blocks with pressure pulsation probes and finally
4) The manufacturing of a special nozzle for the module to deflect the exhaust jet from the fuselage of the test bed.

On the carrier side, two preliminary configurations for the integration of the SCRAM-Jet engine were considered. A single (shown in Fig. 53) and a

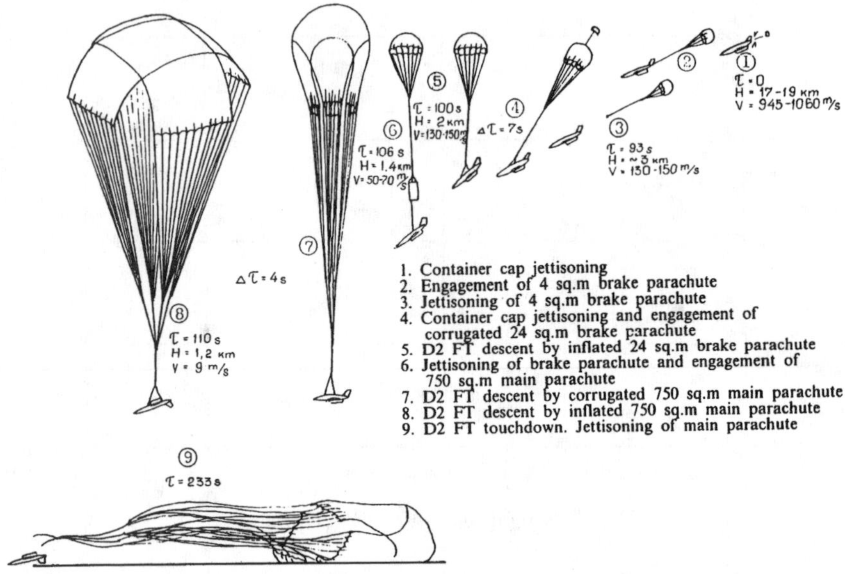

Fig. 52 Three-stage parachute recovery system.

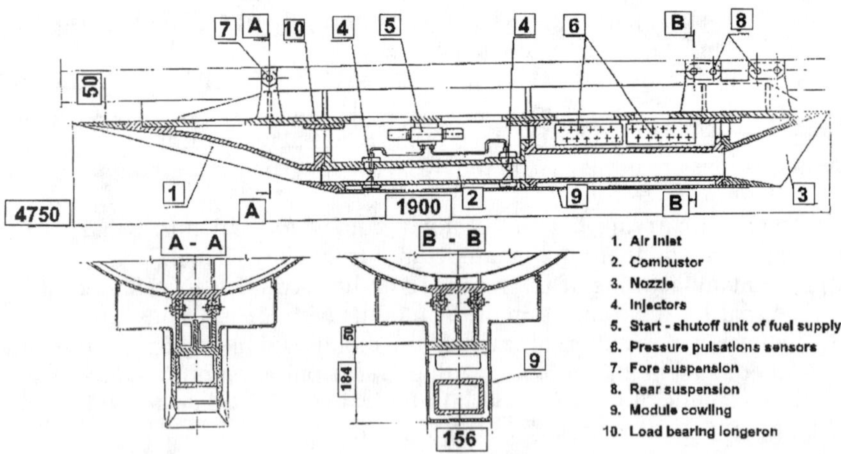

Fig. 53 SCRAM-Jet engine test module for flight testing single engine configuration.

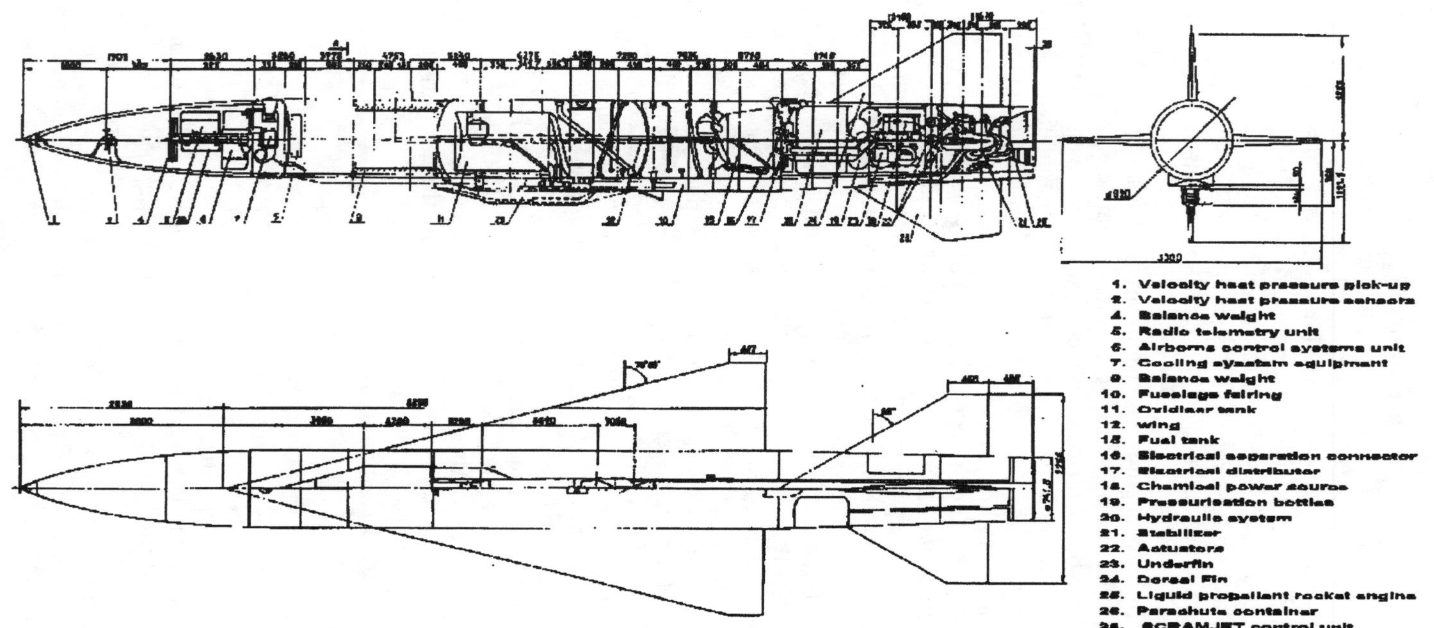

Fig. 54 Attachment of the SCRAM-Jet engine test module to the RADUGA-D2.

double engine mounting was investigated. The complete Flight Test vehicle with integrated SCRAM-Jet module is shown in Fig. 54.

On the RADUGA-D2 three structural elements, influencing the test engine installation are located on the lower side of its fuselage: 1) a triangular cross-section ventral fairing extending from $x=3037$ mm to the fuselage base section, 2) a vertical wedge with a maximum height of 400 mm extending from the section 5490 mm to the section $x=8460$ mm, and 3) the lower fin with a maximum height of 800 mm foldable during flight on the carrier aircraft.

Although the test engine dimensions are significantly less than those of the flying testbed, the installation of the test engine can influence the drone flight characteristics. Consequently, computations using 3-D Euler codes were conducted to investigate the influence of systematic changes for the geometric shape of the fairing in front of the intake on oncoming flow uniformity. These modifications include cases with and without ventral fairing, with the initial shape of the fairing and fairings with sharp forebody, with reduced wedge-shaped slope and fairings with rectangular cross section.

The results have shown that in spite of the relative small dimensions, the ventral fairing has a rather significant influence on the flow field and causes non-uniformity in the area of the assumed location of the test SCRAM-Jet engine intake. If the test module is moved forward by about 3.5 m the fairing stops influencing the flow non-uniformity in the region of the intake position. However the local mean Mach-number was reduced significantly in comparison with M_∞ ($Mx=5.26$ compared to $M_\infty=6.0$). But the results of the CFD flow field calculations and trajectory investigations have shown that the selection of the considered RADUGA-D2 flight profiles and of an optimized appropriate SCRAM-Jet location will guarantee the coincidence of the flight and ground-test flow fields upstream of the inlet.

The "Key Problem" of any future airbreathing high speed propulsion systems is the optimal engine airframe integration in order to obtain a maximum thrust minus drag value for acceleration and cruise. This value is the difference of two nearly equal large figures and both suffer from a rather large uncertainty bandwidth. Therefore, the assessment of thrust plays a most important role and may be last but not least the most urgent driver for flight testing SCRAM-Jet.

Three different thrust-definitions can be specified:

1) Internal thrust (isolated airbreathing engine)
The difference of the gas momentum at the nozzle exit section and the momentum of the stream of undisturbed air entering the inlet.

2) Net thrust (isolated airbreathing engine)
The vector difference between the internal thrust of the engine and the forces due to external drag of the engine nacelle including drag due to spillage

3) Actual thrust (airbreathing vehicle powerplant)
The sum of the engine net thrust P_{net}, the entry momentum determined for the section passing through the top of the first stage of compressor wedge,

the axial component of forces of airframe/nacelle aerodynamic interference forces ($\Sigma\ \Delta Xi$), and the forces of the interaction between airframe elements and the jet (ΔP).

$$P_{act} = P_{net-}\ (\Sigma\ \Delta Xi) + (\Delta P)$$

To determine the actual thrust of the vehicle powerplant, it is necessary to have the dependence of the engine intake entrance Mach-number $M_1 = f(M_\infty)$ and the data of parametric wind tunnel tests of the isolated engine model at various Mach numbers.

Proceeding from the definitions given above, the following methods of defining the thrust on the basis of the results of the flight experiment can be considered.

1) Actual thrust measurement by means of strain-gauges or balance where the nacelle is suspended to the vehicle. This method suggests a direct measurement of the thrust, additional studies are necessary for its technical realization.
2) Determination of the actual thrust of the operating ramjet engine by measuring the value of varying g-load, which acts upon the testbed, if the test engine is started. in this case a preliminary estimation of the drag installed on the test vehicle nacelle including the connections which are necessary for engine in flight operation.
3) Determination of the actual thrust by computing the vector difference between entry and exit, using the results of measuring the parameters (velocity, pressure, temperature) of the flow passing through the engine.
4) Determination of the actual thrust as the difference between the nozzle exit and the undisturbed flow in front of the vehicle. Then net thrust can be obtained by using the results of calculations.

Using the results of computer calculations, bench testing of inlet, combustor, nozzle models and wind tunnel tests of the SCRAM-Jet modules, a mathematical model of the powerplant is derived. The required thrust value can be determined on the basis of the mathematical model by using the flight test data. The experience shows, that the thrust should be determined by different ways simultaneously and the conclusion on the true thrust value can be made only after the comparison of the results obtained by different methods.

As it follows from above, only the actual thrust of the vehicle powerplant can be measured directly in the flight experiment while the net thrust of the isolated engine can be measured in the ground experiment (tests of isolated SCRAM-Jet model in free jet facilities). Therefore for the correct comparison of flight and ground tests data it is necessary to extract all integration effects from the flight test. Neglecting integration effects means inlet momentum $\Rightarrow I_\infty$, interaction forces between the airframe elements and the jet $\Delta P \approx 0$ and the axial component of the aerodynamic interference between airframe and nacelle $\Delta Xi \approx 0$. Under these conditions, the actual thrust is equal to the net thrust.

References

[1]German Federal Ministry for Research and Technology, "Hypersonic Technology Program," Bonn, Germany, 1988.

[2]Hirschel, E. H. "The Technology Development and Verification Concept of the German Hypersonics Technology Program," Dasa Ottobrunn, LME12-HYPAC-STY-0017, Germany, 1995.

[3]Walther, R., "Hyperschalltechnology Propulsion—Executive Summary Report Scramjet-Technology," MTU Munich, B96 EW-0011, Germany, 1996.

[4]Schmatz, M. A. "Hypersonic Three-Dimensional Navier-Stokes Calculations for Equilibrium Air," AIAA Paper 89-2183, 1989.

[5]Eberle, A., Schmatz, M. A., and Bissinger, N. C. "Generalized Flux Vectors for Hypersonic Shock-Capturing," AIAA Paper 90-0390, 1990.

[6]Jürgens B., Sasse S., Schneider A., and Koschel, W., "Scramjet Technology," RWTH Aachen, Inst. for Jet Propulsion and Turboengines, Final Rept., Germany, 1996.

[7]Bissinger, N. C., and Eberle, A., "CFD Contributions During Hypersonic Airplane Intake Design," AGARD CP 510, 1992.

[8]Walther, R., "Hypersonic Technology Program," MTU Munich, Final Rept. on Scramjet-Technology, MTUM-B96EW-0011, Germany, 1996.

[9]Walther, R., Koschel, W., Sabelnikov, V., Korontvit, Y., and Ivanov, V., "Investigations into the Aerodynamic Characteristics of Scramjet Components," International Symposium on Air Breathing Engines, ISABE 97-7085.

[10]Zellner, B., "Investigations of Alternative Concepts and Configurations of Hypersonic Flight Test Vehicles," Dasa-M, Rept. LMLE1-HYPAC-STY-0015, Germany, 1996.

[11]Koschel, W., "Study: Flight Testing SCRAM-Jet," DLR-Lampoldshausen, Final Rept. Doc. Nr. 86456060, Germany, 1996.

Chapter 4

Scramjet Engine Research at the National Aerospace Laboratory in Japan

Nobuo Chinzei,* Tohru Mitani,† and Nobuyuki Yatsuyanagi‡
National Aerospace Laboratory, Kakuda, Miyagi 981-1525, Japan

Nomenclature

C_d = drag coefficient (−)
C_{df} = friction drag coefficient (−)
C_{dp} = pressure drag coefficient (−)
c_f = local skin friction coefficient (−)
D_{int} = internal engine drag, N
F_p = axial component of integrated pressure force over the internal surface of the engine, N
G = minimum flow-path gap between sidewalls or between sidewall and the strut, mm
p_w = wall pressure, Pa
p_0 = air total pressure, Pa
T_{int} = internal engine thrust, N
T_0 = air total temperature, °K
t_b = ignition delay time with radicals, s
t_0 = ignition delay time without radicals, s
X_0 = initial mole fraction of radicals (−)
x = axial distance from the leading edge, mm
ΔF = thrust increment with and without fuel injection, N
ΔF_p = difference of F_ps with and without fuel injection, N
η_c = local combustion efficiency (−)
Φ = total fuel equivalence ratio (−)
φ = local fuel equivalence ratio (−)

*Head, Director, Ramjet Propulsion Research Division, Kakuda Research Center.
†Head, Ramjet Combustion Section, Ramjet Propulsion Research Division, Kakuda Research Center. Member AIAA.
‡Director, Kakuda Research Center.

I. Introduction

THE research activities related to the scramjet in Japan date back to the late 1970s. The activities during that period were limited to fundamental, laboratory-scale experiments on supersonic combustion at universities. Yoshida and Tsuji[1] examined the ignition mechanism of a supersonic diffusion flame by conducting an experiment on a supersonic combustion flowfield with slot injection of gaseous hydrogen transversely into a Mach 1.81 airstream. They concluded that the ignition under these kinds of flowfield occurred in the separated region upstream of the fuel injection. Tsuji and Matsui[2] conducted an experiment on coaxial subsonic fuel flow in a Mach 1.26 flow and examined a flowfield produced by combustion and the structure of the flame formed. A recirculation zone was observed downstream of the fuel injector and was found to have an important role in flame holding. A normal-shock induced combustion of a coflowing fuel jet was experimentally examined by Takeno et al.[3] in a Mach 1.91 high-temperature airstream. The size and location of the normal shock were found to be affected by the exothermic reaction behind it, but reaction quenching was observed downstream. Kimura et al.[4] conducted an experiment on flame stabilization and promotion of combustion by a plasma jet in Mach 2.1 and 2.7 airflows. Its potential for those purposes was found to be quite promising even at very low static temperature.

The decade after the early 1980s was relatively less active in this field except for the experimental work at the Kakuda Research Center of the National Aerospace Laboratory (NAL-KRC), which was stimulated by the studies just mentioned. Details of those studies at the NAL-KRC will be introduced later. Research in this field has once again been active at universities since the early 1990s. Tomioka et al.[5] conducted an experiment on a supersonic combustor with a tangential slot injector in a Mach 2 airstream and examined the effects of combustion on mixing. They found that mixing was remarkably affected by combustion, especially in the region near the injector. With a strut divided streamwise into two parts, and with fuel injection into the intervening gap, Niioka et al.[6] examined the stabilization of a flame formed there in a Mach 1.5 airflow. A fundamental explanation of the flame-holding mechanism and the possibility of a controllable strut for supersonic combustors were proposed. Recently, Takahashi et al.[7] conducted a test for active control of the flame holding in a fixed geometry supersonic combustor at an off-design point and for improving autoignition performance.

Scramjet research at NAL-KRC[8,9] originated with the study on a hydrogen-fueled air-breathing rocket engine in the late 1970s.[10] After several years research on the components of scramjet engines, covering such factors as aerodynamics of the inlet,[11–20] supersonic combustion,[21–43] nozzle performance,[44–51] and cooling structures and refractory materials[52–64] for the engine, was undertaken.

Based on the results of the fundamental research on scramjet engine components, a side wall-compression-type scramjet engine was designed and fabricated between 1991 and 1994 by NAL-KRC to investigate the performance of the scramjet engine and to elucidate the interaction of the components as related to such performance.[65–86]

The scramjet engine model is one of the modules of an airframe-integrated engine. The basic configuration of its air inlet is the sidewall-compression type as in the first model of NASA Langley. It is separable into six parts to allow changing of the components for parametric study. Both water-cooled and heat-sink models with the same internal flow path and interface geometry were fabricated; a liquid-hydrogen (LH_2) cooled air inlet was also fabricated. Other components of the LH_2-cooled engine model are now being fabricated.

NAL-KRC also completed the construction of a Ramjet Engine Test Facility (RJTF) for testing the engine in 1994.[87–95] The RJTF is a free-jet-type hypersonic propulsion wind tunnel, with which it is possible to test the scramjet engine in the range of simulated flight Mach numbers from 4 to 8 at altitudes from 20 to 35 km. The hardware of the facility was completed in 1993, and a series of calibration tests was performed for one year. The first engine test using a storage air heater (SAH) was completed in March 1994, and M8 tests using SAH and a vitiated air heater (VAH) were begun in September 1995. Since March 1994, a total of about 140 tests have been conducted under simulated conditions of flight Mach numbers 4, 6, and 8.

A free-piston high-enthalpy shock tunnel (HIEST), the largest in the world, was completed in November 1997 at NAL-KRC for testing in the aerodynamic study of the HOPE-X model.[96–98] The operating conditions of the HIEST were designed to produce a stagnation enthalpy of up to 25 MJ/kg and a stagnation pressure of up to 150 MPa. The HIEST will be ready for scramjet testing in the range of Mach 8 to 15 by the end of March 1999; this should provide new information on the scramjet in higher velocity ranges and possibly lead to a breakthrough in scramjet engine technology.

This chapter reviews the subjects focused on NAL's research and development projects. The subjects include the scramjet engine models, test facility, measurements and procedures, engine test results, and supplementary studies on engine testing.

II. Engine Model

The engine model is based on one of the modules of a hypothetical airframe-integrated engine as can be seen in Fig. 1. Its major dimensions are shown in Table 1. It consists of an inlet leading edge, a downstream part of the inlet, a fuel injector, and diverging sections of a combustor and a nozzle. They are replaceable for parametric tests by independently changing their geometry. Among these components, to allow airflow spillage, the inlet part is not covered with a cowl. Because its inlet is a sidewall-compression type, the shape is similar to the NASA Langley module. However, most of the dimensions are based on the results of the component tests conducted at the NAL-KRC. Overall length is 2100 mm with equal entrance and exit dimensions of 200 mm in width and 250 mm in height. As will be described later, the projected passage area of the engine entrance was chosen to be less than 20% of the facility nozzle exit section (510 mm square) to ease the wind-tun-

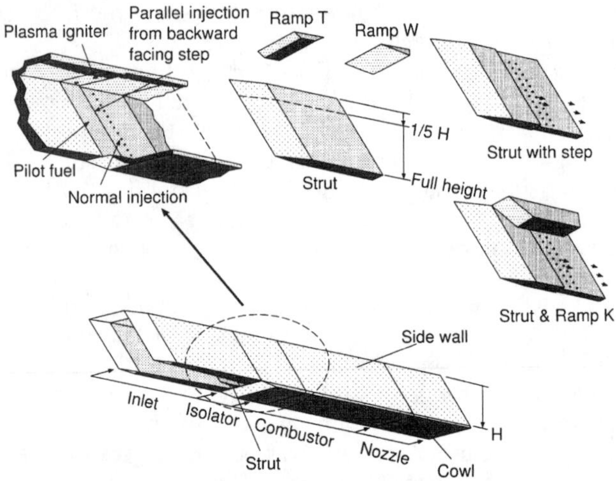

Fig. 1 Scramjet engine model tested.

nel operation. The blockage area, including sidewall fairing and cowl thickness, turned out to be approximately 25% of the latter.

The engine models tested are a heat-sink model, a water-cooled model, and a LH_2-cooled model, all with the same internal flow path except when additional devices such as a strut are attached. The heat-sink model is made of copper (except the cowl and inlet leading edge, which is made of nickel), and the other models are made of zirconia-copper alloy. The former model was also used for the acceptance test of RJTF at flight conditions of Mach 4.[91] It is suitable for force measurement because there are no cooling water supply tubes attached because such tubes might produce undesirable interaction. Each segment of the water-cooled model has independent cooling channels, and the heat-input to each segment can be measured by the enthalpy increment between the inlet and the outlet of the cooling water. Water channels were machined on the wall and then covered by electroforming or electron-beam welding. There are nearly 200 flexible tubes to supply/drain water. It is designed to be capable of a test with a duration of 30 s whereas the heat-sink model can sustain only 10 to 15 s of combustion.

As one of the key technologies necessary for the development of a flight engine, a LH_2-cooled engine has been designed, fabricated, and tested.[82] It was partially fabricated and tested under a M6 condition by replacing some parts of the heat-sink model with its parts. Like the water-cooled model, cooling channels were fabricated by machining and electroforming or electron-beam welding. The procedures of its design will be described later.

In all of the models, the leading edges of the sidewalls, the cowls, and the strut have a radius of 1.0 mm, which was selected as a compromise among cooling requirements, fabrication difficulties, and aerodynamic performance. The mass of the engine with mounts and feed-line interfaces is approxi-

Table 1 Engine Configurations

Dimensions	
Inlet	
Frontal area (height × width), mm²	250 × 200
Throat width (w/o strut or ramp[a]), mm	70
Contraction ratio (w/o strut or ramp[a])	2.9
Sidewall wedge half angle, deg	6
Sidewall leading-edge sweep angle, deg	45
Leading-edge radius, mm	1
Cowl leading-edge location	Inlet end
Isolator	
Length, mm	100 (short), 200 (long)
Height × width (w/o strut or ramp[a]), mm²	250 × 70
Combustor	
Length, mm	800
Constant-area length (w/o strut or ramp[a]), mm	160, 60
Expansion ratio (w/o strut or ramp[a])	1.7
Step height, mm	4 (side), 2 (top), 4 (strut)
Fuel injection orifices	
Main perpendicular injection (MV1, SV1)	
Diameter, mm	1.5 (side), 1.5 (strut)
Location, mm from step	30 (side), 30 (strut)
Number	12 × 2 (side), 12 × 2 (strut)
Main perpendicular injection (MV2,[b] SV2)	
Diameter, mm	1.5 (side), 1.5 (strut)
Location, mm from step	16 (side), 16 (strut)
Number	12 × 2 (side), 12 × 2 (strut)
Main parallel injection (MP,[b] SB)	
Diameter, mm	3.0[c] (side), 3.0[c] (strut)
Location, mm from step	0 (side), 150 (strut)
Number	12 × 2 (side), 12 (strut)
Pilot fuel (SP, TP)	
Diameter, mm	0.5 (side), 0.5 (top)
Location, mm from step	− 50 (side), − 70/− 50[d] (top)
Number	94 (side), 7/6[d] (top)
Nozzle	
Length, mm	330
Expansion ratio	1.7

[a] See Table 2 for models with struts or ramps.
[b] Short isolator model has MV1 and MP whereas long isolator model has MV1 and MV2.
[c] Parallel injection orifices are divergent with their throat diameter of 1.5 mm.
[d] 7 on the center, 3 on both sides.

mately 450 kg. The test conditions simulating flight Mach number of 4, 6 with storage heating, 6 with vitiation heating, and 8 are termed M4, M6S, M6V, and M8, respectively. In the M4, M6S, and M6V tests, the heat-sink model was used, whereas in the M8 tests the water-cooled model was used because the stagnation temperature of the air for these tests is around 2600°K.

A. Inlet

The inlet has a leading edge of the sidewalls with a contraction of 6 deg. The leading edge is swept back at an angle of 45 deg to deflect the airstream for suitable spillage required for starting at low Mach numbers. This action led to all interface sections, rows of fuel injection orifices, and backward-facing steps on the sidewalls being swept back at the same angle in favor of uniform flow properties along this angle. Selection of the sweep angle was based on the results of the inlet tests as described elsewhere,[11,12,14,16] where the inlet models with sweep angles from 60 deg forward to 60 deg backward were tested in a Mach 4 wind tunnel. A 45-deg swept-back model gave good starting characteristics as well as high performances in both flow capture and pressure recovery.

B. Struts and Ramps

In the M4 engine tests struts for final compression of the airstream were not employed, whereas in the M6S, M6V, and M8 tests various types of struts or ramps were attached to the top wall of the model to increase the pressure in the combustor section for better combustion performance. Figure 2 and Table 2 show those struts and ramps. The leading and the trailing edge of the strut are also swept back with the same angle as that of the inlet leading edge. Each strut differs in height (247 or 50 mm), length (393 or 493 mm), thickness (30 or 46 mm), and location (with their leading edges 43 or 143 mm downstream of the swept line passing the cowl leading edge). The 247- and 50-mm high struts are termed full-height and 1/5-height struts, respectively. The full-height strut was specifically selected to have a 3-mm margin from the inner surface of the cowl to avoid collision because of thermal expansion during the run. They are made of stainless steel, copper, or zirconia-copper alloy depending on the specific models. Geometrical contraction ratios of the inlet depend on whether the struts are used or not and on their types and are also shown in Table 2. In the inlet tests under the Mach 4 freestream condition,[14] fairly good performances had been obtained even with a geometrical contraction of 5, which corresponds to the full-height strut model. Among the struts shown, fuel injection orifices are installed only in the water-cooled thin strut (model E), which was used in the M8 tests.

Besides the struts just described, other options, namely, the use of ramps to increase the pressure in the combustor section, were also employed in some runs at M6 and M8 conditions. They are also shown in Fig. 2 and Table 2. In the M6 tests two types of ramps attached to the top wall in the isolator were used, as shown in Fig. 1. The overall length and height of the ramps are 150 and 30 mm, respectively. The wedge angle of the front and rear faces of

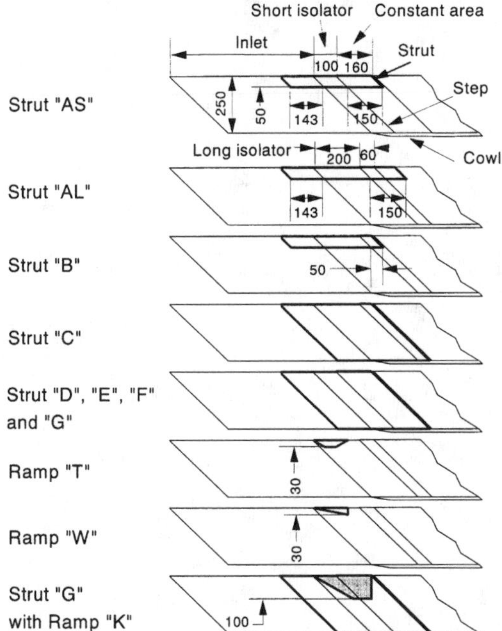

Fig. 2 Struts and ramps examined.

ramp T (30 deg) was determined so that the shock wave emanating from the ramp's front face would impinge on the leading edge of the cowl, assuming an inlet throat Mach number of 4. Such a ramp is expected to be a shock generator and a fence that blocks a boundary-layer separation propagating from the combustor into the inlet. As to the former effect, however, it would be cancelled to some extent by an expansion wave further downstream. In the M8 tests two ramps attached to the channels divided by the thin strut shown in Fig. 2 and Table 2 were used. Their length and height are 250 and 100 mm, respectively. The wedge angle of the front face is equal to that of ramp T (30 deg) and it is also expected to be a shock generator and a fence against the inlet/combustor interaction. Geometrical contraction ratios, when each of these ramps is attached to the isolator section, are also shown in the table.

C. Isolator, Fuel Injector, and Combustor

The fuel injector and the combustor of the engine were designed using the results of the combustor component tests. The fuel injector section has a 4-mm high, swept backward-facing steps on the sidewalls and a 2-mm high, nonswept one on the top wall for flame holding, upstream of which a constant-area section acting as an isolator is attached to mitigate the propagation of combustion-induced disturbances into the inlet. In the M4, M6S, and M6V tests with the heat-sink model, isolators with lengths of 100 and

Table 2 Strut and ramps examined

	Struts								Ramps		
	AS	AL	B	C	D	E	F	G[f]	T	W	K[f]
Length, mm	393	493	393	←	←	←	←	←	150	←	250
Height, mm	50	←	←	247	←	←	←	←	30	←	100
Thickness, mm	30	←	←	←	←	←	46	30	70	←	20 × 2
Leading edge											
Half angle, deg	6.0	←	←	←	←	←	←	←	30	11	30
Location from step, mm	143	243	143	←	←	←	←	←	100	←	←
Cooling[a]	HS	←	←	←	←	WC	←	←	HS	←	WC
Model installed[b]	HS/SI	HS/LI	HS/LI	←	HS/SI	WC/SI	←	←			
When attached to engine											
Throat width, mm	20 × 2, 70	←	←	20 × 2	←	←	12 × 2	20 × 2[f]	70	←	20 × 2[f]
Throat height, mm	250,200	←	←	250	←	←	←	←[f]	220	←	150[f]
Contraction ratio	3.1	←	←	5.3	←	←	8.3	5.3[f]	3.3	←	8.3[f]
Fuel injector[c]	N/A	←	←	←	←	SV1/SV2/SB	N/A	←	←	←	←
Material[d]	SS	←	←	←	←	ZC	CU	←	SS	←	CU
Conditions tested[e]	M6S/V	M6S	M6S/V	←	M6S	M8	←	←[f]	M6V	←	M8[f]

[a] HS: heat-sink, WC: water-cooled.
[b] SI: short-isolator model, LI: long-isolator model, HS: heat-sink model, WC: water-cooled model.
[c] SV1/SV2: perpendicular injection, SB: parallel injection from base.
[d] SS: stainless steel, ZC: zirconia copper, CU: copper.
[e] See Table 3.
[f] In the tests conducted so far, strut G was tested together with ramp K.

200 mm are employed to examine their effects. Because the overall length of the fuel injection section is kept constant in the model, note that the long isolator results in a short constant-area section downstream of the step. The water-cooled and LH_2-cooled models have only the short (100 mm) isolator.

The models with the short isolator were designed and fabricated first, their fuel injector design being based on a model with the 30-mm thick full-height strut. Thus, the throat gap width, G between the strut and the sidewall was the most important scaling factor in designing the combustor. Perpendicular and parallel injectors were employed to control the heat release schedule along the combustor by combining the two injection patterns. Perpendicular injection orifices (termed MV1) with a diameter of 1.5 mm were drilled along a single swept row located 30 mm downstream of the step on both of the sidewalls.

In the heat-sink model with the short isolator, a row of supersonic parallel injection orifices (termed MP) with diameters of 1.5 mm at the throat and 3.0 mm at the exit were also drilled at the base of the step on the sidewall. Because parallel injection contributes thrust through its momentum, supersonic injection is preferred. The models other than the heat-sink model with the short isolator do not have parallel injection orifices because at the time after conducting several runs with the heat-sink model with the short isolator and before designing these models there were no evident effects of parallel injection. Furthermore, arrangement of the perpendicular injection orifices closer to the step was favorable in view of flame-holding characteristics.[67] Therefore, an additional swept-back row of the perpendicular injection orifices (termed MV2) with the same diameter as orifices MV1 were installed in the heat-sink model with the long isolator, instead of parallel injection ones. Their location was chosen to be 16 mm downstream of the step. The same arrangement was also employed in the water-cooled and LH_2 cooled (short isolator) models.

Lateral spacing of the orifices was set equal to G, which has been recommended in the combustor tests at NASA-Langley[99] and was experimentally reconfirmed at NAL-KRC. The perpendicular and parallel injection orifices were located in a staggered manner with each other in the lateral direction to avoid fuel stratification. However, perpendicular injection orifices MV1 and MV2 in the long isolator and water-cooled models were aligned with each other because we expected an interaction between two perpendicular fuel jets to form a large recirculation region behind the step for flame holding. In the combustor tests the step height was $0.1G$, which resulted in successful mixing and combustion efficiencies. This dimension is kept on the top wall, but not on the sidewalls because of the need to install the parallel injection orifices at the base of the step for the heat-sink model with the short isolator. Thus, the height became twice that on the top wall as just described.

As to the fuel-injection strut used in the M8 tests, the same design concept and arrangement of the injection orifices as those for the sidewalls were employed. A 2-mm high backward-facing step and, as on the sidewalls, two rows of perpendicular injection orifices 16 mm (termed as SV2) and 30 mm (termed as SV1) downstream of the step on each side were installed. The supersonic parallel injectors (SB) are on the base of the strut with diameters of 1.5 mm at the throat and 3.0 mm at the exit. They were again located in a staggered manner with perpendicular orifices and parallel ones on the sidewalls to avoid fuel stratification, which led to the arrangement between perpendicular injection orifices on the strut and on the sidewalls also being staggered. In the M8 tests parallel injection from the strut base has not been attempted.

All of the injection orifices just mentioned are for the main fuel. In addition to them, rows of 47 orifices for pilot fuel injection (SP) with a diameter of 0.5 mm were drilled on each of the sidewalls. They were located in the isolator 50 mm upstream of the step. The pilot fuel injection orifices were also installed on the top wall (TP); their locations being 70 and 50 mm upstream of the step for the central seven orifices and three orifices on both sides, respectively. This arrangement for the central pilot fuel injection on the top wall is because of the necessity for locating the igniter on the top wall. When using the strut or the ramp, fuel is not injected from these central pilot orifices on the top wall.

During the initial series in M4 and M6S conditions, the pilot injection from the sidewalls (SP) was tried, but the results were less effective and thrust performance deteriorated in most cases. This may be from the effects of enhancement of the inlet-combustor interaction or blowing out of the flames held at the step. Therefore, these rows of injection orifices were not used in the later runs for these test conditions. In the M8 tests the pilot fuel injection from the sidewalls was again tried in the process of managing to find how to attain the intensive combustion, which will be defined later. Because the pilot fuel from the top wall (TP) was found to be very effective, it was used in almost all of the runs. All of the fuel injection orifices arrangements are shown in Fig. 3.

A twin plasma igniter was attached 20 mm upstream of the step on the top wall to obtain combustion at the two flow passages divided by a strut. Feed-

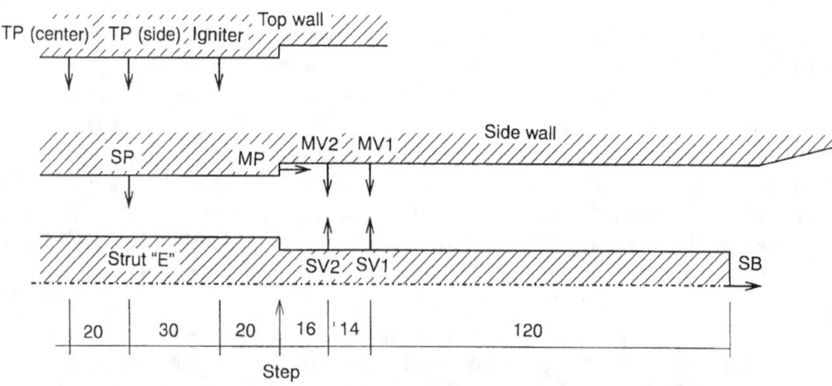

Fig. 3 Arrangement of fuel injection orifices (dimensions in mm).

stock and coolant of the igniter was gaseous oxygen, and input power was 1.0–1.5 kW. Its diameter and length were 21 and 73 mm, respectively. The igniter itself and its arrangement with the pilot fuel injector/backward-facing step were experimentally confirmed to be successful in the combustor component tests.[33] The intention is that the pilot flame be formed with the plasma igniter first and then propagate along the step base to the main fuel injected from the sidewalls. Even when there was fuel injection, the plasma igniter usually continued to be turned on. When using ramp K shown in Table 2 in the M8 tests, the igniter was not operated because it was masked by the ramp.

D. Combustor Downstream Section and Nozzle

Downstream of the fuel injectors, there is a constant-area section to stabilize the combustion. Its length is 160 or 60 mm depending on the

Table 3 Specifications and characteristics of RJTF

Characteristics	Specifications			
Simulated flight conditions				
Flight Mach number	8	6		4
Altitude, km	35	25		20
Static pressure, kPa	0.58	2.55		5.53
Flight dynamic pressure, kPa	26	64		62
Static temperature, °K	237	222		217
Facility nozzle condition	M8	M6S	M6V	M4
Air heater[a]	SAH+VAH	SAH	VAH	SAH
Total temperature, 18K	2600	1690	1560	870
Total pressure, MPa	10.0	5.2	4.9	0.9
Air supply rate, kg/s	6.2	30.9	29.8	45.9
O_2 supply rate, kg/s	2.16	0	5.44	0
H_2 supply rate, kg/s	0.18	0	0.45 0	
Specific heat ratio	1.38	1.40	1.39 1.40	
Nozzle throat dimensions, mm	32.5 × 32.5	12.0 × 510		74.5 × 510
Nozzle exit dimensions, mm		510 × 510		
Unit Reynolds number, m^{-1}	1.4E + 6	6.5E + 6		8.4E + 6
Boundary-layer thickness				
Velocity thickness, mm	70	52		35
Displacement thickness, mm	25	17		11
Test duration, s	30	60		60
Engine conditions				
Inlet Mach number	6.7	5.3	5.2	3.4
Static temperature, °K	324	280		274
Static pressure, kPa	1.6	5.8		12.3
Dynamic pressure, kPa	50	111		100
Inlet air speed, m/s	2570	1780	1740	1130
Inlet air-capture ratio	0.91	0.87		0.72
Stoichiometric H_2 flow rate, g/s	40	140		180

[a] SAH: storage air heater, VAH: vitiation air heater.

isolator length (100 or 200 mm) because, as already stated, the overall length is fixed. A diverging combustor section with a half angle of 1.9 deg follows it. The overall length of the combustor from the step is 800 mm, which corresponds to 40 or 10*G* for the model with or without the full-height strut, respectively. In the direct connect tests of combustor components with the entrance Mach number of 2.5,[30–34] mixing and combustion efficiencies were near to or exceeded 80% for these combustor lengths.

Only a short internal nozzle is incorporated in this engine because of the limited size of the test cell. Because the chosen entrance and exit dimensions are equal, its expansion ratio from the minimum area at the isolator is the same as the inlet contraction and depends on the type of strut employed.

E. LH_2-Cooled Model

The LH_2-cooled model was designed to have the same aerodynamic geometry as the existing heat-sink and water-cooled engines so as to have similar aerodynamic and combustion performance. Also, the LH_2-cooled model consists of five blocks, namely, a leading edge, an inlet, a fuel injector, a combustor, and a nozzle, which are compatible with the heat-sink engine. Each engine block consists of panels with cooling slots. Zirconia-copper alloy was selected as a material because of its high conductivity and strength at high temperature. Electroforming was selected to fabricate the close-out of cooling slots.

In the design of active cooling, the flow condition within the engine duct was estimated based on one-dimensional analysis of the scramjet engine. In the calculation of the gas-side heat-transfer coefficient, a two-dimensional cylinder stagnation point model for the leading edge, a plane flow model for the panels of the inlet and the cowl, and Bartz' equation[100] for the combustor and the nozzle were used. Heat flux based on the one-dimensional flow analysis was verified by the experimental data of the heat-sink engine. The heat transfer coefficient of LH_2 was calculated from Taylor's correlation.[101] Coolant paths were designed by a quasi-one-dimensional method. The orifices with a pressure drop of 1 MPa were installed at the entrance of each coolant path by electroforming. This was to ensure even flow rate to each cooling path, which might be affected by nonuniform heat flux during the run. A maximum temperature of the wall allowed in the quasi-one-dimensional design was limited to 500°K. The stress was calculated by the two-dimensional finite element method from a thermal elastic analysis based on the one-dimensional cooling design.

III. Test Facility

A. Outline

The NAL-completed construction of the RJTF at the KRC in 1994. It is a free-jet-type hypersonic propulsion wind tunnel, with which it is possible to test the scramjet engine in the range of simulated flight Mach numbers of 4, 6, and 8 at altitudes of 20, 25, and 35 km. Environmental pressure corresponding to each altitude can be simulated with a steam-ejector exhaust system. Its major specifications are listed in Table 3. The air heater SAH denotes a storage heater where air flow is heated by passing it through the preheated materials. The VAH is a vitiation heater where heating is attained by lean burning of hydrogen with oxygen and air. The nozzle conditions correspond to the engine inlet conditions behind the bow shock with conical half angles of 12 (Mach 4 flight condition) and 7.5 deg (Mach 6 and Mach 8 flight conditions). At a simulated flight condition of Mach 6, we can operate either the SAH or the VAH independently. Although the VAH is very convenient for supplying hot air, how the combustion products of the vitiation heater affect the engine performance is not well understood. Thus, a comparison of test results using different heaters provides a basis for the analysis of the contamination effects in the VAH mode.[102–104] To generate the temperature conditions of Mach 8 flight, we have to employ cascade heating, i.e., the VAH following the SAH, because the SAH capability is far below the required enthalpy. Therefore, examination of the effects of vitiation at M6 conditions is extremely important in interpreting the M8 test results.

The facility operating envelope, presented in terms of Mach number and altitude, is compared with those of other propulsion test facilities in Fig. 4. Except in the Arc-Heated Scramjet Test Facility at NASA-Langley (AHSTF), vitiation heating is employed in these facilities. This is in contrast to one of the unique features of the RJTF where storage heating of pure air is employed for M4 and M6S tests. However, the test altitude of the RJTF is relatively high compared with those of other facilities, except the AHSTF, because the maximum pressure of the facility nozzle reservoir was chosen to be 10 MPa to ease the structural requirements. As a result of this design restriction, the simulated flight dynamic pressure is 26 kPa for the M8 tests. For the M4, M6S, and M6V tests, however, the supply capability of air is the limiting parameter instead of the maximum pressure. Consequently, the

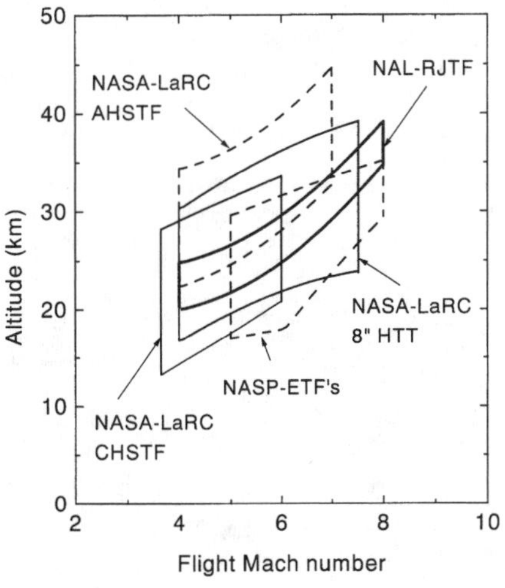

Fig. 4 Comparison of RJTF with other facilities.

flight dynamic pressures are 62 kPa for the M4 test and 64 kPa for the M6S and M6V tests.

We estimated the minimum dimensions of scramjet engines to be about 200 mm square and 2000 mm in length to meet the requirements for the RJTF by considering the scale effects dictating engine combustion phenomena and the difficulties of manufacturing the cooling structure of the models. Considering the blockage ratio of the nozzle flow by the engines, we chose a nozzle exit dimension of 510 mm square. As to the free-jet size, the present facility is approximately twice as large as those of the AHSTF and the Combustion-Heated Scramjet Test Facility at NASA-Langley (CHSTF). However, it is half and 1/5 those of the Engine Test Facilities for National Aerospace Plane (NASP-ETF) and 8-ft High Temperature Tunnel at NASA-Langley (8-ft HTT) respectively. Details of the RJTF are described in Refs. 94 and 95.

B. Components

Figure 5 gives an elevation view of the RJTF. High-pressure air of 23 MPa delivered by a four-stage reciprocating compressor (250 kW) is accumulated in two air reservoirs (15 m^3 each). Their capacities and that of the air evacuation system determine run duration and test frequency, i.e., 60 s for M4 and M6 conditions and 30 s for a M8 condition with a single run a day. The SAH is a pressure vessel with an outer diameter of 1776 mm and a height of 10 m. To reduce dust contamination and bed floating in the heater, cored bricks made of alumina were adopted for thermal storage. Preheating of the SAH bed is done with a propane-air burner installed on the top of the heater. To isolate the combustion gases during the heating of the SAH from critical components downstream, a hot valve is utilized at the outlet of the heater. Constant nozzle stagnation air temperature and pressure conditions during a run are obtained by mixing an appropriate amount of bypassed air with the storage-heated air. During the opening and closing periods of the valve, cold air is supplied through the bypass air line to the downstream area to keep the pressure difference nearly zero so as to protect the valve throat. In the VAH gaseous hydrogen and oxygen are used. Because of the large difference of

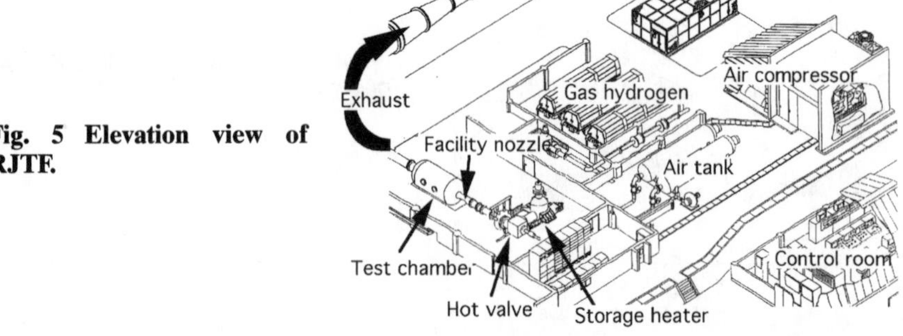

Fig. 5 Elevation view of RJTF.

operating conditions, namely, the difference in the temperature of the air fed to the heater, two separate heaters for M6 and M8 conditions were built.

The facility nozzles for the M4 and M6 conditions are of a two-dimensional design with a slot throat height of 74.4 and 12 mm, respectively. They share a common downstream heat-sink expansion section. The nozzle for the M8 condition is square in cross section at all axial locations with a dimension at the throat of 32.5 mm square. The test cell is a cylindrical steel chamber 3 m in diameter and 5.1 m long as shown in Fig. 6. It houses a nozzle downstream section, the engine to be tested, a force balance (FMS), a diffuser extension, and a diffuser pickup section. To simulate the boundary-layer ingestion, the top wall, i.e., the inner surface of the engine, is aligned with the bottom surface of the nozzle. The downstream sections of the nozzles have an extension cut with an angle of 50 deg for the M4 and M6 nozzles and 15 deg for the M8 nozzle. The cutback prevents the wave initiated at the nozzle lip from impinging on the engine inlet and provides a larger test rhombus. The engine is installed in the test cell through an overhead hatch. A 1/5-scale model test was conducted to determine the proper test-section configuration.[89] Good visibility of the entire engine is facilitated by longitudinally eliminating the portions of cylindrical diffuser pickup as shown in Fig. 6. The engine is set onto the FMS with its top-wall side down as shown in Fig. 7 with fore and aft mounts, the latter having a sliding linear bearing to relax thermal expansion of the engine in the axial direction during the run.

The test cell is exhausted by an air exhaust system consisting of a two-stage steam ejector. It uses 40 kg/s and 120 kg/s of steam for the first and the second stages. It evacuates the test cell pressure to 10 kPa (M4 tests) and 1 kPa (M8 tests) for the facility air of 46 kg/s and 8.3 kg/s, respectively, with the assistance of the diffuser section. Although some afterburning of the engine fuel in the diffuser often caused increased cell pressure, no interference between the facility nozzle flow and engine testing has been found.

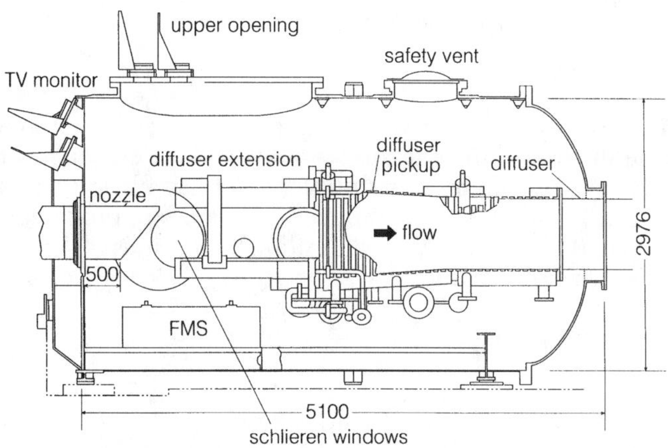

Fig. 6 The test cell evacuated by a two-stage steam ejector system.

Fig. 7 The engine model installed in the test cell.
a) Heat-sink model
b) Water-cooled model

The engine fuel is gaseous hydrogen at room temperature or liquid hydrogen. Gaseous hydrogen to be fed to the VAH and the engine is supplied by three trailers. An oxygen reservoir for the VAH has a storage capacity of 6 m^3 at 20 MPa. Two supply sources of cooling water are prepared. They are pressurized at 10 MPa and 4 MPa with storage capacities of 4 and 20 m^3 for engines and for the facility nozzles, respectively.

C. Calibration of the RJTF

Uniformity of Nozzle Flow

The quality of the core flow of the nozzles was measured by using total temperature and pitot-pressure/gas-sampling rake probes.[93] Because no thermocouples could survive under the M8 test condition, the total temperature was

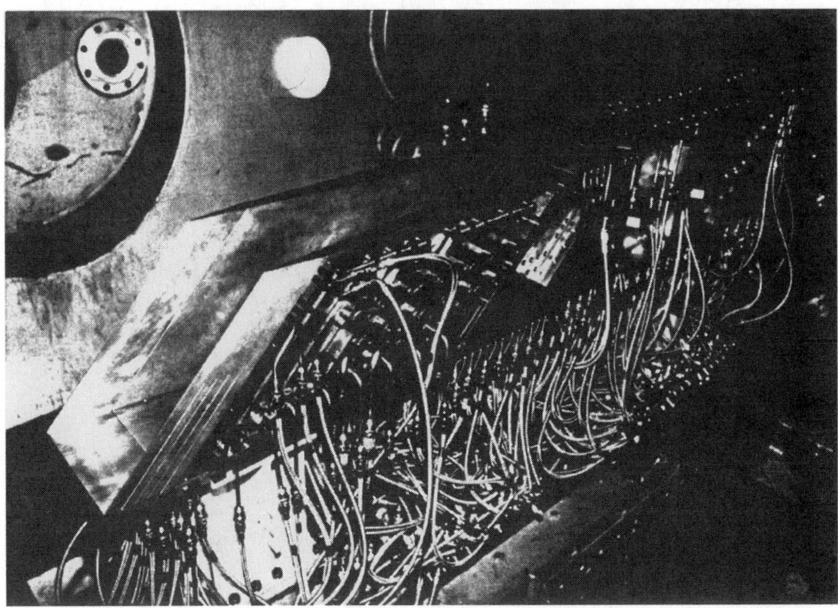

Fig. 7 (continued)

estimated from the chemical equilibrium calculation with gas composition measurements for this condition. The spatial distribution in the M4 and the M6S flow conditions was surveyed at 21 locations in the core regions across the nozzle exits. The M6V and M8 flows were measured at 42 locations in the airflow heated by the VAH because the injection of hydrogen and oxygen might result in poor mixing and deteriorate the uniformity of nozzle flow.

The measurements indicated fairly uniform distributions of total temperature and pitot pressure. The average Mach number across the M4 facility nozzle was evaluated to be 3.46 ± 0.05. Because the test conditions to be simulated are the flowfield behind the bow shock, the Mach numbers at the nozzle exits are lower than those of flight conditions. The total temperature was 890 ± 30 K. In the M6 nozzle there were two peaks of high temperature, 1600°K in the central portion and a low temperature of 1500°K near the corners. However, this distribution does not disturb engine testing because only the central portion of nozzle flow enters the engine. The average Mach numbers were 5.30 ± 0.10 for the M6S case and 5.15 ± 0.15 for the M6V cases because the specific heat ratio between these two cases are different.

Figure 8a shows the total temperature distribution at the exit of the M8 nozzle. The open circles represent the locations measured. The square nozzle exit boundary is shown by the solid bold lines. It has been reported that the corner flow emerged at the central portions of the nozzle exit and deteriorated the uniformity in temperature and velocity fields. However, Fig. 8a does not indicate any distortion in the core flow. A uniform temperature of

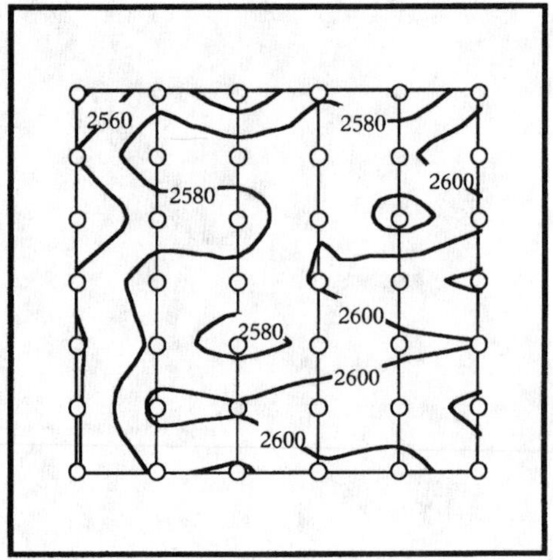

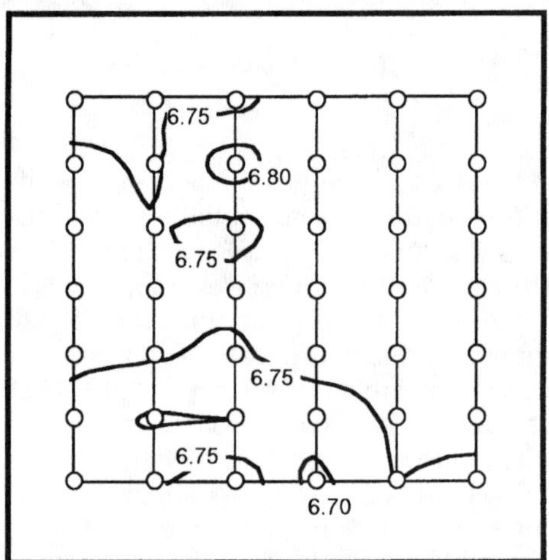

Fig. 8 Uniformity at the M8 nozzle exit.
a) Total-temperature distribution
b) Mach-number distribution

2600 ± 50 K is confirmed. The uniform Mach number around 6.73 ± 0.10 is found in Fig. 8b. The oxygen concentration was measured for the M6V and M8 conditions, and uniform distributions of 20.9 ± 1.0% were assured.

Boundary-Layer Measurement

Airframe-integrated scramjet engines ingest boundary layer developed on the forebody of vehicles, and the engine performance must be affected by the boundary layer. Therefore engines tested in the RJTF are installed aligned with the wet surface of the facility nozzle to simulate the boundary-layer ingestion. Because the relative boundary-layer thickness in the nozzles to the engine heights is an important parameter, their measurements were carried out by water-cooled rakes with 10-mm-pitched pitot probes. The boundary-layer thickness is summarized in Table 3, in which the boundary layers share from 14% (M4 nozzle) to 35% (M8 nozzle) of the half height of the nozzle exits. The displacement thickness varies from 4% (M4 nozzle) to 13% (M8 nozzle) of the half height of the nozzle exits. The relative boundary-layer thickness was found to be 1/7 (M4 nozzle) and 1/3 (M8 nozzle) of the engine height (250 mm).

Dust Concentration in Air from Storage Air Heater

The principal motivation for selecting the cored-brick-type storage air heater is to achieve low dust contamination in the heated air as compared with that of the pebble-type air heater. The dust in test media from the storage heater was sampled using a pitot-type sampling tube and introduced to a two-stage impinger. The dust particles in the sampled gas were counted using a Coulter counter, and the number density was evaluated from the airflow rate to the sampler. The results showed that the mass-averaged particle diameter varied from 2 to 7 μm. A typical number density of particles was found to be $[1.0 \pm 0.4] \times 10^7/\text{m}^3$ in the standard condition. This number density with the average mass of particles gave the mass fraction of dust contained in the airflow, which was from 10^{-7} (diameter 2 μm) to 10^{-5} (7 μm).

Effects from the dust contained in the test media on engine performance are discussed in Section VI.D. The results showed that the heat sink-effect of dust appeared if the mass fraction of dust exceeded 10^{-1}. The analysis also indicated that chemical effect on the dust surface, if any, did not appear in the case for the mass fraction of dust less than 10^{-3}. Because the dust fraction was less than 10^{-5} in the air supplied from the SAH, the airflow in the RJTF was found to be sufficiently clean where the thermal and the chemical effects of dust are negligible.

IV. Measurements

A. General Features

Axial force, lift force, and pitching moment are measured by the FMS, which is capable of resolving the thrust/drag component in the range of 7000/1700 N with an accuracy of ± 0.5% of readings down to 20% of full scale. Lift component is estimated to be 20% of thrust, and the required accuracy is ± 1.0% of readings. Prior to each run, the FMS is calibrated by exerting calibration loads with hydraulic actuators via standard load cells. The engine is equipped with approximately 200 wall-pressure taps connected to strain-gauge-type transducers. Most of their output is electrically scanned and recorded. It is utilized to obtain information on the flowfield within the engine and its operating modes. By integrating the wall-pressure data over the internal surface and comparing the results with the FMS outputs, information on friction also can be obtained.

To estimate the local heat flux to the engine, the engine wall temperature is monitored at 40 stations using thermocouples. They are embedded 1 mm from the inner surface of the heat-sink type engines. Because the Fourier number is sufficiently large, the local heat flux is evaluated from the heat capacity of the wall and the time derivative of the temperature. The wall is approximated to be a slab of a given thickness (typically 20 mm) with an insulated outer surface. Details of the measurement and a survey of the combustion region inside the engines by heat flux are reported by Hiraiwa et al.[75]

Flow rates of gas and liquid are also measured with orifices or turbine meters. These data are recorded with the data acquisition systems (DAS) of the RJTF. The sampling intervals are 5 ms during a crucial period where the combustion might occur, but otherwise 80 ms. Selected high-frequency test parameters are also recorded on FM tapes. Two optical quality windows, 600 mm in diameter, provide schlieren flow visualization and direct photography.

B. Engine Exit Survey

The combustion performance of the scramjet engine was measured by a rake probing in all of the test conditions.[74] Probings were made 5 mm downstream of the engine nozzle exit with three water-cooled, twelve-probe, gas-sampling rakes. Details of the design and calibration studies on quenching of chemical reaction are found in Sec. IV.B. Sampled gas was collected in 40-cc bottles for 1.4 sec, and a gas chromatography was used to measure the gas composition. The local values of the equivalence ratio and the combustion efficiency are calculated from the gas composition in terms of mole fraction ratios by the following equations.

$$\varphi = 1 + \left[\frac{1}{2}\left(\frac{X_{H2}}{X_{N2}}\right) - \left(\frac{X_{O2}}{X_{N2}}\right)\right] \cdot \left(\frac{X_{N2}}{X_{O2}}\right)_0 \tag{1}$$

$$\eta_c = \left[1 - \left(\frac{X_{O2}}{X_{N2}}\right) \cdot \left(\frac{X_{N2}}{X_{O2}}\right)_0\right] / \varphi \qquad (\varphi \le 1) \tag{2}$$

or

$$= 1 - \left(\frac{X_{O2}}{X_{N2}}\right) \cdot \left(\frac{X_{N2}}{X_{O2}}\right)_0 \qquad (\varphi > 1) \tag{3}$$

where X_i denotes the mole fraction of $i = H_2$, O_2, and N_2, and the subscript *0* corresponds to the air composition. Formation of NO_x might cause errors in the N_2 and O_2 concentrations. However, kinetic studies indicate that the maximum NO is as low as 1.3% in the M8 condition employing the storage and vitiation heaters. It does not cause significant errors in Eqs. (1) to (3) because the NO yield affects the denominator and the numerator simultaneously. The rake probing was repeated twice at the same engine condition to obtain the local values of combustion efficiency, fuel equivalence ratio, and Mach number at 72 locations on the engine exit plane. In addition to the measurement during engine operation, pitot pressures under the condition without fuel injection were measured to confirm the accuracy of local mass-flux evaluation. Based on these gas composition data and pitot-pressure distributions, all local flow properties, such as static temperature, velocity, density, etc., can be estimated. In addition to the local properties, the bulk combustion efficiency, weighted by the local flux of H_2, is also informative by comparison with the one-dimensional analysis.

V. 5 Test Results

From April 1994 to April 1998, more than 200 runs were conducted under the simulated conditions of flight Mach numbers 4, 6, and 8, including scramjet and ramjet[105,106] tests. The contents and major results of the scramjet tests are summarized in Table 4.

In the initial several runs at the M4 condition, startability of the test-cell arrangement and reproducibility of the airflow conditions within the engine were confirmed. Then testing procedures of the pilot fuel injection for smooth ignition and of the main fuel injection for steady combustion were checked. During the first and second series of M4 tests, the FMS was clamped to protect against an unexpected load. Over the past four years through experience gained in the test series and analysis of data, various improvements in testing techniques and data reliability have been successfully carried out.

The results of aerodynamic tests with 1/5-models tests[79] (see Section VI.C) were used to obtain a breakdown of drag factors such as internal, external, and engine support/fairing drags. In this paper the engine drag is termed an axial component of the FMS outputs, including those of the engine support and the fairing. The 1/5-models tests also provided capture ratios of the inlet, which were used to estimate stoichiometric fuel flow rates, which are shown in Table 3.

A typical history of the axial component of the FMS output together with the fuel flow rate is shown in Fig. 9. In this paper thrust performance is presented as a thrust increment ΔF, i.e., the difference between the axial component of the FMS outputs with and without fuel injection.

A. General Features of the Engine Operation

Aerodynamic Characteristics

The flowfields within the engine prior to fuel injection were examined by Tani et al.[76] Their major results are as follows. For the models without struts

Table 4 Contents and major results of the tests conducted

Period	Model	Series	Runs	Remarks
March–April 1994	Heat-sink	M4-1	6	Facility/engine shakedown First series of firing tests First successful firing at M4 condition Without strut
May–June 1994	Heat-sink	M4-2	7	Follow-up to M4-1 series
Nov. 1994	Heat-sink	M6S-1	11	First series at M6S condition FMS applied first (the same as in the following series) Wall-pressure taps increased (the same as in the following series) Wall-pressure measuring line shortened to improve response (the same as in the following series) Without strut Firing unsuccessful
Jan.–Feb. 1995	Heat-sink	M4-3	11	Follow-up to M4-2 series
April 1995	Heat-sink	M6S-2	9	Follow-up to M6S-1 with 1/5-height strut to attain combustion First successful firing at M6S condition
June 1995	Heat-sink	M6S-3	11	Follow-up to M6S-1 with recessed wall injector to attain combustion together with oxygen injection into the recessed cavity and fuel injection into the recessed cavity

				Firing partially successful
Sept.–Oct. 1995	Water-cooled	M8-1	5	First series at M8 condition First series with water-cooled model Without strut Firing unsuccessful
March 1996	Heat-sink	M6V-1	7	First series at M6V condition With and without 1/5-height strut Fuel control system modified to change flow rate during a run (the same as in the following series) Gas-sampling/pitot-pressure measurement at the engine exit applied first (the same as in the following series except those with a note)
March–April 1996	Heat-sink	M4-4	9	Follow-up to M4-3 with long isolator
June 1996	Heat-sink	M6S-4	12	Follow-up to M6S-2 with a long isolator, location and length of a strut varied, and 1/5- or full-height strut
Sept. 1996	Heat-sink	M6S-5	7	Follow-up to M6S-4 with a short isolator and 1/5- or full-height strut
Oct. 1996	Heat-sink	M6V-2	10	Follow-up to M6V-1 with a long isolator and 1/5- or full-height strut, or ramps
May 1997	LH2-cooled	M6S-6	3	First series with LH2-cooled inlet with a long isolator and full-height strut Gas-sampling/pitot-pressure measurement at the engine exit not applied
Oct. 1997	Water-cooled	M8-2	8	Follow-up to M8-1 with full-height, fuel-injection strut Firing successful for the first time
March 1998	Water-cooled	M8-3	16	Follow-up to M8-2 with a thick, full-height strut and ramp and normal full-height strut

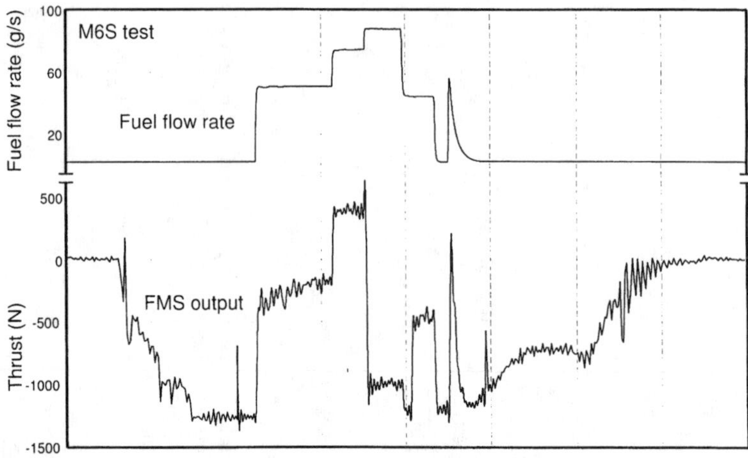

Fig. 9 Typical thrust history with the fuel flow rate sequence (M6S test).

or ramps, the impinging point of the cowl-originated shock wave on the top wall moved downstream as the incoming Mach number to the engine increased. This location gave the approximate Mach number at the cowl leading edge as 2.5 and 3.5 for M4 and M6 cases, respectively. For the M8 case identifying the point of impingement was not possible because it never appeared within the engine because of a shallow angle of the shock wave. The extension of the isolator caused the pressure decrease just upstream of the combustor due to the expansion wave that originated at the inlet exit corner of the opposite sidewall. The flow was then recompressed because of a rather strong reattachment shock wave in the constant-area section downstream of the step. These features are shown in Table 5 for all of the tested models. The length of the isolator must be optimized to achieve a suitable pressure condition for the combustion requirements.

On the top wall the struts kept the pressure level higher in the constant-area combustor but created an expansion wave from its truncated trailing edge. This resulted in a low-pressure region at the entrance of the diverging combustor. The ramp attached to the to wall of the isolator created a low-pressure region just upstream of the combustor, but kept the pressure level unchanged in the diverging combustor. The location of this sort of attachment should be determined with reference to the desired pressure distribution around the combustor. Without struts or ramps the model was able to supply sufficiently compressed air only in the M4 condition. For Mach 6 or higher Mach number flight conditions the additional structures such as full-height struts were indispensable. In the model without any attachments, wall-pressure distribution along the step line was skewed near the cowl because of the cowl-originated shock wave. Except near the cowl region, the pressure variance was within $\pm 30\%$ of the average at all test conditions. The attachment of struts or ramps, especially ramp T, had a

Table 5 Pressures at the critical locations of the model normalized by freestream values

	Test condition							
	M4		M6S		M6V		M8	
	Location (exit of)							
Isolator	inlet	isolator	inlet	isolator	inlet	isolator	inlet	isolator
Short isolator								
W/o strut	4.0	4.9	4.3	3.7	4.7	3.2	3.1	2.8
1/5-height strut	—	—	4.7	3.7	5.6	3.7	—	—
Full-height strut	—	—	9.7	4.4	—	—	—	—
Long isolator	—	—	—	—	—	—	—	—
W/o strut or ramp	3.6	3.9	—	—	—	—	—	—
1/5-height strut	—	—	3.8	7.2	4.0	6.3	—	—
Full-height strut	—	—	3.8	12.8	4.2	13.2	—	—
Ramp T	—	—	—	—	6.7	5.8	—	—
Ramp W	—	—	—	—	4.2	6.3	—	—

negative effect on the pressure uniformity along the swept line. The geometry of the ramp should be more moderate, which would also increase the drag coefficient by the factor of up to 1.3. Optimization of their configuration and location considering the combustion performance will be pursued further.

Combustion Mode

The following common trend appeared in the engine operation mode when fuel was injected, and its flow rate was increased.[68] Namely, when the fuel flow rate was low, a mode (a weak combustion mode) appeared that resulted in low thrust. With a further increase in the fuel flow rate, a mode producing much higher thrust (intensive combustion mode) appeared. Further increase in the flow rate, however, caused the engine to fall into unstart. Threshold values of the fuel flow rate for the transition from the weak to intensive and finally to unstart depended on various test conditions. In the M8 tests, however, the intensive combustion mode was not observed.

To define the intensive and weak combustion modes more clearly with some of the observed experimental evidences would be convenient.[68,74,78] Detailed insight into the intensive mode was done by Kanda.[83]

Wall-pressure and heat-flux distributions. Typical wall-pressure distributions on the top wall in both combustion modes are compared in Fig. 10 for the M6S tests with a short isolator and a 1/5-height (50 mm high) strut. In the intensive mode low pressure behind the step is masked by a large pressure rise initiated upstream, whereas in the weak mode it remains

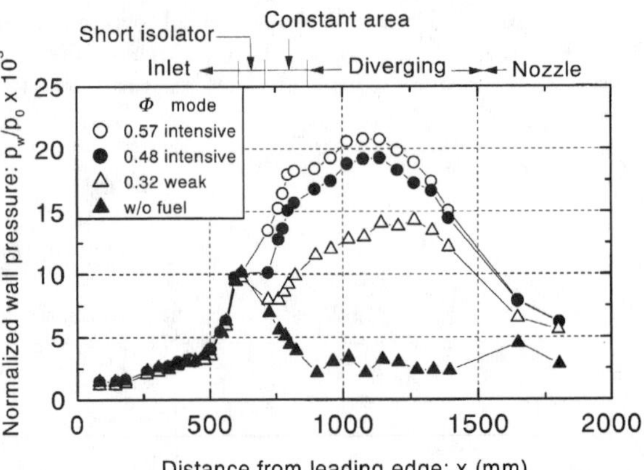

Fig. 10 Wall-pressure comparison on the top wall between intensive and weak combustion mode (M6S test with short isolator and 1/5-height strut).

unchanged. The same tendencies were observed in the distribution on the sidewall. On the sidewall near the cowl, an effect of a shock wave emanating from the leading edge of the cowl caused a much higher pressure rise, and this was because of combustion there that was more intensive than in the other regions. In the intensive mode, therefore, it seems that flames are stabilized at the step, and a large separation bubble appears, whereas in the weak mode combustion begins far downstream of the step.

The development from the weak to the intensive combustion modes can be observed from the variation of heat-flux distribution in the M6S test. The heat-flux distributions along the center line of the sidewall are shown in Fig. 11. In Fig. 11, the triangles represent the heat-flux distribution without fuel injection. With a fuel flow rate of 30 g/s ($\Phi = 0.21$), the first exothermic reaction is detected in the downstream part of the combustor ($x = 1400$ mm) where the shock wave from the cowl leading edge impinges. Note that there is no increase of heating rate behind the step in the combustor from $x = 800$ to 900 mm. As the fuel flow rate increases from 30 to 48 g/s ($\Phi = 0.34$), the heating rate increases, and the maximum point moves upstream.

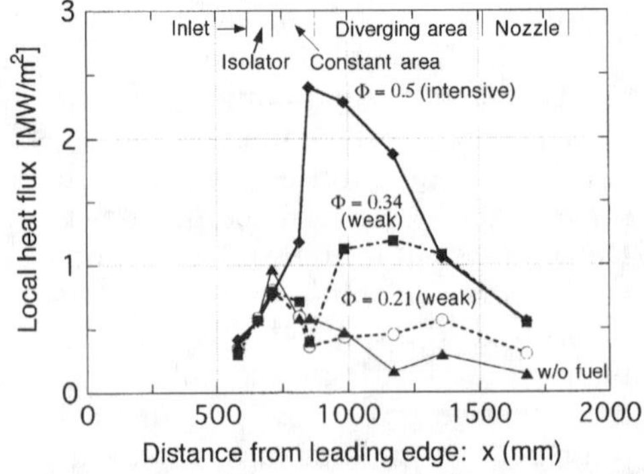

Fig. 11 Heat-flux comparison between intensive and weak combustion modes.

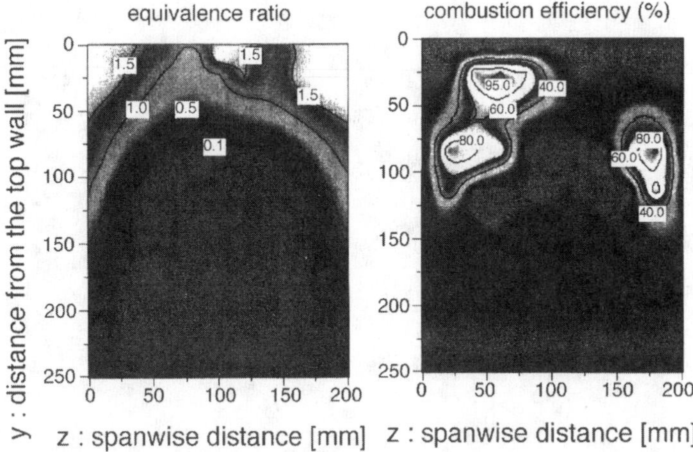

Fig. 12 Distribution of fuel equivalence ratio and combustion efficiency found in the weak combustion mode. Left: fuel equivalence ratio; Right: combustion efficiency.

When intensive combustion is attained at 70 g/s ($\Phi = 0.50$), the combustion is held at the step on the sidewall. The highest heat-flux region moves upstream on the top wall and on the sidewall, especially near the cowl. All of these observations correspond to the wall-pressure distributions shown in Fig. 10. This implies that when increasing fuel flow rate the flame propagates upstream and beyond the step, especially near the cowl.

Cross-sectional distribution of local equivalence ratio and combustion efficiency. Figures 12 and 13 illustrate distributions of local values of the equivalence ratio φ and the combustion efficiency η_c at the engine exit under conditions of the M6 testing.[74] The left figure of Fig. 12 presents the distribution of φ in the weak combustion when only a negligible thrust was observed with a total fuel equivalence ratio Φ of 0.33. Although the fuel is injected from uniformly distributed orifices on the sidewalls, hydrogen is concentrated near the top wall. No fuel exists near the cowl, and the H_2-lean region ($\varphi < 0.1$) covers almost all of the region of the cross section because of the distorted airflow coming from the swept-back inlet. Hydrogen is confined to a narrow region near the sidewalls, especially around the corners with the top wall.

The right figure of Fig. 12 illustrates distribution of local η_c in the weak combustion mode. Confined combustion is found near the sidewalls, the locations coinciding with those of observed faint plumes. Comparison with φ distribution indicates that combustion takes place in the region with $0.5 < \varphi < 1.5$ and that the location of the highest combustion efficiency coincides with the region with $\varphi \sim 1$ in the boundary layer. The bulk combustion efficiency, as an average weighted with the local hydrogen mass flux, was as low as 40%.

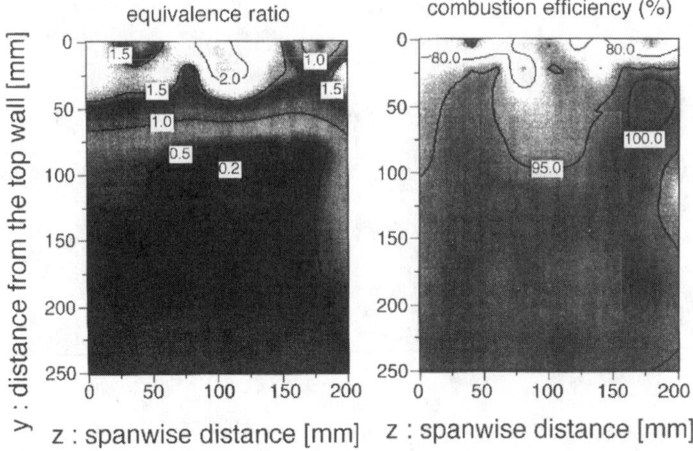

Fig. 13 Distribution of fuel equivalence ratio and combustion efficiency found in the intensive combustion mode. Left: fuel equivalence ratio; Right: combustion efficiency.

The weak combustion mode switched to intensive combustion as the fuel flow rate increased. The left figure of Fig. 13 is the φ distribution in the intensive combustion when the engine delivers ΔF of 1050 N with $\Phi = 0.46$. Although the nonuniform distribution of fuel in the vertical direction still remains, it is flattened, and the region with $\varphi < 0.2$ almost disappears. The span-wise distribution is also improved with the intensive combustion. The right figure of Fig. 13 presents the η_c distribution in the intensive combustion mode. The engine exit is occupied by a region with η_c higher than 80%, and the bulk combustion efficiency was found to be 85%. This improved mixing, and combustion efficiencies produced the higher thrust performance in the intensive combustion mode.

Integrated Wall Pressure

Wall pressures were integrated over the inner and the outer surface of the engine, and their axial components were compared with the thrust component of the FMS outputs to estimate reliability of the FMS and the friction force acting on the outer and the inner surfaces of the engine. Typical results for the M6 tests are compared with the FMS outputs in Fig. 14, where ΔF and ΔF_p are plotted against Φ; they are for the perpendicular fuel injection in a model with a full-height strut and a short isolator. As for the ΔF, ΔF_p is the difference of pressure integrals with and without the fuel injection. When the engine is in the start condition, the difference between ΔF and ΔF_p is small, which implies that the friction forces with and without combustion are not so much different. When in the unstart condition, however, the difference becomes large because of the difference in the freestream condi-

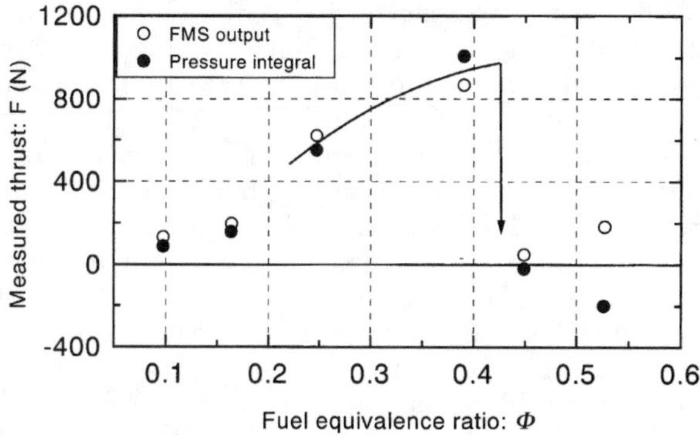

Fig. 14 Comparison of FMS output and pressure integral (M6S test).

tion outside the engine where the external flowfield is assumed to be unchanged in calculating ΔF_P.

B. Mach 4 Tests[9,65–67,71,72,78,81]

The storage heater and only the heat-sink, strutless model were used at this test condition.

Effects of Isolator Length

The most remarkable feature in the results is the effect of the isolator length. Figure 15 shows a variation of ΔF against Φ for two lengths of the isolators and various fuel-injection patterns tested. With the short isolator the thrust monotonically decreases with an increase in the fuel flow rate, which is contrary to the characteristics a thrust-producing machine should

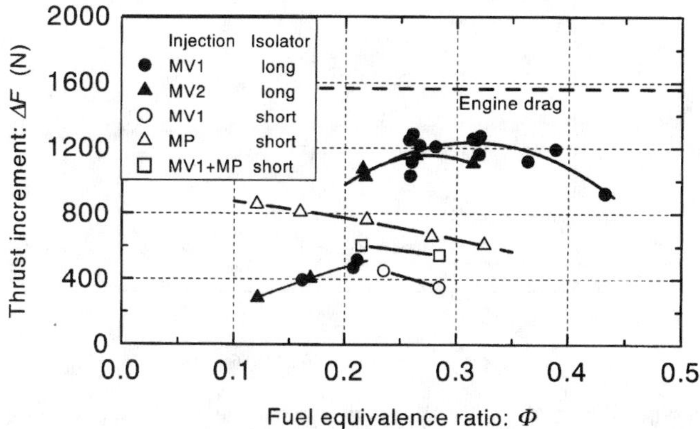

Fig. 15 Variation of thrust increment against total fuel equivalence ratio: M4 tests.

have. The reason for this deficiency is the combustion-induced disturbance propagating beyond the isolator into the inlet, i.e., a combustor/inlet interaction. Its strength grows with an increase in Φ as can be seen in Fig. 16a. In the figure the axial components of the integrated wall pressure over the internal surface of each engine component F_p are plotted against Φ. The injection pattern corresponds to the open-circle symbols ○ in Fig. 15. Although F_ps increase with Φ for the thrust-contributing components, that for the drag-contributing inlet also increases in magnitude even more steeply, resulting in a decrease in their sum.

The deficiency just described was mostly eliminated by employing a long isolator as shown in Figs. 15 and 16b. Figure 16b corresponds to the solid circle symbols ● in Fig. 15. In the figure ΔF increases, but F_p for the inlet never increases with Φ up to its value of 0.3, implying that there is no inlet-combustor interaction. An abrupt increase in ΔF near Φ of 0.2–0.25 is

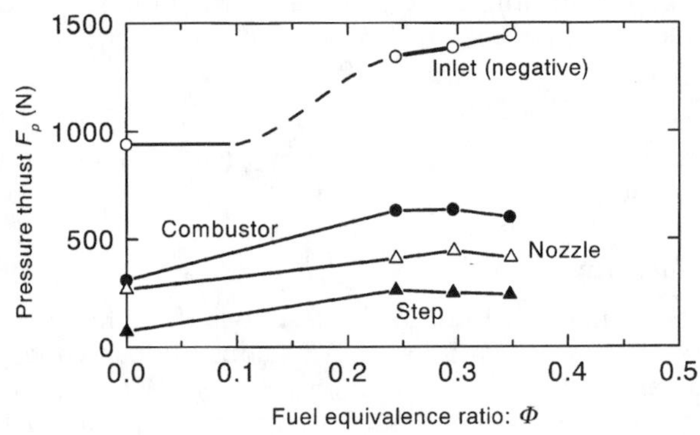

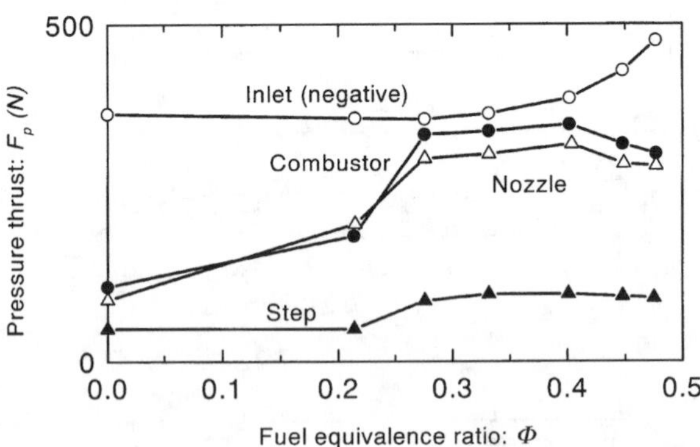

Fig. 16 Pressure thrust of each component against fuel equivalence ratio: M4 tests.
a) Short isolator
b) Long isolator

the phenomenon defined as a transition from the weak to the intensive combustion modes. Even with the long isolator, however, an inlet/combustor interaction appears for Φ beyond 0.3–0.4, as evidenced by an increase in F_p for the inlet. Thus, ΔF never increases with Φ beyond its value of 0.4. Sunami et al.[78] defined the inlet-combustor interaction as engine unstart. However, even at the highest fuel flow rate of $\Phi = 0.67$, a catastrophic unstart condition resulting in negative or nearly zero ΔF was not observed.

Wall-pressure distributions showed the same tendencies as those already described. Namely, the pressure rise caused by the combustion traveled upstream into the inlet for the short isolator model while it remained within the isolator for the long isolator model. The pressure rise profiles were similar in both models when the short isolator model had been shifted downstream by a difference in the isolator lengths (100 mm). In both the short and long isolator cases the pressure on the top wall reached a maximum at the exit of the isolator and remained constant throughout the constant-area section downstream of the step.

As already stated, the long isolator resulted in a short constant-area section downstream of the step. This length has some effects on the combustor performance through suppression of the maximum pressure there[30,34] or by shifting its peak location into the diverging section. However, the report[78] was made that there was little change in the maximum pressures between both models at the same fuel flow rate. Therefore, the effect of the length downstream of the step should be minor in the present model at the M4 test condition.

Combustion Mode and Inlet-Combustor Interaction

Because the low pressure behind the step disappears at least on the top wall in the intensive mode, F_p for the step in this mode should be higher than those in the weak mode. Therefore, all of the data for the short isolator shown in Fig. 15 are in the intensive mode while they are in both modes for the long isolator, as can be judged from Fig. 16b. At the transition range of $\Phi = 0.2$–0.25 in Fig. 15, a steep rise in F_p is also confirmed in Fig. 16, except for the inlet.

Typical wall-pressure profiles along the axial distance on the top wall of the long isolator model are shown in Fig. 17 together with the one without fuel injection. With an increase in Φ, their maximum value increases. Even with the combustion-induced pressure rise traveling upstream, however, it remains nearly constant around $p_w/p_0 = 0.135$. This suggests that simply extending the isolator length has a limit in suppressing the inlet-combustor interaction or the engine unstart. In the intensive mode effects of the combustion-induced pressure rise propagated deeper on the top wall than on the sidewall. This implies that the inlet-combustor interaction was caused by the boundary-layer separation on the top wall. At the maximum fuel flow rate of $\Phi = 0.6$, alternating operation between those with and without the inlet-combustor interaction was observed and will be discussed in detail later.

Based on whether the wall-pressure distributions in Fig. 17 show a wavy pattern or not, subsonic combustion occurs in the intensive mode, at least in

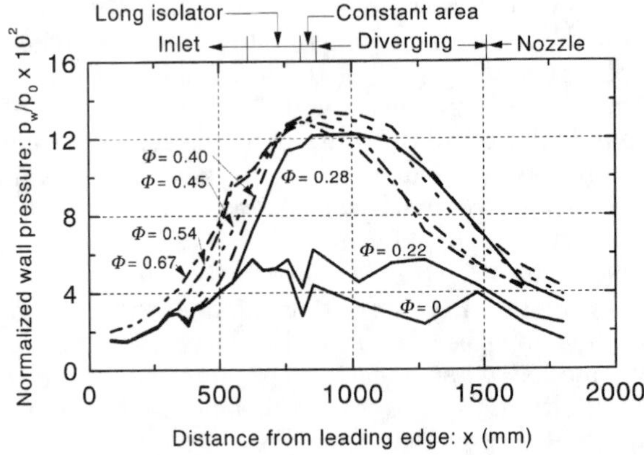

Fig. 17 Pressure distribution on the top wall in the long isolator model: M4 tests.

the constant-area section of the combustor. In the weak mode supersonic flow survives throughout the engine.

Effects of Injection Pattern and Comparison with Engine Drag

Another feature derived from Fig. 15 is the effect of the fuel-injection pattern. For the case of the short isolator, parallel injection from the step base produces the highest thrust. The difference between them changes slightly with a change in Φ. Therefore, parallel injection seems to be effective in mitigating the inlet/combustor interaction although its intensity is still strong. For the long isolator model injecting fuel from the orifices closer to the step (MV2) results in a slightly earlier transition to the intensive mode than that from the downstream orifices. However, the maximum thrust attained is almost the same.

The engine drag measured with the FMS prior to fuel injection is also shown in Fig. 15 and is much higher than the maximum ΔF attained. However, approximately 30% of it was found to be from engine support and fairing drag in the subscale aerodynamic tests[79] (see Section VI.C). This implies that pure engine drag should be around 1000 N in this condition. Thus, the thrust increment exceeded the engine drag in the intensive mode for the long isolator model.

Unsteady Characteristics and Engine Exit Gas Sampling

Unsteady characteristics of the thrust and the wall pressures were examined during the runs.[72,78] The wall pressures were measured with Kulite high-frequency-response transducers at the exit of the inlet. When Φ exceeded 0.4, fluctuations of the inlet pressure and the thrust appeared. The amplitude of the inlet pressure fluctuations increased with increases in Φ. Under this operation the lowest value of the fluctuating pressure was at the same level as the fuel-off pressure. The time-averaged combustor pressure was at the same level as that in the intensive mode without the inlet-combustor interaction (see Fig. 17, which indicated that the precombustion shocks moved between the inlet and the isolator for this range of Φ. These shock motions were also observed in the high-speed schlieren video.

Gas sampling was conducted at the engine exit under the intensive combustion mode. Distributions of φ were nearly uniform in the span-wise direction. They showed relatively high φ near the top wall although nearly 80% of fuel was injected uniformly from MV1, and so this suggests that there was a crossflow transporting the injected fuel from the cowl side to the top-wall side. The distributions of η_c were similar to those of φ. Relatively high η_c, greater than 70%, were achieved in the half region on the top-wall side, whereas almost no combustion was detected on the cowl side. This feature was consistent with the result of heat-transfer distributions, which showed that high heat flux on the sidewall was observed on the top-wall side. The values of the bulk combustion efficiencies estimated by a one-dimensional analysis varied from 55% for the weak mode to 90% for the intensive mode.

C. Mach 6 Tests[9,65–68,70,72,75,77–81,83,85]

The heat-sink model was used with and without the strut under this test condition. As can be seen in Table 4, we have been conducting tests most intensively under this condition. In the initial series of tests with the strutless, short isolator model,[65–67] we were not successful in attaining satisfactory combustion and thrust generation. Their operational modes were later identified as weak. These unsatisfactory results were thought to be because of too low pressure and temperature to attain continuous combustion within the combustor and to the absence of a large recirculation region for flame-holding. The static pressure prior to fuel injection at the combustor entrance was 20 kPa compared with that of 70 kPa under the M4 conditions. To overcome this difficulty, a 1/5-height (strut AS) was attached within the engine. This resulted in successful combustion and thrust generation for the first time in the M6S condition[68]; intensive combustion was also attained. In Fig. 18 ΔF of the perpendicular fuel injection with this model are plotted against Φ with solid circles ●. Beyond Φ of 0.25, the intensive mode appears, and ΔF gradually grows with a further increase in Φ, but suddenly falls to unstart at Φ of around 0.40. Figure 19 shows a photograph of the engine operating in the intensive mode.

Comparison Among Operation Modes

Wall-pressure distributions along the center of the sidewall are shown in Fig. 20 for the three modes of operation, namely, weak, intensive, and unstart modes, together with that one without fuel injection. The distributions in the figure correspond to those shown in Fig. 10 on the top wall. In the intensive mode the level of the wall pressure downstream of the step is much higher than those in the other modes. The combustion-induced pressure rise can be seen to invade into the isolator but not into the inlet, which is in contrast to

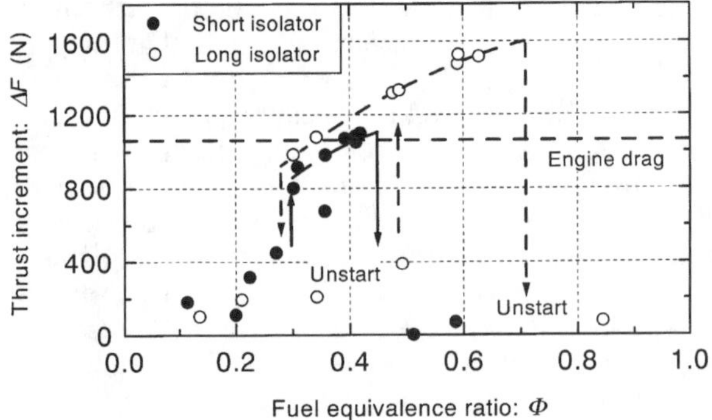

Fig. 18 Thrust increment found in M6S tests with 1/5-height strut.

the M4 test results where the pressure rise was observed in the inlet. In the weak mode the wall pressure is slightly higher than that without fuel injection only in the downstream part of the combustor and in the nozzle. When falling to unstart, the pressure distribution in the inlet changes drastically.

Distributions of the heat flux to the wall show similar conclusions derived from the wall-pressure results. The highest heat flux was observed in the intensive mode while the lowest one was in the weak mode. In the unstart condition the heat flux in the inlet became higher because pressure became

Fig. 19 Engine operating in the intensive combustion mode: M6S tests.

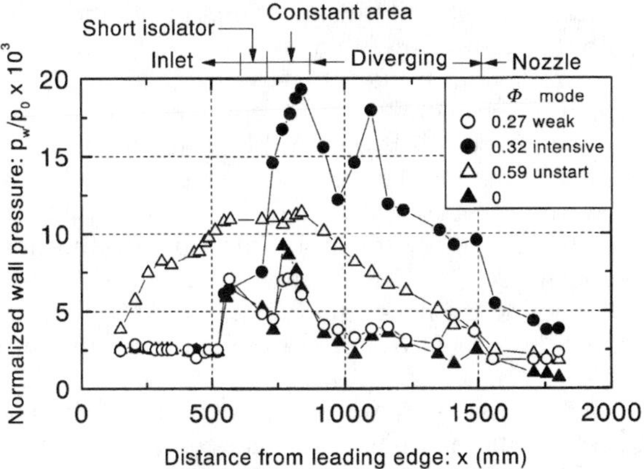

Fig. 20 Effect of combustion mode on the sidewall pressure distribution: M6S tests (short isolator, 1/5-height strut).

higher there. According to the one-dimensional calculation based on the wall pressure-distribution,[68] the bulk combustion efficiencies were 95% for the intensive mode and only 2% for the weak mode. Because the wall-pressure distribution shows a wavy pattern even in the intensive mode, the flow is judged to be supersonic throughout the model except in the unstart condition.

Figure 21 compares the wall pressure distributions on the top wall without fuel injection for this model with that for the strutless model. That with highest fuel flow rate ($\Phi = 1.0$) in the latter model is also shown in Fig. 21. The strut creates a high-pressure region, and the pressure around the step is

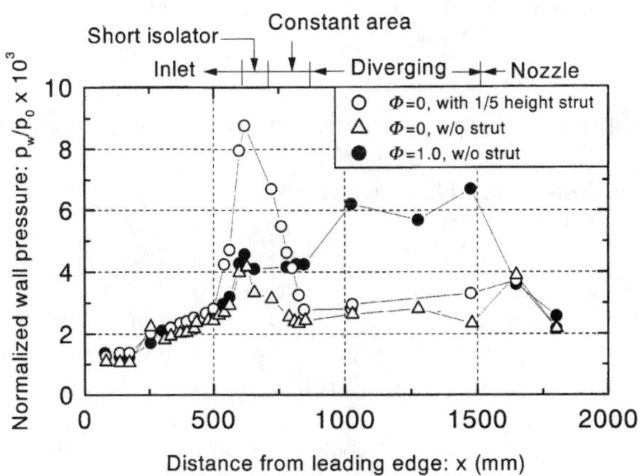

Fig. 21 Effect of 1/5-height strut on the top-wall pressure distribution prior to fuel injection: M6S tests.

more than doubled even without fuel injection. A similar rise in pressure was also observed on the sidewalls. With a large amount of the fuel injected ($\Phi = 1.0$), the pressure level around the step on the sidewall in the strutless model approached or exceeded that in the model with the strut. However, the pressure on the top wall is low with this large amount of fuel, approximately half of that in the latter model. Hence, the region of high pressure seems to be one of the necessary conditions for the initiation of intensive combustion. Factors necessary to attain intensive combustion other than pressure have been proposed by Kanda.[83] He suggests that causing an interaction of the boundary layer with fuel jets and forming a large recirculation region extending downstream of the fuel injection are necessary for this end.

The sudden fall to unstart shown in Fig. 18 was not observed in the M4 tests (Fig. 15). In the latter figure the inlet-combustor interaction grows gradually, and ΔF never falls to near zero. The wall-pressure distributions for the M4 and M6S cases also showed the same characteristics. As can be seen in Fig. 20, change between the start and the unstart conditions is drastic for the M6S case. On the other hand, there was little change for the M4 case even with the pressure at the leading edge departing from that without fuel injection. The state just described for the M4 case was termed partial unstart by Tani et al.[16] in their experiments on inlets. The reasons for this difference can be summarized[68] as follows: namely, the stronger the compression effect of the inlet, the larger may be the change in flow condition there with a slight change in the downstream condition. Similar results were shown in the inlet tests at Mach 4.[16] The change from start to unstart condition was found to have abruptly occurred when the swept angle was small, i.e., strong compression effect. Note that both higher inflow Mach number and attachment of the strut intensify the compression effects in the M6S case compared with the M4 case. The difference in location of the major inlet-combustor interaction, i.e., on the top wall for the M4 case and on the sidewall near the cowl for the M6S case, was thought to be another reason.

Effect of Fuel-Injection Pattern

As another feature of the test results for this short isolator model with the 1/5-height strut, the effects of fuel-injection pattern are summarized as follows:[68] approximately 20% of injection from the pilot fuel orifices on the sidewalls (SP) in addition to MV1 was found to make it possible to attain the transition from the weak to the intensive modes. However, injecting all of the fuel from SP caused the engine to fall to the unstart or to the weak mode. These results suggest that perpendicular fuel injection downstream of the step was indispensable for attaining intensive combustion. The addition of parallel fuel injection from the step base to the perpendicular one caused the thrust to decrease. Decreasing the latter deteriorated the thrust performance further. The wall-pressure distributions showed that the parallel injection caused the combustion mode to change to the weak from the intensive mode, which had been attained with perpendicular injection only. This implies that parallel injection was less effective in attaining the intensive mode.

Encouraged by the successful thrust generation and attainment of intensive combustion in the short isolator model with 1/5-height strut, models with the long isolator were tested with various kinds of struts to achieve further improvement in thrust performance.[77] These results will follow next.

Effect of Isolator Length

The threshold values of the fuel flow rate for the mode transition and the fall to unstart depended on the models used and the fuel injection pattern. In Fig. 18 effects of the isolator length are shown for the 1/5-height strut model. Among the three types of struts employed in the long isolator model, the results for strut B are shown in the figure. The strut chosen for comparison here has the same length and location relative to the step on the sidewall as in the short isolator model. Although the threshold for the transition to the intensive mode is the same ($\Phi = 0.25$), the long isolator model falls to unstart at higher Φ than that with the short isolator.

The wall-pressure distributions on the sidewall for these models with the same fuel flow rate, both in the intensive mode, are compared in Fig. 22. The results with the highest fuel flow rate ($\Phi = 0.74$) with the intensive mode) are also shown for the long isolator model. Because the effects of the combustion-induced pressure rise never reach the inlet for all of the cases, even the short isolator was of sufficient length for the 1/5-height strut model. Thus, unlike in the M4 tests, there should be another major reason for the extension of the intensive mode range because of the lengthening of the isolator. In Fig. 22 the maximum pressure is in the constant-area section downstream of the step for the short isolator model, whereas it is in the diverging section for the long isolator model because of a difference in the length of the constant-area section downstream of the step. This difference caused the long isolator model to mitigate propagation of the upstream

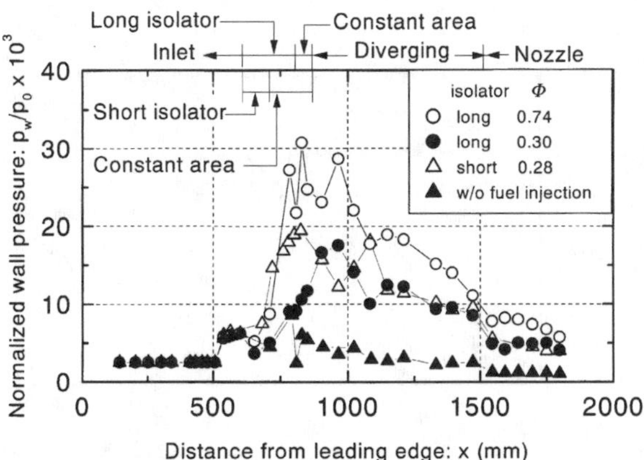

Fig. 22 Effect of isolator length on the sidewall pressure distribution: M6S tests.

influence into the isolator and the inlet and thus contributed to the extension of the intensive mode range.

The same plot as in Fig. 18 is shown for the full-height strut model (struts C and D for the long and short isolator models, respectively) in Fig. 23. Again, the range for the intensive mode is extended by employing the long isolator model. For these models, however, the wall-pressure rise was observed in the inlet section at higher Φ for the short isolator model. The length of the short isolator was insufficient, and its extension functioned as expected.

Effect of Height and Location or Length of the Strut

Effects of the strut height can be seen by comparing Figs. 18 and 23. The transition points from the weak to the intensive mode and to the unstart are not so different between the two models except that the latter point is at lower Φ for the full-height strut model. Attained thrust levels also differ slightly although the maximum value for the full-height strut model is higher than that for the 1/5-height model at the same Φ. Thus, the effects of strut height are not remarkable for the short isolator model in the thrust performances. In the long isolator model, however, the effect appears in thrust levels only at higher Φ and in a value of Φ at the transition to unstart. The engine drag in the full-height strut model is higher than that for the 1/5-height model by 250 N.

With the strut AL attached in the long isolator model, the intensive mode never appeared, but the engine finally fell to unstart. Thus, the intensive mode was limited within a very narrow and higher Φ range with this strut. We never observed the intensive mode in the test series because of limited resolution of the fuel flow rate we set, though. The subtle change in the thrust performance with the geometry or the location of the strut suggests that locations of shock and expansion waves formed by the strut have dominant effects on the attainment of thrust generation.

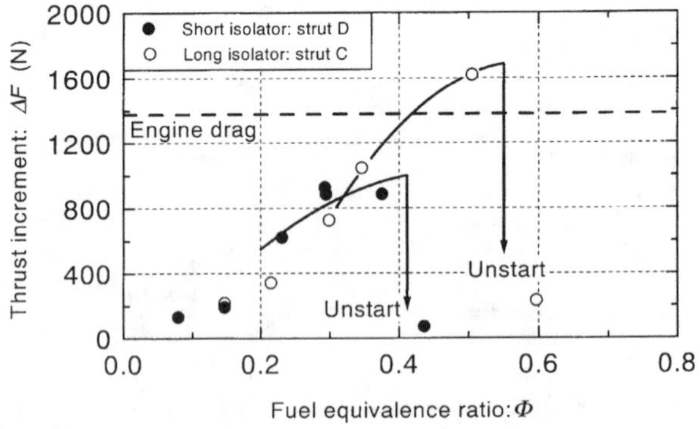

Fig. 23 Thrust increment found in M6S tests with full-height strut.

Effect of Storage/Vitiation Heated Air

The combustion behaviors of the scramjet engine were found to be very sensitive to the test media as shown in Table 6. The hydrogen autoignited without the assistance of igniters in the V mode, whereas autoignition was difficult in the S mode. In addition, flameholding in the V mode experiments was much easier than in the S mode. Thus, the vitiated air may yield higher but false ignition performance in scramjet engine testing. The easier ignition with the vitiated air was interpreted as being because of radicals supplied from the vitiated air heater later.[74]

Thrust performance was also affected by the test air. The thrusts measured by the FMS in the S mode and the V mode experiments are compared in Fig. 24 as functions of Φ. There are two combustion modes found in Fig. 24: the lower thrust mode up to about $\Phi = 0.30$ (weak combustion) and the higher thrust mode (intensive combustion) from $\Phi = 0.30$ to the limit flow rate, which causes engine unstart. Slightly higher thrust is observed in the S mode than in the V mode from $\Phi = 0.30$ to 0.40. The change in burning behavior of the two airflows becomes apparent when the limit fuel flow rates for engine unstart are compared. The engine easily falls into unstart condition in the S mode as shown in Fig. 24. The maximum Φ for keeping an engine at start condition in the S mode was found to be about $\Phi = 0.45$. In the V mode, the engine works with a higher Φ of 0.80 with the parallel injection and overcomes the engine drag. The engine drag was measured to be 1060 N in the S mode air and 1120 N in the V mode airflow.

The sensitivity to the testing media suggests kinetically controlled combustion in the scramjet engine. The comparative studies of scramjet performance conducted at NASA Langley also raise these issues, namely, where the flame is held and whether the combustion is mixing- or kinetic-controlled.[102] In the case of weak combustion, the thrust was low, and the engine exhausted a faint plume from the nozzle exit. The combustion was concluded to be the autoignition of the premixture in the boundary layer developed on the engine wall, which must be kinetically controlled. On the other hand, the flame in the intensive combustion may be held near the backward-facing step. However, there is no way for combustion to be af-

Table 6 Comparison of combustion performance in the storage and the vitiation-heated air

Air modes tested	Storage-heated	Vitiation-heated
Autoignition without preheating	Impossible	Possible
Autoignition with preheating	Marginal	Possible
Flameholding without PJ igniter	Marginal	Good
Without 1/5H-strut	Poor	Good
In weak combustion mode	Marginal	Fair
Combustion with parallel injection	Poor	Good
Fuel equivalence ratio limits	0.45	0.8

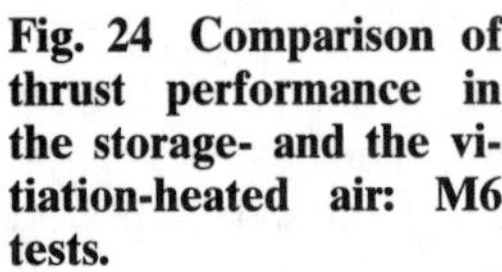

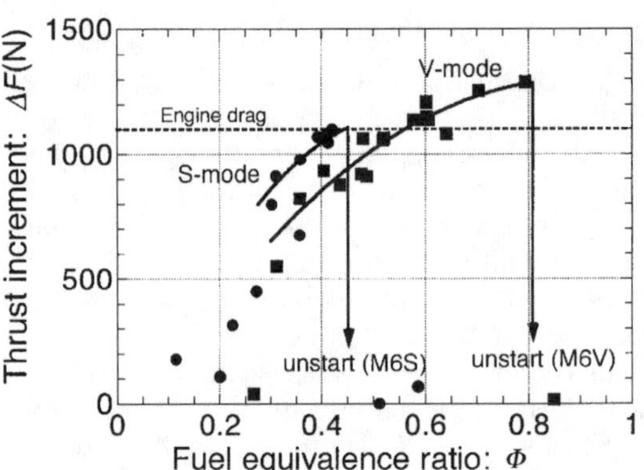

Fig. 24 Comparison of thrust performance in the storage- and the vitiation-heated air: M6 tests.

fected by the chemical reactiveness of test air if the combustion is rate-controlled by mixing process. Physical properties of the test airs with S and V modes do not differ so much that the effects to mixing are negligible. The dependence of thrust on the test air illustrated in Fig. 24 suggests that the flame is not fully mixing-controlled.

Other Results for Thrust Performance

In Figs. 18, 23, and 24, the engine drags measured with the FMS without fuel injection are shown. Some of the ΔF in the intensive mode exceed the engine drag. Considering the fairing and support drags, the positive thrust, the difference between ΔF the engine drag should be much higher.

In the long isolator model a hysteresis in thrust characteristics depending on the direction of the change (increase or decrease) in fuel flow rate can be seen in the Fig. 18. When the fuel flow rate decreases from that causing the intensive mode or even engine unstart, the range of intensive mode extends to the lower flow rate side. This observation implies that once the flow conditions sufficient for the intensive mode are established within the combustor, then they survive to some extent during the reduction process in the fuel flow rate. Similar phenomena were also observed in the M6V tests where the intensive and weak modes appeared alternatively during the constant fuel flow rate operation.

Unsteady Characteristics

As in the M4 tests, unsteady characteristics of the thrust and the wall pressures were examined in the M6 tests.[72,78] The unsteady wall pressures were measured at the exit of the inlet and at the entrance of the diverging combustor. Compared with the oscillations in the start and unstart states, a relatively large amplitude was observed in those around the boundary between both states. In this boundary state the lowest level of the fluctuating pressure in the combustor was the same as that at the unstart state, and their highest was much greater than that at the start state. These behaviors were observed in a narrower region of fuel flow rate ($0.64 < \Phi < 0.70$) than that in the M4 conditions. In the latter case this mode was observed in a Φ range of 0.40–0.52, and the engine thrust decreased gradually with an increase in the fuel flow rate. The phase of the pressure fluctuation of the combustor was opposite that of the fluctuation in the inlet.

Other Options

Besides the model with and without various struts, other options to enhance the thrust performance were attempted. One of them was to attach a recess and auxiliary injectors to the top wall.[70] The recess was to augment pressure and residence time within the recirculation region behind the step base. The results showed that application of the recess was not sufficient to initiate intensive combustion. However, supplemental injection of gaseous oxygen into the region caused intensive combustion. Other options were to attach one of the ramps shown in Fig. 1 and Table 2 to the top wall.[76] Unfortunately, the tests with ramp W we were not successful because of facility and igniter failures. In the model with ramp T, a less oscillatory operation at the maximum Φ and extension of the intensive mode toward lower Φ compared with the strut models were observed.

D. Mach 8 Tests[69,72,74,81,84–86]

The air heating mode is storage heating augmented with vitiation in this test condition. The water-cooled engine was used for the first time, and only the short isolator was used. In the first series of tests, a strutless configuration was tested.[69] Three modes of fuel injection were tested, namely, perpendicular injection, parallel injection, and a combination of the two. However, a partial combustion was attained only in the case of perpendicular injection. Thrust generation was not confirmed, and the flame observed at the engine exit was very faint even at the optimum condition. The combustion efficiency was estimated to be around several percent.[69]

The preceding results imply the same problems as were seen in the M6 case without the strut, i.e., difficulties in fuel mixing with the airflow and in attaining conditions for fast chemistry. Despite these unsatisfactory results as to combustion, the engine's durability in high-temperature airflow was satisfactory. To overcome the difficulties in combustion and thrust generation, a water-cooled strut with fuel injection was applied in the second series, and combustion and thrust generation were attained.[84]

Thrust Performance

The results in the second series are shown in Fig. 25 for ΔF against Φ. Increasing Φ, ΔF increases gradually. With the amount of fuel from the sidewalls beyond $\Phi = 1.6$, however, the engine falls into unstart (symbols ○). A further increase in Φ is possible by partitioning part of the fuel to strut injection (symbols ▲). However, comparison of the thrust level and wall-pressure distributions with those in the M4 and M6 conditions shows that intensive combustion was still not attained. As in the M6 tests, utilization of

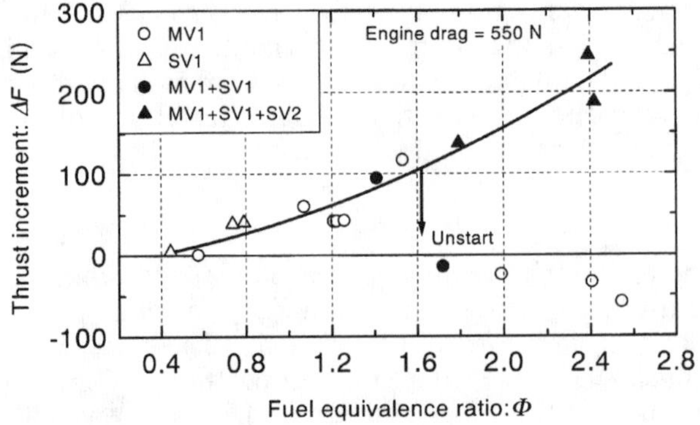

Fig. 25 Thrust increment found in M8 tests with a thin strut.

hysteresis was attempted in this model. A large amount of fuel sufficient for unstart was injected first and then reduced to realize the intensive mode. During the fuel flow rate reduction process after the unstart, the wall temperature and wall pressure increased for a moment. Their levels were large enough to correspond to the intensive mode, but they quickly fell to those in the weak mode. Considering this fact, intensive combustion might have been possible with these models if we had tested a model with a long isolator.

The thrust performance was further improved in the third series by thickening the strut from 30 to 46 mm as can be seen in Fig. 26. In this series the strut with fuel injection was not used, but other fuel-injection patterns from the sidewalls were tried. They were parallel injection from the step base and pilot fuel injection combined from the sidewall. Irrespective of the injection pattern, the thrust performance differs slightly, except the parallel injection from the strut only (symbol △). The engine operates alternatively between

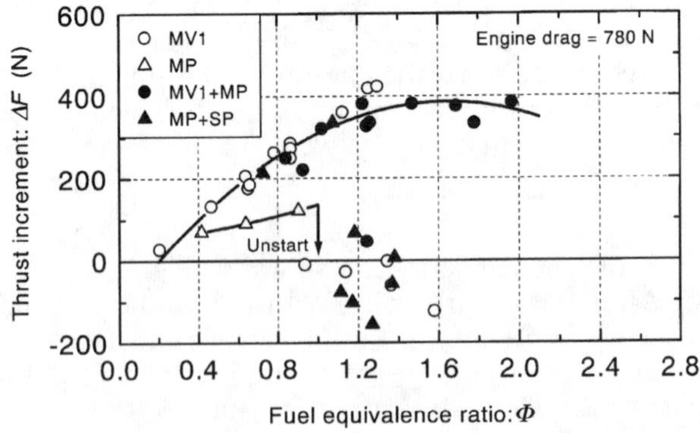

Fig. 26 Thrust increment found in M8 tests with a thick strut.

the start and unstart states for $\Phi = 0.9$–1.3. Beyond this range of Φ, ΔF decrease with an increase in Φ. The generated thrust with parallel injection is because of the momentum force of the fuel jets, not to combustion. Contrary to our expectation, the improved thrust performance with this model is still within the weak mode; this can be judged from the comparison of ΔF levels and wall pressure with those in the M4 and M6 conditions. The same attempt as in the second series to utilize hysteresis for attaining the intensive mode was tried again. However, this was not successful and even the spike-wise jump just mentioned in the wall temperature and pressure was not observed. The engine drag is much higher for this model (approximately 800 N compared with 550 N for the model with a thin strut).

A model attached with both ramp K and the thin strut was also tested. The results were less satisfactory compared with those only with a thick strut. They showed similar trends in thrust performance as those in the thin-strut model.

Gas-Sampling and Pitot-Pressure Survey

Gas-sampling and pitot-pressure survey were carried out using probes with fine 0.3-mm sampling orifices in the M8 engine testing after the probe calibration discussed later in Sec. VI.B. Because disturbances from the installation of the sampling rakes sometimes propagated upstream and interfered with combustion in the engine, the number of rakes installed at the engine exit was reduced from three to two. Because the total temperature of the airflow is high at 2600 K and intrusive probing is difficult in the M8 condition, only a limited number of experiments was successful.

Figure 27 illustrates the distribution of φ measured at 60 locations at the exit of the engine with the thick strut. Bulk hydrogen of $\Phi = 0.6$ (28 g/s) was injected perpendicularly from the sidewalls, and the engine exhibited a

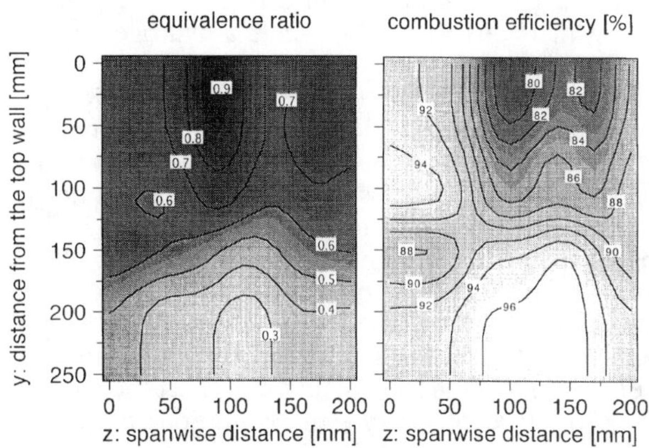

Fig. 27 Gas sampling at the engine exit in the M8 test. Left: distribution of fuel equivalence ratio; Right: distribution of combustion efficiency.

thrust increment of about 200 N. The distribution of φ shows a different feature from that found in Fig. 13a for the M6S case. In the intensive mode of the M6S case, rich H_2 regions are formed in the corners between the top wall and the sidewalls, irrespective of the strut heights and the length of the isolators. Mixing is retarded in supersonic flow, and the influence of combustion on the flowfield must be limited because of the small heat release relative to the enthalpy of incoming air in the M8 condition. Therefore, we expected that the feature found in the weak combustion mode in the M6 testing might be reproduced in the M8 testing.

However, strong interference between the flowfield and the combustion and improved mixing can be seen in Fig. 27. Hydrogen is concentrated in the central part near the top wall in the M8 testing, and this may be caused by the wake accompanied with the thick strut. The contour line of $\varphi = 0.7$ widens to the cowl in the M8 combustion compared with that in the M6 case although the relatively high φ region near the top is the same. Decreased deflection angles at the inlet with the higher Mach number and the thicker strut employed are the causes of this improved mixing.

The contour lines of local combustion efficiency η_c are shown on the right-hand side in Fig. 27. Most of the engine exit is occupied by $\eta_c > 85\%$. A narrow region with $\eta_c < 85\%$ is found at the center of the top wall. The bulk combustion efficiency was found to be 95% in this testing. A lower bulk combustion efficiency of 86% was derived in the $\Phi = 1.3$ testing at which a greater thrust increment of 370 N was observed. The fuel specific impulse was estimated to be 7.5 and 7.2km/s in these $\Phi = 0.6$ and 1.3 cases. Generally speaking, thrust produced in scramjet engines depends on the location of heat addition. The maximum specific impulse attainable in this M8 engine is now being sought by adjusting the mixing and combustion schedules along the streamwise direction in the given engine configuration.

E. Liquid-Hydrogen-Cooled Engine Tests[72,82]

The RJTF has a LH_2 supply system to feed the coolant into the LH_2-cooled engine, which consists of a run tank, a hydrogen-gas pressurization system and a LH_2 line. Heated hydrogen gas after engine cooling will be fueled into the engine combustor in the future. However, hydrogen after cooling is disposed of through a vent stack for the present. The design pressure of the run tank is 10 MPa. A flow rate of 1.5 kg/s is possible under the condition of an upstream pressure of 8.5 MPa and a downstream pressure of 6 MPa at the engine interfaces. The pressure of hydrogen is about five times higher than the critical pressure.

Prior to firing test of the LH_2-cooled engine assembly, preliminary experiments on the inlet block combined with the heat-sink engine were conducted. Objectives of the preliminary experiments were to obtain cryogenic flow characteristic data of the LH_2 facility, to inspect the engine components and the cooling structure under cryogenic conditions, to establish the procedure of the operation, and to verify the safe operation of the engine in high-enthalpy airflow. Following it, a firing test of the engine with LH_2 cooling was conducted in a M6S airflow.

In this test the plasma torch igniter successfully operated, in spite of being chilled completely, and ignited the hydrogen fuel, delivering a thrust. The average heat flux to the inlet was 60 kW/m^2. We confirmed a normal operation under the combined condition of the facility and the engine. As a result, the procedure and the sequence of firing test with the LH_2 cooling were established, and the compatibility of the engine with the facility was verified from the integrated test.

VI. Supplementary Studies for Engine Testing

A. Computational Fluid Dynamics

Numerical Simulations Under Mach 4 and 6 Flight Conditions

To assess the validity and accuracy of computational fluid dynamics (CFD) for the internal flow in scramjet engines, a newly developed, unstructured grid flow solver[20] was used to compute the flowfield. The three-dimensional computational region inside of the engine is discretized by an unstructured hybrid grid for accurate and efficient computations as well as ease of grid generation for complex configuration. The prismatic semistructured grid generated for viscous boundaries on the surfaces allows the control of minimum spacing required for resolving viscous sublayers. This assures the quality of the solution for high Reynolds-number flows without decreasing the advantage of the flexible unstructured grid. The inner region of the flowfield is covered with tetrahedral cells.

The solution algorithm used to compute the compressible Navier–Stokes equations is based on a cell-vertex, upwind, finite-volume scheme for arbitrarily shaped cells. The computational efficiency is drastically improved by the LU-SGS implicit method with a reordering algorithm for the unstructured hybrid grid. The turbulent kinetic viscosity is evaluated by a one-equation turbulence model proposed by Goldberg and Ramakrishnan.[107] The numerical accuracy of the flow solver was validated by several benchmark tests, and the grid size was optimized using the law of wall. A grid system with 0.52 million nodes was employed for the internal flow of the engine.

Figure 28 shows a colored contour map of the local skin friction coefficient c_f, which is calculated using the dynamic pressure in the incoming flow and shown in units of 10^{-3}. A higher value found on the sidewall leading edge is from the undeveloped boundary layer. The friction is large in the diverging combustor section near the cowl because of the flow compressed by the shock wave generated at the cowl. The decelerated flow region in the combustor is occupied by the lower friction coefficient. Because the top wall is covered with an ingested thick boundary layer, the friction coefficient at the leading edge is low.

Wall-pressure distributions derived with the CFD agreed with the experimental results. Engine drag because of the pressure by the internal flow could be estimated by integrating the wall pressure. The resultant drag duplicated the pressure drag measured in experiments within $\pm 2.1\%$ in the M4 flight condition. Thus, the discrepancy between CFD and the experimental findings was small for the pressure force.

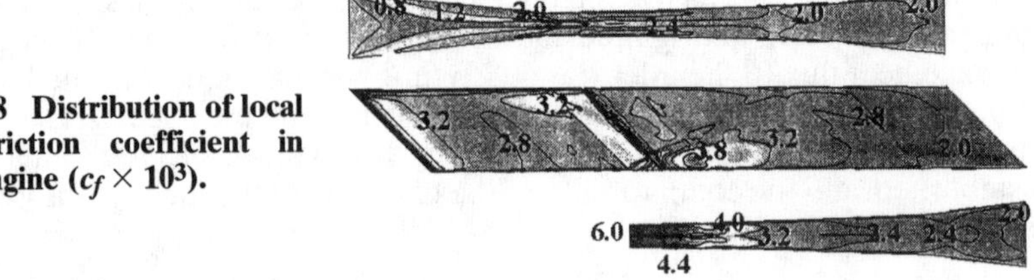

Fig. 28 Distribution of local skin-friction coefficient in the engine ($c_f \times 10^3$).

On the other hand, the CFD value overestimated the friction by 31% in the M4 simulation. Consequently, the discrepancy in the total drag between the CFD and experiments was 15%. Because the discrepancy in the inlet was small and grew in the downstream sections of the model, it might be caused by error accumulated because of the limited grids in the streamwise direction in the engine internal flow. Optimization in grid distributions in streamwise and transverse directions is needed.[79]

Numerical Simulation Under Mach 8 Flight Condition

Numerical simulation for the M8 condition is also being undertaken. The first simulation was conducted for the no-fuel condition assuming calorically perfect gas. The first purpose of the simulation is to validate our simulation methods such as CFD code, turbulence model, grid, etc., through comparison with RJTF experimental results. The second purpose is to confirm the performance of the inlet/strut compression system and resultant isolator/combustor internal flows, such as three-dimensionally distorted flows, shock-boundary-layer interaction, boundary-layer separation, and pressure and temperature fields. The third is to infer how such internal flowfields contribute to fuel/air mixing and ignition.

A multilock Navier–Stokes solver developed and parallelized for the Numerical Space Engine (NSE) was used. The solution algorithm is based on the cell-centered Chakravarthy–Osher total variation diminishing (TVD). The spatial accuracy was of third order, and a steady-state solution was obtained using three-step Runge–Kutta integration with local time steps. A 360 × 120 × 70 structured grid was used. The Baldwin–Lomax's turbulence model was employed for all of the walls and the wake of the strut.

Figure 29a shows the engine cut in half at the symmetric plane, the cross-sectional density contours at various x positions, pressure distributions on the strut surfaces, and the streamlines that start near the leading edges of the top and the sidewall. The three-dimensional inlet and the strut with a 45-deg swept-back angle cause boundary-layer separation on the top wall. They also cause cross-flows toward the cowl, especially in the boundary layers on the sidewall and the strut surfaces, and in the recirculation region behind the backward-facing steps and the strut base. Such crossflows and the resulting strong cowl shock produce locally high-pressure regions and strong heat load near the cowl leading edge. They in turn introduce swirl flows into

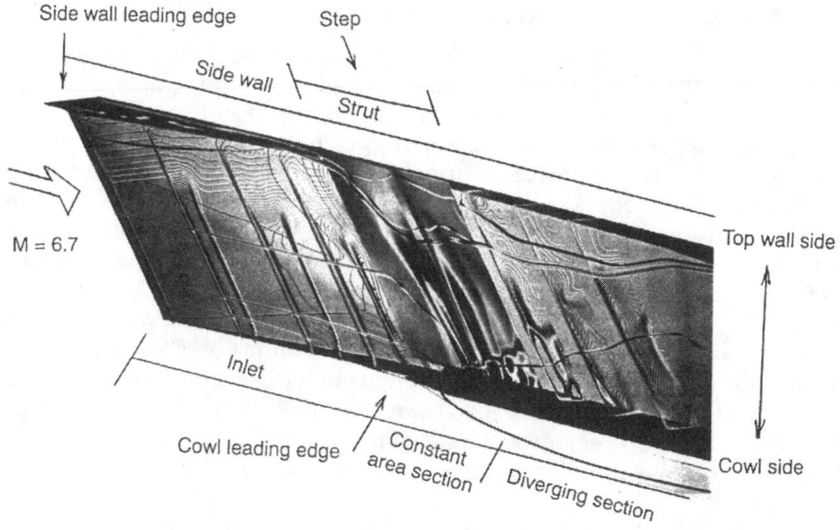

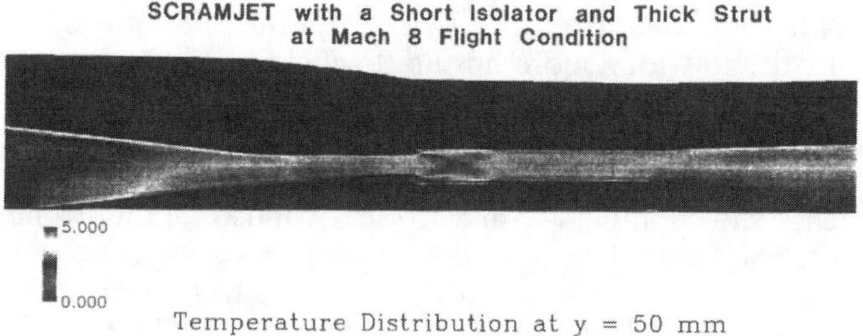

Fig. 29 CFD results for the M8 condition.
a) Density, pressure distribution, and streamlines
b) Temperature distributions near the step

the combustor through their interactions with the inlet shock waves, which are repeatedly reflected between the sidewall and the strut. Such three-dimensional flows make it difficult to apply the results of component tests to the whole engine directly.

Figure 29b shows the temperature distributions near the top wall and around the combustor with a strut and backward-facing steps. Low-speed boundary-layer flow on the top wall (thus the high temperature flow) is entrained into the recirculation region behind the strut. The high-temperature region travels along the strut base toward the cowl and becomes wider on the top wall side. Behind the backward-facing steps high temperature appears in a narrow region. Knowledge of these flow features is necessary for better understanding of the inlet/strut performance, mixing/ignition ability of the combustor, and for the development of new engines.

Numerical Simulation for Reactive Flows

The most prominent features of the scramjet nozzle are the effects of reactions in the high-speed, low-pressure, and nonuniform flow from the combustor. These factors affect lift force as well as axial force delivered by single-expanded ramp nozzles (SERN), in which the underbelly of the space plane is utilized as the expansion surface of the engine. Therefore, the mixing and combustion processes in the external nozzles were studied using a three-dimensional, reactive Navier–Stokes code incorporated with the k-ω turbulence model.[108] The results successfully simulated the unique competition between chemical reactions and mixing in supersonic nozzle flows. This CFD study also made it possible to evaluate the thrust recovery resulting from the combustion of residual fuel from the combustor. For instance, the data of local equivalence ratio and combustion efficiency, derived with gas sampling at the internal nozzle exit, were specified as the upstream boundary conditions for this external nozzle computation. The CFD work will be extended to evaluate performance gain with staged fuel injectors in scramjet combustors.[43]

Simulations of reactive flows in nozzles are relatively easy because the interference between the heat release by combustion and the flowfield is weak in the accelerated flows. Therefore, the reactive nozzle flow is a useful benchmark to calibrate the reactive CFD. After the calibrations the reactive code will be tested with a more hostile flowfield in combustors, where the strong adverse-pressure gradient by supersonic combustion may cause flow separation and the chemical source terms pose the stiffness problem in the reactive flow simulation. The distributions of equivalence ratio and combustion efficiency shown in Figs. 12 and 13 clearly illustrate the strong interaction between the heat release and the flowfield. Recently, the unstructured code introduced in Sec. VI.A was revised to cope with reactive flows. Calibration studies by tracing helium flow in the 1/5-subscale model described in Sec. VI.C are now in progress to assess the supersonic mixing process in the reactive code. With these stepwise validations the reactive CFD codes will be applied to simulate the supersonic combustion inside engines.

B. Chemical Quenching in Gas-Sampling Probes

The combustion performance of the scramjet engine was measured by rake probing. In rake probing gas analysis yields total temperature, and pitot pressure and wall pressure at the engine exit give the local Mach number of the exhaust gas flow. However, the probing becomes difficult in the hypersonic regime because the probe suffers the high heating rate and reaction quenching in the sampled gas, which is questionable.

Governing equations for the gas temperature and the radical concentration in the gas-sampling probes were constructed to evaluate the criterion for reaction quenching in probes. A phase-plane analysis of sampled gas temperature and the radical concentration led to autoignition or quenching of reaction systems depending on whether the temperature of sampled gas was high or low. The criterion, discriminating the quenching and autoignition, could be expressed as the separatrix in the phase plane.[73] The study revealed that reactions in probes could be extinguished if the cooling Damkohler number, defined as the ratio between flow residence time and chemical reaction time, was less than about 10. The analytical studies also revealed the strong dependency of the Damkohler number on the tip diameters of sampling probes.

The analytical results derived assuming a reduced kinetics model were examined by a one-dimensional reaction code with full kinetics including convective heat transfer. Calculations were initiated by specifying the Mach number, static pressure and temperature, gas compositions, and the Nusselt number in probes. The numerical simulations duplicated the characteristic behaviors of autoignition and quenching in sampling probes. The study also revealed that supersonic sampling through the narrow orifice was preferable for quenching of reactions.

Reducing the tip size decreases the Reynolds number, and the viscous effect becomes dominant. It promotes friction choking in the tip and causes a strong shock wave ahead of the tip. Therefore, the flowfield with a Reynolds number of 100 was investigated using a viscous code to study the flow structure around probes. The calculation indicated that sampled gas was inevitably heated by the normal shock in the central part of any sampling probe. However, the flow was reaccelerated to supersonic speed behind the shock if an expanding section followed downstream of the shorter straight section of the tip. This acceleration is favorable for quenching the reaction because the static temperature and the static pressure decrease.[80]

Four kinds of gas-sampling probes, i.e., the freezing-oriented and reaction-oriented probes shown in Fig. 30, were designed based on the preceding studies. The water-cooled, freezing-oriented probes (Figs. 30a and 30b) have a 30-deg half-angle conical tip with a passage on the centerline for pitot-pressure measurement and gas-sampling withdrawal. The conventional probe (Fig. 30a) has a tip diameter of 0.7 mm with a 6-mm straight section following it. The new probe (Fig. 30b) has a minimum passage diameter of 0.3 mm and a 0.6 mm straight section. The static-pressure-type probe (Fig. 30c) has a tip diameter of 1 mm to grasp the external flow across the boundary layer on the probe surface. The effectiveness of the freezing-oriented probes can be examined by comparison with the reaction-oriented probe (Fig. 30d). This probe is pitot-type to promote reactions ahead of and inside the probe. The probe is made of nickel to promote the catalytic effect and is not water-cooled.

A supersonic combustor operating with an airflow at Mach 2.5 and total temperature T_0 up to 2250° K was employed to examine the reaction quenching in probes.[85] The combustor was used as a supersonic flowreactor to calibrate the sampling technique. Variations in gas composition and pitot pressure were measured with multipoint rakes at the exit of the combustion duct. Figure 31 summarizes the T_0 dependence of combustion efficiency η_c measured at the central core region. The reaction-oriented probes show a clear-cut behavior of $\eta_c = 0$ or 1.0 between $T_0 = 875$ and 910 K. That is, all of the H_2 remains unburned for $T_0 < 875°$ K, and the H_2 is consumed

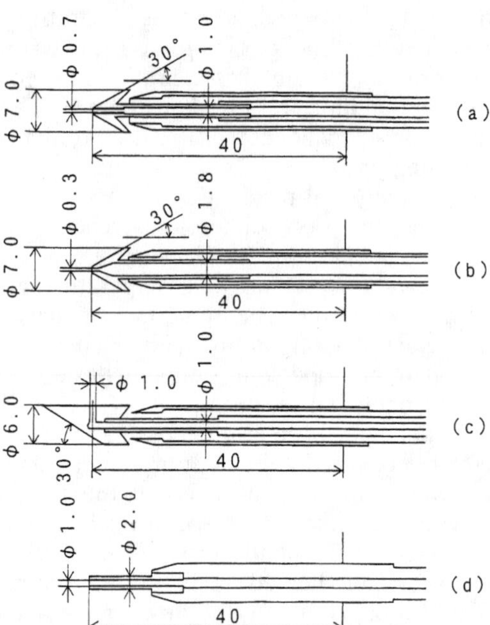

Fig. 30 Gas-sampling probes.
a) Conventional, freezing-oriented probe
b) New, freezing-oriented probe
c) Static-pressure-type probe
d) Reaction-oriented probe

completely if $T_0 > 910$ K. This is autoignition of the H_2-air premixture found by the phase-plane analysis and the full-kinetic code.

Because heating across the shock wave is absent in the static-pressure-type probe shown in Fig. 30c, the results obtained by the static probes must represent the standard of frozen gas compositions. Figure 31 shows that the data obtained with the new probes (shown as 0.3d-pitot in the figure) agree with those by the static probes. The static-type probes and the new probes indicate that combustion was initiated when $T_0 > 1250$ K, at which condition a faint luminous flame was observed at the duct exit. The measurement of pitot pressure also detected this weak, localized combustion of H_2 at T_0 =

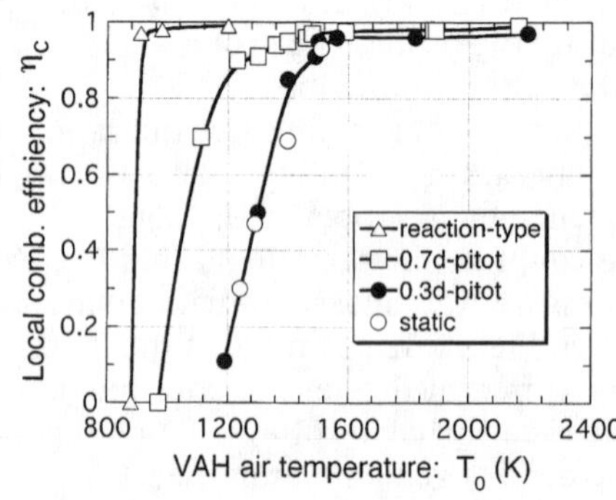

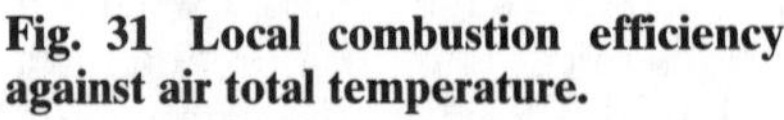
Fig. 31 Local combustion efficiency against air total temperature.

1250 K and supported sufficient quenching of reactions in the new probes. These studies recommended the use of probes with a tip finer than 0.3 mm in diameter.

C. Subscale Wind-Tunnel Testing

Because scramjet engine testing in the RJTF requires extensive manpower and the frequency of experiments is limited, supplementary aerodynamic testing was conducted using a subscale wind tunnel with a nozzle exit of 102 mm square. This wind tunnel is a freejet type at a scale exactly 1/5 of the RJTF. The air can be heated by a storage air heater up to 1000 K, and the maximum nozzle pressure is 10 MPa. Typical unit Reynolds numbers are 5×10^7/m, and those of RJTF can be duplicated by controlling total temperature and pressure.

Three 1/5-scale models (flow-path dimensions of 40 mm wide, 50 mm high, and 420 mm long) of the scramjet engine were constructed. The first model was for measurement of the air-capture ratio in engine inlets. A 2-m long subsonic diffuser section was installed downstream of the inlet. The air mass flux was metered using the pressure choked at a gate valve after establishing the supersonic flow throughout the model. The flow rate choked at the gate valve was separately calibrated. The air-capture ratios of the engine were found to be 0.72 at the M4 and 0.80 at the M6 and M8 conditions.[86] The second model was for wall-pressure measurement, in which the internal flow geometry was reproduced and instrumented with wall-pressure taps at 122 locations. The third one is a force model duplicating the external geometry as well as the internal geometry. The model was designed to divide the support strut, the two sidewalls, the cowl, and the engine strut to permit itemization of the individual drags by a strain-gauge load cell.

Pressure and Frictional Drag in Engines

The wall-pressure distributions on the inner walls of the second model were obtained by measurements at 26 locations on the top wall, 90 on the sidewall, and 6 on the cowl. The pressure drag was obtained by summation of the streamwise component of the force on the area elements. To evaluate the friction on the model, the skin friction coefficient c_f must be known. The c_f can be expressed by the Reynolds number and the Mach number, and the dynamic pressure can be calculated from the local Mach number and p_w if boundary-layer approximation is applied. Because the compression ratio in the models is low, the whole internal flow was approximated to be isentropic. Thus, the local Mach number was evaluated by the isentropic relation between the p_0 and p_w. The frictional drag could be derived by summation of the tangential component of the stress force on wetted surface elements.

The engine drag coefficients in various engine configurations can be obtained by integration of the pressure drag and the frictional drag calculated from the p_w distributions. Figure 32a shows the coefficients of pressure drag C_{dp} and friction drag C_{df} of the internal flow and the sum with the external drag in the M4 flight condition. The total internal drag of 0.121 can

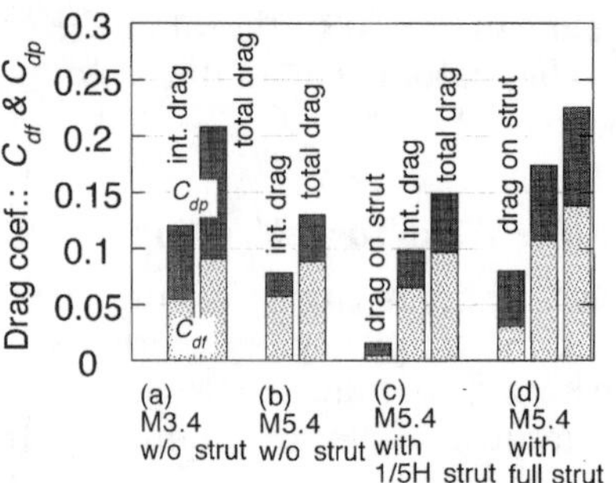

Fig. 32 Coefficients of internal pressure and friction drags and external drag.

be divided into the C_{dp} (0.066) and C_{df} (0.055). The sum with an external drag results in a total engine drag of 0.208. Figure 32b indicates the drags with the same model under the M6 condition. Although the frictional drag coefficient does not change, the pressure drag decreases to 1/3 because the compression ratio decreases at the higher Mach number. The decrease in the internal compression lowers the internal drag to 0.078, and consequently the total drag becomes 0.130.

The attachment of the 1/5-height strut produces a drag of 0.016, which increases the total internal drag to 0.098 in Fig. 32c. The difference between the internal drag and the strut drag denotes the drag produced in the internal flow besides that due to the strut. The larger full-height strut yields a strut drag of 0.080, and the total internal drag increases to 0.174 as shown in Fig. 32d. Because there is no effect on the external flow of engine by installing struts, the total engine drag increases to $C_d = 0.123$ with the 1/5-height strut and $C_d = 0.226$ with the full-height strut. Thus, the wall-pressure measurements provide useful data to evaluate the pressure and frictional drag and to optimize the engine configurations.

Force Measurement and Internal Thrust in Engine Testing

The engine installed on the supporting mount in free-jet-type wind tunnels is subjected to an excessively large external drag. The drag produced by the engine supporting mounts could be separated by comparing the drags with and without the windshield covering the mount, which was disconnected from the balance. The internal and external drags of the engine components were estimated from the increments of drags. For instance, Fig. 32a showed that the sum of the internal and external drag coefficients was 0.208 under the M4 condition. The total drag coefficient including the installation drag of 0.296 was measured by the force balance, which means that the drag coefficient due to the top wall, the supporting mount, and the fairing was 0.088 as the installation drag. The force measurements indicated that the drag indicated by the load cells was able to be divided into approximately three equal parts. They were 1/3 from the supporting mounts with the fairing, 1/3 from the external walls, and 1/3 from the internal walls for the engine without any strut.

Scramjet thrust performance has generally been based on parameter ΔF, which is defined as the thrust increment in the axial force with and without fuel injection. Engine performance is determined by the impulse functions of the inflow stream tube captured by engines. In the M4 engine testing, for instance, only the internal drag $D_{int} = -490$ N out of the engine drag of 1480 N should be included when discussing the engine internal performance because D_{int} should remain unchanged when fuel is injected into the engine. Thus, overall thrust performance is the change in internal thrust T_{int}, which is related to ΔF by $T_{int} = \Delta F + D_{int}$.

The FMS mounting the H_2-fueled scramjet engine in the RJTF indicated outputs with an axial component of −1490 N without fuel injection and −200 N with a fuel injection of 52 g/s under M4 conditions. This implies ΔF = 1290 N. Adding $D_{int} = +\ 490$ N to the ΔF, the internal thrust was found to be $T_{int} = 800$ N. This internal thrust performance corresponds to a specific impulse of about 16 km/s based on the H_2 fuel rate in the M4 condition.

Similar calculations have been carried out for various engine configurations under the M6 and M8 engine test conditions. The installment of the 1/5-height strut improved combustion under the M6 condition. The drag penalty of increased internal drag by the 1/5-height strut was recovered by the increased ΔF of the intensive combustion. The thick full-height strut promoted flameholding, resulting in delivery of an increased thrust of 430 N in the M8 engine testing. However, attachment of the strut also increased the internal drag by 320 N. The increased thrust was almost cancelled out with that in D_{int}. Studies on the tradeoff between D_{int} and ΔF and optimization of geometry of the strut are required.

Loads in Wind-Tunnel Starting and Engine Unstarting

During the initial transient period of the wind tunnel, fluctuating load is frequently imposed on the FMS. It is defined as starting load and is several times larger than steady load. Several methods, such as clamping the FMS until wind-tunnel start, injecting a model into the flow after start, etc., are used to avoid the starting load. It becomes a more serious problem in engine wind tunnels. For example, total pressure of the RJTF in the case of M8 reaches 10 MPa, and model size becomes large. The FMS in the RJTF was, therefore, designed with an onboard clamping system to lock the floating frame during the wind-tunnel starting process.

In addition to wind-tunnel starting, the unstart condition of engines also causes large amplitude load oscillations. In the case of engine unstart, it is impossible to protect the FMS from overloading quickly enough with the methods just mentioned. Therefore, the engine unstart load characteristics and detection of incipient engine unstart should be investigated for effective operation of the facility.[72]

At first, power spectral density functions and probability of amplitudes of pressure drag and loads indicated by the FMS were calculated with the fast Fourier transform (FFT) in the subscale wind tunnel. The expectations of maximum amplitudes in the input and the output were evaluated after confirming the Gaussian process. Then static and dynamic transmission coefficients of the FMS clamp system were measured by hammering tests, and the wind-tunnel starting loads were measured by the clamped FMS. The clamped tests revealed that the magnitude of the wind-tunnel starting load and the engine unstart load were related to the drag coefficient of the models. The powerful two-stage ejector for the RJTF reduced the back pressure for the diffuser in the test cell and lowered the wind-tunnel starting pressure. Consequently, the expected maximum peak load was found to be quite smaller than the designed values allowed for the FMS and the load cells. With these procedures the RJTF is now confidently operated with the FMS under clamp-free conditions to measure the loads of wind-tunnel starting and engine unstarting, and this enables monitoring of the zero shift of load cells to improve the accuracy of force measurement.

Flush-mounted sensors revealed that engine unstart was accompanied by the formation of separation bubbles that traveled up to the inlet region. Before strong engine unstart began, pressure spikes appeared around the inlet area intermittently, and the frequency of these spikes increased with the strength of engine unstart. Consequently, severe engine unstart can be detected by monitoring instantaneous pressure and can be prevented by reducing the fuel supply rate before the occurrence of catastrophic engine conditions.

D. Reaction Kinetic Studies on the Scramjet

Air can be heated by a ceramic storage heater or by a VAH burning fuel with oxygen. However, each of the heating methods has limitations and disadvantages in the production of hot air. For instance, it is difficult for storage heaters using refractory materials to heat air higher than 2000 K. In addition to this limitation, dust produced by the refractory materials and insulator present a serious problem in engine testing. VAHs are very convenient devices for the supply of hot air. Their disadvantage is the contamination by H_2O, NO_x, and radicals in the air supplied to engines. In this section kinetic studies on ignition of H_2 are outlined to elucidate the effects of radicals in engine testing using vitiation heaters. Next, an analysis of thermal and chemical effects of dust particles is briefly presented to assess how clean the air supplied from the cored-brick storage heater should be.[35]

Ignition in Vitiation-Heated Air

The higher ignition performance found in the V-mode engine tests suggests promotion of ignition by the residual radicals from the combustion in the VAH. The shortened ignition delay time has been investigated by using chemical kinetic codes. On the other hand, a linear analysis of the ignition stage of the H_2-O_2 system clarified that the addition of an O atom was 1.5 times more effective than the addition of H for ignition. This result can be applied to plasma-jet igniters often used in scramjet engines.[32]

Using the reduced kinetic model, let us consider an ignition problem consisting of a chain-initiation reaction, $H_2 + O_2 \rightarrow 2OH$ (R_6) and a branching reaction $H + O_2 \rightarrow OH + O$ (R_1). Although the R_6 reaction is eventually overcome by the R_1, initial chain carriers must first be produced by the R_6. The reaction rate of R_6 is slow compared with that of R_1, especially at lower temperatures. The addition of radicals provides a shortcut to accelerate the R_1 reaction.

The calculated shortened ignition times are plotted against the initial radical concentration with various temperatures in Fig. 33, where the ignition delay with an initial condition of $X_0 = 0$ is expressed as t_0 and the ignition time with radical addition ($X_0 \neq 0$) is defined as t_b. For instance, the ignition time at 1500K decreases by ½ with a radical mole fraction of 10^{-4}. The premature ignition effect by radicals can be expressed quantitatively as an equivalent temperature rise in test media because the ignition delay is governed by the R_1 reaction at an activation temperature of 8400 K. Namely, testing with vitiated air at $T_0 = 1500$ K is equivalent to the S mode experiment at $T_0 = 1820$ K if the vitiated air contains an initial radical concentration of 10^{-4} (100 ppm) from the VAH, which is the reason why the engine always autoignited without any ignition source in the V-mode experiments as discussed in Sec. V.C.

Ignition in Storage-Heated Air

The heat-sink effect of dust in ignition was investigated. Effects of heat capacity and size of particles could be expressed in terms of an exponential integral function. The results indicate that larger particles have no thermal inhibition effect and that particles work as an inert gas and delay the ignition at another limit of fine dust particles. This result implies that the maximum delay is determined by the heat capacity and that the heat sink effect appears if the mass fraction of dust becomes greater than 10^{-1}. Therefore, effects of dust in engine testing can be neglected because the dust concentration produced by storage air heaters is much lower than this criterion.

Because dust particles in supplied air may work as powdered flame suppressant in engine combustion, the chemical effect of fine particles was investigated next. Radical quenching on the surface of particles has two opposing roles in the ignition process: the radical termination lowers the concentration of radicals, and this releases heat. Governing equations for radicals and temperature were constructed considering the concurrent effect of surface quenching and gaseous quenching with various reaction

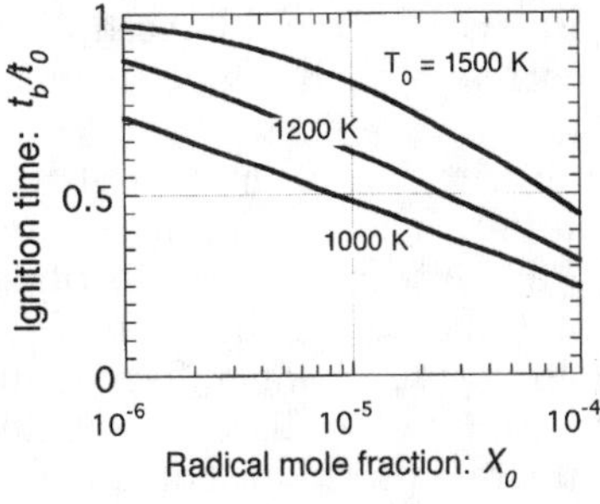

Fig. 33 Premature ignition caused by the addition of radicals.

orders. The termination rate at the surface was controlled by the diffusion rate of radicals at the fast surface reaction limit. The analytical study revealed that the radical termination on the particle surface caused an effect equivalent to heat loss and yielded a criterion for the thermal explosion. The chemical inhibition effect of dust may appear if the mass fraction of dust exceeds 10^{-3} and the dust consists of chemically active particles with a diameter less than about 1 μm. This result implies that the thermal and chemical effects of dust can be neglected in our engine testing because a typical mass fraction in the hot air supplied in the RJTF is estimated to be less than 10^{-5} (see Sec. III.C)

VII. Conclusions and Future Prospects

We have been attempting to test various methods in combustion, thrust generation, and inlet/combustor interaction by attaching a strut, injecting fuel from it, and extending the isolator length. Through these efforts many valuable insights into the phenomena occurring in the model, such as intensive and weak combustion modes and the effects of vitiation heating of the air, have been obtained. Effects of the engine configurations, such as the strut configurations and isolator lengths and those of the fuel injection patterns, were also examined. Furthermore, improvements have been made in test techniques and data acquisition and its reliability. All of them helped us in clarifying the phenomena within the engine and in improving the engine performance. They are, for example, an employment of gas sampling at the engine exit or back-up experiments employing 1/5-models and computational work. An LH_2-cooled model test has been conducted for the first time. Results of the past runs can be summarized as follows:

1) In all of the runs, combustion and thrust generation were observed.

2) There were two types of combustion modes, i.e., weak and intensive modes, depending on the thrust level generated or on the location of the major combustion region.

3) Considering the engine support and fairing drag, the generated thrust exceeded the engine drag at M4 and M6, but never at M8 conditions.

4) Increase in the length of the isolator improved the engine performance by relaxing the inlet/combustor interaction or by avoiding engine unstart.

5) Between M6S and M6V conditions, there was a large difference in the maximum fuel flow rate, which could be injected without causing engine unstart.

Despite the improvements in engine performance, we are still faced with various problems. Major problems with the current models can be summarized as follows:

1) There is a large nonuniformity in the distribution of the fuel at the engine exit. The effects of nonuniformity on the engine performance are most serious within the combustor. Therefore, the relation between distributions at the engine exit and in the combustor should first be examined. Besides this, there should already be a certain degree of nonuniformity in the air mass flux at the combustor entrance (inlet exit). Its level should depend on the flight conditions. We have to tolerate these kinds of nonuni-

formity to some extent. To this end the effects of nonuniformity in the air mass flux at the inlet exit on the engine performance should be clarified. If the effects are serious, some measures to alleviate or compensate for them, such as arrangement optimization of diameter and location and direction of the fuel injection orifices, are necessary.

2) For the M4 and M6 conditions the fuel could not be supplied beyond the equivalence ratio of unity without causing the engine unstart because of an excessive combustion-induced pressure rise in the constant-area section downstream of the step, which leads to engine unstart. The way to avoid this difficulty is to suppress the pressure rise there, which could be made by controlling and distributing the heat release or cross-sectional area in the axial direction in accordance with flight conditions. The former method has already been examined in component tests, which showed fairly good results.

3) To date, we have not attained intensive combustion for the M8 condition, even with an equivalence ratio above unity. This shows that simply inserting the strut with fuel injection or thickening it is not sufficient. Further improvement taking the interaction between combustion and flowfield into consideration is necessary.

Acknowledgments

The authors would like to express their gratitude to Goro Masuya at Tohoku University and to all of the members at NAL-KRC who have been participating in the engine tests. We also would like to thank the Director of NAL-KRC, Hiroshi Miyajima, who has been patiently supporting the construction of the RJTF and the engine tests. The tests presented in this paper were conducted as cooperative research between NAL and Mitsubishi Heavy Industries Co. (MHI). We also would like to thank the staff of Kobe Steel, Ltd., and Ishikawajima Harima Heavy Industries Co. for their support conducting the engine tests.

References

[1]Yoshida, A., and Tsuji, H., "Supersonic Combustion of Hydrogen in Vitiated Airstream Using Transverse Injection," *AIAA Journal,* Vol. 15, No. 4, 1977, pp. 463–464.

[2]Tsuji, H., and Matsui, K., "A Diffusion Flame Formed in a Low Supersonic, High-Temperature, Vitiated Oxidizer Flow," *Combustion Science and Technology,* Vol. 16, 1977, pp. 1–10.

[3]Takeno, T., Uno, T., and Kotani, Y., "An Experimental Study on Shock-Induced Combustion of Hydrogen," *Acta Astronautica,* Vol. 6, No. 7–8, 1979, pp. 891–915.

[4]Kimura, I., Aoki, H., and Kato, M., "The Use of a Plasma Jet for Flame Stabilization and Promotion of Combustion in Supersonic Air Flows," *Combustion and Flame,* Vol. 42, 1981, pp. 297–305.

[5]Tomioka, S., Nagata, H., Segawa, D., Ujiie, Y., and Kono, M., "Effect of Combustion on Mixing Process in a Supersonic Combustor," *Proceedings of the 10th Inter-*

national Symposium on Air Breathing Engines, AIAA, New York, 1991, pp. 886–891 (AIAA 95–2447).

[6]Niioka, T., Terada, K., Kobayashi, H., and Hasegawa, S., "Flame Stabilization Characteristics of Strut Divided into Two Parts in Supersonic Airflow," *Journal of Propulsion and Power,* Vol. 11, No. 1, 1995, pp. 112–116.

[7]Takahashi, S., Sato, M., Tsue, M., and Kono, M., "Control of Flame-Holding in Supersonic Airflow by Secondary Air Injection," *Journal of Propulsion and Power,* Vol. 14, No. 1, 1998, pp. 18–23.

[8]Miyajima, H., Chinzei, N., Mitani, T., Wakamatsu, Y., and Maita, M., "Development Status of the NAL Ramjet Engine Test Facility and Sub-Scale Scramjet Engine," AIAA Paper 92-5094, 1992.

[9]Yatsuyanagi, N., and Chinzei, N., "Status of Scramjet Engine Research at NAL," *Proceedings of the 20th International Symposium on Space Technology and Science,* 1996, pp. 51–57.

[10]Masuya, G. et al., "A Study of Air Breathing Rockets (III)—Supersonic Mode Combustors," National Aerospace Lab., NAL TR-756, Miyagi, Japan, (1983) (in Japanese).

[11]Kanda, T., et al., "Mach 4 Testing of Scramjet Inlet Models," AIAA Paper 89-2680, July 1989; also *Journal of Propulsion and Power,* Vol. 7, No. 2, 1991, pp. 275–280.

[12]Tani, K., et al., "Flow Measurements in Scramjet Inlets," *Proceedings of the 17th International Symposium on Space Technology and Science,* 1990, pp. 831–836.

[13]Ishiguro, T., Ogawa, S., and Wada, W., "Numerical Calculation of Scramjet Inlet Flow," National Aerospace Lab. NAL TR-1174T, Miyagi, Japan, 1992.

[14]Tani, K., et al., "Aerodynamic Performance of Scramjet Inlet Models with a Single Strut," AIAA Paper 93-0741, 1993.

[15]Ito, T., et al., "Wind Tunnel Tests of the Model of Intake-Airframe Integration," *Proceedings of the 11th International Symposium on Air Breathing Engines,* 1993, pp. 1033–1041.

[16]Tani, K., Kanda, T., and Tokunaga, T., "Starting Characteristics of Scramjet Inlets," *Proceedings of the 11th International Symposium on Air Breathing Engines*, 1993, pp. 1071–1080.

[17]Tani, K., and Dolling, D. S., "Fluctuating Wall Pressures in a Mach 5 Crossing Shock Wave/Turbulent Boundary Layer Interaction Including Asymmetric Effects," AIAA Paper 96-0045, 1996.

[18]Kanda, T., et al., "Impulse Function and Drag in Scramjet Inlet Models," *Journal of Propulsion and Power,* Vol. 12, No. 6, 1996, pp. 1181–1183.

[19]Ito, T., Taneda, H., Hozumi, K., and Yoshizawa, A., "Hypersonic Wind Tunnel Test of Sidewall Compression Type Scramjet Inlet," *Proceedings of the 18th International Symposium on Space Technology and Science,* 1997, pp. 163–168.

[20]Kodera, M., Nakahashi, K., Obayashi, S., Kanda, T., and Mitani, T., "Effect of Inflow Boundary Layer Thickness on Scramjet Inlet Flow Field," *Journal of Japan Society for Aeronautical and Space Sciences,* Vol. 45, No. 519, 1997, pp. 216–221 (in Japanese); also AIAA Paper 98-0962, 1998.

[21]Komuro, T., Murakami, A., Kudo, K., Masuya G., and Chinzei, N., "An Experiment on a Cylindrical Scramjet Combustor (I) Simulated Flight Mach Number of 4.4," National Aerospace Lab., NAL TR-918, Miyagi, Japan, 1986 (in Japanese).

[22]Komuro, T., Murakami, A., Kudo, K., Masuya, G., and Chinzei, N., "Experiment on a Cylindrical Scramjet Combustor (II) Simulated Flight Mach Number 6.7," National Aerospace Lab., NAL TR-969, Miyagi, Japan, 1988 (in Japanese).

[23]Sato, Y., et al., "Effectiveness of Plasma Torches for Ignition and Flameholding in Scramjet," AIAA Paper 89-2564, 1989; also *Journal of Propulsion and Power,* Vol. 8, No. 4, 1992, pp. 883–889.

[24]Ishiguro, T., Ogawa, S., Wada, Y., "Numerical Computations of Supersonic Chemically Reacting Flows Using Hydrogen-Air Combustion Models," *International Symposium on Computational Fluid Dynamics, NAGOYA,* 1989.

[25]Sato, Y., et al., "Experimental Study on Autoignition in a Scramjet Combustor," *Journal of Propulsion and Power,* Vol. 7, No. 5, 1991, pp. 657–658.

[26]Kudo, K., Masuya, G., Komuro, T., Murakami, A., and Chinzei, N., "Autoignition and Flameholding in a Cylindrical Scramjet Combustor," National Aerospace Lab., NAL TR-1067, Miyagi, Japan, 1990 (in Japanese).

[27]Komuro, T., et al., "Experiment on Rectangular Cross Section Scramjet Combustor," National Aerospace Lab., NAL TR-1068, Miyagi, Japan, 1990 (in Japanese).

[28]Masuya, G., et al., "Some Governing Parameters of Plasma Torch/Flame-Holder in a Scramjet Combustor," AIAA Paper 90-2098, 1990; also *Journal of Propulsion and Power,* Vol. 9, No. 2, 1993, pp. 176–181.

[29]Kudo, K., et al., "An Experiment on Ignition in a Rectangular Cross Section Scramjet Combustor," National Aerospace Lab., NAL TR-1080, Miyagi, Japan, 1990 (in Japanese).

[30]Chinzei, N., et al., "Effects of Injector Geometry on Scramjet Combustor Performance," *Journal of Propulsion and Power,* Vol. 9, No. 1, 1993, pp. 146–152.

[31]Komuro, T., et al., "Combustion Test and Thermal Analysis of Fuel Injection Struts on a Scramjet Combustor," *Procceedings of the 10th International Symposium on Air Breathing Engines*, 1991, pp. 1228–1233.

[32]Mitani, T., "Chemical Reaction in Scramjet Engine," *Combustion Science and Technology (Japan edition),* Vol. 1, 1992, pp. 19–28 (in Japanese); also National Aerospace Lab., NAL TR-1184, Miyagi, Japan, 1992 (in Japanese).

[33]Masuya, G., et al., "Ignition and Combustion Performance of Scramjet Combustor with Fuel Injection Struts," *Journal of Propulsion and Power,* Vol. 11, No. 2, 1995, pp. 301–307.

[34]Murakami, A., Komuro, T., and Kudo, K., "Experiment on a Rectangular Cross Section Scramjet Combustor (II), Effects of Fuel Injector Geometry," National Aerospace Lab., NAL TR-1220, Miyagi, Japan, 1993 (in Japanese).

[35]Mitani, T., "Ignition Problems in Scramjet Testing," *Combustion and Flame,* Vol. 101, 1995, pp. 347–359.

[36]Tomioka, S., et al., "Auto-Ignition in a Supersonic Combustor with Perpendicular Injection behind Backward-Facing Step," AIAA Paper 97-2889, 1997.

[37]Masuya, G., Uemoto, T., and Wakana, Y., "Performance Evaluation of Scramjet Combustors," *Proceedings of the 13th International Symposium on Air Breathing Engines,* 1997, pp. 952–959.

[38]Boyce, R., et al., "Supersonic Combustion—A Shock Tunnel and Vitiation-heated Blowdown Tunnel Comparison," AIAA Paper 98-0941, 1998.

[39]Matsuura, K., et al., "An Experiment on Scramjet Isolators with Backward-Fac-

ing Step," International Symposium on Space Technology and Science (ISTS) 98-a-1-22, 1998.

[40]Kitadani, H., et al., "Effects of Injector Configuration on Flow-Field in Ignition Region of Scramjet Combustor," International Symposium on Space Technology and Science (ISTS) 98-a-1-24, 1998; also AIAA Paper 98-3127, 1998.

[41]Kobayashi, K., Niioka, T., and Mitani, T., "Asymptotic Analysis of Plasma Jet Igniter," International Symposium on Space Technology and Science (ISTS) 98-a-1-25, 1998.

[42]Sunami, T., Wendt, M. N., and Nishioka, M., "Supersonic Mixing and Combustion Control Using Streamwise Vortices," AIAA Paper 98-3271, 1998.

[43]Tomioka, S., Murakami, A., Kudo, K., and Mitani, T., "Combustion Tests of a Staged Combustion with a Strut," AIAA Paper 98-3273, 1998.

[44]Miyajima, H., et al., "Studies on Scramjet Nozzles (1) Performance of Two Dimensional Nozzles," National Aerospace Lab., NAL TR-1149, Miyagi, Japan, 1992 (in Japanese).

[45]Mitani, T., et al., "Evaluation of Scramjet Nozzle Performance," *Journal of Japan Society for Aeronautical and Space Sciences,* Vol. 40, No. 464, 1992, pp. 515–522 (in Japanese).

[46]Watanabe, S., "A Scramjet Nozzle Experiment with Hypersonic External Flow," AIAA Paper 92-3289, 1992; also *Journal of Propulsion and Power,* Vol. 10, No. 4, 1993, pp. 521–523.

[47]Mitani, T., et al., "Experimental Validation of Scramjet Nozzle Performance," AIAA Paper 92-3290, 1992; also *Journal of Propulsion and Power,* Vol. 9, No. 5, 1993, pp. 725–730.

[48]Hiraiwa, T., et al., "Off-Design Performance of Scramjet," *Proceedings of the 11th International Symposium on Air Breathing Engines,* 1992, pp. 1103–1114.

[49]Ishiguro, T., Mitani, T., Hiraiwa, T., and Takagi, R., "Three-Dimensional Analysis of a Scramjet Nozzle Flow," AIAA Paper 93-5059, 1993; also *Journal of Propulsion and Power,* Vol. 10, No. 4, 1994, pp. 540–545; also *Journal of Japan Society for Aeronautical and Space Sciences,* Vol. 42, No. 488, 1994, pp. 548–555 (in Japanese).

[50]Tomioka, S., Hiraiwa, T., Mitani, T., Matsumoto M., and Yamamoto M., "A Study on the Boundary Layer in a Scramjet Nozzle Operating Under High Enthalpy Conditions," AIAA Paper 94-2820, 1994.

[51]Hiraiwa, T., Mitani, T., Tomioka, S., Izumikawa, M., and Matsumoto, M., "Performance of a Scramjet Nozzle in Hypersonic Flight Condition," AIAA Paper 95-2579, 1995; also *Journal of Propulsion and Power,* Vol. 11, No. 3, 1995, pp. 403–408.

[52]Hirano, T., et al., "An Improvement in Design Accuracy of Functionally Gradient Material for Space Plane Applications," *Proceedings of the 17th International Symposium on Space Technology and Science,* 1990, pp. 501–506.

[53]Ueda, S., et al., "A Study on Heat Transfer in a Scramjet Leading Edge Model," National Aerospace Lda, NAL TR-1187T, Miyagi, Japan, 1992; also *Proceedings of the 18th International Symposium on Space Technology and Science,* 1993, pp. 823–830.

[54]Sohda, Y., et al., "Carbon/Carbon Composites Coated with SiC/C Functionally Gradient Materials," *Ceramic Transactions,* Vol. 34: *Functionally Gradient Materials,* 1992, pp. 125–132.

[55]Wakamatsu, Y., et al., "Development of a Hot Gas Flow Test Device and an

Evaluation of FGM Panel," *Ceramic Transactions*, Vol. 34 *Functionally Gradient Materials*, 1992, pp. 263–270.

[56]Matsuzaki, Y., et al., "Analysis-Assisted Fabrication of TiAl-based Thermal Barrier FGM and Its Performance in a Supersonic Combustion Gas Flow," *Ceramic Transactions*, Vol. 34 *Functionally Gradient Materials*, 1992, pp. 323–330.

[57]Wakamatsu, Y., Saito, T., Ueda, S., and Niino, M., "Development of Thermal Shock Evaluation Device of Functionally Gradient Materials for Aerospace Applications," *Thermal Shock and Thermal Fatigue Behavior of Advanced Ceramic*, 1993, pp. 555–565.

[58]Kumakawa, A., Niino, M., "Thermal Fatigue Characteristics of Functionally Gradient Materials for Aerospace Applications," *Thermal Shock and Thermal Fatigue Behavior of Advanced Ceramic*, 1993, pp. 567–577.

[59]Yamaoka, Y., et al., "Thermal Structural Analyses of SiC/TiC Type Functionally Gradient Material," *Advanced Composites '93*, 1993.

[60]Matsuzaki, Y., Fujioka, J., Ueda, S., and Wakamatsu, Y., "Thermal Barrier Design of γ-TiAl Functionally Gradient Materials (FGMs) for Scramjet," *Journal of Japan Society for Aeronautical and Space Sciences,* Vol. 41, No. 473, 1993, pp. 359–367 (in Japanese).

[61]Saito, T., et al., "Application of Functionally Gradient Materials to Scramjet Engines," *Proceedings of the 11th International Symposium on Air Breathing Engine,* 1993, pp. 662–672.

[62]Saito, T., et al., "Testing of Regeneratively Cooled Light-Weight Panel," AIAA Paper 95-2718, 1995.

[63]Kanda, T., Ono, F., Takahashi, M., Saitoh, T., and Wakamatsu, Y., "Experimental Studies of Supersonic Film Cooling with Shock Wave Interaction," AIAA Paper 95-3141, 1995; also *AIAA Journal,* Vol. 34, No. 2, 1996, pp. 265–271.

[64]Wakamatsu, Y., et al., "Evaluation Test of C/C Composite Coated with SiC/CFGM, under Simulated Condition for Aerospace Application," *Proceedings of 4th International Symposium on Functionally Graded Materials,* 1996, pp. 463–468.

[65]Masuya, G., and Chinzei, N., "Experiment of Scramjet Combustors and their Application to a Sub-Scale Engine," *Proceedings of the 20th International Symposium on Space Technology and Science,* 1994, pp. 9–16.

[66]Yatsuyanagi, N., Chinzei, N., and Miki, Y., "Initial Tests of a Sub-Scale Scramjet Engine," *Proceedings of the 12th International Symposium on Air Breathing Engines,* 1995, pp. 1330–1337.

[67]Masuya, G., and Chinzei, N., "Scramjet Engine Tests at Mach 4 and 6," *Proceedings of the IUTAM Symposium on Combustion in Supersonic Flows,* 1995, pp. 147–162.

[68]Kanda, T., et al., "Mach 6 Testing of a Scramjet Engine Model," AIAA Paper 96-0380, 1996; also *Journal of Propulsion and Power,* Vol. 13, No. 4, 1997, pp. 543–551.

[69]Saito, T., et al., "Mach 8 Testing of a Scramjet Engine Model," *Proceedings of the 20th International Symposium on Space Technology and Science,* 1996, pp. 58–63.

[70]Tomioka, S., Izumikawa, M., Hiraiwa, T., and Sato, S., "Enhancement of Ignition Ability in a Model Scramjet Engine," AIAA Paper 96-3240, 1996.

[71]Chinzei, N., et al., "On the Isolator for the Scramjet Engines," *Proceedings of the 20th International Symposium on Space Technology and Science*, 1996, pp. 83–93.

[72]Shimura, T., Sakuranaka, N., Izumikawa, M., and Mitani, T., "Load Oscillations

Due to Unstart of Engines and Hypersonic Wind Tunnels," AIAA Paper 96-3242, 1996; also *Journal of Propulsion and Power,* Vol. 14, No. 3, 1998, pp. 348–353, or International Symposium on Space Technology and Science (ISTS) 98-a-1-28, 1998.

[73]Mitani, T., "Quenching of Reaction in Gas Sampling Probes to Measure Scramjet Engine Performance," *Proceedings of the 26th Symposium (International) on Combustion,* 1996, pp. 2917–2924.

[74]Mitani, T., Hiraiwa, T., Sato, S., Tomioka, S., Kanda, T., and Tani, K., "Scramjet Engine Testing in Mach 6 Vitiated Air," AIAA Paper 96-4555, 1996; also *Journal of Propulsion and Power,* Vol. 13, No. 5, 1997, pp. 635–642.

[75]Hiraiwa, T., et al., "Testing of a Scramjet Engine Model in Mach 6 Vitiation Air Flow," AIAA Paper 97-0292, 1997.

[76]Tani, K., Kanda, T., Sunami, T., Hiraiwa, T., and Tomioka, S., "Geometrical Effects to Aerodynamic Performance of Scramjet Engine," AIAA Paper 97-3018, 1997.

[77]Sato, S., Izumikawa, M., Tomioka, S., and Mitani, T., "Scramjet Engine Test at the Mach 6 Flight Condition," AIAA Paper 97-3021, 1997.

[78]Sunami, T., Sakuranaka, N., Tani, K., Hiraiwa, T., and Shimura, T., "Mach 4 Tests of a Scramjet Engine—Effect of Isolator," *Proceedings of 13th International Symposium on Air Breathing Engines,* 1997, pp. 615–625.

[79]Igarashi, Y., Kodera, M., Nakahashi, K., Mitani, T., and Shimura, T., "Experimental and Numerical Analysis of Scramjet Internal Flows—Comparative Studies on Engine Drag," AIAA Paper 98-1512, 1998; also Mitani, T., Kanda, T., Hiraiwa, T., Igarashi, Y. and Nakahashi, K., "Drags in Scramjet Engine Testing: Experimental and Computational Fluid Dynamics Studies," *Journal of Propulsion and Power*, Vol. 15, No. 4, 1999, pp. 578–583.

[80]Mitani, T., Takahashi, M., Tomioka, S., Hiraiwa, T., and Tani, K., "Measurements of Scramjet Combustion Properties by Gas Sampling," AIAA Paper 98-1590, 1998; also International Symposium on Space Technology and Science (ISTS) 98-a-1-34, 1998.

[81]Chinzei, N., Mitani, T., Wakamatsu, Y., Shimura, T., Sato, S., and Yatsuyanagi, N., "Sub-Scale Scramjet Engine Tests at NAL-KRC," International Symposium on Space Technology and Science (ISTS) 98-a-1-26V, 1998.

[82]Wakamatsu, Y., et al., "Design and Preliminary Experiments of Liquid Hydrogen Cooled Scramjet Engine," International Symposium on Space Technology and Science (ISTS) 98-a-1-27, 1998.

[83]Kanda, T., "Study of the Intensive Combustion in the Scramjet Engine," AIAA Paper 98-3123, 1998.

[84]Tomioka, S., et al., "Testing of a Scramjet Engine with a Strut at Mach 8 Flight Condition," AIAA Paper 98-3134, 1998.

[85]Mitani, T., Chinzei, N., and Masuya, G., "Experiments on Reaction Quenching in Gas Sampling Probes for Scramjet Engine Testing," *27th Symposium on Combustion,* pp. 2151–2156, 1998.

[86]Kanda, T., et al., "Mach 8 Testing of a Scramjet Engine Model," AIAA Paper 99-0617, 1999).

[87]Sato, S., Kumagai, T., Izumikawa, M., Sakuranaka, N., and Mitani, T., "A Preliminary Study of a Supersonic Wind Tunnel for a RamJet Test Facility," *Proceedings of the 17th International Symposium on Space Technology and Science,* 1990, pp. 825–830.

[88]Miyajima, H., Arai, T., and Hanus, G. J., "Design Features of the NAL Ramjet Engine Test Facility," *Proceedings of the 18th International Symposium on Space Technology and Science,* 1992, pp. 155–161.

[89]Kurosaka, T., Yamamura, T., Iwagami, S., Grunnet, J., Hayakawa, K., and Miyajima, H., "A Model Study on Diffuser Pressure Recovery in NAL Scramjet Test Facility with Simulated Hydrogen Combustion," AIAA Paper 92-3979, 1992.

[90]Mitani, T., "Effects of Dust from Storage Heaters on Ignition in Scramjets," National Aerospace Lab., NAL TR-1234, Miyagi, Japan, 1994 (in Japanese).

[91]Mitani, T., Wakamatsu, Y., Yatsuyanagi, N., Iwagami, S., Endo, M., and Hickel, S., "Acceptance Tests of Ramjet Engine Test Facility," International Symposium on Space Technology and Science (ISTS) 94-a-04, 1994.

[92]Sakuranaka, N., et al., "Starting Loads in Freejet-type Wind Tunnel," *Journal of Japan Society for Aeronautical and Space Sciences,* Vol. 44, No. 508, 1996, pp. 299–305 (in Japanese); also International Symposium on Space Technology and Science (ISTS) 96-a-2-18, 1996.

[93]Hiraiwa, T., Mitani, T., Izumikawa, M., and Ono, H., "Calibration of Nozzle Flow in Ramjet Engine Test Facility," International Symposium on Space Technology and Science (ISTS) 96-d-14, 1996.

[94]Yatsuyanagi, N., et al., "Ramjet Engine Test Facility," National Aerospace Lab. NAL TR-1347, Miyagi, Japan, 1998 (in Japanese).

[95]Yatsuyanagi, N., et al., "Ramjet Engine Test Facility (RJTF) in NAL-KRC, JAPAN," AIAA Paper 98-1511, 1998.

[96]Itoh, K., "Tuned Operation of Free Piston Shock Tunnel," *Proceedings of the 20th International Symposium on Shock Waves,* Vol. I, 1995, pp. 43–51.

[97]Itoh, K., et al., "Design and Construction of HIEST (High Enthalpy Shock Tunnel)," *Proceedings of International Conference on Fluid Engineering,* ICFE'97, Vol. I, 1997, pp. 353–358.

[98]Itoh, K., et al., "Improvement of a Free Piston Driver for a High-Enthalpy Shock Tunnel," *Shock Waves* (to be published).

[99]Bartz, D. R., "Simple Equation for Rapid Estimation of Rocket Nozzle Convective Heat Transfer Coefficients," *Jet Propulsion,* Vol. 27, 1957, pp. 47–51.

[100]Taylor, M. F., "Correlation of Local Heat-Transfer Coefficients for Single-Phase Turbulent Flow of Hydrogen in Tubes with Temperature Ratios to 23," NASA TN D-4332, 1968.

[101]Eggers, J. M., Reagon, P. G., and Gooderum, P. B., "Combustion of Hydrogen in a Two-Dimensional Duct with Step Fuel Injectors," NASA TP-1174, May 1978.

[102]Guy, R. W., Rogers, R. C., Puster, R. L., Rock, K. E. and Diskin, G. S., "The NASA Langley Scramjet Test Complex," AIAA Paper 96-3243 1996.

[103]Northam, G. B., "Report on Combustion in Supersonic Flow," *Proceedings of 21st JANNAF Comb. Meeting,* 1984, pp. 399–410.

[104]Eggers, J. M., "Composition Surveys of Test Gas Produced by a Hydrogen-Oxygen-Air Burner," NASA TM X-71964, 1974.

[105]Ohshima, T., et al., "Experimental Approach to the HYPR Mach 5 Ramjet Propulsion System," AIAA Paper 98-3277, 1998.

[106]Futamura, H., et al., "Freejet Test of Ramjet System for Hypersonic Transport Vehicle," *Proceedings of 13th International Symposium on Air Breathing Engines,* 1997, pp. 573–581.

[107]Goldberg, U. C., and Ramakrishnan, S. V., "A Pointwise Version of Baldwin–Barth Turbulence Model," *Computational Fluid Dynamics,* Vol. 1, 1993, pp. 321–338.

[108]Lee, S.-H., and Mitani, T., "Reactive Flow in Scramjet Nozzles," AIAA Paper 99-0616, 1999.

Chapter 5

Scramjet Research and Development in Russia

Vladimir A. Sabel'nikov* and Vyacheslav I. Penzin[†]
TsAGI, Zhukuovski 140160, Moscow Region, Russia

I. Introduction

THE stages of scramjet working processes, and investigations, developments, and tests of scramjet models, were carried out in parallel in many countries (Russia, USA, France, Germany, Japan) beginning from the moment of scramjet invention: 1957—Russia, E. S. Shchetinkov; 1958—USA, R. J. Weber and J. S. Mackay (1958), R. Dunlap, R. L. Brehm and J. A. Nichols (1958); A. Ferri 1958—France, P. M. Roy (1958).

The scramjet development history in Russia is closely connected with the name of the inventor of this engine, Prof. E. S. Shchetinkov (1907–1976); see his photography in Fig. 1. Scramjet appearance at the scene of airbreathing engines was a natural consequence of progressive development of ramjet engines, the research of which has a wealth of experience in Russia. The base of ramjet engine theory was developed in 1929 by B. S. Stechkin (1929). At the beginning of the 1930s Yu. A. Pobedonostsev tested ramjets installed on artillery projectiles (Yu. A. Pobedonostsev, 1970). The first development and flight tests of a subsonic ramjet were done by I. A. Merkulov in 1939 (I. A. Merkulov, 1965). The engine was attached to the flying aircraft bed and ignited in flight.

The joint work of S. P. Korolev and E. S. Shchetinkov on winged rockets made in GIRD (Group of Propulsion Investigation) in 1936 (M. V. Keldysh, 1980, pp. 395–398) can be considered as a start of activity for flying vehicles using a ramjet engine as a composite part of jet propulsion. The ramjet engine produced by E. S. Shchetinkov was tested in 1941 (B. V. Raushenbakh et al., 1980). For the ensemble of works on ramjet engines he was conferred the scientific degree of candidate of science in 1941. After the Second World War E. S. Shchetinkov in NII-1 (Scientific Re-

*Professor, Chief of Project
[†]Senior Engineer

Fig. 1 Prof. E.S. Shchetinkov (1907–1976), the inventor of scramjet in Russia, 1957.

search Institute No 1; now Keldysh Research Center of Thermal Processes, NII-TP, Moscow) developed the research base for creation of supersonic ramjets for vehicles of different applications. As the head of the Ramjet Theoretical Subdivision from 1946, he participated in all ramjet and scramjet basic problem investigations. His main interest was in engine thermodynamics and combustion. Under the leadership of E. S. Shchetinkov, in 1946–1950 samples of reliably working combustors with combustion efficiency about 90–95% were produced. In the 1950s E. S. Shchetinkov, unified some able young people and created the scientific school that, side by side with successful solutions of actual practical problems of high efficiency combustors, developed and refined their understanding of combustion process theory. Under the supervision of E. S. Shchetinkov more than 20 of his pupils became doctors and candidates of science. E. S. Shchetinkov contributed significantly to the theory of turbulent combustion. At the end of the 1940s E. S. Shchetinkov developed the original microvolume model of turbulent premixed combustion (E. S. Shchetinkov, 1958). He wrote the well-known Russian monograph on physics of gas combustion (E. S. Shchetinkov, 1965).

In the 1950s ramjets were developed successfully in Russia. Groups under the leadership of S. P. Korolev together with academic and industry research institutions carried out scientific and design investigations aimed at intercontinental rocket creation. It was shown that along with rocket propulsion systems it was possible to create a two-stage rocket with ramjet engines (M. V. Keldysh, 1980, p. 17).

In 1944 the work on ramjet development was concentrated at special Design Bureau "Krasnaya Zvezda" headed by Chief Designer S. M. Bondaryuk, the great talented designer–organizer and scientist. He is the author of the excellent book on ramjet theory (S. M. Bondaryuk and A. V. Il'yashenko, 1958). He founded the highly qualified team and scientific and designer's school that was able to develop several effective ramjet systems. Among them were the missile SA-4 GANEF using kerosene ramjet and the "Burya" Mach 3 class large cruise missile. The cruise missile was designed for an intercontinental range and had the total weight of 130,000 kg (see, e.g., V. A. Sosounov, 1993). S. M. Bondaryuk's team took the most active part in the development of the scramjet invented by E. S. Shchetinkov in 1957 (I. V. Bespalov and V. S. Makaron, 1982; V. I. Penzin, 1982). At that time E. S. Shchetinkov worked in NII-1. In his fundamental paper written in 1957 E. S. Shchetinkov calculated scramjet performances over a wide velocity range (up to flight Mach number $M_\infty = 20$), and analyzed scramjet principal advantages over ramjet. It was stated that scramjets can operate in off-designed mode more efficiently. The lower static temperature in a combustor means the less dissociation of the combustion products and the lower static pressure causes less heat transfer as well as lower structural loads and hence ensures less weight. E. S. Shchetinkov proposed to solve the scramjet cooling problem by taking into account fuel type and engine mode of operation. For hydrogen-fueled scramjets it was proposed to use hydrogen "lubrication," i.e., injection of a small quantity of hydrogen through a porous wall, which besides the cooling decreases friction. At very high flight Mach numbers when scramjet thrust could be insufficient for vehicle acceleration, it was proposed to inject into the combustor some excess of substance including fuel, oxygen, water and the like.

In the above-mentioned paper E. S. Shchetinkov also investigated different modes of heat release in scramjet combustor. The combustor had constant and diverging area sections. To avoid the excessive total pressure losses E. S. Shchetinkov proposed to control the heat release zone length and its location depending on flight conditions and scramjet combustion mode. E. S. Shchetinkov showed that inlet contraction ratio must be equal to 0.1 or less to avoid overall excessive total pressure losses in scramjet, and that combustor length must be chosen to reach good air–fuel mixing, since usually the chemical kinetic factor is not the determining one. E. S. Shchetinkov was the first who understood and began the study of the scramjet–vehicle integration problem. He showed that only in the case of the use of scramjet engine–vehicle integration, i.e., when the aircraft forebody is used as the external compression inlet and the aircraft afterbody is used to continue the engine nozzle expansion, it is possible to reach high scramjet performances. E. S. Shchetinkov was the generator of ideas on development of the combined-cycle propulsion systems in which scramjet could be a composite part. This concerns, for example, ejector-rocket (ram-rocket) airbreathing engines with atmospheric air storage, engines using catalytic properties of materials.One can read about modern development of these engines in V. S. Zuev and V. S. Makaron (1971), R. I. Kourziner (1977, 1989), V. G. Gurylev et al. (1996). The team of scientists in NII-1 founded by E. S. Shchetinkov actively developed his ideas. Many scientists from different organizations took an active part in scramjet development thanks to a scientific seminar that was organized and headed by E. S. Shchetinkov. Since the late 1950s and early 1960s scramjet investigations were started at TsAGI (Central Aerohydrodynamic Institute, Zhukovsky, Moscow region), CIAM (Central Institute of Aviation Motors, Moscow), and later at MAI (Moscow Aviation Institute), ITAM [Institute of Theoretical and Applied Mechanics, Siberian Branch of Russian Academy of Sciences (SO RAS), Novosibirsk] and other organizations. Boundaryuk Design Bureau took the most active participation in scramjet development. Collaborators of Boundaryuk Design Bureau together with NII-1 partners directly tested scramjet combustors as well as designed and manufactured test facilities and scramjet modules. E. S. Shchetinkov undertook energetic efforts to attract various design bureaus to work on compound investigations of scramjets integrated with vehicles and to create scramjet flying models. It is necessary to mention Mikoyan Design Bureau, Grushin Design Bureau, Shavyrin Design Bureau, among others.

Unfortunately such energetic activity continued only 7–8 years. In the second half of the 1960s the activities in the field of hypersonic airbreathing propulsion technology in Russia began to slow down after successful launch of the Russian R7 intercontinental ballistic missile. Only some smaller research efforts were maintained in institutions such as CIAM, TsAGI, ITAM, and MAI. Due to this situation E. S. Shchetinkov's team changed the working place and continued its activity at TsAGI from 1968–1969. Only a small part of his team continued working at the NII-1 facilities up to 1972. The passage of E. S. Shchetinkov's team to TsAGI deprived the team for a certain time of facilities. The small facility was transferred to MAI to support scramjet investigations of E. S. Shchetinkov's team during this transitional period. On the base of this facility the special laboratory for scramjet investigations was organized in MAI in 1979. This laboratory was working in close coordination with Propulsion Division of TsAGI.

Considering the great role of E. S. Shchetinkov's and S. M. Bondaryuk's teams in developing the scramjet it is reasonable to distinguish 1957–1972 as a separate period. Because of that the following presentation of historical material is split into two parts. The first part concerns the initial phase of scramjet investigations that gave answers to principal problems connected with the possibility of creation of the scramjet. The second part describes the next phases of investigation in leading scientific and industry institutes focused on the more detailed development of scramjet working processes.

II. Initial Stage of Scramjet Investigations (1957–1972)

First it was necessary to prove the possibility of air–fuel mixing and combustion in supersonic flows within reasonable lengths (about 2m)—the basic processes in scramjet. The team of specialists on turbulent combustion of NII-1 under the leadership of E. S. Shchetinkov and scientists of TsAGI and CIAM started to solve these problems. Already in 1959 the tests carried out by K. P. Vlasov in NII-1 showed the possibility of supersonic combustion in the duct (see, e.g. V. I. Penzin, 1995). M. S. Volynsky (1963) from NII-1, based on the detailed experimental investigations of the mixing of air with different fuels (hydrogen, kerosene, and other hydrocarbon fuels), considered the different fuel supply modes for supersonic combustors to provide effective fuel–air mixing. In particular, he analyzed 1) staged slot fuel injection to offer thermal protection, skin friction reduction, and separation prevention; 2) pylon fuel supply to provide more uniform fuel injection across the combustor section; 3) a combination of these two modes of fuel supply to design combustors with optimum performances at different flight conditions. The numerous investigations of processes of mixing and burning conducted in NII-1 were summarized in the book of A. G. Prudnikov et al. (1971).

In 1960 Nikolaeva and M. P. Samozvantsev (CIAM) investigated combustion of liquid hydrocarbon fuels in supersonic flow in a tube.

In 1961 V. T. Zhdanov and A. A. Semenov (TsAGI) obtained in the 100-mm-diam tube at Mach number $M = 3.0$ principal results concerning the possibility of kerosene autoignition and combustion stabilization with the use of hot chamber walls ($T_w \approx 700$ K) at moderate air stagnation temperature at the duct entry $T_t = 1000 - 1400$ K. Later other hydrocarbon fuels were investigated.

In 1963–1966 M. P. Samozvantsev and V. Ph. Fedyukov (CIAM) investigated combustion of different fuels in the tube of 110-mm. diam. It was shown that combustion of the products of the decomposition of solid propellant, kerosene, and hydrogen could be performed with combustion efficiency (hereinafter combustion efficiency is defined as reacted hydrogen, divided by the total amount of hydrogen) 0.8–0.9 in the combustor of length 0.6–0.9 m at Mach number at the chamber entry $M = 1.8$–2.8 corresponding to flight Mach number $M_\infty = 5$–10.

The deeper investigation of processes occurring in scramjets demanded more sophisticated test facilities. The most important place in E. S. Shchetinkov's team activity was the design and building of the special aerodynamic test facilities. These test facilities had high stagnation pressure P_t and stagnation temperature T_t and were used in direct-connect and freejet tests schemes. Several test beds allowed testing of two-dimensional combustors that were prototypes of chambers that could be integrated with vehicle. In 1962 under E. S. Shchetinkov's supervision the first large hypersonic facility (BMG) was constructed in NII-1. The nozzle diameter was equal to 400 mm, nozzle Mach numbers were in the range 5–8. It was possible to test a relatively large scramjet module with a diameter up to 300 mm and with a model length of 1800 mm. High total pressure allowed scramjet tests in the range of turbulent Re numbers at M_∞ up to 7. The temperature of "vitiated" air in the heater reached $T_t = 2300$ K. This facility was transferred to CIAM in 1973. The same kind of air heating was used for another smaller facility. It had a square 170×170 mm^2, Mach number $M = 5.2$, nozzle. An absolutely unique wind tunnel with a thermal storage air heater (graphite heat exchanger) that allowed a stagnation temperature P_t = 1900 K and the stagnation pressure P_t = 80 MPa was practically constructed in NII-1 in 1968. This facility would allow performance of scramjet components tests in clean air flow at large *Re* numbers in the range of Mach numbers M = 5–8. It could be used as a direct-connect facility and allowed to test combustors in conditions corresponding to M_∞ = 12–15. Unfortunately, the construction of this wind tunnel was not finished.

Using new and modernized facilities the new stage of scramjet investigations aimed to reach conditions existing in real apparatus was begun. All tests listed below were unique and solved important practical problems.

In 1962 I. M. Kuptsov (NII-1) and S. V. Shteiman (S. M. Bondaryuk Design Bureau) carried out tests with products of the decomposition of solid propellant combustion in a two-dimensional diverging combustor, Fig. 2. Mach number at the combustor entry was 2.8. The cross section areas at the entrance and at the exit of the combustor were 16 × 120 mm² and 80 × 120 mm², respectively. It was concluded that the combustion of products of the decomposition of a solid propellant was most similar to combustion of gaseous jets of hydrogen.

In 1964 B. P. Leonov (NII-1) and A. I. Zaslavsky and S. V. Shteiman (S. M. Bondaryuk Design Bureau) investigated supersonic combustion of kerosene in a two-dimensional diverging combustor (Fig. 2) with a jet stabilization system. Jet stabilizers were placed on the narrow duct walls. Mass flow of jets was about 5–6% of the air flow. The stable combustion was received. The combustion efficiency was about 0.8.

In 1964 R. A. Kolyubakin and V. I. Penzin (NII-1) investigated an axisymmetric model of a dual-mode scramjet at free stream Mach number M_∞=6 (Fig. 3). This model was manufactured in the S. M. Bondaryuk Design Bureau. The overall model length was approximately 1.7 m, diameter was 350 mm. The model had a conical shape and was the prototype of the flying model. The conical shape provided stability in flight. Combustion of prod-

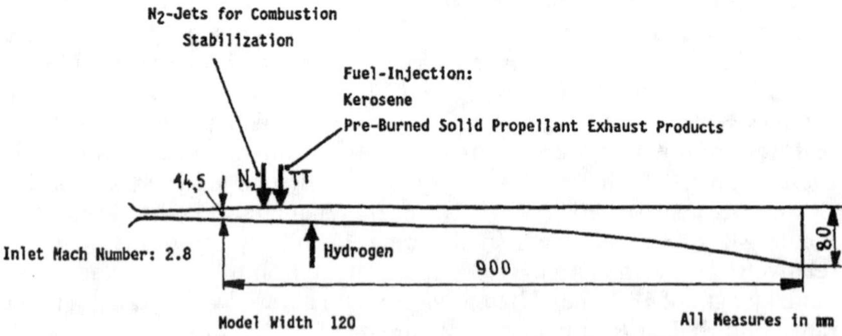

Fig. 2 Scheme of 2-D (plane) diverging-area airframe integrable scramjet combustor connected-pipe tested at NII TP during 1962–1966 by I. M. Kuptsov and S. V. Shteiman; B. P. Leonov, A. I. Zaslavsky, and S. V. Shteiman.

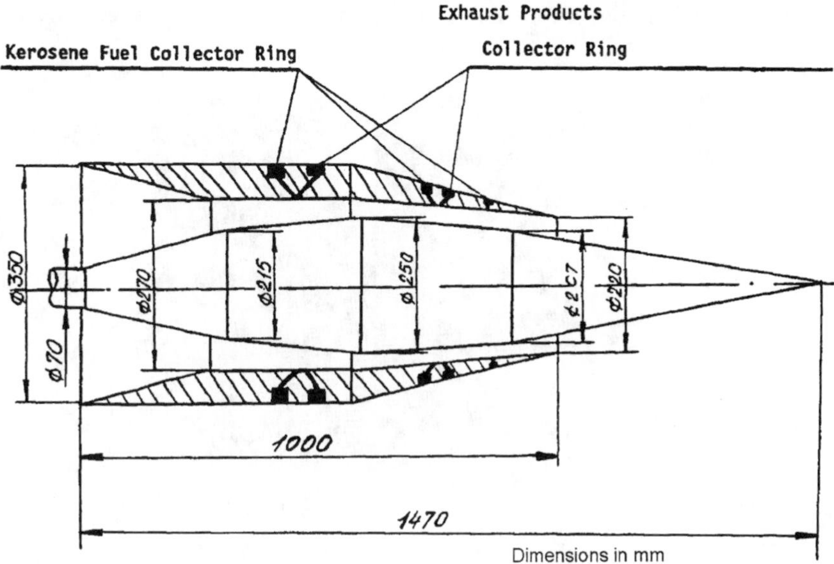

Fig. 3 First Russian scramjet model freejet tested at NII TP by R. A. Kolyubakin and V. I. Penzin in 1964.

ucts of the decomposition of solid propellant in pseudoshock in a diverging duct was realized. The maximum pressure rise in pseudoshock was about 85% of normal shock. The combustion stabilization in supersonic flow in the narrow part of the scramjet passage was obtained with the use of the annular cavity in the central body.

In 1965–1966 A. G. Prudnikov and B. P. Leonov (NII-1) and S. V. Shteiman and A. I. Zaslavsky (S. M. Bondaryuk Design Bureau) investigated the combustion of products of the decomposition of solid propellant, kerosene, and hydrogen in a two-dimensional combustor (Fig. 2). The tests illustrated the possibility of supersonic combustion of different fuels in two-dimensional combustors. Some features of supersonic combustion of nonpremixed gases were analyzed.

In 1966 V. S. Makaron and V. N. Sermanov (NII-1) tested an axisymmetric model of a scramjet at $M_\infty=6$, Fig. 4. The model had a two-ramp external compression inlet with 220-mm-diam at the entrance. The finite ramp angle was $\theta = 20°$. The products of the decomposition of solid propellant were supplied near the inlet throat. The interaction between the inlet and the combustor was investigated. The limiting heat releases were reached.

In 1967 V. N. Strokin and V. Ph. Fedyukov (CIAM) studied the hydrogen ignition and combustion at co-flow and perpendicular fuel supply in direct-connect tests of a two-dimensional combustor. The variation of fuel supply mode was an indispensable condition of dual-mode scramjet combustion control.

In 1967 V. A. Chernov and E. N. Kiseleva (NII-1) investigated combustion of products of the decomposition of a solid propellant in the large constant

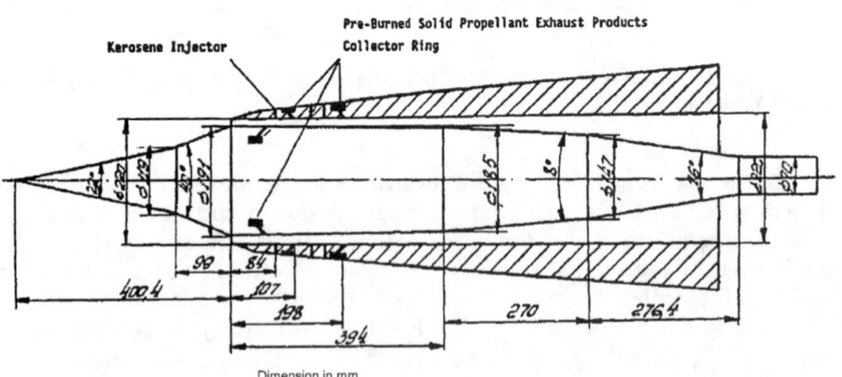

Fig. 4 Scramjet model freejet tested at NII TP by V. S. Makaron and V. N. Sermanov in 1966.

area square combustor (170 × 170 mm²) at entrance Mach number M=3.5 and stagnation temperature T_t =500 K. The increase of the combustor dimensions to the size of real ones did not break down the combustion efficiency of the combustor.

In 1967 O. V. Voloshchenko and E. N. Kiseleva (NII-1) investigated supersonic combustion of products of the decomposition of a solid propellant in the constant area 120 × 170 mm² rectangular burner with a shock system (inlet shock simulation) evoked by the wedge, Fig. 5. Mean Mach numbers at the nozzle exit and combustor entrance were 5.2 and 3.2, respectively. Stagnation temperature was T_t = 800 K. Combustion efficiency was about 80%. The maximum possible heat release (thermal choking) was obtained. It showed the positive influence of shocks on the mixing and combustion processes.

In 1969 V. N. Strokin and S. I. Rozhitsky (CIAM) investigated hydrogen combustion in a conical burner that was a prototype of a diverging dual-mode scramjet combustor.

In 1969 L. I. Gershman, S. V. Shteiman, and A. I. Zaslavsky (S. M. Bondaryuk Design Bureau), A. V. Kulikov and V. L. Zimont (NII-1) investigated a two-dimensional burner with a flow at the entrance which was generated by an oblique nozzle and boundary layer bleed, Fig. 6. The inclination of the nozzle provided the simulation of flow at the final inlet ramp, and the boundary layer bleed provided the simulation of the flow at the inlet cowl. Tests verified the effectiveness of this concept of flight condition simulation. The oblique nozzle provided conditions corresponding to free tests in using direct-connect hardware. This technique was aimed at testing large combustors at low flow rates and at the same time saving the flow structure in the combustor. Visualization of flow in combustor helped to show the great influence of shock systems on mixing and combustion.

In 1969 O. V. Voloshchenko and V. I. Penzin (NII-1) performed freejet tests

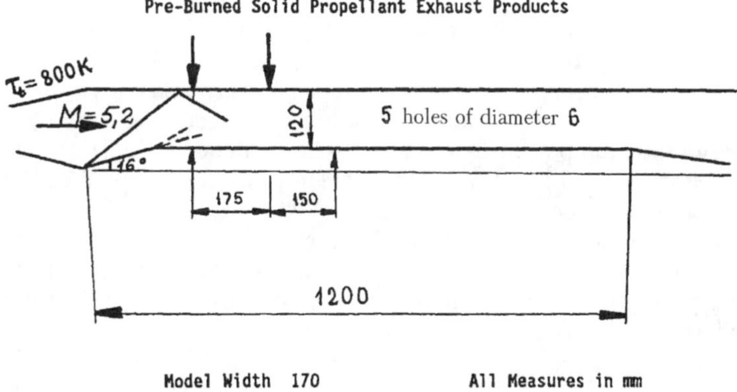

Fig. 5 Scheme of 2-D (plane) combustor including inlet shock wave system simulation connected-pipe tested at NII TP by O. V. Voloshchenko and E. N. Kiseleva in 1967.

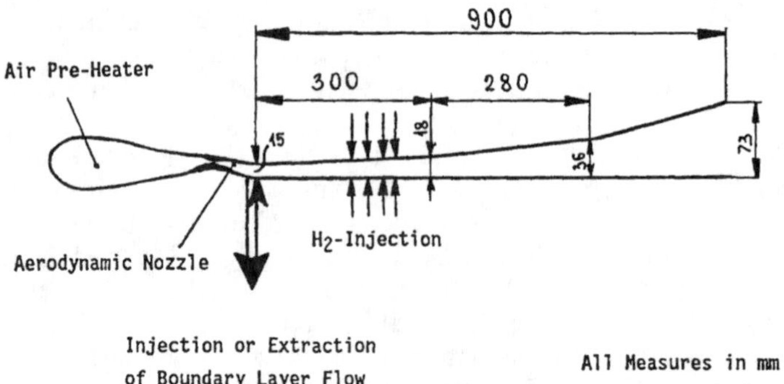

Fig. 6 Scheme of 2-D (plane) combustor including inlet flow and spillage simulation connected-pipe tested at NII TP by L. I. Gershman, S. V. Shteiman, A. I. Zaslavsky, A. V. Kulikov, and V. L. Zimont in 1969.

of a two-dimensional scramjet model at M=5.2, Fig. 7a. The inlet front cross-section was 36×120mm². The model had a one-ramp inlet (θ =15°). The lower wall was inclined by 1°. Three modes of fuel supply (gas-generator products) were tested: normal, parallel to the flow, and the combination of normal and parallel modes. Visualization showed flow structure with and without combustion (Fig. 7b and 7c). The combustion process took place in pseudoshock which was realized in a rectangular duct with aspect ratio of 8. The pressure distributions in scramjet passage correlated to hydraulic diameter showed that the hydraulic diameter concept worked well both for isothermal and for nonisothermal flows. Practically complete fuel burning was obtained on the length of 7–8 hydraulic diameters without disturbance of inlet flow.

In 1969 V. A. Chernov and E. N. Kiseleva investigated the simplest scramjet model—round tube in free stream at Mach number M = 2.5. Combustion of gas-generator gases was carried out without non-start of flow at the entry of the tube.

In 1970–1972 R. A. Kolyubakin and V. N. Sermanov investigated the axisymmetric model of a scramjet fueled by hydrogen over a wide range of flow stagnation temperatures and fuel–air equivalence ratio. The boundaries of hydrogen self-ignition were obtained. They also investigated combustion of fuels containing boron.

E. S. Shchetinkov (1972) presented at the All-Union Combustion Symposium the discussion of the primary problems of supersonic combustion. In 1973 he published a careful analysis of the combustion in pseudoshock (E. S. Shchetinkov, 1973).

All investigations done in NII-1 were supervised by E. S. Shchetinkov. At the same time E. S. Shchetinkov supervised scramjet performance computations and passing them to different design bureaus for more detailed study to assist invention activity. Dual-mode scramjet operation at low M_∞ at a subsonic combustion mode in the wide part of the combustor and at greater

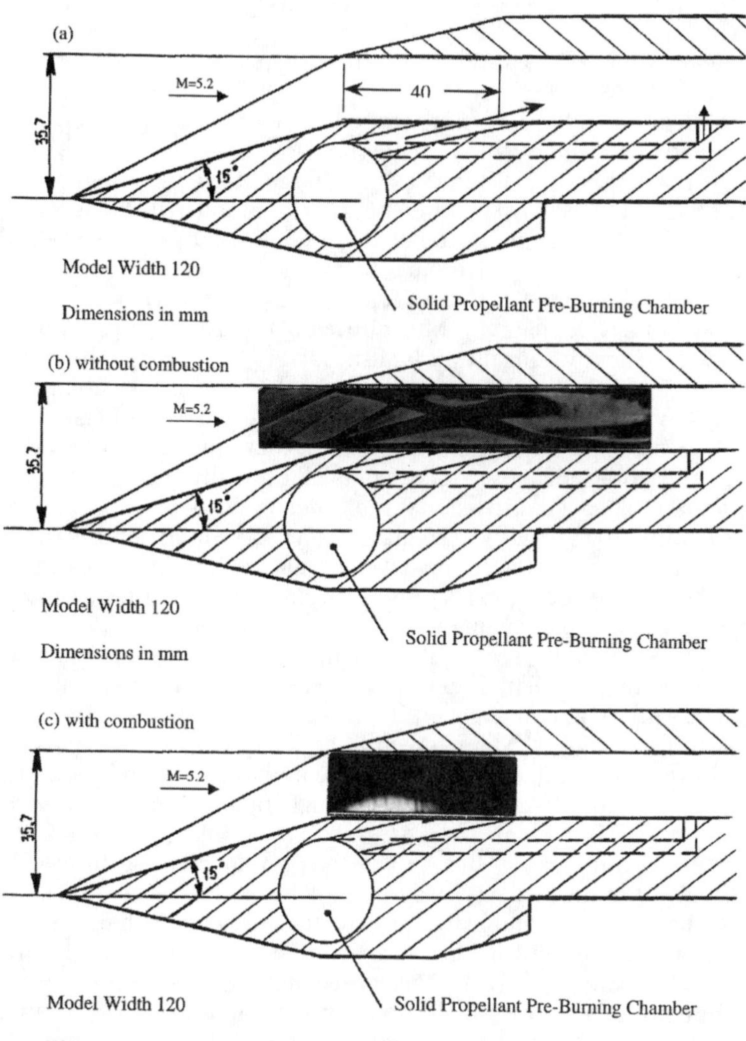

Fig. 7 Investigation into optimum fuel injection modes free-jet tested at NII TP by O. V. Voloshchenko and V. I. Penzin in 1969: a) Scheme of scramjet model, b) flow pattern in scramjet without combustion, and c) flow pattern in scramjet with combustion.

M_∞ at a supersonic mode in the narrow part of the combustor was proposed by V. I. Penzin (patent license 23639 from 8.7.1961). He also proposed an optimum geometry of scramjet passage (patent license 23680 from 8.7.1961), scramjet with passive control of cross-section, and using specially constructed scramjet passage walls that are made of layered material with designed burning of these layers during flight trajectory (patent license 31577 from 30.12.1964).

V. S. Makaron and V. I. Penzin proposed in 1965 a scramjet/rocket engine that extended scramjet velocity range into low Mach number regions and gave the possibility of thrust increasing at large Mach numbers.

NII-1 was at the origin of creation of supersonic and hypersonic inlets (see section IV for more details). G. I. Petrov and E. P. Ukhov (1947) investigated optimum shock systems and applied them for the construction of the most effective inlets for ramjet engines of rockets. This rich experience was used for scramjet inlet development (V. M. Anufriev, S. L. Vishnevetsky) over a wide range of velocities by using the air-ballistic facility. NII-1 had good experience on the nozzle development, too (U. G. Pirumov). V. Ya. Borodachev and A. M. Gubertov were experts in the problem of heat protection. Osminin and Samoilov were experts in dynamics of flight. Alternative types of ram/rocket engines, and storage of air were also studied thoroughly by V. S. Makaron and G. M. Pankov (detailed discussion is presented in the book by V. S. Zuev and V. S. Makaron, 1971, see also V. S. Makaron and V. N Sermanov, 1997).

All this knowledge allowed the development in NII-1 in 1966 of the original single-stage-to-orbit (SSTO) flying airplane. Boosting this vehicle was done with a ram/rocket engine placed in the same passage with the ramjet. The ramjet started at the second stage of flight. Then the scramjet started and finally the liquid propellant rocket started and delivered the vehicle into the orbit. The vehicle parameters were similar to the National Aerospace Plane proposed in the United States much later. The total take-off mass of the vehicle was 150,000–250,000 kg, the payload at the orbit was 6,000–11,000 kg. Scramjet inlet was two-dimensional, the angle of a last compression wedge was 25°. Design Mach number was $M_d = 9$. Relative inlet throat was 0.06. The last section of compression surface was the rotating plane surface, which at $M_\infty \leq 8$ was turned in such a way as to open the entrance of the ramjet passage. The scramjet combustor was a rectangular duct with the height 135 mm and with the width 12 m. There were eight supporting pylons that divided the combustor into separate modules. The length of the combustor was 2 m. Optimum working process in the combustor was provided by combined hydrogen supply. One part of hydrogen was injected by round jets with the use of injectors, the other part of hydrogen was injected from the wall at the small angle. Wall jets interacted between each other and quite soon joined making a plane curtain. Fuel supply was performed at three locations along the combustor length. Such staged fuel injection provided heat protection and diminished friction losses. Hydrogen- staged supply provided effective mixing in the combustor. The combustor was also stable to acoustic pulsation. Hydrogen was used for the cooling of leading edges of vehicle inlet, combustor, and nozzle. The overall hydrogen cooling capacity was enough for the whole flight trajectory. Wall temperature did not exceed 1200 K. Hydrogen temperature in panels did not exceed 700 K.

As mentioned earlier, in the late 1960s both interest in the scramjet and the funding of scramjet projects waned. After E. S. Shchetinkov's death in 1967, scramjet studies were headed by Academician G. G. Chernyi. With decreased interest in scramjets, studies eventually ceased. The investigations carried out in TsAGI, CIAM, ITAM were no longer coordinated. Different aspects of activity by Prof. E. S. Shchetinkov are discussed in a number of articles and reviews (O. M. Belotserkovsky and Yu. A. Shcherbina, 1982; I. V. Bespalov and V. S. Makaron, 1982; R. I. Kourziner, 1982; V. I. Penzin, 1982b, 1995; A. G. Prudnikov and V. A. Frost, 1982).

III. Scramjet Investigations, 1972–1996

A. TsAGI Investigations

In 1969, the TsAGI Propulsion Division focused its efforts on the creation of new test facilities needed for scramjet investigations. The important role in facility designing belonged to the team headed by A. G. Popoviyan. The complex of aerodynamic wind tunnels with air heaters and without them was constructed (see, e.g. description in R. Walther et al., 1993; R. Walther and V. A. Sabel'nikov, 1993). The complex includes:

- Freejet test bed T-131B (Fig. 8). This facility has the fire (kerosene-fueled) heater. To compensate the losses of oxygen consumed by the kerosene combustion, oxygen is added to air flow before it enters the heater. The heater provides stagnation temperatures and pressures up to 2350 K and 10.0 MPa, respectively. The facility is equipped with four aerodynamic nozzles, Mach numbers 5, 6, 7, and 8, with exit area 400 mm in diameter.
- Experimental setup for direct-connect tests T-131V (Fig. 9). The air heater, stagnation temperature, and pressure are the same as for freejet facility T-131B. The aerodynamic nozzles can give the following Mach numbers at the combustor entrance, 2.5, 3.0, 3.5, 4.0
- Small wind tunnel SVV-1. This facility is used for internal flow gasdynamic investigations $M=1\text{–}6$, $P_t \leq 10$ MPa.

These facilities provided wide possibilities for investigations of aerothermodynamics of scramjets.

Until 1969 supersonic combustion and scramjet activity was in the department headed by V. T. Zhdanov. Since 1969, after E. S. Shchetinkov came to TsAGI from NII-1, these investigations were done in a scramjet department headed by him until retirement (1974). The team of V. I. Penzin studied gasdynamics of scramjets and the team of V. L. Zimont investigated the physics of supersonic combustion. Since 1989 supersonic combustion and scramjet activity has been headed by V. A. Sabel'nikov.

Since the 1970s in TsAGI (partly in cooperation with MAI) a great number of investigations of physical processes in the dual-mode scramjet

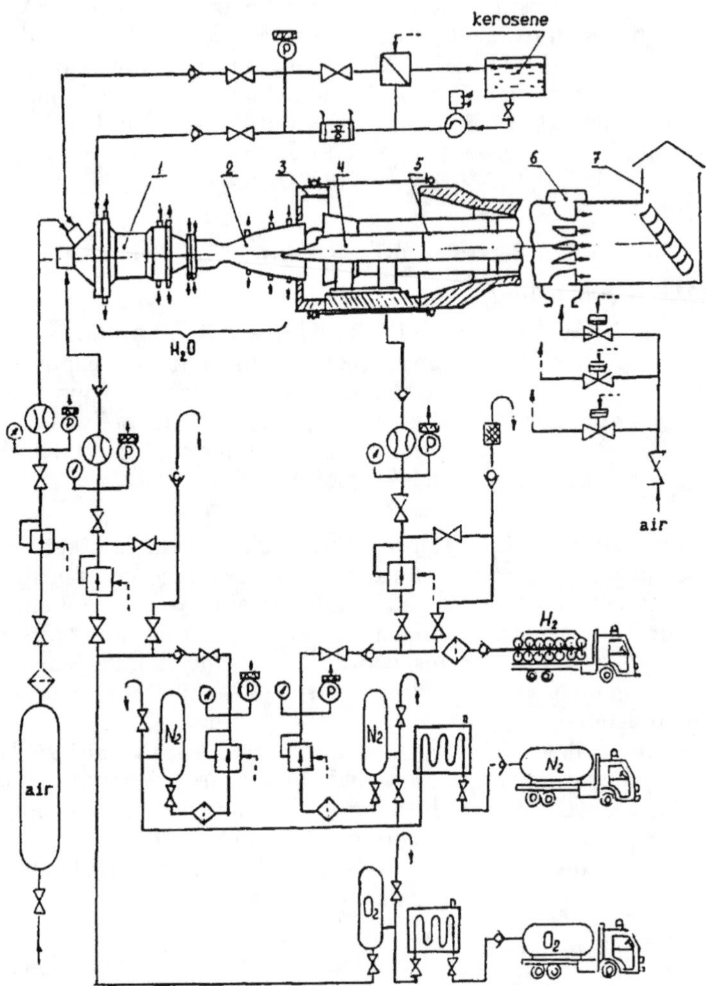

Fig. 8 Scheme of the TsAGI free-jet test facility T-131 B. 1) Kerosene-fueled water cooled, air preheater; 2) aerodynamic nozzle; 3) test chamber; 4) scramjet free-jet model (2-D or axisymmetric models with inlet diameters up to 200 mm and 2300 mm in total length can be mounted); 5) external and internal diffusers; 6) ejector system; 7) exhaust-gas chimney.

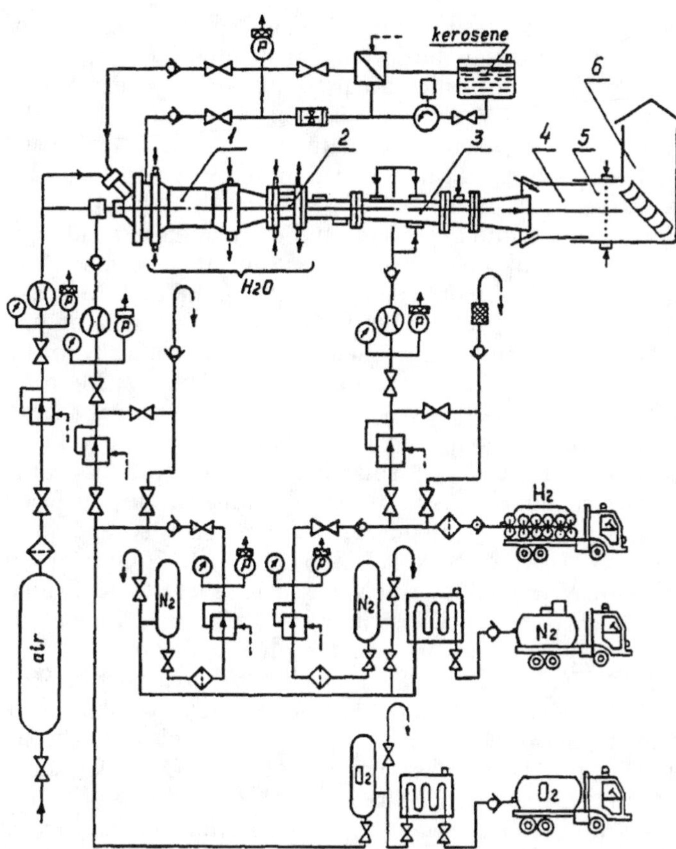

Fig. 9 Scheme of the TsAGI connected-pipe test facility T-131V. 1) kerosene-fueled water cooled air preheater; 2) aerodynamic nozzle; 3) plane or axisymmetric combustor model; 4) adapter tube; 5 exhaust-gas tube with potential water injection; 6) exhaust-gas chimney.

passage, including combustion in supersonic flow and pseudoshock, were carried out.

Supersonic combustion of hydrogen at co-flow injection to air flow, the specific features of combustion of nonpremixed gases, and supersonic heat release efficiency in the combustors of variable geometry were studied in articles of E. A. Meshcheryakov (1987), E. A. Meshcheryakov et al. (1983), E. A. Meshcheryakov and V. A. Sabel'nikov (1981, 1988), V. L. Zimont et al. (1987), E. A. Meshcheryakov et al. (1983), O. M. Kolesnikov (1982). It was found that the essential decreasing of a heat release took place in divergent area combustors (V. L. Zimont et al., 1970). E. A. Meshcheryakov and V. A. Sabel'nikov (1983, 1988) analyzed the causes of such heat release decreasing. They found that in cases when, before the divergent part of the combustor, the combustion is close to thermodynamic equilibrium, the decrease of heat release is mainly due to the decrease of an intensity of turbulent mixing. Main results of this investigation are presented in Appendix A, Part A.I.E. A. Meshcheryakov and V. A. Sabel'nikov considered also the influence of concentration and temperature fluctuations on the supersonic combustion of nonpremixed gases (E. A. Meshcheryakov and V. A. Sabel'nikov, 1985; see also V. L. Zimont et al., 1983).

Combustion stabilization problems and calculation of conditions of the combustion extinction behind the step and in the supersonic flow cavity were analyzed by V. L. Zimont et al. (1972, 1982), and E. A. Meshcheryakov and O. V. Makasheva (1976). The boundaries of self-ignition and flame extinction were determined experimentally and theoretically for hydrogen and hydrocarbon fuels (V. L. Zimont et al., 1972, 1982; E. A. Meshcheryakov and O. V. Makasheva, 1976). It was shown that the cavity has a large region of stable combustion as compared with a step.

The pseudoshock structure in adiabatic flows (i.e. without heat release), methods of calculation of pseudoshock resulting in the satisfactory estimates of wall pressure distributions, pressure recovery, and so on were studied by V. G. Gurylev and S. N. Eliseev (1972), V. G. Gurylev and A. K. Trifonov (1976), V. G. Gurylev and V. I. Penzin (1981), A. K. Trifonov et al. (1977), V. N. Ostras' and V. I. Penzin (1974, 1976), V. L. Zimont and V. N. Ostras' (1974, 1976), V. E. Kozlov and V. A. Sabel'nikov (1982), and V. I. Penzin (1973, 1979, 1983a,b,c, 1985, 1986a,b, 1987a,b,c, 1988a,b, 1993a,b,c).

Combustion in a pseudoshock in a two-dimensional combustor was studied both in the above-mentioned work of O. V. Voloshchenko and V. I. Penzin (see Section II) which was completed in TsAGI in 1969 and by E. S. Shchetinkov (1973) and V. L. Zimont et al. (1978). It was found that a pseudoshock results in a considerable intensification of mixing and changes the character of fuel burning (V. L. Zimont et al., 1978). This conclusion confirmed the early results of O. V. Voloshchenko and V. I. Penzin (1969). It was shown that the combustion in pseudoshock is quite sensitive to changes in flow parameters. For example, the relatively small decrease of stagnation temperature of air or small increase of fuel–air equivalence ratio (ER) results in a large upstream shift of pseudoshock and combustion zone.

Different flow separations typical to those in a dual-mode scramjet passage were analyzed over a wide range of flow parameters. The main interest was to learn the specific features of supersonic flow deceleration, the influence on pressure recovery, and the length of the deceleration region of Mach and Reynolds numbers, boundary layer thickness, and duct cross section shape. All these problems and also the interaction between separations of different types, pulsation characteristics, and possibilities of separation flow control were investigated by V. I. Penzin (1987a,b,c, 1988a,b). Some selected results on deceleration of supersonic flows in smoothly diverging area rectangular ducts are given in Appendix B. Components of a drag force contributed by friction, steps, cavities, and struts were studied by V. N. Ostras and V. I. Penzin (1972), and V. I. Penzin (1976, 1983a,b,c; 1987a,b,c, 1993a,b,c).

The heat exchange in the adiabatic pseudoshock (i.e., without combustion) was studied by O. V. Voloshchenko et al. (1988). The pseudoshock was generated in an axisymmetric duct of internal diameter 146 mm and length of 1500 mm. The Mach number at the entrance of the duct was 3.8, stagnation temperature was in the range 900–1500 K, and stagnation pressure was 1.3 MPa. It was concluded that the good accuracy estimation of the level of the heat flux in the pseudoshock can be obtained from the well-known two-layer model of pseudoshock (see, e.g., E. S. Shchetinkov, 1973). As opposed to the supersonic flow, the heat flux distribution in pseudoshock reaches the maximum level at the distance of 4–5 diameters from the beginning of the pseudoshock.

With connected pipe tests the mechanisms of supersonic combustion, including hydrogen and kerosene fuels, were thoroughly studied (O. V. Voloshchenko et al., 1992). Considerable emphasis was given to stabilization and effective combustion at the lower flight Mach number range 3–6. The problem of combustion stabilization is especially important for liquid hydrocarbon fuel application since the air temperature at the combustor inlet is below the self-ignition temperature of the fuel–air mixture. In addition, Mach numbers within the combustor are low and special attention must be paid to the problem of thermal choking. Numerous investigations were conducted with kerosene-fueled combustors (see their description in O. V. Voloshchenko et al., 1992; R. Walther and V. A. Sabel'nikov, 1993). To provide effective liquid kerosene combustion, stabilization tests with gas-generator were performed. Some selected results of these investigations are presented below. These tests were carried out using the axisymmetric kerosene-fueled combustor with the coaxial gas-generator (Fig. 10). Two cases were investigated. In the first case kerosene was supplied to the gas-generator. Inside of the gas-generator the fuel-rich mixture was premixed, prevaporized, and preburned. The hot and chemically reactive combustion products were injected into the main air flow through a nozzle and six longitudinal slots located at the rear part of the gas-generator shaft. Within the gas-generator the fuel was supplied by a pilot swirl injector at the dump plate of the inlet central body and by a primary fuel supply from the walls using an annular manifold. The flow Mach number at the entrance of the

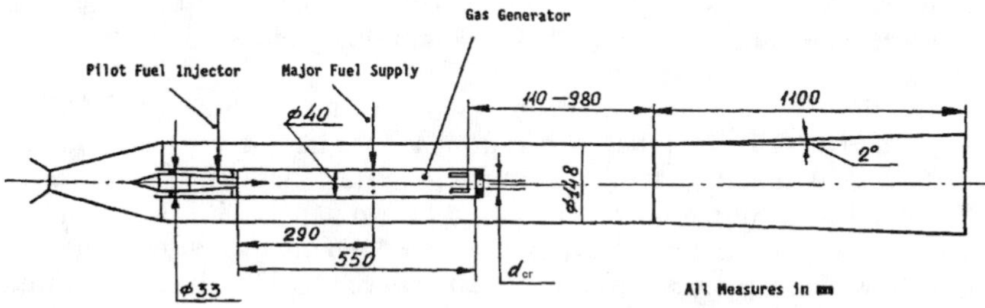

Fig. 10 Axisymmetric scramjet combustor with coaxial gas generator connected-pipe tested at TsAGI (O. V. Voloshchenko et al., 1992; V. A. Sabel'nikov et al., 1993).

combustor was $M = 2.7$, stagnation pressure $P_t = 4.2$ MPa, and stagnation temperature $T_t = 1150 - 1350$ K that corresponds to the flight Mach number range $M_\infty = 4.8 - 5.3$. The relative mass rate of the gas-generator combustion products was 9%. The tests revealed that optimum ignition and combustion within the main combustor occurred at the stagnation temperature of incoming airflow above 1250 K at equivalence ratios in the gas-generator in the range ER = 2.5–3. The combustion efficiency strongly depended on the combustor length. At fuel equivalence ratio in the range ER = 0.17– 0.27 in the main combustor, the combustion efficiencies at combustor lengths of 370 mm and 650 mm reached 74% and 95%, respectively. In the second case only 20% was kerosene supplied into the gas-generator. The rest of the kerosene (80%) was supplied into the main combustor through the annular manifold installed 300 mm upstream of the gas-generator. Tests showed that to reach combustion efficiency at the 95% level the combustor length has to be increased up to 1 m.

In tests with the two-dimensional divergent combustor the injection of kerosene was carried out normally to the main flow through five struts placed in one row. For better fuel–air mixing and to increase penetration of injected jets into main air flow the liquid kerosene was aerated by gas (air). To perform such kerosene aeration the bubbling kerosene injection system developed in MAI by V. N. Avrashkov et al. (1990 a,b) was used (hereinafter bubbling kerosene jet is named barbotaged jet). The air jet throttling at the end of the combustor provided the initial ignition of the fuel. After ignition the throttling was switched out. Stable combustion took place at $T_t \geq 1600$ K and equivalence ratio was in the range ER = 1.5–2.3.

The combustion of kerosene was also studied in the two-dimensional combustor with a sudden expansion. The burner had the entry cross-section of 30 × 100 mm². The cross-section at the location of expansion was 58×100mm². The length of the forward part of the burner was 790 mm and the length of the second one was 510 or 910 mm. The test conditions were M=2.5, P_t = 2.5–3.0 MPa, T_t =1000–1400 K. Kerosene was injected through three diamond-shaped pylons which could be installed at different distances from the combustor entry: 90 mm, 180 mm, or 690 mm. Maximum combustion efficiency reached about 90%. The stable burning was preserved when stagnation temperature dropped to 1000 K. The promising results were also obtained with this combustor fueled by kerosene and using gas-generator exhaust gases for combustion stabilization (V. A. Sabel'nikov et al., 1993b).

The original slot pylons fuel supply concept investigated in TsAGI is shown in Fig. 11. The combustor was designed for large-scale tests and had the cross-section area of 100 × 100 mm². The total length of combustor was 1400 mm. Each slot pylon was composed of two panels with sharp leading edges. The distance between the two installed pylons was 30 mm, the distance between plates was 8 mm, and the blockage was 8%. The tests showed that the injection of hydrogen into the slot passage provided the greater penetration of fuel into the supersonic flow, and more reliable flame holding characteristics as compared with the normal, direct injection of fuel. The

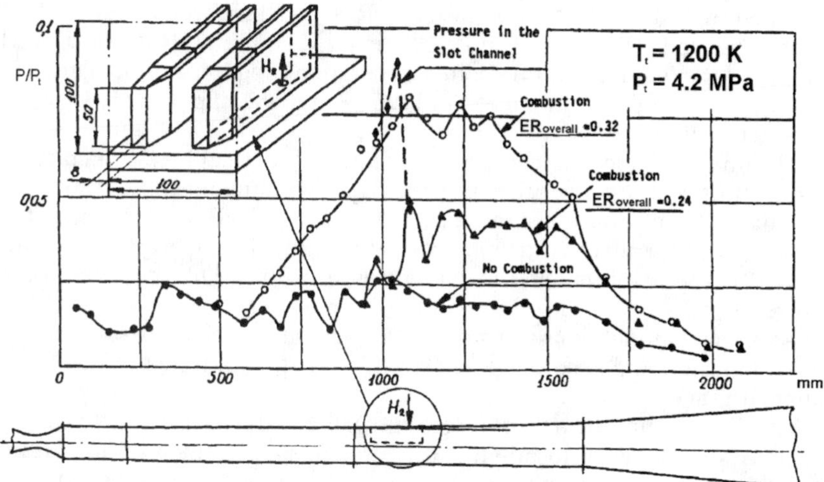

Fig. 11 Fuel injection by slot pylons connected-pipe tested at TsAGI (O. V. Voloshchenko et al., 1992; V. A. Sabel'nikov et al., 1993).

combustion efficiency of 0.75– 0.9 was obtained at M=2.3, ER = 0.2 − 0.32, P_t = 4 MPa, $T_t \approx$ 1200 K.

The tests with two-dimensional combustors, the walls of which were covered with the ceramic heat-resistant materials on the basis of Sic-Sic and Carbon-Carbon with a special protecting coating, showed that the heat-sink chamber worked reliably at $Tt \approx$ 1900 K and P_t = 3.5 MPa during 60 at stoichiometric hydrogen air mixture (ER=1).

In addition, the scramjet engine models fueled by kerosene and hydrogen were free-jet tested at M_∞=5 and 6. The emphasis was placed on the investigation into the gasdynamic/thermodynamic interactions between combustor and inlet. At the end of the 1980s the dual-mode scramjet engine model was realized practically. This module was shown at the following exhibitions: "Aviation engines" (1992) in Moscow and "Airshow" in Zhukovsky (1992, 1995). These works were conducted in close coordination with design bureaus "Raduga" and "Soyuz." The collaborators of the latter, D. D. Gilevich, O. N. Romankov and A. M. Tereshin, took part in the variable geometry combustor (including combustor with a sudden expansion) design and also in tests where the combustor fuel was kerosene barbotaged with air (1989–1991). "Raduga" collaborators I. S. Seleznev, V. D. Savchuk, R. Sh. Khaikin, and V. A. Koval'chuk took part in the development of the dual-mode scramjet module (1985–1986).

The creation of the complex of facilities T-131 allowed performance of large volumes of work with scramjet combustors and modules in the velocity range M_∞=5–7. Part of this investigation was carried out with German research and industry institutes MTU, DLR, DASA, and IABG within the framework of the comprehensive German Hypersonic Technology Program initiated by BMFT—Federal Ministry of Research and Technology in 1988. Motoren- und Turbinen- Union Munich (MTU) and Russian TsAGI launched the first phase of the joint cooperation program on scramjet technology for hypersonic propulsion in 1992 (V. A. Sabel'nikov et al., 1993; R., Walther et al., 1993, 1995, 1997).

The investigations of combustion characteristics and gasdynamics of scramjet combustors over the range of flight Mach numbers 5–7 were carried out in the combustor of the special design, which allowed change to the passage geometry in the wide range of diverging angles and expansion ratios (Fig. 12). The Mach number at the entrance of the combustor was in the range $M = 2.5$–3.5.

A scramjet module consisting of the fixed geometry inlet, variable geometry combustor, and simple nozzle (Fig. 13) was freejet-tested in the Mach number range 5–7.

The direct-connect combustor tests were directed to investigate impact of combustion geometry, location, and hydrogen supply mode on important aerothermodynamic combustor characteristics such as combustion efficiency, total pressure losses, generation of a pressure peak in the vicinity of the fuel supply station, and upstream pressure distortion propagation. Three different fuel injection variants were tested (Fig. 14). Investigations allowed selection of the optimum passage geometry and the fuel delivery law and the choice of one of three tested injector types allow the choice of one of maximum combustion efficiency at minimum drag and stable work of the inlet-combustor system at maximum heat release. Experimental data emphasized strongly the important role that relative heat release played (ratio of heat release to stagnation enthalpy at the entrance of the combustor) on the gasdynamics of a dual-mode scramjet. In fact, it was concluded that the relative heat release is the primary similarity parameter for simulation of real conditions for dual-mode scramjet engine. Some test results are given in Figs. 15–17. The correlation of the wall pressure peak with the fuel–air equivalence ratio for different fuel supply modes is depicted in Fig. 15.

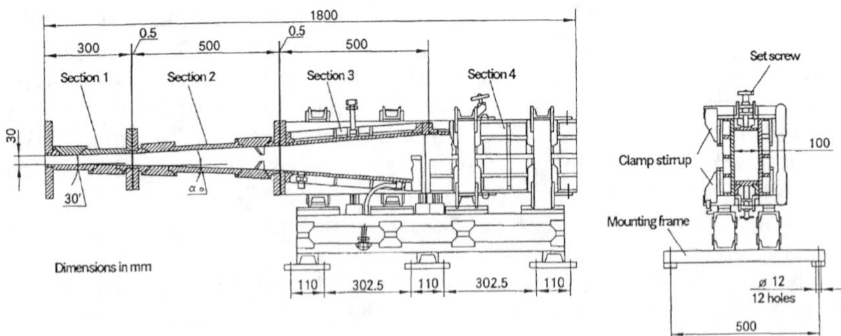

Fig. 12 Geometrically variable combustor model TsAGI-MTU (R. Walther et al., 1995, 1997).

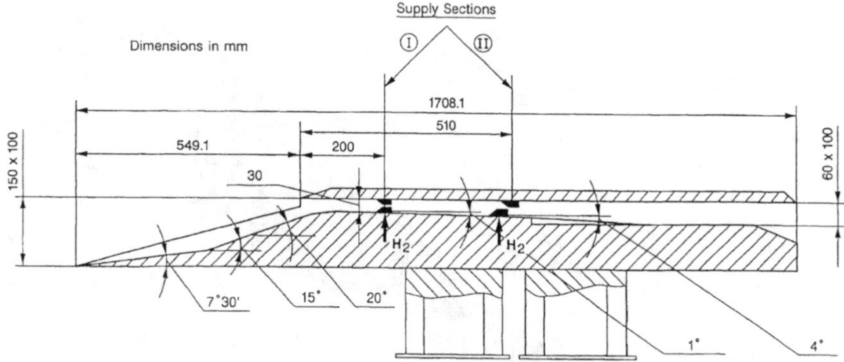

Fig. 13 Scheme of subscale engine model TsAGI-MTU (R. Walther et al., 1995, 1997).

Figure 16 shows the dependence between the length of upstream influence and the pressure peak. The comparison of the efficiencies of different fuel supply modes is presented in Fig. 17. (The specific impulses of diverging-area combustor were used for evaluation.) It is seen that the wedge-shaped fin injectors performed better than the other two configurations (tube injectors and swept wall-mounted ramps). Measurements of forces imposed on the scramjet module fueled by hydrogen and done with the use of six-component balances thrust were found to be in satisfactory agreement with theo-

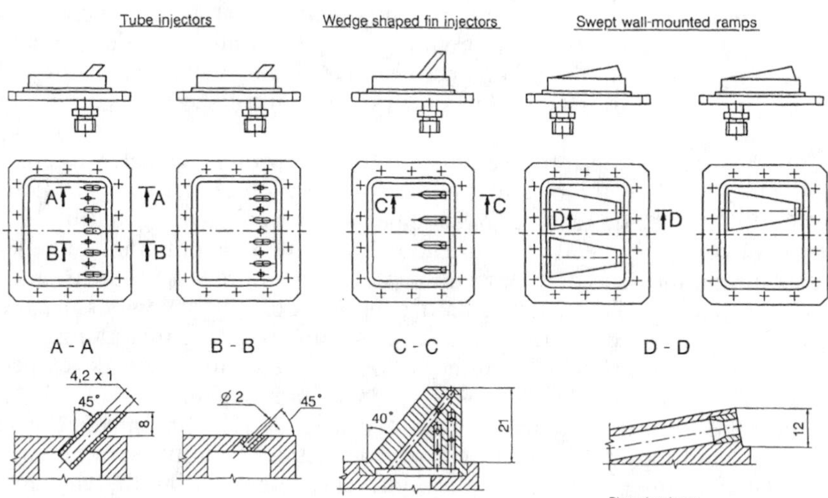

Fig. 14 Various injector configurations used for the combustor model TsAGI-MTU (R. Walther et al., 1995, 1997).

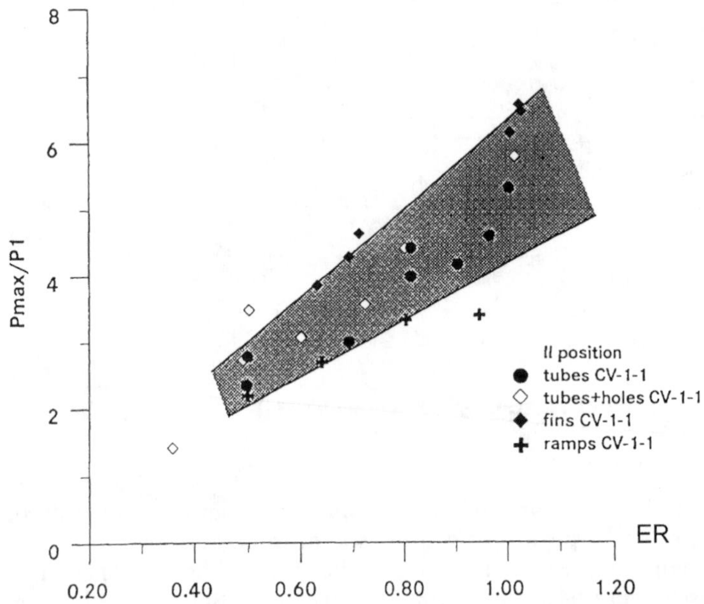

Fig. 15 Impact of fuel supply on static pressure peak; notation CV-1-1 means combustor of variable geometry with angles of divergenge 1 deg and 1 deg for second and third parts of combustor, respectively (R. Walther et al., 1995, 1997).

retical predictions, based on some simplifying assumptions. Combustion efficiency in the scramjet module was about 0.9.

The choice of direction in scramjet investigations was closely linked with analysis of optimum combined propulsion systems which include scramjet. Since the 1960s scramjet performance was investigated by V. T. Zhdanov and his collaborators I. K. Romashkin and A. A. Semenov (see V. G. Gurylev et al., 1996).

The general approach to a scramjet integrated with the vehicle was originated by E. S. Shchetinkov and was developed by V. G. Gurylev et al. (1992, 1996). Some specific features of scramjet inlets were investigated by V. I. Penzin (1993a,b,c; 1994). The method of selection of optimum scramjet geometry on the basis of minimization of total pressure losses due to shock waves and heat release, taking into consideration restrictions associated with fixed geometry of scramjet passages, scramjet–vehicle integration, and the type of combustion in the burner (diffusion, detonation) was developed by V. I. Penzin (1990, 1992) and V. V. Andreev and V. I. Penzin (1992, 1994). To elucidate, in general, there are two types of flames: 1) premixed flame, when reactants are perfectly mixed before reaction; 2) diffusion or nonpremixed flame, when reactants diffuse into each other during the chemical reaction. Detonation combustion the same as deflagration (subsonic combustion wave) takes place in premixed gases but in contradistinction to deflagration, the detonation wave propagates at supersonic speed. Appen-

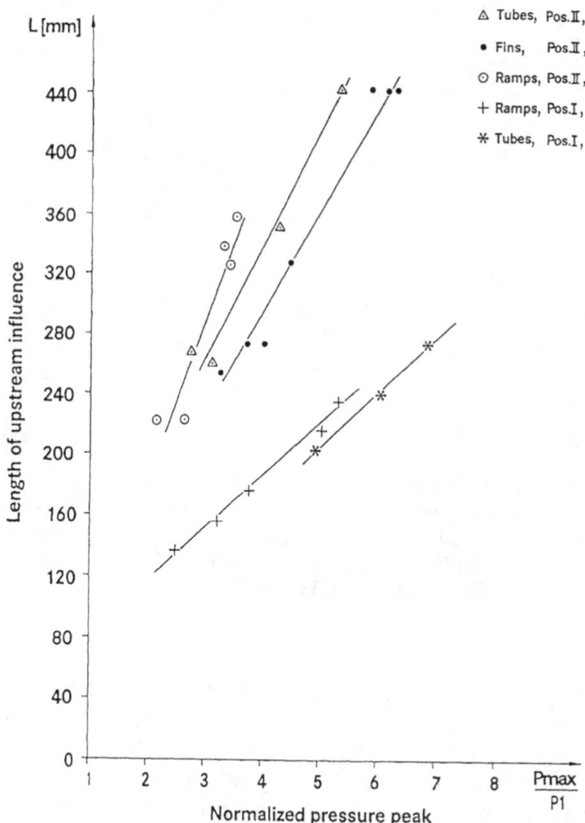

Fig. 16 Upstream influence for various fuel injection (R. Walther et al., 1995, 1997).

dix C contains some aspects of the scramjet–vehicle integration method developed by V. I. Penzin.

For the investigation of complex physics-chemistry processes in the scramjet passage, methods of computational aerodynamics were widely used in TsAGI (O. M. Kolesnikov, 1989; O. M. Kolesnikov et al., 1993; V. A. Sabel'nikov et al., 1990; E. A. Meshcheryakov and V. A. Sabel'nikov, 1981, 1988; V. V. Vlasenko and V. A. Sabel'nikov, 1994). The structure of the flow in a two-dimensional scramjet combustor with tangential hydrogen injection was analyzed (O. M. Kolesnikov, 1989; O. M. Kolesnikov et al., 1993). It was established that in the case of underexpanded jets the shock wave structure strongly influences the ignition and hydrogen burning process (V. A. Sabel'nikov et al., 1990). The impact of hydrogen wall injection on friction and heat fluxes in the combustor and nozzle was analyzed (O. M. Kolesnikov et al., 1993). It was shown that the most effective friction reduction (from the viewpoint of minimum hydrogen mass rate) was obtained when the hydrogen injection was distributed along the combustor. The causes of heat re-

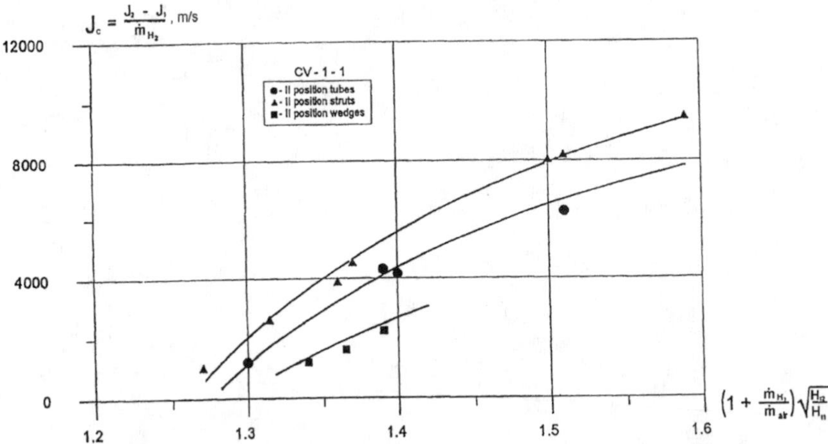

Fig. 17 Combustor-specific impulse for different fuel mode supply as a function of relative heat release. J_2, J_1 are axial impulse function at the entrance and the exit of the combustor, respectively; $\dot{m}_{H2}$ amd $\dot{m}_{air}$ are mass rates of hydrogen and air, respectively; H_{t2} and H_{t1} are stagnation enthalpies at the entrance and the exit of thecombustor (R. Walther et al., 1995, 1997).

lease reduction in the divergent combustors were analyzed (E. A. Meshcheryakov and V. A. Sabel'nikov, 1981, 1983, 1988), and this analysis is reproduced in Appendix A). It was shown that the combustor must have the constant area initial part long enough for good mixing in front of the expansion section. V. V. Vlasenko and V. A. Sabel'nikov (1994) studied the structure of the stabilization zone of a detonation wave on the wedge at finite chemical reaction rates. Great attention was paid to development of mathematical models of combustion in turbulent flows (V. R. Kuznetsov and V. A. Sabel'nikov, 1990; V. A. Sabel'nikov, 1997). The problem of the simulation and the creation of the flow conditions in a scramjet combustor in wind tunnels was studied by V. I. Alfyorov (1993). The values and the distributions of heat fluxes over the surface of a single fuel pylon and over a pylon group were obtained by V. I. Alfyorov in the modified hypersonic wind tunnel and in the hypervelocity wind tunnel with MHD acceleration.

B. CIAM Investigations

Since the middle of the 1970s much work had been done in CIAM on supersonic combustion and gasdynamics of scramjets by several teams of scientists under the supervision of V. N. Strokin, Yu. M. Annushkin, and V. A. Vinogradov. These teams were developing both applied aspects of scramjet combustors and the principal aspects of supersonic combustion problems (thermodynamics of combustion, combustion efficiency, mixing, stabilization and flame holding, injection system arrangement, heat protection with hydrogen cooling, flame extinction, influence of shocks on supersonic combustion, and so forth). For the most part, problems were solved by experimental investigations. To perform them, several experimental facilities with chemical and arc heating were developed and built. CIAM has scramjet test facilities that can provide tests of models with dimensions up to 200 × 200 mm^2 and 300 × 300 mm^2 with the flight Mach number simulation up to 6–8 (Fig. 18). These facilities allowed them to test models both in a free stream flow and in the direct-connect pipe (V. A. Sosounov, 1993a). The supersonic diffusion type of combustion was taken as the main one for scramjet combustors because G. G. Chernyi (1968) and M. P. Samozvantsev (1964) showed that stationary overdriven (strong) detonation waves (which were proposed to be used in detonation wave engines) are unstable and branch off into an adiabatic shock wave and a front of slow combustion.

It is worth mentioning the development of instrumentation for gas sampling in supersonic flow. Swallow shock wedge probes sampling were developed and successfully used in supersonic combustion investigations (S. I. Rozhitsky and V. N. Strokin, 1974; V. N. Strokin and V. M. Khailov, 1974).

Several types of combustors (cylindrical, divergent, two-dimensional with various number of injection nozzles and various types of fuel injection (from wall and from struts) were tested. (See review paper by V. N. Strokin and V. A. Grachev, 1997.)

Some results on autoignition were obtained by V. N. Strokin, 1972. The main result in this field was the necessity of flame holding for both the wall and strut fuel injection up to flight Mach number $M_\infty = 8$. A cavity flame holding system was developed for wall fuel injection (V. A. Grachev and V. N. Strokin et al., 1981; V. N. Strokin and V. A. Grachev, 1994). This flame holding system was found very useful to provide the stable combustion in flight tests of scramjet (see A. S. Roudakov, 1993b; A. S. Roudakov et al., 1993).

The result was obtained under investigation of combustion efficiency. It was found that at the self-ignition conditions (or at the use of the flame stabilization) the increase of combustion efficiency took place with the

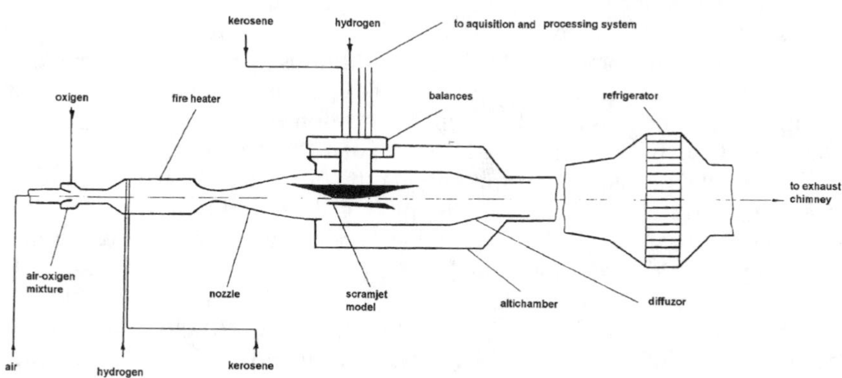

Fig. 18 Principal scheme of CIAM hypersonic test facility (BMG).

increase of equivalence ratio (i.e., with the increase of fuel mass flow rate) S. I. Rozhitsky and V. N. Strokin (1982). This result has the reverse character in comparison with the case of subsonic combustion and possibly is linked with the role played by shock waves in a supersonic combustor. The correlation relationship was found for combustion efficiency at the wall fuel injection. The heat release decrease was found to be very important for divergent combustors. This result was confirmed later during scramjet model tests (V. A. Vinogradov et al., 1990).

It was found that the length of the forward part of the pseudoshock preceding the combustion zone (precombustion shock structure) depends, mainly, on a relative amount of heat release (S. I. Rozhitsky and V. N. Strokin, 1988). For a wall fuel injection in a case of pseudoshock combustion, hydrogen burning was enhanced to a great extent in comparison with supersonic and even subsonic combustion without preceding pseudoshock. It was concluded that for the pseudoshock combustion it was not so difficult to get practically uniform fuel distribution in the combustor using a wall fuel injection. But it was practically impossible for a supersonic mode of combustion. Based on these results it was concluded that the optimum fuel injection system for a dual-mode scramjet has to combine wall and strut modes of fuel injection.

The results of the investigation of operation domain of strut fuel feed system for model two-dimentional scamjet combustor are presented in the paper by V.L. Semenov et al. (1998). The limits of self-ignition and stable combustor with divergence angle up to 7 degrees are determined. An important contribution to ram- and scramjet combustor development was brought by Yu. M. Annushkin (Yu. M. Annushkin, 1981; V. A. Sosounov et al., 1988; Yu. M. Annushkin and E. D. Sverdlov, 1978, 1979, 1984; Yu. M. Annushkin et al., 1983; V. A. Sosounov, 1993c). The problem of stabilization of diffusion hydrogen flames in the air co-flow was solved experimentally. Also, the possibility to shorten the length of the combustor using a greater number of fuel injectors was studied. The minimum critical distance between fuel nozzles was determined experimentally. At that distance the flames do not interact with each other yet and do not merge into the one joint flame. It was found that critical distance decreases with the increase of relative air speed and the fuel nozzle base area. Yu. M. Annushkin, using the vast experimental data, developed the generalized method of calculating combustion efficiency for the case of diffusion combustion. Combustion efficiency was linked with the ratio of combustor length to the length of diffusion flame.

Exceptional persistence in spite of all difficulties and the lack of funding was demonstrated by CIAM in completing the "Kholod" project, which included the successive fulfilment of experimental work on combustors and on scramjet model development for ground- and then flight tests (V. A. Sosounov, 1993a; A. S. Roudakov, 1993b). Pioneered by S. M. Shlyachtenko, E. S. Shchetinkov, and D. A. Ogorodnikov in 1970, the "Kholod" project was carried out by joint efforts of CIAM, TsAGI, Boundaryuk Design Bureau, and LII (Flight Testing Institute, Zhukovsky) up to 1977. Aerodynamic problems of scramjet model accommodation on the rocket, the influence of boundary layer and of blunt leading edges of inlets on scramjet performance, scramjet passage losses and choice of optimum scramjet geometry were studied by E. S. Shchetinkov, V. T. Zhdanov, V. I. Penzin, O. V. Voloshchenko, and S. N. Eliseev of TsAGI. LII was engaged for remote trajectory measurements, data acquisition, and processing. Since 1973 the initiative and responsibility for the "Kholod" project was completely switched over to CIAM.

A great number of organizations were attracted to this project. Among them were design bureaus "Gorizont," "Phakel," "Automatika" and "Soyuz." The last organization manufactured scramjet modules for ground and flight tests. Work was carried out under the guidance of D. A. Ogorodnikov, V. A. Sosounov, R. I. Kourziner and, after 1989, A. S. Roudakov. The main contribution into the project was made by V. S. Kourziner, V. N. Strokin, V. A. Vinogradov, Yu. M. Shichman, G. I. Petrov, and E. D. Sverdlov. It is necessary to stress the activity of the leading designer of CIAM, M. V. Strokin. His input to the project cannot be overestimated. In fact, during more than 20 years he was the informal organizer of this project.

The first scramjet model scheme was a two-dimensional one with a three-shock inlet and divergent combustor. This model was manufactured in CIAM during 1975–1976 and tested in the CIAM free-jet test facility BMG during 1979–1989, in free stream at Mach numbers 5 and 6, and in the direct-connect facility (M=4). In 1979 TMKB "Soyuz" joined the "Kholod" project and manufactured the axisymmetric scramjet module with the annular combustor taking into account the two-dimensional model test results. In addition to supporting the scramjet model concept, the ground-test scramjet model also was manufactured in CIAM in 1983. A series of gasdynamic investigations of the inlets and the engine module passage (with simulation of mass supply and combustion) was carried out in 1986 in cooperation with TsAGI. At the same period CIAM facilities (Ts-101 and BMG) were modernized. BMG characteristics and equipment became: aerodynamic nozzles in the Mach number range 3.5–7.5, the balances, liquid nitrogen for cooling, fire air-heater with a new scheme of heating, a new system of hydrogen and liquid hydrocarbon supply, and a new system of data acquisition and processing. In 1983–1985 a series of tests of the scramjet engine module "57M" (Fig. 19) were carried out in test facilities Ts-101 and BMG in the Mach

Fig. 19 Axisymmetric dual-mode scramjet model "57M" designed and tested by CIAM in the Mach number range 4–6.4.

number range 4–6.4. Stable combustion was obtained both at supersonic and subsonic combustion modes with different methods of fuel injection.

The stable region of the joint inlet/combustor operation was determined. The data that are necessary for the engine control system for flight tests were obtained. The requirements for cooling the system with liquid hydrogen were elaborated. In total, recommendations for the first flight test were given. In 1986 and 1988 successful flight tests of the "cold" scramjet model were carried out. During these tests the launch complex and all operating systems were checked.

To date, CIAM has conducted three rocket-boosted flight tests of axysimmetric dual-mode Mach number 6 design scramjet (Fig. 20). The first flight test took place in November of 1991. The second flight test was carried out in November 1992 together with French specialists. The trajectory segment over which the hydrogen-fueled scramjet worked flawlessly, lasted for 23 s. Stable combustion at hydrogen injection both from the subsonic and supersonic section of fuel manifold was obtained at flight Mach number 5.3. The aims and results of flight tests are described by S. W. Kandebo (1992). The third attempt was also with France in March 1995. It was unsuccessful. The engine failed to operate because of some onboard systems problem (see A. S. Roudakov et al., 1996).

Under a contract with NASA, CIAM and NASA team on 12 February 1998 conducted the fourth successful flight test of dual–mode scamjet at Mach 6.5. Valuable, unique subsonic and supersonic combustion data were obtained from Mach 3.5 to greater than Mach 6.4. Preliminary flight test results are presented by A.S. Roudakov et al. (1998) and by J.W. Hicks (1998).

Flight tests of a dual-mode scramjet became the splendid finale of the "Kholod" project that continued for more than 20 years.

This remarkable success was prepared by intensive ground tests. V. A.

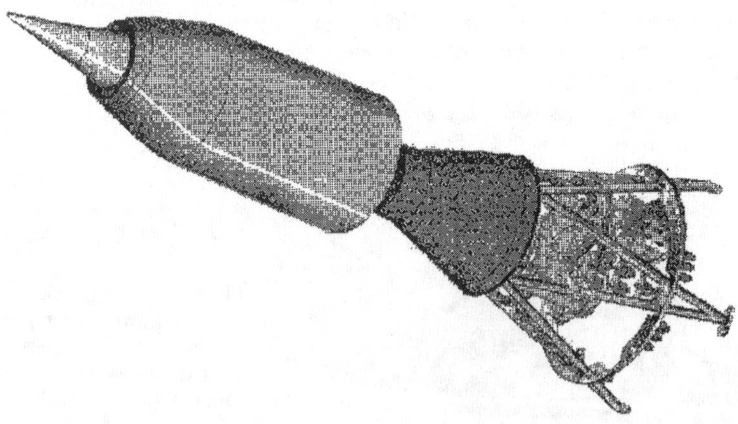

Fig. 20 Axisymmetric dual-mode scramjet flight tested by CIAM in 1991–1992 (from A. S. Roudakov et al., 1996).

Vinogradov et al. (1990) tested the two-dimensional dual-mode scramjet at supersonic, subsonic, and mixed modes of combustion over the range of flight Mach numbers M_∞= 4–6. At M_∞ = 4 tests were carried out using direct-connect scheme. At M_∞ = 5 and 6, tests were done in the free stream. The wall pressure distributions, combustion, and mixing efficiencies were studied. Data on stabilization of combustion, wall heat flux distributions, and joint operation of the combustor and the inlet were obtained. Analogous investigations of an axisymmetric dual-mode scramjet module were carried out at supersonic combustion in the annular combustor by V. A. Vinogradov et al. (1992a). These data were the basis for a preparation of flight tests of the same scheme of scramjet module. The experience of modal dual-mode scamjet creation and the main results of ground and pre-flight tests are analized by D.A. Ogorodnikov, et al., (1998).

Tests with a two-dimensional dual-mode kerosene-fueled scramjet module (Fig. 21) at flight Mach number M_∞=6 and T_t=1500 K were carried out by V. A. Vinogradov et al. (1992b). Several variants of fuel injectors and flame holders were tested. A small amount of hydrogen was used for ignition of the air–kerosene mixture. The conditions were found at which the combustion in the burner (which was composed of the constant cross-section

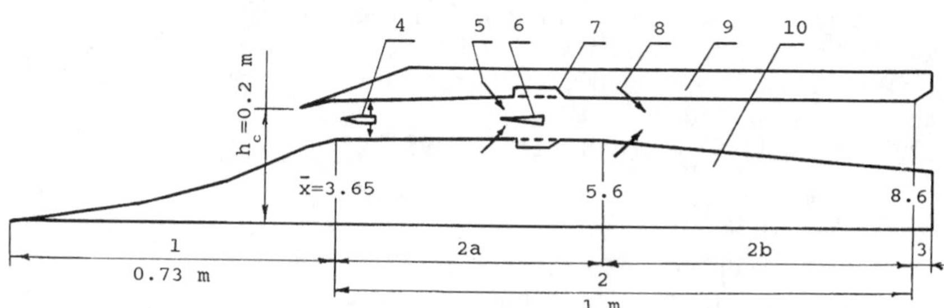

Fig. 21 Two-dimensional scramjet tested at CIAM by V. A. Vinogradov et al., (1992). 1) inlet; 2) combustor (2a is constant area section, 2b is expanding section); 3) nozzle; 4) kerosene injector; 5) hydrogen injectors; 6) hydrogen strut stabilizator; 7) cavity flameholders; 8) hydrogen injectors; 9) cowl (upper wall); 10) bottom wall.

area and divergent section) was continuing after switching off the hydrogen supply.

Problems of kerosene atomization in the wake behind the pylon were analyzed by V. A. Vinogradov and A. G. Prudnikov (1993). The effect of the pylon shape on the fuel stream geometry, degree of atomization, and on the possibility of filling by kerosene droplets of maximum volume of the combustor is analyzed. CIAM investigated the possibility of application of endothermic fuels for hypersonic propulsion systems (including ramjets and scramjets). A review of this effort and obtained results are given in the recent paper by L. S. Ianovski et al. (1997).

Over several years V. I. Kopchenov et al. (1992, 1993a,b) carried out investigations intended to find the means to intensify combustion and mixing processes in the scramjet combustors using methods of mathematical modelling. Visible results were reached with the nonaxisymmetric nozzle, the supersonic part of which was transformed from the circle to the ellipse. At a certain nozzle expansion angle the hydrogen jet disintegrated into two jets resulting in the enhancement of the mixing process. This effect is manifested clearly in cases when the air and hydrogen flow velocities are nearly the same. This effect formed the basis of the method of the combustion intensification in the scramjet combustor. L. V. Bezgin et al. (1995, 1997), also used the method of mathematical modelling for the analysis of scramjet/vehicle integration and the estimation of the scramjet efficiency. Special attention was paid to the influence of the finite chemical reaction rates and viscosity effects. More details about scramjet activity carried out in CIAM can be found in reviews by V. A. Sosounov (1993 a, b, c), A. S. Roudakov (1993 a, b), and V. I. Kopchenov et al. (1993 a, b).

C. ITAM Investigations

The work on supersonic combustion at ITAM (Novosibirsk) were carried out in the department of gasdynamics of combustion under the guidance of V. K. Baev. In 1977 the supersonic combustion laboratory (as part of V. K. Baev's department) was organized and was headed by P. K. Tret'yakov.

The results of investigations were displayed in a number of reviews and books by V. K. Baev et al. (1976b, 1984a, 1987), P. K. Tret'yakov et al. (1993), and P. K. Tret'yakov (1996). Let us limit ourselves only to short notes on these investigations.

A number of fundamental investigations of supersonic combustion were carried out. The articles of V. K. Baev et al. (1974), R. S. Tyulpanov and O. V. Pritsker (1972), and V. Ph. Sokolenko et al. (1971, 1972) studied the structure of supersonic diffusion flame in a duct. The effect of finite chemical reaction rates (V. K. Baev et al., 1974), and temperature on the combustion intensity (R. S. Tyulpanov and O. V. Pritsker, 1972) was investigated. Turbulence characteristics such as intensity and length, and scales of temperature pulsations were measured (V. Ph. Sokolenko et al., 1971, 1974). Optical methods of diagnostics of supersonic combustion were developed (S. S. Vorontsov et al., 1976, 1978; V. K. Baev et al., 1974). The stabilization of combustion in supersonic flows (in particular by recirculation zones) was investigated (V.

K. Baev et al., 1973, 1975, 1976a, 1984a). V. A. Zabaikin and A. M. Lazarev (1986), V. A. Zabaikin et al., (1985, 1986), and P. K. Tret'yakov and A. M. Lazarev (1991) investigated the character of hydrogen combustion at different means of fuel injection into high temperature air flow (Fig. 22). It was found that the shock wave structure of the external supersonic flow strongly affects the combustion process and the combustion stabilization that allows it to control the flame position (Figs. 23 and 24).

Injection and autoignition of liquid hydrocarbon fuels in supersonic air flow were investigated by V. K. Baev et al. (1981a, 1982a). Considerable attention was paid to the development of numerical methods and kinetic models for the description of supersonic combustion. Results of these investigations were summarized by V. K. Baev et al. (1976, 1984a, 1987) and V. I. Dimitrov (1982).

The main efforts were directed to investigating the dual-mode scramjet with stepped combustor. Results of investigations of gasdynamics of a combustor with a sudden expansion were analyzed in detail by V. K. Baev et al. (1984a). The main significance of these works was in the understanding that a combustor with a sudden expansion has a number of advantages, the main ones being the reliable combustion stabilization at rather low temperatures

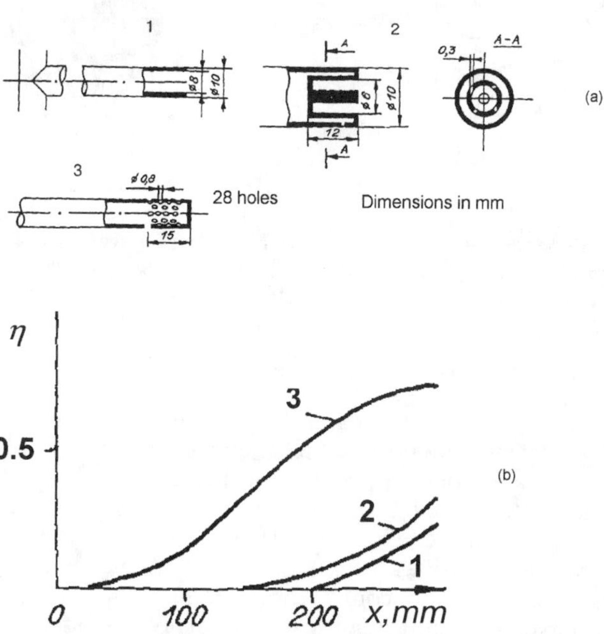

Fig. 22 Fuel injection: a) Different hydrogen injectors tested at ITPM (V. A. Zabaikin and A. M. Lazarev, 1986; V. A. Zabaikin et al., 1985, 1986); 1) co-flow injection; 2) swirling jet injection; 3) radial injection; b) combustion efficiency distribution along diffusion flame at different fuel injection: 1) co-flow injection; 2) swirling jet injection; 3) radial injection (M=1.4 ; T_t=1850 K).

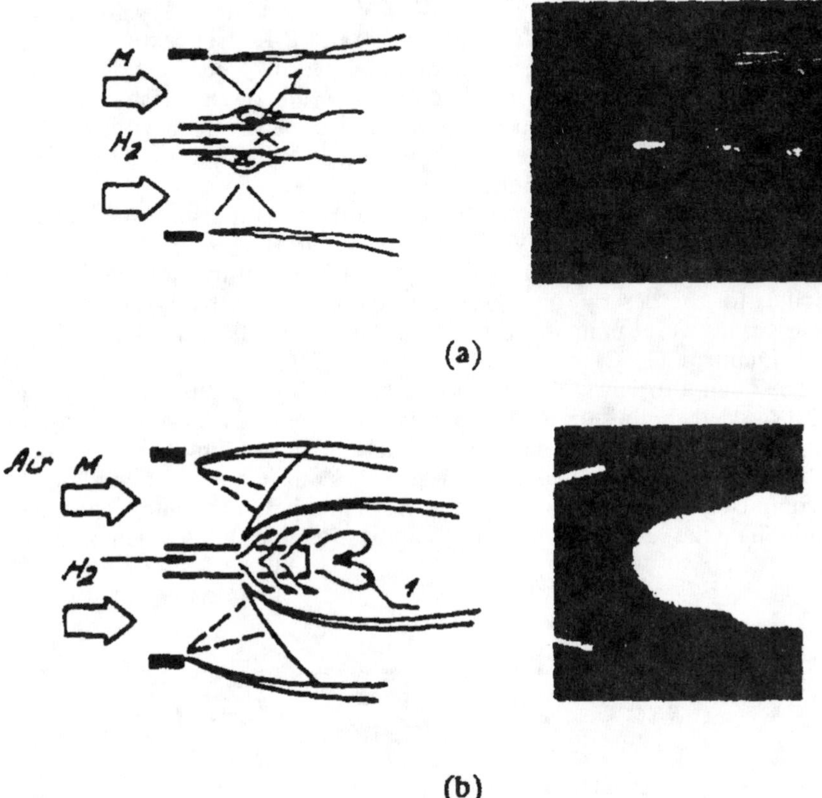

Fig. 23 The dependence of flame structure from shock waves: a) co-flow injection of hydrogen; b) radial injection of hydrogen. M=2.2; p_{jet}/p_{ext} = 0.7, 1 (separation zone). Scheme of flow structure (left), direct flame photo (right) (from P. K. Tret'yakov, 1996).

of incoming flow ($T_t \leq 1000$ K) and the possibility to operate over a wide range of flight Mach numbers and fuel equivalence ratio. These advantages compensate for additional total pressure losses due to a step. Principal similarity parameters of the combustor were also revealed. This knowledge was necessary to control combustor tests. It was found that primary of such parameters was the relative heat release (given that the ignition and stabilization of combustion did not control the process). Stepped combustors were used for the development of small-scale axisymmetric models of dual-mode scramjets (V. K. Baev et al., 1983, 1984b, 1985; V. K. Baev and V. V. Shumskii, 1995). V. K. Baev et al. (1983, 1984b, 1985) tested small-scale (diameter of 23 mm and 72 mm) gasdynamic models of the dual-mode scramjet in high enthalpy air flow at the impulse facility ITAM IT-301. The scheme of the model tested by V. K. Baev et al. (1984b) is given in Fig. 25a. Tests were run

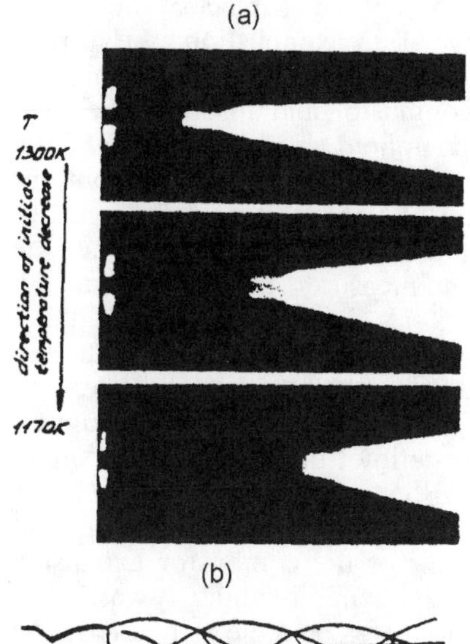

Fig. 24 Flame stabilization of underexpanded hydrogen jet. Influence of a barrel-shock structure: a) direct flame photo; b) scheme of flow structure (from P. K. Tret'yakov, 1996).

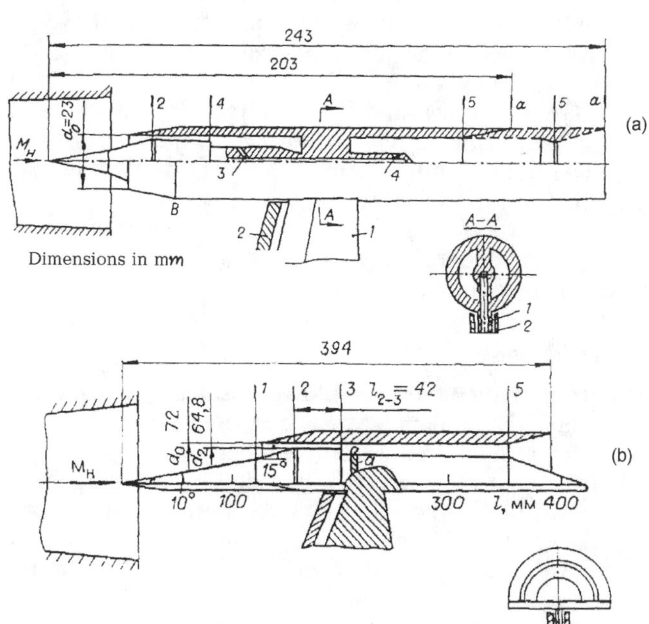

Fig. 25 The schemes of subscale dual-mode scramjet models tested in ITPM: a) V. K. Baev et al., 1984b; (b) - V. K. Baev et al., 1985.

at P_t = 7–60 MPa, T_t = 1000–1850 K, M_∞ = 4.9. The subject of these tests was to study the overall effects of hydrogen combustion: the thrust, pseudoshock movement in the forward section of a combustor, combustion efficiency, the physical picture of flow in combustor, and unstart of the inlet. Calculated scramjet performances were compared with experimental data. V. K. Baev et al. (1985) measured wall pressure and heat flux distributions along another small-scale scramjet model (Fig. 25b). This model was manufactured as a half of the axisymmetric one and was mounted at the side pylon through which hydrogen pipeline and measurement cables were laid. Tests were run at P_t=50–7 MPa, T_t=970–1600 K, M_∞=7.9. Tests showed that at the opposed flow fuel injection it is possible to obtain quite high hydrogen combustion efficiency (0.9–0.95) in pseudoshock at rather short lengths (130–180 mm) without unstart of an inlet. V. K. Baev and V. V. Shumskii (1995) studied the effect of the expansion ratio of the combustor of small-scale ramjet model on engine performance. Tests were performed at M_∞=5; T_t=1200–1500 K; P_t=35–40 MPa and stoichiometric hydrogen-air equivalence ratio (ER=1). Calculated predictions of the combustor expansion ratio on engine performance were confirmed experimentally. It was found that at the pseudoshock combustion mode it was possible to decrease the combustor expansion ratio to the level determined by the theory. Integration of a hypersonic airbreathing vehicle based on an experiment performed by ITAM is presented in the joint paper Dassault Aviation (France)—ITAM (Perrier P. et al., 1995); F. Chalot et al., (1998).

L. N. Pusyrev et al. (1990) discussed the problems that have to be taken into account when developing the small-scale scramjet models for testing during a short time in high enthalpy flows.

P. K. Tret'yakov (1993a,b) made the analysis of some characteristics of heat release in pseudoshock. It was shown that the average heat release correlates well with the length of combustion zone and does not depend on the duct length. The ratio of a local heat release value to its maximum is correlated with the ratio of a combustion zone length to the length of a supersonic part of a pseudoshock. Last ratio does not depend on the duct length or the fuel injection mode.

D. MAI Investigations

The small group that studied in MAI supersonic combustion was organized in 1966 under the guidance of E. L. Solokhin. The laboratory for the investigation of scramjet combustion chambers was organized in MAI in 1979. Head of the laboratory was S. I. Baranovsky and scientific advisor of the laboratory was G. N. Abramovich. Since 1991 the laboratory has been under the guidance of V. N. Levin. Investigations of combustion processes in scramjet combustors and the modernization of test facilities were launched at the same time. Reviews of scramjet research carried out in MAI were made by V. N. Avrashkov et al. (1995).

Investigations of the fundamental processes taking place at supersonic combustion of axysimmetric jets of hydrogen and kerosene were performed by S. I. Baranovsky et al. (1990b). Supersonic and pseudoshock modes of kerosene and hydrogen combustion in ducts were investigated by S. I. Baranovsky et al. (1986, 1992b), and V. N. Avrashkov et al. (1990b). A qualitative physical picture of processes in a combustor was formulated on the basis of the hydrogen combustion study by S. I. Baranovsky and V. M. Levin (1990a, 1991a). The phenomenon of thermal choking and mechanical throttling (in an adiabatic pseudoshock) of the supersonic flow in a combustor was analyzed by S. I. Baranovsky and V. M. Levin (1990a, 1991a). A similarity of precombustion shocks induced by a combustion zone and shocks in the front of a pseudoshock was demonstrated using miniature probes for stagnation and static pressure measurements at thermal and mechanical duct throttling. It was stated that shocks in the front of a pseudoshock are the means of a flame stabilization and the cause of the combustion intensification in supersonic flows in ducts (Fig. 26). The unique transparent combustor (Fig. 27) was constructed in 1985–1991 for the study of a supersonic combustion mechanisms (S. I. Baranovsky et al., 1992c). The combustion chamber side walls have quartz inserts of length 40 mm. This transparent combustor was used for the investigation of an interaction of shocks with a burning mixing layer of fuel and air in the duct using optical methods (laser spectroscopy and fluorescence). One example of supersonic flow visualization obtained by the schlieren method in this combustor is presented in Fig. 28.

Several new methods and devices were proposed to improve the mixing efficiency in gas and two-phase jets at supersonic and subsonic velocities (patent license N309415 from 18.08.88, V. N. Avrashkov et al., 1989). The

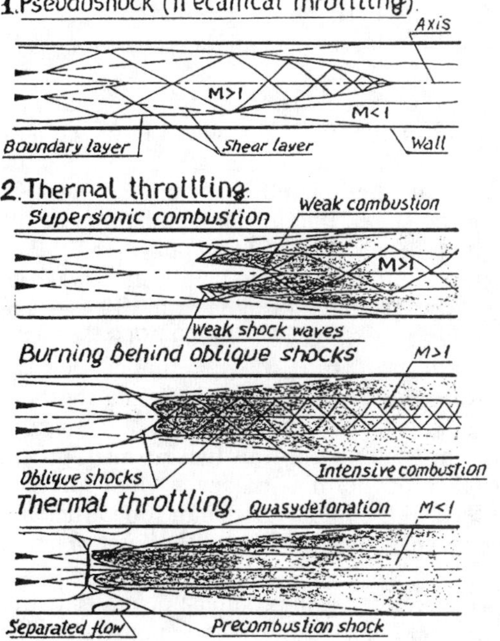

Fig. 26 Physical models of mechanical throttling and thermal throttling and choking of supersonic flow (from from V. N. Avrashkov et al., 1995).

Fig. 27 Transparent combustion chamber test rig of MAI.

injection of liquid fuel (kerosene) jets barbotaged (aerated) by gas (air, hydrogen, and so forth) allowed a sharp increase in mixing efficiency (V. N. Avrashkov et al., 1990a).

The original fuel-supplying tubular micro-pylons were designed and manufactured. The developed fuel-supply manifold resulted in the equally removed jets (V. N. Avrashkov et al., 1988). The use of this manifold and the barbotage system allowed achievement of autoignition and high combustion efficiency of kerosene in the constant area duct of length 600 mm at the following incoming flow parameters: M=2.5; T_t=1600 –1700 K; P=10^5 Pa.

At the next step of investigation stable kerosene combustion was obtained in a diverging-area combustor (V. N. Avrashkov et al., 1990b). Parameters at the combustor entrance were: M=2.5; T_t = 1200–1700 K; P=10^5 Pa; combustor expansion area ratio was 1.6. It was determined that the autoignition took place at fuel–air equivalence ratios exceeding the ones corresponding to the thermal choking of supersonic flow. The system of forced ignition start (pneumatic throttling) of a lean fuel/air mixture was developed to overcome the problem of thermal choking. Quite promising results were obtained during tests of this combustor: combustion efficiency was about 0.85 for the combustor length of 0.45 m, and the total pressure loss coefficient was about 0.35.

The influence of shocks on turbulent mixing and combustion intensity was determined by V. N. Avrashkov et al. (1990b, 1996). It was demonstrated that

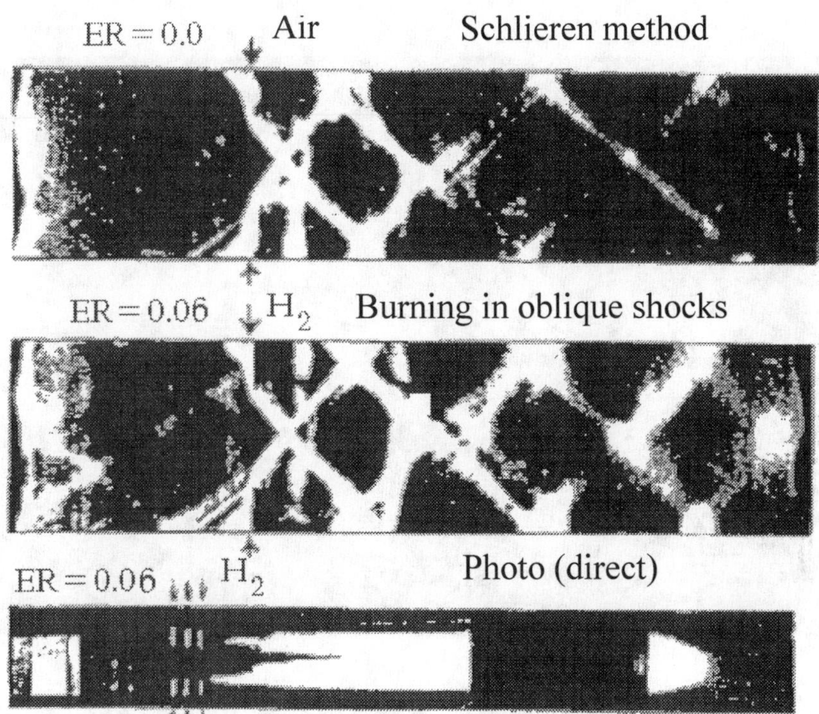

Fig. 28 Supersonic combustion visualization in transparent combustor of MAI (from V. N. Avrashkov et al., 1995).

the shock–mixing layer interaction results in the intensification of mixing up to 1.5–2.0 times.

Until 1987 uncooled heat-sink combustors made of a niobium alloy with a protective antioxidant coating and fueled with hydrogen and kerosene were designed, manufactured, and tested in conditions corresponding to M_∞=6 (Fig. 29a, hydrogen fueled combustor; Figs. 29b and 29c, kerosene fueled combustor). Materials comprising the combustors were developed in MAI at the material department. The forward combustor cross-section was 50 × 100 mm^2. The micropylon fuel injection was used in these combustors. Test conditions for hydrogen fueled combustor were: M=2.5, T_t = 1700 K, P = 0.1 Mpa, ER = 0.5–1.0. Kerosene-fueled combustor entrance parameters were M=2.5, T_t = 1690 K, P = 0.1 Mpa, ER = 0.5–0.9, wall temperature during tests was about $T_W \approx$ 2150 K. Combustion efficiency for the stoichiometric equivalence ratio at the length of 0.45 m exceeded 0.89, and the total pressure loss coefficient was about 0.3.

In 1987 the concept of a wide-range (M_∞= 3–12) ramjet fueled by hydrogen was initiated (see S. I. Baranovsky and V. N. Levin, 1991; S. I. Baranovsky et al., 1992a). The principal geometry modes of operation and the stabilizer

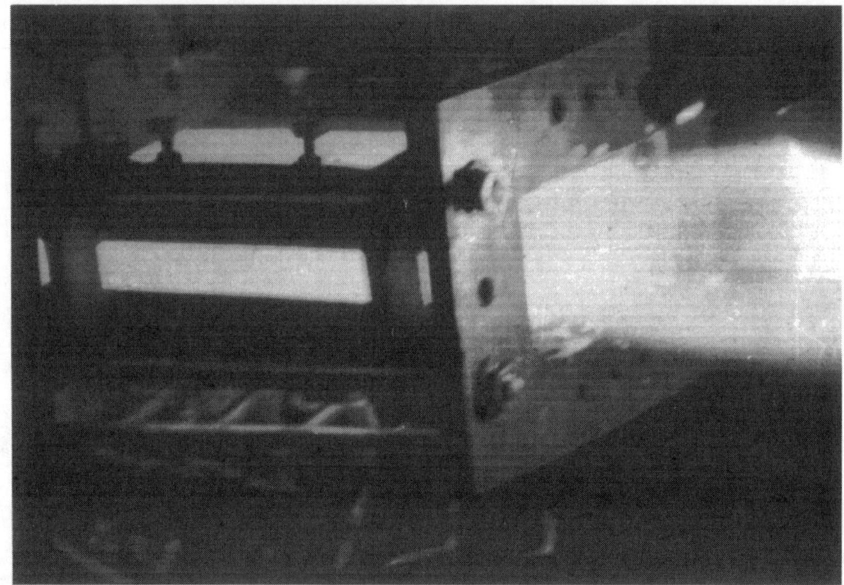

a)

b)

Fig. 29 Uncooled heat-sink hydrogen- and kerosene-fueled scramjet combustors made of niobium alloy with a protective antioxidant coating (from V. N. Avrashkov et al., 1995): a) hydrogen-fueled combustor during direct-connect tests in MAI; b) kerosene-fueled combustor installed at direct-connect test facility.

Fig. 29 (continued) Uncooled heat-sink hydrogen- and kerosene-fueled scramjet combustors made of niobium alloy with a protective antioxidant coating (from V. N. Avrashkov et al., 1995): c) kerosene-fueled combustor during direct-connect tests in MAI.

transformation principle of a wide-range combustion chamber are displayed in Figs. 30a and 30b. The main idea of such a combustor was in joining the tools that are used for ramjet and scramjet combustion process control. It is a common fact that the volume of combustors for ramjets and scramjets differ at M_∞>7 for stoichiometric mixture. The most effective means of combustion stabilization in ramjets are V-gutter stabilizers and gas-generator jets while in scramjet—shocks, cavity, and jet stabilizers. The new concept combustor has a discrete changeable passage geometry comprising the supersonic diffuser aft of the inlet, combustor and nozzle (imitated by the gate-valve), fuel micropylons with a branching space configuration, and turning stabilizers (patent N4883378/06 from 15.10.91). The merit of this scheme consists in the construction of universal stabilizers providing the transformation of V-gutter stabilizers into the cross cascade of thin wedges and flame stabilization in a regular structure of shock waves. Through 1990 this combustor had been successfully tested at conditions corresponding to flight Mach numbers M_∞=3 and M_∞=6 in the subsonic and supersonic combustion regimes, respectively. Tests conditions were: ER=0.5–1.0, T_t =600 K, and P=0.3 MPa for the ramjet regime, T_t=1690 K and P=0.1 MPa for the scramjet regime. The length of the combustor was 400 mm. Tests demonstrated the stable subsonic and supersonic combustion modes with combustion efficiency more than 0.85 and total pressure loss coefficient in the range 0.1–0.3 (Fig. 30c). For the last few years investigations of this concept have been carried out in cooperation with "Aerospatiale" (France), A. Chevalier et al., (1996). Ground tests of a large scale WRR prototype are planned to conduct in Aerospatiale Bourges-Subdray in 1999–2000, F. Falempin et al. (1998).

During 1988–1990 the axysimmetric uncooled kerosene fuel scramjet demonstrator "67M" (Fig. 31a) was designed, manufactured, and tested in

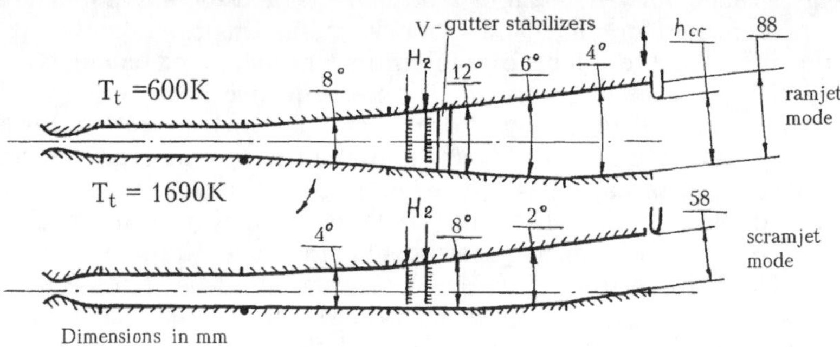

Fig. 30 Wide-range combustion chamber developed and tested in MAI (from V. N. Avrashkov et al., 1995): a) photograph; b) principal scheme.

CIAM (Fig. 31b). The engine was made of heat-resistant materials of the NiAl, FeGrAl, carbon-carbon type with a special protecting coating. The length of the engine was 1280 mm, diameter was 300 mm. Test conditions were M_∞=6.3, T_t =1500–1860 K, and P_t=5.5 MPa.

IV. Short Remarks on Scramjet Inlet and Nozzle Developments

The scramjet inlet development problems are determined by large flight velocities, thick boundary layers (possibly laminar or transitional) on deceleration surfaces, large heat fluxes into the wall, and practically little differ-

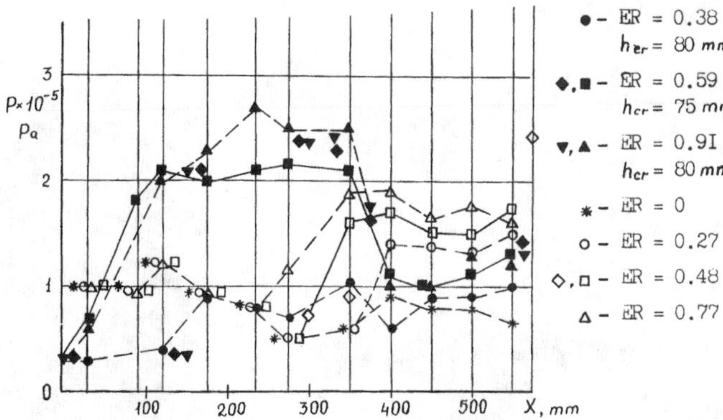

Fig. 30 (continued) Wide-range combustion chamber developed and tested in MAI (from V. N. Avrashkov et al., 1995): c) axial static pressure distributions on the walls of wide-range ramjet combustor.

ence from ramjet inlets destined for use at the same flight conditions. However, because scramjets are intended for use in a larger range of velocities, their inlets have specific features. Among these features are the following: special types of boundary layer separation on the flow deceleration surfaces, a sharp heat flux increase in the region of pressure peaks, blunt inlet leading edges, and so on (see reviews of V. G. Gurylev et al., 1992, 1993a,b, 1996). A

Fig. 31 Scramjet demonstrator: a) Axisymmetric uncooled kerosene fueled scramjet demonstrator "67M" of MAI.

Fig. 31 (continued) Scramjet demonstrator: b) Demonstrator "67M" of MAI in the free-jet test facility of CIAM.

further specific feature of scramjet inlets that does not exist for ramjet inlets is the intrinsic interplay of combustor and inlet. As a consequence the optimal inlet geometry must correspond to the minimum of the overall shock wave and thermal losses. All of these problems were investigated in NII-1, TsAGI, and CIAM.

NII-1 was a pioneer in the field of inlet research for high speed vehicles. The work of G. I. Petrov and E. P. Ukhov (1947) provided a basis for the supersonic and hypersonic inlet design theory. The first tests of axisymmetric inlets at high Mach numbers using aeroballistics and shock tubes were carried out this institute (V. M. Anufriev et al., 1969, 1972). The relationship between the relative inlet throat area and Reynolds number was obtained. At *Re* numbers close to the regime of the turbulent boundary layer the inlet start occured at a relative inlet throat area about 0.12. The flow pattern at the inlet entry at lower Mach numbers was investigated. Earlier unknown types of shock systems and boundary layer separations were discovered. The flow patterns at the flow Mach numbers that are lower, equal to, and greater than design Mach number M_d were also analyzed (M_d = 8). The total pressure recovery coefficients turned out to be near the calculated ones. The essential contribution in inlet investigations of NII-1 was made by S. L. Vishnevetsky and D. A. Mel'nikov (S. L. Vishnevetsky, 1955, 1960; S. L. Vishnevetsky and D. A. Mel'nikov, 1961; V. I. Goryachev, 1963). They studied the different types of inlets, including the annular at Mach numbers up to 6. The problem of inlet shock system optimization was analyzed by S. L. Vishnevetsky (1960).

At TsAGI, high velocity (M ≥ 4) inlet investigations were initiated in the 1960s. The experiment revealed the flow structure in the inlet passage at flight Mach numbers that are lower, equal to, or greater than the design Mach number for the inlet start/unstart and the throttling (P. D. Goncharuk, V. G. Gurylev, E. V. Piotrovich, V. P. Starukhin). The relationship of the inlet throat area from Reynolds number for the laminar, transitionary, and turbulent boundary layers was determined. It was shown that the inlet start and unstart at hypersonic velocities are determined by the stability of the separation zone at the wedge at the entrance of the inlet, but not by the condition of the sonic velocity in the throat, which is typical for supersonic velocities (V. G. Gurylev, A. K. Ivanyushkin, and E. V. Piotrovich, 1973).

In 1964 and 1965, experimental investigations of axisymmetric hypersonic inlets were carried out by V. G. Gurylev, V. V. Zatoloka, Yu. A. Mametiev, and V. P. Starukhin. The boundary layer separation characteristic unique to hypersonic inlets and aggravation of the inlet performance were discovered in these investigations. Since 1965 two-dimensional inlet experimental investigations (V. P. Starukhin, Yu. A. Mametiev, E. V. Piotrovich, A. Ph. Chevagin, V. G. Gurylev, and N. N. Shkirin) were pursued in TsAGI, and inlet performances at M_∞= 4–10, Re = 10^6–10^5 were obtained (Yu. A. Mametiev, E. V. Piotrovich, A. Ph. Chevagin, V. G. Gurylev, N. N. Shkirin). The experimental data and calculations were generalised in V. G. Gurylev's dissertation (TsAGI, 1967). The two-dimensional inlet wedge and side-wall boundary layer interaction with a shock system was examined by V. P. Starukhin and

A. G. Taryshkin (1975, 1982), V. P. Starukhin and A. Ph. Chevagin (1994) of TsAGI, and M. A. Gol'dfeld and V. N. Dolgov (1973) of ITAM. The simplified method of determining inlet–integrated vehicle performance was developed by V. P. Starukhin et al. (1970) and by S. N. Bosnyakov et al. (1994). The two-flowpath inlet for the combined-cycle airbreathing engines with controlled external and internal flaps was proposed by V. G. Gurylev, V. P. Starukhin, and A. Ph. Chevagin in 1977–1988. It allows thrust losses to be diminished at transonic velocities. These results were published by V. G. Gurylev et al. (1992). Experimental results of inlet investigations were generalized in TVF (*Tekhnika Vozdushnogo Flota*) magazine (1977) and later were published in articles (V. G. Gurylev et al., 1992, 1993a,b). Appendix D contains some selected results of experimental investigations on the effect of leading edge bluntness on performance of hypersonic two-dimensional air intake.

The number of computer codes for calculating scramjet two-dimensional and axisymmetric inlets in the range of $M_\infty = 6$–20 taking into account real gas properties, boundary layer, and leading edge bluntness were developed by V. G. Gurylev, Yu. V. Glazkov, S. N. Eliseev, S. B. Mikhailov, G. G. Nersesov, N. N. Shkirin, and N. S. Yatskevich (see, e.g. V. G. Gurylev et al., 1994; E. V. Aleksandrovich, 1994; V. G. Gurylev and S. N. Eliseev, 1994). It was shown in particular that at $M_\infty \geq 10$, thermal flux peaks at the internal inlet surfaces and great net thermal flux may restrict the use of two-dimensional inlets at very high M_∞.

Special among hypersonic inlet investigations are three-dimensional inlets—convergent inlets and Trexler-type (side-wall compression) inlets where the wetted area is lower, and correspondingly they have lower requirements for cooling. Compression surfaces of the convergent inlet are formed from some of the streamlines (of the inviscid flow) behind the shock wave in the flow past the two-dimensional or axisymmetric body. Such construction is based on the idea put forward by G. I. Maikapar (1957) of TsAGI, and named later the convergent air intake. In the same year V. I. Vasiliev (TsAGI) tested an inlet of that type. Further, a rather large series of investigations of convergent inlets was done in ITAM, Novosibirsk in the 1970s by V. V. Zatoloka, B. I. Gutov, Gilyazetdinov, M. A. Gol'dfeld, Kisel', Yu. P. Gun'ko (see review by B. I. Gutov and V. V. Zatoloka, 1983, and detailed list of references in it).

Trexler-type inlets were thoroughly tested both in TsAGI and CIAM (V. A. Vinogradov and V. A. Stepanov of CIAM; Aleksandrovitch and A. Ph. Chevagin of TsAGI) in the 1980s (see, for instance, V. A. Vinogradov et al., 1992c,d). CFD methods were used in CIAM by V. A. Vinogradov et al. (1980, 1982) and V. A. Vinogradov and V. A. Stepanov (1982) (inviscid formulation) and V. I. Vasiliev (1991) (parabolized Navier–Stokes equations). The initial stage of the CIAM inlet investigations was summarized in V. A. Vinogradov's dissertation (1975). Review of experimental and computational research results of 3-D hypersonic inlets is presented in the paper by V.A. Vinogradov et al. (1998).

Several CIAM investigations (V. A. Vinogradov et al., 1974, 1975, 1982; V. A. Vinogradov and V. V. Duganov, 1979) were applied to axisymmetric and two-dimensional inlet configurations. These investigations described the flow pattern at flow Mach numbers greater than the design Mach number. Two-dimensional inlet tests were carried out in wind tunnels with resistance and arc heaters (V. A. Vinogradov and D. A. Ogorodnikov, 1970; V. T. Grin' et al., 1975; N. N. Zakharov, 1974; D. A. Ogorodnikov et al., 1972; D. A. Ogorodnikov, 1993; V. A. Vinogradov et al. (in cooperation with TsAGI, 1992c,d). Investigations determined the impact of the inlet geometry, incoming flow parameters, angle of attack, degree of inlet surface cooling, boundary layer control using suction, injection/bleeding, and flow throttling on start and unstart processes and inlet performances.

Joint CIAM–TsAGI investigations of Trexler-type inlets (V. A. Vinogradov et al., 1992c,d) showed that this type of inlet starts at a lower Mach number range and has lower drag at a higher Mach number range than two-dimensional inlets. However, at a lower Mach number range, two-dimensional inlets have some advantage in drag and pressure recovery. The tangential hydrogen injection in scramjet at high flight velocities was investigated in CIAM for boundary layer control in regions with unfavorable pressure gradients (V. T. Grin', 1967, 1971; V. T. Grin' and N.N. Zakharov, 1971; V. I. Vasiliev et al., 1973; A. A. Betev and N. N. Zakharov, 1992). In some cases the gas injection allows separation into attached flow. Boundary layer control permitted not only improved inlet performance but also a minimal relative inlet throat (V. T. Grin' et al., 1975). It was found that the "isolator" shape and its length have the essential influence on the joint inlet/combustor operation. The combustion efficiency was in turn a function of the flow pattern in the entrance of the combustor (V. A. Vinogradov et al., 1990, 1992a,b). Examples of an inlet geometry design that uses CIAM recommendations were given by O. N. Romankov and F. I. Starostin of "Soyuz" (1993) and V. A. Vinogradov et al. (1993, 1994) of CIAM.

There was no flight flow condition simulation in ground facilities at high flow velocities ($M_\infty \geq 6$), and there were principal difficulties in converting test data into corresponding flight ones. Methods of equivalent models based on the stimulation of flow at the final inlet ramp were developed in TsAGI. They allowed some of the difficulties to be overcome both at flight Mach numbers M_∞ lower than the design Mach number M_d (V. G. Gurylev, A. K. Ivanyushkin and E. V. Piotrovich) and at the more subtle case when $M_\infty > M_d$ (V. G. Gurylev). These methods allowed testing of the scramjet combustors at conditions nearing natural ones. Two-dimensional inlets placed under the vehicle wing (which provides air flow precompression) were analyzed in TsAGI by V. P. Starukhin and A. G. Taryshkin (1982), V. P. Starukhin et al., (1970), V. P. Starukhin and A. Ph. Chevagin (1994), S. N. Bosnyakov et al., (1994), S. A. Bakharev et al., (1990).

For flight Mach number $M_\infty > 5$, heat flux problems begin to be of importance. Tests and calculations that started in TsAGI in the 1970s demonstrated that cooling inlet surfaces prevented boundary layer separation at high positive gradients of pressure (V. G. Gurylev and Yu. A. Mametiev, 1975). The combined use of the thermoindicator paint method that was developed in TsAGI and the method of eroding "soot-oil paints" showed the flow pattern and the heat flux distributions. It was found that the highest heat fluxes occur at the corners of the inlet throat and where shocks interact with the boundary layer at the inlet compression wedge and cowl (V. G. Gurylev and N. N. Shkirin, 1978). Specific requirements of the scramjet passage geometry caused by shock and total thermal loss minimization, optimal shock system selection, diffusion, or combustion detonation mode in combustor and ramjet/vehicle integration were investigated by V. I. Penzin (1966, 1980, 1981, 1982a, 1992).

The scramjet nozzle investigations were started in the 1960s in TsAGI (A. N. Dubov and V. D. Sokolov) and in CIAM (V. M. Khailov and R. Tagirov). Numerical methods were largely used in TsAGI. Calculation procedures for equilibrium and nonequilibrium gas flows in nozzles were developed by A. N. Dubov in 1962 and in 1966–1980, respectively. The analogous calculation codes were developed in CIAM (V. M. Khailov, 1975).

The factors that influence scramjet nozzle loss were determined. It was established that the main impact came from the contour losses and losses connected with the off-design nozzle flow. Fewer losses come from friction and nonequilibrium chemistry in the nozzle. The optimal scramjet nozzle contouring used the method developed in ITPM (Novosibirsk) by A. I. Rylov (1981).

Experimental methods for determining scramjet nozzle performance were developed in TsAGI by A. N. Dubov through 1987. These methods were based on the comparison of the results of tests with the use of different nozzle aft parts.

V. Conclusion

This paper tried as far as possible to give a complete and objective review of the historical scramjet technology development process in Russia. It should, however, be kept in mind that until now some part of the former research and development results were still restricted (classified) and not available for publication in open literature. We ask in advance to accept our apology to those specialists whose names and results turned out to be outside of this paper; such omissions happened unintentionally.

On the basis of the results presented in the paper it can be concluded that Russian research and industry institutions have a wealth of experience in and an excellent understanding of the numerous complex physical phenomena encountered in scramjet engine technology. This know-how has been combined with test facilities offering testing potentials that cannot be found in any other European country.

We sincerely hope that our paper can help to reconstruct some pages of the glorious history of scramjet technology development in Russia and to do justice to the outstanding personalities of this history.

Acknowledgments

We sincerely wish to express our appreciation to our colleagues V. N. Ostras', V. N. Sermanov, O. V. Voloshchenko and V. P. Starukhin of TsAGI, V.

N. Strokin, V. A. Vinogradov and V. I. Kopchenov of CIAM and V. M. Levin and V. N. Avrashkov of MAI who have put some material on the history of scramjet investigations in the above institutions at our disposal. V. A. Vinogradov, V. N. Strokin, E. A. Meshcheryakov, V. P. Starukhin, V. K. Baev and P. K. Tret'yakov made a great number of remarks and advice concerning historical aspects of the problem considered in the paper. These remarks and advice allowed both to refine the subject of the presentation and to improve the paper. The authors are grateful to the above colleagues for their help. But, naturally, all responsibility for the final presentation of the material lies with the authors of the paper.

Bibliography

Albegov, R. V., Kourziner, R. I., Petrov, M.D., and Sheechman, Yu.M., "Working Process Features in a Combined Combustor at Heat Release to Subsonic and Supersonic Flows," Trudy 7-kh Tsanderovkskikh Chtenii, Moscow, 1985 pp. 93–106 (in Russian)

Albegov, R. V., Vinogradov, V. A., Petrov, M. D., and et al., "Experimental Investigation of 2-D Scramjet Analog-Demonstrator at BMG Test Facility at $M = 7$," Trudy Central Inst. of Aviation Motors, No. 700–127, 1993, (in Russian).

Aleksandrovich, E. V., "Drag of Blunt Leading Edges of Hypersonic Inlets in the Mach Number Range M_∞ =0.6–23," Trudy 18-kh Nauchnykh Chtenii po Kosmonavtike, Fazis, Moscow, 1994, pp. 16–18 (in Russian).

Alfyorov, V. I., "On the Problem of Creating the Flow Conditions in Scramjet Combustion Chamber in Wind Tunnels," AIAA Paper 93–1843, 1993.

Andreev, V. V., and Penzin, V. I. "On Integration of a Scramjet with a Plane," TsAGI Preprint No. 67, 1992 (in Russian).

Andreev, V. V., and Penzin, V. I., "On the Choice of a Scramjet Geometry Integrated with a Plane," Trudy 18-kh Nauchnykh Chtenii po Kosmonavtike, Fazis, Moscow, 1994 pp. 5–7 (in Russian).

Annushkin, Yu.M., "Governing Principles of Hydrogen Jets Burn-of in Air Ducts," *Fizika Goreniya i Vzryva,* Vol. 17, No. 4, 1981, pp. 59—71 (in Russian).

Annushkin, Yu.M, and Sverdlov, E. D., "Stability of Submerged Diffussion Flames in Subsonic and Underexpanded Supersonic Gas-Fuel Streams," *Combustion, Explosion and Shock Waves,* Vol. 14, No. 5, 1979, pp. 597–606.

Annushkin, Yu.M., and Maslov, G.Ph., "Experimental Investigations of Hydrogen-Kerosene Fuel in Ramjet," *Fizika Goreniya i Vzryva,* Vol. 18 No. 2, 1982, pp. 30–36 (in Russian).

Annushkin, Yu.M., Maslov, G.Ph. and Sverdlov, E. D., "Stabilization of Hydrogen Diffusion Flame in Co-Stream of Air," *Fizika Goreniya i Vzryva,* Vol. 19, No. 6 1983, pp. 14–20 (in Russian).

Annushkin, Yu.M., and Sverdlov, E. D., "On the Rules of Length Change of Diffusion Flames in Co-Stream of Air," *Fizika Goreniya i Vzryva,* Vol. 20, No. 3, 1984, pp. 46–51 (in Russian).

Anufriev, V. M., Kozlov, G. I., and Roitenburg, G. I., "Using of the Shock Tube for Super-Sonic Inlet Investigations," *Izvestiya AN SSSR, Mekhanika Zhidkosti i Gaza,* No. 2, 1969, 33–36 (in Russian).

Anufriev, V. M., Kozlov, G. I., and Roitenburg, G. I., "Investigation of Inlet Characteristics in Shock Tube," *Izvestiya AN SSSR, Mekhanika Zhidkosti i Gaza,* No. 1 1972 pp. 156–161 (in Russian).

Avrashkov, V. N., Baranovsky, S. I., and Levin, V. M., "Experimental Investigation of Ignition and Combustion of a Liquid Kerosene in Supersonic Flow in a Plane Channel," *Trudy MAI: Raschetnye i experimentalnye Issledovaniya Vozdushno-Reaktivnykh Dvigatelei,* Izdatel'stvo MAI, Moscow, 1988, pp. 22–27 (in Russian).

Avrashkov, V. N., Baranovsky, S. I., and Davidenko, D. M., "On Fuel in Supersonic Combustors at Injection of Liquid Fuel from Wall," *Izvestiya VUZov, Aviatsionnaya Tekhnika,* No. 3, 1989, pp. 72–74 (in Russian).

Avrashkov, V. N., Baranovsky, S. I., and Davidenko, D. M. "Penetration Height of a Liquid Jet Saturated by Gas Bubbles," *Izvestiya VUZov, Aviatsionnaya Tekhnika,* No. 4, 1990a, pp. 96–98 (in Russian).

Avrashkov, V. N., Baranovsky, S. I., and Levin, V. M., "Gasdynamic Features of Supersonic Kerosene, Combustion in Model Combustion Chamber," AIAA Paper 90–5268, 1990b.

Avrashkov, V. N., Grigor'ev, S. V., Davidenko, D. M., and Levin, V. M., "Peculiarities of Experimental Investigations Methodology of a Working Process in Combustion Chambers of Ramjet in MAI," *Teorya Vozdushno-Reaktivnykh Dvigatelei,* Izdatel'stvo MAI, Moscow, 1995, pp. 194–206 (in Russian).

Avrashkov, V. N., Mayinger, F., and Grüning, C., "Influence of Shock Waves on Mixing Processes in Supersonic Hydrogen-Air Flames," Tech. Rept. INTAS-94-0079, 1996.

Baev, V. K., Lokotko, A. V., and Tret'yakov, P. K., "Influence of the Structure of the External Supersonic Flow on the Stabilization of Flame Behind of Axisymmetric Body," *Fizika Goreniya i Vzryva,* Vol. 9, No. 5, 1973, pp. 721–724 (in Russian).

Baev, V. K., Golovichev, V. I., Dimitrov, V. I., and Yasakov, V. A., "Calculation of Auto-Ignition and Combustion of a Hydrogen Jet Air with Finite Chemical Reaction Rates," *Acta Astronautica,* Vol. 1, pp. 1974, 1227–1238.

Baev, V. K., and Yasakov, V. A., "Experimental Investigations of Combustion of Axisymmetric Hydrogen Jet in Channel with Constant Cross Area Section," *Fizika Goreniya i Vzryva,* Vol. 11, No. 5, 1975, pp. 687–693 (in Russian).

Baev, V. K., Garanin, A.Ph., and Tret'yakov, P. K., "Investigation of Flow Structure Behind of Axisymmetric Body in Supersonic Flow at Injection of Inert and Reactive Gases," *Fizika Goreniya i Vzryva,* Vol. 11, No. 6, 1975, pp. 859–963 (in Russian).

Baev, V. K., Garanin, A.Ph., and Tret'yakov, P. K., "Flame Stabilization in Supersonic Flow on Opposed Hydrogen Jet," *Phizicheskaya Gazodinamica,* ITPM SO AN SSSR, 1976a, pp. 52, 53 (in Russian).

Baev, V. K., Golovichev, V. I., and Yasakov, V. A., *Two-Dimensional Turbulent Flows of Reactive Gases,* Nauka, Novosibirsk, 1976b (in Russian).

Baev, V. K., Vorontsov, S. S., Zabaikin, V. A., and Konstantinovskii, V. A., "Using of the Method of Resonant Absorption for Determination of Residence Time in a Recirculation Zone," *Fizika Goreniya i Vzryva,* Vol. 15, No. 6, 1979a, pp. 83–86 (in Russian).

Baev, V. K., Konstantinovskii, V. A., and Tret'yakov, P. K., "Simulation of a Ramjet Combustor," *Dinamika Goreniya v Sverkhzvukovykh Potokakh,* ITPM SO AN SSSR, Novosibirsk, 1979b, pp. 3–26 (in Russian).

Baev, V. K., Boshenyatov, B. V., Pronin, Yu.A., and Shumskii, V. V., "Investigation of a Liquid Injection into Supersonic Flow of High-Enthalpy Gas," *Fizika Goreniya i Vzryva,* Vol. 17, No. 3, 1981a, pp. 72–76 (in Russian).

Baev, V. K., Tret'yakov, P. K., and Konstantinovskii, V. A., "The Study of Hydrogen Combustion in a Ducted Supersonic Flow with a Sudden Expansion," *Archivum Combustion,* Vol. 1, No. ¾ 1981b, pp. 251–259.

Baev, V. K., Pronin, Yu.A, and Shumskii, V. V., "Auto-Ignition of Liquid Fuels in Supersonic Flow of Air," *Fizika Goreniya i Vzryva,* Vol. 18, No. 4, 1982a, pp. 22–26 (in Russian).

Baev, V. K., Shumskii, V. V., and Yaroslavtsev, M. I., "Investigation of Dual-Mode Scramjet Combustor at Subsonic Heat Release," *Gasodinamika Techenii v Soplakh i Diffusorakh,* ITPM SO AN SSSR, Novosibirsk, 1982b pp. 86–105 (in Russian).

Baev, V. K., Shumskii, V. V., and Yaroslavtsev, M. I., "Investigation of Gasdynamics of the Model with Combustion in Pulse Facility," *Prickladnaya Mechanika i Tekhnicheskay Physicka,* No. 6, 1983, pp. 58–66 (in Russian).

Baev, V. K., Golovichev, V. I., and Tret'yakov, P. K., *Combustion in Supersonic Flow,* Nauka, Novosibirsk, 1984a (in Russian).

Baev, V. K., Shumskii, V. V., and Yaroslavtsev, M. I., "Investigation of Power Characteristics and Flow Parameters in the Channel of a Model with Combustion," *Prickladnaya Mechanika i Tekhnicheskay Physicka,* No. 1, 1984b, pp. 103–109 (in Russian).

Baev, V. K., Shumskii, V. V., and Yaroslavtsev, M. I., "Investigation of the Pressure Distribution and Heat Transfer in Gasdynamic Model with Combustion in a High-Enthalpy Air Stream," *Prickladnaya Mechanika i Tekhnicheskay Physicka,* No. 5, 1985, pp. 56–65 (in Russian).

Baev, V. K., Golovichev, V. I., and Tret'yakov, P. K., "Combustion in Supersonic Flow," *Fizika Goreniya i Vzryva,* Vol. 23, No. 5, 1987, pp. 5–15 (in Russian).

Baev, V. K., Shumskii, V. V., and Yaroslavtsev, M. I., "Study of Combustion and Heat Exchange Processes in High-Enthalpy Short-Duration Facilities," *High-Speed Flight Propulsion Systems,* edited by S. N. B. Murthy and E. T. Curran, Progress in Astronautics and Aeronautics, Vol. 137, Washington, DC, 1991, Chap. 8, pp. 457–481.

Baev, V. K., and Shumskii, V. V., "Effect of the Gasdynamics of a Two-Regime Combustor on the Power Characteristics of a Model with Combustion," *Combustion, Explosion and Shock Waves,* Vol. 31, No. 6, 1995, pp. 661–670.

Bakharev, S. A., Gurylev, V. G., and Kosykh, A. P., "Aerodynamic Characteristics of Slender Sharp Triangular Wings with Air Extraction Through Inlet at Hypersonic Velocities," *Uchenye Zapiski TsAGI,* Vol. 21, No. 6, 1990, pp. 67–77 (in Russian).

Baranovsky, S. I., Levin, V. M., and Turishchev, A. I., "Experimental Investigation of Turbulent Kerosene Spray Flame," *Fizika Goreniya I Vzryva,* Vol. 21, No. 6, 1985 (in Russian).

Baranovsky, S. I., Levin, V. M., and Turishchev, A. I., "Supersonic Combustion of Kerosene in a Cylindric Duct," *Structure of Gas-Phase Flames. Pt. 1,* Novosibirsk, 1986, pp. 114–120 (in Russian).

Baranovsky, S. I., and Levin, V. M., "Gas Dynamics of Flow Structure in a Channel Under Thermal and Mechanical Throttling," *1st Symposium (Int.) on Experimental and Computational Aerothermodynamics of Internal Flows,* 1990, pp. 763–767.

Baranovsky, S. I., Levin, V. M., Nadvorsky, A. S., and Turishchev, A. I., "Heat

Transfer in Supersonic Coaxial Reacting Jets," *International Journal of Heat Mass Transfer,* Vol. 33, No. 4, 1990, pp. 641–648.

Baranovsky, S. I., and Levin, V. M., "Precombustion Shock Wave as a Means of the Working Process Control in a Supersonic Combustion Chamber," *International Symposium on Air Breathing Engines,* 1991a.

Baranovsky, S. I., and Levin, V. M., "Wide Range Combustion Chamber of Ramjet," AIAA Paper 91–5094, 1991b.

Baranovsky, S. I., Davidenko, D. M., and Levin, V. M., "Combustion Chamber of Ramjet for Aerospace Plane," *9th World Hydrogen Conference Paris,* 1992a, pp. 1583–1591.

Baranovsky, S. I., Davidenko, D. M., Konovalov, I. V., and Levin, V. M., "Experimental Study of the Hydrogen Supersonic Combustion," *9th World Hydrogen Conference Paris,* 1992b, pp. 699–1708.

Baranovsky, S. I., Davidenko, D. M., and Levin, V. M., "Test Facility for the Flow Structure Study in the Supersonic Combustion Chamber," *6th International Symposium on Flow Vizualization,* 1992c.

Belotserkovskii, O. M., and Shcherbina, Yu. Ya., "About Scientific and Pedagogical Activity of E. S.Shchetinkov," *Iz Istorii Aviatsii i Kosmonavtiki, Vypusk 46,* AN SSSR, Moscow, 1982, pp. 41–43 (in Russian).

Bespalov, I. V., and Makaron, N. S., "Life and Activity of Prof. E. S.Shchetinkov," *Iz Istorii Aviatsii i Kosmonavtiki, Vypusk 46,* AN SSSR, Moscow, 1982, pp. 3–12 (in Russian).

Betev, A. A., and Zakharov, N. N., "Control of a Separation Flow by Using a Tangential Injection in Channels of Supersonic Energetic Facilities," *Sbornick Statei CIAM, Vypusk 3,* 1992, pp. 12–29 (in Russian).

Bezgin, L. V., Ganzhelo, A. N., Gouskov, O. V., Kopchenov, V. I., Laskin, I., and Lomkov, K., "Numerical Simulation of Supersonic Flows Applied to Scramjet Duct," *12th International Symposium on Airbreathing Engines,* 1995, pp. 895–905.

Bezgin, L. V., Ganzhelo, A. N., Gouskov, O. V., and Kopchenov, V. I., "Numerical Simulation of Viscous Non-Equlibrium Flows in Scramjet Elements," *13th International Symposium Air Breathing Engines,* 1997, pp. 976–986.

Bondaryuk, S. M., and Il'yashenko, A. V., *Ramjets,* Oborongiz, Moscow, 1958 (in Russian).

Bosnyakov, S. M., Starukhin, V. P., and Chevagin, A.Ph., "Influence of Wing on Characteristics of a Underwing Two-Dimensional Inlet," *Uchenye Zapiski TsAGI,* Vol. 24, Nos. 1–2, 1994, pp. 67–76 (in Russian).

Chalot, F., Rostand, Ph., Perrier, P., Goun'ko, Yu. P., Kharitonov, A. M., Laypov, A. F., Mazhul I. I. & Yaroslavtsev, M. I. 1998. Validation of gobal aeropropulsive characteristics of integrated configurations. AIAA paper 98-1624, 8p. Norfolk, USA, 8th International Spaces and Hypersonic Systems and Technologies Conference, April 1998.

Chernyi, G. G., "Supersonic Flow Past Bodies with Generation of Detonation and Slow Combustion Fronts," *Astronautica Acta,* Vol. 13, 1968, pp. 467–480.

Chernov, V. A., and Kiseleva, E. N., *Kinetics and Aerodynamics of Fuel Combustion Processes,* Nauka, Moscow, 1969 (in Russian).

Chevalier, A., Levin, V. M., Bouchez, M, and Davidenko, D. M., "French-Russian Partnership on Hypersonic Wide Range Ramjets," AIAA Paper 96-4554-CP, 1996.

Dimitrov, V. I., *Reduced Kinetics*, Nauka, Novosibirsk, 1982 (in Russian).

Dunlap, R., Brehm, R. L., and Nickolls, J.A., "A Preliminary Study of the Application of Steady-State Detonative Combustion to a Reaction Engine," *Jet Propulsion,* Vol. 28, 1958, p. 451.

Falempin, F., Levin, V., Avrashkov, V., Davidenko, D. & Bouchez, M. 1998. French-Russian Partnership on Hypersonic Wide Ramjets: Status in 1998. AIAA paper 98-1545, 8p. Norfork, USA, 8th International Spaces and Hypersonic Systems and Technologies Conference, April 1998.

Ferri, A. 1960. Possible Directions of Future Research in Airbreathing Engines. In: Fourth AGARD Colloquium, Milan, Italy; also 1961: Combstion and Propulsion—High Mach Number Air Breathing Engines, Edited by Jaumotte, Rothrock and Lefebvre, Pergamon Press, New York, pp 3–15.

Gol'dfeld, M. A., and Dolgov, V. N., "Experimental Investigation of Turbulent Boundary Layer on Triangular Plate with Wedge," *Izvestiya SO AN SSSR, Seriya Technich. Nauk,* Vyp. 2, No. 8 1973, pp. 16–22 (in Russian).

Goryachev, V. I., "Supersonic Flow over Two-Shock Round Inlets," NII-1, Rept. No. 1963, Moscow, p.63.

Grin', V. T., "Experimental Investigation of Control of Boundary Layer Using Injection on a Plane Plate at Mach Number M=2.5," *Izvestiya AN SSSR, Mekhanika Zhidkosti i Gaza,* No. 6, 1967, pp. 115–117 (in Russian).

Grin', V. T., "Tangential Injection into Supersonic Boundary Layer," *Sbornick statei "Pogranichnyi sloi i teploobmen,"* Trudy Central Inst. of Aviation Motors, No. 507, 1971, pp. 58–69 (in Russian).

Grin', V. T., and Zakharov, N. N., "Experimental Investigation of Influence of a Tangential Injection and Cooling of Wall on Flow with a Separation," *Izvestiya AN SSSR, Mekhanika Zhidkosti i Gaza,* No. 6, 1971, pp. 144–147 (in Russian).

Grin', V. T., Zakharov, N. N., and Ogorodnikov, D. A., "Investigation of Flow in Channels of Scramjet Inlets," Trudy 3-kh Chtenii, Posvyashchennykh Tsanderu, 1975, pp. 98–112 (in Russian).

Gromov, V. G., Larin, O. B., and Levin, V. A., "On Application of the 'Unmixedness' Model for the Calculation of a Turbulent Wall-Jet of Hydrogen in Co-Flowing Supersonic Air Stream," *Khimicheskaya Physica,* Vol. 3, No. 8, 1984, pp. 1190–1195 (in Russian).

Gurylev, V. G., and Eliseev, S. N. "About Theory of Pseudo-Shock at Entrance of a Channel," *Uchenye Zapiski TsAGI,* Vol. 3, No. 3, 1972, pp. 25–35 (in Russian).

Gurylev, V. G., Ivanyushkin, A. K., and Piotrovich, E. V., "Experimental Investigation of Reynolds Number Influence on Start of Inlets at High Supersonic Velocities of Flow," *Uchenye Zapiski TsAGI,* Vol. 4, No. 1, 1973, pp. 33–44 (in Russian).

Gurylev, V. G., and Mamet'ev, Yu.A., "Influence of Cooling of Central Body on Start and Breakup of Flow at the Entrance and Throttling Characteristics of Inlets at Supersonic and Hypersonic Velocities," *Uchenye Zapiski TsAGI,* Vol. 6, No. 2, 1975, pp. 139–146 (in Russian).

Gurylev, V. G., and Trifonov, A. K., "Pseudo-Shock in the Simple Inlet as a Cylindric Duct," *Uchenye Zapiski TsAGI,* Vol. 7, No. 1, 1976, pp. 80–89 (in Russian).

Gurylev, V. G., and Shkirin, N. N., "Heat Fluxes in Hypersonic Inlets with Turbulence Generator and Bluntness of Central Body," *Uchenye Zapiski TsAGI,* Vol. 9, No. 4, 1978, pp. 24–34 (in Russian).

Gurylev, V. G., and Penzin, V. I., "Transition of Supersonic Flow into Subsonic in Channel with Short Contractor and Cylindrical Entrance Part," 7 Chteniya Tsandera, pp. 84–92 (in Russian).

Gurylev, V. G., Starukhin, V. P., Zhdanov, V. T., Chevagin, A.Ph., Glazkov, Yu.V., and Eliseev, S. N., "Aerodynamics Problems of Inlets of Combined Propulsion Engines of Hypersonic Vehicles with Air-Breathing Engines," *"1st International Aviation-Space Conference 'Chelovek-Zemlya-Kosmos',"* Moscow, 1992, pp. 510–521 (in Russian).

Gurylev, V. G., Starukhin, V. P., and Chevagin, A.Ph., "Problems of Aerogasdynamics for Hypersonic Inlets of Combined Propulsion System", *International Conference on Powerplant Aincraft Aerogazodinamics,* Zhukovsky, Russia, 1993a, p. 29 (in Russian).

Gurylev, V. G., Starukhin, V. P., Chevagin, A.Ph., Zhdanov, V. T., Eliseev, S. N., and Glazkov, Yu.V., "Aerogasdynamics of Inlets of Combined Propulsion Engines of Air-Space Planes with Air-Breathing Engines," 17 Nauchnye Chteniya po Kosmonavtike, Fazis, Moscow, 1993b, pp. 72–74 (in Russian).

Gurylev, V. G., and Eliseev, S. N., "Controlled Two-Dimensional Inlets of Scramjet for Mach Number Range $M_\infty = 5–15$," 18 Nauchnye Chteniya po Kosmonavtike, Fasis, Moscow, 1994, pp. 19, 20 (in Russian).

Gurylev, V. G., Nersesov, G. G., and Kupriyanova, T. V., "Two-Dimensional Fixed Geometry Scramjet Inlets with Blunt Leading Edges at High Flight Mach Numbers ($M_\infty = 6,10,16$)," 18 Nauchnye Chteniya po Kosmonavtike, Fasis, Moscow, 1994, pp. 14, 15 (in Russian).

Gurylev, V. G., Dubov, N. A., Sabel'nikov, V. A., and Starukhin, V. P., "Aerodynamics of Propulsion Systems of Hypersonic Vehicles with Air-Breathing Engines," *TsAGI-Osnovnye Etapy Nauchnoi Deyatel' Nosti 1968–1993,* edited by G.S. Byushgens, Nauka, Moskow 1996, pp. 208–216 (in Russian).

Gusarov, G. P., and Repnikov, A. A., "Theoretical Investigation of Different Means of Scramjet Forcing," *Trudy 2-kh Chtenii Tsandera,* 1972, pp. 35–45 (in Russian).

Gutov, B. I. and Zatoloka, V. V., "Computational and Experimental Investigation of New Configuration of Convergent Inlets with 3-D Flows," ITPM SO AN SSSR, Novosobirsk, Preprint No. 30–83, 1983 (in Russian).

Hicks, J. W. 1998. International efforts scram into flight. *Aerospace America,* June 1998, pp. 28–33.

Ianovski, L. S., Sosounov, V. A., and Shikhman, Yu. M., "The Application of Endothermic Fuels for High Speed Propulsion Systems," 13th International Symposium on Air Breathing Engines 1997, pp. 59–69.

Janovsky, R., Weinss, W. & Kovaltschuk, V. 1998. The Hypersonic Technology Demonstrator D2. AIAA Paper 98-1645, 9p., Norfork, USA, 8th International Spaces and Hypersonic Systems and Technologies Conference, April 1998.

Kandebo, S. W., "Russians Want US to Join Scramjet Tests," *Aviation Week and Space Technology,* 1992 March 30, pp. 18–20.

Keldysh, M. V. (ed.), *Tvorcheskoe Nasledie Akademika Koroleva,* Nauka, Moscow, 1980 (in Russian).

Khailov, V. M., *Chemical Relaxation in Jet Engine Nozzles,* Mashinostroenie, Moscow, 1975, (in Russian).

Kolesnikov, O. M., "Ignition of Underexpanded near Wall Hydrogen Jet in a

Supersonic Stream," *3-D International Seminar on Flame Structure, Book of Abstracts,* ITPM, Novosibirsk, 1989 (in Russian).

Kolesnikov, O. M., Makarov, I. G., Meshcheryakov, E. A., and Sabel'nikov, V. A., "Numerical Modeling of Supersonic Hydrogen Combustion," TsAGI, Preprint No. 63, 1992 (in Russian).

Kopchenov, V. I., and Lomkov, K., "The Enhancement of the Mixing and Combustion Processes Applied to Scramjet-Engine," AIAA Paper 92-3428, 1992.

Kopchenov, V., Lomkov, K., Miller, L., Krjukov, V., Rulev, I., Vinogradov, V., Stepanov, V., Zakharov, N., Tagirov, R., and Aukin, M., "Scramjet CFD Methods and Analysis," *Part 1. Scramjet CFD Methods,* AGARD Lecture Series 194, 1993a, pp. 4.1–4.20.

Kopchenov, V., Lomkov, K., Zaitsev, S., and Borisov, I., "Scramjet CFD Methods and Analysis," *Part 2. Scramjet CFD Analysis,* AGARD Lecture Series 194, 1993b, pp. 8.1–8.30.

Kourziner, R. I., *Jet Engines for High Supersonic Velocities of Flight (Theory Fundamentals),* Mashinostroenie, Moscow, 1977 (in Russian).

Kourziner, R. I., "About Activity of Prof. E. S. Shchetinkov," *Iz Istorii Aviatsii i Kosmonavtiki, Vypusk 46,* AN SSSR, Moscow, 1982, pp. 13–18 (in Russian).

Kourziner, R. I., *Jet Engines for High Supersonic Velocities of Flight, 2nd ed.* Mashinostroenie, Moscow, 1989 (in Russian).

Kozlov, V. E., and Sabel'nikov, V. A., "Computation of Deceleration of Viscous Supersonic Flow in Channels," *Izvestiya AN SSSR,* Mekhanika Zhidkosti i Gaza, No. 2, 1982, pp. 162–166 (in Russian).

Kuznetsov, V. R., and Sabel'nikov, V. A., *Turbulence and Combustion,* Hemisphere, New York, 1990.

Leonov, B. P., Shteiman, S. V. and Kulikov, A. V., "Methods of Calculation Combustion Efficiency in Supersonic Flows," *Fizika Goreniya i Vzryva,* Vol. 7, No. 4, 1971, pp. 572–577 (in Russian).

Maikapar, G. I., "On the Wave Dray of Axisymmetric Bodies at Supersonic Speeds," *Prikladnaya Matematika i Mekhanika,* Vol. 23, 1959, p. 528 (in Russian).

Makaron, V. S., and Sermanov, V. N., "Rocket-Ramjet Engine of Air Liquifaction Cycle (LACRRE) Performance Analysis and Experimental Investigations," *13th International Symposium on Air Breathing Engines,* 1997, pp. 1259–1265.

Merkulov, I. A., "First Experimental Tests of Ramjets Developed in GIRD," *Iz Istorii Aviatsii i Kosmonavtiki Vypusk 3,* AN SSSR. Moscow, 1965, pp. 22–32 (in Russian).

Meshcheryakov, E. A., and Makasheva, O. V., "Calculation of Condition for Extinction of Combustion Behind of a Step and in a Cavity in Supersonic Flow of Hydrogen-Air Mixture," *Fizika Goreniya i Vzryva,* Vol. 12, No. 6, 1976, pp. 871–879 (in Russian).

Meshcheryakov, E. A., and Sabl'nikov, V. A., "Combustion of Hydrogen in Supersonic Flow in Channel at Co-Flow Supply of Fuel to Air Flow," *Fizika Goreniya i Vzryva,* Vol. 17 No. 2, 1981, pp. 55–64 (in Russian).

Meshcheryakov, E. A., Levin, V. M., and Sabel'nikov, V. A., "Numeric and Experimental Investigation of Combustion of Hydrogen Jet in Co-Flow Air Stream in Channel," *Trudy TsAGI,* Vypusk 2193, 1983 (in Russian).

Meshcheryakov, E. A., and Sabel'nikov, V. A., "About Influence of Concentration

Fluctuations on Supersonic Combustion of nonpremixed Gases in Ducts," *Trudy 7-kh Tsanderovskikh Chtenii,* Moscow, 1985, pp. 116–125 (in Russian).

Meshcheryakov, E. A., "On Heat Release to Supersonic Flow in Channel," *Uchenye Zapiski TsAGI,* Vol. 18 No. 2, 1987, pp. 125–129 (in Russian).

Meshcheryakov, E. A., and Sabel'nikov, V. A., "Interplay of Mixing and Kinetics in Decreasing of Heat Release at Supersonic Combustion of Nonpremixed Gases in Diverging Channels," *Fizika Goreniya vi Vzryva,* Vol. 24, No. 5, 1988, pp. 23–32 (in Russian).

Obraztsov, I.Ph. (ed.), "The Development of Aviation Science and Technology in USSR," *History-Tekhnic Essay,* Inst. Istorii Estestvoznaniya i Tekhniki AN SSSR, Moscow 1980, p. 192.

Ogorodnikov, D. A., "Control of Supersonic Boundary by Suction or Bleed," *Sbornick Statei "Pogranichnyi Sloi i Teploobmen,"* Trudy Central Inst. of Aviation Motors, No. 507, 1971, pp. 42–57 (in Russian).

Ogorodnikov, D. A., Grin', V. T., and Zakharov, N. N., "Boundary Layer Control of Hypersonic Inlets," NASA TTF-1397, 1972.

Ogorodnikov, D. A., "CIAM Hypersonic Investigations and Capabilities," AIAA Paper 93–5093, 1993.

Ogorodnikov, D. A., Vinogradov, V. A., Shikhman, Ju. M. & Strokin, V. N. 1998. Design and research Russian program of experimental hydrogen fueled dual mode scamjet: choice of conception and results of pre-flights tests. AIAA Paper 98-1586. Norfolk, USA, 8th International Spaces and Hypersonic Systems and Technologies Conference, April 1998, pp. 724–734.

Ostras', V. N., and Penzin, V. I., "Experimental Investigation of Force to Internal Surface of Cylindric Tube at Flowing Through It of Nonhomogeneous Supersonic Flow Generated by Conical Nozzles," *Uchenye Zapiski TsAGI,* Vol. 3, No. 4, 1972, pp. 29, 30 (in Russian).

Ostras', V. N., and Penzin, V. I., "Experimental Investigation of Friction Force in Channel with Pseudo-Shock," *Uchenye Zapiski TsAGI,* Vol. 5, No. 3, 1974, pp. 151–155 (in Russian).

Ostras', V. N., and Penzin, V. I., "About Change of Characteristics of Separated Flow Generated by Throttling of Supersonic Flow in Channel," *Uchenye Zapiski TsAGI,* Vol. 7, No. 3, 1976, pp. 39–46 (in Russian).

Penzin, V. I., "About Conditions of an Optimum of Supersonic Flows with Oblique Shock Systems and Subsequent Heat Release," *Aviatsionnaya Tekhnika,* No. 4, 1966, pp. 114–120 (in Russian).

Penzin, V. I., "Experimental Investigation of Normal Injection to Supersonic Flow," *Uchenye Zapiski TsAGI,* Vol. 4, No. 3, 1973, pp. 112–118 (in Russian).

Penzin, V. I., "Experimental Investigation of a Separation of Supersonic Boundary Layer in Axisymmetric Channel," *Uchenye Zapiski TsAGI,* Vol. 5, No. 4, 1974, pp. 106–112 (in Russian).

Penzin, V. I., "Separated Flow in an Annular Cavity," *Uchenye Zapiski TsAGI,* Vol. 7, No. 6, 1976, pp. 124–130 (in Russian).

Penzin, V. I., "On Choice of Maximum Angle of Inlet Wedge of Scramjet," *3* Korolevskie Chteniya, Moscow, 1981, pp. 70–80 (in Russian).

Penzin, V. I., "On Optimum System of Shocks of Scramjet Inlet," 4 Korolevskie Chteniya, Moscow, 1982a, pp. 16–28 (in Russian).

Penzin, V. I., "E. S.Shchetinkov—Innovator of Scramjet," *Iz Istorii Aviatsii i Kosmonavtiki.* Vypusk 46, AN SSSR Moscow, 1982b, pp. 27–33 (in Russian).

Penzin, V. I., "Interaction of Pseudo-Shock with a Obstacle," *Uchenye Zapiski TsAGI,* Vol. 14, No. 5, 1983a, pp. 39–46 (in Russian).

Penzin, V. I., "An Estimation of a Drag of a Channel with Installed Inside of It Wedge-Like Bodies at Supersonic Flow," *Idei F. A.Tsandera i Razvitie Raketno-Kosmicheskoi Nauki,* Nauka, Moscow, 1983b, pp. 123–130 (in Russian).

Penzin, V. I., "Influence of Gasdynamics and Geometrical Parameters of Flow in Duct with Sudden Expansion upon Base Pressure," *Uchenye Zapiski TsAGI,* Vol. 14, No. 6, 1983c, pp. 113–120 (in Russian).

Penzin, V. I., "Deceleration of Supersonic Flow in Channels with Annular and Circular Shapes of Cross Section," *Trudy 7-kh Nauchnykh Chtenii po Kosmonavtike,* Moscow, 1985, pp. 72–77 (in Russian).

Pensin, V. I., "Deceleration of Supersonic Flow in Tube with Sudden Expansion," *Trudy 8-kh Nauchnykh Chtenii po Kosmonavtike,* Moscow 1986a, pp. 73–80 (in Russian).

Penzin, V. I., "Using of Hydraulic Diameter Conception at Estimation of Deceleration Parameters," *Trudy 8kh Nauchnykh Chtenii po Kosmonavtike,* Moscow 1986b, pp. 55–60 (in Russian).

Penzin, V. I., "Influence of Initial Nonhomogeneouty of Flow on Pseudo-Shock," *Mezhvuzovskii Sbornik: Voprosy Teorii i Rascheta Rabochikh Protsessov Teplovykh Dvigatelei,* UAI, pp. 28–44 (in Russian).

Penzin, V. I., "Rebuilding of Separated Flow into Pseudo-Shock Inside of Rectangular Channel," *Mezhvuzovskii Sbornik: Voprosy Teorii i Rascheta Rabochikh Protsessov Teplovykh Dvigatelei,* UAI, 1987b, pp. 14–27 (in Russian).

Penzin, V. I., "Dependence of Pressure at Step from Shape of Cross-Section of Channel with Sudden Expansion," *Uchenye Zapiski TsAGI,* Vol. 18, No. 1, 1987c, pp. 65–72 (in Russian).

Penzin, V. I., "Pseudo-Shock and Separated Flow in Rectangular Channels," *Uchenye Zapiski TsAGI,* Vol. 19, No. 3, 1988a, pp. 105–112 (in Russian).

Penzin, V. I., "Influence of Shape of Cross-Section of Direct Channel of Deceleration of Supersonic Flow," *Uchenye Zapiski TsAGI,* Vol. 19, No. 3, 1988b, pp. 55–59 (in Russian).

Penzin, V. I., "On Optimization of a Shape of an Internal Duct of Dual-Mode Scramjet," Preprint TsAGI, Vypusk 12, 1990 (in Russian).

Penzin, V. I., "About Place of Detonation Scramjet in a Family of Ramjet Engines," Preprint TsAGI, Vypusk 59, 1992 (in Russian).

Penzin, V. I., "Influence of Differences in the Means of Generation of a Back-Pressure on Separation of Supersonic Flow in Channel of Constant Cross-Section," *Trudy 16-kh Nauchnych Chtenii po Kosmonavtike,* Moscow, 1993a, pp. 71–73 (in Russian).

Penzin, V. I., "Experimental Investigation of a Deceleration of Supersonic Flow in Divergent Rectangular Channels," Preprint TsAGI, Vypusk 80, 1993b (in Russian).

Penzin, V. I., "Experimental Investigation of Drag of a Channel with Installed Inside of It Pylon Cascade," *Trudy 17-kh Nauchnych Chtenii po Kosmonavtike,* Moscow, 1993c, pp. 45–49 (in Russian).

Penzin, V. I., "Pressure Pulsitions Behind of Pylon Cascade in Model Scramjet

Combustion Chamber," *Trudy 18-kh Nauchnych Chtenii po Kosmonavtike,* Moscow, 1994, pp. 21–23 (in Russian).

Penzin, V. I., "E. S. Shchetinkov (1917–1976) and Initial Stage of Researches on Scramjet in Russia," *Korolevskie Chteniya,* Moscow, 1995 (in Russian).

Perrier, P., Rostand, P., Stoufflet, B., Baev, V. K., Latypov, A. F., Shumsky, V. V. and Yaroslavtsev, M. I., "Integration of an Hypersonic Airbreathing Vehicle: Assessment of Overall Aerodynamic Performances and of Uncertainities," AIAA Paper 95-6100, 1995.

Petrov, G. I., and Ukhov, E. P., "Computation of the Total Pressure Recovery Coefficient for Deceleration of Supersonic Flow to Subsonic Flow for Different Systems of Plane Shock Waves," *Gostekhizdat,* Moscow, 1947, (in Russian).

Pobedonostsev, Yu.A., "First Flight-Tests of Ramjets," *Iz Istorii Aviatsii i Raketnoi Tekhniki,* AN SSSR, Moscow, 1970, pp. 109–121 (in Russian).

Prudnikov, A. G., and Frost, V. A., "Works of E. S. Shchetinkov on Study of Turbulent Combustion," *Iz Istorii Aviatsii i Kosmonavtiki, Vypusk 46,* AN SSSR, Moscow, 1982, pp. 19–26 (in Russian).

Prudnikov, A. G., Volynskii, M. S., and Sagalovich, V. N., *Processes of Mixing and Combustion in Air-Ramjet Engines,* Mashinostroenie, Moscow, 1971 (in Russian).

Puzyrev, L. N., Shumskii, V. V., and Yaroslavtsev, M. I., "Principles of Design of Gasdynamic Models with Combustion for Tests in High-Enthalpy Short-Duration Facilities," ITPM SO AN SSSR, Novosibirsk, Preprint 7-20, 1990 (in Russian).

Raushenbakh, B. V. (ed.), "Development of the Aviation Science and Technique in USSR," *Istoriko-Tekhnicheskie Ocherki,* Nauka, Moscow, 1980, p. 192 (in Russian).

Romankov, O. N., and Starostin, F.I., "Design and Investigation of the Stand and Flying Scramjet Models. Conceptions and Results of Experiments," AIAA Paper 93-2447, 1993.

Roudakov, A. S., "Some Problems of Scramjet Propulsion for Aerospace Planes," *Part 1—Scramjet: Aim and Features,* AGARD Lecture Series 194, 1993a, pp. 3.1–3.20.

Roudakov, A. S., "Some Problems of Scramjet Propulsion for Aerospace Planes," *Part 2—Scramjet: Development and Test Problems,* AGARD Lecture Series 194, 1993b, pp. 9.1–9.26 .

Roudakov, A. S., Schickhman, Ya., Semenov, V., Novelli, Ph, and Fourt, O., "Flight Testing an Axisymmetric Scramjet: Russian Recent Advances," 44th Congress of the International Astronautical Federation, 1993.

Roudakov, A. S., Semenov, V. L., Kopchenov, V. I., and Hicks, J. W., "Future Flight Test Plans of an Axisymmetric Hydrogen-Fueled Scramjet Engine on the Hypersonic Flying Laboratory," AIAA Paper No. 96-4572 CP, 1996.

Roudakov, A. S., Semenov, V. L., Kopchenov, V. I. & Hicks, J. W. 1998. Recent Flight Test Results of the Joint CIAM–NASA Mach 6.5 Scramjet Flight Program. AIAA Ppaer No. 98-1643, 10p., Norfork, USA, 8th International Spaces and Hypersonic Systems and Technologies Conference, April 1998.

Rozhitskii, S. I., and Strokin, V. N., "On Method of Taking a Probe of Combustion Products in Supersonic Reactive Flow," *Fizika Goreniya i Vzryva,* Vol. 10, No. 4, 1974, pp. 492–498 (in Russian).

Rozhitskii, S. I., and Strokin, V. N., "Deceleration of Supersonic Flow in Channel at Combustion," *Trudy po Kosmonavtike,* Moscow, 1988, pp. 57–60 (in Russian).

Rozhitskii, S. I., and Strokin, V. N., "On Peculiarities of Mixing and Combustion in Supersonic Flow," *Gorenie Geterogennykh i Gazovykh Smesei,* 1982 (in Russian).

Roy, P. M., "Propulsion Supersonique par Turboreacteurs et par Statoreacteurs," 1st International Congress on Aeronautical Sciences under the Auspices of the International Council of the Aeronautical Sciences, 1958.

Rylov, A. I., "Designing of Compact Nonsymmetrical Nozzles with Maximum Thrust at Given Lift Force," *Izvestiya AN SSSR. Mechanika Zhidkosti i Gaza,* No. 6, 1981, pp. 132–136 (in Russian).

Sabel'nikov, V. A., "Supersonic Turbulent Combustion of Nonpremixed Gases-Status and Perspectives," *International Colloquium on Advanced Computation and Analysis of Combustion,* edited by G. D. Roy, S. M. Frolov, and P. Givi, ENAS Publishers, Moscow, 1997.

Sabel'nikov, V. A., Vorozhtsov, I. I., and Yumashev, V. L., "Numerical Modelling of Supersonic Hydrogen Combustion Process in Co-Flowing Air Stream on the Basis of Parabolized Navier-Stokes Equations," *1-st Asian-Pacific International Symposium on Combustion and Energy Utilization,* 1990, pp. 173–177.

Sabel'nikov, V. A., Voloshchenko, O. V., Ostras', V. N., Sermanov, V. N., and Walter, R., "Gasdynamics of Hydrogen-Fueled Scramjet Combustors," AIAA Paper 93-2145, 1993a.

Sabel'nikov, V. A., Voloshchenko, O. V., Kolesnikov, O. M., Meshcheryakov, E. A., Ostras', V. N., and Sermanov, V. N., "Gasdynamics of Scramjet Combustors," *Trudy 16kh Korolevskikh Chtenii,* Fasis, Moscow, 1993b, pp. 28–49 (in Russian).

Samozvantsev, M. P., "About Stabilization of Detonation Waves Using Blunt Bodies," *Prickladnaya Mechanika i Tekhnicheskay Physicka,* No. 4, 1964, pp. 126–129 (in Russian).

Semenov, V. L. & Romankov, O. N. 1998. The investigation of operation domain of strut fuel feed system for model scramjet combustor. AIAA Paper 98-1514. Norfork, USA, 8th International Spaces and Hypersonic Systems and Technologies Conference, April 1998. A collection of technical papers, pp. 83–92.

Sermanov, V. N., "E. S. Shchetinkov and Problem of Experimental Investigations of Scramjet," *Iz Istorii Aviatsii i Kosmonavtiki. Vypusk 46,* AN SSSR, Moscow, 1982, pp. 34–40 (in Russian).

Shchetinkov, E. S., "Calculation of Flame Velocity in Turbulent Stream," *7th Sympssium (International) on Combustion,* Combustion Inst., Pittsburgh, PA, 1958, pp. 583–589.

Shchetinkov, E. S., *Physics of Gas Combustion,* Nauka, Moskow, 1965 (in Russian).

Shchetinkov, E. S., "On Problems of Supersonic Combustion," *Gorenie i Vzryv,* Nauka, Moscow, 1972, pp. 276–281 (in Russian).

Shchetinkov, E. S., "On Piece-Wise One-Dimensional Models of Supersonic Combustion and Pseudo-Shock in a Duct Combustion," *Explosion and Shock Waves,* Vol. 9, No. 4, 1973, pp. 409–417.

Sokolenko, V.Ph., Tyulpanov, R. S., and Ignatenko, Yu. V., "On Structure of Diffusion Flames," *Fizika Goreniya i Vzryva,* Vol. 7, No. 4, 1971, pp. 566–571 (in Russian).

Sokolenko, V.Ph., Tyulpanov, R. S., Morin, O. V., and Ignatenko, Yu. V., "Interaction of Characteristics of a Turbulent Field and Hydrogen Diffusion Flame in Closed Channel," *Fizika Goreniya i Vzryva,* Vol. 10, No. 2, 1974, pp. 240–244 (in Russian).

Sosounov, V. A., Annushkin, Yu.M., Sverdlov, E. D., and Pagy, D. G., "Investigation of Hydrogen Diffusion Flames in Direct-Flow Combustors," *Hydrogen Energy Progress,* Vol. 3, Pergamon, New York, 1988, pp. 2009–2024.

Sosounov, V. A., *Introduction and Overview,* AGARD Lecture Series 194, 1993a, pp. 1.1–1.20.

Sosounov, V. A., *Research and Developments of Ramjets/Ramrockets. Part II—Integral Liquid Fuel Ramjets,* AGARD Lecture Series 194, 1993b, pp. 5.1–5.23.

Sosounov, V. A., *Research and Developments of Ramjets/Ramrockets. Part III—the Study of Gaseous Hydrogen Ram Combustors,* AGARD Lecture Series 194, 1993c, pp. 6.1–6.6.

Starukhin, V. P., Aleksandrovich, E. V., Glushatov, A. A., and Trifonov, A. K., "Experimental Investigation of Flow Parameters at the Entrance of the Inlet Installed Under Triangular Wing," Trudy TsAGI, 1970 (in Russian).

Starukhin, V. P., and Taryshkin, A. G., "Experimental Investigation of Turbulent Boundary Layer in Supersonic Flow on Plane Compression Surfaces with Sharp Corners," *Uchenye Zapiski TsAGI,* Vol. 6, No. 4, 1975, pp. 95–100 (in Russian).

Starukhin, V. P., and Taryshkin, A. G., "Investigation of Boundary Layer Parameters in Front of Two-Dimensional Supersonic Inlet, Installed Under Triangular Plate," *Uchenye Zapiski TsAGI,* Vol. 13, No. 2, 1982, pp. 69–77 (in Russian).

Starukhin, V. P., and Chevagin, A.Ph., "Influence of Bluntness of Leading Edges on Characteristics of Underwing Inlets," *Uchenye Zapiski TsAGI,* Vol. 25, No. 1–2, 1994, pp. 89–100 (in Russian).

Stechkin, B. S., *Theory of Air Ram Engine,* Tekhnika Vozdushnogo Flota, Moscow 1929 (in Russian).

Strokin, V. N., "Analysis of Auto-Ignition of Turbulent Fuel Jet in an Oxidant Stream," *Inzhenerno-Phizicheskii Zhyrnal,* Vol. 22, No. 3, 1972a, pp. 480–487 (in Russian).

Strokin, V. N., "Process of Auto-Ignition and Hydrogen Combustion in Supersonic Flow," *Gorenie i Vzryv,* Nayka, Moscow, 1972b, pp. 386–391 (in Russian).

Strokin, V. N., and Khailov, V. M., "On an influence of NO on Induction Delay of Hydrogen Auto-Ignition in Air," *Fizika Goreniya i Vzryva,* Vol. 10, No. 2, 1974, pp. 230–236 (in Russian).

Strokin, V. N., and Grachev, V. A., "Possible Scheme of Flame Holding in Hydrogen Fueled Scramjet Combustor," *Proceedings of International Aerospace Congress,* Vol. 1, Moscow, 1994, pp. 630–632.

Strokin, V. N., and Grachev, V. A., "The Peculiarities of Hydrogen Combustion in Model Scramjet Combustors," *13th International Symposium on Air Breathing Engines,* 1997, pp. 374–384.

Tret'yakov, P. K., "Determination of Heat Fluxes for Flow with a Pseudo-Shock in a Duct," *Fizika Goreniya i Vzryva,* Vol. 29, No. 3, 1993a, pp. 71–77 (in Russian).

Tret'yakov, P. K., "Pseudo-Shock Regime of Combustion," *Fizika Goreniya i Vzryva,* Vol. 29, No. 6, 1993b, pp. 33–39 (in Russian).

Tret'yakov, P. K., "The Study of Supersonic Combustion for a Scramjet," *Experimentation, Modeling and Computation in Flow, Turbulence and Combustion, Vol. 1,* edited by Désideri et al., Wiley, New York, 1996, pp. 319–336.

Tret'yakov, P. K., and Lazarev, A. M., "Hydrogen Combustion Flame Holding by Using Wave Structure of Supersonic Air Flow," 13th ICDERS Meeting, 1991.

Tret'yakov, P. K., Golovitchev, V. I., and Bruno, C., "Supersonic Combustion: a Russian View 1967–1993," Joint Meeting of the Italian and Spanish Sections of the Combustion Institute, 1993.

Trifonov, A. K., Bogdanov, V. V., and Gurylev, V. G., "Total Pressure Pulsations in Flow Behind of Pseudo-Shock at the Entrance of a Simple Inlet Like Cylindric Duct," *Uchenye Zapiski TsAGI,* Vol. 8, No. 3, 1977, pp. 64–74 (in Russian).

Tyulpanov, R. S., and Pritsker, O. V., "Impact of Temperature on the Burning of the Diffusion Hydrogen Flame in Supersonic Flow in Closed Duct," *Fizika Goreniya i Vzryva,* Vol. 8, No. 1, 1972, pp. 77–82 (in Russian).

Vasiliev, V. I., Stepanov, V. A., and Zakotenko, S. N., "Numerical Investigation of Mixing Process in Hypersonic Inlets," *Uchenye Zapiski TsAGI,* Vol. 22, No. 6, 1991, pp. 57–67 (in Russian).

Vishnevetskii, S. L., "Shaping of Supersonic Inlets of Airbreathing Engines," *Trudy NII-1,* No. 7, 1955 (in Russian).

Vishnevetskii, S. L., "Some Properties of Systems of Plane Pressure Shocks," *Izvestiya AN SSSR, Otdelenie Tekhn. Nauk, Mekhanika i Mashinostroenie,* No. 6, 1960, pp. 15–18 (in Russian).

Vishnevetskii, S. L., and Mel'nikov, D. A., "Improving of Characteristics of Annular Inlets," *Trudy NII-1,* No. 39, 1961 (in Russian).

Vinogradov, V. A., and Ogorodnikov, D. A., "Experimental Investigation of Two Dimensional Scramjet Inlet at Mach Numbers M=9–13," Trudy Central Inst. of Aviation Motors, No. 480, 1970 (in Russian).

Vinogradov, V. A., Ogorodnikov, D. A. & Stepanov, V. A.1998. Experimental and computational researches of the spatial (3-D) scheme hypersonic inlets. AIAA paper 98-1527, 8p. Norfork, USA, 8th International Spaces and Hypersonic Systems and Technologies Conference, April 1998. A collection of technical papers, pp. 159–168.

Vinogradov, V. A., Zakharov, N. N., and Ivanov, M. Ya., "Computation of Deceleration of Two-Dimensional Flow of Inviscid Gas in Duct and Possibility of Realization of Such Flow," *Uchenye Zapiski TsAGI,* Vol. 6, No. 2, 1975, pp. 161–166 (in Russian).

Vinogradov, V. A., and Duganov, V. V., "Computation of Flow in Supersonic Inlets Taking into Account Boundary Layer on the Washed Surfaces," *Uchenye Zapiski TsAGI,* Vol. 10, No. 5, 1979, pp. 29–34 (in Russian).

Vinogradov, V. A., Makaron, V. E., and Stepanov, V. A., "Numerical Investigation of Three-Dimensional Flow in Inlet of a Scramjet Module," *Trudy Nauchnykh Chtenii Pionerov,* Kosmosa 1980, pp. 191–201 (in Russian).

Vinogradov, V. A., and Stepanov, V. A., "Numerical Investigation of a Scramjet Inlet with Fixed Geometry with Three-Dimensional Deceleration of Flow at Mach Numbers M=5–7," *Uchenye Zapiski TsAGI,* Vol. 13, No. 4, 1982a, pp. 81–89 (in Russian).

Vinogradov, V. A., Duganov, V. V., and Stepanov, V. A., "Application of Numerical Methods for Calculations of Characteristics of Supersonic and Hypersonic Inlets for Airbreathing Engines," *Uchenye Zapiski TsAGI,* Vol. 13, No. 2, 1982b, pp. 62–68 (in Russian).

Vinogradov, V., Grachev, V., Petrov, M., and Sheekhman, J., "Experimental Investigation of 2-D Dual-Mode Scramjet with Hydrogen Fuel at Mach 4–6," AIAA Paper 90-5269, 1990.

Vinogradov, V. A., Albegov, R. V., and Petrov, M. D., "Experimental Investigation of Hydrogen Burning and Heat Transfer in Annular Duct at Supersonic Velocity," *International Council of the Aeronautical Sciences,* 1992a, pp. 737–743.

Vinogradov, V., Kobigsky, S., and Petrov, M., "Experimental Investigation of Liquid Carbon-Hydrogen Fuel Combustion in Channel at Supersonic Velocities," AIAA Paper 92-3429, 1992b.

Vinogradov, V. A., Stepanov, V. A., and Alexandrovich, E. V., "Numerical and Experimental Investigation of Air Frame—Integrated Inlet for High Velocities," *Journal of Propulsion and Power,* Vol. 8, No. 1, 1992c, pp. 151–157.

Vinogradov, V. A., Evdokimov, S. V., Stepanov, V. A., and Chevagin, A.Ph., "Peculiarities of Internal Characteristics of Three-Dimensional Hypersonic Inlets," *Tekhnika Vozdushnogo Flota,* No. 2, 1992d, pp. 33–38 (in Russian).

Vinogradov, V. A., and Prudnikov, A. G., "Injection of Liquid into the Strut Shadow at Supersonic Velocities," Society of Automotive Engineers, 93-1455, 1993.

Vinogradov, V. A., Semenov, V. L., Romankov, O. N., and Vedeshkin, G. K., "Bench Test Development Methodology of Experimental Scramjet Module for HFL Flight Test," *International Aerospace Congress,* Vol. 1 1994, pp. 625–629.

Vinogradov, V. A., Stepanov, V. A., and Goldfeld, M. A., "Experimental and Numerical Investigation of Two Concepts of the Hypersonic Inlet," AIAA Paper 95-2721, 1995.

Vlasenko, V. V., and Sabel'nikov, V. A., "Numerical Simulation of Inviscid Flows with Hydrogen Combustion after Shock Waves," AIAA Paper 94-3177, 1994.

Voloshchenko, O. V., Ostras', V. N., and Sermanov. V. N., "Investigation of Heat Transfer in Pseudo-Shock," *Pionery Osvoeniva Kosmosa I Sovremennost'.* Nauka, Moscow 1988, pp. 62–67 (in Russian).

Voloshchenko, O. V., Kolesnikov, O. M., Meshcheryakov, E. A., Ostras', V. N., Sabel'nikov, V. A., and Sermanov, V. N., "Supersonic Combustion and Gasdynamics of Scramjet," *International Council of the Aeronautical Sciences,* 1992, pp. 693–702.

Volynskii, M. S., "Injection of Liquid into Supersonic Flow," *Izvestiya An SSSR. Mekhanika i Mashinostroenie,* No. 2, 1963, pp. 20–27 (in Russian).

Vorontsov, S. S., Konstantinovskii, V. A., and Tret'yakov, P. K., "Determination of Combustion Efficiency of Hydrogen in Supersonic Flow by Optical Method," *Fizicheskaya Gazodinamika,* ITPM SO AN SSSR, Novosibirsk, 1976, pp. 69–72 (in Russian).

Vorontsov, S. S., Garanin, A.Ph., and Pikalov, V. V., "Investigation of Axisymmetric Hydrogen Flame by Optical Method," *Invesiya Abelya i eie Obobshchenie,* Novosibirsk, ITPM SO AN SSSR, 1978, pp. 244–251 (in Russian).

Walther, R., and Sabel'nikov, V. A., "Russian Scramjet Technology Development : From the First Steps to the Current Status," *Space Course on Low Earth Orbit Transportation,* Vol. 2, TU, München, 1993, pp. 33.1–33.31.

Walther, R., Sabel'nikov, V. A., Korontsvit, Yu, Ph., Voloshchenko, O. V., and Sermanov, V. N., "Progress in the Joint German - Russian Scramjet Technology Programme," *International Symposium on Air Breathing Engines,* 1995, pp. 1217–1329.

Walther, R., Koschel, W., Sabel'nikov, V. A., Korontsvit, Yu. Ph., and Ivanov, V. V., "Investigations into the Aerothermodynamic Characteristics of Scramjet Components," *International Symposium on Air Breathing Engines,* 1997, pp. 598–606.

Weber, R. J., and Mackay, J. S., "An Analysis of Ramjet Engines Using Supersonic Combustion," NACA TN 4386, 1958.

Zabaikin, V. A., Lazarev, A. M., and Tret'yakov, P. K., "Experimental Investigation of Influence of Gasdynamic Structure and Flame in Air Flow," *International Work-Seminar "Protsessy Turbulentnogo Perenosa v Reagiruyushchikh Sistemakh,"* Minsk, 1985, pp. 94–99 (in Russian).

Zabaikin, V. A., and Lazarev, A. M., "Influence of Different Means of Hydrogen Injection on Its Burning in Supersonic Air Flow," *Izvestiya SO AN SSSR, Seriya Tekh. Nauk,* Vol. 4 No. 4, pp. 1986, 44–49 (in Russian).

Zabaikin, V. A., Lazarev, A. M., Solovova, E. A., and Tret'yakov, P. K., "Gasdynamics of Supersonic Flow in a Channel of Variable Cross-Section at Heat Release." *Vestnik AN BSSR, Ser. Fiz.-Energ. Nauk,* No. 3, 1986, pp. 102–106 (in Russian).

Zakharov, N. N., "Influence of Heat-Transfer on Separation of a Turbulent Boundary Layer," *Sbornick Statei "Pogranichnyi Sloi i Teploobmen,"* Trudy Central Inst. of Aviation Motors, No. 507, 1971, pp. 70–84 (in Russian).

Zakharov, N., Tagirov, R. & Aukin, M. 1993a. Scramjet CFD methods and analysis. Part 1. Scramjet CFD Methods. In: AGARD Lecture Series 194, pp. 4.1–4.20.

Zimont, V. L., Ivanov, V. K., and Oganesyan, S.Kh., "Autoignition and Extinction of Combustion in a Recirculation Zone Behind of Step or in a Cavity for Supersonic Flow of Fuel Mixture," *Gorenie i Vzryv,* Nauka, Moscow, pp. 386–391 (in Russian).

Zimont, V. L., and Ostras', V. N., "Calculation of Pseudo-Shock in a Cylindric Channel," *Uchenye Zapiski TsAGI,* Vol. 5, No. 3, pp. 1974, 40–48 (in Russian).

Zimont, V. L., Ivanov, V. I., Mironenko, V. A., and Solokhin, E. L., "Experimental Investigation of a Combustion Mechanism in Supersonic Flow at Co-Flows of Fuel and Oxidant." *Gorenie i Vzryv,* Nauka, Moscow, 1977, pp. 388–393 (in Russian).

Zimont, V. L., and Osras', V. N. "Deceleration in Pseudo-Shock at Supersonic Flow in Channel." *Trudy 4kh Chtenii Tsandera,* Moscow, 1978, pp. 37–54 (in Russian).

Zimont, V. L., Levin, V. M., and Meshcheryakov, E. A., "Hydrogen Combustion in Supersonic Flow in a Channel in Pseudo-Shock," *Fizika Goreniya i Vzryva,* Vol. 14, No. 4 1978, pp. 23–36 (in Russian).

Zimont, V. L., Levin, V. M., and Meshcheryakov, E. A., "On Combustion Stabilization in Supersonic Flow," *Fizika Goreniya i Vzryva,* Vol. 18, No. 3, 1982, pp. 40–43 (in Russian).

Zimont, V. L., Levin, V. M., Meshcheryakov, E. A., and Sabel'nikov, V. A., "Peculiarities of Supersonic Combustion of Nonpremixed Gases in Channels," *Fizika Goreniya i Vzryva,* Vol. 19, No. 4, 1983, pp. 75–78 (in Russian).

Zimont, V. L., and Meshcheryakov, E. A., "Processes of Turbulent Transfer and Combustion in Supersonic Flows in Channels," *International Work-Seminar "Protsessy Turbulentnogo Perenosa v Reagiruyushchikh Sistemakh,"* Minsk, pp. 3–18 (in Russian).

Zuev, V. S., and Makaron, V. S., *Theory of Ramjet and Ram-Rocket Engines,* Mashinostroenie, Moscow, 1971 (in Russian).

Appendix A: Three Problems in Supersonic Combustion

Vladimir A. Sabel'nikov
TsAGI, Zhukovsky, Moscow Region, 140160, Russia

These parts of the present appendix address three problem of combustion in supersonic flows. The first one is the retardation of the heat release in supersonic diverging-area combustor. The theoretical analysis of the reasons of such a retardation (given in the first part of the appendix) follows closely the article of E. A. Meshcheryakov and V. A. Sabel'nikov (1986, see the bibliography in the main text of the chapter). The second problem involves the combustion stabilization in supersonic flow using a free recirculation bubble that is obtained by the interaction of a concentrated vortex with a normal shock wave. Results of the experimental investigation of a such a flow configuration for combustion stabilization of liquid kerosene in supersonic flow performed by V. A. Sabel'nikov et al. (1998a, see the bibliography in the main text of the chapter) are presented in the second part of the appendix. The third problem we consider in the last part of the appendix is the enhancement of supersonic combustion. This part contains the results of the experiments conducted by V. A. Sabel'nikov et al. (1998b, see the bibliography in the main text of the chapter) on supersonic mixing and combustion enhancement in a scramjet combustor using effervescent (i.e. aerated by hydrogen or air) liquid kerosene jets injected through elliptic nozzles.

I. Retardation of Heat Release in Supersonic Diverging-Area Combustor

A. Introduction

The design of a high efficiency supersonic combustor needs elucidating the physics of the noticeable retardation of heat release of supersonic combustion of nonpremixed gases in diverging-area channels (see, e.g., Refs. 1–6), and clarifying which of the two factors, mixing or finite chemical reaction rates (i.e. kinetics) is responsible first and foremost for this retardation, and under what conditions. There are considerable difficulties in solving this problem by experimental methods because it is virtually impossible to separate these two factors and examine their effects in pure form. In this appendix the retardation problem is tackled by numerical simulation. Boundary layer approximation has been used to simplify the conservation equations for a turbulent multicomponent chemically reacting gas flow (an axisymmetric supersonic/subsonic hydrogen jet in co-flow supersonic air stream in a combustor). Zeldovich's theory of the extinction of nonpremixed laminar flame[7] is invoked for a qualitative analysis of the retardation phenomenon.

B. Formulation

We consider a turbulent supersonic/subsonic hydrogen jet exhausting from an axisymmetric nozzle into a co-flow supersonic air (oxidizer) stream in an axisymmetric combined combustor, i.e., a combustor that has both constant-area and diverging-area parts (see Fig. A.I.1). The fuel jet is supposed to be matched; i.e., the static pressures at the nozzle exit and in co-flow air stream are the same, $P_1 = P_2 = P_{in}$ (subscripts 1 and 2 refer to jet and co-flow parameters, respectively). The turbulent effects on the mixing and combustion are incorporated simply by replacing in the conservation equations (for the mean quantities) the molecular transport coefficients by their turbulent counterparts. This approach is often named the quasilaminar model of turbulent combustion.[8–10] The mean conservation equations for mass, momentum, and energy for a multicomponent axisymmetric chemically reacting gas flow applying the boundary layer approximation[9] can be written as follows:

$$\frac{\partial \rho u y}{\partial x} + \frac{\partial \rho v y}{\partial y} = 0 \qquad \text{(A.I.1)}$$

$$\rho u \frac{\partial u}{\partial x} + \rho v \frac{\partial u}{\partial y} = -\frac{\mathrm{d}P}{\mathrm{d}x} + \frac{1}{y}\frac{\partial}{\partial y}\left(\rho v_t y \frac{\partial u}{\partial y}\right) \qquad \text{(A.I.2)}$$

$$\rho u \frac{\partial Y_i}{\partial x} + \rho v \frac{\partial Y_i}{\partial y} = \frac{1}{y}\frac{\partial}{\partial y}\left(\frac{\rho v_t y}{sc_t}\frac{\partial Y_i}{\partial y}\right) + \omega_i \qquad \text{(A.I.3)}$$

$$\rho u \frac{\partial H}{\partial x} + \rho v \frac{\partial H}{\partial y} = \frac{1}{y}\frac{\partial}{\partial y}\left(\frac{\rho v_t y}{Pr_t}\frac{\partial H}{\partial y} + \rho v_t y \sum_i \frac{Le_t - 1}{Pr_t} h_i \frac{\partial Y_i}{\partial y} \rho v_t y u \frac{\partial u}{\partial y}\right) \qquad \text{(A.I.4)}$$

$$P = \rho R^0 \sum_{i=1}^{N} \frac{Yi}{\mu_i} T$$
$$H = \sum_{i=1}^{N} h_i Y_i + \frac{u^2}{2},\ h_i = h_i^0 + \int_{T_0}^{T} Cp_i(T)\,\mathrm{d}T,\ i = 1,\ldots,N \qquad \text{(A.I.5)}$$

Here x and y are the longitudinal and transverse coordinates; ρ is density;

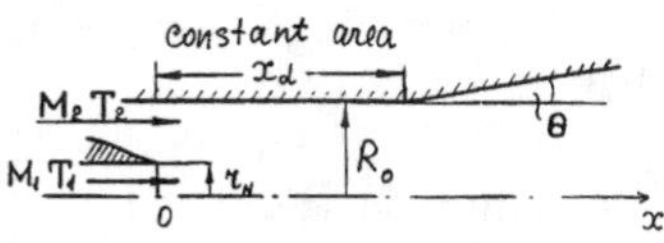

Fig. A.I.1 Scheme of a co-flow fuel supply in a combined (constant-area + diverging-area) supersonic axisymmetric combustor; 1) fuel jet, 2) co-flow air (oxidizer) stream; x_d is the length of constant-area combustor part; R_0 and r_N are radii of constant-area combustor part and fuel nozzle, respectively; Θ is the semi-angle of combustor expansion; subscripts 1 and 2 refer to jet and co-flow parameters, respectively

u and v the velocity components in the x and y directions; P is the static pressure; $h_i(T)$, h_i^0, and Cp_i are the specific enthalpy, heat of formation, and specific heat of species i, respectively; Y_i, μ_i, and ω_i are the mass function, molecular mass, and rate of formation of species i, respectively; R^0 is the universal gas constant; T is the absolute temperature; N is the number of components; Pr_t, Sc_t, Le_t are the turbulent Prandtl, Schmidt, and Lewis numbers, respectively, values of which were taken in the calculations at $Pr = Sc_t = 0.8$, $Le_t = 1.0$; v_t is the turbulent viscosity.

A phenomenological equation[10] was used for turbulent viscosity to close Eqs. (A.I.1–A.I.4)

$$\rho u \frac{\partial v_t}{\partial x} + \rho v \frac{\partial v_t}{\partial y} = \rho v_t k_0 f(M) \left| \frac{\partial u}{\partial y} \right| + \frac{1}{y} \frac{\partial}{\partial y} \left(\frac{\rho v_t y}{\sigma} \frac{\partial v_t}{\partial y} \right) + \xi v_t \left(u \frac{\partial \rho}{\partial x} + v \frac{\partial \rho}{\partial y} \right) \tag{A.I.6}$$

$$f(M) = \begin{cases} 1, & M \le 1 \\ 1/M, & M > 1 \end{cases}$$

Here $k_0 = 0.2$, $\sigma = 0.5$, and $\zeta = 2/3$ are empirical constants; function $f(M)$ in Eq. (A.I.6) has been introduced to incorporate the experimentally observed reduction of the thickness of supersonic mixing layer with increasing Mach number (the effects of compressibility on turbulence). The last term in the right-hand-side of Eq. (A.I.6):

$$\text{Effect of pressure gradient } = \xi v_t \left(u \frac{\partial \rho}{\partial x} + v \frac{\partial \rho}{\partial y} \right) \tag{A.I.7}$$

describes the influence of the axial pressure gradient on the turbulent mixing (in the diverging-area part of the supersonic combustor the pressure gradient is usually favorable (i.e., negative) even at relatively small expansion angle).

The chemical scheme we have used involves eight active species H, O, OH, H_2O, HO_2, H_2O_2, O_2, and H_2, inert nitrogen N_2 and includes 20 elementary reactions.[9,13,14]

Existing measurements were used to specify the initial profiles of the parameters; when none were available, the initial profiles at the nozzle exit cross-section and in the co-flow were taken as piecewise uniform functions. Boundary layers were assumed at the lip of the fuel nozzle and on the combustor walls. The profiles of the axial velocity in boundary layers were approximated by power law with exponent 1/7. The turbulent viscosity profile is usually not measured in experiments, and the initial profile of it was chosen from some preliminary estimates and curve fitting calculations of experimental measurements (see, e.g., Ref. 10).

There are certain difficulties in specifying the initial mass fractions of active chemical radicals. It is known[15] that the induction time period is strongly dependent on the initial level of these radicals in the inflow. Unfortunately, the gas composition of the inflow is usually not known in such detail, particularly when the air co-flow stream is preheated in a combustion heater (so called "vitiated" air). We determined the initial O, H, and OH mass fractions from the partial equilibrium conditions (exceptions are mentioned especially). The mass fractions of other intermediate radicals were taken as 10^{-12} (reduction in that constant had virtually no effect on the final results).

For the sake of simplicity and to reduce the computational time, instead of non-slip boundary conditions on the channel walls, boundary conditions were posed in the constant shear sublayer (logarithmic layer), i.e. the so-called "law of the wall" was used.[9,10] The wall heat fluxes were calculated from the Reynolds analogy. The value of the local skin friction factor was taken in accordance with Ref. 16.

Eqs. (A.I.1–A.I.6) were integrated numerically using the finite difference method described in Ref. 10. The recommendations of Ref. 9 were used for integrating the stiff kinetic equations. Most of the calculations were performed with 54 nodes in the transverse direction (check calculations were done with 100 nodes). The axial integration step was $\Delta x = (0.5 - 2.5) \cdot 0.4 \cdot 10^{-2} R_0$ (R_0 is the combustor radius at the initial cross-section); the value of Δx was decreased as P_{in} increased.

C. Results

At first, the validity of the above formulated model was verified comparing the calculations with measurements in a diverging-area supersonic combustor.[17] Combustor geometry is given in Fig. A.I.2a, $R_0 = 25$ mm, $r_N = 5.5$ mm, $x_d = 125$ mm. Experimental conditions were: $M_2 = 1.98$, $P_{in} = 0.09$ MPa, total temperatures of "vitiated" air and hydrogen jet $T_{t2} = 1870$ K and $T_{t1} = 350$ K, respectively. Experiments were carried out at three different values of fuel equivalence ratio: ER = 0.434 ($M_1 = 1.0$), ER = 0.333 ($M_1 = 0.8$) and ER = 0.227 ($M_1 = 0.57$). Calculations were done at the following conditions: $v_{t1} = 10^{-4} U_2 R_0$, combustor wall temperature = 720 K; initial boundary layer thickness $0.1R_0$, $0.02\ R_0$ at combustor wall, at external and internal wall sides of fuel nozzle, respectively; local skin friction factor $1.5 \cdot 10^{-3}$. Experimental data and calculated results are presented in Fig. A.I.2b. It is seen in that figure that self-ignition took place practically at the nozzle exit section. Calculated results were in accordance with experimental data if it was assumed that there was a small amount of super-equilibrium atomic oxygen in the incoming "vitiated" air flow, namely $Y_O = 2.5 \cdot 10^{-5}$ (equilibrium O mass fraction is $1.46 \cdot 10^{-7}$). "Vitiated" air in Ref. 17 was composed of water vapor, O, OH, and H mass fractions that facilitate rapid combustion, but Ref. 17 does not contain any information about the values of the above-mentioned active radical mass fractions.

Fig. A.I.2b shows also the calculated combustion efficiency η, using Eq. (A.I.8) given below (combustion efficiency η was not measured in Ref. 17). It is seen that the combustion efficiency is virtually the same and is quite small for three values of fuel equivalence ratio ER (combustion efficiency $\eta = 0.24$ at the combustor exit). The low combustor performance in Ref. 17 can be explained by three factors:

1) The comparatively low turbulent viscosity at the combustor entrance,

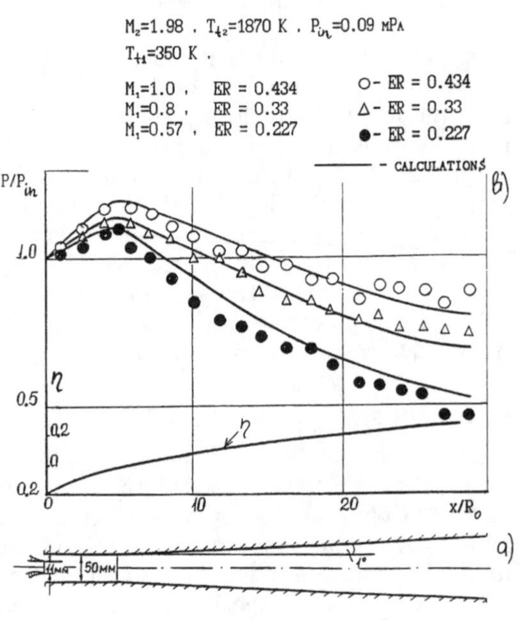

Fig. A.I.2 Comparison of the experimental data[17] with numerical results: a) Scheme of the axisymmetric combined (constant- area + diverging-area) supersonic combustor tested in Ref. 17; R_0 = 25 mm, r_N = 5.5 mm, x_d = 125 mm. b) Impact of fuel equivalence ratio ER on the axial static pressure and combustion efficiency distributions in the combined combustor; symbols and curves are the experimental data; Ref. 17 and calculated results, respectively: 1) 1, ○- ER = 0.434, M_1 = 1.0; 2) 2, Δ - ER = 0.33, M_1 = 0.8; 3) 3, ● - ER = 0.227, M_1 = 0.57; M_2 = 1.98, T_{22} = 1870 K, $P_1 = P_2 = P_{in}$ =0.09 MPa, T_{tl} = 350 K. Calculations were done at the following conditions: $\nu_{t1} = 10^{-4} U_2 R_0$, $\nu_{t2} = 0.6 \cdot 10^{-3} U_2 R_0$, combustor wall temperature 720 K. Initial boundary layer thicknesses were: 1) 0.1 R_0 at combustor wall, 2) 0.02 R_0 at external wall side of nozzle, 3) 0.08 R_0 at internal wall side of nozzle; local skin friction factor is 1.5 $\cdot 10^{-3}$.

$\nu t_2 = 0.6 \cdot 10^{-3} U_2 R_0$ (compare with the value $\nu t_2 = 5.2 \cdot 10^{-3} U_2 R_0$ which, was used in [A.I.10] to describe similar experimental data obtained in the joint work of MAI-TsAGI);

2) The small difference in the jet and coflow velocities (the velocity ratio U_1/U_2 was in the range 0.8–0.9) and, as a consequence, the turbulence generation was negligibly small in combustor;

3) The expansion of combustor, which is liable for decreasing of the turbulent viscosity (see analysis below, Fig. A.I.6).

For the purpose of gaining insight into the reasons of the heat release decrease in the diverging-area combustor, we carried out parametrical calculations for the hydrogen co-flow jet combustion in the constant and diverging-area combustors at the following inflow conditions: $M_1 = 1.0$, $M_2 = 2.6$, $T_1 = 300 - 1000$ K, $T_2 = 900 - 1500$ K, $P_1 = P_2 = P_{in} = 0.01 - 0.4$ Mpa, $\nu t_1 = 10^{-4} \cdot U_2 R^0$, $\nu_{t2} = (3{-}5) \cdot 10^{-3} \cdot U_2 R_0$. The fuel jet was a hydrogen–nitrogen mixture with the mass fraction of the latter enabling one to vary independently the fuel equivalence ratio ER. The jet nozzle radius was equal to $r_N/R_0 = 0.3$. The greater part of the calculations was performed for $R_0 =$ 50 mm. The combustor was a combination of the constant-area part of length $x_d = 5R_0$ and diverging-area conical part. The semi angle of the conical part was varied over the following range: $\theta = 0 - 5° 40'$ ($\tan\theta = 0 -$ 1).

Figures A.I.3, A.I.4, and A.I.5 show the impacts of the initial static pressure magnitude and of the angle of combustor divergence on the axial static combustion efficiency distributions for constant-area. (Fig. A.I.3) and combined combustors (Fig. A.I.4) and combined combustors (Figs. A.I.4 and A.I.5). The efficiency η was calculated using the following relationship:

$$\eta = \frac{I(x) - I(0)}{G^{in}_{H_2}Q} \tag{A.I.8}$$

where

$$I(x) = \sum_{i=1}^{8} G_i h_i$$

is the chemical energy flux; G_i is the mass flow rate of species i; $G^{in}_{H_2}$ is the mass flow rate of hydrogen at combustor entry;

$$Q = h^0_{H_2} + h^0_{O_2}L_0 - (1 + L_0)\,h^0_{H_2O}$$

is the specific release of the global chemical reaction $H_2 + 1/2\,O_2 = H_2O$; L_0 is the stoichiometric coefficient. Figures A.I.3a,b indicate that with decreasing the initial static pressure, finite chemical reaction rates play a continuously more important role as evidenced by the increase of the ignition delay length and departure from thermodynamic equilibrium in the diffusion flame. As P_{in} falls, so does the combustor performance (see the curve for the combustion efficiency η in Fig. A.I.3b), and for $P_{in} \leq 0.05$ MPA it is impracticable to obtain effective combustion even in a constant-area combustor.

We draw the conclusion from Figs. A.I.4 and A.I.5 that combustor expansion reduces considerably the combustion efficiency both in the thermodynamic equilibrium and finite chemical reaction rates cases, and it does more so the larger combustor expansion angle and the smaller the length of the constant-area combustor part (not shown here). At $\tan\theta \geq 0.06$ the combustion efficiency levels off, i.e., heat release almost ceases. Particular attention is deserved by the cases in which the combustion is already stabilized and flame is close to thermodynamic equilibrium in the constant-area part of combustor (a rough estimation of condition for that occurrence is $P_{in} \geq 0.03$ MPa, see below), i.e., combustion is mainly controlled by mixing. In such cases the following rule is applied: The relative reduction of the combustion

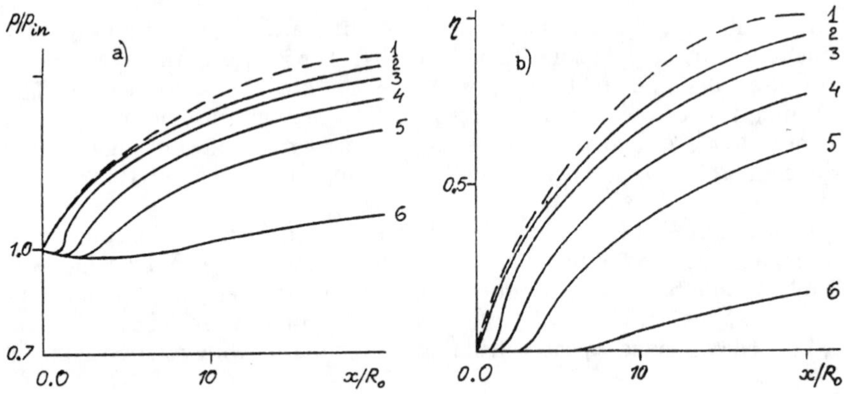

Fig. A.I.3 Impact of initial static pressure magnitude in constant-area supersonic combustor on a) axial static pressure distributions and b) axial combustion efficiency distributions.
1) thermodynamics equilibrium; 2) P_{in} = 0.02 MPa; 3) P_{in} = 0.01 MPa; 4) P_{in} = 0.05 MPa; 5) P_{in} = 0.03 MPa 6) P_{in} = 0.01 MPa; M_1 = 1.0, M_2 = 2.6, T_1 = 1000K, T_2 = 1500 K, $v_{t1} = 10^{-4} U_2 R_0$, $v_{t2} = 3.5 \cdot 10^{-3} U_2 R_0$, ER = 0.23.

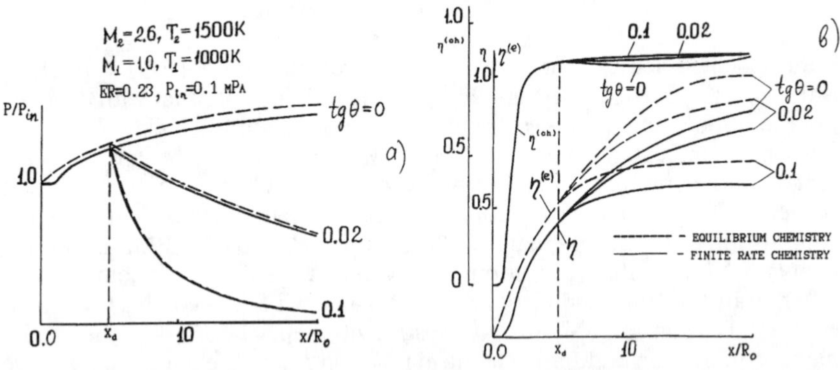

Fig. A.I.4 Impact of the combustor expansion angle in the combined (constant-area + diverging-area) combustor on a) axial static pressure distribution and b) axial combustion efficiency, equilibrium combustion efficiency, and the chemical combustion efficiency solid and dashed curves are for finite chemical reaction rates and thermodynamic equilibrium cases, respectively: M_1 = 1.0, M_2 = 2.6, T_1 = 1000 K, T_2 = 1500 K, $V_{t1} = 10^{-4} U_2 R_0$, $v_{t2} = 3.5 \cdot 10^{23} U_2 R_0$, ER = 0.23, P_{in} = 0.1 MPa.

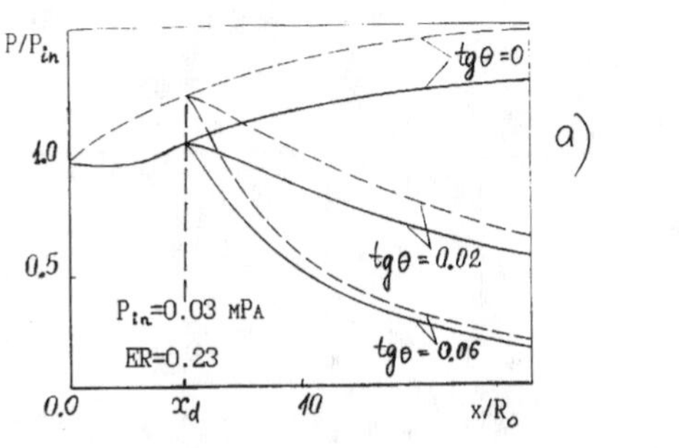

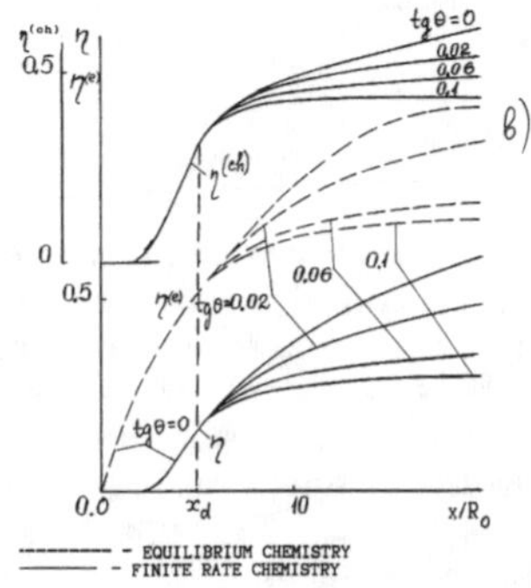

Figure A.I.5. Impact of the combustor expansion angle in the combined (constant-area + diverging-area) combustor on
a) axial static pressure distribution
b) axial combustion efficiency, equilibrium combustion efficiency and the chemical combustion efficiency solid and dashed curves are for finite chemical reaction rates and thermodynamic equilibrium cases, respectively
$M_1 = 1.0$, $M_2 = 2.6$, $T_1 = 1000$ K, $T_2 = 1500$ K, $\nu_{t1} = 10^{-4} U_2 R_0$, $\nu_{t2} = 3.5 \cdot 10^{-3} U_2 R_0$, ER = 0.23, $P_{in} = 0.03$ MPa

efficiency (in comparison to the case of thermodynamic equilibrium combustion) due to the finite chemical reaction rates is almost the same for the constant-area and combined combustor (constant-area followed by diverging-area). For example, for $P_{in} = 0.1$ MPa, the relative reduction of combustion efficiency η at $x/R_0 = 5.0$ is equal to 16%, whereas at the combustor exit ($x/R_0 = 20.0$) the reduction is equal to 13%, no matter what the combustor expansion angle is (Figs. A.I.4a,b). This result shows that for diffusion flame stabilized before diverging-area combustor part, the reduction of the heat release in the diverging-area part of combustor (under the above stated conditions) is mainly due to the reduced mixing. The mixing reduction is illustrated in Fig. A.I.6 where the axial maximum turbulent viscosity distribution is plotted. The reduction of turbulent viscosity in the divergent part of the combustor, which is seen in Fig. A.I.6, is explained by the negative impact of the favorable pressure gradient in the diverging-area combustor part (so-called relaminarization effect, e.g., Ref. 12). In the frame of the current approach, the relaminarization effect is described by the last term [see Eq. (A.I.7)] referring to the right-hand-side of Eq. (A.I.6). Without this term [i.e., equating constant ζ in Eq. (A.I.6) to zero, $\zeta = 0$] Eq. (A.I.5) would result in a monotonously weakly increasing axial maximum turbulent viscosity distribution in the diverging-area part of the combustor.

One can clearly differentiate the impacts of mixing and finite chemical reaction rates on the heat release retardation by factorization of the relationship for the combustion efficiency η Eq, (A.I.8) into two factors:

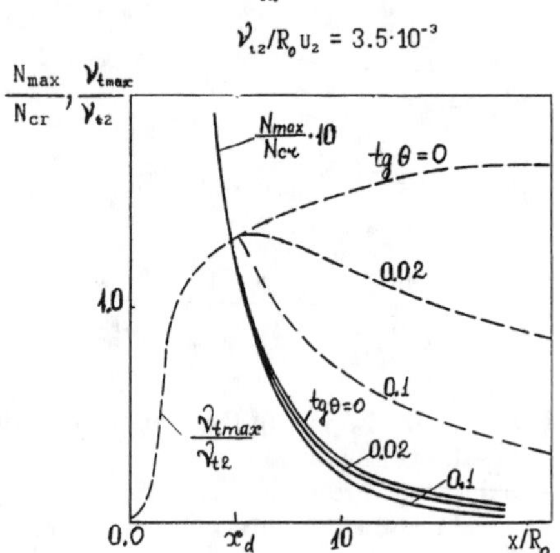

Fig. A.I.6 Axial maximum scalar dissipation to critical scalar dissipation ratio and maximum turbulent viscosity to inflow turbulent viscosity ratio in the combined (constant-area + diverging-area) supersonic combustor $M_1 = 1.0$, $M_2 = 2.6$, $T_1 = 1000$ K, $T_2 = 1500$ K, $\nu_{t1} = 10^{-4}U_2R_0$, $\nu_{t2} = 3.5 \cdot 10^{-3}U_2R_0$, ER = 0.23, $P_{\text{in}} = 0.1$ MPa.

$$\eta = \eta^{(e)}\eta^{(ch)} = \frac{I^{(e)}(x) - I(0)}{G_{\text{H}_2}^{\text{in}}Q}\,\frac{I(x) - I(0)}{I^{(e)}(x) - I(0)} \tag{A.I.9}$$

The first factor $\eta^{(e)}$

$$\eta^{(e)} = \frac{I^{(e)}(x) - I(0)}{G_{\text{H}_2}^{\text{in}}Q} \tag{A.I.10}$$

is under chemical equilibrium, or mixing combustion efficiency that is dependent only on mixing. For the case of the global chemical reaction $H_2 + 1/2O_2 = H_2O$, Eq. (A.I.10) for $\eta^{(e)}$ is reduced to

$$\eta^{(e)} = 1 = \frac{G_{\text{H}_2}^{\text{in}}(x)}{G_{\text{H}_2}^{\text{in}}} \tag{A.I.11}$$

The second factor $\eta^{(ch)}$

$$\eta^{(ch)} = \frac{I(x) - I(0)}{I^{(e)}(x) - I(0)} \tag{A.I.12}$$

is chemical combustion efficiency; it is determined by the finite chemical reaction rates.

Calculated equilibrium and chemical combustion efficiencies are plotted in Figs. A.I.4b and A.I.5b for two values of P_{in}, 0.1 MPa and 0.03 MPa, respectively. It is seen that chemical combustion efficiency $\eta^{(ch)}$ is weakly dependent on the expansion angle at $P_{\text{in}} = 0.1$ MPa (Fig. A.I.4b) and substantially at $P_{in} = 0.03$ MPa (Fig. A.I.5b). The slight increase of the chemical combustion efficiency at $P_{in} = 0.1$ MPa as expansion angle increases can be explained by the reduction of the oxidizer flux to the combustion zone (as the result of the decreasing of the turbulent viscosity) and as a consequence a closer approach of the flame to thermodynamic equilibrium state (see the analysis below). Calculations showed that at the $P_{in} \leq 0.05$ MPa the combustion does not stabilize in the constant-area combustor part of length $x_d = 5\ R_0$, Figure A.I.5b illustrates that in such cases both combustion efficiencies $\eta^{(e)}$ and $\eta^{(ch)}$ fall in the diverging-area combustor part, as expansion angle increases. Thus we can conclude that in these cases both mixing and finite chemical reaction rates retard the heat release, particularly for $P_{in} \leq 0.03$ MPa.

D. Qualitative Analysis of Results

There is a qualitative explanation for the effects of the combustor expansion on the combustion efficiency when combustion stabilization takes place before the divergent combustor part. This explanation is based on the theory[17] that establishes the extinction limit for the laminar combustion of nonpremixed gases. The analysis[18,19] is based on the proposition that chemical processes in a turbulent flame are mostly confined in the thin nonstationary zones, located in the vicinity of the stoichiometric surfaces. The instantaneous structure of these zones, which are named flamelets, is similar to that in a laminar diffusion flame. Flamelets move randomly in the space because of the velocity fluctuations. That fine structure vanishes if one averages the instantaneous picture. The main result of Ref. 17 is that diffusion combustion is quenched if the fuel and, correspondingly, oxidizer flux to the flame front (frame load) exceed some critical value. This condition can be written in the following form, which is convenient for practical application:

$$N < N_{cr} = A\left(\frac{T_2}{293}\right)^2\left(\frac{P}{P_0}\right) \tag{A.I.13}$$

Here constant $A = 200$ or $28\ s^{-1}$ for the hydrogen–air and propane–air mixtures, respectively, $P_0 = 0.1$ MPa, Z is the mixture fraction (see, e.g., Ref. 18), $N = \langle D(\partial Z/\partial x_i)^2 \rangle$ is the mean mixture fraction dissipation, D is the molecular diffusion coefficient, and the exponent n describes the dependence of the normal laminar velocity on the pressure; $n = 0-0.2$ for hydrogen.[20,21] For the estimations below the intermediate value $n = 0.1$ is used. Scalar dissipation can be found from the following semi-empirical relationship (see, e.g., Refs. 18, and 22).

$$N = b\frac{E}{\nu_t}\sigma^2 \tag{A.I.14}$$

where $b = 0.19$ is an empirical constant; E is the turbulent kinetic energy; and σ^2 is the variance of mixture fraction. The values of E and σ^2 were calculated from semi-empirical balance equations.[18,22] Figure A.I.6 shows axial distributions of the ratio N_{max}/N_{cr} in the combustor for various expansion angles; N_{max} is the maximum value of scalar dissipation in the cross-section at x. It is seen that 1) ratio of $N_{max}/N_{cr} \ll 1$ for $x/R_0 \geq 5$; and 2) value of this ratio decrease with increasing the expansion angle. Thus we can conclude that flame extinction does not take place in the diverging-area combustor (under the above stated inflow conditions), and the increase of the expansion angle may even reduce the departure from thermodynamic equilibrium. This qualitative conclusion agrees completely with the one obtained on the basis of numerical integration of Eqs. (A.I.1–A.I.6) and presented earlier. It confirms the conclusion that was drawn in the preceding paragraph that the main reason for the heat release retardation in the diverging-area combustor (for the cases when combustion is stabilized in constant-area combustor part) is the reduction of the turbulent mixing.

If nitrogen in the fuel jet was replaced by air (all other things being equal) the above-stated conclusion is not altered, although some combustion heat release enhancement was observed in this case (not shown here).

E. Conclusions

1) It was found that finite chemical reaction rate factors can have a pronounced impact on heat release in a supersonic combustor, for example, at inflow static pressure $P_{in} < .03$ MPa it is impossible to deliver efficient heat release even in the constant-area combustor.

2) If the combustion is already stabilized before the diverging-area combustor part (e.g., at $P_{in} \geq 0.1$ MPa) the heat release retardation in the diverging-area part is mainly due to reducing the intensity of turbulent mixing. At lower inflow static pressure levels ($P_{in} \leq 0.03$ MPa) both the reduced mixing and finite chemical reaction rates play comparable roles in the retardation of heat release in a diverging-area combustor.

The practical significance of the above analysis is that several ways can be proposed to overcome the retardation of heat release in diverging-area supersonic combustors: First, proving a sufficient length of the forward constant-area combustor part, which should not be less than the ignition delay length; second, enhancing the turbulent mixing using additional turbulization of flow that can be done with various types of devices or by producing shear layers in the flow (an alternative simple solution is to use a backward-stepped combustor, e.g., Ref. 6; and third, to counteract the destructive role of the favorable axial pressure gradient in the diverging-area combustor part using, for example, additional distributed (staged) fuel injection in the diverging-area part. This latter idea, in fact, is used in adopting the thermal-compression principle.[23,24]

Despite the fact that the analysis was based on the boundary layer approximation, it is supposed that the qualitative conclusions of the study were not altered if a more complete analysis would be performed (e.g., based on the parabolized Navier–Stokes equations), since shock waves and local recirculating zones that may be uncovered in such analysis will facilitate combustion.[25]

References

[1]Zimont, V. L., Ivanov, V. I., and Mironenko, V. A., et al. *Combustion and Explosion,* Nauka, Moscow, 1997 (In Russian)

[2]Cookson, R. A., and Isaac, J. J. *Astron. Acta,* No. 6, 1979, p. 531.

[3]Albegov, R. V., Kurziner, R. I., and Petrov, M. D., et al. *Proceed. 7th Tsander Chtenii. Theory of Flying Vehicle,* Moscow, 1984 (In Russian).

[4]Guy, R. W., and Mackey, E. A. AIAA Paper No. 79-7045, 1979.

[5]Stalker, R. J., and Morgan, R. G. *Combust. Flame,* 57, No. 1, 1984, p. 55.

[6]Zabaikin, V. A., Lazarev, A. M., Solovova, E. A., et al. *Vestnik AN BSSR, Ser. Fiz.-Energ. Nauk,* No. 3, p. 103 (In Russian)

[7]Zel'dovich, Ya. B.,. *Zh. Tekh. Fiz.,* 19, No. 10, 1949, p. 1199 (In Russian).

[8]Zimont, V. L., Levin, V. M., Meshcheryakov, E. A., and Sabel'nikov, V. A., *Fiz. Goreniya I Vzryva,* 19, No. 4, 1983, p. 75 (In Russian).

[9]Baev, V. K., Golovichev, V. I., and Yasakov, V. A. *Two-dimensional Turbulent Reacting Gas Flows*, Nauka, Novosibirsk (In Russian), 1976.

[10]Meshcheryakov, E. A., and Sabel'nikov, V. A., *Fiz. Goreniya i Vzryva,* 16, No. 2, 1981, p. 55 (In Russian).

[11]Abramovich, G. N., Krasheninnikov, S. Yu., and Secundov, A. N., *Turbulent Flows Subjected to a Bulk Forces and Non Self-Similar Behaviour,* Mashinostroenie, Moscow, 1975 (In Russian).

[12]Golfeld, M. A., and Tyutina, E. G., Preprint No. 12-82. ITPM, Novosibirsk, 1982 (In Russian).

[13]Dimitrov, V. I., *Simple Kinetics*, Nauka, Novosibirsk, 1982 (In Russian).

[14]Baulch, D. L., Drysdale, D. D., Horne, D. G., et al., *Evaluated Kinetic Data for High-Temperature Reactions,* Vol. 1, 1972, London.

[15]Baev, V. K., Golovichev, V. I., Domitrov, V. I., et al., *Fiz. Goreniya i Vzryva, 9,* No. 6, p. 823, 1973 (In Russian).

[16]Orth, R. C., Billig, F. S., and Grenleski, S. E., *Instrumentation for Air-Breathing Propulsion,* Vol. 34, 1974.

[17]Cookson, R. A., Flanagan, P., and Penny, G. S., *12th Symp. (Int.) on Combustion,* 1969.

[18]Kuznetsov, V. R., and Sabel'nikov, V. A., *Turbulence and Combustion,* Nauka, Moscow, 1986 (In Russian).

[19]Kuznetsov, V. R., Trudy CIAM No. 1086, Moscow, 1983 (In Russian).

[20]Shchetinkov, E. S., *Gas Combustion Physics,* Nauka, Moscow, 1965 (In Russian).

[21]Warnatz, J., *Combust. Sci. Technol.,* 26, No. 5, 6, 1981, p.203

[22]Zimont, V. L., Meshcheryakov, E. A., and Sabel'nikov, V. A., *Fiz. Goreniya i Vzryva,* 14, No. 3, p. 55, 1978 (In Russian).

[23]Ferri, A., AIAA Paper No. 66-826, 1966.

[24] Billig, F. S., Orth, R. C., and Lasky, M. J., *J. Spacecraft,* 5, No. 9, 1968, p. 1076.

[25]Zabaikin, V. A., and Lazarev, A. M., *Simulating Processes in Hydrogasdynamics and Power Engineering,* Novosibirsk, 1985 (In Russian).

II. Combustion Stabilization in Supersonic Flow Using Free Recirculating Bubble

A. Introduction

Experimental investigations[1–8] revealed that the interaction of a concentrated vortex with a shock wave at excess of critical conditions results in breakup of the vortex and generation of a free recirculating zone (bubble). Vortices in Refs 1–3 were generated using different types of generators, e.g., semispan wing having a diamond shape airfoil section that was installed at the angle of attack to the incoming flow. This kind of generator allowed the vortex intensity to be varied easily by the variation of the angle of attack. Shock waves in Refs. 1–3 were generated by 1) supersonic two-dimensional inlets, 2) blunt bodies, 3) wedges, and 4) axisymmetric diffusers (which were similar to supersonic nozzle). In Refs. 5 and 6 the phenomenon of free recirculating bubble generation was studied at interaction of the vortex with a central shock wave (Mach stem) in overexpanded jet. Already in Ref. 3 it was concluded that breakup of the vortex and generation of a free recirculating bubble is determined mainly by intensities of vortex and shock wave and does not depend on the type of vortex and shock generators.

It is a very attractive idea to use a free recirculating bubble for combustion stabilization in supersonic flow. High efficiency of such a kind of combustion stabilization in supersonic flow was demonstrated in Ref. 9 at studying H_2–air combustion (with a forced ignition). Free recirculating bubble in Ref. 9 was generated by the interaction of shock waves with base wake behind of cylindrical model of diameter 20 mm in supersonic flow with $M = 2.1$ and $T_t = 375$ K.

In Refs. 10 and 11, using the numerical simulation, the self-ignition and combustion stabilization of methane was demonstrated in free recirculating bubble (with lateral dimension of the latter of about 5 cm) in supersonic flow with $M = 3.0$ and $T_t = 1400$ K.

The objective of the present experimental study was to study self-ignition and combustion stabilization of aerated liquid kerosene (hereinafter called barbotaged kerosene) in supersonic flow using a free recirculating bubble that was generated by the interaction of concentrated vortex with normal shock wave. The investigation was conducted in the hypersonic facility T-131B of TsAGI. Tests were performed in nearly matched supersonic jet at the vicinity of the exit of two-dimensional channel at $M = 2.0$–2.8 at $T_t =$ 1200–1400 K.

B. Estimation of Minimum Dimension of Recirculating Bubble Needed for Self-Ignition and Combustion Stabilization

For evaluation of the minimal necessary size of a free recirculating bubble that is required for self-ignition and stabilization of combustion we use the criterion of combustion stabilization behind a bluff body in subsonic flow of homogeneous fuel mixture (see, e.g., Refs. 9, 12, and 14:

$$\tau_{res}/\tau_{ind} \approx 1 \tag{A.II.1}$$

where τ_{res} and τ_{ind} are residence and induction times, respectively. In experiments[9] it was shown that this criterion is applicable both for a supersonic flow and a nonpremixed combustion. Residence time can be roughly estimated by the equation[9,12–14]

$$\tau_{res} = k \cdot h/u_e \tag{A.II.2}$$

where k = 30–50, h is maximum thickness of the recirculating bubble, u_e is velocity at the boundary of the recirculating bubble (i.e., behind the conical shock). Estimation for T_t = 1200–1400 K and M = 2.5 gives $u_e \approx$ 1000 m/s. The value k = 50 was obtained in Ref. 3, where residence time was measured in a base flow region of an axisymmetric body at M = 4.1.

For the estimation of the induction time we assume the static temperature inside the recirculating bubble is nearly total temperature. Another assumption is that static pressure inside the recirculating bubble is equal to atmospheric pressure. For T_t = 1200 K and 1400 K at atmospheric pressure induction times for stoichiometric kerosene–air mixture are[12,14] $\tau_{ind} \approx 5 \cdot 10^{-3}$ s and $3 \cdot 10^{-4}$ s, respectively (for the hydrogen–air mixture $\tau_{ind} \approx 5 \cdot 10^{-5}$ s and $1 \cdot 10^{-5}$ s, respectively). Hence, as one can obtain from Eqs. A.II.1, and A.II.2, for the stabilization of combustion of kerosene–air mixture in the recirculating bubble it is necessary that the thickness of recirculating bubble were greater than

$$h > h_{cr} = u_e \tau_{ind}/k \tag{A.II.3}$$

where $h \approx$10–16 cm at T_t = 1200 K $h_{cr} \approx$ 0.6–1 cm at T_t = 1400 K.

In our experiment the entrance diameter of the diffuser (shock-generator) was d = 25 mm. It is known that the thickness of recirculating bubble h is almost the same as d (Refs. 1–3). Thus we can expect the self-ignition and stabilization of combustion in our tests will occur only if $Tt \geq$ 1400 K.

C. Scheme of the Experiment: Facility and Test Conditions

Experiments were conducted in the hypersonic facility T-131B of TsAGI, which is used for scramjet combustor connected pipe tests. "Vitiated" air parameters in the kerosene pre-heater were T_t = 1200–1400 K, $P_t \leq$ 4.0 MPa. Scheme of the experiment is presented in Fig. A.II.1. Supersonic nearly matched jet was exhausted into still atmosphere from the exit of a rectangular divergent channel (3), which is connected to a rectangular nozzle (1) with exit cross section 30×100 mm^2 and designed for M = 2.5. The free recirculating bubble was generated due to interaction between vortex (11) generated by vortex generator (4) mounted on the channel wall and bow shock wave in front of the diffuser (5), which is placed downstream the flow.

D. Experimental Model

The experimental model includes channel (3), injectors (2), pylon-vortex generator (4), diffuser (shock-generator) (5), transversing equipment (6) (Fig. A.II.2). The channel (3) consists of two sections. Front section has length of 300 mm and entrance cross-section 30 × 100 mm^2, it diverges along the bottom wall with a half-deg angle. Second section has a length of 500 mm, it diverges along the bottom and top walls with an angle of 2 deg. Exit cross-section of the channel is 67 × 100 mm^2.

Aerated by air, liquid kerosene was injected perpendicular to the main flow through holes drilled in walls of four tube-injectors (2), which were installed at 70-mm distance from the channel entrance. Injectors were mounted on the lid of a hatch across the channel. Each of the four injectors had three holes (of the diameter d = 0.6 mm) on each side, Fig. A.II.3. The distance between injectors and channel exit was about 1 m, thus the kerosene–air mixture was nearly homogeneous at the channel exit.

Barbotage of kerosene was carried out in the specially designed mixing device,[16] which provides a kerosene–air mixture with mass rate ≤0.25 kg/s and gass mass fraction in the mixture less than 5%, Fig. A.II.4. Using barbotage allowed to provide better mixing of kerosene with supersonic air flow.

The vortex generator (1) in Fig. A.II.1 was a semispan wing having a diamond-shape airfoil section with a chord length of 15 mm, a span of 32 mm, a half-angle of 15 deg, and angle of attack in the range ±180 deg, Fig. A.II.5. Vortex generator was mounted on the wall of the channel at 50-mm upstream of the exit. The construction of the vortex generator had a cooling duct (3) to cool generator during start-up of the test facility. This duct was also used to supply hydrogen (or another fuel) into the recirculating bubble. To provide the durability of the generator under high temperatures and pressures, it was manufactured from a strong alloy steel.

A schematic plot of the diffuser (shock-generator) is given in Fig. A.II.6. As mentioned above the dimensions of the diffuser were chosen to get the free recirculating bubble, emerging as the result of interaction between vortex and bow shock in front of the diffuser, sufficient to sustain self-ignition and combustion stabilization. The diffuser was fixed in a special holder (3) which is mounted on the stand of transversing equipment (4). Transversing equipment allowed the holder to move with the diffuser in and out of the desired region of the flow during a test run. The diffuser had a sharp edge lip. Relation between throat area entrance and diffuser area was chosen according to the recommendations of work.[17] Fillment of this relationship assured the start of the diffuser in supersonic flow with M >2.1 and when the vortex goes into the diffuser. Throttling hollow (5) was connected to the diffuser throat by six holes (6). The holes were spread evenly across the circle of the channel's cross-section and were 1.5 mm in diameter. To achieve the throttling of the diffuser, gas was supplied into the throttling hollow through the connecting pipe (7). Gas-dynamic throttling allowed the ability to vary effective throat area of the diffuser, if necessary. This gave a possibility to change continuously from start to unstart regimes of the diffuser. During test runs static pressure along the diffuser wall was measured in three points. To

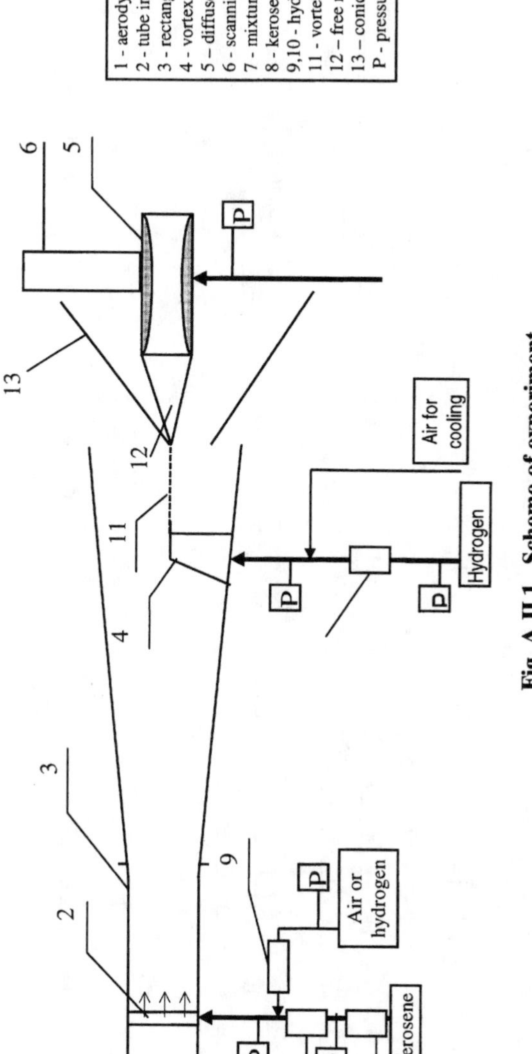

Fig. A.II.1 Scheme of experiment.

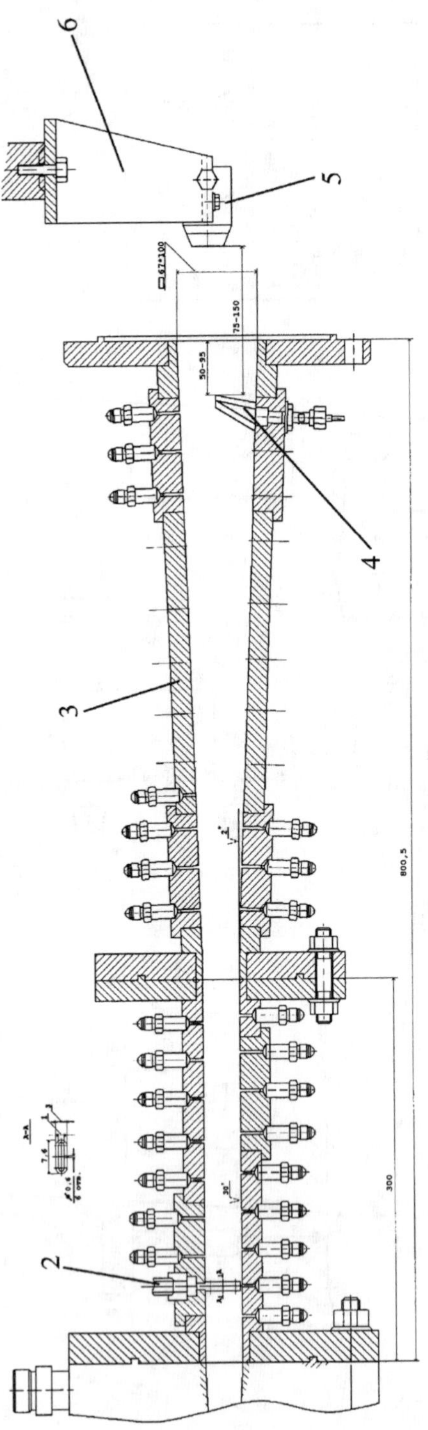

Fig. A.II.2 Scheme of the experimental model (all dimensions in millimeters).

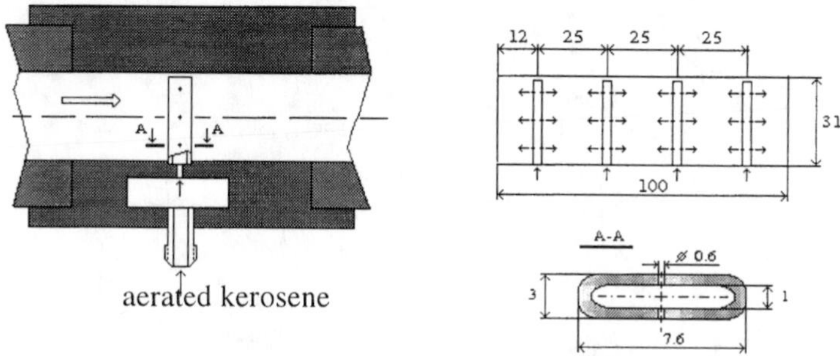

Fig. A.II.3 Geometry and location of tube-injectors (all dimensions in millimeters).

ensure operation under high temperatures and loads the holder and constructions of the diffuser were made of endurable steel, and its lips are from heat-resistant steel.

E. Tests Methodology and Measurements

At first, the conditions of the generation of the free recirculating bubble were found. These runs were done at the total temperature $T_t = 1200$ K for two cases: 1) with injectors; 2) without injectors. In the first case the fuel was not supplied into the injectors. The following parameters were varied during tests: Angle of attack of the vortex generator, distance between the vortex generator and the diffuser, and throttling intensity. Besides, the influence of air or hydrogen injection into vortex (throughout the vortex generator) on the generation of the free recirculating bubble was investigated.

Investigation of self-ignition and combustion stabilization was conducted after the conditions of the appearance of free recirculating bubble were established. The test runs were done at $T_t = 1200$–1400 K. During the test runs the following measurements were done:

1) Axial static pressure distributions on the top and bottom walls of the channel.
2) Axial static pressure distribution on the wall of diffuser.
3) Pre-heater pressure P_t.

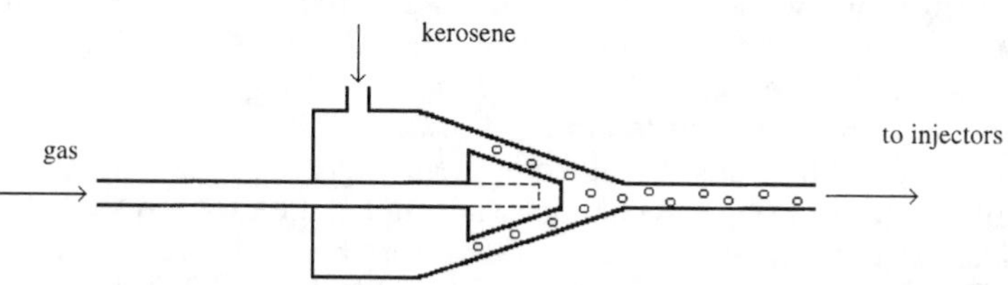

Fig. A.II.4 Scheme of the mixture system (barbotage) device.

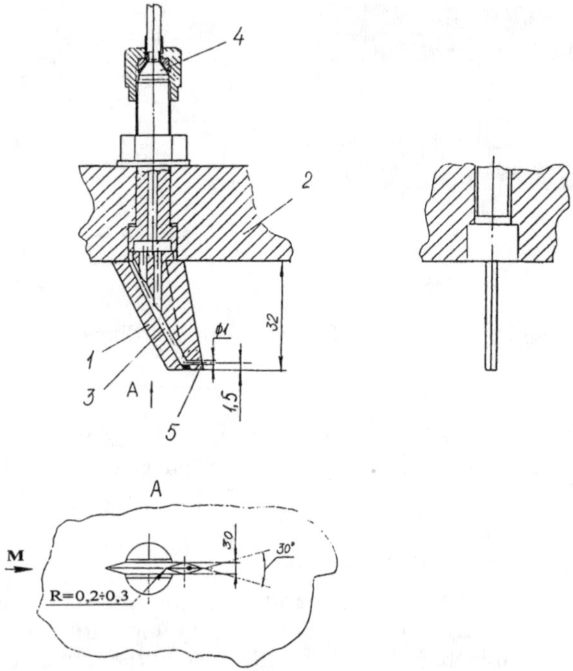

Fig. A.II.5 Scheme of the vortex-generator (all dimensions in millimeters).

4) Air, hydrogen, and oxygen pressures and temperatures before their flow-meter nozzles.

5) Pressure and mass rate of kerosene before the barbotage-mixer device.

6) Fuel mixture pressure before the injectors.

7) Pressure of hydrogen (or air) supply into the vortex generator.

8) Pressure of throttling air in the diffuser.

9) Videotaping of the flow in the vicinity of the exit of the rectangular channel made using schlieren system.

F. Experimental Results

A list of test runs and conditions at which they were conducted is given in Table A.II.1. Stagnation pressure fields at the vicinity of the channel exit with injectors and without injectors were measured at section $x = 18$ mm and $x = 50$ mm from channel exit in runs no. 1 and 2, respectively. Parameters at the channel entrance were: $M_{\text{ent}} = 2.5$, $P_t = 4.0$ MPa, $T_t = 1400$ K. Stagnation pressure was measured by scanning horizontal 10-point rake, which was vertically moved by transversing equipment. Fig. A.II.7 presents static pressure distributions along the channel walls and stagnation pressure field for run No. 2. Two conclusions can be drawn from Fig. A.II.7: 1) static pressure on walls close to channel exit was within range 0.8–1.1 bar, i.e., jet was matched to atmospheric pressure; 2) stagnation pressure variation inside of

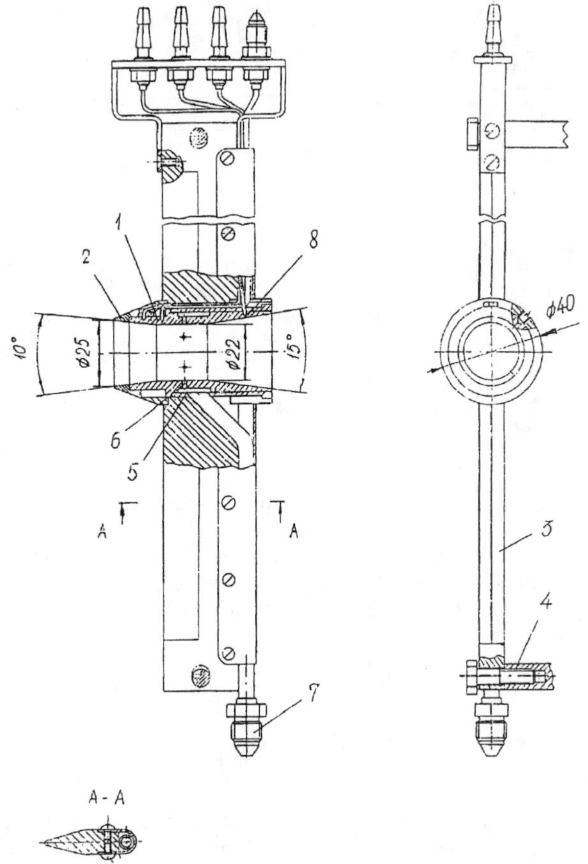

Figure A.II.6. Scheme of the disffuser (shock-generator, all dimensions in millimeters).

jet in vertical direction within range 19 mm $< y <$ 48 mm (see reference system in Fig. A.II.7) was negligible and so the practically uniform supersonic flow core thickness was about 30 mm. The Mach numbers, calculated on the base of measured stagnation pressure and static pressure equal 0.1 MPa were within range M = 2.5–2.8 and M = 2.57–2.85 at distances 18 mm and 50 mm, respectively.

Beginning from run no. 5 four tube injectors were installed at the channel (Fig. A.II.3). Installation of injectors resulted in a 1.5 time increase of static pressure at the exit of channel in comparison with runs no. 1 and 2 in which injectors were absent. Stagnation pressure fields at the vicinity of channel exit for runs no. 5–11 were not measured, but rough estimation shows that injection installation resulted in a decrease of Mach number down to M = 2.0 at distance x = 18 mm from exit. For matched condition to be obtained

Table A.II.1 Test Conditions and Results

Run number	P_t, MPa	T_t, K	M_{ent}	ER	Objective and results
No. 1	3.85	1418	2.5	0	
	3.81	1408	2.5	0	
	3.79	1403	2.5	0	
	3.81	1402	2.5	0	
	3.83	1407	2.5	0	Stagnation pressure fields at supersonic jet at distance $x = 18$ mm and $x = 50$ mm from the channel exit were measured.
No. 2	3.87	1433	2.5	0	
	3.90	1441	2.5	0	
	3.87	1425	2.5	0	
	3.89	1429	2.5	0	
	3.87	1424	2.5	0	
No. 3	3.95	1197	2.5	0	Bow shock in front of the shock wave-generator was obtained.
	3.93	1194	2.5	0	
No. 4	3.91	1187	2.5	0	Free recirculation bubble in front of the shock wave–generator as a result of vortex and bow shock interaction was obtained.
	3.91	1192	2.5	0	
	3.88	1185	2.5	0	
No. 5	2.75	1220	2.5	0	
	2.76	1210	2.5	1.23	No self-ignition and combustion stabilization of barbotage kerosene in free recirculation bubble.
No. 6	2.75	1280	2.5	0	
	2.72	1275	2.5	0.77	
No. 7	3.80	1400	2.5	0	Self-ignition and combustion stabilization of barbotaged kerosene in free recirculation bubble was obtained. Outward propagation of combustion took place.
	3.82	1407	2.5	0	
	3.80	1401	2.5	0	
	3.79	1396	2.5	0.7	
No. 8	3.93	1433	2.5	0	
	3.94	1436	2.5	0	
	3.92	1423	2.5	1.46	Self-ignition and combustion stabilization of kerosene in free recirculation bubble was obtained. No outward propagation of combustion.
No. 9	3.97	1461	2.5	0	
	3.96	1453	2.5	0	
	3.97	1460	2.5	1.56	
No. 10	2.53	1190	2.5	0	Role played alone by bow shock in combustion stabilization was studied. It was shown that self-ignition was absent.
No. 11	2.66	1384	2.5	0	
	2.69	1406	2.5	0.9	

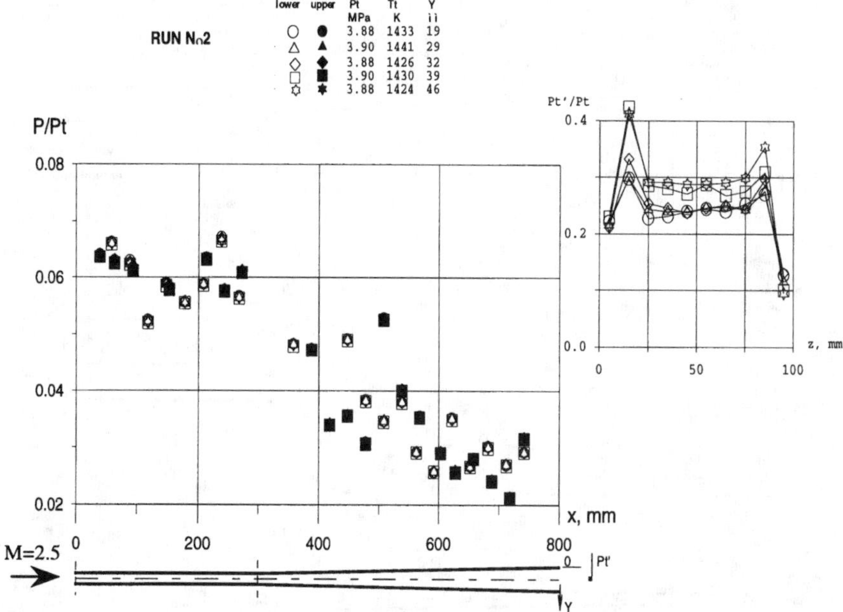

Fig. A.II.7 Axial static pressure distribution along channel walls and pitot pressure fields in run no. 2 at section x = 50 mm from the channel exit.

at the channel exit in this case, it was needed to decrease pressure in the pre-heater to P_t = 2.7 MPa.

In run no. 3 methodology of bow shock wave generation was refined. Run was performed without injectors and vortex generator. Diffuser was located at a distance 50 mm downstream of channel exit. Flow parameters at the channel entrance were: M_{ent} = 2.5, P_t = 3.9 MPa, T_t = 1200 K. Fig. A.II.8 shows axial pressure distributions on channel walls and also on diffuser wall for two cases: 1) without diffuser throttling and 2) with diffuser throttling using air jets in the diffuser throat. It is seen from Fig. A.II.8 that the flow was supersonic inside of the diffuser for the case without throttling. Bow shock was generated in front of the diffuser for the case with throttling. Pressure rise roughly corresponded to M = 2.5. Subsonic flow behind bow shock wave acclerated and reached sonic velocity at the throttling section.

Generation of a free recirculating bubble was performed in run no. 4. Flow parameters at the channel entrance were: M_{ent} = 2.5, P_t = 3.9 MPa, T_t = 1200 K. A test was done without injectors but with vortex generator and diffuser, with throttling of the latter. Angle of attack of vortex generator was 15 deg. It was installed at a distance 50-mm upstream from channel exit. Diffuser was installed at a distance 50-mm downstream from channel exit. Two cases were considered: 1) with air supply into the vortex (through generator); and 2) without air supply into the vortex. Schlieren pictures are

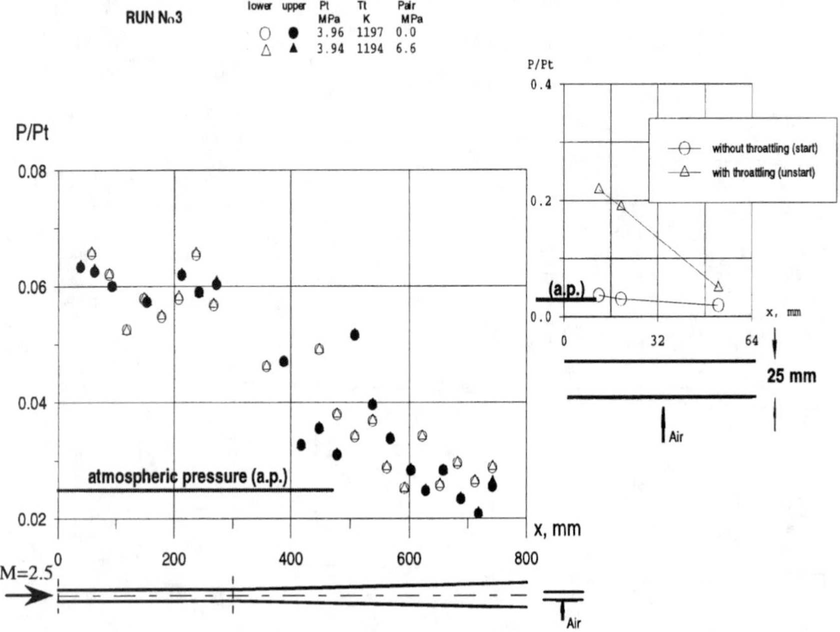

Fig. A.II.8 Axial static pressure distribution along channel walls and diffuser wall in run no. 3.

presented in Figs. A.II.9a and A.II.9b, respectively. In the first case air mass rate was 1.5 g/s (pressure supply was 1.0 MPa). It is seen from Figs. A.II.9a,b that in both cases free recirculating bubble arose in front of the diffuser and that recirculating bubble is larger in the second case. Thus, this test shows that injection of high pressure gas into the vortex results in the decrease of vortex intensity, the decrease of dimension of the recirculating bubble is a consequence of that. It can be assumed that some critical value (dependent of course, on shock intensity and vortex structure) of air mass rate exists, when breakup of vortex does not take place. In this case interaction of vortex with shock is referred as weak (see, e.g., Ref. 1.) and only bending of shock is realized during such interference. Unfortunately, this interesting problem was outside of our investigation.

Self-ignition and combustion stabilization of barbotaged kerosene was obtained in run no. 7 at $T_t = 1400$ K. Time development of the self-ignition process was run as follows. At first, free recirculating bubble was organized in supersonic flow, Fig. A.II.10a. Afterwards supply of barbotaged kerosene throughout was performed and hydrogen was injected throughout the vortex generator. Fuel equivalence ratio was ER = 0.7, mass rate of H_2 was 0.5 g/s. Suddenly in some region of the free recirculating bubble, self-ignition arose followed by propagation of flame throughout free recirculating bubble and at final stage outward of zone to supersonic flow of kerosene–air

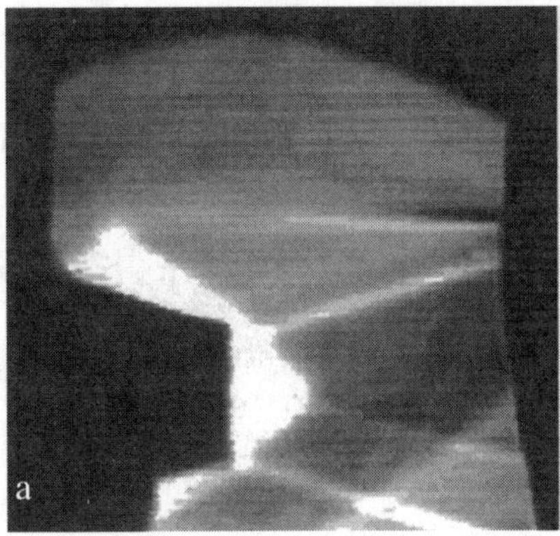

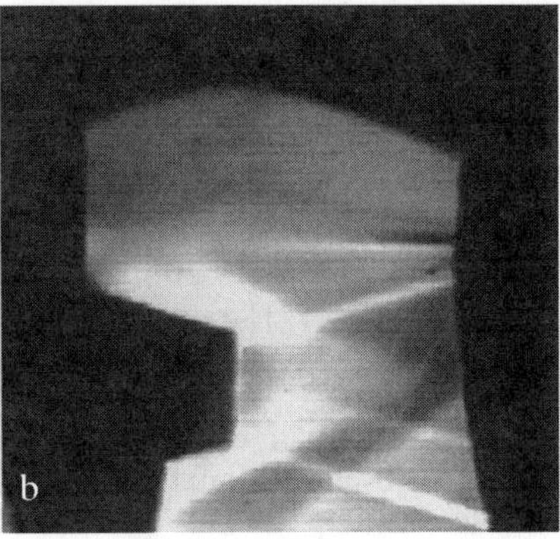

Fig. A.II.9 Schlieren picture of flowfield in run no. 4 a) with air supply through vortex generator and b) without air supply through vortex generator.

mixture, Fig. A.II.10b. Upstream propagation of the flame was not observed. Approximately 3 s after the beginning of kerosene supply, diffuser and holder were thermally damaged due to intensive heat release in front of and in vicinity of diffuser.

In runs no. 8 and 9 efforts were undertaken to obtain self-ignition of pure liquid kerosene, at the same time temperature $T_t = 1400$ K as in run no. 7. Fuel equivalence ratio was ER $\approx$1.4. The diffuser was installed at the distance 25-mm downstream of the channel exit. In run no. 8 air was supplied throughout the vortex generator. In run no. 9 after some time air supply was switched off, and H_2 was injected throughout vortex generator. The radiation of the light was observed mainly from recirculating bubble, i.e., the burning in external supersonic flow was practically absent. A possible reason of such a decreasing of the intensity of combustion, as compared to the run

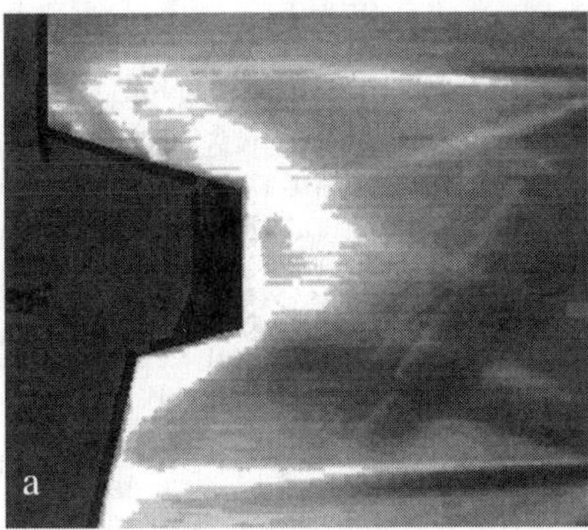

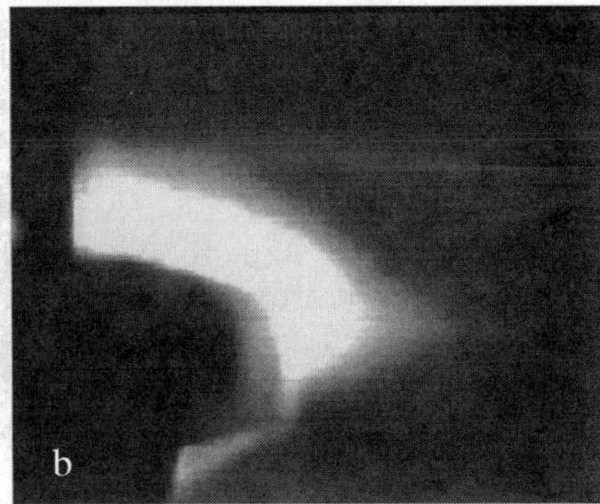

Fig. A.II.10 Schlieren picture of flowfield in run no. 7 a) just before self-ignition and b) just after self-ignition.

no. 7, is the worsening of mixing between kerosene and air without barbotage of the former.

The last two runs, nos. 10 and 11, were directed to learn the role of shock waves in the combustion stabilization. To this end these runs were conducted without a vortex generator and, consequently without free recirculating bubble, not only with the bow shock in front of the diffuser. Tests were performed at two stagnation temperatures $T_t = 1200$ K (run no. 10) and $T_t = 1400$ K (run no. 11). Tests showed the absence of self-ignition of barbotaged kerosene in the vicinity of the diffuser for start and unstart regimes of the diffuser work. At $T_t = 1400$ K self-ignition was observed in the far wake region downstream of the diffuser. Thus, it can be concluded that self-ignition of kerosene–air mixture was realized only due to the free recirculating bubble in front of the diffuser

G. Conclusions

Experiments were carried out to study the self-ignition and combustion stabilization of aerated by gas liquid kerosene in supersonic flow using a free recirculating bubble that was generated by the interaction of a concentrated vortex with a normal shock wave. The following results were obtained:

1) Scheme of combustion stabilization in supersonic flow using free recirculating bubble is proposed.

2) Aerodynamic model for experimental confirmation of combustion stabilization scheme for aerated by air (barbotaged) kerosene in supersonic high-entalphy flow in the hypersonic facility T-131B of TsAGI was designed and manufactured.

3) Dimensions of the aerodynamic model were chosen to realize self-ignition and combustion stabilization. Criterion of stabilization of the flame in the wake of bluff body was used. It was estimated that the dimensions of the facility T-131B permit obtaining the self-ignition and combustion stabilization of barbotaged kerosene at stagnation temperature $T_t = 1400$K.

4) Experiments confirmed principles on which model design was performed and dimensions of it. At $T_t = 1400$ self-ignition and combustion stabilization of barbotaged kerosene was obtained in nearly matched supersonic jet by using free recirculating bubble generated by the interaction of concentrated vortex and bow shock. Outward combustion propagation into supersonic flow was obtained (without upstream propagation). Self-ignition of pure liquid kerosene took place only inside of free recirculating bubble without outward propagation into supersonic flow.

5) Self-ignition and combustion stabilization of barbotaged kerosene in free recirculating bubble at $T_t = 1200$ K–1300 K was not achieved.

6) It was shown that injection of gas into the vortex results in weakening of the concentrated vortex and, as a consequence, decrease of dimensions of free recirculating bubble. A hypothesis was put forward that a limiting value of injection mass rate exists and at excess of which the free recirculating bubble disappears.

Acknowledgements

This work was supported by US Air Force Office of Scientific Research (AFMC), EOARD, contract No. SPC-96-4043. We sincerely wish to express our appreciation to W. L. Bain III USAF Wright Laboratory, WPAFB, OH for his valuable support of this study. Thanks to O. V. Voloschenko, V. N. Sermanov, and Yu. Korotkov for their valuable contribution in running the tests.

References

[1]Zatoloka, V. V., Ivanyushkin, A. K., and Nikolaev, A. V., "Interference of Concentrated Vortices with Shock Wave in Inlet," *Breakup of Vortices,* Uchenye Zapiski TsAGi, 6, No. 2, pp. 134–138, 1975 (In Russian).

[2]Zatoloka, V. V., Ivanyushkin, A. K., and Nikolaev, A. V., "Interference of Vortexes with Shocks in Aircoops, Dissipation of Vortexes," *Fluid Mechanics-Soviet Research,* Vol. 7, No. 4, July–August, pp. 153–158, 1978.

[3]Ivayushkin, A. K., Korotkov, Yu.V., and Nikolaev, A. V., "Some Peculiarities of Interference of Shock Waves with Aerodynamic Wake Behind Body," *Uchenye Zapiski TsAGI,* Vol. 10, No. 5, pp. 33–42, 1989 (In Russian).

[4]Delery, J., Horovitz, E., Leuchter, O., and Solingac, J., "Etudes Fondamentales sur les Écoulements Tourbillonnaires," *La Recherche Aérospatiale,* No. 2, pp. 81–104, 1984.

[5]Mettwally, O., Settles, G., and Horsman, C., "An Experimental Study of Shock Wave/Vortex Interaction," AIAA Paper 89-0082, 12 p., 1989.

[6]Glotov, G.Ph., "Interference of Concentrated Vortex with Shock Waves in Free Flow and Nondesigned Jets," *Uchenye Zapiski TsAGI,* Vol. 10, No. 5, pp. 21–32, 1989 (In Russian).

[7]Michael, K., Smart, K. M., and Kalkhoran, I. M., "Effect of Shock Strength on Oblique Shock-Wave/Vortex Interaction," *AIAA Journal,* Vol. 33, No. 11, pp. 2137–2143, 1995.

[8]Kalkhoran, I. M., Smart, K. M., and Betti, A., "Interaction of Supersonic Wing-tip Vortices with a Normal Shock," *AIAA Journal*, Vol. 34, No.9, pp. 1855–1861, 1996.

[9]Winterfeld, G., "On the Burning Limits of Flame-Holder-Stabilized Flames in Supersonic Flow," *AGARD, IX,* Vol. 2, No. 34, p. 12, 1968.

[10]Sabel'nikov, V. A., "Supersonic Turbulent Combustion of Nonpremixed Gases-Status and Perspectives," *Advanced Computation and Analysis of Combustion,* ENAS Publishers, Moscow, pp. 208–237, 1997.

[11]Figueira da Silva, L. F., Sabel'nikov, V. A., and Deshaies, B., "The Stabilization of Supersonic Combustion by a Free Recirculating Bubble: A Numerical Study," *AIAA Journal,* Vol. 35, No. 11, pp. 1782–1784, 1997.

[12]Shchetinkov, E. S., *Physics of Gas Combustion,* Nauka, Moscow, 1965 (In Russian).

[13]Zakkay, V., and Sinha, R., "Residence Time Within a Wake Recirculation Region in an Axisymmetric Supersonic Flow," AIAA Paper 70-111, 1970.

[14]Baev, V. K., Golovichev, V. I., Tret'yakov, P. K., et al., *Combustion in Supersonic Flow,* Nauka, Moscow, Novosibirsk, 1984 (In Russian).

[15]Westbrook, C. K., and Dryer, F. L., "Chemical Kinetic Modeling of Hydrocarbon Combustion," *Progress in Energy and Combustion Science,* Vol. 10, pp. 1–57, 1984.

[16]Avrashkov, V. N., Baranovsky, S. I., and Levin, V. M., "Gasdynamic Features of Supersonic Kerosene Combustion in Model Combustion Chamber," AIAA Paper 90-5268, 1990.

[17]Ivanyushkin, A. K., and Korotkov, Yu.V., "Influence of Flow Vortocity on Diffuser Start at Supersonic and Hypersonic Speeds," *TsAGI Workshop-School "Fluid Mechanics": Research in Hypersonic Flows and Hypersonic Technologies,* TsAGI, pp. 27–28, 1995.

III. The Enhancement of Liquid Hydrocarbon Supersonic Combustion using Effervescent Sprays and Injectors with Noncircular Nozzles

A. Introduction

Combustion in a supersonic combustor is considerably dependent (along with the kinetics) on the intensity of turbulent mixing. The factors that give the methods of supersonic mixing enhancement a special significance are as follows: 1) the decrease of mixing intensity in supersonic flows; 2) the small residence time due to the length of the combustor that does not exceed 3 m and a flow speed which is in the range 1–2 km/s. Among enhancement techniques we can mention (see also Refs. 1–4): 1) the interaction between fuel jets, shock, and expansion waves; 2) the use of injectors which geometry favors the generation of intense longitudinal vortices (e.g., NASA swept wedges); 3) the use of a noncircular nozzle geometry (e.g., elliptic nozzles) for the fuel supply.

Up to now, above listed techniques of mixing intensification were used basically to accelerate the gaseous fuel jet mixing. The main mechanism of mixing intensification in a gaseous jet is vortex-induced and is related to the excitation of large-scale modes of instability. Opportunities of such a mechanism are apparently limited for jets of liquid fuel (e.g., kerosene, the promising fuel for small-dimension hypersonic vehicles). Hence, the idea of the aeration (hereinafter referred as to barbotage) of liquid fuel jets by the gas (effervescent sprays) is considered to be appealing. Investigations conducted at MAI (see, e.g. Ref. 5) and at TsAGI showed that effervescent sprays by their expansion angle are close to gaseous jets.

The main objective of the present investigation was to study the potential possibilities of supersonic mixing combustion enhancement by using gas-aerated (hydrogen or air) liquid kerosene and noncircular nozzles. Fuel was injected through elliptic nozzles from injectors of two geometry's: 1) tube-micropylons and 2) fin-pylons. Tests were conducted under scramjet combustor conditions. Flow parameters at the combustor entrance were $M = 2.5$ and $T_t = 1650–1800$ K. This work presents ignition delay characteristics, axial pressure distributions, combustion efficiencies, and pressure-area integrals for elliptic and round nozzles.

B. Experimental Facility: Test Methodology

Tests were conducted using the scramjet combustor and hypersonic facility of MAI equipped with kerosene-fueled pre-heater (vitiated air). Oxygen mass fraction in the vitiated air $Y^0_{O_2}$ was slightly lower than in the atmospheric air. $Y^0_{O_2}$ values for each test run can be found in Table A.III.1. With an oxygen mass fraction in the atmospheric air of 0.232 the kerosene equivalence ratio (ER) in vitiated air is determined by the following relation:

$$ER = (0.232/Y^0_{O_2})L_0G_{ker}/G_1$$

where $L_0 = 14.7$ is the stoichiometric coefficient for kerosene combustion in atmospheric air, G_{ker} and G_1 are mass rates of kerosene and vitiated air, respectively. Total flow parameters and other parameters characterizing the facility and combustor operation regimes are given in Table A.III.1. Fig. A.III.1a depicts the schematic view of combustor. The combustor has four sections: 1) a 150-mm-length section with a constant cross-area 52×104 mm² (height h = 52 mm and width 2 = 2h); 2) a 150-mm-length section with a divergence angle of 6.85 degree along upper wall leading to an exit cross-area of 70×104 mm²; 3) a 300-mm-length section with divergence angle of 1.9 deg along upper wall leading to an exit cross-area of 80×104 mm²; 4) a 570-mm-length section with a constant cross-area. Thus, the total length of the combustor is 1050 mm with an area-expansion ratio of 1.7. Flow from the combustor was exhausted into a still atmosphere. Axial pressure distributions were measured by taps placed on the upper and lower combustor walls.

Kerosene jets were aerated (barbotated) with hydrogen or air. Barbotage device scheme is shown in Fig. A.III.2. Mass fraction of gas used for aeration was small enough: indeed, while the kerosene mass rate was 6–130 g/s, hydrogen mass rate was about 1 g/s and air mass rate was about 10 g/s. Mixture pressure in the barbotage device was in the range 1.5–2.5 MPa. The volume fractions of kerosene and gas at the nozzle exit of the injectors were of the same order of magnitude. Injection of effervescent kerosene sprays into the flow with a much lower pressure level causes the explosion of the jet that promotes the vaporization and mixing of liquid kerosene.[5] Fuel was injected into the combustor in two ways: 1) at the angle of 45 deg relative to the mainstream air flow direction throughout tube-micropylons; 2) in the co-flow direction with mainstream flow throughout the fin-pylons. The injectors were mounted in rows of four pieces on the upper and lower combustor walls. The distance between the combustor entrance and the injectors' location was 105 mm. Fuel injection for both injector types was performed through either round nozzles of diameter 1.2 mm or elliptic nozzles with dimensions of principal axis 0.6 mm and 1.9 mm. Injector geometries are given in Fig. A.III.1b (tube-micropylons) and Fig. A.III.1c (fin-pylons); their placement scheme is given in Fig. A.III.1d.

Test runs nos. 1–3 were done with hydrogen barbotaged kerosene at fixed fuel equivalence ratios. The four other test runs nos. 9–12 where kerosene was barbotated with air were done in the following sequence: after reaching the desired combustor entrance conditions the fuel was injected during 20–30 s. During this time interval the magnitude of ER was gradually decreased and changed from values nearly stoichiometric to values at which

Table A.III.1 Test Parameters

RUN	Nozzles	P_t, MPa	T_t, K	P_m, MPa	G_{air}, kg/s	G_{02} kg/s	$G_{h,ker}$, kg/s	G_{ker}, kg/s	G_{bg}/G_{ker}	$Y^0_{O_2}$	ER
Tube-micropylons injectors, barbotage by hydrogen											
1	Round	1.44	1690	—[a]	2.125	0.522	0.103	0	—[a]	0.16	0
1	Round	1.44	1650	1.5	2.125	0.522	0.103	0.132	~0.015	0.16	1.06
2	Elliptic	1.46	1790	—[a]	2.185	0.257	0.101	0	—[a]	0.1548	0
2	Elliptic	1.45	1780	2.18	2.185	0.257	0.101	0.08	~0.01	0.1548	0.69
3	Elliptic	1.41	1780	—[a]	2.097	0.257	0.106	0	—[a]	0.1539	0
3	Elliptic	1.42	1750	1.6	2.097	0.57	0.106	0.07	~0.01	0.1547	0.62
Tube-micropylons injectors, barbotage by air											
9	Elliptic	1.46	1754	—[a]	2.175	0.257	0.1189	0	—[a]	0.1354	0
9	Elliptic	1.45	1793	2.29	2.125	0.2514	0.1182	0.094	0.16	0.1347	0.945
9	Ellptic	1.44	1756	1.96	2.115	0.2514	0.1187	0.075	0.18	0.1335	0.763
9	Elliptic	1.44	1756	1.86	2.116	0.2514	0.1184	0.07	0.19	0.1341	0.709
10	Round	1.43	1771	—[a]	2.08	0.257	0.1254	0	—[a]	0.1224	0
10	Round	1.45	1736	2.39	2.155	0.2514	0.1220	0.094	0.16	0.1304	0.962
10	Round	1.46	1775	2.16	2.137	0.2514	0.1212	0.075	0.18	0.1308	0.771
10	Round	1.44	1727	2.15	2.144	0.2514	0.1217	0.07	0.19	0.1304	0.720
Fin-pylons injectors, barbotage by air											
11	Round	1.43	1765	—[a]	2.109	0.257	0.1226	0	—[a]	0.1278	0
11	Round	1.45	1745	2.45	2.045	0.2514	0.1231	0.1011	0.16	0.1272	1.1
11	Round	1.42	1788	2.08	2.064	0.2514	0.1226	0.0787	0.18	0.1259	0.87
11	Round	1.43	1807	2.05	2.064	0.2514	0.1223	0.0782	0.19	0.1263	0.86
12	Elliptic	1.41	1732	—[a]	2.077	0.257	0.1216	0	—[a]	0.1286	0
12	Elliptic	1.42	1755	2.19	2.087	0.2536	0.1218	0.0984	0.16	0.1288	1.05
12	Elliptic	1.42	1764	1.95	2.082	0.2536	0.1216	0.0817	0.18	0.1288	0.87
12	Elliptic	1.42	1762	1.91	2.080	0.2536	0.1217	0.0795	0.19	0.1286	0.85

[a]Air supply for cooling of injectors, no fuel supply; G_{air}—air mass rate through pre-heater, kg/s; G_{02}—oxygen mass rate through pre-heater, kg/s; $G_{h,ker}$—kerosene mass rate through pre-heater, kg/s; G_{ker}—kerosene mass rate through combustor model; P_m—total pressure of mixture in injectors; G_{bg}/G_{ker}—barbotage gass mass rate to kerosene mass rate ratio.

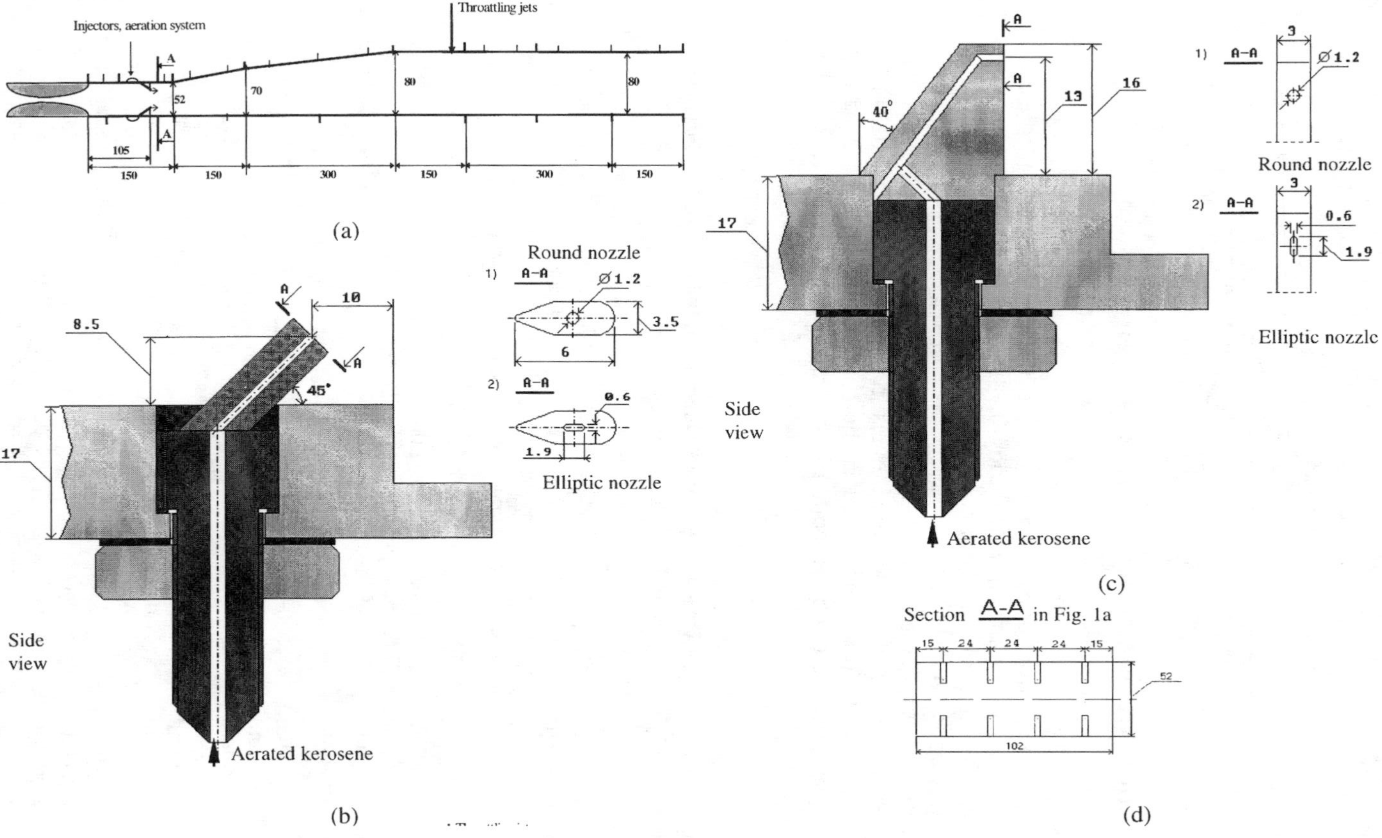

Fig. A.III.1 Combustor: (a) scheme of combustor, (b) geometry of tube-micropylon, (c) geometry of fin-pylon, (d) scheme of injectors location; all dimensions in millimeters.

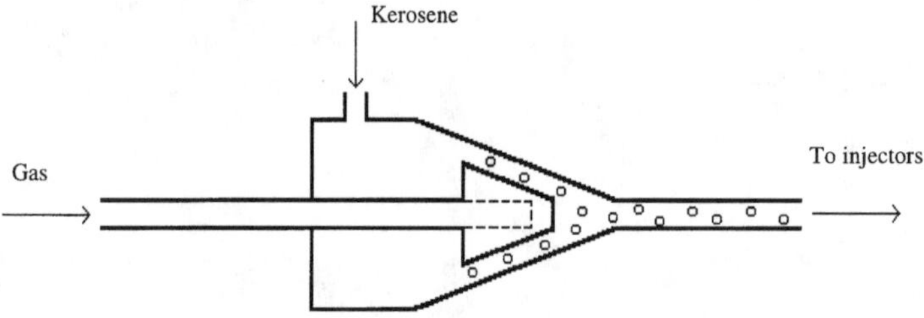

Fig. A.III.2 Scheme of the mixture system (barbotage) device.

combustion blowout took place. To ignite the combustor, high pressure throttling air jets were injected during 0.5–1.0 s in the section located at distance 780 mm from the combustor entrance. After ignition the air throttling jets were switched off.

During the tests, the axial pressure distributions on the upper and lower walls of combustor were measured. In test run no. 2 (see Table A.III.1) the total pressure field was measured in the combustor exit plane. Measurements were carried out by 10-point transversing rake. In the other test runs,

Table A.III.2 Flow Parameters at x = 900 mm from the Combustor Entrance

Run	Nozzles	ER	M	η combustion efficiency	σ total pressure recovery coefficient
	Tube-micropylons injectors, barbotage by hydrogen				
1	Round	1.06	0.98	1	0.368
2	Elliptic	0.69	1.02	1	0.354
3	Elliptic	0.62	1.08	1	0.345
	Tube-micropylons injectors, barbotage by air				
9	elliptic	0.945	1.06	1	0.344
9	elliptic	0.763	1.05	1	0.334
9	elliptic	0.079	1.5	0.7	0.352
10	round	0.962	1.1	0.98	0.338
10	round	0.771	1.33	0.8	0.342
10	round	0.720	1.9	0.4	0.322
	Fin-pylon injectors, barbotage by air				
11	round	1.1	1.1	1	0.336
11	round	0.87	1.1	0.93	0.33
11	round	0.86	1.2	0.79	0.34
12	elliptic	1.05	1.1	1	0.337
12	elliptic	0.87	1.1	0.95	0.333
12	elliptic	0.85	1.2	0.89	0.331

the total and static pressures were measured in a single point at the combustor exit plane.

Experimental data were analyzed using a one-dimensional (1-D) method. This method is based on the solution of the conservation equations of the energy, mass and impulse at known (from experiment) axial pressure distributions on the walls of the combustor (pressure is assumed constant over cross-sections of the combustor). The 1-D calculation results for the section at distance 900 mm from combustor entrance are given in Table A.III.2.

C. Test Results

Fig. A.III.3 compares axial normalized static pressure (static presssure P divided by the pressure P_t in the pre-heater) distributions on the combustor walls for the runs nos. 9 (elliptic nozzles) and 10 (round nozzles) with practically the same values of ER for elliptic and round nozzles. Kerosene was barbotated with air. Fuel was injected at the angle of 45 deg relative to the mainstream air flow direction throughout tube-micropylons. The flow in the combustor remained supersonic in test runs nos. 9 and 10 over the range of ER given in Table A.III.2. It can be seen from Fig. A.III.3 that for the tests with the combustion the values of the pressure along the length of the combustor are almost everywhere higher in the case of the elliptic nozzles, i.e., the enhancement of the supersonic combustion took place when kerosene was injected through elliptic nozzles. Fig. A.III.3 shows that at the aft

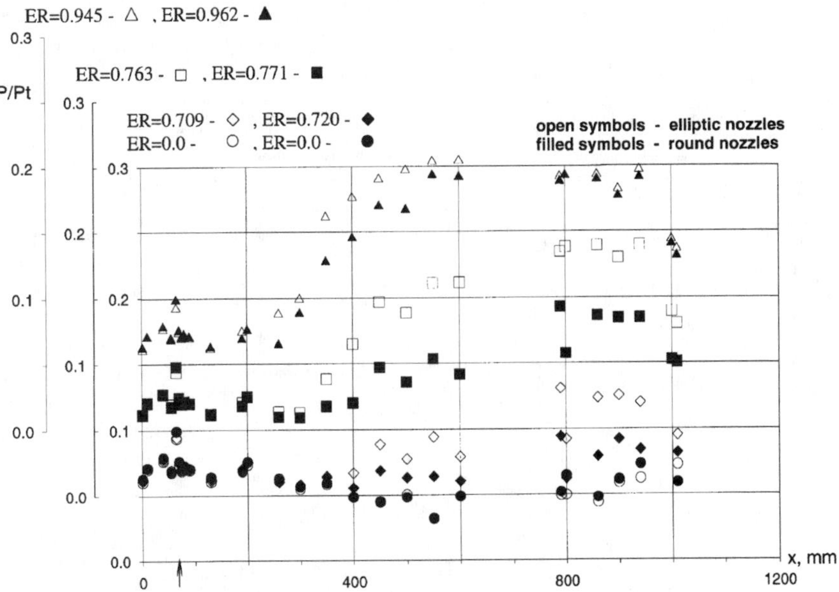

Fig. A.III.3 Axial static pressure distrubutions on combustor wall for tube-micropylons. Open and filled symbols for elliptic and round nozzles, respectively; arrow shows the place of fuel supply.

of the combustor, a flow separation occurred for the case without combustion, i.e., at ER = 0 (due to the overexpansion of the flow). It is also seen from Fig. A.III.3 that after some ignition delay (for the combustion cases) pressure increased monotonously along the combustor (with the exception of the aft of the combustor). Fig. A.III.4 shows the dependence of ignition delay length on ER. It can be concluded that ignition delay length was shorter for elliptic nozzles.

Supersonic combustion enhancement can be analyzed using local characteristic-normalized difference of pressure rise due to combustion for elliptic and round nozzles, i.e.:

$$\Delta\overline{P} = (P_{\text{ell}} - P_{\text{round}})/(P_{\text{round}} - P_{\text{no combustion}}) \qquad \text{(A.III.1)}$$

The results of calculation of $\Delta\overline{P}$ using data in Fig. A.III.3 are presented in Fig. A.III.5. It can be seen that $\Delta\overline{P} > 0$ along the length of the combustor, i.e. the elliptic nozzles provide better combustion performance than round nozzles. Better indicator of enhancement of supersonic mixing and combustion is obtained from the analysis of the impact of the fuel supply mode on the integral characteristic–pressure-area integral for the diverging-area supersonic combustor (see, e.g., Refs. 6 and 7). The combustion-induced pressure-area integral for 2-D combustor (Fig. A.III.1a) were calculated from

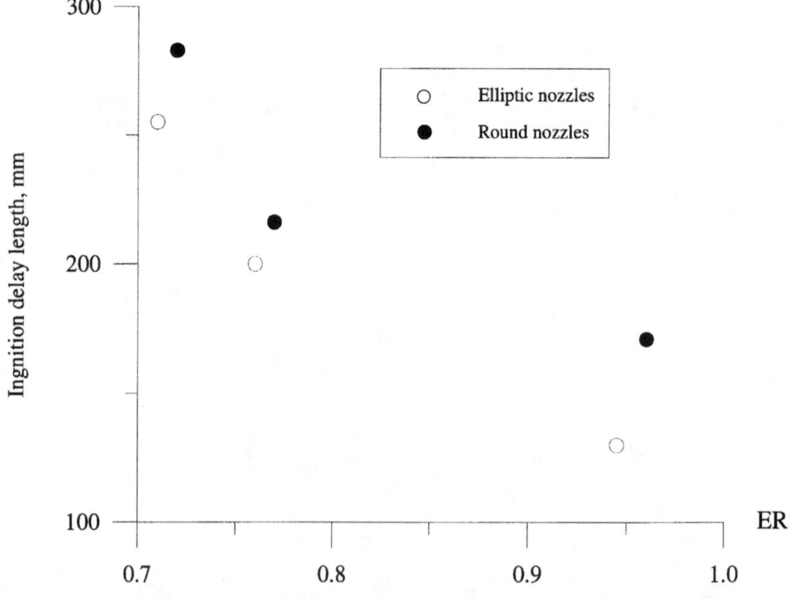

Fig. A.III.4 Ignition delay lengths for tube-micropylons; open and filled circles for elliptic and round nozzles, respectively.

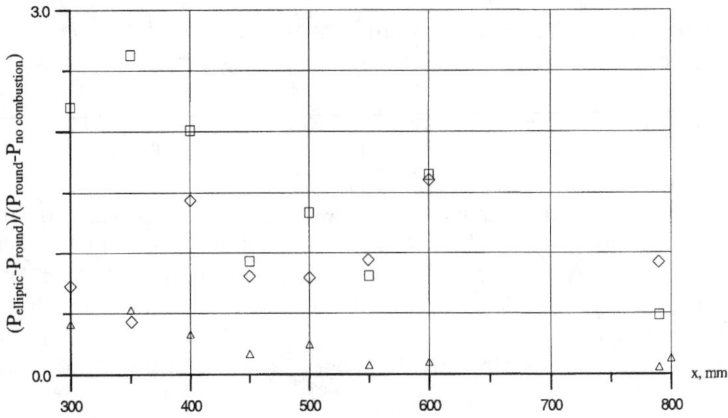

Fig. A.III.5 Normalized difference of combustion-induced pressure rises for elliptic and round nozzles drilled at tube-micropylons; Δ- ER = 0.945, ER = 0.962 for round and elliptic nozzle, respectively, □- ER = 0.763, ER = 0.771 for round and elliptic nozzles, respectively, ◇- ER = 0.709, ER = 0.72 for round and elliptic nozzles, respectively.

the measured axial wall pressure distributions from the following relationship (see, e.g., Ref. 7):

$$\Delta\overline{F} = w\int (P_{\text{combustion}} - P_{\text{no combustion}})\, tg\theta dx \qquad \text{(A.III.2)}$$

where θ is the local wall angle with respect to flow and x is the axial coordinate. Fig. A.III.6 shows normalized combustion induced pressure-area integral $\Delta\overline{F} = \Delta F/I_1$, where $I_1 = (P + \rho u^2)_1 hw$ is the axial impulse function

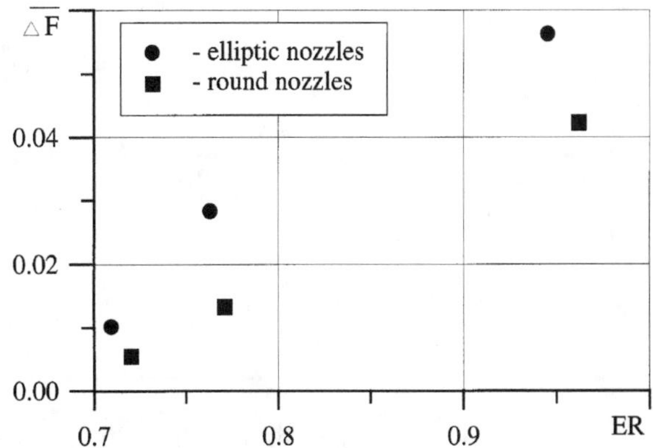

Fig. A.III.6 Comparison of normalized combustion-induced pressure-area integrals for tube-micropylons with elliptic and round nozzles.

at the combustor entrance. It is seen that the magnitude of $\Delta\overline{F}$ increases with increasing of ER and is clearly higher for elliptic nozzles than for the round nozzles.

Hydrogen was used for the aeration of kerosene in test runs nos. 1–3 (see Table A.III.1). The influence of the type of gas used for aeration on the magnitude of $\Delta\overline{F}$ is illustrated in Fig. A.III.7. One can conclude that barbotage of kerosene with hydrogen provided higher mixing and combustion enhancement than barbotage with air. The possible reasons of greater hydrogen barbotage efficiency are the following: 1) greater specific work capacity of hydrogen during expansion compared to that of the air; 2) favorable influence of hydrogen on combustion kinetics of kerosene. The last factor is hardly possible since hydrogen fraction in the mixture is quite low (about 1%).

Figure A.III.8 compares axial pressure distributions (with practically the same values of ER for elliptic and round nozzles) for two tests for which air-barbotated kerosene was injected through round (test run no. 11) and elliptic nozzles (test run no. 12) located at the base of fin-pylons in the co-flow direction to the mainstream flow. One can conclude from Fig. A.III.8 that the combustion-induced pressure rises for elliptic and round nozzles are nearly the same; i.e., mixing and combustion efficiencies practically coincide for both types of nozzles. This conclusion is confirmed by the calculation of the combustion-induced pressure-area integrals for both types of nozzles (Fig. A.III.9). It follows from comparison of Figs. A.III.6 and A.III.9 that tube-micropylons with injection at angle of 45 deg relative to the mainstream air flow direction provide better performance than fin-pylons with co-flow injection.

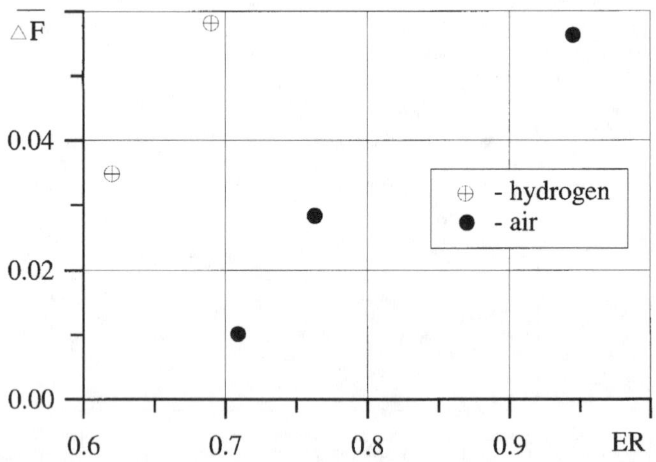

Fig. A.III.7 Impact of type of the gas for kerosene aeration normalized combustion-induced pressure-area integral for tube-micropylons with elliptic nozzles.

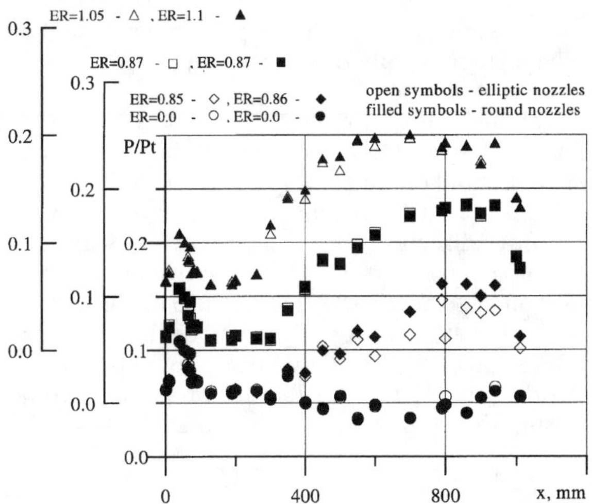

Fig. A.III.8 Axial static pressure distributions on combustor wall for fin-pylons; open and filled symbols for elliptic and round nozzles, respectively.

D. Conclusions

An experimental study was carried out to study the supersonic mixing and combustion enhancement in scramjet combustor using aerated by gas liquid kerosene jets (effervescent sprays) injected through elliptic nozzles from tube-micropylons and fin-pylons. The following results were obtained:

1) Elliptic nozzles provided greater mixing and combustion efficiencies in comparison with round nozzles for the cases when barbotated kerosene was injected from tube-micropylons at the angle of 45 deg relative to the main-stream air flow direction.

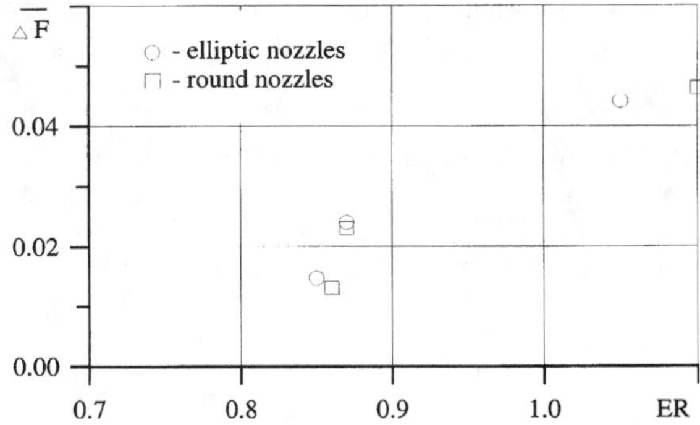

Fig. A.III.9 Comparison of normalized combustion-induced pressure-area integral for fin-pylons with elliptic and round nozzles.

2) Barbotage of kerosene with hydrogen provided higher mixing and combustion enhancement compared to barbotage with air at injection from tube-micropylons.

3) Test results obtained for fin-pylons with co-flow injection of barbotated kerosene did not show noticeable difference in mixing and combustion efficiencies for round and elliptic nozzles.

4) Injection from tube-micropylons at an angle of 45 deg relative to the mainstream air flow direction provided greater mixing and combustion efficiencies in comparison to co-flow injection from fin-pylons.

The investigation showed clearly that the use of the effervescent sprays and elliptic nozzles for the injection of the fuel enable to realize the efficient supersonic combustion of the liquid hydrocarbon fuels. It would be interesting in future work to study the possibilities of the supersonic combustion enhancement using the elliptic nozzles drilled at the base of NASA swept wedges.

Acknowledgments

This work was supported by the U.S. Navy, Office of Naval Research contract no. N00014-96-1-0869, with Gabriel Roy as the technical monitor. We sincerely with to express our appreciation to V. M. Levin and V. N. Avrashkov of MAI for their valuable contribution in running the tests without which this study could not have been conducted.

References

[1]Gutmark, E. J., Schadow, K. S., and Y. K. H. "Mixing Enhancement in Supersonic Free Shear Flows," *Ann. Rev. Fluid Mech.,* Vol. 27, 1995, pp. 375-417.

[2]Haimovitch, Y., Gartenberg, E., and Roberts, A. S. Jr., "Investigation of Rump Injector for Supersonic Mixing Enhancement," NASA-CR-4634, 1994.

[3]Haimovitch, Y., Gartenberg, E., Roberts, A. S. Jr., and Northam, G. B. "Effects of Internal Nozzle Geometry on Compression-Ramp Mixing in Supersonic Flow," *AIAA Journal,* Vol. 35, No. 4, 1997, pp. 663-670.

[4]Kopchenov, V. I., and Lomkov, K. "The Enhancement of the Mixing and Combustion Processes Applied to Scramjet-Engine," AIAA Paper 92-3428, 1992.

[5]Avrashkov, V. N., Baranovsky, S. I., and Davidenko, D. M. "Penetration Height of a Liquid Jet Saturated by Gas Bubbles," *Izvestiya Vuzov, Aviatsionnaya Tekhnika,* No. 4, pp. 96-98, 1990 (In Russian).

[6]Kay, I. W., Peschke, W. T., and Guile, R. N. "Hydrocarbons-Fuelled Scramjet Combustor Investigation," AIAA Paper 90-2337, 1990.

[7]Stouffer, S. D., Vandsburger, U., and Northam, G. B. "Comparison of Wall Mixing Concepts for Scramjet Combustors," AIAA Paper 94-0587, 1994.

Appendix B: Deceleration of Supersonic Flows in Smoothly Diverging-Area Rectangular Ducts

Vyacheslav I. Penzin
TsAGI, 140160, Moskow Region, Russia

Experimental investigations of supersonic flow deceleration to subsonic velocities in straight ducts with oblong cross-sections at various back pressures showed that characteristic properties of the flow essentially depend on the entrance flow Mach number M and the cros- section aspect ratio $\bar{b}$, $\bar{b} = b/h$; b and h are duct width and height respectively (V. I. Penzin, 1997 and 1988a,b). At certain values of these parameters there is a transition from a pseudo-shock, where no region of appreciable stall exists, to flow with long boundary layer separations. The transition to this flow pattern leads to a decrease in the pressure recovery and to an increase of the supersonic/subsonic transition region length. The pseudo-shock length versus changes of parameters predicted by computations is in good agreement with experiments. There are no such results in the case of ducts with oblong cross sections.

Detailed measurements of static pressure distribution, total pressure profiles and wall static pressure pulsations characteristics, made in ducts with different oblong cross-sections, allowed the dependence of M on $\bar{b}$ to be determined, which establishes the line of division between separated and attached flows. Flow will separate from the duct surface if the required conditions are met there. The correlation of pressure recovery, distance necessary to complete supersonic/subsonic transition, and parameters M and $\bar{b}$ were also analyzed. Besides, it was shown that if the hydraulic diameter d_h (d_h=4S/P, S and P are duct cross-section area and perimeter, respectively) which was used as a scaling factor is decreased below a certain limit, pressure recovery diminishes substantially (combined effects of the Reynolds number and the displacement thickness of boundary layer). The principal cause of the flow transition from attached to separation mode in the ducts with oblong cross-section is associated with non-symmetric separations. The flow is allowed to separate over a portion of the side wall surface and then reattaches. Owing to the reverse flow, the boundary layer thickens considerably. These separations are not suppressed by an opposite wall at high aspect ratios. In practical devices, for instance, in scramjets, an air passage can include a rectangular duct section with one of the wide walls inclined to the axis. The back pressure can be the combustion pressure of a scramjet engine. Such diverging-area ducts provide more stable flow at various back pressures but lower the pressure recovery due to larger Mach numbers at which deceleration occurs. However, the diverging of the rectangular duct leads to a decrease in the aspect ratio that can partly compensate for the drop in the pressure recovery.

As a continuous effort of duct flow study the following issues will be addressed in this appendix : 1) an investigation of the characteristic properties of the flow in diverging-area rectangular ducts at various back pressures,

2) definition of integral characteristics of flow (such as pressure recovery and supersonic/subsonic transition region length) as a function of wall inclination angle, area expansion ratio and hydraulic diameter at the duct entrance, and 3) determination of separation movement along the duct as a function of the back pressure that characterizes flow stability.

The experimental facility consisted of replaceable contoured nozzles with exit diameter d = 81.4 mm and a test chamber connected to an exhaust ejector. The nozzle Mach numbers M = 2.6, 3.2, and 3.8 were calculated using the nozzle critical to exit area ratio. All the test ducts were formed from one basic 12.5 × 37.5 mm^2 cross-section area and 600-mm length duct on the lower wall of which were mounted wedge-like inserts. The constant cross-section basic duct became a combined one with three different sections having constant, variable, and again constant-area. A schematic side view of the duct is sketched in Fig. B.1. This configuration is a typical one for an airframe-integrated scramjet. The first section was 50-mm long for all variants and the third one had a stream cross-section of 12.5 by 37.5 mm. The angle of inclination of the lower wall changed in the range of η = 0.5 to 3 deg. The duct geometries are given in Table B.1.

For the purpose of comparison there were investigations of flow in constant-area planar ducts with cross section configurations corresponding to those cited in Table B.1. These ducts were formed from duct 1 by mounting into it plates of various thickness. Duct inner wall roughness in all cases was equal to 3 μm. A throttling valve at the duct outlet allowed the pressure build-up in the duct to be controlled. The duct entrance sections were located at the nozzle outlet in the uniform free stream. The earlier investigations of static pressure distributions are found to be particularly useful in flow character diagnostics. Since flow integral characteristics were the main

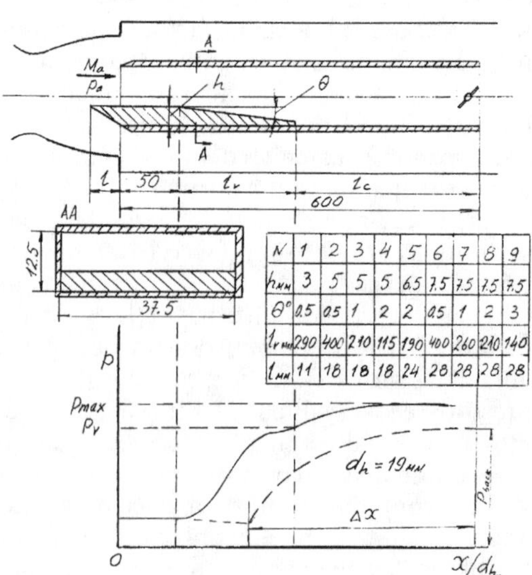

N	1	2	3	4	5	6	7	8	9
h мм	3	5	5	5	6.5	7.5	7.5	7.5	7.5
θ°	0.5	0.5	1	2	2	0.5	1	2	3
lv мм	290	400	210	115	190	400	260	210	140
l мм	11	18	18	18	24	28	28	28	28

Fig. B.1 Schematics of diverging rectangular ducts.

Table B.1 Duct Geometries

Duct number	Inlet cross section, mm2	Aspect ratio, $\bar{b}$	Wall inclination θ, deg	Area expansion ratio, f	Entrance hydraulic diameter, mm
1	12.5 × 37.5	30	0	1	19
2	9.5 × 37.5	39	0.5	1.37	15
3	7.5 × 37.5	50	0.5	1.67	12.5
4	7.5 × 37.5	50	1.0	1.67	12.5
5	7.5 × 37.5	50	2.0	1.67	12.5
6	6.2 × 37.5	60	2.0	2.03	10.6
7	5.0 × 37.5	74	0.5	2.5	8.8
8	5.0 × 37.5	74	1.0	2.5	8.8
9	5.0 × 37.5	74	2.0	2.5	8.8
10	5.0 × 37.5	74	3.0	2.5	8.8

purpose of this investigation, static pressure measurements were the only kind of measurements taken. Static pressure taps in duct 1 were located along the middle of the upper wide flat wall with the pitch of a half of the hydraulic diameter and along the narrow side wall (at the distance of ¼ of duct height from the upper wall) with the pitch equal to the hydraulic diameter. There were no measurements along the wedge-like inserts. The pressure was measured by special instrumentation and normalized by the plenum pressure. The relative pressure measurement accuracy was ± 1%. The air stagnation temperature was $T_t = 250 - 260$ K. The flow Reynolds number range was $0.5 \times 10^6 < Re < 2 \times 10^6$ based upon the average velocity at the entrance section of the duct and hydraulic diameter. Longitudinal dimensions were normalized by duct 1 hydraulic diameter $d_h = 19$ mm.

Pressure distribution along models 9 and 3 at different M and opened throttle are presented in Fig. B.2. It is seen that supersonic flow in diverging and constant-area sections of the duct are non-uniform, especially behind the diverging part. At the flow Mach number 2.6, as it was shown earlier, the pseudo-shock–type flow can be established in all the above ducts. At Mach 3.2 and 3.8 there is a greater possibility of flow separation, pressure recovery drop and an increase in the required duct length. There are two factors of the duct diverging-area effect on the flow character. On one hand, hydraulic diameter increases and aspect ratio decreases tends to restrict the separation and on the other hand, the duct expansion makes the flow similar to the separation flow in an over-expanded nozzle. The details involving these two factors are discussed later. Attached to separation flow transition is accompanied by certain changes in the static pressure distributions. For instance, in the pressure curve segments with small gradient, some kind of plateau appears and longitudinal static pressure distributions along the narrow and the wide walls do not coincide. A plateau in the pseudo-shock does not exist and the static pressure over the circumference of the cross-section of a duct is constant. General behavior of the pressure distribution subject to various

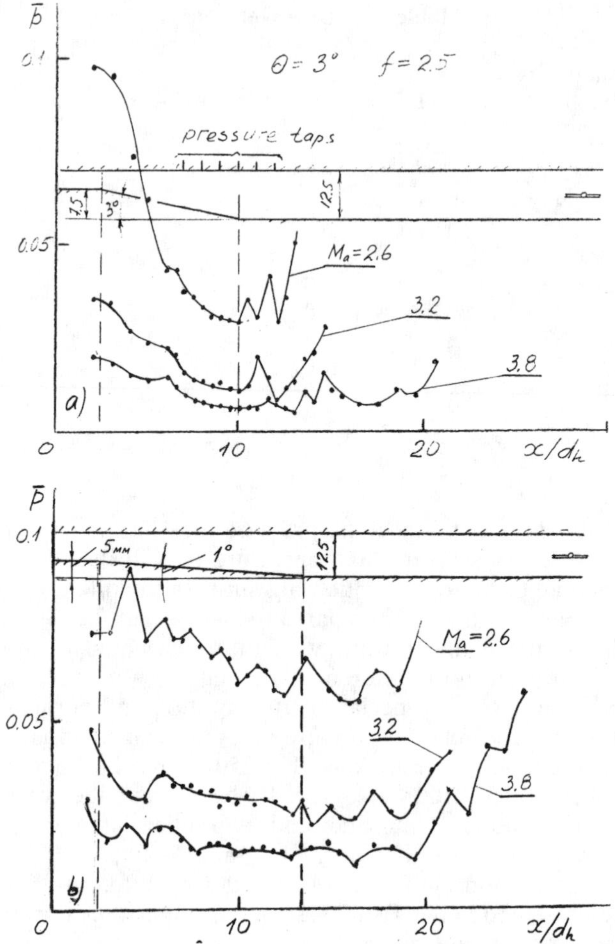

Fig. B.2 Supersonic flow pressure distribution in rectangular ducts.

back pressures is depicted in Fig. B.3. Typical internal wall pressure distributions in several ducts are shown for different Mach numbers. The upper diagram (duct 7, M = 2.6, θ = 0.5 deg, f = 2.5, $\bar{b}$=7.4) describes attached flows. The static pressure distributions along the wide (filled symbols) and the narrow (open symbols) walls coincide, the pressure curves are characteristic for a pseudo-shock.

The middle diagram (duct 5, M= 3.2, θ = 2 deg, f = 1.67, $\bar{b}$ = 3) include pressure curves characteristic of attached and separation flows. At low back pressure the pressure recovery region is located in the duct constant-area section. The pressure curves are similar to those in a pseudo-shock (curves 1, 2), the transition region length is about 10 duct length hydraulic diameters. The curve with maximal pressure build-up is near to that of pseudo-shock

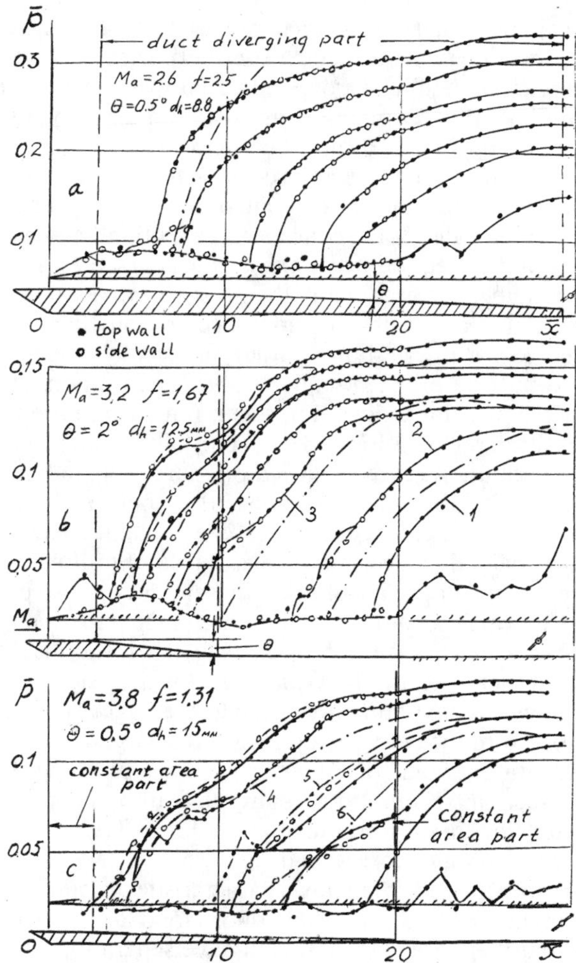

Fig. B.3 Pressure distributions in rectangular diverging ducts at different back pressures, 1-6 flow regimes.

(with average Mach 3.7 at the outlet of the diverging section). The dash-dotted lines correspond to duct with constant-area pressure build-ups (pseudo-shock) at Mach 3.8 and they are similar to curves 1 and 2 in the combined duct. If the pressure recovery region approaches the diverging section (curve 2), the Mach number before this region grows and, if it moves along this section (curve 3) the aspect ratio grows and hydraulic diameter decreases. It is now observed that these factors assist in flow separation because there are plateaus in the pressure curves and they do not coincide. Dashed lines have been placed through the open-symbol data (narrow side wall) and solid lines through the shaded-symbol data (wide wall). The pressure build-up along the narrow side wall moves upstream further then along the wide one. This phenomenon can be considered as the appearance of the separation tongues along the narrow walls. That would make physical sense considering the fact that a three-dimensional boundary layer separation at the duct corner occurs at the lower pressure ratio.

Now it is necessary to note one more characteristic property of the flow in a constant-area section of the combined duct. If a supersonic/subsonic transition region is located in the constant-area section (curves 1 and 2) then the pressure build-up curves have the same character as pseudo-shock curves: there is a negative gradient in the pressure curves after reaching the maximum, which indicates subsonic flow acceleration. Such a pressure distribution character, as it was shown earlier, practically does not depend on flow non-uniformity before the pseudo-shock. The flow non-uniformity created by the two-dimensional diverging duct is not an exception in this sense. But if the transition region is located even partially in the diverging section, the pressure distribution (curve 3) has a pressure plateau and no pressure maximum notwithstanding the supersonic flow before the constant-area duct section. Such curve characteristics can be explained partially as a consequence of the back flow occurring. Thus, this kind of flow can be considered as a specific kind of pseudo-shock. This tendency is also pronounced in stepped tubes in which separation regions are present invariably.

Mach number increasing up to $M = 3.8$ (Fig. B.3, lower diagram, duct 2, $\theta = 0.5°$, $f = 1.31$) leads to the flow pattern at which the separation starts immediately downstream of the throat due to the large aspect ratio (greater than 3). The pressure recovery region length reaches 20 and more caliber's of the duct length hydraulic diameter. It occupies both the diverging and constant-area sections. There are pressure plateau segments in the curves. The dash-dotted lines in the lower diagram of Fig. B.3 (curves 4, 5 and 6) are given for comparison and correspond to a 9.5×37.5 mm^2 constant-area duct. One can note the following flow characteristic properties: the pressure distributions in the front part of the diverging (solid lines) and constant-area duct sections (curve 4) are very close and have separation character. However, pressure recovery in the combined duct is higher than in the constant-area duct due to a higher hydraulic diameter and lower aspect ratio. If the pressure recovery regions are at the aft part of the duct (curve 6) the picture is reversed; pressure recovery is higher in the constant-area duct. It is explained by the higher Mach number before the pressure recovery region in the combined duct and consequently higher-pressure losses occur.

Absence of the distinct pressure maximum in axial pressure distribution $p(x)$ in the case of combined duct makes it difficult to answer the question whether a diverging-area section permits the shortening of the required duct length for supersonic/subsonic process completion. However, as it follows from the earlier statement, the initial pressure gradient grows if a pseudo-shock is fixed and its length decreases. Solid (combined duct) and dash-dotted (constant-area duct) curves in Fig. B.3 show that using the diverging-area sections does not lead to growth in the pressure gradient and consequently the supposed duct shortening can not be expected. Only at a high Mach number and equal pressure recoveries (compare with curve 4 for the constant-area duct) is there a small gain in the duct length due to lesser $\bar{b}$ and greater d_h. Maximal pressure build-ups for all the ducts at the three Mach numbers are outlined in Figs. B.4–B.6. Maximal mass rate through the ducts existed while the static pressure at the duct entrance was independent on the back pressure. It can be supposed that maximal back pressure could be raised a little if static taps were located somewhat nearer to the duct entrance (nearer than 38 mm). Experiments, however, show that this increment can not surpass 5–7%. A normal shock in front of the duct entrance arises at greater back pressures. The extreme complexity of the flow, availability of separations, and back flows at the beginning of the diverging-area section at the maximal back pressures is testified to by many investigations.

The curves in Figs. B.4–B.6 are grouped in such a way that one can trace the influence of the wall inclination angle at an identical area expansion ratio f, and the effect of duct expansion ratio f at an indentical θ on the axial pressure distribution. For the first two groups of curves $f = 2.5$ and 1.67 and for the second groups $\theta = 0.5$ and 2.0 deg. Figure B.4 demonstrates the test results for $M = 2.6$ at which a pseudo-shock occurs at all aspect ratios. Examination of the top curve group (scale I, $f = 2.5$, $d_h = 8.8$mm, $\bar{b} = 7.4$) allows the following conclusions to be made: pressure recovery in the combined duct depends substantially on the wall inclination angle, its increase from 0.5 to 3.0 deg decreases the pressure recovery by 1.5 times. For Mach 3.2 and 3.8 this decrease is approximately 1.35–1.3 times. A Mach number increase tends to decrease the flow stability as displayed in the top pressure build-up position scatter at $f = 2.5$. At $f = 1.67$ the curves tend to converge. At $\theta = 2$ and 3 deg the curve character testifies to the appearance of separation flow at the frontal part of the diverging duct. At $\theta = 0.5$ and 1.0 deg there is no such kind of flow. Thus, a decrease in $\bar{b}$ and an increase in d_h along the diverging-area duct (at $\theta > 1$ deg) was insufficient to check the separation stimulated by the expansion itself due to a Mach number increase. Figs. B.5 and B.6 (Mach 3.2 and 3.8) show that separations take place at all θ. Decreasing of the duct area expansion ratio does not cause a change in the flow character.

Let us look now at the duct area expansion ratio effect on flow character at identical wall inclination angles. The curve groups in Figs. B.4–6 for various f at identical θ are located in the lower diagram parts (scales III and IV). Analysis of the curves indicates that the duct expansion ratio change carries no essential alteration in the axial pressure distribution along the initial part of the duct. It is defined by the $M(\bar{b})$ relationship. Pressure recovery at various f differ substantially: however, to define this effect quantitatively was not possible. As was noted earlier there is the combined effect of f, d_h and $\bar{b}$ on the axial pressure distribution. Influence of the hydraulic diameter should be noted separately. As experiments with a number of constant-area rectangular ducts, showed pressure recovery substantially decreases if d_h is less than the value of hydraulic diameter of the baseline variant (d_h=19 mm) due to change of Reynolds number and relative boundary layer displacement thickness. The hydraulic diameters at the entrance and outlet sections of the combined ducts are $d_h < 19$ mm and $d_h = 19$ mm respectively. It is possible that pressure recovery can be dependent on some average hydraulic diameter. Maximal/minimal possible pressure recovery ratio can scarcely be larger than the theoretical ratio of pressures after normal shocks located at the entrance and the exit of the diverging duct, respectively.

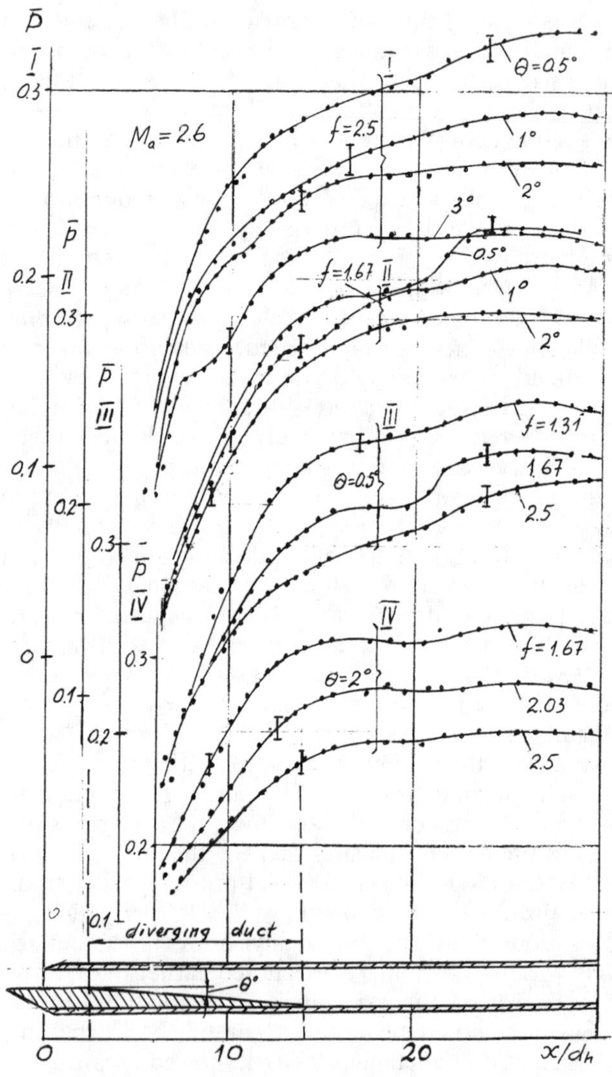

Fig. B.4 Limiting pressure distributions in different rectangular diverging ducts.

Figures B.5 and B.6 present as well curves related to the constant-area duct (dashed lines) having the same cross section as that at the entrance to the combined ducts. Curve comparisons (Fig. B.5, scale III; Fig. B.6, scale I) allows the conclusion that pressure recovery change in the combined duct by changing f can be explained mainly by the d_h influence, being consistent with previous results. The variation of θ effects only small changes in pressure distribution. At $d_h > 15$ mm the pressure recovery in both ducts are close (the upper dash-dotted line in Fig. B.5 relates to duct with $d_h = 19$ mm). Analogous result takes place at the rest of the Mach numbers. Dependence of $p_{\max}$ for all the ducts on f is presented in Fig. B.7. Curves have regular

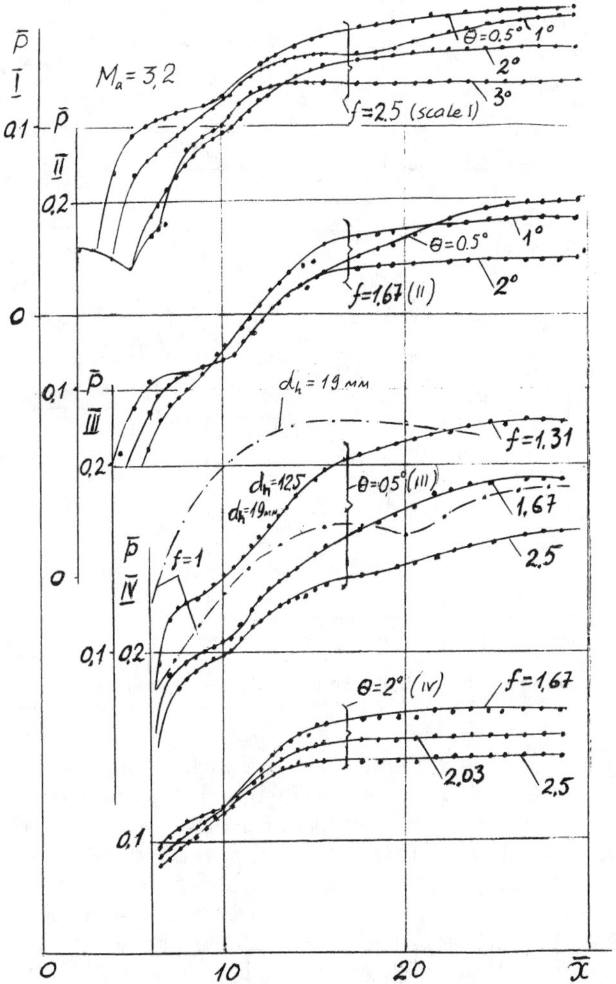

Fig. B.5 Limiting pressure distributions in different rectangular diverging ducts.

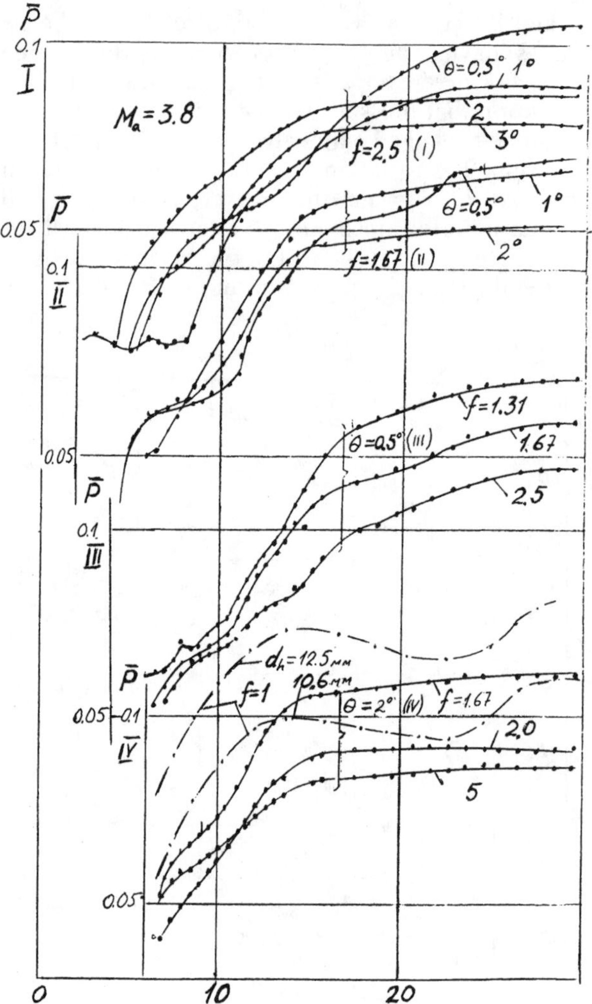

Fig. B.6 Limiting pressure distributions in different rectangular diverging ducts.

character. Curiously, it was found that curve inclination depends little on M and θ in the observed range of parameters. Let us pay attention to the definition of the internal characteristics of the flow, such as the pressure recovery. To evaluate the effects of θ and f on the pressure recovery it is necessary to exclude the effect of d_h as much as possible. Let us denote the maximal static pressure at the end of the diverging section and at the end of the constant-area duct as p_v and p_c, respectively. Figure B.8 presents p_v/p_c ratio as a function of f for all θ and M combinations. As it follows from Fig. B.8 there is no possibility to establish the effect of M on p_v/p_c. This is due to a large scatter of data caused by a finite number of axial pressure distribu-

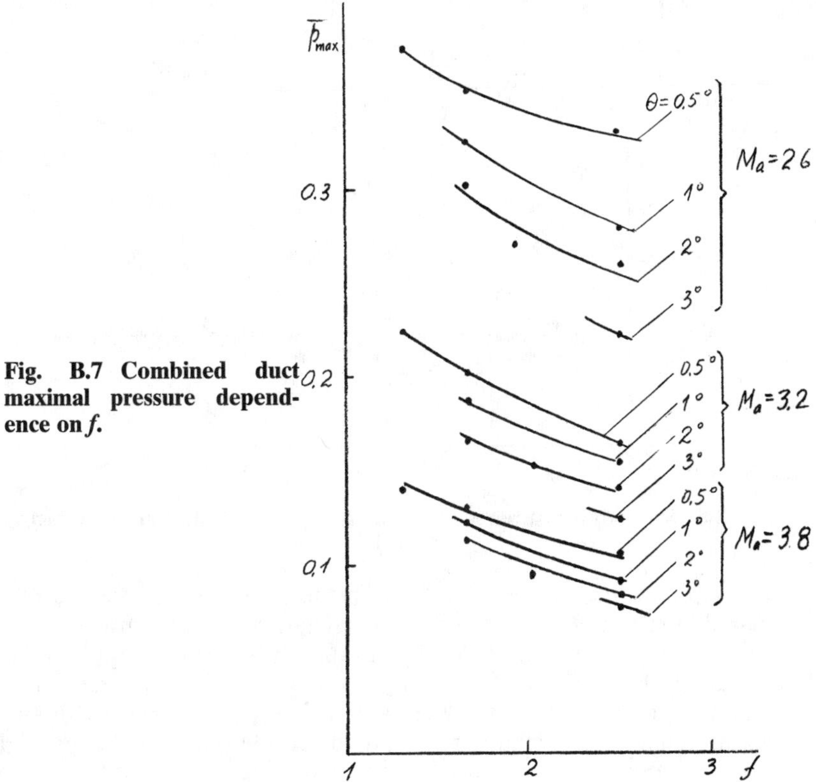

Fig. B.7 Combined duct maximal pressure dependence on *f*.

tion curves used for this relation determination, and also by the duct inlet geometry difference. At the same time there is an obvious influence of f on p_v/p_c noted. At $\theta = 0.5$ deg and $f < 2$ the value of p_v/p_c is greater then 1, i.e., demonstrating performance rivaling the constant-area duct. It depends on the flow mode in the ducts with long cross sections. A small wall inclination angle provides sufficient duct length for flow deceleration occurring at larger average d_h and lower average $\bar{b}$, than in constant-area ducts with the same inlet configurations, which promotes pressure recovery rise.

At greater angles θ the pressure recovery in diverging sections yields essentially to that in constant-area ducts. The length of the diverging section is not enough to complete deceleration and the flow at that section exit is supersonic. The pressure recovery can be even lower than that after a normal shock stationed at the diverging section exit. For rough estimations of p_v/p_c an empirical relation can be used:

$$\frac{p_v}{p_c} = 1.1 - 0.2 \cdot \theta + (f - 2) \cdot (0.07 + 0.1 \cdot \theta)$$

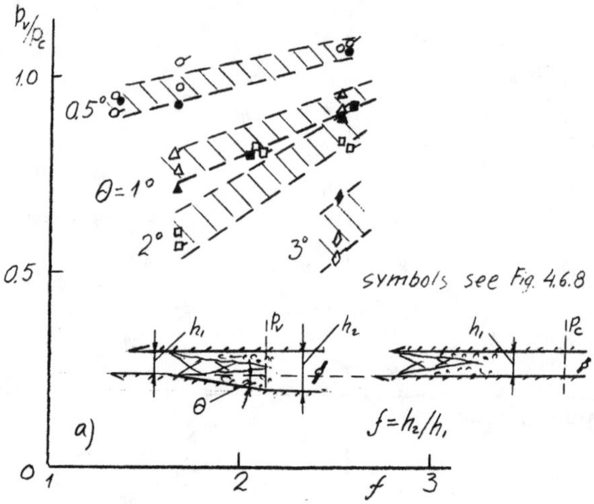

Fig. B.8 Ratio of pressure recoveries in diverging and constant-area ducts.

Let us denote the maximal pressure in the constant-area section of the combined duct $p_{\max}$ Then the ratio $p_v/p_{\max}$ characterizes the fraction of the summary pressure recovery belonging to the diverging section of the combined duct. In Fig. B.9 values of $p_v/p_{\max}$ are plotted as a function of f for all combinations of θ and M. It can be seen that $p_v/p_{\max}$ is less than 1 for all the above ducts and consequently in not one diverging section is the deceleration process completed. To do these greater expansion ratios f (about

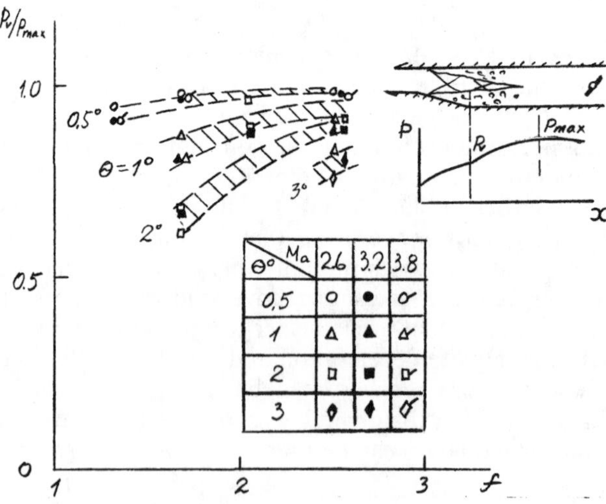

θ° \ Ma	2.6	3.2	3.8
0.5	○	●	○ (flagged)
1	△	▲	△ (flagged)
2	□	■	□ (flagged)
3	◊	♦	◊ (flagged)

Fig. B.9 Ratio of maximal pressure recoveries in diverging and combined ducts.

2.5–3.0) for ducts with $\theta = 0.5$–3.0 deg are needed. An empirical relation for p_v/p_{max} intended for rough e stimations is

$$\frac{p_v}{p_{max}} = 1 - 0.15 \cdot (3 - f) \cdot \theta$$

Consider now the total effectiveness of the flow deceleration in the combined duct. The figure of merit for this effectiveness used here is p_{max}/p_d, p_d is the maximum expected calculated pressure recovery successively provided by the normal shock and by an ideal subsonic diffuser with a fixed expansion ratio equal to each f of the combined duct from Table. In Fig. B.10 the values of p_{max}/p_d are plotted as a function of f for $\theta = 0.5$ and 2.0 deg and various M. Analysis of Fig. B.10 permits the conclusion that the relative pressure recoveries in the examined diverging ducts are very low and at $f = 2.5$ ($d_h = 8.8$mm $\bar{b} = 7.4$) can be lower then 50%. In "classical" pseudo-shock this value is about 95%. The main cause of such low effectiveness in the supersonic flow deceleration is small hydraulic diameter d_h (small Re) and large cross-sectional aspect ratio $\bar{b}$.

Effect of the wall inclination angle on p_{max}/p_d can be seen in Fig. B.11. For approximate estimation an empirical relation can be used:

$$\frac{p_{max}}{p_d} = \left(\frac{p_{max}}{p_d}\right)_{A=est} \cdot (0.55 - 0.05 \cdot \theta)$$

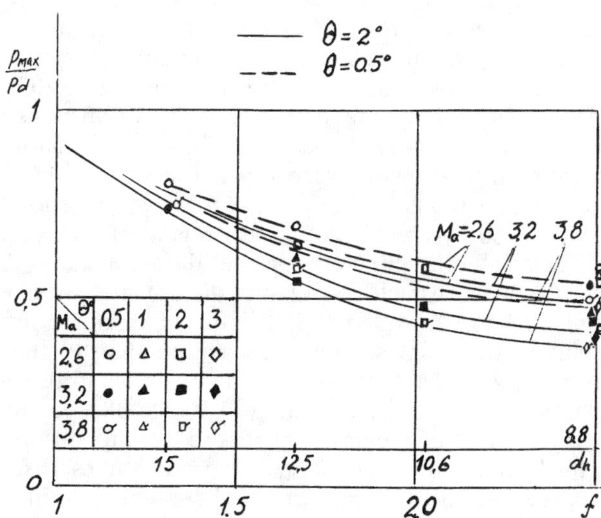

Fig. B.10 Dependence of ratio of pressure recoveries in combined duct (experimental and theoretical maximal) on expansion ratio.

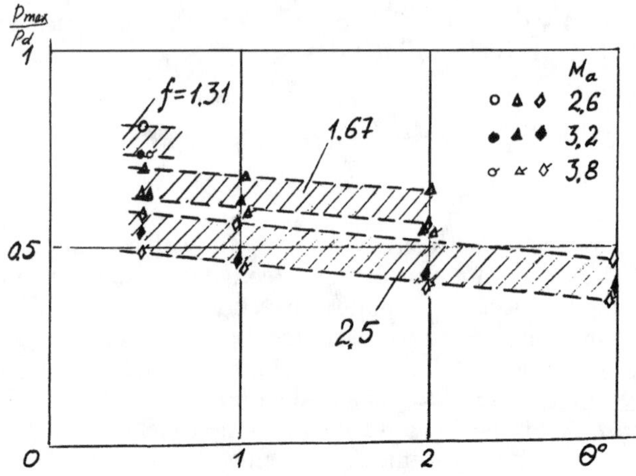

Fig. B.11 Dependence of ratio of pressure recoveries in combined duct (experimental and theoretical maximal) on divergence angle.

Here $(p_{\max}/p_d)_{A=\mathrm{cst}}$ is the relative pressure recovery in a constant-area duct having a cross-section which is equal to the entrance cross-section of the corresponding combined duct. In analyzing supersonic flow deceleration in ducts containing diverging parts with long cross sections it is possible to state that the pressure recovery in them is lower than in ducts with the same exit to entrance area ratio and consisting of constant-area + subsonic diffuser sections. In conclusion let us consider the conditions of propagation of pressure disturbances upstream in the combined duct due to back pressure rise. It can be connected with the flow stability, say in a scramjet passage. As was stated above the developed pseudo-shock in a constant-area duct is very sensitive to pressure rise. A small back pressure change leads to the essential transition of the pseudo-shock that creates a danger of duct unstart.

In technical applications the flow stability in special situations may be more important than the pressure recovery, but in all cases pressure losses must be as minimal as possible. The pseudo-shock propagation in a round tube can be restricted by creation of a local region of high pressure. This local region was created by circular ledges of different shape. A base in a stepped duct also is an obstacle to pressure disturbances, which do not go upstream while the pressure at the step is lower than the pressure that occurs at incipient turbulent separation, corresponding to the flat plate-ramp (two-dimensional). Methods of creating obstacles to pressure disturbance propagation are specific. Mounting of wedge-like bodies along the circumference of a duct of rectangular cross-section, for instance, did not bring a pseudo-shock fixation as took place for round tubes. Disturbances in this case propagate primarily along corners where there is a thick boundary layer and shock waves interact three-dimensionally. The back flow is not eliminated or reduced. In order to achieve an optimum design the vanes should be arranged in the vicinity of the inlet. In Figs. B.4, B.6, B.12a,b,c the distances of upstream pressure influences $\Delta\bar{x}$ measured from the combined duct exit are plotted as a function of the back pressure. Curves correspond to $M = 2.6$, 3.2 and 3.8, $f = 1.67$ and 2.5 and $\theta = 0.5$–3 deg. Open symbols are pressure data obtained along the middle of the wide wall and filled symbols are along the narrow side walk. The dashed and dash-dotted lines correspond to constant-area duct with 12.5 and 7.5 by 37 mm² cross sections. Horizontal dash-dotted lines indicate the start of the diverging section and horizontal dashes on each curve are the end of it.

Inspection of Fig. B.12 indicates that separation region movement along the constant-area section of combined ducts resembles the movement along constant-area ducts. This result corresponds to the above conclusion about the similarity of pressure distributions for these two cases. The separation region moving into a diverging section leads to a decrease in the gradient $d(\Delta\bar{x})/dp$, in some cases to zero. That can be interpreted as a fixation of the

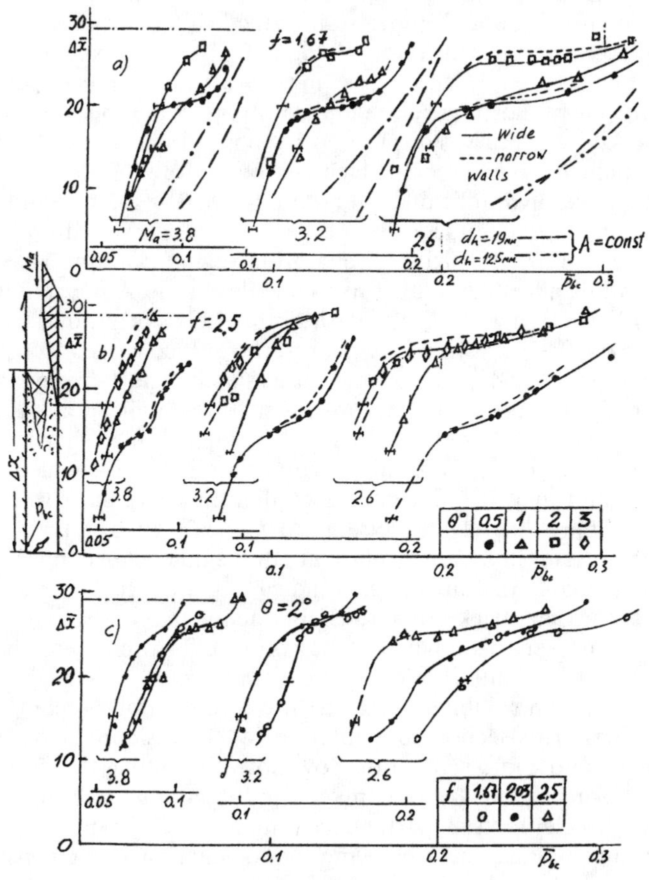

Fig. B.12

separation at the upper flat wall surface. The greater the length of a plateau in curve $\Delta\bar{x}(p)$ the more stable the flow is. In this sense the duct with $\theta = 0.5$ deg may seem most prospective. However, the plateau in this case is located rather far from the diverging section inlet, which reduces the maximal pressure recovery. Hence, it is expected that the most favorite divergence angles lies between $\theta = 0.5$ deg and 2 deg. It is possible that the separation region fixation at the lower sloped wall could exist in a greater range of back pressure change as it takes place at the base of a stepped duct. Further back pressure rise leads to a quicker upstream separation movement and then to flow disruption. As mentioned above the pressure disturbance front is irregular, there are tongues of separation regions along the sidewalls but their advance is not as great. As it can be seen from Figs. B.4, B.6, and B.12 the distance between incipient separations along side (small dashed lines) and upper (solid lines) walls do not exceed one or two duct heights. It should be noted that an increase in the wall inclination angle makes the separation front more uniform and in the proximity of the wall, break line curves corresponding to side and upper walls, practically coincide and even can change places. From an analysis of Figs. B.12a and 12b it is possible to conclude that, at $f = 1.67$, an increase in the wall inclination angle promotes plateau movement upstream close to the wall break line, but pressure recovery drops. At $f = 2.5$ curves displace to regions of lower back pressures that can be explained by the greater Mach number before the separation region and by the lower hydraulic diameter. At $f = 2.5$, $\theta = 1$–3 deg and maximal back pressures, curves practically merge, only the curve with $\theta = 0.5$ deg moves. At $f = 1.67$, $\theta = 0.5$ deg separation region can approach the wall breakline, while at $f = 2.5$ stable supersonic flow aside. A flow failure occurs earlier. In Figs. B12a and 12b the dash-dotted lines correspond to duct 1 ($f = 1.31$ and $\theta = 0.5$ deg), which are near to the dashed curves (constant-area duct 12.5×37.5 mm^2). That means that at small θ and f diverging sections of combined ducts practically do not assist flow stabilization - there is no plateau in curves, and gradient d $(\Delta\bar{x})/dp$ is high. Some other conclusions could be made from Fig. B12c where curves are plotted for ducts with $\theta = 2°$ and f, greater than 1.31. It can be seen that the curve parts with plateaus are located approximately at the same $\Delta\bar{x}$ values. However, the greater the f, the greater these curves are displaced to the region of lesser back pressures (at Mach 3.2 and 2.6). Thus the data indicate that the highest performance in terms of maximal back pressure is obtained with a duct having a small f ($f = 1.67$). However, this conclusion may prove to be premature because the compared ducts had different hydraulic diameters.

We deduce from exhibited data that the diverging of a rectangular duct with an oblong cross-section does not remove the possibility of transforming separationless flow into separation flow, notwithstanding the decreasing of $\bar{b}$ and increasing d_h, i.e., factors preventing the boundary layer separation. Additional deceleration of, partly decelerated vortex flow, in the diverging section of a combined duct, following supersonic flow in the constant-area portion takes place downstream practically at constant pressure which makes it different from pseudo-shock flow type.

Pressure recovery in rectangular diverging ducts with oblong cross-sections can exceed somewhat the pressure recovery in a constant-area duct with cross-section equal to the entrance section of the diverging-area duct. It occurs at $\theta = 0.5$ deg due to greater d_h and lower $\bar{b}$. At $\theta > .5$ deg the diverging-area duct leads to a greater decrease, in comparison with a constant-area duct, in efficiency the greater θ is, due to the M increase at which the deceleration take place. Pressure recovery in divergence sections of the ducts with a given length depends little on θ if $\theta < 3$ deg. The obtained dependence of pressure recovery on diverging angle depends little on M in the examined range. The length of the rectangular diverging duct necessary for supersonic flow deceleration into subsonic flow in the examined range of $\bar{b}$ is $f = 2.5$–3 deg. In all tested combined ducts, the concluding flow deceleration took part in the constant-area duct. The pressure recovery in the examined combined ducts decreases substantially as f increases and can be less than 5% of the pressure recovery in a normal shock. The basic reason for this is the fact that as f increases, the value of d_h can reach values at which deceleration effectiveness of all duct types decreases. The presence of the diverging-area section in the rectangular duct assists in flow stabilization, decreases pressure upstream influence, and helps pseudo-shock movement fixation. The nearest pseudo-shock position to the entrance takes place at $\theta = 2$ deg ($f > 1.6$). At $f < 1.3$ there is no separation fixation.

Bibliography

Penzin, V. I., "Dependence of Pressure at Step from Shape of Cross-Section of Channel with Sudden Expansion," *Uchenye Zapiski TsAGI,* Vol. 18, No. 1, 1987 pp. 65–72 (in Russian).

Penzin, V. I., "Pseudo-Shock and Separated Flow in Rectangular Channels," *Uchenye Zapiski TsAGI,* Vol. 19, No. 3, 1988a, pp.105–112 (in Russian).

Penzin, V. I., "Influence of Shape of Cross-Section of Direct Channel of Deceleration of Supersonic Flow," *Uchenye Zapiski TsAGI,* Vol. 19, No. 3, 1988b, pp. 55–59 (in Russian).

Appendix C: Some Aspects of Scramjet-Vehicle Integration

Vladimir V. Andreev and Vyacheslav I. Penzin

TsAGI, 140160, Moskow Region, Russia

Aerodynamically the integration of an airbreathing engine and a vehicle implies an arrangement of the engine whereby the lower surface of the forebody acts as part of the inlet for preliminary compression and the afterbody is used for enlarging the nozzle size and increasing the degree over-expansion of exhaust products. In the case of a two-dimensional engine the body may have a flat surface used as a first panel of a two-dimensional multi-staged inlet. The shape, the angle of setting and the length of this surface should allow for a minimum of wave loss and air flow spillage before the inlet. The solution of this problem requires a change of the body shape, the mounting of the side walls and so on. All these changes will have an effect on the vehicle performance, which, in its turn, should result in a change of the optimal geometry of an airbreathing engine. In other words, optimal geometry of an airbreathing engine and a vehicle should be selected by means of successive approximations.

The most complicated and closely integrated engine/vehicle interaction is observed in the case of a propulsion system with a scramjet. The problem here is not only that the scramjet passage may occupy the entire lower surface of the vehicle body and, consequently, specify aerodynamic characteristics to a great extent but also that the optimal geometry of the scramjet engine passage, in contrast to the other types of air breathing, is determined by its operation mode and heat supply. To each value of fuel/air equivalence ratio (ER) corresponds a certain optimal geometry of the inlet and the combustor. The variation of ER is defined by the required thrust and the vehicle drag. The problem of matching the operation mode and the geometry of an idealized engine has been discussed in several papers. The influence of the operating conditions of the engine on the optimal geometry of the scramjet passage at M_∞ equal to the inlet design Mach number M_d was investigated in Refs. 1–3. Variations of the final compression ramp inclination angle θ_f of the two-dimensional inlet and the inlet throat (combustor height) H_{th} with respect to M_∞ and fuel/air equivalence ratio ER were determined on the basis of minimizing the total wave and thermal losses. Furthermore, the influence of the design Mach number M_d the throat height H_{th}, the number of the inlet ramps N, the methods of contouring the compression wedge and the cowl on the optimal geometry were investigated in the above-mentioned papers.

The variation laws for the scramjet geometry when a divergent combustion chamber (cycles M=const, p=const) is used are considered in Ref. 4. The next phase of the study was to consider the booster type vehicle. The fixed geometry scramjet cannot be optimal throughout the flight Mach number range M_∞, so there is a need for a compromise geometry. The parametric analysis of effective characteristics of a fixed geometry scramjet in the flight Mach number range from 5 to 16 has been made.in Ref. 5 for geometry optimization of booster type scramjets. It was shown that the maximum range of engine operation with respect to M_∞ is achieved if the final ramp inclination angle of the two-dimensional inlet θ_f is equal to 10–15 degrees. The optimal design Mach numbers, the relative inlet height H_i with respect to θ_f and the required scramjet effective thrust coefficient at the start flight Mach number are also specified. At low values of θ_f the pressure level in the engine passage becomes lower, the non-uniformity of the flow behind the inlet throat with a cowl inclination angle equal to 0 deg decreases, and the range of Mach numbers in which the heat supply can occur in the constant cross-sectional area combustor is extended. According to calculations,[5] the values of θ_f decrease with the reduction of the throat height (in the variable at the start inlet) and the shortening of its side walls (resulting in a flow coefficient f decrease). Thrust augmentation due to the extra fuel supply to the combustor (as compared with a stoichiometric mixture) results in optimal θ_f increase that may reach 20 deg. It was suggested in Ref. 5 that the preliminary flow compression occurs under the flat screen, whose effect on the flow at a given length and angle of setting is close to the effect of a delta wing. However, the selection of the optimal geometry of the scramjet passage based only on the analysis of thrust-efficiency engine performance[5,6] can be regarded as one of the preliminary phases of engine selection. The point is that the law of angle of attack variation along the trajectory defined mainly by aerodynamic characteristics of the vehicle is not known in advance, and the same applies to the flow rate and engine thrust.

According to Ref. 5, it also seems impossible to evaluate the influence of the flight trajectory constraints (on dynamic pressure, for example) and the differences in engine thrust-efficiency characteristics associated with the use of different approximation techniques on the optimal geometry of the scramjet. High values of the specific impulse at high values of M_∞ cannot guarantee the effective acceleration, since the scramjet thrust coefficients may appear to be insufficient and acceleration may be too slow. The selection of the scramjet geometry merely for reasons of achieving maximum characteristics at high values of M_∞ may result in considerable decrease of characteristics at low values of M_∞ and make acceleration slow as well. Thus, it was necessary to perform parametric trajectory analysis of a vehicle acceleration using this engine in order to select the optimal scramjet passage geometry with respect to the maximum mass of a vehicle (minimum fuel consumption) at the point of a trajectory corresponding to the scramjet propulsion end.

The purpose of the present work is to solve two problems based on trajectory analysis: 1. The revision of the range of the passage geometry variation for idealized scramjet suggested in Ref. 5 as applied to concrete aerodynamic characteristics of the vehicle. 2. The evaluation of the influence of aerodynamic characteristics, flight conditions (trajectory restrictions) and the differences in the technique of determining the thrust-efficiency characteristics of the scramjet on its optimal geometry. A two-dimensional scramjet with multi-staged external compression inlet was considered. Configurations of a vehicle lifting body and scramjet passage are shown in Fig. C.1. The scramjet passage geometry was defined by the following inlet parameters: the final ramp inclination angle θ_f, the design Mach number M_d, and the number of the ramps N. The inlet throat height, H_{th}, equal to the height of a constant cross sectional area combustor H_c was defined by propulsion start conditions at the starting Mach number. The nozzle geometry was not optimized, and the nozzle flow was assumed to be one-dimensional. A scramjet is a component of a propulsion system consisting of a ramjet in parallel arrangement with a scramjet and a liquid rocket engine. A liquid rocket engine operates in the ranges of M_∞ from 0.8 to 2.5 and from 15 to 29. A ramjet and scramjet operate in the ranges of M_∞ from 2.5 to 6(7) and from 6 to 15, respectively. It was assumed that ramjet and scramjet inlet panels are set at a required position before start at the starting Mach number and then,

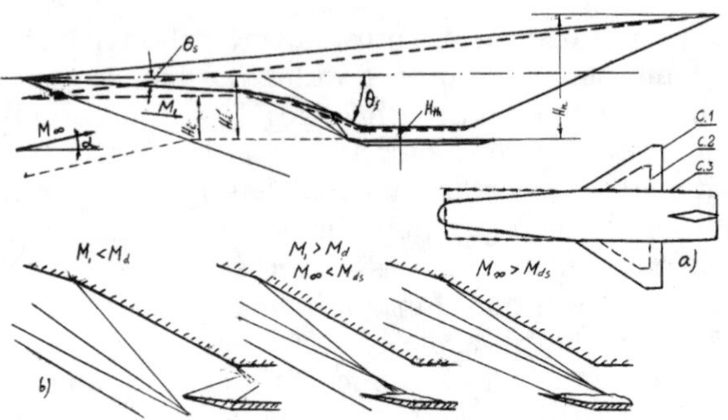

Fig. C.1 Airframe -integrated scramjet concept.

with increasing M_∞ the passage geometry remains fixed. Kinematics of panel displacement was not considered in this appendix. The scramjet and ramjet were located under the vehicle body, the forebody was considered to be the first inlet panel and the afterbody - the nozzle surface. The forebody lower surface was flat. Operational ranges of the ramjet and scramjet were selected approximately from particular considerations. Thus, for a ramjet the starting M_∞=2.5 was minimal. At lower values of M_∞ the acceleration was not possible because of the thrust deficiency. Maximum combustor height was taken to be 80% of the inlet entry height. The minimum starting M_∞ for the scramjet with a constant cross sectional area combustor (M_∞=6) was also defined by the condition providing the vehicle's acceleration. The maximum operation Mach number M_∞ for the scramjet was taken to be equal to 15. At this value of M_∞ the scramjet characteristics drop off. The same value of M_∞ was taken in Ref. 6 because of excessive thermal loads in the scramjet passage at high values of M_∞. Strictly speaking, each scramjet passage should have its own value of $M_{\infty max}$. However, according to estimations, the vehicle mass put in orbit depends only slightly on the individual value of $M_{\infty max}$ for the five passage geometry variations under discussion.

At present many techniques exist for the analysis of three-dimensional gas flow in the engine passage regarding the real gas effects, fuel mixing and combustion processes, and the nozzle flow characteristics. These techniques make it possible to define the forces affecting the engine and the vehicle as a whole. However, it is not worthwhile to use these techniques directly for parametric flight trajectory analysis aimed at preliminary determination of the vehicle concept on account of computer slow response and high cost of calculations. It is also not worthwhile to use these techniques for calculating a set of grids for the scramjet thrust-efficiency performance because of the large number of flight parameters. The purpose of the present analysis is the same as that of Ref. 5 that is not to determine the scramjet thrust-efficiency performance in the range of M_∞, but to select the optimal geometry of the engine passage using approximate characteristics determined in the process of the trajectory analysis. This approach simplifies the task. In this case some factors improving the scramjet performance, such as hydrogen heating on the vehicle exterior surfaces before it enters the combustor, can be ignored or taken into account crudely. In the scramjet performance analysis requirements for the account of friction forces are reduced. In the present analysis as well as in Ref. 5 an approximate technique was used for calculating the characteristics of the two-dimensional inlet, the combustor, the nozzle and the engine on a whole that for the sake of simplicity can be applied to each flight trajectory integration step. Gas real properties on the assumption that the flow is in equilibrium were taken into account according to technique [7] resulting from I-S diagrams processing. The combustor and the nozzle flow were considered to be one-dimensional. Losses by friction along the gas-air passage were accounted for approximately in proportion with the length of each panel and the friction coefficient was defined by accounting for compressibility and the temperature factor. In parametric analysis the inlet final ramp inclination angle θ_f varied from 10 to 20 deg, the design Mach number M_d, from M_d = 6–11. The number of inlet ramps with angles of inclination chosen at $M_\infty = M_d$ under the condition of equal intensity of shocks, was taken to be equal to 3. The nozzle expansion ratio $\bar{H}_n = H_n/H_i$ as set to be equal to 3.3. Two values of the dynamic pressure along the flight trajectory in the region of airbreathing engine operation were considered: q=77 kPa and q=52 kPa. In the analysis, regardless of operating mode and the passage geometry the hydrogen combustion completeness in the combustor was assumed to be 0.9 and impulse losses in the nozzle 3%. To evaluate the tendency of the scramjet passage geometry variation with respect to the level of its thrust-efficiency performance, the latter was defined under two assumption: 1) cold hydrogen enters the combustor (as in Ref. 5); 2) the hydrogen is heated on the exterior surfaces of the vehicle up to 1000 K which increases its enthalpy and the impulse of hydrogen jets injected in the combustor parallel to the main flow. Thrust coefficient increase due to hydrogen heating was calculated by using a correction $C_t = C_{tcold}(0.9+0.33)$ $M_\infty)$ obtained by processing the previously performed calculations.

Since the purpose of this work was to determine the laws of the scramjet geometry variations, the parameters and operation modes of liquid rocket engine and ramjet were not optimized. The specific impulse of a liquid rocket engine at the first and the second leg was I_1=415 s and I_2=475 s, respectively. The starting thrust loading n was assumed to be equal to one and at M_∞= 15, n = 1–1.15. The ramjet inlet geometry was defined by the condition that θ_f=20° and M_d = 4. The height of the nozzle critical section throughout the entire range of ramjet operation was supposed to vary so as to ensure the maximum engine thrust. In the general case the optimal geometry of the scramjet passage used as a component of the propulsion system of a vehicle put into orbit vehicle should be selected by the maximum vehicle mass m_{max} at the end flight point. In the particular case, when the flight is limited by the given dynamic pressure at the trajectory leg of airbreathing engine operation, the selection is made by the value of m_{max} at the scramjet fixed propulsion end. Since the required engine thrust is defined by the aerodynamic characteristics of a vehicle these have both a direct influence on these selection of the scramjet optimal geometry and operation mode and an indirect influence, through the angle of attack variations and the related variations of flow rate and total pressure losses.

It is convenient to evaluate the influence of aerodynamic characteristics variation on the scramjet geometry caused by the use of different configurations when the relation of flow rate-to angle of attack is constant. This is feasible if the vehicle body for all configurations remains the same, capable of providing for the preliminary air compression and if aerodynamics variation is made at the expense of the wing. For example, it may be supposed that cantilevers turn around the longitudinal axis. In the case of the lifted cantilevers the drag is reduced. The drag reduction can be achieved in the swing wing configuration as well. The similar change of aerodynamic characteristics can be achieved through the assumption that configurations for simplicity of analysis differ only by the wing area. Three configurations have been selected: 1) full area wing cantilever configuration; 2) half-area wing cantilever configuration; 3) all-body configuration.

For parametric trajectory analysis aerodynamic characteristics of lifting body configurations were used as calculated by I.Ya. Grechko (TsAGI) at M_∞>3. At lower values of M_∞ characteristics of liquid rocket engine operation aerodynamic performance was taken approximately by analogy with those available in literature. Characteristics used in the analysis are shown in Fig. C.1 as related to a full area wing. The assumption made in selecting the characteristics at M_∞<3 has no effect on the results of comparative study of the scramjet optimal geometry, it will take effect only on the mass of the vehicle in orbit. Variation in the vehicle mass related to variation of the wing area was neglected. The landing problem of the vehicle was not considered in the present work. The results of the vehicle trajectory analysis (configuration 3) with a scramjet inlet θ_f= 15 deg and M_d=8 are shown in Fig. C.2. The vehicle relative mass m (for the three configurations), angle of attack a,

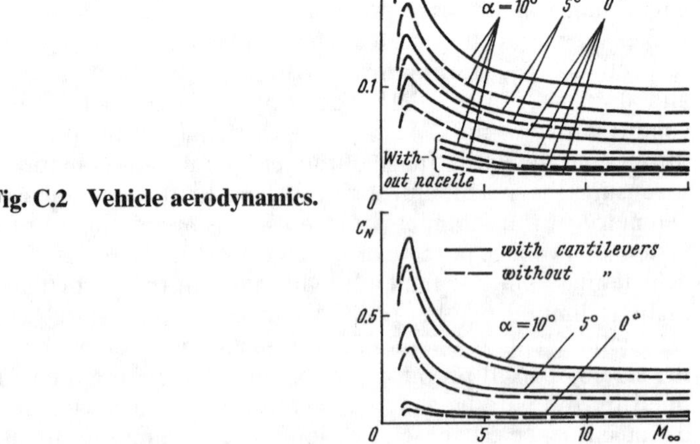

Fig. C.2 Vehicle aerodynamics.

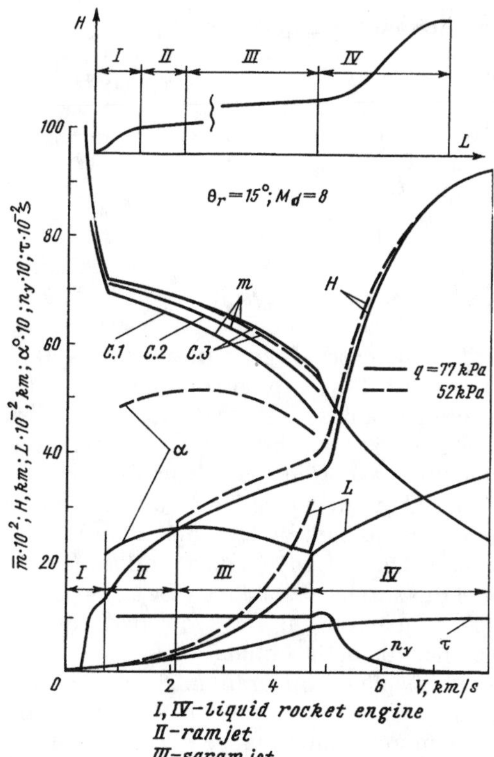

Fig. C.3 Vehicle trajectory parameters.

distance L, altitude H, time τ, and overload n_y are presented with respect to the flight speed. V. The flight with operating ramjet and scramjet corresponds to the dynamic pressure $_q$ of 52 kPa and 77 kPa. As follows from Fig. C.3, the vehicle lifting body has a great lift. Exclusion of a wing cantilever (configuration 3) does not cause considerable growth of induced drag, as a result the vehicle relative mass $\overline{m}$ at M_∞= 15 corresponding to the scramjet propulsion end, increases by $\Delta\overline{m}$ = 0.075 as compared with configuration 1. In orbit the gain in m is less and equals $\Delta\overline{m} \pm 0.03$. In changing for a dynamic pressure of 52 kPa (dashed line), the vehicle mass is reduced a little at M_∞=15, because the angle of attack in this case increases nearly twice as much (from a = 2.5 deg to a = 5 deg). The flow rate (f) and effective thrust (C_t) grow accordingly, providing for sufficiently effective vehicle acceleration.

To make clear the details of trajectory flight of a vehicle powered by optimal geometry scramjet at different dynamic pressure let us consider the relative fuel capacity μ_f in all typical legs of the flight:

As can be seen from Table C.1, ramjet and scramjet fuel consumptions during the flight at a dynamic pressure of 52 kPa are higher than at q=77 kPa. However, fuel total consumption in orbit due to the higher altitude of

Table C.1 Ramjet and Scramjet Fuel Consumptions

Dynamic pressure	μ_f liquid rocket engine	μ_f ramjet	μ_f scramjet	μ_f liquid rocket engine
$q = 77$ kPa	0.275	0.038	0.138	0.290
$q = 52$ kPa	0.270	0.040	0.155	0.274

flight at q=52 kPa was found to be the same. This result may be attributed to the effect of vehicle integration with an engine whose thrust increases with increasing angle of attack. The increase of angle of attack a with a corresponding increase in engine thrust allows for retaining the fuel–air equivalence ratio ER equal to 1 at high values of M_∞ and, thereby, to raise the engine efficiency which decreases in case the thrust increase is achieved by over-enriching the air–fuel mixture. According to calculations the dynamic pressure variation has not caused the change in flow coefficient f. This is due to the fact that the value of f is defined by the angle between the free-stream velocity vector and the vehicle's body (screen) surface (in the present work the angle of setting of this surface θ_s is taken to be 6 deg). Thus, during the change from a =2.5 to a =5 deg the angle $a + \theta_s$ varied from 8.5 to 11 deg, that is only by 30%. The selection of the optimal angle is a separate problem closely related to the scramjet/vehicle integration. The decrease of θ_s should result in a total pressure recovery increase in the air stream incoming to the scramjet and in the more sudden change of coefficient f in changing the angle of attack. But in this case the drag of the body section surrounded by the flow that does not enter the engine, may increase. The scramjet thrust coefficient C_t at q = 52 kPa is a little lower than C_t at q=77 kPa owing to the larger angle of attack, the specific impulse being decreasing owing to the smaller nozzle expansion ratio.

We now return to the problems of the optimization of the selected configuration of the scramjet passage geometry. The vehicle relative mass variation (configuration 2) with respect to the design Mach number for the three values of θ_f is shown in Fig. C.4a. Since apart from θ_f and M_d the scramjet passage geometry is determined by the inlet throat height H_{th} (combustor height) the variation of this value with respect to θ_f and M_d is also shown in Fig. C.4a. As follows from Fig. C.4a the maximum value of m is obtained at θ_f=15 deg. It is concluded in Ref. 5 that the optimal value of θ_f ranges from 10 deg to 15 deg. The obtained refinment is related to the fact that at low values of the thrust coefficient at low M_d decreases which result in slow acceleration. The plots illustrate that the optimal value of M_d increases with decreasing θ_f and is equal to 8.2, 8.5 and 9.0 for values equal to 20, 15, and 10 deg, respectively. For the optimal value of θ_{f} the optimal design Mach number exceeds the initial free-stream Mach number of the scramjet propulsion start by ΔM = 1.5. The relative mass variation with respect to $\overline{m}$ (M_d) for the three configurations at θ_f = 15 deg is shown in Fig. C.4b. It can be seen that the optimal value of M_d at equal values of the dynamic pressure is

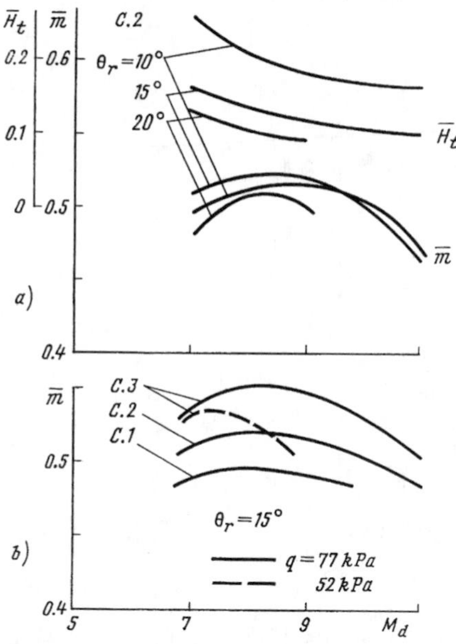

Fig. C.4 Variation of inlet throat height and relative vehicle mass with flight Mach number for two vehicle configurations.

relay the same for all three configurations and equals $M_d = M_{st} + 1.5$. The tendency of M_d to increase with increasing the wing area is only implied. This results from the fact that there was a small increase in angles of attack with decreasing wing area. The angle of attack α increases up to 5 deg with decreasing the dynamic pressure to 52 kPa. This caused a considerable change in the plot of $\overline{m}(M_d)$. The optimal value M_d decreases and approaches the starting $M_\infty = 0.5$.

Angles of attack corresponding to the scramjet propulsion start and end (M_∞.=7 and 15, respectively) are shown in Fig. C.5 for the same configuration and the same values of . θ_f as is shown in Fig. C4. As follows from Fig. C.5a the value of α is practically independent of θ_f. and M_d at M_∞=15 variation of α with respect to θ_f and M_d is large as well. Flow coefficient variations for the same conditions are presented in Fig. C.6. The previously made conclusion about a small variation in f for all configurations is confirmed when considering the plots. In this respect the vehicle's body cannot be regarded suitable for investigation of configurations operating at substantially different angles of attack. It is hoped that the selection of another vehicle body having lower drag and lower lift at a greater extension may ensure more effective use of vehicle/engine integration. But in this case uncertainties remain concerning the possibility of retaining the flow characteristics because of the flow spillage before the inlet.

Plots of $\overline{m}$ versus M_d are shown in Fig. C.7a for configuration 2 at θ_f =15 deg for the scramjet propulsion start at M_∞.=6. It can be seen that the higher value of $\overline{m}$ is achieved at M_{st} =7. As for the optimal value of M_d at M_{st} =6 it tends to lower values such that the value of remains approximately the

Fig. C.5 Variation of vehicle angle of attack with designed Mach number for different vehicle configurations and final ramp angle.

same as in the case with M_{st}=7. Plots of $\overline{m}$ versus M_d are compared in Fig. C.7b for the two design conditions of the scramjet thrust-efficiency characteristics with and without considering the hydrogen heating on the vehicle's exterior surfaces. It can be concluded from Fig. C.7b that accounting for hydrogen heating results in a slightly increased optimal value of M_d. Considering the fact that the scramjet performance with account for hydrogen heating was somewhat exaggerated it may be concluded that the noted above differences in the technique of determining C_t and I_t do not result in

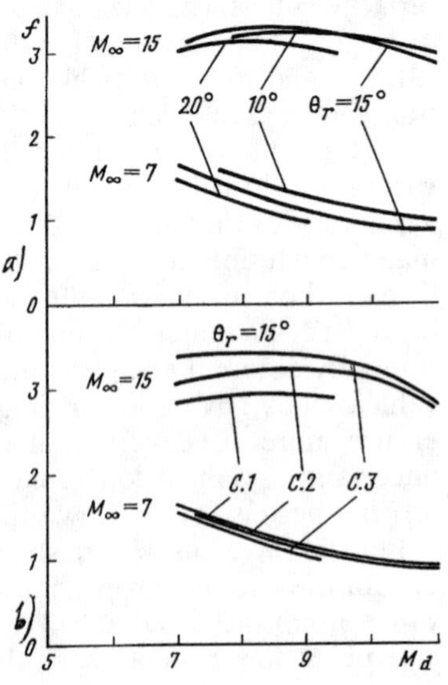

Fig. C.6 Variation of inlet mass flow coefficient with designed Mach number for different flight Mach numbers, final ramp angles, and vehicle configurations.

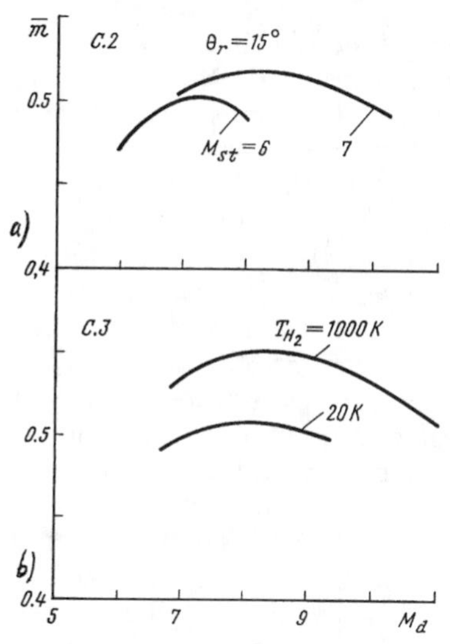

Fig. C.7 Variation of vehicle relative mass with designed Mach number for different initial Mach number and hydrogen temperature.

noticeable changes of the scramjet optimal geometry. The solution of this and other problems requires further investigations, design studies included. Among these are the problems of considering the effects of the nozzle geometry and operation mode, the moment characteristics of the vehicle and other effects on the optimal shape of the scramjet passage. The flight trajectory analysis of different versions of the idealized vehicle powered by two-dimensional fixed geometry scramjet operating in the Mach number range from 6 to 15 showed that: 1. The conclusion previously made in Ref. 5 about the advantage of using the external compression inlet with small flow turning angles is confirmed. At the same time certain corrections should be made: The optimal value of the inlet final ramp angle is found to be 15 degrees for all vehicle configurations considered. The optimal inlet design Mach number exceeds the starting Mach number by $\Delta M = 0.5–2.0$. The change of aerodynamic characteristics of a vehicle with a lifting body did not result in noticeable variations in the scramjet passage geometry, and this is attributed to a large body lift and a corresponding small change in the angle of attack and engine flow rate. The change of flight conditions, for example, the imposing of restrictions for the dynamic pressure along the trajectory results in noticeable variations of the scramjet passage geometry. Thus, the transition from the dynamic pressure of 77kPa to 52 kPa results in an almost two-fold increase in the angle of attack. This corresponds to the flow turning angle increase and consequently to the decrease in optimal design Mach number approximately by 1. The improvement of the scramjet thrust efficiency characteristics by accounting for the hydrogen heating on the vehicle exterior surface results in insignificant changes in the optimal geometry of the scramjet passage. This result confirms the possibility of using different approximate techniques for the scramjet performance analysis to evaluate the optimal engine geometry.

All previous calculations were carried out at the screen setting angle θ_s equal to 6 deg. This angle was selected rather arbitrary. Now we try to select optimal value of it by varying this angle. Two types of the screen were taken into consideration: delta wing with sweptback angle $\chi = 80$ deg and rectangular screen (two-dimensional flow). It was assumed that aerodynamics is the same for both cases. The influence of wedge bluntness on flow rate and total pressure loss was disregarded. Using numerical results from Refs. 8 and 9 three-dimensional data averaged flow parameters in front of inlet entrance depending on M and α are taken. These averaged parameters are: Mach number, total pressure recovery coefficient ν, and flow contraction coefficient K. Two-dimensional screen is more convenient from the viewpoint of vividness and the ease of presentation, for example, of screen length influence on engine performance. As in the previous section, three flow modes were considered (Fig. C.1). In the case of delta screen the third flow mode at which shocks go under the cowl was not considered as well as friction forces. The engine passage with θ_s=0 deg was taken to be initial geometry (dashed lines in Fig. C.1). The length of engine passage and the nozzle exit height H_n were assumed to be invariant for each screen designed Mach number M_{des} that changed in the range M_{des}=15–M$_d$. The right boundary corresponds to the inlet designed Mach number (without screen-isolated engine). The increase in θ_s involves an increase in inlet entrance height H_i and correspondingly a decrease in ratio H_n/H_i that leads to engine performance drop. This ratio in the initial variant is taken to be 3.3. In Fig. C.8 all the inlet dimensions are related to inlet height. Parameters θ_f, M_d and N were taken to be equal, in accordance with previous recommendations, 15 deg, 8.5 and 3, respectively. Varying screen parameters were: M_{des} = 15, 12 and 8.5 (no screen); θ_s = 0, 3, 6, and 9 deg. Since θ_f is assumed to be invariant and the screen is assumed to be the first inlet ramp the increase in θ_s means other ramps angle decrease and inlet nearing to 1-shock inlet. The increase in θ_s as is seen from Fig. C.11 also decreases the upper body surface inclination and its aerodynamic drag. Besides variants with θ_f = 15 deg calculations with another θ_f (9–21 deg) were carried out.

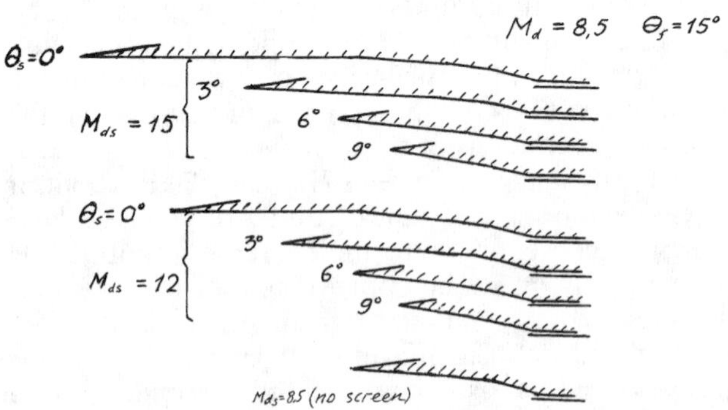

Fig. C.8 Inlet ramp configuration.

The variation of the inlet throat height range corresponds to the assumed earlier. All other calculation conditions are also the same. Scramjet starting Mach number was assumed to be 7 to simplify calculations (constant area combustor). It was shown earlier that starting Mach decreasing down to 5–6 does not change inlet optimal geometry substantially.

The vehicle was assumed to be a material point that moves in a vertical plane at constant dynamic pressure with wings level and without sideslip. Fig. C.9 illustrates trajectory calculation results. We covered variants with delta and rectangular screen ($M_{\text{des}} = 15, \theta_f = 15$ deg, $M_d = 8.5, \theta_s = 4$ deg) at two values of dynamic pressure $q = 50$ and 75 kPa. The plots of fuel fraction μ necessary for vehicle boosting from M=7, flight range L, angle of attack a, engine specific impulse I_{sp} and air rate coefficient f versus M are presented. Some preliminary conclusions can be done. 1) The use of delta screen instead of rectangular does not change engine and vehicle characteristics much. So it is possible to use rectangular screen instead delta one for qualitative analysis of engine performances. 2) The transition from q = 75 kPa to q = 50 kPa dynamic pressure brings out two-fold increase in a and substantial increase in flow rate. However this does not compensate fully the drop in engine performance and fuel rate increases. 3) At

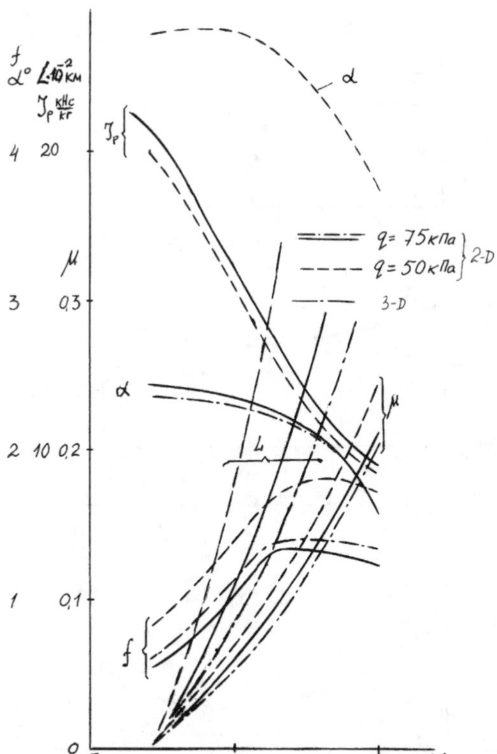

Fig. C.9 Scramjet/vehicle trajectory parameters.

large M the flow rate changes little. This is connected with a drop due to vehicle getting lighter and shocks going under the cowl (flow stream tube limiting). The most strong growth in fuel consumption takes place at M = 11–13. The thrust drop brings about pace of boosting decrease and sharp increase in the flight range.

Plots of μ versus θ_s for different M_f and two types of screens are presented in Fig. C.10. The engine variant without screen was analyzed as with and without accounting of the drag force exerted to the «back» of the vehicle (upper surface). Cross-hatched bands indicates these variants at $M_f = 9$ and 11. There is not upper boundary at $M_f = 13$ because in this case the vehicle stops boosting at $M = 12$. At $q = 50$ kPa minimal fuel rate is reached at θ_s = 0–2 deg. Greater θ_s corresponds to greater M_f because at large M angle of attack is lower and optimal wave and thermal loss is reached at greater θ_s. The increase of dynamic pressure up to 75 kPa resulted in optimal θ_s increase up to 4 deg due to lower a. It is interesting to note that the angle of meeting of velocity vector and screen surface $\theta_\Sigma = \theta_s + a$ for both dynamic pressures is approximately the same, namely 6 deg. This fact possibly can be used at θ_s selecting at intermediate dynamic pressure. The relationship $\mu(\theta_s)$

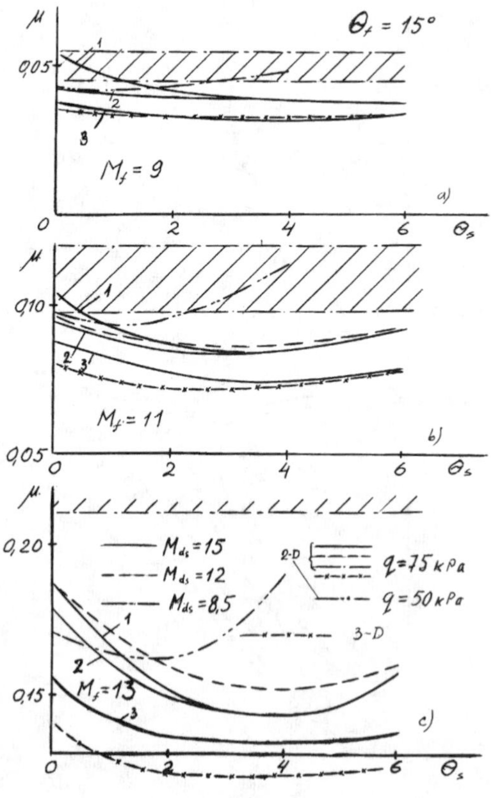

Fig. C.10 Variation of fuel fraction required to accelerate to velocity with screen setting angle for different final Mach numbers, rectangular and delta screen configurations and designed Mach numbers.

for delta and rectangular screen is of similar character. The lesser μ for delta screen is explained by lesser wave loss and also by friction force neglecting. The comparison of curves corresponding to $M_{\rm des} = 15$ and 12 at $q = 75$ kPa shows that minimal μ is reached at approximately the same θ_s and screen shortening at that results in 10% μ drop. The absence of screen ($M_{\rm des} = M_d = 8.5$) at $M_f = 9$ and 11 leads to 30–40% μ increase. The following figures illustrate the engine–vehicle integration effect. The vehicle boosting can somewhat differ depending on starting thrust coefficient. The curves marked by Fig. C.11 and 12 correspond to starting ($M_{\rm st} = 7$) $C_t = 0.2$ and 0.4. It is seen that difference in C_t did not influence the optimal θ_s selection. Curves 3 correspond to the case in which the throat and combustor heights are assumed to be adjustable along the whole trajectory (see earlier). The influence of inlet throat control on inlet final ramp angle was analyzed earlier. It was shown there that inlet height change does not practically change optimal θ_f. The influence of inlet throat on optimal angle of screen inclination θ_s is also small. There is only small tendency of θ_f decreasing with inlet throat diminishing that brings about some total loss decrease.

Figure C.11 illustrates the influence of $M_{\rm des}$ on μ. It is seen that screen

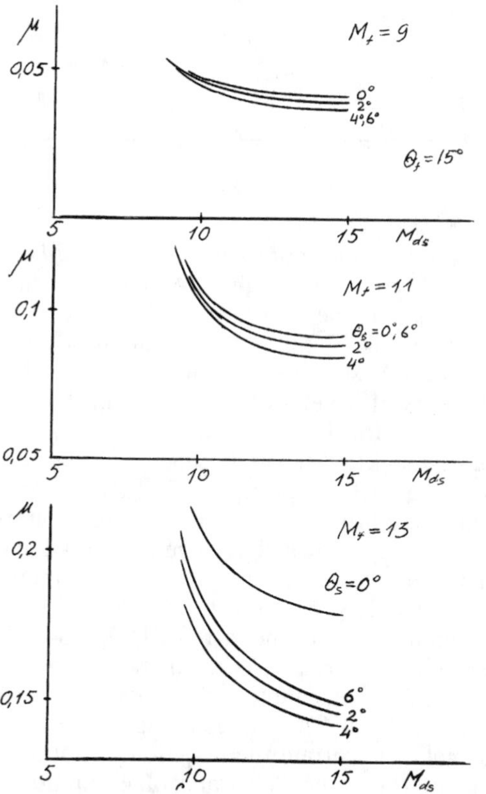

Fig. C.11 Variation of fuel fraction required to accelerate to velocity with screen designed Mach number for different M_f and θ_s.

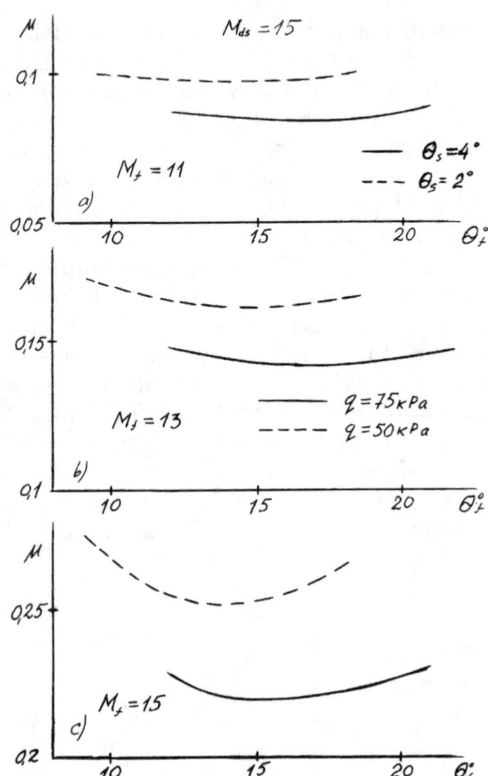

Fig. C.12 Variation of fuel fraction required to accelerate to velocity with final ramp angle for different dynamic pressures and screen setting angles.

length increasing involves a decrease in μ the more noted the more M_f is. The increase in M_{des} over M_f decreases μ only slightly. At large M_f the dependence of μ on θ_f increases. It was obtained earlier that at fixed $\theta_s = 6$ deg and $M_f = 15$ the optimal $\theta_{\mathrm{f}} = 15$ deg. Taking into consideration that optimal θ_s is smaller than 6 deg one can suppose that optimal θ_f will be greater than 15 deg. In Fig. C.12 plots of μ versus θ_{f} for different M_f are presented. These curves were calculated for θ_s near to optimal ($\theta_s = 4$ deg for $q = 75$ kPa and $\theta_s = 2$ deg for $q = 50$ kPa). It can be concluded that the increment in optimal θ_f does not surpass 1–2 deg at $q = 75$ kPa. As one would expect at $q = 50$ kPa optimal θ_f is a little less, 13–14 deg due to greater angle of attack and wave loss. It is well known that at moderate Mach numbers the increase in number of oblique shocks over 3 does not practically increase engine thrust. At hypersonic speeds it can be expected that optimal number of shocks will increase. However the increase in M leads to an decrease in optimal θ_f that assist in N decrease. As calculations showed the increase in N over 3 does not save fuel consumption more than 1%. The influence of N on throat height is also small. The curves in Figs. 10 and 12 have sloping minimum in μ that makes uncertain the selection of optimal parameters. Other factors, for example pressure level and Mach number, determining the combustion process and structural strength can impact the engine geometry selection. It is clear that these parameters must be accounted in real scramjet designing. Therefore the results of present calculations have qualitative character and can be used only on the initial step of scramjet passage selection.

References

[1]Penzin, V. I., "On Optimum Conditions for Supersonic Oblique Shock Flows Followed by Heat Supply," *Izvestiya Vysshikh Uchebnykh Zavedeniy, Seriya "Aviatsionnaya Tekhnika,"* No.4, 1966 (in Russian).

[2]Penzin, V.I., "The Problem of Selection of the Maximum Ramp Angle of the Scramjet Inlet," *Trudy III Nauchnykh Chteniy po Kosmonavtike,* Izd. ANSSSR, Moscow, 1981 (in Russian).

[3]Penzin, V. I., "On the Problem of Optimum System of Scramjet Inlet Shocks," *Trudy IV Nauchnykh Chteniy po Kosmonavtike,* Izd. AN SSSR, Moscow, 1982 (in Russian).

[4]Penzin, V. I., "On Optimum System Oblique Shocks + Heat Supply for Off-Design Modes," *Tsanderovskiye Chteniya,* Isd. AN SSSR, Kharkov, 1983 (in Russian).

[5]Penzin, V. I., "Optimization of Passage Shape for Dual Hypersonic Ramjet," Preprint TsAGI, No. 12, 1990 (in Russian).

[6]Ikawa, H., "Rapid Methodology for Design and Performance Prediction of Integrated Scramjet/Hypersonic Vehicle," AIAA Paper 89-2682, 1989.

[7]Varshavsky, G. A., Guber, E. Ya., and Kiselev, A. P., "On Thermodynamics of Equilibrium Gas Mixture Flows Formed by C,H,N,O Species," *Trudy TsAGI,* Issue 978, 1966 (in Russian).

[8]Bosnyakov, S. M., et al., "Investigations of Three-Dimensional Flow and Aerodynamic Performances of Plane Inlets with Different Inlet Shape and Side Walls Dimensions," *Uchenie Zapiski TsAGI,* Vol. 14, No.3, 1983 (in Russian).

[9]Bosnyakov, S. M., and Zlenko, N. A., "Calculations of Three-Dimensional Coefficients of Flow near the Wedge Surfaces with Finite Width," *TsAGI Transactions,* No. 2187, 1983 (in Russian).

Appendix D: Leading-Edge Bluntness Effect on Performance of Hypersonic Two-Dimensional Air Intakes

Vladimir P. Starukhin and Alexander Ph. Chevagin
TsAGI, 140160, Moskow Region, Russia

Introduction

Due to intense aerodynamic heating of outer surface of hypersonic vehicles the wing and intake leading edges must be blunted. Shock waves with local subsonic zones in front of of the blunt leading edges change external and internal flows as compared with the sharp edge design. This may change both the external characteristics (such as drag and lift coefficients) and internal characteristics (the mass flow and total pressure recovery coefficients) of an intake and ,consequently, aircraft aerodynamics and thrust-fuel consumption characteristics of a propulsion system. Therefore designers of hypersonic intakes should be aware of how to evaluate the possible degradation of intake performances due to bluntness of leading edges of the compression wedge, the side wall and the cowl lip, and the wing leading edges for the underwing intakes. It is necessary to learn what are the possibilities for compromising between radii required as a successful countermeasure against the aerodynamic heating and the allowable leading edge radii which can adopted to retain the intake performance at an acceptable level. At present there are no rather reliable calculation methods nor sufficiently extensive systematic experimental data which could be used to evaluate the influence of an intake leading edge radius on intake performances. Hereinafter data of such experimental studies are presented which make it possible to partly answer a number of questions that arise in connection with the necessity to blunt the inlet leading edge. Experimental investigations employed two models:

1) the model of an isolated two-dimensional intake (with a design Mach number of 6) whose scheme is presented in Fig. D.1.
2) the model of an underwing two-dimensional intake (with a design mach number of 5.3) whose scheme is presented in Fig. D.2.

During the investigation the configuration and radius of leading edges of a wing, compression wedge, cowl lip and side walls were varied by replacing

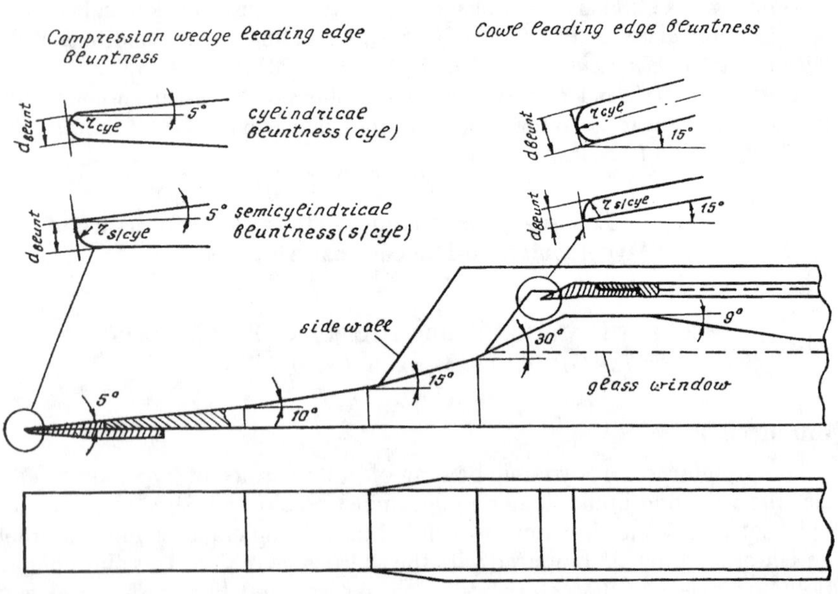

Fig. D.1 Scheme of isolated two-dimensional intake; design Mach number is 6.

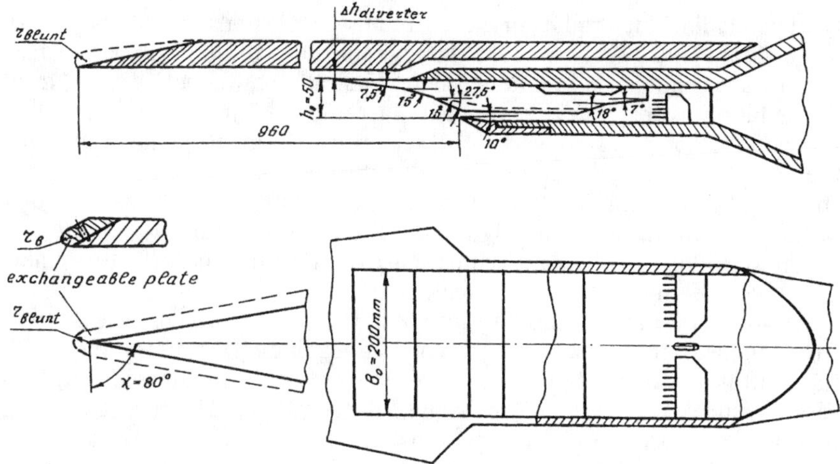

Fig. D.2 Scheme of underwing two-dimensional intake; design Mach number is 5.3.

the parts. The objective of tests was to evaluate the influence of inlet leading edge bluntness on flow patterns in the entry section of the intake and its total internal characteristics (i.e., the mass flow and total pressure recovery coefficients).

Isolated Two-Dimensional Intake

The first stage of tests was carried out using the isolated intake model over a Mach number range M=4–5.5, when $M < M_d$ (M_d is the design Mach number). First of all, it has been revealed that the influences of the blunted cowl lip and compression wedge are substantially different: the cowl lip blunted to a considerable extent (r_{lip}=0.025h_0) does not cause the intake unstart, nor notable variation of mass flow and total pressure recovery coefficients, but blunting the compression wedge leading edge, to the same radius may result in intake unstart and sudden degradation of internal characteristics of the intake.

The bluntness effect on the flow patterns in the intake entry section considerably depends on how shock wave from the last ramp of the compression wedge interacts with the shock wave that exists in front of the blunt cowl lip ,and so it substantially depends on both the free stream Mach number and the leading edge radius. When M = 4-4.5, i.e., when the Mach number is noticeably less than M_d, the shock wave from the last ramp of the wedge passes noticeably over the cowl lip leading edge and may be said to interact not with the central part of the bow shock wave (in front of the blunt edge) but with its continuation which transforms into a weaker oblique shock wave of the same direction as the shock wave from the last ramp of the compression wedge. In this case, according to the shock wave interference classification by Edney,[1] the type VI interaction appears. Increasing the cowl lip leading edge radius in these circumstances spoils the inflow pattern in the intake entry section and, at larger radii, does result in intake unstart. Two reasons should be mentioned here. First, the bow shock wave in front of the bluntness begins, as its intensity increases to interact with boundary layer on the compression wedge and separates the boundary layer. Second, the increase of the cowl lip leading edge thickness expands the area of the stream tube entering the channel, because of a shift of the stagnation point on the cowl lip leading edge. When the throat area is close to a minimum required to start-up an intake with a sharp cowl lip, the above factors make, of course, the intake start-up difficult.

At M=5, i.e., in flight modes that are closer to $M = M_d$, the shock wave from the last ramp passes near the cowl lip leading edge and interacts with the central part of the bow shock wave in front of the leading edge. In this case the shock wave interaction is close to type IV interference, according to Edney's classification.[1] The flow pattern in the entry section does not experience substantial variations when the cowl lip leading edge is made more blunted. Even at a radius of $0.025h_0$ the inflow pattern is close to that for a sharp leading edge of cowl lip. When the cowl lip radius is increased, the bow shock wave would be generated in front of a blunt body and would interact with a shock wave from the last ramp so as to show type IV interference; but the flow pattern photographs in Fig. D.3 show that, instead, a special system of oblique shocks forms, such that the blunt edge is as if a sharp one, and the shock wave reflected from the cowl lip leading edge into the intake, only slightly increases its intensity in comparison with the design with a sharp edge. This can be explained by the influence of the inlet fences and, in particular, by interaction of their boundary layer with shock waves at the intake entry. When the shock wave from the last ramp interacts with the bow shock wave in front of bluntness, the boundary layer in corners formed by fences and cowl lip separates along the cowl lip leading edge and "sharpness" the cowl lip on some proportion of its width (Fig. D.4). In this case, the separated flow similar to that over a cylindrical obstacle on a plate[2] seems to appear. At a certain ratio of a distance between side walls and sizes of the separation zone, the entire blunt cowl lip may be flowed as if sharp.

The comparison in Fig. D.5 of the total pressure recovery coefficient ν_{max} and the mass flow coefficient f_{max} for different leading edge radii makes it possible to quantify the influence of cowl lip leading edge bluntness. In this case it should be remembered that mass flow coefficients for all versions were calculated using the capture area F_0 for the design with sharp leading edges. The maximum levels of these coefficients for the different versions differ quite insignificantly. Though, the following trend should be noted: the mass flow coefficient slightly increases(within 4–5%) with the radius, which is clearly due to stagnation point shift on the cowl lip leading edge. The total pressure recovery coefficient variation in this case is quite within the accuracy of experiments.

The effects of cylindrical and semi-cylindrical cowl lips with equal thickness of leading edges are presented in Fig. D.5. Replacing the semi-cylindrical leading edge with the cylindrical one slightly increases the mass flow

$M_\infty = 5 \quad d_{wedge} = 0$

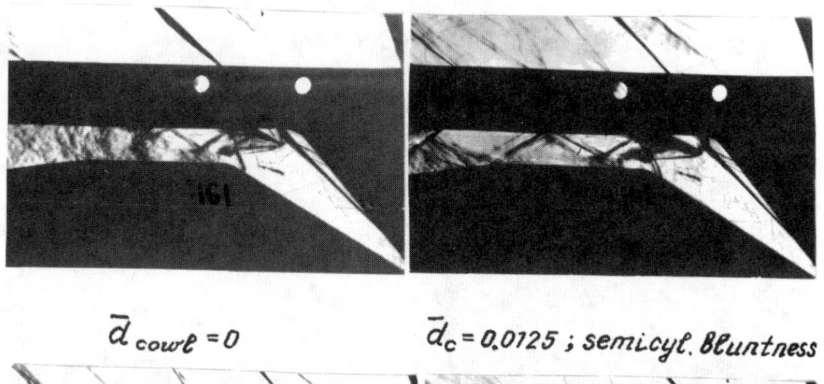

$\bar{d}_{cowl} = 0$ | $\bar{d}_c = 0.0125$; semicyl. bluntness

$\bar{d}_c = 0.025$; semicyl. bluntness | $\bar{d}_c = 0.025$; cyl. bluntness

$\bar{d}_c = 0.05$; semicyl. bluntness | $\bar{d}_c = 0.05$; cyl. bluntness

Fig. D.3 Flow pattern pictures for isolated two-dimensional intake; Mach number is 5; $\bar{d}_{cowl} = d_{cowl}/h_0$.

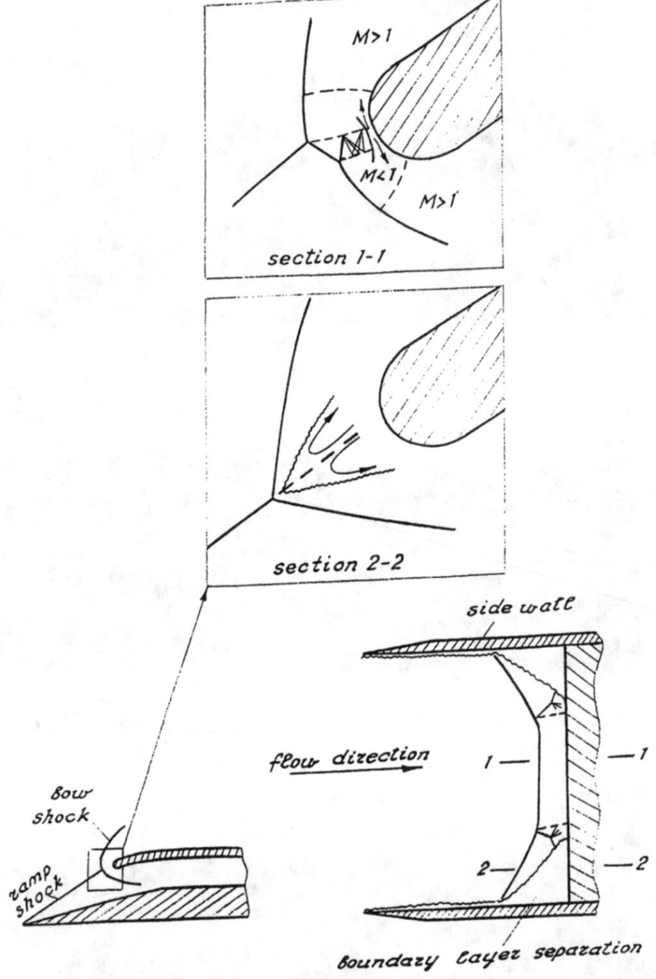

Fig. D.4 Entry flow scheme for isolated two-dimensional intake.

coefficient $f_{\max}$ (by up to 4%). This is likely to be due to a larger shift of the cowl lip leading edge stagnation point to the external side of the cowl lip. In this case the recovery coefficient either remains almost invariable or slightly increases by 3–4% as well. Summarizing, we may say that blunting a cowl lip leading edge of a two-dimensional intake flown with $M < M_d$, insignificantly varies the intake mass flow and total pressure recovery coefficients.

Compression wedge leading edge bluntness considerably influences the internal intake characteristics as distinct from cowl lip bluntness effects. The comparison of $f_{\max}$ and $\nu_{\max}$ at different radii of semi-cylindrical compression wedge is presented in Fig. D.6. It is seen that increasing the com-

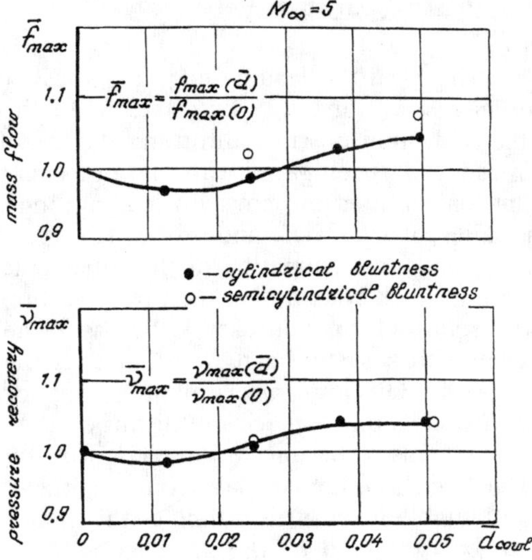

Fig. D.5 Impact of cowl leading edge bluntness on total pressure recovery and mass flow coefficients for isolated two-dimensional intake; $\bar{d}_{cowl} = d_{cowl}/h_0$.

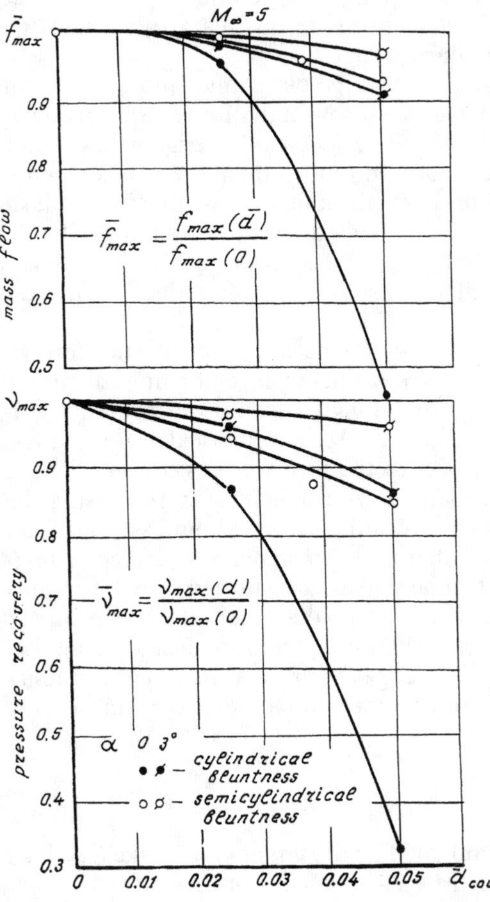

Fig. D.6 Impact of compression wedge leading edge bluntness on total pressure recovery and mass flow coefficients for isolated two-dimensional intake; $\bar{d}_{cowl} = d_{cowl}/h_0$.

pression wedge radius to $r_w = 0.05h_0$ (and the leading edge thickness to $d_w = 0.05h_0$) reduces the coefficients f_{max} and ν_{max} by 10–15%. In this case it should be noted that the maximum possible throttling degree of the intake with a blunt wedge is reduced as compared with that of a sharp wedge irrespective of how blunt, or sharp, is the cowl lip leading edge. For example, a wedge with a radius of 0.0375 h_0 and $0.05h_0$ (to d_w = 0.0375 h_0 and 0.05 h_0 respectively) leads to termination of the supersonic flow regime in a throat if the closing shock system is in the throat section (immediately upstream of the diverging subsonic diffuser), whereas in the case of a sharp wedge (the closing shock system can well-join with the shock wave reflected from the cowl lip. On flow pattern photographs for the inlet with a blunt compression wedge, one can see the disturbed flow layer (over the wedge surface) that seems to be due to interaction of the entropy layer on the blunt wedge with the boundary layer on its surface. The reduction of the maximum attainable level of throttling of the intake with the blunt wedge is likely to be explained by the presence of such disturbed layer penetrating the intake channel. When throttling the intake, such layer seems to transfer the backpressure upstream to much larger distances than in the case of a sharp wedge.

The profile of the blunt wedge leading edge plays much more significant role (see Fig. D.6). For example, the mass-flow and total pressure recovery coefficients of an intake with a semi cylindrical leading edge with a radius $r_w = 0.025h_0$ ($d_w = 0.025h_0$) reduce by no more than 1–2% whereas the cylindrical leading edge with the same thickness $d_w = 0.025h_0$ leads to reduction as a great as 8–10%. A semi-cylindrical leading edge with the thickness $d_w = 0.05h_0$ reduces both f_{max} and ν_{max} by 10–15%, whereas a cylindrical leading edge causes intake unstart and large reduction of these coefficients of course.

It is obvious that, due to variation of stagnation point location on the cylindrical leading edge (see Fig. D.7), a thicker layer of flow disturbed by bluntness comes to the intake entry along the compression wedge surface-thicker in comparison with that on a semi-cylindrical leading edge; this layer is less resistant with respect to the positive pressure gradient in a shock wave at the entry and separates from the surface. Increasing the angle of attack moderates the negative influence of bluntness (Fig. D.6). This seems to come from the fact that, at higher angles of attack, a stagnation point shifts (Fig. D.8) and some part of a disturbed flow layer generated by the blunted structure is moved to the external surface of the compression wedge and consequently, does not penetrate inside the intake. Thus, blunting the wedge leading edge with a rather large radius ($r_{wedge} \approx 0.025\ h_0$) may lead to unstart and a notable deterioration of total intake characteristics.

Underwriting Intake

The second stage of tests was conducted on the underwing intake (see Fig. D.2) over the Mach number range M = 4–10 which corresponds to intake

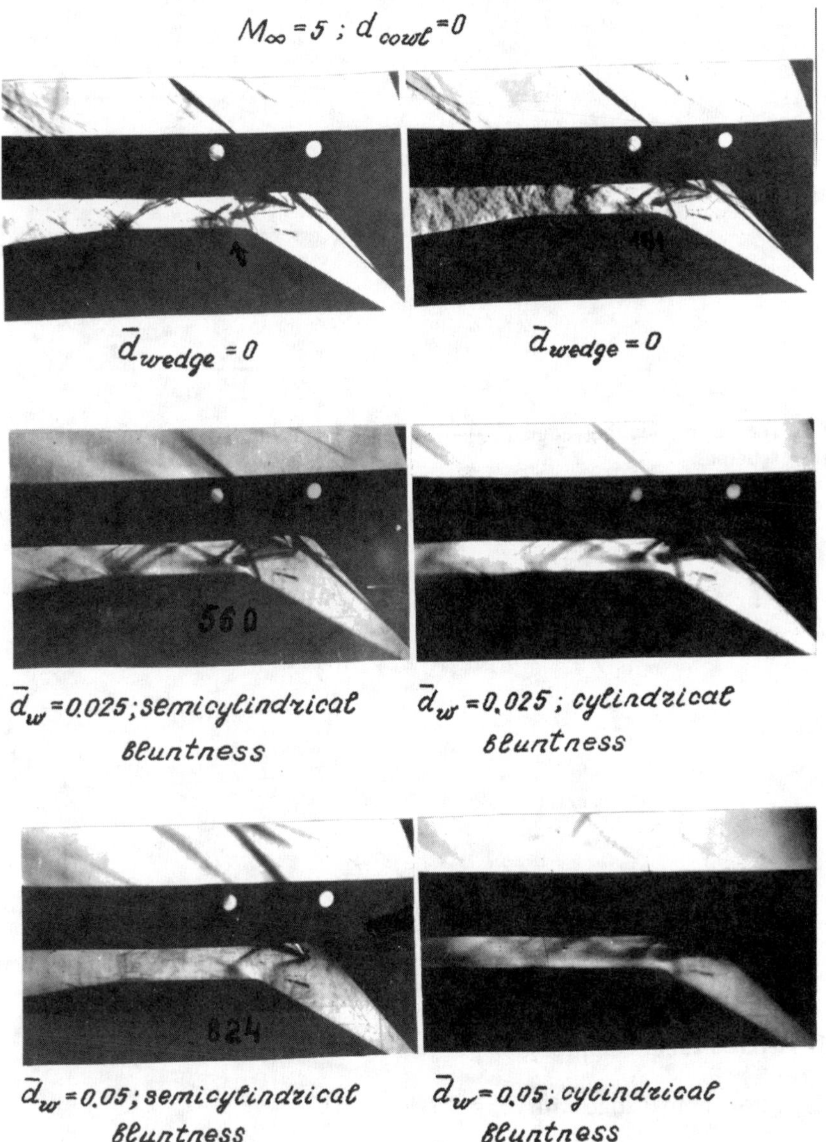

Fig. D.7 Flow pattern pictures for isolated two-dimensional intake; Mach number is 5; $\bar{d}_{wedge} = d_{wedge}/h_0$.

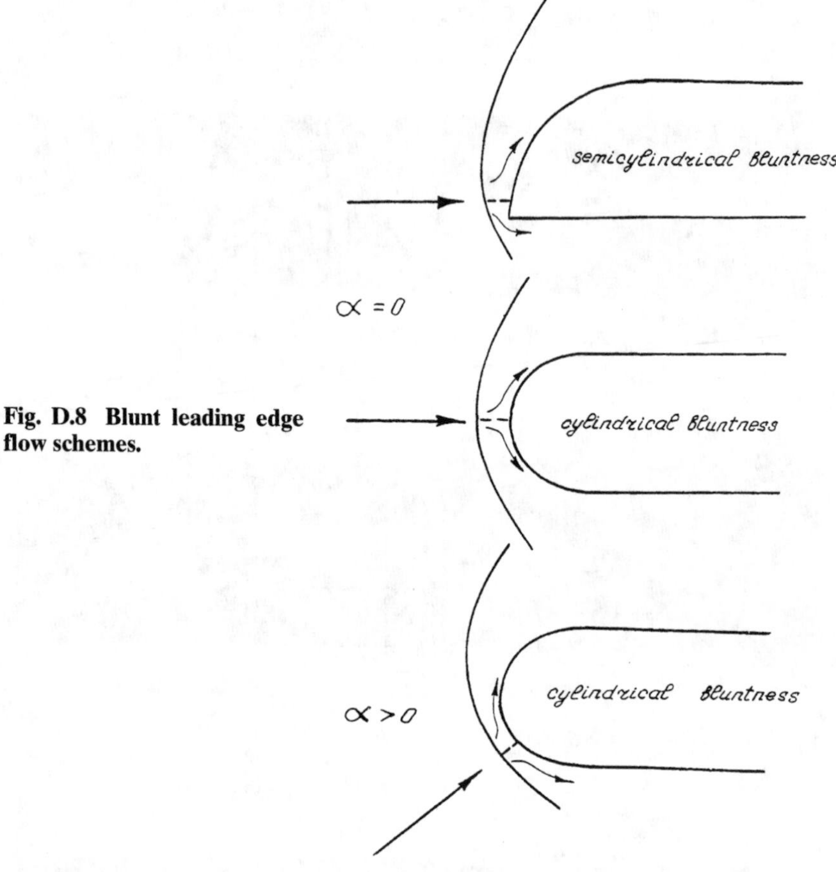

Fig. D.8 Blunt leading edge flow schemes.

operating regimes with both $M < M_d$ and $M > M_d$. Tested at $M \leq 7$ were following intakes:

1) A three-ramp compression wedge with final inlet ramp angle $\theta\Sigma = 27.5$ deg.
2) A two-ramp wedge with $\theta\Sigma = 15$ deg.

Whereas the tests at $M > 7$ involved only wedges with $\theta\Sigma = 15$ deg, the intake performances at Mach numbers $M \geq 6$ were also investigated on regimes with supersonic flow in the throat (i.e., $M_{throat} > 1$).

The tests of the underwing intake model show also (see Fig. D.9) that when the wing leading edge radius is increased, we have the following:

Both the maximum mass flow coefficient f_{max} and the total pressure recovery coefficient for subsonic flow in the intake throat ν_{max} gradually reduce; the coefficient ν_{max} reduces more significantly as the leading edge radius increases in comparison with coefficient f_{max}. This can be explained

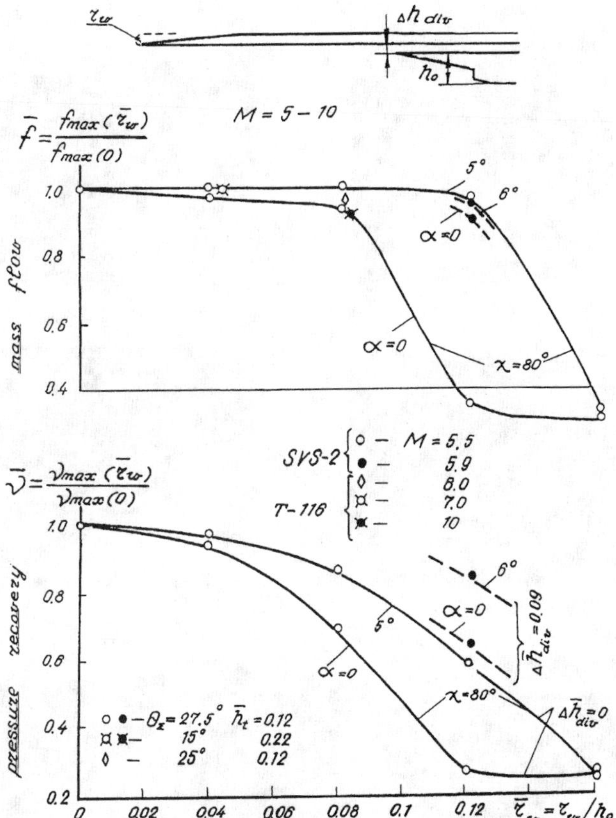

Fig. D.9 Impact of wing leading edge bluntness on total pressure recovery and mass flow coefficients for underwing two-dimensional intake; $\bar{r}_w = r_w/h_0$.

by the decrease of the intake throttling capability. In intakes with large radii the coefficients f_{max} and ν_{max} are small due to both destruction of inflow pattern and flow separation in front of the intake entry, see Fig. D.10.

The negative effect of bluntness is also elevated when the angle of attack increases. For example, at a=0 deg the coefficients f_{max} and ν_{max} notably decrease when $r_{wing} > 0.08\ h_0$, whereas at a=5 deg the coefficients decrease only when $r_{wing} > 0.12\ h_0$. The effect of angle of attack seems also to be due to the fact that at a>0° some part of the entropy layer on the blunt edge moves to the upper surface of the wing. The wing bluntness influences similarly the total pressure recovery coefficient when $M_{throat} > 1$.

Of interest is it to compare effects of a blunt delta wing (which is in fact the first ramp of the compression surface of the underwing intake with swept leading edges, χ=80 deg) and effects of a blunt first ramp of the compression

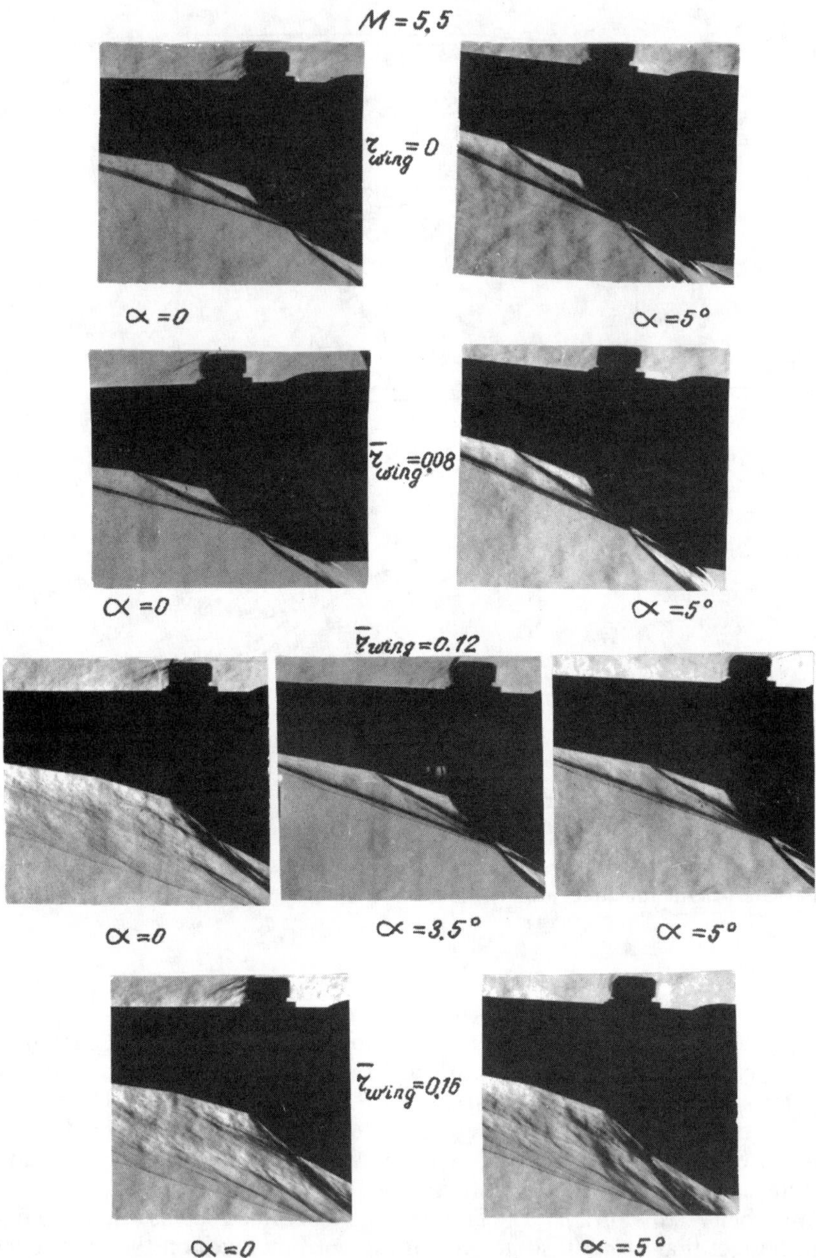

Fig. D.10 Flow pattern pictures for underwing two-dimensional intake; Mach number is 5.5: $\bar{r}_{\text{wing}} = r_{\text{wing}}/h_0$.

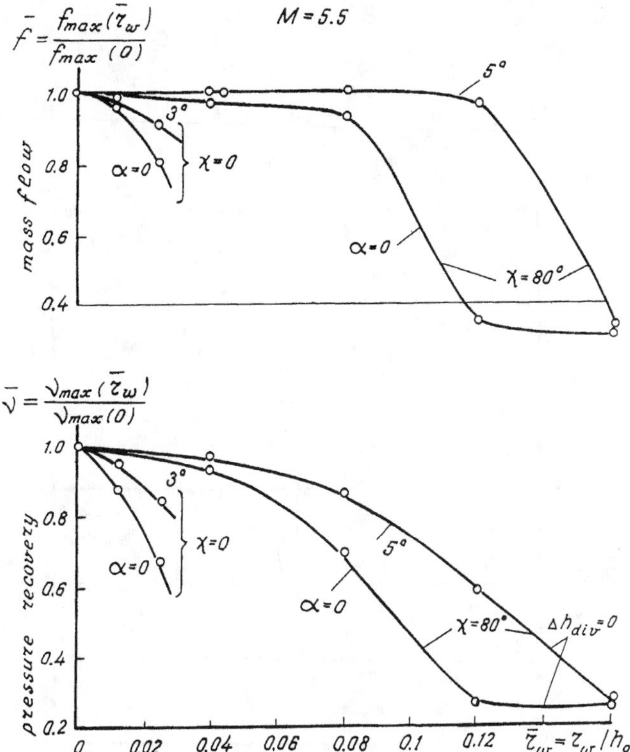

Fig. D.11 Impact of leading edge sweep angle on total pressure recovery and mass flow coefficients for underwing two-dimensional intake; $\bar{r}_w = r_w/h_0$.

wedge in the isolated two-dimensional intake with an unswept leading wedge. This comparison makes it possible to evaluate the influence of sweep on the compression surface leading edge bluntness effect. Fig. D.11 shows that the increase of the sweep angle substantially reduces the negative influence of blunt leading edge. In the present case the sweep reduces the bluntness effect by a factor of more than three for the coefficient ν_{max} and of more than five for the coefficient f_{max}. Blunting the cowl lip and side wall leading edges may by said not to worsen characteristics, see Fig. D.12).

The main question which arises during investigation of bluntness effect, is that on the admissible radius which insignificantly degrades intake internal characteristics. The test results lead to the following conclusions:

1) For the compression wedge of an isolated two-dimensional intake with an unswept leading edge, it is permissible to increase the leading edge radius to 0.01 h_0, and for a two-dimensional intake under the delta wing with the sweep angle χ=80 deg, the wing leading edge radius may be increased to

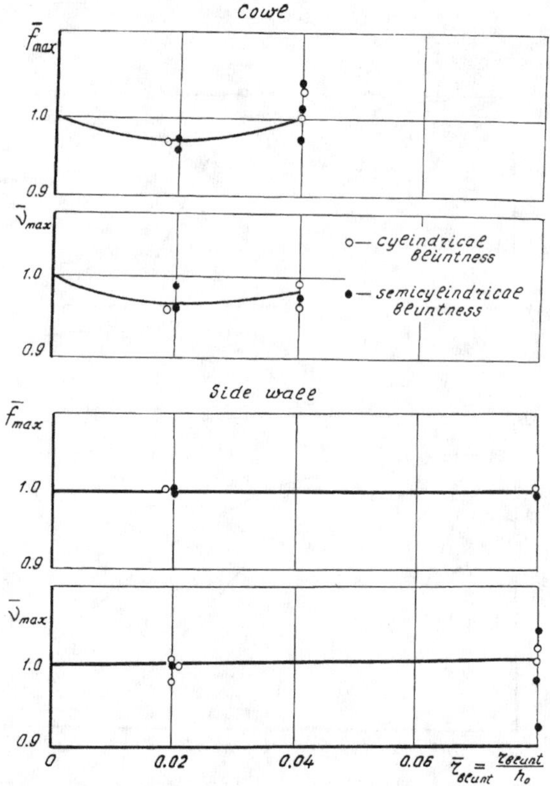

Fig. D.12 Impact of cowl and side-wall leading edge bluntness on total pressure recovery and mass flow coefficients for underwing two-dimensional intake; $\bar{r}_{blunt}$ =.

0.04–0.05 h_0 without notable negative effect on internal characteristics of the intake operating with $M < M_d$.

2) Cowl lip and side wall leading edge radii for $M \leq M_d$ may be greater.

However, it should be noted that in the final decision on the admissible radius one must also take into account the increased external drag of the intake. For example, Fig. D.13 demonstrates how the drag of an engine nacelle of an aircraft with flight M =6 increases due to cowl lip leading edge bluntness. It is likely that taking into account of the external drag increment, the admissible cowl lip radii will be substantially lesser than the ones allowed by internal characteristics, since at cowl lip radius r_{cowl} = 0.04 h_0 the drag increment due to blutness exceeds the total nacelle drag.

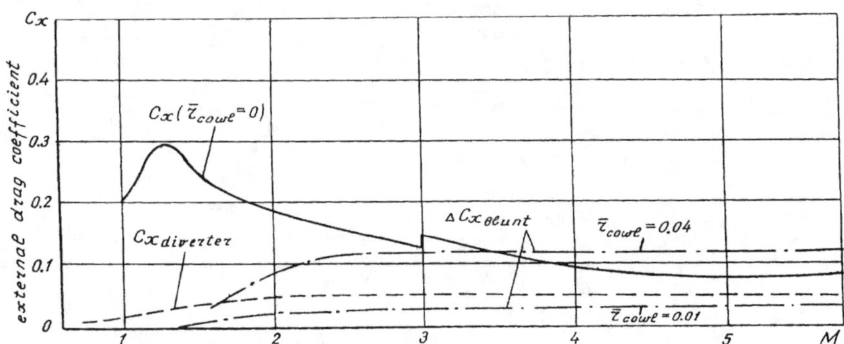

Fig. D.13 Impact of cowl leading edge bluntness on external drag of air-intake; $\bar{r}_{cowl} = r_{cowl}/h_0$.

References

[1]Edney, B. E., "Effects of Shock Impingement on the Heat Transfer Around Blunt Bodies," *AIAA Journal,* Vol. 6, No. 1, 1968, pp. 16–24.

[2]Borovoy, V. Ya., "*Gas Flow and Heat Exchange in Regions of Shock-Wave/Boundary Layer Interaction,*" Mashionostroenie, Moscow, 1983 (in Russian).

Chapter 6

Scramjet Performance

Griffin Y. Anderson,* Charles R. McClinton,† and John P. Weidner‡
NASA Langley Research Center, Hampton, Virginia

Introduction

INTEREST at NASA (then NACA) in technology relevant to hypersonic propulsion and to scramjets in particular extends back into the late 1940s and early 1950s. Various theoretical[1,2] and experimental[3,4] studies were reported, including an early comprehensive ideal gas cycle analysis to assess the potential performance of scramjets in 1958 (Ref. 5). The period from the mid-1950s to mid-1960s can be appropriately thought of as Aerospace Plane I, where a substantial investment was made, primarily by the U.S. Air Force, in system studies and technology relevant to air-breathing propulsion for orbital transportation. With the creation of NASA in 1958 and the subsequent focus of U.S. aerospace resources on expendable rocket propulsion for lunar exploration, the termination and/or limitation of most hypersonic airbreathing research inevitably resulted. The NASA Hypersonic Research Engine (HRE) Project was a notable exception that achieved the demonstration of high levels of dual-mode scramjet engine performance in ground tests at Mach 5 to 7 conditions, and also successfully demonstrated fabrication techniques, aerothermodynamic performance, and robust survival of regeneratively cooled flight-type engine structures in ground tests at Mach 7 conditions.[6] While completing the HRE Program, NASA structured a modest research effort at the Langley Research Center to address issues relevant to airframe-integrated scramjet propulsion systems.[7] The culmination of this effort was the demonstration in ground tests of fixed geometry engine performance sufficient to accelerate a candidate manned research

* Retired Head, Hypersonic Airbreathing Propulsion Branch.
† Technology Manager, Hyper-X Program Office.
‡ Retired, Senior Propulsion Engineer, Hypersonic Airbreathing Propulsion Branch.

airplane concept through the Mach 4 to 7 speed range. This successful demonstration of a ground-based technology extending into the hypersonic speed range became part of the supporting data for embarking on the National Aero-Space Plane (NASP) Program in the 1980s. The NASP Program marked the first allocation of significant resources to hypersonic air-breathing flight technology since the 1960s. NASP was narrowly focused on a single-stage-to-orbit (SSTO) concept; and thus, from an historical perspective NASP may be appropriately thought of as Aerospace Plane II. In the 1980s and 1990s the Aerospace Plane II effort has centered around the development of an airplane, the X-30, with orbital capability, which was intended to reduce the cost and increase the operational flexibility of inserting small high-value payloads into low Earth orbit on demand, thereby making space more readily accessible for a variety of missions.

In general, any hypersonic air-breathing vehicle must maintain a flight trajectory within the atmosphere that adheres to structural heating and load limits while delivering enough airflow to the propulsion system to maintain an adequate level of net thrust for acceleration. Turbine inlet temperature is the limiting factor for the turbojet engine in achieving high specific thrust (classical Brayton cycle). At flight speeds exceeding approximately Mach 3, the compressor is no longer required to achieve sufficient cycle-pressure ratio, and the ramjet becomes the propulsion cycle of choice. To achieve hypersonic speeds (Mach number greater than 5), active cooling is required, which utilizes the capacity of the fuel for heat sink in a regenerative fashion; however, the performance penalty associated with cooling large internal surface areas of the engine is not trivial, and cooling becomes a serious design consideration. As flight speed is further increased, the elevated stagnation temperature and pressure preclude use of the ramjet cycle. The thermal management task is further complicated by increasing component heat load, and cryogenic (hydrogen) fuel technology is generally considered necessary. As flight speed is still further increased, the supersonic combustion ramjet (scramjet) cycle becomes the engine cycle of choice because it does not require excessive speed reduction of the high-enthalpy mainstream airflow. Therefore, operation over a wide flight-speed range dictates an engine capable of operating as both a ramjet and as a scramjet, and this operation typically requires variable geometry.

The fundamental differences between the ramjet and scramjet cycle should be examined before proceeding to a more detailed treatment of the scramjet propulsion system operating at hypervelocity speeds. (Hypervelocity as used herein refers to combustor operation at hypersonic, Mach number greater than 5, conditions). These differences are highlighted in Fig. 1. The normal or strong shock system required in the operation of a ramjet with subsonic combustion located downstream of the inlet throat is stabilized by back pressure generated by choking the flow in the throat of the exhaust nozzle. To maximize ramjet performance, the Mach number at the beginning of the shock system must be in the range of Mach 1.4 to 1.8 to minimize total-pressure loss across the normal shock wave. This requires a relatively large inlet contraction ratio, and boundary-layer bleed may also be needed to prevent inlet un-

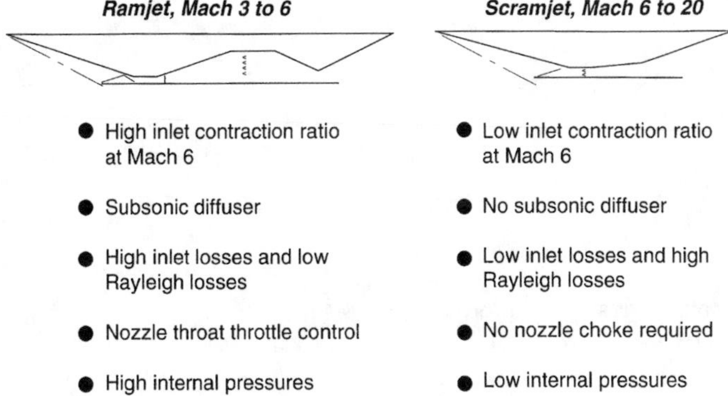

Fig. 1 Hypersonic engine cycles.

start and/or to stabilize the shock system. The net result is a larger inlet total-pressure loss compared to that generated by the scramjet cycle, which avoids the normal shock and operates at a lower inlet contraction ratio. The major advantage of the ramjet cycle is comparatively small Rayleigh losses associated with the combustion process, whereas the scramjet incurs higher Rayleigh losses associated with combustion at supersonic speeds. This trade-off between inlet pressure recovery and combustion pressure losses governs the relative fuel specific impulse of the ramjet and scramjet propulsion cycles. Fuel specific impulse, presented in Fig. 2, is one performance discriminator between the two cycles and typically indicates a benefit for operating the ramjet to Mach 6 or 7. However, the internal static pressure is another discriminator (illustrated in Fig. 3) that is directly related to installed engine weight. For

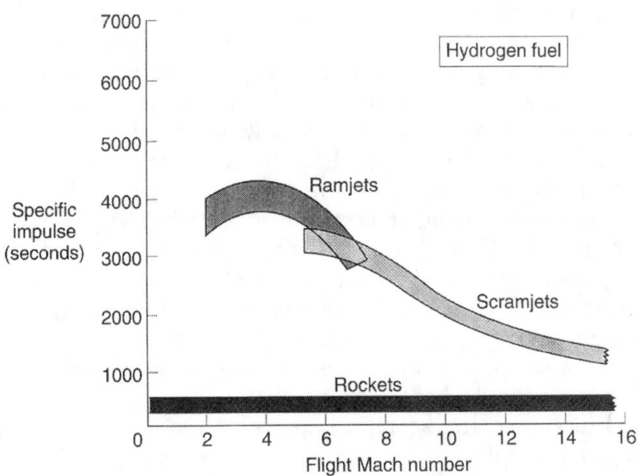

Fig. 2 Engine performance.

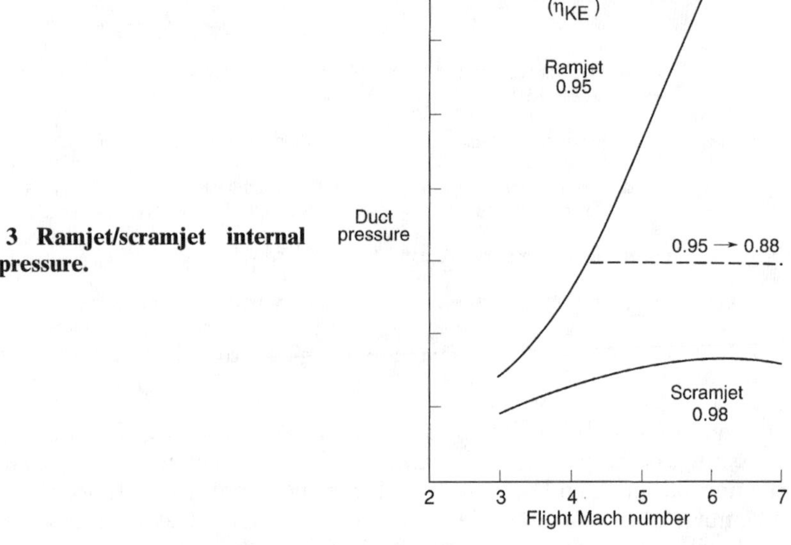

Fig. 3 Ramjet/scramjet internal duct pressure.

the data presented in Fig. 3, the ramjet duct pressure within the subsonic diffuser was calculated assuming an inlet kinetic energy efficiency (η_{KE}) of 0.95 and is seen to increase rapidly as a function of Mach number. Additionally, accelerator vehicles, such as those envisioned for space launch applications, tend to optimize at high flight-dynamic pressures typically on the order of 2000 psf or more, further dictating high values for the internal static pressure. Rather than degrading ramjet performance through reduced pressure recovery, a change in the mode of engine operation to the scramjet cycle would effectively reduce engine internal static pressure by maintaining supersonic flow throughout the engine. A dual-mode engine, operating as a ramjet at low Mach numbers and transitioning to a scramjet cycle as flight speed is increased, would satisfy these general criteria. Compromises in ramjet performance of the dual-mode engine could be potentially offset by a shorter diffuser and use of a thermal choke rather then a mechanical choke at the nozzle throat. The development and demonstration of such an engine was the objective of the NASA HRE Project started in the 1960s (Refs. 6 and 8). In ground facility tests the HRE successfully demonstrated dual-mode ramjet and scramjet cycle operation at Mach 5, 6, and 7 and, in addition, successfully demonstrated flight-type regeneratively cooled structure at Mach 7 flight conditions. More contemporary examples of dual-mode scramjet engines are described in other chapters of this volume.

As speed increases, the specific impulse of the scramjet decreases and finally approaches the performance level of a rocket (Fig. 2). At or somewhat before this point the airbreathing engine has no advantage, and the vehicle would transition to rocket propulsion to accelerate out of the atmosphere into orbit. Performance of the airbreathing vehicle is improved over a rocket-powered vehicle by extending the useful operation of the scramjet engine to as high a flight speed as possible. Extending operation of a hypersonic vehicle to hypervelocity speeds puts extreme demands on the propulsion system. This system must survive the severe thermal environment and operate at a high level of efficiency while producing enough net thrust to adequately accelerate the vehicle. At high-speed component efficiency becomes critical as a consequence of the increasing kinetic energy in the airflow processed by the engine (Fig. 4). For example, at hypervelocity (Mach 16+) the level of energy in the airflow processed by the engine becomes much larger than the energy added by burning hydrogen fuel; therefore, the net thrust (the difference between gross nozzle thrust and the ram drag) becomes only a small fraction of the stream thrust of the airflow entering the engine. Specifically, at Mach 6 the energy associated with heat release of a stoichiometric hydrogen fuel-air mixture is about twice that of the kinetic energy contained in the airflow, whereas at Mach 16 this value is only about one-fourth. At the same time additional factors contribute to increased performance sensitivity: the high inlet area contraction ratio (required to compress the air at high altitudes), high temperatures (to be contained within the engine), and high flow velocities. Consequently, performance sensitivity to component efficiency rapidly increases with increasing vehicle speed throughout the hypersonic flight regime. The following sections of this chapter examine some of the important loss mechanisms and design challenges of the hypersonic air-breathing propulsion system.

Cycle Considerations

The overall cycle performance of the scramjet propulsion system is a result of individual component performances and interactions among the various components, as shown schematically in Fig. 4. Starting at the front of the engine, the forebody/inlet configuration must efficiently compress the air to a suitable pressure level to allow for both ignition and robust combus-

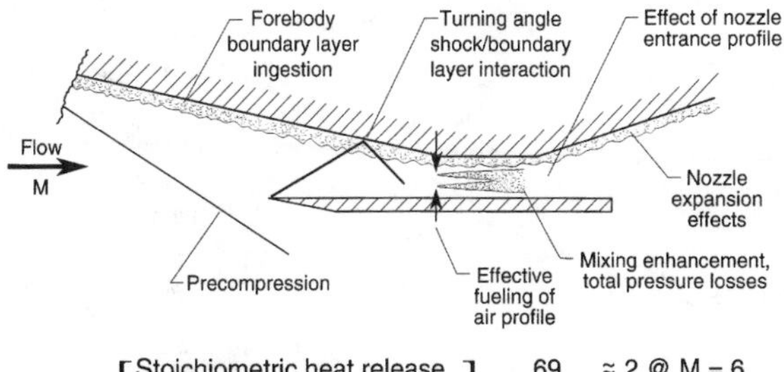

$$\left[\frac{\text{Stoichiometric heat release}}{\text{Kinetic energy of airstream}}\right] \approx \frac{69}{M^2}; \quad \approx 2 \text{ @ } M = 6; \quad \approx 1/4 \text{ @ } M = 16$$

- Scramjet performance is most sensitive to component efficiencies at high Mach numbers
- High energy level of airstream required for combustion research

Fig. 4 Hypervelocity scramjet design challenge.

tion of the fuel. Additionally, the inlet influences cycle performance in several ways: the degree of flow uniformity delivered across the width of the vehicle, the level of compression accomplished via the external and internal portion of the inlet, combustor entrance flowfield distortion caused by shocks and expansions, and total-pressure losses resulting from shock waves, wall friction, and shock/viscous interactions. An additional impact of inlet efficiency on cycle performance at high flight Mach numbers is the associated static temperature rise that accompanies inefficiency in the inlet compression process. This effect can limit scramjet performance because higher static temperature can result in excessive chemical dissociation, which significantly reduces the combustion heat release. As illustrated in Fig. 5, achieving a given static-pressure level utilizing an inlet with high efficiency where compression is comprised mainly of isentropic turning and weak shock waves, results in a much lower static temperature than that generated by a less efficient inlet employing stronger shock waves. By entering the combustor at a lower static temperature (for a given static pressure), a correspondingly lower combustion temperature results; and less chemical dissociation is created. Ideally, an equilibrium nozzle expansion process might allow the dissociated species to recombine and recover the dissociation energy. However, optimized hypersonic vehicle designs generally require rapid area expansion in the initial nozzle contour, which generates a correspondingly rapid decrease in static pressure that results in a frozen chemical process. Thus, the dissociation energy is lost from the propulsion cycle, and, consequently, the net thrust is reduced.

Having achieved the desired fluid dynamic and thermodynamic state in the forebody/inlet compression process, the inlet airstream is mixed with fuel and burned to yield heat release. Achieving effective fuel distribution at high flight speeds in reasonable combustor length while matching the local

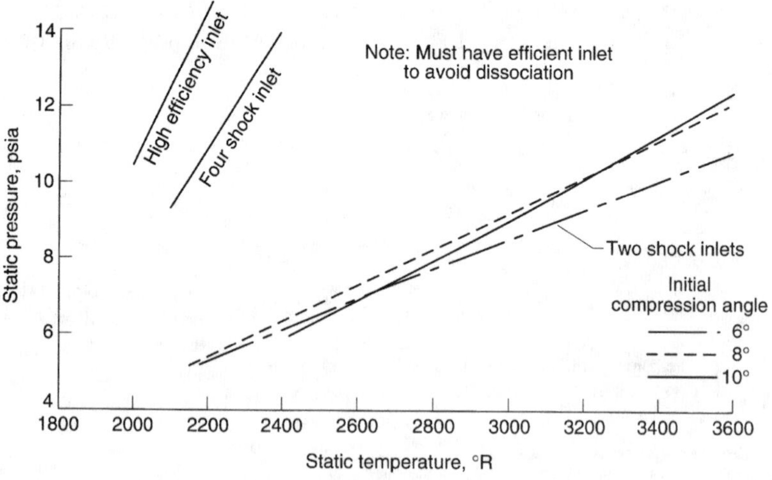

Fig. 5 Static conditions at inlet throat at a flight $M = 14$.

inlet airflow profile is exceedingly difficult with fuel injection only from the walls of the combustor. Thus, consideration of intrusive fuel injectors (struts and the like) is often required in spite of serious design concerns that they introduce, such as drag and cooling requirements. Hence, an obvious design trade exists that requires a balance between intrusive injector losses and reduced combustor length. Typically, the fuel injector/combustor design goal for an accelerator engine is the consumption of all of the oxygen in the airstream within a reasonable length (i.e., reasonable weight). Also, balancing the overall engine-cooling requirement with the available fuel-cooling capacity is another engine design trade that is influenced by many choices in the engine cycle selection. Classical limiting processes, such as constant-area or constant-pressure heat addition, are often used for simplicity to represent the combustor analytically. The actual three-dimensional process is much more complex; but, as pointed out by Ferri,[9] the overall cycle process is probably best considered in terms of the peak pressure generated in the engine, rather than the pressure generated only by the inlet. Thus, a diverging combustor with increasing area as a function of axial distance may be appropriate to control peak pressure and heat transfer as the fuel mixes and burns, even at some penalty in reduced thrust performance.[7] Again, a complex trade between engine design and operating parameters is required to achieve high levels of overall system performance.

For a typical hypersonic scramjet the optimum exhaust area, inclusive of both the internal nozzle and the vehicle afterbody, is approximately 1.5 to 2.0 times the freestream engine capture area. Because the exhaust flow is always highly underexpanded due to the losses incurred during the compression and combustion processes, an optimization of the nozzle/afterbody surfaces, inclusive of the truncation of the nozzle contours, is required to simultaneously achieve a high net thrust value, proper vehicle trim characteristics, and low system weight. Again, there are numerous trades required to optimize overall system performance, particularly among the amount and rate of nozzle expansion and the length of the vehicle afterbody. In summary, throughout the entire flight regime nozzle/afterbody performance effects are very important; and a comprehensive study addressing these issues is one of the most complex and challenging design considerations involved in achieving high overall system performance. The following sections address each component process of the scramjet engine and the associated performance issues in more detail.

Flow Nonuniformity and Cycle Performance

Interpreting multidimensional fluid-dynamic phenomena within a one-dimensional context is frequently required to assess integrated engine-vehicle performance and is particularly acute for off-axis engine design and analysis. Yet, at present, no singular design methodology is widely applied by the industrial or scientific community. Historically, integral boundary-layer techniques[10] have provided much of the focus for this analytic reduction process, with the major emphasis directed toward the characterization of the mass, momentum, and energy defects within the distorted region of the flow. Other specialized fluid-dynamic implementations have been explored, with particularly notable successes in the design of diffusers[11] and ramjet isolators.[12,13] Additionally, the concept of applying integral-distortion techniques to engine cycle performance analysis is not novel[14] and is required to address the inherent limitations of purely one-dimensional cycle performance analysis techniques.[15]

A simple frictionless adiabatic constant-area flow that transitions from supersonic to subsonic Mach number serves to illustrate this point. Given that at the inflow plane, the following are uniquely specified: total enthalpy flux, mass-flow rate, cross-section area, and stream thrust, then only two one-dimensional fluid states are compatible with the specific inflow conditions, i.e., the subsonic and supersonic solutions associated with the normal shock states. Hence, a purely one-dimensional methodology is only relevant to the two end states of this simplified ramjet engine isolator process, and therefore excludes all other pressure rises associated with the separated nonuniform fluid flowfields typically observed in experiments.[16] Commonly, the one-dimensional engineering approach to performance analysis is to match the integrated values of the mass, momentum, and energy fluxes within the control volume while concurrently employing an equation of state consistent with an equilibrium thermodynamic chemistry assumption. Unfortunately this approach, as does each purely one-dimensional methodology, creates an unknown bias, because the flowfield nonuniformity and angularity are absorbed within the definition of the associated pseudothermodynamic states. Individual component thrust and drag accounting can be tautologically satisfied; however, no unique cycle or process efficiency can be deduced because the one-dimensional methodology characterizes numerous multidimensional flowfields. Historically, this entire matter has not been aggressively pursued in any Mach-number regime because the presumption of minimal performance effects; however, with the maturing of hypersonic engine design technology, flow nonuniformity effects, inclusive of angularity effects, must be reassessed and, hence, incorporated into cycle performance analysis in a more complete and comprehensive manner.

To help illustrate the usefulness of a distortion-based methodology, a time-independent spatially one-dimensional integral formulation of the fluid-equation set is presented in flux-conserving form[17] with arbitary source terms:

$$\frac{\partial}{\partial x}\left\{\begin{matrix} \rho u A \eta_A \eta^*{}_A \\ PA\eta_P^* + \rho u^2 A\, \eta_F \eta_F^* \\ \rho u A \eta_A \eta_A^* \, \overline{H}_T \end{matrix}\right\} = \left\{\begin{matrix} \dot{S}_1 \\ \dot{S}_2 \\ \dot{S}_3 \end{matrix}\right\}$$

where P is the static pressure, ρ is the density, u is the axial velocity, A is the geometric cross-sectional area, $\bar{H}_T$ is the bulk total enthalpy, $\dot{S}_1$ is the mass source terms, $\dot{S}_2$ is the momentum source terms, $\dot{S}_3$ is the energy source term, $\eta_A \eta_A^*$ are the area distortion parameters (core and boundary layer) $= 1 - \delta/A$ where δ equals the area defect of the flowfield, $\eta_F \eta_F^*$ are the momentum distortion parameters (core and boundary layer) $= 1 - \delta/A - \theta/A$ where θ

equals the momentum defect of the flowfield, η_p^* is the pressure distortion parameter $\int PdA/PA$, and x is the spatial coordinate.

The spatial forms quantify the mass, stream thrust, and total enthalpy fluxes defined by the corresponding control surface. This formulation utilizes numerous distortion parameters to quantify the spatial nonuniformity, and each is analogous to the defect parameters utilized by integral boundary-layer methodologies. The justification for this added complexity is demonstrated via the application of this equation set to the prior example of the simplified ramjet isolator problem, i.e., the frictionless adiabatic constant-area flowfield transitioning from supersonic to subsonic Mach number. Modeling the separated zone-pressure variation vs length as a function of local dynamic pressure,[12], and assuming that the mass and momentum defects are identical, yields a subset of equations that have a closed-form solution[18]:

$$\xi = \gamma \frac{C_a^2}{C_b}\left[\frac{\tilde{P}-1}{(C_b-\tilde{P})(C_b-1)} - \frac{1}{C_b}\,\ell n\left|\frac{C_b-\tilde{P}}{(C_b-1)\tilde{P}}\right|\right] - \frac{\gamma-1}{2\gamma}\,\ell n\left|\tilde{P}\right|$$

where γ is the ratio of specific heats, R is the universal gas constant, T_t is the total temperature, F is the stream thrust, $\dot{m}$ is the mass-flow rate, D_H is the hydraulic diameter, $C_{f_{\text{entrance}}}$ is the coefficient of friction at the entrance, P_{entrance} is the static pressure at the entrance, $\tilde{P} = P/P_{\text{entrance}}$, $\xi = (44.5\, C_{f_{\text{entrance}}}/D_H)\, x$, $C_a = (\dot{m}/A)\, 1/P_{\text{entrance}}\, RT_t/\gamma$, and $C_b = (F/A)\, 1/P_{\text{entrance}}$.

C_a and C_b characterize the flowfield total enthalpy and stream thrust, and ξ is proportional to the axial position associated with the corresponding pressure ratio $\tilde{P}$. This resulting distortion-based solution is in good agreement with the existing ramjet isolator data for Mach number 2.23, summarized in Ref. 12, detailing a pressure-ratio factor of 3.8 achieved in 7.80 hydraulic diameters vs a prediction of 7.77 derived from this distortion methodology. Hence, not only is the distortion methodology consistent with the limiting value, i.e., the experimentally obtained normal shock recovery value, but also it is consistent with the inherent spatial scales of the problem. This enhanced ability to assess and predict flowfield evolution is required to realistically assess hypersonic scramjet cycle performance. It is recommended that a more complete cycle methodology, including distortion parameters of the form defined in the prior text, be adopted by the community at large.

Inlet

Sidewall Compression Concepts

Inlet research at NASA Langley Research Center (LaRC) over the last three decades has emphasized three-dimensional designs that have the potential to optimize both aerodynamic and structural performance. The airframe-integrated propulsion system may be made up of a group of rectangular propulsion modules attached to the bottom surface of the vehicle, as illustrated in Fig. 6. The forebody serves as a portion of the external inlet, and the afterbody completes the nozzle expansion process. The inlet illustrated in Fig. 6 is one of a class of three-dimensional inlets that utilize lateral compression from the inlet sidewalls to complete the vertical compression initiated by the vehicle forebody. The bottom surface (cowl) leading edge is typically located aft near the inlet throat to allow air to spill at low speeds, thus making possible designs with fixed geometry that can operate over a wide speed range (see Ref. 19).

Performance characteristics of this inlet class change with the sweep angle of the sidewall compression surfaces (Fig. 7). This latter feature is particularly desirable from a structural and heating perspective. When flow is compressed by a swept sidewall, a flow component is generated in the direction of the sweep, which turns the flow down toward the cowl in the case of the aft-swept leading edge. Flow is allowed to spill ahead of the cowl, making the inlet easier to start, that is, to pass the normal shock and establish supersonic flow inside the inlet. The downward component of flow is reduced at higher speeds, which reduces spillage and results in a significant increase in captured airflow.[19] However, the downward turning of the flow and spillage caused by sweep also cause a shock wave from the cowl leading edge inside the inlet, and this shock can increase distortion ahead of the combustor. Reversing the sweep of the sidewall leading edge causes flow to be turned upward ahead of the cowl and reduces spillage and flow distortion. Inherently this design makes the inlet harder to start, and consequently variable geometry may be required. Starting characteristics of this class of inlet have been studied,[20] and some results are shown for inlets with aft sweep in Fig. 8. Note that starting characteristics are a function of the aspect ratio of the inlet and that inlets having a low aspect ratio can be started with

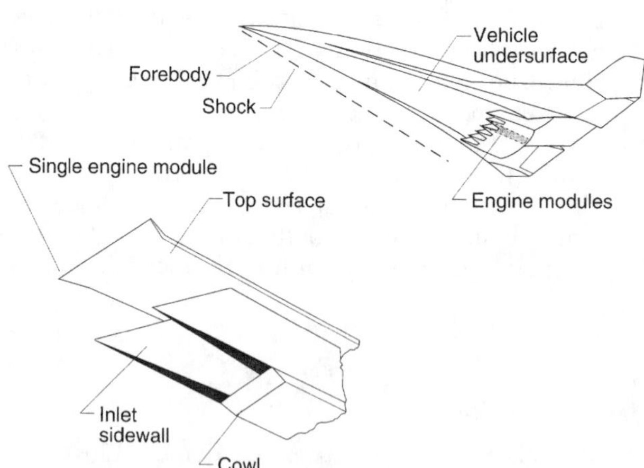

Fig. 6 Airframe-propulsion system integration.

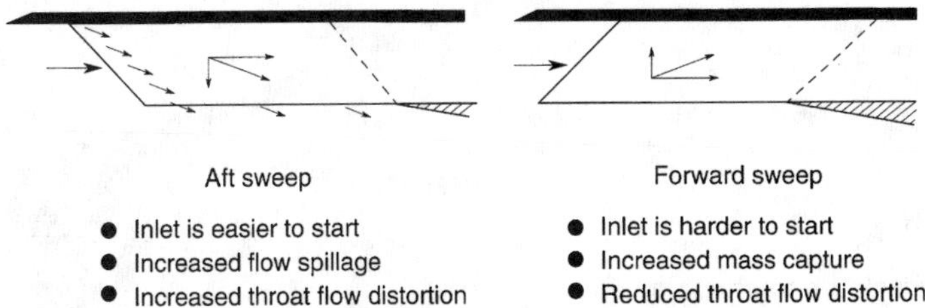

Fig. 7 Aft-sweep vs forward-sweep inlets.

a fairly high contraction ratio. Some experience has also been gained with inlets having a forward sweep,[21] and these inlets have proven to be much harder to star without variable geometry. Recent work has examined other sidewall shapes that would retain the desired starting characteristics of inlets with aft sweep while improving mass capture, pressure recovery, and throat flow distortion.[22]

The interaction of shock waves with the boundary layers formed within the inlet and ingested from the vehicle forebody are a first-order concern in the design of the hypersonic inlet. A particularly troublesome shock wave/viscous interaction exists in a sidewall compression inlet that is similar to the interaction of a shock wave generated by a fin on a surface with a thick boundary layer, and this interaction has been studied by several researchers.[23,24] In this interaction (Fig. 9) the incoming boundary layer separates for all but the weakest interactions and rolls up into a strong vortex near the base of the compression surface. A number of studies have also

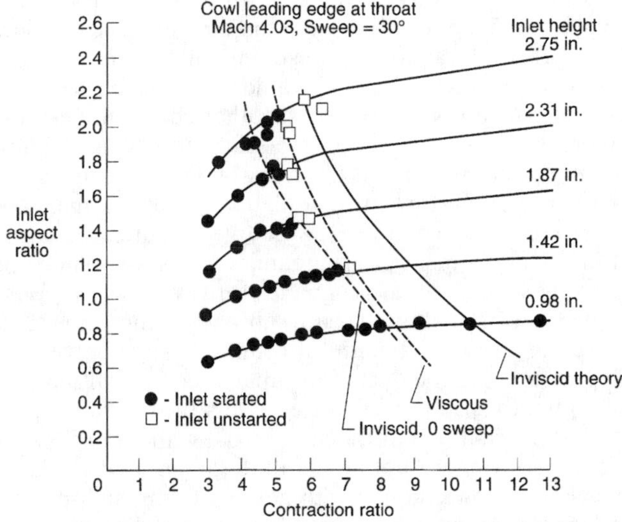

Fig. 8 Sidewall compression inlet starting predictions compared with data.

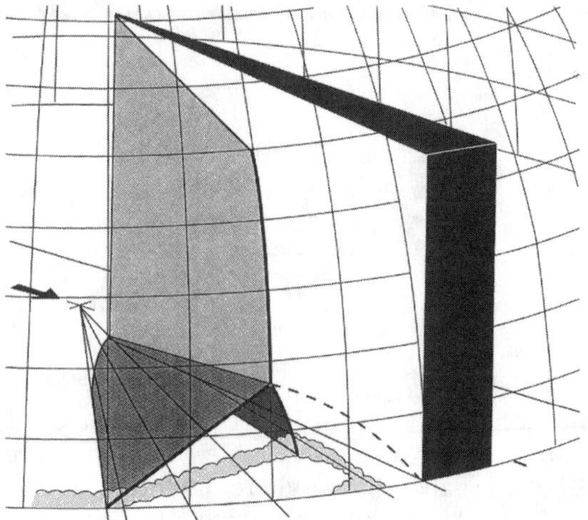

Fig. 9 Single fin interaction flowfield projected onto a conical cross plane.

been conducted to examine the crossing shock-wave interaction formed by two fins, which is again similar to the compression system generated by a sidewall compression inlet.[25–27] These investigations find a strong vortex-vortex interaction occurring along the flowfield centerline next to the body surface, which creates a core of low total-pressure air. Computational studies of this flowfield[28] illustrate this low energy core (Fig. 10), which penetrates deep into the inviscid flow. This figure illustrates the viscous nature of the flow and suggests that a full three-dimensional Navier–Stokes solution is required to capture the details of the interaction. Recent research has examined the consequences of this complex flow on inlet performance, as illustrated by computational efforts[22] to examine the influence of sidewall shape on the vortex-vortex interaction and resulting inlet performance. The problems associated with this interaction would be expected to be reduced at higher hypersonic speeds, as is typical of other shock-viscous interactions. However, there has been little work to study this type of interaction at hypersonic speeds (an exception is the work by Frank Lu in Ref. 29 and 30), so that its influence on performance at higher speeds is largely unknown.

A significant consequence of the low-energy air core at low speeds is the resultant effect on performance during ramjet operation. When operating as a ramjet, a strong shock system progresses forward toward the inlet throat as a consequence of the downstream subsonic combustion process. The low-energy core of air acts as a path for communication between the downstream high pressure and the upstream supersonic flow within the inlet and may cause the inlet to unstart prematurely. Thus, the maximum pressure that can be sustained downstream of the inlet throat without unstarting the inlet is a measure of performance potential in the ramjet mode. This problem of a supersonic viscous flow coupled with a downstream high-pressure region is difficult to study computationally but has been studied experimentally in Ref. 31. Back pressure was applied to the inlet with a throttling device to

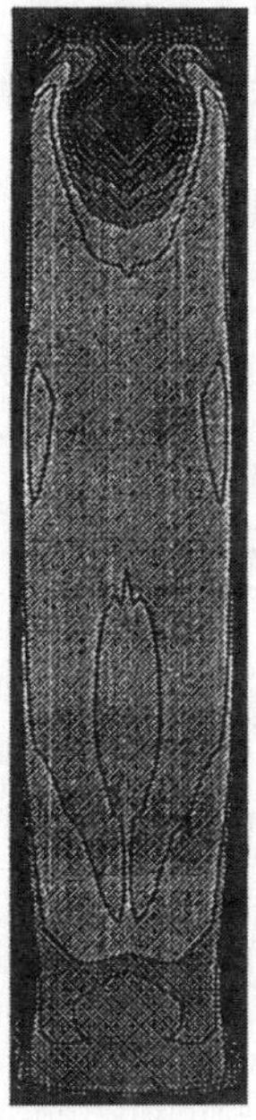

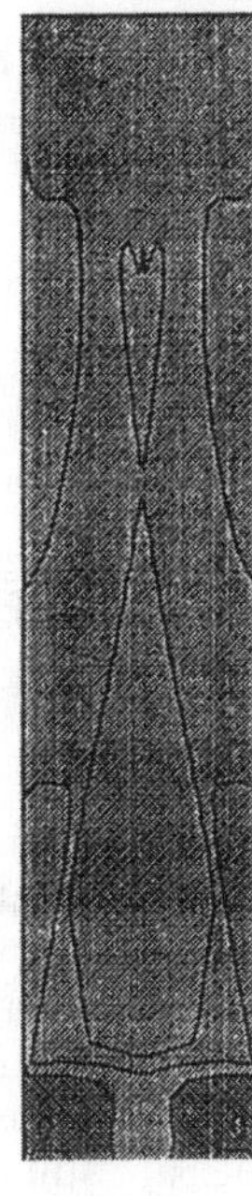

Fig. 10 Mach-number profiles at inlet throat from CFD.

simulate a subsonic combustion process. The back pressure was increased during the test until the inlet was forced to unstart. The objective of the study was to alter the body-side geometry to reduce the effect of the vortex interaction region and, thereby, allow a higher back pressure to be obtained before unstart. Some results that indicate a benefit of adding a compression ramp to the body-side surface and an increase in the level of maximum back pressure achieved with increasing ramp angle are presented in Fig. 11. These results are encouraging, but more work is needed to understand the complex flow mechanism and to determine how to optimize performance of this type of inlet.

Interactive Inlet Design

In general, computationally based interactive and/or automated design algorithms that are relevant to ramjet and scramjet inlet configurations and address realistic three-dimensional viscous gas dynamics are prohibitively costly to implement because of the large computational resource requirements. By necessity this limitation, long recognized by the design community, led to numerous uniquely ingenious inlet design solutions and insights, a prominent early example being the nearly equal strength multiple-shock solution of Oswatitsch.[32] Yet, given the historical trend of increasingly more powerful computational hardware, numerous researchers in associated and parallel disciplines continue to address the inherent technical limiting factors

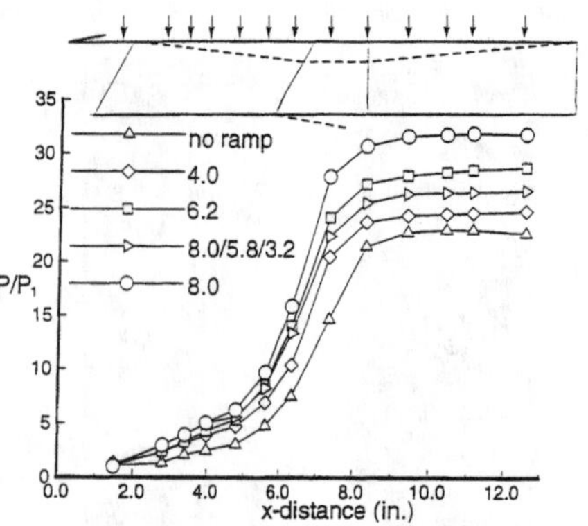

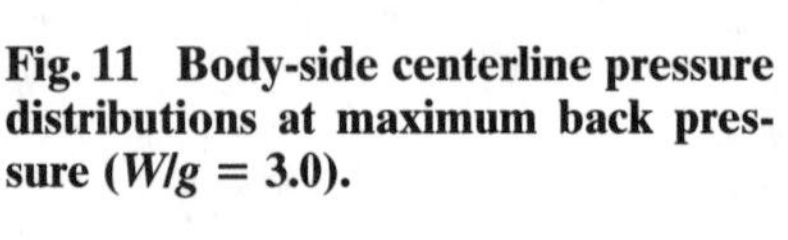

Fig. 11 Body-side centerline pressure distributions at maximum back pressure ($W/g = 3.0$).

of interactive analytic design, with notable progress being achieved in nonlinear optimization algorithms[33,34] and wall-function methodologies.[35,36] The former addresses the generation of an optimal solution without invoking the typical iterative methodologies of constrained optimization,[37] whereas the latter addresses circumventing the inherent stability and accuracy restrictions of numerically generated solutions to the equations governing viscous fluid phenomena. Albeit that these techniques are still theoretically incomplete, each contributes to the ultimate goal of producing a useful interactive inlet design tool. Also to this end, Korte et al.[22,38] utilized a parabolized Navier–Stokes fluid-dynamic spatial marching algorithm to examine design issues relevant to laminar two-dimensional flight-type scramjet inlet configurations, as well as turbulent three-dimensional scramjet inlet geometry, and concluded that optimal inlet designs must invoke contoured nonplanar wall shapes.

In summary, clearly the relevant constraints and the exact performance criteria to be optimized are design specific and, hence, complicate the generalization of any encompassing procedure. Yet, the need to ultimately enhance the relevant technology remains, and, therefore, future research efforts to address the definition of optimized three-dimensional forebody/inlet compression systems are appropriate.

Inlet/Isolator Interactions

The inlet is coupled aerodynamically to the combustor through both the viscous flow that originates within the inlet and shock and expansion waves that continue into the combustor. Interactions between fuel injectors located in the combustor with the flow structure from the inlet may result in flow separation and affect inlet/combustor performance. At lower speeds, when the engine is operating in a subsonic combustion ramjet mode, performance is most sensitive to the aerodynamic coupling between these two components as a result of the large pressure rise from combustion in the downstream portion of the combustor that forces the shock train in the isolator to reduce flow velocity to subsonic speeds. The level of the pressure rise that can be sustained in the combustor without causing inlet unstart is indicative of the performance level that can be achieved by the ramjet engine. Before moving to the combustor, we will examine performance characteristics of a constant-area isolator section intended to aerodynamically isolate the supersonic inlet from the downstream subsonic combustion. Ideally, the downstream pressure rise would approach that which would exist behind a normal shock wave at the inlet throat. However, viscous interactions within the isolator, which are influenced by distortion from the inlet, will inhibit the ability to sustain a high downstream pressure level and make the pressure rise achievable a function of the inlet and isolator design. Most of the previous isolator work has been based on tests conducted in a constant-area duct with a uniform inflow profile. The flowfield within the constant-area duct represents conditions at the inlet throat, and the length of duct that is affected by the pressure rise is a measure of the length of isolator required to support a given pressure rise. Although this is an accurate method to develop isolator performance maps, the method assumes that the flow from the inlet is uniform or that inlet flow uniformity does not substantially affect isolator performance.

Tests have been conducted at Langley Research Center in a Mach 4 blow-down tunnel to address inlet/isolator performance when representative inlet flowfields are included.[38a] The model used for these tests is described in Fig. 12 and consists of an inlet, isolator, diverging section, and downstream choke mechanism. The inlet has a foreplate of variable length to vary the thickness of the boundary layer entering the inlet and a rotating cowl to start the inlet and to change the inlet contraction ratio. Two cowl lengths were provided to change the flowfield within the inlet and the resulting flow profile entering the isolator. Note that the long cowl produces

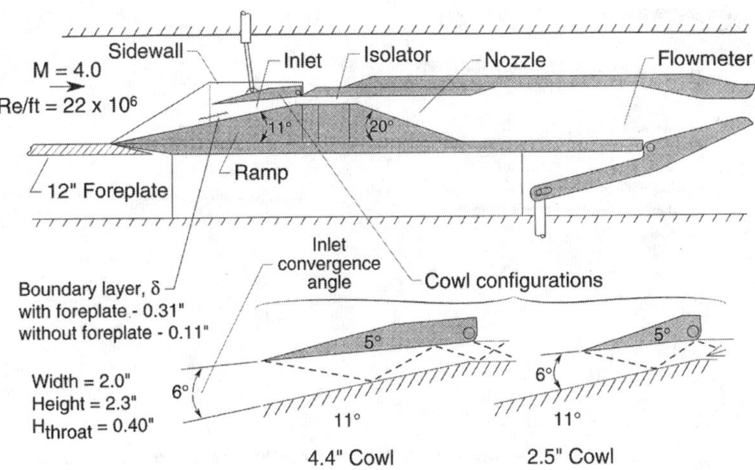

Fig. 12 Two-dimensional inlet/isolator parametric model.

a shock system that nearly cancels on the body-side shoulder to minimize distortion while the short cowl produces a shock system that reflects from the cowl side opposite the shoulder to produce a much higher level of distortion. The test procedure was to start the tunnel and inlet with the cowl deflected down. Once the inlet is started, the inlet cowl is rotated to a predetermined position, and the flow choking device is slowly closed until the inlet unstarts. Data channels are scanned continuously during this process, which enables static pressure to be plotted through the model at successive times, as illustrated in Fig. 13. Figure 13 slows the large pressure fluctuations (denoted by the three indicated pressure distributions) within the isolator before inlet unstart occurs. This fluctuation is presumably because of distortion entering the inlet throat and the increase in downstream pressure up to the point of inlet unstart.

Tests were conducted over a range of inlet-cowl rotation angles for the two inlet cowls illustrated in Fig. 12 with and without the forward boundary-layer generating plate. Maximum back pressure achieved before inlet unstart is summarized in Fig. 14 for an isolator length that is 4.7 times the inlet throat height. Maximum back pressure is plotted against inlet convergence angle, which is the difference between the body ramp angle and the cowl angle. Back pressure generally increases with convergence angle and is much higher for the longer cowl. Several interesting points can be observed from this figure. First, a higher inlet convergence angle is achievable with a thinner incoming boundary layer, and the maximum convergence achieved is independent of cowl length. This latter observation suggests that inlet unstart caused by increasing convergence angle is not from overcompressing the flow, but is caused by the strength of the shock/boundary-layer interaction ahead of the throat because the compression level between the two cases is considerably different. It is not surprising that the long cowl produces higher pressure because back pres-

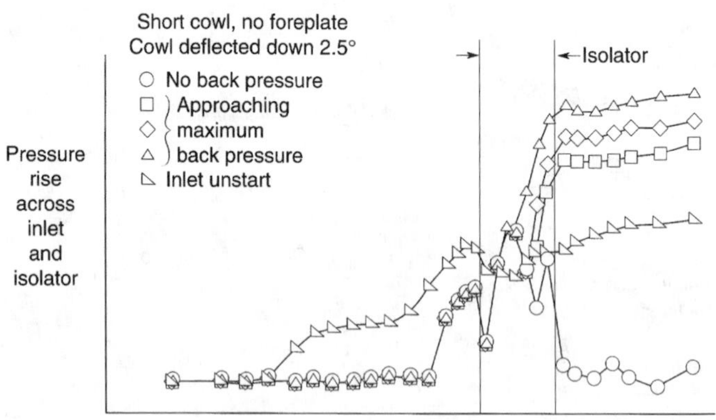

Fig. 13 Isolator back-pressure characteristics.

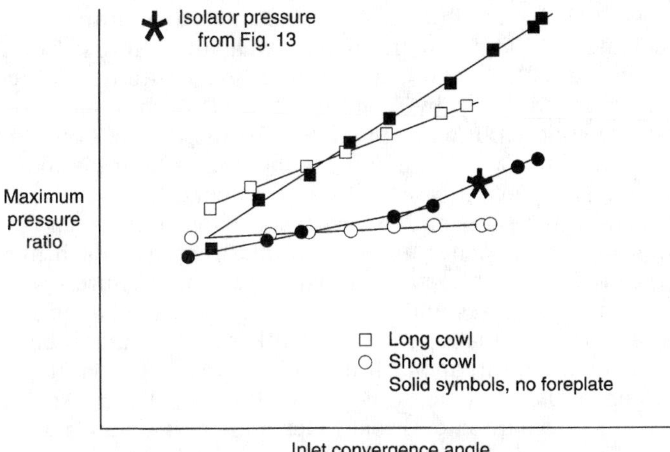

Fig. 14 Maximum pressure ratio achieved across inlet and isolator.

sure is a product of the pressure rise within the isolator and the inlet pressure ratio. Because of its larger area contraction ratio, the pressure ratio generated by the longer cowl is greater than that generated by the short cowl. Also note that all of the data converge at low inlet convergence angles where there is only a small contraction ratio from the cowl lip to the inlet throat.

Data from Fig. 14 are given in a different form in Fig. 15 where only the isolator pressure rise is plotted against inlet contraction ratio to separate isolator performance from inlet performance. The Mach 4 flow is com-

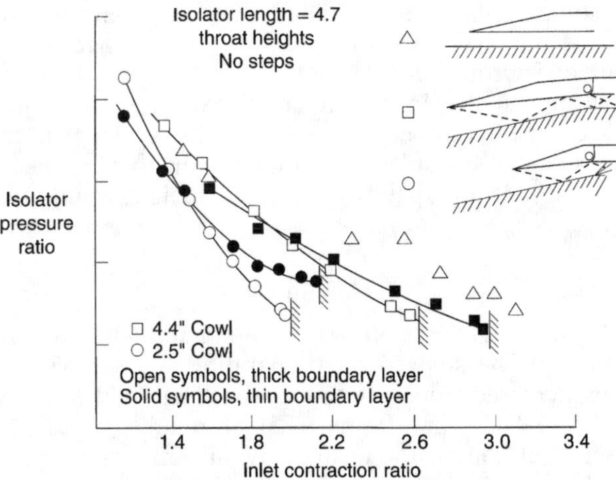

Fig. 15 Isolator maximum pressure rise.

pressed by the 11-deg forebody to Mach 3.2 and is then further compressed by the inlet internal contraction from the cowl lip to the throat. Comparisons are made with data from the constant-area isolator study reported in another chapter in this book by relating the local Mach number within the isolator to the contraction ratio required to compress the flow from Mach 3.2 to that Mach number which exists within the isolator but ahead of the pressure rise. Thus, for all cases with a given contraction ratio, the Mach number is about the same and is increased to approach Mach 3.2 as inlet contraction ratio approaches unity. Note that as throat Mach number increases with reduced contraction ratio, the isolator pressure rise from the cases considered increases and tends to converge. Thus, when the inlet is operating at a low contraction ratio, very little pressure ratio is being generated by the inlet, the throat Mach number is fairly high, and isolator performance approaches that measured in a constant-area duct. At higher inlet contraction ratios where inlet throat Mach number is significantly reduced, the effect of inlet configuration and resulting distortion entering the isolator becomes obvious, particularly when a thick body-side boundary layer that is consistent with a flight vehicle is considered. At this incoming Mach number with maximum inlet contraction ratio, the isolator behind the short cowl, which produces a shock pattern that causes a very high distortion level, yields only about half the pressure rise found for an isolator operating on the flow from a constant-area duct without distortion.

To summarize, boundary-layer thickness, inlet contraction ratio, throat Mach number, and the nature of the compressive shock pattern within the inlet all contribute to a level of distortion ahead of an isolator that has been shown to have a direct impact on isolator performance. This discussion of inlet/combustor isolators has emphasized the low Mach-number portion of flight regime where subsonic combustion is expected, a large pressure rise is associated with the subsonic diffusion and combustion process, and the isolator section is necessary to contain that high pressure. At higher speeds where supersonic combustion is initiated, some isolation from the inlet may still be required as a result of a close proximity of fuel injectors and resulting local regions of subsonic flow near the inlet throat. At high hypersonic speeds a high supersonic Mach number will exist at the inlet throat so that an isolator section will not be required. Thus, the challenge in the design of a dual-mode scramjet capable of operation over a wide speed range is to provide an isolator at low speeds without penalizing scramjet performance at high hypersonic speeds.

Combustor

The next section highlights efforts to extend and improve scramjet combustor technology. The demonstrated operating range of scramjet combustors has been increased from a status approaching flight Mach number less than 10 in the 1975–85 time frame to Mach number 15 to 20 today. The quality of scramjet combustor technology, in terms of the level of detail addressed and the accuracy of predictions, has also been dramatically improved during this time frame with the development and application of laser diagnostic measurement techniques and three-dimensional reacting flow computational fluid dynamics (CFD). This improvement was driven by the small thrust margin at hypervelocity (see Fig. 2). Methods have been implemented to quantify the effect of losses on overall vehicle performance. Today, optimized combustor designs are based, as a minimum, on engine thrust potential. Integrated design tools are now available that are capable of tracking all combustor losses and their impact on vehicle performance over the vehicle operating range. Combustor designers can account for losses from fuel mixing, nonequilibrium chemical kinetics, injector drag, combustor flow distortion, combustor wall shear and heat flux, overall thermal balance, etc. The large gain in scramjet combustor technology obtained over the past decade has resulted from synergistic application of experimental, analytical, and numerical methods.

The status of supersonic combustor technology in the 1985 time frame is summarized by Northam and Anderson.[39] Up to that time NASA combustor technology was focused on a Mach 4–8 research aircraft. Scramjet combustor design tools in this era were based largely on experimental studies, and designs were evaluated both in direct connect combustor and in freejet engine experiments (see Ref. 39). Performance levels for both purely supersonic and dual mode (mixed supersonic and subsonic combustion) were established. Some of the critical problems encountered in this low supersonic combustion speed regime were addressed, including fuel mixing, ignition, flame holding, mode transition, and combustor/inlet interactions. The scramjet concepts investigated were shown capable of thermally balanced operation with sufficient thrust and efficiency to accelerate a research aircraft.

Hypersonic Combustion Physics

Single-stage-to-orbit air-breathing concepts, such as NASP, provide new challenges to combustor designers. As flight speed increases, the combustor environment becomes truly hypersonic. At such speeds the kinetic energy of the freestream air entering the scramjet propulsion cycle is large compared to the energy released by reaction of the oxygen content of air with hydrogen fuel. Thus, the effects of reaction at Mach 25 speeds where heat release from combustion may be 10% of the total enthalpy of the working fluid will be small compared to Mach 8 flight where the kinetic energy of the air and the potential combustion heat release are roughly equal. Hypersonic combustion, perhaps, corresponds closely to the gradual diffusive mixing and burning process described by Ferri. Flow deflections caused by heat release are small—a few degrees at most—and flow boundaries are conceived as contoured to control the pressure rise at the location of the flame and eliminate the possibility of strong shock formation.

To the contrary, at speeds of Mach 8 and below combustion in ducted flows can generate large local pressure rise, flow deflection, and separation. The behavior of this upstream interaction has been studied extensively by Billig[42] and is characteristic of supersonic combustion in constant-area channels below flight speeds of about Mach 8. In such flows involving local separation and a bulk Mach number near one, local-wall static pressure is representative of the pressure across the entire flow at a given axial station. Therefore, a one-dimensional approximation to the flow can provide a reasonable description of the flow behavior. In hypersonic combustion, however, local Mach number remains high throughout the combustor flowfield and a significant variation in static pressure across the combustor flow is likely to occur. Typically, a fully three-dimensional representation of the flowfield will be required; and as pointed out by Stalker[40], special care must be taken to properly relate the implications of one-dimensional calculations to fully three-dimensional experimental data in a meaningful way.

So hypersonic combustion flows differ from supersonic combustion flows in that the flow remains hypersonic throughout in a bulk sense, the effects of heat release are smaller, and the pressure field is fully three-dimensional. Common features of hypersonic and supersonic combustion flows include real-gas effects, nonadiabatic wall boundaries, finite-strength shock waves, dissimilar gas injection, turbulent mixing, finite-rate chemical reaction, flow separation, etc. Also, because aerodynamic, fluid mixing, and chemical-rate processes are all expected to be important, experimental simulation requires near full-size hardware and duplication of flight conditions. Simulation requirements are discussed in more detail in the next section.

Hypersonic combustion raises some additional uncertainty and concerns. First, the effects of (extreme) compressibility on turbulence generation and mixing are not well known or understood. Second, at about Mach 12 the velocity of the injected (hydrogen) fuel stream equals the velocity of the combustor airstream, and at higher flight speeds the air velocity exceeds the fuel velocity. The behavior of fuel-air mixing under these conditions is also not well known. In fact, compared to flight at Mach 8 and below, there is limited data on which to base confidence in our understanding of or ability to model hypersonic combustion flows.

Simulation Requirements

The primitive variables available that describe the flow in a hypersonic combustor are P, pressure (or ρ density); T, temperature; u, velocity; L, model length; and ν_i, gas composition.

In principle, the composition can be manipulated in any fashion that results in duplication of the flight values of certain well-known dimensionless groups, i.e., simulation parameters. The first-order simulation parameters are M, Mach number; Re, Reynolds number; St, Stanton number; D_1, Damkohler's first number; D_2, Damkohler's second number; and GW, wall enthalpy ratio. To these may be added certain second-order parameters, such as Pr, Prandtl number and Sc, Schmidt number.

The physical interpretation of the first three fluid dynamics and heat-transfer parameters is also well known, i.e., the Mach number represents the ratio of kinetic to thermal energy, the Reynolds number the ratio of inertial to viscous forces, and the Stanton number the ratio of heat flux to inviscid energy flux. However, the interpretation of the last three is less well known as they are more specific to hypersonic reacting flows. These three parameters are the ratios of flow transit time through the combustor to chemical reaction time D_1, the ratio of heat added by reaction to the stagnation enthalpy of the inviscid flow D_2, and the ratio of enthalpy at the wall temperature to stagnation enthalpy of the inviscid flow GW. The second-order parameters represent gas properties, e.g., the ratio of viscosity to thermal conductivity and the ratio of viscosity to diffusivity. However, to the extent that these parameters reflect turbulence characteristics, they are more dependent on the first-order simulation parameters than on molecular properties of the gas. The first-order parameters can, in turn, be related to the primitive variable as follows:

$$M \sim \frac{u}{\sqrt{T}} \tag{1}$$

$$Re \sim \frac{\rho u L}{\sqrt{T}} \sim \rho L M \tag{2}$$

$$St \sim \frac{q_w}{\rho u H} \tag{3}$$

$$D_1 \sim \frac{L}{u t_c} \tag{4}$$

$$D_2 \sim \frac{\eta_c \Delta h_c}{c_p T + u^2/2} \tag{5}$$

$$GW \sim \frac{c_p Tw}{c_p T + u^2/2} \tag{6}$$

where t_c is a characteristic combustion time, q_w is the heat flux to the wall, η_c is a combustion efficiency, Δh_c is the heat of combustion, c_p is a characteristic specific heat, and T_w is the temperature of the combustor wall. The overall reaction time for a typical combustion process is generally proportional to both a function of pressure (or density) with an exponential dependency of approximately 1.75 and a function of temperature with an exponential dependency of unity. In certain restricted conditions where only binary (two-body) reactions occur, the combustion time is linear in density, leading to a direct relationship between Reynolds number and Damkohler's first number:

$$D_1 \sim \rho L/u \exp(-T)$$

$$\sim Re \frac{\sqrt{T}}{u^2} \exp(-T) \tag{7}$$

Hence, if velocity and temperature were to be duplicated, then simulation of Reynolds number would also satisfy the requirements for simulation of binary reaction time and vice versa. However, even in the simplest of situations, manipulating the temperature, velocity and model length in a fashion that will simultaneously preserve the values of Mach number, Reynolds number, and Damkohler's numbers is virtually impossible.

Therefore, it is clear that, in general, in hypersonic combustion experiments one needs to duplicate the primitive variables, including model length and gas composition, to ensure a faithful representation of the coupled chemical and flow processes. This automatically satisfies all of the simulation parameter requirements, with the possible exception of wall temperature and wall reactivity simulation.

Experimental Simulation

As previously stated, combustion heat release in air produces about the same energy increment as the kinetic energy of flight at Mach 8. Thus, the experimental simulation of supersonic combustion flow conditions for propulsion studies in ground facilities frequently utilizes so-called direct-combustion heating with oxygen replenishment or vitiation heating as a means of generating a test environment. Essentially, the test is conducted in (fuel-lean) combustion products where the amount of free oxygen is adjusted to represent the free-oxygen content of air. Of course, the remaining nonoxygen content of the test gas is not principally nitrogen as in air but will consist of some carbon dioxide and/or water vapor, etc., depending on the choice of fuel and oxidizer. The NASP Engine Test Facility (ETF) at Aerojet is an example of a Mach 8 vitiated propulsion facility based on a storable propellant (nitrogen tetroxide/ monomethylhydrazine) rocket gas generator.[44] The NASA Langley 8-Foot High-Temperature Tunnel (8′ HTT) is an example of a Mach 7 air/methane/oxygen direct-combustion heated propulsion facility.[45]

Contamination of the test media with combustion products is, of course, a concern because the contaminants are not inert and do not have the same thermodynamic properties as the nitrogen they replaced. Suffice it to say that interpretation of the test data from vitiated facilities (and from any facility, in fact) requires detailed consideration of the actual test media as it affects the results in comparison with free flight in air. As the flow speed of the ground-test simulation increases, in addition to the turbulence (velocity fluctuation) intensity of the test flow, other (turbulence) factors like temperature and oxygen-concentration fluctuation levels, chemical contamination, the degree of dissociation of contaminants and of molecular oxygen, etc., all may become issues that must be addressed. A recently completed joint Russian/USA flight project, discussed in Appendix A, was undertaken in part to address the effects of these type of facility contaminants on scramjet performance.

Other sources of energy such as storage heaters or electric-arc heaters can also provide high-temperature test conditions. Contamination (with the addition of contamination from either dust particles or NO_x formation, respectively) is still an issue that must be addressed. To their credit, arc heaters have the potential of reaching energy levels corresponding to much higher flight speeds than combustion-heated facilities and, in fact, have been used successfully to simulate the orbital re-entry flow environment. However, for propulsion flow simulation conventional arc heaters lack the ability to operate at stagnation pressures adequate for hypersonic combustion simulation. Typically, arc heaters can achieve operating pressures of 20 to 50 atm where pressures of 2000 to 5000 atm and higher are desired to duplicate the flight conditions.

The enthalpy requirements for hypersonic combustion simulation are summarized in Fig. 16. The sensible total enthalpy of flight is shown plotted against Mach number. The right-hand-most curve shows the freestream Mach number along the flight path. The forebody flowfield and inlet compression process reduce the local Mach number and raise the flow static pressure along a nearly constant total enthalpy path, as indicated by the arrow. The shaded band represents the range of Mach number and total enthalpy representative of combustor entrance conditions. The local static temperature, stagnation temperature, and stagnation pressure representative of the combustor entrance condition are shown at the points indicated for flight Mach numbers of 10, 15, 20, and 25. The static pressure is typically 0.5 to 1 atm. Note that the static temperature remains in the range of 3000° to 4000° R while the stagnation temperature quickly reaches levels where significant dissociation will occur. Also note that the stagnation pressure rises exponentially with Mach number from achievable levels (100 atm) at Mach 10 to extreme levels exceeding a million psi at Mach 25. The scale at the far right shows a level of oxygen dissociation expected in a facility that adds energy to the test gas at rest (such as an arc heater or a reflected-shock tunnel) as a function of stagnation enthalpy. Pulse facilities, which generate high pressure and hypervelocity flows for a short time,

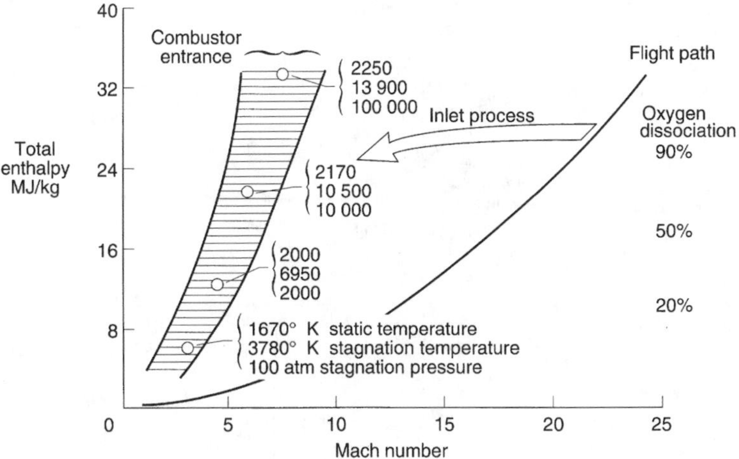

Fig. 16 Total-enthalpy requirements for hypersonic combustion simulation.

represent the only presently proven means to approach simulation of these hypersonic combustion flow conditions on the ground.

Figure 17 shows the total-enthalpy simulation capability of selected pulse facilities on the same coordinates used in Fig. 16 (data from Ref. 46). Aerodynamic simulation of Mach number and Reynolds number is achieved in the Calspan reflected-shock tunnels with ambient temperature or moderately heated (600°F) helium as a driver gas, but duplication is limited to a flight Mach number of about 10 for a test duration of 1 ms. The reflected-shock tunnels T4 and T5 take advantage of the higher temperature achieved by free-piston compression of helium driver gas to achieve energy approaching orbital velocity. The Ames 16-in. shock tunnel is intermediate in energy simulation capability with a hydrogen combustion-heated driver. The expansion tube (HYPULSE), which is essentially two shock tubes in tandem, can achieve Mach 16+ energy with an ambient temperature helium driver.

Several high-energy driver methods have been investigated to extend the capability of the HYPULSE expansion tube to duplicate Mach 16+ combustor flows. These include free-piston and detonation-driven expansion tubes. These approaches will provide the potential capability to duplicate both total enthalpy and total pressure, as well as velocity and gas composition, above about Mach 16.[47] This capability has been demonstrated in a pilot-scale (1 ½-in. inside diameter) facility.[48] Addition of a free-piston driver to the 6-in. inside diameter NASA expansion tube (HYPULSE) located in Ronkonkoma, Long Island, New York, and operated by GASL has been assessed in detail and could increase the operating envelope as indicated in Fig. 17. (Recent HYPULSE facility upgrades are discussed in Appendix B).

The total pressure required for hypersonic combustion simulation is shown in Fig. 18. The shaded bands show pressure requirements for either

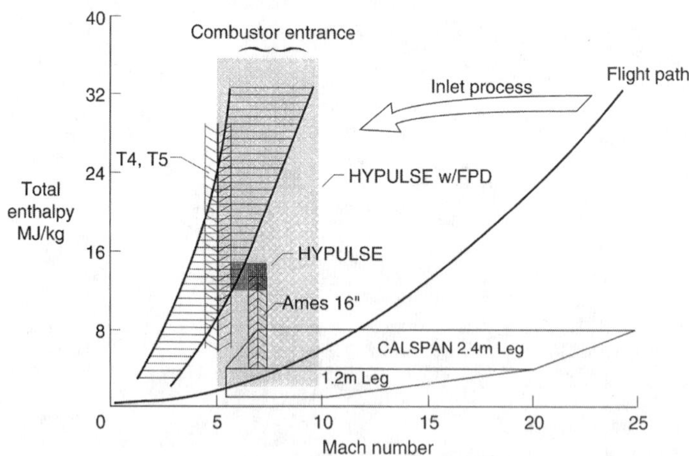

Fig. 17 Total-enthalpy capability of selected pulse facilities for hypersonic combustion simulation.

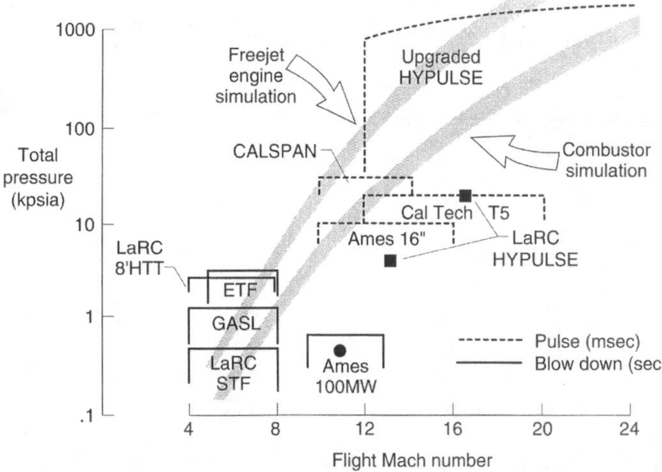

Fig. 18 Total-pressure capability of selected facilities for hypersonic combustion simulation.

freestream engine simulation or internal combustor flow simulation along with the capabilities of the selected facilities. Blow-down facilities with run times on the order of hundreds of seconds like the NASA Langley 8′ HTT and the NASP ETF's are limited to 2 to 3 kpsi. Higher enthalpy facilities like the NASA Ames 100 MW Arc-Heated Facility have even less pressure capability because of the facility nozzle throat-cooling problem. Pulse facilities like the Calspan, NASA Ames 16-Inch and Cal Tech T5 reflected-shock tunnels[46] can produce and contain significantly higher pressures (up to 30 kpsi) largely because of shorter flow times on the order of milliseconds. Note that at Mach numbers above 12, the total-pressure requirement approaches a million psi, and only the expansion tube is capable of producing those pressures. The unique capability of the expansion tube is a result of the fact that the acceleration process in the expansion tube adds velocity directly to the flow without stagnating it. This means that the facility need not withstand the stagnation pressure of the flow that it generates, and this feature is a considerable advantage when duplication of flight conditions is required. (Scramjet engine test capabilities in the HYPULSE facility, down to Mach 7, are discussed in Appendix B).

The fundamental difficulty in generating hypersonic flows in a ground facility is related to putting the energy into the proper mode in the test gas that is generated. Figure 19 is an attempt to explain this in more detail. The energy required for flight along a constant dynamic-pressure trajectory of 1000 psf is taken as a reference with the generation of a combustor pressure of 0.5 atm and the assumption of an inlet process with a kinetic energy efficiency of 98%. The facility process assumes stagnation heating with isentropic expansion from the stagnation conditions to the required combustor entrance condition and chemical freezing of the gas composition at

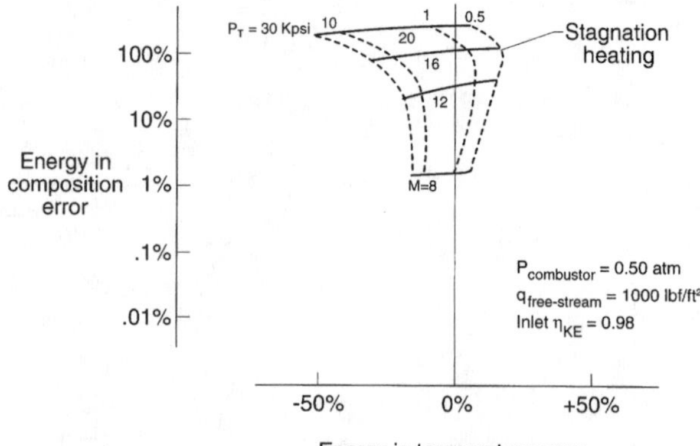

Fig. 19 Ground simulation of scramjet flight combustor conditions with stagnation heating.

the throat of the facility nozzle. The horizontal axis represents the error in energy in static temperature compared to flight of this particular facility stagnation pressure and total enthalpy where the energy scale has been made dimensionless with respect to the amount of energy in stoichiometric combustion of hydrogen and air. Similarly, the vertical axis is the error in energy because of composition. That is, if the gas is dissociated at the stagnation energy level, this dissociation persists in expansion through the nozzle, and energy is tied up in NO and atomic oxygen, which would not be present in clean air in flight. Various flight Mach numbers are shown by the solid lines, and various total-pressure levels are indicated by the dashed lines. Because flight in clean air is taken as the reference, in Fig. 19 goodness is toward the center and bottom of the figure with a small energy error due to static temperature mismatch and a small energy error in dissociation. Note that as Mach number increases, the amount of energy error in composition increases substantially. Also, as Mach number increases, generally the required stagnation pressure for simulation of flight significantly increases. Figure 20 adds a similar carpet plot of energy error for a facility like the expansion tube, which generates velocity directly without heating at stagnation conditions. As can be seen in Fig. 20, the level of energy in composition error is approximately an order of magnitude smaller for the expansion tube or nonstagnating facility compared to a stagnation heating facility.

Another way to assess facility contamination level is in terms of the amount of molecular oxygen in the test flow compared to the oxygen content of clean air in flight. The performance of a number of facilities is shown in Fig. 21 as a function of flight Mach number. The hatched band shows the performance typical of reflected-shock tunnels with particular facilities such as T5 and the Ames 16-Inch Tunnel shown by symbols on the figure. The expansion tube HYPULSE is shown by the circular symbols. As Mach

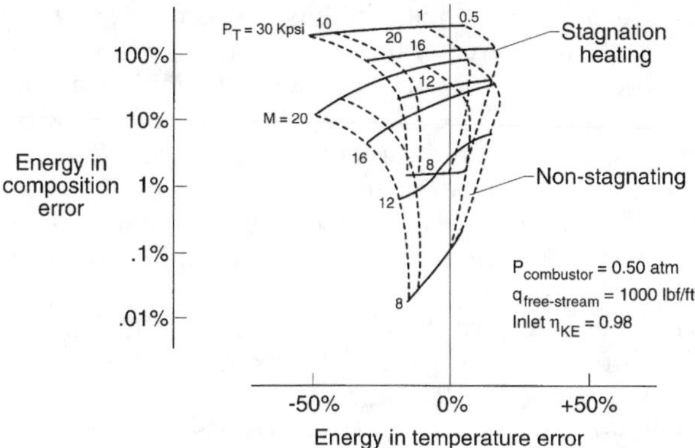

Fig. 20 Ground simulation of scramjet flight combustor conditions comparing stagnation and nonstagnating heating.

number increases beyond Mach 12, the expansion tube facility shows an increasing advantage over the reflected-shock tunnels in terms of available oxygen to simulate flight in clean air.

The comparatively moderate investment required to generate the desired test conditions in pulse facilities is achieved at the expense of flow duration. Producing adequate flow duration to establish a sensibly steady flow representative of the real steady flow becomes an issue, which (like the contamination issue) must be addressed in facility design and in the interpretation of data. However, short flow duration has an advantage in high-energy flow tests in that the model cooling requirement can be met by simple heat-sink approaches. The short flow duration also makes some types of measurements, like local heat flux and optical measurements requiring windows in

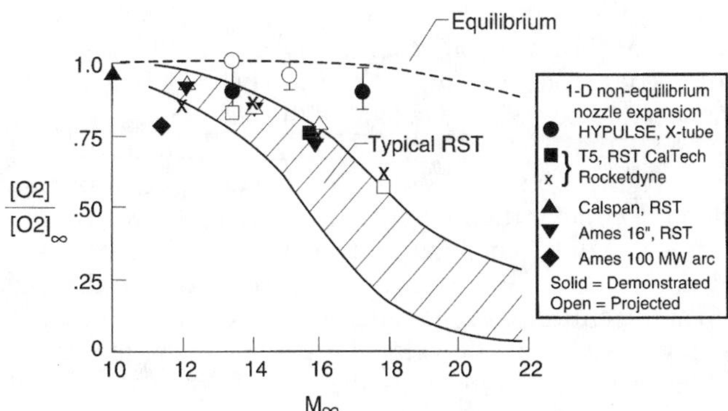

Fig. 21 Test gas oxygen dissociation level at simulated combustor entrance conditions.

the combustor, much easier to make in pulse flows than in steady flows. On the other hand, all measurements must have very high frequency response—on the order of 10^6 Hz—to provide meaningful data from test times of 10^{-3} or less. In this regard we may note that hypersonic tests conducted in a pulse facility almost inevitably involve an unheated (room temperature) model. Hence the wall enthalpy ratio GW is typically on the order of 1/10 or less, whereas in flight it might be as high as 2/10. Fortunately in point of fact, the heat-transfer and skin-friction coefficients both become virtually independent of GW for values less than about 3/10.

Of greater or equal concern than duplication of wall temperature is the simulation of wall reactivity. This consists of both the tendency of the wall to catalyze gas-phase reactions and the extent to which the gas-phase reactions are thermally quenched (or promoted) at the wall. The catalytic efficiency is primarily dependent on certain gross characteristics of the wall materials, i.e., metals tend to be fully catalytic and glassy materials (e.g., quartz) tend to be noncatalytic. Thus, use of a model composed of metal walls with glass windows may present a potential local disruption of wall catalytic effects. However, the extent of the problem is easily determined by using metal blanks in place of windows. The effects of thermal quenching at a model wall temperature that is substantially less than exists in flight is more difficult to determine experimentally, as wall-mounted instrumentation may not function at the true wall temperature. However, this would appear to be an instance where CFD simulations could be trusted to correct the data for wall-temperature effects. If necessary, the model could be preheated (e.g., radiatively), and nonintrusive or noncontacting instrumentation (e.g., exit plane surveys) used to assess wall-temperature effects.

Comparison of Combustion Data

This section describes an experimental assessment of the effects of the level contamination by dissociation of test gas on a mixing and combustion experiment at Mach 17 flight energy conditions. The intention is only to show the effects of the state of the test gas produced by different facilities on the experimental results and not to imply that one facility type should be pursued to the exclusion of another. In fact, as noted earlier, increased contamination goes along with higher energy (Fig. 16). In general, all hypervelocity data need to be assessed in light of the test media involved including the initial composition and finite-rate chemistry, which may be important in the flowfield. Figure 22 shows the experimental apparatus. Basically the configuration is a circular tube 1½ in. in inside diameter with a sharp leading edge. The tube is about 36 in. long, and an annular injector is located about 7 in. from the entrance of the tube to inject hydrogen into the high velocity air captured at the entrance of the tube. Hydrogen is supplied to the injector by a fast-acting valve from a Ludwig tube, and the hydrogen flow is initiated such that hydrogen is flowing into the model when the facility flow is established and initiates the combustion simulation.

Two identical models were constructed. One was tested in the NASA HYPULSE facility at Mach 17 flight conditions. The other was tested in the

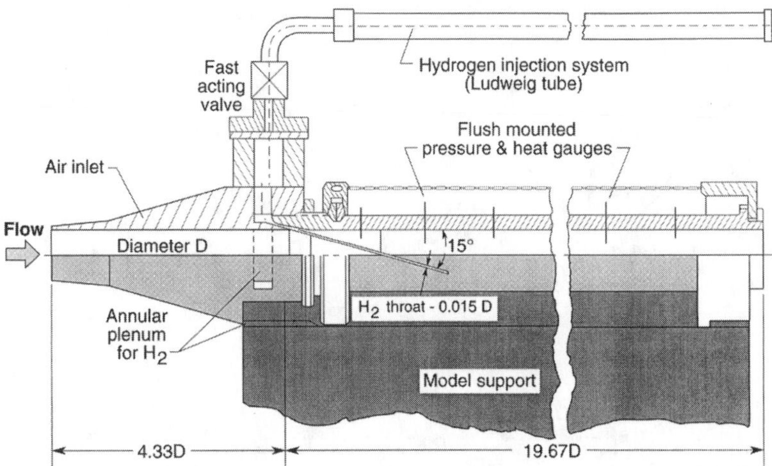

Fig. 22 Mach 17 combustion experimental apparatus—axisymmetric model.

reflected-shock tunnel T4 at the University of Queensland in Brisbane, Australia. Typical flight conditions and the wind-tunnel test conditions are shown in Table 1.

The deficiencies of simulation are nonequilibrium chemical contamination (O and NO) and low total pressure and Reynolds number. Other potential differences between facilities are turbulence intensity and contamination from diaphragm material, etc. Note that in flight only molecular oxygen would be expected to be present in the test stream, but at Mach 17 energy level the reflected-shock tunnel T4 has significant dissociation compared to the clean-flight condition. Wall pressure measured in the tube downstream of the injection point in these tests is shown in Fig. 23. The combustion pressure rise parameter shown is the pressure with injection into air minus the pressure for injection into nitrogen divided by the pressure with injection into nitrogen. Presenting the data in this way shows the maximum effect of combustion on wall pressure and tends to remove some of the wave structure that otherwise produces scatter in the data. The open symbols show data from the reflected-shock tunnel T4, and the solid symbols show data from the expansion tube HYPULSE. Clearly, in Fig. 23 considerably more pressure rise is generated in the reflected-shock tunnel T4, which

Table 1 Hypersonic Combustion Conditions: Mach 17 flight simulation

Model	*P, psia*	*T*, °R	*V, ft/s*	O_2, Mass fraction	Dissociation, %	*M*
HYPULSE	3.29	3758	16660	0225	1.5	5.75
T4	2.11	3717	15450	0108	48	5.17
Typical flight	7–20	3600	17000	0237	0	7

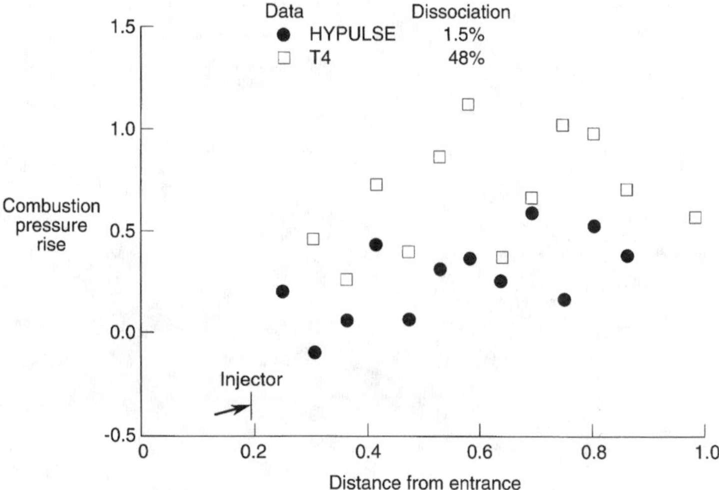

Fig. 23 Comparison of combustion pressure rise at Mach 17 for fuel equivalence ratio of 3.

has a level of dissociation approximately equal to half of the molecular oxygen compared to freestream undissociated air.

Reference 51 presents an analysis to understand the reasons for this difference in measured pressure. The flow in the tube is modeled with a simple one-dimensional flow conservation equation that includes three temperatures for the mixing/reacting fuel and air flowing down the tube. Initial temperature for the fuel and air are defined by the initial conditions in the experiment, and a mixed stream of fuel and air with an independent temperature generated by a finite-rate kinetic computation follows development of the flow. Wall friction, shock losses, and the rate of mixing in the tube are modeled based on the expansion tube wall-pressure data and are held constant with the different initial composition of the reflected-shock tunnel flow (Table 1) to derive the predicted pressure distribution that is compared with the data in Fig. 24. Excellent agreement between the prediction for the reflected-shock tunnel T4 is achieved with data using this three-temperature scheme with the finite-rate kinetic model. Apparently, even though a significant amount of the molecular oxygen is withheld from reaction with hydrogen fuel as NO, the initial level of atomic oxygen present in the dissociated test gas increases heat release and the amount of reaction resulting in the greater pressure rise.

Figure 25 shows results from the same computation in terms of the energy yield predicted in the experiment. Results for the expansion tube are shown by the solid lines, and results for the reflected-shock tunnel are shown by the dashed lines. The lowest pair of curves shown is for the actual pressure and temperature of the experiment. Because of the initial dissociation in the reflected-shock tunnel, the energy yield is somewhat higher than for the

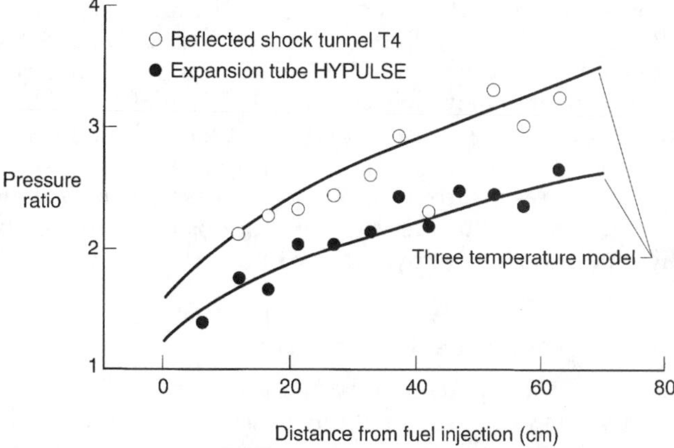

Fig. 24 Assessment of experimental Mach 17 combustion pressure rise.

expansion tube. If the pressure were raised to a level of 1 atm, which is more representative of a typical flight trajectory, an even greater difference would be observed; and if the temperature were also reduced to a level of 1200 K (again, more representative of flight), a still greater energy yield would be achieved with a larger difference between the reflected-shock tunnel and expansion tube resulting. The top curves show the computed result if the flow were in local equilibrium as it mixed along the length of the tube. The energy level is considerably above the actual measured result for the pressure and temperature of the experiment, indicating that finite-rate chemistry plays a dominant role in the results for the conditions of the test. Again, the intent of this comparison is to emphasize the need to account for the composition of the test media in hypervelocity combustion simulations in

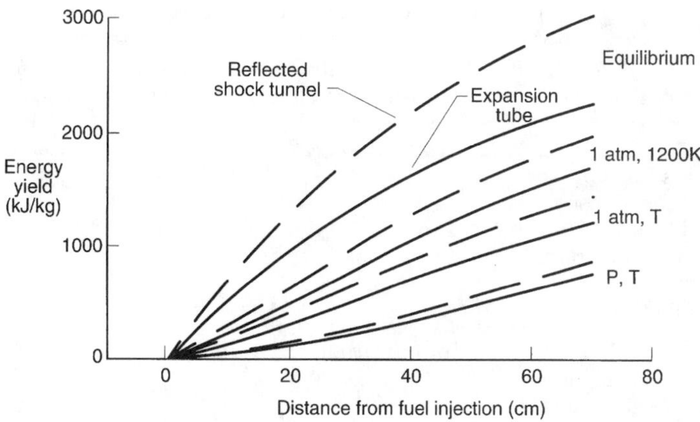

Fig. 25 Computed energy yield for Mach 17 combustion experiment.

order to understand the implications of ground simulations on flight performance. Bakos et al.[52–55] and Bélanger et al.[56,57] have continued studies of the effects of contamination in impulse facilities on hypersonic combustion simulation. In particular, they have continued to exploit comparison of data in identical hardware tested at similar energy levels in both reflected-shock tunnels and expansion tubes. The interested reader is urged to consult the cited references and results from continuing studies that are establishing the detailed data needed to interpret the implications of pulse facility simulations for flight performance of hypersonic scramjets.

Instrumentation/Measurement Requirements

Measurement of scramjet combustor performance is difficult to achieve in hypervelocity flows because the influence of combustion is small on the easily measured quantities, such as wall pressure. An in-depth study of measurement requirements was performed by Bittner.[58] This study showed that for reasonable hypervelocity combustor designs, fuel mixing and combustion efficiency are the most important combustor performance parameters (i.e., engine thrust is about proportional to combustion efficiency). Considering the measurement uncertainty and sensitivity of indirect measurements (not including uncertainty of assumptions required to deduce performance from these indirect measurements) for deducing mixing and combustion efficiency, Bittner concludes that these performance parameters must be directly measured. Table 2 illustrates the performance uncertainty expected based on several measurement approaches that could be used to determine combustion and mixing efficiency. The measurement uncertainty column indicates the assumed accuracy of the measurement technique: performance sensitivity to measurement is the relative change in the measurement to the performance parameter, and the performance uncertainty is the

Table 2 Combustor performance uncertainty at Mach 15 based on various measurement approaches

Measurement	Perform. sensitivity to measurement	Measurement uncertainty, %	Performance uncertainty, %
	For combustion efficiency		
Water mass fraction	1.0	± 5	± 5
Water number density	0.8	± 5	± 51
Total wall heat load	3.4	± 3	± 10
Oxygen mass fraction	0.6	± 80	± 48
Average pressure	3.4	± 3	± 10
Oxygen utilization	1.0	± 80	± 80
	For mixing efficiency		
Peak fuel mass fraction	0.5	± 5	± 25
Peak fuel number density	0.9	± 5	± 51
Comb. exit min. temperature	1.1	± 80	± 95

product of the measurement uncertainty and performance sensitivity. Even without considering uncertainty introduced in deducing performance from indirect measurements, direct measurements (such as water for combustion efficiency) are the most accurate approach. This study also shows the limitation of many planar measurements, which are only order-of-magnitude accurate (± 80 %).

The best experimental measurement for determining combustion efficiency is combustor-exit water mass fraction (determined by line-of-sight laser absorption), and for mixing efficiency is fuel mass fraction distribution. Bittner demonstrated, by evaluation of typical CFD solutions, that 3 or 4 ½-mm diameter laser absorption paths for a 1-inch-high combustor are adequate to resolve the combustion efficiency to ± 5 % uncertainty, at the combustor exit, as illustrated in Fig. 26. This measurement technique has been applied to combustion tests both in the NASA Ames 16-Inch Reflected Shock Tunnel and the NASA LaRC HYPULSE expansion tube at GASL.

Measurements of fuel-mixing efficiency and fuel distribution are being accomplished using an approach, which-will be described in the following paragraphs, by tracing the injected fluid with particles. In this technique the hydrogen fuel to be injected has a small fraction, say 3% of silane (SiH_4) added. Silane is very reactive and burns on contact with air or oxygen even at room temperature and in very dilute mixtures. Immediately prior to injection into the airflow, oxygen is added to this silane/hydrogen mixture sufficient to burn the silane present and produce silicon dioxide particles. These particles, typically on the order of 0.2μ in diameter, are small enough to follow the fuel as it is injected, mixed, and burned in the flow. Illumination of a cross plane in the flow with laser light can then produce a Mie scattering image of the particles, which can be recorded with a fast electronic camera, and thus visualize the fuel distribution in the experiment. This process is shown schematically in Fig. 27 for a typical combustor duct with ramp-type fuel injectors.[59]

The process of correcting the raw image and processing it to produce a true cross plane in the flow is presented in Fig. 28. Both an image with injection in the flow and a reference background luminosity image are recorded. These images are substracted and then stretched to represent a

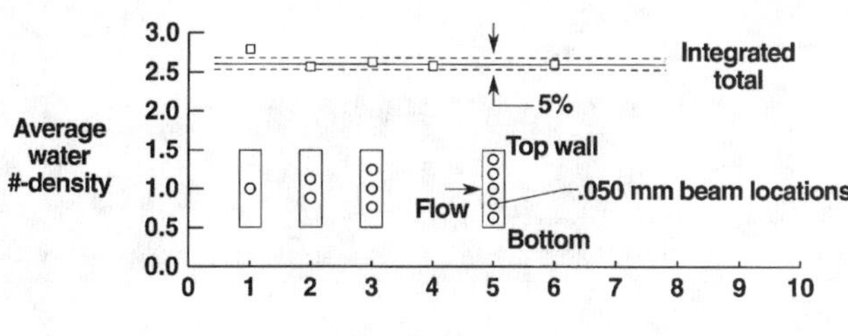

Fig. 26 Indicated performance sensitivity to number of measurements.

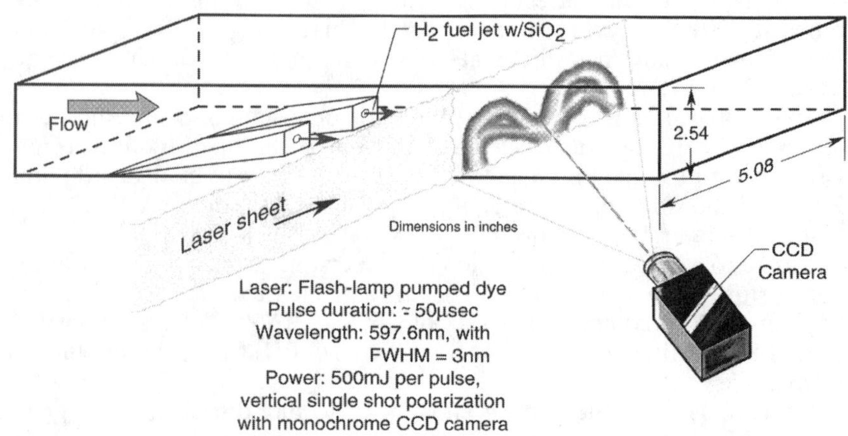

Fig. 27 Schematic of fuel plume imaging in HYPULSE.

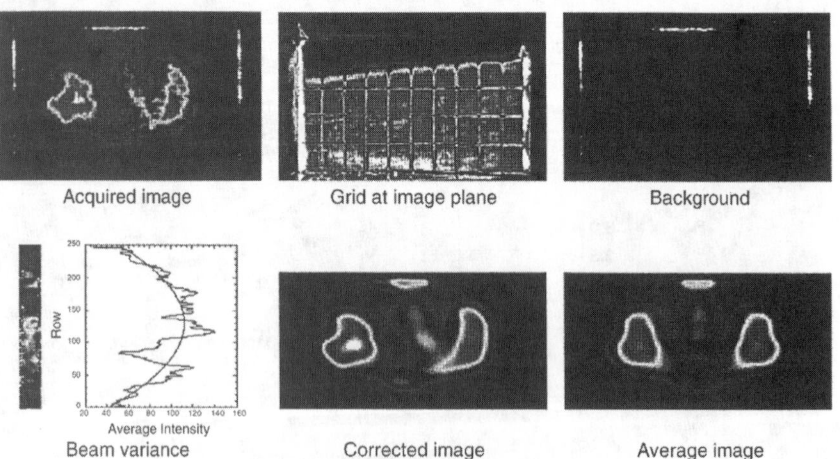

Fig. 28 Fuel plume image processing procedure.

true cross plane in the flow. This image is further corrected to reflect the actual intensity profile of the laser sheet illumination to provide an accurate picture of particle concentration in a cross plane in the flow. If particles are neither created nor destroyed in the mixing/reacting flow as it proceeds downstream, then the total amount of scattered light in the corrected image from each plane should be approximately the same. Thus, the local amount of reflected light in each image, nondimensionalized by the average reflected light in each plane, becomes a measure of the local fuel concentration relative to the average concentration in the overall bulk flow. Thus, this technique allows a quantitative measure of the fuel distribution to be determined with successive pictures at successive planes downstream from the injection location in the duct. Data obtained in this way provide a direct, experimental measurement of fuel-mixing efficiency and can be used to calibrate computations of the type described in the following section. The overall intent is to use this approach to provide a calibration of CFD codes to determine the level of hypersonic combustion performance achievable in flight.

Computational Simulation

Much has been written about the computer age and the impact of modern CFD on engineering in general and on aeronautics in particular. CFD has an important role to play in the understanding of hypersonic combustion. However, one must recognize at the outset that some of the important physics of hypersonic combustion flows are either not included or at best only crudely modeled in even the most advanced computer codes. For instance, chemical reaction occurs at the molecular level, but CFD treats the flow as a continuous fluid. Further, turbulence is generally modeled by some physical analogy rather than being computed directly, and any interaction between turbulence and chemical reaction is generally ignored. Also, the physical modeling of turbulence is based entirely on empiricism derived from velocity fluctuation measurements in incompressible flows; but compressibility introduces some effects that tend to suppress the generation of turbulence, and the assumption of simple analogous behavior of mass, momentum, and energy transport is inadequate for combustion flows.

In spite of these factors, CFD predictions of complex fully three-dimensional hypersonic combustion flowfields are possible today on a routine basis. Examination of three-dimensional CFD solutions can provide valuable insight to complex hypersonic combustion flows. If these computational results can be calibrated with data from experimental simulations of hypersonic combustion at the appropriate conditions, then in spite of incomplete physics in the analysis the code results can become an important design tool. Once a code is benchmarked against data, it can be used to conduct parametric variations in design variables to help improve and refine a given configuration.

One must recognize that because the actual physical processes going on in a hypersonic combustion flow are much more complex than current CFD tools can model, the most complex CFD methods may not be the tool of first choice for analysis. For instance, the added computational complexity of finite-rate chemistry may not be justified in a hypersonic combustion flowfield where reactions may not begin to proceed until the scale of the turbulence has reached a level where adequate molecular collisions of fuel and oxidizer can occur, as suggested by Swithenbank.[60] Following this, local chemical equilibrium may prevail, as in the diffusive burning model for supersonic combustion envisioned by Ferri.[41] A computational approach employing a combination of chemically frozen flow and a simple complete reaction model, much like the sudden freezing model developed by Bray[61] in the 1960s for the reverse process in nozzle flows, could be more accurate than a very complex finite-rate reaction scheme that models turbulence-chemistry interactions as augmented laminar diffusion. The point is that the complexity of the various parts of the computational model should be balanced to achieve efficient engineering results, as well as consistency with the real physical processes.

Computational Methods

Today, a multilevel analysis approach is used by most numerical application organizations. This multiple-level approach uses simplified methods to screen problems, followed by more sophisticated and costly approaches for selected problems. Hypervelocity scramjet combustor flows feature predominantly supersonic flow with only small separation regions associated with predominantly axial injection, small if any base regions, little ignition delay (caused by high static temperature), and extremely cold walls—tending to stabilize the boundary layer. These characteristics allow credible solutions with lower-level approximate approaches such as parabolized Navier–Stokes formulation (PNS). During the NASP era, CFD application of scramjet combustors at NASA primarily relied on three codes: SPARK,[62,63] SHIP,[64] and GASP.[65] These codes (Table 3) provide flexibility in evaluation of combustor flow characteristics and performance. The SHIP PNS code, which utilizes a boundary condition to allow PNS modeling of separated flow regions, is very fast and used for both screening and trade studies (such as demonstrated in Ref.

Table 3 NASA LaRC combustor CFD codes

Code	Formulation	Turbulence	Chemistry	Relative solution time
SPARK	Complete PNS	B-L	General finite-rate chemistry	1.0 0.2
SHIP	PNS	B-L, k-ε, q-ω, comp. correct.	Complete reaction	0.02
GASP	Complete space marching	B-L, k-ε	General finite-rate chemistry	1.0 0.2

66). This code utilizes simplified one-step $H_2 + O_2 \rightarrow H_2O$ complete reaction chemistry modeling, but include several turbulence models. SHIP combustor solutions generally use the q-ω two-equation incompressible turbulence model. (The relative time shown is based on that model usage.) Other turbulence models in the code include Baldwin–Lomax (B-L) and k-ε. In addition, compressibility corrections are available for the q-ω turbulence model. GASP or SPARK, on the other hand, are used for in-depth evaluation of flow details[67] and verification of the screening results.

All of these methods have been demonstrated capable of predicting scramjet fuel mixing and combustion efficiency. Because of the large grid requirements to resolve the combustor wall and injected fuel mixing and combustion simultaneously, only the SHIP PNS approach is used to resolve the entire combustor flowfield. The GASP code has been shown to be capable of predicting separated flow regions associated with shock/boundary-layer interaction and flow over steps, using the Goldberg-corrected Baldwin–Lomax turbulence model.[68]

Confidence in CFD's predictive capability has been enhanced by comparison with experimental cold-flow mixing simulations of high-speed scramjet combustors, including transverse hydrogen injection into a Mach 4 airflow,[69–74] helium into Mach 3 and 6 airflow,[75–78] and flush wall and ramp injection of air into a Mach 2 airflow.[79–81]. Generally, agreement with fuel distribution, peak injectant concentration decay, and fuel-mixing efficiency are acceptable[66,67,82,83] in the combustor far field (i.e., at lengths required for nearly complete mixing). Typical examples for the three codes are presented in the following paragraphs.

Comparison of GASP predictions with experimentally measured injectant mole fraction for staged injection of air transverse to a Mach 2 airstream are presented in Fig. 29 (Ref. 83). Penetration, spreading, and decay of the maximum concentration of injectant are accurately modeled, indicating that the GASP code represents the flow physics at an acceptable level to reliably predict complex combustor fuel injection and mixing flowfields. Comparison of SPARK predictions and experimentally measured injectant mole fraction for low-angle wall injection of helium into a Mach 3 airstream are presented in Fig. 30 (Ref. 84). Excellent agreement with the vertical centerline helium mass-fraction contours indicate that the SPARK code can accurately model fuel-mixing flowfields. In practice, the best figure of merit for these validation studies is not fuel contour replication, it is the bulk parameter fuel-mixing efficiency used in design analysis.

Most low-enthalpy experimental tests provide details of the flowfield, but not sufficient definition to calculate the fuel-mixing efficiency used in engine cycle analysis to model scramjet performance. Historically, mixing efficiency η_m is defined[70] as that fraction of the least-available reactant (i.e., oxygen or fuel), which would be consumed if the fuel-air mixture were reacted everywhere to completion without additional mixing. Thus, in fuel-rich regions all of the local oxygen would be consumed and is considered mixed so that it can react, whereas in fuel-lean regions all of the fuel is mixed. Thus, the

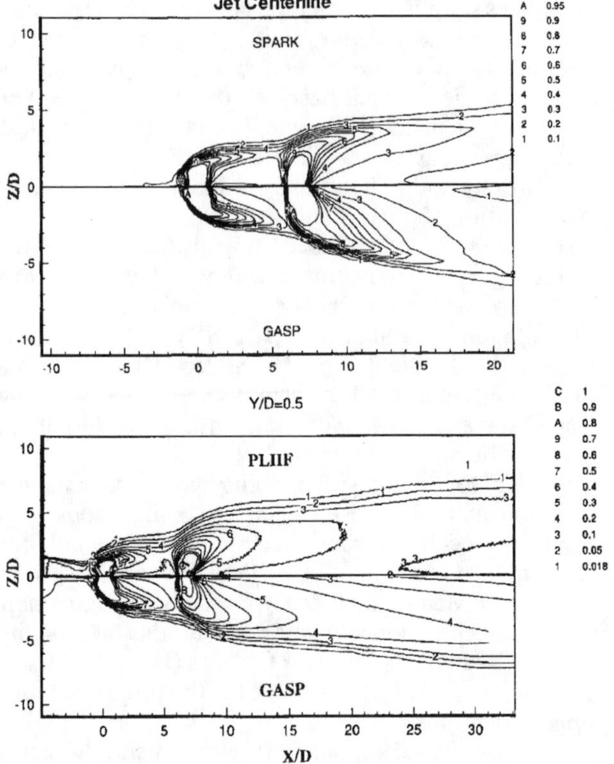

Fig. 29 Comparison of injectant mole fraction contours with PLIIF image in the streamwise planes (*Y/D* = 0.0 *Y/D* = 0.5).

definition of fuel fraction mixed so that it can react α_R, depends on the local equivalence ratio, as follows:

$$\alpha_R = \begin{cases} \alpha & \text{for } \alpha < \alpha_s \\ (1-\alpha)\,\alpha/(1-\alpha_s) & \text{for } \alpha > \alpha_s \end{cases}$$

and mixing efficiency is calculated from

$$\eta_m = \frac{m_{\mathrm{H}_2,\mathrm{mix}}}{m_{\mathrm{H}_2,\mathrm{total}}} = \frac{\int_{A_{\infty=0}} \alpha_R \rho u \,\mathrm{d}A}{\int_{A_{\alpha=0}} \alpha \rho u \,\mathrm{d}A}$$

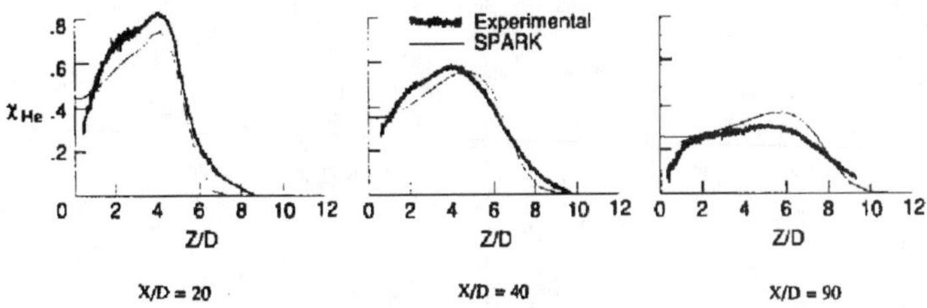

a. Vertical centerline helium mole fraction profiles.

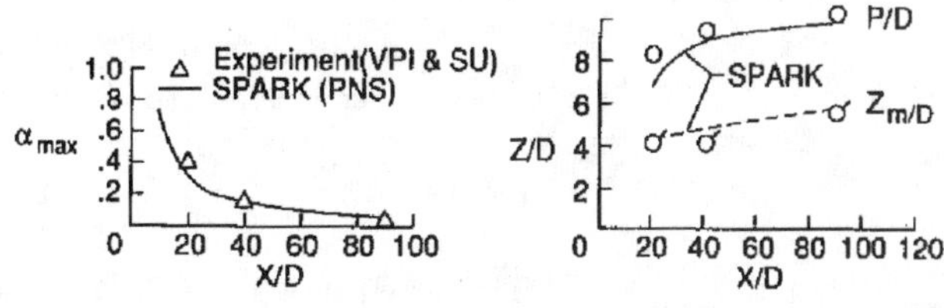

b. Maximum helium mass fraction c. Helium penetration

Fig. 30 Comparison of experimental data and numerical simulation of a 30-deg flush wall He injector into a Mach 4 airflow.

where a is hydrogen (fuel) mass fraction, a_s is the H_2 stoichiometric mass fraction (0.0285), $a_{s=0}$ is the area enclosed by zero H_2 contour defining the extent of mixing region, $m_{H_2,mix}$ is mixed H_2 mass-flow rate, and $m_{H_2,total}$ is total flow rate from the flowfield integration when $a_{max} \leq a_s$ and η_m equals 1.0.

Comparison of CFD predictions with experimentally measured fuel-mixing efficiency is presented in Fig. 31, which illustrates a comparison of the SHIP three-dimensional PNS calculation[66] for transverse hydrogen injection into a Mach 4 airflow.[73] Results of the SHIP calculations for these six injector configurations are within ± 10% of the experimental correlation, which was developed and presented in Ref. 73. In fact, SHIP predictions are in better agreement and exhibit the same trends as the experimental data. This comparison illustrates that the PNS code can accurately model the fuel mixing, even for transverse injectors, which have small upstream regions of flow recirculation. Comparisons with ramp injectors and low-angled flush wall injection[64] exhibit similar accuracy in modeling fuel-injector flow physics and fuel-mixing efficiency.

Recent advances in pulse-tunnel combustor technology have allowed direct quantitative measurement of fuel mixing in hypervelocity combustor flowfields using the fuel plume imaging already discussed. These images are integrated to provide fuel-mixing efficiency. A typical experimental fuel plume image is presented in the right half of Fig. 32. This image was obtained at flight Mach 13.5 simulation in the HYPULSE expansion tube facility. Fuel was injected ($\varphi = 1.0$) 3 in. upstream of this image, using a swept-ramp injector photographically scaled from that tested by Northam.[85] This image

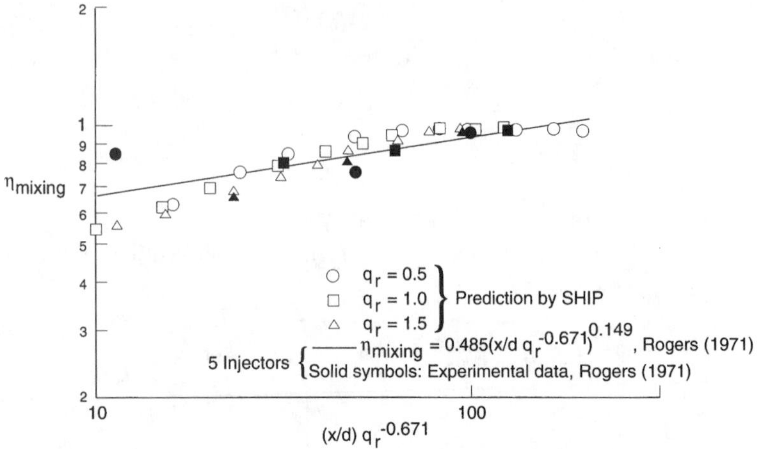

Fig. 31 Predicted and measured fuel mixing for normal sonic H_2 injection into a Mach 4 airflow.

illustrates the effect of vortical stretching of the fuel plume and is integrated to provide fuel-mixing efficiency. Numerical simulation using the SPARK code, presented on the left, closely approximates the fuel plume shape, penetration and spreading, peak fuel mass fraction, and integrated fuel-mixing efficiency. This (and other) comparison demonstrates acceptable simulation of the fuel-mixing process in these hyper-velocity combustor conditions and justifies the use of CFD for design studies. Additional validation of CFD for these flows is required as data become available.

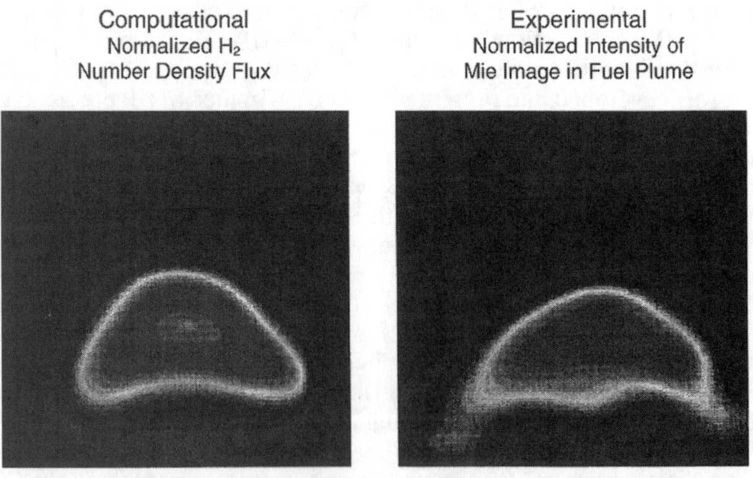

Fig. 32 CFD data comparison of fuel plume.

Combustor Performance Index—Thrust Potential

Use of CFD for combustor design provides more information than generally available from experimental studies. Trends from this mass of information can be effectively used to improve the combustor and engine design, but only by selection of appropriate parameters for determination of design goodness. One measure of the combustor performance is energy availability,[86,87] defined as the potential of the combustor-exit flow to generate vehicle thrust. Available energy is destroyed in the burner by injector blockage, mixing, wall heat transfer, and frictional drag on the injector and combustor walls. Generally, it is increased by fuel momentum, the axial wall-pressure integral, and the release of energy into the flow by exothermic chemical reaction. The combustor-exit flow expands in the nozzle and along the afterbody where the bulk of the work potential added in the combustor is realized by increased wall pressure and, hence, the generation of vehicle thrust. The ultimate engineering significance of any combustor analysis, experiment, or numerical simulation must be measured by its success in increasing understanding of the ability of the engine to produce thrust. The goal of such studies must lead to understanding of and accurate prediction of the thrust potential for use in vehicle design efforts.

Earlier work on flow losses (or thrust potential) has been performed by many workers, notably by Swithenbank[86] who identified concerns with mixing enhancement strategies that could entail greater flow losses than performance gains recovered from the additional mixing. Czysz and Murthy[87] present treatment of useful work availability (or exergy) in high-speed propulsion systems. Kamath and McClinton[88] among others, have used an inverse approach, using an entropy based approach to the description of flow losses (effectively 'lost' work rather than 'available' work). The performance parameter presented herein, thrust potential, was developed by Riggins and McClinton.[89,90]

The approach taken to analyze thrust potential is twofold. First, the three-dimensional CFD generated flowfield is one-dimensionalized,[90] using a scheme that conserves all mass fluxes (including individual species mass fluxes), momentum fluxes, and energy fluxes between the three-dimensional solution and the one-dimensional representation of the solution. Next, this flux-equivalent one-dimensionalized flow is expanded (from any or all cross-sectional plane in the combustor) in an ideal, or reference, nozzle. This ideal expansion to a referenced area provides a net combustor/nozzle thrust or the net thrust potential. This net thrust potential is then either nondimensionalized by an ideal thrust to form a combustor effectiveness parameter,[90] or the actual net thrust is used directly either for comparative purposes or to build more sophisticated design models. This latter approach is most useful in combustor trade studies on a specific engine (with specified inlet and nozzle) or flight vehicle configuration. More sophisticated engine or vehicle design models use this combustor/nozzle thrust potential in conjunction with inlet performance and engine/vehicle weight models and incorporate thermal balance to specify minimum fuel equivalence ratio requirements.

The result of applying this technique to post process, a typical scramjet combustor CFD solution, is illustrated in Fig. 33. This figure presents a typical distribution of thrust potential through a combustor. For this example, the inflow plane A has some thrust-generating potential (as inlet drag was not considered in this case). Upstream of the fuel injection B, frictional drag and heat loss to both the combustor walls and to any intrusive injector surfaces, injector generated shock waves and pressure drag, reduce the thrust potential. The region of the flow between B and C is dominated by the injection of fuel; the sharp increase in thrust potential is caused by the fuel momentum.

Exothermic reaction occurs in region C to E and is responsible for increasing the thrust potential; in the region C to D, the benefit of energy release associated with the mixing and combustion overcomes the losses due to wall friction, heat transfer to the walls, and shocks. Finally, although combustion continues in the region between D and E, the losses overcome the gain. The designer's task is to find injector and combustor designs that optimize the peak thrust potential with acceptable peak heat flux, total heat load, and engine length (weight).

The greatest limitation of this method of quantifying combustor thrust potential is associated with the one-dimensionalization process. Irreversible loss in the available thrust potential, which is analogous to that occurring in a multistream mixing process, can be seen both in an increase in integrated entropy or in a corresponding decrease in thrust potential. Evaluation of the magnitude of the mixing loss resulting from the one-dimensionalization process has been studied by individually expanding the flow in each cell in the combustor CFD exit flow plane and summing the resulting thrust potential. The thrust potential obtained without averaging is approximately the same as that computed by using the one-dimensional method. The difference found is less than 2% for most cases examined, except in the injection near-field where the flow is highly nonuniform.

Application of the thrust-potential model to compare competing fuel injector concepts is demonstrated in Fig. 34 (Ref. 90). This thrust-potential assessment is for simulation of a flight Mach 7 scramjet with expanding combustor, comparing two fuel injector designs: a 30-deg flush-wall sonic injector, and a 10-deg swept-ramp injector. Both injectors provide comparable fuel mixing and combustion efficiency. Differences in the thrust potential are seen ahead of injection where ramp drag reduces the thrust potential of

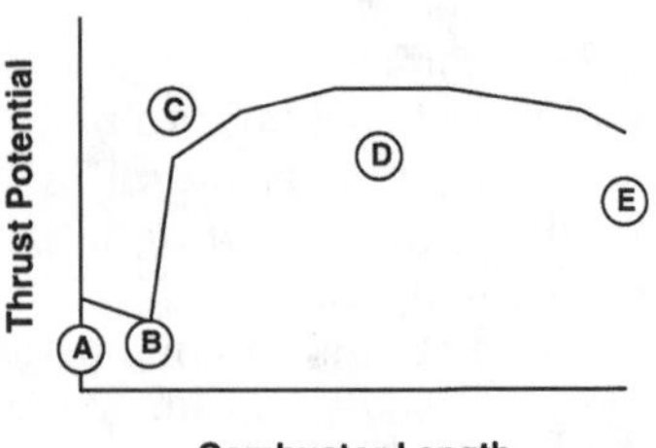

Fig. 33 Typical distribution of thrust potential through a combustor.

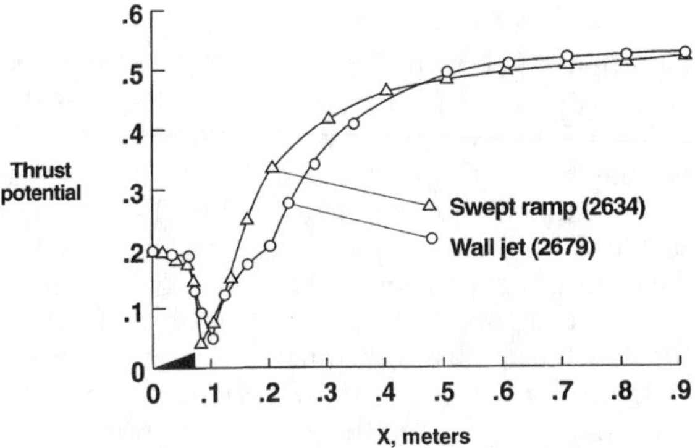

Fig. 34 Combustor effectiveness vs axial length for swept ramp and wall jet.

the ramp injector. In the region downstream of injection, axial jet momentum imparts a large increase in thrust potential. In the near field $0.15 < x < 0.3$, the ramp injector both mixes and reacts fuel faster than the flush-wall injector, hence the rapid rise in thrust potential. In the far field mixing for the wall injector catches up with the ramp injector case, and the thrust potential exceeds that of the ramp. The difference between the two cases is about equal to the ramp injector drag. The small differences observed in Fig. 34 can become very significant in the Mach 15 regime, where, for example, combustor shear increases dramatically to 25–50% of the engine net thrust and net thrust becomes a small difference between large forces. In the Mach 15 regime all losses become a larger fraction of the net thrust, and the fraction of energy availability added to the flow by combustion becomes smaller.

The purpose of this section is to present an assessment of scramjet combustor technology following eight years of extensive effort to expand this technology for speeds approaching Mach 20, which are required for an SSTO air breather. Facilities and test techniques have been brought on line to provide combustor simulation at Mach 14–20 flight energy, albeit at flight dynamic pressure and scale somewhat below those desired. Methods for extending these to full simulation were discussed, and plans are under way to accomplish this extension in the near future. Some of these enhancements are referenced in the discussions in Appendix B. Advances in instrumentation for pulse facility testing currently provide accurate measurement of scramjet fuel mixing, wall pressure, and heat flux; and work is progressing to resolve both combustion efficiency and combustor wall shear. Advances in CFD application have provided solutions, which, although lacking in some physics, provide significant insight into this new, complex high-speed combustor environment. Nevertheless, comparisons with existing data provide surprising similarity. Improvements in design methods are based on the experimentally anchored CFD methods, such as the use of thrust potential. These design methods provide a consistent methodology for improving the combustor design.

Nozzle

The nozzle completes the propulsion flow path and has the job of expanding the high pressure and temperature gas mixture generated within the inlet and combustor into a high velocity exhaust with greater momentum than the captured airflow, thus generating net thrust. In the hypersonic vehicle this is accomplished by expanding the combustor-exit flow starting within the engine module and continuing the expansion over a large portion of the afterbody of the vehicle. In this expansion process potential energy is changed to kinetic energy, and the shape of the afterbody gives a direction to the propulsion flow path, which establishes the angle of the gross thrust vector relative to the vehicle's flight direction. These two factors must be given primary consideration in the design of the hypersonic nozzle: the efficient generation of thrust and the aerodynamic balance of the vehicle. At hypersonic speeds these factors are critical because the nozzle is working on the total flow through the engine, and net propulsive thrust is a small difference between two large numbers, i.e., the gross thrust exiting the nozzle and the ram drag of the stream tube entering the inlet. The nozzle thrust direction dominates the trim of the vehicle at hypersonic speeds. The impact of these factors on the design of the combined vehicle and propulsion system is discussed in the next section along with other nozzle issues that must be considered at lower speeds.

Performance of the nozzle is a result of the upstream flow process as well as the flow process that occurs within the nozzle. There are five principal loss mechanisms that will be discussed: flow profile at the nozzle entrance, failure to recombine dissociated species, skin friction, flow divergence, and underexpansion losses. Flow profiles at the nozzle entrance are a result of boundary-layer growth from the body side and cowl surfaces, inlet shock waves that were not canceled at the throat and passed through the combustor, and the fuel injection, mixing and combustion process. The effect of these profiles is not intuitively obvious, given the complex nature of their interaction with the expanding flow and their effects on the other loss mechanisms. There has been some thought of altering the flow profile with speed by changing the way fuel is proportioned between body side and cowl fuel injectors, thereby altering the gross thrust vector angle and improving the aerodynamic balance of the vehicle over a range of Mach number.

The effects of the other flow mechanisms on nozzle performance are somewhat better understood, and always result in a loss to the scramjet flow path. A typical measure of nozzle performance is CFG, which is defined as gross thrust of the stream tube at the exit of the nozzle referenced to the equilibrium ideal thrust of the flow-path stream tube expanded to the freestream static pressure. A good nozzle would have a value of CFG of about 0.96, with around 0.97 representing a practical upper limit. The magnitude of these losses as well as their relative importance is a strong function of the design approach for the propulsion system. Perhaps the most complex of these is the dissociation losses, which are those losses resulting from the flow freezing in the rapid expansion process within the nozzle, so that energy is not generated by the recombination of free radicals produced by the combustion process. This is not a significant problem at low speeds because dissociation within the combustor is a strong function of static temperature and pressure, and at low speeds pressure tends to be high and static temperature is relatively low. Thus, features and performance levels within the engine that affect static temperature at high Mach numbers also affect dissociation and the potential for losses through freezing the flow in the nozzle expansion process. Figure 35 gives the potential for losses in CFG caused by freezing dissociated flow for propulsion parameters that include combustor fuel equivalence ratio, inlet contraction ratio, and inlet kinetic energy efficiency for a Mach 14 flight condition. Dissociation losses are maximized at a fuel equivalence ratio of unity corresponding to the highest combustor-exit temperatures. Note that the losses are particularly sensitive to the inlet design parameters. A high contraction ratio is desirable at high Mach numbers to increase static pressure and temperature for rapid ignition and combustion; but, as can be seen, high contraction must be limited to avoid high losses caused by kinetic freezing affects. In addition, chemical kinetic losses are particularly sensitive to reduced inlet kinetic energy efficiency. Analysis of finite-rate chemical kinetics of the expansion process generally leads to the conclusion that the flow is very close to a frozen process. Efforts to minimize this loss must rely on revising component design parameters to limit static temperature and dissociation at the combustor exit (such as some combustor divergence) or find ways to catalyze reaction in the nozzle flow and drive the expansion process further toward equilibrium.

The effects of friction within the nozzle on nozzle CFG are illustrated in Fig. 36, with friction coefficient given in terms of local nozzle surface flow conditions. Friction is reduced as the flow expands at the nozzle throat, but because the nozzle is so large the total drag becomes significant. A larger portion of the friction is generated in the first half of the nozzle where pressures are higher so that the behavior of the boundary layer in the initial portions of the nozzle is most important in controlling friction and heat transfer. Efforts have been made to evaluate the possibility of the boundary layer relaminarizing just downstream of the nozzle throat.[91] The forming of laminar flow within the nozzle would significantly reduce friction drag and heat transfer.

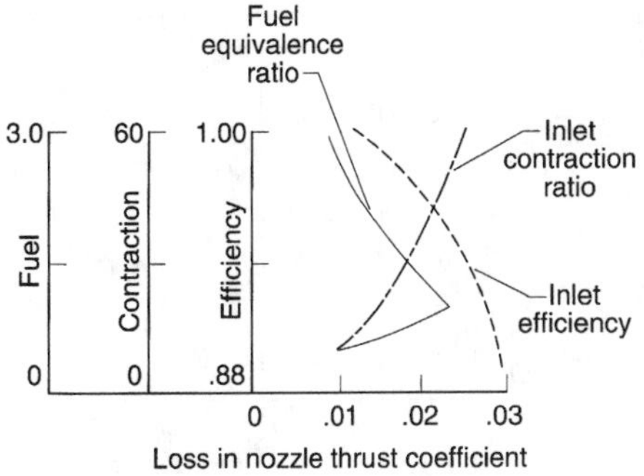

Fig. 35 Effect of inlet and combustor parameters on nozzle thrust coefficient for frozen flow chemistry compared to equilibrium flow chemistry at Mach 14.

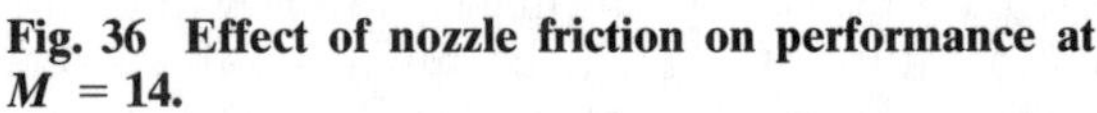

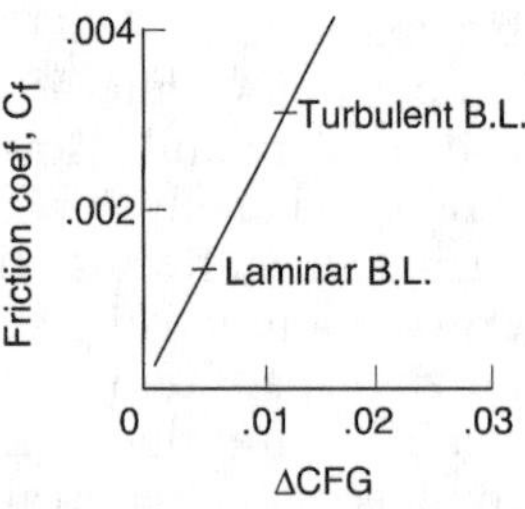

Fig. 36 Effect of nozzle friction on performance at M = 14.

The final two losses that will be considered are caused by flow divergence and under expansion (Fig. 37). Flow divergence is a result of streamlines at the nozzle exit being at different angles relative to that of the flight path. Internal flow divergence is a result of the nozzle shape and three-dimensional features and can be analyzed using CFD calibrated against test results.[92] Under-expansion losses are a result of the nozzle not being large enough to fully expand the flow to the ambient pressure and are constrained by the physical size of the airframe afterbody. Note that nozzle performance can be very sensitive to both of these nozzle design parameters.

Engine/Vehicle System Integration

Forebody/Inlet

The vehicle forebody serves as the external compression portion of the inlet. In designing the hypersonic inlet, to maximizing the external forebody compression to minimize internal inlet surface area and heat load is desirable. In addition to providing inlet compression, the forebody must provide a uniform distribution of air across the vehicle ahead of the propulsion module. Figure 38 illustrates a poor forebody design in that the static-pressure distribution ahead of the propulsion modules results in a large accumulation of boundary layer in the center of the forebody. Such an airflow distribution would cause an unacceptably thick boundary layer and airflow loss in the center propulsion module. The importance of finite-rate chemistry in calculating lateral airflow distribution as well as flowfield profiles between the body and cowl is also illustrated in this figure. Studies have

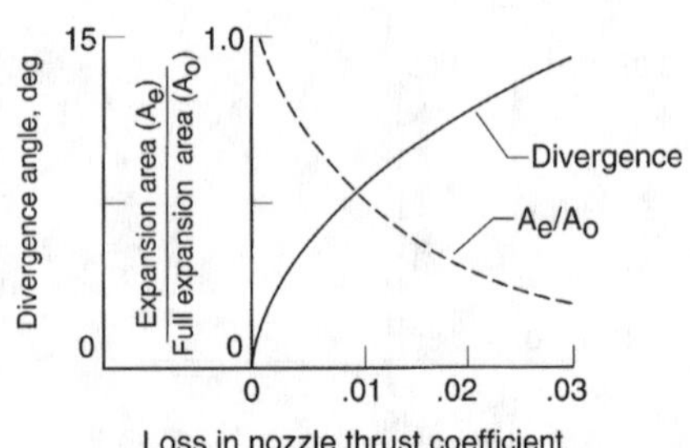

Fig. 37 Effect of divergence and under expansion on nozzle performance at Mach 14.

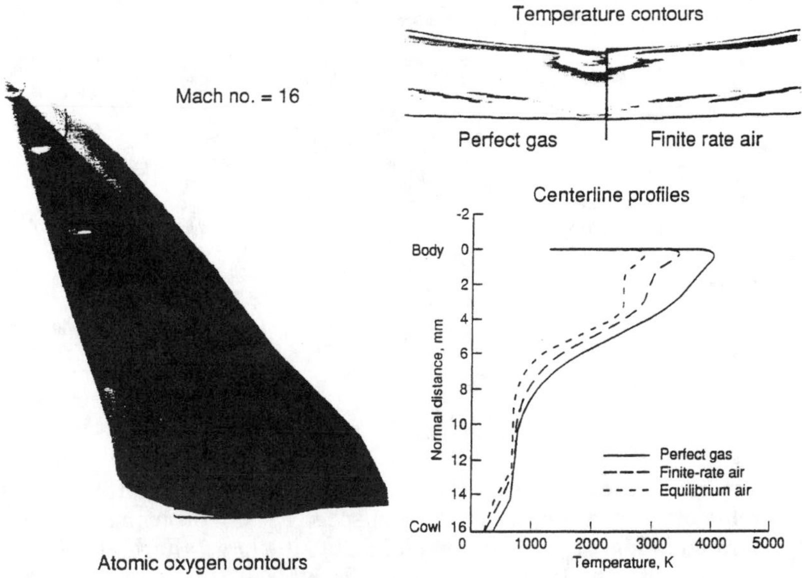

Fig. 38 Forebody analysis on a typical SSTO vehicle.

been conducted[93] to determine forebody shapes that maximize compressive performance and minimize airflow distortion.

In addition to airflow compression and distribution, forebody shape also plays an important part in determining the point of boundary-layer transition. A delayed transition from a laminar to a turbulent boundary layer reduces friction drag on the forebody and has a significant effect on overall vehicle drag at hypersonic speeds. Also, a reduced rate of boundary-layer growth results in thinning the boundary layer entering the propulsion module, thereby increasing the airflow processed through the propulsion system. The result would be a significant increase in payload for an SSTO vehicle (Fig. 39). Unfortunately, most aerodynamic research is conducted in noisy facilities that have turbulence levels sufficient to affect boundary-layer transition. Research in noisy tunnels would predict a much earlier transition point on the forebody as indicated by a comparison with quiet tunnel data at supersonic speeds. In addition, a higher transition Reynolds number Re_T would be predicted for a cone in a noisy tunnel as opposed to a flat plate, although linear theory and quiet tunnel data would give the opposite result. Confirming these trends at high hypersonic Mach numbers will be important to define the forebody shape that would be most desirable to maximize vehicle performance.

Nozzle/Afterbody

A unique feature of the hypersonic airframe-integrated propulsion system is a propulsion module separating large forebody and afterbody surfaces (Fig. 40). The detailed design of these surfaces is dictated by propulsion

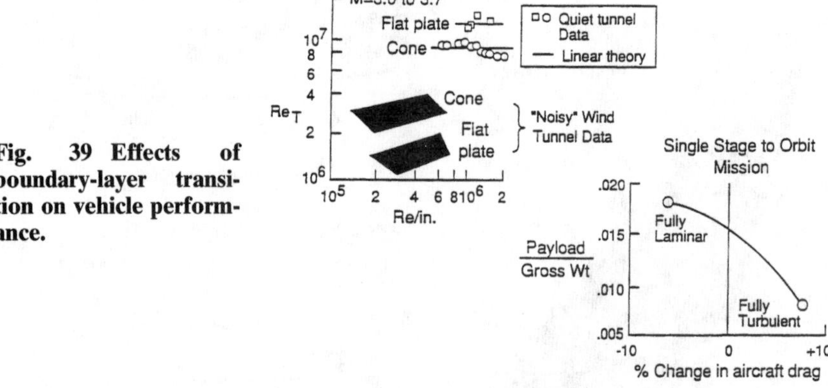

Fig. 39 Effects of boundary-layer transition on vehicle performance.

requirements at the high end of the vehicle's design speed range, resulting in extreme off-design conditions at lower speeds. These problems start early, as illustrated in Fig. 41. At rotation on takeoff, the large afterbody becomes nearly parallel to the ground plane and is subject to a reduced pressure resulting in a loss of lift when propulsion exhaust is simulated. These effects, which are influenced by elevon, wing, and nozzle geometry,[94] could result in a serious design penalty, given the inherently poor aerodynamic performance of hypersonic vehicle shapes at subsonic speeds.

Another consequence of the large afterbody is high transonic drag. The nozzle must be designed for a very high-pressure ratio at hypersonic flight conditions, leaving a highly overexpanded nozzle at transonic speeds. Figure 41 summarizes results from nozzle tests conducted in the 16-Foot Transonic Tunnel at LaRC. Results show that pressure drops below atmospheric pressure immediately downstream of the cowl exit and then recovers back to atmospheric pressure at approximately half of the nozzle length. The downstream pressure recovery back to atmospheric pressure results from the overexpansion shock originating from the cowl lip as well as outside airflow filling in the large base area. Nozzle sidewall fences inhibited pressure recovery,

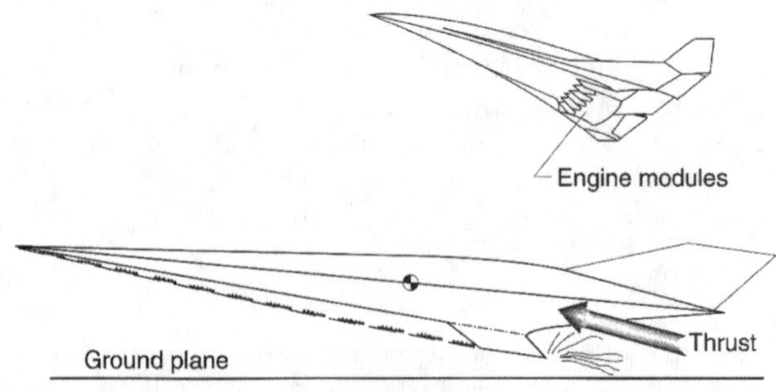

Fig. 40 Hypersonic vehicle at takeoff.

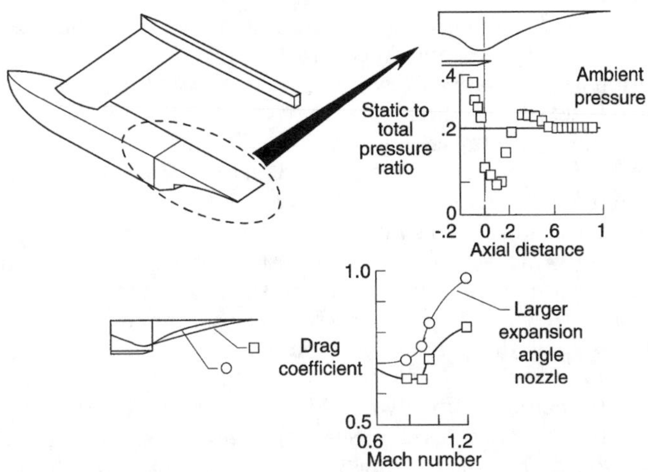

Fig. 41 Aft-body performance at transonic speeds.

which leads to a much higher level of base drag. Note in the lower part of the figure that drag maximizes at Mach 1.2 rather than Mach 1 and is a function of nozzle afterbody expansion angle. Tests have also been conducted on nozzle afterbodies[95,96] in the Langley 20-Inch Mach 6 Tunnel (Fig. 42) to determine performance characteristics. Parametric tests included the nozzle sidewall fence and air or simulant gas to represent the nozzle exhaust flow. The simulant gas was a cold mixture of gases intended to properly reproduce the engine exhaust flow ratio of specific heats throughout the nozzle expansion process. Note that measured nozzle forces are increased when the exhaust flow is simulated as compared to results using air. In addition, increases in nozzle thrust and lift occur when a flow fence is installed because the nozzle is

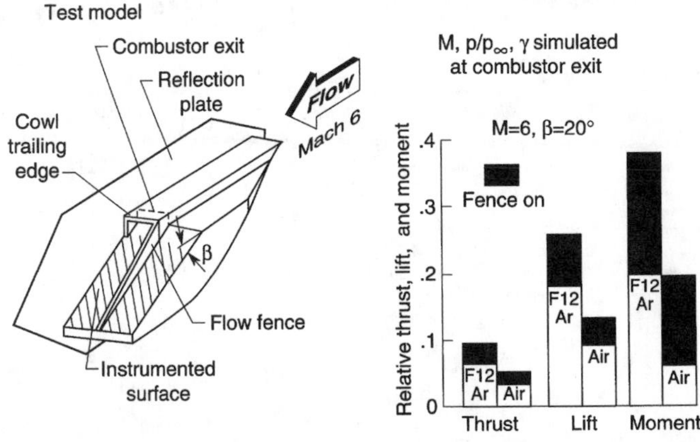

Fig. 42 Hypersonic nozzle exhaust simulation; effect of flow fence and simulant gas.

not overexpanded and exhaust flow containment within the nozzle maximizes thrust at higher speeds. In contrast, at transonic speeds (Fig. 41) the configuration without sidewalls would have less base drag.

As would be expected, the nozzle afterbody has the potential for producing large lift forces and resulting pitching moments that would be imposed on the vehicle. Previous studies conducted on research airplane concepts[97,98] have shown the magnitude of the lift force to be a function of the afterbody expansion angle, the vehicle shape, and the axial location of the propulsion module on the vehicle. Effects of propulsion forces as a function of axial engine location are given in Fig. 43. The results shown in this figure assume shock-on-lip at Mach 10, so that airflow and resulting thrust at the combustor exit increase as the propulsion module is moved aft and the distance between the body and forebody shock wave becomes greater. Nozzle expansion angle also increases along with a decrease in nozzle length, both acting to reduce lift. The result is a nose-up pitching moment that requires a positive elevon deflection to trim the vehicle (Fig. 44), which in turn produces a drag and reduces thrust margin. At hypersonic speeds elevon trim drag can dominate overall vehicle drag leading to a large loss of installed propulsive performance. A challenge of the hypersonic vehicle designer is to deal with a balance between thrust, lift, and pitching moment over a wide speed range and at different propulsive power settings that may impose additional constraints on the design of the integrated propulsion system.

Concluding Remarks

This chapter has presented an overview with depth added in selected parts of the 30-plus years of research involvement with airframe-integrated ramjets and scramjets at the NASA Langley Research Center. Following the resurgence of interest in hypersonic air-breathing propulsion for SSTO vehicles revolving around the NASP Program, which is perhaps more appropriately termed Aerospace Plane II the industry and the country are at a decision point for hypersonic propulsion technology. Because the NASP Program has now been concluded in part because of an inability to successfully demon-

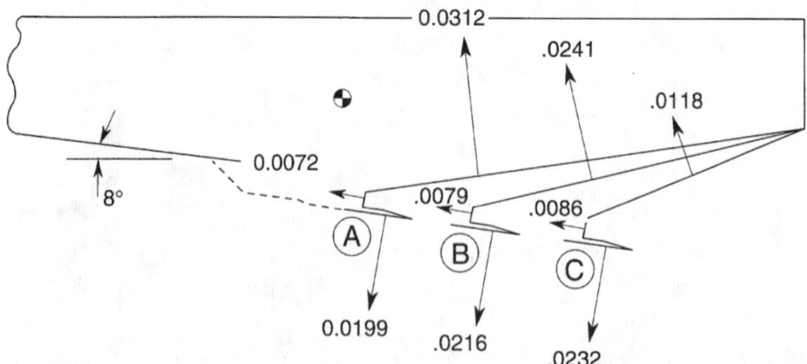

Fig. 43 Propulsion force coefficients as a function of scramjet location at Mach 10.

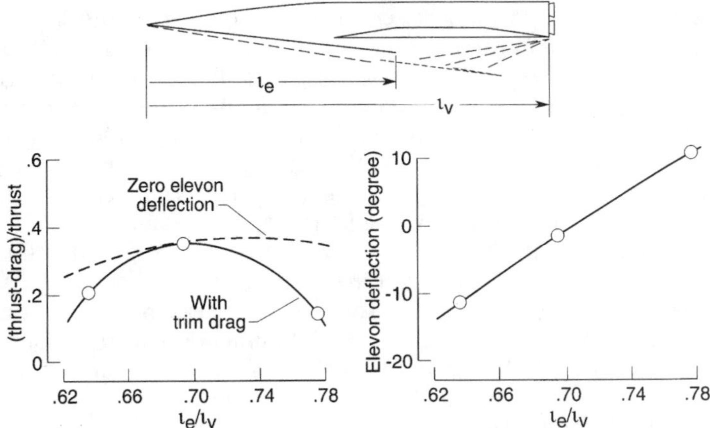

Fig. 44 Engine axial location effects at Mach 10.

strate adequate dual-mode scramjet performance, the prudent path would seem to follow a continued effort to exploit the NASP investment in infrastructure and technology by continuing a highly focused program of refinement and innovation in hypervelocity propulsion technology. Given the changing world political situation and reordered national priorities of the 1990s, the logical culmination of this propulsion technology program might be a rocket-boosted demonstration of a specially designed high-performance scramjet engine flight package rather than a manned airplane capable of SSTO (i.e., the X-30), which was the goal of the NASP Program. In fact, NASA has started such a program, the Hyper-X (Appendix B). The primary objective of this program is to elevate scramjet technology readiness by acquiring flight data for an airframe-integrated hypersonic vehicle.

Although many challenging and interesting issues and problems remain in the Mach number less-than-8 speed range, the real proof of principal for (air-breathing) scramjet application to orbital transportation is in the level of performance achievable in flight at Mach 10 and above. The NASP investment has provided the tools (impulse facilities, CFD codes, measurement techniques, trained people and infrastructure) to pursue and demonstrate the required propulsion technology. The challenge remains to apply these tools to bring the Aerospace Plane II era to a successful conclusion in the 1990s–2000s and make high-performance hypervelocity propulsion available as a viable option for future space transportation systems.

Appendix A: Central Institute of Aviation Motors—NASA MACH 6.5 Scramjet Flight Test

Introduction

On February 12, 1998, the Central Institute of Aviation Motors (CIAM) performed the highest speed, longest duration dual-mode scramjet flight test yet conducted. An axisymmetric scramjet was flown on the nose of a modified SA-5 surface-to-air missile launched from the Sary Shagan test range in the Republic of Kazakhstan. It achieved 77 s of liquid hydrogen regeneratively cooled and fueled engine data at Mach numbers from 3.5 to 6.5. NASA contracted with CIAM in November 1994 to perform this flight test and a companion set of ground tests of the CIAM designed scramjet. Previously CIAM had conducted three flight tests of a similar scramjet.[99,100] The first achieved a peak Mach number of about 5.5. The second and third tests, conducted jointly with the French,[100] reached Mach 5.35 and 5.8, respectively. However, on the third flight the scramjet failed to operate because of an onboard power system problem. The NASA contract provided for ground and flight tests at Mach 6.5 of a modified scramjet design. The ground tests were performed by CIAM in the CIAM C-16V/K facility at Tureavo. The overall program was designed to provide the first ever flight demonstration of supersonic combustion and ground-to-flight comparisons for scramjet engine design tool methodology verification. On previous flights the dual-mode scramjet combustor operated in the subsonic combustion mode.

Experimental Apparatus and Test Conditions

For the CIAM/NASA flight test the previous scramjet test article was redesigned for the higher Mach 6.5 heating environment and to ensure that the combustor remained supersonic (see geometry in Fig. A1a and A1b). These changes included expanding combustor sections to assure scramjet operation and an improved thermal/structural design, including an improved cooling liner design and modifications of the inlet cowl leading-edge material. A cross-section view of the modified annular combustor cooling liner structure is presented in Fig. A1b. Most of the cooling liner uses copper alloy material on the hot side and steel on the cold side. However, all steel liners are utilized at selected locations to meet structural strength requirements. The modified thermal/structural design is schematically illustrated in Fig. A2. Details of the pretest analyses that led to these design changes are included in Ref. 101.

The flight tests were conducted using the CIAM designed Hypersonic Flying Laboratory (HFL) KHOLOD.[100] The HFL is an experimental support unit (fuel, controls, instrumentation) specifically designed to support captive carry tests of these engines on the nose of an SA-5. A photograph of the scramjet, HFL, and SA-5 missile during launch preparations is presented in Fig. A3.

Flight and Ground-Test Results

The flight test occurred just after 2:00 PM Thursday, February 12, 1998 (Ref. 100). The flight trajectory parameters of altitude (H), dynamic pressure (Qbar), Mach number, and relative fuel flow rate vs time are illustrated in Fig. A4. The flight data indicate that there was fuel flow to the scramjet for about 77 s, starting at a Mach number of approximately 3.5 (38 s into the flight), continuing through the maximum velocity of 1830 m/s at 21.4 km

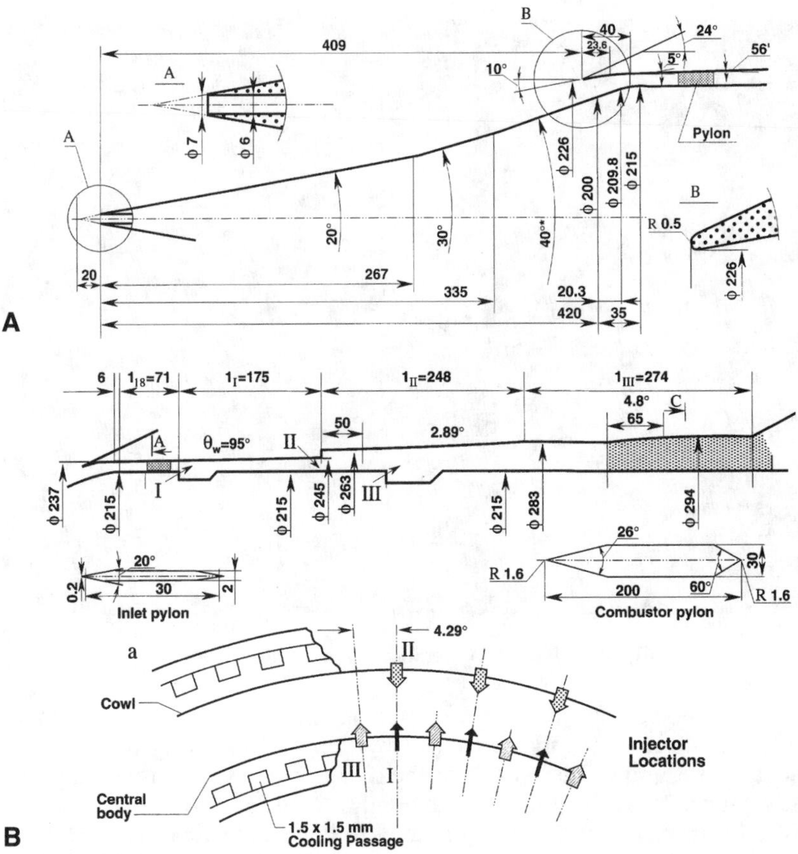

Fig. A1 Inlet and combustor geometry details of scramjet engine (units, mm; φ signifies diameter): a) Inlet details; b) Combustor details.

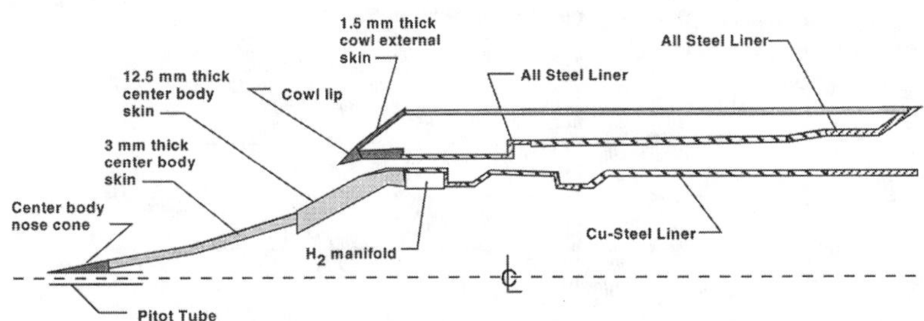

Fig. A2 Scramjet engine structural and material details.

Fig. A3 CIAM Hypersonic Flying Laboratory.

altitude at booster burnout (56.5 seconds), which corresponds to a Mach number of 6.48 given the measured (by weather balloon) static temperature, static pressure, and winds. Note that the dynamic pressure is relatively constant through this 18.5-s portion of the flight. After Mach 6.48 is achieved at booster burnout, the missile and scramjet follow a ballistic trajectory. After missile burnout the scramjet gradually slows to Mach 5.8, and dynamic pres-

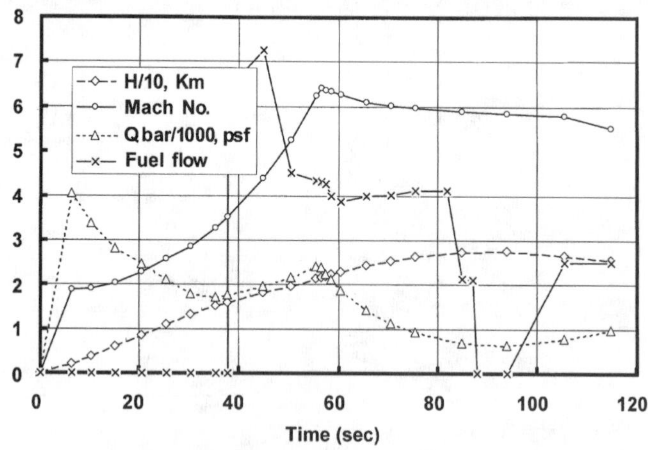

Fig. A4 Flight-test trajectory.

sure decreases to the maximum altitude (27 km) condition at 90 s, then increases until the flight-test termination. Fuel flow continued through the ballistic portion of the flight (except for a few seconds at about 90 s) until a flight termination device was activated at 115 s. The scramjet was located after the flight and recovered, dented but intact. CIAM is presently conducting a post-flight inspection of the recovered engine (Fig. A5), but preliminary indications are that the structure and cooling system performed as predicted.

Several anomalies appear to have occurred during the test. First, the missile flew a little lower than anticipated. The altitude at maximum velocity was 21.3 km rather than 24 km. Second, the inlet appears to have unstarted when fuel was first injected and remained unstarted for about the first 10 s of fueled operation. The inlet appears to have restarted and then remained started as the vehicle accelerated to Mach 5. The third anomaly was that while the aft two fuel-injection stages operated as designed from inlet start, the first-stage fuel did not come on when it was commanded to 53 s into the flight at about Mach 5.5. This resulted in this dual-mode scramjet engine operating in the subsonic combustion mode, rather than the desired supersonic combustion mode.

Ground tests of another copy of the same engine were run by CIAM in their C-16 V/K facility just after the flight test.[103] These tests were run at simulated Mach 6.5 test conditions with fuel injection from the second-and third-stage injectors only as in the flight test. A schematic of the test facility with the engine installed is presented in Fig. A6. This ground-test engine had the same

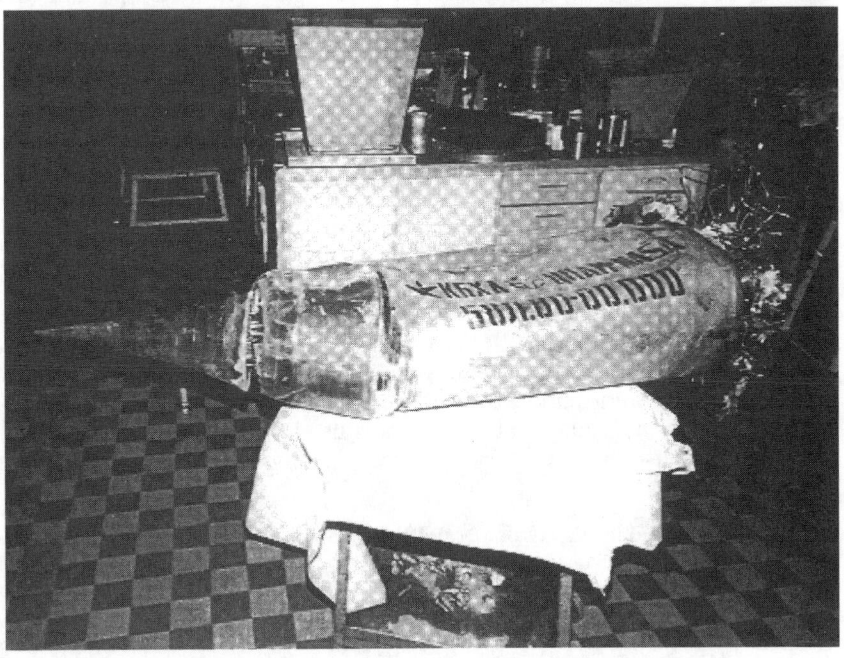

Fig. A5 Recovered scramjet model after flight.

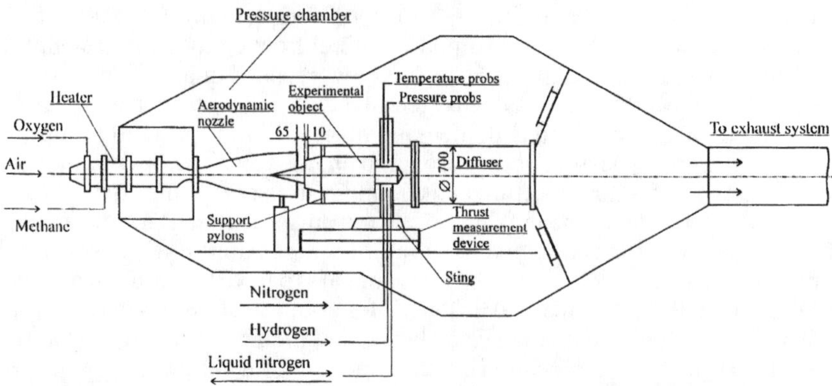

Fig. A6 Axisymmetric scramjet engine installed in CIAM V/K facility, Mach 6.

instrumentation as the flight engine plus the addition of a force measurement system to measure the engine's thrust. A comparison of the flight-and ground-test conditions is presented in Table A1. The flight condition selected is that with the highest enthalpy, which is the closest to the wind-tunnel value. Initial comparisons of the pressure measurements show small differences between ground and flight combustor normalized pressure, as illustrated in Fig. A7. These results were obtained at the same test gas total enthalpy and fuel equivalence ratio (phi) but different total pressure and Mach number (see Table A1). Detailed analysis of these results provide useful demonstration of facility contaminant effects vis -à-vis flight scaling of ground-test data.[103a]

Appendix B: NASA'S Hyper-X Program

Introduction

NASA initiated the Hyper-X Program in 1996 to mature hypersonic air-breathing propulsion and related technology.[104,105] Hyper-X is a joint program executed by the NASA Langley Research Center (LaRC) and Dryden Flight Research Center (DFRC). The program is designed to move hypersonic air-breathing vehicle technology from the laboratory to the flight environment, the last stage preceding prototype development. The program goal is to verify and demonstrate the experimental techniques, computational methods, and analytical design tools used for hypersonic, hydrogen-fueled, scramjet-powered aircraft. Accomplishing this goal requires data

Table A1 Ground-and flight-test gas conditions

Test point	M	q, kPa (psf)	h_t, MJ/kg (Btu/lbm)
Ground	5.96	55.5 (1160)	1.91 (820)
$t = 56.5$ s	6.40	11.4 (2376)	1.88 (808)

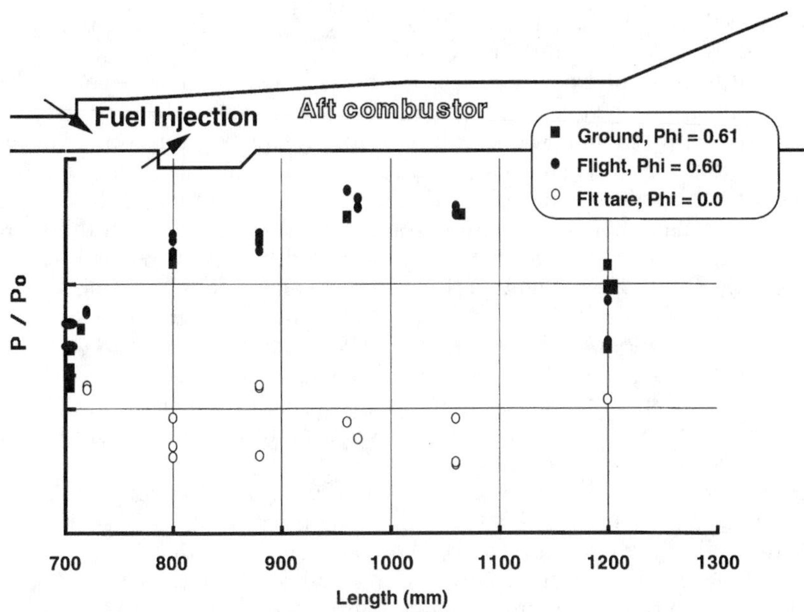

Fig. A7 Comparison of ground and flight combustor data (P_0 is test gas freestream total pressure.).

from a successful scramjet-powered vehicle in flight. Because of the integrated nature of scramjet-powered vehicles, the complete vehicle must be developed and tested: propulsion verification cannot be separated from other hypersonic technologies. A significant driver in the Hyper-X program is the modest budget and the challenge to perform better, faster, cheaper. For this reason, the flight-test vehicle is small to minimize vehicle development cost plus the cost of achieving boost-to-test condition. In addition, the vehicle is based on an existing configuration (specifically designed for a Mach 10 cruise, global-reach mission) and the large National Aerospace Plane (NASP) database to reduce design costs. Flight tests are intended to demonstrate the design cruise condition, plus the Mach 7 accelerating, off-design condition. Additional ground testing is planed for Mach 5, the scramjet take-over condition.

Flight-Test Vehicle Design and Fabrication

The conceptual design for the Hyper-X was completed in May 1995,[106] and the preliminary design (called the government candidate design) was completed in October 1996. This included basic structural design, thermal protection system selection, identification of most system and subsystem components and potential vendors, preliminary packaging, power requirements, stage separation details, booster integration, and flight-test planning. The Hyper-X research vehicle was essentially scaled photographically from a vision vehicle developed in a previous study. The vision vehicle is a Mach 10 cruise global reach concept. Scaling the configuration external lines allowed utilization of existing databases, as well as rapid convergence to a controllable flight-test vehicle with low trim drag penalty. The scramjet flow path, on the other hand, was reoptimized for engine operability and vehicle acceleration, accounting for scale, wall-temperature effects, etc. For example, the inlet contraction, fuel injector detail, and combustor length have been modified, rather than simply photographically scaled. Although there are differences between the vision and Hyper-X flight test vehicles, demonstration of the Hyper-X predicted performance is validation of the design process.[107] Figure B1 illustrates results from the highest level methods used for design. This Reynolds-averaged Navier–Stokes solution, produced using the General Aerodynamic Simulation Program (GASP) code,[108], provides a complete solution of the flowfield of the flight research vehicle at the Mach 7, scramjet-powered test condition and was used to verify the performance based on simpler design methods.

Wind-tunnel testing commenced in early 1996 to verify the engine design, develop and demonstrate flight-test engine controls, develop experimental aerodynamic databases for control law and trajectory development, and support the flight research activities, Mach 7 engine performance and operability were verified in reduced dynamic-pressure tests of the DFX (dual-fuel experimental) engine in the NASA LaRC Arc-Heated Scramjet Test Facility (AHSTF, Ref. 109). Figure B2 shows the full-scale, partial-width DFX flowpath model in the AHSTF test cabin. Preliminary experimental results for the Mach 5 and 10 scramjet combustor designs were obtained using the direct-connect combustor module rig and an existing semi-direct connect combustor model in the Hypersonic Pulse Facility (HYPULSE) operating in the reflected shock mode.[110] To protect the engine during boost, the scramjet inlet design includes an articulating cowl door. Tests were performed early in the program to verify inlet starting with this configuration, using the parametric inlet model illustrated in Fig. 12.

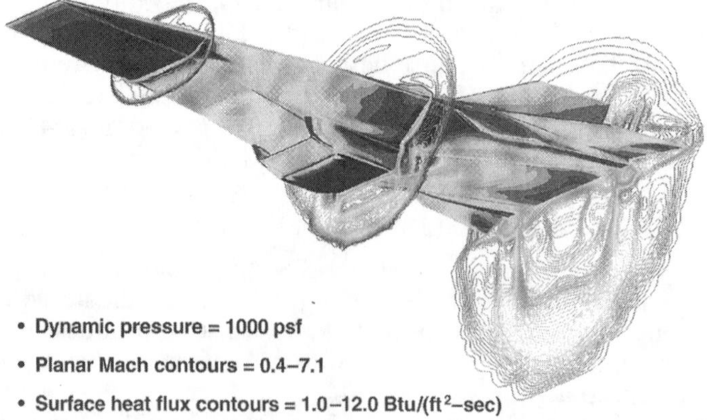

Fig. B1 Hyper-X Mach 7 powered CFD solution.

Fig. B2 DFX engine in AHSTF.

A preliminary aerodynamic database was developed in 1996 from results of quick-look experimental programs on 11 separate wind-tunnel models using over 1000 wind-tunnel runs. These tests were performed using 8.3 and 3.0% scale models of the Hyper-X research vehicle (HXRV) and the Hyper-X launch vehicle (HXLV) at Mach numbers of 0.8–6, 6 and 10. The aerodynamic database includes boost, stage separation, research vehicle powered flight (with propulsion data) and unpowered flight back to subsonic speeds. The aerodynamic database is being filled in with additional wind-tunnel tests as discussed under "Risk Reduction." Figure B3 presents the 8.3%-scale HXRV model in the 20 Inch Mach 6 Tunnel at LaRC.

Other work leading up to the HXRV development contract included preliminary control law development, preliminary evaluation of trajectories, development of aerothermal loads, and preliminary design of engine thermal structure. Preliminary control laws[111] were developed for feasibility studies. The trajectory evaluation included some Monte Carlo uncertainty analysis using the methods demonstrated in Ref. 112. For the scramjet-powered part of the trajectory, longitudinal and lateral control laws were developed for angle-of-attack (AOA) and sideslip control. These include AOA and sideslip estimators that utilize motion data, aerodynamic data, and atmospheric and flight condition data. Preliminary assessments of flight trajectories and stability margins for the longitudinal control laws demonstrated that the vehicle meets the flight-test requirements, using conservative structural bending mode filters.

Aerothermal load development activities covered boost, separation, and the HXRV in free flight. The vehicle structural design and preliminary sys-

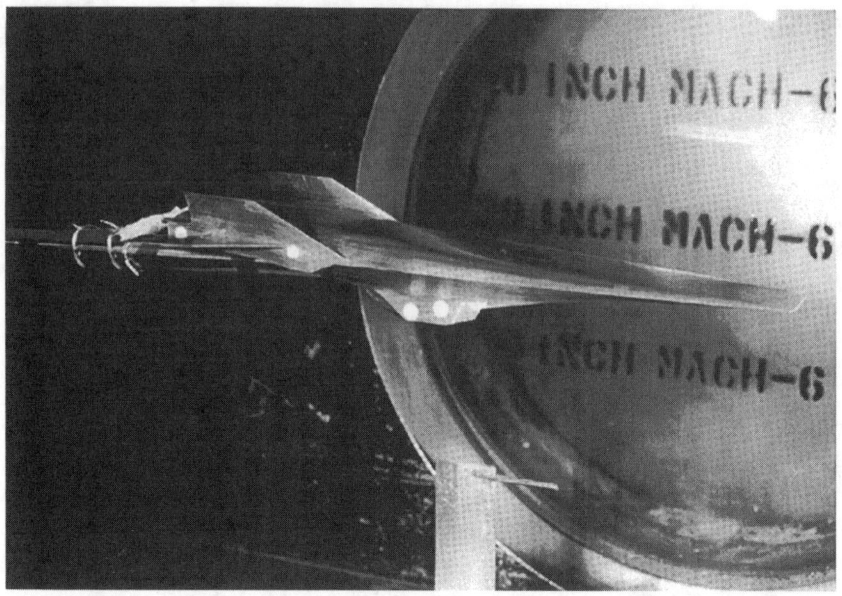

Fig. B3 The 8.3% scale HXRV model in the NASA Langley 20-Inch Mach 6 Tunnel.

tems layout are presented in Fig. B4. Vehicle structure utilizes metallic (largely aluminum) keels, bulkheads, and skins. Thermal protection consists of alumina-enhanced thermal barrier tiles, which have been fully characterized for the Space Shuttle, and carbon-carbon wing, tail, and forebody nose leading edges. The majority of the wings and tails are constructed from high-temperature steel. The flight engine is a robust design similar to wind-tunnel engine models. Most of the engine structure is copper, heat-sink cooled for the short duration scramjet test, but limited water cooling is included for the cowl and sidewall leading edges. An articulating cowl leading-edge section closes the flow path, protecting the engine internal surfaces during boost and descent after the scramjet test. High-pressure gaseous hydrogen (fuel), silane (scramjet ignition), and helium (fuel system and internal cavity purge) are contained in off-the-shelf fiber-wound aluminum tanks. Instrumentation, flight and engine-cowl control actuators and controllers, and the flight-control computer are all either off-the-shelf units or derivatives thereof. A high-pressure water system is included for engine-cowl cooling.

Measurements and instrumentation for the Hyper-X flight test are selected to 1) determine overall vehicle performance, 2) evaluate local flow phenomena, 3) validate design methods (propulsion, aerodynamic, thermal, structures and controls), 4) monitor vehicle systems for safety, and 5) identify failure modes. The instrumentation approach is to utilize only a limited quantity of proven reliable measurements (pressure, temperature, and strain gauge) to ensure program schedule and cost goals. Off-the-shelf data system

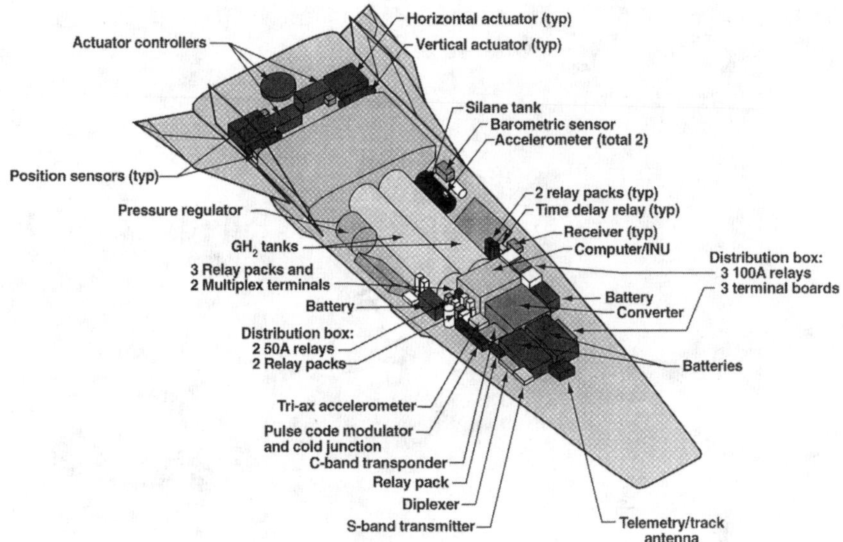

Fig. B4 Hyper-X equipment layout.

components are utilized to process and telemeter measurements. The primary measurement is vehicle acceleration. Of the 503 measurements, 194 are pressure, 107 are temperature, and 13 are strain gauge. About 160 of the total surface measurements provide propulsion flow-path-related data and 100 provide aerodynamic data.

Flight-Test Plans

For the flight itself the NASA DFRC B-52 will carry the HXLV-mounted research vehicle to launch altitude (see Fig. B5). The desired HXRV test condition at Mach 7 and 10 is a dynamic pressure of 1000 psf. For the Mach 10 tests launch from the B-52 will take place at about 40,000 ft. For the Mach 7 tests, however, booster launch at lower altitudes (about 20,000 ft) will restrain the HXLV from overaccelerating the HXRV. For this same reason, in fact, the booster assembly will also incorporate up to 5 tons of ballast. The research vehicle will be boosted to approximately 95,000 ft for Mach 7 and 110,000 ft for Mach 10 tests. Fully autonomous, these vehicles will fly preprogrammed, 700 to 1000-mile due-west routes in the Western Test Range off the California coast, while telemetering the test data.

Nominal flight sequence for the Mach 7 flight test is illustrated in Fig. B6. Following drop from the B-52 and boost to a predetermined stage separation point, the research vehicle will be ejected from the booster stack (see Fig. B7, Ref. 113) and will start the programmed flight test. Once separated from the booster, the research vehicle will establish unpowered, controlled flight. The flight controller will also determine the true flight conditions,

Fig. B5 HXLV on B-52.

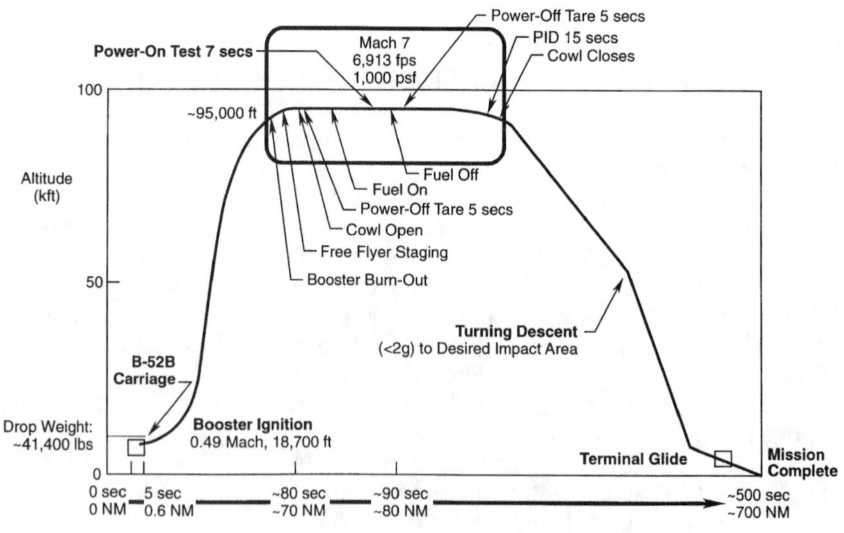

Fig. B6 Nominal Hyper-X flight trajectory.

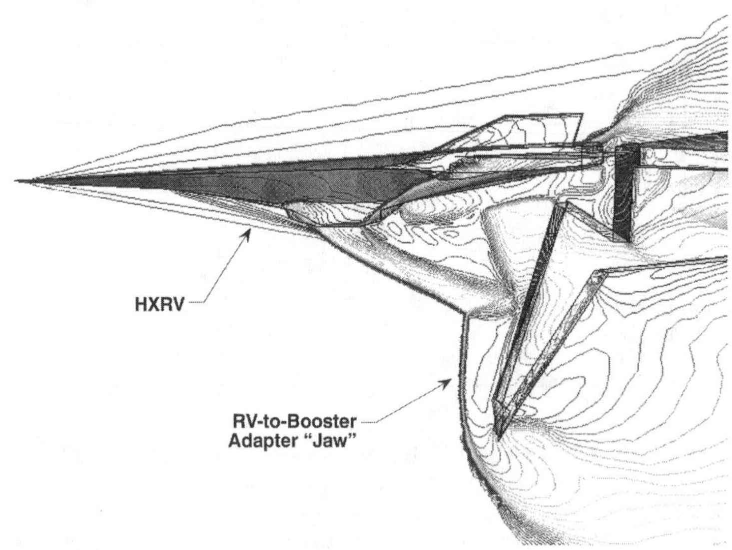

Fig. B7 Typical CFD solution for RV stage separation.

which are required to correctly program the fuel system. Following the powered engine test (discussed next) and 15 s of aerodynamic parameter identification maneuvers,[114,115] the cowl door will be closed. The vehicle will then fly a controlled deceleration trajectory to low subsonic speed. In the process short-duration, programmed test inputs will be superimposed on the control surface motions to aid in the identification of aerodynamic parameters. However, it is important to realize that mission success only requires demonstration of scramjet-powered vehicle acceleration. Thus, less preparation is expended on this lower priority segment of the flight.

Additional details of the scramjet-powered segment are illustrated in Fig. B8, using results from preliminary simulation models. Figure B8 presents elevator position and AOA as functions of time, from stage separation through cowl closure. Initially the elevator controls are locked, and the vehicle is assumed to be at the launch vehicle stage separation condition of 0-deg AOA. Aerodynamic and separation forces drive the vehicle nose down initially toward negative AOA. When active control is established, the control system pulls the elevator (wing) up to regain 0-deg AOA. At 0.5 s from separation, the flight controls switch to the powered flight attitude, which in this early simulation was 2-deg AOA. For this simulation the cowl door opens at 2 s. Tare operation (i.e., with no fuel injection) is maintained for 5 s. At 8 s after separation, ignitor (silane) and fuel flow are initiated. The ignitor is turned off at about 9.5 s as the fuel is ramped up to full power. Full power—design fuel flow rate—is maintained from about 11 to 13.5 s in this simulation. Combustion blowout and fuel ramp-down is complete at 14 and 14.5 s, respectively. Five seconds of engine tare data and 15 s of performance

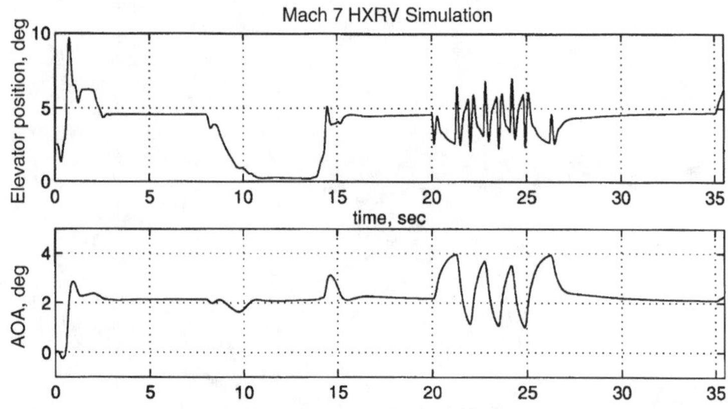

Fig. B8 Flight-control evaluation (elevator and AOA history).

identification maneuvers are completed before the cowl flap is closed 35 s after stage separation. During this process, the elevator excursions are within reasonable limits, and vehicle response is adequate for the flight test.

Hyper-X Technology

The Hyper-X program technology (the application of science to a practical purpose) development is concentrated on four main objectives required for practical hypersonic flight: 1) vehicle design and flight-test risk reduction, 2) flight validation of design methods, 3) method enhancements, and 4) future vehicle systems.

Flight-Test Risk Reduction

Risk reduction includes detailed analysis and testing to ensure that the flight test is successful, and it includes all phases of the flight and all disciplines. Flight phases are HXLV captive carry and separation from the B-52, the HXLV boost phase, stage separation, and the HXRV powered and unpowered flight. Analysis and testing include propulsion, aerodynamic, thermal-structural, guidance-trajectory-control, systems, etc. This activity serves to hone and can lead to significant improvements in the design tools. Scramjet design tools have never before been required for a real airframe-integrated flight vehicle. These tools include analytical, numerical, and experimental methods synergistically applied to reduce flight performance and operability uncertainty. Successful demonstration of the scramjet-powered vehicle's predicted performance will validate the use of these tools in the design process. Experimental propulsion flow path, aerothermal, and propulsion integration experimental tests play a key role in this activity and are highlighted in the following paragraphs.

Aerodynamic facilities, models and flight scaling. The primary aerodynamic wind tunnels used for testing the HXRV, HXLV, and stage separation include the facilities listed in Table B1. Details of these facilities are presented in Refs. 116–118., and discussion of the use of these facilities in the Hyper-X program is discussed in Ref. 113 and 119. A large range of models are being tested (see Table B1). The HXRV models include 12-in. (8.3% of full scale) and 18-in. (12.5%) steel models of the research vehicle and a 33%-scaled model of the research vehicle forebody. These HXRV models include variable control surfaces, including the all-moving horizontal tails and twin vertical rudder surfaces. The 12-in. model was used for preliminary and quick-look data (designed to scope, i.e., develop a course database for preliminary validation of the basic configuration). These tests were performed with rudder δ_r, aileron δ_a, and elevon δ_e, control-surface deflections from -20 to $+20$ deg, -20 to $+20$ deg, and -10 to $+10$ deg, respectively in 10-deg increments. The 18-in. HXRV model is used for detailed (i.e., to define control effectiveness over the entire expected flight operating range) and benchmarked (fine enough to accurately identify the nonlinear characteristics) tests. This wind-tunnel model has smaller control-surface deflection increments (2.5 deg). Similar tests were performed for the HXLV and for stage separation. Results from these aerodynamic wind-tunnel tests form the basis of the design aerodynamic database used in determination of scramjet thrust requirements and for flight simulation and control law development. The aerodynamic databse development process uses both engineering and CFD methods to account for other factors not addressed in the experiments. These include Mach number, test gas enthalpy, wall temperature and Reynolds number differences between wind tunnel and flight, sting- and/or blade-mounting interference, differences between cowl open and closed, and the powered portion of the scramjet engine test.[119] These factors are accounted for incrementally, and the resulting deltas are added

Table B1 Aerodynamic wind-tunnel tests (model scale in % of full scale)

Facility	HXLV, %	Stage separation, %	HXRV, %	BL control, %	FADS calibration, %
LaRC 31″ Mach 10	3[d]	8.3[b]	83[d] 125[d]	33[d]	—
LaRC 20″ Mach 6	3[d]	8.3[b]	83[d] 12.5[d]	33[d]	—
AEDC					
VKF-B Mach 6	8.3[d]	8.3[c]	83[d]	—	80[b]
Polysonic $0.4<M<4.6$	6[d]	—	16.7[b]	—	80[b]
BNA Mach 0.2	—	—	12.5[a]	—	—

[a]Preliminary/similar config.;
[b]quick look;
[c]detailed;
[d]benchmarked.

to the wind-tunnel results to develop the flight (design) aerodynamic database. In the process wind-tunnel aerodynamic data is used to anchor or validate the CFD and analytical methods, that is, the CFD and analysis are performed at the wind-tunnel condition and compared with the measured results before being used for generation of increments. In addition, the final aerodynamic database model is verified by comparison with CFD predictions for the flight conditions. This process decreases the design aerodynamic database uncertainty. The extensive database also provides background for quick validation of design methods and identification of error sources if the flight data do not agree with the design database.

Scramjet testing facilities, models, and flight scaling. Facilities utilized by the Hyper-X Program for scramjet engine flow path and propulsion-airframe integration development and flight-test risk reduction are summarized in Table B2. Details of these facilities and use in wind-tunnel tests are discussed in Refs. 109, 110, 113, 120, and 121. Unlike aerodynamic tests, scramjet tests require high-temperature (enthalpy) and high-pressure air or test gas. In these facilities air is heated by several means: electric arc, combustion, or reflected shock. For each of these facilities, the test gas contain some contamination relative to the flight environment. The impact of the contamination on performance adds some uncertainty to the predicted flight performance and engine operability (ignition, flameholding, inlet starting, and inlet isolator effectiveness). In addition, some of these facilities do not fully simulate the flight dynamic pressure (listed in Table B2). Shock tunnels, normally used for higher speed testing, are also used for Hyper-X Mach 7 engine tests. The NASA HYPULSE facility[110] at GASL, operating in the reflected-shock tunnel mode, provides clean-air test gas at dynamic pressures in excess of 1000 psf for Mach 7 full-scale flight simulation (albeit with short test times). Testing the engine at Mach 7 also provides a nearly direct comparison with long duration tests to verify the HYPULSE reflected-shock tunnel testing methods before the facility is used to determine the Mach number 10 scramjet engine performance. Using a shock-induced detonation driver, the HYPULSE facility provides true Mach 10 simulation for freejet scramjet module tests to simulated flight dynamic pressure of about 800 psf and to nearly 2000 psf for combustor testing.

For the Mach 7 flight tests propulsion risk reduction is accomplished by a series of wind-tunnel tests and related analysis. The test matrix, presented in Table B2, includes both engine and complete vehicle flow path as well as propulsion-airframe integration tests. The DFX engine (Fig. B2) was used to support engine flow-path design and help establish flow-path performance and operability. This engine simulates the full-scale, partial-width (44%) engine flow-path, as illustrated in Fig. B9. This simulation includes the correct cowl leading-edge radius and all flow-path lines from the vehicle midforebody to midaftbody stations, as denoted by the shaded region in Fig. B9. This entirely heat-sink cooled engine incorporated adequate parametric capability for preliminary feasibility studies. Two additional partial-width engines have been built for parametric testing over the Mach 5 to Mach 10 speed range: the Hyper-X Engine Model (HXEM) and the HYPULSE Scramjet

Table B2 Propulsion test facility statistics

Facility	Primary use	Flow energizing method (Max $T_{t,\infty}$, °R)	Simulated flight Mach no.	Nozzle exit Mach no.	Nozzle exit size, in.	Test section dimensions, ft	Dynamic Pressure psf
Direct-connect module GASL	Combustor tests	H_2/O_2/Air combustion (3800)	4.0 to 7.5	2.2	4.71 × 6.69	—	>4000 $M5$ and 7
Combustion-heated scramjet test facility (CHSTF)	Engine tests	H_2/O_2/Air combustion (3000)	3.5 to 5.0 4.7 to 6.0	3.5 4.7	13.26 × 13.26	2.5 W × 3.5 H × 8 L	1500 at $M5$
Arc-heated scramjet test facility (AHSTF)	Engine tests	Linde (N = 3) arc heater (5200)	4.7 to 5.5 6.0 to 8.0	4.7 6.0	11.17 × 11.17 10.89 × 10.89	4 diameter × 11 L	800 600
8-ft high temperature tunnel (8′ HTT)	Engine tests	CH_4/O_2/Air combustion (3560)	4.0 5.0 6.8	4.0 5.0 6.8	96 diameter	8 diameter × 12 L (26 diam chamber)	1500 at $M7$
Hypersonic pulse facility (HYPULSE)	Engine and combustor tests	RST (15 550)	7 10	6.5	24 diameter	7 diameter	>2000 at $M7$ ~800 at $M10$
GASL Leg IV	Engine tests	Pebble-bed + H_2/O_2/Air combustion (5200)	5 7	4.7 6.0	13.26 × 13.26	2.5 W × 3.5 H × 8 L	1200 1700

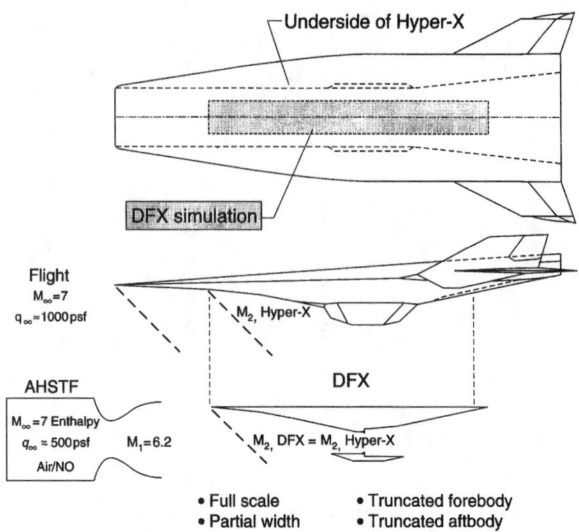

Fig. B9 DFX simulation compared to Hyper-X.

Model (HSM). The partially water-cooled HXEM is designed for testing in long duration, full enthalpy and pressure wind tunnels at Mach numbers of 5 and 7 (see Ref. 122). Mach 7 tests of this engine will be performed in the AHSTF, the GASL Leg IV, and the 8-ft. HTT (see Table B3). Tests in the AHSTF provide a direct comparison with DFX results. Tests in the GASL Leg IV at full enthalpy and at full or partial pressure provide comparisons of performance and operability for high-to-low test gas dynamic pressure and for facility contaminant (i.e., H_2-Air-O_2 combustion-heated facility to arc-heated facility). Tests in the methane-oxygen-air heated 8-ft. HTT also provide full-pressure simulation, with CO and H_2O contaminated test gas. In addition, the engine flow-path can be tested in the 8-ft HTT with both partial and full-length forebody. This allows quantification of the effect of full-length boundary-layer development on engine performance and operability. In addition, testing the partial-length forebody provides data for direct comparison with results from the smaller AHSTF and GASL Leg IV. More importantly, these tests provide a benchmark of the 8-ft. HTT facility effects on engine performance before testing either the full-width flight engine (HXFE) or the Hyper-X research vehicle (see Table B3).

The uncooled HSM is designed for testing in the HYPULSE reflected-shock tunnel at Mach numbers of 7 to 10 and possibly future tests at Mach numbers greater than 14 in the HYPULSE expansion tunnel. For each test the engine is mounted in the 24-in. diameter, Mach 6.5 facility nozzle using the approach illustrated for the DFX (Fig. B2).

Flight scaling of performance and operability, determined in wind-tunnel tests of the Hyper-X engine, is accomplished using a different approach than used for aerodynamic database development. Design methods, including

Table B3 Mach 7 scramjet tests

	Scramjet model / width / length					
	DFX	HXEM Part.		HFE	HSM	HXRV
Facility (contaminants)	Part.	Part.	Full	Full	Part.	Full
AHSTF			—	—	—	—
(NO^-)	500	500				
GASL	—		—	—	—	—
Leg IV		500 1000				
(H_2O)						
8-ft HTT	—	600	600	600	—	
(H_2O & CO_2)		1000	1000	1000		1000
					500	
HYPULSE	—	—	—	—	1000	—
(none)						
					2000	
Flight						1000
(none)						

Notation:
dynamic pressure, psf;
partial/full width/length simulation

analytical and CFD-based methods (see Ref. 107) can model, to some degree, both the wind-tunnel test gas and the flight environments. These prediction methods are verified by comparison with multiple ground tests of the HXRV engines. Flight scaling is accomplished simply by using these methods to predict the flight performance for the projected flight conditions.[119]

Flight Validation of Design Methods

A primary program goal is flight verification of scramjet propulsion, aerodynamic, and propulsion-airframe integration design methods. This is required to develop confidence in future hypersonic vehicle designs. The hypersonic vehicle aerodynamic and scramjet engine design methods have been successfully verified using wind-tunnel data. This verification includes comparision with and understanding of the differences and/or uncertainty between predicted and experimental results. The experiments include unit, component, and complete system tests (some of this activity is discussed in Chap. 6). Unit problems include boundary-layer heat transfer and friction; shock boundary-layer interaction, shock-shock interaction heating, and corner flows; simple film and fuel injection and mixing (Figs. 30 and 33); inlet isolator shock-train pressure rise (Fig. 13), and expansion corner flow phenomena (Ref. 92). Component problems, such as vehicle forebody, inlets, combustor (Fig. 32) or nozzle (Refs. 91 and 92) components, include multiple

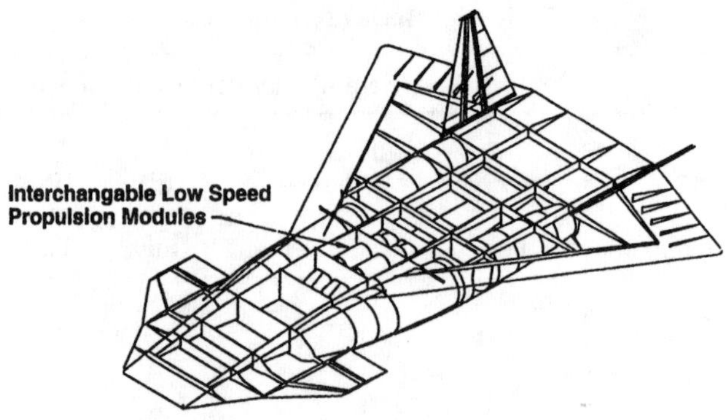

Fig. B10 HySID canard-wing configuration.

unit effects. System database includes complete unpowered vehicles (Fig. B3) and engine-flow-path modules (Fig. B2).

Propulsion-airframe integration design methods, unlike aerodynamic and propulsion methods, have not been adequately verified because of the limited nature of appropriate data. For example, data reported in Ref. 96 include a simulation gas mixture for powered nozzle effects, but does not incorporate an inlet. This limitation will be corrected with validation studies of wind-tunnel tests of the HXFE and/or HXRV itself in the 8-ft. HTT (Table B3). These tests of the engine and vehicle (simulated or real) will provide the first data set for a fully integrated scramjet-powered vehicle and will likely include some challenging modeling of wind-tunnel effects. The next step will be to validate the design methods with the HXRV flight data. Some issues expected in flight validation include 1) low freestream turbulence effects on fuel mixing, shock-induced boundary-layer separation, and boundary-layer transition control; 2) full total enthalpy effects on slender-body, hypersonic, wind-tunnel based aerodynamic performance; 3) clean-air test gas effects on ignition, flameholding, and flame propagation; and 4) unknowns in propulsion-airframe integration.

Method Enhancements

Scramjet and scramjet-powered vehicle design requires a matrix of highly integrated design tools encompassing engineering and higher-order CFD-based analysis methods[107] and specialized experimental facilities and measurements.[123] Successful development of hypersonic air-breathing engine-powered vehicles requires continued refinement of design tools. As discussed in Chap. 6, current methods are tailored to a limited class of engines (predominantly two-dimensional). Part of the challenge is expanding this capability to truly three-dimensional designs.[124] Other challenges lie in reduced turn-around time required for the vehicle design process, in refinement and implementation of low-speed aeropropulsion integration methods, and in development and automation of multidiscipline design processes.

Hyper-X Phase II and Beyond (Future Vehicle Design)

This technology area represents the long-term look at future systems. The Hyper-X program, as discussed in detail in Ref. 125, is a two-phase program. Phase 1 emphasis is on the Mach 5–10, dual-mode scramjet operating speed range. Phase II is not funded, but studies leading to a Phase II are progressing. Phase II is intended to provide flight validation of critical technologies for hypersonic aircraft or access-to-space vehicles operating from takeoff up to Mach 7 into the scramjet operating speed range. This includes operation on and transition from the low-speed engine to the dual-mode scramjet. Phase II is currently envisioned to include a reusable flight-test vehicle. One candidate for the flight vehicle, referred to as the Hyper System Integration Demonstration (HySID) vehicle,[126] is illustrated in Fig. B10.

Other longer-term technology development is being directed toward the following: 1) hypervelocity scramjet engine technology with Mach numbers of 14–20; 2) alternate engine cycles, including rocket-based combined cycle (RBCC), pulse detonation engine (PDE), and pulse detonation rocket (PDR)[127,128]; 3) plasma aero, magneto-hydrodynamics, virtual inlet power generation, etc.[129]; and 4) system studies to refine existing or to identify new concepts and missions for hypersonic air-breathing reusable vehicles.

Acknowledgments

The authors are pleased to acknowledge the numerous contributions of their colleagues in NASA, other government labs, and the aerospace industry to the ideas and accomplishments described in this paper. Although all are recognized at least in part by direct reference where possible, specific additional mention is due to Aaron Auslender for his contributions to the sections on "Flow Nonuniformity . . . " and on "Interactive Inlet Design." Also, credit for the idea to use smoke and mirrors (Mie scattering) in fuel plume imaging is due in part to Clay Rogers at Langley, and for successful implementation of the technique, credit is due to the HYPULSE team at GASL including at least John Erdos, Jose Tamagno, Rich Trucco, Robert Bakos, and the dedicated staff who support them.

References

[1]Becker, J. V., and Baals, D. D., "The Aerodynamic Effects of Heat and Compressibility in the Internal Flow Systems of Aircraft," NACA ACR, Sept. 1942; also NACA TR 773.

[2]Hicks, B. L., "Addition of Heat to a Compressible Fluid in Motion," NACA ACR E5A29, Feb. 1945.

[3]Perchonok, E., and Wilcox, F. A., "Investigation of Ramjet Afterburning as a Means of Varying Effective Exhaust Nozzle Area," NACA RM E52H27, Nov. 1952.

[4]Dorsch, R. G., Serafini, J. S., and Fletcher, E. A., "A Preliminary Investigation of Static-Pressure Changes Associated with Combustion of Aluminum Borohydride in a Supersonic Wind Tunnel," NACA RM E55F07, 1955.

[5]Weber, R. J., and MacKay, J. S., "An Analysis of Ramjet Engines Using Supersonic Combustion," NACA TN 4386, Sept. 1958.

[6]Andrews, E. H., and Mackley, E. A., "Review of NASA's Hypersonic Research Engine Project," AIAA Paper 93-2323, June 1993.

[7]Henry, J. R., and Anderson, G. Y., "Design Considerations for the Airframe-Integrated Scramjet," NASA TM X-2895, 1973.

[8]"Hypersonic Research Engine Project—Phase II. Aerothermodynamic Integration Model Test Report," AiResearch Manufacturing Co. of California, Document AP-74-10784, NASA Paper CR-132655.

[9]Ferri, A., "Review of SCRAMJET Propulsion Technology," *Journal of Aircraft,* Vol. 5, No. 1, 1968.

[10]Schlichting H. *Boundary Layer Theory,* McGraw–Hill, New York, 1977.

[11]Ratekin G. H., "Vaned Diffuser Development Program," Final Rep. Air Force Weapons Lab., AFWL-TR-77-57, June 1977.

[12]Ortwerth, P. J., "A Generalized Distortion Theory of Internal Flow," Air Force Weapons Lab., AFWL-TR-77-118.

[13]Pinckney, S. Z., "Isolator Modeling for Ramjet/Scramjet Transition," *1993 National Aero-Space Plane Technology Review* (Monterey, CA), April 1993, (Paper 171)

[14]McLafferty, G., "A Generalized Approach to the Definition of Average Flow Quantities in Nonuniform Streams," United Aircraft Corp. Research Dep. Rep. SR-13534-9, Dec. 1955.

[15]Wyatt, D. D., "Analysis of Errors Introduced by Several Methods of Weighting Nonuniform Duct Flows," NACA Technical Note 3400.

[16]Pratt, D. T., and Heiser, W. H., "Isolator-Combustor Interaction in a Dual-Mode Scramjet Engine," AIAA Paper 93-0358, 1993.

[17]Ortwerth, P. J., private communications.

[18]Quan, V., private communications.

[19]Trexler, C. A., and Souders, S. W., "Design and Performance at a Local Mach Number of 6 of an Inlet for an Integrated Scramjet Concept," NASA TN D-7944, April 1975.

[20]Trexler, C. A., "Inlet Starting Predictions for Sidewall-Compression Scramjet Inlets," AIAA Paper 88-3257, 1988.

[21]Hudgens, J. A., and Trexler, C. A., "Operating Characteristics at Mach 4 of an Inlet Having Forward-Swept, Sidewall-Compression Surfaces," AIAA Paper 92-3101, July 1992.

[22]Korte, J. J., Singh, D. J., Kumar, A., and Auslender, A. H., "Numerical Study of the Performance of Swept, Curved Compression Surface Scramjet Inlets," AIAA Paper 93-1837, 1993.

[23]Settles, G. S., and Dolling, D. S., "Swept Shock Wave/Boundary Layer Interactions," *Tactical Missile Aerodynamics,* edited by J. Nielsen and M. Hemsch, AIAA Progress in Astronautics and Aeronautics, AIAA, New York, 1986.

[24]Settles, G. S., and Dolling, D. S., "Swept Shock/Boundary Layer Interactions—Tutorial and Update," AIAA Paper 90-0375, Jan. 1990.

[25]Narayanswami, N., Knight, D., Bogdonoff, S. M., and Horstman, C. C., "Crossing Shock Wave-Turbulent Boundary Layer Interactions," AIAA Paper 91-0649, Jan. 1991.

[26]Garrison, T. J., Settles, G. S., Narayanswami, N., and Knight, D. D., "Structure of Crossing-Shock Wave/Turbulent Boundary Layer Interactions," AIAA Paper 92-3670, July 1992.

[27]Bodgonoff, S. M., and Stokes, W. L., "Crossing Shock Wave Turbulent Boundary Layer Interactions—Variable Angle and Shock Generator Length Geometry Effects at Mach 3," AIAA Paper 92-0636, Jan. 1992.

[28]Kumar, A., Singh, D. J., and Trexler, C. A., "A Numerical Study of the Effects of Reverse Sweep on a Scramjet Inlet Performance," AIAA Paper 90-2218, July 1990.

[29]Pace, E. G., and Lu, F. K., "On the Scale of Surface Features in Hypersonic Shock Boundary Layer Interactions," AIAA Paper 91-1769, June 1991.

[30]Chung, K., and Lu, F. K., "Hypersonic Turbulent Expansion-Corner Flow with Shock Impingement," AIAA Paper 92-5101, Dec. 1992.

[31]Rodi, P. E., "An Experimental Study of the Effects of Bodyside Compression on Forward Swept Sidewall Compression Inlets Ingesting a Turbulent Boundary Layer," AIAA Paper 93-3125, July 1993.

[32]Oswatitsch, K., "Pressure Recovery for Missiles With Reaction Propulsion at High Speeds (the Efficiency of Shock Diffusers)," NACA TM-1140, June 1947.

[33]Ta'asan, S., Kuruvila, G., and Salas, M. D., "Aerodynamic Design and Optimization in One Shot," AIAA 92-0025, 1992.

[34]Iollo, A., Salas, M. D., and Ta'asan, S., "Shape Optimization Governed by the Euler Equations Using an Adjoint Method," NASA C R No. 191555, ICASE Report No. 93-78.

[35]Wahls, R. A., "Development of a Defect Stream Function, Law of the Wall/Wake Method for Compressible Turbulent Boundary Layers," NASA C R 4286, March 1990.

[36]Barnwell, R. A., "Nonadiabatic and Three-Dimensional Effects in Compressible Turbulent Boundary Layers," *AIAA Journal,* Vol. 30, N. 4, 1992, pp. 897–904.

[37]Scales, L. E., *Introduction to Non-Linear Optimization,* MacMillan, London, 1985, pp. 110–136.

[38]Korte, J. J., and Auslender, A. H., "Optimization of Contoured Hypersonic Scramjet Inlets with a Least-Squares Parabolized Procedure," *Computing Systems in Engineering,* Vol. 4, No. 1 1993, pp. 13–26.

[38a]Emani, S., Trexler, C.A., Auslender, A.H., and Weidner, J.P., "Experimental Investigation of Inlet-Combustor Isolators for a Dual-Mode Scramjet at Mach 4," NASA-TP-350, May 1995.

[39]Northam, G. B., and Anderson, G. Y., "Supersonic Combustion Ramjet Research at Langley," AIAA Paper 86-0159, Jan. 1986.

[40]Stalker, R. J., and Morgan, R. G., "Free-Piston Shock Tunnel T4—Initial Operation and Preliminary Calibration," Univ. of Queensland, Dept. of Mech. Eng. Research Rept., Queensland, Australia, Jan. 1988; also NASP CR-1923, Aug. 1988; also NASA CR-181721, Sept. 1988.

[41]Ferri, A.; "Mixing Controlled Supersonic Combustion," *Annual Review of Fluid Mechanics* Vol. 5, 1973.

[42]Billig, F. S., and Dugger, G. L., "The Interaction of Shock Waves and Heat Addition in the Design of Supersonic Combustors," *Twelfth Symposium (International) on Combustion,* The Combustion Inst., Pittsburgh, PA, 1969, pp. 1125–1134.

[43]Stalker, R. J., "Thermodynamics and Wave Processes in High Mach Number Propulsive Ducts," AIAA Paper 89-0261, Jan. 1989.

[44]Hooper, W. G., "Mach 5 and Mach 8 Hypersonic Test Facility," AIAA Paper 89-2297, July 1989.

[45]Puster, R. L., Rebush, D. E., and Kelly, H. N., "Modification to the Langley 8′ High Temperature Tunnel for Hypersonic Propulsion Testing," AIAA Paper 87-1887, July 1987.

[46]Rogers, R. C. (ed.), "Workshop on the Application of Pulse Facilities to Hypervelocity Combustion Simulation," *Eighth NASP Technology Symposium*(Monterey, CA), March 1990 (NASP WP-1008).

[47]Tamagno, J., Bakos, R., Pulsonetti, M., and Erdos, J., "Hypervelocity Real Gas Capabilities of GASL's Expansion Tube (HYPULSE) Facility," AIAA Paper 90-1390, June 1990.

[48]Paull, A., Stalker, R., and Stringer, I., "Experiments on an Expansion Tube with a Free Piston Driver," *AIAA 15th Aerodynamic Testing Conference*(San Diego, CA), May 1988.

[49]Auslender, A. H., private communication, NASA Langley Research Center, Hampton, VA.

[50]Private communication, Facility data from Reference 7 were used to compute test section conditions with the NENZF computer code.

[51]Jachimowski, C. J., "An Analysis of Combustion Studies in Shock Expansion Tunnels and Reflected Shock Tunnels," NASA TP-3224, July 1992.

[52]Bakos, R. J., and Morgan, R. G., "Effects of Oxygen Dissociation on Hypervelocity Combustion Experiments," AIAA Paper 92-3964, 1992.

[53]Bakos, R. J., and Morgan, R. G., "Chemical Recombination in an Expansion Tube," *AIAA Journal,*Vol. 32, No. 6, pp. 1316–1319.

[54]Bakos, R. J., and Morgan, R. G., "Scramjet Combustion and Thrust Measurements in a Reflected Shock Tunnel with Varied Test Gas Atomic Oxygen Contamination," AIAA Paper 94-2520, June 1994.

[55]Bakos, R. J., "An Investigation of Test Flow Nonequilibrium Effects on Scramjet Combustion," Ph.D. Dissertation, Dept. of Mechanical Engineering, Univ. of Queensland, Australia, June 1994.

[56]Bélanger, J., "Studies of Mixing and Combustion in Hypervelocity Flows with Hot Hydrogen Injection," Ph.D. Dissertation, California Inst. of Technology, CA, April 1993.

[57]Bélanger, J., and Hornung, H. G., "Transverse Jet Mixing and Combustion Experiments in the Hypervelocity Shock Tunnel T5," AIAA Paper 94-2517, June 1994.

[58]Bittner, R. D., private communications, Nyma, Hampton, VA.

[59]Rogers, R. C., Weidner, E. H., and Bittner, R. D., "Quantification of Scramjet Mixing in Hypervelocity Flow of a Pulse Facility," AIAA Paper 94-2518, June 1994.

[60]Swithenbank, J., Eames, I., Chin, S., Ewan, B., Yang, Z., Cao, J., and Zhao, X., "Turbulent Mixing in Supersonic Combustion Systems," AIAA Paper 89-0260, Jan. 1989.

[61]Bray, K. N. C., "Chemical Reactions in Supersonic Nozzle Flows," *Ninth Symposium (International) on Combustion,*Academic, 1963, p. 770.

[62]Drummond, J. P., Rogers, R. C., and Hussaini, M. Y., "A Detailed Numerical Model of a Supersonic Reacting Mixing Layer," AIAA Paper 86-1427, 1986.

[63]Kamath, H., "Parabolized Navier Stokes Algorithm for Chemically Reacting Flows," AIAA Paper 89-0386, Jan. 1989.

[64]Kamath, P., Mao, M., and McClinton, C., "Scramjet Combustor Analysis with the SHIP3D PNS Code," AIAA Paper 92-5090, 1992.

[65]Walters, R. W., Slack, D. C., Cinella, P., Applebaum, M., and Frost, C., "A Users Guide to GASP," NASA Langley Research Center/Virginia Polytechnic Inst. and State Univ., Rev. 0, Hampton, VA/Blacksburg, VA, 1990.

[66]Kamath, P., Hawkins, R. W., and McClinton, C., "A Highly Efficient Engineering Tool for 3-D Scramjet Flowfield and Heat Transfer Computations," NASA CP-10045, April 1990.

[67]Riggins, D. W., and McClinton, C. R., "A Computational Investigation of Mixing and Reacting Flows in Supersonic Combustors," AIAA Paper 92-0626, Jan. 1992.

[68]Mekkes, G. L., "Computational Analysis of Hypersonic Shock Wave/Wall Jet Interactions," AIAA Paper 93-0604, Jan. 1993.

[69]Torrence, M. G., "Concentration Measurements of an Injected Gas in a Supersonic Stream," NASA TN D-3860, 1967.

[70]Rogers, R. C., "A Study of the Mixing of Hydrogen Injected Normal to a Supersonic Airstream," NASA TN D-6114, 1971.

[71]McClinton, C. R., "Effect of Ratio of Wall Boundary Layer Thickness to Jet Diameter on Mixing of Normal Hydrogen Jet in a Supersonic Stream," NASA TM X-3030, 1974.

[72]Orth, R. C., Schetz, J. A., and Billig, F. S., "The Interaction and Penetration of Gaseous Jets in Supersonic Flows," NASA CR-1386, 1969.

[73]Rogers, R. C., "Mixing of Hydrogen from Multiple Injectors Normal to a Supersonic Airstream," NASA TN D-6476, 1971.

[74]McClinton, C. R., "The Effect of Injection Angle on the Interaction Between Sonic Secondary Jets and a Supersonic Free Stream," NASA TM D-6669, Feb. 1972.

[75]Mays, R. B., Thomas, R. H., and Schetz, J. A., "Low Angled Injection into a Supersonic Flow," AIAA Paper 89-2461, July 1989.

[76]Ng, W. F., Kwok, F. T., and Ninnermann, T. A., "A Concentration Probe for the Study of Mixing in Supersonic Shear Flows," AIAA Paper 89-2459, July 1989.

[77]Fuller, E. J., Mays, R. B., Thomas, R. H., and Schetz, J. A., "Mixing Studies of Helium in Air at Mach 6," AIAA Paper 91-2268, June 1991.

[78]Fuller, E. J., Thomas, R. H., and Schetz, J. A., "Effects of Yaw on Low Angled Injection into a Supersonic Flow," AIAA Paper 91-014, Jan. 1991.

[79]McDaniel, J. C., and Graves, J., "A Laser-Induced-Fluorescence Visualization Study of Transverse, Sonic Fuel Injection in a Non-Reacting Supersonic Combustor," AIAA Paper 86-0507, Jan. 1986.

[80]McDaniel, J. C., Fletcher, D., Hartfield, R., Jr., and Hollo, S., "Staged Transverse Injection into Mach 2 Flow Behind a Rearward Facing Step," AIAA 91-5071, Dec. 1991.

[81]Donohue, J. M., McDaniel, J. C., and Haj-Hariri, H., "Experimental and Numerical Study of Swept Ramp Injection into a Supersonic Flow Field," AIAA Paper 93-2445, June 1993.

[82]Vitt, P. H., Riggins, D. W., and McClinton, C. R., "Validation and Application of Numerical Modeling to Supersonic Mixing and Reacting Flows," AIAA Paper 93-0606, Jan. 1993.

[83]Srinivasan, S., Bittner, R., and Bobskill, G., "Summary of GASP Code Application & Evaluation for Scramjet Combustor Flow Fields," AIAA Paper 93-1973, 1993.

[84]Mao, M., Riggins, D. W., and McClinton, C. R., "Numerical Simulation of Transverse Fuel Injection," NASA CP-10045, April 1990.

[85]Northam, G. B., Greenburg, I., and Byington, C. S., "Evaluation of Parallel Injector Configurations for Supersonic Combustion," AIAA Paper 89-2525, July 1989.

[86]Swithenbank, J., Eames, I., Chin, S., Ewan, B., Yang, Z, Cao, J., and Zhao, X., "Turbulence Mixing in Supersonic Combustion Systems," AIAA Paper 89-0260, Jan. 1989.

[87]Czysz, P., and Murthy, S. N. B., "Energy Analysis of High-Speed Flight Systems," *High-Speed Flight Propulsion Systems,* edited by S. N. B. Murthy and E. T. Curran Progress in Astronautics and Aeronautics, AIAA, Washington, DC, 1991, pp. 143–235.

[88]Kamath, P., and McClinton, C. R., "Computation of Losses in a Scramjet Combustor," AIAA Paper 92-0635, Jan. 1992.

[89]Riggins, D. W., and McClinton, C. R., "A Computation Investigation of Flow Losses in a Supersonic Combustor," AIAA Paper 90-2093, July 1990.

[90]Riggins, D. W., and McClinton, C. R., "Analysis of Losses in Supersonic Mixing and Reacting Flows," AIAA Paper 91-2266, July 1991.

[91]Jentink, T. N., "An Evaluation of Nozzle Relaminarization Using Low Reynolds Number *k*-e Turbulence Models," AIAA Paper 93-0610, Jan. 1993.

[92]Jentink, T., "CFD Code Validation for Nozzle Flowfields," AIAA Paper 91-2565, June 1991.

[93]Boppe, C. W., and Davis, W. H., "Hypersonic Forebody Lift-Induced Drag," SAE Paper TP 892345, Sept. 1989.

[94]Gatlin, G. M., "Ground Effects on the Low-Speed Aerodynamics of a Powered, Generic Hypersonic Configuration," NASA TP-3092, May 1991.

[95]Tatum, K., Monta, W., Witte, D., and Walters, R., "Analysis of Generic Scramjet External Nozzle Flowfields Employing Simulant Gases," AIAA Paper 90-5242, Oct. 1990.

[96]Heubner, L. D., and Tatum, K. E., "Computational and Experimental Afterbody Flow Fields for Hypersonic, Airbreathing Configurations with Scramjet Exhaust Flow Simulation," AIAA Paper 91-1709, June 1991.

[97]Small, W. J., Weidner, J. P., and Johnston, P. J., "Scramjet Nozzle Design and Analysis as Applied to a Highly Integrated Hypersonic Research Airplane," NASA Paper TN D-8334, Nov. 1976.

[98]Weidner, J. P., Small, W. J., and Penland, J. A., "Scramjet Integration on Hypersonic Research Airplane Concepts," *Journal of Aircraft,* Vol. 14, No. 5, 1977, pp. 460–466.

[99]Roudakov, A., Schickhman, J., Semenov, V., Novelli, P., and Fourt, O., "Flight-testing an Axisymmetrical Scramjet—Russian Recent Advances," *44th Congress of the IAF* (Graz, Austria), Oct. 1993.

[100]Roudakov, A., Semenov, V., and Hicks, J., "Recent Flight Test Results of the Joint CIAM-NASA Mach 6.5 Scramjet Flight Program," AIAA Paper 98-1643, April 1998.

[101]McClinton, C., Roudakov, A., Semenov, V., and Kopehenov, V., "Comparative Flow Path Analysis and Design Assessment of an Axisymmetric Hydrogen Fueled Scramjet Flight Test Engine at a Mach Number of 6.5," AIAA Paper 96-4571, Nov. 1996.

[102]Roudakov, A., Semenov, V., Kopehenov, V., and Hicks, J., "Future Flight Test Plans of an Axisymmetric Hydrogen-Fueled Scramjet Engine on the Hypersonic Flying Laboratory," AIAA Paper 96-4572, Nov. 1996.

[103]Alexandrov, V., Prokhorov, A., Roudakov, A., Belykh, S., Vedeshkin, G., Shutov, A., Yurin, V., and Hicks, J. "Support and Realization of Tests of Axisymmetric Scramjet on Test Cell C16VK CIAM RTC," *9th International Conference on the Methods of Aerophysical Research* (Novosibirsk, Russia), July 1998.

[103a]Voland, R.T., Auslender, A.H., Smart, S.M., Roudakov, A., and Semenov, V., "CIAM/NASA Mach 6.5 Scramjet Flight and Ground Experiments," AIAA 99-4848, Oct. 1999.

[104]Rausch, V. L., McClinton, C. R., and Crawford, J. L., "Hyper-X: Flight Validation of Hypersonic Airbreathing Technology," (Chattanooga, TN), Sept. 1997 (ISABE 97-7024).

[105]Rausch, V. L., McClinton, C. R., and Hicks, J. W., "NASA Scramjet Flights to Breath New Life into Hypersonics," *Aerospace America,*July 1997.

[106]"Air Launched Flight Experiment Final Report," McDonnell Douglas Aerospace, June 1995.

[107]Hunt, J. L., and McClinton, C. R., "Scramjet Engine/Airframe Integration Methodology," *AGARD Future Aerospace Technology Conference,*(Palaiseau, France), April 1997 (Paper C35).

[108]Godfrey, A. G., "GASP Version 3 User's Manual," Aerosoft, Inc., May 1996.

[109]Guy, R. W., Rogers, R. C. Pulster, R. L., Rock, K. E., and Diskin, G. S., "The NASA Langley Scramjet Test Complex," AIAA 96-3243, July 1996.

[110]Bakos, R. J., Castrogiovanni, A., Calleja, J.F., Nucci, L., and Erdos, J. I., "Expansion of the Scramjet Ground Test Envelope of the HYPULSE Facility," AIAA 96-4506, Nov. 1996.

[111]Lallman, F., "Preliminary Control Laws," NASA, Hyper-X Document H7006-M7-01-GCT, March 24, 1997.

[112]Desai, P. N., Braun, R. D., Powell, R. W., Engelund, W. C., and Tartabini, P. V., "Six Degree-of-Freedom Entry Dispersion Analysis for METEOR Recovery Analysis," *Journal of Spacecraft and Rockets,*Vol. 34, No. 3, 1997, pp. 334–340.

[113]McClinton, C. R., Holland, S. D., Rock, K. E., Engelund, W. C., Voland, R. T., Huebner, L. D., and Rogers, R. C., "Hyper-X Wind Tunnel Program," AIAA Paper 98–0553, Jan. 1998.

[114]Iliff, K. W., and Shafer, M. F., "Extraction of Stability and Control Derivatives from Orbiter Flight Data," NASA CP 3248, April 1995.

[115]Morelli, E. A., "Flight Test Validation of Optimal Input Design and Comparison to Conventional Inputs," AIAA Paper 97-3711, Aug. 1997.

[116]Miller, C. G., III, "Langley Hypersonic Aerodynamic/Aerothermodynamic Testing Capabilities—Present and Future," AIAA CP-90-1376, June 1990.

[117]Pirrello C. J., "An Inventory of Aeronautical Ground Research Facilities. Vol. 1—Wind Tunnels," NASA CR 1874, Nov. 1971.

[118]Penaranda, F. E., and Freda, M. S., "Aeronautical Facilities Catalogue, Vol. 1—Wind Tunnels," NASA RP-1132, Jan. 1985.

[119]McClinton, C. R., Voland, R. T., Holland, S. D., Engelund, W. C., White, J. T., and Pahle, J. W., "Wind Tunnel Testing, Flight Scaling and Flight Validation with Hyper-X," AIAA Paper 98-2866, June 1998.

[120]Huebner, L. D., Rock, K. E., Voland, R. T., and Wieting, A. R., "Calibration of the Langley 8-Foot High Temperature Tunnel for Hypersonic Airbreathing Propulsion Testing," AIAA CP-96-2197, June 1996.

[121]Roffe, G., Bakos, R. J., Erdos J. I., and Swartwout, W., "The Propulsion Test Complex at GASL," ISABE 97-7096, Sept. 1997.

[122]Voland, R. T, Rock, K. E., Huebner, L. D., Witte, D. W., Fisher, K. E., and McClinton, C. R., "Hyper-X Engine Design and Ground Test Program," AIAA Paper 98-1532, April 1998.

[123]Rogers, R. C., Capriotti, D. P., and Guy, R. W., "Experimental Supersonic Combustion Research at NASA LaRC." AIAA Paper 98-2506, July, 1998.

[124]Smart, M. K., "Design of Three-Dimensional Hypersonic Inlets with Rectangular to Elliptic Shape Transition," AIAA Paper 98-Reno, Jan. 1998.

[125]Hunt, J. L., and Couch, L. M., "Beyond Hyper-X," *Space '98. Sixth ASCE Specialty Conference,* April 1998.

[126]Hunt, J. L., and Rausch, V. L., "Airbreathing Hypersonic System Focus at NASA Langley Research Center," AIAA Paper 98-1641, April 1998.

[127]Pegg, R. J., Couch, B. D., and Hunter, L. G., "Pulse Detonation Engine Air Induction System Analysis," AIAA Paper 96-2918, July 1996.

[128]Bussing, T. R. A., and Pappas, G., "An Introduction to Pulse Detonation Engines," AIAA Paper 94-0263, Jan. 1994.

[129]Gurijanov, E. P., and Harsha, T., "Ajax: New Directions in Hypersonic Technology," AIAA Paper 96-4609, Nov. 1996.

Chapter 7

Scramjet Inlets

David M. Van Wie*
The Johns Hopkins University, Applied Physics Laboratory, Laurel, Maryland

Nomenclature

A_i = inlet reference area
A_0 = area of captured stream tube in the freestream
C_d = drag coefficient, discharge coefficient
C_h = Stanton number
C_l = lift coefficient
C_p = specific heat at constant pressure
C_v = specific heat at constant volume
C_0 = $\rho_w \mu_w / \rho_e \mu_e$
D = drag
d = blunt leading-edge diameter
F = stream thrust
G_θ = Goertler number
H_t = total enthalpy
h = static enthalpy
I_{sp} = specific impulse
M = Mach number
$\dot{m}$ = mass flow
P = pressure
Pr = Prandtl number
$\dot{Q}$ = heat loss
$\dot{Q}'$ = fraction heat loss
q = dynamic pressure
$\dot{q}$ = heat flux
R = specific gas constant, radius of curvature, spherical coordinate

*Supervisor, Aeronautical Sciences and Technology Group.

Re = Reynolds number
r = nonequilibrium index
S = entropy
T = temperature
U = axial component of velocity
V = transverse component of velocity
W = crossflow component of velocity
W_{H2O} = water flow rate
X, Y, Z = Cartesian coordinates
X_n = virtual wedge offset
α = incipient separation angle
β_f = fuel-injection angle
γ = C_p/C_v
Δ = shock standoff distance, difference operator
δ = turning angle, boundary layer thickness
ε = convergence criteria
η_B = compression efficiency
η_c = combustion efficiency
η_i = isentropic efficiency
η_{KD} = process efficiency
η_{KE} = kinetic energy efficiency
$\eta_{KE_{ad}}$ = adiabatic kinetic-energy efficiency
η_N = nozzle efficiency
η_{P_t} = total pressure recovery
η_R = static pressure efficiency
η_S = entropy rise efficiency
θ = angular coordinate, momentum thickness
θ_i = incipient separation convergence angle
μ = viscosity
ρ = density
τ = shear stress
$\bar{\chi}$ = viscous interaction parameter
$\bar{\chi}^*$ = cold-wall viscous interaction parameter

Subscripts

add = additive
ave = average condition
b = body
cowl = cowl
e = boundary-layer edge
ex = exhaust condition
extra = outside flow path
f = fuel
in = entering condition
inlet = inlet property
Isentropic = isentropic property
Kantrowitz = Kantrowitz limit

n	= nose
out	= exiting condition
pl	= plenum
s	= shock, separation
$stag$	= stagnation condition
t	= total condition
tr	= transition
w	= wall property
x,y	= Cartesian coordinates
0	= freestream
2	= properties at cowl lip
4	= throat properties
4′	= properties downstream of precombustion shock

I. Introduction

THE primary purpose of an inlet (also referred to as an intake or diffuser) for any airbreathing propulsion system is to capture and compress air for processing by the remaining portions of the engine. In a conventional jet engine the inlet works in combination with a mechanical compressor to provide the proper compression for the entire engine. For vehicles flying at high supersonic ($3 < M_0 < 5$) or hypersonic ($M_0 > 5$) speeds, adequate compression can be achieved without a mechanical compressor. Because the airflow and compression ratio for these engines are provided entirely by the inlet, an efficient design of an inlet is crucial to the success of the engine operation.

The goal in the design of any hypersonic inlet is to define a minimum weight geometry that provides an efficient compression process, generates low drag, produces nearly uniform flow entering the combustor, and provides these characteristics over a wide range of flight and engine operating conditions. The design of hypersonic inlets is complicated by the many constraints, both aerodynamic and mechanical, that are imposed on the inlet. Examples of aerodynamic constraints include starting limits, boundary-layer separation limits, and constraints on combustor entrance flow profiles. Examples of mechanical constraints include limits on leading-edge radii, variable geometry flexibility, and cooling system limits.

The features of a scramjet inlet tend to be different from those of external compression ramjet or turbojet inlets, as illustrated in Fig. 1. The ramjet or turbojet inlet captures, compresses, and diffuses the airstream to low subsonic speeds. Most of the compression occurs on the external portion of the inlet with little to no internal contraction. These inlets require a large amount of turning to achieve the desired compression ratio, which creates a tradeoff between the external cowl drag and the internal aerodynamic performance of the inlet. Along with the high level of total turning, the inlets often use boundary-layer bleed to increase pressure recovery, stabilize shock-wave/boundary-layer interactions, and serve as a trap for a terminal shock system.

The design of scramjet inlets is influenced greatly by vehicle and flight

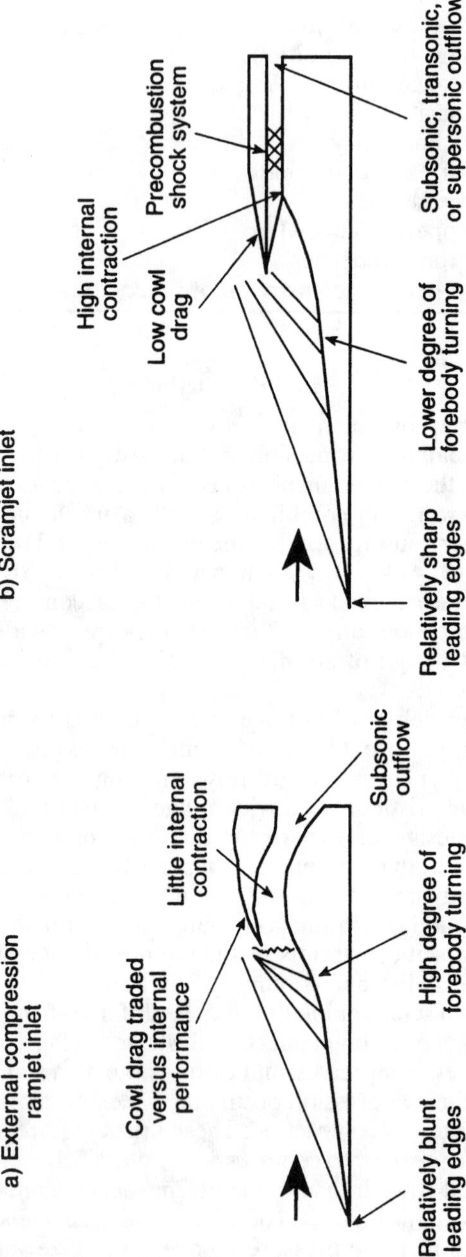

Fig. 1 Comparison of external compression ramjet and scramjet models.

constraints. Because scramjets operate at higher speeds than ramjets or turbojets, the desired compression ratios for scramjets can be achieved with smaller amounts of turning compared to ramjet or turbojet inlets. In most scramjet inlet designs the compression is split between the external and internal portions of the inlet, so that high internal contraction ratios are common. Because these engines are designed to fly at very high speeds, optimum designs possess very low external cowl drag. Although bleed is common in a ramjet or turbojet inlet, the use of boundary-layer bleed is uncommon in a scramjet inlet because of the high air temperatures encountered and the inherent resistance to separation of hypersonic boundary layers. At high speeds and/or low levels of heat release, the scramjet inlet will operate with supersonic flow throughout. At lower speeds and/or higher values of heat release, a precombustion shock system forms in the inlet throat as a result of the thermal blockage of the combustion process. Depending on the strength of this precombustion shock system, the flow exiting the inlet can be supersonic or subsonic.

At the most general level scramjet inlets are designed to provide supersonic airflow to the combustion process. A vast array of geometries has been tested in attempting to accomplish this seemingly simple task. Several examples of scramjet inlets are shown in Fig. 2, illustrating a diversity of designs.[2–9] The inlets can be categorized in terms of two-dimensional planar, two-dimensional axisymmetric, or three-dimensional designs. Each of these inlets attempts to satisfy certain design requirements in different ways. A description of the advantages and disadvantages of various concepts will be presented in later sections.

In the next section various figures of merit for evaluating the performance of scramjet inlets are discussed along with their relative merits. In Sec. III, a brief description of the many aerodynamic phenomena that are encountered in scramjet inlets is presented. The state of the art in the understanding and modeling of these phenomena is discussed. In Sec. IV, cycle calculations that illustrate the significance of efficient inlet operation on the performance of a sample scramjet engine are presented. Various issues associated with the measurement of scramjet inlet performance are discussed in Sec. V. In Sec. VI, several classes of inlet designs are discussed, and the available performance database is summarized. In the last section recommendations are provided for areas in which improvements can be made in the understanding of scramjet-inlet operating characteristics.

II. Definitions of Performance Parameters

Within the existing literature a variety of performance parameters are used to describe the operating characteristics of scramjet inlets. These parameters have been developed over time by various investigators. Although no single parameter can be used to uniquely define the performance of an inlet, certain parameters are more useful than others for illustrating the effects of particular phenomena. In this section techniques for calculating the most prominent inlet performance parameters are presented.

The numbering scheme illustrated in Fig. 3 is used in the following dis-

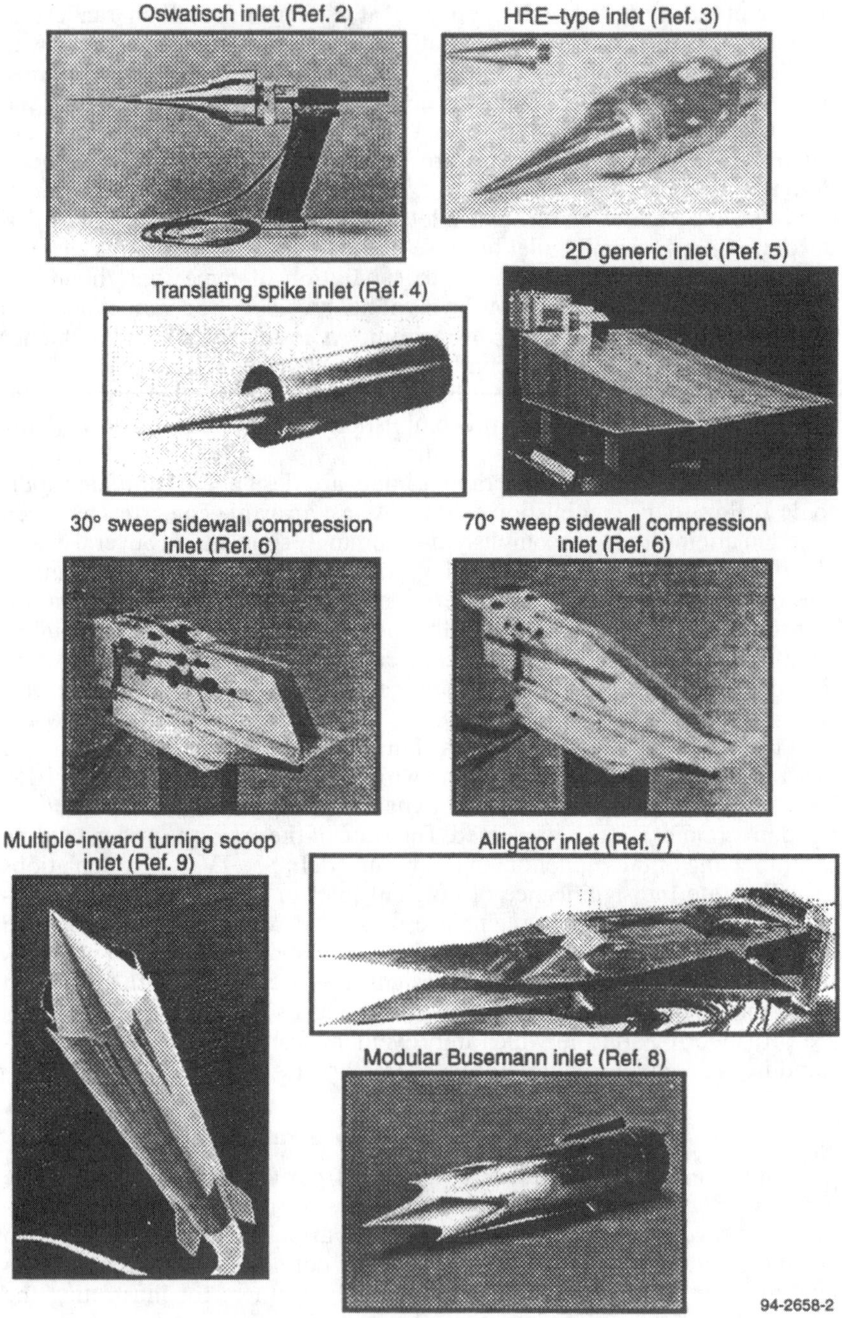

Oswatisch inlet (Ref. 2)

HRE–type inlet (Ref. 3)

2D generic inlet (Ref. 5)

Translating spike inlet (Ref. 4)

30° sweep sidewall compression inlet (Ref. 6)

70° sweep sidewall compression inlet (Ref. 6)

Multiple-inward turning scoop inlet (Ref. 9)

Alligator inlet (Ref. 7)

Modular Busemann inlet (Ref. 8)

94-2658-2

Fig. 2 Examples of scramjet inlets.

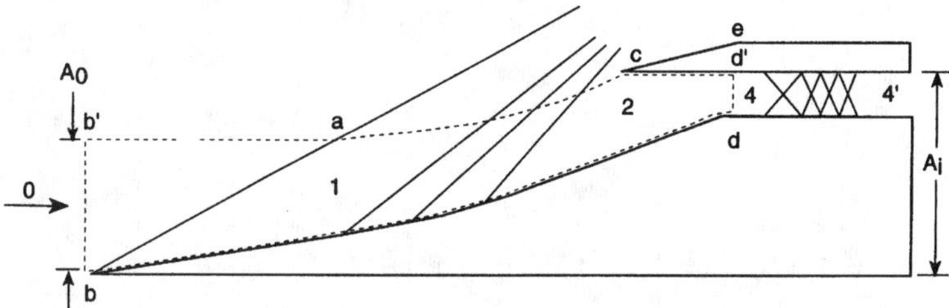

Fig. 3 Scramjet inlet station definitions.

cussions. Station number 0 will refer to the conditions in the freestream ahead of any portion of the inlet compression process. Station number 2 refers to the properties at the entrance to the internal portion of the inlet. Station 4 refers to the properties at the inlet throat (i.e., minimum area), and station 4′ refers to properties downstream of any precombustion shock system.

The operating characteristics of a scramjet inlet can be defined in terms of the inlet geometry together with the mass, momentum, and energy of the flow at the beginning and end of the compression process. Rather than dealing with dimensional quantities, these characteristics are usually described through a series of nondimensional quantities. The mass flow through the inlet is characterized by the air capture ratio A_0/A_i, which is defined as the cross-sectional area of the captured stream tube in the freestream A_0 referenced to the projected frontal area of the inlet A_i.

The energy of the flow at the inlet throat is specified by the total enthalpy at the inlet throat H_{t4}, which can be specified either directly or indirectly through the heat loss to the portions of the inlet wetted by the captured streamtube $\dot{Q}$. These quantities are often specified in nondimensional terms of either the fractional heat loss $\dot{Q}' = \dot{Q}/\dot{m}H_{t0}$ or the total enthalpy ratio H_{t4}/H_{t0}.

The stream thrust at the inlet throat is often determined using the overall drag coefficient of the captured streamtube C_d, which is defined as follows:

$$C_d = \frac{F_0 - F_4}{q_0 A_i} \tag{1}$$

where F_0 and F_4 are the stream thrusts of the captured stream tube in the freestream and at the inlet throat. This drag coefficient can be broken down into an additive drag coefficient $C_{d_{\text{add}}}$ and a drag coefficient for the forces on the actual inlet surfaces $C_{d_{\text{inlet}}}$ as follows:

$$C_{d_{\text{add}}} = \int_a^c (P - P_0) dA_y \tag{2}$$

and

$$C_{d_{\text{inlet}}} = \int_b^d P dA_y + \int_b^d \tau dA_x \int_c^{d'} P dA_y + \int_c^{d'} \tau dA_x \tag{3}$$

where dA_x and dA_y are the axial and vertical projections of the surface area, respectively. An additional drag term that is often of interest is the external cowl drag coefficient $C_{d_{\text{cowl}}}$ defined as follows:

$$C_{d_{\text{cowl}}} = \frac{\int_c^e P dA_y + \int_c^e \tau dA_x}{q_0 A_i} \tag{4}$$

For a two-dimensional inlet the normal force coefficient is also needed in specifying the total force on the captured stream tube. The overall lift coefficient for the captured streamtube C_l is defined as follows:

$$C_l = \left\{ \int_{bdd'cb'b} P dA_x + \int_{bdd'cb'b} \tau dA_y \right\} / q_0 A_i \tag{5}$$

This term can be split into an additive lift coefficient $C_{l_{\text{add}}}$ and an inlet normal force coefficient $C_{l_{\text{inlet}}}$, which are defined as follows:

$$C_{l_{\text{add}}} = \int_{b'}^c P dA_x \,/\, q_0 A_i \tag{6}$$

and

$$C_{l_{\text{inlet}}} = \left\{ \int_b^d P dA_x + \int_b^d \tau dA_y - \int_c^{d'} P dA_x - \int_c^{d'} \tau dA_y \right\} / \, q_0 A_i \tag{7}$$

Given the freestream conditions and the nondimensional performance parameters, the mass flow, stream thrust, and energy flow at the inlet throat can be calculated as follows:

$$\dot{m} = \rho_0 U_0 \frac{A_0}{A_i} A_i \tag{8}$$

$$F_{x_4} = (P_0 + \rho_0 U_0^2) A_0 - C_d\, q_0 A_i \tag{9}$$

$$F_{y_4} = C_l q_0 A_i \tag{10}$$

and

$$H_{t_4} = (1 - \dot{Q}') H_{t_0} \tag{11}$$

Knowing the mass flow, stream thrust, and energy of the flow in the inlet throat, a one-dimensional set of flow properties can be defined from the following equation set:

Mass:

$$\dot{m} = \rho_4 U_4 A_4 \tag{12}$$

Axial momentum:

$$F_{x_4} = (P_4 + \rho_4 U_4^2) A_4 \tag{13}$$

Normal momentum:

$$F_{y_4} = \rho_4 U_4 V_4 A_4 \tag{14}$$

Energy:

$$H_{t_4} = h_4 \frac{1}{2}\left(U_4^2 + V_4^2\right) \tag{15}$$

State:

$$h_4 = h(P_4, \rho_4) \tag{16}$$

Assuming an ideal gas, the equation of state can be simply stated as follows:

$$h = \frac{\gamma}{\gamma - 1} \frac{P}{\rho} \tag{17}$$

For nonideal gases this functional relationship is more general and is usually provided in terms of a computer subroutine (for examples, see Ref. 10 and 11).

For an ideal gas Eqs. (12–16) can be solved directly to provide the throat Mach number as follows:

$$M_4 = \left(\frac{-B + \sqrt{B^2 - 4AC}}{2A} \right)^{1/2} \tag{18}$$

where

$$C = (\gamma - 1)H_{t4}\left(\frac{\dot{m}}{\gamma F_4}\right)^2; \qquad B = 2\gamma C - 1\,; \qquad A = \gamma^2 C - \frac{\gamma - 1}{2}$$

With the throat Mach number the other throat flow properties are easily obtained, for example;

$$P_4 = \frac{F_4}{(1 + \gamma M_4^2)A_4} \tag{19}$$

$$T_{t4} = H_{t4}\,\frac{\gamma - 1}{\gamma R} \tag{20}$$

$$T_4 = T_{t4}\left(1 + \frac{\gamma - 1}{2}\,M_4^2\right)^{-1} \tag{21}$$

$$\rho_4 = \frac{P_4}{RT_4} \tag{22}$$

When the gas is nonideal, an iterative procedure must be used in solving Eqs. (13–17). One possible solution procedure is as follows:

1) Guess h_4.
2) $U_4 = \sqrt{2(H_{t4} - h_4)}$.
3) $\rho_4 = \dfrac{\dot{m}}{U_4 A_4}$
4) $P_4 = \dfrac{F_4}{A_4} - \rho_4 U_4^2$
5) $h'_4 = h(P_4, \rho_4)$
6) If $|h_4 - h'_4| > \varepsilon$, then repeat steps 1–6.

In the preceding procedure ε is some acceptable convergence criterion. Any standard numerical solution scheme, such as the secant method, can be used in converging on the proper choice for h_4. The one-dimensional flow properties derived from this procedure are referred to as stream-thrust-averaged conditions.[12]

For the purposes of describing the efficiency of the inlet compression process, the state variables at the inlet throat can be combined to obtain several different efficiency parameters.[13,14] The definitions of the efficiency parameters can be easily visualized using the Mollier diagram shown in Fig. 4. The more prevalent efficiency parameters include the following:

Total pressure efficiency:

$$\eta_{P_t} = \frac{P_{t4}}{P_{t0}} \tag{23}$$

Kinetic energy efficiency:

$$\eta_{\mathrm{KE}} = \frac{h_{t4} - h(P_0, s_4)}{h_{t0} - h_0} \tag{24}$$

Adiabatic η_{KE}:

$$\eta_{\mathrm{KEad}} = \frac{h_{t0} - h(P_0, s_4)}{h_{t0} - h_0} \tag{25}$$

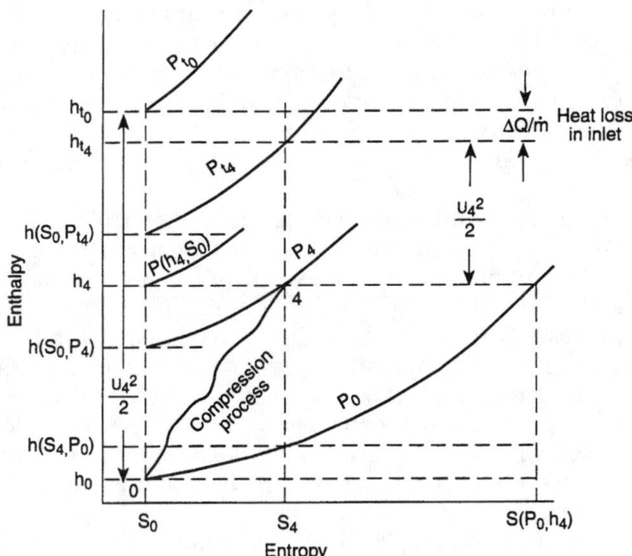

Fig. 4 Inlet compression process illustrated in enthalpy-entropy plane.

Isentropic efficiency:

$$\eta_i = \frac{h(P_{t4}, s_0) - h_0}{h_{t0} - h_0} \tag{26}$$

Static pressure efficiency:

$$\eta_R = \frac{P_4}{P(h_4, s_0)} \tag{27}$$

Dimensionless entropy rise:

$$\frac{\Delta s}{R} = \frac{s_4 - s_0}{R} \tag{28}$$

Entropy rise efficiency:

$$\eta_s = 1 - \frac{s_4 - s_0}{s(P_0, h_4) - s_0} \tag{29}$$

Process efficiency:

$$\eta_{\text{KD}} = \frac{h_4 - h(P_0, s_4)}{h_4 - h_0} \tag{30}$$

Compression efficiency:

$$\eta_B = \frac{h(P_4, s_0) - h_0}{h_4 - h_0} \tag{31}$$

For an ideal gas closed-form relationships between the various efficiency parameters can be derived.[13,14] These relationships are provided for reference in Table 1. For ideal gas problems η_{P_t} or η_{KE} are used most often whereas η_{KE} is quoted most often for nonideal gas problems. Each of these efficiency parameters offers advantages and disadvantages, especially with respect to the modeling of inlet performance. One should keep in mind though that the use of a single efficiency parameter is insufficient to completely specify the performance of an inlet.

In Ref. 13 arguments are presented that the compression ratio P_4/P_0 and contraction ratio A_0/A_4 should be used as the fundamental figures of merit inlet for inlet performance. Given these quantities together with the heat loss, the stream-thrust-averaged conditions at the inlet throat can be calculated. Sample results are presented in Fig. 5 for inlets at Mach 5, 10, and 15. Results are presented for both perfect gas and air in chemical equilibrium for an inlet flying along a $q_0 = 1000$-psf trajectory. In the pressure-area plane lines of constant adiabatic kinetic energy efficiency and throat Mach number are shown. Note that increasing the inlet efficiency at a given contraction ratio results in a decrease in the compression ratio and an increase in the throat Mach number.

The results provided in Fig. 5 can be used to assess the importance of high-temperature gas effects at high speeds. For each Mach number a sample inlet is noted in the pressure-area domain. In comparing results for the sample

	Known Parameter				
	η_B Compression Efficiency	η_{KD} Process Efficiency	η_{KE} Kinetic Energy Efficiency	η_{PT} Total Pressure Recovery	η_R Static Pressure Recovery
η_B		$1-(1-\eta_{KD})(P_4/P_0)^{(\gamma-1)/\gamma}$	$\frac{\left(\frac{P_4}{P_0}\right)^{(\gamma-1)/\gamma}-1}{\left(\frac{P_4}{P_0}\right)^{(\gamma-1)/\gamma}\left[\frac{T_{t4}}{T_{t0}}\left(1+\frac{\gamma-1}{2}M_0^2\right)-\left(\frac{\gamma-1}{2}\right)M_0^2\eta_{KE}\right]-1}$	$\frac{\left(\frac{P_4}{P_0}\right)^{(\gamma-1)/\gamma}-1}{\frac{T_{t4}}{T_{t0}}\left(\frac{P_4/P_0}{\eta_{PT}}\right)^{(\gamma-1)/\gamma}-1}$	$\frac{\left(\frac{P_4}{P_0}\right)^{(\gamma-1)/\gamma}-1}{\left(\frac{P_4}{P_0\eta_R}\right)^{(\gamma-1)/\gamma}-1}$
η_{KD}	$1-\frac{(1-\eta_B)}{(P_4/P_0)^{(\gamma-1)/\gamma}}$		$\frac{\left[\left(\frac{P_4}{P_0}\right)^{(\gamma-1)/\gamma}-1\right]\left[\frac{T_{t4}}{T_{t0}}\left(1+\frac{\gamma-1}{2}M_0^2\right)-\left(\frac{\gamma-1}{2}\right)M_0^2\eta_{KE}\right]}{\left(\frac{P_4}{P_0}\right)^{(\gamma-1)/\gamma}\left[\frac{T_{t4}}{T_{t0}}\left(1+\frac{\gamma-1}{2}M_0^2\right)-\left(\frac{\gamma-1}{2}\right)M_0^2\eta_{KE}\right]-1}$	$\frac{\left(\frac{P_4}{P_0}\right)^{(\gamma-1)/\gamma}-1}{\left(\frac{P_4}{P_0}\right)^{(\gamma-1)/\gamma}-\eta_{PT}^{(\gamma-1)/\gamma}(T_{t4}/T_{t0})}$	$\frac{\left(\frac{P_4}{P_0}\right)^{(\gamma-1)/\gamma}-1}{\left(\left(\frac{P_4}{P_0}\right)^{(\gamma-1)/\gamma}-\eta_R^{(\gamma-1)/\gamma}\right)}$
η_{KE}	$\frac{2}{(\gamma-1)M_0^2}\left[\frac{(1/\eta_B-1)}{(P_4/P_0)^{(\gamma-1)/\gamma}}+\frac{T_{t4}}{T_{t0}}\left(1+\frac{\gamma-1}{2}M_0^2\right)-\frac{1}{\eta_B}\right]$	$\frac{\frac{T_{t4}}{T_{t0}}\left(1+\frac{\gamma-1}{2}M_0^2\right)+\frac{1}{\frac{1}{\eta_{KD}}\left[\left(\frac{P_4}{P_0}\right)^{(\gamma-1)/\gamma}-1\right]-\left(\frac{P_4}{P_0}\right)^{(\gamma-1)/\gamma}}}{(\gamma-1)M_0^2/2}$		$\frac{2}{(\gamma-1)M_0^2}\left(\frac{T_{t4}}{T_{t0}}\right)\left[1+\frac{\gamma-1}{2}M_0^2-\eta_{PT}^{(1-\gamma)/\gamma}\right]$	$\frac{T_{t4}}{T_{t0}}-\frac{2}{(\gamma-1)M_0^2}\left[\eta_R^{(1-\gamma)/\gamma}-\frac{T_{t4}}{T_{t0}}\right]$
η_{PT}	$\frac{\left(\frac{P_4}{P_0}\right)\left(\frac{T_{t4}}{T_{t0}}\right)^{\gamma/(\gamma-1)}}{\left\{1+\frac{1}{\eta_B}\left[\left(\frac{P_4}{P_0}\right)^{(\gamma-1)/\gamma}-1\right]\right\}^{\gamma/(\gamma-1)}}$	$\left\{\left[\left(\frac{P_4}{P_0}\right)^{(\gamma-1)/\gamma}-\frac{\left(\frac{P_4}{P_0}\right)^{(\gamma-1)/\gamma}-1}{\eta_{KD}}\right]\frac{T_{t4}}{T_{t0}}\right\}^{\gamma/(\gamma-1)}$	$\left[\left(1+\frac{\gamma-2}{2}M_0^2\right)-\frac{(\gamma-1)M_0^2\eta_{KE}}{2T_{t4}/T_{t0}}\right]^{\gamma/(1-\gamma)}$		$\eta_R\left(\frac{T_{t4}}{T_{t0}}\right)^{\gamma/(\gamma-1)}$
η_R	$\frac{P_4/P_0}{\left\{1+\frac{1}{\eta_B}\left[\left(\frac{P_4}{P_0}\right)^{(\gamma-1)/\gamma}-1\right]\right\}^{\gamma/(\gamma-1)}}$	$\left[\left(\frac{P_4}{P_0}\right)^{(\gamma-1)/\gamma}\left(1-\frac{1}{\eta_{KD}}\right)+\frac{1}{\eta_{KD}}\right]^{\gamma/(\gamma-1)}$	$\left[\left(\frac{T_{t4}}{T_{t0}}\right)\left(1+\frac{\gamma-1}{2}M_0^2\right)-\frac{\gamma-1}{2}M_0^2\eta_{KE}\right]^{\gamma/(1-\gamma)}$	$\eta_{PT}\left(\frac{T_{t4}}{T_{t0}}\right)^{\gamma/(1-\gamma)}$	

Table 1 Relationships of inlet efficiency parameters

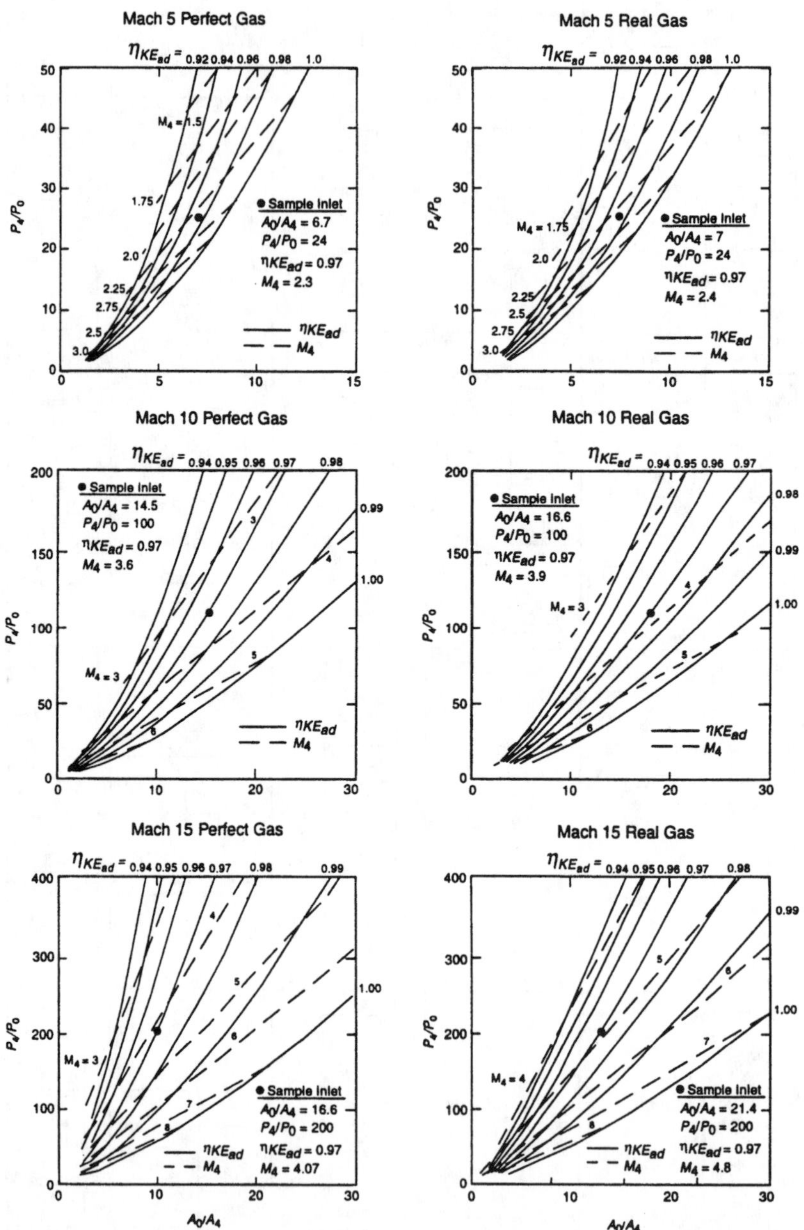

Fig. 5 Illustration of the use of the pressure-area domain in specifying inlet performance.

inlets, the inlet efficiency and compression ratio held constant. At Mach 5 a contraction ratio of 6.7 and a throat Mach number of 2.3 result for an assumed compression ratio of 24 and an η_{KE} of 0.97. For the real-gas case the contraction ratio is 7 and the throat Mach number is 2.4 for the same compression ratio of 24 and η_{KE} of 0.97. Note that the differences between the perfect gas assumption and the assumption of air in chemical equilibrium are small.

At Mach 15 with a perfect gas assumption, a contraction ratio of 16.6 and a throat Mach number of 4.07 result for the assumed compression ratio of 200 and η_{KE} of 0.97. For air in chemical equilibrium, the contraction ratio is 21.4 and the throat Mach number is 4.8 for the same compression ratio of 200 and η_{KE} of 0.97. Note that for this example, a difference in contraction ratio of 28% is required to maintain the same compression ratio between the perfect gas case and the case with air in chemical equilibrium. This example illustrates the fact that high-temperature gas dynamics can play an important role in the design and operation of hypersonic inlets. Other aspects of high-temperature gas dynamics will be presented in a later section.

The discussion up to this point has focused on the use of stream-thrust-averaged quantities in the inlet throat. These quantities are preferred because they maintain the mass, momentum, and energy of the true flowfield, and they can be measured experimentally. The disadvantage of the stream-thrust-averaged flow quantities is that they do not accurately represent the entropy in the true flow. In the development of the stream-thrust-average quantities, the assumption is made that the flow is mixed to a uniform, one-dimensional state. This mixing process introduces an entropy rise that does not exist in the actual flow. Consequently, the stream-thrust quantities represent a lower bound on the true performance of an inlet.

To avoid the entropy rise associated with mixing, mass-average flow quantities are often stated. These flow quantities are useful in defining an average flow quantity (such as average Mach number) that may be required for the analysis of some downstream flow phenomena. Care must be taken in using mass-averaged flow quantities in performance estimates because the mass, momentum, and energy of the true flow may not be accurately represented. Additional information on various averaging techniques is presented in Ref. 1.

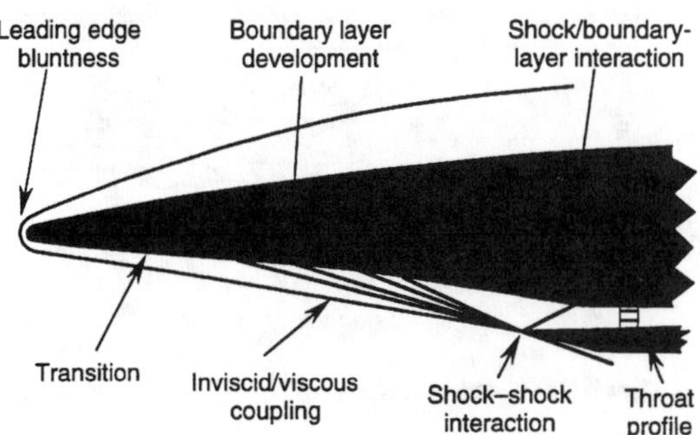

Fig. 6 Schematic of the inlet flowfield features that affect inlet performance and operability.

III. Inlet Design Issues

In the design of scramjet inlets for hypersonic flight, a variety of aerothermodynamic phenomena are encountered. As illustrated in Fig. 6, these phenomena include blunt leading-edge effects, boundary-layer development issues, transition, inviscid/viscous coupling, shock-shock interactions, shock/boundary-layer interactions, and flow profile effects. For inlets that are designed to operate within a narrow Mach number/altitude envelope, an understanding of a few of these phenomena might be required. For inlets designed to operate over very large ranges of speed and altitude (such as an aerospace plane), nearly all of these aerodynamic issues must be addressed. In this section several predominant flowfield phenomena are discussed.

A. Starting and Contraction Limits

For proper operation all scramjet inlets must operate in a started mode. The term started is used to denote operation under conditions where flow phenomena in the internal portion of the inlet do not alter the capture characteristics of the inlet. An inlet can be unstarted either by overcontracting to the point where the flow chokes at the inlet throat or by raising the back pressure beyond the level that can be sustained by the inlet.

The flowfield of an unstarted hypersonic scramjet inlet can be quite different from that found in a conventional external compression turbojet or ramjet inlet, as illustrated in Fig. 7. In a ramjet or turbojet inlet the ratio of the boundary-layer height to inlet height at the cowl-lip plane is usually small. When these inlets unstart, a normal shock is expelled, and flow is spilled subsonically. For hypersonic scramjet inlets a substantial portion of the flow at the cowl lip can be boundary layer. When these inlets unstart, the expelled shock system is sufficiently strong to separate the boundary layer, creating the flowfield illustrated in Fig. 7. The flowfield is characterized by a large separated region and supersonic flow spillage. In this condition the portion of the flow that is captured by the unstarted inlet can be predominately supersonic. In general, an unstarted inlet captures less airflow with

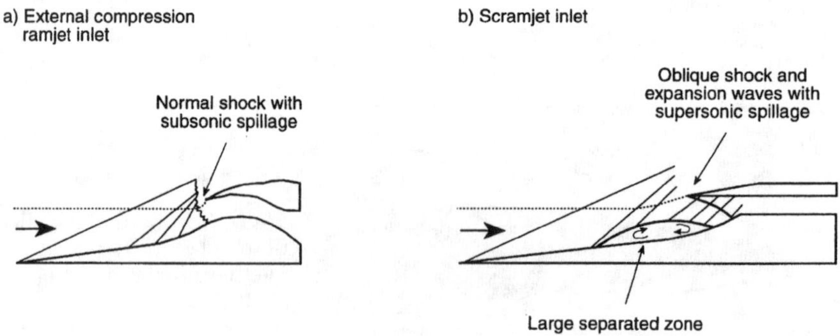

Fig. 7 Comparison of unstarted flowfield characteristics of ramjet and scramjet inlets.

lower efficiency and higher aerodynamic and thermal loads compared to a started inlet.

Inlet starting has been studied extensively and has been found to be dependent on the local Mach number, internal contraction ratio, diffuser flowfield pressure recovery, and the time-dependent details of the starting process.[3,15–21] For a specific inlet geometry and a set of freestream conditions, an inlet will start at a certain internal contraction. After the inlet starts variable geometry features of the inlet, if present, can be adjusted to increase the contraction ratio up to a maximum contraction ratio, at which point further increases would cause the inlet to unstart. Preliminary estimates of the internal contraction ratio that will self-start can be obtained from the Kantrowitz limit.[20] This limit is determined by assuming a normal shock at the beginning of the internal contraction and calculating the one-dimensional, isentropic, internal area ratio that will produce sonic flow at the throat. For a perfect gas the Kantrowitz limit can be calculated as follows:

$$\left(\frac{A_2}{A_4}\right)_{\text{KANTROWITZ}} = \frac{1}{M_2}\left[\frac{(\gamma+1)M_2^2}{(\gamma-1)M_2^2+2}\right]^{\frac{\gamma}{\gamma-1}}\left[\frac{\gamma+1}{2\gamma M_2^2-(\gamma-1)}\right]^{\frac{1}{\gamma-1}} \left[\frac{1+\gamma-1/2\, M_2^2}{\gamma+1/2}\right]^{\frac{\gamma+1}{2(\gamma-1)}} \tag{32}$$

In Eq. (32) the assumption is made that the inlet is internally contracting from station 2 to the throat at station 4. If the internal contraction begins at a different point in the compression process, the Mach number at that point would be used instead of M_2.

A summary of published results is presented in Fig. 8 for both inlet starting and maximum contraction ratios.[3–5,15–32] Data are also shown several inlets for which an operating contraction ratio (rather than the maximum contraction ratio) has been reported. These operating contraction ratios were obtained from reported tests in which no explicit attempt was made to determine a maximum contraction. In Fig. 8 the data are plotted in terms of the inverse of the contraction ratio (i.e., A_4/A_2 and A_4/A_0). Also shown in this figure is the Kantrowitz limit [Eq. (32)] and the isentropic contraction limit, which can be calculated as follows:

$$\left(\frac{A_4}{A_0}\right)_{\text{ISENTROPIC}} = M_0\left(\frac{\gamma+1}{2}\right)^{\frac{\gamma+1}{2(\gamma-1)}}\left(1+\frac{\gamma-1}{2}M_0^2\right)^{-\frac{\gamma+1}{2(\gamma-1)}} \tag{33}$$

The allowable starting and maximum contraction ratios both increase as the Mach number increases, but the experimental data show a significant variation at higher speeds. This variation is a result of differences in inlet geometry, Reynolds number, and wall-to-freestream temperature ratio. The data also suggest that the Kantrowitz starting limit becomes conservative at high speeds as a result of the assumption of a single normal shock. Because the shock system of an unstarted hypersonic inlet has a higher pressure recovery compared to a normal shock, internal contraction ratios higher than that indi-

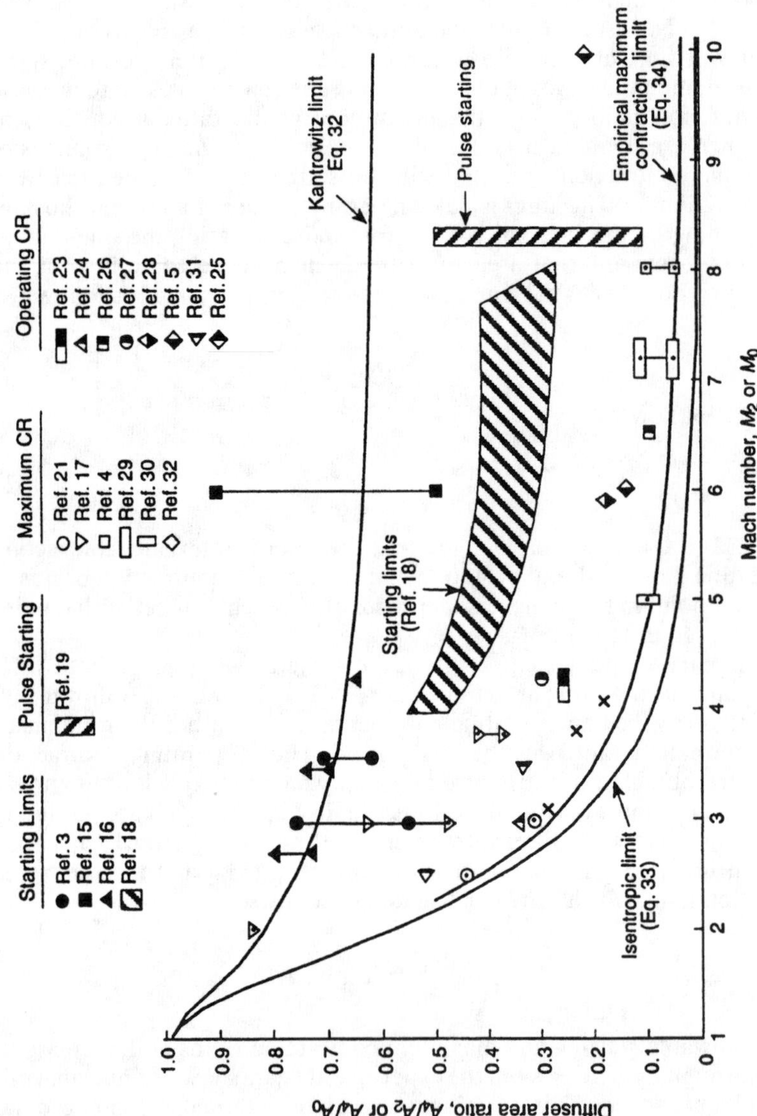

Fig. 8 Starting and maximum contraction ratio (CR) limits.

cated by the Kantrowitz limit can be tolerated. Allowable started internal contraction ratios as high as 2–3 are obtained for some inlet geometries.

For inlet designs in which large internal contractions are desirable, several approaches to achieving high contraction ratios have been attempted, including swept-back sidewalls in two-dimensional configurations,[9] open cowls in sidewall compression configurations,[33] and bypass doors and wall perforations in various types of configurations.[18,34] These approaches all aim at increasing the available spill area during the starting process.

The maximum contraction ratio data shown in Fig. 8 show that higher contraction ratios can be generated as the Mach number increases. Obviously, the limiting contraction ratio is lower than that for the corresponding isentropic contraction because of the effects of shock waves and viscous losses. The data imply a practical limit on the maximum contraction ratio. An empirical fit to the data for this limit is as follows:

$$\frac{A_4}{A_2} = 0.05 - \frac{0.52}{M_0} + \frac{3.65}{M_0^2} \qquad (2.5 < M_0 < 10) \tag{34}$$

Because this line lies below the observed maximum contraction data, this equation provides an estimate of the limit on the achievable contraction ratio.

One factor that affects the maximum contraction ratio for an inlet is the degree of flow nonuniformity in the flowfield at the inlet throat. In Ref. 17 the effect of throat distortion on the minimum average Mach number that could be produced without inlet unstart was investigated for an axisymmetric inlet. In this work the throat flow nonuniformity was characterized in terms of a distortion parameter D, defined as follows:

$$D = \frac{\displaystyle\int_A \left| (\rho U^2) - (\rho U^2)_{\text{ave}} \right| dA}{(\rho U^2)_{\text{ave}} A} \tag{35}$$

The effect of the distortion on the minimum achievable throat Mach number is shown in Fig. 9. As the throat distortion increases, the minimum throat Mach number that can be produced without unstart also increases. For high levels of throat distortion, the minimum Mach number at which the flow would choke could be as high as 2.6. These results show that the inlet throat profile should be as uniform as possible if it is desirable to operate at high contraction ratios.

Inlets with very high internal contraction have been found to start in pulse facilities.[16,19,35] The impulsive nature of the facility starting process creates a situation where the flowfield pictured in Fig. 7 never establishes, so that the Kantrowitz limit does not apply. Some evidence exists that even overcontracted inlets will initially start in a pulse facility and then will unstart as the boundary layers fully develop.[16] In Ref. 19 the starting process for simple two-dimensional inlets was investigated at Mach 8.3. Using a gun tunnel, starting contraction ratios were measured for various geometries, boundary-layer thicknesses, Reynolds numbers, and initial pressure ratios across the

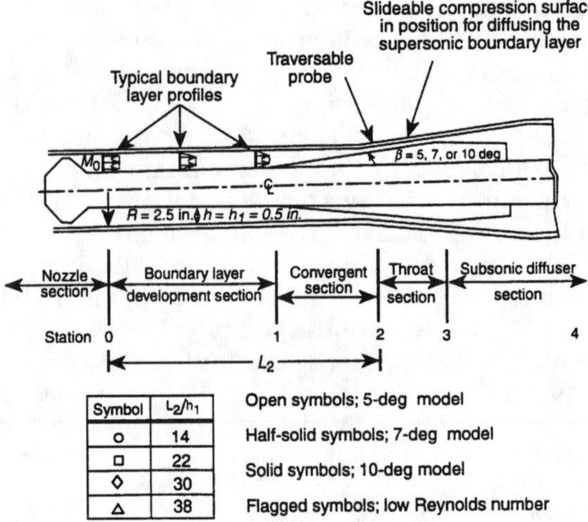

Symbol	L_2/h_1
○	14
□	22
◇	30
△	38

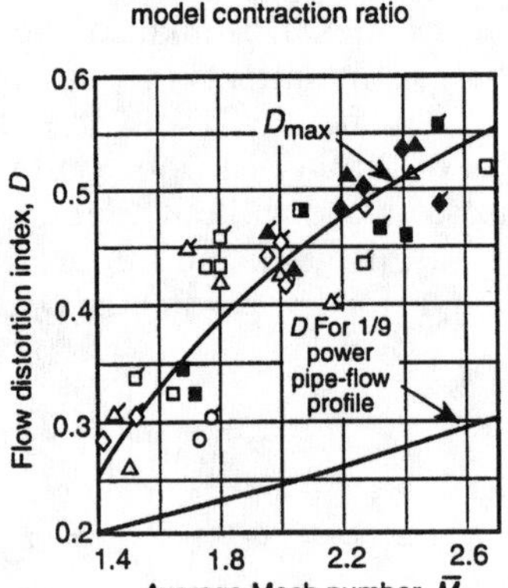

Fig. 9 Variation of the distortion index with average Mach number (from Ref. 17).

diffuser. Typical results are shown in Fig. 10 where the starting contraction ratio is shown as a function of the cowl angle and initial pressure in the test section. In these tests the ratio of the facility total pressure to the initial pressure in the test section before facility startup was shown to play a dominant role in the starting process. In general, the contraction ratio limit observed in pulse facilities falls between the Kantrowitz limit and the maximum steady-flow contraction limit.

B. High-Temperature Effects

One issue associated with hypersonic flight of a scramjet inlet concerns the high-temperature gas effects that are present in the flowfield and their impact on the operating characteristics of the inlet. At high speeds a large amount of kinetic energy is contained in the flow, and the temperature of

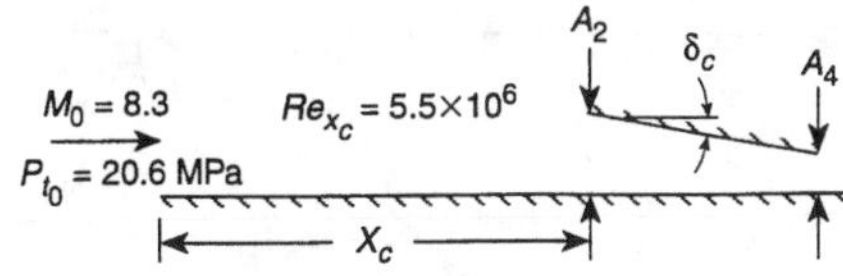

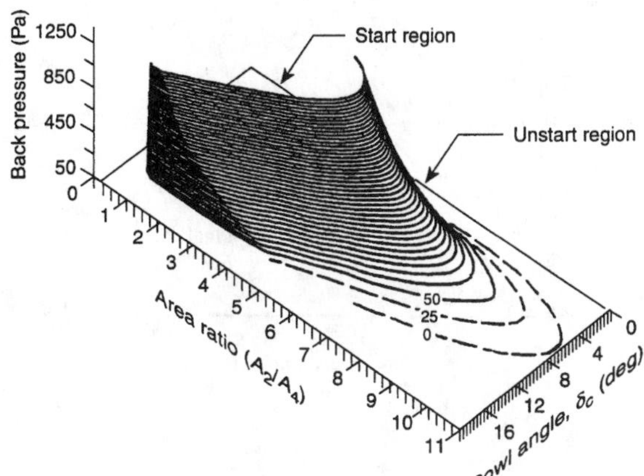

Fig. 10 Effect of initial test cabin pressure on starting characteristics of a two-dimensional inlet (from Ref. 19).

the captured airstream rises as it is slowed either through the compression process or because of viscous effects. For temperatures below approximately 600 K, air can be modeled as a perfect gas, but at higher temperatures the effects of vibrational excitation, dissociation, and ionization can become important. These high-temperature gas effects are often erroneously called real-gas effects. The term real gas actually refers to effects caused by intermolecular forces between gas molecules and the finite volume of each molecule. Actual real-gas effects occur at very high densities for air and generally are not important in the operation of scramjet inlets.

Some idea of the importance of air chemistry on inlet operation is seen in Fig. 11, where results are presented for optimum inviscid four-shock inlets at Mach 15 on a $q_0 = 1000$-psf trajectory. Plotted are the contraction ratio and total turning of the optimum inlets as a function of compression ratio for both an ideal gas case and a case that assumes air in chemical equilibrium. These results show that the effect of the gas chemistry on the total turning is very small. For example, at a compression ratio of 250, only 0.2-deg difference in total turning exists between the ideal gas and air in chemical equilibrium cases. The results show that the gas chemistry affects the contraction ratio of the optimum inlets, and this effect becomes more important as compression ratio is increased. For example, a 13% difference in contraction ratio exists between the two cases at a compression ratio of 250. This difference is driven by the changes in the shock angles and density ratios between the two different air models.

The potential influence of high temperatures on an inlet is seen in Fig. 12, where the temperatures at various locations throughout an inlet flowfield

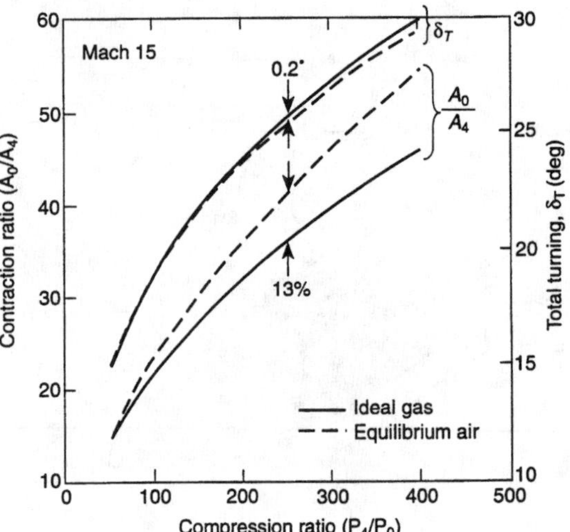

Fig. 11 Effect of gas chemistry on total turning and CR for optimum inviscid four-shock scramjet inlets at Mach 15.

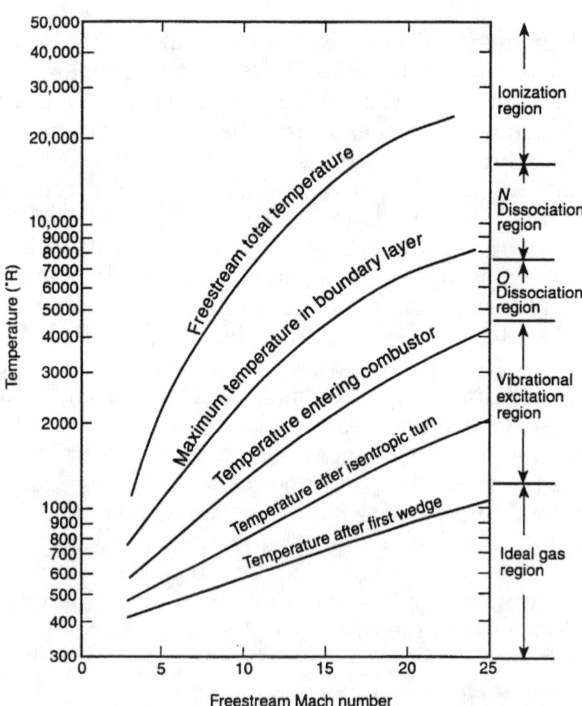

Fig. 12 Temperatures at various locations in a typical inlet (from Ref. 36).

that is in chemical equilibrium are shown. The entire flow, except for the stagnation regions, can be represented by an ideal gas at speeds below Mach 5. For speeds up to Mach 12, the inviscid portion of the flowfield can be represented by an ideal gas, but vibrational effects become important in the boundary layers. At speeds above Mach 12, vibrational effects are important in the inviscid flow, and dissociation effects become important in the boundary layers. The effects of gas ionization are limited to the stagnation regions at the highest speeds.

The consideration of the gas model (i.e., ideal gas, equilibrium mixture, etc.) is important in the design of hypersonic inlets because the positioning and strengths of shocks can be influenced by high-temperature effects. If these effects are neglected, the flow structure and resulting inlet performance can be significantly affected. This issue is particularly important with regard to the testing of hypersonic inlets in ground-test facilities where a complete simulation is not possible.

At very high Mach numbers and high altitudes, the partition of energy between various states may not remain in equilibrium when the flow properties are rapidly changing. For very high-speed inlet designs ($M_0 > 15$) these nonequilibrium effects are most important in the stagnation regions and in the high-temperature portions of the boundary layers. An example of the types of differences that can exist at Mach 25 is shown in Fig. 13 for an axisymmetric inlet.[36] In this example the temperature distribution through the shock layer at the cowl-lip plane is shown for operation at an altitude of 47.5 km. The temperature distributions were calculated with PNS codes, and results are shown for three air models: ideal, equilibrium, and nonequilibrium. These results show that the temperatures in the inviscid core are nearly the same for the three gas models. Within the high-temperature region of the boundary layer, significant differences in the temperature distribution can result with the nonequilibrium calculations producing the

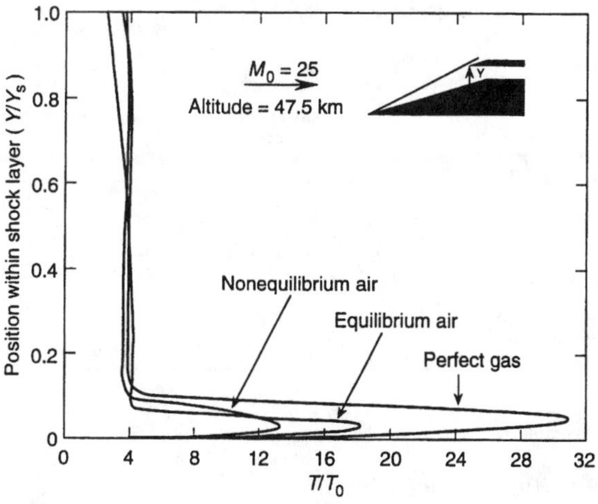

Fig. 13 Effect of gas model on the temperature profile at the cowl-lip plane of a typical inlet (from Ref. 36).

lowest local temperatures. Because the nonequilibrium phenomena are limited to a narrow region of the boundary layer, cheaper equilibrium-based design tools are often used in the design process.

C. Blunt Leading-Edge Effects

One critical design issue for scramjet inlets concerns the blunting of leading edges on surfaces such as the forebody and cowl lip. These surfaces must be blunted to obtain acceptable heating levels at hypersonic speeds. Unfortunately, the bluntness generally degrades the performance of an inlet. As shown schematically in Fig. 14, blunt leading edges cause curved bow shocks that generate large entropy layers. These entropy layers modify the downstream boundary-layer development, including transition, and can create significant changes in the inviscid flowfield such as the shock positioning and air-capture characteristics.

An illustration of the potential effects of forebody nose bluntness is provided in Fig. 15, where schlieren photographs of the forebody flowfield of a two-dimensional inlet at Mach 13.1 are shown.[37] This inlet forebody was 36 in. long and consisted of a 5-deg blunted wedge followed by an additional 5 deg of isentropic compression. In the schlieren photograph on the top, a sharp (R_n = 0.005 in.) leading edge was used, and the expected forebody flowfield was obtained. The forebody bow shock, isentropic compression field, and forebody boundary layer are clearly visible. In the schlieren photograph on the bottom, a blunter (R_n = 0.100 in.) leading edge was used, and the outward displacements of the bow shock and isentropic compression field are readily seen. The existence of several oblique shocks upstream of the cowl lip suggests that the inlet is unstarted. For this inlet design the change in the nose bluntness from 0.005 to 0.100 in. is the difference between the inlet operating and not operating.

Because blunt leading edges can significantly alter the shock structure and pressure distribution in an hypersonic inlet, engineering estimates of these effects are needed. A correlation for the shock shape around a two-dimensional blunted leading edge is given in Ref. 38 as follows:

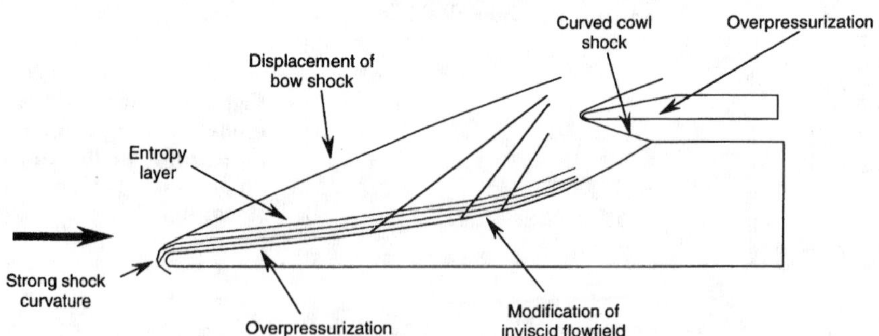

Fig. 14 Blunt leading-edge effects.

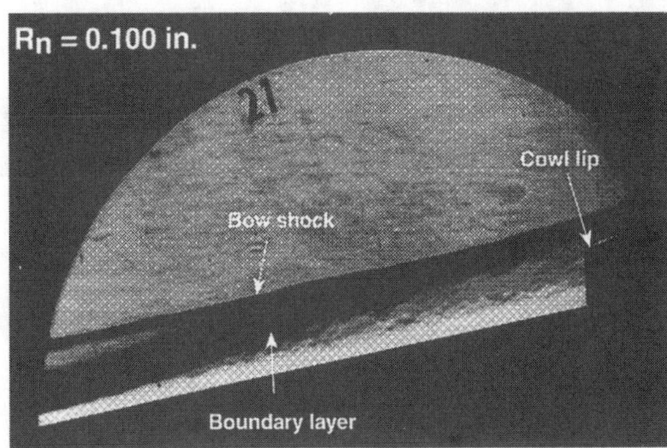

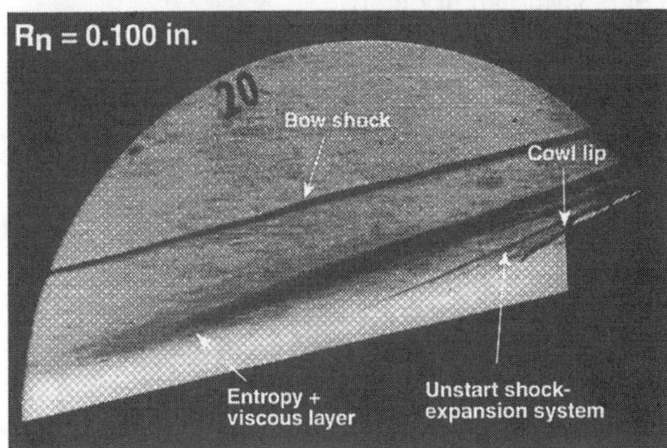

Fig. 15 Schlieren photographs showing the effect of R_n on the inlet flowfield at M_0 = 13.1, Re = 1.3 × 10^6/ft and T_w/T_0 = 6.1 (from Ref. 37).

$$x = R + \Delta - R_c \operatorname{cotan}^2\theta\left[\left(1 + \frac{y^2\tan^2\theta}{R_c^2}\right)^{1/2} - 1\right] \tag{36}$$

where x and y are the Cartesian coordinates of the shock shape, R is the radius of curvature of the leading edge, R_c is the radius of curvature of the shock at the vertex, and Δ is the shock standoff distance. Correlations for Δ/R and R_c/R are provided in Ref. 38 as follows:

$$\frac{\Delta}{R} = 0.386 \exp\left(\frac{4.67}{M_0^2}\right) \tag{37}$$

and

$$\frac{R_c}{R} = 1.386 \exp\left[\frac{1.8}{(M_0 - 1)^{0.75}}\right] \tag{38}$$

The effects of blunt leading edges on the pressure distributions downstream of the leading edge on flat plates and cylinders have been investigated extensively, and theoretical approaches have been developed using the hypersonic equivalence principal. The pressure distribution on a flat plate and the bow-shock shape downstream of the blunted leading edge have been correlated as a function of distance, freestream Mach number, and the drag coefficient of the leading edge as follows[39,40]:

$$\frac{P}{P_0} = 0.117\, M_0^2 C_d^{2/3}\left(\frac{x}{d} + \frac{2}{3}\right)^{-2/3} + 0.732 \tag{39}$$

A comparison of this correlation with results from the Method of Characteristics and experimental data is shown in Fig. 16.

The extent of the blunt leading-edge interaction can be estimated using Eq. (39). For a leading-edge drag coefficient of 1.2, the length of flat plate at which the induced pressure is 50% above the freestream pressure (i.e., $P/P_0 = 1.5$) can be calculated as follows:

$$\frac{x}{d} = 0.714 M_0^3 \tag{40}$$

For freestream Mach numbers of 5, 10, 15, and 20, the leading-edge interaction would extend to $x/d = 8.9, 71.4, 241$, and 571, respectively. One can see that this interaction can extend over significant portions of the inlet at high speeds even for relatively sharp leading edges.

The shock shape created by the blunt leading edge has also been correlated as follows[39,40]:

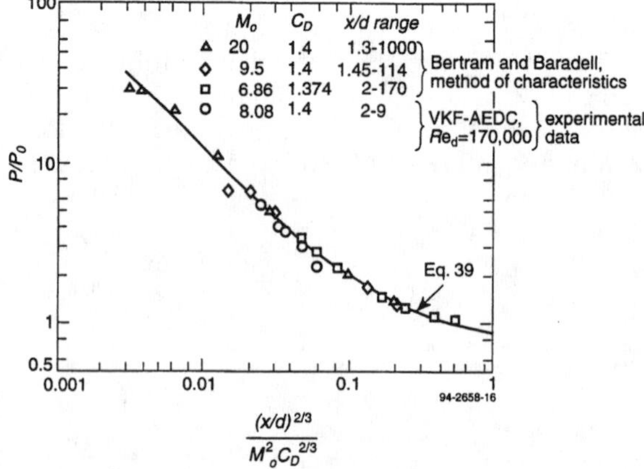

Fig. 16 Correlation of pressure distribution for a blunt-nosed flat plate (from Ref. 40).

$$\frac{r}{d} = \frac{0.77\, M_0^2 C_d}{M_0^2 \left(\frac{C_d}{x/d}\right)^{2/3} - 1.09} \tag{41}$$

where r is the vertical height of the shock above the stagnation point and d is the diameter of the blunt leading edge. The validity of this correlation is illustrated in Fig. 17, where a comparison between Eq. (41) and results from experiments and Method of Characteristics calculations is shown. For the purposes of preliminary design, this correlation is useful in predicting the displacement of the bow-shock shape caused by the blunt leading edge.

The results presented thus for the pressure distribution and shock shape have been for blunted flat plates. These same results can apply (to first order) for a blunted wedge where the pressure distribution and shock shape given by Eqs. (39) and (41) represent changes to the pressure and shock shape generated by a sharp wedge.

A second blunt leading-edge phenomenon that must be addressed is the situation that exists when a shock wave intersects a blunt leading-edge flowfield. This flow situation is of interest because of the tremendous pressure and heat-transfer amplifications that can result. This so-called shock-shock interaction has been studied extensively for both spherical leading edges and blunted cylinders. Edney has identified six different types of interactions depending on the location of the incident shock relative to the bow shock.[41] These six interactions are shown schematically in Fig. 18. Type I, II, and V interactions result in shock-wave/boundary-layer interactions. The Type III interaction produces a shear layer that may attach to the blunt leading edge, causing high heat-transfer regions typical of reattaching separated boundary layers. In a Type IV interaction a supersonic jet is produced that grazes or impinges on the leading edge. The Type VI interaction results in an interaction between an expansion fan and the boundary layer.

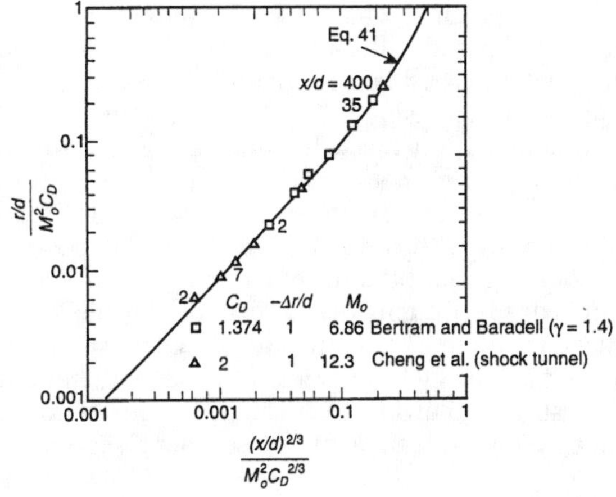

Fig. 17 Correlation of shock-wave shapes and blunt-nosed flat plate (from Ref. 40).

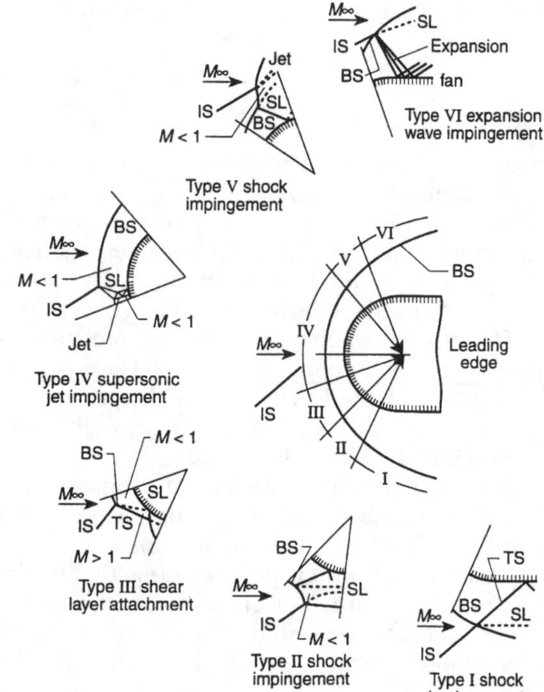

Fig. 18 Six types of shock-wave interference patterns (from Ref. 42): BS, bow shock; IS, oblique impinging shock; SL, shear layer; TS, transmitted shock.

The Type IV interaction produces the most significant increases in pressure and heat transfer.[41–48] This interaction occurs when an oblique shock intersects the nearly normal portion of the bow wave. This shock-shock interaction results in the formation of a supersonic jet in which the flow is efficiently compressed before impingement on the leading edge. The amplification of the pressure and heat transfer in this impingement region can be a factor of 20–40 times higher than that encountered with an isolated leading edge. The amplification factors increase with increasing freestream Mach number, increasing incidence shock angle, and decreasing ratio of specific heats.

In Ref. 44 the effect of sweepback on the leading edge was shown to reduce the amplification factors. Peak pressure amplification factors were shown to be proportional to $\cos^4\lambda$, where λ is the sweep-back angle. Peak heat-transfer amplification factors were shown to be proportional to $\cos^{2.2}\lambda$.

Because of the high local heating rates, the development of an efficient cooling system for the leading edge represents a significant design challenge for hypersonic scramjet inlets. Preliminary attempts at using transpiration cooling on the leading edge with a Type IV interaction have met with limited success.[45] Using high levels of surface blowing, the peak heat transfer could only be reduced 8%, although the overall heat transfer level could be substantially reduced.

In Ref. 46 the effects of two impinging oblique shocks on a leading edge (typical of a scramjet inlet) were investigated. Using a two-dimensional forebody consisting of a 7.5-deg wedge followed by a 5-deg turn, circumferential measurements of pressure and heat transfer were obtained at Mach 8. Results presented in Fig. 19 show that the heat-transfer amplification (local heat transfer referenced to the undisturbed stagnation point heat transfer) for this case was approximately 38, and the pressure amplification (local pressure referenced to the undisturbed pitot pressure) was approximately 14. Although the amplifications occur over a narrow portion of the leading edge, these tremendous heat-transfer rates pose a serious problem with regard to the design of a survivable leading edge.

Examples of the magnitude of the heat-transfer levels are shown in Fig. 20 for operation at Mach numbers between 7 and 25 (Ref. 36). Maximum heat-transfer levels are shown for a forebody leading-edge radius of 50 mm, a cowl-lip radius of 2.5 mm, and a cowl with a 2.5-mm leading-edge radius with a shock interaction. These results assume a heat-transfer amplification factor of 20. Shown on the right-hand side of this figure is the time that would be required to melt the surface of an uncooled, semi-infinite slab of copper. (Obviously, the leading edges at hypersonic speeds would be cooled, so this example is purely for illustrative purposes.) At the highest heat-transfer levels the surface material can be raised to the melting point in millisec-

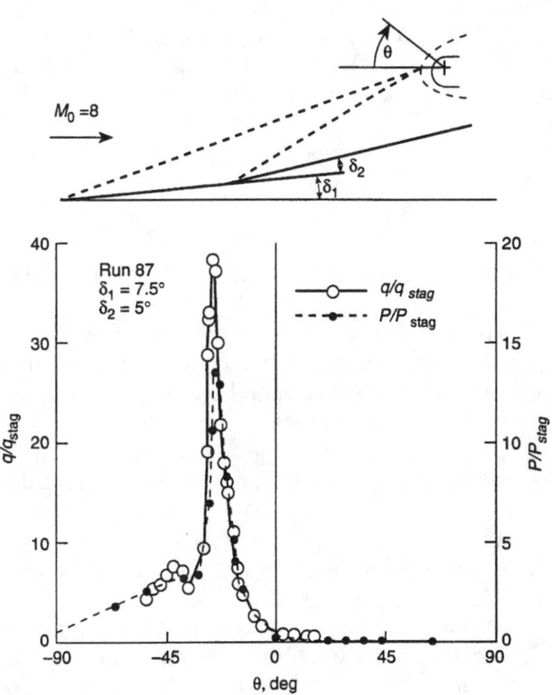

Fig. 19 Schlieren photograph and heat-transfer rate and pressure distribution for dual incident shock: L = 53.47 cm (from Ref. 46).

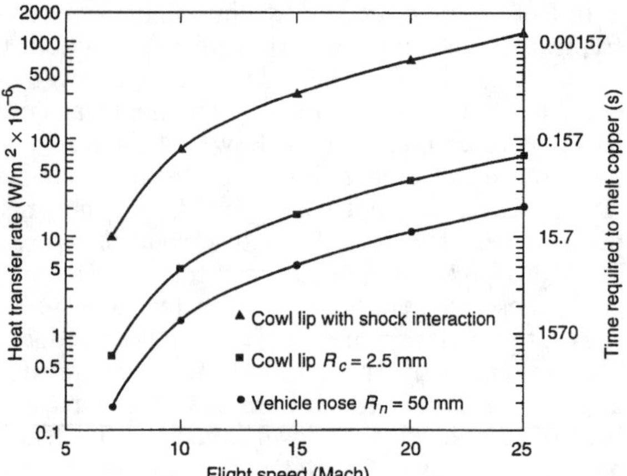

Fig. 20 Stagnation point heat-transfer rates for a typical ascent trajectory (from Ref. 36).

onds. The high stresses created by these high-temperature gradients can significantly reduce the life of the materials.

Two approaches to predicting the magnitude of the pressure and heat-transfer loads at the leading edge have been attempted for the shock-shock interactions. The simpler approach involves modeling of the supersonic jet that is formed.[41,49,50] By calculating the flowfield through the shocks and expansions within the jet that lead to a final normal shock, the pressure in the impingement region can be estimated. Once the pressure is obtained, the heat transfer is found through a relationship of the type:

$$\frac{q}{q_{\text{stag}}} = 1.4\left(\frac{r}{w}\frac{p}{p_{\text{stag}}}\right)^{0.5} \tag{42}$$

where r is the radius of the leading edge and w is the width of the supersonic jet.

The alternate approach at solving the shock-shock interaction flowfield is through the use of computational fluid dynamics.[51–57] Accurate prediction of the heat transfer within the interaction region depends principally on adequacy of the grid resolution and turbulence modeling.

In Ref. 57 a computational analysis has shown that the flowfield within a Type IV interaction can be unsteady under certain conditions. This time-dependent nature of the flow was traced to an unsteady vortex production within the shear layers of the supersonic jet. For an interaction at Mach 8.1 with an 18-deg incident shock angle, the pressure oscillation with a frequency of approximately 1400Hz was encountered.

In summary, the effects of blunt leading edges on scramjet inlet flowfields must be considered in the design. Reducing the bluntness and sweeping the leading edges tend to reduce the effects. For shock-shock interactions, weakening the strength of the impingement shocks and sweeping the leading edges helps to reduce the peak pressure and heat-transfer loads.

D. Viscous Phenomena

At hypersonic speeds the development of the boundary layer within an inlet has a major influence on the performance and operability of the inlet. This influence arises because the growth of the boundary layer adds to the effective compression of the captured flow. In regions of high adverse-pressure gradients, the separation of the boundary layer must be considered and may lead to design constraints on the inlet. Furthermore, the losses as a result of friction within the boundary layer represent the single largest loss mechanism for hypersonic inlets.

As illustrated in Fig. 21, the effect of boundary-layer growth on the pressure distribution along a flat plate can be correlated for laminar flows in terms of a viscous interaction parameter $\overline{\chi}$:

$$\overline{\chi} = M_0^3 \sqrt{\frac{C_0}{Re_x}} \tag{43}$$

where

$$C_0 = \frac{\rho_w \mu_w}{\rho_e \mu_e}$$

The results are divided into a weak interaction regime $\overline{\chi} < 3$ and a strong

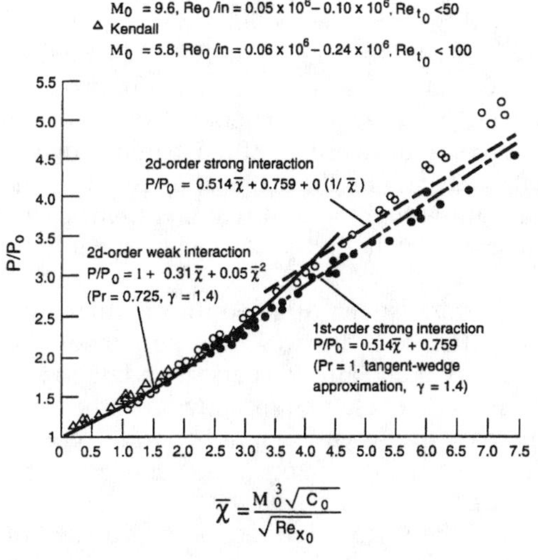

Fig. 21 Correlation of induced pressures on a flat plate as a result of boundary-layer growth (from Ref. 58).

interaction region $\bar{\chi} > 3$ (Ref. 58). In the weak interaction region the boundary-layer displacement growth can be used in calculating the resulting inviscid flowfield. In the strong interaction zone the viscous and inviscid portions of the flowfield must be solved simultaneously. The pressure distributions for the weak and strong interactions have been correlated as follows:

Weak interaction:

$$\frac{P}{P_0} = 1 + 0.31\bar{\chi} + 0.05\bar{\chi}^2 \qquad \bar{\chi} < 3 \tag{44}$$

Strong interaction:

$$\frac{P}{P_0} = 0.514\bar{\chi} + 0.759 \qquad \bar{\chi} < 3 \tag{45}$$

These correlations have been developed for flat-plate boundary layers, but they also can be applied (to first order) for two-dimensional wedges if $\bar{\chi}$ is evaluated behind the oblique shock.

Some feel that for the importance of the boundary layer growth can be obtained by considering a simple two-dimensional inlet consisting of two 5-deg external shocks followed by a single 10-deg internal shock. An estimate of the significance of the displacement thickness can be obtained by calculating the Mach number and altitude where $\bar{\chi} = 2$, which corresponds to a condition where the pressure induced by the boundary layer is approximately 1.8 times the inviscid pressure. In these calculations $\bar{\chi}$ is evaluated at a point one ft downstream of the leading edge.

In Fig. 22 the Mach number and altitude at which $\bar{\chi} = 2$ are shown for a flat plate and a 5-deg wedge. The flat plate would correspond to the cowl in an oversped configuration, and the 5-deg wedge corresponds to the initial forebody compression. Also overlaid on the plot are lines of constant dynamic pressure between 500 and 2000 psf that roughly define the practical air-breathing corridor. One can see that $\bar{\chi} = 2$ will exist at a point one ft downstream of a flat plate leading edge for Mach numbers as low as 10. For the 5-deg wedge $\bar{\chi} = 2$ would exist one ft downstream of the leading edge at Mach 13 for a 500-psf trajectory. These results show that the viscous compression at leading edges is significant within the air-breathing flight corridor. The viscous interaction is most significant on the cowl when the inlet is operated in an oversped mode.

The degree of viscous interaction was also investigated for the inlet configuration just discussed, assuming the cowl lip was downstream of the forebody compression. The Mach number and altitude corresponding to $\chi = 1$ on the cowl are also shown in Fig. 22. These results show that the inlet compression process serves to increase the Reynolds number and decrease the Mach number such that the viscous interaction is not as significant compared to that already presented.

Another example of the effects of the viscous interaction is provided in

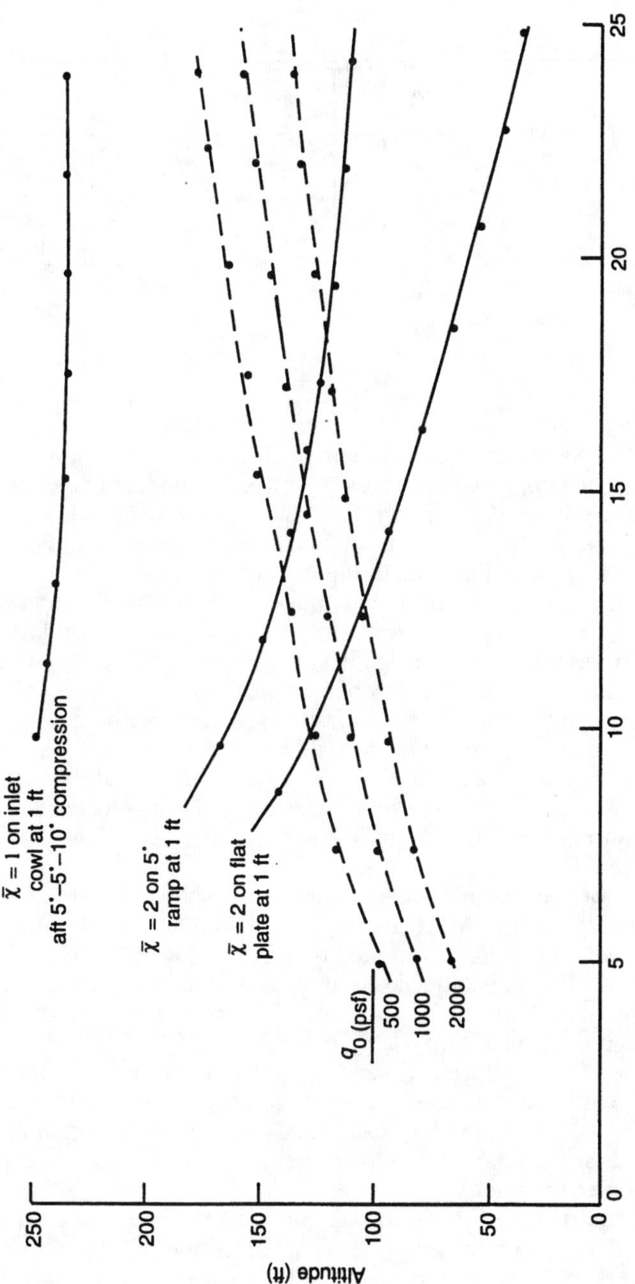

Fig. 22 Regimes of viscous interaction in an inlet.

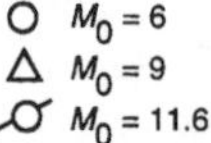

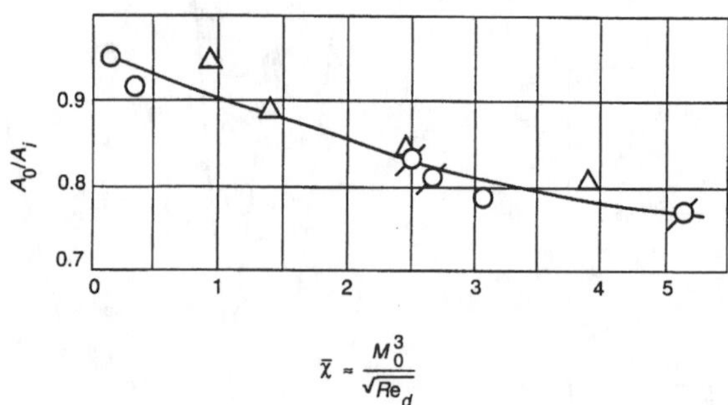

Fig. 23 Effect of $\bar{\chi}$ on air capture for an axisymmetric inlet (from Ref. 59).

Fig. 23 (from Ref. 59) where measurement of air-capture ratio for an axisymmetric inlet has been correlated versus $\bar{\chi}$ for Mach numbers between 6 and 11.6. Not only are these types of correlations important in correlating measured performance, but they also extremely useful in projecting the performance measured in ground facilities to flight conditions.

Having reviewed viscous interactions that result with laminar boundary layers, transition is the next phenomenon to be discussed. Boundary-layer transition is important in inlet design because it can significantly affect the performance and operability. As shown in the next section, turbulent boundary layers can sustain much larger pressure gradients without separation compared to laminar boundary layers, so that turbulent boundary layers can extend the operational capabilities of the inlet. Unfortunately, turbulent boundary layers also increase the heat loads and frictional losses within an inlet. Ideally, an inlet would contain just enough turbulent flow to prevent separation yet minimize viscous losses.

An example of the importance of transition is shown in Fig. 24, where results are provided from a Mach 10.4 test of a two-dimensional inlet.[37] The external forebody of this inlet consisted of a 5-deg blunted wedge followed by an additional 5 deg of isentropic compression. Both the surface pressure and Stanton number distribution are shown. The results show that the heat transfer through the transition zone increases by nearly a factor of 10. Also note that the length of the transition zone can be an appreciable portion of the forebody length. For the example shown in Fig. 24, the transition zone is approximately 16% of the external forebody length. In this example the forebody boundary layer was fully turbulent before the entrance to the internal portion of the inlet.

Given the importance of transition on inlet performance and operability, being able to design an inlet knowing when transition will occur is highly desirable. Several different modes of transition exist. (See Ref. 60 and 61 for a detailed overview of hypersonic transition technology). First-mode, second-mode, Goertler vortices, and crossflow instabilities are the predomi-

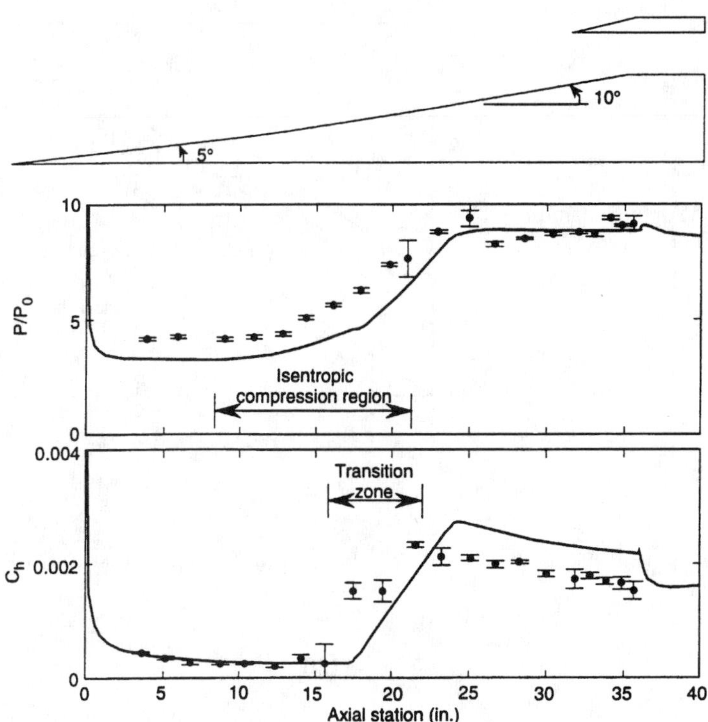

Fig. 24 Comparison of test results with PNS predictions of the forebody pressure and heat-transfer distribution at M_0 = 10.4, *Re* = 5.7 × 106 ft, T_w/T_0 = 6.1, and R_n = 0.005 in. (from Ref. 37).

nant mechanisms for transition. In the first mode the generation of Tollmien–Schlichting waves is responsible for transition. In zero-pressure gradient situations this mode can dominate for speeds as high as Mach 7 for adiabatic walls. For cold-wall cases above Mach 4, the second-mode instabilities, which are destabilized by wall cooling, can dominate. In the situation of flow over a compressive concave corner, the development of Goertler vortices may be the driving mechanism for transition. The final transition mechanism is cross-flow instabilities that can exist in three- dimensional flowfields.

The use of linear stability theory has been invoked extensively in the prediction of transition with reasonable success. Unfortunately, the application of linear stability theory for complex shapes can be computationally expensive, and so several rules of thumb have been developed. The simplest model is a correlation of transition vs the ratio of Reynolds number based on momentum thickness to the boundary-layer edge Mach number (Re_θ/M_e). As shown in Fig. 25 (from Ref. 62), this parameter has been used to correlate transition data obtained from flight experiments for zero-pres-

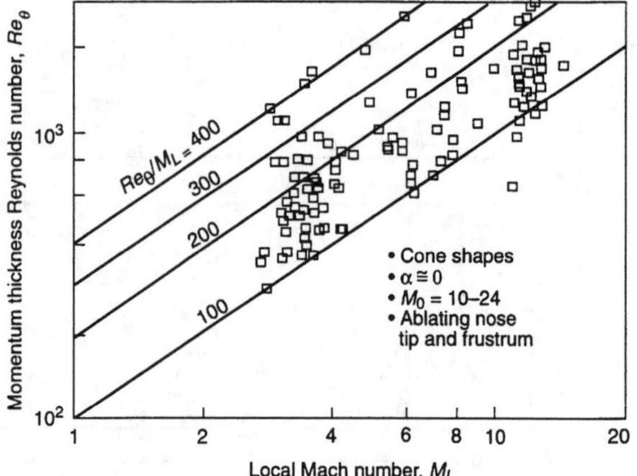

Fig. 25 Flight transition correlation (from Ref. 62).

sure gradient flows. A conservative rule of thumb is that transition will occur when

$$\frac{Re_\theta}{M_e} > 150 \tag{46}$$

For flow in which transition is caused by the generation of Goertler vortices, another rule of thumb is that transition will occur for Goertler numbers greater than 8 or 9 as follows:

$$G_\theta = Re_\theta \sqrt{\frac{\theta}{R_c}} = 8 - 9 \tag{47}$$

where θ is the boundary-layer momentum thickness and R_c is the radius of curvature of the compression surface. For flow in which a crossflow instability triggers transition, the rule of thumb for crossflow transition is as follows:

$$Re_{\text{CF}} = \frac{\rho W \delta}{\mu} = 250 \tag{48}$$

where W is the crossflow velocity and δ is the boundary-layer thickness. These simple rules of thumb are useful for first-order results in a preliminary design process. Significant uncertainties in the transition location will result with the use of these correlations, so that either the design must be made robust or more refined analysis must be conducted.

Besides the classic transition instabilities, several bypass transition mechanisms exist. These mechanisms create large disturbances in the boundary layer that directly excite nonlinear instabilities. These bypass mechanisms include the effects of acoustics, roughness, and vibration. Very little detailed information on these bypass mechanisms is available for hypersonic speeds.

With a model for the transition location, modeling of the transition process itself becomes important. Several overviews of transition modeling are available.[63–65] In most of these models, the local total viscosity is assumed to consist of the laminar viscosity plus some fraction of the turbulent viscosity. By modeling the development of the viscosity between the fully laminar and fully turbulent ends of the transition zone, the transition process itself can be simulated.

The major areas of uncertainty with respect to transition modeling include the effects of adverse-pressure gradient, roughness, and bluntness. Only through the development of additional experimental data and analysis techniques will this issue be resolved.

E. Boundary-Layer Separation

One critical issue associated with the design of scramjet inlets concerns the separation of boundary layers. Because of the desire to compress the captured airstream in a short length, the boundary layers are subjected to high adverse-pressure gradients. Boundary-layer separation is undesirable, and limits are often placed on the design of an inlet such that separation does not exist.

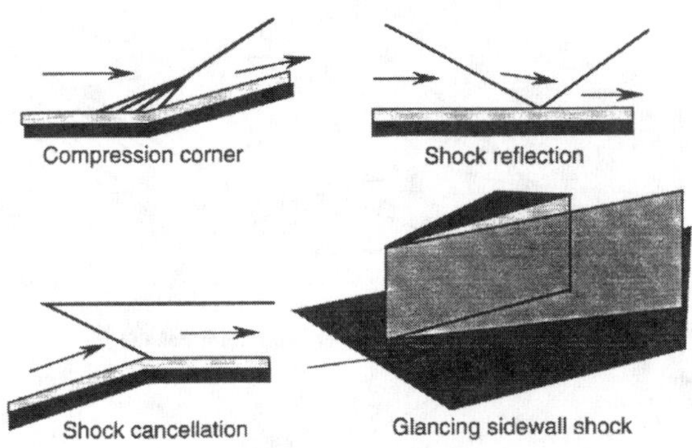

Fig. 26 Possible types of shock-wave/boundary-layer interactions.

Some possible shock-wave/boundary-layer interactions that can be encountered in scramjet inlets are shown in Fig. 26. These interactions include nominally two-dimensional interactions, such as those caused by a compression corner or shock reflection, and three-dimensional interactions caused by glancing shock waves or swept compression surfaces. In each of these interactions, an incipient separation pressure rise can be defined as the highest pressure rise that can be sustained without evidence of significant separation. The corresponding turning angle is called the incipient turning angle. (Although a minuscle amount of separation may exist for small pressure rises, this type of interaction does not affect inlet design.)

The separated flowfields for two-dimensional[66] and three-dimensional[67] shock-wave/boundary-layer interactions are illustrated in Fig. 27. The two-dimensional interaction is characterized by a separation shock followed by an expansion around the separation bubble and a recompression shock as the flow is again turned parallel to the wall. The three-dimensional interaction results in a complicated flow structure that includes the roll up of a vortex that generates a zone of high heat transfer.[68,69] This vortex roll up

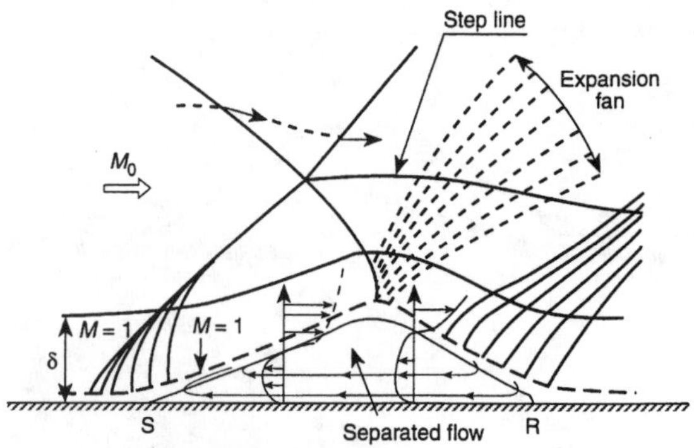

a) Two-dimensional interaction (from Ref. 66).

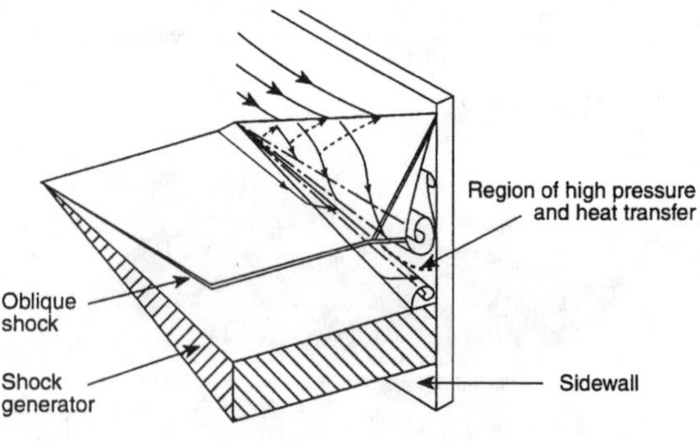

b) Three-dimensional interaction (from Ref. 67).

Fig. 27 Two- and three-dimensional shock-wave/boundary-layer interactions. a) Two-dimensional interaction (from Ref. 66) b) Three-dimensional interaction (from Ref. 67)

occurs in both sidewall compression inlets[70,71] and along the sidewalls of two-dimensional inlets.[72–74]

The existence of separated flow in a scramjet inlet is undesirable because of 1) the creation of additional shock waves that may not have been designed as part of the compression process, 2) the losses associated with compression-expansion-recompression of the flow; 3) the existence of a zone of high heat transfer near the reattachment point; 4) the generation of unsteady waves within the inlet that create large acoustic loads; 5) the weakening of the boundary layer such that downstream influences are more easily propagated upstream; and 6) the generation of an aerodynamic contraction that may cause the inlet to unstart.

Given that it is undesirable to operate inlets with substantial separation, one must define the limits under which separation may be expected. This problem has been studied extensively, and several good survey papers are available.[75–77] For two-dimensional interactions a first-order approximation is that the source of the interaction does not affect the incipient pressure rise. The pressure rise can be caused by either a ramp or an incident shock, and the incident shock can be reflected or canceled. In each case the incipient separation pressure rise is nearly the same.

For two-dimensional laminar boundary layers Holden[78,79] has found that the incipient separation angle a in a compression corner can be correlated as follows:

$$M_0 a = 4.32(\bar{\chi}^*)^{-1/2} \tag{49}$$

where

$$\bar{\chi}^* = \frac{\gamma - 1}{\gamma + 1}\left(0.664 + 1.73\,\frac{T_w}{T_0}\right)\bar{\chi}$$

For interactions in which blunt leading-edge effects are important, the incipient separation angle was correlated as follows:

$$M_0 a_i\left(\frac{\kappa}{\bar{\chi}^{*2}}\right) = 4.34\left(\frac{\bar{\chi}^*}{\kappa^{2/3}}\right)^{-7/5} \tag{50}$$

where

$$\kappa = \frac{(\gamma - 1/\gamma + 1)k M_0^3 t}{L}$$

and

$$k = \frac{D_n}{\frac{1}{2}\rho_0 U_0^2 t}.$$

The parameter $\bar{\chi}^*/\kappa^{2/3}$ is called the Cheng parameter and is used to characterize flow with both displacement and bluntness effects. In Ref. 79 the correlation provided in Eq. (50) was shown to hold at conditions ranging from displacement dominated flows to bluntness dominated flows (i.e., $10 > \bar{\chi}^*/\kappa^{2/3} > 0.08$).

In general, wall cooling increases the incipient separation pressure for a laminar boundary layer.[79–81] In situations where the boundary layer does separate, wall cooling can dramatically reduce the size of the separated zone. This effect of wall temperature on separation may partially explain the sensitivity of inlet starting to wall temperature that has been reported.[3,16]

For turbulent boundary layers the pressure rise required to separate a boundary layer is nearly five times that required for a laminar boundary layer.[76] The incipient separation ramp angle is a function of Mach number and Reynolds number.[82–92] A compilation of existing incipient separation data is shown in Fig. 28 (from Ref. 82). The results indicate that the incipient separation ramp angle increases for increasing Mach number. At low Reynolds number the incipient separation angle decreases with increasing Reynolds number. At higher Reynolds number the incipient separation angles becomes insensitive to Reynolds number.

In Ref. 93 and 94 Korkegi has correlated existing two- and three-dimensional turbulent interaction data in terms of the incipient pressure rise. As shown in Fig. 29, the incipient pressure rise for two-dimensional interactions can be correlated as follows:

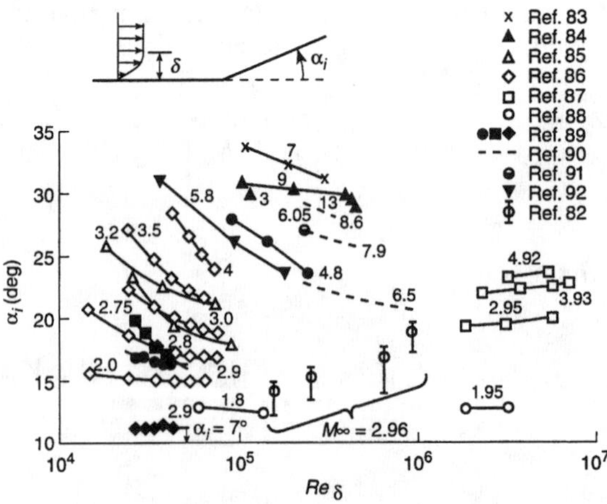

Fig. 28 Comparison of experimentally determined incipient separation angle data for turbulent interactions (from Ref. 82).

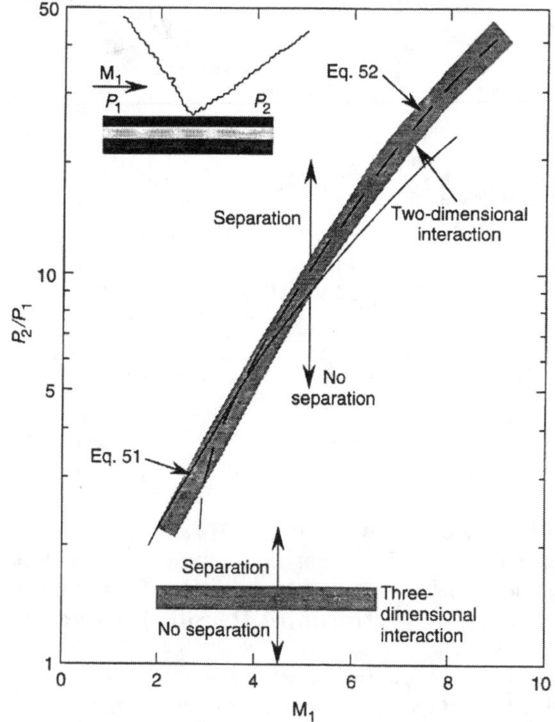

Fig. 29 Pressure rise required to separate a boundary layer (from Ref. 94).

$$\frac{P_2}{P_1} = 1 + 0.3M^2 \qquad \text{for} \quad M < 4.5 \qquad (51)$$

and

$$\frac{P_2}{P_1} = 0.17M^{2.5} \qquad \text{for} \quad M > 4.5 \qquad (52)$$

Similarly, the incipient separation turning angle for a wedge mounted normal to the flow with a glancing shock interaction was correlated as follows:

$$M\alpha_i = 0.3 \text{ rad} \qquad (53)$$

In Ref. 95 the issue of separation for a swept-shock interaction was considered. In this work the recommendation was made that the incipient pressure rise for a two-dimensional interaction could be used for the swept-shock interaction if the Mach number normal to the swept shock is used in the pressure correlations.

The incipient turning angle of an incident shock is also dependent on the

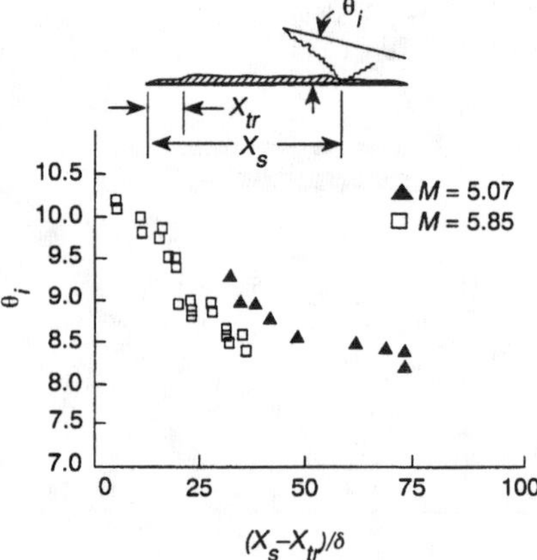

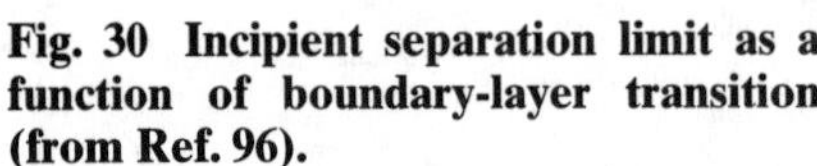
Fig. 30 Incipient separation limit as a function of boundary-layer transition (from Ref. 96).

distance of the interaction downstream of transition.[96] As shown in Fig. 30, interactions that occur closer to the end of transition can sustain a higher pressure rise without separation. Based on the data from Ref. 96, the interaction must be approximately 50δ downstream of transition before this effect disappears.

One interaction of particular interest in inlet design concerns the interaction of a shock wave with a boundary layer near a convex corner. Initial investigations of this phenomenon have shown that the flowfield changes significantly with changes in the positioning of the incident shock relative to the corner.[97–100] The corner influences the interaction when the shock impinges within 6–7δ of the corner. For interactions within 1δ of the corner, the incident shock is nearly canceled. The degree of cancellation of the incident shock has been shown to be a function of the incoming boundary-layer height.[5]

Given the sensitivity of inlet operation to boundary-layer separation, inlet designs without separation are pursued. In cases where separation is unavoidable, attempts have been made to minimize or control the interaction. The two prominent techniques for controlling shock-wave/boundary-layer interactions are through either bleeding or blowing. With a bleed system the low-momentum flow near the wall is removed such that the remaining flow can negotiate subsequent pressure rises. The disadvantage of a bleed system lies in the substantial drag associated with the bleed air. At hypersonic speeds an additional disadvantage lies in the high temperature of the bleed air, which complicates the design of the bleed system. A review of the bleed system database is provided in Ref. 101.

The alternate to a bleed system is a blowing system. In a blowing system flow is injected axially in an attempt to energize the boundary layer such that subsequent adverse-pressure gradients can be negotiated.[102–105] Blowing systems avoid the high drag and internal ducting of a bleed system, but a high-pressure source for the injected fluid must be provided.

F. Isolators/Supersonic Diffusers

As was illustrated in Fig. 1, scramjet inlets may be required to operate with a precombustion shock system in the inlet throat. This shock system is forced into the constant area throat section by the thermal blockage of the heat release in the combustor at low speeds and/or high levels of heat release. The portion of the engine where this precombustion pressure system is stabilized is called the isolator or supersonic diffuser. Because of the relatively thick boundary layers present in scramjet inlets, a long isolator may be required to sustain the required pressure rise. If the combined pressure recovery of the inlet and isolator is not sufficient to supply the required pressure to the combustor, the inlet will unstart.

G. Combustor Entrance Profiles

The airflow captured and compressed by the inlet ultimately must be fueled within the combustor. As the flow profiles at the end of the inlet become more skewed or as they change between one operating condition and another, difficulties in fueling are created. Furthermore, large flow profiles at the combustor entrance may create intolerable flowfield situations within the combustor with respect to heat loads or combustor operability.

Although it is expected that the flow profiles at the entrance to the combustor can play an important role in the performance and operability of the engine, little information exists on what constitutes a good profile. For most design systems an attempt is made to make the profile as uniform as possible, which should help in fueling the combustor.

IV. Engine Cycle Calculations

Sample engine-cycle calculations are used to illustrate the sensitivity of a scramjet engine to the inlet operating characteristics. Three generic scramjet engine configurations were analyzed using the Johns Hopkins University/Applied Physics Laboratory (JHU/APL) RamJet Performance Analysis (RJPA) code.[106] The sample configurations are meant to be typical of hydrogen-fueled scramjet engines operating at Mach 5, 10, and 20. A sketch illustrating the engine nomenclature is shown in Fig. 31, and the characteristics of the engines are provided in Table 2. The sample Mach 5 engine operates as a dual-mode engine with a normal shock in the inlet isolator. The Mach 10 and 20 engines operate as pure scramjets.

Using the sample engines, the sensitivity of the engine specific impulse I_{sp} to the inlet performance was calculated using the cycle analysis, and the results are provided in Fig. 32. The results show that the sensitivity of engine I_{sp} to air capture changes by a factor of approximately two over the Mach number range investigated. At Mach 5 a 10% variation in the inlet mass capture results in a 4% variation in the engine I_{sp}. At Mach 10 a 10% variation in the mass flow causes between a 4 and 10% variation in the engine I_{sp}.

The results provided in Fig. 32 also show that the sensitivity of engine I_{sp}

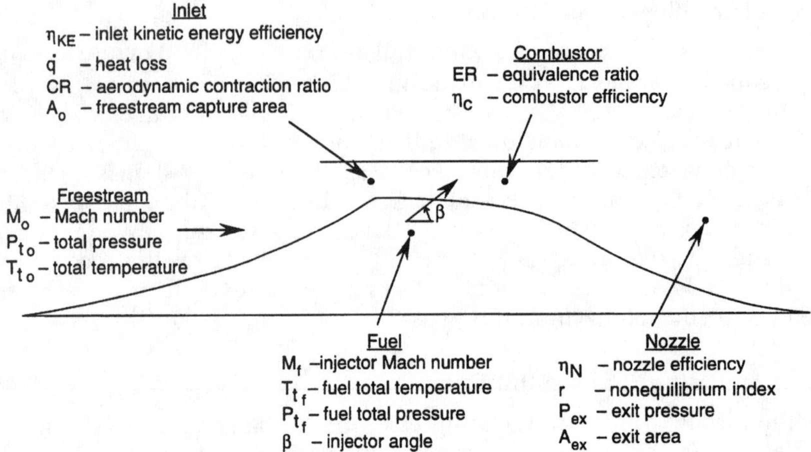

Fig. 31 Variables used in engine-cycle calculations for a sample engine.

to heat loss and inlet efficiency increases significantly as the freestream Mach number increases. For example, 2% variation in η_{KE} at Mach 5 creates about a 1 to 2% variation in engine I_{sp}. At Mach 20 a 2% variation in η_{KE} creates a 50% variation in engine I_{sp}. This tremendous increase in the sensitivity of engine performance to inlet performance at high speeds cre-

Table 2 Sample engine characteristics

	Mach 5	Mach 10	Mach 20
η_{KE}	0.97	0.97	0.97
A_0/A_4	7	15	35
q	1%	1%	1%
Combustion Mode	Dual Mode	Scramjet	Scramjet
Combustor Area Ratio	>1	1	1
Equivalence Ratio	1	1	1
Fuel T_t (°R)	1000	2000	2000
Fuel P_t (psi)	1000	1000	360
M_f	1	3.9	3
β_f	90°	0°	0°
η_c	100%	100%	100%
η_N	98%	99%	98%
r	1.0	0.33	0.33
Expansion to:	$P_{ex}=P$	$A_{ex}=1.7A_0$	$A_{ex}=1.7A_0$

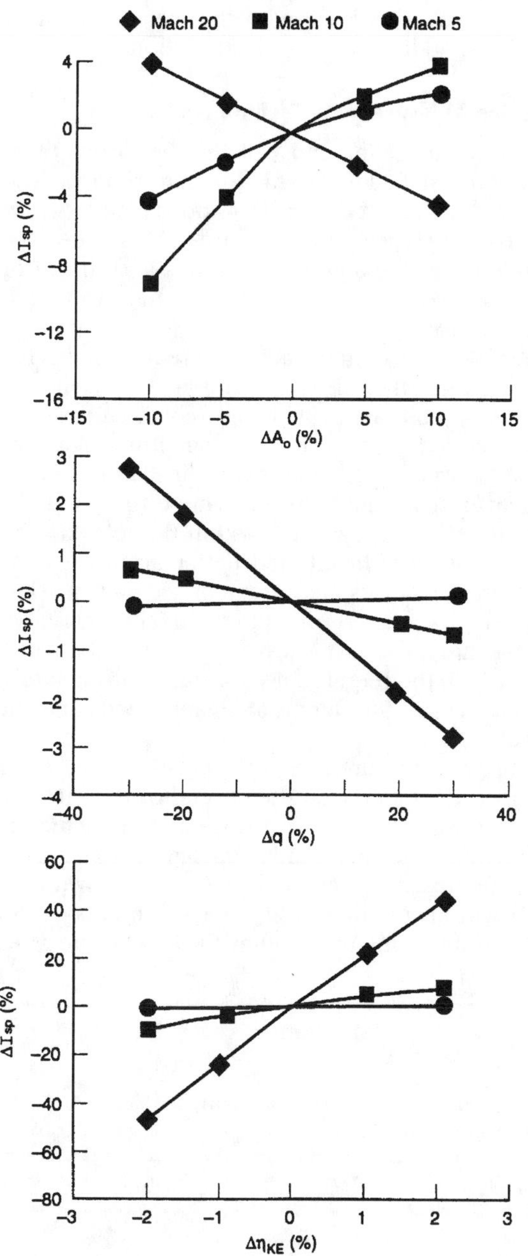

Fig. 32 Sensitivity of sample engine specific impulse to uncertainties in air capture, heat loss, and efficiency.

ates the requirement for a very refined, high-performance inlet design and creates challenges with respect to the accuracy required in the measurement of inlet performance. Specific issues associated with the measurement of scramjet inlet performance will be addressed in the next section.

V. Performance Measurement Techniques

Because of the high cost of flight testing, most of the development of scramjet inlets occurs in ground-test facilities with subscale models. The combination of small models and high inlet contraction ratios result in very small internal flowfield dimensions (typically on the order of 0.1–1.0 in.). These small dimensions, coupled with the high temperatures and short run times encountered in most hypersonic facilities, greatly complicate the measurement of scramjet inlet performance.

In Ref. 107 estimates were made regarding the accuracy requirements for measurements of the air-capture ratio, inlet heat loss, and inlet efficiency as a function of Mach number; these results are provided in Table 3. Across the Mach number range the accuracy requirement for an air-capture measurement is between 1 and 3%. At lower hypersonic speeds heat loss is not a major factor, and so large measurement uncertainties can be tolerated. As the Mach number increases, the effect of heat loss within the inlet on the engine performance becomes more significant, and better accuracies are required. At Mach 20 the required heat loss measurement accuracy is 10 to 20%. The accuracy requirement for an η_{KE} measurement also increases significantly with increasing Mach number. At Mach 5 an accuracy between 1 and 2% is acceptable. At Mach 20 the required accuracy is approximately 0.05%. These very stringent accuracy requirements at higher speeds present many measurement challenges.

The standard technique for the measurement of the air-capture ratio involves incorporation of a mass flow meter on the aft end of the inlet, as shown in Fig. 33. This flow-metering device consists of a plenum, which is used to mix and diffuse the incoming airstream, and a sonic nozzle. The sonic nozzle is calibrated such that the discharge coefficient $C_{d\text{pl}}$ is known as a function of a characteristic Reynolds number. Using measurements of the total pressure and total temperature within the plenum, the air-capture ratio can be obtained as follows:

Table 3 Measurement acuracies required to maintain 1% uncertainty in engine Isp

	Mach 5	Mach 10	Mach 20
Air Capture	1–2%	2–3%	2–3%
Heat Loss	≈100%	30–50%	10–20%
η_{KE}	1–2%	0.2–0.5%	≈0.05%

$$\frac{A_0}{A_i} = C_{d_{\text{pl}}} \frac{1}{M_0} \left[\frac{1 + (\gamma - 1/2)M_0^2}{(\gamma + 1/2)} \right]^{\frac{\gamma+1}{2(\gamma-1)}} \frac{P_{t_{\text{pl}}} A_{\text{pl}}^*}{P_{t0} A_i} \sqrt{\frac{T_{t0}}{T_{t_{\text{pl}}}}} \tag{54}$$

The principal difficulty with this technique in long-duration facilities involves adequately mixing the captured stream so that uniform pressures and temperatures exist at the entrance to the exhaust nozzle. The plenum usually incorporates striker plates, baffles, and aerogrids to aid in this mixing. In Ref. 108 the use of a heat exchanger within a mass flow meter was discussed. This heat exchanger is used to reduce the temperature of the airstream to more measurable levels and to make the temperature uniform. With a high-quality mass flow meter accuracies of approximately 0.5% are achievable.

In pulse facilities adequate run time may not exist to establish steady flow through a conventional mass flow meter. In Ref. 109 and 110 a technique was investigated in which a sealed plenum was attached to the inlet. By measuring the rate of pressure rise and heat loss within the plenum, the captured mass flow can be calculated as follows:

$$\dot{m} = \frac{V}{\gamma R T_t} \left[\frac{dP_{t_{\text{pl}}}}{dt} + (\gamma - 1) \frac{\dot{Q}}{V} \right] \tag{55}$$

where V is the plenum volume and $\dot{Q}$ is the heat loss within the plenum. Measurement accuracies of approximately 5% were demonstrated using a plenum without a heat exchanger in pulse facilities with run times as short as 2–3 ms.

Two techniques are available for measurement of the heat loss to the inlet. The first technique involves measurement of the heat-transfer distribution using either discrete heat-transfer gauges or an optical thermal mapping technique. With measurement of the heat-transfer distribution, standard numerical integration techniques are used to obtain the global inlet heat loss. The accuracy of this technique depends principally on the accuracy of the heat-transfer measurements and the adequacy of the resolution of the heat-transfer distribution. Accuracies on the order of 10 to 20% are achievable.

A second technique for the measurement of heat loss is available for cooled models tested in long-duration wind tunnels. As illustrated in Fig. 34,

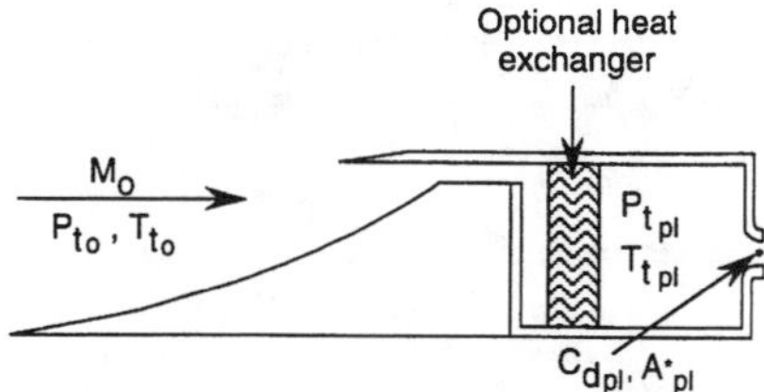

Fig. 33 Air-capture measurement using a mass flow meter.

a measurement of the temperature rise in the cooling water can be used to provide a global measurement of the heat loss if the cooling lines that affect the captured stream tube can be isolated. If the heat added to the cooling water is split into the inlet heat loss $\dot{Q}$ and the heat loss from portions of the model outside of the captured stream tube $\dot{Q}_{extra}$, the inlet heat loss can be determined as follows:

$$\dot{Q} = Cp_{H_2O}\dot{W}_{H_2O}(T_{out} - T_{in}) - \dot{Q}_{extra} \tag{56}$$

If the extra heat loss can be minimized or accurately estimated, this technique can be used with accuracies approaching 3 to 5%. This technique was demonstrated in Ref. 16.

Finally, measurements are required that allow determination of the inlet efficiency. Because inlet efficiency cannot be measured directly, it must be inferred from other measurements. At supersonic and low hypersonic speeds the detailed measurement of the static and pitot pressure distributions in the inlet throat can be used to estimate η_{KE}. As the freestream Mach number increases above 5 or 6, higher inlet contraction ratios, higher inlet heat losses, and tighter measurement accuracy requirements combine to preclude the use of throat flowfield measurements as a viable means of determining the inlet efficiency.

The only technique for the accurate measurement of inlet performance at hypersonic speeds involves determination of the drag on the portions of the model wetted by the captured airstream.[5,107,111–113] Both one-step and two-step force measurement schemes have been developed depending on whether the captured stream tube can be identified (see Ref. 107 for details on the two-step technique). As shown in Fig. 35, the one-step force measurement technique involves the determination of the drag on the captured stream tube. By subtracting this drag from the captured freestream stream thrust, the throat stream thrust can be determined. The throat stream thrust, captured mass flow, and inlet heat loss then can be combined to determine the stream-thrust-averaged, one-dimensional set of flow properties at the inlet throat using the procedures described in Sec. II.

The measurement of the drag on the captured stream tube can involve a combination of discrete pressure, skin friction, and heat-transfer measure-

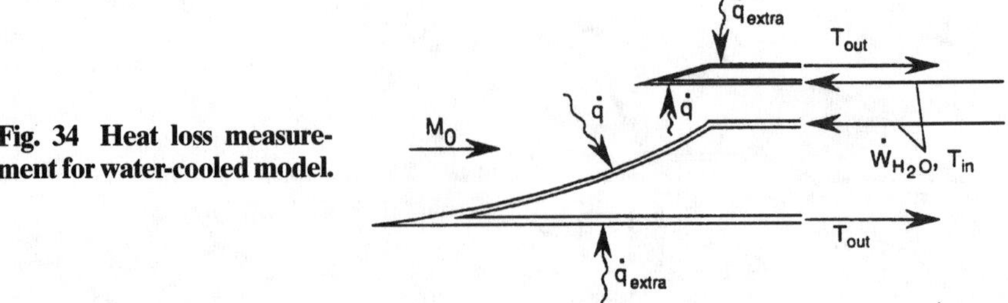

Fig. 34 Heat loss measurement for water-cooled model.

ments together with direct force measurements and some analytical estimates. For example, the configuration illustrated in Fig. 35 uses a direct force measurement to obtain the innerbody drag, discrete heat-transfer measurements coupled with Reynolds analogy to determine the cowl drag, and theoretical estimates to obtain the additive drag.

By using a well-instrumented inlet model with standard instrumentation, inlet efficiency can be determined using this drag measurement technique with the required accuracies for Mach numbers as high as 16 to 17. By using a high-accuracy, direct force measurement system, the required accuracies can be met at Mach numbers above 20.

In summary, techniques are available for the accurate measurement of hypersonic inlet performance. As Mach number increases, the required accuracies become more difficult to achieve, and so careful attention must be paid to the details of the measurement implementation.

VI. Design and Performance of Scramjet Inlets

As pointed out in the opening section, many different inlet designs have been developed in an attempt to provide high performance without violating given constraints. The choice of a design for a particular application will result from tradeoffs between inlet performance, weight, survivability, and integration with other engine and vehicle components. In this section the design procedures for various two-dimensional planar, two-dimensional axisymmetric, and three-dimensional inlet concepts are discussed together with a summary of available performance information.

A. Two-Dimensional Planar Designs

Two-dimensional planar inlets use a series of oblique shocks and isentropic compression regions arranged in a manner such that the desired compression and high performance will be achieved. Although a totally isentropic inviscid compression process can be designed, these inlets tend to be very long with substantial viscous losses. As a compromise between inviscid and viscous losses, discrete oblique shocks are often used in the compression process. Numerous investigations of the operating characteristics of two-dimensional planar inlets have been performed.[5,15,32,59,72–74,114–119]

Optimum inviscid inlet performance can easily be determined for an inlet consisting of any number of discrete oblique shocks.[120,121] Consider the

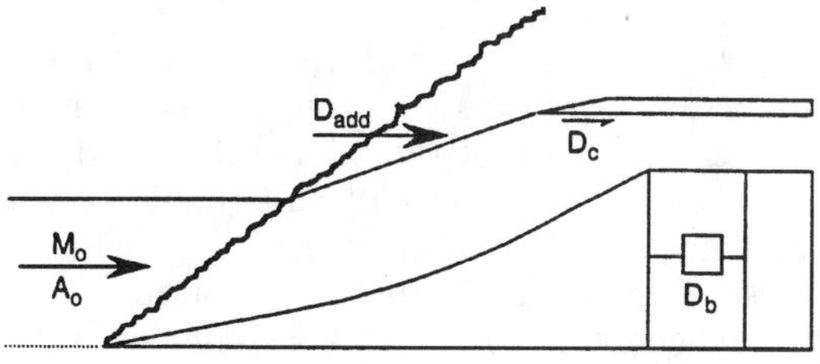

Fig. 35 Inlet efficiency measurement using forebody force measurement.

design of an inlet consisting of four oblique shocks generated by four discrete turns of angles δ_1 through δ_4. If the freestream conditions and overall compression ratio are specified, a single combination of δ_1 through δ_4 can be determined that will provide the highest inlet efficiency. For a perfect gas the maximum total pressure recovery will occur when the total pressure ratio across each shock is identical (i.e., the normal component of Mach number is the same for all shocks). For a nonideal gas the optimum performance must be determined using numerical optimization techniques, but the optimum performance is still obtained when the normal Mach number for each shock is nearly the same.

With the primary purpose of a scramjet inlet to capture and compress air for use in the combustor, a natural question arises concerning the optimum degree of compression. The compression must be sufficiently high that combustion can be sustained, but also sufficiently low that large nonequilibrium chemical losses are not incurred within the combustor and nozzle. This issue can be investigated with the two-dimensional planar designs. To provide insight into the requirement for providing adequate compression for combustion, again consider the design of an inlet consisting of four oblique shocks generated by four discrete turns of angles δ_1 through δ_4. In Fig. 36 the amount of turning in optimum four-shock inlets is shown for a range of compression ratios for Mach numbers between 5 and 25, assuming air in chemical equilibrium. The amounts of turning in optimum two-shock, optimum three-shock, and isentropic inlets are also shown. Overlaid on these curves is a band of compression ratio vs freestream Mach number that would be required to produce a combustor entrance pressure of 0.5 atm for flight at dynamic pressures between 500 and 2000 psf. (A combustor entrance pressure of 0.5 atm is a rough rule of thumb on the minimum pressure required to sustain combustion in a scramjet engine.) These results show that the total amount of turning required to produce a given compression ratio in a two-dimensional design does not strongly depend on the efficiency of the compression process. (As shown earlier, the amount of turning also does not depend strongly on whether an ideal gas or equilibrium air assumption is used.) Furthermore, the results show that higher amounts of turning are required to achieve the required compression ratio at speeds between Mach 3 and 5 than are required at speeds above Mach 10. For an inlet required to operate over a wide range of Mach numbers, this significant variation in required total turning complicates the design of the inlet. Only through the selective use of variable geometry and multiple shock reflections can this required turning be provided over the entire Mach number spectrum.

After using inviscid tools to construct the basic inlet layout, compensation for the boundary layer must be made. As illustrated in Fig. 37, the inviscid flow pattern is generated by the combination of the inlet surface and the boundary-layer displacement thickness. The growth of the boundary layer provides compression that is in addition to the turning provided by the inlet surface. The boundary layer also modifies the apparent location of the inviscid shocks relative to the actual geometry. Through development of the

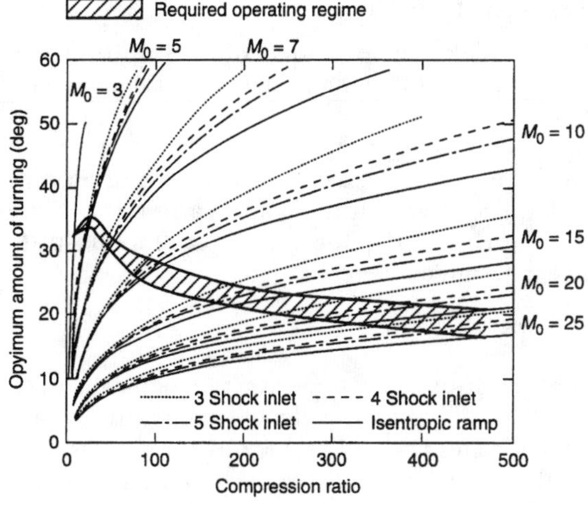

Fig. 36 Optimum inviscid scramjet inlets.

proper correlations, the inviscid surface can be directly modified to account for the boundary-layer displacement. A coupled inviscid/viscous analysis is then used to verify the inlet design.

Planar inlet designs offer the advantage of ease of design in that two-dimensional analysis tools can be used. As in all inlets, three-dimensional aspects of the flow exist, but the predominant flow characteristics can be made to be two-dimensional. Two-dimensional planar designs also offer the advantage of simple incorporation of variable geometry features such as moveable ramps and cowls, so that the inlets can be made to function over a wide range of flight and engine operating conditions.

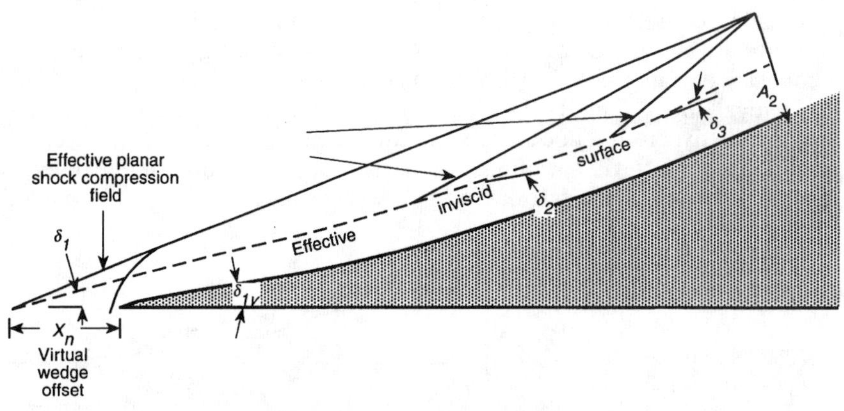

Fig. 37 Design of external compression field with viscous-inviscid interaction.

B. Two-Dimensional Axisymmetric Designs

Two-dimensional axisymmetric inlets can be divided into outward-turning and inward-turning designs. The outward-turning designs are usually developed using the same approach as the two-dimensional planar designs although slightly more turning is required to achieve the same compression ratio. The Oswatitisch inlet will be used to provide an example of an outward-turning axisymmetric inlet. The inward-turning inlets are substantially different from planar designs. The Busemann inlet will be used to illustrate an inward-turning axisymmetric design concept.

Oswatitisch Inlet

As illustrated in Fig. 38, the Oswatitisch inlet consists of a conical shock, isentropic compression field, and a single cowl-reflected shock that is canceled on an innerbody shoulder.[2] The inviscid surface of the Oswatitisch inlet is designed using the Method of Characteristics such that the cowl-reflected shock is straight and uniform parallel flow is provided to the combustor. By varying the initial cone angle and degree of isentropic compression, a wide range of inlet designs can be generated. After development of the inviscid design, the surface geometry is modified to account for the boundary-layer growth using a coupled inviscid-viscous analysis. The basic operating characteristics of an Oswatitisch inlet were demonstrated in Ref. 2.

The principal disadvantages of the Oswatitisch inlet are the small internal throat heights, which lead to large viscous losses, high cowl-lip drag, and a significant sensitivity to angle of attack. These disadvantages are inherent in all outward-turning axisymmetric designs.

Busemann Inlet

The Busemann inlet is an inward-turning axisymmetric design that provides uniform parallel inviscid flow to a combustor.[122–124] As shown in Fig. 39, the Busemann inlet consists of an inward-turning isentropic compression that leads to a free-standing conical shock wave that is canceled at a shoulder. Numerically, the inlet designs are generated by starting with the desired combustor entrance conditions and specifying the strength of the freestanding conical shock. The Taylor–McColl equations are then solved for the flowfield between the shock wave and the freestream. Because the strength of the shock wave is specified, the inviscid performance of the inlet is directly known. Having calculated the inviscid shape of the inlet, a coupled

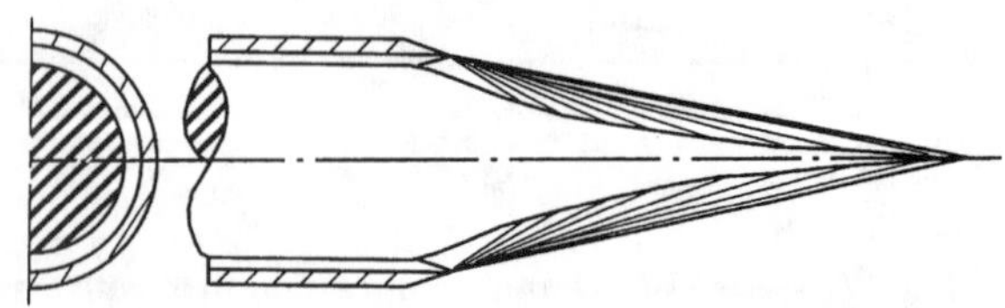

Fig. 38 Oswatitsch inlet design (from Ref. 2).

inviscid-viscous analysis can be used to modify the geometry to account for the boundary-layer growth. The basic operating characteristics of the Busemann inlet were verified in Ref. 2, 123, and 124.

Busemann inlets offer the advantage of high performance because the loss-producing shock occurs after a substantial portion of isentropic compression. Furthermore, the design is such that the large surface areas are in the forward region of the inlet where the boundary layer is likely to be laminar. In the aft region of the inlet where the boundary layer may be turbulent, the surface area is small.

The primary disadvantages of the Busemann inlet are that it will not start under steady flow conditions, and it does not integrate well with variable geometry requirements. Streamline tracing through the basic Busemann inlet flow is used to overcome these disadvantages as discussed in the following section.

C. Three-Dimensional Inlet Designs

Besides the two-dimensional planar and axisymmetric designs, several three-dimensional inlet concepts have been developed. In each case these inlets use a two-dimensional design as the starting point. Examples of this type of inlet are the sidewall-compression inlet and the modular Busemann inlet.

Sidewall-Compression Inlet

As the name implies, the sidewall-compression inlet uses lateral compression provided by the inlet sidewalls. As illustrated in Fig. 40, the compressive sidewalls are usually swept aft to weaken the strength of the shocks. In a plane normal to the swept leading edges, the inviscid compression process looks like a two-dimensional planar inlet. The sweep of the sidewalls creates a vertical component to the flow that is helpful in starting the inlets. The operating characteristics of the sidewall-compression inlets have been investigated for a wide range of sidewall sweep, including reverse sweep, entrance boundary-layer characteristics, gas chemistry, and body-side compression shapes.[6,25,70,125–134]

The primary disadvantages of the sidewall-compression inlets are related

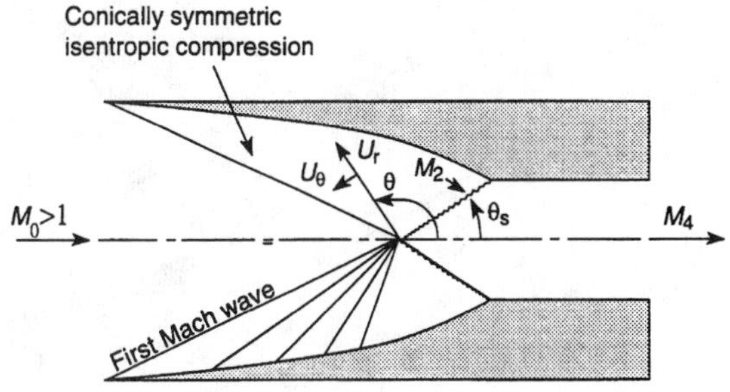

Fig. 39 Schematic of the Busemann inlet flowfield.

to viscous effects and spillage drag. The sidewall-compression inlet generates swept, glancing shocks that create a pair of vortices that roll up and propagate into the inlet throat. The combination of the vortex structure coupled with the shock created from the cowl creates an extremely complex flowfield that is difficult to both analyze and fuel. Spillage drag of these inlets also tends to be high because the spillage flow is forced out through the narrow opening near the cowl leading edge. This narrow opening forces the flow to spill at large angles, creating large spillage drags.

Modular Busemann Inlets

The basic Busemann inlet suffers from the large internal contraction ratio that cannot be started under steady flow operating conditions. Streamline tracing within the basic Busemann inlet flowfield can be used to overcome this problem.[124] As shown in Fig. 41, stream tubes can be traced within the basic flow such that low internal contractions result when the walls are cut back along the initial compression waves. These modular Busemann inlets provide the same inviscid flow characteristics as the basic Busemann inlet designs. One challenge with this design is the modification of the inviscid geometry to account for the boundary-layer growth. The operating characteristics of the modular Busemann inlet have been demonstrated in Ref. 135–137.

The modular Busemann inlet offers the advantage of high performance, highly swept leading edges, and the flexibility to choose a stream-tube shape that will allow easy fueling within the combustor. The primary disadvantages lie in the complicated boundary-layer compensation process and the high spillage drag at off-design conditions.

D. Performance Characteristics

The performance characteristics of scramjet inlets have been investigated and reported for a variety of inlet concepts. Comparison of the results is

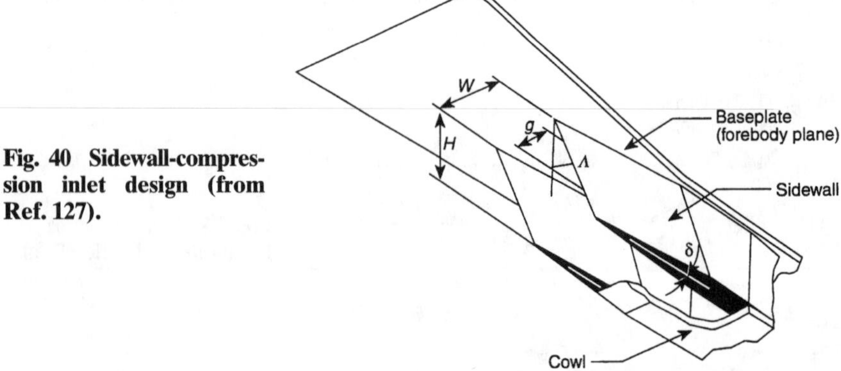

Fig. 40 Sidewall-compression inlet design (from Ref. 127).

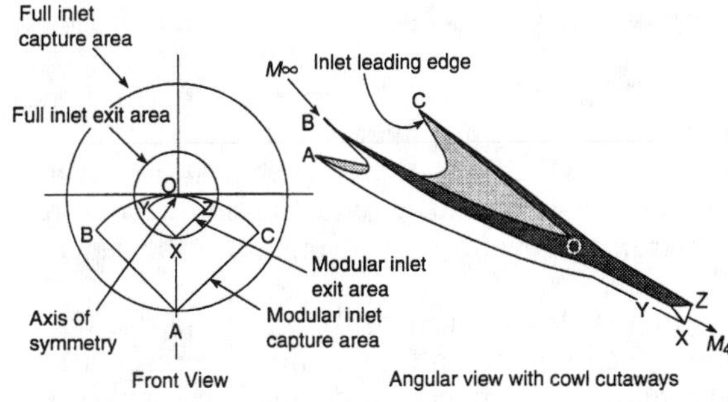

Fig. 41 Streamline tracing technique for the design of a modular Busemann inlet (from Ref. 124).

complicated by the diversity of measurement techniques employed, but some general trends can be observed.

The utility of the pressure-area domain for discussing inlet performance is demonstrated in Fig. 42, where measured performance from a two-dimensional planar inlet at Mach 10.4 is provided. The inlet model used in this configuration was a mixed external-internal compression with a forebody consisting of a 5-deg blunted wedge and an additional 5-deg of isentropic compression.[5] A single cowl shock was used to turn the flow back to a horizontal direction. Test results are shown for both sharp and blunt forebody nose configurations. In these tests the inlet cowl was translated vertically to provide a range of compression ratios. At the shock-on-shoulder operating point for the sharp-nosed configuration, the measured inlet efficiency was $\eta_{KE} = 0.978$. As the cowl was lowered to provide higher contraction ratios, the compression ratio increased, and the inlet efficiency

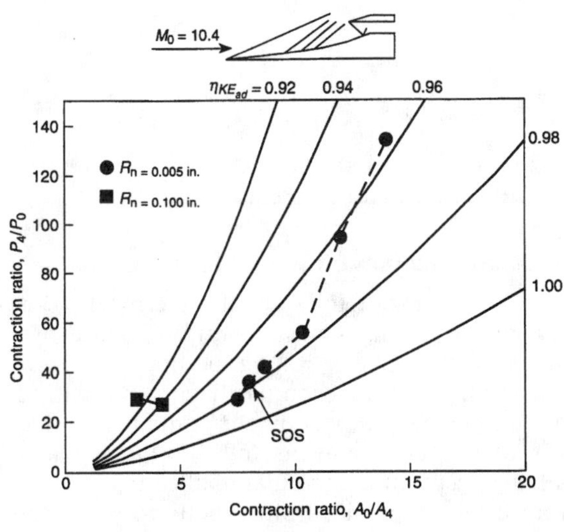

Fig. 42 Effect of CR and nose bluntness on the performance of a two-dimensional inlet at Mach 10.4 (from Ref. 5).

decreased. At the highest compression ratios generated, the inlet efficiency had dropped by nearly 2%.

The results shown in Fig. 42 also illustrate the potential performance decrement created by blunt leading edges at hypersonic speeds. With the blunt leading edge the forebody flowfield was altered such that the aerodynamic contraction ratio was cut in half with little reduction in the compression ratio. This resulted in extremely low inlet performance in the range of η_{KE} equal to 0.915 to 0.945.

While the pressure-area domain is useful in picturing the performance results at a given freestream Mach number, alternate data representations are helpful when trying to develop performance modeling. For example, the inlet performance was correlated as a function of the throat-to-freestream Mach number ratio in Ref. 138 as follows:

$$\eta_{\mathrm{KE}} = 1 - 0.4\left(1 - \frac{M_4}{M_0}\right)^4 \tag{57}$$

In Fig. 43 a summary of reported inlet performance is provided in terms of the M_4/M_0 together with the correlation provided by Eq. (57). The results show that as the inlet contraction ratio increases and M_4/M_0 decreases, the inlet efficiency drops. The data also show that Eq. (57) provides a first-order accurate model for the inlet performance with most two-dimensional designs at high Mach numbers performing slightly better and most three-dimensional designs at lower Mach numbers performing slightly worse. The available data are not sufficient to determine whether this effect is caused by differences in the freestream Mach number or differences in the inlet designs.

In Ref. 130, a detailed investigation of one particular sidewall-compression inlet was presented with the recommendation that Eq. (57) be modified to represent the test results better. A best-fit correlation to their data was provided as follows:

$$\eta_{\mathrm{KE}} = 1 - 0.528\left(1 - \frac{M_4}{M_0}\right)^{3.63} \tag{58}$$

This correlation is also provided in Fig. 43 for reference.

VII. Summary and Recommendations for Future Investigations

The development of a high-performance inlet system is critical to the success of a scramjet engine concept. This inlet must operate efficiently over the entire range of expected flight and engine operating conditions. Over the past 30 years a substantial amount of information has been generated related to scramjet inlets and the fundamental phenomena contained in their flowfields. This information provides the framework to begin the engineering task of developing an inlet concept for a particular application.

A significant amount of uncertainty remains in the ability to accurately

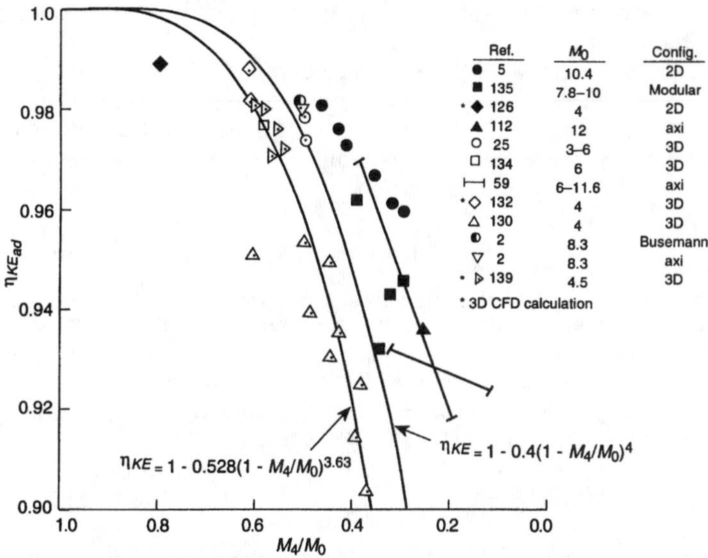

Fig. 43 Compilation of reported inlet efficiencies.

predict inlet performance and operability. The issues creating the uncertainty in the performance predictions include: 1) knowledge of boundary-layer transition location, length, and characteristics; 2) detailed effects caused by inviscid-viscous coupling; 3) high-temperature gas effects on the boundary-layer development and subsequent separation criteria; 4) knowledge of the losses associated with three-dimensional aspects of the flowfield; and 5) details of the losses created in the development of turbulent boundary layers in regions of high-pressure gradients.

Concerning the operability of the inlet, the principal areas of uncertainty involve 1) the adequacy of the separation models used in defining allowable shock-wave/boundary-layer interactions and 2) the reliability with which inlet starting and unstarting can be predicted or calculated. At the present time nearly all information related to these two operability issues has been developed experimentally. Development of reliable analysis techniques that could be used to address these issues would greatly aid in the inlet development process.

One particular issue for which little information exists concerns the allowable flow profiles at the end of the inlet compression process. Because the inlet flow must ultimately be fueled, constraints on the profiles must exist, but no rational method exists at the present time to specify these constraints. In many current design systems attempts are made to make the flow entering the combustor as uniform as possible, but this may not be the optimum way of coupling the inlet and combustor. Development of guidance on allowable or preferred combustor entrance profiles will be essential in the development of optimum scramjet engines.

Finally, the design of creative variable geometry concepts will be required for vehicles that are to fly over wide ranges of flight conditions and engine operating characteristics. A definite art exists in the development of an inlet that can meet performance goals without violating the many separate constraints over an entire flight spectrum.

References

[1]McLafferty, G.H., "A Generalized Approach to the Definition of Average Flow Quantities in Nonuniform Streams," United Aircraft Corporation Research Department Report SR-13534-9, December 1955.

[2]Molder, S., McGregor, R. J., and Paisley, T. W., "A Comparison of Three Hypersonic Air Inlets," *Investigations in the Fluid Dynamics of Scramjet Inlets,*Ryerson Polytechnical Univ. and Univ. of Toronto, Canada, July 1992.

[3]Andrews, E. H., McClinton, C. R., and Pinckney, S. Z., "Flowfield Starting Characteristics of an Axisymmetric Mixed-Compression Inlet," NASA TM-X-2072, Jan. 1971.

[4]Karanian, A. J., and Kepler, C. E., "Experimental Hypersonic Inlet Investigation with Application to Dual Mode Scramjet," AIAA Paper 65-588, June 1965.

[5]Van Wie, D. M., and Ault, D. A., "Internal Flowfield Characteristics of a Two-Dimensional Inlet at Mach 10," *Journal of Propulsion and Power,*Vol. 12, No. 1, 1996, pp. 158–164.

[6]Holland, S., and Perkins, J., "Mach 6 Test of Two Generic Three-Dimensional Sidewall Compression Scramjet Inlets in Tetraflouromethane," AIAA Paper 90-0530, Jan. 1990.

[7]Waltrup, P. J., Anderson, G. Y., and Stull, F. D., "Supersonic Combustion Ramjet (SCRAMJET) Engine Development in the United States," *3rd International Symposium on Air Breathing Engines,*(Munich, Germany), March 1976.

[8]Rejeske, J. V., and Sharp, B. M., "SCRAM Inlet Studies Final Report," Feb. 1971.

[9]White, M. E., Stevens, J. R., Van Wie, D. M., Mattes, L. A., and Keirsey, J. L., "Investigation of Cowl Vent Slots for Supercritical Stability Enhancement of Dual-Mode Ramjet Inlets," *Journal of Propulsion and Power,*Vol. 26, No. 3, 1990, pp. 225–226.

[10]Tannehill, J. C., and Mugge, P. H., "Improved Curve Fits for the Thermodynamic Properties of Equilibrium Air Suitable for Numerical Computation Using Time-Dependent or Shock-Capturing Methods," NASA CR-2470, Oct. 1974.

[11]Lewis, C. H., and Burgess, E. G., "Empirical Equations for the Properties of Air and Nitrogen to 15,000°K," AEDC-TDR-63,138, July 1963.

[12]McLafferty, G. H., "A Generalized Approach to the Definition of Average Flow Quantities in Non-Uniform Streams," United Aircraft, Research Dep. Rep. SR-13534-9, Dec. 1955.

[13]Billig, F. S., and Van Wie, D. M., "Efficiency Parameters for Inlets Operating at Hypersonic Speed," *1987 International Society of Airbreathing Engines Symposium*(Cincinnati, OH), June 1987.

[14]Curran, E. T., and Bergsten, M. B., "Inlet Efficiency Parameters for Supersonic Combustion Ramjet Engines," APL-TDR-64-61, June 1964.

[15]Goldberg, T. J., and Hefner, J. N., "Starting Phenomena for Hypersonic Inlets with Thick Boundary Layers at Mach 6," NASA TN-D6280, Aug. 1971.

[16]Gurylev, V. G., and Mamet'yev, Yu. A., "Effect of Cooling of the Central Body on Start-Up Separation of the Flow at the Intake and the Throttling Characteristics of Air Scoops at Supersonic and Hypervelocity Velocities," *Fluid Mechanics—Soviet Research,* Vol. 7, No. 3, 1978.

[17]Cnossen, J. W., and O'Brien, X. X., "Investigation of the Diffusion Characteristics of Supersonic Streams Composed Mainly of Boundary Layers," *Journal of J Aircraft,* Vol. 2, No. 6, 1965.

[18]Mahoney, J. J., *Inlets for Supersonic Missiles,* AIAA Education Series, AIAA Washington, DC, 1990.

[19]McGregor, R. J., Molder, S., and Paisley, T. W., "Hypersonic Inlet Flow Starting in the Ryerson/University of Toronto Gun Tunnel," *Investigations in the Fluid Dynamics of Scramjet Inlets,* Ryerson Polytechnical Univ. and Univ. of Toronto, Canada, July 1992.

[20]Kantrowitz, A., and Donaldson, C., "Preliminary Investigation of Supersonic Differs," NACA WRL-713, 1945.

[21]Van Wie, D. M., Kwok, F. T., and Walsh, R. F., "Starting Characteristics of Supersonic Inlets," AIAA Paper 96-2914, July 1996.

[22]Watson, E. C., "An Experimental Investigation at Mach Numbers From 2.1 to 3.0 of Circular Internal Compression Inlets Having Translatable Centerbodies and Provisions for Boundary Layer Removal," NASA TM-X-156, Jan. 1960.

[23]Vahl, W. A., and Oehman, W. I., "Internal Flow Characteristics of a Fixed-Geometry Induction System Having Axial Symmetry at Mach Numbers From 3.8 to 4.2," NASA TM-X-759, Jan. 1963.

[24]Sakata, K., Yanagi, R., Murakami, A., Shindo, S., Honami, S., Shizaua, T., Sakamoto, K, Shiraisha, K., and Omi, J., "An Experimental Study of Supersonic Air-Intake With 5-Shock System at Mach 3," AIAA, Paper 93-2305, 1993.

[25]Trexler, C. A., and Souders, S. W., "Design and Performance at a Local Mach Number of 6 of an Inlet for an Integrated Scramjet Concept," NASA TN-D-7944, Aug. 1975.

[26]Demarest, P. E., "High Temperature Tests of an External-Plus-Internal Compression Inlet at Mach 6.5," ASD-TDR-62-280, Sept. 1962.

[27]Karanian, A. J., and DeBlois, X. X., "Investigation of Precompression Devices for Downward Turning Inlets," United Airlines Research Lab., Rep. M911356-10, April 1973.

[28]Heins, A. E., Reed, G. J., and Woodgrift, K. E., "Hydrocarbon Scramjet Feasibility Program. Part III—Freejet Engine Design and Performance," AFAPL-TR-70-74, Jan. 1971.

[29]"Hypersonic Ramjet Program for the Period 1 March to 31 December 1964, Vol. 1: 18-in. Engine Program," AFAPL-TR-65-36, Vol. 1, May 1965.

[30]McFarlin, D. J., and Kepler, C. A., "Mach 5 Test of Hydrogen-Fueled Variable Geometry Scramjet," AFAPL-TR-68-116, Oct. 1968.

[31]Ragsdale, W. C., "An Investigation of Leeward Aft-Entry Ramjet Inlet Performance," NOL-TR-73-25, Feb. 1973.

[32]Gunther, F., "Development of a Two-Dimensional Adjustable Supersonic Inlet," JPL/CIT Report 20-247, 1954.

[33]Trexler, C. A., "Inlet Starting Predictions for Sidewall Compression Scramjet Inlets," AIAA Paper 88-3257, July 1988.

[34]McLafferty, G. H., "Tests of Perforated Convergent-Divergent Diffusers for Multi-Unit Ramjet Application," United Aircraft Corp. Research Dep., Rep. R-53133-19, June 1950.

[35]Lushkov, A. I., and Nikol'skii, X. X., "Shock Starting of a Supersonic Diffuser," *Inzhenernvi Zuhrnal,*Vol. 2, No. 1, 1966, pp. 11–16.

[36]Van Wie, D. M., White, M. E., and Corpening, G. P., "NASP Inlet Design and Testing Issues," *Johns Hopkins APL Technical Digest,*Vol. 11, Nos. 3 and 4, 1990, pp. 353–361.

[37]Ault, D. A., and Van Wie, D. M., "Comparison of Experimental Results and Computational Analysis for the External Flowfield of a Scramjet Inlet at Mach 10 and 13," (*Journal of Propulsion and Power,*Vol. 10, No. 4, 1994, pp. 533–539).

[38]Billig, F. S., "Shock-Wave Shapes Around Spherical- and Cylindrical-Nosed Bodies," *Journal of Spacecraft and Rockets,*Vol. 4, No. 6, 1967, pp. 822–823.

[39]Anderson, J. D., Jr., *Hypersonic and High Temperature Gas Dynamics,*McGraw–Hill, New York, 1989.

[40]Lukasiewicz, J., "Blast-Wave Hypersonic Flow Analogy—Theory and Application," *American Rocket Society Journal,*Vol. 32, No. 9, 1962, pp. 1341–1346.

[41]Edney, B. E., "Effects of Shock Impingement on the Heat Transfer Around Blunt Bodies," *AIAA Journal,*Vol. 6, No. 1, 1986, pp. 15–21.

[42]Wieting, A. R., and Holden, M. S., "Experimental Study of the Shock Wave Interference Heating on a Cylindrical Leading Edge at Mach 6 and 8," *AIAA Journal,*Vol. 27, No. 11, 1989, pp. 1557–1565.

[43]Hains, F. D., and Keys, J. W., "Shock Interference Heating in Hypersonic Flows," *AIAA Journal,*Vol. 10, No. 11, 1972, pp. 1441–1447.

[44]Wieting, A., "Shock Interference Heating in Scramjet Engines," AIAA Paper 90-5238, Oct. 1990.

[45]Nowak, R., Holden, M., and Wieting, A., "Shock/Shock Interference on a Transpiration Cooled Hemispherical Model," AIAA Paper 90-1643, June 1990.

[46]Wieting, A. R., "Multiple Shock-Shock Interference on a Cylindrical Leading Edge," *AIAA Journal,*Vol. 30, No. 8, 1992, pp. 2073–2079.

[47]Holden, M., Sweet, S., Kolly, J., and Smolinski, G., "A Review of the Aerothermal Characteristics of Laminar, Transitional and Turbulent Shock/Shock Interaction Regions in Hypersonic Flows," AIAA Paper 98-0899, Jan. 1998.

[48]Carl, M., Hannemann, V., and Eitelberg, G., "Shock/Shock Interaction Experiments in the High Enthalpy Shock Tunnel Gottingen," AIAA Paper 98-0775, Jan. 1998.

[49]Bramlette, T. T., "Simple Technique for Predicting Type III and IV Shock Interference," *AIAA Journal,*Vol. 12, No. 8, 1974, pp. 1151–1152.

[50]Lutz, S. A., "Correlation of Type III Turbulent Shock Interaction Heating Data on a Hemisphere," *AIAA Journal,*Vol. 30, No. 12, 1992, pp. 2973–2974.

[51]Tannehill, J. C., Holst, T. L., and Rakich, J. V., "Numerical Computation of Two-Dimensional Flows with an Impinging Shock," *AIAA Journal,*Vol. 14, No. 2, 1976., pp. 204–211.

[52]Tannehill, J. C., Vigneron, Y. C., and Rakich, J. V., "Numerical Simulation of Two-Dimensional Blunt Body Flows with Impinging Shocks," *AIAA Journal,*Vol. 17, No. 12, 1979, pp. 1289–1290.

[53]Singh, D. J., Kumar, A., and Tiwari, S. N., "Influence of the Shock-Shock Interaction on the Blunt Body Flowfield at Hypersonic Flight Speeds," AIAA Paper 89-2184-CP, July 1989.

[54]Klopfer, G., and Yee, H., "Viscous Hypersonic Shock-on-Lip Interaction on Blunt Cowl Lips," AIAA Paper 87-0233, Jan. 1987.

[55]Moon, Y., and Holt, M., "Interaction of an Oblique Shock Wave with Turbulent Blunt Body Flows," AIAA Paper 89-0272, Jan. 1989.

[56]Prabu, R., Stewart, J., and Thareja, R., "Shock Interference Studies on a Circular Cylinder at Mach 16," AIAA Paper 90-0606, Jan. 1990.

[57]Lind, C. A., and Lewis, M. J., "A Numerical Study of the Unsteady Processes Associated with a Type IV Shock Interaction," AIAA Paper 93-2479, June 1993.

[58]Hayes, W. D., and Probstein, R. F., *Hypersonic Flow Theory,* Academic, New York, 1959.

[59]Orgorodnikov, D., "CIAM Hypersonic Investigations and Capabilities," AIAA Paper 93-5093, Dec. 1993.

[60]Malik, M. R., "Prediction and Control of Transition in Supersonic and Hypersonic Boundary Layers," *AIAA Journal,* Vol. 27, No. 11, 1989, pp. 1487–1493.

[61]Malik, M. R., Zang, T., and Bushnell, D., "Boundary Layer Transition at Hypersonic Speeds," AIAA Paper 90-5232, Oct. 1990.

[62]Elias, T. I., and Eisworth, E. A., "Stability Studies of Planar Transition in Supersonic Flows," AIAA Paper 90-5233, Oct. 1990.

[63]Narasimha, R., "The Laminar-Turbulent Transition Zone in the Boundary Layer," *Progress in Aerospace Sciences,* Vol. 22, 1985, pp. 29–80.

[64]Arnal, D., "Laminar-Turbulent Transition Problems at Hypersonic Speeds," *The Second Joint Europe/U.S. Short Course on Hypersonics,* 1989.

[65]Narasimha, R., and Dey, J., "Transition Zone Models for Two-Dimensional Boundary Layers: A Review," *Acad. Proc. in Eng. Sci.,* Vol. 14, 1989, pp. 93–120.

[66]Delery, J. M., "Shock-Wave/Turbulent Boundary Layer Interaction and Its Control," *Progress in Aerospace Sciences,* Vol. 22, 1985, pp. 209–280.

[67]Stollery, J. L., "Some Aspects of Shock-Wave/Boundary-Layer Interaction Relevant to Intake Flows," *Hypersonic Combined Cycle Propulsion,* AGARD-CP-479, June 1990.

[68]Kubota, H., and Stollery, J. L., "An Experimental Study of the Interaction Between a Glancing Shock Wave and a Turbulent Boundary Layer," *Journal of Fluid Mechanics,* Vol. 116, 1982.

[69]Alvi, F. S., and Settles, G. S., "Physical Model of the Swept Shock Wave/Boundary Layer Interaction Flowfield," *AIAA Journal,* Vol. 30, No. 9, 1992, pp. 2252–2258.

[70]Tani, K., and Kanda, T., "3D Shock and Boundary Layer Interaction In Scramjet Inlets," *Shock Wave Symposium,* (Japan), Dec. 1990.

[71]Murakami, A., Yanagi, R., Shindo, S., Sakato, K., Honami, S., Tanako, A., and Shiraishi, K., "Mach 3 Wind Tunnel Test of a Mixed Compression Supersonic Inlet," AIAA Paper 92-3625, July 1992.

[72]Weir, L. J., Reddy, D. R., and Rupp, G. D., "Mach 5 Inlet CFD Study and Experimental Results," AIAA Paper 89-2355, July 1989.

[73]Fisher, S. A., "Three Dimensional Flow Effects in a Two-Dimensional Air Intake," *Journal of Propulsion and Power,* Vol. 2, No. 6, 1986, pp. 546–551.

[74]Reddy, D. R., and Weir, L. J., "Three-Dimensional Viscous Analysis of a Mach 5

Inlet and Comparison with Experimental Data," *Journal of Propulsion and Power,* Vol. 8, No. 2, 1992, pp. 432–440.

[75]Delerey, J., and Marvin, J., "Shock-Wave/Boundary-Layer Interactions," AGARDograph No. 280, Feb. 1986.

[76]Delerey, J., "Shock-Shock and Shock-Wave/Boundary-Layer Interactions in Hypersonic Flows," AGARD-FDP-VKI Special Course on Aerothermodynamics of Hypersonic Vehicles, May 1988.

[77]Settles, G. S., and Dolling, D. S., "Swept Shock Wave/Boundary Layer Interactions," *Tactical Missile Aerodynamics,* Vol. 104, AIAA, Progress in Astronautics and Aeronautics, 1986.

[78]Holden, M. S., "Boundary-Layer Displacement and Leading Edge Bluntness Effects on Attached and Separated Laminar Boundary Layers in a Compression Corner. Part I: Theoretical Study," *AIAA Journal,* Vol. 8, No. 12, 1970, pp. 2179–2188.

[79]Holden, M. S., "Boundary-Layer Displacement and Leading-Edge Bluntness Effects on Attached and Separated Boundary Layers in a Compression Corner. Part II: Experimental Study," *AIAA Journal,* Vol. 9, No. 1, 1971, pp. 84–93.

[80]Gray, J. D., and Rhudy, R. W., "Effects of Blunting and Cooling on Separation of Laminar Supersonic Flow," *AIAA Journal,* Vol. 11, No. 9, 1973, pp. 1296–1301.

[81]Lewis, J. E., Kubota, T., and Lees, L., "Experimental Investigation of Supersonic Laminar Boundary-Layer Separation in a Compression Corner with and without Cooling," *AIAA Journal,* Vol. 6, No. 1, 1968, pp. 7–14.

[82]Law, C. H., "Supersonic Turbulent Boundary-Layer Separation," *AIAA Journal,* Vol. 12, No. 6, 1974, pp. 794–797.

[83]Batham, J. P., "An Experimental Study of Turbulent Separating and Reattaching Flows at High Mach Number," *Journal of Fluid Mechanics,* Vol. 52, Pt. 3, 1972, pp. 425–435.

[84]Elfstrom, G. M., "Turbulent Separation in Hypersonic Flow," Imperial College, Aero. Rep. 71—16, Univ. of London, England, UK, 1971.

[85]Kessler, W. C., Reilly, J. F., and Mackayetris, L. J., "Supersonic Turbulent Boundary Layer Interaction with an Expansion Ramp and Compression Corner," McDonnell Douglas, MDCEO264, St. Louis, MO, 1970.

[86]Kuehn, D. M., "Experimental Investigation of the Pressure Rise Required for the Incipient Separation of Turbulent Boundary Layers in Two-Dimensional Supersonic Flow," NASA Memo 1-21-59a, 1959.

[87]Thomke, G. J., and Roshko, A., "Incipient Separation of a Turbulent Boundary Layer at High Reynolds Number in Two-Dimensional Supersonic Flow over a Compression Corner," McDonnell Douglas Astro Co., Western Div. DAC 59819, Huntington Beach, CA, 1969.

[88]Drougge, G., "An Experimental Investigation of the Influence of Strong Adverse Pressure Gradients on Turbulent Boundary Layer at Supersonic Speeds," FFA Rept. 47, Stockholm, Sweden, 1953.

[89]Spaid, F. W., and Frishett, J. C., "Incipient Separation of a Supersonic Turbulent Boundary Layer, Including the Effects of Heat Transfer," *AIAA Journal,* Vol. 10, No. 7, 1972, pp. 915–922.

[90]Holden, M. S., "Shock Wave-Turbulent Boundary Layer Interaction in Hypersonic Flow," AIAA Paper 72–74, 1972.

[91]Gray, J. D., and Rhudy, R. W., "Investigation of Flat-Plate Aspect Ratio Effects on Ramp-Induced, Adiabatic, Boundary Layer Separation at Supersonic and Hypersonic Speeds," AEDC-TR-70, 1971.

[92]Sterret, J. R, and Emery, J. C., "Experimental Separation Studies for Two-Dimensional Wedges and Curved Surfaces at M=4.8 to 6.2," NASA TN-D-1014, 1962.

[93]Korkegi, R. H., "A Simple Correlation for Incipient Turbulent Boundary-Layer Separation due to a Skewed Shock Wave," *AIAA Journal,* Vol. 11, No. 11, 1973, pp. 1578–1579.

[94]Korkegi, R. H., "Comparison of Shock-Induced Two- and Three-Dimensional Incipient Turbulent Separation," *AIAA Journal,* Vol. 13, No. 4, 1975, pp. 534,535.

[95]Korkegi, R. H., "A Lower Bound for Three-Dimensional Turbulent Separation in Supersonic Flow," *AIAA Journal,* Vol. 23, No. 3, 1985, pp. 475,476.

[96]Frew, D., Galassi, L., Stava, D., and Azevedo, D., "A Study of Incipient Separation Limits for Shock-Induced Boundary Layer Separation for Mach 6 High Reynolds Number Flows," AIAA Paper 93-2481, June 1993.

[97]Hawbolt, R. J., Sullivan, P. A., and Gottlieb, J. J., "Experimental Study of Shock Wave and Boundary Layer Interaction Near a Convex Corner," AIAA Paper 93-2980, July 1993.

[98]Lu, F. H., and Chung, K. M., "Exploratory Study of Shock Reflection Near an Expansion Corner," AIAA Paper 93-3132, July 1993.

[99]Chew, Y. T., "Shockwave and Boundary Layer Interaction in the Presence of an Expansion Corner," *Aeronautical Quarterly,* Vol. 30, 1979, pp. 506–527.

[100]White, M. E., "Expansion Corner Effects on Hypersonic Shock Wave/Turbulent Boundary Layer Interactions," AIAA Paper 96-4542, Nov. 1996.

[101]Hamed, A., and Shang, J. S., "Survey of Validation Database for Shockwave Boundary-Layer Interactions in Supersonic Inlets," *Journal of Propulsion and Power,* Vol. 7, No. 4, 1991, pp. 617–624.

[102]White, M. E., Lee, R. E., Thompson, M. W., Carpenter, A., and Yanta, W. J., "Tangential Mass Addition for Shock/Boundary-Layer Interaction Control in Scramjet Inlets," *Journal of Propulsion and Power,* Vol. 7, No. 6, 1991, pp. 1023–1029.

[103]Alzner, E., and Zakkay, V., "Turbulent Boundary Layer Shock Interaction with and without Injection," *AIAA Journal,* Vol. 9, No. 9, 1971, pp. 1769–1776.

[104]Orgorodnikov, D. A., Grin, V. T., and Zakharov, N. N., "Controlling the Boundary Layer in Hypersonic Air Intakes," *1st International Symposium on Air Breathing Engines,* (Marseille, France), June 1972.

[105]Tindell, R. H., and Willis, B. P., "Experimental Investigation of Blowing for Controlling Oblique Shock/Boundary Layer Interactions," AIAA Paper 97-2642, July 1997.

[106]Pandolfini, P. P., Billig, F. S., Corpening, G. P., Corda, S., and Friedman, M. A., "Analyzing Hypersonic Engines Using the Ramjet Performance Analysis Code," *APL Technical Review,* Vol. 2, No. 1, 1990, pp. 118–126.

[107]Van Wie, D. M., "Techniques for the Measurement of Scramjet Inlet Performance at Hypersonic Speeds," AIAA Paper 92-5104, Dec. 1992.

[108]Van Wie, D. M., Walsh, R. F., and McLafferty, G. H., "Measurement of Mass Flow in Hypersonic Facilities," AIAA Paper 92-3904, July 1992.

[109]Van Wie, D. M., Corpening, G. P., Mattes, L. A., Carpenter, D. A., Molder, S., and McGregor, R., "An Experimental Technique for the Measurement of Mass Flow of

Scramjet Inlets Tested in Hypersonic Pulse Facilities," AIAA Paper 89-2331, July 1989.

[110]Corpening, G. P., Van Wie, D. M., and Mattes, L. A., "Further Assessment of a Scramjet Inlet Mass Flow Measurement Technique for Use in Hypersonic Pulse Facilities," AIAA Paper 91-0551, Jan. 1991.

[111]Williams, R. L., "Application of Pulse Facilities to Inlet Testing," *Journal of Aircraft,* Vol. 1, No. 10, 1964, pp. 236–241.

[112]Kutschenreuter, P. H., and Balent, R. L., "Hypersonic Inlet Performance from Direct Force Measurements," *Journal of Spacecraft,* Vol. 2, No. 2, 1965, pp. 192–199.

[113]Keirsey, J. L., "A Study of the Aerodynamics of Scramjet Inlets," Johns Hopkins Univ., APL TG-732, Sept. 1963.

[114]Marquart, E. J., "Predictions and Measurements of Internal and External Flowfields of a Generic Hypersonic Inlet," AIAA Paper 91-3320-CP, Sept. 1991.

[115]Minucci, M. A. S., and Nagamatsu, H. T., "Experimental Study of a Two-Dimensional Scramjet Intake, $M_\infty = 10.1\text{-}25.1$," *Journal of Propulsion and Power,* Vol. 8, No. 3, 1992, pp. 680–686.

[116]Vinogradov, V., Grachev, V., Petrov, M., and Sheechman, J., "Experimental Investigation of 2D Dual-Mode Scramjet with Hydrogen Fuel at Mach 4–6," AIAA Paper 90-5269, Oct. 1991.

[117]Minucci, M. A. S., and Nagamatsu, H. T., "Investigation of a Two-Dimensional Scramjet Inlet, $M_\infty = 8\text{-}118$ and $T_0 = 4100$ K," *Journal of Propulsion and Power,* Vol. 9, No. 1, 1993, pp. 139–145.

[118]Gnos, A. V., Watson, E. C., Seebaugh, W. R., Sanator, R. J., and DeCarlo, J. P., "Investigation of Flow Fields with Large Scale Hypersonic Inlet Models," NASA TN-D-7150, April 1973.

[119]Bissinger, N., and Schmitz, D., "Design and Wind Tunnel Testing of Intakes for Hypersonic Vehicles," AIAA Paper 93-5042, Dec. 1993.

[120]Van Wie, D. M., "An Application of Computational Fluid Dynamics to the Design of Optimum Ramjet Powered Missile Components," Ph.D. Dissertation, Univ. of Maryland, University Microfilms, 1986.

[121]Van Wie, D. M., White, M. E., and Waltrup, P. J., "Application of Computational Design Techniques in the Development of Scramjet Engines," AIAA Paper 87-1420, June 1987.

[122]Busemann, A., "Die Achsensymmetrische Kegelizeueber-Schallstromung," *Luftfahtforschung,* Vol. 19, 1942, pp. 137–144.

[123]Molder, S., and Szpiro, E. J., "Busemann Inlet For Hypersonic Speeds," *Journal of Spacecraft,* Vol. 3, No. 8, 1966.

[124]Van Wie, D. M., and Molder, S., "Applications of Busemann Inlet Designs for Flight at Hypersonic Speeds," AIAA Paper 92-1210, Feb. 1992.

[125]Tani, K., Kanda, T., Komuro, T., Murakami, A., Kudou, K., Wakamatsu, Y., Masuya, G., and Chinzei, N., "Flow Measurements in Scramjet Inlets," *Proceedings of the 17th International Symposium on Space Technology and Science*(Tokyo), 1990.

[126]Cozart, A. B., Holland, S. D., Trexler, C. A., and Perkins, J. N., "Leading Edge Sweep Effects in Generic Three-Dimensional Sidewall Compression Scramjet Inlets," AIAA Paper 92-0674, Jan. 1992.

[127]Holland, S. D., and Perkins, J. N., "Contraction Ratio Effects in a Generic 3D

Sidewall Compression Scramjet Inlet: A Computational and Experimental Investigation," AIAA 91-1708, June 1991.

[128]Holland, S. D., "Reynolds Number and Cowl Position Effects for a Generic Sidewall Compression Scramjet Inlet at Mach 10: A Computational and Experimental Investigation," AIAA 92-4026, June 1992.

[129]Holland, S. D., and Perkins, J. N., "Internal Shock Interaction in Propulsion/Airframe Integrated 3D Sidewall Compression Scramjet Inlets," AIAA Paper 92-3099, July 1992.

[130]Tani, K., Kanda, T., Kudou, K., Murakami, A., Komuro, T., and Itoh, K., "Aerodynamic Performance of Scramjet Inlet Models with a Single Strut," AIAA Paper 93-0741, Jan. 1993.

[131]Holland, S. D., and Murphy, K. J., "An Experimental Parametric Study of Geometric, Reynolds Number, and Ratio of Specific Heats Effects in Three-Dimensional Sidewall Compression Scramjet Inlets at Mach 6," AIAA Paper 93-0740, Jan. 1993.

[132]Korte, J. J., Singh, D., Kumar, A., and Auslender, A., "Numerical Study of the Performance of Swept Curved Compression Surface Scramjet Inlets," AIAA Paper 93-1837, June 1993.

[133]Kanda, T., Komuro, T., Masuya, G., Kudo, K., Murakomi, A., Tani, K., Wakamatsu, Y., and Chinzei, N., "Mach 4 Testing of Scramjet Inlet Models," *Journal of Propulsion and Power,* Vol. 7, No. 2, 1991, pp. 275–280.

[134]Trexler, C. A., "Inlet Performance of the Integrated Langley Scramjet Module (Mach 2.3 to 7.6)," AIAA Paper 75-1212, Oct. 1975.

[135]Sharp, B. M., and Hollweger, D. J., "Four Module Scramjet Inlet Development Studies Final Report," Johns Hopkins Applied Physics Lab. Laurel, MD, March 1969.

[136]Hartill, W. R., "Analytical and Experimental Investigations of a Scramjet Inlet of Quadriform Shape," AFAPL-TR-65-74, Aug. 1965.

[137]Billig, F. S., "SCRAM-A Supersonic Combustion Ramjet Missile," AIAA Paper 93-2329, June 1993.

[138]Waltrup, P. J., Billig, F. S., and Stockbridge, R. D., "Engine Sizing and Integration Requirements for Hypersonic Airbreathing Missile Applications," AGARD-CP-307, No. 8, March 1982.

[139]Kumar, A., Singh, D. J., and Trexler, C. A., "Numerical Study of the Effects of Reverse Sweep on Scramjet Inlet Performance," *Journal of Propulsion and Power,* Vol. 8, No. 3, 1992, pp. 714–719.

Chapter 8

Supersonic Flow Combustors

P. Kutschenreuter*
GE Aircraft Engines

Nomenclature

A = area
a = speed of sound
BL = boundary layer
C_D = drag coefficient
CFD = computational fluid dynamics
C_{FG} = nozzle gross thrust coefficient (actual/ideal for actual nozzle area ratio)
CR = contraction ratio
D = diameter
ER = fuel equivalence ratio
F = thrust
F/A = fuel-air ratio
g = gravitational constant, 32.17 ft/s^2
ΔH_C = net heat release
h = enthalphy
I_{SP} = fuel specific impulse, s
J = ratio of work unit to heat unit, 778.16 ft lb/Btu
K = constant
L/W = sum of injector configuration axial length/width
M = Mach number
P = pressure
psf = lbs/ft^2
Q_{LOSS} = heat loss
q = dynamic pressure
R = gas constant
Re = Reynold's number

---

*Retired.

S = entropy
T = temperature
U = axial velocity
V = velocity
w = flow rate, lbm/s
x/w = axial downstream distance from fuel jet/fuel jet lateral spacing
Y = vertical distance from combustor wall or fuel jet
β = fuel injection angle measured relative to axial
γ = ratio of specific heats
Δ = difference
η = efficiency
ρ = density
ϕ = stream thrust

Subscripts and Superscripts

A = air
C = core flow or convective
F = fuel
G = gross
J = jet
M = mixing
N = net
P = projected
R = ratio
REF = reference
SEP = boundary-layer separation
STEP = average value for combustor step
X = upstream of normal shock
Y = downstream of normal shock
0 = freestream
1 = inlet capture station
1.5 = area ratio of combustor including any isentropic expansion/ $A_2 = 1.5$
2 = inlet exit or combustor entrance station
2A = isolator entrance station
2.8 = area ratio of combustor including any isentropic expansion/ $A_2 = 2.8$
3 = combustor exit or nozzle entrance station
4 = nozzle exit station
* = sonic condition

I. Introduction

BEFORE getting into details, we will answer the question, "Why *supersonic* combustion ramjets?" The simple answer to this question is that the use of supersonic combustion can result in higher levels of performance at increased flight Mach number than can be achieved with subsonic combustion. This is illustrated in the Fig. 1 from Ref. 1, which shows specific impulse trends for each type of engine.

Understanding these trends can be a help in designing a better supersonic combustor. To develop this understanding, let us first take a look at the implications of *not* using supersonic combustion, i.e., higher freestream Mach number operation with subsonic combustion. Figure 2 illustrates some important trends. The upper plot can be thought of as a key to such subsonic combustor performance implications, whereas the lower plot provides insight into the corresponding subsonic combustor thermal protection and weight implications. As indicated in Fig. 2, a 1000-psf flight trajectory ($q_0 = \gamma_0 P_0 M_0^2/2$) has been used.

Figure 2a typifies the significant losses in freestream total pressure because of viscous and shock-wave losses incurred by the inlet in decelerating the captured flow from increasingly higher freestream Mach numbers to the low subsonic Mach numbers required for stable and efficient subsonic combustion. Also indicated is the rapid rise to total temperatures in excess of 2000° R. With corresponding static temperatures approaching these total temperature levels, molecular dissociation occurs, and the net combustion kinetics include not only the heat-releasing exothermic reactions but also heat-absorbing endothermic reactions. As a consequence of these competing reactions, with increasing temperature (as well as with decreasing pressure) a decreasing proportion of the thermochemical energy can be converted into kinetic energy.

Figure 2b illustrates the rapid increase in subsonic combustor entrance pressure differential relative to ambient (in excess of 30 atm for the Mach 9 freestream conditions). In addition, to keep the wall temperature from exceeding the assumed 2200° R material limit, thermal protection of the inlet subsonic diffuser is required beginning at about Mach 5. Clearly then, such subsonic combustors operating at high freestream Mach numbers must be designed to contain relatively high internal pressures while absorbing significant heat loads to maintain required structural integrity; not necessarily a recipe for light weight.

Recognition of the potential use of supersonic combustion to achieve improved ramjet engine performance at higher flight Mach numbers gave birth to the scramjet (*s*upersonic *c*ombustion *ramjet*). By the late 1960s U.S. Air Force, NASA, and U.S. Navy sponsored development resulted in ground testing of complete scramjet engines[1] designed and built by Garrett Air

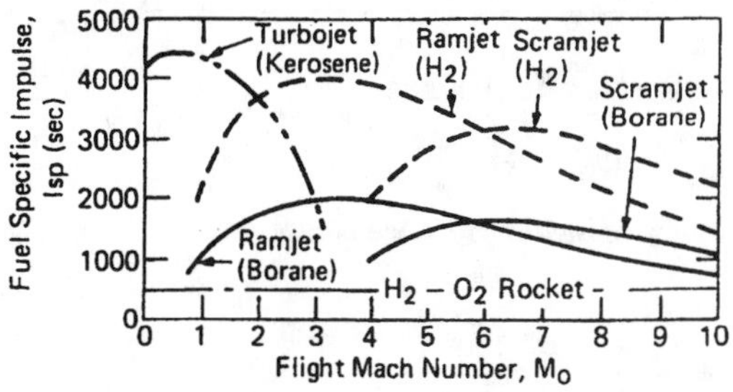

Fig. 1 Typical engine fuel specific impulse characteristics.

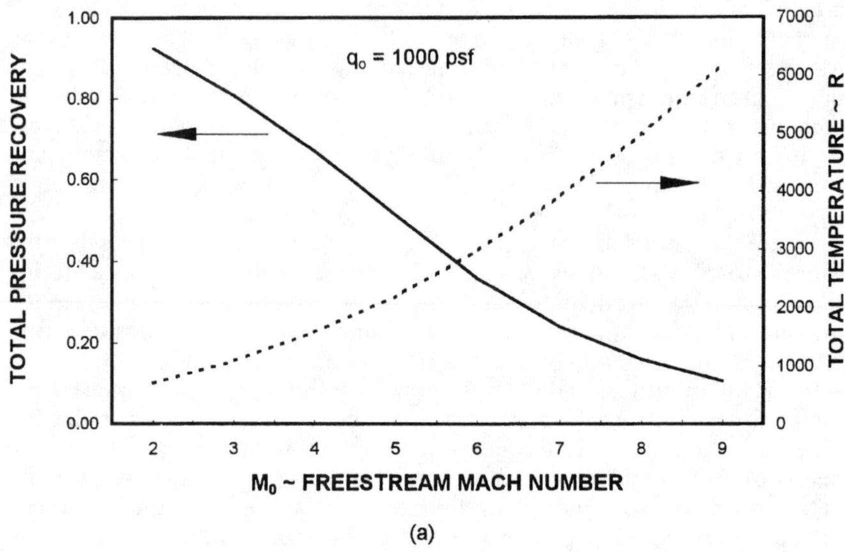

(a)

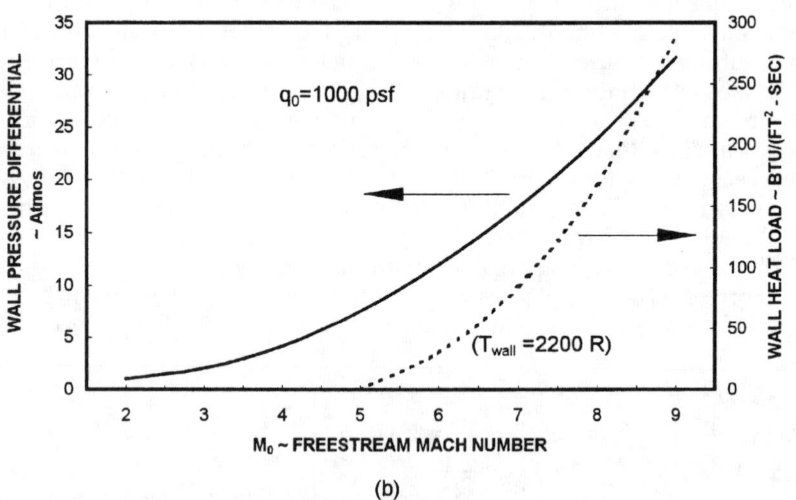

(b)

Fig. 2 Typical subsonic combustor entrance conditions.

Research, GE Aircraft Engines, General Applied Science Laboratory, Johns Hopkins Applied Physics Laboratory, United Aircraft Research Laboratory, and Marquardt. Some of these engines were dual mode, i.e., operated with subsonic combustion at low freestream Mach numbers, switching to supersonic combustion at the higher freestream Mach numbers. Hydrogen as well as hydrocarbon fuels were burned. Such tests were generally limited to nominal freestream Mach numbers of 8 because of steady-state free-jet facility limitations; however, steady-state direct connect supersonic combustion research was conducted up to freestream enthalpy levels corresponding to nominally Mach 12 flight conditions.[2] More recently, maturation of higher energy, short-duration pulse facilities[3] has permitted significant expansion of the ground-test facility envelope for scramjet research.

Application of supersonic combustor technology to scramjet engine design requires not just an understanding of supersonic combustion phenomena, but must also involve inlet and nozzle considerations in order to permit achievement of adequate engine cycle performance and a reasonable mechanical design. As an aid to subsequent discussion, Fig. 3 defines the scramjet inlet, combustor, and nozzle component interface station notation used in this chapter.

II. Phenomenological Considerations

A basic understanding of some of the unique phenomena and implications associated with supersonic combustion is a prerequisite to the intelligent design of a supersonic combustor for a given application. Supersonic combustor considerations essentially dictate many of the inlet requirements, and the resulting combustor exit flow chemistry can also limit key exhaust nozzle options. The approach taken in this section is to illustrate phenomena and trends through simple (one-dimensional), but representative, examples. Some specific results from more sophisticated three-dimensional, finite-rate, viscous flow computational fluid dynamics (CFD) solutions are presented in the higher freestream Mach number material in Sec. V.

A. Inlet Flow

For a subsonic combustion ramjet the captured supersonic airflow is first decelerated in a supersonic diffuser to a lower supersonic Mach number, then

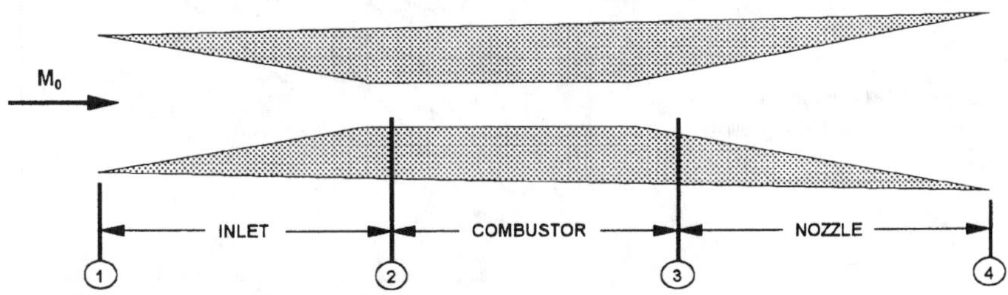

Fig. 3 Scramjet component interface station notation.

brought to subsonic conditions through a normal shock system, and finally diffused further to a lower subsonic Mach number. This action is typically accomplished by a supersonic diffuser employing a number of oblique shocks and a throat section followed by a subsonic diffuser. Obviously for a scramjet inlet, only a supersonic diffuser is required. Consequently, the question of how much supersonic diffusion or inlet compression should be provided at the entrance of the supersonic combustor becomes an important issue. The answer involves integration of the design of the inlet, combustor, and nozzle such that the required engine performance is achieved at acceptable weight, operational reliability, and cost.

Figure 4 illustrates parametric continuity and energy solution approach (Appendix A) results which are used in establishing appropriate supersonic combustor entrance conditions for a subsequent Mach 8 flight example. Combustor entrance static pressure and static temperature are shown for lines of constant inlet contraction ratio (CR) A_0/A_2 where A_2 is the combustor entrance area and A_0 is the inlet freestream tube area of the captured mass flow. Note that lines of constant combustor entrance Mach number coincide with lines of constant entrance temperature for the thermally perfect inlet airflow at these conditions. Cycle performance initially favors high CR (It permits larger nozzle expansion area ratio and thus higher exit momentum.) typically until thermal choking in the combustor limits the required fuel equivalence ratio (ER) or the higher inlet losses (increased entropy representing higher temperature at a given pressure) resulting from the increased contraction ratio cause significant dissociation that limits net heat release.

Insight into the realities of how high a CR to select can be obtained from Fig. 4 by examining the lines of constant inlet entropy increase. Because entropy increase relates the amount of temperature increase to the amount of pressure increase, it is a natural choice as an inlet efficiency parameter, particularly in that the objective of inlet compression is essentially the maximum pressure rise for a particular temperature rise across the inlet. Because increased CR implies increased supersonic diffusion, then inlet losses also increase with contraction ratio. Note in Fig. 4 that at a given

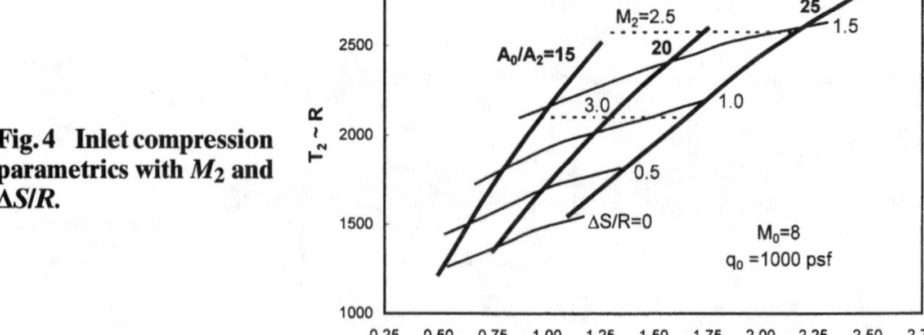

Fig. 4 Inlet compression parametrics with M_2 and $\Delta S/R$.

contraction ratio increased P_2 is the result of additional inlet losses (entropy increase). Achieving a given P_2 at a lower A_0/A_2 results in a lower T_2. Lower T_2 at a given P_2 implies reduced dissociation and higher net heat release. The impact of these trends on net heat release increases with increased flight Mach number. Thus it is implied that lower levels of P_2 and T_2 during low flight Mach-number scramjet operation are likely to permit improvements in higher flight Mach-number performance levels. This raises the question of how low? A 1960's rule of thumb derived from the then available kinetics data was that a scramjet inlet should provide at least a temperature (T_2) of 2000°R and a pressure (P_2) greater than ½ an atmosphere to ensure auto-ignition and sustained supersonic combustion of hydrogen. Test results quickly established that fuel injection details could have a significant influence on auto-ignition limits as subsequently documented in Ref. 4.

With hydrogen as the fuel, Fig. 5 (calculated from the specie mol fractions and heats of formation from Ref. 5 equilibrium data for $H_2O, H_2, N_2, A, O, OH, H, O_2, NO, N, NH_3$, and NH) provides additional insight. The net heat release (ΔH_C) decreases with decreased pressure and increased temperature because of dissociation and endothermic reactions. Thus inlet temperature levels that produce combustor temperatures of nominally 4500°R or greater, can result in significant reduction in net heat release depending on the pressure level. As a consequence, note that even when flight enthalpy is duplicated in a ground-test facility, if facility pressure limitations result in reduced operating pressures compared to flight, then lower ground-test net heat release levels can be expected. If determination of scramjet performance is the test objective, then results interpretation under such circumstances can be difficult.

The magnitude of dissociation and endothermic reaction problems for scramjet combustors is largely driven by increases in flight Mach number (similar to subsonic combustion ramjets, but for scramjets it is delayed to higher flight Mach numbers). Typically along an accelerating flight path altitude increase with increasing flight Mach number results in a significant reduction in ambient pressure at relatively constant ambient temperature. Figure 6 illustrates typical atmospheric characteristics along a flight path of 1000-psf dynamic pressure. Note that across the flight Mach number range

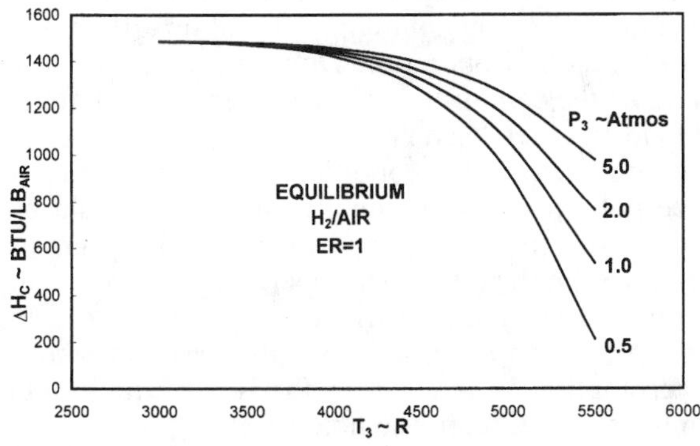

Fig. 5 Equilibrium net heat release trends.

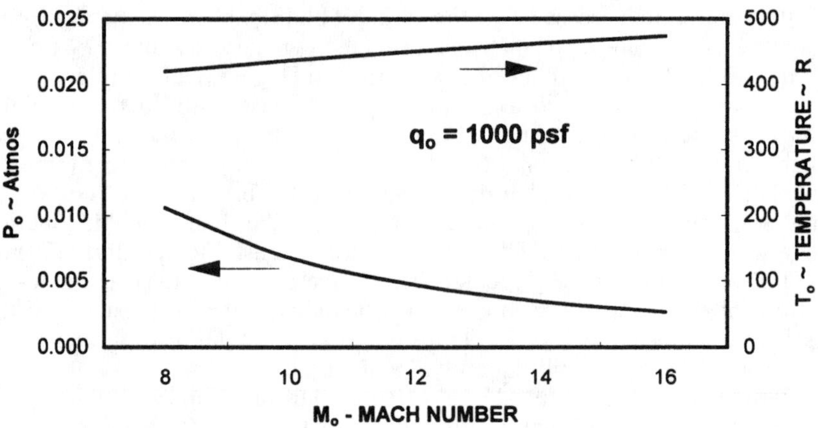

Fig. 6 Flight-path ambient characteristics.

the ambient pressure decreases by about a factor of four whereas the temperature increases by about 10%. Consequently, if the inlet provides increased pressure ratio to compensate (increasing A_0 or decreasing A_2 with increasing M_0), it may then produce too high a temperature ratio. To illustrate the nature of the resulting problem at the combustor entrance, a fixed geometry inlet with a contraction ratio of 25 that provides the combustor entrance Mach number schedule illustrated in Fig. 7a was assumed. From Fig. 7b continuity and energy solution results one can that at $M_0 = 16$ combustor entrance pressure decreases to nominally half an atmosphere, and the combustor entrance temperature is almost 4000°R. Recalling Fig. 5 and allowing for nominal increases in temperature and pressure, a reduction in net combustor heat release approaching $M_0 = 16$ can be anticipated. Note that a more efficient inlet than the one used here, while providing lower T_2, would further decrease P_2 (at the same CR). Increased CR at $M_0 = 16$ coupled with inlet efficiency improvement would be a better thermodynamic solution, but would likely require variable inlet geometry in order to provide the CR reduction required at say $M_0 = 8$ for maintaining full capture flow and avoiding thermal choke or reduced ER operation. Doubling the flight-path dynamic pressure would double the combustor entrance pressure, but at the expense of increased heat load and fuel delivery pressure.

Clearly selection of the combustor inlet conditions can have a major impact on the performance potential of a scramjet engine and should be iteratively established considering inlet, cycle, and mechanical design, as well as combustor performance implications. The operating environment typically requires treating the thermodynamic variables as functions of both pressure and temperature. (All of the illustrative one-dimensional results presented in this chapter accomplish the latter using the Ref. 5 equilibrium thermodynamics and chemistry unless otherwise noted.)

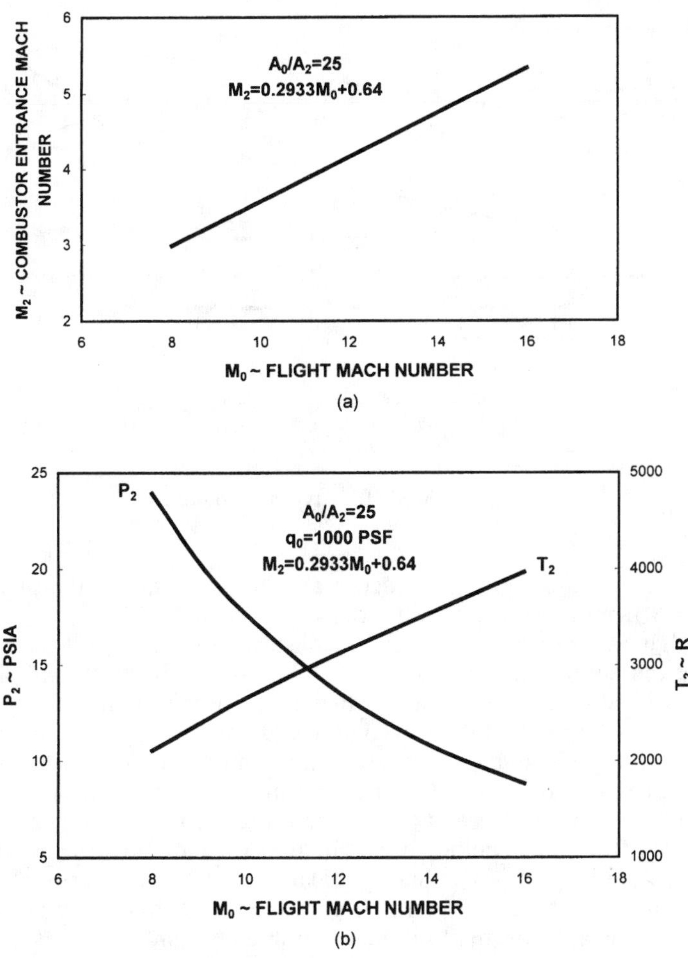

Fig. 7 Example fixed geometry inlet trends.

B. Combustor Flow

Figure 8 illustrates several flow representations on the enthalphy-entropy diagram useful in simple modeling of scramjet constant-area combustor flow processes: the well known Rayleigh line for simple heating and the Fanno line for flow with friction (Ref. 6), plus a variant introduced here as profile line flow, which is explained later. These perfect gas calculations with constant ratio of specific heats were made for the uniform one-dimensional Mach 2 combustor inlet profile conditions summarized in the lower right corner of Fig. 8 and plotted as point *X*. Sonic points on each of the three curves are indicated by the square symbols. Distinguishing features for each flow type are summarized in Table 1. For all three flow types static pressure

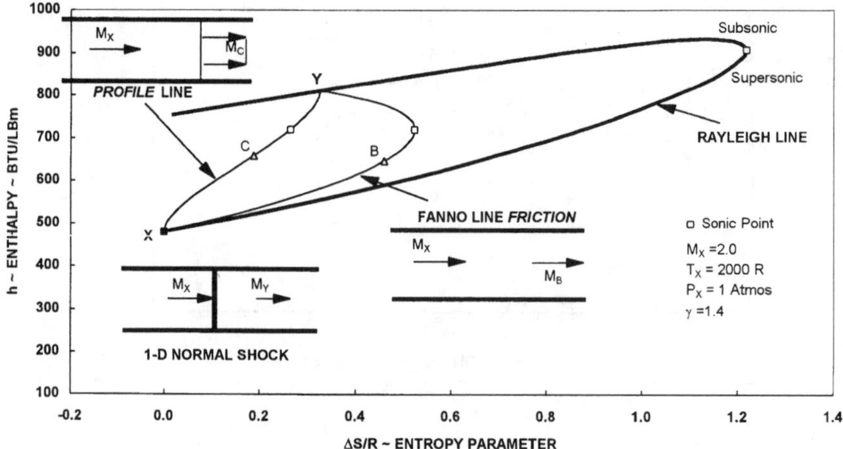

Fig. 8 Simple flow types example.

is assumed constant across the duct in all planes normal to the duct walls ($\delta P/\delta Y = 0$), but the static pressure does vary axially.

Rayleigh lines are simply curves for constant flow per unit duct area and constant momentum per unit duct area, plotted in entropy-vs-enthalpy coordinates. Heat addition (increasing stagnation enthalpy) for both the upper (subsonic flow) branch of the Rayleigh line and the lower (supersonic flow) branch of the Rayleigh line increases to the sonic limit. Thus, although heat addition increases the duct Mach number in ramjets, it decreases the duct Mach number in scramjets and can also result in thermal choking of scramet combustors. If thermal choking occurs before stoichiometric fuel to air ratio (F/A) is achieved (ER = 1), then significant thrust loss can result.

In a somewhat analogous fashion conventional Fanno lines are simply curves for constant flow per unit duct area and constant energy (stagnation enthalpy) per unit duct area, also plotted in entropy-vs-enthalpy coordinates. Momentum loss (such as from wall friction) for both the upper subsonic branch and the lower supersonic branch drives the one-dimensional duct Mach number to the sonic limit. Thus, friction, or for that matter any

Table 1 Simple flow characteristics summary

	Constant area duct	Constant mass flow	Constant momentum	Constant energy	$dP/dY = 0$	Uniform full duct profile
Rayleigh line	Yes	Yes	Yes	No	Yes	Yes
Fanno line	Yes	Yes	No	Yes	Yes	Yes
Profile line	Yes	Yes	Yes	Yes	Yes	No

source of momentum loss (such as in-stream fuel injectors), couples with the heat addition effect in decreasing the scramjet combustor Mach number and leading to possible thermal choke.

When a point on the supersonic branch of either a Rayleigh or Fanno line has the same mass flow, momentum and energy as a point on its subsonic branch, a normal shock solution results (points X and Y in Fig. 8).

Profile lines are simply conservation-equation solution curves for constant mass, momentum, and energy for uniform upstream one-dimensional flow but with a nonuniform downstream flow profile (Appendix B). Such profiles are useful in a simple inviscid representation of separated flow and flow with a thick enough boundary layer (BL) that the flow per unit area near the duct wall is small with respect to that of the mean flow. As observed from Fig. 8, the degree of profile distortion must diminish on either side of the sonic point to become fully one-dimensionally uniform across the entire duct at the normal shock solution points X and Y coincident with the Rayleigh and Fanno lines. For the Fig. 8 calculations the profiles consist of uniform flow with streamtube areas decreasing from that of the duct area at X and Y to a minimum value determined by the duct Mach number and entropy increase. An important attribute is that while the mass, momentum, and energy for the profile flow (such as represented by C in Fig. 8 or any other point on the profile line) is identical to that of the one-dimensional uniform flow at X and Y; flow entropy for a C profile is always higher than that for the uniform supersonic X flow, and in the limit flow entropy subsonically approaches that of the normal shock solution at Y. Note also that whereas constant-area mixing of subsonic profile flows to a uniform flow with the same mass flow, momentum, and energy results in an entropy increase the opposite occurs for supersonic profile flows; mixing out a profile to a uniform duct flow gives the inappropriate mathematical result of an entropy decrease.

Expanding on the Rayleigh line thermal choking observations, Fig. 9 (Ref. 5 thermodynamics and equilibrium chemistry) based on the Fig. 7a combustor entrance Mach number schedule suggests that constant-area duct thermal choking at ER $= 1$ is likely to be encountered at freestream Mach numbers less than 8. For example, the same calculations at $M_0 = 6$ indicate that an ER < 1 is required to avoid thermal choking in a constant-area combustor. Low freestream Mach-number scramjet operation typically requires increased or divergent-area combustors to avoid thermal choking at the higher ER levels typically required for rapid flight vehicle acceleration. Because exit momentum determination for divergent-area combustors is a coupled function of axial heat release rate and rate of area increase, the calculation process is generally more complex. Reference 7 develops divergent combustor solution approaches based on Crocco processes and provides numerous results. Note also from Fig. 9b that the net heat release (ΔH_C) at $M_0 = 16$, even with equilibrium chemistry, is about half that of the $M_0 = 8$ value. This is simply a reflection of the Fig. 5 thermodynamics and chemistry coupled with the Fig. 7 combustor entrance conditions provided by the inlet.

The corresponding static pressure ratio characteristics across the example constant area combustor are summarized in Fig. 10 for two different stream-

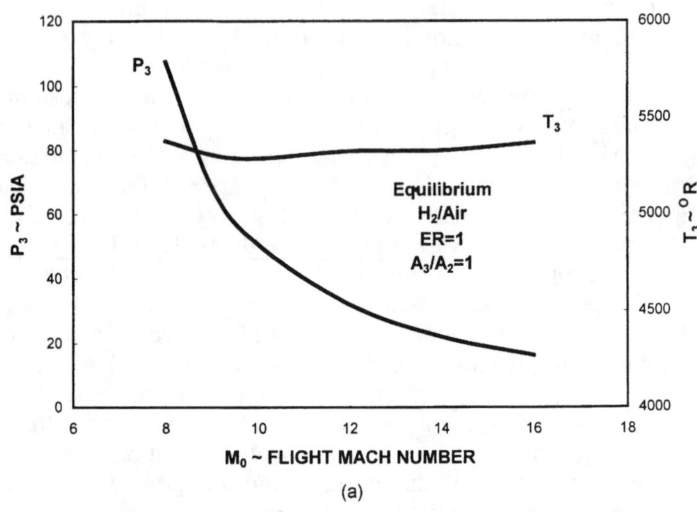

(a)

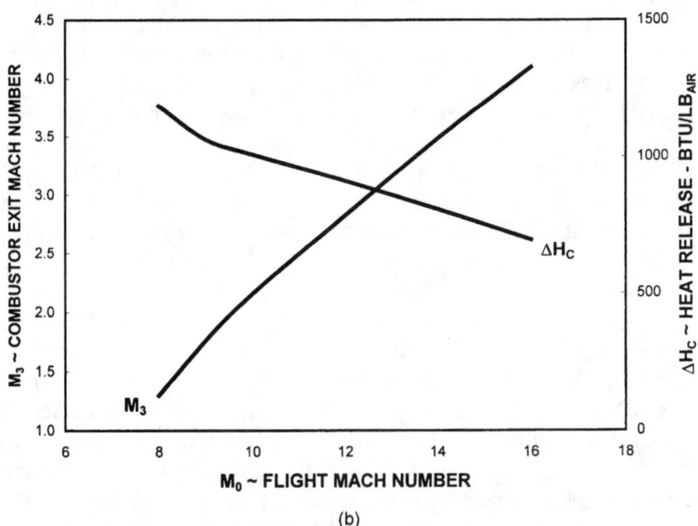

(b)

Fig. 9 Constant-area combustor solutions.

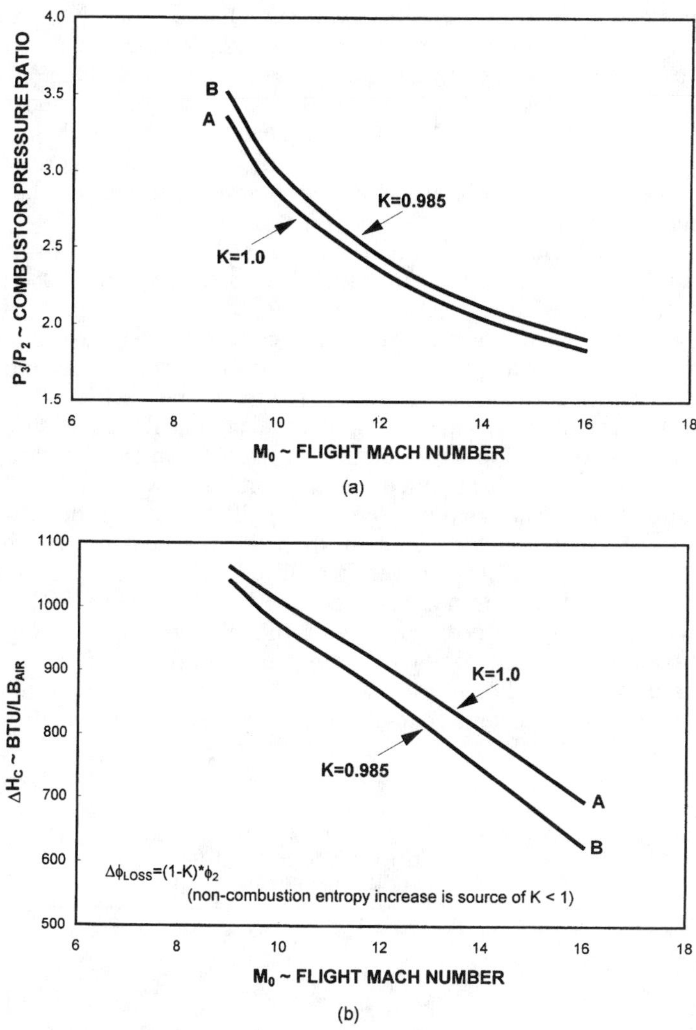

Fig. 10 Combustor stream-thrust loss impact.

thrust [$\phi = PA[1 + V^2/(RT)]$] exit condition assumptions for $K = \phi_3/\phi_2$. For $K = 1$ only those entropy increases directly associated with the combustion process are included. $K = 0.985$ reflects a 1.5% exit stream-thrust loss from additional entropy increases as might be induced by shock waves, friction, in-stream fuel injectors, etc. In both cases the decrease in combustor pressure ratio with increasing freestream Mach number is related to the decreasing ratio of net heat release to freestream stagnation enthalpy. Related Fig. 10 numerical results (not shown) also indicate that when the constant-area combustor exit stream thrust (ϕ_3) is lower than ϕ_2 then M_3 is indeed reduced in accordance with Fanno line observations. Thus entropy increase from other causes (wall friction, in-stream fuel injectors, etc.) than combustor heat release can lead to earlier (over shorter length) thermal choking. From the $K = 1$ and $K = 0.985$ results in Fig. 10, losses not only reduce M_3 as mentioned, but also increase P_3. Recall from Fanno and Rayleigh line observations that losses and heat release both represent entropy increases, and thus static pressure rise across a scramjet combustor should not be totally attributed to heat release.

All of the preceding example results are based on assumptions of 100% mixing at ER = 1 and equilibrium flow. The assumption of 100% mixing implies that provided adequate combustor length ER = 1 mixing of the fuel and air is achieved homogeneously in the combustor. The assumption of equilibrium chemistry implies that for steady flow the specie concentrations are not (or no longer) a function of time (forward and backward reactions occur at the same rate). Because the individual chemical reactions require a finite time for completion as a function of pressure and temperature, the preceding calculations also implied that the combustor was long enough for the finite rate chemical reactions to reach equilibrium. Consequently, if the combustor length is not long enough, then the net heat release for stoichiometric combustion of hydrogen is reduced below the levels in Fig. 5. Reaction-rate completion times for hydrogen (reproduced from Ref. 8) are shown in Fig. 11. Superimposing on Fig. 11 the inlet conditions of Fig. 7 and assuming the axial velocity of the fuel is approximately that of the air results in estimates of required combustor length of nominally only several inches. Consequently, an important conclusion here is that scramjet hydrogen/air

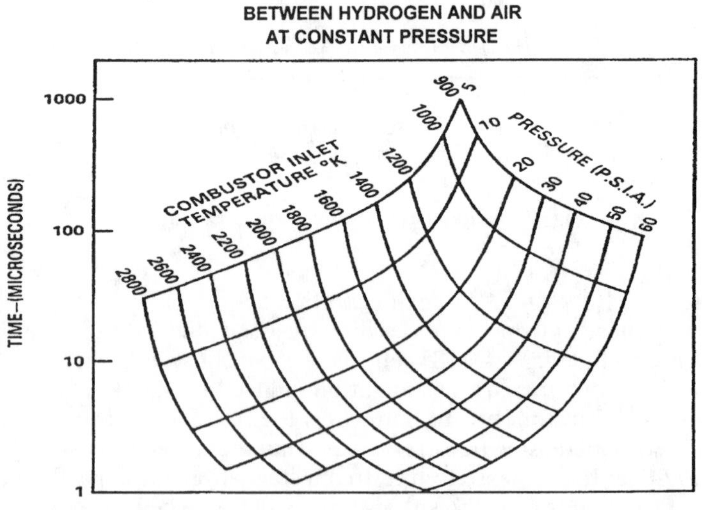

Fig. 11 Hydrogen/air reaction completion characteristics.

combustion at these conditions should be expected to be mixing-rate limited rather than reaction-rate limited.

III. Design Approach Implications

Whereas the preceding one-dimensional inviscid examples were intended to provide an initial introduction to supersonic combustion phenomena as applied to scramjets, selected results are useful in providing a more detailed understanding of design approach implications. Given the almost instantaneous heat release rate of hydrogen over a wide range of pressure and temperature, often further accelerated by the presence of local shock waves from nonaxial fuel injection into the airstream, the actual combustion pressure rise can be expected to be rather abrupt with an initially high adverse pressure gradient rapidly followed by an asymptotic approach to the final level. This sharp pressure gradient coupled with the impracticality of scramjet inlet BL removal for such high-temperature flow raises the concern of BL separation. For simplicity, Fig. 12 based on a curve fit [$P_{SEP}/P_2 = 0.9018(2.074^{M2})M_2^{-.9898}$] to selected data from Ref. 9 is used here to typify BL separation limits in terms of only M_2 and an abruptly reached pressure ratio increase P/P_2. When the Fig. 12 approximation is combined with the combustor pressure ratio results from Fig. 10, Fig. 13 shows that there is a high probability of BL separation (and potentially inlet unstart) at the lower hypersonic flight Mach numbers for constant-area combustion at ER = 1.

A. Step Combustors

One design approach to the BL separation problem is the use of a step combustor as schematically illustrated in Fig. 14. The resulting $A_3/A_2 > 1$ provides a basis for accommodating the combustor pressure rise. Fuel is typically injected in close proximity to the base of the step, which can also

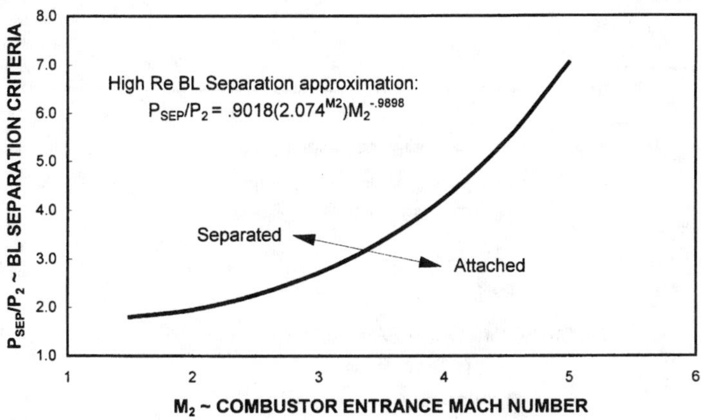

Fig. 12 Boundary-layer separation approximation.

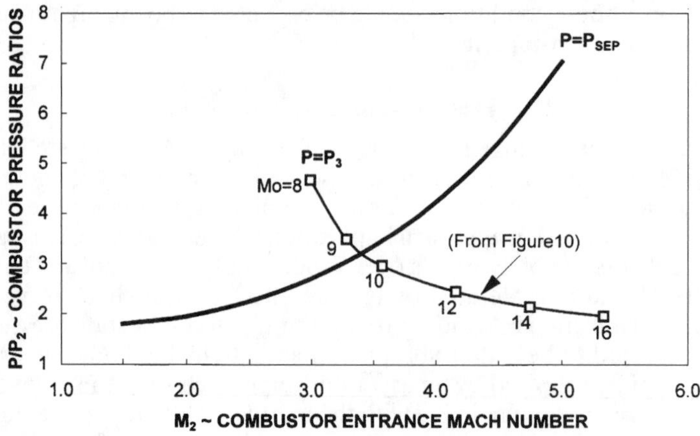

Fig. 13 Combustor flow separation implications.

provide small subsonic recirculation regions for ignition and flame holding. Boundary-layer separation occurs downstream of the station 2 interface but is only a concern in that the resulting fuel injection and combustion-induced step pressure must not subject the station 2 BL to such a high back pressure that significant upstream separation results. The flow characteristics downstream of the step are very complex and generally not amenable to insightful one-dimensional approximations except possibly beginning at station 3, the combustor exit, where by definition the fuel/air mixing and combustion processes are complete. The station 3 stream-thrust change over that of the inlet air flow at station 2 is a function of the station 2 to 3 wall friction, the

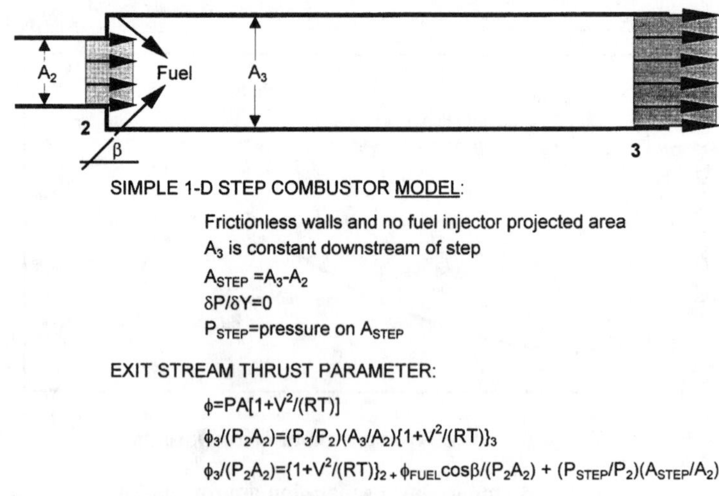

Fig. 14 Step combustor schematic.

axial component of the fuel stream thrust and the step wall force. For simplicity, the latter is characterized here using P_{STEP} as a one-dimensional value of step pressure, which when multiplied by the step area equals the step wall force. Assuming for simplicity of illustration that the friction force is small relative to the other stream-thrust components, then determining the step wall force becomes the issue.

All subsequent illustrations in this section are based on the $M_0 = 8$ inlet conditions from Fig. 7 as the initial combustor entrance conditions for complete combustion of hydrogen at ER = 1.

Parametric results for step combustor pressure ratio at three values of A_3/A_2 are summarized in Fig. 15. As is clear from the stream-thrust equation in Fig. 14, the best performance results from operation at the highest step pressure level. These simple one-dimensional conservation-equation solution results require step pressure level bounds to permit rational approximation of likely operating points for such combustor geometry. Such results can then be used as an aid in preliminary combustor geometry selection as a starting point for three-dimensional, viscous CFD analyses. One obvious bound is the BL separation limit, plotted as the dashed line in Fig. 15, permitting us to exclude solution results at higher step pressure ratios. Perhaps a less obvious bound is the line of step pressure equal to the combustor exit pressure. Consistent with anticipating an initially abrupt then somewhat asymptotic combustor axial pressure distribution, we assumed in a constant area duct that the step pressure should not exceed the combustor exit pressure. A third and very important bound is that the step pressure not be so high that a violation of the Second Law of Thermodynamics is implied. This bound will be discussed next; however, as noted in Fig. 15 the entropy limits (Appendix C) as indicated by the square symbols in Fig. 15 represent bounds in step pressure ratio even higher than either the separation limit or the exit pressure limit and thus do not impact this example. In conducting such studies, in general, all three bounds should be evaluated to determine the lowest upper bound on the step pressure. An implication of the latter is that the maximum potential performance points

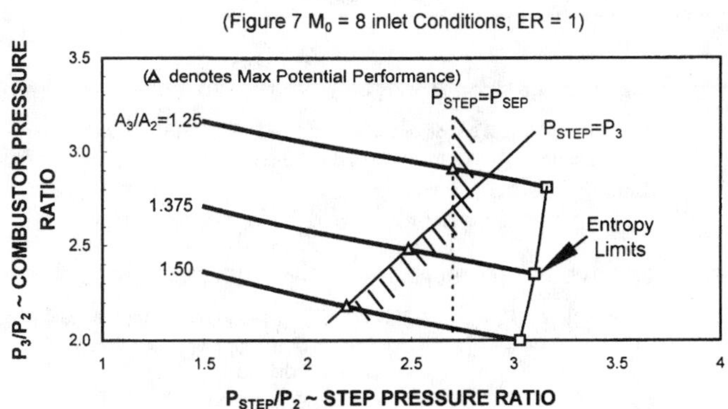

Fig. 15 Step combustor pressure ratio solutions.

determined by the separation limit and the exit pressure limit (plotted as the triangular symbols in Fig. 15 and subsequent figures) reflect entropy penalties associated with the use of the step that reduce combustor performance levels compared to the earlier inviscid constant-area results.

For the Fig. 15 parametric results a maximum potential performance point was identified for each of the three step area ratios, but which step area ratio is likely to provide the highest performance is yet to be determined. This rather complex issue is first addressed by attempting to assess the relative thrust potential (Appendix D) of the three combustor step area ratios, and then examining factors associated with achieving such potential. To accomplish the former, use is made of a one-dimensional reference combustor as typified in Fig. 16. For the reference combustor entrance air flow conditions at station 2, the fuel ER, and the fuel axial component of momentum are identical to those of the step combustors being evaluated. For the reference combustor an inviscid constant-area combustor is used from station 2 to station 3 to establish a reference combustor exit entropy level. The station 3 flow of the reference combustor is then isentropically expanded to an area at station 4 corresponding to the exit area of each step combustor being evaluated in order to establish a reference level of performance. This permits the exit stream-thrust values for a reference and each step combustor to be compared at identical overall A_4/A_2. Step pressure level (P_{STEP}) assumptions that result in stream-thrust levels higher than the reference combustor at the same A_4 have lower station 4 entropy levels and are considered Second Law (entropy limit) violations. Thus the value of P_{STEP} resulting in the same A_4 stream thrust as that of the reference combustor is considered to represent the entropy limit value. Figure 17 provides such pressure ratio comparative results over a wide range of A_4/A_2. The lower curve represents an isentropic expansion from constant-area reference combustor ($A_3 = A_2$) results. The upper curve represents the entropy limit levels of P_{STEP}/P_2 for

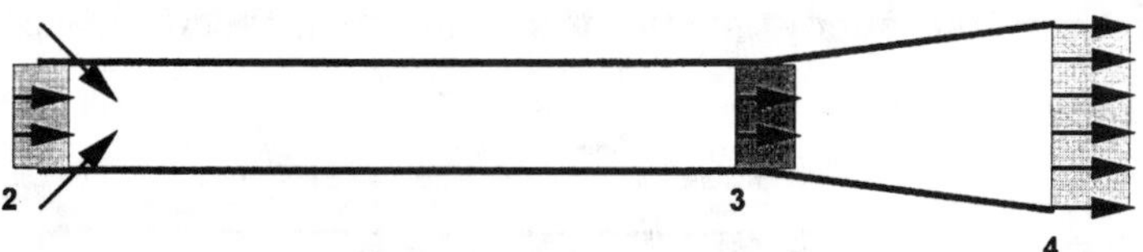

1-D REFERENCE COMBUSTOR MODEL:

Entrance air & fuel conditions identical to STEP COMBUSTOR
Constant area (A_3=A_2) frictionless combustion
Isentropic Expansion from A_3 to A_4
A_4 of REFERENCE COMBUSTOR=A_3 of STEP COMBUSTOR

REFERENCE COMBUSTOR STREAM THRUST PARAMETER:

$[\phi/(P_2A_2)]_{REF} = (P_4/P_2)(A_4/A_2)\{1+V^2/(RT)\}_4$ used to represent the maximum exit stream thrust for given air & fuel entrance conditions

Fig. 16 Thrust potential criteria.

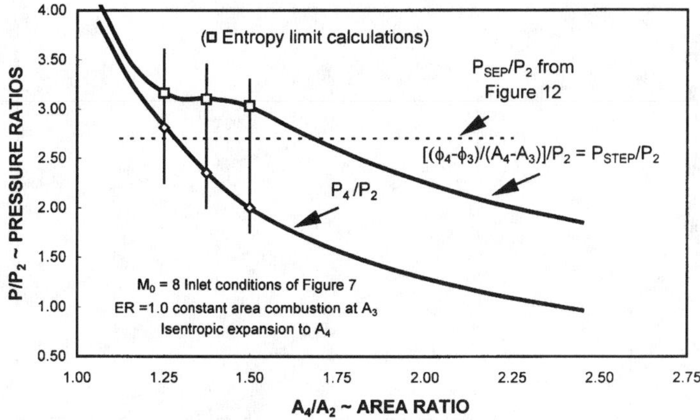

Fig. 17 Reference combustor parametric results.

step combustors. Thus, a step combustor at A_3/A_2 with a step wall force equal to the product of the A_4/A_2 one-dimensional pressure value and the step area (A_3/A_2) will have the same exit stream thrust as the reference combustor. Second Law (entropy limit) operating points for the Fig. 15 combustor area ratios are indicated by the square symbols. From superposition of the BL separation limit, we can again see for these step combustors at these operating conditions that the entropy limit step pressures are higher than permitted by the BL separation limit.

Figure 18 compares the exit stream-thrust levels for each of the three step combustor area ratios parametrically as a function of P_3/P_{STEP} and P_{STEP}/P_2. The $A_3/A_2 = 1.25$ and 1.375 step combustor calculation results were isentropically expanded to $A_4/A_2 = 1.5$ to permit comparison at the same overall area ratio of the largest area ratio ($A_3/A_2 = 1.5$) step combustor. Exit stream-thrust values for each of the three step combustors at the common overall area ratio of 1.5 ($\varphi_{1.5}$) were nondimensionalized by the exit stream thrust of the constant-area reference combustor after isentropic expansion to $A_4/A_2 =$ 1.5 ($\varphi_{1.5\ REF}$). Note from Fig. 18a that high values of P/P_{STEP} are indicators of performance loss and that increasing step area ratio (A_3/A_2) results in increased loss. Thus use of steps larger than required to avoid significant combustion-induced BL separation upstream of station 2 results in unnecessary stream-thrust losses. Figure 18b quantifies for these conditions the corollary that increased step pressure represents increased stream thrust as was immediately obvious in Fig. 14 from examination of the stream-thrust equation.

The stream-thrust results of Fig. 18 (at the common overall area ratio of 1.5) are replotted as a function of step combustor exit entropy (S_{STEP}) relative to the constant-area reference combustor exit entropy (S_{REF}) in Fig. 19b, illustrating a consistent loss in stream thrust with entropy increase. Whereas minimum combustor exit momentum loss (~1%) is achieved with the lowest step area ratio ($A_3/A_2 = 1.25$), recall from Fig. 15 that the lowest

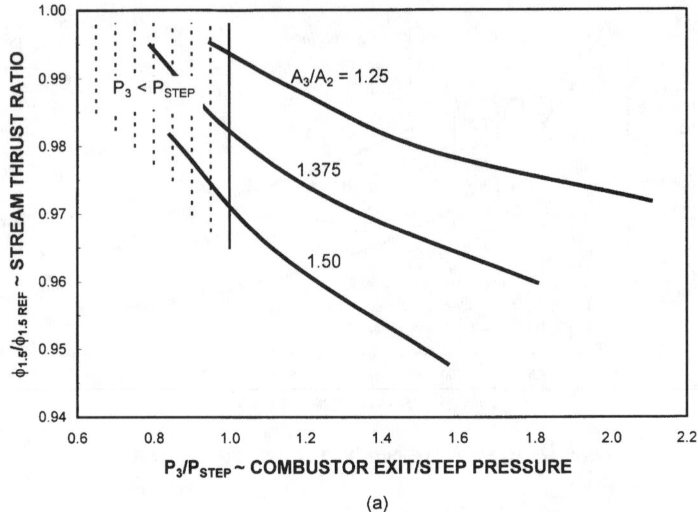

(a)

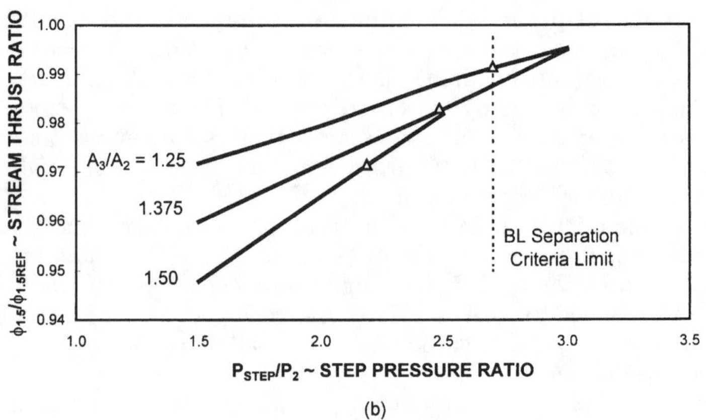

(b)

Fig. 18 Step combustor exit stream-thrust solutions.

step area ratio is also the most likely to separate the BL upstream of station 2 as a result of step pressure at the separation limit (no separation margin). Consequently, selection of $A_3/A_2 = 1.375$ at less than roughly another 1% in combustor stream-thrust loss achieves nominally 9% in step pressure separation ratio margin (Compare the potential operating point indicated by the triangular symbol in Fig. 15 on the curve of $A_3/A_2 = 1.375$ instead of the potential operating point indicated by the triangular symbol on the curve of $A_3/A_2 = 1.25$.). Fig. 19a relates the entropy penalty as a function of step pressure ratio or the corresponding combustor pressure ratio. As before, for a particular A_3/A_2 the most probable operating conditions are considered to

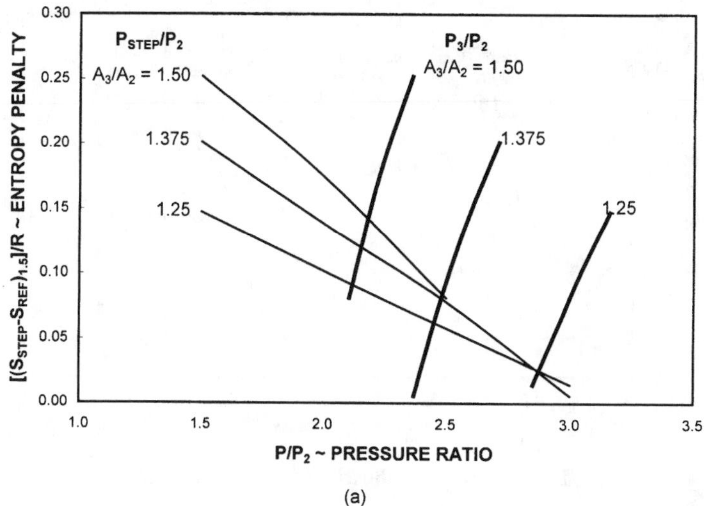

(a)

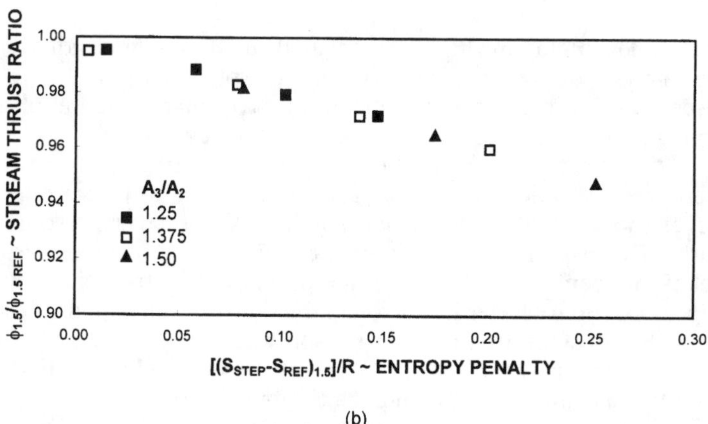

(b)

Fig. 19 Step combustor entropy penalty.

be above the intersection of these two pressure ratio lines (above $P_{STEP} = P_3$) for those P_{STEP} levels not exceeding the BL separation limit P_{SEP}.

Thus, the step combustor can be a design solution when BL separation is likely to occur in a constant-area combustor. The step area increase provides 1) a mechanism for accommodating wall forces resulting from fuel injection and combustion pressures $P_{STEP}>P_2$ without separating the inlet BL upstream of station 2 and 2) relief in reducing the overall combustion induced pressure rise. The resulting step force is a source of combustor exit momentum increase, but the reduced overall combustor pressure level represents reduced net heat release coupled with entropy increase. The objective is to determine the minimum step size that avoids BL separation upstream of

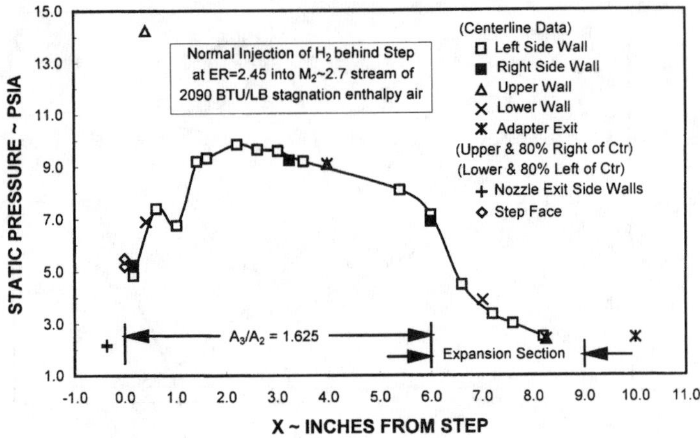

Fig. 20 Step combustor $P(x)$ data sample.

station 2 while maintaining high exit stream thrust. One-dimensional control volume analysis, although not providing an explicit solution for the step pressure, is useful in initial assessments of performance potential and in identifying preliminary geometry for more detailed evaluation.

Figure 20 is a sample step combustor pressure distribution from Ref. 10. The step pressure ratio is nominally 2.4, and the combustor entrance pressure level (nozzle exit sidewall data) indicates normal supersonic entry conditions. Downstream of the immediate region of the fuel injectors, there is excellent agreement of the centerline pressure data from all four walls. The supersonic flow accelerates through the downstream expansion section and exits the constant-area adapter section without being back pressured. The fuel normally injected ($\beta = 90°$) and at the relatively high ER can be expected to contribute to the high peak combustor pressure rise as a consequence of increased initial fuel/shock interactions immediately downstream of the step, while also assuring near instantaneous penetration into the air stream.

Reasonably high step pressure ratios have been experimentally demonstrated over a wide range of operating conditions. Figure 21 is a sample of data from Ref. 2, 10, and 11. The highest A_3/A_2 data set represents a circular duct, the smallest a two-dimensional duct with a step on one wall, and the middle two sets a two-dimensional duct with stepped upper and lower walls. In all cases combustor heat loss from the water cooling and use of cold hydrogen can be expected to have resulted in lower exit stagnation enthalpy than to be expected at flight conditions with a regeneratively fuel cooled combustor. Furthermore, it is important that the highest values of pressure ratio for each data set in Fig. 21 not be misconstrued as limiting values. In all of these cases station 2 combustor entrance pressure levels indicated typical supersonic entry conditions.

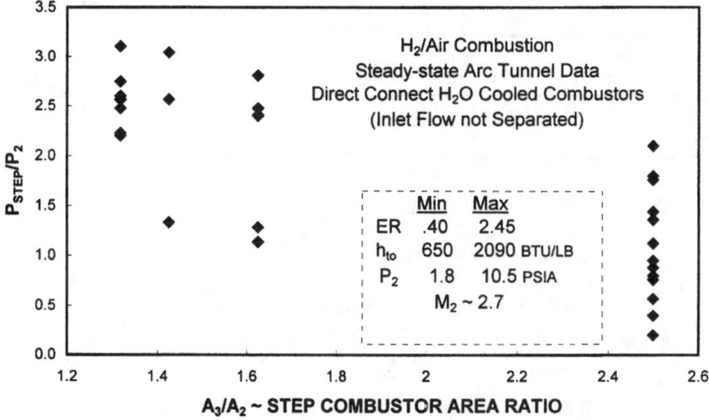

Fig. 21 Step pressure ratio data samples.

B. Isolator Combustors

Another design solution when BL separation is likely to occur in a constant-area combustor is the use of an inlet isolator[12] in conjunction with an increased area combustor. The increased area combustor could be a step combustor, a divergent-area wall combustor, or a combination. Whereas divergent-area combustors need not be of the constant pressure type, such combustors are used here for illustration because of their simplicity of calculation. Figure 22 shows a simplified variation of such an approach. The isolator section is intended to provide the length required for precombustion increase of the inlet pressure to the combustor pressure level without unstarting the inlet. Typically, such backpressuring of the viscous inlet flow is accomplished by a shock train (close-coupled series of shock-wave BL interactions). The isolator is designed to contain this shock train, thus preventing it from extending further upstream into the inlet compression system. The higher the combustor back pressure, the greater the throttling required, and consequently the smaller is the flow area A_{2A} required to satisfy the conservation equations at the isolator exit.

Figure 23 provides isolator/combustor matching results based on the simple model described in Fig. 22. Required isolator stream-tube and matching combustor exit area ratios are plotted as a function of overall isolator/combustor pressure ratio. These results, and all subsequent isolator/combustor calculations in this chapter, use the same $M_0 = 8$ (Fig. 7) inlet conditions (station 2) and ER = 1 fuel injection conditions that were used for the step combustor. The reduction in isolator stream-tube area ratio with increasing combustor pressure ratio implies increasing profile distortion, which recalling the profile line flow characteristics depicted in Fig. 8 also implies entropy increase. Although isolator losses will be less at the lower levels of combustor pressure ratio, recall from Fig. 5 that such reduced pressure operation can be expected to reduce the net heat release.

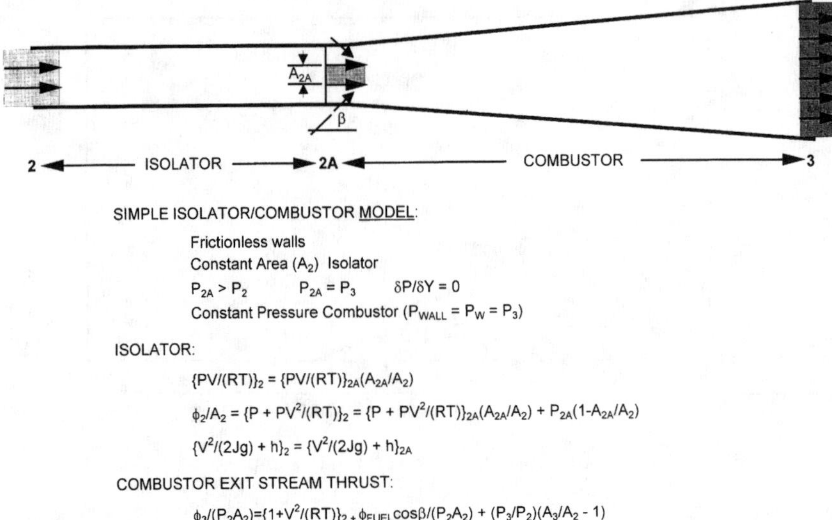

Fig. 22 Isolator/combustor schematic.

The isolator-induced increase in combustor entrance flow entropy as a function of combustor pressure ratio is illustrated in Fig. 24. At the higher combustor pressure ratios the entropy increase from the isolator represents nominally a 38% increase over the inlet station 2 level at no benefit in CR increase. Although it is required in the constant-area frictionless isolator model of Fig. 20 that station 2 and station 2A stream thrust be constant, note that for the resulting simple exit profile at station 2A the $P_{2A}(A_2\text{-}A_{2A})$ component is not recoverable as a wall force in the combustion process as

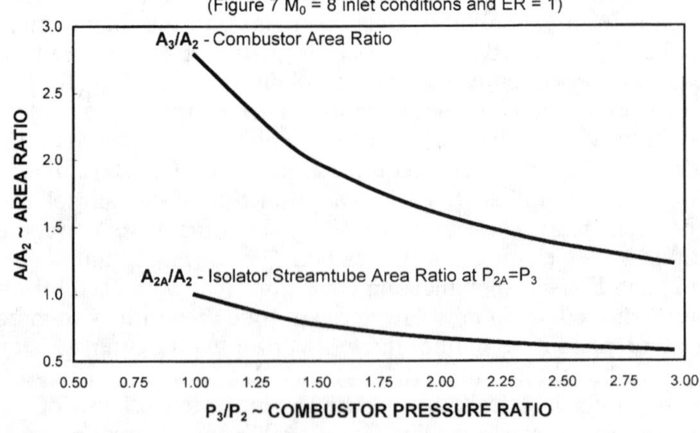

Fig. 23 Isolator/constant pressure combustor matching.

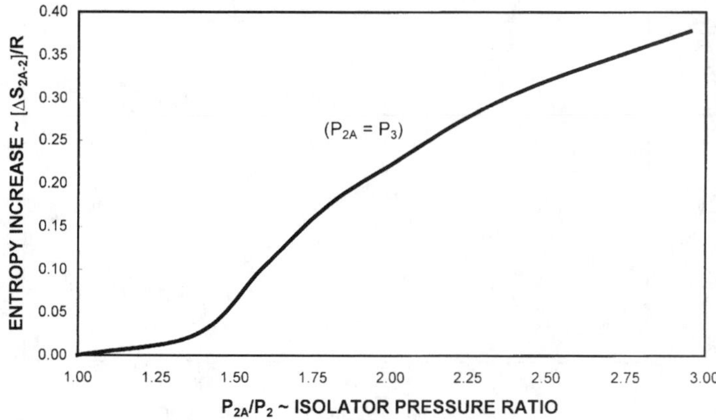

Fig. 24 Isolator entropy characteristics.

would be the case had there been hard wall contraction from A_2 to A_{2A}. The amount of this inferred wall force stream-thrust loss $\Delta\phi = P_{2A}(A_3\text{-}A_{2A})$ nondimensionalized by the station 2 stream thrust is depicted in Fig. 25 as a function of the combustor pressure ratio.

Isolators with constant pressure combustor performance results are summarized in Fig. 26 and 27, again using the reference combustor approach. For operation at low values of isolator precombustion pressure rise, the resulting low-pressure combustion process results in significantly higher entropy levels than that of the inviscid constant-area reference combustor (identical station 2 airflow conditions and identical fuel injection). With increasing isolator precombustion pressure rise, the combustor exit entropy level approaches that of the reference combustor, suggesting that the isolator flow

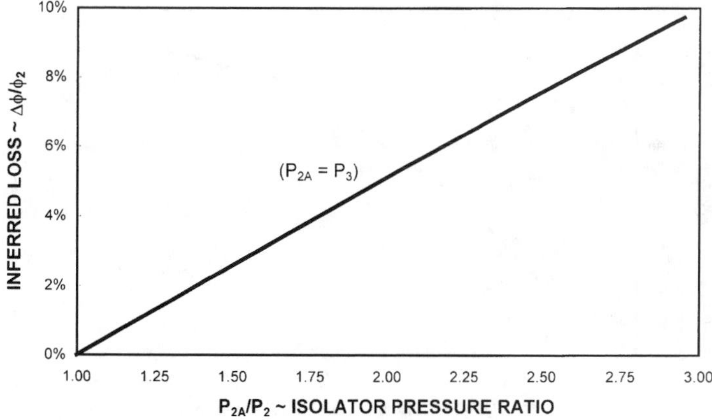

Fig. 25 Isolator stream-thrust loss approximation.

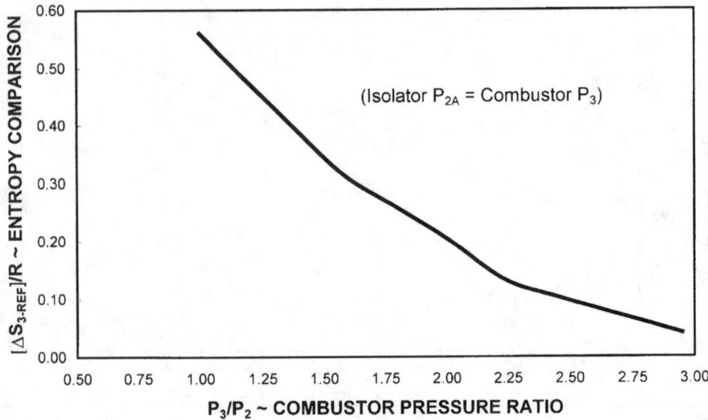

Fig. 26 Constant pressure entropy penalty.

entropy increase was an effective tradeoff relative to overall combustor performance. The latter is more evident in the Fig. 27 stream-thrust ratio comparison. Because the largest A_3/A_2 for the isolator with constant pressure combustion was 2.8 (Fig. 23), the combustor exit flows for the lower combustor area ratios (represented by the lower combustor exit pressure ratios) were isentropically expanded to $A_4/A_2 = 2.8$. The resulting exit stream-thrust values were then compared against that of the reference combustor, also isentropically expanded to $A_4/A_2 = 2.8$. Consistent with the entropy increase results of the preceding figure, increased precombustion pressure rise permits operation at combustor exit stream-thrust levels within a few percent of the reference combustor.

Step combustor and isolator with constant pressure combustion results are compared in Fig. 28. For this comparison the Fig. 18 combustor results

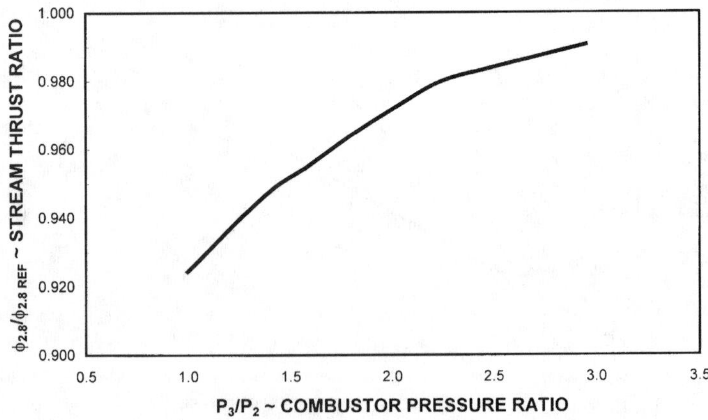

Fig. 27 Isolator constant pressure stream-thrust penalty.

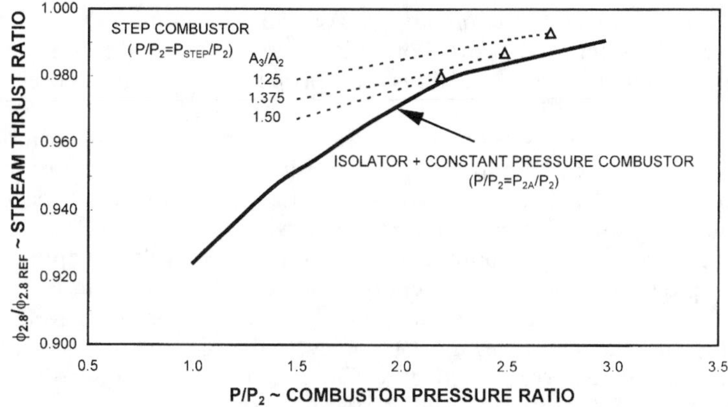

Fig. 28 Combustor concept stream-thrust comparison.

were isentropically expanded to the $A_4/A_2 = 2.8$ value for comparison with the Fig. 23 isolator with constant pressure combustion results. (The abscissa in Fig. 28 is P_{STEP}/P_2 for the step combustor and P_3/P_2 for the isolator with constant pressure combustor.)

Whereas the Fig. 28 comparison may suggest a slightly higher potential performance for the step combustor at these example conditions, such results by themselves are generally not adequate for selecting one approach over the other for a particular application. Such simple calculations, however, can be useful for quickly narrowing down the geometry and operating conditions for more detailed study and higher level system tradeoffs. These introductory calculations have not dealt with the realities of achieving the different fuel injector characteristics required for either the step combustor or the isolator with constant pressure combustion that provide the assumed penetration and mixing (100% as assumed in these calculations). In addition, for the case of the step combustor approach, the level of step wall force actually achieved (key to its performance) is an issue requiring more work. Similarly, in the case of the isolator with constant pressure combustion, the realities of the controlled heat release and area distribution that achieve the constant pressure characteristics need to be examined. Reference 12 provides a procedure for estimating isolator length requirements based on a comprehensive test database, which can then be factored into engine thermal protection and weight trades. Furthermore, for both approaches the impact of such geometry and potential concept selection on engine performance at other flight conditions requires evaluation.

IV. Fuel Injection Basics

To a large extent, for given inlet conditions the net heat release achieved in a scramjet combustor is driven by the efficiency and effectiveness of the fuel injection. Efficiency is reflected in the degree of fuel/air mixing achieved; effectiveness is associated with minimization of the combustor exit stream-thrust losses incurred in the mixing process and the extent of the additional wall cooling or thermal protection risk associated with the fuel injector concept. Scramjet cycle calculations must properly represent the thermodynamics of both the efficiency and effectiveness of the fuel injector concept. Earlier discussion focused on stream-thrust balance across the combustor assuming 100% mixing and complete combustion not to complicate the fundamental issues then being addressed. In this section the focus is on understanding some fundamental implications of the fuel injection process and advantages and disadvantages of typical fuel injector design approaches.

A simple scramjet combustor energy balance is illustrated in Fig. 29. Of particular concern is the fuel enthalpy and the combustor heat loss through the walls. In a balanced scramjet engine cycle the fuel temperature is typically a result of the regenerative cooling required throughout the entire engine to ensure that no material temperature limits are exceeded. Consequently, the fuel enthalpy term in the combustor energy balance can be greater than the heat loss term through the combustor walls such that in the terminology of Fig. 29, $w_F h_{TF} - Q_{LOSS} > 0$. Thus proper simulation in a direct-connect scramjet combustor test requires that the fuel temperature be high enough for consistency with such a combustor energy imbalance. On the other hand, arbitrarily increasing the injected fuel temperature in a scramjet engine cycle simulation can lead to apparent free energy and thus higher than possible calculated performance levels when the injected fuel temperature used is higher than the maximum value set by materials property limits. For simplicity in the examples of the preceding sections, we assumed that $w_F h_{TF} - Q_{LOSS} = 0$. Such an approach, however, does not eliminate the need for specifying a fuel temperature when the fuel is injected with an axial component of velocity. Recall from Fig. 14 the term $\varphi_{FUEL} \cos \beta/(P_2 A_2)$ in the combustor stream-thrust balance. A fuel injection angle of 60 deg was used in the examples presented, and to permit the required stream-thrust determination, the further assumption was made that the gaseous H_2 at ER $= 1$ was injected at nominal conditions of 2000°R and $M = 2$. The net implication then is that such fuel injection conditions reflect the combustor heat loss through

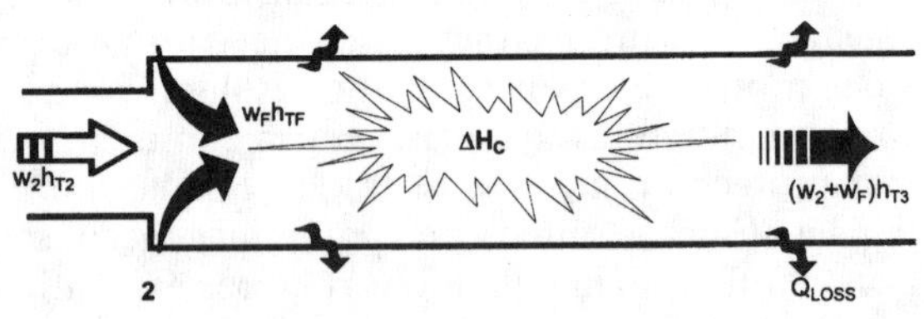

Fig. 29 Combustor simple energy balance.

$$w_2 h_{T2} + w_F h_{TF} + w_2 \Delta H_C - Q_{LOSS} = (w_2 + w_F) h_{T3}$$

if $w_F h_{TF} - Q_{LOSS} = 0$, then:

$$h_{T3} = (h_{T2} + \Delta H_C)/(1 + F/A)$$

the walls. The point is that establishing the correct fuel injection temperature typically requires a system energy balance.

Proper penetration of the injected fuel into the airstream is a necessary but not sufficient prerequisite for achieving adequate mixing. Typical fuel injector concepts for accomplishing this can be classified into two general types: 1) wall jets and 2) in-stream injectors, but there are many variants of both types. Hypermixers can be considered to be a subset of in-stream injectors. Each type has its unique advantages and disadvantages. Figure 30 provides generic illustrations.

A. Wall Jets

Wall jets, although minimizing intrusion of the combustor flow path, result in a relatively complex flow structure in the immediate vicinity of the jets as characterized in Fig. 31 for normal injection through a round hole on a flat plate. The resulting regions of high-pressure gradient on the wall near the jet are also a source of increased local heat flux. (Reference 13 provides infrared thermal images of surface temperature levels and variations resulting from normal sonic injection of hydrogen into high enthalpy M = 3 airflow.) Many simplified models have been proposed for approximating the complex flow structure depicted in Fig. 31. Such models can be a source of understanding the principles of the basic flow phenomena, which in turn can provide a basis for formulating design approaches for increased penetration,

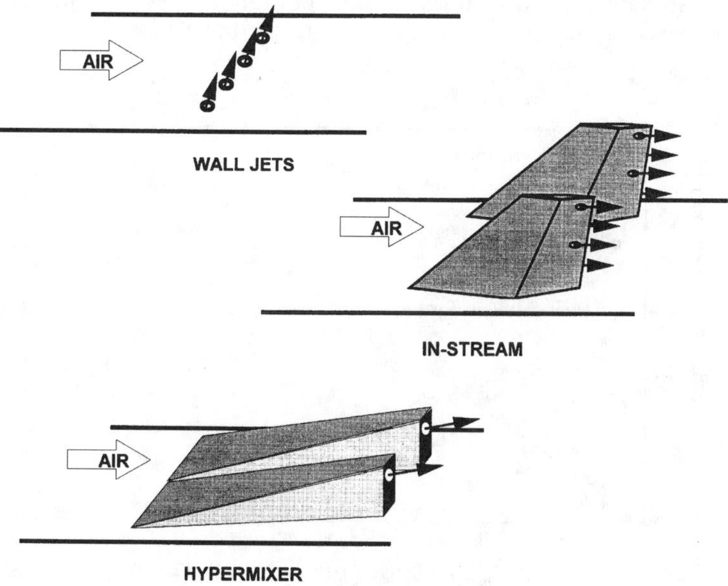

Fig. 30 Fuel injector concepts.

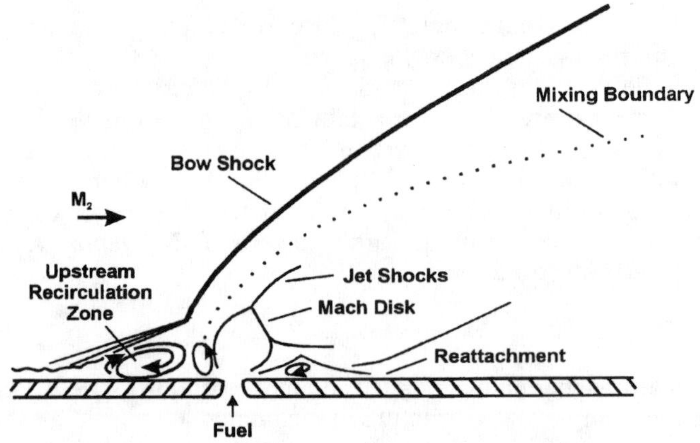

Fig. 31 Wall jet flow structure characterization.

prioritization of test variables in parametric experiments, related design tradeoffs, and initial design geometry for subsequent CFD analyses.

Experimental data correlate reasonably well when plotted in the $(Y/D^*)^2 = Kq_R$ formulation as illustrated in Fig. 32 using the schlieren-determined penetration height results of Ref. 11 ($q_R = q_{\text{FUEL}}/q_{\text{AIR}} = [\rho V^2]_{\text{FUEL}}/[\rho V^2]_{\text{AIR}}$ and for any injector geometry D^* is the sonic hole diameter based on $D^* = (4\Sigma A^*/\pi)^{0.5}$, wherein for an injector configuration with multiple holes, the total area of all of the holes is used.) The Fig. 32 data points were obtained with a flat plate aligned with $M = 3.25$ wind-tunnel flow. Gaseous helium, air, and gaseous nitrogen were injected through holes in replaceable insert

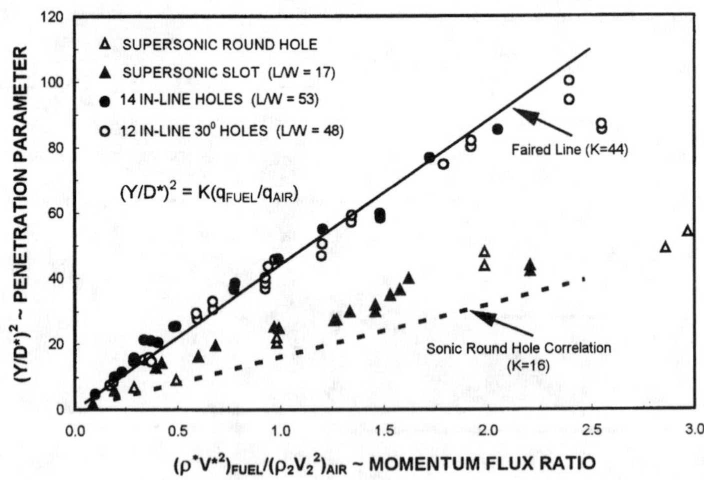

Fig. 32 Wall jet penetration characteristics.

plates. The dashed Sonic Round Hole Correlation line in Fig. 32 is used as a reference for comparing results obtained with different hole geometry. In addition to the normal sonic hole data of Ref. 11, this reference line is also based on the normal sonic hole concentration profile data of Ref. 14 and 15. In spite of differences in defining and determining penetration height, the penetration characteristics for normal injection with round sonic holes from all three sources were found in Ref. 11 to agree well with the simple expression $(Y/D^*)^2 = 16q_R$. Thus for round sonic holes penetration increases with q_R and/or D^* (increased fuel injection per hole).

For the same amount of fuel injection per hole, the supersonic round hole data in Fig. 32 indicate increased penetration over sonic injection. Potential contributing factors to such improvement are illustrated in Fig. 33 based on a $\gamma = 1.4$ injectant. For hole exit Mach numbers (M_J) less than about 1.48, the exit momentum per unit area is slightly higher than that for sonic flow (increase in effective q_R). For a given injectant flow rate (fixed D^* and q^*/q_R) different values of exit static pressure (P_J) are available depending on M_J. A lower exit static pressure for the same q^*/q_2 may require less shock-induced increase in air pressure and less external injectant expansion to achieve a common static pressure. In addition, isentropic expansion inside the injector should result in a smaller flow diameter than for the same static pressure achieved in external expansion because the latter typically involves over expansion shocks and is not isentropic. Thus supersonic expansion inside the injector should result in a smaller A_P for the same static pressure, which should then result in a higher value of K. In attempting to match the injectant pressure to a representative pressure in the disturbed flow (thus

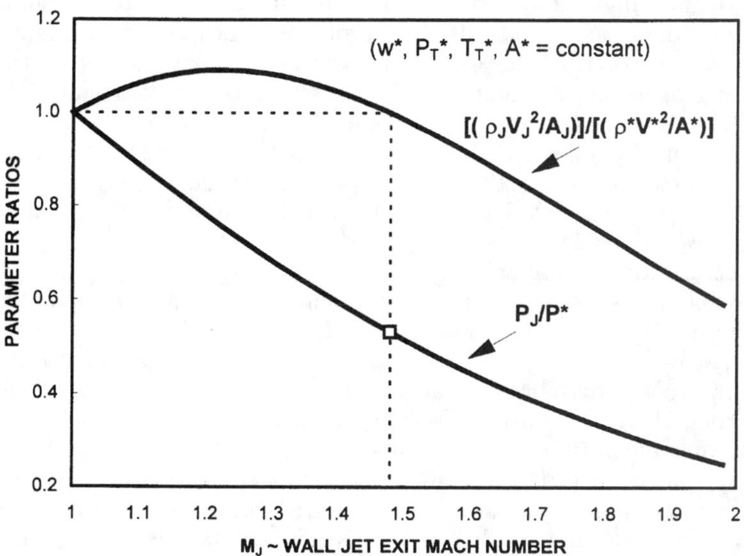

Fig. 33 Wall jet supersonic exit flow characteristics.

minimizing over and under expansion effects), Ref. 8 suggests approximating the effective back pressure as the static pressure behind a normal shock in the undisturbed flow when the BL momentum thickness is small relative to the jet diameter and as the pressure required to separate the BL when the momentum thickness is larger than the jet diameter.

Another approach for increasing the mass flux per fuel stream tube frontal area is illustrated in the Fig. 32 example. Based on the sonic round hole correlation at constant q_R, if the hole diameter is doubled then the penetration should increase by $2^{0.5}$ at the expense of four times the fuel injected per hole. If, instead, four holes of the undoubled diameter are axially aligned, then for the same four times the small single hole fuel flow there is no increase in the fuel stream tube frontal area. To a first approximation, this might be considered analogous to achieving either four times the effective q_R with essentially the small hole diameter or a factor of two reduction in the frontal area for the same amount of flow. Either way, the penetration would be expected to increase by a factor of two (instead of $2^{0.5}$). The implied generalization is then $(Y/D^*)^2 = 16nq_R$ where n is the number of holes axially aligned.

On this basis for the 14 in-line hole (1/16-in.-diam jets on 1/4-in. centers) results in Fig. 32, the predicted K value would be nominally 60 compared to the faired data result of $K = 44$. Although the 14 in-line sonic hole result of $K = 44$ represents almost a factor of two improvement over a single sonic round hole, it falls 23% below the simple reasoning value of $K = 60$. A contributor to this shortfall (aside from overly simplistic reasoning) may involve hole spacing. When in-line jets are spaced too far apart, the benefit of increased momentum flux per unit frontal area is likely to diminish, approaching single hole performance in the limit. On the other hand, too close an axial spacing could also be a problem in terms of constraining fore and aft expansion (high-pressure jet proximity) at the expense of the frontal area increasing lateral expansion. For the Ref. 11 sonic hole data, $P^*/P_2 = (3.25/1)^2\ q_R$ or nominally $10.6q_R$ (where P^* is the sonic injectant static pressure and P_2 is the undisturbed $M = 3.25$ air static pressure). Thus at a q_R of approximately 1.15, the injectant pressure P^* meets the Ref. 8 criteria of matching an effective back pressure corresponding to the normal shock pressure at $M = 3.25$.

Necessity for increased expansion to the side may be a contributing factor in the Fig. 32 supersonic slot results not being that much better than that of the single supersonic round hole. Such slots represent the limit in close axial hole spacing, i.e., none at all. Thus it appears that optimization of in-line hole spacing could potentially provide additional improvements over the reference round hole penetration capability reflected in the Fig. 32 data.

Probably the most impressive results in Fig. 32 are those for the 12 in-line 30 deg sonic holes (1/16-in.-diameter on 1/4-in. centers). Whereas the indicated penetration is comparable to the high level of the 14 in-line normal sonic holes, they additionally provide a significant axial momentum contribution that becomes particularly important for high Mach-number scramjet performance. The sweep effect of the inclined holes probably reduces the force on the fuel jet stream tube (lower C_D), while also increasing the time and distance for achieving a given penetration height compared to normal injection. The latter, however, could induce an earlier initiation of lateral mixing at the expense of penetration. Thus, again, there appears to be an opportunity for improved wall jet penetration over the Fig. 32 results through optimization of both hole inclination and in-line spacing; perhaps also coupled with the right amount of supersonic internal expansion.

B. In-Stream Injectors

From a purely penetration standpoint in-stream injectors have to be the right choice. Fuel jets can be vertically spaced as required to achieve desired local F/A distribution across the entire combustor height, matching even nonuniform flow air profiles. For fuel injected from the sidewalls or out of the top of in-stream injectors, much of the previous discussion of wall jets applies. For fuel injected axially downstream from the trailing edge, full axial fuel momentum is also achieved but at what can be a significant mixing rate penalty. At high combustor velocities such axial injection can lead to increased distance (combustor length) to achieve adequate mixing. The foregoing benefits of in-stream injectors must be evaluated against the drag on the injector struts and the ability/risk and complexity/cost of meeting their challenging cooling requirements.

Reference 14 provides force-balance-determined drag levels for nine injector strut configurations. Geometry variables include leading-and trailing-edge sweep, leading-edge radius, thickness ratio, and percent chord at maximum thickness ratio. Figure 34 schematically typifies injector installation in the wind-tunnel test facility. Available access panels on the wind-tunnel sidewalls were used for installing the different strut configurations, resulting in a 90-deg plane-of-rotation reference shift relative to typical strut installation on an inlet with horizontal surface ramps. Cold flow data were obtained at $M = 2$, 2.5, and 3 over a range of dynamic pressures from nominally 800 to 1000 psf. The base of the 5-in. span struts were immersed in the sidewall BL from the 12 by 12 in. wind-tunnel cross section. Top and

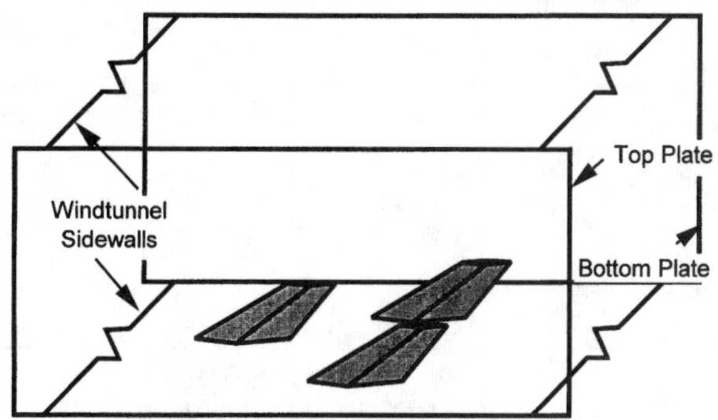

Fig. 34 In-stream injector schematic.

bottom plates of a 9-in. span permitted bypassing the top and bottom wall wind-tunnel BL to minimize sidewall effects on isolated injector drag data and on multiple injector interaction.

Figure 35 is from Ref. 14 force measurements with six injectors installed and converted to drag coefficients based on total strut projected area and 1008-psf wind-tunnel dynamic pressure. Such results are in reasonable agreement with the linear wing theory for the 7.5% thickness ratio (t/c) geometry but are a factor of two to three higher for the 10% and 16% t/c injectors. Tests conducted with only two injectors installed indicated essentially the same drag per injector. For six injectors installed, based on the Fig. 35 drag coefficients, combustor exit stream-thrust losses ($\Delta\phi/\phi_2$) of approximately 3.4, 1.6, and 1.2%, respectively, can be calculated for the 16, 10, and 7.5% t/c injectors. With helium injected through the fuel injector orifices at the rate required for a stoichiometric mixture of hydrogen, measured drag levels of the lower t/c struts were reduced by about 8 to 9%.

Speculating on the impact on injector drag with hydrogen injection from similar struts but with aft surfaces redesigned providing attributes similar to a step combustor as discussed in Sec. III. A is informative. Consider the Fig. 35 results to be comprised of forward and aft surface pressure forces and friction. Assuming combustion effects on the friction drag are small with respect to the total strut drag, then when the average combustion back pressure P_B on such step-type aft surfaces equals the average forward surface pressure P_F the strut drag is clearly just the friction drag. When P_B is greater than P_F, the net strut drag is less than its friction drag. Assuming strut aft surfaces with combustion emulate the characteristics of a step combustor, then $P_B/P_F > 1$ should be achievable, and the net strut drag should be near zero or less. Thus although strut drag is a concern with such in-stream injectors, it may not have to be a major issue.

The major issues then become those associated with thermal protection

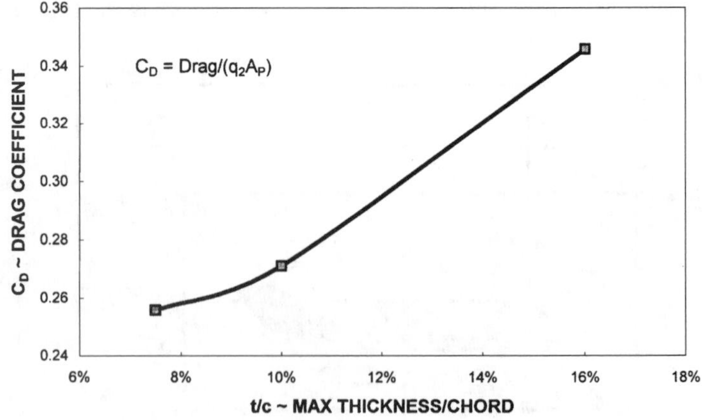

Fig. 35 In-stream injector drag characteristics.

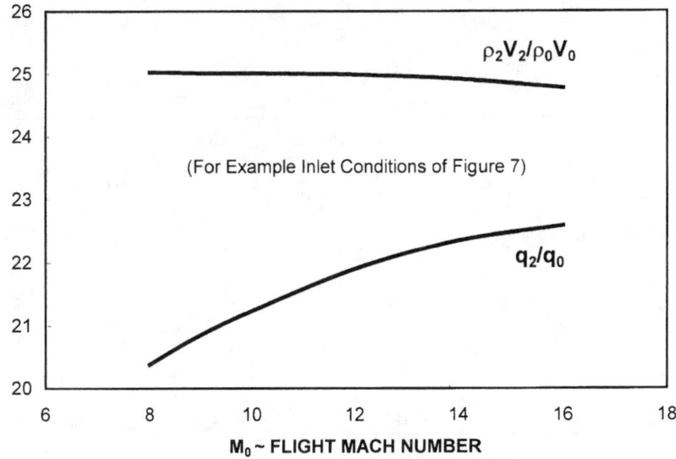

Fig. 36 Injector drag and heat flux parameters.

and structural integrity. Figure 36, based on the inlet characteristics used in Sec. II.B indicates q_2/q_0 dynamic pressure ratios in excess of 20 (which for the 1000-psf trajectory used means $q_2 > 20{,}000$ psf) and $\rho_2 V_2/\rho_0 V_0$ values of 25. Interpreted in terms of the challenge normally associated with providing adequate thermal protection for the scramjet inlet cowl leading edges, these are rather formidable numbers. With the exception of whatever benefit is actually achieved from the swept leading edges, for the same material and leading-edge diameter of the inlet cowl this can result in 25 times greater heat flux. Relief by increasing the diameter of the strut injector leading edge can lead to significant drag increase. Aside from concern about ability to cool drag acceptable injector strut leading-edge diameters, the total fore and aft injector surface area cooling requirements, while at significantly reduced heat flux levels compared to the leading edge, can be a source of increased fuel cooling ER requirements.

C. Hypermixers

Figure 37 depicts hypermixers based on the fuel injector designs tested in Ref. 15. In contrast to the in-stream strut type fuel injectors of Ref. 14 discussed in Sec. IV.B, the low 10.3-deg ramp angle and increased ramp surface area rather than the nominally 60-deg swept sharp leading edges reduce the stagnation heat flux problem. The axial downstream fuel injection permits maximum momentum recovery from the fuel (the importance of which increases with flight Mach number). Vortex shedding from the corners, step-type recirculation behind the aft surfaces, and impingement of the reflected ramp shock just aft of the injectors are relied upon to enhance the otherwise relatively slow mixing of the axially injected M = 1.7 hydrogen in Ref. 15. Maximum blockage of the duct was 11.6%. The upstream airflow was $M = 2.0$ at controlled total temperatures from 2500 to 4000°R.

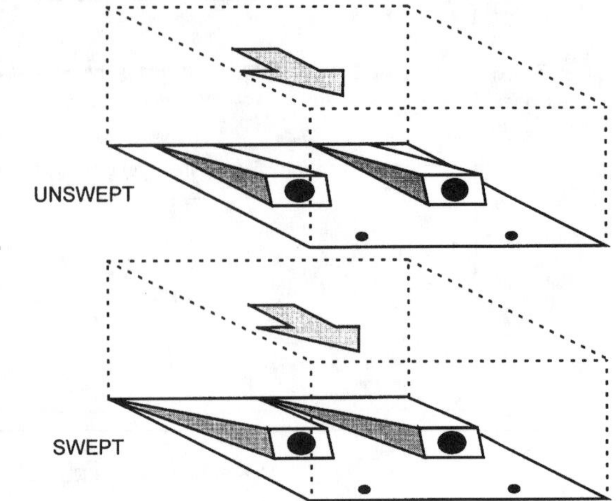

Fig. 37 Hypermixer geometry.

To limit upstream propagation of downstream disturbances, a constant-area section (isolator) of length 2.65 times the duct height was added upstream of the injectors. A divergent section was added downstream of the injectors; the divergent section was 31.8 duct heights in length. Reference 15 provides measured wall-pressure distributions over a wide range of ER and corresponding calculated combustion efficiency levels based on one-dimensional analyses. Combustion efficiencies calculated for the swept ramp injector of Fig. 37 were higher than those calculated for the unswept ramp injector. Furthermore, combustion performance achieved when perpendicular wall jet injection was added downstream and combined with the swept injector was reported to nearly equal the mixing performance projected for a normal wall jet based on an empirical mixing correlation.

The fuel injector ramp and reflected shocks are a source of noncombustion entropy increase and unlike well-designed inlet compression ramps do not provide permanent increased inlet contraction for the losses incurred. Intentional creation of vortices for mixing enhancement also introduces additional axial momentum losses. With lateral spacing between the aft step-type surfaces, drag offsetting combustion-induced pressure rise can spread laterally to lower pressure regions and thus can be expected to be less than with conventional step-type combustors. Thus the challenge of the hypermixer concept is to achieve adequate mixing of near axially injected fuel in reasonable combustor lengths without offsetting drag increases.

D. Mixing

Simplistically, penetration positions the fuel from each injector port into the airstream such that it can subsequently mix within its assigned spatial control volume to the intended homogeneous F/A. Viewed somewhat more realistically, portions of the penetration and mixing processes overlap each other. Heat release is an ongoing process as hydrogen and oxygen molecules at adequate temperature and pressure find each other (mix) as a function of time and distance, and their resulting reaction spatially changes the pressure and temperature for the unmixed hydrogen and oxygen molecules. Such mixing processes, particularly for supersonic flow, are not likely to be predicted well by simple analytical models. Consideration of the latter, however, can be instructive in understanding trends and characteristics.

Analytical and experimental incompressible flow results are presented in Ref. 16 beginning on p. 496 for the wake far downstream of a row of round bars, indicating that significant length is required for such shear-layer mixing. The wake velocity profiles for the latter results are very similar to those measured in free shear-layer flow.[17] The growth rate (mixing) of free shear-layer compressible flow when nondimensionalized by the incompressible value has been found to correlate well with the convective Mach number, M_C.[18–20]

$$M_C = (U_2 - U_c)/a_2$$

where

$$U_C = (a_2 U_{\mathrm{FUEL}} + a_{\mathrm{FUEL}}\, U_2)/(a_2 + a_{\mathrm{FUEL}})$$

with a denoting the speed of sound and assuming $U_{\mathrm{FUEL}} < U_2$. (The initial rationale for these equations is provided in Ref. 21.) Reference 17 experimental results are typical of other results indicating that as M_C increases from 0 the compressible shear-layer growth rate rapidly decays from the incompressible value. Typically, at M_C of nominally 1.0 or more, the compressible shear flow growth rate can be on the order of only 30% of the incompressible flow value. Consequently, scramjet combustors primarily dependent on free shear-layer mixing (such as unenhanced simple axial fuel injection) can be expected to require an unacceptably long length to achieve acceptable mixing.

Fuel injector designs that introduce curved shock waves and local BL separation increase the rotationality of the flow and thus for the losses induced provide near-field mixing increases over that of free shear flow. Combustion in the near and far field, along with induced vortices, also increases mixing relative to free shear flow. Referring back to step combustor results of Fig. 20 based on the measured combustor peak pressure level, one might conclude that mixing was either essentially complete or at the stage where further combustion of any unmixed fuel was dependent on free shear flow mixing and thus impractical at nominally 3 to 4 in. downstream of the step. Measured from the step, this corresponds to an axial distance nondimensionalized by fuel jet lateral spacing (x/w) of 12 to 16.

Currently, previous test results and CFD analyses provide the best basis for evaluating fuel injector effectiveness and the implications of different combustor mixing length.

V. High Mach-Number Implications

At the higher flight Mach numbers, many of the problems precipitating the switch from ramjets to scramjets return as major issues limiting scramjet practicality. The increased flow-path velocities associated with increased flight Mach number coupled with no-slip wall conditions result in significantly increased heat transfer levels. Consequently, fuel cooling ERs can be considerably higher than required for stoichiometric combustion, resulting in reduced I_{SP} levels and increasing fuel volume requirements. Thus once again, ability to thermally protect, and the weight and cost consequences of such thermal protection become limiting factors.

In addition, net thrust per unit airflow is reduced by the diminishing ratio of net heat release from combustion relative to freestream stagnation enthalpy as illustrated in Fig. 38 from Ref. 22. As a consequence, the net thrust becomes a small difference between two large numbers; the gross thrust and the ram drag.

$$F_N = F_G - F_R = (w_A + w_F)V_4/g + (P_4 - P_0)A_4 - w_A V_0/g$$

which, assuming $(P_4 - P_0)A_4$ is small relative to the other terms at hypervelocity conditions, yields a net specific thrust of

$$F_N/w_A \sim [(1 + F/A)V_4 - V_0]/g$$

From the energy equation

$$(h_{T4}/h_{T0})\,(1 + F/A) = 1 + \Delta H_C/h_{T0}$$

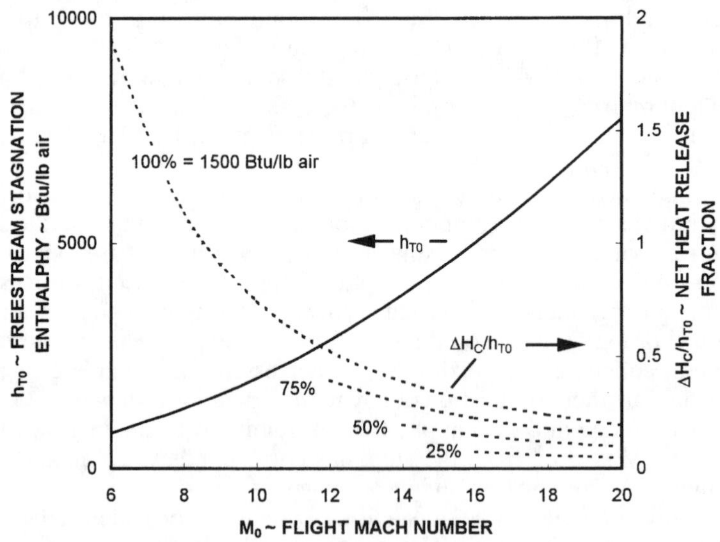

Fig. 38 Diminishing impact of heat addition.

Also assuming $h << h_T$ at hypervelocity conditions, then

$$V_4 \sim [2Jg(h_{T4})]^{0.5} \quad \text{and} \quad V_0 \sim [2Jg(h_{T0})]^{0.5}$$

resulting in the approximation

$$F_N/w_A \sim (V_0/g)[[(1 + F/A)(\Delta H_C/h_{T0} + 1)]^{0.5} - 1]$$

Using a nominal maximum net heat release (ΔH_C) of 1500 Btu/lb of air and a nominal F/A of 0.03, Fig. 39 illustrates a significant lapse rate of F_N/w_A with increasing hypervelocity (which also approximates the lapse rate of I_{SP}).

An important implication of Fig. 39 is the requirement for increased inlet capture area (A_1), with increasing flight velocity. Assuming the vehicle drag coefficient and external drag projected area are essentially constant at hypervelocities, then for cruise at a particular V_0

$$A_1 = (C_D A_P/2)(V_0/g)/(F_N/w_A) => A_1 \sim V_0/(F_N/w_A)$$

where A_P is the vehicle external drag projected frontal area. Figure 40 results based on this approximation imply that substantial increases in inlet capture area are required to provide scramjet thrust levels high enough to meet cruise drag requirements at increasing hypervelocities.

Figures 38–40 have included parametric results for four levels of net heat release. High sensitivity of F_N/w_A and A_1 to reduction in ΔH_C at hypervelocities are indicated. At such conditions reductions in ΔH_C are typically because

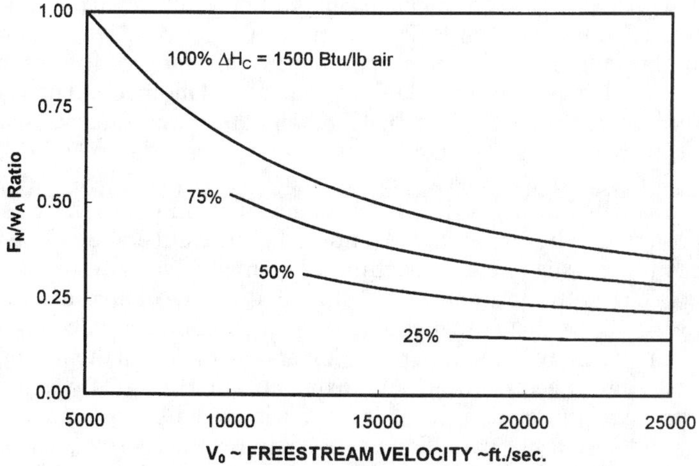

Fig. 39 Net thrust per unit airflow characteristics.

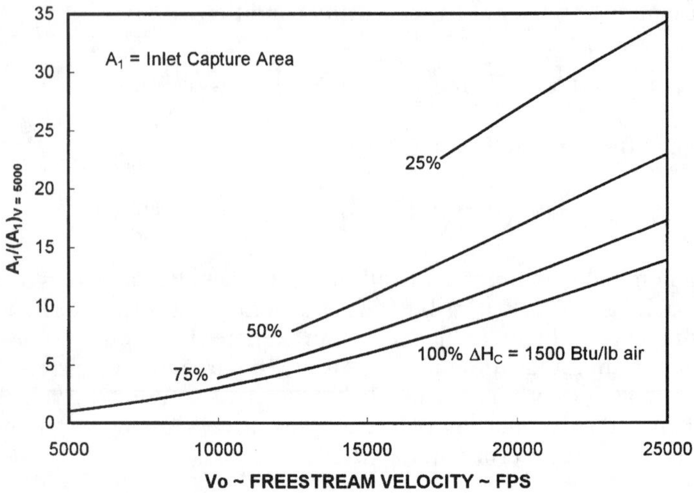

Fig. 40 Inlet capture area growth trends.

of 1) increasing difficulty in achieving adequate F/A mixing (with low fuel injector losses and a significant axial component of fuel momentum) and 2) increased endothermic (heat absorbing) reactions in the combustor.

A. Mixing

As in the case of lower flight Mach numbers, proper penetration by the fuel across the combustor remains a prerequisite for good mixing. At hyper-velocity conditions the degree of fuel axial momentum associated with achieving good penetration can have a major impact on engine performance as illustrated in Fig. 41 from Ref. 22 wherein 100% mixing was assumed for such equilibrium H_2/air chemistry parametric calculations. The top dashed and three solid lines quantify the effect of axial fuel momentum on I_{SP} when hydrogen fuel is injected 0 (axial), 45, 60 and 90 (normal) degrees with no intrusive injector drag loss. Under such conditions axial injection almost doubles the I_{SP} compared to normal injection. If good penetration and mixing of axially injected fuel is to be achieved, an intrusive device such as a strut is likely to be required. Depending upon strut geometry and blockage, calculations suggest the resulting drag might typically be at a level comparable to between 1 and 2% of the ram drag. Assuming such to be the case, Fig. 41 shows that if fuel could be injected nonintrusively at 45 or 60 deg and could achieve similar penetration and mixing, then the incentive for using instream injectors to achieve axial injection is greatly diminished. Furthermore, if the avoidance of instream injectors permits an increased dynamic pressure flight path, finite rate combustor and nozzle calculations such as those shown in Fig. 42, which use the kinetics of Ref. 23, suggest that recombination kinetics are likely to provide enough incremental I_{SP} to make

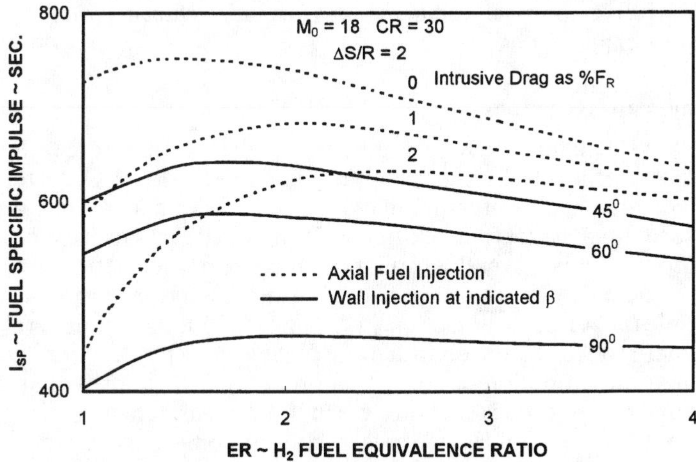

Fig. 41 Fuel injector drag and injection angle impact.

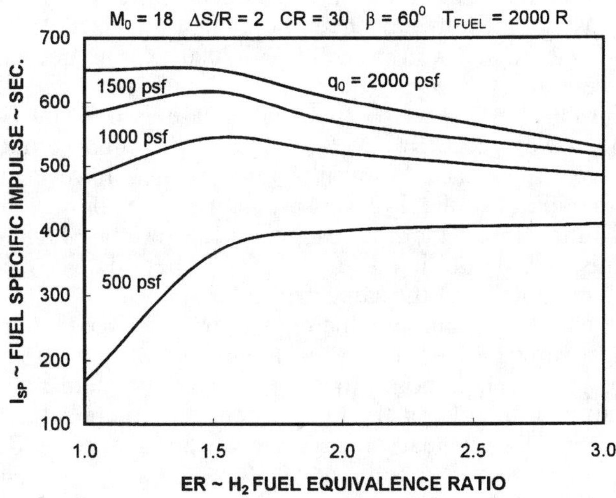

Fig. 42 Finite rate kinetics impact on Mach 18 I_{SP}.

the nonintrusive approach the higher performance choice as well as having the least mechanical design risk.

B. Combustor Reactions

At relatively high pressure (~ 1 atm) and modest temperature (~ less than 3000 °R), the product of H_2/air combustion is essentially H_2O. As temperature is increased and/or pressure decreased such as with increasing scramjet flight Mach numbers (refer back to Fig. 7), molecular dissociation of H_2, O_2, and N_2 occur, followed by the formation of other products such as OH and NO. Consequently, not only is less hydrogen and less oxygen available for the exothermic (heat release) reactions forming H_2O, but the net heat release is also further reduced by the endothermic (heat absorbing) reactions of dissociation and the formation of additional products. (The impact of such exothermic and endothermic reactions was initially illustrated in Fig. 5.) At such conditions the amount of H_2O formed relative to some maximum theoretical value is not a good indicator of combustion efficiency because it does not account for the total impact of the endothermic reactions. Net heat release itself can be used in place of a combustion efficiency; absolute values in terms of Btu/lb H_2 or Btu/lb air work well. Units of Btu/lb air have been used throughout this section; part of the reasoning being that with ER > 1 as typical for thermal protection (regenerative cooling) at hypervelocity flight conditions, values of the net heat release per pound of H2 can vary considerably with noncombustion related ER changes even if the associated variations in absolute net heat release are small.

Following the established heat of reaction sign convention of negative for exothermic and positive for endothermic, Table 2 based on the thermodynamics and chemistry of Ref. 5 illustrates contributions of the heats of formation for the resulting combustion products to the net heat release. For ease of comparison, results have been nondimensionalized with respect to the net heat release at 1 atm pressure and 2000 °R. The net values in the right-hand column are simply the sum for the six tabulated combustion products. Thus, at $P_3 = 1$ atm ΔH_C (net heat release) is 32.7% at $T_3 = 5500$°R of the value at $T_3 = 2000$°R. For $T_3 < 4000$°R the endothermic reactions generally have little impact. As was illustrated in Fig. 7 however, at hypervelocity flight conditions inlet T_2 levels are likely to be approaching 4000°R, implying significant endothermic reaction at the resulting higher T_3 levels. Other chemistry and additional reactions will effect details such as those summarized in Table 2 but the same principles apply.

As flight vehicle Mach number increases, flow residence time in a particular sramjet combustor length may no longer be adequate to permit the reactions to go to completion, prompting the use of finite rate chemistry instead of equilibrium chemistry. One-dimensional finite rate calculations provide the Fig. 43 net heat release rate results at three different inlet pressure levels and inlet conditions of $M_2 = 6.0$ and $T_2 = 3474$°R. The freestream stagnation enthalpy level corresponds to $M_0 = 17$. Complete mixing was assumed for all three combustor entrance pressure levels; thus the Fig. 43 results are reaction-rate limited. For $P_2 = 1$ atm, nominally 95%

Table 2 Relative net heat release parametrics (~% ΔH_C)

T_3 ~ R	H_2O%	H%	O%	N%	OH%	NO%	Net%
			P_3 = 1 atmos				
2000	−100.0	0.0	0.0	0.0	0.0	0.0	−100.0
3000	−99.8	0.0	0.0	0.0	0.0	0.0	−99.8
4000	−96.4	0.2	0.0	0.0	0.2	0.2	−95.8
5000	−76.2	4.2	1.7	0.0	1.2	0.9	−68.2
5500	−55.3	13.1	5.8	0.0	2.2	1.5	−32.7
			P_3 = 0.5 atmos				
2000	−100.0	0.0	0.0	0.0	0.0	0.0	−100.0
3000	−99.8	0.0	0.0	0.0	0.0	0.0	−99.8
4000	−95.5	0.2	0.1	0.0	0.2	0.2	−94.8
5000	−70.4	6.5	2.7	0.0	1.5	1.0	−58.7
5500	−46.0	19.5	8.8	0.0	2.5	1.6	−13.6
			P_3 = 0.2 atmos				
2000	−100.0	0.0	0.0	0.0	0.0	0.0	−100.0
3000	−99.7	0.0	0.0	0.0	0.0	0.0	−99.7
4000	−93.9	0.4	0.1	0.0	0.3	0.2	−92.9
5000	−60.8	11.3	4.8	0.0	1.8	1.1	−41.8
5500	−32.5	31.7	14.6	0.0	2.7	1.7	18.2

Equilibrium H_2/air ~ ER = 1.

of the net heat release is achieved approximately 8.0 in. downstream of fuel injection. At the lower combustor inlet pressures the slower reaction rates require significantly longer combustor lengths to achieve 95% of their lower net heat release levels.

C. CFD Solution Results

The simple design approaches of the preceding sections were used to establish the geometry for the different Mach 17 scramjet combustors of this section. CFD was used in evaluating the performance of these combustors to permit attainment of more detailed results. The CFD solutions were obtained using the RPLUS3D code of Ref. 24 modified to include an algebraic turbulence model, specified wall-temperature boundary conditions, and the kinetics of Ref. 23.

With ER levels in excess of stoichiometric for Mach 17 scramjet combustors, we now mention how mixing efficiency is defined. Excess oxygen and global mixing definitions (Ref. 25) are two useful concepts when evaluating CFD results.

The excess oxygen definition results in 100% mixing efficiency when that amount of oxygen (in all forms) required to burn a stoichiometric amount of fuel is mixed with that amount of fuel. Mixing of any excess fuel above the stoichiometric level is ignored in this definition, which is intended to be

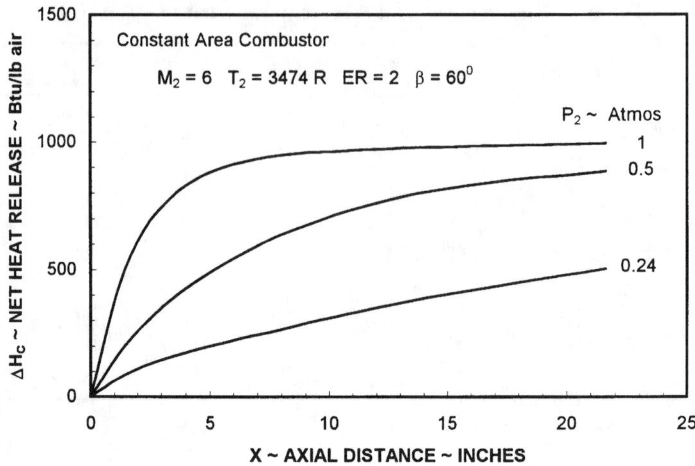

Fig. 43 Finite rate kinetics impact on net heat release.

an indicator of the degree of mixing for only the amount of oxygen that can potentially react with the fuel.

$$\eta_{M,\mathrm{EO}} = 1.0 - \text{Excess Oxygen/Oxygen Present}$$

Excess oxygen is that oxygen (in all forms) that has not yet mixed with any fuel. At a given axial plane in the combustor, the mixing efficiencies within each computational node are mass averaged.

Global mixing is based on the more conventional way of evaluating mixing; 100% mixing is not achieved until all of the fuel and all of the air become a homogeneous mixture. Thus if fuel is injected in the amount of ER = 2, then 100% global mixing requires ER = 2 at every computational node.

Different definitions of the F/A boundary are also frequently used. Such differences can become important in comparing results from different investigators. Fundamentally, the boundary should be based on what is considered to be the smallest amount of fuel that constitutes a meaningful presence of that fuel. Sometimes the lean flammability limit is used. The theoretical lean flammability limit for hydrogen and air is 4% by volume, which translates into a mass fraction of 0.288%. From propulsive thrust considerations a somewhat higher mass fraction may be appropriate. A value of 1% mass fraction (roughly ER = 0.34) of hydrogen in all forms (H_2O, OH, H, etc.) has been used in this section.

CFD-determined F/A boundaries can also be grid-size dependent. Even with the same F/A boundary definition a coarser grid will generally overpredict fuel penetration and mixing whereas tighter mesh for the same conditions should be closer to the truth for a particular CFD code.

Combustor geometry and initial flow conditions used for subsequent step combustor wall jet calculations comparing round vs slot wall jet injector

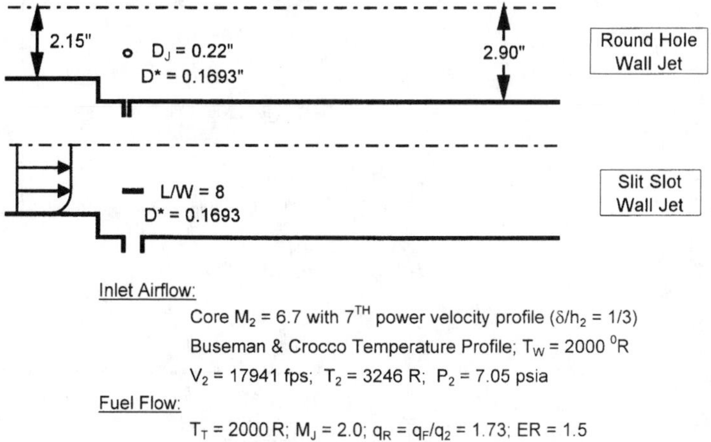

Fig. 44 Study combustor geometry and inlet conditions.

results are summarized in Fig. 44. Excess oxygen mixing efficiency is seen in Fig. 45 favor the slot injector over the round hole by about 18%. Comparing the reacting and nonreacting slot injector results of Fig. 46 implies that combustion enhanced the nonreacting mixing by about 17% (comparable calculations indicated about 11% for the circular hole). Net heat release and contributing products are summarized for the slot wall injector in Fig. 47. Consistent with Table 2, endothermic reactions resulting in the dissociation of H_2 and O_2 significantly offset the exothermic reactions leading to the formation of H_2O, resulting in relatively low net heat release at these condi-

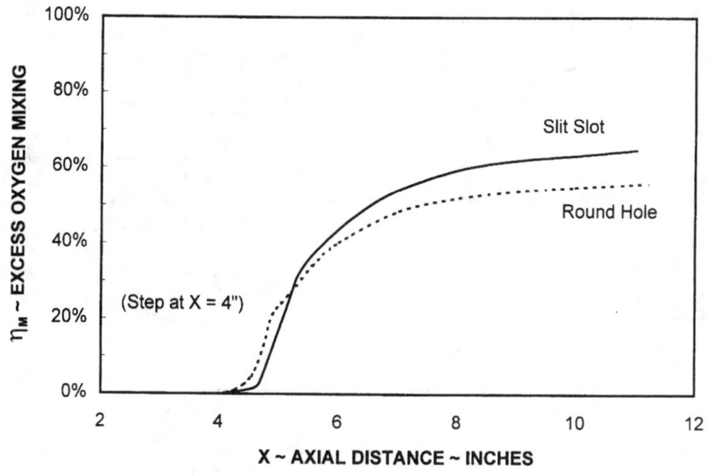

Fig. 45 Mixing efficiency comparison.

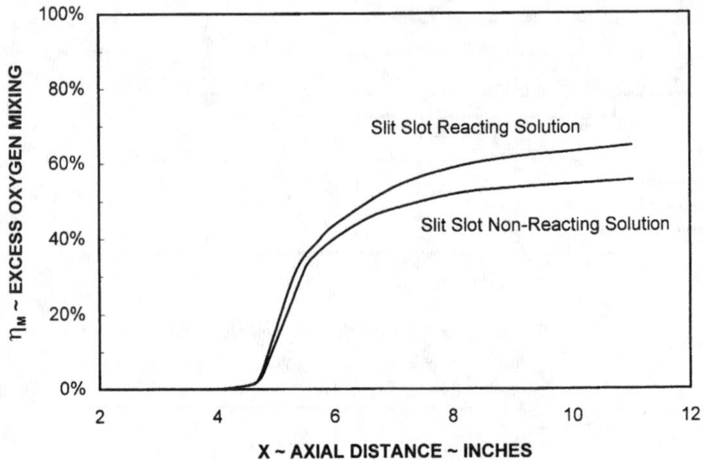

Fig. 46 Reaction impact on mixing.

tions. The equilibrium chemistry solution for this same problem is compared with the finite rate chemistry result in Fig. 48, indicating a substantial decrease in the net heat release because of the finite rate chemistry at these conditions. Returning to finite rate chemistry results, Fig. 49 compares the impact of slot injector angularity on excess oxygen mixing efficiency. With a 60° injection the same mixing efficiency as with a normal injection is achieved in the same distance (but at a lower rate) while also providing 50% axial fuel momentum. Axial (tangential) injection while providing 100%

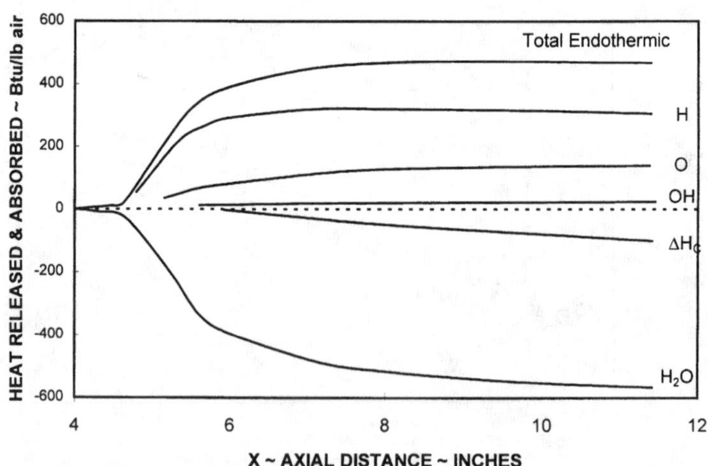

Fig. 47 Slit-slot net heat release summary.

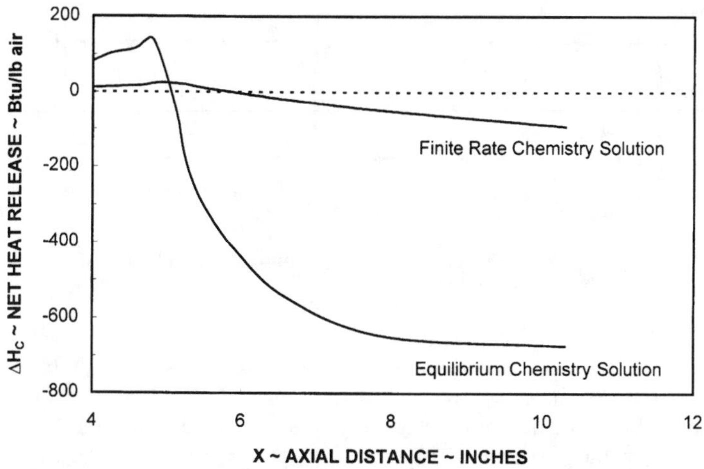

Fig. 48 Slit-slot net heat release comparison.

axial fuel momentum achieves only 75% of the 60° injector mixing efficiency while requiring twice the axial distance.

Fuel injector and combustor geometry from Ref. 26 are depicted in Fig. 50. The combustor step contains a slanted section permitting angled injection of cold hydrogen (T = 360°R) at M = 2.0 through slit-slot configured wall jets. A constant area extends 6.4 in. upstream of the step, providing BL growth for nominally M_2 = 6.0 and T_3 = 3474°R, Mach 17 stagnation enthalpy combustor entrance flow. One percent H_2 mass fraction contour-boundary-solution results are summarized in Fig. 51. (Note that axial station numbers represent distance from the leading edge of the constant-area duct

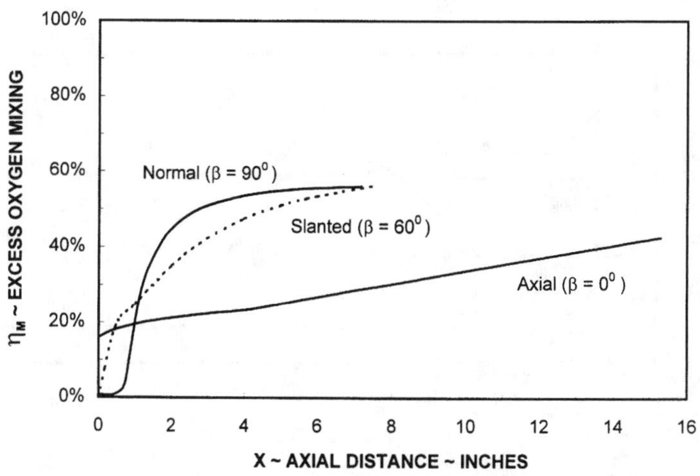

Fig. 49 Slit-slot angled injection mixing comparison.

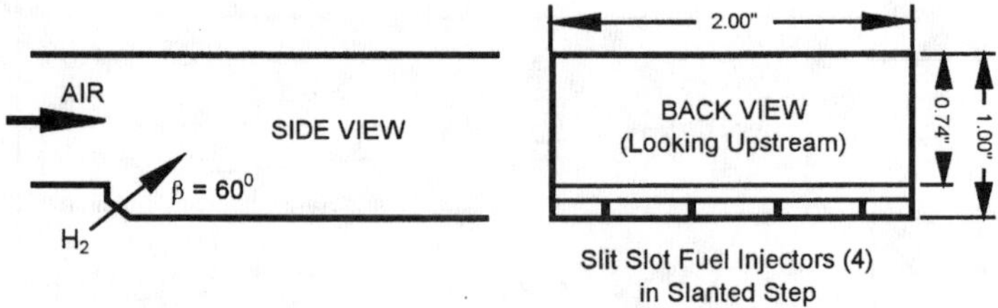

Fig. 50 Slanted slit-slot geometry schematic.

upstream of the step, not distance downstream of the step.) In an axial distance of approximately two combustor heights (2 in.) downstream of the step, about 90% of the combustor height is fueled to at least 1% hydrogen mass fraction along the fuel injector centerline compared to about 70% midway between adjacent fuel jets. Excess oxygen and global mixing efficiencies are compared in Fig. 52. At a distance of 10 combustor heights (10 in.) downstream of the step, the calculated excess oxygen mixing efficiency reaches 90%, compared to 85% at an axial distance of only six combustor heights (6 in.) downstream of the step.

These results suggest that although there may be design solutions for achieving good mixing efficiency and conserving a reasonable axial component of the fuel momentum without resorting to instream fuel injectors, flow chemistry remains a major barrier to acceptable high Mach number scramjet performance. The atmospheric properties of temperature and pressure tend to limit hypervelocity combustor inlet conditions to those with a propensity for endothermic reactions and significant finite rates of reaction. Thus with increasing flight velocity the chemical energy available from supersonic combustion of hydrogen is further reduced relative to

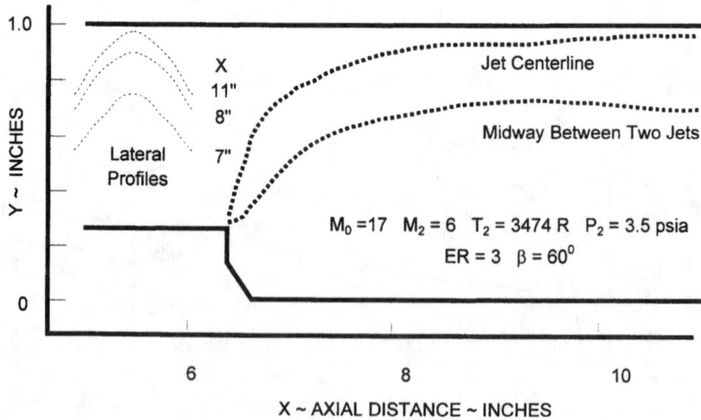

Fig. 51 Finite rate 1% H_2 mass fraction boundaries.

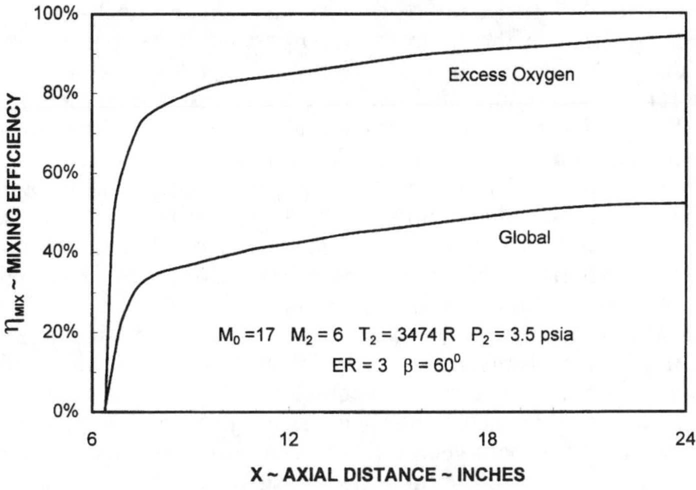

Fig. 52 Finite rate mixing efficiencies.

increasing values of freestream stagnation air enthalpy and scramjets are no longer as attractive an airbreathing propulsion solution as at lower flight Mach numbers.

D. Design Philosophy

Reflecting on the many challenges of high Mach number scramjet design implied in the preceding figures and text, the fundamental design philosophy and design approach suggestions from Ref. 26 are presented. The design philosophy is summarized in the following:

low inlet $\Delta S/R$
+ CR based on large A_4/A_2 vs dissociation losses
+ mixed = burn kinetics
+ high *net* axial fuel momentum with good mixing
+ large nozzle with high C_{FG}
= high efficiency components carefully matched!

While such a summation may seem intuitively obvious or overly simplistic, it provides a basis for identifying and developing specific design approaches.

Based on the resulting design philosophy, we suggest some specific design approaches: thin and swept leading edges; external ramp shocks inside cowl lip; weak shocks traded against viscous losses; integrated inlet and combustor; CR $\approx$ 30; $P_2 = 7$ psia or greater; $\beta_{\text{FUEL}} = 45$ deg and nonintrusive; high ER determined by cooling requirements; integrated combustor and nozzle; low angularity loss nozzle; and nozzle BL relaminarization. The importance of a low loss inlet cannot be over emphasized. In addition to all of the lower flight Mach-number reasons for high inlet performance, the impact on combustor and nozzle kinetics can be substantial at higher Mach-number flight conditions as seen in the finite rate combustion and expansion scramjet cycle calculation results of Fig. 53. Highly swept sharp cowl leading edges may be required to reduce inlet leading-edge losses at acceptable heat flux levels. With initial inlet ramp shock impingement inside the cowl lip, less cooling is required for such thinner leading edges. Such reduced cooling requirements might be translated into a permissible increase in flight-path dynamic pressure to reduce combustor endothermic reactions and enhance kinetic recombination in the nozzle. The increased ER levels required for regenerative cooling provide an opportunity to achieve adequate fuel penetration and mixing without resorting to in-stream fuel injectors. The use of swept cowl leading edges suggests consideration of using a similar sweep in the plane of the fuel injectors to help integrate the inlet and combustor. Caret-type compression surfaces as depicted in Fig. 54 would then seem to offer high potential for achieving such a swept propulsion design with reasonable scramjet flowpath geometry. Inlet capture geometry could be configured in a variety of different ways (using the same caret compression surfaces) to accomplish required vehicle/propulsion integration for a specific mission application. Nozzle geometry might be similarly swept to both aerothermally and geometrically integrate the combustor and nozzle. Such a highly swept and integrated flow path is much less likely to have crossflow-induced performance problems at high Mach-number flight conditions than possibly in the Mach 6 to 8 range; and if so, advantage should be taken of this as well as any other high-speed flow phenomena. Relaminarization of the BL nozzle flow may also fall into this latter category and could play a critical role in permitting high enough C_{FG} levels to be achieved with low exit flow angularity.

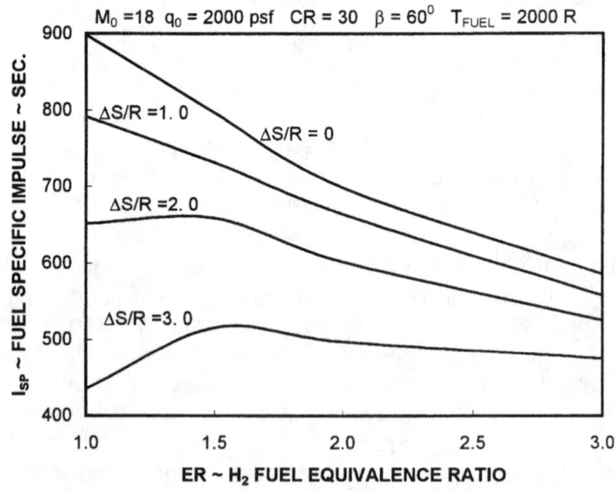

Fig. 53 Inlet efficiency impact on finite rate kinetics.

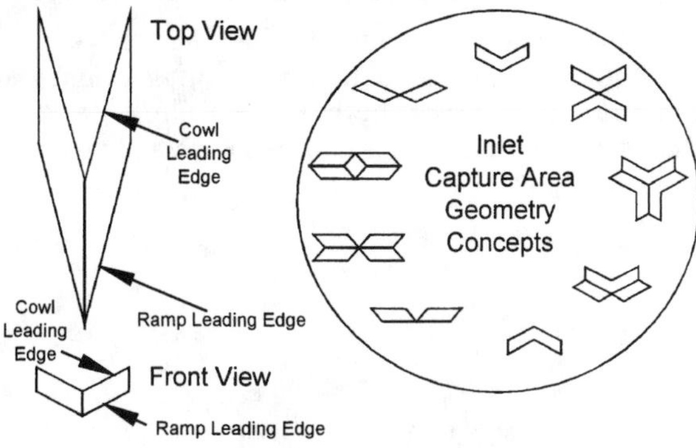

Fig. 54 Swept scramjet propulsion concept.

Figure 55 attempts to summarize Mach 18 scramjet performance potential based on these concepts. Selected CR, ER, nozzle area ratio, and fuel injection angle values used are consistent with the design philosophy and design approaches already discussed. Assuming a reasonable degree of success in implementing most of these design approaches, something on the order of 600 in I_{SP} is suggested as a realistic challenge on the fringes of the state of the art. Vehicle/propulsion integration constraints, operational requirements down to Mach 6 to 8, etc., would be expected to compromise high Mach-number scramjet design and thus its performance potential. Increased optimism, discovery, and invention will lead to higher projected values.

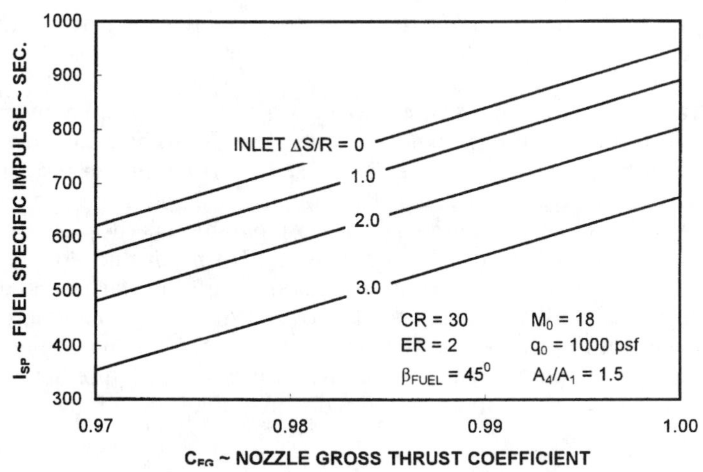

Fig. 55 Projected Mach 18 scramjet performance potential.

Appendix A: Inlet One-Dimensional Continuity and Energy Flow Solution

For simplicity, the assumption is made that the inlet captures full mass flow and that the inlet flow is adiabatic (radiation and heat loss due to wall cooling are ignored). Then using the main text Fig. 3 station notation, the required two conservation equations can be expressed as

Continuity:

$$\rho_2 V_2 A_2 = \rho_0 V_0 A_0$$

Energy:

$$h_T = V_0^2/(2Jg) + h_0 = V_2^2/(2Jg) + h_2$$

Combining the continuity equality with the equation of state provides

$$P_2/P_0 = (A_0/A_2)(R_2/R_0)(V_0/V_2)(T_2/T_0)$$

permitting the solution for the given M_0, P_0, T_0 as follows:

1) At a selected T_2, guess P_2 providing h_2, R_2 and a_2 from thermodynamic properties.

2) $V_2 = [2Jg(h_T\text{-}h_2)]^{0.5}$

3) At a selected A_0/A_2, check $P_2/P_0 = (A_0/A_2)(R_2/R_0)(V_0/V_2)(T_2/T_0)$ against guessed P_2 and iterate with new P_2 guess until solution is achieved.

4) $M_2 = V_2/a_2$

5) At T_0, P_0 and T_2, P_2, determine $\Delta S/R$ from thermodynamic properties.

Appendix B: Profile Flow Solution

Simple profile flow as indicated in Table 1 of the main text, although required to meet the condition $\delta P/\delta Y = 0$, provides the flexibility of not requiring uniform velocity across the duct. Such flexibility can be used to permit approximation of distorted flow profiles ranging from conventional boundary layers to the extreme of separated flow. Classical power law profiles such as $u/V = (y/\delta)^{1/N}$ can be used to approximate viscous flow characteristics of typical attached BL flow, or as an alternative a uniform core flow using the conventional BL integral parameters (displacement and momentum thicknesses) can be used. (Ref. 27) The profile flow example results in Fig. 8 in the main text are based on a uniform Mach-number core flow of area less than or equal to that of the duct. For adiabatic flow of perfect gas with constant specific heat ratio of 1.4, the one-dimensional conservation equations applied to a constant-area duct require between the uniform upstream condition X and the downstream profile flow C

Continuity:

$$P_X A_X M_X T_X^{-0.5} = P_C A_C M_C T_C^{-0.5}$$

Momentum:

$$P_X A_X(1 + 1.4M_X^2) = P_C A_C(1 + 1.4M_C^2) + P_C(A_X - A_C)$$

Energy:

$$T_X(1 + M_X^2/5) = T_C(1 + M_C^2/5)$$

which combined to provide the expression:

$$A_C/A_X = \left\{(1 + 1.4M_X^2)(M_C/M_X)\left[(1 + M_C^2/5)/(1 + M_X^2/5)\right]^{0.5} - 1.4M_C^2\right\}^{-1}$$

where A_C is the downstream core flow area in the constant duct area A_X. The entropy increase from flow condition X to flow condition C is then

$$\Delta S/R = 3.5ln(T_C/T_X) - ln(P_C/P_X)$$

Numerical results for the $M_X = 2$ inlet conditions of Fig. 8 in the main text are plotted here in Fig. B.1 as a function of M_C. An important consequence of $\Delta S/R$ being a monotonic function of M_C is that unlike Fanno and Rayleigh Line flow processes that can begin with either subsonic or supersonic flow and progress as far as the limiting sonic point, profile flow processes must always progress from a higher initial Mach number to a lower value to avoid Second Law of Thermodynamic's violation. Thus supersonic profile flows should not be "mixed out" to an "equivalent" one-dimensional uniform supersonic flow with the same mass flow, momentum, and energy. Also note

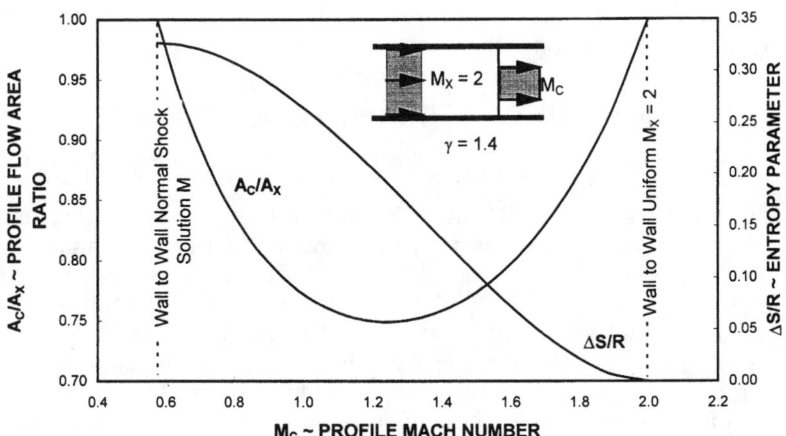

Fig. B.1 Example profile characteristics.

in Fig. B.1 that A_C/A_X is not a minimum at $M_C = 1.0$ because of the variation in $\Delta S/R$ with M_C.

Such profile flow approximations are not restricted to a perfect gas with constant ratio of specific heat as in the Fig. 8 and B.1 examples. Subsequent figures in the main text dealing with isolator flow are based on the thermodynamics and chemistry of Ref. 5 and the recast conservation equations as provided in Fig. 22.

Appendix C: Entropy Limit Concept

The lowest entropy increase process (ideal) for a given set of initial conditions as used in the main text has been characterized as that resulting from one-dimensional, inviscid flow in a constant-area duct with complete F/A mixing and fully reacted combustion. Thus for the same given set of initial conditions (regardless of A_3/A_2) flow with a lower entropy level implies a process more efficient (unlikely) than resulting from one-dimensional, inviscid flow in a constant-area duct with complete F/A mixing and fully reacted combustion.

Two ways for determining if an input wall force approximation has violated such a characterized limit are (1) direct entropy level comparison and (2) exit stream-thrust comparison at the same A_3/A_2. In the former, if the entropy level at A_3 for the combustor being analyzed is greater than the level of the characterized ideal, then no Second Law violation is considered to have occurred. The alternative of comparing the exit stream thrust can be accomplished by isentropically expanding the characterized ideal combustor exit flow to the same A_3 as that of the combustor being analyzed. If the exit stream thrust at the same exit area of the expanded ideal flow is greater than that of the combustor being analyzed, then no Second Law Violation is considered to have occurred. For simplicity of illustration, such ideal expansions for the examples in the main text were based on equilibrium flow.

The concept of the reference combustor (Fig. 16 of the main text) as used in evaluating thrust potential (Appendix D) is a natural outgrowth of entropy limit considerations.

Appendix D: Combustor Thrust Potential Concept

Simply stated, the best measure of scramjet combustor performance is the engine thrust subsequently produced for given combustor entrance airflow conditions at a specified ER. The combustor thrust potential as used here is a comparative value of combustor exit momentum based on common inlet inflow conditions and common fuel ER evaluated at a common exit area ratio greater than the station 2 inlet area. Its purpose is to provide an initial basis for ranking the potential of different combustor designs relative to resulting engine thrust level. Clearly, specific inlet airflow conditions can compromise combustor contribution to engine thrust potential as can the exhaust nozzle expansion process, underscoring the need for engine system performance component trade studies. Furthermore, combustor length, weight, and thermal protection requirements can also be important system level discriminators. Thus, the thrust potential of the scramjet combustor, although an important discriminator, is not the only factor in selecting one design over the other.

For the one-dimensional examples in the main text, evaluation of thrust potential is relatively straightforward. Inlet conditions are prescribed at station 2, the ER is set at 1.0 and H_2 at a given temperature is injected at 60 deg. The problem then, given these resources, is determining combustor geometry likely to permit the highest engine performance without unstarting the inlet or at least ranking those studied relative to achieving the latter. The defined one-dimensional combustor exit flow conditions for these examples greatly simplify the calculation of ideal equilibrium expansion to the common area ratio required for thrust potential comparisons of different combustor geometries. Note, however, that differences in even such idealized combustor exit flow conditions influence the split between equilibrium and frozen expansion in the exhaust nozzle. The latter can have a significant impact on exhaust nozzle gross thrust. Furthermore, the reality of non-uniform combustor exit profile expansion processes can also significantly impact the exhaust nozzle gross thrust, particularly for different profiles with close to the same integrated momentum.

Thus the concept of thrust potential, although simple in principle, can be quite complex when properly applied to the three-dimensional, viscous, finite rate chemistry environment of the scramjet engine.

References

[1]Waltrup, P. J., Anderson, G. Y., and Stull, F. D., "Supersonic Combustion Ramjet (SCRAMJET) Engine Development in the United States," *The 3rd International Symposium on Air Breathing Engines,* (Munich, Germany) March 1976.

[2]Kenworthy, M. J., Stanforth, C. M., and Colley, W. C., "Investigation of Instrumentation and Simulation Techniques for the Supersonic Combustion Process," AFAPL-TR-66-76, Oct. 1966.

[3]Anderson, G., Kumar, A., and Erdos, J., "Progress in Hypersonic Combustion Technology with Computation and Experiment," AIAA-90-5254, Oct. 1990.

[4]Huber, P. W., Schexnayder, C. J., Jr., and McClinton, C. R., "Criteria for Self-Ignition of Supersonic Hydrogen-Air Mixtures," NASA TP1457, Aug. 1979.

[5]Browne, W. G., and Warlick, D. L., "Properties of Combustion Gases; System: H2 - Air," General Electric, GE Rep. R62FPD-366, Nov. 1962.

[6]Shapiro, A. J., "The Dynamics and Thermodynamics of Compressible Fluid Flow, Vol. I," 1954, pp. 190–203.

[7]Billig, F. S., "Combustion Processes in Supersonic Flow," *Journal of Propulsion and Power,* Vol. 4, May-June 1988, pp. 209–216.

[8]Billig, F. S., "Key Issues Regarding the Aerothermochemistry of Turbulent Combustion in the Design and Development of Hypersonic Propulsion Systems," *Eleventh Sandia Cooperative Group Meeting on Turbulent Combustion,* March 29-30, 1988.

[9]Nestler, D. E., and Daywitt, J. E., "Boundary-Layer Separation Correlations for Hypersonic Inlet Analyses," General Electric, GE Report AMJ-86-01, Nov. 1986.

[10]Elkins, R.T., "Phenomenology of Supersonic Combustion," Wright Lab., WL-TR-2012, June 1991.

[11]"Analytical and Experimental Evaluation of the Supersonic Combustion Ramjet Engine," AF APL-TR-65-103, *Component Evaluation,* Vol. II, Dec. 1965.

[12]Billig, F. S., "Research on Supersonic Combustion," AIAA-92-0001, Jan. 1992.

[13]Byington, C. S., Northam, B. G., and Capriotti, D. P., "Transpiration Cooling in the Locality of a Transverse Fuel Jet for Supersonic Combustors," AIAA 90-2341, July 1990.

[14]Povinelli, L. A., "Aerodynamic Drag and Fuel Spreading Measurements in a Simulated Scramjet Combustion Module," NASA TN D-7674, May 1974.

[15]Northam, B. G., Greenberg, I., and Byington, C. S., "Evaluation of Parallel Injector Configurations for Supersonic Combustion," AIAA-89-2525, July 1989.

[16]Schlichting, H., "Boundary Layer Theory," Pergamon, 1955.

[17]Samimy, M., Erwin, D. E., and Elliott, G. S., "Compressibility and Shock Wave Interaction Effects on Free Shear Layers," AIAA-89-2460, July 1989.

[18]Papamoschou, D., "Structure of the Compressible Shear Layer," AIAA Paper 89-0126, 1989.

[19]Samimy, M., and Elliott, G. S., "Effects of Compressibility on the Structure of Free Shear Layers," AIAA Paper 88-3045A, 1988.

[20]Gutmark, K. E., Schadow, K. C., and Wilson, K. J., "Mixing Enhancement in Coaxial Supersonic Jets," AIAA Paper 89-1812, 1989.

[21]Bogdanoff, D. W., "Compressibility Effects in Turbulent Shear Layers," *AIAA Journal,* Vol. 21, No. 6.

[22]Kutschenreuter, P. H., Subramanian, S. V., Gaeta, R. J., Jr., Hickey, P. K., and Davis, J. A., "A Design Approach to High Mach Number Scramjet Performance," AIAA Paper 92-4248, 1992.

[23]Gaeta, R. J., Jr., "Techniques for Scramjet Engine Performance Analysis Using 3-D CFD Solutions," GE Aircraft Engines TM, TM92-182, Sept. 1992.

[24]Yu, S. T., Tsai, Y. L., and Shuen, J. S., "Three-Dimensional Calculation of Supersonic Reacting Flows Using an LU Scheme," AIAA Paper 89-0391, Jan. 1989.

[25]Oldenborg, R. et al., "Hypersonic Combustion Kinetics," NASA TM 1107, 1990.

[26]Kutschenreuter, P. H., Subramanian, S. V., and Gaeta, R. J., "Slanted Step Slit-Slot Scramjet Fuel Injector Mixing & Combustion Studies at Hypervelocity Test Conditions," Unpublished GE Aircraft Engines Rep., March 1993.

[27]Kutschenreuter, P. H., Davis, J. A., and Gaeta, R. J., "Hypersonic Inlet Efficiency Revisited," *JANNAF Propulsion Meeting,* Feb. 1992.

Chapter 9

Aerothermodynamics of the Dual-Mode Combustion System

W. H. Heiser*
U.S. Air Force Academy, Colorado
D. T. Pratt†
University of Washington, Washington

Nomenclature

A = through-flow area
C_p = specific heat at constant pressure
f = mass-basis fuel/air ratio
H = dimensionless static enthalpy
h = specific enthalpy
I = impulse function, $= pA + \dot{m}u$
K = dimensionless kinetic energy
M = Mach number
$\dot{m}$ = mass-flow rate
n = polytropic process exponent
p = absolute pressure
R = engineering or universal gas constant
T = absolute temperature
u = magnitude of the axial velocity
x = axial direction; axial location
γ = ratio of specific heats
τ = total temperature ratio
Φ = dimensionless stream thrust function
χ = dimensionless axial position

*Professor Emeritus, Department of Aeronautics.
†Professor Emeritus, Department of Mechanical Engineering.

Subscripts

a = location on H-K diagram
b = combustion (burner) process; burner exit; location on the H-K diagram
c = compression process; critical or sonic point; isolator exit core flow
d = fuel injection position downstream
e = exit or exhaust; expansion process
f = fuel
i = initial; injection conditions; inlet or entry
m = location on H-K diagram
o = overall
p = location on H-K diagram
s = reattachment point (end of separation)
t = throat; total or stagnation condition
u = separation point (beginning of separation) upstream
x = axial direction; axial location
0 = undisturbed freestream conditions as seen from the reference frame of the airbreathing engine or vehicle
1–4 = airbreathing engine reference station numbers (Fig.5)
10 = expansion nozzle exit station
* = choking or critical (sonic) point
& = sonic point (not choked)

I. Introduction

A LARGE and growing interest in the behavior of chemically reacting compressible flows is being stimulated by the desire to understand and control the operation of "dual-mode" ramjet/scramjet combustion systems for hypersonic airbreathing propulsion engines. Briefly, these devices coordinate carefully tailored distributions of chemical energy release and flowpath throughflow area in order to produce good thrust performance over a very wide range of flight Mach numbers (perhaps as much as $3 < M < 20$). The defining feature of dual-mode combustors is the substitution of "thermal choking" for a physical nozzle throat during ramjet operation.

The purpose of this article is to provide the reader with the physical insight and mathematical tools necessary to comprehend and analyze the complex aerothermodynamics of these devices. This will be done primarily through the use of one-dimensional compressible flow analysis, which has been found to provide clarity and insight into the important physical phenomena at work.

This article is based primarily on material presented in our recent AIAA Education Series textbook *Hypersonic Airbreathing Propulsion,*[1] which contains a great deal of background information for interested readers. In general, it is an implied reference for *everything that follows,* although specific citations are absent in order to avoid excessive repetition. In particular, the nomenclature and engine station designations are taken from that reference. The computations described here can also be carried out using software provided with the textbook.

The combustion system is the key to achieving successful hypersonic airbreathing engine performance. Referring to Fig. 1, the combustor or burner is downstream of a compression system (commonly called the inlet or diffuser) designed to decelerate the relative freestream flow. The constant area duct located between the inlet and combustor is known as the isolator because its function is to contain disturbances originating in the combustor that could cause unstart if they reached the inlet. The combustor is followed by an expansion system, or nozzle, designed to accelerate the combustion products to the surrounding or ambient static pressure with a minimum of aerodynamic losses and deviation from the desired thrust direction.

Unless otherwise stated, all analyses are based on the classical steady, one-dimensional equations for the flow in ducts of compressible, calorically perfect gases having constant ratios of specific heats and gas constants.[2] All flow properties may be interpreted as their mass flow weighted averages. The gas properties may be selected to reflect the most likely averages for the interval under consideration. The chemical energy released by reactions in the combustor or burner will, in accordance with firmly established tradition, be treated as "heat addition" or "total temperature increase" without mass addition.

II. H-K Diagram

Portraying the behavior of internal flows in airbreathing engines in a simple but revealing way is a challenging task, because there are several types of aerothermodynamic processes involved, and because there are many physical properties that one would like to follow. Authors through the years have therefore carefully tailored their graphical presentations to the particular situation at hand, often with mixed results. In our pursuit of understanding and explaining the intricacies of ramjet and scramjet engines, it has been found that representation of processes on dimensionless enthalpy-kinetic energy coordinates is singularly helpful. For brevity, this rep-

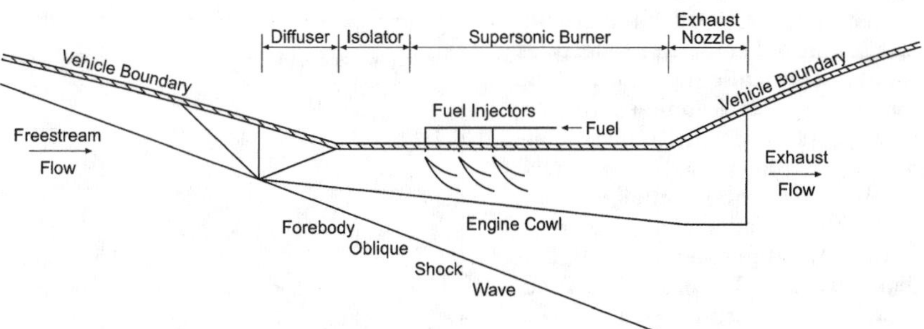

Figure 1. Schematic diagram of the dual-mode ramjet/scramjet airbreathing engine.

resentation will be called the "H-K diagram," as shown in Fig. 2. The local static enthalpy and kinetic energy are most conveniently made dimensionless by dividing by the total enthalpy of the undisturbed relative freestream flow.

One especially attractive feature of the H-K diagram is that the paths of many simple processes appear as straight lines, as seen in Fig. 2. Isolines of constant static enthalpy, static temperature, kinetic energy, velocity, Mach number, and total enthalpy all fit this description. In particular, the adiabat (constant total enthalpy or total temperature isoline) is given by the diagonal straight line

$$H + K = T_t/T_{to} \equiv \tau = \text{constant} \tag{1}$$

From the definition of the Mach number, isolines of constant M are given by the radial straight lines

$$\left(\frac{\gamma - 1}{2}\right)M^2H - K = 0 \tag{2}$$

Close examination of the one-dimensional equations of motion reveals that frictionless constant static pressure flow with heating or cooling corresponds to flow with constant velocity or kinetic energy, so that constant static pressure isolines are also vertical straight lines.

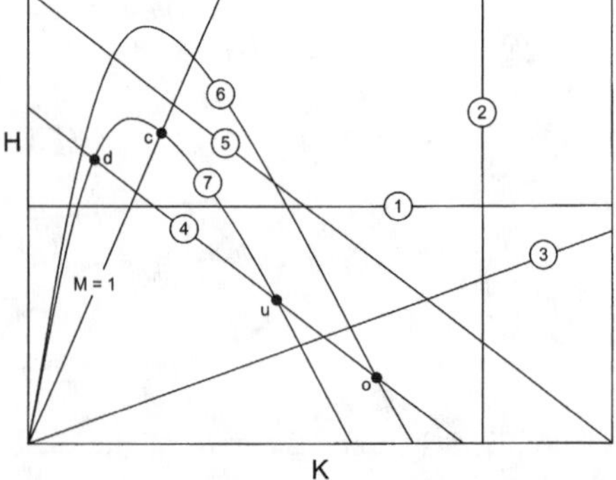

Figure 2. The H-K diagram, depicting representative constant-property isolines. Key: 0 = freestream reference state. Point c = choked condition at constant impulse function. Points u and d denote end states of normal shock. Circled numbers denote isolines of constant property, as follows: (1) static enthalpy, static temperature. (2) Kinetic energy, velocity, pressure (for frictionless heating or cooling only). (3) Mach Number. (4) total enthalpy, total temperature (adiabat), case τ = T_t/T_0 = 1. (5) post-heat release adiabat, case $\tau > 1$. (6) impulse function/stream thrust, area (for frictionless flow with heating or cooling only), case $\Phi = \Phi_0$. (7) impulse function, case $\Phi < \Phi_0$.

The single important exception to straight-line representation on H-K coordinates is the isoline of constant impulse function I (equal to $pA + mu$), which can be written in dimensionless form as

$$\Phi \equiv \frac{I}{\dot{m}\sqrt{C_p T_{to}}} = \left(\frac{\gamma - 1}{\gamma}\right)\frac{H}{\sqrt{2K}} + \sqrt{2K} = \text{constant} \tag{3}$$

Note that the non-dimensional impulse function at any point on the H-K diagram does not have some arbitrary value, but is uniquely related to H and K by Eq. (3) and to the operational variables M and $\tau = T_t/T_{t0}$ by

$$\Phi = \left(1 + \frac{1}{\gamma M^2}\right)\sqrt{\frac{\tau(\gamma - 1)M^2}{1 + \frac{\gamma - 1}{2}M^2}} \tag{4}$$

Another virtue of the H-K diagram is that all of the familiar "simple types" of compressible flows[2] are easily visualized:

1) First, the constant total enthalpy or τ isoline (adiabat) represents the locus of states for adiabatic flows with or without friction and shock waves and with or without area change, including isentropic flows in ducts of varying area and constant energy flows in ducts of constant area (Fanno flow).
2) The constant impulse function Φ isoline represents the locus of states for diabatic, frictionless flow in ducts of constant area (Rayleigh flow).
3) The two intersections of a constant τ adiabat with a constant Φ isoline are the end states for normal shock waves (Rankine-Hugoniot "jump" conditions) labeled u for upstream and d for downstream in Fig. 2.
4) The point of tangency between a constant τ adiabat and a constant Φ isoline, denoted by c in Fig. 2, corresponds to sonic or choked flow, which is the condition of maximum heat addition for a given value of impulse function or stream thrust.

For completeness, Table 1 summarizes the most important H-K diagram mathematical relationships. Note, in particular, that in every case the slopes of the resulting curves depend only upon the local Mach number and specific heat ratio γ, and not upon the magnitude of the constant associated with any one curve. The Mach number can again be seen to be the most important determinant of compressible flow behavior.

A. Scramjet and Ramjet H-K Diagrams

The value of the H-K diagram for airbreathing propulsion applications can be illustrated by using it to visualize the internal flow processes within scramjet and ramjet engines.

Figure 3 is an H-K diagram for air being processed through a scramjet with burner heat addition corresponding to an exit total temperature ratio

Table 1 Summary of H-K diagram relationships

Dimensionless quantity	H-K equation	d*H*/d*K* equation
Total enthalpy, $\tau \equiv T_t/T_{t0}$	$H + K = \tau$	-1
Mach number, $M \equiv u/\sqrt{\gamma RT}$	$K = \left(\frac{\gamma - 1}{2}\right) M^2 H$	$\frac{2}{(\gamma - 1)M^2}$
Impulse function, $\Phi \equiv I/\dot{m}\sqrt{C_p T_{to}}$	$\left(\frac{\gamma - 1}{\gamma}\right)\frac{H}{\sqrt{2K}} + \sqrt{2K} = \Phi$	$\left(\frac{1}{\gamma - 1}\right)\left(\frac{1}{M^2} - \gamma\right)$

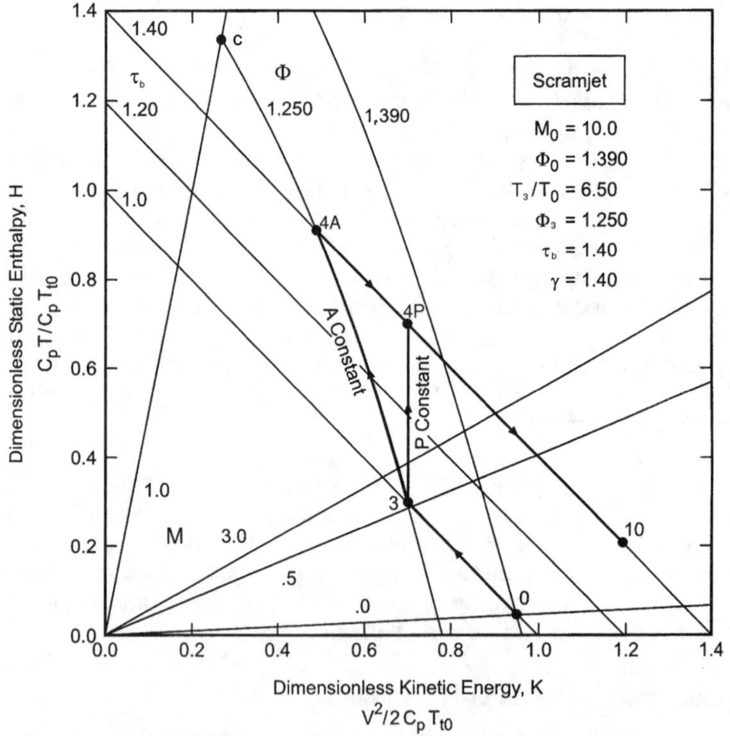

Figure 3. The H-K diagram for a scramjet. Inlet compression is represented by the path 0–3. Constant-area and constant-pressure combustion are represented by paths 3-4A and 3-4P, respectively. Expansion through the thrust nozzle is represented by path 4A-10 or 4P-10.

$\tau_b = T_t/T_{t0} = 1.40$. The engine is powering a vehicle at a freestream Mach number $M_0 = 10$, so that $\Phi_0 = 1.39$ from Eq. (4). The air is first adiabatically decelerated from the freestream condition (point 0) to the burner entry condition (point 3) by a combination of isentropic compression and oblique shock waves. The air is then "heated" in a combustion process in which the chemical energy of the fuel is released as sensible thermal energy in the combustion product stream. The precise path of this process depends on design choices made for the burner, and two of the many possible burner designs are depicted in Fig. 3. The first, path 3-4A, is frictionless, constant area heating, which is a Rayleigh line of constant Φ. The second, path 3-4P, is frictionless, constant static pressure heating, which is a line of constant K. For the scenario shown here, there is clearly no danger of encountering point c or thermal choking.

Since vertical lines of constant K are also lines of constant static pressure for frictionless flow, Fig. 3 reveals that constant area heat release imposes a positive (adverse) axial pressure gradient due to the accompanying thermal blockage or occlusion, which could cause boundary layer separation. Of course, the burner could be designed with increasing throughflow area in the axial direction in order to accommodate the dilation of the flow due to combustion heat release. However, the reduced static pressures and temperatures will suppress chemical reaction rates and thus combustion efficiency, as well as reducing the thermodynamic cycle efficiency because of a reduction of the mean temperature of heat addition.

The heated air is then accelerated and expanded to the freestream static pressure (Point 10). Because there are total pressure losses within the scramjet, the Mach number there can never be as large as M_0, but M_{10} is large enough that the kinetic energy and velocity at Point 10 exceed those of Point 0, which means that the scramjet produces net thrust.

Figure 4 shows process paths on the H-K diagram for air being processed by a ramjet, also with $\tau_b = 1.40$. In this case, the ramjet is powering a vehicle at a freestream Mach number of $M_0 = 3$, so that $\Phi_0 = 1.224$. The air must be first decelerated and compressed from the freestream condition (Point 0) to near-stagnation conditions in order to maintain approximately the same burner entry static temperature (Point 3) as previously described for the scramjet. In fact, for $M_0 = 3$, the largest possible total-to-static temperature ratio T_{t0}/T_0 is only 2.8. Experience shows that the required inlet compression is best accomplished by a combination of isentropic compression and oblique shock waves from Point 0 to Point u, followed by a normal shock wave from Point u to Point d, and then by subsonic diffusion from Point d to Point 3. The normal shock wave is ordinarily produced and stabilized within a divergent passage known as a transsection, which is capable of providing a range of normal shock Mach numbers depending on the axial location of the shock wave within the duct.

The combustion "heating" of the air is again portrayed as being frictionless, either constant area, path 3-4A, or constant static pressure, path 3-4P. Since the Mach number is very small in either case, Points 4A and 4P lie very close together. Note that a constant area heating process starting from Point

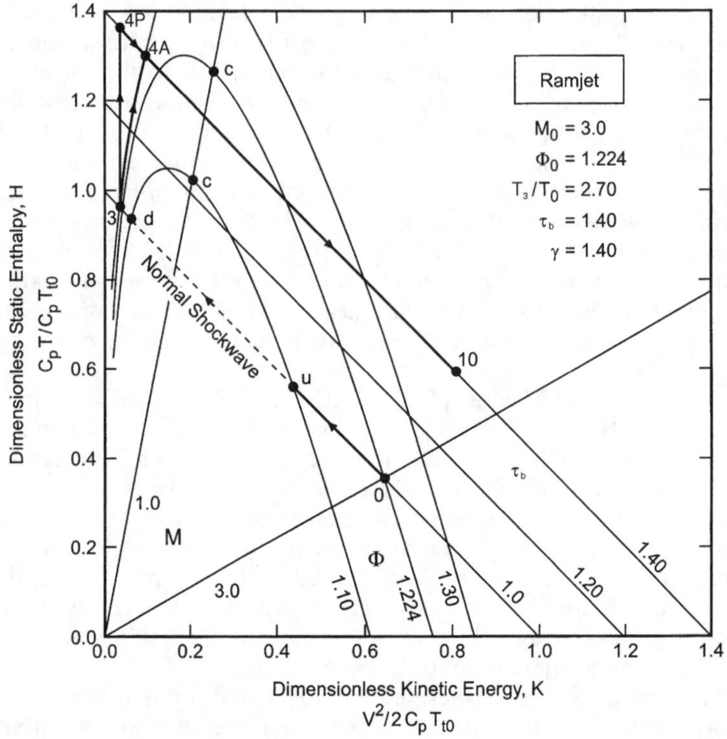

Figure 4. The H-K diagram for a ramjet. Inlet compression, including a normal shock from u to d, is represented by path 0–3. Constant-area and constant-pressure combustion are represented by paths 3-4A and 3-4P, respectively. Expansion through the convergent-divergent thrust nozzle is represented by path 4A-10 or 4P-10.

d would have reached Point c and thermal choking long before the desired amount of energy could have been added. Fortunately, further deceleration such as path d-3 can always be provided in order to lower the burner entry Mach number and raise Φ_3 enough that the desired energy can be added without thermal choking.

The heated air is then accelerated and expanded to the freestream static pressure (Point 10) in order to produce net thrust. In moving along the burner exit adiabat from Point 4A or 4P to full nozzle expansion at Point 10, the process path must again cross the $M = 1$ isoline. This can only be accomplished with a convergent-divergent nozzle having a choked throat. In fact, the minimum throughflow area (or blockage) of the nozzle throat determines the burner exit or back pressure, and thereby the axial location and strength of the trans-section normal shock wave. This also determines the overall total pressure loss of the flow through the ramjet and the ultimate performance of the engine.

B. H-K Diagram Closure

The H-K diagram, for all its virtues, is *not* a thermodynamic state diagram, because only H is a thermodynamic property of the flow, while K is a mechanical property. Consequently, there is no direct relationship between a point on the H-K diagram and other intensive thermodynamic properties of the fluid such a static pressure or entropy. However, when certain properties are known or remain constant along isolines or process paths, the H-K diagram will provide the right amount and type of information to both illustrate and analyze the important combustion and flow phenomena for the design and analysis of the propulsion system. These advantages will become apparent as the more complex interactions between the components of the dual-mode combustion system are described.

III. Dual-Mode Combustion System

The combustion system will be considered as consisting of two components, the inlet isolator and the combustor or burner, as illustrated in the schematic Fig. 5. Since fuel and air are still mixing and burning as they flow supersonically out of the burner and into the expansion system, the thrust nozzle could also logically be regarded as part of the combustion system. However, the viewpoint adopted here is that the function of the combustion system is to cause the fuel and air streams to begin to mix and start to burn, and that the design of the expansion system must take into account the degree of incompleteness of mixing and reaction in the gas stream exiting the burner, as a too-rapid drop in pressure will inhibit further mixing and "freeze" the reaction chemistry. Conversely, while thrust is generated in any burner with an expanding area ratio, the burner is not usually regarded as part of the expansion system.

In Fig. 5 engine reference stations 3 and 4 designate burner entry and exit, respectively. Station 2 will be used to designate entry to the isolator. Station 3, which designates both isolator exit and burner entry, is defined as the axial location *of the most upstream fuel injector.* Stations u and d designate the upstream and downstream limits or "ends" of a positive or adverse axial pressure gradient, respectively, and station s designates the upstream "end" of the negative or favorable pressure gradient which extends through the remainder of the burner and right on through the expansion system. As will be shown presently, station s is also approximately the location of the lowest Mach number in the combustion system.

Before getting deeply into analytical detail, it will be helpful to review certain relevant material, introduce some new concepts, define some new terms, and establish some cause-and-effect relationships which will help guide the reader through some rather tricky phenomena.

A. Dual-Mode Concept

The conflicting requirements of high cycle thermal efficiency and dissociation of the working fluid at excessively high static temperatures dictate that the combustion process must be subsonic (ramjet) for flight Mach

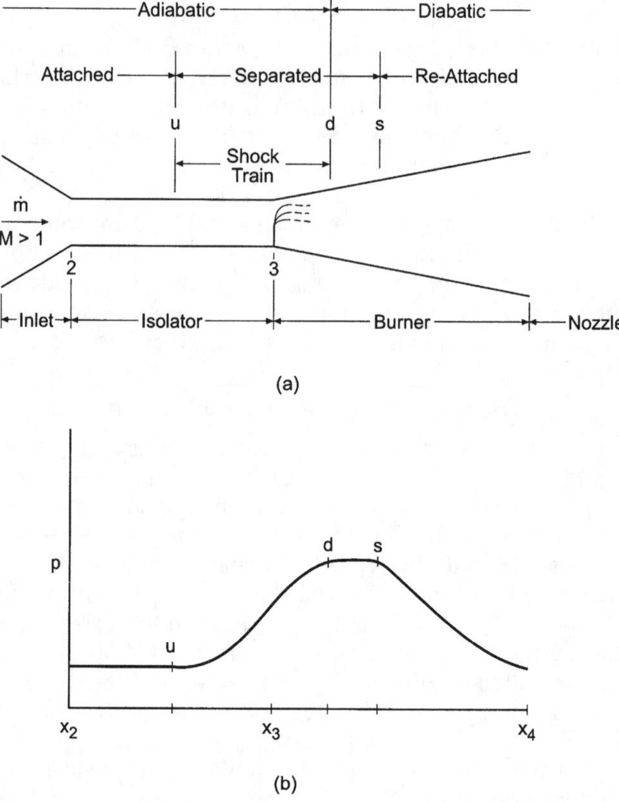

Figure 5. (a) Designation of axial locations for combustion system geometry (stations 2, 3 and 4) and axial variation of static pressure within the combustion system (stations u, d and s.) (b) Typical axial distribution of wall pressure for scramjet mode operation.

numbers less than about 5, and supersonic (scramjet) for M_0 greater than about 7.

A pure ramjet engine, which operates at supersonic flight speeds but with subsonic combustion, requires two area constrictions (physical throats). The first throat, at the outlet from the inlet diffuser, is required to stabilize the final, normal shock wave in the area expansion downstream of the throat (a supersonic-to-subsonic diffuser, referred to as the *transition section* or *trans-section*), in order to deliver subsonic flow to the burner. The second throat, downstream of the burner, is required to accelerate the subsonic flow to supersonic velocity in the expansion nozzle.

Unlike the ramjet engine depicted in Fig. 4, the scramjet engine of Fig. 3, with entirely supersonic throughflow has no physical throat. Since the Mach number never drops to or below unity in a pure scramjet, there is no need for either an upstream or a downstream throat.

To avoid having to carry two different engines for ramjet and scramjet operation, it would be desireable to operate the engine duct in *either* ramjet or scramjet mode using only the no-throat geometry of the scramjet. In other words, it is desired to be able to have subsonic flow in the burner *without area constriction either upstream or downstream of the burner,* as such constrictions would limit the mass flow rate at higher flight Mach numbers, when supersonic combustion is required. Of course, while it is conceivable to design "rubber" or variable-throat inlet compression and expansion system geometries to accomplish this, it is obviously difficult to do this.

To satisfy this design goal, Curran and Stull proposed in 1963, and patented in 1969, the concept of a *dual mode combustion system,*[3] in which both subsonic and supersonic combustion can be made to occur within the same scramjet engine geometry. The first experimental demonstration of ramjet mode operation in the open literature was reported by Billig in 1966.[4,5]

B. Ramjet Mode (Subsonic Combustion)

In ramjet mode the flow must be subsonic at the burner entry. In the dual-mode engine, the transition from supersonic flow to subsonic flow is accomplished by means of a constant-area diffuser called an *isolator,* which is capable of providing burner entry Mach number and static pressure conditions anywhere between those at Station 2 and those corresponding to a normal shock wave at the Station 2 conditions. In order that the burner entry flow be subsonic, the flow must be *choked* ($M = 1$) somewhere downstream, which causes a large back pressure p_3 at burner entry. This back pressure causes a *normal shock train* to form in the isolator, just upstream of station 3. As long as the back pressure p_3 does not exceed the isolator's ability to contain the normal shock train, the isolator performs the same functions as the divergent diffuser or trans-section. The function of the second ramjet throat — to choke the flow and thereby fix the burner entry back pressure p_3 — is provided for in the dual-mode burner by means of a *choked thermal throat,* which is brought about by choosing the right combination of area distribution $A(x)$ and fuel-air mixing and combustion, as represented by the total temperature distribution $T_t(x)$.[2,6] It will be shown in some detail in Sec. IV.C. how this is accomplished.

Although not shown in Fig. 5, an asterisk will be used to designate the axial location of the choked thermal throat, whenever one exists.

C. Scramjet Mode (Supersonic Combustion)

In scramjet mode, since there is no need for a physical throat either upstream or downstream of the burner, and the flow is supersonic at burner entry, there is apparently no need for an inlet isolator. However, even though the flow is *ideally* neither choked nor subsonic anywhere within the engine, it frequently happens that if the area increase in the burner is not sufficient to relieve the *thermal occlusion* resulting from heat addition to a supersonic stream, an adverse pressure gradient arises. This effect has been seen for frictionless constant-area heat addition (Rayleigh flow) in Fig. 3. If the pressure rises too abruptly within the burner, the boundary layer will separate. The resulting pressure rise propagates freely upstream through the separated boundary layer, even though the *confined core flow* remains supersonic.[7] Unless the upstream migration of the shock train is contained within the isolator, the engine inlet will unstart. Happily, it is possible to get double duty from the isolator, which not only can contain the normal shock train required for subsonic burner entry in ramjet mode, but also can contain an *oblique shock train* with a *supersonic* confined core outflow, which provides the necessary adiabatic pressure rise in the confined core flow to match the pressure rise resulting from heat addition in the burner, and thereby prevents unstart of the engine inlet.

D. Transition from scramjet to ramjet mode

An example process path for shock-free supersonic combustion on H-K coordinates is shown in Fig. 6. This example process path was calculated for a variable-area scramjet process, using analytical methods to be described in Sec. IV.

The shape of the burner process path in Fig. 6 is interesting. At burner entry, where the heat addition rate (dT_t/dx) is greatest, the process path is close to the desirable constant-*P*/constant-*K* process path, as the pressure

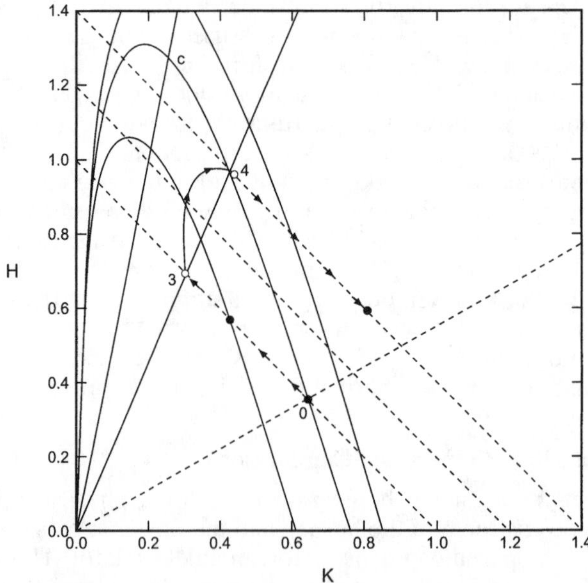

Figure 6. Supersonic combustion process path on H-K coordinates. Example case illustrated: Scramjet mode with shock-free isolator, $M_2 = M_3 = 1.5$. $A(x)/A_3 = 1+\chi$, $T_t(x)/T_{t2} = 1+0.4\chi(2-\chi)$, where $\chi \equiv (x-x_3)/(x_4-x_3)$.

rise due to heat addition is counteracted by the relief due to increasing area. However, as dT_t/dx tapers off toward burner exit, the relieving effect of increasing area now dominates the occlusion due to heat addition, so the pressure starts to decrease (K increases), and the burner process path approaches in turn a constant-M path, then a constant-T path, and finally approaches a constant-T_t (adiabatic) process path near burner exit. Note that, in this example, the burner Mach number passes through a minimum $M_s = 1.33$, which is well above Mach one. The axial locus of the minimum Mach number, denoted station s in Fig. 5, is called the *thermal throat* of the burner, by analogy to the physical throat of a converging-diverging nozzle. Also note that, just as with a physical throat, it is possible for a thermal throat to exist without being choked.

In the burner process path illustrated in Fig. 6, the minimum Mach number $M_s = 1.33$ is not sufficiently less than $M_3 = 1.5$ to cause the boundary layer to separate, so the isolator is shock-free. However, note that the burner area ratio $A_4/A_3 = 2$ is too great to maintain the exit pressure p_4 close to p_3. This drop-off in static pressure is accompanied by a decrease in static temperature and an increase in Mach number, compared to the ideal constant-P heat addition process. By reducing the mean temperature at which heat is added in the burner, the cycle thermal efficiency is reduced. In addition, supersonic mixing is inhibited at higher Mach numbers, and chemical kinetic rates are strongly proportional to both static pressure and temperature. Thus for the heat addition process shown in Fig. 6, compared to the desirable constant-P process, burner residence time is decreased, the fuel-air mixing rate is depressed, combustion reactions tend to "freeze" before the desired amount of heat has been released, and the mean temperature at which heat is added is reduced, even for the same $\tau_b = T_{t4}/T_{t3}$. Clearly, it is very important to maintain the design pressure in the combustion system whenever possible.

In order to make the burner outlet pressure p_4 closer to the inlet value p_3, the burner area ratio could be reduced from 2 to, say, 1.73, and the calculation repeated. The result is plotted as process path B on Fig. 7. In this case, the Mach number at the thermal throat, $M_s = 1.19$, is still not sufficiently less than $M_3 = 1.5$ to separate the boundary layer. However, it is getting closer to unity. Also, the exit pressure p_4 hasn't been raised that much, and is still less than p_3.

As the burner area ratio is further decreased to 1.57, resulting in process path C in Fig. 7, the thermal throat moves increasingly closer to the $M = 1$ ray, as now $M_s = 1.03$. However, the flow is still not quite choked at the thermal throat, and p_4 is still less than p_3.

A further decrease in the area ratio to $A_4/A_3 = 1.55$ *does* cause the flow to choke at the thermal throat. As a result, the flow must now be subsonic at burner entry, so that M_3 and Φ_3 can be made sufficiently low to allow more room for the desired heat addition, as described in Sec. II.A. A normal shock train forms in the isolator to provide the required pressure and subsonic flow at burner entry. The resulting heat addition process is represented as path D in Fig. 7. The dual-mode combustion system is now oper-

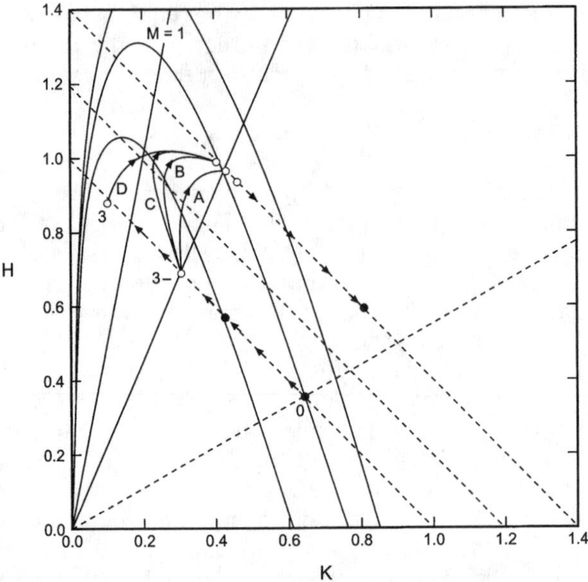

Figure 7. Scramjet-to-ramjet mode transition by reducing burner area ratio A_4/A_3, for same $T_t(x)$ as in Fig. 6. Path A is that illustrated in Fig. 6 for $A_4/A_3 = 2.0$, path B for $A_4/A_3 = 1.73$, path C for $A_4/A_3 = 1.57$, path D for $A_4/A_3 = 1.55$.

ating in ramjet mode, with *subsonic* flow into and *supersonic* flow out of the burner.

There are other ways besides changing $A(x)$, as described above, to transition from scramjet to ramjet mode and back. For example, by keeping $A(x)$ fixed while increasing the fuel flow rate to "add" more heat, thus changing $T_t(x)$ until the increase in τ_b is sufficient to cause thermal choking. This process can be reversed by reducing τ_b until the flow un-chokes and supersonic flow is re-established at burner entry. This process is demonstrated experimentally in Ref. 7.

IV. One-Dimensional Flow Analysis of the Isolator-Burner System

Understanding the behavior of the isolator-burner system is critical to both designing dual-mode scramjets and interpreting their experimental data. We begin with an examination of the operation of the isolator, a deceptively simple and benign component that serves a critical purpose.

The material in the following Sec. IV and V is a drastically reduced version of that originally published in Ref. 1. Readers interested in delving more deeply will find an abundance of valuable information there. Furthermore, the emphasis of Sec. IV and V is on the analysis (or design "from the inside out") of dual-mode scramjet combustion systems. In contrast, Sec. VI is dedicated to the interpretation of experimental data ("from the outside in").

A. Control Volume Analysis of the Isolator

The isolator can produce any static pressure at station 3 between the static pressure of the supersonic flow at station 2 (i.e., $p_{3\min} = p_2$) and that corresponding to a normal shock at the conditions of station 2 (i.e., $p_{3\max} = p_{2y}$). A back pressure p_3 greater than p_{2y} will cause the compression system to unstart.

Since the isolator has constant area, and friction and heat transfer may be neglected, it is a straightforward matter to apply the one-dimensional conservation laws to the control volume and determine the flow conditions at station 3 for a given p_3. The analysis is based upon the reasonable assumption that the isolator exit flow consists of a confined, uniform core flow of area $A_{3c} < A_3$ and velocity u_3 surrounded by a dead region of separated flow,[7] as shown in Fig. 8.

The most important results of the analysis for our purposes are

$$M_3 = \left\{ \frac{\gamma_b^2 M_2^2 \left(1 + \left(\frac{\gamma_b - 1}{2}\right) M_2^2\right)}{\left(1 + \gamma_b M_2^2 - (p_3/p_2)\right)^2} - \left(\frac{\gamma_b - 1}{2}\right) \right\}^{-1/2} \tag{5}$$

which gives the resulting core M_3 as a function of the isolator entrance Mach number M_2 and the required back pressure ratio p_3/p_2, and

$$\frac{A_{3c}}{A_2} = \frac{1}{\gamma_b M_3^2} \left[\frac{p_2}{p_3} \left(1 + \gamma_b M_2^2\right) - 1 \right] \tag{6}$$

which gives the ratio $(A_{3c}/A_2) < 1$ as a function of M_2, M_3, and p_3/p_2, for $(p_{3\max}/p_2) > (p_3/p_2) > 1$. Equation (6) is plotted on Fig. 9 for an isolator entry Mach number $M_2 = 2$ and $\gamma_b = 1.4$. Note that A_{3c}/A_2 is indeed less than one, and that $p_{3\max}/p_2$ is that of a normal shock wave. There is evidence from CFD modeling[9] and experiments[10] that the scramjet mode (i.e., $M_3 > 1$) gives rise

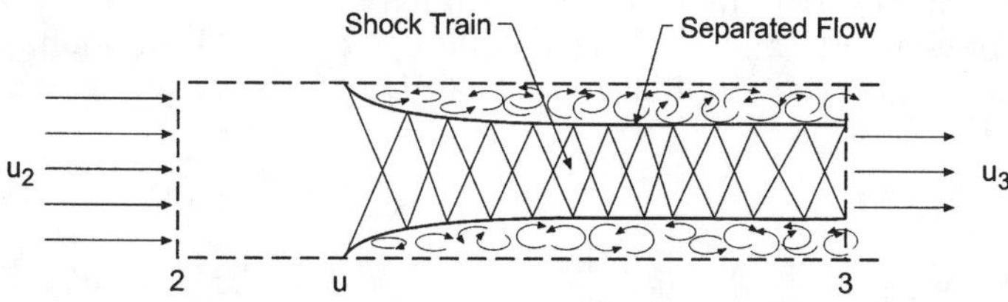

Figure 8. Control volume for analysis of the isolator.

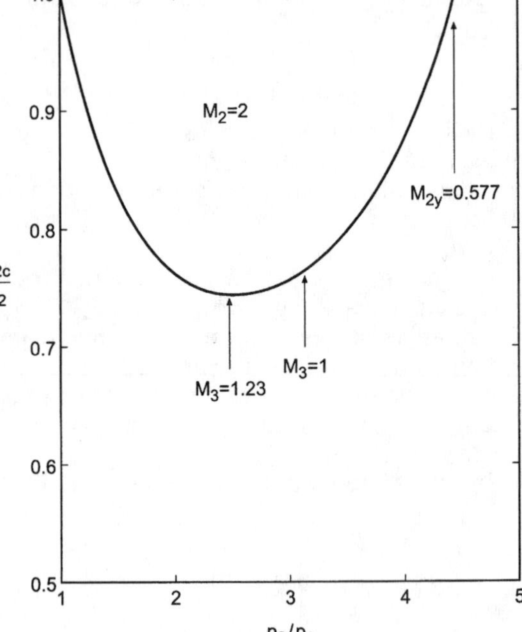

Figure 9. Variation of the confined flow area fraction A_{3c}/A_2 with imposed back pressure ratio p_3/p_2, for isolator entry Mach number $M_2 = 2$ and $\gamma_b = 1.4$.

to an oblique-shock train in the isolator, while the ramjet mode ($M_3 < 1$) gives rise to a normal-shock train. The length of the shock trains can be fairly estimated using an experimental correlation developed by Waltrup and Billig.[11]

B. One-Dimensional Flow Analysis of the Burner

We now shift focus to the active component of the combustion system, namely the burner, in order to analyze and quantify the processes that determine the magnitude of the back pressure p_3. This analysis is based directly on the classical "generalized one-dimensional flow" methods for ideal gases with constant gas properties of Refs. 2 and 6. Because combustion reactions are occurring within the burner, the reader must exercise good judgement in selecting gas properties used in the equations.

For the case of frictionless flow without mass addition but with change in throughflow area A and total temperature T_t due to "heat addition," the governing ordinary differential equation (ODE) for axial variation of Mach number is

$$\frac{\mathrm{d}M}{\mathrm{d}x} = M\left(\frac{1+\left(\frac{\gamma_b - 1}{2}\right)M^2}{1-M^2}\right)\left\{-\left(\frac{1}{A}\frac{\mathrm{d}A}{\mathrm{d}x}\right)+\frac{(1+\gamma_b M^2)}{2}\left(\frac{1}{T_t}\frac{\mathrm{d}T_t}{\mathrm{d}x}\right)\right\} \tag{7}$$

For present purposes, it is assumed that $A(x)$ and $T_t(x)$ are prescribed or given a priori, and are treated as independent variables. Of course, while $A(x)$ is relatively easily controlled, one cannot know in advance precisely what $T_t(x)$ will arise because it ultimately depends on the finite-rate processes of mixing and chemical reaction. Nevertheless, for a wide variety of scramjet mixers, $T_t(x)$ can be usefully represented in non-dimensional form by rational functions. As a result, in the spirit of preliminary design, $T_t(x)$ can be varied systematically within realistic limits in order to explore the effects of various fuel injection and mixing strategies and devices on dual-mode scramjet burner performance.

To determine a unique solution to Eq. (7), it is necessary to specify the burner entry Mach number M_3, as well as the forcing functions $A(x)$ and $T_t(x)$. The ODE Eq. (7) can then be numerically integrated to find $M(x)$ from burner entry to exit. The remaining desired physical properties $p(x)$, $p_t(x)$, $T(x)$, and $u(x)$ can be determined directly from $A(x)$, $T_t(x)$, and $M(x)$ using the convenient "useful integral relations" of Ref. 2:

$$T(x) = T_2 \frac{T_t(x)}{T_{t2}} \left[\frac{1 + \left(\frac{\gamma_b - 1}{2}\right) M_2^2}{1 + \left(\frac{\gamma_b - 1}{2}\right) M^2(x)} \right] \tag{8a}$$

$$p(x) = p_2 \frac{A_2}{A_c(x)} \frac{M_2}{M(x)} \sqrt{\frac{T(x)}{T_2}} \tag{8b}$$

$$p_t(x) = p_{t2} \frac{p(x)}{p_2} \left[\frac{T_2}{T(x)} \frac{T_t(x)}{T_{t2}} \right]^{\frac{\gamma_b}{\gamma_b - 1}} \tag{8c}$$

$$u(x) = u_2 \frac{M(x)}{M_2} \sqrt{\frac{T(x)}{T_2}} \tag{8d}$$

Finally, the following equations are useful for plotting the burner process path on H-K coordinates

$$H(x) = \frac{T(x)}{T_{t2}} \quad \text{and} \quad K(x) = \left(\frac{\gamma_b - 1}{2}\right) M^2(x) H(x) \tag{9}$$

C. Establishing a Choked Thermal Throat

Since there is no physical minimum area throat in the dual-mode scramjet, choking in the divergent combustor passage is brought about by tailoring $T_t(x)$ and $A(x)$ so that the former drives the local Mach number up to 1 and the latter continues the acceleration to supersonic velocities. This process is known as thermal choking, and the location of the sonic flow is known as the choked thermal throat. Thermal choking is the fundamental physical phenomenon that makes dual-mode scramjet operation possible.

Equation (7) reveals that thermal choking *can* occur for any pair of functions $A(x)$ and $T_t(x)$ at the critical axial location x^* where $M = 1$ and the sum of the terms inside the right hand brackets is zero. Note that this is a necessary but not sufficient condition, and does not guarantee that thermal choking *will* occur. This must be determined by other means.

Having located the position of choking x^* from Eq. (7), if it exists, Eq. (7) is integrated backward from x^* to find conditions at the burner entry. The initial slopes $(dM/dx)^*$ are also determined from Eq. (7) using l'Hopital's rule (see Ref. 2), and can be either positive or negative, just as is the case for isentropic nozzle flow. The resulting unique subsonic entry Mach number is designated M_{3p} (for plus root) and the supersonic M_{3m} (for minus root). When a choked thermal throat is known or assumed to exist, the corresponding burner entry pressures p_{3p} and p_{3m} that the isolator "sees" as back pressures are determined by substituting M_{3p} and M_{3m} into Eq. (5) and solving for p_{3p}/p_2 and p_{3m}/p_2. Note that the confined core area A_{3c}/A_2 at burner entry is given by Eq. (6) and that all other properties of the confined core flow are determined as usual by Eq. (3) and the "useful integral relations".[2] Finally, conditions from x^* to x_4 can be obtained by integrating Eq. (7) in the downstream direction.

V. System Analysis of Isolator-Burner Interaction

Having developed the necessary analytical tools for calculating the aerothermodynamic behavior of the isolator and burner, it is now possible to analyze the interaction between these two components of the dual-mode combustion system, for each of the three cases in Table 2 and described in Sec. III.

Table 2 Relative axial location of rising (*u-d*) and falling (*s-4*) pressure gradients, for different combustion system modes and type of shock train. (Axial station designations are as illustrated in Fig. 5 and as described in text. $M = 1$ at stations & and *.)

System mode	Shock train	Axial location of pressure gradients	Separated	Subsonic	Adiabatic
Scramjet	None	2, 3, u, $d=s$, 4	No	No	2-u
	Oblique	2, u, 3, d, s, 4	u-s		2-d
Ramjet	Normal	2, u, &, $d=3$, s, *, 4		&-*	

A. Scramjet with Shock-Free Isolator

In the first scramjet case shown in Table 2, there is no interaction between the burner and the isolator, and the state of the airflow is unaltered between stations 2 and 3. All properties between stations 3 and 4 are found by integrating Eq. (7) and using Eq. (3) and the "useful integral relations" Eqs. (8).

B. Scramjet with Oblique Shock Train

Heat addition to a supersonic flow causes the static pressure to increase which, in turn, can cause the wall boundary layer to separate and an oblique shock train to arise in the isolator. If the flow is predicted to separate, the state of the confined core at the burner entry must be determined. This is accomplished in two steps:

1) Assume that the flow in the isolator is undisturbed, so that $M_3 = M_2$, and integrate Eq. (7) for the given $A(x)$ and $T_t(x)$ to find $M(x)$. Identify the axial location where the smallest $M(x)$ [and greatest $p(x)$] occurs, designated station s in Fig. 5 and Table 2. If the Mach number ratio M_s/M_2 satisfies the approximate empirical criterion $(M_s/M_2) < 0.762$ the wall boundary layer is assumed to separate. If the boundary layer remains attached, the burner is operating in the shock-free isolator scramjet mode as described in Sec. V. A.

2) If separation is indicated, the flow is assumed to internally adjust itself in such a way that the heat added to the separated core flow in the burner (process *d-s* in Fig. 5 and Table 2) occurs at constant pressure equal to the maximum pressure $p_s = p(x_s)$, as determined in step 1. The back pressure at station d, where the pressure rise through the isolator oblique shock train matches the pressure of the separated flow, is therefore determined by the requirement $p_d = p_s$. The complete thermodynamic state of the separated core flow at station *d*, as well as the variation of all properties between stations *d* and *s*, is obtained from the constant pressure solution for $M(x)$

$$M(x) = \frac{M_3}{\sqrt{\tau(x)\left(1 + \left(\frac{\gamma_b - 1}{2}\right)M_3^2\right) - \left(\frac{\gamma_b - 1}{2}\right)M_3^2}} \tag{10}$$

together with Eq. (3) and Eqs. (8). The confined core flow does not conform to the burner area $A(x)$ between stations *d* and *s*, but rather forms its own axial variation $A_c(x)$, as given by

$$A(x) = A_3\left[\tau(x)\left(1 + \frac{\gamma_b - 1}{2}M_3^2\right) - \left(\frac{\gamma_b - 1}{2}\right)M_3^2\right] \tag{11}$$

The confirmed core flow is assumed to re-attach immediately downstream of the axial location of the maximum pressure at station s, due to the establishment of a favorable pressure gradient there.

C. Scramjet with Normal Shock Train

As the amount of heat addition to the supersonic stream is gradually increased, $M(x^*)$ as calculated by the methods of Sec. IV.C (with no separation, and $M_3 = M_2$) will gradually approach and finally reach 1 (i.e., $M_3 = M_{3m}$ and $p_3 = p_{3m}$). When the slightest additional amount of heat is added, thermal choking occurs, which forces an abrupt change to a normal shock train in the isolator and a fully subsonic burner entry flow. Since the heat is now added to a subsonic flow, a favorable pressure gradient is established and the boundary layer remains attached throughout the burner. The corresponding conditions at the isolator exit and burner entry are determined from the subsonic branch of $M(x)$ and, provided that the required p_3 is less than or equal to the available $p_{3\max} = p_{2y}$, the burner will operate in the ramjet mode. If p_{3p} exceeds $p_{3p\max}$, the isolator will "unstart" (i.e., require a change to its entry conditions) and overall engine performance is usually greatly diminished.

VI. Interpretation of Experimental Data

In the preceding Sec. IV and V, a procedure was developed for designing an isolator-burner system "from the inside out", by assuming various algebraic equation models for $A(x)$ and $T_t(x)$, then numerically solving Eq. (7) together with Eqs. (8) for $M(x)$ and related properties. However, in order to interpret experimentally determined wall pressures measured during scramjet burner tests, it is necessary to analyze the isolator-burner system "from the outside in."

Figure 5b can be thought of as an idealized plot of measured wall static pressures $p(x)$ at various axial stations in a scramjet burner. The flow and thermodynamic state of the gas at entry is known, as is the geometric area distribution $A(x)$. It is desired to determine the axial distributions of $M(x)$ and $T_t(x)$ from the known $A(x)$ and measured $p(x)$ data, in order to determine both the overall heat release and the axial distribution of heat release.

The first step is to inspect the $p(x)$ data to identify the axial location of adverse or favorable pressure gradients with respect to station 3, where the fuel is first injected, and identify which of the three cases of Table 2 is indicated by the data. Choices are then made for the axial station locations x_u, x_d and x_s, defined as appropriate to the case identified from Table 2 and accompanying text. It is emphasized that the data reduction analysis which follows is only meaningful if the axial location of regions of separated flow are recognized as such and separation is taken into account.

The next step is to "smooth" the $p(x)$ data by curve-fitting in the sense of least-squares. Following the recommendation of Waltrup and Billig,[11] $p(x)$ between u and d can be represented by a cubic polynomial,

$$\frac{p(x)}{p_u} = 1 + \left(\frac{p_d}{p_u} - 1\right)\chi^2(3 - 2\chi) \quad \text{where} \quad \chi \equiv \frac{x - x_u}{x_d - x_u} \tag{12}$$

In the interval $[x_d, x_s]$, $p(x) = p_s = p_d =$ constant.

Whereas any smoothly decreasing function *could* be used to fit the $p(x)$ data from x_s to x_4, Billig[4,5] recommends a particularly useful function, the "polytropic" relationship

$$pA^n = \text{constant} \tag{13}$$

where the exponent n is determined from the pressure data on the interval $[x_s, x_4]$ by

$$n = -\frac{\ell n\left[p(x_s)/p(x_4)\right]}{\ell n\left[A(x_s)/A(x_4)\right]} \tag{14}$$

It is certainly not obvious that Eq. (13) will fit any arbitrary set of $p(x)$ data, but experience shows that this is very often so, although it may be necessary to adjust the end-state values $p(x_s)$ and $p(x_4)$ in Eq. (13) to obtain the best fit in the sense of least-squares to all of the intermediate $p(x)$ data.

Since it is assumed that only pressure forces act on the duct walls, a differential change in stream thrust function is given by $\mathrm{d}I = p\mathrm{d}A$, so that between any two axial locations x_i and x_e in the entire data range $[x_2, x_4]$,

$$I(x_e) = I(x_i) + \int_{x_i}^{x_e} p(x') \frac{\mathrm{d}A(x')}{\mathrm{d}x'} \mathrm{d}x' \tag{15}$$

It is important to recognize that, since Eq. (15) is based on pressure measured at the walls, Eq. (15) is valid whether the flow is separated or attached. It is in this sense that the analysis is being carried out "from the outside in."

The evaluation of the definite integral in Eq. (15) is straightforward:

1.) If $A_e = A_i$, then $I_e = I_i$ follows at once.
2.) If $p(x)$ is locally fitted by the cubic Eq. (12) and $A(x)$ is linear, then $\mathrm{d}A/\mathrm{d}x$ is constant and factors, leaving just the cubic polynomial integrand which integrates to a quartic expression.
3.) If $A(x)$ is quadratic in x (as in a straight-walled conical geometry), then $\mathrm{d}A/\mathrm{d}x$ is linear, and the integrand is a fourth-order polynomial which integrates to a fifth-order polynomial for $I(x)$.
4.) If $A(x)$ is variable in the range $[x_d, x_s]$, then the constant pressure term factors, and the definite integral is evaluated as $p_d[A(x_e)\text{-}A(x_i)]$.
5.) In the range $[x_s, x_4]$, because of the choice of curvefit function Eq. (13) with n determined from Eq. (14), the definite integral in Eq. (15) is evaluated as

$$\frac{p(x_e)A(x_e) - p(x_i)A(x_i)}{1-n}, \; n \neq 1 \tag{16}$$

With $I(x)$ thus determined for all x, the Mach number $M(x)$ is obtained from the definition of the stream thrust function, $I \equiv pA(1 + \gamma_b M^2)$, as

$$M(x) = \sqrt{\frac{1}{\gamma_b}\left[\frac{I(x)}{p(x)A(x)} - 1\right]\frac{A(x)}{A_c(x)}} \tag{17}$$

Once again, $M(x)$ from Eq. (17) must be interpreted as the Mach number *of the separated core flow* within the separated flow range $[x_u, x_s]$.

All other properties of interest may be determined, for the appropriate case, from Eqs. (8). In the scramjet with shock-free isolator case, as the flow is attached everywhere, all pressure rise and fall is due to the interaction of heat "addition" and area increase, so that $T_t(x)$ is determinate immediately from Eqs. (8a) and (8b). In the two cases with shock trains originating in the isolator, the data treatment is different in regions of separated and attached flow. In the *adiabatic, separated* flow interval $[x_u, x_d]$, $T_t = T_{t2}$ is constant and, so that the confined core area $A_c(x)$ is evaluated from Eqs. (8a) and (8b). In the *diabatic, attached* flow interval $[x_s, x_4]$, $A(x)$ is given, and $T_t(x)$ is evaluated from Eqs. (8a) and (8b). Thus all properties are determinate from $p(x)$ and the known state of the air at isolator entry station 2, everywhere *except in the subinterval* $[x_d, x_s]$, wherein the flow is both separated and diabatic, so that both $A_c(x)$ and $T_t(x)$ are unknown. In this interval, any simple smooth function could be used to patch $T_t(x_d) = T_{t2}$ to $T_t(x_s)$, from which the remaining properties could be determined *approximately* within $[x_d, x_s]$.

A. Billig's Experimental Wall-Pressure Measurements

Figures 10 and 11 present experimental pressure measurements from hydrogen-air combustion in laboratory scramjet burners, reported by Billig in Refs. 12 and 13.

From the schematic Fig. 10a, it is apparent that a divergent duct diffuser was used instead of a constant-area isolator. It is apparent from the relative axial location of the constant-pressure plateau in Fig. 10b that the combustion system is in scramjet mode with an oblique shock train. Application of Eqs. (12) through (17), together with Eqs. (3) and (8), gives results summarized in Table 3.

Note in Table 3 the rise in static temperature in the pre-combustion shock train at burner entry [2–3 or u–d]. An H-K diagram of the analyzed process is shown in Fig. 10c. Note that the process path 2–3 in the isolator is constant-P heat "addition" in the confined flow within the constant-A portion of the burner, just as represented by Fig. 5 and the second case of Table 2. In the variable-area portion of the burner, from station s to station 4, the H-K process path asymptotes rapidly to the $\tau_b = 1.65$ adiabat. While the burner satisfies the design requirement $p_4 \sim p_2$, 95 percent of the heat "addition" occurs at constant-P along path 3-s, where the pressure is more than three times greater than p_2 or p_4.

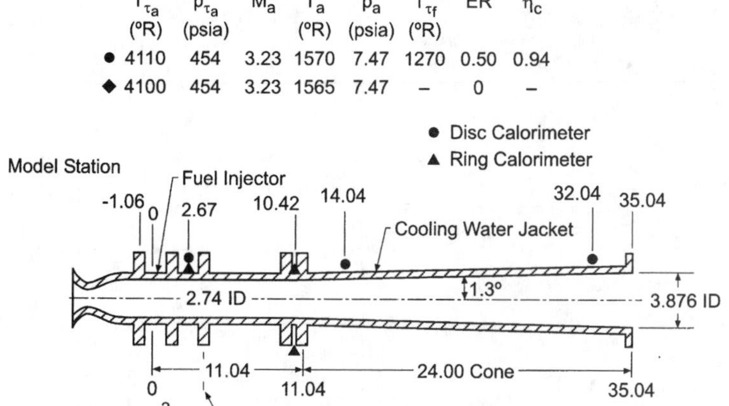

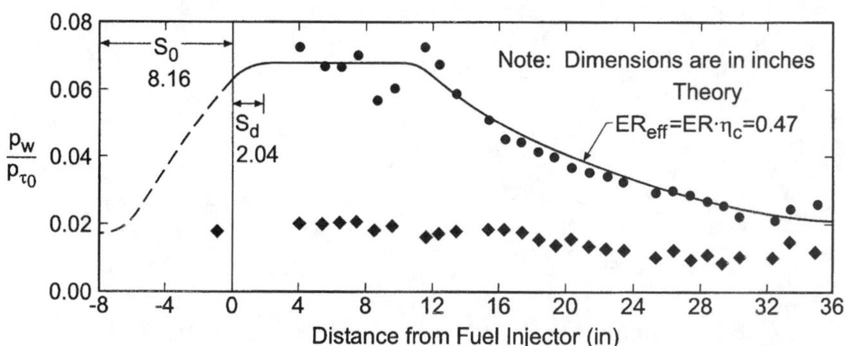

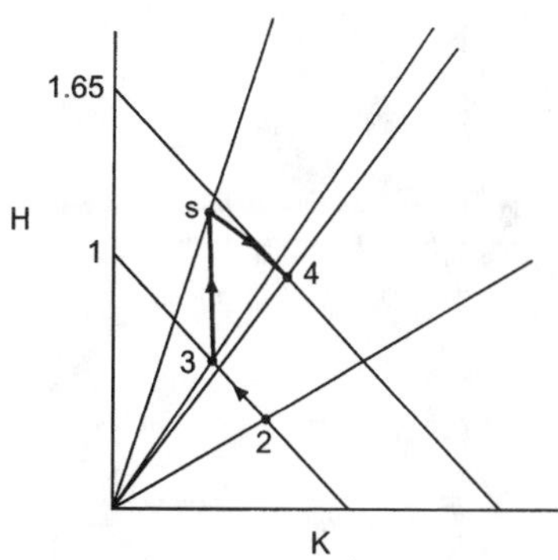

Figure 10. Experimental wall pressure measurements in a cylinder-cone H2-air scramjet combustor. (Fig. 8 of Ref. 12.) a) schematic of burner, b) axial pressure distribution, and c) H-K diagram of process path within burner.

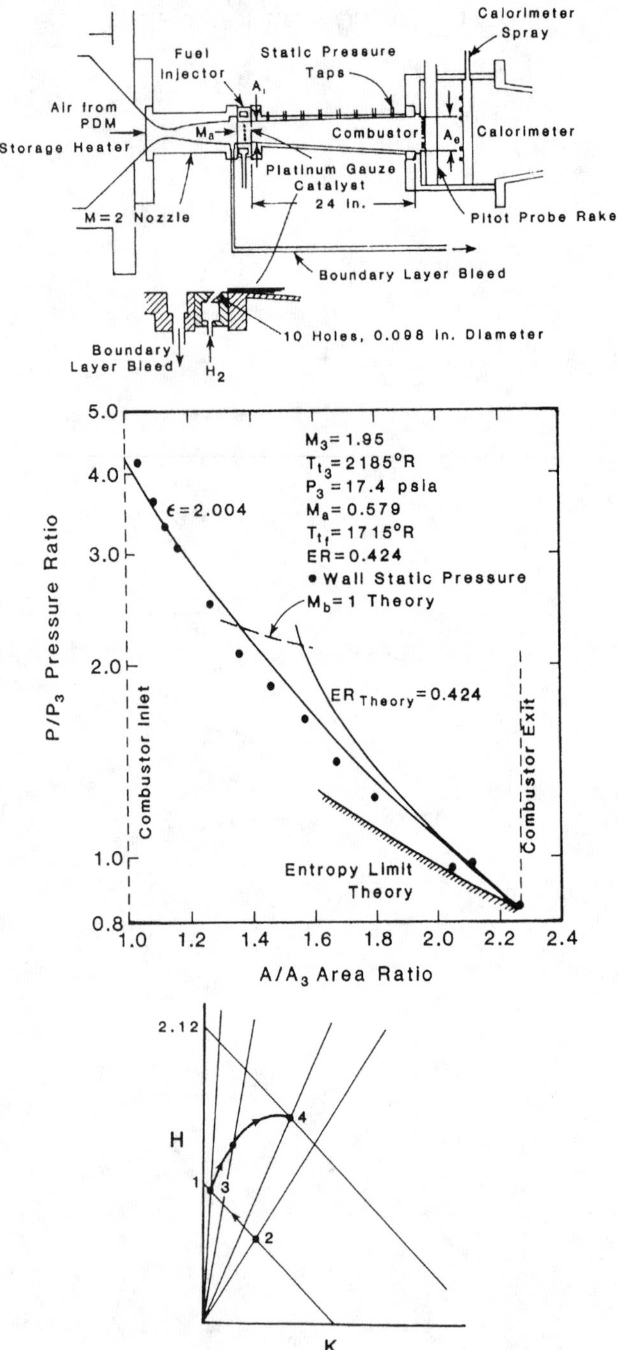

Figure 11. Experimental wall pressure data in a hydrogen combustor test apparatus. (Figs. 7 and 8 of Ref. 13). a) schematic of burner, b) axial pressure distribution, and c) H-K diagram of process path within burner.

Table 3 Analysis of wall-pressure data of Fig. 10. (overall H_2-air equivalence ratio $\varphi_0 = 0.50$, assumed $\gamma_b = 1.31$.)

Station	x, in.	T_t, °R	T, °R	u, ft/s	M
$2 = u$	−8.16	4109	1570	6081	3.23
3	0.0	4109	2510	4822	2.03
d	2.04	4109	2562	4822	1.97
s	12.0	6644	5098	4743	1.40
4	35.0	6813	3934	6476	2.17

Table 4 Analysis of wall-pressure data of Fig. 11. (overall H_2-air equivalence ratio $\varphi_0 = 0.424$, assumed $\gamma_b = 1.31$.)

Station	x, in.	T_t, °R	T, °R	u, ft/s	M
2	0.0	2186	1320	3410	1.95
$3 = d$	11.0	2186	2066	1272	0.58
s	12.0	2304	2183	1272	0.57
*	18.55	4112	3506	2851	1.00
4	34.0	4752	3245	4501	1.64

Figure 11 is the test data from what is believed to be the first experimental demonstration of ramjet mode operation in a dual-mode combustion system.[5,13] Results from applying Eqs. (12) through (17) to this data are summarized in Table 4.

There are some very interesting features in the data and analyzed results summarized in Table 4. First, note that the burner entry state corresponds to state $2y$, the normal shock limit for inlet unstart. These data were collected as part of a series of experiments during which Billig recognized the need for an inlet isolator.[5,13] In the case represented in Fig. 11, boundary layer bleed was used, in the absence of an isolator, to stabilize the normal shock at burner entry. The region of flow separation is very short, as the pressure begins to drop immediately downstream of the stabilized normal shock wave at burner entry. Note also that the flow passes smoothly through the critical point as it accelerates from subsonic entry to supersonic flow at burner exit.

VII. Closure

For preliminary design of a scramjet combustion system, the "generalized one-dimensional flow analysis" method of Sections IV and V is a straightforward *synthesis* procedure for finding the right combination of $A(X)$ and $T_t(x)$ "from the inside out" to achieve a desired burner performance. The method is especially advantageous for designing dualmode burners, because of its capability of independently locating a potential critical point prior to solving the ordinary differential equation for M, thus avoiding iteration of the numerical solution to locate the critical or sonic point. When the design calculations show that flow separation is expected, adjustments can be made to the required burner entry states, which can then be compared with the admissible range of isolator exit states to see whether or not the isolator can contain the pre-combustion shock train.

For determining burner performance "from the outside in" from experimental data, the integral analysis method of Sec. VI is an easily applied, powerful *analysis* method. In particular, when a critical point occurs in dual-mode combustion, the integral analysis method locates the choked thermal throat explicitly, whereas an attempt to replicate the $p(x)$ data by trial-and-error application of the generalized one-dimensional flow analysis method would lead to numerical difficulties owing to the $(1\text{-}M^2)$ term in the denominator of Eq. (7).

It is clear that the puzzle of optimal design of scramjet and dual-mode combustion systems has not yet been solved. It is hoped that this brief introduction will inspire the reader to join the search.

References

[1]Heiser, W. H., and Pratt, D. T., *Hypersonic Airbreathing Propulsion,* AIAA Education Series, AIAA Washington, DC, 1994.

[2]Shapiro, A. H., *The Dynamics and Thermodynamics of Compressible Fluid Flow, Vol. I,* Ronald Press, New York, 1953.

[3]Curran, E. T., and Stull, F. D., "The Utilization of Supersonic Combustion Ramjet Systems at Low Mach Numbers," Aero Propulsion Lab., RTD-TDR-63-4097, Jan. 1964.

[4]Billig, F. S., "Combustion Processes in Supersonic Flow," *Journal of Propulsion and Power,* Vol. 4, No. 3, 1988, pp. 209–216.

[5]Billig, F. S., "Research on Supersonic Combustion," *Journal of Propulsion and Power,* Vol. 9, No. 4, 1993, pp. 499–514.

[6]Zucrow, M. A., and Hoffman, J. D., *Gas Dynamics, Vol. I,* Wiley, New York, 1976.

[7]Sullins, G. A., "Results of Mach 8 Direct-Connect Scramjet Combustor Tests," Paper AL-93-P112, 1992.

[8]Shchetinkov, E. S., "Piecewise-One-Dimensional Models of Supersonic Combustion and Psuedo Shock in a Duct," *Combustion, Explosion, and Shock Waves,* Vol. 9, No. 4, 1975.

[9]Lin, P., Rao, G. V. R., and O'Connor, G. M., "Numerical Investigation on Shock Wave/Boundary Layer Interactions in a Constant Area Diffuser at Mach 3," AIAA Paper 91-1766, 1991.

[10]Elmquist, A. R., "Evaluation of a CFD Code for Analysis of Normal-Shock Trains," AIAA Paper 93-0292, 1993.

[11]Waltrup, P. J., and Billig, F. S., "Prediction of Precompression Wall Pressure Distributions in Scramjet Engines," *Journal of Spacecraft and Rockets,* Vol. 10, No. 9, 1973, pp. 620–622.

[12]Billig, F. S., Dugger, G. L., and Waltrup, P. J., "Inlet-Combustor Interface Problems in Scramjet Engines," *Proceedings of the 1st International Symposium on Airbeathing Engines,* 1972.

[13]Billig, F. S., "Design of Supersonic Combustors Based on Pressure-Area Fields," *Eleventh Symposium (International) on Combustion,* Combustion Inst., 1967.

Chapter 10

Basic Performance Assessment of Scram Combustors

S. N. B. Murthy*
Purdue University, West Lafayette, Indiana

I. Introduction

THREE problems of interest in the design and operation of scram combustors are 1) fuel combustion and energizing of air, the primary working fluid, 2) integration of the combustor into the propulsion system, and then the overall vehicle (internal and external) flowpaths, and 3) physical scaling of the combustor for testing and operation. The current chapter addresses these problems from a generic point of view to examine the extent to which relatively simple approaches and tools can provide a basis for conceptual and some aspects of preliminary design.

The first of those problems, the one related to energy utilization, is generally considered in practice in terms of realizing the best specific impulse I_{sp} and propulsive efficiency, η_{PR}. Those parameters are in fact overall engine parameters, including the performance of other components such as the inlet and the nozzle. However, assuming that all such other components operate under ideal conditions, I_{sp} and η_{PR} can also be considered as combustor performance parameters with respect to a chosen set of reference

---

*Senior Researcher.

Dr. R. Meenakshisundaram, Post-doctoral Fellow, assisted in the computational work presented.

The researchers were supported during the preparation of much of the material that is included in this chapter by a grant form the Air Force Aero-Propulsion and Power Directorate, the NASA Langley Research Center, and the NASA Glenn at Lewis Research Center; a number of very useful discussions were held with R. Buckley, C.R. McClinton, D. Palac and S. Thomas. The SCRAM-3L code was made available by the NASA Langley Research Center in its original form. The reports dealing with the Hypersonic Research Engine were provided by Earl H. Andrews. We are deeply indebted to each of them. Finally, Dr. E.T. Curran must be mentioned, who has continuously supported advances in scramjets and demanded a thorough assessment of each advance at a fundamental level; his review of this chapter has been most valuable.

ambient or environment conditions, and hence also as indicative of the efficacy of energy utilization.

Several noteworthy attempts at analysis of those problems have been made on various simplified bases in the past. These also include extensive, multi-dimensional, and occasionally time-dependent calculations with attention to complexities in geometry, fuel injection, and thermal management; however, they are not under consideration here, although it is important to note that such calculations are essential in detailed design of the system and the components.

In the current volume, the three basic problems of flowpath layout, energy analysis, and scaling have been discussed by A. Paull and R.J. Stalker (Chapter I.1), G.Y. Anderson, C.R. McClinton, and J.P. Wiedner in Chapter I.6, P. Kutschenreuter (Chapter III.9), and P.J. Ortwerth (Chapter IV.15), with different emphasis on individual problems. The following chapter (by A. Siebenhaar, et al.) also deals with flowpath analysis in the considerably more complex case of a combined cycle engine. The current chapter considers only the case of the scramjet combustor and then with very simple geometry and fuel injection schemes, while treating the entry conditions to the combustor, friction and heat transfer losses at the walls, and mixing schedule and efficiency as parameters. This chapter may thus provide a breather to reflect on the developments reported by others mentioned above.

A significant consideration in energy analysis is the selection of state parameters for the working fluid flow. Considering only gaseous fluids, an equation of state can be assumed, supplemented with such additional equations as needed to account for high temperature and chemical reaction phenomena. In any practical process there is a gain of entropy, and entropy thus becomes an important parameter. The flow speeds of interest in hypersonic propulsion are high, and there are continuous, although, in comparison with a gas turbine jet engine, for example, relatively small, changes in kinetic energy along the flowpath, as well as substantial changes in enthalpy. The processes along the flowpath may therefore be assessed by a parameter made up of the sum of the two. The use of such a parameter has been illustrated throughout Chapter III.9 of W. Heiser and D. Pratt. In the current chapter, for various reasons as will become clear later, stagnation pressure and exergy or energy availability are also considered major and necessary parameters for performance analysis. Energy availability, it may be pointed out, is simply the part of the total energy of the fluid that is available for conversion to work; loss of energy availability is irrecoverable as far as the working fluid is concerned. Further details can be found in Ref. 1.

The flowpath layout is a part of engine design. The objective of energy analysis is establishment of the effectiveness of the engine in converting the energy in the supply fluids, consisting of air breathed in and the fuel, into useful work throughout the flight path for lift and thrust action. The effectiveness is rationalized on the basis of the First and Second law of thermodynamics in terms of efficiencies. A small set of efficiencies must then be selected that together reflect how well a component or the system as a whole is performing in itself and in meeting the overall desired output of the engine. Accuracy, comprehensiveness, and clarity should provide the basis for the selection of the set. These are particularly important in scaling design parameters and test results.

It has now become well established that hypersonic systems must be fully integrated through a combined consideration of the internal flowpath with the external flow past the vehicle. The interactions between the two arise in air mass flow distribution and through forces that affect both flows at various locations along the flowpath. These interactions must be taken into account also in undertaking energy analysis, for example in evolving design convergence and eventual vehicle optimization. In all such studies, it has also become clear that the combustor, where the working fluid receives fuel combustion energy, is the appropriate component relative to which the flowpath as a whole can be examined. In this manner, the combustor becomes the central component of the flowpath, the inlet and the isolator serving the purpose of taking in air of the right quality and without disturbances from downstream, and the nozzle converting the energy of the working fluid at combustor exit to thrust work during vehicle motion (Fig. 1). In considering the inlet and the nozzle, the external flow past the vehicle needs accounting, and in the case of the combustor, the interactive flow processes at the cowl. However, in the current chapter, the main emphasis is on combustor analysis, while some remarks are made at the end on inlet selection and nozzle optimization for system design convergence.

It may be interesting to note that the choice of combustor as the basic element of the engine or the internal flowpath is not due to the importance of magnitude of energy addition to the air. Such combustion-generated energy becomes less than half of the kinetic energy of air even with an inlet Mach number of 10. However, what are of crucial importance are (a) the loss of available energy of air and fuel supplied due to the inefficiencies of the flow and combustion process and the wall heat loss, and (b) the physical size, particularly the length and the depth, of the combustor, which along with fuel density has a direct effect on vehicle volume, wetted area, and drag. A minimum sized combustor with maximum possible gain in energy availability is (at least) a main requirement for a hypersonic flight vehicle. These matters have been discussed by P. Czysz and S.N.B. Murthy in Ref. 2.

We begin with a discussion of scram combustor effectiveness.

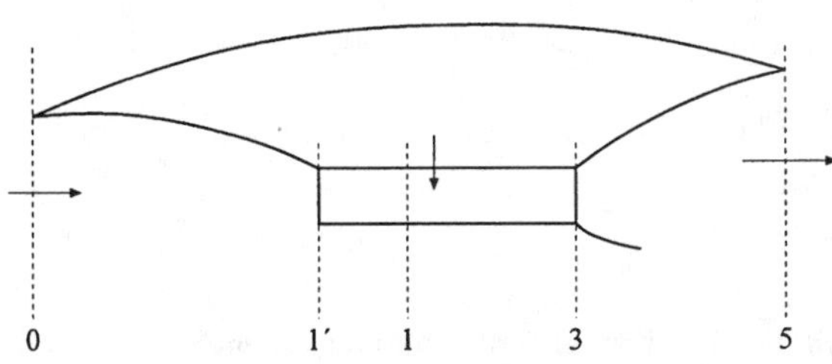

Fig. 1. Schematic of a hypersonic Engine with Inlet (01'), Isolator (1'1), Combustor (13), and Nozzle (35).

II. Scram-Combustor Effectiveness

It is often of interest, especially in assessing design and operation, to identify changes in various flow and state properties along the combustor, in particular in the form of efficiencies. It can be argued that a set of efficiencies based on the combined considerations of 1) kinetic energy, 2) energy availability, and 3) stagnation pressure is needed and adequate for assessing the performance of a scram combustor. This proposal is described in this section.

A traditional starting point for energy analysis is the determination of efficiencies based on energy flux and the First and Second law of thermodynamics. In the first case, referring to Fig. 2, an efficiency can be defined for the combustor as follows:

$$\eta_{\mathrm{FL}} = \frac{E_3(1+\varphi)}{E_1 + E_{f1}\cdot\varphi} \tag{1}$$

where, considering a unit mass of air-fuel mixture, E_1, E_3, and E_f represent, respectively, the sum total of kinetic energy and enthalpy of air at 1, products of combustion at 3, and fuel, the latter including the heat released by combustion under stoichiometric conditions, Q^*. The fuel to air ratio is denoted by φ.

Based on Second law, an efficiency can be defined as

$$\eta_{\mathrm{SL}} = \frac{(E_3 - T_3\cdot\Delta s_3)(1+\varphi)}{\left(E_1 - T_1\Delta s_1\right) + \left(E_f - T_{f1}\cdot\Delta s_1\right)\varphi} \tag{2}$$

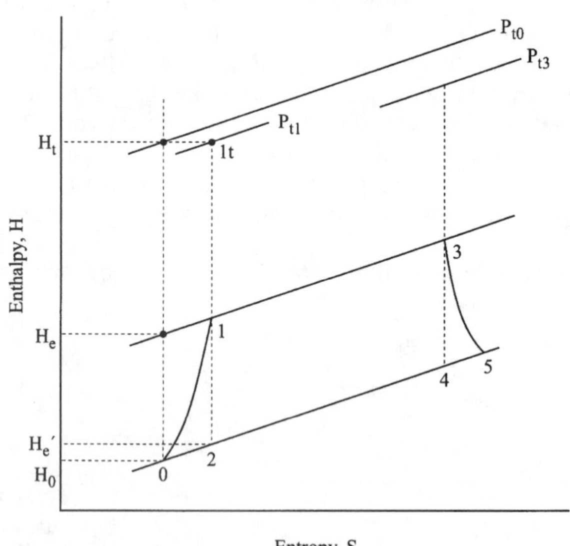

Fig. 2. Thermodynamic Diagram for a Scramjet Engine.

where Δs denotes the entropy gain relative to zero of absolute pressure and temperature at the stations considered. In general $(E - T \,.\, \Delta s)$ represents what is called energy availability, the part of total energy available for performing work, also often called exergy of the substance. Further discussion follows later. Here it is adequate to note that E and Δs can be expressed in terms of state parameters governed by the equation of state.

It may be observed that definitions (1) and (2), refer to the ratios of energy and exergy realized at the combustor exit, respectively, with respect to their values at entry. Each ratio is less than unity due to the loss processes between the entry and exit. The losses occurring in exergy include, in part, irrecoverable losses due to entropy gain.

Before proceeding further, it may be of interest to note that a set of efficiencies were proposed in Ref. 3 for examining the air inlet as a component of a hypersonic propulsion system. These have been described in Chapter II.8 of this volume, along with an additional efficiency parameter (Ref. 8 of that chapter) namely the so-called compression efficiency. All of these efficiencies are convertible into one another for the case of an ideal gas with constant ratio of specific heats, and Appendix A of this chapter provides a note on such conversion.

A. Kinetic Energy Efficiency

The concept of kinetic energy efficiency, η_{KE}, expounded in Refs. 3–5, can be used to obtain the following definition with reference to Fig. 2:

$$\eta_{\mathrm{KE}} = \frac{H_{\mathrm{te}} - H_e'}{H_{t0} - H_0} = \frac{V_e^2}{V_0^2} \tag{3}$$

where H, H_t and V refer to static enthalpy, stagnation enthalpy, and velocity, respectively, and subscript e refers to any station, other than the reference ambient state denoted by subscript O, along the flowpath, H'_e denotes the value of enthalpy at the end of isentropic expansion from H_{te} to ambient pressure. The definition has been extended to include the change in stagnation enthalpy between stations and 0 and e, by writing

$$\eta_{\mathrm{KE}} = \frac{V_e^2}{V_0^2} \cdot \frac{H_{t0}}{H_{\mathrm{te}}} \tag{4}$$

It will be noted that Eq. (4) does not address the cause of the stagnation enthalpy change between 0 and e. The change may be due in part to heat transfer through the walls and, in a combustor, due to heat release between 0 and e. The change in stagnation enthalpy can be related to the change in stagnation pressure and also in static pressure for given conditions at 0. Such changes are determined directly when one assumes that the working fluid can be considered everywhere a perfect gas with constant specific heats. In general, combustion may involve nonequilibrium chemistry with various complexities in reaction paths and products, and heat of formation and thermal properties of different substances.

In the case of an inlet, Eq. (4) is often applied, noting that, in general, 1) the mass flow does not change, and 2) any change in stagnation enthalpy may only be due to heat transfer or change in internal state of the fluid. There is, of course, friction and shock loss, leading to changes in stagnation pressure, which, however, can be included explicitly in the kinetic energy definition. Thus, η_{KE} represents a reasonable measure of the performance of an inlet.

In the case of a scram combustor, the process of heat addition can vary widely in different cases, and it is usual to undertake thermodynamic analysis in terms of processes with constant pressure (or, equivalent, constant velocity, as in the well-known Brayton cycle), constant Mach number, or constant area of cross-section. Furthermore, addition of fuel changes the mass flow, while combustion heat release changes the stagnation pressure even in the absence of friction and shock losses. These matters are discussed in Appendix B of this chapter. Here, it is adequate to note the following:

- The kinetic energy efficiency given in Eq. (4) involves V_0^2. It can be seen that η_{KEC}, the kinetic energy efficiency for a combustor, becomes for the case of an isentropic inlet directly proportional to P_{t3}/P_0 and inversely proportional to M_0^2, noting

$$\eta_{KEC} = \frac{1-(P_0/P_{te})^{\gamma-1/\gamma}}{1-(P_0/P_{t0})^{\gamma-1/\gamma}} \tag{3a}$$

Thus, the kinetic energy efficiency becomes a function of the reference ambient conditions chosen. If the inlet efficiency is assumed to be unity. η_{KEC} is still a function of flight or inlet entry Mach number.

In the case of an inlet (and also, provided the expansion is complete to ambient conditions, a nozzle) the component interacts directly with the ambient atmospheric condition, and there is no ambiguity about the definition of η_{KE}. However, the combustor entry conditions may only be related indirectly to the ambient atmosphere. In a connected mode test, thus, the kinetic energy efficiency for a combustor may only be defined with respect to a chosen set of reference conditions (M_0, P_0), and therefore is of value only in comparing combustors in the same type of tests with the same reference conditions, or the same combustor under different connected mode test conditions but the same reference conditions. Thus, also, η_{KEC} is a parameter of limited usefulness for comparing different combustors in connected mode tests under different reference conditions.

- The kinetic energy definition of Eq. (3) when applied to an inlet and a nozzle respectively, is not the same as the inlet and nozzle efficiency introduced in his analysis by Builder (see discussion in Ref. 1). Thus, the optimum diffusion, or, equivalent, the thermal ratio (or thermal compression ratio, as it is sometimes referred to), is given in the case of constant pressure combustion by

$$\psi_{\mathrm{OPT}} = \left[\frac{\eta_I \cdot \eta_N}{1 - \eta_I \cdot \eta_N} \cdot \frac{Q}{H_0} \right]^{1/2} \tag{5}$$

where the subscripts I and N refer to inlet and nozzle, respectively, and Q denotes the heat supplied per unit mass or airflow.

$$\eta_{\mathrm{I}} = \left[\eta_{\mathrm{KE\ inlet}} + \left(\frac{V_1}{V_0} \right)^2 \right] \Big/ \left[1 - \left(\frac{V_1}{V_0} \right)^2 \right] \tag{6a}$$

and

$$\eta_N = \left[\eta_{\mathrm{KE\ nozzle}} + \left(\frac{V_3}{V_5} \right)^2 \right] \Big/ \left[1 - \left(\frac{V_3}{V_5} \right)^2 \right] \tag{6b}$$

It is possible that (V_1/V_0) is smaller than (V_3/V_5); then, the nozzle performance may be more significant than that of the inlet in choosing ψ.

The performance of the combustor enters into the determination of ψ_{OPT} through Q, and there is no simple relation between Q and η_{KEC}.

- The numerical value of η_{KE} for any component is nearly equal to unity, and therefore does not provide a readily visible distinguishing parameter. Noting this fact and also that estimates of the parameter are hard to come by in available experimental data, an attempt is made, quite successfully, in Ref. 4 to introduce kinetic energy efficiency of various components along the flowpath to determine the specific impulse of the engine, which can be written as a function of overall engine kinetic energy efficiency, η_{KEO}, given by

$$\eta_{\mathrm{KEO}} = \eta_{\mathrm{KEinlet}} \cdot \eta_{\mathrm{KEcombustor}} \cdot \eta_{\mathrm{nozzle}} \tag{6}$$

in the case of a simple scramjet engine. The expression for specific impulse, following Ref. 5, is

$$I_{\mathrm{SP}} = \frac{V_0}{\varphi g} \left\{ \left[\eta_{\mathrm{KEO}}(1+\varphi) \left(1 + \frac{\varphi Q^*}{H_{t0}} \right) \right]^{1/2} - 1 \right\} \tag{7}$$

for a flight velocity V_0, when the fuel to air ratio is φ and the heat added by unit mass of the fuel is Q^*. Figure 3 displaying the variation of I_{sp} as a

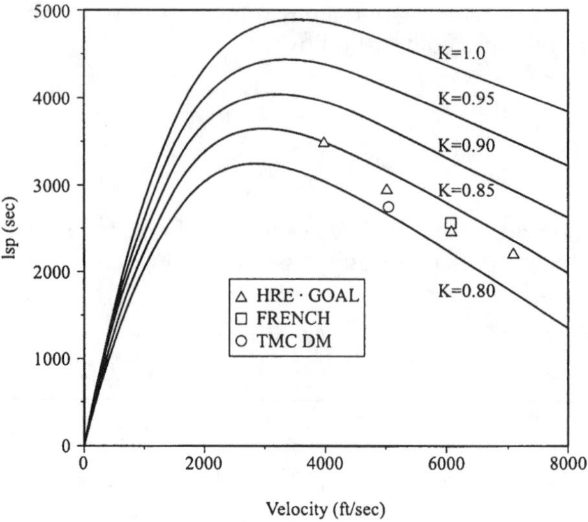

Fig. 3. Performance of Scramjet Engines with Hydrogen Fuel, Stoichiometric Combustion.

function of V_0 using η_{KEO} as a parameter is reproduced from Ref. 4; some available experimental data were also included in that figure.

- Real gas effects and nonequilibrium chemical reaction effects are extremely important in combustors and also nozzles. The kinetic energy efficiency should be determined accounting for these effects; an estimate based on assuming a constant value of ratio of specific heats is not, in general, valid.

B. Energy Availability Efficiency

It is a natural law that the total energy of a substance in a given state is available for performing work only with respect to its current state. Any gain in energy for a substance at a given state, however brought about, is again available in total for performing work only in its current state. An unambiguous parameter for representing the state of a substance is its energy availability with reference to absolute zero of state conditions or another such reference state. It is, however, a comprehensive parameter, and thus is both advantageous in comparing the overall performance of one system with another, and, at the same time, however, not entirely and directly useful in design or scaling. In such tasks as improving design and designing a size-based variant, one obviously needs a knowledge of primitive parameters (such as pressure, enthalpy, velocity, etc.), and energy availability does not provide explicitly the effect of changes in such parameters. At the same time, for a working fluid on the whole, energy availability is the single, unambiguous and comprehensive parameter representative of its ability to perform work.

It is now possible to construct two efficiency parameters based on the use of energy availability as follows:

1. Component-Wise Exergy Efficiency

Referring to Figs. 1 and 2, one can define an efficiency based on energy availability change across the combustor as follows.

$$\eta_{\varepsilon 1} = [\varepsilon_3 (1 + \varphi) - \varepsilon_1] / \varepsilon_{1f} \tag{8}$$

where

$$\varepsilon_1 = \varepsilon_{1a} + \varphi \cdot \varepsilon_{1f} \tag{9}$$

and

$$\varepsilon_{1f} = \varepsilon_{1fP} + \varepsilon_{1fC} \tag{10}$$

the subscripts *a* and *f* referring to air and fuel, and the additional subsripts *P* and *C* referring to the physical and chemical components. Physical exergy is based on kinetic energy and enthalpy of fuel; and chemical exergy is based on heat generated by combustion of fuel under stoichiometric conditions at the supply pressure and temperature. Thus, $\eta_{\varepsilon 1}$ provides a measure of the efficiency with which the combustor is able to convert the total energy of fuel supplied to the given air stream into exergy gain. It will be observed that $\eta_{\varepsilon 1}$ is not the same as $(1 - \eta_{SL})$ in the definitions provided here.

Between stations 1 and 3, corresponding to combustor inlet and exit, the change in enthalpy should account for the heat generated by combustion, the loss of heat to the walls due to work done on wall friction, and radiative transfer, if any, and changes in the molecular state of the working fluid (due to dissociation and ionization). The heat transferred may, in some part, become re-introduced into the combustor as fuel enthalpy or kinetic energy; this, however, would be accounted for in the energy supplied by the fuel.

The loss of energy availability is thus due to 1) entropy gain depending on the type of heat addition process, 2) energy entrapment in the internal degrees of freedom of the molecular, atomic, and ionic states, 3) friction work at the wall, 4) mixing work between the fuel and air streams, and 5) heat transfer to the wall. The extent of the loss of energy availability in the combustor is shown by the value of $\eta_{\varepsilon 1}$.

2. System-Wise Exergy Efficiency of Combustor

Referring again to Figs. 1 and 2, one can define a second efficiency parameter, namely

$$\eta_{\varepsilon 2} = (\varepsilon - \varepsilon_4)(1 + \varphi) / (\varepsilon_{1a} + \varepsilon_{1f} \cdot \varphi) \tag{11}$$

where ε_4 is the value of exergy per unit mass of products of combustion expanded to ambient pressure. It may be pointed out that $\eta_{\varepsilon 2}$ is thus determined with respect to a flight altitude and flight velocity. The efficiency parameter provides a measure of the best work output realizable at a given flight velocity by expansion of products of combustion to ambient pressure, relative to the exergy input with air and fuel. The exergy values are considered with reference to ambient temperature and pressure.

The best work output feasible may be looked upon as the best thrust work obtainable and therefore as a representation of best thrust realizable. $\eta_{\varepsilon 2}$ is thus similar to the thrust coefficient, θ, of Builder, but expressed as a ratio of energy availability with reference to ambient pressure and the input energy availability of air and fuel. In any case, $\eta_{\varepsilon 2}$ is in the nature of an engineering parameter to depict overall engine performance.

It may be pointed out that the expansion to atmospheric ambient conditions invoked in the definition is not arbitrary; at any location along the flight trajectory, the return of the combustion products to the end state of equilibrium with the reference condition of the atmosphere represents the work potential of the products. However, the selection of the flight velocity value in Eq. (11) is arbitrary insofar as the combustor is concerned, although the flight velocity has to be established with care while choosing the trajectory.

It will be observed that the parameter $\eta_{\varepsilon 2}$ can be utilized to compare 1) the performance of a given combustor operating under different initial conditions and also different values of φ, and 2) two different combustors operating with the same initial conditions and φ, in both cases, however, relative to fixed flight conditions.

The two efficiency parameters $\eta_{\varepsilon 1}$ and $\eta_{\varepsilon 2}$ serve two different purposes, and therefore both are necessary when the combustor is considered as part of a flight propulsion system.

C. Stagnation Pressure Efficiency

In Ref. 3 an efficiency was defined based on the loss of stagnation pressure during supersonic flow diffusion in an inlet. Since the primary interest in high speed flow is the extent of diffusion and the inefficiency of that process, and as Mollier charts for air were not available over the pressure range of interest, this efficiency parameter was not pursued. However, across a scram combustor, the loss of stagnation pressure is of fundamental interest, and one may introduce an efficiency based on such loss as follows:

$$\eta_{PT} = P_{T3} / P_{T1} \tag{12}$$

It is also possible to see from Eq. (3a) how η_{PT} and η_{KE} can then be related under various assumptions. Also

$$\eta_{PT} = \frac{P_{T3}}{P_{T0}} \cdot \frac{P_{T0}}{P_{T1}} \tag{13}$$

the second term on the right hand side representing the stagnation pressure ratio across the inlet.

D. Combustion Process

The two main fuels utilized in scram combustors are hydrogen and hydrocarbons, with various types of additives for increasing chemical reactivity and density. In the following only hydrogen fuel is considered. Two overall parameters depicting effectiveness of fuel combustion over a duration, and equivalently, a length of combustor or combustion are 1) the heat released, and 2) the amount of a reference element such as hydrogen that is converted to water indicating completion of combustion of that element. The latter may also be expressed in terms of the amount of hydrogen that remains in any form other than that of water at a location of interest in the combustor in relation to the amount of hydrogen made available initially.

We may therefore define three efficiencies as follows referring to the combustion process in each case considering a unit mass of fuel added:

$$\eta_{C1} = \frac{\text{heat released}}{\text{heat releasable by hydrogen supplied}} \tag{14}$$

$$\eta_{C2} = \frac{\text{amount of hydrogen reacted}}{\text{amount of hydrogen supplied}} \tag{15}$$

and

$$\eta_{C3} = \frac{\text{amount of hydrogen converted to water}}{\text{amount of hydrogen supplied}} \tag{16}$$

It will be observed that the three efficiency parameters implicitly include 1) the fuel to air ratio used and 2) the effects of mixing of air and fuel in initially non-premixed cases. At the same time, the efficiency parameters do not refer explicitly to the effects of ignition process or changes in combustion rate due to any fluid mechanical or transport causes.

E. Set of Efficiencies

In order to undertake thermodynamic analysis of a gaseous flow, it is necessary to define the state of the fluid in terms of two state variables and the kinetic energy of gas motion. A set of efficiencies based on those parameters or some equivalent of those would also serve to define the efficiency of a process involving the gas flow. Thus, we recommend the use of kinetic energy efficiency, the component-wise exergy efficiency, and the efficiency based on loss of stagnation pressure as a set for the determination of the efficiency of a scram combustor, comprehensively, uniquely, and unambiguously. All three efficiencies are important, both in comparing the performance of a combustor under different conditions, and comparing one combustor with another.

It may be pointed out here that conversion of one efficiency parameter to another is possible and simple only for a perfect gas with constant ratio of specific heats. Since scram engines do not operate under such conditions, conversion is not feasible in general. Thus each of the efficiencies becomes unique. At the same time, each definition of efficiency leaves out of account changes in one or more parameters across the combustor. Finally, the three efficiencies together account for all state and fluid dynamic variables, including all chemical reaction and combustion processes.

In general, the various efficiencies are unlikely to become the highest at a single location along the combustor for given geometry, fuel injection, and entry conditions. From the point of view of energy analysis, $\eta_{\varepsilon 1}$ may be selected as the dominant or reference performance parameter. Under given conditions of operation, $\eta_{\varepsilon 1}$ varies as a function of length along the combustor. Noting the importance of combustor length in vehicle configuration synthesis, as has been discussed in this volume in Chapter IV.17, the location along the length where $\eta_{\varepsilon 1}$ attains a peak value may be chosen as the location of interest, and other efficiencies, namely $\eta_{\mathrm{FL}}, \eta_{\mathrm{SL}}, \eta_{\mathrm{KE}}, \eta_{\varepsilon 2}, \eta_{\mathrm{PT}}, \eta_{C1}, \eta_{C2}$, and η_{C3}, determined at that location to depict details of performance as they become of interest. Ordinarily $\eta_{\varepsilon 1}$, η_{KE}, and η_{PT} are adequate to undertake thermodynamic estimates of performance. Next, η_{FL}, η_{C1}, η_{C2}, and η_{C3} provide, as a set, measures for the combustion and heat release process. Finally η_{SL} and $\eta_{\varepsilon 2}$ can be utilized to depict performance in terms of parameters of engineering interest.

An example may be useful here.

1. Enhanced Combustion and Detonation

Both the processes are assumed to occur under the same supersonic flow conditions. Enhanced combustion is considered as deflagration of initially unmixed reactants when a small volume has attained a degree of mixing over a short length of travel of reactants and combustion is initiated either by self-ignition or supply of necessary energy by external means. The ignition process may be represented by equilibrium chemical reaction. Combustion then continues along the supersonic flow, which undergoes changes by internal heat addition, geometry of the duct, and friction and heat transfer losses.

In the case of detonation, the reactants are assumed to be fully mixed, and ignition is expected to set in by the increase in temperature arising on account of the formation of a shockwave. Combustion is expected to be completed in a negligibly short period of time and over a short distance. Considering an oblique shockwave for generality, a stable detonation requires that V_{2n}, the normal component of velocity downstream of the shockwave, should be equal to or less than the local acoustic velocity, since a combustion wave propagation is governed by molecular processes even as an acoustic wave is. The important requirement in a detonation is the coupling between combustion and shock process, although in reality the (short, as specified earlier) characteristic length for combustion is several orders of magnitude larger than that of the shock process. Combustion occurs in the reduced supersonic speed of flow following passage through the shockwave; the flow speed then increases by the action of heat release to some higher value, the increase depending on the fuel-to-air ratio and heat released by combustion. These matters are discussed in Refs. 6–8. Appendix B provides the thermo-gas dynamic relations of interest, governing the two processes for ready reference.

Here the interest is in comparing the two processes on the basis of η_{KEC}, and $\eta_{\varepsilon 1}$, and η_{PT}, and further showing that the three together provide a basis for comparison. In order to proceed it is necessary to specify the heat addition process further, for example, by considering combustion in ducts of 1) constant area, 2) constant flow Mach number, and 3) divergent area following a constant area section.

As shown in Appendix B, the C-J detonation wave is the best means of heat addition, and the ODW is the next in order. In deflagration, constant pressure combustion is better than constant Mach number combustion. In a diverging area combustor, the Mach number always increases, as opposed to a constant area combustor with a decrease in Mach number. Accordingly the losses are always greater in a diverging area combustor than in a constant area combustor. The losses must include the stagnation pressure loss explicitly, since stagnation pressure change is not a deducible parameter given the change in stagnation enthalpy and the change in static pressure and static enthalpy.

It is often considered in practice that the specific impulse or the thrust work that can be obtained from a combustor in relation to the fuel supplied is an adequate performance parameter for engineering analysis. However, such a parameter does not provide either a means of comparing different combustors or of modifying a combustor in a particular manner. Even within a class of combustors, for example a constant area duct followed by a divergent section, a detailed comparison would require the change in stagnation pressure, static temperature, and entropy gain before modifications can be considered. Entropy gain is a comprehensive parameter, it is true; therefore one also needs to know how the static pressure gain for a given heat addition is modified in different cases.

III. Computational Tool and Limitations

The efficiencies that together represent the effectiveness of a combustor have been defined with respect to unit mass flow of air at entry to the combustor. When the flowfield is considered on an integral basis, for example under the assumption of one-dimensional flow, the efficiencies can thus be readily determined. The one-dimensional approximation can be appropriately modified, when necessary, with a non-uniformity coefficient to account for crosswise variations in state or flow properties; however, the variation of such coefficients along the flow will have to be obtained by other means. For example, flow predictions can be carried out along with fuel injection, but no chemical reaction, in the given geometry of the combustor. With given initial flow conditions on an appropriate multi-dimensional basis, one-dimensional averaged values of flow and state variables can be established along the flow. Changes in such variables can then be expressed in the form of changes in nonuniformity coefficients along the flow for use in one-dimensional energy analysis.

Also, a form of one-dimensional procedure can be applied to multiple streams that together make up a given stream. An example of the latter approach is the division of a flow into a core inviscid stream with adjacent viscous layers at the flow duct walls. One approach to flowfield analysis based on such an artifice has been discussed in Chapter II.15 of this volume.

However, a scram combustor flowfield is fully three-dimensional with imbedded shocks, often including recirculations, local regions of concentrated vorticity, and highly complex shear layers. There is also a possibility of local subsonic regions, if only on a transient basis. These complexities arise variously depending on inlet discharge flowfield, possible effects of the isolator, combustor geometry, fuel admission features, ignition of the mixture and flame propagation dynamics, and the general progress of combustion. It is, therefore, very difficult to establish *a priori* what flow features to de-emphasize in any flowfield prediction scheme to set up a simplified model. In recognition of such difficulties, considerable effort has been made for determining the flowfield on a fully three-dimensional basis. These efforts have yielded greatly interesting results and insights; however, they also have several limitations:

1) Predictions with complex inflow and boundary conditions are difficult and require substantial computational resources and time. It is extremely difficult, if not well nigh impossible, to verify the details of the predictions through measurement in well-defined test cases. At this stage various codes can yield useful results only in respect of the overall features of the flow along with one or two selected specific features.
2) In undertaking energy analysis, there are considerable ambiguities in assigning such integral parameters as mass flow over the grid, and estimates of energy and exergy flux over a cross-sectional plane become erroneous in unknown extent. Dense grids do not necessarily yield better accuracy.
3) Some processes such as three-dimensional flow separation and, during combustion, the occurrence of local or global extinction (due to long reaction times or instability) are extremely difficult to capture in current codes, except through an externally imposed physical criterion.
4) Experience so far has been that detailed three-dimensional calculations are too expensive in preliminary design. Their use is most effective in improving a design based on test results with respect to specific features.

At the same time, one-dimensional, that is integrated, flow calculations can be variously improved with some knowledge of fully three-dimensional flow predictions. For example, integrated flow calculations require the following in a parametric form: 1) mixing schedule and efficiency, 2) wall friction coefficient, 3) shock formation and losses, and 4) specific features of the flow arising from complexities in geometry or other mechanical causes, including fuel injection effects. Such features can be obtained through calculations of the flowfield under chemically non-reactive conditions. The results have the virtue of nearly detailed verification with current-day measurement feasibilities even in short-duration facilities. Not enough attention appears to have been paid to this aspect of prediction, at least as observed from the lack of adequate correlations for those parameters. In any case, such correlations may also be derived from predictions. Meanwhile, any set of one-dimensional flowfield calculations can be improved with some predictions from three-dimensional non-reactive flowfield predictions.

A one-dimensional code has been utilized here, the initial version for which was provided by the NASA Langley Research Center (McClinton, C.R., private communication, NASA Langley Research Center, Nov. 1996). The output desired is the performance of the combustor including the changes in the state properties of the working fluid and the set of efficiencies everywhere along the combustor. The working fluid consists of the air and fuel supplied.

A. One-Dimensional Calculation Scheme

The one-dimensional calculation scheme is based, as shown in Fig. 4, on recognizing three streams in the combustor, namely the air stream, the fuel stream, and the product stream, each on a one-dimensional basis, and the three together also on a one-dimensional basis. The mass flow in each stream is determined along the combustor as the mixing and combustion between air and fuel proceed. The mass averaged properties of the three streams are utilized to establish mean values which determine the chemical reaction, friction loss, and heat transfer.

Chemical reaction between the fuel and air is assumed to begin in a small volume of mixture that is estimated based on the fuel equivalence ratio, and the reaction rate is specified on the lines of the Arrhenius reaction rule or another as desired. The thermal properties of various reactants and products are taken in standard polynomial form relative to temperature.

The mixing rule between the streams may be chosen as desired, but must be specified. Friction loss and heat transfer loss also may be chosen and specified as desired.

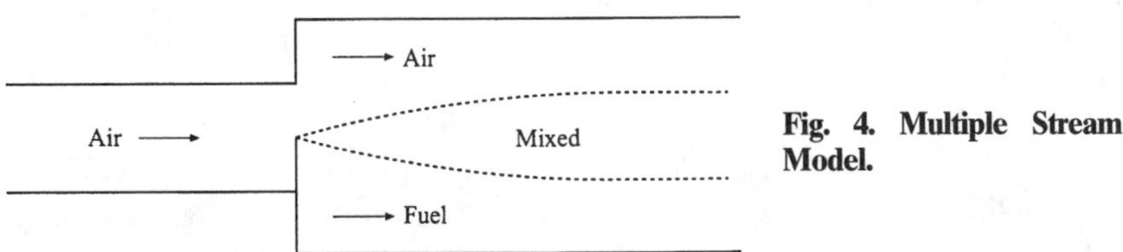

Fig. 4. Multiple Stream Model.

Some other possible variations included in the calculation procedure are as follows:

1) The air stream composition and properties may be chosen as desired.
2) The fuel stream composition and properties, as well as the injection angle of fuel relative to the air stream, may be chosen as desired. When the fuel stream is not admitted parallel to the air stream, the momentum of the fuel can be taken into account in an appropriate part along the air stream.
3) The location of commencement of chemical reaction and combustion may be chose as desired, either with no external ignition energy addition or with the required amount of such addition.
4) Chemical reaction may be specified in as much detail as necessary. For example, the initial reactions on ignition may be treated as equilibrium reactions, and later reactions can be assumed to follow non-equilibrium reaction schemes.

The combustor geometry along the flow can be specified as necessary; however on a one-dimensional basis only the cross-sectional area and the wall surface wetted perimeter are required in the calculations. Other details of the geometry cannot be taken into account, for example the geometry of backward facing steps or the presence of cavities. However, discrete changes in area variation can be introduced at desired locations along the flow, with multiple injection locations for fuel admission.

By the same token it is also possible to introduce discrete fuel admission at several locations along the flow, for example coincidentally with cross-sectional area changes. In each such case of fuel admission both the angle of injection relative to the air stream and the mixing rules can be chosen as desired.

1. Mixing Rules

A set of mixing rules that may be adopted consists in specifying 1) the rate of mixing, 2) the length of mixing, and 3) a mixing efficiency at the end of mixing. Other sets of rules may also be specified. Chapters I.6, III.9 and IV.15 provide examples of such rules.

2. Friction and Heat Transfer

A coefficient of friction may be specified directly for the combustor, or one based on local Reynolds number, e.g., the coefficient is proportional to $(\ln Re)^{-n}$, or $(Re)^{-n}$ depending on the velocity distribution function chosen for the boundary layer and the turbulent characteristics of the boundary layer.

Heat transfer is generally calculated based on Reynolds analogy, but another rule may be specified.

3. Air and Fuel Composition

In addition to standard air, air vitiated with products of combustion (as may happen during combustion-generated heating), atomic oxygen (as may happen during arc-heating of air) or humidity (again due to combustion-generated heating or unavailability of driers) may also be specified.

The calculation procedure has no means of accounting for formation of shockwaves or expansions along the flow. If details are available on such processes, for example from experimental data or other means, one can introduce such concentrated changes locally in the form of property changes relative to the calculated one-dimensional values.

In a similar fashion, any nonuniformity in flow and property variations at entry may be expressed in the form of nonuniformity coefficients for mass, momentum, and energy. However, this begs the question of variation of such nonuniformity coefficients along the combustor, and one needs this information from another source.

IV. General Illustrative Studies

The illustrative cases are divided into two groups: 1) parametric studies, and 2) studies on selected variations of operating conditions and observed effects.

In all cases, the fuel is assumed to be hydrogen. In certain cases silane has been added to increase chemical reactivity at low temperatures. The reaction schemes and rates are obtained from Refs. 9 and 10, and included for ready reference in Appendix C.

A. Parametric Studies

The parametric studies have been conducted with 1) two basic geometry variations, 2) two injection schemes, and 3) variations of various other parameters.

Two generic geometry variations, shown in Figs. 5a and 5b, are chosen for the studies conducted, namely, 1) constant cross-section duct with a step, and 2) a duct with initially constant cross-section followed by a series of panels to obtain varying area geometry.

The injection schemes are assumed to be one or more injections through the wall at different locations and at different angles to the air stream. It is assumed that the fuel momentum component at injection that is not in the general flow direction of air is lost as part of mixing, and only the component of momentum in the direction of air flow is accounted for in subsequent analysis. This then is the means adopted to distinguish normal, parallel, and other types of injection schemes.

The operational flow parameters of interest are 1) air input conditions: pressure, temperature, velocity, and concentration of oxygen, and other gases, if any; 2) fuel input conditions: temperature, velocity, and equivalence ratio; 3) additive to improve ignition and reactivity: silane concentration; 4) mixing of air and fuel; 5) wall friction; 6) cooling and thermal management; and 7) ignition schemes.

Mixing of air and fuel is based on specifying 1) rate of mixing, 2) length of mixing, and 3) maximum efficiency of mixing at the length chosen.

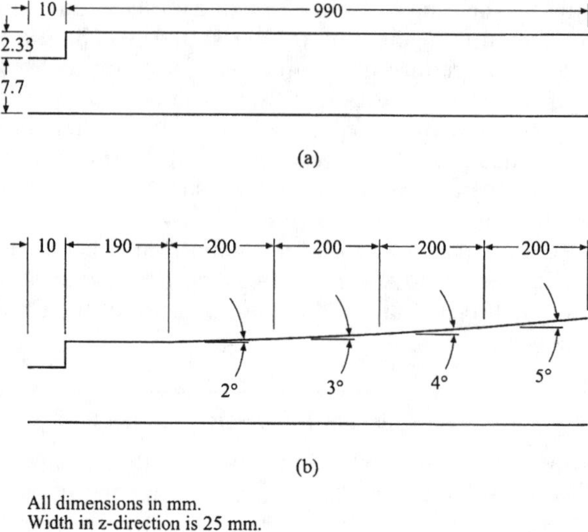

Fig. 5. Schematic of Combustor Geometry. (a) One-dimensional 3-Stream Model. (b) Typical Combustor Schematic - 1. (c) Typical Combustor Schematic - 2.

The exponential "mixing rule" employed is as follows. The exponential rule is employed for the calculation of mixing efficiency at any location.

$$\eta_m = 1 - e^{-cc \cdot x} \tag{17}$$

where x = axial location, $cc = \ell n\,(1 - \eta_{max}) / L_{mix}$, L_{mix} = specified mixing length, and η_{max} = maximum mixing efficiency at the end of mixing length.

In parametric studies, mixing is varied in different cases. One type of mixing assumed is referred to as "typical mixing" in which case the following are specified: 1) mixing length, and 2) maximum mixing efficiency at the end of mixing length. For example, a mixing length of 20 cm. and maximum mixing efficiency of 95% may constitute a typical mixing definition. In other cases, mixing is specified in terms of mixing lengths for different values of mixing efficiency.

Ignition is assumed to occur instantaneously when the initial mixed volume is at or above the ignition temperature for the fuel. The initial mixed volume is a small volume in the vicinity of fuel injection where the fuel and air are assumed to be mixed instantly to attain the maximum efficiency value selected for the case under consideration.

When the initial mixed volume temperature is below the ignition temperature of fuel, ignition is assumed to be 1) delayed until the mixed volume temperature becomes equal to the ignition temperature by the action of a wall friction, or 2) initiated by the supply of required energy from an external source to bring the initial mixed volume temperature to the ignition level. (In the latter case, it may be noted that the external energy supplied needs to be accounted for in the final energy balance calculations.)

At ignition, the chemical reactions may be assumed 1) to proceed as under non-equilibrium conditions, or 2) to be initiated under equilibrium conditions with corresponding large changes in pressure and temperature, to be followed by a form of relaxation and non-equilibrium reactions.

Next, the friction loss in the combustor depends on the flow in the vicinity of the boundary wall and hence the flow Reynolds number, the wall surface roughness (which changes over time due to deposits), the action of shock-waves embedded in the flowfield, and the heat transfer to the wall, which affects local turbulence characteristics and also the shock-boundary layer interaction processes. The determination of friction coefficient and the associated heat transfer losses may be specified by adopting 1) a so-called "typical friction coefficient" of 0.003 and 2) Reynolds analogy for heat transfer. The combustor may be cooled convectively with fuel that may be utilized for combustion, when the fuel entry conditions into the combustor need to be chosen appropriately.

The *output from the calculations* follows. The performance of the combustor with hydrogen fuel is derived in terms of the resulting variations along the combustor of 1) pressure and temperature, 2) Mach number and flow velocity, 3) mass fraction of water produced, and 4) mass fraction of total hydrogen that remains unreacted.

In each case, the output is post-processed to obtain the following: 1) the local values of various efficiencies along the combustor from entry up to a specified length of the combustor; 2) the location along the combustor where the exergy-based efficiency, $\eta_{\varepsilon I}$, is a maximum, and the values of the other efficiencies at that location; and 3) any other features of interest.

In addition the specific impulse that can be realized by expansion of combustion products at the location with the best value of $\eta_{\varepsilon 1}$ to a set of realistic, albeit arbitrary, reference ambient conditions has been calculated. It is possible that in certain cases there may be no possibility of realizing thrust with reference to chosen ambient conditions.

B. Results

1. Parametric Studies

The parameters varied in different combinations in these studies are the following: geometry, straight duct and duct with multiple areas of cross-section (i.e. with panels); air inlet pressure P_{1a}, 2.0 and 0.4 atm; air inlet temperature T_{1a}, 900, 1100 and 1671 °K; air inlet Mach number M_{1a}, 3.0, 3.5, and 4.8; fuel inlet pressure P_{1f}, 4.0 and 1.2 atm; mixing efficiency η_m, 0.9, 0.7, and 0.5; and fuel equivalence ratio φ 0.7, 0.9. and 1.1. In all cases the fuel inlet temperature and Mach number are assumed to be 166.7°K and 2.0, respectively.

It is interesting to compare the performance in the following cases.

Details of all of the test cases studied and the predicted results are given in Ref. 10. An interested reader will find in that reference a discussion of the effect of the following on combustor performance: 1) air inlet temperature, 2) air inlet Mach number, 3) introducing several changes in area of cross-section along the combustor, 4) distributed injection, and 5) different mixing efficiencies under different injection conditions.

Here we confine attention to a small number of studies described in Tables 1 and 2; the predicted results for these cases are given in Figs. 6a–6f, each of which is in two parts. The efficiencies deduced for the location in the chamber where $\eta_\varepsilon{}^1$ is the maximum are given in Table 3.

Case 2.1. Case 2.1 deals with a straight duct combustor, Fig. 5b, operated with air and hydrogen. The initial mixing of air and fuel results in a reduction in temperature in the volume of the mixed fluid. The average temperature across the combustor increases along the flow due to the initial chemical reactions and wall friction. When the temperature becomes larger than the ignition temperature, the mixture becomes ignited and thereafter combustion occurs along the combustor. Figure 6a provides the performance parameters along the combustor corresponding to operation with three different values of fuel equivalence ratio: pressure, temperature, Mach number, mass fraction of water formed, and mass fraction of hydrogen unreacted in the mixture. It is found that the Mach number in all cases tends to a value of unity between 20 cm and 27 cm from the location of fuel injection; no further predictions are possible beyond that flow choking station.

Figure 6b shows the variation of six efficiencies along the combustor for operation at different values of fuel equivalence ratio. Table 3 provides the values of different efficiencies at the location where the exergy-based effi-

Table 1 Operational conditions in selected parameter variation cases: principal features

Case no.	Geometry	Main features	Figure no.
2.1	Straight duct	φ: 0.7, 0.9, and 1.1	6a,6b
6.1	Multiple section geometry	φ: 0.7, 0.9, and 1.1	6c
8.1	Multiple section geometry	Multiple fuel injection; $\varphi_1 = 0.5$; $\varphi_2 = 0.3$; $\varphi_3 = 0.2$; $\eta_{max} = 0.95$; $T_{1a} = 900°K$	
8.2	Multiple section geometry	Multiple fuel injection; $\varphi_1 = 0.5$; $\varphi_2 = 0.3$; $\varphi_3 = 0.2$; $\eta_{max} = 0.95$; $T_{1a} = 1100°K$	
9.1	Multiple section geometry	Multiple fuel injection; $T_{1a} = 900°K$ $\varphi_1 = 0.5$; $\eta_{1max} = 0.95$; $\varphi_2 = 0.3$; $\eta_{1max} = 0.7$; $\varphi_3 = 0.2$; $\eta_{1max} = 0.5$	6d–6f
9.2	Multiple section geometry	Multiple fuel injection; $T_{1a} = 1100°K$ $\varphi_1 = 0.5$; $\eta_{1max} = 0.95$; $\varphi_2 = 0.3$; $\eta_{1max} = 0.7$; $\varphi_3 = 0.2$; $\eta_{1max} = 0.5$	

Table 2 Operational conditions in selected parameter variation cases

Case no.	P_{1a}, atm	T_{1a}, °K	M_{1a}	P_{1f}, atm	T_{1f}, °K	M_{1f}	Mixing	Friction and heat loss
2.1	2.0	900.0	3.5	4.0	166.7	2.0	Typical; Lmix = 20 cm	Typical
6.1	2.0	900.0	3.5	4.0	166.7	2.0	Typical; Lmix = 20 cm	Typical
8.1	2.0	900.0	3.5	4.0	166.7	2.0	Typical; Lmix = 20 cm	Typical
8.2	2.0	1100.0	3.5	4.0	166.7	2.0	Typical; Lmix = 20 cm	Typical
9.1	2.0	900.0	3.5	4.0	166.7	2.0	Etam = 0.95, 0.7 & 0.5	Typical

ciency, $\eta_{\varepsilon 1}$, is a maximum, that is, for example with $\varphi = 0.70$, at a location 32.6 cm from the fuel injection station, the maximum efficiency value being 63.8%. Also, at that location, the highest value of static pressure is obtained and equal to about 14.0e5.0 Pascals, noting that the entry air pressure is about 2.03e5.0 Pascals.

The combustor performance at any fuel equivalence ratio is given, as stated earlier, by the combination of $\eta_{\varepsilon 1}$, η_{KE}, and η_{PT}. The combination of these has been obtained, as can be seen in Table 3, when the fuel equivalence ratio is 0.7, 0.9, and 1.1.

Case 6.1. Case 6.1 deals with a combustor that consists of a straight duct followed by one with a diverging wall; the wall is assumed to be diverging at 2°, 3°, 4°, and 5° from the horizontal beginning at 20 cms from the location of injection, over 20 cms at each angle. The combustor is supposed to be operated with three different values of fuel equivalence ratio in three different cases; the entire fuel is injected at one location, at the beginning of the straight duct part of the combustor.

Figure 6b provides the performance of the combustor with respect to various parameters. It is found that, only when the equivalence ratio is 0.7, the combustion continues up to the combustor exit; however the static pressure drops to the value of entry pressure at about 52 cm from entry and then continues to decrease. The static temperature also decreases after a substantial gain, following ignition, over about 20 cms. When the fuel equivalence ratio is raised to 0.9 and 1.1, the flow chokes in the first area expansion section.

Figure 6c provides the variation of six efficiencies along the combustor. The best values of efficiencies are obtained when the fuel equivalence ratio is 0.7; the additional length of the combustor provides adequate time for combustion completion. The wall divergence also may have had an effect; more on this aspect in Section 5.5 to follow.

Table 3 provides the estimated values of efficiencies in this case as in Case 6.1.

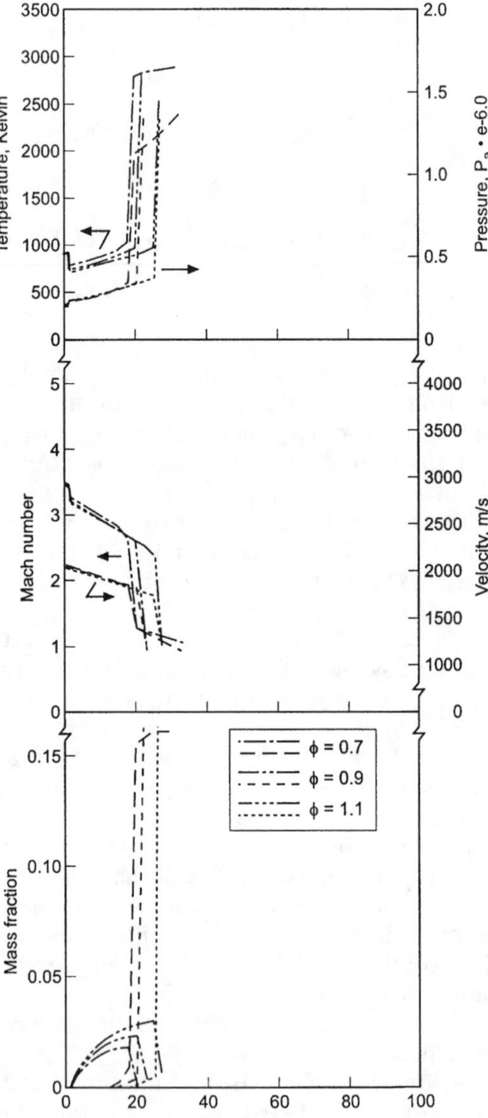

Fig. 6a Performance in Case 2.1. See Table 1 for operating conditions.

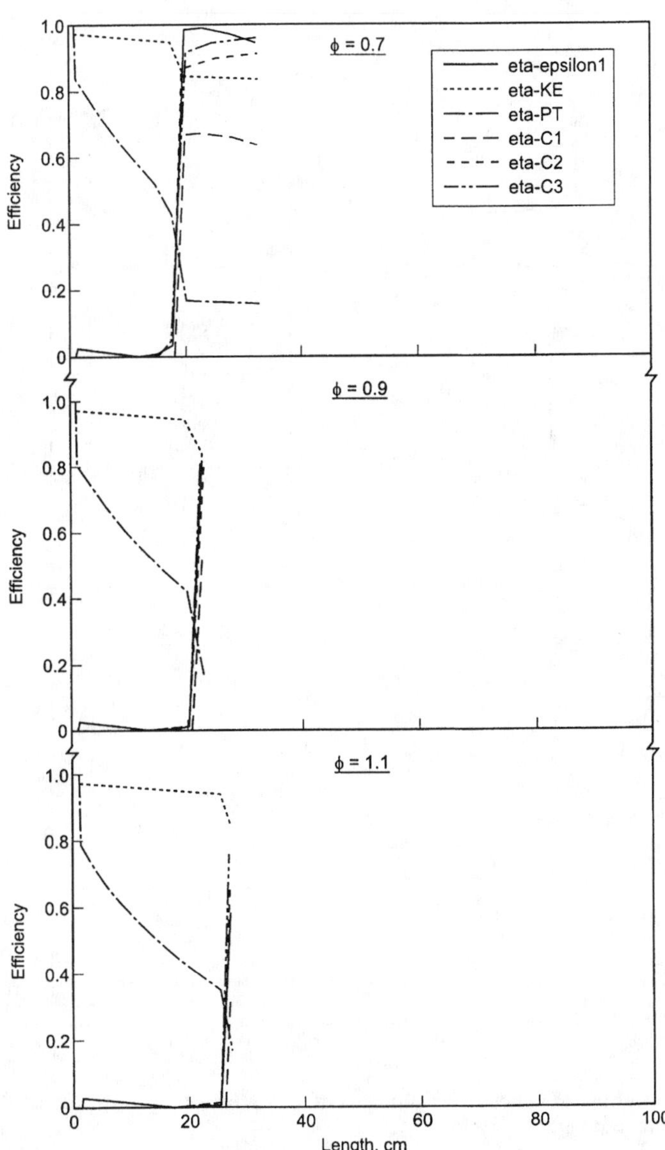

Fig. 6b Variation of efficiencies along combustor length in Case 2.1 at different values of fuel equivalence ratio, φ.

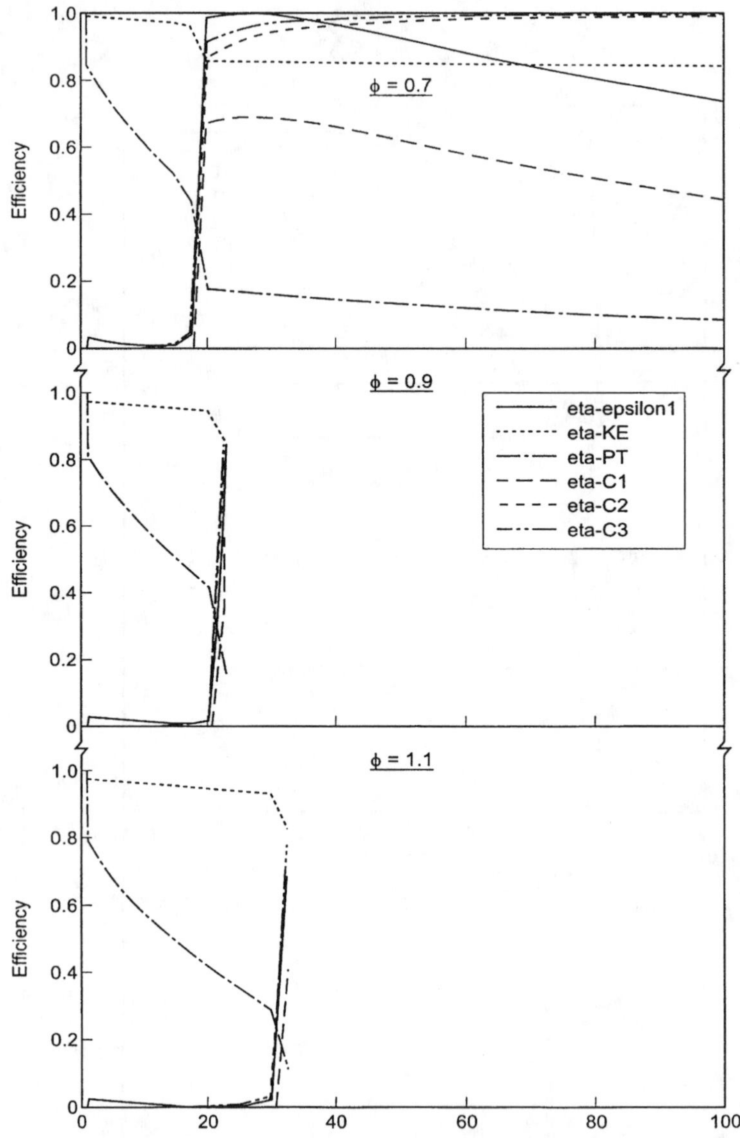

Fig. 6c Variation of efficiencies along combustor length in Case 6.1 at different values of fuel equivalence ratio, φ.

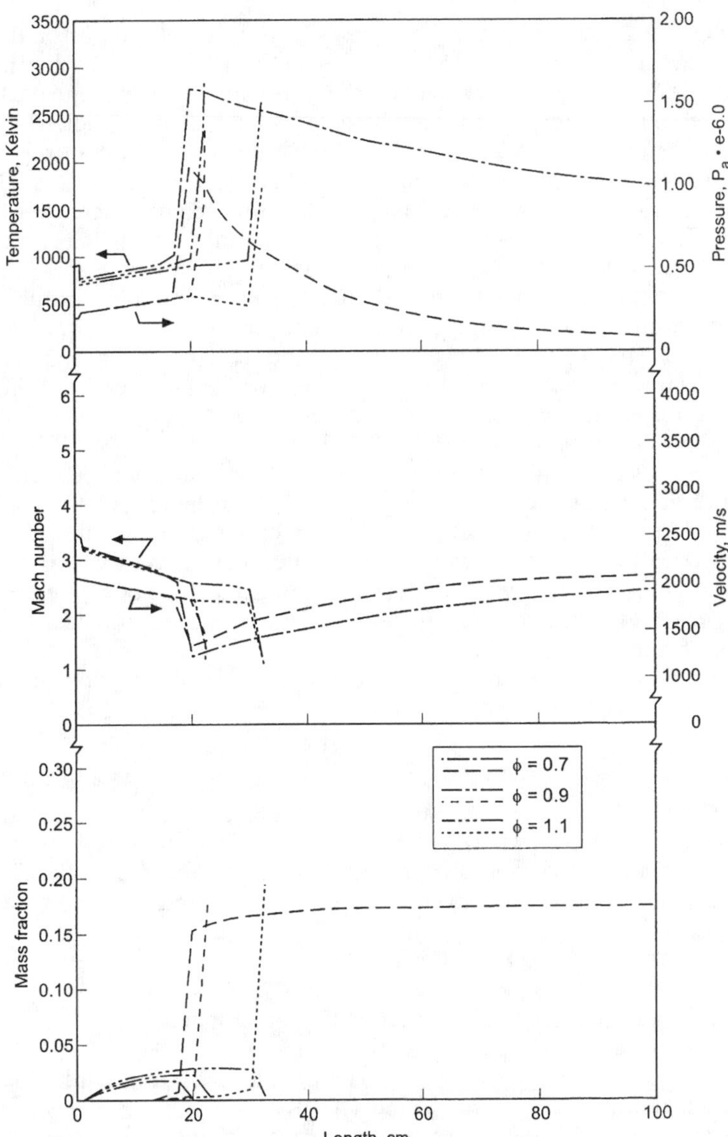

Fig. 6d Performance in Cases 8.1 and 8.2. See Table 1 for operating conditions; note multiple fuel injection at three area change locations, while η_m is the same for each injection of fuel; T_{1a} = 900°K and 1100°K.

Cases 8.1, 8.2, 9.1, and 9.2. These cases pertain to a combustor that is configured the same as in Case 6.1; however, in Case 8.1 fuel is supplied at entry to the combustor and at two turns of the wall, namely the first and the third turn, in amounts corresponding to different fuel equivalence ratios, while the supply air temperature is 900°K; Case 8.2 is the same as Case 8.1 except the supply air temperature is 1100°K; and Cases 9.1 and 9.2 correspond to Cases 8.1 and 8.2, respectively, with the maximum mixing efficiency decreasing from 0.95 to 0.7, and then to 0.5, serially, for the three amounts of fuel injected.

Figures 6d and 6e provide the performance parameters in Cases 8.1 and 8.2, and 9.1 and 9.2, respectively. Although the overall amount of fuel supplied corresponds to fuel equivalence ratio of one, one can see, in comparison with Fig. 6b no flow choking occurs. The effect of entry air temperature is minimal insofar as the combustor exit conditions are concerned, although there are differences in the changes in pressure and temperature immediately following ignition.

Somewhat surprisingly there are no large differences between the performance depicted in Figs. 6d and 6e; in other words, the reduction in mixing efficiency for the fuel admitted in the later sections of the combustor does not seem to affect the performance to any noticeable extent. Even as in Case 2.1, it is not possible to say what effect was produced (see Fig. 6f) by the wall divergence and the associated increase in area, relative to the effect of mixing efficiency.

Table 3 provides the efficiencies in the four cases at the location where $\eta_{\varepsilon 1}$ is the maximum.

2. *Studies on Selected Observed Effects*

Based on the large number of test results reported in the literature, a few cases have been chosen where specific features have been introduced in the operation of a scram combustor, for undertaking predictions. The operating conditions in these cases are given in Tables 4 and 5, along with the numbers of figures where the performance is shown.

These cases pertain to the following:

1) Case 1: presence of atomic oxygen in the air supplied.
2) Case 2: presence of water vapor in the air supplied.
3) Cases 3 and 4: mixing length reduced while operating with different equivalence ratios at T_{1a} = 900°K and 1100°K.
4) Cases 5, 6 and 7: mixing length varied while operating at different M_{1a} and fixed T_{1a} = 700°K.
5) Case 8: effect of multiple section geometry while operating with quick mixing at T_{1a} = 900°K and 1100°K.
6) Cases 9, 10, and 11 with silane added equal to 0%, and 1%: effect of operation with silane added to fuel at injection in different amounts.

The results for these cases are given in Figs. 7–17 and in Table 6. The latter provides the efficiencies again at the location where $\eta_{\varepsilon 1}$ is the maximum along the chamber.

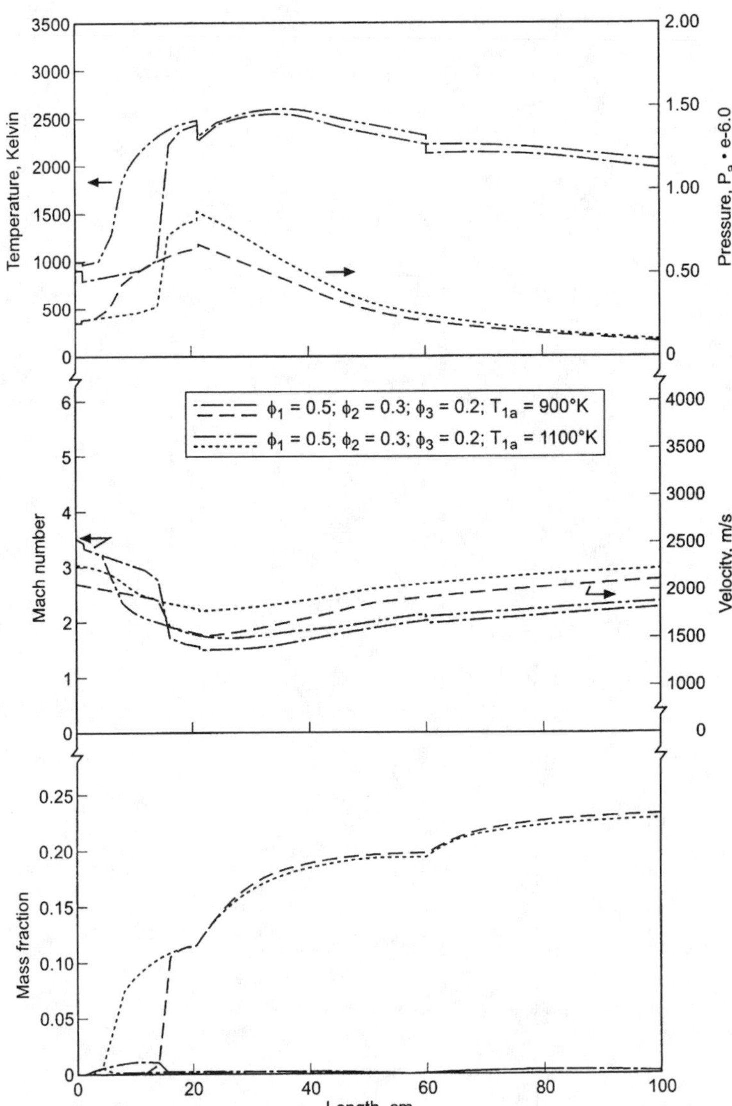

Fig. 6e Performance in Case 9.1 and 9.2. See Table 1 for operating conditions; note multiple injection ($\varphi = 0.5$ with $\eta_m = 0.95$; $\varphi = 0.30$ with $\eta_m = 0.70$; and $\varphi = 0.20$ with $\eta_m = 0.50$ at three successive area change locations); T_{1a} = 900°K and 1100°K.

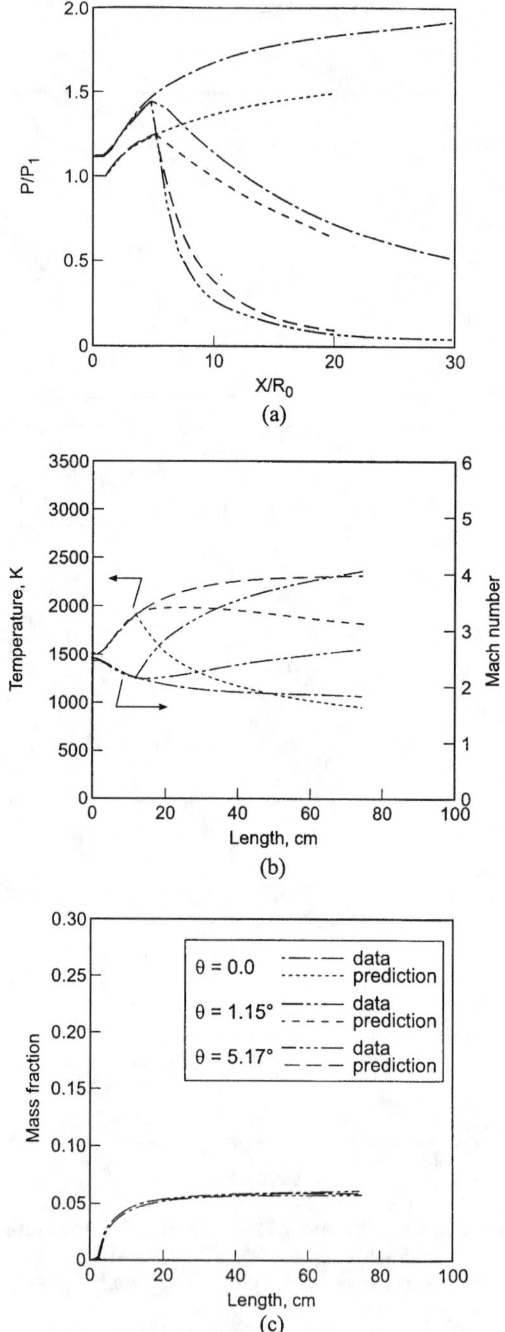

Fig. 6f Variation of Efficiencies along Combustor Length in Case 8.1 and 9.1.

Table 3 Efficiencies for cases given in Tables 1 and 2

Case no./ fig. no.	Length at $\eta_{I\max.}$, cm	η_I	η_{II}	η_{FL}	η_{SL}	η_{KE}	η_B	η_{C1}	η_{C2}	η_{C3}	I_{sp}, s
2.1/6 a,b											
$\varphi = 0.7$	32.59	0.638	0.077	0.995	0.884	0.833	0.161	0.944	0.909	0.958	700.1
$\varphi = 0.9$	22.50	0.531	0.923	0.800	0.101	0.838	0.165	0.806	0.764	0.848	804.6
$\varphi = 1.1$	26.84	0.343	0.063	0.791	0.660	0.848	0.164	0.585	0.649	0.762	455.0
6.1/6 c											
$\varphi = 0.7$	100.0	0.439	—	0.901	0.786	0.821	0.082	0.734	0.989	0.996	—
$\varphi = 0.9$	22.71	0.568	0.114	0.945	0.820	0.833	0.158	0.847	0.785	0.860	903.7
$\varphi = 1.1$	32.46	0.410	0.083	0.835	0.701	0.827	0.120	0.664	0.690	0.787	598.5
8/6 d–f											
$T_{1a} = 900°K$	100.0	0.473	0.088	0.878	0.751	0.798	0.069	0.732	0.950	0.929	658.1
$T_{1a} = 1100°K$	100.0	0.396	0.104	0.0852	0.737	0.789	0.069	0.653	0.935	0.919	860.1
9/6 d–f											
$T_{1a} = 900°K$	100.0	0.441	0.059	0.833	0.732	0.806	0.698	0.648	1.000	0.856	441.7
$T_{1a} = 900°K$	100.0	0.370	0.081	0.814	0.723	0.797	0.070	0.575	1.000	0.850	672.8

The following observations can be made from the results:

1) The presence of atomic oxygen influences only the initial chemical reactions. Once the atomic oxygen undergoes a reaction, the reactions proceed according to the scheme specified. There is, of course, some effect on ignition, which can be considered by comparing, for example, Cases 2.2 in Table 2 and Case 1 in Table 5.
2) Water vapor affects the performance only through changes in thermodynamic properties due to composition change. There is however some change in chemical reactions due to water acting as a third body in the initial length of the combustor. In terms of specific impulse, a substantial reduction arises with increase of water vapor content to 15%.
3) A reduction in mixing length is helpful only when the air temperature is sufficiently high to give rise to early ignition. When the air temperature is small, there is delay in setting in of chemical reaction, and the flow tends to choke early along the combustor. All of the efficiencies and the specific impulse improve with a higher air inlet temperature. There is also an effect of fuel equivalence ratio and it is found that performance improves with increase in fuel equivalence ratio.
4) Variation of M_{1a} at fixed T_{1a} implies variation in flow velocity and

Table 4 Operational conditions in special cases: principal features

Case no.	Geometry	Main features	Fig. no.
1	Straight duct	Atomic oxygen in air: 1% and 3%	7
2	Straight duct	Water vapor in air: 5% and 15%	8
3	Straight duct	Instant mixing; $T_{1a} = 900°K$; φ varied.	9
4	Straight duct	Instant mixing; $T_{1a} = 1100°K$; φ varied.	10
5	Straight duct	Fuel accumulation before ignition; $T_{1a} = 900°K$; M_{1a} varied.	11
6	Straight duct	Fuel accumulation before ignition; $T_{1a} = 1100°K$; M_{1a} varied.	12
7	Straight duct	Fuel accumulation before ignition; mixing length reduced to 0.1 cm; M_{1a} varied.	13
8	Multiple section geometry	Multiple sections and fuel injections; $\varphi_1 = 0.5, 0.3$; and 0.2; in successive locations; $T_{1a} = 900°K$ and 1100°; φ varied.	14
9	Straight duct	Silane added: 1% of entry air; $M_{1a} = 3.0$; φ varied.	15
10	Straight duct	Silane added: 1% of entry air; $M_{1a} = 3.5$; φ varied.	16
11	Straight duct	Silane added: 1% of entry air; $M_{1a} = 4.8$; φ varied.	17

Table 5 Operational Conditions in Special Cases

Case no.	P_{1a}, atm.	T_{1a}, °K	M_{1a}	P_{1f}, atm.	T_{1f}, °K	M_{1f}	Mixing	Friction and heat loss
1	2.0	1100.0	3.5	4.0	166.7	2.0	Typical; Lmix = 20 cm	Typical
2	2.0	1100.0	3.5	4.0	166.7	2.0	Typical; Lmix = 20 cm	Typical
3	2.0	900.0	3.5	4.0	166.7	2.0	Lmix = 0.1 cm	Typical
4	2.0	1100.0	3.5	4.0	166.7	2.0	Lmix = 0.1 cm	Typical
5	2.0	700.0	3.5	4.0	166.7	2.0	Typical; Lmix = 20 cm	Typical
6	2.0	700.0	3.5	4.0	166.7	2.0	Lmix = 99 cm	Typical
7	2.0	700.0	3.5	4.0	166.7	2.0	Lmix = 0.1 cm	Typical
8	2.0	900.0	3.5	4.0	166.7	2.0	Lmix = 0.1 cm	Typical
9	2.0	700.0	3.0	4.0	166.7	6.0	Typical; Lmix = 20 cm	Typical
10	2.0	700.0	3.5	4.0	166.7	6.0	Typical; Lmix = 20 cm	Typical
11	2.0	700.0	4.8	4.0	166.7	6.0	Typical; Lmix = 20 cm	Typical

residence time in the combustor. As the Mach number and the flow velocity at entry decrease, there is a tendency towards early choking. The chemical reactions show no change. The performance drops as the entry Mach number is reduced.

5) When a combustor is configured with several sections and fuel injection at the start of each section, it is found that, even as stated under 3) above, there is improvement in performance as T_{1a} is increased.

6) The addition of silane for improving reactivity of the mixture is effective most when M_{1a} is small and φ is also small.

V. Specific Illustrative Studies

A number of experimental studies have been reported in literature on testing of scramjet combustors, generally in a connected mode insofar as air supply is concerned, and with some form of expanding area ducting added to the combustor for discharging the products of combustion. A few of these cases have been reported with adequate detail on the combustor geometry, air and fuel supply, ignition and other details, and these lend themselves to prediction based on the one-dimensional procedure undertaken here. Broadly they fall into two categories, one, where the combustor cross-sectional area is constant with some form of flame-holding device at entry, and two, where the combustor area is varied along the length at discrete locations. Different types of fuel injection schemes have been employed in various cases. It is generally difficult to categorize them in other ways, and, therefore,

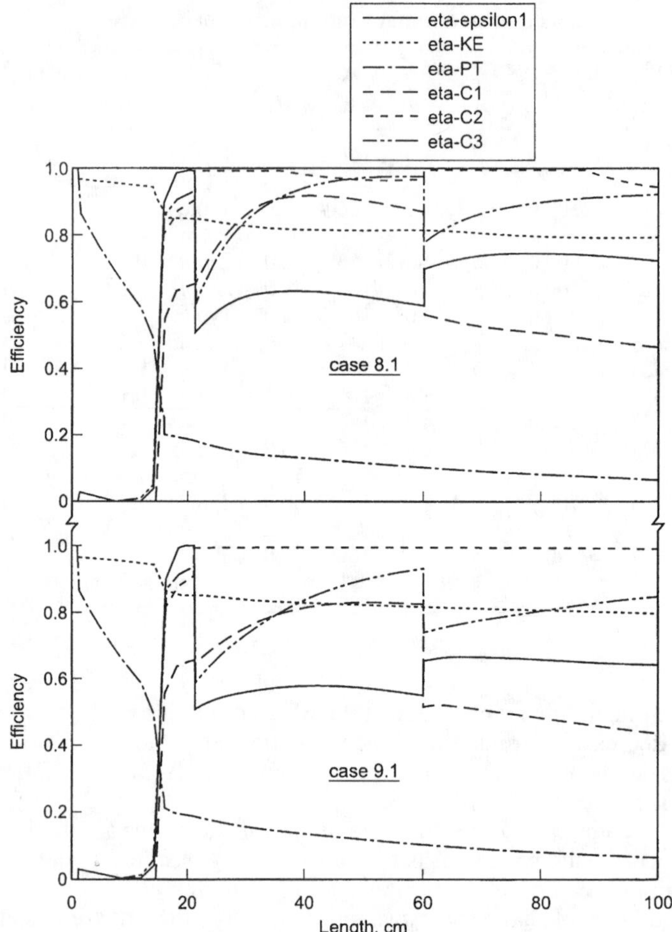

Fig. 7. Performance of Case 1. See Table 5 for operating conditions; note air supplied contains 1% atomic oxygen and 3% atomic oxygen.

in the following each case is referred to by the primary investigator and an appropriate reference where further details may be found on the tests.

In all of the selected cases the fuel utilized is hydrogen.

Generally no details are available on friction and heat transfer losses. Therefore, for purposes of performance prediction, the friction coefficient has been assumed to be the typical friction coefficient, and heat transfer is estimated based on Reynolds analogy. No account is taken of the possible effects of the presence of internal shockwaves, or flow separation in the vicinity of fuel injection devices.

Ignition and the associated chemical reaction are dealt with in the same fashion as in the case of parametric studies.

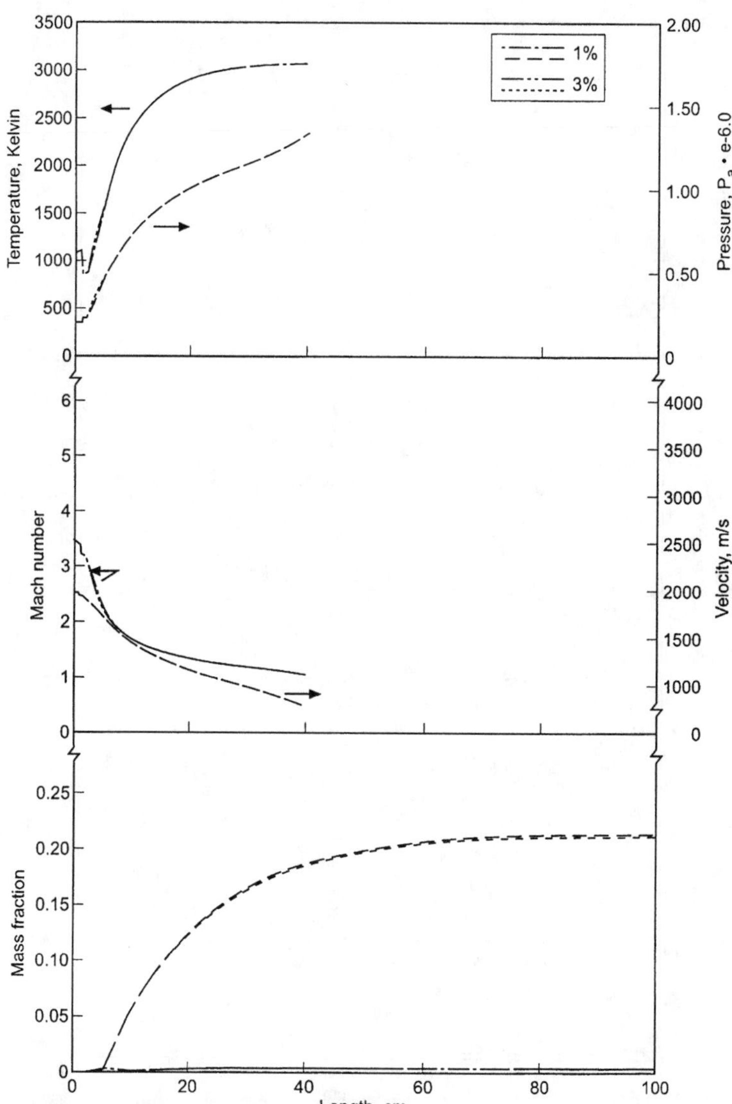

Fig. 8. Performance in Case 2. See Table 5 for operating conditions; note air supplied is humid with 5% water vapor and 15% water vapor.

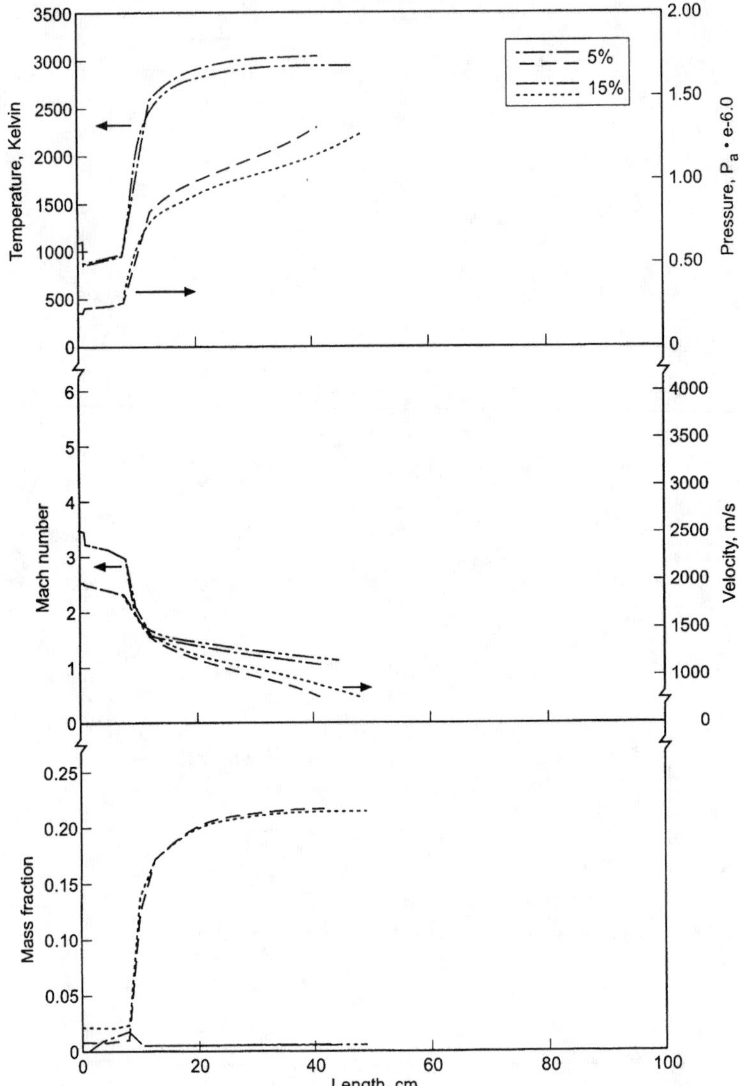

Fig. 9. Performance in Case 3. See Table 5 for operational conditions; note instant mixing of air and fuel; T_{1a} = 900°K.

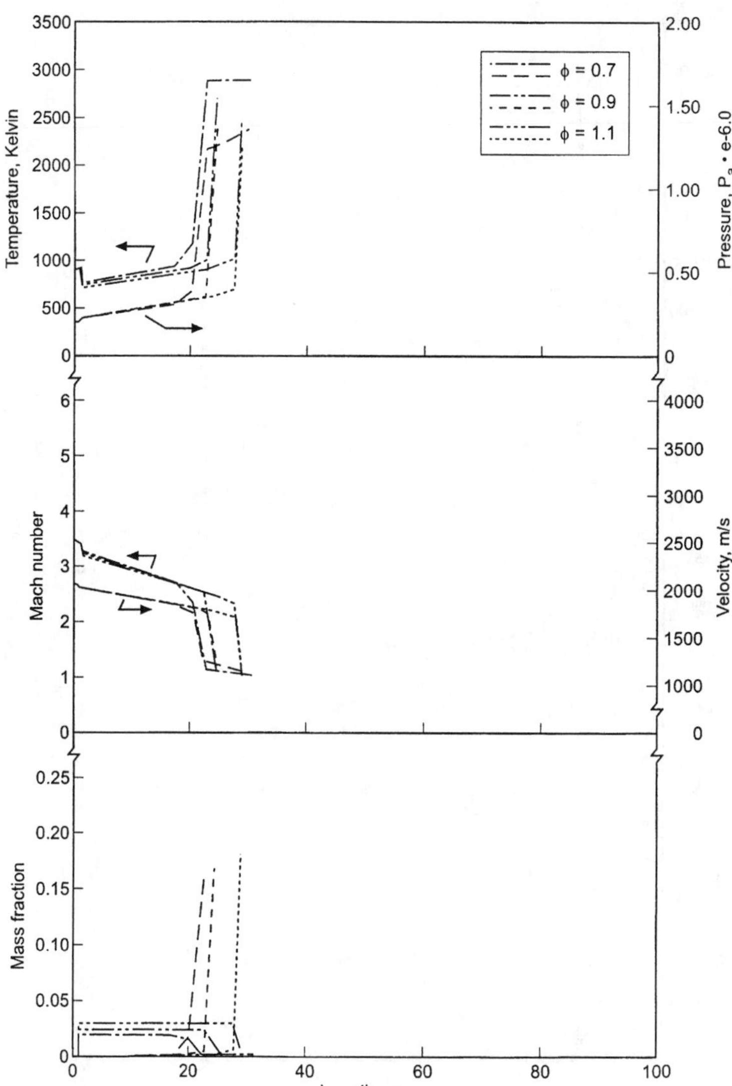

Fig. 10. Performance in Case 4. See Table 5 for operational conditions; note instant mixing of air and fuel; T_{1a} = 1100°K.

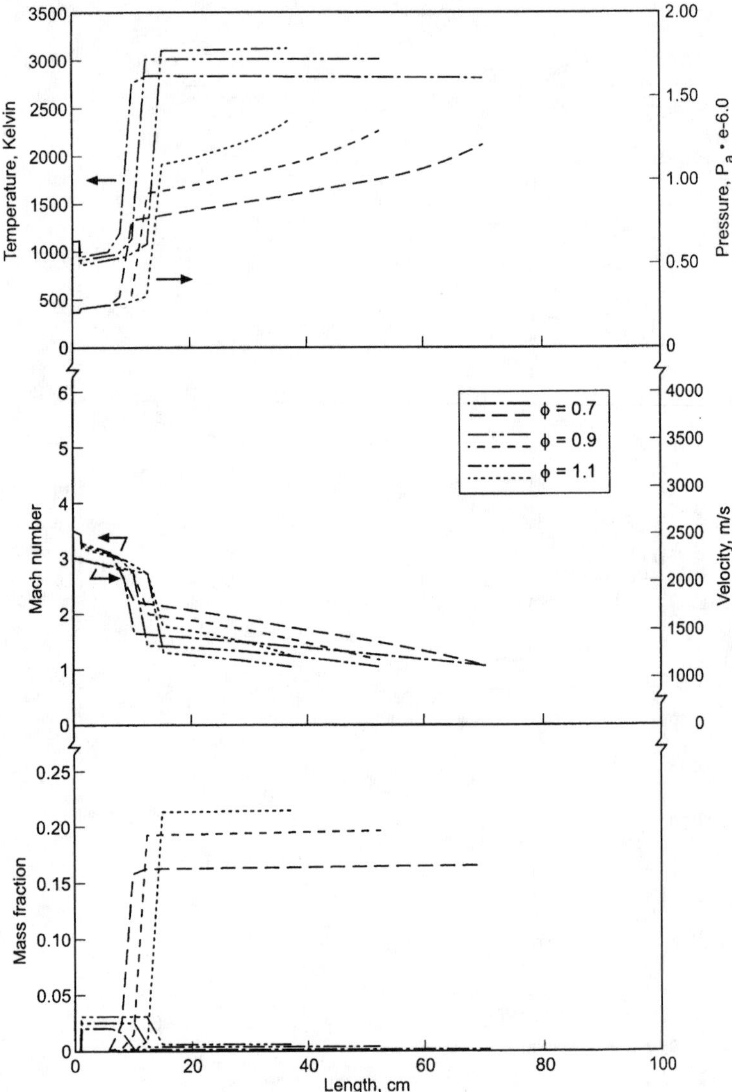

Fig. 11. Performance in Case 5. See Table 5 for operational conditions; note fuel accumulation before ignition; mixing length = 20 cm; M_{1a} = 3.0, 3.5, and 4.8.

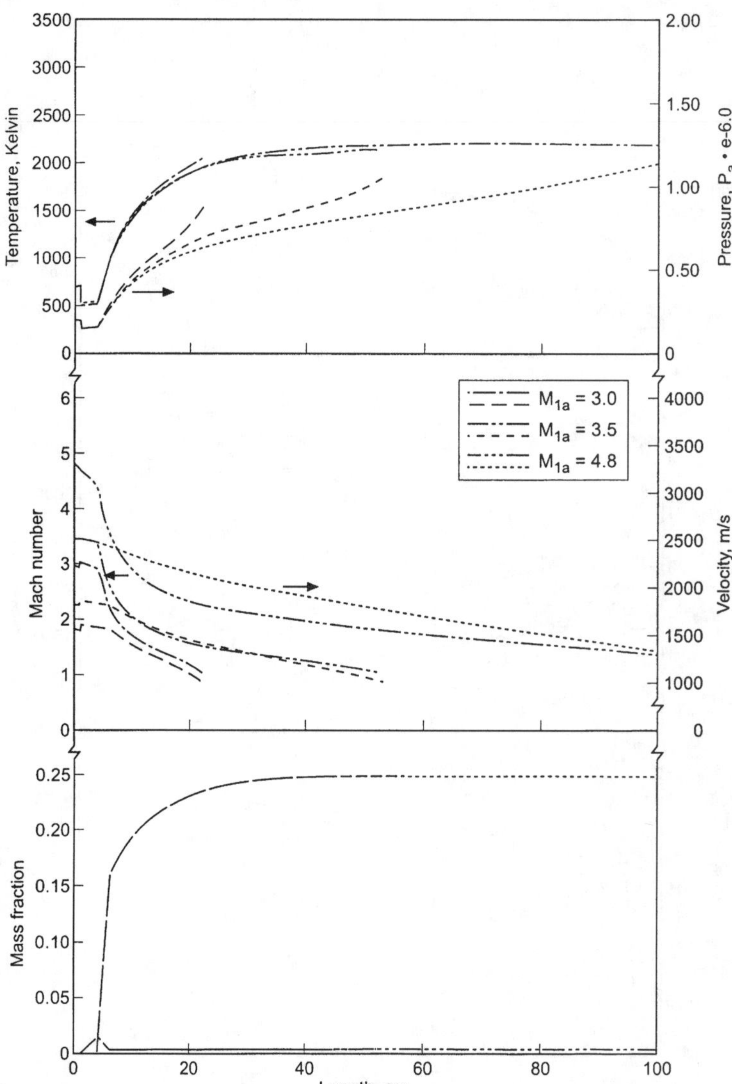

Fig. 12. Performance in Case 6. See Table 5 for operational conditions; note fuel accumulation before ignition; mixing length chosen = 90 cm; M_{1a} = 3.0, 3.5, and 3.8.

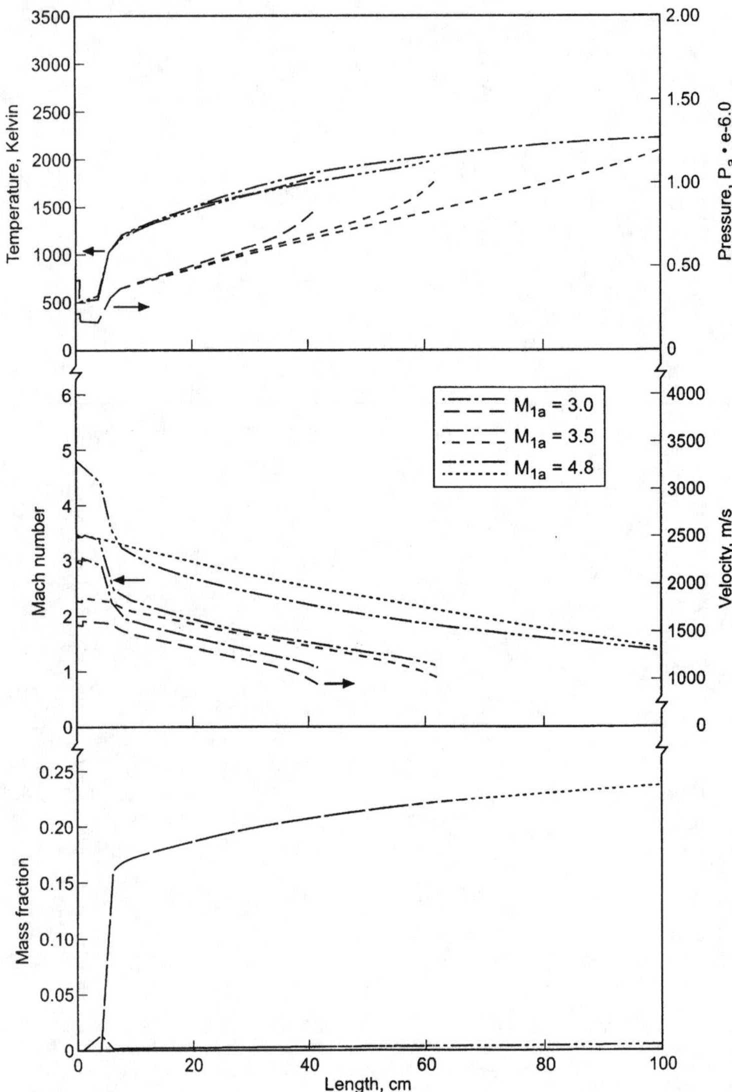

Fig. 13. Performance in Case 7. See Table 5 for operational conditions; note fuel accumulation before ignition; mixing length = 0.1cm; M_{1a} = 3.0, 3.5, and 3.8.

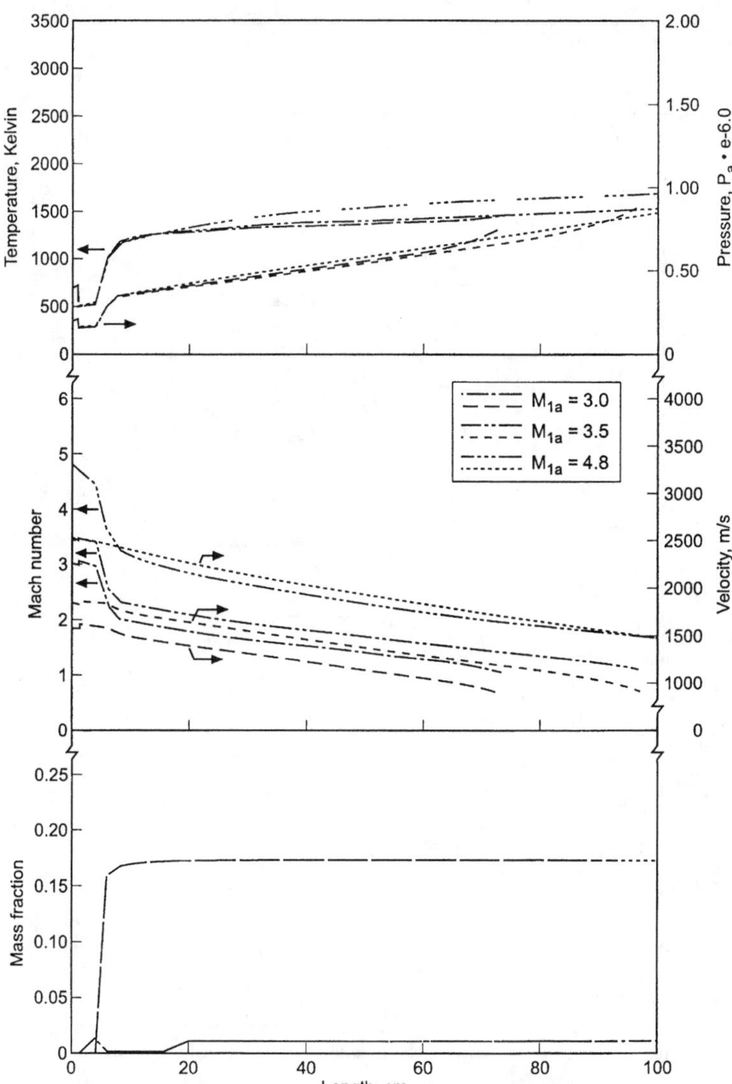

Fig. 14. Performance in Case 8. See Table 5 for operational conditions; note multiple fuel injection (φ = 0.5, 0.3, and 0.2 successively at three area change locations, with η_m identical for all injections); T_{1a} = 900°K and 1100°K.

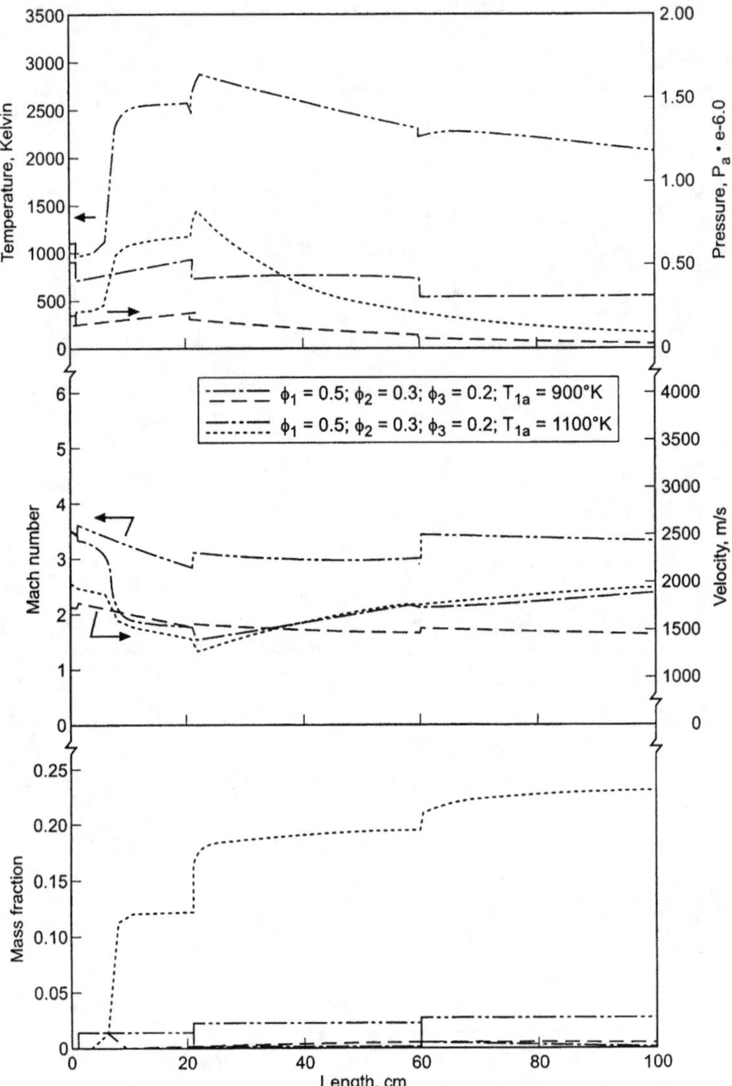

Fig. 15. Performance in Case 9. See Table 5 for operational conditions; note fuel mixed with silane at injection; $\varphi = 0.7, 0.9$, and 1.1; $M_{1a} = 3.0$.

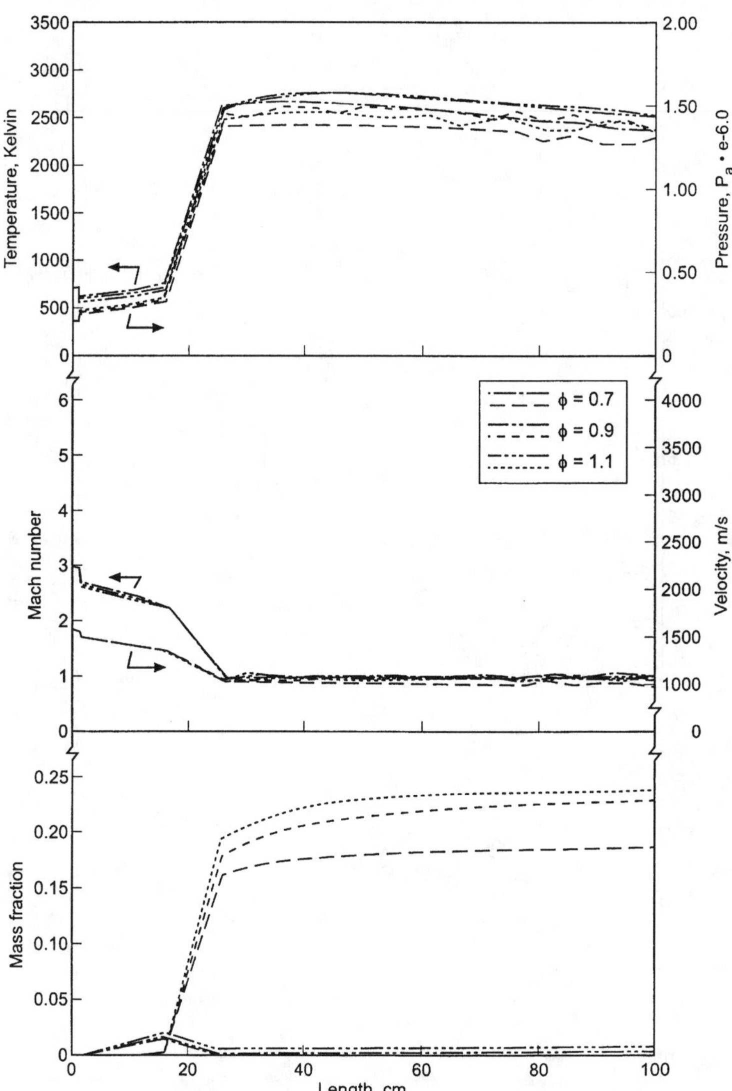

Fig. 16. Performance in Case 10. See Table 5 for operational conditions; note fuel mixed with silane at injection; φ = 0.7, 0.9, and 1.1; M_{1a} = 3.5.

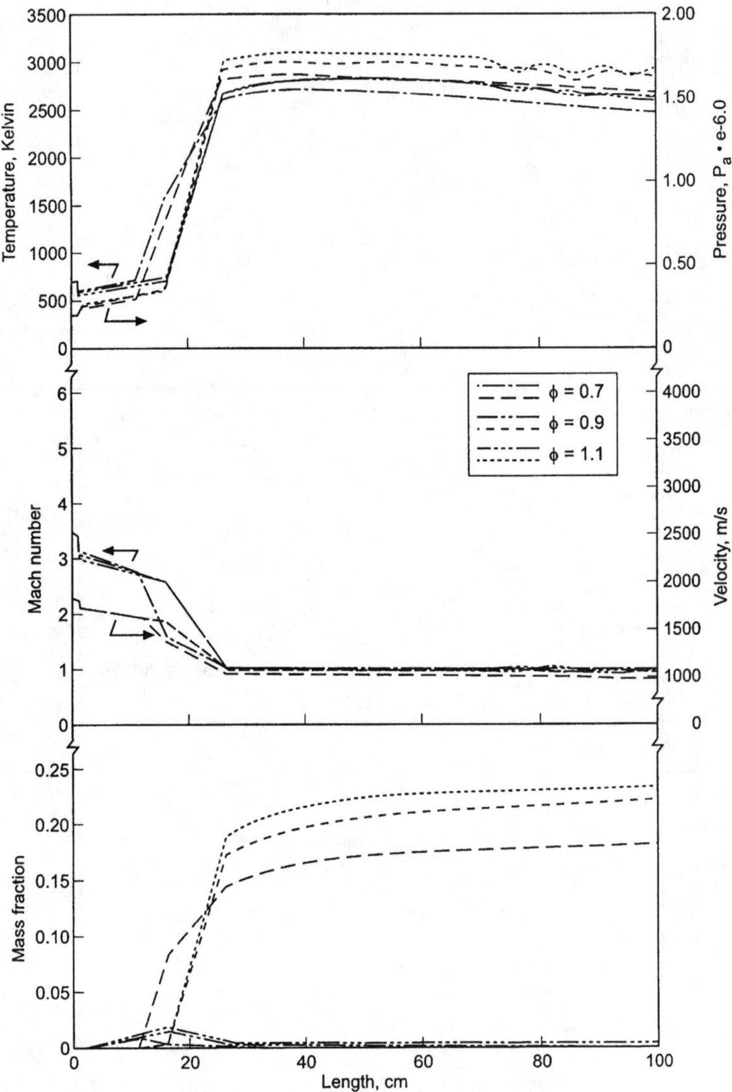

Fig. 17. Performance in Case 11. See Table 5 for operational conditions; note fuel mixed with silane at injection; $\varphi = 0.7, 0.9$, and 1.1; $M_{1a} = 4.8$.

As in the description of parametric studies, the details of the operating conditions in various cases are summarized in Tables 7 and 8, which also lists the numbers of figures in which the performance parameters are shown. The various cases are discussed in the following.

A. Hypersonic Research Engine

The Hypersonic Research Engine (HRE) was initiated at NASA Langley Research Center in 1967; the program was terminated in 1976. A full-scale (18.0 in. diameter cowl and 87.0 in. length) HRE concept engine, designated the Aerothermodynamic Integration Model (AIM), was tested in 1974 at the NASA Plum Brook Station Hypersonic Test Facility at Mach numbers 5, 6, and 7. Test data from those tests have been presented, along with analysis and predictions carried out at that time, in Refs. 10 and 11. The current study is based on some of the data sets taken from those references.

The overall geometry of the combustor, along with the locations of fuel injection, for the HRE is shown in Fig. 18. It will be observed that the combustor follows the forebody and the inlet section, is of varying cross-sectional area over the length, has four injections in two pairs through the wall, and a final section with struts. Our interest is in the region between the first set of injectors and the struts; thus no analysis has been attempted in the region of the struts and beyond.

The HRE used hydrogen as fuel in all of the tests.

The HRE was operated in a number of modes with different objectives in various cases. Three of those modes have been chosen for further consideration here, namely, 1) auto-ignition, 2) supersonic combustion, and 3) optimized supersonic combustion.

We will discuss these further in the following paragraphs.

In all cases, the combustor may be considered in two parts: 1) beginning with the first set of injectors up to a location approximately 17.0 cm downstream and 2) from the latter location up to the location of the struts further downstream. In 1) there is supersonic combustion beginning with the entry Mach number to a value of Mach number equal to a near-unity value. Thus one obtains the entry conditions to 2). With the second set of injectors at entry to 2) one then examines the possibility of auto-ignition or supersonic combustion as desired. The total equivalence ratio of fuel in the first set of injectors is small, of the order of 0.2 to 0.48, while that in the second set is related to the specific tests under consideration.

The measured data in all cases consisted of engine entry conditions, wall pressure measurement along the combustor duct length, and the thrust force generated. Other parameters along the combustor were deduced using a simple one-dimensional model.

In Sec. I (Fig. 18) the data seem to indicate that heat release by combustion occurred in conjunction with gas dynamic oblique shock processes. This has been confirmed using the current one-dimensional model while allowing fuel accumulation over a small length of combustor following injection and intense combustion with equilibrium chemistry.

Table 6 Efficiencies for cases given in Table 4 and 5

Case no./ figure No.	Length at $\eta_{\varepsilon 1 max.}$	η_I	η_{II}	η_{FL}	η_{SL}	η_{KE}	η_{PT}	η_{C1}	η_{C2}	η_{C3}	Isp, s
1/7											
Atomic oxygen = 1%	25.0	0.517	0.171	0.905	0.790	0.821	0.168	0.770	0.750	0.807	1356.0
Atomic oxygen = 3%	25.0	0.521	0.171	0.904	0.790	0.822	0.169	0.768	0.742	0.807	1354.0
2/8											
Water = 5%	25.0	0.490	0.162	0.895	0.782	0.821	0.168	0.750	0.747	0.815	1296.2
Water = 15%	25.0	0.435	0.143	0.870	0.764	0.822	0.173	0.701	0.740	0.837	1156.3
3/9											
$\varphi = 0.7$	30.27	0.659	0.085	1.000	0.894	0.833	0.162	0.967	0.911	0.964	774.1
$\varphi = 0.9$	24.49	0.499	0.088	0.905	0.782	0.840	0.164	0.771	0.764	0.860	699.4
$\varphi = 1.1$	28.70	0.320	0.522	0.776	0.646	0.849	0.163	0.560	0.654	0.779	337.4
4/10											
$\varphi = 0.7$	70.17	0.339	0.026	0.866	0.771	0.818	0.140	0.617	0.925	0.970	262.9
$\varphi = 0.9$	52.31	0.451	0.105	0.889	0.783	0.815	0.148	0.714	0.861	0.915	926.4
$\varphi = 1.1$	37.16	0.487	0.155	0.889	0.775	0.816	0.156	0.739	0.773	0.829	1227.6
5/11											
$M_{1a} = 3.0$	22.29	0.299	—	0.708	0.600	0.866	0.271	0.521	0.824	0.860	—
$M_{1a} = 3.5$	30.0	0.289	—	0.721	0.621	0.862	0.185	0.510	0.848	0.883	—
$M_{1a} = 4.8$	27.5	0.233	—	0.746	0.665	0.878	0.108	0.453	0.839	0.876	—

(Table 6 continued)

6/12											
$M_{1a} = 3.0$	41.49	0.168	—	0.606	0.501	0.878	0.255	0.371	0.7308	0.764	—
$M_{1a} = 3.5$	62.56	0.152	—	0.625	0.525	0.861	0.155	0.356	0.775	0.809	—
$M_{1a} = 4.8$	37.5	0.075	—	0.656	0.577	0.893	0.109	0.275	0.713	0.750	—
7/13											
$M_{1a} = 3.0$	73.03	0.006	—	0.474	0.380	0.896	0.231	0.175	0.606	0.638	—
$M_{1a} = 3.5$	97.39	—	—	0.468	0.379	0.882	0.138	0.106	0.606	0.636	—
$M_{1a} = 4.8$	100.0	—	—	0.508	0.433	0.876	0.552	—	0.606	0.638	—
8/14											
$T_{1a} = 900°K$	100.0	—	—	0.379	0.267	0.931	0.111	—	0.023	0.030	—
$T_{1a} = 1100°K$	100.0	0.408	0.109	0.859	0.744	0.790	0.071	0.667	1.908	0.928	906.9
9/15											
$\varphi = 0.7$	36.05	0.470	0.057	0.827	0.721	0.855	0.333	0.836	0.929	0.966	565.3
$\varphi = 0.9$	46.05	0.470	0.111	0.812	0.700	0.850	0.347	0.815	0.892	0.934	968.0
$\varphi = 1.1$	46.05	0.436	0.138	0.774	0.657	0.852	0.357	0.763	0.792	0.846	1088.4
10/16											
$\varphi = 0.7$	41.05	0.396	0.048	0.793	0.692	0.849	0.187	0.754	0.904	0.952	509.5
$\varphi = 0.9$	46.05	0.418	0.106	0.790	0.683	0.844	0.195	0.759	0.873	0.924	982.1
$\varphi = 1.1$	46.05	0.395	0.133	0.759	0.647	0.845	0.201	0.719	0.783	0.844	1111.4
11/17											
$\varphi = 0.7$	31.05	0.303	0.111	0.788	0.702	0.854	0.061	0.657	0.862	0.934	1465.5
$\varphi = 0.9$	36.05	0.273	0.120	0.749	0.656	0.847	0.055	0.604	0.796	0.879	1349.3
$\varphi = 1.1$	36.05	0.254	0.132	0.718	0.619	0.845	0.053	0.568	0.719	0.807	1314.7

Table 7 Cases supported by test data: description of cases supported by test data

Case no.	Identification	Description	Fig. no.
12.1.1	HRE ( )	Supersonic combustion; $\varphi_1{}^a = 0.24$; $\varphi^b = 0.59$; freestream conditions: $M_0 = 6$; $p_0 = 0.402$ psi; $T_0 = 420$ R; $V_0 = 5993$ ft/s.	—
12.1.2	HRE ( )	Supersonic combustion; $\varphi_1 = 0.27$; $\varphi = 0.76$; freestream conditions: $M_0 = 6$; $p_0 = 0.401$ psi; $T_0 = 419$ R; $V_0 = 5985$ ft/s.	—
12.2.1	HRE ( )	Supersonic combustion; $\varphi_1 = 0.19$; $\varphi = 0.83$; freestream conditions: $M_0 = 5$; $p_0 = 0.384$ psi; $T_0 = 511$ R; $V_0 = 5731$ ft/s.	—
12.2.2	HRE ( )	Supersonic combustion; $\varphi_1 = 0.18$; $\varphi = 0.7$; freestream conditions: $M_0 = 5$; $p_0 = 0.42$ psi; $T_0 = 485$ R; $V_0 = 5585$ ft/s.	—
12.3.1	HRE ( )	Supersonic combustion; $\varphi_1 = 0.48$; $\varphi = 0.34$; freestream conditions: $M_0 = 7$; $p_0 = 0.156$ psi; $T_0 = 271$ R; $V_0 = 5881$ ft/s.	—
12.3.2	HRE ( )	Supersonic combustion; $\varphi_1 = 0.47$; $\varphi = 0.55$; freestream conditions: $M_0 = 7$; $p_0 = 0.157$ psi; $T_0 = 287$ R; $V_0 = 6053$ ft/s.	—
12.4.1	HRE ( )	Optimized performance; $\varphi_1 = 0.21$; $\varphi = 0.73$; freestream conditions: $M_0 = 6$; $p_0 = 0.399$ psi; $T_0 = 421$ R; $V_0 = 5986$ ft/s.	—
12.4.2	HRE ( )	Optimized performance; $\varphi_1 = 0.21$; $\varphi = 0.36$; freestream conditions: $M_0 = 6$; $p_0 = 0.385$ psi; $T_0 = 402$ R; $V_0 = 5896$ ft/s.	—
12.5.1	HRE ( )	Optimized performance; $\varphi_1 = 0.48$; $\varphi = 0.0$; freestream conditions: $M_0 = 7$; $p_0 = 0.158$ psi; $T_0 = 278$ R; $V_0 = 5921$ ft/s.	—
12.5.2	HRE ( )	Optimized performance; $\varphi_1 = 0.26$; $\varphi = 0.51$; freestream conditions: $M_0 = 7$; $p_0 = 0.155$ psi; $T_0 = 279$ R; $V_0 = 5955$ ft/s.	—
13.1	Billig (1973)	Conical combustor; $\varphi = 0.8$;	20
13.2	Billig (1973)	Short cylinder - cone combustor; T_{1a} and T_{1f} varied; φ: 0.78 and 0.49.	21
13.3	Billig (1973)	Step cylinder - cone combustor; T_{1a} and T_{1f} varied; φ: 0.93 and 0.51.	22
14.1	Northam (1989)	Configuration III geometry; Unswept ramp injector; parallel fuel injection; φ: 0.6 and 0.9, and 1.23.	24
14.2	Northam (1989)	Configuration III geometry; Unswept ramp injector; parallel + perpendicular fuel injection; φ: 0.93, 1.12, and 1.57.	25
15.1	Paull (1993)	Low pressure run; $T_{1a} = 700°$K φ: 0.9 and 0.42.	27
15.2	Paull (1993)	Low pressure run; $T_{1a} = 1230°$K φ: 1.0 and 0.55.	28
15.3	Paull (1993)	High pressure run; $T_{1a} = 630°$K φ: 1.0 and 0.5.	29

Table 7 (continued)

Case no.	Identification	Description	Fig. no.
15.4	Paull (1993)	High pressure run; T_{1a} = 1270°K φ: 0.9 and 0.5.	30
16.1	Sabel'nikov (1993)	KSR combustor; P_{1a}, T_{1a} and M_{1a} varied; φ: 0.53 and 0.41; fuel injection: 50% parallel & 50% perpendicular.	32
16.2	Sabel'nikov (1993)	KSS combustor; P_{1a}, T_{1a} and M_{1a} varied; φ: 0.54 and 1.0; perpendicular fuel injection.	33
16.3	Sabel'nikov (1999)	φ: 0.434, 0.333, and 0.227; M_{1f} varied.	34
16.4	Sabel'nikov (1999)	straight duct: wall angle = + 1.15 ° and + 5.17 ° φ = 0.23.	35

[a] φ_1 = equivalence ratio for first set of fuel injectors in HRE.
[b] φ = equivalence ratio for test section.

1. *Auto-Ignition Tests*

These tests were conducted to establish auto-ignition of fuel supplied with the second set of injectors. Autoignition is confirmed in the tests by the measured variation of pressure along the combustor and the use of the flow-combustion model and code. The current one-dimensional code with the equilibrium chemistry model does not provide any further details on ignition. The pressure rise due to ignition is found to be smaller than that indicated by the test data; however the predictions do not account for any shockwave affects. The temperature rise is comparable.

2. *Supersonic Combustion Tests*

The objectives in these tests were to establish that combustion can continue in the second part under supersonic flow conditions and to optimize the performance with variation of fuel equivalence ratio in the second set of injectors.

In these cases, the various thermodynamic efficiencies have been calculated corresponding to the test data available.

The cases considered are listed in Table 7 and the operating conditions for those cases are given in Table 8. It must be repeated that the operating conditions are reproduced from Ref. 12, which include measured data as well as predictions using the original model and code.

The predicted efficiencies in these cases have been given in Table 9, utilizing the test data as reported. It will be observed that Table 9 does not provide any of the efficiencies pertaining to combustion and chemical reactions. This is because there have been no test data available on the chemical composition of the combustor gases. An attempt was made to simulate the combustor flowfield including the chemical state of the gases, but was not successful.

Table 8 Operational conditions in cases supported by test data

Case no.	P_{1a}, atm	T_{1a}, °K	M_{1a}	P_{1f}, atm	T_{1f}, °K	M_{1f}
12.1.1	0.9671	796.1	2.5	2.0	833.3	1.0
12.1.2	1.0321	811.7	2.5			
12.2.1	0.7187	908.9	2.1	2.0	833.3	1.0
12.2.2	0.9022	880.6	2.1			
12.3.1	1.0270	817.8	2.4	2.0	833.3	1.0
12.3.2	0.8419	732.8	2.6			
12.4.1	0.8694	803.9	2.5	2.0	833.3	1.0
12.4.2	0.8830	794.4	2.5			
12.5.1	0.9545	768.3	2.4	2.0	833.3	1.0
12.5.2	0.9738	820.0	2.4			
13.1	0.5008	841.7	3.2	1.0	236.9	1.0
13.2	0.5117	878.3	3.2	1.0	536.1	1.0
					537.0	
13.3	0.4981	806.7	3.2	1.0	521.3	1.0
					548.6	
14.1	1.0	926.0	2.0	1.21	166.7	1.7
				1.90		
				2.77		
14.2	1.0	926.0	2.0	4.325	166.7	1.7
						1.0
15.1	0.3454	700	4.5	2.57	243.8	1.0
				1.20		
15.2	0.2961	1230	4.4	1.70	243.8	1.0
				0.94		
15.3	0.7106	630	4.6	5.79	243.8	1.0
				2.90		
15.4	0.8586	1270	4.4	4.08	243.8	1.0
				2.27		
16.1	1.0720	592.9	3.0	14.55	280.0	1.0
	1.7098	720.0	2.5	16.16		
16.2	1.4729	516.4	2.5	20.13	280.0	1.0
	0.9968	585.7	3.0	28.83		
16.3	0.8882	1048.2	2.0	0.8882	289.3	1.0
						0.8
						0.57
16.4	0.9869	1500	2.6	0.9869	1000.0	1.0

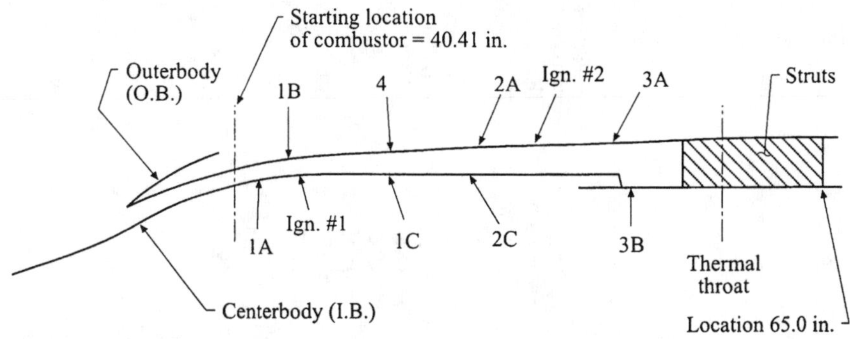

Injector Parameters

Injector	Number of Injectors	Diameter, in.	Injection Angle[a], deg.	S/d	x, in.	Location
1A	37	0.119	90	13.1	40.5	I.B.
1B	37	0.119	90	13.9	41.25	O.B.
1C	37	0.119	106	13.5	44.5	I.B.
4	37	0.119	90	14.2	44.5	O.B.
2A	60	0.095	67	11.4	48.5	O.B.
2C	60	0.095	119	10.6	46.5	I.B.
3A	114	0.090	65	7.0	53.75	O.B.
3B	102	0.095	90	6.3	55.9	I.B.

Ignitor Parameters

Ignitor	x, in.	Circumferential locations						Injection Angle[a], deg.	Location
1	42.00	55	110	165	230	290	350	94.5	I.B.
2	50.98	40	100	–	220	240	280	60.0	O.B.

a. With respect to Aerothermodynamic Integration Model

All Dimensions in Inches

Fig. 18. Schematic of Hypersonic Research Engine.

Table 9 Efficiencies for HRE cases (from HRE data)

Case Details	η_I	η_{II}	η_{FL}	η_{SL}	η_{PT}	η_{C1}	η_{KE} (computed)	η_{KE} (HRE data)	Combustor effectiveness (HRE data)
Case 12.1.1	0.266	0.508	0.771	0.629	0.148	0.641	0.843	0.917	0.698
Case 12.1.2	0.449	0.597	0.768	0.671	0.173	0.731	0.915	0.887	0.869
Case 12.2.1	0.190	0.458	0.668	0.475	0.289	0.576	0.862	0.920	0.674
Case 12.2.2	0.487	0.618	0.840	0.698	0.287	0.844	0.912	0.812	0.890
Case 12.3.1	0.005	0.372	0.576	0.468	0.010	0.326	0.838	0.925	0.487
Case 12.3.2	0.200	0.416	0.581	0.489	0.009	0.446	0.836	0.959	0.664
Case 12.4.1	0.381	0.573	0.771	0.634	0.173	0.724	0.885	0.872	0.748
Case 12.4.2	0.451	0.605	0.864	0.733	0.170	0.834	0.865	0.898	0.795
Case 12.5.1	0.112	0.438	0.695	0.581	0.111	0.439	0.847	0.973	0.558
Case 12.5.2	0.191	0.450	0.656	0.545	0.010	0.488	0.839	0.942	0.572

It will also be seen that Table 9 lists the kinetic energy efficiency and the so-called combustor effectiveness given as part of the test data in the references cited above. The value of η_{KE} as determined from the test data appear to be different from those cited in the original references. This may be due to the fact that the original values may have been calculated using ratio of specific heats at some unspecified location. Combustor effectiveness is defined in the reports of the test data as the ratio of actual total momentum change from entrance to exit of the combustor to the total momentum that can be generated by expanding the combustion products to ambient static pressure under isentropic conditions. The change in momentum across the combustor is a result of the combustion process, but it reflects only a part of the change in energy or exergy.

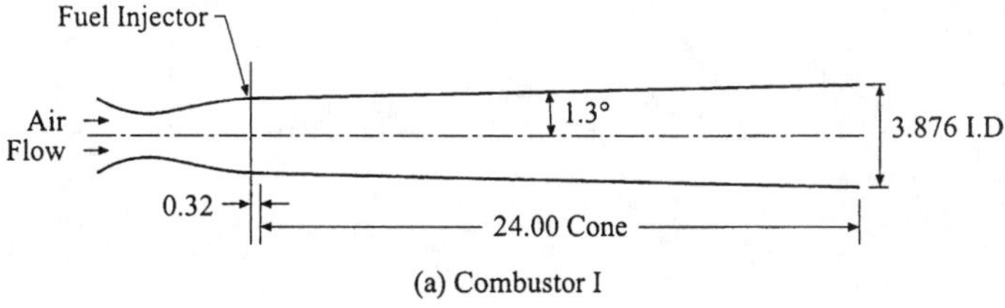

(a) Combustor I

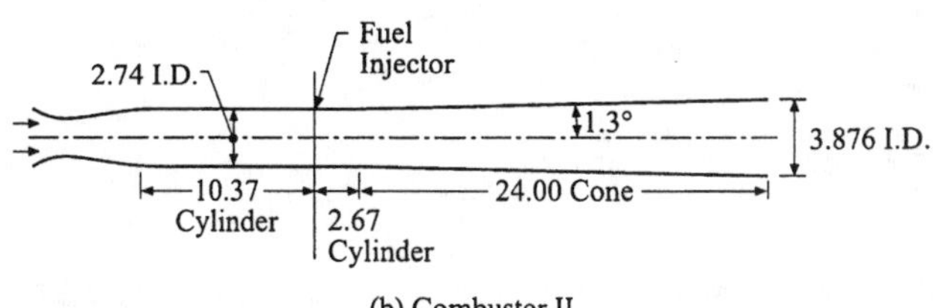

(b) Combustor II

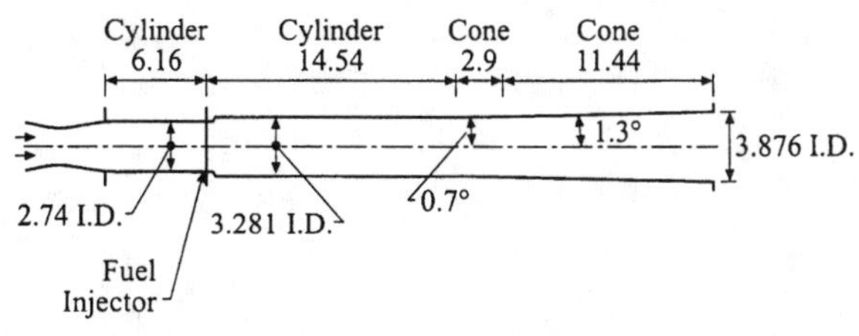

(c) Combustor III

All Dimensions in Inches

Fig. 19. Schematic of Combustors used by Waltrup and Billig.

B. Direct Connect Combustion Tests due to Waltrup and Billig (1973)

The direct-connect-combustor tests with hydrogen fuel are reported in Refs. 12 and 13. The main interest in the original tests was in the distribution of pressure in the pre-combustion zone; however, some test data have been made available on the combustor sections.

Three combustor geometries were employed, viz., conical, short cylinder-cone, and stepcylinder-cone, which are shown schematically in Fig. 19. The conical combustor, Fig. 19a, is close-coupled to the nozzle, whereas in the other two cases, Figs. 19b and 19c, the constant area sections are coupled. The conical combustor was operated with an equivalence ratio of 0.8, the short cylinder-cone with equivalence ratios of 0.78 and 0.49, and the step-cylinder-cone combustor with equivalence ratios of 0.93 and 0.51. In all cases, the entry diameter and the entry-to-exit area ratio were 2.74 in. and 2.0, respectively, and the entry Mach number was 3.23.

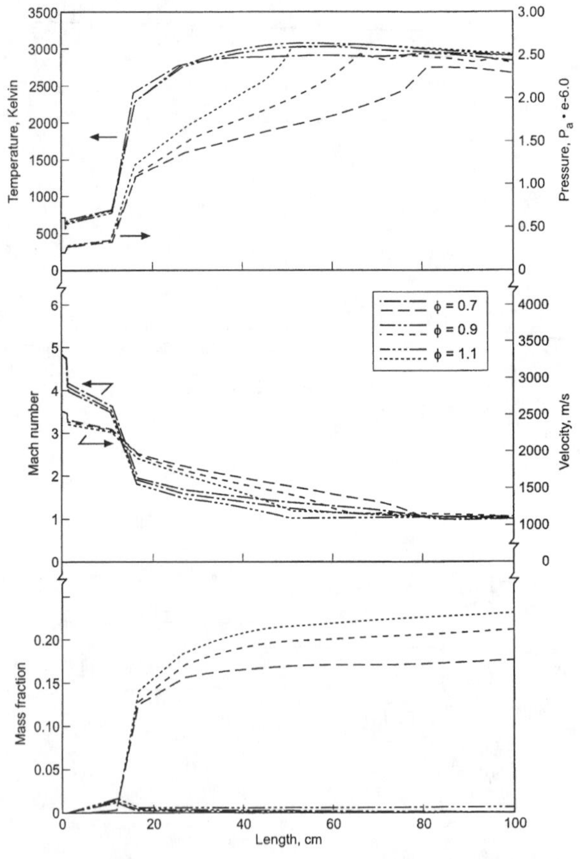

Fig. 20. Performance along the combustor in Fig. 19(a); see Case 13.1 of Table 7.

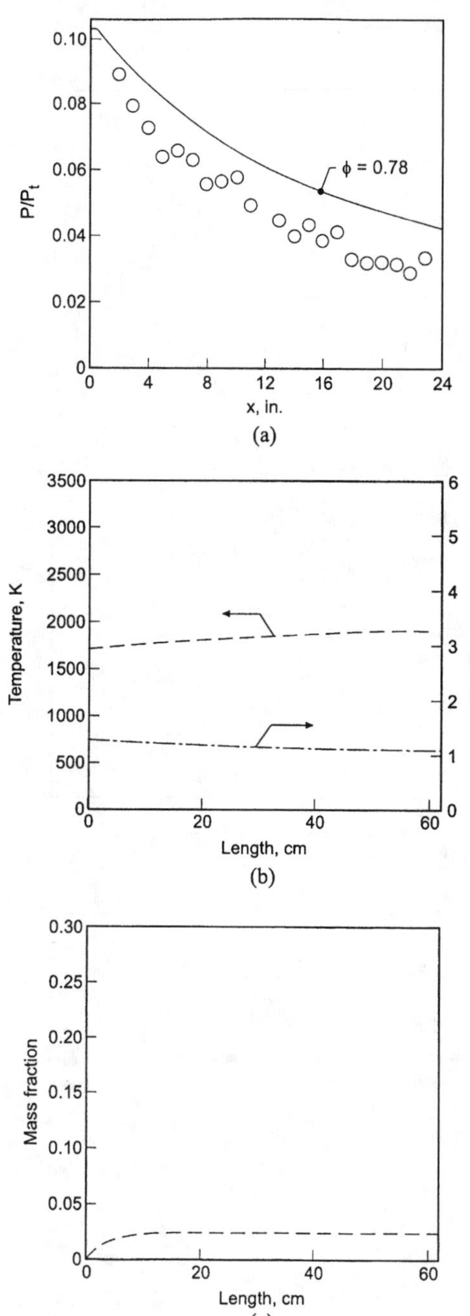

Fig. 21. Performance along the combustor Fig. 19(b); see Case 13.2 of Table 7.

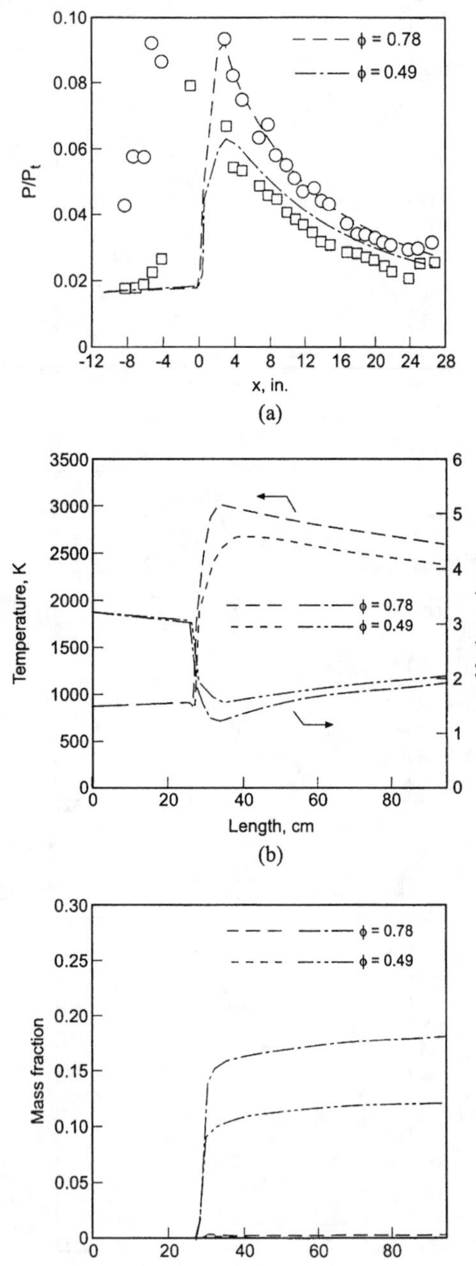

Fig. 22. Performance along the combustor in Fig. 19(c); see Case 13.3 of Table 7.

Table 10 Efficiencies for cases 13–16 supported by test data

Case no fig. no.	η_I max. length at $\eta_{I\max}$, cm	η_I	η_{II}	η_{FL}	η_{SL}	η_{KE}	η_B	η_{C1}	η_{C2}	η_{C3}	Isp, s
Case 12.1.1	40.31	0.562	0.238	0.903	0.712	0.697	0.378	0.947	0.801	0.879	2478.7
Case 12.1.2	42.16	—	0.128	0.472	0.261	0.843	0.413	0.220	0.004	—	1414.1
Case 12.2.1	42.16	—	0.146	0.585	0.414	0.845	0.374	0.265	7.8e-8	—	1817.3
Case 12.2.2	42.16	—	0.128	0.467	0.292	0.836	0.313	0.221	9.6e-6	—	1338.0
Case 12.3.1	42.16	—	0.136	0.483	0.304	0.846	0.308	0.224	1.9e-5	—	1431.9
Case 12.3.2	42.16	—	0.158	0.598	0.429	0.857	0.370	0.275	6.2e-8	—	2018.2
Case 12.4.1	42.16	—	0.143	0.511	0.335	0.857	0.322	0.233	9.8e-7	—	1571.0
Case 12.4.2	42.16	—	0.138	0.462	0.288	0.851	0.309	0.217	0.005	—	1413.2
Case 12.5.1	42.16	0.036	0.196	0.619	0.508	0.900	0.476	0.249	4.6e-8	—	2707.5
Case 12.5.2	42.16	0.002	0.183	0.524	0.403	0.882	0.352	0.235	1.7e-8	—	2106.3
Case 12.6.1	42.16	—	0.180	0.497	0.383	0.879	0.388	0.205	1.3e-6	—	2029.1
Case 12.6.2	42.16	—	0.164	0.464	0.343	0.882	0.315	0.209	2.2e-6	—	1695.7
Case 13.1/20											
$\varphi = 0.80$	50.0	0.968	—	1.000	0.994	0.784	0.300	1.000	0.936	0.949	—
Case 13.2/21											
$\varphi = 0.78$	94.0	—	—	0.480	0.390	0.913	0.702	—	8.7e-7	—	—
$\varphi = 0.49$	94.0	—	—	0.585	0.501	0.916	0.682	—	3.2e-7	—	—
Case 13.3/22											
$\varphi = 0.93$	73.6	—	—	0.299	0.192	0.814	0.184	—	2.0e-8	—	—
$\varphi = 0.51$	73.6	—	—	0.385	0.283	0.804	0.155	—	4.0e-5	—	—

(continued)

Table 10 (continued)

Case no fig. no.	η_I max. length at $\eta_{I\max}$, cm	η_I	η_{II}	η_{FL}	η_{SL}	η_{KE}	η_B	η_{C1}	η_{C2}	η_{C3}	Isp, s
Case 14.1/24											
$\varphi = 0.6$	70.0	0.483	—	0.841	0.743	0.837	0.625	0.664	1.000	0.979	—
$\varphi = 0.9$	75.0	0.443	—	0.767	0.675	0.815	0.621	0.597	0.986	0.970	—
$\varphi = 1.23$	25.3	0.203	—	0.538	0.431	0.856	0.875	0.316	0.576	0.576	—
Case 14.2/25											
$\varphi = 0.93$	84.4	0.554	—	0.864	0.719	0.805	0.564	0.743	0.966	0.962	—
$\varphi = 1.12$	79.4	0.473	—	0.786	0.642	0.806	0.591	0.647	0.896	0.891	—
$\varphi = 1.57$	24.9	0.126	—	0.452	0.312	0.865	0.852	0.249	0.531	0.530	—
Case 15.1/27											
$\varphi = 0.9$	45.0	0.542	—	0.915	0.836	0.799	0.170	0.0791	0.823	0.934	—
$\varphi = 0.5$	130.0	0.508	—	0.993	0.898	0.804	0.072	0.936	0.997	0.984	—
Case 15.2/28											
$\varphi = 1.0$	65.0	0.177	—	0.819	0.744	0.776	0.156	0.482	0.729	0.818	—
$\varphi = 0.55$	100.0	—	0.792	0.712	0.773	0.083	0.122	0.854	0.918	—	
Case 15.3/29											
$\varphi = 1.0$	40.0	0.654	—	0.943	0.856	0.824	0.146	0.853	0.913	0.916	—
$\varphi = 0.5$	130.0	0.455	—	0.905	0.807	0.846	0.078	0.691	1.000	0.998	—

(continued)

Table 10 (continued)

Case no fig. no.	η_I max. length at $\eta_{I\max}$., cm	η_I	$\eta_{\rm II}$	$\eta_{\rm FL}$	$\eta_{\rm SL}$	$\eta_{\rm KE}$	η_B	η_{C1}	η_{C2}	η_{C3}	Isp, s
Case 15.4/30											
$\varphi = 0.9$	30.0	0.494	—	0.923	0.854	0.826	0.205	0.741	0.795	0.855	—
$\varphi = 0.5$	40.0	0.476	—	0.954	0.887	0.848	0.214	0.759	0.912	0.951	—
Case 16.1/32											
$M_{1a} = 3.0$	62.0	0.932	0.293	1.000	0.982	0.811	0.292	1.000	0.992	0.994	3517.3
$M_{1a} = 2.5$	52.0	0.878	0.229	1.000	0.945	0.839	0.408	1.000	0.985	0.988	3081.8
$M_{1a} = 2.5$	141.2	—	0.377	0.240	1.000	0.743	—	—			
$M_{1a} = 3.0$	141.2	—	—	0.316	0.185	0.977	0.733	—	—	—	—
Case 16.2/33	120.0	0.496	—	0.910	0.829	0.770	0.151	0.751	0.466	0.638	—
	23.3	0.176	—	0.752	0.675	0.822	0.156	0.436	0.628	0.665	—
Case 16.3/34											
$\varphi = 0.434$	75.0	—	—	0.562	0.447	0.898	0.797	0.014	—	—	
$\varphi = 0.33$	37.45	0.613	—	0.989	0.877	0.802	0.531	0.937	0.934	0.963	—
$\varphi = 0.227$	47.5	0.710	—	1.000	0.883	0.817	0.551	1.000	0.996	0.994	—
Case 16.4/35											
$\theta = 0°$	52.5	0.627	—	0.973	0.894	0.820	0.587	0.990	0.986	0.990	—
$\theta = 1.15°$	45.0	0.517	—	0.953	0.872	0.830	0.581	0.871	0.972	0.981	—
$\theta = 5.17°$	15.0	0.285	—	0.907	0.826	0.847	0.702	0.600	0.850	0.902	—

The main measurement was that of pressure distribution along the combustor.

The performance of the three combustor configurations, with operating conditions as in Tables 7 and 8, is shown in Figs. 20–22. A comparison of the measured pressure distribution with that predicted is shown in Figs. 20a, 21a, and 22a for the three combustors. The predicted performance with respect to efficiencies is summarized in Tables 9 and 10.

C. NASA Langley Direct-Connect Tests due to Northam, Greenberg, and Byington (1989)

The direct-connect tests with hydrogen fuel are reported in Ref. 14. The main emphasis in the tests was on exploring enhanced mixing and its effects with parallel injection of fuel.

One of the configurations with two successive wall inclinations is shown schematically in Fig. 23; thus, a duct with two successive increases in area is under consideration.

Two test cases (Tables 7 and 8) are considered, one with parallel injection of fuel, and the second with parallel injection of fuel for combustion followed by perpendicular injection of fuel further downstream, the latter for establishing a gain in thrust due to such injection of fuel.

The main measurement was that of wall static pressure along the combustor.

The predicted performance of the combustor is given in Figs. 24 and 25, where a comparison with measured pressure values along the combustor is given in Figs. 24a and 25a. Figure 24 pertains to the first case with parallel injection of fuel with fuel equivalence ratio of 0.6, 0.9, and 1.23. It is found that the pressure variation along the combustor is nearly the same as the measured values, although slightly smaller, in the high values of φ; however there is substantial difference in the case $\varphi = 0.6$, both in the peak value of pressure reached and the location of that peak along the combustor. The data seem to indicate two peak values of pressure, separated by about 10 in. The prediction yields one peak value at the first peak value location in the data.

Figure 25 refers to the second case with equal amounts of parallel and perpendicular fuel injection, but only at the higher fuel equivalence ratio values of 0.93 and 1.12. In this case the data are predicted satisfactorily.

Unfortunately there are no other measured data. The predictions show

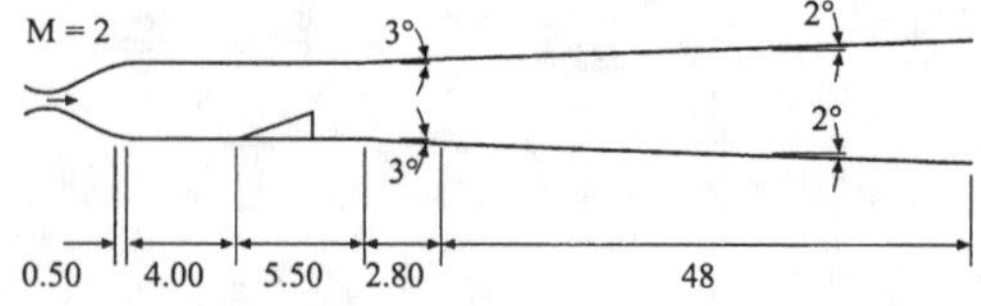

Fig. 23. Schematic of Combustor used by Northam et al.

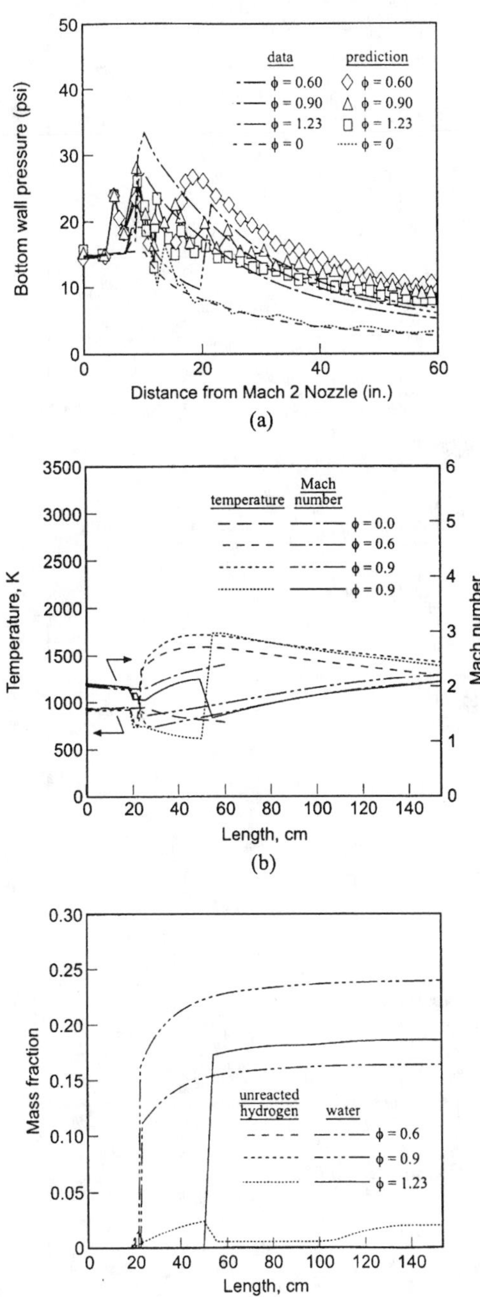

Fig. 24. Performance along the Combustor Length; see Case 14.1 of Table 7.

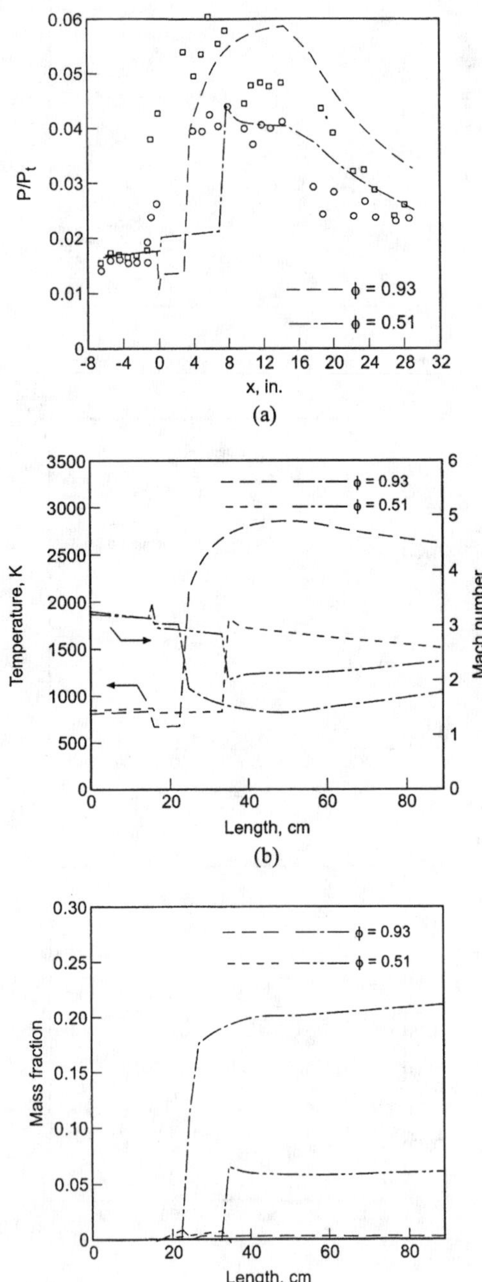

Fig. 25. Performance along the Combustor Length; see Case 14.2 of Table 7.

that in the first case (Fig. 24), the temperature rise and mass fraction of reactants vary quite differently for the case $\varphi = 0.6$ from the manner of their variation for the case of high values of φ.

D. Free Piston Shock Tunnel Experiments due to Paull (1993)

The details of the free-piston shock tunnel and the testing undertaken on scram combustors with hydrogen fuel utilizing the tunnel are given in Ref. 15.

A schematic of the scram combustor and so-called thrust surface added to the combustor is given in Fig. 26. The figure also shows the 5 mm fuel injector that was located in the center of the combustor duct.

The Mach number at entry to the combustor was nominally 4.5 and the static pressure ranged from 30 to 120 kPa. The experimental results are said to have shown that higher specific impulse was obtained with long combustors and at increased values of operating pressure.

The predicted performance of the combustor in four cases (Tables 7 and 8) is presented in Figs. 27–30. The predicted efficiencies are summarized in Tables 9 and 10.

E. Test Data due to (1) Sabel'nikov, Voloschenko, Ostras, and Sermanov, and Walther (1993) and (2) Mescheryakov and Sabel'nikov (1981).

Two sets of test data are considered, one reported in 1993 (Ref. 16), and the other in 1981 (Ref. 17).

1. 1993 Test Data

The details of the direct-connect testing conducted in TsAGI hypersonic facility T-131A utilizing hydrogen as fuel are given in Ref. 16.

A schematic of the two scram combustor ducts employed is given in Figs. 31a and 31b. In the combustor, referred to as the KSR combustor, shown in Fig. 31a, the first and fourth sections have constant area sections, while the second and third sections have expansion angles of 3.0 and 2.33 degrees, respectively. In the other combustor, referred to as the KSS combustor and shown in Fig. 31b, the first two sections have an identical, small expansion angle of 0.5 degree; the third section has an expansion angle of 5 degree; and

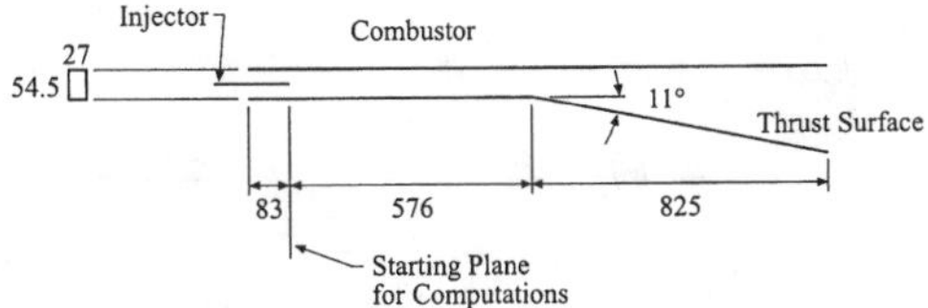

Fig. 26. Schematic of Combustor Configuration used by Paull.

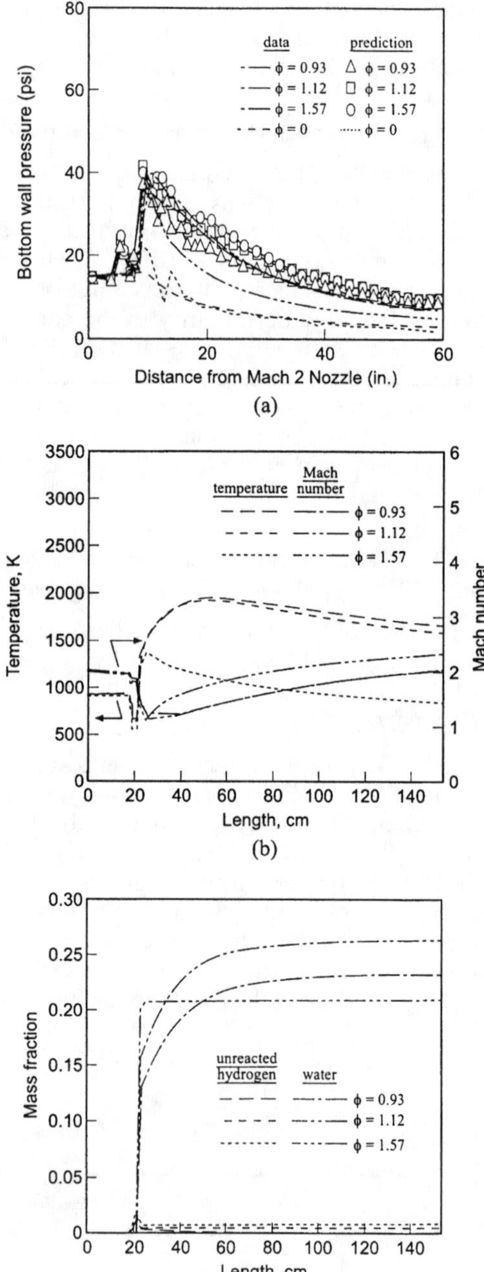

Fig. 27. Performance along the Combustor Length; see Case 15.1 of Table 7.

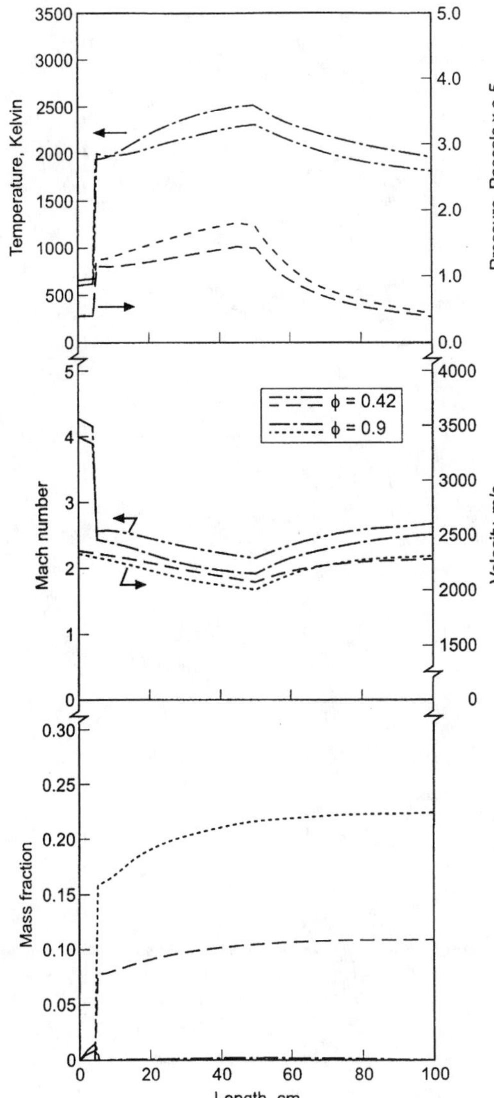

Fig. 28. Predicted Performance along Combustor Length. See Case 15.2 of Table 7.

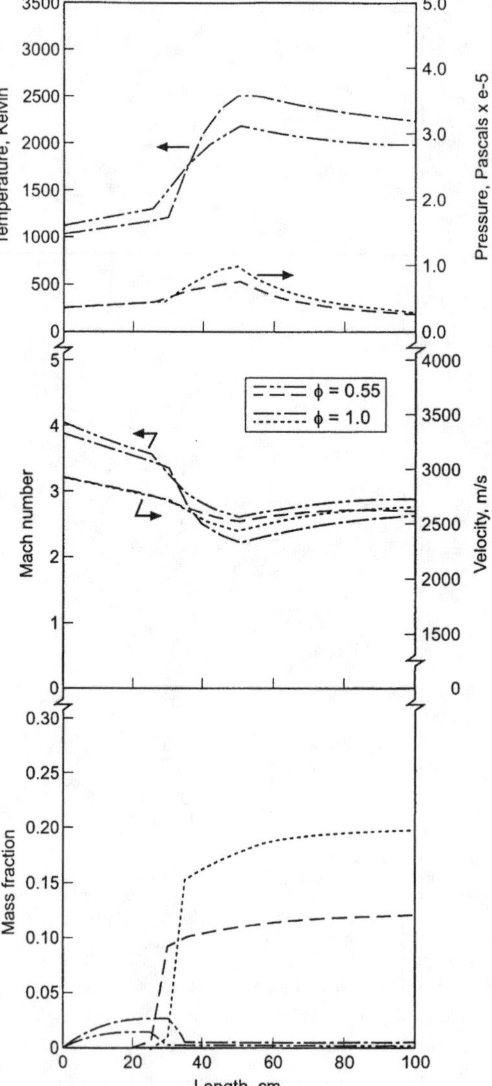

Fig. 29. Performance along Combustor Length. See Case 15.3 of Table 7.

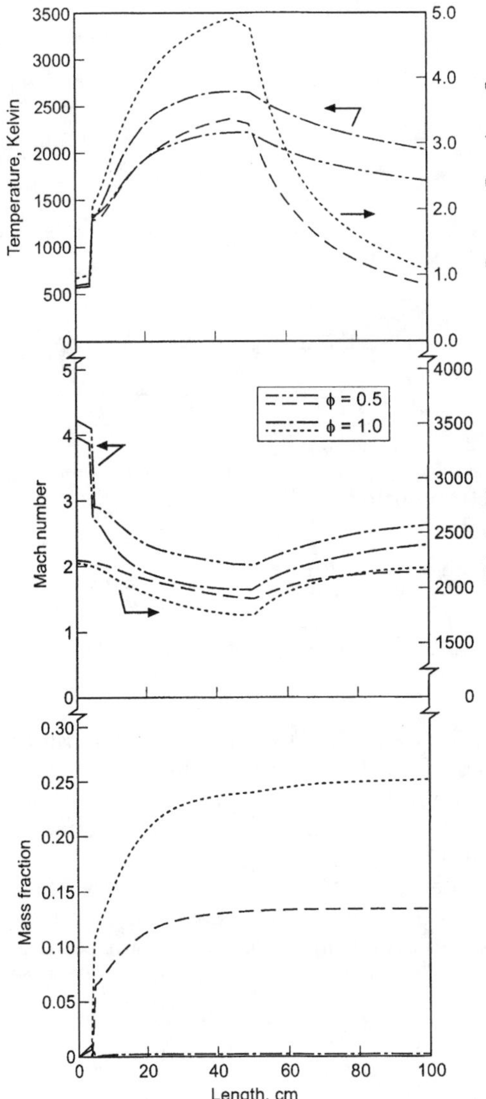

Fig. 30. Performance along Combustor Length. See Case 15.4 of Table 7.

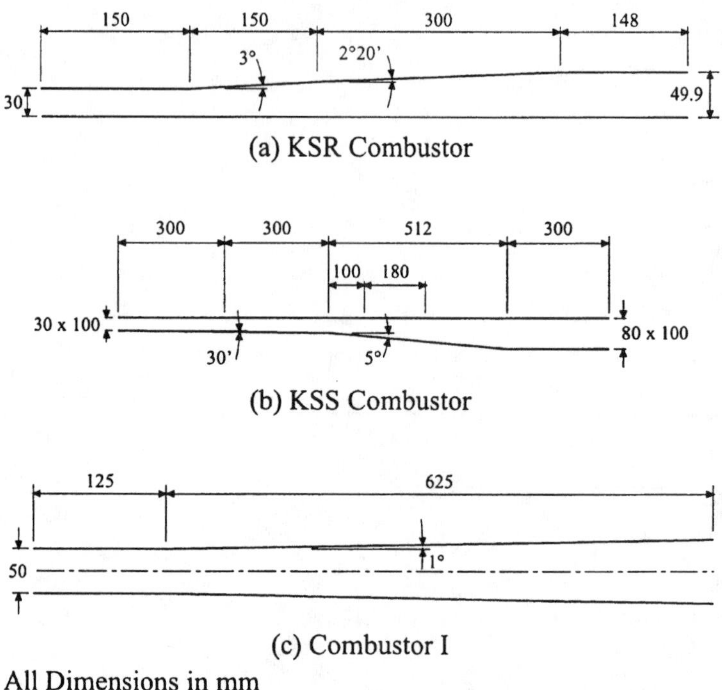

Fig. 31. Schematic of Combustor used by Sabel'nikov et al.

the last section is of constant area. The interest in the study is in establishing the influence of geometry, and location and mode of fuel injection.

The test cases considered are given in Tables 7 and 8. Two test cases, Case 1 and Case 2, with different entry conditions have been considered in the tests with the KSR and KSS combustor.

The predicted performance for the two test cases (one at entry Mach number 2.5 and the other of 3.0) are given in Figs. 32 and 33. The measured pressure distribution along the combustor wall is compared with the predicted values in Figs. 32a and 33a.

2. *1981 Test Data*

Details of the testing are provided in Ref. 18.

A schematic of the combustor duct employed is shown in Fig. 31c. The duct consists of a straight section followed by a diverging wall.

Four test cases have been considered as shown in Tables 7 and 8. In the first case, the angle of divergence of the wall is held constant at 1 degree and the fuel equivalence ratio is varied over the range of 0.227 to 0.434. In the remaining three cases, the fuel equivalence ratio is held constant at 0.223

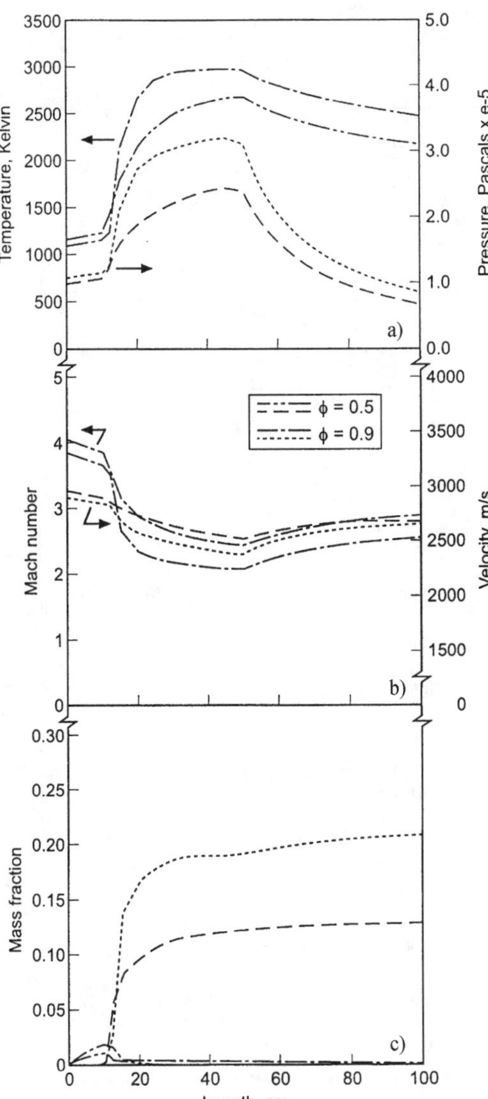

Fig. 32. Predicted Performance along Combustor Length. See Case 16.1 of Table 7.

while the angle of divergence of the wall is set at 0, 1.15, and 5.17 degrees in different cases.

The predicted performance in the first case shown in Fig. 32, where 32a provides a comparison of predicted and measured pressure distribution along the wall. The predicted performance in the other three cases is shown in Figs. 33–35.

In Appendix A of Chapter I.4 of the current volume, Sabelnikov discusses

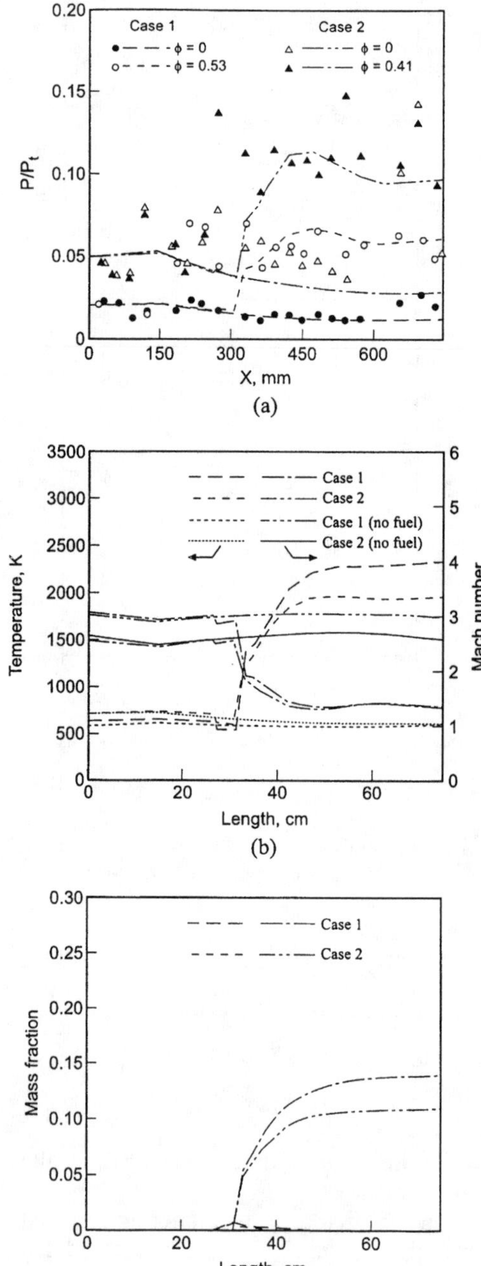

Fig. 33. Predicted Performance along Combustor Length. See Case 16.2 of Table 7.

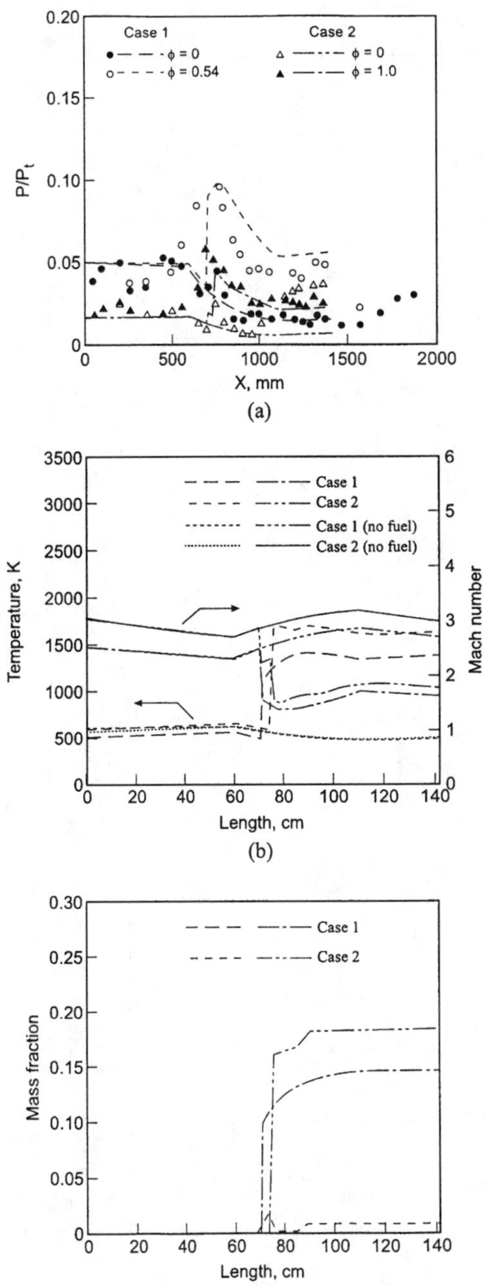

Fig. 34. Predicted Performance along Combustor Length. See Case 16.3 of Table 7; note 1° inclined wall.

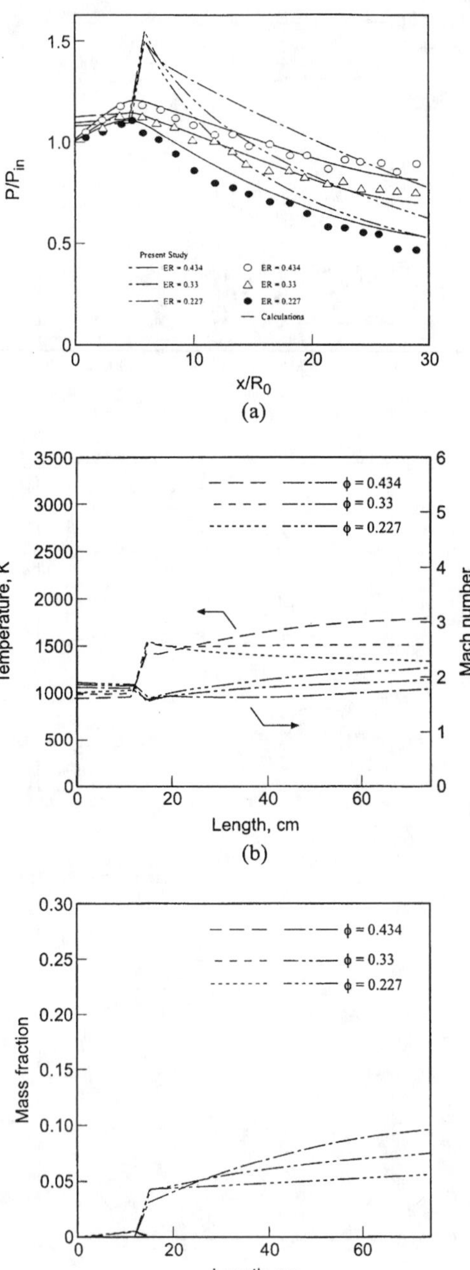

Fig. 35. Predicted Performance along Combustor Length. See Case 16.4 of Table 7; note wall inclination = 0°, 1.15°, and 5.17°.

the controversy regarding the possible cause for decreased combustor performance in combustor ducts of diverging area of section. He invokes Zeldovich's theory of the extinction of nonpremixed laminar flame for a quantitative estimate of the reduction in combustor efficiency through a change in chemical reaction rates; references for a discussion of Zeldovich's theory can be found in Sabelnikov's chapter in this volume.

Figure 35a shows the change in static pressure along the combustor from the test data and current predictions; the nature of changes are generally in agreement. The predicted temperature distribution along the length, given in Fig. 35b, also displays the right trend in different cases. Considering the progress of chemical reactions in Fig. 35c, it appears that wall divergence has negligible effect on the formation of water and, also, on the hydrogen that remains unreacted in the mixed stream. Based on the current estimates, it is thus difficult to conclude that chemical reactions are substantially modified by the reduction in pressure and residence time caused by the increase in area along the combustor.

The current predictions, however, have been carried out with 1) a global "typical" mixing rule, without considering the details of turbulent motion and diffusion, and 2) instantaneous reaction of mixed fuel under non-equilibrium conditions following ignition. It seems therefore unfair to question the applicability of the Zeldovich model on the basis of current results. In the experiments reported by Sabel'nikov there were no measurements related to the composition of the combustion gas. The controversy on the effect of wall divergence remains until more detailed measurements become feasible.

VI. Scaling Performance and Geometry

The scaling of scram combustors is of interest in regard to 1) geometrical features and 2) operational parameters. Thus, the performance of a given combustor under different operating conditions, as discussed in Sec. IV, may also be interpreted in terms of scaling with respect to selected operational parameters. Of greater interest is the manner in which operational parameters need to be adjusted when a combustor size is altered. In particular, in view of the available test facilities and the cost of undertaking flight tests, the combustors that can be tested tend to be small in cross-sectional area and in length. The natural questions then are, if the size of the combustor is required to be larger in practical application, what would be the appropriate operating conditions for realizing approximately the same type of performance and the same order of efficiencies as in the case of the small combustor that may be optimized, through testing, for nearly best performance under a given set of operating conditions; and, *vice versa,* for a large combustor that may be required in the design of a scramjet engine for operating over a range of flight conditions with assumed values of performance, what are the test conditions under which a scaled down of the combustor should be tested to ensure the validity of assumptions in design.

Scaling of a combustor thus involves setting up a verifying inter-relations governing the changes in input operation required for changes in design, in particular the size, of the combustor. In practice the inter-relations, which take the form of scaling rules, may be set up only from the point of view of specific aspects of performance; a group of such scaling rules can then, in principle, may serve to examine the essential aspects of performance, provided no conflicts are found. However, verification through testing imposes constraints natural to the test facilities that are available. The alternative of undertaking verification by recourse to modeling and computation also has its constraints. The subject of scaling of scramjets and combustors is largely in its infancy.

In practical application, the thrust required to be generated by a scramjet depends on the vehicle parameters, the acceleration or cruise desired, and the flight trajectory in altitude-speed space. An important parameter of the trajectory is the changing ambient pressure. In order to operate the combustor at a desired value of air entry pressure, the compression in the inlet needs to be adjusted along the flight trajectory, with some attention to the optimum compression for best thrust generation, discussed earlier in Sec. I. The conclusion is that the inlet and the combustor need to be considered together in scaling from one size to another in thrust generation and hence in size. The problem, of course, is further complicated by the need often felt to separate the combustor operation from the inlet operation with an isolator between the two components. Here, however, attention is focussed on a combustor, which is supplied with air and fuel under certain conditions from reservoirs.

With the given air and fuel supply, several processes are of interest in a scram combustor, each of them being quite complex: 1) mixing of air and fuel, 2) initiation of chemical reactions and ignition, 3) progress of chemical reactions and combustion, and 4) friction and heat transfer along the flow, accompanied by formation of shockwaves and their interaction with wall layers. In each case the governing parameters need to be identified, and scaling rules have to be set up relating the parameters to design variables.

The four processes together may be considered as leading to the thrust potential of the combustor, which is a function of the static pressure and enthalpy of combustion products in relation to the external ambient conditions. Considering two combustors of different physical size and nearly the same geometry, design, and operational modes, it is thus of interest to examine the changes in pressure and temperature along the combustor wall, including the exit, when operating under input conditions that are adjusted in the two combustors based on scaling rules relevant to the four processes identified.

One set of tests conducted specifically to establish the implications of such scaling rules is due to Pulsonetti (Refs. 18 and 19); a brief mention of this work may also be found in Chapter I.1 of this volume, This study has been chosen for further analysis here.

The experimental work was carried out utilizing the impulse, short-duration tunnel, the T-4 of The University of Queensland. A schematic of the combustor is given in Fig. 36. The operating conditions are included in Table 11 for the so-called small combustor and large combustor, the scaling factor

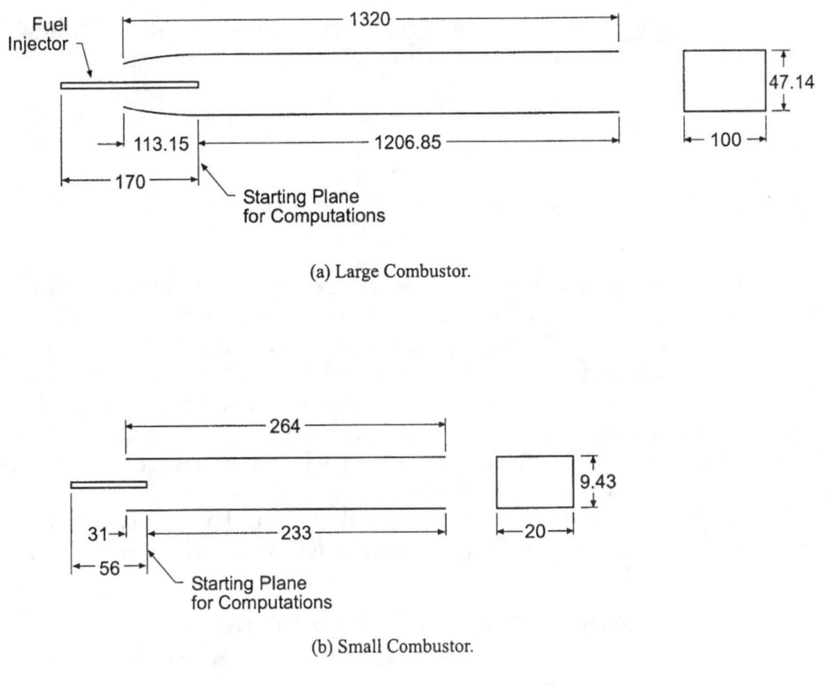

Fig. 36. Schematic of combustors of The University of Queensland.

being 5; that is the large combustor is five times larger than the small combustor in all dimensions including the cross-section and the length. The small combustor was 20 mm in width, 9.43 mm in height, and 264 mm in length. The fuel was injected with a central injector which was 2.016 mm thick. In the large combustor the fuel injector was also centrally placed and was 10.0 mm thick. As reported, there was some difficulty in realizing the desired scaling with respect to the length of the injector and the location of the injector tip relative to the model tip; this, as expected, would have changed the free flow area at the tunnel nozzle exit, for example decreased it compared to the linear scaling value. However, the observed results, it is correctly surmised, are unlikely to have been affected to any significant extent by such deficiencies.

Both combustors were operated with hydrogen gas as fuel.

The combustors were fitted with pressure transducers and heat flux gages for pressure and heat flux measurement at the wall. No measurements were made in the flow; nothing is known about the progress of chemical reactions and combustion. The occurrence of ignition was sought to be determined based on the observed rise in wall pressure.

In order to appreciate the basis for the selection for the operating parameters during the tests, given in Table 11 and also the results to follow, it is necessary to understand the approach and basis for scaling adopted by Pulsonetti.

A. Approach

Several noteworthy features in the approach are as follows:

1) From various points of view, the tests were conducted in two groups: one, with constant stagnation pressure and different stagnation enthalpy values (Cases 17.1 through 17.4 and 18.1 through 18.4) and two, with constant stagnation enthalpy and different values of stagnation pressure (the remaining test cases in Table 11 for the large and the small combustor). The tunnel employed was particularly suitable for doing this. The results could be correlated with respect to these parameters, which may be looked upon as valid for a particular flight trajectory.
2) The scaling factor 5 between the small and the large combustor was chosen on the basis of several reasons connected with tunnel operation and

Table 11 Cases from Pulsonetti's test data

Case no.	Description	P_{1a}, atm	T_{1a}, °K	M_{1a}	P_{1f}, atm	T_{1f}, °K	M_{1f}
17.1	Large combustor; $\varphi = 1.23$	0.1678	1100	4.42	0.1678	128	2.56
17.2	Large combustor; $\varphi = 1.41$	0.1796	1520	4.37	0.1796	130	2.53
17.3	Large combustor; $\varphi = 1.26$	0.1382	1640	4.38	0.1382	136	2.43
17.4	Large combustor; $\varphi = 1.32$	0.2063	2080	4.26	0.2063	140	2.36
17.5	Large combustor; $\varphi = 1.23$	0.1401	1150	4.43	0.1401	129	2.55
17.6	Large combustor; $\varphi = 1.53$	0.0774	1010	4.47	0.0774	118	2.74
18.1	Small combustor; $\varphi = 1.39$	0.7678	1100	4.47	0.7678	109	2.85
18.2	Small combustor; $\varphi = 1.24$	0.9050	1530	4.29	0.9050	118	2.64
18.3	Small combustor; $\varphi = 1.62$	0.8201	1760	4.17	0.8201	114	2.75
18.4	Small combustor; $\varphi = 1.35$	0.9830	2400	3.96	0.9830	128	2.50
18.5	Small combustor; $\varphi = 1.23$	0.6553	1060	4.48	0.6553	108	2.85
18.6	Small combustor; $\varphi = 1.31$	0.4234	1080	4.47	0.4234	107	2.80

also, realization of acceptable values of stagnation enthalpy, static pressure, and static temperature.

3) Two factors were considered significant in regard to the compressible turbulent mixing that could be expected to occur between air and hydrogen fuel supplied: one, the distance from the point of contact of fuel and air to the location where adequate micromixing would have occurred to permit the notion of a distance for start of the mixing process, and two, a distance for the completion of mixing based on the notion of a mixing efficiency defined as the local ratio of the two fluids (air and fuel) undergoing mixing relative to the stoichiometric ratio. These lengths, it can be argued, vary inversely as pressure, and thus a pressure-length scaling should apply where the areas of cross-sections have been scaled also with respect to pressure.

4) The criteria for the onset of ignition generally emphasize reaction rates for carefully selected reactions, in the current case, of hydrogen and air, to establish ignition time as a function of pressure and temperature. The influence of temperature is considered significant in the case of slow chain carrying reactions. It is then argued heuristically that ignition time should vary inversely as pressure and pressure-length scaling is valid for the ignition process.

5) Considering the progress of combustion, in this case of hydrogen and air, the reaction time is that required for the production of 95% of heat produced by the formation of water, reckoned from the instant of ignition. Again selecting a set of reactions with various intermediate products and considering reaction rates, it can be shown that reaction time is inversely proportional to pressure raised to the power 2 or 1.7, depending on assumptions concerning reactions and reaction rates. Thus p^n - L (where P is pressure raised to the power n, equal to 2 or 1.7, and L is length) scaling should apply.

6) Finally, considering friction and heat transfer processes, in the simplest case viscosity and turbulence effects have to be examined. In more complicated cases, the formation of shockwaves in the flow lead to shock-boundary layer interaction and friction and heat transfer processes become modified. In general, these processes depend on wall boundary layer characteristics, which can be expressed as a function of the inverse of flow Reynolds number. It is then possible to adopt P - L scaling in this case also.

The foregoing brief and admittedly terse discussion should serve the purpose of showing the bases for developing an approach to the rather complicated problem of scaling scram combustors. An interested reader should study the foundational texts of fluid mechanics, combustion theory, and chemical reaction theory to obtain a thorough understanding of the issues involved in various background subjects. Many assumptions have been introduced throughout the discussion, and their validity in different cases is uncertain in detail. It will have been observed that density of the gas has been replaced by pressure throughout for different reasons in different contexts: temperature having negligible effect or temperature variation being small, and so on. Such assumptions are indeed subject to further analysis.

Except in the progress of reaction and combustion, following ignition, the pressure-length scaling can be taken to be applicable to the processes in a combustor. This is equivalent to assuming similarity with linear scaling of geometry and inverse linear scaling of pressure. Pulsonetti adopted this approach and used the scaling rule

$$PD = \text{constant} \tag{18}$$

Where D is a length dimension. This means that if all of the dimensions of the combustor are increased by a certain factor, the operating pressure must be decreased by the same factor, while retaining the temperature to be the same.

Thus, for operation of a small combustor and a large combustor with a linear scaling factor S, at the same entry Mach number and static temperature value, the mass flux has to be

$$\dot{m}_L = \dot{m}_S \cdot S \tag{19}$$

and the stagnation and static pressure at entry become reduced by the factor S while the stagnation enthalpy at entry remains the same. To realize five times the thrust of the small combustor by correct expansion of combustion products to the same ambient pressure conditions, while the fuel equivalence ratio and combustion efficiency are the same in the two cases, the large combustor exit stagnation pressure has to increase by a factor of $(S)^{(\gamma-1)/\gamma}$, assuming that γ, the ratio of specific heats, and the exit static temperature remain equal in the two combustor flows. If the large combustor is operated in ambient conditions where the static pressure is smaller by a factor 5 relative to that for the small combustor, then the thrust of the large combustor becomes five times that of the small combustor provided the stagnation pressure and the static temperature at the exit of the two combustors is the same. One can notice how such demands necessitate completion of combustion to equal extents, with equal heat release and losses, in the lengths provided; the longer length of the large combustor may involve an increase in friction loss, although the change in its hydraulic diameter is compensated by the increase in length. A variety of other types of operation can be examined on the same basis. The operational parameters in the test cases given in Table 11 may also be analyzed on the same basis, noting that in the tests S was set equal to 5.

B. Ignition Delay Estimate

In carrying out the combustion experiments an attempt was made to establish ignition location and time in different cases. The estimate was based on the rise in static pressure along the wall of the combustor as observed in experimental data. No clear trends were observable among the small and large combustors in view of 1) ambiguity in entry conditions and wall pressure measurements, and 2) uncertainty in the occurrence of mixing-limited and reaction-limited combustion in different cases. The same observation was made for global ignition delay times that were calculated using the estimated temperature of air-fuel mixture in the case of high and low enthalpy air and fuel.

Thus the testing at this stage may be said to have yielded mainly comparisons of pressure rise along the combustor.

C. Pressure Rise Along Combustor

The results of performance calculations are presented in Figs. 37–42, along with relevant test data for the following parameters: 1) static pressure variation along the combustor for given entry pressure under specific operating conditions in the large and the small combustor tests (Fig. 37); 2) static pressure ratio across the combustor as a function of stagnation enthalpy in the large and the small combustor tests under specific operating conditions (Fig. 38); 3) ratio of static pressure ratio across the small and the large

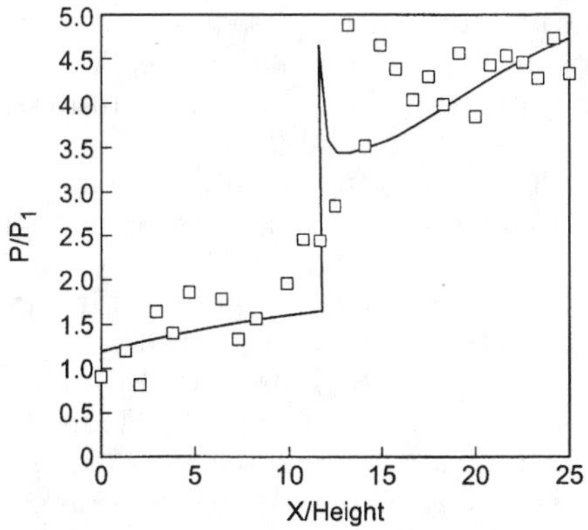

(a) Large Scramjet; See Test Condition in Table I.4.

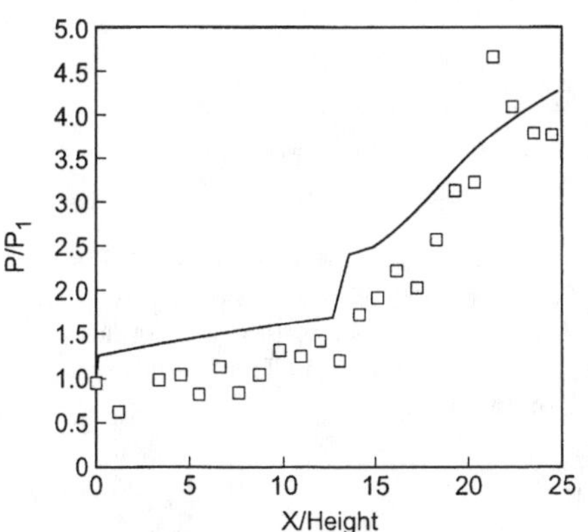

(b) Small Scramjet; See Test Condition in Table I.4.

Fig. 37. Pressure distribution along the combustor: a) large combustor for case 17.1 of Table 4. b) small combustor for Case 18.1 of Table i.3.

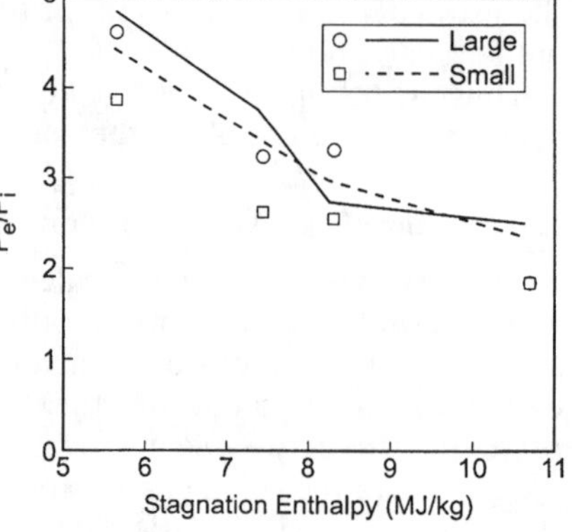

Fig. 38. Variation of Ratio of Final to Initial Pressure Ratio as a Function of Stagnation Enthalpy Input in Large and Small Combustors.

combustor as a function of stagnation enthalpy addition under specific operating conditions (Fig. 39); 4) static pressure ratio across the combustor as a function of nozzle stagnation pressure in the large and the small combustor tests under specific operating conditions (Fig. 40); and 5) ratio of static pressure ratio across the small and the large combustor as a function of nozzle stagnation pressure under specific operating conditions (Fig. 41). In all cases the measured pressure has been nondimensionalized with pressure at the point of fuel injection.

Figure 37 shows that the variation of pressure along the combustor length is predicted in agreement with data, provided, as done in the calculations, the location of ignition is chosen to be that given by Pulsonetti. Figure 38 also establishes the general trends of values processed from data. Figure 39 presents the same ratio of static pressure ratio across the small and the large combustor, but as a function of stagnation enthalpy; at entry to the combustor. Here there seems to be a difference in the value of stagnation enthalpy, about 15%, at which the ratio of pressure ratio increases; however the

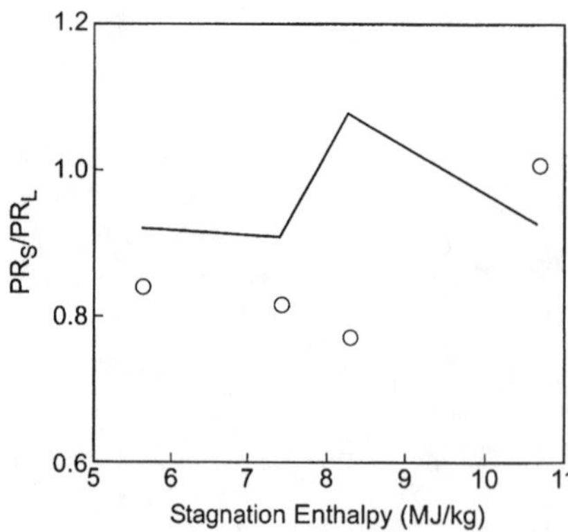

Fig. 39. Variation of Ratio of Final Pressure to Pressure at Point of Injection Ratio as a Function of Stagnation Enthalpy at Entry in Large and Small Combustors.

general trend of variation is captured in the predictions. The difference is probably due to the basis of chemical reaction assumed.

Figures 40 and 41 show the ratio of final to initial pressure in the large and the small combustor, one with respect to stagnation pressure at entry and the other with respect to stagnation enthalpy. The predictions yield slightly optimistic values.

One other set of values have been deduced from the preductions, namely the enthalpy rise across the combustor, non-dimensionalized with respect to entry stagnation enthalpy. These values are shown as a function of entry stagnation pressure for the case of the large and the small combustors in Fig. 42. It is clearly observed that the small combustors do not perform as well as the large combustors. Pulsonetti recognizes this based on pressure rise scaling in the large and the small combustors, although no results are presented on the enthalpy rise. Considering reaction-limited combustion as the basis in small combustors, Pulsonetti refers to the reaction associated with the production of water, a chain-breaking reaction involving a third body, and argues that this may have depleted the concentration of free radicals needed for intense combustion to occur in the small combustor operating at the scaled-up value of static pressure. This problem requires further study. Meanwhile the prediction scheme could not be used effectively to establish the progress of chemical reactions since enough data could not be found on the parameters needed for such predictions.

In general, the predicted results match the measurements in all cases. The main difference is in the region of combustion initiation. It may be noted that the calculations have been performed assuming that ignition occurs where the experimental data indicates the setting in of ignition; that is, the combustion-generated pressure rise is assumed to begin at the same location as in experiments. Thus in the calculations, the fuel injected is assumed to undergo mixing and accumulate up to the location of combustion. In any case, the conclusion is that the prodiction schemes confirms the applicability of the scaling rule, Eq. (18) for the cases under consideration. It is also of interest to note that experimental data of the type generated in pulsed test

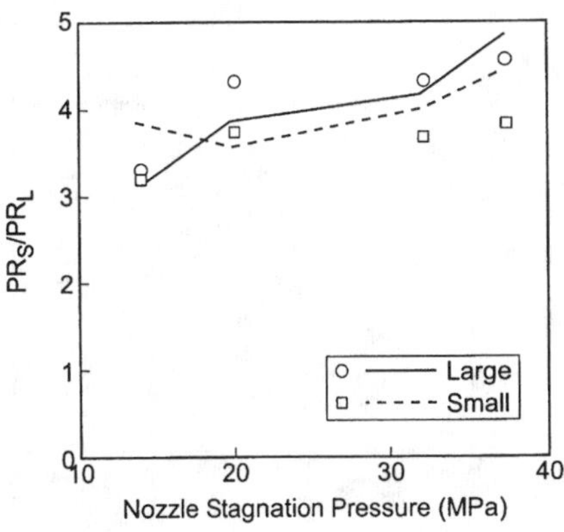

Fig. 40. Variation of Ratio of Final Pressure to Pressure at Point of Injection Ratio as a Function of Stagnation Pressure at Entry in Large and Small Combustors.

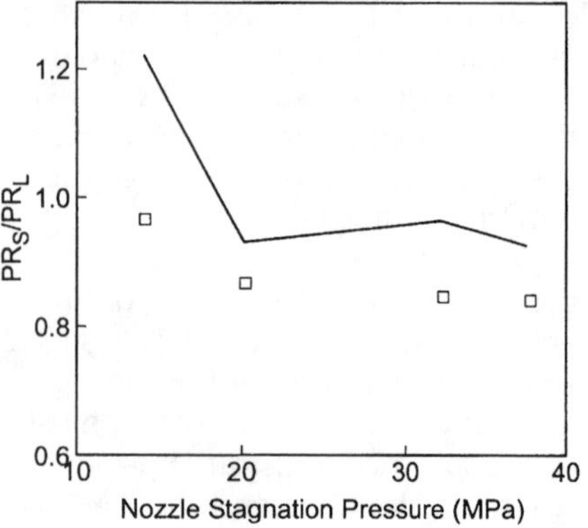

Fig. 41. Variation of Ratio of Final Pressure to the Pressure at Point of Injection as a Function of Entry Stagnation Pressure in Large and Small Combustors.

facilities can be generated using the relatively simple prediction scheme utilized here. Other aspects of scaling and its effects need more detailed measurements and more sophisticated analytical-computational tools for verification.

Based on the predictions presented in Sec. V and the current section in various practical cases, it may be said that the one-dimensional three-stream code provides a reasonable basis for establishing the basic overall performance of a scram combustor, including some details on combustion, when the combustor consists of a simple duct and the fuel injection scheme can be represented by the mass and momentum of the fuel in relation to the mass and momentum of the air stream. There is also some scope for examining specific features of performance in various cases in terms of the influence of mixing schedule, friction, and heat transfer, as well as combination of equilibrium and nonequilibrium chemical kinetics in the setting in and progress of chemical reaction and combustion. The code thus provides a simple and effective tool for a configuration layout of the combustor, which can be used in the layout of the overall engine flowpath, including the inlet and the

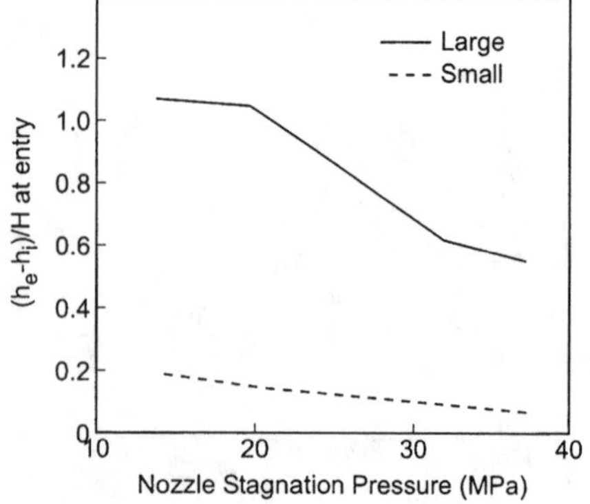

Fig. 42. Variation of the Ratio of Enthalpy Increase Across Combustor Normalized with Respect to Enthalpy at Entry as a Function of Entry Stagnation pressure in Large and Small Combustors.

nozzle. The tool obviously can also be utilized to arrive at optimum values of the parameters included in the code for best performance of a given type of configuration. However, the tool is of use only when the duct geometry and the fuel injection scheme permit an integrated approach to the flowpath with no local, disturbing processes.

VII. Combustor-Based System Integration

In Section 2.2, the combustor was considered as a part of the scramjet engine as a system, and a system-wise exergy efficiency was defined in Eq. (11). The interest was in the best value of energy availability that could be recovered as thrust work, by expansion of the products of combustion to the ambient pressure, in relation to the total energy availability input to the engine through supply of air and fuel. In practice the thrust nozzle expansion would also suffer losses and the actual energy availability recovered as thrust work would be less than the ideal value.

From the point of view of the vehicle system, the thrust generated must be equal to drag of the vehicle for a cruise mission, and greater than the drag if the vehicle is to accelerate or undertake maneuver. As stated earlier, Ref. 2 shows how the vehicle configuration and drag depend on the details of the propulsion system geometry, and, in particular, the size of the intake frontal cross-section, the combustor cross-section and length, the fuel volume at take-off, and the nozzle length. Considering the combustor as the central element, its size and efficiency become dominant parameters in the thrust-drag balance. The combustor can be considered in itself and also, accounting for interactions with the inlet and the nozzle (Fig. 1). In fact the system-based combustor efficiency can account for such combined interactions, as can be seen from Fig. 2b.

In order to undertake the determination of the propulsion system internal flowpath and the vehicle system external flowpath in analysis of configuration convergence, the interactions between the two can be accounted for in terms of the following: 1) vehicle external flow on thrust nozzle exit flow: 2) air mass flow and momentum flux split between the two streams; and 3) thermal management of the total vehicle system. The latter is often dealt with separately from the first two, although thermal management, in addition to affecting size and mass of the vehicle, also affects the external surface treatment of the vehicle and cooling of sharp leading edge as well as fuel supply parameters. It is becoming increasingly clear that thermal management needs to be dealt with integrally with other aspects of vehicle design and operation. At the same time, drag and mass flow split can be established without all of the details of thermal management.

A. Inlet and Nozzle Efficiency

It was stated earlier that the optimal thermal compression ratio [Eq. (5)] for best thrust work output is a function of inlet and nozzle efficiency. Considering the magnitude of energy loss, the largest loss is in the thermal and kinetic energy of the exhaust stream. For best propulsive efficiency, the kinetic energy of the exhaust stream should differ as little as permissible with the available mass flow, from the kinetic energy of the inlet air and fuel on admission to the engine. For a given value of nozzle pressure ratio with respect to the ambient atmosphere, the kinetic energy obtained through expansion depends on the nozzle entry stagnation enthalpy. Thus, unless part of the enthalpy of the nozzle working fluid can be recovered by another means of thermal management, one always finds a certain large loss of enthalpy in the exhaust stream corresponding to the expansion pressure ratio employed. In turn this implies that the nozzle itself should operate at, and, therefore, has to be designed for the best realizable efficiency. Thus the inlet performance (as depicted by the efficiency with which compression is realized) is critical for the best performance of the engine, as well as the nozzle efficiency, the latter playing a more direct and dominant role.

B. Inlet Layout

The configuration design of the air inlet has been discussed in detail in Chapter IV.15 of this volume. Both the internal flowpath design as well as interaction with the external flowpath have been have been dealt with.

Other developments on inlet internal flowpath determination have been reported in Ref. 20. These developments are particularly significant since they are based on establishing the flowpath from the point of view of setting up the diffusion process through the selection of flow contours, while the flow contours in turn are derived in terms of specific shapes in the longitudinal and the transverse planes. Such a procedure yields the greatest choice in designing the flowpath. For example, continuous and discrete (obtained through weak shockwaves) changes in flow deceleration and increase of static pressure can be realized with the least losses.

The analysis provides a means of obtaining streamline traces through axisymmetric and planar flows. A family of inlet shapes can be generated ranging from outward turning axisymmetric ones, realized with aerodynamic spikes, to fully inward turning ones of the well-known Busemann type with shock cancellation. In the latter case, at the design Mach number, Mach waves from the curved inward turning surface coalesce at the apex of the internal shock. The conical isentropic compression is governed by the Taylor-Maccoll equation for axisymmetric flow.

In the general case the formulation is based on the so-called radial deviation parameter (RDP), which varies from negative unity value for outward turning flows to positive unity value for inward turning flow. The radial deviation parameter is introduced into the flow with a core uniform flow or a core body, such that the desired radial flow can be obtained.

The computational methodology is based on the use of the method of characteristics with weak shock fitting as necessary, and with a subsequent correction for viscous displacement thickness. An inlet cross-section can be specified at entry or exit, the latter corresponding to combustor entry, for example. Then the streamlines can be traced downstream of entry or upstream of exit as necessary to obtain the internal compression surfaces.

C. Nozzle Layout

In an overall nose-to-tail vehicle configuration study, for the internal flowpath the thrust generated can be established when the nozzle and the nozzle internal flow and the external flow affecting the nozzle flow are established. The methodology for three-dimensional nozzle design and performance estimation is described in Refs. 21 and 22. The direct optimization of such nozzles can follow the technique described in Ref. 23. Appendix D provides a summary of the approach.

References

[1]Czysz, P., and Murthy, S. N. B., "Energy Analysis of High Speed Flight Systems," *High Speed Flight Propulsion Systems, Progress in Aeronautics and Astronautics,* Vol. 137, AIAA, Washington, DC, 1991.

[2]Czysz, P., and Murthy, S. N. B., "Energy Management and Vehicle Synthesis," *in Developments in High-Speed Vehicle Propulsion Systems,* Progress in Aeronautics and Astronautics, Vol. 165, AIAA, Reston, VA, 1996.

[3]Curran, E. T., and Bergsten, M. B., "Inlet Efficiency Parameters for Supersonic Combustion Ramjet Engines," AF Aero Propulsion Lab. Rep. APL TDR 64-61, June 1964.

[4]Curran, E. T., Leingang, J. L., Carriero, L. R., and Petters, D. P., "A Review of Kinetic Energy Methods in High Speed Engine Cycle Analysis," International Symposium on Air Breathing Engines, Paper No. ISABE 91-10.5(L), AIAA, 1991.

[5]Curran, E. T., Leingang, J., Carriero, L. R., and Petters, D., "Further Studies of Kinetic Energy Methods in High Speed Ramjet Cycle Analysis," AIAA Paper 92-3805, July 1992.

[6]Strehlow, R. A., *Fundamentals of Combustion,* International Textbook Co., 1968.

[7]Pratt, D. T., Humphrey, J. W., and Glenn, D. E., "Morphology of a Standing Oblique Detonation Wave," AIAA Paper No. 87-1785, June–July 1987.

[8]Sislian, J. P., Chapter V.17 in the current volume.

[9]Meenaksh, R., and Murthy, S. N. B., "One-Dimensional Parametric Studies on Scram Combustors," Purdue Univ. Internal Rept. No. M/00/R-1, 1999.

[10]Andrews, E. H., and Mackley, E. A., "Hypersonic Research Engine/Aerothermodynamic Integration Model—Experimental Results," NASA Rept. TMX-72824, Vol. I–V, April 1976.

[11]Andrews, E. H., and Mackley, E. A., "Analysis of Experimental Results of the Inlet for the NASA Hypersonic Research Engine Aerothermodynamic Integration Model," NASA, June 1976.

[12]Billig, F. S., Dugger, G. L., and Waltrup, P. J., "Inlet-Combustor Interface Problems in Scramjet Engines," *The First International Symposium on AirBreathing Engines,* June 1972.

[13]Waltrup, P. J., and Billig, F. S., "Prediction of Precombustion Wall Pressure Distributions in Scramjet Engines," *Journal of Spacecraft,* Vol. 10, No. 9, 1973, pp. xx–xx.

[14]Northam, B. G., "Evaluation of Parallel Injector Configurations for Supersonic Combustion," AIAA Paper No. 89-2525, July 1989.

[15]Paull, A., "Hypersonic Ignition and Thrust Production in a Scramjet," AIAA Paper No. 93-2444, June 1993.

[16]Sabel'nikov, V. A., Voloshenko, O. V., Ostras, V. N., Shermanov, V. N., and Walter, R. "Gasdynamics of Hydrogen-Fueled Scramjet Combustors," AIAA Paper No. 93-2145, June 1993.

[17]Sabel'nikov, V. A., and Penzin, V. I., "Scramjet Research and Developments in Russia," Chapter I.5 in the current volume. (See Appendix A).

[18]Pulsonetti, M. V., "Scaling Laws for Scramjets," Ph.D. Dissertation, Univ. of Queensland, June 1995.

[19]Pulsonetti, M. V., and Stalker, R., "A Study of Scramjet Scaling," AIAA Paper No. 96-4533, Nov. 1996.

[20]Billig, F. S., and Kothari, A. P., "Streamline Tracing, A Technique for Designing Hypersonic Vehicles," *Proceedings of the XIII,* AIAA, International Symposium on Air Breathing Engines, Reston, VA, 1997.

[21]Meenaksh, R., Hoffman, J. D., and Murthy, S. N. B., "Design and Performance Computations in Complex 3-D Nozzles," AIAA Paper No. 99-0882, Jan. 1999.

[22]Meenaksh, R., Hoffman, J. D., and Murthy, S. N. B., "Three-Dimensional Nozzles: Direct Optimization and Integration," AIAA 9th International Space Planes and Hypersonics Conference, Nov. 1999.

[23]Allman, J.G., and Hoffman, J.D., "Design of Maximum Thrust Nozzle Contours by Direct Optimization Methods," *AIAA Journal,* Vol. 19, No. 6, 1981, pp. 750, 751.

Appendix A: Efficiency Relations

Simple Efficiency Interrelations

Direct efficiency conversions or interrelations can be established assuming that the working fluid is a perfect gas with a constant ratio of specific heats, following Refs. 3 through 5.

A generalization of kinetic energy efficiency to account for mass flow changes has been attempted in Ref. 1 in this Appendix. On this basis, the kinetic energy efficiency for a combustor may be written as

$$\eta_{\text{KEC gen}} = \frac{H_{t3p}(1+f) - H'_{ep}(1+f)}{H_{t1a}\left[1 + f\left(H_{t1f}/H_{t1a}\right)\right] - H'_{2a}\left[1 + f\left(H_{t1f}/H_{t1a}\right)\right]} \tag{A1}$$

$$= \frac{(1+f)}{\left[1 + f\left(H_{t1f}/H_{t1a}\right)\right]} \cdot \frac{H_{t3p} - H'_{ep}}{H_{t1a} - H'_{2a}} \tag{A2}$$

where f is the fuel to air ratio.

Normalizing each of the enthalpy changes of Eq. (A2) with the corresponding initial values of enthalpy, one can write

$$\eta_{\text{KEC gen}} = \frac{1+f}{\left[1 + f\left(H_{t1f}/H_{t1a}\right)\right]} \cdot \frac{V_e^2}{V_1^2} \cdot \frac{H_{t1}}{H_{t3}} \tag{A3}$$

The stagnation pressure loss efficiency is expressed by Eq. (12) as

$$\begin{aligned}\eta_{PT} &= P_{te}/P_{t1}\\ &= \frac{P_{te}}{P_e}\cdot\frac{P_o}{P_{t1}}\\ &= \left(\frac{H_{te}}{H_e}\right)^{\frac{\gamma}{\gamma-1}}\cdot\left(\frac{Ho}{H_{t1}}\right)^{\gamma/(\gamma-1)}\cdot\left(\frac{P_{to}}{P_{t1}}\right)\\ &= \left[\frac{1+(V_e^2/2JH_e)}{1+(V_1^2/2JH_1)}\right]^{\gamma/(\gamma-1)}\cdot\left(\frac{P_{to}}{P_{t1}}\right)\end{aligned} \tag{A4}$$

For an ideal inlet, by definition

$$\frac{P_{to}}{P_{t1}} = 1 \tag{A5}$$

For a constant pressure combustor with an ideal inlet,

$$\eta_{PT} = \left[\frac{1+(r-1/2)M_e^2}{1+(r-1/2)M_o^2}\right]^{\gamma/\gamma-1} \tag{A6}$$

Referring to Eq. (A4), it follows that

$$\frac{V_e^2}{V_1^2} = \left[\eta_{PT}\left(1+\frac{V_1^2}{2JH_o}\right)^{\gamma/(\gamma-1)} - 1\right]^{\frac{(\gamma-1)}{\gamma}}\left(\frac{H_e}{H_o}\right)\cdot\frac{2JH_o}{V_o^2}\cdot\left(\frac{V_o^2}{V_1^2}\right) \tag{A7}$$

Hence, combining Eqs. (A3) and (A5), one has a relation between $\eta_{KEC.gen}$ and η_{PT}.

If one assumes f is small compared to unity and $H_{t1a} = H_{t1f}$, one obtains the simple relation of Ref. 5 in the main text, namely

$$\eta_{KEC} = 1 - \frac{2JH_o}{V_o^2}\left[\left(\frac{1}{\eta_{PT}}\right)^{\frac{\gamma-1}{\gamma}} - 1\right] \tag{A8}$$

where, it is assumed that the inlet operates with no stagnation pressure loss.

References

A1 Hoose, K. V., "A Newly Defined KE Parameter and Its Application in the Design of Scramjet," AIAA Paper 96-4608, 1996.

Appendix B: Heat Addition to a Supersonic Gas Flow

We will assume throughout this analysis frictionless adiabatic flow (that is, the walls being insulated and hence permitting no heat exchange with the flow) and deal with it on a one-dimensional basis.

I. Constant Pressure Heat Addition in a Duct

In a Brayton cycle, referring to Fig. 2, the heat addition across the heat addition zone or the combustor per unit mass of air is given by

$$Q = H_{t3}\left(1+\frac{f}{a}\right) - H_{t1a}\left(1+\frac{f}{a}\cdot\frac{H_{t1f}}{H_{t1a}}\right) \tag{B1}$$

$$= H_3\left(1+\frac{f}{a}\right) - H_1\left(1+\frac{f}{a}\cdot\frac{H_{t1f}}{H_{t1a}}\right) \tag{B2}$$

since $V_3 = V_1$. Over an element of length Δx along the flow, the heat addition may be written as

$$\Delta Q_x = H_x\left(1+\frac{f}{a}\right) - H_1\left(1+\frac{f}{a}\cdot\frac{H_{t1f}}{H_{t1a}}\right) \tag{B3}$$

where subscript x denotes the location in the combustor. The corresponding change in flow area is given by

$$\frac{\Delta A}{A_1} = \frac{H_x}{H_1} - 1 \tag{B4}$$

assuming a constant ratio of specific heats, γ. And, the change in stagnation pressure over Δx may be written as

$$\frac{\Delta P_{tx}}{P_{t1}} = 1 - \frac{\left(1+(\gamma-1/2)M_x^2\right)^{\gamma/r-1}}{\left(1+(\gamma-1/2)M_1^2\right)^{\gamma/r-1}} \tag{B5}$$

where

$$M_x^2/M_1^2 = H_1/H_x \tag{B6}$$

Finally the exergy gain between stations 1 and x is given by

$$\varepsilon_x - \varepsilon_1 = H_{tx} - H_{t1} - T_1(s_x - s_1) \tag{B7}$$

II. Constant Mach Number Heat Addition in a Duct

The heat addition per unit mass of air is again given in this case by Eqs. (B1) and (B2), since $M_1 = M_3$. The relation (B3) also holds in this case.

The corresponding change in flow area over an elemental length Δx along the flow is given by

$$\frac{\Delta A}{A_1} = \frac{P_1}{P_x} \cdot \left(\frac{T_x}{T_1}\right)^{1/2} \tag{B8}$$

And the change in stagnation pressure over Δx may be written as

$$\frac{\Delta P_{tx}}{P_{t1}} = 1 - \frac{P_x}{P_1} \tag{B9}$$

$$= 1 - \left[\left(1 + \frac{\Delta A}{A_1}\right)^{-1} \cdot \left(1 + \frac{\Delta Q}{H_{01}}\right)\right] \tag{B10}$$

Finally the exergy gain between stations 1 and x is given by the same relation as Eq. (B7).

III. Heat Addition in a Constant Area Duct

The heat addition per unit mass of air is given by Eq. (B1). The heat addition over an element of length Δx is given by

$$\Delta Q_x = H_{tx} - H_{t1} \tag{B11}$$

The change in static pressure in this case over Δx is given by

$$\frac{\Delta Px}{P_1} = 1 + \frac{1 + \gamma M_1^2}{1 + \gamma M_x^2} \tag{B12}$$

The corresponding change in stagnation pressure is then obtained as

$$\frac{\Delta P_{tx}}{P_{t1}} = 1 - \left[\left(1 + P\frac{\Delta P_x}{p_1}\right)\frac{\left[1 + (\gamma - 1/2)M_x^2\right]^{\gamma/\gamma-1}}{\left[1 + (\gamma - 1/2)M_1^2\right]^{\gamma/\gamma-1}}\right] \tag{B13}$$

The Mach number ratio may be written as follows:

$$\frac{M_x^2}{M_1^2} = \frac{V_x}{V_1} \cdot \left[1 - \gamma M_1^2\left(\frac{V_x}{V_1} - 1\right)\right]^{-1} \tag{B14}$$

$$= \frac{V_x}{V_1} \cdot \frac{1 + \gamma M_x^2}{1 + \gamma M_1^2} \tag{B15}$$

Finally, the exergy gain from 1 to x *is* given by the same relation as Eq. (B7).

IV. Heat Addition in a General Diverging Area Duct

The heat addition per unit mass flow of air is given by Eq. (B1). the heat addition over an element Δx can be written as in Eq. (B11).

The change in static pressure in this case over a length Δx from station 1 is given by the following:

$$\frac{\Delta P_x}{P_1} = 1 - \left[\left(1 + \frac{\Delta A}{A_1}\right)^{-1} \cdot \frac{1 + \gamma M_1^2}{1 + \gamma M_x^2}\right] \tag{B16}$$

The corresponding change in stagnation pressure is then given by the same expression as Eq. (B13)

The Mach number ratio is given by

$$\frac{M_x^2}{M_1^2} = \frac{V_x}{V_1} \cdot \frac{A_1 P_1}{A_2 P_x} \tag{B17}$$

where

$$\frac{V_x}{V_1} = \frac{A_1 P_1}{A_x P_x} \cdot \frac{P_{tx}}{T_{t1}} \tag{B18}$$

The corresponding change in stagnation pressure ratio is given by Eq. (B5). Finally the gain in exergy over the length Δx is given by Eq. (B7).

V. Heat Addition Following a Shockwave

The shockwave is a process occurring with no change in stagnation enthalpy. There is a gain in static enthalpy at the expense of kinetic energy and such gain can act as an ignition source for a combustible mixture. The largest gain in static enthalpy and the accompanying increases in gains in static pressure and, also, exergy occur in the case of a normal shock.

When the reactants, such as fuel and air, are fully mixed ahead of the presence of the shockwave, one may obtain a detonation type combustion. However, both in unmixed and mixed propellants, the combustion, which can always be treated as a wave, can occur in the slower deflagration mode. The deflagration type combustion depends on the type of conditions under which it takes place, for example according to types discussed in I–IV.

A detonation occurs when there is a strong coupling between the shockwave and the heat release, and the combustion is completed in a relatively small distance, although larger than the shock thickness. Detonative combustion is known to be stable so long as the normal component of velocity following the shockwave is such that its Mach number is equal to or less than unity. The case when the Mach number is unity is a limiting case, and since then the heat added is the largest, it is recognized as an ideal situation, and is commonly referred to as the Chapman-Jouget (C-J) limit or type of detonation. This ideal corresponds to heat addition in deflagration to the choking limit starting from a (subsonic or) supersonic condition; the C-J detonation is a supersonic flow process.

It is often useful to consider combustion as a wave with a propagating speed in a mixture. One can then ask what the flow velocity of the mixture is when an amount of heat Q released would render the wave propagation velocity become equal to that of sound in the medium. V. Levich in Ref. 1 in this Appendix provides an expression for the Mach number of flow:

$$M_1^{*2} = \frac{1 + 2\overline{Q}}{1 + 4\,(\gamma - 1/\gamma + 1)\overline{Q}}\left[1 \pm \left[\frac{1 + 4\,(\gamma - 1/\gamma + 1)\overline{Q}}{(1 + 2\overline{Q})^2}\right]^{1/2}\right] \tag{B19}$$

where $\overline{Q} = Q/H_1$, and

$$M_1^* = \frac{V_1}{a^*} \tag{B19a}$$

$$a^* = (\gamma R T_1^*)^{1/2} = \left[\gamma R\left(\frac{2T_{t1}}{\gamma + 1}\right)\right]^{1/2} \tag{B19b}$$

Here a^* is the critical acoustic velocity corresponding to the critical temperature T^* or M equal to unity case, and M^* is the Mach number based on a^*. The positive sign in the radical of Eq. (B.19) corresponds to supersonic velocity of flow, which is the primary interest here.

It may be pointed out that M^*_1 in supersonic flow is in fact the minimum Mach number at which the addition of heat equal to $\overline{Q}$ will cause the Mach number to reduce to unity. In this case, an approximate form of Eq. (B.19) for an estimate of M^*_1 is given as

$$M_1^* = \frac{2 + 4\overline{Q}}{1 + 4\,(\gamma - 1/\gamma + 1)\overline{Q}} \tag{B20}$$

Now, a detonation wave can be considered as a combination of a shockwave and a flame front. The flame front following the shockwave occurs in the reduced velocity field, and finally at the end of the combustion zone the local M^* attains the value of unity. As pointed out in Ref. 7 in the main text, it is necessary to note that neither deflagration nor detonation can propagate at a rate greater than M^* equal to unity.

In the case of an oblique shockwave, since there is no change in the tangential component of velocity across a shockwave, the constraint on the propagation velocity applies to the normal component of velocity; hence, the velocity reached at the end of combustion is such that M_n^*, the normal component, is unity or less. When it is unity, the limiting value, one obtains the C-J detonation as stated earlier.

The terminology introduced by D.T. Pratt (Ref. 7) is particularly useful here: in the case of oblique waves, *weak underdriven oblique detonation waves* are distinguished from *weak and strong overdriven oblique detonation waves* by whether M_{2n}, the normal component of Mach number on the downstream side of an oblique shockwave following heat addition is greater than or less than unity, respectively. Thus underdriven ODW cannot occur, and the limit to overdriven ODW is the C-J detonation condition.

A. *Oblique Detonation Wave*

To proceed, we will consider the case of an oblique shock with detonation following it in the under-driven mode. Referring to Fig. A1 and using the terminology of Ref. 7, it can be shown that the ratio of static pressure across the detonation process is given by the relation,

$$\frac{P_3}{P_1} = 1 + \gamma M_1^2 \sin^2 \beta (1 - \chi) \tag{B21}$$

where $\chi = \rho_2/\rho_1$, the density ratio, and the static temperature ratio is given by

$$\frac{T_3}{T_1} = 1 + \overline{Q} + \left(\frac{\gamma - 1}{2}\right) M_1^2 \sin^2 \beta (1 - \chi^2) \tag{B22}$$

In Eqs. (B21) and (B22) the condition at the end of combustion is referred to as condition 3 to be consistent with the terminology in Secs. I–IV. The density ratio, noting that there is no change in geometry over the short distance of the detonation process and assuming a constant area duct, is given by

$$\chi = \frac{V_{s2n}}{V_{\sin}} = \frac{\tan(\beta - \theta)}{\tan \beta} \tag{B23}$$

where V_{sin} and V_{s2n} refer to the normal components of velocity on the two sides of the shockwave; $V_{\text{sin}} = V_{1n}$.

The heat added can be written as

$$Q = H_{t3} - H_{t1} \tag{B24}$$

Here the subscript 3 refers to the end of combustion process, as stated above.

The stagnation pressure ratio across the detonation process is obtained readily from Eq. (B21) using the Mach number ratio for the process, namely,

$$\frac{M_{2n}^2}{M_{1n}^2} = \frac{\left[1 + \sin(\beta - \theta)^2\right]}{(4 \sin\beta)^2} \cdot \frac{T_1}{T_2} \tag{B25}$$

The gain in exergy across the detonation process is again given by Eq. (B7).

B. C-J Detonation Case

In the limiting case with M_{2n} becoming equal to unity with the release of heat equal to $\overline{Q}$, the density ratio can be expressed for the case of a normal shock by

$$\chi = \frac{1 + \gamma M_{1n}^2}{(1 + \gamma) M_{1n}^2} \tag{B26}$$

The corresponding ratios of static pressure and static temperature can be obtained from Eqs. (B21) and (B22). Similarly one can obtain the ratio of stagnation pressure, noting that

$$M_{1n}^2 = \left[1 + \overline{Q}(\gamma + 1)\right] + \left\{\left[1 + \overline{Q}(\gamma + 1)\right]^2 - 1^{1/2}\right\} \tag{B27}$$

It will be recalled that the lowest value of M_{1n} possible is given in Eqs. (B19) and (B20) for the case of a normal shockwave.

A significant feature of the C-J detonation process is that the entropy rise is the minimum. The relevant algebra can be found in Ref. 6 in main text and in this Appendix. The implication is that in a constant area combustor the C-J detonation provides the largest static pressure rise for the largest heat addition that can be supplied to a given Mach number supersonic flow. Thus, the exergy gain is the highest under the C-J detonation condition among not only all detonation processes but also in comparison with any of the processes of heat addition is discussed in Secs. I–IV.

C. Time for Setting in of Detonation

As stated earlier, the fetch of the reactive medium for reaction on-set and completion, however small, is still finite. In other words, there is a finite length of combustor between the beginning of the shock process and the end of combustion. Reference 2 in this Appendix deals with an estimate of the time required, and the estimate for the so-called induction time is provided in the following.

$$\ell n\, \tau = \ell n\, \tau^* + A^*/(T - T^*) \tag{B28}$$

where, for stoichiometric concentrations of an air and hydrogen flame, in the ranges

$$1500\ k < T < 500\ k$$

and

$$0.05 < P < 100\ \text{atm}$$

$$\ell n\, \tau^* = -\ 14.54 + 1.30\ Z^2/(6.09 + Z^2)$$

$$A^* = 4100 - 3850\ Z^2/(7.33 + Z^2),$$

and

$$T^* = 300 - 1350\ Z/(18.75 + Z)$$

Here Z denotes a function of P; for the given velocity flow, one can then calculate the distance or fetch required.

The maximum distance for combustion initiation, when one can still assume coupling with the shockwave is a moot point; Ref. 3 in this Appendix assumes that any calculated distance under one foot may be considered as leading to detonation, or, equivalently, orders of milliseconds in time.

VI. Efficiencies in Heat Addition

The kinetic energy in all cases I–V is given by

$$\eta_{\text{KEC}} = \frac{V_4^2}{V_2^2} \cdot \frac{H_{t1}}{H_{t3}} \tag{29}$$

where V_4 and V_2 are obtained by isentropic expansion from P_{t3} and P_{t1}, respectively, to P_O, the ambient pressure.

The exergy-based efficiencies are in each case those given by Eqs. (8) and (11). Determining these requires values for Δs, the change in entropy, which vary in different cases I–V.

The stagnation pressure ratio across the combustor with reference to the entry stagnation pressure can be found starting with the static pressure at the exit and the local Mach number.

A detailed knowledge of the thermodynamic parameters at the combustor exit for given conditions is obtainable starting with the foregoing three efficiency parameters. It can also be seen that they are necessary for a complete description of combustor performance. The ideal value of enthalpy at the combustor exit is the sum of the entry value of air kinetic energy and enthalpy, fuel kinetic energy and enthalpy, and the heat releasable by stoichiometric combustion of fuel added to ideally saturated products. However, it is not possible to determine what the change in static pressure and entropy are without reference to the specific type of combustion process.

If combustors are to be compared on the basis of the practically significant value of specific impulse realizable, a reasoning of the causes leading to changes in specific impulse would require a knowledge of the type of combustion process and the parameters entering into the definition of specific impulse. These parameters may be chosen variously in different cases, and here the three efficiency parameters η_{KEC}, $\eta_{\varepsilon 1}$, and η_{PT} are chosen as adequate.

References

[B1]Williams, F., *Combustion Theory*, Academic Press, New York, 1985.

[B2]Oran, E. S., Boris, J. P., Young, T., Flanagan, M., Burks, T., and Picone, "Numerical Simulations of Detonations in Hydrogen-Air and Methane-Air Mixtures," *Eighteenth Symposium (International) on Combustion,*Combustion Inst., 1981.

[B3]Ostrander, M. J., Hyde, J. C., Young, M. F., Kissinger, R. D., and Pratt, D. T., "Standing Oblique Detonation Wave Engine Performance," AIAA Paper No. 87-2002, June–July 1987.

Appendix C: Hydrogen Combustion Scheme

The reaction scheme utilized is given in Table C1. The table also includes reaction rate coefficients *A, B,* and *E* in the expression for reaction rate, k_f:

$$k_f = A\,T^B \exp\,(-\,E/T) \tag{C1}$$

Silane Reaction Mechanism: In the cases where silane is added, the reaction scheme adopted and the rate coefficients are given in Table C2. Here, the rate coefficient is defined by

$$k = A\,\exp(-\,E/RT) \tag{C2}$$

where A is in s^{-1} for unimolecular reactions, cm^3 / mole-s for bimolecular reactions, and cm^3 / $mole^2$-s for termolecular reactions. The activation energy E is in cal/mole.

I. Thermodynamic Properties

The thermodynamic functions are calculated by polynomial equations for C_p, h, S, and G as a function of temperature. Seven coefficients $A_1, A_2, \ldots,$ A_7 are used for each species in the following equations, where T is the temperature in Kelvin. Two sets of coefficients are given for each species. The first set was obtained for the temperature range 1000 to 5000 K, and the second set for 300 to 1000 K. These coefficients calculated by Gordon and McBride[1] are stored in the thermodynamic database.

$$\frac{C_p}{R} = A_1 + A_2T + A_3T^2 + A_4T^3 + A_5T^4 \tag{C3a}$$

$$\frac{h}{RT} = A_1 + \left(\frac{A_2}{2}\right)T + \left(\frac{A_3}{3}\right)T^2 + \left(\frac{A_4}{4}\right)T^3 + \left(\frac{A_5}{5}\right)T^4 + \frac{A_6}{T} \tag{C3b}$$

$$\frac{S}{R} = A_1 \ln T + A_2T + \left(\frac{A_3}{2}\right)T^2 + \left(\frac{A_4}{3}\right)T^3 + \left(\frac{A_5}{4}\right)T^4 + A_7 \tag{C3c}$$

$$\frac{G}{RT} = A_1(1 - \ln T) - \left(\frac{A_2}{2}\right)T - \left(\frac{A_3}{6}\right)T^2 - \left(\frac{A_4}{12}\right)T^3 - \left(\frac{A_5}{20}\right)T^4 + \frac{A_6}{T} - A_7 \tag{C3d}$$

II. Equilibrium and Nonequilibrium Combustion

The initiation of combustion is determined by the entry air temperature. If the air temperature is lower than the ignition temperature of the fuel, it is assumed that external energy is supplied to raise the temperature of air to the ignition temperature. The fuel is assumed to accumulate for the specified distance and the whole set of reactions are solved for equilibrium chemistry at the ignition point, which is a reaction-limited event; the properties along the rest of the combustor are calculated using non-equilibrium chemistry. If the air temperature is higher than the ignition temperature of fuel, it is assumed that ignition occurs readily and instantaneously, and non-equilibrium chemistry is used, since combustion then is mixing-limited. The chemical reactions occurring in the system are symbolized by

$$\sum_m a_{mr}\, x_m \rightleftarrows \sum_m b_{mr}\, x_m \tag{C4}$$

Table C1 Hydrogen reaction scheme

Head	A	B	E
H2 + O2 → HO2 + H	7.00e + 13	0.0	56800.
H + O2 → OH + O	2.20e + 14	0.0	16800.
O + H2 → OH + H	5.06e + 04	2.67	6290.
OH + H2 → H2O + H	2.16e + 08	1.51	3430.
OH + OH → H2O + O	2.50e + 09	1.14	0.
H + OH → H2O + M	8.62e + 21	−2.0	0.
H + H → H2 + M	7.30e + 17	−2.0	0.
H + O → OH + M	2.60e + 16	−0.6	0.
O + O → O2 + M	1.10e + 17	−1.0	0.
H + O2 → HO2 + M	2.30e + 18	−1.0	0.
HO2 + H → OH + OH	1.50e + 14	0.0	1000.
HO2 + O → O2 + OH	2.00e + 13	0.0	0.
HO2 + OH → H2O + O2	2.00e + 13	0.0	0.
HO2 + HO2 → H202 + O2	2.00e + 12	0.0	0.
H + H2O2 → H2 + HO2	1.70e + 12	0.0	3780.
H + H2O2 → OH + H2O	2.00e + 13	0.0	3580.
O + H2O2 → OH + HO2	2.80e + 13	0.0	6400.
OH + H2O2 → H2O + HO2	7.00e + 12	0.0	1430.
OH + OH → H2O2 + M	1.60e + 22	−2.0	0.
N + N → N2 + M	2.80e + 17	−0.8	0.
N + O2 → O + NO	6.40e + 09	1.0	6300.
N + NO → N2 + O	1.60e + 13	0.0	0.
N + OH → NO + H	6.30e + 11	0.5	0.
H + NO → HNO + M	5.40e + 15	0.0	−600.
H + HNO → H2 + NO	4.80e + 12	0.0	0.
O + HNO → OH + NO	5.00e + 11	0.5	0.
OH + HNO → H2O + NO	3.60e + 12	0.0	0.
HO2 + HNO → H2O2 + NO	2.00e + 12	0.0	0.
HO2 + NO → NO2 + OH	3.40e + 12	0.0	−260.
HO2 + NO → HNO + O2	2.00e + 11	0.0	1000.
H + NO2 → NO + OH	3.50e + 14	0.0	1500.
O + NO2 → NO + O2	1.00e + 13	0.0	600.
M + NO2 → NO + O	1.16e + 16	0.0	66000.

Table C2 Silane reaction mechanism

Reaction	Rate coefficient
$SiH_4 \rightarrow SiH_2 + H_2$	$6.0 \times 10^{13} \exp(-54960/RT)$
$SiH_4 + O_2 \rightarrow SiH_3 + HO_2$	$2.0 \times 10^{11} \exp(-44000/RT)$
$H + SiH_4 \rightarrow H_2 + SiH_3$	$1.5 \times 10^{13} \exp(-2500/RT)$
$O + SiH_4 \rightarrow OH + SiH_3$	$4.2 \times 10^{12} \exp(-1600/RT)$
$OH + SiH_4 \rightarrow H_2O + SiH_3$	$8.4 \times 10^{12} \exp(-100/RT)$
$H + SiH_3 \rightarrow SiH_2 + H_2$	$1.5 \times 10^{13} \exp(-2500/RT)$
$O + SiH_3 \rightarrow SiH_2O + H$	$1.3 \times 10^{14} \exp(-2000/RT)$
$OH + SiH_3 \rightarrow SiH_2O + H_2$	5.0×10^{12}
$SiH_3 + O_2 \rightarrow SiH_2O + OH$	$8.6 \times 10^{14} \exp(-11400/RT)$
$SiH_2 + O_2 \rightarrow HSiO + OH$	$1.0 \times 10^{14} \exp(-3700/RT)$
$H + SiH_2O \rightarrow H_2 + HSiO$	$3.3 \times 10^{14} \exp(-10500/RT)$
$O + SiH_2O \rightarrow OH + HSiO$	$1.8 \times 10^{13} \exp(-3080/RT)$
$OH + SiH_2O \rightarrow H_2O + HSiO$	$7.5 \times 10^{12} \exp(-170/RT)$
$H + HSiO \rightarrow H_2 + SiO$	2.0×10^{14}
$O + HSiO \rightarrow OH + SiO$	1.0×10^{14}
$OH + HSiO \rightarrow H_2O + SiO$	1.0×10^{14}
$HSiO + \neq M \rightarrow H + SiO + M$[a]	$5.0 \times 10^{14} \exp(-29000/RT)$
$HSiO + O_2 \rightarrow SiO + HO_2$	3.0×10^{12}
$SiH_2O + HO_2 \rightarrow HSiO + H_2O_2$	$1.0 \times 10^{12} \exp(-8000/RT)$
$SiO + O + M \rightarrow SiO_2 + M$	$2.5 \times 10^{15} \exp(-4370/RT)$
$SiO + OH \rightarrow SiO_2 + H$	$4.0 \times 10^{12} \exp(-5700/RT)$
$SiO + O_2 \rightarrow SiO_2 + O$	$1.0 \times 10^{13} \exp(-6500/RT)$

[a]+ M represents a third body.

where x_m represents one mole of species m and a_{mr} and b_{mr} are integral stoichiometric coefficients for reaction r.

The non-equilibrium reactions proceed at a rate $\dot{w}_r$ given by

$$\dot{w}_r = k_f \, \Pi \left(\frac{\rho_m}{W_m} \right) a'_{mr} - k_b \, \Pi \left(\frac{\rho_m}{W_m} \right) b'_{mr} \qquad \text{(C5)}$$

where W_m is the molecular weight of species m. Here the reaction orders a'_{mr} and b'_{mr} need not equal a_{mr} and b_{mr}, so that empirical reaction orders can be used. The coefficients k_f and k_b are assumed to be of a generalized Arrhenius form

$$k_f = A\, T^B \cdot \exp(-E/RT) \qquad \text{(C6)}$$

However, the rates of equilibrium reactions are implicitly determined by the constraint conditions

$$\prod_m \left(\frac{\rho_m}{W_m} \right)^{b_{mr} - a_{mr}} = k_c \qquad \text{(C7)}$$

The equilibrium constant is expressed as

$$k_c = e^{-\Delta G^o/\mathrm{RT}} (RT)^{b_{\mathrm{mr}} - a_{\mathrm{mr}}} \tag{C8}$$

where ΔG^o is the standard molal free energy change for the reaction.

Appendix D: Three-Dimensional Nozzles—Design and Integration

Complex three-dimensional geometries for nozzles are a means for obtaining optimal integration with the engine and the external flowpath, and realizing special effects, such as reduced plume noise and signature. In the case of scramjet engines, the nozzle follows the combustor, and may also be utilizing an integrated thermal management scheme with the combustor. The nozzle entry section, if independent of the combustor exit, would require a transition or connecting section. As a scram combustor exit flow is usually supersonic, the nozzle is a supersonic flow device throughout its length.

The shape and integration of the nozzle give rise not only to the generation of thrust force but, in general, also to lift and side force. At given flight conditions along a trajectory, the control and management of those forces and the associated moments and torque for a dynamically stable flight is a major concern in the design of the vehicle.

The nozzle entry conditions may in general involve chemically reactive flows, and then the calculation of the nozzle flowfield and its interaction with the external flow may require accounting for chemical reaction effects. However, the speed of flow may permit an assumption of chemically frozen conditions, which when they arise in practice reduce the realizable thrust due to gas expansion.

The general problems of nozzle design and installation may be divided into 1) establishing the internal flowpath and 2) determining the interaction with the external flowpath. Details on calculation procedures can be found in Refs. 21 and 22. Here a brief summary follows.

I. Internal Flowpath

The internal flowpath may be considered in terms of development of the shape and geometry in longitudinal planes along the general direction of flow expansion, and in a number of transverse planes. One thus is able to accommodate a given pressure gradient along the nozzle flow and also a shape in the transverse plane. Referring to Fig. D1, this would be equivalent to the selection of 1) the shapes of a number of beams in a number of longitudinal xy planes between the entry and the exit planes where the nozzle shapes and the area ratio are given, and 2) the shapes of the connecting curves between the beams in the transverse plane zy and their development along the x direction. Considerable choice exists in the selection of these planar curves in the xy and the yz planes. For example, the beams may be chosen as of parabolic shape, and the connecting curves may be chosen as super-ellipses or other planar curves. If there are constraints on the choice

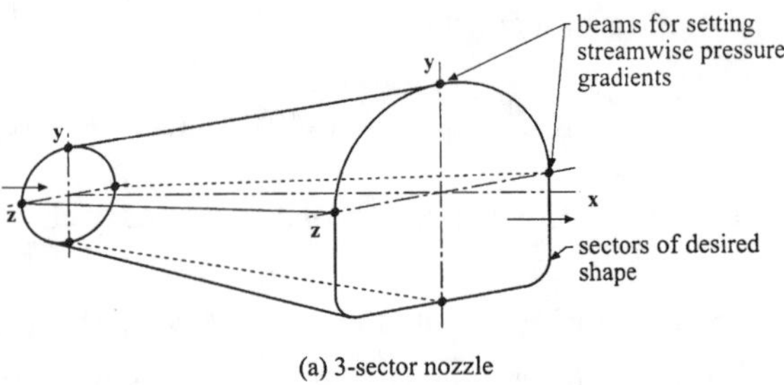

(a) 3-sector nozzle

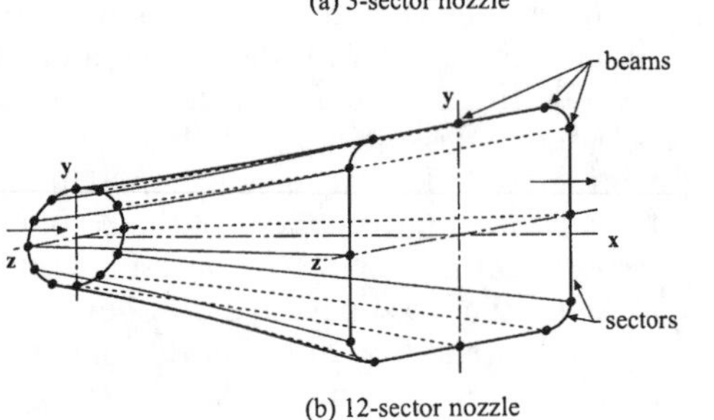

(b) 12-sector nozzle

Fig. D1 Three-Dimensional Nozzle Contours with Multiple-Sector Wall.

of these curves, such as specific shapes and turns, they can be introduced in the analytical curves employed.

In the foregoing, it is assumed that the length of the nozzle is given. If there is some choice, then one can establish the nozzle geometry for a length that is related specifically to the minimum length nozzle, for example, for maximum thrust or minimum mass of the nozzle. In this case, rather than following the traditional method of calculation of the minimum length or surface area through the application of calculus of variations, one can adopt a direct optimization procedure as described in Ref. 24 in main text.

Throughout the foregoing analysis, considering the primary interest in configuration design, one can adopt an inviscid flow approximation, and utilize the three-dimensional method of characteristics for determining the nozzle flowfield. It will be observed that the procedure lays out the nozzle as well as determines the performance, for given entry and exit cross-sections, length, and other constraints on the flowpath.

II. Integration with the Vehicle External Flow

The internal flowpath of the engine and the external flowpath of the vehicle are part of the same flow considering the ambient atmosphere that is utilized and becomes affected by the flight vehicle from a location ahead of the vehicle to a location behind the vehicle wake. There is also some interaction between the two flowpaths specifically at the forebody, the leading edge of the inlet, and the engine cowl.

The primary interest in the aft-body of the vehicle, where the nozzle is integrated, is in the interaction between the nozzle discharge plume and the external flow, and the resulting three-dimensional forces and moments (relative to the center of gravity of the vehicle, for instance), and the torque, if any, on the body. Such interaction therefore needs to be established under various flight and operational conditions.

During configuration design, the nozzle flowpath and plume as well as the external flow may be considered in the inviscid flow approximation, while the external flow is represented in a desired nonuniform fashion. The three-dimensional method of characteristics can be utilized for establishing the interactive flowfield with shock fitting by a standard procedure. However, viscous effects are very important in the flow interaction, and crucial guidance is urgently required from detailed observations and measurements before one can be confident about predicted results. There is also considerable scope here for virtual design and testing techniques.

Chapter 11

Strutjet Rocket-Based Combined-Cycle Engine

A. Siebenhaar* and M. J. Bulman†
GenCorp Aerojet, Sacramento, California
and
D. K. Bonnar‡
Boeing Company, Huntington Beach, California

I. Introduction

THE multi-stage chemical rocket has become established over many years as the propulsion system for space transportation vehicles while there is increasing concern about its continued affordability and rather involved reusability. Two broad approaches to addressing this overall launch cost problem consist in one, the further development of the rocket motor, and two, the use of air-breathing propulsion to the maximum extent possible as a complement to the limited use of a conventional rocket. In both cases a single-stage-to-orbit (SSTO) vehicle is considered a desirable goal. However, neither the all-rocket nor the all-air-breathing approach seems realizable and workable in practice without appreciable advances in materials and manufacturing. An affordable system must be reusable with minimal refurbishing on-ground and large mean time between overhauls, and thus with high margins in design. The suggestion has been made that one may use different engine cycles, some rocket and others air-breathing, in a combination over a flight trajectory, but this approach does not lead to a converged solution with thrust-to-mass, specific impulse, and other performance and operational characteristics that can be obtained in the different engines. The reason is this type of engine is simply a combination of different engines with no commonality of gas flow path or components, and therefore tends to have the deficiencies of each of the combined engines. A further development in this approach is a truly combined cycle that incorporates a series of

* Development Team Leader
† Development Engineer
‡ Development Engineer

cycles for different modes of propulsion along a flight path with multiple use of a set of components and an essentially single gas flow path through the engine. This integrated approach is based on realizing the benefits of both a rocket engine and an air-breathing engine in various combinations by a systematic functional integration of components in an engine class usually referred to as a rocket-based combined-cycle (RBCC) engine.

RBCC engines exhibit a high potential for lowering the operating cost of launching payloads into orbit. Two sources of cost reductions can be identified. First, RBCC-powered vehicles require only 20% takeoff thrust compared to conventional rockets, thereby lowering the thrust requirements and the replacement cost of the engines. Second, because of the higher structural and thermal margins achievable with RBCC engines coupled with a higher degree of subsystem redundancy, lower maintenance and operating costs are obtainable. Both of these reductions result from the increased specific impulse of RBCC engines and the reduction in takeoff and dry mass.

The strutjet engine is described in detail in Sec. II, along with the delineation of five modes of engine operation, optimization of the propulsion system, and the resulting engine architecture. Section III is devoted to integration of the engine with the flight vehicle and also addresses the performance of the engine as a propulsion system and in relation to various types of flight vehicles. In addition, the concept of robustness space is utilized to assess the design of the engine and its ability to reduce operations cost. Section IV deals with the testing accomplished to date and future plans for on-ground and flight vehicle testing. Section V addresses the maturity of various elements of the overall RBCC technology. Finally, the applicability and superior potential of the strutjet engine, as a leading rocket-based combined-cycle engine for potentially low operating cost launch vehicles, is summarized in Section VI.

II. Strutjet Engine

The aerojet strutjet engine is a member of the RBCC class of engines with several new technologies and innovations. Many of these technologies are also of far-reaching interest in other propulsion schemes. The Strutjet engine operates all of the cycles with liquid hydrogen (LH2) for fuel and liquid oxygen (LOX) and atmospheric air as necessary for combustion of fuel.

The name *strutjet* is derived from the use of a series of struts in the front part of the engine flow path. The struts serve a number of functions in the engine, including compression of incoming air, isolation of combustion from air inlet (that is, as an isolator), fuel distribution and injection, ram/scram combustion, and rocket-thruster integration in the different modes of engine operation. The struts finally provide efficient structural support for the engine. The struts thus form a key element in the engine.

The strutjet engine is in principle a single engine configuration with three propulsion elements, namely rocket, ramjet, and scramjet in its five mode operation. The elements are highly integrated in design and function. Engines of related conceptual designs have been considered for similar applications by Escher and Flornes[1] and also by Billig and Van Wie[2] in a U.S. patent for an RBCC-type engine. The Billig–Van Wie engine has a separate inlet for compression and then uses struts downstream for the purpose of distributing various fuel injectors across the flow path. The difference between such earlier concepts and the aerojet strutjet is found in the higher degree of functional integration of various strutjet engine components, which results in a shorter, higher thrust-to-weight engine with good performance. The specific impulse of the strutjet engine is characteristic of other known air-breathing engines without the thrust-to-mass penalty of having separate propulsion systems for different flight conditions.

The strutjet engine concept is founded on obtaining specific impulse I_{sp} higher than that of a rocket, a better thrust-to-mass ratio (F/m_e) than an all air-breathing engine, and a substantial reduction in vehicle gross takeoff weight (GTOW). Thus the mean I_{sp} for the strutjet is estimated to be 586 s, while the conventional hydrogen-oxygen rocket provides an I_{sp} of 425 s. In comparison, the mean I_{sp} for an all air-breather such as the U.S. National AeroSpace Plane (NASP) was estimated to be 755 s. A SSTO-type all-rocket may yield F/m_e equal to 80 lbf/lbm and a NASP-type SSTO all air-breather, about 6 lbf/lbm. Using current state-of-the-art technology, a strutjet can be built with F/m_e of 35 lbf/lbm; however, as shown later, for the sake of increased engine robustness F/m_e values as low as 25 lbf/lbm are employed in the analysis. The GTOW, made up of the empty weight, propellant, and payload, becomes reduced for a strutjet because of the reduction in the amount of oxygen that needs to be carried on board at vehicle launch. We expect that the propellant mass fraction can be maintained at the present state-of-the-art levels, about 85%, compared to the required 90% value for an all-rocket SSTO system. At the same time the engine and the launch vehicle are expected to have substantially higher structural reliability margins than the all-rocket and the all-air-breathing engine system. Finally, the thermal management of these vehicles also can be accomplished within the current state-of-the-art because high-speed atmospheric air-breathing operation is limited to a relatively low Mach number such as 10. This avoids the severe vehicle heating conditions that are known to have imposed serious engineering challenges, typical for hypersonic atmospheric flight in the regime above Mach 10 in NASP-type vehicles. The strutjet vehicle trajectory is chosen such that the ascent thermal loads are equal to or less than the thermal loads that a reusable vehicle would experience during re-entry.

A. Flow-Path Description

Figure 1 provides a schematic of the strutjet and identifies the different elements in the combined-cycle engine. The propulsion subsystems are integrated into a single engine using common propellant feed lines, cooling systems, and controls. The air inlet along with the struts, the combustion sections, and the nozzle make up the main engine flow path. An isometric view of a typical RBCC strutjet engine is given in Fig. 2.

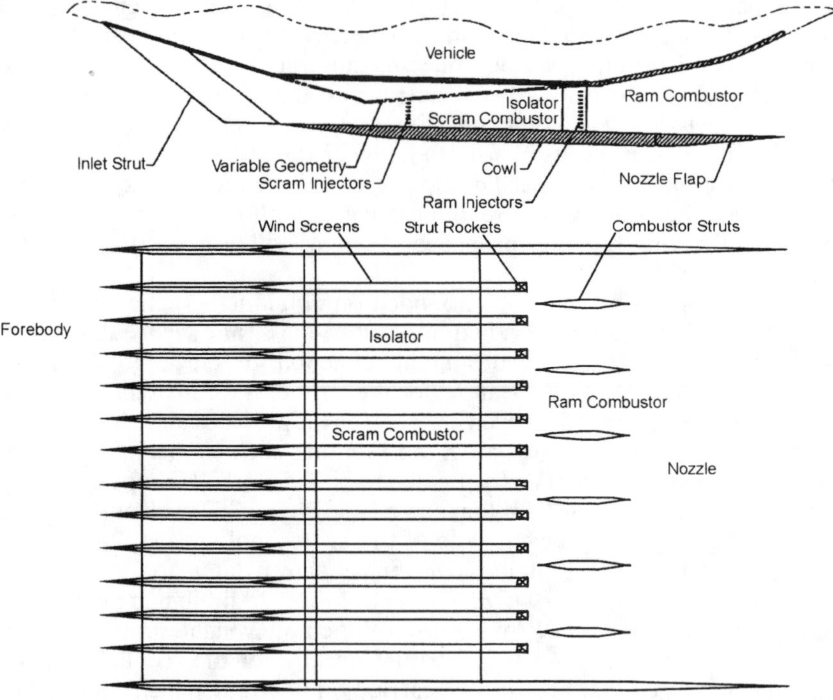

Fig. 1 Strutjet engine has clean unobstructed flow path and simple two-dimensional for inlet and nozzle variable geometry.

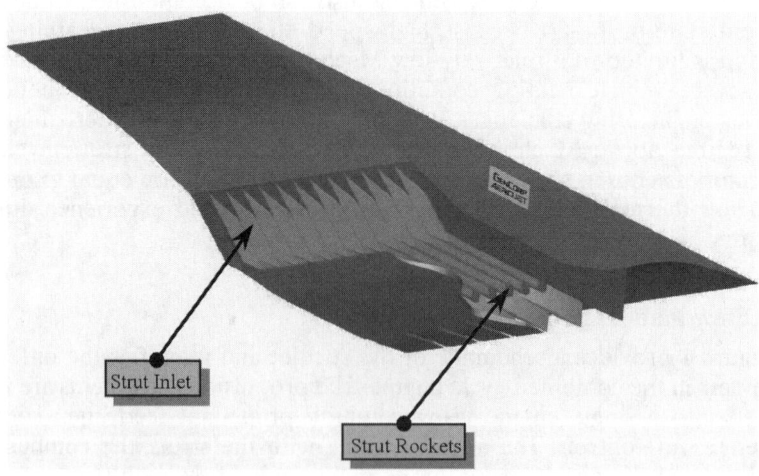

Fig. 2 Strutjet engine concept.

B. Engine Architecture

Flow-Path Elements

The variable geometry inlet incorporates two engine ramps that maximize air capture and control compression as required by the engine. The inlet combines effective forebody precompression with strut compression. This results in soft start, low spill drag, and good capture and recovery efficiencies. The soft start is a result of the increased openness of the inlet on the cowl side, which causes a gradual decrease in spillage with increasing Mach number. This, in turn, provides smooth increases in captured air mas flow and pressure recovery.

The inlet geometry and the changes in contraction ratio as a function of flight Mach number are shown in Fig. 3. The struts or flow dividers, which extend into the inlet, channel the flow into discrete, narrow flow paths that diffuse the flow over the shortest length possible. In the forward part of the inlet where the struts are designated as windscreens, the air capture is enhanced by locating the cowl lip near the minimum flow-area section (variable geometry wide open); this also facilitates inlet start at low supersonic speeds. The struts are integrated in the inlet such that each flow passage between two adjoining struts behaves like a sidewall compression inlet and provides what may be called strut compression. Past the cowl lip, at low speeds the flow area between two adjacent struts remains constant; the strut section serves as an inlet combustor isolator during ducted rocket and ramjet mode. At higher speeds ($M\approx 5$) the inlet ramps are deployed to increase the engine contraction and performance.

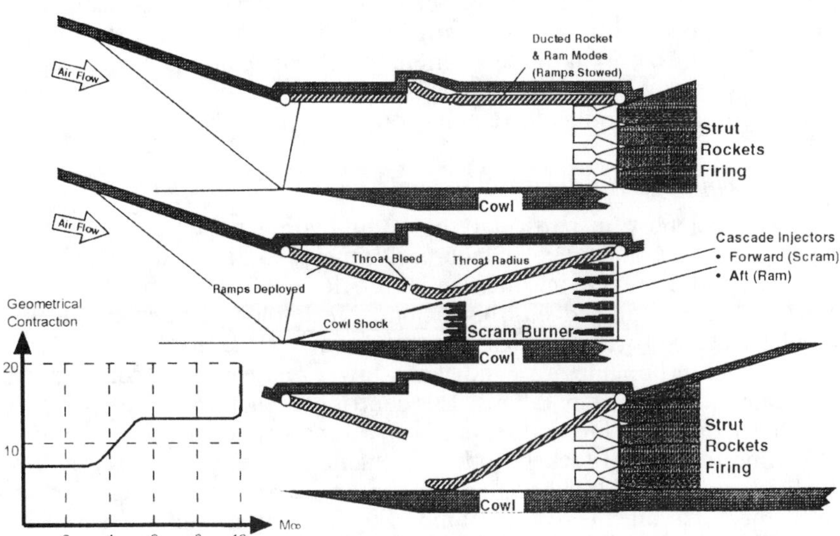

Fig. 3 Simple two-dimensional variable geometry inlet allows for geometrical contraction variation.

Above Mach 6 the scram mode is the most efficient, and the diverging isolator duct between the fully contracted inlet throat and the rockets is used as the scram combustor. During this mode of operation, the variable nozzle geometry adjusts the scram combustor flow path into a continuously diverging configuration.

Maximum engine performance is achieved when the fuel is injected as far forward as possible without degrading the inlet air capture capability. In general, this requires the fuel injection point to start well back in the flow path and move continuously forward as the vehicle accelerates. Ideally, this would imply an infinite number of injectors that would be turned on or off as the Mach number changes. The strutjet flow path employs three stages of fuel injection to cover this range. These injectors are referred to as the base axial, aft, and forward injectors. The base axial injectors delay the heat release at low speeds until the flow path area is large enough to tolerate it. The rockets used in the boost phase of the mission and the ascent phase are housed in the trailing end of the main struts. These strut rockets also integrate two of the air-breathing injector stages. At intermediate speeds the aft (ram) injectors are employed in combination with one of the other injectors to control the effective heat-release location as required in order to maximize the engine performance. At high scram speeds the fuel is injected from the forward (scram) injectors to complete the combustion before the fuel can leave the engine.

The ram combustor used at speeds up to Mach 6 is located aft of the strut rockets. This combustor provides enough area to permit stoichiometric subsonic combustion at low speeds.

The strutjet variable geometry nozzle is a simple flap used to control subsonic combustion pressure to create the optimum thrust as dictated by operating mode and flight Mach number. This nozzle flap is the principal control in the transition to the scram mode. By opening the nozzle flap up at approximately Mach 6, the combustor pressure drops, and the flow remains supersonic through the combustor.

Turbomachinery, Propellant Supply, and Thermal Management

Turbomachinery, in particular the liquid-hydrogen fuel pump, may be considered the Achilles heel of high-performance rocket engines. Unlike the conventional rocket engine, the strutjet provides high performance without relying on the cutting-edge turbomachinery technology, particularly the use of advanced, high-strength, high-temperature materials. The demands on the strutjet fuel turbopump are significantly lower relative to the conventional rocket as summarized in Table 1. This contributes significantly to increased reliability, extended life, and reduced cost of the strutjet engine operation.

During the ducted-rocket mode, the engine fuel supply is powered by a fuel-rich gas generator exhausting into the ram-scram duct through the base axial injectors while the oxygen supply is powered by an oxidizer-rich staged combustion cycle. In contrast to the hydrogen side, stage combustion on the oxygen side does not impose a technical challenge because oxygen, being a high-density fluid, can be pumped to the required pressure levels at rela-

Table 1 Comparison of rocket and strutjet turbopumps

Parameter	Rocket	Strutjet	Reduction, %
Chamber pressure	3000 psia	2000 psia	33
Fuel pump			
Discharge pressure	7000 psi	3000 psi	57
Turbine inlet	1700 °R	1000 °R	41
Temperature	2000 rpm	1700 rpm	15
Turbine tip speed	100%	72%	28
Rotational stress			

tively benign shaft speed. Oxygen-rich preburner technology provides advantages of reduced turbine temperature and elimination of interpropellant seals needed for fuel-rich gas-driven oxygen pumps. Figure 4 illustrates the engine cycle during the ducted-rocket mode of operation. During pure air-breathing modes, the entire oxygen feed system is, of course, inactive. As shown in Fig. 5, the fuel side operates now in a simple expander cycle mode using heat available from the required cooling of the engine structure. In the scram-rocket mode the fuel remains operating in the expander cycle, and the oxidizer circuit is restarted in the staged combustion cycle, as shown in Fig. 6. Figure 7 illustrates the location of these various operating modes on the fuel pump operation map.

Engine Cycle

The overall strutjet engine propellant flow is illustrated in the previously shown Fig. 4. As displayed, there are three subsystems: 1) the hydrogen and oxygen fuel tanks, turbopumps feed system and powerhead, 2) the strutrocket and fuel-injection assembly, and 3) the engine structure and cooling system. Both the strut rocket and the engine structure are operating in the thermal and combustion gas dynamic environment.

Fuel-rich gases generated in the fuel gas generator (FGG) drive the hydrogen turbine and are subsequently injected into the engine internal airstream at selected locations through base-axial, aft, and forward injectors. The selection depends on the engine operating mode and is accomplished through appropriate valving. Hydrogen gas is also used to cool the rocket chambers (the figure including representationally only one) and the engine structure before injection into the combustor section. In the expander cycle hydrogen heated by the engine structure bypasses the preburner and drives the turbine in an expander cycle.

The oxygen side of the propellant system is only active during rocket operation and operates (always) in a stage-combustion cycle. The oxygen-rich turbine drive gases are generated in the preburner (OPB) through the burning of a small amount of hydrogen and all of the oxygen-flow, prior to injection into the rocket chambers.

The table shown in Figure 4 includes four attributes for each of the five

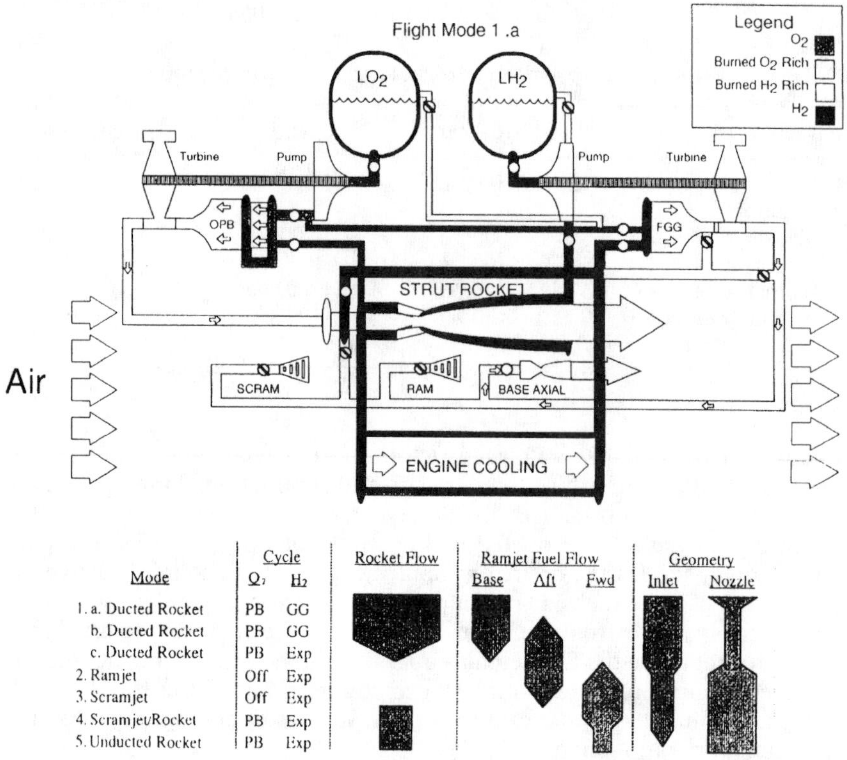

Mode	Cycle O_2	Cycle H_2
1. a. Ducted Rocket	PB	GG
b. Ducted Rocket	PB	GG
c. Ducted Rocket	PB	Exp
2. Ramjet	Off	Exp
3. Scramjet	Off	Exp
4. Scramjet/Rocket	PB	Exp
5. Unducted Rocket	PB	Exp

Fig. 4 Strutjet engine cycles and operation.

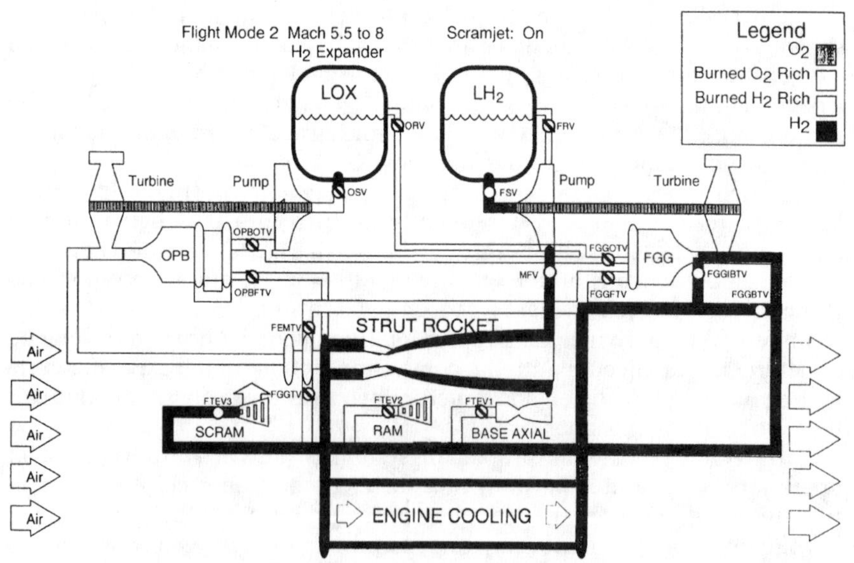

Fig. 5 Hydrogen expander cycle during ram and scramjet modes.

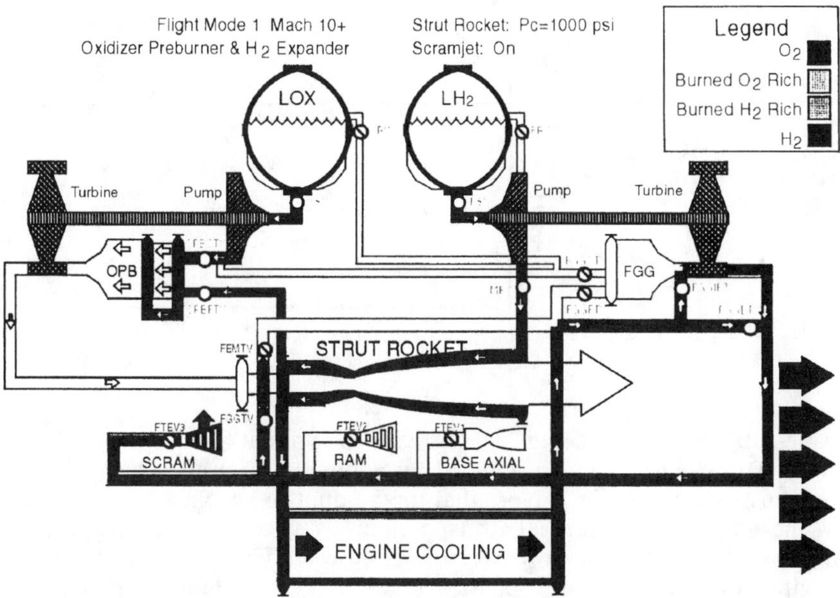

Fig. 6 Oxydizer preburner and hydrogen expander cycles during scramjet/rocket and ascent-rocket modes.

operating modes: 1) selected power cycle in the oxygen and hydrogen supplied, 2) amount of rocket propellant flow, 3) amount of hydrogen injected into the airstream, and 4) settings of the inlet and nozzle variable geometry.

Structural Concept

The struts provide a very efficient means for a number of processes. First, the struts are an important structural element of the engine. They reduce the unsupported spans and reduce weight and enhance fuel injection, mixing, and combustion at supersonic speeds. They also incorporate the strut rockets. These compact rockets serve to induce airflow and to assist in air-fuel mixing in the ducted-rocket mode. When operative, they provide the bulk of the thrust.

The current design employs additional aft struts in the aft end of the combustion zone, which gives rise to a substantial engine panel mass reduction. These engine panels are essentially two-layer laminates comprised of a structural and a thermal protection layer. Minimization of the total panel weight over the entire engine is accomplished by optimizing the design of each of these layers. In the case of the structural layer, substantial mass savings can also be realized by incorporating, where possible, supporting structures to reduce the unsupported panel span. In the inlet isolator the main engine struts inherently serve as supporting structures. However, the strut rockets define the end of the main struts, leaving a large, unsupported

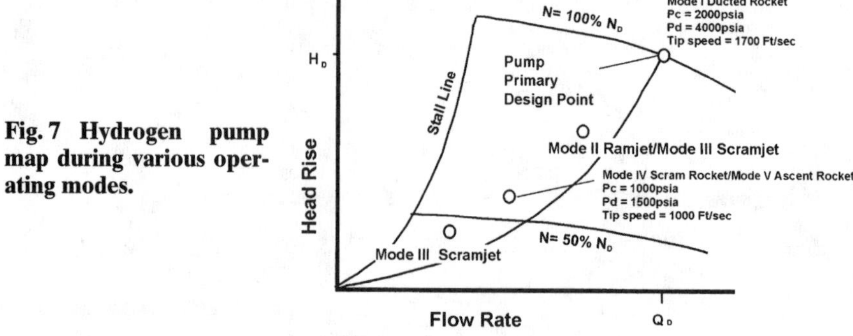

Fig. 7 Hydrogen pump map during various operating modes.

panel. Structure weight-trade studies have shown that the panel mass in the combustion region can be reduced by adding several aft struts into this region, as shown in Fig. 8. Note that the amount of weight savings, relative to the number of additional struts, diminishes as the number of aft struts increases because of their own weight. As shown in Fig. 1, the present strutjet engine design incorporates six aft struts, with a ram-combustor panel mass saving of nearly 80% compared to a configuration with no aft struts. Based on symmetric design considerations, a set of 11 aft struts would be the next possible configuration, but this would substantially increase internal engine drag and heat load on the cooling system without a corresponding payoff in

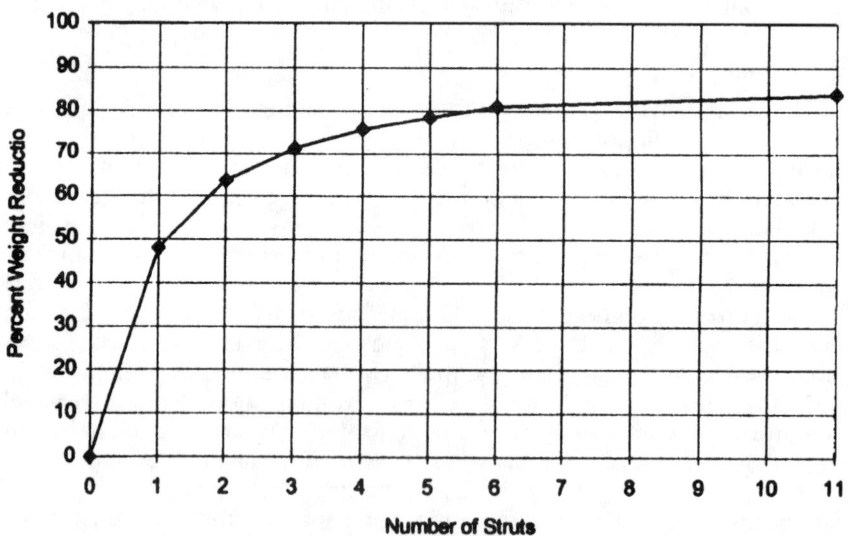

Fig. 8 Combustor weight reduces with increased number of aft struts.

structural weight savings. The achievable engine thrust-to-mass using advanced technology is as high as 35 lbf/lbm.

C. Strutjet Operating Modes

Along a typical SSTO ascent trajectory, a strutjet operates in five modes with smooth transition between modes: 1) ducted-rocket operation for takeoff and acceleration through the transonic speed regime into the supersonic region, 2) ramjet operation from Mach 2.5 to about 6, 3) scramjet operation from Mach 6 into the hypersonic speed range up to Mach 10, 4) scram /rocket operation from Mach 10 to low vacuum conditions, and 5) ascent rocket from low vacuum operation up to orbital speeds.

Ducted-Rocket Mode

Before discussing the ejector ramjet concept employed in the strutjet, revisiting two known, classical ejector ramjet concepts will be helpful.[3] The most efficient of the two is the socalled diffuse and afterburn (DAB) concept, which employs a sequential series of the ejector processes. This is more thermodynamically efficient but has substantial adverse consequences. To prevent premature combustion during the ejector pumping (driven by shear mixing between primary and secondary flows) and diffusion, stoichiometric or oxygen-rich rocket mixture ratios are needed. This reduces the rocket I_{sp} and is a challenge for the cooling system. Because no available fuel is entrained in the rocket exhaust, separate fuel injectors are required in the large area of the ram burner. This approach adds significant length and weight to the engine.

The second type of ejector ramjet is called the simultaneous mix and combust (SMC). In this type ejector ramjet the processes are allowed to occur in parallel in a shorter duct. The fuel for afterburning comes from the conventionally fuel-rich rocket exhaust. After burning begins immediately, a diverging duct is needed to keep from drastically reducing the air induction caused by the thermal occlusion. Although shorter and lighter, this approach loses up to 30% of the thrust of the DAB ejector scramjet at takeoff.

The strutjet combines the best of each of these historic ejector ramjets. Here, the duct is as short as the SMC but with the performance close to the DAB type. To realize a short duct length, the flow path is divided up into many narrow gaps by the struts. The ejector pumping occurs between the struts and is complete much faster than without struts. The diffusion occurs in the other plane by the divergence of the body wall. The rockets are operated at a high but still fuel-rich mixture ratio. The body-wall divergence provides the needed tolerance to the incidental combustion of this fuel. The bulk of the fuel (provided in the fuel-rich turbine exhaust) for afterburning is injected from the base of the struts. To maintain high performance, the amount of air the strut rockets can induct into the ram combustor must be maximized. The artifice employed is to reduce the rate of early heat release. Figure 9 shows the performance loss because of reduced air injection as a

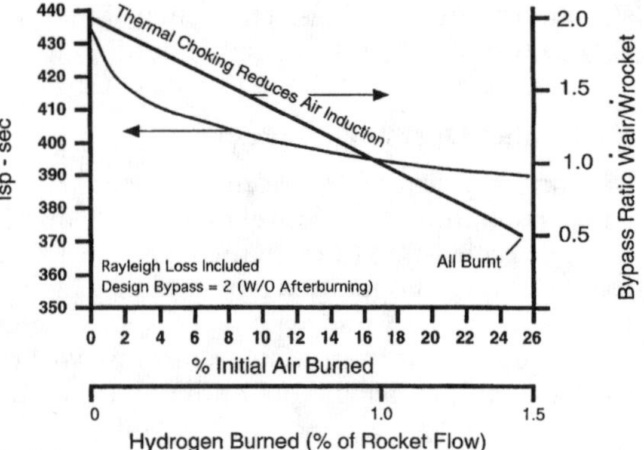

Fig. 9 Thermal choking causes air reduction reducing specific impulse.

function of the amount of air burned at the front of the ram combustion zone. To prevent the performance loss from early heat release in the ram combustion, two strategies are employed. One, the amount of free hydrogen in the rocket exhaust is minimized thereby minimizing the amount of fuel that can come in contact with air in the front part of the ram combustion zone; the strut rockets are designed to operate near stoichiometric conditions with only a small amount of fuel film cooling. The second strategy consists in delaying the afterburning of the turbine exhaust. The fuel-rich turbine exhaust is injected into the ram combustion chamber from the base of the struts between the rocket nozzles as illustrated in Fig. 10. This fuel-rich turbine exhaust is buried inside the rocket plumes. After the diffusion is complete, when the heat release can be tolerated the turbine exhaust emerges from the rocket plumes and burns with the air in the ramburner. Complex auxiliary injectors and flame holders are thus avoided. This concept is illustrated in Fig. 11. In this mode the engine inlet is wide open, and the nozzle is in a partially closed position that is chosen to maximize the engine thrust.

The ducted-rocket mode provides some added specific impulse over an all-rocket engine up to Mach 2.5, where the ramjet mode takes over.

Mode Ia. During the ducted-rocket mode, the variable geometry inlet is wide open. Figure 10 provides an upstream view into the combustion zone. As can be seen, three base axial injectors are located at the strut base between four rocket thrusters. Under takeoff conditions when maximum

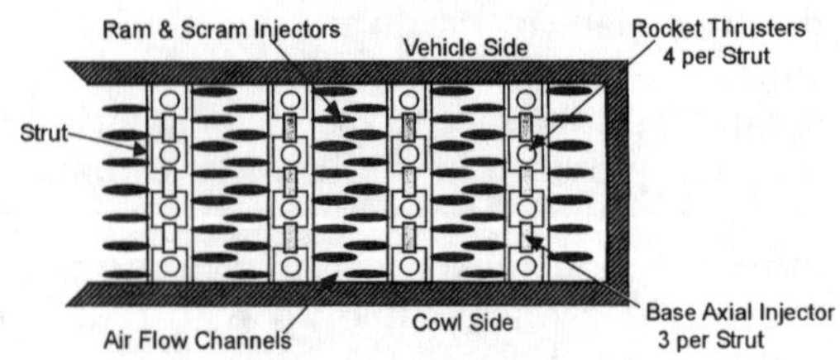

Fig. 10 Struts provide mounting platform for three-dimensional fuel distribution.

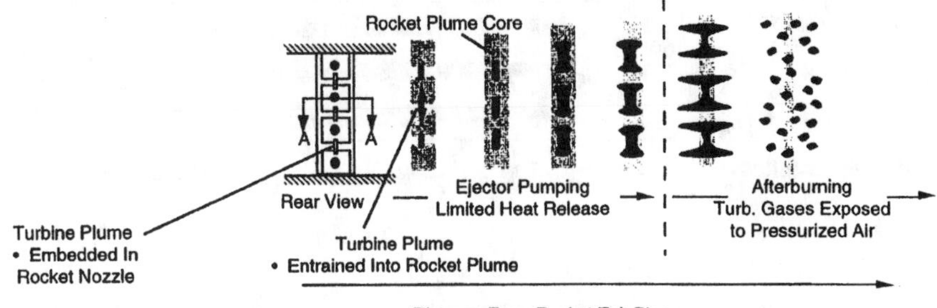

Fig. 11 Embedding fuel in rocket plume delays afterburning.

thrust is required, the strut rockets are operated at maximum chamber pressure. Because operation in the ducted-rocket mode consumes a significant proportion of the propellant on board the vehicle, increasing performance with induced air (thrust augmentation) is considered essential and contributes to the high mission-average specific impulse required by Earth-to-orbit vehicles.

Mode Ib. At higher flight speeds in ducted-rocket mode, additional fuel is injected into the combustion zone through the aft injectors. Figure 10 shows the location of these injectors and also illustrates how the fuel distribution can be tailored in a direction normal to airflow in the combustion zone, so as to match the fuel flow to the locally available airflow.

Ramjet Mode

Struts provide an ideal mounting place for ram and scram injectors. Combustion efficiencies up to 95% have been demonstrated with both hydrocarbon and hydrogen fuels at Mach 8 conditions at high fuel equivalence ratios, at which it is most difficult to burn completely. Figure 12 illustrates how strut-mounted injectors reduce the mixing gap, thus allowing a significant reduction in combustor length and weight. The strutjet's shorter combustion zone has lower internal drag and reduced heat load. In the case of hydrogen, a cascade fuel injector has been specifically designed for the injection of gaseous hydrogen in the scramjet. A cascade injector[4] delivers the fuel in the form of a low drag wedge-shaped fuel plume. The fuel is injected normally at supersonic speed into the supersonic airstream, the injectors being tailored such that the injected gas is expanded to a level close to the ambient static pressure in the combustor, thereby realizing a low drag shape and increased momentum. In addition to its superior penetration and mixing, shown in Fig. 13, a cascade injector reduces the wall heat flux near the injector by a factor of five or more by avoiding the classic separation bubble found in front of a high drag normal fuel jet. The reduction of these injection-induced hot spots reduces the cooling system complexity and pressure drop.

The predicted air-breathing I_{sp} performance can only be obtained in a practical engine if high combustion efficiency is achieved. Whereas the effi-

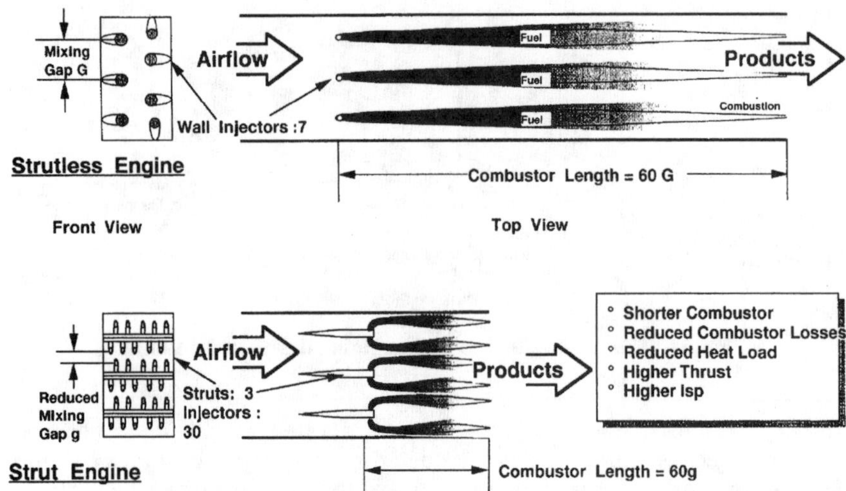

Fig. 12 Strut reduce mixing gap.

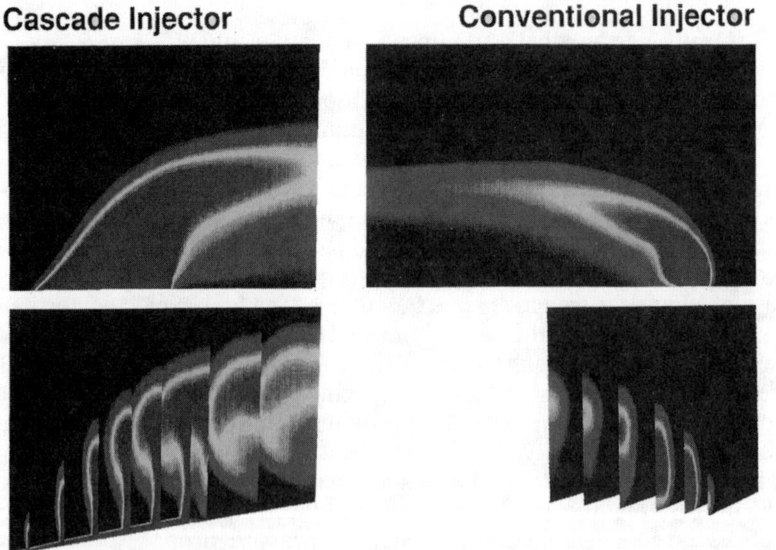

Fig. 13 Cascade injector penetrates deeper than conventional injector.

cient generation of thrust, even in large-area rocket engines, is well understood, the maximization of ram-scram combustion efficiency and thrust generation, being primarily a function of the fuel-injection scheme, is a technical challenge; however, a practical strategy is now well in hand. The strutjet fuel-injection schemes provide the required flexibility of timing and location of the heat release without significant total-pressure loss.

Modes Ic and II. Prior to, and during transition, and in the ramjet mode, engine cooling provides enough energy to the hydrogen in the cooling circuit, and the gas generator can be turned off; the rocket then operates on the fuel side in an expander cycle. When full ramjet operation is achieved, the rocket fuel and oxidizer circuits are shut down, and the fuel to the ram and scram injectors is controlled such that combustion moves forward as Mach number increases. Ramjet takeover occurs at Mach 2.5. This low Mach number is because of the high capture efficiency and low design point of the strutjet engine. At flight speeds beyond Mach 3, the inlet ramps are deployed to gradually increase contraction, eventually reaching full contraction before scram transition.

In low ramjet mode the strutjet variable geometry remains open to maximize air capture and thrust; however, the nozzle opening is dictated by the desired ramjet burner pressure. The strut rockets are turned off. As described earlier, no turbine exhaust gas is available in this mode. The desired combustion location (location of highest pressure) is downstream from the struts in the ram combustion area. For this case two injection sites, one normal injector at the aft end of the isolator section (see top of Fig. 1), and a second axial injector in the strut rocket base (see Fig. 10), are provided. The available injection sites permit heat-release optimization through a division of fuel flow rates between the two principal injection sites, referred to as the aft and base axial injectors.

Scramjet Mode

Mode III. During transition from ramjet to scram mode, the nozzle is opened fully, and some of the fuel is injected from the forward injectors. In scram mode the geometry in the engine internal flow path is similar to that in the ram mode except now the inlet is fully contracted and the combustion nozzle is wide open. The flow through the engine is supersonic, the highest pressure location being in the flow isolator section between the struts. The combustion control options are similar to those of the ram mode with the injection shifted forward. Heat release is controlled by regulating the fuel flow between the forward and aft injectors. At the high scram Mach numbers of 7 to 10, the aft injectors are turned off, and all of the fuel is injected through the forward injectors. These are located downstream of the inlet throat as seen in Fig. 1 and are vertically arranged as indicated in Fig. 10.

Scram/Rocket and Ascent-Rocket Modes

The end of strutjet air-breathing operation begins with a pitch-up maneuver taking the vehicle outside the sensible atmosphere. This maneuver is initiated with the strut rockets reignited while the scram engine is still producing significant thrust. The rocket operation gives rise to scramjet performance benefits from the additional contraction of the incoming air caused by the displacement of the rocket plume gases. At Mach 10 such combined operation results in roughly twice the engine thrust as either element operating alone. This synergy is very beneficial at this point in flight time, as large thrust is needed to overcome the increased drag associated with the pitch-up maneuver.

Modes IV and V. During transition from scram to scram-rocket and then to ascent rocket mode, the rocket operates in the same cycle as during the transition to the ram mode. When the flight dynamic pressure falls below internal engine pressure, the inlet is closed off completely to prevent the rocket gases from escaping out the front. The scram injectors operate at a small flow rate to provide some bleed flow, which cools the strutjet inlet and isolator sections forward of the rockets and which also minimizes attachment shocks of the rocket exhaust plumes. During this operation, the nozzle is kept wide open, as stated earlier, to provide the maximum rocket gas expansion possible. With the inlet closed off this results in a very large nozzle area-ratio nozzle that expands its exhaust products over the boat tail of the vehicle. This action generates the highest possible rocket specific impulse. A very important parameter in the design of an RBCC vehicle is the selected transition Mach number M_t, which impacts the mission average specific impulse, engine/vehicle weights, and correspondingly the vehicle GTOW. For the sake of system robustness and associated increased system reusability, a lower transition Mach number $M_t \leq 10$ is advantageous because it exposes the vehicle and engine to lower heat loads. For the vehicle discussed in this chapter, a transition Mach number of 10 has been baselined. Beyond this, the rocket is reignited and during this submode, referred to as scram/rocket mode, the vehicle begins its pull-up out of the atmosphere; the scram/rocket operation provides an addition of 5 to the mission average I_{sp}. At approximately Mach 12 (and a dynamic pressure about 25 psf) the inlet is closed off, and the engine operates beyond that point in the ascent/rocket mode as a high area-ratio rocket.

D. Optimal Propulsion System Selection

An optimal propulsion system may be considered as one yielding minimum dry mass. To perform the mission, the vehicle required a propulsion system that can operate over the complete speed and altitude range. An RBCC propulsion system with its multiple modes is capable of performing over the required range. The question is whether the RBCC is the correct propulsion system, and, if it is, how is it designed to maximize performance. Several alternative strategies may be employed in this connection. For example, one may consider that mission-averaged I_{sp} may be improved if, in place of a ducted rocket, a high I_{sp} boost engine is employed, further reducing the amount of onboard oxidizer that must be burned with fuel to accelerate the vehicle to ramjet takeover speed. Alternately, or in addition, air-breathing operation can be extended into the high Mach-number scram operation region. Unfortunately, the mass of the resulting conglomerate of engines and decrease in propellant bulk density have a diminishing impact on the vehicle mass fraction that in most cases will more than offset the gains of higher I_{sp} performance. The system parameters are highly interrelated as can be seen in the following:

Boost-Mode Selection

If an engine such as an advanced turbojet or other low-speed air-breathing engine is added, improved performance may appear to result. The effective F/m_e ration of an engine conglomerate is determined by the ratio of the sum of the thrust of the active engines at takeoff divided by the sum of the mass of all engines, active or not. A separate boost-mode engine, used up to ramjet takeover (Mach 2.5), becomes dead mass for the rest of the mission. During boost, such a high I_{sp} boost-mode engine may save 10 mass units of propellant for each mass unit of added engine mass, but to accelerate its greater mass after shutdown to orbit approximately 4 lbm of propellant per unit engine mass must be burned. After accounting for the added main engine, tank, and thermal protection system masses, the propellant required to accelerate the inactive boost-mode engine over the rest of the trajectory significantly exceeded the propellant savings during boost. The addition of a boost-mode air-breathing engine also increases the complexity of the vehicle requiring integration of additional ducting with diverters to isolate the boost-mode engine from the high enthalpy flow occurring later in the flight. The associated overall engine development, production, and operation costs are also expected to increase. Because a rocket is needed in the final phase of flight, little weight penalty is experienced with the ducted-rocket boost engine. The fully integrated ducted rocket, yielding a lower vehicle dry weight and a simpler, less costly engine, therefore becomes the preferred choice for the boost phase.

Engine Design Point Selection

The design of the ram/scram mode is critical because it is the only mode that can add significantly to the mission effective I_{sp}. The thrust produced by an air-breathing engine is directly related to the mass of air processed. This air is captured by the inlet and compressed to raise the pressure for combustion and subsequent expansion. The net accelerating force is the difference between the gross thrust and total vehicle drag (including the spill drag). This total drag is highest at low speeds when the gross thrust is lowest. When the net accelerating force is low, most of the fuel burned is wasted overcoming the vehicle drag. Higher thrust is necessary to perform the mission. One method is to leave the rockets on longer, but this results in much higher propellant consumption. A better method is to increase the air-breathing thrust. The thrust of a ram/scramjet can be expressed in terms of specific impulse ($F/W\text{dot}_f$), fuel-to-air ratio ($W\text{dot}_f/W\text{dot}_a$), and the captured air-weight flow ($W\text{dot}_a$), yielding

$$F = (F/Wdot_f) * (Wdot_f/Wdot_a) * Wdot_a$$

We can show that under the conditions that $M > 5$, $Wdot_f/Wdot_a$ scheduled to provide minimum fuel requirements for acceleration up to $M > 10$, and flight at constant dynamic pressure, the thrust is proportional to the product $Fct(M) * \eta_{cap}$ in which $Fct(M)$ is a function decreasing with increasing Mach number and η_{cap} is the inlet capture efficiency. Although $Fct(M)$ is to some degree dependent on the efficiency of the overall ram/scram engine design and its integration into the vehicle, the strongest influence on thrust can be materialized through appropriate manipulation of η_{cap}. The inlet capture efficiency depends on the inlet type and the inlet design Mach number. The main determination of the inlet design point is the speed at which the bow shock sweeps back to the cowl lip, a condition referred to as shock-on-lip. Typically, an inlet will achieve full or nearly full capture at its design Mach number. At lower Mach numbers the inlet will spill increasing amounts of the air; this reduces the thrust produced by the engine and creates spill drag. For a given capture area thrust can only be increased by selecting a lower inlet design point. Figure 14 shows the effect of inlet design point on the capture efficiency. A high η_{cap} at low speeds gives twice the thrust at a time when it is the most needed.

Figure 15 shows the thrust resulting from two different design points. The engine with the Mach 6 design point produces more than twice the thrust of the Mach 12 design at Mach 6 in the middle of the air-breathing acceleration. The thrust does not cross over until after Mach 8, then it is only slightly better than at the Mach 6 design. This higher thrust reduces the drag integral as well as the thermal soak time. The reason that the Mach 12 design outperforms the Mach 6 design above Mach 8 is higher inlet pressure recovery and greater contraction. These are second-order thermodynamic parameters and yield less than 8% improvement in the thrust production. The objective of a space access vehicle is to accelerate the vehicle to orbital

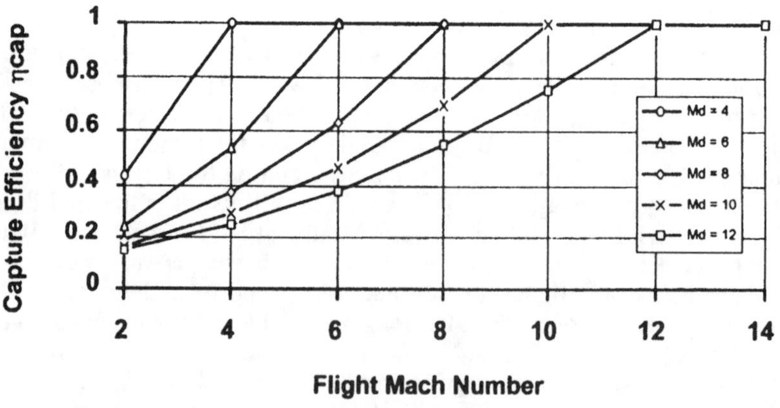

Fig. 14 Lower *M* inlet design points capture more air.

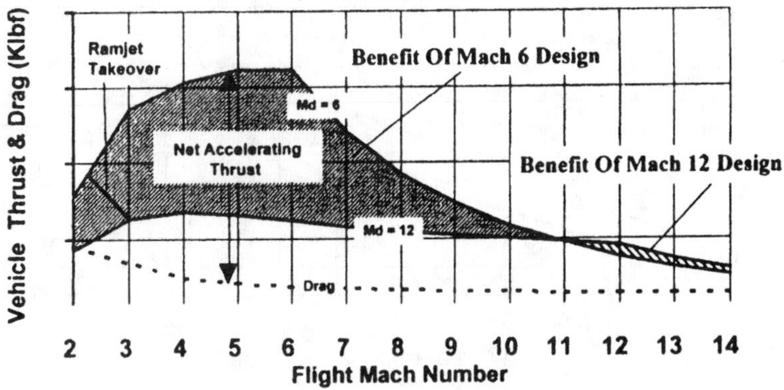

Fig. 15 Mach 6 design produces twice the thrust of Mach 12 design.

velocity and altitude. Only the net thrust, gross thrust minus drag, is providing this acceleration. Net I_{sp} is based on net thrust. Figure 16 illustrates this net I_{sp} as a function of flight Mach number. Figure 17 illustrates the impact of the higher net thrust and I_{sp} on the mission average I_{sp}. The gains in I_{sp} start at the lowest speed with a substantial improvement caused solely to the earlier ramjet takeover. The gains continue to improve all the way up to about Mach 8 where the slop of the Mach 6 curve begins to decrease faster than the Mach 12 design. Although performing better at high speed, the Mach 12 design never recovers the large gains made by the Mach 6 design.

The Mach 6 design point engine appears an obvious choice. A critical design issue concerns the ability of the inlet to operate beyond its design point at overspeed conditions. The decline in the thermodynamic efficiency has already been discussed and does not change the design point selection. Typically, this results in the vehicle bow shock falling inside the inlet cowl. If

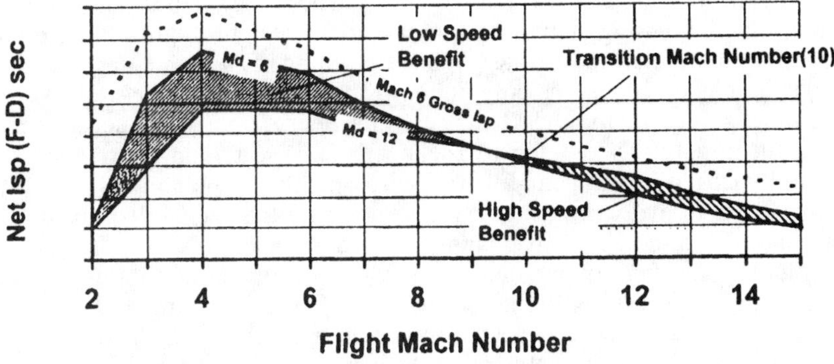

Fig. 16 Mach 6 design produces higher net specific impulse almost up to the transition Mach number.

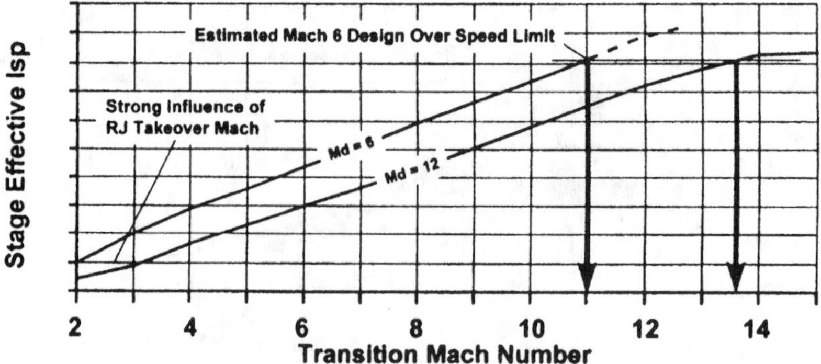

Fig. 17 Vehicles designed for Mach 6 outperform vehicles with higher design points.

this causes sufficient disruption of the inlet operation, the engine thrust can decrease to a point where it equals the drag. In that case the higher performance gained at lower speeds is of little use if the engine stops accelerating the vehicle. The inlet designer must build in sufficient overspeed capability into the inlet to permit acceleration past the point of the transition to ascent mode. In Fig. 17 the assumption is made that the Mach 6 inlet can be designed to operate to at least Mach 10.

Ascent-Rocket Transition Point Selection

Up to around Mach 15, a scramjet has higher I_{sp} than a rocket. The issue is at what point is the rising mission average I_{sp} offset by the rising penalties. High Mach-number operation increases the thermal loads that must be dealt with. Stagnation temperature increases roughly as the square of the Mach number, requiring more active cooling of the engine and vehicle and thicker passive thermal protection. Higher speed operation requires a more complex engine with greater variable geometry. These factors add significantly to the vehicle dry weight, and any increase in dry weight requires additional propellant to accelerate the mass. While operating in any pure air-breathing mode, the engine consumes only onboard fuel, which, in the case of hydrogen, weighs only 4.5 lbm/ft^3. If the scramjet operation is extended beyond Mach 10, more of this low-density fuel needs to be stored on-board, increasing the hydrogen fraction and further decreasing the propellant bulk density. The increase in hydrogen usage has also several other negative impacts on vehicle performance. The propellant tank mass is generally proportional to its volume. Low drag hypersonic-shape factors produce nonoptimal structural shapes contributing to higher tank mass. The larger tank adds volume and surface to the thermal protection system mass that must be applied to the larger exterior of the vehicle. At the same time the larger frontal area, increased wetted surface, and extended operation in the atmosphere add to the vehicle drag losses. Figure 18 illustrates the effect of air-breathing opera-

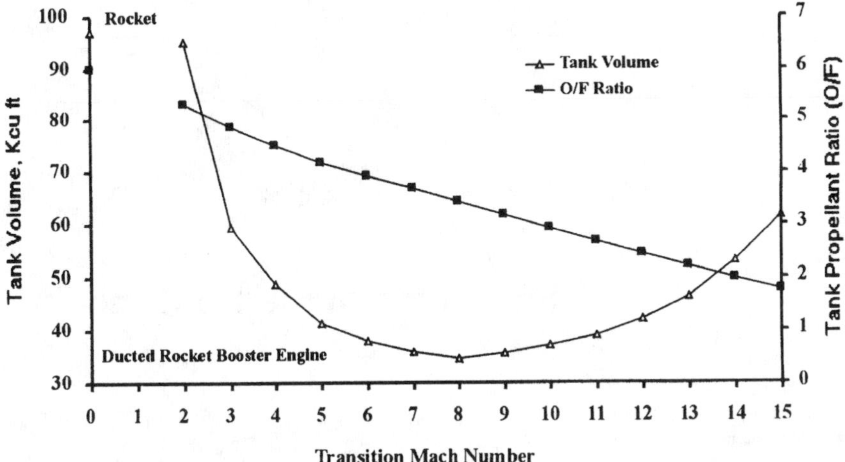

Fig. 18 While O/F ratio will decreases with transition Mach number, tank volume shows a minimum.

tion quantified by the choice of transition Mach number on the tanked oxidizer-to-fuel (O/F) ratio and the resulting tank volume for the strutjet-powered vehicle, assuming that the low speed system is a ducted rocket. The tank volume initially decreases because of the increasing I_{sp}. However, as the stored O/F ratio continues to decrease because of greater air-breathing operation, the fuel tank volume begins to increase. At a Mach number of about 10 and a stored O/F ratio of about 3.0, the minimum dry weight is obtained. Selecting the transition Mach number to be 10 not only reduces the vehicle dry weight but also reduces the engine technology level requirements.

III. Strutjet Engine/Vehicle Integration

A. Strutjet Reference Mission

An extensive analysis was conducted to determine which launch vehicle configurations have the highest potential for meeting ground takeoff requirements for a launch system delivering 8 klbm to geosynchronous transfer orbit (GTO). Another vehicle was added to this family that would deliver 25 klbm to the international space station (ISS) orbit, proving that strutjet propulsion offers versatility in overall vehicle design.

Figure 19 shows two options for integrating this type of engine into a vehicle. In the option shown on the left, the strutjet engine is integrated on the underside of a semi-axisymmetric vehicle. This configuration maximizes the vehicle net thrust value and also provides an efficient structural shape. The figure on the right is a two-dimensional engine configuration.

Of particular interest for two-stage-to-orbit (TSTO) vehicles is the ramjet

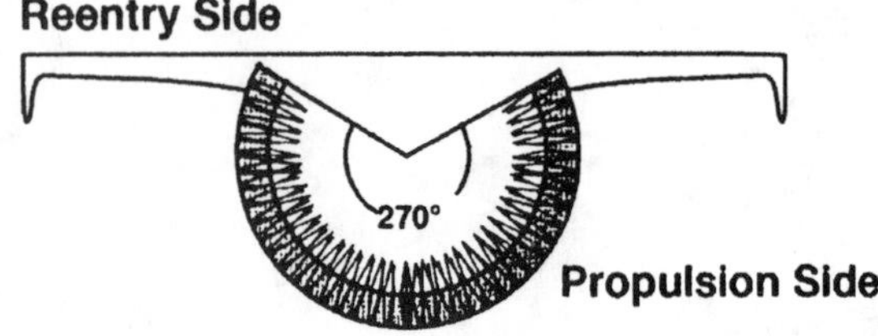

Semi-Axisymmetric Concept

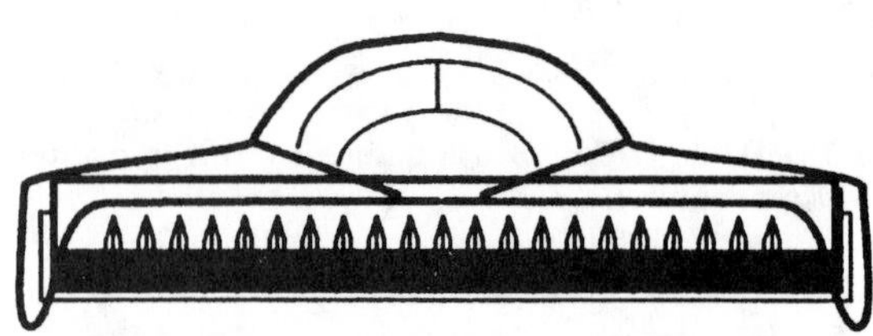

2-D Lifting Body Concept

Fig. 19 Strutjet can be integrated in semi-axisymmetric and two-dimensional lifting body vehicle concepts.

mode of the strutjet, which enables the vehicle to cruise in the atmosphere like an airplane. This allows omniazimuthal launching from a single launch site even in the presence of territorial overflight constraints. The vehicle would simply fly in the highly fuel-efficient ramjet mode to a geographical location from where the launch into orbits with inclinations different from the launch site latitude are permitted. This capability also provides the promising prospect of a new, efficiently configured, effectively operated, and conveniently located future spaceport. The cruise mode of operation also provides more flexibility for the return of the reusable first-stage vehicle to the launch site after separation of the second stage.

These vehicle studies led to the following conclusions: 1) both horizontal and vertical takeoff are feasible and practical; 2) horizontal landing is required; 3) a ducted rocket provides good takeoff and transition thrust; 4) TSTO systems would stage at 16 kft/s with the first stage returning to launch site; 5) SSTO systems are feasible to low-Earth orbit (LEO) and ISSO; and 6) the potential exists for a single launch site for all orbital inclinations.

During these studies, it was recognized that it is possible to account for the synergistic benefit when the rocket is reignited, but the vehicle is still in the upper atmosphere. This effect, as stated earlier, amounts to about 5

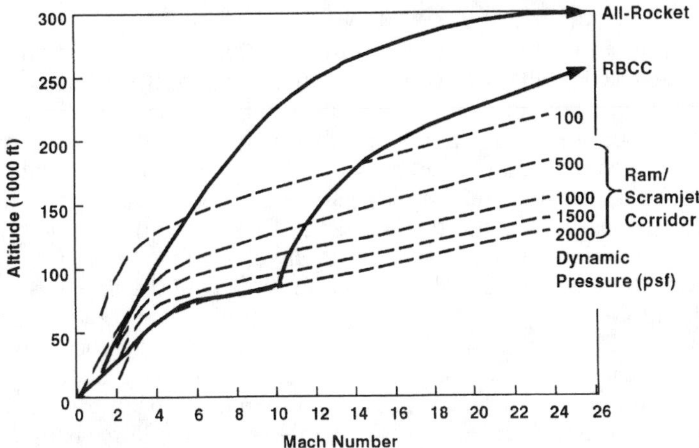

Fig. 20 Typical strutjet vehicle flies at higher dynamic pressure than all-rocket vehicle.

in I_{sp}. Also, the performance of the ascent rocket is critical to the increase in I_{sp}. For example, the mission average I_{sp} is reduced by 4 when the mission ends at the ISSO, instead of LEO, because of the longer operation required in the relatively low performance ascent-rocket mode. Some noticeable attributes of the strutjet can be summarized as follows: 1) substantially lower takeoff and dry weights than conventional rocket vehicles for all strutjet-powered vehicles considered; 2) a high dynamic pressure air-breathing operation to only Mach 10 that has manageable aeroheating environments and is testable with existing facility capabilities; 3) available material technologies for engine and vehicle; 4) large potential for future payload growth and cost reductions with an I_{sp} increase through air-breathing operation at higher than Mach 8, a mass fraction improvement through advanced materials, a robustness increase through higher margins and redundancy; and 5) a single propellant combination for all operation modes. LOX/LH_2.

A typical RBCC vehicle trajectory is shown in Fig. 20 compared to an all-rocket vehicle trajectory. The baseline RBCC SSTO-HTHL vehicle flies at constant dynamic pressure of 2000 psf until the transition Mach number is reached. Then, the ascent-rocket mode takes over to fly to orbit. The vehicle, having a burnout altitude of about 250,000 ft, is flown into an initial elliptical orbit. The final desired circular orbit is obtained by a reaction control system engine (burn phase) to reach low-Earth or higher orbits. For the ascent/rocket case the vehicle nominally burns out at about 250,000 ft, and then performs a circularization burn for the final orbit injection. The vehicle masses used here are based on actual vehicle and engine layouts and simulated flights along trajectories that are optimized either for all-rocket or RBCC propulsion.

B. Engine-Vehicle Considerations

Frequently, when comparing propulsion systems, the focus is on the engine characteristics with insufficient attention given to the vehicle contribution to the propulsion process. When concentrating on the engine only, its I_{sp} and F/m_e appear to be the only discriminators between propulsion systems. However, the vehicle stores and provides propellants to the engine, and in the case of air-breathing engines also carries the inlet to supply the ambient air and provides part of the thrust nozzle. These essential propulsion features have significant impact on the overall vehicle mass and performance. The low bulk density of a hydrogen-powered vehicle requires large, heavy tanks. In addition, a strutjet-powered vehicle flies, when operating in air-breathing modes, more depressed trajectories than typical rocket-powered vehicles to ingest the necessary air for propulsion. These factors increase both drag and heat load on the airframe. When comparing propulsion systems, one must give full consideration to these vehicle impacts of the propulsion choice.

C. Vehicle Pitching Moment

The strutjet configuration as illustrated in Fig. 1 is most effectively integrated on the underside of a vehicle. During atmospheric flight, the forebody and aft-body pitching moments caused by the compression of air and the expansion of combustion gases are balanced with vehicle body flaps. Similarly, exoatmospheric control of the pitching moment resulting now only from the expansion of combustion gases can be achieved with a vehicle trailing-edge flap as shown in Fig. 21. Pitch trim, i.e., net thrust vector directed through vehicle center of mass, can be maintained as the vehicle burns off its ascent mode propellant by deflecting the trailing-edge flap by 5 deg or less. The increase in mass and complexity for the mechanical

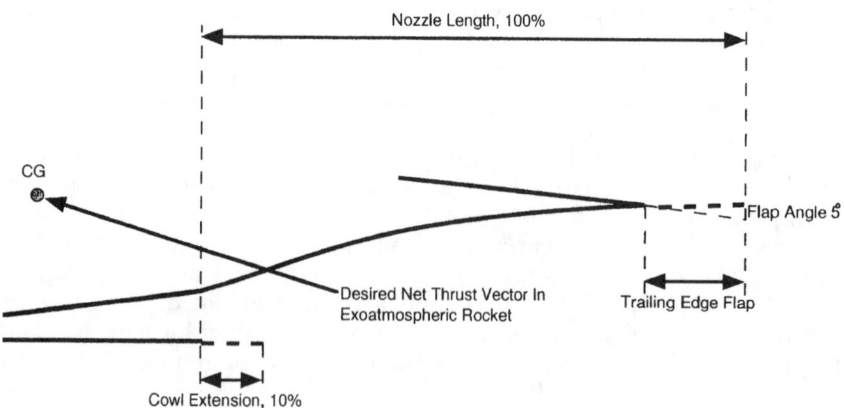

Fig. 21 Exit nozzle can be shaped such that net thrust vector goes through CG during exoatmospheric operation.

integration and actuation of the trailing-edge flap is less than the incorporation of additional rocket thrusters to control the vehicle during its exoatmospheric operation.

D. Engine Performance

The air drawn into the engine by ejector action at subsonic vehicle speeds and by ram effect at higher speeds provides significant (100% and possibly higher) thrust augmentation during boost. The ramjet contribution starting at about Mach 1 increases to full takeover at Mach 2.5. This transition provides the full benefit of a ramjet mode of operation with gross I_{sp} values approaching 3800 s. The specific impulses obtained in the ramjet and in the subsequent scramjet mode are shown in Fig. 22; these values, which are obtained at a nominal angle of attack of zero degree and variable geometry of both inlet and nozzle, are corrected for vehicle drag losses corresponding to varying angle of attack during a typical space launch mission.

Figure 23 presents the thrust generated as a function of Mach number for a typical (single) RBCC engine for various operating modes and for a dynamic pressure of 1500 psf. The capture area is 200 ft^2 per engine designed at a shock-on-lip of Mach 6. The net specific impulse generated by the engine is also provided in Fig. 22, assuming that the rocket is ignited at flight Mach of 10. It is significant to observe here 1) the importance of sizing the engine for transonic flight, 2) the need for high thrust in the vehicle ascent phase, and 3) the effect of the choice of rocket reignition Mach number.

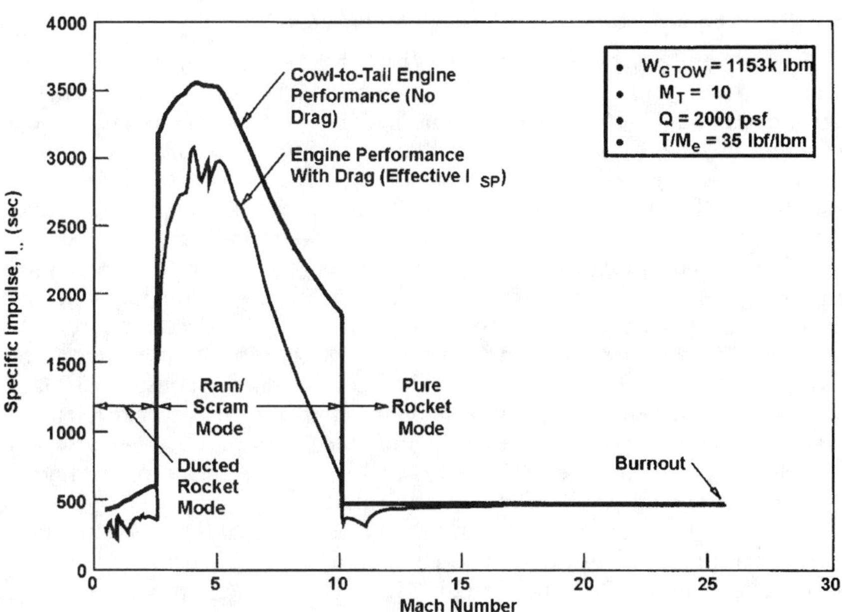

Fig. 22 Strutjet gross and net specific impulse.

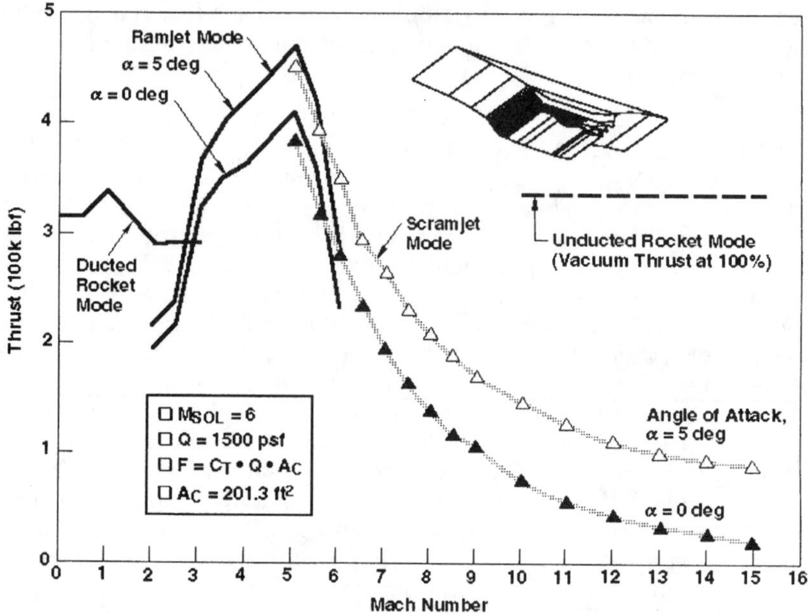

Fig. 23 Strutjet thrust during various operating modes.

E. Reduced Operating Cost Through Robustness

From several points of view, we can see that the strutjet engine technology leads to lower operating cost for Earth-to-orbit transportation systems. Referring to Fig. 24, there are two reasons for this: 1) lower engine replacement cost because the thrust required is reduced to about 70 to 80%, and 2) lower maintenance and operating cost because of higher structural margins and redundancy. Both of these savings result from the increased I_{sp} of the strutjet RBCC engine and the associated reduction in takeoff and dry masses.

RBCC vehicle design trades that consider engine F/m_e and M_T are summarized in Fig. 25, showing the gross mass at ignition as a function of M_T for various values of engine F/m_e. The data show that beyond a M_T of about 10 the vehicle gross mass flattens out. In the generation of these data, the assumption was made that the thermal protection mass is constant at Mach numbers greater than 10 regardless of the higher heat load experienced. Variations in these two key design parameters, F/m_e and M_T, show higher vehicle dry mass for lower M_T and lower gross takeoff mass for the higher M_T values, even as the F/m_e ratios are lowered.

RBCC SSTO/HTHL vehicle thrust-to-mass ratio and transition Mach number are mapped in Fig. 26 and compared to the all-rocket case with a gross takeoff mass of about 2.7 million lbm. The available thrust and structure design margins for the RBCC design are shown in the shaded area for the two baseline cases, depending on the key RBCC design parameters

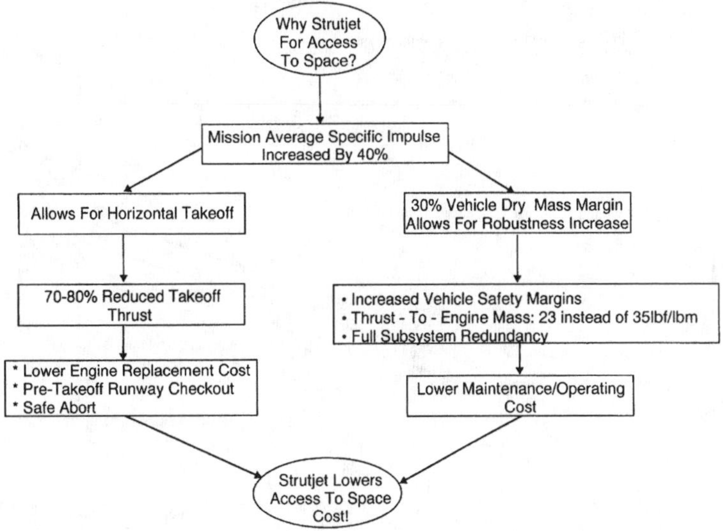

Fig. 24 Strutjet concept provides potential for lower cost access to space operation.

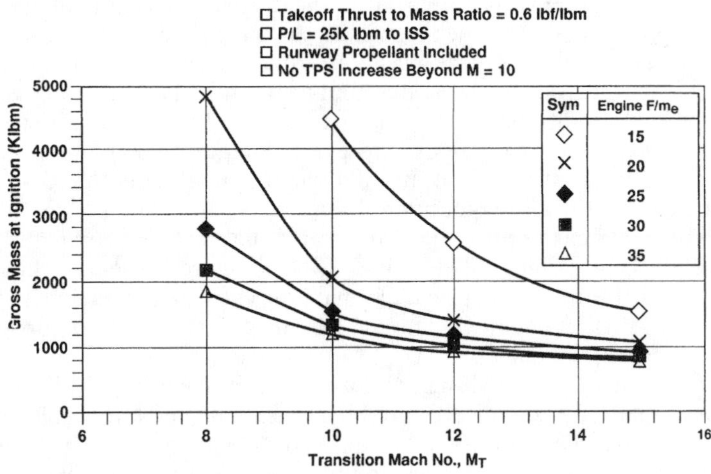

Fig. 25 Low transition Mach number reduces thermal protection requirements but increases vehicle mass at takeoff.

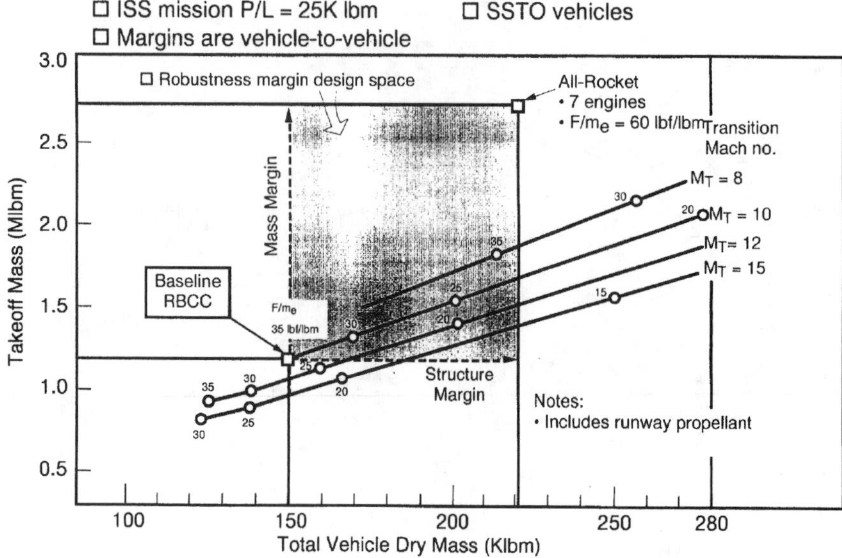

Fig. 26 Shaded area indicates available thrust and structural design margins.

selected. The baseline RBCC vehicle with a gross takeoff mass of 1.12 million lbm is shown at an F/m_e of 35:1 and an M_T of 10.

The strutjet SSTO/HTHL vehicle design trades are compared to two all-rocket SSTO/VTHL designs in Fig. 27. The gross takeoff mass is shown as a function of total vehicle dry mass for the two design cases. Again, the shaded area represents the robustness design space margin for the RBCC baseline vehicle relative to the all-rocket vehicle. For the all-rocket case with an abort capability, the gross mass difference is about 1.5 million lbm, representing a significant thrust margin between the two design cases. The vehicle dry mass difference is about 71,000 lbm, representing a structure margin between the two designs. These data include 45,000 lbm of propellant used by the RBCC on the runway for horizontal takeoff and an abort landing gear for the all-rocket design.

The strutjet vehicle shows a positive design margin when compared to an all-rocket SSTO, as illustrated in Fig. 28. This figure presents vehicle dry mass as a function of total vehicle dry mass for various subsystems: 1) prime structure, 2) strutjet engine, 3) landing system, 4) TPS, and 5) other subsystems (e.g., power, avionics).

Figure 29 illustrates the design philosophy employed to gain high robustness for the RBCC system leading to high reusability and affordable operating cost. The design point of the strutjet vehicle is determined by assuming that the state-of-the-art materials and processes, design and analysis techniques, and producibility and fabrication technologies are equivalent to an advanced all-rocket SSTO vehicle design. The structural robustness factor is

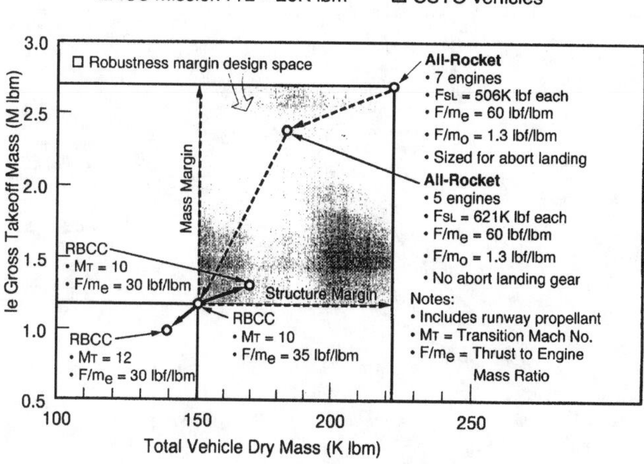

Fig. 27 Strutjet HTHL designs have significant robustness margins relative to all-rocket vehicle designs.

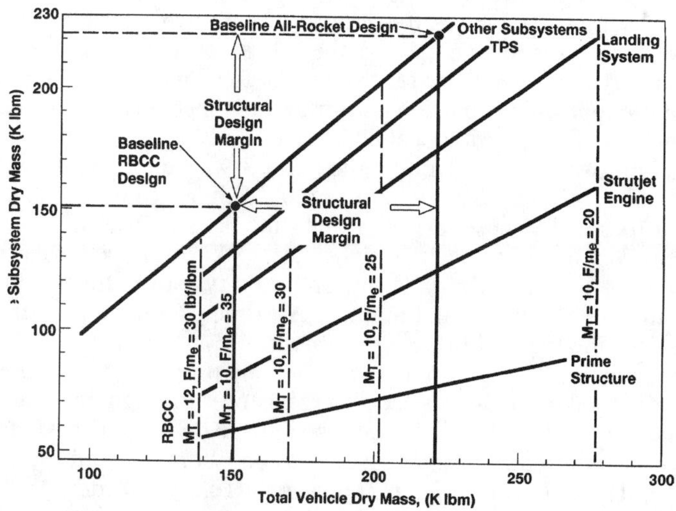

Fig. 28 Strutjet vehicle shows positive structural design margin.

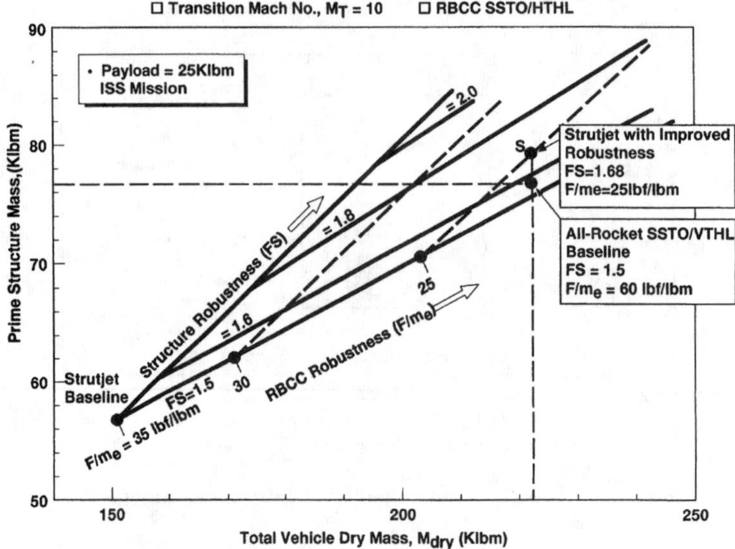

Fig. 29 Strutjet provides significant robustness space.

represented by a tank design factor of safety (FS), and the corresponding propulsion robustness parameter for the RBCC engine is its F/m_e. The two baseline vehicle designs have a tank factor of safety of 1.5. As can be seen in Fig. 29, increasing the RBCC tank FS to 1.95 provides the same prime structure mass as the all-rocket vehicle, and decreasing the RBCC engine F/m_e ratio to about 23:1 increases the total dry mass to the same level as the all-rocket case. As is to be expected, the robustness curves emanating from the baseline RBCC design are far from orthogonal to each other, indicating the strong coupling of propulsion and structure, which is typical for Earth-to-orbit launch vehicles. A practical combination of structural and engine robustness improvements is given by the point S, with the same dry weight as the all-rocket vehicle but FS equal to 1.68 and F/m_e ratio equal to 25lbm/lbm.

The all-rocket dry mass serves in this study as a reference and a design limit. However, in future designs/higher dry mass RBCC vehicles could be considered, as long as robustness and reusability can be further increased and the mission objectives maintained. A prime candidate for further improvement is the thermal protection system.

As noted in Fig. 28, the primary structure and propulsion have the largest share of the total dry mass, and adding robustness through mass increase to these elements would have the largest impact on overall dry mass. The landing gear is already designed for mission abort loads; during routine takeoffs and landings, its design margins are large, and no additional robustness is required.

Auxiliary propulsion and other dry mass items make up less than 10% of the total. Robustness increase through redundancy of these items does not have a significant effect on the system dry mass.

The reduced takeoff mass of the strutjet vehicle, as shown in Fig. 30, allows a corresponding reduction in takeoff thrust. Typically, thrust-to-mass ratios of 1.3 lbf/lbm for vertical and 0.6 lbf/lbm for horizontal takeoff is required to provide adequate vehicle acceleration and engine out capability. A reduction in takeoff thrust reduces the required engine size, which lowers both replacement and maintenance and operation cost. Relative to an all-rocket design, a thrust reduction of 80% for the strutjet baseline or about 73% for a strutjet vehicle design with increased structural and engine robustness is possible, as indicated by the point S in Fig. 30.

Tables 2 and 3 provide, respectively, candidates for engine and summary of vehicle design features leading to more robustness, but at the cost of increased engine and vehicle mass. Of all of the features listed, the powerhead/valving deserves a special comment. The powerhead/valving scheme previously illustrated in Fig. 4 provides the lowest weight system. However, the required mode switching and the various operating conditions of the fuel turbine lead to the surmise that separate turbopumps may represent a more robust approach.

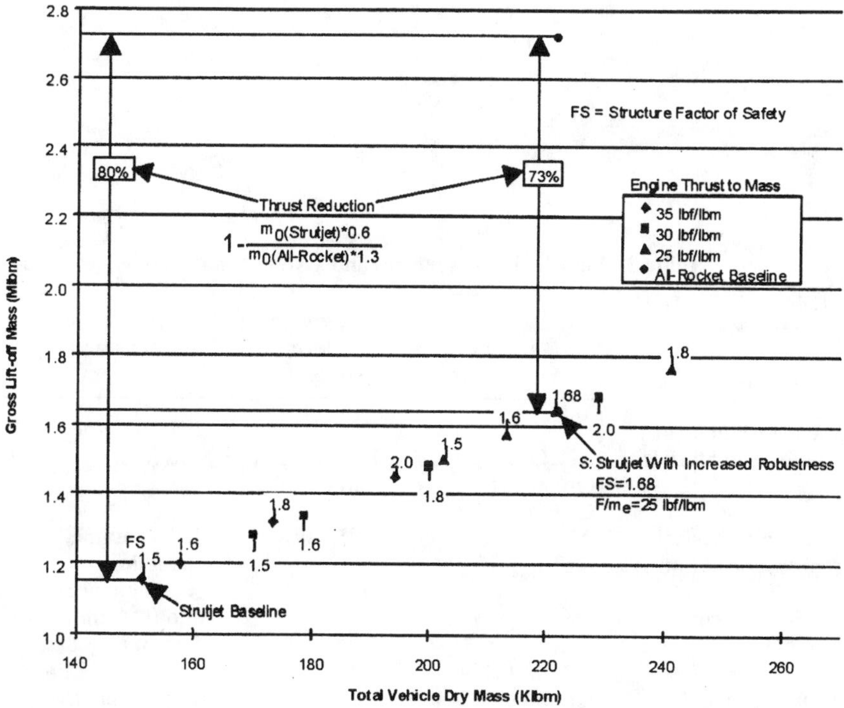

Fig. 30 Strutjet vehicle takeoff thrust is significantly higher than all-rocket takeoff thrust.

Table 2 Sensitivity to Engine Robustness

Feature providing engine robustness	Sensitivity to engine robustness increase
General	—
Engine thrust-to-mass	High
Reduced transition Mach number	High
Specific	—
Redundant ignition	Low
Health monitoring	Low
TPA housing stiffness	Medium
Reduced turbine speeds/temperature	Medium
Reduced chamber pressure	Medium
Separate fuel TPA's	High
ducted-rocket mode	
ram/scam and unducted-rocket modes	

Table 3 Vehicle design features and system robustness

Design approach	Nominal condition	→	Growth level potential	Impact on robustness/ system
Increased tank (FS)	1.5	→	>2.0	High (lifetime)
Choice of materials	Al LO_2 Gr/Ep LH_2 Gr/Gp Str.	→	Develop new Materials	Med (cost)
Materials physical properties (Tanks)	Al = 80 ksi Gr/Ep = 75 ksi	→	+25–50%	High (lifetime, handling, maintenance)
Design margins	0	→	+25%	Med (lifetime)
Handling	Min gauge 0.100–0.200	→	×2.0	High (maintenance)
Fabrication and design	Optimum thickness	→	Thicker skins and web isogrid	Med (lifetime, handling)

F. Vehicle Comparisons

To proceed further, the baseline all-rocket and RBCC vehicle designs are compared to the current Space Shuttle launch vehicle in Fig. 31. The nominal all-rocket SSTO vehicle is designed with a large forward hydrogen tank, a 15 by 30 ft payload bay, and a single, large rear LOX tank. This arrangement leaves a small trim control margin because of the heavy masses of LOX and engines in the rear. Forward canards may be required, increasing the gross liftoff mass further. The all-rocket SSTO vehicle is about 230 ft long and the RBCC SSTO vehicle about 210 ft. The RBCC strutjet vehicle has two forward and two aft hydrogen tanks and twin LOX tanks on each side of the payload for optimum center of gravity control. Prior to adding increased robustness, the RBCC baseline gross mass is about 44%, and the RBCC has a dry mass of about 68% of the all-rocket case.

A layout of a SSTO all-rocket vehicle is shown in Fig. 32. This vehicle is assumed to have five rocket engines with 708,000 lbf thrust each (or 7 engines at 506,000 lb each) at a F/m_e of 60:1 The vehicle, with a liftoff thrust-to-mass ratio of 1.3, is sized for abort landing gear with partially loaded propellant tanks. The tanks are designed with a FS of 1.5 using a large graphite-epoxy hydrogen tank and a single, large aluminum-lithium LOX tank. The mass breakdown for this all-rocket vehicle is provided in Table 4. A layout of a SSTO RBCC vehicle is shown in Fig. 33, and its mass breakdown is summarized in Table 5. Comparative design parameters of both vehicles are provided in Table 6.

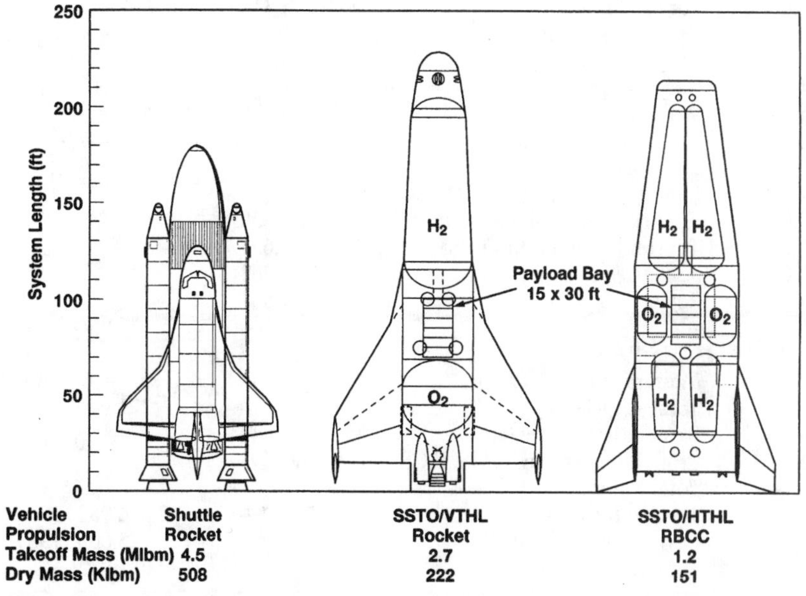

Fig. 31 Strutjet and all-rocket vehicles for delivering 25 klbm to ISSO.

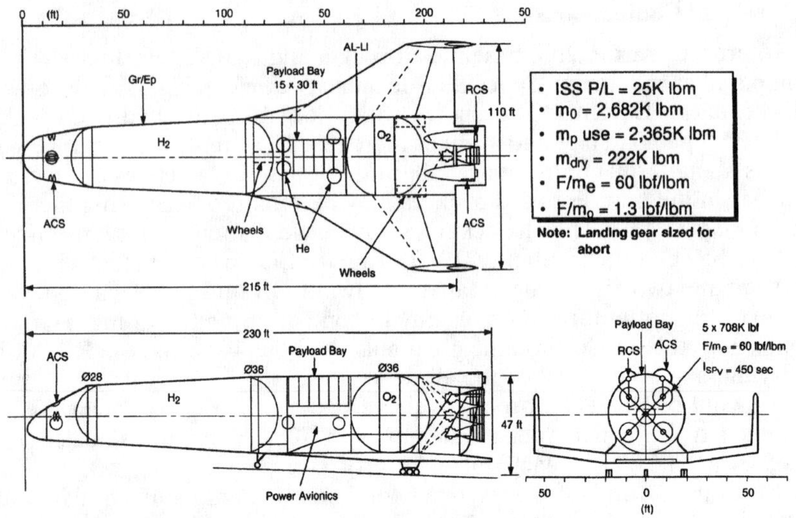

Fig. 32 All-rocket SSTO-VTHL vehicle design concept.

Table 4 All-rocket vehicle mass breakdown

Module	Component	Mass, (lbm)	Mass, (lbm)
1.0	Forward nose	3,347	—
2.0	LH_2 tanks + intertank	27,773	—
3.0	LO_2 tanks + intertank	15,021	—
4.0	Cargo airframe	11,062	—
5.0	Wing/tail/structure	16,776	—
6.0	Boat-tail structure	2,969	—
—	Total primary structure[a]	—	76,948
7.0	Landing system	24,093	—
8.0	Electrical/mechanical/actuation	5,450	—
9.0	Thermal protection (TPS)	26,633	—
10.0	Avionics/environment	2,263	—
11.0	Power	567	—
—	—	—	135,954[a]
12.0	OMS/RCS (dry)	4,346	—
13.0	Main propulsion (dry)	81,640	(F/m_e = 60 lbf/lbm)
—	Total vehicle dry[a]	—	221,940
14.0	Residuals/reserves/He	39,644	—
15.0	Propellants (main, OMS, RCS)	2,395,398	—
—	Total liquids[a]	—	2,435,042
—	Total vehicle (no P/L)[a]	—	2,656,982
—	Payload[a]	—	25,000 (ISS)
—	Total ignition[a]	—	2,681,982

[a]F/m_0 = 1.3 lbf/lbm at liftoff.

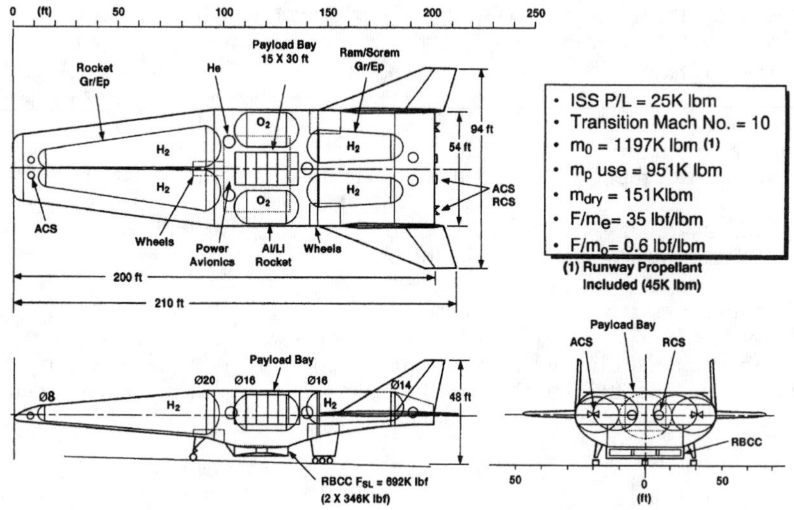

Fig. 33 Strutjet SSTO-HTHL vehicle design concept.

Table 5 Strutjet vehicle mass breakdown

Module	Component	Mass, lbm	Mass, lbm
1.0	Forward nose	2,477	—
2.0	LH_2 tanks + intertank	25,089	—
3.0	LO_2 tanks + intertank	7,509	—
4.0	Cargo airframe	7,015	—
5.0	Wing/tail/structure	11,388	—
6.0	Boat-tail structure	3,456	—
——	Total primary structure[a]	—	56,933
7.0	Landing system	34,896	—
8.0	Electrical/mechanical/ actuation	5,450	—
9.0	Thermal protection (TPS)	18,174	—
10.0	Avionics/environment	2,263	—
11.0	Power	567	—
——	Total dry (less prop)[a]	—	118,283
12.0	OMS/RCS (dry)	3,690	—
13.0	Main propulsion (dry)	33,166	(RBCC F/m_e = 35:1)
——	Total vehicle dry[a]	—	151,449
14.0	Residuals/reserves/He	19,692	—
15.0	Propellants (main, OMS, RCS)	972,061	(m_{runway} = 44,629 lbm)
——	Total liquids[a]	—	991,753
——	Total vehicle (no P/L)[a]	—	1,143,202
——	Payload[a]	—	25,000 (ISS)
——	Total ignition[a]	—	1,168,202

[a] M_T = 10.0; T/m_o = 0.6 lbf/lbm at liftoff.

Table 6. Design parameters for the all-rocket and the RBCC-SSTO vehicle

Parameter/vehicle	All-rocket	Strutjet RBCC
Vehicle length	230 ft	210 ft
Wing span (tip to tip)	110 ft	94 ft
Number/thrust per engine	5 at 708,000 lbf or 7 at 506,000 lbf	2 at 346,000 lbf
Engine thrust-to-mass	60 lbf/lbm	35 lbf/lbm
Liftoff thrust-to-mass ratio	1.3 lbf/lbm	0.6 lbf/lbm
Tanks		
Operating pressure	45 psi	45 psi
Factor of safety	1.5	1.5
Number/material LH_2 tank	1 ea graphite/epoxy	4 ea graphite/epoxy
Number/material LO_2 tank	1 ea aluminum-lithium	2 ea aluminum-lithium
Oxidizer-to-fuel loading ratio	6.0:1	3.4:1
Runway propellant	None	45,000 lbm
Propellant ullage	3%	3%
Propellant residuals	1% for the main 3% for the OMS/RCS	1% for the main 3% for the OMS/RCS
Vacuum specific impulse for all onboard propulsion systems	450 s	450 s
Vehicle prime structure mass	77,000 lbm	57,000 lbm
Total dry (less propulsion) mass	136,000 lbm	118,000 lbm
Total vehicle dry mass	222,000 lbm	151,000 lbm
Total useable propellants	2,365,000 lbm	951,000 lbm
Useful mass fraction	0.8992	0.8461
Landing gear mass	1% of mass at ignition (Note: Vehicle does not reach ISS orbit with higher landing gear mass.)	3% of mass at ignition
Mass at ignition	2,682,000 lbm	1,168,000 lbm
Robustness margin	Reference: 0k lbm	71,000 lbm
Thrust reduction	Reference: 0%	80% prior to, 70% after robustness increase to no robustness margin

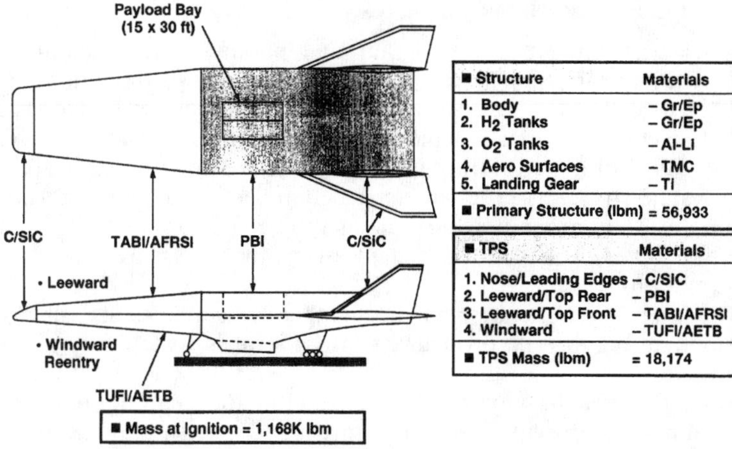

Fig. 34 SSTO all-rocket and RBCC vehicles use similar structure and TPS technology.

Both the SSTO all-rocket and the SSTO RBCC vehicle use similar structure and TPS technology, as shown in Fig. 34, for the strutjet vehicle. The TPS is designed for a transition Mach number of 10 and for re-entry thermal conditions. The body structure is graphite-epoxy along with the hydrogen tanks; the LOX tanks are aluminum-lithium; the aerosurfaces are titanium matrix composite (TMC), and the landing gear is titanium. The TPS materials are similar, having carbon-silicon carbide (C-SiC) nose and leading edges and the polybenzimidezole blanket insulation (PBI) on the top leeward side; Nextel-tailored advanced blanket insulation (TABI) or advanced flexible reusable surface insulation (AFRS) is used on the top leeward side (similar to Shuttle FRSI); and tiles of toughened unipiece fibrous insulation (TUFI) and alumina-enhanced thermal blanket (AETB) on the re-entry lower windward side. The windward side could also have advanced internal multilayer insulation (IMI) but at a much higher cost. The TPS unit mass for the current design over the whole vehicle surface is estimated to be about 1.0 lb/ft^2. This represents a slightly less areal density than current technology (Shuttle is about 1.5 lb/ft^2 average for the whole vehicle).

IV. Available Hydrocarbon and Hydrogen Test Data and Planned Future Test Activities

The overall test program, accomplished to date or planned for the near future, may be divided into three groups:

A. Storable Hydrocarbon System Tests
B. Gaseous Hydrogen System Tests, and
C. Planned Flight Tests

A. Storable Hydrocarbon System Tests

Aerojet performed a test program that parametrically examined the RBCC strutjet propulsion system from Mach 0 to 8 over the altitude range from 0 to 100,000 ft. The tests were carried out in the context of a long-range missile, two configurations of which are shown in Figs. 35 and 36. However, the test results obtained during this campaign are equally applicable to a launch system strujet like the one described in Sec. II. In over 1,000 hot fire and inlet tests a number of achievements were realized:

1) The strut inlet provides excellent air capture, pressure recovery, and unstart margin.

2) The integration of compact high chamber pressure rockets using gelled hypergolic and cryogenic propellants into a strut is structurally and thermally feasible.

3) A fixed-engine flow-path geometry suitable for all modes of operation can be established providing adequate thrust and specific impulse to accomplish mission objectives.

4) Static sea-level thrust augmentation of 13% can be achieved because of the interaction of air ingested with the fuel-rich rocket plume.

5) The ducted-rocket thrust increases with increased flight Mach number. At Mach 2.85 and altitude 20,000 ft the thrust increase is over 100%. And, at Mach 3.9 and altitude 40,000 ft the ramjet thrust exceeds the rocket sea-level thrust by 19%.

6) Dual-mode operation of the ram-scram combustor with a thermally choked nozzle is feasible.

7) Efficient combustion at high altitude and with short combustors is possible with hypergolic pilots. Combustion efficiency of 90% can be demonstrated with a combustor only 30 in. long at Mach 8 conditions.

Inlet Tests

As part of the first strutjet inlet test program, a subscale inlet model, shown in Figs. 37 and 38, was constructed to evaluate design options for the freejet engine inlet design. Prior to this testing, engine/vehicle performance was based on extrapolating the performance of the inlet from the literature

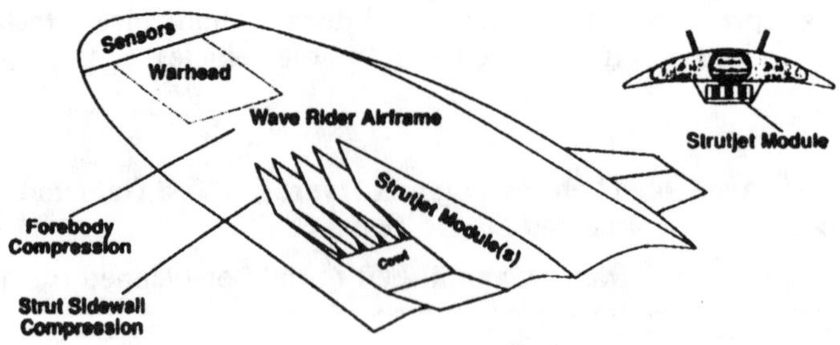

Fig. 35 Wave-rider-type Mach 8 cruise missile provides significant forebody compression.

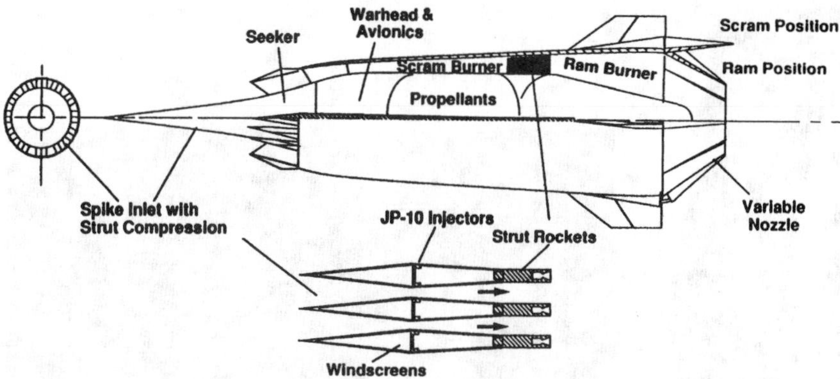

Fig. 36 Axisymmetric Mach 8 strutjet engine facilitates engine-vehicle integration and provides centerline thrust vector.

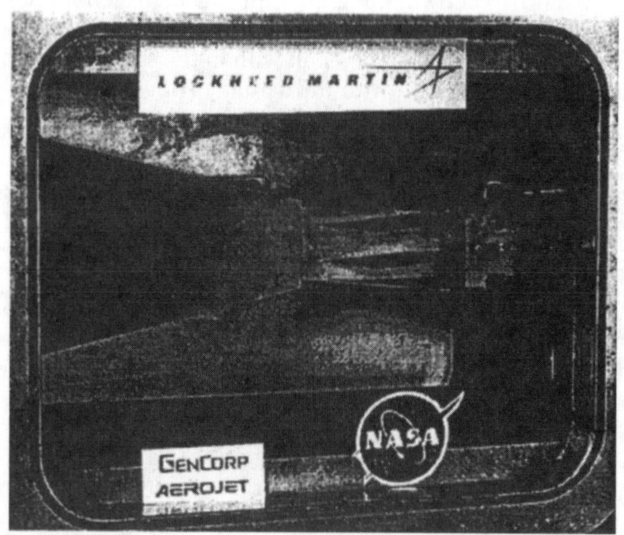

Fig. 37 Subscale strutjet inlet test hardware mounted in wind tunnel.

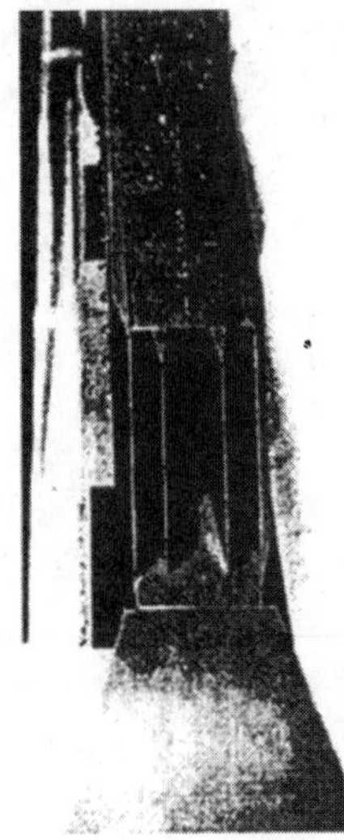
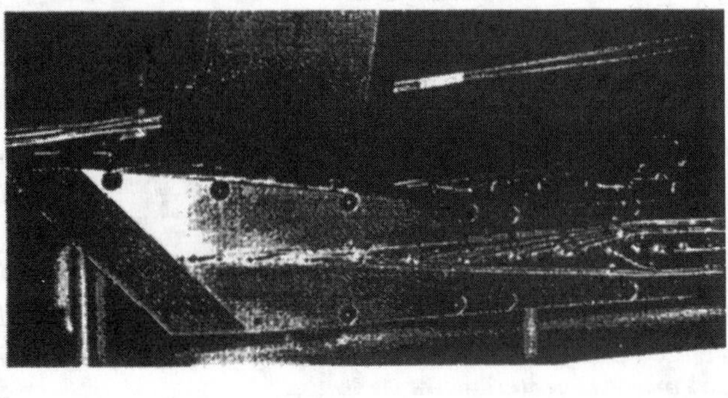

Fig. 38 Subscale strutjet inlet test hardware.

(mostly the work of NASA LaRC). The objectives of the inlet development were to define a missile-like inlet that would interface with the hydrocarbon combustor that up to this point had only been tested in a direct-connect configuration. A very conservative design was selected that had no internal contraction to ensure starting. This inlet was found to start at all tested Mach numbers (4–6), and, as shown in Fig. 39, it produced excellent pressure rise. As predicted, it exhibited low energy flow near the body side of the flow path, a phenomenon that had to be dealt with in the subsequent freejet tests via matching the fuel injection with the actual airflow distribution.

Ducted-Rocket Tests

In these tests each strut contained three water-cooled gelled IRFNA and MMH propellant rockets, as shown in Fig. 40. The injector pattern consisted of 36 pairs of fuel and oxidizer elements arranged in concentric rings, the outermost ring providing fuel film cooling to the chamber. The chamber geometry is shown in Fig. 41. The characteristic parameters of the injector are summarized in Table 7.

As illustrated in Fig. 42, the rig representing the strutjet engine was designed as a sandwich with hinged side-wall sections. The duct section housing the strut rocket had a fixed geometry of 4.0 by 6.6 in. The isolator section in front of the strut duct could be connected to either a bell mouth for a static test or a hydrogen-fueled vitiated air heater. Two struts were mounted in the strut duct, dividing the flow path to the inlet into three channels.

In the sea-level static ducted-rocket tests, the isolator section in front of the strut duct was fitted with a calibrated bell mouth. The duct geometry was

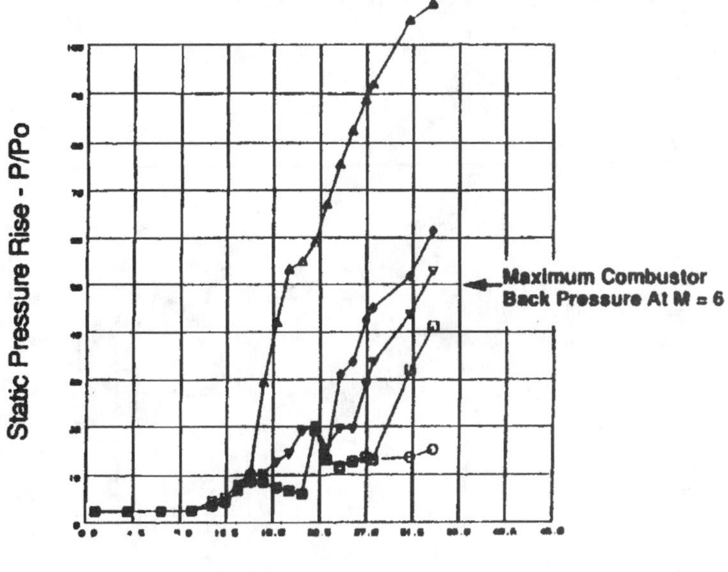

Fig. 39 First strutjet inlet generated high-pressure ratios.

Fig. 40 Missile size strut rockets.

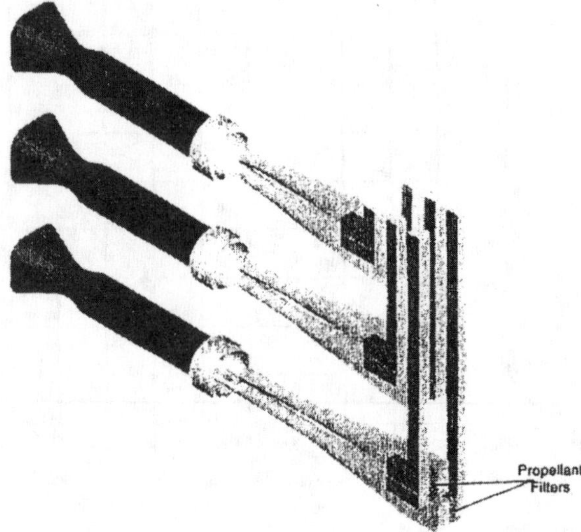

Fig. 41 Strut-rocket chamber geometry and propellant feeds for test article.

varied to determine the configuration yielding the maximum thrust. Sensitivities of rocket chamber pressure, mixture ratio, and rocket nozzle expansion ratio were also established. As shown in Fig. 43, thrust enhancement was a strong function of the ram-burner throat area and a somewhat weaker function of the ram-burner geometry. With a duct geometry of 3–3 deg, 13% more thrust was obtained than for the reference rocket in a particular test with a throat area of 32 in^2. Data analysis indicated that oxygen content of the inducted air was completely consumed in approximately 8 in. from the rocket baseline. Considering the influence of chamber pressure, operation at 2000 psia generated more thrust than operation at 1600 psia; however, airflow and thrust augmentation were reduced by 19 and 3%, respectively. The area ratio of 11:1 generated 12% higher induced airflow and 6% more thrust than the lower area ratio of 5:1. Finally, the observation was made that greater airflow and higher thrust result from operation at higher mixture ratios because of reduced thermal choking resulting from the afterburning scheme.

Table 7 Strut-Rocket design parameters

Parameter	Design	Nominal operation
Mixture ratio	1.6	1.4
Chamber pressure, psia	2500	1600–2000
Thrust, lbf	1000	600–700
Expansion ratio	5.1 and 11.1	5.1 and 11.1

Fig. 42 Strutjet uncooled test engine with adjustable two-dimensional combustor geometry.

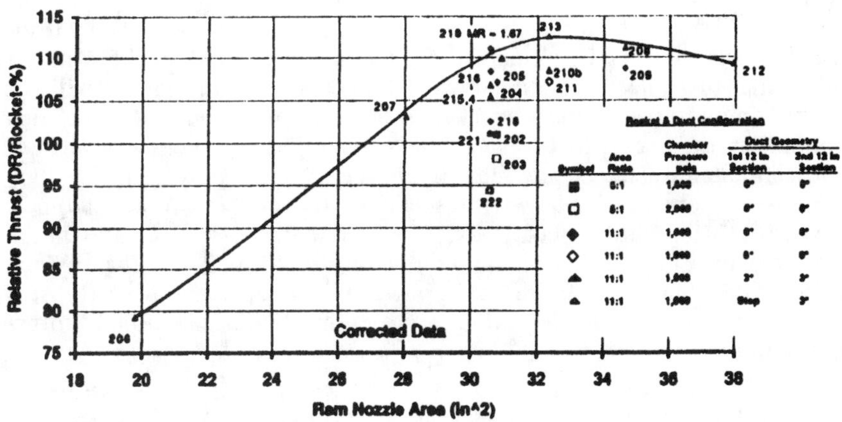

Fig. 43 Direct-connect ducted-rocket test data show peak thrust enhancement of 13%.

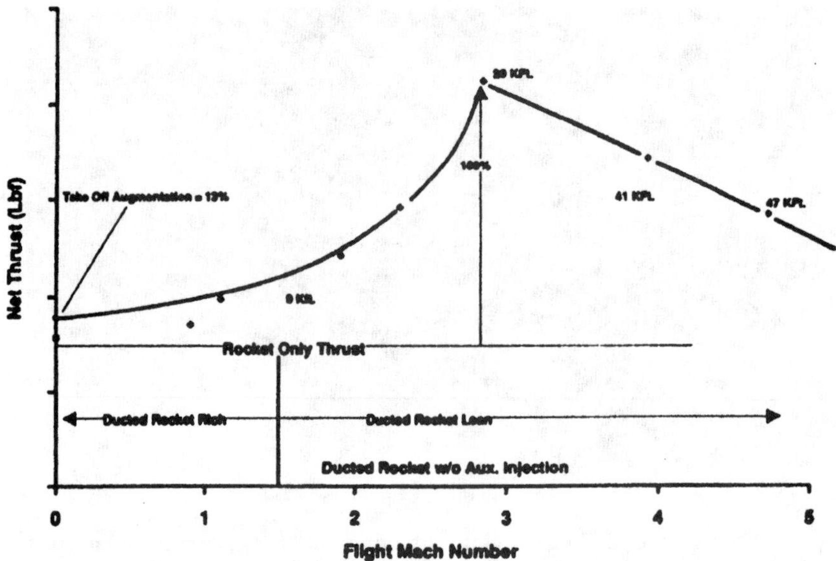

Fig. 44 Fuel-rich rocket doubles thrust in ducted-rocket mode.

In the direct-connect ducted-rocket tests that allow evaluation of the engine under flight trajectory conditions, the isolator section in front of the strut duct was connected to a hydrogen vitiated air heater with a Mach 2 nozzle. The duct geometry, which in the static tests provided the maximum takeoff thrust augmentation, was maintained in these direct-connect tests. Measuring the duct pressure and assuming a particular inlet performance allows for the determination of flight altitude and Mach number simulated in a given test.

Figure 44 shows the thrust obtained in the ducted-rocket and ramjet tests. The left branch of the figure depicts the thrust of the ducted rocket without additional fuel injection, and the right branch that of the ramjet without rocket operation. The ducted-rocket tests were conducted under fuel-rich conditions; the excess fuel was sufficient to support 10 lbm/see of airflow. The simulated trajectory provides 10 lbm/see of air at approximately Mach 1.5. Tests beyond Mach 1.5 were thus lean on an overall engine stoichiometric basis. Auxiliary fuel injectors could be used to increase the engine thrust. The peak thrust is seen to occur at a simulated altitude of 23,000 ft and Mach 2.85 with 31 lbm/see of air being supplied to the engine. The peak thrust is over twice the bare rocket value, representing better than 100% thrust augmentation.

Direct-Connect Ram and Scramjet Tests

The ramjet tests were conducted at Mach numbers of 2 and higher without rocket operation to optimize ramjet injector performance. The primary

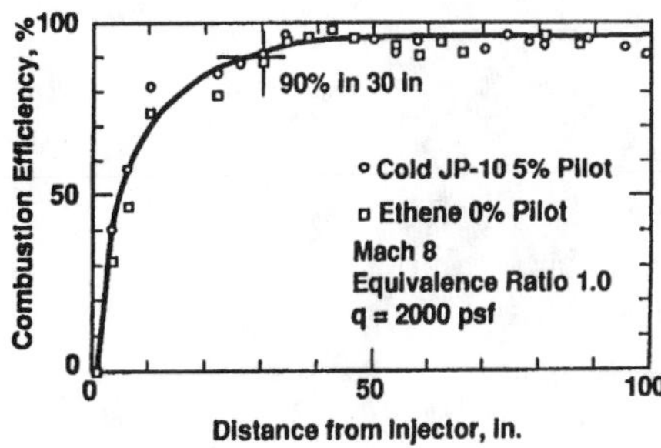

Fig. 45 Both gaseous ethene and pilot vaporized cold JP-10 yield high combustion efficiency.

ramjet test variable, other than the injector parameters, was the fuel of choice. In support of the strategy for minimizing heat load and hot spots, the main emphasis was on achieving a short combustor length.

In the scramjet operating regime three test series were conducted. In the first tests the scramjet geometry was explored, and high combustion efficiency with a fixed geometry over the Mach number range of 2 to 8 were demonstrated. Mach 2 and 4 tests were conducted with JP-10 fuel. The Mach 8 tests simulated the effect of regeneratively heated fuel by using ethane instead of JP-10.

In the second test series the tests conducted with ethane were repeated with JP-10 fuel, using a slightly modified duct geometry, namely, the first 12 in. downstream being of constant area followed by a 2-deg double-sided expansion over the remaining duct length. Autoignition was not achieved; however, when pilots were utilized for ignition and flame sustaining, stable combustion at 95% efficiency was observed. As shown in Fig. 45, the test duplicated, in essence, the performance previously achieved with ethene, with only the slight change in duct geometry. The strutjet design used in these tests provided for a contact pilot at each injection point of the hydrocarbon ramjet fuel. This pilot derives its energy from the combustion of small amounts of the gelled rocket propellants, which are injected and burned upstream of the hydrocarbon injection. Because of the hypergolic nature of the employed rocket propellants, the pilots act initially as igniters and subsequently as flame sustainers, allowing flight at high Mach numbers and high altitudes. The demonstration of this feature is verified by the data presented in Fig. 46. At a simulated flight condition of Mach 4 and 40,000 ft of altitude, JP-10 was ignited by the pilot, resulting immediately in a thrust increase of about 2000 lbf. Combustion and thrust production were sustained as long as the pilot stayed on. When turned off, the combustion ceases and thrust collapses.

The third test series was in support of the first freejet tests of the strutjet engine to be conducted by NASA Lewis Research Center (LeRc) at their Hypersonic Test Facility (HTF) in Plum Brook. This facility has the capability to run simulated freejet flight conditions at Mach numbers of 5, 6, and 7 with a dynamic pressure of 1000 psf. All previous strutjet direct-connect tests were conducted at a dynamic pressure of 2000 psf or higher. In addition the strutjet

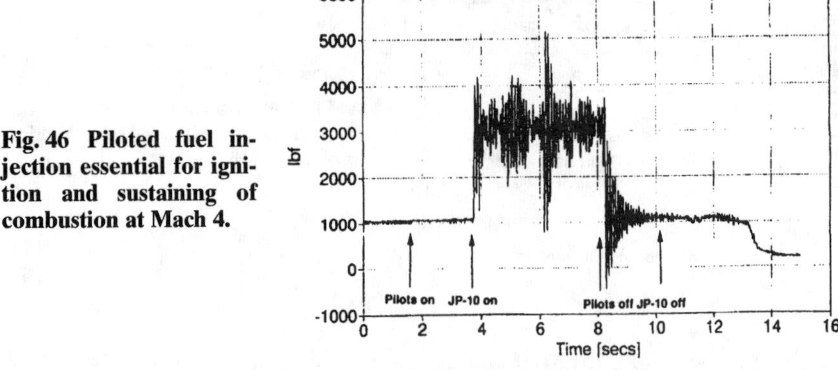

Fig. 46 Piloted fuel injection essential for ignition and sustaining of combustion at Mach 4.

testing had only been conducted at simulated flight conditions of $M = 0$–4 and 8. For the sake of risk reduction, additional direct-connect tests were conducted at Mach 6 and 7 at the reduced dynamic pressure. Figure 47 shows the test configuration and also the duct pressures achieved. These tests used the initial combustor divergence found efficient in the ducted-rocket/ramjet test series. By properly staging the pilot and the unheated liquid JP-10 injection, good combustion efficiency was achieved without reducing the duct divergence. This was a significant accomplishment because the employed engine geometry proved to be satisfactory for operation from Mach 0 to 8.

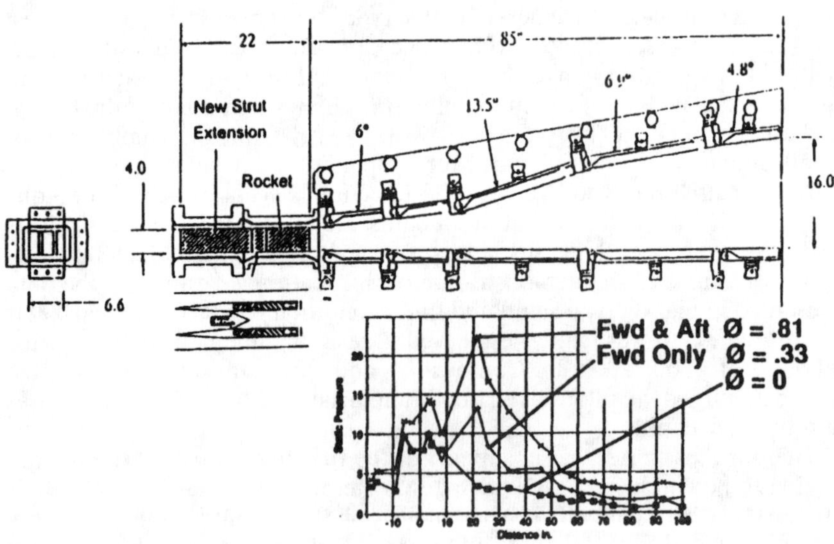

Fig. 47 Mach 7 combustion achieved high efficiency with rapidly expanding geometry.

Fig. 48 Freejet engine installed in wind tunnel.

Freejet Tests

The freejet engine shown in Fig. 48 was designed and constructed by Aerojet, then delivered to and tested at Plum Brook. Tests were conducted with the identical fuel-injection strategy used in the Mach 7 direct-connect tests already discussed. These tests demonstrated inlet starting, fueled unstart with a large forward-fueling split followed by inlet restart with shifting the fuel aft and substantial thrust increase. Figure 49 shows the internal pressure profiles and compares the freejet to the inlet data. Figure 50 shows the differential thrust produced as a function of the fuel flow. The slope break at an equivalence ratio of 0.55 is notable. This is the expected result

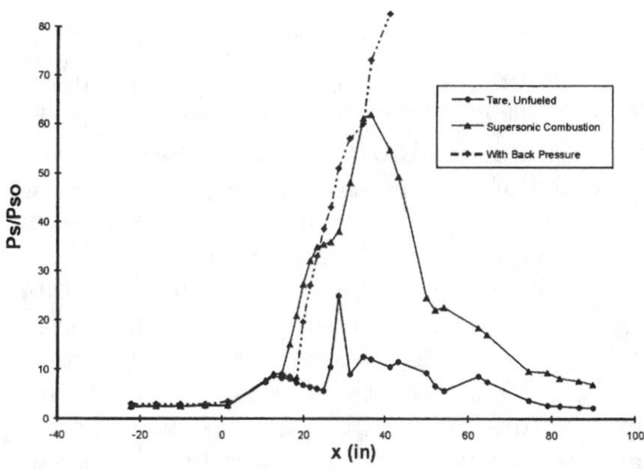

Fig. 49 Good agreement obtained between inlet only and freejet internal pressure measurements.

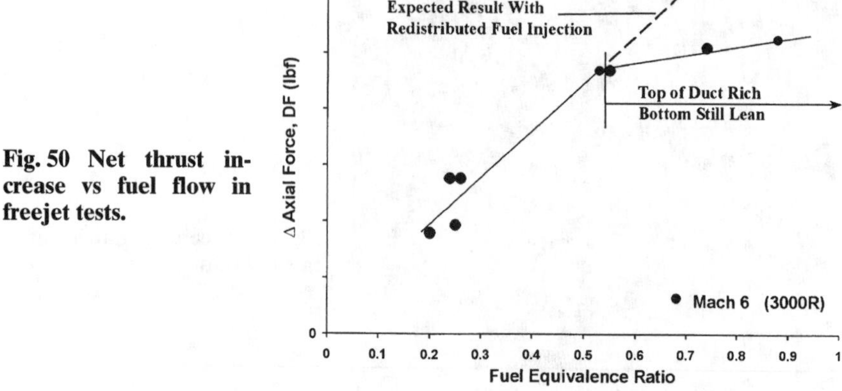

Fig. 50 Net thrust increase vs fuel flow in freejet tests.

of the nonuniform airflow distribution in the isolator. In the final test at HTF, the fuel-injection distribution was shifted to better match the airflow. Unfortunately the facility experienced a hot isolation valve failure that prematurely ended the test campaign.

B. Gaseous Hydrogen System Tests

All hydrogen systems considered for the strutjet engine use cryogenic hydrogen. This hydrogen is used to cool the engine regeneratively. During this cooling process, the hydrogen converts from a cryogen to a gas. All combustion-related processes of the strutjet engine will then use gaseous hydrogen as a fuel. Therefore, all combustion-related tests were conducted with gaseous hydrogen. The tests described here were executed as part of the Advanced Reusable Technology program sponsored by NASA Marshall Space Flight Center (MSFC) and supported by Aerojet with the objective to demonstrate the technology of the hydrogen-fueled RBCC strutjet engine previously described in Sec. II.

Inlet Tests

Figure 4 illustrates schematically the airflow path on the vehicle underbody. Forebody compression reduces the inlet approach Mach number. For example, if the freestream Mach number is 6.0, then the approach Mach number is reduced to 4.2. A model of the inlet-isolator test article is shown in Fig. 51. The inlet model accurately simulates the inlet from just upstream of the struts all the way through to the ram combustor. Geometrical similarity between the full-scale SSTO vehicle, capable of delivering 25,000 lbm payload to the International Space Station, and the subscale test article is maintained. The model is 6.8% scale of the full-size inlet. The inlet is preceded by a plate simulating the vehicle forebody boundary layer. Whereas the full-scale engine contains 8 to 16 struts and two sidewalls, the test article is composed of two struts and two sidewalls. In the model the sidewalls are

Fig. 51 Strutjet inlet in wind tunnel.

positioned to represent the symmetry plane between struts, thereby fully simulating flow around and between struts. The full-scale engine has two bleed locations, a forebody, and a throat bleed. To adjust for the nonlinear scale effects, the inlet subscale model has an additional strut bleed that removes excess boundary-layer buildup on the sidewalls. A throat plug is used to simulate combustion pressure increase.

The testing of this inlet in the NASA LeRC supersonic wind tunnel provided excellent results. Tests were performed over simulated flight Mach numbers from 3.6 to 8.1. The inlet started at all Mach numbers and exhibited unstart margins of over 20% in ramjet and 100% in scramjet modes. With forebody spill excluded engine capture efficiency exceeded 90%, which is remarkable, considering the wide operating range and large unstart margin realized with this inlet. The inlet started easily at all Mach numbers and generated excellent pressure rise, as shown in Fig. 52.

Strut-Rocket Tests

Aerojet and NASA MSFC completed proof-of-concept testing on a new strut-rocket injector element developed to enable operations under the

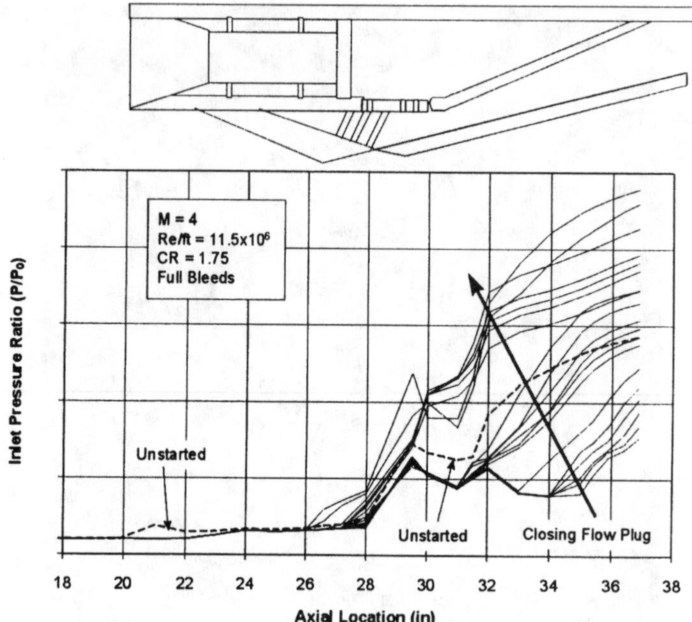

Fig. 52 Test data exhibit large unstart margin.

unique strut-rocket conditions. This element has been incorporated into the design of six subscale strut rockets for ducted-rocket testing.

The primary objective for a part of these tests was to demonstrate the performance and durability of a new injector that had been designed to maximize thruster efficiency while minimizing thruster length. The injector design ensures efficient mixing of the propellants within the chamber by employing an impinging element design and by utilizing a refined element pattern: 18 elements on the 0.5-in-diam. injector face. Figure 53 shows the injector firing in a single chamber in tests at Aerojet. Thermocouple data verified that the temperature of the injector face was within the limits predicted for the design. Post-test visual inspections of the test articles indicate that the injectors suffered virtually no erosive or other damage related to excessive face temperatures. Scanning electron microscope images of single elements and single orifices produced at MSFC are particularly encouraging. The data obtained during 16 hot-fire tests extend over the ranges indicated in Table 8.

Subscale strut rockets were fabricated and check-out tests were completed. In these tests the propellant, coolant, and ignition sequences required for RBCC operation were developed, and unaugmented rocket thrust was determined to establish a reference for subsequent combined-cycle testing. Aerojet is also currently fabricating a full-scale strut rocket using the same injector element. During the testing of this test article, laser

Fig. 53 Single rocket-chamber ignition test.

diagnostics will be used to measure the fuel distribution in the rocket exhaust for various design and operating conditions.

Direct-Connect Ram and Scramjet Tests

Aerojet successfully demonstrated the efficiency of this fuel-injection strategy at Mach 6 and 8. Figure 54 shows one of the test strut assembly for the direct connect campaign. Two struts are shown installed in the test rig in Fig. 55.

Three test series were conducted: ram and scram tests at stimulated Mach 6 flight conditions and scram-mode tests at Mach 8 conditions. High performance was achieved in all three test series in only 27 tests. A key to this success was the use of the cascade scram injectors in the forward location. The fuel split for each mode and Mach number was determined by the inlet tolerance to the combustion pressure rise. Excellent performance at stoichiometric conditions was demonstrated at each Mach number and mode tested by simple adjustment of the fuel flow to each of the three injectors. The data indicated that the performance of the cascade injector was even better than expected. Evidence of overpenetration suggested additional improvement can be made by increasing the number of cascades from four per strut side to five or more; this can be expected to increase the combustion rate and permit even shorter, lighter engine designs.

Table 8 Evaluation ranges of subscale strut-rocket injector

Chamber pressure, psia	Mixture ratio	Test duration, s
100–1800	4.76–7.0	1–5

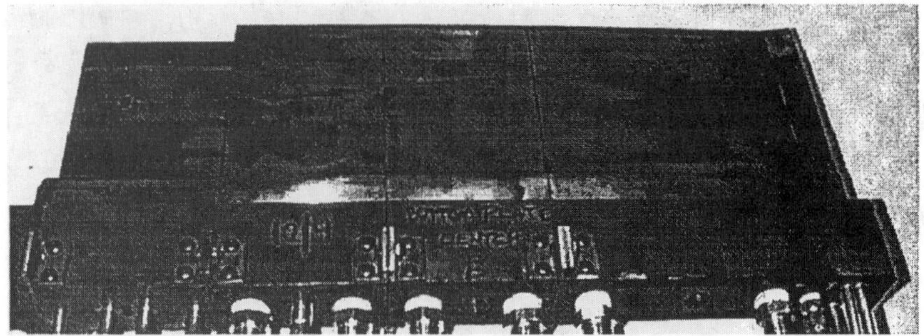

Fig. 54 Individual direct-connect strut.

Planned Tests

In the first two tests strut rockets, shown in Fig. 56, will be installed in the direct-connect duct to explore the performance in the scram/rocket and ascent/rocket modes.

A second pair of strut rockets will be installed in the new freejet engine and tested first under sea-level static conditions and then in the ducted-rocket and ramjet modes. Of particular interest is the transition from ducted-rocket to ramjet mode at a flight Mach number of about 2.5. An external view of the free et test hardware is shown in Fig. 57. To accommodate aerodynamic flow conditions inside the freejet test facility, the test hardware nozzle flap shown in Fig. 3 is integrated on the body side of the engine; in the flight engine the nozzle flap will be on the opposing cowl side.

The contemplated test facility has the capability to provide accelerating test conditions and this allows tests to begin at one Mach number and sweep

Fig. 55 Two struts in direct-connect rig.

Fig. 56 Strut rocket integrated into strut base.

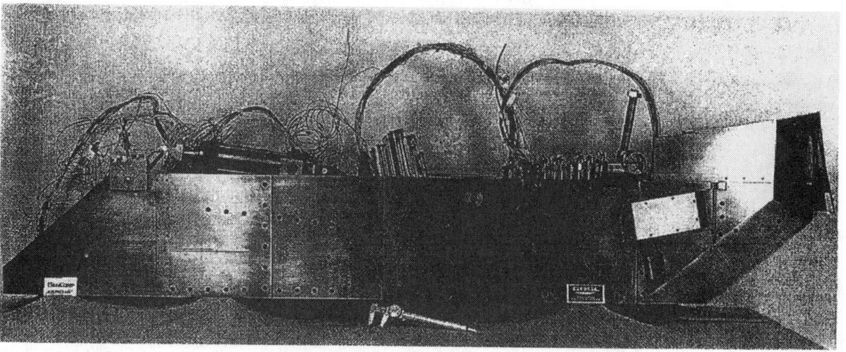

Fig. 57 Freejet engine test hardware.

continuously to a higher one while simultaneously matching pressure and enthalpy. Using this facility, it will be possible to demonstrate the ducted-rocket to ramjet and the ram to scram transitions.

C. Planned Flight Tests

The development of a strutjet-powered vehicle poses challenges above and beyond that of classical rocket vehicles. The most significant difference is that for RCC propulsion vehicle, an engine must be developed simultaneously with the flight vehicle. Flight demonstrations are essential because simulated flight conditions are costly or not at all obtainable on the ground. To reduce the development risk of such vehicles, we proposed the advancement of the technology readiness levels incrementally using two demonstration vehicles and engines as defined in Tables 9 and 10. With a length of about 35 ft, the first test vehicle TV-1 is about 1/16th scale relative to the full-scale SSTO vehicle already presented in Fig. 33.

Test Vehicle TV-1

A typical, small 35-ft RBCC X-plane is shown in Fig. 58. Its dry mass is about 24,000 lbm. This vehicle could be air launched from a B-52 or ground launched to test out all RBCC engine operating modes. The air-launched vehicle is sized for a maximum burnout Mach number of 10. A small payload bay (2 × 4 ft) is available for instrumentation. The configuration layout is

Table 9 Flight-test vehicle attributes

Attribute	Test vehicle: TV - 1	Test vehicle: TV - 2
Airframe demo objective	Propulsion and vehicle system technology test bed Vehicle system development Mid/high-speed controllability Propulsion integration	Flight weight engine Flight weight structure All vehicle systems Aircraft-like operations
Length	Small scale (30–40 ft)	Large scale (~200 ft)
Speed range	Ground launched: $M = 0$–7.8 Air dropped launched: $M = 0.8$–10.0	$M = 0$–10.0
Vehicle structure	Inexpensive Robust Advanced materials and structures	Flight weight materials and structures
Thermal protection	Robust baseline TPS Actively cooled leading edges and inlet ramps	Active and passive lightweight TPS
Demonstrations	Cryogenic tankage and propellant feed system Power generation Actuation systems Avionics	All flight weight systems Redundancy Reliability/maintainability

Table 10 Flight-test engine attributes

Attribute	Test engine: TE - 1	Test engine: TE - 2
Propulsion demo objective	Performance all modes all transitions Flight weight structure Controllability/operability	Performance All modes all transitions Flight weight structure Controllability/operability Reliability/maintainability
Rocket propellants	LOX/LH2	LOX/LH2
Ram/scram propellants	LH2	LH2
Engine structure	Lightweight Inexpensive Composite structure Advanced materials	Flight weight materials and structure
Engine thermal mgmt	Water cooled	Hydrogen cooled
Rocket	Uncooled	Hydrogen cooled
Leading edge/ inlet duct	Water cooled	Hydrogen cooled
Ram combustor		
Powerhead		
Ox preburner	Modified existing design	New full-scale design using hydrostatic bearings
Ox TPA	New subscale design using ball bearings	
Fu PB	Modified IPD design	
Rocket fuel TPA	New subscale design using ball bearings	
Ram/scram fuel TPA	New subscale design using ball bearings	
Other subsystems	bread broad controller Existing flight-type valves Electromagnetic valve and VG actuation Instrumentation as required	All components reflect high degree of life and robustness Reliability Maintainability

similar to the full-scale SSTO vehicle. There are two RBCC engines, each having a sea-level thrust of 9800 lbf, for a total ignition thrust of 19,600 lbf.

The two trajectories for the X-plane flights from a B-52 at 40,000-ft altitude and ground launch are compared in Fig. 59. The air-launched mission profile is flown at a maximum dynamic pressure of 1500 psf and achieves a burnout Mach number of about 10 at about 100,000-altitude. In contrast, the ground-launched mission profile is flown at a maximum dynamic pressure of 2000 psf and achieves a burnout Mach number of about 7.8 at 83,000-ft altitude. These two trajectories were computed using the OTIS (OPTIMAL TRAJECTORIES by IMPLICIT SIMULATION) code developed by The Boeing Company.

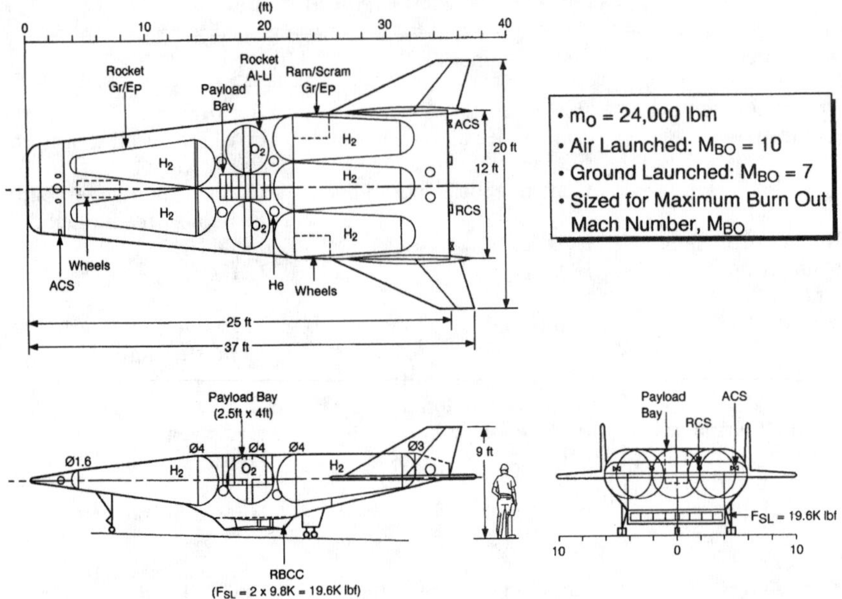

Fig. 58 TV-1 X-Plane preliminary layout.

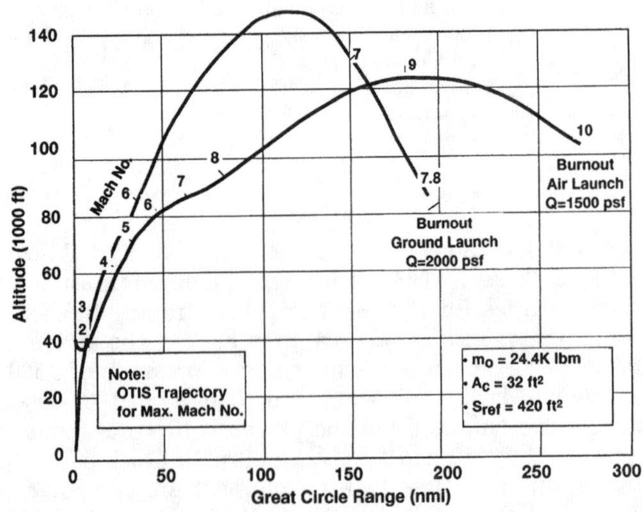

Fig. 59 TV-1 trajectories optimized for maximum Mach number.

The engines required for this 35-ft vehicle will focus on thrust, specific impulse performance, and controller characteristics. They will be lightweight; however, because of their small size the demonstration of thrust-to-mass ratio and life capability are not the objective of these tests. Control functions to be developed involve variable geometry for inlet and nozzle, inlet bleed valve control, and rocket and ram-scram propellant flow control.

Test Vehicle TV-2

This vehicle can be similar in length to the full-scale SSTO vehicle shown in Fig. 32, except it will only be powered by one full-size strutjet engine with 346,000 lbf of thrust, which allows the vehicle to be about 40% narrower from tip to tip. Because the purpose of this vehicle is to demonstrate reusability and low maintenance operation, it will be structurally very similar to the final vehicle.

V. Maturity of Required Strutjet Technologies

The maturity of RBCC engines is best defined in terms of a Technology Readyness Level (TRL) as defined by NASA. The following assessment is as of Summer 1998 relative to a TRL of 6, which requires technology demonstration in a relevant environment be it simulated on the ground or actually flown. Table 11 categorizes various aspects of an RBCC engine into the degree of maturity relative to TRL 6.

The table shows that a large amount of development must be accomplished before a TRL of 6 is achieved for the strutjet RBCC engine. With this assessment in mind, it may not be prudent to risk at this point in time the commitment of a large amount of resources toward the exclusive RBCC approach to achieve low cost access to space. Alternate approaches, like further maturation of all rocket propulsion, should be employed. However, the potential of RBCC engines is so overwhelming that it is also not prudent to casually dismiss the opportunity to exploit this option.

VI. Summary and Conclusions

Strutjet propulsion has the potential to provide substantial engine robustness design margins over an all-rocket system. These design margins can be used to reduce operating costs through reduced maintenance activities of the highly reusable system and the reduction in engine replacement cost possible through a significantly smaller vehicle and horizontal takeoff. Using advanced technologies, the achievable thrust-to-mass ratio of the engine is about 35 lbf/lbm. Making use of the available robustness design margin, this number can be reduced to about 23 lbf/lbm. Runway takeoff provides a reliable preairborne propulsion health verification coupled with an operational attractive abort mode.

A strutjet-powered SSTO HTHL design has the potential to also offer structural robustness design margins over an all-rocket SSTO VTHL vehicle. Results show that, prior to adding robustness, the RBCC vehicle gross takeoff mass is about 44% of the all-rocket case, and the RBCC vehicle dry

Table 11 Degree of technology maturity relative to TRL 6

Top-level technology item	Subtier technology item	Degree of technology maturity: Low	Medium	High
Flowpath integration	Design point selection	—	x	—
	Balanced performance along Flow path	—	x	—
Fuel injection	Penetration, mixing, vaporization	—	x	—
	Stable combustion	—	x	—
	Controlled heat release	—	x	—
Lightweight structure	High-temperature materials	—	—	x
	Radiation-cooled structures	x	—	—
	Regeneratively cooled structures	x	—	—
	Endothermic fuel reactions	—	x	—
	Closed-loop cooling	x	—	—
Strut rockets	High thrust-to-mass	—	—	x
	Long life	—	x	—
Turbopumps	Low weight	x	—	—
	Long life bearings	—	x	—
	Low-temperature turbines	—	x	—
	Multimode operation	x	—	—
Engine controls	For all flight modes	x	—	—
Testing	Ground-test subscale	—	x	—
	Ground-test full scale	x	—	—
	Captive flight-of-engine module	x	—	—
	Self-Powered flight-of-flight type engine	x	—	—

mass is about 68% of the all-rocket case. Adding structural robustness to the RBCC vehicle, by for example increasing the factor of safety from 1.5 to 1.68 and using an engine with a thrust-to-mass ratio of 25 lbf/lbm, the resulting dry mass does not exceed that of an all-rocket vehicle. The gross takeoff mass (1.7 million lbm) increases to only about 63% of the all-rocket level (2.7 million lbm) while keeping the transition Mach number at 10. Because the required thrust-to-mass ratio of the strutjet HTHL vehicle is only 0.6 vs the corresponding ratio of 1.3 for the all-rocket vehicle, the strutjet vehicle with its increased robustness benefits from a thrust reduction of 73%.

If the all-rocket vehicle landing system were designed for a total liftoff abort mass with fully loaded tanks, even greater robustness could be built into the RBCC vehicle before dry-mass equivalence is reached. Also, the RBCC robustness design margins would increase slightly for higher transition Mach numbers and higher engine thrust-to-mass ratios, provided that no added cooling requirements are needed for the TPS subsystems. Therefore, the RBCC vehicle design has added structure and engine robustness design margins available to become equivalent to an all-rocket vehicle's total dry mass under the same design ground rules.

A. Hydrogen and Hydrocarbon Strutjet Engines

These engines, particularly the hydrogen-fueled one, have made considerable progress, and test data established to date verify their fundamental feasibility and support earlier performance predictions. Whereas the current NASA Advanced Reusable Technology (ART) Program as well as Aerojet-sponsored test activities provided additional data during 1998 and early 1999, no further design or test activities are currently planned for storable hydrocarbon RBCC engines.

B. Strutjet Technology Maturity

Although significant achievements have been made toward a Technology readiness Level of 6 demonstration in a relevant environment, most of the effort to date is focused on flow-path development and performance assessment. Future work must be done on engine structure, thermal management, and propellant feed system. Ground tests of flight-type engines are mandatory before committing these advanced engines to in-flight, captive-carry or self-powered, evaluation.

C. Overall Recommendation

It has been shown that strutjet propulsion, with its superior robustness, has the potential to reduce maintenance, operations, and engine replacement cost. However, at this point in time, no specific cost figures are obtainable because of the unavailability of sufficiently refined engine and vehicle designs, the lack of detailed maintenance procedures and operational scenario definitions, and above all, the absence of a representative, generally accepted operational cost model. Current models use the vehicle dry mass as the figure of merit for cost-operating cost determinations. For the RBCC case this method would be invalid because the RBCC system with its intentionally increased dry mass is substantially more operations cost effective than the reference all-rocket system with equal mass. The establishment of a representative and credible cost model is needed before the cost benefits of robust RBCC propulsion can be quantified.

References

[1] Escher, W. J. D., and Flornes, B. J., A Study of Composite Propulsion Systems for Advanced Launch Vehicle Applications, Main Technical Report, Volume 2, NAS7-377, April 1967

[2] Billig, F.S., and Van Wie, D.M., "Translating Cowl Inlet with Retractable Propellant Injection Struts," Johns Hopkins University, U.S. Patent No. 5,214,914, June 1, 1993.

[3] Flornes, B.J., and Stroup, K.E. "1964 Advanced Ramjet Concepts Program, Volume 1, Advanced Jet Compression Engine Concepts," USAF Aero Propulsion Laboratory Report No. AFAPL-TR-65-32, WPAFB, Dayton, OH, May 1965.

[4] Bulman M.J., "Scramjet Injector," Aerojet, U.S. Patent No. 5,220,787, June 22, 1993.

Chapter 12

Liquid Hydrocarbon Fuels for Hypersonic Propulsion

Lourdes Maurice* and Tim Edwards†
Air Force Research Laboratory,
Wright-Patterson Air Force Base, Ohio
and
John Griffiths‡
University of Leeds, Leeds, England, United Kingdom

Nomenclature

A = collision frequency factor
C_p = heat capacity at constant pressure
E = activation energy
i- = iso isomer
I_{sp} = specific impulse
K = equilibrium constant
k = rate constant
k_f = forward-rate constant
k_r = reverse-rate constant
M = passive collision partner
n = temperature dependence exponent
n- = normal isomer
PAH = polycyclic aromatic hydrocarbon
R = ideal-gas constant
t_1 = timescale from chain branching
t_2 = timescale from chain branching
v = reaction rate
x = fraction of fuel reacted
τ = residence time
φ = net branching ratio

---

*Air Force Deputy for Propulsion.
† Senior Chemical Engineer.
‡ Professor.

I. Introduction

The composition of aircraft hydrocarbon fuels is generally determined by specifications that are primarily based upon operational requirements. The specifications control seven general characteristics: volatility, fluidity, composition, combustion, corrosion, stability, and contamination levels. These properties influence important operational considerations such as range potential, operability, system maintenance requirements, and safety. Moreover, cost and availability are always of paramount importance in the selection of a jet fuel. Selected properties and cost of typical aviation fuels are shown in Table 1.

The evolution of jet fuels from the early days of the gas turbine engine to the present era of JP-8 and JP-5 for military, and Jet A for commercial aviation has been traced in an informative review article by Lander (1997). The author recounts that the first jet fuel was aviation gasoline (avgas). However, researchers realized early on that avgas would have to be altered to meet the requirements of jet-powered aircraft. The high-volatility gasoline caused fuel-vapor loss from vented tanks at high altitudes and pump cavitation problems. The light components of avgas also contributed to poor lubrication, which led to increased wear and reduced the life of fuel metering pumps. Also, octane-enhancing antiknock additives, which contained lead, eroded the turbine blades and caused maintenance problems in the engine hot section. It became increasingly evident that the heavier fraction of petroleum, not used for aviation gasoline, resulted in very attractive jet fuels. Heavier crude fractions offered increased volumetric heating value and consequently increased flight range. Moreover, the fast-igniting and clean-burning heavy paraffins found in the kerosene or middle distillate fraction of petroleum were ideally suited for the gas turbines. Also, the higher hydrogen content resulted in less radiant flames and, thus, increased combustor life. Thereby, the basic characteristic of jet fuels were evolved; detailed specifications and properties may be found in sources such as the *Aviation Fuel Properties Handbook* (Coordinating Research Council, 1983).

The primary purpose of jet fuel is clearly to provide propulsive energy. However, a secondary use for jet fuel soon becomes apparent. The need to cool hot parts in the engine has been evident from the invention of the internal combustion engine. As engine performance goals and flight speeds increased, the need for thermal management became apparent. Compressor bleed air is commonly used to provide cooling of current aircraft subsystems. The compressor bleed air is naturally hotter than ambient air because of mechanical compression. However, air stagnation temperature increases as a function of flight speed because of associated aerodynamic heating. Therefore, neither compressor bleed nor inlet air is an effective coolant at hypersonic flight speeds [e.g., Keirsey, 1992; Wiese, 1992; Ianovski, 1993; Sobel and Spadaccini, 1997]. The fuel used for propulsion is an attractive coolant and can provide a heat sink through sensible heating and latent phase change. Unfortunately, *specific heat,* as well as *thermal stability* considerations, limit its usefulness as a coolant. The specific heat of a fuel is the amount of heat energy transferred into or out of a unit mass of the fuel when increasing or

Table 1 Properties of typical aviation fuels[a]

Fuel	Formula	Freeze point C	Flash point C	Net heating value, kJ/m^3	Viscosity at −40C, mm^2/s	Cost,[b] $/liter	Heat sink
Hydrogen	H_2	−259.2	Gaseous	8,133	0.0084 (0°C)	16.33	high
JP-7	$C_{12}H_{24}$ (avg)	−44	63	34,423	17	0.8	↓
JP-8	$C_{12}H_{22}$ (avg)	−50	38	34,674	10	0.24	↓
JP-10	$C_{10}H_{16}$	−79	54	39,441	19	2.94	low
		↑ operability	↑ safety	↑ range	↑ operability	↑ economics	↑ cooling

[a]Hydrogen included for comparison. [b]Price paid by U.S. Air Force in 1998.

decreasing its temperature. Thermal stability of a fuel is described as its tendency to resist chemical decomposition at high temperatures, which leads to the formation of undesirable products [Coordinating Research Council, 1979; Hazlett, 1991]. The conclusion has been made that the limiting operational bulk temperature for (any) liquid hydrocarbon in aircraft was 435 K (325°F) (Croswell and Biddle, 1992). This design temperature limit was set in the fuel systems of all commercial and military jet aircraft. The temperature barrier was broken in the 1960s with the development of the low-volatility JP-7, a heavily refined fuel developed specifically for the SR-71 aircraft. JP-7 has a thermal stability limit of 560 K (550°F) (Croswell and Biddle, 1992), which translates to a flight Mach-number limit of about 4.6 (Wiese, 1992) at a typical operating flight altitude. A thermally stable kerosene jet fuel [jet propellant thermally stable (JPTS)] with a bulk temperature capability of 492 K (425°F) has also been developed for the U-2 aircraft. The cost of both JP-7 and JPTS is substantially higher than that of JP-8. Hence, to meet the thermal management needs of the present and near-term fleets while maintaining the current availability at minimal cost, the U.S. Air Force embarked on a program to enhance the thermal stability of JP-8 via the development of an additive package. JP-8 with the additive package is referred to as JP-8 + 100 [e.g., Heneghan et al., 1996a]. The additive package allows the bulk fuel temperature to increase from 435 to 492 K (325 to 425°F) without generating deleterious fuel system deposits, providing an increase in heat sink capacity of 50%. Hundreds of additives have been tested and a package formulated that contains a dispersant, an antioxidant, and a metal deactivator. The efficacy of the package has been demonstrated in multiple fuels, test rigs, full-scale component tests, and engine tests, and the packages have been evaluated for fuel-material compatibility. Several base-level demonstrations have documented maintenance cost reductions, and fleet-wide conversion to JP-8+100 is underway.

Since the 1960s, the promise of hypersonic flight has both challenged and evaded researchers. Both reusable launch vehicles and reusable high-speed cruise air vehicles capable of speeds greater than Mach 5 may provide attractive commercial and military payoffs. Moreover, the development of high-speed missiles is recognized as a fundamental military advantage (Henderson, 1997). Typical hypersonic application classes include expendable and reusable air vehicles and transatmospheric vehicles. Billig (1991) has extensively discussed these three classes of hypersonic vehicles, and it is apparent that propulsion choices for these systems are dependent on many factors. The performance of potential candidates, i.e., turbojets, ramjets, and scramjets, varies markedly as a function of flight speed (Fig. 1). Furthermore, engine thrust-to-weight ratio is also an important performance consideration, and attractive propulsion solutions for hypersonic systems may require combined-cycle systems (Curran, 1991). Consequently, viable fuel options must consider the requirements of this broad range of propulsion systems.

A cryogenic fuel such as hydrogen is an attractive propellant for hypersonic systems both from energetic and vehicle coolant perspectives. The constant-pressure specific heat C_p of typical jet fuels at 344 K (160°F) lies in

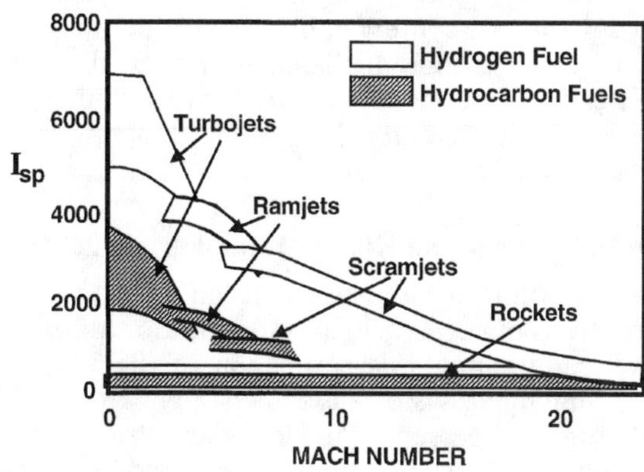

Fig. 1 High-speed propulsion operating regime.

the range from 2.15 to 2.30 kJ/kg K (0.51 to 0.55 Btu/lb °F). By contrast the constant-pressure specific heat of hydrogen at 343 K (158 °F) is 14.38 kJ/kg K (3.44 Btu/lb °F). However, hydrogen's low energy density may result in aerodynamically unattractive vehicles. Cryogenic propellants also pose logistic and safety concerns, as evidenced by ground personnel requirements of cryogenic vs solid propellant rockets. Moreover, availability and cost considerations (Table 1) presently limit the use of hydrogen to specialty applications. Hydrocarbon fuels have been shown to be competitive with cryogenic hydrogen for hypersonic cruise and two-stage-to-orbit (TSTO) air-breathing applications because of their much higher density and logistical benefits (e.g., Lewis and Gupta, 1997; Palaszewski et al., 1997). Furthermore, all-hydrocarbon orbital launch vehicle concepts might potentially provide payload fractions in the 3–4% range, a marked improvement over present shuttle class vehicles (Leingang et al., 1996), albeit with an associated increase in complexity.

Despite the recognized benefits of hydrocarbon fuels for hypersonic applications, these fuels clearly exhibit poorer specific impulse (I_{sp}, see Fig. 1), heat sink, combustion properties, and thermal stability than hydrogen. Therefore, improvement of the three latter capabilities, as well as addressing materials compatibility issues, are crucial for enabling the use of hydrocarbon fuels for hypersonic applications.

In the rest of this chapter, the authors' intent is to focus on fuel issues for hypersonic vehicles that are primarily inherent to hydrocarbon fuel chemistry. We also seek to acquaint the reader briefly with the key operational issues and provide an assessment of the current state of the art of implementation of liquid hydrocarbon fuels in hypersonic propulsion systems. It is our goal to provide a first resource that will lead the interested student to more detailed references. Firstly, the fuel heat-sink requirements of hypersonic vehicles are discussed, and the concept of endothermic fuels is introduced. A detailed discussion of the characteristics of these fuels is provided. Fuel system challenges are subsequently addressed, focusing on fuel thermal stability considerations. Finally, the fundamental combustion chemistry of hydrocarbon fuels is discussed within the context of hypersonic fuel systems, and combustor development considerations are addressed from the perspective of the fuel system designer. An appendix discussing issues inherent to chemical kinetic models for complex multi-dimensional reacting flow computations is also included.

II. Fuel Heat-sink Requirements and the Role of Endothermic Fuels

The heat-sink requirements of a fuel have long been recognized, and there is generally qualitative agreement between various system trade studies. A plot of typical fuel heat-sink requirements as a function of flight Mach number for both manned and unmanned systems is shown in Fig. 2 (Lander and Nixon, 1971). A similar estimate is presented in Heiser and Pratt (1994) for *only* engine cooling demand, and consequently their required heat sink at a given Mach number is somewhat lower than shown in Fig. 2. The fuel heat sink required is dependent upon the engine design and size, cooling scheme design, structural materials selected, and cooling required by the airframe and other subsystems (such as avionics and flight controls). The heat capacity C_p and thus the sensible heat sink potential ($C_p\Delta T$, where ΔT is the available rise in bulk fuel temperature) of liquid hydrocarbon fuels computed is significantly lower than that offered by hydrogen as already discussed. However, it was recognized in the late 1960s and early 1970s that heat sink in a fuel could be derived from two sources: the traditional *sensible heat sink* derived from heating of the fuel, as well as a *chemical heat sink* induced via fuel *endothermic* reactions (Lander and Nixon, 1971).

At first glance there appears to be a contradiction between the terms *endothermic* and *fuels.* The outcome of a chemical reaction is either absorption or release of energy, and the latter is clearly essential to generate propulsive thrust. However, in the context of jet fuels, *endothermic* pertains to a class of reactions that take place in the fuel system *prior* to ignition. Endothermic fuels feature an additional heat sink over that of conventional fuels by undergoing heat-absorbing chemical reactions that are supported by energy extracted from heated air, (i.e., aerodynamically heated inlet air or compressor bleed). The extent of the necessary endothermic reaction can

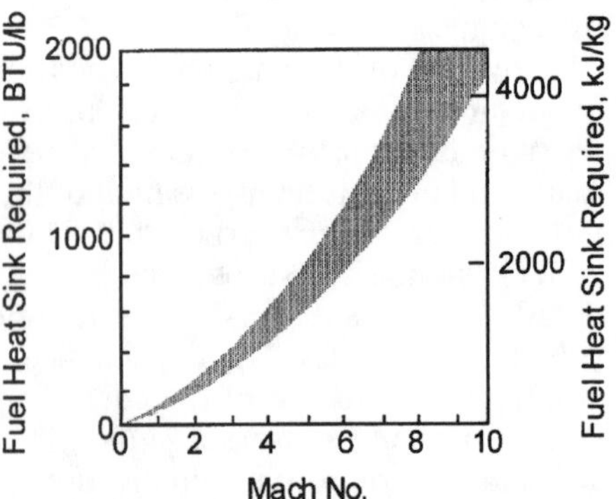

Fig. 2 Heat sink required as a function of Mach number (Lander and Nixon, 1971). Lower bound—missiles and upper bound—aircraft.

vary from 0 to nearly 100% over the flight envelope as a function of cooling load requirements.

Endothermic fuel reactions can enable the use of air-breathing propulsion at hypersonic flight speeds without resorting to cryogenic propellants (Lander and Nixon, 1971; Nixon, 1986, Wiese, 1992; Ianovski, 1993; Sobel and Spadaccini, 1997; Edwards, 1996b). Endothermic fuels were also being considered for application to gas turbines for the High Speed Civil Transport (Glickstein and Spadaccini, 1997). The relationship between fuel temperature and heat sink for thermally cracking of a kerosene-class hydrocarbon is shown schematically in Fig. 3 (Sobel and Spadaccini, 1997; Lander and Nixon, 1971). Typical kerosene-class fuels yield a total sensible heat sink of approximately 2300 kJ/kg (1000 Btu/lb) for fuels heated from ambient to 1000 K (1340°F) (Lander and Nixon, 1971). By contrast, the heat sink for hydrogen under comparable conditions is 10,000 kJ/kg (4300 Btu/lb).

A. Characteristics of Endothermic Fuels

The chemical compositions of endothermic fuels have not been precisely defined to date. Nevertheless, it is apparent that these specialty fuels will comprise saturated linear and cyclic hydrocarbon molecules featuring at least seven carbon atoms. There are several options in selecting an endothermic fuel, as summarized in Table 2. Moreover, if heat sinks of candidate fuels are composed in terms of propulsive energy (heat of combustion), hydrocarbons clearly became competitive with hydrogen (Table 3). This data is consistent with that of Favorskii and Kurziner (1990), who report comparable ratios of heat sink to combustion at 1073 K (1471°F) for liquid H_2 (0.127) and hydrocarbon fuels (0.097). The thermal decomposition reactions of hydrocarbon fuels feature a substantial activation energy barrier and hence offer a potential heat sink. Moreover, the resulting heat-of-combustion of decomposition products remain essentially unchanged or slightly increased. Many types of fuel decomposition reactions have been considered for endothermic applications, including dehydrogenation, dehydrocyclization, dedimerization, cracking, and steam reforming. Reaction classes and potential endotherms are defined in Table 2.

These reactions may be effected via either catalytic or thermal means. Whereas thermally driven endothermic reactions reduce fuel system complexity, catalytically driven reactions offer enhanced reaction product selectivity and lower operational temperatures (Lander and Nixon, 1971; Sobel

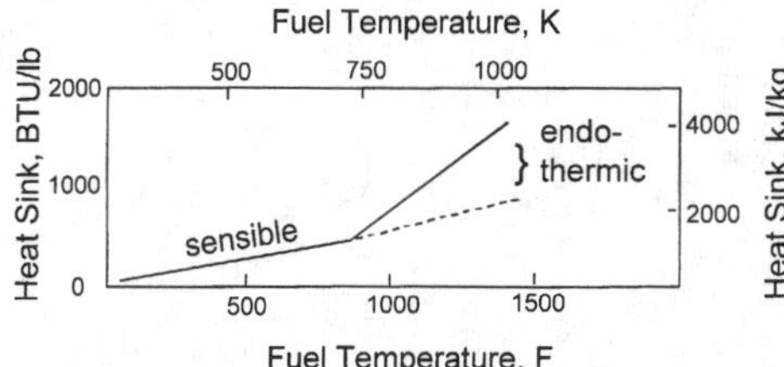

Fig. 3 Notional diagram of fuel heat sink as a function of temperature.

Table 2 Chemical heat-sink values for various endothermic fuels (Heat sink for kerosene calculated using dodecane heat of formation as an approximation.)

Endothermic reaction	Reaction types	Theoretical chemical heat sink, kJ/kg fuel reacted	Calculated heat of combustion of endothermic products, kJ/kg fuel reacted
C_7H_{14} (methylcyclohexane) $\rightarrow$ C_7H_8 (toluene) + 3 H_2	Dehydrogenation	2,190	45,800
C_7H_{16} (*n*-heptane) $\rightarrow$ C_7H_8 (toluene) + 4 H_2	Dehydrocyclization	2,350	47,300
$C_{10}H_{10}$ (dicyclopentadiene) $\rightarrow$ 2 c-C_5H_5	Dedimerization	621	43,630
$C_{12}H_{24}$ (kerosene) $\rightarrow$ C_2H_4 (ideal)	Cracking	3560	47,200
$C_{12}H_{24}$ $\rightarrow$ CH_4, C_2H_4, C_2H_6, etc. (actual)	Cracking	<<3,500	
$C_{12}H_{24}$ + 6 H_2O $\rightarrow$ 9 CH_4 + 3 CO_2 (AJAX stage 1)	Steam reforming		
CH_4 + H_2O $\rightarrow$ CO + 3 H_2 (AJAX stage 2)	Steam reforming	net: (stage 1 + 2)	net: (stage 1 + 2)
Net: $C_{12}H_{24}$ + 15 H_2O $\rightarrow$ 9 CO + 3 CO_2 + 27 H_2		5,490[a]	21,240
$2NH_3$ $\rightarrow$ N_2 + 3 H_2	Dehydrogenation	2,720	19,280
CH_3OH $\rightarrow$ CO + 2 H_2	Dehydrogenation	4,000	20,420
2 CH_4 $\rightarrow$ C_2H_2 + 3 H_2	Addition-dehydrogenation	11,765	62,860
Benzene (C_6H_6) $\rightarrow$ 3 C_2H_2	Ring fracture of aromatics	7,650	48,280
Decalin ($C_{10}H_{18}$) $\rightarrow$ napthalene ($C_{10}H_8$) + 5 H_2	Dehydrogenation	2,210	40,700

[a] Heat sink for steam reforming on total propellants (fuel + water).

Table 3 Cooling capacities of candidate endothermic fuels as a function of propulsive efficiency (heat of combustion)

Fuel	Heat of combustion kJ/kg	Total heat sink (chemical + physical) at 1000 K (1340°F), kJ/kg	Ratio of heat sink/heat of combustion
H_2	128,823	13,931	0.108
Norpar 12	43,654	3,950	0.090
C_7H_{14}(MCH)	45,800	5,074	0.110
CH_3OH	20,420	5,897	0.289

and Spadaccini, 1997). *Dehydrogenation* and *cracking* have received the most attention by far both in the U.S. and Russia (Lander and Nixon, 1971; Ianovski, 1993). Consequently, these reactions are discussed next in some detail.

Naphthenic (cycloparaffin) fuels can be dehydrogenated to an aromatic and hydrogen. The classic example of this type of reaction is the catalytic dehydrogenation of methylcyclohexane (MCH, c-C_7H^{14}) to toluene (C_7H_8) and hydrogen (Lander and Nixon, 1971; Petley and Jones, 1992).

$$c\text{-}C_7H_{14} \rightarrow C_7H_8 + 3H_2$$

This reaction provides an endothermic heat sink of 2190 kJ/kg (940 Btu/lb).

Supported platinum catalysts in packed beds have been used to demonstrate chemical heat sinks near the theoretical value of 2190 kJ/kg (940 Btu/lb) for dehydrogenation of methylcyclohexane, and heat fluxes up to 170 W/cm^2 (150 Btu/ft^2-s) have been achieved. The reaction yield could approach near 100%, dependent upon the amount of heating supplied. The U.S. Air Force sponsored large efforts at the Shell Development Corporation in the 1960s and early 1970s and Allied-Signal Aerospace in the 1980s and early 1990s to study catalytic dehydrogenation of fuels for advanced systems (Lander and Nixon, 1971 and references therein; Lipinski and White, 1992; Johnson et al., 1992). The Allied-Signal work culminated in the testing of a coupled endothermic fuel/air heat-exchanger reactor (see Fig. 4) and combustor (Lipinski and White, 1992).

Dehydrogenation reactions are favored at temperatures lower than those required for cracking, thus allowing the use of lighter, more common structural materials. However, dehydrogenation reactions are usually carried out in the presence of supported platinum catalysts, which are readily subject to poisoning, and the reactions generally result in the formation of aromatics. Aromatics are well known to augment soot formation and increase radiative heating of the combustor liner. Hence, the use of dehydrogenation is presently receiving less emphasis.

Another plausible reaction that affords a high endothermic heat sink is

Fig. 4 Endothermic fuel/air heat exchanger/reactor (Lipinski and White, 1992).

cracking of liquid hydrocarbon fuels (Lander and Nixon, 1971; Ianovski, 1993; Sobel and Spadaccini, 1997). The U.S. Air Force recognized early the potential of thermal cracking as an aircraft heat sink, sponsoring work at Monsanto Co., Inc., to determine fundamental cracking rates for pure hydrocarbons under pressures, temperatures, and flow rates typical of aircraft fuel systems (Fabuss et al., 1962; Fabuss et al., 1964). The cracking reaction will naturally occur at higher temperatures than dehydrogenation, or it may be enhanced through the use of catalysts or chemical initiators. The endothermicity of this reaction depends strongly on the products produced. For example, the cracking reaction of *n*-decane may proceed via a number of paths producing hydrogen and a variety of smaller hydrocarbons. The realizable theoretical endothermicity for *n*-decane cracking is dependent on the products of reaction. The production of acetylene affords the maximum theoretical endothermic heat sink of 9800 kJ/kg (4200 BTU/lb).

$$n\text{-}C_{10}H_{22} \rightarrow 5C_2H_2 + 6H_2$$

Unfortunately, acetylene production requires reaction temperatures above 1000 K (1340°F), which exceed reactor and catalyst materials limits, and demands unrealizable catalyst product selectivity (the ability of the catalyst to enhance the formation of desired products) (Sobel and Spadaccini, 1997).

The production of ethylene, a more saturated compound, reduces the theoretical endothermic heat sink to 3600 kJ/kg (1545 Btu/lb).

$$n\text{-}C_{10}H_{22} \rightarrow 5C_2H_4 + H_2$$

The temperature necessary for equivalent conversions is lower, ensuring the survival of the heat exchangers and catalysts, but catalyst product selectivity remains problematic. Finally, production of saturated compounds, which feature negative heats of formation, further decreases the endothermic heat sink.

In reality, cracking produces a mix of products from both endothermic and exothermic reactions. The products are dependent both on catalyst product selectivity and reactor operating conditions such as temperature, pressure, and residence time. The endothermic cracking of a normal paraffin blend with average carbon number 12 (Norpar 12) in the presence of inexpensive, commercially available zeolite catalysts at 1000 K (1340°F) has been investigated experimentally by Sobel and Spadaccini (1997). The product efflux consisted primarily of low molecular weight alkenes and alkanes (ethylene, propene, propane, ethane, and methane) and hydrogen. An actual total heat sink of approximately 3950 kJ/kg (1700 Btu/lb) at 1000 K (1340 °F) was measured; the chemical heat sink was 1740 kJ/kg (750 Btu/lb). At higher temperatures both increased sensible and chemical heat sinks can be obtained. Ianovski (1993) showed a total heat sink of 5000 kJ/kg (2150 Btu/lb) at 1170 K (1650°F) for cracking of endothermic fuel T-15, a fuel that has properties that "are not distinguished from the standard jet fuels such as Russian T-6 or American JP-7." The resulting products from all experiments featured attractive combustion characteristics.

Thermal cracking is generally regarded as less endothermic (less selective to endothermic products) than catalytic cracking (Lander and Nixon, 1971; Sobel and Spadaccini, 1997), although the net heat sink is a strong function of reaction conditions, especially residence (reaction) time. Endothermic cracking reactions are favored by short residence times. For example, thermal cracking of jet fuels in static reactors produces primarily saturated products (methane, ethane, propane) (Lai et al., 1992) with the result that the reactions are net exothermic. Thermal cracking of hydrocarbons to produce ethylene is carried out commercially on an enormous scale (Crynes and Albright, 1987). In the commercial processes, operating at near-atmospheric pressures with large tubes and substantial steam dilution maximizes ethylene yield. Regrettably, these options are not feasible in hypersonic fuel systems.

Other alternatives have been presented to the standard cracking and dehydrogenation endothermic reactions. For example, Gurijanov and Harsha (1996) have discussed the Russian AJAX concept of steam reforming of liquid hydrocarbon fuels for hypersonic vehicles, as depicted in Fig. 5 and outlined in Table 2. The concept is essentially a two-stage steam reforming process with individual optimization of catalysts and temperatures for each stage. The first stage is steam cracking of the kerosene fuel to form primarily methane; the second stage is the highly endothermic steam reforming of methane to form CO and H_2. Steam reforming offers a very large heat-sink potential at the expense of combustor heat release. However, there is also a vehicle range penalty caused by the water that must be carried onboard.

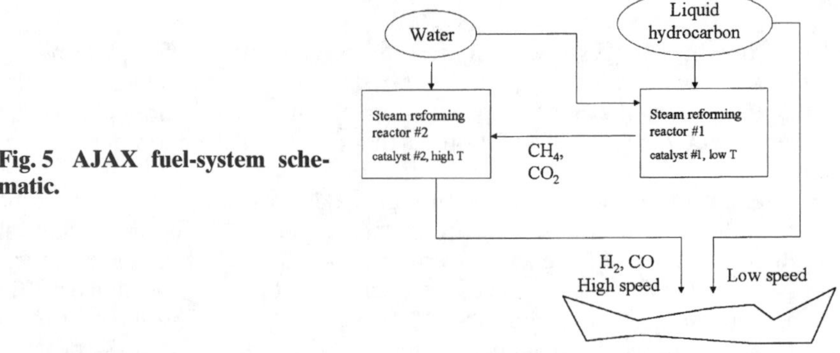

Fig. 5 AJAX fuel-system schematic.

Other nonhydrocarbon fuels, such as methanol and ammonia, could conceivably be used for applications where heat-sink requirements are more pressing than heat release. Another example of a highly endothermic fuel is the reaction of methane to form acetylene and hydrogen (Zubrin, 1992). However, one must recognize that these reactions will not proceed to 100% completion as written in Table 2, but instead yield other intermediate products with associated lower heat sinks. The endothermic reaction selection for the first fielded liquid hydrocarbon-fueled hypersonic vehicle will most likely be either cracking or dehydrogenation. Whichever approach is ultimately chosen may well depend on the combustion properties of the resulting reaction products, as discussed below.

B. Fundamental Considerations of Heat Removal

In the present context the rate of heat removal through fuel reactions is also of interest. Generally, thermal cracking is approximated as first order, where conversion (reaction) is related to residence time via

$$kt = ln\,[1/(1-x)] \tag{1}$$

where k is the rate constant [function of temperature, $k = Ae^{-(E/RT)}$] and t is the residence time. There is a substantial body of rate data in the literature for thermal cracking because of its industrial importance. An endothermic fuel system designer has to balance the heat-sink capability of the fuel with the heat-transfer requirements of the system. The heat-transfer coefficient to the fuel can be modeled by the standard Nusselt number correlation. As the heat flux to be absorbed increases, the fuel velocity must also increase with a corresponding decrease in residence time.

For catalytically driven fuel reactions the Arrhenius-rate expressions differ from those governing thermal cracking. The reactions are modeled to occur at the surface of catalyst particles or catalyst-coated surfaces with the surface temperature replacing the fuel temperature in the Arrhenius expression for the rate constant (Lander and Nixon, 1971; Sobel and Spadaccini,

1997; Zhou and Krishnan, 1996). Naturally, at high fuel temperatures both thermal and catalytic reactions occur with surface (catalytic) reactions promoted by high surface temperatures. Consequently, catalyst-coated heat-transfer surfaces are preferred to packed beds because wall-coated catalysts expose the fuel to the highest temperature at the catalyst surface; whereas in packed beds, the fuel transfers heat to the catalyst, so that the fuel catalyst interface is the coolest surface in the system. It may also be noted that chemical initiators can be used to enhance the fuel reaction rate (Ianovski, 1993; Chen, et al. 1998).

III. Fuel System Challenges

As already discussed, hypersonic vehicles will face severe thermal environments, particularly in the engine. At Mach 8, for example, uncooled combustor surfaces will surpass 3000 K (5000°F), far in excess of known structural material capability (Newman, 1993). There are many barriers to achieving the high levels of hydrocarbon fuel heat sink required for hypersonic flight, including: 1) deposition from high temperature reactions in the fuel (fuel thermal stability, fuel/material compatibility), 2) structural and heat-transfer performance, and 3) fuel system integration and control.

The intent of the present chapter is to focus on issues that are inherent to hydrocarbon fuel chemistry; only fuel thermal stability and fuel/material compatibility are addressed in specific detail. Structural and heat-transfer performance, and associated heat transfer and flow instabilities, are also briefly discussed. The fuel-system integration and control issues resulting from chemical modification of the fuel in its passage through the fuel system have been previously addressed elsewhere (Van Griethuysen et al., 1996). Only a brief summary is included herein for completeness.

A schematic diagram of a regeneratively cooled hypersonic engine is shown in Fig. 6 (Chen et al., 1998), and a basic fuel system circuit incorporating endothermic reactor is shown in Fig. 7 (Van Griethuysen et al., 1996). Clearly, the implementation of endothermic fuels results in increased fuel-system complexity. A complex fuel control system will be required to distribute fuel to the various engine structure panels to be cooled and to collect and distribute the hot fuel to the various fuel injectors. Temperatures typical of endothermic fuel systems are substantially higher than encountered in currently fielded vehicles where bulk fuel temperatures do not exceed 436 K (325°F). These temperatures coupled with operating pressures above critical [~34 bar (500 psia)] will necessitate the use of robust fuel-system construction materials. An important point, which is also often overlooked by researchers, is the complexity of facilities for ground testing endothermic-fueled engines. Such a facility must be capable of preconditioning the fuel to the appropriate composition, temperature, and pressure, which requires roughly 2 MW of power per kg/s of feed fuel. The use of slave combustors as the heat source is often favored over electrical heaters. Additionally, provisions must be made for post-treatment of the test fuels, and contingencies put in place to handle failure modes. These issues necessitate the use of condensers and flare stacks to dispose of endothermic reaction products in

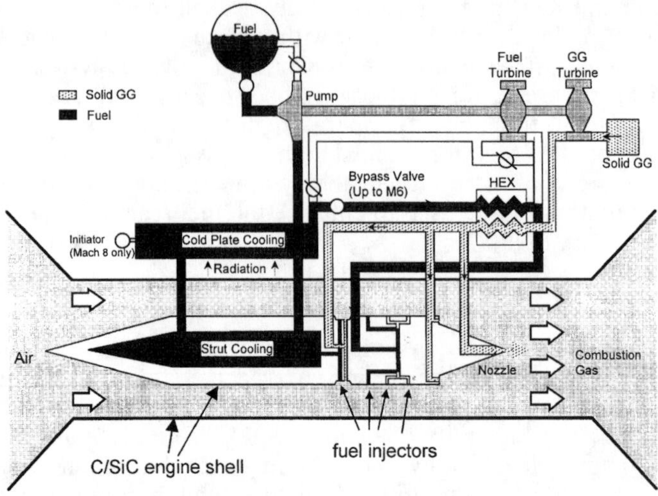

Fig. 6 Notional schematic of a fuel system for a hypersonic air-breathing missile (Chen et al., 1998).

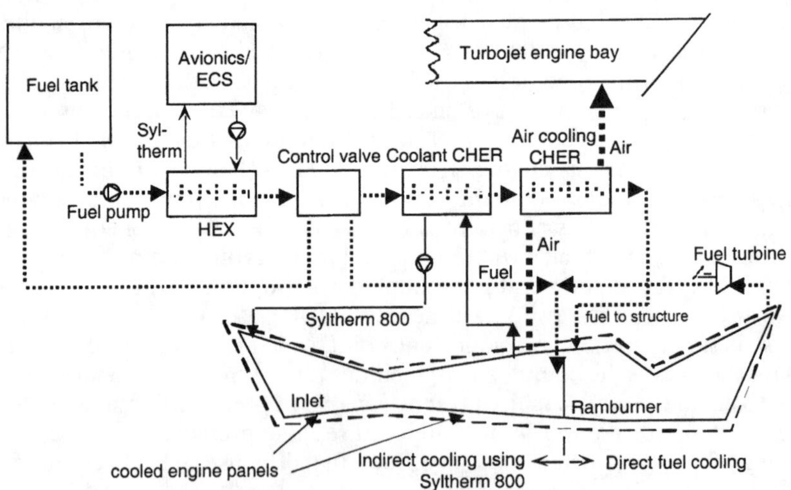

Fig. 7 Schematic of an endothermic fuel system featuring catalytic heat exchanger/ reactors (after Van Griethuysen et al., 1996). CHER denotes catalytic heat exchanger/ reactor. Syltherm is a synthetic heat-transfer fluid.

Fig. 8 Engine ground-test facility at the Air Force Research Laboratory featuring endothermic fuels.

ground-test facilities. A sample of a fuel system comprising an endothermic reactor for a relatively modest engine test facility [13.6 kg/s (30 lb/s) airflow] at the Propulsion Directorate of the Air Force Research Laboratory is shown in Fig. 8. The complexity of such facilities is also illustrated in Ianovsky (1997).

A. Thermal Stability

Fuel thermal stability is arguably *the* greatest challenge facing the use of high-temperature hydrocarbon fuels for aircraft systems where system lives of 2000 hours or more are desired. For missiles or other short-lived systems, with lifetimes measured in minutes, this is probably not the case. The thermal stability of the fuel is usually characterized by the amount of deposits that a particular fuel forms at a specific temperature in a particular test device. Translated to the vehicle, a thermally stable fuel would, by definition, create fewer deposits in the fuel system than an unstable fuel. Deposits can be created on heat-exchanger surfaces, filters, injectors, and control valves. Depending upon the temperature of the fuel, two types of deposits are found: thermal-oxidative and pyrolytic (Fig. 9). In the absence of fuel tem-

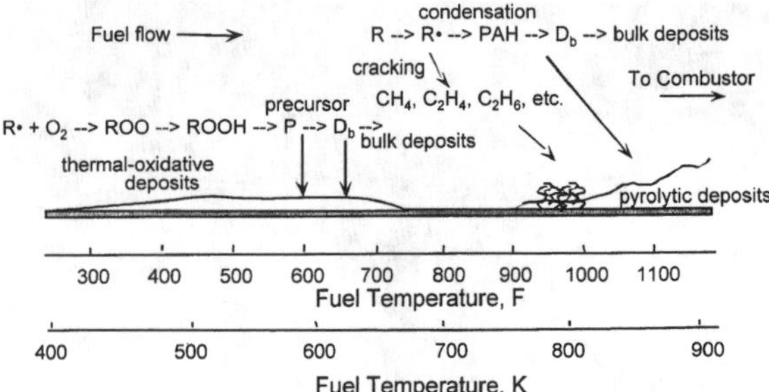

Fig. 9 Fuel deposition regimes.

perature information, the two types of deposits can often be distinguished by chemical or morphological analysis (Hazlett, 1991; Eser, 1996). Deposits are often described as *coking* or *fouling* without a consistent definition of these terms. In the hydrocarbon processing literature coking is usually confined to deposits arising from pyrolytic reactions. This classical definition is adopted in the present chapter.

Thermal-Oxidative Deposition

Thermal-oxidative deposits result from fuel molecule reactions with the oxygen dissolved in fuel (~70 ppm); these deposits begin to occur at temperature on the order of 400 K (300°F). The thermal-oxidative stability of conventional Jet A/JP-8 jet fuels has been the subject of much study (Hazlett, 1991). The type of thermal-oxidative deposition behavior seen in Fig. 9 (nonmonotonic deposition vs temperature) is often observed in high-temperature fuel thermal stability tests (Taylor, 1974; Marteney and Spadaccini, 1986; TeVelde and Glickstein, 1983; Edwards and Zabarnick, 1993; Hazlett et al., 1977). Arguably, the thermal-oxidative deposition reaches a peak and then decreases because of complete consumption of the dissolved oxygen in the fuel (Edwards and Zabarnick, 1993; Jones et al., 1996). Note that thermal-oxidative deposition is essentially eliminated by fuel deoxygenation (Hazlett, 1991). The challenge for JP-8 fuels lies in the variability of fuel thermal stability within the specification. For blended hydrocarbon fuels with stricter specifications than those for typical aviation kerosene such as JP-7, and for single component fuels such as JP-10 or MCH, this variability in thermal stability is much less; all JP-7s should have roughly similar thermal stability. Typical thermal-oxidative deposition levels for various fuels are shown in Table 4 (Edwards and Liberio, 1994; Edwards and Krieger, 1995). The MCH, JP-10, and decalin data are more limited; thus the deposition rates are only estimates. The relatively large thermal-oxidative deposition observed from the pure hydrocarbon MCH is consistent with earlier data (Bradley and Martel, 1975) and has not been adequately explained to date.

Many mitigation measures have been studied for thermal-oxidative fouling. In general, these measures fall into three categories: fuel deoxygenation,

Table 4 Thermal-oxidative surface deposition rates (complete oxygen consumption); fuels heated to 480 C.

Fuel	Total surface deposition rate, ppm
JP-8 (Jet A)	0.8–1.6
JP-8+100	0.08–0.2
JPTS	0.12
JP-7	0.07
MCH	~0.5
JP-10	~0.1
Decalin	~0.08

fuel additives, and surface modification of fuel-system materials. Fuel deoxygenation has been found to be very effective in reducing thermal-oxidative deposition (Hazlett, 1991; Taylor, 1974; Edwards and Liberio, 1994). Unfortunately, a flight-weight system for performing fuel deoxygenation has not been developed (Darrah, 1988). Fuel tanks for aircraft are usually vented to equalize pressure as the aircraft climbs and descends. Thus, the fuel is usually exposed to air, picking up ~70 ppm of dissolved oxygen at sea level. A chemical deoxygenation scheme would use a reactive additive species to react with the dissolved oxygen before it attacks the fuel molecules. One possibility, triphenyl phosphine, originally suggested by Beaver et al. (1994) was studied at the Air Force Research Laboratory (Heneghan et al., 1996b). The studies showed that many other undesirable fuel-additive reactions occured as the fuel was heated, preventing the additive from acting purely as an oxygen scavenger. This is not suprising, considering the wide variety of chemical species present in Jet A/JP-8 type fuels.

The most effective means for reducing thermal-oxidative fouling appears to be the use of detergent/dispersant/metal deactivator additives (Heneghan et al., 1996a). Detergent/dispersant additives act to prevent agglomeration (growth) of oxidatively created particles in the bulk flow, which keeps the particles very small (100–200 nm) and minimizes deposition (Vilimpoc et al., 1996). These additives have demonstrated reductions in deposition in JP-8 fuels to JPTS levels with the addition of additive packages at the 100-ppm concentration level. Reductions in surface deposition approached 90% for the JP-8+100 additive package devised by Betz, designated 8Q462, which is presently undergoing fleetwide conversion as already discussed. Changes in surface composition and finish have been found to affect the induction time, but not the subsequent deposition rate (Hazlett, 1991). This is only reasonable because the surface is covered with a thin deposit layer after the induction time and thus should not be interacting with the fuel. In current aircraft where operational practices limit fuel temperatures to 436 (325 °F), the dissolved oxygen in air-saturated fuel (~70 ppm) is typically only partially consumed. Tests show that as temperatures exceed ~644 K (700°F), however, the oxygen is completely consumed for any physically realistic residence time. Thus, all of the high heat-sink fuels described in this chapter will be operating under conditions of complete oxygen consumption during hypersonic flight. This simplifies the study of the thermal-oxidative stability of fuels because a variety of diagnostic tests give similar results for a given fuel under conditions of complete oxygen consumption (Edwards and Krieger, 1995). The deposition rates in Table 4 were determined by running a series of experiments at varying test times, then plotting the total thermal-oxidative deposition (as measured by carbon burnoff) as a function of test time. The slope of the deposition vs test time line gives the deposition rate (μg/h), which is usually divided by the fuel flow rate (g/h) to yield a deposition rate in ppm (as shown in Table 4). Appreciable induction times, where the deposition initially is not measurable, are common. If consideration of this induction time is neglected, the calculation of deposition rates may be significantly affected for fuels of high thermal stability (Edwards and Krieger, 1995). In a number of tests with widely varying flow rates and complete oxygen consumption, the deposition rate (in ppm) was found to be relatively constant (Edwards and Krieger, 1995). Note that the deposition rate expressed in $\mu g/cm^2$-h (an often-used unit) would not be constant, but would increase as flow rate increases. It is important to understand that the oxygen is completely consumed in all of these tests. As flow rate and residence time vary, the fuel temperature range over which the oxidative deposition occurs changes, but the overall deposit amount is relatively constant. In reality the deposition is shifted to higher temperatures as the flow rate increases. In lower flow rate (~1 ml/min) tests Jones et al. (1996) found the deposition rate dropped as tube diameter increased, implying that some part of the deposition process is diffusion-controlled. Recent work sponsored by the Air Force Research Laboratory has developed chemical kinetic models of the fuel oxidation and deposition processes (Katta et al., 1997; Ervin and Zabarnick, 1998). The fuel oxidation and deposition reactions are reduced to approximately 10 global steps with rate constants determined from data produced in various laboratory thermal stability test devices. The thermal stability or quality of the fuel is characterized by the sulfur level, which can vary from essentially zero to the specification limit (0.3 wt%) for JP-8. The current models incorporate only one fuel-quality-dependent parameter, which is proportional to the fuel sulfur level. Incorporating an additional parameter to account for the presence of a dispersant additive into the deposition model will be required for advanced fuels such as JP-8+100 because of the dramatic reductions in deposition as just described (Heneghan et al., 1996a).

Pyrolytic Deposition

At temperatures above approximately 755 K (900 °F) in flowing systems, thermal cracking and other reactions of the base hydrocarbons begin to occur. When these reactions are deliberate (because of the extension of the heat-absorbing capability of the fuel), the fuel is termed endothermic (as already described). Whether deliberate or not, thermal reactions of the bulk fuel often lead to deposition on surfaces. Typically, pyrolytic deposits would occur in the highest temperature components of the fuel system. In a typical high-temperature fuel system, the fuel would pass through a series of heat exchangers, pumps, and control devices (see Fig. 6 and 7). Thermal-oxidative deposits could be a problem for lower temperature heat exchangers and components, e.g., airframe heat exchangers and fuel controls and pumps. Pyrolytic deposits might be found in the hottest heat exchangers and in the fuel injectors. Note that in any given fuel system component the fuel temperature and pressure will vary throughout a vehicle's mission. It is conceivable that a heat exchanger, for example, would be subject to thermal-oxidative deposition during one part of a mission and pyrolytic deposition during another part of the flight.

A large body of literature exists on thermal cracking of hydrocarbons (e.g., Crynes and Albright, 1987). A generic coking mechanism from the industrial hydrocarbon pyrolysis literature is shown in Fig. 10. However, aircraft fuels at high temperatures experience a number of environmental conditions that are significantly different than typical industrial fuel applications. Industrial cracking typically occurs at near-atmospheric pressure with steam-diluted hydrocarbons in large [>2.5 cm (1 in.) internal diameter (ID)] tubes. In contrast, typical aircraft fuel system pressures are 35–70 bar (500–1000 psia), the fuel is undiluted, and the passage size is much smaller, on the order of millimeters (Ronald, 1995; Beal et al., 1991). Thus, it might be anticipated that high-temperature aircraft fuels might show some differences in behavior from industrial experience. In particular, coking rates might be expected to be much higher in aircraft fuel systems because of the higher hydrocarbon concentrations and the larger surface-volume ratios.

As already noted, pyrolytic deposition from aircraft fuels begins to occur

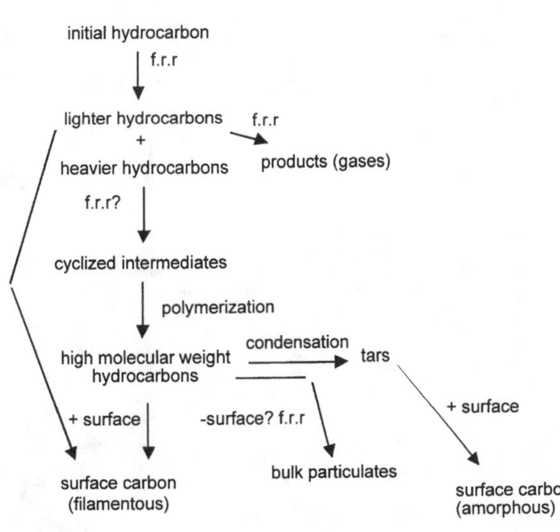

Fig. 10 Coking mechanism adapted from Trimm (1983), Baker et al. (1978, 1982, 1989, 1996), and Albright and Marek (1988).

at fuel temperatures on the order of 750 K (900°F), depending upon residence time. In general, this type of deposition appears to be directly related to thermal cracking of the fuel. The cracking rates of different hydrocarbons vary depending upon their structure, with paraffins and isoparaffins being the least stable to cracking, followed by cycloparaffins (naphthenes) and aromatics. The authors have avoided the use of thermal stability in this context because thermal stability in this paper is used to denote the stability relative to deposit formation. This is another illustration that care must be taken with the terms thermal stability and coking, as defined in the present context, because they are used to describe very different processes. In addition to the propensity to crack, we have found that some fuel structures are more prone to pyrolytic deposition than others. Thus, the coking potential of various fuels at a given temperature consists of two parameters: 1) the amount of cracking of the fuel and 2) the tendency of a fuel to form deposits once cracking has occurred. It is easy to imagine a situation where a fuel that is stable to cracking might form significant deposits once it does begin to crack. This fuel should then be limited to conditions where cracking does not occur. For example, JP-7 and JP-8 fuels are mostly paraffinic and crack much more readily at a given temperature than decalin or JP-10, which are naphthenes. However, in flowing tests decalin and JP-10 tend to form more pyrolytic deposits than JP-7 and JP-8 at a given temperature, despite the lower amount of cracking (Edwards and Atria, 1995; Edwards, 1996b). Thus, decalin could have a lower use temperature than JP-7 or JP-8, despite its greater stability to cracking. The pyrolytic and thermal-oxidative deposition for several fuels at a maximum fuel temperature of 920 K (1200°F) is illustrated in Fig. 11 (Edwards and Atria, 1995). The difficulties of using the term thermal stability in this context are evident.

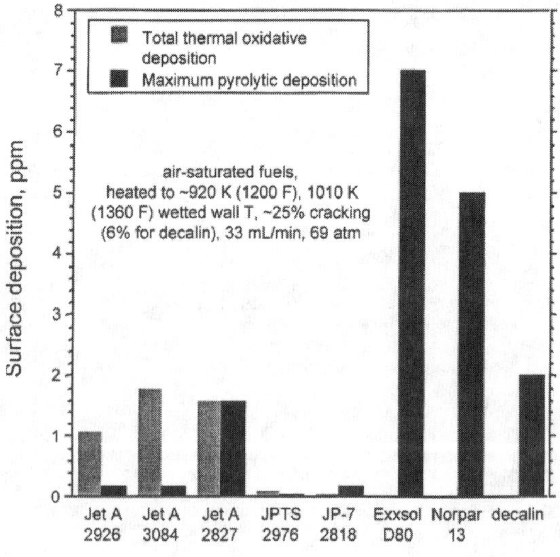

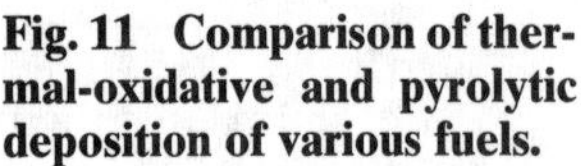
Fig. 11 Comparison of thermal-oxidative and pyrolytic deposition of various fuels.

The formation of polycyclic aromatic hydrocarbon (PAH) species has long been linked to soot nucleation and mass growth. An endothermic fuel pyrolytic reactor is arguably analogous to a combustion system in which the equivalence ratio approaches infinity. Furthermore, the formation of saturated and aromatic species leads to a significant decrease in the endothermicity of the reaction. Finally, the reacting timescales of benzene and PAH molecules are generally slower than those of aliphatic molecules; thus, significant formation of aromatics can reduce overall system efficiency in kinetically limited hypersonic combustors. The formation of PAH species within the fuel system is an issue that must be addressed. The formation of benzene and polyaromatic hydrocarbon (PAH) soot precursors within the endothermic thermal cracking reactors used in hydrocarbon-fueled high-speed vehicles may be viewed as analogous to PAH formation in conventional combustion systems (Fig. 12). Experimental observations have indeed confirmed the formation of aromatic compounds in the endothermic cracking process of dodecane and JP-7 (Edwards and Anderson, 1993; Striebich and Rubey, 1998). Stewart et al. (1998) have also observed PAH formation in the pyrolysis of MCH and toluene/*n*-heptane fuels at temperatures up to 840 K (1050°F). Aromatic formation appears to precede conditions where deposit formation rates are high. In direct analogy to combustion systems, naphthalene and indene are commonly observed PAH species. Moreover, bibenzyl formation is substantial in toluene/*n*-heptane pyrolysis. Cyclic formation in *n*-dodecane pyrolysis is also experimentally observed by the pre-

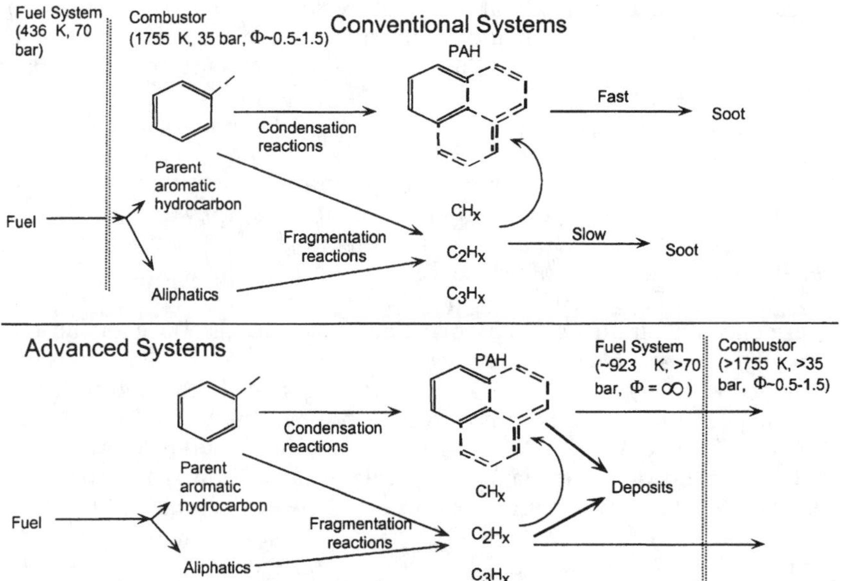

Fig. 12 Soot-fuel system deposit analogy.

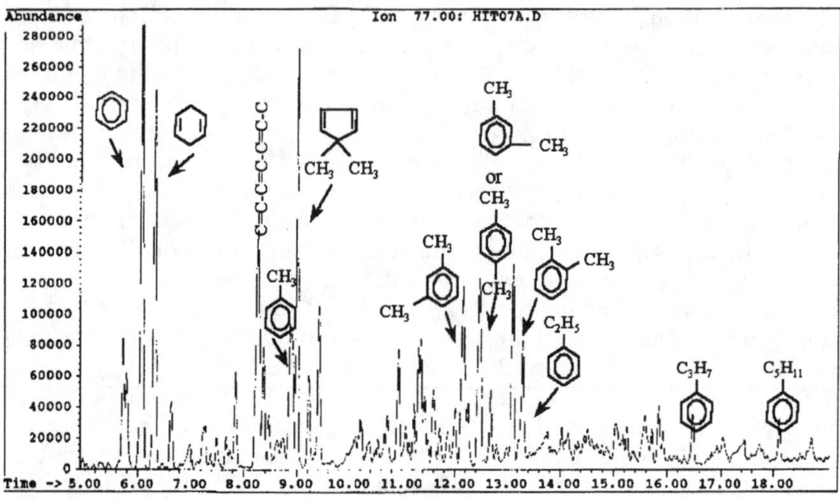

Fig. 13 Cyclic formation in *n*-dodecane pyrolysis at 973 K, 34 bar in the System for Thermal Diagnostic Studies.

sent authors (Fig. 13) (Maurice et al., 1998a) in the System for Thermal Diagnostic Studies (STDS) (Striebich and Rubey, 1990, 1996). Exponential growth of benzene, naphthalene, and indene yields is noted as a function of temperature, and reactor residence time is shown to be a key parameter influencing molecular growth at temperatures typical of endothermic pyrolytic reactors. Therefore, formation of undesirable benzene and PAH species might be mitigated by controlling endothermic reactor residence time (Maurice et al., 1998a,b).

Although there is much less information available for pyrolytic deposition than for thermal-oxidative deposition, some generalizations can be made. For a given fuel at a given flow rate (or given residence time), the level of deposition is roughly exponential in temperature, as shown in Fig. 14 (see also Fabuss et al., 1962, and Edwards and Liberio, 1994). This is a very strong dependence! The effect of residence time on deposition is less clear, although comparing the 20% cracking points in Fig. 14 indicates that there is a residence time influence on pyrolytic deposition levels above and beyond the influence associated with conversion or cracking level. Thus, deposition increases with increases in temperature (conversion) for a fixed residence time; deposition decreases with increasing temperature for a fixed conversion level. In other words, if the heat loads on a vehicle require a certain level of fuel conversion, it appears that pyrolytic deposition is minimized by reducing residence time. Firm relationships between coking and temperature/residence time are not possible from this data because of the nonisothermality of the test. Modeling efforts are underway to develop this understanding (Sheu et al., 1998). In any event pyrolytic deposition is a primary barrier to the use of endothermic fuels in aircraft.

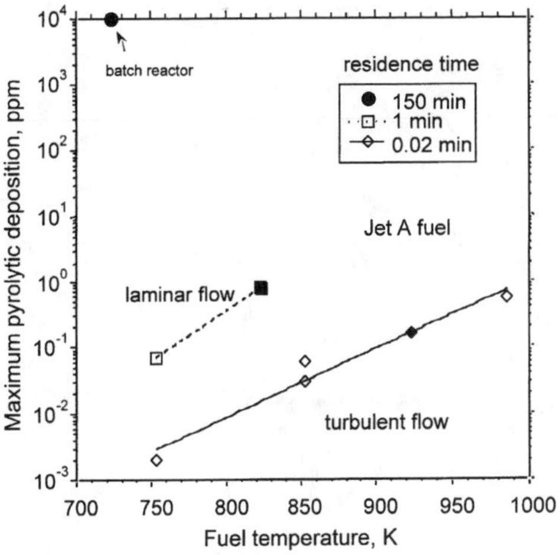

Fig. 14 Pyrolytic deposition as a function of maximum fuel temperature and residence time. Data from Edwards and Atria (1995), Lai et al. (1992), and Atria and Edwards (1996). Solid symbols represents 20% conversion.

Three distinct types of pyrolytic deposits have been recently identified in the literature based upon their morphology: filamentous, amorphous, and graphitic (Baker et al., 1978, 1982, 1989, 1996; Albright and Marek, 1988; Trimm, 1983). An amalgamation of several mechanisms cited in the literature for the various types of coking is shown in Fig. 10. Filamentous and amorphous carbon have been identified for the first time under high-pressure fuel-system conditions (Atria et al., 1996; Edwards, 1996b). Identifying the high-pressure conditions that control the formation of the various types of coke deposits remains desirable. Filamentous carbon is the most deleterious form of coke because it involves the removal of small pieces of structural metal, weakening the material and reducing ductility. Iron-, nickel-, and cobalt-containing alloys are most susceptible to attack, whereas other materials (such as rhenium) are resistant. This type of coking is predominant at 920 K (1200°F) under steam-cracking conditions, which involve run times of months in superalloy furnace tubes (Baker, 1996). By contrast, in two-hour tests (relevant to an expendable aircraft) with 1030 K (1400°F) fuel-tube surface temperatures, the discovery was made that very little filamentous carbon was formed from high-pressure jet fuels on typical superalloys (Edwards, 1996b). No weakening of the thin (0.4-mm/0.015-in. wall thickness) superalloy tubes was found. Hence superalloys may be acceptable for short fuel-system lifetimes.

Several mitigation measures for pyrolytic deposition have been studied. Among the general approaches examined were fuel deoxygenation, additives, and surface treatments. Note the similarity to the mitigation approaches for thermal-oxidative fouling. However, pyrolytic behavior is quite different from thermal-oxidative behavior. For example, fuel deoxygenation often increases pyrolytic deposition (Edwards and Liberio, 1994; Edwards and Atria, 1995; Atria and Edwards, 1996). Thus, it appears that the oxidation products act as a deposit suppresser, at least for pure hydrocarbons. Alter-

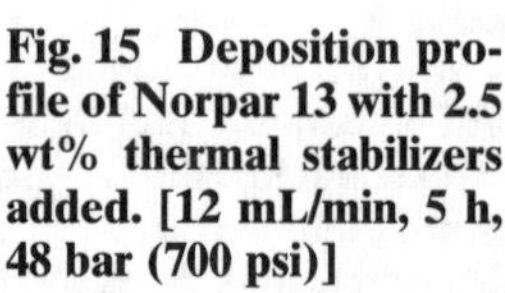

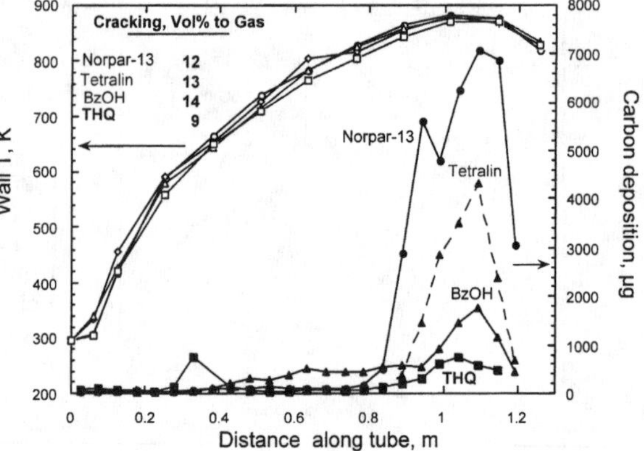

Fig. 15 Deposition profile of Norpar 13 with 2.5 wt% thermal stabilizers added. [12 mL/min, 5 h, 48 bar (700 psi)]

natively, some pyrolytic deposit-promoting species may be consumed by oxidation reactions. A number of high-temperature pyrolysis-suppressing additives have been studied in batch reactors at 720 K (840°F) (Yoon et al., 1996). The most effective additives in reducing both conversion and deposition were benzyl alcohol (BzOH) and tetra-hydro quinoline (THQ). At the ~2.5% concentration level these hydrogen-donating additives are effective in reducing deposition (but not cracking/conversion) in a flow reactor, as shown in Fig. 15 (Atria and Edwards, 1996). THQ is also effective in reducing pyrolytic deposits in JP-8 (Minus and Corporan, 1998). It would be preferable to use additives that were effective at lower concentration; however, those additives have not yet been identified.

The most effective means of preventing pyrolytic coking was found to be the use of inert coatings or surface treatments on the metal tube surfaces (Ianovski, 1993; Atria et al., 1996) These types of coatings have been found to be effective in industrial applications in reducing filamentous carbon formation (Baker et al., 1982, Szechy et al., 1992). As shown in Fig. 16, a

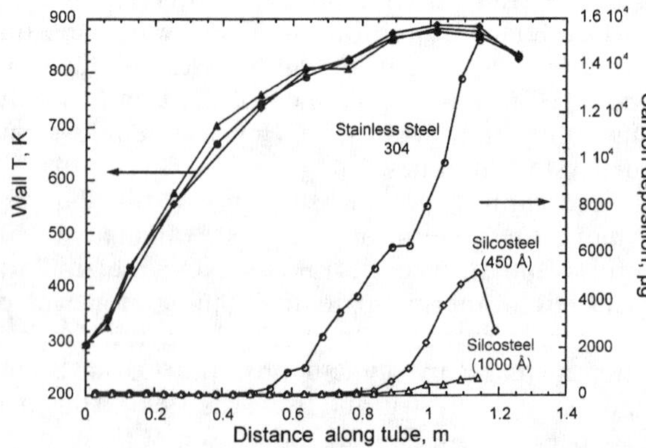

Fig. 16 Deposition profile comparing uncoated to Silcosteel-coated stainless-steel tubes.

deposited silica coating (Silcosteel®) is very effective at preventing coking under fuel system conditions, although the type of pyrolytic deposit (filamentous, amorphous, or graphitic) present under these conditions was not identified. The coating apparently acts as a barrier to the formation of catalytic (filamentous) carbon, as is seen industrially in steam cracking (Crynes and Albright, 1987; Baker et al., 1982). Earlier U.S. Air Force results showing that the Silcosteel coating was ineffective in reducing pyrolytic fouling (Edwards and Atria, 1995) were apparently from the use of the thinner 450 - Angstrom coating. A strong dependence on coating thickness is shown in Fig. 16 (Atria et al., 1996). Electrolytic/plasma processing of surfaces is also reported to be very effective at reducing pyrolytic deposition (Ianovski, 1993).

In summary, two distinct routes of deposit formation could be important in hydrocarbon-fueled hypersonic vehicles: thermal-oxidative and pyrolytic. The mechanism of thermal-oxidative fouling is fairly well understood, but control is difficult because of the wide variety of fuel species present in distillate aircraft fuels. Detergent/dispersant additives, often in combination with a metal deactivator, offer the best means for minimizing thermal-oxidative fouling. In contrast, the mechanism of pyrolytic coking is less well understood under aircraft conditions, especially as it relates to the various forms of surface carbon (filamentous, amorphous, graphitic). The most effective means of minimizing pyrolytic coking to date appears to be the use of inert (e.g., silica) surface coatings.

B. Structural and Heat-Transfer Considerations

Practical application of storable fuels for scramjets will have to rely initially on the cooling capability demonstrated primarily in single-tube heat-exchanger tests and subsequently extend such data to fuel-cooled structures (Fig. 6). For this application one expects that fuel-cooled combustors will resemble regeneratively cooled liquid rocket engines with coolant passages in the combustor and nozzle walls. AiResearch Manufacturing Company designed, fabricated, and tested a JP-5 cooled test combustor using a platefin sandwich structure (Harris et al. 1973). A water-cooled version of such combustor was also built and extensively tested in a freejet engine (Fig. 17) by Kaiser Marquardt (Jensen and Braendlein, 1996). A Mach 8 scramjet would exhibit uncooled combustor wall temperature levels similar to a cooled RP-1/LOX (liquid oxygen) rocket engine ~ on the order of 3000 K (5000°F) (Heiser and Pratt, 1994; Newman, 1993). One major difference exists: liquid rocket engine fuel flows are much larger than scramjet fuel flows for similar size combustors. This leads to typical maximum bulk fuel temperatures of ~420 K (300°F) for RP-1-fueled rockets (Van Huff, 1972), whereas Mach 8 scramjet fuel temperatures are expected to exceed 800 K (1000°F) (Lander and Nixon, 1971). For RP-1 (similar to JP-7 and JP-8) the allowable maximum wall temperature is limited, by coking, to 730 K (850°F) (Van Huff, 1972). Significant degradation in heat transfer from coking can be apparent under these conditions in as little as 1000 s (Rosenberg et al., 1991, 1992). In this regard, fuel-materials interactions can also be important

Fig. 17 Kaiser Marquardt MA194 Dual-Mode Hydrocarbon-Fueled Scramjet Engine.

(Roback et al., 1983; Giovanetti et al., 1985). High-energy additives such as azides have been found to enhance coking under rocket cooling conditions (Linne and Munsch, 1995). JP-7 has recently been compared to RP-1 under these conditions and found to be more thermally stable (Linne and Munsch, 1995) in contrast to earlier results (Roback et al., 1983). Catalysts coated on the heat exchanger passage walls would enhance endothermic reactions. As already, discussed, catalysts promote both overall conversion and better product distribution at a given temperature, as compared to thermal cracking. Moreover, wall-coated catalysts have significant advantages in heat transfer and pressure drop over packed beds.

The fabrication of fuel-cooled structures, including manifolds, is expected to be a significant challenge. At high fuel temperatures and heat fluxes heat-transfer instabilities are often encountered (Hines and Wolf, 1962; Faith et al., 1971; Harris et al., 1973; Hitch and Karpuk, 1997; Linne et al., 1997). The mechanism of these instabilities is unclear, although the theory has been stated that the large property variations between subcritical and supercritical fuels created the oscillations, which were flow instabilities termed pseudoboiling. Buoyancy effects have also been found to be important in heat transfer involving supercritical fluids such as CO_2 and H_2O (Polyakov, 1992). Under some conditions flow relaminarization can occur, significantly degrading heat transfer and causing a "heat-transfer crisis" (Ianovski and Sapgir, 1996; Yanovskii and Kamenetskii, 1993). The use of inserts to augment swirl in the flow can enhance the heat transfer under supercritical conditions (Dreitser et al., 1993; Hitch and Karpuk, 1998). Similarly, stratification of the flow at low velocities can cause heat-transfer failure (Dutton, 1960). The heat-flux distribution inside scramjet combustors can often be quite uneven with shock impingement on walls (caused by injectors, pilots, etc.) causing significant hot spots. These hot spots are expected to exceed

570 W/cm^2 (500 Btu/ft^2-s) in some cases. These high heat fluxes will require high coolant velocities for cooling with Reynolds numbers in excess of 150,000. The average heat flux in the combustor at Mach 8 has been estimated to be about 170 W/cm^2 (150 Btu/ft^2-s) (Heiser and Pratt, 1994). In a multiple-passage fuel system measures must be taken to ensure that shock impingement does not cause the affected passage to become fuel-starved because of increased pressure losses (ΔP). Measures must also be taken to ensure that the cooling system is dynamically stable.

C. Fuel-System Integration and Control

For a hydrocarbon-fueled hypersonic vehicle the fuel system becomes one of the key integration challenges for the vehicle designer. In the missile system, as illustrated in Fig. 6, the fuel is expected to serve as a coolant as well as a source of energy during combustion. In a typical Mach 8 missile mission, two flight points are expected to be especially challenging: 1) transition from boost to ramjet operation, nominally at ~ Mach 4, and 2) Mach 8 cruise. At ramjet takeover the fuel is a liquid with relatively poor combustion properties at the low air total temperatures (~ 870 K/1100°F) corresponding to Mach 4 flight speeds. At Mach 8 in a fixed geometry engine the thrust can only be reduced for cruise by reducing combustor equivalence ratio, e.g., reducing the fuel (and thus the coolant) flow. Thus, high Mach cruise is the most thermally challenging part of the mission: low fuel (coolant) flows and high heat loads. Intermediate flight Mach numbers present primarily control and operability challenges for the fuel system: ensuring that the cooling and combustion are both satisfied in a controlled manner. This is also a significant challenge, especially if one considers that the combustor fuel control in Figs. 6 and 7 is controlling high-temperature fuel. The design of an integrated thermal management system for hypersonic vehicles is beyond the scope of this chapter, but many studies have been performed (e.g., Petley and Jones, 1992; Bergholz and Hitch, 1992). These studies generally indicate maximum Mach numbers of 6–8 for endothermic storable-fueled vehicles with regenerative cooling required above about Mach 6–6.5 (Waltrup, 1997).

In addition to coking issues (just discussed), the high-temperature fuel has significantly different fuel properties from the cold fuel. As illustrated in Figs. 18 and 19, the fuel density and viscosity change significantly with temperature, especially near the fuel critical temperature (Edwards, 1993). The critical temperature and pressure of Jet A/JP-8/JP-7 fuels are approximately 660–670 K (725–750°F) and 20–25 bar (300–370 psia), respectively (Yu and Eser, 1995; Edwards, 1993; Guisinger and Rippen, 1989; Steele and Chirico, 1989). The control of the fuel system when the fuel properties vary so significantly will be a challenge. Simple fuel controls used in aircraft (fuel pressure is used to control flow) will not work properly when the fuel properties change significantly. The mass flow through a simple valve will be quite different for a 311 K (100°F) fuel and a 980 K (1300°F) fuel at a given pressure. Similar concerns apply to fixed-orifice injectors.

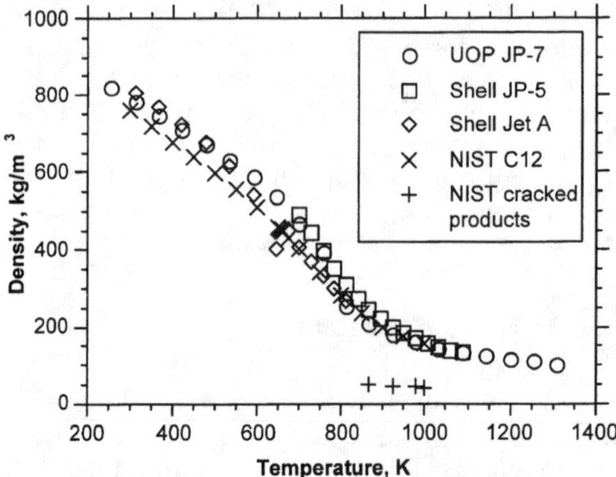

Fig. 18 Literature data for fuels (Edwards, 1993; TeVelde and Glickstein, 1983) and calculated (Ely and Huber, 1990) densities for *n*-dodecane and cracked products as a function of temperature, 68 bar (1000 psia). Cracked products are dodecane-0.03 mole fraction, toluene-0.14, benzene-0.08, methane-0.15, ethane-0.15, ethylene-0.11, propane-0.11, propylene-0.11, and butane-0.12. These were simulated JP-7 thermal cracking products (Edwards, 1993).

IV. Combustion Challenges

Combustion in dual-mode scramjet engines featuring combustion in both subsonic and supersonic flow (see Fig. 1) is a complex function of injection, mixing, kinetics, piloting, and heat-release phenomena. An important issue is the effect of the fuel preprocessing as a function of Mach number in the fuel system before combustion. As discussed in the preceding sections, this processing can range from preheating to partial reaction. Thus, the typical liquid fuel combustion process, injection→ atomization→ vaporization→ pyrolysis→ mixing→ oxidation, is modified to the extent that all but the mixing and oxidation steps might occur in the fuel system.

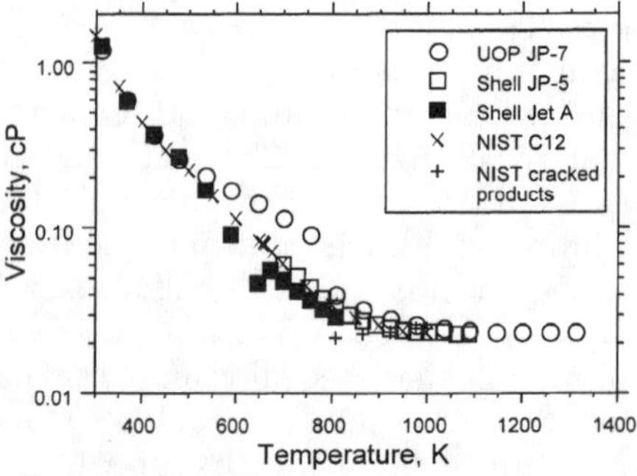

Fig. 19 Literature data for fuels and calculated viscosities for *n*-dodecane and cracked products as a function of temperature, 68 bar (1000 psia).

Well-stirred reactor (WSR) tests have been used to determine the effect on blowout parameters of the endothermic fuel reactions. The experimental apparatus and procedures are as described by Blust et al. (1997). Classically, blowout limits are widest (best) for hydrogen, followed by ethylene, then propane and liquid fuels (Northam, 1985; Beach, 1992). Experiments were conducted to determine if fuel prevaporization and reaction will increase the liquid fuel stability limits to those of pure ethylene. Simulated endothermic products (comprising 13% methane, 22% ethane, 52% ethylene, and 13% toluene by volume) for the thermal decomposition of a JP-type fuel were tested in the WSR to compare the blowout limit of this fuel mixture to those of pure hydrocarbons and Jet A. Blowout equivalence ratios of several fuels, including an endothermic simulant, at a loading parameter (Longwell and Weiss, 1955) of ~ 1.3 g-mol/s/l/atm$^{1.75}$ were compared, as shown in Fig. 20. Reactor residence time is approximately 7.3 ms, and blowout temperature ranges from 1300 K (1880°F) (Jet A) to 1500 K (2240°F) (methane). Liquid fuels are preheated to provide adequate vaporization, and the equivalence ratios of liquid fuels are corrected to an inlet temperature of 297 K. Also shown for reference are the (partial) stability loops of methane and ethane. As shown, the endothermic simulant displays a modestly improved blowout limit compared to other fuels.

Partial fuel conversion in the fuel system improves the combustion kinetics of the fuel (reduces ignition delay/increases blowout limits), but not to the level of ethylene, a major fuel reaction product. This is in direct agreement with recent experimental observations in MCH/toluene blends (Zep-

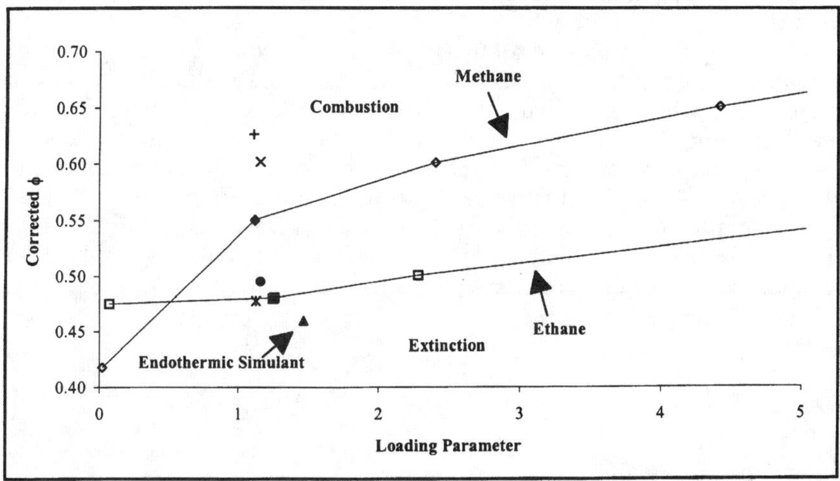

Fig. 20 Corrected LBO equivalence ratio vs loading parameter for several hydrocarbons: methane (diamond), ethane (square), cyclohexane (circle), *n*-dodecane (X), toluene (star), Jet A (+), and endothermic simulant (triangle). Also shown (solid lines) are calculated methane and ethane stability loops.

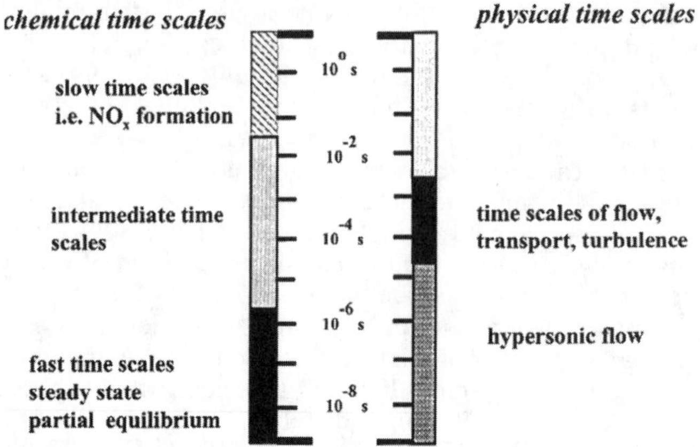

Fig. 21 Structure of timescales representing chemical and physical processes in combustion (after Maas and Pope, 1992).

pieri et al., 1997), which imply that each fuel component follows its own individual oxidation mechanism. The present results are also similar to those observed in shock tubes for MCH dehydrogenation products: the ignition delay for a toluene/hydrogen mixture is intermediate between hydrogen and toluene (and similar to MCH) (Hawthorn and Nixon, 1966). Thus, despite an enhancement provided by the common radical pool, the combustion kinetics behavior of a fuel mixture cannot approach that of the most reactive component. Hence, ignition aids will still be necessary as discussed in the following.

The combustion properties of hydrocarbons have long been recognized as a significant limitation to successful operation of storable-fueled hypersonic vehicles (Hawthorn and Nixon, 1966; Northam, 1985; and Beach, 1992). With the exception of nitric oxides, polycyclic aromatic hydrocarbons, and soot formation, the physical flow time scales of conventional combustors generally exceed the chemical timescales (Fig. 21). However, hypersonic combustors are limited to residence times of the order of fractions of a millisecond (Fig. 22), which are commensurate with a broader range of combustion reaction timescales (1 μs to 10 ms; Griffiths and Barnard, 1995).

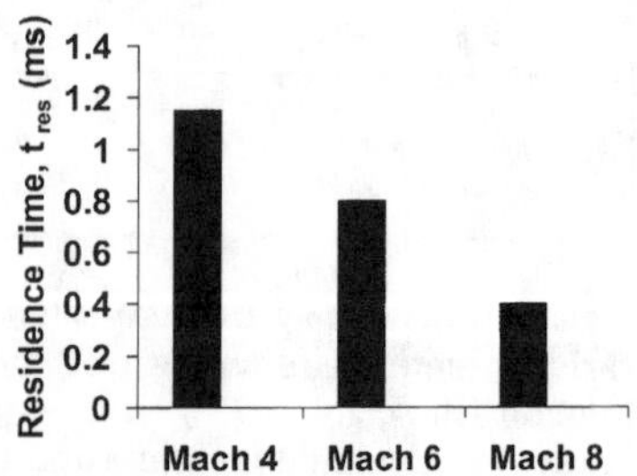

Fig. 22 Typical scramjet combustor residence time.

Table 5 Typical combustor entrance conditions

Freestream condition→	Mach 4	Mach 6	Mach 8
Mach number	2	2.9	3.5
Static temperature, K (R)	500 (900)	600 (1100)	830 (1500)
Total temperature, K (R)	900 (1600)	1700 (3000)	2500 (4500)

A key hypersonic vehicle combustion challenge is achieving ignition in the short residence times available. The relevant temperatures are shown in Table 5 for Mach 4, 6, and 8 (Billig, 1993; Heiser and Pratt, 1994; Newman, 1993; Van Griethuysen et al., 1996). These temperatures are a function of dynamic pressure (trajectory) and vehicle design, but the values shown in Table 5 are typical. The static conditions are the most challenging in which to achieve ignition; recirculation zones, boundary layers, and shock areas can achieve temperatures approaching the total temperature or the stagnation temperature and are thus promising ignition zones, at the expense of increasing pressure loss (Beach, 1992).

It appears evident, as shown in Fig. 23, that the ignition delays of conventional hydrocarbon fuels must be significantly reduced for successful supersonic combustion. Additionally, it is also necessary to increase the flame speeds and broaden the blowout limits of conventional hydrocarbon fuels for successful combustor development (Edwards, 1996a; Tishkoff et al., 1997; Morris et al., 1998). Storable hydrocarbon development efforts for hypersonic applications have generally focused on heat sink and thermal stability considerations. However, the development of highly reactive fuels and an accompanying understanding of combustion chemistry are clearly also necessary steps toward addressing the requirements of 21st century hypersonic propulsion systems. The ability to meet this challenge is predicated on an understanding of the basic kinetic foundations of hydrocarbon oxidation, as discussed next. The present understanding of jet fuel combustion kinetics

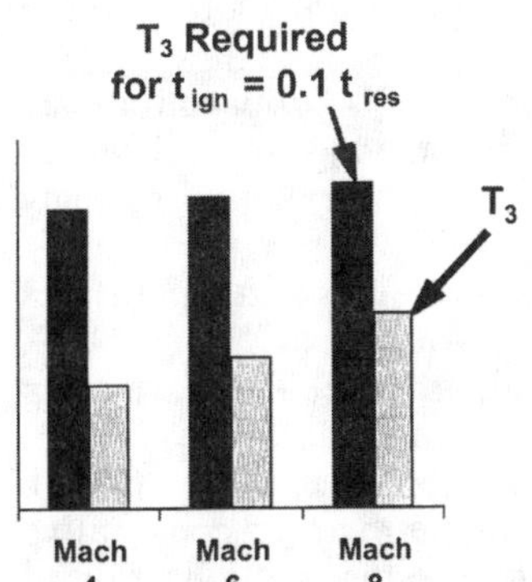

Fig. 23 Mach-number influence on ignition delay (Weber, 1997). T_3 denotes combustor-entrance temperature.

and potential approaches for enhancing hydrocarbon fuel reactivity are subsequently discussed.

A. Chemical Kinetic Foundations

Combustion processes are normally driven by free radical chain reactions, comprising initiation, propagation, branching, and termination stages. An overview of the progress to complete oxidation of a hydrocarbon fuel molecule (RH) may be regarded to involve the sequence of molecular species

$$RH \rightarrow CH_2O \rightarrow CO \rightarrow CO_2 \tag{2}$$

Formaldehyde (CH_2O) is known to be a molecular intermediate of all hydrocarbon flame (or other high temperature) processes involving C and H-containing fuels, and it is involved at the final stage of the conversion to carbon oxides. Virtually all carbon dioxide (CO_2) is formed via the oxidation of carbon monoxide (CO). Water (H_2O) is formed at each stage, principally by an H-atom abstraction in a propagation process of the form

$$RH + OH \rightarrow R + H_2O \tag{3}$$

where RH represents any reactant or intermediate molecule.

The main source of heat release during the combustion of hydrocarbons comes from the formation of the final products of combustion, CO_2 and H_2O, with significant contributions also from the recombination of propagating free radicals in termination processes. CO formation is also accompanied by heat release, and it may be a final product during inefficient combustion regimes, such as in very fuel-rich conditions.

The main propagating species of the chain reactions at temperatures above 1000 K (1340°F) are H, O, OH, HO_2, R, and RO, where R represents a fragment of a hydrocarbon fuel molecule (such as the alkyl radical). For reasons to be discussed, the fuel fragments tend to be of low molecular mass, containing fewer carbon atoms than the primary fuel molecules. With the exception of hydrogen and carbon monoxide combustion, at temperatures below 1000 K (1340°F) oxygen atoms make very little contribution to the reaction chain propagation.

As the temperature is decreased toward 750 K (890°F), propagation may be supplemented by the reactions of RO_2 species (alkylperoxy radicals, in the case of alkanes), and the organic fragments (R) become increasingly representative of the primary fuel.

A more detailed overview for the combustion of alkanes at temperatures in excess of 1000 K (1340°F) was constructed by Warnatz (1984) (Fig. 24), which demonstrates the important interrelationships between the formation and removal of a variety of low mass intermediate and molecular species. How reaction develops, and the time taken to ignition (the ignition delay), depends upon this type of complex sequence of competitive and consecutive reaction steps, including also chain branching, that is, the multiplication of

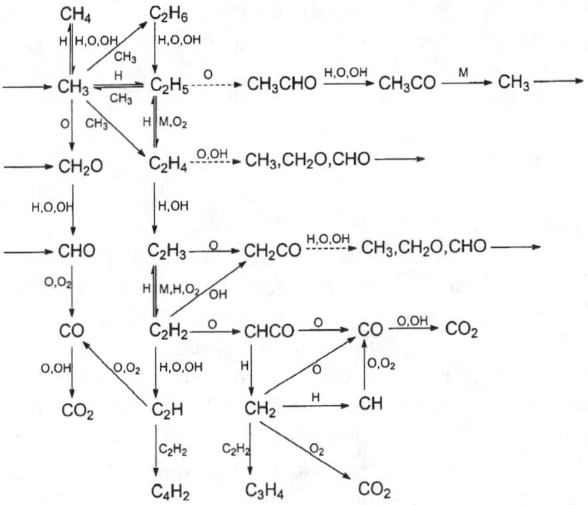

Fig. 24 Combustion of alkanes (after Warnatz, 1984).

propagating species. Throughout much of the temperature range, the most important of the branching reactions is

$$H + O_2 \rightarrow OH + O \tag{4}$$

This extremely important role of H atoms can be restricted by the reaction conditions, as discussed later. The diagrammatic form of the Warnatz scheme (1984) includes solid lines to represent single-step reactions and broken lines to represent more complex sequences from one stage to the next. The inclusion of forward and backward reactions in some cases is not intended here as an indication of chemical equilibrium and does not imply that other sequences are not allowed to proceed in either direction. Whether or not there is the possibility of reversibility is a consequence of the temperature dependence of the elementary reactions involved.

Temperature Dependence of Elementary Reactions and Competition Between Reactions

Each of the elementary reactions involved in the mechanism exhibits some dependence on temperature, and this is encapsulated in the rate constant k for the reaction. The common representation for the temperature dependence (which is usually very satisfactory over a limited temperature range) is given by the Arrhenius expression

$$k = A\, e^{(-E/RT)} \tag{5}$$

where E/R represents the temperature coefficient, comprising the activation energy (E/J mol^{-1}) or (E/ Btu mol^{-1}) and the universal gas constant R (= 8.314 J mol^{-1} K^{-1}). The two-parameter form of the Arrhenius equation (5) is not always adequate to represent the temperature dependence over very wide temperature ranges, as is commonly encountered in combustion systems. The dependence is then usually represented in a three-parameter expression of the form

$$k = A\,T^n\,e^{(-E/RT)} \tag{6}$$

in which *A, n, and E* become the independent variables.

Typically $0 < E < 100$ kJ mol^{-1} (94.86 Btu mol^{-1}) in chain propagation. Some of these reactions have activation energies close to zero, so that the rate constant is virtually independent of temperature. Even at $E >> 0$, but at temperatures such that T is of the order of E/R [for example, when flame temperatures are reached at $T > 2000$ K (3140°F)], the rate constant becomes progressively less sensitive to temperature change. Consequently, the rate constant is then governed mainly by the magnitude of the collision frequency A factor. By contrast, at $E/R >> T$ there is a highly nonlinear response to temperature change. Thus competitive reactions that have significantly different reaction rates at low temperatures may become comparable in rate at high temperatures. The importance of these effects is discussed in next sections.

Chemical Equilibrium in Flame and Postflame Gases

The exponential form of the temperature dependence of elementary reactions establishes how active the combustion system has become by virtue of the intensity of the free radical pool available for propagation of the reaction and generation of heat. It also determines whether or not both the forward and backward reactions have to be taken into account as shown in the following paragraphs.

The intensity of the radical pool in the reaction zone is a function of the rates of termination reactions. For example, an important termination process, associated with the major propagating reactions, is the recombination

$$H + OH + M \leftrightarrow H_2O + M \tag{7}$$

in which M plays the part of a chaperone species (or third body) and represents any molecule in the system that is capable of contributing kinetic energy or taking up excess energy during a collision. If the rate constants for the forward and reverse of Eq. (7) are considered to be $k_{7f} = 1.0 \times 10^{16}$ $e^{(900/T)}$ cm^6 mol^{-2} s^{-1} (i.e., virtually temperature independent) and $k_{7r} = 1.0 \times 10^{17}\ e^{(-58650/T)}$ cm^3 mol^{-1} s^{-1}, then the equilibrium k_{7f} / k_{7r} $(= K_7)$ takes values ranging from 1.4×10^{25} mol^{-1} cm^3 at 1000 K (1340°F) to 2.2×10^9 mol^{-1} cm^3 at 2500 K (4040°F). Thus at low temperatures the forward reaction acts as an extremely effective radical sink insofar that the recombination effectively locks up the propagating species as H_2O. In such circumstances the reverse reaction has virtually no part to play. However, at much higher temperatures the equilibrium is displaced sufficiently far to the left-hand side of the elementary reaction (7) that the H and OH radicals are maintained at a higher concentration [by a factor of 10^8 at 2500 K (4040°F)], arising from the enhanced dissociation of water.

While only the H and OH concentrations are directly affected by the shift of the equilibrium of reaction (7), there is a knock-on effect to other species via other equilibria. For example, at equilibrium reaction (4) is represented by

$$K_4 = (k_{4f} / k_{4r}) = ([OH][O] / [H][O_2]) \tag{8}$$

with rate constants for the forward and reverse processes given by $k_{4f} = 2.00 \times 10^{14}\, e^{(-8455/T)}$ cm^3 mol^{-1} s^{-1} and $k_{4r} = 1.46 \times 10^{13}\, e^{(-252/T)}$ cm^3 mol^{-1} s^{-1}. The equilibrium constant K_4 takes the values 3.75×10^{-3} to 0.5 over the temperature range 1000–2500 K (1340–4040°F). An enhancement of [H] and [OH], in Eq. (7), cause [O] also to be raised when Eq. (4) is equilibrated. At flame and postflame gas temperatures the rate constant for the forward reaction is sufficiently high that an equilibrium between the free radical concentrations for O, H, and OH is rapidly established.

There is an additional effect of Eq. (7) in the reaction zones of flames or combustion systems for which residence time in the reaction zone is exceedingly short. The radical recombination rate in Eq. (7) is comparatively slow because it involves a third body. (It is also strongly pressure-dependent.) This means that, at very short times in the reaction zone, there may not be sufficient time for the equilibrium of Eq. (7) to be established. The radicals H and OH are then maintained at concentrations higher than those that would exist at equilibrium (i.e., super-equilibrium concentrations). Because Eq. (4) involves only bimolecular interactions in its forward and reverse steps, this is readily maintained at equilibrium throughout the reaction zone. Thus the O-atom concentration is also forced higher than that for true chemical equilibrium if [H] and [OH] are in excess. There is sufficient time of residence in the post-reaction zone for the super equilibrium concentrations to relax to their appropriate equilibrium values, which are governed by the prevailing gas temperature. These remarks apply also to all other free radicals whose concentrations are established in bimolecular equilibria of the type exemplified by Eq. (8).

With regard to interpretation of Fig. 25, at temperatures below 1000 K (1340°F) free radical concentrations are low. As a consequence of the powerful radical sink (7), reverse reactions involving the bimolecular interactions of free radicals may be disregarded, such as that of OH and O expressed in Eq. (8) as the reverse of reaction (4). However, at high temperature the reverse of reaction (4), and others like it, play a significant part and must be taken into account in any kinetic analysis.

Relative Rates of Oxidation and Degradation of the Primary Fuel

The propagation reactions are the principal means of reactant consumption by abstraction reactions, for example,

$$RH + OH \rightarrow R + H_2O \tag{9}$$

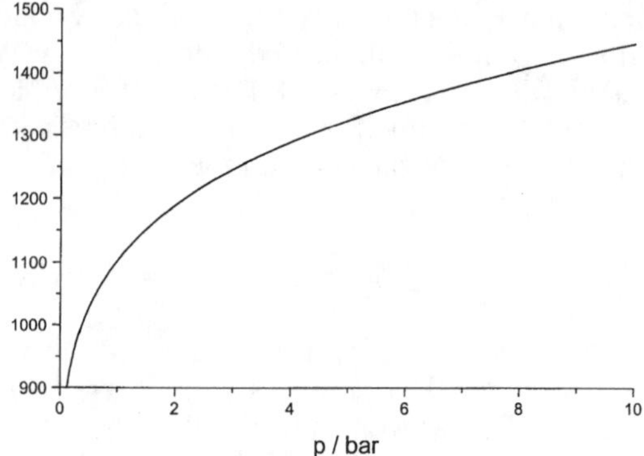

Fig. 25 Pressure vs temperature (K) at which the condition $v_4/v_{16} = 1$ is satisfied.

where R would represent an alkyl radical generated from an alkane. Whether or not this species decomposes or oxidizes depends upon the temperature and the concentration of oxygen. Consider a component of kerosene, such as dodecane ($C_{12}H_{26}$). The dodecyl radical may undergo the competitive reactions

$$\underset{\text{alkyl radical}}{C_{12}H_{25}} \rightarrow \underset{\text{alkene}}{C_6H_{12}} + \underset{\text{lower alkyl radical}}{C_6H_{13}}, \quad k_{10} = 2.5 \times 10^{13}\, e^{(-14433/T)} \text{s}^{-1} \tag{10}$$

$$\underset{\text{alkyl radical}}{C_{12}H_{25}} + O_2 \rightarrow \underset{\text{alkene}}{C_{12}H_{24}} + \underset{\text{hydroperoxy radical}}{HO_2}, \quad k_{11} = 1.0 \times 10^{12}\, e^{(-1000/T)} \text{mol}^{-1}\text{cm}^3\text{s}^{-1} \tag{11}$$

and the relative rates are given by

$$v_{10}/v_{11} = k_{10} \,/\, k_{11}\, [O_2] \tag{12}$$

The relative magnitudes of the rate constants and the relative reaction rates at an oxygen molar density of 2.6×10^{-6} mol cm^{-3} (typical of atmospheric conditions), over the temperature range 750–1500 K (890–2240°F), are given in Table 6.

Table 6 Relative rates of dodecyl radical consumption reactions at an oxygen molar density of 2.6×10^{-6} mole cm^{-3} (typical of atmospheric conditions)

T / K	750	1000	1250	1500
(k_{10}/k_{11}) /mol^{-1} cm^3	4.75×10^{-7}	4.05×10^{-5}	5.83×10^{-4}	3.45×10^{-3}
v_{10}/v_{11}	0.18	15.58	224	1327

The significance of these results is that there is an increasing tendency for the carbon structure of the fuel to be degraded as the temperature increases. Consequently, the chemistry at temperatures below 1000 K (1340°F) tends to be specific to the primary fuel structure, whereas at higher temperatures the reactions that take place are common to a very wide range of fuel molecules. In fact, a species such as the hexyl radical shown as a product of reaction (10) is also capable of decomposing in a similar fashion, as will its further products. Hence, the generality of the Warnatz (1984) mechanism (Fig. 24) for the oxidation of alkanes at temperatures above 1000 K is well supported. The relative rates of reactions (10) and (11) are susceptible to the prevailing oxygen concentration, such that decomposition (10) would be favored in very fuel-rich conditions. The predominance of oxidation would be maintained to a higher temperature during combustion at reaction pressures above ambient, even with air as the oxidizing medium.

Relative Rates of Initiation of the Primary Fuel

The spontaneous initiation of reaction may occur by processes that are similar to those cited for the primary propagation, although the activation energies involved are considerably greater. Thus, from the very start, the mechanism may be predisposed to be very similar for a wide range of fuel components, but the trend is strongly temperature dependent, as seen from the following competition for dodecane consumption:

$$\underset{\text{alkane}}{C_{12}H_{26}} \rightarrow \underbrace{C_6H_{13} + C_6H_{13}}_{\text{lower alkyl radicals}}, \qquad k_{13} = 4.0 \times 10^{16}\, e^{(-44743/T)}\text{s}^{-1} \tag{13}$$

$$\underset{\text{alkane}}{C_{12}H_{26}} + O_2 \rightarrow \underset{\text{alkyl radical}}{C_{12}H_{25}} + \underset{\text{hydro-peroxy radical}}{HO_2}, \tag{14}$$

$$k_{14} = 1.8 \times 10^{14}\, e^{(-25000/T)}\text{mol}^{-1}\text{cm}^3\text{s}^{-1}$$

and the relative rates are given by

$$v_{13}/v_{14} = k_{13} / k_{14}\,[O_2] \tag{15}$$

The data in Table 7 were derived in a manner similar to those in Table 6. Although there is a qualitative similarity to the switch of the propagation reactions, at initiation the degradation of the primary fuel molecule becomes

Table 7 Relative rates of dodecane consumption reactions at an oxygen molar density of 2.6×10^{-6} mole cm^{-3} (typical of atmospheric conditions)

T / K	750	1000	1250	1500
(k_{13}/k_{14})/mol^{-1} cm^3	8.21×10^{-10}	5.92×10^{-7}	3.07×10^{-5}	4.28×10^{-4}
v_{13}/v_{14}	3.16×10^{-4}	0.23	11.8	165

predominant at a higher temperature than that during propagation. The absolute rates of either of the initiation steps are considerably slower than those of the corresponding propagation reactions. Not only do reactions (13) and (14) have very different activation energies, but also they have significantly different endothermicities. Whereas reaction (14), and reactions like it of other similar hydrocarbons, is about 200 kJ mol^{-1} endothermic, typically, the endothermicity of reaction (13) exceeds 350 kJ mol^{-1}.

Chain Branching

The self-sustaining nature of combustion processes arises mainly from the autocatalysis of the chemistry through chain branching, coupled to physical transport processes within the combustion system. Reaction (4) has a particularly important role in this respect because H atoms diffuse so readily. Competition with the branching process (4) arises in the kinetics from the recombination reaction

$$H + O_2 + M \rightarrow HO_2 + M, \quad k_{16} = 2.3 \times 10^{18}\, T^{-0.8}\, \text{mol}^{-2}\, \text{cm}^6\, \text{s}^{-1} \tag{16}$$

The relative reaction rates are given by

$$v_4 / v_{16} = k_4 / k_{16}[\text{M}] \tag{17}$$

Thus, whether or not chain branching predominates is strongly dependent on temperature, as shown in Table 8, where the total species concentration [M] is 1.22×10^{-5} mol^{-1} cm^3 [corresponding to air at 1 bar and 1000 K (14.7 psia and 1340°F)].

Chain branching is the more important step of the two at temperatures above ~1200 K (1700°F) at atmospheric pressure because there is an inverse dependence on the total concentration of species in the system. However, the opportunity for chain branching through this competition becomes progressively less accessible as the pressure is raised. The shift in balance may be illustrated with respect to the pressure vs. temperature at which the condition $v_4/v_{16} = 1$ is satisfied (Fig. 25). Once the temperature has risen above about 1400 K (> 2000°F), chain branching via Eq. (4) is inevitable at the pressures associated with virtually all combustion applications. Consequently exceedingly vigorous and strongly exothermic reaction is easily sustained. That is, combustion not subject to kinetic constraints and fluid

Table 8 Relative rates of primary chain branching and recombination reactions corresponding to air at 1000 K and 1 bar

T / K	750	1000	1250	1500
(k_4/k_{16}) /mol^{-1} cm^3	1.09×10^{-7}	2.33×10^{-6}	1.49×10^{-5}	5.3×10^{-5}
v_4/v_{16}	8.9×10^{-3}	0.19	1.22	4.34

mechanical considerations may become efficiency controlling. However, kinetic considerations may still be relevant in circumstances for which pollutant formation is a major issue.

When reaction (16) is important, H, O, and OH radical propagation is displaced by HO_2 (hydroperoxy) radical propagation. This restricts that rate of propagation because the activation energies of its molecular reaction are relatively high. That is, reactions of the kind

$$RH + HO_2 \rightarrow R + H_2O_2 \tag{18}$$

may have activation energies in the range 50–80 kJ mol^{-1}, according to whether the H atom that is abstracted is from a tertiary, secondary, or primary site in the alkane fuel (RH). Even if the attack is made on an intermediate molecular species with a labile H atom, such as formaldehyde

$$CH_2O + HO_2 \rightarrow CHO + H_2O_2 \tag{19}$$

the activation energy exceeds 40 kJ mol^{-1}. An alternative reaction for HO_2 involves its recombination

$$HO_2 + HO_2 \rightarrow H_2O_2 + O_2 \tag{20}$$

for which there is no activation energy, but its rate has a second-order (i.e., squared) dependence on the free radical concentration.

The combination of reactions represented by Eqs. (18) and (11) gives an overall stoichiometry (based on $C_{12}H_{26}$) of the form

$$C_{12}H_{26} + O_2 = C_{12}H_{24} + H_2O_2 \tag{21}$$

which is virtually thermoneutral. Thus, if the reaction is driven into conditions that are dominated by HO_2 propagation, not only is the reaction relatively slow but also the rate of heat release is very modest.

However, hydrogen peroxide (H_2O_2) is relatively unstable and undergoes the decomposition

$$H_2O_2 + M \rightarrow 2\,OH + M, \qquad k_{22} = 3 \times 10^{17}\, e^{(-23635/T)}\ cm^3\ mol^{-1}\ s^{-1} \tag{22}$$

to yield OH radicals. At atmospheric pressure the half-life for this reaction is less than 1 ms at $T > 1000$ K (1340°F) but increases to more than 40 ms if the temperature is lower than 800 K ($T < 1000$°F). Thus, there is a potential for regeneration of OH radicals as the main propagating species, but when residence times are very short, it is viable only in higher temperature ranges. The reaction is not strictly one of chain branching if reaction (20) is the main route to hydrogen peroxide formation, but the conversion of HO_2 to OH via Eqs. (20) and (22) is important, nevertheless. The reaction constitutes chain branching if the decomposition of hydrogen peroxide by Eq. (22) is preceded by its formation through reactions like Eqs. (18) or (19).

Relationship of Ignition Delay to Chemical Timescales and Chain-Branching Factors

The rate of development of spontaneous ignition, and the ignition delay, is a complicated function of the kinetic development and thermal feedback from the exothermic reaction. Clearly, the physical configuration of the engine is also important, but beyond the present scope. The dependence of the duration of the ignition delay on temperature and pressure can be interpreted from the timescales associated with the chain-branching kinetics. The characteristic timescales are obtained from the net branching factor φ. If reactions (4) and (16) are considered to represent the competition between the branching and nonbranching reaction channels, then the net branching factor (φ_1 / s^{-1}) is defined as

$$\varphi_1 = 2k_4 + k_{16}[M] \tag{23}$$

According to Eq. (23), the reaction can accelerate exponentially at all conditions for which φ_1 is greater than zero, and the exponential growth is a function of the rate constants k_4 and k_{16} and the prevailing pressure. A nonbranching reaction would appear to occur for $\varphi_1 < 0$ although in practice there may be another opportunity for chain branching governed by

$$\varphi_2 = k_{22}[M] \tag{24}$$

Nevertheless, there are quite different timescales associated with each of these branching factors, defined respectively as $t_1 = 1 / (\varphi_1[O_2])$ and $t_2 = (1 / \varphi_2)$. The temperature dependence of each of t_1 and t_2 are shown over a range of pressures up to 5 bar (73.5 psia) in Fig. 26. The ordinate is cut off at 0.4 ms in Fig. 26 because each of the curves is already close to its asymptotic approach to a limiting temperature at which $t \rightarrow \infty$. These timescales give a guide to the typical duration of ignition delays insofar that they represent the time taken for the chemical reaction rate to multiply by a factor of 2.7 (i.e., a factor of e^1) as a result of chain branching.

Whereas t_2 decreases with increasing pressure at any given temperature,

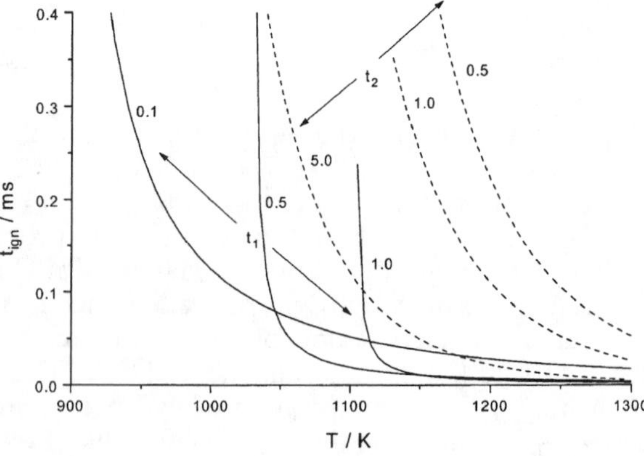

Fig. 26 Temperature dependence of timescales associated with t_1 and t_2 chain-branching factors. Labels are pressure in bar.

t_1 exhibits a more complex pressure and temperature dependence, as can be seen from the crossover of its curves at different pressures. For pressures up to 1 bar and temperatures up to 1300 K (14.7 psia and 1880°F), t_1 is always considerably shorter than t_2, and it will be the principal controlling factor in the determination of the development of spontaneous ignition. However, at higher pressures within this temperature range, t_2 may be the shorter of the two [for example, $\varphi_1 > 0$ is not accessible below 1300 K at p = 5 bar (1880°F and 73.5 psia)], and therefore t_2 controls the development of spontaneous ignition under these conditions.

Although not implicit in Fig. 26, spontaneous gas phase ignition of hydrocarbon fuels does occur at temperatures below 900 K (1160°F) because supplementary chain reactions involving alkylperoxy radicals (RO_2) occur, and there is also thermal feedback from the exothermicity of reaction. Simple relationships such as Eqs. (23) and (24) cannot be derived for predicting the rate of development of reaction in the lower temperature region. Numerical simulation, involving detailed kinetic models coupled to thermal feedback, has to be used to predict the response of the system. This is also the case for a fully quantitative interpretation of the behavior over the ranges of temperature and pressure already discussed. Numerical approaches are reviewed in a later section.

Kinetic Consequences of the Degradation of Endothermic Fuels Prior to Combustion

A repercussion of using the fuel as a heat sink is that the composition entering the combustion zone contains proportions of degraded products. That is, some of the pyrolytic stages of the combustion process that lead to the initial breakdown of the fuel to smaller species occur in the endothermic reactor, and so the chemistry in the main reaction zone is dominated by the combustion chemistry of smaller molecular species. The kinetic implications may be illustrated by reference to the reaction pathways of *n*-heptane in diffusion flames (Fig. 27), although normally the free radicals would not be able to survive sufficiently long for them to be carried through from the endothermic reactor to the reaction zone. Nevertheless, as discussed earlier the energetics of the combustion process may be increased by these transformations, and the reactivity may be enhanced by the presence of more reactive fuel components, such as hydrogen.

B. Present State of Chemical Kinetics

The preceding discussion underscores the importance of understanding the combustion chemical kinetics of hydrocarbons for application to hypersonic systems that are limited by ignition issues. However, studies concerned with larger alkane molecules commonly found in jet fuels are scarce. The rate constants for the thermal decomposition of such molecules have been determined experimentally (Rumyantsev et al., 1980). A detailed chemical kinetic mechanism for the combustion of endothermic fuels must include higher-order normal paraffin molecules. Experimental and modeling studies

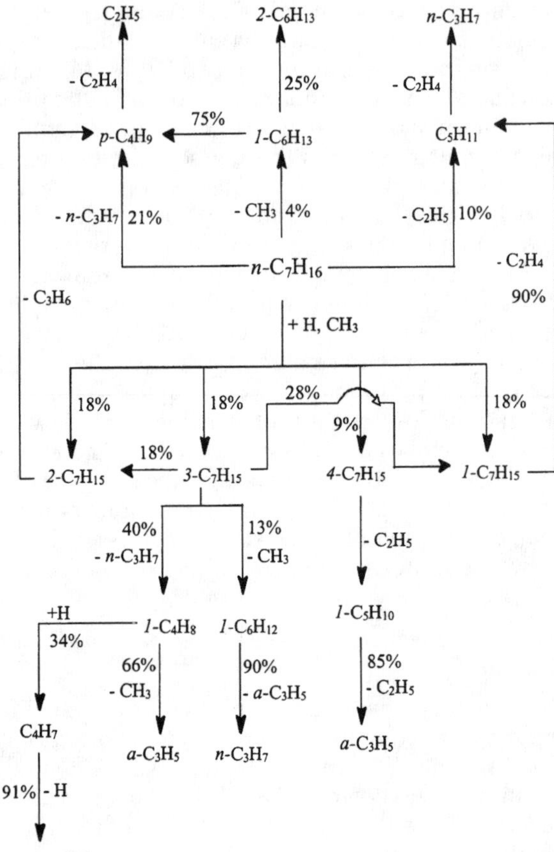

Fig. 27 Consumption paths of *n*-heptane in diffusion flames (Maurice, 1996).

of the combustion kinetics of normal paraffins generally have been confined to molecules containing less than eight carbon atoms. The oxidation of *n*-decane has also been studied in jet-stirred reactors at pressures from 0.06 – 40 bar (.88 – 588 psia) and low to intermediate temperatures [≤ 1150 K (1610°F)] by Balès-Guéret et al. (1992) and Dagaut et al. (1994). Finally, the chemical structures of subatmospheric (Delfau et al., 1990) and atmospheric pressure (Douté et al., 1995) *n*-decane laminar premixed flames have been investigated experimentally. Detailed kinetic modeling of the low-pressure flames focused primarily on benzene formation (Vovelle et al., 1991, 1994).

The ability of the hierarchically constructed detailed mechanism of Lindstedt and co-workers to capture the combustion kinetics of the products of the endothermic reaction has been established (e.g., Leung and Lindstedt, 1995; Leung, 1996; Lindstedt and Skevis, 1997; Skevis, 1996). The extension of the mechanism to *n*-decane has resulted in a predictive tool applicable to

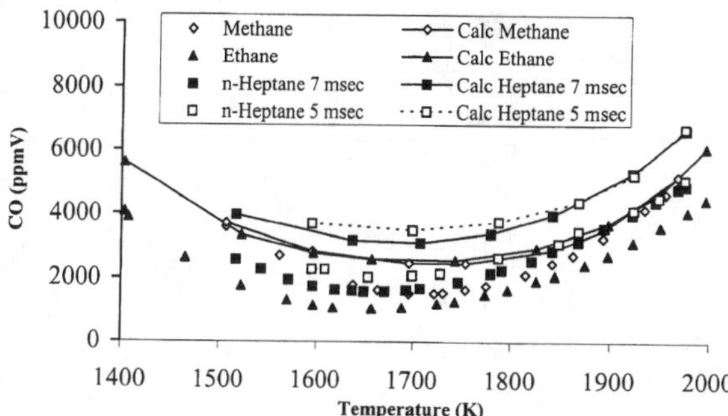

Fig. 28 Computed and measured CO emissions from alkanes vs reactor temperature. Residence times are ~ 7.3 ms for solid symbols and lines and ~ 5.3 ms for hollow symbols and dotted lines. Temperature variations are affected by varying equivalence ratio.

the parent endothermic fuel as well as to the products of the endothermic reforming reaction.

The mechanism is applied to predict the combustion chemistry of a variety of endothermic fuel components in the well-stirred reactor. Sample carbon monoxide concentration predictions, indicative of combustion efficiency, for pure fuels and fuel blends are shown in Figs. 28 to 30. Theoretical predictions of fuel consumption, carbon dioxide, carbon monoxide, and nitrogen oxides for hydrocarbons ranging from methane to ethylbenzene are generally within levels of experimental uncertainties. The mechanism can hence be applied to analytically study and predict the ignition behavior of practical aviation fuels and help guide fuel formulation efforts.

A fundamental understanding and a predictive capability for benzene, PAH, and soot formation is of considerable fundamental and practical significance in endothermic fuel reactors as well as flames. The highly integrated nature of the fuel and propulsion systems of high-speed vehicles, where the fuel serves as a coolant as well as the primary energy source, underscores the need for a fuel model that can be used from tank to combustor over a broad range of fuel temperatures and pressures. Moreover, reacting timescales for benzene and PAH molecules are slower than for aliphatic molecules; hence, aromatic formation may be detrimental to

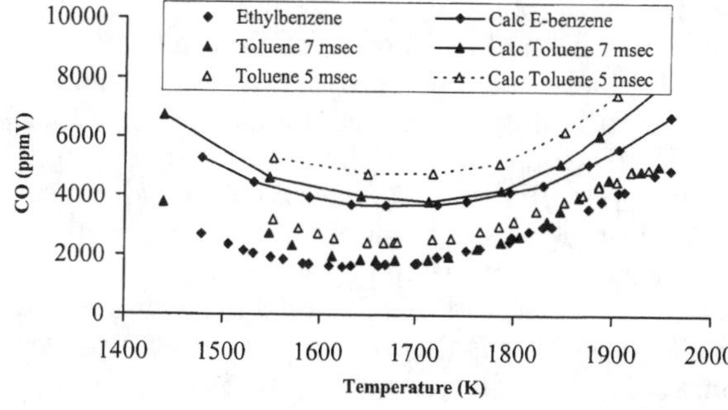

Fig. 29 Computed and measured CO emissions from aromatics vs reactor temperature. Residence times are ~ 7.3 ms for solid symbols and lines and ~ 5.3 ms for hollow symbols and dotted lines. Varying equivalence ratio affects temperature variations.

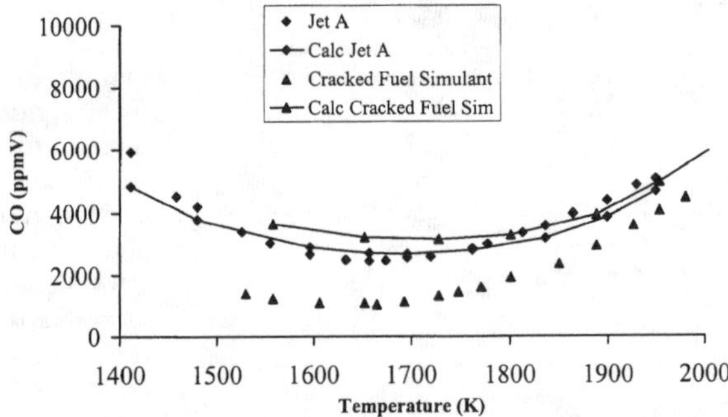

Fig. 30 Computed and measured CO emissions from hydrocarbons mixes vs reactor temperature at $\tau \sim 7.3$ ms. Temperature variations are affected by varying equivalence ratio. Symbols denote experimental data, and lines denote computations. Jet A is computationally represented by a surrogate model comprising 78% *n*-decane and 22% ethylbenzene by volume.

combustor performance. Finally, if soot is a substantial accompanying product of the combustion product, there is an accompanying loss of energy, and hence performance.

C. Combustor Development Considerations

In the past successful supersonic combustion of hydrocarbon fuels has involved severe chemical enhancement via pyrophoric (and generally toxic) materials such as silane or boranes, injection of a very reactive oxidizer (i.e., chlorine trifluoride), hydrogen enhancement, the use of catalysts within a pilot, or the use of pilots in which a portion (or all) of the hydrocarbon fuel was partially reacted (Billig, 1993; Jensen and Braendlein, 1996; Waltrup, 1997). Alternatives that are presently being explored include 1) the use of various kinds of pilots/flameholders, including plasma ignitors; 2) ionization of the fuel or airstream (Gurijanov and Harsha, 1996; Morris et al., 1998); and 3) other modifications of the fuel composition to yield highly reactive fuels. The latter may invoke the use of nontoxic additives, such as dialkyl ethers, as ignition aids. Fundamentally, the energy barriers associated with the decomposition of oxygenated compounds such as ethers are considerably lower than those associated with hydrocarbon depletion initiation reactions. Moreover, OH radical attack reactions on hydrocarbons generally feature lower energy barriers than H abstraction via H and O atom attack and pyrolysis reactions (Table 9). Consequently, the use of ethers might provide an enhanced OH radical pool that could decrease the overall timescales of hydrocarbon consumption. Moreover, endothermic fuel reactions can be expected to improve the fuel's combustion properties, if for no other reason than that the fuel enters in a hot, vaporized state. Furthermore, the mixing length of a supersonic combustor is theoretically inversely proportional to the gas diffusion coefficient (Heiser and Pratt, 1994). The gas

Table 9 Typical *n*-heptane consumption reactions (Maurice, 1996)

Reaction	*A*, $(kmol/m^3)^{1-m}/s$, *m* is reaction order	*n*	*Ea*, kJ/kmole
$C_7H_{16} \leftrightarrow n\text{-}C_3H_7 + p\text{-}C_4H_9$	2.500*E* + 13	0.00	261.90
$C_7H_{16} + H \leftrightarrow 2\text{-}C_7H_{15} + H_2$	1.820*E* + 04	2.00	20.92
$C_7H_{16} + O \leftrightarrow 2\text{-}C_7H_{15} + OH$	6.400*E* + 02	2.50	20.92
$C_7H_{16} + OH \leftrightarrow 2\text{-}C_7H_{15} + H_2O$	2.350*E* + 04	1.61	0.00

diffusion coefficient is in turn inversely proportional to the square root of the molecular weight of the diffusing components (*The Chemical Engineers Handbook,* 1973). Therefore, the large amounts of hydrogen generated from endothermic reactions such as reforming might also be expected to improve the fuel combustion characteristics by lowering the average fuel molecular weight. Finally, techniques developed to enhance the combustion characteristics of liquid hydrocarbon fuels may also be applicable to solid and slurry fuels (which feature increased energy density, hence increased vehicle range), thereby providing further potential performance enhancements for future hypersonic systems.

D. Prospects for Modeling Large Kinetic Systems

Kinetic mechanisms generally employed to model chemical reactions in combustion processes can be divided into four classes: detailed, skeleton, reduced, and global. A detailed reaction mechanism is a verified and validated comprehensive kinetic mechanism that contains sufficient detail to reproduce a wide range of observed trends. A skeleton reaction mechanism is a subset of a detailed kinetic mechanism that contains all of the relevant information necessary to reproduce a particular observed feature. A reduced reaction mechanism is a form of the skeleton mechanism from which the independent scalars have been removed on the basis of rigorous analysis. A global reaction mechanism is a series of empirically derived reaction step(s) fitted to a specific data set. The mechanism is generally only valid to reproduce the fitted data. Although detailed chemical kinetic modeling of practical scramjet combustors is not computationally feasible at present, detailed kinetic models can be used to address such questions as the optimum fuel constituents for minimum ignition delay or maximum reactivity. Modeling can also point out key radicals or ions that most affect the combustion chemistry.

Successful numerical modeling as a predictive mode to the performance of practical combustion systems requires excellence on a number of different fronts. At the computational end there is a major issue of the coupling of the thermokinetic model to the fluid dynamic code. The description of the physics itself (fluid dynamics and heat transfer), as applied to air-breathing hypersonic combustors, is outside the scope of this review. Our emphasis is on the chemical input, which involves the following components: 1) an accurate knowledge of thermochemical and kinetic data or reliable predictive models for such data, 2) a detailed kinetic and thermochemical model constructed to represent the combustion chemistry over wide temperature and pressure ranges and validated by comparison with experiment, and 3) formal methods for mechanism reduction to an acceptable level of tractability while retaining accuracy of the global predictions to be used in concert with the fluid dynamic model.

Detailed reaction mechanisms comprising around hundreds of species and thousands of elementary reactions are presently too complex for direct application to multidimensional fluid dynamics problems. *Simplified* reduced kinetics reaction mechanisms must be developed. However, reduced kinetic mechanisms are only as reliable as the *detailed* mechanism upon which they are based. It is therefore essential that the detailed mechanism feature the capability to address the fundamental science of the practical problems. It is also important to understand the limitations of the detailed kinetic mechanism. These important subjects are extensively treated in the Appendix.

V. Summary

This chapter has addressed the issues associated with the application of hydrocarbon fuels to hypersonic missiles and hypersonic cruise and space access vehicles. Endothermic fuels have been identified as the critical technology to enable the use of hydrocarbons at hypersonic speeds. The salient challenges to adopting endothermic fuels from a chemistry perspective are recognized as deposition from high-temperature effects on the fuel (fuel thermal stability, fuel/material compatibility) and fuel reactivity. Approaches to extend fuel heat-sink capability, improvement of hydrocarbon combustion properties, and fuel-system fouling mitigation were discussed, and the reader is referred to the appropriate references. Combustion chemical kinetics are identified as a critical issue for hydrocarbon-fueled scramjets. Finally, a review of the fundamental steps necessary to adopt chemical kinetic mechanisms into multidimensional computations is provided in the Appendix.

Acknowledgments

The authors gratefully acknowledge the support of the Air Force Office of Scientific Research (Julian Tishkoff), the HyTech Office (Bob Mercier), and the University of Leeds Chemistry Department. We are also indebted to the editors of this volume for their helpful comments.

Bibliography

Albright, L. F., and Marek, J. C., "Mechanistic Model for Formation of Coke in Pyrolysis Units Producing Ethylene," *Ind. Eng. Chem. Res.*, Vol. 27, 1998, pp. 755–759.

Atria, J. V., and Edwards, T., "High Temperature Cracking and Deposition Behavior of an *n*-Alkane Mixture," *ACS Petroleum Chemistry Div. Permits,* Vol. 41, No. 2, 1996, pp. 498–501.

Atria, J. V., Cermignani, W., and Schobert, H. H., "Nature of High Temperature Deposits from *n*-Alkanes in Flow Reactor Tubes," *ACS Petroleum Chemistry Division Preprints,* Vol. 41, No. 2, 1996, pp. 493–497.

Baker, R. T. K., and Harris, P. S., "The Formation of Filamentous Carbon," *Chemistry and Physics of Carbon,* edited by P. L. Walker, Vol. 14, 1978, pp. 83–165.

Baker, R. T. K., Yates, D. J. C., and Dumesic, J. A., "Filamentous Carbon Formation over Iron Surfaces," *Coke Formation on Metal Surfaces,* edited by Albright and Baker, ACS Symposium Series 202, 1982, pp. 1–21.

Baker, R. T. K., "Catalytic Growth of Carbon Filaments," *Carbon,* Vol. 27, No. 3, 1989, pp. 315–323.

Baker, R. T. K., "Coking Problems Associated with Hydrocarbon Conversion Processes," *ACS Fuel Chemistry Division Preprints,* Vol. 41, No. 2, 1996, pp. 521–524.

Balès-Guéret, C., Cathonnet, M., Boettner, J. C., and Gaillard, F., "Experimental Study and Kinetic Modeling of Higher Hydrocarbon Oxidation in a Jet-Stirred Reactor," *Energy and Fuels,* Vol. 6, 1992, pp. 189–194.

Baulch, D. L., Cobos, C. J., Cox, R. A., Esser, C., Frank, P., Just, Th., Kerr, J. A., Pilling, M. J., Troe, J., Walker, R. W., and Warnatz, J. J., "Evaluated Kinetic Data for Combustion Modelling," *Journal of Physical Chemistry,* Ref. Data, Vol. 21, 1992, pp. 411–734.

Baulch, D. L., Cobos, C., Cox, R. A., Frank, P., Just, Th., Kerr, J. A., Murrells, T., Pilling, M. J., Troe, J., Walker, R. W., and Warnatz, J. J., "Evaluated Kinetic Data for Combustion Modelling," *Journal of Physical Chemistry,* Ref. Data, Vol. 23, 1994, pp. 847–1033.

Baulch, D. L, "Kinetic Databases," *Comprehensive Chemical Kinetics,* Low-Temperature Combustion and Ignition, edited by M.J. Pilling, Vol. 35, Elsevier, Amsterdam, 1992 pp. 235–292.

Beach, H. L., "Supersonic Combustion Status and Issues," *Major Research Topics in Combustion,* edited by Hussaini, Kumar, and Voigt, Springer–Verlag, New York, 1992.

Beal, E. J., Hardy, D. R., and Burnett, J. C., "Results and Evaluation of a Jet Fuel Thermal Stability Flow Device Which Employs Direct Gravimetric Analysis of Both Surface and Fuel Insoluble Deposits," *4th International Conference on Stability and Handling of Liquid Fuels,* 1991, pp. 245–259 (DOE/CONF 911102).

Beaver, B. D., Demunshi, R., Sharief, V., Tian, D., and Teng, Y., "Development of Oxygen Scavenger Additives for Jet Fuels," *5th International Conference on Stability and Handling of Liquid Fuels,* 1994, pp. 241–254 (DOE/CONF 941022).

Benson, S. W., *Thermochemical Kinetics,* 2nd., Wiley, New York, 1976.

Bergholz, R. F., and Hitch, B. D., "Thermal Management Systems for High Mach Airbreathing Propulsion," AIAA Paper 92-0515, Jan. 1992.

Billig, F. S., "Propulsion Systems from Takeoff to High-Speed Flight," *High-Speed Flight Propulsion Systems,* Vol. 137, Progress in Astronautics and Aeronautics, edited by S.N.B. Murthy and E.T. Curran, AIAA, Washington, DC, 1991.

Billig, F. S., "Research on Supersonic Combustion," *Journal of Propulsion and Power,* Vol, 9, No. 4, 1993, pp. 499–514.

Blust, J. W., Ballal, D. R., and Sturgess, G. J., "Emissions Characteristics of Liquid Hydrocarbons in a Well Stirred Reactor," AIAA Paper 97-2710, July 1997.

Bradley, R. P., and Martel, C. R., "Effect of Test Pressure on Fuel Thermal Stability Test Methods," AFAPL-TR-74-81, April 1975.

Brezinsky, K., "The High-Temperature Oxidation of Aromatic Hydrocarbons," *Progress in Energy and Combustion Science,* Vol. 12, 1986, pp. 1–24.

Burcat, A., and McBride, B. J., "1994 Ideal Gas Thermodynamic Data for Combustion and Air Pollution Use," TAE697, 1994.

Burdette, G. W., Lander, H. R., and McCoy, J. R., "High-Energy Fuels for Cruise Missiles," *Journal of Energy,* Vol. 2, No. 5, 1978, pp. 289–292.

Castaldi, M. J., Marinov, N. M., Melius, C. F., Huang, J., Senkan, S. M., Pitz, W. J., and Westbrook, C. K., "Experimental and Modeling Investigation of Aromatic and Polycyclic Aromatic Hydrocarbon Formation in a Premixed Ethylene Flame," *Twenty-Sixth Symposium (International) on Combustion,* Vol. 1, Combustion Inst., 1996, pp. 693–702.

Chakir, A., Belliman, M., Boettner, J. C., and Cathonnet, M., "Kinetic Study of *n*-Heptane Oxidation," *International Journal Chem. Kin.,* Vol. 24, 1992, pp. 385–140.

Chen, F., Tam, W. F., Shimp, N. R., and Norris, R. B., "An Innovative Thermal Management System for a Mach 4 to Mach 8 Scramjet Engine," AIAA Paper 98-3734, July 1998.

Chevalier, C., Pitz, W. J., Warnatz, J., Westbrook, C. K., and Melenk, H., "Hydrocarbon Ignition: Automatic Generation of Reaction Mechanisms and Application to Modeling of Engine Knock," *Twenty-Fourth Symposium (International) on Combustion,* Combustion Inst., Pittsburgh PA, 1992, pp. 93–101.

Chin, L. P., and Katta, V. R., "Numerical Modeling of Deposition in Fuel-Injection Nozzles," AIAA Paper 95-0497, Jan. 1995.

Chinnick, S. J., Baulch, D. L., and Ayscough, P. B., "An Expert System for Hydrocarbon Pyrolysis Reactions," *Chemometrics and Intelligent Laboratory Systems,* Vol. 5, 1988, pp. 39–52.

Cohen, N., and Westberg, K. R., "Evaluation and Compilation of Chemical Kinetic Data," *Journal of Physical Chemistry,* Vol. 83, 1979, pp. 46–50.

Cohen, N., and Westberg, K. R., "Chemical Kinetic Data Sheets for High Temperature Reactions, Part II," *Journal of Physical Chemistry,* Ref. Data, Vol. 20, 1991, pp. 1211–1311.

Come, G. M., Warth, V., Glaude, P. A., Fournet, R., Battin-Leclerc, and Scacchi, G., "Computer-Aided Design of Gas-Phase Oxidation Mechanisms: Application to the Modeling of *n*-Heptane and *i*-Octane Oxidation," *Twenty-Sixth Symposium (International) on Combustion,* Combustion Inst., Pittsburgh PA, 1996, pp. 755–762.

Coordinating Research Council, "CRC Literature Survey on the Thermal Oxidation Stability of Jet Fuel," CRC Rept. No. 509, April 1979.

Coordinating Research Council, "Handbook of Aviation Fuel Properties," CRC Rept. No. 530, 1983.

Croswell, B. M., and Biddle, T. B., "High Temperature Fuel Requirements and Payoffs," *Aviation Fuel: Thermal Stability Requirements,* edited by Perry W. Kirklin and Peter David, American Society for Testing and Materials, Philadelphia, 1992 (ASTM STP 1138).

Crynes, B. L., and Albright, L. F., "Thermal Cracking," *Encyclopedia of Physical Science and Technology,* Vol. 3, Academic, 1987, pp. 768–785.

Curran, E. T., "Introduction," *High-Speed Flight Propulsion Systems,* edited by S. N. B. Murthy and E. T. Curran, Vol. 137, Progress in Astronautics and Aeronautics, AIAA, Washington, DC, 1991.

Curran, H. J., Gaffuri, P., Pitz, W. J., Westbrook, C. K., and Leppard, W. R., "Autoignition Chemistry in a Motored Engine: An Experimental and Kinetic Modelling Study," *Twenty-Sixth Symposium (International) on Combustion* Vol. 2 Combustion Inst., Pittsburgh, PA, 1996, pp. 2669–2677.

Dagaut, P., Reuillon, M., and Cathonnet, M., "High Pressure Oxidation of Liquid Fuels from Low to High Temperature. 3. *n*-Decane," *Combustion Science and Technology,* Vol. 103, 1994, pp. 349–359.

Darrah, S., "Jet Fuel Deoxygenation," Air Force Wright Aeronautical Lab. AFWAL-TR-88-2081, 1988.

Delfau, J. L., Bouhria, M., Reuillon, M., Sanogo, O., Akrich, R., and Vovelle, C., "Experimental and Computational Investigation of the Structure of a Sooting Decane-O_2-Ar Flame," *Twenty-Third Symposium (International) on Combustion,* Combustion Inst., Pittsburgh, PA, 1990, pp. 1567–1572.

Dougherty, E. P., and Rabitz, H. J., "Computational Kinetics and Sensitivity Analysis of Hydrogen-Oxygen Combustion," *Chemical Physics,* Vol. 72, 1980, pp. 6571–6586.

Douté, C., Delfau, J. L., Akrich, R., and Vovelle, C., "Chemical Structure of Atmospheric Pressure Premixed *n*-Decane and Kerosene Flames," *Combustion Science and Technology,* Vol. 106, 1995, pp. 327–344.

Dreitser, G. A. et al., "Integrated Study of Scientific and Applied Problems on Heat Transfer Enhancement in Tubular Heat Transfer Apparatuses," *Journal of Engineering Physics and Thermophysics,* Vol. 65, No. 1993, pp. 638–643.

Dryer, F. L., "The Phenomenology of Modelling Combustion Chemistry," *Fossil Fuel Combustion,* edited by W. Bartok, and A. F. Sarofim, Wiley, New York, 1989, p. 121.

Dutton, R. A., "An Experimental Investigation of the Suitability of JP-4 Fuel for the Regenerative Cooling of a Hypersonic Ramjet Engine," Marquardt Rep. S-154, March 1960.

Edwards, T., "USAF Supercritical Fuels Interests," AIAA Paper 93-0807, Jan. 1993.

Edwards, T., and Anderson, S., "JP-7 Thermal Cracking Assessment," AIAA Paper 93-0806, Jan. 1993.

Edwards, T., and Zabarnick, S., "Supercritical Fuel Deposition Mechanisms," *Industrial and Engineering Chemistry Research,* Vol. 32 1993, pp. 3117–3122.

Edwards, T., and Krieger, J., "The Thermal Stability of Fuels at 480 C. Effect of Test Time, Flow Rate, and Additives," *ASME Turbo Expo '95* (Houston, TX), 195 (ASME 95-GT-68) 1995.

Edwards, T., and Liberio, P., "The Thermal-Oxidative Stability of Fuels at 480 C (900 F)," *ACS Petroleum Chemistry Division Preprints,* Vol. 39, No. 1, 1994, pp. 86–91.

Edwards, T., and Atria, J., "Deposition from High Temperature Jet Fuels," *ACS Petroleum Chemistry Division Preprints,* Vol. 40 No. 41, 1995, pp. 649–654.

Edwards, T., "Combustion Challenges of High Temperature Jet Fuels," *AIChE Annual Meeting* (Chicago, IL), Nov. 1996 (Paper 204a).

Edwards, T., "Research in Hydrocarbon Fuels for Hypersonics," *1996 JANNAF Airbreathing Propulsion Subcommittee Meeting* Vol. I, 1996, pp. 17–26 (CPIA Publication 654).

Ervin, J. S., and Zabarnick, S., "Computational Fluid Dynamics Simulation of Jet

Fuel Oxidation Incorporating Pseudo-Detailed Chemical Kinetics," *Energy and Fuels,* Vol. 12, 1998, pp. 344–352.

Ervin, J., Williams, T., and Hartman, G., "Flowing Studies of an Endothermic Fuel at Supercritical Conditions," AIAA Paper 98-3760, July 1998.

Eser, S., "Mesophase and Pyrolytic Carbon Formation in Aircraft Fuel Lines," *Carbon,* Vol. 34, No. 4 1996, pp. 539–547,

Fabuss, B. M., Smith, J. O., Lait, R. I., Borsanyi, A. S., and Satterfield, C. N., "Rapid Thermal Cracking of *n*-Hexadecane at Elevated Pressures," *I&EC Process Design and Development,* Vol. 1, No. 4, 1962, pp. 293–299,

Fabuss, B. M., Smith, J. O., Lait, R. I., Fabuss, M. A., and Satterfield, C. N., "Kinetics of Thermal Cracking of Paraffinic and Naphthenic Fuels," *I&EC Process Design and Development,* Vol. 3, No. 1, 1964, pp. 33–37,

Faith, L. E., Ackerman, G. H., and Henderson, H. T., "Heat Sink Capabilities of Jet A Fuel: Heat Transfer and Coking Studies," NASA CR-72951, July 1971.

Favorskii, O. N., and Kurziner, R. I., "Development of Air-Breathing Engines for High-Speed Aviation by Combining Advances in Various Areas of Science and Engineering," *High Temperature,* Vol. 28, 1990, pp. 606–614.

Giovanetti, A. J., Spadaccini, L. J., and Szetela, E. J., "Deposit Formation and Heat-Transfer Characteristics of Hydrocarbon Rocket Fuels," *Journal of Spacecraft,* Vol. 22, No. 5, 1985, pp. 574–580.

Glickstein, M. R., and Spadaccini, L. J., "Applications of Endothermic Reaction Technology to the High Speed Civil Transport," *JANNAF Airbreathing Propulsion Subcommittee Meeting,* Vol. I, 1997, pp. 99–106 (CPIA Publication 666),

Griffiths, J. F., "Reduced Kinetic Models and Their Application to Practical Combustion Systems," *Progress in Energy and Combustion Science,* Vol. 21, 1995, pp. 25–107.

Griffiths, J. F., and Barnard, J. A., *Flame and Combustion,* Chapman and Hall, London, 1995.

Guisinger, S. J., and Rippen, M. E., "Critical Property Determination of Advanced Fuels," *ACS Petroleum Chemistry Division Preprints,* Vol. 34, No. 4, 1989, pp. 885–896.

Gurijanov, E. P. and Harsha, P. T., "Ajax: New Directions in Hypersonic Technology," AIAA Paper 96-4609, Nov. 1996.

Hanson, R. K., and Salimian, "Rate Coefficients in the N/H/O System," *Combustion Chemistry,* edited by W.C. Gardiner, Springer-Verlag, New York, 1984.

Harris, E. N., Buchmann, O. A., Chessmore, G. L., Sun, Y. H. and Vuigner, A. A., "Design and Test of a Regeneratively Cooled, Hydrocarbon Fueled Combustor," AFAPL-TR-73-14, Feb. 1973.

Hawthorn, R. D., and Nixon, A. C., "Shock Tube Ignition Delay Studies of Endothermic Fuels," *AIAA Journal,* Vol. 4, 1966, pp. 513–520.

Hazlett, R. N., Hall, J. M., and Matson, M. "Reactions of Aerated *n*-Dodecane Liquid Flowing Over Heated Metal Tubes," *Ind. Eng. Chem. Prod. Res. Dev.,* Vol. 16, 1977, pp. 171–177,

Hazlett, R. N., "Thermal Oxidation Stability of Aviation Turbine Fuels," American Society for Testing and Materials, ASTM Monograph 1, Philadelphia, PA, 1991.

Heiser, W. H., and Pratt, D. T., *Hypersonic Airbreathing Propulsion,* AIAA, Washington, DC, 1994.

Henderson, R. E. (ed.), "Hypersonic Air Breathing Missile," *Propulsion and Energy Issues for the 21st Century,* AGARD Report 824, March 1997, Chap. 1.

Heneghan, S. P., Zabarnick, S., Ballal, D. R., and Harrison, W. E., "JP-8 + 100: The Development of High Thermal Stability Jet Fuel," AIAA Paper 96-0403, Jan. 1996.

Heneghan, S. P., Williams, T., Whitacre, S. D., and Ervin, J. S. "The Effects of Oxygen Scavenging on Jet Fuel Thermal Stability," *ACS Petroleum Chemistry Division Preprints,* Vol. 41, No. 2, 1996, pp. 469–473.

Hines, W. S., and Wolf, H., "Pressure Oscillations Associated with Heat Transfer to Hydrocarbon Fluids at Supercritical Pressures and Temperatures," *ARS Journal,* March 1962, pp. 361–366.

Hitch, B. D., and Karpuk, M. E., "Experimental Investigation of Heat Transfer and Flow Instabilities in Supercritical Fuels," AIAA Paper 97-3043, July 1997.

Hitch, B., and Karpuk, M., "Enhancement of Heat Transfer and Elimination of Flow Oscillations in Supercritical Fuels," AIAA Paper 98-3759, July 1998.

Ianovski (Yanovskii), L. S., "Endothermic Fuels for Hypersonic Aviation," *AGARD Conference on Fuels and Combustion Technology for Advanced Aircraft Engines.* 1993, pp. 44-1 to 44-8 (AGARD CP-536)

Ianovski, (Yanovskii) L. S., and Sapgir, G., "Heat and Mass Transfer to Hydrocarbon Fuels at Thermal Decomposition in Channels of Engines," AIAA Paper 96-2683, July 1996.

Ianovski, (Yanovskii) L. S., Sosounov, V. A., and Shikhman, Y. M., "The Application of Endothermic Fuels for High Speed Propulsion Systems," *Proceedings of the 13th International Symposium on Airbreathing Engines,* Vol. 1, AIAA Reston, VA, 1997, pp. 59–69.

Jensen, J., and Braendlein, B., "Review of the Marquardt Dual Mode Mach 8 Scramjet Development," AIAA Paper 96-3037, July 1996.

Johnson, R. W., Zackro, W. C., Germanas, D., and Keenan, S., "Development of High Density Endothermic Fuel Systems," *1992 JANNAF Propulsion Meeting,* Vol. III, 1992, pp. 139–150 (CPIA Publication 580).

Jones, E. G., and Balster, W. J., "Phenomenological Study of the Formation of Insolubles in Jet-A Fuel," *Energy & Fuels,* Vol. 7, 1993, pp. 968–977.

Jones, E. G., Rubey, W. A., and Balster, W. J., "Fouling of Stainless-Steel and Silcosteel Surfaces During Aviation Fuel Autoxidation," *ACS Petroleum Chemistry Division Preprints,* Vol. 40 No. 4, 1995, pp. 655–659.

Jones, E. G., Balster, W., and Pickard, J. M., "Surface Fouling in Aviation Fuels: An Isothermal Chemical Study," *Journal of Engineering for Gas Turbines and Power,* Vol. 118, 1996, pp. 286–291.

Katta, V. R., Jones, E. G., and Roquemore, W. M., "Modeling of Deposition Process in Liquid Fuels," AIAA Paper 97-3040, July 1997.

Kee, R. J., Grocar, J. F., Smooke, M. D., and Miller, J. A., "PREMIX: A Fortran Program for Modeling Steady-State, Laminar, One-Dimensional Premixed Flames," Sandia Lab. Rep, SAND 85-8240, 1985.

Kee, R. J., Rupley, F. M., and Miller, J. A., "Chemkin II: A Fortran Package for the Analysis of Gas-Phase Chemical Kinetics," Sandia Lab. Rep, SAND 89-80098, 1993.

Keirsey, J. L., "Airbreathing Propulsion for Defense of the Surface Fleet," *Johns Hopkins Applied Physics Laboratory Technical Digest,* Vol. 13, No. 1, 1992, p. 57.

Lai, W.-C., Song, C., Schobert, H., and Arumugam, R., "Pyrolytic Degradation of

Coal- and Petroleum-Derived Aviation Jet Fuels and Middle Distillates," *ACS Fuel Chemistry Division Preprints,* Vol. 37, No. 4, 1992, pp. 1671–1680.

Lam, S. H., and Goussis, D. A., "Understanding Complex Kinetics with Computational Singular Perturbation," *Twenty-Second Symposium (International) on Combustion,* Combustion Inst. Pittsburgh, PA, 1988, pp. 931–940.

Lam, S. H., and Goussis, D. A., "The CSP Method for Simplifying Kinetics," *International Journal Chem. Kin.,* Vol. 26, 1994, pp. 461–486.

Lander, H., and Nixon, A. C., "Endothermic Fuels for Hypersonic Vehicles," *Journal of Aircraft,* Vol. 8, No. 4 1971, pp. 200–207.

Lander, H., "Endothermic Fuels: A Historical Perspective," *JANNAF Airbreathing Propulsion Subcommittee Meeting,* Vol. I, 1997 pp. 45–60 (CPIA Publication 666).

Lefebvre, A. H., "Investigation of Flame Speeds of Endothermic Fuels," Air Force Wright Aeronautical Lab., AFWAL-TR-88–2027, 1988.

Leingang, J. L., Maurice, L.Q., and Carreiro, L. R., "In-Flight Oxidizer Collection Systems for Airbreathing Space Boosters," *Developments in High-Speed-Vehicle Propulsion Systems,* edited by S. N. B. Murthy and E. T. Curran, *Progress in Astronautics and Aeronautics,* AIAA, Reston, VA, Vol. 165, 1996.

Leung, K. M., and Lindstedt, R. P., "Detailed Kinetic Modelling of C_1-C_3 Alkane Diffusion Flames," *Combustion and Flame,* Vol. 102, 1995, pp. 129–160.

Leung, K. M., "Kinetic Modelling of Hydrocarbon Flames Using Detailed and Systematically Reduced Chemistry," Ph.D. Dissertation, Univ. of London, 1996.

Lewis, M. J., and Gupta, A. K., "Impact of Fuel Selection on Hypersonic Vehicle Optimization," Sept. 1997, (ISABE 97–7200).

Lindstedt, R. P., and Maurice, L. Q., "Detailed Kinetic Modelling of *n*-Heptane Combustion," *Combustion Science and Technology,* Vol. 107, 1995, pp. 317–353.

Lindstedt, R. P., and Maurice, L. Q., "Detailed Kinetic Modelling of Toluene Combustion," *Combustion Science and Technology,* Vol. 120, 1996, pp. 119–167.

Lindstedt, R. P., and Maurice, L. Q., "A Detailed Chemical Kinetic Model for Aviation Fuels," AIAA Paper 97–2836, July 1997.

Lindstedt, R. P., "Systematically Reduced Kinetic Models: A European Perspective," AIAA Paper 97–3368, July 1997.

Lindstedt, R. P., and Skevis, G., "A Study of Acetylene Chemistry in Flames," *Combustion Science and Technology,* Vol. 125, 1997, pp. 73–137.

Lindstedt, R. P., "Systematically Reduced Kinetic Models: A European Perspective," AIAA Paper 97–3368, July 1997.

Linne, D. L., and Munsch, W., "Comparison of Coking and Heat Transfer Characteristics of Three Hydrocarbon Fuels in Heated Tubes," *JANNAF Combustion Subcommittee Meeting,* Vol. II, 1995, pp. 95–101

Linne, D. L., Meyer, M. L., Edwards, T., and Eitman, D. A., "Evaluation of Heat Transfer and Thermal Stability of Supercritical JP-7 Fuel," AIAA Paper 97–3041, July 1997.

Lipinski, J. J., and White, C., "Testing of an Endothermic Heat Exchanger/Reactor for a Mach 4 Turbojet," *1992 JANNAF Propulsion Meeting,* Vol. III, 1992, pp. 161–170 (CPIA Publication 580).

Longwell, J. P., and Weiss, M. A., "High Temperature Reaction Rates in Hydrocarbon Combustion," *Industrial and Engineering Chemistry,* Vol. 47 No. 8, 1955, pp. 1634–1643.

Maas, U., and Pope, S. B., "Simplifying Chemical Kinetics: Intrinsic Low-Dimensional Manifolds in Composition Space," *Combustion and Flame,* Vol. 88, 1992, pp. 239–264.

Maas, U., "Coupling of Chemical Reaction with Flow and Molecular Transport," *Applied Mathematical Modeling,* Vol. 40, 1996, pp. 249–266.

Marteney, P. J., and Spadaccini, L. J., "Thermal Decomposition of Aircraft Fuel," *Journal of Engineering for Gas Turbines and Power,* Vol. 108, 1986, p. 648.

Maurice, L. Q., "Detailed Chemical Kinetic Models for Aviation Fuels," Ph.D. Dissertation, Univ. of London, 1996.

Maurice, L. Q., Striebich, R. C., and Edwards, T., "The Analogy of Cyclic Compound Formation in the Gas-Phase and Supercritical Fuel Systems of Hydrocarbon Fueled High Speed Vehicles," *ACS Petroleum Chemistry Division Preprints,* Vol. 43, No. 3, 1998, pp. 423–427.

Maurice, L.Q., Striebich, R. C., and Edwards, T., "Formation of Cyclic Compounds in the Fuel Systems of Hydrocarbon Fueled High Speed Vehicles," AIAA Paper 98-3534, July 1998.

Mendez, F., and Trevino, C., "Ignition in a Vertical Wall in Contact with a Combustible Gas: Catalytic Reaction in one Surface of the Plate," *Combustion Theory and Modeling,* Vol. 1, No. 2, 1997, pp. 167–182.

Minus, D. K., and Corporan, E., "Deposition Effect of Radical Stabilizing Additives in JP-8 Fuel," *ACS Petroleum Chemistry Division Preprints,* Vol. 43, No. 3, 1998, pp. 360–363.

Morris R. A., Arnold, S. T., Viggiano, A. A., Maurice, L. Q., Carter C., and Sutton, E., "Investigation of the Effects of Ionization on Hydrocarbon-Air Combustion Chemistry," *Proceedings of the 2nd Weakly Ionized Gas Workshop,* 1998, pp. 163–176.

Muller, C., Scacchi, G., and Come, G. M., "THERGAS: A Computer Program for the Evaluation of Thermochemical Data of Molecules and Free Radicals in the Gas Phase," *Journal of Chemical Physics,* Vol. 1995, pp. 1154–1178.

Nagashima, T., Kitamura, H., and Obata, S., "Supersonic Combustion of Hydrogen in Tandem Transverse Injection with Oxygen Radicals," Sept. 1997 (ISABE 97-7055).

Nau, M., Neef, W., Maas, U., Gutheil, E., and Warnatz, J., "Computational and Experimental Investigations of a Non-Premixed Methane Flame," *Twenty-Sixth Symposium (International) on Combustion,* Combustion Inst., Pittsburgh, PA, 1996, pp. 83–89.

Nehse, M., Warnatz, J., and Chevalier, C., "Kinetic Modeling of the Oxidation of Large Aliphatic Hydrocarbons," *Twenty-Sixth Symposium (International) on Combustion,* Combustion Inst., Pittsburgh, PA, 1996, pp. 773–780.

Newman, R. W., "Oxidation-Resistant High-Temperature Materials," *Johns Hopkins Applied Physics Laboratory Technical Digest,* Vol. 14, No. 1, 1993.

Nixon, A. C., "A Study on Endothermic and High Energy Fuels for Airbreathing Engines," U.S. Air Force, AFSC/ASD, Air Force Contract F33615-84-C-2410 Final Rep., 1986.

Northam, G. B. (ed.), "Workshop Report: Combustion in Supersonic Flow," *22nd JANNAF Combustion Meeting Proceedings,* Vol. 1, 1985 (CPIA Publication 432).

Paczko, G., Lefdal, P. M., and Peters, N., "Reduced Reaction Schemes for Methane, Methanol and Propane Flames," *Twenty-First Symposium (International) on Combustion,* Combustion Inst., Pittsburgh, PA, 1986, pp. 739–748.

Palaszewski, B., Ianovski (Yanovskii) L. S., and Carrick, P., "Propellant Technologies: A Persuasive Wave of Future Propulsion Benefits," *3rd International Symposium on Space Propulsion,* (Beijing, China), Aug. 1997.

Perry, R. H., and Chilton, C. H. (eds.), *Chemical Engineers Handbook,* McGraw–Hill, New York, 1973.

Peters, N., "Flame Calculations with Reduced Mechanisms—an Outline," *Reduced Kinetic Mechanisms for Application in Combustion Systems,* edited by N. Peters and B. Rogg, Springer-Verlag, Berlin, 1993, pp. 3–13.

Petley, D. H., and Jones, S. C., "Thermal Management for a Mach 5 Cruise Aircraft Using Endothermic Fuel," *Journal of Aircraft,* Vol. No. 3 1992, pp. 384–389.

Pitsch, H., Peters, N. and Seshadri, K., "Numerical and Asymptotic Studies of the Structure of Premixed Iso-Octane Flames," *Twenty-Sixth Symposium, (International) on Combustion*, The Combustion Institute, Pittsburgh, pp. 763-771, 1996.

Polyakov, A. F., "Heat Transfer under Supercritical Pressures," in *Advances in Heat Transfer*, Vol. 21, pp. 1-53, 1992.

Richards, G.A. and Lefebvre, A.H., "Turbulent Flame Speeds of Hydrocarbon Fuel Droplets in Air," *Combust. Flame*, Vol. 78, pp. 299-332, 1989.

Richards, G.A. and Sojka, P.E., "A Model of H_2 Enhanced Spray Combustion," *Combust. Flame*, Vol. 79, pp. 319-332, 1990.

Ritter, E.R., "THERM: Thermodynamic Property Estimation for Gas Phase Radicals and Molecules," NJIT, Newark, 1989.

Roback, R., Szetela, E. J. and Spadaccini, L. J., "Deposit Formation in Hydrocarbon Fuels," *Journal of Engineering for Power, Vol. 105, pp. 59-65, 1983.*

Ronald, T., "Status and Applications of Materials Developed for NASP," AIAA 95-6131, April 1995.

Rosenberg, S. D., Gage, M. L., Homer, G. D. and Franklin, J. E., "Hydrocarbon-Fuel/Copper Combustion Chamber Liner Compatibility, Corrosion Prevention, and Refurbishment," *Journal of Propulsion and Power*, Vol. 8(6), pp. 1200-1207, 1992.

Rosenberg, S. D. and Gage, M. L., "Compatibility of Hydrocarbon Fuels with Booster Engine Combustion Chamber Liners," *Journal of Propulsion and Power*, Vol. 7(6), pp. 922-928, 1991.

Rumyantsev, A.N., Shevel'kova, L.V., Sokolova, V.M. and Nametkin, N.S., "Dependence of the Decomposition Rate Constant of Higher *n*-Paraffin Hydrocarbons on Their Molecular Weight," *Neftekhim*, Vol. 20 pp. 212-217, 1980.

Salooja, K.C., "Studies of Combustion Processes Leading to Ignition in Hydrocarbons," *Combust. Flame*, Vol. 4, pp. 117-136, 1960.

Salooja, K.C., "Effects of Temperature on the Ignition Characteristics of Hydrocarbons," *Combust. Flame*, Vol. 5, pp. 243-247, 1961.

Salooja, K.C., "Studies of Combustion Processes Leading to Ignition of Isomeric Hexanes," *Combust. Flame*, Vol. 6, pp. 275-285, 1962.

Schmidt, D., Maas, U., Segatz, J., Riedel, U. and Warnatz, J., "Simulation of Laminar Methane-Air Flames Using Automatically Simplified Chemical Kinetics," *Combust. Sci. Tech.*, Vol. 113, pp. 3-16, 1996.

Seaton, W.H., Freedman, E. and Treweek, D.N., "CHEETAH: the ASTM Chemical Thermodynamic and Energy Release Evaluation Program," ASTM DS 51, American Society for Testing Materials, 1974.

Schinke, R., Keller, H.M., Flothmann, H., Stumpf, M., Beck, C., Mordant, D.H.,

Dobbyn, A.J., Rice, S.A., Marcus, R.A., Troe, J., Neumark, D.M. and Koltman, M.C., "Resonances in Unimolecular Dissociation: from Mode-Specific to Statistical Behavior," *Advances in Chemical Physics*, 1997, Vol. 101, pp. 745-787

Seshadri, K., "Multistep Asymptotic Analysis of Flame Structures," *Twenty-Sixth Symposium (International) on Combustion*, The Combustion Institute, Pittsburgh, pp. 831-847, 1996.

Sheu, J.-C., Jones and E. G., Katta, V., "Thermal Cracking and Fouling of Norpar-13 Fuel Under Near-Critical and Supercritical Conditions," AIAA 98-3758, July 1998.

Skevis, G. (1996), "Soot Precursor Chemistry in Laminar Premixed Flames," Ph.D. thesis, University of London.

Sobel, D. R. and Spadaccini, L. J., "Hydrocarbon Fuel Cooling Technologies for Advanced Propulsion," ASME *Journal of Engineering for Gas Turbines and Power*, Vol. 119, pp. 344-351, 1997.

Steele, W. V. and Chirico, R. D., "Accurate Measurement of Thermochemical and Thermophysical Properties of Future Jet Fuels," *ACS Petroleum Chemistry Division Preprints*, Vol. 34 (4), pp. 876-884, 1989.

Stein, S.E., Lias, S.G., Leibman, J.F., Levin, R.D. and Kafafi, S.A. (1994) "NIST Structures and Properties, " NIST Standard Reference Database 25, Version 2.0.

Stewart, J.F., Brezinsky, K. and Glassman, I., "Supercritical Pyrolysis of Methylcyclohexane," *ACS Petroleum Chemistry Division Preprints*, Vol. 43(3), 433-437, 1998.

Striebich, R.C. and Rubey, W.A., "A System for Thermal Diagnostics Studies," American Laboratory, pp. 64-67, 1990.

Striebich, R. C. and Rubey, W. A., "High-pressure, high-temperature pyrolysis reactions in the condensed phase with in-line chemical analysis," *ACS Petroleum Chemistry Division Preprints*, Vol. 43(3), 378-381, 1998.

Szechy, G., Luan, T.-C., Albright, L. F., "Pretreatment of High-Alloy Steels to Minimize Coking in Ethylene Furnaces," pp. 341-359 in *Novel Production Methods for Ethylene, Light Hydrocarbons, and Aromatics*, Albright, L. F., Crynes, B. L., Nowak, S., eds, Marcel Dekker, 1992.

Taylor, W.F., "Deposit Formation from Deoxygenated Hydrocarbons. 1. General Features," Ind. Eng. Chem. Prod. Res. Dev., Vol 13(2), pp. 133-138, 1974.

TeVelde, J. A. and Glickstein, M. R. "Heat Transfer and Thermal Stability of Alternative Aircraft Fuels, Vol. I," NAPC Report NAPC-PE-87C, AD A137 404, November 1983.

Tishkoff, J. M., Drummond, J. P., Edwards, T. and Nejad, A. S., "Future Directions of Supersonic Combustion Research: Air Force/NASA Workshop on Supersonic Combustion," AIAA 97-1017, January 1997.

Tomlin, A.S., Turanyi, T. and Pilling, M.J., "Chapter 4, Mathematical Tools for the Construction, Investigation and Reduction of Combustion Mechanisms," in *Comprehensive Chemical Kinetics* Volume 35, Low-Temperature Combustion and Ignition, M.J. Pilling (ed.), Elsevier, Amsterdam, pp. 293-437, 1997

Trimm, D. L., "Fundamental Aspects of the Formation and Gasification of Coke," in *Pyrolysis, Theory and Industrial Practice*, Albright, Crynes, and Corcoran, eds. Marcel Dekker, pp. 203-232, 1983.

Troe, J., "Rigidity Factors in Unimolecular Reactions," Berichte der Bunsen-Gusellschaft fur Physikalische Chemie, 101(3), pp. 438-444, 1997.

Tsang, W., "Chemical Kinetic Database for Combustion Chemistry. Part 2. Methanol," *J. Phys. Chem. Ref. Data*, Vol. 16, pp. 471-508, 1987.

Tsang, W., "Chemical Kinetic Database for Combustion Chemistry. Part 3. Propane," *J. Phys. Chem. Ref. Data*, Vol. 17, pp. 887-951, 1988.

Tsang, W., "Chemical Kinetic Database for Combustion Chemistry. Part 4. Isobutane," *J. Phys. Chem. Ref. Data*, Vol. 19, pp. 1-68, 1990.

Tsang, W., "Chemical Kinetic Database for Combustion Chemistry. Part 5. Propene," *J. Phys. Chem. Ref. Data*, Vol. 20, pp. 221-273, 1991.

Tsang, W., Bedanov, V. and Zachariah, M.R., "Master Equation Analyses of Thermal Activation Reactions: Energy Transfer Constraints on Falloff Behavior in the Decomposition of Reaction Intermediates with Low Thresholds," *J. Phys. Chem.*, Vol. 100, pp. 4011-4018, 1996.

Turanyi, T., "Application of Repro-Modeling for the Reduction of Combustion Mechanisms," *Twenty-Fifth Symposium (International) on Combustion*, The Combustion Institute, Pittsburgh, pp. 948-955, 1994.

Van Griethuysen, V.J., Glickstein, M.R., Petley, D.H., Gladden, D.H. and Kubik, D.L., "High-Speed Flight Thermal Management," in *Developments in High-Speed-Vehicle Propulsion Systems, Progress in Astronautics and Aeronautics*, Vol. 165, eds. S.N.B. Murthy and E.T. Curran, 1996.

Van Huff, N. E., "Liquid Rocket Engine Fluid-Cooled Combustion Chambers," NASA SP-8087, April 1972.

Vilimpoc, V., Sarka, B., Weaver, W.L., Gord, J.R., and Anderson, S., "Techniques to Characterize the Thermal-Oxidative Stability of Jet Fuels and the Effects of Additives," ASME 96-GT-44, Intenational Gas Turbine and Aeroengine Congress & Exhibition, Birmingham, UK, June 1996.

Vovelle, C., Delfau, J.L., Reuillon, M., Akrich, R., Bouhria, M. and Sanogo, O., "Comparison of Aromatic Formation in Decane and Kerosene Flames," *ACS Division of Fuel Chemistry Preprints*, New York, Vol. 36(4) pp. 1456-1463, 1991.

Vovelle, C., Delfau, J.L. and Reuillion, M., "Formation of Aromatic Hydrocarbons in Decane and Kerosene Flames at Reduced Pressures," in *Soot Formation in Combustion: Mechanisms and Models,* (H. Bockhorn, Ed.), Springer-Verlag, 1994.

Walker, R.W. and Morley, C., "Chapter 1, Basic Chemistry of Combustion," in *Comprehensive Chemical Kinetics,* Vol. 35, Low-Temperature Combustion and Ignition, M.J. Pilling (ed.), Elsevier, Amsterdam, pp. 1-120, 1997.

Waltrup, P.J., "Hypersonic Airbreathing Missile Propulsion," AGARD-CP-600 Vol. 3, pp. C3-21, 1997.

Warnatz J., "Chemistry of High Temperature Combustion of Alkanes up to Octane," *Twentieth Symposium (International) on Combustion*, The Combustion Institute, Pittsburgh, pp. 845-856, 1984.

Warnatz, J., "Rate Coefficients in the C/H/O System," in *Combustion Chemistry,* W.C. Gardiner (ed), Springer-Verlag, New York, 1984.

Warth, V., Stef, N., Glaude, P.A., Battin-Leclerc, F., Scacchi, G. and Come, G.M., "Computer Aided Derivation of Gas-Phase Oxidation Mechanisms: Application to the Modeling of the Oxidation of *n*-Butane," *Combust. Flame*, Vol. 14, pp. 81-102, 1998.

Westbrook, C.K. and Dryer, F.L., "Simplified Reaction Mechanisms for the Oxidation of Hydrocarbon Fuels in Flames," *Comb. Sci. Tech.*, Vol. 27, pp. 31-43, 1981.

Weber, J., personal communication, 1997.

Wiese, D.E. "Thermal Management of Hypersonic Aircraft Using Noncryogenic Fuels," *SAE Transactions*, Vol. 100(1), p. 1313, 1992.

Yanovskii, (Ianovski) L. S. and Kamenetskii, B. Y., "Heat Exchange During the Forced Flow of Hydrocarbon Fuels at Supercritical Pressures in Heated Tubes," *Journal of Engineering Physics and Thermophysics* (translation of Inzhenerno-Fizicheskii Zhurnal), Vol. 60(1), pp. 38-42, 1991.

Yoon, E. M., Selvaraj, L., Song, C., Stallman, J. B., and Coleman, M. M., "High-Temperature Stabilizers for Jet Fuels and Similar Hydrocarbon Mixtures. 1. Comparative Studies of Hydrogen Donors," *Energy and Fuels*, Vol. 10, pp. 806-811, 1996.

Yu, J., and Eser, S., "Determination of Critical Properties (T_c, P_c) of Some Jet Fuels," *Ind. Eng. Chem. Res.*, Vol. 34, pp. 404-409, 1995.

Zeppieri, S., Brezinsky, K., and Glassman, I., "Pyrolysis Studies of Methylcyclohexane and Oxidation Studies of Methylcyclohexane and Methylcyclohexane/Toluene Blends," *Combust. Flame*, Vol. 108, pp. 266-286, 1997.

Zhou, N., and Krishnan, A, "A Numerical Model for Endothermic Fuel Flows with Heterogeneous Catalysis," AIAA 96-0650, January 1996.

Zubrin, R. M., "The Methane-Acetylene Cycle Aerospace Plane: A Promising Candidate for Earth to Orbit Transportation," AIAA-92-0688, January 1992.

Addendum—Recent Work

In the time since this manuscript was submitted, several relevant papers have been published. Two papers have shown that endothermic (partially converted) fuels have reduced ignition delays, as compared to the unconverted fuel [Ad. 1, Ad. 2]. Reference Ad. 1 also contains a detailed description and characterization of a state-of-the-art facility reactor for endothermic fuel combustion studies. The potential for enhanced hydrocarbon fuel combustion through ionization (as in the "AJAX" concept) has been investigated [Ad. 3]. Also, chemical additives have been shown to enhance ignition of complex JP fuels [Ad. 4]. The second stage of the two-stage AJAX steam reforming process has been experimentally investigated [Ad. 5]. More details have been published on the mode of action of endothermic initiators [Ad. 6]. Further studies of fuel pyrolysis and hydrogen donor additives for suppressing pyrolytic deposition have been reported [Ad. 7, Ad. 8]. A study on the effects of various types of cooling passage geometry on deposition has recently been published [Ad. 9]. It was shown that geometry can exert a strong influence on the amount and location of deposition. Representative hydrocarbon-fuel-cooled panel tests have been reported [Ad. 10, and Ad. 11]. Lifetimes consistent with expendable vehicles have been demonstrated. Simpler "surrogates" for complex kerosene fuels have been identified for modeling and experimental purposes [Ad. 12].

[Ad. 1]. Yanovskii, L.S., Sapgir, G.B., Strokin, V.N., Ivanov, V.F., "Endothermic Fuels: Some Aspects of Fuel Decomposition and Combustion at Air Flows," ISABE 99-7067, Sept 1999.

[Ad. 2]. Colket, M.B., and Spadaccini, L.J., "Scramjet Fuels Autoignition Study," ISABE 99-7069m Sept. 1999.

[Ad. 3]. Williams, S., Midley, A.J., Arnold, S.T., Bench, P.M., Viggiano, A.A., Morris, R.A., Maurice, L.Q., and Carter, C.D.,"Progress on the Investigation of the Effects of Ionization on Hydrocarbon/Air Combustion Chemistry," AIAA 99-4907, Nov. 1999.

[Ad. 4]. Sidhu S., Graham J. L., Kirk, D. C., and Maurice, L., "Investigation of Effect of Additives on Ignition Characteristics of Jet Fuels: JP-7 and JP-8," 22nd International Symposium on Shock Waves, July 1999.

[Ad. 5]. Korabelnikov, A., and Kuranov, A., "Thermochemical Conversion of Hydrocarbon Fuel Under the Concept 'AJAX'," AIAA-99-4921, Nov. 1999.

[Ad. 6]. Wickham, D.T., Engle, J.R., Hitch, B.D., and Karpuk, M.E., "Initiators for Endothermic Fuels," ISABE 99-7068, Sept. 1999.

[Ad. 7]. Maurice, L.Q., Corporan, E., Minus, D., Mantz, R., Edwards, T., Wohlwend, K., Harrison, W.E., Striebich, R. C., Sidhu, S., Graham, J., Hitch, B., Wickham, D., and Karpuk, M., "Smart Fuels: Controlled Chemically Reacting," AIAA 99-4916, Nov. 1999.

[Ad. 8]. Minus, D.M., and Corporan, E., "Thermal Stabilizing Tendencies of Hydrogen Donor Compounds in JP-8+100 Fuel," AIAA 99-2214, July 1999.

[Ad. 9]. Spadaccini, L.J., Sobel, D.R., and Huang, H., "Deposit Formation and Mitigation in Aircraft Fuels," ASME 99-GT-217.

[Ad. 10]. Siebenhaar, A., Chen, F., Karpuk, M., Hitch, B., and Edwards, T., "Engineering Scale Titanium Endothermic Fuel Reactor Demonstration for Hypersonic Scramjet Engine," AIAA 99-4909, Nov. 1999.

[Ad. 11]. Faulkner, R.F., and Weber, J.W., "Hydrocarbon Scramjet Propulsion System Development, Demonstration, and Application," AIAA 99-4922, Nov. 1999.

[Ad. 12]. Edwards, T., and Maurice, L.Q., "Surrogate Mixtures to Represent Complex Aviation and Rocket Fuels," AIAA Paper 99-2217, July 1999.

Appendix: Basic Elements of Chemical Kinetic Mechanisms

Thermochemical and Kinetic Databases

The origins of thermochemical and kinetic data reside in experimental measurement involving simple species or simple chemical systems, but the diversity and complexity of the fuels involved in most practical combustion systems is such that it is necessary to extend the scope of tabulations by predictive methods (Baulch, 1997). There are a number of existing, and highly respected databases from which information is usually drawn (e.g., Burcat and McBride, 1994). Examples of computational packages that may be used for the estimation of thermochemical data include CHETAH (Seaton et al., 1974), THERM (Ritter, 1989), NIST DB 25 (Stein et al., 1994), and THERGAS (Muller et al., 1995). These procedures are based on the group additivity rules devised by Benson and co-workers (Benson, 1976).

The major sources of kinetic data for elementary reactions involving C, H, and O are the critically evaluated data sets published by the CEC group (Baulch et al., 1992, 1994) and NIST (Tsang, 1987, 1988, 1990, 1991). In addition, there are a number of data sheets (e.g., Cohen and Westberg, 1991) and reviews with reference to hydrocarbon combustion at low temperatures below 1200 K (1700°F) (e.g., Walker and Morley, 1997) and at higher temperatures (e.g., Warnatz, 1984). High-temperature reactions involving N atoms and N-containing species have been discussed by Hanson and Salimian (1984).

These sources of quantitative information do not cover the full range of reactions that are required in a comprehensive representation of the combustion chemistry of higher hydrocarbons. Thus, it is necessary at the present time to derive the appropriate kinetic parameters for many reactions from carefully derived estimates. Confidence in the numbers may be gained by analogy to the (known) data for similar reactions within a particular class. Fortunately, the hierarchical nature of the mechanisms of hydrocarbon combustion, particularly at temperatures above 1200 K (1700°F), permits quite extensive generalizations to be made.

The pre-exponential factors of bimolecular reactions can be predicted relatively easily. The main issues relate to the accuracy of the temperature dependencies of rate constants (i.e., their activation energies) if wide temperature ranges have to be taken into account. The pressure dependencies of the rate constants for unimolecular decomposition reactions are also very important, but they are particularly difficult to quantify. There are successful methods for quantitative interpretation that are applicable to single reactions (Schinke et al., 1997), but simplifications are essential if pressure dependencies are to be taken into account for many reactions of a complex kinetic scheme (Tsang et al., 1996). This requirement has yet to be widely recognized in combustion modeling where validation of the model has been made at pressures that are far removed from those of the combustion application. Fortuitously, the present range of model validation (< 2 bar) is particularly suited to scramjet application. However, pressures in the fuel systems of hypersonic vehicles are on the order of 34–68 bar (500–1000 psia), underscoring the need to take into account the pressure dependencies of elementary reactions in comprehensive kinetic models applicable to the complete tank-to-engine exit nozzle propulsion system.

Construction and Validation of Comprehensive Combustion Models

The most extensively developed model for combustion chemistry of hydrocarbons applies to alkanes containing up to eight carbon atoms, the primary objective being the understanding of gasoline combustion in spark ignition engines (e.g., Curran et al., 1996). The model originated more than 15 years ago and examined hydrogen, carbon monoxide, and methane as the primary fuels reacting at temperatures above 1200 K (1700°F). The model has gradually evolved from the reactions of single carbon atom components to those relevant to C_8 alkanes and incorporates extensive oxidation chemistry of the numerous unsaturated and partially oxygenated molecular components that are involved. The model includes thousands of elementary reactions. The range of validity has also been extended by adding classes of reactions that are known to take place below 1000 K so that combustion above about 600 K can be investigated numerically.

A largely empirical route to the estimation of rate parameters has been adopted, although this is backed up by formal determination of the thermochemistry and kinetic constants of the reaction (Kee et al., 1993). Validation of the models is normally based on experimental studies of combustion reactions in controlled conditions, from which chemical analysis of the intermediate and final molecular products is made (Dryer, 1989). Supplementary tests also emerge from the comparison of ignition delays in shock tubes (high temperature) and rapid compression machines (low temperature); also, certain one-dimensional laminar flame properties such as burning velocities and CO, CO_2, and H_2 mole fractions can also be predicted and compared with experiment (Kee et al., 1985).

The original goals did not include the prediction of soot formation or other pollutant emissions, and only recently has attention been paid to other classes of hydrocarbons or other organic compounds, such as the aromatics (Castaldi et al., 1996) and ethers (Curran et al., 1996). Even with access to the most powerful computers, a model such as this can be applied only to zero-dimensional or, in limited circumstances, one-dimensional modeling. Coupling of full kinetic schemes to complex fluid dynamic codes or involving direct numerical simulation of the Navier–Stokes equations is impossible at present. Such an application would require computing power beyond that which is foreseeable in the near future.

Warnatz (1984) also began the development of comprehensive kinetic models at about the same time as the development of the spark ignition engine models already discussed, and more recently the two sources have been drawn together (Chevalier et al., 1992). There have also been other developments of comprehensive kinetic models that are applicable to alkane combustion (e.g., Dagaut et al., 1994). These have been validated against chemical analyses made in a high-pressure, well-stirred flow reactor. There have also been very important developments of programs dedicated largely to the detailed understanding of diffusion and premixed flames. The work has placed particular emphasis on the chemical complexities that emerge in very fuel-rich conditions, such as formation of PAH and soot (Leung and Lindstedt, 1995; Lindstedt and Skevis, 1997), with extensions to the components of kerosene (Maurice, 1996; Lindstedt and Maurice, 1995, 1996; Lindstedt and Maurice, 1997). There are also numerical models of intermediate detail that have been created to help the interpretation of the global behavior of a wide range of combustion systems. They are far too numerous to cite, and they have very limited general value because the scope of their test and application is so restrictive. The way forward to creation of models of limited complexity but broad application must be through formal methods of reduction of comprehensive kinetic schemes, as discussed in the next subsection.

Part of the modern culture to the development of comprehensive kinetic models is the automatic generation of reaction schemes using formal chemical rules without prejudicial input from the modeler, i.e., the development of an expert system (Griffiths, 1995; Tomlin et al., 1997). The first numerical experiments in which an expert system was developed were performed on hydrocarbon pyrolysis in the range C_1 to C_4 (Chinnick et al., 1988). The automatic generation yielded a scheme for C_4H_{10} pyrolysis comprising 76 species in 179 reactions. There have been subsequent developments, using similar techniques, applied to alkane oxidation up to heptane, which include 1200 species in 7000 reactions (Chevalier et al., 1992) and by Nehse et al. (1996) for alkanes up to *n*-decane. However, these mechanisms have been validated solely against measured ignition delays in a shock tube, which is not a sufficiently sensitive or exhaustive test of validity. Certainly it is hardly adequate to distinguish the quality of the automatic generation from schemes that have been deduced wholly from a subjective, human perspective.

At the present time the most substantial progress in the development of expert systems has been achieved by the Départment de Chimie-Physique de Réactions combustion research group in Nancy, France (Come et al., 1996; Warth et al., 1998). In work by Come et al. (1996), models for the oxidation of *n*-heptane and *i*-octane have been developed. The models were validated against chemical analyses obtained in a well-stirred flow reactor at temperatures in the range 950–1150 K at 1 bar and at 600–850 K at 10 bar (Chakir et al., 1992).

Formal Routes to Sensitivity Analyses and Mechanism Reduction

The relevance of mechanism reduction is the requirement for much abbreviated reaction schemes and, more importantly, considerably fewer species (< 20) to be involved if detailed kinetics are to be incorporated in multidimensional fluid dynamic codes (see Table A.1). Nevertheless the global properties of the combustion process, and the way in which they are modified by changes of the prevailing conditions, must be retained. A review of formal routes to sensitivity analysis can be found in Lindstedt (1997).

The crux of this issue is that certain reactions exert a strong, even dominant control of the overall behavior to the extent that many reactions may be disregarded and the behavior of many species may be lumped into one group. The latter ensures that the identity of the class of compounds is not lost because they contribute ultimately to the overall heat generation and final product formation, but computational power is not exhausted in unnecessarily detailed calculations. The problem is that neither the same species nor the same reactions are likely to be dominant throughout all ranges of conditions.

A common mathematical procedure that is used to distinguish the most important features is that of sensitivity analysis (Dougherty and Rabitz, 1980; Griffiths, 1995; Tomlin et al., 1997). In this analysis the sensitivity of the combustion behavior to each component and each reaction is quantified through normalized sensitivity coefficients obtained in a sensitivity matrix. Thus, the relative importance of components can be distinguished by determining the normalized sensitivity coefficient for the rate of production of product species j with respect to the concentration of each intermediate species i. Similarly, the relative importance of reactions can be distinguished by determining the normalized sensitivity coefficient for the rate of production of product species j with respect to the rate coefficient for each of the elementary reactions. Proprietary programs are available for performing a sensitivity analysis [e.g., KINALC (Turanyi, 1994)]. The sensitivity of other

Table A.1 Number of independent scalars for various kinetic problems

	Full CFD two- and three-dimensional with chemistry at each grid point	Flamelet models one-dimensional chemistry
H_2 - C_3H_8 flames	2–10 step systematic reduction or ILDM[a]	Reduction unnecessary
C_mH_n flames, m = 4-12	Between 6 and 10 steps systematic reduction or ILDM	Reduction unnecessary
Ignition C_mH_n, M = 4-12	16–34 step necessary systematic reduction or ILDM	Unknown presently
PAH and soot	Between 10 and 20 steps systematic reduction or ILDM	Unknown presently
NO_x	Between 7 and 15 steps systematic reduction or ILDM	Reduction unnecessary

[a] Note that the Intrinsic Low Dimensional Manifolds (ILDM) technique is presently restricted to 3–6 degrees of freedom.

properties of the system to the variation of the rate constants may also be determined, such as the temperature change, ignition delay, or burning velocity (e.g., Fig. A.1). Alternative methods for mechanism reduction, which are related to sensitivity analysis, are principal component analysis and uncertainty analysis (Tomlin et al., 1997).

Three classes of species are defined for the purpose of the procedure for mechanism reduction. These are 1) the *important species,* which are those about which information may be required, such as the concentration profiles for the primary reactants or the final products; 2) the *necessary species,* which are the ones that have to be retained in the reduced model in order to satisfactorily reproduce the overall behavior that is predicted by the full kinetic model; and 3) the *redundant species,* which are those that may be eliminated from the model without significant effect on the output of the model.

The first step of mechanism reduction is to ensure that there are no superfluous variables in the model. In this respect the redundant species are determined from the first of the normalized sensitivity parameters just described (Tomlin et al., 1997). The next step is to reduce the number of reactions by taking out those that show negligible sensitivity with respect to the rate of product yield (or other parameter of interest). The third step is to classify as many species as possible within a more limited number of groups, each of the groups being represented by a single variable (species lumping). The new lumped variables are related to the original variables by a function, typically called the *lumping function.* This function will contain the concentrations of individual species and also other parameters, such as reaction rate constants. Lumping functions are derived numerically (Tomlin et al., 1997).

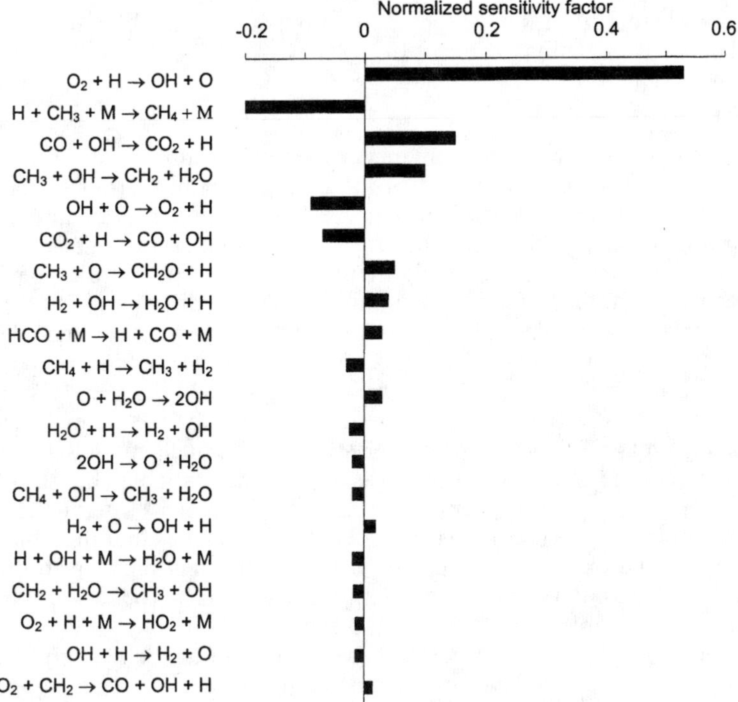

Fig. A.1 Typical sensitivity analyses.

An alternative approach to mechanism reduction used in chemical kinetics is a procedure based on the investigation of timescales (Fig. 21). An example of this procedure is the relationship of free radical or atom concentrations that can be derived using the quasistationary state approximation. Other methods that bear some relation to this are the computational singular perturbation method (Lam and Goussis, 1988, 1994) and the slow manifold approach (Maas and Pope, 1992). The physical basis for these methods is that certain species will have reached a stationary state (or even chemical equilibrium) on a timescale that is orders of magnitude shorter than other variables of the system. A graphic description of this approach is the term *slow manifold,* as applied to the slowest changing. A simple illustration is the control of superequilibrium concentration of all radical species by the slow, three-body termination processes in the reaction zones of flames, as described earlier. Applications of the intrinsic low-dimensional manifold (ILDM) approach in which there is a coupling of chemical reaction with flow and molecular transport are discussed by Maas (1996) and others (Schmidt et al., 1996; Nau et al., 1996). It is noteworthy that these applications are, as yet, confined to methane combustion chemistry. However, the Air Force Research Laboratory is pursuing extending these techniques to typical endothermic fuels. Finally, a recent alternative approach to the reduction of the numbers of variables is that of fitting algebraic equations to detailed kinetic models, as used in repro-modeling (Turanyi, 1994).

Skeletal Models

These mathematical developments are extremely important for the accurate, quantitative representation of the combustion properties in terms of simplified chemical models. Unfortunately, the experience to date is that, according to the mathematics, the irreducible kinetic representation of hydrocarbon combustion over a wide temperature and pressure range is still too complicated for coupling to computational fluid dynamics (CFD) representations of turbulent flows. Consequently, a variety of skeleton kinetic schemes has been generated to represent the overall behavior of hydrocarbons with particular reference to ignition delays (Griffiths, 1995). There are also abbreviated schemes to represent burning velocities of hydrocarbon flames (Peters, 1993). The potential strengths of such models lie in the restriction of the chemistry to as few as four species variables, but the choice of appropriate rate parameters in the associated reactions usually then rests on an empirical match to experimental data. This means that the range and validity of the application must be treated with caution. An attempt at quantifying the number of independent scalars (species) for various types of reacting flow problems is shown in Table A.1.

As an example, certain aspects of hydrocarbon flame chemistry, such as the temperature dependence of the inner zone on pressure and composition and the unstretched laminar burning velocity dependence on pressure, can, in the case of methane, be captured within a kinetic structure of the following (Paczko et al., 1986; Seshadri, K., 1996). However, it must be noted that the scheme is not applicable to the determination of ignition delays, as it does not consider extensive initiation chemistry via oxygen attack.

$$CH_4 + 2H + H_2O \leftrightarrow CO + 4H_2 \tag{A.1}$$

$$CO + H_2O \leftrightarrow CO_2 + H_2 \tag{A.2}$$

$$H + H + M \leftrightarrow 2H + M \tag{A.3}$$

$$O_2 + 3H_2 \leftrightarrow 2H + 2H_2 \tag{A.4}$$

The formal reduction to obtain reactions (A.1–A.4) is based largely on the quasistationary state and partial equilibrium approaches to free radical concentrations. This yields H as the sole remaining propagating free radical. Consequently, the chain branching reaction (4), which is found to have the highest normalized sensitivity factor (Fig. 29), is subsumed into Eq. (A.4). Similarly, the oxidation of CO, which would be determined mainly by reaction with OH radicals as

$$CO + OH \leftrightarrow CO_2 + H \tag{A.5}$$

is encapsulated in the water-gas shift reaction (A.2). The other principal elementary steps that are incorporated within this model are

$$H + O_2 + M \rightarrow HO_2 + M \tag{16}$$

and

$$CH_4 + H \rightarrow CH_3 + H_2 \tag{A.6}$$

and the rate constants of these elementary reactions are used to represent the global terms. It is relevant that Eqs. (A.1–A.4) include the three elementary reactions showing the highest sensitivity factor in Fig. A.1. The fourth [reaction (16)] has a low sensitivity in low-pressure flames but will be of much greater importance at higher pressures (Fig. 23).

There are features of scheme (A.1–A.4) also in common with much earlier empirical approaches to global mechanisms (Westbrook and Dryer, 1981). For these the argument was made that two-step chemistry, represented by fuel oxidation to CO followed by its oxidation to CO_2, was a reasonable global foundation for flame chemistry, reflecting the relatively slow rate of oxidation of CO to CO_2.

The scheme (A.1–A.4) has also been adapted for *n*-heptane flames (Seshadri et al., 1996) and *i*-octane flames (Pitsch et al., 1996). For the latter, to reactions (A.2–A.4) have been added

$$C_3H_4 + O_2 + H_2O \rightarrow 3H_2 + 3CO \tag{A.7}$$

$$i\text{-}C_8H_{18} + 2H + H_2O \rightarrow C_3H_4 + 4H_2 + CO \tag{A.8}$$

$$i\text{-}C_4H_{18} \rightarrow 2\ i\text{-}C_4H_8 + H_2 \tag{A.9}$$

The scheme comprising Eqs. (A.2–A.4) and (A.7–A.9) was derived by the formal reduction of a 967 reaction scheme involving 109 species (Pitsch et al., 1996). It has been used to investigate numerically, and by asymptotic analysis, the structure of premixed *i*-octane flames. It is possible that this type of reaction scheme might form a basis for the interpretation of CFD modeling of turbulent combustion systems.

Rather different approaches to the development of abbreviated forms of kinetic schemes have been applied to determine ignition delays (Griffiths, 1995). However, a reduced scheme for methane ignition (Mendez and Trevino 1997), which is somewhat related to the reactions (23), (24), (A.1), and (A.2), comprises

$$5CH_4 + 2O_2 \rightarrow CO + 3H_2O + 4CH_3 + H_2 \tag{A.10}$$

$$CH_4 + CH_4 \rightarrow C_2H_6 + H_2 \tag{A.11}$$

$$2CH_4 + O_2 + H_2 \rightarrow C_2H_6 + 2H_2O \tag{A.12}$$

$$C_2H_6 \rightarrow C_2H_4 + H_2 \tag{A.13}$$

$$C_2H_4 \rightarrow C_2H_2 + H_2 \tag{A.14}$$

$$C_2H_2 + O_2 + 0.5H_2O \rightarrow CO + 0.5\,CO_2 + 1.5H_2 \tag{A.15}$$

This scheme was tested only against shock-tube measurements at temperatures at about 1300 K (2860°F), and so its validity in any other circumstances is not known.

Chapter 13

Detonation-Wave Ramjets

Jean P. Sislian*
University of Toronto, Toronto, Ontario, Canada

Introduction

A DETONATION-wave ramjet is a mode of scramjet propulsion where heat addition to a supersonic *premixed* combustible mixture at below its ignition temperature is accomplished through a conveniently located shock wave that raises the temperature and pressure of the gas mixture to its ignition point. If ignition occurs far enough downstream so that the ensuing combustion process does not influence the preceding shock wave, the heat addition is said to be shock-induced. If ignition occurs close to the preceding shock wave, the combustion process couples with the shock and generates a detonation wave. The shock wave inducing combustion or the detonation wave may be normal or oblique to the oncoming flow, planar or axisymmetric, and may be generated by gasdynamic processes, by a wedge, cone, or blunt body. The detonation wave may also be laser-initiated. Several configurations for the use of shock-induced combustion in ramjets were advanced, as shown in Fig. 1. In Fig. 1a the detonation wave is normal to the oncoming freestream, and there is no precompression, except for the shock component of the detonation wave. The detonation Mach number equals that of the flight Mach number, and combustion behind the normal shock is not supersonic. In Fig. 1b a certain amount of precompression in the inlet is employed ahead of the normal to the oncoming flow detonation wave. Therefore, the detonation-wave Mach number is less than the flight Mach number, but combustion is still not supersonic. In configurations Figs. 1c and 1d, oblique detonation waves are employed at flight Mach number and at below flight Mach number, respectively, and flow velocities remain supersonic throughout the propulsion stream tube. A possible configuration that uses a blunt body, a bump, to generate detonative or shock-induced combus-

*Professor. Associate Fellow AIAA.

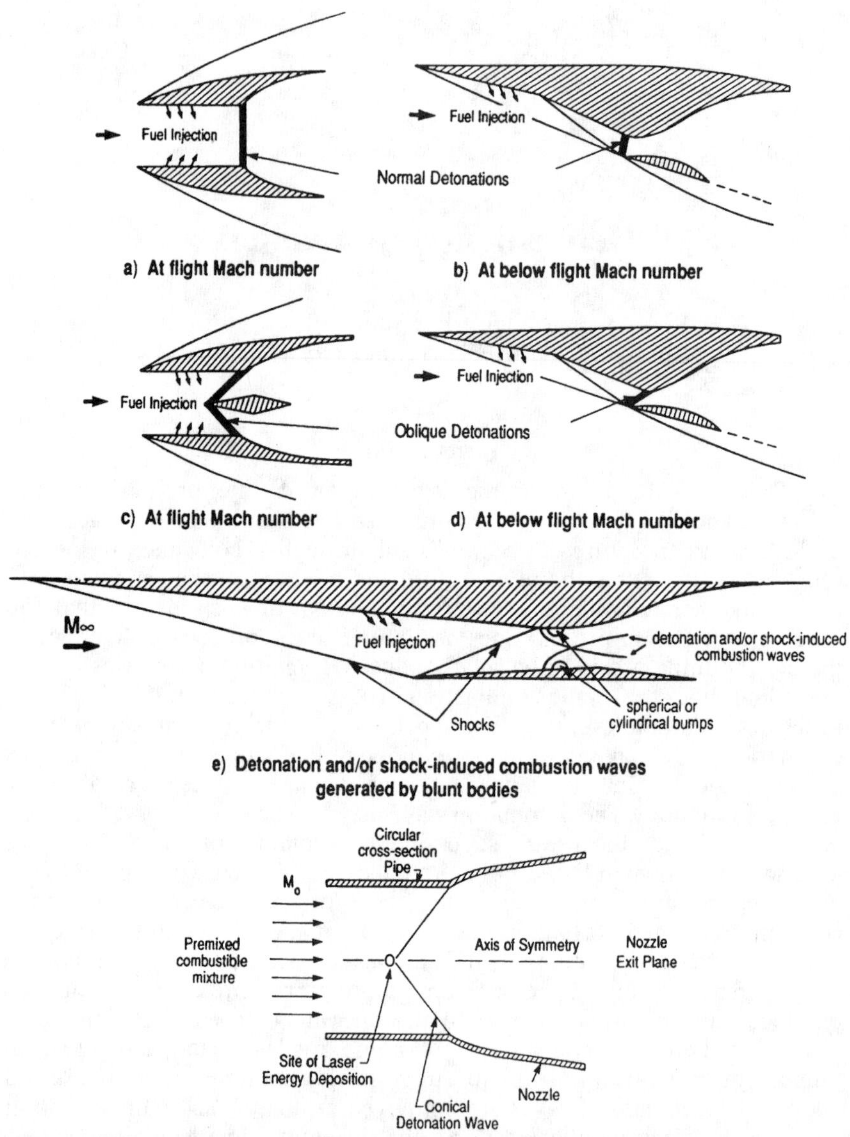

Fig. 1 Detonation-wave ramjet configurations: a) at flight Mach number; b) at below flight Mach number; c) at flight Mach number; d) at below flight Mach number; e) detonation and/or shock-induced combustion waves generated by blunt bodies; and f) conical detonation wave generated by laser energy deposition at point 0 (Ref. 21).

tion is shown in Fig. 1e. In Fig. 1f the ignition device is nonintrusive, and the conical detonation wave is stabilized by using a train of energy-depositing, rapidly repeated laser pulses. Each pulse initiates at a fixed site *O* relative to the combustor, a spherical detonation wave in the combustible mixture that flows faster than its Chapman–Jouguet detonation wave speed. As the fixed, finite frequency of detonations is increased to a large value, a stabilized conical detonation wave with negligible corrugations arises.

There are many advantages of the detonation-wave ramjet over the scramjet. The shock component of the detonation or shock-induced combustion process provides the additional large compression and the corresponding high temperatures required for rapid combustion. Consequently, the compression process in the inlet of the detonation wave ramjet is much lower than that for the scramjet, and, therefore, the losses involved in flow deceleration in the inlet are less than those for the scramjet. In addition, with the proper choice of fluid dynamic and geometric parameters of the detonation wave ramjet configuration for given flight conditions, the shock-induced combustion process will be very rapid and will occur over a very short distance from the shock, resulting in a short combustor length, less combustor cooling load, and an overall shorter and lighter-weight engine system than for the scramjet. Considering that for acceleration missions, as the single-stage-to-orbit vehicle, the engine weight is a strong concern; these advantages are not insignificant. As is the case for scramjets, the main disadvantage of a detonation-wave ramjet is the lack of static thrust.

However, the practical implementation of detonation-wave ramjets poses serious research and technical challenges. A number of formidable tasks must be tackled to achieve satisfactory operation of such ramjets.

1) Experimental and theoretical evidence of the stability of detonation waves, and conditions under which such waves are possible, must be established at all required combustor inlet values of velocity, pressure, temperature, and fuel/air ratio.
2) The injected fuel must be more or less homogeneously mixed with the high Mach number vehicle forebody airflow upstream of the intended detonation wave. Accurate estimates of losses accruing from these processes must be established.
3) Ignition of the hot injected fuel at the injection point, in the forebody flow, and in the forebody wall boundary layer must be avoided, or minimized.
4) Boundary-layer separation caused by shock-induced combustion or detonation waves must be controlled.
5) Realistic propulsive performance estimates of detonation wave ramjets must be obtained.

The concept of using detonation waves for propulsion purposes was advanced as early as 1946 in France by Roy.[1] However, this concept has received little attention in the past and, consequently, has not yet been explored in a systematic manner. The earliest works in this area were those by Dunlap et al.[2] and Sargent and Gross.[3] They presented engine performance characteristics (specific thrust, specific fuel consumption) in the flight Mach number range $3 \leq M_\infty \leq 10$ based on simplified one-dimensional analysis. In Ref. 4, Townend studied a more elaborate hypersonic detonation-wave ramjet configuration consisting of multishock diffusers, optimized by the Oswatitsch criterion, i.e., having shocks of equal strength, matched to strong or Chapman–Jouguet detonation waves, normal or oblique, in such a way as to minimize the net total pressure loss in the inlet and heat addition processes. He derived an analytical expression for this optimum condition. He concluded that heat addition by Chapman–Jouguet detonation is competitive with heat addition at constant pressure and that the problems just mentioned of detonation-wave ramjets may, in some cases, be avoided. He emphasized "the need for research on oblique rather than normal detonation and conical rather than plane waves."

Morrison[5,6] published two reports on oblique detonation wave ramjets, in which he studied in some detail multishock external and internal compression diffusers, fuel-injection losses, combustion-chamber configurations, effects of chemistry on oblique detonation-wave ramjets, and of real-gas effects on nozzle expansion. Estimates of the propulsive performance of external and internal compression ramjets were also reported based on simplified one-dimensional analysis. He concluded that "the oblique detonation wave ramjet offers a great potential as an airbreathing propulsor to extend the useful range of ramjet flight Mach numbers from 6 to 16 and above. Specific impulses and thrust coefficients that would be obtainable in the above flight range would exceed 70 percent of ideal." Detonation-wave ramjets were also investigated by Billig at the Johns Hopkins University Applied Physics Laboratory (e.g., see Refs. 7 and 8) within the context of a wider study of external burning in supersonic streams.

The concept of heat addition to a supersonic flow by shock-induced combustion has been actively developed in the former USSR (e.g., see Refs. 9 and 10).

A revival of interest has occurred lately in detonation-wave ramjet propulsion for hypersonic flight. In Ref. 11 the oblique detonation wave engine in combination with a dual-fuel, dual-expander rocket engine has been proposed as a propulsion device for a single-stage Earth-to-orbit vehicle. In Ref. 12 the analysis performed by Morrison in Ref. 6 has been taken a step further by assuming more realistic working gas properties and considering a more elaborate detonation-wave model. The estimated performance parameters (fuel-specific impulse, thrust-per-unit inlet area, etc.) are essentially based on a one-dimensional cycle-type analysis rather than on a specific vehicle geometry. Computational fluid dynamic methods together with a Chapman–Jouguet detonation-wave model or a shock-induced finite-rate hydrogen/air combustion process were used in Refs. 13–19 to determine the entire flowfield of a class of hypersonic planar and axisymmetric, external or internal compression, detonation-wave ramjet models and to assess their aerodynamic and propulsive performance characteristics. Three-dimensional wave-rider configurations derived from these studies were also investigated. An analysis of the performance of a conceptual transatmospheric vehicle powered by an oblique detonation-wave engine is given in Ref. 20. The thrust-to-drag ratio of an air-breathing combustor based on a stabilized, conically configured, laser-driven detonation-wave is analyzed in Ref. 21.

Another interesting application of this propulsion mode is the ram accelerator.[22] In this device a projectile-shaped body is fired into a tube filled with gaseous propellants at high pressure, which consists of premixed fuel and oxidizer, such as hydrogen/oxygen, methane/oxygen, or acetylene/oxygen, and possibly a diluent. The pressure and temperature of the combustible mixture increase as the flow passes through a series of oblique and normal shocks until it reaches the ignition temperature. The combustion process travels with the projectile, generating a pressure distribution that produces forward thrust on the projectile. Several modes of ram-accelerator operation have been proposed. Figure 2a shows the thermally choked ram-accelerator mode, which operates at velocities below the Chapman–Jouguet detonation speed of the combustible mixture by stabilizing a subsonic combustion zone behind the base of the projectile. An oblique detonation ram-accelerator mode, which can operate only at speeds greater than the Chapman–Jouguet detonation speed by igniting the propellant mixture with a reflected oblique shock, is shown in Fig. 2b. Conceptually, the ram accelerator can be considered as a simpler version of the detonation-wave ramjet because the upstream fuel/air mixing problem is circumvented and the projectile shape is simpler than configurations used in hypersonic detonation-wave ramjets.

The pulsed detonation-wave engine is another propulsive device employing unsteady, intermittent, gaseous detonative combustion as a thrust-producing mechanism. It can be visualized as a cylindrical tube of finite length with an ignition device located at the closed end while the other end is open. After introduction of the gaseous detonatable mixture, the ignition device initiates a flame front at the closed end that subsequently develops into a detonation wave, propagates down the tube, and reaches its open end, generating a reflected rarefaction wave. Gradual exhausting of the burned

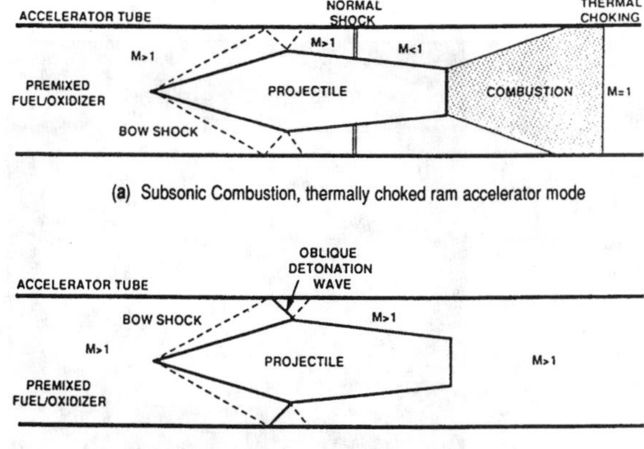

Fig. 2 Ram-accelerator operation: a) subsonic combustion, thermally choked ram-accelerator mode and b) oblique detonation ram-accelerator mode.

gas occurs as the rarefaction wave moves upstream. Eventually, the tube reaches subatmospheric pressures, and a fresh charge is introduced for the next cycle. An intermittent detonation engine offers many advantages, one of the important being the capability of static thrust as well as efficient operation at supersonic velocities. A comprehensive review of propulsion applications of the pulsed detonation engine concept is given in Ref. 23. Ram accelerators and pulsed detonation engines are being developed extensively as separate research areas, and, therefore, will not be considered further herein.

Experimental Evidence of Standing Detonation Waves

Central to the development of detonation-wave ramjets is the actual possibility of generating standing, stable, oblique detonation waves with acceptable losses in a more or less homogeneous combustible mixture at all required ramjet combustor-inlet conditions of pressure, temperature, Mach number, and fuel/air ratio. Therefore, soon after Roy's proposal,[1] researchers attempted to experimentally stabilize detonation waves in the laboratory environment. The most cited experiments in these early stages are those by Gross[24] and Gross and Chinitz.[25] Standing, stable, normal, and oblique shock-induced combustion or detonation waves were produced in a supersonic, $M = 3.15$, wind tunnel of rectangular, 75×15 cm test cross-section in hydrogen/air or methane/air combustible mixtures. Fuel was injected from a small orifice axially located a short distance upstream of the tunnel nozzle throat. Two appropriately chosen side-wall wedges in the test section generated a Mach reflected-shock pattern that caused ignition behind the normal wave. A schlieren photograph of this wave pattern is shown in Fig. 3a. An increase in hydrogen flow caused the normal wave to increase in size, move upstream to a new stable position, and the downstream total temperature to rise. Their shape, size, and axial location in the test section are sketched in Fig. 3b for various amounts of hydrogen fuel flow, quantified by $\overline{Q} = Q/c_pT$, which is the ratio of the heat added to the flow to the initial,

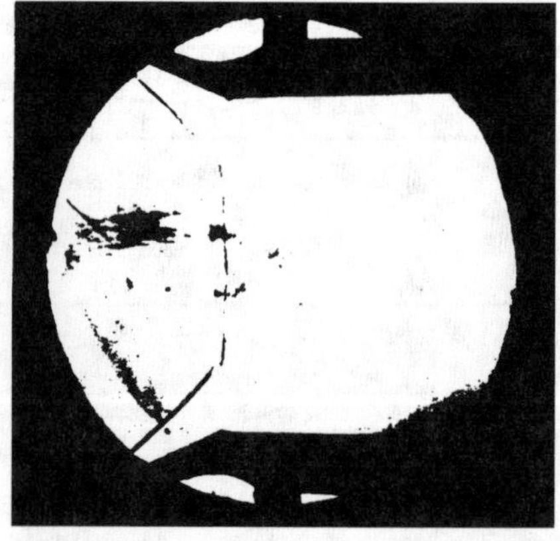

Fig. 3a Schlieren photograph of a hydrogen/air normal detonation wave. The flow is from left to right. $M_1 = 3.15$ and $\overline{Q} = 3.75$. The dark smudges are dirt on the window.[24]

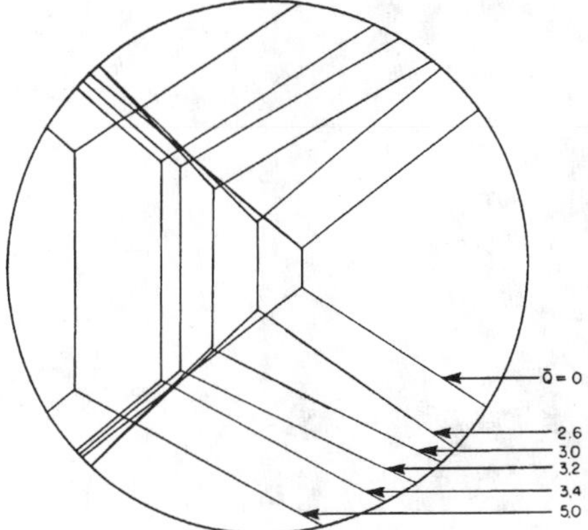

Fig. 3b Detonation-wave position and size in function of added heat. $\overline{Q} = Q/c_pT$ (Ref. 24).

prewave enthalpy of the flow. Axial traverses with a small probe through the normal wave indicated that the wave thickness is on the order of 2.5 mm. There appeared to be no measurable separation between the normal shock wave and the onset of chemical reaction. Total temperatures and pressures measured a short distance and parallel to the wave front agreed rather well with those predicted by detonation theory. Attempts also have been made to generate oblique detonation waves by placing single wedges in the supersonic test section, taking care not to wet the wedge by the injected fuel. The oblique shock wave formed by the wedge, upon extending into the combustible flow region, steepened, and formed a detonation wave (see Fig. 4). An increase in fuel flow increased the wave angle. These normal and oblique detonations were steady and reproducible and had remarkably plain fronts for the highly nonuniform fuel/air distribution ahead of the waves. However, these results were considered controversial. For instance, Fletcher[26] argues that the fuel, in Gross' experiments, was not at all burning at the shock front but rather at some distance behind it and that "the phenomena he is observing are the result of interactions between reflected shock waves and the subsonic heated region through which pressure disturbances can be fed back upstream to the normal shock front." More recently, Pratt et al.[27] analyzed Gross' data on oblique detonation waves and showed that "the flow was very near to thermal choking and that no solutions to the governing equations exist anywhere near the stated conditions." They then argued that the combined thermal flow area occlusions (caused by the wedge) caused an increase in back pressure, forcing a nonreacting strong oblique shock to be stabilized on the wedge. One should also note the effect of the observed fuel combustion at the fuel injector on Gross' results. At about the same time, at the University of Michigan, Nicholls et al.[28] succeeded in stabilizing normal shock-induced combustion in the Mach reflected-shock zone of an underexpanded freejet. Experimental attempts to stabilize oblique detonation waves on wedges in hydrogen/air mixtures were also reported in the early 1960s in Refs. 29 and 30. The experiments were conducted in a Mach 3 wind

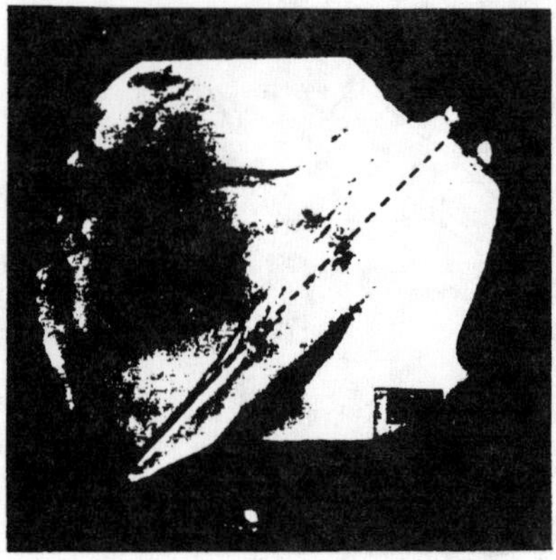

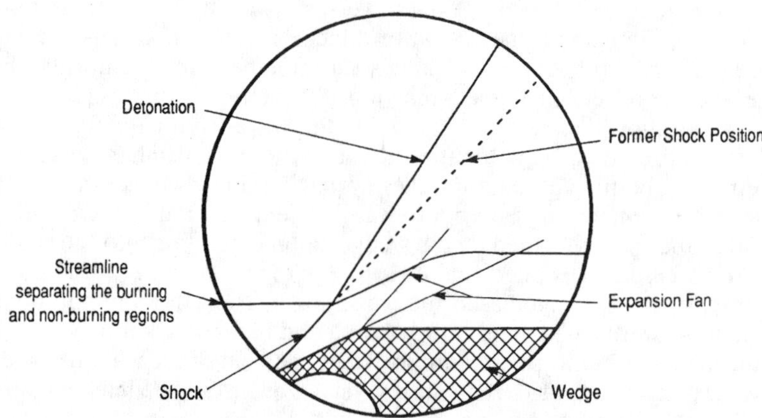

Fig. 4 Schlieren photograph and its corresponding sketch of an oblique detonation wave. $M_1 = 3.0$, $\bar{Q} = 1.0$, and wedge angle $\delta = 27$ deg (Ref. 24).

tunnel, and the fuel injected from the trailing edge of a thin double-wedge strut was placed in the wind-tunnel nozzle. A typical schlieren photograph of the oblique shock pattern and the shock-induced combustion aft of the oblique shock is shown in Fig. 5. Unfortunately, the results of this set of early experiments were inconclusive, as experimental limitations permitted only low approach Mach numbers and, correspondingly, limited substoichiometric heat addition to the flow.

More recently, an interesting experimental facility was built in Germany to investigate freejet flames that are matched to the ambient pressure.[31] The experimental principle for the generation of combustion by oblique shock waves in a premixed fuel/air mixture is apparent in Fig. 6. Compressed and

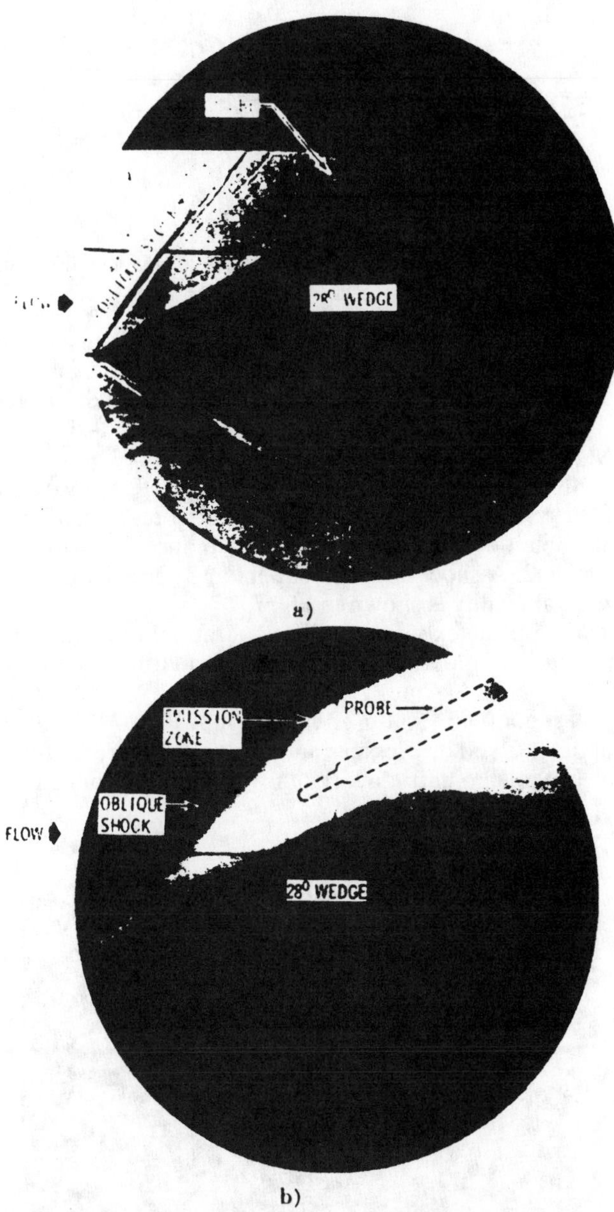

a)

b)

Fig. 5 Shock-induced combustion aft of an oblique shock in a Mach 3 stream: a) schlieren photograph of flow without combustion and b) combined schlieren and emission photograph with combustion.[29]

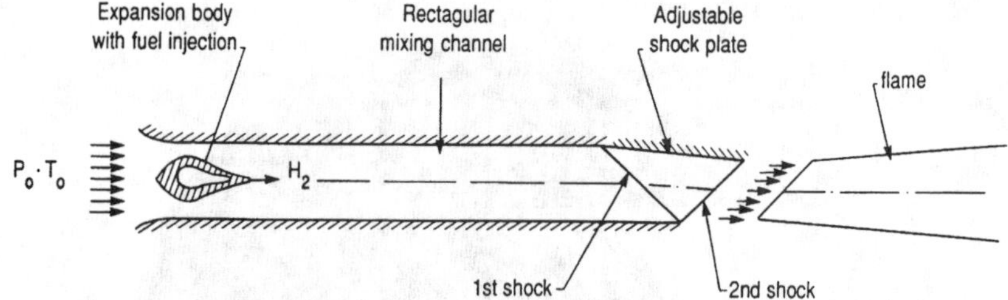

Fig. 6 Generation of supersonic premixed flames in a shock tunnel.[31]

heated air is expanded in a rectangular channel containing a central body of Laval shape and thus brought to a supersonic velocity. Care was taken to introduce hydrogen into the flow from this central body in a shock-free manner by carefully designing the body contour and adjusting the hydrogen pressure. At the end of the mixing channel, two oblique shocks can be generated with an adjustable flap. With appropriate adjustment the flow after the shock system will leave the channel horizontally. Unfortunately, the highest achievable flow Mach number in the induction region of the flame is approximately 2. A shock-induced supersonic flame by the first shock in this experimental facility is shown in Fig. 7.

A recent attempt at experimentally generating and stabilizing oblique detonation waves in supersonic combustible mixtures is the experiment performed in the NASA-Ames Arcjet Hypersonic Wind Tunnel.[32] A schematic of the test configuration used is shown in Fig. 8. The test flow Mach number is about 4.6, and its pressure and temperature are 0.016 atm and 840 K, respectively. Stoichiometric hydrogen was injected from struts placed at

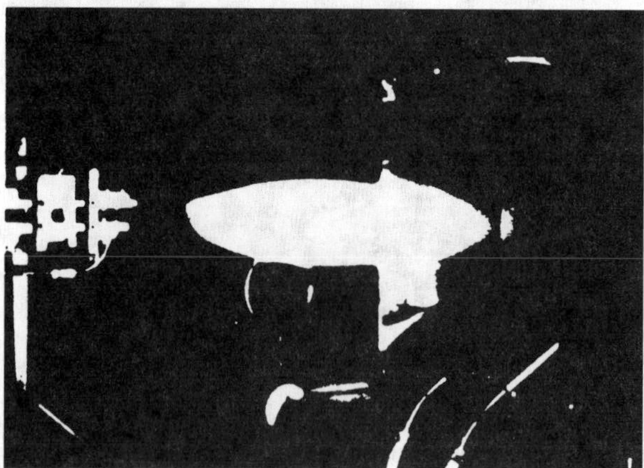

Fig. 7 Shock-induced combustion in hydrogen/air combustible mixture by the first shock.[31]

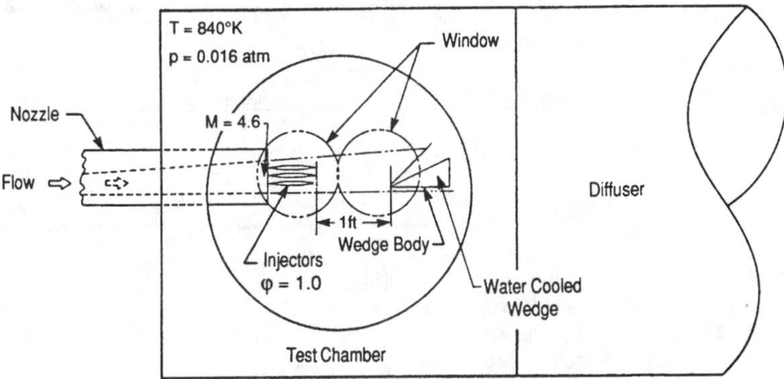

Fig. 8 Test setup in NASA Arcjet Hypersonic Wind Tunnel.[32]

the exit of the nozzle. Unfortunately, published schlieren photographs do not indicate clearly the presence of the detonation wave. The only evidence given by the authors of its presence is the detected pressure rise on the wedge surface when hydrogen was injected.[33]

An interesting and relatively simple technique to generate and stabilize oblique detonation waves is presented in Refs. 34 and 35. It is based on the fact that when a detonation wave propagating in a (primary) layer of combustible mixture comes into contact with an adjacent (secondary) layer of another combustible mixture an oblique shock or detonation wave may be generated in the secondary layer. Experimentally, the phenomenon is realized in two adjacent square cross-sectional shock tubes, each of which is filled by a different combustible mixture of gases. At the test section the two mixtures are separated by a very thin (cellulose, aluminum, etc.) foil, which does not interfere with the resulting detonation/shock interaction. A schematic of typical interactions is shown in Fig. 9. The propagating (normal) detonation wave in the primary layer is followed by an expansion wave. Downstream of this expansion wave, the interface between the combustion products of the primary detonation and the secondary gas mixture acts similar to a wedge or ramp at angle δ_3 (see Fig. 9) to the horizontal, moving supersonically, with the velocity of the primary detonation relative to the second layer. In coordinates fixed to this primary detonation, the second combustible gas layer appears as a supersonic premixed fuel/air or oxidizer stream passing through an oblique detonation or shock wave as necessary. The resulting interaction depends strongly on the properties of both the primary and secondary combustible layers. To establish an oblique detonation in the secondary layer, the detonation velocity in the primary layer must be high enough, and the induced oblique discontinuity angle high enough so that the normal component of the Mach number in the secondary gas layer is equal to the Chapman–Jouguet Mach number or higher. Using this technique, one can extend the range of initial conditions that have been used to date to experimentally generate oblique detonation waves.

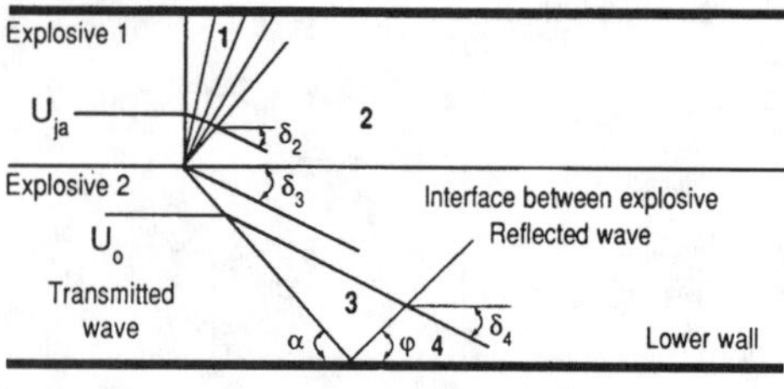

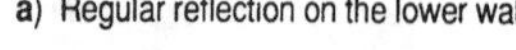
a) Regular reflection on the lower wall

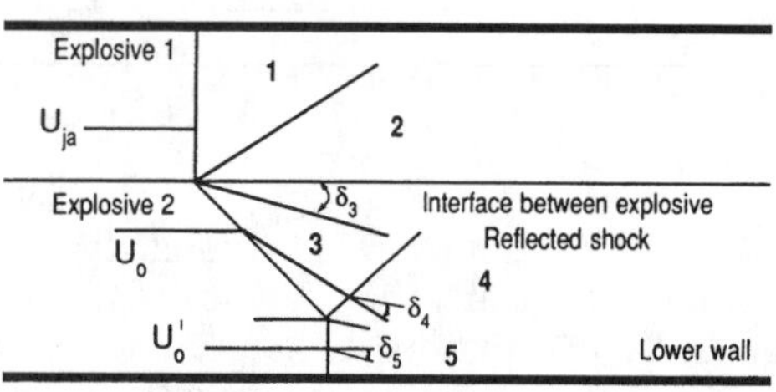

b) Mach reflection on the lower wall

Fig. 9 Glancing interaction between detonation and explosive boundary: a) regular reflection on the lower wall and b) Mach reflection on the lower wall.[35]

From the preceding, one can see that there is currently little experimental evidence concerning the flame holding and stability of oblique detonation waves, especially at high-approach flow Mach numbers. Experimental difficulties associated with generating the correct flow conditions for stabilizing an oblique detonation wave are caused mainly by the high Mach numbers of detonations, the required relatively high test flow pressures and temperatures, the preparation of homogeneous fuel/air mixtures, and to possible premature ignition of the combustible mixture at the injection port.

An appropriate tool to experimentally generate and stabilize detonation or shock-induced combustion waves is the recently developed ram-accelerator facility when operated in the superdetonative mode. This facility offers the possibility of tailoring the conditions of the test gas to match closely the detonation-wave ramjet combustor-inlet flow parameters of pressure, equivalence ratio, and Mach number.[36] Such an attempt at experimentally verifying the stability of oblique detonation waves and assessing their applicability to hypersonic propulsion is made in Ref. 37.

Operating Envelope of Standing Detonation Waves

As with the case of shock waves, stabilizing a detonation wave on a wedge or cone for given oncoming flow parameters is not always possible. Therefore, it is of interest to determine the envelope of operating parameters within which a stable oblique or conical detonation wave can be realized by a wedge or cone for the required conditions of pressure, temperature, velocity, and fuel/air ratio of the combustible mixture at the combustor inlet section of a hypersonic detonation-wave ramjet. At the high temperatures and pressures involved accurate analysis requires the use of temperature-dependent specific heats and equilibrium or finite-rate chemistry calculations. However, a qualitative analysis can be performed within the framework of perfect gas theory and a Chapman–Jouguet detonation-wave model wherein the detonation wave is considered as a discontinuity across which a certain amount of heat Q per unit mass of mixture is released as a result of combustion.[38–46] That is, the conservation of energy across the discontinuity is written as

$$h_1 + Q + \frac{u_1^2}{2} = h_2 + \frac{u_2^2}{2} \tag{1}$$

where h_1 and h_2 are the specific enthalpies and u_1 and u_2 are normal to the detonation-wave velocity components ahead of and behind the wave. Together with the conservation laws of mass and momentum across the discontinuity, and the equation of state, a closed-form solution can be obtained for the flow parameters behind the detonation wave that can be presented and interpreted in a number of ways (see aforementioned references). The resulting dependence of the detonation wave angle β to the oncoming flow on the wedge angle θ, i.e.,

$$\theta = \beta - \tan^{-1}\left[\frac{(1 + \gamma M_{1n}^2 \pm \sqrt{(M_{1n}^2 - 1)^2 - 2\,(\gamma + 1)\, M_{1n}^2 \overline{Q}}}{(\gamma + 1) M_{1n}^2 \sqrt{(M_1/M_{1n}^2)^2 - 1}}\right] \tag{2}$$

where $M_{1n} = M_1 \sin\beta$ and $\overline{Q} = Q/c_p T_1$, is instructive (see Fig. 10). For $\overline{Q} = 0$ Eq. (2) has one nontrivial solution that is the well-known result for oblique shock waves.[47] In this case for any given M_1 and wedge angle θ, there are two angles β that satisfy the conservation equations, the smaller β corresponding to the weak branch and the larger β to the strong branch of the oblique shock-wave solution. These two branches are separated by the maximum value of the flow deflection angle θ_{max} (see Fig. 10).

For $\overline{Q} > 0$ Eq. (2), as plotted in Fig. 10, has two nontrivial solutions, i.e., for given M_1, T_1, $\overline{Q}$, and β there exists two distinct values of θ satisfying the conservation equations. Of particular importance is the limiting case when both solutions coincide, i.e., when the expression under the radical in Eq. (2) vanishes:

$$(M_{1n}^2 - 1)^2 - 2\,(\gamma + 1)\, M_{1n}^2 \overline{Q} = 0 \tag{3}$$

Under this condition the conservation and state equations show that the normal component of the Mach number downstream of the wave M_{2n} is equal to unity: $M_{2n} = 1$. This is the so-called Chapman–Jouguet state, and, as

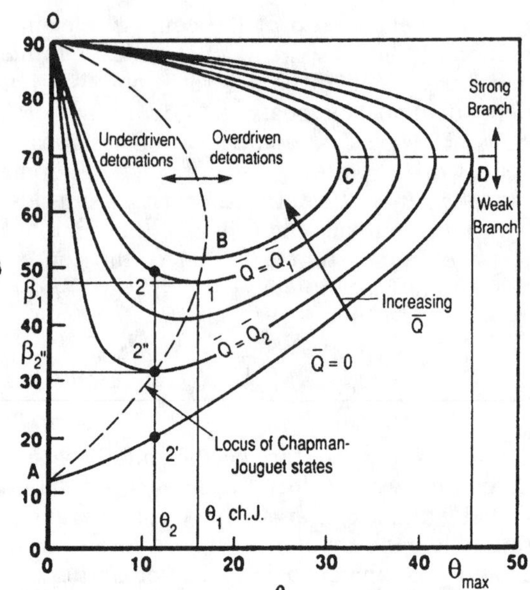

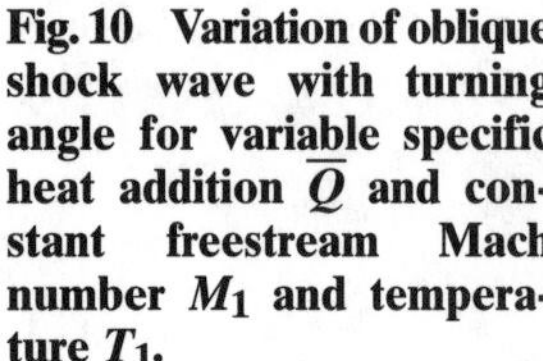
Fig. 10 Variation of oblique shock wave with turning angle for variable specific heat addition $\overline{Q}$ and constant freestream Mach number M_1 and temperature T_1.

is apparent from Fig. 10, it is the point of minimum detonation-wave angle β on each locus of states for given $\overline{Q} > 0$ and, therefore, corresponds to the minimum total-pressure loss for given approach conditions. Curve ABO in Fig. 10 is the locus of these Chapman–Jouguet states. The two nontrivial solutions of Eq. (2) can then be classified by whether M_{2n} is less or greater than unity. To avoid confusion with the terms *weak* and *strong* as applied to oblique shock waves (lower and upper branches, respectively, of the $\overline{Q} = 0$ curve in Fig. 10), Pratt et al.[27] recommend to call the solution of Eq. (2) corresponding to $\overline{Q} > 0$ and $M_{2n} > 1$ *weak underdriven oblique detonation waves* (the curves in Fig. 10 to the left of ABO) and those corresponding to $\overline{Q} > 0$ and $M_{2n} < 1$ *weak overdriven oblique detonation waves* (the curves in Fig. 10 to the right of ABO up to the maximum turning angle $\theta_{\max}$ beyond which an oblique detonation wave will detach). Finally, strong oblique shock waves with $\overline{Q} > 0$ and for which $M_{2n} < 1$ always are called simply *strong detonation waves* (the upper curves in region OCD in Fig. 10). It can be shown (e.g., see Refs. 9, 39, 40, 44, and 27) that underdriven weak detonation waves are not physically meaningful solutions. Therefore, for given M_1, T_1, and $\overline{Q}$, physically meaningful, attached oblique detonation waves are represented only by the curves in region ABCD in Fig. 10. One can see that the effect of heat addition is to reduce the range of possible flow turning angles for attached oblique detonation waves.

The Chapman–Jouguet state is of critical significance for applications in detonation-wave ramjets. As mentioned earlier, the Chapman–Jouguet flow turning angle $\theta_{\text{Ch.J.}}$ corresponds to the minimum total-pressure loss for given oncoming flow conditions; moreover, in the vicinity of the Chapman–Jouguet state, the detonation-wave angle and, hence, the total-pressure loss through the wave are insensitive to the flow turning angle θ (see Fig. 10). Consequently, small amounts of detonation overdrive near the Chapman–Jouguet point could be employed (say, for detonation-wave stability purposes) in a detonation-wave ramjet with little penalty in its performance. The flow deflection angle corresponding to the Chapman–Jouguet condition can be obtained from Eq. (2) as

$$\theta_{\text{Ch.J.}} = \beta_{\text{Ch.J.}} - \tan^{-1}\left[\frac{1+\gamma M_{1n\text{Ch.J.}}^2}{(\gamma+1)M_{1n\text{Ch.J.}}^2\sqrt{(M_1/M_{1n\text{Ch.J.}})^2-1}}\right] \tag{4}$$

where $M_{1n\text{Ch.J.}}$ is related to the amount of heat released by Eq. (3), and $M_{1n\text{Ch.J.}} = M_1 \sin\beta_{\text{Ch.J.}}$. Figure 11 from Ref. 27 depicts how $\theta_{\text{Ch.J.}}$ varies with M_1 for a given reactant mixture and also shows the variation with Mach number of the maximum turning angle $\theta_{\max}$, corresponding to wave detachment from the wedge. It represents the range of permissible wedge angles for attached oblique detonation waves for a particular set of approach conditions.[27]

The effects of oncoming flow parameters on the relationship between the wedge angle and the detonation-wave angle are shown in Fig. 12. These plots were obtained from calculations based on variable specific heats and chemical equilibrium of the burned gas. The heat-release parameter is represented by the fuel/air equivalence ratio φ. One can readily see that the effect of increasing the combustor-inlet temperature and Mach number is to increase the capacity of the oblique detonation wave to accept full stoichiometric heat release.[27]

As mentioned earlier, weak-attached underdriven oblique detonation waves are not physically meaningful. Therefore, the question arises on the real nature of the flow when the flow turning angle is less than the minimum turning angle $\theta_{\text{Ch.J.}}$ for stable oblique detonation waves. Rutkowski and Nicholls,[40] Woolard,[43] Chernyi,[9,44] and Bartlmä[46] argued that because the normal component of the Mach number behind the detonation wave is

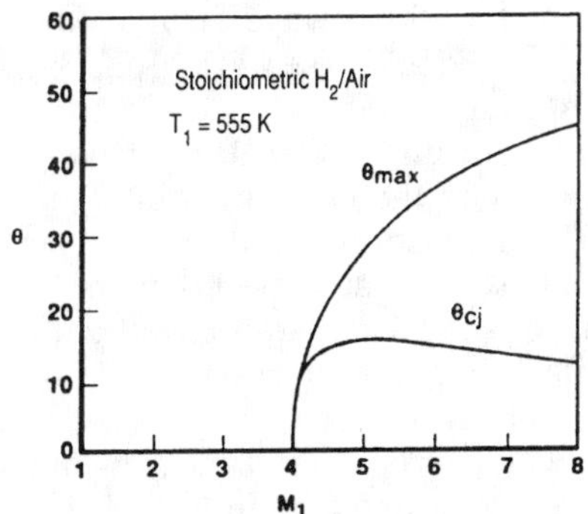

Fig. 11 Variation of maximum and minimum turning angles with freestream Mach number for stable oblique detonation waves for stoichiometric hydrogen/air.[27]

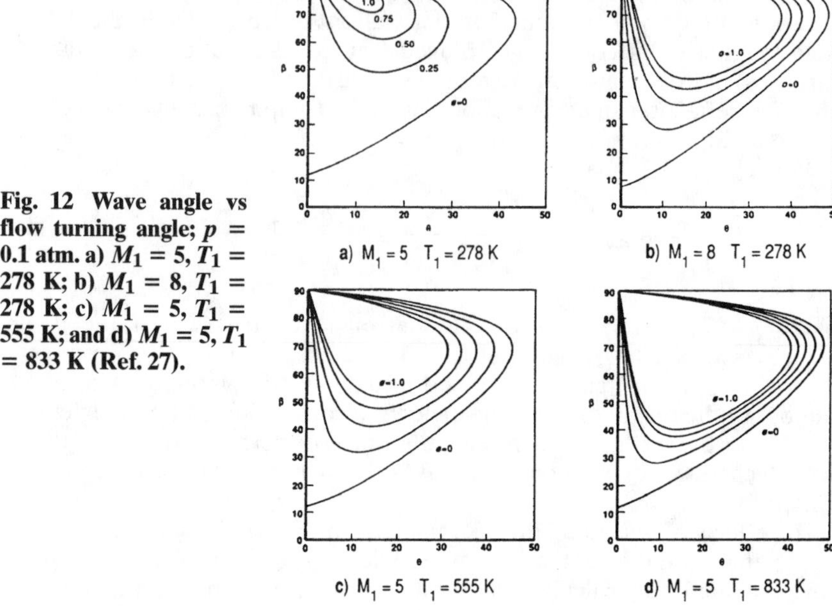

Fig. 12 Wave angle vs flow turning angle; p = 0.1 atm. a) $M_1 = 5$, T_1 = 278 K; b) $M_1 = 8$, T_1 = 278 K; c) $M_1 = 5$, T_1 = 555 K; and d) $M_1 = 5$, T_1 = 833 K (Ref. 27).

unity, i.e., the detonation wave is parallel to the acoustic characteristics behind the wave, and because $\theta_{\mathrm{Ch.J.}}$ is the smallest angle through which the flow can be deflected for a given amount of heat release $\overline{Q}$, the flow behind the fixed Chapman–Jouguet detonation adjusts itself to the smaller than $\theta_{\mathrm{Ch.J.}}$ wedge angle θ_w through a centered Prandtl–Meyer expansion starting immediately behind the detonation wave (Fig. 13), which turns the flow isentropically parallel to the wedge surface. However, Pratt et al.[27] state that because, for underdriven weak waves $M_{2n} > 1$, the Mach waves (characteristics) behind the detonation wave diverge from the latter (see Fig. 13, flow region to the right of characteristic OI) and "since the presence of the wall cannot be transmitted to an underdriven oblique wave, such a wave cannot be attached to and supported by a wedge"[27] (see also Ref. 39). Consequently, they advanced the argument that "as point 2 (see Fig. 10) is not stable and the flow is still required to turn through the angle θ_2, the only stable endpoints possible are those lying between the limits of point 2$'$, no detonation (i.e., at most shock-induced combustion), and incomplete oblique detonation waves with partial heat release at most $\overline{Q}_2$, point 2$''$, determined by the condition $\theta_{\mathrm{Ch.J.}} = \theta_2''$"[27]; i.e., the Chapman–Jouguet detonation wave of angle β_1 in state 1 adjusts itself to a new position $\beta_2'' < \beta_1$ at Chapman–Jouguet point 2$''$ corresponding to heat release $\overline{Q}_2$.

Recently, the present author and his colleagues considered the flow of a stoichiometric hydrogen/air mixture over a wedge depicted in Fig. 14. The

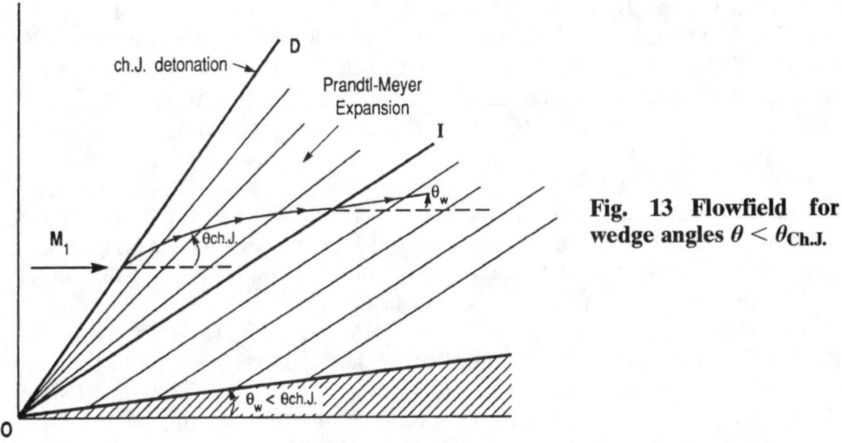

Fig. 13 Flowfield for wedge angles $\theta < \theta_{Ch.J.}$

oncoming flow with $M_1 = 5.4$, $T_1 = 900$ K, and $p_1 = 0.22$ atm (22 kPa) makes an angle $\theta_1 = 20$ deg with the horizontal (i.e., for the horizontal wedge surface the flow turning angle is 20 deg). The finite-rate shock-induced combustion process is described by 33 reactions between 13 species (H_2, O_2, H, O, OH, H_2O, HO_2, H_2O_2, N, NO, HNO, N_2, and NO_2) taken from Ref. 48. The resulting flowfield was computed by using the numerical scheme developed in Refs. 49 and 50. The computed flowfield for a wedge surface tilted 5 deg upwards from the horizontal, i.e., for a flow deflection angle $\theta = 15$

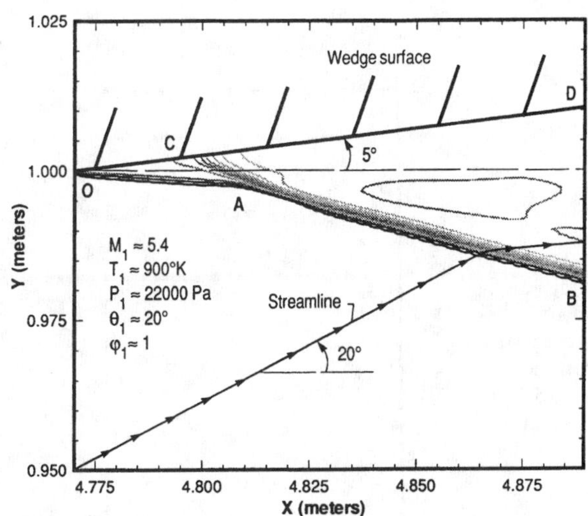

Fig. 14 Wedge flow of a stoichiometric hydrogen/air mixture with finite-rate combustion chemistry. Flow deflection angle $\theta = 15$ deg. Static temperature contours.

deg, is shown in Fig. 14. Because of the ignition delay behind the oblique shock OA, the combustion starts in the vicinity of point C on the wedge surface. The combustion generates a set of deflagration waves that coalesce with the oblique shock wave and steepen into detonation wave AB (Fig. 14). Deflection angles 7.5 deg $\leq \theta \leq$ 25 deg were considered and the angle of the detonation wave AB numerically determined. Figure 15, which is a plot of the calculated detonation-wave angle vs the deflection angle, shows that detonation-wave angle β remains constant for deflection angles $\theta \leq$ 15 deg; the flow behind the detonation-wave (see Fig. 16) for θ = 7.5 deg expands immediately downstream of the detonation wave, with the pressure matching the pressure induced behind the oblique shock wave OA (Fig. 14). This resulting flowfield is very similar to that depicted in Fig. 13 for the case when the combustion process is in equilibrium. The Chapman–Jouguet point for the case considered is $\theta_{\text{Ch.J.}} \approx$ 15 deg and $\beta_{\text{Ch.J.}} \approx$ 31 deg. This result seems to support the theory advanced in Refs. 9, 40, 43, 44, and 46 by earlier investigators. Off-Chapman–Jouguet flow situations over a wedge may occur in the combustor region of detonation-wave ramjets at design and off-design flight conditions. Therefore, it is of great interest to investigate their structural stability similar to that performed in Ref. 51 for overdriven oblique detonation waves. Experimental evidence of standing detonation waves for wedge angles, smaller than the corresponding Chapman–Jouguet flow turning angle, is lacking.

For wedge angles larger than $\theta_{\max}$ the oblique detonation wave detaches, and either a normal detonation wave or normal shock-induced combustion will occur in the vicinity of the wedge vertex. Away from the axis of symmetry, these waves decay and bifurcate in a number of ways, exhibiting, under certain conditions, a periodic structure. For example, see Refs. 9 and 52–57 for details.

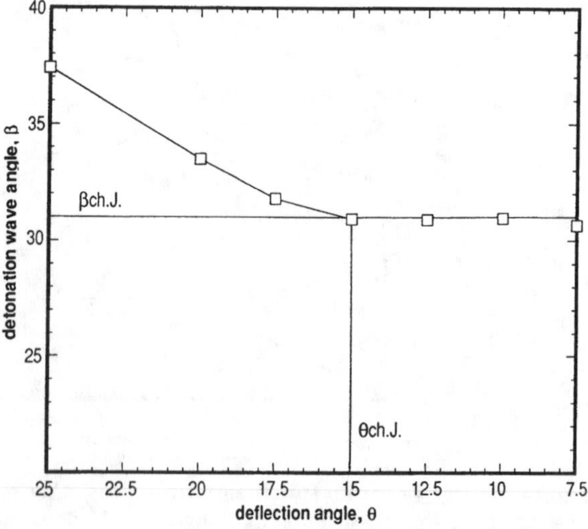

Fig. 15 Detonation-wave angle vs turning angle for a stoichiometric hydrogen/air mixture with finite-rate combustion chemistry: M_1 = 5.4, T_1 = 900 K, p_1 = 0.2 atm.

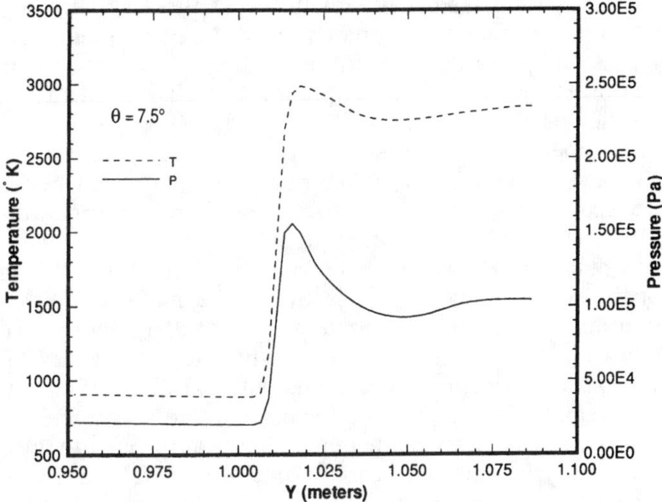

Fig. 16 Profiles of static pressure and temperature at section *BD* (Fig. 14) for a deflection angle θ = 7.5 deg.

Fuel/Air Premixing Process

Injecting and premixing the fuel in the high Mach number forebody (inlet) airflow is the most difficult and challenging task in the practical implementation of detonation-wave ramjets. Fuel must be injected into the airstream at a location where the static temperature of the fuel/air mixture is below its ignition temperature. Heat recycling from the airframe and engine will lead to high injected-fuel total temperatures, on the order of 1000 K or above. High total airflow temperatures, well above the ignition temperature of the combustible mixture, will preclude oblique or normal to the airflow fuel injection; the induced shocks could easily cause premature ignition in the forebody flow. Therefore, fuel must be injected parallel to the oncoming airflow at matched flow conditions at the fuel-injector nozzle-exit plane to avoid the presence of shock waves that could be generated in its vicinity; parallel injection will, however, produce long mixing lengths. Even with parallel injection, ignition of the combustible mixture may occur in the boundary layers in the vicinity of the nozzle exit and next to the forebody wall where the maximum static temperatures may exceed the fuel/air ignition temperature. Finally, the use of in-stream fuel-injection struts, if possible, should be avoided, as it would add drag and cooling problems[32] and, most importantly, would create hot spots and cause premature ignition.

When parallel-wall slot fuel injection is used to perform the premixing, the resulting mixing lengths can be estimated by an approximate analysis[58] based on the following assumptions: 1) the initial boundary layers and the wall boundary layer adjacent to the mixing zone are neglected; 2) the Prandtl and Lewis numbers are unity; 3) the turbulent shear stress is determined by Prandtl's mixing length relation and the empirical constant is based on incompressible experimental data; 4) the nondimensional velocity profile is independent of distance from the mixing origin; 5) the mixing is isobaric; and 6) the mixing similarity parameter is independent of distance from the mixing origin.

When the preceding approximate analysis is applied to slot injection in a two-dimensional ramjet inlet with an air-capture area of ~ 1 m radius, one obtains the mixing lengths, calculated to reach stoichiometric concentrations on the wall, and the width of the mixture profile with a stoichiometric peak for various slot heights shown in Fig. 17. Depending on the slot width mixing lengths for this single-slot wall fuel injection in the inlet vary from 3–8 m for low Mach number to about 2–3 m at very high Mach numbers. An example of the application of these plots to a wedge-type inlet at Mach 12 is shown in Fig. 18. Here wall injection is combined with parallel fuel injection from a slotted in-stream wedge to produce an approximate stoichiometric fuel/air profile (indicated by the dotted line in the figure) at the inlet to the combustor. A similar analysis used in the case of one- or two-wall slot fuel injection in the internal compression channel of the inlet is shown in Fig. 19. Mixing lengths for this case vary from ~ 3–7 m for low flight Mach numbers to ~ 2–3.5 m for very high Mach numbers. Mixing distances to achieve stoichiometric concentration predicted by the preceding analysis are not prohibitively long for realistic detonation-wave ramjet dimensions. However, it should be kept in mind that the analysis is approximate and requires experimental verification.

As mentioned earlier, the amount of compression in the inlet of the detonation-wave ramjet is less than that for the scramjets, and, therefore, the flow Mach numbers are higher. With mixing taking place in this high Mach number inlet flow, the detonation-wave ramjet will be penalized severely by fuel-injection losses. An exact estimate of these losses would be prohibitively complicated, although computational fluid dynamics codes, incorpo-

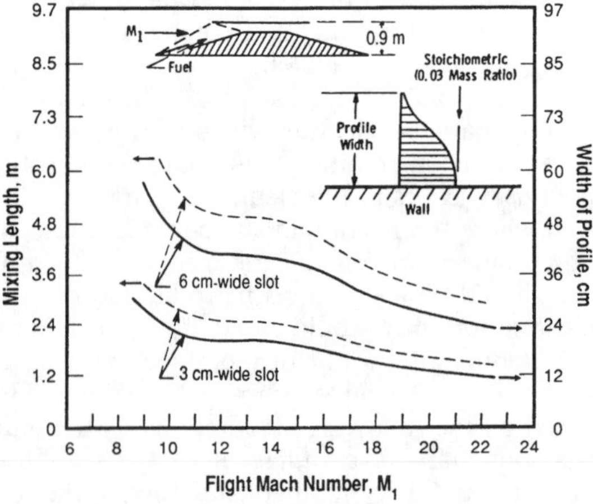

Fig. 17 Theoretical mixing length and fuel profile width in engine inlet for slot hydrogen injection to 0.03 mass ratio at peak. V_{fuel}/V_{air} = 0.8 and $T_{t_{fuel}} / T_{t_{air}}$ = 0.1(Ref. 58).

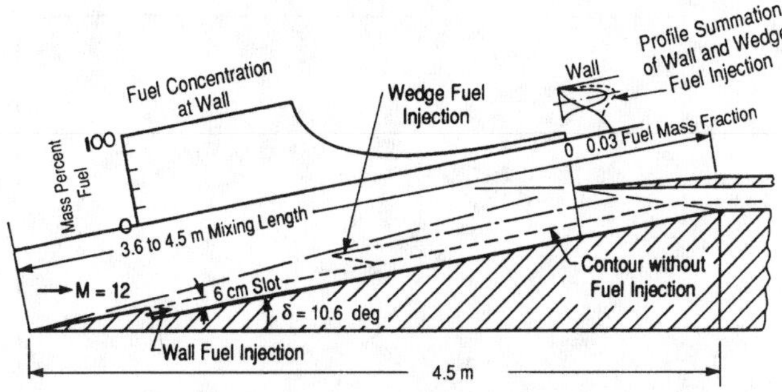

Fig. 18 Mach 12 two-shock inlet with fuel injection.[58]

rating validated turbulent mixing models, could provide realistic estimates.[59] However, an approximate estimate of the magnitude of these losses for parallel injection in a high Mach number flow can be obtained[2] if one assumes that the two mixing gases (air and fuel), at different temperatures and velocities at station 1, are inviscid perfect gases with average values of the specific heats in the temperature range considered, heat and other losses are absent, and that the gases are fully mixed at a downstream station 2. Conservation of mass, momentum, and energy applied at stations 1 and 2 provide, among other flow parameters, an expression for the total-pressure ratio at stations 2 and 1, which is plotted in Fig. 20, as a function of airflow Mach number at the initial mixing station 1 for a stoichiometric hydrogen/air mixture for cases when mixing occurs at constant pressure or constant area. The mixture temperature at station 2 is prescribed, $T = 1100$ K, to keep it below its ignition temperature. Curves for two different fuel-inlet conditions are shown. In the Mach number range $2 \leq M_{1a} \leq 5$ the least loss in total pressure is obtained by injecting the fuel at a low energy level

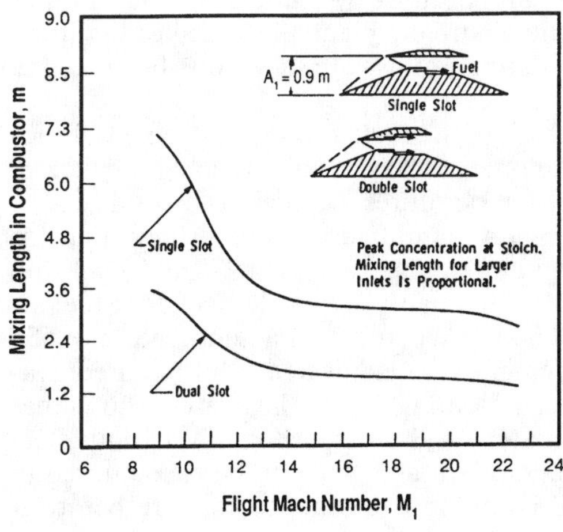

Fig. 19 Theoretical compression chamber mixing length for single- and double-slot injection; inlet area = 0.9 m high. V_{fuel}/V_{air} = 0.8 and $T_{t_{fuel}} / T_{t_{air}}$ = 0.1 (Ref. 58).

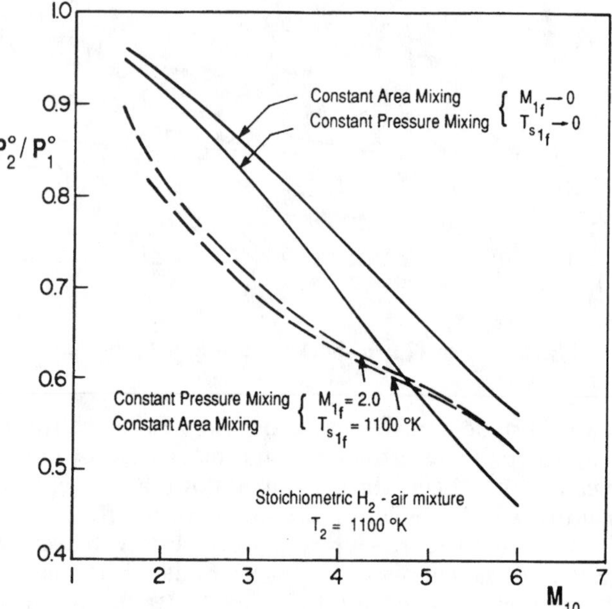

Fig. 20 Total-pressure ratio across mixing zone.[2]

because more thermal energy is transferred from the air to the fuel in a frictionless heat extraction process during which the total pressure of the gas increases.[2] As the Mach number of the air increases, the air total temperature becomes so large that the effect of the difference in fuel-inlet conditions becomes less important. Mixing losses depend on the fuel/air equivalence ratio, as well as fuel/air velocity and temperature ratios. Large fuel/air ratios induce higher mixing losses; this can be offset by parallel fuel injection at velocities approaching those of the airstream. Figure 20 indicates that mixing losses for high Mach number, detonation-wave ramjet inlet flows can be significant.

The problem of premature ignition of the fuel/air mixture during wall slot injection in the forebody of the detonation-wave ramjet is a major concern, especially when the injected-fuel temperatures are high as a result of engine/airframe cooling requirements. A simplified analysis[12] shows that for flight Mach numbers above ~ 7 and wall temperatures of 278 and 555 K the maximum temperature in a turbulent boundary layer will exceed the fuel/air ignition temperature (assumed 1000 K). When fuel at a temperature of 555 K (equal to the assumed wall temperature) is injected to form an equivalence ratio of 0.1 (the adopted lean limit for ignition), the analysis shows that preignition of the combustible mixture will occur at flight Mach numbers greater than 10.5. For higher fuel temperatures likely to be encountered in practical applications, premature ignition of the fuel/air mixture before it reaches the detonation wave seems to be unavoidable. Calculations performed in Ref. 12 suggest that preignition can be suppressed by injecting a relatively cool inert gas between the fuel and air (a buffer gas), e.g., nitrogen or water, up to flight Mach numbers of 16. Further research into this problem may reveal other efficient solutions to this crucial process for the operation of a detonation-wave ramjet. State-of-the-art computational-fluid -dynamics methods incorporating turbulent mixing models and finite-rate chemistry, and subsequent experimentation in hypersonic high-enthalpy tunnels, represent important tools for the analysis of fuel injection and the mixing process and of the possibility of premature ignition in the forebody flow.[32,33,59]

Because of the impracticality of achieving a homogeneous fuel/air mixture in hypersonic flows, it is of great interest to assess the effects of mixing inhomogeneities on detonation or shock-induced combustion wave structure, form, and stability. Results of such an analysis[59] are shown in Figs. 21 and 22. In Fig. 21 the stoichiometric parameter φ in the freestream is artificially modulated by a pseudosinusoidal function in the interval $0.1 \leq \varphi \leq 1.9$, with $\varphi = 1$ in the center region of the flow. The shock wave in the nonreacting case is curved because of Mach-number variations caused by molecular weight variations. In the reacting case the curvature of the detonation wave is amplified clearly. Nevertheless, the combustion is complete,

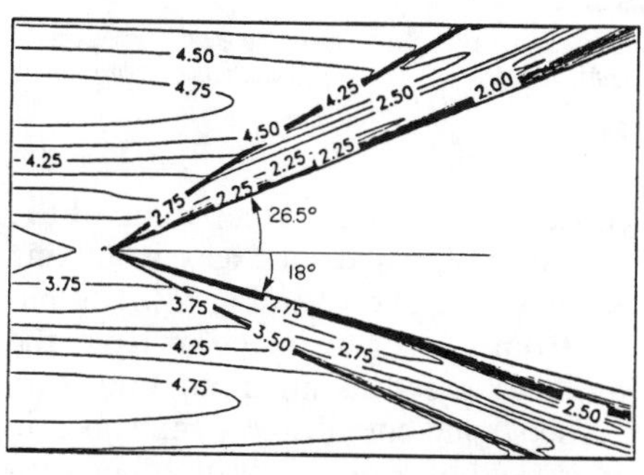

a) Mach number for non-reacting case, variable φ

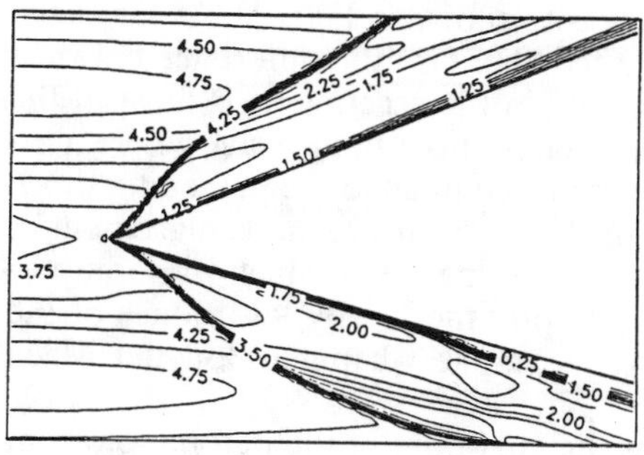

b) Mach number for reacting case, variable φ

Fig. 21 Variable stoichiometric parameter φ: $0.1 < \varphi < 1.9$. Oncoming flow Mach number $M_\infty \approx 4.2$ near center region. a) Mach number for nonreacting case, variable φ and b) Mach number for reacting case, variable φ (see Ref. 59).

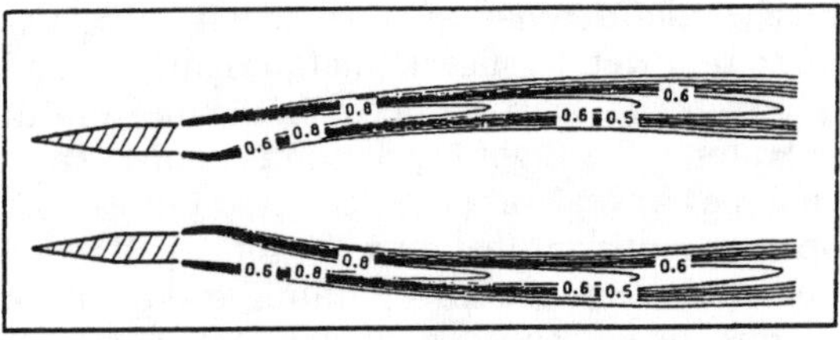

a) H_2 concentration for fuel-injection

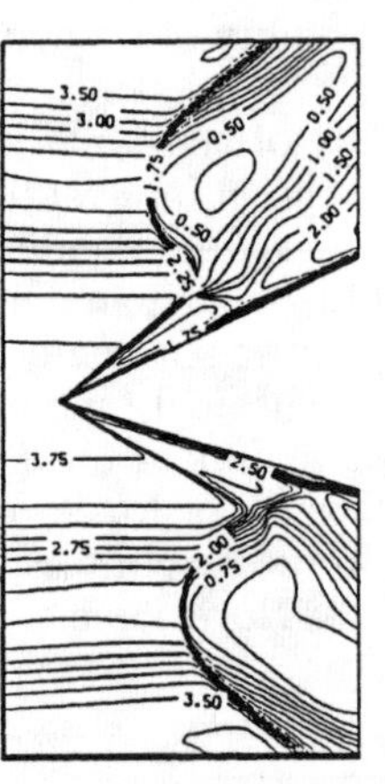

b) Mach number, non-reacting case; input from strut flow-field

c) Temperature, reacting case; input from strut flow-field

Fig. 22 Effect of mixing inhomogeneities on detonation-wave structure: a) H_2 concentration for fuel injection; b) Mach number, nonreacting case; input from strut flowfield; and c) temperature, reacting case, input from strut flowfield (see Ref. 59).

i.e., all of the hydrogen or oxygen is consumed,[59] and the steepening of the wave becomes mixing-dependent. The effects of an extreme case of mixing inhomogeneity on detonation-wave behavior are shown in Fig. 22. The numerically simulated flowfield resulting from hydrogen injection from the base of a two-strut configuration (Fig. 22a) is fed as an input boundary condition to the flow over the wedge. The strong curvature of the shock in the nonreacting case (Fig. 22b) is again caused by large Mach-number variations of the oncoming flow caused by molecular weight changes. Of interest is the temperature field in the reacting case shown in Fig. 22c. As far as the locations of discontinuities are concerned, there is little difference between nonreacting and reacting cases. However, in the reacting case, most of the combustion occurs in the two highly curved portions of the wave where most of the fuel is located. The flame seems to be coupled to the preceding shock in that region, which is responsible for the slight upstream displacement of the wave front (compared with the nonreacting case). Much more importantly, the detonation front is stable, despite the locally low values of the Mach number, estimated to be lower than the Chapman–Jouguet Mach number.[59]

In these poorest mixing conditions the curvature of the detonation front is dominated by the fuel/air mixing pattern instead of heat-release distributions. Curved shocks may be used as a flame holder in hypersonic detonation-wave ramjets, whether or not the flame is attached to the shock. However, this phenomenon needs further investigation.

The fuel/air mixing process in very high Mach-number flows is poorly understood. Further extensive research is needed to obtain numerically and experimentally reliable results, especially in the turbulence modeling area, for application to detonation-wave ramjet design.

Performance Analysis

Because the detonation-wave angle to the oncoming flow is minimum at the Chapman–Jouguet condition (see Figs. 12 and 15), total-pressure loss or entropy increase across the detonation wave clearly can be minimized by the attainment of this condition at the lowest possible (normal to the wave) Chapman–Jouguet detonation Mach number. This Mach number is proportional to the amount of heat released, i.e., the fuel/air equivalence ratio φ, and inversely proportional to the square root of the static temperature ahead of the detonation wave.[5] Therefore, for best performance the static temperature of the unreacted combustible mixture must be allowed to approach its limiting ignition value. The effect of the predetonation wave Mach number (and, hence, predetonation wave temperature) on optimum air-specific impulse is shown in Fig. 23 for a hypersonic ramjet employing a normal detonation wave with specified amounts of losses in the inlet and nozzle.[3] The shift in optimum prewave Mach number results from the interplay of prewave temperature and detonation-wave losses.

Optimum performance of a detonation-wave ramjet may be obtained by properly balancing the inlet shock losses against those of the oblique detonation wave, provided the heated flow is expanded isentropically in the nozzle. Within the framework of one-dimensional inviscid perfect gas theory,

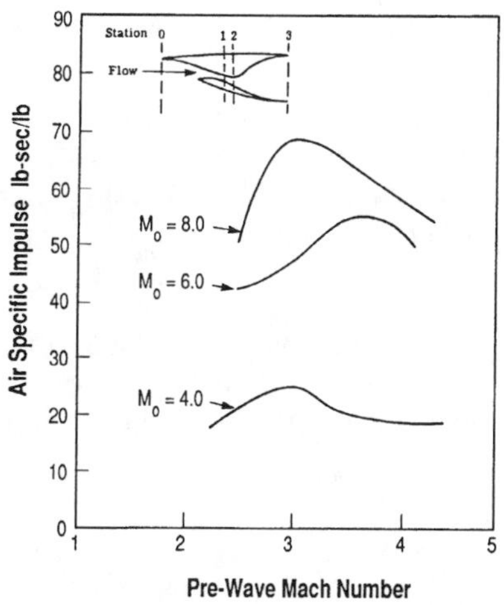

Fig. 23 Air-specific impulse vs prewave Mach number. Normal detonation-wave ramjet.[3]

the velocity ratio across any ramjet is shown to be given by (e.g., see Refs. 4 and 5)

$$\frac{V_e}{V_o} = \left\{ \left[1 + \frac{2}{(\gamma-1)M_o^2} \right] \frac{T_{\rm pd}^o}{T_o^o} \left[1 - \left(\frac{p_e/p_o}{p_{\rm pd}^o/p_o^o} \right)^{\frac{\gamma-1}{\gamma}} \frac{1}{1+(\gamma-1/2)M_o^2} \right] \right\}^{1/2} \tag{5}$$

Here subscripts o, e, and pd denote the oncoming, exit, and postdetonation or postcombustion flow parameters, and superscript o the stagnation values. Assuming the combustion products are fully expanded, i.e., $p_{\rm e} = p_{\rm o}$, the ramjet thrust

$$Th = \dot{m}_a V_o \left(\beta \frac{V_e}{V_o} - 1 \right) \quad ; \quad \beta = 1 + \frac{\dot{m}_{\rm fuel}}{\dot{m}_a} \tag{6}$$

is maximum if V_e/V_o is maximum or $p_{\rm pd}^o/p_o^o$ is maximum for given $\dot{m}_{\rm fuel}/\dot{m}_a$, flight Mach number M_o, and total temperature ratio $T_{\rm pd}^o/T_o^o$ across the detonation wave. For a detonation-wave ramjet having N equal strength shocks $M_{n1}, \ldots, M_{nN} = M_{n_s}$, where M_{n_s} is normal to the shock-wave component of the Mach number (Oswatitsch's criterion for minimum entropy rise through the inlet[60]) and a detonation wave characterized by M_{nd} (normal to the wave component of the oncoming flow Mach number), Townend[4] has derived the following analytical relationship for maximum $p_{\rm pd}^o/p_o^o$ for given values of M_o and $T_{\rm pd}^o/T_o^o$

Condition of maximum $p_{\rm pd}^o/p_o^o$ for given M_o and $T_{\rm pd}^o/T_o^o$

$$M_{n_s}^2 = \left[1 \pm \sqrt{1-(1-\chi_F)(1-\gamma\chi_F)} \right] / (1-\gamma\chi_F) = f_1(F, M_{nd}) \tag{7}$$

where

$$\chi_F = \left(\frac{M_{nd}^2 - 1}{M_{nd}^2 + 1} \right) \frac{F[M_{nd}^4(\gamma+1-F) - FM_{nd}^2(\gamma-1) + \gamma(F-1) - 1]}{[\gamma+1+\gamma F(M_{nd}^2-1)][F+(\gamma+1-F)M_{nd}^2]} \tag{8}$$

and F is a dimensionless detonation-wave classification parameter (see Refs. 5 and 45): $F = 1$ for Chapman–Jouguet detonations, $1 < F < 2$ for overdriven detonation waves, and $F = 2$ for shock waves. Another relationship between n, F, M_{n_s} and M_{nd} is[4]

$$\frac{T_{\rm pd}^o}{T_o^o} = 1 + f_3(n, M_{n_s}) \frac{F(2-F)(M_{nd}-1)^2}{(\gamma+1)[2+(\gamma-1)M_o^2]M_{nd}^2} = f_2(n, F, M_{n_s}, M_{nd}) \tag{9}$$

$$= 1 + \frac{q}{c_p T_o^o} = \text{given quantity } (q - \text{heat added per unit mass}) \quad (10)$$

Here

$$f_3(n, M_{n_s}) = \frac{T_{\text{comp}}}{T_o} = \left\{ \frac{\left[2\gamma M_{n_s}^2 - (\gamma - 1)\right]\left[2 + (\gamma - 1) M_{n_s}^2\right]}{(\gamma + 1)^2 M_{n_s}^2} \right\}^n \quad (11)$$

T_{comp} being the temperature at the end of the inlet compression process and ahead of the detonation wave. Finally, the (maximum) engine cycle static temperature attained on the discharge side of the detonation wave is[45]

$$\frac{T_{\text{max}}}{T_o} = \frac{T_{\text{pd}}}{T_o} = f_3(n, M_{n_s}) \frac{\left[\gamma + 1 + \gamma F(M_{nd}^2 - 1)\right]\left[F + (\gamma + 1 - F) M_{nd}^2\right]}{(\gamma + 1)^2 M_{nd}^2} \quad (12)$$

$$= f_4(n, F, M_{n_s}, M_{nd}) \quad (13)$$

The preceding relationships can be used to study various cases of optimum (maximum thrust) detonation-wave ramjets for a given number of n equal-strength inlet shocks.

1) If F is specified, i.e., the detonation-wave type is chosen (Chapman–Jouguet or overdriven), then Eqs. (5) and (6) determine M_{n_s} and M_{nd} required for a maximum thrust detonation wave ramjet. For a given flight Mach number, all of the flow parameters of the propulsive cycle are thereby determined. Note that in this case no control is exercised over T_{comp} and T_{max}.
2) If the static temperature at the end of the inlet compression is specified (this temperature should be less than the fuel/air ignition temperature at flow conditions at that location), i.e., if T_{comp}/T_o is specified, then Eq. (7) determines M_{n_s}, and Eqs. (5) and (6) are used to determine F and M_{nd}, required for a maximum thrust detonation-wave ramjet with controlled inlet discharge flow temperature. In this case no control is exercised over F and T_{max}.
3) If a given maximum engine cycle temperature is specified, i.e., $T_{\text{max}}/T_o = T_{pd}/T_o$ is given, then Eqs. (5), (6), and (8) will determine M_{n_s}, F, and M_{nd} of a maximum thrust detonation-wave ramjet. In this case the inlet discharge flow temperature T_{comp} and F are obtained from the resulting solution.

The preceding analysis, although simplified, may serve as a guide to more elaborate methods for determining the optimum performance of detonation-wave ramjets.

A more recent performance analysis of detonation-wave ramjets was conducted by Ostrander et al.[12] In that study the engine cycle was divided into four parts: inlet, fuel/air premixing process, detonation wave (combustor), and nozzle expansion. A sketch of the considered inlet is shown in Fig. 24a. The forebody angle δ is specified, and at design conditions the shock generated by the forebody intersects the leading edge of the cowl. This shock is subsequently reflected from the upper surface of the cowl and the lower surface of the ramjet and ends at the detonation-wave wedge (point D). The inclination of the upper surface of the cowl is determined by the condition that the first two shocks be of equal strength. For the axisymmetric case the forebody shock is conical, and it is assumed that subsequent oblique shock reflections in the inlet occur in a planar flow. The gas in the inlet flow is treated as calorically imperfect, and its composition is approximated as follows (Fig. 24b): air is composed of three species—O_2, N_2, and Ar; starting from the fuel-injection station and extending a certain distance downstream is a zone of pure hydrogen fuel. In this zone fuel is added by parallel injection, and the injection station is well upstream of the detonation wave where the static temperature of the air is low. Finally, a perfectly mixed fuel/air mixture is assumed downstream of this zone and up to the detonation wave. The mixing of hydrogen and air is assumed to occur in a constant area. The static and total temperatures of hydrogen are specified,

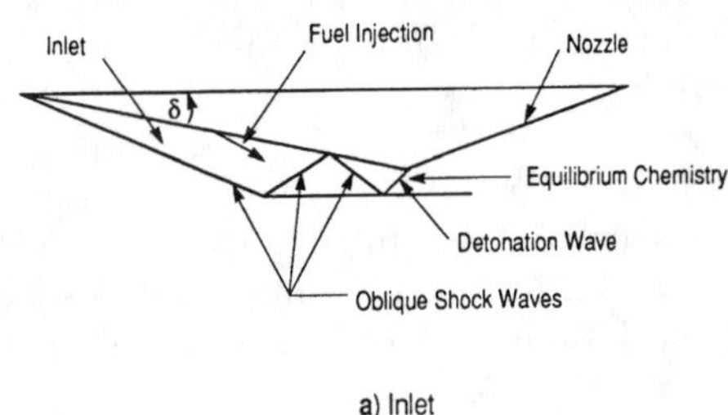

a) Inlet

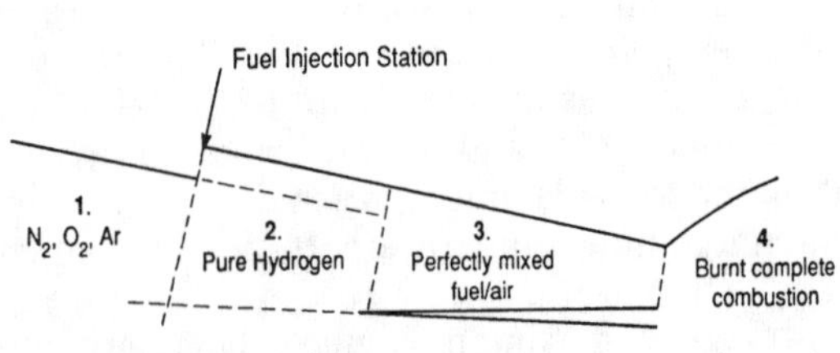

b) Gas composition zones assumed in the inlet

Fig. 24 Oblique detonation-wave engine sketch: a) inlet and b) gas composition zones assumed in the inlet.[12]

and its pressure at the point of injection is equal to that of the airstream at that point.

The detonation wave is modeled as an oblique shock wave with an additional term in the energy equation to account for the heat released by the combustion of hydrogen as the mixture passes through the shock wave. The induction times for the hydrogen/air mixture near stoichiometric concentrations were predicted by the simplified procedure suggested in Ref. 61, and complete combustion is assumed after the detonation wave. The assumption is made that if downstream of the oblique shock the calculated induction distance is less than 1 ft, the deflagration front couples with the oblique shock front, which steepens and forms a detonation wave. The burnt gas is then expanded for the conditions just behind the detonation wave to a point where the exit area is equal to the inlet area, resulting in an underexpanded nozzle. The expansion process is assumed to be one-dimensional, isentropic, and chemically frozen.

Based on the preceding model of a detonation-wave ramjet, Ostrander et al.[12] calculated some of its performance parameters in the flight Mach-number range $5 \leq M_\infty \leq 20$ and for two constant dynamic pressure trajectories $q = 500$ and 2000 psf. For given freestream conditions and cowl height (inlet capture area), the given forebody angle is varied until a geometry is obtained (with the first two shocks being of equal strength) such that the induction distance is less than 1 ft. This gives a geometry and performance for one Mach number on a mission trajectory.

Figure 25 shows the estimated fuel-specific impulse vs flight Mach number, each point on the curve representing a different geometry. Of interest is the lower Mach-number portion of this plot. It shows that a detonation-wave ramjet has a definite lower limit in Mach number at which it can provide a significant amount of thrust. A detailed plot of this low Mach-number region is shown in Fig. 26, where each point is labeled with the maximum equivalence ratio that could be accommodated. At low Mach numbers the amount of energy that can be added at the oblique detonation

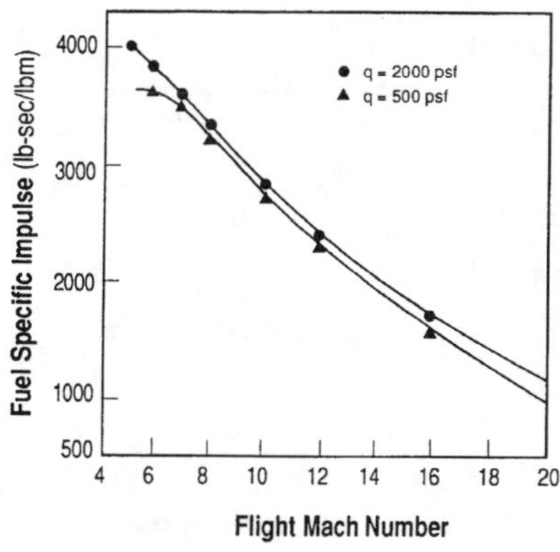

Fig. 25 Fuel-specific impulse vs flight Mach number for two values of freestream dynamic pressure.[12]

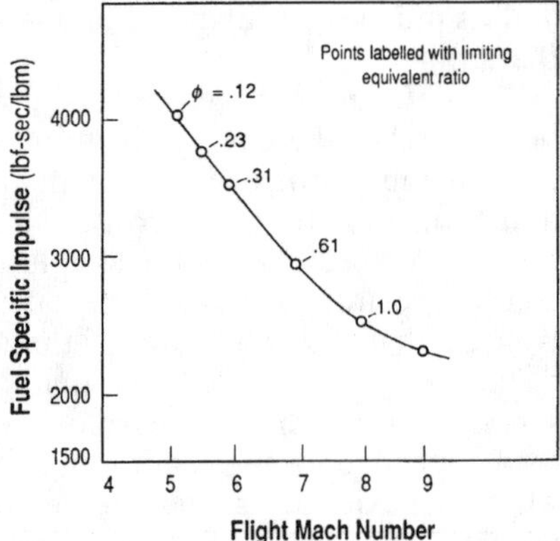

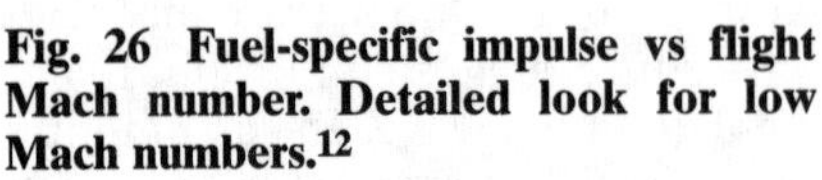
Fig. 26 Fuel-specific impulse vs flight Mach number. Detailed look for low Mach numbers.[12]

wave is limited. The thrust-per-unit inlet capture area in Fig. 27 shows a strong dependence on flight dynamic pressure. Note that the maximum amount of thrust is obtained near $M = 8$.

The performance of the detonation-wave ramjet is compared in Fig. 28 to the performance of a scramjet taken from Ref. 62. Data for the scramjet take into account viscous losses and are shown for three different types of nozzle-expansion calculations: frozen, finite-rate, and equilibrium chemistry. The detonation-wave ramjet data do not take into account viscous losses and correspond to a frozen nozzle-expansion process. Ostrander et al.[12] conclude that "with all the loss mechanisms taken into account, the scramjet will probably perform better at low Mach numbers, but at higher Mach numbers it seems that the oblique detonation wave engine might do better."

As mentioned earlier, each point on the performance curves represents a particular vehicle/engine configuration for a given flight Mach number.

Fig. 27 Net thrust-per-unit inlet area vs flight Mach number for two values of freestream dynamic pressure.[12]

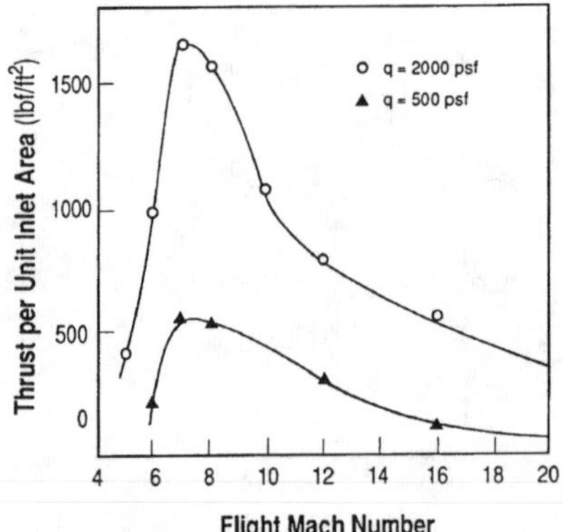

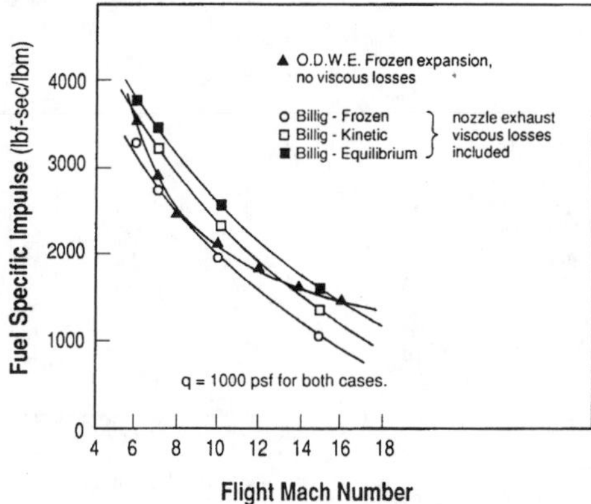

Fig. 28 Fuel-specific impulse vs flight Mach number for a scramjet[62] and an oblique detonation-wave engine (O.D.W.E.).[12]

It is of interest to determine the performance of a detonation-wave ramjet, designed for a specific flight Mach number, at flight Mach numbers other than those for which it was designed. The results of such a study performed in Ref. 12 are shown in Fig. 29 for $q = 1000$ psf for two configurations designed for $M = 10$ and 12. The dotted curve represents the performance at design conditions. They indicate that in the vicinity of the design point the detonation-wave ramjet performs better at off-design conditions than at design conditions. The corresponding curves for the thrust-per-unit inlet area are shown in Fig. 30. Note the change in slope of the curves in the vicinity of design flight Mach numbers. At lower flight Mach numbers the inlet flow spillage combined with the limitation on heat addition decreases the thrust significantly (see Ref. 12). Ostrander et al. draw the general conclusion that as a result of the limitation of heat addition at low Mach numbers (Fig. 25) the thrust potential of the detonation-wave ramjet ap-

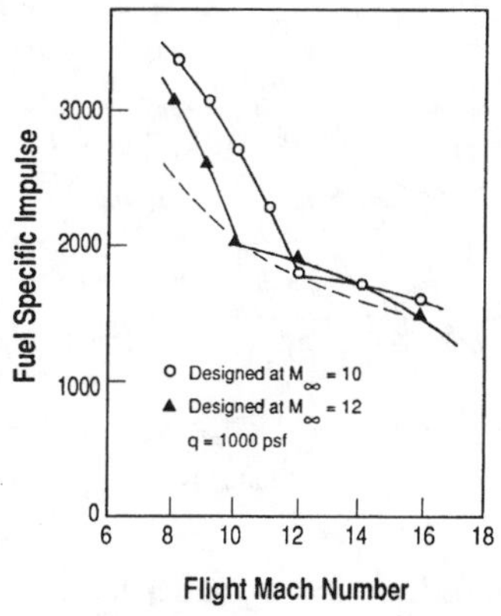

Fig. 29 Fuel-specific impulse vs flight Mach number; performance of two configurations at off-design conditions.[12]

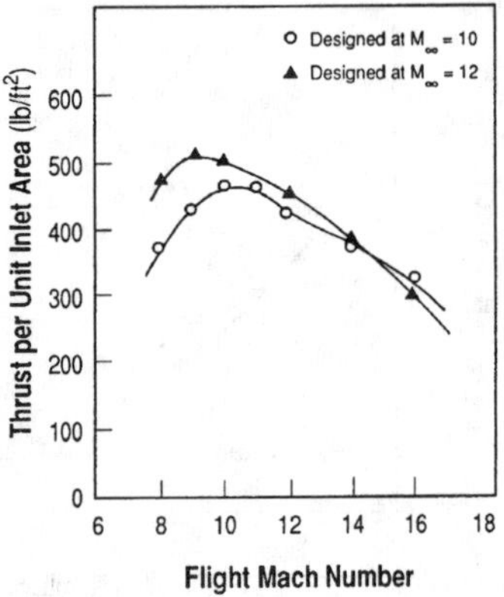

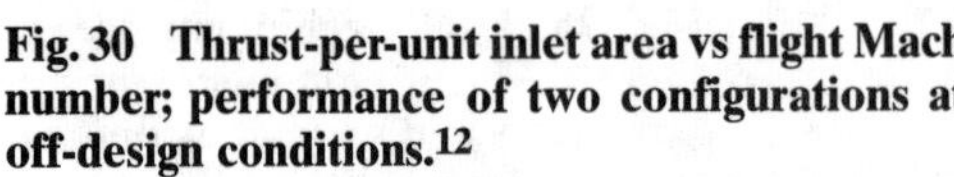

Fig. 30 Thrust-per-unit inlet area vs flight Mach number; performance of two configurations at off-design conditions.[12]

pears to be more limited than that of the scramjet in this lower end of flight Mach numbers. However, it seems that in the high flight Mach-number range the detonation wave is a more efficient supersonic combustion process.

Another approach to obtain detonation-wave ramjet performance estimates was taken in Refs. 14–19. Computational fluid dynamic methods were used to determine the entire inviscid hypersonic flowfield around a planar or axisymmetric shocked-combustion ramjet or shcramjet model. The shock (discontinuity) tracking capability of the computational scheme was used to accurately determine shock or detonation-wave and wall intersection points and construct a series of wall contours resulting in a body geometry for given freestream conditions. Pressures acting on these planar and axisymmetric bodies were calculated and used to determine various

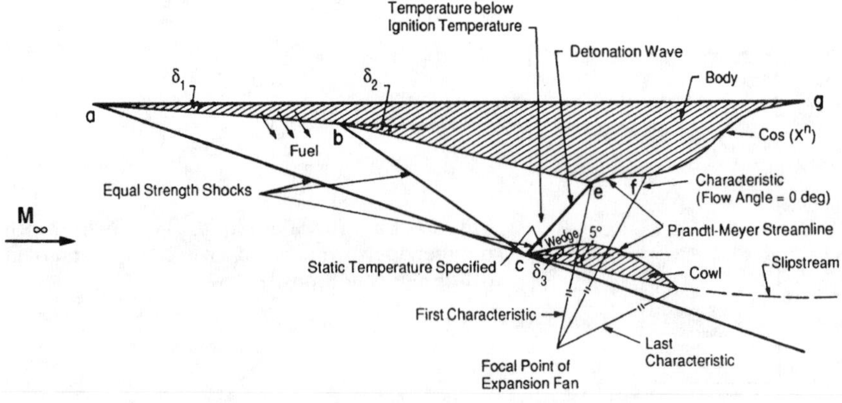

Fig. 31 External compression shcramjet model (planar or axisymmetric).

aerodynamic and propulsive characteristics over a range of flight Mach numbers.

The shcramjet model considered is shown in Fig. 31. It requires that the fuel injected somewhere in the inlet be mixed homogeneously with the airflow before reaching the region in the vicinity of the cowl lip. To circumvent this formidable task, it is assumed that the portion of the oncoming flow captured by the inlet is a homogeneous hydrogen/air mixture at a prescribed equivalence ratio. The inlet, which is a biconic or double wedge, compresses the oncoming mixture through two shocks *ac* and *bc* (Fig. 31), thus raising its pressure and temperature. The flow is then deflected back toward the centerbody by the upper surface of the cowl placed at the point of intersection *c* of shocks *ac* and *bc* (on-design conditions), which generates a third shock *ce*. The temperature T_c behind this third shock is prescribed, effectively controlling the amount of compression in the inlet. This prescribed temperature should not exceed a value for which the temperature of the fuel/air mixture before the third shock would be greater than its ignition temperature, yet should be sufficiently high to ignite the mixture after shock *ce*. The two angles of the biconic, or double wedge, δ_1 and δ_2 and the cowl deflection angle δ_3 are determined from the condition that these three shocks be of equal strength at their points of inception (*a*, *b*, and *c* in Fig. 31). The cowl has a flat lower surface and forms a 5-deg (arbitrarily chosen minimum value) leading-edge angle. The upper surface of the cowl consists of an initial straight portion followed by a contoured section that corresponds to a streamline in a centered Prandtl–Meyer expansion flow until it intersects the lower surface of the cowl. The transition point is determined by the condition that the head (first characteristic) of the Prandtl–Meyer expansion does not intercept the detonation wave *de* (typically, the head intercepts 0.01 m downstream of the detonation-wave interception point *e*, see Fig. 31). In the case of shock-induced combustion, this transition is behind the point where the temperature of the combustion products on the lower surface of the body reaches a maximum value. This ensures that the cowl upper surface does not weaken the detonation wave before it reaches the centerbody or quenches the combustion process. The cowl length is varied by controlling the position of the focal point along the head of the centered Prandtl–Meyer expansion.

The combustion products are expanded in a half-open nozzle, which consists of a surface corresponding to a streamline in a centered Prandtl–Meyer expansion to point *f*, where the slope is zero. From this point to exit point *g*, the nozzle wall is of the form $\cos(x^n)$. The cowl and nozzle designs are by no means optimum but have been used in the study as a convenient design alternative.

Combustion of the fuel/air mixture is either shock-induced or by a detonation wave. In Fig. 32a the third shock wave from the upper surface of the cowl (*ce*) raises the temperature of the hydrogen/air mixture. Combustion begins downstream of the shock but does not initially affect the oblique shock wave. As the combustion begins on the surface of the cowl, it gener-

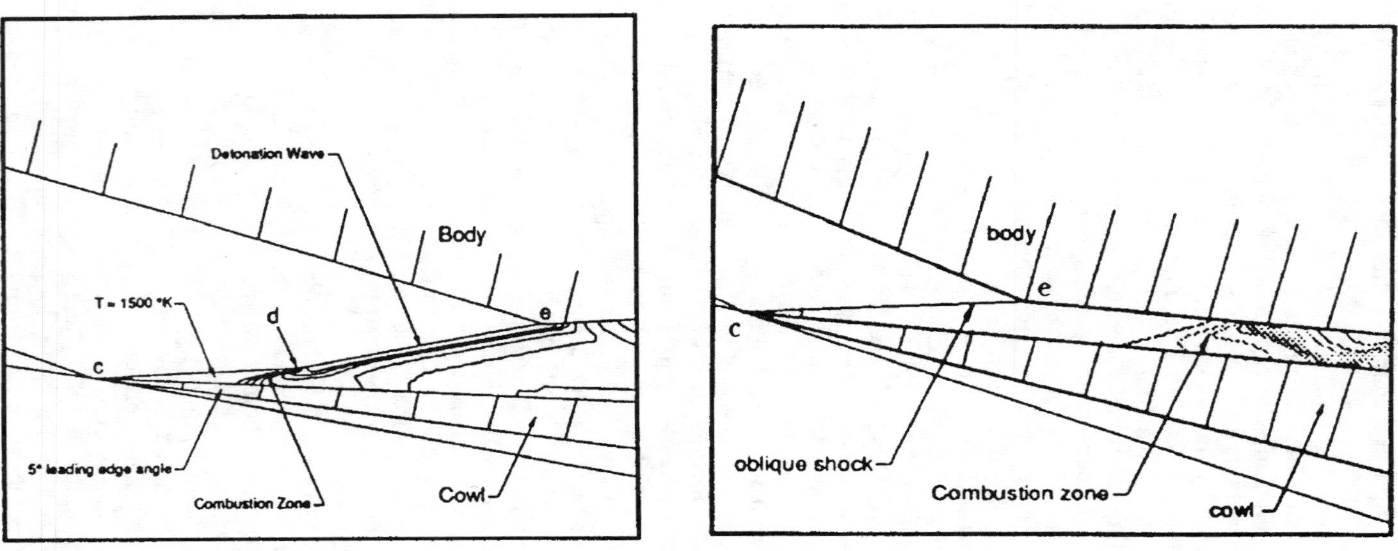

Fig. 32 Various shock combustion processes: a) detonative combustion and b) shock-induced combustion.

ates a series of compression waves that coalesce and impinge on the cowl leading-edge shock wave (*ce*) creating a kink at point *d*. From this point to point *e*, the wave, which began as an oblique shock wave, now behaves as a detonation wave. For other flight conditions the shock-induced combustion occurs far downstream of shock *ce* and does not affect the shock. Such a case is shown in Fig. 32b, where the shock-induced combustion process is sluggish and, therefore, takes a *sizable* distance to completion. In all cases, because of the discontinuity tracking capability of the computational method, the shock and combustion processes are well delineated.

The system of equations describing the steady axisymmetric flow of an inviscid, nonheat-conducting and reacting gas is used to generate the shcramjet geometry and determine its flowfield variables:

$$\frac{\partial \boldsymbol{a}}{\partial x} + \frac{\partial \boldsymbol{b}}{\partial r} = \boldsymbol{f} \tag{14}$$

where

$$\boldsymbol{a} = \begin{bmatrix} \rho u \\ P + \rho u^2 \\ \rho uv \\ \rho u H^o \\ \rho u a_i \end{bmatrix}, \quad \boldsymbol{b} = \begin{bmatrix} \rho v \\ \rho uv \\ P + \rho v^2 \\ \rho v H^o \\ \rho v a_i \end{bmatrix}, \quad \boldsymbol{f} = \nu \begin{bmatrix} \rho v/r \\ r uv/r \\ \rho v^2/r \\ \rho v H^o/r \\ \rho v a_i/r - \dot{w}_i \end{bmatrix} \tag{15}$$

with $\nu = 0$ for planar and 1 for axisymmetric flows. The variables u, v, P, r, H_o, a_i, and $\dot{w}_i$ are the velocity components in the x and r directions, pressure, density, total enthalpy, species mass fraction, and production, respectively. This system of equations is closed by the following equation of state:

$$P = \rho \mathcal{R} T / \mu(\boldsymbol{a}) \tag{16}$$

where $\mathcal{R}$ is the universal gas constant and $\mu(\boldsymbol{a})$ the molecular weight of the gas mixture. The finite-rate shocked-combustion process of the hydrogen/air mixture is described by 33 reactions between 13 species (H_2, O_2, H, O, OH, H_2O, HO_2, H_2O_2, N, NO, HNO, N_2, NO_2) taken from Ref. 48. The rate coefficients used for the forward reactions are those given in Ref. 48. The rate coefficients for the reverse reactions were calculated from the forward-rate coefficients and the appropriate equilibrium constants. All thermochemical data for the hydrogen, oxygen, and nitrogen species were taken from the JANNAF (1971) tables.

A first-order Godunov computational scheme[63] employing the preceding combustion model and capable of discontinuity tracking was used to solve the chemical kinetics and gasdynamic equations in a coupled manner. Details of the numerical solution procedure are given in Ref. 17. This method was chosen because it allows, if desired, the tracking of discontinuities (i.e., shock waves, slipstreams, etc.) in the flow with resulting sharp, discontinuous profiles in the flow variables across them. Thus, in the shock-induced combustion process, the shock and post-shock combustion processes are well delineated. This capability is essential for the design of the shcramjet, as it allows the exact determination of the point of intersection of shock and detonation waves with walls or other shock waves, as well as the position of the slipstream separating the nozzle from the outer flow, and any shocks created or destroyed in the flowfield.

Estimated performance parameters of shcramjets were determined for accelerator vehicles on constant flight dynamic pressures P_{dyn}, trajectories for two values of $P_{dyn} =$ 1000 and 2000 psf, and for two inlet compression ratios characterized by T_c = 1300 and 1700 K, where T_c is the temperature just behind the third shock, generated by the cowl (Fig. 31). The thrust coefficient C_T and propulsive efficiency η_p are defined as

$$C_T = \frac{Th}{\dot{m}_o V_o} \tag{17}$$

$$\eta_p = \frac{Th \times V_o}{Q} \tag{18}$$

where Th is the net generated thrust, V_o the freestream velocity, $\dot{m}_o$ the captured air mass flow rate, and Q the total heat added to the flow.

For a given flight Mach number the net thrust produced by shcramjets subject to the just-described design methodology is controlled by varying the cowl length. As this length increases, the net thrust increases to a maximum value and then decreases. All geometric dimensions are referred to the distance of the cowl lip to the axis of symmetry (i.e., the cowl radius is fixed at unity). Each point on the performance curves represents a specific on-design maximum thrust configuration for given flight conditions. Results presented are for a stoichiometric air/fuel ratio.

For the cowl and nozzle design methodologies adopted, the variation of performance parameters depended on the type of combustion processes induced by the cowl shock. In the planar case, for P_{dyn} = 1000 psf and T_c = 1300 K, the combustion process gradually changes from detonative to shock-induced. Flowfield computations show that this transition occurs at low T_c values. For the planar case this occurs in the flight Mach number range $14 \leq M \leq 17$ (Fig. 33) and for the axisymmetric case in the $12 \leq M \leq 14$ and $16 \leq M \leq 18$ ranges (see Figs. 34–36). This transition results in an increase of performance parameters. In addition, the effect of increasing the inlet compression does not seem to be significant in the high flight Mach-number range considered. All performance parameters are based on the net thrust, i.e., the net resulting force in the flight direction.

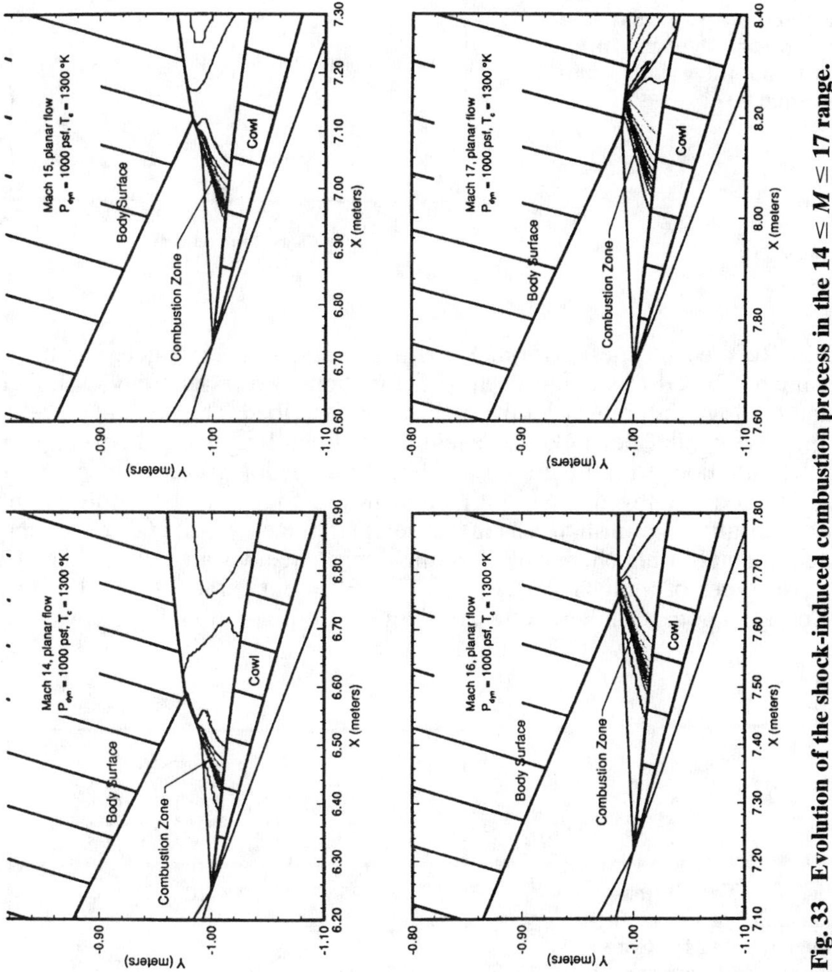

Fig. 33 Evolution of the shock-induced combustion process in the $14 \leq M \leq 17$ range.

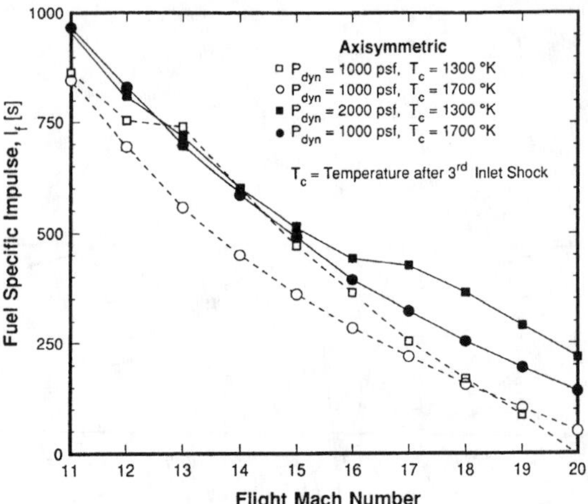

Fig. 34 Net fuel-specific impulse vs flight Mach number for two values of freestream dynamic pressure and three inlet compression levels.

In Refs. 64–66 the performance of detonation-wave ramjets has been evaluated based on an approach to detonation-wave ramjet model design methodology different from the model just described. The inviscid detonation-wave ramjet flowfield was again determined by solving the foregoing Euler equations with the finite-rate H_2/air combustion process. The numerical method employed was a fully implicit and coupled, Newton-iteration, total variation diminishing scheme, developed in Refs. 67 and 68, for solving the nonequilibrium chemically reacting flow at steady state.

Two types of multishock, external (Fig. 37, Configuration I) and mixed (Configuration II) inlets, were considered. Configuration I employs two

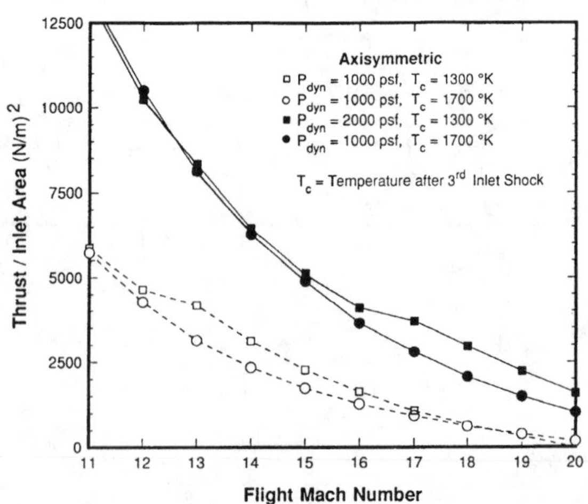

Fig. 35 Net thrust-per-unit inlet area vs flight Mach number for two values of freestream dynamic pressure and three inlet compression levels.

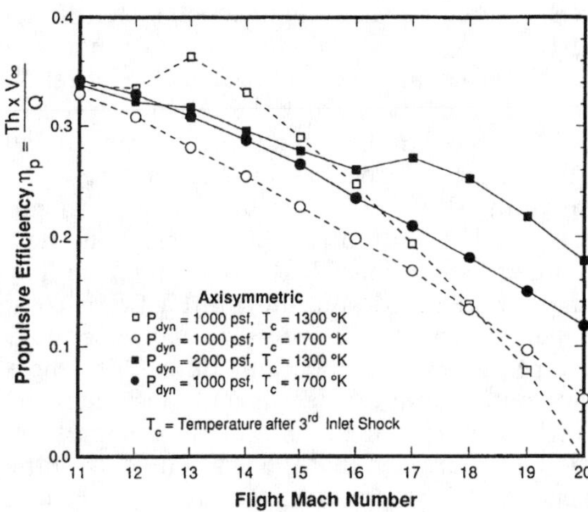

Fig. 36 Propulsive efficiency vs flight Mach number for two values of freestream dynamic pressure and three inlet compression levels.

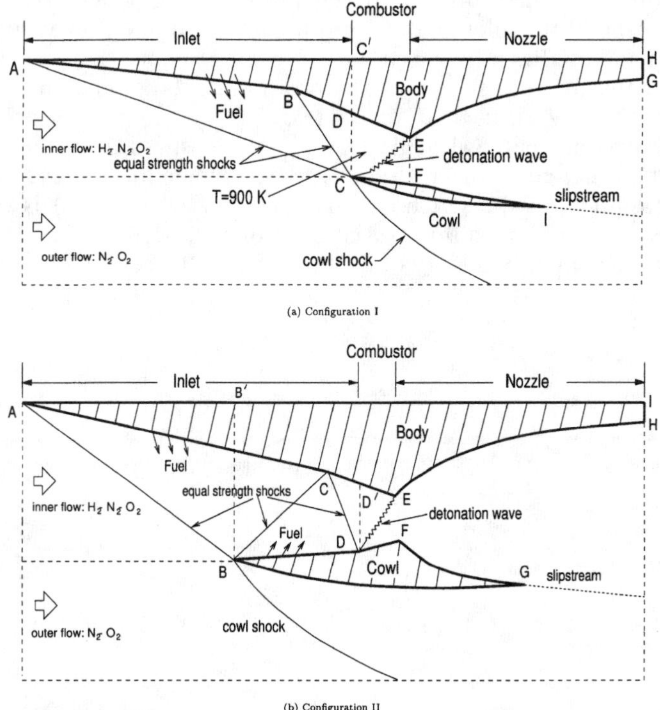

(a) Configuration I

(b) Configuration II

Fig. 37 Shcramjet model (axisymmetric or planar).

equal-strength shocks, whereas a three equal-strength shock system is assumed for Configuration II. At the design flight condition the two shocks in Configuration I are assumed to intersect at the cowl tip *C*; in Configuration II the cowl tip *B* is always situated on the bow shock *AB*. As in the preceding, it is also assumed that fuel is injected in the forebody/inlet flow parallel to the oncoming airflow and that a homogeneous fuel/air mixture at an equivalence ratio of one at the combustor-entrance section (*DC* in Configuration I and *DD′* in Configuration II). This portion of the oncoming flow above the cowl lip is labeled inner-flow in Fig. 37. The portion of the flow below the leading edge of the cowl, labeled outer-flow in Fig. 37, is comprised of air only. We assume that the temperature, pressure, and velocity are identical for the inner and outer flows. However, because of the differences in composition of the inner and outer flows, the density and Mach numbers differ. The temperature of the inner flow at the end of the compression process must be sufficiently low to guarantee that the ignition delay of the combustible mixture is long enough to prevent burning in the inlet. Hence, the maximum compression temperature at the combustor inlet (*DC* or *DD′*, Fig. 37) is set, rather arbitrarily, at 900 K, ensuring that no premature ignition occurs. The inlet length (axial distances *AC′* or *AB′*, Fig. 37) is fixed at 15 m. These requirements uniquely determine the planar or axisymmetric inlet surface geometry. The inlet flowfield is solved iteratively using a non-reacting version of the Euler code. The dotted lines *CD* or *DD′* (Fig. 37) represent the exit plane of the inlet and the grid points where the flow variables are extracted and used as inflow to the combustion system.

The combustor consists essentially of an oblique detonation wave generated by the cowl surface *CF* or *DF* (Fig. 37). The combustor design proceeds by varying the cowl angle δ (Fig. 38) measured from the axis and solving the Euler equations with the LUSGS scheme for the resulting flowfield. The inflow at the left boundary *CD* or *DD′* (Fig. 37) is obtained from the distribution of flow variables along these lines from the inlet flowfield. The upper boundary of the combustor (the central body surface),

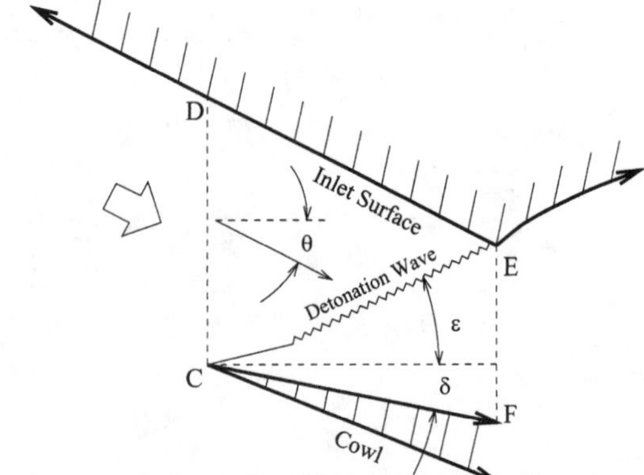

Fig. 38 Combustor design.

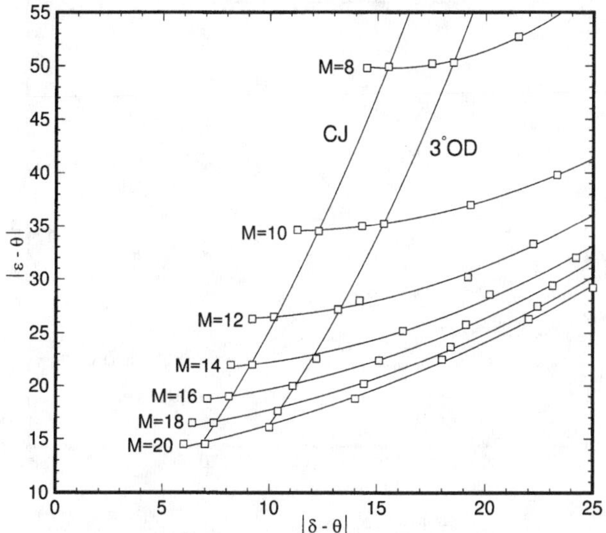

Fig. 39 Detonation-wave angle vs deflection angle for planar shcramjets, $8 \leq M_{\text{inner}} \leq 20$ and locus of Chapman–Jouguet and 3-deg overdriven angles.

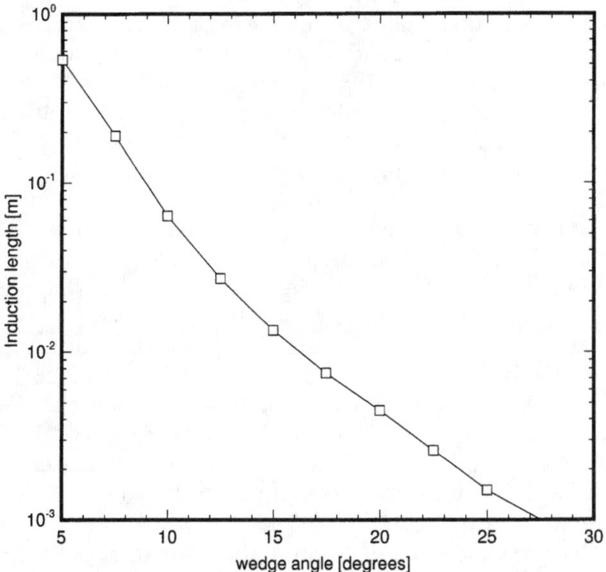

Fig. 40 Ignition-delay distance along wedge surface as a function of wedge angle (δ_w) for shock-induced combustion wedge for $\varphi = 1$, $P_\infty = 23000$ Pa, $T_\infty = 900$ K.

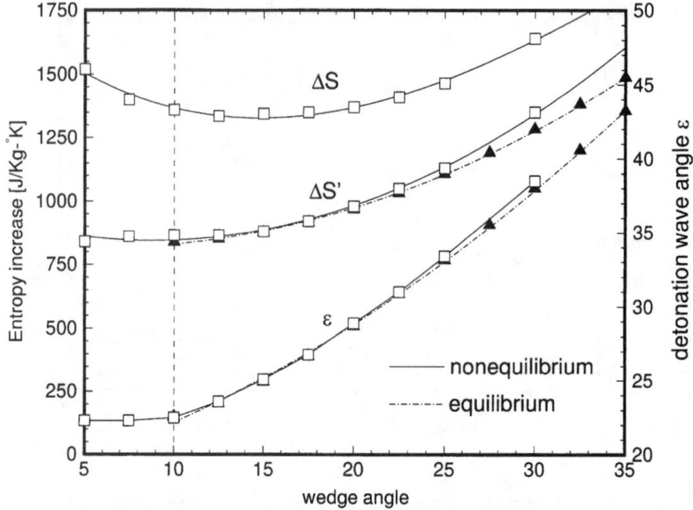

Fig. 41 Total entropy (ΔS), entropy when chemical reactions are in equilibrium ($\Delta S'$), and detonation wave angle (ε) as a function of wedge angle (δ_w) for $\varphi = 1$, $P_\infty = 23000$ Pa, $T_\infty = 900$ K, and $M_\infty = 7$. Comparison of results with equilibrium chemically reacting flows.[36]

lines DE or $D'E$ (Figs. 37 and 38), are determined by tracking the streamlines, starting at the coordinates of points D or D' (inlet exit point on the body surface), through the combustor section until the detonation wave is intercepted. To determine the Chapman–Jouguet and, therefore, the minimum entropy angle, the wedge angle is decreased until the detonation-wave angle ceases to decrease (see Fig. 15). The net deflection and detonation-wave angles are plotted in Fig. 39 for the planar flow case for the range of inner flow Mach numbers of interest. Generally, the weaker the detonation wave, the longer the ignition delay (see Fig. 40); hence, at the Chapman–Jouguet point the ignition delay becomes exceedingly long. Noting that the entropy curve is rather shallow near the Chapman–Jouguet point (see Fig. 41), the 3-deg overdriven case was chosen for the combustor cowl angle, i.e., $\delta_{\text{combustor}} = \delta_{\text{CJ}} - 3°$, for all flight Mach numbers (Fig. 38), thereby considerably shortening the ignition-delay distance and accordingly the combustor length with no penalty in entropy rise. We can also see why the inlet length was chosen to be 15 m. For this inlet length and geometrically scaled inlet area at the combustor, a shorter inlet length would lead to a shock-induced combustion process and not to a detonation wave. The main objective of this detonation wave ramjet model was to have a combustor consisting of a detonation wave working at very near Chapman–Jouguet condition for all flight Mach numbers considered. The second objective was to design a more realistic nozzle than in the previous model (Fig. 31). For purposes of determining the nozzle surfaces (walls), the flow in the nozzle is assumed to be chemically frozen and the frozen method of characteristics used to build the central body and cowl inner surfaces (nozzle walls; *EG* and *FI*, Fig. 37) by specifying a uniform flow in the axial (oncoming flow) direction at the exit plane of the nozzle. The nozzle-exit pressure is iteratively determined to expand the flow to the axis of the vehicle. The initial values of the flow variables are extracted from the combustor-exit section (line *EF*, Fig. 38) and averaged. Furthermore, average specific heats, total pressure, and temperature are calculated along this initial line distribution, which is kept constant throughout the method of characteristics calculations. The false-wall technique is used for Configuration I, whereas a dual-wall technique was found to be more appropriate for Configuration II (see Ref. 69) (The false-wall technique would produce negative thrust for this configuration.) Because both the central body and cowl inner nozzle walls obtained were exceedingly long, they were cut off to provide 96% of the overall thrust to reduce frictional drag. The outer surface of the cowl (*CI*, Fig. 37) was designed by prescribing a polynomial, matching the locations of the cowl lip *C*, and the trailing edge *I*, of the inner surface. At the leading edge the cowl angle was prescribed to be 5 deg and at the trailing edge is set to be zero (parallel to the local flow). Thus, the integrated detonation-wave ramjet model was obtained by assembling all surfaces generated for each component and the entire ramjet flowfield, from tip to tail, determined by using the LUSGS technique for the Euler equations describing the nonequilibrium reacting flows of H_2/air combustible mixtures. At the trailing edge *I* of the cowl, a contact surface is formed between the inner (propulsive stream tube) and outer (freestream deflected by the cowl shock) flows. It is important to approximately match the pressure across this interface as the creation of a shock into the propulsive stream tube of the nozzle (as would occur if the pressure is less in the nozzle than in the outer flow) would adversely affect the detonation-wave performance. Figure 42 shows the pressure distributions at lateral sections 3 and 4, i.e., at the cowl trailing-edge plane and the nozzle-exit plane. A sharp drop in pressure can be observed at $y \approx 3.3$ m because of the resultant shock formed between the inner and outer flows. One can see that the pressure mismatch is not severe; in any event, no shock enters the propulsive stream tube from the cowl trailing edge (see also Fig. 44). The nonuniformity of the pressure distribution at the exit plane 4 is not significant and is greater than the freestream pressure. Other flowfield features of interest are the distributions of H_2O mass fraction and (nonequilibrium) entropy in the combustor outflow plane 2 (Fig. 44). Clearly there is surplus production of H_2O, and the entropy rise is less near the cowl surface, corresponding to the shock-induced combustion region. Therefore, the shock-induced combustion is apparently more efficient than heat addition by a detonation wave. This is also confirmed by the performance characteristics obtained from the previous ramjet model considered (see Figs. 33–36). Hence, a scramjet model employing solely shock-induced combustion may deserve further consideration. Of particular interest also is the sizable temperature in-

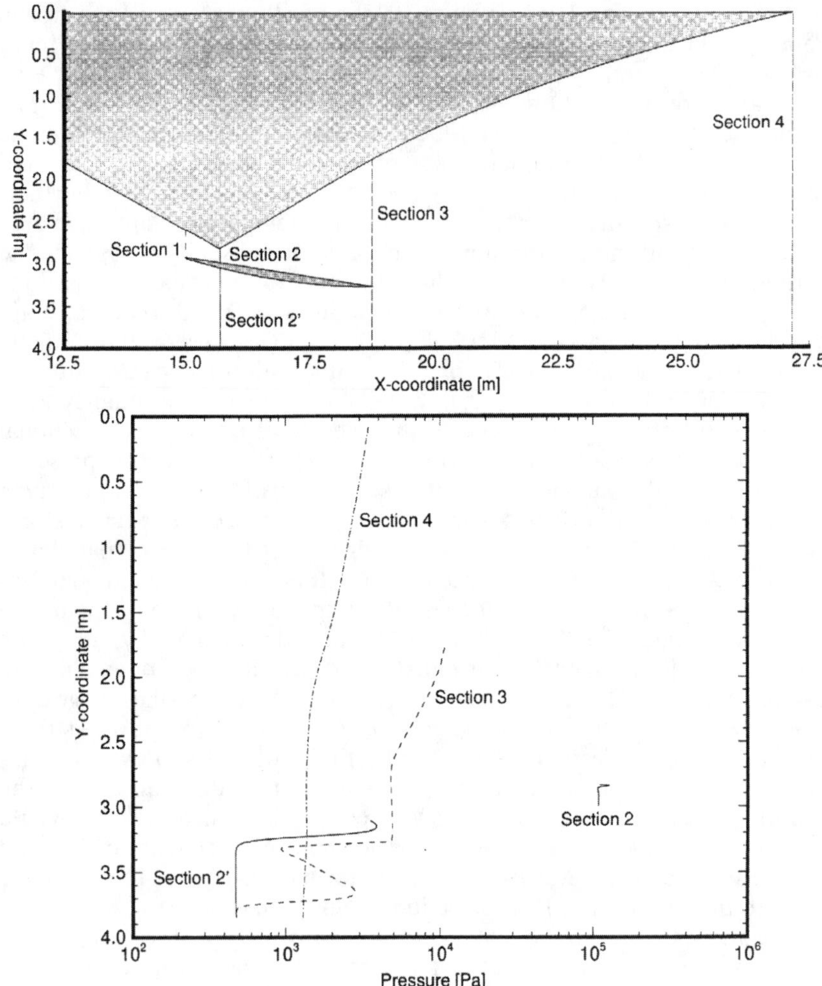

Fig. 42 Pressure along three vertical extraction lines in a Mach 14; planar shcramjet at a flight dynamic pressure of 1400 psf.

crease along the body nozzle surface because of H_2O recombination in the forward section of the nozzle depicted in Fig. 44. A detailed analysis of planar detonation-wave ramjet flowfields for both Configurations I and II for Mach 14, equivalence ratio $\varphi = 1$, and for a flight trajectory at constant dynamic pressure $q=1400$ psf is given in Ref. 66.

Several pertinent performance parameters of interest are shown in Figs. 45–49 for both Configurations I and II. Figure 45 contains the net fuel specific impulse I_{sp} for planar flow configurations. The magnitude of I_{sp} is greater for Configuration II by an average of 22–23% over the entire flight

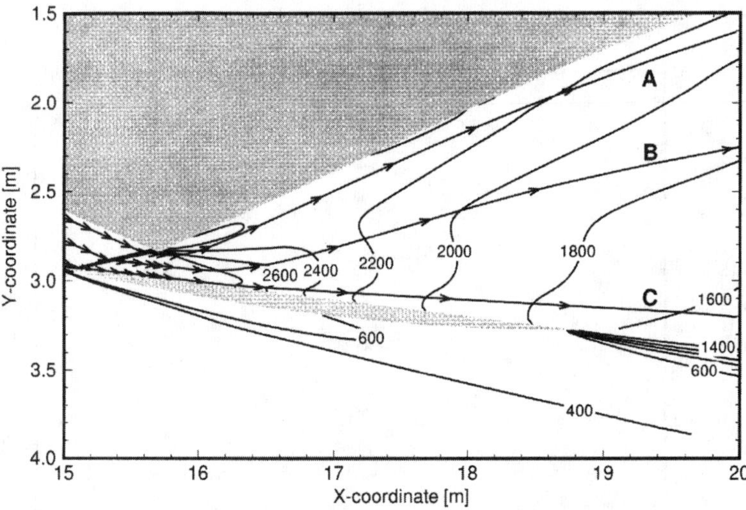

Fig. 43 Temperature contours at the start of the nozzle 58for Mach 14, planar configuration shcramjet. Flight dynamic pressure $q = 1400$ psf.

Mach-number range. A major consequence of Fig. 45 is that the detonation-wave ramjet I_{sp} is comparable to that of the rocket up to flight Mach numbers 20 ~ 22. These first-approach results show that the detonation-wave ramjet can be a viable means of hypersonic propulsion. However, much additional work is needed to definitely substantiate this claim. The nozzle-exit area to inlet-capture area ratio for an axisymmetric propulsive stream tube is shown in Fig. 46. This ratio is larger for Configuration II and over the entire flight Mach-number range varied by ~ 33% for Configuration I and by ~ 25% for Configuration II. Note that for both cases this ratio is close to unity, especially at high Mach numbers. The overall efficiency of the planar detonation-wave ramjet, which is the product of thermal and propulsive efficiencies, is given in Fig. 47. Because the propulsive efficiency is very close to unity, especially at very high flight Mach numbers, the depicted trend of the overall efficiency is virtually identical to the thermal overall efficiency. Clearly, the mixed compression ramjet exhibits a higher overall efficiency over the entire flight Mach-number range. Of interest is the percentage contribution to the overall thrust of each ramjet component shown in Fig. 48. A trend is observed for external compression ramjets whereby the thrust contribution by the cowl component (cowl combustor and nozzle surfaces and the outer cowl surface) diminishes as the flight Mach number increases. At $M = 24$ the cowl contribution is seen to be minimal. This trend corresponds to the fact that the 3-deg overdriven Chapman–Jouguet condition reduces the cowl combustor angle to be parallel to the freestream at high flight Mach numbers. For Configuration II the same condition produces negative thrust (drag) by the combustor wall of the

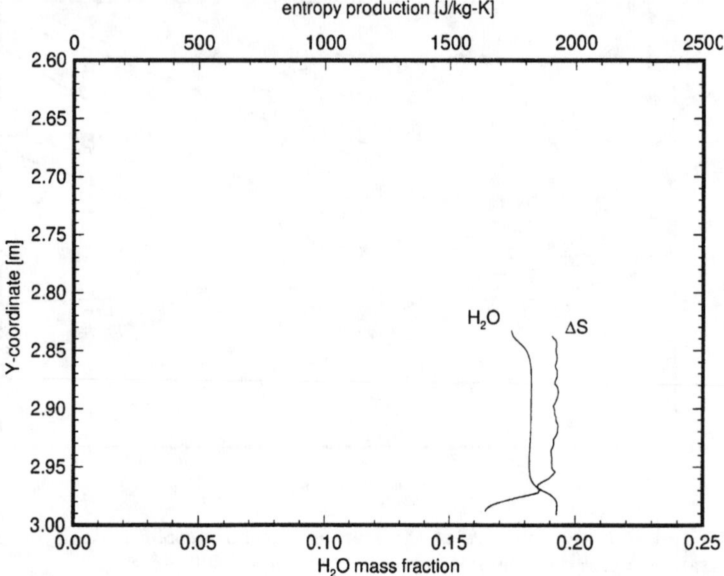

(a) Entropy and H_2O mass fraction at combustor outflow, section 2

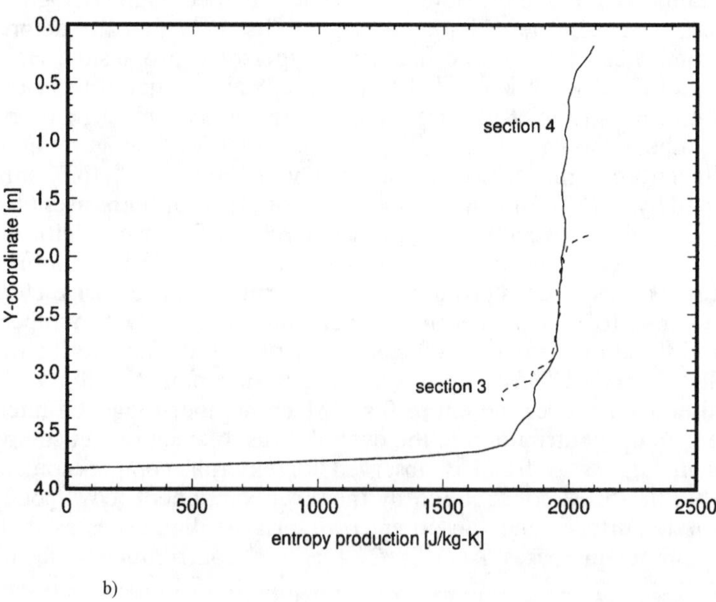

b)

Fig. 44 Entropy and H_2O mass fraction along vertical extraction lines in a Mach 14; planar external compression shcramjet at a flight dynamic pressure of 1400 psf.

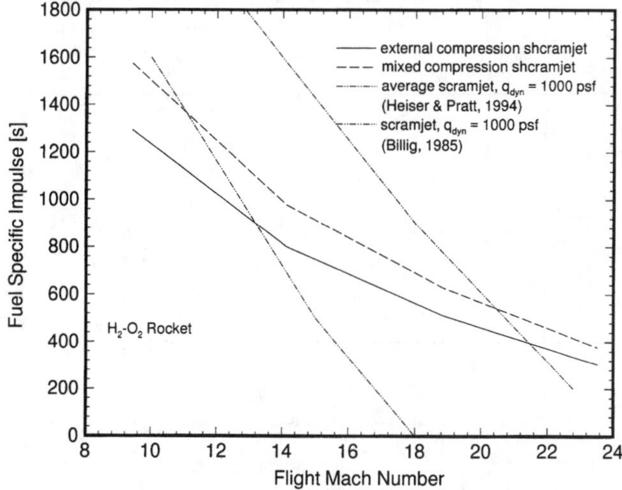

Fig. 45 Mixed vs external compression planar shcramjet fuel specific impulse comparison, at a flight dynamic pressure of 1400 psf; equivalence ratio $\theta = 1$

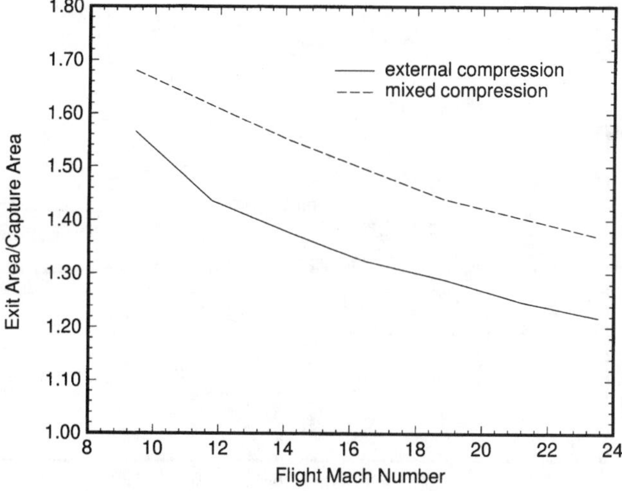

Fig. 46 Stream-tube exit area to inlet-capture area ratio for axisymmetric configuration shcramjets (external and mixed-compression) at a flight dynamic pressure of 1400 psf.

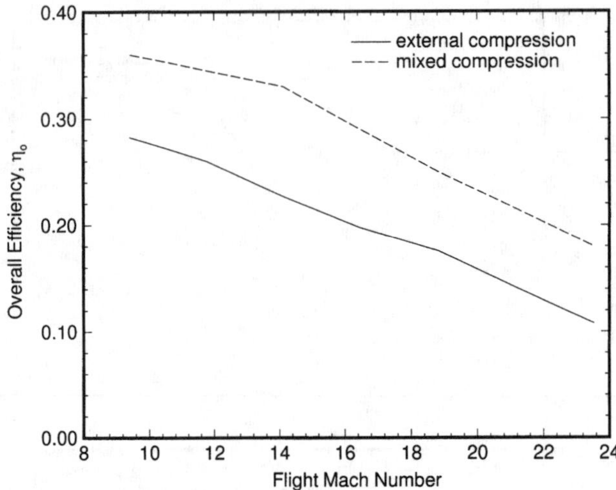

Fig. 47 Overall efficiency η_o for planar configuration shcramjets (external and mixed-compression) at a flight dynamic pressure of 1400 psf.

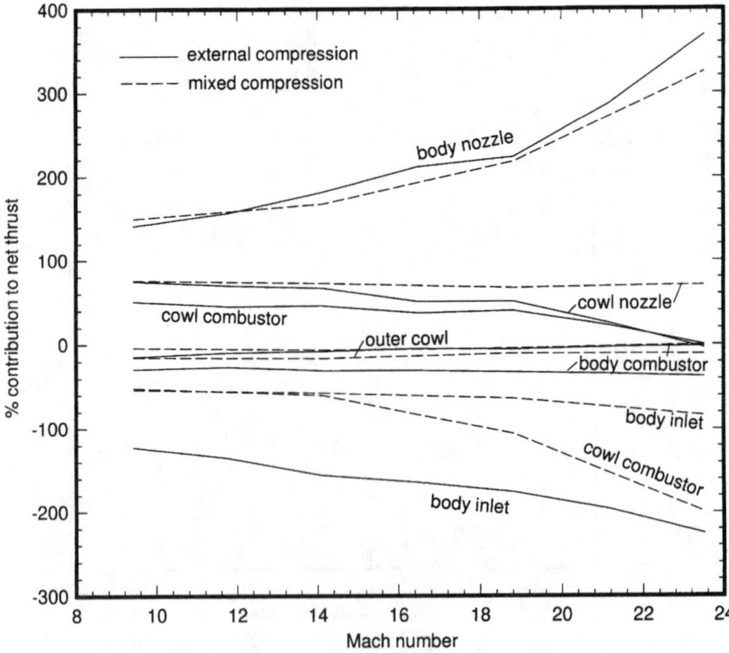

Fig. 48 Component contribution to overall thrust for planar configuration shcramjets (external and mixed-compression) at a flight dynamic pressure of 1400 psf.

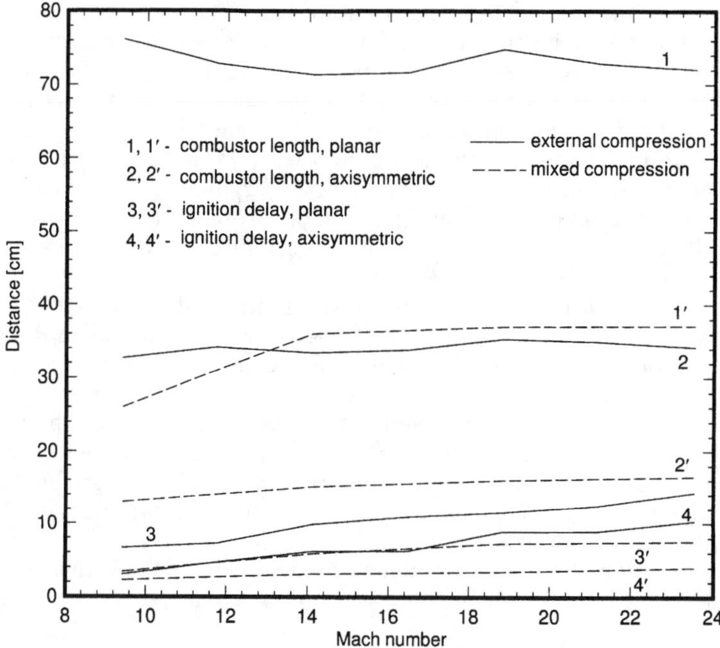

Fig. 49 Ignition-delay distance in the combustor for planar and axisymmetric configuration shcramjets (external and mixed-compression) at a flight dynamic pressure of 1400 psf.

cowl. As expected, the inlet and combustor drag along the centerbody is lower for the mixed compression configuration. The body nozzle contribution to overall thrust increases as the flight Mach number increases for both configurations.

Of particular interest for detonation-wave ramjets is the ignition-delay distance along the cowl surface in the combustor section and the total length of combustor (*CF* of *DF*, Figs. 37 and 38) over the considered flight Mach-number range. The combustor length is governed by a number of factors, including the specification of the cowl angle and the 900 K temperature limit at the end of the compression process, which in combination dictate the ignition-delay distance, the detonation-wave angle, and hence the relative geometry of the cowl and body surfaces in the combustor. The combustor length is then obtained from the intersection of the detonation wave and the centerbody surface. From Fig. 49 it is evident that in order to satisfy the condition that the detonation wave operate at 3-deg overdriven Chapman–Jouguet condition, the ignition-delay distance increases substantially as the flight Mach number increases. Remarkably, for both planar and axisymmetric ramjets, the combustor length remains relatively unchanged at very high Mach numbers. This property might be of interest because an accelerating, variable geometry, detonation-wave ramjet would not have to alter the combustor length, but only the cowl angle. The combustor length is lower for the axisymmetric configuration by virtue of the fact that the combustor flow area is narrower than for planar detonation-wave ramjets. One can also see that ignition delay distances and combustor lengths are shorter for mixed compression detonation-wave ramjets. Results also show that planar detonation-wave ramjets exhibit marginally better performance characteristics than the axisymmetric ramjets throughout the entire range of flight Mach numbers. More details on performance characteristics are provided in Ref. 66.

Two studies concerning off-design and nonoptimal flow conditions for the just-described detonation-wave ramjet models were performed in Ref. 70 to determine deviations from on-design performance parameters already presented. Specifically, the impact of nonuniform fuel/air mixing, arising from practical difficulties of fuel injection into a hypersonic flow, on the propulsive characteristics of the detonation-wave ramjet was considered, and an investigation of detonation-wave ramjets operating at off-design flight Mach numbers, i.e., at flight Mach numbers different from those used for on-design operation, has been carried out.

Consideration of nonuniform fuel/air mixing is simulated with a Gaussian distribution of the equivalence ratio φ as inflow conditions (Fig. 50), i.e., this

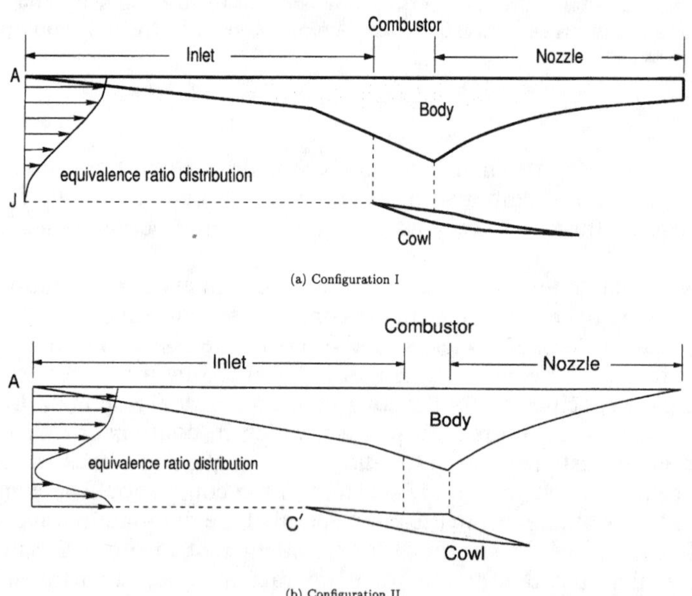

Fig. 50 Nonuniform equivalence ratio distribution: a) tangential fuel injection along the inlet wall, Configuration I and b) tangential fuel injection along the inlet and cowl upper walls, Configuration II.

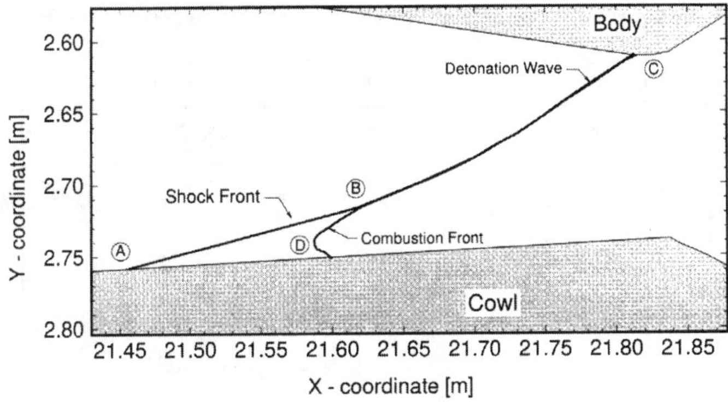

(a) Planar mixed compression shcramjet at $M_\infty = 9$, $q_{\rm dyn} = 1400$ psf

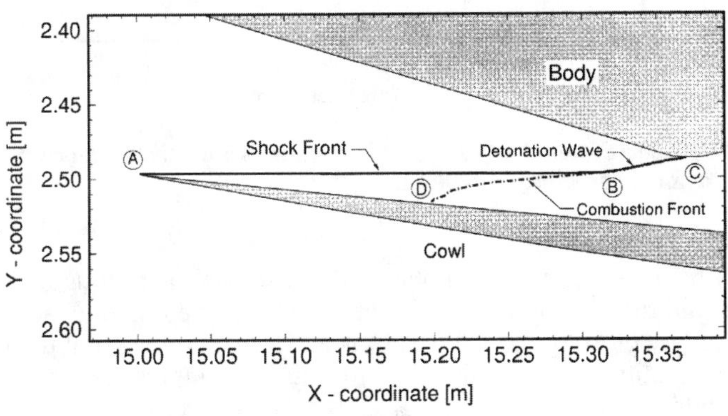

(b) Axisymmetic external compression shcramjet at $M_\infty = 14$, $q_{\rm dyn} = 1400$ psf

Fig. 51 Shock-induced combustion detonation-wave configurations generated by nonuniform fuel/air distributions of Fig. 50.

distribution, which reflects the fuel/air concentration field in the vicinity of the wall injector exit, is assumed to persist in the flow up to the combustor entry. A tangential fuel injection is assumed to take place along the inlet wall for the external compression detonation-wave ramjet (Fig. 50a), whereas for the mixed compression case the fuel is assumed to be injected tangentially along the centerbody inlet wall and the cowl upper surface (Fig. 50b). For both cases the overall equivalence ratio is unity, i.e., the amount of fuel injected is the same as in the homogeneously mixed case. The equivalence ratio distribution ranges from $\varphi \approx 2.4$ at the inlet surface to $\varphi \approx 0.02$ at the furthest point from the body for Configuration I and $0.02 \leq \varphi \leq 3.6$ for both upper and lower distributions for Configuration II. Figure 51 depicts extensive regions of distinct combustion processes both shock-induced and deto-

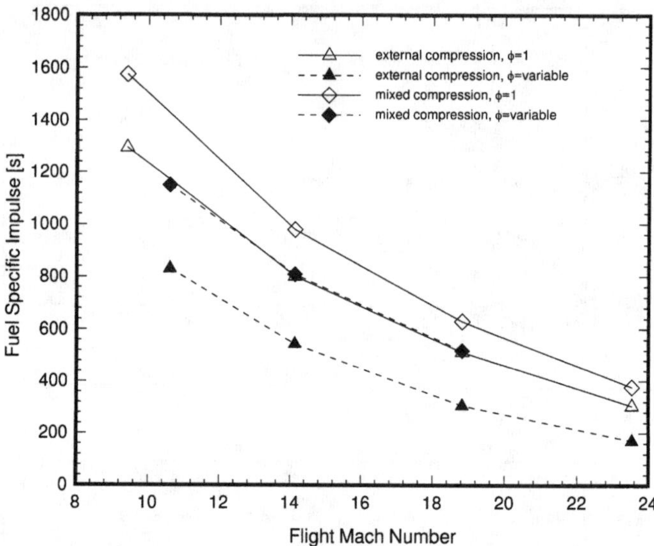

Fig. 52 Fuel-specific impulse vs Mach number for various shcramjet configurations at a flight dynamic pressure of 1400 psf.

native with curved fronts. In this case a distinct Chapman–Jouguet wave did not exist, and the maximum overall net thrust for given flight conditions was determined by plotting the net thrust as a function of the combustor-wedge angle and locating its peak value. The fuel specific impulse determined from such considerations is shown in Fig. 52 as a function of flight Mach number. It characterizes the range of detonation-wave ramjet performance deterioration because of fuel/air mixing degradation from the ideal homogeneous to highly incomplete.

One of the most important issues related to the feasibility of the detonation-wave ramjet concept is the off-design performance of the engine. The design of a detonation-wave engine is inextricably linked to the flight conditions under which the engine is to operate to the extent that, ideally, the engine geometry varies with varying flight conditions. The engine will perform differently in off-design regimes, i.e., under flight conditions for which it was not designed, and an assessment of performance deviation from on-design performance must be made. To date, only a few studies of off-design detonation-wave ramjet operation have been performed.[6,12,15] Although each study has made useful contributions, each has its own limitations and deficiencies. The analysis in Ref. 6 was based on a simple ideal gas model with constant specific heats and completely disregarded the hydrogen-air chemistry of the shock-induced combustion. Ostrander et al.[12] also employed an ideal gas model but modeled the hydrogen combustion with empirical formulas. The applicability of the modeling approach they used and the validity of their results is questionable considering that they obtained better performance for the engine during off-design operation than during on-design operation. Also, their design procedure did not necessarily produce an engine geometry that minimized losses or maximized performance. A two-dimensional numerical simulation in the investigation of off-design detonation-wave ramjet performance was employed in Ref. 15. However, the detonation wave was modeled by simple heat addition, meaning that again the hydrogen-air reaction was not taken into consideration. In fact, the amount of heat addition was varied independently of the flight Mach number. Because in reality the amount of heat released depends on the strength of the detonation wave and hence on flight Mach number, this aspect of the study was clearly not realistic. Finally, the engine geometry was not designed to provide Chapman–Jouguet or near-Chapman–Jouguet detonation.

A physically more realistic model of detonation-wave ramjets that addresses the deficiencies of previous studies was considered in Ref. 70 and a detailed analysis was performed of the off-design propulsive characteristics of mixed and external compression oblique detonation-wave ramjets (Configurations I and II) designed according to the methodology adopted in Ref. 66 for operation at flight Mach numbers $12 < M_\infty < 20$. The detonation-wave ramjet flowfields were simulated by means of a fully implicit, fully coupled numerical method that solves the Euler equations of motion for a compressible reactive mixture of hydrogen and air in chemical nonequilibrium. As in Ref. 66 the combustion of the hydrogen in air was modeled by means of a 13-species, 33-reaction model.[48] Figure 53 depicts the fuel specific impulse as a function of the flight Mach number for planar mixed and external compression detonation-wave ramjets. The solid curve represents the locus of points corresponding to on-design operation, whereas data points not lying on the curve (square symbols) are the performance values for off-design operation. Performance values for a single engine geometry operating at lower-than-design flight Mach number, at flight Mach number, and higher-than-design Mach number are joined by dotted lines. One can see that fuel specific impulse deteriorates in the off-design operating regime and more so for mixed compression detonation-wave ramjets than for external compression ramjets. In general, this deterioration is more acute at lower Mach numbers but moderate in all other cases. In a similar way, the net specific thrust at off-design conditions is plotted in Fig. 54. The general trend for the net specific thrust during off-design operation is similar to that of fuel specific impulse in that performance is degraded with respect to on-design operation and more so for Configuration I than for Configuration II. Note the significant drop in performance of the Mach $\sim$ 14 on-design external compression ramjet operating at Mach $\sim$ 12. Figure 55 shows that whereas for some off-design cases the thermal efficiency deteriorates, in some cases it even exceeds that for on-design operation. In general, Configuration II produces more net thrust than Configuration I when operating at off-design flight conditions. For both configurations the relative loss of net thrust at off-design conditions decreases as the flight Mach number increases, but Con-

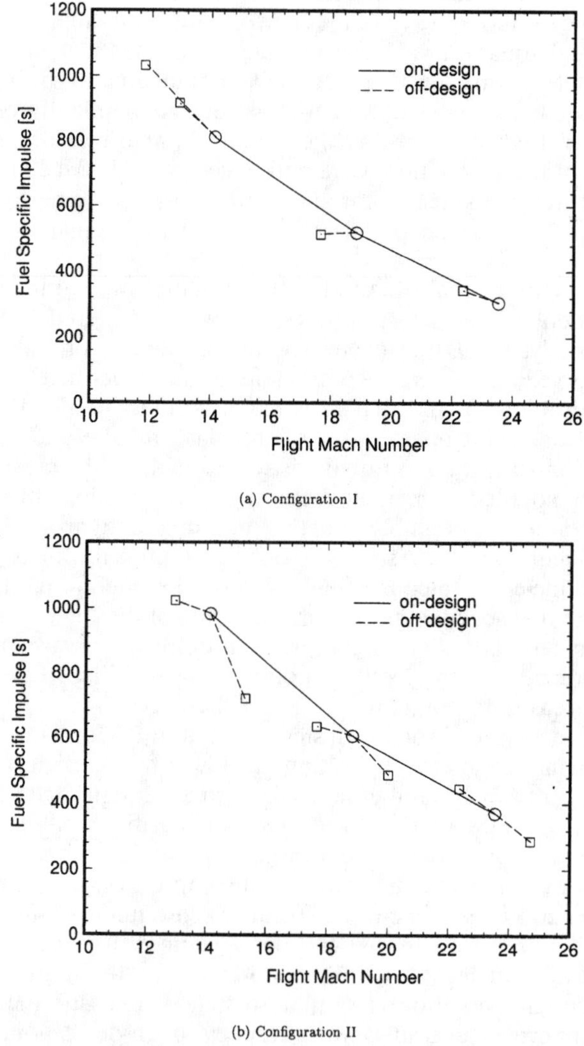

(a) Configuration I

(b) Configuration II

Fig. 53 Net fuel-specific impulse for an on- and off-design planar detonation-wave ramjet at q_{dyn} = 1400 psf.

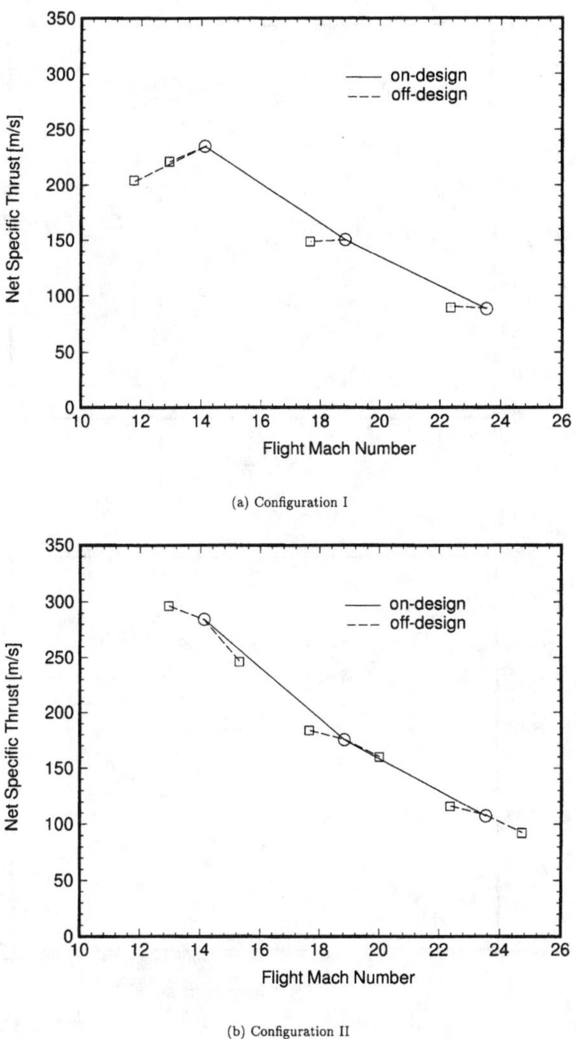

(a) Configuration I

(b) Configuration II

Fig. 54 Net specific thrust for an on- and off-design planar detonation-wave ramjet at q_{dyn} = 1400 psf.

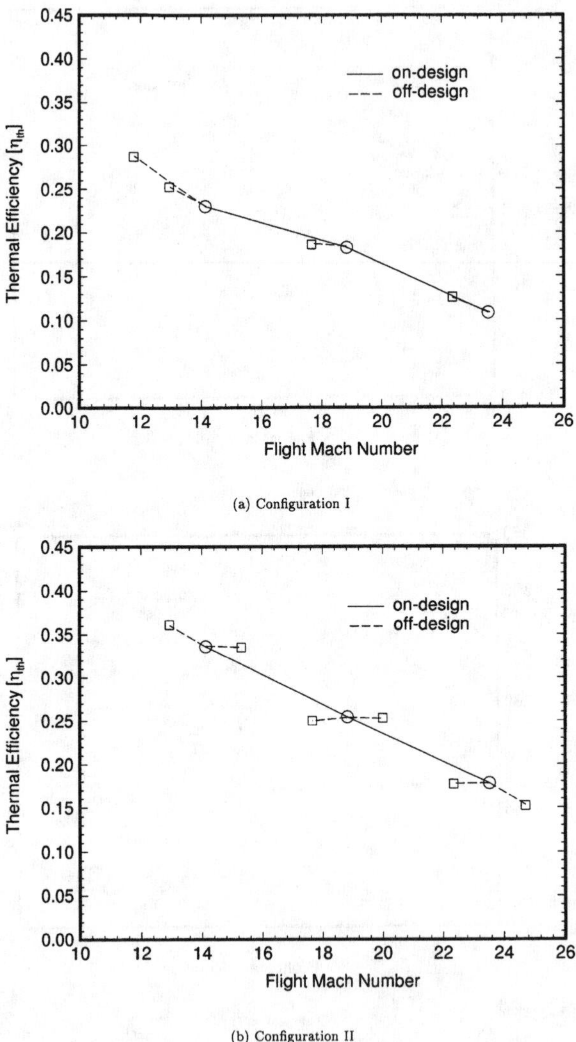

(a) Configuration I

(b) Configuration II

Fig. 55 Thermal efficiency for an on- and off-design planar detonation-wave ramjet at q_{dyn} = 1400 psf.

figuration I is more sensitive to off-design operation than Configuration II. It is, therefore, clear that in the absence of changes to engine geometry thrust production of the engine will deteriorate at off-design operation. Most effective modification in the engine's geometry is a shift in the nozzle throat to ensure that expansion of the flow always takes place immediately downstream of the detonation wave.

Although the aforementioned geometry change would likely effect more of a recovery of thrust than any other change, a number of other possibilities for improving thrust production exist. Shifting the leading edge of the cowl to ensure that no spillage of intake flow occurs would recover thrust lost from a reduction in ingested fuel and air, especially for Configuration I. Shifting the position of the leading edge of the detonation wedge to ensure that it is always located at the point of intersection of the final inlet shock with the cowl surface could also provide a small gain in thrust, especially for the higher-than-design Mach numbers where the shock intersection would thereby be avoided. Investigating the effect of these changes on off-design performance offers much potential for further study. More details on off-design propulsive characteristics of detonation-wave ramjets can be found in Ref. 70.

Shcramjet/Airframe-Integrated Wave Rider

By now it is well established that there is a strong interaction between the propulsion aspects and airframe aerodynamics of a hypersonic air-breathing vehicle. To yield high thrust margins and high lift force, the engine and the airframe must be highly integrated in a vehicle with high volume efficiency. One attractive means for studying the aerodynamic, propulsive, and geometric characteristics of such highly integrated vehicles is to utilize the wave-rider concept. At their design conditions the shape of wave-rider vehicles can be generated from known, accurately determined planar or axisymmetric flowfields. Such an approach was taken in Refs. 17 and 18 to determine the aerodynamic and propulsive performance characteristics of a class of shcramjet-powered engine/airframe-integrated vehicles derived from the planar and axisymmetric shcramjet model flowfields just described.

The three-dimensional wave-rider configuration consists of four surfaces: upper surface, lower surface, base area, and cowl surface. The leading edge is defined by specifying the vehicle capture area in the $x = 0$ plane and projecting it onto the bow-shock surface of the shcramjet. The trailing-edge contour may take any form desired, provided that it begins and ends on the streamlines from the leading edge and that the Mach number normal to the trailing-edge contour is greater than or equal to 1. For simplicity this contour is chosen in the plane $x = L$, where L is the length of the wave rider. Because the shock wave is attached to the leading edge at the design condition, the upper and lower surfaces of the configuration have independent flowfields and can thus be designed separately. With the leading- and trailing-edge contours specified, the lower surface of the wave rider is generated by replacing the stream surfaces with solid surfaces. The upper surface is

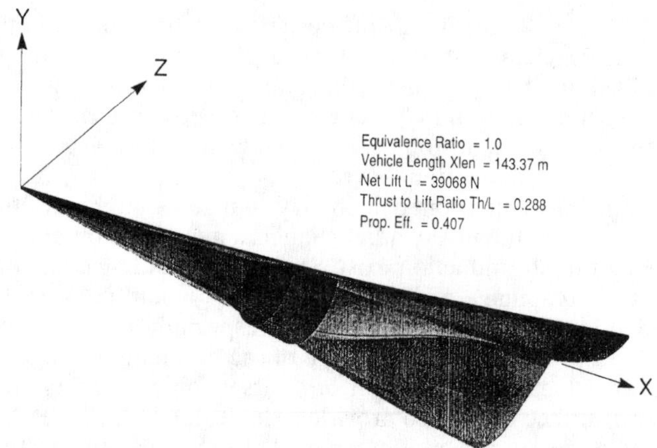

Fig. 56 Detonation-wave ramjet/airframe integrated wave rider (M_o = 14 and anhedral angle = 30 deg).

aligned with the freestream flow, starting at the leading edge and ending at the $x = L$ plane. The area between the upper and lower surfaces of the configuration at $x = L$ is the vehicle base area. The cowl, located at point c (Fig. 31) on the leading edge, is embedded in the flowfield and spans from one side of the body to the other.

A typical example of a wave rider with such an air-capture area generated from a Mach 14 axisymmetric shcramjet flowfield is shown in Fig. 56. The vehicle capture area in the y-z plane at $x = 0$ is defined by a circular arc of radius R, tangent to a line drawn at an angle θ_a (anhedral angle) from the z axis at the origin (Fig. 57). The purpose of designing the capture area curve OA concave upwards is to obtain a wing at a positive angle of attack. The pressure acting on the base area, and hence the thrust contribution of this surface, was assumed to be zero (i.e., $p \equiv 0$) because the flowfield behind the body is no longer two-dimensional. This assumption results in a conservative estimate of the propulsive force of the body because the pressure acting on this surface is probably of the same, or slightly smaller, order of magnitude as the freestream pressure. In any case, it is small compared with pressures acting on the lower surface.

The lower surface of the body of the wave rider consists of the main body of revolution and wings (formed from stream surfaces). The flow over the main body is axisymmetric, thus the x and y components of the force acting on it, and the surface area of the body can be expressed as

$$F_{x,\text{body}} = 2\int_0^{r(L)} \int_{-\frac{x}{2}}^{-\theta_a} p r_b \, \mathrm{d}\theta \, \mathrm{d}r \tag{19}$$

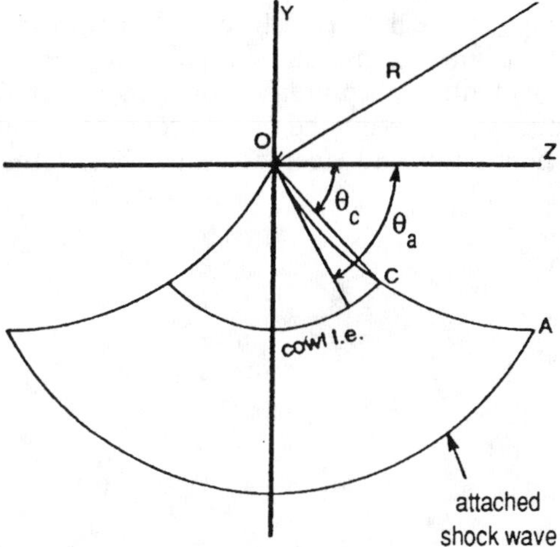

Fig. 57 Vehicle air-capture area or projection of leading-edge contour on $x = O$ plane and θ_a = anhedral angle (negative for $y > 0$).

$$= 2\left(\frac{\pi}{2} - \theta_a\right)\int_0^{r(L)} pr_b \, \mathrm{d}r \tag{20}$$

$$F_{y,\text{body}} = 2\int_0^L pr_b \cos(\theta_a) pr_b \, \mathrm{d}x \tag{21}$$

$$S_{\text{body}} = 2\int_0^{S(L)}\int_{-\frac{x}{2}}^{-\theta_a} r_b \, \mathrm{d}\theta \, \mathrm{d}S \tag{22}$$

$$= 2\left(\frac{\pi}{2} - \theta_a\right)\int_0^L r_b\sqrt{1 + \left(\frac{\mathrm{d}r}{\mathrm{d}x}\right)^2}\,\mathrm{d}x \tag{23}$$

where r_b is the radial coordinate of the main body surface. Similar expressions are used to calculate the forces on the upper and lower surfaces of the cowl by replacing θ_a by θ_c in Fig. 57, r_b by $r^u{}_c$ by $r^l{}_c$—the cowl upper and lower surface coordinates, respectively, and L by l—the cowl length.

The force acting on the lower surface of the wings is determined by dividing this surface into longitudinal strips by adjacent streamlines and subdividing the elementary strip areas into triangles. Knowing the coordinates of the triangle vertices situated on adjacent streamlines and the pressures acting on them, the x and y components of the force are determined from

$$F_x = \sum_{i=i}^{n_e} p_i \left(\frac{n_x}{|n|} \right)_i S_i \tag{24}$$

$$F_y = \sum_{i=1}^{n_e} p_i \left(\frac{n_y}{|n|} \right)_i S_i \tag{25}$$

where S_i is the area of the ith triangle and n_e the number of elemental triangles.

The wave-rider design is derived from the maximum net thrust axisymmetric shcramjet flowfield for stoichiometric hydrogen/air ratios at on-design conditions. The vehicle capture area depends on two parameters: the anhedral angle θ_a and the radius R of the circular arc. Some of the vehicle's performance parameters are presented in Figs. 58 and 59 for two values of the freestream dynamic pressure, $P_{\text{dyn}} = 1000$ psf and 2000 psf, and for one inlet compression ratio characterized by $T_c = 1500$ K after the third shock cd (Fig. 31). The thrust-to-lift ratio is strongly dependent on the anhedral angle and changes drastically when θ_a varies between -30 and $+30$ deg (Fig. 58), reaching a value of ~ 0.3 for $M_o = 11$ and $\theta_a = 60$ deg. The largest Th/L ratio is attained for $\theta_a = -30$ deg, low circular arc radius, low M_o and large P_{dyn}. This ratio is not very sensitive to dynamic pressure, but the influence of R is quite appreciable in the $\theta_a = -30$ deg case. The variation of volumetric

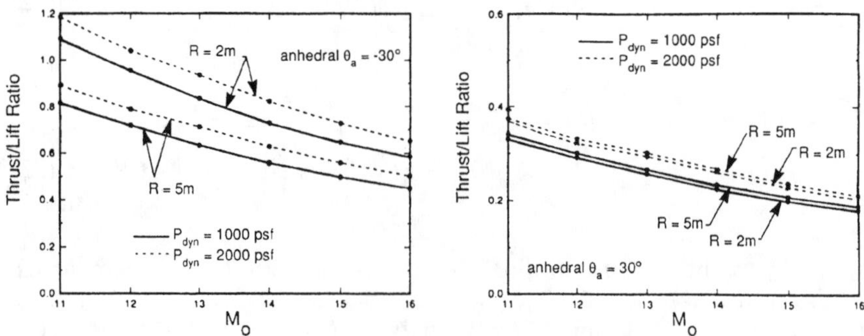

Fig. 58 Thrust-to-lift ratio.

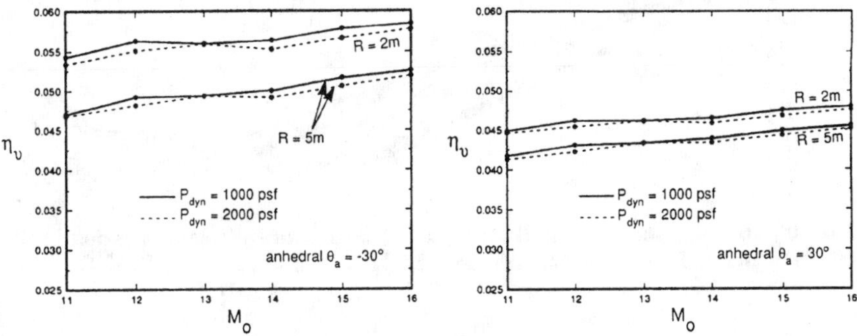

Fig. 59 Volumetric efficiency, $\eta_v = V_{2/3}/S$.

efficiency η_v defined as the 2/3 power of the total volume of the vehicle to its total surfaces, is shown in Fig. 59. It increases steadily with M_o, is insensitive to P_{dyn} and shows a slight dependence on R but an appreciable dependence on θ_a. For given M_o, P_{dyn}, and R, it attains the largest value at $\theta_a = -30$ deg and $R = 2$ m.

Concluding Remarks

Recent renewed research effort in detonation-wave ramjets, although limited in scope, has generated some revealing results concerning, particularly, the propulsive characteristics of such engines. There is now some evidence that detonation-wave ramjets could perform better than scramjets in the very high flight Mach-number range. However, much more additional work is needed to definitely substantiate this claim. Realistic estimates of fuel/air mixing losses in the high-temperature, high Mach-number vehicle forebody flow, effects of fuel/air mixing inhomogeneities, the optimum choice (from the point of view of entropy consideration) between inlet shock intensities and that of the oblique detonation wave, the viscous losses in the inlet, nozzle, and, particularly, viscous effects on the detonation-wave behavior in the narrow passage between the cowl and the central body of the ramjet on the propulsive characteristics of such ramjets are but a few of the problems yet to be tackled.

Preliminary and partial answers to these problems were recently presented in Ref. 71 where it is found that the boundary layer on the forebody (inlet) of the engine walls can be a crucial factor for the realization of shock-induced combustion. In Figs. 60 and 61 taken from this reference, hydrogen is injected near the external edge of the boundary layer on the vehicle surface (inlet). As a result of hydrogen jet and wall boundary-layer interaction, the hydrogen jet is ignited at a small distance from the point

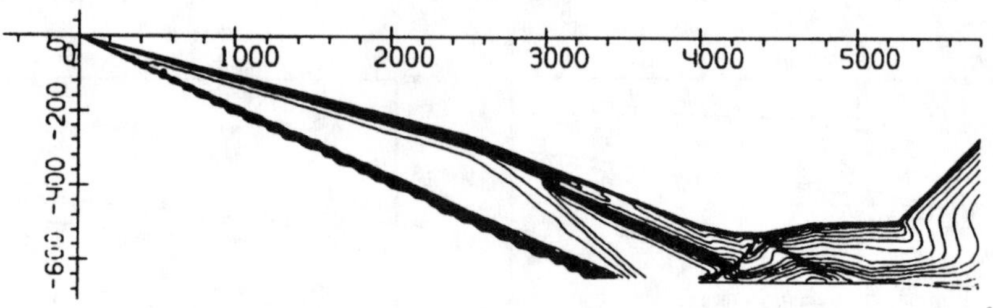

Fig. 60 Mach-number field in the propulsive duct with hydrogen injection in the inlet. Oncoming flow M_∞ = 12; hydrogen jet, M_{H_2} = 2.45, T_{H_2} = 450 K (Ref. 71).

of injection, and diffusive combustion is realized upstream of the combustor (See Fig. 61 for H_2O mass fraction distribution in the inlet and engine duct.)

Of utmost interest is also the realistic estimate of its off-design performance characteristics, which can serve as an indication to the degree of variable geometry needed, and, hence, the mechanical complexities involved, to achieve satisfactory performance in practice. Experimental evi-

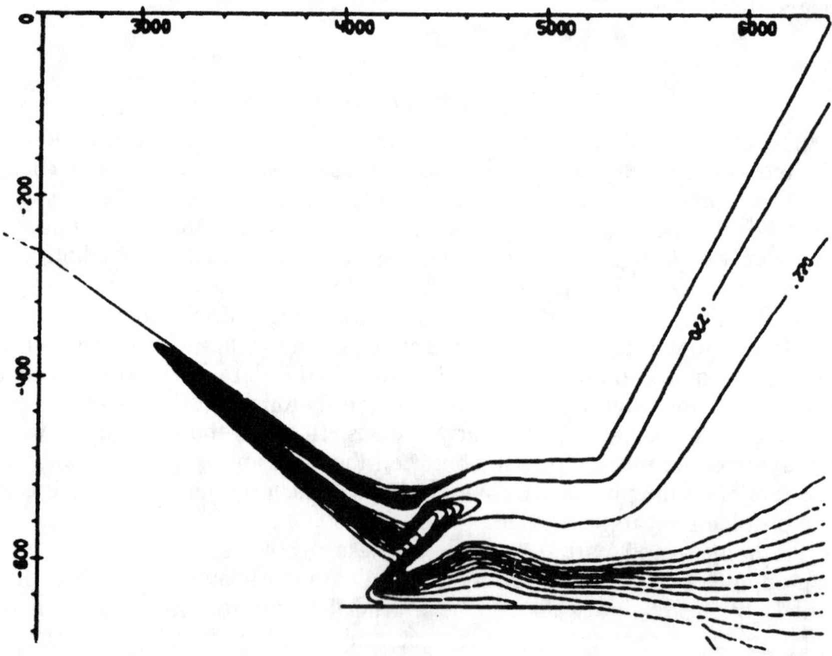

Fig. 61 H_2O mass fraction contours in the propulsive duct. Oncoming flow M_∞ = 12; hydrogen jet, M_{H_2} = 2.45 and T_{H_2} = 450 K *(Ref. 71).*

dence of oblique detonation-wave stability for combustor-inlet flow conditions encountered in actual flight trajectories has yet to be established. But most important for the technical realization of the detonation-wave ramjet is the feasibility of quasihomogeneous fuel/air mixing in the relatively hot high Mach-number forebody airflow without premature ignition upstream of the detonation wave and along the forebody boundary layer. This is a formidable technical challenge. Extensive computational-fluid-dynamic studies of this problem supported by experimental data and innovative mixing and cooling techniques are needed to solve this difficult task, if detonation-wave ramjets are likely to be a technically feasible alternative to scramjets.

Acknowledgment

The author would like to thank Research Assistant J. Schumacher for his comments and careful reading of the material and for his valuable assistance in the preparation of the manuscript.

References

[1]Roy, M.M., "Moteur Thermiques," *Comptes Rendus de l'Academie des Sciences*, Vol. 222, 1947.

[2]Dunlap, R., Brehm, R. L., and Nicholls, J. A., "A Preliminary Study of the Application of Steady-State Detonation Combustion to a Reaction Engine," *Journal of Jet Propulsion*, Vol. 28, No. 6, 1958, pp. 451–456.

[3]Sargeant, W. H., and Gross, R. A., "A Detonation Wave Hypersonic Ramjet," Air Force Office of Scientific Research, N 589, 1959.

[4]Townend, L. H., "Detonation Ramjets for Hypersonic Aircraft," Royal Aircraft Establishment TRept. 70218, 1970.

[5]Morrison, R. B., "Evaluation of the Oblique Detonation Wave Ramjet," NASA CR 145358, 1978.

[6]Morrison, R. B., "Oblique Detonation Wave Ramjet," NASA CR 159192, 1980.

[7]Billig, F. S., "External Burning in Supersonic Streams," *Proceedings of the 18th International Astronautical Congress*, Vol. 3, 1968, pp. 23–54.

[8]Billig, F. S., "Combustion Process in Supersonic Flow," *Proceedings of the 7th International Symposium on Airbreathing Engines*, 1985, pp. 245–256.

[9]Chernyi, G. G., "Problems of Hydrodynamics and Continuum Mechanics," *Supersonic Flow Past Bodies with Formation of Detonation and Combustion Fronts*, Society for Industrial and Applied Mathematics, Philadelphia, PA, 969, pp. 145–169.

[10]Chushkin, P. I., "Combustion in Supersonic Flows Past Various Bodies," *Journal of Comp. Math. and Math. Phys.*, Vol. 1, No. 6, 1969, pp. 1367–1377.

[11]O'Brien, C. J., and Kobayashi, A. C., "Advanced Earth-to-Orbit Propulsion Concepts," AIAA Paper 86-1386, 1986.

[12]Ostrander, M. J., Hyde, J. C., Young, M. F., and Kissinger, R. D., "Standing Oblique Detonation Wave Engine Performance," AIAA Paper 87-2002, 1987.

[13]Sheng, Y., and Sislian, J. P., "A Model of a Hypersonic Two-Dimensional Oblique Detonation Wave Ramjet," UTIAS, TN 257, 1985.

[14]Sislian, J. P., and Atamanchuk, T. M., "Aerodynamic and Propulsive Performance of Hypersonic Detonation Wave Ramjets," *Proceedings of the Ninth International Symposium on Air-Breathing Engines*, 1989, pp. 1026–1035.

[15]Atamanchuk, T. M., and Sislian, J. P., "Performance Characteristic of Hypersonic Detonation Wave Ramjets," *Proceedings of the Seventy-fifth AGARD Symposium on Air-Breathing Engines*, 1990.

[16]Atamanchuk, T. M., and Sislian, J. P., "On-and-Off Design Performance Analysis of Hypersonic Detonation Wave Ramjets," AIAA Paper 90-2473, July 1990.

[17]Atamanchuk, T. M., and Sislian, J. P., "Hypersonic Detonation Wave Powered Lifting-Propulsive Bodies," AIAA Paper 91-5010, Dec. 1991.

[18]Atamanchuk, T. M., Sislian, J. P., and Dudebout, R., "An Aerospace Plane as a Detonation Wave Ramjet/Airframe Integrated Waverider," AIAA Paper 92-5022, Dec. 1992.

[19]Sislian, J. P., and Dudebout, R., "Hypersonic Shock-Induced Combustion Ramjet Performance Analysis," *Proceedings of the Eleventh International Symposium on Air-Breathing Engines*, 1993, pp. 413–420.

[20]Menees, G. P., Adelman, H. G., Cambier, J-L, Bowles, and J. V., "Wave Combustors for Trans-Atmospheric Vehicles," *9th International Symposium on Airbreathing Engines*, 1989.

[21]Fendell, F., Mitchell, J., McGregor, R., and Sheffield, M., "Laser-Initiated Conical Detonation Wave for Supersonic Combustion. II," *Journal of Propulsion and Power,* Vol. 2, No. 2, 1993, pp. 182–190.

[22]Hertzberg, A., Bruckner, A. P., and Bogdanoff, D. W., "Ram Accelerator: A New Chemical Method for Accelerating Projectiles to Ultrahigh Velocities," *AIAA Journal*, Vol. 26, No. 2, 1988, pp. 195–203.

[23]Eidelman, S., Grossmann, W., and Lottari, I., "Review of Propulsion Applications and Numerical Simulations of Pulsed Detonation Engine Concept," *Journal of Propulsion and Power*, Vol. 7, No. 6, 1991, pp. 857–865.

[24]Gross, R. A., "Exploratory Studies of Combustion in Supersonic Flow," Air Force Office of Scientific Research, TRept. 59–587, 1959.

[25]Gross, R. A., and Chinitz, W., "A Study of Supersonic Combustion," *Journal of Aerospace Sciences*, Vol. 27, No. 7, 1960, pp. 517–524.

[26]Fletcher, E. A., "Stationary Detonation Waves," *Combustion and Propulsion, Fourth AGARD Colloquium: High Mach Number Air-Breathing Engines*, edited by A. L. Jaumotte, H. Lefebvre, and A. M. Rothrock, Pergamon, New York, 1961, pp. 169–177.

[27]Pratt, D. T., Humphrey, J. W., and Glenn, D. E., "Morphology of Standing Oblique Detonation Waves," *AIAA Journal*, Vol. 7, No. 5, 1991, pp. 837–845.

[28]Nicholls, J. A., Debora, E. K., and Gealer, R. I., "Studies in Connection with Stabilized Gaseous Detonation Waves," *Seventh International Symposium on Combustion*, Oxford Univ. Press, Oxford, England, UK 1958, pp. 766–772.

[29]Rubins, P. M., and Rhodes, R. P., Jr., "Shock-Induced Supersonic Combustion in a Constant-Area Duct," *AIAA Journal*, Vol. 1, No. 12, 1963, pp. 2778–2789.

[30]Rubins, P. M., and Cunningham, T. H. M., "Shock-Induced Combustion in a Constant-Area Duct," *Journal of Spacecraft and Rockets*, Vol. 2, No. 2, 1965, pp. 199–205.

[31]Algermissen, J., "Investigation of the Ignition Behavior of Hydrogen/Air Mix-

tures by Oblique Shock-Induced Combustion," *Zeitschrift für Flugwissenshaften und Weltranmforschung*, Vol. 10, No. 2, 1986, pp. 73–81 (in German).

[32]Adelman, H. G., Cambier, J. -L., Menees, G. P., and Balboni, J. A., "Analytical and Experimental Validation of the Oblique Detonation Wave Engine Concept," AIAA Paper 88-0097, 1988.

[33]Menees, G. P., Adelman, H. G., and Cambier, J-L., "Analytical and Experimental Investigations of the Oblique Detonation Wave Engine Concept," *AGARD Propulsion and Energetics Panel, Seventy-fifty Symposium, Hypersonic Combined Cycle Propulsion*, Paper No. 26, 1990.

[34]Debora, E. K., and Wagner, H. G., "A Technique for Establishing Oblique Detonations," Max-Plank Inst. for Flow Research, TRept. 102/1988, Goettingen, Germany, 1988.

[35]Fan, B. C., Sichel, M., and Kauffman, C. W., "Analysis for Oblique Shock-Detonation Wave Interaction in the Supersonic Flow of Combustible Medium," AIAA Paper 88-0441, 1988.

[36]Bruckner, A. P., Knowlen, C., and Hertzberg, A., "Applications of the Ram Accelerator to Hypervelocity Aerothermodynamic Testing," AIAA Paper 92-3949, 1992.

[37]Takashi, N., Schuh, M. J., Randall, D. S., Dahm, D. T., and Pratt, J., "Demonstration of Oblique Detonation Wave for Hypersonic Propulsion," Astronautical Research and Engineering, TRept. Final Rep., Sunnyvale, CA, 1989.

[38]Samaras, D. G., "Gas Dynamic Treatment of Exothermic and Endothermic Discontinuities," *Canadian Journal of Research*, Series A, Vol. 26, No. 1, 1948, pp. 1–21.

[39]Siestrunck, R., Fabri, J., and Le Grives, E., "Some Properties of Stationary Detonation Waves," *Fourth Symposium on Internal Combustion*, 1953, pp. 498–501.

[40]Rutkowski, J., and Nicholls, J. A., "Considerations for the Attainment of a Standing Detonation Wave," *Proceedings of the Gas Dynamics Symposium in Aerothermochemistry*, 1955, pp. 243–253.

[41]Chinitz, W., Bohrer, L. C., and Foreman, K. M., "Properties of Oblique Detonation Waves," Air Force Office of Scientific Research, TRept. 5-59-462, April 1959.

[42]Gross, R. A., "Oblique Detonation Waves," *AIAA Journal*, Vol. 1, No. 1, 1963, pp. 1225–1227.

[43]Woolard, H. W., "Analytical Approximations for Stationary Conical Detonations and Deflagrations in Supersonic Flow," Applied Physics Laboratory, Johns Hopkins Univ., TRept. TG-446, Baltimore, MD, May 1963.

[44]Chernyi, G. G., "Self-Similar Flows of Combustible Gas Mixtures," *Mekhanika Jhidkosti i Gaza,* Vol. 16, 1966, pp. 10–24 (in Russian).

[45]Townend, L. H., "An Analysis of Oblique and Normal Detonation Waves," RAE, TRept. 66081, 1966.

[46]Bartlmä, F., "Randbedingungen bei Schiefen Reaktionsfronten in Uberschallströmung," *Zeitschrift für Flugwissenshaften*, Vol. 16, No. 2, 1968, pp. 437–444.

[47]Liepmann, H. W., and Roshko, A., *Elements of Gasdynamics*, Wiley, New York, 1957.

[48]Jachimowsky, C. J., "An Analytical Study of the Hydrogen-Air Reaction Mechanism with Application to Scramjet Combustion," NASA TP-2791, 1988.

[49]Yee, H. C., "Upwind and Symmetric Shock-Capturing Schemes," NASA TM-89464, 1987.

[50]Shuen, J. S., and Yoon, S., "Numerical Study of Chemically Reacting Flows Using a Lower-Upper Symmetric Successive Overrelaxation Scheme," *AIAA Journal*, Vol. 27, No. 12, 1989, pp. 1752–1760.

[51]Singh, D. J., Carpenter, M. H., and Kumar, A., "Numerical Simulation of Shock-Induced Combustion/Detonation in a Premixed h_2-Air Mixture Using Navier-Stokes Equations," AIAA Paper 91-3359, June 1991.

[52]Chernyi, G. G., and Gilinskii, S. M., "High Velocity Motion of Solid Bodies in Combustible Gas Mixtures," *Astronautica Acta*, Vol. 15, No. 5, 1970, pp. 539–545.

[53]Ruegg, F. W., "A Missile Technique for the Study of Detonation Waves," *Journal of Research of the National Bureau of Standards, Section C: Engineering and Instrumentation*, Vol. 66, 1962, pp. 51–62.

[54]Alpert, R. L., and Toong, T. Y., "Periodicity in Exothermic Hypersonic Flows about Blunt Projectiles," *Astronautica Acta*, Vol. 17, Nos. 4 and 5, 1972, pp. 539–560.

[55]Lehr, H. F., "Experiments on Shock-Induced Combustion," *Astronautica Acta*, Vol. 17, 1972, pp. 589–597.

[56]Fujiwara, T., Matsuo, A., and Nomoto, H., "A Two-Dimensional Detonation Supported by a Blunt Body or a Wedge," AIAA Paper 88-0098, Jan. 1988.

[57]Wang, Y., Fujiwara, T., Aoki, T., Arakawa, H., and Ishiguro, T., "Three-Dimensional Standing Oblique Detonation Wave in a Hypersonic Flow," AIAA Paper 88-0478, Jan. 1988.

[58]Rubins, P. M., and Bauer, R. C., "A Hypersonic Ramjet Analysis with Premixed Fuel Combustion," AIAA Paper 66-648, 1966.

[59]Cambier, J.-L., Adelman, H., and Menees, G. P., "Numerical Simulation of an Oblique Detonation Wave Engine," AIAA Paper 88-0063, Jan. 1988.

[60]Oswatitsch, K., "Pressure Recovery for Missiles with Reaction Propulsion at High Supersonic Speeds. (The Efficiency of Shock Diffusers)," NASA TM-1140, 1947

[61]Oran, E. S., Boris, J. P., Young, T., Flanigan, M., and Picone, M., "Numerical Simulations of Detonations in Hydrogen-Air and Methane-Air Mixtures," *Eighteenth International Symposium on Combustion*, The Combustion Inst., Waterloo, Ontario, Canada, 1981, pp. 1641–1649.

[62]Billig, F. S., "Proposed Supplement to Propulsion System Management Support Plan," DARPA, TRept., 1985.

[63]Godunov, S. K., *Numerical Solutions of Multidimensional Problems of Gasdynamics*, Nauka, Moscow, 1976.

[64]Dudebout, R., and Sislian, J. P., "Numerical Simulation of Hypersonic Shock-Induced Combustion Ramjet Flowfields," AIAA Paper 94-3098, 1994.

[65]Sislian, J. P., Dudebout, R., Schumacher, J., Islam, M., and Oppitz, R., "Inviscid Propulsive Characteristics of Hypersonic Shcramjets," AIAA Paper 96-4535, Nov. 1996.

[66]Sislian, J. P., Dudebout, R., and Oppitz, R., "Inviscid On-Design Propulsive Characteristics of Hypersonic Shock-Induced Combustion Ramjets," Univ. of Toronto Inst. for Aerospace Studies, TRept. 352, 1997, Toronto, Canada.

[67]Jameson, A., and Yoon, S., "Lower-Upper Implicit Scheme with Multiple Grids for the Euler Equations," *AIA Journal*, Vol. 25, No. 7, 1987, pp. 929–935.

[68]Yee, H. C., Klopfer, G. H., and Montagnè, "High-Resolution Shock-Capturing Schemes for Inviscid and Viscous Hypersonic Flows," *Journal of Computational Physics*, Vol. 88, 1990, pp. 31–61.

[69]Park, H-K., "Model of an Aero-Space Plane Based on an Idealized Cone-Derived Waverider Forebody," Ph.D. dissertation, Univ. of Oklahoma, OK, 1990.

[70]Sislian, J. P., Dudebout, R., Schumacher, J., Islam, M., and Redford, T., "Inviscid Off-Design Propulsive Characteristics of Hypersonic Shock-Induced Combustion Ramjets," Univ. of Toronto Inst. for Aerospace Studies, TRept. 354, 1997, Toronto, Canada.

[71]Bezgin, I., Ganxhelo, A., Gouskov, O., and Kopchenov, V., "Some Numerical Investigation Results of Shock-Induced Combustion," AIAA Paper 98-1513, April 1998.

Chapter 14

Problem of Hypersonic Flow Deceleration by Magnetic Field

A. B. Vatazhin* and V. I. Kopchenov†
Central Institute of Aviation Motors, Moscow, Russia

Introduction

THE present article is devoted to an analysis of various aspects of magnetohydrodynamic (MHD) control over supersonic and hypersonic flows. The important application of this magnetogasdynamics division is connected with development of aerospace vehicles powered by propulsion systems including the supersonic ramjet engine (scramjet).

All problems considered in this article are formulated as independent scientific problems. Nevertheless, the authors kept in mind the possibility of using obtained results in hypersonic technologies.

The introduction includes the following: a brief analysis of peculiarities of MHD control over gasdynamic flows, a review of available MHD applications to modern hypersonic technologies, and a structure of the present article.

Peculiarities of MHD Control

It is very important that MHD force $\boldsymbol{f} = \boldsymbol{j} \times \boldsymbol{B}$ ($\boldsymbol{j}$ and $\boldsymbol{B}$ are vectors of electric current density and of magnetic field) acts on unit *volume* (in contrast, for example, to gravity force density, acting on unit *mass*). Therefore, an influence of MHD force on gasdynamic flow increases with diminution of gas density on retention of all other parameters. Really, a parameter

---

*Professor.

†Chief of Project.

This work was supported by the Russian Foundation for Fundamental Studies (project N 98-01-00923) and by grant of EOARD. Some results of this article will be published in the collected papers devoted to 90th Anniversary of Academician L.I. Sedov. Authors wish to thank colleagues who took part in carrying out this investigation, especially E. Kholshchevnikova, V. Likhter, and O. Gouskov.

of MHD interaction is defined as $S = jBL/\rho u^2$, where ρ is gas density, L is length scale, and u is the characteristic velocity.

The direction of force $\boldsymbol{f}$, in many cases, is determined by directions of external magnetic $\boldsymbol{B}$ and electric $\boldsymbol{E}$ fields. Therefore, a possibility to point the force $\boldsymbol{f}$ to the prescribed direction exists. This circumstance is very attractive for gasdynamic flow control.

The term $\boldsymbol{jE}$ constituting an input of electric power to unit volume appears in the energy equation written for a change of kinetic and inner energy sum per unit volume. If $\boldsymbol{jE} < 0$, electric power is extracted from a flow, i.e. generator regime is realized. If $\boldsymbol{jE} > 0$, electric power is introduced into a flow and the flow acceleration is possible.

The realization of generator regime, as well as of accelerator one, may be controlled with confidence. It allows electric power extraction from any gasdynamic duct portion and introduction of this power or its part into another duct portion. As will be shown below, such MHD energy bypass opens an attractive possibility to solve some problems of scramjet operation.

The above-mentioned well-known peculiarities of MHD control will be discussed in the present article in relation to their possible use when applied to hypersonic propulsion system. Other peculiarities are analyzed in numerous monographs and publications.

There are some serious restrictions on the possibilities of using MHD methods. The main limitations are connected with the problem of creating technical systems for production of external magnetic and electric fields (*aspect A*) and for the raising of electroconductivity (*aspect B*) and also with the problem of high irreversible losses in MHD flows (*aspect C*).

Aspect A has a history in relation, for example, to development of the MHD generator, and there is appreciable progress in this field.

Aspect B is the main restriction on MHD use for a flight vehicle. Although various facilities for a rise in conductivity have been produced, their application to flight conditions is a very hard task. This problem is principally a technical–economical one, and will not be considered in this article.

Aspect C is very important in obtaining a correct assessment of MHD control possibilities. These possibilities are frequently overestimated. What actually happens is that high losses appear in MHD flows. This circumstance can result in the deterioration of MHD control efficiency (even though problem B will be successfully decided). That is why the problem of irreversible losses in MHD flows will be discussed in detail.

Review of Proposals to Use MHD Control

The research for improving the characteristics and the operational range of hypersonic systems of the aerospace plane type (with the propulsion system, including scramjet using non-traditional technologies) has received much attention in the past few years.[1–4] These new concepts are based on some new technologies. One of them is connected with a creation of weakly ionized plasma and with the use of effects arising in the flow of weakly ionized gas adjacent to the vehicle. The essential constituent of the new concept is also the magneto-plasma-chemical engine.[1–4] Not dwelling on all aspects, which are called to improve the characteristics of such systems, it must be emphasized that a significant part in the increase of their efficiency is assigned to MHD means of the flow control.[5–8]

In accordance with the available proposals, the MHD-generator, established onboard of the flight vehicle, generates at hypersonic flow deceleration of the electrical energy that is used partially for the flow ionization. These ionization is necessary for the MHD system functioning. The submitted assessments [3,4,6] testify that it is possible to realize a self-sustaining operational mode of the MHD system, when the energy, produced by generator, certainly exceeds the consumptions on the forced flow ionization. Moreover the electrical energy excess can be used, in the authors' [3,6] opinion, for other needs, for example, for creation of plasma in the flow over the vehicle with the purpose of the vehicle drag reduction. This aspect, associated with effects in weakly ionized plasma flow over bodies, was widely discussed in Ref. 9 and these questions will not be touched upon in this review.

The second important proposal consists of the use for the characteristics improvement of the scramjet proper.[1,3,5–8] The results show the possibility of a specific impulse increase in the case of the MHD system use in comparison with conventional scramjet that are obtained in the framework of a one-dimensional model.[6] It is possible to distinguish the following factors allowing an increase of scramjet efficiency at the expense of MHD methods application for the flow control.

The proposal is made to use MHD control for the improvement of the air intake characteristics.[1,3,5] With the MHD system use, the characteristics of the scramjet air intake are supposed to be improved in the wide range of flight Mach numbers without mechanical regulation. So, it is shown [3,6] on an example of the 2-D air intake designed on flight Mach number $M_f = 10$ that MHD control allows an increase in the flow rate on exit from the air intake at the flight regime $M_f = 6$. The MHD control of the inlet flow also provides increase of specific impulse and thrust of the scramjet in comparison with the one without MHD control. The estimations are performed using 2-D Euler equations.

The important factor,[5] which MHD energy conversion provides, is the possibility to bypass a portion of the energy from the air intake exit or combustor entry to the combustor exit or nozzle in order to optimize the engine operation through the combustion process improvement, inlet and nozzle flow field transformation. The MHD generator is installed at the inlet exit before the combustor for energy extraction from the flow ahead of the combustor entry. The electric energy produced by MHD generator can be expended partially on the ionizer and onboard equipment operation, and also on the additional acceleration of the combustion products in the MHD accelerator installed downstream from the combustor in front of the nozzle. The analysis was performed[6] in 1-D approach for ideal gas (inviscid and nonheat conducting) in the case when all the energy produced by MHD generator is transmitted to the MHD accelerator and the flow in the combustor receives all the energy potentially enclosed in the fuel. Presented results of specific impulse estimations indicate that the advantage of an MHD-controlled system, in comparison with usual scramjet, can be attained at the proper deciding of the guiding parameters of MHD- system at the conditions of identical inlet, combustor and nozzle geometry. The physical aspects providing the merits of the MHD-controlled system are not presented in cited papers.[5,6]

The concept of the hypersonic system,[1] including the possibility[5,6] to bypass the part of flow energy from the inlet exit to the nozzle entry, has been studied extensively in the paper.[4] The analysis performed by the authors[4] and their colleagues gave them ground to believe that energy bypass increases the value of flight Mach number at which the transition from subsonic combustion mode to the supersonic one must be realized.

The interesting proposal to use the part of the scramjet duct at once as the scramjet combustor and as MHD-generator to provide the MHD control of combustion process was formulated in Ref. 7. The scramjet combustor–MHD generator scheme with thin electric conducting layers (T-layers) is analyzed. The local plasma regions are generated with definite frequency at the exit cross section of air intake. The deceleration of this plasma layer due to interaction with external magnetic field results in the generation of unsteady flow structure including shock compressed zone and T-layer. It is proposed to realize the combustion process in the shock compressed zone. The simplified mathematical model for the scramjet performances evaluation was developed. The estimation of the specific impulse averaged in time in the case of MHD control for the regime corresponding to $M_f = 10$ for hydrogen combustion shows that this characteristic increases appreciably in comparison with usual scramjet. The main reason for the performance improvement of the MHD-controlled engine, in the authors' opinion,[7] is combustion efficiency increase due to the higher temperature and lower velocity in the shock compressed region in comparison with usual scramjet.

The important general question on the scramjet operation at high flight Mach numbers will be considered here before the discussion of proposal,[8] which is a further development of the concept.[7] It is necessary to note that one of the principal problems at high hypersonic flight Mach number is the efficient supply of energy contained in the fuel at its combustion to the high enthalpy flow corresponding to high M_f. On the one hand, the problem is the realization of the mixing of fuel and air at high velocities in the combustor on the acceptable length. Low mixing efficiency prevents effective combustion. On the other hand, the flow deceleration to smaller velocities in inlet (to provide better conditions for mixing) is followed by temperature rise at the combustor entry to an inadmissible level at which the effective realization of combustion is once again a difficult task, but for another reason. The general problem lies in the fact that under the conditions, corresponding high M_f, temperature levels in the combustor are too high to realize the effective heat release. This is due to the fact that chemical equilibrium is shifted to the formation of intermediate components with large part of radicals in the products of combustion. This has a strong negative effect on the heat release during the combustion process. Even if high mixing efficiency is reached, the last circumstance results in the rather low combustion efficiency estimated on the released energy (see, for example, Refs. 10 and 11). Moreover, this negative tendency can occur in the scramjet combustor even at moderate hypersonic M_f when it is possible to realize high mixing efficiency and to initiate intensive combustion process in the presence of shock system in the combustor duct. Despite moderate temperature at the combustor entry, an unacceptable high temperature and high level of radicals in combustion products are then detected at the exit section of the combustor. This negative effect of combustion efficiency losses is enhanced by the nozzle processes. The high energy radicals are exhausted through the nozzle without recombination that is followed by the thrust losses. The special scramjet duct design can be proposed as one way of overcoming this difficulty.[10,11] Another way is concerned with the addition of special chemical substances to increase the level of recombination.[12]

It seems that MHD flow control can be considered as an alternative tool for scramjet efficiency increase at high flight Mach numbers. In connection with this, the following scheme[8] of process in the scramjet can be proposed to provide the optimal conditions for fuel combustion. The deceleration of hypersonic flow is performed simultaneously with energy extraction from the flow up to the creation of conditions optimal for fuel combustion. It is desirable to return at least part of the energy extracted at the deceleration to the flow where this will have a positive effect, for example, in the accelerated duct in order to compensate partly for the impulse losses at the energy extraction. The MHD conversion of flow energy into the electric energy on the external load and vice versa is proposed for this goal. The energy is extracted from the flow in the block combustor—MHD generator and is transformed to electrical energy. The accelerating block is the nozzle combined with the MHD accelerator (the nozzle walls are electrodes) and the part of electrical energy is returned to the flow due to MHD accelerator. In general, various MHD schemes can be proposed for this idea realization. The principal scheme of MHD generator in Ref. 8 is analagous to this one proposed in Ref. 7 with T-layer initiation. Conditions of T-layer initiation and sustaining for different points of vehicle trajectory, and some results of numerical simulation of MHD generator flow are presented in Ref. 8. Unfortunately, the quantitative estimations of influence of the proposed energy bypass scheme using an MHD system on the combustion efficiency and the scramjet performances are not presented.

From our point of view, the proposal to increase the combustion efficiency and scramjet performances at high M_f using energy bypass with the aid of MHD control may be of interest for extending the scramjet application range. This proposal must be carefully verified in future investigations.

Besides, MHD methods can be used actively for control of forces acting on flight vehicle or its control devices. In particular, it is noted[5] that it is possible to influence the flow structure, including shock wave formation, with an MHD control system. Also it is pointed out, for example, in Ref. 5

that application of a magnetic system like dipole reduces the heat flux strength. It is proposed to control the ionized gas flow by MHD in the region of a forebody of a flight vehicle up to the engine entrance and also in the engine duct.[6] Considerable attention is given to the prevention of boundary layer separation in the engine duct as a potential source of engine performance improvement.[6] Unfortunately, this last possibility is not illustrated by quantitative results and these estimations still remain to be performed in the future. At the same time, the principal capability of MHD control on the boundary layer is confirmed by the well-known solution of classical Hartmann problem (see, for example, Ref. 13). Besides, the investigation of MHD action on the boundary layer separation was carried out, for example, in Ref. 4. It is necessary to point out the widespread program of computational and experimental investigations of MHD control on hypersonic flow.[15]

All aforementioned factors give promise that the positive effect can be obtained at the use of MHD tool onboard the vehicle. Of course, the weight and cost estimations are necessary. Only comprehensive analysis of advantages and costs will allow a final conclusion about expediency of offered control means. The attempt of such versatile assessment is made in Ref. 4.

The special question is connected with creation of medium conductivity. This question is considered in Ref. 5. The performed assessments show that the appreciable level of natural conductivity arises behind the vehicle bow shock on high-flight altitudes at flight Mach numbers exceeding 15. At the same time, at lower altitudes and Mach numbers the MHD system function demands artificial ionization with the help of special devices placed onboard a flight vehicle. Therefore it is interesting to note that, in accordance with estimations, [8] the natural conductivity in T-layer can be self-maintaining at the acceptable values of applied magnetic field.

It is necessary to note the following factor which has not been taken essentially into account at the efficiency estimations of magneto-plasma-chemical engine. This question is related to the influence of plasma on chemical processes in the combustion chamber. At the same time, some questions of the plasma source application for supersonic combustion enhancement and mechanisms of this enhancement were investigated in Ref. 16. The ionization effectiveness in the ignition promotion and in the combustion enhancement of hydrocarbon fuels was the subject of investigation.[17] The modification of a detailed chemical kinetics scheme for complex hydrocarbon fuels concerned with the presence of charged particles was performed[17] on the basis of kinetics experiments.

To conclude this brief review it is necessary to note that the analysis of MHD control efficiency applied to scramjet is limited by comparatively simple mathematical models (one-dimensional for the engine duct or 2-D inviscid flow for scramjet inlet). At the same time, it was shown earlier that role of viscous effects appears rather essential at the consideration of flows in conventional MHD generators.[18] The boundary layer separation is accompanied by an essential change in the flow structure and by significant losses in extracted electrical power. Therefore the necessity to develop more detailed and realistic mathematical models for the description of

MHD flows in the scramjet duct arises. These questions are of prime importance in the investigations of super- and hypersonic flow deceleration using MHD-control. This problem was studied in CIAM in the last few years.

Contents of the Present Article

The following will be discussed:

MHD Parameters Behind Normal Shock Wave

The main objective is to determine the conditions when the appreciable level of natural conductivity behind normal shock is reached. It is shown that a desirable effect takes place at large flight Mach numbers and at high altitudes only ($M > 15$, $H > 30$ km). All MHD dimensionless parameters behind shock wave are estimated.

MHD Equations System

Full Navier-Stokes equations with additional MHD terms and electrodynamic equations system in the case of small magnetic Reynolds numbers are written. Various MHD regimes, corresponding dimensionless parameters and parameters determining irreversible losses, asymptotic behavior of solution of 1-D MHD equations at $M \to \infty$, computational code for the abovementioned system of MHD equations are discussed in detail.

MHD Control on Boundary Layer

The relationship for the parameter of MHD boundary layer separation is found. The situations are considered with the magnetic field alone and when both the magnetic and electric fields influence the boundary layer. The possibility of effective MHD control on boundary layer is proved.

Deceleration of Supersonic Flow by Axisymmetric Magnetic Field

The main purpose is to evaluate total pressure losses in such flow. The complicated flow pattern (shock waves, boundary layer separation, recirculation zones) is analyzed with the aid of numerical solution of 2-D Navier-Stokes equations for laminar and turbulent flows. Total pressure losses for a wide range of MHD interaction parameters are calculated.

Deceleration of Supersonic Two-Dimensional Flow in MHD Generator Regime

Performances of deceleration, total pressure losses, and efficiency of MHD energy conversion are determined for laminar and turbulent flows.

Relative Value of MHD Effects in Hypersonic Airflows

Electroconductivity of Air and Dimensionless MHD Parameters behind a normal shock wave in a hypersonic flow

Behind strong normal shock wave, pressure p and enthalpy h are determined by following approximate formulas

$$p = \rho_H v_H^2, \qquad h = 1/2\, v_H^2 \tag{1}$$

Here ρ_H and v_H are density and velocity of air ahead of shock wave.

Values of p and h allow us to find all thermodynamic parameters behind shock wave at thermodynamic equilibrium with the aid of tabulated data for air (see, for example, Ref. 19).

The conductivity of air is determined by the following relationship

$$\sigma = \frac{e^2 n_e}{m_e \nu_{\text{eff}}}, \quad \nu_{\text{eff}} = c_e \sum_a n_a Q_{ea}, \quad ce = \left(\frac{8kT_e}{\pi m_e}\right)^{1/2} \tag{2}$$

Here e, m_e, n_e, T_e, and C_e are charge, mass, concentration, temperature, and "heat" velocity of electrons, Q_{ea} is electron collision cross section (including electron-ion collision cross-section Q_{ei}).

Air is assumed to be electrical neutral and concentrations of ions and electrons are equal.

Functions $Q_{ea}(T_e)$ and $Q_{ei}(T_e)$ are known.[20]

Formula (2) is written for the general case when temperature T_e of electron component and temperature T of heavy components are different.

First, the equilibrium case is considered ($T_e = T$). Then, electron concentration n_e is given in Ref. 19. Nevertheless, with the aim of further generalization to nonequilibrium case, it is desirable to obtain an analytical relationship for n_e. Analyses of ionization processes in air at temperatures $T \leq 6000$ K show that production of ions and electrons is, mainly, a result of an NO ionization. Then, using Saha formula for NO equilibrium ionization and known physical constants, the following relationship may be found

$$x_{\text{NO}^+}^2 = 0.333\, \frac{2g_{\text{NO}^+}}{g_{\text{NO}}} \frac{x_{\text{NO}} T}{p}\, T^{3/2} \exp\left(-\frac{11600\varphi_i}{T}\right) \tag{3}$$

$$x_{\text{NO}} = \frac{n_{\text{NO}}}{\sum_a n_a}, \quad x_{\text{NO}^+} = \frac{n_{\text{NO}^+}}{\sum_a n_a}, \quad p = kT\sum_a n_a, \quad x_{\text{NO}^+}$$

$$([T] = \text{K}, \;\; [\varphi_i] = \text{V}, \;\; k = 1.38 \times 10^{-16}\ \text{Erg/K}, \;\; [n] = \text{cm}^3)$$

Here a is number of component, n_a and X_a are numerical and molar concentrations of a-component, p is pressure of air, g_{NO} and g_{NO^+} are inner statistic sums for NO and NO^+, X_e is electrons concentration, which is equal to concentration of NO^+. The following values were used in calculations [20, 21]

$$\frac{g_{NO^+}}{g_{NO}} = 0.185, \qquad \varphi_i = 9.5\,\mathrm{V}$$

The physical model used is an approximate one. It is assumed that NO gives the main contribution to ionization process, twice ionization and formation of negative ions are not accounted for. Nevertheless, the calculations of electron molar concentration on the base of (3) and the tabulated data[19] coincide with high accuracy up to $T = 5000$ K.

Gasdynamic parameters, electroconductivity, MHD interaction parameter S and Hall parameter β behind strong normal shock wave were calculated with the aid of (1)–(3). Parameters S and β are determined as follows:

$$S = \frac{\sigma B^2 L}{\rho_H v_H}, \qquad \beta = \frac{eB}{m_e v_{\mathrm{eff}}} \tag{4}$$

Estimates of these parameters are calculated when characteristics length L and magnetic field B are equal to 10 cm and 10^4 Gs, correspondingly.

Some results of performed calculations are presented in Table 1.

Data in Table 1 were obtained for possible characteristic trajectory points of hypersonic vehicle with Scramjet. Two parameters—altitude H and Mach number M of vehicle flight—determine all gasdynamic values ahead shock wave. Values behind shock wave were calculated with the aid of (1–3).

As evident from Table 1, air conductivity and MHD interaction parameter S are very small for $M \approx 9$; therefore, magnetic field influence on gasdynamic flow is impossible. Air conductivity for $M \approx 14$ is much larger and the value of MHD interaction parameter becomes sufficient for MHD effects.

Therefore, special methods are necessary for conductivity enhancement with the aim of MHD control realization.

Evaluation of Capabilities of Conductivity Increase in Pure Air

Electron concentration in air, when there is electric field E^* (in moving coordinate system) and therefore electron temperature T_e is larger than temperature T of heavy components, will be obtained under assumptions that only component NO is ionized and molar electron concentration $x_e = x_{NO^+}$ remains so small that value of x_{NO} (under given air pressure p and temperature T) is the same one as for equilibrium ionization. Modified Saha formula (3) in the case $T_e \neq T$ is written as

$$x_{NO^+}^2 = 0.123\,\frac{x_{NO}T}{p}\,T_e^{3/2}\exp\left(-\frac{11600\varphi_i}{T_e}\right) \tag{5}$$

Table 1

Head	Head	$H = 36.3$ km, $M = 9.7$	$H = 42.0$ km, $M = 14$	$H = 32.1$ km, $M = 9.9$	$H = 37.3$ km, $M = 14.4$
		Equilibrium conductivity			
Temperature, K	T	3100	4900	3163	4981
Pressure, N/m^2	p	6×10^4	6×10^4	1.2×10^5	1.2×10^5
Collision frequency of electrons, s^{-1}	ν_{eff}	2.98×10^{10}	3.34×10^{10}	5.78×10^{10}	6.57×10^{10}
Electron concentration, cm^{-3}	n_e	5.9×10^{10}	2.71×10^{13}	1.24×10^{11}	5.27×10^{13}
Conductivity, mo/cm	σ	5.6×10^{-4}	0.230	6.04×10^{-4}	0.226
MHD interaction parameter $S = \sigma B^2 L/(\rho_H \upsilon_H)$ ($B = 10^4$ Gs, $L = 10$ cm)	S	2.8×10^{-4}	0.17	1.5×10^{-4}	8.5×10^{-2}
Hall parameter $\beta = eB/(m_e \nu_{eff})$	β	5.9	5.3	3.04	2.68
		Nonequilibrium conductivity ($E^* = 10$ V/cm)			
Temperature, K	T_e	—	—	4550	—
Collision frequency of electrons, s^{-1}	ν^*_{eff}	—	—	7.47×10^{10}	—
Electron concentration, cm^{-3}	n^*_e	—	—	3.32×10^{13}	—
Conductivity, mo/cm	σ^*	—	—	0.125	—
MHD interaction parameter	S^*	—	—	3.14×10^{-2}	—
Hall parameter	β^*	—	—	2.35	—

Estimates of the electron temperature increase in the presence of electric field $\boldsymbol{E}^* = \boldsymbol{E} + \boldsymbol{v} \times \boldsymbol{B}$ may be done on the basis of energy equation for electron gas in which there are only two terms: energy input into electron gas and energy losses as a result of electron collisions with other particles. This approximate equation has the following form:

$$\boldsymbol{j}\boldsymbol{E}^* = 3n_e m_e k(T_e - T)\sum_a \left(\nu_{ea}\frac{\delta_a}{m_a}\right) \tag{6}$$

Here $\boldsymbol{j}$ is electric current density, ν_{ea} is frequency of electron collisions with particles of a-component (including electron-ion collisions), $\delta_a > 1$ is coefficient of nonelastic electron collisions losses [in general case, $\delta_a = \delta_a(T_e)$], m_a is mass of a-particle.

Estimate of temperature T_e will be performed further without Hall effect. In this case

$$\boldsymbol{j} = \sigma\boldsymbol{E}^*, \quad \sigma = \frac{e^2 n_e}{m_e \nu_{\text{eff}}}, \quad \nu_{\text{eff}} = c_e\sum_a n_a Q_{ea}, \quad c_e = \left(\frac{8kT_e}{\pi m_e}\right)^{1/2}, \tag{7}$$
$$Q_{ea} = Q_{ea}(T_e)$$

Substitution of Eq. (7) into Eq. (6) results in the following equation:

$$T_e = T + \beta, \quad \beta = \frac{1}{3k}\left(\frac{eE^*}{m_e c_e}\right)^2\left[\left(\sum_a n_a Q_{ea}\right)\left(\sum_a \delta_a Q_{ea}\frac{n_a}{m_a}\right)\right]^{-1} \tag{8}$$

Concentration n_{NO^+} is determined with the aid of Eq. (5) and consequently is a function on unknown variable T_e. Other concentrations n_a are found under given p and T using tabulated data[19] for equilibrium state of air. Values of δ_a are estimated following Ref. 20. Therefore, relationship (8) is an equation for the determination of T_e (quantities p, T and E^* are given parameters). Solution of Eq. (8) has been performed using iteration procedure.

Calculations were carried out under flight conditions $H = 32.1$ km, $M = 9.9$, and $\boldsymbol{E}^* = 10$ V/cm. Results are presented in Table 1. It is evident from this table that the nonequilibrium effect enhances conductivity and MHD interaction parameter appreciably. However, the obtained value of nonequilibrium effect is overestimated. (In particular, in energy equation for electrons, convection and diffusion terms are not accounted for.)

Equations of Magnetic Gas Dynamics at Small Magnetic Reynolds Numbers. Main Parameters. Methods of Numerical Analysis.

Equations of Magnetic Gasdynamics and Main Dimensionless Parameters

The used system of MHD equations includes the equation of continuity, the equation of momentum containing the electromagnetic force $\boldsymbol{f} = \boldsymbol{j} \times \boldsymbol{B}$, the equation of energy containing the term $q = \boldsymbol{j}\boldsymbol{E}$, the equation of state (all above-mentioned equations will be written later) and the following electromagnetic equations (in the stationary case) (see Ref. 13)

$$\boldsymbol{j} = \sigma(-\nabla\varphi + \boldsymbol{v} \times \boldsymbol{B}) - a\boldsymbol{j} \times \boldsymbol{B}, \quad \operatorname{div} \boldsymbol{j} = 0, \quad \boldsymbol{E} = -\nabla\varphi \qquad (9)$$
$$\operatorname{div} \boldsymbol{B} = 0, \quad \operatorname{rot} \boldsymbol{B} = 0 \quad [a = e/m_e \nu_{\text{eff}}]$$

In the first equation of (9), the Hall effect for electrons only is accounted for.

Here $\boldsymbol{j}$, $\boldsymbol{B}$, and $\boldsymbol{E}$ are the vectors of electric current, magnetic and electric fields; σ and φ are electric conductivity and electric potential, a is the Hall parameter for electrons divided by $|\boldsymbol{B}|$ [see Eqs. (2) and (4)], y is the gas velocity. The induced magnetic field is supposed to be insignificant (see Table 1), the external current sources are arranged outside the flow. In this case, $\boldsymbol{B}$ is the known external magnetic field. Equations (9) intended for determination of quantities $\boldsymbol{j}$ and φ must be solved simultaneously with gasdynamic equations.

Gasdynamic equations for two-dimensional and axisymmetrical flows can be written in the form

$$\frac{\partial \boldsymbol{U}}{\partial t} + \frac{\partial \boldsymbol{F}}{\partial x} + \frac{\partial \boldsymbol{G}}{\partial y} + \frac{\boldsymbol{R}}{y} = \boldsymbol{Q} \qquad (10)$$

$$\boldsymbol{F} = \begin{pmatrix} \rho u \\ \rho u^2 + p - \tau_{xx} \\ \rho u v - \tau_{xy} \\ \rho u h^* - u\tau_{xx} - v\tau_{xy} + q_{hx} \end{pmatrix}, \quad \boldsymbol{G} = \begin{pmatrix} \rho v \\ \rho u v - \tau_{yx} \\ \rho v^2 + p - \tau_{yy} \\ \rho v h^* - u\tau_{yx} - v\tau_{yy} + q_{hy} \end{pmatrix}$$

$$\boldsymbol{Q} = \begin{pmatrix} 0 \\ f_x \\ f_y \\ q \end{pmatrix}, \quad \boldsymbol{R} = \begin{pmatrix} \rho v \\ \rho u v - \tau_{yx} \\ \rho v^2 + \tau_{yy} + \tau_{\theta\theta} \\ \rho v h^* - u\tau_{yx} - v\tau_{yy} + q_{hy} \end{pmatrix}, \boldsymbol{U} = \begin{pmatrix} \rho \\ \rho u \\ \rho v \\ \rho e^* \end{pmatrix} \qquad (11)$$

$$h^* = h + \frac{1}{2}(u^2 + v^2), \quad e^* = e + \frac{1}{2}(u^2 + v^2) \qquad (12)$$

$$q_{hx} = -\frac{\mu}{\Pr}\frac{\partial h}{\partial x}, \quad q_{hy} = -\frac{\mu}{\Pr}\frac{\partial h}{\partial y} \qquad (13)$$

Here $\boldsymbol{U}$ is a vector of conservative variables, $\boldsymbol{F}$ and $\boldsymbol{G}$ are generalized fluxes including viscous and heat terms (τ_{ik} are the components of viscous stresses tensor), $\boldsymbol{F}$, $\boldsymbol{G}$ and $\boldsymbol{R}$ are functions of $\boldsymbol{U}$.

In these equations, x and y are axial and radial coordinates, ρ, p, e, and h are density, pressure, internal energy and enthalpy of gas, u and v are axial and radial velocities, q_{hx} and q_{hy} are axial and radial components of heat flux vector $\boldsymbol{q}_h$, Pr is Prandtl number.

Tensor of viscous stresses and tensor of deformation velocities are connected by a well-known linear relation with a proportionality coefficient that is equal to coefficient of dynamic viscosity μ.

In the case of turbulent flows, after averaging of full Navier-Stokes equations and using Boussinesq hypothesis, the equations transform to the same form as Eqs. (10–13) but, in tensor of viscous stresses, the coefficient μ is replaced by $\mu_e = \mu + \rho v_t$ and the coefficient μ/Pr in relation (13) is replaced by $(\mu/Pr + \rho v_t/Pr_t)$, where Pr_t is turbulent Prandtl number. Various turbulent models must be used for v_t determination. The turbulent model [22] was used in our calculations

Equations (10–13) are closed by the equation of state. In the simplest case, this equation is

$$h = \frac{\gamma}{\gamma - 1}\frac{p}{\rho}, \quad p = \rho \Re T; \quad \gamma = \text{const}, \quad \Re = \text{const} \tag{14}$$

Here γ is the ratio of specific heat capacities, $\Re$ is the gas constant.

Transport coefficients σ, μ and others are given functions of any two thermodynamic parameters.

Equations (10–14) are written in the form that is very convenient for numerical solution.

Some remarks concerning electrodynamic Eqs. (9) may be made.

If the first equation of Eq. (9) is solved with respect to vector $\boldsymbol{j}$ components, the direct linear relationship between these components and the electric potential derivatives is obtained. With the aid of the second equation of Eq. (9), the elliptic equation for potential φ is found. The boundary conditions for the last equation are: $\varphi = \text{const}$ on the electrodes, the normal component $j_n = 0$ on the nonconductive surfaces, $j_n = j_n(\Sigma)$ or $\varphi = \varphi(\Sigma)$ are the given functions on the wall with the sectional electrodes (Σ consists of wall points). The conditions at the entrance and exit channel cross-sections must reflect the features of considered MHD devices.

Without the Hall effect ($\beta = 0$) the equation in φ is

$$\Delta\varphi + (\nabla \ln\sigma)\nabla\varphi = (\nabla \ln\sigma)\,(\mathbf{v} \times \mathbf{B}) + \mathbf{B}\ \text{rot}\ \mathbf{v} \tag{15}$$

For the case when $\sigma = \text{const}$ the following equation is obtained:

$$\Delta\varphi = \mathbf{B}\ \text{rot}\ \mathbf{v} \tag{16}$$

A set of dimensionless parameters for the overall MHD equation system (9–14) includes traditional gasdynamic parameters (Mach and Reynolds numbers, etc.) and the new (MHD) parameters:

$$S = \frac{\sigma B^2 L}{\rho u}, \quad K = \frac{\delta\varphi}{uBL}, \quad \beta = \frac{eB}{m_e v_{\text{eff}}} \tag{17}$$

In these parameters the characteristic values are used, L and $\delta\varphi$ are the characteristic length and electric voltage, β is the Hall parameter, S is the MHD interaction parameter, K is the load parameter. The well-known parameter—magnetic Reynolds number—is small and it is not introduced into the set (17). Another well-known parameter—Hartmann number—is expressed as

$$Ha^2 = \frac{\sigma B^2 L^2}{\mu} = S \cdot Re\,, \quad Re = \frac{\rho u L}{\mu} \tag{18}$$

Parameters Describing Irreversible Losses in MHD Flows

The MHD flow in each its point may be realized in regime with extraction of electric power (generator regime, $q = \boldsymbol{jE} < 0$) or with input of electric power (accelerator regime, $q > 0$). The generator and accelerator regimes for the whole MHD device are determined by conditions:

$$\int_D q\,\mathrm{d}D < 0 \text{ and } \int_D q\,\mathrm{d}D > 0 \tag{19}$$

Here D is MHD device volume.

The main integral characteristics of MHD devices are introduced with the aid of the following relationships:

$$\int_D \boldsymbol{jE}\,\mathrm{d}D = \int_D \boldsymbol{j}\left(\frac{\boldsymbol{j}}{\sigma} - \boldsymbol{v} \times \boldsymbol{B} + \frac{a}{\sigma}\,\boldsymbol{j} \times \boldsymbol{B}\right)\mathrm{d}D = Q - A$$

$$Q = \int_D \frac{j^2}{\sigma}\,\mathrm{d}D, \quad A = -\int_D \boldsymbol{v}\,(\boldsymbol{j} \times \boldsymbol{B})\,\mathrm{d}D \tag{20}$$

Here Q is the Joule dissipation in D region (Q is always positive) and A is the gas work (per second) in magnetic field.

According to the relationship

$$\int_D \boldsymbol{jE}\,\mathrm{d}D = -\int_D \mathrm{div}\,(\varphi \boldsymbol{j})\,\mathrm{d}D = -N, \quad N = \oint_\Sigma \varphi j_n \mathrm{d}\Sigma \tag{21}$$

where N is the power extracted from the MHD device and Σ is the closed surface of volume D, the following efficiencies may be introduced:

$$\eta = \frac{N}{A} = \frac{N}{N+Q} \text{ (generator regime: } A = N + Q, N > 0, A > 0)$$

$$\eta = \frac{(-A)}{(-N)} = \frac{(-A)}{Q+(-A)} \Big[\text{accelerator regime:}$$

$$(-N) = Q + (-A), N < 0, A < 0\Big] \tag{22}$$

Two methods will be used to estimate irreversible losses in MHD flow. The first method is related to determination of losses of total pressure p^*. In the case of 3-D flow, the pressure p^* should be averaged over the cross-section of a channel. Total pressure losses will be determined by the ratio $\sigma' = p_e^*/p_0^*$, where p_e^* is the total pressure at exit cross-section of a channel of the equivalent one-dimensional flow, which is characterized by the same values of flow rate, total enthalpy flux and longitudinal impulse flux, as ones in the real flow; p_0^* is the total pressure average by the same way at the channel entrance.

The irreversible losses in the MHD channel can also be estimated from the change of entropy. The equation for entropy s (another form of the energy equation) in the stationary case is as follows:

$$\operatorname{div} \rho \mathbf{v} s = \frac{1}{T}(-\operatorname{div} \boldsymbol{q}_h + \tau_{1k}\frac{\partial v_i}{\partial x_k} + \frac{\boldsymbol{j}}{\sigma}) \tag{23}$$

The summation is performed over twice-repeating subscripts (i, $k = 1, 2, 3$). The second term in round parentheses represents a viscous dissipation, the third one represents the Joule dissipation. Both these terms are positive. Letting $\boldsymbol{q}_h = -\lambda\nabla T$ (λ is the heat conductivity coefficient) and integrating Eq. (23) over the closed region D with surface Σ and external normal $\boldsymbol{n}$, the following integral relation is obtained:

$$\oint_{\Sigma} \rho v_n s \, d\Sigma = -Q_h + \Gamma,$$

$$Q_h = \oint_{\Sigma} \frac{(\boldsymbol{q}_h)_n}{T} d\Sigma, \quad \Gamma = \int_D \left[\frac{\lambda(\nabla T)^2}{T^2} + \frac{1}{T}\tau_{ik}\frac{\partial v_i}{\partial x_k} + \frac{j^2}{\sigma T}\right] dD > 0 \tag{24}$$

For the channel with non-perforated walls one can find from Eq. (24):

$$G[< s_e > - < s_0 >] = -Q_h + \Gamma, \quad < s > = G^{-1}\int_F \rho u s \, dF \tag{25}$$

Here subscripts e and 0 correspond to exit and entrance cross-sections of the channel, respectively; F is the area of channel cross-section, G is the mass flow rate of gas. According to Eq. (25), the variation of the entropy, averaged over mass flow rate, consists of the component $(-Q_h)$, which may be either positive or negative, and the irreversible positive component Γ.

When calculations are carried out on the base of Euler equations, then $Q_h = 0$. The left-hand side of relation (25) and the third term in expression for Γ are calculated directly. This, in turn, allows to find the sum of first and second terms in Γ. This sum represents irreversible losses in shock waves in this case.

MHD Deceleration of a Hypersonic Flow in One-Dimensional Approach

In many cases, a one-dimensional approach is a very useful instrument for analysis of MHD flow peculiarities. The one-dimensional flow is considered below when the perfect gas with constant specific heat capacities c_p and c_v moves in the presence of following magnetic and electric fields:

$$\mathbf{B} = [0, 0, B(x)], \quad \mathbf{E} = [0, E(x), 0]; [B > 0, \quad E > 0 \tag{26}$$

$B(x)$ and $E(x)$ are known functions. This model and its applications were discussed in Refs. 13 and 23. The governing equations for isotropic electric conductivity σ are:

$$\rho u F = \text{const}$$

$$\rho u u' + p' = \tau_{\text{w}} + \sigma(E - uB)B$$

$$\rho u(c_{\text{p}} T' + u u') = -q_{\text{w}} + \sigma(E - uB)E$$

$$p = \rho \Re T \tag{27}$$

Here u is the longitudinal velocity, M is the Mach number, F is cross-section area, τ_w is the projection of wall shear stress vector on the x axis, q_w is the density of wall heat flux extracted from the channel, $\Re$ is the gas constant. Index "$'$" indicates the x derivative.

Distributions of F, τ_w, q_w, E, and B along the channel may be considered as control functions. Then, the gasdynamic variables p, ρ T, and u may be found from the system (27).

When τ_w and q_w do not contain the derivatives of gasdynamic variables with respect to x, the equations for u', p', T', etc. may be obtained [13,23] from the system (27). These equations determine the qualitative influence of control functions on the gas-dynamic values, i.e., to find the conditions when the flow accelerates or decelerates in each channel cross-section. It is significant that the right parts of these equations are proportional to the term $(M^2 - 1)^{-1}$ and, consequently, the influence of control functions changes its sign at sonic velocity transition. Therefore, the continuous sonic velocity transition is possible only when the sign of total influence in the right parts of these equations is changed at the point $M=1$.

The simpler case corresponds to the situation when $q_w=0$, $\tau_w=0$ and $F=$ const. The most interest in this case is connected with asymptotic behavior of flow under large Mach numbers. The one-dimensional equations for $M \to \infty$ are written as:

$$\frac{u'}{u} = -(\gamma - 1)S(1 - K)\left(\frac{\gamma}{\gamma - 1} - K\right)$$

$$\frac{T'}{T} = (\gamma - 1)\Lambda, \quad \frac{p'}{p} = (\gamma - 1)\Lambda, \quad \frac{M'}{M} = -\frac{\gamma - 1}{2}\Lambda, \quad \frac{(p^*)'}{p^*} = -\Lambda$$

$$\frac{s'}{c_v} = (\gamma - 1)\Lambda, \quad \frac{(T^*)'}{T^*} = -2SK(1 - K)$$

$$\Lambda = \gamma M^2 S(K - 1)^2 > 0, \quad S = \frac{\sigma B^2 h}{\rho u}, \quad K = \frac{E}{uB} \tag{28}$$

Here h is the channel height, K is the local load parameter, S is the local MHD interaction parameter; p^* and T^* are the total pressure and temperature, $\gamma = c_p/c_v$; longitudinal coordinate x is changed by x/h.

From Eq. (28) the following significant result is obtained:

$$\frac{T'}{T} = \frac{p'}{p} = \frac{s'}{c_v} = -(\gamma - 1)\frac{(p^*)'}{p^*} \tag{29}$$

The result of integration of Eq. (29) is

$$\frac{p_0^*}{p^*} = \left(\frac{p}{p_0}\right)^{\frac{1}{\gamma - 1}} \tag{30}$$

According to Eq. (30), for the flow with large Mach numbers, the relative decrease of total pressure is greater than the relative increase of the static pressure, i.e., gasdynamic losses grow more rapidly than static pressure grows.

One should emphasize that the asymptotic relation (30) is valid for arbitrary distributions of electric and magnetic fields in a channel.

Consider now specific examples of hypersonic flow deceleration in constant electric and magnetic fields. Note that, depending on the initial conditions, flow with input of electric power ($K > 1$) and flow with extraction of electric power ($K < 1$) may be realized. The results of some calculations are presented in Tables 2 and 3.

Here K_0 is the load parameter at the initial cross section of a channel, $X = \int_0^x (\sigma B^2/\rho u)dx$ is the dimensionless length, on which the deceleration from M_0 to M_e takes place. Tables 2 and 3 are presented for the flows with electric power output and input, respectively. Of special interest is the regime of Table 3. In this case, the input of electric power is mainly transformed into the heat delivered to a flow. Just this heat provides the supersonic flow deceleration.

Numerical Method for Solution of MHD Equation System

The aforementioned system of equations for inviscid (Euler equations) and viscous flows (full Navier-Stokes equations averaged in turbulent case) was solved numerically using FNAS2D code.[24] This code is based on the time relaxation procedure for the modified implicit version[25] of Godunov scheme.[26] The modified scheme provides the second order accuracy of the steady state solution on regular, near uniform grid and approximation on arbitrary nonuniform grid. The presented finite volume method is realized on the grid constructed in physical space without transformation to the computational space. The scheme is fully implicit. This means that parameters on the new time level are used to approximate the convective and viscous fluxes through cell boundaries. The implicit approximation was used also for the sources terms [$\boldsymbol{Q}$ and $\boldsymbol{R}$ terms in Eqs. (10) and (11)].

The system of conservation equations is written for each cell using increments in time[27] of main dependent variables: density, velocity components, pressure, turbulent viscosity, etc. The arbitrary discontinuity breakdown problem (Riemann problem) is used in modified form to include it into the implicit scheme. The Riemann problem solution for the new time level is

Table 2 $M_e = 1.3, K_0 = 0.4$

M_0	X	P_e/P_0	u_e/U	T_e/T_0	σ'
5	1.17	7.2	0.484	3.5	0.0375
10	1.78	25.5	0.431	11.0	0.00166

Table 3 $M_e = 1.3, K_0 = 4$

M_0	X	P_e/P_0	u_e/U	T_e/T_0	σ'
5	0.082	13.3	0.898	11.9	0.070
10	0.087	52.9	0.893	47.2	0.0035

considered in linear approach. It is supposed that the configuration of discontinuities realized after breakdown on the new time level is identical to this one obtained on the old time level. Therefore, identical relations exist between parameters on the cell boundaries and parameters in the neighboring cells before breakdown on the old time level and between parameters increments in time on the boundaries and in the cells centers. If the pressures ratio on the shock and/or on the expansion wave is a large value, then the iterative procedure is used to obtain the accurate solution of the nonlinear Riemann problem on the old time level. These corrected values are used for convective fluxes approximation on cell boundaries on the old time level. The linear problem solution is not corrected for parameters increments in time.

To provide the higher order accuracy for the steady state solutions the piecewise linear parameters distributions within the cell are assumed and the parameters in the middle point of each cell edge are estimated for adjacent cells on the old time level. These parameters are used as the initial data for the Riemann problem. These initial data are evaluated with the aid of a minimal derivatives principle[28] modified to arbitrary nonregular grids in Ref. 29. The minimal derivatives or minimal space increments principles provide the monotonicity condition[30] of the higher order accuracy scheme. As to the parameters increments in time, their values in cell centers are chosen as the initial data for the Riemann problem. It is acceptable without loss of steady state solution accuracy.

The viscous stresses and thermal fluxes on the cell boundaries on the old time level are approximated with the aid of central differences generalization appropriate for arbitrary grids. The contribution of viscous terms into finite difference operator on the new time level is taken into account approximately because the contribution of some grid points is dropped to provide the five-diagonal structure of the algebraic system of equations for increments of the aforementioned dependent variables in time.

The resulting system of linear algebraic equations for the main parameters increments in time is solved using an iteration method (internal iterations on each time step). Sequential downstream and upstream sweeping in Gauss-Seidel method are performed for each time step. The number of internal iterations (in Gauss-Seidel method) on each time step is chosen depending on the global residual level and Courant number.

The possibility to use the adapted grid is introduced into this code. The grid adaptation is based on the so-called spring analogy.[31] If one had to choose some parameter for adaptation, this principle will provide construction of the grid with almost uniform distribution of parameter increment on neighborhood cells.

Boundary-Layer Separation Parameter in Magnetogasdynamics

Parameter of Boundary-Layer Separation in the Case of Nonconducting Wall

The deceleration of gasdynamic flow by magnetic field is very often accompanied by boundary layer separation. In gasdynamics, the prediction of separation regime is usually carried out by introducing the separation parameter ξ_{cr} that is a value of dimensionless parameter $\xi = p'\delta^*/\rho_0 u_0^2$ at the separation cross-section. Here p' is a pressure gradient in an external flow, δ^* is a boundary layer displacement thickness, ρ_0 and u_0 are density and velocity on a boundary of a boundary layer. All listed variables are functions of coordinate x counted along a surface. Parameter ξ_{cr} is determined with the aid of experimental and theoretical investigations using the hypothesis that the flow in the cross-section of a boundary layer separation depends only on the flow in the nearest vicinity of this cross-section (hypothesis of local influence).[32] Therefore, the parameter ξ_{cr} has universal character for each of boundary layer classes. If $\xi < \xi_{cr}$, the boundary layer separation does not happen.

The information on the separation parameter ξ_{cr} allows, on the one hand, one to predict the boundary layer separation and, on the other hand, to realize non-separated flows. In gas dynamics the turbulent boundary layer separation parameters were determined in papers.[32,33] Quantity ξ_{cr} for a laminar MHD boundary layer was found in Ref. 14. The extension of this result to the MHD turbulent boundary layer of compressible gas is given below.

Two-dimensional MHD boundary layer on a nonconducting surface is considered when the magnetic field lies in a plane of flow and the electrical field is perpendicular to this plane. It is assumed that the boundary layer thickness δ is less than the characteristic length of change of the external magnetic field, the induced magnetic field is insignificant, and the Hall effect is absent. Then longitudinal MHD force in a boundary layer is written as

$$f_x = \sigma B(E - uB) \tag{31}$$

where $B = B(x)$ is an external magnetic field component perpendicular to a surface, $E = E(x)$ is an electrical field perpendicular to a flow plane. $B(x)$ and $E(x)$ are calculated in points of a surface and are the given functions.

The momentum equation in x direction has a form

$$I \equiv \rho u \frac{\partial u}{\partial x} + \rho v \frac{\partial u}{\partial y} + p' - \frac{\partial \tau}{\partial y} - f_x = 0 \tag{32}$$

$$\tau = \tau_l + \tau_t \tag{33}$$

$$\tau_l = \mu \frac{\partial u}{\partial y}, \quad \tau_t = \rho l^2 \left(\frac{\partial u}{\partial y}\right)^2 \tag{34}$$

Here y is a coordinate perpendicular to a surface, τ is a friction stress including laminar τ_l and turbulent τ_t components, l is a mixing path length. For τ_t, the simplest approximation[34] is used.

The complete system of equations of MHD boundary layer is not presented here for the sake of brevity.

Determination of ξ_{cr} is performed with the aid of an approximate expression for a profile of τ at the cross-section of a boundary layer separation, where, by definition, $\tau = 0$ at $y = 0$. The τ-profile is approximated by a fourth-power polynomial. The following conditions serve to obtain the polynomial coefficients

$$x = x_s, y = 0: \quad \tau = 0, \quad \frac{\partial \tau}{\partial y} = p' - \sigma BE, \quad \frac{\partial^2 \tau}{\partial y^2} = -BE\frac{\partial \sigma}{\partial y} \tag{35}$$

$$\frac{\partial^3 \tau}{\partial y^3} = \frac{\sigma B^2}{\mu}(p' - \sigma BE) - BE\frac{\partial^2 \sigma}{\partial y^2}$$

$$x = x_s, \quad y = \delta: s = 0 \tag{36}$$

Here δ is a boundary layer thickness.

For determination of second, third, and fourth conditions in Eq. (35), the relationships (33), (34) and the relationships $I = 0$, $\partial I/\partial y = 0$, $\partial^2 I/\partial y^2 = 0$, taken in a point $x = x_s$, $y = 0$, were used. Expression (36) is a usual asymptotic condition on the external boundary of a boundary layer.

The obtained expression for a polynomial should be equated with τ. It is essential, that the laminar friction can be neglected at the cross-section of a boundary layer separation: τ_l is zero at the point $x = x_s, y = 0$, it is small near a wall, and always $\tau_l << \tau_t$ far from a wall. Therefore, $\tau \approx \tau_t$.

The mixing path length in the separation cross-section can be approximately represented by the formula

$$l = \delta m\psi(S), \quad S = \frac{\sigma_w B^2 \delta^*}{\rho_0 u_0}, \quad m = \text{const}, \quad \psi(0) = 1 \tag{37}$$

Here S is an MHD interaction parameter, the function $\psi(S)$ takes into account possible influence of magnetic field on a turbulence. The subscripts indexes 0 and w correspond to parameters on the boundary of a boundary layer and on a surface, respectively.

Taking into account the aforesaid, we receive the following differential equation:

$$\bar{\rho}\left(\frac{\partial \bar{u}}{\partial \bar{y}}\right)^2 = F(\bar{y}, \zeta, H^2, \beta, \beta_1), \quad \bar{u}(0) = 0, \quad \bar{u}(1) = 1$$

$$F = \zeta\left[\bar{y} - \bar{y}^4 + H^2(\bar{y}^3 - \bar{y}^4)\right] - \beta(\bar{y}^2 - \bar{y}^4) - \beta_1(\bar{y}^3 - \bar{y}^4) \tag{38}$$

$$\zeta = \frac{(p' - \sigma_w BE)\delta}{\rho_0 u_0^2 m^2 \psi^2}, \quad \beta = \frac{BE\delta^2}{2\rho_0 u_0^2 m^2 \psi^2}\left(\frac{\partial \sigma}{\partial y}\right)_w$$

$$H^2 = \frac{\sigma_w B^2 \delta^2}{6\mu}, \quad \beta_1 = \frac{BE\delta^3}{6\rho_0 u_0^2 m^2 \psi^2}\left(\frac{\partial^2 \sigma}{\partial y^2}\right)_w$$

$$\overline{y} = \frac{y}{\delta}, \overline{\rho} = \frac{\rho}{\rho_0}, \quad \overline{u} = \frac{u}{u_0} \tag{39}$$

Here H is proportional to Hartmann number, the parameters β and β_1 reflect the influence of MHD force (factor BE) on the flow and effect of conductivity change on a wall. The value of $(-\,p' + \sigma_w BE)$ (see parameter ζ) represents a sum of projections on χ axis, both pressure force and electromagnetic force. At $B = 0$, a boundary layer separation is caused by increase of pressure. The force $\sigma_w BE$, depending of B and E orientations, allows control of boundary layer separation.

For determination of η profile at the cross-section of a boundary layer separation, it is necessary to invoke an energy equation. However, in the first approximation, it is assumed that velocity u and enthalpy h are related by Crocco integral

$$\overline{h} = \overline{h}(\overline{u}) = \overline{h}_w\,(1 + a) + \overline{u}(1 - \overline{h}_w)(1 + a) - a\overline{u}^2$$

$$\overline{h} = \frac{h}{h_0}, \quad \overline{h}_w = \frac{h_w}{h_0 + 1/2\,u_0^2}, \quad a = \frac{u_0^2}{2h_0} \tag{40}$$

[This integral is exactly executed only at $p' = 0, B = 0, E = 0, h(x,0) = h_w = \text{const}$ and at laminar and turbulent Prandtl numbers equal to unit. It is known that the use of Crocco integral in traditional gasdynamics is possible also at the violation of indicated conditions, especially when hereinafter the problem is decided by an integral method. The special calculations have shown that at the use of an integral method in magnetohydrodynamics, the introduction of relation (40) between h and u is quite allowed.]

As the pressure across a boundary layer does not vary, that $\rho = \rho(\overline{h}) = \rho[\overline{h}(\overline{u})] = \rho(\overline{u})$, and the Eq. (38) is transformed into the following ordinary differential equation for determination of a function $\overline{u}(\overline{y})$

$$\frac{d\overline{u}}{d\overline{y}} = \sqrt{\frac{F(\overline{y}; \zeta, H^2, \beta, \beta_1}{\overline{\rho}(\overline{u}; a, \overline{h}_w)}} \tag{41}$$

Two boundary conditions in Eq. (38) find the function $\overline{u}(\overline{y})$ and the parameter ζ, as functions of $a, \overline{h}_w, H^2, \beta$ and β_1. Also, the value δ^*/δ becomes a function of the same parameters. As a result, the following functional relationship is obtained

$$\xi_{cr} = SK + m^2\psi^2(S)\Re\,(a, \overline{h}_w, H^{*2}, \beta^*, \beta_1^*)$$

$$S = \frac{\sigma_w B^2 \delta^*}{\rho_0 u_0}, \quad K = \frac{E}{u_0 B}, \quad H^{*2} = \frac{\sigma_w B^2 \delta^*}{\mu}$$

$$\beta^* = \beta\left(\frac{\delta^*}{\delta}\right)^2, \quad \beta_1^* = \beta_1\left(\frac{\delta^*}{\delta}\right)^3 \tag{42}$$

Quantities β^* and β_1^* are received from β and β_1 by replacement δ on δ^*.

The value of m is found from the condition, that for incompressible fluid, at the absence of electric and magnetic fields and of heat transfer ($a = 0, \overline{h}_w = 1, H^* = 0, \beta^* = 0, \beta_1^* = 0$), parameter $\xi_{cr} = 0{,}015$. Using this circumstance, we obtain the final formula

$$\xi_{\mathrm{cr}} = SK + \psi^2(S)\Gamma(a, \overline{h}_w, H^{*2}, \beta^*, \beta_1^*)$$

$$\Gamma = 0{,}015\frac{\Re(a, \overline{h}_w, H^{*2}, \beta^*, \beta_1^*}{\Re(0{,}1{,}0{,}0{,}0)}$$

$$\left(m^2 = \frac{0{,}015}{\Re(0{,}1{,}0{,}0{,}0} = 0{,}0077\right) \tag{43}$$

Various particular cases of formula (43) are considered below.

1) Magnetic field is absent ($B = 0$). It is an ordinary gasdynamic case:

$$\xi_{\mathrm{cr}} = \Gamma\ (a, \overline{h}_w, 0, 0, 0) = \xi_{\mathrm{cr}}\ (a, \overline{h}_{\mathrm{w}})$$

Function $\xi_{\mathrm{cr}}\ (a, \overline{h}_w)$ has been determined in Ref. 35.

Parameter a in Eq. (43) accounts for an influence of gas compressibility on ξ_{cr}. [In the case of the perfect gas with constant heat capacities, $a = 1/2\ (\gamma - 1)\ M_0^2$.] Quantity $\overline{h}_w$ ($0 < \overline{h}_w < 1$) in Eq. (43) is connected with heat transfer (wall heat flux is absent at $\overline{h}_w = 1$ and increases when $\overline{h}_w$ diminishes). Parameter ξ_{cr} increases and boundary layer separation is delayed when a and $\overline{h}_w$ decrease.

2) Electric field is absent, but magnetic field is applied to the flow. When σ = const and influence of magnetic field on turbulence is unsubstantial [$\psi^2(S) = 1$], separation parameter becomes

$$\xi_{\mathrm{cr}} = \Gamma\ (a, \overline{h}_w, H^{*2}, 0, 0) = Q(H^{*2}; a, \overline{h}_{\mathrm{w}}) \tag{44}$$

The function Q is presented in Fig. 1. Curves 1, 2, 3 correspond to values $M_0 = 0, \overline{h}_w = 0.1$; $M_0 = 8.165, \overline{h}_w = 0.1$; $M_0 = 8.165, \overline{h}_w = 1$, respectively. (Here $\gamma = 1.3$.)

In this case, magnetic field decelerates the flow, separation parameter decreases and boundary layer separation begins earlier.

Fig. 1

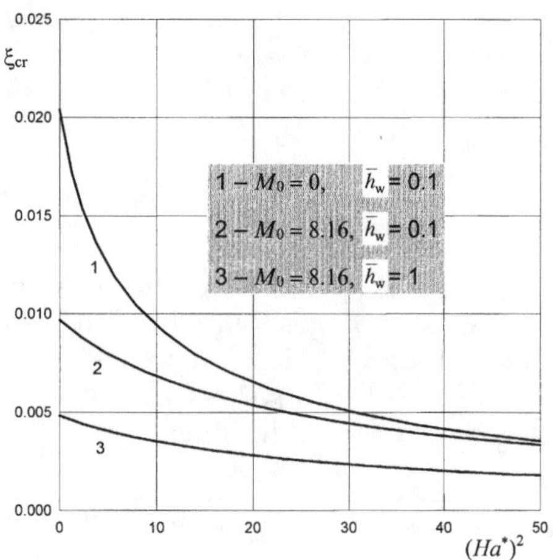

3) Magnetic and electric fields act jointly on the flow. If σ = const and $\psi^2(S)$=1, then ξ_{cr} has the following form:

$$\text{gin}_{cr} = SK + Q \tag{45}$$

The first term in Eq. (45) accounts for the influence of both B and E on ξ_{cr}.

If $E > 0$ ($K > 0$, $SK > 0$) and value of SK is sufficiently large, then ξ_{cr} is greater than its value at $B = 0$ despite the fact that Q decreases with increase of B (see Fig. 1).

If $E < 0$, then always ξ_{cr} is less than its value at $B = 0$ and boundary layer separation begins earlier.

4) There are electric and magnetic fields but the wall is cold and $\sigma_w = 0$. In this case, ξ_{cr} is determined by formula

$$\xi_{cr} = \Gamma\,(a, \bar{h}_w, 0, \beta^*, \beta_1^*) \tag{46}$$

Dependencies of ξ_{cr} on β^* at a =5,4 ($M_0 = 6$ when $\gamma = 1.3$), $\bar{h}_w = 0.16$ are shown in Fig. 2 for $\beta_1 = 4; 0; -4$. Values of β_1^* vary along each of these curves. Averaged values of β_1^* for these curves are: 1.1; 0; −1.3 respectively. Data in Fig. 2 correspond to the case when $\beta^* > 0$, i.e. $BE > 0$. (In the considered case, the wall conductivity gradient is a positive value).

Parameter of Boundary-Layer Separation in the Case of Conducting Wall

When MHD boundary layer is developed on the wall that is electrode, force f_x in the momentum Eq. (32) is presented as

$$f_x = f_x(x) = j_y B_z \tag{47}$$

This situation models MHD flow when external magnetic field is perpendicular to the flow plane (x,y). In the boundary layer approximation, electric current j_y does not change across layer and is the known function of coordi-

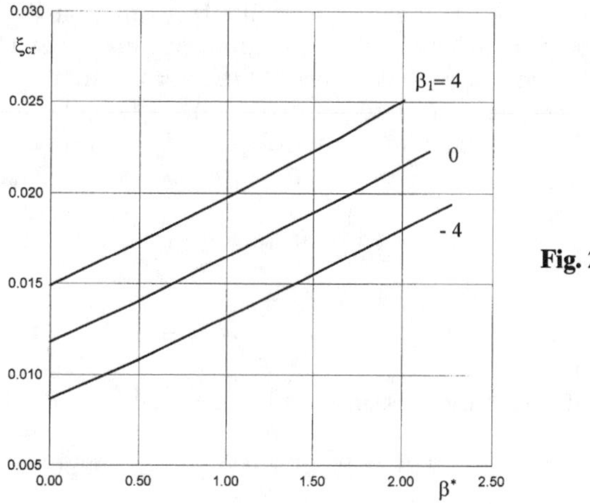

Fig. 2

nate x. If, as in section 7, the length of external magnetic field changing is much more than boundary layer thickness, then B_z is a given function of x-coordinate on surface. Therefore, force f_x, like the pressure gradient, is the function of x only and determination of boundary layer separation parameter is severely simplified and is reduced to the ordinary boundary layer case.

Using the method developed in the preceding section, the following relationship for ξ_{cr} can be found:

$$\xi_{cr} = \frac{j_y B_z \delta^*}{\rho_0 u_0^2} + \Gamma(a, \bar{h}_w, 0,0,0) \quad (48)$$

Function Γ in Eq. (48) is determined by formula (44). The possible influence of magnetic field on turbulence in the boundary layer is not accounted for.

In the absence of MHD, effects, $\xi_{cr} = \Gamma$. As it was shown above, value of ξ_{cr} in this case was calculated in Refs. 33 and 35. The control of boundary layer separation is effected by the first complex in formula (48). As a rule, this complex is negative for generator regime, so ξ_{cr} decreases and the earlier separation of the boundary layer takes place. When regime with electric power input to the flow is realized, this complex is positive in many cases, so ξ_{cr} increases and boundary layer separation is delayed.

Deceleration of a Supersonic Flow in a Circular Nonconducting Tube by an Axisymmetric Magnetic Field

Flow Deceleration in a Circular Tube by Magnetic Field of a Single Current Loop

Formulation of a Problem

An axisymmetric flow of conducting gas in the axisymmetric magnetic field along the circular tube with non-conducting walls is considered. To find out the features of MHD deceleration of a hypersonic flow, the model (9) at constant gas conductivity and in the absence of the Hall effect will be used. Under these assumptions, if an azimuth velocity at the tube entrance is zero, then a velocity vector lies in a meridian plane everywhere over the tube.

The right part of Eq. (15) becomes zero because rot $\boldsymbol{v}$ is perpendicular to meridian plane and vector $\boldsymbol{B}$ lies in this plane. The solution of equation $\Delta\varphi = 0$ with the boundary condition $\partial\varphi/\partial n = 0$ on all tube surfaces is $\varphi = \text{const}$, i.e., $\boldsymbol{E}=0$.

MHD source terms $\boldsymbol{f}$ and q in accordance with the velocity, magnetic, and electric fields structure are

$$f_x = -\sigma B_y(uB_y - vB_x), \quad f_y = \sigma B_x(uB_y - vB_x); \quad q = 0 \tag{49}$$

Here x and y are axial and radial coordinates, and u and υ are axial and radial velocity components.

According to equations div $\boldsymbol{B} = 0$, rot $\boldsymbol{B} = 0$, an axisymmetric magnetic field may be presented (see Ref. 13, p. 469) as

$$B_x = B_* \sum_{k=0}^{\infty} a_{2k} r^{2k}, \quad B_y = B_* \sum_{k=0}^{\infty} a_{2k+1} r^{2k+1}$$

$$a_{2k+1} = (-1)^{k+1} \frac{\psi^{(2k+1)}(\xi)}{2^{2k+1} k!(k+1)!}$$

$$a_{2k} = (-1)^k \frac{\psi^{(2k)}(\xi)}{2^{2k}(k!)^2}$$

$$\psi(\xi) = \frac{B_x(x,0)}{B_*}, \quad \xi = \frac{x}{H}, \quad r = \frac{y}{H} \tag{50}$$

Here B^* and H are the characteristic magnetic field and radius of external current loops generating the magnetic field. The range of the series convergence is $0 \le r < r^*$, $r^* = y^*/H > R/H$, where R is the tube radius, y^* is the radius of loop nearest to the tube.

Formulas (50) determine the external magnetic field distribution using the axial distribution $B_x(x, 0)$ only.

Function $\psi(\xi)$ for the magnetic field of a single current loop positioned at the $x = 0$ cross section has the form

$$\psi(\xi) = (1 + \xi^2)^{-3/2} \tag{51}$$

Magnetic field lines for this case are illustrated in Fig. 3.

The flow at the duct entry is supposed to be supersonic and uniform. All parameters distributions must be given for supersonic flow at the duct entry. The slip velocity conditions are posed on tube walls in inviscid case and no-slip conditions—in viscous case. In viscous case the wall temperature is supposed

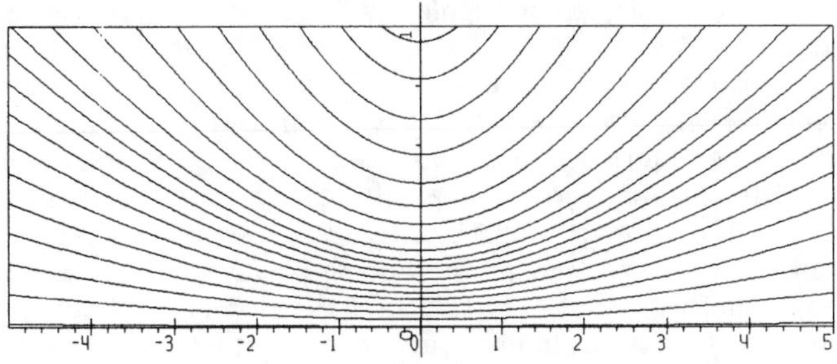

Fig. 3

to be equal to free stream temperature. In supersonic inviscid flow, the additional boundary conditions are not required on the exit computational boundary. In viscous case, the so-called drift boundary conditions (when normal derivatives of all parameters on the duct exit are assumed to be equal to derivatives determined in the nearest nodes of computational region) are posed both in the supersonic core region and in the subsonic part of the wall boundary layer.

The dimensionless parameters for the considered problem in inviscid case are:

$$\gamma,\ M_0,\ S=\frac{\sigma B_*^2 R}{\rho_0 U},\ \frac{H}{R} \tag{52}$$

where U is the entrance velocity.

Numerical Simulation of Flow in Euler Approach (Inviscid Gas)

Calculations were performed at M_o=5, γ = 5/3,H/R = 2, and at different values of S.

In the case of $S << 1$, force f, in accordance with Eq. (49), is approximated as

$$f = (f_x, f_y),\quad f_x \approx -\sigma U B_y^2,\quad f_y \approx \sigma U B_x B_y$$

$$B_x \geq 0;\ \ B_y < 0 \text{ at } x < 0,\ \ B_y = 0 \text{ at } x = 0,\ \ B_y > 0 \text{ at } x > 0 \tag{53}$$

Therefore, the force f_x is always negative and the force f_y is negative in the region x<0 and positive in the region $x > 0$. Such MHD force orientation admits the possibility for the flow to detach from the wall ahead of the cross section $x = 0$ and then to attach to the wall downstream from the cross section $x = 0$.

The flow velocity field at $S \geq 1$ is deformed significantly but the aforementioned tendency must be retained. Real calculations were performed using relations (49).

Some results of steady flow calculations obtained by time relaxation are presented below. The steady flow field illustrated by M = const lines (plotted with interval $\Delta M = 0.1$) for S=3 is shown in Fig. 4 (the last pattern). The calculated region is: $|x| \leq 5, 0 \leq y \leq 1$ (all coordinates are referred to duct radius). Characteristic features of the flow structure are: the existence of a cavity with extremely low gas velocities, the generation of shock in the $x > 0$ zone due to flow compression by transverse force f_y; the presence of shock in the $x < 0$ zone, generated on the leading edge of a cavity. Fig. 4 also shows the history of flow structure development in conditional relaxation time, beginning from uniform supersonic flow at the magnetic field switching on.

Results of numerical calculations of ratios of main averaged parameters at exit and entrance cross-sections as well as exit Mach numbers are presented in Table 4 (inviscid cases).

Numerical Simulation of Viscous Laminar Flow

The full Navier-Stokes system was integrated with the aid of numerical code described in the “Numerical Method” section. Calculations were performed at $M_0 = 5, \gamma = 5/3, H/R = 2, Pr = 0.72, Pr_t = 0.9$ and at different values of MHD interaction parameter S and of Reynolds number Re.

Figure 5 shows the field of Mach numbers for deceleration of a viscous laminar flow in the case of $S = 1.5$, $Re = 2\times10^5$. The calculations were carried out in the region: $-15 \leq x \leq 5, 0 \leq y \leq 1$. The boundary layer thickness at $x = -15$ was supposed to be zero. The flow is characterized by the extensive zone of boundary layer separation. The separation is initiated by pressure rise near the wall downstream from $x = 0$ (that is concerned with the shock induced by transverse MHD force) and by upstream propagation of disturbances through the subsonic near-wall region. The development of separation is also promoted by longitudinal MHD force directed against the flow and by transverse MHD force acting at $x < 0$ in the magnetic field zone and directed from the wall. The separation zone extends far upstream from the zone of real magnetic field influence. In this case, the whole extensive separation zone with low velocities is excluded from the MHD interaction. In the near-axis region, MHD interaction is weak because of very small transverse component of magnetic field. These circumstances give rise to two anomalies: 1) as S increases, the resulting decrease of supersonic flow deceleration is possible, and 2) at transition from an inviscid flow calculation to a viscous laminar flow calculation at the same S, MHD flow deceleration can also decrease (see Table 4).

Numerical results for main averaged parameters at exit cross-section are presented in Table 4 (its middle part).

Numerical Simulation of Turbulent Flow

Calculations were carried out on the base of full averaged Navier-Stokes equations and with the aid of computer code described in the “Numerical Method” Section.

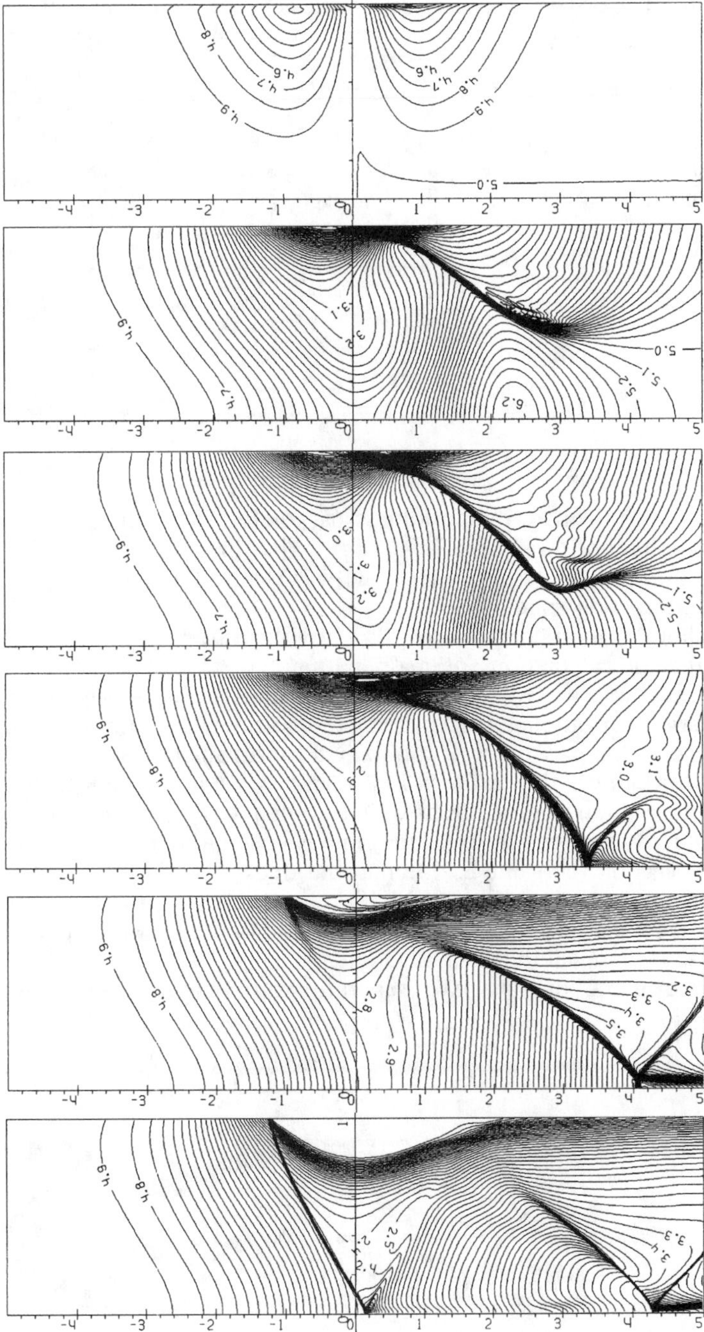

Fig 4.

Table 4 Calculation results for magnetic field of a current loop ($M_0 = 5$)

Re	S	P_e^*/P_0^*	P_e/P_0	M_e	T_e/T_0
Inviscid	0.75	0.296	2.62	2.95	2.39
Inviscid	1.5	0.255	3.0	2.73	2.68
Inviscid	3	0.247	3.07	2.68	2.74
Inviscid	5	0.219	3.47	2.5	3.02
2×10^4	0.75	0.390	1.99	3.41	1.84
2×10^4	1.5	0.297	2.52	2.98	2.26
2×10^5	0.75	0.388	2.06	3.37	1.93
2×10^5	1.5	0.429	1.89	3.54	1.79
2×10^6	0	0.665	1.33	4.24	1.29
2×10^6	0.75	0.284	2.67	2.90	2.40
2×10^6	3	0.234	3.17	2.62	2.78

The values of main dimensionless parameters were the same as in the cases of laminar flow except Reynolds number that was 2×10^6. The transition of laminar flow to turbulent one was realized at this Re upstream of MHD interaction zone.

The lines of M = const for $S = 0.75$ are illustrated in Fig. 6.

Results of calculations of main parameters at exit cross-sections are shown in Table 4 (its lower part). Note that the presence of turbulent boundary layer on duct walls provides some flow deceleration, even in the case $S = 0$. Comparison of data obtained for turbulent flow at S=0.75 with results for laminar flow at $Re = 2\times10^5$ and 2×10^4 at the same S shows that flow deceleration is more intensive in turbulent case. As follows from Fig. 6, in turbulent case for $S = 0.75$, the flow is unseparated in contradiction to laminar case at the same S although the thickness of turbulent boundary layer increases as the result of MHD interaction after the flow entry into magnetic field. The shock and wave structure in the flow core in the turbulent case is identical to the inviscid one.

The turbulent boundary layer separation originates at larger S (see Fig. 7 for $S = 3$).

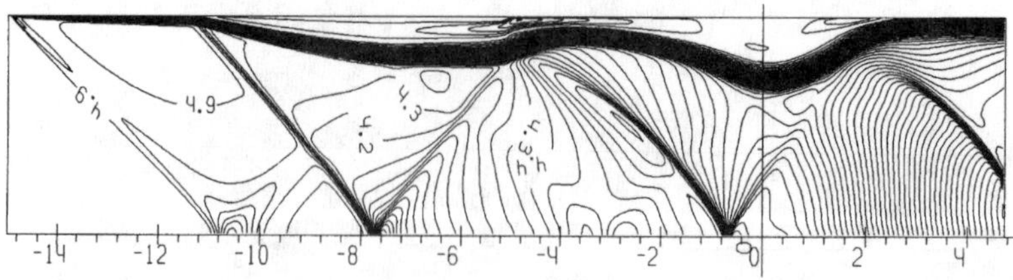

Fig. 5

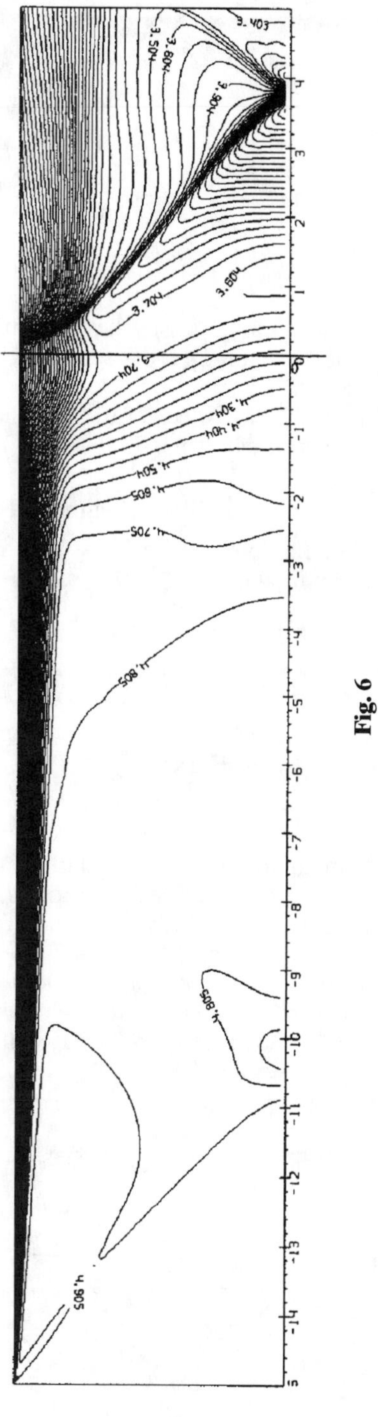
3.403
3.504
3.604
3.904
3.704
3.604
3.704
4.304
4.404
4.504
4.605
4.705
4.805
4.805
4.905
4
3
2
1
0
-1
-2
-3
-4
-5
-6
-7
-8
-9
-10
-11
-12
-13
-14

Fig. 6

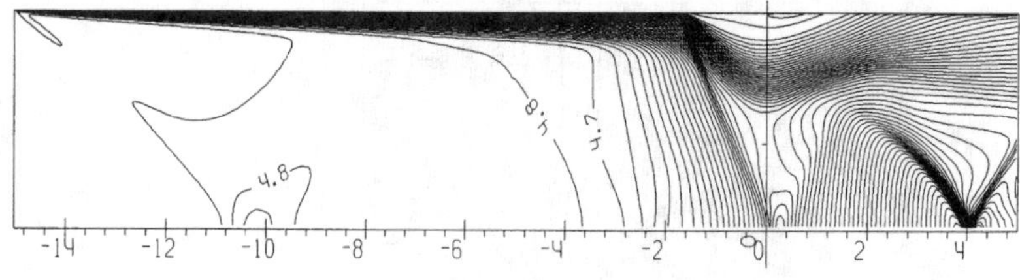

Fig. 7

In contradiction to the laminar case, the separation region in Fig. 7 is located in the vicinity of current loop cross-section. This situation is identical to one observed in the inviscid case.

According to Table 4, the flow deceleration in the turbulent case is enhanced if parameter S increases from 0.75 to 3.

Flow Deceleration in a Circular Tube by Magnetic Field of a Solenoid

The flow in the magnetic field of a solenoid is considered when distribution of $B_c(X, 0)$ is determined as

$$\psi(\xi) = \frac{1}{2}\left[\frac{\xi}{\sqrt{1+\xi^2}} - \frac{\xi - \Lambda}{\sqrt{1+(\xi-\Lambda)^2}}\right], \quad \Lambda = 5$$

$$\psi(\xi) = \frac{B_x(x,0)}{B_*}, \quad \xi = \frac{x}{H} \tag{54}$$

The distribution of magnetic field in the whole channel is found with the aid of Eqs. (50) and (54). In this case, the solenoid radius is twice greater than duct radius. The solenoid entry is located in the cross section $x = 0$. The solenoid exit cross section is located $x = 10$ and the right boundary of calculation region is at $x = 15$. The magnetic field lines are shown in Fig. 8.

The length of the duct from the entry cross-section to the entrance to the solenoid is equal to 5 and 15 duct radii in the cases of inviscid and of viscous

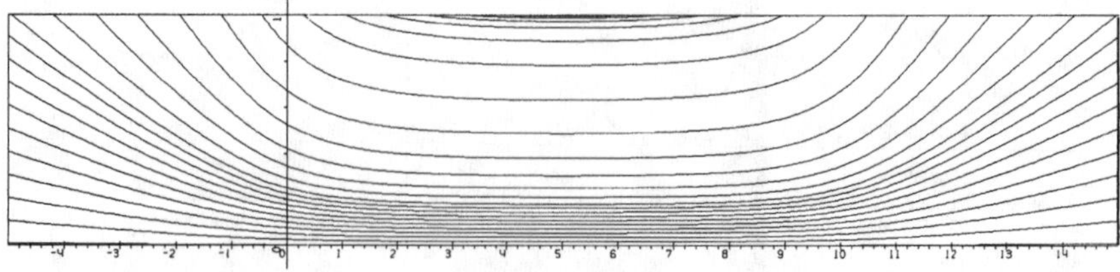

Fig. 8

flows, correspondingly. Thus, the duct geometry upstream and downstream from the solenoid is the same as in the case of a current loop. The total duct length in these two cases differs only by solenoid length.

MHD interaction for inviscid flow (Euler equations) and for laminar and turbulent viscous flows at different Reynolds numbers (full Navier-Stokes equations) was analyzed. The calculations were performed at the same boundary conditions and at the same values of the main dimensionless parameters as in the case of magnetic field generated by the current loop (see preceding section). Results of these calculations are presented in Table 5.

Mach number lines at $S = 1.5$ are presented in Fig. 9a (inviscid flow), in Fig. 9b (viscous laminar flow, $Re = 2{\times}10^5$) and in Fig. 9c (viscous laminar flow, $Re = 2{\times}10^4$). The cavity is absent in Fig. 9a, the separation zone is extensive and closed in Fig. 9b, and it is absent in Fig. 9c.

Analysis of these and other calculation results allows to make the following conclusions:

In inviscid flow, cavity appears, in the solenoid case, at the larger S in comparison with current loop case.

In viscous laminar flow, the separation zone occurring at $S = 1.5$ and at relatively large Reynolds number ($Re = 2{\times}10^5$) (see Fig. 9b) is closed in the solenoid case: flow attaches to the wall in the vicinity of the solenois end. At lower Reynolds number ($Re = 2{\times}10^4$) and at the same S (see Fig. 9c), the separation zone is absent in contradiction to current loop case.

In turbulent flow, the separation zone in the magnetic field of solenoid appears only at sufficiently large values of S. So, boundary layer separation is absent in Fig. 10 where Mach numbers field at $S = 5$ is presented for turbulent flow. Thick turbulent boundary layer contributes to overall flow deceleration. Smaller dimensions of separation zone in the case of magnetic field of solenoid result in vanishing of some anomalies mentioned in the preceding section.

Table 5 Calculation results for magnetic field of a solenoid ($M_0 = 5$)

Re	S	P_e^*/P_0^*	P_e/P_0	M_e	T_e/T_0
Inviscid	0.75	0.459	1.81	3.63	1.73
Inviscid	1.5	0.303	2.56	2.99	2.35
Inviscid	3	0.244	3.07	2.68	2.77
Inviscid	5	0.227	3.31	2.56	2.93
2×10^4	0	0.411	1.92	3.48	1.75
2×10^4	0.75	0.328	2.29	3.14	2.06
2×10^4	1.5	0.290	2.52	2.96	2.26
2×10^5	0	0.734	1.24	4.44	1.21
2×10^5	0.75	0.432	1.88	3.54	1.77
2×10^5	1.5	0.386	2.06	3.365	1.92
2×10^6	0	0.624	1.35	4.18	1.34
2×10^6	3	0.191	3.83	2.33	3.21
2×10^6	5	0.165	4.45	2.12	3.63

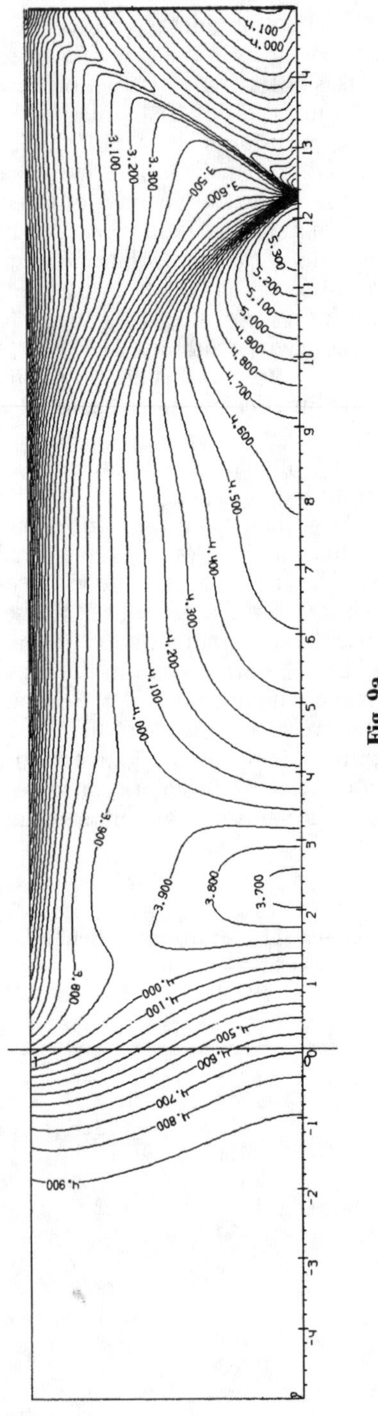
4.100
4.000
3.100
3.200
3.300
3.500
3.600
5.200
5.100
5.000
4.900
4.800
4.700
4.600
4.500
4.400
4.300
4.200
4.100
4.000
3.900
3.900
3.800
3.700
3.800
4.000
4.100
4.500
4.600
4.700
4.800
4.900
14
13
12
11
10
9
8
7
6
5
4
3
2
1
0
-1
-2
-3
-4

Fig. 9a

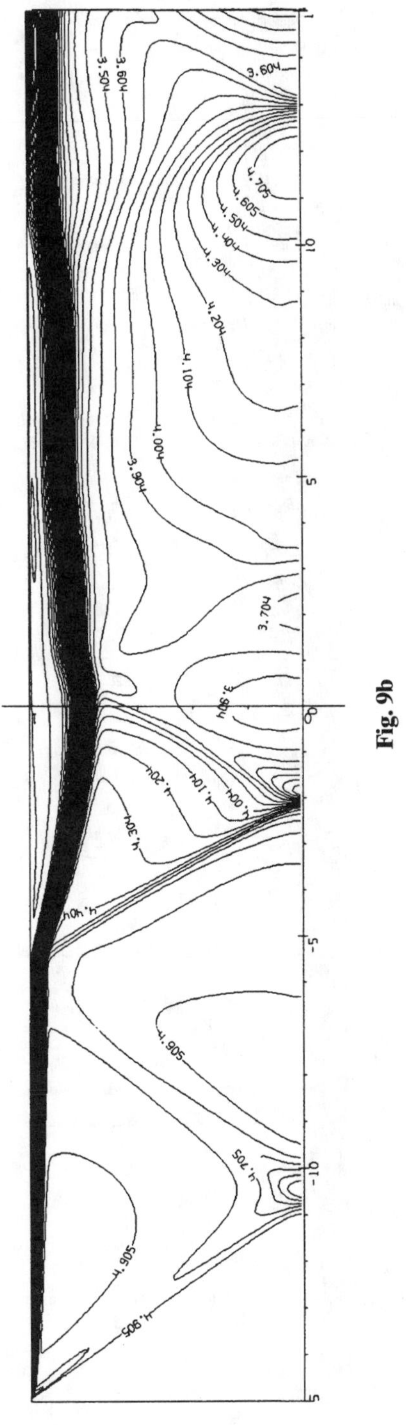
3.604
3.504
3.604
4.705
4.605
4.504
4.404
4.304
4.204
4.104
4.004
3.904
3.704
3.904
4.204
4.104
4.004
4.304
4.404
4.905
4.705
4.905
4.905

Fig. 9b

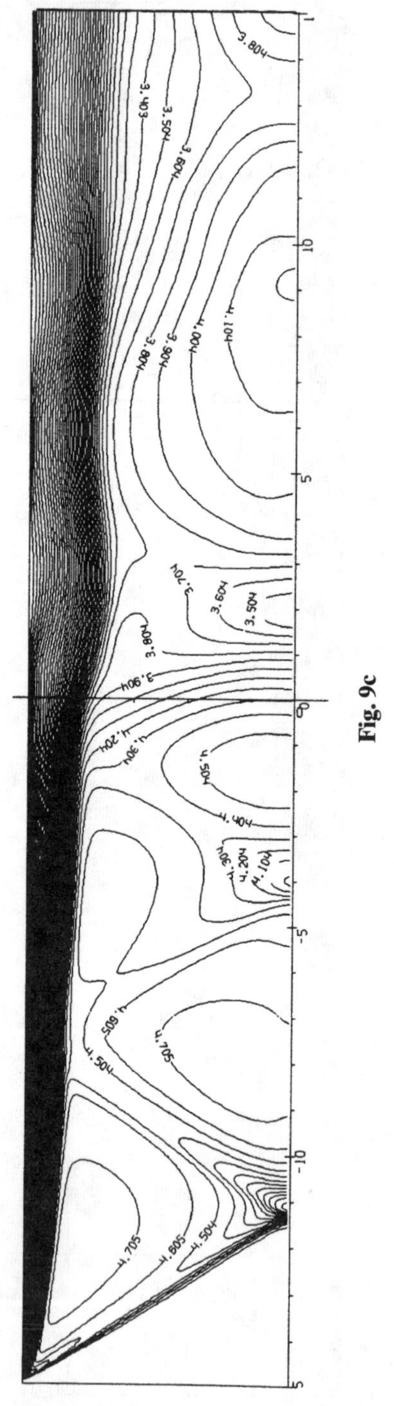
3.804
3.403
3.504
3.604
3.804
3.904
4.004
4.104
3.704
3.604
3.504
3.804
3.904
4.204
4.304
4.504
4.404
4.304
4.204
4.104
4.605
4.504
4.705
4.705
4.605
4.504
10
5
0
-5
-10

Fig. 9c

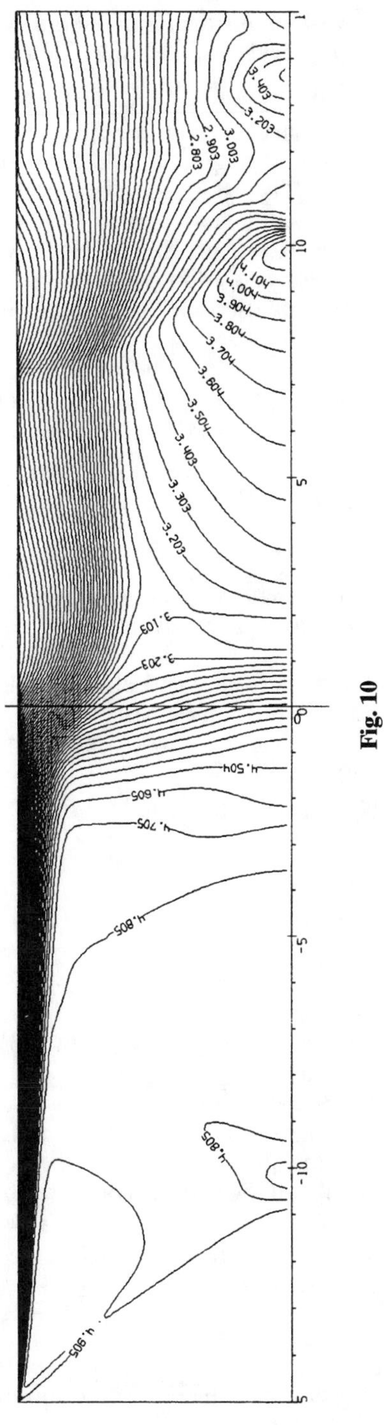
3.403
3.203
3.003
2.903
2.803
4.104
4.004
3.904
3.804
3.704
3.604
3.504
3.403
3.303
3.203
3.103
3.203
4.504
4.605
4.705
4.805
4.805
4.905
10
5
0
-5
-10

Fig. 10

Deceleration of Two-Dimensional Supersonic Flow in Channels by Magnetic Field Perpendicular to a Flow Plane In Generator Regime

Formulation of a Problem

Let's consider a stationary MHD flow in a channel when "working" component B_z of an external magnetic field $\boldsymbol{B}$ (x,z) is perpendicular to a flow plane (x,y). The field $\boldsymbol{B}$ is produced by magnetic system positioned out of channel and has a longitudinal component B_x (x,z). Therefore, in general case, the flow is a 3-D one. But, averaging MHD flow over coordinate z and using some assumptions considered in detail in Ref. 13 transform the original 3-D problem to 2-D equations with the following structure of gasdynamic and electromagnetic fields:

$$\mathbf{v} = [u(x,y), v(x,y), 0], \quad \mathbf{B} = [0, 0, B(x)], \quad B(x) = < B_z(x,z) >_z$$

$$\boldsymbol{j} = [j_x(x,y,), j_y(x,y), 0], \quad \varphi = \varphi(x,y), \quad \boldsymbol{E} = \left(-\frac{\partial\varphi}{\partial x}, -\frac{\partial\varphi}{\partial y}, 0\right) \tag{55}$$

The symbols "< >" designate an averaging over z. An induced magnetic field is not accounted for. $B(x)$ is the given function determined by external magnetic system.

Distributions of electric current $\boldsymbol{j}$ and electric field $\boldsymbol{E}$ are obtained with the aid of Eqs. (9) and (55) and can be written as:

$$j_x = \sigma_\beta\left[-\frac{\partial\varphi}{\partial x} + vB + \beta\left(\frac{\partial\varphi}{\partial y} + uB\right)\right]$$

$$j_y = \sigma_\beta\left[-\frac{\partial\varphi}{\partial y} - uB + \beta\left(-\frac{\partial\varphi}{\partial x} + vB\right)\right] \tag{56}$$

$$\frac{\partial j_x}{\partial x} + \frac{\partial j_y}{\partial y} = 0 \quad \left(\sigma_\beta = \frac{\sigma}{1+\beta^2}, \beta = aB\right) \tag{57}$$

Here β is the Hall parameter, conductivity σ and quantity a are given functions of thermodynamic parameters.

The following elliptical equation for φ is obtained from Eqs. (56) and (57)

$$\Delta\varphi + \frac{\partial\varphi}{\partial x}\left[\frac{\partial \ln\sigma_\beta}{\partial x} + \beta\frac{\partial \ln(\beta\sigma_\beta)}{\partial y}\right] + \frac{\partial\varphi}{\partial y}\left[\frac{\partial \ln\sigma_\beta}{\partial y} - \beta\frac{\partial \ln(\beta\sigma_\beta)}{\partial x}\right]$$

$$= \sigma_\beta^{-1}\left\{\frac{\partial}{\partial x}\left[B\sigma_\beta(v + \beta u\right] - \frac{\partial}{\partial y}\left[B\sigma_\beta(u - \beta v)\right]\right\} \tag{58}$$

The forces f_x, f_y and local electric power input q to gas are defined as

$$f_x = j_y B,\ f_y = -j_x B,\ q = -j_x \frac{\partial \varphi}{\partial x} - j_y \frac{\partial \varphi}{\partial y} \tag{59}$$

The solution of Eq. (58) requires the formulation of the boundary conditions on the channel walls and at the entry and exit duct cross-sections. The formulation of these boundary conditions depends on the physical problem.

It is assumed that magnetic field $B(x)$ and potential difference $\delta\varphi$ between channel walls are concentrated in the finite region of the channel and decrease rapidly to the entrance and to the exit cross-sections. Therefore, in these cross-sections, boundary condition $j_x = 0$ is assumed. If the wall is an electrode, then $\varphi_w = \text{const}$. If the wall is the sectioned electrode, then φ_w is the given function of coordinate along the wall. (In the last case, the distribution of the normal component j_n on the wall may be given instead of φ_w distribution.) If the wall consists of electrodes and isolators, then boundary conditions are: $\varphi_w = \text{const}$ on electrodes, and $j_n = 0$ on isolators.

For generator regime of flow, in accordance with Eqs. (20–22), efficiency η is introduced as the ratio of extracted electric power to a gas work in magnetic field.

The another coefficient that characterizes the MHD generator is the ratio of extracted electric power N [see relationship (21)] to the flux of total enthalpy at the entrance cross-section F_0:

$$\eta_N = N \Big/ \int_{F_0} \rho u h^* \, dy \tag{60}$$

Quantity h^* is determined by Eq. (12).

Gasdynamic equations were presented in Eqs. (10–14). For considered 2-D flow, the last term in the left-hand side of Eq. (10) must be omitted.

Boundary conditions on the channel walls and conditions at the entrance and exit cross-sections are the same as in the "Deceleration of a Supersonic Flow" section.

The main dimensionless parameters for the problem are

$$M_0 = \sqrt{\frac{\rho_0 V_0^2}{\gamma p_0}},\quad Re = \rho_0 V_0 h_0/\mu_0,\quad S = \sigma_0 B_*^2 h_0/\rho_0 V_0,\quad K_0 = \frac{(\delta\varphi)_*}{V_0 B_* h_0}$$

$$\mathrm{PR},\quad \mathrm{Pr_t},\quad \gamma, \beta_0 = a_0 B^* \tag{61}$$

where subscript "0" corresponds to the core flow at the entrance cross-section, B^* and $(\delta\varphi)^*$ are the characteristic values of magnetic field and electric potential difference, h is the channel height. New dimensionless parameters in Eq. (61), are the load parameter K_0 and the Hall parameter β_0.

All numerical calculations were carried out at $M_0 = 5, S = 0.1, K_0 = 0.3, \gamma = 1.4, Pr=0.72, Pr_t = 0.9$. Laminar flows of viscous gas were calculated at $Re = 5\times10^3$, turbulent flows at $Re = 10^6$. Quantities σ and a were constant. Hall parameter was equal to 0 or to 1.

Quasi-One-Dimensional Approximation for Electrical Variables

Using the results of Ref. 13, one can obtain the one exact solution of Eqs. (56–58). Let's assume magnetic field is uniform, channel height h is constant, quantities $\xi = (u, v, \sigma_\beta, \beta)$ are functions of transverse coordinate y only, and boundary conditions for electrical variables are not changed along longitudinal coordinate x. Under these assumptions, the solution of Eqs. (56–58) has the following structure: $\varphi = Ax + \kappa(y), j_y = \text{const}, A = \text{const}$ and is written as

$$j_y = C - \frac{A < \beta >}{\langle \sigma_\beta^{-1} \rangle}, \quad C = \frac{B\langle u - \beta v\rangle - (\delta\varphi/)}{\langle \sigma_\beta^{-1} \rangle}$$

$$j_x = \sigma(-A + vB) - \beta j_y, \quad E_x = -A, \quad E_y = uB + (j_y - \beta_{j_x})/\sigma$$

$$< j_x >= -A < \sigma > +B < \sigma v > - < \beta > j_y, \quad \delta\varphi = \varphi(0) - \varphi(h) = \varphi_- - \varphi_+$$

$$< \xi >= h^{-1} \int_0^h \xi \, dy \tag{62}$$

This solution depends on two parameters: A and $\delta\varphi/h$. If potentials of walls are constant, then A=0. If channel walls are sectioned electrodes and mean longitudinal current $< j_x >$=0, then this condition determines A and, consequently, the solution (62) depends on quantity $\delta\varphi/h$ only.

Now let's assume that quantities B, h and $\delta\varphi$ are not constant ones but they are weak functions of x. Also, variables ξ are supposed to depend on x as weak functions.

Under these conditions, it may be assumed that distributions of $\boldsymbol{j}$ and $\boldsymbol{E}$ are determined by formulas (62), but now they depend not only on y but also on x as a result of x dependence of B, h, $\delta\varphi$ and ξ. (Analogous approach is often used in usual hydrodynamics. For example, an analysis of fully developed viscous flow in a channel with variable cross-section is performed with the aid of Poiseuille formula that is exact solution only for a channel with constant cross-section.)

Attractiveness of such approach consists in its local nature. The solution of complex system (56–58) is not needed at the use of this approach.

Certainly, the various modifications of relations (62), were used pre-

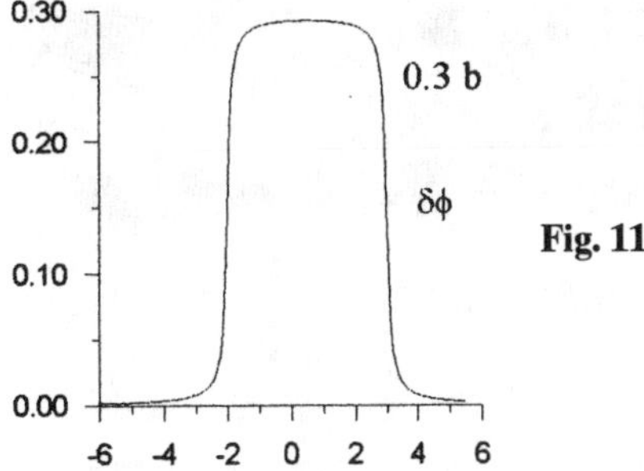

Fig. 11

viously in engineering calculations of MHD devices. Therefore, it is necessary above all to compare the results of MHD flow calculation at the use of Navier-Stokes equations and Eqs. (56–58) with results of the use of Navier-Stokes equations and approximate relations (62).

Let's assume distributions of magnetic field $B = B^*b(x)$ and load parameter $K(x) = \delta\varphi/(B^*V_0h)$ are similar to one another as shown in Fig. 11 ($\varphi_- = -\varphi_+, \delta\varphi = 2\varphi_-$), $A = 0$ and the channel height $h =$ const. Calculation region is supposed to be $-6 \leq x \leq 5.5, 0 \leq y \leq 1$ (here coordinates x and y are referred to the channel height h). All parameters required for calculations were indicated at the end of the preceding section.

For these conditions, base calculation of viscous laminar MHD flow with the aid of full Navier-Stokes equations and Eqs. (56) and (57) for electric field was performed. Calculation results are indicated in Table 6 as Fig. 12.

The corresponding Mach number field is shown in Fig. 12. The main peculiarity of this field is the absence of flow symmetry, in spite of the fact that the flow with isotropic conductivity is considered. (This peculiarity will be discussed below.) The flow is characterized by availability of a developed cavity on the lower wall and small cavity on the upper wall. The cavities are formed as a result of boundary layer separation due to an action of MHD

Table 6

Fig	Re	β_0	P_e/P_0	P_e^*/P_0^*	M_e	T_e/T_0	η	η_N
12	5.10^3	0	2.34	0.150	2.95	1.91	0.462	0.048
14	5.10^3	0	1.57	0.351	3.81	1.44	—	—
13	5.10^3	0	2.35	0.148	2.93	1.91	0.504	0.0539
15	5.10^3	0	2.35	0.148	2.93	1.91	0.504	0.0539
16	5.10^3	0	2.29	0.156	2.99	1.88	0.516	0.0561
17	5.10^3	1	2.27	0.163	3.02	1.88	0.433	0.0434
18	10^6	0	1.26	0.577	4.36	1.21	—	—
19	10^6	0	2.84	0.114	2.64	2.26	0.474	0.0622
20	10^6	0	3.61	0.0784	2.25	2.63	0.411	0.0856
21	10^6	1	2.88	0.113	2.62	2.28	0.369	0.0618

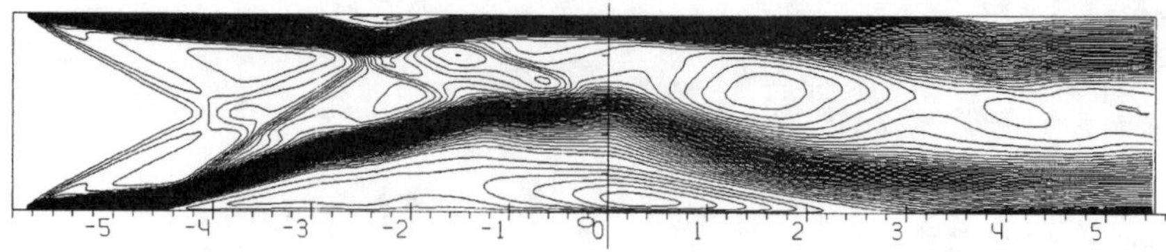

Fig. 12

decelerating force $f_x > 0$ and upstream disturbances propagation. The resultant longitudinal dimension of a large cavity exceeds a length of an effective MHD interaction zone.

The results of MHD flow calculation under the same initial and boundary (for steady problem) conditions, but with the use of approximate model (62) instead of Eqs. (56) and (57), are submitted in Table 6 as Fig. 13. Mach number field is shown in Fig. 13. The flow patterns in Figs. 12 and 13 coincide with each other in most details. Resultant flow characteristics also differ insignificantly (see Table 6). This proves the validity of approximate model (62), despite apparently unfavorable conditions in this case owing to large gradients of $b(x)$ and $\delta\varphi(x)$. Therefore, the set of Navier-Stokes equations together with Eq. (62) will be used hereinafter in all calculations.

Numerical Analysis of Laminar and Turbulent Flows

Laminar Flows

Results of calculations are shown in Table 6 (Figs. 13–17) and Mach number fields.

Fig. 14 corresponds to flow at $B = 0$. Some deceleration in this is caused by boundary layer development on the channel walls. The results presented in Figs. 13 and 15 were obtained at the use of the gasdynamic field shown in Fig. 14, as initial conditions for the solution of Navier-Stokes non-stationary equations in the presence of magnetic field. A very interesting result was found, namely, the dependence of equation solution on time relaxation process. Thus, if the condition $T = T_w = T_0$ on the lower wall was substituted for adiabatic condition at the beginning of relaxation process, the pattern shown in Fig. 13 was realized. If the temperatures on both walls in the whole relaxa-

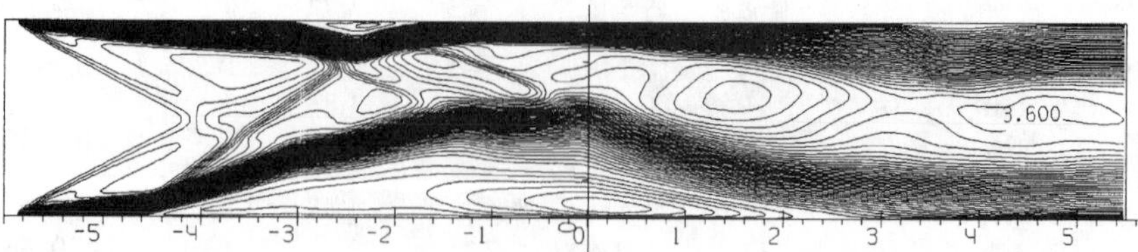

Fig. 13

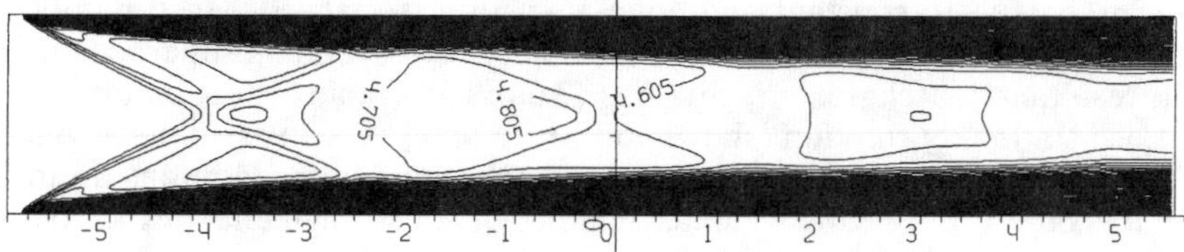

Fig. 14

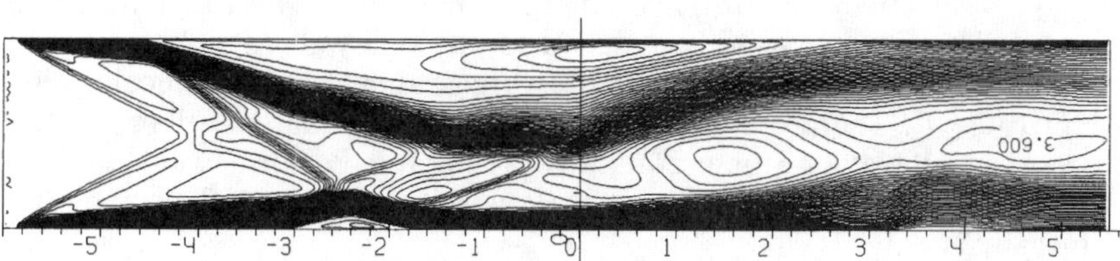

Fig. 15

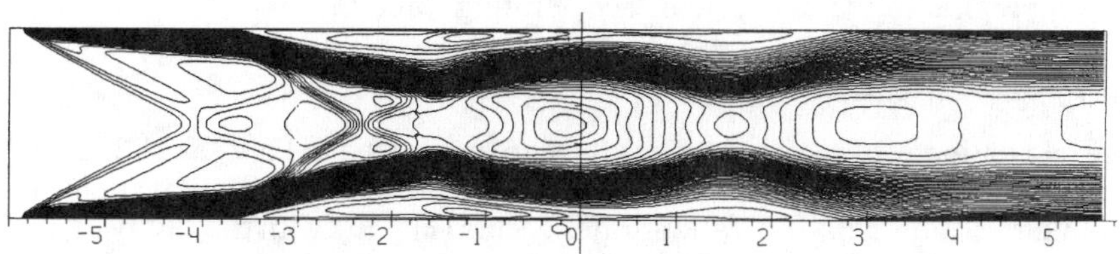

Fig. 16

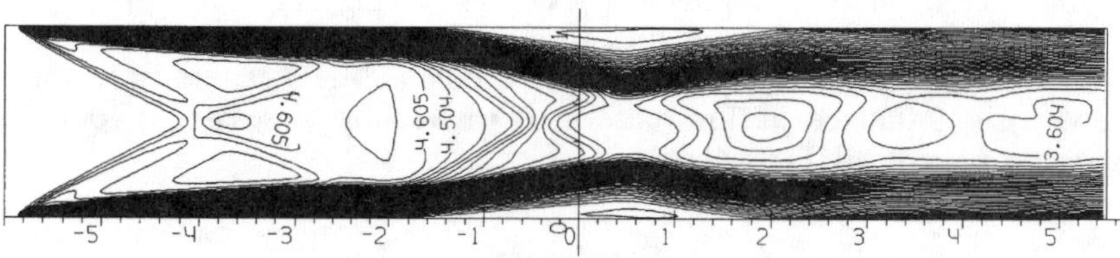

Fig. 17

tion process were constant and equal, Fig. 15 was realized. Mach number fields (Figs. 13 and 15) completely coincide at corresponding superposition. The resultant flow characteristics (Figs. 13 and 15 in Table 6) are identical.

Each stationary solution (shown in Figs. 13 and 15) is an asymmetric one. It may be explained, apart from asymmetry of boundary conditions during the relaxation process, by weak asymmetry of some fragments of the computational procedure.

It is useful to note that in gasdynamics, asymmetric flows may be realized when boundary conditions are symmetric ones. For example, it is shown in Ref. 36, that at throttling of a supersonic flow channel, asymmetric system of shock waves can arise. There is also some analogy with subsonic flow in the expanding diffuser. Even in a diffuser with weak expansion, the flow, being very sensitive to a positive gradient of pressure, begins to separate and the arising separation zone is attached to any one wall of a channel.[37] In considered flow, the role of positive pressure gradient plays the decelerating MHD force f_x. Therefore, more careful numerical analysis of relaxation process for full Navier-Stokes equations is required. Performed calculations assume that the stationary symmetric flow is unstable to small asymmetric disturbances.

The stationary symmetric flow (Fig. 16) is received by considering only half of a channel with the fixed symmetry conditions on a channel center line.

Hall effect influence on flow is illustrated by Fig. 17. In the relaxation process, the temperatures on both walls were fixed and equal. The flow in Fig. 17 should be considered as a transformation of the flow in Fig. 15 due to action of the force F_y stipulated by Hall effect and directed to the lower wall, having regard to the decrease of decelerating force f_x.

Turbulent Flows

Calculation results for turbulent flow are presented in Figs. 18–21. Flow in Fig. 18 corresponds to the case of $B = 0$. In the presence of magnetic field, an asymmetric flow with separation zones occurs (Fig. 19), with the cavity on the upper wall. The flow qualitatively coincides with one in Fig. 15 except that the boundary layer separation on the lower wall in the case of turbulent flow is absent. The data in Fig. 20 correspond to symmetrical flow, and in Fig. 21 they correspond to the flow with Hall effect. This effect, as well as in laminar case, reduces an asymmetry of a flow: the separation zone on the upper wall is reduced as a result of the diminishing of deceleration force f_x. Hall effect also results in a reduction of output electric power (see Fig. 21 in Table 6)

The research performed in this article shows that in the process of a hypersonic flow deceleration by a magnetic field in MHD generator regime, the viscosity influence on flow pattern, on irreversible losses and on output characteristics of MHD devices.

Conclusions

1) The review of available proposals devoted to use of MHD control meth-

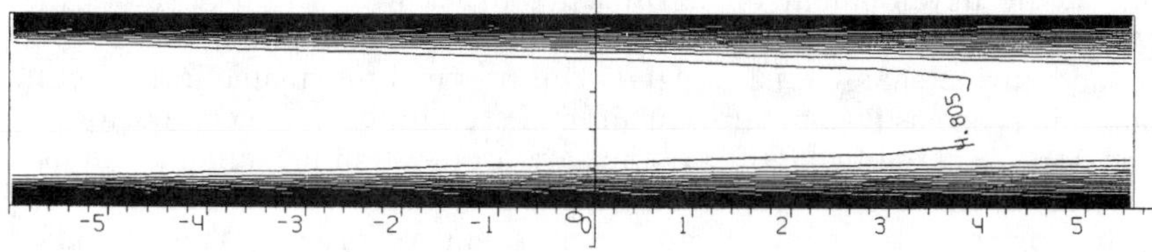

Fig. 18

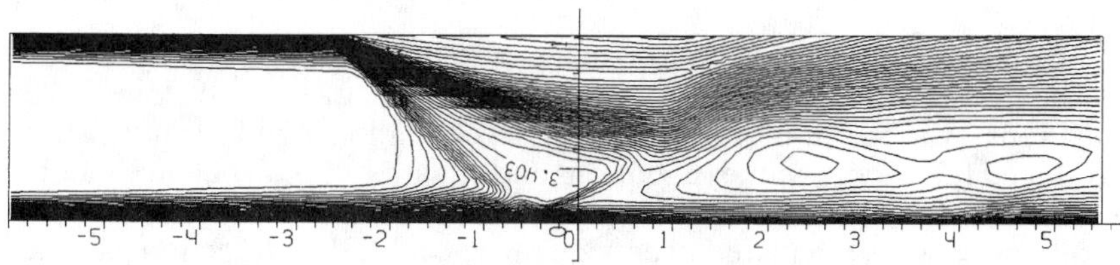

Fig. 19

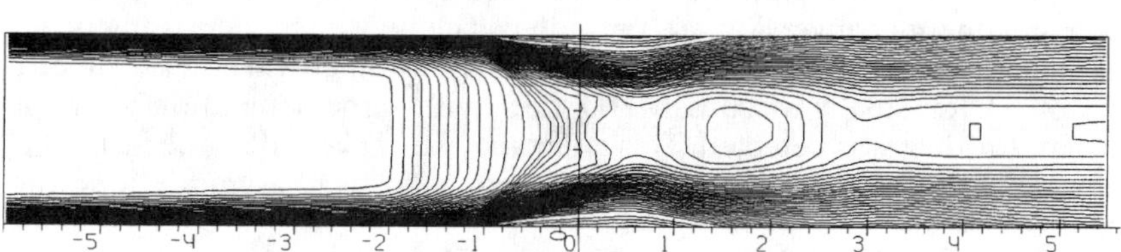

Fig. 20

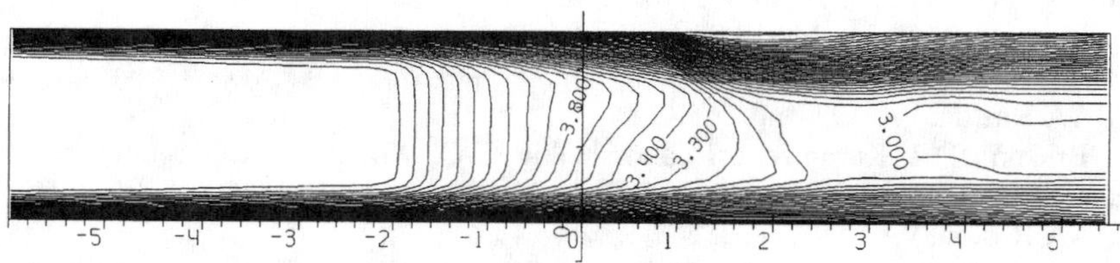

Fig. 21

ods at the development of aerospace vehicles with scramjet shows that MHD control may be considered an effective way of improvement in vehicle and engine characteristics. MHD control on aerodynamic forces acting on vehicle, on gasdynamic flow in air intake, on processes in combustor and also MHD energy bypass for optimization of system including air intake, combustor and nozzle are discussed as a new effective tool in hypersonic technologies. Current investigations are often performed using simplified flow models (1-D or inviscid 2-D flows) and therefore more careful analysis of MHD effects is needed.

2) The essentials of MHD equations systems as applied to supersonic flows were considered. The main dimensionless MHD parameters, parameters characterizing irreversible losses, peculiarities of 1-D MHD flow at large Mach numbers, and MHD characteristics of flow behind normal shock are presented.

3) The numerical investigation of supersonic flow deceleration by axisymmetric magnetic field was performed. Calculations of inviscid flow (Euler equations), various flow (full Navier-Stokes equations) and turbulent flow (averaged full Navier-Stokes equations and equations for turbulence model) were made. The total pressure losses and efficiency of MHD supersonic flow deceleration were calculated over wide ranges of MHD interaction parameter and Reynolds number. The strong interaction of MHD and viscous effects followed by boundary layer separation and by the appearance of recirculation zones was detected and nontrivial qualitative results were obtained.

4) The numerical investigation of supersonic flow deceleration in 2-D duct in MHD generator regime was made. Data for efficiencies of flow deceleration and energy conversion are presented. The new effect consisting of the existence of two different numerical solutions (symmetrical and asymmetrical ones) for steady viscous MHD flow with large separation zones in symmetric duct was detected. This effect may be explained by the instability of symmetric flow with large separation zones to small asymmetric disturbances.

5) Analysis performed and calculation results suggest that estimation of MHD control efficiency for hypersonic technologies requires accounting for viscous and spatial effects.

References

[1]Gurijanov, Å. P., and Harsha, P. T.,"AJAX: New Directions in Hypersonic Technology," AIAA Paper 96-4609, 1996.

[2]Kuranov, A. L., Kuchinsky, V. V., Sepman, V.Yu., et al.,"Flow about Power Control for the AJAX Project," *Proceedings of the 2nd Weakly Ionized Gases Workshop*, AIAA Reston, VA, 1998, pp. 263–268.

[3]Brichkin, D. I., Kuranov, A. L., and Sheikin, E. G.,"MHD-Control Technology for Hypersonic Vehicle," *Proceedings of the 2nd Weakly Ionized Gases Workshop,* AIAA Reston, VA, 1998, pp. 251–261.

[4]Bruno, C., and Csysz, P. A.,"An Electro-Magnetic-Chemical Hypersonic Propulsion System," AIAA Paper 98-1582, 1998.

[5]Bityurin, V. A., Zeigarnic, V. A., and Kuranov, A. L.,"On a Perspective of MHD Technology in Aerospace Applications," AIAA Paper 96-2355, 1996.

[6]Brichkin, D. I., Kuranov, A. L., and Sheikin, E. G.,"MHD Technology for Scramjet Control," AIAA Paper 98-1642, 1998.

[7]Vasilyev, E. N., Derevyanko, V. A., and Latypov, A. F.,"Mathematical Simulation of Gas in Hypersonic Ramjet with MHD-Control," *Proceedings of the International Conference on the Methods of Aerophysical Research,* 7, Pt. 2, Novosibirsk, 1994, pp. 229–235.

[8]Kaznelson, S. S., and Zagorsky, À.Â.,"Magnetohydrodynamic Control of the Flow in the Scramjet Duct," *Thermophysics and Aeromechanics,* Vol. 4, No. 1, 1997, pp. 41–46.

[9]*Proceedings of the 2nd Weakly Ionized Gases Workshop*, AIAA, Reston, VA, 1998.

[10]Bezgin, L., Ganzhelo, A., Gouskov, O., Kopchenov, V., Laskin, I., and Lomkov, R.,"Numerical Simulation of Supersonic Flows Applied to Scramjet Duct," *International Symposium on Air Breathing Engines,* Symposium Papers, Vol. II, 1995, pp. 895–905.

[11]Kopchenov, V. I., and Lomkov, K. E.,"Numerical Simulation of the Supersonic Combustion Enhancement and 3-D Scramjet Combustor Design with Account of Nonequilibrium Chemical Reactions," *High Speed Aerodynamics,* Vol. 1, No. 1, 1997, pp. 43–53 (published in Russia).

[12]Singh, D. J., Carpenter, M. H., and Drummond, J. P.,"Thrust Enhancement in Hypervelocity Nozzles by Chemical Catalysis," *Journal of Propulsion and Power,* Vol. 13, No. 4, 1997, pp. 574–576.

[13]Vatazhin, G. A., Lyubimov, S. A., and Regirer, Magnetohydrodynamic Flows in Channels,Nauka, Moscow, 1970. (in Russian).

[14]Vatazhin, A. B., "On the Magnetohydrodynamic Boundary Layer Separation," *Prikladraya Matematika y Mekanika*, No. 2, 1963.

[15]Miller, J. H., Kashuba, R. J., Kelley, J. D., Vogel, P., Smereczniak, P., and Chadwick, K.,"Recent Progress in Plasma Aerodynamics," *2nd Weakly Ionized Gas Workshop,* pp. 127–143.

[16]Nagashima, T., Kitamura, H., and Obata, S.,"Supersonic Combustion of Hydrogen in Tandem Transverse Injection with Oxygen Radicals," *International Symposium on Air Breathing Engines*, 97-7057, 1997.

[17]Morris, R. A., Arnold, S. T., Viggiano, D. A., Maurice, L. Q., Carter, C., and Sutton, E. A.,"Investigation of the Effect of Ionization on Hydrocarbon-Air Combustion Chemistry," *Proceedings of the 2nd Weakly Ionized Gases Workshop,* AIAA, Reston, VA, 1998, pp. 163–176.

[18]Ivanov, V. A.,"A Method of Calculation MHD-Current with a Separation of a Boundary Layer," Vol. 32, No. 6, 1994, pp. 909–919.

[19]Predvoditelev, A. S., et al., Tables of Thermodynamic Functions of Air (for Temperature Between 200 and 6000 K and Pressure Between 0.00001 and 100 atm), Moscow, 1962 (in Russian).

[20]Sutton, G. W., and Sherman, A., *Engineering Magnetohydrodynamics,* McGraw-Hill, New York, 1965.

[21]Glushko, V. P., et al., Thermodynamic Properties of Individual Substances, Nauka, Moscow, 1978 (in Russian).

[22]Gulyaev, A. N., Kozlov, V. E., and Sekundov, A. N.,"On the Development of a Universal One-Parametric Model for Turbulent Viscosity," *Izvestiya, Academy of Sciences, USSR, Mekh. Zhidk.i Gaza,* No. 4, 1993, pp. 69–81.

[23]Resler, E. L., and Sears, W. R.,"The Prospects for Magnetohydrodynamics," *JAS,* Vol. 25, No. 4, 1958 pp. 235–245.

[24]Gouskov, O., Kopchenov, V., and Nikiforov, D.,"Flow Numerical Simulation in the Propulsion Elements of Aviation-Space Systems Within Full Navier-Stokes Equations," *Procedures of the International Conference on the Methods of Aerophysical Research,* Pt. 1, Novosibirsk, 1994, pp. 104–109.

[25]Topekha, E., and Kopchenov V.,"Implicit Relaxation Finite-Difference Scheme for Navier-Stokes Equations," *Proceedings of the Methods for Investigations of Hypersonic Vehicles,* Pt. 3, TsAGI, 1994, pp. 9.1–9.10.

[26]Godunov, S. K., Zabrodin, A. V., Ivanov, M.Ja., Kraiko, A. N., and Prokopov, G. P., *Numerical Solution of Multidimensional Gas Dynamics Problems*, Nauka, Moscow, 1976.

[27]Beam, R. M., and Warming, R. F.,"An Implicit Factored Scheme for Compressible Navier-Stokes Equations," *AIAA Journal,* Vol. 16, No. 4, 1978, pp. 393–402.

[28]Kolgan, V. P.,"Minimal Derivatives Principle Using at the Development of Finite-Difference Schemes for Calculations of Gasdynamic Discontinuous Solutions," *TsAGI Scientific Notes,* Vol. 3, No. 6, 1972, pp. 68–77.

[29]Tillyaeva, N. I.,"Generalization of Modified Godunov's Scheme on Space-Nonuniform Grids," *TsAGI Scientific Note,* Vol. 12, No. 2, 1986, pp. 19–26.

[30]Godunov, S. K.,"Finite Difference Method for the Numerical Calculations of Discontinuous Solutions of Hydrodynamic Equations," *Mathematical Proceedings,* Vol. 47, No. 3, 1959, pp. 271–306.

[31]Baruzzi, G.,"Structured Mesh Grid Adapting Based on a Spring Analogy," *Proceedings of the Conference of CFD Society of Canada,* 1993, pp. 425–436.

[32]Bam-Zelikovich, G. M.,"Calculation of Boundary Layer Separation," *Izvestiya, Academy of Sciences, USSR, OTN,* No. 12, 1954.

[33]Abramovich, G. N., Applied Gas Dynamics, Nauka, Moscow, 1976.

[34]Schlichting, H., Grenzschicht-Theorie,Verlag G. Braun, Karlsruhe, 1965.

[35]Zakharov, N. N., "Influence of Heat Transfer on the Turbulent Boundary Layer Separation," Central Inst. of Aviation Motors, Moscow, Trudy No. 507, p. 70.

[36]Ikui, T., Matsuo, K., and Nagai, M.,"The Mechanism of Pseudo-Shock Waves," *Bulletin of the Japan Society of Mechanical Engineers,* Vol. 17, June 1974, pp. 731–739.

[37]Chang, P. K., *Separation of Flow*, Pergamon, New York, 1969.

Chapter 15

Rudiments and Methodology for Design and Analysis of Hypersonic Air-breathing Vehicles

James L. Hunt and John G. Martin

Introduction

HYPERSONIC, air-breathing, horizontal takeoff and landing vehicles are highly integrated systems with unprecedented levels of interdisciplinary interactions involving a broad spectrum of technologies. The intensity of couplings among vehicle systems such as propulsion, propellant feed, auxiliary power, thermal management, airframe, controls, and especially the coupling between aerodynamic and propulsive flowfields requires these vehicles to be more highly integrated than previously attempted. This increased level of integration will be substantially greater than that required for current high-performance aircraft, but the added effort will be the key to greater performance potential for these vehicles.

In today's hypersonic scenarios both cruise- and accelerator-type vehicles are of interest. The commonality shared by these vehicles is that all operate within an air-breathing corridor; they will be powered by air-breathing engines—a subsidiary cycle for low-speed acceleration (turbojets, ejector ramjets, etc.), ramjets to Mach 5, and scramjets to potentially Mach 20—and they take off and land horizontally on standard runways. Some will be designed to ascend to cruise at hypersonic speeds, Mach 5 to 12, 20 or more miles above the ground; others (aerospace planes) will continue to accelerate upward through an air-breathing corridor to Mach 25 and, with minimal rocket power, transition to a low-Earth orbit, 100 miles up.

This hypersonic air-breathing technology represents the capability to cruise and maneuver into and out of the atmosphere, to provide rapid response for low-Earth-orbit missions, or to attain very rapid transport service between remote Earth destinations. But there are differences between configurations dedicated for cruise and those that accelerate to orbit. The accelerator must have a much bigger inlet area relative to body cross-section area than the cruiser to facilitate sufficient thrust margin, and thus acceleration, to reach orbital speed. On the other hand, the cruiser requires no thrust margin at the design cruise speed. A primary aerodynamic/propulsion design challenge for the accelerator is to maximize inlet capture per unit drag across the Mach range at trimmed angles of attack, whereas maximizing configuration lift-to-drag ratio at the design point is a goal for cruisers. Because the accelerator design encompasses the spectrum of flight dynamics and aerodynamic-propulsion coupling issues, and to some extent embodies both capabilities, the aerospace-plane-configuration design space will be used herein to delineate the hypersonic air-breathing vehicle force and moment matrix and the engine/airframe integration/design methodology.

The aerospace-plane matrix encompasses a broad spectrum of configurations ranging from wing-fuselage, to lifting bodies, to conical arrangements, as depicted in Fig. 1. The matrix is bracketed on one extreme (left in Fig. 1) by an ogive-cylinder wing-body design where the engine integration is constrained by the wing. Because of inadequate precompression and air processing ability of the ogive cylinder and an engine size constrained by the wing, this design reflects quasi-cruise characteristics that limit its acceleration potential. On the other extreme (right in Fig. 1) is a conical configuration with engines wrapped completely around the fuselage. This arrangement provides the ultimate in capture-per-unit parasitic drag and thus acceleration capability. However, with current heavy engine weights the fuel fraction available for this conical arrangement becomes unattractive for single-stage-to-orbit (SSTO) missions. Moreover, this vehicle must fly very close to zero angle of attack from takeoff to near Mach 24, which is difficult to attain.

The configurations of primary interest for aerospace-plane applications are concentrated at the center of the matrix of Fig. 1. These involve 1) high fineness-ratio conical winged bodies with partially wrapped engines, 2) mod-

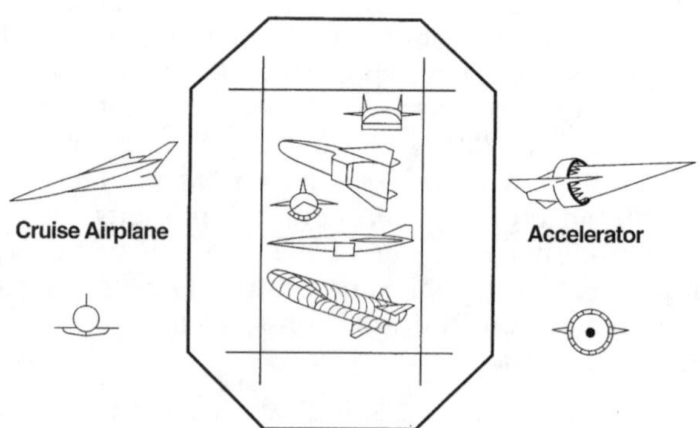

Fig. 1 Aerospace plane matrix.

erate fineness-ratio lifting bodies with rotating wings (horizontal control surfaces) with quasi-two-dimensional propulsion flow paths considered as a reference herein, and 3) inward-turning inlet/nozzle configurations that may offer more installed specific impulse. This chapter focuses on the flight dynamics of the reference aerospace-plane configurations to illustrate the trim setup (forces and moments) for hypersonic air-breathing vehicles and analysis/design methodology being used at Langley Research Center (LaRC) to resolve the matrix.

Rudiments of Design

Coordinate System

The right-handed set of coordinates that will be used in the discussions that follow is shown in Fig. 2 (Ref. 1). The body axes are fixed at the center of gravity relative to the body and, thus, move with the body. The x axis points forward, y to the right, and z downward. The wind axis is also shown in Fig. 2 and is distinguished from the body axis with the subscript a. The aerodynamic forces (lift and drag) and propulsion force (resultant thrust vector) are resolved in the wind axis. The propulsion force is calculated in the body axis and resolved in the wind axis. Forces are usually projected in the body axis for calculating moments and resolving trim; however, the resultant moment depends only on the reference point and is independent of axis. All forces are resolved in the wind axis for trajectory analysis.

In addressing the stability and control of an air vehicle, its motion in the plane of symmetry (x-z plane or pitch plane) can be uncoupled from the motion of the plane of symmetry. The motion in the x-z plane is referred to as longitudinal motion and treats linear motion along the x and z axes and rotation about the y axis. Motion of the plane on symmetry, known as lateral or lateral-directional motion; deals with linear motion along the y axis and angular rotations about the x and z axes.

For aerospace planes with underslung engine integration architectures, the resultant thrust vector couples with the resultant aeroforces primarily in

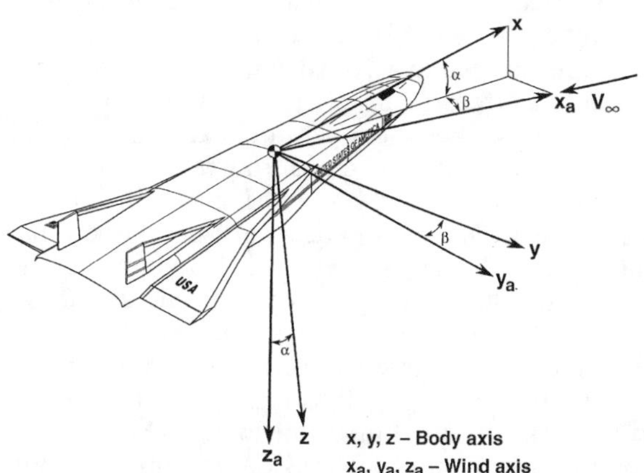

Fig. 2 Coordinate systems.

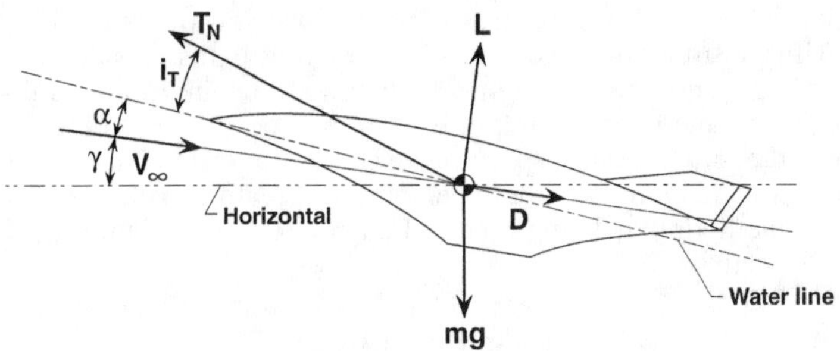

Fig. 3 Plane-of-symmetry coordinate system.

the *x-z* plane. This chapter, therefore, will focus on longitudinal motion and trim in the pitch plane. The pitch plane (plane of symmetry) coordinate system is shown in Fig. 3. It introduces a third reference system—the horizontal/vertical from which flight-path angle is measured and weight is projected, respectively.

Force Accounting System

The force accounting system defines the aerodynamic and propulsion force accounting responsibilities for both power-on and power-off conditions. The aerodynamic/propulsion interface is used to divide the pressure and shear forces acting on the vehicle surface into those charged to propulsion and those charged to aerodynamics.

For an underslung ramjet/scramjet airframe-integrated vehicle in which the vehicle lower forebody acts as a precompression surface and the vehicle lower aft body acts as a high expansion ratio nozzle, the entire undersurface of the vehicle is a propulsion flow path. This propulsion flow path is defined by those surfaces that are wetted by air that flows through the engine nacelle, and the forces acting thereon are charged to propulsion (Fig. 4a). This includes the lower external forebody, the interior nacelle, and the exterior nozzle aft body. Forces on all other exterior surfaces including the exteriors of the engine cowl and sidewall are charged to aerodynamics. This classic force accounting system is referred to as freestream-to-freestream or more commonly nose-to-tail. When the main engine is not operating, exterior forebody and nozzle forces are charged to aerodynamics (e.g., during re-entry).

In a similar manner forces on an Orbital and Ascent Maneuvering System (OAMS) must be accounted for. The OAMS is a linear aerospike rocket that is integrated at the trailing edge of the aerospace plane that augments the air-breathing propulsion system during takeoff, through transonic, and above Mach 15 and also serves as the orbital maneuvering system. Forces acting on the upper OAMS nozzle surfaces are charged to aerodynamics when the OAMS is not operating. When it is operating, those forces are charged to propulsion (see blowup in Fig. 4a). Forces acting on the lower OAMS nozzle surfaces are charged to propulsion for both OAMS on and off condition

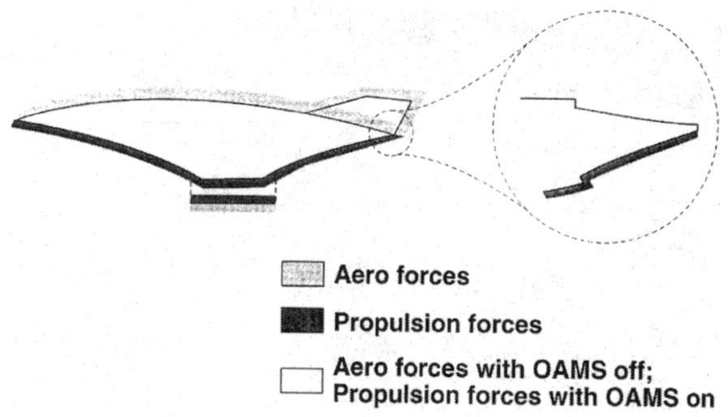

Fig. 4a Nose-to-tail force accounting system.

because these surfaces are contiguous with the aft-body nozzle and form its trailing edge.

A second force accounting system, known as cowl-to-tail, is shown in Fig. 4b. Here, the propulsion accounting begins at the cowl lip rather than the apex of the vehicle and proceeds through the engine and out the aft-body nozzle. This approach usurps the need to trace streamlines forward from the cowl lip to the freestream to define the forebody control volume in the nose-to-tail accounting system and gives the aerodynamics a more conventional role that now includes the lower forebody.

Other force accounting systems that have been employed, such as ramp-to-tail, fall in between the nose-to-tail and cowl-to-tail systems already discussed. Many of these systems were constructed to facilitate propulsion performance computations rather than to provide appropriate accounting for aero and propulsion.

If a control-volume (momentum balance) cycle-analysis approach is used to resolve the forces in the propulsion flow path as depicted in Fig. 5a (Ref. 2), then additional propulsion-related forces should be designated that do not represent actual forces acting on the vehicle propulsion flow-path surfaces but rather are a result of the way in which control volumes are defined. These are 1) spillage drag caused by shock losses associated with uncaptured spilled air (Fig. 5a); 2) plume drag, which is a fictitious drag captured by the control volume at the external nozzle-flow interface with the freestream flow (a virtual surface, Fig. 5a) and thus must be added back into the force accounting; 3) ram drag, which is the stream thrust at the forward control

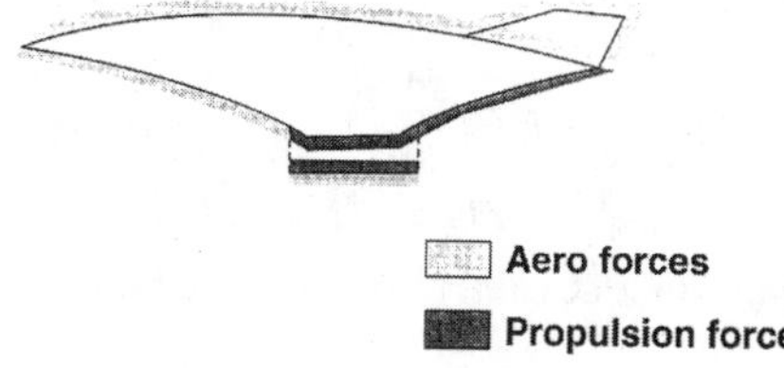

Fig. 4b Cowl-to-tail force accounting system.

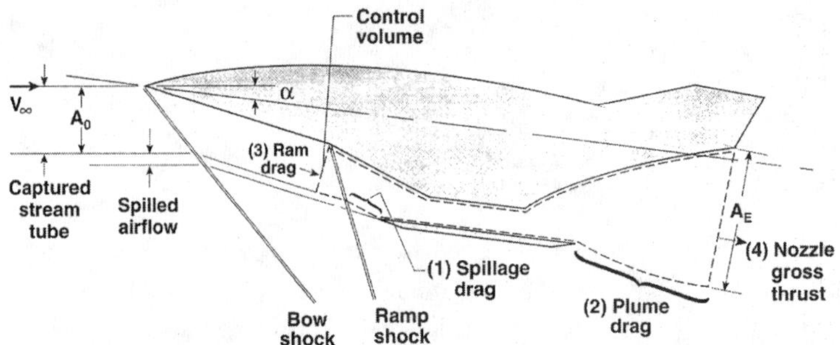

Fig. 5a Two-dimensional propulsion flow path, model, control volume, and vectoral relationship—traditional approach.

volume interface with the forebody flow (subsequently captured) that is required to perform the momentum balance across the combustor; and 4) nozzle gross thrust, which is the stream thrust at the nozzle-exit control-volume interface. At the time that this approach was first implemented in hypersonic propulsion cycle analysis, only forces in the flight direction were of interest; effective I_{sp} was the primary focus. The control-volume approach could not adequately predict propulsive lift and pitching moment, so an improved method was needed.

In propulsion cycle-analysis methods that integrate the pressures on the propulsion surfaces in contrast to the control-volume and momentum-balance approach, none of the preceding corrections are required. This also applies to hybrid schemes[3] in which a control volume is used only for the combustor force resolution, and wall pressure plus skin-friction integrations are used to resolve the forces on the remainder of the flow-path surfaces. Consequently, hybrid schemes lend themselves very well to propulsive lift and pitching-moment computations. Figure 5b illustrates a hybrid scheme. This approach has become more practical in recent history because of improvements in computational technology. Thus, in general, forces resolved

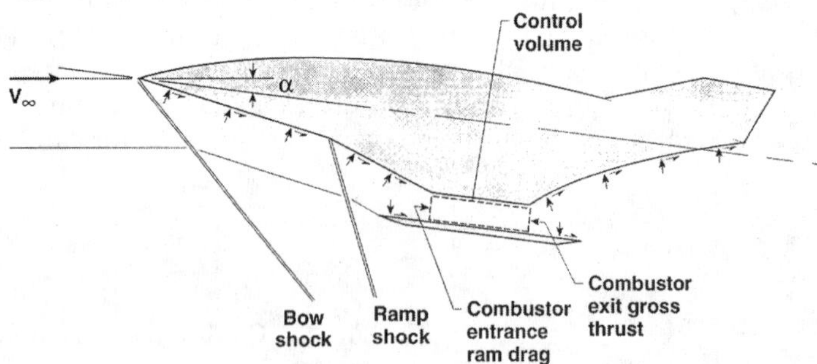

Fig. 5b Two-dimensional propulsion flow path, model, control volume, and vectoral relationship—hybrid approach.

from control volumes confined to interior surfaces require no virtual interface corrections.

Force accounting systems also have to address anomalies. For example, the main engine nozzle surface is charged to propulsion, independent of the state of the nozzle plume. Propulsion is responsible for all nozzle expansion surfaces even if the exhaust is overexpanded and not just the entire expansion surface wetted by the engine exhaust flow. If the exhaust is under expanded and the plume spills over onto adjacent aerodynamic surfaces (such as the sides of the aft body), the incremental effects of spillage on these aerodynamic surfaces are treated as jet effects and are book-kept by aerodynamics.

Details of hypersonic propulsion system force-accounting methods and the accompanying component-force summation equations are given in Refs. 4–8.

Nominal SSTO Vehicle/Trajectory

To illustrate the force, moment, and trim setup for a SSTO hypersonic air-breathing vehicle, it is necessary to define a nominal vehicle design flying a nominal trajectory. Because the vehicle design is predominantly defined by ascent loads, an ascent trajectory will be the focus.

The vehicle design selected for illustration is a moderate fineness-ratio lifting body. The configuration is quasi-two-dimensional with a spatular apex to provide high air-mass capture-per-unit parasitic drag to the underslung propulsion system. The forebody acts as a precompression surface to the quasi-two-dimensional ramped inlet system that transitions to a throat section, then to a constant-area combustor section, and then to an internal nozzle, all housed in a quasi-two-dimensional nacelle. The propulsion flow path terminates in a high expansion ratio aft-body nozzle with a portion of the trailing edge of the vehicle being made up of a linear array of aerospike rocket nozzles. A generic representation of the SSTO vehicle is presented in Fig. 6. It has twin verticals with trailing-edge rudders for directional control and fixed wings with elevons (or rotating wings) for pitch and roll control as well as body flaps situated outboard of the linear rocket nozzles.

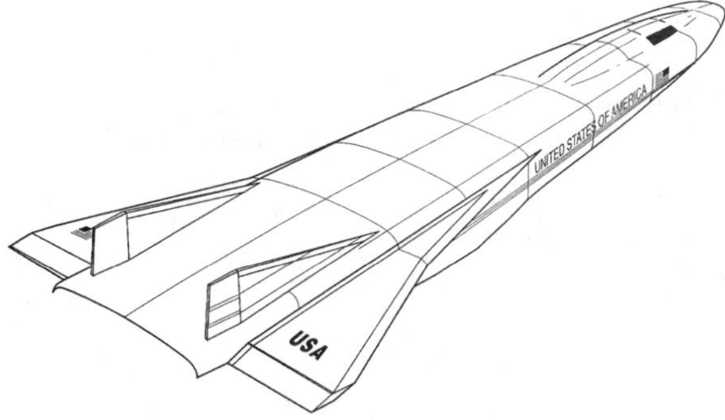

Fig. 6 Generic representation of nominal SSTO vehicle (moderate fineness-ratio lifting body).

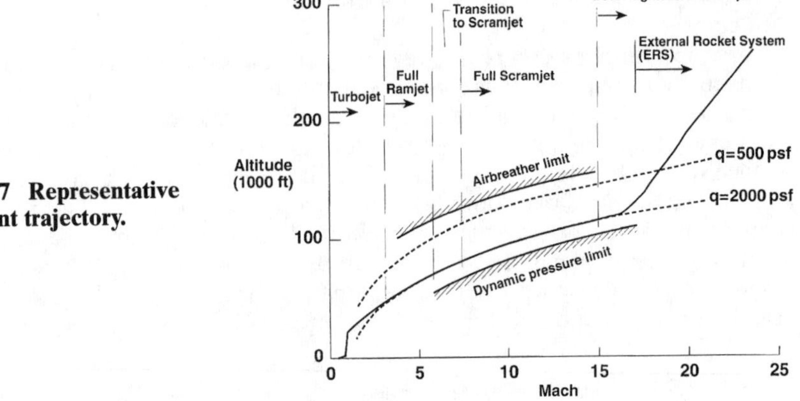

Fig. 7 Representative ascent trajectory.

The nominal ascent trajectory is presented in Fig. 7. The hypersonic air-breathing portion of interest here is the scramjet-powered segment above Mach 6. The dynamic pressure is 2000 psf to Mach 15+ where a pull-up is initiated and the external rocket is ignited for thrust augmentation. A nominal center-of-gravity schedule for the ascent trajectory is shown in Fig. 8. Liquid oxygen (LOX) is located in both fore and aft tanks and the c.g. moves forward from takeoff to Mach 3 because the LOX in the rear tank is used first. This allows a more rearward c.g. location for rotation at takeoff while providing the forward c.g. required for high-speed stability.

Loads

Aerodynamic and propulsion forces impart the external loading on a hypersonic vehicle. They are reduced to fuselage bending loads and superimposed on each other along with inertia loads to provide a net bending

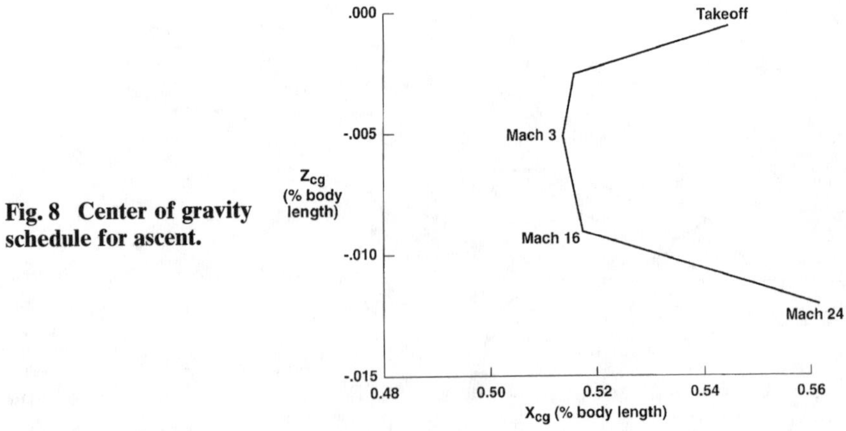

Fig. 8 Center of gravity schedule for ascent.

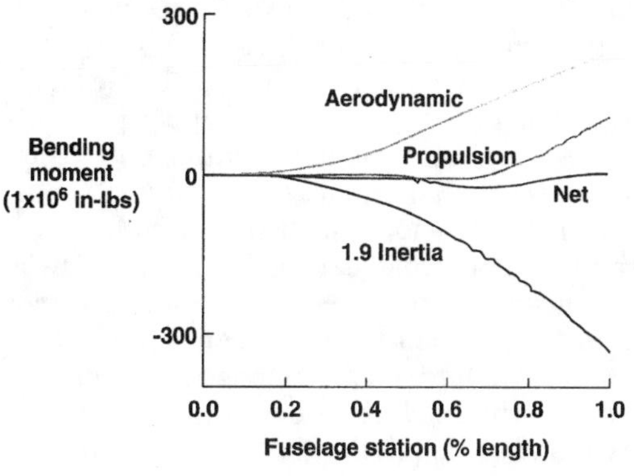

Fig. 9 Mach 10 load distribution.

moment on the airframe fuselage. These loads and their superposition for a Mach 10 flight condition are shown in Fig. 9 for the nominal configuration. The airframe structure must be designed to withstand these Mach 10 fuselage bending moments as well as all other flight-induced moments (i.e., thermal stress, vibration, propellant slosh, etc.). Therefore, the net maximum positive and negative bending moments for the nominal SSTO vehicle at each trajectory event are combined, as a function of fuselage station, to produce a loads envelope as shown in Fig. 10. For example, at a fuselage station of 20% (relative to vehicle length) the structure is designed by a Mach 0.9 pull-up bending moment of 15×10^6 in./lb and a Mach 1.2 pull-down bending moment of 20×10^6 in./lb; at 60% of length the structure is designed by a Mach 1.2 pull-up bending moment of 65×10^6 in./lb and a taxi bending moment of -60×10^6 in./lb. Note that at 80% of vehicle length the Mach 10 pull-up is a controlling load case and its moment of -15×10^6 in./lb appears in both Fig. 9 and Fig. 10.

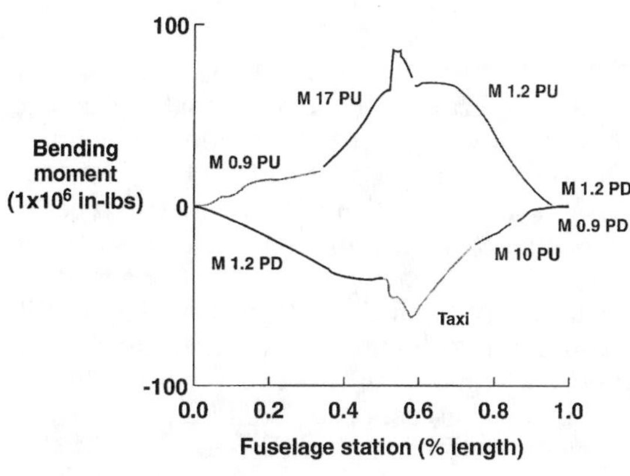

Fig. 10 Airframe design loads envelope.

External fuselage bending moments are resolved with finite element analysis (FEA) into internal running axial loads in the surface panels. A pull-up maneuver causes compression in the upper surface and tension in the lower surface. A push-down maneuver causes just the opposite. Those panel loads and other FEA-computed internal loads provide a composite design-to, mechanical loads envelope (Fig. 10) for structural component sizing. Mechanical loads must be superimposed with thermal loads produced simultaneously by material expansion of the structural shell from applied temperature increases. The resulting thermomechanical loads environment is complex. Panel shapes, sizes, and thicknesses must be optimized to obtain a minimum weight vehicle design for thermomechanical load environments.[9]

For maneuvers that cause the vehicle to have an angular acceleration such as pull-ups and push-downs, the induced inertia forces on the distributed mass are quantified about the vehicle c.g. and are applied with the vehicle external aerodynamic and propulsion loads. Vehicle equilibrium requires that the external loads be balanced at all flight conditions. As a result, net bending moments are zero at the fuselage forward and aft ends and are highest usually near the vehicle c.g. (Fig. 9). Aerodynamic, propulsion, and inertia loads are balanced with control-surface loads to maintain controlled flight.

The level of longitudinal static instability that can be managed by the control system is dependent upon the vehicle's thermoelastic stiffness. This is because control-surface actuation rate and frequency are limited by the airframe's structural elastic response. Formulation of the vehicle's thermoelastic stiffness must consider both types of temperature gradients: in-plane and through-the-thickness (see Ref. 10). Those temperature gradients reduce the vehicle's thermoelastic stiffness, which in turn lowers the vehicle's fundamental bending-mode frequencies. At hypersonic speeds during ascent, these gradients become severe; this is also a critical region of flight for vehicle stability and will be discussed in the next section. Bending-mode natural frequencies are driven down to the point where they begin to interfere with the reaction rate of the controls, impairing the vehicle's ability to manage some important stability behavior. The thermoelastic responses of a hypersonic vehicle can be quantified with coarse-meshed finite element models and optimized to meet aeroservo stability and control criteria.

Stability and Control

Airplane stability and control is concerned with the rotational motion and moments about an aircraft's c.g. and how to control this motion to maintain stable flight. This rotational motion is defined about the three body axes x, y, and z that are called roll, pitch, and yaw, respectively. There are two types of stability: static stability and dynamic stability. Static stability deals with the instantaneous tendencies of rotation whereas dynamic stability addresses the time-dependent rotational motion of a vehicle in flight. This chapter will only focus on pitch-plane motion (motion about the y axis) and static stability. Dynamic stability is beyond the scope of this text and will not be addressed, with the exception of limited discussion on some unstable static control issues that also have dynamic stability implications.

To be statically stable in pitch, a vehicle, upon being disturbed, must return by itself to its original angle of attack without the aid of the control system. Static stability is achieved by positioning the c.g. ahead of the neutral point. The neutral point is defined as the location on the vehicle where the total vehicle pitching moment (aerodynamic plus propulsive) remains constant as angle of attack varies, $(C_{m_\alpha} = 0)$.

Hypersonic air-breathing vehicles, and in particular accelerators, must be designed to fly through a broad Mach-number range. This range includes subsonic, transonic, supersonic, low hypersonic, and high hypersonic (Mach 15 +) speeds. These requirements place conflicting demands on achieving static stability. The neutral point will vary greatly as the vehicle accelerates through the Mach-number range. In addition, the c.g. location will vary as the propellants are consumed as was shown in Fig. 8. The level of stability therefore varies significantly as the vehicle traverses the Mach range as shown in Fig. 11. Here, static margin is plotted vs Mach number where the vehicle is nominally stable below Mach 5 and statically unstable above Mach 5.

At hypersonic speeds, static stability typically requires a very forward c.g. position, which is usually beyond the limits of the vehicle's c.g. envelope. Furthermore, this forward position usually exceeds what can be designed into a vehicle while trying to maintain an aft-ward c.g. position necessary for takeoff. This problem becomes exacerbated when the main engines are not operating, either during re-entry, or more significantly during ascent, if abort capability is desired at high Mach numbers. When the nozzle is unfilled with exhaust flow (unpressurized), the neutral point shifts forward. This leads to the conclusion that some level of static instability will be present at hypersonic speeds. To determine how much static instability can exist and remain manageable by the vehicle's controls, we will need to gain additional insight into some specific vehicle characteristics.

Static instability can be expressed in terms of time-to-double-amplitude. Time-to-double-amplitude, or time-to-double (T_2) in units of seconds, is the time required for the angle of attack of a vehicle to double for a given pitch disturbance. T_2 is a function of the stick-fixed rate of change of total-vehicle

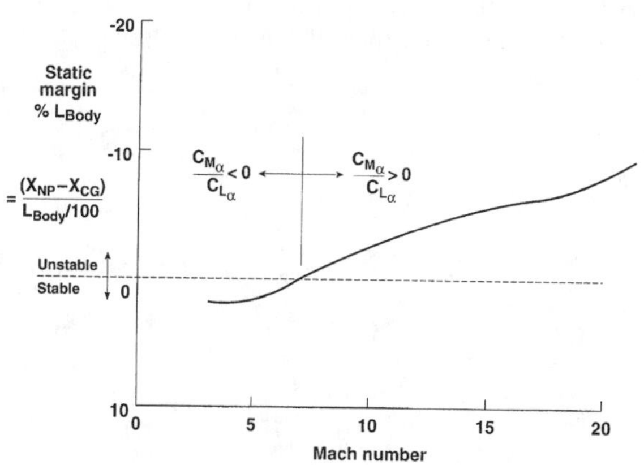

Fig. 11 Typical static margin variation along trajectory.

pitching moment M_a and pitch-plane moment of inertia I_{yy} and is shown in Eq. (1).

$$T_2 = \ell n(2)/\sqrt{M_a/I_{yy}} = \ell n(2)/\sqrt{\bar{q}S\bar{c}C_{m_a}/I_{yy}} \tag{1}$$

As M_a increases and I_{yy} decreases, T_2 decreases, and the more unstable the vehicle will be. Note that M_a or C_{m_a} is positive in this example because of the vehicle being unstable. Minimum acceptable values of T_2 are determined by the responsiveness of the control system and effectiveness of the control surfaces and will require six-degree-of-freedom (6-DOF) analysis to quantify. T_2 limits also can be influenced by vehicle flexibility if the fundamental bending-mode frequency of the fuselage gets low enough. In this case the rate at which pitch control can be applied will be restricted by this bending-mode frequency. Once minimum time to double has been established, an acceptable c.g. envelope can be defined that will satisfy all constraints.

Another approach to improving the high-speed stability problem is to increase I_{yy} by locating liquid oxygen (LOX) tanks at both ends of the fuselage. A significant amount of LOX usually remains in the tanks during the high hypersonic air-breathing phase that is reserved for the terminal boost, insertion, and orbital phases. Arranging the LOX tankage in this manner also provides the potential for expanded c.g. control by pumping LOX from one end of the vehicle to the other.

Representative Forces and Moments

Component forces acting on the reference SSTO configuration at Mach 7 are presented in Fig. 12 for an angle of attack of zero degrees and a flight dynamic pressure of 2000 psf. The forces are calculated for a vehicle sized at approximately 140 ft long with a reference planform area of 3400 ft^2, which does not necessarily represent closure for an SSTO mission. The accounting system is nose-to-tail; the forebody force is separated from the aerodynamic force vector, which is acting at the vehicle c.g. with a nose-down moment of 1.2×10^6 ft/lbf. Obviously, from the almost downward slant of the aerodynamic force vector, the negative lift acting on the lee surface and the wings/controls is on the order of four times that of the drag. The resultant thrust from the sum of the component forces on the propulsion flow path is

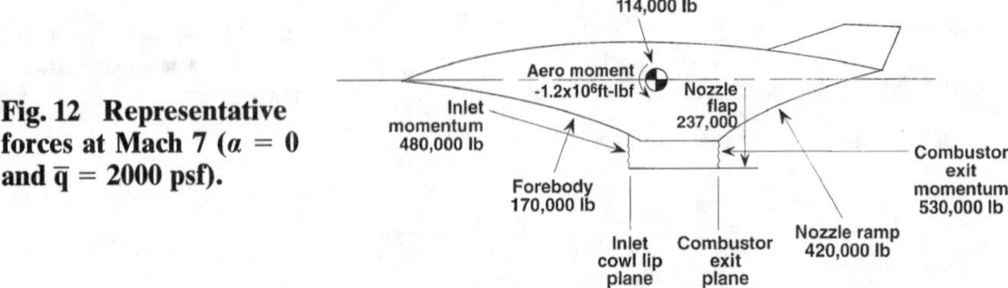

Fig. 12 Representative forces at Mach 7 ($\alpha = 0$ and $\bar{q} = 2000$ psf).

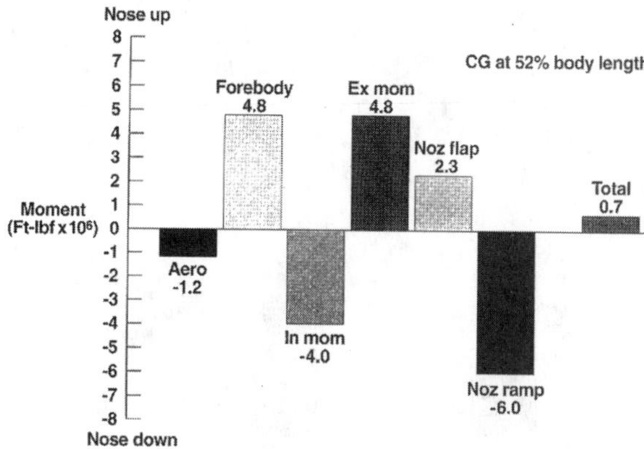

Fig. 13 Pitching-moment breakdown at Mach 7 (α = 0 and $\overline{q}$ = 2000 psf).

approximately 200,000 lb. The aft-body nozzle ramp is a very important contributor to the resultant thrust given that the difference between the inlet momentum and combustor-exit momentum parallel to the body axis coordinate is only 70,000 lb. The nozzle ramp contribution will become even more important at higher Mach numbers.

The pitching moment breakdown for this Mach 7 case ($a = 0, q = 2000$ psf) with a c.g. at 52% of body length is shown in Fig. 13. As indicated earlier in Fig. 13, the aerocontribution is shown as -1.2×10^6 ft/lbf of nose-down moment. In the propulsion flow path the forebody, combustor-exit momentum, and nozzle flap provide nose-up moment while the inlet momentum and nozzle ramp provide nose-down moment. The resulting moment is slightly nose up at 0.7×10^6 ft/lbf. To trim the vehicle, this moment will have to be counteracted using the body flaps and wing elevons (or rotating wings) depending on the choice of controls and will result in some additional drag (trim drag).

Component forces acting on the nominal SSTO configuration at Mach 15 are presented in Fig. 14, still for an angle of attack of zero degrees and a flight dynamic pressure of 2000 psf. The aerodynamic force is 35% lower than at Mach 7 (Fig. 12). However, the inlet and combustor-exit momentum forces are much higher than at Mach 7, but the difference in the two are much lower so that the vehicle must depend more on the nozzle ramp for thrust margin as the Mach number increases.

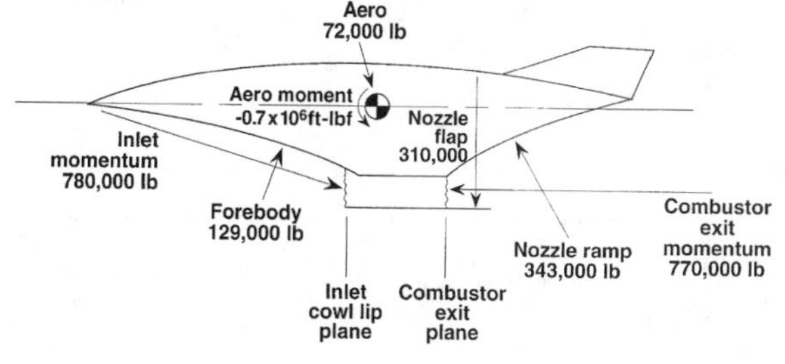

Fig. 14 Representative forces at Mach 15 (α = 0 and $\overline{q}$ = 2000 psf).

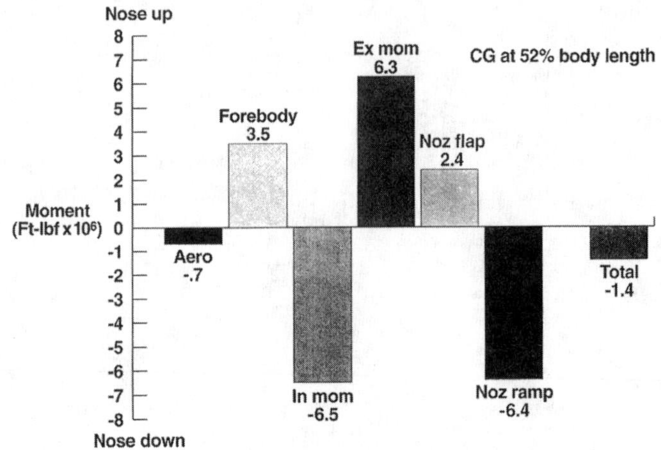

Fig. 15 Pitching-moment breakdown at Mach 15 ($\alpha = 0$ and $\overline{q} = 2000$ psf).

The pitching-moment breakdown for this Mach 15 case is presented in Fig. 15. Compared to Mach 7, the aero and forebody moments are smaller, but the inlet moment, combustor-exit moment, nozzle flap, and nozzle ramp moments are larger. The important fact to note here is that the resultant moment is nose down, the reverse of the Mach 7 case.

Impact of Propulsion Lift on Aerodynamics

Engine/airframe integration is a major driver in shaping an air-breathing SSTO configuration. For the reference configuration focused on here, the coupled impact of the propulsion integration on vehicle aerodynamic performance is examined through a series of drag polars at Mach 22 (Figs. 16–20).

A conventional untrimmed drag polar at Mach 22 is shown in Fig. 16 for the nominal SSTO vehicle using a cowl-to-tail accounting system. Aerodynamic forces are determined by the pressure integral around the body plus

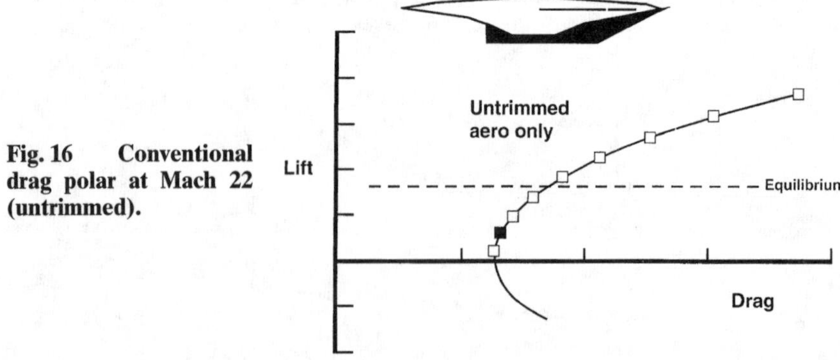

Fig. 16 Conventional drag polar at Mach 22 (untrimmed).

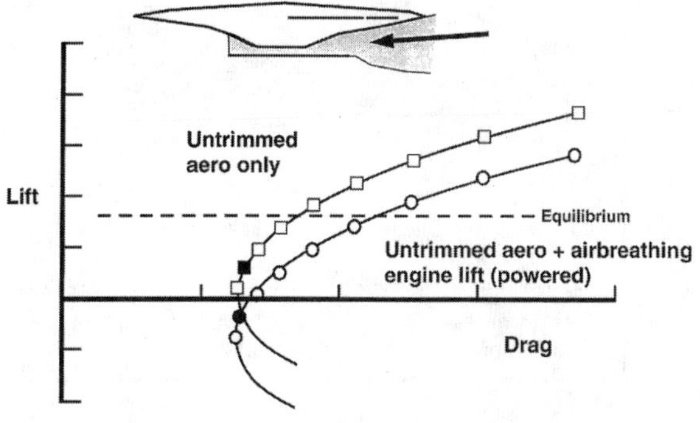

Fig. 17 Powered drag polar at Mach 22 (untrimmed).

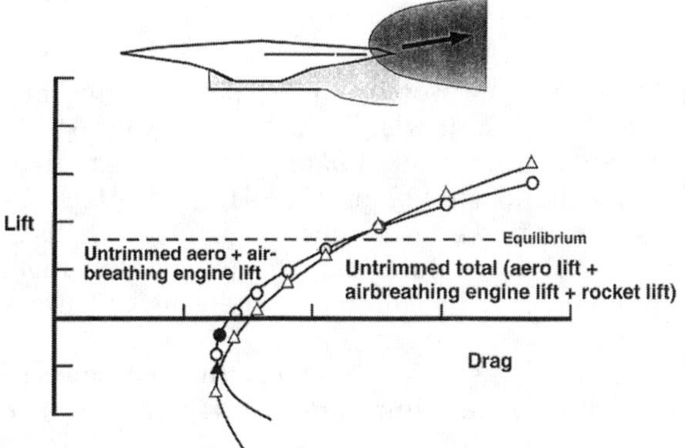

Fig. 18 Powered drag polar at Mach 22 including OAMS (untrimmed).

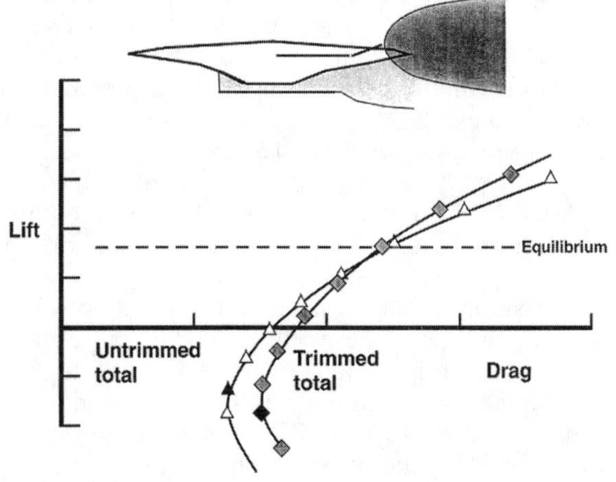

Fig. 19 Powered drag polar at Mach 22 including OAMS (trimmed).

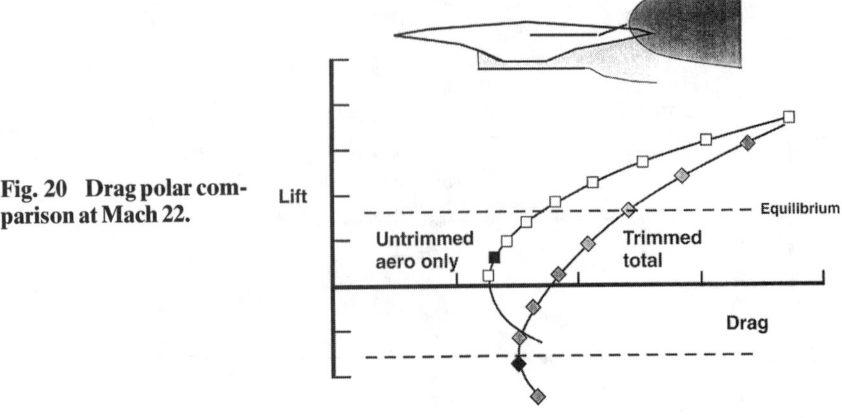

Fig. 20 Drag polar comparison at Mach 22.

skin-friction terms except for the internal propulsion flow path from the inlet cowl to the trailing edge of the aft-body nozzle. The external cowl is part of the aerodynamics. The drag polar shows symbol data points at angle-of-attack intervals of one degree; the darkened symbol is zero angle of attack.

For the purposes of this illustration, the assumption is made that the weight of the nominal SSTO vehicle is such that the equilibrium flight condition (lift equals weight minus centrifugal force) is at an angle of attack of 2.5 deg (Fig. 16).

In the next sequence of these figures (Fig. 17), the untrimmed powered drag polar is presented along with the untrimmed drag polar. In the powered polar the lift of the air-breathing propulsion system is added to the aerodynamic lift, but the net thrust parallel to the wind axis is not included. The line with the circle data points shows the impact of including the air-breathing engine lift. Note the large decrease in lift, which comes from the higher integrated pressures internal to the nacelle on the cowl side rather than on the body side. The forebody turns the propulsion flow downward, and the cowl must turn it back parallel to the axis of vehicle, thus creating substantial negative lift. Converting the exit momentum into a single vector provides an equal and opposite force on the vehicle as shown in the figure—it has a slightly downward slope indicative of the negative propulsive lift associated with the thrust of the air-breathing propulsion at Mach 22. Note that the loss in lift caused by including the scramjet propulsion flow path would require an increase in angle of attack from 2.5 to 4.5 deg to maintain equilibrium conditions.

In Fig. 18 the OAMS lift forces are included in the untrimmed powered drag polar. For this aft-rocket orientation, one would normally expect that its inclusion would result in higher drag because of the associated higher angle of attack required to maintain the desired lift. In this case the drag at zero lift is slightly lower than with the aft rocket off because of the thrust-drag accounting system; when the aft rocket is off, the base of the rocket is a small drag penalty at this high Mach number. The biggest difference is caused by the large lift changes with angle of attack. As the angle of attack increases beyond the initial installation angle, the rocket lift loss can turn into a lift benefit. Note that trimming the rocket nose-up moment can also offer a lift benefit, but also potentially a trim-drag penalty depending on the location of the c.g.

The final step in establishing the powered drag polar is to include the vehicle's control-surface trim effects. These effects can be a penalty or an advantage. In this case trim is a small advantage because of a slight reduction in drag for the same lift as shown in Fig. 19. Although the minimum trimmed drag is higher than the minimum untrimmed drag, the slope of the drag polar is more favorable, providing lower drag at the equilibrium flight condition.

In the last figure of this sequence (Fig. 20), the trimmed powered drag is compared to the untrimmed conventional aerodynamic drag polar. The shape is much different because of air-breathing propulsion (scramjet), the aft rocket, and the hypersonic vehicle's trim characteristics. We note a desensitivity to angle of attack in this geometry and some interesting characteristics. As shown by the negative-lift dashed line, this vehicle would have lower drag at the equilibrium condition if it flew upside down. A task for the hypersonic propulsion/airframe integrator is to determine whether this upside-down advantage was the result of poor integration or a consequence of designing the vehicle for best overall performance across the Mach range.

There are other aspects to this upside-down discussion than just drag at the equilibrium flight conditions at Mach 22. The powered drag polar does not include the net thrust parallel to the wind axis and that the upside-down equilibrium condition is at an angle of attack 0.3 deg. At this condition the forebody propulsion flow path is experiencing less compression, which would reduce the inlet capture and probably the net thrust, and, in turn, the effective specific impulse of the vehicle, which is the figure of merit in determining the fuel fraction required to perform the mission. (Net thrust is characterized with a probable reduction with less compression in this upside-down example because the nozzle-exit area relative to the capture area has been increased relative to the vehicle's upright equilibrium position. The installed specific impulse could be higher.)

The challenge facing the vehicle designer for this moderate fineness-ratio air-breathing SSTO lifting-body class of vehicles is to optimize configuration shape, propulsion integration, internal vehicle arrangement (for optimum c.g. travel), and control surfaces, so that the vehicle achieves the maximum effective specific impulse at each point along the ascent trajectory from takeoff to orbit. This design must then be capable of re-entering, descending, and landing unpowered in a safe and reliable fashion.

A nominal re-entry/descent trajectory is shown in Fig. 21. It was tailored to remain within the temperature capability of the thermal protection system and not to exceed a total acceleration of 1.5 gs. Re-entry is performed by flying at low dynamic pressure and nearly constant low angle of attack (~6 deg). Banking is used as a control to regulate vertical lift during re-entry in order to maintain an acceptable dynamic pressure level.

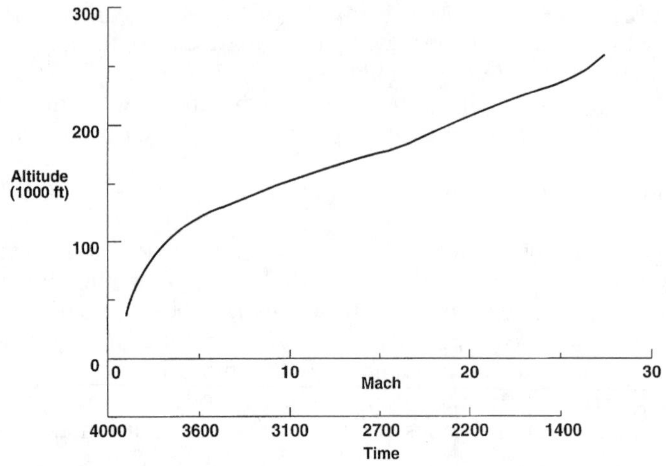

Fig. 21 Representative vehicle descent trajectory.

The trim issues for re-entry are less challenging than for ascent because the propulsion is off, the vehicle is very lightly loaded (no propellant left), and the c.g. is constant. The concerns here are to stay above the altitude of the thermal limits while remaining low enough to maintain aerodynamic control. This trade becomes near critical around Mach 10 where the use of reaction control jets for a short time may be necessary.

Engine/Airframe Integration Methodology

To accelerate the air-breathing corridor to Mach 10 or above, a hypersonic air-breathing vehicle's engine and airframe must be totally integrated so that within the propulsion flow path the forebody acts as a precompression surface for the dual-mode scramjet inlet and the aft body acts as a high expansion-ratio nozzle. More than any other activity, it is the high-speed engine/airframe integration that forges the design and shape of the vehicle. Thus, the methodology discussion will instinctively focus on the scramjet engine/airframe integration. [The low-speed engine system whether an over-positioned turbojet/turboramjet (two-duct) or an ejector ramjet (single-duct) has much less influence on shaping the vehicle.]

Scramjet engine/airframe integration methodology can be classified into four levels (Fig. 22, Ref. 11). Level 1 uses analytical methods and generally includes iteration on closed-form solutions, which are coded into fast-running computer programs. Level 2 makes the transition to numerical analysis and includes finite difference/element/volume inviscid (Euler) flowfield analysis and heat conduction/transfer codes. Also included in Level 2 are the integral boundary-layer codes and finite element stress analysis codes. Levels 1 and 2 constitute the engineering methodology category because they are used extensively in conceptual/preliminary design and performance

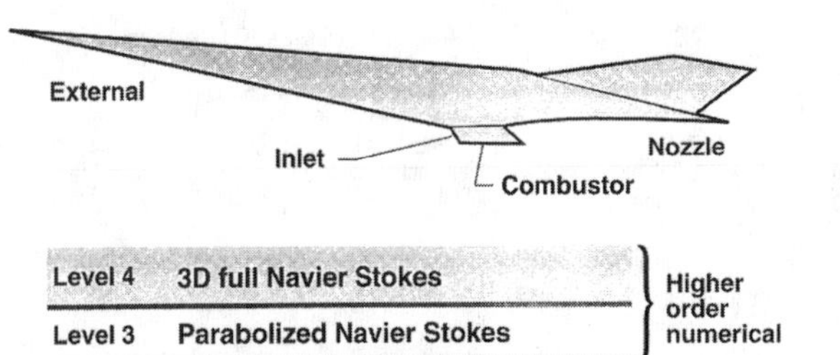

Fig. 22 Methodology classification levels.

tasks. Level 2 also includes hybrid methods that combine and integrate methodologies across the fluid-structural-thermal disciplines.

Level 3 consists of the parabolized Navier–Stokes (PNS) finite difference/volume codes that are used for parabolic problems. These flows generally consist of large supersonic regions with only embedded subsonic pockets. Level 4 is the highest level of analysis and consists of time-averaged Navier–Stokes (TANS) codes. These can be full Navier–Stokes (FNS) codes or Navier–Stokes solutions using the thin-layer approximation (TLNS). These are used for flows that are viscous dominated and elliptic in nature, i.e., downstream pressure feedback effects are included. The Navier–Stokes (NS) codes allow shear stress and heat transfer to be computed directly. Also included in Level 4 are the new coupled multidisciplinary codes that include significant interaction among the fluid-structure-thermal effects.

Do not be confused by the levels. More is not always better. PNS is completely appropriate for flows that have no separation, whereas FNS would provide the same results at greater cost. Engineering methods are also best for preliminary trade studies because they are rapid and use few resources.

Engineering Methods

Engineering methods constitute the Level 1 and 2 classes of methodologies. They are used primarily in conceptual and preliminary design and performance tasks.

Cycle Analysis

The ramjet/scramjet cycle code used for characterizing performance as well as refining flow-path design for highly integrated engine/airframe configurations at LaRC is SRGULL (Fig. 23, Ref. 3). It accurately resolves the net propulsive thrust of an air-breathing vehicle as a small difference between the combustor/nozzle thrust and the forebody/inlet drag. The forebody flowfield properties and the inlet mass capture that SRGULL predicts are critical in resolving the net thrust.

SRGULL uses a two-dimensional or axisymmetric Euler (finite difference,

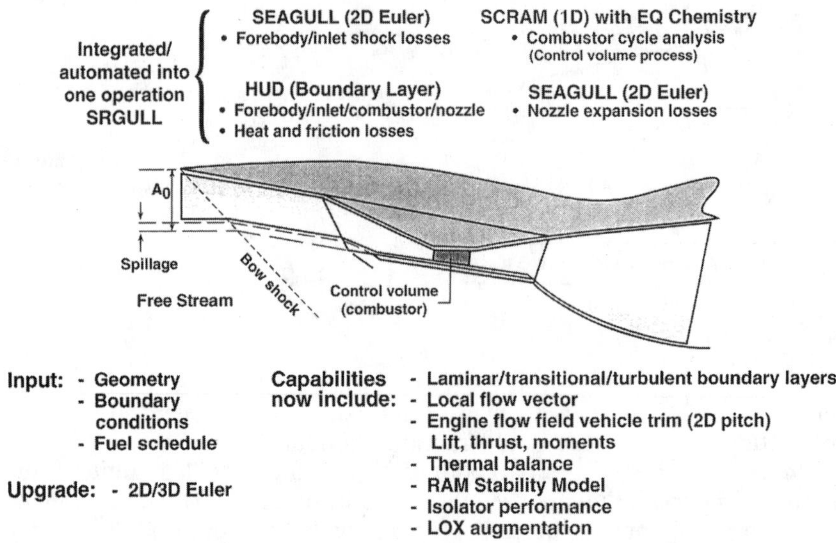

Fig. 23 Tip-to-tail scramjet/ramjet cycle analysis, SRGULL.

shock fitting) algorithm on the forebody and inlet, coupled with a boundary-layer solution, to predict the forebody/inlet drag and the flow properties entering the engine. The ramjet/scramjet solution is then completed using a one-dimensional cycle analysis with equilibrium chemistry and multiple steps through the combustor. A fuel-mixing distribution as a function of length is required input. Finally, the nozzle forces are resolved using the two-dimensional or axisymmetric Euler and boundary-layer codes. A three-dimensional Euler capability is now being implemented into the code.

Capabilities in the SRGULL code include the analysis of laminar, transitional, and turbulent boundary layers; integration of engine flow-path forces such as lift, thrust, and moments; and computation of LOX augmentation of the scramjet, which consists of small rocket motors in the flow path firing parallel to the flow just downstream of the throat either at stoichiometric, fuel-rich, or fuel-lean conditions. To first order, a thermal balance also can be accomplished. Given the wall temperature, heat flux to the walls (calculated by the code), and the fuel injection temperature, the amount of fuel flow required to actively cool the vehicle is then determined. This fuel flow rate is then used to predict the net thrust for a thermally balanced system. Particularly at high hypersonic flight Mach numbers, the increased fuel flow rate, which is generally above an equivalence ratio of one, will increase thrust and decrease engine-specific impulse, which to a point can actually increase the effective specific impulse. The prediction of coolant fuel flow rate is further refined in the thermal management analysis as described in the upcoming section.

SRGULL also has the capability to predict engine unstart,[12] which is

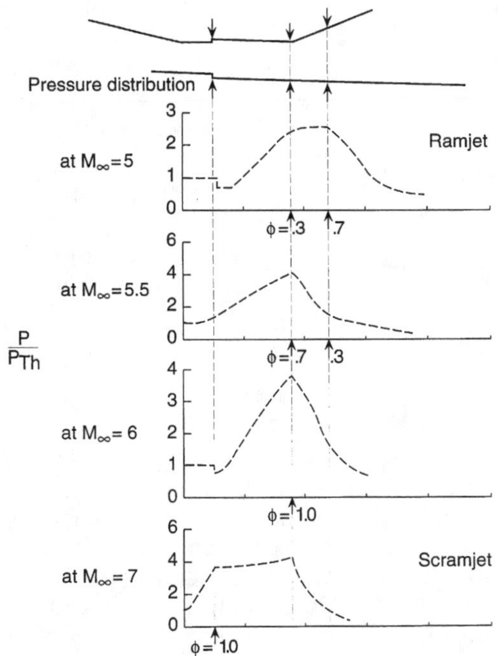

Fig. 24 Ramjet-to-scramjet mode transition with SRGULL.

another unique feature of this cycle code. Figure 24 (Ref. 12) shows an isolator/ramjet/scramjet keelline at the top. The arrows mark points where fuel can be injected. The four plots show the pressure distribution through the engine as a function of distance along the engine for various freestream Mach numbers where transition between pure ramjet and pure scramjet occurs. Note that in the top plot, fuel is being injected from the forward injector at an equivalence ratio of 0.3 and from the downstream injectors at an equivalence ratio of 0.7. Also note the rise in pressure that occurs upstream of the forward fuel injector. If more fuel were to be added at this fuel injector, the pressure rise would be pushed farther and farther upstream, until at some point an engine unstart occurs. Note that as the freestream Mach number increases, the fuel can be injected farther upstream without causing unstart.

Figure 25 (Ref. 12) shows an experiment run in a Langley tunnel to study the effects of geometry changes on isolator flowfield characteristics. As shown, SRGULL accurately predicts the pressure distribution in the isolator.

Inviscid Flow

Euler codes are used to approximate inviscid flows in support of flowpath design and performance analysis of external forebodies and inlets and aft-body nozzles in which forebodies precompress the air entering the inlet and aft bodies provide combustor flow expansion surfaces.

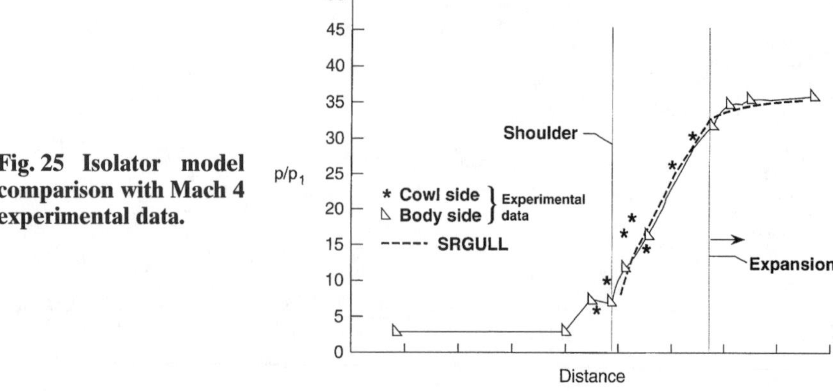

Fig. 25 Isolator model comparison with Mach 4 experimental data.

The two-dimensional/axisymmetric Euler code used in the Systems Analysis Office and Numerical Applications Office at NASA Langley is SEAGULL.[13] It was developed by Manual Salas at Langley in the mid 1970s. It is a floating shock-fitting technique in which second-order difference formulas are used for the computation of discontinuities. A procedure, based on the coalescence of characteristics, is used to detect the formation of shock waves. Mesh points that are crossed by discontinuities are recomputed. The technique provides resolution for two-dimensional external or internal flows with an arbitrary number of shock waves and contact surfaces. An example solution for the inviscid flow internal to a two-dimensional scramjet is presented in Fig. 26.

To resolve three-dimensional inviscid flows, an unstructured, adaptive mesh Euler code (SAMflow, Ref. 14) has been implemented at LaRC. The unstructured, adaptive mesh methodology[15] was selected to provide resolution of shocks in a capturing technique with minimum griding effort by the analyst.

The spatial discretization is accomplished via finite element techniques on unstructured tetrahedral grids. To achieve high execution speeds, edge-based data structures are used. Either central or upwind flux (Van Leer,

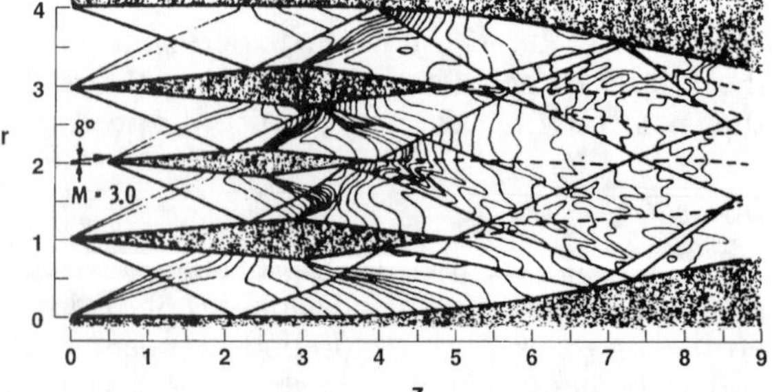

Fig. 26 Flowfield for a simulated scramjet, showing shock waves, vortex sheets, and isobars.

Roe) formulations can be used. For the temporal discretization both Taylor–Galerkin and Runge–Kutta time-integration schemes are available. Monotonicity of the solution may be achieved through a blend of second- and fourth-order dissipation, flux-corrected transport (FCT), or classic total variational dimensioning (TVD) limiters. The equations of state supported by SAMflow include ideal gas, polytropic gas, and real air table look-ups.

A variety of boundary conditions can be prescribed to simulate engineering flows: subsonic, transonic, and supersonic inflow/outflow boundary conditions; total-pressure inflow boundary conditions; static pressure, Mach number, and normal flux outflow boundary conditions; and porous walls and periodic boundary conditions.

An example application is shown in Fig. 27 in which the SAMflow code is used to resolve the three-dimensional nose-to-tail inviscid flow on a Mach 10, lifting-body airplane. These calculations were used to quantify the three-dimensional inlet and nozzle flows in a dual-fuel lifting body configuration development study.[16]

Also, the methodology[15] includes the capability for treating moving boundaries with prescribed motion or moving rigid bodies with motion computed from six degree-of-freedom mechanics using the computed aerodynamic forces that are then linked back to the flow solver. To this end, the equations that constitute SAMflow are solved in the arbitrary Lagrangean–Eulerian (ALE) frame. It is from this perspective, in addition to steady-state solutions, that SAMflow is being used to assist in resolving the Hyper-X stage separation flow dynamics and resulting aerodynamic forces.

Boundary Layer

Boundary-layer calculations are required in the propulsion flow path in conjunction with inviscid flow predictions to quantify heat transfer, skin friction (shear), and displacement thicknesses. For engineering calculations the integral method is a reliable option.

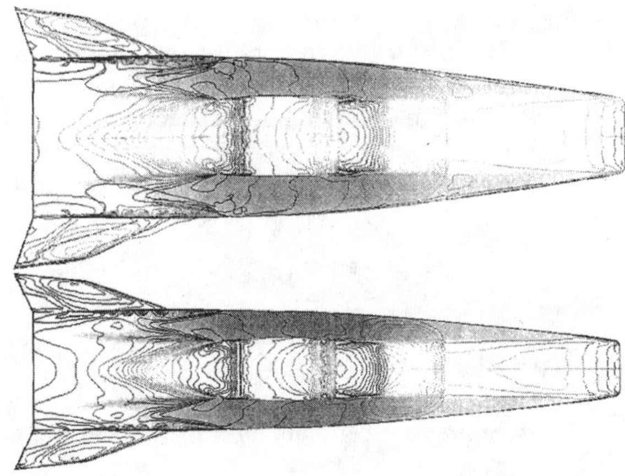

Fig. 27 Three-dimensional inviscid pressure contours on lifting-body cruise configuration at Mach 10 (design point) for two fineness ratios.

In the cycle calculation SRGULL[2] the basic integral method used[17] is applicable to the prediction of axisymmetric and two-dimensional laminar and turbulent boundary layers. It requires the simultaneous solution of the integral momentum, moment of momentum, and energy equations. To obtain this simultaneous solution, auxiliary relations are used for the boundary-layer velocity and enthalphy profiles, the shear distribution across the boundary layer, and the local surface friction and heat transfer, all of which are derived to be a function of the local pressure gradient and the total heat removed from the boundary layer forward of the local station. These relations are derived using modified flat-plate log-log type velocity profiles for pressure gradients as a basis of departure from flat-plate solutions,[17] modified flat-plate Crocco-type enthalpy-velocity profile[18] to account for the total heat removed from the boundary layer, and flat-plate friction and heat-transfer methods (Reynolds analogy). For laminar boundary layers the flat-plate friction correlation method used is a combination of the Blasius incompressible friction coefficient correlation[19] and Eckert's reference temperature method for the compressibility correction.[20] For turbulent boundary layers the flat-plate friction correlation method used is the modified Spalding–Chi method of Neal and Bertram.[21] For the heat transfer the flat-plate method is the modified Reynolds analogy of Colburn.[22]

For more general applications boundary-layer predictions are calculated with a boundary-layer integral matrix procedure (BLIMP, Ref. 23). This well-known/widely used code was developed through U.S. Air Force funding to compute viscous boundary-layer effects over two-dimensional axisymmetric or planar conditions as inputs. Results from the Euler solver SAMflow[14] are used to provide boundary-layer edge conditions to BLIMP. The edge conditions are supplied along inviscid streamlines along which the integral BLIMP procedure parabolically marches. This provides a reasonable merging of the accuracy of SAMflow for three-dimensional inviscid flowfield computations and the reliability of BLIMP for viscous computations. In this manner boundary layers on three-dimensional configurations (propulsion flow path or aerodynamic surfaces) can be approximated; streamline divergence is included but without boundary-layer crossflow. An example of the coupled SAMflow-BLIMP software application on a hypersonic configuration in terms of pressure contours and heat transfer/shear stress distribution at Mach 2 is given in Fig. 28a and 28b, respectively.

Fig. 28a SAMFLOW-generated surface pressure contours for a hypersonic cruise vehicle design.

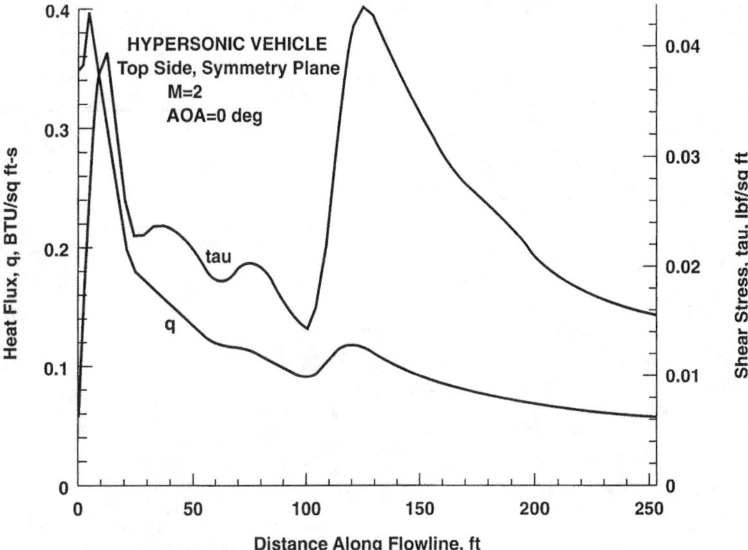

Fig. 28b BLIMP-generated heat flux and shear stress along top centerline for a hypersonic cruise vehicle design.

Thermal Management

The thermal management approach used for hypersonic air-breathing vehicle analysis at LaRC[24] is based on a three-dimensional transient thermal analyzer (SINDA-85, Ref. 25). It has been deemed the Integrated Numerical Methods for Hypersonic Aircraft Cooling Systems Analysis and includes capability for Thermal Protection System (TPS) sizing.[26] The focus here is the propulsion flow path only.

Generally it is known a priori that the engine flow path requires active cooling. An example of a coolant routing along the keel line of the inlet, combustor, and nozzle on the body side of the propulsion flow path is shown in the upper left-hand corner of Fig. 29. Schematically, the active cooling network is shown in the middle of the figure. Inputs to the network analysis include the initial coolant system architecture, propulsion heat loads, flow-path geometry, coolant supply temperature, coolant and material properties, and the total-pressure drop through the network based on the pumping system and the desired fuel-injection pressure. From this the coolant mass flow, temperature, and pressure distribution, along with the panel temperature distribution are determined. The panel temperatures are checked to ensure that they remain below the material temperature limits. Also, panel stresses are calculated, and minimum thicknesses are computed to meet structural requirements. For example, if a hole is punctured in one of the cooling-panel walls, the stress on that wall must not be allowed high enough to cause the panel to unzip. The network architecture and panel designs are

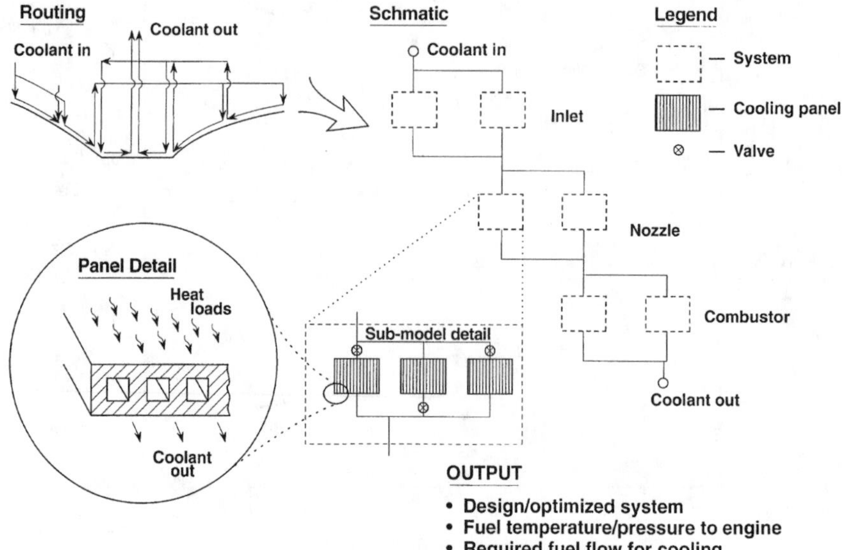

Fig. 29 Cooling system design/analysis.

modified until the overall cooling system weight and coolant flow rate are minimized, while meeting the preceding constraints. As noted in the propulsion section, the coolant flow rate and the fuel-injection properties have a significant impact on the net propulsive thrust and specific impulse.

As an example, consider the cooling network design for the Access to Space air-breathing/rocket SSTO vehicle.[27] Slush hydrogen was stored in the tank at 20 psig and 25° R. It was pumped to 5500 psi and 60° R before circulating through the cooling panels, then through a turbine to drive the pump, back into the cooling network again, and out into the combustor. The heat exchangers were sized at Mach 15 conditions where the heating rates were the greatest. The cooling-panel network was designed to deliver hot hydrogen to the injectors. Detailed thermal and fluid analysis was conducted on the cooling panels to determine the channel dimensions, pressure drop across each panel, and material selection.

Structures

Hypersonic vehicle structures are characterized by thermal stresses that are as high as the mechanical stresses; for portions of the propulsion flow path, the thermal stresses can be even higher than the mechanical stresses. Because of the design sensitivities inherent in air-breathing hypersonic vehicles, it is necessary to accurately predict structural weight, as well as the aerothermoelastic flight response of the vehicle even at the conceptual/preliminary design level. Some of the codes used for this endeavor include Pro/ENGINEER[28] for computer-aided design, MSC/NASTRAN,[29], P3

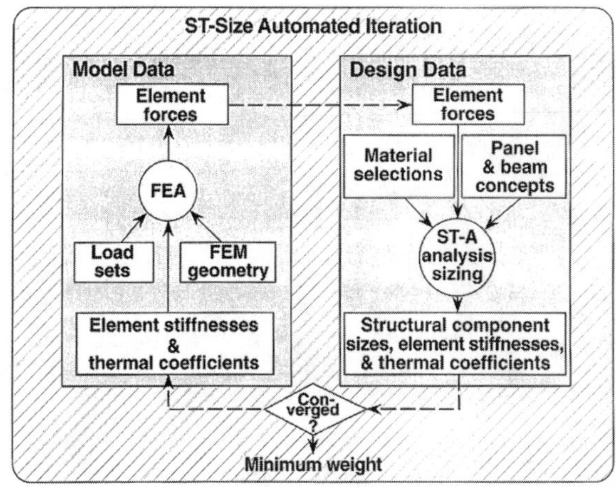

Fig. 30 Structural sizing process.

PATRAN,[39] Pro/MECHANICA[3] for finite element analysis to predict element loads, and an in-house developed software packaged called ST-SIZE[32] to perform panel-failure mode analysis and panel sizing.

The automated structural design process[32] is shown schematically in Fig. 30. This figure illustrates how a structural panel is sized in ST-SIZE.[32]. Starting on the lefthand side of the figure, initial element stiffnesses, thermal coefficients, thermal and mechanical loads, and the finite element geometry are input into the finite element analysis code. Forces on each of the elements are then determined. Moving to the right of the figure, the element forces, material selections, and panel and beam concepts are input to the ST-SIZE code. Here up to 30 failure mode analyses in strength and 26 failure mode analyses in stability are performed, and the panel is sized to meet all failure modes. Given the new panel design, the element stiffnesses and thermal coefficients change, and the FEA must recalculate the element forces. This iterative process continues until convergence is achieved. The net result is the minimum panel weight, resulting from a fully stressed panel that satisfies all of the failure mode tests, all within the margin-of-safety.

In general, structural panels of air-breathing hypersonic vehicles are unsymmetric—geometrically and/or thermally. As a result, traditional two-dimensional panel methods, which do not account for panel asymmetry, can predict inaccurate panel sizes. In contrast, an enhanced version of ST-SIZE, developed in SAO,[33], models the panel asymmetry. This is accomplished by calculating the membrane bending coupling in the two-dimensional element. The methods of ST-SIZE are the basis for the HyperSizer™code that is a commercial product of Collier Research and Development Corporation.[33]

The unit weights of the engine primary structure for the Access to Space air-breathing SSTO vehicle[27] were the result of finite element method FEM analysis and automated structural design using the structural/thermal sizing

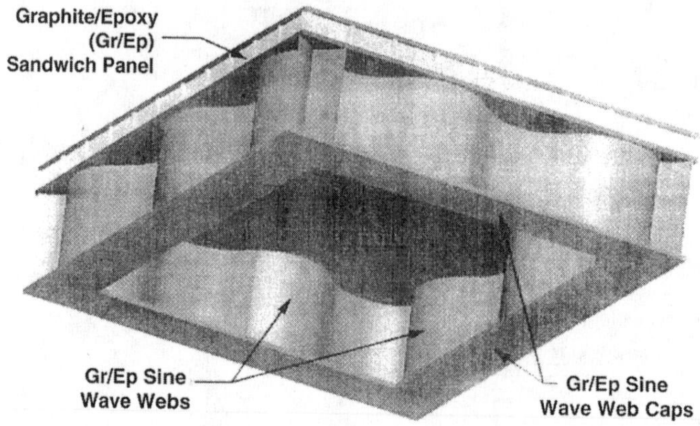

Fig. 31 Primary structure concept for engine of SSTO vehicle.

code ST-SIZE. The primary structure for supporting the propulsion flow path operating pressure loads was a system of honeycomb panels, backed by integrally attached stiffening beams made up of sine-wave webs and flat caps, as shown in Fig. 31. This arrangement transfers the engine forces to the trusses, which are directly attached to the integral tank structure of the airframe. These trusses also provide stiffness to the airframe and naturally invoke some load sharing. The primary structure of the engine is isolated from the hot gas in the flow path by nonintegral heat exchangers that transmit the pressure forces through to the honeycomb panels.

Higher-Order Numerical Methods

Resolution of the scramjet propulsion flow path in preliminary and final design activities, and especially the resolution of mixing and combustion in the combustor, requires the most sophisticated, computationally intensive and less stable numerical methodologies of Levels 3 and 4. These high-fidelity approaches with suitable modelings of turbulence, viscous effects, and chemistry are the full NS (elliptic) and PNS (marching) codes that capture both the inviscid and viscous flow characteristics simultaneously.

Full Navier–Stokes

The code most relied on in NAO to resolve the most complex problems in the flow path from three-dimensional shock/boundary-layer interaction in the inlet to fuel injection and mixing modeling in the combustor to three-dimensional expansion and possible relaminarization in a chemically reacting nozzle is the GASP code (General Aerodynamic Simulation Program, Refs. 34 and 35). It was developed to provide generalized numerical predictions of flows over aerodynamic and propulsion flow-path surfaces at the high level of fidelity and detail that hypersonic air-breathing vehicles require.

GASP is a finite volume, upwind-biased code that can solve one-dimensional, two-dimensional, axisymmetric, and fully three-dimensional flows.[36] It has various chemical and thermodynamic models for solving (single or multiple species) perfect gas flows, flow in chemical equilibrium, chemically frozen flows, and flows with finite chemical reactions. It can be run in the space-marching (time-dependent PNS) or elliptic mode, either implicit or explicit, with Euler, TLNS, and FNS terms. Turbulence is modeled by either the standard algebraic Baldwin–Lomax model, a high Reynolds-number model for shear flows, or a choice of either the Lam–Bremhorst[37] or Chien's[38] two-equation turbulent models that integrate completely through the boundary layer.

The GASP code is versatile[39] because of its multiblock and multizone features and is convenient to use for solving complex flowfields. The ability to switch from solving the FNS equations (elliptical) to the PNS equations (marching mode) at any streamwise location in the computational domain makes it very convenient and efficient to use.

GASP is routinely used for the analysis of scramjet component and engine flow-path performance. Figure 32 illustrates one such solution for a powered wind-tunnel model tested at NASA LaRC. This type of analysis provides comparison with experimental data. Correlation with the experimental data provides confidence in the ability to predict flight vehicle engine performance. The GASP code has also been compared with simple unit inlet, combustor, and nozzle experimental databases. Figure 33 represents calibration[36] of the GASP turbulence model for nozzle heat transfer. This study demonstrated the requirement for a two-equation turbulence model for nozzle relaminarization effects on heat transfer. Similar studies have illustrated turbulence-modeling requirements for the inlet shock boundary-layer interactions,[40], including flow separation modeling and grid resolution for the solution of shock-shock interaction heat flux on cowl leading edges. The GASP code has also been extensively verified for combustor analysis.[39,42–44]

Parabolic Navier–Stokes

PNS or space-marching solutions are adequate for much of the scramjet flow path, including large regions of the forebody and all of the nozzle.

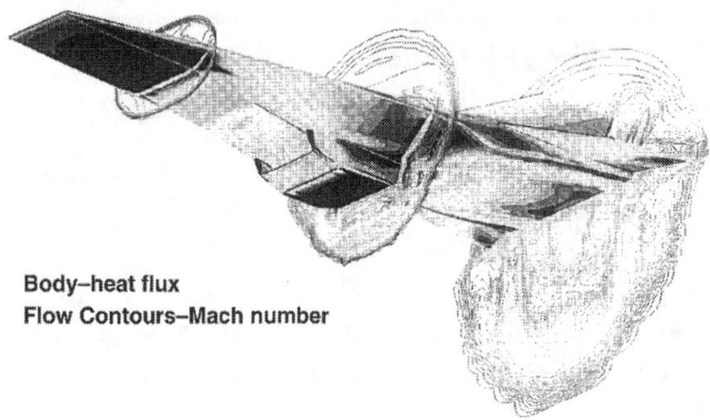

Fig. 32 Powered hypersonic (Mach 7) vehicle CFD solution.

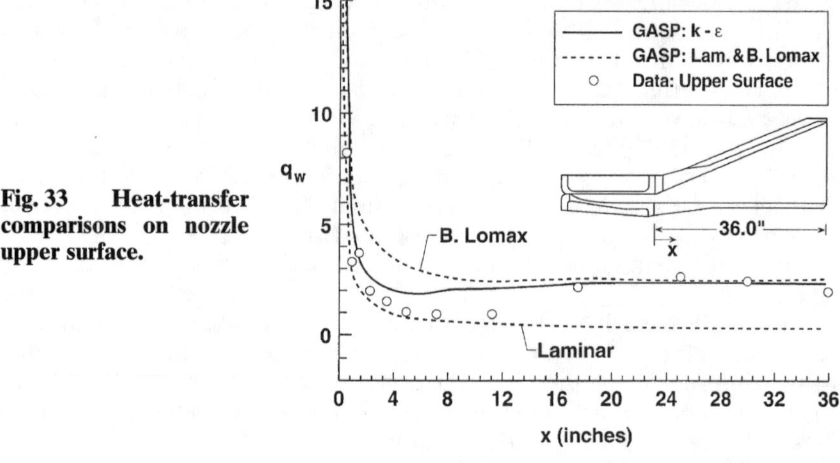

Fig. 33 Heat-transfer comparisons on nozzle upper surface.

Design and analysis of scramjet fuel injection, mixing and combustion is one of the best uses for three-dimensional CFD methods. This process cannot be modeled with simpler methods, as the flow will always be three-dimensional. Effective design evaluation of scramjet combustor performance requires a rapid, approximate method for screening concepts. The Supersonic Hydrogen Injection Program (SHIP) was developed for that purpose. The scramjet combustor, being predominantly supersonic flow, can be approximated using either space-marching (GASP) or PNS SHIP solutions. The small subsonic regions are approximated by wakes, established by forcing the flow downstream, as described in Ref. 45. The SHIP three-dimensional code solves the parabolized, Favra-averaged equations for the conservation of mass, momentum, total energy, total fuel, and turbulence fields in a variable area domain of rectangular cross section.[46] Turbulence resolution is at the two-equation level, with one of several high-*Re* or low-*Re* models, including corrections for compressibility. The governing transport equations are solved by the SIMPLEC pressure correction algorithm[47] extended to compressible flow.

Vehicle Design Methodology

Hypersonic air-breathing vehicles are predominantly shaped by the integration of the propulsion flow path with the airframe, as discussed in the proceeding section. However, the emphasis in this section is on the entire vehicle design. Although the propulsion flow-path design shares some design disciplines and analytical tools with the airframe (i.e., engine thermal management and structures as already described), there are some disciplines yet to be covered. These include aerodynamics and aerothermodynamics, airframe structures and weights, airframe thermal management (i.e., TPS sizing and propellant boil-off estimates), trajectory analysis, and vehicle synthesis and sizing.

Aerodynamics/Aerothermodynamics

All speed regimes must be considered in the aerodynamic analysis of a hypersonic vehicle. For aerothermal analysis in which thermal loads are generated, only supersonic and hypersonic speeds need be considered. APAS[48] is used for engineering analysis suited for the conceptual design phase. It is an interactive code that features the Unified Distributed Panel (UDP) Method for subsonic, transonic, and low supersonic analysis, and the Hypersonic Arbitrary Body Program (HABP) for hypersonic analysis.

HABP can calculate surface pressure and skin-friction coefficients, radiation-equilibrium wall temperature or heat flux for a given wall temperature, for each surface element of an analysis model. These properties can either be mapped onto a structural or a thermal analysis model as input or integrated for aerodynamic force and moment coefficients to be used in vehicle performance analysis. HABP can also compute aerodynamic stability derivatives for stability and control analysis. Tangent-wedge and tangent-cone analysis techniques are used in HABP to predict pressure forces. Shear forces are predicted using either the Reference, Temperature, Reference Enthalpy, or the Spalding–Chi empirical correlation methods.[20–22]

Traditional panel methods used for subsonic and supersonic aerodynamic analysis are well suited for wing-body configurations that derive a majority of their lift from the wings. These methods often lack resolution, both in magnitude and occasionally in trends, for hypersonic air-breathing configurations that derive a majority of their lift from the body. To achieve accurate results, three-dimensional Euler with integral boundary-layer solutions[14,23] are used to anchor the low-speed aerodynamic database.

Structures/TPS Sizing

Typical air-breathing hypersonic airframe structures can be categorized into two basic load carrying classes: hot shell structure with nonintegral, insulated tanks where the structure is directly exposed to the external aerodynamic and propulsion heating; or cold integral structure in which the tanks carry the load and are insulated from external heating by a TPS.

Analytical examinations of the two types of structures were conducted on an air-breathing SSTO vehicle[27] and a Mach 10, dual-fuel global-reach airplane[49] using finite element, failure mode analysis with appropriate load cases for each mission. Results show that for these mission scenarios the cold integral-tank structural approach, using graphite/epoxy tanks, was lighter weight and more viable than the titanium matrix composite (TMC) hot shell structure. The structural analysis tools used in these analyses were similar to the ST-SIZE (Hypersizer™,[33]) code profiled in the engine integration section.

Sizing the TPS to accommodate the heating environment and operational

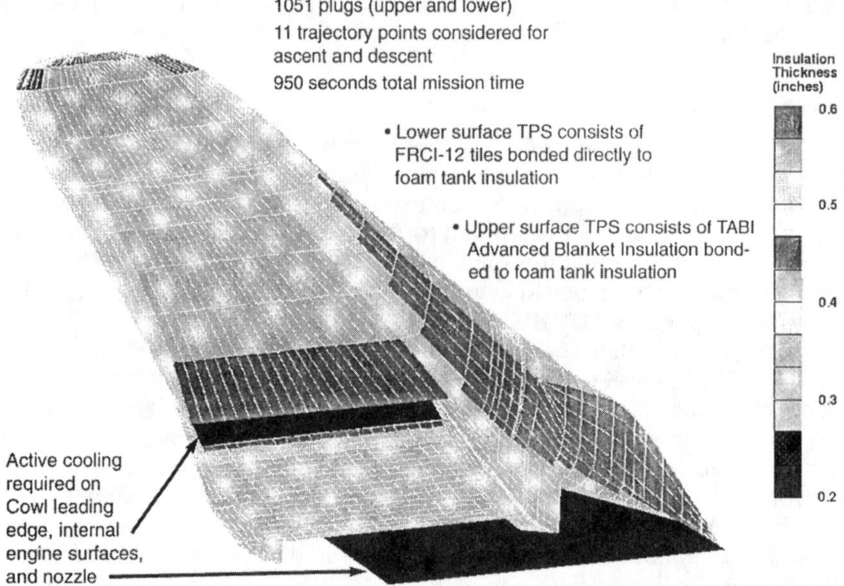

Fig. 34 Thermal protection system thickness distribution.

considerations such as cryo-pumping and frosting during ground hold, and boil-off during flight, is performed by the same software tools that are used in the thermal management analysis of the engine as already described in the thermal management section. Required inputs are aerodynamic and propulsion heating at multiple points along the trajectory. In conceptual and preliminary design engineering codes such as APAS[48] can be used. Improved resolution of the heating environment can be obtained using MINIVER.[50] Navier–Stokes codes can be used for even higher levels of accuracy.

Figure 34 shows the results of a transient thermal analysis completed using the SINDA code[26] on an SSTO configuration with a cold integral structure. Note that the surface is divided into more than a thousand panels. The color coding represents the appropriate thermal protection thickness as predicted for each plug (panel). The transient analysis starts with the predicted heat flux into each panel (in this case using the APAS code) at several points along the trajectory. Figure 35 shows a representative cross-sectional view of one panel, or plug. Node 1 is the exposed surfaced of the vehicle. The TPS is located between Node 1 and Node 7 where Node 7 represents the bond between the TPS and the fuel tank insulation. Below Node 9 is the structural panel and is typical for structural panels that are integral with the tank as described in the preceding section. For a nonintegral structure where the structural shell is separate from the tank, the foam insulation is on the inside of the structural shell, and the TPS is attached directly to the structure.

1002 plugs, 18 ascent and descent trajectory points analyzed

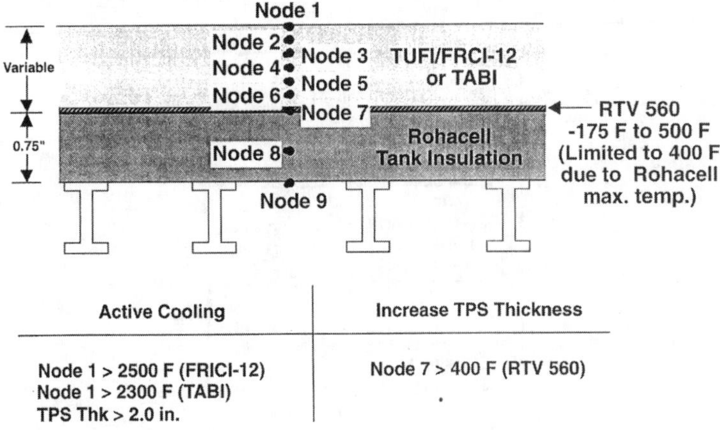

Fig. 35 Access-to-Space SSTO thermal analysis plug model.

Closure

Closure is the process by which a vehicle's size (both weight and scale) that will satisfy the mission requirements is determined. The closure process is carried out by first analyzing a baseline (as drawn) vehicle. Vehicle performance is computed to determine the propellant required to perform the mission and is typically expressed as propellant fraction required (PFR). Propellant fraction is simply the propellant weight divided by the takeoff gross weight (TOGW). The propellant fraction available (PFA) for the as-drawn vehicle is determined from the vehicle synthesis process. If the PFA is less than the PFR, the vehicle must be scaled up, PFR and PFA are recomputed, and the process is repeated until PFA equals PFR, thus closing the vehicle. The caption in the upper right-hand corner of Fig. 36 illustrates the closure process.

Vehicle Performance

Performance analysis begins with a set of mission requirements and a nominal mission definition. Required input data include vehicle design limits and trajectory constraints, vehicle mass properties and c.g. schedule, propulsion mode schedule (i.e., turbojet, ejector ramjet, ramjet, scramjet, etc.), throttle schedules, and propulsion and aerodynamic force and moment databases.

A representative air-breathing SSTO vehicle accent trajectory is given in Fig. 7 with representative engine mode changes and air-breathing corridor limits. The trajectory analysis methods typically used in quantifying vehicle performance (PFR) in order of increasing accuracy and computational time are 1) energy-state,[51] 2) three-DOF, and 3) six-DOF methods. The most com-

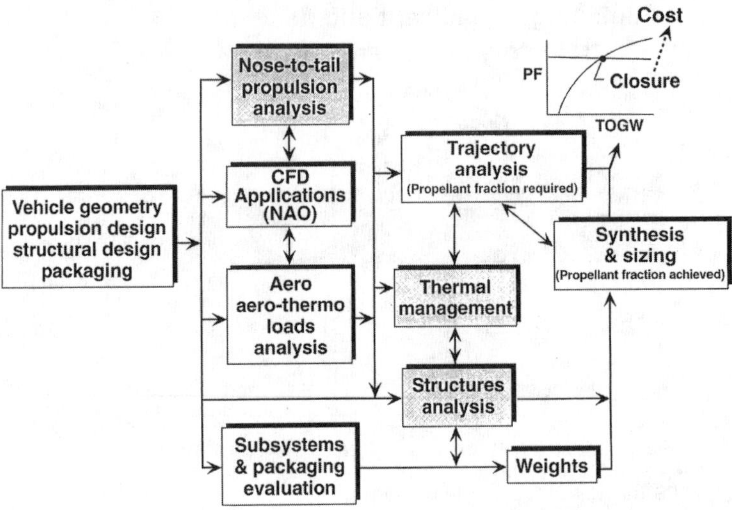

Fig. 36 Vehicle design/analysis process.

monly used method at Langley in analyzing conceptual and preliminary designs is the three-DOF Program to Optimize Simulated Trajectories (POST, Ref. 52) code.

Synthesis/Sizing

In conceptual design the methods used to match a vehicle design to a given mission is referred to as synthesis and sizing. Synthesis is the process by which component math models (weight and volume) are combined to provide a parametric model of a vehicle's PFA as a function of vehicle size. Typical vehicle components include structure, TPS, engines, tanks, empannages, landing gear, propellant feed system, etc. Sizing is the process by which the parametric math model of the vehicle is used to determine the vehicle's size to yield a desired PFA. At Langley, this process is currently performed on a PC spreadsheet.

Design Automation/Optimization

Hypersonic air-breathing vehicle design involves numerous variables, many with strong sensitivities and with a multitude of couplings and as such represents a complex and laborious process. The interdependence among the disciplines is depicted in Fig. 37. For example, the aerodynamics discipline 1) interacts with the propulsion discipline in defining the external vehicle shape, 2) provides heating rates to the thermal management analysis, 3) provides pressure and temperature profiles for structural analysis, and 4) provides forces and moments for trajectory analysis and stability and control assessment.

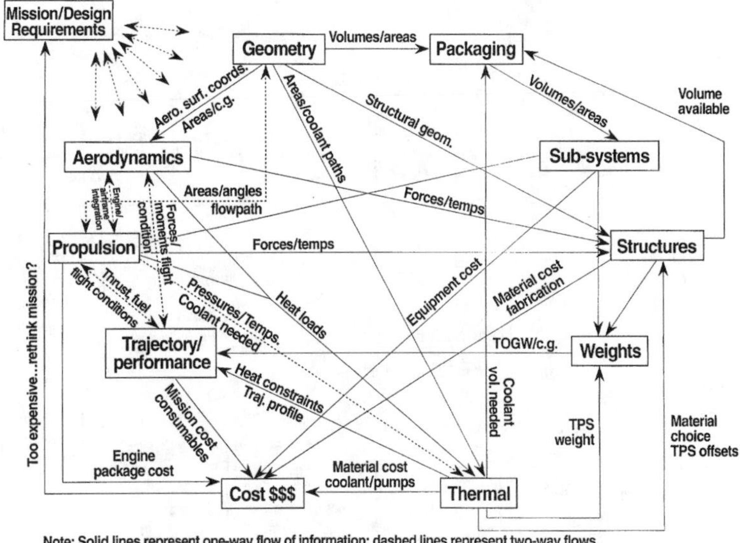

Fig. 37 Discipline interdependence.

To automate the design process and to be sure to capture all interactions, a working environment for multidisciplinary design, analysis, and optimization of air-breathing hypersonic vehicles (HOLIST) is being developed at Langley[53] in part through a contract with Boeing.[54] HOLIST will help eliminate disconnects between disciplines, enable rapid multidisciplinary parametrics, allow the evaluation of design sensitivities, and will enable the optimization of the vehicle design and trajectory. Currently, a Pro/ENGINEER[28] parametric geometry model interface is to be incorporated into HOLIST. This will enable the entire vehicle configuration to be represented with a limited number of design variables that can be optimized. HOLIST is modular in construction such that when improvements are made with any of the discipline tools, or new tools become available, they can be easily incorporated. A user-friendly optimizer, Optdes-X,[55] has been integrated into the HOLIST environment, and the entire system is set up to run on workstations, complete with graphical user interfaces.

Figure 38 is a simplified flowchart illustrating how an optimization proceeds in HOLIST. In the upper left-hand corner the process setup includes defining the design variables, objective function, constraints, and convergence criteria for a run. The baseline vehicle geometry and packaging models, together with a definition of the mass and thermal properties, is set up next. Analysis of the configuration proceeds with aerodynamics, propulsion, etc. (Note that for simplification of the diagram several disciplines are not represented here, including structures and thermal management, for example.) The analysis can either be performed by running discipline analysis

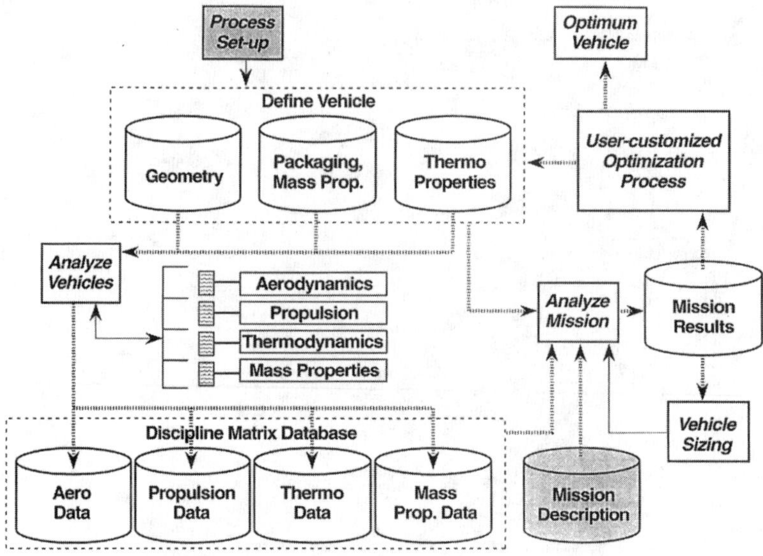

Fig. 38 HOLIST design optimization.

codes real time or by accessing a database to obtain the discipline results. It is important to note that there is more than just one discipline analysis result being passed through this flowchart at any given time. In other words, because the vehicle will fly some trajectory, matrices of aerodynamic and propulsion data representing the coefficients of lift, drag, and thrust, and fuel flow rate, for example, at appropriate values of angle of attack and Mach number, must be passed through the loop. In addition, the propulsion flow-path geometry may vary along the trajectory requiring multiple geometry definitions.

Once the analyses are completed, the vehicle is flown as represented by the Analyze Mission box. From the mission results the vehicle is sized. (It is also possible to define a scaling factor as a variable and use |IPFR-PFA| $\leq$.1 as a constraint. This would eliminate the need to perform the sizing process in the extra loop.) At this point if only a single vehicle analysis were required, the process would be complete. However, if optimizing the vehicle is desired, the optimization process begins. Finite differences are used to calculate the derivatives of the objective function with respect to each of the design variables. Thus, for the perturbation of each design variable, one pass through the loop is made. Based on the derivative information, the vehicle design for the next iteration is defined. The objective function for the new design is evaluated, the derivatives at the new point in the design space are determined, and the process continues with the vehicle definition for the next iteration. Iterations continue until the convergence criteria and all of the constraints are satisfied, yielding the optimum vehicle configuration.

Summary

The rudiments and basic principles of hypersonic air-breathing vehicle design and performance optimization presented in this chapter are critically important in achieving viable, attractive, and compelling vehicle design concepts. Many design variable couplings and first-order design drivers were discussed. Primary emphasis is placed on the subject of engine-airframe integration as being the key element and that the high-speed operation is the dominant driver. However, poor low-speed performance can severely degrade the best high-speed design, demanding that serious attention be paid to the low-speed flight regime.

The development of the scramjet engine-airframe integration methodology has progressed to a degree that allows resolution of hypersonic air-breathing vehicle designs for space access vehicles, cruise airplanes, and missiles for the dual-mode ramjet flow-path element of the design. The challenges that lie ahead are 1) reducing the turn-around time required for application of this methodology in the hypersonic air-breathing vehicle design process, 2) implementation and refinement of low-speed aeropropulsion integration methods, and 3) the development and automation of multidiscipline design process. As these design processes mature, viable space access and hypersonic cruise air-breathing vehicle designs will evolve.

Acknowledgments

The authors would like to thank Fred M. Robertson of Lockheed-Martin and Joe G. Nagy of Boeing for their contributions.

References

[1]Anon., "Recommended Practice: Atmospheric and Space Flight Vehicle Coordinate Systems," ANSI/AIAA R-004-1992, 1992.

[2]Pinckney, S. Z., "Internal Performance Predictions for Langley Scramjet Engine Module," NASA TM X-74038, Jan. 1978.

[3]Pinckney, S. Z., and Walton, J. T., "Program SRGULL: An Advanced Engineering Model for the Prediction of Airframe-Integrated Subsonic/Supersonic Hydrogen Combustion Ramjet Cycle Performance," NASA TM 1120, Jan. 1991.

[4]Billig, F. S., and Sullins, G. A., "A Generalized Method of Force Accounting," *24th JANNAF Combustion Meeting,* Oct. 1987.

[5]Numbers, K., "Hypersonic Propulsion System Force Accounting," AIAA Paper 91-0228, Jan. 1991.

[6]Perkins, S. C., Jr., and Dillenius, M. F. E., "Estimation of Additive Forces and Moments for Supersonic Inlet," AIAA Paper 91-0712, Jan. 1991.

[7]Lehrach, R. P. C., "Thrust/Drag Accounting for Aerospace Plane Vehicles," AIAA Paper 87-1966, June–July 1987.

[8]Sullins, G. A., and Billig, F. S., "Force Account for Airframe Integrated Engines," AIAA Paper 87-1965, June–July 1987.

[9]Collier, C. S., "Structural Analysis and Sizing of Stiffened, Metal Matrix Composite Panels for Hypersonic Vehicles," AIAA Paper 92-51015, Dec. 1992.

[10]Collier, C. S., "Thermoelastic Formulation of Stiffened, Unsymmetric Composite

Panels for Finite Element Analysis of High Speed Aircraft," AIAA Paper 94-1579, April 1994.

[11]Hunt, J. L., and McClinton, C. R., "Scramjet Engine/Airframe Integration Methodology," AGARD Paper C35, April 1997.

[12]Pinckney, S. Z., "Isolator Modeling for Ramjet and Ramjet/Scramjet Transition," *National Aero-Space Plane Technology Review,* Paper No. 171, 1993.

[13]Salas, M. D., "Shock Fitting Method for Complicated Two-Dimensional Supersonic Flows," *AIAA Journal,* Vol. 14, No. 5, 1976.

[14]Spradley, L. W., Robertson, S. J., Mahaffey, W. A., and Lohner, R., "The Solution Adaptive Modeling (SAM) CFD Software System," RSI 96-002, Sept. 1996.

[15]Lohner, R., et. al., "Fluid-Structure Interaction Using a Loose Coupling Algorithm and Adaptive Unstructured Grids," AIAA Paper 95-2259, June 1995.

[16]Bogar, T. J., Alberico, J. F., Johnson, D. B., Espinosa, A. M., and Lockwood, M. K., "Dual-Fuel Lifting Body Configuration Development," AIAA Paper 96-4592, Nov. 1996.

[17]Pinckney, S. Z., "Turbulent Heat-Transfer Prediction Method for Application to Scramjet Engines," NASA TN D-7810, Nov. 1974.

[18]Pinckney, S. Z., "Flat-Plate Compressible Turbulent Boundary-Layer Static Temperature Distribution with Heat Transfer," AIAA Paper, Jan. 1967.

[19]Schlichting, H., *Boundary Layer Theory,* Pergamon Press, 1955.

[20]Nestler, D. E., "Survey of Theoretical and Experimental Determinations of Skin Friction in Compressible Boundary Layers: Part I. The Laminar Boundary Layer on a Flat Plate," Thermodynamics TM No. 100, Sept. 29, 1958.

[21]Neal, L., Jr., and Bertram, M. H., "Turbulent-Skin-Friction and Heat-Transfer Charts Adapted from the Spalding and Chi Method," NASA TN D-3969, 1967.

[22]Nestler, D. E., "Survey of Theoretical and Experimental Determinations of Skin Friction in Compressible Boundary Layers: Part II. The Turbulent Boundary Layer on a Flat Plate," Thermodynamics TM No. 109, Jan. 29, 1959.

[23]Murry, A. L., "Further Enhancements of the BLIMP Computer Code and User's Guide," Air Force Wright Aeronautical Labs, AFWAL-TR-88-3010, June 1988.

[24]Petley, D. H., Jones, S. C., and Dziedzic W. M., "Integrated Numerical Methods for Hypersonic Aircraft Cooling Systems Analysis," AIAA Paper 92-0254, Jan. 1992.

[25]Anon., "SINDA '85/FLUINT User's Manual," Denver Aerospace Div. of Martin Marietta Corp., Denver, CO, Nov. 1987.

[26]Petley, D. H., and Yarrington, P., "Design and Analysis of the Thermal Protection System for a Mach 10 Cruise Vehicle," *1996 JANNAF Propulsion and Joint Subcommittee Meeting* (Albuquerque, NM), Dec. 1996.

[27]Hunt, J. L., "Airbreathing/Rocket Single-Stage-to-Orbit Design Matrix," AIAA Paper 95-6011, April 1995.

[28]Parametric Technology Corp., "Pro/ENGINEER Fundamentals," Release 17.0.

[29]Reymond, M., and Miller, M., of the McNeal–Schwendler Corp., "MSC/NASTRAN—Quick Reference Guide," Version 69.

[30]PDA Engineering, PATRAN Div., "P3/PATRAN User Manual. Vol. 1, Part 1: Introduction to PATRAN 3 and Part 2: Basic Fundamentals," Publication No. 903000, 1993.

[31]Parametric Technology Corp., "Pro/MECHANICA™ Installation Guide," Release 17.0, Aug. 1996.

[32]Collier, C., "Thermoelastic Formulation of Stiffened, Unsymmetric Composite Panels for Finite Element Analysis of High Speed Aircraft," AIAA Paper 94-1579, 1994.

[33]Collier, C., "Structural Analysis and Sizing of Stiffened, Metal Matrix Composite Panels for Hypersonic Vehicles," AIAA Paper 92-5015, 1992.

[34]Walters, R. W., Cinnella, P., Stack, D. C., and Halt, D., "Characteristics Based Algorithms for Flows in Thermo-Chemical Nonequilibrium," AIAA Paper 90-0393, Jan. 1990.

[35]Walters, R. W., "GASP—A Production Level Navier–Stokes Code Including Finite-Rate Chemistry," *Proceedings of the Fourth Annual Meeting of the Center for Turbomachinery and Propulsion Research,* April 1990.

[36]Jentink, T. N., "An Evaluation of Nozzle Relaminarization Using Low Reynolds Number *k*-e Turbulence Models," AIAA Paper 93-0610, Reno, Nevada, Jan. 1993.

[37]Lam, C. K. G., and Bremhorst, K., "A Modified Form of the *k*-e Model for Predicting Wall Turbulence," *Transactions of the ASME,* Vol. 103, Sept. 1981, pp. 456–460.

[38]Chien, K., "Predictions of Channel and Boundary-Layer Flows with a Low-Reynolds Number Turbulence Model," *AIAA Journal,* No. 1, 1982, pp. 33–38.

[39]Srinivasan, S., Bittner, R. D., and Bobskill, G. J., "Summary of the GASP Code Application and Evaluation Effort for Scramjet Combustor Flowfields," AIAA Paper 93-1973, June 1993.

[40]Mekkes, G. L., "Computational Analysis of Hypersonic Shock Wave/Wall Jet Interaction," AIAA Paper 93-0604, Jan. 1993.

[41]Srinivasan, S., and Erickson, W. D., "Influence of Test Gas Vitiation on Mixing and Combustion at Mach 7 Flight Conditions," AIAA Paper 94-2816, June 1994.

[42]MacDaniel, J. M., Fletcher, D., Hartfield, R., and Hollo, S., "Staged Transverse Injection into Mach 2 Flow Behind a Rearward Facing Step: A 3-D Compressible Test Case for Hypersonic Combustion Code Validation," AIAA Paper 91-5071, Dec. 1991.

[43]Rogers, R. C., et. al., "Quantification of Scramjet Mixing in the Hypervelocity Flow of a Pulse Facility," AIAA Paper 94-2518, June 1994.

[44]McClinton, C. R., "Comparative Flowpath Analysis and Design Assessment of an Axisymmetric Hydrogen Fueled Scramjet Flight Test Engine at a Mach Number of 6.5," AIAA Paper 96-4571, Nov. 1996.

[45]Kamath, P. S., Mao, M., and McClinton, C. R., "Scramjet Combustor Analysis with the SHIP3D PNS Code," AIAA Paper 91-5090, 1991.

[46]Kamath, P. S., and Mao, M., "Computation of Transverse Injection Into a Supersonic Flow with the SHIP3D PNS Code," AIAA Paper 92-5062, 1992.

[47]Braaten, M. E., "Development and Evaluation of Direct and Iterative Methods for the Solution of Equations Governing Recirculating Flows," Ph.D. Dissertation, Univ. of Minnesota, MN, 1985.

[48]Bonner, E., Clever, W., and Dunn, K., "Aerodynamic Preliminary Analysis System, Part 1 Theory," NASA CR-165628, 1981.

[49]Hunt, J. L., and Eiswirth, Ed. A., "NASA's Dual-Fuel Airbreathing Hypersonic Vehicle Study," AIAA Paper 96-4591, Nov. 1996.

[50]Engle, C. D., and Praharaj, S. C., "MINIVER Upgrade for the AVID System, Vol. I: LANMIN User's Manual," NASA CR-172212, 1983.

[51]Jackson, C. M., Jr., "Estimation of Flight Performance with Closed-Form Approximations to the Equations of Motion," NASA TR R-228, Jan. 1966.

[52]Brauer, G. L., et. al., "Program to Optimize Simulated Trajectories (POST)," NASA Contract NAS1-18147, Sept. 1989.

[53]Lockwood, M. K., Petley, D. H., Hunt, J. L., and Martin, J. G., "Airbreathing Hypersonic Vehicle Design and Analysis Methods," AIAA Paper 96-0381, Jan. 1996.

[54]Alberico, J., "The Development of an Interactive Computer Tool for Synthesis and Optimization of Hypersonic Airbreathing Vehicles," AIAA Paper 92-5076, Dec. 1992.

[55]Design Synthesis, Inc., "OptdesX—A Software System for Optimal Engineering Design Users Manual," Release 2.0.3, 1994,

Chapter 16

Transatmospheric Launcher Sizing

Paul Czysz*
Parks College, Saint Louis University, St. Louis, Missouri
and
Jean Vandenkerckhove†
VDK System, Brussels, Belgium

Nomenclature

A = Küchemann L/D parameter
A_C = inlet air capture area
A_{cap} = geometric capture area
A_{cowl} = airbreather cowl area
A_{LE} = air, liquid enriched from separator
A_{OP} = air, oxygen poor
b = span spatular vehicle
b_w = span of pointed vehicle of same length as spatular vehicle
B = Küchemann L/D parameter
c = spatular width
C_{sys} = constant system weight
C_D = total drag coefficient
C_{D0} = zero lift-drag coefficient
$C_D S$ = drag area
C_{un} = unmanned system weight
D = drag
$?E$ = energy increment
ΔI_{SPab} = change of I_{SP} in airbreathing mode
ΔI_{SPc} = change of I_{SP} with A_{LE}
ΔI_{SR} = change of I_{SP} in rocket mode

---

*Oliver L. Parks Endorsed Chair in Aerospace Engineering, Aerospace and Mechanical Engineering Department.
†Consultant Jean Vandenkerckhove passed away before this manuscript was completed.

E_{aero} = aerodynamic efficiency, L/D
E_{prop} = energy conversion efficiency, $\frac{T \cdot V}{Q_c \cdot \dot{w}_{fuel}} = \frac{T \cdot V}{Q \cdot \dot{w}_{air}}$
E_{TW} = engine thrust-to-weight ration, sea level static (SLS)
ff = fuel fraction, W_{fuel}/W_{GTO}
f_{col} = fraction collected
f_{crw} = crew member specific weight
f_{mnd} = crew system specific weight
f_{cprv} = provisions to accommodate crew
f_s = stoichiometic f/A
f_{sys} = variable system weight coefficient
F_i = thrust per unit span
I_p = propulsion index, $\rho_{ppl}/(WR - 1$
I_{str} = structural index, W_{str}/S_{wet}
I_{SP} = specific impulse, $T/\dot{w}_{ppl}$
I_{SPab} = specific impulse, $T_{ab}/(\dot{w}_{ppl})_{ab}$
I_{SPC} = specific impulse with LEA
I_{SPE} = effective specific impulse, $(T - D)/\dot{w}_{ppl}$
I_{SPf} = fuel specific impulse, $T/\dot{w}_{fuel}$
K_D = drag as fraction of kinetic energy
K_{FB} = fuselage bending moment parameter
K_w = S_{wet}/S_{pln}
k_{crw} = crew member volume
k_{mix} = fuel-air mixing losses in combustor, as a fraction of the freestream kinetic energy
k_{mnd} = crew member provision volume
k_{sup1} = support weight coefficient
k_{uc2} = undercarriage relief coefficient
k_{ve} = engine volume coefficient
k_{vs} = system volume coefficient
k_{vv} = void volume coefficient
KE = kinetic energy, $V^2/2$
K_v = scaled propellant volume fraction, $\left(V_{ppl}/V_{total}\right) \cdot S_{pln}^{-0.0717}$
L = length of vehicle
L_{1d} = planform loading at landing
L_{to} = planform loading at takeoff
M_{tr} = airbreather to rocket transition Mach number
M_{Wi} = ramjet engine module width
N_{crw} = number of piloting crew
P_{yO} = dry weight payload ratio (W_{pay}/W_{EO})
P_{yG} = gross weight payload ratio, (W_{pay}/W_{GTO})
Q_c = fuel heat of combustion
Q = Brayton cycle heat addition, $Q_c \bullet$ (f/A
R_{c1} = collection ratio, also Γ
$r_{F/A}$ = fuel/air ratio
$r_{O/F}$ = oxidizer/fuel ratio
r_{pay} = payload to empty weight ratio, W_{pay}/W_{EO}

r_{ppl} = propellant to empty weight ratio, W_{ppl}/W_{EO}
R_{str} = structure to empty weight ratio, W_{str}/W_{EO}
r_{sys} = system to empty weight ratio, W_{sys}/W_{EO}
r_{use} = useful load to empty weight ratio, W_{use}/W_{EO}
s = $S_{wet}/V_{tot}^{0.667} = K_w \cdot t$
sfc = specific fuel consumption, $3600/I_{SPf}$
spc = specific propellant consumption, $3600/I_{SP}$
S_{pln} = planform area
S_{wet} = wetted area
S_{pmx} = collection plant maximum, A_{LE} flow rate capacity
t = $S_{pln}/V_{tot}^{0.667}$
T = thrust
T_j = thrust of airbreather
T_{jc} = thrust of airbreather with collection
T_{SP} = specific thrust, $T/\dot{w}_{air}$
TW_0 = take off thrust to weight ratio
v_{col} = collection plant specific volume
V = flight velocity
V_0 = flight velocity
V_C = combustor gas velocity
V_{crw} = volume for each crew member
V_{pcrw} = crew provisions volume
V_{col} = collection plant volume
V_{fix} = fixed system volume
V_{pay} = payload volume
V_{ppl} = propellant volume
V_{tr} = velocity at transition from air breather to rocket
V_{sys} = system volume
V_{tot} = total vehicle volume
V_{un} = unmanned fixed sys. volume
V_{void} = void volume
w_{col} = collection plant specific weight
$\overline{W}$ = average width of spatular vehicle
W_{col} = collection plant weight
W_{cprv} = crew provisions weight
W_{crw} = crew and consumable weight
W_{dry} = dry weight ~ W_{EO}, W_{EO} − trapped fluids − crew consumables
W_{eng} = engine weight
W_{EO} = $W_{OE} - W_{pay} - W_{crew} \sim W_{dry}$
W_{OE} = $W_{GTO} - W_{ppl}$
W_{GTO} = gross takeoff weight
W_{pay} = payload weight
W_{fuel} = fuel weight
W_{oxid} = oxidizer weight
W_{ppl} = propellant weight, $W_{fuel} + W_{oxi}$
W_{str} = structure weight

W_{sup} = support weight
W_{sys} = system weight
W_{use} = useful load, $W_{pay} + W_{crw}$
W_R = weight ratio, W_{GTO} / W_{OE}
$\dot{w}_{hj}$ = hydrogen flow rate
$\dot{w}_{oj}$ = oxygen flow rate
$\dot{w}_{hjc}$ = hydrogen flow rate with collection
$\dot{w}_{ojc}$ = oxygen flow rate with collection
X_c = length from nose to engine cowl inlet lip
X_{O2} = oxygen concentration in A_{LE}
X_{N2} = nitrogen concentration in A_{LE}
Y_{N2} = nitrogen concentration in A_{OP}
Y_{O2} = oxygen concentration in A_{OP}
Γ = collection ratio, also R_{cl}
Λ = leading edge wing sweep angle
φ = equivalence ratio
μ_a = margin on inert weight
ρ_{pay} = payload density
ρ_{ppl} = propellant density
η_{carnot} = Brayton cycle Carnot efficiency
η_{sep} = separation efficiency
γ = flight path angle
σ = bow shock angle with respect to vehicle centerline
σ = $V_{tot}/S_{wet}^{1.5}$, $= \tau/K_w^{1.5}$
τ = $\tau = v_{tot}/s_{p\,ln}^{1.5}$ (Küchemann's parameter)

I. Introduction

THE major driver, in the development of launch vehicles for the twenty-first century, is reducing the cost of payload to orbit. This focuses vehicle characteristics toward a continuous use basis with the capability to recover fully operational the vehicle and payload if forced to abort the mission and reduction of launch time and resources required. Such requirements become variously qualified and constrained in each nation by its state and commercial policies, geography, and other considerations. There is a fundamental need to rethink the basic approach to conceptual design in terms of the technical requirements for meeting mission goals. This chapter provides an approach and systematic method for applying the approach for evolving various types of vehicle systems.

A. Theme

An approach to the conceptual design of such vehicles is still a matter of debate. Although several design synthesis methods have been developed[1–3] the difficulty is in rationalizing needs, capabilities, and opportunities. While it is fully recognized that airbreathing propulsion has a crucial role in meeting the stated goals for launch vehicles, and that the vehicle needs to be fully integrated in design, functions, and operation, the difficulty is whether reasonable estimates of available and required industrial capabilities can be made and a conceptual design rationalized. However, there are invariably ambiguities and controversies associated with estimates of available and required industrial capabilities, whether propulsion-propellant schemes, configuration geometries, materials and structures, flight management, or controls are considered, individually or collectively. The approach taken in this report is directed toward clarifying and through the use of simple and direct, basic principles and estimates, overcoming some of the ambiguities. The outcome is what may be referred to as a sizing code that addresses the convergence of a vehicle system in terms of a set of design parameters to meet a given mission.

B. Objectives

The authors' objectives in the development and use of the sizing programs were the following:

1) Provide a quantitative sizing model based on simple and direct, principles and estimates, to assess SSTO characteristics that account for system weight and volume as well as for an explicit margin on inert weight.
2) Provide simplified input requirements for screening parametric studies based on engineering experience that represents current and future manufacturing capabilities. Specifically, the authors identify a "CURRENT" set of volume and weight assumptions considered within today's industrial manufacturing and materials capabilities, and a "FUTURE" set which results from application of the current on-going R&D in Europe. These two sets bound the possible "design space."
3) Apply the model to assess the Single-Stage-to-Orbit (SSTO) performance sensitivity to changes in assumptions and interaction between these assumptions.
4) Extend the sizing model to Two-Stage-to-Orbit (TSTO) and perform sensitivity analysis as for SSTO.
5) Compare SSTO and TSTO performance.
6) Assess the potential of LOX collection for both SSTO and TSTO.

II. Vehicle Sizing Approach

In the development of atmospheric flight vehicles at lower speeds, it was accepted practice to adopt variants of a methodology, developed about a quarter of a century ago[4,5] for conceptual design. The method is illustrated in Ref. 4 and Fig. 1 of Ref. 5. The method is based on historical data on design, test results, and operational experience. In the case of hypersonic vehicles the total operational experience is small. However, for historical design and test data that is not necessarily so. One author's (PAC) career in hypersonic vehicles is based on the approach pioneered in the Mercury and Gemini reentry vehicles. That is a conventional, cold, load carrying structure protected by relatively smooth radiation shingles.[6–8] This approach could and did yield statistically weighted correlations for evolving an optimum concept that weighed less for high lift-to-drag lifting bodies than comparable low performance bodies.[9] Propulsion systems integrated into the vehicles spanned a broad spectrum of engines, ranging from turboramjets[10–12] to scramjets[13–16] during that time period. This led directly to the NASA-sponsored Hypersonic Research Facilities Studies.[17] Reference 17 describes 102 Hypersonic Research Objectives required to achieve Mach 12 flight, and the Hypersonic Research Facilities performance and cost to achieve some significant fraction of those research objectives. To put the study into perspective, the Ground Research Facilities represented about one-eighth of that effort and Flight Research Facilities represented about seven-eighths of the effort, so the study was primarily a research aircraft, not a ground facility effort. The objective of this chapter is to document a constant performance, volume and mass converged sizing procedure.

A. Approach

When both authors began their careers in aerospace, the standard practice was to begin design by drawing constant wing area or constant weight concept aircraft. Each system component was independently sized, designed, and assembled. Each concept was manually redrawn and iterated to approximately the same mission radius/range; however, performance could differ significantly between concepts. This was not a very satisfactory method for high performance aircraft.

Sizing aircraft concepts to both mission distance and maneuver performance produced a change in how concepts were evaluated.[18–21] Decisions could now be made on equal performance aircraft of differing size and weight rather than vice versa. This aircraft sizing approach sized an aircraft configuration to mission performance requirements, then iterated on the system weight until assumed and computed were equal.[22] This is the approach taken in this chapter. The significant difference between a conventional aircraft and a hypersonic aircraft/space launcher is the propellant volume. For conventional commercial aircraft, the significant volume is the passenger volume.

If commercial airliners have passenger volume that approaches 80% of the total volume, then space launchers can have a propellant volume that approaches 80% of the total aircraft volume.[23] Although updated in subsequent references, this observation was reported in early studies.[9] Volume limitations of were recognized early on as forcing a balance between aerodynamic performance and usable mission volume. As for aircraft, space launcher sizing programs size for constant performance, then consider both volume and mass in its convergence criteria. The sizing procedures converge on system volume and weight. The mass ratio for the mission was determined independently by trajectory analysis. The volume of the vehicle was iterated until volume available equaled volume required and the mass ratio equaled the mass ratio required.[17,24] The interdependence of aerodynamics, propulsion and structure required this approach to consider the vehicle as a single system, not an assembly of separate systems. Both authors have always used this approach for hypersonic aircraft; that is, considering a constant performance vehicle system sized to mission weight ratio and volume requirements.

A significant number of critical conditions have to be met at the higher speeds. As with all high performance vehicles, there are overriding demands with respect to industrial capabilities in propulsion and materials and structures. For whatever reasons, launch vehicle design has continued in its present form with all rocket schemes that include limited recovery, and reuse capability with refurbishment. This is also the reason payload to orbit cost has not been significantly reduced. It is clear that the approach to conceptual design of hypersonic vehicles needs to focus on payload to orbit cost and long duration use.[25,26] Achievement of a conceptual vehicle rests on 1) whatever data and projections can be established, including results of capabilities that may become available from preliminary study[27,28] and 2) recognition of the fact that the most significant gains may only be realized from propulsion-propellant capabilities. These represent the principal challenge.

Based on this reasoning, an approach to determine launch vehicle size emphasizes energy management and propulsion as the principal elements, given the available industrial capability and the required freedom in selecting configuration geometries repetition. The objective of this chapter is to provide a parameter set that represents the industrial capabilities (technology) available today to fabricate a vehicle launcher system. These are based on personal observations of both authors and private communications with industry representatives responsible for the industrial capability. Based on earlier work, a methodology is developed for the rational synthesis of vehicles based on utilization of available data, projections, and characteristics of different configuration concepts. The methodology is applied to SSTO, and TSTO with various limits for airbreathing propulsion, and for SSTO and TSTO with air collection and air collection with separation. The methodology applies to both aircraft and launch vehicles.

B. Sizing Methodology

The approach described was applied to three vehicle classes: 1) the Douglas Aircraft Company Phase I systems studies of NASA-sponsored High Speed Civil Transport (HSCT), which determined the Phase II configurations, sizes and weights.[29,30] Over 30 airframe/propulsion system/fuel combinations were analyzed in Phase I, and three were selected for further study in Phase II; 2) the recoverable vertical launch vehicle[31] for a government-sponsored McDonnell Douglas Astronautics Company (later named the Delta Clipper); 3) and to size demonstrator/prototype vehicles[32–34] and launch vehicles for research papers.

This approach was implemented in the early 1980s, by one of the authors (VDK), as three separate computer sizing programs, namely SIZING, ABSSTO and ABTSTO, which were used to generate the data for this report. The development of the sizing programs began with the methodology described in "Hypersonic Convergence,"[35] where we begin with the fundamental equation that defines the weight ratio to orbit.

$$W_R = \frac{W_{\text{GTO}}}{W_{\text{OE}}} = \frac{W_{\text{OE}} + W_{\text{ppl}}}{W_{\text{OE}}} = 1\frac{W_{\text{ppl}}}{W_{\text{OE}}} = 1(1 + r_{\text{O/F}}) \cdot \frac{W_{\text{fuel}}}{W_{\text{OE}}} \tag{1}$$

The oxidizer-to-fuel ratio ($r_{\text{O/F}}$) is averaged over the trajectory and is equal to ($W_{\text{oxid}}/W_{\text{fuel}}$. For a given fuel and dry weight fuel fraction, the weight ratio is driven by the oxidizer-to-fuel ratio. So whatever the fuel, the weight ratio can be minimized if the oxidizer-to-fuel ratio can be minimized. The weight ratio may also be expressed in terms of the effective specific impulse (I_{SPE}) and:

$$W_R = \exp\left(\frac{\Delta V}{G \cdot I_{\text{SPE}}}\right) \tag{2}$$

Thus the weight ratio (W_R) and effective specific impulse (I_{SPE}) are functions of the oxidizer-to-fuel ratio for a given fuel. Rearranging the above equations, we arrive at two fundamental equations on which this sizing approach is built.

$$W_{\text{OE}} = \frac{V_{\text{ppl}}}{S_{\text{pln}}} \cdot \frac{\rho_{ppl}}{WR - 1} \cdot S_{\text{pln}} = \frac{V_{\text{ppl}}}{V_{\text{tot}}} \cdot \frac{V_{\text{tot}}}{S_{\text{pln}}^{1.5}} \cdot I_p \cdot S_{\text{pln}}^{1.5} \tag{3}$$

$$I_p = \frac{\rho_{\text{ppl}}}{W_R - 1} = \left[\frac{\rho_{\text{fuel}} \cdot (1 + r_{\text{O/F}})}{1 + r_{\text{O/F}} \cdot \dfrac{\rho_{\text{fuel}}}{\rho_{\text{oxid}}}}\right] \cdot \left\{\exp\left[\frac{\Delta V \cdot \dfrac{T}{D}}{g \cdot I_{\text{SP}} \cdot \left(\dfrac{T}{D} - 1 - \sin\gamma\right)}\right] - 1\right\}^{-1} \tag{4}$$

The operational weight empty (W_{OE}) is a product of three terms. The first term $V_{\text{ppl}}/S_{\text{pln}}$ is determined by geometry, the second $\rho_{\text{ppl}}/WR - 1$ by the aero-thermo-propulsion system, and the third (S_{pln}) by size.

The propulsion index I_P is expressed by the product of two terms. The first term is a function of the density of the propellants chosen and their oxidizer-to-fuel ratio. The second term is more complex. It is a function of the propellant and engine selection; engine size, excess thrust over drag, and climb angle for a given increment in velocity. The propulsion index (I_p) can be evaluated along a trajectory or used to represent an index for a given propulsion not/but-propellant system for the entire trajectory. The magnitude of the propulsion index (I_p) is a function of maximum sustained speed of the vehicle and not a significant function of the specific propulsion type. In the authors' analyses for SSTO space launchers, based on SSME class of turbopumps and operating pressures, the propulsion index, spans the spectrum from an all rocket SSTO to an all airbreather SSTO, which is 4.0 ± 0.5. For, any given vehicle speed, the larger the propulsion index, the smaller and lighter the vehicle. The mean value of the propulsion index as a function of the maximum sustained Mach number of the vehicle is

$$I_p = 107.6 \cdot 10^{-0.081 \cdot M} \tag{5}$$

The scatter around the mean is about ± 10% from a subsonic cruise fighter with supersonic dash capability to a SSTO vehicle.

C. Fundamental Sizing Relationships

The non-dimensional volume index, τ, introduced by Küchemann[36] and attributed to him by J. Collingbourne from an unpublished reference relates volume to planform area. The W_{OE} can now be related to vehicle design parameters. Although Küchemann introduces tτu as a volume parameter, it can indeed be considered a slenderness parameter. This is clearly illustrated in Fig. 1 for a long-range, hypersonic aircraft sized with three different fuels: JP/kerosene (752 kg/m³, 47 lb/ft³), sub-cooled liquid methane (464 kg/m³, 29 lb/ft³), and sub-cooled liquid hydrogen (74.6 kg/m³, 4.66 lb/ft³). This is an order of magnitude range in fuel density. For a kerosene-fueled low volume-per-unit-planform-area slender aircraft like a SST, $\tau = 0.03$. As fuel density decreases, the value of τ increases to 0.039 to 0.147. For a high volume-per-unit-planform-area vehicle like a hydrogen-oxygen propellant-combined cycle-powered space launcher, τ can be in the 0.18 to 0.20 range. (Also see Fig. 13.) Introducing τ, Eq. (3) becomes:

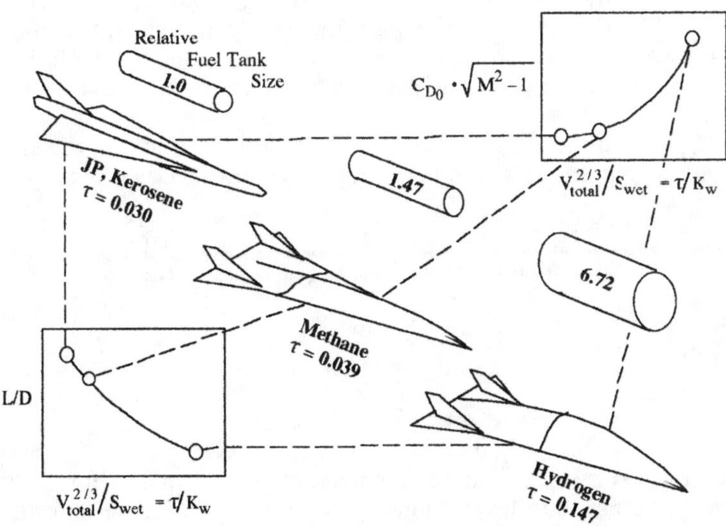

Fig. 1 Propellant Density Drives Configuration Concept and Slenderness.

$$W_{OE} = \frac{\rho_{ppl}}{WR-1} \cdot \left(\frac{V_{ppl}}{V_{tot}}\right) \cdot \tau \cdot S_{pln}^{1.5} = I_P \cdot \left(\frac{V_{ppl}}{V_{tot}}\right) \cdot \tau \cdot S_{pln}^{1.5} \tag{6}$$

where

$$\tau = \frac{V_{tot}}{S_{pln}^{1.5}} \tag{7}$$

Recalling that $W_{EO} = W_{OE} - W_{pay} - W_{crew} \sim W_{dry}$, it follows that

$$\begin{aligned} W_{EO} &= \left(\frac{\rho_{ppl}}{WR-1}\right) \cdot \left(\frac{V_{ppl}}{V_{tot}}\right) \cdot \tau \cdot S_{pln}^{1.5} - W_{pay} - W_{crew} \\ &= \left(\frac{\rho_{ppl}}{WR-1}\right) \cdot \left(\frac{V_{ppl}}{V_{tot}}\right) \cdot \frac{\tau \cdot S_{pln}^{1.5}}{(1+r_{use})} \end{aligned} \tag{8}$$

We now have the design variables related directly to the dry weight. However, a word of caution: the three items in Eq. (8) *are not* independent variables. They are related through the propellant and propulsion system. From Eq. (6) it might seem that a low value of the propulsion index is desirable. In fact, for the combined volume and weight convergence point, the higher the propulsion index, the less the operational empty weight. This is because the other two parameter groups *are not* independent of the value of the propulsion index. As pointed out by Froning and Leingang[37], r_{pay} is essentially a constant for most launch vehicles. Thus, Fig. 2 shows that the payload-to-gross weight ratio is only an artifact of the weight ratio to orbit. A much more meaningful ratio is the payload-to-empty weight ratio. This ratio is essentially constant with the airbreathing speed increment. The data for the comparison are for the payload only. The vehicles forming the database were manned, so adding a value to r_{py} to represent the ratio of the crew weight to W_{OE} (r_{crw}) provides a value for the useful payload ratio, to r_{use}.

Using one additional definition for structural fraction, r_{str}, the series of fundamental equations is complete with the following equation:

$$\frac{W_{str}}{S_{wet}} = \left(\frac{\rho_{ppl}}{WR-1}\right) \cdot \left(\frac{V_{ppl}}{V_{tot}}\right) \cdot \frac{(r_{str})}{(1+r_{use})} \cdot \frac{\tau \cdot S_{pln}^{1.5}}{K_W} \tag{9}$$

where

$$W_{str} = OEW \cdot r_{str} \tag{10}$$

$$K_w = S_{wet}/S_{pln} \tag{11}$$

Equation (9) now directly relates geometry-based parameters with the material/structure- and propulsion-based parameters. Please note that the propulsion index, the propellant volume ratio, and the geometric terms directly affect the required structural weight per unit wetted (surface) area.

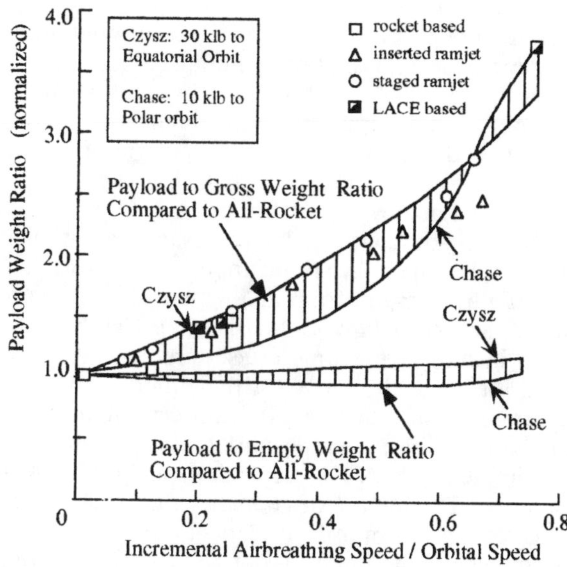

Fig. 2 Payload Weight Ratios Show Empty Weight Ratio as Constant.

The greater the propulsion-propellant system performance (i.e., the greater the value of I_p), the heavier the structural weight allowed for convergence becomes, and therefore, the less technology is required. The corollary is that poor propulsion performance always demands structural/ material/fabrication breakthroughs.

D. Effect of τ on Configuration Concepts

To visualize the effect of Küchemann's τ on configuration, the blended body configuration in Fig. 3 represents blended body configurations from very slender to very stout, and their associated value of τ, the ratio of wetted to planform surfaces (K_w) and the maximum lift-to-drag ratio at Mach 12. The minimum size configuration is the minimum volume vehicle consisting of only the propulsion-configured compression side of the vehicle and a flat upper surface. The stout vehicle represents the limits of transonic drag for a practical propulsion system to obtain a high value of thrust minus drag.

E. Parametric Sizing Interactions

The relationship of τ and K_w[28] is dependent on the configuration concept. The premise for the sizing approach of the Hypersonic Convergence is that families of geometries can be used to represent the characteristics of hypersonic vehicles rather than detailed point designs. Given the propulsion system characteristics and the industrial capability to manufacture hypersonic vehicles, the result is a configuration concept derived from the value of these geometric parameters that permits convergence within the technology limits set by the structural and propulsion indices. Thus, the configuration concept is a result of a parametric analysis, not an initial assumption.

Figure 4 shows the range of τ and K_w for a number of families of hypersonic configuration concepts appropriate for launchers, all with 78° sweep angle. (See Appendix B for the full range of configuration concepts). Also shown as a reference point is the vertical launch rocket configuration with an aft wing, the NASA Langley WB-004 configuration. The three propulsion integrated launchers (blended body, wing-body, and Nonweiler waverider) are from converged design studies that supported the work in Refs. 39 and 40. The other configurations are from mathematical models for the surface area and volume (see Appendix B for details). Combined cycle engine launchers (which include hypersonic cruise aircraft) are powered by air-breathing propulsion over all or part of their flight path. The hypersonic glider configurations (with blunt bases) are ascent vehicles that return to earth unpowered and are based on the work at the Air Force Flight Dynamics Laboratory[41] in the 1960s. All of the vehicles include, in the wetted area, control surface areas. The impact of geometry on the size and weight of a launch aircraft is clearly shown in Ref. 42, pages 631 to 637.

In Fig. 1, the correlating parameter is not τ, but $V_{\text{total}}^{2/3}/S_{\text{wet}}$. In one author's (PAC) work experience in advanced design, the aerodynamic correlating parameters based on volume and area were referenced *area,* not *volume.*[9] In Ref. 28 both correlation parameters s and t are presented where parameter s is called the "volumetric efficiency factor" and parameter t is called the "shape efficiency factor." So the same variables were used, but in different combinations. The original material is presented in this chapter, as generated, so the transformations shown below can be helpful.

Küchemann's convention:

$$\tau = \frac{V_{\text{total}}}{S_{\text{pln}}^{1.5}} = t^{-1.5}$$

U.S. industry convention:

$$t = \frac{S_{\text{pln}}}{V_{\text{total}}^{0.667}} = \tau^{-0.667} \tag{12}$$

$$\sigma = \frac{V_{\text{total}}}{S_{\text{wet}}^{1.5}} = \frac{V_{\text{total}}}{(K_W \cdot S_{\text{pln}})^{1.5}} = s^{-1.5} \qquad s = \frac{S_{\text{wet}}}{V_{\text{total}}^{2/3}} = \frac{K_W \cdot S_{\text{pln}}}{V_{\text{total}}^{2/3}} = \sigma^{-0.667} \tag{13}$$

F. Summary of Parameter Groups

Parameter groups that dominate the sizing process are listed in the following discussion. The variables within these parameters are interrelated, so a change in one can result in a change in the magnitude of some of the other parameters. This means that the sizing process is very interdependent and interactive among propulsion, propellant, geometry-size, materials, and struc-

From reference 37, volume 3

	Minimum	Slender	Nominal	Stout	Circular Cone
tau	0.032	0.063	0.104	0.229	0.393
K_W	2.44	2.51	2.64	3.39	3.61
$\beta \bullet C_{D0}$	0.0613	0.0639	0.0574	0.0809	0.0980
L/D M = 12	4.82	4.71	4.57	4.13	3.70

Fig. 3 The Blended Body Has a 7 to 1 Volume Range by Upper Body Shaping.

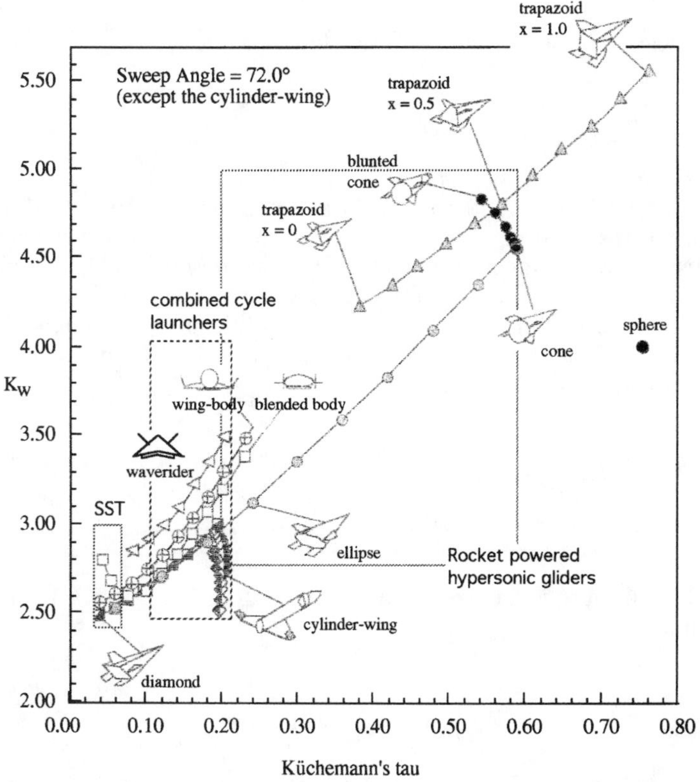

Fig. 4 Surface and Volume Characteristics of Hypersonic Configuration Concepts Span a Broad Range.

tural concept. A later discussion about sizing of high-speed aircraft will include discussions about propulsion, propellants, aerodynamics, and geometry. One observation is that the weight ratio is a function of oxidizer to fuel ratio, Eq. (1), as is the resulting configuration characteristics, so the configuration concept is a result of a parametric analysis and not the input.

$$\frac{\rho_{ppl}}{(WR-1)} = I_p \propto \text{propulsion concept, propellant, aerodynamics energy}$$

$$\frac{W_{str}}{S_{wet}} = I_{str} \quad \propto \text{materials, structural concept, manufacturing capability}$$

$$\frac{V_{ppl}}{V_{tot}} \quad \propto \text{size, fineness ratio (tau), geometry}$$

$$\frac{W_{str}}{OEW} = r_{str} \quad \propto \text{materials, size, fineness ratio (tau), geometry}$$

$$\frac{W_{pay}}{OEW} = r_{pay} \quad \propto \text{approximately constant}$$

$$\frac{S_{wet}}{S_{pln}} = K_W \quad \propto \text{size, fineness ratio (tau), geometry}$$

G. External Aerodynamics

The sizing program included a parametric solution technique that provides the vehicle size and weight as a function of τ. Vehicle drag, and therefore, thrust-to-drag ratio must be determined to correct weight ratio for thrust-to-drag changes as a function of τ. As presented on pages 670 and 671 of Ref. 27, this is accomplished via an empirical correlation. These correlations were done by Dwight Taylor while at McDonnell Douglas Corporation in the 1960s (private communication, Dwight Taylor, 1993). Briefly, Taylor's original correlation parameter was:

$$\sqrt{\left(\frac{V_{tot}^{0.667}}{S_{pln}}\right) \cdot \left(\frac{S_{wet}}{S_{pln}}\right)^{1.5}} = \tau^{0.333} \cdot K_W^{0.75} = F \tag{14}$$

From Ref. 36 there is a correlation for lift-to-drag ratio by Küchemann, of the format:

$$\left(\frac{L}{D}\right)_{max} = \frac{A}{M} \cdot (M + B) \tag{15}$$

the constants A and B are as defined by Küchemann and the authors for slender aircraft:

1959 SOA	Future SOA	This chapter database
$A = 3$	$A = 4$	$A = 3.063$
$B = 3$	$B = 3$	$B = 3$

The aerodynamic correlations for drag and lift-to-drag ratio are then:

$$\left(\frac{L}{D}\right)_{\max} = \frac{3.063}{M} \cdot (M+3) \cdot (1.11238 - 0.1866\dot{\theta} \cdot F) \tag{16}$$

$$\beta C_{D0} = 0.05772 \cdot \exp(0.4076 \cdot F) \tag{17}$$

The zero lift drag coefficient is a function of relative volume, relative wetted area and Mach number. It is not necessary to do a complete drag build-up to determine total drag. The total drag can then be estimated using the approach of Vinh.[42] Then

$$\beta \cdot C_d \cdot S_{\text{pln}} = \beta \cdot C_{d0} \cdot (1+B) \cdot S_{\text{pln}} \tag{18}$$

For L/D maximum, B is equal to 2. That is the classical case where the optimum induced drag for a symmetrical airfoil section is equal to the zero lift drag. As developed by Vinh, the values for $(1+B)$ are:

Acceleration, $C_L \sim 0.1 \cdot (C_L)_{\text{L/D max}}$ and $(1 + B) = 1.075$:
Minimum fuel flow Cruise, $C_L \sim 0.82 \cdot (C_L)_{\text{L/D max}}$ and $(1 + B) = 1.75$:
$(L/D)_{\max}$ Glide: $C_L \sim (C_L)_{\text{L/D max}}$ and $(1 + B) = 2.0$:

Given a reference configuration and drag, then the thrust to drag along the trajectory can be corrected for total volume. That is:

$$\left(\frac{T}{D}\right)_{\tau} = \left(\frac{T}{D}\right)_{\tau\,\text{ref}} \cdot \frac{(\beta \cdot C_{d0})_{\tau\text{ref}}}{(\beta - C_{d0})_{\tau}} \tag{19}$$

$$(ISPE)_{\tau} = (ISPE)_{\tau\,\text{ref}} \cdot \frac{(1 - T/D)_{\tau}}{(1 - T/D)_{\tau\,\text{ref}}} \tag{20}$$

$$WR = (WR)_{\tau\,\text{ref}} \cdot \exp\left(\frac{ISPE_{\tau}}{ISPE_{\tau\,\text{ref}}}\right) \tag{21}$$

So from the trajectory analysis the drag corrected propulsion index (I_p) can be determined using Eq. (4)

$$I_p = \frac{\rho_{\text{ppl}}}{WR - 1} \qquad \text{in density units } i \tag{22}$$

The foregoing equations apply to an accelerating space launcher vehicle.

For long-range cruise applications the correction must be made on the range equation not the rocket acceleration equation.[43]

H. Technology Maturity Determination

One result of the "Hypersonic Convergence" work was the definition of a primary structure and propulsion interaction that controlled the size and weight of the aircraft, derived from Eq. (9). This evolved into the Industrial Capability Index (ICI) as a measure of the practicality of the vehicle under consideration, in terms of the industrial materials/fabrication-not/propulsion capability available. The concept of an index is that there is a quantity that can represent the relative measure of one level of technical maturity with another. The overall technical maturity obviously involves capability in a number of areas, starting with propulsion and progressing through aerodynamics, materials, manufacturing, and vehicle integration, as well as others. The technical maturity is obviously an engineering capability to meet a specified goal. A suggested relation for Industrial Capability Index is

$$ICI = \text{Industrial Capability Index} = 10 \cdot \left(\frac{\rho_{\text{ppl}}/(WR - 1)}{W_{\text{str}}/S_{\text{wet}}} \right) = 10 \cdot \frac{I_p}{I_{\text{str}}} \tag{23}$$

Figure (5) clearly shows that the enabling capabilities are the propulsion system and the structural weight per unit surface area. And these are interdependent. If the Structural Index is increased (industrial technology decreased), and if the Propulsion Index is not correspondingly increased (industrial technology increased), the size and geometry of the possible vehicle must change (become larger and stouter). The opposite is true if the Propulsion Index is improved. It enables a converged vehicle with higher structural weight per unit surface area. The technologies applicable to each

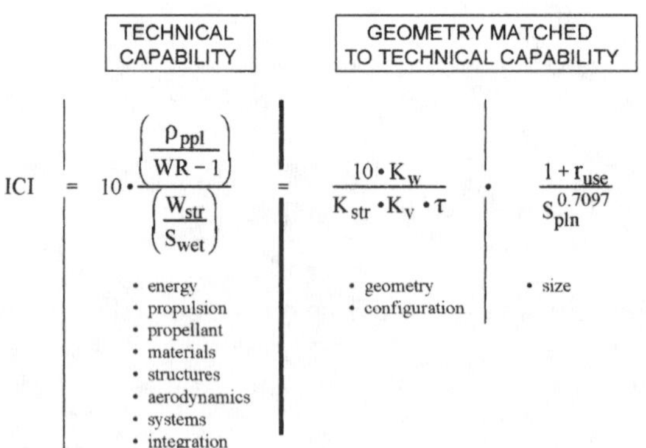

Fig. 5 Industrial Capability Index: Technology Required Equates to Size and Geometry of Configuration Concept.

side of the equation are indicated. The structural index is readily determined from current or projected industry achievements and manufactured hardware. The lower the "technology" in the materials and structures, the higher the value of the structural index, i.e., the heavier the structure per unit surface area.

The Propulsion Index is more an index of the propulsion system hardware (turbopumps, heat exchangers, etc.) than is the Propulsion Index. Determined from current hardware, the Industrial Capability Index can be established the thermodynamic cycle. Given the SSME engine hardware as a reference, the Propulsion Index from all-airbreather to all-rocket varies less than 15% for application of the SSME hardware to other propulsion cycles (see Ref. 3). For the SSME case it will be found that the Propulsion Index is 57.0 ± 10 kg/m^3 (3.56 lb/ft^3 ± 0.5) and the Structural Index is 21 kg/m^2 (4.3 lb/ft^2) resulting in a value of 10· ICI of 27.1 ± 5 m^{-1} (8.26 ± 1.5 ft^{-1}).

Equation (23), as presented in Fig. 5, implies that for a given ICI there is a minimum-sized vehicle for each combination of geometric parameters. That is, the geometric solution can be less than the ICI in magnitude but not greater, and that the greater the magnitude of ICI, the greater the technology required. If a smaller-sized vehicle is desired, then either the Structural Index must be reduced or the Propulsion Index increased. Considering the demonstrated expander cycle of ISAS in Tokyo, Japan, it will be found that the Propulsion Index is 64.0 ± 10 kg/m^3 (4.02 lb/ft^3 ± 0.5) and the Structural Index is 19.5 kg/m^2 (4.0 lb/ft^2), resulting in a value of 10 · ICI of 32.1 ± 5 m^{-1} (10.0 ± 1.2 ft^{-1}). When the same ICI is desired, the Structural Index can increase to 23.7 kg/m^2 (4.86 lb/ft^2) without any change in vehicle size, or the vehicle can be 87% of the planform area with the SSME industrial capability. A maximum index of 10 · ICI of 37.7 ± 5 m^{-1} (11.5 ± 1.2 ft^{-1}) appears possible using the values in Ref. 65.

Equation (23) can be mapped to show the available design space for a selected configuration.[44] It is important to recognize that Eq. (23) shows that the smaller vehicles are technologically challenging, not the larger. The most costly and technically challenging is a small demonstrator with a zero payload, not a larger vehicle. The design space map for the blended body is given in Fig. 6. The technical capability is what was judged to be available in the 1994 time frame. The small circle symbols are the authors evaluation of the 1994 Industrial Capability Index (available in Europe). One author (VDK) focused on the maximum margin and minimum technology solutions that were the least slender (i.e., stout). The other author (PAC) focused on the solutions at the current industrial capability boundary.

The sizing process defined up to this point provides an indication of the possible design space, dependent on mission, configuration, propulsion and propellant. The Structural Index is straightforward. For non-space launchers, weight ratio is not the measure of propellant load, but of fuel fraction. For an aircraft application, Propulsion Index is:

$$I_p = \left(\frac{\rho_{\text{ppl}}}{WR - 1}\right) = \frac{(1 - ff) \cdot \rho_{\text{ppl}}}{ff} \tag{24}$$

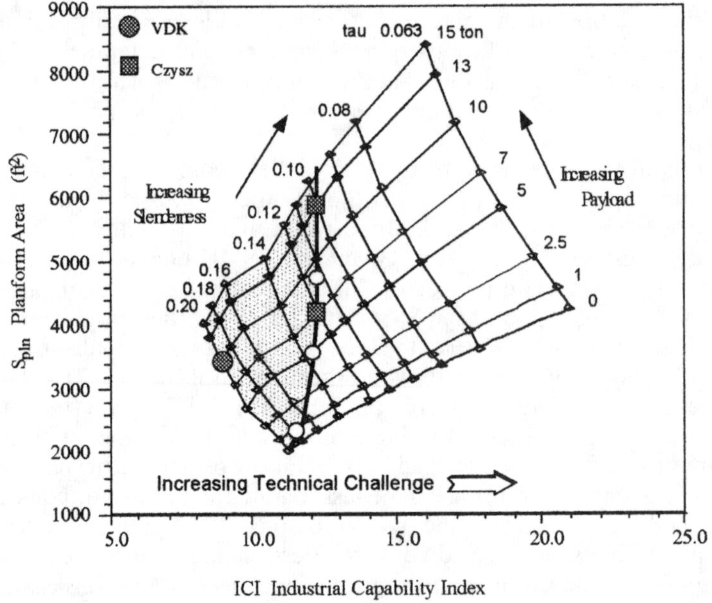

Fig. 6 "Design Space" is Bounded by Realities of Technology and Geometry.

where

$$WR - 1 = \frac{ff}{1 - ff} \qquad ff = \frac{W_{\text{fuel}}}{W_{\text{GTO}}}$$

As previously stated, the Propulsion Index is a function of maximum sustained Mach number, and this sizing technique is not limited to space launchers. As applied to a HSCT problem, the Propulsion Index for kerosene fuel was 609 kg/m^3 (38.0 lb/ft^3) and 350 kg/m^3 (21.8 lb/ft^3) for liquid methane. That resulted in an ICI of 356.2 m^{-1} (108.6 ft^{-1}) for kerosene fuel. When design space evaluation was executed for the HSCT, the result was not like that of Fig. 6. The minimum size and weight for a wing-body transport configuration with advanced variable bypass turbofan engines and hydrocarbon fuel was, for τu ~ 0.035, not 0.20 as in Fig. 6. This method thus provides a logical starting point for configuration development not based on conjecture or tradition but fundamental physical relationships. Thus, much less time is needed to find a configuration that will converge.

III. Propulsion Systems

Airbreathing propulsion can be beneficial over a part of the flight trajectory. Historically there are three broad categories of airbreathing propulsion:

1) a combination of individual engines operating separately (sometimes in parallel, sometimes sequentially) that can include a rocket engine[45]
2) an individual engine (usually a rocket engine) operating in conjunction with an engine that can operate in more than one cycle mode,[46–48] or a combined cycle engine
3) a single, combined cycle engine that operates in all of the cycle modes required, over the entire flight trajectory.[49,50]

For the single, combined cycle concept, the engineering challenge is transitioning from one cycle to the next within a single engine. The transition from one engine cycle operation to another must be made efficient (on First Law basis that means the total energy losses must be minimized) and effective (on Second Law basis that means that when the energy is available for recovery as useful work, the energy conversion must be accomplished immediately or it becomes unrecoverable).[51,52] A category (*c*) engine is designed for minimum entropy rise across the cycle. The scope and limitations of these engines are discussed in detail in Refs. 37 and 38, and several advantages to such a scheme have been identified private communication. (Escher, W. D., "More Power for next Generation Space Transports: Combined Airbreathing + Rocket Propulsion," March, 1994). [53,54]

A. Performance Characteristics of Airbreathing Engines

The performance of an airbreathing engine is governed principally by the state properties of air and vehicle characteristics that include: the captured inlet air mass flow, the entry air kinetic energy, the energy released to the cycle by combustion of the fuel, and the internal drag and energy losses through the engine flowpath.[55] Evaluating these factors permits the establishment of performance boundaries based on first principles. The result is an altitude-speed representation of performance potential and constraints for Brayton cycle airbreathing engines defined by an altitude (equivalency, entropy state of exhaust gas) boundary and a velocity (equivalently, air kinetic energy to combustion energy ratio) boundary.

The first boundary is a function of the entropy of the gas exiting the propulsion system nozzle. Since the freestream entropy increases with altitude, for a fixed entropy rise engine cycle, the entropy of the exhaust nozzle gas also increases with altitude. The second boundary is a function of the kinetic energy of the freestream flow. At higher speeds the air kinetic energy can significantly exceed the Brayton cycle heat addition to the airflow by combustion of a fuel. The ratio of the maximum combustion energy per unit mass of air to the kinetic energy per unit mass of air is:

$$\frac{Q_{\text{net}}}{KE} = \frac{2 \cdot Q \cdot \eta_{\text{carnot}}}{V^2} \tag{25}$$

Remember the Carnot Cycle loss is the unrecoverable energy loss because the atmosphere (the receiver) is not at absolute zero temperature. A reasonable value for η_{carnot} is 0.79. The Brayton cycle heat addition (Q) for hydrogen is 1503 Btu/lb and for most hydrocarbons 1280 $\pm$ 20 Btu/lb.[35] From hydrocarbons to hydrogen, the Brayton cycle heat addition equals the air kinetic energy between 7,100 ft/sec to 7,700 ft/sec. As the speed increases, the combustion energy added to the airstream becomes a smaller fraction of the freestream kinetic energy. For hydrocarbons to hydrogen, between 14,200 ft/sec and 15,400 ft/sec, the Brayton cycle heat addition is 25% of the freestream kinetic energy. For hydrocarbons to hydrogen, between 21,300 ft/sec to 23,100 ft/sec the Brayton cycle heat addition is 11% of the freestream kinetic energy. Energy input from combustion must overcome the losses that result from the external drag of the vehicle, energy losses associated with the internal engine flow, and irreversible losses in the thermodynamic cycle, as well as supply the energy required for acceleration to orbital speed. So the energy available to overcome drag and provide acceleration is reduced by 4 every time the flight speed is doubled. The losses to be overcome, however, are not a strong function of speed; so there is a vehicle speed where available energy just equals the drag energy; that is maximum airbreathing speed. For example, various losses may be expressed in the form of energy losses non-dimensionalizing with respect to kinetic energy of coming air.

Combustor drag losses:

$$\left(\frac{\Delta E}{KE}\right)_{\text{comb}} = -\left(\frac{v_c}{v_0}\right)^2\left(\frac{C_D S}{A_{\text{cowl}}}\right)_{\text{eng}} \tag{26a}$$

Fuel mixing losses:

$$\left(\frac{\Delta E}{KE}\right)_{\text{mix}} = -k_{\text{mix}} \cdot \left(\frac{v_C}{v_0}\right)^2 \tag{26b}$$

Vehicle drag losses:

$$\left(\frac{\Delta E}{KE}\right)_{\text{vehicle}} = -\left(\frac{C_D S}{A_c}\right)_{\text{vehicle}} \tag{26c}$$

Fuel injection energy:

$$\left(\frac{\Delta E}{KE}\right)_{\text{fuel}} = +\,\varphi \cdot f_s \cdot \left(\frac{V_{\text{fuel}}}{V_0}\right)^2 \tag{26d}$$

Energy to accelerate:

$$\left(\frac{\Delta E}{KE}\right)_{\text{accel}} = -\left(\frac{T}{D}\right)\cdot\left(\frac{C_D S}{A_C}\right)_{\text{vehicle}} \tag{26e}$$

The only term that adds to the available energy of the working fluid is the injected fuel energy. If the temperature of the fuel (in this case hydrogen) is scheduled so that the injected fuel velocity is equal to the flight speed, and the fuel injection angle is about 6°, then the injected fuel energy to kinetic energy ratio is $0.0292 \cdot \varphi$. For an equivalence ratio of six, this provides an energy addition of 17.5%, or the equal of the maximum available combustion energy from fuel at 18,400 ft/sec. So recovering normally discarded energy as thrust is as critical as burning fuel in the engine. This is reflected in Fig. 7.

As the speed increases, the engine performance becomes more an energy conservation engine than a chemical combustion engine.[56] The result is a spectrum of operation over the speed regime that was developed by Czysz and Murthy[57] and is shown in Fig. 7. The figure illustrates the extent to which the kinetic energy of freestream air entering the vehicle inlet capture area and the fuel mass and internal energy become gradually more significant and critical as the flight speed increases. Thus, the operating limits of the airbreather can be clearly identified.

Examining Fig. 7, it should be clear to the reader that airbreathing propulsion is limited in both speed and altitude. The speed regime to the right of the energy ratio 4 line is questionable for an operational vehicle. It's possible for a research vehicle to investigate this area, but as we shall see at the energy ratio 4 boundary, the airbreathing vehicle has achieved a significant

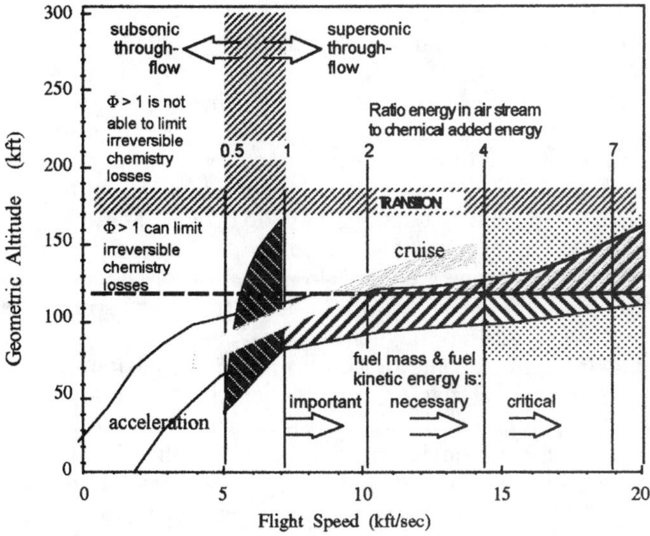

Fig. 7 Brayton Cycle Operation Is Increasingly Dependent on Energy Conservation, not Fuel Combustion as Flight Speed Increases.

fraction of the benefits from incorporating airbreathing. So from an energy viewpoint, a practical maximum airbreathing speed is 14,200 ft/sec (4.33 km/sec). To the right of this line the payoff achieved compared to the resources required yields diminishing returns. The authors' contribution early on established a practical maximum for operational airbreathing launchers[58] at 3.9 km/sec (12,700 ft/sec) with the possibility to reach 14,000 ft/sec (4.27 km/sec) from a vehicle sizing, compression side materials, and minimum dry weight approach.[59] The altitude regime above 120,000 ft produces a degradation of thrust because increasing entropy levels limit the internal molecular energy that can be converted into kinetic energy (exhaust gas velocity). Excess hydrogen provides abundant third bodies for the dissociated air molecules to recombine with—up to a flight altitude, about 170,000 ft. Above that altitude it is improbable a Brayton cycle engine can produce sufficient thrust. If excess fuel is used in Brayton cycle engines below 120,000 feet and less than 14,500 ft/sec it is to convert a fraction of the aerodynamic heating into net thrust via injection of the hydrogen at high velocity into the engine (such as velocity corresponding to flight speed). Note that cruise engines operate at greater cycle entropy levels than acceleration engines.

Thus, up to this point, we have used first principles to establish that the vehicle will be stout, and not too small if it is to be built from available industrial capability, Fig. 6. We have also established it is not practical for an operational vehicle to exceed 14,200 ft/sec in airbreathing mode, and apparently 12,700 ft/sec would be less challenging while retaining the benefits of airbreather operation.

B. Major Sequence of Propulsion Cycles

There are a significant number of propulsion system options that have been studied. The authors have focused on those that are applicable to transatmospheric vehicles. The intent is to define the single-stage-to-orbit weight ratio and the onboard oxygen ratio carried by the vehicle. The less the weight ratio and the oxygen-to-fuel ratio, the smaller the size and gross weight of the vehicle. In terms of these parameters, the authors examined four principal propulsion categories with hydrogen as fuel, as shown in Fig. 8. The first category is rocket-derived, air-augmented propulsion where the primary propulsion element is a rocket motor. The second category is airbreathing rocket-derived propulsion where the propulsion elements are a rocket motor and an air/fuel heat exchanger. The third category is the thermally integrated, combined cycle engine propulsion where the principal element is a rocket-ejector ramjet where the rocket-ejector provides both thrust and compression.[60,61] The fourth category is the thermally integrated combined cycle engine propulsion where the thermally processed air is separated into nearly pure liquefied oxygen and oxygen-poor nitrogen; the liquid enriched air is stored for later use in the rocket engine. Thermal integration means that the fuel passes through both rocket and the scramjet to scavenge rejected heat and convert it into useful work before entering the combustion chambers, increasing the specific impulse.

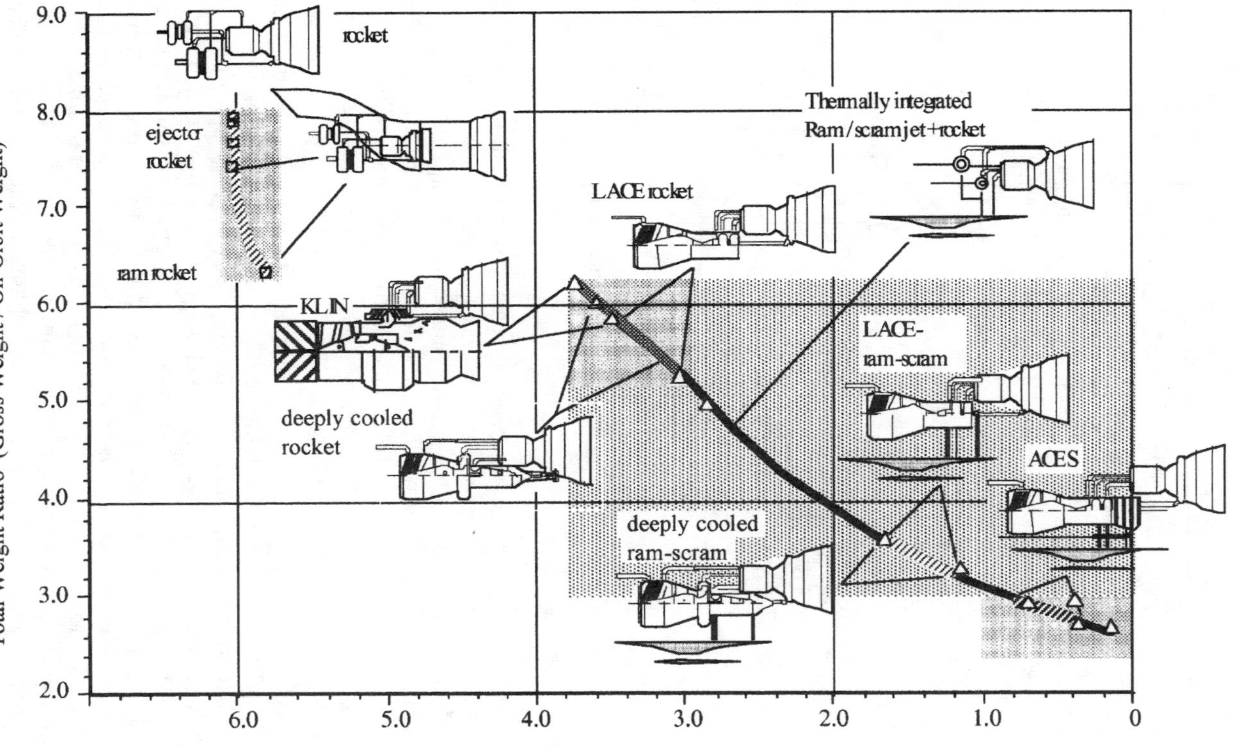

Fig. 8 Propulsion Cycles Determine Carried Oxidizer and To-Orbit Weight Ratio.

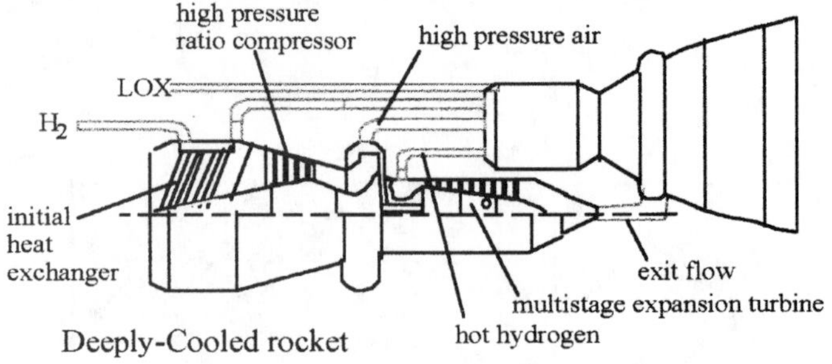

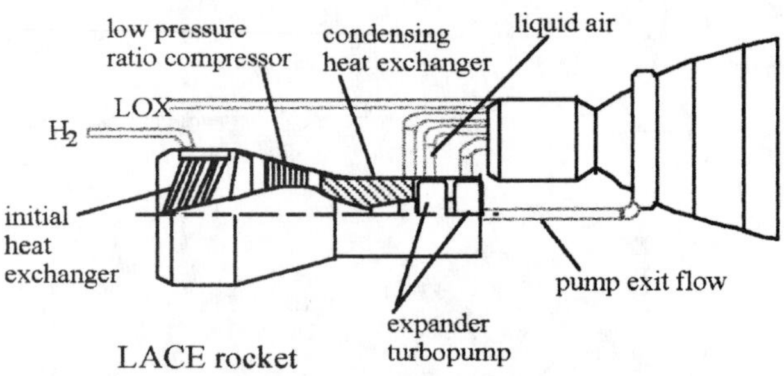

Fig. 9 Thermally Integrated Airbreathing Rockets.

The combined cycle concept dates back 40 years[62] to the Marquardt Company.[39,63] Marquardt had a propulsion concept that could go hypersonic in a single engine (Ref. 16). One of the Marquardt Company's concepts incorporated folding rotating machinery[48] into their cycle; however, it is still a single engine that can go from takeoff to hypersonic speed.

1. Rocket Derived Propulsion

Rocket derived propulsion systems generally operate up to Mach 6 or less because of pressure and temperature limits of the air induction system. At Mach 6 inlet diffuser static pressures can typically equal 20 atmospheres and 3,000°R (1,666 K). Although no rocket derived propulsion systems are evaluated in this chapter, they are included for completeness in the comparisons. These propulsion systems can offer major advantages when applied to existing rocket launchers.[64] As shown in Fig. 8, rocket derived systems occupy the upper left-hand corner of the parameter space. The weight ratio to orbit is reduced proportionally to the thrust augmentation of the airbreathing system, but there is little change in the carried oxygen to fuel ratio. Examples of the rocket derived air-augmented propulsion are the following:

1) Air-augmented rockets employ the rocket motor as a primary ejec-

tor[60,65,66] so that some of the external airstream can be mixed with the rocket exhaust to increase mass flow and thrust at lower Mach numbers ($M < 6$), thus increasing the specific impulse. The rocket motor operates at its normal oxidizer to fuel ratio, The reduction of the mass-averaged exhaust velocity increases propulsion efficiency. This concept is not designed to burn the oxygen in the entrained air. The weight ratio is reduced to 7.5. The external air inlet system does add empty weight. But with a mass ratio reduction of one-half, the system weighs less if the inlet system is less than 6.7% of the dry weight.

2) Ram rocket is an air-augmented rocket cycle where the rocket is operated at a richer-than-normal oxidizer-to-fuel ratio so that the oxygen in the entrained air can burn the excess fuel at the normal airbreathing air/fuel ratios for the fuel used.[67] The external airstream is mixed with the rocket exhaust to increase mass flow and with the combustion of the excess fuel, increases thrust and specific impulse at lower Mach numbers ($M < 6$). The weight ratio is reduced to 6.3, and the fuel-rich rocket operation reduces the oxygen-to-fuel ratio slightly. This is the best operational mode for the air-augmented rocket.

2. *Airbreathing Rocket Propulsion*

Airbreathing rocket-derived propulsion systems generally operate up to Mach 6 or less because of pressure and temperature limits of the air induction system. At Mach 6 inlet diffuser static pressures can typically equal 20 atmospheres and 3,000°R (1,666 K). airbreathing rocket propulsion concepts employ a method to reduce the temperature of air entering the inlet system so it can be compressed to rocket chamber operating pressures with reduced power requirements. There are two options: One option is to deeply cool the air just short of saturation and use a turbocompressor to pump the gaseous air into the rocket chamber. The second option is to liquefy the air and use a turbopump to pump the liquid air into the rocket chamber, see Fig. 9. The rocket motor operates at nearly normal oxygen-to-fuel ratios, except that there is now a large mass of nitrogen also introduced into the combustion chamber. Again, the mass average exhaust velocity is reduced and the total mass flow increased, thus increasing thrust and propulsion efficiency. These propulsion systems are the darker shaded rectangle at the upper left-hand part of the shaded area in Fig. 8:

1) The deeply cooled rocket is an expander cycle rocket developed by Rudakov and Balepin at CIAM[68] and Alan Bond for HOTOL. In Fig. 9, a more detailed view of the two airbreathing rocket cycles is shown. In the deeply cooled cycle there is a hydrogen/air heat exchanger in the air inlet to capture the inlet air kinetic energy. This controls the air temperature entering the compressor, and limits the work of compression and the compressor-corrected speed. The warmed hydrogen then enters the rocket combustion chamber to recover additional energy. The total thermal energy collected from the incoming air and hydrogen combustion chamber is then used to drive an expansion turbine which in turn drives a turbocompressor that

compresses the cooled inlet air. That air can be cooled to nearly saturation by the hydrogen flow, then compressed to rocket operating pressures and introduced into the combustion chamber. A rocket motor combustion chamber heat exchanger is necessary to provide sufficient energy to drive the turbomachinery. In effect, the rocket becomes an airbreathing rocket for Mach numbers less than 6. In this concept there is no other airbreathing engine. This cycle reduces the mass ratio to the 5.2 to 6 range and the oxygen-to-fuel ratio to about 3.2.

2) LACE rocket is the rocket part of the Aerospace Plane propulsion concept developed by the Marquardt Company in the mid-to-late 1950s. LACE is *L*iquid *A*ir *C*ycle *E*ngine. It was examined in Russia,[69,70] Japan,[71–73] and India.[74] As depicted in Fig. 9, this cycle, as with the deeply cooled, employs a hydrogen/air heat exchanger in the air inlet to capture the inlet air kinetic energy from the incoming air and cool it to nearly saturation. The cooled air is then pressurized to a few atmospheres and then flows into the pressurized liquefying heat exchanger. The total thermal energy collected from the incoming air and hydrogen combustion chamber is used to drive an expansion turbine which in turn drives a turbopumps liquefied air into the rocket motor. A heat exchanger in the rocket motor's combustion chamber is necessary to provide sufficient energy to drive the turbomachinery. In effect, the rocket becomes an airbreathing rocket for Mach numbers less than 6. In this concept there are no other airbreathing engines. This cycle reduces the mass ratio to the 5 to 5.8 range and the oxygen-to-fuel ratio to about 3.

3. *Thermally Integrated Combined Cycle Propulsion*

The fundamental element of the combined cycle engine concept is a rocket ejector ram-rocket-ramjet, which is thermally integrated into a rocket propulsion system.[75] In this section, propulsion systems eight and nine are the propulsion systems employed for the vehicle sizing studies. In the class of integrated ejector ram-scramjet propulsion, the integral rocket ejectors provide both thrust and compression at lower Mach numbers.[64,76] The combination of ramjet and turbojet results in a poor acceleration combination. However, the introduction of a deeply cooled turbojet that is thermally integrated with an expander rocket (KLIN cycle[77]) becomes analogous to the rocket ejector ram-rocket-ramjet, with the additional benefit of excellent low-speed performance. Examples of the thermally integrated engine's combined cycle propulsion are the following:

1) Deeply cooled turbojet-rocket (KLIN cycle) is an adaptation of Rudakov and Balepin's deeply cooled rocket ramjet to a deeply cooled turbojet. The turbojet and rocket are thermally integrated. Unlike the ramjet, the precooler on the turbojet keeps the compressor air inlet temperature low to reduce required compressor work and to increase mass flow and thrust. With the precooler, the turbojet does not see the inlet temperature associated with higher Mach number flight, so it appears to be at lower flight speed. The pre-cooled turbojet provides a significant increase in transonic thrust. The

pre-cooled turbojet provides operation from takeoff to Mach 5.5 with rocket thrust augmentation when required, such as in the transonic region. Above Mach 5.5 turbomachinery is shut down and the rocket operates as a conventional cryogenic rocket. The KLIN cycle is equivalent to cycles 3 and 4 in that the mass ratio is reduced to the 5.5 to 6 range and the oxygen-to-fuel ratio to about 3.4, but it is not like these cycles in that it produces fuel efficient, low-speed thrust.

2) Deeply cooled rocket-ram-scramjet is the integration of the deeply cooled cycle developed by Rudakov and Balepin at CIAM[66] and Alan Bond for HOTOL with a subsonic through-flow ramjet. In this cycle, the thermal energy from the incoming air and hydrogen combustion in both the rocket and ramjet is used to drive an expansion turbine, which in turn drives a turbo-compressor. The incoming inlet air is cooled to nearly saturation in an air-hydrogen heat exchanger, and then compressed to rocket operating pressures by the turbocompressor so it can be introduced into the rocket combustion chambers. A heat exchanger in the rocket motor's combustion chamber is necessary to provide sufficient energy to drive the turbomachinery. After leaving the expansion turbine, the hydrogen is introduced into the ramjet combustion chamber. At Mach 6 or less, the rocket is essentially an airbreathing rocket operating in parallel with a ramjet. Above Mach 6, the rocket is not used, and the ramjet operates as a supersonic through-flow ramjet (scramjet). After scramjet shutdown, the rocket operates as a conventional cryogenic rocket. The operational line is represented by the heavy line traversing the shaded area. For airbreather operation to the 12,000 to 14,000 ft/sec range, this cycle can achieve weight ratios in the 3 to 4 range with oxygen-to-fuel ratios less than one.

3) LACE Rocket-Ram-Scramjet is a Liquid Air Cycle Engine. It is like the Aerospace Plane propulsion concept developed by John Ahern at the Marquardt Company in the late 1950s. It was examined in the 1990s by Russia,[67,68] Japan,[69–71] and India[72] In this cycle, the thermal energy from the incoming air and hydrogen combustion is used to drive an expansion turbine, which in turn drives a turbopump. The inlet air is cooled to nearly saturation by an air-hydrogen heat exchanger, then pressurized to a few atmospheres. It then flows into the pressurized liquefying heat exchanger. The turbopump pressurizes the liquid air to rocket operating pressures so it can be introduced into the rocket combustion chamber. A rocket motor combustion chamber heat exchanger is necessary to provide sufficient energy to drive the turbomachinery. After exiting the turbomachinery, the hydrogen is introduced into the ramjet combustion chamber. At Mach 6 or less, the rocket is essentially an airbreathing rocket operating in parallel with a ramjet. The ramjet can convert to supersonic through-flow (scramjet) at Mach 6. Above Mach 6, the rocket is not used when the scramjet is operating. After scramjet shutdown the rocket operates as a conventional cryogenic rocket. The operational line is represented by the heavy line traversing the shaded area. For airbreather operation to the 12,000 to 14,000 ft/sec range, this cycle can achieve weight ratios in the 3 to 4 range with oxygen to fuel ratios less than one. Cycles 5, 6, and 7 can achieve specific impulse in the

4,500 second range in the Mach 6 to 3 range. Thermal integration provides about 1,500 seconds of the 4,500 second I_{SP}.

4) Ejector Ram-Scramjet-Rocket is an ejector ramjet thermally integrated with a rocket.[78,79] The ejector may be a hot gas ejector and/or a rocket ejector. Remember, if the ramjet is a subsonic through-flow engine, then the scramjet is simply a supersonic through-flow engine. The maximum air-breathing speed can be selected to be from Mach 6 to at least Mach 14.5. At Mach = 6, the system is an ejector ramjet analogous to system #2, except the rocket ejectors are distributed in the struts inside the ramjet engine module.[80] Subsonic thrust is generated in the same manner as system #2. Above Mach 6 it is a conventional scramjet engine with variable configuration injectors to minimize internal drag (ref. HC). The operational line is represented by the heavy line traversing the shaded area. This cycle can produce weight ratios from 6 to 3 depending on the maximum airbreathing speed. It is simpler than cycles 7 and 8, but lacks the lower speed (M <6) high-specific impulse.

4. *Thermally Integrated Enriched Air Combined Cycle Propulsion*

These cycles are thermally integrated combined cycle propulsion analogous to cycles 7 and 8 except the thermally processed air is separated into nearly pure liquefied oxygen and oxygen-poor nitrogen. The liquid-enriched air is stored for use in the rocket engine during the ascent portion of the rocket's trajectory. The oxygen-poor nitrogen is introduced into the ramjet, creating the equivalent of a mixed-flow bypass turbofan. That is, the mass averaged exhaust velocity is reduced but the specific impulse, engine mass flow and thrust are increased. Thermal integration means that the fuel passes through both rocket and the scramjet to scavenge rejected heat and convert it into useful work before entering the combustion chambers, thus increasing the specific impulse. Examples of thermally integrated, enriched, air-combined cycle propulsion follow:

1) ACES-LACE Ejector Ram-Scramjet-Rocket is *A*ir *C*ollection and *E*nrichment *S*ystem.[81] ACES is an option added to system #7. The liquid air is not pumped to the rocket immediately, but passed through a fractionating system to separate the oxygen component as liquid-enriched air (LEA contains 80 to 90% oxygen) and nitrogen component as liquid oxygen-poor air (OPA contains from 2 to 5% oxygen).[71,82] The oxygen component is then stored for use in the rocket's ascent portion of the flight. The oxygen-poor nitrogen component is injected into the ramjet to create a hypersonic bypass engine that increases engine mass flow, thrust, and reduces the mass averaged exhaust velocity. At takeoff this can significantly reduce the takeoff perceived noise. It is done for the same reasons a conventional mixed flow bypass gas turbine was invented. It was originally proposed for the space plane of the late 1950s and the subject of intense investigation in the 1960 to 1967 time period.[83] For airbreather operation to the 12,000 to 14,000 ft/sec range, this cycle can achieve weight ratios less than 3 with oxygen-to-fuel ratios approaching one-half.

2) ACES-Deeply Cooled Ejector Ram-Scramjet-Rocket is Air Collection and Enrichment System. ACES is an option added to system #8. The deeply cooled gaseous air is not pumped to the rocket immediately, but passed first through a vortex tube initial separator (at this stage the LEA contains about 50% oxygen), and then into a cryogenic magnetic oxygen separator. The oxygen component is then liquefied as (LEA contains 80 to 90% oxygen) stored for use in the rocket's ascent portion of the flight. The gaseous component of oxygen-poor air (OPA contains from 2 to 5% oxygen). The oxygen-poor nitrogen component is injected into the ramjet, to create a hypersonic bypass engine that increases engine mass flow, thrust, and reduces the mass averaged exhaust velocity. At takeoff this can significantly reduce the takeoff perceived noise. It is done for the same reasons a conventional mixed-flow bypass gas turbine was invented. This system is in laboratory testing and studies[84] but has not as yet been developed as propulsion hardware. For airbreather operation to the 12,000 to 14,000 ft/sec range, this cycle can achieve weight ratios less than 3 with oxygen-to-fuel ratios approaching one-half.

C. Cycle Comparison

When these propulsion systems are compared to the rocket, a number of observations are possible. The first of these regards the weight-ratio-to-orbit. Figure 10 presents the weight-ratio-to-orbit for the four categories discussed in 3.2. The first two categories merge into a rocket-derived curve. The inserted ramjet and staged ramjet are integrated ejector ramscramjet and rocket propulsion systems. The former have an airbreather inserted between two rocket operations (one from takeoff, the other from airbreather shutdown), while the latter have an airbreather function from takeoff followed by a rocket operation. The weight ratio does not include propellant for orbital operations. If a nominal quantity were included, the weight ratio would be as indicated in the upper right hand corner of the figure. The curve in the upper left indicates the region of applicability for rocket-derived propulsion systems. The other curve indicates the region of applicability for thermally integrated combined cycle propulsion. The lower boundary of that area represents the maximum speed for airbreathing operation developed in Fig. 10. That limits achieves 88% of the maximum benefit realized by airbreathing to the minimum weight ratio (about 22,000 ft/sec or 6.7 km/sec). The technical, hardware and economic challenges to achieve the last 12% of the weight ratio benefit by flying some 8,000 ft/sec faster, probably exceeds the benefits in the authors' opinion.

The sizing studies reported in this chapter focus on the shaded area, that is airbreathing speeds between 6,000 ft/sec (1.83 km/sec) and 12,000 ft/sec (3.96 km/sec). There is an area where the rocket-derived propulsion and combined cycle propulsion are equivalent in the 5,000 to 6,000 ft/sec (1.52 to 1.83 km/sec) region. Reference 70 addresses this area. What Fig. 10 implies is that if an all-rocket gross weight is 7.5 times W_{OE} then a thermally integrated combined cycle powered vehicle will be from 5.5 to 3 times W_{OE}, depending on the maximum airbreathing speed. As Fig. 3 showed, the ratio of W_{pay} to

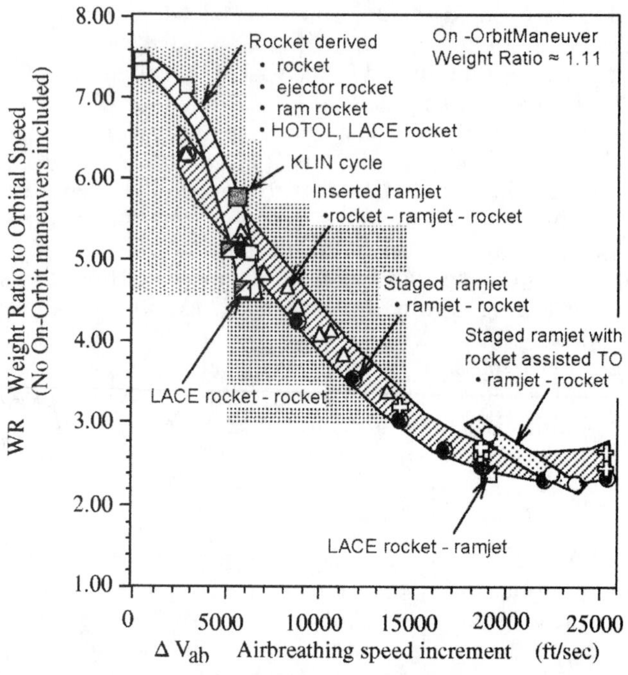

Fig. 10 Weight Ratio Reduction at 14,500 ft/sec is 88% of Maximum.

W_{OE} is essentially constant with airbreathing speed, so the combined cycle propulsion reduces the gross weight by 2 to 4.5 times the W_{OE}!

1. Takeoff Gross Weight and Takeoff Mode

In reality, horizontal or vertical takeoff, like the configuration concept, is less a choice than a result of the propulsion concepts selected. Figure 11 shows the impact of assuming vertical or horizontal takeoff for sized configuration for the same payload weight as a function of weight-ratio-to-orbit. Three different takeoff wing loadings were evaluated. VTOHL takeoff thrust-to-weight ratio is 1.35. HTOL takeoff thrust-to-weight ratio is 0.75. Prior work suggested the nominal takeoff thrust to weight ratios, and no attempt was made to find an optimum takeoff thrust to weight ratio each case. If the HTOL gross weight exceeds the VTOHL gross weight, then the lighter vehicle is a vertical takeoff mode. If thrust vectoring is available for nose wheel lift off, then the 200 lb/ft^2 (976 kg/m^2) is acceptable,[17] although the takeoff speed is rather high (about 344 knots). The VTOHL/HTOL boundary for 200 lb/ft^2 is weight ratio 5.2, or an airbreathing speed of about 7,000 ± 1,000 ft/sec. For a takeoff wing loading of 125 lb/ft^2 (610 kg/m^2), the takeoff speed is 291 knots, the VTOHL/HTOL boundary is now a weight ratio 4.3, or an airbreathing speed of 10,000 ± 1,000 ft/sec. This wing loading is also correct to air launch horizontal landing (ALHL) in the Mach

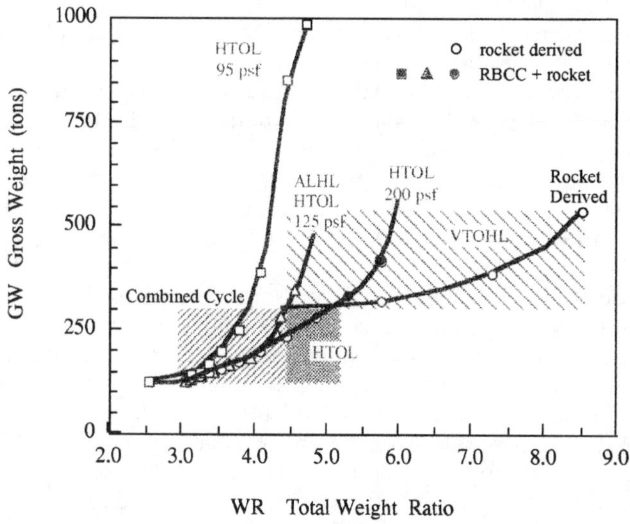

Fig. 11 Wing Loading and Weight Ratio Determine Gross Weight.

0.72 at 35,000 ft region. For a takeoff wing loading of 95 lb/ft^2 (464 kg/m^2) takeoff wing loading and a takeoff speed of 254 knots, only the maximum airbreathing speed would permit horizontal takeoff. This wing loading is in fact too low to be practical for launchers as it drives the gross weight to unacceptable levels. The conclusion is, that if the weight ratio is greater than 4.3, the vehicle is best a vertical takeoff configuration or an air-launched configuration. For all of the vehicles in this report, the landing mode is horizontal.

Choosing the 125 lb/ft^2 (610 kg/m^2) takeoff wing loading means that only launchers with airbreathing speeds over 10,000 ft/sec will be considered for horizontal takeoff. Thus, like configuration concept, takeoff mode is a result of engineering decisions, not an arbitrary selection. In terms of configuration concept selection, the choice is based on whether or not the airbreather is a rocket derived or thermally integrated combined cycle. Landing wing loading is equivalent to a combat fighter, less than 45 lb/ft^2 (220 kg/m^2).

2. *Configuration Concept*

Given the space infrastructure of the 21st century, it is important to recall that rescue and supply of the manned space facilities requires the ability to land in a major ground-based facility at any time from any orbit and orbital location. The cross and down range needed to return to a base of choice also requires high aerodynamic performance. For the rocket derived propulsion concepts that are limited to Mach 6 or less, an acceptable inlet can be integrated into the vehicle configuration derived, for example, from the FDL-7 series of hypersonic gliders developed by the Flight Dynamics Laboratory[42] and the work of the McDonnell Douglas Astronauatics Company. The thermally integrated combined cycle configuration concept is derived from the McDonnell Douglas Advanced Design organization in St. Louis. This is a family of rocket-accelerated hypersonic airbreathers.[34] They can take off horizontally, vertically or be air launched. In its initial 1960s propulsion configuration, the vehicle was accelerated by a main rocket in the aft end of the body. Today, it can retain this concept or use combined cycle propulsion. In any case, rockets are usually mounted in the aft body for space propulsion.

This hypersonic glider and the hypersonic aircraft in Fig. 12 both have hypersonic lift to drag ratios in excess of 2.7. That means unpowered cross ranges in excess of 4,500 nautical miles and down ranges on the order of the circumference of the earth. So these two craft can depart from any low altitude orbit in any location and land in the continental United States (CONUS). Both are stable over the entire glide regime. The zero lift drag can be reduced in both by adding a constant width section to create a spatular configuration. The maximum width of this section is generally the pointed body half-span. The pointed configurations are shown in Fig. 10. No winged-cylindrical body configurations were considered.

3. *On-Board (Carried) Oxidizer*

The question is, why all the trouble about airbreathers? Is not a rocket good enough? Perhaps for ballistic missiles, but not for vehicles that need to achieve airline flight frequency and durability. The key to reducing size and weight, so the vehicle can abort at launch with vehicle and payload surviving, in a fail operational state, is to reduce of the onboard propellant and oxidizer. The rocket derived propulsion reduces the weight-ratio-to orbit but does not significantly affect the carried oxygen-to-fuel ratio. Both airbreathing rocket derived propulsion and the thermally integrated engine combined cycle engine reduce weight ratio and carried oxygen-to-fuel ratio. The air collection and enrichment system provides the greatest reduction in both weight ratio and oxygen to fuel into. Airbreathing rocket cycles (i.e, LACE or Deeply Cooled) can eliminate about 40% of the oxidizer from the launcher, so that for every 100,000 pounds of hydrogen there is

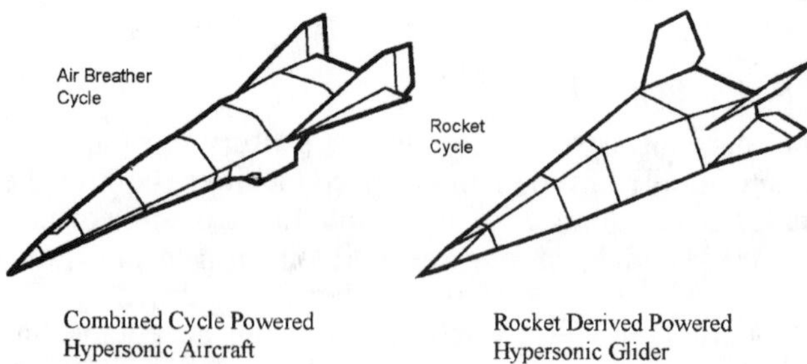

Fig. 12 Propulsion Cycle Determines Configuration Concept.

about 36,000 pounds of liquid oxygen carried on-board instead of 600,000 pounds for the pure rocket. For the thermally integrated combined cycle propulsion, the liquid oxygen load can be only 200,000 ob. For the air collection and enrichment system propulsion it might be possible to reduce the liquid oxygen load to 100,000 lb or less. The result is smaller, lighter vehicles that have better abort capability and have the potential of affordable sustained operations, with scheduled maintenance[85]

Ashford and Emanuel have compared ejector ramjet to the Oblique Detonation Wave Engine (ODWE). The ODWE can be one operating regime of a combined cycle propulsion system[86] when internal drag of the engine module becomes so large as to significantly diminish a thrust to drag ratio at high hypersonic speeds.[28]

IV. Sizing Code

Traditionally, the aircraft companies used constant gross weight analyses and photographic scaling as tool of the design trade. Herbst[19,87] introduced to McDonnell Aircraft Company and one of the authors (PAC) a scaling approach based on requirements, not fixed weight. In the requirements sizing approach each component is sized iteratively until the entire system meets all of the requirements in the sizing criteria. Formerly, each configuration concept of the same weight had a different performance. Each sized configuration concept now has the same performance with different size and mass. Performed with a computer-aided design program, this approach was revolutionary. Cycle time to evaluate a configuration concept was drastically reduced. With the sizing program the system meets the specifications, but each component is not "the optimum within its own application" but what is necessary for the system. Component performance is just sufficient to meet the system specifications.

The hypersonic sizing program is both mass and volume controlled. Space launchers and passenger-carrying aircraft offer the additional volume problem of a payload net density approximately that of hydrogen. The general approach taken in the sizing program is to specify the payload and propulsion system performance. An initial estimate is made for the planform area. The resultant iterations continue until volume available equals volume required. Two approaches are presented in Sec. IV.B and IV.C for applying the relationships developed in Sec. II, depending on the information available to the user.

A. Hypersonic Convergence Sizing Code

The first approach is that used in the Hypersonic Text for Parks College of Engineering and Aviation, "Hypersonic Convergence," by P.A. Czysz[36]. The approach correlates geometric data from references including Refs. 17, 18, 23, 32, 33. For a fixed payload and crew weight, this historical database was used to correlate the maximum propellant volume available for hydrogen-fueled aircraft as Kv.

$$K_v = \frac{V_{\text{ppl}}}{V_{\text{total}}} \cdot S_{\text{pln}}^{-0.0717} \tag{27}$$

$$\frac{V_{\text{ppl}}}{V_{\text{total}}} = K_v \cdot S_{\text{pln}}^{-0.0717} \tag{28}$$

Correlations for four configuration concepts are given in Ref. 36. As this correlation represents a maximum propellant volume ratio with high density electronic payloads, corrections for low density payloads are given in Ref. 36.

In order to determine the allowable structural weight per unit surface area, an estimate of the structural fraction was necessary. The initial correlation was based on one author's (PAC) hypersonic aircraft experience. When this sizing approach was employed for the HSCT study, the Douglas Aircraft Company data overlaid the hypersonic aircraft data (Refs. 30 and 31). Other aircraft data indicate that this approach gave an acceptable first order value that was consistent with initial estimates. That correlation is:

$$K_{\text{str}} = 0.228^{\pm 0.035} \cdot \tau^{0.20} \tag{29}$$

B. Final Hypersonic Convergence Relationships

This approach does not integrate an engine design/performance program or a trajectory analysis. These are done on a separate Microsoft Excel spreadsheet. The adequacy of this approach is documented in Section 5 of Ref. 36. Equations (1) through (17) then become as follows. These equations yield dimensional values, so the units must be consistent with the units of the other parameters. W_{OE} is an American term that is the dry weight plus trapped fluids and crew consumables. It is slightly greater than the European W_{dry}. With respect to parametric screening, the differences between W_{OE} and W_{dry} are inconsequential. The result then, is two equations that give the Operational Empty Weight and the structural index required for convergence.

$$W_{\text{OE}} = K_V \cdot \tau \cdot \left(\frac{\rho_{\text{ppl}}}{WR-1}\right) \cdot S_{\text{pln}}^{1.5717} - W_{\text{pay}} - W_{\text{crew}} \tag{30}$$

$$\frac{W_{\text{str}}}{S_{\text{wet}}} = \frac{K_{\text{str}} \cdot K_v \cdot \tau}{K_W} \cdot \left(\frac{\rho_{\text{ppl}}}{WR-1}\right) \cdot \frac{S_{\text{pln}}^{1.5717}}{(1+r_{\text{use}})} \tag{31}$$

$$I_{\text{str}} = \frac{W_{\text{str}}}{S_{\text{wet}}} \quad \text{Structural Index (weight/length}^2\text{)} \tag{32}$$

Equation (31) clearly shows that for the same propulsion index and geometry, the smaller the planform area the less the structural weight per unit surface area for solution. In order to compensate, and keep the structural index constant, the geometric parameter must increase accordingly, i.e., become stouter.

If the configurations presented in Fig. 4 and in Appendix B are used with the assumptions of this approach, the geometrical term in Eq. (31) collapses into a single function, as given in Eq. (33).

$$\left(\frac{K_w}{\tau}\right)\cdot\left(\frac{1}{K_{str}\cdot K_v}\right)=\left(\frac{K_w}{\tau}\right)\cdot\frac{11.35^{\pm 2.29}}{\tau^{0.206}} \tag{33}$$

$$\frac{K_w}{\tau}=\exp\left\{0.081\left[\ell n(\tau)\right]^2-0.461\cdot\ell n(\tau)+1.738\right\} \tag{34}$$

In Fig. 13 the value (K_w/tau is presented for all of the configurations given in Appendix B. As indicated from the data in Fig. 4, the range of τ spans the complete spectrum of aircraft configurations from the SST wing body con-

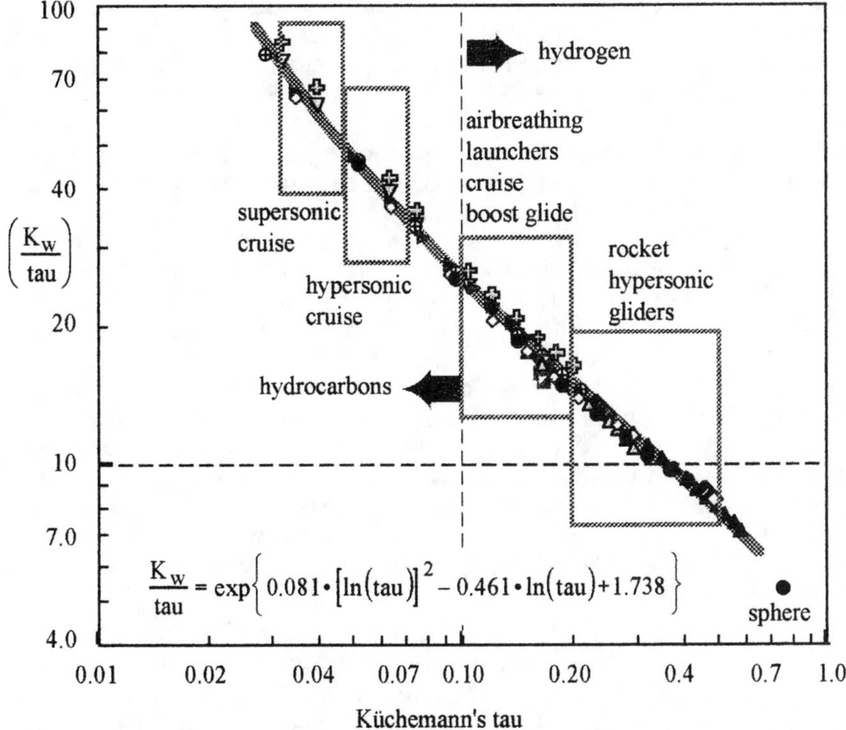

Fig. 13 Geometric Parameter Spans Complete Spectrum of Aircraft Configurations.

figuration with a τ of 0.03 to a sphere with a τ of 0.75. Equation (34) represents the curve through the data. This means that given the propulsion and structural indices, the first order vehicle size can be readily estimated as a function of tau and a configuration concept. Thus:

$$S_{\text{pln}} = \left[\frac{\rho_{\text{ppl}}/(WR-1)}{W_{\text{str}}/S_{\text{wet}}} \cdot \left(\frac{K_w}{\tau} \right) \cdot \left(\frac{1}{K_v \cdot K_{\text{str}}} \right) \cdot \left(1 + \frac{W_{\text{pay}}}{OEW} \right) \right]^{1.409} \tag{35}$$

The late Jean Vandenkerckhove thought this approach had merit. He used the approach as a screening tool and termed his adaptation of this program "SIZING." He did not think the approximations, the separate Excell trajectory determination, separate ramjet/scramjet size, and performance determination were acceptable for his applications. So a more detailed approach was undertaken to evaluate Eq. (1) through (17).

C. Vandenkerckhove Sizing Code

Vandenkerckhove (VDK) began his adventure into airbreathing after an encounter with the co-author at a conference in London in 1983. VDK set out to show that only rockets had a future in launcher development. The approach used by Vandenkerckhove [83,84,88,89] was to use an existing European-developed trajectory code to which he added the vehicle characteristic information: from reference 23, from a number of European references and from information gained in personal discussions with European Aerospace engineers. For the propulsion performance he constructed a one-dimensional ram-scramjet, nose-to-tail energy-based performance code (HYPERJET Mk #3),[90] similar to those developed by Dr. Frederick Billig,[91] formely of APL/JHU. The final programs were identified as ABSSTO and ABTSTO for airbreathing single-stage-to-orbit and two-stage-to-orbit respectively. The result obtained was just the opposite of that anticipated. In the mid 1980s the two co-authors began collaboration on airbreathing launchers. The first step was for VDK to incorporate the sizing routines from Ref. 36 into his code so that constant gross weight solution could be avoided. The payload and crew weight was fixed. Rather than use a separate trajectory code to establish the required weight ratio, in VDK's code, the sizing code, engine design/performance code and trajectory code were integrated into a single program. The code was re-iterated until the assumed initial values matched the output from the code within a small tolerance. The dry weight was determined by solving the weight and volume equations simultaneously, as developed in Chapter 5.

V. VDK Sizing Approach

The intent of this chapter is to show the basis for the single-stage-to-orbit sizing code (ABSSTO) and the two-stage-to-orbit sizing code (ABTSTO). In the process, the principal parameter inter-relations that drive space launcher concepts will be illustrated and the inter-relationship between parameters that shows a priori selection of parameter values is not prudent. System sizing a transatmospheric vehicle is a very interactive process, considering the different system elements. Each sizing is the product of a large number of iterations to converge each set of selected sizing inputs. In the late 1960s a sizing procedure was introduced that differed from the traditional constant gross weight and photographic scaling. This new process involved inputting the requirements to yield an aircraft sized to those specific requirements, as presented in Sec. 70. Comparisons could now be done between aircraft of different sizes and weights with the same performance/same payload that meet the same requirements. Without such a tool, the sizing process of a transatmospheric vehicle is a series of size and weight guesses. It is like finding one's way in a labyrinth, with too many options, too few reference points and frequent blind alleys. The availability of practical pathfinding tools becomes a necessity. The sizing tool presented in this chapter readily provides the system consequences for each choice of variables. Since Jean Vandenkerckhove's passing, the sizing programs described in this chapter have continued to be developed and can be obtained from Patrick Hendrick, Ecole Royale Militaire, ERM-KMS (MAPP), Renaissance Avenue, 30 B-1000 Brussels, Belgium.

The first application deals with single-stage-to-orbit vehicles (this chapter) and the second application is the adaptation to two-stage-to-orbit vehicles, (chapter 7). The single-stage-to-orbit vehicles are an especially challenging application because the weight ratio (WR) to orbital speed is a direct consequence of the oxidizer carried on-board, as is demonstrated by repeating Eq. (1).

$$W_R = 1 + \frac{W_{\text{ppl}}}{W_{\text{OE}}} = 1 + (1 + r_{O/F}) \cdot \frac{W_{\text{fuel}}}{W_{\text{OE}}}$$

$$\frac{W_{\text{fuel}}}{W_{\text{OE}}} = \frac{WR - 1}{1 + r_{O/F}}$$

$$\frac{W_{\text{oxidizer}}}{W_{\text{OE}}} = \frac{r_{O/F}}{1 + r_{O/F}} \cdot (WR - 1) \tag{36}$$

There is some variation in the fuel to operational empty weight ratio ($W_{\text{fuel}}/W_{\text{OE}}$) for different propulsion systems, but for a given set of variables, the direct driver of the magnitude of the weight ratio is the on-board oxidizer to fuel ratio ($r_{\text{O/F}}$). Reviewing 22 different propulsion configurations from "Hypersonic Convergence" (Ref. 35) from all rocket to airbreathing rocket/ejector ramjet to 22,000 ft/sec, the mean value of ($W_{\text{fuel}}/W_{\text{OE}}$) is 1.063 with a standard deviation of $\pm$ 0.159. The largest value of $W_{\text{fuel}} / W_{\text{OE}}$ = 1.62 for an all airbreather to orbital speed (25,659 ft/sec). This is a result of the poor thrust minus drag at high Mach numbers. So the fuel fraction ($W_{\text{fuel}} / W_{\text{OE}}$), Eq. (2.2a), remains almost constant, as oxidizer to fuel ratio decreases with airbreathing speed increases,[35] that is, the carried oxidizer fraction is decreasing. The gamut of parameters is a challenging one, i.e., the weight ratio spans a range from two to ten or greater. That means the propellant mass fraction can range from 50% to over 90% compared to 30% to 50% for aircraft. Because of the net propellant densities and their relationship to the weight ratio, the propellant volume is essentially constant (70% or greater) regardless of the propulsion system concept. The propellant volume can be a very dominant factor. The individual propellant densities can vary from 70 kg/m³ to 1300 kg/m³ (normal boiling point hydrogen to triple point LOX). So volume becomes the dominant factor. The hypersonic sizing program is then volume dominated. The next two parts present the weight and volume budgets that were developed by the authors.

A. Weight Budget

The equation for $W_{dry} \sim W_{OE}$, with margin is

$$W_{dry} \approx (1+\mu_a)\cdot\left(W_{str}+W_{eng}+W_{sys}+W_{\substack{crew\\provisions}}\right)=W_{EO} \tag{37}$$

where the equations, with their nominal range of parameter values for large operational systems are, as follows:

$$W_{str} = I_{str}\cdot K_w\cdot S_{pln} + W_{cprv} \qquad 17 \le I_{str} \le 23\ \text{kg/m}^2$$

$$W_{sys} = C_{sys} + f_{sys}\cdot W_{dry} \qquad 0.16 \le f_{sys} \le 0.24\ \text{ton/ton}$$

$$C_{sys} = C_{un} + f_{mnd}\cdot N_{crw} \qquad 1.9 \le C_{un} \le 2.1\ \text{ton}$$

$$1.45 \le f_{mnd} \le 1.05\ \text{ton/person}$$

$$W_{eng} = \frac{TW_0\cdot W_R}{E_{TW}}\cdot(W_{dry}+W_{pay}+W_{crw} \qquad 10 \le E_{TW} \le 25\ \text{kg thrust/kg weight}$$

$$W_{cprv} = f_{cprv}\cdot N_{crw} \qquad 0.45 \le f_{cprv} \le 0.50\ \text{ton/person}$$

$$W_{crw}\cdot N_{crw} \qquad 0.14 \le f_{crw} \le 0.15\ \text{ton/person}$$

If $N_{crew} = 0$, the vehicle is piloted by an automatic control system and has no provisions for a light crew, although the payload may include people. The values for nominal range of values for the sizing parameters came from personal discussions with European Aerospace engineers. Solving Eq. (37) for the dry weight (W_{OE}), we have

$$W_{\text{dry}} = \frac{\left[I_{\text{str}} \cdot K_w \cdot S_{\text{pln}} + C_{\text{sys}} + W_{\text{cprv}} + TW_0 \cdot W_R/E_{\text{TW}} \cdot (W_{\text{pay}} + W_{\text{crw}}\right]}{\left[(1/1 + \mu_a) - f_{\text{sys}} - TW_0 \cdot W_R/E_{TW}\right]} \tag{38}$$

The denominator is a difference term that can dominate the magnitude of the dry weight. The propulsion system performance is a critical factor in determining dry weight as it appears in both the numerator and denominator. This is an important observation, as it may not be the structural technology that is producing heavy dry weights but a poor performance propulsion system. For nominal values of the dry weight margin and system weight parameter,

$$0.63 \leq \left(\frac{1}{1 + \mu_a} - f_{\text{sys}} \right) \leq 0.71 \tag{39}$$

This means that for the denominator to increase the numerator by less than a factor of two, the value of the propulsion parameter should be in the range:

$$\begin{aligned} 4.8 &< \frac{E_{\text{TW}}}{TW_0 \cdot W_R} < 7.7 \\ 0.16 &< f_{\text{sys}} < 0.24 \end{aligned} \tag{40}$$

For transition to horizontal takeoff at weight ratios less than 4.3, the minimum and maximum engine thrust to weight ratios for increasing the numerator of Eq. (38) by less than two are for the nominal value of system dry weight ratio are:

The greater the system weight fraction, the greater the demands on the propulsion system to converge the dry weight with minimum growth. Or, stated another way, the greater the propulsion system performance, the more tolerance there is in the structural index and system weight fraction.

Table 1 Nominal range for takeoff engine thrust-to-weight ratio

W_R	T_{W0}	E_{TW}	E_{TW}
9.0	1.35	57	94
8.0	1.35	51	83
7.0	1.35	44	73
6.0	1.35	38	62
5.0	1.35	32	52
4.0	0.75	14	23
3.0	0.75	11	17
		$f_{\text{sys}} = 0.16$	$f_{\text{sys}} = 0.24$

Note that the minimum values for the engine thrust to weight ratios are not inconsistent with values already achieved in past experimental programs across the entire weight ratio range.

B. Volume Budget

The volume budget is as important as the weight budget, as it is the volume to planform area ratio that determines the configuration concept alternatives acceptable for the system under consideration.

The total volume relationship is

$$V_{tot} = V_{ppl} + V_{sys} + V_{eng} + V_{void} + V_{pay} + V_{crew} \tag{41}$$

where the equations with their nominal range of parameter values for large operational systems are as follows:

$$V_{tot} = \tau \cdot S_{pln}^{1.5} \qquad 0.032 \leq \tau\ 0.20$$

$$V_{ppl} = W_{OE} \cdot \frac{(WR-1)}{\rho_{ppl}} \qquad 5.0 \leq V_{un} \leq 7.0\ \text{m}^3$$

$$V_{fix} = V_{un} + f_{crw} \cdot N_{crw} \qquad 11.0 \leq f_{crw} \leq 12\ \text{m}^3\ /\ \text{person}$$

$$V_{sys} = V_{fix} + K_{vs} \cdot V_{tot} \qquad 0.02 \leq k_{vs} \leq 0.04\ \text{m}^3/\text{m}^3$$

$$V_{eng} = k_{ve} \cdot TW_0 \cdot W_R \cdot W_{OE} \qquad 0.25 \leq k_{ve} \leq 0.75\ \text{m}^3/\text{ton thrust}$$

$$V_{void} = k_{vv} \cdot V_{tot} \qquad 0.10 \leq k_{vv} \leq 0.20\ \text{m}^3/\text{m}^3$$

$$V_{pay} = W_{pay}/\rho_{pay} \qquad 48 \leq \rho_{pay}\ 130\ \text{kg/m}^3$$

$$V_{crw} = (V_{pcrv} + k_{crw}) \cdot N_{crw} \qquad 0.9 \leq k_{crw} \leq 2.0\ \text{m}^3/\text{person}$$

$$6.0 \leq V_{pcrv} \leq 5.0\ \text{m}^3/\text{person}$$

The values for the sizing parameters came from personal discussions with European Aerospace engineers. The payload density is characteristic of satellites in payload bay and payload shrouds. In the case of an all electronic internal payload, the payload density would be three times greater. Summing up the individual volumes, and solving for W_{OE}, we have

$$W_{OE} = \frac{\tau \cdot S_{pln}^{1.5} \cdot (1 - k_{vv} - k_{vs}) - (v_{pcrw} - k_{crw}) \cdot N_{crw} - W_{pay}/\rho_{pay}}{(WR-1)/\rho_{ppl} + k_{ve} \cdot TW_0 \cdot W_R} \tag{42}$$

$$W_{dry} = W_{OE} - W_{pay} - f_{crw} \cdot N_{crw}$$

Again, as in Eq. (38) we see that propulsion performance directly affects the dry weight by influencing the magnitude of the denominator. Equations (38) and (47) have τ and planform area as variables and can be solved simultaneously. A first estimate for the planform area with which to begin the sizing iterations can be obtained from either of two relationships, depending on whether landing or takeoff is critical:

$$\frac{1.1 \cdot W_{OE}}{L_{ld}} \leq S_{pln} \leq \frac{W_{GW}}{L_{to}} \tag{43}$$

Propulsion parameters continue to dominate the empty weight. Combining Eqs. (38), (42), and (43) provides a direct solution for the required volume, vehicle weight and slenderness.

Thus, selection of values for these sizing parameters is neither arbitrary nor independent of each other. With the trajectory program providing weight ratio and propellant bulk density, the three equations can be solved for the dry weight, τ and planform area. For vertical takeoff, the maximum landing planform loading results in the highest value of τ for an acceptable landing speed. The configuration concept considered must have a τ range that includes the τ value from the solution. An alternate approach is to establish a range of τ, and solve for the W_{OE} as a function of tau. The planform area and loading are then a fallout, and the vehicle selection can be made on consideration of the system characteristics. Considering the propulsion term in Eq. (38), $E_{TW}/(TW_0 \cdot W_R$, as the denominator increases (when moving from airbreather to rocket) the numerator increases at about the same rate. As a result, as reported by Froning and Leingang,[92] dry weight is essentially constant with the transition Mach number to rocket power, M_{tr}.

C. Input Values Assumptions

Two sets of assumptions can be considered to bound the "design space" of potential transatmospheric vehicles. The CURRENT and the FUTURE assumption sets are given in Table 2. Wherever applicable, manned and unmanned control are differentiated. These values were developed by J. Vandenkerckhove from discussions with the leading European Aerospace companies. The FUTURE set results from application of current European research and development that in 1994 had not yet been prepared for production application. Note that in Fig. 14, the 1970 projections for 1985 state-of-the-art hypersonic cruise vehicles are comparable to the FUTURE assumptions. The representative structural concept (Fig. 14), is based on the 1985 industrial capability, and had a I_{str} less than 18.0 kg/m^2 for a non-metallic composite structure protected by high temperature metal composite shingle with insulation. The 1970 state-of-the-art hypersonic cruise vehicles[17] based for an aluminum structure protected by high temperature refractory metal shingle with insulation is slightly greater than current assumption. It might be that considering Refs. 93 and 94 FUTURE should be near future.

Table 2 SSTO program inputs based on industrial capability

Head	CURRENT	FUTURE	Unit
Structural index I_{str}	21.0	18.0	kg/m^2
Engine thrust to weight ratio, E_{TW}	12.5	17.5	kg/kg
Fixed system weight C_{sys} unmanned	1900	1600	kg
manned	4400	3800	kg
Fixed system volume V_{fix} unmanned	6.00	5.00	m^3
manned	40.0	38.0	m^3
Variable system weight fraction f_{sys}	18.0	16.0	%
Takeoff planform loading L_{to}	400	440	kg/m^2
Engine specific volume k_{ve}	0.40	0.30	m^3/ton
System volume fraction k_{vs}	5.0	4.0	%
Void fraction k_{vv}	15.0	12.0	—
Crew & consumables W_{crew} manned only	640	600	kg
Hydrogen condition	Sub-cooled	50% slush	
	66.4	82.1	kg/m^3
Configuration concept	Blended body		
I_{sp} of main rocket mode Isp_1	460		sec
Mixture ratio of main rocket $(O/F)_1$	6.5		—
I_{sp} of OMS rocket I_{sp2}	440		sec
Mixture ratio of OMS rocket $(O/F)_2$	6.0		—
I_{sp} of ACS rocket I_{sp3}	400		sec
Mixture ratio of ACS rocket $(O/F)_3$	5.0		—
Payload density in bay ρ_{pay}	50.0		kg/m^3

In this chapter, SSTO is always assumed to be long duration, continuous use vehicle, regardless of the propulsion system under consideration. The vehicle can be under either unmanned or manned control with a crew of two, for a 10-day design mission. The term reusable is not used because the concept is to re-use an essentially disposable vehicle. Lindley and Penn (Ref. 26) show that for low launch rate ballistic vehicles, over 90% of the payload launch costs are from propellant and infrastructure. The payload launch cost is still dominated by propellant cost, but the infrastructure cost is minimal. In neither case is the production cost of the vehicle a critical cost item. Commercial aircraft have become more complex, larger, and more expensive while the cost per passenger trip is falling. The reason is flight rate. The vehicles in this chapter are intended to achieve a high flight rate.

D. Volume and Weight Assumptions

CURRENT set provides volume and weight assumptions considered realistic within Europe's industrial capabilities:

1) These are comparable but somewhat conservative when compared with published European aerospace industry results.
2) The takeoff thrust-to-weight ratio used is $E_{TW} = 16.5$ kg thrust/kg. This

value is very conservative for a rocket ejector combined cycle engine (RBCC). Rudakov and Balepin give a value for E_{TW} of 22.4 kg thrust/kg weight.[69,75,95] The values used in Table reflect a more conservative approach to the weights.

3) The system weights are comparable but somewhat conservative when compared with published European aerospace industry results.

4) Planform loading at takeoff of 400 kg/m² (82 lb/ft²) is adequate to get a liftoff velocity of about 420 km/hr (261mph), which is very conservative. As shown in Fig. 4, this will penalize HTOL in favor of VTOHL. For HTOL to gain the advantage, the planform loading should exceed 610 kg/m² (125 lb/ft²)

E. Aerodynamics

When a trajectory program is used, a database is needed to provide aerodynamic coefficients C_{D0}, C_D and C_{La} as functions of Mach number M_0 that takes vehicle shape into account. In the absence of specific data a "generic" aerodynamic database is used.[36] Either a blended body or body-wing configuration can be assumed. The vehicle total volume and planform area are accounted for in calculating C_{D0}, which explicitly depends on Küchemann's τ.

F. Propulsion

When a trajectory program is used, it needs to access a database[91,96–99] providing engine performance, net thrust per unit span and net specific impulse, as functions of Mach number M_0, dynamic pressure q_0 and angle of attack a_0. Unless otherwise indicated, the use of compact and lightweight rocket ejector combined cycle engines[58,76,78,100] capable of operating successively in air/rocket-ejector, ramjet, scramjet and rocket modes is assumed. The length of the combustor is consistent with pertinent experiments and computations.[101,102] for rocket ejector mixing-combustion sections. These are the characteristics of the HYPERJET Mk #3 engine dimensioned for airbreathing operation up to Mach 15 as described in Ref. 105. The propulsion performance is computed along the trajectory by the HYPERJET program[107]. This is a baseline reference engine. The number of modules required to provide the necessary thrust is computed using a not-to-exceed width of 1 meter per engine module. The number of modules cannot exceed the vehicle width divided by the module width (1 m): static thrust; 234 kN (52,600 lb) per engine; width; 1.00 m per engine module; length; 3.9 m; and weight (without margin); 1450 kg (3197 lb) per engine.

The baseline sea level static engine thrust to weight ratio, $E_{TW} = 16.5$ kg thrust/kg weight, was conservatively estimated for this engine. The width of the engine is a variable (not to exceed a width of 1 meter per engine module) so the vehicle thrust can be matched to the trajectory requirements. That is, a minimum total engine width is determined that can meet the performance required along the entire trajectory, which determines the number of one meter modules required. Since the underside of the vehicle is the propulsion

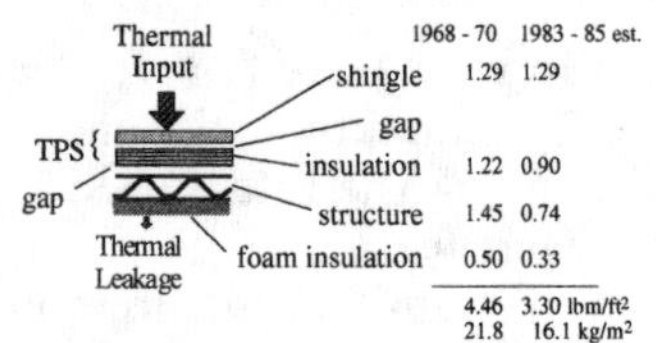

Fig. 14 Structural Concept for Hypersonic Vehicles, Ref. 110.

system, the freestream capture area, total engine cowl area, and cowl height are determined by the vehicle length. The reference baseline engine provides a point of departure to determine the thrust for each vehicle's unique propulsion system.

G. Trajectory

For the trajectory program, the following are trajectory segments. After a takeoff the trajectory assumes a steep subsonic climb followed by a low flight path angle acceleration through the transonic regime, followed by a climb/acceleration at constant dynamic pressure, followed by a cruise with a constant heading roughly perpendicular to the target orbital plane but somewhat "anticipating" the subsequent turn which aligns the vehicle into the orbital plane. After another acceleration at constant q, a pull-up maneuver is initiated just prior to transition from the scramjet to rocket during which the trajectory slope increases to reduce drag losses during the subsequent rocket acceleration. After transition to rocket mode, the SSTO continues the pull-up for a few seconds before accelerating and climbing along a so-called "linear-tangent" trajectory toward injection into transfer orbit at a specified altitude (for instance 90 km to keep kinetic heating low) and flight path angle. Although it is for TSTO, Fig. 15 contains all the elements of the SSTO trajectory. Included in the TSTO trajectory are the pull-up and staging that are not a part of the SSTO trajectory.

VI. SSTO Launcher Sizing

The SSTO launcher is sized to an SSTO mission that is launched from Istres in Southern Europe. There are two missions considered: delivery of a payload to a Low Earth Orbit (LEO) and delivery of a payload to a sun-synchronous mission (passes over the launch site at the same time every 24 hours). The requirements are 1) unmanned control SSTO mission, 2) horizontal takeoff and landing (HTOL), 3) blended body configuration, 4) takeoff planform loading, 400 kg/m^2, 5) liquid hydrogen/oxygen propellant, 6) current industrial capability rocket motor, 7) 8000 kg payload to 463 km / 28.5 degrees (design mission), 8) 8000 kg payload to 200 km / 98 degrees (alternate mission), 9) in-bay payload density (payload weight/payload bay volume) = 50 kg/m^3 10) based in Europe (Istres), 11) rendezvous and docking propellant allowance (100 m/sec), and 12) dry mass margin: $\mu = 15\%$.

The launchers were sized to requirements using ABSSTO for the two levels of industrial capability assumptions (Table 1). The design parameters assumed for the two SSTO cases produce different vehicle sizes and weights, as shown in Table 3.

Unless otherwise noted, the sizing program determined the minimum dry weight vehicle that met the takeoff planform loading requirements and that determined the Küchemann volume parameter, tau. The low takeoff planform loading forces a dramatic increase in the dry weight payload fraction with the application of the CURRENT assumption set. With a larger takeoff planform loading the planform area would be at least 200 m^2 less. Even though the volume for the FUTURE assumption set is reduced, the planform area has been reduced even more so that tau has substantially increased.

In the authors' judgment either one of these vehicles could be built today. That does not imply a lack of engineering challenge. What it does imply is that with a creative, innovative approach and a dedicated, small engineering team, these vehicles are possible. In many aspects, the vehicles represent a lesser challenge than the A-12/SR-71 did 35 years ago.

A. Determination of Vehicle Length

A spatular nosed vehicle can be characterized by an average width-to-length ratio. This permits construction of Fig. 16 that shows vehicle length as a function of landing weight (W_{OE}) for five values of the average width-to-length ratio from range 14 to 22%. For instance, an average width-to-length ratio of 16% yields a length of 55.3 m for W_{OE} = 66.2 ton for CURRENT assumption set, and 44.2 m and W_{OE} = 43.8 ton for the FUTURE assumption set.

Figure 16 implies that the engine capture ratio is increasing to the right as average width-to-length increases. For a given W_{OE}, increasing the nose width reduces the length significantly. For a W_{OE} of 70 ton, the length decreases from 61 m for a pointed configuration to 49 m for an average width-to-length ratio of 22%. More importantly, for the pointed vehicle, each size requires a different engine module width. That makes engine developers very uneasy, as the tested engine module may not be exactly the engine module for the final sized vehicle. However, the spatular nosed vehicle solves that problem because rather than making the vehicle longer, thereby changing the size of each engine module, the same length vehicle can be made wider using an integer number of the same size engine module. This holds true over a wide range of planform areas. The specific relationships that are employed follow from the basic geometry given in Fig. 17.[103]

$$L = \sqrt{\frac{S_{pf} \cdot \tan \Lambda}{1 + c/b}} \tag{44}$$

$$\frac{\overline{W}}{L} = \frac{1 + c/b}{(1 - c/b) \cdot \tan \Lambda} \tag{45}$$

Table 3 SSTO vehicle from the sizing program spans the design space

Head	CURRENT	FUTURE	Unit
Mass ratio WR	3.24	3.30	——
Propellant bulk density ρ_{ppl}	134.5	128.0	kg/m^3
Küchemann volume parameter τ	0.132	0.213	
Planform area S_{pln}	536.6	329.9	m^2
Wetted (surface) area S_{wet}	1469.7	1068.6	m^2
Dry weight W_{dry} (OEW)	58.2	35.8	ton
Gross mass GW (TOGW)	214.6	144.7	ton
Gross weight payload fraction P_{yg}	3.7	5.5	%
Dry Weight payload fraction P_{yo}	13.8	22.4	%
Landing planform loading L_{lnd}	123.4	133.2	kg/m^2
Total Volume V_{tot}	1635	1273	m^3
Propellant Volume V_{ppl}	1142	890.0	m^3
Propellant volume ratio	69.8	70.0	%
Ratio structure weight to dry weight	61.0	61.8	%
Ratio engine weight to dry weight	13.0	11.1	%
Ratio system weight to dry weight	26.0	27.1	%

$$\frac{c}{b} = \frac{(\overline{W}/L) \cdot \tan \Lambda - 1}{(\overline{W}/L) \cdot \tan \Lambda + 1} \tag{46}$$

$$\frac{A_{capture}}{S_{pln}} = \tan \sigma \cdot \left(\frac{X_c}{L_{ssto}}\right)^2 \cdot \left\{\frac{\left[1 + 2 \cdot c/b/(X_c/L_{ssto}) \cdot (1 - c/b)\right]}{1 + \left[2 \cdot c/b/(1 - c/b\right]}\right\} \tag{47}$$

Table 4 shows the increase in capture area and decrease in length for the spatular noses. This class of vehicle has been investigated by a Krieger,[104] Pike [105] and Townend.[106] Departing from the pointed nose shape permits a greater planform area and greater capture area for a given length.

Küchemann's tau reflects the vertical dimension of the vehicle for a given

Table 4 Spatular characteristics for a 72° sweep blended body

x/b	0.0	0.065	0.123	0.175	0.220
A_{cap} / S_{pln} $\sigma = 8°$	0.0558	0.0598	0.0630	0.0655	0.0676
A_{cap} / S_{pln} $\sigma = 10°$	0.0700	0.0750	0.0790	0.0822	0.0848
$\overline{W}/L$	14%	16%	18%	20%	22%

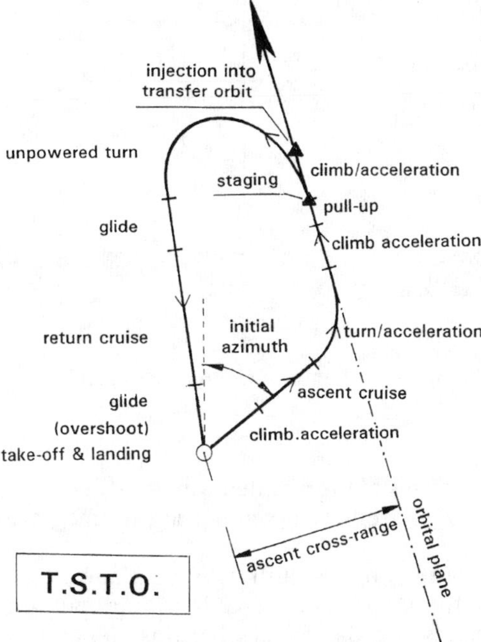

Fig. 15 Two-Stage-to-Orbit (T.S.T.O.) Flight Path Showing Ascent Cross Range and Orbital Launch Plane.

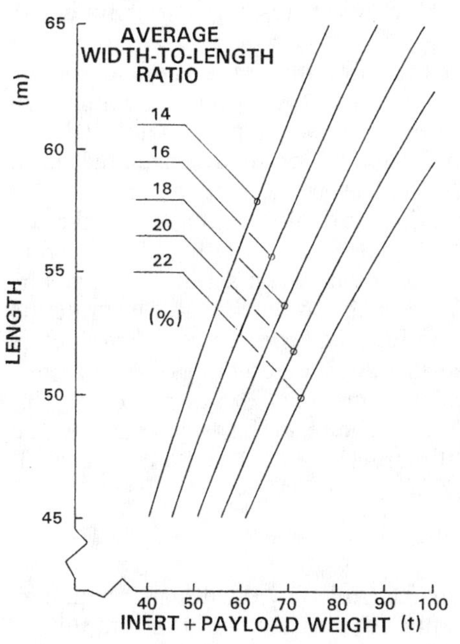

Fig. 16 Vehicle Length as a Function of Dry + Payload Weight (OWE) for Different Spatular Widths, Expressed as an Average Width-to-Length Ratio.

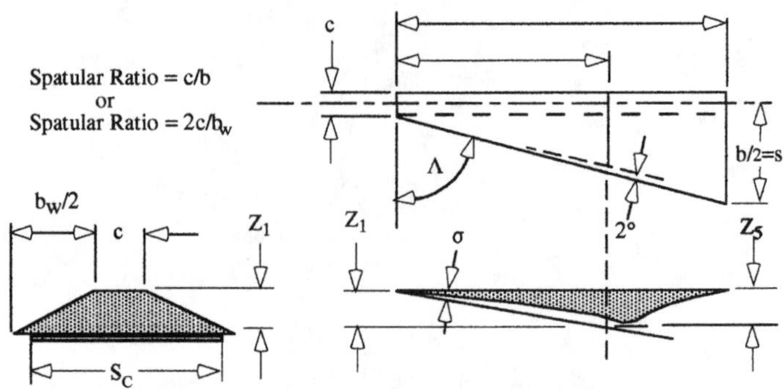

Fig. 17 Spatular Planform Configuration Relationships.

planform shape. In terms of the blended wing-body configuration, the range of possible values of tau represent geometric extremes, namely:

1) When the top of the body is flush with the plane containing the leading edges, then the only volume is the aero-thermo-propulsion configured propulsion system that is also the vehicle. For this minimum volume case, $\tau =$ 0.032 and Mach 15 lift-to-drag ratio is about 4.5
2) When the top curvature of the body is arched sufficiently so that the volume to planform area approaches that of a right circular cone of the same leading edge sweep, then the volume is at a practical maximum. For this maximum volume case, $\tau = 0.23$ and Mach 15 lift-to-drag ratio is about 2.0.

So, for the blended wing-body of the same planform size, the volume spans a total volume ratio of 7.18 from minimum to maximum τ. Volume can be penalizing from both an empty weight and a drag viewpoint. It is not unusual that so much effort has been exerted to densify hydrogen fuel, even slightly, as it has a strong affect on the vehicle size, weight and drag.

The discussion of parametric sensitivities is divided into three categories, Design (VI.B), Mission (VI.C), and Geometry (VI.D). Using the trajectory code, ABSSTO representative relationships between selected parameters was determined. These results provide a coherent database to illustrate the principal parameter interactions. The influence of the important parameters was assessed using a baseline vehicle sized to the design mission of launch from southern Europe (Istres) to achieve a 463 km / 28.5 degree orbit. This requires a cruise phase of about 1000 km. The result baseline vehicle was a 240 ton vehicle, with 15% dry weight margin, $\tau = 0.10$, and a takeoff planform loading of $L_{to} = 400$ kg/m^2.

B. Design

Design represents those parameters controlled by the designer who may change them in order to seek an acceptable sized configuration concept.

1. Influence of Transition Mach & Takeoff Thrust/Weight

Figure 18 shows the variation in payload weight, (W_{pay}) as a function of takeoff thrust-to-weight ratio (TW_0) for nine combinations of M_{tr} and engine thrust/weight (ETW). For each combination, an optimum value of TW_0 was found. Increasing TW_0 past this maximum resulted in a rapidly decreasing payload. The locus of the maxima is indicated for each transition Mach number. This figure points out that too large an engine is just as debilitating as too small an engine. However, in terms of engine thrust-to-weight ratio, greater performance yields greater payloads. The same is true for transition Mach numbers. Increasing transition Mach numbers result in an increasing payload. The maximum payload points define a narrow range of vehicle takeoff thrust-to-weight ratios. The three sets of curves for the three transition Mach numbers show the dramatic increase in payload with increasing transition Mach numbers. Parametric studies by the authors (not shown here) have shown that as transition Mach numbers are increased beyond Mach 15, the increase in payload diminishes while the takeoff thrust-to-weight ratio continues to increase. The same studies show for a transition

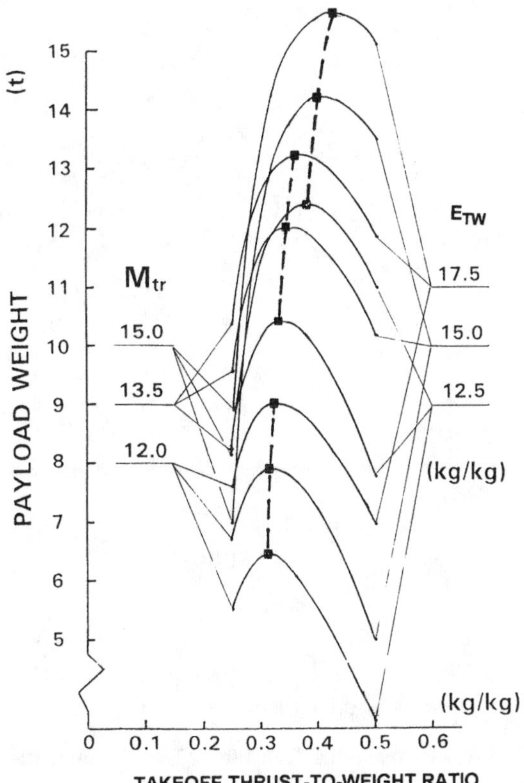

Fig. 18 Payload Increases with Increasing Transition Mach Number and Takeoff Thrust-to-Weight Ratio.

Mach number of eight, with a low engine thrust-to-weight ratio, it very is possible for the payload to be zero.

2. *Takeoff Thrust-To-Weight Ratio For Maximum Payload*

Plotting the maximum payload points from Fig. 18 results in Fig. 19. This figure shows the variation of the optimum take-off thrust-to-weight ratio (TW_0) as a function of transition Mach number M_{tr} for three values of engine thrust to weight ratio, 12.5, 15.0 and 17.5 kg thrust / kg weight. It is seen that the optimum value of TW_0 increases significantly with transition Mach number. The TW_0 approximately doubles between Mach 10 and 20. Engine thrust-to-weight ratio has less effect on TW_0 than transition Mach number. From the definition of the propulsion energy conversion efficiency (E_{prop}) both specific impulse (I_{SP}) and specific thrust (T_{SP}) must decrease with increasing speed, so the higher the transition Mach number the larger the engine required to maintain the desired acceleration. In spite of this, at least over the transition Mach numbers analyzed, the payload continues to increase. As discussed in 5.6, there is a limit to the engine size increase that

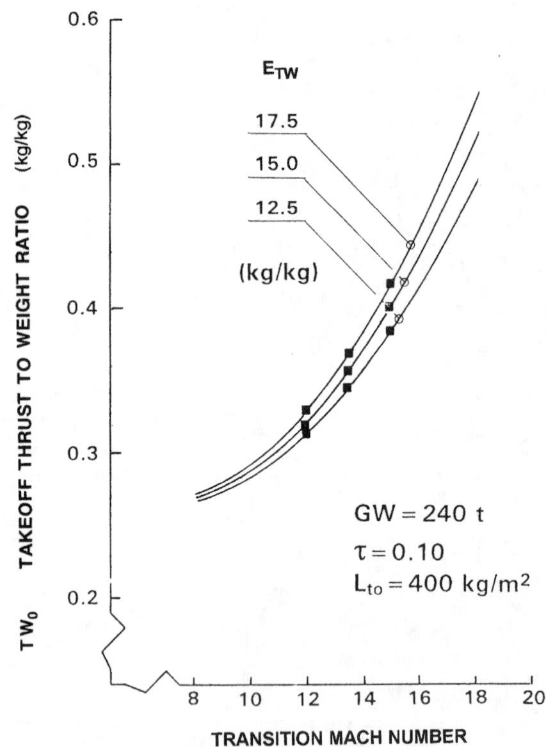

Fig. 19 Takeoff Thrust to Weight Ratio Increases as Transition Mach Number Increases

can occur. The width of the engine, consisting of n one meter modules, is determined by the takeoff thrust required. The maximum engine width engine width, of n modules, possible is the width of the vehicle at the engine cowl axial location (X_c). As shown in Fig. 17, the cowl width usually is set away from the vehicle leading edge by 2 degrees. Careful design of a spatular vehicle can eliminate this constraint by minimizing the streamline divergence near the leading edge. Whether for a spatular design, or a pointed design (c/b = 0) it is the leading edge flow that determines the proximity of the cowl edge. When the total engine width equals this vehicle width, the thrust cannot be increased further by making the engine wider, so the flight speed is at this point, it is the maximum airbreathing speed. That determination has to be made for each configuration. This consideration is usually not a critical factor in that other factors, as discussed in section 3, usually limit the maximum airbreathing speed first.[107]

In the present case, TW_0 is driven by the thrust minus drag "pinch" at the transition Mach number as the propulsion transitions from scramjet to rocket. This may not always be the case, however. If a LOX-LH2 rocket ejector is used at low speed with an I_{SPE} ~ 400 sec.to obtain a higher value of TW_0, perhaps 0.80 to 1.00, the transition Mach number "pinch" is minimized and it permits transition to ejector ramjet mode well below Mach 2.

3. *Planform Loading* L_{to}

Figure 20 is calculated for the reference condition of M_{tr} = 15, E_{TW} = 15 kg thrust / kg weight and τ = 0.1. Bulk propellant density increases and weight ratio decreases as takeoff wing loading is increased. Increasing the planform loading at takeoff reduces the vehicle size and total drag, thus requiring less fuel and a lower weight ratio. Thus, a higher wing loading at takeoff is an advantage (refer to Fig. 11 and the accompanying discussion). At 440 kg/m^2 (90 lb/ft^2) the weight ratio curve has not yet reached a minimum. Drastically reducing the takeoff wing loading can have a detrimental effect on the launcher convergence, another reason for having the correct takeoff thrust-to-weight ratio and corresponding takeoff wing loading. Again the selection is not arbitrary, but determined by the assumptions made. The fact that the weight ratio decreases with increasing wing loading shows the problem is dominated by zero lift drag area ($C_{D0} \cdot S_{pln}$), not by induced drag area. The typical acceleration flight path dynamic pressure is high (47.8 kPa or 998 lb/ft^2 or greater) and the vehicle flies at a low angle of attack, near zero lift drag. All above results are for "50% slush" hydrogen at 82.2 kg/m^3

4. *Influence of Transition Mach Number* M_{tr}

The results presented in Figs. 21 and 22 are always subject to skeptical criticism by those who do not believe future launchers can be anything other than rocket powered,[91]. These results reflect the trends from other airbreathing propulsion system studies,[108] and Refs. 57, 58, 109–112. All of these references show minimum gross weight and dry weight occur at a transition

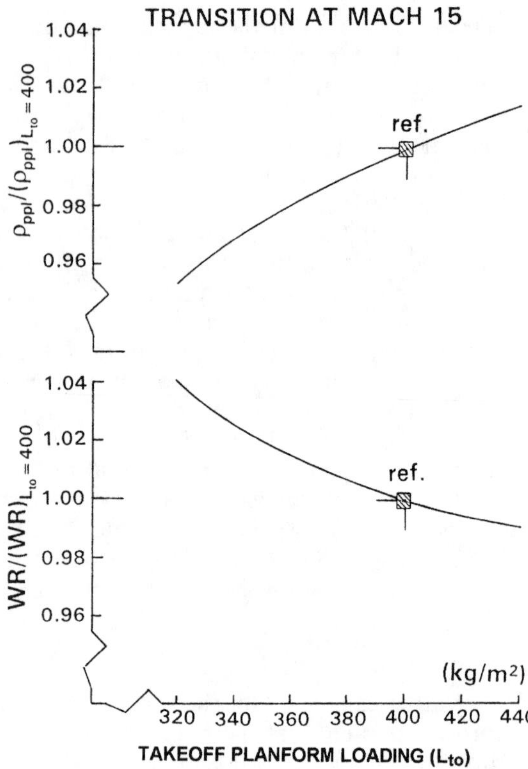

Fig. 20 Increasing Takeoff Platform Loading Decreases the Weight Ratio-to-Orbit and Increases Bulk Propellant Density.

Mach number between 12 and 16, depending on specific assumptions. This is within the range of practical airbreathing propulsion, as illustrated in Fig. 7, with a maximum air kinetic energy to Brayton cycle combustion energy ratio of 5. This is just one unit higher than the energy ratio 4 shown in Fig. 7 and is possible but slightly more challenging. Reference 112 shows the results of the Strutjet® sizing investigations as having a lower dry weight and gross weight than NASP.

Figures 21 and 22 show the change of gross weight and of dry weight as functions of transition Mach numbers. Both the LEO 463 km / 28.5 deg mission and the sun-synchronous 200 km/98.0 deg mission and for CURRENT and FUTURE systems assumptions are shown. These results clarify the importance of specifying the design mission and payload as an aircraft designer would do. With the sizing program the results clearly show the result of increasing transition Mach numbers. The minimum weights are reached in the Mach 14 to 17 region. Note that the minimum gross takeoff weight and dry weight occur at different transition Mach numbers. For transition Mach numbers less than the minimum the: (1) weight ratio increases and (2) the rocket propellant load increases (Figs. 23 and 24) (also see Figs. 29 and 30) result in increasing dry and gross weights. Thus, selecting a lower transition

Mach number, such as Mach 8 or 10, will not result in a lower weight solution but a higher weight solution. For a transition Mach number greater than the minimum, the combined result of (1) higher thrust-to-weight ratio and (2) a larger propellant tank to accommodate the increasing tank volume as bulk propellant density decreases (Figs. 21 and 22). The impact of the CURRENT assumption set of industrial capability, that is generally judged lower risk, is to actually increase the risk because of increased weight growth. The FUTURE assumptions are much more tolerant to changes in transition Mach numbers. As pointed out earlier, the FUTURE assumptions are based on those judged available in 1983. Note that the minimum gross takeoff weight and dry weight occur at different transition Mach numbers.

The difference between CURRENT and FUTURE system assumptions is rather dramatic. Most of this dramatic difference results from the selected takeoff planform loading (L_{to}). For transition at Mach 15, the gross weight decreases from 214.6 ton down to 144.7 ton (−32.6%) and dry weight from 58.2 ton down to 35.8ton (−38.5%) for the LEO mission. The gross weight decreases from 291.5 ton down to 171.3 ton (−41.2%) and dry weight from 76.3 ton to 40.5 ton (−46.2%) for the sun-synchronous mission. For the FUTURE system assumptions, the τ is near the maximum for the blended wing-body (0.23) and therefore the planform area is near its minimum size.

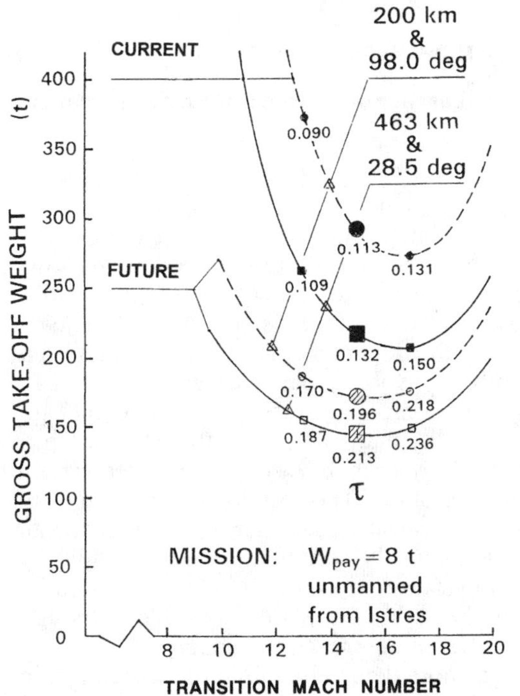

Fig. 21 Takeoff Gross Weight (W_{GTO}) as a Function of Transition Mach Number for Two Orbital Altitude and Inclinations.

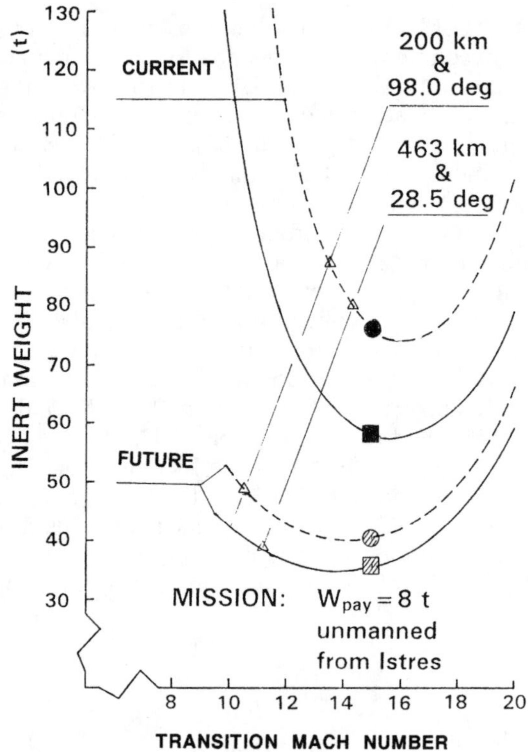

Fig. 22 Inert (dry) Weight (W_{EO}) as a Function of Transition Mach Number for Two Orbital Altitudes and Inclinations.

For CURRENT assumptions the tau is near the minimum τ for the blended wing-body (0.09) and represents a large planform area forced by the value of L_{to}. For CURRENT assumptions, transition to rocket below Mach 11 to 13 result in high gross weights for exceeding 400 tons (nearly that of a 747 airliner) and dry weights over twice the minimum. For FUTURE assumptions, transition to rocket below Mach 10 is possible, albeit at a significant increase in dry and gross weight. These are compelling reasons to implement improvements in the industrial capability and to improve the application of current industrial capability. As industrial capability improves, it reduces the difference between the mission descriptions. The desired design mission and payload should be specified and the sensitivities determined. Rather than assume a fixed gross weight and lament the results, an acceptable solution can be found through creative application of known industrial capabilities, i.e., technology. In many cases, in the authors' experience, problems thought impossible were solved when a different approach was taken.

The sun-synchronous mission is more demanding than the LEO mission. For the CURRENT assumptions, the gross weight is 36% heavier and dry weight is 31% heavier. Note that the minimum dry weight occurs at a lower transition Mach number than the gross weight. Again, improved industrial capability makes the system easier, not harder. Improving the industrial capability has greater payoff than increasing the engine thrust-to-weight ratio.

Once transition to a supersonic through-flow engine is made (about Mach 6) there is no penalty for flying faster until a transition Mach number of about 15 is reached (Refs. 52, 58, 109–113). As discussed in Sec. III, there are other considerations that can affect the maximum value of the transition Mach number. Note also that the lower industrial capability causes convergence at lower values of tau and higher transition Mach numbers. Also shown are values of the Küchemann's tau that yielded minimum weights. The values of tau for the minimum weight values for LEO orbit approaches the blended body configuration concept constraints if tau > 0.229 (see Fig. 3).

5. *Influence of LH2 state and Küchemann's Volume Coefficient*

Figures 23 and 24 show the variation of gross weight and of dry weight as a function of transition Mach number, for CURRENT system assumptions and

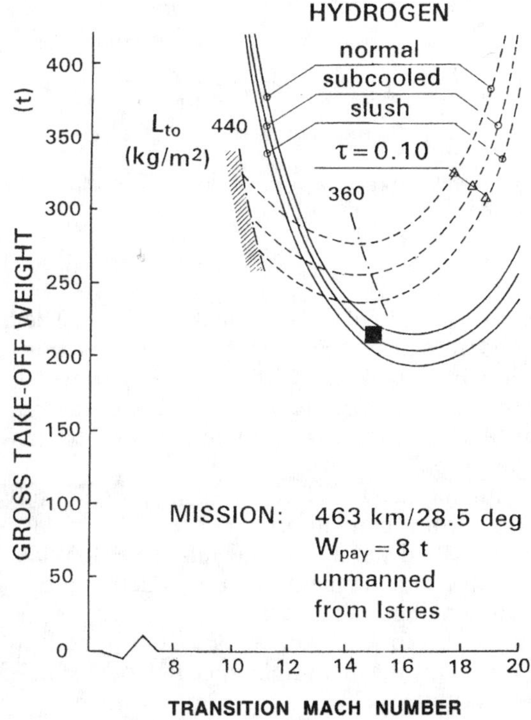

Fig. 23 Take Off Gross Weight (W_{GTO}) as a Function of Transition Mach Number for Variable and Fixed Küchemann's Tau.

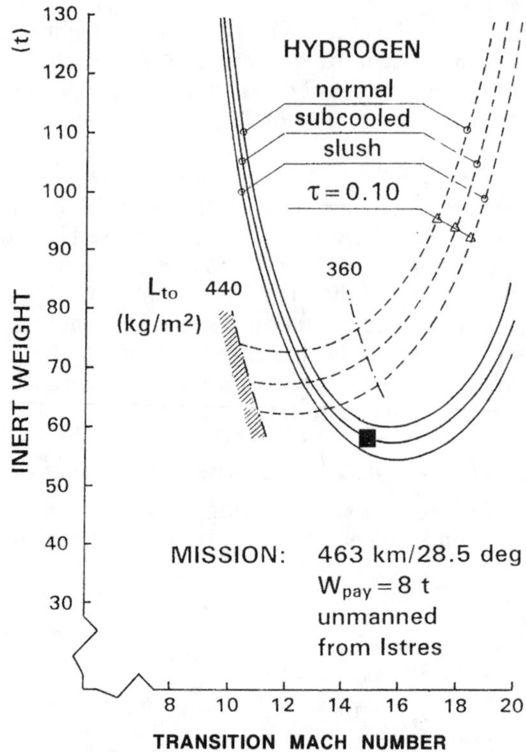

Fig. 24 Inert (dry) Weight (W_{EO}) as a Function of Transition Mach Number for Variable and Fixed Küchemann's Tau

for *normal*, *sub-cooled* and *50% slush* liquid hydrogen. The hydrogen density has only a moderate influence on these weights. For transition to rocket at Mach 15, a shift from *normal* to *50% slush* liquid hydrogen lowers the gross weight from 224.1 to 205.4 tons (−8.3%) and the inert weight from 61.0 to 55.5 tons (9.0%). For dry weight that is a − 0.54% in dry weight per percent change in propellant density.

As normally implemented by VDK, the sizing program solves for volume and determines tau from the planform area specified by planform loading required by the specified landing speed (solid line). When a solution is determined for a fixed tau, the planform area is determined by tau (dashed line) and planform loading is a fallout.

The minimum weights for constant tau occur at distinctly lower transition Mach numbers, especially the dry weight, specifically Mach 12.5 compared to Mach 16.3. The constant τ solutions are also more sensitive to changes in propellant density. The dry weight sensitivity is −1.08% change in dry weight per percent change in propellant density, twice the variable τ solutions. Because the fixed τ cases result in a more slender vehicle than the τ for minimum weight (see Fig. 21), the minimum weights for the fixed τ solutions are 12.7% greater.

Figure 15 in Ref. 7 shows the weight penalty for increasing slenderness and increased cross-range. In early studies even though long glide range systems could de-orbit from any arbitrary orbit and orbital location for return to CONUS (continental United States) with no waiting, these were usually rejected as too costly in terms of vehicle weight (Ref. 33). It is out of the scope of this study to discuss the requirement for cross-range. In previous experience, a good hypersonic glider (Lift-to-Drag ratio of 2.1 to 2.3 at Mach 22) with just enough propellant for one missed landing was sufficient.[93,108]

6. *Influence of Weight Margin*

Figures 25 and 26 represent the variation of gross weight and of inert weight as functions of dry weight margin, (μ_a) for CURRENT system assumptions and for transition to rocket at Mach 15. These results indicate that rather large margins can be used without incurring an excessive performance penalty. The selected value for the takeoff planform loading again shows its influence. The Küchemann's τ varies with μ_a as the sizing program solves for the slenderness for minimum dry weight because the planform

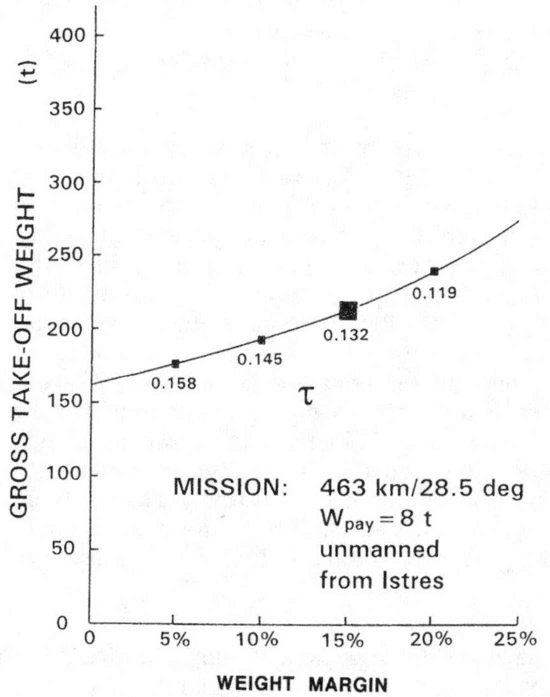

Fig. 25 Takeoff Gross Weight (W_{GTO}) as a Function of Dry Weight Margin.

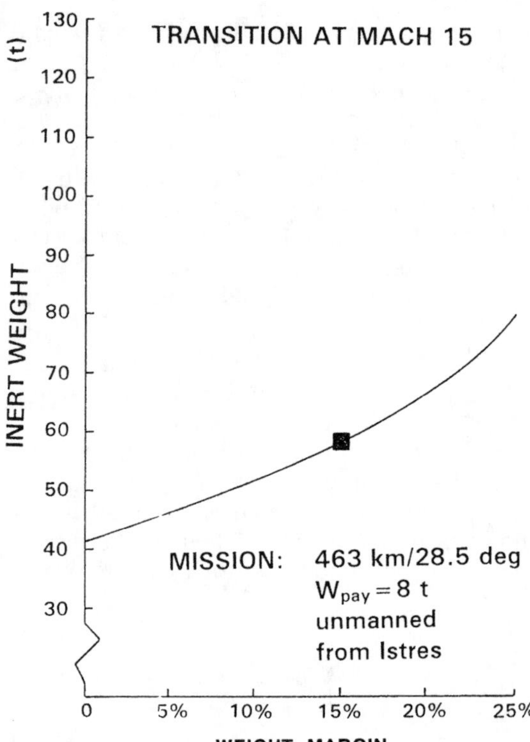

Fig. 26 Inert (dry) Weight (W_{EO}) as a Function of Dry Weight Margin

area is increasing faster than the total volume as the gross takeoff weight is increasing. If the takeoff planform loading were greater, the change in τ would not be as great. Should the margin not be fully used for coping with development problems, then the vehicle will have extra payload capability. When margin is changed in a sizing program, the result is a change in the dry mass and can be changed in geometry, depending on the takeoff planform loading. Again, for the L_{to} assumed, the key system attribute is dry weight driven primarily by the increasing planform area. For a 20% increase in dry mass margin, the vehicle becomes more slender. The dry weight at 20% dry weight margin is 63% greater than the dry weight for zero margin. For the L_{to} assumed, the linear sensitivity for a 1% increase in dry weight margin increases the dry weight by 3.2%. As implied in the results presented in Figs. 25 and 26, for the FUTURE system assumptions, the sensitivity is less. So there is a significant but tolerable dry and gross weight increase for modest weight margins that depend on the L_{to} assumed and the system assumptions.

7. *Influence of I_{str}, E_{TW}, and L_{to}*

Figures 27 and 28 show the change in gross weight and of dry weight as functions of structure index I_{str}, engine thrust-to-weight ratio, E_{TW}, and

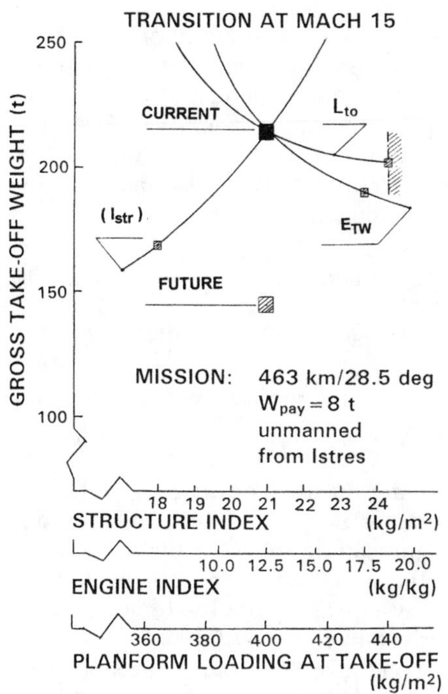

Fig. 27 Takeoff Gross Weight (W_{GTO}) Sensitivities With Respect to Basic Design Parameters, CURRENT Assumption Set.

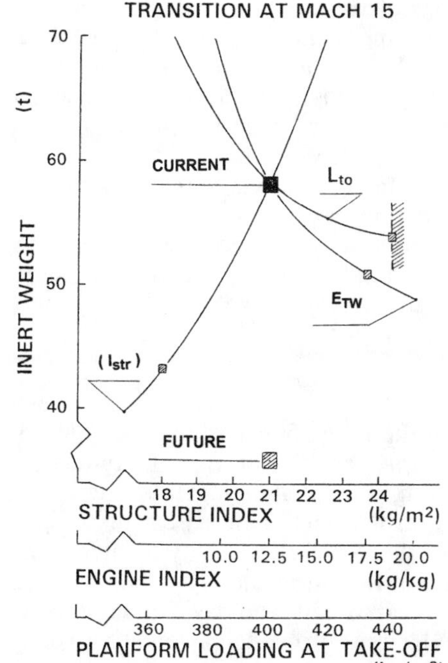

Fig. 28 Inert (dry) Weight (W_{EO})) Sensitivities With Respect to Basic Design Parameters, CURRENT Assumption Set.

takeoff planform loading, L_{to} for CURRENT system assumptions and for a transition to rocket at Mach 15. Sensitivities are non-linear, but the slopes at the reference point were determined in terms of percent change in the dependent variable per percent change in the independent variable. CURRENT system assumptions sensitivity is greater than FUTURE system assumptions, as implied in Figs. 21 and 22. For CURRENT system assumptions, the percent change is dry mass per percent change in parameter:

1) Structural index is + 2.20, i.e., a 1% reduction in I_{str} results in a 2.20% reduction in dry weight.
2) Engine index is −0.80, i.e., a 1% increase in E_{TW} results in a 0.80% reduction in dry weight.
3) Takeoff planform loading is −1.35, i.e., a 1% increase in L_{to} results in a 1.35% reduction in dry weight

In summary, a decrease in $I_{str} <$ kg/m², an increase in $E_{TW} > 12.5$ kg thrust/kg or to a lesser degree an increase in $L_{to} > 400$ kg/m² improves performance significantly. The most significant influence is that of I_{str} since for the CURRENT system assumptions the structure weight represents more than 60% inert weight than for FUTURE system assumptions. Remember that the increase in dry weight is as much driven by the takeoff planform loading as is the structural index. Concentrating on the material, structure, and fabrication industrial capability is an important investment as is increasing the takeoff wing loading. Reference 17 uses a takeoff planform loading as high as 976.3 kg/m² (200 lb/ft²) for the hypersonic research aircraft. With a high takeoff thrust-to-weight ratio and a vectoring (gimbaled) rocket engine nose wheel rotation can be accomplished at lower flight speeds and the climb initiated using thrust support. Increasing the engine thrust per unit mass is the least beneficial of the parameters investigated. Considering the importance of I_{str} a review of the historical values from Refs. 93, 108, and 114 is appropriate as shown below in Table 5, for operational vehicles with one example for a hypersonic demonstrator.

The proposed hypersonic research aircraft could accommodate such a large I_{str} because of a very small propellant fraction resulting from being designed for only 5 minutes at Mach 12 (see Appendix A and Ref. 17 for a discussion of these hypersonic research aircraft).

8. *Influence of Payload Density*

The influence of payload density in the range $50 < \rho_{pay} < 100$ kg/m³ is shown in Table 6 for assumption Set A and transition at Mach 15. It demonstrates why payload net densities are very low when diameter is fixed.

A 50% reduction in payload density results in a 6% increase in dry weight and 5.6% increase in gross weight. So the vehicle sensitivity to significant changes in payload density is primarily reflected in the propellant volume ratio. Correlation of then-available payload weights and payload bay dimensions for non-military satellites yielded a most probable payload (bay) density of 80 to 70 kg/m³ (5.0 to 4.4 lb/ft³) or about the density range of liquid

Table 5 Structural index based on estimated industrial capability

Source	Structural index, kg/m^2	Structural index, lb/ft^2
Refs. 93 and 108 1970 projection to 1985	17.1	350
VDK, CURRENT	21.0	368
VDK, FUTURE	18.0	430
Refs. 93 and 108 1970 industrial capability	22.0	450
NASA, circa 1990 [114]	15.1	320
Refs. 93 and 108 low cost 1970 hypersonic demonstrator	29.3	600

hydrogen. Since the payload shroud, enclosing the payload, for the Arianespace Ariane 5 and NASA Shuttle cargo bay are 4.5 m in diameter the payload density should reflect the weight and length of the longest anticipated satellite that a 4.5 m diameter payload bay must accommodate.

C. Mission

Mission relates to those parameters which are determined by the trajectory, transition Mach number, orbital altitude, inclination, and the orbital maneuver propellant required.

1. *Influence of Mission Requirements*

Correction factors for alternate missions are provided in the following:

1) Keeping the transition at Mach number of 15, launch from Kourou instead of Istres to the same 463 km / 28.5 deg orbit, no cruise leg to the orbital plane is required, so the increments are:

Table 6 Impact of payload density on vehicle size for 8-ton payload

Head	Head	Head	Head	Unit
Payload density	50	75	100	kg/m^3
Bay volume	160	107	80	m^3
Bay length for a 4.5-diameter	10.0	6.7	5.0	m
Planform area	536.6	517.8	507.9	m^2
V_{ppl} / V_{tot}	69.8	72.5	74.0	%
Gross weight	214.6	207.1	203.2	ton
Dry weight	58.2	56.0	54.9	ton

$$\Delta W_R/W_R = -0.036 \quad \text{and} \quad \Delta\rho_{ppl}/\rho_{ppl} = +0.0480$$

2) Keeping the transition at Mach 15, and launching into a sun-synchronous orbit 200 km/98.0 degrees produces a rather long launch window of several hours per day, depending on base latitude. It is made possible by the cruise capability, results in the following increments:

$$\Delta W_R/W_R = +0.077 \quad \text{and} \quad \Delta\rho_{ppl}/\rho_{ppl} = -0.0030$$

3) Adding orbital maneuver propellant and provision for an additional 100 m/sec for in-orbit maneuvers (orbital maneuver LH_2-LOX engines with I_{sp} = 420 sec) has these increments:

$$\Delta W_R/W_R = +0.025 \quad \text{and} \quad \Delta\rho_{ppl}/\rho_{pp1} = -0.0075$$

2. *Weight Ratio to Orbit vs Transition Mach*

Figure 29 shows the evolution of weight ratio to orbit (W_R) as a function of transition Mach number, M_{tr} for three engine thrust-to-weight ratios (E_{TW}) = 12.5, 15.0, and 17.5 kg thrust/kg weight and the corresponding

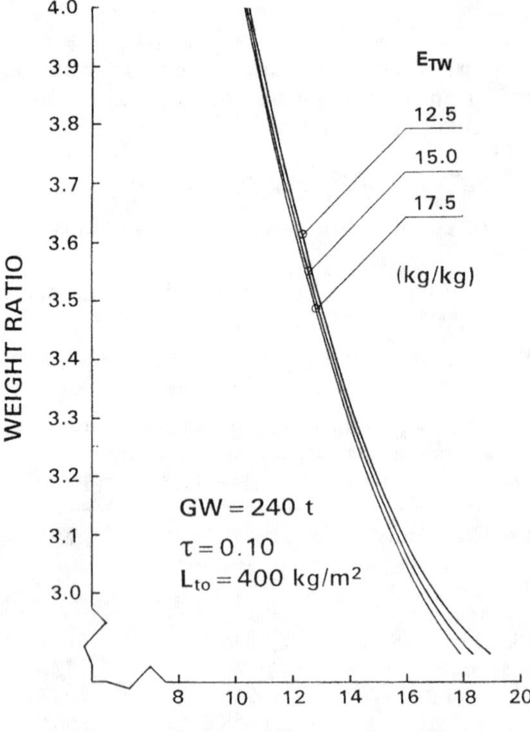

Fig. 29 Weight Ratio to Orbit Decreases as Transition Mach Number Increases.

optimum thrust-to-weight ratio at takeoff, TW_0. W_R is seen to decrease rapidly with increasing M_{tr} but only little with improving E_{TW} as the flight time to orbit gets shorter (not as strong a driver as often perceived). The fuel mass in terms of multiples of the W_{OE} is essentially constant but the oxidizer mass is continuously decreasing as the carried oxidizer to fuel ratio decreases. The result shows the fuel (hydrogen) fraction of the propellant increasing and the bulk density decreasing.

3. *Bulk Propellant Density vs Transition Mach*

The density of 50% solid slush hydrogen is 82.2 kg/m^3. The bulk density of 6:1 oxygen/hydrogen for a rocket motor is about 378 kg/m^3. So, as the transition Mach number increases and the quantity of rocket propellant decreases, the bulk density decreases. Figure 30, calculated for "slush" hydrogen, shows the evolution of bulk propellant density ρ_{ppl} as a function of transition Mach number M_{tr} for three E_{TW} = 12.5, 15.0 and 17.5 kg thrust / kg weight for the optimum takeoff thrust-to-weight ratio (TW_0). Propellant bulk density decreases rapidly with increasing M_{tr} but only very little with improving E_{TW}. This figure is calculated for "50% slush" hydrogen at 82.2

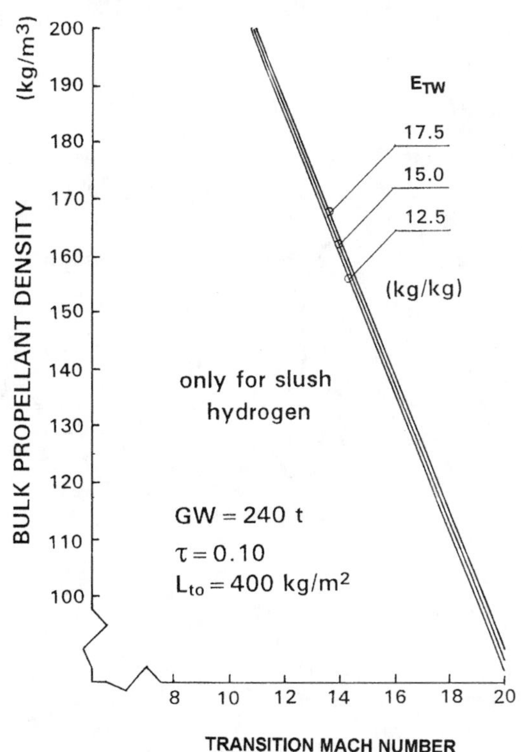

Fig. 30 Propellant Bulk Density Decreases as Transition Mach Number Increases

kg/m³. For sub-cooled liquid hydrogen the density is 75.9 kg/m³ and for Normal Boiling Point (NBP) hydrogen the density is 70.5 kg/m³. That 16.6% increase in density does impact the final size of the vehicle. The engine thrust-to-weight ratio has little impact on these results. Since weight ratio and propellant bulk density decrease at similar rates, the propulsion Index I_p remains constant (Ref. 104).

4. *Influence of Crew and of Payload Weight*

Figures 31 and 32 show the change in gross weight and of dry weight as functions of payload weight W_{pay} for CURRENT system assumptions and FUTURE system assumptions, and for both manned and unmanned vehicles with transition to rocket at Mach number 15. Beyond the normal albeit moderate increase in gross take-off weight and inert weight associated with heavier payloads, the following is seen:

1) Crew presence increases GW by 63.9 ton, i.e., 29.8% and W_{dry} by 20.4 ton, i.e., 35.0% for CURRENT system assumptions, if W_{pay} is kept equal at 8 ton.

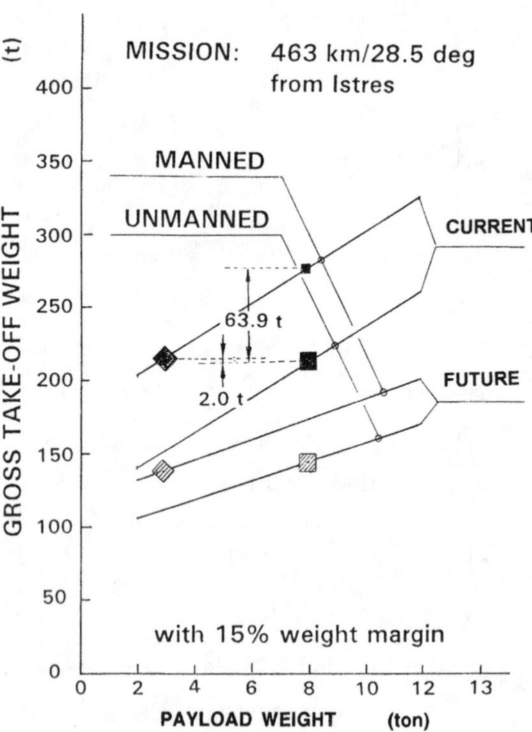

Fig. 31 Takeoff Gross Weight (W_{GTO}) Has a Greater Growth Factor for Manned Compared to Unmanned Operation.

2) Crew presence only increases GW by 2.0 ton, i.e., 0.9% and W_{dry} by 5.9 ton, i.e., 10.0% for CURRENT system assumptions, if W_{pay} is lowered from 8 down to 3 ton in the manned version (including crew and associated consumables),

3) The difference between manned and unmanned vehicles decreases as technology improves as shown by the comparison of the curves for CURRENT system assumptions and FUTURE system assumptions.

D. Geometry

1. Küchemann's Parameter τ

Figure 33 is calculated for the reference condition of $M_{tr} = 15$, $E_{TW} = 15$ kg thrust/kg weight, $L_{to} = 400$ kg/m^2 and shows the influence of τ with respect to this condition. As the vehicles become stouter, that is, increasing τ, the weight ratio required to achieve orbital speed increases, the drag increases, and the volume of the vehicle is increased. The propellant density decreases as the hydrogen fraction is increasing. The reason for this behavior is the increase in zero lift drag as slenderness decreases.

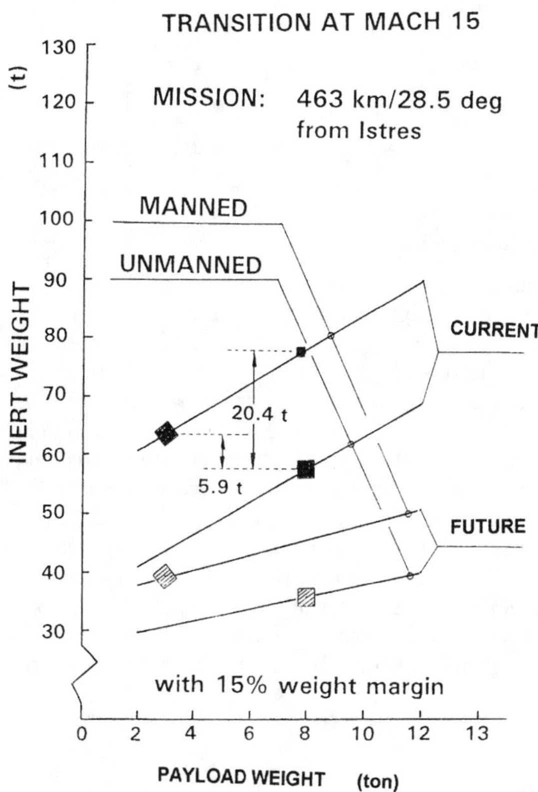

Fig. 32 Inert (dry) Weight (W_{EO})) Has a Greater Growth Factor for Manned Compared to Unmanned Operation.

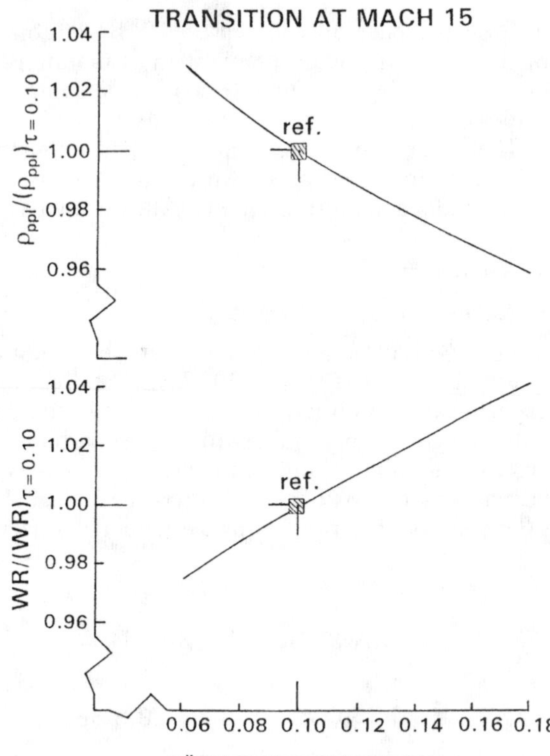

Fig. 33 Increasing Küchemann's tau Increases the Weight Ratio to Orbit and Reduces Bulk Propellant Density.

2. *Slenderness and Fuselage Bending Strength*

An example of cross-coupling is the interaction between Küchemann's τ and structure index I_{str}. An a priori judgment might be that the structure is heavier if the configuration is shouter, i.e., a larger τ. But, this is not the case. Indeed, when Küchemann's τ increases, the relative cross-section of the integral tanks gets larger with a corresponding increase in section moment of inertia available to carry bending loads. This increased section inertia can be taken to advantage to lower structural index I_{str}.[94] K_{FB} is a fuselage bending correction for the increase in cross-section inertia as tau is increased. It is an approximate relationship, but a reasonable representation of the bending stiffness increase with increasing based on aircraft experience. The first parameter, K_{FB}, establishes a correction for the skin weight as a function of τ with $\tau = 0.10$ as the reference point ($K_{FB} = 1.0$). The second relationship applies this correction to the structural index, I_{str}. The relationship between τ and I_{str} is through the fuselage bending parameter, K_{FB} are:

$$K_{FB} = 1.25 \cdot 10^{-0.97 \cdot \tau} \tag{48a}$$

$$I_{str} = K_{FB} \cdot (I_{str})_{\tau=0.10} \tag{48b}$$

Figure 34 shows the variation of I_{str} as a function of τ for four values of $(I_{str})_{\tau=0.10}$ from 18 to 21 kg/m², which cover the range between FUTURE system assumptions and CURRENT system assumption. The comparison between a *nominal* vehicle characterized by $\tau = 0.10$ and $I_{str} = 21$ and one with $\tau = 0.14$ and $I_{str} = 18$ shows that 60% of the improvement in I_{str} is due to the stouter configuration and 40% due to technology. The result is that the more slender the configuration the greater the required structural index to carry the bending loads. There appears to be a dichotomy in this result. The greater the structural index, the less advanced the industrial capability required. The lesser the structural index, the more advanced the industrial capability required. In this case, however, it should be interpreted not as a requirement but a measure of margin. For a given level of industrial capability, the stouter vehicle has more margin between available and required. When material advances are available they reduce the structural index, and can be incorporated without significant redesign to permit payload increase.

3. *Slenderness and Thrust Minus Drag*

Fig. 35 shows the tentative variation of propulsion index E_{TW} (kg static thrust per unit kg weight) as a function of Küchemann's τ for five values of

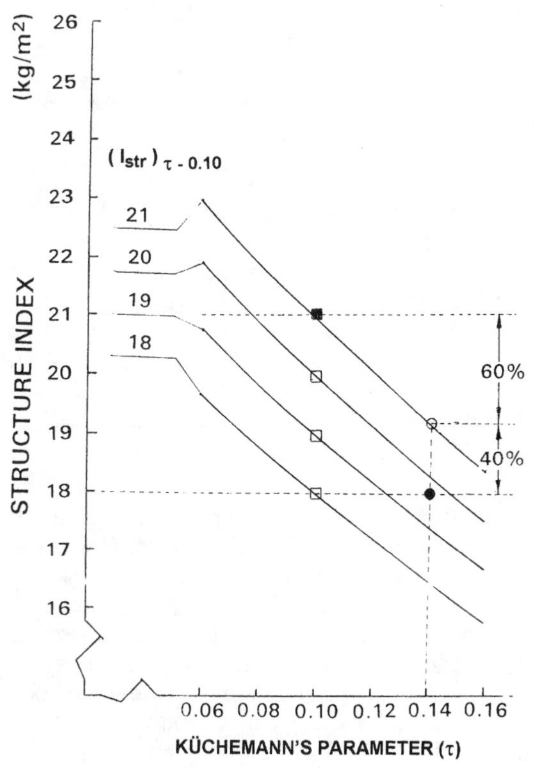

Fig. 34 Structural Index Required for Convergence Decreases as Küchemann's Tau Increases.

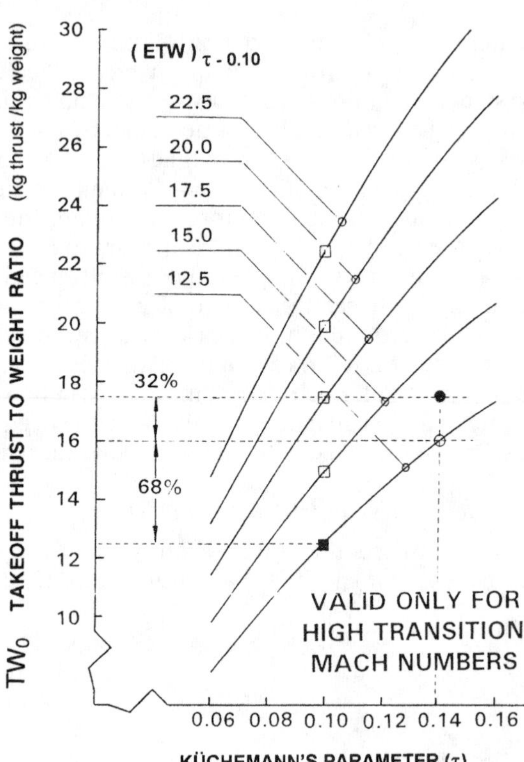

Fig. 35 Propulsion Index Required for Convergence Increases as Küchemann's Tau Increases.

$(E_{TW})_{\tau=0.1}$. These span the range between FUTURE system assumptions and CURRENT system assumptions. Drag is a function of τ as defined in Eqs. (12) through (18). So as τ is increased the drag is increased, and additional thrust is required to provide the required trajectory acceleration. The comparison between a nominal vehicle characterized by $\tau = 0.10$ and $E_{TW} = 12.5$ and a vehicle with $\tau = 0.14$ and $E_{TW} = 17.5$ requires a 68% increase in E_{TW} due to the stouter configuration and 32% to technology. This relation between E_{TW} and τ applies within limits for higher transition Mach numbers from scramjet to rocket, on the order of 15.

4. *Influence of Vehicle Shape*

Figure 36 indicates three of the possible shapes for hypersonic airbreathing configurations. The spatular nosed waverider configuration (from the University of Maryland) has essentially the same characteristics as the blended body (private communication, Mark Lewis, Univ. of Maryland, College Park, MD, Aug. 1997). The principal difference is that the waverider leading edge captures the leading edge shock and the blended body does not. The spatular nose waverider intersects the shock cone offset from the shock apex. The Nonwieler waverider intersects the shock cone at the apex, and has the least volume potential (see Ref. 115). The blended body and wing-body

	Blended Body	Wing – Body	From reference 29, page 667 Wave Rider Nonweiler type
tau	0.104	0.0948	0.1081
K_W	2.639	2.703	2.950
L/D	4.573	4.594	4.627
$\beta \cdot C_{D0}$	0.0674	0.0668	0.0686
I_{str}	22.3 kg/m^2	20.1 kg/m^2	18.3 kg/m^2
V_{tot}	1258 m^3	1308 m^3	1508 m^3
S_{plan}	525 m^2	575 m^2	579 m^2

Fig. 36 Three Hypersonic Configurations Sized for Constant Propellant Volume.

(designed to integrate hypersonic airbreathing propulsion) are two configuration concepts considered. The three configuration concepts in Fig. 36 are sized for the same propellant volume. Note that as the wetted area increases, the value of required I_{str} decreases (increasing industrial capability), the lift to drag ratio decreases, and the planform area (size) increases.

The influence of geometric characteristics on mission sized vehicles is illustrated by comparing a blended body configuration concept with a wing-body concept for CURRENT system assumptions, and transition to rocket at Mach number 15, as given in Table 7.

The influence of vehicle shape is the same as the influence of hydrogen fuel density and payload bay density. The wing-body configuration concept increases gross liftoff weight by 6.5% and inert weight by 7.6%, with respect to the blended body configuration concept. Figure 33 has comparisons between different configuration concepts.

5. *Inlet Capture Area Interaction*

Another example is the influence of increasing forebody turning angle (VDK has used the term forebody asymmetry to signify it is not a symmet-

Table 7. Comparison of configuration concepts

Configuration concept	Blended body	Wing-body	Unit
Planform area S_{pln}	536.6	571.3	m^2
V_{ppl} / V_{tot}	69.8	70.5	%
Gross weight	214.6	228.5	ton
Dry weight	58.2	62.6	ton

rical body or cylinder) to improve engine specific impulse, thrust per unit span and weight per unit thrust at the price of some increase in drag (the curves $\Delta\delta_0 = 1$ deg on Figs. 35 and 36) show the benefits derived from the improvement in engine ISP and thrust resulting from a one degree increase in flow turning by the forebody beyond the 10 degree already assumed.

E. Strong Parameter Cross-Couplings

These analyses have indicated where multiple parameter interactions can occur. In Figs. 37 and 38, these multiple parameter interactions are shown for a 5 ton payload to the sun-synchronous orbit. The changes in Gross Weight and Dry Weight are represented as functions of transition Mach number. Three cases are presented: the CURRENT system assumptions, modified CURRENT system assumptions for $I_{str} = K_{FB} \langle (I_{str})_{\tau=0.10}$ and FUTURE system assumptions. Comparison between the first two cases shows that the cross-coupling improves performance by a few percentage points (4.3% for W_{GTO} at $M_{tr} = 15$) and shifts the optimum transition Mach number from 14.8 to 15.1. The interaction sharply increases the penalty for a lower value of τ as the transition Mach number decreases. The values of τ are repre-

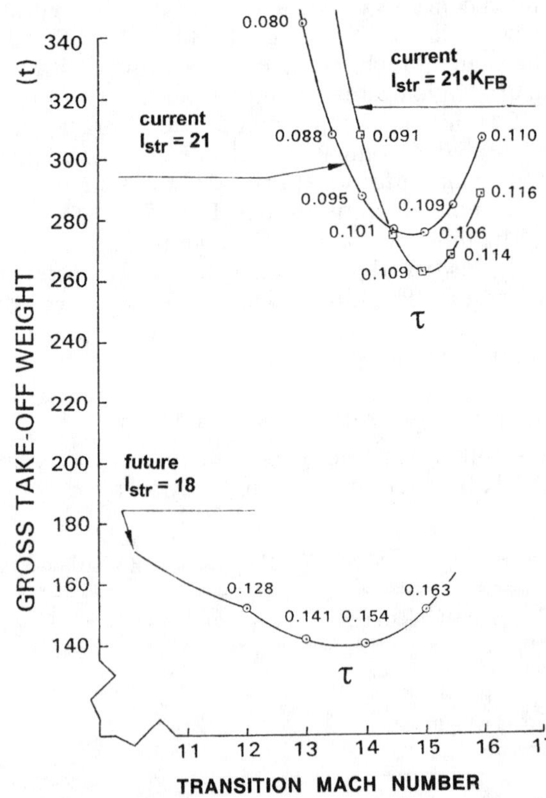

Fig. 37 Takeoff Gross Weight (W_{GTO}) as a Function of Transition Mach Showing "Snowball" Effect With Respect to μ_{str}.

sented for a number of points on Figs. 37 and 38 and the corresponding values of $I_{str} = K_{FB} \langle (I_{str})_{\tau=0\cdot;\cdot10}$ on Fig. 34. If the change in stiffness and thrust with τ are taken into account, then the weight versus transition Mach number is less broad and the minimum weight point moves to an increased transition Mach number. It is also clear that the sensitivity to transition Mach number for the FUTURE system assumptions is decreasing. The two sets of assumptions bound the design space of potential interest. The principal driver behind the difference between CURRENT system assumptions and FUTURE system assumptions is the chosen value of the takeoff planform loading (L_{to}).

1. Effects of Inadequate Thrust-Minus-Drag

Appendix B discusses the four hypersonic research aircraft shown in Fig. 39. Since a launcher flies an acceleration trajectory, any significant reduction in the average acceleration significantly affects vehicle size. There are significant differences in size between vehicles resulting from choice of low speed accelerator propulsion. Figure 39 represents four possible low speed propulsion configuration concepts for an experimental horizontal takeoff, Mach 12 aircraft designed for a five-minute test time at Mach 12.

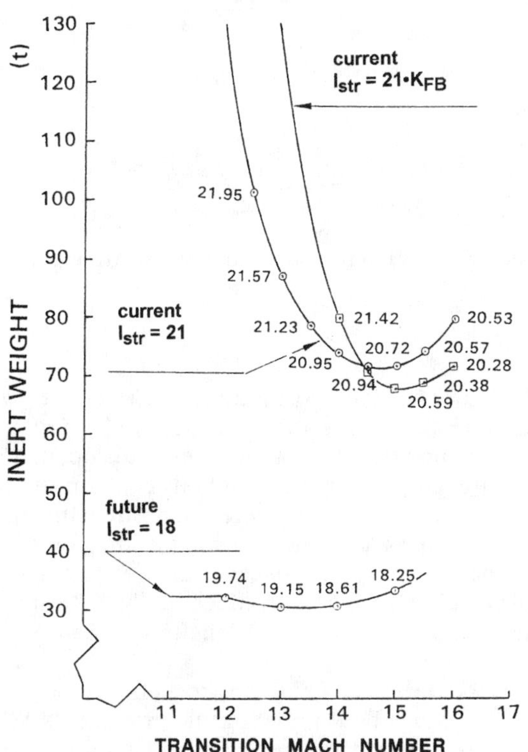

Fig. 38 Inert (dry) Weight (WEO) as a of Transition Mach Showing "Snowball" Effect With Respect to μ_{str}.

Turbojet Accelerator: Wing-Body, Horizontal Take Off and Landing.

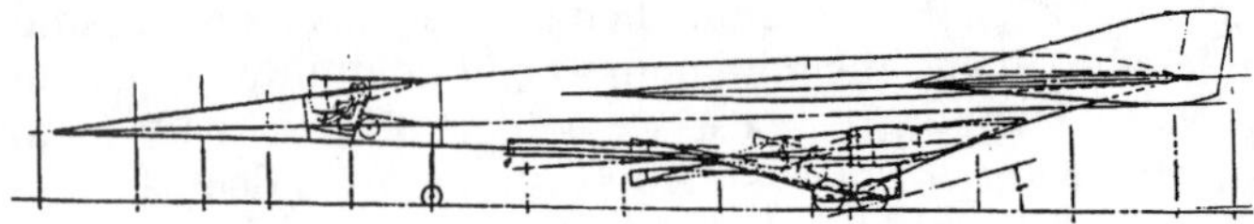

Rocket Accelerator: Wing-Body, Horizontal Take Off and Landing.

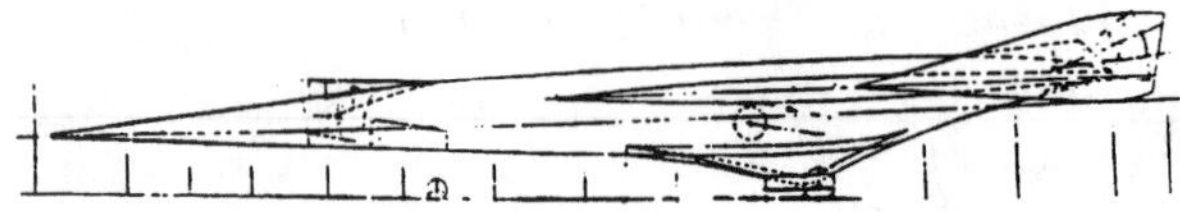

Rocket Accelerator: All-Body, Horizontal Take Off and Landing.

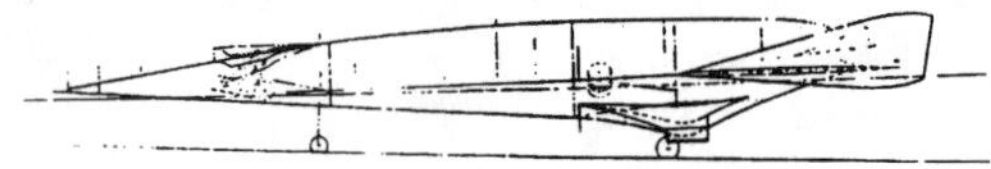

Rocket Accelerator, All-Body, Air Launched Horizontal Landing

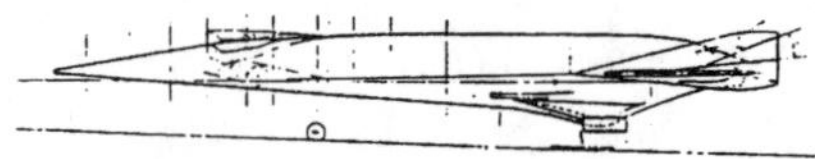

Fig. 39 Comparison of Propulsion System Sized Research Aircraft: The Greater the Transonic Thrust, minus Drag, the Smaller The Vehicle.

The top configuration in the figure has the poorest transonic acceleration and is the largest and heaviest airframe. It has the least LOX to hydrogen ratio and therefore has the least propellant weight, but greatest propellant volume. Because the transonic thrust is so poor, the wing-body configuration concept is required and a low value of τ (0.114) so there is a positive thrust, minus drag, with turbojets. This produces a low takeoff wing loading or 201 kg/m^2. This is the classic combination of engine concept, turbojet power to Mach 2, then powered by convertible scramjets to Mach 12: Specifically; W_{GTO} = 53.5 tons, W_{dry} = 36.3 tons, I_{str} = 26.1 kg/m^2, length = 43.6 m, S_{pln} = 265.7 m^2, and K_w = 2.70.

The second configuration has the turbojets replace the rockets but retains the wing-body configuration. The wings hold no cryogenic propellants, so the volumetric efficiency of this and the previous vehicle is very poor. There still remains a low value of τ (0.107). The takeoff wing loading is 393 kg/m^2. This uses a rocket to Mach 3, then is powered by convertible scramjets to Mach 12. It is smaller than the turbojet aircraft but weights more because the fuel/oxidizer propellant load is greater. Specifically; W_{GTO} = 80.0 tons, W_{dry} = 19.9 tons, I_{str} = 23.6 kg/ m^2, length = 37.5 m, S_{pln} = 203.5 m^2, and K_w = 2.69.

The third configuration is a rocket accelerated blended-body configuration with a TW_0 greater than one. There are no wings so the volumetric efficiency is very good. Because it is a blended body, τ is 0.160. The takeoff wing loading is 497 kg/m^2. This is a high, but practical takeoff wing loading with the rocket acceleration provided by a thrust to weight ratio in excess of one. Performance is very good. This is rocket powered to Mach 3, then powered by convertible scramjets to Mach 12. It is significantly smaller than the wing-body aircraft. Specifically; W_{GTO} = 61.3 tons, W_{dry} = 15.2 tons, I_{str} = 28.0 kg/m^2, Length = 29.6 m, S_{pln} = 123.4 m^2, and K_w = 2.68.

The fourth configuration shown is a rocket accelerated blended-body configuration. It was designed to be air launched from an under-wing pylon on a C-5 or could be launched atop an Antonov An-225. There are no wings so the volumetric efficiency is very good. Because it is a blended body, τ is 0.157. The launch wing loading is 400 kg/m^2. This is all rocket powered from air launch speed to Mach 3, then powered by convertible scramjets to Mach 12. It is smaller than the runway takeoff aircraft. Specifically; W_{GTO} = 31.6 tons, W_{dry} = 9.72 tons, I_{str} = 26.2 kg/m^2, Length = 24.3 m, S_{pln} = 79.0 m^2, and K_w = 2.75.

The length of the third configuration has been shortened to 67.9% of the length of the first configuration. The dry weight of the third configuration has been reduced to 41.9% of the first configuration. The air launch option reduces the length another 17.9% and the dry weight another 36%. Details of the sized configurations in this figure are to be found in Appendix B. Note that the low speed accelerator thrust minus drag has a significant effect on vehicle size. The classic turbojet/fan is a poor launcher accelerator. However a deeply cooled, thermally integrated turbomachinery-rocket is an excellent accelerator (see Refs. 31 and 115). If the latter two configurations were to have been accelerated by an RBCC engine instead of a pure rocket, then the volume and dry weight would have been similar, but the propellant weight would have been reduced by about 60%. The weight reduction would have come from a reduction of the on-board oxygen, and the volume would have remained the same because of the higher hydrogen volume fraction.

VII. TSTO Launcher Sizing

The second application deals with Two-Stage-to-Orbit vehicles. This is a two-part problem. The weight ratio (W_R) to staging speed is a direct function of the oxidizer carried on-board the first stage. The fuel fraction (W_{fuel}/GW) remains almost constant, with oxidizer fraction ($W_{oxidizer}/GW$) decreasing as airbreathing speed increases.[36] The second stage is an all rocket powered vehicle; one way to affect weight ratio from staging speed to orbital speed is a change in propellants. As in the case for the SSTO sizing, the thrust to weight ratio was varied to achieve a minimum weight vehicle. The propellant volume can be a very dominant factor. The individual propellant densities can vary from 70 kg/m³ to 1300 kg/m³ (normal boiling point hydrogen to triple point LOX). So volume becomes the dominant factor. The hypersonic sizing program is thus volume dominated.

The launchers were sized to requirements using ABTSTO for the two levels of industrial capability assumptions (Table 8). The hypersonic sizing program is both mass and volume controlled. Space launchers and passenger carrying aircraft offer the additional volume problem of a payload net density approximately that of hydrogen. For the two stage to orbit vehicle there are two vehicles to be sized. The second stage is an all rocket vehicle whose size and weight are determined by the staging Mach number. The greater the staging Mach number the larger the first stage and the smaller the second stage. As in the SSTO studies, there is a staging Mach number for a minimum weight system (stage 1 + stage 2). An initial estimate is made for the first and second stages planform area based on landing wing loading. The resultant iterations continue until available volume equals required volume for both stages.

A. Assumptions

Results are directly driven by the assumptions made. The assumptions used in this section build on what has been accomplished within the aerospace industry and are therefore state-of-the-art. In the view of the authors,

Table 8. TSTO program inputs based on industrial capability, 1st stage

Head	CURRENT	FUTURE	Unit
Structural index I_{str1}	21.0	18.0	kg/m²
Support mass coefficient k_{sup1}	0.06	0.05	kg/kg
Engine thrust to weight ratio, E_{TW1}	6.0 min 15.0 max	8.0 min 20.0 max	kg/kg
Fixed system weight C_{sys1} unmanned	2100	1900	kg
manned	4400	3800	kg
Fixed system volume V_{fix1} unmanned	8.00	6.00	m³
manned	32.0	28.0	m³
Variable system weight fraction f_{sys1}	0.20	0.18	kg/kg
Takeoff planform loading L_{to1}	250	300	kg/m²
Engine specific volume k_{ve1}	0.25 min 0.30 max	0.25 min 0.30 max	m³/ton
System volume fraction k_{vs1}	0.05	0.04	m³/m³
Void fraction k_{vv1}	0.15	0.12	m³/m³
Crew & consumables W_{crew1} manned only	640	600	kg
Hydrogen condition	subcooled 66.4	50% slush 82.1	kg/m³
Configuration concept	Blended body		——
Payload	Second stage		——

the assumptions are conservative but provide a realistic, achievable, aircraft system. Very rational, traditional aircraft assumptions can, and have in the past doomed many airbreathing system studies to failure. On the other hand, too optimistic assumptions have led to unrealistic results. Although derived from the SSTO, the weight and volume assumptions for the TSTO are slightly different, as discussed below.

B. Volume and Weights

CURRENT set provides volume and weight assumptions considered realistic within Europe's current industrial capabilities. The parameters with the greatest affect on the results follow:

1) Structure index I_{str} = 21.0 kg/m^2: the value chosen is nearly state-of-the-art.
2) As for the SSTO, the sea level static engine thrust-to-weight ratio, E_{TW} = 16.5 kg thrust / kg. The value chosen is very conservative for rocket based, combined cycle engines (RBCC). Rudakov and Balepin give a value of 22.4 kg thrust/kg weight.[70,75,116] The values used in Table 8 reflect a more conservative approach to the weights.
3) The system weights are comparable but somewhat conservative when compared with published European aerospace industry results.
4) Planform loading at takeoff of 400 kg/m^2 (82 1b/ft^2) is adequate to get a liftoff velocity of about 420 km/hr (261 mph), which is very conservative. As shown in Fig. 9, this will penalize HTOL in favor of VTOHL.

With respect to the structural index, note that in 1970 these were considered state-of-the-art for hypersonic cruise vehicles.[95] The I_{str} value selected represents a 1970 assessment of industrial capability with light alloy structure protected by high temperature alloy and refractory metal, insulated heat shields, as shown in Fig. 14.

FUTURE set results arise from application of current European research and Development that may not yet be prepared for production application. Note again that in 1970 the projections for 1985 state-of-the-art for hypersonic cruise vehicles were comparable.[17] The representative mixed composite/metal structural concept had a I_{str} less than 18.0 kg/m^2 based on the assessment of the 1985 industrial capability with non-metallic composite structure protected by high temperature metal composite, insulated heat shields.

C. Propulsion

Database for HYPERJET Mk #3 (see Sec. V.F) can also be used for the first stage of a TSTO or the second stage if it is an airbreather.

D. Aerodynamics

In the absence of specific data the same generic aerodynamic database used for the SSTO (see Eqs. 25 through 28) can be used, provided that the total volume of the two stages is used prior to staging. If the stages are separated, each is handled individually as in the case for SSTO. For the purposes of these studies the aerodynamic estimates are judged to be adequate.

E. Trajectory

The trajectory used is shown in Fig. 15 for the TSTO mission. It is essentially the same as the SSTO mission up to initiation of the pull-up maneuver prior to staging, during which the vehicle first continues to accelerate but soon reaches its maximum speed in airbreathing and may even slightly decelerate to reach a favorable slope at the staging point. A climb angle of 20 degrees at Mach 6.5 can reduce the ?V to orbit from 6.4 km/sec to 5.9 km/sec.[117] After staging, the first stage returns to base (with the capability to perform an overshoot at landing if necessary) and the second stage continues the pull-up for a few seconds before accelerating and climbing along a so-called *linear-tangent* trajectory before insertion into transfer orbit at the specified altitude and slope. When a trajectory program is used, a trajectory must be specified. The magnitude of the speed may actually decrease in order to reach the desired flight path angle for staging.

F. First Stage

The first stage of the TSTO is a long operational life, continuous use vehicle. Propulsion is always assumed to be rocket ejector combined cycle engine (except as noted in alternative engine evaluation section) and manned with a crew of two, with missions of less than 2 hours. The crew provisions are not greater than a military combat aircraft. Whenever applicable, the same sets of technical characteristics are assumed for TSTO first stage as for the SSTO, except as noted:

1) Introduction of second stage support coefficient μ_{sup1} (which also accounts for the extra mass of the undercarriage for the first stage),

$$W_{\text{sup1}} = k_{\text{sup1}} \cdot GW_2$$

$$W_{\text{sys1}} = C_{\text{sys}} + f_{\text{sys}} \cdot W_{\text{dry}} + k_{\text{sup1}} \cdot GW_2 \tag{49}$$

2) Introduction of a second stage under carriage relief coefficient, μ_{uc2}

$$\Delta\, W_{\text{sys2}} = -k_{\text{uc2}} \cdot (GW_2 - GW_1)$$

$$W_{\text{sys2}} = C_{\text{sys}} + f_{\text{sys}} \cdot W_{\text{dry}} - k_{\text{uc2}} \cdot (GW_2 - GW_1) \tag{50}$$

3) Selection of a slightly less favorable system weight fraction to account for the separation mechanism,
4) Selection of a slightly lower takeoff planform loading.

The following two sets of assumptions can be considered to bound the "design space" of potential interest. The CURRENT and the FUTURE sets of assumptions are given in Table 8.

Weight breakdown is an adaptation of the general sizing equations for TSTO, and the equation set is modified as follows; where the equations with their nominal range of parameter values for large operational systems are as follows:

$$W_{\text{str1}} = I_{\text{str1}} \cdot S_{\text{wet1}} \qquad 17 \leq I_{\text{str1}} \leq 23 \text{ kg/m}^2$$

$$C_{\text{sys1}} = C_{\text{un1}} + f_{\text{mnd1}} \cdot N_{\text{crw1}} \qquad 1.9 \leq C_{\text{un1}} \leq 2.1 \text{ ton}$$

$$1.15 \leq \text{f}_{\text{mnd1}} \leq 0.95 \text{ ton/person}$$

$$W_{\text{sys1}} = C_{\text{sys1}} + f_{\text{sys1}} \cdot W_{\text{dry1}} + k_{\text{sup1}} \cdot W_{G2} \qquad 0.16 \leq f_{\text{sys1}} \leq 0.24 \text{ ton/ton}$$

$$0.06 \leq k_{\text{sup1}} \leq 0.05 \text{ ton/person}$$

$$W_{\text{eng}} = \frac{TW_{01} \cdot W_{R1}}{E_{\text{TW1}}} \cdot (W_{\text{OE1}} + W_{G2}) \qquad 6.0 \leq 20. \text{ kgthrust/kg weight}$$

$$W_{\text{crw prv1}} = f_{\text{prv1}} \cdot N_{\text{crw1}} \qquad 0.21 \leq f_{\text{prv}} \leq 0.22 \text{ ton/person}$$

$$W_{\text{crw}} = f_{\text{crw1}} \cdot N_{\text{crw1}} \qquad 0.14 \leq f_{\text{crw}} \leq 0.15 \text{ ton/person}$$

Volume breakdown is an adaptation of the general sizing equations for TSTO and the equation set is modified as follows; where the equations with their nominal range of parameter values for large operational systems are as follows:

$$(V_{\text{ppl}}) = \frac{(W_{\text{OE1}} + W_{\text{G2}}) \cdot (W_{R1} - 1)}{(\rho_{\text{ppl}})_1}$$

$$V_{\text{fix1}} = V_{\text{un1}} + f_{\text{crw1}} \cdot N_{\text{crw1}} \qquad 6.0 \leq V_{\text{un1}} \leq 8.0\text{m}^3$$

$$11.0 \leq f_{\text{crw1}} \leq 12.0 \text{ m}^3\text{/person}$$

$$V_{\text{sys1}} = V_{\text{fix1}} + k_{\text{vs1}} \bullet V_{\text{tot1}} \qquad 0.018 \leq k_{\text{vs1}} \leq 0.02 \text{ m}^3\text{/m}^3$$

$$V_{\text{eng1}} = k_{\text{ve1}} \bullet TW_{01} \bullet W_{R1} \bullet W_{\text{OE1}} \qquad 0.30 \leq k_{\text{ve1}} \leq 2.0 \text{ m}^3\text{/ton thrust}$$

$$V_{\text{void1}} = k_{vv1} \bullet V_{\text{tot1}} \qquad 0.12 \leq k_{vv} \leq 0.15 \text{ m}^3\text{/m}^3$$

$$V_{\text{pa1}} = k_{\text{pa1}} \bullet V_{\text{tot1}} \qquad -0.10 \leq k_{\text{pa1}} \leq 0.0$$

$$V_{\text{crw1}} = (V_{\text{pcrv}} + k_{\text{crw}}) \bullet N_{\text{crw}} \qquad 0.9 \leq k_{\text{crw1}} \leq 2.0 \text{ m}^3\text{/person}$$

The payload accommodation parameter (V_{pa}) accounts for any depression in the first stage that reduces the total volume of the first stage to partially submerge the second stage in the first stage.

G. Second Stage

The TSTO reusable second stage is either unmanned or manned with a crew of two, with 10-day emissions. The same sets of technical characteristics can be assumed for the sustained use TSTO second stage as for SSTO except 1) introduction of undercarriage relief coefficient μ_{uc2}, and 2) much lighter rocket engines $E_{TW2} > 60$ kg thrust/kg) with their large nozzles mounted outside the vehicle moldline so the volume per unit thrust is less.

The following two sets of assumptions can be considered to bound the "design space" of potential interest. The CURRENT and the FUTURE sets of assumptions are given in Table 9.

Weight breakdown is an adaptation of the general sizing equations for TSTO.

Equations with their nominal parameter range are as follows:

$$W_{str2} = I_{str2} \cdot S_{wet2} \qquad 18 \leq I_{str2\ 1}\# \ 21 \text{ kg/m}^2$$

Table 9 TSTO program inputs based on industrial capability, 2nd stage

Head	CURRENT	FUTURE	Unit
Structural index I_{str2}	21.0	18.0	kg/m^2
Engine thrust to weight ratio, E_{TW2}	65.0	70.0	kg/kg
Undercarriage relief coeff. k_{uc2}	0.03	0.028	kg/kg
Fixed system weight C_{sys2} unmanned	1900	1600	kg
manned	4400	3800	kg
Fixed system volume V_{fix2} unmanned	6.00	5.00	m^3
manned	40.0	38.0	m^3
Variable system weight fraction f_{sys2}	18.0	16.0	kg/kg
Launch planform loading L_{to2}	400	440	kg/m^2
Engine specific volume k_{ve2}	0.25	0.20	m^3/ton
System volume fraction k_{vs}	0.05	0.04	m^3/m^3
Void fraction k_{vv2}	0.15	0.12	m^3/m^3
Crew & consumables W_{crew2} manned only	640	600	kg
Hydrogen condition	Subcooled 66.4	50% slush 82.1	kg/m^3
Configuration concept	Blended body		——
Isp of main rocket mode I_{sp1}	460		se
Mixture ratio of main rocket $(r_{O/F})_1$	6.5		——
Isp of OMS rocket I_{sp2}	440		sec
Mixture ratio of OMS rocket $(r_{O/F})_2$	6.0		——
Isp of ACS rocket I_{sp3}	400		sec
Mixture ratio of ACS rocket $(r_{O/F})_3$	5.0		——
Payload density in bay	50.0		kg/m^3

$$C_{sys2} = C_{un2} + f_{mnd2} \cdot N_{crw2} \qquad 1.6 \le C_{un2} \le 1.9 \text{ ton}$$

$$1.25 \le f_{mnd2} \le 1.10 \text{ ton/person}$$

$$W_{sys2} = C_{sys2} + f_{sys2} \cdot W_{dry2} \qquad 0.16 \le f_{sys2} \le 0.18$$

$$-k_{uc2} \cdot (W_{GR2} - W_{GR2}) \qquad 0.01 \le k_{uc2} \le 0.03 \text{ ton/ton}$$

$$W_{crpv} = f_{prv2} \cdot N_{crw} \qquad 0.45 \le f_{prv2} \le 0.50 \text{ ton/person}$$

$$W_{crw2} = f_{crw2} \cdot N_{crw} \qquad 0.14 \le f_{crw2} \le 0.15 \text{ ton/person}$$

Volume breakdown is an adaptation of the general sizing equations for TSTO where the equations, with their nominal range of parameter values for large operational systems, are as follows:

$$V_{fix2} = V_{un2} + k_{crw} \cdot N_{crw} \qquad 5.0 \le V_{un2} \le 7.0 \text{ m}^3$$

$$11.0 \le k_{crw} \le 12.0 \text{m}^3$$

$$V_{sys2} = V_{fix2} + k_{vs2} \cdot V_{tot2} \qquad 0.02 \le k_{vs2} \le 0.04 \text{ m}^3/\text{m}^3$$

$$V_{eng2} = k_{ve2} \cdot TW_{O2} \cdot W_{R2} \cdot W_{OE2} \qquad 0.25 \le k_{ve2} \le 0.75 \text{ m}^3/\text{ton thrust}$$

$$V_{void2} = k_{vv2} \cdot V_{tot2} \qquad 0.10 \le k_{vv2} \le 0.20 \text{ m}^3/\text{m}^3$$

$$V_{pay2} = \frac{W_{pay}}{\rho_{pay}} \qquad 48.0 \le \rho_{pay} \le 130 \text{ kg/m}^3$$

$$V_{crw} = k_{crw} \cdot N_{crw} \qquad 0.9 \le k_{crw} \le 2.3 \text{ m}^3/\text{person}$$

H. TSTO Sizing Results

The application of the sizing model code ABTSTO to the TSTO problem examined a number of pertinent TSTO alternatives and questions. The TSTO launcher is sized to a TSTO mission that is launched from Istres. The mission considered a payload to a Low Earth Orbit (LEO) will deliver. The requirements are 1) manned TSTO first stage, 2) unmanned TSTO second stage, 3) horizontal takeoff and landing (HTOL), 4) blended body configuration, 5) takeoff planform loading, 250 kg/m2, 6) liquid hydrogen/oxygen propellant, 7) current industrial capability rocket motor, 8) 7000 kg payload to 463 km / 28.5 degrees (design mission), 9) in-bay payload density = 50 kg/m3, 10) based in Istres, i.e., a cruise of some 1000 km, 11) rendezvous and docking propellant allowance (100 m/sec), and 12) dry mass margin : μ_a = 10% for first stage.

The 2nd stage weight is selected through iteration in order to keep the margin on the 1st stage inert weight constant (10%). Unless otherwise noted, the sizing program determined the minimum dry weight vehicle that met the takeoff planform loading requirements that determined the Küchemann volume parameter, tau.

I. Influence of First Stage Propulsion Concept

Most of the TSTO systems employ first stage propulsion systems other than ejector ramjets. To evaluate the difference between turbomachinery and ejector ramjet propulsion, three different cases were examined. The three cases differing only by their 1st stage engines, are compared with fully coherent weight, aerodynamic and trajectory assumptions: turboramjets (E_{TW} = 6 kg thrust/kg) ejector-ramjets with nominal thrust and Isp (E_{TW} = 15 kg thrust/kg), and ejector-ramjets degraded by 15% in thrust$_i$ and I_{sp} (E_{TW} = 13 kg thrust/kg)

The degraded ejector-ramjet with 15% less net thrust per unit span, net specific impulse and static thrust-to-weight ratio, has been considered to account for the scarcity of published data on this type of engines,[118,119] Turbo-ramjets, with known successful flight operations is the and the Pratt & Whitney J-58 on the SR-71. Other turboramjets with known tests are the Republic XF-103, and the the Marquardt Company super-charged ejector's ramjet engine, SERJ,[120] that was tested in 1966 at true temperature and pressure at the engine face-up to Mach 5.[40] Conceptual engines that may have been tested are the wrap-around ramjets engine series JZ8 and JZ6C by General Electric.[10,11] Other ram compression engines are a Mach 6 subsonic throughflow engine, the MA-145 dual mode ramjet,[13] and McDonnell Aircraft's[15] fully integrated two-dimensional scramjet the MA-188.[14]

As can be seen in Table 10, for a staging Mach number of seven, the real culprit is the turbomachinery in the first stage. Its high weight to acceleration characteristics essentially doubles the gross and dry weight of the first stage (see Ref. 43, p. 10, Fig. 3). Even a poor performance ejector ramjet is a better choice. As Rudakov clearly points out (Ref. 115) a thermally integrated RBCC ejector ramjet with LH_2/LOX injection can simultaneously have the thrust of the rocket and the I_{sp} of turbomachinery (see Ref. 68).

J. Discussion of Results

The nominal ejector-ramjet significantly lowers takeoff gross weight by 33.6% and the total dry weight of the two stages by 45.6% (from 219.7 tons down to 119.6 tons). The engine weight is lower by 87.9%, i.e., approximately 1/8th of turboramjet.

The "degraded" ejector-ramjet with net thrust per unit span and net specific impulse lowered by 15%, but 15% heavier than the "nominal baseline" increases gross takeoff weight by 13.2% and total dry weight (of the two stages) by 20.1% (from 109.6 tons up to 132.4 tons). In this case, the penalty which results from degraded engine performances is moderate since engine weight increases only by 30% (i.e., 15% for the lower thrust per unit span and 15% for the lower thrust to weight ratio).

The improvement resulting from the ejector-ramjet is primarily due to its lower weight, resulting in a much lighter first stage. The time to staging is

Table 10 Performance comparison for two types of engines

Head	Turbo-ramjet	Ejector-ramjets Nominal	Degraded
Takeoff weight, ton	393.0	261.0	295.5
Time to staging	56m 20 sec	55m 51sec	60m 10sec
1st stage			
Static thrust/weight	84.0	38.0	38.0
Structure weight, ton	95.6	66.3	74.3
Engine weight, ton	60.5	7.3	9.5
System weight, ton	43.3	22.5	25.0
Dry weight, ton	199.5	96.1	108.8
Propellant weight, ton	83.2	45.5	67.0
Stage weight, ton	282.7	141.6	175.8
2nd stage			
Static thrust/weight	1.15	1.15	1.15
Structure weight, ton	14.6	17.2	17.3
Engine weight, ton	2.1	2.3	2.3
System weight, ton	3.6	4.0	4.0
Dry weight, ton	20.3	23.5	23.6
Payload weight, ton	7.0	7.0	7.0
OWE, ton	27.3	30.5	30.6
Propellant weight, ton	81.6	87.9	88.4
Stage weight, ton	108.9	118.4	119.0

almost the same for the turbo-ramjets and the "nominal" ejector-ramjets, and only 10% higher for the degraded ejector-ramjets. Another major improvement resulting from the ejector-ramjet is the thrust to weight ratio at takeoff. For the turbo-ramjets it is 0.84 and for ejector-ramjet it is 0.38. With turbo-ramjets the 84% results from optimizing the payload weight. With ejector-ramjets 38% is imposed by the acceleration during the turn to accelerate (see Fig. 15) so the end Mach number is 6.7 pull-up is initiation. The turbo-ramjet thrust at staging speed is only 20% to 25% of the static thrust and transonic thrust. However, the ejector-ramjet thrust is nearly constant from transonic speeds to staging speeds. For the TSTO the static thrust is determined predominantly by the need to keep a reasonable acceleration at staging. Then the turbo-ramjet will have a greater sea level static thrust than the ejector-ramjet, since the former has a greater thrust decrease with speed.

VIII. Comparison Between SSTO And TSTO

Airbreather powered launcher size and weight determined predominantly by maximum airbreathing Mach numbers and available industrial capability for both SSTO and TSTO. The impact of these elements can be illustrated via parametric analyses. This comparison was computed with ABSSTO and ABTSTO sizing codes for a 5-ton payload to the demanding sun-synchronous orbit with daily launch windows of one hour (which requires cruise and turn capability). Figures 40 and 41 provide the variation of takeoff gross weight (W_{GTO}) and total inert (dry) weight (W_{dry}) as functions of transition to rocket Mach number (M_{tr}). To ensure comparability, fully comparable assumptions are used for the SSTO and the TSTO, with respect to technology, weight and volume models, aerodynamic database and engine performance. Unless otherwise stated, a 15% margin on all inert weight is systematically applied. The SSTO and second TSTO stage is unmanned but the TSTO first stage is manned.

The first comparison uses the CURRENT assumption set weight and volume assumptions for the sun-synchronous mission. The observations follow.

Firstly, for the same payload weight, the total dry weight of the SSTO and TSTO vehicles is comparable, with the minimums occurring at different transition Mach numbers. The TSTO minimum occurs about four Mach numbers lower than the SSTO minimum. With the vehicle mass plus payload that achieves orbital speed of 30.5 tons for TSTO and 66.2 tons for SSTO, it is not unexpected that the takeoff gross weight of the TSTO is considerably less than the SSTO (150 tons compared to 280 tons). This implies that on a "cost per pound basis" the airframe costs might be comparable while the long-term operational costs could favor the TSTO (less propellant). The

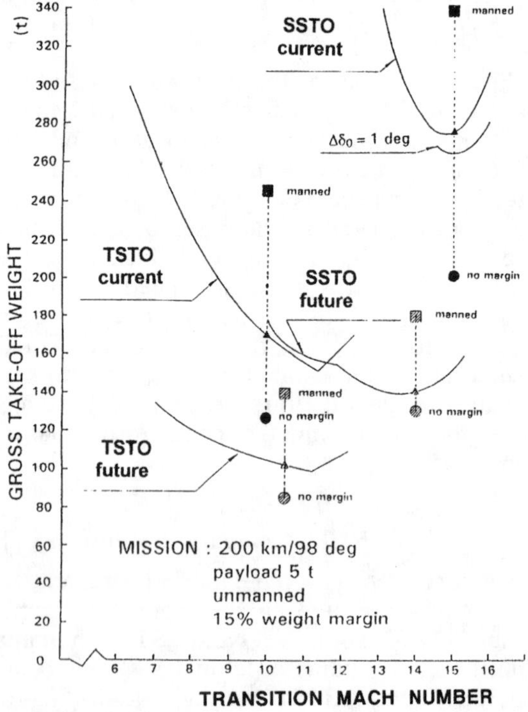

Fig. 40 Takeoff Gross Weight (W_{GTO}) as a Function of Transition Mach Number for "Reference" and "Advanced" Assumptions.

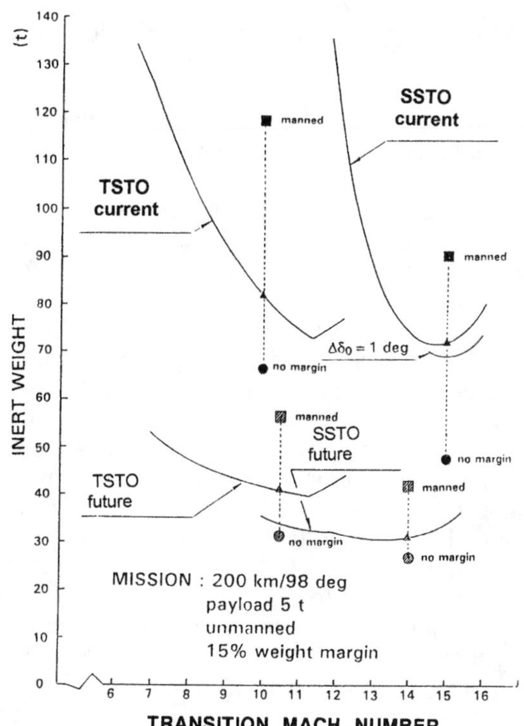

Fig. 41 Inert (dry) Weight (WEO) as a Function of Transition Mach Number for "Reference" and "Advanced" Assumptions.

TSTO has approximately the same total dry weight as the single SSTO vehicle. The difference is that for the TSTO, only the upper stage (about 20% of total dry weight) must endure an atmospheric entry at orbital speed. So the TSTO may be more expensive in the short term, but perhaps is equal for a long operational life vehicle considering maintenance and propellant costs.

Secondly, the TSTO gross weight and total dry weight both show sharp minimums at a staging Mach number of 11.4. A major improvement over ramjets (limited to Mach number 6) is obtained by using scramjets. When staging increases from Mach number 6 up to 11, the gross weight decreases by 44.5% (from 280.0 to 155.5 tons) and the total dry weight decreases by 42.7% (from 130 to 74.5 tons). The minimum weight for a direct ascent would be about Mach number 12 due to the decrease in second stage weight with increasing staging Mach numbers. A point is arrived at when the second stage is essentially a minimum sized vehicle based on aerodynamic consideration, and further reduction in weight ratio does not reduce its size (see ref. 141 for similar results). For the TSTO, the return to base introduces an additional constraint that results in a sharp minimum. For staging Mach numbers greater than 11.4, the return cruise leg distance goes to zero (see Fig. 15), which forces a shortened glide phase and an increase in the vehicle weights. Selection of different trajectories or further optimization may somewhat alter the comparison.

The SSTO gross weight and total inert (dry) weight both show sharp minimums at transition Mach number Mach 15. These results parallel the results in Figs. 21 and 22. This optimum is due to the increase in tank volume and consequent vehicle drag, and to the need to install more and more thrust to overcome the hypersonic thrust minus drag pinch. At the transition point, the effective specific impulse, EI_{sp} available to accelerate equals the specific impulse in rocket mode, therefore, the rocket is more effective beyond that point.

Thirdly, Figs. 40 and 41 show that for the same five ton payload, the minimum TSTO gross weight is about 55% of the minimum SSTO gross weight. However, the minimum TSTO total dry weight (stages one and two added together) nearly equals that of the minimum SSTO dry weight. The SSTO uses much more propellant to launch a given payload since its weight in orbit is much higher, but it is a single vehicle and does not require a separation. At minimum weight, both SSTO and TSTO require scramjet propulsion. The most important difference is that the minimum weight TSTO requires airbreathing operation only to Mach 11 but the minimum weight SSTO requires airbreathing operation to Mach 15.

Fourth, with the FUTURE assumption set the airbreathing Mach number for minimum weight varies only slightly for the TSTO but decreases from 15 to 13.5 for the SSTO. If scramjet operation is limited to the Mach 10 to 12 region, then the SSTO will experience a large weight growth for the CURRENT assumption set, and a much smaller weight growth for the FUTURE assumption set. Perhaps one of the more important observations is that the gross weight of the CURRENT assumption set TSTO is equal to the FUTURE assumption set SSTO in the Mach 10 to 12 region. However, the dry weights are significantly different, 70 tons for the CURRENT TSTO and 30 tons for the FUTURE SSTO.

One question to be answered is the effect of manned flight control versus unmanned flight control, and the impact of margin. Figures 40 and 41 show the very significant impact on gross weight and total dry weight at constant payload, near the airbreathing Mach number for minimum weight, from requiring a manned flight control or accepting a zero margin on inert weight. The impact is significantly greater for the CURRENT assumption set than for the FUTURE assumption set, i.e., less impact as the weight is less. Requiring manned flight control for SSTO vehicles increases the dry weight by 29% and the gross weight by 21%. The percentages are larger for TSTO vehicles, 47% and 49% respectively. Elimination of the margin will reduce the dry weight, especially the CURRENT assumption set SSTO vehicles (34% reduction in dry weight and 29% in gross weight). For the FUTURE assumption set the impact of margin is much less. However, this mean a vehicle without margin to accommodate unexpected weight increases and without growth potential might not be prudent for a long duration operation vehicle. The two sets of assumptions probably bound the design space of potential interest.

IX. Air Liquefaction and LOX Collection

Up to this point, all of the rocket ejector ramjet-rocket integrated propulsion cycles have not utilized the available energy stored in the liquid hydrogen. It is possible to recover a significant portion of the energy required to liquefy the hydrogen if a suitable thermodynamic system is employed.[56] A fundamental change in the operation of airbreathing launchers is possible with the incorporation of such a thermodynamic cycle. The first step was initiated by the Marquardt Company in the 1960s as the Liquid Air Cycle Engine (LACE). Reference 121 discusses this early effort. For a LACE system, the thermal energy in a portion of the atmospheric air entering airbreathing inlet is used to boil the liquid hydrogen, and that process liquefies the air. This liquefied air can then be pumped into the rocket to create an "airbreathing" rocket. Application to launchers is covered in Refs. 75, and 79–81. There has been a standing concern about atmospheric water condensing on the heat exchanger coils and blocking the heat exchanger. References by Tanatsugu with respect to the ATREX engine show a wide operational range possible without icing occurring. In fact, if the transition through low altitude is rapid, there is little chance of icing. In 1988 one author witnessed Mitsubishi HI run a heat exchanger (a cube, approximately one meter per side) in 38 C, 95% relative humidity air without any water ice accumulation. Water, carbon dioxide, and argon ice accumulation in the liquid air was checked, but it is not a program-stopping element.

The next step also began in the 1960s not only to liquefy the atmospheric air, but separate and store the oxygen subsequently required in rocket mode by a cryogenic fractional distillation. The separated nitrogen was injected into the ramjet to create a ramjet engine that is equivalent to a low bypass ratio turbojet with improved I_{sp} (see references by Rudakov of CIAM). This cycle was called ACES (Air Collection Enrichment System) and was investigated up to ground testing in the United States. Recent references with historical information are Refs. 86, 109–112, 122, 123. Recently LACE and ACES have been the object of much renewed interest in Japan,[114,121], Russia[20,69,70,113,124,125] and India.[115,126]. See Sec. III for brief discussions of the different engine cycles. As we shall see in this chapter, in-flight air collection to liquefy, separate and store the oxygen subsequently required in rocket mode improves performance of both the SSTO and the TSTO

A. Propulsion System Configuration

The thermally integrated LACE and/or ACES propulsion systems are essentially rocket ejector ramjet-rockets with additional equipment. Repeated from Sec. III for clarity, we have the following.

The rocket ejector ram-scramjet-rocket is an dual-mode rocket ejector ramjet thermally integrated with a rocket.[78,81] The maximum airbreathing speed can be selected from Mach 6 to at least Mach 14.5. At Mach 6, the system is rocket ejector ramjet analogous to system # 7 in Sec. III. This is illustrated in Fig. 42. Note that in order to recover the maximum available

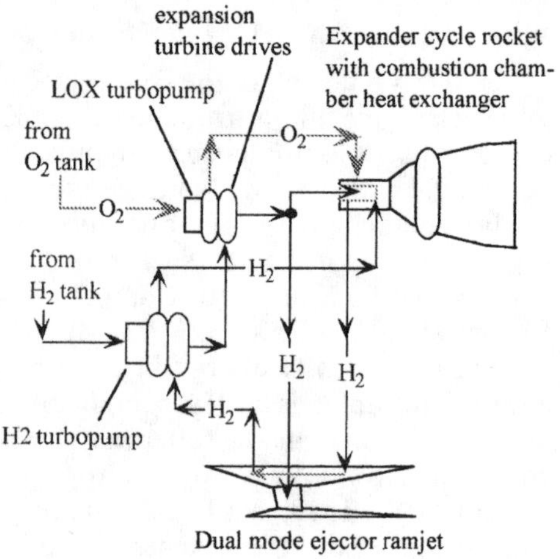

Fig. 42 Ejector Dual-Mode Ramjet Rocket System Schematic.

energy the hydrogen flows from the hydrogen turbopump to the dual-mode ramjet and rocket before flowing to the two expansion turbines, and finally into the rocket/dual-mode ramjet for combustion. As quantified by Rudakov (ref. SS), this thermal integration provides an additional 1500 seconds I_{SP}. For adequate engine performance, a heat exchanger should be installed in the rocket combustion chamber.[127]

We begin with a conventional LACE system. The measure of LACE system performance is the collection ratio (Γ) that is the weight of air liquefied per unit weight of liquid hydrogen. In the LACE cycle, atmospheric air flows into the auxiliary inlet and into a precooler where the air temperature is reduced to just above saturation. The deeply cooled air is then pressurized to several atmospheres by a compressor and then flows into the liquefying heat exchanger. The liquid air is then pressurized in an expansion turbine-driven turbopump and introduced into the rocket engine similarly to the LOX.

We then move to ACES (Air Collection and Enrichment System).[128] ACES system is a dual-mode rocket ejector ramjet thermally integrated with a rocket and an air liquefaction and separation system. This is illustrated in Fig. 43. In the ACES system, the liquid air flows into a cryogenic fractionating system (based industrial air plant hardware) to separate a portion of the oxygen from the air. The oxygen component emerges as liquid-enriched air (LEA contains over 95% oxygen; the remainder is nitrogen) and nitrogen component as liquid oxygen-poor air (OPA contains just a few percent oxygen; it's mostly nitrogen). The separation efficiency is the fraction of the liquid air that is actually separated into LEA and OPA. The balance of the unseparated air emerges from the separator as liquid air which is mixed with the liquid OPA[86,129] and injected into the ramjet to create a hypersonic low bypass ratio engine. The LEA is stored for use in the rocket ascent portion of the flight and any on-orbit maneuvering. ACES was

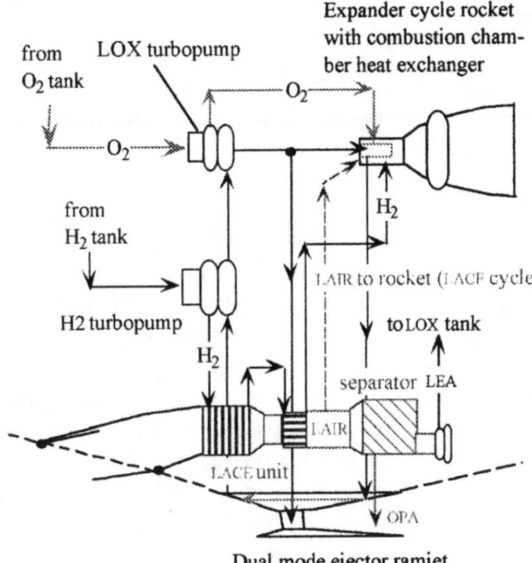

Fig. 43 LACE and ACES with Ejector Dual-Mode Ramjet Rocket System Schematic.

originally proposed for the space plane of the late 1950s and was the object of intense investigation from 1960 to 1967 (Ref. 83).

A unit of air contains 21% oxygen and 79% nitrogen. What emerges from the separator is 18.2% LEA, which contains 98% oxygen and 2% nitrogen; and 81.8% OPA that contains 3.9% oxygen and 96.1% nitrogen. If the collection ratio is about 6, then six units of LEA are collected for each unit of liquid hydrogen. The OPA flow is then 26.9 units for each unit of liquid hydrogen. If the ramjet is operating at a 35 to 1 air-to-fuel ratio, then the OPA injected into the ramjet produces the equivalent of a 0.77 bypass ratio engine.

Because the thermodynamic process is different and the vehicle's weight history of the vehicle is different from conventional vehicles (during the outbound leg, liquid-enriched air is collected so the vehicle weight increases, the takeoff weight is not the maximum weight but some point along the outbound trajectory). The sizing program was modified to reflect this characteristic.

B. Sizing Model Modifications and Assumptions

These modifications apply to both the SSTO and TSTO first stage sizing programs. The first item that must be added is the weight of the air collection and separator plant. The plant weight is proportional to the maximum weight flow rate capacity of the plant, S_{pmx} and the plant specific weight W_{col}. Thus,

$$W_{\text{col}} = S_{\text{pmx}} \cdot w_{\text{col}} \tag{51}$$

The addition of the collection and separation equipment adds to the internal volume occupied by non-propellants and must be added to the volume equation. Thus,

$$V_{\text{col}} = S_{\text{pmx}} \cdot v_{\text{col}} \tag{52}$$

There are two specific impulse corrections required. The first is that the main propulsion motors (1), the in-orbit maneuver motors (2) and the altitude control motors (3) need to be corrected for the nitrogen content of the liquid oxygen. In the outbound flight, the fraction of the total oxygen needed for the mission (after transition to rocket propulsion, collected by the ACES system) is the fraction collection parameter (f_{col}). The increment in rocket specific impulse due to nitrogen in the liquid oxygen is ΔI_{SR}. If only LEA is used in the rocket motor, the ΔI_{SR} increment is negative. Then,

$$I_{\text{SPC}} = I_{\text{SP}} + f_{\text{col}} \cdot \Delta I_{\text{SR}} \tag{53}$$

The second is the correction to the airbreathing propulsion for the added OPA and the added inlet and ram drag associated with the air collection system.

$$I_{\text{Sabc}} = I_{\text{Sab}} + \Delta I_{\text{Sab}} \tag{54}$$

The oxygen depleted air (OPA) injected into the ramjet can offset the inlet and ram drag associated with the air collection system and the I_{SP} increment can be zero or positive.

Because of the changes in mass flows through the system, the equation set within the sizing model must be changed to reflect collection, thus the terms for hydrogen flow rate, LEA flow rate, ramjet thrust, and specific impulse are the following:

Without collection:

$$\dot{w}_{\text{hj}} = \frac{T_{\text{ab}}}{I_{\text{SPab}}}$$

$$\dot{w}_{\text{oj}} = 0.0$$

$$T_{\text{j}} = F_{\text{i}} \cdot M_{\text{wi}}$$

$$I_{\text{SP}} = \frac{T_j}{\dot{w}_{\text{hj}}}$$

With collection:

$$\dot{w}_{hjc} = \frac{T_{ab}}{I_{SPab} + \Delta I_{SPab}}$$

$$\dot{w}_{ojc} = \Gamma \cdot \dot{w}_{hj}$$

$$T_{jc} = F_i \cdot M_{wi} - 4.321 \cdot \dot{w}_{ojc} \cdot V$$

$$I_{SPc} = \frac{T_{jc}}{\dot{w}_{hjc}}$$

Even though oxygen (LEA) is collected, it does not enter into the specific impulse determination as it is stored and not a propellant for this phase of the flight. The selected values of the essential parameters needed for LOX collection are summarized in Table 11.

C. Application to SSTO

Using the current assumption set and the assumptions in Table 11, four parametric sets are investigated for a 5-ton payload to the demanding sun-synchronous orbit. 1) The influence of planform loading, 2) The influence of the transition Mach number from airbreather to rocket, the influence of collection ratio, and the influence of future assumption set. The first parametric shows that in-flight oxygen collection simultaneously improves both takeoff gross weight and total inert weight while it lowers the Mach number for minimum weight. Figures 44 and 45 shows an SSTO with collection is 28% lighter in gross weight and 14% in dry weight than its 276-ton gross weight, 72-ton dry weight counterpart without collection (Figs. 21 and 22). This is despite a major reduction in transition Mach number from 15 to 11. The collection ratio for this particular case is 6 kg of liquid enriched air per kg of hydrogen. Because of the outbound cruise leg chosen, only 74% of the oxygen subsequently required in rocket mode was collected. As for prior sizing results, the greater the takeoff planform loading, the less the dry and

Table 11 Assumptions for in-flight LOX collection

Parameter	Assumed value	Unit
Γ collection ratio	6.0	kg LEA/kg H2
collection Mach number range	$3 < M < 5$	
w_{col} specific collection plant weight	32.0	kg/kg LEA/sec
v_{col} specific collection plant volume	0.50	m^3/kg LEA/sec
η_{sep} separation efficiency fraction of the air separated	85%	kg sep / kg in
X_{O2} oxygen concentration in LEA	98%	——
?I_{SPR} change is rocket ISP due to LEA	−4	sec
?I_{SPab} change in airbreather ISP due to collection	0	sec

gross weights. In this case, the actual minimum was not reached by a small increment.

Tables 12–15 shows the parametric cases in Figs. 44 and 45, adjusted for cruise length, so the collected fraction is 95%, using the same mission (5 ton to 200 km/98 degrees the current assumption set (except for vertical takeoff, slush hydrogen, and a landing gear sized for horizontal landing). Table 12 shows the same trend as in the referenced figures; the weights decrease as wing loading is increased, as does the weight of the collected LEA. Note that the minimum plant capacity and weight occurs at a planform loading of 280 kg/m2. So selection of parametric set as a baseline vehicle depends on the requirements and program goals, and is not a straightforward task based on a single criterion.

Table 14 shows the sensitivity to transition Mach number. The planform loading was selected for minimum collection and separation plant weight. There is no strong advantage to proceed to a transition Mach number of 10. There is only a 5% reduction in both dry and gross weight for a 10% percent increase in transition Mach number.

Without exceeding Mach 8 in the airbreathing mode the takeoff gross weight is lowered to 164.8 tons and the dry weight to 58.4 tons. With collection and separation there is little incentive to exceed Mach 8 transition.

The sensitivity to collection ratio is also high, as illustrated by Table 15. The practical limit for collection ratio is just over 7. That is enabled with lightweight heat exchangers and cruise speeds at a moderate Mach number

Table 12 Influence of planform loading

Head	Head	Head	Head	Unit
Payload weight	5.0	5.0	5.0	ton
Planform loading	**260**	**280**	**300**	**kg/m2**
Transition Mach no.	8.0	8.0	8.0	——
Fraction collected[a]	95.0	95.0	95.0	%
Collection ratio	6.0	6.0	6.0	——
Start Mach no.	3.0	3.0	3.0	——
End Mach no.	5.0	5.0	5.0	——
Collection plant weight	32.0	32.0	32.0	kg/(kg LEA/sec)
Collection plant volume	0.50	0.50	0.50	m3/(kg LEA/sec)
Change in rocket I_{SP}	−4	−4	−4	Sec
Change in airbreather I_{SP}	0	0	0	Sec
Takeoff gross weight	196.2	164.8	151.1	Ton
Dry weight	71.0	58.4	52.5	Ton
Küchemann's tau	0.097	0.119	0.139	
LEA collected	198.6	166.4	152.1	Ton
Ascent cross range	26.7	22.9	19.8	Degree
Collection capacity	80.5	76.8	78.8	kg LEA/sec
Collection plant weight	2.96	2.83	2.90	Ton

[a]Without accounting LOX consumption at low speed (applies to Tables 12–15).

Table 13 Influence of transition Mach number

Head	Head	Head	Head	Unit
Payload weight	5.0	5.0	5.0	Ton
Planform loading	280	280	280	g/m2
Transition Mach no.	**8.0**	**9.0**	**10.0**	—
Fraction collected*	95.0	95.0	95.0	%
Collection ratio	6.0	6.0	6.0	—
Start Mach no.	3.0	3.0	3.0	—
End Mach no.	5.0	5.0	5.0	—
Collection plant weight	32.0	32.0	32.0	kg/(kg LEA/sec)
Collection plant volume	0.50	0.50	0.50	m3/(kg LEA/sec)
Change in rocket I_{SP}	−4	−4	−4	Sec
Change in airbreather I_{SP}	0	0	0	Sec
Takeoff gross weight	164.8	150.9	144.3	Ton
Dry Weight	58.4	53.7	51.3	Ton
Küchemann's tau	0.119	0.124	0.126	—
LEA collected	155.4	139.3	120.4	Ton
Ascent cross range	22.9	20.3	18.2	degree
Collection capacity	76.8	71.9	68.7	kg LEA/sec
Collection plant weight	2.83	2.64	2.53	Ton

Table 14 Influence of collection ratio

Head	Head	Head	Head	Head	Unit
Payload weight	5.0	5.0	5.0	5.0	ton
Planform loading	280	280	280	280	kg/m^2
Transition Mach no.	8.0	8.0	8.0	8.0	—
Fraction collected*	95.0	95.0	95.0	95.0	%
Collection ratio	**4.0**	**6.0**	**8.0**	**10.0**	—
Start Mach no.	3.0	3.0	3.0	3.0	—
End Mach no.	5.0	5.0	5.0	5.0	—
Collection plant weight	32.0	32.0	32.0	32.0	kg/(kg LEA/sec)
Collection plant volume	0.5	0.5	0.5	0.5	m^3/(kg LEA/sec)
Change in rocket I_{SP}	−4	−4	−4	−4	sec
Change in airbreather I_{SP}	0	0	0	0	sec
Takeoff gross weight	319.8	164.8	135.8	127.7	ton
Dry weight	104.1	58.4	50.7	48.9	ton
Küchemann's tau	0.090	0.119	0.128	0.132	—
LEA collected	286.3	166.4	146.9	142.1	ton
Ascent cross range	37.0	22.9	14.9	10.5	degree
Collection capacity	79.5	76.8	94.0	115.7	kg LEA/sec
Collection plant weight	2.93	2.83	3.46	4.26	ton

Table 15. Influence of combined improvements

Head	Head	Head	Head	Unit
Payload weight	5.0	5.0	5.0	Ton
Planform loading	280	**280**	**280**	g/m^2
Transition Mach no.	8.0	**8.0**	**7.0**	——
Fraction collected[a]	95.0	**96.0**	**96.0**	%
Collection ratio	6.0	**7.0**	**7.0**	——
Start Mach no.	3.0	**2.0**	**2.0**	——
End Mach no.	5.0	5.0	5.0	——
Collection plant weight	32.0	32.0	32.0	kg/(kg LEA/sec)
Collection plant volume	0.50	0.50	0.50	m^3/(kg LEA/sec)
Change in rocket I_{SP}	−4	−4	−4	Sec
Change in airbreather I_{SP}	0	0	0	Sec
Takeoff gross weight	164.8	85.3	86.8	Ton
Dry weight	58.4	27.6	28.3	Ton
Küchemann's tau	0.119	0.164	0.163	——
LEA collected	166.4	87.8	99.5	Ton
Ascent cross range	22.9	9.7	11.1	degree
Collection capacity	76.8	67.2	69.0	kg LEA/sec
Collection plant weight	2.83	2.47	2.54	Ton

[a]Without accounting LOX consumption at low speed.

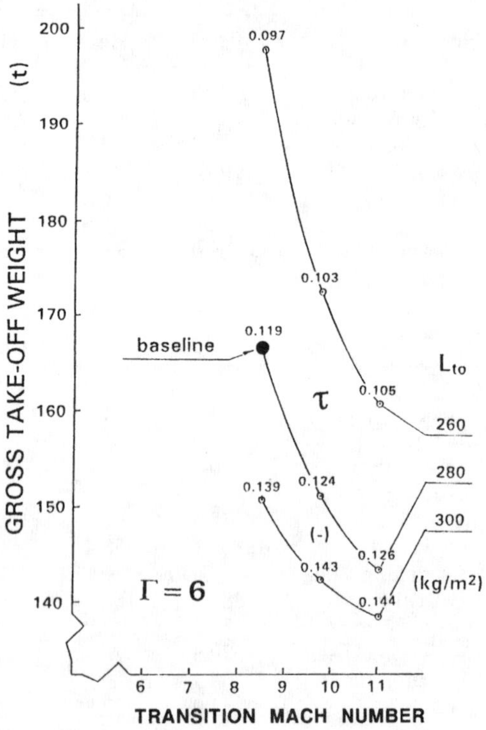

Fig. 44 Takeoff Gross Weight (W_{GTO}) as a Function of Transition Mach Number for "Reference" Assumptions and S.S.T.O. with LACE and Collection.

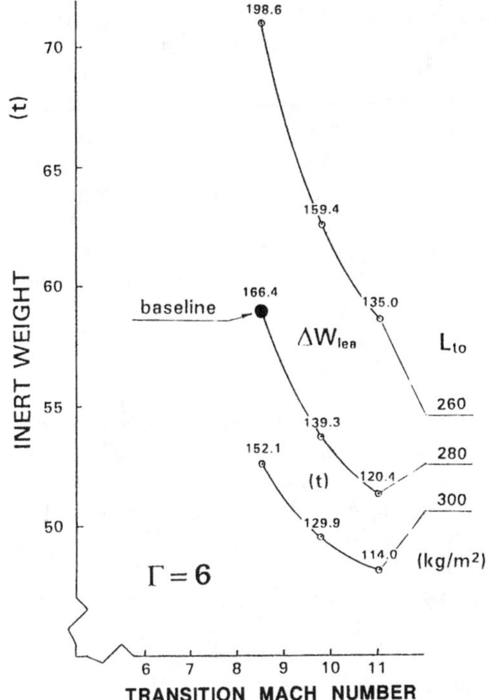

Fig. 45 Inert (dry) Weight (W_{EO}) as a Function of Transition Mach Number for "Reference" Assumption and S.S.T.O. with LACE and collection.

of 3.2 with a supercharged condenser, para to ortho hydrogen conversion and recirculation to the tanks initially filled with slush hydrogen. Examining the LACE references from Japan will show evaluations of such systems. Sensitivities to other collection characteristics, in particular the specific weight of the LOX collection plant, and decrease of specific impulse in rocket mode from the LEA are smaller.

To investigate the limits of the design space, Table 15 compares the performance of the baseline (CURRENT assumptions set) and of two very good cases for FUTURE assumptions. The sizing program results show that when the FUTURE assumptions set is employed several key characteristics change.

A higher Küchemann tau, in excess of 0.16 resulted that gives the possibility to lower the structure weight by taking advantage of the higher cross-section moment of inertia. The fuselage bending moment parameter was set to unity for these cases. A higher collection ratio of 7 was assumed as well as easy improvements of starting collection at Mach 2 instead of 3. The result was that the fraction collected equal to 96% instead of 95%.

The first case assumes transition to rocket mode at Mach 8 and the second at Mach 7, with the 15% margin kept unchanged. In both cases the performance improves dramatically with both takeoff gross weight and dry weight down nearly 50% of the baseline and a dry weight payload fraction in excess of 18%. The difference between the two cases is minimal.

An STTO of less than 100 tons gross takeoff weight and a dry weight payload fraction in excess of 18% does not appear to be an unrealistic goal. A Russian experimental aircraft, the Tu-2000, is credited with collection and having a gross takeoff between 70 and 90 tons with a crew of two, an unspecified payload to an unspecified orbit.[130] With in-flight air separation and collection, the use of scramjets may not be mandatory for the SSTO. If the FUTURE assumptions (Table 15) are applied to a vehicle, then the penalty for a transition at Mach 7 instead of 8 is minimal, so a ramjet may be the engine of choice. This, however, does not imply that the ramjet is not without risk, as the internal pressure can exceed 1.03 MPa (150 psia) above Mach 5.5, and a lightweight, thermally survivable design is a challenge.

D. Application to TSTO

LOX collection can also be harnessed for a TSTO launcher. The collection/separation equipment is installed on the first stage and the LEA transferred to the rocket propelled second stage. As in the case of the four conventional TSTO, both vehicles are long-life, frequent use vehicles. Two staging Mach numbers are evaluated, Mach 10.0 and Mach 6.5. The results for two cases with and without collection are summarized in Table 15.

For the 5-ton payload and staging at Mach 10, the addition of LOX collection and minor trajectory changes (cruise at Mach 3.5 instead of 4.5) decreases the TSTO gross weight by 10.3% from 170.5 to 153 tons. The sum of the dry weights is essentially the same for both cases. Considering that 96.3% of the LOX required by the second stage has been collected, that is a rather modest improvement. The daily launch window does increase from 1 to 3.5 hours, and this is obtained without jeopardizing second stage size

For the 5-ton payload with staging at Mach 6.5 the addition of LOX collection decreases the TSTO gross weight by 10.2% from 288.5 to 259 tons. Again, the sum of the dry weights is essentially the same for both cases. Considering that 95% of the LOX required by the second stage has been collected, that is again a rather modest improvement. The daily launch window increases from 1 to nearly 6 hours, and this is obtained without jeopardizing second stage size. Thus separation and collection can result in a factor of 6 increase in the launch window and obtain a 10% reduction in gross takeoff weight.

The results of J. Leingang et al.[38,41,42] for staging at Mach 5 on a similar mission estimates a much more significant improvement in takeoff gross weight; that is, a 39.3% reduction from 375.1 to 228.1 tons. The larger gain results primarily from a low staging Mach and circling during collection instead of extending the outbound cruise and fixing the inbound return leg.

E. Summary

A major advantage of LOX collection is that it takes place at relatively low Mach numbers, below 5 to 7. That means the system can be ground tested in suitable Mach 6 true temperature facilities. The air collection and separation system as well as the low speed cycle of the combined engine can be validated on a supersonic instead of hypersonic demonstrator. Duplication of Mach 8 flight requires a zirconium dioxide cored brick heater for uncontaminated air or an aluminum oxide cored brick heater with a vitiated topping heater. With proper operation either can provide test times at constant temperatures in for about 60 seconds. For test times of several minutes, the hydrogen/nitrous oxide/oxygen combustion test developed by Dr. Peter Kramer[131] at the Daimler–Chysler Ottobrunn facility is very acceptable.

One alternative is to incorporate the LACE system with collection and separation into a very large subsonic transport, and incorporate the RBCC propulsion system for a fully reusable aircraft launched off the subsonic carrier. The orbital vehicle is not burdened with the weight of the liquid collection and separation system. The launcher takes off with only hydrogen on board, and the LOX tanks are filled with LEA as the subsonic transport flies to the launch point. This provides a suitable solution with only subsonic staging (such a system can be viewed as a mobile-based SSTO[132]) This system has the advantage of the lowest testing Mach number requirements. The LACE system testing does not challenge the existing ground test facility capability in the United States or Europe

The results of Sec. IX have been added to results presented in Figs. 40, 41, 46 and 47. Gross weight is of interest for the long-life aircraft operation, because it is the propellant and operational costs that drive the cost of payload to orbit. The planform loading envelope parametric from Fig. 45 is plotted on Fig. 47.

TSTO reduces the gross weight (less propellant), but the sum of the dry weights are very much the same. So the choice is to build and develop one or two vehicles of the same aggregate dry weight. That should drive the interest to SSTO rather than TSTO for launchers. Note that for TSTO, collection has very little impact on size and weight (Fig. 46), although the launch window can be expanded up to a factor of six. However, for SSTO the impact is quite dramatic. The single dashed line for SSTO with collection, on the gross weight (Fig. 46), is for a takeoff wing loading of 280. Using combined improvements (Table 16), the SSTO with collection has significant advantages, labeled SSTO FUTURE collection in Figs. 45 and 46. The potential exists for a less than 100-ton takeoff gross weight launcher with existing technology, although it is not ready for production application.

Are these results reasonable? With no operational SSTO or TSTO vehicles in existence, only study results can be compared. Where all-rocket results are presented, these can be compared to existing rockets, only if the study rockets are expendable. Zagaynov[133] provided a method to do so. Figure 48 shows the original Zagaynov figure with results added from this chapter, and from Richard Nau and from references in the chapter. The NASP data is from the original Zagaynov figure. There are a number of observations. One fully reusable, long-life all-rocket launchers weigh much more than expendables. Two, launchers incorporating airbreather propulsion have a lower growth rate with payload than all-rocket propulsion, for all the different airbreathing speeds reported. Three, launchers incorporat-

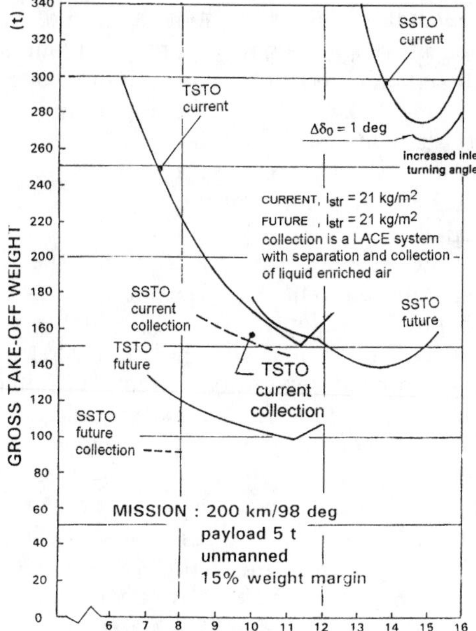

Fig. 46 Takeoff Gross Weight (W_{GTO}) Summary of Results as a Function of Transition Mach Number for SSTO and Maximum Airbreathing Mach Number for TSTO.

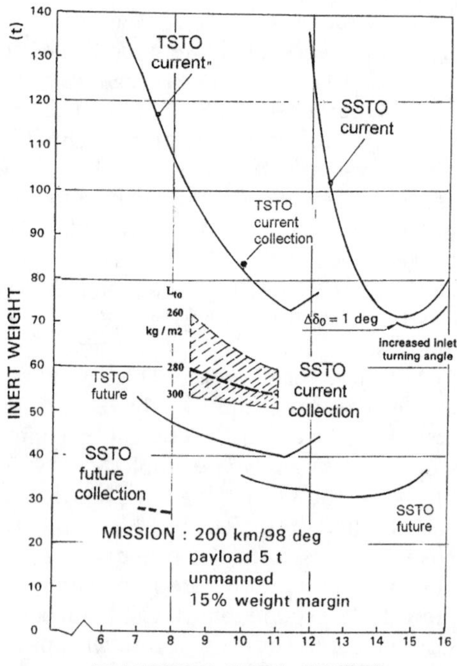

Fig. 47 Inert (dry) Weight (W_{EO}) Summary of Results as a Function of Transition Mach Number for SSTO and Maximum Airbreathing Mach Number for TSTO.

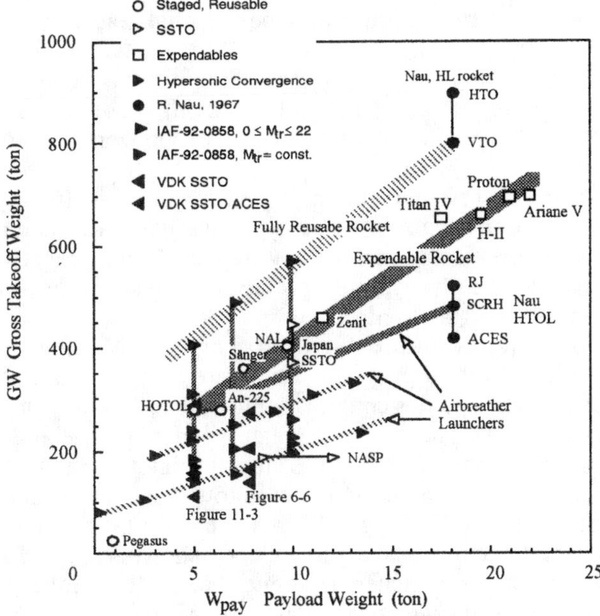

Fig. 48 These Results Are Consistent with the Published Results.

ing airbreathing propulsion weigh much less than launchers with all-rocket propulsion. Four, in sustained operations with high flight rates it is the propellant weight and operational costs that drive cost to orbit, not acquisition cost.

X. Conclusions

1) In the authors' judgment, the real issue for the space launcher organizations is not technology, but creating an operationally affordable vehicle that can be as reliable and frequent in delivering cargo to orbit as aircraft are in delivering cargo to another city.[134] The launcher that is at least partially airbreathing can meet the needs of frequent flights to orbit. That potential is not recognized by the space organizations of the world, and typical of their positions is[126,135] " . . . the only propulsion system for the 21st century is rocket. . . . ." Less frequently flying heavy lift vehicles to LEO are a different matter, and vehicles designed for eventual full reuse, such as ENERGIYA, are appropriate.

2) A vehicle sizing approach is presented that integrates simultaneous volume and weight sizing solutions that are a function of configuration concept, propulsion concept, propulsion-aerodynamic-structural-energy efficiency, and trajectory. A method of describing the total parametric design space and the design space where solutions are possible is described. The parameter

interactions are such that a priori judgments often lead to non-converged results.

3) A broad spectrum of potential airbreathing propulsion systems are described and their impact on weight ratio, takeoff mode and size are presented to show the impact of airbreathing propulsion and some of the choices available.

4) Vehicle sizing is presented that permits sizing to specific requirements. This provides greater insight into the hypersonic aircraft systems' interactions than does constant size exercises. With size constant negative payloads can and do result. The physical interpretation of negative payload and the volume of that payload is obscure.

5) The key to creating an affordable, flexible and reliable launcher is a lightweight high-thrust propulsion system. There are a large number of different engine cycles discussed in this chapter. Some of these employ turbo machinery as part of the cycle. The need for a high specific impulse and high thrust leads to the thermally integrated LACE ejector ramjet concept. The desire to reduce carried oxidizer to a minimum leads to the collection and separation adaptation of the thermally integrated LACE ejector ramjet concept. As the HyFac Flight and Ground Test Facilities clearly showed, as good as turbojets are as fighters, the poor launcher transonic acceleration make separate gas turbine engines an expensive price to pay for familiar conventionality.

6) The TSTO or mobile-based SSTO with the Liquid Air Cycle Engine (LACE) incorporated into the subsonic carrier for collection purposes can already provide a flexible, fully reusable concept with only subsonic staging.

7) For the SSTO without in-flight collection, the use of scramjets is mandatory. To represent it as an unachievable device in the near term is to discredit the pioneers of the late 1950s and 1960s who built and successfully tested these engines up to at least Mach 8 inlet diffuser, exit, duplicated conditions. With air collection, the use of scramjets is not essential, provided advanced lightweight, high internal pressure and temperature ramjets are available. With subsequent upgrades scramjets with increased payloads have a significant growth potential.

References

[1]Johnson, D., "Beyond the X-30: Incorporating Mission Capability," AIAA Paper 91-5078, 1991.

[2]Plokhikh, V. P., "Sensitivity Analysis of SSTO Reusable Vehicle Parameters," International Astronautical Federation, 89-223, Oct. 1989.

[3]Schindel, L., "Design of High Performance Ramjet or Scramjet Powered Vehicles," AIAAA Paper 89-0379, Jan. 1989.

[4]Frederick, J., Sutton, R., and Martens, R., "Turbine Engine Cycle Selection Procedure," *Proceedings of the 3rd International Symposium on Airbreathing Engines,* DGLR-Fachbuch, No. 6, 1976.

[5]Czysz, P., and Murthy, S. N. B., "Energy Management and Vehicle Synthesis," AIAA Paper 95-6101, April 1995.

[6]"Hypersonic Scramjet Vehicle Study," Vol. III, *Structures and Weights,* (U), McDonnell Douglas Corp., MAC Rept. F666, Oct. 1967 (Secret).

[7]Taylor, R. J., "High Temperature Airframe Weight Estimation," SAWE Technical Chap. 479, May 1965.

[8]"Structural Weight Estimation," Vol. I, MCAIR Rept. 747, circa 1965.

[9]"Mission Requirements of Lifting Systems (U), Summary of Significant Results & Figures," McDonnell Aircraft Rept. B947, prepared by NASA NAS 9-3562, Aug. 1965 (declassified 1970).

[10]"GE5/JZ6 Study, Wrap-Around Turboramjet, Turbo-Accelerator Propulsion System Study Data," (U) General Electric, Rept. SS-65-2, Dec. 1965 (confidential).

[11]"Variable Cycle Turboramjet GE14/JZ8," Preliminary Performance and Installation Manual (U), General Electric Rept., U.S. Air Force Contract AF33615-69-1245, Spoat 1969 (confidential).

[12]"Comparative Propulsion Systems Concepts Study (U)," U.S. Air Force Rept. AFRPS-TR-69-19, Sept. 1969 (secret).

[13]"Hypersonic Ramjet Propulsion Program, Engine Performance, (U)," The Marquardt Co., Marquardt Rept. 6112, Aug. 1966 (confidential).

[14]"MA188-XAB Baseline Dual Mode Scramjet for the McDonnell/Douglas Reusable Launch Vehicle Application, (U)," Marquardt Letter to MCAIR, 1966, (confidential).

[15]"Hypersonic Scramjet Vehicle Study, (U)" McDonnell Douglas Corp., MCAIR Rept. F666, Oct. 1967 (secret).

[16]"A Study of Advanced Airbreathing Launch Vehicles with Cruise Capability," Lockheed Aircraft Corp., Lockheed Rept. IR 21042, circa 1967.

[17]"Hypersonic Research Facilities Study, Volumes I through VI," NASA CR 114322–114331, Contract NAS2-5458, OART, Oct. 1970 (declassified 1972).

[18]Tjonneland, E., "Survey of Integration Problems, Methods of Solutions, and Applications," Purdue Univ. Short Course on Engine-Airframe Integration, West Lafayette, IN, July 1988.

[19]Herbst, W., and Ross, H., "The Systems Approach to Systems Engineering," ASPR Inst. Dec. 1969.

[20]Czysz, P., Glaser, F. C., and LaFavor, S. A., "Potential Payoffs of Variable Geometry Engines in Fighter Aircraft," *Journal of Aircraft,* Vol. 10. No. 6, 1973, pp. 342–349.

[21]Plokhikh, V.P., "Problems of Creating Reusable Aerospace Transporting Systems," International Astronautical Federation, 95-V.4.05, Oct. 1995.

[22]Czysz, P., Dighton, R.D., and Murden, W.P., "Designing for Air Superiority," AIAA Paper, Aug. 1973.

[23] Billig, F.S., "Hypersonic Vehicles II," *Proceedings of the Short Course of Engine Airframe Integration,* School of Mechanical Engineering, Purdue Univ., July 1989.

[24]Krieger, R.J., "A Summary of Features and Design Issues for Single-State-to-Orbit Vehicles," July 1990.

[25]Koelle, H.H., "Lunar Space Transportation Cutting Costs of Logistics," International Academy of Astronautics; 95-IAA.1.1.08, Oct. 1995.

[26]Lindley, C., and Penn, J., "Requirements and Approach for a Space Tourism Launch System," International Academy of Astronautics, 97-IAA.12.08, Oct. 1997.

[27]Murthy, S. N. B., and Czysz, P., "Energy Management and Vehicle Synthesis," *Developments in High-Speed-Vehicle Propulsion Systems,* edited by, S. N. B. Murthy and E.T. Curran, *Progress in Astronautics, Vol. 165,* AIAA, Washington, DC, 1996.

[28]Barrére, M., and Vandenkerckhove, J., "Energy Management," *11th International Symposium on Airbreathing Engines,* edited by J.D. Escher, The Synerjet SAE Progress in Technology Series, SAE PT-54, 1997.

[29]Page G. S., "Vehicle Synthesis Program (VSP)," Douglas Aircraft Memorandum C1-E82—ACAP-86-1381, Dec. 1986

[30]Page, G. S., "HSCT Configurations for Phase II Analysis," Douglas Aircraft Memorandum AVI-ACAP-AAP-HAH5-108, Aug. 1987

[31]Czysz, P. A., Unpublished work on Delta Clipper Sizing Program for McDonnell Douglas Astronautics Co., Huntington Beach, CA, May 1991.

[32] Czysz, P. A., and Murthy, S. N. B., "SSTO Launcher Demonstrator for Flight Test," AIAAA Paper 96-4574, Nov. 1996.

[33]Czysz, P. A., Froning, H. D., and Longstaff, R., "A Concept for an International Project to Develop a Hypersonic Flight Test Vehicle," *Proceedings of the International Workshop on Spaceplanes/RLV Technology Demonstrators,* 1997.

[34]Czysz, P. A., and Froning, H. D., "A SSTO Launcher/Demonstrator Concept for International Development for a Flight Test Vehicle," *The International Workshop on Space Plane/RLV Technology Demonstrators,* March 1997.

[35]Czysz, P., "Hypersonic Convergence: Volumes 1 through 10," Saint Louis Univ., Parks College, Aerospace Engineering Dept., Course AE-P493-50, 1992–93: also Purdue Univ. Short Course, "Integration of Winged Flight Vehicles," West Lafayette, IN, 1989.

[36]Küchemann, D., *The Aerodynamic Design of Aircraft,* Pergamon, New York, 1978, p. 214.

[37]Froning, H. D., Jr., and Leingang, J. L., "Impact of Aerospace Advancements on Capabilities of Earth-to-Orbit Ships," International Astronautical Federation, Oct. 1990.

[38]Czysz, P., "Advanced Propulsion Concepts for the XXIst Century," International Academy of Astronautics Workshop on Advanced Propulsion Concepts, The Aerospace Corp., El Segundo, CA, Jan. 1998.

[39]Escher, W. J. D., "Rocket-Based Combined Cycle (RBCC) Powered Spaceliner Class Vehicles Can Advantageously Employ Vertical Takeoff and Landing (VTOL)," AIAA Paper 95-6145, April 1996.

[40]Escher, W. J. D., and Czysz, P. A., "Rocket-Based Combined-Cycle Powered Spaceliner Concept," International Astronautical Federation, IAF-93-S.4.478, Oct. 1993.

[41]Buck, M. L., Zima, W. P., Kirkham, F. S., and Jones, R. A., "Joint USAF/NASA Hypersion Research Aircraft Study," AIAA Paper 75-1039, Aug. 1975.

[42]Vinh, N. X., *Flight Mechanics of High-Performance Aircraft,* Cambridge Aerospace Series 4, Cambridge Univ. Press, 1993.

[43]Czysz, P. A. "Propulsion Concepts and Technology Approaches that Enable a Launcher for the XXIst Century," *Proceeding of the 5th International Symposium La Propulsion dans les Transports Spatiaux,* 1996, p. 15.7.

[44]Czysz, P. A., "Interaction of Propulsion Performance with the Available Design

Space," *Proceedings of the XII International Symposium on Airbreathing Engines,* 1995.

[45]"Single-Stage-to-Orbit Concept Comparison," Aerospace Corp., 86-2602-301-ADA, Oct. 1985.

[46]Tanatsugu, N., Inatani, Y., Makino, T., and Hiroki, T., "Analytical Study of Space Plane Powered by Air-Turbo Ramjet with Intake Air Cooler," International Astronautical Federation, 87-264, Oct. 1987.

[47]Nouse, H., Minoda, M., et al., "Conceptual Study of Turbo-Engines for Horizontal Take-Off and Landing Space Plane," International Astronautical Federation, 88-253, Oct. 1988.

[48]Balepin, V. V., Maita, M., Tanatsugu, N., and Murthy, S. N. B., "Deep-Cooled Turbojet Augmentation with Oxygen—Cryojet for an SSTO Launch Vehicle," AIAA Paper 96-3036, July 1996.

[49]Maita, M., Ohkami Y., and Mori T., "Conceptual Study of Space Plane Powered by Hypersonic Airbreathing Propulsion System," AIAA Paper, Oct. 1990.

[50]Yugov, O. K., et al., "Optimal Control Programs for Airbreathing Propulsion System or Single-Stage-to-Orbit Vehicles," International Astronautical Federation, 89-308, Oct. 1989.

[51]Curran, E. T., "The Potential and Practicality of High Speed Combined Cycle Engines," *Hypersonic Combined Cycle Propulsion,* AGARD Conference Proceeding No. 479, AGARD, 1993, pp. K1–9.

[52]Billig, F. S., "The Integration of the Rocket with a Ram-Scramjet as a Viable Transatmospheric Accelerator," *Proceedings of the XI ISABE,* AIAA, Washington, DC, 1993, pp. 173–187.

[53]Czysz, P., and Little M., "Rocket-Based Combined Cycle Engine (RBCC)—A Propulsion System for the 21st Century," AIAA Paper 93-5096, 1993.

[54]Czysz, P., "Rocket Based Combined Cycle (RBCC) Propulsion Systems Offer Additional Options," *Proceedings of the XI ISABE,* AIAA, Washington, DC, 1993, pp. 119–137.

[55]Yugov, O. K., Dulepov, N. P., and Harchenvnikova, G. D., "The Analysis of Hypersonic and Combined Cycle Engines in the Propulsion System of the SSTO Vehicles," International Astronautical Federation, Oct. 1990.

[56]Ahern, J. E., "Thermal Management of Air-Breathing Propulsion Systems," AIAA Paper 92-0514, Jan. 1992.

[57]Czysz, P., and Murthy, S. N. B., "Energy Analysis of High-Speed Flight Systems," *High-Speed Flight Propulsion Systems,* Vol. 137, edited by Curran and S.N.B. Murthy, Progress in Astronautics and Aeronautics, AIAA, Washington, DC, 1991, pp. 183–186.

[58]Czysz, P. A., "Space Transportation Systems Requirements Derived from the Propulsion Performance Reported in the *Hypersonic and Combined Cycle Propulsion Session* at the 1991 IAF Congress," International Astronautical Federation, 92-0858, Sept. 1992.

[59]Czysz, P. A., "Interaction of Propulsion Performance with the Available Design Space," *Proceedings of the XII International Symposium on Airbreathing Engines,* 1995.

[60]Nicholas, T. M. T., et al., "Mixing Pressure-Rise Parameter for Effect of Nozzle

Geometry in Diffuser-Ejectors," *Journal of Propulsion and Power,* Vol.12, No. 2, 1996, pp. 431–433.

[61]Der, J., Jr., "Characterizing Ejector Pumping Performance," *Journal of Propulsion and Power,* Vol. X, No. 3, 1991.

[62]Escher, W. J. D., "A History of RBCC Propulsion in the U.S.—A Personal Recounting," White Paper, Kaiser Marquardt, Van Nuys, CA, June 1998.

[63]Escher, W. J. D., "A Winning Combination For Tomorrow's Spaceliners," *Aerospace America,* Feb. 1996 pp. 38–43.

[64]Czysz, P. A., and Richards, M. J., "Benefits from Incorporation of Combined Cycle Propulsion," AIAA Paper 98-S.5.10, Oct. 1998.

[65]Mossman, E. A., Rozycki, R. C., and Holle, G. F., "A Summary of Research on a Nozzle-Ejector-System," Martin Denver Research Rept., R-60-31, Oct. 1960.

[66]Harper, R. E., and Zimmerman, J. H., "An Investigation of Rocket Engine Thrust Augmentation with a Nozzle-Ejector System," Arnold Engineering Development Center Rept. TRD-62-42, March 1942.

[67]Scherrer, D., "Evaluation du Concept de Fusee-Statoreacteur Pour la Propulsion Hypersonique," ONERA Activities 1988, ONERA, Paris, April 1988.

[68]Rudakov, A. S, and Balepin, V. V., "Propulsion Systems with Air Precooling for Aerospaceplane," Society of Automotive Engineers, 911182, April 1991.

[69]Rudakov, A. S, Gatin, R. Y., Dulepov, N. P., Koralnik, B. N., Harchevnikova, G. D., and Yugov, O. K., "Analysis of Efficiency of Systems with Oxidizer Liquefaction and Accumulation for Improvement of Spaceplane Performance," International Astronautical Federation, IAF-91-270 Oct. 1991.

[70]Balepin, V. V., Harchermikova, G. D., Tjurikov, E. V., and Avramenko, A.Ju., "Flight Liquid Oxygen Plants for Aerospace Plane: Thermodynamic and Integration Aspects," Society of Automotive Engineers, April 1993.

[71]Aoki, T., Ito, T., et al., "A Concept of LACE for SSTO Space Plane," AIAA Paper 91-5011, Dec. 1991.

[72]Miki, Y., Taguchi, H., and Aoki, H., "Status and Future Planning of LACE Development," AIAA Paper 93-5124, Nov. 1993.

[73]Ogawara, A., and Nishiwaki, T., "The Cycle Evaluation of the Advanced LACE Performance," International Astronautical Federation, 89-313, Oct. 1989.

[74]"Hyperplane," International Astronautical Federation, Oct. 1988.

[75]Lashin, A. I., Kovalevski, M. M., Romankov, O. N., and Tjurikov, E. V.,"Combined Propulsion System for Advanced Multipurpose Aerospace Plane (ASP)," International Astronautical Federation, IAF-93-S.4.479, Oct. 1993.

[76]Bulman, M., and Siebenhaar, A., "The Strutjet: The Overlooked Option for Space Launch," AIAA Paper 95-3124, July 1995.

[77]Balepin, V. V., and Hendrick, P., "Application of the KLIN Cycle to Vertical Take-Off Lifting Body Launcher," AIAA Paper 98-1503, 1998.

[78]Bulman, M., and Siebenhaar, A., "The Strutjet: Exploding the Myths Surrounding High Speed Airbreathing Propulsion," AIAA Paper 95-2475, July 1995.

[79]Vandenkerckhove, J. A., "A Peep Beyond SSTO Mass Marginality," International Astronautical Federation, 92-0656, Sept. 1992.

[80]Stroup, K. E., and Pontez, R. W., "Ejector Ramjet Systems Demonstration," The Marquardt Corp. final report under USAF Contract AF33(615)-3734, Rept. AFAPL-TR-67-118, May 1968.

[81]Hendrick, Ir. P., "SSTO & TSTO LOX Collection System Performances: Influence of LOX Plant Architecture," International Council of the Aeronautical Sciences, 96-3.8.3.

[82]Leingang, J. L., Maurice, L. Q., and Carreiro, L. R., "Space Launch Systems Using Collection and Storage," International Astronautical Federation, IAF 92-0664, Sept. 1992.

[83]Vandenkerckhove, J. A., "A Peep Beyond SSTO Mass Marginality" International Astronautical Federation, 92-0656, Oct. 1992.

[84]Vandenkerckhove, J. A., "SSTO Configuration Assessment," VDK System S.A., WLC Phase 5, WP 260, Chap. 2, revision 1, Brussels, Belgium, Aug. 1992.

[85]Czysz, P., and Froning, H. D., "A Propulsion Technology Challenge—Abortable, Continuous Use Vehicles," International Astronautical Federation, 95-S.2.03, Oct. 1995.

[86]Townend, L., and Vandenkerckhove, J., "External Afterburning and Shock-Confined Combustion in Supersonic Flow," APECS-VDK 001/94, European Space Agency contract 120285, May 1994.

[87]Herbst, W., and Ross, H., "Application of Computer Aided Design Programs for the Technical Management of Complex Fighter Developments & Projects," AIAA Paper 70-364, March 1970.

[88]Vandenkerckhove, J. A., "A First Assessment of Scramjet-Propelled Single-Stage-to-Orbit (SSTO) Vehicles," VDK System S.A., WLC Phase 5, WP 260, Chap. 1, Brussels, Belgium, Feb. 1991.

[89]Vandenkerckhove, J. A., "Further Assessment of Scramjet-Propelled Single-Stage-to-Orbit (SSTO) Vehicles" VDK System S.A., WLC Phase 5, WP 260, Chap. 1, revision 1, Brussels, Belgium, Oct. 1991.

[90]Vandenkerckhove, J., "HYPERJET Mk #3, a Rocket Derived Combined Engine," VDK System Rept., April 1993.

[91]Billig, F. S., "Propulsion Systems from Take-Off to High-Speed Flight," *High-Speed Propulsion Systems,* edited by S.N.B. Murthy and E.T. Curran, Progress in Astronautics and Aeronautics, Vol. 137, Chap. 1.

[92]Froning, H. D., Jr., and Leingang, J. L., "Impact of Aerospace Advancements on Capabilities of Earth-to-Orbit Ships," International Astronautical Federation, Oct. 1990.

[93]Pegg, R. J., Hunt, J. L., et al., "Design of a Hypersonic Waverider-Derived Airplane," AIAA Paper 93-0401, Jan. 1993.

[94]Hunt, L., and Martin, A., "Aero-Space Plane Figures of Merit," AIAA Paper 92-5058, Dec. 1992.

[95]Sosounov, V., "Study of Propulsion for High Velocity Flight," International Sysposium on Airbreathing Engines, Sept. 1991.

[96]Rudakov, A. S., "Cryogenic Propellant Rocket Engine Problems; Liquid Air Rocket Engines & Some Perspective of Engines," unpublished, Jan. 1993.

[97]Aoki, T., and Ogawara, A., "Study of LACE Cycle for SSTO Space Plane," International Astronautical Federation, 88-252, Oct. 1988.

[98]Miki, Y., Togawa, M., Tokunaga, T., Eguchi, K., and Yamanaka, T., "Advanced SCRAM-LACE System Concept for SSTO Space Plane," International Astronautical Federation, IAF-91-272, Oct. 1991.

[99]Togawa, M., Aoki, T., and Hirakoso, H., "A Concept of LACE for SSSTO," International Astronautical Federation, 91-5011, Dec. 1991.

[100]Czysz, P., "Implications of Propulsion System Integration on the Configuration of High Speed Vehicles," Contribution to WLC Phase 5 WP-260, April 1992.

[101]Saber, A. J., and Chen, X., "H2/Air Subsystem Combustion Kinetics in Aerospace Powerplants," International Astronautical Federation, 91-267, Oct. 1991.

[102]Swithenbank, J., and Chigier, N. A., "Vortex Mixing for Supersonic Combustion," *Proceedings of the 12th Symposium (International) on Combustion,* Combustion Inst., Pittsburgh, PA, 1969, pp. 1153–1182.

[103]Nelson, K., "Thesis," M. S. Thesis, Parks College, St. Louis Univ., MO, 1993.

[104]Krieger, R. J., "Summary of Design and Performance Characteristics of Aerodynamic Configures Missiles," AIAA Paper 81-0286, 1981.

[105]Pike, J., *The Aeronautical Quarterly,* Nov. 1972.

[106]Townend, L., "Base Pressure Control for Air Breathing Launchers," AIAA Paper 90-1936, July 1990.

[107]Czysz, P., and Murthy, S. N. B., "Energy Management and Vehicle Systems," Progress in Astronautics and Aeronautics Series, Vol.—, edited by E. T. Curran and S. N. B, Murthy, AIAA, Washington, DC, 1996.

[108]Czysz, P., "Maximize the Energy Conserved, Minimize the Energy Expended Yields Excellent Return on Investment (ROI)," European Space Agency, Feb., 1992.

[109]Escher, W. J. D., "Cryogenic Hydrogen-Induced Air-Liquefaction Technologies," *Hypersonic Combined Cycle Propulsion,* AGARD-CP-479, June 1990.

[110]Froning, H. D., and Czysz, P., "Impact of Emerging Technologies on Manned Transportation Between Earth and Space," International Astronautical Federation, 91-196, Oct. 1991.

[111]Maurice, L. Q., Leingang, J. L., and Carreiro, L. R., "Airbreathing Space Boosters Using In-Flight Oxidizer Collection," *Journal of Propulsion and Power,* July 1991, pp. xx–xx.

[112]Maurice, L. Q., Carreiro, L. R., and Leingang, J. L., "The Benefits of In-Flight LOX Collection for Air-Breathing Space Boosters," AIAA Paper 92-3499, Dec. 1992.

[113]Rudakov, A. S., Gatin, R. Yu., et al., "Analysis of Efficiency of Systems with Oxidizer Liquefaction and Accumulation for Improvement of Aerospace Plane Performance," International Astronautical Federation, 91-279, Oct. 1991.

[114]Togawa, M., Aoki, T., and Ito, T., "On LACE Research," AIAA Paper 92-5023, Dec. 1992.

[115]Gopalaswami, R., Gollakota, S., Venugolapan, P., Nagarathinam, M., and Sivathanu, P. A., "Concept Definition and Design of a Single-Stage-to-Orbit Launch Vehicle HYPERPLANE," International Astronautical Federation, 88-194, Oct. 1990.

[116]Sosounov, V., "Study of Propulsion for High Velocity Flight," Symposium on Air Breathing Engines, Sept. 1991.

[117]Tanatsugu, N., et al., "A Study on Two-Stage-to-Orbit Launcher with Air-Breathing Propulsion," *Proceedings, of AAR-JRS Joint Symposium,* 1985.

[118]Vandenkerckhove, J., "Comparison Between Ejector-Ramjets & Turbo-Ramjets for TSTO Propulsion," AIAA Paper 93-5095, Nov. 1993.

[119]Doublier, M., Pouliquen, M., and Scherrer, D., "Combined Engines for Advanced European launchers," International Astronautical Federation, 88-251, Oct. 1988.

[120]Escher, W. J. D., and Flornes, B. J., "A Study of Composite Propulsion Systems for Advanced Launch Vehicle Applications," The Marquardt Corp. Rept. 25, 194, NASA Contract NAS7-377, Van Nuys, CA, Sept. 1966.

[121]Maita, M., Miyajima, H., and Mori, T., "System Studies on Space Plane Powered by SCRAM/LACE Propulsion System," AIAA Paper 92-5024, Dec. 1992.

[122]Leingang, J. L., Carreiro, L. R., and Maurice, L. Q., "On-Demand Reusable Space Launch Systems That Use In-Flight Oxidizer Collection," Society of Automotive Engineers, TPS 931451, April 1993.

[123]Bond, W. H., and Yi, A. C., "Prospects for Utilization of Air Liquefaction and Enrichment System (ALES) Propulsion in Fully Reusable Launch Vehicles," AIAA Paper 93-2025, June 1993.

[124]Balepin, V. V., and Tjurikov, E. V., "Integrated Air Separation and Propulsion System of Aerospace Plane with Atmospheric Oxygen Collection," Society of Automotive Engineers, 92-0974, April 1992.

[125]Balepin, V. V., Dulepov, Folomeev, E. A., et al., "Flight Liquid Oxygen Plant for Aerospace Plane Thermodynamic and Integration Aspects," April 1993.

[126]Nagappa, R., Subramanyan, J. D. A., and Shilen, S., "Development of Airbreathing Launch Vehicles," National Conference on Air Breathing Engines & Aerospace Propulsion, 1992.

[127]Tanatsugu, N., et al., "Test Results of the Expander Cycle Air Turbo Ramjet for a Future Space Plane," International Astronautical Federation, 91-271, Oct. 1991.

[128]Hendrick, Ir. P., "SSTO & TSTO LOX Collection System Performances: Influence of LOX Plant Architecture," International Council of the Aeronautical Sciences, 96-3.8.3.

[129]Ogawara, A., and Nishiwaki, T., "The Cycle Evaluation of the Advanced LACE Performance," International Astronautical Federation, 89-313, Oct. 1989.

[130]Anifinmov, N. A., "In Searching for an Optimal Concept of Future Russian Reusable Space Transportation system," *Proceedings of the International Workshop on Spaceplane/RLV Technology Demonstrators,* pp. 67–96.

[131]Kramer, P., "Ramjet Combustion Test Facility," European Space Agency, Feb. 1992.

[132]Czysz, P. A., and Little, M. J., "The Rocket Based Combined Cycle Engine (RBCC)—A Propulsion System for the 21st Century," 1993.

[133]Zagaynov, G. I., and Plokhikh, V. P., "USSR Aerospace Plane Program," AIAA Paper 91-5103, Dec. 1991.

[134]Penn, J., and Lindley, C. A., "Spaceplane Design and Technology Considerations over a Broad Range of Mission Applications," Aerospace Corp., El Segundo, CA, 1998.

[135]Freeman, D. C., Talay, T. A., and Austin, R. E., "Single-Stage-to-Orbit—Meeting the Challenge," International Astronautical Federation, 95-V.5.07, Oct. 1995.

[136]Vandenkerckhove, J., "HYPERJET Mk #3, a Rocket Derived Combined Engine," VDK System Rept., April 1993.

Appendix A: Hypersonic Configuration Geometric Characteristics

This appendix presents the fundamental sizing relationships from Hypersonic Convergence and then develops the geometrical relationships that are inherent in the approach. The fundamental premise of the approach was that the geometry of hypersonic vehicles related to volume and area could be approached parametrically rather than single point designs. Ten families of hypersonic configurations are in the Hypersonic Convergence database, and all the configurations have delta planforms with the wing apex beginning at the nose. A NASA Langley cylinder-wing configuration (WB-004) was added as a reusable all-rocket reference point. There are two different scaling modes. One is to fix the sweep angle and vary the volume by changing the maximum cross-sectional area. The other is to fix the cross section and vary the wing sweep through 72 to 80 degrees. The configurations and the pertinent equations are included in this appendix, so the reader can develop whatever relationships desired. For this report the authors used a fixed sweep angle of 78 degrees for all configurations. All the curves shown in the appendix are for pointed bodies.

To begin, the Hypersonic Convergence is briefly reviewed to show the derivation of the three principal size-determining elements and where the geometric characteristics of the configuration play a role. The three principal elements are 1) ratio propellant volume to planform area, 2) ratio propellant density to weight ratio minus one, and 3) magnitude of planform area. The geometry of the configuration will be of first order importance in the first and third elements. Configuration will play a role in the second term, but only as a correction to the weight ratio term for thrust to drag ratio. Beginning with the definition of weight ratio, we have,

$$WR = \frac{TOGW}{OWE} = \frac{OWE + W_{\text{ppl}}}{OWE} = 1 + \frac{W_{\text{ppl}}}{OWE} \tag{A1}$$

the fundamental definition of Operational Weight Empty.

$$OWE = \frac{W_{\text{ppl}}}{WR - 1} = \frac{V_{\text{ppl}}}{S_{\text{pln}}} \cdot \frac{\rho_{\text{ppl}}}{WR - 1} \cdot S_{\text{pln}} = OEW + W_{\text{pay}} + W_{\text{crew}} \tag{A2}$$

Incorporating Küchemann's volume parameter, we have

$$\tau = \frac{V_{\text{tot}}}{S_{\text{pln}}^{1.5}} \text{ (after Küchemann)}$$

$$OWE = \frac{\rho_{\text{ppl}}}{WR - 1} \cdot \frac{V_{\text{ppl}}}{V_{\text{tot}}} \cdot \tau \cdot S_{\text{pln}}^{1.5} \tag{A3}$$

Introducing a geometric parameter, the ratio of wetted (surface) area to planform area, and a correlation for the structure weight fraction with respect to the OEW, we have

$$K_w = S_{wet}/S_{pln} \tag{A4}$$

$$W_{str}/OEW = K_{str} \cdot S_{pln}^{0.138}$$

$$K_{str} = 0.228^{\pm 0.035} \cdot \tau^{0.206} \tag{A5}$$

now a relationship for the technology of the airframe structure (including thermal protection) as related to the propulsion-propellant technology and geometry.

$$\frac{W_{str}}{S_{wet}} = \left(\frac{\rho_{ppl}}{WR-1}\right) \cdot \left(\frac{V_{ppl}}{V_{tot}}\right) \cdot \frac{W_{str}/OEW}{1+(W_{use}/OEW)} \cdot \frac{\tau \cdot S_{pln}^{1.5}}{K_w} \tag{A6}$$

With respect to the propellant volume fraction, the correlation from a series of detailed design hypersonic cruise vehicle from a F-15 weight class to an AN-225 weight class provided the database. Because the correlation parameter, K_v, is dimensional, the two versions are given for both unit systems. The original correlations were for an all-electronic, high density payload, so the initial value, K_{vo}, is scaled with respect the bulk density of the payload. This is the payload weight divided by the payload bay volume.

$$\frac{V_{ppl}}{V_{tot}} \approx K_v \cdot S_{pln}^{0.0717} \text{ (English)}$$

$$\approx 1.1857 \cdot K_v \cdot S_{pln}^{0.0717} \text{ (Metric)} \tag{A7}$$

$$K_v = K_{vo} - 6.867E-3 \cdot \tau^{-1} + 8.2777E-4 \cdot \tau^{-2} - 2.811E-5 \cdot \tau^{-3}$$

$$K_{vo} = 0.40 \cdot \left(\frac{\rho_{pay}}{5.0}\right)^{0.123} \qquad 0.40 \cdot \left(\frac{\rho_{pay}}{176.5}\right)^{0.123} \tag{A8}$$

The payload fraction was correlated for two classes of vehicles. The top equation of (A9) is for the propulsion integrated configuration concepts with a body-integrated inlet ramp system and exhaust nozzle. The bottom equation of (A9) is for the blunt base rocket powered hypersonic glider configuration concepts. Note that the payload fraction is a function of both the geometrical slenderness and the absolute value of payload. The payload must be in metric tons for Eq. (A9).

$$W_{pay}\big/OEW = \frac{\exp(2.10 \cdot \tau)}{24.79} \cdot \exp[0.71 \cdot \ell n(W_{pay})]$$

$$W_{pay}\big/OEW = \frac{\exp(1.29 \cdot \tau)}{25.4} \cdot \exp[0.71 \cdot \ell n\left(W_{pay}\right) \tag{A9}$$

Equation (A6) can then be written as

$$\frac{W_{\text{str}}}{S_{\text{wet}}} = \frac{K_{\text{str}} \cdot K_v \cdot \tau}{K_w} \cdot \left(\frac{\rho_{\text{ppl}}}{WR - 1}\right) \cdot \frac{S_{\text{pln}}^{0.7097}}{1 + \left(W_{\text{pay}}/OEW\right)} \tag{A10}$$

This equation can be rearranged to yield a first order estimate of the vehicle planform area based on the available industrial capability (technology), payload faction, and configuration geometry.

$$S_{\text{pln}} = \left[\left(\frac{K_w}{\tau}\right) \cdot K_{\text{str}} \cdot K_v \cdot \frac{\rho_{\text{ppl}}/(WR-1)}{W_{\text{str}}/S_{\text{wet}}} \cdot \left(1 + \frac{W_{\text{pay}}}{OEW}\right)\right]^{1.409} \tag{A11}$$

The three primary terms are then

1) $\dfrac{\rho_{\text{ppl}}/(WR-1)}{W_{\text{str}}/S_{\text{wet}}} = \dfrac{I_p}{I_{\text{str}}}$

2) $\left(1 + \dfrac{W_{\text{pay}}}{OEW}\right) = 1 + \dfrac{\exp(2.10 \cdot \tau)}{24.79} \cdot \exp\left[0.71 \cdot \ell n\left(W_{\text{pay}}\right)\right.$

3) $\left(\dfrac{K_w}{\tau}\right) \cdot K_{\text{str}} \cdot K_v = \left(\dfrac{K_w}{\tau}\right) \cdot \dfrac{0.093^{\pm 0.017}}{\tau^{0.794}}$

A most likely value for K_v of $0.40^{\pm 0.02}$ is assumed for the last term. For the (K_w/τ term, ten families of hypersonic configurations from the Hypersonic Convergence database are given. All configurations have delta planforms with the wing apex beginning at the nose. The NASA Langley cylinder-wing configuration (WB-004) was added as a reusable all-rocket reference point. There were two different scaling modes. One was to fix the sweep angle and vary the volume by changing the maximum cross-sectional area. The other was to fix the cross section and vary the wing sweep through 72 to 80 degrees. For this report the authors used a fixed sweep angle of 78 degrees for all configurations. The configurations and pertinent equations are given herein, so readers can generate their own scaling models. All the curves shown in this reference are for pointed bodies. A spatular body fixes the length, and adds width and volume. Since the length determines the engine module height and length, a fixed length, but wider, vehicle can incorporate an increased number of the same module configurations. This eases the concerns of the propulsion community with respect to engine module certification. The spatular nosed waverider from the University of Maryland has essentially the same characteristics as the blended body (Lewis, Mark, private communication, June 1997).

The ten configuration families were fox fixed 78 degree sweep angle with variable cross-section shape. Rocket derived propulsion includes LACE and deeply cooled rocket cycles.

Airbreathing propulsion integrated configurations are 1) blended body, 2) wing-body (not cylinder-wing), and 3) Nonweiler waverider.

Rocket derived hypersonic gliders are the following: 1) diamond cross section. Base height to width 0.1 to 1.0; 2) elliptical cross section. Base height to width from 0.1 to 1.0; 3) trapezoidal cross section. Base top width to bottom width from 0 to 1.0; 4) blunted right circular cone. Nose to base diameter ratio from 0 to 0.3; 5) half-diamond cross section. Base height to width from 0.05 to 0.5; 6) half-elliptical cross section. Base height to width from 0.05 to 0.5; and 7) blunted half-right circular cone. The nose to base diameter ratio ranges from 0 to 0.3.

The eleventh configuration was the NASA Langley cylinder-wing (WB-004) configuration used as a vertical launch, recoverable rocket vehicle reference. The exposed wing area and diameter of the tank were held constant. The volume changed by varying the length to diameter of the cylinder. This configuration was not used in this report.

Figure A1 shows the wetted area to planform area ratio versus tau for configurations that include aerodynamic control surfaces based on the configurations shown in Fig. 10. These are possible candidates for space launchers having values of tau less than 0.20 and lower values of wetted area to planform area ratio. The wing-body and Nonweiler waverider have larger values of wetted area to planform area ratio than integral wing-body configurations. The WB-004 configuration has very different geometric properties than the highly swept integral wing-body configurations. It is essentially

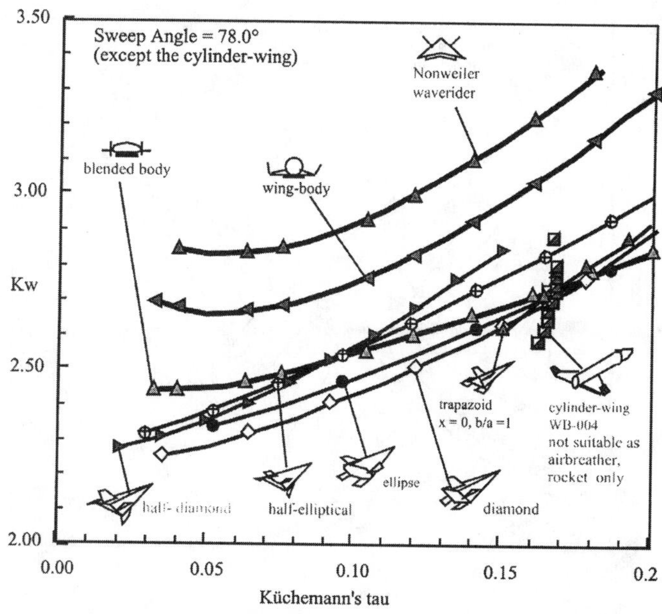

Fig. A1 Potential Space Launcher Geometric Design Space.

a constant tau configuration (0.162 to 0.167) over a 2 to 1 volume ratio. The full range of hypersonic shapes extend beyond tau = 0.20. Figure A2 shows the range of configuration characteristics to tau =0.5. That is about the limit for a reasonable lift to drag ratio for an acceptable cross range and down range. Not shown is the sphere with a tau = 0.752 and a K_w = 4.00.

The elliptical cone spans the widest range of tau and K_w as it progresses from an ellipse with a height 10% of its width to a circle. The diamond and trapezoid shapes span similar ranges. There are two trapezoidal shapes. One with *(b/a)* = 1 has a height equal to the half width. The other with *(b/a)* = 2 has a height equal to the width. The parameter in the sizing equation is the ratio of tau to K_w. That ratio is plotted in Fig. A3, for all of the configurations shown in Fig. A2. The result is the collapse of the geometric characteristics into nearly a single line (Fig. A3). In this graph the sphere is shown, and it has the lowest value of the $(K_w/\tau$ term. So the sphere has the lowest OEW if not the lowest drag.

The different class of vehicles and the propellants can be differentiated on this figure. The denser the propellant, the less the propellant volume, the more slender the shape and the larger the planform area with respect to the propellant volume. The important conclusion is that as a first order estimate, only tau need be known. After the first order estimate, then the refinement of the estimate using different geometries can proceed. The primary determinant then is the propulsion index that results from a trajectory or cruise

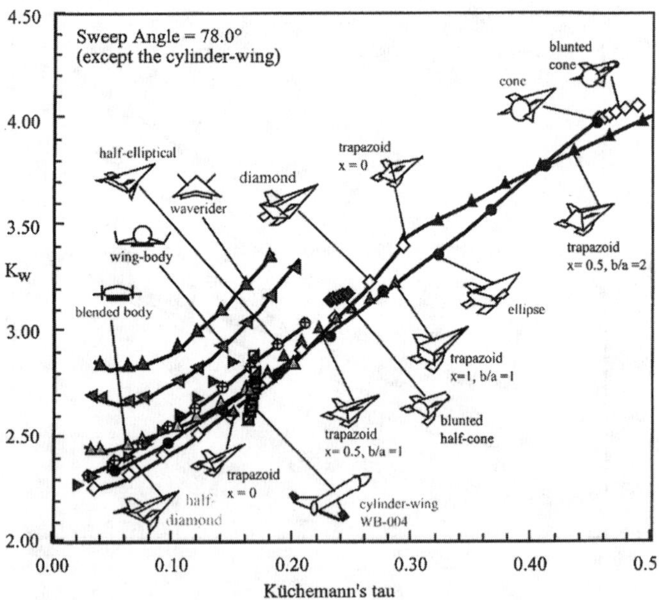

Fig. A2 Geometric Design Space for Hypersonic Vehicles

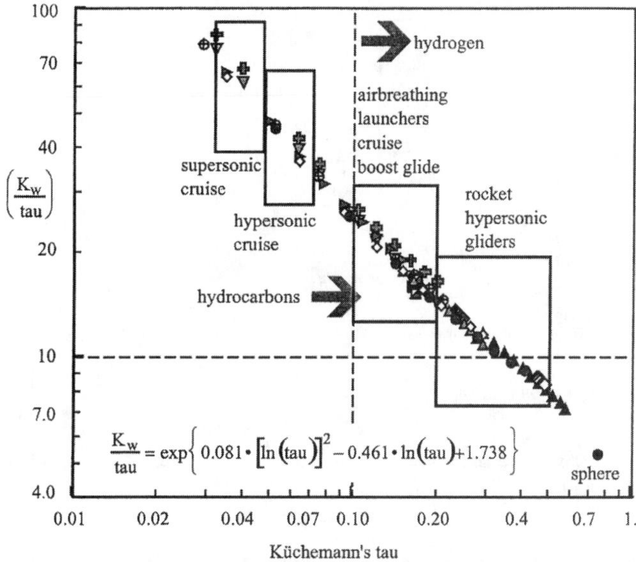

Fig. A3 Sizing Geometry Parameter, (K_w/τ is Determined by Tau.

performance analysis. The remainder of the appendix gives the configuration concepts and the description of the geometric properties.

Figure A2 shows the wetted area to planform area ratio (K_w versus τ for configurations that includes aerodynamic control surfaces. Configurations with values of τ less than 0.20 are possible candidates for space launchers, and have lower values of wetted area to planform area ratio. The wing-body and Nonweiler waverider have larger values of wetted area to planform area ratio than the blended body configurations. The WB-004 configuration has very different geometric properties than the highly swept integral wing-body configurations. It is essentially a constant tau configuration (0.162 to 0.167), which means volume changes require planform area changes. The full range of hypersonic shapes extend beyond tau = 0.20. Figure A3 shows a broader range of configuration concepts that encompasses most hypersonic aircraft, launchers, and gliders. The complete figure that includes the sphere is Fig. 4.

The elliptical cone spans the widest range of τ and K_w as it progresses from an ellipse with a height 10% of its width to a circle. The diamond and trapezoid shapes span similar ranges. There are two trapezoidal shapes. One with $(b/a) = 1$ has a height equal to the half width. The other with $(b/a) = 2$ has a height equal to the width. The parameter in the sizing equation is the ratio of τ/K_w. The ratio is plotted in (Fig. A4), for all of the configurations shown in (Fig. A3). The results is the collapse of the geometric characteristics into nearly a single line (Fig. A4). In this graph the sphere is shown, and it has the lowest value of the (K_w/τ) term. So the sphere has

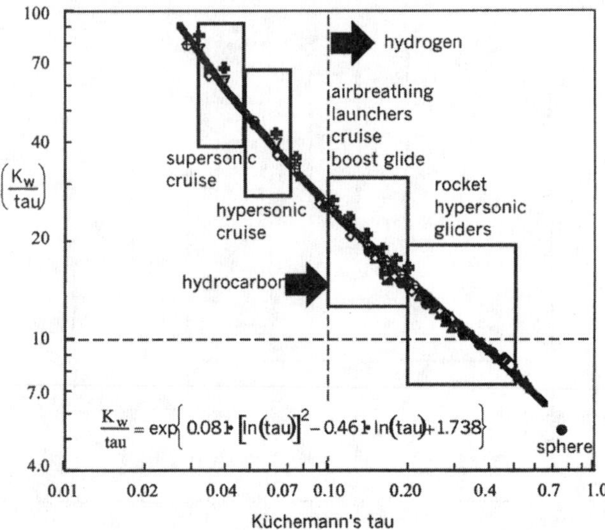

Fig. A4. Sizing Geometry Parameter, (K_w/τ) is Determined by τ.

potentially the lowest W_{OE} and the highest drag, making it a simple ballistic vehicle.

The different class of vehicles and the propellants can be differentiated on this figure. The denser the propellant, the less the propellant and payload volume; the more slender the shape, the larger the planform area is with respect to the propellant volume. Thus, the configuration concept is not an arbitrary selection but a consequence of the role and mission. The important conclusion is that for a first order estimate, only the role/mission and τ need be known. After the first order estimate, then refining the estimate using different geometries can proceed. The primary determinant then is the propulsion index that results from a trajectory or cruise performance analysis. The remainder of the appendix gives the configuration concepts and the description of the geometric properties.

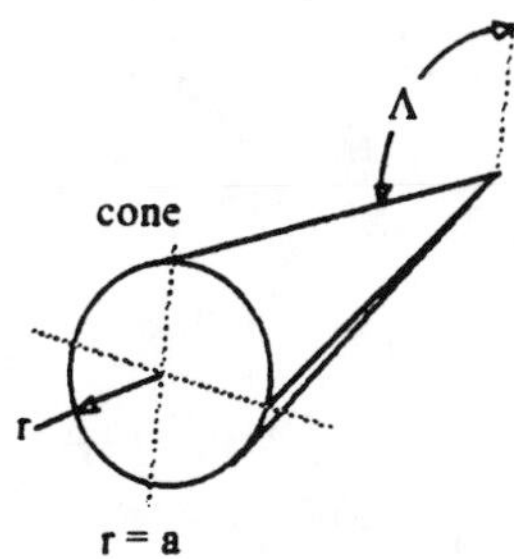

$$A_{\text{base}} = \pi r^2$$

$$S_{\text{wet}} = \pi r^2\left(1 + \frac{1}{\cos\Lambda}\right)$$

$$V_{\text{tot}} = \frac{\pi r^3}{3}\tan\Lambda$$

$$K_w = \pi\left(\frac{1}{\tan\Lambda} + \frac{1}{\sin\Lambda}\right)$$

$$\tau = \frac{\pi}{\sqrt[3]{\tan\Lambda}}$$

blunted cone

r_n

r_b

$r_b = a$

$0.00 \leq r_b / r_n \leq 0.30$

$$A_{\text{base}} = \frac{\pi r_b^2}{2}$$

$$S_{\text{plan}} = r_b^2 \tan\Lambda\left[1 - (r_n/r_b)^2\right] + \frac{\pi r_b^2}{2}\left(\frac{r_n}{r_b}\right)^2$$

$$S_{\text{wet}} = \pi r_b^2\left[1 + \frac{1 + (r_n/r_b)^2}{\cos\Lambda} + 2\left(\frac{r_n}{r_b}\right)^2\right]$$

$$V_{\text{tot}} = \frac{\pi r_b^3}{3}\left(1 - \frac{r_n}{r_b}\right)\left[1 + \frac{r_n}{r_b} + \left(\frac{r_n}{r_b}\right)^2\right]\tan\Lambda + \frac{2\pi r_b^3}{3}\left(\frac{r_n}{r_b}\right)^3$$

$$(Kw)_{78^\circ} = 4.600 \cdot \left(\frac{r_n}{r_b}\right)^2 - 2.350 \cdot \frac{r_n}{r_b} + 4.111$$

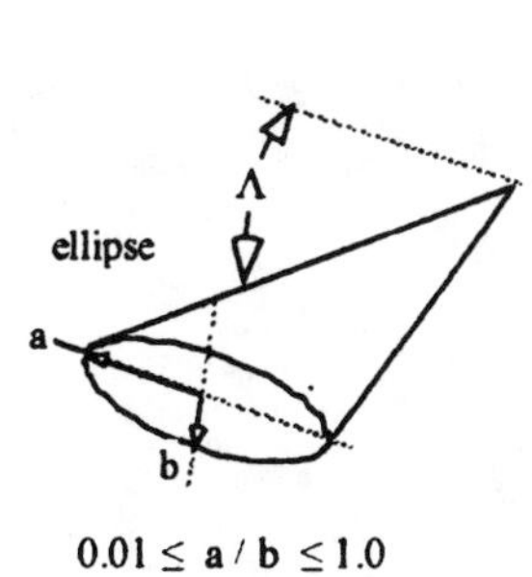

$0.01 \leq a / b \leq 1.0$

$$A_{\text{base}} = \pi a^2 e$$

$$S_{\text{plan}} = a^2 \tan\Lambda$$

$$S_{\text{wet}} = \pi a^2 \frac{(1+e)}{\cos\Lambda}\left(1 + \frac{R^2}{4} + \frac{R^4}{64} + \frac{R^6}{256}\right) + \pi a^2 e$$

$$V_{\text{tot}} = \frac{\pi a^3 e}{3}\tan\Lambda$$

$$e = b/a \qquad R = \frac{1-e}{1+e}$$

$$(Kw)_{78^\circ} = 2.404 \cdot \tau^2 + 2.920 \cdot \tau + 2.174$$

$$\tau = 0.449 \cdot \left(\frac{b}{a}\right) + 0.007$$

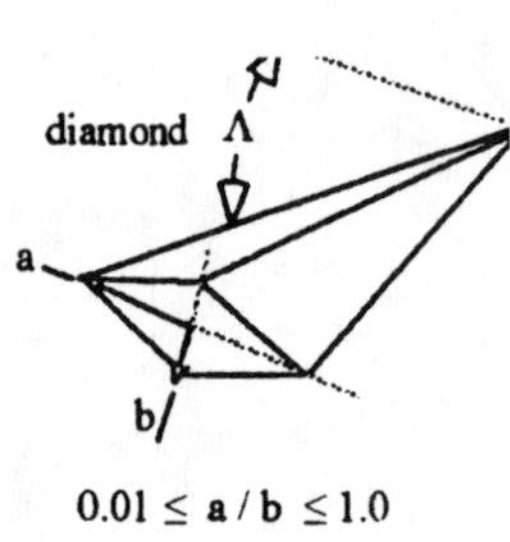

$$A_{\text{base}} = 2a^2e$$

$$S_{\text{plan}} = a^2 \tan\Lambda$$

$$S_{\text{wet}} = 2a^2\sqrt{1+e^2}\,\tan\Lambda + 2a^2e$$

$$V_{\text{tot}} = \frac{2a^3e}{3}\tan\Lambda$$

$$e = b/a$$

$$(Kw)_{78^\circ} = 8.023\cdot\tau^2 + 1.872\cdot\tau + 2.173$$

$$\tau = 0.296\cdot\left(\frac{b}{a}\right) + 0.007$$

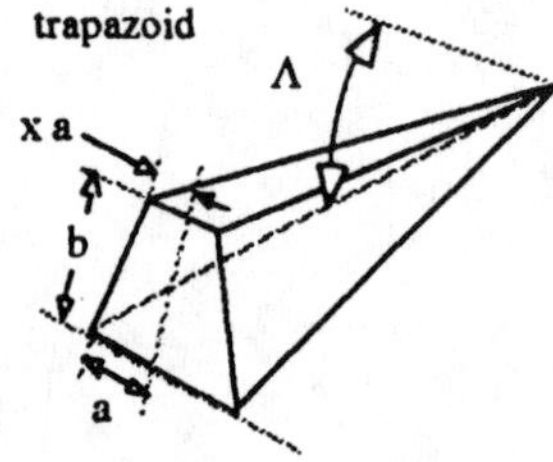

$$A_{\text{base}} = a^2e(1+X)$$

$$S_{\text{plan}} = a^2 \tan\Lambda$$

$$S_{\text{wet}} = a^2\left[(1+x) + \frac{\sqrt{e^2+(1+x)^2}}{\cos\Lambda}\right]$$

$$V_{\text{tot}} = \frac{a^3(1+x)e}{3}\tan\Lambda$$

$e = b/a \qquad x = \text{top width/bottom width}$

$$(Kw)_{78^\circ} = -22.447\cdot\tau^2 + 13.850\cdot\tau + 1.053$$

$$\tau = 0.143\cdot x + .150 \qquad b/a = 1$$

$$(Kw)_{78^\circ} = -2.894\cdot\tau^2 + 5.053\cdot\tau + 2.199$$

$$\tau = 0.286\cdot x + .292 \qquad b/a = 2$$

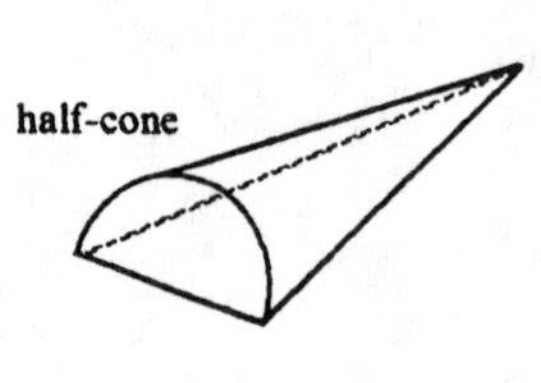

$$A_{\text{base}} = \frac{\pi r^2}{2}$$

$$S_{\text{plan}} = r^2 \tan\Lambda$$

$$S_{\text{wet}} = \frac{\pi r^2}{2}\left(1 + \frac{1}{\cos\Lambda}\right) + r^2\tan\Lambda$$

$$V_{\text{tot}} = \frac{\pi r^3}{6}\tan\Lambda$$

$$K_w = \frac{\pi}{2}\left(\frac{1}{\tan\Lambda} + \frac{1}{\sin\Lambda}\right) + 1$$

$$\tau = \frac{\pi}{6\sqrt{\tan\Lambda}}$$

blunted half-cone

$$A_{\text{base}} = \frac{\pi r_b^2}{4}$$

$$S_{\text{plan}} = r_b^2 \tan\Lambda\left[1-(r_n/r_b)^2\right] + \frac{\pi r_b^2}{2}\left(\frac{r_n}{r_b}\right)^2$$

$$S_{\text{wet}} = \frac{\pi r_b^2}{2}\left[1+\frac{1+(r_n/r_b)^2}{\cos\Lambda}+2\left(\frac{r_n}{r_b}\right)^2\right] + S_{\text{plan}}$$

$$V_{\text{tot}} = \frac{\pi r_b^3}{6}\left[\left(1-\frac{r_n}{r_b}\right)\left[1+\frac{r_n}{r_b}+\left(\frac{r_n}{r_b}\right)^2\right]\tan\Lambda + 2\left(\frac{r_n}{r_b}\right)^3\right]$$

$$(K_w)_{78^\circ} = 58.592\left(\frac{r_n}{r_b}\right)^2 - 25.755\frac{r_n}{r_b} + 5.970$$

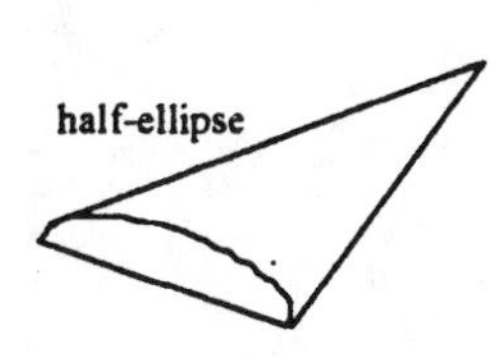

$$A_{\text{base}} = \pi a^2 e/2$$

$$S_{\text{plan}} = a^2 \tan\Lambda$$

$$S_{\text{wet}} = \frac{\pi a^2}{2}\left\{\frac{(1+e)}{\cos\Lambda}\left(1+\frac{R^2}{4}+\frac{R^4}{64}+\frac{R^6}{256}\right)+e\right\} + S_{\text{plan}}$$

$$V_{\text{tot}} = \frac{\pi a^3 e}{6}\tan\Lambda$$

$$e = b/a \qquad R = \frac{1-e}{1+e}$$

$$(K_w)_{78^\circ} = 4.689\cdot\tau^2 + 2.917\cdot\tau + 2.226$$

$$\tau = 0.224\cdot\left(\frac{b}{a}\right) + 0.007$$

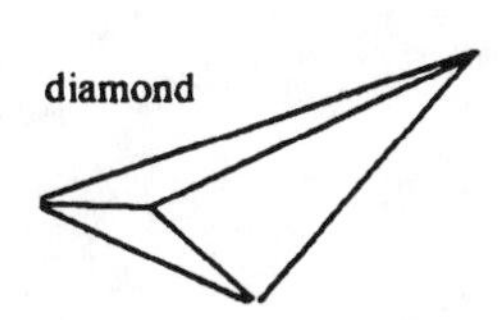

$$A_{\text{base}} = a^2 e$$

$$S_{\text{plan}} = a^2 \tan\Lambda$$

$$S_{\text{wet}} = a^2\left(\sqrt{1+e^2}\tan\Lambda + e\right) + S_{\text{plan}}$$

$$V_{\text{tot}} = \frac{a^3 e}{6}\tan\Lambda$$

$$e = b/a$$

$$(K_w)_{78^\circ} = 15.387\,\tau^2 + 1.865\,\tau + 2.228$$

$$\tau = 0.143\left(\frac{b}{a}\right) + 0.007$$

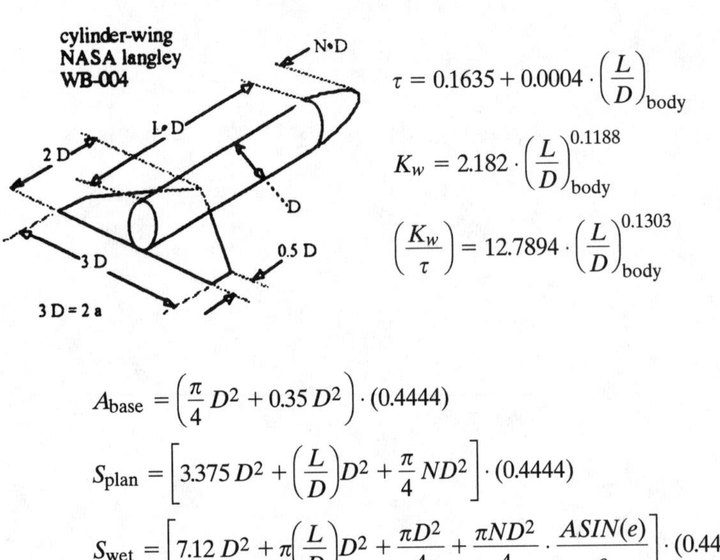

$$\tau = 0.1635 + 0.0004 \cdot \left(\frac{L}{D}\right)_{\text{body}}$$

$$K_w = 2.182 \cdot \left(\frac{L}{D}\right)_{\text{body}}^{0.1188}$$

$$\left(\frac{K_w}{\tau}\right) = 12.7894 \cdot \left(\frac{L}{D}\right)_{\text{body}}^{0.1303}$$

$$A_{\text{base}} = \left(\frac{\pi}{4} D^2 + 0.35\, D^2\right) \cdot (0.4444)$$

$$S_{\text{plan}} = \left[3.375\, D^2 + \left(\frac{L}{D}\right) D^2 + \frac{\pi}{4} N D^2\right] \cdot (0.4444)$$

$$S_{\text{wet}} = \left[7.12\, D^2 + \pi\left(\frac{L}{D}\right) D^2 + \frac{\pi D^2}{4} + \frac{\pi N D^2}{4} \cdot \frac{ASIN(e)}{e}\right] \cdot (0.4444)$$

$$V_{\text{tot}} = 0.27\, D^3 + \frac{\pi D^3}{4}\left(\frac{L}{D}\right) + \frac{\pi N D^3}{6}$$

$$e = \frac{\sqrt{2N - 1}}{N} \quad \text{nose length} = N \cdot D$$

$$\frac{V_{\text{ppl}}}{V_{\text{tot}}} = \frac{\frac{L/D}{4} + \frac{N}{6} - \frac{3}{4} + \frac{1}{3}}{\frac{L/D}{4} + \frac{N}{6}}$$

Propulsion Integrated Configurations:

Blended Body $K_W = -62.217 \cdot \tau^3 + 29.904 \cdot \tau^2 - 1.581 \cdot \tau + 2.469$

Wing-Body $K_W = -93.831 \cdot \tau^3 + 59.920 \cdot \tau^2 - 5.648 \cdot \tau + 2.821$

Nonweiler Waverider $K_W = -533.451 \cdot \tau^3 + 220.302 \cdot \tau^2 - 22.167 \cdot \tau + 3.425$

Appendix B: Impact of Lower Speed Thrust Minus Drag

There are a number of hypersonic studies[1–4] that provide a basis for hypersonic configuration concepts, as shown in Fig. B1. Please note: Refs. 1–4 were both a hypersonic flight and ground research facilities study, with a majority of the resources going to the former. For hypersonic speeds, a freestream inlet/nacelle (i.e., the Mach number of the inlet flow field is the same as the flight Mach number) is on the same order of length as the fuselage. Except for the lowest hypersonic Mach numbers, the preferred installation is the airframe integrated propulsion system in which the airframe is part of both the inlet and nozzle. The engine module inlet is then immersed in a compression flow field where the local Mach number is less than the flight Mach number. For such installations, the engine cowl Mach number approaches a limiting value; thus, the performance of the cowl/engine system is relatively fixed for high Mach numbers. The front underside of the vehicle is the inlet. The vertical dimension of the underside depends on the distance the engine module cowl is from the nose. (See Figs. 17 and B10). As the vehicle size is iterated, both the vertical and longitudinal distances change. It is only the converged solution that provides the correct cowl Mach number. The flowfield Mach number ahead of the cowl is

$$M_{\text{cowl}} = 2.37 \cdot Ln(M_{\text{flight}}) - 0.43 \tag{B1}$$

At flight Mach numbers above 2.7, the Mach number at the inlet cowl lip is two or greater, and there is a shock attached at the cowl lip entering the engine module inlet, (see Fig. B10). Airframe configurations that can incorporate integrated inlets system are wing body, waverider, all body, and blended body.

There can be interactions between the propulsion system configuration and the airframe configuration that mutually exclude some possible combi-

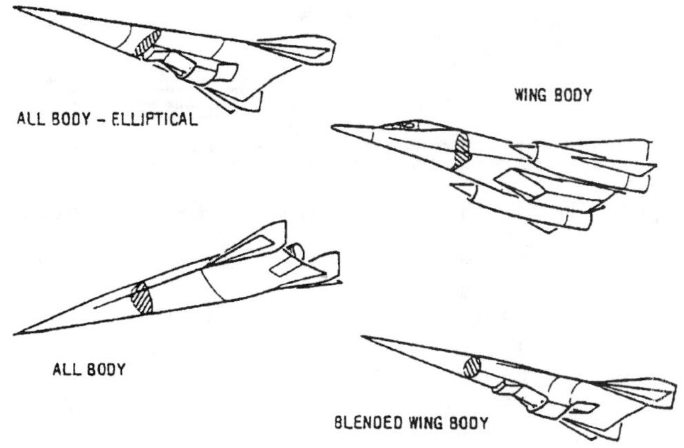

Fig. B1 Representative Hypersonic Configurations

nations. To have a consistent comparison base, the examples shown are for research aircraft that must achieve a 5-minute test time at either Mach 6 or Mach 12. These are not propellant-dominated designs as would be the case for an operational cruiser or SSTO launcher. However, the strong interaction between airframe and propulsion is still evident.

Figure B2 shows a representative air-launched, rocket-accelerated integrated ramjet Mach 6 research vehicle. Even though the inlet-nacelle is within the bow shock field to provide some initial compression, the propulsion package is out of proportion (too long with respect to vehicle length) compared to an operational sized vehicle. This research aircraft can be the basis of research, but not for determining size trends for full-scale vehicles.

Figure B3 shows a representative air-launched, rocket-accelerated, scramjet-powered Mach 12 research vehicle. This vehicle exhibits a more representative proportion between the engine module and the airframe. This is an air-launched concept, with an unpowered landing capability. The Mach 12 class of research vehicles was selected to illustrate the interaction between the propulsion concept and the airframe concept because of the better correspondence with operational sized vehicles.

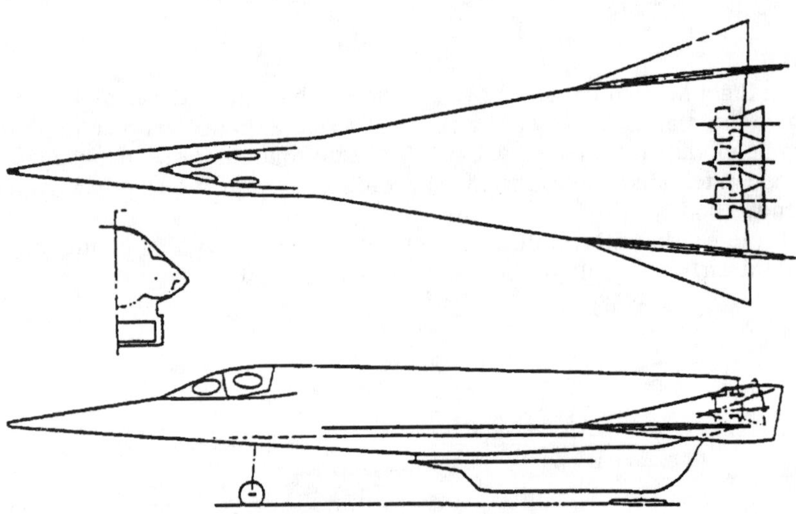

Fig. B2 Air-Launched Mach 6 Hypersonic Research Aircraft.

Wing Sweep	**80°**
Length	**19.3 meters**
Span	**7.1 m**
Planform area	**48.3 m²**
Propellant Volume	**28.5 m³**
τ	**0.201**
W_{TOG}	**< 18 ton**
W_{OE}	**< 11 ton**

Correct size with respect to Mach 12 manned research aircraft in Fig B3.

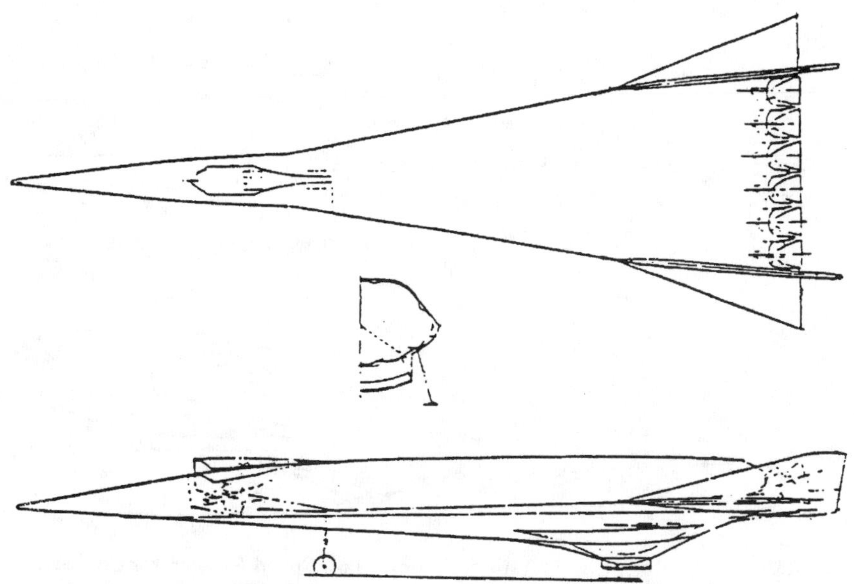

Fig. B3 Air-Launched Mach 12 Hypersonic Research Aircraft.

Wing Sweep	**80°**
Length	**25.5 meter**
Span	**10.7 m**
Planform area	**110 m²**
Propellant Volume	**100 m³**
τ	**0.147**
W_{TOG}	**< 36 ton**
W_{OE}	**< 12 ton**

Correct size with respect to Mach 6 manned research aircraft in Fig. B2.

Although the Mach 12 vehicle has twice the speed and is larger in size than the Mach 6 vehicle, the W_{OE} is only 1 metric ton greater. It is also more slender (τ is less). The integrated ramjet propulsion requires a much smaller engine-shorter nacelle/module than the Mach 6 aircraft. The absence of a takeoff landing gear is a major contributor in that it eliminates concentrated landing gear loads and the landing gear weight. Perhaps this is an indication that an air-launched operational system is not unreasonable.

Propulsion Airframe Strong Interactions

There is not much surviving data from hypersonic studies of 40 years ago, but there are enough fragments that some insight into the interactions between propulsion and airframe concept can be illustrated. The materials systems that were available from 1965 to 1970 from used refractory metals to accommodate the highest temperatures. The airframe weight per unit

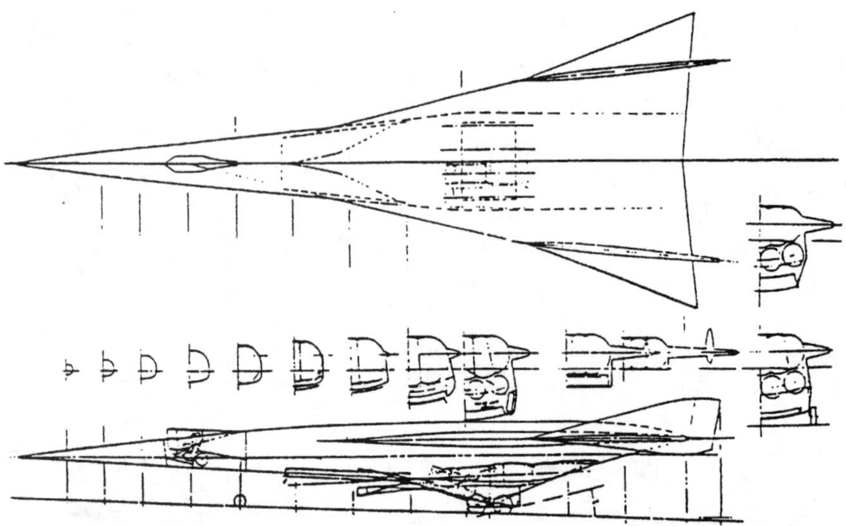

Fig. B4 A Wing-Body Configuration, Turbojet Low Speed Accelerator Convertible Scramjet Test Vehicle. Horizontal Take-off and Landing (HTOL) 5 minute test time at Mach 12.

Length	**43.6 meters**
Payload	**453.6 kg**
$S_{planform}$	**265.7 m^2**
τ	**0.114**
S_{wet}	**717.1 m^2**
L_{to}	**201 kg/m^2**
Wstr/Swet	**26.1 kg/m^2**
Wtps/Swet	**5.06 kg/m^2**
Structure	**15.1 tons**
TPS	**3.63 tons**
Equipment	**2.59 tons**
Propulsion	**15.7 tons**
Empty Weight	**36.3 tons**
W_{OE}	**37.9 tons**
Propellant	**16.0 tons**
Gross Weight	**53.5 tons**

surface area ranged from 23 to 27 kg/m^2 in most cases. The configurations from Refs. 1–4 and the four illustrated in Figs. B4–B7 were all sized to provide 5 minutes of test time at Mach 12 at altitudes that corresponded to potential operational aircraft. All four aircraft shown incorporated a ramjet that transitioned from subsonic throughflow to supersonic throughflow (i.e., dual-mode scramjet) as part of the research objectives.

For the four examples shown for two acceleration power plants, the hydrogen/oxygen rocket, and the hydrogen fueled afterburning turbojet. There

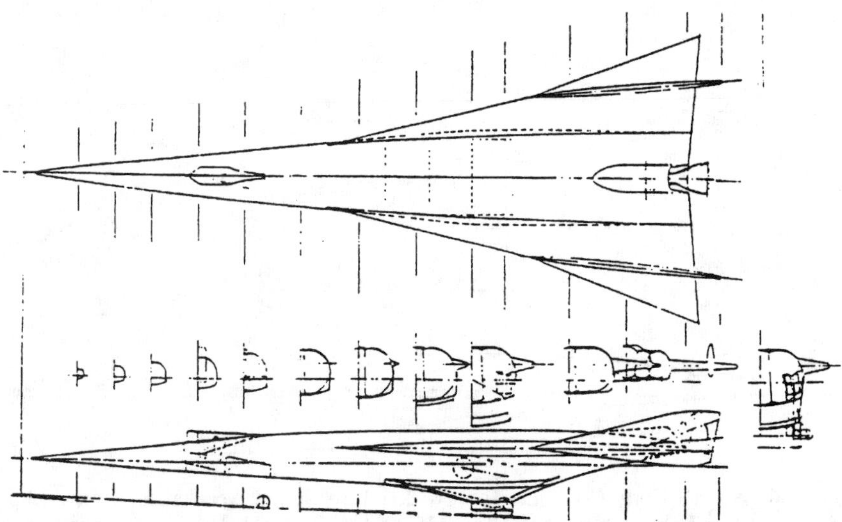

Fig. B5 A Wing-Body Configuration, Rocket Low Speed Accelerator Convertible Scramjet Test Vehicle. Horizontal Takeoff and Landing (HTOL) 5 minutes test time at Mach 12.

Length	**37.5 meters**
Payload	**453.6 kg**
$S_{planform}$	**203.5 m^2**
τ	**0.107**
S_{wet}	**547.4 m^2**
L_{to}	**393 kg/m^2**
Wstr/Swet	**23.6 kg/m^2**
Wtps/Swet	**3.75 kg/m^2**
Structure	**12.9 tons**
TPS	**2.06 tons**
Equipment	**2.42 tons**
Propulsion	**6.01 tons**
Empty Weight	**19.9 tons**
W_{OE}	**21.7 tons**
Propellant	**58.2 tons**
Gross Weight	**80.0 tons**

are two airframe configurations, the all-body with the rocket accelerator, and the wing-body with afterburning turbojet accelerator.

One combination failed to converge for this study—the all-body with an afterburning turbojet accelerator. The afterburning turbojet provides only a marginally positive thrust minus drag for the wing-body configuration (Fig. B4), and negative thrust minus drag for the all-body configuration.

The turbo-ramjet with a conventional 2-D overhead ramp inlet and C-D nozzle was satisfactory to produce an acceptable Mach 6 research aircraft.[3]

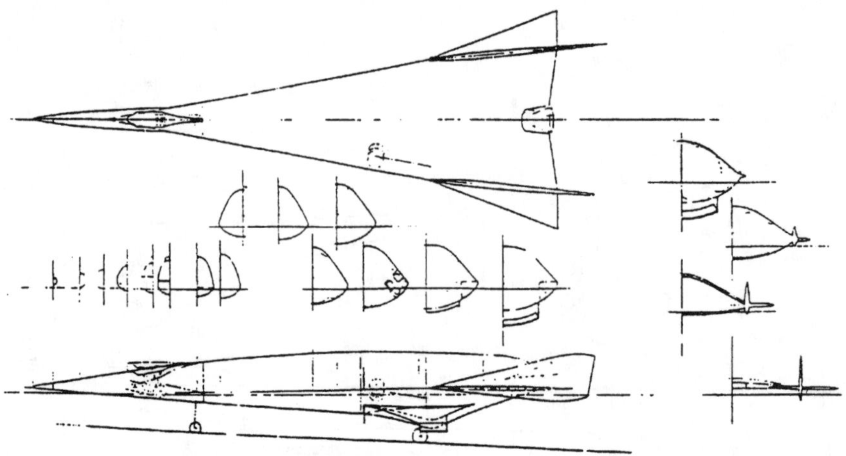

Fig. B6 An All-Body Configuration, Rocket Low Speed Accelerator Convertible Scramjet Test Vehicle. Horizontal Takeoff and Landing (HTOL) 5 minutes test time at Mach 12.

Length	**29.6 meters**
Payload	**453.6 kg**
$S_{planform}$	**123.4 m²**
τ	**0.160**
S_{wet}	**331.3 m²**
L_{to}	**497 kg/m²**
Wstr/Swet	**28.05 kg/m²**
Wtps/Swet	**4.74 kg/m²**
Structure	**7.9 tons**
TPS	**1.6 tons**
Equipment	**2.19 tons**
Propulsion	**4.18 tons**
Empty Weight	**15.2 tons**
W_{OE}	**16.5 tons**
Propellant	**44.8 tons**
Gross Weight	**61.3 tons**

The resulting aircraft rather resembled the Canadian CF-106 Arrow with very large inlets and did not offer a significant research value with regard to fully integrated propulsion systems. It had a W_{OE} of 21.98 tons and a gross weight of 48.5 tons. By comparison the selected Mach 12 air-launched, rocket accelerated, dual-mode scramjet research vehicle had a W_{OE} of 13.97 tons and a gross weight of 23.7 tons.

The following four figures, B4–B7, show the progressive reduction in research vehicle size as transonic and low supersonic acceleration is improved. Examination of the trends in these four figures can give insight as to the difficulty in arriving at an airbreathing launcher if the propulsion system is selected a priori. In the figures, *W*tps/*S*wet is the weight of the

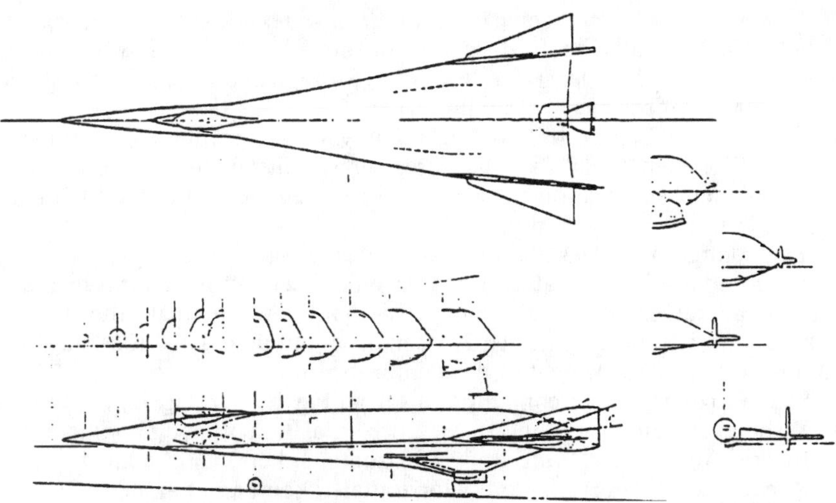

Fig. B7 An All-Body Configuration, Rocket Low Speed Accelerator Rocket Boosted, Convertible Scramjet Test Vehicle. Air-Launched with Horizontal Landing (ALHL) 5 minute test time at Mach 12.

Length	**24.3 meters**
Payload	**453.6 kg**
$S_{planform}$	**79.0 m^2**
τ	**0.157**
S_{wet}	**217.4 m^2**
L_{to}	**400 kg/m^2**
Wstr/Swet	**26.15 kg/m^2**
Wtps/Swet	**4.81 kg/m^2**
Structure	**4,72 ton**
TPS	**1.61 ton**
Equipment	**2.34 ton**
Propulsion	**2.59 ton**
Empty Weight	**9.72 ton**
W_{OE}	**10.8 ton**
Propellant	**20.8 ton**
Gross Weight	**31.6 ton**

thermal protection system mounted over the aluminum/composite structure per unit wetted area, and *W*str/*S*wet is the weight of the structure per unit wetted area.

This configuration is the largest, heaviest airframe with the least propellant quantity. The only afterburning turbojet accelerated configuration to converge and had the lowest acceleration margin of any of the configuration concepts in the transonic regime. This propulsion configuration is very sensitive to drag changes. Effective I_{SP} (EI_{SP}) or acceleration I_{SP} based on thrust minus drag was very poor despite high ISP turbojet propulsion.

Absence of usable cryogenic fuel volume in wings adds to size and weight. This configuration is the most slender for the volume, i.e., it has the lowest Küchemann τ and the least usable volume. Storage of low density, cryogenic hydrogen is not practical within thin wings.

The wing loading on this configuration was the maximum for a runway takeoff. The high thrust rocket assures runway liftoff (aids in nose wheel rotation) and acceleration through transonic regime, but carried oxidizer weight is a heavy burden at takeoff.

This configuration was the smallest, lightest airbreathing configuration. It represented a solution that assures transonic acceleration and the smallest, lightest airframe. Perhaps it is not unreasonable to eliminate the runway takeoff and air launch. The An-225 can accommodate a 300-ton launcher, almost ten times the weight of this demonstrator.

Summarizing the four configurations from Fig. B4–B7, Fig. B8 follows.

Turbojet Accelerator: Wing-Body, Horizontal Takeoff and Landing.

Rocket Accelerator: Wing-Body, Horizontal Takeoff and Landing.

Rocket Accelerator: All-Body, Horizontal Takeoff and Landing.

Rocket Accelerator, All-Body, Air-Launched Horizontal Landing

There is one vehicle-propulsion concept combination not shown. That is the all-body configuration with turbojet accelerator. The wing-body configuration has a transonic acceleration margin of just over 0.1 g. (0.98 m/sec^2). The all-body acceleration is minus 0.6 g (-5.88 m/sec^2) at Mach 1.2. Thrust minus drag is zero at Mach number 0.98. The rocket-boosted all-body has twice that in positive acceleration, 1.2 g, or 11.7 m/sec^2. An investigation into the transonic acceleration will identify the shortcoming of high speed turbojet aircraft in the transonic flight regime.[4]

All the afterburning turbojet/fans used in the HyFac study were conventional installations where the airflow to the compressor increased in temperature and pressure. At about Mach number 1.8, most compressors reach the point where the corrected speed for design efficiency equals the mechanical rotational speed limit of the compressor. At Mach numbers above this point the mechanical speed is constant and the corrected compressor speed $\left(N/\sqrt{T/288K}\right)$ decreases. If there is a cryogenic heat exchange in the inlet between the inlet exit and compressor entrance, then the temperature of the air entering the compressor can be kept at that for best corrected speed, up to the Mach number limit where the heat exchanger can no longer keep the temperature within limits. When such a turbojet is thermally integrated with a rocket, there is no longer a transonic acceleration deficiency. This is not a classic definition of a turbojet.

The underside of the airbreathing vehicle is an integrated propulsion system. The beginning of the inlet is the nose of the vehicle. At the inlet design Mach number, the bow shock is on the cowl lip (shock-on-lip). In principle, the maximum operating Mach number is the shock-on-lip Mach number. Equation (B1) defines the Cowl Mach Number, that is the flow that defines the cowl flow. At a Cowl Mach number of 2 or greater, the cowl lip shock should be attached to the cowl lip.

Turbojet Accelerator: Wing-Body, Horizontal Take Off and Landing.

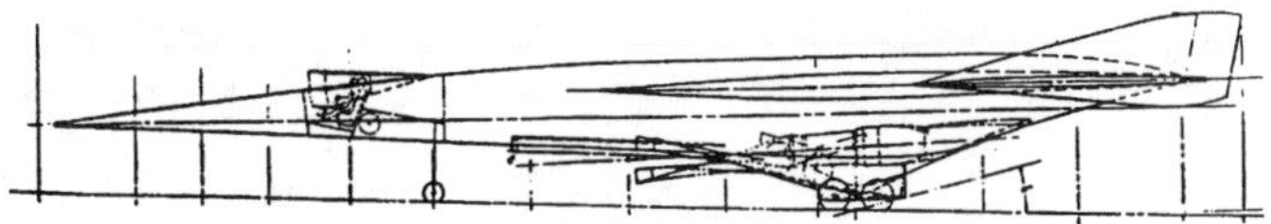

Rocket Accelerator: Wing-Body, Horizontal Take Off and Landing.

Rocket Accelerator: All-Body, Horizontal Take Off and Landing.

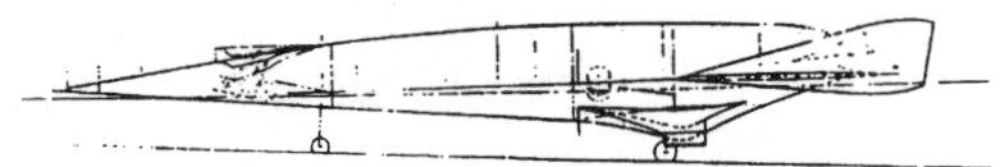

Rocket Accelerator, All-Body, Air Launched Horizontal Landing

Fig. B8 Comparison of Propulsion Sized Aircraft.

References

[1]"Hypersonic Research Facilities Studies," Vol. II, Phase I, *Preliminary Studies,* Pt. 2, "Flight Vehicle Synthesis," NASA Contract, NAS2-5458, NACA CR 114324, Oct. 1970 (declassified Oct. 1982).

[2]"Hypersonic Research Facilities Studies," Vol. II, Phase II, *Parametric Studies,* Pt. 2, "Flight Vehicle Synthesis," NASA Contract, NAS2-5458, NACA CR 114326, Oct. 1970 (declassified Oct. 1982).

[3]"Hypersonic Research Facilities Studies," Vol. IV, Phase III, *Final Studies,* OART 1, "Flight Research Facilities," NASA Contract, NAS2-5458, NACA CR 114327, Oct. 1970 (declassified Oct. 1982).

[4]Townend, L., "Base Pressure Control for Air-Breathing Launchers," AIAA Paper 90-1936, July 1990.

Chapter 17

Scramjet Flowpath Integration

Paul J. Ortwerth*
Boeing Airplane Company, Rocketdyne Propulsion and Power, Albuquerque, New Mexico

I. Background

A. Scramjet-Powered Vehicles

1. Philosophy

SCRAMJET-POWERED vehicles are effectively surfing on a high-pressure combustion wave. This chapter will deal with the wave nature of scramjet and flowpath integration from nose to tail of the vehicle. The power needed is not free, as it is for the surf board but is generated by the combustion of the onboard fuel supply with captured air generating a supersonic combustion wave system. The first priority will be to study the mechanics of the scramjet flowpath and the efficiency of thrust generation. The engine is only part of the total vehicle flowpath and one must be concerned with the drag of the nonpropulsive flowpath. The ratio of thrust to drag directly determines effective specific impulse, and sizing the engines is crucial. If all the air of the total flowpath were captured based on frontal area of the bow wave at the cowl plane, then only cowl drag, drag due to lift, and trim drag would remain. At first sight, if engines did not weigh any more than the airframe, all vehicles would prefer maximum capture. Even then the optimal vehicle may use fewer engines because smaller engines obtain larger expansion ratio and efficiency.

Thus, minimization of propellant fraction required leads to selection of slender, minimum drag shapes for the rest of the vehicle. However, this can lead to low volumetric efficiency or high surface area per unit volume concomitant high specific weight per unit volume, and therefore, low propellant fraction available. Thus, closure of the integration process may yield a larger vehicle to match propellant fraction required to propellant fraction available at low drag than a more volumetrically efficient, higher drag vehicle.

Increased fuel density can offset low volumetric efficiency at a cost of lower cycle fuel efficiency since the thrust per pound of air capture is primarily determined by the oxygen content of air. Another feature of fuel selection is the fuel's coolant capacity. The engine flowpath must be so thermally balanced so that all surface heat is completely transferred to the fuel as in regeneratively cooled rocket engines. Thus, fuel selection has an impact on engine flowpath cycle optimization and/or the upper Mach limit of stoichiometric operation through thermal balance constraint.

A minimum size vehicle is desired for the required payload, range, time of flight. Scaling laws are required for performance, structural thermal protection of the aircraft's weight and the engine weight. Sizing the vehicle and its propulsion system are overall goals of integration.

Thus, integration is an expert search process for an optimal set of airframe and propulsion configuration specifications: *I*sp, air capture, fuel, drag, lift-to-drag (L/D), trim drag, volumetric efficiency, propellant fraction, and thermal balance to size the overall system. The process begins by drawing the future system's configuration lines and then walking around in that design space. The need for a level playing field is paramount when comparing different concepts. A direct comparison of size and performance interactions directly hooked to configuration specifications is required to select the preferred concept.

The goal presented in the following work is to outline an approach to directly relate the geometric flowpath lines to propulsion cycle performance, volumetric efficiency, weight and propellant fraction available, aerodynamic efficiency and trim, and, finally, the thermal loads and thermal protection system requirements. To keep the study scope tractable, major emphasis is placed on relating the configuration lines to the propulsion cycle performance. The work is best served by using explicit formulations with an appropriate process-based ratio of specific heats.

This way, major configuration problems that arise with the flowpath can be identified early and solved in the concept evaluation phase of the design process. Because some applications will require takeoff and vacuum thrust, the internal flowpath is assumed to be a single propulsive duct operating a rocket-based, combined cycle (RBCC) engine (Ratekin (1999), Siebenhaar (1999)).

Historically, advanced vehicle design has been enabled by advanced propulsion. Early in aviation history, this was well understood and companies were vertically integrated with operations, air frames, and propulsion in the same house. This was considered monopolistic and eventually disintegrated. The current practice of separating propulsion and airframe design is not well-suited to scramjet integration and returning to an integrated company may be necessary.

2. *Flowpath of an Integrated Scramjet Engine*

To help define terms, a flowpath is shown in Fig. 1 in which the scramjet is integrated to the vehicle fuselage inside a cowl located somewhere aft of the midplane. The vehicle forebody and aftbody are major parts of the compression and expansion surfaces of the propulsion cycle. The flowfield could be three- or two-dimensional (3D or 2D), depending on whether the fuselage cross section is flat or round, or perhaps of a wave-rider shape. While the forebody may have multiple turns for added external compression, for simplicity only a single ramp or cone is depicted in Fig. 1. The same comment applies to the nozzle specifications.

We assume a large bluntness at the fuselage nose rather than a sharp nose in order to emphasize that the flowfield and shock position are determined by distance measured by nose diameters on the forebody. Bluntness also has a strong affect on transition, boundary-layer heating, and displacement of the forebody flow entering the cowl. Thus, one sees that even the vehicle's nose affects the cycle performance shown in Fig. 1. Many interactive configuration specifications are depicted in Fig. 1. The goal of the unified integration approach is to include the effects of all their specifications.

The engine's internal flowpath, depicted in the five internal components, is 3D for modular sidewall, compression engine types and 2D designs when flow splitters are installed. A boundary-layer control bypass system is not used for the scramjet, but is employed at lower speeds to control the isolator standoff pressure.

The 76 specifications included relative to Fig. 1 are geometric except for the material life-cycle design temperatures for the various components. These temperatures are very important as one can only fly as fast as radiation and fuel heat transfer can help cool the external and internal engine flowpath. Some specifications directly define the cycle compression ratio, combustion process, and the expansion ratio. Others are directly related to the operability and efficiency of flowpath components. Of course, all affect vehicle aerodynamics, packaging, and engine sizing. With so many specifications, "finding the long poles in the tent" is clearly important. Selecting a main design specification can be aided by traditional cycle analysis.

A few remarks regarding these geometric specifications is in order to help orient the reader. In particular, one should recognize that A_O is measured with respect to the water line of the configuration, and if the vehicle flight altitude is such that the water line is at angle of attack relative to the wind axis, then the capture area for shock-on-lip condition is different from A_O. The same remarks apply to the aftbody's projected exit area A_E; the overall expansion ratio is different from ε_O. The inlet spill flow through area A_{SPILL} generates drag and lift forces and both the spill mass flow rate and spill angle are needed for the vehicle's force balance. Thus, the internal flowpath generates normal forces and moments and should be accounted for in the vehicle trim accounting, not just as the force on the external surfaces.

3. *Integration Requirements*

The task is to establish the mission's performance requirements for the integrated system and the propellant mass fraction. The analysis should include both the cruise and the acceleration phases of the flight.

The starting point for the analysis is the dynamic equation for the moving

a)

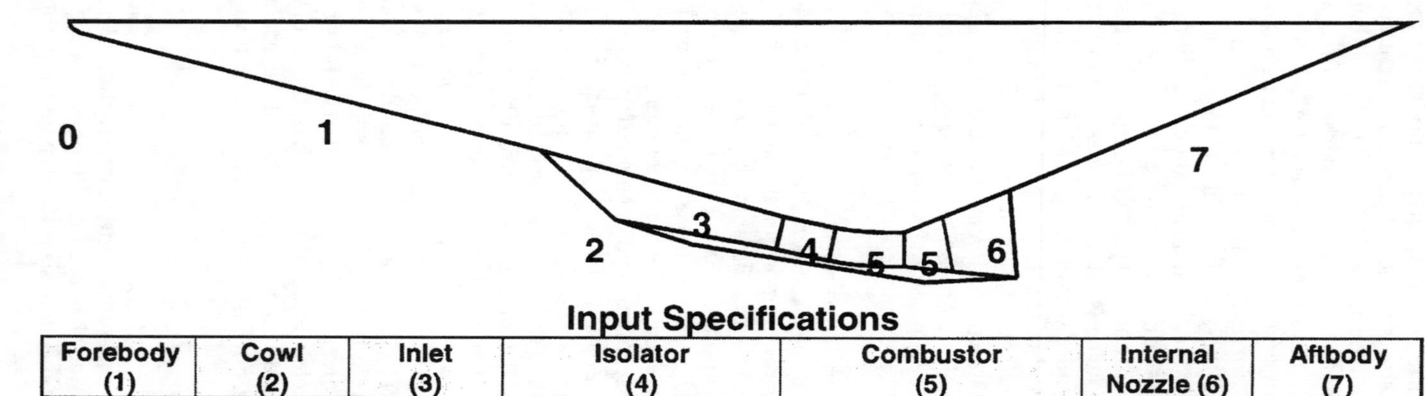

Fig. 1a Integrated Scramjet Vehicle Flowpath and Components

Fig. 1b Boeing Rocketdyne RBCC Engine Module Asembled for Free Jet Test, circa 1997 sponsored by NASA, MSFC.

vehicle, which is the instantaneous force balance equation. Details of the analysis are presented in Appendix A.

For the cruise flight, as discussed in Appendix A, the well-known Breguet Eq. (A3) is obtained by equating the propulsive thrust required overcome the drag. The thrust is generated by combustion of a fuel intrinsic to the vehicle; the drag corresponds to the value at the flight altitude, and speed under level flight conditions. Also, the vehicle's altitude is assumed to increase as fuel is burned off so that the L/D ratio is a constant optimum value. Defining overall flight efficiency as $(V \cdot I_{sp})$, where V and I_{sp} refer to flight speed and fuel specific impulse, respectively, the flowpath has to be designed for efficiency.

It will be shown later that $V \cdot I_{sp}$ is nearly constant over a wide range of flight speeds. Thus, the optimum L/D, which is a function of vehicle volume and speed, becomes the basis for selecting the cruise speed. Of course, one must recognize that in a well-integrated scramjet much of the vehicle drag is canceled at the shock on cowl lip-operating speed. In this connection, a waverider vehicle with a high L/D ratio of 7 to 8, and hydrocarbon fuel with a high propellant mass fraction could be possible, noting the high density of the fuel. Though the waverider has poor volumetric efficiency, the product $[(L/D) \cdot \text{In}\,(W_o\,/\,W_f)]$ would be close to the maximum, where subscripts o and f denote the initial (takeoff) and final weights.

Considering accelerating flight (analysis presented in Secs. B–D, of Appendix A), the propellant mass fraction for a rocket motor with nearly constant I_{sp} is given by Eq. (A8), while that for an engine such as the scramjet, with nearly constant $(V \cdot I_{sp})$, the mass fraction is given by Eq. (A16). These two results permit a nominal comparison of a single-stage-to-orbit (SSTO) vertical takeoff rocket with a scramjet. However, a scramjet engine requires vehicle acceleration to scramjet takeover speed, and, in general, a rocket would be needed for orbit insertion when the scramjet is no longer efficient at high altitudes. Such low speed and orbital insertion requirements would suggest the use of an RBCC, which combines the static thrust and orbital insertion capability of a rocket, with high thrust-to-weight ratio, and the fuel efficiency of a scramjet.

Using the results of Appendix A, one can construct an illustration (as in Fig. 2) to compare the weight history of vehicles with rocket accelerators and vehicles with an RBCC engine accelerator that includes a scramjet during some part of the flight. Such comparisons are always based on the flight Mach regime of scramjet operation and the performance of the rocket motor on either side of scramjet usage. However, the results presented in Fig. 2 lead to some important aspects of engine flowpath design.

Based on Fig. 2, the SSTO mass fraction required for both the rocket and the scramjet are a challenge to successful vehicle-propulsion integration. The rocket-only reusable SSTO, using an average I_{sp} of 378 sec, requires a dry weight fraction of 10%, which must include recovery wings, landing gear, propellant residuals for landing (if vertical takeoff and landing are adopted), and payload. The system's feasibility is enhanced only by very lightweight, advanced, state-of-the-art structural and thermal protection design. The scramjet, on the other hand, requires the same system for reusability; but, owing to the higher fuel efficiency, allows a higher dry weight fraction, about 20%, and the feasibility of the system is enhanced. The gains are due to the high RBCC scramjet propulsive efficiency in the range of 0.45. Yet another

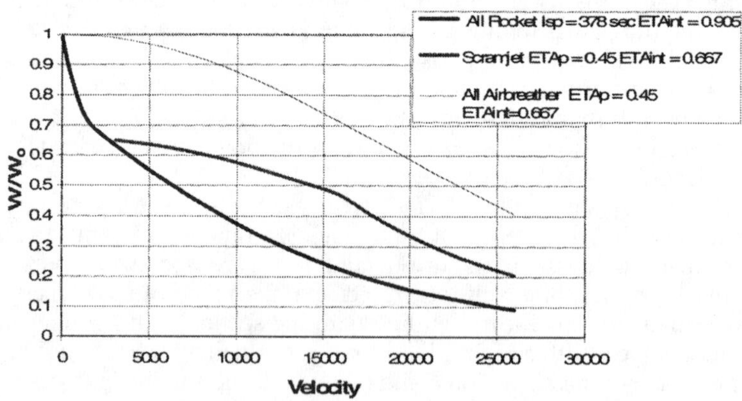

Fig. 2 SSTO Vehicle Mass History Comparison for Power by Rocket and RBCC Engines.

advantage occurs because (compared to the rocket-only mode) the RBCC enjoys a substantial advantage in nozzle expansion ratio up to $\varepsilon = 600$ and an added 20-sec I_{sp}.

Integration of a scramjet for an SSTO still requires aggressive structural design, use of lightweight materials, and propulsion performance advances. This is seen by considering sizing an engine for a fixed vehicle airframe, i.e., examining thrust-drag ratio as it is controlled by thrust and directly affects effective I_{sp} through integration efficiency, η_{int}. Increasing the number of engine modules increases thrust, weight, and expense, thus reducing available propellant fraction while propellant fraction decreases, which raises costs and, hopefully, payload. However, when one considers payload, fixed improvements in both I_{sp} and η_{int} reduce the total vehicle size, therefore an even larger savings in cost is achieved by overall reduction in takeoff weight and dry weight of aircraft and propulsion. Payload is an important design requirement.

Returning to the issue of successfully executing propulsion-vehicle integration, one must address issues inherent in the practice of dividing the design responsibility for propulsion and vehicle manufacture to two different companies. This tends to divide the flowpath into two parts: the external vehicle flowpath and the internal propulsion flowpath. The flowpath design is assigned to two separate companies. Generally, the airframe company is responsible for the drag and engine size or mass capture from the atmosphere, and the propulsion company is responsible for propulsive thrust generation and I_{sp}.

There is, however, a difference between low speed flight vehicle integration and hypersonic flight vehicle integration: With a turbojet, the airframe company is responsible for the inlet; the propulsion company is responsible for the engine and the thrust nozzle. With a scramjet, the propulsion company is responsible for all components involved in the internal flowpath, from inlet entry to nozzle exit. However, sizing the engine is the responsibility of the vehicle company, which is the system integrator. The internal flowpath and the (external) vehicle flowpath must be fully compatible for successful vehicle integration.

The division of the flowpath into propulsion and airframe parts in a scramjet is also of interest in regard to force accounting. Thus, broadly, all aspects of vehicle drag become the responsibility of the airframe company and all aspects of propulsion become the responsibility of the propulsion company. Such division provides some clarity in responsibilities, but also necessitates a careful division of efforts; e,g., an electronically collocated team is needed. In particular, two main difficulties are that 1) while the inlet internal flowpath is the responsibility of the propulsion team, the internal turn in the flowpath may cancel a part of the forebody drag associated with the forebody downturn; and 2) while the aftbody completes the expansion part of the propulsion cycle, it generates significant lift and moments. Note that this is substantially affected substantially by thrust nozzle exhaust plume interaction with the external vehicle flowpath.

The propulsion integration team is authorized to provide pitchplane 2D force and moment accounting. Attempting force accounting while searching for closure, it is important to watch the drag per unit mass capture as well as the thrust per unit mass capture. These accounting parameters provide the visibility for keeping the integration efficiency $\eta_{int} = (1 - D/T)$ high, and also the direction of propellant fraction required, λ_{pr} and the incurred cost of propellant fraction available.

In summary, establishing the performance of an integrated flowpath is a critical task for both the airframe and the propulsion designers and it is extremely important for the two groups to perform the task of integration together.

Another integration requirement in a flight vehicle with pilot, crew, and passengers is adequate cooling capacity to cool the personnel and vehicle environment with an adequate margin. It is important that the vehicle have a hypersonic mission abort capability to enable a return to base. This probably requires a zoom capability to exo-atmospheric conditions for flight system safety and an auxiliary rocket engine and fuel supply to allow deceleration before a cool, safe decent trajectory. This will have an effect on cruise speed, and necessarily on size or range for a given payload, but as long as the overall vehicle efficiency given by $[(L/D)\ V \cdot I_{sp}]$ remains constant, this would be feasible.

B. Flowpath Optimization

1. *Classical Analysis*

An important foundation starting point for flowpath optimization was established by Builder (1964). It pertains to the adiabatic Brayton cycle illustrated in Fig. 3 in entropy-enthalpy space (Mollier diagram). Some aspects of this cycle's analysis are presented in Appendix B.

The chief parameters in the Brayton cycle are the thermal compression ratio, $\psi = H_3 / H_o$, the heat energy supplied by the combustion of fuel, $\Delta \psi_c = \Delta H_c / H_o$, and the efficiency of the inlet compression and thrust nozzle expansion process. The air specific impulse, I_A, can be expressed by those parameters for given flight conditions (Mach number and altitude) by the relation Eq. (B3). Note that the second term under the square root is very small when the flight Mach number is high. Thus, by series expansion, one can write the following for specific power:

$$I_A \cdot V_o = (\gamma R T_o)\left(\frac{\psi - 1}{\gamma - 1}\right)(\eta_c\,\eta_e)\left[\left(1 + \frac{\Delta\psi_c}{\psi}\right) - 1\right] \tag{1}$$

which then becomes a function of T_o or altitude. If γ, η_c, and η_e can be assumed to be constant, $(I_A \cdot V_o)$ also may be considered to be a constant. Eq. (1) thus becomes an important scaling law for the cruise operation of scramjet engines at high speeds. The scaling law may be stated as follows: for a scramjet engine operating at given (V_o, T_o) conditions, the air specific impulse is a function of ψ and $(\Delta \psi_c / \psi)$.

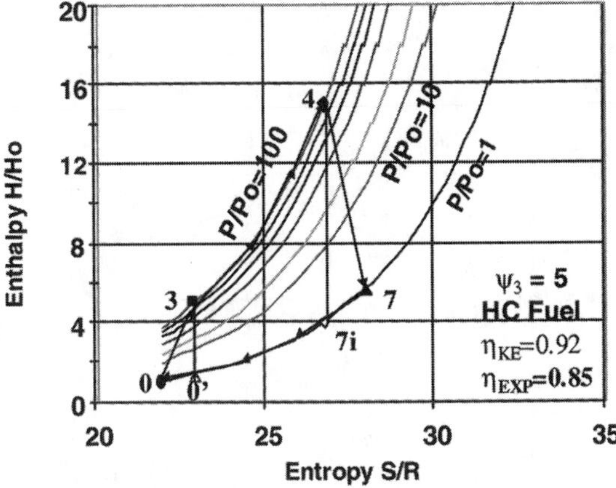

Fig. 3 Adiabatic Brayton Cycle Scramjet (ABCS) on Mollier Chart.

2. *Ideal Adiabatic Brayton Cycle Performance*

Several interesting aspects of cycle performance can be illustrated using the Builder analysis. The Builder analysis, however, suffers from a number of built-in assumptions: 1) constant value of ratio of specific heats, 2) neglect of fuel mass, and 3) constancy of diffusion and expansion efficiency. Some deficiencies can be corrected. Meanwhile, within the framework of the simple analysis, important conclusions can be arrived at that are of fundamental importance in the conceptual design of a flowpath; several examples follow.

Assuming that the thermal ratio of air diffusion is given by the ratio of stagnation to static temperature, that is,

$$\psi = H_{To} / H_o \tag{2}$$

a limiting case of Builder, one can establish the ideal upper limit air specific impulse as a function of flight Mach number for a given combustion energy input; for example, considering hydrogen and jet fuel as the fuels in two different cases, one obtains the predictions illustrated in Fig. 4. However, the variation of air specific impulse would strictly apply only to a ramjet, although the variation with Mach number is typical of a dual-mode ram/scram engine; both the peak value of I_A and the decay of I_A with respect to Mach number are similar to what is realizable with a ramjet followed by a scramjet. One can also show that the increase of thrust per unit atmosphere per unit area, F/P_0A_0 [see Eq. (B6)], with respect to Mach number, as shown in Fig. 5, also is similar. Finally, Fig. 6 shows the manner in which the chemical conversion efficiency, defined by the ratio of thrust work generated to the heat of combustion of fuel [see Eq. (B7)], also increases with Mach number. Above about Mach 8, the chemical conversion efficiency of the adiabatic Brayton cycle scramjet is superior to that of any other cycle, and this is an important argument for the development of scram combustors and scramjets.

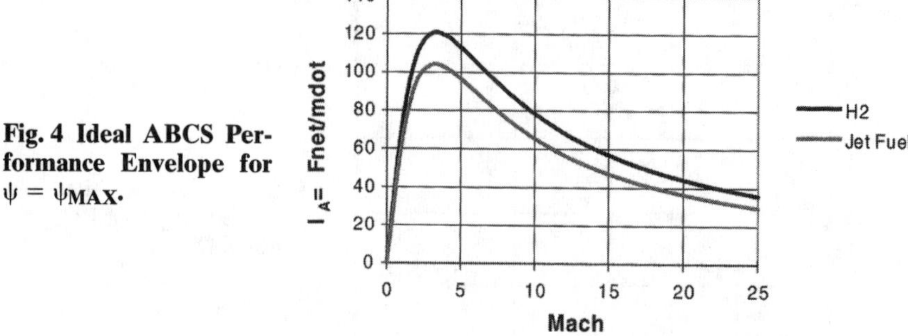

Fig. 4 Ideal ABCS Performance Envelope for $\psi = \psi_{MAX}$.

Practical Brayton cycle. The foregoing results of the analysis of the ideal adiabatic cycle scramjet offer a conceptual cornerstone for scramjet engine integration. Unfortunately, the formal elegance of these results will have to be left behind as more true-to-life processes are considered. The problem with establishing the efficiencies η_c and η_e, or their equivalent, is an example. The efficiencies are related, through aerodynamic analysis, typically 3D Navier–Stokes computational fluid dynamics methods, to geometry, Mach number, and Reynolds number, as a function of flight altitude. Very high convergence accuracy, on the order of 10E-5 for mass flow and 10E-6 for momentum, are required for accurate post-processing of the solution outputs from computations. This is a resource hurdle for most conceptual design studies, which often is bypassed with the result that the outcome of the studies are of rather low value. Therefore, a different and more direct approach is needed, one that is more reliable and at the same time includes the major physics of scramjets.

In reality, the losses in the compression (inlet) and expansion (nozzle) components of the engine consist mainly of shock and under-expansion losses. This can be seen in Fig. 7, where I_A is shown as a function of inlet exit pressure (or, equivalently thermal ratio, ψ) at flight Mach number 10, assuming $\eta_c = 0.92$ and $\eta_e = 0.95$. It can be observed that I_A is zero at two values

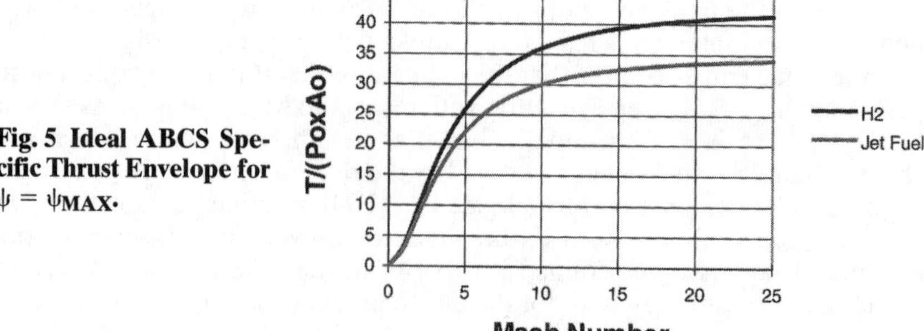

Fig. 5 Ideal ABCS Specific Thrust Envelope for $\psi = \psi_{MAX}$.

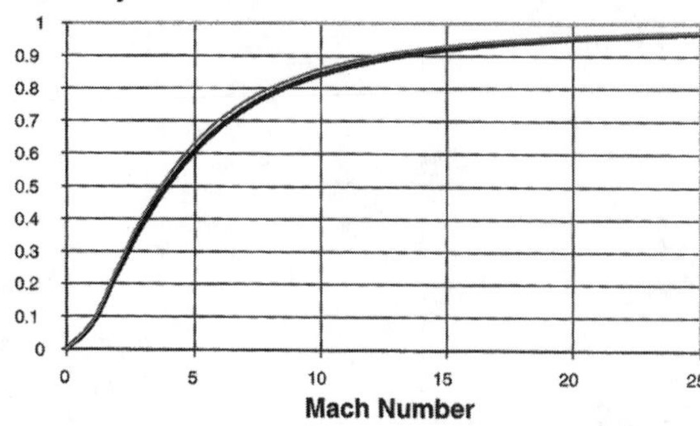

Fig. 6 Ideal ABCS Chemical Energy Conversion Efficiency for $\psi = \psi_{MAX}$ Negligible Fuel Dependence.

of ψ, one at ψ equal to unity and the other at ψ equal to $(\psi_{OPT})^2$ where ψ_{OPT} corresponds to ψ for maximum I_A. There are two curves in Fig. 7. The upper curve is based on equilibrium chemistry and a fixed value of $\Delta\psi_c$. The maximum value of I_A occurs in the upper curve at ψ nearly equal to the low value of 4, a value also obtained from Builder's analysis. The lower curve presents the value of I_A realized with ψ equal to ψ_{OPT} required for maximum I_A, however, including combustion with equilibrium chemistry and thrust nozzle expansion with frozen chemistry. The optimum value of ψ in this case is even lower.

The optimum value of ψ with component and chemical losses is much lower than the full stagnation compression of the ideal engine [see Eq. (2)]. The optimum value is a function of flight Mach number and altitude with a given fuel. At the lower supersonic Mach numbers, during subsonic operation of the combustor or in the ramjet mode, normal shock losses occur, and this leads to an optimum value for ψ. As the Mach number increases, the losses, including the chemical losses, become too large and one needs

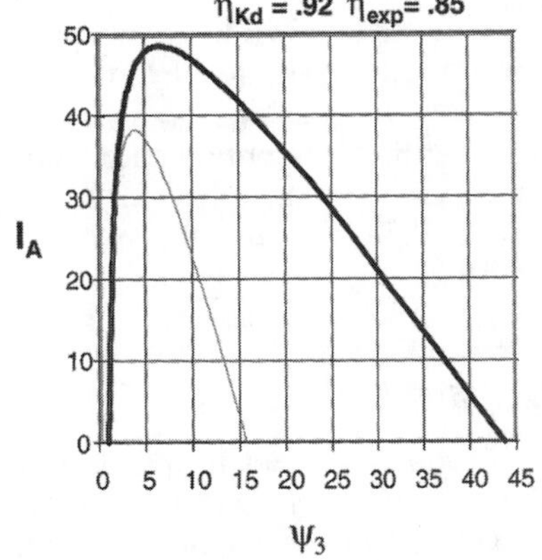

Fig. 7 Practical ABCS Cycle Shows Performance Maximum at $\psi = \psi_{OPTIMUM}$. Lower Curve With Frozen Nozzle Expansion.

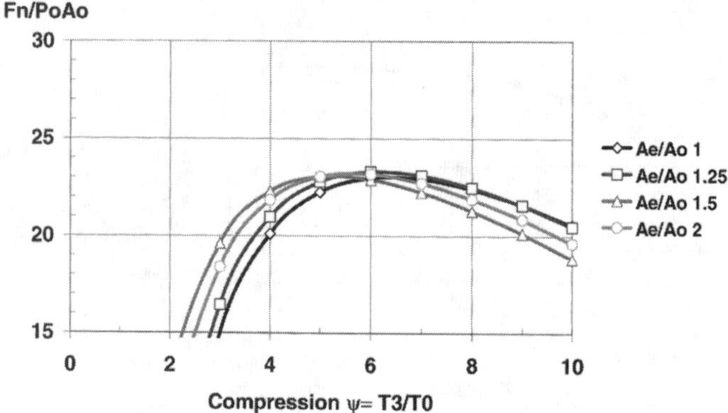

Fig. 8 Optimum Thermal Compression ψ_{OPT} is a Function of overall Geometric Expansion.

to retain the flow in the combustor at supersonic speeds, that is, transition to scramjet operation. Thus, the transitioning from subsonic to supersonic conditions in the combustor is determined entirely by consideration of realizable efficiency. It is important, throughout the operation, to maintain the highest effective combustion heat. This leads to the natural transition from ramjet to scramjet modes in the Mach 6 to 8 range. The optimal value of ψ, in turn, becomes determined by such considerations.

Limitations of Brayton cycle. A closer inspection of the Brayton Cycle Mollier diagram in Fig. 3 reveals a disturbing fact about the expansion losses called underexpansion losses. If one traces the 100 atm combustor pressure to ambient and then stops at the 10 atm bar, the area inside the loop decreases about 40%. In real hardware, the high combustor pressure ratio is still appropriate but the cycle expansion is limited by geometry to a pressure ratio about 10 times ambient. If the cycle is truncated at that expansion, a very large loss of available work would occur (Fig. 8). This limitation is addressed by changing the heat addition process to recoup the expansion ratio, but as referenced to $\Delta\psi$, available in expansion for overall area expansion ratio limits of practice. This is an important feature of the current view in the community of flowpath designers.

This introduction covers some important issues in scramjet flowpath technology, particularly the desirability of finding the optimum thermal ratio; setting specific performance limits for practical engines; and the need to carefully develop the combustion cycle to offset underexpansion and the friction drag not yet covered that can further degrade performance.

II. Energy Analysis

There is a fundamental difference in the dynamics of jet propulsion at lower flight speeds and at hypersonic flight speeds, Ortwerth (1991). At lower speeds, large changes arise in relative velocity in the engine flowpath, while at hypersonic speeds, they are of second order. This difference in flow velocity changes profoundly affects the manner in which compression and expansion processes are viewed under hypersonic flight conditions. One may say, by graphical analogy, that at lower speeds, compression is arranged along the flowpath, or longitudinally, while at hypersonic speeds, it is arranged transversely. This can be seen by considering the relation between velocity and area changes for supersonic flow under adiabatic conditions:

$$\frac{\mathrm{d}V}{V} = \frac{1}{M_o^2 - 1} \cdot \frac{\mathrm{d}A}{A} \tag{3}$$

Clearly for M_0^2 near unity, large changes occur in V for (small) area changes. The reverse happens at hypersonic speeds; as $M_0 \to \infty$, area changes produce virtually no changes in velocity.

Considering one-dimensional approximation for the flowpath, one can make several interesting observations. For instance, considering mass conservation, one can write

$$\frac{\mathrm{d}\rho}{\rho} + \frac{\mathrm{d}A}{A} + \frac{\mathrm{d}V}{V} = 0 \tag{4}$$

At low speeds, $\mathrm{d}A/A \to 0$ yields the relation

$$\rho_2 / \rho_1 = V_1 / V_2 \tag{5}$$

At hypersonic speeds, $\mathrm{d}V/V \to 0$ yields the relation

$$\rho_2 / \rho_1 = A_1 / A_2 \tag{6}$$

and, thus, compression is entirely by change of geometry. These differences have immediate changes in propulsion paradigm or viewpoint, as illustrated in Fig. 9, where compression of the capture stream can be considered as changing from the longitudinal to the transverse mode in the low and high speed regimes.

One useful approach from the preceding is to separate the velocity changes, in examining drag or thrust work done on or by the fluid (air) in the propulsion process, from changes in internal energy. This may be viewed as energy partitioning into kinetic energy and internal energy. To develop such an analysis, consider the following:

Momentum conservation:

$$\int_A \left(\tilde{\tilde{P}} + \rho \boldsymbol{V}\boldsymbol{V} \right) \cdot \boldsymbol{n} \cdot \mathrm{d}s = \int_\Sigma \left(\tilde{\tilde{P}}_w + \tau \right) \cdot \boldsymbol{n} \cdot \mathrm{d}s \tag{7}$$

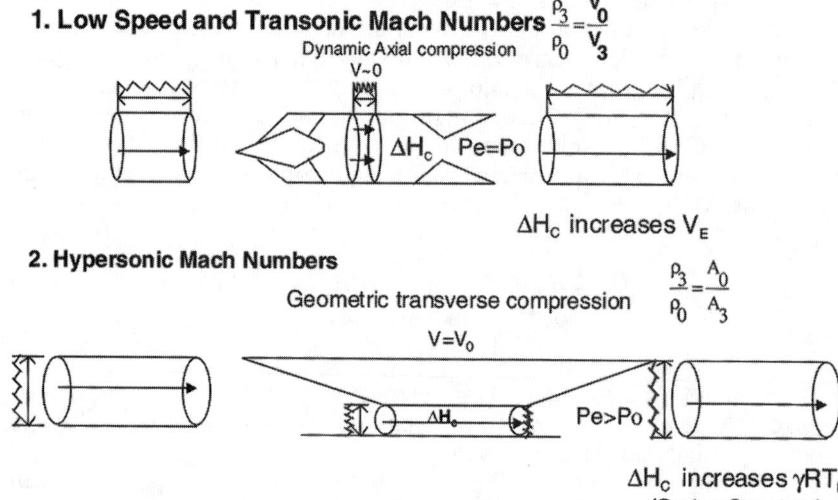

Fig. 9 Jet Propulsion Paradigm Shifts with Mach Number.

Energy conservation:

$$\int_A \rho \boldsymbol{V} H_t \cdot \boldsymbol{n} \cdot \mathrm{ds} = \int_\Sigma \boldsymbol{q}_w \boldsymbol{n} \cdot \mathrm{ds} \tag{8}$$

where

$$H_T = H + \frac{V^2}{2} \tag{9}$$

Equation (7) gives the relation for determining the stream thrust.

It is noted that the integration of the momentum equation is not universal, since the flow surfaces of the inlet and the exit of the flowpath need not be planar or even contiguous; this is the reason that Eqs. (7) and (8) are left in dyadic form. However, in order to proceed, one may consider that the stream momentum dyadic trace with the surface normal is vector differential, which defines the normal surface at any flowpath station by aligning the surface normal with the integrated mass velocity vector. This procedure will be adopted in the following discussion. Thus, one may write

$$\int \rho \boldsymbol{V}\boldsymbol{V} \cdot \boldsymbol{n}\, \mathrm{ds} = \boldsymbol{V} \frac{\mathrm{d}m}{\mathrm{d}t} = \boldsymbol{V} \cdot \dot{m} \tag{10}$$

This relation formally includes three components of stream thrust. Thus, considering the pitch plane forces, one can write the following for differences of stream thrust, along the coordinates normal and parallel to the original wind axes, in terms of lift and drag forces:

$$F_2 \sin \theta_2 - F_1 \sin \theta_1 = L \tag{11}$$

$$F_2 \cos \theta_2 - F_1 \cos \theta_1 = D \tag{12}$$

where

$$F = PA + \dot{m}V \tag{13}$$

and subscripts 1 and 2 refer to two stations along the flowpath. The flow angles θ_1 and θ_2 should be chosen appropriately. Considering on-axis flow only, the change in stream thrust is equal to the drag on the walls of the engine components. There is a companion equation for conservation of flux of angular momentum and applied torque $\vec{T}$, derived from equation 7.

$$\vec{T} = \vec{R}_2 \times \vec{F}_2 - \vec{R}_1 \times \vec{F}_1 \tag{7a}$$

One can use equation (7a) to find flowpath moments from the component solutions for F and station area centroid R for 1st order results.

To examine thrust work, one recalls the energy balance equation, Eq. (8), and rewrites it in the form

$$\overline{H}_{t2} - \overline{H}_{t1} = \dot{Q}/\dot{m} \tag{14}$$

A. Hypersonic Energy Partitioning

Assuming that

$$V = (V_1 + V_2)/2 \tag{15}$$

and noting that the velocity changes little along the flowpath, one can multiply the stream thrust equation by V and obtain the thrust work from the energy balance equation.

It is observed that

$$V_2V - V_1V = \frac{V_2^2}{2} - \frac{V_1^2}{2} \tag{16}$$

And since $V_1 / V_0 < 1$, $V_1 / V_2 \simeq 1$, $V_2 / V_{exit} < 1$, and $\dot{m}_1 = \dot{m}_2 = \dot{m}$, say, it follows that

$$\left[\rho_2 A_2 V\left(\frac{P_2}{\rho_2}\right)+\dot{m}\frac{V_2^2}{2}\right]-\left[\rho_1 A_1 V\left(\frac{P_1}{\rho_1}\right)+\dot{m}\frac{V_1^2}{2}\right]=D\cdot V \tag{17}$$

Now one can subtract the energy equation from the above, and write the stream enthalpy as the sum of internal energy and flow work. We obtain a surprising result for compressible flow, namely that the kinetic energy terms can be canceled in Eq. (17), and thus

$$e_2-e_1=(-\dot{m}h_2+\dot{m}\frac{P_2}{\rho_2}+\dot{m}h_1-\dot{m}\frac{P_2}{\rho_1})=D\cdot V-\dot{Q} \tag{18}$$

This shows that the total power needed to drive the inlet compression, less the recovered energy, goes directly into internal energy of the working fluid in the cycle and the engine coolant, if any.

In the same fashion, considering the thrust nozzle, under the same approximations as given above, one can write

$$e_2-e_1=\frac{T\cdot V}{\dot{m}}-\frac{\dot{Q}}{\dot{m}} \tag{19}$$

where the thrust is taken to be negative drag, and e represents enthalpy.

The importance of the foregoing relation is that the scramjet cycle can be viewed very much the same as the reciprocating internal combustion cycle in a stationary ground-reference coordinate system; the energy conservation law is the same for the two cycles. Equation (19) may now be rewritten as follows.

$$\frac{D}{P_1A_1}=\frac{\psi-1}{\gamma-1}+\frac{\dot{Q}}{\dot{m}RT} \tag{20}$$

This expression for drag is correct to 1 part in 10^4, and represents what will be referred to as the *hypersonic energy partitioning* (HEP) *principle.*

The HEP Eq. (20) is similar to the unsteady flow analogy useful in external hypersonic problems, or the hypersonic blast wave theory. This equation, it can be stated, is based on the central assumption of velocity changes along a hypersonic flowpath being small, by design and by theory. Other assumptions are generally no different from those of Builder. The HEP is foundational in that, if the pressure and friction forces are calculated for any component through aerodynamic analysis, based on initial conditions and geometry, then the state of the gas at the exit of that component is determined; and those conditions then become the initial conditions for the following component. Component efficiencies are then calculated for reference, but they are not needed *a priori* for flowpath performance analysis. The approach to analysis based on HEP, thus, is a radical departure from the classical cycle analysis methods currently used, and offers a way of avoiding specification of component efficiencies, generally required to high orders of accuracy, which is necessary for accurate prediction of scramjet performance, however, it is well nigh impractical.

It may be of interest to explain the reasoning behind the approach to analysis based on the HEP. In this connection, one can establish a close connection between the classical cycle analysis and the determination of forces along the engine flowpath. The following will illustrate this connection. The flowpath is assumed to be thermally balanced. The thermal loads can be neglected explicitly since they sum to zero.

The inlet drag can be written based on the HEP as follows:

$$\frac{F_{\text{inlet}}}{P_o A_o} = -\frac{(\psi - 1)}{(\gamma - 1)} \tag{21}$$

The thrust of the combustor is expressed by

$$\frac{F_{\text{comb}}}{P_o A_o} = \frac{F_{\text{comb}}}{P_3 A_3} \cdot \frac{P_3 A_3}{P_o A_o} \tag{22}$$

$$= \frac{P_3(A_4 - A_3)}{P_3 A_3} \cdot \frac{P_3 A_3}{P_o A_o} \tag{23}$$

$$= \frac{P_3 A_3 (T_4/T_3 - 1)}{P_3 A_3} \cdot \frac{P_3 A_3}{P_o A_o} \tag{24}$$

$$= \frac{\Delta H_c}{H_o} = \Delta \psi_c \tag{25}$$

The thrust of the nozzle is similarly given by

$$\frac{F_{\text{noz}}}{P_o A_o} = \frac{F_{\text{noz}}}{P_4 A_4} \cdot \frac{P_4 A_4}{P_o A_o} \tag{26}$$

$$= \frac{(\psi - 1)}{(\gamma - 1)} \cdot \eta_c\, \eta_e \left(1 + \frac{\Delta \psi_c}{\psi}\right) \tag{27}$$

The static drag due to atmospheric pressure on the cowl may be written as

$$\frac{F_{\text{cowl}}}{P_o A_o} = (\psi_E - 1) \tag{28}$$

$$= (\psi - 1) - \Delta \psi_c + (\psi - 1)\eta_c\, \eta_e \left(1 + \frac{\Delta \psi_c}{\psi}\right) \tag{29}$$

where $\psi_E = T_7 T_0$.

On adding the forces associated with the different components, one obtains the same net thrust as in the classical cycle analysis approach, namely Eq. (B6) of Appendix B:

$$\frac{F_{\text{net}}}{P_o A_o} = \frac{\gamma}{(\gamma - 1)} \cdot (\psi - 1)\left[\eta_c\ \eta_e\left(1 + \frac{\Delta\psi_c}{\psi}\right) - 1\right] \tag{30}$$

In subsequent sections, the net thrust will be related to the sum of wall pressure integrals, and wall friction integrals, and thus a formal connection will be established between the geometry of the flowpath and the aerothermodynamic parameters. This leads to the design of flowpath geometry under given conditions for specific requirements.

B. Summary and Statement of the Design Problem

The section on flowpath optimization and the current section serve to introduce the concept of optimal cycle parameters, the methodology of hypersonic energy partitioning, and the identification of component design principles. These, then, become the basis for integrating the propulsion system and the vehicle for a given mission. The focus is on integrated flowpath performance analysis and an approach to relate cycle thermodynamic parameters to forces in the different components. The thrust and drag are calculated from aerodynamic analysis of pressure and viscous forces over the geometry of the flowpath.

This approach, therefore, requires an identification of geometry and the extent of changes in geometry at the beginning of the analysis. In terms of geometrical lines and surfaces making up the flowpath, the simplest choice is to start with straight lines and planar surfaces for connecting entry and exit sections of all components. One can then proceed to make modifications to the lines and surfaces for best performance. The initial choice of straight lines and sections is not restrictive, it will be realized, since the final results show small deviations in angles and the curvature needed is thus small. However, it will be useful to remember that curvature may be crucially important locally, for example, in regions of shock coalescence. The identification of geometry and its changes may thus be looked upon as the first task under this approach.

The other step in this approach is to define the parts to be analyzed. It is important that none of the essential flight vehicle system parts are left out. Referring to Fig. 1, starting with the nose of the flight vehicle, one needs to proceed to the tail in a contiguous and systematic fashion, covering each of the following components.

The second step is to define the parts to be analyzed. As stated before parts or component integration requires not leaving anything out that is essential. Starting with the nose of the aircraft or missile one can proceed to the tail in an orderly way and define the following needed components:

1) Forebody of the vehicle. The forebody provides the initial external compression of the propulsion cycle. This is an external flow component which contributes to the lift, drag, and moments of the vehicle. This is a large surface containing a significant volume for fuel or payload. The forebody is radiation or ablatively cooled to reduce the forebody weight and the critical cooling heat load on the fuel.
2) Internal Inlet. The internal inlet provides final compression of the propulsion cycle. This is an internal flow inlet which is actively cooled instead of radiation cooled as are the other internal flow components. A proper tradeoff between external forebody and internal inlet compression is critical of good integration of the scramjet and vehicle.
3) Isolator. The isolator provides the aerodynamic isolation of combustor pressure rise from adversely affecting inlet flow, particularly for low mid-speed operation where a pseudo or normal shock is supported by combustion and is not required above about Mach 8. The isolator is usually a constant area section and may require boundary layer bleed at low speed to minimize length, which is important in keeping the wetted and cooled area small.
4) Combustor. This is the principal component of the engine flowpath operating at the highest pressure and therefore drag and heat transfer rates in the propulsion flowpath. The combustor must provide mixing, flame holding, and combustion heat release over a wide speed and throttle range. Minimization of combustor drag and heat transfer is critical to good performance. A trade of drag and combustion efficiency with length is in principle required. The combustion heat addition effectiveness in the cycle also is a function of the inlet compression ratio through real gas effects, principally through the dissociation level of the equilibrium products.
5) Combustor nozzle transition section or second stage combustor. This section is provided to allow for low speed scramjet or dual mode scramjet operation. At lower speed the heat addition is robust and can unstart the flowpath by thermal choking. Thus, a divergent heat addition section is required, but not at the rapid expansion rate of the internal nozzle, which would quench the second stage mixing and combustion. At high speed, $M > 8$, all combustion is in the main combustor and this transition section is simply an expansion. This lower expansion rate can aid in achieving equilibrium during the initial phase of the expansion.
6) Internal Nozzle. The internal nozzle initiates efficient total expansion of the propulsion flowpath and is designed, along with the transition and external expansion, for some design Mach number. The internal nozzle generally has lower than ideal pressure and thrust at high speed due to the rapid Prandtl Meyer expansions in the internal nozzle. Since this is an active cooled component, the rapid expansion results in lower internal heat load.
7) External Nozzle. The external nozzle provides final expansion and thrust of the propulsion cycle; and since an external flow also generates lift and moment on the vehicle. The external nozzle like the forebody is desired to be radiation cooled for lightweight and reduced cooling fuel flow. During integration the effects of the cowl on vehicle force and moment balance

considerations should be included in the trimmed vehicle performance. Note that in scramjet integration, the effects of the cowl on force and moment balance considerations should be included in trimmed vehicle performance. Two noteworthy factors in this connection are a) the cowl provides significant external lift and drag, and b) the drag rises sharply with deflection angle at high speed and, therefore, engine exit area. Cowl drag rise thus sets a limit to nozzle expansion ratio.

In the following sections of this chapter, the individual elements of the scramjet, namely the inlet, forebody, cowl, isolator, combustor, and nozzle are each examined as part of the engine internal flowpath. One can then address integration and modular design of the propulsion system.

In the analysis of the various parts of the flowpath of a scramjet, it is obviously essential to account for viscosity effects in terms of friction and heat losses. As stated earlier, a general viscous compressible flow analysis may not be effective in conceptual design. Therefore, the analysis is based on Prandtl's boundary layer theory, with all viscous effects confined to the boundary layer while the core flow is treated as inviscid. This suggests separating the viscous forces and heat transfer from pressure forces. One then has a propulsion flowpath with two parts: one part associated with a classical heat addition cycle, consisting of compression, combustion heat addition, and expansion, and a second part associated with a cycle, that may be called the friction cycle, which accounts for friction work, heat transfer, and mass addition, all of them occurring in the wall layer of the propulsion flowpath. The concept of the friction cycle and its implications are reviewed briefly at the end of Sec. A in Appendix C. The propulsion flowpath is then represented and dealt with as a sum of the two parts. Such division is fully compatible with the HEP principle, and leads to a variety of simplifications in the analysis.

III. Inlet

The goal here is to relate aerothermal performance directly to flowpath geometry; in particular, in the case of the inlet, the objective is to establish a connection between optimum thermal ratio, ψ_{OPT}, and the geometric contraction, and thus, the boundary of the flowpath. Such connection would eventually provide a means of determining the effective pressure distribution corresponding to the engine cycle; thereby providing a means of establishing the work done in different components along the flowpath. The effective pressure distribution must obviously account for waves and fluid viscosity effects, as well as their interactions in the different parts of the flowpath, so that through appropriate integration one can obtain the overall friction forces and drag.

The approach is based on exploiting the fact that every area distribution can be represented by a one-dimensional (1D) equivalent pressure distribution, which can be related to the distribution of thermodynamic pressure along the flowpath. To do this accurately, one needs to undertake a 3D flowfield computation, Hsia (1991), that is generally impractical during conceptual and preliminary design. A simpler and more direct approach seems indicated, Ortwerth (1996). Such an approach should still yield approximate values of wall forces and heat transfer along the flowpath, that can be matched, following integration in 1D, with values generally obtained in practice. This will provide both a validation of the approach as well as the results needed in integration.

The approach may be divided into two parts: 1) establishing direct relations among the principal design parameters, namely geometry, thermal ratio, velocity ratio, contraction ratio, and thermodynamic performance, and 2) determining geometry of the inlet flowpath.

A. Some Useful Direct Relations

The first part of the approach is based on introducing what may be called a pseudo pressure coefficient, K_{WP}, defined by

$$K_{WP} = \frac{\overline{P_W - P}}{P} \tag{31}$$

Here P is the average pressure over the cross-section of the duct, and P_W is the wall pressure at any location; assuming that a wall layer, in which the viscous and the shock losses are confined, separates the core inviscid flow, P_W at the wall becomes different from P. The overbar over the pressure difference represents an average value over a length of the flowpath; considering an inlet, K_{WP} may refer to any part of the inlet flowpath length, including the overall length.

Details on the analysis of the pressure coefficient are presented in Appendix C, Sec. A. It is shown that K_{WP} can be related directly to the contraction ratio of an inlet and the entropy gain over the length of the inlet flowpath, viz.,

$$K_{WP} = \frac{\Delta S}{R} / \ell n\, CR \tag{32}$$

The change in entropy ΔS is due to the losses in the inlet. A visualization of the linear distribution of entropy production in a single shock inlet is shown in Fig. 10. Note that Eq. (32) provides a powerful connection in terms of K_{WP} between the thermodynamic change and the geometrical change. With a knowledge of pressure distribution as a function of cross-sectional area along the flowpath, as well as those of other thermodynamic properties, the flowfield can be determined explicitly using the mass conservation equation.

Proceeding in the same fashion, it has further been shown in Appendix C [see Eqs. (C20) and (C21) in Sec. B.2] that the thermal ratio ψ can be related directly to the pressure coefficient:

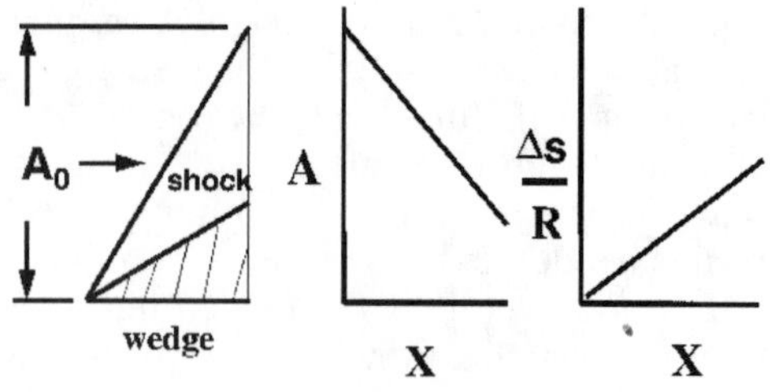

Fig. 10 A Simple Wedge Flow Illustrates how Average Entropy is Directly Related to Area.

$$\psi = \tilde{A}^{-(\gamma-1)(1+K_{WP})}\ \tilde{V}^{\,-(\gamma-1)} \tag{33}$$

$$\psi = \tilde{A}^{-(a-1)}\left[\frac{\beta-\psi}{\beta-1}\right]^{-(\gamma-1)/2} \tag{34}$$

Where $\tilde{A}$ and $\tilde{V}$ represent the area ratio and velocity ratio, respectively, across the inlet. β is a function of γ and Mach number, and a is a function of γ and K_{WP}. They are defined in Appendix C.

Other properties such as $\tilde{P}$ and $\tilde{\rho}$ also are provided in Appendix C in terms of $\tilde{A}$, $\tilde{V}$, β, and ψ.

Equations (32–34) thus provide all the details pertaining to the inlet in terms of relating the contraction ratio, thermal ratio, velocity ratio, and area ratio, by using the pressure coefficient.

Within the hypersonic approximation, one may write

$$\beta = 1 + \frac{\gamma-1}{2}M^2 \Rightarrow \infty \tag{35}$$

It then follows (see Appendix C, Sec. B) that various quantities of interest become a function of K_{WP} and area ratio only. One can then use the following approximate relations over any length of the inlet flowpath starting from the inlet entry plane:

$$\psi = \tilde{A}^{-(a-1)} \tag{36}$$

$$\tilde{P} = \tilde{A}^{-a} \tag{37}$$

$$\tilde{\rho} = \tilde{A}^{-1} \tag{38}$$

$$\tilde{H} = \tilde{A}^{-(a-1)} \tag{39}$$

$$\tilde{V} = 1 \tag{40}$$

$$M = M_O / \sqrt{\psi} \tag{41}$$

It may be worth emphasizing that two important gains have been made through the foregoing analysis using the definition of pressure coefficient, K_{WP}:

1) Various quantities of interest along the flowpath have been related in a simple and readily applicable form to thermodynamic and geometrical parameters and the thermal ratio.
2) A set of relations have been established that, following integration, readily yield pressure and shear force over the inlet flowpath surface area.

It will be recalled that the latter is the goal in the approach being developed for integration, noting that the integral of pressure and viscous forces may be looked on as net thrust or drag of the inlet with reference to the internal flowpath.

B. Flowfield in the Inlet Flowpath Introduces Distortion Parameters

It is possible next to determine the flowfield in the inlet flowpath using the integral drag value discussed previously. However, in the same approach based on integral values, one also needs the wall displacement area due to viscosity effects before the flowfield can be determined. One therefore needs a link between the inviscid core and the viscous wall layer, and this can be easily obtained based on Prandtl's boundary layer theory. The flux conservation analysis approach is revisited to introduce this important flow feature.

It is adequate to consider the three flux variables, mass, momentum, and energy, and the cross-sectional area of the flowpath to determine the flow variables. The analysis may be summarized as follows.

Based on Foa (1958), the so-called normal flow function may be introduced, viz.,

$$N = \dot{m}\left[\frac{RT_T}{\gamma}\right]^{1/2} / F \tag{42}$$

The function N only varies from 0.456435 to 0.319438 as Mach number varies from 1 to infinity!

This function can then be expressed in terms of core flow Mach number in the flowpath and two factors, which are in the nature of efficiency or correction (Ortwerth 1977 and McLafferty 1955), one for the presence of the displacement area in the flowpath due to viscous action, namely

$$\eta_A = 1 - (\delta * / A) \tag{43}$$

where δ^* is the displacement thickness, and the other for the momentum defect, namely

$$\eta_M = \eta_A - (\theta / A) \tag{44}$$

Where θ is the momentum thickness. One can then write

$$N = \eta_A \cdot M \cdot \left[1 + \frac{\gamma - 1}{2} M^2\right]^{1/2} / \left[1 + \gamma M^2 \eta_M\right] \tag{45}$$

Now, from the expressions in Eqs. (42) and (45), one can write

$$\frac{\dot{m}(RT_T/\gamma)^{1/2}}{F} = \frac{\eta_A M\left[1 + (\gamma - 1/2)M^2\right]^{1/2}}{(1 + \gamma M^2 \eta_M)} \tag{46}$$

This equation then provides a means of obtaining any of the flowfield variables if the mass flux, thrust, and the factors η_A and η_m are known.

Thus, the pressure can be obtained from Eq. (42) by writing

$$\frac{P = (F/\ A)}{1 + \gamma \eta_m M^2} \tag{47}$$

Other state variables may be found the same way.

Hence, it has been shown that starting with freestream values of mass flow, stream thrust, and total enthalpy, and with knowledge of drag and distortion factors η_A and η_m, the state variables can be calculated at any location along the flowpath.

Noting that the inlet exit corresponds to the combustor entry section in a scramjet engine, and as the inlet exit Mach number is a function of the thermal ratio only, the search for optimum ψ or flowpath is related directly to the viscous flow solution throughout. The emphasis on cycle efficiency through Ψ requires accurate M, which requires the combined viscous interaction solution of the flowpath [see Auslender (1997)].

1. Combined Viscous Interaction Solution

The difficulty with flux conservation analysis is clearly revealed by the normal function N for which 5 or 6 place accuracy is required. An alternate approach can be used to complete a combined solution simultaneously through K_{WP}. This is fundamental in Chapter 8. The main result is the δ^* formula [Eq. (C40)], which is repeated here for emphasis.

$$\tilde{\delta}* = \tilde{A}^{a(\tilde{H}-1)}\left(\tilde{\delta}_i^* + \frac{\tilde{\Sigma}}{\tilde{m}_\sigma}\left\{\frac{\left[H\,(c_f/2)\right]_{\text{ref}}}{(1+K_{wP})^{1-n}}\right\}\cdot\frac{1-\tilde{A}\left[a(\tilde{H}-n)+1\right]}{\left[a(\tilde{H}-n)+1\right]}\right) \tag{48}$$

Thus, if K_{WP} and geometry are known, the solution is obtained from nose to tail.

C. Determination of K_{WP}

The main unknown throughout the analysis is K_{WP}. It is clear that K_{WP} needs to be determined through an *a priori* analysis.

1. Experimental Correlation of K_{WP}

A first step is to establish a database of reliable data based on a variety of inlets. This approach is presented in Appendix C, Sec. D, and consists, in summary, of establishing through tests, the stagnation pressure ratio across a large body of carefully selected inlets, and then analyzing the test data in terms of 1) ψ, the thermal ratio; and 2) a work parameter W, which represents the work done when the velocity change across the inlet is assumed to be small—a generally valid assumption, as had been made earlier in hypersonic flow

$$W = \left(\frac{(\gamma-1)}{\gamma} - \frac{1}{M_o^2}\right)$$

and 3) the similarity rule for affine bodies in hypersonic flow, viz.,

$$K_\theta = \left(M_o^2 - 1\right)^{1/2}\cdot\theta \tag{49}$$

The result of the analysis is a correlation given by the simple formula:

$$\begin{aligned} K_{\text{WP}} &= \frac{WK_\theta^2}{\psi^{1/2}} \\ &= \frac{WK_\theta^2}{(CR^{a-1})^{1/2}} \end{aligned} \tag{50}$$

where ψ is rewritten using Eq. (36).

The correlation pertains only to a 2D internal flowpath. However, other effects are added to this correlation for the forebody, taking into account the leading edge bluntness of the forebody and inlet leading edges. Also observe that the inlet internal flowpath enters the isolator-combustor at the inlet exit, as seen in Fig. 1. Thus, at the inlet downstream end, there is interaction between the internal flowpath of the inlet due to the turning of the flow from the inlet into the isolator-combustor. These two interactions, one at the forebody leading edge and the other at the turning in the cowl leading edge region, are critical in assessing the internal flowpath in the inlet in hypersonic air intakes. Both the mass flow and its distribution with respect to state variables become affected, and inadequate attention to these details can lead to a serious loss of engine thrust.

Carefully read Appendix C.

D. Inlet Testing and Determination of K_{wp}

The parameter K_{WP} is one of the easiest inlet performance measures to extract from experimental data of all those commonly used. The alternatives all involve some form of the mass and stream thrust averaged 1D analysis by inversion of Eq. (45), described above. Some difficulties with the approaches are discussed below. Determination of K_{WP} first is an accurate way of determining the other inlet performance parameters.

Experimentally, it is practically impossible to determine the drag accurately and precisely for resolution of the total stream thrust and mass flow to solve for mass-averaged mach numbers at hypersonic speed. The pitot rake data density are difficult to afford at the entrance and exit of the inlet. Further uncertainty under test due to lack of resolution and uncertainty of the stream static pressure and the pitot probe location. Since the Mach ratio and η_{KE} are determined by the small differences in the integrated stream thrust at the entrance and exit, the results are unreliable. Further, the cost of accurately determining the viscous layer displacement at the throat is also practically impossible. Thus, the assumptions that the drag and distortion are known is not sustained and inlet data reduction at high speed is problematic. The inlet is usually installed with some form of forebody simulation to determine the effects of initial boundary layer at the inlet entrance and its thickness must be measured to find the effective inlet entrance area $\eta_A A_1$.

The accurate data available are usually the wall pressures and heat transfer. The heat transfer data are used to determine the location of transition. The wall pressure data are usually distributed around the perimeter of the throat at several locations and an accurate average wall static pressure in the inlet throat is available. The use of the combined interaction solution Eq. (C40) in Appendix C enables the inlet performance to be determined from the heat transfer and throat static pressure. The analysis provides the inlet process K_{WP} and η_A. The solution for the aerodynamic contraction and the thermal ratio ψ of the inlet is determined after K_{WP} and η_A. The solution process is illustrated by inspecting Figs. 11–13 for a Mach 10 test case. Note that the unknown K_{WP} determines the inlet distortion and thermal ratio. The thermal ratio ψ converts into the pressure drag and can be compared to the experimental integral of wall pressure for consistency if the wall tap density is high and there are taps in the corners, this is independent of any knowledge of the total stream thrust.

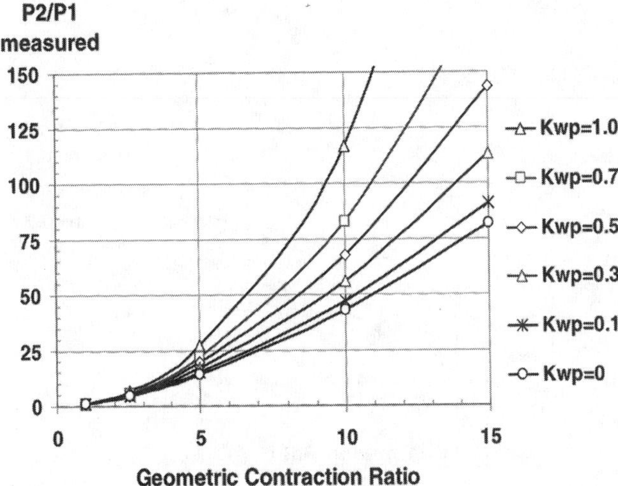

Fig. 11 Illustration for Determining Kwp From Inlet Contraction and Test Data, P_2/P_3, M = 10, REX = 10^7, Cold Wall.

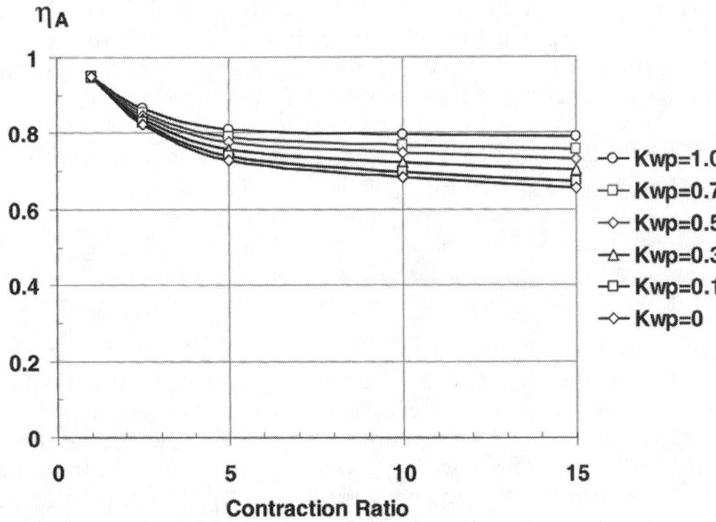

Fig. 12 Illustration for Experimental Determination of ηA from Inlet Contraction and Kwp, M = 10, REX = 10^7, Cold Wall.

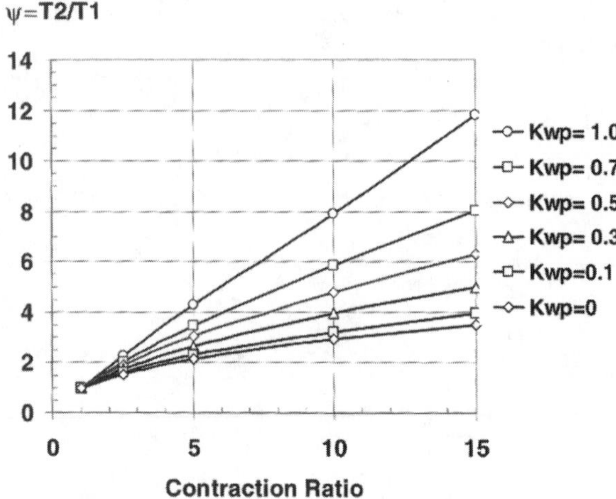

Fig. 13 Illustration for Experimental Determination of Inlet Thermal Compression ψ from Contraction and Kwp, M = 10, REX = 10^7, Cold Wall.

E. Implications of Thermodynamic Analysis to Design of Inlets

The thermodynamics of the inlet process can be inferred from Fig. 10 in terms of the change in enthalpy and that in entropy of the working fluid of the inlet internal flowpath. It becomes apparent by inspecting the figure that the inlet performance can be assessed in a number of ways using thermodynamic parameters and kinetic energy. These matters are discussed in Appendix C, Sec. F and also in an earlier chapter in this book by VanWie.

It is of particular interest to consider kinetic energy efficiency, η_{KE} and the diffusion coefficient K_D. In the analysis of Brayton cycle presented earlier on flowpath optimization (Sec. I.B), the efficiency of the inlet process was defined by η_c. Also, the optimal thermal ratio based on η_c, η_e, and $\Delta H_c / H_o$ was denoted by ψ_{OPT} given by

$$\psi_{OPT} = \left[\frac{\eta_c\ \eta_e}{1 - \eta_c\ \eta_e} \frac{\Delta H_c}{H_o} \right] \tag{51}$$

It is shown in Appendix C, Sec. E, that 1) η_c for adiabatic flow diffusion to stagnation conditions is the same as η_{KE}; and 2) η_c can also be related, for adiabatic diffusion to supersonic conditions, to K_D, although the definition of K_D applies to non-adiabatic flow as well as adiabatic flow. In particular, when V_2 tends to zero, η_{KE} becomes the same as K_D. Thus, the conclusion is that ψ_{OPT} can be obtained using Eq. (51) by substituting, as convenient, η_{KE} or an expression [see Eq. (C74) in Appendix C] involving K_D. Since by definition $\eta_c \rightarrow 1$ at high Mach number the issue needs study.

1. *Useful Drag Efficiency*

Another efficiency parameter for an inlet is the so-called useful drag efficiency η_{UD}, which is defined based on the relation between drag and change in internal energy given in Eq. (19). The definition and the relation between η_{UD} and K_D are given in detail in Appendix C, Sec. F [see Eq. (C77)]. That relation involves the thermal ratio. Thus, the thermal ratio can be said to be a function of useful drag efficiency and diffusion coefficient, albeit a complicated one.

Ordinarily in testing, one measures the total pressure recovery in an inlet, although there is renewed interest in the use of wind tunnels and drag balances for inlet model testing. Precise measurement of inlet performance using high speed wind tunnel tests is difficult. Typical required measurements include freestream pitot survey, wall static pressure, wall heat flux, wall temperature, and throat pitot survey. Internal total drag and total heat flux (which requires calorimetric data) are not usually measured due to prohibitive costs. The measurement of overall mass capture is feasible, but again, complicated.

If one used CFD methodology to determine inlet performance in a consistent fashion, one again faces various difficulties. For example, specifying the mass flux through the throat becomes problematical unless accurate throat viscous layer blockage can be established with accuracy. In many investigations, a mass-averaged total pressure recovery is all that can be obtained.

Some relations for stagnation pressure recovery may be written as follows, based on the analysis of Appendix C.

Adiabatic case.

$$\left(\frac{\overline{P_{T2}}}{P_{To}}\right) = CR^{-K_{WP}} = \psi^{\left(\frac{K}{\gamma-1}\right)} \tag{52}$$

where

$$K = \frac{a - K_{wp}}{a - 1} = \left(\frac{K_{wp}}{1 + K_{wp}}\right) \tag{53}$$

Nonadiabatic case.

$$\left(\frac{\overline{P_{T2}}}{P_{To}}\right) = \psi^{\left(\frac{\gamma}{\gamma-1}\right)k}\left[1 - \frac{Q_{total}}{\dot{m}H_t}\right]^{\gamma/(\gamma-1)} \tag{54}$$

where Q_{total} is the amount of heat transfer.

In summary, the foregoing discussion presents the different ways one can specify η_c in the determination of ψ_{OPT} and the difficulties and ambiguities associated with each method.

IV. Forebody

A. Forebody Design

The installation of a scramjet engine invariably means that the forebody becomes part of the propulsion flowpath. Therefore, it is necessary to discuss some aspects of forebody design and their effects on the internal flowpath. However, details of hypersonic external flowpath will not be covered. The primary interest is to focus on drag per unit mass capture as a characteristic (discriminating) parameter in optimum forebody shapes.

An excellent review of optimum aerodynamic shapes for hypersonic forebodies can be found in Miele (1965), who shows that under various constraints of, for example, length and diameter, length and volume, or length and base area, one can obtain various optimal shapes, depending on the speed range of interest. Thus, in supersonic flow, with length and diameter constrained, one obtains, using linearized theory, the von Karman ogive. At hypersonic speeds, using Newtonian or modified Newtonian flow, power law bodies are obtained. When base area and length are constraints, the result is a star shape, and also a waverider shapes for inviscid flow; however, when viscous effects are included, the number of star points is reduced and the "valleys" become enlarged, tending to give rise to a conical shape. A conical shape also is obtained when one imposes minimum surface area and length requirements, and also in the case of power law bodies when the index is set to unity.

Observe that optimum shapes with constant volume become, when viscous effects are included, those for which friction drag and pressure drag are equal. This can be explained as follows: slender bodies have low pressure drag and large friction drag; as thickness increases, surface area is reduced and the pressure increases; thus, pressure drag is dominant for thick bodies; and the optimum shape with minimum total drag is realized when friction drag and pressure drag become equal. The same result has been obtained in the case of waveriders, Bowcutt (1999). It becomes evident in the following that the same result is obtained in the case of optimized inlets.

The discussion in Miele on optimum bodies including a thrust cowl is of direct relevance to propulsion installation. In the case of such a body, under Newtonian theory, all the thrust due to turning the thrust vector is recovered, and the optimal shape is a wedge in 2D flow and a cone in 3D flow. However, no attempt was made in the original work to establish the total pressure recovery or the drag as a function of captured mass. These are important in installation, and will be discussed in the following paragraphs.

The properties of wedges and cones in hypersonic flow are available in Chernyi (1961), and Rasmussen (1994). However, their relation to internal flow needs to be established. In this connection, it is necessary to review elements of the hypersonic slender body theory with specific application to forebodies and internal flows. A brief discussion of relevant aspects is presented in Appendix D, with particular reference to wedges and cones.

The main results of that analysis are as follows:

1) On the basis of linearized similarity theory, one can establish the pressure coefficient as a function of the product, κ, of upstream flow Mach number M_∞ and the shape parameter, t / ℓ, with t as thickness and ℓ, a characteristic body length.
2) The pressure coefficient can then be related to the contraction ratio.
3) Next, the pressure coefficient K_{WP} can be determined as

$$K_{WP} = \frac{\ell n\left[\left(\frac{2\gamma}{\gamma-1}k_s^2 - \frac{\gamma-1}{\gamma+1}\right)\cdot\left(\frac{2}{\gamma-1}\frac{1}{k_s^2} - \frac{\gamma-1}{\gamma+1}\right)^{\gamma}\right]}{\ell n\, CR^{(\gamma-1)}} \tag{55}$$

where

$$k_s = k\cdot\frac{\tan\delta}{\tan\theta} \tag{56}$$

and it is assumed that $\tan\theta = \sin\theta = \theta$. Observe that K_{WP} as given in Eq. (55) shows the influence of k, or the thickness, and is different from the approximate expression, Eq. (C64), given in Appendix C.

It is then possible to establish the drag per unit mass flow captured in different cases:

Wedge:

$$\frac{D}{\dot m} = \frac{V_\infty}{2g}\left[C_{\mathrm{pw}}\left(\frac{\theta}{\delta}\right)\right] \tag{57}$$

Cone:

$$\frac{D}{\dot m} = \frac{V_\infty}{2g}\left[C_{\mathrm{pw}}\left(\frac{\theta}{\delta}\right)\right]^2 \tag{58}$$

The results are compared in Fig. 14, where drag in nondimensional form is shown as a function of $(1/k)$; for this choice of coordinates, the nondimensional drag for a wedge is unity.

Note that comparison between the wedge and the cone is only partial, since they have different types of angles and shock-on-lip Mach numbers. However, at this level of analysis, the cone is found to be better than the wedge at the same initial angle over a range of the similarity parameter. This difference may result in advantageous surface area per unit volume or performance.

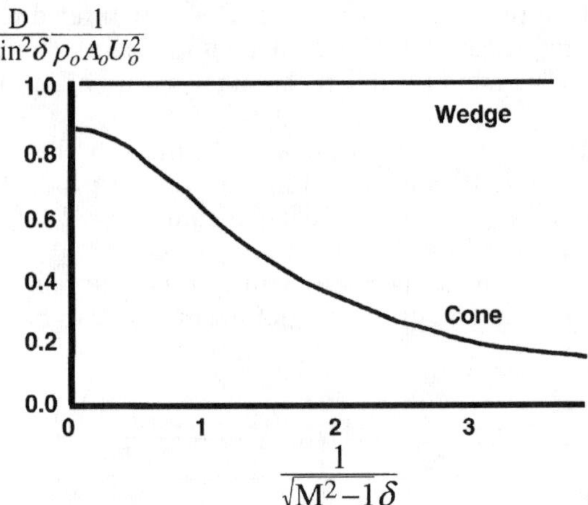

Fig. 14 Drag per Unit Mass Capture for Wedge and Cone Plotted Against Similarity Parameter.

The leading edge of all elements in hypersonic flow must be, to some extend, blunt, otherwise the heat load on the element at that location becomes impractically high for survival of most materials. The bluntness, of course, causes a bow shock, the effect of which stretches far beyond the leading edge region. The flow through the shock gives rise to an entropy layer of hot, low density gas over the surfaces making up the leading edge, and this affects the stand-off distance of the shockwave. The resulting flowfield and heat transfer are substantially different in 2D and 3D flows. For wedges, the total drag is increased by a large amount over that of a wedge with a sharp leading edge. On the other hand, for a cone, there actually arises a reduction in drag with bluntness for certain ranges of bluntness. Figure 14 (see Appendix D, Sec. B) illustrates this. The reason for this lies in the over-expansion occurring at the junction between the nose and the cone.

On this basis, one can proceed to find a correction for the drag in the case of a wedge and a cone; these matters are discussed in Appendix D, Sec. B.

The important conclusion is that leading edge interaction with the flowpath is quite large. For example, for the drag to drop below 10% due to bluntness, the ℓ/d in the case of a wedge needs to be over 4,500. Thus, a 0.5 in. diameter cylinder for the leading edge would increase the drag by 10% on a 189-ft-long forebody of 5-deg-wedge shape. A 50% drag increment arises for $\ell/d = 1{,}340$ or, in the wedge case, with 58 ft length when bluntness or d/D is about 4%, where D is the nominal cross-dimension. Thus, efficient active cooling of leading edges of wedge (2D) shape is of the highest priority in the design of hypersonic flight vehicles.

On the other hand, the problem is not serious in the case of a conical-shaped leading edge with a spherical nose. For example, a 10% drag reduction on a 5-deg cone is realized on an 85-ft-long forebody, when a 1-ft diameter spherical nose with a bluntness of $d/D = 0.067$.

The implications of these estimates for the design of the intake internal flowpath are clear and noteworthy.

B. Inlet Forebody Integration

The integration of the forebody and inlet requires a consideration of geometric specifications that can lead to unmanageable consequences. Presented below are the results of an integration study that illustrates this feature of geometrically constrained integrated performance.

The geometry study included a sidewall inlet installed on a conical forebody and a 2D inlet installed on a 2D forebody that started as a wedge and continued external compression to a specific total turn angle of 12 deg (Fig. 15). The boundary layer was laminar on the conical forebody and inlet transition was fixed at the first shock intersection on the sidewalls. The boundary layer was laminar on the 2D forebody until the first ramp, the inlet cowl was laminar until the shock reflection. Both geometries include nose and inlet bluntness. Thus, all the important features of the combined interaction analysis in Appendix C were included as fairly as possible [see Ortwerth (1996)].

Results are presented in Figs. 16–20 for Mach numbers increasing from 6 to 14. The initial compression is traded with internal contraction producing performance maps for each approach.

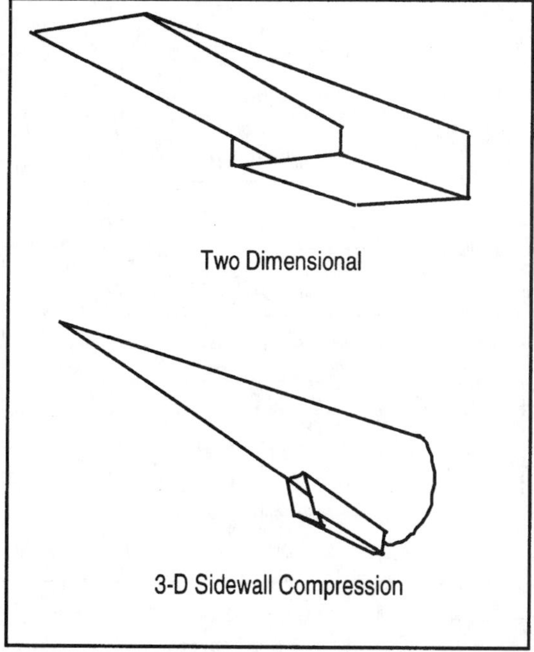

Fig. 15 Forebody Inlet Integration Types for Study Comparison.

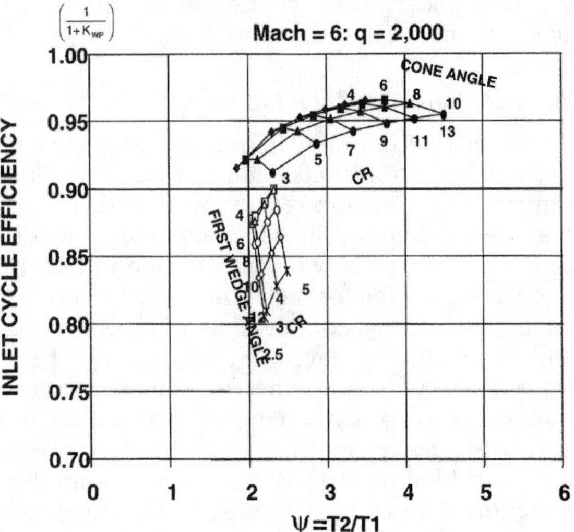

Fig. 16 M = 6 Inlet Forebody Integration Performance Maps.

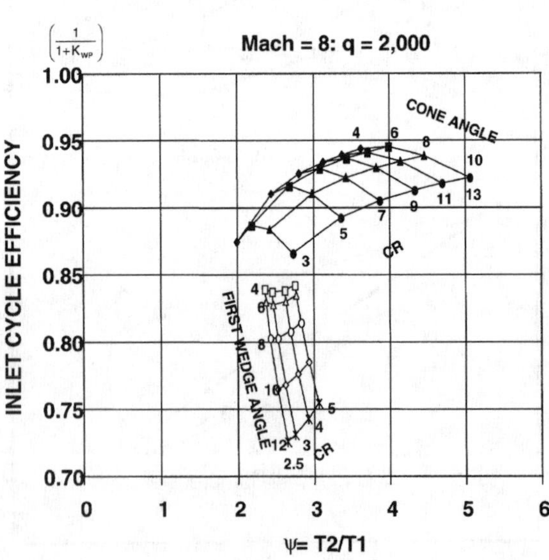

Fig. 17 M = 8 Inlet Forebody Integration Performance Results Maps.

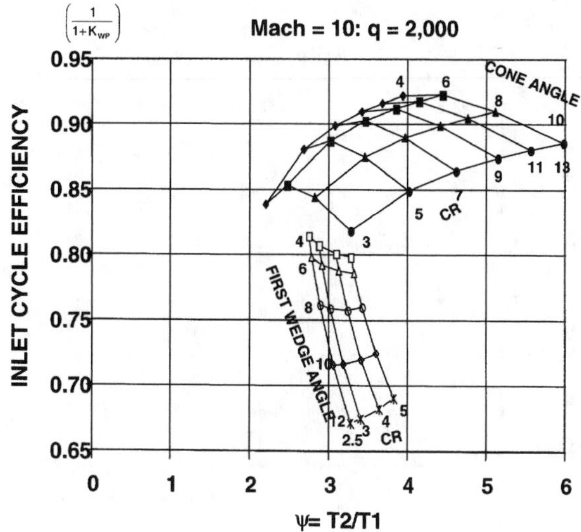

Fig. 18 M = 10 Inlet Forebody Integration Performance Maps.

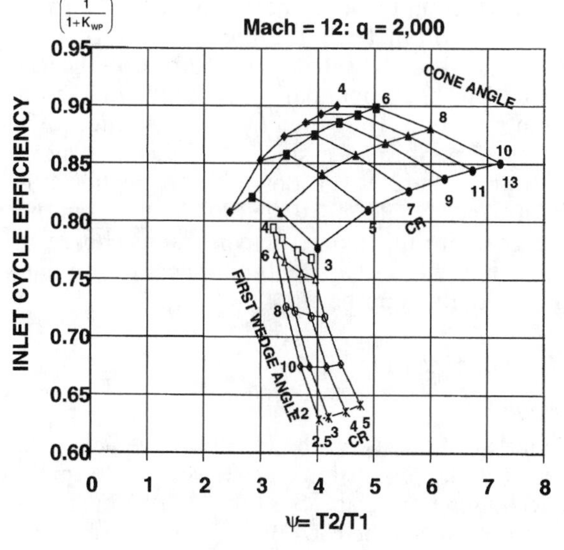

Fig. 19 M = 12 Inlet Forebody Integration Performance Maps.

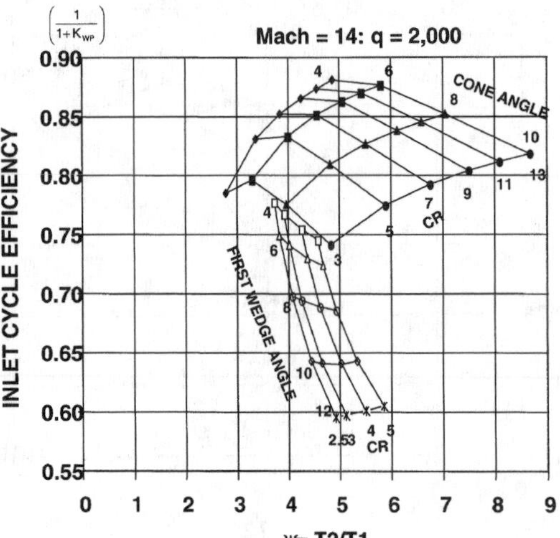

Fig. 20 M = 14 Inlet Forebody Integration Performance Maps.

There are no overlapping regions and the conical 3D is consistently outperforming the 2D integrated forebody inlet. General principles are found common to both 2D and 3D types. First, the performance of each configuration decreases with increasing Mach number. Second, when designing for a specified ψ at a desired Mach number starting with lower forebody losses, that is lower angle, is better than trying to recover later in the process as it never happens over the parameter ranges in this study.

The 3D integration enjoys a wider and higher operational thermal compression range. This is due to an unfortunate choice of external cowl angle constraint, naming zero degrees, for the 2D inlets. With zero turn, the 2D inlet is operationally restricted by shock boundary layer separation to a maximum internal contraction ratio of 5. Thus, the 3D forebody-inlet combination has a much better cycle performance since it operates at more superior conditions thermally than the 2D. The possibility of higher efficiency and higher cycle performance would make the 3D combination the choice of many flowpath designers for the geometric constraints chosen in this example. Integration is rather subjective and proponents of any approach will probably continue with that approach.

V. Force Accounting

A. Force Accounting Viewpoint

We touch briefly on accounting of forces on the integrated vehicle with regard to the interactions between the propulsion and the vehicle flowpath. The difficulties with force accounting are traceable to the discipline's cultural problems. The problem roots are in the 6 DOF viewpoint, or culture, of aerodynamics and the 1D thermodynamic viewpoint of propulsion engineering. Hopefully, the present approach of direct integration of pressure area and viscous force can offer some unifying impetus. We will concentrate on the specific topic of accounting for the drag of the forebody, cowl, and nozzle, with a numerical example.

The analysis presented previously has separated the shear and pressure forces for on-axis flowpaths. The pressure forces on the thrust and contraction surfaces include the isentropic ideal expansion or compression resulting from flowpath area changes and also the wave drag or non-ideal effects through K_{WP}. The present propulsion engineering approach would normally use the analysis for complete nose-to-tail performance estimation within the framework of the present state of development. It is necessary to reconcile the two viewpoints by accounting for turning the flow off-axis, which has the effect of increasing forebody drag from the on-axis analysis value without changing the state of the flow from the on-axis solution. Vehicle engineering ordinarily places design requirements on propulsion that often are difficult with forebody turning, in addition to the obvious task of flowpath compression. A recommendation of force accounting should be able to clarify the implications of meeting such requirements.

As the enters the cowl of the internal propulsion flowpath, it is usually turned back along the vehicle waterline before the combustor exit station. The manner of turning may be abrupt as through a sharp turn at constant area without drag recovery, or the flow may be turned gradually obtaining a drag recovery. The turn sometimes is avoided until the internal nozzle expansion. These approaches result in different flow states at the entrance to the heat addition component. In both cases, the cowl cancels the forebody's normal force due to the captured flow.

The situation becomes clear by way of an example. A wedge of 5 degrees turning is considered in a Mach 12 flow. The solution from shock theory gives the following data: $M_0 = 12, \delta = 5$ deg, $\theta = 8.68$ deg, $P_1/P_0 = 3.6596$, $P_{T1}/P_{T0} = 0.80779$, and $T/T_0 = 1.5398$.

The normalized wedge pressure drag is obtained by the exact solution from the pressure ratio and frontal area of wedge and shock capture area:

$$\frac{D}{P_0 A_0} = \frac{P_1}{P_0} \frac{\left[1 + (\tan\theta - \tan\delta)\sin\delta\cos\delta\right]\tan\delta}{\tan\theta}$$

$$D/P_0 A_0 = 2.1091$$

From the geometry of the wedge and shock angle, the contraction ratio is obtained. The flow area after the shock is taken normal to the wedge surface.

$$CR = \frac{\tan\theta}{(\tan\theta - \tan\delta)\cos\delta}$$

$$CR = 2.3513$$

With the contraction ratio and total pressure ratio, the pressure coefficient K_{WP} is

$$K_{WP} = 0.24966$$

The adiabatic internal flow exponent is obtained from Eq. (C23).

$$a = 1.4999$$

Using approximate formulas to compare the pressure ratio, thermodynamic compression, and drag from the hypersonic flowpath gives the results as

$$\frac{P_1}{P_0} = CR^a = 3.6055$$

showing pressure percent accuracy to within 1.5% and

$$\psi = \frac{T_1}{T_0} = CR^{a-1} = 1.5334$$

is 0.4% accurate, so the on-axis drag formula, which is hooked directly to cycle compression ψ as follows:

$$\frac{D}{P_0 A_0} = \frac{\psi - 1}{\gamma - 1} = 1.3334$$

shows a large error and is 36% low.

The compression compares well with the shock solution; however, there is a difference in drag of 36% that needs to be accounted for. The on-axis internal flow solution for drag and the pressure area integration solution differ because the turning effects are not included in the drag formula of the internal flow solution, which contains the effects of compression and wave losses only. To correctly account for all the forces in the K_{WP} analysis, the drag of the turn must be added without a change in the adiabatic exponent. The turn drag to good accuracy is simply the cosine loss of freestream momentum.

$$\frac{D_{TURN}}{P_0 A_0} = \gamma M_0^2 \left(\frac{V_2}{V_0}\right)(1 - \cos\delta) = 0.7671 \tag{59}$$

Adding the turn drag to the compression and wave effects of the K_{WP} theory, the comparison of total pressure drag is now accurate to 0.4%. Thus, it is observed that cycle compression ψ accounts for the drag due to contraction and wave drag but not turn drag. Propulsion engineering should now recognize and work with the fact that force balance is not hooked formally to the state of the flow as it is implied in the 1D domain.

Each forebody turn is increased by 5 degrees of isentropic turn to a total of 10 degrees. The continued analysis shows drag of turning increases to 53% of total drag as turning angle is increased by 5 degrees of Prandtl–Meyer compression.

The contraction is increased until nearly proportional to the total turning:

$$CR = 5.0558 \tag{60}$$

The wave losses remain the same as for the wedge alone.

$$\frac{P_{T2}}{P_{T0}} = 0.80779 \tag{61}$$

Therefore, the wall pressure factor is decreased, since some compression is isentropic.

$$K_{\mathrm{WP}} = 0.13172 \tag{62}$$

The adiabatic exponent is also nearer to the isentropic exponent value.

$$a = 1.4527 \tag{63}$$

The inlet pressure ratio and cycle compression ratios are increased due to added contraction.

$$\frac{P_2}{P_0} = 10.83 \tag{64a}$$

$$\mathrm{w} = 2.0988 \tag{64b}$$

The drag of the flowpath area contraction and wave drag are slightly more than doubled and follow the geometric contraction ratio somewhat fortuitously.

$$\left(\frac{D}{P_0 A_0}\right)_{1D} = 2.747 \tag{65}$$

The big change is due to the turn drag, which has increased by a factor of four following the square of the turn angle ratio, which is in fact the true dependence on the angle.

$$\frac{D_{\mathrm{TURN}}}{P_0 A_0} = 3.063 \tag{66}$$

Thus, the total drag has increased by nearly a factor of three.

$$\frac{D_{\mathrm{TOTAL}}}{P_0 A_0} = 5.8098 \tag{67}$$

The important point to make is that when flowpath analysis is done on-axis or 1D from nose to tail, then force accounting and thermodynamics can be disconnected. In fact, errors can be made that are serious to actual integrated performance. Sorting out these effects needs to be done with a coordinated analysis accounting system with vehicle aerodynamics.

With explicit design, the internal flow can be turned back on-axis smoothly and the turn generates a thrust that can be equal to the forebody turn drag, and, therefore, no change in flow-state properties are required. In general, the cycle compression ψ will be increased according to the amount of unrecovered drag in the internal flow after realignment, to balance the total drag. This effect on the propulsion cycle should be addressed during engine flowpath integration or the engine performance will be computed at the wrong compression ratio and with a wrong internal thrust balance. The propulsion flowpath force balance accounting sheet must include an entry for flow angle at each station to enable calculation of lift, drag, and moment as well as flow-state properties.

1. Turn Comparison

The manner of turn will be considered with further examples. Three integration schemes will be examined. The cases shown in Fig. 21 are three ways to deal with the forebody turn in the integrated flowpath. The same expansion ratio of freestream capture area to exhaust plume exit area is considered. The same thermodynamic compression ψ is achieved. The compression ψ is assumed to be optimum. The combustion process is constant so that ψ_c and M_c are the same in every case.

The first case is the sudden realignment through a sharp constant area turn, which is a feature of straight line junctions between the inlet and isolator. This type of turn provides no recovery of lost momentum of the forebody turn. Thus, the compression ψ will be increased accordingly to

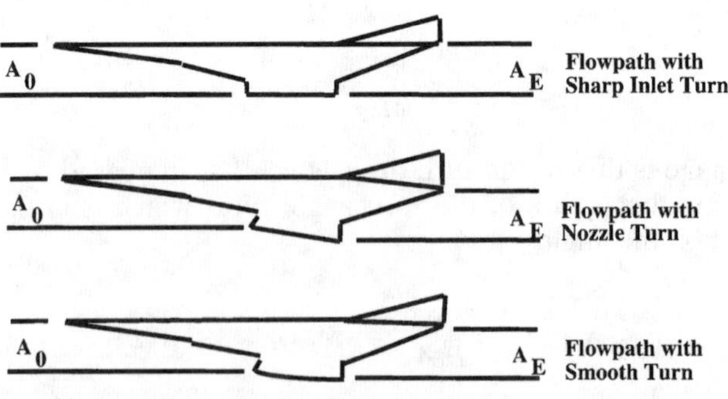

Fig. 21 Engine Vehicle Cowl Turn Integration Cases.

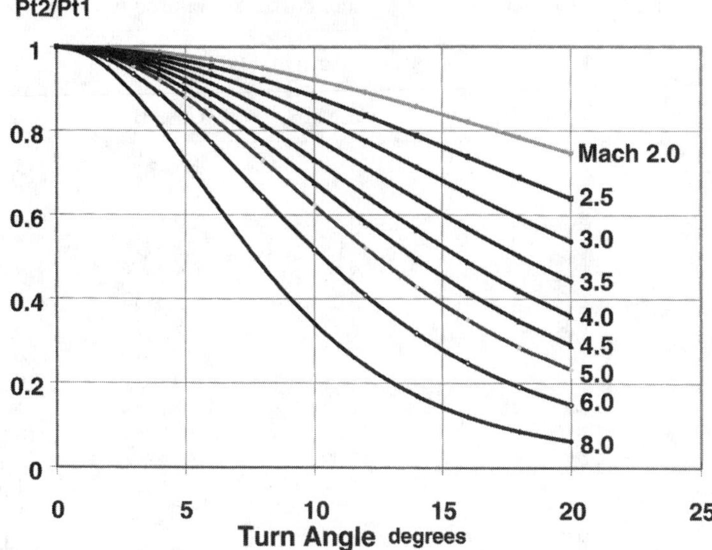

Fig. 22 Inlet Cowl Turn Loss for Sharp Turn of Constant Area.

account for the same area, mass flow, and enthalpy, but lower total stream thrust. A graphic representation of sharp turn losses as total pressure ratio across the turn is shown in Fig. 22. For low angles and low Mach numbers, the effect is small, while for high angles of 10 degrees or more and Mach numbers of 4 or greater, the sharp turn is a large loss. Because propulsion seeks to design for optimum ψ, the state will be obtained at lower contraction than a flowpath without turn losses. The nozzle expansion ratio will be similarly reduced at constant ψ.

The sharp inlet turn case suffers from loss of nozzle expansion compared with the other two due to increased inlet losses and smaller contraction ratio to achieve the same desired compression. An alternative to turning at the inlet cowl or through the internal flowpath is to assume the forebody turn drag can be recovered in the nozzle expansion turning contour. This requires providing a surface for generating a side force normal to the thrust axis to turn the stream parallel to the axis. The only way to achieve this presently is to turn the flow with the cowl. Cowl design requires added length, however, added length is undesirable for engineering thermal structural penalty reasons, also for trim reasons nozzle lift is desired. However, the turn will coalesce into a shock if the cowl is short. Thus, the nozzle design for this case is unknown because of the design complexity at this time and the area needs research. The contour is assumed unavailable since, in principle, it requires the expansion contour to generate more thrust than ideal expansion to the same overall expansion area ratio; thus, the ideal thrust is used based on available area ratio. Results of the comparison are shown in Table 1.

Table 1 Flowpath Integration and Force Accounting Example

	Parameter	Sharp Turn at Inlet Cowl Case A	Turn Delayed Until Nozzle Case B	Smooth Internal Turn Case C
Forebody	δ_{shock}	5	5	5
	$\delta_{isentropic}$	5	5	5
	δ_{total}	10	10	10
	ψ	2.099	2.099	2.099
	D/P_0A_0	5.8098	5.8098	5.8098
	$CR_{forebody}$	5.0558	5.0558	5.0558
Inlet	CR_{inlet}	2.735	5	5
	$\Delta\psi_{sharp}$	1.228	0	0
	δ	0	10	10
	ψ	4.7	4.7	4.7
	P/P_0	65.05	118.85	118.85
	CR_{total}	13.83	25.28	25.28
Combustor	$\varepsilon_c=A_c/A_{tr}$	1.5	1.5	1.5
	ψ	15.5	15.5	15.5
	$\Delta F_{comb.}/P_0A_0$	4.67	4.67	4.67
	$\Delta F_{turn}/P_0A_0$	0	0	3.07
	δ	0	10	0
Nozzle	$\varepsilon_0=A_E/A_0$	1.5	1.5	1.5
	$\varepsilon_N=A_E/A_c$	13.83	25.28	25.28
	$\Delta F_{Nozzle}/P_0A_0$	31.67	36.9	36.9
	δ	0	10	0
Flowpath	$\Delta F_{internal}/P_0A_0$	27.09	29.26	32.33
Static drag	$D/P_0A_0=\varepsilon_0-1$	0.5	0.5	0.5
Lift	L/P_0A_0	0	33.0	0
Installed thrust	$\Delta F_{Net}/P_0A_0$	26.59	28.76	31.83
Loss%		16.5	9.6	

Differences are clearly noticeable at 17 and 10% performance loss due to integration effects compared to the smooth turn reference case.

Integration of the flowpath with the vehicle is a highly subjective process and is not easily resolved without counting lift, volume, and moments. In case B with the body aligned, cowl will have lift. Converting lost thrust to a lift-to-drag (L/D) ratio by dividing lift to lost thrust $(L/D)_{flowpath} = 11$, which is a very high effective L/D for hypersonic wings and equals the L/D of an ideal flat plate airfoil at 5 degrees angle of attack. Emphasizing this result cannot be overdone since the good wings at high C_L have L/D values in the 4.5 range. Aero design usually tries to integrate the wing area to obtain minimum drag not maximum L/D; more typical L/Ds are 2.5. Therefore, depending on the trajectory dynamic pressure and required angle of attack, case B is in the race with case C for the overall installed $I_{sp_{EFF}}$ winner when drag due to lift is included for all vehicles.

Case A loses to case C since canceled lift and wing lift and drag due to lift requirements would be the same for both, however, case C has higher propulsion performance. Thus, depending on vehicle requirements, some case between B and C is best and the integrated flowpath should be designed for flowpath lift with its high aerodynamic efficiency.

2. *Smooth Turns*

The required radius of curvature and length of the arc of the smooth turn required for the total recovery of thrust available is needed for flowpath integration. The 2D constant area geometry is the configuration choice for analysis, since it will keep the overall contraction and compression on design. The geometry is shown in Fig. 23.

The fluid dynamics is that of Prandtl–Meyer turning with compression on the cowl side and expansion on the body side. Pressure is just a function of the flow angle if the arc is short and waves of one family are present on each wall without interaction. The thrust is a path function and the required area for the pressure integral to act on is needed. A general integral formula for thrust shows that it can be set equal to the turn drag fraction desired. The hypersonic approximation for the Prandtl–Meyer function gives the formula for pressure ratio in a turn.

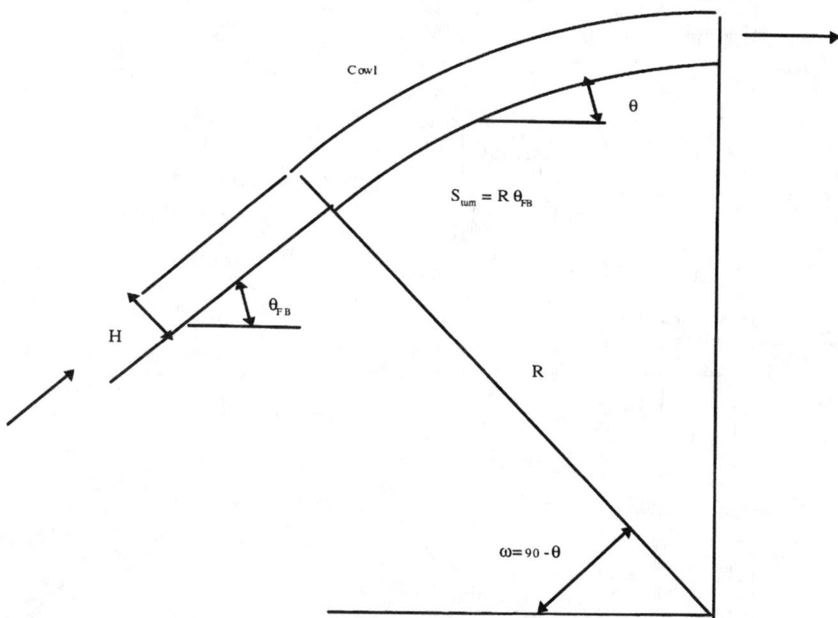

Fig. 23 Geometry for Smooth Constant Area Turn Analysis.

$$\frac{P}{P_1} = \left[1 \pm \left(\frac{\gamma-1}{2}\right) M_1 \lambda\right]^{\left(\frac{2\gamma}{\gamma-1}\right)} \tag{68a}$$

The plus sign is for compression and the minus sign is for expansion from Mach M_1, pressure P_1 through angle λ. The form leads to easy integration with the small angle approximation for sine in the area differential.

$$T = \frac{P_1(R+H)\theta^2}{K_\theta^2}\left\{\frac{K_\theta\left[1+\left(\frac{\gamma-1}{2}\right)K_\lambda\right]^{\frac{3\gamma-1}{\gamma-1}}}{\left(\frac{3\gamma-1}{\gamma-1}\right)\left(\frac{\gamma-1}{2}\right)} - \frac{\left[1+\left(\frac{\gamma-1}{2}\right)K_\lambda\right]^{\frac{4\gamma-2}{\gamma-1}}}{\left(\frac{4\gamma-2}{\gamma-1}\right)\left(\frac{\gamma-1}{2}\right)^2} + \frac{\left[1+\left(\frac{\gamma-1}{2}\right)K_\lambda\right]^{\frac{3\gamma-1}{\gamma-1}}}{\left(\frac{3\gamma-1}{\gamma-1}\right)\left(\frac{\gamma-1}{2}\right)^2} - \left[\frac{K_\theta}{\left(\frac{3\gamma-1}{\gamma-1}\right)\left(\frac{\gamma-1}{2}\right)} - \frac{1}{\left(\frac{4\gamma-2}{\gamma-1}\right)\left(\frac{\gamma-1}{2}\right)^2} + \frac{1}{\left(\frac{3\gamma-1}{\gamma-1}\right)\left(\frac{\gamma-1}{2}\right)^2}\right]\right\} \tag{68b}$$

$$D = \frac{P_1 R\theta^2}{K_\theta^2}\left\{-\frac{K_\theta\left[1-\left(\frac{\gamma-1}{2}\right)K_\lambda\right]^{\frac{3\gamma-1}{\gamma-1}}}{\left(\frac{3\gamma-1}{\gamma-1}\right)\left(\frac{\gamma-1}{2}\right)} - \frac{\left[1-\left(\frac{\gamma-1}{2}\right)K_\lambda\right]^{\frac{4\gamma-2}{\gamma-1}}}{\left(\frac{4\gamma-2}{\gamma-1}\right)\left(\frac{\gamma-1}{2}\right)^2} + \frac{\left[1-\left(\frac{\gamma-1}{2}\right)K_\lambda\right]^{\frac{3\gamma-1}{\gamma-1}}}{\left(\frac{3\gamma-1}{\gamma-1}\right)\left(\frac{\gamma-1}{2}\right)^2} - \left[\frac{-K_\theta}{\left(\frac{3\gamma-1}{\gamma-1}\right)\left(\frac{\gamma-1}{2}\right)} - \frac{1}{\left(\frac{4\gamma-2}{\gamma-1}\right)\left(\frac{\gamma-1}{2}\right)^2} + \frac{1}{\left(\frac{3\gamma-1}{\gamma-1}\right)\left(\frac{\gamma-1}{2}\right)^2}\right]\right\} \tag{68c}$$

The maximum thrust recovery available is written in the hypersonic variable notation and is used to normalize the results for plotting.

$$\Delta F_{\max} = \frac{P_1 H \ \gamma \ K_\theta^2}{2} \tag{68d}$$

The maximum available net thrust occurs when λ is equal to the forebody angle θ_{FB}. The radius of curvature can be found for net thrust. T-D is equal to the forebody cosine loss.

The length found is well approximated by the first term of a series expansion of the *R/H* factors such that a quick solution for arc length and radius of curvature is approximately found by the equation below as a function of turn entrance Mach and forebody angle.

$$\frac{R\theta_{FB}}{H} = \frac{3M}{2} \tag{69}$$

Fig. 24 shows a view of the solution as a function of turning, with the length normalized by the channel height divided by the tangent of the Mach angle. The results show that the assumption that the duct was short enough for the wall to be noninteracting, $H/L < 1/M$, is violated and the last third of the duct will interact and the efficiency of the turn in the last portion will be reduced, since in this region every degree of turn is canceled by a wave of the opposite family. A correction using constant pressure in this region is shown for the Mach 8, 10-degree curve. The effect is not large and only reduces the thrust recovery by less than 3%. The curves are easily recognized because

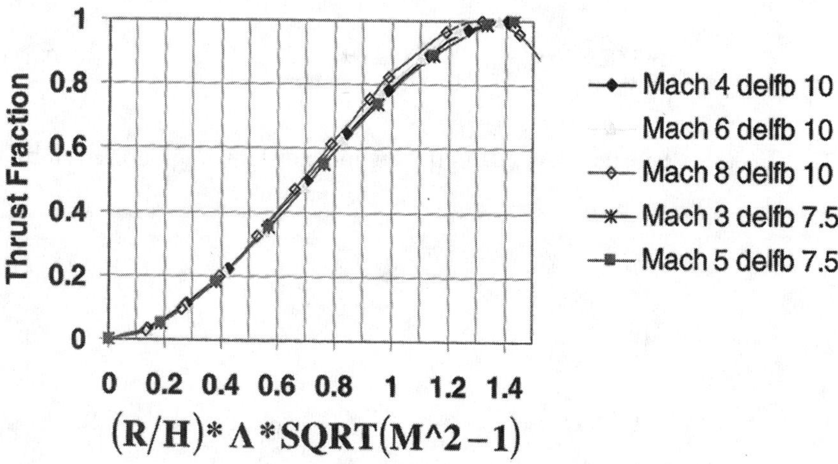

Fig. 24 Thrust Recovery of Smooth Turn in Flowpath.

these curves were overturned and the recovery decreases after the maximum. Of interest is that the different solutions are nearly collapsed to a single curve in the modified coordinates.

A caution is that the cost of the thrust turn is not free. The internal flowpath is required to operate with a pressure gradient from ramp to cowl (which is a strong function of turn angle), that is the power of 7 for $\gamma = 1.4$.

A high-pressure ratio from cowl to body will require specific attention from the flowpath designer dealing with operation of internal flow components, and specifically, the combustor fuel distribution. The turn radius of curvature Mach number dependency indicates that the minimum radius is obtained with the turn in the combustor.

B. Lift Drag

The discussion of nozzle performance introduces the topic of accounting for the lift and drag losses due to induced drag. This question directly relates to selecting the wing size and, therefore, to determine the trade of $\mathrm{Isp_{eff}}$ and dry weight. The basic assumption in this analysis is that lift equals weight during the airbreathing portion of the trajectory. In the following discussion, the small angle approximation is used. One can write the aerodynamic lift normalized by the initial takeoff weight W_0 as follows:

$$\frac{L_{\mathrm{AERO}}}{W_0} = \frac{W}{W_0}\left[1 - \left(\frac{V_0}{26000}\right)^2\right] - \frac{L_{\mathrm{PROP}}}{W_0} \tag{70a}$$

The lift generated by the propulsion flowpath due to vehicle angle of attack (AOA) at any capture A_0 is given by

$$\frac{L_{\mathrm{PROP}}}{W_0} = K_{\mathrm{TRAJ}}\left(\frac{A_0}{A_C}\right)\left(1 + \frac{K_{\mathrm{ENG}} + 2}{\gamma M_0^2}\right) AOA \tag{70b}$$

The trajectory constant K_{TRAJ} and the engine constant K_{ENG} are given by

$$K_{\mathrm{TRAJ}} = \frac{2q}{W_0/A_C} \tag{71}$$

and

$$K_{\mathrm{ENG}} = \frac{T}{P_0 A_0} \tag{72}$$

A drag breakdown will include loss in the flight path direction as previously accounted. The aero drag of the nonpropulsion flowpath consists of a value at zero angle of attack plus an induced drag:

$$\frac{D}{W_0} = \frac{D_0}{W_0} + \frac{L_{\mathrm{AERO}} AOA}{W_0} + \frac{T}{W_0} \frac{AOA^2}{2} \tag{73}$$

One can write

$$\frac{T}{W_0} = \frac{K_{\mathrm{ENG}} K_{\mathrm{TRAJ}} (A_0/A_C)}{\gamma M_0^2} \tag{74}$$

Thus, the ratio of drag and thrust is

$$\frac{D}{T} = \left[\frac{D_0}{W_0} + \frac{L_{\mathrm{AERO}}}{W_0} AOA + \frac{T}{W_0} \frac{AOA^2}{2} \right] \bigg/ \frac{T}{W_0} \tag{75}$$

The lift curve slope in the linear range of AOA is as follows:

$$C_L' = 2\delta_W \left[\frac{2}{K_\delta} + \left(\frac{\gamma + 1}{6} \right) K_\delta \left(3 + \frac{AOA^2}{\delta_W^2} \right) \right] \tag{76}$$

The C_{DO} for a double wedge airfoil is given by

$$C_{D0} = \frac{4}{M_0} \delta_W^2 + \frac{\gamma + 1}{3} M_0 \delta_W^4 \tag{77}$$

To solve the set of equations, one must find the angle of attack by solving the lift in terms of lift curve slope with the required wing area, S_W, as follows:

$$\begin{aligned} AOA\, C_L'\, K_{\mathrm{TRAJ}} K_W &= \frac{W}{W_0} \left[1 - \left(\frac{V_0}{26000} \right)^2 \right] \\ &\quad - K_{\mathrm{TRAJ}} \frac{A_0}{A_C} \left(1 + \frac{K_{\mathrm{ENG}} + 2}{\gamma M_0^2} \right) AOA \end{aligned} \tag{78}$$

The wing area includes the cowl shadow, fuselage nonpropulsion planform area, the area assigned to the forebody spill flow at speeds below the shock-on-lip (SOL) Mach number, and the wetted wing area external to the fuselage. A useful reference notation for the wing area is the freestream projected geometric capture area. The only reason that K_W is not constant is the variation of the forebody spill area in the total; this is a small term at hypersonic speed, viz.,

$$K_W = \frac{S_W}{2A_C} \tag{79}$$

The final result is given by

$$AOA = \frac{W/W_0\,[1-(V_0/26000)^2]}{C_L'\,K_{\mathrm{TRAJ}}\,K_W + K_{\mathrm{TRAJ}}(A_0/A_C)(1 + K_{\mathrm{ENG}} + 2/\gamma M_0^2} \tag{80}$$

The set of equations for angle of attack involves determining the effective I_{sp} as well as the drag on the trajectory flight path.

A series of charts for the angle of attack, lift, drag, and integration efficiency are found for an 8-degree half angle conical forebody using mass ratio along the trajectory given in Figs. 25–28. The AOA just reaches 3 degrees at $M = 5$, but falls below 2 degrees at Mach 12. The wing could be made smaller, which counterintuitively reduces induced drag even though the AOA increases. That is because the engine lift is twice as efficient as the wing and also a smaller wing reduces the zero lift drag. Thus, reducing the wing area to nothing is strongly desirable, the lifting body concept emerging as the vehicle's shape. The merits and complexities of a lifting body HTOHL SSTO system is a separate subject.

In the current example, the vehicle SOL Mach number is 12. Fig. 28 shows the integration efficiency, η_{PdA}, along the trajectory; the value obtained in the example is quite good, being above 90%.

C. Flow Turning and Overall Design of Inlet

In Secs. III and IV, the aerothermal design of the inlet internal flowpath and the design and analysis of forebodies were presented. The inlet flowpath then enters the isolator-combustor element, and undergoes a turning at the junction between the inlet and the isolator-combustor. The turning of the flow, therefore, has an important bearing on the overall design of the inlet. This section serves to discuss the aerothermodynamics of the flow turning process.

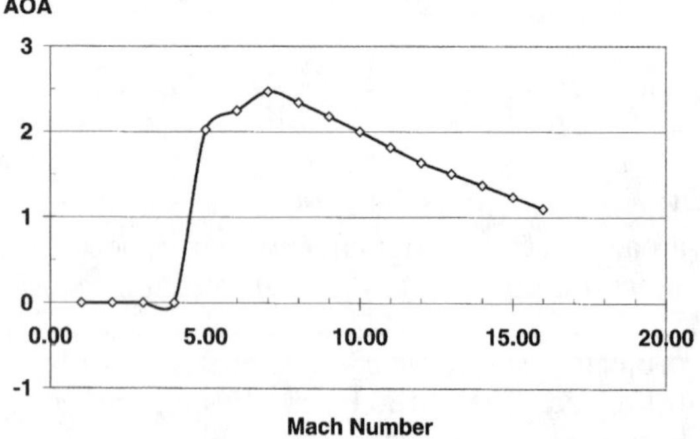

Fig. 25 VTOHL Vehicle Angle of Attack with Propulsion Lift Included.

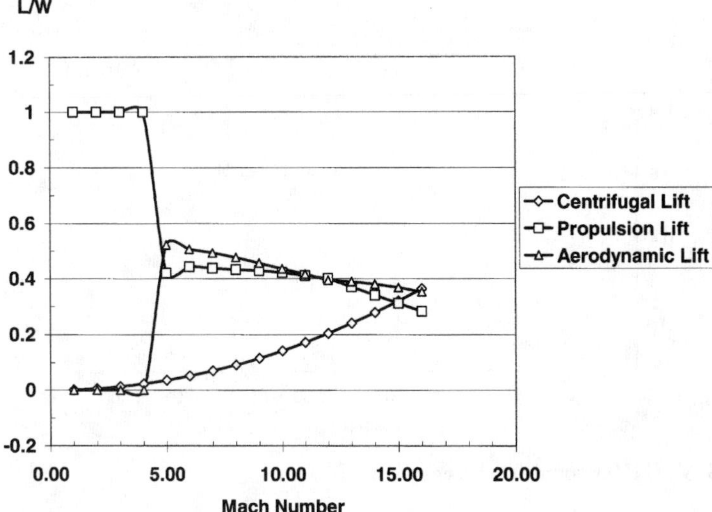

Fig. 26 VTOHL Vehicle Lift Decomposition for Propulsion, Wing, and Centrifugal Lift.

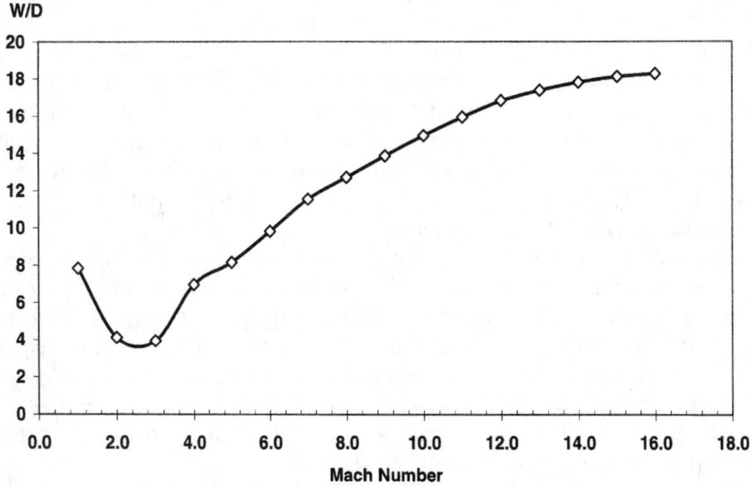

Fig. 27 VTOHL Vehicle Effective Lift to Drag Ratio.

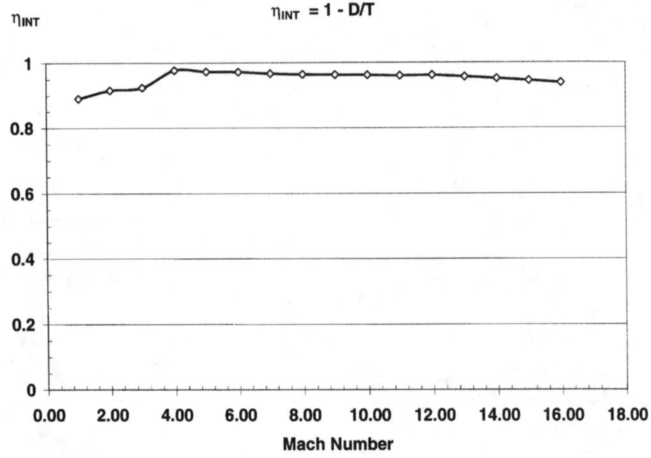

Fig. 28 VTOHL Propulsion Integration Efficiency.

Also note the effect of the flow turning on the forces generated in the inlet. An earlier analysis introduced a separation of pressure forces from shear forces, assuming an on-axis flowpath. The pressure forces included those on the contraction surfaces of an inlet (and, by implication, also on the expansion surfaces of a thrust nozzle) due to area changes, and wave drag as well as viscous drag; the latter were accounted for by using the pressure coefficient K_{WP}. Now, during flow turning, without area change, there arises an increase in forebody drag compared to the value calculated with an on-axis flow, the force being generated entirely as a result of flow turning.

Note that, on one hand, there is an additional force due to flow turning, and, on the other hand, the internal flowpath, from the point of view of flowpath analysis, has suffered only friction and wave drag losses, if any, due to the turning under a 1D flow assumption. Also, from an overall vehicle system point of view, the design requirements in propulsion demand both flow compression with change of area as well as flow turning following the area change; thus, the design problem, becomes even more complex. The overall design of the inlet must include the internal flowpath, the forebody, and the flow turning at the cowl section.

As the flow enters the cowl section, it is usually turned back along the flow waterline, either at entry to the combustor or by the location of the combustor exit section; that is, the manner of turning at the cowl may be abrupt with a sharp constant area turn without drag recovery, or the flow may be turned gradually, resulting in a drag recovery. These two methods of turning result in different entry conditions to the isolator-combustor element. In both cases, the turning at the cowl causes a cancellation of the forebody's normal force component due to the captured flow.

A discussion on how these considerations affect force accounting is presented in Sec. V.A.

Meanwhile, it is of interest to analyze the internal flowpath changes during flow turning.

1. Analysis of Flow Turning

The object here is to determinate the radius (or curvature) and length of arc of a smooth turn, which is required for the total recovery of thrust available over the flowpath. The flowpath of the turn is chosen as a two-dimensional, constant area duct; the area or the contraction ratio of the inlet is thus maintained at the original value. While the analysis is relatively simple for two-dimensional geometry, the turn may in fact be three-dimensional in practice.

The two main design parameters are determined by the following:

1) Radius of turn is established for net thrust, $(T - D)$, to become equal to the forebody cosine loss.
2) Length is found simply by the first term of a series expansion of R/H (see Fig. 24) as a function of forebody angle and the entrance Mach number.

A note of caution: thrust recovery with the turn is not entirely free. The internal flowpath is necessarily required to function with a pressure gradient from ramp to cowl; it is a strong function of turning angle. This is illustrated in Fig. 22, where the pressure ratio is given as a function of M_1 and λ, where M_1 is the entry Mach number and λ is the turning angle.

2. Overall Design of Inlet

Overall design of an inlet must, in general, include the forebody with a leading edge and a cowl entry section.

Considering the combination of a forebody (which may consist of a wedge or a cone, and an inlet flowpath), which may be 2D or 3D, and noting the two performance parameters (thrust and drag), one needs to consider eight combinations in analyzing the minimum value of K_{WP} as a function of ψ or *CR*. Obviously, the effective η_e and ΔH_c are dependent on ψ. In general, there are important considerations of viscous drag and heat transfer, as well as real gas effects, that may well cause the flowpath integrator to choose a cycle not designed for ψ_{OPT}, based on the adiabatic Brayton cycle, but rather the cycle with the lowest fuel equivalence ratio over a stoichiometric one so as to realize the highest I_{sp} (as long as it is adequate to meet the cooling needs). However, the problem of finding the best combination of ψ and minimum K_W remains, even under such other constraints. One should recall here the discussion on determining K_{WP} in the earlier section on inlets.

Complexities in the forebody-inlet combinations are 1) the blunt leading edge, 2) 3D nature of many inlets, 3) presence of unequal angles, and 4) possible operation of the inlet in the so-called overspeed mode. Here, a summary is presented on dealing with such complexities.

Drag increment due to blunt leading edge. With the case of the sharp leading edge as the basis, note that entropy changes are additive along successive parts of the flowpath, one can write the corrected drag for the case of a blunt leading edge in the form of corrections. Thus,

$$C_D = C_{D_i} + \Delta C_{D_{\mathrm{BLE}}} + \Delta C_{C_{\mathrm{INTER}}} \tag{81}$$

where the subscripts *i*, BLE, and INTER refer to the sharp leading edge, the blunt leading edge, and interference between the internal flowpath and the leading edge, respectively. The latter is large and unfavorable in the case of a wedge, and favorable for slightly blunted cones. Then, considering pressure drag, one can write

$$K_{\mathrm{WP}} = \left[ln(1-\gamma)C_D \cdot (\gamma-1)/2 \cdot M_\infty^2\right]/[(\gamma-1)ln\,CR] \tag{82}$$

One can then proceed to determine the viscous drag integral.

Drag correction for three-dimensional inlets. An interesting case is 3D compression, where a sidewall inlet is complemented with the addition of cowl and ramp compression. One then assumes that each wave source or wall adds its own contribution to the drag and that such contributions are linearly superposable. One can then write the ratio of the final corrected total pressure to the initial total pressure as

$$\frac{P_{\mathrm{Tf}}}{P_{\mathrm{Ti}}} = \frac{P_{\mathrm{tr}}}{P_{\mathrm{T1}}} \cdot \frac{P_{\mathrm{Tc}}}{P_{\mathrm{Ti}}} \cdot \frac{P_{\mathrm{Tsf}}}{P_{\mathrm{T1}}} \cdot \frac{P_{\mathrm{Ts2}}}{P_2} \tag{83}$$

where the four terms on the right side represent the ramp loss factor, the cowl loss factor, sidewall one loss factor, and sidewall two loss factor, respectively. The latter is the same as sidewall one loss factor unless the inlet has a yaw angle; in that case, differences arise in effective compression angles between sidewalls one and two.

This allows the overall wall pressure coefficient to be written as follows.

$$K_{\mathrm{WP\ overall}} = \sum \frac{K_{\mathrm{WP}i}\,\ell n\,CR_i}{\ell n\,CR_T} \tag{84}$$

This may seem somewhat arbitrary since CR_i have not been uniquely defined. This is especially true in a 3D sidewall inlet where waves can be found to travel in all directions, not just up and down and across. However, much success has occurred by using Eq. (62), probably because the overall CR remains the same as that being considered whatever the CR_i are. The form of the expression for overall K_{WP} indicates that the average compression angle is a weighted rms value of the angle.

Also note that in the special case of isentropic flow turning, K_{WP} simply becomes zero.

Two cases of interest are 1) contraction on four walls, top, bottom, and two sides; and 2) compression in one direction over various angles.

Two reasons for using unequal angles are 1) design or packaging may require a larger angle on one wall compared to the other, and the interest is in the penalty resulting from the choice of such angles; and 2) operational requirements, for example, from penalties due to an angle of attack or yaw, which would effectively change the apparent angle of compression relative to the oncoming stream.

Details on analysis of the effects of unequal angles can be found in Appendix D, Sec. C.

At a high enough flight Mach number, forebody shock can be expected to intercept the cowl lip. Further increases in Mach number will cause the shock to move inside the cowl region. When this occurs, the inlet is said to be operating in an *overspeed mode.* Air entering the inlet flows in two directions and has two values of Mach number.

Applying the K_{WP} formula here causes additional difficulty because of the need to account for two apparently adjoining flows. The problem is made tractable by introducing a slip line (with flow direction and pressure continuous across it) between the two (interacting) parts of the flow. The methodology for dealing with the overspeed situation of an inlet is discussed in Appendix D, Sec. D.

At low speed the primary issue of the hypersonic inlet is starting and mass capture. The starting characteristics of inlets are discussed in Trexler (1988), and Anderson (1991). Useful recipes are presented for starting contraction ratio with Mach number as well as estimates of the mass capture.

3. Review of inlet compression process. The compression process may be considered from the point of view of drag work. Several sources of drag are identified: 1) D_{ideal} — frictionless adiabatic (isentropic) compression drag for compression to the minimum cross-section (throat) of the engine flowpath; 2) D_{wave} — drag due to shockwaves; 3) D_{BLE} — drag due to entropy layer production at the leading edge of the forebody; 4) D_{vis} — drag due to viscosity effect; this leads to additional work by the fluid in the form of heat, which can be recovered in a fuel friction cycle in part; and 5) D_{turn} — drag in turning the flow from the flight path freestream direction.

In searching for a good design, an optimum design requires considering all of the above. Often in practice, one may include some amount of isentropic ramp compression to reduce wave losses with a constant cross-sectional area; however, one then must account for drag due to turn, unless appropriate measures are taken to recover smoothly the associated wave losses.

With all the pressure drag components accounted for, an overall K_{WP} can then be determined, suitable for viscous flow calculations.

Based on the foregoing design procedure, one can proceed to assess the best technology for inlet compression. As expected, the 3D inlet on the cone with the thrust cowl appears to be the best choice, and, thus, becomes the starting point for designers considering modifications and improvements. The flowlines established in this procedure also become a suitable metric for comparison of various designs, and also flowpaths, that may be established through more sophisticated fluid mechanical calculations.

D. Force Accounting Approaches

Two approaches, one useful for the propulsion group and a second that is useful for dealing with the vehicle group, are discussed in Appendix F.

VI. Combustor

This section presents a discussion from two points of view: 1) cycle performance and 2) practical operability. The heat addition cycle offers many more options in a supersonic combustion environment than the constant pressure process in the adiabatic Brayton cycle scramjet. As expected, the heat addition process affects overall cycle performance and the flowpath optimization. Thus, the effects of heat addition gas dynamics on cycle performance will be addressed first. The concerns of friction drag and heat transfer are practical factors in selecting the combustor approach. These considerations will lead to investigating the friction cycle side of the performance equation.

The practical operability aspects include ignition delay and reaction time, mixing, base pressure and flame holding. Some major aspects are covered after cycle considerations.

Heat addition cycles: The scramjet offers many modes of heat addition. In the 1D analytical view, several possibilities exist for conservation analysis to be closed. A partial list includes 1) constant area, 2) constant density, 3) step combustor, 4) constant pressure, 5) constant Mach number, and 6) constant temperature. The processes listed exclude any cases in which the combustor area decreases. The reasons are twofold. First is the belief that area contraction belongs to the domain of the compression part of the cycle and for a given contraction, it would be better to start heat addition after all contraction. The second reason is that all 1D treatment of processes above admit jump solutions and do not require a heat addition distribution to be known, which is required for heat addition in a contraction. See Billig and Dugger (1963).

However, the 1D restrictions can be circumvented by considering 2D wave dynamics for shocks and flames from the Hugoniot equations. Such a wave dynamics approach is used to search for the optimal combustion process, including combustors with contraction. A. Ferri (1969), was a pioneer in the application of combustion wave dynamics in the design of scramjets. Stalker (1984), emphasized the importance of wave dynamics in the interpretation of experimental data in his experiments.

The wide range of speed operation has an impact on the type of combustor selected when avoiding inlet unstart due to heat choking is considered. The previously examined inlet, thermal compression range, $\psi = 4 - 5$, provides mixed fuel air gas temperature below the autoignition temperature of hydrogen, Huber (1979). Reliable combustor operation requires continuous ignition to effect flame holding. Huellmantel (1977) introduced a low drag cavity flame holder applicable to scramjet combustors which has been applied with success by Vinogradov (1990). Typically, flame holding is provided by a recirculation zone, such as the wake of a strut, step, cavity, or in the case of hydrogen, the separation around a fuel jet. Morrison (1997), has surveyed the flame holding aspects of high speed combustion for common flame holder types. This consideration, along with thermal management requirements, prompts the evaluation of the combustor area as one of the primary design specifications of the combustor.

This brief discussion leads to the question of presenting the cycle performance results on a chart with the parameter space that unifies the results and allows ready interpretation of the cost of a combustor feature.

A. Isolator

The isolator is an important component of the dual-mode scramjet or the conventional mode ramjet because it enables the inlet to remain started when the combustion heat chokes the flow. This isolation is needed for the conventional Brayton cycle ramjet also D. Pratt and W. Hieser (1993), have treated the isolator in a very satisfactorily much the same as this work. The physics of operation is the same as that of the pseudo shock diffuser, which finds application in supersonic wind tunnel diffusers, supersonic ejectors, supersonic aircraft inlets, and subsonic mode ramjets. The cross-sectional area may be constant and provides the highest pressure recovery, normal shock efficiency, except at low entrance Mach numbers. In the latter flow process, the normal shock Mach number downstream is very near unity. The flow downstream is then choked by the distortion generated in the separation and reattachment process in the real flow after compression, but not upstream in the supersonic flow. Operating in this regime is always unsteady due to large oscillations in duct pressure. To obtain nil to low total pressure losses in the normal shock, supersonic aircraft were designed to operate in this region. To operate smoothly, many bleed holes are involved in the upstream region and in the throat. The wind tunnel avoids the oscillation by slightly diverging the isolator wall by a small angle. The earliest visualization of the process was in Shapiro (1954). Ironically, it has taken a long time to understand and model the process in a constant area channel.

The work of Ortwerth (1977b), is briefly outlined and is one of several papers published on distortion theory in that issue. The outline of the solution is presented here with the notation being the same as this text and some of the physics is made more explicit. The supersonic solution is obtained first.

1. Scramjet Isolator

For a scramjet engine operating above the dual-mode range, some purpose for the isolator other than a source of drag and heat is needed. The isolator is available for various geometry mechanisms to continue the contraction of the inlet. The isolator is the preferred location for ramp injectors.

A design formula for the drag and heat load of the attached supersonic flow is needed. Although the combined interaction solution Eq. (C40) is capable of solving the flow in principle, it is not now formulated for the isolator. Thus, another more direct solution is necessary to avoid reformulation.

Based on the analytical approach adopted in the discussion of inlets, one can write the following equation for the entropy change in the isolator duct, in a fashion similar to Eq. (C18):

$$\frac{\mathrm{d}s}{R} = \left(\frac{P_W - P}{PA}\right)\mathrm{d}A + f_D \frac{\tau_W\,\mathrm{d}\sigma}{PA} \tag{85}$$

where the first term on the right refers to the pressure drag and the second to the viscous drag.

Denoting the pressure drag and the associated drag coefficient by D_p and C_{D_p}, respectively, one can write

$$\frac{\mathrm{d}D_p}{\mathrm{d}\sigma} = \frac{C_{D_p}\frac{1}{2}\rho V^2 A_{\mathrm{iso}}}{\Sigma_{\mathrm{iso}}} \tag{86}$$

Here Σ is the wetted surface area and A the isolator cross-section area

$$\frac{\tau}{\tau_{\mathrm{ref}}} = \left(\frac{P}{P_{\mathrm{ref}}}\right)^n = 1 + n(\tilde{P} - 1) \tag{87}$$

where the reversible heat transfer is expressed in terms of the pressure ratio.

It then follows that the pressure ratio in the isolator can be expressed by the relation

$$\frac{P}{P_2} - 1 = \tilde{P} - 1 = \frac{a}{b}\left\{\exp\left[(\gamma - 1)\,b\left(\frac{\sigma}{\Sigma_{\mathrm{iso}}}\right)\right] - 1\right\} \tag{88}$$

where

$$a = \tilde{D}'_P + f_D\,\tilde{D}'_V \tag{89}$$

and

$$\mathrm{b} = \mathrm{n}f_{\mathrm{D}}\,\tilde{D}_{\mathrm{v}} \tag{90}$$

Here the viscous drag is denoted by D_v.

Integrating Eq. (86) and introducing the relation in Eq. (88), the drag of the isolator is given by

$$\left(\tilde{D}_V\right)_{\text{ISO}} = \left(\frac{\tau}{P}\right)\left(\frac{\Sigma_{\text{iso}}}{A_{\text{iso}}}\right)\left\{1 + n\frac{a}{b}\left[\frac{(e^b - 1)}{b} - 1\right]\right\} \tag{91}$$

Also, using the temperature relation in Eq. (91), the heat transfer can be expressed by

$$\frac{Q}{\dot{m}h_o} = f_q \frac{D_v}{\dot{m}h_0} \tag{92}$$

Finally, one can write the combined pressure-viscosity effect in the isolator in terms of

$$\Delta\psi = \psi_3 - \psi_2 = (\tilde{P} - 1)\,\psi_2 \tag{93}$$

where ψ is, as before, the local thermal ratio.

B. Dual-Mode Combustor Isolator

In the ramjet mode, a serious problem of interaction with the inlet happens when the amount of heat addition is such as to cause the flow to choke and the inlet to unstart. An isolator, in the true sense of the word, is then essential. The isolator does not remove the choking limit but enables the engine to operate at the limit if desired.

The basic flow contains three regions, as shown in Fig. 29. Region I is a core flow that experiences a pressure gradient in the supersonic region by increasing area contraction and in the subsonic by expansion of the area; and the separation region III balances the pressure gradient by shear stress on the outer boundary, which is region II. Since the pressure gradient in the core flow must equal the pressure gradient, that shear can support in the separation region flow an interaction equation is established. The pressure ratio $\tilde{P}$ over a length Δx can be shown to vary as follows:

$$\frac{\mathrm{d}P}{\mathrm{d}X} = 4K'(\gamma PM^2) \tag{94}$$

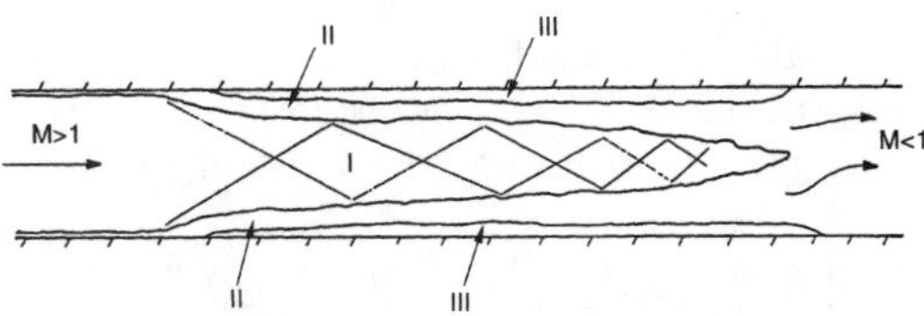

Fig. 29 Flow Model for Normal Shock in Duct.

where

$$4\,K' = (44.5C_{fo}) \tag{95}$$

A solution of this equation, along with the relevant conservation equations, determines the interaction length required for any pressure. The solution in normalized coordinates is

$$\tilde{\ell} = \frac{g_1^2}{\gamma f_1}\left[\frac{\tilde{P}-1}{(f_1-\tilde{P})(f_1-1)} + \frac{1}{f_1}\,\ell n\,\frac{\tilde{P}(f_1-1)}{(f_1-\tilde{P})}\right] + \frac{\gamma-1}{2\gamma}\,\ell n\tilde{P} \tag{96}$$

where $\tilde{P} = P/P_1$ pressure ratio.

The length is related to actual distance by the definition,

$$\tilde{\ell} = 4K'\left(\frac{X}{D_H}\right)$$

$$D_H = \text{hydraulic diameter} = \frac{4A}{\text{perimeter}}$$

$$f_1 = \frac{F_1}{P_1 A_1}$$

where F_1 is the stream thrust at start of interaction,

$$g_1 = \frac{\dot{m}a_T}{P_1 A_1}$$

where $\dot{m}$ is the mass flow,

$$a_T = \sqrt{(\gamma-1)H_T}$$

where H_T is the total enthalpy,

$$4\,K' = 44.5C_{f1}$$

where C_{f1} is the initial friction coefficient.

In the author's experience, the foregoing relation presents the best estimate over a wide range of Mach numbers, ratio of specific heats, and cross-sectional shapes of common use.

Comparing experimental data for overall recovery length to achieve the full pressure rise available, is shown in Fig. 30 from the original reference. The comparison of overall isolator length is comprehensive, covering rectan-

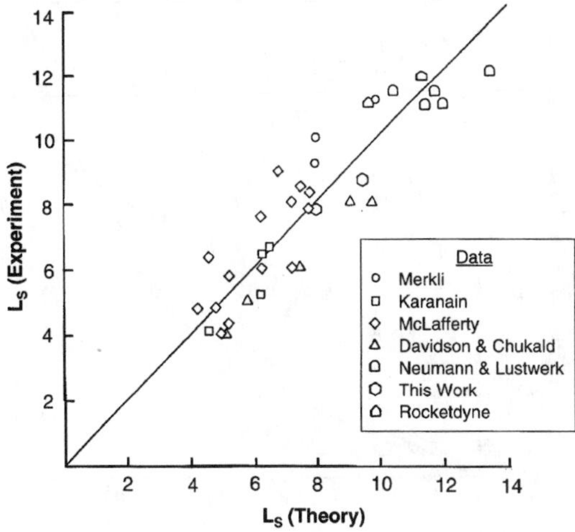

Fig. 30 Goodness of Fit for Shock Length to Maximum Pressure Rise in Circular and Rectangular Ducts.

gular, circular, and square channels for a range Reynolds number that differed by a factor of 40, Mach number from 1.5 to 5. Most experiments were carried out in air. The required length depends on a large number of initial conditions and fluid properties though Reynolds number and the isolator entrance friction coefficient law. This dependence was extracted by varying each parameter separately.

1. Generalized Results

The length of the shock train in an isolator can be expressed by the six factors that follow:

$$\mathrm{L_{sh}} = f_S \cdot f_M \cdot f_\theta \cdot f_\gamma \cdot f_\mathrm{W} \cdot f_\eta \cdot 15 \tag{97}$$

These factors are presented in Figs. 31–36. The subscripts stand for the following: S is the shape of P distribution with distance, M is the Mach-number dependence, θ is the initial Reynolds number based on momentum thickness, γ is the ratio of specific heats, $W = T_W/T_{\mathrm{AW}}$, and η is the normal shock efficiency $= 1 - \delta^*{}_\mathrm{A}$. The shock structure factor, f_S indicates how performance of an isolator varies with length, and the considerable difficulty in accurately determining shock region length in experiments, since 90% of the pressure rise occurs in 70% of the length. The other factors show the influence of various initial conditions.

A final comparison between predictions and experimental data is shown in Fig. 37. The data were obtained in experiments conducted with a rocket located in a diffuser under altitude simulation conditions; the initial conditions of air were far removed from the ambient atmospheric conditions. The

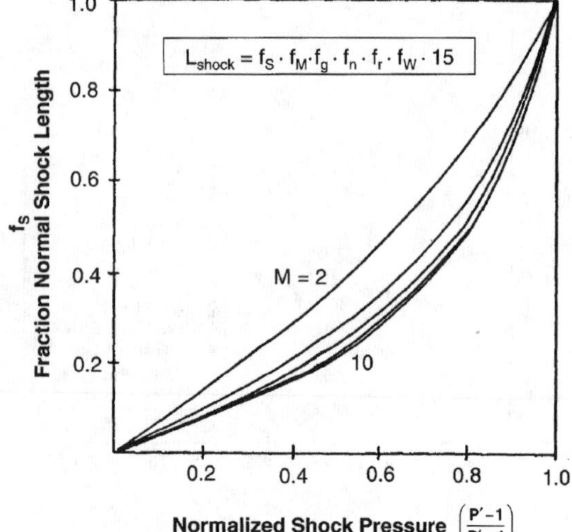

Fig. 31 Length or Shape Factor f_s for Normal Shock Length Recipe as Function of Design Pressure Rise.

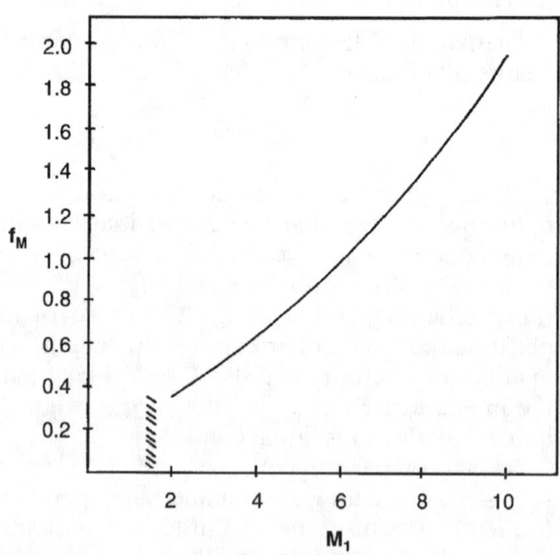

Fig. 32 Mach Number Factor f_m for Normal Shock Length.

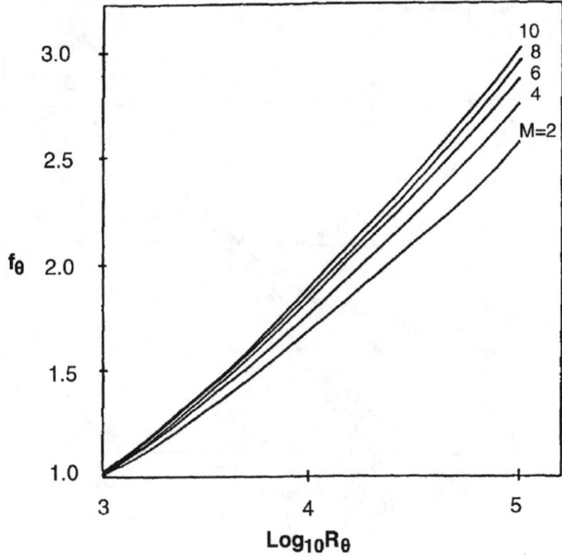

Fig. 33 Initial Boundary Layer Factor f_θ for Normal Shock Length.

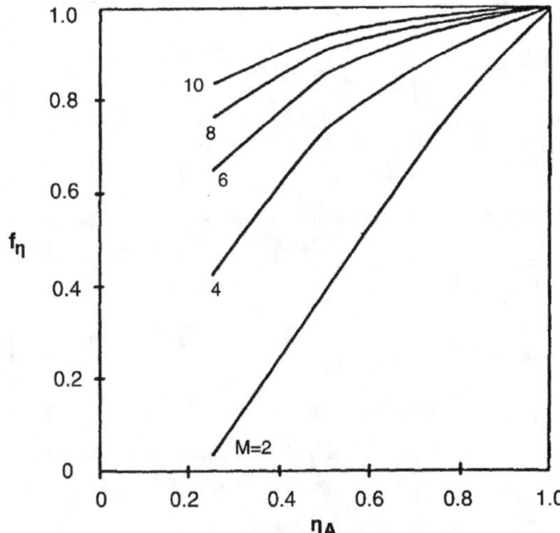

Fig. 34 Initlal Distortion Factor f_η for Normal Shock Length Mach Number is Parameter.

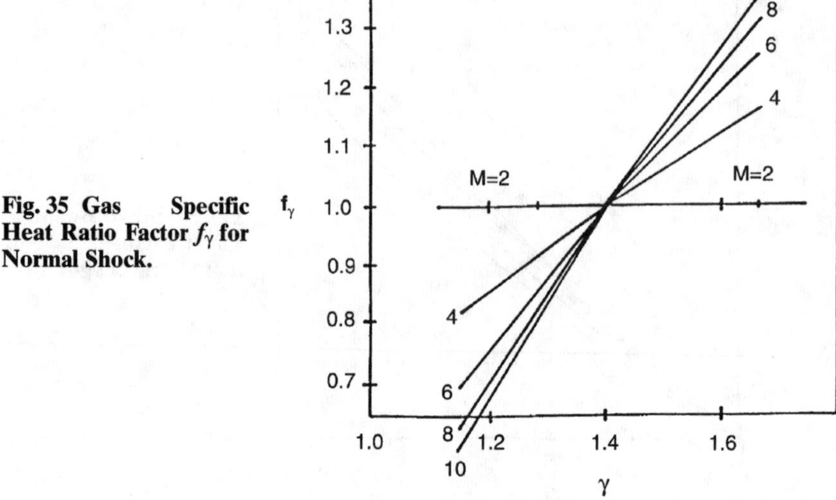

Fig. 35 Gas Specific Heat Ratio Factor f_γ for Normal Shock.

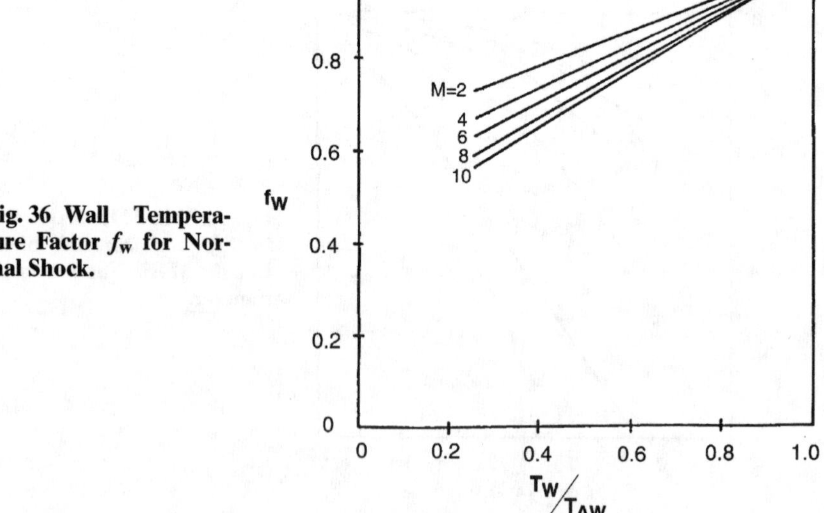

Fig. 36 Wall Temperature Factor f_w for Normal Shock.

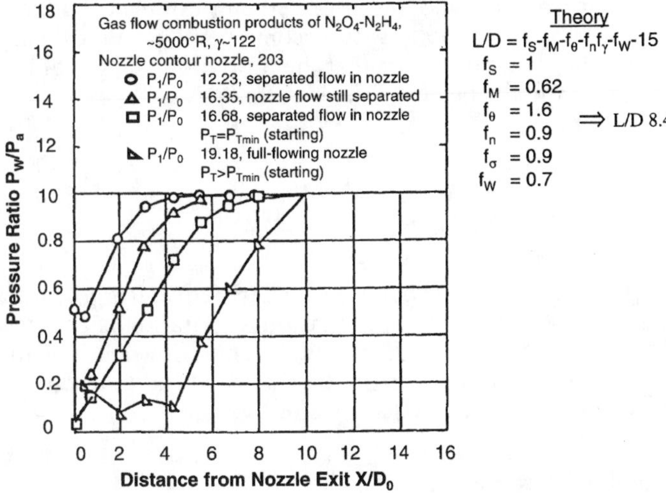

Fig. 37 Comparison of Normal Shock Recipe with JPL Data.

agreement between predictions and data is excellent in the variation of pressure ratio as a function of length.

Thus, the formula or the graphic results can be used to scale subscale tests to flight scale.

2. *Some Complexities*

However, various scramjet features can cause problems and, therefore, need to be accounted for. One concern is that nonuniformity conditions at isolator exit cause distorted inflow conditions in the combustor. Such distorted flow occurs when, for example, the entrance boundary layer is rather thick on one side (or wall) of the isolator. The boundary layer on that wall will then undergo separation and the other walls apparently become reflection walls for the waves created. This can have a dramatic effect on the isolator by reducing the pressure gradient and length required for a desired pressure rise. This is a situation where an inlet ingests the forebody boundary layer, doubling the hydraulic diameter. The remedy is to introduce a large boundary layer bleed at the entrance to the isolator for the wall with the large boundary-layer build-up.

Another problematic scramjet feature is non-axial flow at the entrance to the isolator. The isolator can affect the flow only when the flow is parallel to one of its walls. The flow requires some distance to become parallel to a wall when it is initially at an angle to it. This results in what is called a slick spot or slip region at the entrance. Its occurrence can be explained, for example, in the presence of a compression ramp as follows: the flow undergoes separation at the corner of the ramp and isolator in a direction parallel to the ramp, which immediately produces a large displacement of the slipline and no pressure rise results; the displacement, however, interacts with a shockwave, which causes the flow to become parallel to the wall. For geometrical considerations, it follows that the ratio of the length of the separation region, ℓ_{sep}, and the height of the isolator, h, is given by

$$\frac{\ell_{sep}}{h} = \frac{2}{\tan\delta_{sep} + \tan\delta_{sh}} \tag{98}$$

where δ_{sep}, the angle of flow separation, is nearly equal to θ_{ramp}, and δ_{sh} is the shock angle. It may be pointed out that this separation is an inviscid flow phenomenon, caused by the angularity of the entry flow. Küchemann (1978), discussed this as "separation of the first kind." It is quite different from "ordinary separation" due to viscosity effect.

C. Detonation Wave Engine

The detonation wave structure as first proposed by von Neuman consists of two parts or processes, first a normal shock front solved as a nonreacting Hugoniot jump, followed by a chemical reaction and heat release zone solved as a subsonic deflagration front. This is based on the consideration that at ordinary densities, the fluid time scale through a shock is at most a few times 10^{-11} sec, while the kinetic time scale is a few times 10^{-6} sec or longer as such a clear separation of processes can be expected. This leads to the utility of a shock tube for physical chemistry studies of reaction times. The stable convection speed of a denotation wave is found to be easily calculated with the aid of the Chapman-Jouget (C-J) stability condition for speed of burned gases relative to the shock wave being exactly equal to the speed of sound in the reacted gases. This condition is also known to be the condition of minimum entropy of the burned gas state, Hirschfelder (1954). This is a guide for an optimum combustion cycle search.

Consider the normal detonation wave fixed in the combustor. At some critical flight speeds, the flow velocity entering the combustor equals the C-J detonation velocity, and a detonation wave can form in the engine minimum area section. For analysis, a constant area combustor calculation will yield the same answer as the detonation model, since the area terms cancel and the dynamic equations become the same as the Hugoniot equations.

For flight above this stable C-J speed, the detonation wave will blow out of the exit of the combustor. This condition can be undone if the detonation is stabilized by an inclination of correct angle to result in the normal velocity to the wave corresponding to the C-J velocity. This turn angle can be found from the solution of the Hugoniot equations for two dimensions and is called an oblique detonation wave. If the engine continues to accelerate, the wave angles become shallower and the wedge must move upstream. The stabilization angle of the wedge and combustor opposite wall also become steeper. The combustor cross-sectional area always contracts for the oblique detonation solution above the C-J flight speed. Historically, such engines are proposed by designers to operate in this mode by premixing the fuel and air in the flow upstream of the combustor in the inlet. Such engines are given the appellation oblique detonation wave engines (ODWE).

Below this C-J flight speed, the detonation wave will run out of the front of the combustor into the inlet. This condition can be stabilized by providing an area expansion after the shock and transferring fuel injection downstream to a larger area to increase the total stream thrust to the critical value and permit the deflagration wave to accelerate the flow to Mach one. Thus, the single detonation wave, supersonic combustion model has now been separated into two separate waves for gas dynamic control reasons, while retaining the C-J stability condition. When first proposed, this was thought to be beyond the state of the art, however, it was demonstrated successfully in the early 1960s, which resulted in a cycle patented by Curran and Stull called the thermally choked dual-mode ramjet (DMRJ). As the engine accelerates, the choke plane can be moved back up to the flowpath minimum area just behind the shock, which is assumed stationary during the entire process by means of variable geometry and/or proper fueling techniques.

Another way to operate in the dual-mode region is to provide a two-stage combustion scheme with a constant or step constant area combustor where enough fuel and heat are provided to choke the combustor at less than stoichiometric fueling and then directly inject fuel downstream to burn the remaining fuel at a constant Mach number. As the vehicle accelerates, the fuel in the first stage combustor can be increased while turning it down in the second stage which requires no variable geometry and a simpler injector control scheme. This is called the constant mach cycle DMRJ. These cycles are called dual-mode ramjets because they hold a normal shock in the engine throat until the speed increases above the C-J flight speed. Above C-J speed, the cycle becomes a true scramjet, with combustor exit Mach numbers above one and a full pseudo-shock no longer rests in the isolator.

D. Application to a Dual-Mode Combustor

The discussion of dual-mode combustor operation forms the preamble of the scramjet combustor section and is important in its own right. In favor of its understanding is the fact that the Mach numbers are low enough for the flux conservation analysis to be easily interpretable and accurate. The complexities of dual-mode operation are high, however, and some thought will be expected to follow the development. The introduction of the dual-mode cycle is due to E.T Curran and F.D. Stull of the Air Force Aero-Propulsion Laboratory in the 1960s in a classified patent. The ideal operation included establishing a plane combustion wave with subsonic entrance flow and sonic exit conditions. In practice distributed heat addition must be accounted for and the ideal cycle envisioned by Curran and Stull is only approximated in practice. Even so all thermally choked engine cycles are termed dual-mode today. In certain Russian texts, Baev (1986), the term "psuedo shock combustion" is one the author finds descriptive.

The isolator is a key element in a dual-mode scramjet, and determines both the efficiency and the fuel schedule over the transition from subsonic flow operation to supersonic flow operation. The cycle parameters that one chooses for the low speed operation are important both from an operational point of view as well as in determining the performance of the engine by theoretical considerations. Lopez (1970), experimentally optimized staged combustors in dual-mode operation during the NASA HRE program, in the 1960s. He concluded that the best performance was obtained with a cycle in which combustion occurred at a constant Mach number close to unity, before the inlet interaction caused spillage and thrust loss. Since then, it is understood theoretically that this type of cycle is the best that can be achieved in practice.

As stated earlier, the three components, namely the isolator, the combustor section following it, and the divergent section of the combustor, form the three main elements of the scramjet engine with which the inlet and the nozzle are to be integrated. For generality, the combustor is anticipated to include a step at the entrance for the incorporation of mixers or flame holder elements. The length of the isolator and the area ratio of the combustor step are linked. The isolator in a dual-mode scramjet determines both cycle efficiency and the fuel schedule. Also of importance is the connection with the scramjet. Kenworthy (1965), was one of the first to show that a step combustor can suffer no performance loss in the scram mode, depending on base pressure. Thus, the study of dual-mode with a step combustor is on sound ground theoretically.

An analysis can be carried out to determine engine performance dependence on isolator pressure ratio in continuation of what has been presented in the preceding section. The details are found in Appendix C. The issue addressed is whether a step area change that allows isolator exit pressure to act on a base area can be optimized to find the best step area to choose, based on engine performance and fixed normal shock efficiency.

The combustion is always completed at a stoichiometric fuel air ratio and exit plane Mach number of 1. Thus, the only unknown is the area or pressure at the combustion exit since the product of pressure and area is a constant. This follows from the total stream thrust being constant at the burn plane. The total temperature after combustion is known from φ and ΔH_C. The above remarks can be deduced from simple inspection of the N function at Mach 1.

$$\frac{F^*_{\text{Comb}}}{P_0 A_0} = \frac{\dot{m}}{N^* P_0 A_0}\sqrt{\frac{RT_{\text{Tc}}}{\gamma}}$$

$$= \frac{PA}{P_0 A_0}(\gamma + 1) \tag{99}$$

The stream thrust is related to the area and isolator pressure rise so that for a specified geometry the maximum heating can be obtained at any flight Mach number. The maximum heating in a step combustor can be found by noting that the stream thrust can be written

$$\frac{F^*}{P_0 A_0} = \eta_{NS} \frac{P_{NS}}{P_0} \frac{(A/A_{ISO} - 1)}{CR} - \frac{CR^{(a-1)} - 1}{(\gamma - 1)} + \gamma M_0^2 + 1 \qquad (100)$$

The equation relates the heating to the step area and the isolator pressure rise.

An analysis can be carried out for the performance of the isolator as a part of the engine, in continuation of what has been presented in the immediately preceding section. Such analysis may be carried out under two sets of conditions: 1) determination of heat input in a constant cross-section combustor, such that the flow Mach number becomes close to unity or choked; and 2) determination of heat input at constant Mach number near unity.

The pimary result from the analysis relates to the drag suffered as a function of initial conditions and length of the isolator when the heat input follows a specific fuel schedule as a function of either (1) flow Mach number at entry and exit from the combustor or (2) the near constant flow Mach number at which the combustor is operated.

1. *Case 1*

In case 1, although choking limits smooth heat input in the combustor, to approach this condition in practice, it appears that one should choose to operate at a Mach number greater than unity for stability margin or throttling the engine to meet mission thrust demand needs for best overall performance.

2. *Case 2*

This is the more interesting case in which the isolator, the first stage of the combustor (with constant area), and the second stage (with near constant flow Mach number) are considered together. Fuel supply into the first stage and combustion cause the pressure to rise and the Mach number to low value, but larger than unity. Additional fuel in the second stage then burns, causing a reduction in pressure while the Mach number tends to increase.

Application of the dual-mode scramjet is summarized by an example. Anticipating the results in Sec. VIII, a fixed geometry engine was analyzed. The first result was the ideal step combustor expansion required to complete stoichiometric combustion in stage I as a function of the flight Mach number and the normal shock recovery (Fig. 38). The data show, as expected, that the area ratio at Mach 2 and 3 is rather large and undesirable for integration with the scram mode. An expansion, $\varepsilon_c = 2$, was selected as a reasonable compromise for the step expansion, based on the very low freestream capture at low speed resulting in low installed thrust under best conditions and a compression ratio for robust combustion at $M = 12$ in scram mode. Below a freestream Mach number 4, the system requires staged combustion type I and type II. The installed thrust, based on freestream ambient pressure and the cowl projected area, is shown in Fig. 39. The comparison then shows that

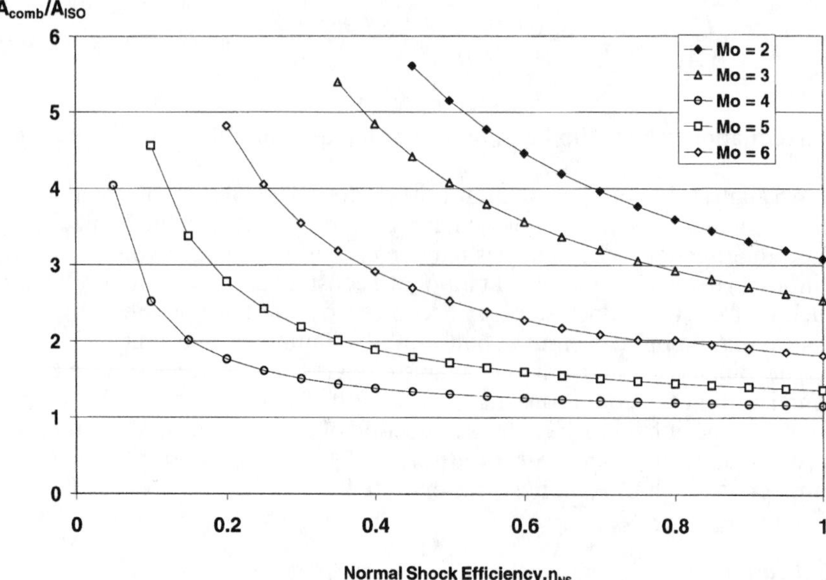

Fig. 38 Dual-Mode Scramjet, Single Stage Step Combustor Area Ratio, ϵ_c, Required as a Function of Isolator Shock Efficiency μ0 Mach Number.

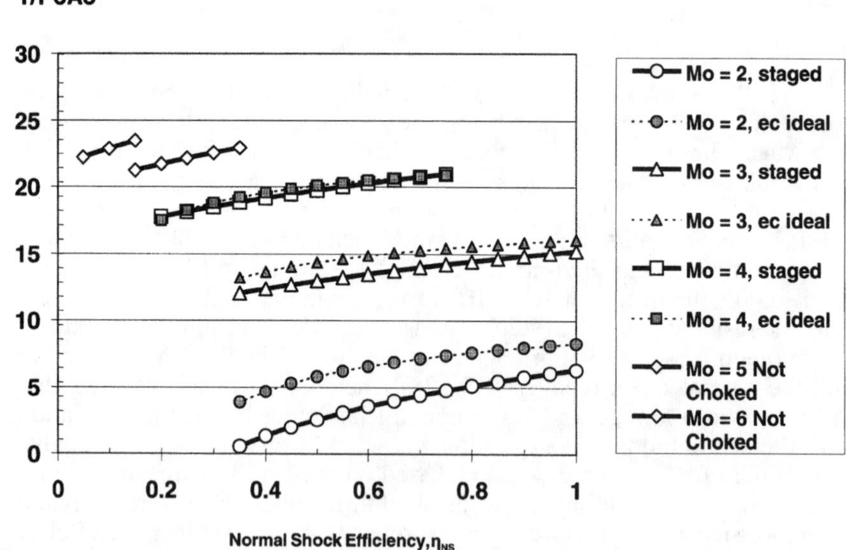

Fig. 39 Dual-Mode Scramjet Specific Thrust for Two-Stage Combustion, ϵ_c = 2.0 Fixed, Compared to Ideal Single-Stage Combustor as a Function of η_{NS}, Mo.

the loss of thrust for the staged combustion mode compared to the ideal complete stage I type operation is slight at Mach 3 and above.

3. *Test with Type II Combustion*

It is obviously of interest to examine all test data when available. In a specific series of tests, on an engine with this type of combustion (reported in Ortwerth and Bowcutt (1995), the pressure profiles obtained as a function of overall fuel-to-air ratio in the unit are presented in Fig. 40. It can be observed that the isolator and the first stage combustor show a smooth pressure rise, while the second stage and operated in the type III mode only as an extreme example of the performance expectation. The diverging combustor was operated in a close approximation of a type II combustion cycle by piloting the main fuel for stable operation. The result is surprisingly good and was helped by the extra length of the combustor, which provided added isolator pressure rise before combustion.

The maximum pressure shows a direct dependence on φ, on the fuel equivalence ratio, which was controlled by second stage fueling after the pilot ignition in the first stage was established. Thus, even though the overall flow was behaving like one of constant Mach number, the inlet isolator seems to adjust to the overall fuel level, and thrust increased nearly linearly with φ. In this respect, the engine was behaving like a ramjet where the upstream pressure is increased by heat addition, and, thus, provided proof of the concept of dual-mode scramjet operation.

From the analysis of the combustor presented in Appendix H, Sec. A, an expression is presented in Eq. (H37) for an approximate value of total thrust obtained by the three-part isolator-combustor unit. This is based on assuming a variation in pressure in the second stage of the combustor, given by an area-based rule:

$$P = A^{-m} \tag{101}$$

and a power-law similarity rule for the ratio of wall shear stress to wall pressure.

Normalizing the total thrust generated by the thrust generated at the exit of the first stage combustor, a relation is obtained as shown in Eq. (H40). This ratio is found to increase monotonically with equivalence ratio, which determined the stagnation temperature ratio, $T_{t\mathrm{II}} / T_{t\mathrm{I}}$. Remember that this ratio corresponds to the smallest value of A_{II} and the largest value of P_{II} for the burn Mach number in the unit. The ratio is entirely consistent with data. Thus, the conditions and the thrust ratio provide a benchmark in setting operating conditions and in design.

4. *Test with Type I Combustion*

Results of freejet combustion tests using both hydrocarbon and hydrogen fuels in a dual-mode integrated scramjet engine with step combustor for operational flame stability without a pilot at simulated Mach ~ 3 – 3 flight

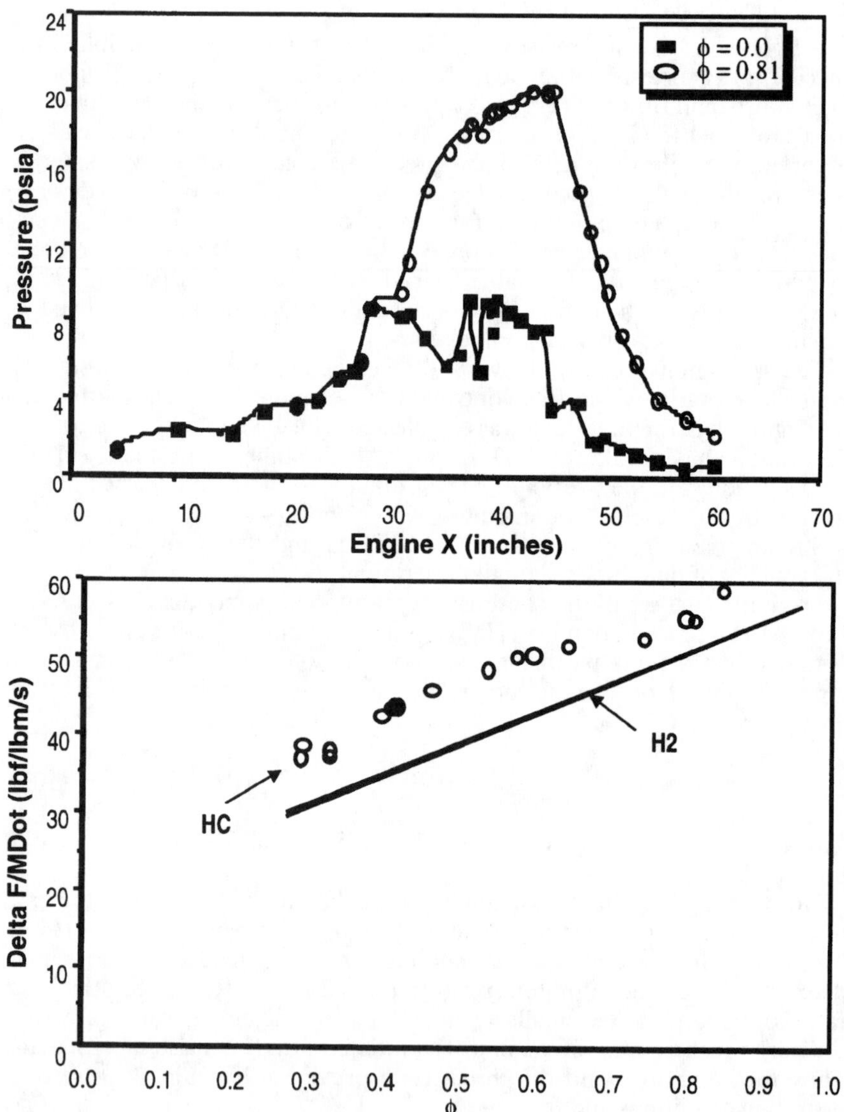

Fig. 40 Dual-Mode Scramjet Pressure and Performance with Constant Mach Combustion.

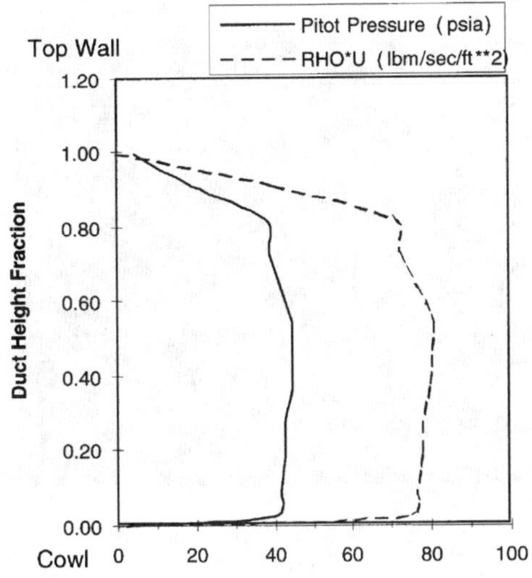

Fig. 41 Engine Throat Pitot Pressure and Mass Flow Profiles Entering Combustor for Test at M_0 = 4.0.

conditions are presented and discussed directly following the paper, Wiles (1987). Heated JP-10 and gaseous hydrogen were the fuel types tested. Thrust measurement and wall pressure data obtained during the test show that the flowpath performed equally well for both fuel types. Pilot survey results for mass flux distribution at the combustor entrance plane show that the 3D sidewall compression inlet produced a highly uniform profile over $< 85\%$ of the duct height. Axial fuel injection tailoring from forward combustor to mid-combustor stations provided a 21% gain in overall thrust performance. The thrust improvement was a result of the capability to increase the amount of fuel injected by 49% before isolator shock movement forward into the inlet. The tests were conducted in heated clean air at the GASL Leg 4 Aero/Thermo Test Facility.

The experimental model used in this freejet test program is shown in Fig. 42. It was a heat sink OFHC copper plate design. It incorporated a 3D sidewall compression inlet with swept leading edges and a contraction ratio of approximately 5 to capture and direct air into the transition section. These tests were conducted with no cowl on the inlet bottom. The transition section functioned as a shock isolator during ramjet operation as well as a housing for the high-speed ramp injectors. The 2:1 area ratio dump combustor featured a large base area for thrust production, management of thermal choking, and localized fuel injector stations.

Fuel injection stations existed at the combustor entrance plane and at three axial locations along the combustor wall. The wall injectors were primarily a single-row sonic orifice-type oriented normal to the wall. The high-speed ramp injectors delivered fuel both axially and normally relative to the ramp. A series of axial sonic injectors were located in the dump

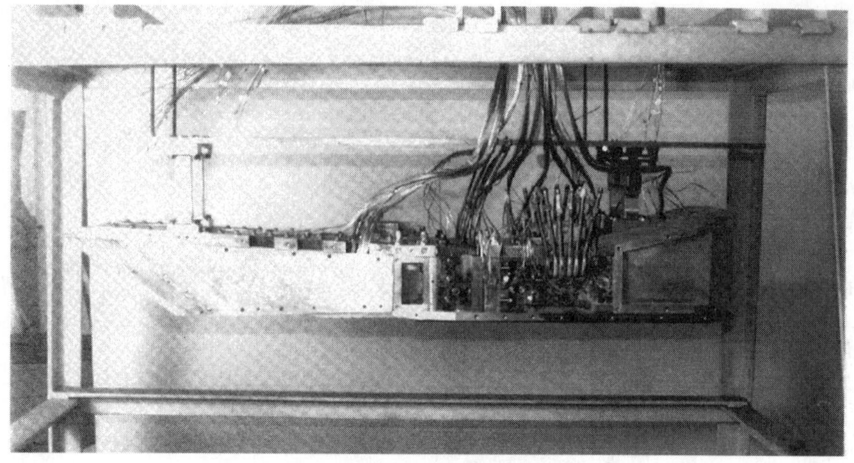

Fig. 42 Test Engine Assembly (Dual Mode).

combustor base to fuel the recirculation zone and generate pressure on the step face.

The test series consisted of two facility shakedown runs with no fuel, one Pilot survey run with no fuel, one hydrogen fuel run, and seven heated JP-10 fuel runs.

5. Results

The step in the combustor provides for enhancement of engine operability (i.e., the control of forward propagation of the thermal choke-induced back pressure shock system) are shown in Fig. 43.

At the maximum pressure condition shown, the shock compression leading edge is contained within the engine. As the η is increased further, air spills from the inlet and the isolator pressure decreased.

Wall pressure data obtained at the combustor exit show a similar trend to that seen in Fig. 44a. It was observed that the engine η at which the peak combustor pressure exists, coincides with the η, which generates the maximum shock isolator pressure in the region of the engine cowl leading edge.

Combustion efficiency is ~95% at low η and reduces to ~50% at the maximum operable engine condition. These efficiencies, while low, were in a range that demonstrated the principles of operation of the type I combustion cycle. Some problems with direct injection of large amounts of H2 fuel directly in a base region are discussed in Ortwerth (1999).

Thrust performance of the engine operating with heated JP-10 fuel is shown in Fig. 44b. Included is the 1D theoretical prediction as a function of η and combustion efficiency. Comparison of the test data to prediction shows combustion efficiencies of ~70% were achieved by injection of fuel only into the base region.

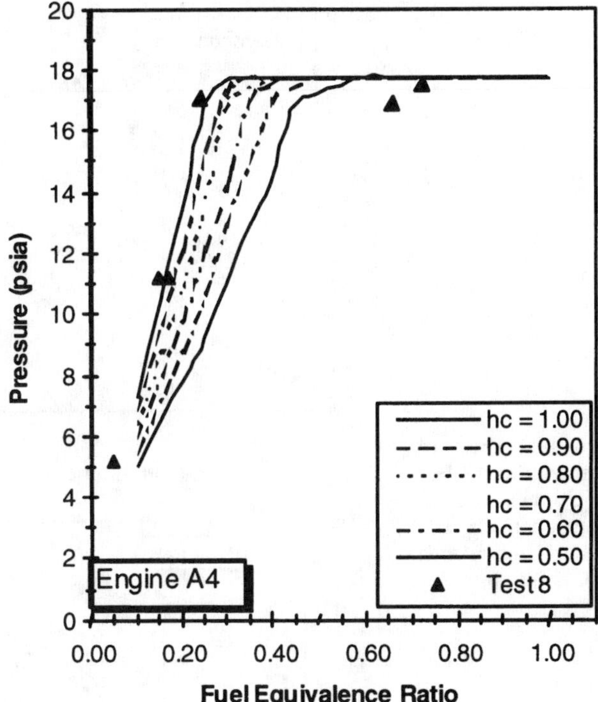

Fig. 43 Dual-Mode Engine Test Data Step Combustor Only.

6. *Drag and Heat Transfer*

At these Mach numbers, drag and heat transfer are sufficiently uncoupled from the heat addition process, and they may each be determined independently of combustion. Also, under factorable pressure conditions, it appears that the friction and heat transfer laws do not need to be modified over wide ranges of Mach numbers. Finally, K_{WP} is set at zero for the low Mach numbers of interest in the combustors.

On this basis, expressions for estimating drag and heat transfer are given at the end of Appendix H, Sec. A.

E. Scramjet

The two-wave detonation model can be generalized for any speed in the supersonic combustion regime into a two-wave supersonic combustion model as shown in Fig. 45. An oblique shock and a deflagration wave are propagating from a corner origin and are not allowed to be parallel. The wave must be subsonic relative to the flow after the shock in order for the final burned state to push the shock to a stable condition. Since the downstream may be expanding at any angle, the exit velocity normal to the deflagration wave is assumed to

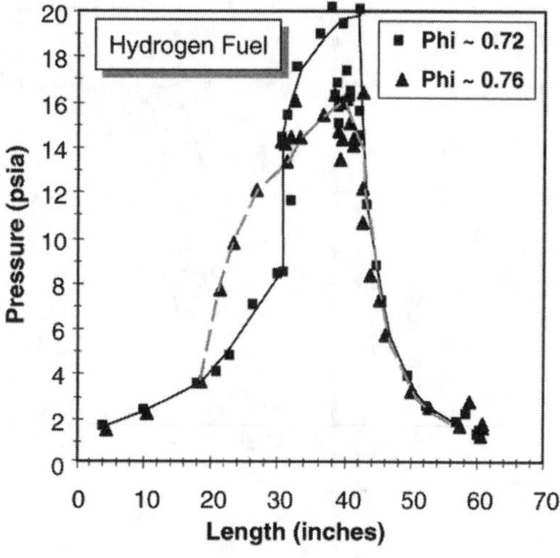

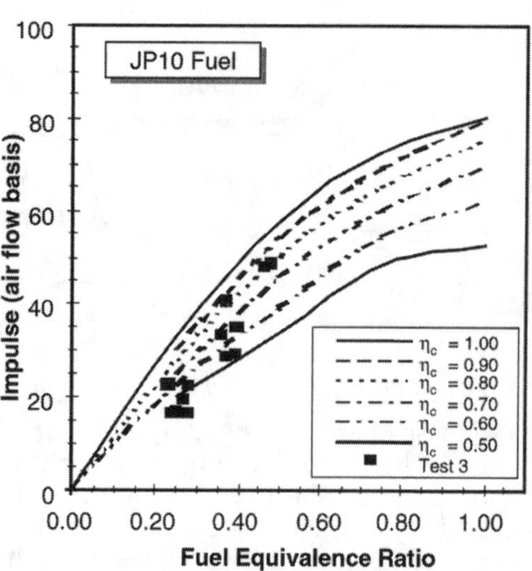

Fig. 44 Engine Test Pressure at Max Performance and After Inlet Unstart (Upper). Engine Performance on HC Fuel up to Unstart Equivalence Ratio (Lower).

be exactly sonic. This latter condition is a simplification since all that is really required is that the total velocity be at least sonic.

In addition to a change in flow direction and flow cross-section or stream tube area change, there exists a pressure differential between the wall with shock, and the wall with flame gases. Thus, the thrust wall has the von Neuman spike pressure and the drag wall has the deflagration pressure, which is exactly what is desired. Thus, contrary to 1D models, a constant area combustor, with a turn, can generate a thrust. There are solutions for which not only the area but the density or pressure after the deflagration wave are constant, similar to the 1D solution, but yielding a higher overall performance. The

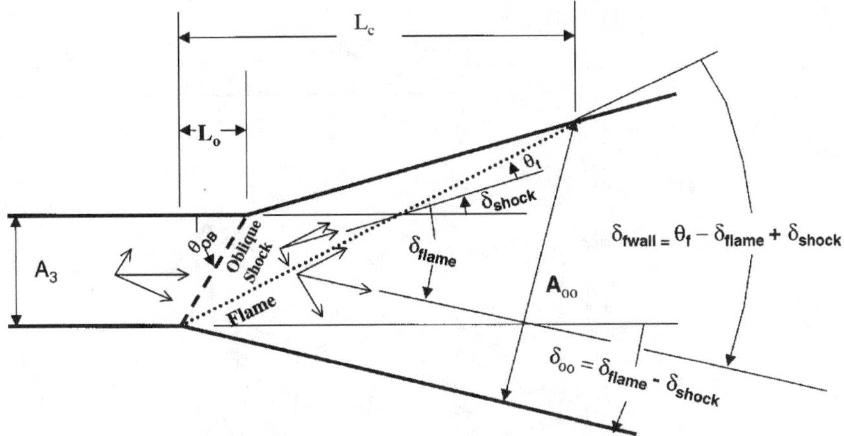

Fig. 45 Flow Diagram for Two-Wave Combustor Analysis.

two-wave model has a C-J solution for many area ratios across the combustor, including some contracting areas. In the following discussion, a supersonic combustion solution map is presented and the ideal frictionless combustor cycle performance evaluated; reference should be made to Fig. 46.

Selection of the abscissa of the combustor map is guided by the step constant area solution requirement to specify outside the conservation solution a value for the base pressure to calculate the thrust increment in the combustor. Since engine thrust is generated by a flowpath process, including the expansion work that the gas does on the nozzle, the best ordinate is the total thrust available in the combination path of the combustor, followed by ideal expansion to a fixed exit area. With the exit area fixed, the effects of different nozzle expansion ratios for each case are included.

The next parameter of practical importance is the selected area ratio. The plot shows overall performance increases monotonically with combustor wall pressure. The question is, how far can the pressure be increased? Lines of constant pressure relative to combustor exit pressure are shown crossing the lines of constant area. The practical realization of base pressure will be discussed in the following section.

The upper performance limit is found from the two-wave C-J model for each area ratio and the limit-effective wall pressure is higher than the combustor exit pressure. A locus of maximum C-J performance is thus obtained. Higher performance is found for combustion in the smallest area, and values for the converging combustor are shown to be 10% higher than the 1D constant area case. The ODWE is plotted as a stand-alone point on the converging area or drag side of the abscissa. The ODWE is about 2% higher than the constant area point. The 1D constant area, density, and pressure solution points are found interior to the map for reference. Ranking of the 1D combustors shows constant area to be the preferred type, followed closely by constant density and lastly by constant pressure. The result is

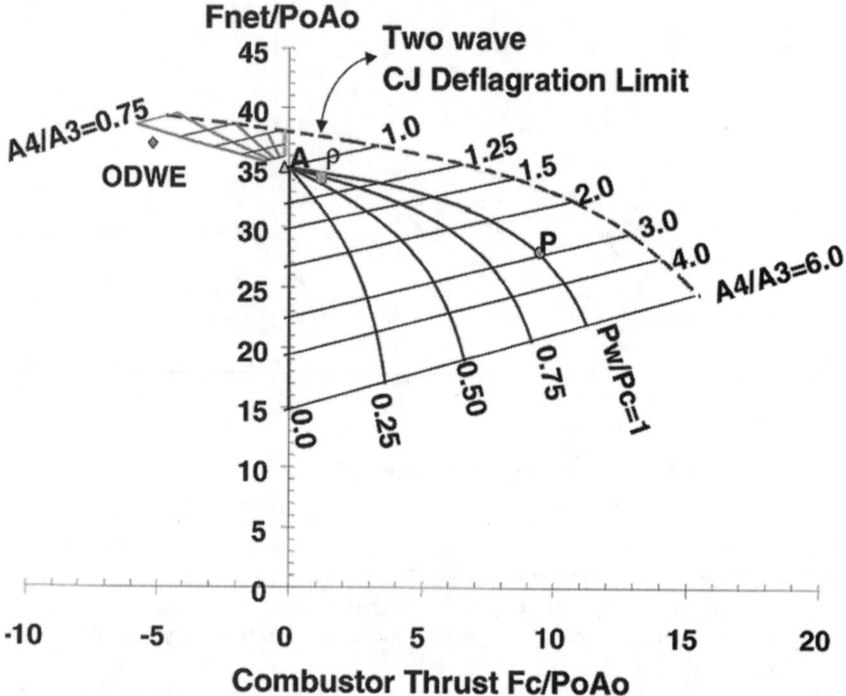

Fig. 46 This Combustor Map Shows Heat Addition Cycle has Large Effects on Specific Thrust $M_0 = 10$, $\phi = 5$, $\phi = 1$.

POINTS
◇Oblique Detonation Wave
A Constant Area Combustor
$\mathcal{S}$ Constant Density Combustor
P Constant Pressure Combustor

somewhat always designer subjective because if maximum cycle pressure were held fixed, the constant pressure point would fall much closer to the performance of the constant area case. Thus, the need to understand the choice of thermal compression ratio is again raised. The thermal compression ratio, ψ, was held fixed for this comparison. The constant area, density, and pressure solutions for the two-wave model are found on the C-J boundary.

The two-wave model contains a natural combustor length as part of the solution that is of academic interest for a reference length and leads naturally to selecting the combustor approach, based on the combined friction and heat addition cycle, which is directly related to the combustor wetted area, and, therefore, length combustor L/D (Fig. 47).

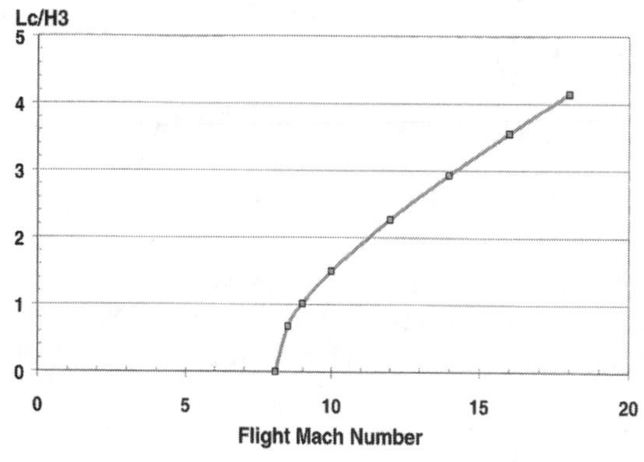

Fig. 47 Required Combustor Length for Oblique Detonation Wave (Reference).

F. Friction Cycle

Analysis of the complete engine cycle necessarily includes the heat load delivered to the coolant, usually the fuel, and the accounting of the friction drag losses. These viscous effects also interact directly through their displacement effects on the outer core flow process of the cycle. Philosophically, it is useful to separate the viscous drag forces and the thrust of the fuel nozzles from the inviscid cycle dynamics. This separation is theoretically provided nicely by the well-known Reynolds analogy between friction drag work and the wall heat transfer. Some features of the friction cycle are examined with the aid of this well-known principle.

The ambient pressure is very much smaller than the fuel nozzle supply pressure. For the purpose of simplicity, the fuel is assumed to expand to ambient pressure with no interaction with the air flow. Conservation of force and energy yield the following relations for thrust drag and efficiency.

$$\text{Viscous drag} = D_v \tag{102}$$

$$\text{Energy dissipated} = D_v Vo \tag{103}$$

$$\text{Energy transferred to propellant} = f_q\, D_v\, Vo \tag{104}$$

For negligible energy of the fuel in the tank, the fuel energy after expansion is equal to the energy

$$\dot{m}_f V_f^2 = f_q D_v V_o \tag{105}$$

transferred. Thrust of the fuel expansion is found for expansion to ambient pressure.

$$T = \dot{m}_f V_f = E_f \frac{2}{V_f} = f_q D_v V_o \frac{2}{V_f} \tag{106}$$

Thus, the thrust to drag of the cycle is simply a function of the friction factor and the velocity ratio.

$$\frac{T}{D_v} = 2f_q \frac{V_o}{V_f} \tag{107}$$

For the hypersonic case, $V_o \sim V$ and f_q is found as a simple relation

$$f_q = \frac{C_h}{C_f} \frac{(H_{\text{aw}} - H_w)}{V_o^2/2} \tag{108}$$

Evaluating this relation with assumption of a cold wall yields values greater than 0.5 and may be greater than 1 when combustion effects are included.

The effective specific impulse (I_{sp}) of the cycle is written in terms of velocity as follows:

$$I_{\text{SPeff}} = \frac{T - D}{m_k} = \frac{V_f}{g}\left(1 - \frac{1}{2f_q}\frac{V_f}{V_o}\right) \tag{109}$$

This equation clearly shows a maximum at a velocity ratio less than one.

$$\left(\frac{V_f}{V_o}\right)_{\text{opt}} = f_q \tag{110}$$

The maximum I_{sp} is simply a linear function of flight velocity and is positive, showing the cycle is theoretically capable of producing net thrust.

$$(I_{\text{SPeff}})_{\max} = \frac{f_q}{2}\frac{V_o}{g} \tag{111}$$

For flight in the atmosphere limited by orbital speed, the maximum value obtained is 250 see. The problem is that the mass flow is much larger than that corresponding to the stoichiometric fuel air ratio. Yet the value is a useful lower bound on the complete scramjet cycle and expresses the fact that the scramjet has an orbital speed capability. Note that a velocity twice the optimum produces a zero value of $I\text{sp}_{\text{eff}}$. Using fuel blends and thermal cracking in place of fuel velocity improves the overall engine efficiency, in addition to improving the fuel's cooling capacity.

In our engine's internal flow accounting system, the total friction drag for a zero-effective friction cycle Isp can be found for the same assumptions.

$$\left(\frac{D_v}{P_o A_0}\right)_{I=0} = \varphi_{\text{cool}} \frac{M_o^2}{20} \tag{112}$$

This relation may be used to form a figure of merit for the as-built or proposed flowpath, which typically exceeds this value at equivalence ratio one. A figure that measures the number of times this drag is exceeded by a given design is

$$K_{fc} = \frac{20}{\varphi_{\text{cool}} M_o^2}\left(\frac{D_v}{P_o A_o}\right)_{\text{Real}} \tag{113}$$

The K_{fc} values for some designs have been on the order of 1.5, and means to achieve such results are discussed next. In the author's experience, the values of K_{fc} are surprisingly constant over a reasonable Mach range around the shock-on-cowl lip design point. As such, it is possible to gain a rapid estimate of the net thrust by combining the previous results into a general performance relationship where the thermal cycle and friction cycle are separated.

$$\frac{T_{net}}{P_o A_o} = K_{Tc} - \frac{(K_{fc} - 1)\varphi M_o^2}{20}$$

$$K_{Tc}(\psi, \varphi) = 33 \qquad \text{at } \psi = 5, \varphi = 1$$

$$K_{fc}(\psi, \varphi) = 1.5 \qquad \text{at } \psi = 5, \varphi = 1 \tag{114}$$

The thermal cycle thrust K_{Tc} is dependent on ψ and K_{wp}, but not explicitly on M_o. The effects of spillage and increased forebody contraction with M_o have been discussed earlier. The form of the equation also suggests finding the geometric sensitivities of the terms K_{Tc} and K_{fc} to set up an optimization technique.

The design goal is to operate the regeneratively cooled engine flowpath at or below a required cooling equivalence ratio of one and it can be determined by writing the energy balance in terms of the normalized viscous drag work.

$$\left(\frac{Q}{\dot{m}h_o}\right) = \frac{f_q}{(\gamma - 1)}\left(\frac{D_v}{P_o A_o}\right) \tag{115}$$

This relation then determines the overall fuel-air ratio required.

$$\left(\frac{f}{a}\right) = \frac{(Q/\dot{m}h_o)}{(H_{Tf}/h_o)} = \frac{\frac{f_q}{(\gamma/\gamma - 1)}(D_v/P_o A_o)}{(H_{Tf}/h_o)} \tag{116}$$

The above relation can be solved for limiting drag and heat transfer by inserting the appropriate conditions for the type of fuel used and an assumed high conductivity metal cooling panel wall. A nominal heat transfer factor of $f_q = 0.5$ is used for the Mach limit estimates shown in Table 2.

These Mach numbers are somewhat flexible at this level of design since they are dependent on total surface area per unit frontal area. However, these numbers represent a reasonable guideline for different fuels. The hydrocarbon fuels are assumed to be thermally dissociated or cracked in the brief flow residence time through the cooling panels to lighter hydrocarbons such as ethylene, methane, and hydrogen. Some hydrocarbon cooling schemes involve recirculating hot fuel into the tank and coking of the fuel is a problem. Also the products of cracking may have a dramatic effect on the chemical erosion of copper heat exchangers particularly methane.

Table 2 Title

Fuel	Viscous drag	Heat load	Mach limit at $\varphi = 1$
H_2	13.1	1.885	12.0–14.0
HC	6.1	0.88	9.0–11.0

1. Practical Combustor Design

The need to reduce heat load on combustion walls is important and will be addressed by starting with the technical considerations of the required mixing length and the means of reducing the pressure and wetted surface are of the combustor through the use of separated bubbles downstream of steps. The mixing length is a boundary condition for the separation and reattachment process and thus will be covered first. For most operating conditions, a step has been found to be more than adequate to anchor the flame propagation in the combustor with hydrogen and hydrocarbon fuels at Mach numbers 3 and over. The flame holding is most difficult at low Mach numbers where the recirculation zone temperatures are of course lowest being proportional to Mach number squared.

2. Mixing Length

Mixing of fuel and air in the combustor is crucial to the effective design of the scramjet engine flowpath. The appearance of design recipes based on sound research is critical to the maturity of the process. No flights except the Russian scramjet vehicle have been reported; thus, the appellation mature mixing for a recipe is perhaps awaiting more confirmation.

3. Mixing Limited Combustor Operation and Integration

The mixing rates of fuel and air are much lower than the reaction rate and heat release rate in most experimental scramjet combustion studies. The mixing process is a major limiting design technology.

The first published design formulas on combustion efficiency in supersonic combustion in this country are Griffin Anderson (1971), and also in Northam (1989), and are based on research done at the NASA Langley Research Center. The working formulas are presented as summary research results and commonly are used for sidewall injection design and also with axial concurrent jet injection.

Optimum geometry of injector element spacing occurs when the distance between the injector elements is equal to the half-width of the isolator with injectors installed on both sidewalls. This result is based on a downstream capture area per injector element, with an aspect ratio of two to one. The mixing recipe for normal and coaxial injection is as follows:

For $\varphi = 1$ define a length,

$$X_L = 60\ G \tag{117}$$

to be used with an injector spacing,

$$S = G/2 = W_{iso}/2 \tag{118}$$

For normal injection from both walls, the length for complete mixing X_φ is found for $\varphi \geq 1$

$$X\varphi = 3.333e^{-1.204}{}_{\varphi} \tag{119}$$

$\varphi \leq 1$

$$X\varphi = 0.179e^{1.72}{}_{\varphi} \tag{120}$$

The mixing efficiency η_m is defined to be

$$\eta_{mix} = \frac{\varphi_{mix}}{\varphi} \qquad \text{for} \quad \varphi \leq 1 \tag{121}$$

$$\eta\ mix = \varphi\ mix \qquad \text{for } \varphi \geq 1 \tag{122}$$

The mixing efficiency for perpendicular injection is

$$\eta_\mu = 1.01 + .176\ \ell n\left(\frac{X}{X_\varphi}\right) \tag{123}$$

For parallel injection, mixing efficiency is

$$\eta_m = \frac{X}{X_\varphi} \tag{124}$$

It is recommended that linear interpolation of the mixing rules for injection angles between zero and 90 degrees to the stream. Using multiple struts to span the combustor is effective in reducing the mixing length.

4. *Ramp Mixing and Combustion*

Presently, the author prefers ramp mixing using coaxial jets to recover friction drag and heat as thrust work from fuel momentum, Ortwerth (1996).

Ramp mixing offers a great advantage, namely unhooking mixing rate from fuel momentum effects, which control penetration and shear in side-wall or coaxial injection. Using ramp injectors, the mixing length is controlled by axial vortex entrainment and is designed into the flowpath to obtain the desired mixing length; it is not dependent on variable dynamic conditions on the flight trajectory.

A second significant advantage is that the blunt base of the ramp becomes a flame holder. This is a flight-proven point since flame holders were used in the Russian scramjet engine flight tests. The ramp injector flame holder gives the designer freedom to choose contraction ratio and combustor inlet temperature unhooked from autoignition fuel temperature. As previously shown, the contraction ratio must be optimized for efficiency.

In the following, flowpath integration requires a determination of the relationship of drag and mixing rate. The basic elements of the analysis closely follow the ideas of Swithenbank (1969), who was a pioneer in vortex-generated mixing. In this approach, we will follow the growth in the area of fuel stream as vortex stirring continues downstream of the generator. Once this area fills the combustor, it is assumed that some additional length is required to complete homogenization and micro-mixing.

5. *Ramp Mixing Analysis*

Two counter-rotating vortices are generated by the ramp (Fig. 48). The planar laser-induced fluorescence (PLIF) data show growth of the mixed gas region. These vortices will entrain air at a rate proportional to the strength of the vortex core and the space between the two cores. These vortices are persistent, Donohue (1995). Assume the vortex entrainment to be the only macro-mixing mechanism. The large scale energy-containing structures in the flowfield are only 1D in the azimuthal coordinate and, thus, very efficient in stirring. Small scales internal to the vortex pair complete the micro-mixing or homogenization of the mixture. The analysis approach is to relate the vortex strength to the ramp drag theoretically, through energy balance. Entrainment rate per unit length is found by multiplying the entrainment velocity, density, and gap span.

Begin by determining the vortex velocity as a function of drag. Consider one side of the ramp only. The stirring power supplied by the ramp is drag times velocity.

$$D_e V = \frac{1}{2} C_{\mathrm{De}} \rho V^3 A_R \tag{125}$$

where C_{De} is the effective ramp pressure drag coefficient, which is some fraction of the total pressure drag coefficient of the ramp.

By energy partition approximation, this drag power is related to the change of internal energy of the mass flow in the vortex.

$$\dot{m}\Delta e = D_e V \tag{126}$$

Using the Swithenbank approximation that this final internal energy rise is exactly that due to dissipation of vortex kinetic energy

$$\dot{m}\Delta e = \frac{1}{2} \dot{m}_{V'} 2 \tag{127}$$

where V' is the mass-average transverse vortex-induced velocity. The ramp scale that represents the mixing transport scale is of the same order as the

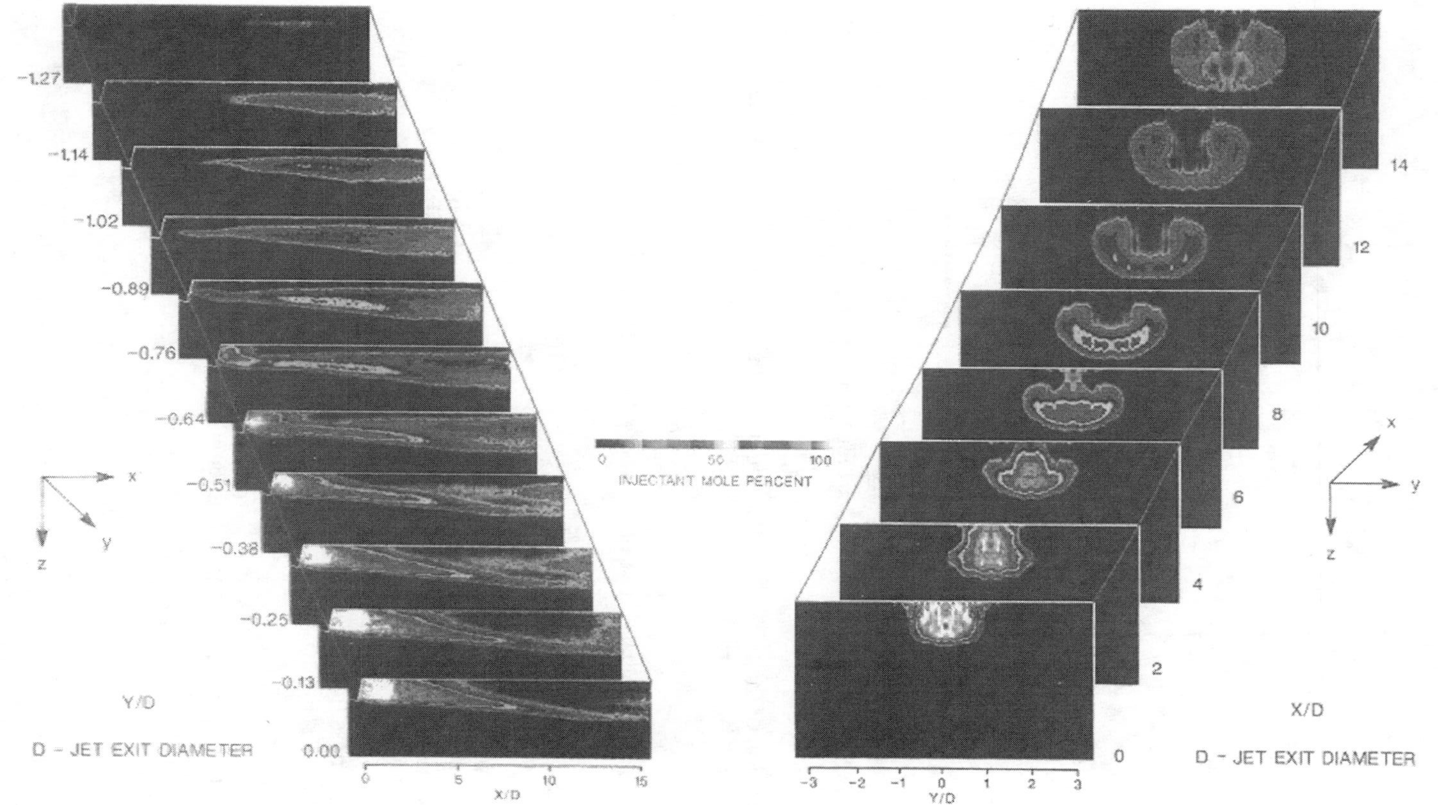

Fig. 48a Ramp Mixing PLIF Data at M = 2.2, Stream Wize Laser Sheet (Left), Normal to Stream Laser Sheet (Right).

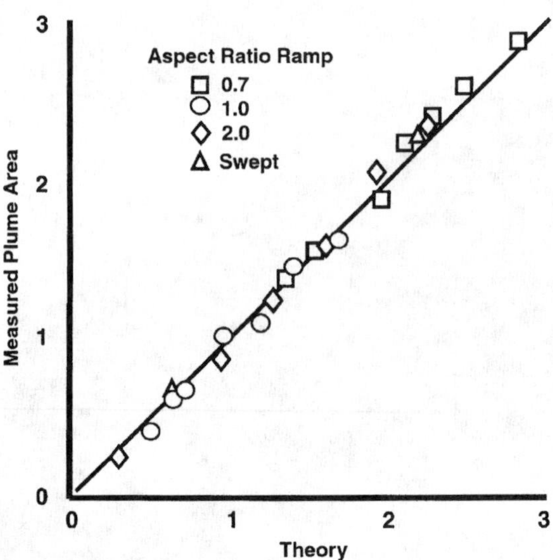

Fig. 48b Goodness of Fit Chart for Ramp Mixing Plume Area Eq. 134.

dimension of the mixer. Assume the initial mass flow captured by the vortex at the end of the ramp is

$$\dot{m} = \rho\, V_{AR} \tag{128}$$

where A_R is the base area of the ramp. One can solve for the swirl velocity in a few steps,

$$\left(\frac{V'}{V}\right)^2 = C_{D_e} \tag{129}$$

Based on the excellent flowfield data of Donohue and McDaniel (1995), write a mixing field conservation equation based on air entrainment in between the two counter-rotating vortices. An important detail to remember is that the fuel is already injected through the base of the key.

The mass flow rate in the axial direction in the plume is

$$\dot{m} = \rho V h w \tag{130}$$

the rate of entrainment per unit length is

$$\varepsilon_{,} = \rho\, g\, V' \tag{131}$$

Thus,

$$\rho V = \frac{\mathrm{d}hw}{\mathrm{d}x} = \rho g V' \tag{132}$$

Integrating above, the solution of the mixing as a function of length is found.

$$(hW)_X - (hW)_0 = C_{D_e} g x \tag{133}$$

Normalize this result by dividing by the ramp reference area for generality.

$$\frac{A_{MIX}}{A_{REF}} = \frac{A_o}{A_{REF}} = \sqrt{C_{D_e}}\left(\frac{g}{X_{REF}}\right)\left(\frac{x}{X_{REF}}\right) = \frac{A_o}{A_R} + \sqrt{C_{D_e}}\left(\frac{g}{W_k}\right)\left(\frac{X}{H_k}\right) \tag{134}$$

where

$$A_{REF} = H_k W_k \tag{135}$$

$$X_{REF} = \sqrt{A_{REF}} \tag{136}$$

The data reveal the same linear dependence of box area as the solution. Thus, the data can be used to find the dependence of g and A_o as a function of the only geometric parameter of the ramp not used in the theory. Summarized data are shown in Table 3.

The solution admits the determination of the gap g and intercept A_O with a smooth function of aspect ratio. The result is for the intercept

$$\sqrt{\frac{A_o}{A_{REF}}} = \left(\frac{1}{b_o} - \frac{1}{2}\right) \tag{137}$$

Where the ramp aspect ratio is

$$b_o = \frac{H_k}{W_k} \tag{138}$$

The gap function contains two factors. The first factor is a square root of aspect ratio. The residual function is a quadratic that varies surprisingly little over the range tested, varying only 20% for W_k/H_k, varying from 0.5 to 1.4.

Table 3

Ramp	Case	*CDe*	$b = Hk/Wk$	Slope	*AO*	$g = slope/\sqrt{C_D}$
Straight	1	0.0523	0.714	0.266	0.8	1.16
Straight	2	0.0374	1.01	0.233	0.25	1.20
Straight	3	0.0187	2.0	0.222	0.0	1.6
Straight	4	0.1230	0.58	0.350	0.05	1.0

Thus,

$$\frac{g}{\sqrt{A_{\text{REF}}}} = \left(1.264 - \frac{.407}{b_e} + \frac{.349}{b_e^2}\right)\sqrt{b_e} = \left(\frac{g}{w}\right)\frac{1}{\sqrt{b_e}} \tag{139}$$

The modified or effective ramp base aspect ratio b_e is formed to include the swept ramp data. From the given base geometry, the base aspect ratio is modified to account for the projected based area of the swept sides

$$b_{\text{SR}} \equiv \frac{H_k^2}{A_{\text{BASE}}} \tag{140}$$

and then averaged with the geometric base aspect ratio to form the effective swept ramp aspect ratio

$$b_e = \frac{b_o + b_r}{2} \tag{141}$$

A composite plot showing the tightness of the data used as compared with the semi-empirical fit is shown in Fig. 48. The aspect ratio $W_{\text{mix}}/H_{\text{mix}}$ is shown in Fig. 49.

The soundness of the recipe can be checked by using it in a predictive fashion for data outside the original set. An orthogonal set was obtained by varying ramp angle at constant aspect ratio. In these data only, the distance to a specified mass fraction on the centerline was determined and the relative length reported for an aspect ratio $b = .714$. The ramp angles tested were 8, 10, and 12 deg, for which C_{De} was .0300, .0374, and .0448 at $x/H_k = 8.2$. The result is reproduced in Fig. 50, and the agreement is good over the angles tested.

The recipe book for combustor mixing length for the common types of injectors is complete.

G. Step Combustor

This section has relevance for the ramp injection scheme just analyzed, wherein a step area change is added at the base of the ramp. A step may also be designed into the combustor to augment performance by reducing friction and heating. Such performance improvements depend to a large extent on the level of base pressure, which produces thrust on the step area of the base. As previously mentioned, Kenworthy (1965), first noted the potential of the step burner. However, he was unable to achieve the pressure level he projected due largely to expecting nature to do it without attention to either base injection requirements on the rate of heat release or the step area. Orth (1975), investigated the flow in the near field of the step. Well now work on the dual problems of base heat release and base injection to achieve high base pressure and reduced friction drag and heat transfer.

The critical problem is determining the pressure at the step in the presence of heat addition in the supersonic freestream. Fortunately, a significant amount of research has been done on the base pressure problem, which can

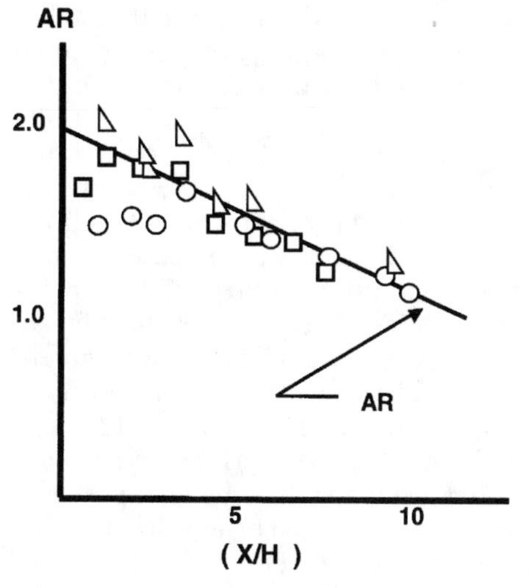

Fig. 49 Ramp Mixing Plume Aspect Ratio Correlation Symbols Same as Fig. 48.

be directly applied to the combustor's internal flow geometry. Details follow the Chapman–Korst (Chapman (1958) and Korst (1956) models and the associated integral profile theories. They can be found in Appendix IQ. The theory can be used to solve the base pressure to isolator pressure ratio and the reattachment length to step height ratio. The only modification required in the original theory is to determine the pressure at any length from the known mixing rate, controlled heat release and the turn angle required at reattachment. The dynamics of base combustion are comprehensively treated in the compendium edited by Murthy (1974).

The physics of the aerodynamic base pressure problem may not be well known to internal flow specialists. However, the base phenomena are routinely encountered in scramjet engine testing. There are three interesting situations in base phenomena for the combustor expert. The first is observed

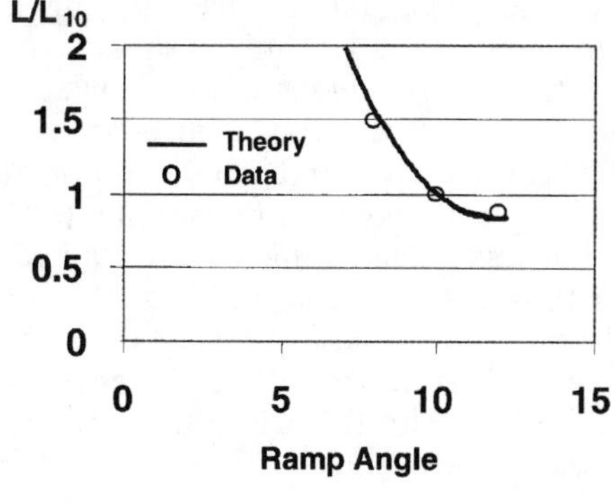

Fig. 50 Comparison of Ramp Mixing with Data Outside Empirical Base.

typically at low enthalpies around the $M_0 = 6$ speed range. Then thermal or heat choking effects are dominant and the upstream base pressure is equal to the combustor exit pressure or higher. The second case is when the step pressure is still much higher than the isolator pressure and is equal to the pressure determined by the efficiency of freestream combustion or pressure at the reattachment location. These are two cases where the base pressure is dependent on the combustor pressure. The third case corresponds to flows where the combustion pressure rise is spread out due to increased mixing length and reduced pressure rise due to combustion at high inlet temperatures. Here, apparently, the base flow is not as closely coupled to the wall pressure at reattachment and is lower than the pressure at the reattachment location. This last case is quite controversial and not well understood because no test facilities operate at high enthalpies in the $M_0 = 12$ range except for pulse direct-connect-type rigs or shock tunnels. The results are scattered and the base pressure is often much lower than the inlet pressure. For such complex interacting flows, the test time is often blamed as the cause of low base pressure, measured using the argument that the flow interactions are not established. The $M_0 = 12$-range cases are a serious cause for concern in projecting actual flight operations, where performance may be degraded (Fig. 46). This is a problem should be well-suited to the external aerodynamic approach. The Chapman–Korst theory is the subject of the base pressure analysis in Appendix I.

In the external aerodynamics work, there are two classes of base flows. The first are base flows that are free of any control effects. These flows fall under the appellation of the free interaction case. There is one main parameter and that is the upstream Mach number at the base separation point or lip Mach number. A second order parameter is the ratio lip boundary layer displacement thickness to step height or base diameter. This presentation of data and its correlation with lip pressure rather than the theoretically correct downstream or reattachment pressure, is the usual practice because the downstream pressure is equal to the lip pressure and no distinction can be found (Fig. 51).

The second case involves base flows with active control input. Three kinds of controls are encountered, the first actively seeks to reduce the Mach number and total pressure on the reattaching streamline. This is done by injecting inert gas or reacting gas Townend (1964), into the base entrained into the shear layer and flows downstream after reattaching the shear layer over the separation bubble. Results of the experimental study of Townend shown in Fig 52. As anticipated, the flows with hydrogen injection and combustion are the most efficient, that is, they have the highest base pressure and the lowest injectant mass flow. By counting drag reduction due to an equivalent thrust from the propulsion system, an Isp can be determined, which is quite high. The difficulty for those interested in traditional thrust generation is that base burning only generates a base pressure equal to ambient or lip pressure and a positive net thrust that could actually accelerate the body to a higher speed is not possible. This first case can therefore be called the weak interaction base pressure problem.

The search for higher than lip pressure then leads to the strong interaction base pressure problem. Approaches to this problem involve using extra means to raise the downstream pressure at the reattachment point. The first technique employs an inward turn or ramp at the reattachment location. This was one of the original experiments of Chapman that led to the discovery of the downstream pressure being the dominant parameter, not the lip pressure. The second approach to downstream pressurization is similar to the first. In this case, a shock is generated externally by a freestream body, Townend (1986), which impinges on the recirculation bubble upstream of reattachment (Fig. 52). These successful controls led to the experiments with external combustion, with research largely aimed at powering artillery shells, Strahle (1982).

The current study is interested in determining the flows above Mach 8. Here the problem strictly falls in the domain of the scramjet, where thermal choking is not possible and the interaction must be with the local burning rate. A series of experiment funded internally by Boeing at the Siberian research center ITAM, are shown in Fig. 53. The data show decreasing base pressure with increasing total temperature at a constant isolator Mach number. The profile theory compares well with the experiment when a suitably chosen combustion efficiency is input as a parameter. The combustion efficiency was determined from the ramp-mixing recipe. The effects of base injection are well captured by theory as second order. All data indicate pressures above the lip pressure except at very early test times when the mainstream combustion was rather weak. This period lasted for 20 to 30 milliseconds and points out a difficulty in testing scramjet combustors in pulse facilities when complex interactions are occurring between base stabilized flames and the pressure at reattachment. This is not the correct procedure for presenting the data since the combustor pressure also is increasing

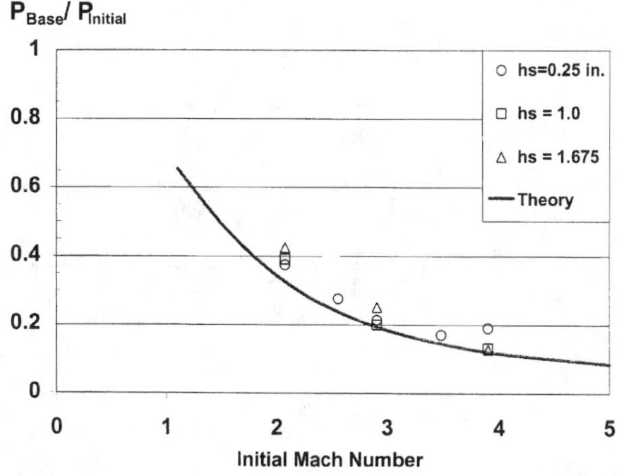

Fig. 51 Supersonic Step Base Pressure for Free Interaction.

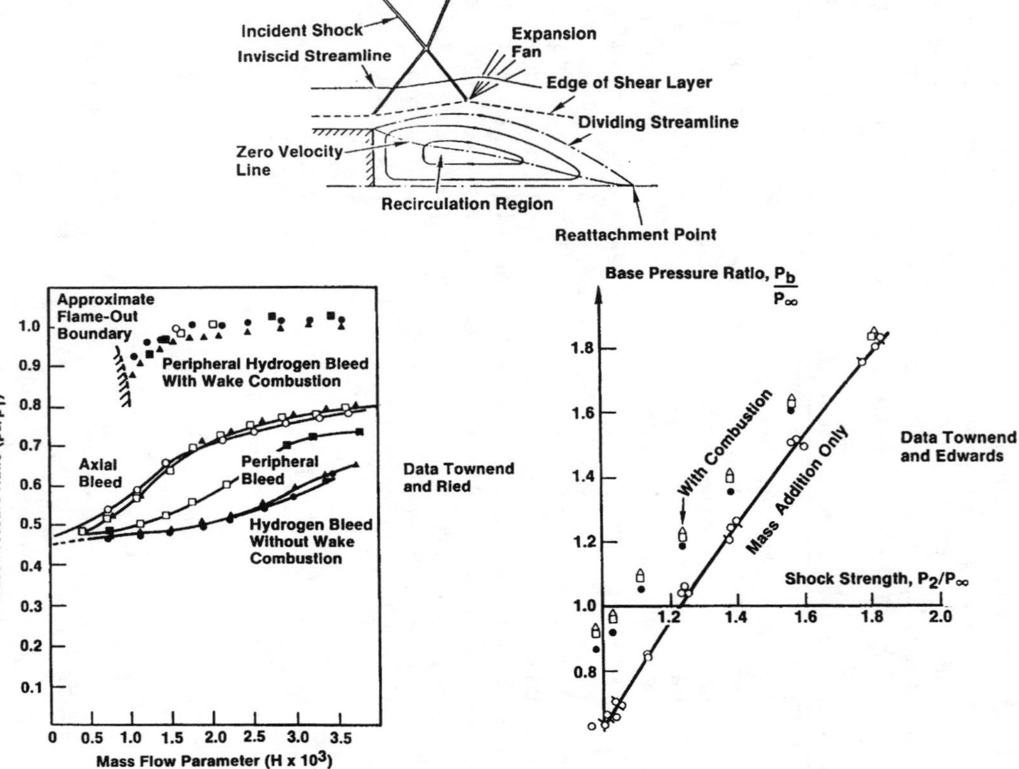

Fig. 52 Base Pressure Flowfield (upper), Base Pressure Control by Base Injection and with Combustion (left), Base Pressure Control Effects, Including Downstream Pressurization (right).

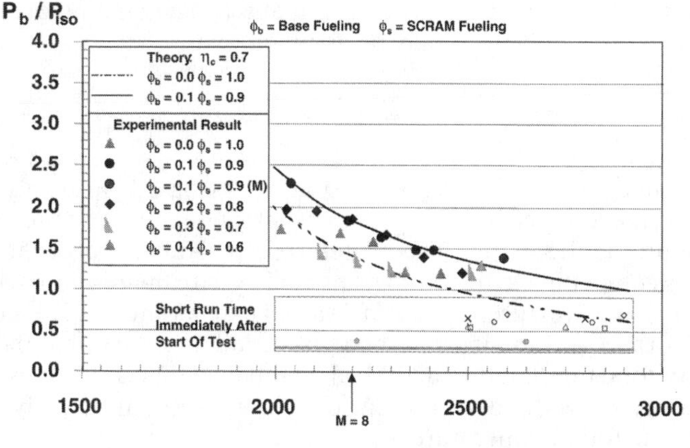

Fig. 53 Base Pressure in Step Combustor $\epsilon_c = 2.0$.

as total temperature is falling. A better approach is to normalize the base pressure with combustor exit pressure (Fig. 54). Here, the cases for steady supply conditions were used and the pressure is again weakly dependent on base bleed once equilibrium pressure is established. The theoretical curve at initial compression of $\psi_{iso} = 4.66$ was computed at freestream Mach of 8 and 16 at constant compression and the resulting curve was virtually identical.

An important observation is that for this flow, the main flow combustion rate dominated the results. It was initially assumed that higher base fuel rate would result in large changes in base pressure by forcing the reattachment point to move downstream to a higher pressure in the combustor. The results indicate that the effect of the rate of combustion is first order and the analysis shows only modest movement, being almost constant at about 5 step heights. Until more definitive experimental data are obtained that simulate flight speeds up to the Mach 12 range, the expectation is that base to

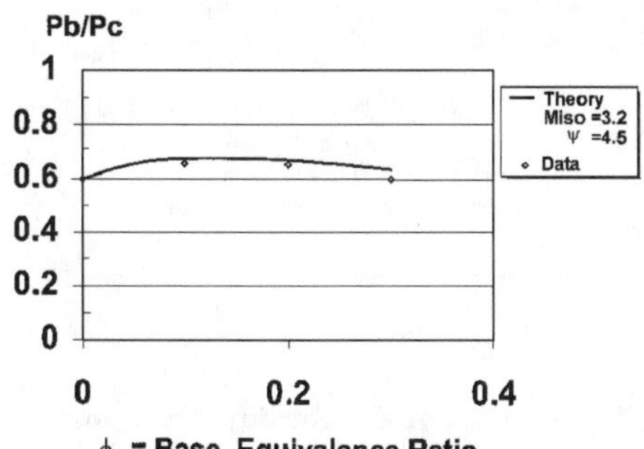

Fig. 54 Base Pressure in Step Combustor Shows Weak Dependence on Base Mass Flow and High Correlation with Max Downstram Pressure P_c.

combustor exit pressure ratio is a function of ψ_{iso}. Thus, until more definitive experiments are available, it may be assumed that

$$\eta_b = \frac{Pb}{Pc} = 0.7 \qquad \text{at} \quad \psi_{iso} \approx 4.66 \pm 0.5 \quad (142)$$

The experimental base pressure ratio above is based on combustor exit pressure is recommended though somewhat aggressive. For the more cautious, the profile theory provides a preferred analytical approach.

There is compelling logic to experimentally verify the results of the profile theory for base pressure, in lieu of Mach 12 experiments. The combustor expansion ratio and mixing rate may be varied as orthogonal parameters, along with total temperature and Mach number so that ψ_{iso} = constant lines can be obtained with very different conditions without actually requiring constant true total temperature.

H. Summary

Combustor design is a highly subjective art. The choice of combustor expansion is such a topic. A comparison of performance as a function of area ratio, ε_c, is shown in Fig. 55. The results at Mach 12 show increased performance as combustor area increases, because at low area $\varphi_{cool} > 1$ until the area ratio reaches ~ 1.75. The data at $M = 8$ decrease monotonically as might be expected from the wave analysis. For an accelerator application, heat transfer reduction by combustor expansion is probably preferred.

VII. Nozzle Component Losses

The nozzle performance is usually not considered a problem. The usual approach is to divide nozzle losses into certain categories and analyze the

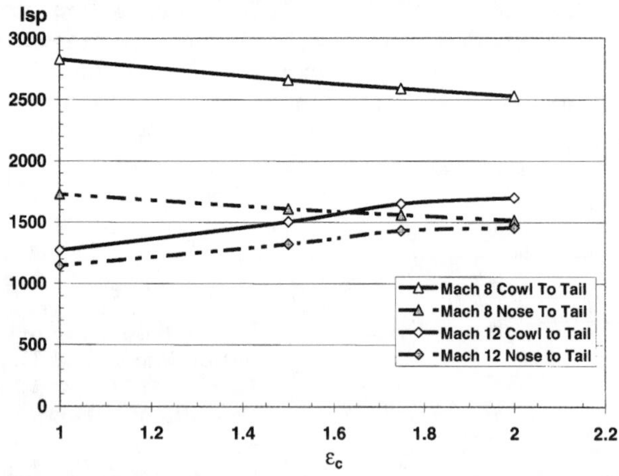

Fig. 55 Engine Practical Performance Dependence on Combustor Expansion ϵ_c Benefits at Higher Mach Number due to Reduced Friction Drag.

loss factors individually. The fully integrated 3D chemically reacting viscous flow with plume and external flow interaction is a tractable, but very expensive (time and effort) problem. Because one expects reasonable performance, nozzle analysis is often given low priority. However, much more needs to be known for such an important component where nearly all the thrust occurs.

A. Standard Loss Categories

The various categories of losses are as follows: 1) underexpansion, 2) thrust vector, 3) divergence, 4) viscous, 5) chemical kinetics, 6) shocks, 7) plume, and 8) trim drag. The items listed will be briefly discussed. The underexpansion loss is the major thermodynamic loss and is simply the ideal internal thrust computed for expansion to ambient pressure, less the ideal thrust computed for expansion to the given nozzle exit area ratio. It is used automatically to trade with cowl drag for net thrust optimization.

The thrust vector loss is simply the cosine loss factor. This factor should be a design parameter since the lift generated by the exit flow vector trades with the wing L/D-ratio as shown in the discussion of force system accounting.

The first loss associated with a non-ideal parallel flow contour is the divergence loss. It is traditionally associated with the rocket engine conical nozzle compared to the isentropic bell. This loss is important for the scramjet because the inlet Mach number is high and the ideal bell contour is extremely long. Presented below is a recipe for divergence loss of straightwall 2D.

Viscous losses are unavoidable. Minimizing the viscous loss is found by minimizing the wetted surface area and, to some degree, the pressure profile. Scramjet integration with a vehicle aftbody normally produces a plug configuration that is geometrically efficient since it has minimum surface for a given expansion ratio.

Chemical kinetic losses are important but to date are largely uncontrollable. Some studies to enhance recombination rates are often initiated. To date, few additives have a third body catalytic efficiency higher than water, which is already present in large amounts as a combustion product. An effective thermodynamic control is already employed to control the thermal compression, ψ, to a level that minimizes the energy in dissociated combustion products. This is why the optimal ψ often appears in the range of 5. Normally, the SSTO size vehicle kinetic studies indicate very early freezing of the gas composition so that minimizing combustor dissociation is critical.

Shock waves and entropy production will reduce the amount of energy available in the flow expansion. They are usually weak and often arise at interfaces between engine modules and the aftbody. For short nozzles, these intrinsic waves are often thought to be beneficial since they can be used to manage the pressure profile. A definitive study of this area has not yet been done.

Plume interaction occurs for all speeds less than that for which the last cowl characteristic just glazes the tail cone. Plume expansion normally reduces the pressure on the aftbody thrust surface and performance. The pitching moments are of course affected and the effect is generated at a large moment arm. Again, no recipes are available and the result is often ignored until later. Cowl length is the primary control of plume losses. A very important flight regime is the transonic one where the opposite problem of negative pressure coefficient occurs because not enough propulsion flow is available to fill the base. Base drag is often the total cause of the transonic drag rise. Here the requirement is to fill the base with something to eliminate the external flow base overexpansion. The most impressive results from base burning to fill the base with total base drag elimination always need mentioning.

Trim drag is a most disagreeable topic simply because it reminds one that it could have somehow been done better. This is an unrealistic expectation, in some respects, because it is a natural phenomenon. However, plume drag should always be part of the design-analysis process. However, our desires in this respect are often frustrated because, among other considerations, the recipe for the treatment of nozzle plume effect is missing . The improvement of nozzle analysis design tools is one of the most important areas remaining for the future. In principle, every plume expansion requires a trim flap correction. The trim problem of minimizing the taxes of the plume and thrust vector can usually has a good outcome.

B. Expansion Process Physics

In principle, the expansion process is well understood from many past studies and the method of analysis characteristic of the midcentury. An excellent review of hypersonic nozzle shaping is found in Doty (1989). The CFD-based tool has been employed with success in this area. Basically, an initial turn sets up a wave system similar to the 2D Prandlt–Meyer expansion wave at a corner, even in 3D designs, Ortwerth (1995). The system of expansion waves is dispersive and the waves are not concentrated, but widen as they propagate, thus, they are hard to track. As the waves are reflected, the pressure distribution has a bouncing ball appearance rather than a smooth monotonic character. The simple straight wall does not relax to a source flow except very slowly and the effect is amplified by supersonic entrance conditions.

A CFD result for a 3D nozzle is shown in Fig. 56. The Prandtl–Meyer step at the corner is evident also as the resulting wave reflections. The Prandtl–Meyer step is not flat, due to combustor flow exit, static pressure gradient from cowl to body; this, in turn, is due to a combustor turn introduced to align the exit flow to the freestream direction, as previously discussed in the section on flow tuning and inlet integration. The area distribution (Fig. 57) shows a change in slope as the internal scramjet flowpath modules interface with the aftbody. The resulting compression is beneficial as the wall pressure integrals catch up to the ideal expansion integral (Fig. 58).

As in the compression process, the low-speed efficiency factor, C_{fg}, will

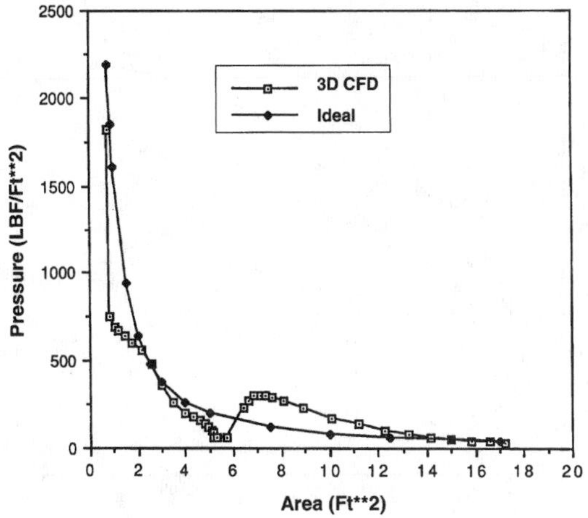

Fig. 56 3-Dimensional Nozzle Integration Pressure Profile Comparison with 1-Dimensional Shows Internal Loss and External Gain Due to Waves in Flow.

tend to unity no matter how poorly the nozzle performs. A simple remedy is to use the factor of the actual integral of wall pressure and area compared to the ideal 1D isentropic pressure area integral, η_{PdA}. This efficiency is equal to unity minus a nozzle K_{wp}; however, because the process is isentropic, there is no change in the adiabatic expansion coefficient. However, it should be used in the shear stress and heat flux, because the effect is to reduce drag and heat transfer. The use of K_{wp} is reserved for nozzle shock loss effects. This need not be discussed further since the formulation is the same. A

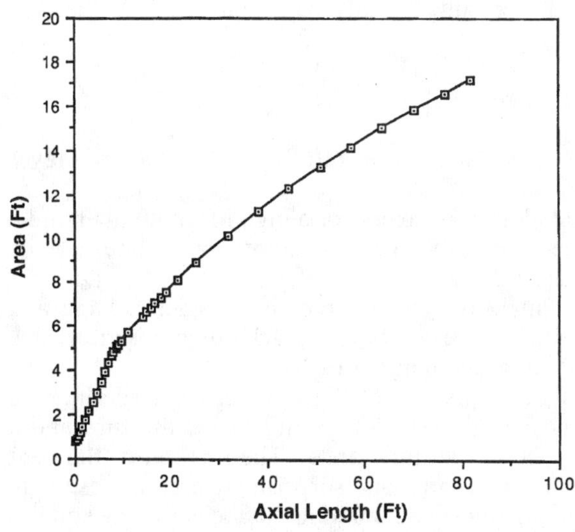

Fig. 57 1-Dimensional Area Distribution for Fig. 56.

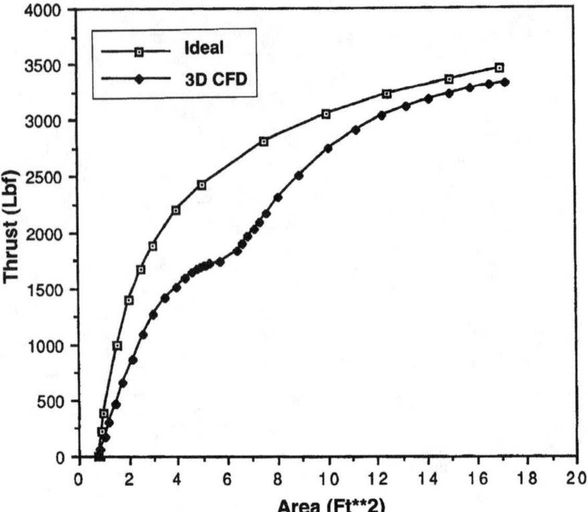

Fig. 58 Thrust Generation in 3D Nozzle.

caution, however, is that the K_{wp} is negative, since entropy loss is distributed in an expanding area, not a contraction.

A simple recipe for straight wall nozzle expansion losses was obtained by correlating the results of the characteristic method for λ over the range 1.4 to 1.15 and entrance Mach number to 4 for a range of expansion angle up to 20 degrees (Ref. 30). The result is shown in Fig. 57. The correlation formula, based on asymptotic value after the Prandtl–Meyer step is given by,

$$\eta\text{PdA} = 1 - 0.3\,\gamma\,M_C\,\theta^2 \tag{143}$$

The expansion angle for a 3D expansion is found in the area expansion rate.

The recipe for η_{PdA} in the step zone is analyzed as follows:

$$\eta_{\text{Pda}} = \left(\frac{P_\theta}{P_c}\right)\frac{(\gamma-1)(\varepsilon-1)\varepsilon^{(\gamma-1)}}{\left[\varepsilon^{(\gamma-1)}-1\right]} \tag{144}$$

where ε is the local area expansion ratio and P^θ is the Prandtl–Meyer expansion pressure.

The recipe indicates that all is well except for a high Mach number and a large expansion angle. The dependence on a Mach number is linear, while the angle effect is quadratic.

An effective expansion angle over 10 degrees is considered large and should be carefully studied when the combustor Mach number is increasing from near unity above a flight Mach number of 10.

It is important to calculate the length of the step region. Experimental observation shows that the heat transfer drops quickly at the turn and a desire to enhance this reduction naturally arises. The integral of the heat transfer is needed, however, and a rapid and surprisingly accurate estimate of the length of this region can quickly be found in the recipes if one equates

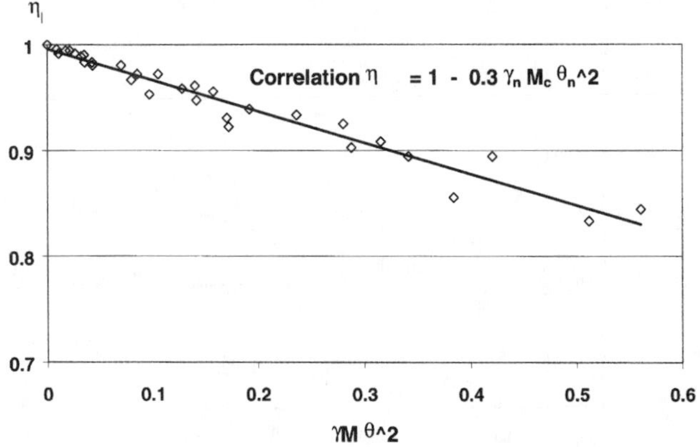

Fig. 59 Assymtotic Downstream Nozzle Efficiency Correlation a Function of Chordal Angle, Mach Number, and ζ

the step zone η_{PdA} to the asymptotic η_{PdA} value. Equation (C49) must be used for quickly nozzle expansion process since β and ψ are same order.

Heat transfer can be mitigated by using film cooling as reported in Baker (1993). Added fuel requirements occur with film cooling because good effectiveness precludes good mixing and the fuel may not efficiently be counted as part of combustion. Film cooling is part of the friction cycle.

VIII. Integration Results

A few results for a conical SSTO are presented. The integration of the flowpath is discussed in three parts. Because the propulsion system occupies such a large fraction of the total area and dry weight of the system, the question of partitioning the scramjet into modules is dealt with first. The second part deals with the detailed analysis of the scramjet module for pressure, viscous forces, heat load, and thrust. The third part is devoted to selecting the vehicle size for the SSTO mission.

A. Partitioning of Internal Flowpath

Partitioning of the internal flowpath is always needed because the cowl has to be attached to the vehicle, and the question (which is tractable) then posed is how many partitions are optimum? The study is constrained by the geometric L/D ratio scaling of flow processes, structural mechanics of panels, and back-up structure for loads as a function of span and area. A significant variable is the direction of compression and expansion. Two orientations are considered: the vertical and the horizontal. The cases do not consider 3D effects such as sweep in counting areas or lengths. The horizontal compression and expansion cases are often referred to as 2D configurations and on flat bodies can include variable geometry. The vertical geometry is referred to as the sidewall compression case. The sidewall configuration includes

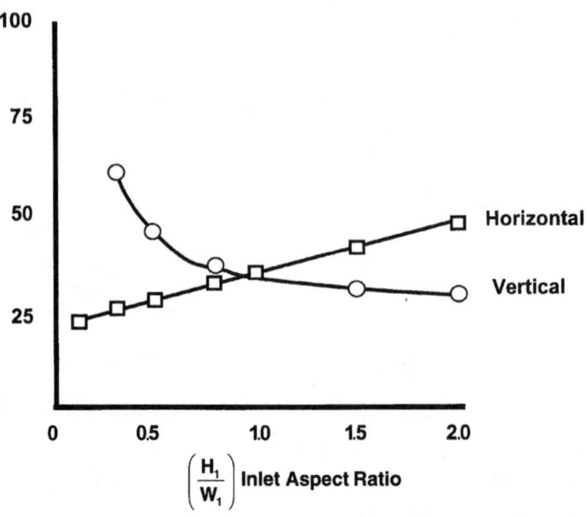

Fig. 60 Effect of Partions on Engine Surface Area Depends on Combustor Orientation and the Number of Partitions Measured by Module Inlet Aspect Ratio.

variable geometry on round or flat bodies, because the moving internal panels will bind together if the containing walls are not parallel. The sidewall engine length becomes shorter with more partitions while the length in the 2D case remains fixed because the flow scaling is based on the cowl height at the inlet plane; even though there are significant benefits from adding structural panels. The details are presented in Appendix G.

Significant results for wetted area and weight are found in Figs. 60 and 61. The more partitions that can be used, the better it is for the sidewall configuration. The limit is determined by the maximum inlet aspect ratio, which is practical for starting. Aspect ratio above 1.5 is achievable, however, a value of $AR=1.5$ is adopted for the sums that follow. The sidewall engine is two-thirds the weight and surface area of the 2D engine. The author considers this to be a general result as long as the internal scales do not become so small that the engine flame blows out.

B. Engine Module Flowpath Integration

A comparison of the design values obtained by use of the recipes in this chapter has been made with experimental data by parts, and a comparison for the integrated flowpath of the engine is needed. Data from an example engine are shown in Fig. 62. The comparison is surprisingly good except for the isolator that contained some strong reflections about the theoretical line so that details are good only on average. The wall pressure distribution is a critical test of the geometry-based combined interaction or direct-integration-over walls approach adopted here, and compares well with the use of traditional flux conservation, 1D methodology. There are no free parameters after the measured mass capture, fuel flow, and wall temperatures are input. The mix-

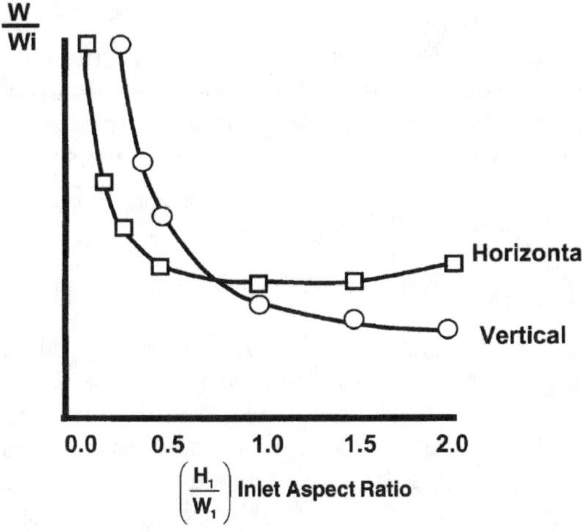

Fig. 61 Effect of Number of Partitions on Engine Weight, Including Structural Efficiency.

ing efficiency recipe was computed to be 1.0, and no corrections were made for less than ideal combustion efficiency. This is a first comparison and one is encouraged enough by its validity to adopt the method for application studies. The total heat load in the isolator and combustor was measured in fuel on and fuel on cases because the engine had flight-type removable panels. The cal-

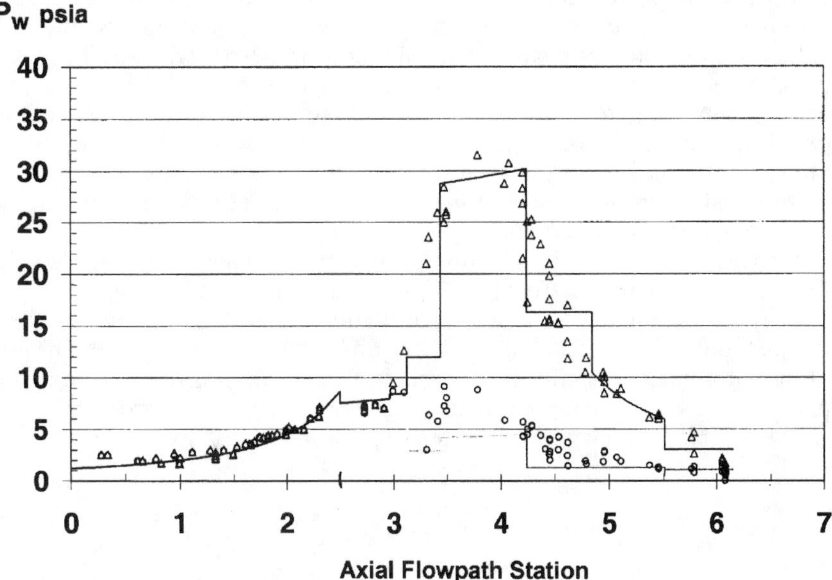

Fig. 62 RBCC Engine Flowpath Pressure Distribution Compared with Theory for Fuel-on and Fuel-off.

orimetry results were equally encouraging, being 1 or 2% higher than theory. Force balance was employed for thrust measurement. The measured value was 1% higher than that given by theory. Perhaps such close agreement in the dynamic environment is fortuitous. The standard 1D flux conservation cycle analysis has difficulties with some details of the flowpath in a dynamic environment; however, they are well known and mature. The results of the present approach (in a limited number of test comparisons) are, however, encouraging. This lends confidence to the approach for applications.

C. Scramjet Integration and Example

A conical SSTO-type vehicle was selected to illustrate engine vehicle integration in this chapter. Sizing a vehicle for a given payload means sizing a vehicle to contain enough fuel to accelerate the payload and vehicle to orbit, plus some residual amount for the return to base. The process involves a search and is dependent on the fact that drag and mass must follow the square-cube law in regard to geometry. For example, thrust and drag are functions of scale squared, while fuel mass increases as scale cubed. Thus, a size is searched so that the fuel required equals the fuel available. Once the size is found, the search process is closed. The mass-fraction available path is discussed first; then the mass-fraction required path.

D. Vehicle Mass Properties

During hypersonic SSTO studies of the mass properties expected in real vehicles, a recipe for mass property scaling was compiled and it became possible to establish a correlation in which everything depends on the fuel volume. Thus, quick estimates of scaling mass properties with size becomes possible. In undertaking such preliminary design estimates, it is necessary to depend on recipes, ad hoc rules, and correlations as well as the availability of an adequate database. Personal preferences also can be expected to influence the choices. In the following, a particular set of choices illustrates the methodology. Other choices lead to a different set of values for various quantities; however, the methodology remains valid.

A significant variable in the correlation is the average propellant density. This density is computed from the thrust, I_{sp}, and mixture ratio along the flight path by calculating the amount of propellants (oxidizer and fuel). This after-the-fact solution is acceptable to start with as it may be adjusted later. In view of the formula's simplicity, the propellant density also may be considered a parameter; then a map is generated, which is a key part of the process.

The vehicle, propulsion, and systems weight algorithm is

$$
\begin{aligned}
GLOW = {} & \rho_{\mathrm{prop}} V_P (1 + C_R) + \left(\frac{V_P}{P_{\mathrm{PFUS}}}\right)^{\frac{2}{3}} (W_S S_{\mathrm{FS}} + W_{\mathrm{TK}} S_{TK} \\
& \times \left[1 + C_{\mathrm{VC}} + C_{\mathrm{OMS}} + C_{\mathrm{ENG}} \theta H P_{\mathrm{ENG}} (1 + C_{\mathrm{ESYS}})(1 + C_{\mathrm{EC}})\right] (1 + X_G) \\
& + GLOW (C_W P_{\mathrm{WING}} + C_T P_{\mathrm{TAIL}} + C_G S_{\mathrm{FG}} + C_{\mathrm{VEHSYS}}) + C_{\mathrm{PAY}}
\end{aligned}
\tag{145}
$$

A designer can adjust the weight growth factor, X_G, to suit the specific need. The input starts by specifying the payload and any known mission constants, C_{PAY}. One has the option of selecting any guess, engineering or otherwise, for volume and engine size based on q and H. Next, the conceptual designer estimates the flight propulsion system cycle as a function of velocity to determine propellant density. This can be automated on a spreadsheet. The result is usually displayed for propellant fraction available with GLOW. Table 4 is a set of representative values for the mass property factors for a vertical takeoff horizontal landing (VTHL) vehicle.

The volume is the only explicit geometric input in the weight equation. This is another serious shortfall of this statistical approach. Various implicit discriminators in the packing factors can be changed by studying specific features. The fuselage-packing factor is too low for this type of conical configuration due to the packaging efficiency of the cone-shaped tanks and the reduced landing gear requirements. The engine factor may be a little high.

The engineering behind this correlation occurs by computing the surface-to-volume ratio raised to 2/3. Thus, the fuselage skin surface shape factor used, $S_{FS} = 10$, is also too high; it should be 6.74 for an 8-deg cone, which has approximately 1.4 times the surface area of a sphere for the same volume. However, there are many bulkheads and stringers to consider. The density factor involves the engineering disciplines of design, structures, thermal analysis, and materials. This is subjective and based on statistics. It is not easy to change a factor without risk, because many parts are involved and design constraints must not be violated. Thus, designs evolve slowly along a learning curve.

The engine has four size factors, which may all scale directly with the number of engines, q, the absolute size to a reference H, and a basic factor that has a specific SOL Mach number or a known surface area to fuselage surface area.

E. Mass Fraction Required

Determination of the required mass fraction is found by applying the results of Appendix A. The first input is a trajectory specified by altitude or dynamic pressure with velocity (Fig. 63). Along the trajectory, engine performance and drag are computed, including the drag due to lift and spill drag to determine an effective, Fuel Specific Impulse, I_sP_{eff}, along the trajectory. The scramjet data are supplemented by estimates of performance for the rocket modes. The flowpath performance and operation will be reviewed first. The engine was sized to be on design at Mach 12 as determined by the shock angle at Mach 12 and a specified cowl height. A complete 360-deg engine wrap was used as a starting point since vertical take off is assumed for the SSTO mission.

The vehicle can be briefly described as 210 ft in overall length with a fuselage volume of 53,650 ft^3. The outer cowl diameter is 36 ft at the internal nozzle exit and about 30 ft at the cowl inlet lip. The vehicle body diameter at the inlet face is 24 ft. The forebody cone angle is 8 degrees, with length of

Table 4 SSTO Mass Property Factors with Gear and Wing Factors Reflecting VTHL

SSTO Mass Property Factors Equation Symbol	Component Description	Value
PFUS	Fuselage packing factor	0.75
C_R	Residuals	0.02
W_S	Skin density	2.8
S_{FS}	Fuselage skin area factor	10
W_{TK}	Tank density	1
S_{TK}	Tank area factor	10
C_{VC}	Vehicle consumables	.02
C_{OMS}	OMS RCS system factors	.07
C_{ENG}	Engine factor	.45
θ	Engine wrap factor	$\theta/2\pi$
H	Engine Height factor	1
P_{ENG}	Engine packing factor	1
C_{ESYS}	Engine systems factor	.51
C_{EC}	Engine consumables, fluids	.033
C_W	Wing factor	.015
P_{WING}	Wing packing factor	1
C_T	Tail factor	.0037
P_{TAIL}	Tail packing factor	1
C_G	Landing gear (VTHL)	.0071
S_{FG}	Gear scaling factor	1
X_G	Weight growth factor	.05
CP_{AY}	Payload	25,000
	Payload bay	2000
	Crew and systems	0

88 ft and the tail cone angle is 10 degrees with length also 88 ft. The inlet forebody compression data used are found in the previous section for the Mach range 6 to 14. The engine length is about 34 ft. The gross liftoff weight (GLOW) of the vehicle is about 550,000 lb and depends on the propellant density. The frontal area of the cowl is 736 ft^2 and the flowpath surface area is 15,700 ft^2. The internal flowpath is about half of the total.

The I_{sp} and installed thrust used for the captured mass fraction estimated in Fig. 64 are given in Figs. 65 and 66. The internal contraction of the flowpath was constant, CR geometric ~5. The overall contraction varied with Mach number because of forebody shock wave movement. The thermal compression ratio at Mach 12 was ψ = 4.68 in a sizable range of overall expansion.

F. Closure

The standard practice used to obtain the required mass fraction is to perform a trajectory analysis using a 6 DOF or a 3 DOF trajectory code

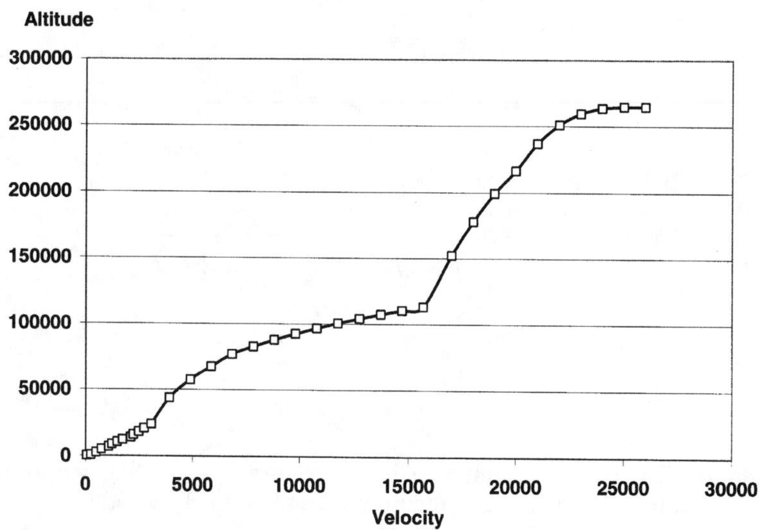

Fig. 63 VTOHL SSTO Trajectory with RBCC Engine.

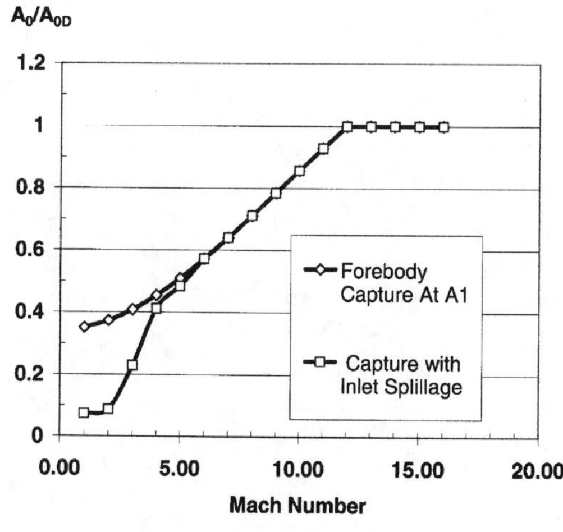

Fig. 64 RBCC Capture Area Characteristics Used in Application Study.

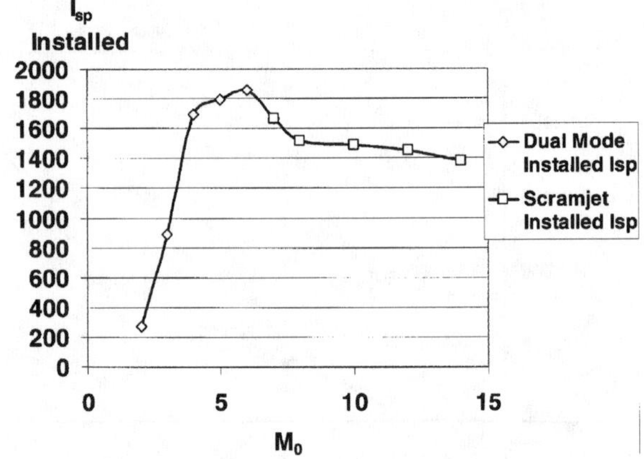

Fig. 65 VTOHL Installed I_{SP} for Angle of Attack. See Fig. 25 Definitions, Appendix F.

simulation. The propulsion and aero data are input. A trajectory optimizer often is employed. Very good agreement with the trajectory code is obtained by using a simple spreadsheet method.

A formally exact solution to a vehicle dynamic flight path is

$$\frac{W^{1,2}\mathrm{PROP}}{W_1} = 1 - e^{-(\Delta V_{\mathrm{LOSS}} + g_{\mathrm{LOSS}})} \tag{146}$$

The propellant burned is exponentially dependent on the trajectory by two factors. The first depends on the energy needed to change the kinetic energy of the vehicle and to overcome drag work, ΔV_{LOSS}. The second factor results from the increase or decrease of potential energy or work against gravity, g_{LOSS}.

$$\Delta V_{\mathrm{LOSS}} = \int_{V_1}^{V_2} \frac{\mathrm{d}V}{g I_{\mathrm{SPeff}}} \tag{147}$$

and

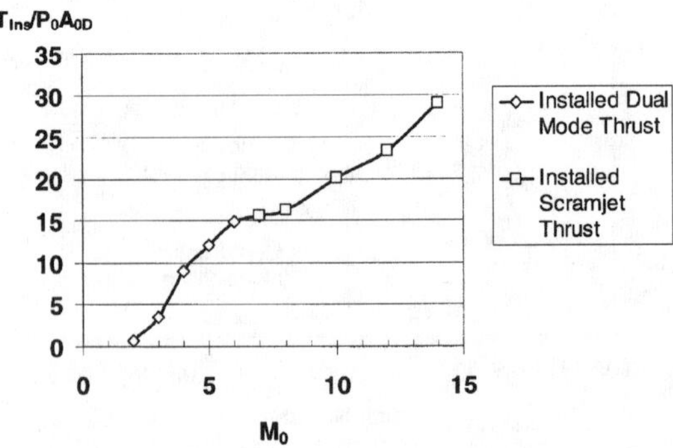

Fig. 66 VTOHL Installed Specific Thrust Scramjet Modes Only, with Out Rocket Operation. For Definitions, See Appendix F.

$$g_{\text{LOSS}} = \int_{V_1}^{V_2} \frac{(\mathrm{d}h/\mathrm{d}V)\mathrm{d}V}{VI_{\text{SPeff}}} \tag{148}$$

The problem immediately becomes tractable since, in principle, the trajectory velocity and h geometric altitude are known from the rocket accelerator take-off solution in Appendix A and the airbreathing q, V specification and an altitude-property atlas. A curve fit routine or linear interpolation formula can be used to find the rate of altitude change with velocity on the flight path. The propulsion solutions provide the needed I_{sp}, thrust, capture area (A_o), and mixture ratio (MR), also for selected points. A curve fit or interpolation formula allows the filling in of ~100 rows in a spreadsheet format where the columns contain the energy work loss integrals and the intermediate data. The columns for thrust and drag permit the effective I_{sp} to be filled in.

By definition

$$I_{\text{SPeff}} \equiv I_{\text{SP}}\left(1 - \frac{D}{T}\right) \tag{149}$$

The drag is found from the vehicle L/D. For low thrust systems, a high degree of precision and accuracy of (T/D) is needed. This has a small effect on the VTHL. The thrust is high and the full cowl fuselage has a very high L/D at full capture. To be accurate, the net thrust vector cosine loss, based on angle of attack, for the symmetric conical vehicle counts as drag since the thrust is not aligned with the wind axis. The lift is the sine of the angle of attack multiplied by the full exit momentum and is huge by comparison. In the analysis, all forces on the flowpath are counted, including the external cowl drag at zero angle of attack and forebody spill drag. Thus, drag due to lift of the wing area external to the fuselage, the cowl planform area needs to be accounted for as normal aerodynamic surfaces for angle-of-attack-generated drag. A C_{Do} applies only to the frontal area of the wing and tail. At this level of analysis, trim drag is neglected because it is small and is estimated inside the uncertainty band of the terms counted for the axisymmetric flow.

Thus, the mass of fuel burned is tracked at each point on the trajectory. To determine the average propellant density, two added calculations are entered for the split of oxidizer and fuel and each total is tracked along the trajectory. With the known mixture ratio,

$$MR = \frac{\dot{m}_O}{\dot{m}_F} \tag{150}$$

each increment of fuel and oxidizer is found.

$$W_F^{1,2} = \frac{W^{1,2}{}_{\text{PROP}}}{MR + 1} \tag{151}$$

and

$$W_O^{1,2} = MR\, W_F^{1,2} \tag{152}$$

The average propellant density then becomes

$$\rho_{\mathrm{AVG}} = \frac{W_F + W_O}{W_F/\rho_F + W_O/\rho_O} \tag{153}$$

In summary for each pull-up Mach number, a different ΔV_{LOSS} and g_{LOSS} is obtained, which is shown in Figs. 67 and 68.

The average propellant density as a function of pull-up Mach number is shown in Fig. 69. Thus, using the known mass fraction and propellant density, the locus of closure is found on the GLOW map on Fig. 70. The region to the right of and inside the closure locus represents available design space. The dashed lines are for constant tank volume and, thus, rise steeply with increasing propellant density. The solid lines are for constant propellant density and, thus, the low slope is obtained owing to the square-cube law. The sensitive parameters in the analysis are the scramjet engine performance, propellant density, and the rocket thrust augmentation on both ends of the flight spectrum. At the performance level found in the Mach 4 to 12 range, sensitivity to scramjet improvement is low. The scramjet was just at the φ=1 limit of high conductivity material temperature cooling at the Mach 12 condition, and the stoichiometric condition was forced without restricting fuel temperature at Mach 14 and 16. Above Mach 12, the flowpath aerothermo design is limiting. Film cooling is a technique used to minimize combustor and nozzle heating. However, high film effectiveness implies low combustion efficiency, and it is not practical to offload main propellant to wall film cooling, however, it must be an added cost.

The main result is finding the minimum GLOW at ~ 435 klb at about pull-up Mach number 12. The vehicle is only 33 ft in diameter and 202 ft long. The minimum size vehicle is the same for Mach 10 or 12 pull-up, at

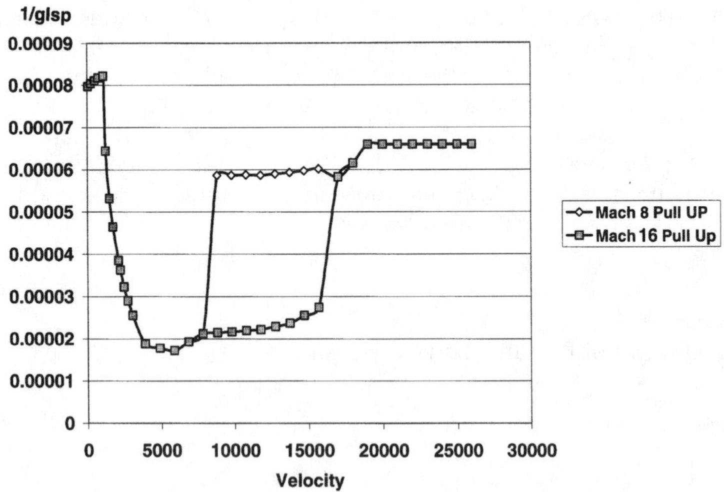

Fig. 67 Effect of Pull-Up Mach Number on Acceleration Loss Integrand Eq. 147.

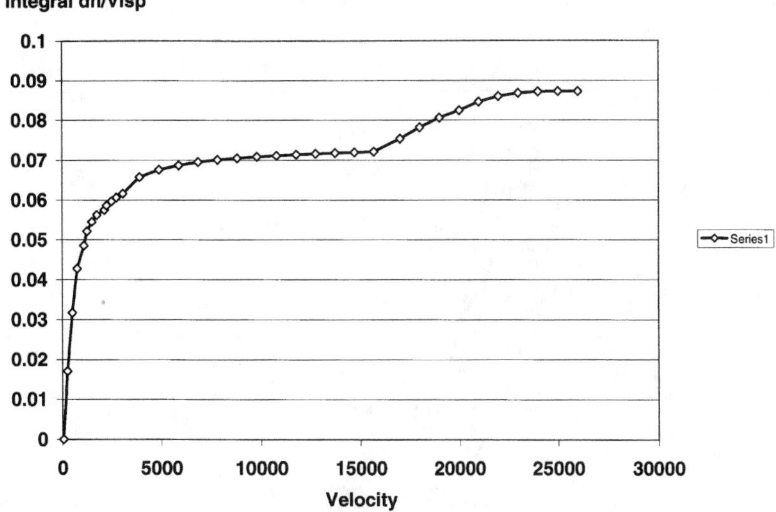

Fig. 68 Gravity Mass Loss Integral Eq. 148.

about 24,000 ft^3, as can be seen from the lines of constant volume. The design study does show a payload-to-orbit fraction of nearly 6%, which is why there is a fairly sustained level of interest by the space transportation community in the rocket-based combined cycle (RBCC) engine technology. The author is cautious about saying that payload fraction is a constant independent of system size; this is a flowpath integration study and as such the absolute size at closure is moot.

The mission design flexibility is robust since a constant vehicle volume

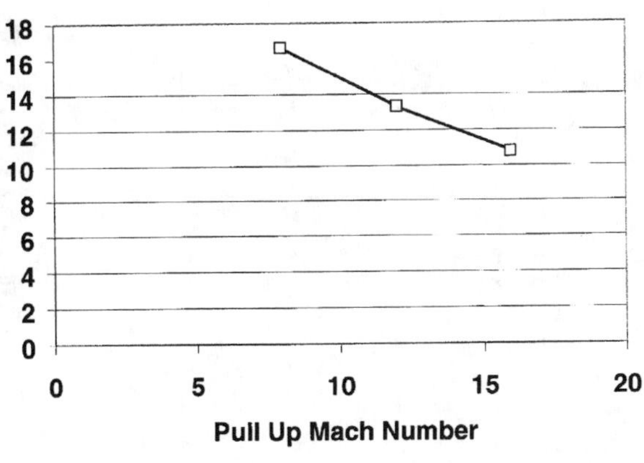

Fig. 69 Effect of Pull-up Mach Number on Bulk Propellant Density Eq. 153.

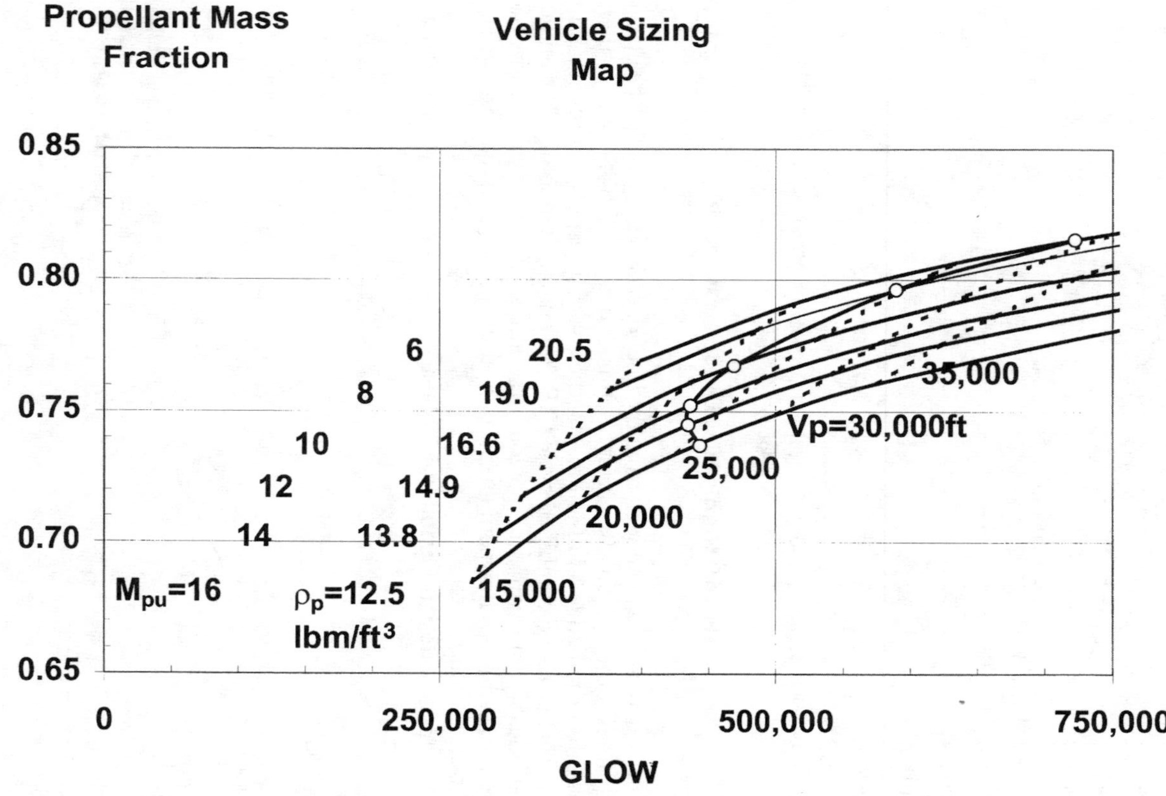

Fig. 70 VTOHL Vehicle Closure Map. Constant Payload 25,000 lb, LHz, LOX Propellants. Minimum Glow for M = 12 Pull-up Mach Number.

closes over pull-up Mach number from 6 to 16 even though the GLOW increases by 50%. The same vehicle volume does not imply that the same vehicle could fly all the missions. The engines may not be adequate to lift off for example, and the wing and landing gear size and weights are not the same since they are proportional to GLOW to keep wing and tire loading constant. At a minimum, the GLOW is constant, thus, there does seem to be a region where the design falls in a bucket and where the specific design sensitivity to any major parameter is small.

One of the primary goals of the sizing analysis is finding a bucket. The existence of the bucket is fundamental to the physics of the analysis. The scramjet SSTO is a stored chemical energy system and the bucket is a function of the fuel's energy density. The results show that the advantage of high I_{sp} is offset by the propellant density decrease.

IX. Summary and Recommendations

This chapter presents a hypersonic theory for the scramjet flowpath integration. A physical model was constructed for each component in the flowpath to enable direct integration of the wall forces and heat transfer. The inviscid core flow and viscous wall layers are solved together in a combined interactive solution. To accomplish this, a new internal wall pressure coefficient, called K_{wp}, was found based on theoretical consideration separating entropy generation sources in a way that could keep the inviscid flow separate from the wall layers. This pressure coefficient provides a way to count the entropy generated in the engine's shock wave system as distinct from the shear work and heat transfer at the wall. Further, the physics of limiting hypersonic flow emerge in the form of a so-called hypersonic energy partitioning (HEP) principle that enables aerodynamic drag to directly relate to gas state property changes separated from gas kinetic energy changes.

The approach admits a further separation of forces into parts, a wave part, an isentropic compression or expansion part, and a part due to pure 2D aerodynamics or turning part. The last feature enables force accounting in the normal and transverse planes, lift and drag, and, in principle, also moment generated in each component. This is very important in force accounting and in the design layout of the flowpath. Internal forces and moments directly affect the overall performance, L/D, of the vehicle.

The physical model package for forebody-inlet integration has been used to compare two classes of flowpath, 2D and 3D, and the discriminating design features and limitations are identified.

To divide the processes into parts that lead to explicit aerodynamic analysis and solutions, such quantities as the gas properties are required for each process, with appropriate attention paid to the chemical state. The virtue of this approach is that clarity is retained, which is obscured in the usual 1D-cycle programs using equilibrium thermochemical programs. More work is required in this area to account for real gas effects. Currently, the full capability of the method is somewhat immature for routine application in design exercises.

However, a successful comparison of the pressure distribution, local and integrated heat load by calorimetry (with experimental test of an integrated flowpath) has verified the soundness of the method. The theoretical recipes used in this chapter and programmed in an Excel spreadsheet, also matched the thrust and *I*sp obtained in tests to well within engineering accuracy.

The value of this approach to the future of scramjet vehicle technology is that performance design and analysis code for propulsion are hooked directly to flowpath geometry. This puts the propulsion analysis in the same genre as the APAS external aerodynamic analysis code. Thus, it is hoped that the scramjet vehicle will achieve the same conceptual design confidence as rocket vehicles and jet aircraft.

In conclusion, a few philosophical points of practical interest in the future of scramjet technology are offered. The soundness of engineering design recipes needs to be matured at the discipline level. No one can send hundreds of engineers to the vehicle or propulsion industry to work the detailed designs without a complete set of working guidelines and recipes. Such rules must be based on sound fundamental knowledge and theories available in the literature.

To develop a scramjet spacecraft such as the SSTO discussed in the last section of this chapter, there is a need for a series of scramjet research vehicles of increasing size to full-scale prototype demonstrations to ensure the future strength of engineering design recipes, the discipline level must mature. In the author's vision, the first test series must address the problem of putting the required design recipes into place. The vehicle size is not daunting for an unmanned SSTO, the dry weight of which is about the same as a large jet transport. What is daunting is that the physics of the propulsion flowpath are different in a scramjet, and are not as well known. The design of a successful scramjet requires that design margins are carefully established and applied to the limit, just as for the 747 aircraft. The mystery needs to be understood and tamed. Confidence is currently low because this is an esoteric field when compared with helicopters, for example. But there are few incentives in current market forces for scramjet technology. Making space transportation affordable is necessary but not high priority. Thus, inexpensive high-value programs are needed and an approach is suggested, as follows.

The primary view is that for about 3 years, at least 10, maybe 20, flight tests are needed each year! If one flight per year is all we do, it will take 60 years.

The economics are such that these tests require an outlay of only $10 million or less. Admittedly, this is a radically counter-cultural vision as our best government and industry program managers think a realistic estimate for program starts at $1 billion. The first milestone, then, is to build and successfully operate an inexpensive test vehicle. The second milestone is to recruit the best available minds and support them and their findings.

Arguing design data at true application-based Reynolds numbers suggests this. Recall that on a full-scale vehicle, it has been found that the mass flow in the boundary layers is half the total. Fig. 71 shows a vehicle with a dry weight of a few hundred pounds. This is an example of a reasonable

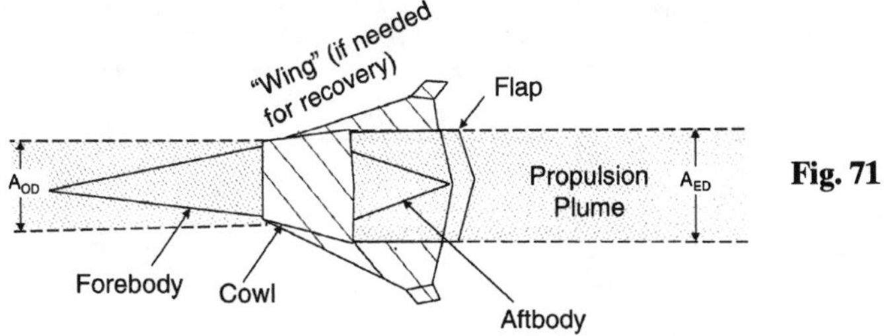

Fig. 71

starting point in flight testing. It meets the cost requirement and is self-powered from Mach 3 to 12. Flight time is short and the range requirement is not thousands, but hundreds of miles. The idea of thinking small and short range is that full simulation is achieved at very high dynamic pressure, 9,000 ~ 20,000 lb/ft^2, with, of course, very high acceleration (20 to 40 gs) and, thus, a 1/10th-scale vehicle achieves full-scale simulation in one's backyard.

The table of scaling parameters covering those needed for valid hypersonic scramjet vehicle design formulas is shown in Table 5—labeled q scaling. The q scaling as proposed works with respect to most pressures of interest in the integrated vehicle.

Fig. 72 shows the pressure distribution calculated for the STTO vehicle example. Flight verification of this supersonic combustion wave is required.

Table 5 Scaling Flight Demonstrator Data to Another Size Application at Constant Mach Number and Reynolds Numbers

Parameter (size of new engine)	
Length	L/L_1 = *Given*
Velocity	$V/V_1 = 1$
Pressure	$P/P_1 = (L/L_1)^{-1}$
Temperature	$T/T_1 = 1$
Dynamic Pressure	$q\ /q\ _1 = (L/L_1)^{-1}$
Friction coefficient (Stanton number)	$C_f/C_{f1} = 1$
Boundary layer thickness	$\delta^*/\delta^*_1 = (L/L_1)$
Mass capture	$(\dot{m}/\dot{m}_1) = (L/L_1)$
Heat flux per unit mass	$(\dot{Q}/\dot{m})/(\dot{Q}/\dot{m})1 = 1$
Thrust per unit mass	$(T/\dot{m})/(T/\dot{m})1 = 1$
Specific impulse	$(I_{SP})/(I_{sp})_1 = 1$
Equivalence ratio	$\varphi/\varphi_1 = 1$
Ignition delay time	$\tau_1/\tau_1 = (L/L_1)$
Recombination time	$\tau_R/\tau_{R_1} = (L/L_1)^n \quad n \sim 2$

Note: Subscript 1 denotes flight data or quantity derived from flight data.

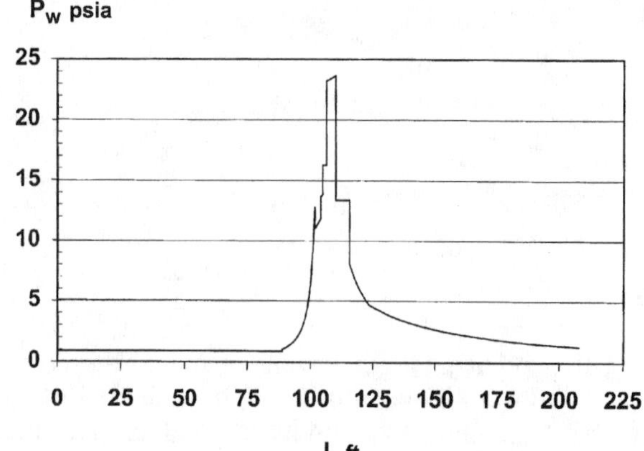

Fig. 72 Flow Path Wave Structure

Bibliography

Anderson, G. Y., and Clayton-Rogers, R., "A Comparison of Experimental Supersonic Combustor Performance with an Empirical Correlation on Non-Reactive Mixing Results," NASA TMX-2429, 1971.

Anderson, M., and Ortwerth, P., "Three Dimensional Hypersonic Inlets — Low Speed Performance," AIAA Paper 91-5021, 1991.

Auslender, A. H., "An Application of Distortion Analysis to Scramjet-Combustor Performance Assessment," JANNAF Workshop on Scramjet Combustor Performance Final Report, 1996.

Baker, N. R., Northam, G. B., Stouffer, S. D., and Capriotti, D. P., "Evaluation of Scramjet Nozzle Configurations and Film Cooling for Reduction of Wall Heating," AIAA Paper 93-0744, 1993.

Baev, V. K., Golovchev, V. I., Tret'yakov, P. K., Garanin, A. F., Konstantinovskiy, V. A., and Yasakov, V. A., "Combustion in Supersonic Flow," NASA TM-77822, 1986.

Billig, F. S., and Dugger, G. L., "The Interaction of Shock Waves and Heat Addition in the Design of Supersonic Combustors," *Twelfth International Symposium on Combustion,* Combustion Inst., Pittsburgh, PA, 1963.

Bowcutt, K. G., "Multidisciplinary Optimization of Airbreathing Hypersonic Vehicles Considering Aerodynamics, Propulsion, Control and Mass Property Effects," *International Symposium on Airbreathing Engines*, 99-7089, Sept. 1999.

Builder, C. H., "On the Thermodynamic Spectrum of Air Breathing Propulsion," AIAA Paper 64-243, June–July 1964.

Chapman, D. R., Kuehn, D. M., and Larson, H. K., "Investigation of Separated Flows in Supersonic and Subsonic Streams with Emphasis on the Effect of Transition," NACA R 1356, 1958.

Chernyi, G. G., *Introduction to Hypersonic Flow,* Academic Press, New York, 1961.

Donohue, J. M., and McDaniel, J. C., Jr., "Complete 3-D Multi-Parameter Mapping of the Supersonic Mixing Flowfield of a Ramp Fuel Injector," AIAA Paper 95-0519, 1995.

Doty, J. H., Thompson, H. D., and Hoffman, J. D., "Optimum Thrust Two-Dimensional NASP Nozzle Study," NASP 1069, 1989.

Edney, B., "Anomalous Heat Transfer and Pressure Distributions on Blunt Bodies at Hypersonic Speeds in the Presence of an Impinging Shock," Aeronautical Research Inst. of Sweden, Rept. 115, Stockholm, 1968.

Ferri, A., "Mixing Controlled Supersonic Combustion," *Twelfth International Symposium on Combustion,* Combustion Inst., Pittsburgh, PA, 1969.

Foa, J. V., "Theory of Propulsion in Air," Aeronautical Engineering Dept., Rensselaer Polytechnic Inst., 1958.

Hirschfelder, Curtiss, Bird, *Molecular Theory or Gases and Liquids,* Wiley, New York, 1954.

Hsia, Y. C., Gross, B. J., and Ortwerth, P. J., "Inviscid Analysis of a Dual Mode Scramjet Inlet," *Journal of Propulsion and Power,* Vol. 7, No. 6, 1991, pp. 1030–1035.

Huber, P. W., Schexnayder, C. J., Jr., and McClinton, C. R., "Criteria for Self-Ignition of Supersonic Hydrogen and Air Mixtures," NASA T Pr 1457, 1979.

Huellmantel, L. W., Ziemer, R. W., and Alibulent, C., "Stabilization of Premixed Propane Air Flames in Recessed Ducts," *Jet Propulsion,* 1977, pp. 31–43.

Kenworthy, M. J., "Analytical and Experimental Evaluation of the Supersonic Combustion Ramjet Engine," General Electric Corp. AF APL-TR-65-103, Evendale, OH, 1965.

Korst, H. H., "A Theory for Base Pressures in Transonic and Supersonic Flow," *Journal of Applied Mechanics,* Vol. 23, 1956, pp. 593–600.

Küchemann, D., *The Aerodynamic Design of Aircraft,* Pergamon, New York, 1978.

Laderman, A. J., "Pressure Gradient Effects on Supersonic Boundary Layer Turbulence," AFFDL-TR-79-3005, Final Rept., 1979.

Lopez, H., "HREP-Phase II - Combustor Program," Final Technical Data Rept., Airesearch Report No. AP-70-6054, NASA CR 66932, 1970.

McLafferty, G., "A Generalized Approach to the Definition of Average Flow Quantities in Non-Uniform Streams," United Aircraft Corp. Rept. SR-13534-9, 1955.

Miele, A., *Theory of Optimum Aerodynamic Shapes,* Academic Press, New York, 1965.

Morrison, C. Q., Campbell, R. L., Edelman, R. B., and Jaul, W. K., "Hydrocarbon Fueled Dual-Mode Ramjet/Scramjet Concept Evaluation," AIAA Paper Sept. 1997.

Murthy, S. N. B., and Osborn, J. R. (eds.), *Aerody of Base Combustion,* Vol. 40, *Progress in Aeronautics and Astronautics*, AIAA, New York, 1974.

Northam, G. B., Greenberg, I., and Byington, C. S., "Evaluation of Parallel Injector Configurations for Supersonic Combustion," AIAA paper, July 1989.

Orth, R. C., and Cameron, J. M., "Flow Immediately Behind a Step in a Simulated Supersonic Combustor," *AIAA Journal,* Vol. 13, No. 19, 1975, pp. xx–xx.

Ortwerth, P. J., "A Generalized Distortion Theory of Internal Flow,". Air Force Weapons Lab. AFWL-TR-77-118, 1977.

Ortwerth, P. J., "Results of the Distortion Theory of Normal Shock Diffusers, Part II: Shock Length Correlations for Normal Shocks in Ducts," Air Force Weapons Labs-tr-77-118, 1977b.

Ortwerth, P. J., Mathur, A. B., and Brown, C. J., "Flow Path Optimization for Hypersonic Vehicles," AIAA, Paper 91-5043, 1991.

Ortwerth, P. J., and Bowcutt, K. G., "Advanced Airbreathing Propulsion System for Long Range Atmospheric Interceptors," AIAA Paper 05-0B3, July 1995.

Ortwerth, P. J., and Goldman, A. L., "Forebody Inlet Comparison for Hypersonic Vehicles," AIAA Paper 96-3039, July 1996.

Ortwerth, P. J., Mathur, A. B., Segal, C., Owens, M. G., and Mullagilli, S., "Combustion Stability Limits of Hydrogen in a Non-Premixed Supersonic Flow," *International Symposium on Airbreathing Engines*, Paper 99-7136, Sept. 1999.

Ortwerth, P., Mathur, A., Vinogradov, V., Grin, V., Goldfeld, M., and Starov, A., "Experimental and Numerical Investigation of Hydrogen and Ethylene Combustion in Mach 3–5 Channel with a Single Injector," AIAA Paper 96-3245, July 1996.

Pratt, D. T., and Heiser, W. H., "Isolator-Combustor Interaction in a Dual-Mode Scramjet Engine," AIAA Paper 93-0358, 1993.

Rasmussen, M., *Hypersonic Flow,* Wiley-Interscience, New York, 1994.

Ratekin, G., Goldman, A., and Ortwerth, P., "Rocketdyne RBCC Engine Concept Development," *International Symposium on Airbreathing Engines,* Paper 99-7179, 1999.

Shapiro, A. H., *The Dynamics and Thermodynamics of Compressible Fluid Flow,* Vol. II, Ronald, New York, 1954.

Stalker, R. J., and Morgan, R. G., "Supersonic Hydrogen Combustion with a Short Thrust Nozzle," *Combustion and Flame* Vol. 57, 1994, pp. 55–70.

Strahle, W. C., Hubbartt, J. E., and Walterick, R., "Base Burning Performance at Mach 3," *AIAA Journal,* Vol. 20, No. 7, 1982, pp. xx–xx.

Swithenbank, J., and Chigier, N. A., "Vortex Mixing for Supersonic Combustion," *Twelfth International Symposium on Combustion,* Combustion Inst., Pittsburgh, PA, 1969.

Townend, L. H., and Reid, J., "Some Effects of Stable Combustion in Wakes Formed in a Supersonic Stream," *Supersonic Flow, Chemical Processes and Radiative Transfer,* edited by Olfe and Zakkay, Pergamon, New York, 1964.

Townend, L. H., and Edwards, J. A., "Base Pressure Control for Aerospace Vehicles," *Proceedings of the Symposium Fluid Dynamics and Space VKI,* June 1986.

Trexler, C. A., "Inlet Starting Predictions for Sidewall-Compression Scramjet Inlets," AIAA Paper 88-3257, July 1988.

White, F. M., *Viscous Fluid Flow,* McGraw-Hill, New York, 1974.

Wiles, K. J., Ketchum, A. C., Emanuel, M. A., Mathur, A. B., and Ortwerth, P. J., "Preliminary Tests of a Dual Fuel-Dual Mode Integrated Scramjet Engine," AIAA, 1987.

Appendix A: Dynamics of a Flight Vehicle

A. Cruise Flight

Considering steady, level flight of a flight vehicle, the mass of the vehicle must be balanced by the lift and drag of the vehicle by propulsive thrust. Then, one can write

$$-gI_{\text{sp}}\frac{\mathrm{d}M}{\mathrm{d}t} = \frac{W}{(L/D)} \tag{A1}$$

where dM is the change in the mass of the vehicle, due to fuel consumption, over a period dt. Multiplying both sides by the flight velocity, V, and rearranging, one obtains the following for the element of range covered in a time interval Δt:

$$dR = V \cdot dt = -I_{sp} V \left(\frac{L}{D} \right) \frac{dW}{W} \tag{A2}$$

Integrating this over a change of W from W_o to W_f, while treating (L/D) as a constant, one obtains the well-known expression for the Breguet range:

$$R = I_{sp} V \left(\frac{L}{D} \right) \ell n \left(\frac{W_o}{W_f} \right) \tag{A3}$$

In practice, to maintain L/D constant over the flight, while fuel consumption reduces the vehicle mass, the flight altitude needs to be increased; the optimum value of $V \cdot I_{sp}$ or best efficiency then determines the flight path. In the alternative, $V \cdot I_{sp}$ may be held nearly constant and the flight speed selected for optimal L/D over the flight path.

B. Accelerated Flight

The instantaneous dynamic equation of motion during acceleration is given by

$$dV = gI_{sp} \left(1 - \frac{D}{T} \right) \frac{dM}{M} \tag{A4}$$

Integrating, one can write

$$\Delta V = gI_{sp\ eff}\ \ell n \left(\frac{M_f}{M} \right) \tag{A5}$$

where the effective I_{sp} is given by

$$I_{sp\ eff} = I_{sp} \left(1 - \frac{D}{T} \right) \tag{A6}$$

Note that D, T, and $I_{sp_{EFF}}$ are all functions of V.

From Eq. (A5) the propellant fraction (PFR) required for acceleration can be stated as

$$PFR = \frac{M_{prop}}{M_o} = 1 - \exp \left(\frac{\Delta V}{g\, I_{sp\ eff}} \right) \tag{A7}$$

It follows that PFR does not depend on acceleration *per se* but on the drag of the thrust ratio during flight, and on I_{sp}, which depends on the flight, the fuel, and, for an airbreathing engine, on the flight path conditions.

C. Application to a Constant I_{sp} Engine

A rocket motor is an outstanding example of an engine with a nearly constant I_{sp}. For a single-stage-to-orbit (SSTO) rocket with constant thrust, one can consider a flight plan made up of two parts: 1) liftoff and vertical acceleration to a desired vertical velocity, and 2) tangential acceleration to orbit. The weight of the rocket is a function of time until burnout, and can be represented by the relation

$$\frac{W}{W_o} = 1 - \frac{a_o t}{I_{sp}} \tag{A8}$$

where a_o represents the thrust-to-weight ratio. Denoting the flight path angle by θ, it is clear that in the first part of the trajectory $\theta = \pi / 2$, and in the second part, with dynamical equilibrium

$$\sin\theta = \frac{W}{T} \tag{A9}$$

For the first or vertical acceleration part, the acceleration is simply thrust-to-weight ratio minus one, and the velocity at the beginning of the gimbals on leg is given by

$$V_{GO} = \frac{g\, I_{sp}}{a_o}\left[a_o\, \ell n\left(\frac{W_{GO}}{W_O}\right)^{-1} - \left(1 - \frac{W_{GO}}{W_O}\right)\right] \tag{A10}$$

For the second part, the tangential acceleration may be written as follows:

$$\frac{dV_T}{dt} = \frac{g\, I_{sp}}{a_o}\left[\left(\frac{T}{W}\right)^2 - 1\right]^{1/2} \tag{A11}$$

Integrating, the tangential velocity can be written as

$$V_T = \frac{g\, I_{sp}}{a_o}\,[a_o \ell n\, F_1]^{-1} - \left\{\left[a_o^2 - \left(\frac{W_f}{W_o}\right)^2\right]^{1/2} - \left[a_o^2 - \left(\frac{W_{GO}}{W_O}\right)^2\right]^{1/2}\right\} \tag{A12}$$

where

$$F_1 = a_o + \left[a_o^2 - \left(\frac{W_f}{W_o}\right)^2\right]^{1/2} \Bigg/ \left(\left(\frac{W_f}{W_{GO}}\right)\left\{a_o + \left[a_o^2 - \left(\frac{W_{GO}}{W_O}\right)^2\right]^{1/2}\right\}\right) \tag{A12a}$$

D. Application to a Constant $V{\cdot}I_{sp}$ Engine

The scramjet engine operates with a nearly constant $V{\cdot}I_{sp}$ value. The product $V{\cdot}I_{sp}$ is closely related to propulsive efficiency, the efficiency of conversion of chemical reaction heat to thrust power, that is,

$$\eta_P = \frac{TV}{\dot{W}\,\Delta H_{\text{comb}}} \tag{A13}$$

Noting that

$$\sin\theta = \frac{1}{V}\frac{\mathrm{d}z}{\mathrm{d}t} \tag{A14}$$

and writing

$$K = \frac{V^2}{2} + gz \tag{A15}$$

one can write in this case that

$$\frac{W_f}{W_o} = \exp\left(\frac{K_f - K_o}{\eta_P \cdot \varsigma \cdot gJ \cdot \Delta H_f}\right) \tag{A16}$$

where ζ is the integration parameter equal to $(1 - D/T)$. This equation can be obtained by starting with Eq. (A5), multiplying it by V, rearranging, and integration following substitution of Eqs. (A14) and (A15)

Appendix B: Brayton Cycle Scramjet

The ideal Brayton cycle on an H-(S/R) diagram is illustrated in Fig. 3.

Following Builder (1964) and within the framework of the assumptions introduced by him, one can write the expansion velocity in the thrust nozzle as

$$V_7^2 = V_o^2 + 2H_o(\psi - 1)\left[\eta_c\eta_e\left(\frac{\Delta H_c/H_o}{\psi}\right) - 1\right] \tag{B1}$$

where the inlet contraction ratio is represented by

$$\psi = H_3 / H_o \tag{B2}$$

One can then write the following for the air-specific impulse of the engine:

$$I_A = F_{\text{net}} \Big/ \dot{m}_a = \frac{V_o}{g}\left(\left\{1 + \frac{\psi - 1}{[\gamma - 1/2]\,M_o^2}\left[\eta_c\eta_e\left(1 + \frac{\Delta\psi_c}{\psi}\right) - 1\right]\right\}^{1/2} - 1\right) \tag{B3}$$

where

$$\Delta\psi_c = \frac{\Delta H_c}{H_o} \tag{B4}$$

Next, the power available from air per unit mass flow rate can be written as

$$\frac{power}{\dot{m}} = \frac{F_{\text{net}} \cdot V_o}{\dot{m}} = \frac{F_{\text{net}}}{P_o A_o} \cdot RT_o \tag{B5}$$

Thus, the power per unit mass flux of air becomes the power per unit area per atmosphere, when P_o is expressed in atmosphere units.

The thrust per unit area per atmosphere is given by

$$\frac{F_{\text{net}}}{P_o A_o} = \left(\frac{\gamma}{\gamma - 1}\right)(\psi - 1)\left[\eta_c \eta_e \left(1 + \frac{\Delta\psi_c}{\psi}\right) - 1\right] \tag{B6}$$

Lastly, the efficiency of thrust generation from chemical conversion of heat can be written as follows, using Eqs. (B4) and (B5):

$$\eta_{\text{chem}} = \frac{F_{\text{net}} \cdot V_o}{\dot{m}\,\Delta\,H_c} \tag{B7}$$

$$= \frac{F_{\text{net}}}{P_o A_o} \cdot \left(\frac{\gamma}{\gamma - 1}\right)^{-1} \cdot \frac{1}{\Delta\psi_c} \tag{B8}$$

$$= \frac{1}{\Delta\psi_c} \cdot (\psi - 1) \cdot \left[\eta_c \eta_e \left(1 + \frac{\Delta\psi_c}{\psi}\right) - 1\right] \tag{B9}$$

Appendix C: Aerothermodynamics of Scramjet Engine

A. Pressure Coefficient

The engine flowpath involves two important features: waves and viscous interaction at the boundary wall. To analyze the flowfield on a one-dimensional (1D) basis, one can proceed by distinguishing between the thermodynamic pressure P, and a corresponding value of wall pressure P_w, at each location along the flowpath. The wall pressure accounts for the effects of shock waves in addition to those of wall friction.

The following 1D differential equations can then be constructed for momentum and energy balance:

$$\frac{1}{\rho}\frac{\mathrm{d}P}{\mathrm{d}x} + V\frac{\mathrm{d}V}{\mathrm{d}x} = \frac{(P_w - P)}{\rho} \cdot \frac{1}{A}\frac{\mathrm{d}A}{\mathrm{d}x} - \frac{\tau_w}{\rho}\frac{1}{A}\frac{\mathrm{d}\sigma}{\mathrm{d}x} \tag{C1}$$

and

$$\frac{dH}{dx} + V\frac{dV}{dx} = -\frac{q_w}{\dot{m}}\frac{d\sigma}{dx} \tag{C2}$$

where $\dot{m}$ is the mass flux over an area of cross-section A. The wall shear stress and heat transfer per unit surface area are denoted by τ_w and q_w, respectively. The element of surface area over the length dx is denoted by $d\sigma$. Equation (C1) can also be derived from the stream thrust integral equation.

Eliminating the velocity terms in Eqs. (C1) and (C2), one obtains

$$dH - \frac{dP}{\rho} = -\left(\frac{P_w - P}{\rho}\right)\frac{dA}{A} + \left(\frac{\tau_w}{\rho} - \frac{q_w}{\rho V}\right)\frac{d\sigma}{A} \tag{C3}$$

Introducing the entropy relationship

$$T\,ds = dH - \frac{dP}{\rho} \tag{C4}$$

it follows that

$$\frac{ds}{R} = -\left(\frac{P_w - P}{P}\right)\frac{dA}{A} + \left(\frac{\tau_w}{P} - \frac{q_w}{PV}\right)\frac{d\sigma}{A} \tag{C5}$$

Considering the engine duct, which may involve a contraction (as in the inlet) or an expansion (as in a nozzle), an average value can be assigned for the so-called pressure coefficient by writing

$$K_{WP} \equiv \overline{\frac{P_w - P}{P}} \tag{C6}$$

The overbar denotes a suitably averaged value. Three major observations on K_{WP} are the following:

1) K_{WP} depends on the process occurring in the duct; thus, the flow deceleration or diffusion process in an inlet, for instance, is an independent variable in determining K_{WP}.
2) As an aerodynamic function of the initial conditions and geometry, K_{WP} should be explicitly dependent on the entry Mach number squared, and the shape parameters or angle related to hypersonic similarity parameters.
3) When K_{WP}, along with viscosity effects, is zero, the internal flow is governed by Euler equations and is isentropic.

When introducing K_{WP}, one can readily integrate pressure, shear stress, and heat transfer on the wall.

Returning to Eq. (C4), noting that the entropy as well as the area of cross-section are both functions of length along the flow, one obtains, after integration, the relation

$$\frac{\Delta S}{R} = K_{WP}\,\ell n\,CR \tag{C7}$$

for an inlet with a contraction ratio CR. K_{WP} is then the pressure coefficient corresponding to an increase in entropy ΔS due to shock waves in an inlet of contraction ratio CR; this provides a definition for K_{WP} in an inlet. Integration is based on considering a specified entry flow Mach number and assuming that the shock waves formed in the flow are a function of that Mach number. For a 2D wedge flow with an attached shock wave, Fig. 10 shows that the linear relationship of Eq. (C7) indeed applies.

In connection with the introduction of the wall pressure, P_w, as a different variable than the thermodynamic pressure, P, a brief review on the concept of the friction cycle may be useful.

Considering a small length Δx at $x = x_1$ along the flowpath, with n change of area, A to $A + (\mathrm{d}A/\mathrm{d}x)$. Δx, and assuming the applicability of Prandtl's boundary layer theory, one can consider the wall layer thickness, δ, at $x = x_1$ growing to $\delta + (\mathrm{d}\delta/\mathrm{d}x) \cdot \Delta x$ over the length Δx. Due to entrainment, the mass flux $\dot{m}$ at $x = x_1$ increases to $\dot{m} + (\mathrm{d}\dot{m}/\mathrm{d}x) \cdot \Delta x$ over length Δx. Considering wall friction over Δx, one should account for a loss of pressure as well as a gain in internal energy of the fluid in the wall layer. Based on the concepts of a control volume between x_1 and $x_1 + \Delta x$ and a thermodynamic cycle in the engineering sense, one is extracting kinetic energy from the fluid and converting it into internal energy through a change in pressure by the action of wall friction, which is simply due to viscosity in the case of a smooth wall. If the thermodynamic pressure is P over Δx in the core flow, one can write the loss in pressure in the wall layer as $(P - P_W)$, where P_W is the wall pressure, noting this difference in pressure acts over the area $(\mathrm{d}A/\mathrm{d}x) \cdot \Delta x$.

Equation (C1) may then be interpreted as follows:

$$\rho AV \frac{\mathrm{d}V}{dx} = -A\frac{\mathrm{d}P}{\mathrm{d}x} + P_W \frac{\mathrm{d}A}{\mathrm{d}x} - P\frac{\mathrm{d}A}{\mathrm{d}x} - \tau_W \frac{\mathrm{d}\sigma}{\mathrm{d}x}$$

$$= -\frac{\mathrm{d}(PA)}{\mathrm{d}x} + P_W \frac{\mathrm{d}A}{\mathrm{d}x} - \tau_W \frac{\mathrm{d}\sigma}{\mathrm{d}x}$$

B. Engine Cycle Thermodynamic Functions

1. Velocity Change and Thermal Ratio

The relation between velocity and ψ, the compression parameter, can be obtained starting with the energy balance equation, namely

$$\frac{H_3}{H_o} = \frac{\beta_o}{\beta_3} \tag{C8}$$

and

$$\beta = 1 + \frac{\gamma - 1}{2} M^2 \tag{C9}$$

The velocity ratio is

$$\left(\frac{V_3}{V_o}\right)^2 = \frac{\beta_3 - \psi}{\beta_o - 1} \tag{C10}$$

where

$$\beta_3 = \frac{H_{to}}{H_3} = \frac{H_{to}}{(H_o \psi)} \tag{C11}$$

and

$$\beta_o = \frac{H_{to}}{H_o} \tag{C12}$$

2. *Thermal Ratio and Geometric Contraction Ratio*

Starting with the entropy relationship

$$T\,\mathrm{d}s = \mathrm{d}H - \frac{\mathrm{d}p}{\rho} \tag{C13}$$

One can write

$$\frac{\mathrm{d}\rho}{\rho} = \frac{1}{\gamma - 1}\frac{\mathrm{d}T}{T} - \frac{\mathrm{d}s}{R} \tag{C14}$$

and

$$\frac{\mathrm{d}\rho}{P} = \frac{1}{\gamma - 1}\frac{\mathrm{d}T}{T} - \frac{\mathrm{d}s}{R} \tag{C15}$$

From the conservation of mass, one can write on a 1D basis

$$\frac{\mathrm{d}\rho}{\rho} = -\frac{\mathrm{d}A}{A} - \frac{dV}{V} \tag{C16}$$

However, from Eq. (C7)

$$\frac{\mathrm{d}s}{R} = \frac{\mathrm{d}A}{A} K_{WP} \tag{C17}$$

Combining Eqs. (C14), (C16), and (C17), one finally obtains the enthalpy

$$\frac{\mathrm{d}H}{H} = -(\gamma - 1)\left[(1 + K_{WP})\frac{\mathrm{d}A}{A} + \frac{dV}{V}\right] \tag{C18}$$

Integrating this equation, an exact thermodynamic relation between area and thermal compression ψ is

$$\psi = \tilde{A}^{-(a-1)}\left[\frac{\beta - \psi}{\beta - 1}\right]^{-\left(\frac{\gamma-1}{2}\right)} \tag{C19}$$

Where area ratio and velocity ratio are designated by $\tilde{A}$ and $\tilde{V}$, respectively, and the adiabatic exponent is

$$a = \gamma + (\gamma - 1)K_{\mathrm{WP}} \tag{C20}$$

$$= 1 + (\gamma - 1)(1 + K_{\mathrm{WP}}) \tag{C21}$$

$$= \gamma(1 + K_{\mathrm{WP}}) - K_{\mathrm{WP}} \tag{C22}$$

In a similar manner, formal expressions, as follows, can be obtained in closed form for $\tilde{P}$ and $\tilde{\rho}$, the pressure and density ratio, respectively.

$$\tilde{P} = \tilde{A}^{-a}\left(\frac{\beta - \psi}{\beta - 1}\right)^{-\left(\frac{\gamma}{2}\right)} = \tilde{A}^{\frac{-a}{\tilde{V}^{\gamma}}} \tag{C23}$$

$$\tilde{\rho} = \tilde{A}^{-1}\left(\frac{\beta - \psi}{\beta - 1}\right)^{-(1/2)} = \tilde{A}^{\frac{-1}{\tilde{V}}} \tag{C24}$$

The foregoing expressions for the various ratios can be further simplified by assuming $1/\beta$ tends to zero as M becomes large, as in hypersonic flow.

C. Boundary-Layer Influence

While the foregoing analysis is 1D based on the assumption of uniform velocity and thermodynamic properties over the cross-section of the flow-path, in general, one needs to account for fluid viscosity effects. An analysis is presented in the following for a viscous flow based on the usual boundary layer approximation.

1. Preliminaries

Within the hypersonic flow approximation, the thermal and momentum thickness of a boundary layer may be set equal. Defining the momentum defect thickness as

$$\theta = \frac{\int_o^\delta \rho V(V_e - V)\,\mathrm{d}y}{\rho_e V_e^2} \tag{C25}$$

and the energy defect thickness as

$$\Delta = \frac{\int_o^\delta \rho V(H_e - H)\,\mathrm{d}y}{\rho_e V_e (H_{aw} - H_w)} \tag{C26}$$

where the conditions at the freestream edge are denoted by subscript *e,* and subscript aw refers to the adiabatic wall (and, thus, $H_{\rm aw} = H_e$), and equating the two thicknesses, it follows that

$$\frac{H - H_w}{H_e - H_w} = \frac{V}{V_e} \tag{C27}$$

This is the well-known Crocco integral-relating enthalpy and velocity difference.

Next, the isobaric condition within the boundary layer can be used to write

$$\frac{\rho}{\rho_e} = \left(\frac{H}{H_e}\right)^{-1} \tag{C28}$$

$$= \left[\beta \frac{H_w}{H_e} + \beta\left(1 - \frac{H_w}{H_e}\right)\tilde{V} - (\beta - 1)\tilde{V}2\right]^{-1} \tag{C29}$$

which provides a relation for density variation within the boundary layer.

Lastly, a relation can be established for the shape factor

$$H = \delta^*/\theta \tag{C30}$$

where δ^* is the displacement thickness of the boundary layer. The velocity profile for laminar and turbulent (1/7th power profile) flow has been integrated for flows up to Mach number 12 for adiabatic and cold walls. Through empirical curve fit of predictions, H is found to be linear both in temperature ratio and (β - 1), that is,

$$H = A\left(\frac{T_w}{T_e}\right) + B\left(\frac{\gamma-1}{2}\right)M_e^2 \tag{C31}$$

where A and B have the following values: for laminar flow $A = 2.6$ and $B = 1.0$; for turbulent flow $A = 1.29$ and $B = 1.0$.

Shear stress. Now, the von Karman momentum integral equation may be written as follows:

$$\frac{\mathrm{d}}{\mathrm{d}x}\left(\rho_e V_e^2 \theta\right) = \tau_w + \delta^* \frac{\mathrm{d}\,P_w}{\mathrm{d}x} \tag{C32}$$

where

$$\rho_e V_e^2 \theta = \frac{\gamma P_e M_e^2 \delta^*}{A(T_e/T_w) + B(\gamma - 1/2)M_e^2} \tag{C33}$$

Here, it turns out in hypersonic flow that a transformed shape factor, namely,

$$\tilde{H} = \frac{\delta^*}{\theta} \cdot \frac{1}{\gamma M_e^2} = \frac{H}{\gamma_e^2} \tag{C34}$$

is constant along the flowpath, at least under the condition T_W = constant, since

$$\tilde{H} = \left[A\frac{T_w}{T_e} + B\left(\frac{\gamma-1}{2}M_e^2\right)\right] \Big/ \gamma M_e^2 \tag{C35}$$

In terms of this transformed shape factor, Eq. (32) can be rewritten as follows;

$$\frac{\mathrm{d}\delta^*}{\mathrm{d}x} + \delta^*(1 - \tilde{H})\left(\frac{1}{P_w}\frac{\mathrm{d}P_w}{\mathrm{d}x}\right) = \tilde{H}\left(\frac{\tau_w}{P_w}\right) \tag{C36}$$

Observe that the shear stress term on the right side of Eq. (36) can be written as

$$H\frac{\tau_w}{P_w} = \frac{\left[H(c_f/2)\right]_{\text{ref}}}{(1 + K_{\text{WP}})} \cdot \tilde{p}^{a-1} \tag{C37}$$

which shows that the shear stress is only a function of pressure. If the variation of pressure or equivalently the cross-sectional area is known, one can determine the variation of shear stress.

Now it is possible to integrate Eq. (C36) and obtain the relation

$$\tilde{\delta}^* = \tilde{A}^{a(\tilde{H}-1)}\left(\tilde{\delta}_i^* + \frac{\tilde{\Sigma}}{\tilde{m}_\sigma}\left\{\frac{\left[H(c_f/2)\right]_{\text{ref}}}{(1+K_{\text{wP}})^{1-n}}\right\}\cdot\frac{1-\tilde{A}^{[a(\tilde{H}-n)+1]}}{\left[a(\tilde{H}-n)+1\right]}\right) \tag{C38}$$

where $\tilde{\delta}^*$, $\tilde{\Sigma}$, $\tilde{m}^\sigma$, and $\tilde{A}$, the ratio of displacement area over the reference area, the ratio of surface area over the reference area, the ratio of the change of cross-section over the initial cross-section, and the area ratio, respectively, are given by

$$\tilde{\delta}^* = \delta^*/A_{\text{ref}}$$

$$\tilde{\Sigma} = \Sigma/A_{\text{ref}}$$

$$\tilde{m}\sigma = 1 - \frac{A_f}{A_i}$$

$$\tilde{A} = A/A_{\text{ref}}$$

while $\tilde{m}^\sigma$ can be expressed by the relation

$$\tilde{m}^\sigma = 1 - \tilde{A}^\sigma + \tilde{\delta}^*$$

and $n = 6/7$.

Since $\tilde{\delta}^*$ is a function of $\tilde{A}$ and $\tilde{m}^\sigma$, the solution needs to be obtained recursively.

Heat transfer. Consistent with the Crocco relation given in Eq. (C27), one can invoke equality of the momentum defect and the total enthalpy defect; that is,

$$\Delta = \theta = \delta^*/H \tag{C39}$$

Then the total heat transfer between two stations along the flowpath is given by

$$\dot{Q}_{2-1} = (\rho_2 V_2 \Delta_2 - \rho_1 V_1 \Delta_1)(H_T - H_w) \tag{C40}$$

where H_r is the recovery temperature at the wall.

2. *Application to the Inlet Flowpath*

Wall pressure force. Based on the boundary layer assumption, it is possible to identify a core or potential flow area A_P as a part of the gross cross-sectional area A_G through the definition

$$\text{A}_\text{P} = \text{A}_\text{G} - \delta^* \tag{C41}$$

Therefore, the net wall pressure force over a length L can be written as

$$\text{net pressure force} = \int_{o}^{L} P_w \, \mathrm{d}A_G \tag{C42}$$

$$= \int_{o}^{L} P_w \cdot \mathrm{d}Ap + \int_{o}^{L} P_w \cdot \mathrm{d}\delta^* \tag{C43}$$

Here

$$P_w \, \mathrm{d}A = \frac{P_{\text{ref}} A_{\text{ref}}(1 + K_{wP}) \, \mathrm{d}\tilde{A}}{\tilde{A}^a \tilde{V}^\gamma} \tag{C44}$$

Noting that

$$\psi = \tilde{A}^{(a-1)} \left(\frac{\beta - 1}{\beta - \psi} \right)^{\frac{\gamma - 1}{2}} \tag{C45}$$

$$\tilde{P} = \tilde{A}^a V_\gamma \tag{C46}$$

$$\tilde{V} = \left(\frac{\beta - \psi}{\beta - 1} \right)^{\frac{1}{2}} \tag{C47}$$

$$\frac{\mathrm{d}\tilde{A}}{\tilde{A}^a} = \frac{\mathrm{d}(\psi \tilde{V}^\gamma)}{(a - 1)} \tag{C48}$$

The pressure force integral over the core flow is

$$\frac{\int P_w \, \mathrm{d}A}{P_1 A_1} = \frac{(\gamma + 1)}{(\gamma - 1)} (\beta - 1)^{1/2} \left[\left(\frac{\beta - \psi}{\beta - 1} \right)^{1/2} - 1 \right] + \beta \left[\left(\frac{\beta - 1}{\beta - \psi} \right)^{1/2} - 1 \right] \tag{C49}$$

The pressure force integral is a function of ψ only. The integral is positive for $\tilde{A} > 1$ or an area expansion along the flow, thus indicating generation of thrust. It is negative for $\tilde{A} < 1$ or an area contraction, indicating creation of a drag force.

Note that Eq. (C49) provides a ID solution. Further corrections need to be introduced under actual 3D and burning flow conditions in inlets and nozzles.

Next, it is necessary to account for the pressure force over the wall layer displacement area. To do this, one starts with the von Karman equation

$$d(\rho_e V_e^2 \theta) = \tau_w \cdot d\sigma + \delta^* \cdot dP_w \tag{C50}$$

and writes the following:

$$P_w \cdot d\delta^* = \tau_w \cdot d\sigma \, d\left(\rho_e V_e^2 \theta\right) + d\left(P_w \delta^*\right) \tag{C51}$$

Integrating, and introducing the definition of pressure force integral

$$\Delta F_{x\delta} = \int_o^L P_w \cdot d\delta^* \tag{C52}$$

it follows that

$$\Delta F_{xV} = \int_o^L \tau_w \, d\sigma - \rho_e V_o^2 \theta \Big|_o^L + P_w \delta^* \Big|_o^L \tag{C53}$$

where the friction drag, taken as negative, is given by

$$\Delta F_{xV} = -\int_o^L \tau_w \, d\sigma \tag{C54}$$

The net reaction force on the wall between any two stations 0 and L can then be written using Eq. (C43) as

$$\Delta F_{xw} = \int_o^L P_w \, d A_p - \rho_e V_e^2 \theta \Big|_o^L + P_w \delta^* \Big|_o^L \tag{C55}$$

Here the friction drag does not appear explicitly. Equation (C53) provides an expression for the sum of the potential and viscous pressure integrals.

Introducing the shape factor H from Eq. (E32)

$$\Delta F_{xV} = \int_{o}^{L} P_w \, \mathrm{d} A_p + \left(1 - \frac{1}{H}\right)\left[(P_w\delta^*)_2 - (P_w\delta^*)_1\right] \tag{C56}$$

Friction force and heat transfer. Using Eqs. (C44) to (C47), one can write the net pressure force as follows:

$$\frac{\int_{o}^{L} P_w \, \mathrm{d} A_p}{P_{\mathrm{ref}} \cdot A_{\mathrm{ref}}} = (1 + K_{wP})\left[\frac{\tilde{A}^{-(a-1)} - 1}{a - 1}\right] \tag{C57}$$

$$= \frac{\psi - 1}{\gamma - 1} \tag{C58}$$

The friction drag force is

$$\begin{aligned} D_V &= \int \tau_w \, \mathrm{d}\sigma \\ &= -\tau_{w\,\mathrm{ref}} \frac{\tilde{\Sigma}}{\tilde{m}} (1 + K_{wP})^n \left[\frac{\tilde{A}^{(1-an)} - 1}{an - 1}\right] \end{aligned} \tag{C59}$$

Finally, the heat transfer over the length L can be written as

$$Q_L = \int q_w \, \mathrm{d}\sigma$$

$$= -q_{w\,\mathrm{ref}} \frac{\tilde{\Sigma}}{\tilde{m}} (1 + K_{\mathrm{wP}})^n \left[\frac{\tilde{A}^{(1-an)} - 1}{an - 1}\right] \tag{C60}$$

Note that Eqs. (C58) through (C60) involve the geometrical parameter $\tilde{A}$, the thermal ratio ψ, and the pressure factor K_{wP}.

Other aspects of scaling of skin friction force and heat transfer along the internal flowpath are discussed in Appendix E.

D. Experimental Determination of Inlet K_{wP}

Guided by considerations of hypersonic flow theory, an attempt has been made to determine K_{wP} as a function of compression ratio, Mach number, ratio of specific heats, and compression angle by exact theory, and then to correlate the theoretical database using hypersonic similarity parameters. The methodology is described in the following paragraphs.

The inlet of a scramjet can be of several types: external, mixed external-internal, or internal compression. Currently, attention has been concentrated on internal compression inlets, where all the compression is generated by physical surfaces enclosing the inlet flowpath. The compression flowfield may be two- or three-dimensional (2D or 3D) depending on design choice, including partitioning of the flowpath as needed by weight, cooling, and other considerations. However, here a parametric series of 2D inlets was chosen for analysis, along with compression angles, Mach numbers, and contraction ratios.

The final database was generated by considering the oblique shock train of a set of 92 straight wall, adiabatic, 2D, multishock inlets over the ranges of mach number 4 to 18, contraction ratio 2 to 20, and wedge angles of 4, 5, and 6 degrees.

The data were correlated by first determining the total pressure ratio of each inlet, and then calculating K_{wP} from the following relation:

$$\frac{P_{T2}}{P_{T1}} = CR^{K_{\mathrm{wP}}} \tag{C61}$$

This body of data was then analyzed using the hypersonic similarity parameter (Ref. 12),

$$K_\theta = \left(M_\infty^2 - 1\right)^{1/2} \cdot \theta \tag{C62}$$

the thermal ratio ψ and a work parameter w given by

$$w = \left(\frac{\gamma - 1}{\gamma} \cdot \frac{1}{M_\infty^2}\right) \tag{C63}$$

The result of the correlation using the data from the set of 92 inlets is

$$K_{\mathrm{wP}} = \frac{wK_\theta^2}{\psi^{1/2}} \tag{C64}$$

The work parameter w is a measure of compression work when the velocity change is small, an assumption utilized everywhere in hypersonic flow.

Equation (C64) provides a simple formula for K_{wP}. It is implicit in contraction ratio since

$$\psi = \frac{H_3}{H_1} = CR^{a-1} \tag{C65}$$

Correlations of data are presented in Figs. C1–C5.

The theory and the correlation can be verified by comparing the exact

Fig. C1

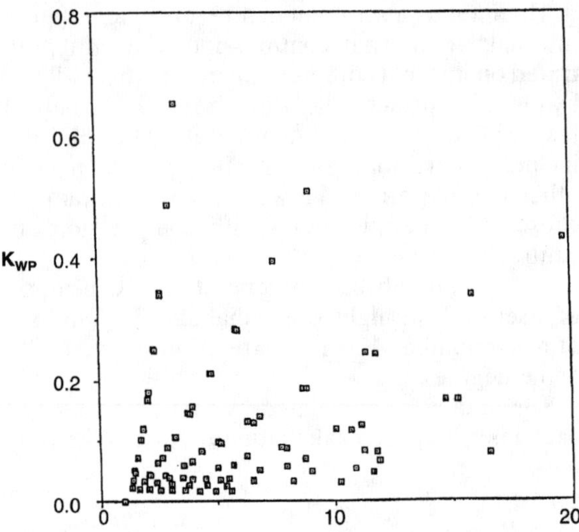
K_{WP}
0.8
0.6
0.4
0.2
0.0
0
10
20
CR

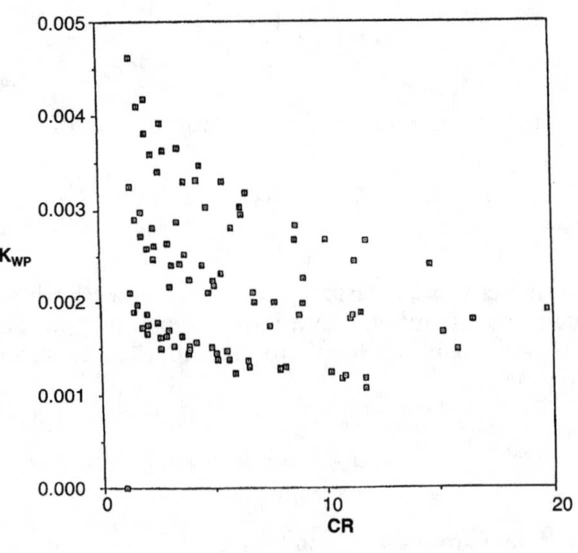
K_{WP}
0.005
0.004
0.003
0.002
0.001
0.000
0
10
20
CR

Fig. C2

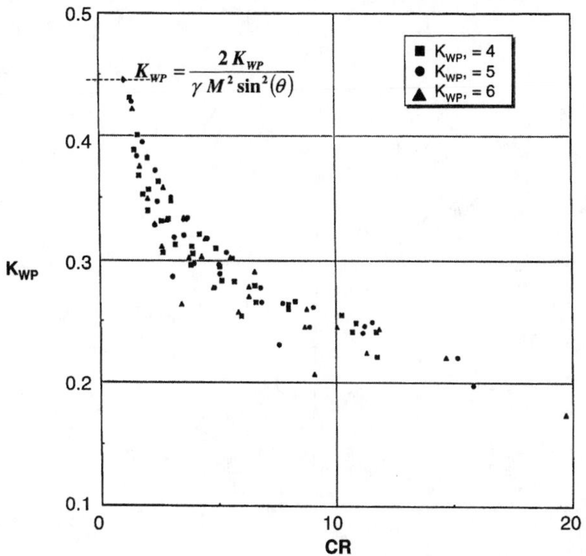

$K_{WP} = \frac{2 K_{WP}}{\gamma M^2 \sin^2(\theta)}$
$K_{WP'}$ = 4
$K_{WP'}$ = 5
$K_{WP'}$ = 6
K_{WP}
CR
0.5
0.4
0.3
0.2
0.1
0
10
20

Fig. C3

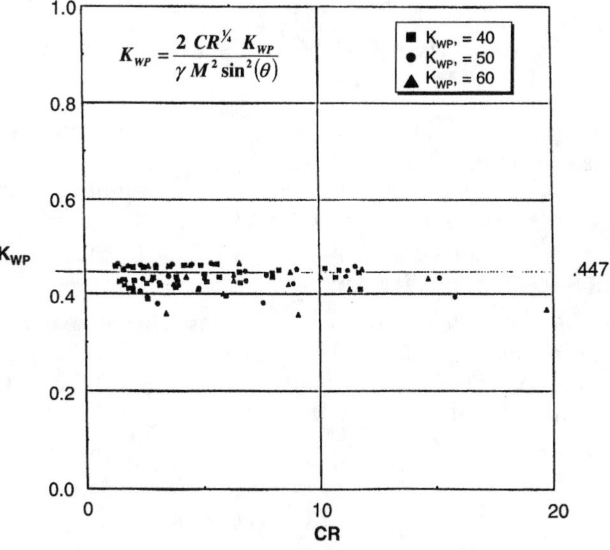

$K_{WP} = \frac{2\ CR^{1/4}\ K_{WP}}{\gamma M^2 \sin^2(\theta)}$
$K_{WP'}$ = 40
$K_{WP'}$ = 50
$K_{WP'}$ = 60
K_{WP}
.447
CR
1.0
0.8
0.6
0.4
0.2
0.0
0
10
20

Fig. C4

Fig. C5

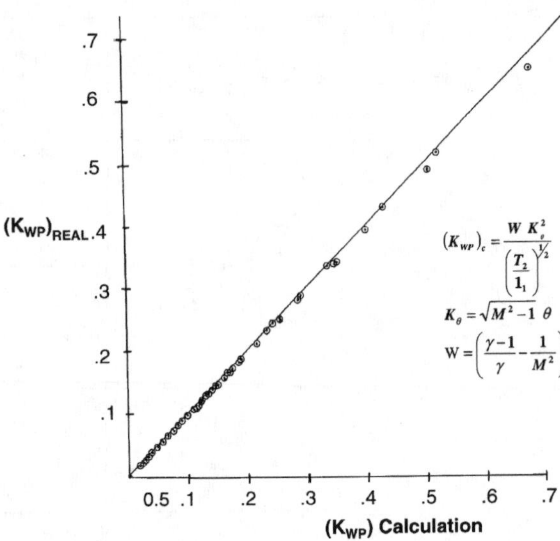

drag data with the drag obtained with the correlation. All computed drag coefficients are compared with the exact value derived from the set of data

$$\int P_w \cdot \mathrm{d}A$$

integral, and the drag coefficient

$$C_D = \frac{(CR^{a-1} - 1)}{(\gamma -)\left[(\gamma/2)M_\infty^2\right]} = \frac{\psi - 1}{(\gamma -)\frac{\gamma}{2}M_\infty^2} \tag{C66}$$

The data comparisons are shown in Fig. C6a–6c, and one can see that agreement is excellent.

Various geometrical features such as leading edge bluntness will affect the correlations. These issues are discussed in the section on forebody.

E. Ratio of Specific Heats for Air

The compression process ψ based formulas can be extended to high values by the simple expedient of replacing the γ with the average value from the sum of the beginning value and the value of g at the end. The value of γ can be written by simple curve fit,

$$\gamma = 1.4 \qquad 0 < \psi < 2$$

$$\gamma = 1.4 - .025\,(\psi - 2) \qquad 2 < \psi < 5$$

$$\gamma = 1.325 - .0125(\psi - 5) \qquad 5 < \Psi < 9$$

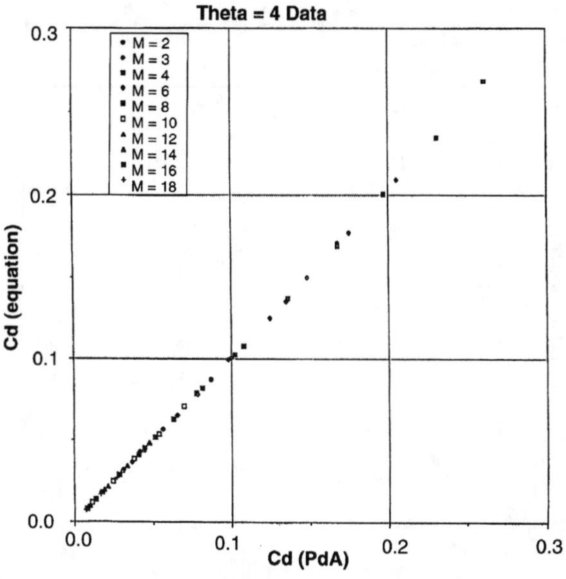

Fig. C6a

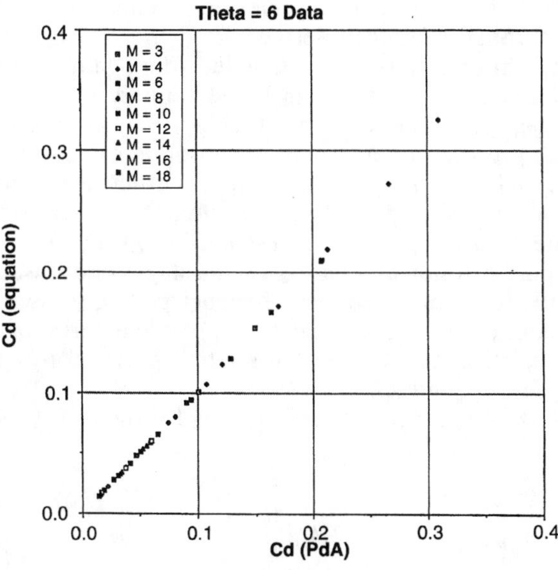

Fig. C6b

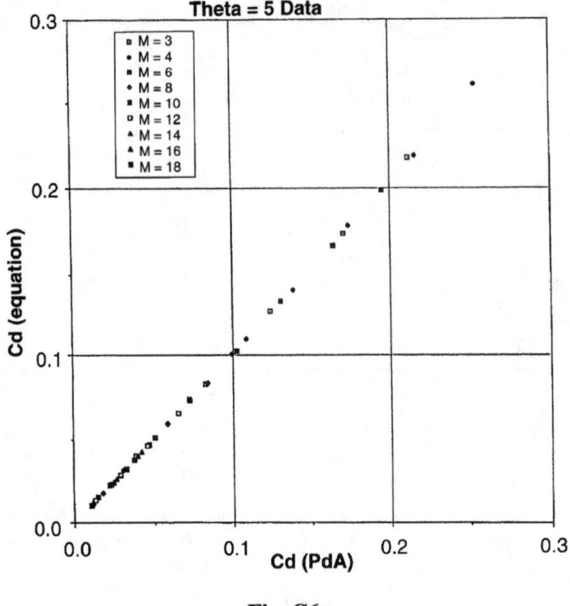

Fig. C6c

F. Further Analysis of Thermal Ratio

Referring to Fig. 3, note that the inlet performance can be assessed in several ways through efficiency derived from thermodynamic considerations: 1) change in thermodynamic state, including gain in entropy, represented by effectiveness; 2) change in kinetic energy relative to the ideal, represented by η_{KE}, the kinetic energy efficiency; 3) change in kinetic energy in the diffusion process, rather than in the absolute value of kinetic energy, represented by K_D, which is called the process efficiency; 4) total pressure recovery in the diffusion process, represented by $\eta\Sigma$; 5) static pressure recovery in the diffusion process, represented by η_R; and 6) polytropic efficiency, based on the notion of a series of small stages of diffusion, each stage efficiency represented by η_{EL}. These efficiencies are discussed by Van Wie later in this book. The efficiencies can be expressed in terms of one another.

Here, the objectives to examine the relation between the thermal ratio and, in particular, η_{KE} and K_D.

Using Eq. (B1), one can write the following for the thrust coefficient:

$$\frac{C_F}{2} = \frac{1}{\dot{m}V_o} \cdot \left\{1 - \frac{\psi - 1}{(\gamma - 1/2)M_o^2}\left[\eta_c\,\eta_e\left(1 + \frac{\Delta H_c/H_o}{\psi}\right) - 1\right]\right\}^{1/2} \tag{C67}$$

By using Eq. (B3), the air specific impulse then becomes

$$I_A = \frac{1}{\dot{m}}\left(\frac{V_o}{g}\right)\left(\frac{C_F}{2}\right) \tag{C68}$$

The cycle applies to a ramjet when air is stagnated before combustion and ψ equals the stagnation to static temperature ratio, β. It can then be shown, noting that V_3 is very small or zero, that η_c is equal to η_{KE} defined by

$$\eta_{\mathrm{c}}\ \eta_{\mathrm{KE}} = \frac{H_{To} - H_{o'}}{H_{To} - H_o} = \frac{V_{o'}^2}{V_o^2} \tag{C69}$$

where o' is indicated in Fig. 3. This efficiency can be rewritten as follows:

$$\eta_{\mathrm{KE}} = 1 - \frac{(\psi^k - 1)}{(\gamma - 1/2)M_o^2} \tag{C70}$$

where

$$k = \frac{(a - \gamma_e)}{(a - 1)} \cdot \frac{1}{\gamma_e} \tag{C71}$$

γ_e is the effective value of γ for expansion, and a is given by Eqs. (C20) through (C22).

By inspection the formula shows that for scramjets, when $\psi \ll \beta_{\mathrm{o}}$, where β_{o} is given by Eq. (C12), η_{KE} will tend to unity independent of losses as the Mach number increases. For ramjets, η_{KE} varies from about 0.92 at $M_{\mathrm{o}} = 2.5$ to 0.8 at $M_{\mathrm{o}} = 5.0$. The diffuser then includes a normal shock.

The inlet equivalent efficiency is the diffusion coefficient K_D, mentioned above, and is given by

$$\eta_{\mathrm{c}} = K_D = \frac{H_3 - H_{o'}}{H_3 - H_o} \tag{C72}$$

$$= (\psi - \psi^k)\ /\ (\psi - 1) \tag{C73}$$

Note that K_D is valid for adiabatic and non-adiabatic flow, while η_{KE} has been defined for adiabatic flow. However, considering adiabatic flow

$$\eta_{\mathrm{KE}} = 1 - (1 - K_D)\left(1 - \frac{V_3^2}{V_o^2}\right) \tag{C74}$$

Note that when V_3 goes to zero, η_{KE} is equal to K_D, again considering adiabatic flow in both cases.

One may define here an efficiency, η_{UD}, for useful drag, that is valid for both adiabatic and non-adiabatic flow. Noting that

$$\frac{P_3}{P_o} = \psi^{a/(a-1)} \tag{C75}$$

one can write

$$\eta_{UD} = \frac{e_{31} - e_o}{e_3 - e_o} = \frac{\psi^{(1-k)} - 1}{\psi - 1} \tag{C76}$$

then it follows that

$$\eta_{UD} = K_D/\psi^k \approx \frac{1}{1 + k_{wp}} \tag{C77}$$

This last relation is most useful.

1. Relation Between ψ_{OPT} and η_e

For an adiabatic Brayton cycle with constant values of η_c and η_e, it has been shown in Ref. 25 that

$$\psi_{OPT} = \left[\frac{\eta_c\,\eta_e}{1 - \eta_c\,\eta_e}\left(\frac{\Delta H_c}{H_o}\right)\right]^{1/2} \tag{C78}$$

This can be rewritten as

$$\psi_{OPT}^{2-k} = \frac{\eta_e}{1 - \eta_e}\left[(1-k)\frac{\Delta H_c}{H_o} - k\psi_{OPT}\right] \tag{C79}$$

after substituting K_D for η_c under the condition V_3 tends to zero.

Appendix D: Hypersonic Slender Body Theory Applied to Forebodies and Leading Edges

A. Forebodies

The basic assumption of hypersonic flow past slender bodies is that axial velocity perturbations are negligible. The equation of motion thus reduces to an unsteady flow in the transverse plane, and a similarity parameter can be derived as follows:

$$\mathrm{k} = \mathrm{M}_\infty\,\tau \tag{D1}$$

where M_∞ is the flow Mach number and τ is the fineness parameter given by

$$\tau = \mathrm{t} / \ell$$

$$= \theta, \text{ approximately}$$

such that, referring to Fig. D1, the pressure coefficient and drag are the same for different values of θ and Mach number when k is held constant.

To determine the pressure distribution, drag per unit volume, and drag per unit mass capture for forebodies of different thicknesses, or τ, one can proceed as follows.

For a wedge, the pressure coefficient and shock angle are given by (Ref. 12)

$$\frac{c_p}{\tan^2\theta} = \frac{\gamma+1}{2} + \left[\left(\frac{\gamma+1}{2}\right)^2 + \cdot \frac{4}{k^2}\right]^{1/2} \tag{D2}$$

and

$$\frac{k_s}{k_\delta} = \frac{\tan\delta}{\tan\theta} = \frac{\gamma+1}{4} + \left[\left(\frac{\gamma+1}{4}\right)^2 + \frac{1}{k^2}\right] \tag{D3}$$

The agreement with exact solution is excellent, as can be seen in Refs. 7 and 12.

For a cone, the shock angle, given in explicit form, is adequate for the

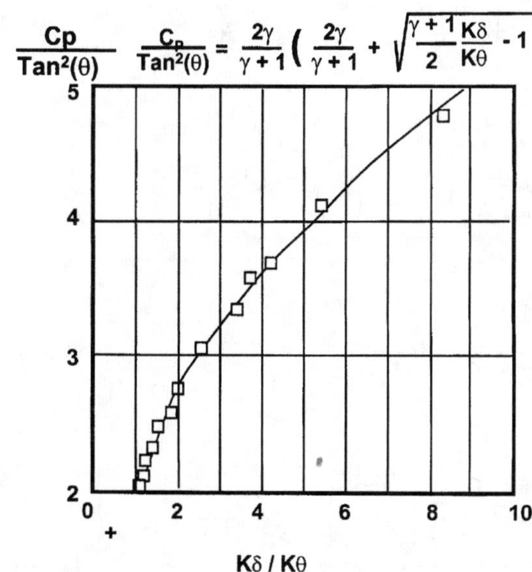

Fig. D1

current purpose. A somewhat simpler expression has been deduced by the author:

$$\frac{k_s}{k_\delta} = \left(\frac{\gamma+1}{2} + \frac{1}{k^2}\right)^{1/2} \tag{D4}$$

1. Relation to Contraction Ratio

Relations given by Eqs. (D3) and (D4) may then be used to form expressions for the forebody contraction ratio:

Wedge:

$$(CR)_w = \frac{(\gamma+1)k + \left[(\gamma+1)^2k^2 + 16\right]^{1/2}}{\left\{\left[(\gamma+1) - 4\right] - k + \left[(\gamma+1)k^2 + 16\right]^{1/2}\right\}} \tag{D5}$$

Cone:

$$(CR)_c = \frac{(\gamma+1)k^2 + 2}{(\gamma+1)k^2 + 2} \tag{D6}$$

2. Relation to K_{WP}

An exact analytical expression can be found for K_{WP} in the case of forebodies, as follows, from the shock angle relation for the pressure ratio and density ratio. Since

$$\frac{\Delta S}{R} = \frac{1}{\gamma-1} \cdot \ell n\left[\frac{P_2/P_1}{(\rho_2/\rho_1)^\gamma}\right] = K_{WP} \cdot \ell n\, CR \tag{D7}$$

where

$$P_2/P_1 = \left[\frac{2\gamma}{(\gamma-1)}\right] M^2 \sin^2\delta - \left(\frac{\gamma-1}{\gamma+1}\right) \tag{D8}$$

and

$$P_2/P_1 = \frac{\gamma-1}{\gamma+1} + \frac{2}{\gamma+1}\frac{1}{M^2\sin^2\delta_s} \tag{D9}$$

one can write

$$K_{WP} = \frac{\ell n\left[\left(\frac{2\gamma}{\gamma+1}k_s^2 - \frac{\gamma-1}{\gamma+1}\right)\left(\frac{\gamma-1}{\gamma+1} + \frac{1}{k_s^2}\frac{2}{\gamma+1}\right)^\gamma\right]}{\ell n(CR)^{\gamma-1}} \tag{D10}$$

Here, it is assumed that $\tan\theta = \sin\theta = \theta$ for small angles.

B. Leading Edges

To reduce the effects of heat transfer, the leading edges, in general, are blunt. The bluntness leads to an additional drag caused by formation of a bow shock and adverse flow interactions downstream of the leading edge. The flow interactions are substantially different in 2D and 3D cases. Thus, in the case of a wedge, the total drag becomes increased in the case of a blunt leading edge, compared with the case of a sharp leading edge.

On the other hand, in the case of a cone, there is a reduction in the total drag with certain values of bluntness, due to the phenomenon of overexpansion at the corner. In this connection, optimum bluntness for a cone is given by the relation

$$\left(\frac{\ell}{\delta}\right)_{\text{opt}} = \frac{0.96}{\tan^2\theta_c}\frac{C_{\text{D BLE}}^{0.5}}{2} \tag{D11}$$

This relation yields minimum drag at 12.5% bluntness, which is in fair agreement with data.

For wedges, the author suggests using the theory of hypersonic plane sections (Ref. 7). Expressing the effect of bluntness as a correction to the drag in the case of a wedge with a sharp leading edge, one can write

$$C_D = (C_{\text{DW}} - C_{\text{D BLE}})\,\text{r} \tag{D12}$$

where r is an interaction parameter. For the case of the limiting Mach number $M \to \infty$,

$$r_\infty = \left[1+\gamma+\frac{2\gamma}{t}\right] \Big/ \left[1+\gamma+\frac{2}{t}\right] \tag{D13}$$

where t, the ratio of wedge drag to drag of wedge with blunt leading edge is given, from Eq. (D12), by

$$C_D = C_{\text{DW}}\left(1+\frac{1}{t}\right)r = \left(\frac{4\tan^3\theta}{C_{\text{D BLE}}}\right)\frac{L}{d_{\text{BLE}}} \tag{D14}$$

Since the shock is displaced outward from the wedge, one can write

$$y_{shock} = x\tan\delta_s - \Delta_s \tag{D15}$$

where

$$\Delta_s = \frac{\gamma-1}{4}\frac{C_{\text{D BLE}}}{\tan^3\theta_w} \tag{D16}$$

The displacement is constant and is observed far from the leading edge.

In the case of a blunt cone, the shock becomes identical with that of a sharp cone far downstream of the leading edge.

Finally, it may be pointed out that the drag coefficient for a blunt leading edge shows a slight Mach number dependence. Thus, for a cylinder

$$C_{D\,BLE} = 1.23 - 0.5/M^2 \tag{D17}$$

and for a sphere

$$C_{D\,BLE} = 0.92 - 0.38/M^2 \tag{D18}$$

C. Case of Unequal Angles

A contraction is considered for an inlet with four walls, two side walls, and top and bottom walls. One can write

$$CR = \frac{h_1 W_1}{h_2 W_2} = \frac{h_1}{h_2} \cdot \frac{W_1}{W_2} = CR_{top} \cdot CR_{side} \tag{D19}$$

In the special case where compression angles are different for walls compressing in two different directions, but the same for walls compressing in the same direction,

$$CR_T = CR_B \quad \text{and} \quad CR_{S1} = CR_{S2}$$

T and B refer to top and bottom walls and $S1$ and $S2$ to the sidewalls. The total effective wall pressure coefficient then becomes

$$K_{WPT} = \frac{\Sigma K_{WPi}\,\ell n\, CR_i}{\ell n\, CR_T} \tag{D20}$$

where the subscript T denotes total and the subscript i refers to any wall. K_{WPi} is given by

$$K_{WPi} = \frac{WK_{\theta i}^2}{\psi_r^{1/2}} \tag{D21}$$

with Ψ_T being the total enthalpy ratio. It is consistent with the assumption that in each direction there is a part of the total loss; that is,

$$(\bar{P}_T)_{CR} = (\bar{P}_T)_{CR1} \cdot (\bar{P}_T)_{CR2} \tag{D22}$$

where 1 and 2 refer to the two walls compressing in the same direction.

Now, we consider compression in one direction only but with different angles for the two walls. In particular, the case with the same value of overall contraction and length is considered. In such an asymmetrical case, one can write, using functional analysis, the mean square compression angle as follows:

$$\bar{\theta}^2 = \frac{\theta_G^2}{\ell n\, CR}\left[\ell n\left(\frac{1-\bar{L}\tan\theta_L}{1-\bar{L}\tan\theta_G}\right)^{1/2} \cdot CR - \frac{\theta_L^2}{\theta_G^2}\,\ell n\left(\frac{1-\bar{L}\tan\theta_G}{1-\bar{L}\tan\theta_L}\right)^{1/2} \cdot CR\right] \tag{D23}$$

where θ_G and θ_L represent the greater and the lesser values, and $\bar{L}$ is L normalized by H.

Note that Eq. (D23) does not require the length to be the same for the two sides. However, there is greater emphasis on the antisymmetric side, and on the intuitive reasoning that for small deviations from symmetry, the larger losses on the larger angle side may be nearly offset by the lower losses on the smaller angle side.

One other factor of interest here is that a yawed inlet will turn the flow from the yaw condition to alignment with the inlet flow axis. The additional drag associated with this turn is given by

$$\Delta\, \mathrm{D}_{\mathrm{turn}} = \dot{m}\, \mathrm{V}\,(1 - \cos\beta) \tag{D24}$$

For small yaw angles, the turn may be neglected in the estimation of K_{WP} or the wave losses.

D. Overspeed Situation in an Inlet

The phenomenon being considered is one wherein the forebody shock moves into the inlet as the flight Mach number exceeds the value at which the shock is located at the cowl lip. It is then necessary to examine the two streams at two different Mach numbers by the introduction of a slipline (in fact, a slip surface) with the governing conditions that the pressure and the flow direction should be continuous across the slipline. The forebody flow is assumed to be known so that the intersection of the forebody shock and the inlet sidewall is known. The slipline originates at this intersection.

These conditions may now be applied to the case of parallel contraction of the two flows, for illustration. The variable of interest is the angle θ_{SL} of the slipline. It defines both the contraction ratio of the two streams and the pressure coefficient K_{WP}, based on the average angle defined by Eq. (D23) in the preceding section. It may be noted that the following analysis applies to the case of distinguishing the individual contraction ratio of the two streams and not to the overall contraction ratio of the two streams together; the two streams need to be considered separately, since, for example, the two streams may have undergone different entropy changes before capture.

Figure D2 provides an illustration of the flowfield.

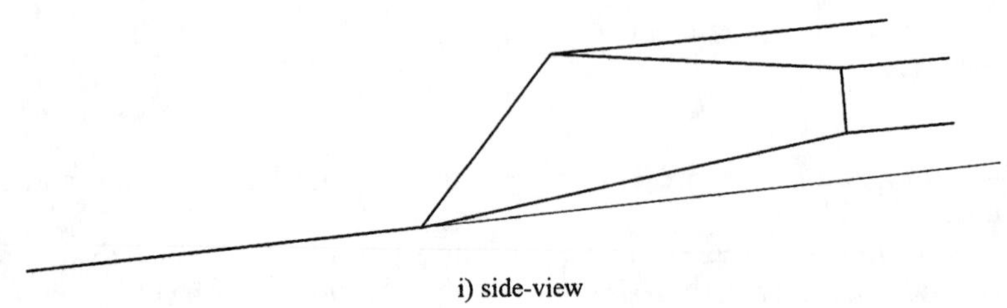

i) side-view

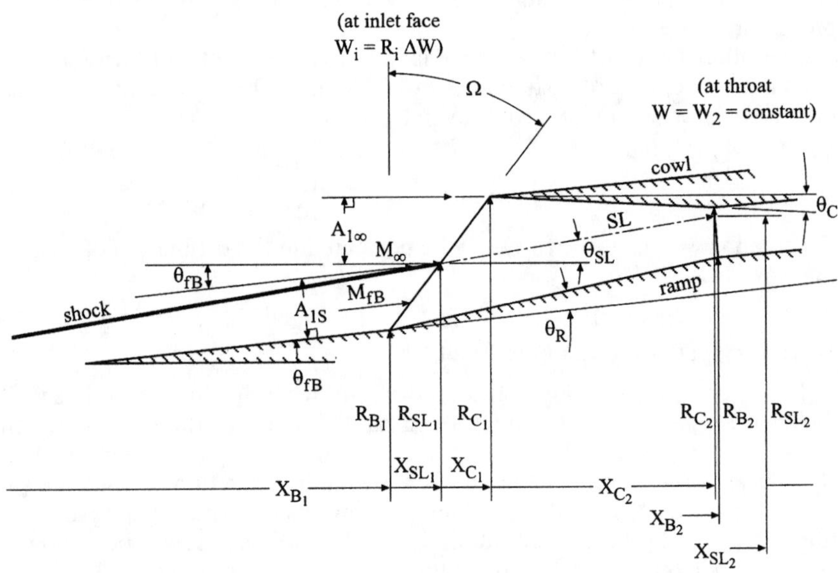

ii) nomenclature

Fig. D2

1. Pressure Ratio Relation

For the over-speed contraction side

$$\left(\frac{P_2}{P_0}\right) = (CR_{OS})^{a_{OS}} \tag{D25}$$

where

$$a_{OS} = \gamma + (\gamma - 1)(K_{WP})_{OS} \tag{D26}$$

For the forebody shock + internal contraction side

$$\left(\frac{P_2}{P_o}\right) = \left(\frac{P_{FB}}{P_o}\right)\left(\frac{P_2}{P_{FB}}\right) \tag{D27}$$

$$= (CR_{FB})^{a_{FB}} (CR_{INT})^{a_{INT}} \tag{D28}$$

Equating the two pressure ratios, one obtains a compatibility relation

$$(CR_{FB})^{a_{OS}} = (CR_{FB})^{a_{FB}} (CR_{INT})^{a_{INT}} \tag{D29}$$

The contraction ratio of the two streams is unknown as are the compression pressure coefficients. Sorting these variables out will become clear when the geometry of the interaction is completed.

2. Geometry Relation

The geometry of the inlet can now be analyzed to determine CR_{OS}, the freestream to throat area ratio on the over-speed side, and CR_{INT}, the internal contraction ratio on the forebody shock side.

The shock defines the geometry of a swept leading edge inlet or a two-dimensional inlet. The problem is to determine CR_{OS} and CR_{INT} as a function of θ_{SL}. The overall contraction has to account for the sidewall compression in the direction normal to the plane of the figure for the three-dimensional inlet.

In the following, angles and lengths are measured from the freestream direction.

Then, one can write the following relations for the geometry of the over-speed side,

$$CR_{OS} = \frac{(R_{C1} - R_{SL1})(W_{OS}/W_2)}{\left[(R_{C1} - R_{SL1}) - (L_C \tan\theta_C - L_{SL}\tan\theta_{SL})\right]\cos\theta_{FB}}$$

$$A_{1OS} = (R_{C1} - R_{SL1})W_{OS}$$

$$W_{OS} = \frac{1}{2}(W_{C1} + W_{SL1})$$

$$A_{2\mathrm{OS}} = (R_{C2} - R_{\mathrm{SL2}})\cos\theta_{\mathrm{FB}} W_2$$

$$L_C = X_{C2} - X_{C1}; \qquad L_{\mathrm{SL}} = X_{\mathrm{SL2}} - X_{\mathrm{SL1}}$$

$$R_{C2} = R_{C1} - L_C \tan\theta_C; \qquad R_{\mathrm{SL2}} = R_{\mathrm{SL1}}(1 + L_{\mathrm{SL}} \tan\theta_{\mathrm{SL}}) \qquad \text{(D30)}$$

For the forebody shock and internal contraction side

$$CR_{\mathrm{INT}} = \frac{A_{1\mathrm{INT}}}{A_{2\mathrm{INT}}} = \frac{(R_{\mathrm{SL1}} - R_{B1})\, W_{\mathrm{INT}} \cos\theta_{\mathrm{FB}}}{(R_{\mathrm{SL2}} - R_{B2})\, W_2 \cos\theta_{\mathrm{FB}}} \qquad \text{(D31)}$$

and

$$CR_{\mathrm{INT}} = \frac{(R_{\mathrm{SL1}} - R_{B1})(W_{\mathrm{INT}}/W_2)}{\left\{(R_{\mathrm{SL1}} - R_{B1}) - \left[L_B \tan(\theta_R) - L_{\mathrm{SL}} \tan\theta_{\mathrm{SL}}\right]\right\}}$$

$$A_{\mathrm{INT}} = (R_{\mathrm{SL1}} - R_{B1})\, W_{\mathrm{INT}} \cos\theta_{\mathrm{FB}}$$

$$W_{\mathrm{INT}} = \tfrac{1}{2}(W_{\mathrm{SL1}} + W_{B1})$$

$$A_{2\mathrm{INT}} = (R_{\mathrm{SL2}} - R_{B2})\, W_2 \cos\theta_{\mathrm{FB}}$$

$$R_{\mathrm{SL2}} = R_{\mathrm{SL1}} + L_{\mathrm{SL}} \tan\theta_{\mathrm{SL}}$$

$$R_{B2} = R_{B1} + L_B \tan(\theta_R) \qquad \text{(D32)}$$

We note that in the equations for CR_{OS} and CR_{INT}, the only unknown is θ_{SL}, and assume θ_{SL} is parametrically specified for now. The angles to use for an average θ^2 for determining K_{WP} and a value of (a) are also found in the previous section on unequal angles.

Then, on the over-speed side

$$\theta_{\mathrm{G}} = \max(\theta_{\mathrm{SW}}, \theta_{\mathrm{C}} - \theta_{\mathrm{SL}}) \qquad \text{(D33a)}$$

$$\theta_{\mathrm{L}} = \min(\theta_{\mathrm{SW}}, \theta_{\mathrm{C}} - \theta_{\mathrm{SL}}) \qquad \text{(D33b)}$$

And, on the shock side

$$\theta_{\mathrm{G}} = \max(\theta_{\mathrm{SW}}, \theta_{\mathrm{SL}} - \theta_{\mathrm{R}}) \qquad \text{(D34a)}$$

$$\theta_{\mathrm{L}} = \min(\theta_{\mathrm{SW}}, \theta_{\mathrm{SL}} - \theta_{\mathrm{R}}) \qquad \text{(D34b)}$$

Now, with the average compression angle from Eq. (D23), and inlet face Mach number known for both sides, the corresponding K_{WP} can be found. The value of θ_{SL} is varied until Eq. (D29) is satisfied. The overall pressure calculation can be used to define an overall K_{WP} and the appropriate total drag, average core Mach number, and ψ found.

Appendix E: Scaling Drag and Heat Transfer

A. Skin-Friction Coefficient

The scram engine internal flowpath is considered as a duct with friction and heat transfer, in which heat is released by combustion. Within the framework of boundary layer theory, the freestream is distinguished from a wall boundary layer. In general, even assuming that the velocity at the outer edge of the boundary layer, under the conditions of high speed flow in the combustor, is nearly constant, it is important to recognize the variation of pressure along the flow, the pressure ratio being on the order of 100.

Based on the analysis of compressible boundary layer and using the notation adopted there, one can write the following for the variation of skin-friction coefficient with velocity, velocity gradients, and pressure gradient:

$$(G - 3\,a\,H)\frac{\mathrm{d}\lambda}{\mathrm{d}x^*} + \frac{V'}{V}\lambda\,(\lambda'\delta^+ - G) - \lambda^4\,\frac{M\,(1/V)^n}{Re_L}\,Re_L V \tag{E1}$$

where the skin-friction parameter λ is equal to $\sqrt{2/C_f}$, a is the pressure gradient parameter, and V and V' are the local velocity normalized by the far upstream velocity, and its gradient, respectively. x^* is the length along the flow normalized by overall length.

Other parameters in Eq. (E1) are as follows: G and H are the shape parameters; and Re_L, the Reynolds number is defined by

$$Re_L = \frac{U_\infty L}{\nu_e}\cdot\left(\frac{M_e}{M_\infty}\right)\left(\frac{T_e}{T_\infty}\right)^{1/2} \tag{E2}$$

where U is velocity and subscript ∞ refers to far upstream conditions.

For a flat plate with a pressure gradient, a can be written as follows:

$$a = \frac{V_e}{\tau_w V^*}\cdot\frac{\mathrm{d}P_e}{\mathrm{d}x} \tag{E3}$$

where subscript w refers to the wall. It is possible to show that for the scram combustor with high speed flow, the pressure gradient parameter given in Eq. (E3) is in fact a fair approximation; the reasoning is as follows.

Denoting the pressure gradient by

$$\frac{\mathrm{d}P}{\mathrm{d}x} = \frac{P_{\max} - P}{L} \tag{E4}$$

and

$$P_{\max} - P = c_{\mathrm{p\,max}}\left(\frac{\gamma P_\infty M_\infty^2}{2}\right) \tag{E5}$$

with $c_{p\,max}$ of order unity, one can write

$$\frac{dP_e}{dx} = c_{p\,max}\left(\frac{\gamma P_\infty M_\infty^2}{2}\right) \tag{E6}$$

and

$$a = \frac{c_{p\,max}}{c_f L^*} \tag{E7}$$

with

$$c_f = \frac{\tau_w}{\left(\frac{1}{2}\rho U^2\right)} \tag{E8}$$

and

$$L^+ = \frac{LU^+}{\nu_\infty} \tag{E9}$$

Some numerical estimates may then be made. The skin friction factor c_w may be taken (from experience) as 0.001. For a turbulent boundary layer

$$\delta^+ + \delta_e \frac{U^+}{\nu_\infty} = 2{,}000 \tag{E10}$$

The reference length may be taken as the wetted length of the cowl, such that

$$\frac{L}{\delta} = 10^3 \text{ to } 10^4$$

It then follows that $L^+ \geq 2.10^5$, and $a \leq 5.10^{-4}$. This order of magnitude analysis shows that the effect of pressure gradient is indeed small, and in practice one may adopt the skin-friction coefficient for a flat plate with zero pressure gradient; in any case, it is clear that the skin-friction coefficient tends to this value even as the boundary layer relaxes to the flat plate behavior.

The conclusion is that for a scram combustor, within the approximation of a high speed flow in which the freestream velocity can be considered nearly constant, the skin-friction drag relaxes quickly to that of a flat plate with zero pressure gradient flow.

In view of the well-known presence of shock waves in a scram combustor, it is necessary to account for their presence through an appropriate modification of the foregoing analysis and estimate. In the following, an extremely simple engineering approach is presented.

First, the flow length can no longer be taken as the geometrical length of the combustor in view of the presence of shockwaves. It needs to be increased by several times the boundary layer thickness to account for the effect of shocks and their interaction with the boundary layer. Second, the pressure gradient parameter should account for the jump in pressure across the shockwaves. One can write approximately

$$a_{\mathrm{shock}} = \frac{c_{\mathrm{p\,shock}}}{2} \tag{E11}$$

For strong shocks, a is of the order of 0.1 to 1.0.

A complexity that may arise in the presence of shock waves is flow separation due to sudden pressure rise. (At such locations, it is also of interest to note that local hot spots may arise, affecting the transport processes.) In any case, to utilize the solution for a flat plate with zero pressure gradient, one used the following obtained from Eq. (E1):

$$G\frac{\mathrm{d}\lambda}{\mathrm{d}x^*} = Re_L \tag{E12}$$

where the density and viscosity are assumed to correspond to the local edge boundary conditions. Since Eq. (E12) is a relaxation equation, some time (or space) lag can be expected for the flat plate solution to apply in the presence of a pressure gradient. However, the lag is not significant. Meanwhile, the pressure gradient along the combustor must be known, either from experimental data correlation or a reliable theoretical method.

A skin-friction coefficient reference should also be known. In this connection, Ref. 11 has shown that the friction law for incompressible flow can be transformed to that of a compressible flow by modifying the Reynolds number by accounting for kinetic heating and wall temperature, and introducing a correction coefficient for compressibility as follows:

$$c_{\mathrm{f\,comb}} = c_{\mathrm{f\,inc}} \cdot \frac{1}{F_c} \cdot (Re_x F_{\mathrm{Re}}) \tag{E13}$$

where, assuming a turbulent boundary layer, one can write

$$c_{\mathrm{f\,inc}} = 0.026\,Re_x^{1/7} \tag{E14}$$

This value of friction coefficient is shown in Ref. (E1) to be within 3% of the theoretical value for compressible flow. However, since the foregoing relations depend on density, it is useful to modify them so that only pressure is involved. For example, one can write

$$F_{\mathrm{RE}} Re_x = P_e \left(\frac{U_e x}{M_w} \right) \frac{(\sin^{-1} A + \sin^{-1} B)}{(T_{Aw} \cdot T_w)^{1/2}} \tag{E15}$$

$$a = \left(\frac{U_e^2}{2C_p T_w} \right)^{\frac{1}{2}} \qquad b = \left(\frac{T_{\mathrm{aw}}}{T_w} - 1 \right)$$

$$A = \frac{2a^2 - b}{(b^2 - 4a^2)^{\frac{1}{2}}} \qquad B = \frac{b}{(b^2 - 4a^2)^{\frac{1}{2}}}$$

where A and B denote the Van Driest functions, which are repeated here for the reader's convenience.

Note that all the terms in Eq. (E15) are constant except P_e and x. Thus, it follows that

$$c_f = 0.026 \left(\frac{T_e}{T_w} \right) (P_e x)^{1/7} \cdot \text{constant} \tag{E16}$$

and, accordingly,

$$\tau_w = P_e^{6/7} U_e^2 x^{-1/7} \text{ constant} \tag{E17}$$

Therefore, knowing the pressure distribution along the combustor, one can write

$$\frac{\tau_W}{\tau_{W\,\mathrm{ref}}} = \left(\frac{P_e}{P_{\mathrm{ref}}} \right)^{6/7} \left(\frac{U_e}{U_{\mathrm{ref}}} \right)^2 \left(\frac{x}{L} \right)^{-1/7} \tag{E18}$$

This then provides the value of wall shear stress along the flowpath. The velocity and x/L dependence is weak and can be neglected.

B. Heat Transfer

The heat transfer can be determined simply by invoking the Reynolds analogy, and writing

$$\frac{q_W}{q_{W\,\mathrm{ref}}} = \left(\frac{P_e}{P_{\mathrm{ref}}} \right)^{6/7} \left(\frac{U_e}{U_{\mathrm{ref}}} \right) \left(\frac{x}{L} \right)^{-1/7} \frac{(H_{\mathrm{aw}} - H_w)}{(H_{\mathrm{aw}} - H_w)_{\mathrm{ref}}} \tag{E19}$$

where the last term represents the enthalpy ratio. This term allows one to account for both heat release due to combustion (which may be large), and wall temperature variation along the flow.

Some data on curved ramps with a constant pressure gradient parameter, A. Laderman (1979), are shown in Figs. E1 and E2. The quick pressure rise

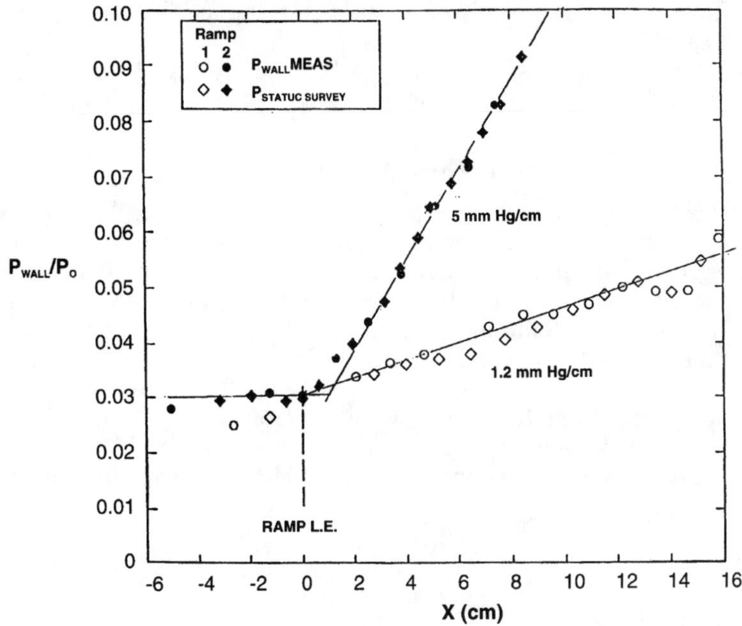
Ramp
1 2
P WALL MEAS
P STATUC SURVEY
5 mm Hg/cm
1.2 mm Hg/cm
P WALL/P o
RAMP L.E.
X (cm)

Fig. E1

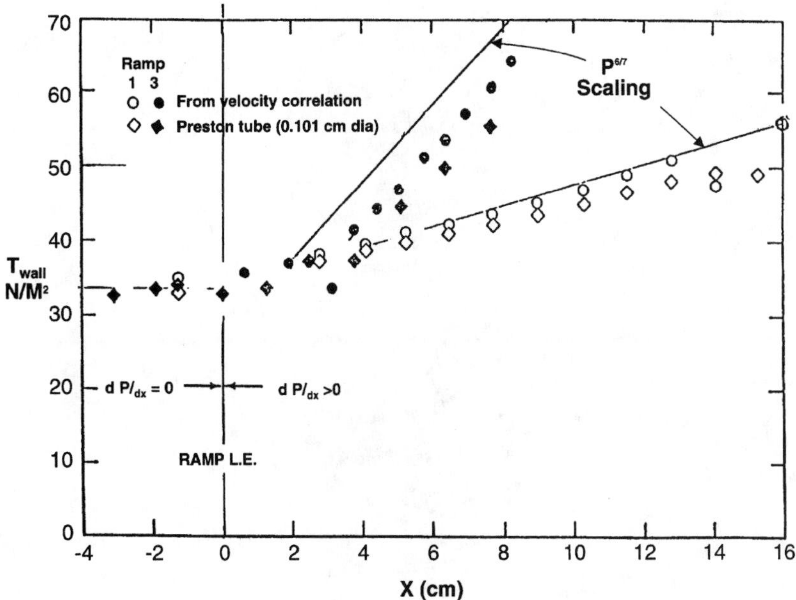
Ramp
1 3
From velocity correlation
Preston tube (0.101 cm dia)
P 6/7
Scaling
T wall
N/M 2
d P/dx = 0
d P/dx >0
RAMP L.E.
X (cm)

Fig. E2

case shows some lag, as anticipated; however, the agreement with predictions is satisfactory.

That these scaling laws are also applicable when strong shocks are present in the flow.

Appendix F: Force Accounting Procedures

This section presents some recipes used by the author for thrust and drag accounting. Two systems are used for interfacing with the propulsion and vehicle groups. The first is called freestream-to-freestream pressure accounting, primarily used for propulsion group; the second is the cowl inlet to aftbody exit area accounting system most often used to communicate installed thrust data to the vehicle group.

The geometry used to define terms is shown in Figs. F1a, F1b, and F2 in which fore, aft, and planform views are shown, respectively. From the figures, let us extract some important areas. The fore view allows us to define the forebody base area at the cowl inlet face used for forebody drag determination.

$$A_{\mathrm{FB}} = \pi R_{\mathrm{BI}}^2 \tag{F1}$$

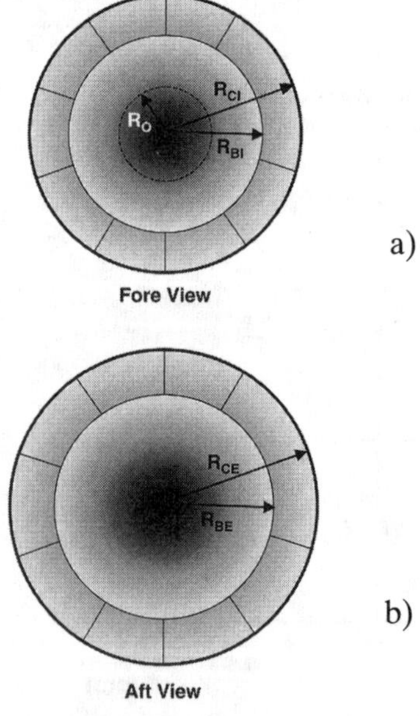

Fig. F1

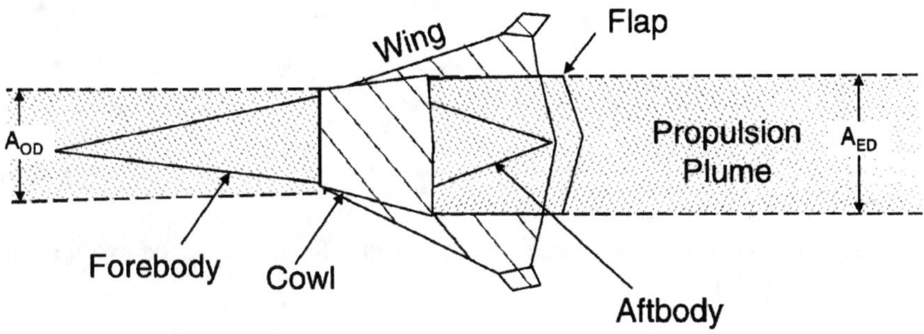

Fig. F2

The forebody cowl projected area and maximum zero angle of attack free-stream capture area is

$$A_{OD} = \pi R_{CI}^2 \tag{F2}$$

The inlet area is the difference of these reference areas

$$A_1 = A_{OD} - A_{FB} \tag{F3}$$

The design forebody contraction ratio is

$$CR_{FBD} = \frac{A_{OD}}{A_1} \tag{F4}$$

Based on forebody aerodynamics the freestream capture area is

$$A_O = \frac{\rho_1 A_1 V_1}{\rho_O V_O} \tag{F5}$$

Thus, at any speed less than shock-on-lip, the forebody contraction ratio is

$$CR_{FB} = \frac{A_O}{A_1} = \frac{\rho_1 V_1}{\rho_O V_O} \tag{F6}$$

From the aft view, the aftbody base area is

$$A_{AB} = \pi R_{BE}^2 \tag{F7}$$

The aftbody cowl projected area, which is useful for defining the propulsion flowpath design exit area, is

$$A_{ED} = \pi R_{CE}^2 \tag{F8}$$

The internal flowpath cowl or internal nozzle design exit area is

$$A_6 = A_{ED} - A_{AB} \tag{F9}$$

The overall design flowpath expansion ratio is

$$\varepsilon_{OD} = \frac{A_{ED}}{A_{OD}} \tag{F10}$$

The actual engine exhaust plume area may be larger or smaller than the design geometric aftbody exit area

$$\varepsilon_O = \frac{A_E}{A_O} \tag{F11}$$

The aftbody design expansion ratio is

$$\varepsilon_{ABD} = \frac{A_{ED}}{A_6} \tag{F12}$$

The plume are when expanded to ambient pressure is computed from the flowpath analysis and the aftbody expansion ratio

$$\varepsilon_{AB} = \frac{A_E}{A_6} \tag{F13}$$

This completes the preliminary definitions commonly used for area ratio nomenclature.

A. Freestream Force Accounting

The area assigned to the vehicle accounting system for forebody drag is

$$A_{FBV} = A_1(CR_{FBD} - CR_{FB}) \tag{F14}$$

The projected aftbody surface area assigned to the vehicle is

$$A_{ABV} = A_6(\varepsilon_{ABD} - \varepsilon_{AB}) \tag{F15}$$

The available information for the data transfer to the vehicle company is readily available from flowpath analysis, however, it is problematic for two reasons: The first reason is the inconvenience to the vehicle house; the second reason is we don't know what to do when the vehicle base area negative.

The force accounting in this system gives the net thrust as

$$T_{NET} = F_E - F_O - P_O(A_E - A_O) \tag{F16}$$

The installed thrust then includes the flowpath surfaces wetted by the captured air and is written

$$T_{INSTALLED} = T_{NET} - C_{PFB}q_O A_{FBV} + C_{PABV}q_O A_{ABV} - C_{PCOWL}q_O A_{COWL} \tag{F17}$$

The forebody term and cowl terms are straightforward but the aftbody term is problematic

B. Cowl-to-Tail Accounting

In principle, the cowl-to-tail accounting has a simple division of forces, because the vehicle forebody gets fixed and the cowl and engine get fixed-internal flow from cowl entrance and the fixed aftbody.

$$T_{NET} = F_{ED} - F_1 - P_O(A_{ED} - A_1) + T_{TURN}$$

$$T_{TURN} = P_O A_O(1 + \gamma M_O^2)[1 - \cos(\delta_{FB})] \tag{F18}$$

This is much better defined in the cowl-to-tail system and the turn thrust term is explicit to the point where it is not visible in the freestream system. The installed thrust is

$$T_{INSTALLED} = T_{NET} - C_{PFB}q_O A_{FB} - C_{PCOWL}q_O A_{COWL} \tag{F19}$$

C. Lift Effects

Turning to the planform for definition of the lift-producing surfaces are two accounting purposes. The wing is defined as the planform, not including the projected design flowpath, and the propulsion lift is defined by the internal flow capture area. Thus, the drag of the wing, tail, and drag due to lift can be written

$$D_{\mathrm{WING}} = C_{\mathrm{DO}q_O} S_{\mathrm{Wing}} + L_{\mathrm{WING}} \tan(AOA) + L_{\mathrm{PROP}} \frac{\tan(AOA)}{2}$$

$$C_{\mathrm{DO}} = C_{\mathrm{DOWING}} + C_{\mathrm{DOTAIL}} + C_{\mathrm{DTRIM}} \tag{F20}$$

The propulsion lift is written as

$$L_{\mathrm{PROP}} = F_{\mathrm{ED}} \sin(AOA) \tag{F21a}$$

$$D_{\mathrm{PROP}} = F_{\mathrm{ED}}[1 - \cos(AOA)] \tag{F21b}$$

This completes a brief summary of the common force accounting entries usually employed.

Appendix G: Geometry and Mass of Integrated Vehicle

A. Geometry

The system is considered in terms of 1) forebody, 2) inlet, 3) combustor, and 4) nozzle.

1. Forebody

Two basic geometries may be considered: the wedge and the cone. These may be compared under different assumptions, as follows: 1) equal in volume, half angle, and capture area; 2) equal in drag, volume, and capture area; and 3) partial capture area. The latter case considers the ratio of geometric surface area to volume, raised to the 2/3 power for various aspect ratio wedges, and various smile angle of conical forebodies as a function of half angle with respect to the waterline.

Case 1. For purposes of illustration, a cone of half angle equal to 5 degrees is considered in Mach 16 flight.

For the wedge, the volume (Vol), capture area (S_{cap}), and wetted surface area S_{wet} are given by the following:

$$Vol = 2\left(\frac{1}{2} L_w^2 \cdot \tan\theta\right) W \tag{G1}$$

$$S_{\mathrm{cap}} = 2\ W\ L_w\ (\tan\theta_w)\ (\delta/\tau) \tag{G2}$$

$$S_{\mathrm{wet}} = \frac{2L_w}{\cos\theta_w} \cdot 2L_y^2 \cdot \tan\theta_w \tag{G3}$$

where $\tau = \tan\theta, \theta = 5$ deg by assumption, and $k = 1.395$. The wetted surface area-to-volume ratio can be written as

$$\frac{S_{wet}}{Vol^{2/3}} = 2\left(\frac{1}{f}\cdot\tan\theta_w\right)\Big/ f\,(\tan\theta_w\;/\;f)^{2/3} \tag{G4}$$

Next, for the cone, Vol, S_{cap} and S_{wet} can be written as

$$Vol = \Pi/_3 \cdot L_c^3 \cdot \tan^2\theta \tag{G5}$$

$$S_{cap} = \Pi\,(L_c \tan\theta)^2\,(\delta_c/\tau)^2 \tag{G6}$$

$$S_{wet} = \Pi L_e^2 \frac{\tan\theta_c}{\cos\theta_c} \tag{G7}$$

Also, the cone, the shock angle, forebody pressure, and pressure recovery are given by

$$\left(\frac{\delta_i}{\tau}\right)_c = 1.329$$

$$(P_T)_W = 0.65$$

$$c_{pc} = 0.0166$$

For equal volume and capture area, for the case of the wedge one then obtains

$$\left(\frac{\delta_W}{\tau}\right)_w = 1.535$$

$$c_{pw} = 0.023$$

$$c_{pw} = 0.0235$$

Also,

$$\frac{L_c}{L_w} = \frac{3}{2}\frac{(\delta/\tau)_c^2}{(\delta/\tau)_w} = 1.75$$

Thus, the conical vehicle will be about 70% longer than the wedge-shaped vehicle.

Since

$$\frac{W}{L_w} = \frac{\Pi}{3}\left(\frac{L_w}{L_c}\right)^2 \tan\theta \tag{G8}$$

One can determine the fineness or aspect ratio of the wedge forebody as

$$f = \frac{L_c}{W} = 2.12$$

Finally, the ratio of surface area of the wedge and the cone is given by

$$\frac{S_w}{S_c} = 1.36$$

Case 1. In this case, with equal drag as a requirement, the effect on the geometrical lines for the wedge would be to reduce the angle relative to that of the cone and, correspondingly, to increase the length of the wedge forebody. The reduction in the wedge angle required is given by

$$\delta_w = \delta\left[\frac{(D/\dot{m})_c}{(D/\dot{m})_w}\right] \tag{G9}$$

so that, for the specific example,

$$\delta_w = 5(2/3)^{0.5} = 4.8°$$

Now k^{-1} for the wedge increases and the wedge shock function becomes

$$(\delta\tau)_w = 1.663$$

Interestingly, in this case of equal drag, the total pressure ratio, as well as the surface pressure, become the same as that for the cone:

$$k = 1.136$$

$$\frac{\delta}{\tau} = 1.66$$

$$C_{pw} = 0.169; \qquad C_{pc} = 0.166$$

$$(P_{\tau 1})_w = 0.78; \qquad (P_{\tau 1})_c = 0.81$$

$$\frac{L_c}{L_w} = 1.60$$

$$f_w = \frac{L_w}{W} = 3.30$$

Thus,

$$\frac{S}{V^{2/3}} = \frac{2(1/f \cdot \tan\theta_w)}{(\tan\theta_w/f)^{2/3}} = 9.65$$

and

$$\frac{S_w}{S_c} = -\ 9.65/6.89 = 1.40$$

which is very close to that obtained in case 1. However, in case 2, the drag and total pressure recovery remain the same for the wedge and the cone.

Case 2. It has been pointed out in Ref. G1 that for large SSTO flight vehicles, the preferred configuration is the one with the engine in the bottom of the vehicle with partial capture area for air.

Referring to Fig. G1, in the case of partial wedges, leaving the forebody fineness free for generality, one can write

$$\frac{S}{V^{2/3}} = \frac{(2/f \cdot \tan\theta_w \cdot (2f)^{2/3}}{(\tan\theta_w)^{2/3}} \tag{G10}$$

For partial cones, specifying the same freedom about the fineness ratio as in the 2D case, one can write

$$\frac{S}{V^{2/3}} = 6^{2/3}\left(1 - \frac{\theta_s}{2\cos\theta_c}\right)\Big/(\tan\theta_c)^{1/3} \cdot (\theta_c)^{2/3} \tag{G11}$$

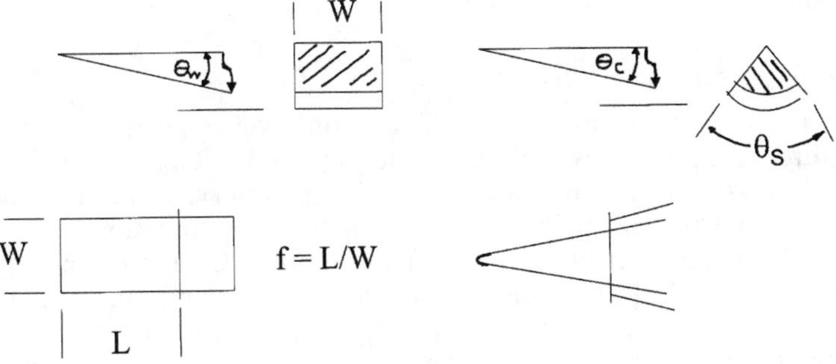

Fig. G1

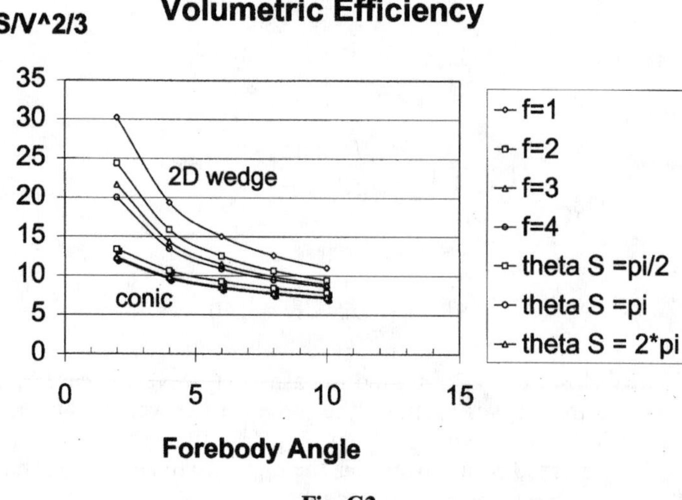

Fig. G2

These results lead to the surface per unit volume for partial cones becoming less than that for half wedges. Also, wedges become more efficient as fineness of angles increase. For the 3D bodies, increased efficiency is obtained with increasing smile angle and half angle.

Fig. G2 shows a convenient chart for wedges and cones.

B. Weight Analysis

Weights are determined for different surfaces—the surface from ramp to cowl (denoted by subscript RC) being distinguished from the side wall surface (denoted by subscript SW). The case of horizontal throat area is to be distinguished from that of vertical throat area.

1. *Surface Area Estimation*

The propulsion package for installation on a vehicle is divided by partitions into a multiple flow path engine module in Fig. G3. A horizontal and vertical throat configuration are depicted. It is observed that the length of the horizontal throat engine can be expressed as a function of h and is, therefore independent of the number of modules. On the other hand, the length of the vertical throat engine is a function of the number of partitions in the engine, because length is dependent on the span of the engine.

In the following, the overall expansion ratio is assumed to be unity.

Surface areas for the vertical throat case.

Inlet:

$$A_{RC} = 2 \cdot \left(\frac{L}{h}\right)_I \cdot h \cdot s \qquad \text{(G12a)}$$

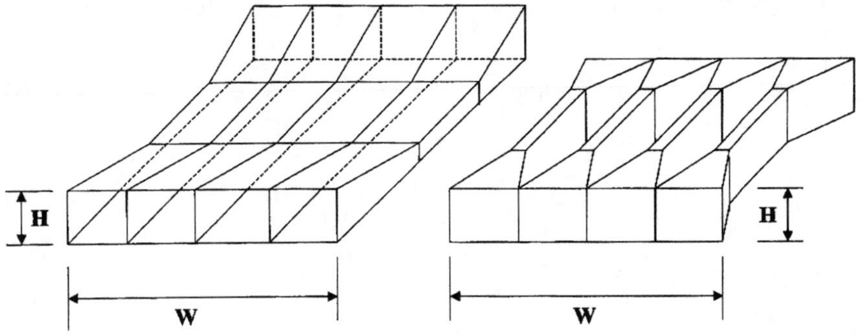

Fig. G3

$$A_{SW} = 2 \cdot \frac{h}{2}\left(1 - \frac{1}{CR}\right)\left(\frac{L}{h}\right)_I \cdot h \tag{G12b}$$

Combustor:

$$A_{RC} = 2 \cdot \left(\frac{L}{h}\right)_c \cdot \frac{h}{CR} \cdot S \tag{G13a}$$

$$A_{SW} = 2 \cdot \frac{h}{CR} \cdot \left(\frac{L}{h}\right)_c \cdot \frac{h}{CR} \tag{G13b}$$

Nozzle:

$$A_{RC} = 2\left(\frac{L}{h}\right)_N \cdot \frac{h}{CR} \cdot S \tag{G14a}$$

$$A_{SW} = 2\frac{h}{CR} \cdot \left(\frac{L}{h}\right)_N \cdot h \cdot \frac{h}{CR}\left(1 - \frac{1}{CR}\right) \tag{G14b}$$

Surface areas for the horizontal throat case.
Inlet:

$$A_{CR} = 2\left(\frac{L}{h}\right)_I \cdot h \cdot \frac{S}{2}\left(1 - \frac{1}{CR}\right) \tag{G15a}$$

$$A_{SW} = 2\left(\frac{L}{h}\right)_I \cdot h \cdot S \tag{G15b}$$

Combustor:

$$A_{CR} = 2\left(\frac{L}{h}\right)_c \cdot h \cdot \frac{h}{CR} \tag{G16a}$$

$$A_{SW} = 2\left(\frac{L}{h}\right)_c \cdot h \cdot \frac{S}{CR} \tag{G16b}$$

Nozzle:

$$A_{CR} = 2 \cdot \left(\frac{L}{h}\right)_N \cdot h \cdot \frac{h}{2}\left(1 - \frac{1}{CR}\right) \tag{G17a}$$

$$A_{SW} = 2 \cdot \left(\frac{L}{h}\right)_N \cdot h \cdot S \tag{G17b}$$

If the overall width of the engine is W, and there are N modules in the engine

$$S = \frac{W}{N} \tag{G18}$$

Now, considering an engine configuration for which $S = h$ and, therefore, the engine cross-section is square, the surface areas in this case for horizontal and vertical throat are equal. Thus,

$$N_{S=h} = \frac{W}{h} \tag{G19}$$

and this number may be assumed to be an integer. Then, in general, the span to height ratio may be written as

$$\frac{S}{h} = \frac{N_{S=h}}{N} \tag{G20}$$

For the square engine, the total length of the engine can be expressed by

$$L = \left[\left(\frac{L}{h}\right)_I + \left(\frac{L}{h}\right)_c \cdot \frac{1}{CR} + \left(\frac{L}{h}\right)_N\right] \cdot S \tag{G21}$$

or, in general,

$$\frac{L}{h} = \left[\left(\frac{L}{h}\right)_I + \left(\frac{L}{h}\right)_c \cdot \frac{1}{CR} + \left(\frac{L}{h}\right)_N\right] \frac{N_{S=h}}{N} \tag{G22}$$

2. *Weights Estimation*

The individual component areas can now be multiplied by the appropriate structural density and heat exchange density factors to obtain the weights and the heat loads, respectively.

The side wall panels weigh less per unit area than the ramp to cowl panels because they do not have unbalanced loads.

A reference density may be selected for normalizing the weight of the engine module. Note that the bending load cannot be used as reference since it is nonlinear, and the pressure load is not an adequate reference. Therefore, ρ_{ref} is expressed in terms of a reference area and P_{ref} as follows:

$$S_{ref} = 1 \text{ ft}$$

$$P_{ref} = 14.7_{psia}$$

Thus,

$$\varsigma_{ref} = 0.5328 \cdot (14.7^2)^{0.4}$$

$$= 1.56 \text{ lbm/ft}^2$$

Then, one can write

$$\frac{\rho_{RO}}{\rho_{ref}} = \left(\overline{P}_D\right)^{0.4}\left(\frac{S}{h}\right)^{0.8}\left(\frac{h}{h_{ref}}\right)^{0.8} \tag{G23}$$

$$= \left(\overline{P}_D\right)^{0.4}\left(\frac{N_{S=h}}{N}\right)^{0.8}\overline{h}^{0.8} \tag{G24}$$

where $\overline{P}_D$ is the design load in atmospheres and $\overline{h}$ is the cowl height in feet.

The surface areas for the two cases can then be determined, and by multiplying by the surface density factor appropriate for each case, the weight-scaling relations established.

3. *Weight-Scaling Relations*

The following presents an illustration of setting up weight-scaling relations.

A generic truss core structure was assumed, using graphite polyamide material for structural parts and a copper alloy for heat exchanger panels. Parametric studies were performed with span and pressure load varied. The heat exchanger panels are taken to be governed by the heat load and coolant pressure, in this case, gaseous hydrogen. Thus, the results are useful for scaling weights of fixed geometry engines with regenerative cooling. In this study, heat exchanger manifolds, attachments, and scales were included.

The heat exchanger panel weights for inlet, combustor, and nozzle were the same for the two basic concepts of horizontal and vertical throat.

For panels with unbalanced loads, as in the ramp, cowl, and end walls, the structural density was expressed as

$$\rho_{RC} = 0.5328\,(PS^2)^{0.4}\ \text{lb/ft}^2$$

where P is the design pressure in psi and S is the span in ft. For panels without unbalanced loads, such as sidewalls, it was assumed that

$$\rho_{SW} = 0.375\,P_D^{0.3}\ \text{lb/ft}^2$$

where P_D is the design pressure in psi. The heat exchanger panel weights were taken to be in the form of an external load and span independent:

$$\rho_{HEX} = 2.7\ \text{lb/ft}^2$$

The structural designs have a safety factor of 1.5 or greater on ultimate strength; the reference area is the engine frontal area at the cowl; and the reference weight is reference area times the reference density. Thus,

$$W_{ref} = 1.56\,h\,W$$

Horizontal throat weight relations.
Inlet:
For each module

$$\left(\frac{W_{RC}}{W_{ref}}\right)_N = \frac{2}{N}\left(\frac{L}{h}\right)_I (\bar{P}_D)^{0.4}\left(\frac{N_{S=h}}{N}\right)^{0.8}(\bar{h})^{0.8} \tag{G25}$$

and

$$\left(\frac{W_{SW}}{W_{ref}}\right)_N = \frac{1}{N_{S=h}}\left(\frac{L}{h}\right)_I\left(1-\frac{1}{CR}\right)\cdot 0.5384\,(\bar{P}_D)^{0.3} \tag{G26}$$

For the total of N modules,

$$\left(\frac{W_{RC}}{W_{ref}}\right) = 2\left(\frac{L}{h}\right)_I (\bar{P}_D)^{0.4}\left(\frac{N_{S=h}}{N}\right)^{0.8}(\bar{h})^{0.8} \tag{G27}$$

and

$$\left(\frac{W_{SW}}{W_{ref}}\right) = \left(\frac{L}{h}\right)_I\left(1-\frac{1}{CR}\right)\cdot 0.5384\,(\bar{P}_D)^{0.3}\cdot\frac{N}{N_{S=h}} \tag{G28}$$

Combustor:
For each module

$$\left(\frac{W_{\text{RC}}}{W_{\text{ref}}}\right) = \frac{2}{N}\cdot\frac{1}{CR}\cdot\left(\frac{L}{h}\right)_C (\bar{P}_D)^{0.4}\left(\frac{N_{S=h}}{N}\right)^{0.8}(\bar{h})^{0.8} \tag{G29}$$

and

$$\left(\frac{W_{\text{SW}}}{W_{\text{ref}}}\right) = \frac{2}{N_{S=h}}\cdot\frac{1}{CR^2}\cdot\left(\frac{L}{h}\right)_C \cdot 0.5384\,(\bar{P}_D)^{0.3} \tag{G30}$$

For N modules

$$\left(\frac{W_{\text{RC}}}{W_{\text{ref}}}\right) = \frac{2}{CR}\cdot\left(\frac{L}{h}\right)_C \cdot(\bar{P}_D)^{0.4}\left(\frac{N_{S=h}}{N}\right)^{0.8}(\bar{h})^{0.8} \tag{G31}$$

and

$$\left(\frac{W_{\text{SW}}}{W_{\text{ref}}}\right) = \frac{2}{CR^2}\cdot\left(\frac{L}{h}\right)_C \cdot 0.5384\,(\bar{P}_D)^{0.3}\left(\frac{N}{N_{S=h}}\right) \tag{G32}$$

Nozzle:

$$\left(\frac{W_{\text{RC}}}{W_{\text{ref}}}\right) = \left(\frac{2}{N}\right)\left(\frac{L}{h}\right)_N (\bar{P}_D)^{0.4}\left(\frac{N_{S=h}}{N}\right)^{0.8}(\bar{h})^{0.8} \tag{G33}$$

and

$$\left(\frac{W_{\text{SW}}}{W_{\text{ref}}}\right) = \frac{1}{N_{S=h}}\left(\frac{L}{h}\right)_N\left(1-\frac{1}{CR}\right)\cdot 0.5384\cdot(\bar{P}_D)^{0.3} \tag{G34}$$

For N modules

$$\frac{W_{\text{RO}}}{W_{\text{ref}}} = 2\left(\frac{L}{h}\right)_N (\bar{P}_D)^{0.4}\left(\frac{N_{S=h}}{N}\right)^{0.8}(\bar{h})^{0.8} \tag{G35}$$

and

$$\left(\frac{W_{\text{SW}}}{W_{\text{ref}}}\right) = \left(\frac{L}{h}\right)_N\left(1-\frac{1}{CR}\right)\cdot 0.5384\,(\bar{P}_D)^{0.3}\left(\frac{N}{N_{S=h}}\right) \tag{G36}$$

Equations (G35) and (G36) show that the ramp-cowl panel weight decreases with increasing N, while the side wall weight increases linearly with N. Thus, an optimum number of modules exists for lowest weight.

Vertical throat weight relations. In the following, the weight relations are given per module. The relations for N modules can be obtained simply by multiplying the expression on the right of each relation by N.

Inlet:

For each module

$$\frac{W_{\mathrm{RC}}}{W_{\mathrm{ref}}} = \frac{1}{N}\cdot\left(\frac{L}{h}\right)_I\left(1-\frac{1}{CR}\right)(\bar{P}_D)^{0.4}\left(\frac{N_{S=h}}{N}\right)^{1.8}(\bar{h})^{0.8} \tag{G37}$$

and

$$\frac{W_{\mathrm{SW}}}{W_{\mathrm{ref}}} = \frac{2}{N}\cdot\left(\frac{L}{h}\right)_I\cdot 0.5384\,(\bar{P}_D)^{0.3} \tag{G38}$$

Combustor:

For each module

$$\frac{W_{\mathrm{RC}}}{W_{\mathrm{ref}}} = \frac{1}{N}\frac{2}{CR^2}\left(\frac{L}{h}\right)_C(\bar{P}_D)^{0.4}\left(\frac{N_{S=h}}{N}\right)^{1.8}(\bar{h})^{0.8} \tag{G39}$$

and

$$\frac{W_{\mathrm{SW}}}{W_{\mathrm{ref}}} = \frac{1}{N}\cdot\frac{2}{CR}\cdot\left(\frac{L}{h}\right)_C\cdot 0.5384\,(\bar{P}_D)^{0.3} \tag{G40}$$

Nozzle:

For each module

$$\frac{W_{\mathrm{RC}}}{W_{\mathrm{ref}}} = \frac{S}{W}\cdot\frac{S}{h}\left(1-\frac{1}{CR}\right)\left(\frac{L}{h}\right)_N\frac{\rho_{\mathrm{RC}}}{\rho_{\mathrm{ref}}} \tag{G41}$$

$$= \frac{1}{\mathrm{N}}\left(\frac{L}{h}\right)_N\left(1-\frac{1}{CR}\right)(\bar{P}_D)^{0.4}\left(\frac{N_{S=h}}{N}\right)^{1.8}(\bar{h})^{0.8} \tag{G42}$$

and

$$\frac{W_{\mathrm{SW}}}{W_{\mathrm{ref}}} = \frac{2}{N}\left(\frac{L}{h}\right)_N\cdot(0.5384)\,(\bar{P}_D)^{0.3} \tag{G43}$$

4. *Heat Exchanger Panel Weight Relations*

The heat exchanger panel mass density is assumed to be constant; thus, ρ_{HEX} is taken to be 2.7 lb/ft^2.

The general weight relation is written as

$$\frac{W_{\text{HEX}}}{W_{\text{ref}}} = \frac{\text{total surface area}}{\text{frontal area}} = \frac{2.7}{1.56} \tag{G44}$$

Horizontal throat engine:

$$\frac{W_{\text{HEX}}}{W_{\text{ref}}} = 1.73\left[\frac{106}{5} - \left(\frac{286}{25}\right)\frac{N}{N_{S=h}}\right] \tag{G45}$$

Vertical throat engine:

$$\frac{W_{\text{HEX}}}{W_{\text{ref}}} = 1.73\left[\left(\frac{286}{25}\right)\frac{N_{S=h}}{N} - \frac{106}{5}\right] \tag{G46}$$

The panel weight in this case is also scale-dependent as in the case of the panels joining the ramp and cowl. This was not taken into account in Ref. G2. To include it, one has to make ($\rho_{\text{HEX}} / \rho_{\text{ref}}$) of the same form in this case, as in the case of ramp and cowl, by including the use of internal insulation.

Appendix H: Two-Wave Combustion Model for Optimal Supersonic Combustion Performance

A. Heat Addition in a Dual-Mode Combustor

In the following, two cases of heat addition are considered: 1) the constant area combustor is thermally choked by heat addition; and 2) heat addition at constant Mach number.

1. *Constant Area Combustor with Choking due to Heat Addition*

The flow function now becomes the following:

$$N = \dot{m}\left(\frac{RT_t}{\gamma m}\right)^{1/2} = \frac{M\left[1 + (\gamma - 1/2)M^2\right]^{1/2}}{1 + \gamma M^2} \tag{H1}$$

Now, the stream thrust ratio at the isolator inlet can be expressed in terms of Ψ, the engine compression ratio, by writing

$$\frac{F_2}{F_1} = 1 - \frac{(\psi - 1)(\gamma - 1)}{F_1} \tag{H2}$$

Then considering the constant area combustor section following the isolator as the first stage of the combustor, the stream thrust at the exit of that section is given by

$$F_1 = F_2 + P_3\,(A_1 - A_3) \tag{H3}$$

where P_3 denotes the pressure at the exit of the constant area combustor, and, therefore,

$$P_3 = P_2 \cdot (P_3/P_2)_{\text{isolator}} \tag{H4}$$

$(P_3/P_2)_{\text{isolator}}$ can be found from the isolator analysis.

Since stream thrust is known, one can solve for the maximum fueling, which would cause choking of the engine, by noting that at choking $N = N^* = [2\,(\gamma + 1)]^{-0.5}$ is a maximum. Then

$$(1 + \mathrm{f}\ell a)_\ell \frac{(T_{t1}/M)}{(T_{t1}/M)} = \left(\frac{N^*}{N_2}\right)\left[1 + \frac{\overline{P}_{\text{iso}}(\varepsilon_c - 1)}{1 + \gamma M_2^2}\right] \tag{H5}$$

where $\psi = \beta_2 / \beta_1$, and

$$\frac{T_{t1}/M}{T_{t2}/M} = 1 + \frac{\Delta\psi_c}{\beta_1} \tag{H6}$$

$$\Delta\,\psi_c = f\,(\psi_c, \mathrm{f}\ell a) \tag{H7}$$

Note that M_2 is a function of β_1 and ψ, such that

$$M_2^2 = \frac{2(\psi\beta_1 - 1)}{\gamma - 1} \tag{H8}$$

Also, since

$$N_2 = N_1 \cdot (F_2 / F_1) \tag{H9}$$

one can write the following relation for maximum fueling:

$$(1 + f\ell a_1)\left(1 + \frac{\Delta\psi_c}{\beta_1}\right)^{1/2}_{M=1.0} = \frac{N^*}{N_1}\left[1 - \frac{\psi - 1}{(\gamma - 1)\overline{F}_1}\right] \times\left[1 + \frac{\overline{P}_{\text{iso}}(\varepsilon_c - 2)}{1 + \dfrac{2\gamma}{\gamma - 1}\cdot(\psi\beta_1 - 1)}\right] \tag{H10}$$

Here the left side is a function of equivalence ratio, φ, only. This expression shows how the isolator and the first stage of the combustor work, using the term $[P_{\mathrm{ISO}}\,(\varepsilon_c - 2)]$.

Flow distortion may be taken into account in the flow function of Eq. (H1), for example,

$$N_{\mathrm{distortion}} = \eta_A M\left(1 + \frac{\gamma - 1}{2} M^2\right)^{1/2} \Big/ (1 + m_T M^2) \qquad \text{(H11)}$$

Considering a large distortion such as $\eta_H = \eta_A = \eta_F = 0.8$, $(M_{\mathrm{crit}})_{10}$, and $N_{\mathrm{crit}} = 0.4170$, the fuel deficit, ~20%, should be added. Thus, geometric blockage trades about one to one for that with heat addition blockage and Mach number.

Next, heat transfer and drag can be determined from the stream thrust, based on reference values. First, solving for pressure

$$(P_c)_{\mathrm{critical}} = (F_1/A_1)/[1 + \gamma\eta] M^2_{\mathrm{critical}} \qquad \text{(H12)}$$

Next, the wetted area of the combustor can be taken as the area of the end walls, neglecting the side walls on the basis of flow separation over those:

$$\Sigma_c = 2\ell\, c W_3 \frac{(\varepsilon_c + 1)}{2} \qquad \text{(H13)}$$

Then, the heat transfer is given by

$$\dot{Q}_{\mathrm{ct}} = q_c \Sigma_c = \dot{q}\left(\frac{P_c}{P_1}\right)^{6/7} \Sigma_c \frac{(T_{t3} - T_w)}{(T_{t1} - T_w)} \qquad \text{(H14)}$$

Similarly, the drag can be written as

$$(D_{V1})_c = \tau_1 \left(\frac{P_c}{P_1}\right)^{6/7} \Sigma_c \qquad \text{(H15)}$$

2. *Case of Constant Mach Number Flow Combustor*

Considering low Mach number, K_{WP} is set equal to zero. Further, in view of the favorable pressure gradient in this flow, Reynolds analogy is assumed to hold. Also, wall shear and heat transfer are considered negligible in all separated regions. Finally, fuel momentum is neglected, assuming crosswise injection. The analysis is based, as earlier, on 1D approximation, and the interest is in the isolator, the first stage of combustion (in the constant area section), and the second stage of combustion with constant Mach number.

Analysis. In this case, the flow function

$$\mathrm{N} = \frac{M\sqrt{\beta}}{(1 + \gamma M^2)} \tag{H16}$$

is a constant. Hence, writing

$$\mathrm{d}F = (P\,\mathrm{d}A + A\,\mathrm{d}p)\,(1 + \gamma\,M^2) \tag{H17}$$

and noting, through momentum balance, that

$$\mathrm{d}F = P\,\mathrm{d}A - \tau_{\mathrm{W}}\,d\sigma \tag{H18}$$

it follows that

$$\mathrm{d}F/F = \frac{1}{2}\,\mathrm{d}\frac{(T_t/M)}{(T_t/M)} \tag{H19}$$

and

$$(1 + \gamma M^2)\,A\mathrm{d}P = -\gamma M^2\,P\mathrm{dA} + \tau_{\mathrm{w}}\left(\frac{\mathrm{d}\sigma}{\mathrm{d}A}\right)\mathrm{d}A \tag{H20}$$

It further follows, on combining $(1 + \gamma\,M^2)\,\gamma\,M^2$, that

$$\frac{\mathrm{d}P}{P} = -m\frac{\mathrm{d}A}{A}\left(1 - \frac{\tau_{\mathrm{W}}}{\gamma_p M^2}\frac{\mathrm{d}\sigma}{\mathrm{d}A}\right) \tag{H21}$$

where

$$\frac{\mathrm{d}\sigma}{\mathrm{d}A} = \frac{\Sigma_{\Pi}}{A_I}\left(\frac{A_{\Pi}}{A_I} - 1\right)$$

$$m = \frac{\gamma M^2}{1 + \gamma M^2} \tag{H22}$$

On non-dimensionalizing, with respect to reference conditions,

$$\frac{\mathrm{d}\widetilde{P}}{\widetilde{P}} = -\frac{\mathrm{d}A}{A}\left[1 - \left(\frac{\tau_w}{P_w \gamma M^2}\right)\frac{\widetilde{\Sigma}_{\Pi}}{m_{\Pi}}\right] \tag{H23}$$

Writing

$$\frac{\tau_w}{P_w} = \left(\frac{\tau_w}{\tau_\gamma}\right)\left(\frac{\tau_\gamma}{P_\gamma}\right) \cdot \tilde{P}^{-1} = \left(\frac{\tau_\gamma}{P_\gamma}\right) \tilde{P}^{-1} \tag{H24}$$

and, through linearizations,

$$\tilde{P}^{n-1} = 1 + (n-1)(\tilde{P} - 1) \tag{H25}$$

for $n = 6/7$, it follows that

$$1 - \left(\frac{\tau_w}{P_w}\right)\left(\frac{1}{\gamma M^2}\right)\frac{\tilde{\Sigma}}{\tilde{m}} = 1 - \left(\frac{\tau_\gamma}{P_\gamma}\right)\frac{(8/7 - \tilde{P}/7)}{M^2} \cdot \frac{\tilde{\Sigma}}{\tilde{m}} \tag{H26}$$

Then, Eq. (H23) may be written as follows, by assuming the variation of pressure is given by

$$\frac{\mathrm{d}P}{P} = -m\frac{\mathrm{d}A}{A}[(1 - 8f) + fP] \tag{H27}$$

where

$$f = \frac{\tau_r}{7P_r\delta M^2} \cdot \frac{\tilde{\Sigma}}{\tilde{m}} \tag{H28}$$

Integrating Eq. (H27), one obtains

$$P\,[\,(1 - 7f)\,A^{m(1-8f)} - f] = (1 - 8f) \tag{H29}$$

or

$$P = \frac{(1 - 8f)}{(1 - 7f)A^{m(1-8f)} - f} \tag{H30}$$

$$= \frac{(1 - 8f)A^{m(1-8f)}}{(1 - 7f) - fA^{-m(1-8f)}} \tag{H31}$$

It is found that f is much less than unity. Thus, approximately,

$$\frac{P_{II}}{P_I} = \left(\frac{A_{II}}{A_I}\right)^{\frac{-\gamma M^2}{1+\gamma M^2}} \tag{H32}$$

and

$$\frac{P_{II}A_{II}}{P_I A_I} = \left(\frac{A_{II}}{A_I}\right)^{\frac{1}{1+\gamma M^2}} = \left(\frac{T_{T\,II}}{T_{T\,I}}\right)^{1/2} \tag{H33}$$

The total thrust obtained at end of combustion and expansion is given by

$$\int_{A_I}^{A_E} P\,\mathrm{d}A = \int_{A_I}^{A_{II}} P\,\mathrm{d}A + \int_{A_{II}}^{A_E} P\,\mathrm{d}A \tag{H34}$$

$$= (F_{II} - F_I) + \frac{F_{II}}{1+\gamma M_{II}^2}\,\frac{\left[1-(A_E/A_{II})^{-(\gamma-1)}\right]}{(\gamma-1)} \tag{H35}$$

Normalizing the total thrust with respect to the thrust with the first stage combustor,

$$\int_{A_I}^{A_E} P\,\mathrm{d}A/F_I = \left[\left(\frac{T_{tII}}{T_{tI}}\right)^{1/2} - 1\right] + \frac{T_{tII}^{1/2}/T_{tI}\left[1-\left(A_E/A_{II}\right)^{-(\gamma-1)}\right]}{(1+\gamma M_{II}^2)(\gamma-1)} \tag{H36}$$

Drag and heat transfer. The pressure integral can be expressed by

$$\int \mathrm{P^n}\,\mathrm{d}A = P_1\,A_1 \int A^{-n\Pi}\,\mathrm{d}\tilde{A} \tag{H37}$$

$$= P_1 A_1 \left[\frac{\tilde{A}^{(1-n_{II})} - 1}{1 - m_{II}}\right] \tag{H38}$$

where

$$m = n\,\gamma\,M^2 \,/\, (1 + \gamma\,M^2) \tag{H39}$$

Then the drag can be expressed by the relation

$$\frac{D_{V\,II}}{P_1 A_1} = \left(\frac{\tau_1}{P_1}\right)\left[\frac{\tilde{A}_{II}^{(1-m_{II})} - 1}{1 - m_{II}}\right]\frac{\tilde{\Sigma}_{II}}{(1-\tilde{A}_{II})} \tag{H40}$$

The heat transfer, by similarity, is given then by

$$\dot{Q}_{II} = \dot{q}w_1 \left[\frac{\tilde{\Sigma}_{II}}{(1-\tilde{A}_{II})}\right]\left[\frac{\tilde{A}_{II}^{(1-m_{II})}-1}{1-m_I}\right] \tag{H41}$$

B. Scramjet Two-Wave Combustor

The dynamics of a shock supported by a Chapman–Jouget deflagration wave can be studied by using Hugoniot equations.

The Hugoniot equations are the integral conservation equations for mass, normal momentum, tangential momentum, and energy for a small stream tube crossing a discontinuity surface. The interesting solution results in an equation of state relating the pressure and density.

$$m = \rho_1 u_1 = \rho_2 u_2 \tag{H42}$$

$$U = (u, v)$$

$$m u_1 + p_1 = m u_2 + p_2 \tag{H43}$$

$$v_1 = v_2 \tag{H44}$$

$$h_1 + \frac{1}{2}u_1^2 + Q = h_2 + \frac{1}{2}u_2^2 \tag{H45}$$

The terms follow standard notation for state variables and velocities normal and tangential to the wave. The equations are general and apply to shocks and flames depending on the heat of combustion Q being set at either zero or a value depending on fuel and mixture ratio. Figure 45 defines the notation for the solution.

The set of equations defines a locus of solutions in the p, ρ plane. The basic Hugoniot equation for the relation of pressure and density is hyperbolic in shape. In the notation of Fig. 45, the state of the unburned gas in front of the flame is station **0** and the burned gas behind the flame is station 00.

$$\frac{p_{00}}{p_0} = \frac{(\gamma + 1/\gamma - 1) + 2Q(\rho_0/p_0 - \rho_0/\rho_{00})}{(\gamma + 1/\gamma - 1)\dfrac{\rho_0}{\rho_{00}} - 1} \tag{H46}$$

A solution for a hydrogen air detonation is shown in Fig. H1.

The slope of the line from the initial state point to the final or solution point has a slope equal to minus the square of the mass flux density -$\mathbf{m}^2$ [Eq. (H42)]. Thus, only states with a negative slope are allowed; otherwise, the

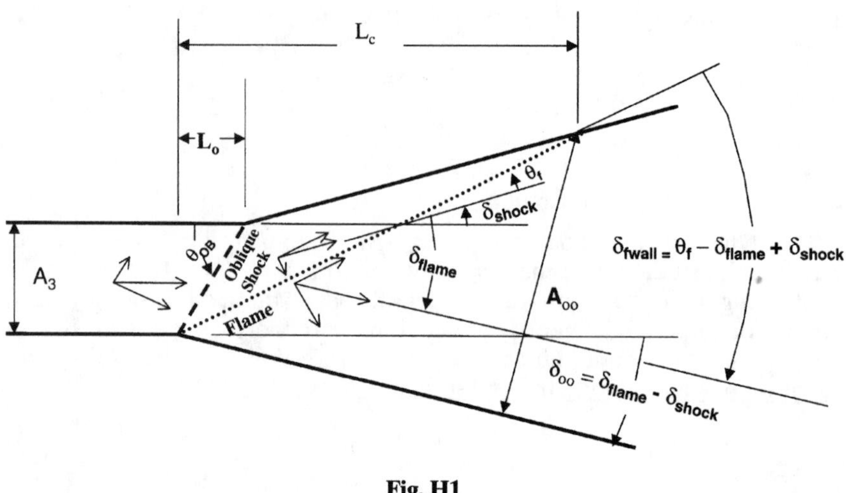

Fig. H1

solution corresponds to imaginary events. This fact neatly divides the diagram into supersonic or detonation branch and the subsonic or deflagration branch. Only the supersonic branch is shown. The shock solution goes through the initial point but the final burned state is displaced by the heat addition Q. The path of the solution is shown where the shock state is calculated first and then the deflagration state is found for the C-J condition.

$$\frac{\mathrm{d}p_{00}}{\mathrm{d}(1/\rho_{00})} = \frac{(p_{00} - p_0)}{(1/\rho_{00} - 1/\rho_0} \tag{H47}$$

In the two-wave process depicted in Fig. 45, the state o corresponds to the state behind an oblique shock and the slope of the line joining the initial point to the burned gas curve is the C-J slope, which is the minimum slope to the solution Hugoniot. Thus, to solve the oblique detonation wave, only the initial state pressure and density are required. The shock pressure rise allows the determination of the oblique shock angle since the Mach number is known for state 3 in Fig. 45. For every initial state o, the C-J condition allows for the unique C-J points to be found when the Hugoniot pressure density equation derivative is set equal to the right side of the C-J condition above.

$$\frac{\rho_0}{\rho_{00}} = 1 + \frac{(\gamma-1)}{\gamma}\frac{Q\rho_0}{p_0} \mp \frac{(\gamma-1)}{\gamma}\frac{Q\rho_0}{p_0}\sqrt{1 + \frac{2\gamma}{(\gamma^2-1)}\frac{p_0}{Q\rho_0}} \tag{H48}$$

Thus, for the ODWE, the solution is found in one step. Solutions for the combustor area ratio and wave angles are shown in Figs. H2 and H3. The

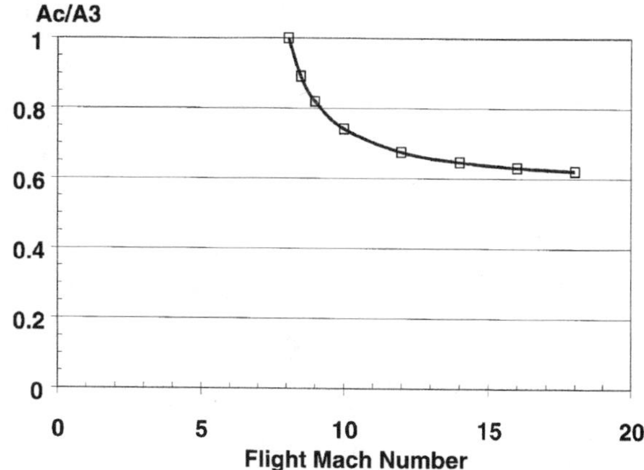

Fig. H2

inlet thermal compression ratio is $\psi = 5$ for H2 fuel and the wave angles and area are the functions of flight Mach number. The curves indicate that variable combustor geometry is required for flight from 8 to 15. Above 15, the area variations are minimal.

The solutions plotted were calculated using approximate curve fits for the ratio of specific heats ε and the thermal ratio ψ after combustion as follows.

First in the Hugoniot solutions, only differences in enthalpy are required and not the isentropic exponent. The heat addition Q is normalized by P/ρ and temperature or molecular weight do not appear. Thus, a correlation of ψ change due to combustion is required. For H2 and a generic hydrocarbon fuel, the following formulas are useful for representative calculations:

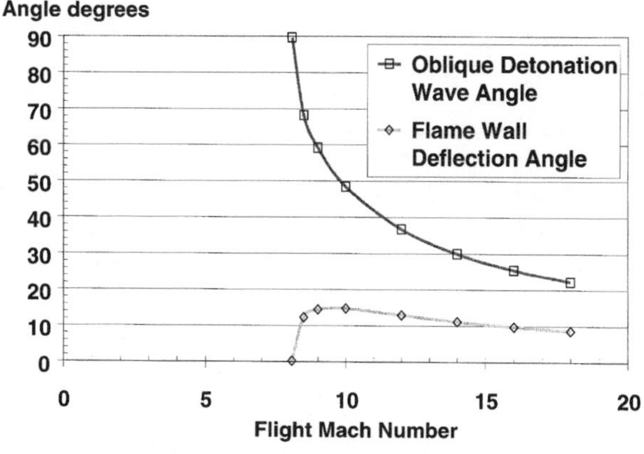

Fig. H3

Hydrogen:

$$\left(\frac{f}{a}\right)_{\text{st}} = 0.0298$$

$$\Delta\psi_{H2} = 12.2 \tag{H49}$$

$$\left(\frac{Q}{RT}\right)_{H2} = 55.0885 \tag{H50}$$

$$\delta_{H2} = 1.0 \tag{H51}$$

Hydrocarbon jet fuel:

$$\Delta\psi_{\text{HC}} = 10 \tag{H52}$$

$$\left(\frac{Q}{RT}\right)_{HC} = 46.69 \tag{H53}$$

$$\delta_{\text{HC}} = 1.7 \tag{H54}$$

The change of ψ depends on mixture ratio φ, initial value ψ_3, and absolute pressure P_3. The above fuel specific formulas are then applied in a set of easy steps as follows to find the ψ after combustion from the given φ,ψ,p.

$$g(\varphi) = \min\left\{\left(\varphi + \frac{\sin(\pi\varphi)}{\Delta\psi_f}\right), \left[1.05 + \frac{\sin(\pi 1.05)}{\Delta\psi_f} + \delta_f(\varphi - 1.05\right]\right\} \tag{H55}$$

$$\Delta\psi(1,p) = \Delta\psi_f + \frac{\ell n\left[p_{(\text{atm.})}\right]}{29}$$

$$\Pi(1,p) = \frac{1 + \ell n\left[p_{(\text{atm.})}\right]}{29} \tag{H56}$$

$$\mu(\varphi,p) = g(\varphi)\Pi(1,p) \tag{H57}$$

$$\Delta\psi(\varphi,p) = g(\varphi)\Delta\psi(1,p) \tag{H58}$$

$$\Delta\psi_c(\varphi,p) = \Delta\psi(\varphi,p) - \mu(\varphi,p)\psi_{oo} \tag{H59}$$

$$\psi_c = \psi_{oo} + \Delta\psi_c(\varphi, p) = \frac{(R_u T_c / Mw_c)}{(R_u T_o / Mw_o)} \tag{H60}$$

The ratio of specific heats is then found by solving the enthalpy difference due to combustion and the state change found above.

$$\frac{\gamma_c}{(\gamma_c - 1)} = \frac{(Q/RT)_{\text{fuel}} \min(1, \varphi)}{(1 + f/a)_{\text{St.}} (\psi_c - \psi_3)} \tag{H61}$$

Note that these thermodynamic relations are accurate for Hugoniot solutions and are not accurate for isentropic expansions in nozzles.

Some useful solution relations for angles are given for reference. The solution of the density ratio given shock pressure ratio allows for solution of the velocity normal to the oblique shock. This enables the shock wave angle and deflection angle to be obtained easily. The relation also is useful for the constant area supersonic combustor cycle solution, since u_0 is given.

$$\frac{U_0^2}{(p_0/\rho_0)} = \left(\frac{p_{00}/p_0 - 1}{1 - \rho_0/\rho_{00}}\right) \tag{H62}$$

When the wave is a C-J wave, the normal velocity U_{00} also is known as follows.

$$\sin(\theta) = \frac{\rho_{00} u_{00}}{\rho_0 U_0} \tag{H63}$$

$$\tan(\theta - \delta) = \frac{u_{00}}{v_{00}} = \frac{\rho_0}{\rho_0} \tan(\theta) \tag{H64}$$

$$\tan(\theta - \delta) = \frac{u_{00}}{v_{00}} = \frac{\rho_0}{\rho_0} \tan(\theta) \tag{H65}$$

$$u_{00}^2 = a_{00}^2 = \gamma \frac{p_{00}}{\rho_{00}} \tag{H66}$$

$$M_{00}^2 = 1 + \frac{v_{00}^2}{u_{00}^2} = 1 + \left[\frac{\rho_{00}}{\rho_0} \frac{1}{\tan(\theta)}\right]^2 \tag{H67}$$

The relations for combustor lengths are easily found from the above angles.

$$\left(\frac{U_{00}}{U_0}\right)^2 = \left(\frac{\rho_0}{\rho_{00}}\right)^2 \sin^2(\theta) + \cos^2(\theta) \tag{H68}$$

$$\frac{A_{00}}{A_3} = \frac{\rho_3}{\rho_{00}}\frac{U_3}{U_{00}} \tag{H69}$$

$$\frac{L_0}{A_3} = \frac{1}{\tan(\theta)} \tag{H70}$$

$$\frac{L_c}{A_3} = \frac{A_{00}}{A_3}\frac{1}{\tan(\theta_{00} - \delta_{00})} \tag{H71 }$$

This concludes the brief review of Hugoniot equations and detonation wave combustion cycles. The detonation cycles are considered advanced and practically impossible to develop since no facility is available for developing the flowpath. The requirements would have to include sufficient size to contain an inlet, mixing scheme, and combustor for testing at $M = 12$ to 15 enthalpy with long test times relative to most pulse facilities capable of reaching the pressure and enthalpy levels required.

Appendix I: Base Pressure Estimate

A. Required Pressure at Reattachment

The first task is to relate rate of combustion and combustion efficiency to the reattachment angle and pressure required to turn the core flow parallel to the combustor wall (Appendix C, Fig. C6).

At the beginning, the pressure in the base is unknown, therefore the core flow is assumed to be expanded or compressed, depending on whether P_b/P_{iso} is less than or greater than one. A useful reference state for the core flow is the constant pressure mixing solution of the main fuel and air at any pressure P_b in an area A_{mix}.

Then for mixing the perfect gases, assuming that U_{mix} and $h_{,\text{ix}}$ are known from the mixing conservation sums, the next step is to find the burn state at constant pressure, P_b as follows:

$$U_{\text{mix}} = \left[\frac{U_{\text{iso}} + (f/a)U_f}{1 + f/a}\right] \tag{I1a}$$

$$h_{\text{mix}} = \left(\frac{H_{\text{To}} + (f/a)H_{\text{Tf}}}{1 + f/a}\right) - \frac{U_{\text{mix}}^2}{2} \tag{I1b}$$

$$\psi_{\text{mix}} = \frac{h_{\text{mix}}}{h_o}\frac{(\gamma/\gamma - 1)_o}{(\gamma / \gamma - 1)_{\text{mix}}} \tag{I1c}$$

The burn gas, ψ_c, can be found by following the procedure given in Appendix H for Hugoniot equation solutions, with

$$\left(\frac{P}{\rho}\right)_{\text{BURN}} = \left(\frac{RT}{M_{\text{WT}}}\right)_{\text{BURN}} \tag{I2}$$

The burn area for the assumed P_b is solved from mass conservation,

$$A_{\text{BURN},P_B} = \frac{\dot{m}\left(1+\frac{f}{a}\right)}{V_{\text{mix}}P_B}\left(\frac{RT}{M}\right)_{\text{BURN}} \tag{I3}$$

One should compare the calculated combustor area, A_{burn}, with the assured P_b with the given combustor area for a check, thereby establishing an upper limit on P_b or the assumed value of P_b should be reduced.

Let us define combustion efficiency for the constant pressure and velocity zone over the separation bubble.

$$\eta_{\text{BC}} \equiv \frac{A - A_{\text{MIX}}}{A_{\text{BURN}} - A_{\text{MIX}}} \tag{I4}$$

Perhaps surprisingly, the combustion efficiency in the constant pressure region becomes known at reattachment where $A = A_c$.

$$\eta_{\text{BC},R} = \frac{A_c - A_{\text{MIX}}}{A_{\text{BURN}} - A_{\text{MIX}}} \tag{I5}$$

From the mixing results, the combustion mass fraction also is a function of length. Thus,

$$\eta_{\text{BC}} = \frac{\left(\frac{A_o}{A_{\text{Ref}}}\right) + S\left(\frac{x}{H_k}\right)}{\varphi^j C} \tag{I6}$$

Where S is the slope of the mixing rate recipe and C is the capture area per ramp.

$$S = \sqrt{C_{\text{De}}}\left(\frac{g}{W_k}\right) \tag{I7}$$

and

$$C = \frac{A_{\mathrm{COMB}}}{n_R A_{\mathrm{RAMP}}} \tag{I8}$$

In Eq. (I6)

$$j = 1 \qquad \text{for } \varphi < 1 \text{ and zero for } \varphi > 1$$

Thus, the length of the separation region or reattachment length is found by equating $\eta_{c\mathrm{R}}$ and η_c, and

$$\frac{X_R}{H_k} = \frac{\varphi^j \eta_{\mathrm{BC},R} C - A_o/A_{\mathrm{ref}}}{S} \tag{I9}$$

The reattachment length of the shear layer over the separation bubble is important for calculating the thickness and mass flows in the shear layer.

The pressure rise at reattachment is found from the flow turning angle of the core flow at the wall. The flow angle is

$$\frac{\mathrm{d}A_{\mathrm{burn}}}{\mathrm{d}x} = W_c \tan\theta_R \tag{I10}$$

The rate of change of η_c is found from the geometric relation

$$\left(\frac{\mathrm{d}\eta_{\mathrm{BC}}}{\mathrm{d}x}\right)_{\mathrm{geo}} = \frac{W_c \tan(\theta_R)}{A_{\mathrm{burn}} - A_{\mathrm{mix}}} \tag{I11}$$

The rate of change of η_{BC} from the mixing recipe is

$$\left(\frac{\mathrm{d}\eta_{\mathrm{BC}}}{\mathrm{d}x}\right)_{\mathrm{mix}} = \frac{S}{\varphi^j C}\frac{1}{H_k} \tag{I12}$$

The reattachment flow turning angle is then given by

$$\theta_R = \arctan\left[\frac{S(A_{\mathrm{burn}} - A_{\mathrm{mix}})}{\varphi^j C H_k W_c}\right] \tag{I13}$$

The pressure rise at reattachment is found from the Prandtl–Meyer compression at the turn angle θ_R and core Mach number. The core velocity is the same as the mix gas velocity so that Mach number will be determined by finding the speed of sound. Using the completion of heat release

$$\left(\frac{P}{\rho}\right)_R = \left(\frac{RT}{M}\right)_R = \left(\frac{P}{\rho}\right)_{\text{MIX}} + \eta_{cR}\left[\left(\frac{P}{\rho}\right)_{\text{BURN}} - \left(\frac{P}{\rho}\right)_{\text{MIX}}\right] \tag{I14}$$

$$M_R = \frac{U_{\text{MIX}}}{\sqrt{\gamma\left(\frac{P}{\rho}\right)_R}} \tag{I15}$$

Thus, M and θ_R, which are needed for determining the pressure ratio required at reattachment, are known.

For simplicity, the Shapiro series expansion for Prandtl-Meyer pressure coefficient is presented here to illustrate that the pressure ratio required is closed for any P_b since the Prandtl-Meyer solution is implicit.

$$\frac{P_R}{P_B} = \left(\frac{\gamma}{2} K_R^2\right)\left(\frac{2}{K_R} + \frac{\gamma+1}{2} + \frac{\gamma+1}{6} K_R\right) + 1 \tag{I16}$$

where

$$K_R = \sqrt{M_R^2 - 1}\tan\theta_R$$

and θ_R = negative in expansion.

With the procedure for determining the pressure ratio required at the location where reattachment is complete, the following section will help determine the pressure ratio available from the total pressure ratio available on the reattaching streamline.

In the following section, using shear layer convention, $U_1 = U_{\text{MIX}}$, $\rho_1 = \rho_{\text{BURN}}$, and $\gamma_1, = \gamma_{\text{BURN}}$.

The Chapman–Korst model, where one ignores any viscous effects on the reattachment streamline in the region of recompression, the pressure at reattachment is equal to the stagnation pressure of the reattaching streamline, viz.,

$$\frac{P_R}{P_B} = \left[1 + \left(\frac{\gamma-1}{2}\right)M_{\text{RS}}^2\right]^{\frac{\gamma}{\gamma-1}} \tag{I17}$$

where the Mach number on the reattaching streamline in the shear layer is

$$M_{\text{RS}} = \left[\sqrt{\frac{\rho_{\text{RS}}}{\rho_R}}\sqrt{\frac{\gamma_R}{\gamma_{\text{RS}}}}\left(\frac{U_{\text{RS}}}{U_1}\right)\right]M_R \tag{I18}$$

The shear layer is modeled as a self-similar constant pressure shear layer developing without initial boundary layer effects. The velocity profile is taken to be the same as the wake function

$$\frac{U}{U_1} = \frac{1}{2}\left(1 + \cos\frac{\pi y}{\delta}\right) \tag{I19}$$

where y is measured from the outer edge, $U = U_{\text{mix}}$ toward the recirculation eddy region, and $U\delta = 0, \delta$ being the shear layer thickness.

The dividing streamline is established first. Note that by definition all high-speed mass entrained in the shear layer is contained between the edge and Y_{DSL}.

$$\int_o^{Y_{\text{DSL}}} \rho U \,\mathrm{d}y = \dot{m}_1 \tag{I20}$$

Furthermore, all momentum in the shear layer comes from the high-speed side. Total momentum flux in shear layer

$$\int_0^{Y_{\text{DSL}}} \rho U^2 \mathrm{d}y = \dot{m}_1 U_1 \tag{I21}$$

From shear layer momentum conservation,

$$\int_o^{\eta_{\text{DSL}}} \tilde{\rho}\tilde{U}\,\mathrm{d}_\eta = \int_o^1 \tilde{\rho}\tilde{U}^2\,\mathrm{d}\eta \tag{I22}$$

where

$$\tilde{\rho} = \rho/\rho_{\text{BURN}} = \rho/\rho_1 \tag{I23}$$

$$\tilde{U} = U/U_1 \tag{I24}$$

$$\eta = \pi y/\delta \tag{I25}$$

In Eq. (I22), η_{DSL} thus specifies the location of the dividing streamline.

The determination of the dividing streamline just performed also is the method for finding other important shear layer coordinates from known conservation laws. These flux conservation conditions rely on profile integrals, hence, the name for the theory is profile theory. The goal of finding the pressure rise available is achieved when the value of the reattaching streamline is found. The total pressure is found as a property of the flow at that location. The task is to find the low speed or eddy stagnation conditions to actualize the properties at that point in the shear layer. There are enough conservation conditions to find these low-speed boundary conditions.

The high-speed mass also is equal to the integral of the total mass flux, times the high-speed fluid mass fraction across the shear layer. We would identify with the high-speed mass fraction. While it is not proof, it is consistent with a Crocco integral of the species conservation for unity turbulent Prandtl and Scmidt number. So for later use write,

$$Y_1 = \tilde{U}$$

$$Y_2 = 1 - \tilde{U}$$

The shear layer thickness δ can be found from the shear layer length and low-speed momentum since all momentum on the low-speed side is equal to the shear stress on the dividing streamline line (DSL) times length of the DSL.

$$\int_{Y_{DSL}}^{\delta} \rho U^2 \, dy = \tau_{DSL} X_R \tag{I26}$$

By normalizing shear stress at reattachment, the shear layer thickness, δ_R, which is a constant dependent on X_R linearly, and, therefore, P_B linearly.

$$\delta_R = \frac{\left(\tau_{DSL}/\rho_1 U_1^2\right) X_R}{\int_{\eta_{DSL}}^{1} \tilde{\rho}\tilde{U}^2 \, d\eta} \tag{I27}$$

$$\frac{\tau_{DSL}}{\rho_1 U_1^2} = K_o \chi$$

$$K_o = .01234$$

$$\chi = 1 - .415M \qquad 0 \le M \le 1$$

$$\chi = .585/M \qquad M \ge 1$$

Normalizing forms an implicit solution for η_R, which is a strong function of P_B/P_{TR} since density and δ_R give us a second power dependence on P_B.

The mass fraction of fuel in the base can be solved. The reattaching streamline can be found by conserving injected mass flow that must be contained, by definition, between the dividing streamline and the reattaching streamline.

$$\int_{Y_{\mathrm{DSL}}}^{Y_R} \rho U \,\mathrm{d}y = \dot{m}_{fB} \tag{I28}$$

$$\int_{\eta_{\mathrm{DSL}}}^{\eta_R} \tilde{\rho}\tilde{U} \,\mathrm{d}\eta = \frac{\dot{m}_{fB}}{\rho_1 U_1 \delta_R} \tag{I29}$$

Once η_R is known, the problem is closed and the pressure rise available can be found. However, the solution is implicit and depends on the as yet unknown total base enthalpy.

The fluid in the base contains direct base injection fluid, and the fluid from the reentrant jet from the reattaching streamline Y_{DSL} to η_R. The reentrant jet contains high-speed fluid and base-fluid injectant. Thus, the base-eddy flow contains high-speed combustion gas and base-injection fluid. The low-speed mass fraction $Y2$ at any point is part base fuel and part high-speed combustion gas, while the high-speed mass fraction $Y1$ contains only speed gas. Note that the mass fraction of base flow *above* the reattaching streamline is

$$Y_f = Y_{\mathrm{fB}}\, Y_2 \tag{I30}$$

Y_{fB} is found in the mass balance of base injection fluid by using the relation between the velocity and the low-speed mass fraction in the shear layer.

$$Y_{\mathrm{fB}} = \frac{\int_{\eta_{\mathrm{DSL}}}^{\eta_R} \tilde{\rho}\tilde{U} \,\mathrm{d}\eta}{\int_o^{\eta_R} \tilde{\rho}\tilde{U}(1-\tilde{U}) \,\mathrm{d}\eta} \tag{I31}$$

The global energy balance for the base flow and shear layer is used to solve for the recirculation zone total enthalpy. Write conservation of the flow into the shear layer that flows downstream after the reattachment point from the high-speed side and the base fuel injected below the dividing streamline.

Mass balance:

$$\dot{m}_{\mathrm{OUT}} = \dot{m}_{\mathrm{fB}} + \dot{m}_1 = \int_o^{Y_R} \rho U \,\mathrm{d}y \tag{I32}$$

Energy balance:

$$\dot{m}_{\mathrm{OUT}}\, H_{\mathrm{OUT}} = \dot{m}_{\mathrm{fB}}\, H_f + \dot{m}_1\, H_1 - \dot{Q}_w \tag{I33}$$

and

$$\dot{m}_{\mathrm{OUT}}\, H_{\mathrm{OUT}} = \int_{o}^{Y_R} \rho U H \,\mathrm{d}y \tag{I34}$$

Crocco profile relation of the total enthalpy is needed to evaluate the integrals.

$$\tilde{H} = \frac{H}{H_1} = \frac{H_B}{H_1} + \frac{(H_1 - H_B)}{H_1}\tilde{U} \tag{I35}$$

The integral of total enthalpy enables a formalism to obtain the implicit solutions in a few steps.

$$H_{\mathrm{OUT}} = H_B + \frac{(H_1 - H_B)\rho_1 U_1 \delta_R}{\dot{m}_1 + \dot{m}_{fB}} \int_{o}^{\eta R} \tilde{\rho} U^2 \,\mathrm{d}\eta \tag{I36}$$

And, as defined, it is true that

$$\rho_1 U_1 \delta_R = \frac{\dot{m}_1}{\displaystyle\int_{o}^{1} \tilde{\rho}\tilde{U}^2 \,\mathrm{d}\eta} \tag{I37}$$

defines a loss function by inspection of the enthalpy integral.

$$\Delta = 1 - \frac{\displaystyle\int_{o}^{\eta} \tilde{\rho}\tilde{U}^2 \,\mathrm{d}\eta}{\displaystyle\int_{o}^{1} \tilde{\rho}\tilde{U}^2 \,\mathrm{d}\eta} \tag{I38}$$

Note from this definition that delta is a function of Y_{fB}.

$$\Delta = \frac{\dot{m}_{\mathrm{fB}}}{\dot{m}_1} \frac{(1 - Y_{\mathrm{fB}})}{Y_{\mathrm{fB}}} \tag{I39}$$

Solve the energy balance for H_B:

$$H_B = \frac{\dot{m}_{\mathrm{fB}} H_f + \dot{m}_1 \Delta H_1 - \dot{Q}_w}{\dot{m}_f + \dot{m}_1 \Delta} \tag{I40}$$

This form can be converted to mass fraction dependence. Inserting the Δ function, find

$$H_B = Y_{\text{fB}} H_f + (1 - Y_{\text{fB}}) H_1 - \frac{\dot{Q}_w Y_{\text{fB}}}{\dot{m}_{\text{fB}}} \tag{I41}$$

To evaluate these formulas, the auxiliary relations for molecular weight of base gas are

$$Mw_B = \left(\frac{Y_{\text{fB}}}{Mw_f} + \frac{(1 - Y_{\text{fB}})}{Mw_1} \right)^{-1} \tag{I42}$$

and the ratio of specific heats in base,

$$\frac{\gamma_B}{\gamma_B - 1} = Y_{\text{fB}} \left(\frac{Mw_B}{Mw_f} \right) \left(\frac{\gamma}{\gamma - 1} \right)_f + (1 - Y_{\text{fB}}) \left(\frac{Mw_B}{Mw_1} \right) \left(\frac{\gamma}{\gamma - 1} \right)_1 \tag{I43}$$

The local specific heat at any point in the shear layer is a function of U and thus η

$$\left(\frac{\gamma}{\gamma - 1} \right) = \frac{(\gamma/\gamma - 1)_1 (\tilde{U}/Mw_1) + (\gamma/\gamma - 1)_B [1 - \tilde{U}/MwB]}{\tilde{U}/Mw_1 + [1 - \tilde{U}/MwB]} \tag{I44}$$

The local density at any point is also just a function of U

$$\tilde{\rho} = \frac{(\gamma/\gamma - 1)}{(\gamma/\gamma - 1)_1} \left[1 \Big/ \frac{h_B}{h_1} + \frac{(H_{\text{MIX}} - H_B)}{h_1} \tilde{U} - \frac{(\gamma - 1)}{2} M_R^2 \tilde{U}^2 \right] \tag{I45}$$

This set of equations can best be solved iteratively by guessing H_{B} or Y_{fB}. All the properties can be evaluated. Check H_B from the energy balance and adjust the guess H_B; this procedure converges quickly.

The subject of heat transfer must now be treated to close the problem, since Q is part of the energy balance. Eckert measured base heat transfer in an experiment in which he showed the base heat transfer was really unhooked from H_B. This was surprising, since Chapman closed by setting $H_B = Hw$. Eckert showed that $H_B = Hw$ closure was a large error. Instead, he found $H_B = H_{\text{MIX}}$. Base heating was very low and approximately the same as a laminar boundary layer with a stagnation point at R with total pressure equal to P_R. Further, he showed that a small base mass flow reduced even this low level to zero. The mass flow needed to reduce Q to zero is far less than practical base pressure control flow rates and thus, Q_w is negligible as far as the eddy energy balance is concerned.

On the attached flow side of the reattachment point, the heat transfer is proportional to P_R to the 0.85th power. This is the normal scaling employed throughout the flowpath.

B. Closure

The problem is closed when the pressure available from the shear layer calculation matches the pressure required to turn the flow parallel in the wall. In the analysis, base pressure is the only free parameter for a given base bleed flow.

NOMENCLATURE FOR FLOWPATH COMPONENT SPECIFICATIONs

Section 1 Forebody:

Rn = vehicle nose radius
$(\theta_{FB})\ 1$ = initial forebody ramp or cone angle
$(\theta_{FB})\ 1.n$ = intermediate angles
$(\theta_{FB})\ 2$ = final forebody angle at inlet face
$L_{FB.n}$ = forebody length from nose to turn n
(Σ_{FB}) = forebody wetted surface area
T_{WFB} = forebody wall material maximum operating temperature

Section 2 Cowl:

AO = projected frontal area of forebody, including the cowl, often equal to the maximum freestream capture area of the flowpath at zero angle of attack
A_{SPILL} = open area of cowl viewed from bottom due to shortened cowl or notch
R_{CT} = radius of curvature of cowl turn
θ_{CT} = cowl turn angle
R_{LEC} = cowl leading edge radius
Λ_{CLE} = cowl leading edge sweep angle viewed from bottom
θ_{CLE} = cowl leading edge external turn angle
$\theta_{C.n}$ = cowl external angle intermediate to leading edge and start of cowl turn
θ_{CTE} = cowl trailing edge angle (often equal to $\theta_{C.n} - \theta_{CT}$)
A_{FC} = cowl frontal area (often equal to $A_E - A_o$)
S_{COWL} = cowl external surface area
T_{WCOWL} = cowl wall material maximum operating temperature

Section 3 Inlet:

Λ_{SW} = inlet sidewall leading edge sweep angle
R_{ILE} = inlet leading edge radius
θ_{ISW} = inlet sidewall compression angle
θ_{ICW} = inlet internal cowl compression angle
θ_{IRW} = inlet internal ramp compression angle
θ_{IE} = inlet exit wall angle
CR = inlet contraction ratio
H_1 = cowl height at inlet face normal to forebody
W_1 = inlet width
S_{INLET} = inlet wetted surface area
T_{WINLET} = inlet wall material maximum operating temperature

Section 4 Isolator: (In an RBCC engine, the isolator may contain some geometric specifications for combustor primary rocket thrusters and scramjet stirring and fuel injectors.)

X_{BLB} = mass fraction of isolator flow diverted to boundary-layer bleed system
θ_{BLDB} = boundary layer diverter wall angle relative to isolator axis
AR_2 = isolator aspect ratio equal to inlet exit aspect ratio
$(L/W)_{ISO}$ = isolator length to width ratio
N_R = number of ramps or primary rockets
θ = compression angle of ramps
b_R = ramp aspect ratio equal to the ramp height/ramp width
B_R = ramp virtual blockage equal to N_R* ramp height* ramp width/A_{ISO}
N_{SCRAM} = number of scram injectors
A_{ESCRAM} = exit area of scram injectors
$A_{EROCKET}$ = exit area of primary rockets
ε_R = expansion ratio of rocket thruster nozzles
ε_{SCRAM} = expansion ratio of scram injector nozzles
T_{WISO} = isolator wall material maximum operating temperature

Section 5: Combustor [For combined cycle engines, there are two general combustor stages to accommodate heat choking at low speed; the first stage is assumed to be a step and constant area section (CAC), and the second stage is a divergent section (CMC)].

Section 5a CAC Combustor:

ε_{CAC} = combustor step expansion ratio equal to A_C/A_{ISO} (often equal to combustor width/isolator width)
AR_C = combustor aspect ratio equal to combustor height/combustor width
$(L/W)_C$ = combustor length/combustor width
N_{BASE} = number of base fuel injectors
ε_{BASE} = expansion ratio of base fuel nozzles
A_{BASE} = base fuel injector area
A_{FILM} = film cooling injector area
ε_{FILM} = expansion ratio of film coolant nozzles
S_{CAC} = surface area of CAC combustor
T_{WCAC} = CAC combustor wall material maximum operating temperature

Section 5b CMC Combustor:

ε_{CMC} = divergent combustor expansion ratio equal to A_5/A_{CAC}
AR_{CM} = combustor aspect ratio equal to combustor height/combustor width
$(L/W)_C$ = combustor length/combustor width
θ_{CMC} = effective expansion angle of CMC combustor

$(\theta_R)_{\text{CMC}}$ = ramp wall expansion angle of CMC combustor
$(\theta_{\text{SW}})_{\text{CMC}}$ = sidewall expansion angle of CMC combustor
$(\theta_C)_{\text{CMC}}$ = cowl wall expansion angle of CMC combustor
S_{CMC} = surface area of CMC combustor
T_{WCMC} = CMC combustor wall material maximum operating temperature

Section 6 Internal Nozzle:

ε_N = internal nozzle expansion ratio A_E/A_O
$(\theta_R)_N$ = internal nozzle ramp wall angle
$(\theta_{\text{SW}})_N$ = internal nozzle sidewall angle
$(\theta_C)_N$ = internal nozzle cowl wall angle
S_N = surface area of internal nozzle
T_{WN} = internal nozzle wall material maximum operating temperature

Section 7 Aftbody Nozzle:

ε_{O} = external nozzle overall expansion ratio A_E/A_O
ε_{AB} = external nozzle expansion ratio A_E/A_6
$(\theta_R)_{\text{ABI}}$ = external nozzle initial ramp wall angle
$(\theta_R)_{\text{ABE}}$ = external nozzle final ramp wall angle
S_N = surface area of internal nozzle
T_{WAB} = aftbody nozzle wall material maximum operating temperature

CR_{OS} The freestream to inlet exit contraction ratio on the over-speed side
CR_{INT} The forebody to inlet exit contraction ratio on the forebody and internal contraction side
CR_{FB} The freestream to forebody contraction ratio
A_{1OS} The inlet face area on the overspeed side
A_{2OS} The flow area at the inlet exit on the over-speed side
A_{1INT} The inlet face area on the forebody and internal contraction side
A_{2INT} The flow area at the inlet exit on the forebody and internal contraction side
R_{B1} The normal distance to the forebody at the inlet face
R_{C1} The normal distance to the cowl at the inlet face
R_{SL1} The normal distance to the slipline at the inlet face
R_{B2} The normal distance to the ramp at the inlet exit
R_{C2} The normal distance to the cowl at the inlet exit
R_{SL2} The normal distance to the slipline at the inlet exit
W_{B1} The width of the inlet on the forebody
W_{C1} The width of the inlet at the cowl
W_{SL1} The width of the inlet at the slipline
W_2 The width of the inlet at the inlet exit
W_{OS} The average width of the inlet face on the over-speed side

W_{1INT}	The average width of the inlet face on the forebody and internal contraction side
X_{B1}	The length to the inlet face on the forebody
X_{C1}	The length to the inlet face at the cowl
X_{SL1}	The length to the inlet face slipline
X_{SL2}	The length of the slipline
L_C	The cowl length
L_{SL}	The cowl length
θ_{FB}	The forebody angle relative to the freestream
θ_R	The inlet ramp angle relative to the freestream
θ_C	The cowl internal angle relative to the freestream
θ_{SL}	The slipline angle relative to the freestream
θ_{SW}	The inlet sidewall angle relative to the freestream

PROGRESS IN ASTRONAUTICS AND AERONAUTICS SERIES VOLUMES

***1. Solid Propellant Rocket Research (1960)**
Martin Summerfield
Princeton University

***2. Liquid Rockets and Propellants (1960)**
Loren E. Bollinger
Ohio State University
Martin Goldsmith
The Rand Corp.
Alexis W. Lemmon Jr.
Battelle Memorial Institute

***3. Energy Conversion for Space Power (1961)**
Nathan W. Snyder
Institute for Defense Analyses

***4. Space Power Systems (1961)**
Nathan W. Snyder
Institute for Defense Analyses

***5. Electrostatic Propulsion (1961)**
David B. Langmuir
Space Technology Laboratories, Inc.
Ernst Stuhlinger
NASA George C. Marshall Space Flight Center
J. M. Sellen Jr.
Space Technology Laboratories, Inc.

***6. Detonation and Two-Phase Flow (1962)**
S. S. Penner
California Institute of Technology
F. A. Williams
Harvard University

***7. Hypersonic Flow Research (1962)**
Frederick R. Riddell
AVCO Corp.

***8. Guidance and Control (1962)**
Robert E. Roberson
Consultant
James S. Farrior
Lockheed Missiles and Space Co.

***9. Electric Propulsion Development (1963)**
Ernst Stuhlinger
NASA George C. Marshall Space Flight Center

***10. Technology of Lunar Exploration (1963)**
Clifford I. Cumming
Harold R. Lawrence
Jet Propulsion Laboratory

***11. Power Systems for Space Flight (1963)**
Morris A. Zipkin
Russell N. Edwards
General Electric Co.

***12. Ionization in High-Temperature Gases (1963)**
Kurt E. Shuler, Editor
National Bureau of Standards
John B. Fenn, Associate Editor
Princeton University

***13. Guidance and Control–II (1964)**
Robert C. Langford
General Precision Inc.
Charles J. Mundo
Institute of Naval Studies

***14. Celestial Mechanics and Astrodynamics (1964)**
Victor G. Szebehely
Yale University Observatory

***15. Heterogeneous Combustion (1964)**
Hans G. Wolfhard
Institute for Defense Analyses
Irvin Glassman
Princeton University
Leon Green Jr.
Air Force Systems Command

***16. Space Power Systems Engineering (1966)**
George C. Szego
Institute for Defense Analyses
J. Edward Taylor
TRW Inc.

***17. Methods in Astrodynamics and Celestial Mechanics (1966)**
Raynor L. Duncombe
U.S. Naval Observatory
Victor G. Szebehely
Yale University Observatory

***18. Thermophysics and Temperature Control of Spacecraft and Entry Vehicles (1966)**
Gerhard B. Heller
NASA George C. Marshall Space Flight Center

***19. Communication Satellite Systems Technology (1966)**
Richard B. Marsten
Radio Corporation of America

---
*Out of print.

***20. Thermophysics of Spacecraft and Planetary Bodies: Radiation Properties of Solids and the Electromagnetic Radiation Environment in Space (1967)**
Gerhard B. Heller
NASA George C. Marshall Space Flight Center

***21. Thermal Design Principles of Spacecraft and Entry Bodies (1969)**
Jerry T. Bevans
TRW Systems

***22. Stratospheric Circulation (1969)**
Willis L. Webb
Atmospheric Sciences Laboratory, White Sands, and University of Texas at El Paso

***23. Thermophysics: Applications to Thermal Design of Spacecraft (1970)**
Jerry T. Bevans
TRW Systems

***24. Heat Transfer and Spacecraft Thermal Control (1971)**
John W. Lucas
Jet Propulsion Laboratory

25. Communication Satellites for the 70's: Technology (1971)
Nathaniel E. Feldman
The Rand Corp.
Charles M. Kelly
The Aerospace Corp.

26. Communication Satellites for the 70's: Systems (1971)
Nathaniel E. Feldman
The Rand Corp.
Charles M. Kelly
The Aerospace Corp.

27. Thermospheric Circulation (1972)
Willis L. Webb
Atmospheric Sciences Laboratory, White Sands, and University of Texas at El Paso

28. Thermal Characteristics of the Moon (1972)
John W. Lucas
Jet Propulsion Laboratory

***29. Fundamentals of Spacecraft Thermal Design (1972)**
John W. Lucas
Jet Propulsion Laboratory

***30. Solar Activity Observations and Predictions (1972)**
Patrick S. McIntosh
Murray Dryer
Environmental Research Laboratories, National Oceanic and Atmospheric Administration

***31. Thermal Control and Radiation (1973)**
Chang-Lin Tien
University of California at Berkeley

***32. Communications Satellite Systems (1974)**
P. L. Bargellini
COMSAT Laboratories

***33. Communications Satellite Technology (1974)**
P. L. Bargellini
COMSAT Laboratories

***34. Instrumentation for Airbreathing Propulsion (1974)**
Allen E. Fuhs
Naval Postgraduate School
Marshall Kingery
Arnold Engineering Development Center

***35. Thermophysics and Spacecraft Thermal Control (1974)**
Robert G. Hering
University of Iowa

36. Thermal Pollution Analysis (1975)
Joseph A. Schetz
Virginia Polytechnic Institute
ISBN 0-915928-00-0

***37. Aeroacoustics: Jet and Combustion Noise; Duct Acoustics (1975)**
Henry T. Nagamatsu, Editor
General Electric Research and Development Center
Jack V. O'Keefe, Associate Editor
The Boeing Co.
Ira R. Schwartz, Associate Editor
NASA Ames Research Center
ISBN 0-915928-01-9

***38. Aeroacoustics: Fan, STOL, and Boundary Layer Noise; Sonic Boom; Aeroacoustics Instrumentation (1975)**
Henry T. Nagamatsu, Editor
General Electric Research and Development Center
Jack V. O'Keefe, Associate Editor
The Boeing Co.
Ira R. Schwartz, Associate Editor
NASA Ames Research Center
ISBN 0-915928-02-7

***39. Heat Transfer with Thermal Control Applications (1975)**
M. Michael Yovanovich
University of Waterloo
ISBN 0-915928-03-5

*Out of print.

***40. Aerodynamics of Base Combustion (1976)**
S. N. B. Murthy, Editor
J. R. Osborn,
Associate Editor
Purdue University
A. W. Barrows
J. R. Ward,
Associate Editors
Ballistics Research Laboratories
ISBN 0-915928-04-3

***41. Communications Satellite Developments: Systems (1976)**
Gilbert E. LaVean
Defense Communications Agency
William G. Schmidt
CML Satellite Corp.
ISBN 0-915928-05-1

***42. Communications Satellite Developments: Technology (1976)**
William G. Schmidt
CML Satellite Corp.
Gilbert E. LaVean
Defense Communications Agency
ISBN 0-915928-06-X

***43. Aeroacoustics: Jet Noise, Combustion and Core Engine Noise (1976)**
Ira R. Schwartz, Editor
NASA Ames Research Center
Henry T. Nagamatsu,
Associate Editor
General Electric Research and Development Center
Warren C. Strahle,
Associate Editor
Georgia Institute of Technology
ISBN 0-915928-07-8

***44. Aeroacoustics: Fan Noise and Control; Duct Acoustics; Rotor Noise (1976)**
Ira R. Schwartz, Editor
NASA Ames Research Center
Henry T. Nagamatsu,
Associate Editor
General Electric Research and Development Center
Warren C. Strahle,
Associate Editor
Georgia Institute of Technology
ISBN 0-915928-08-6

***45. Aeroacoustics: STOL Noise; Airframe and Airfoil Noise (1976)**
Ira R. Schwartz, Editor
NASA Ames Research Center
Henry T. Nagamatsu,
Associate Editor
General Electric Research and Development Center
Warren C. Strahle,
Associate Editor
Georgia Institute of Technology
ISBN 0-915928-09-4

***46. Aeroacoustics: Acoustic Wave Propagation; Aircraft Noise Prediction; Aeroacoustic Instrumentation (1976)**
Ira R. Schwartz, Editor
NASA Ames Research Center
Henry T. Nagamatsu,
Associate Editor
General Electric Research and Development Center
Warren C. Strahle,
Associate Editor
Georgia Institute of Technology
ISBN 0-915928-10-8

***47. Spacecraft Charging by Magnetospheric Plasmas (1976)**
Alan Rosen
TRW Inc.
ISBN 0-915928-11-6

***48. Scientific Investigations on the Skylab Satellite (1976)**
Marion I. Kent
Ernst Stuhlinger
NASA George C. Marshall Space Flight Center
Shi-Tsan Wu
University of Alabama
ISBN 0-915928-12-4

***49. Radiative Transfer and Thermal Control (1976)**
Allie M. Smith
ARO Inc.
ISBN 0-915928-13-2

***50. Exploration of the Outer Solar System (1976)**
Eugene W. Greenstadt
TRW Inc.
Murray Dryer
National Oceanic and Atmospheric Administration
Devrie S. Intriligator
University of Southern California
ISBN 0-915928-14-0

***51. Rarefied Gas Dynamics, Parts I and II (two volumes) (1977)**
J. Leith Potter
ARO Inc.
ISBN 0-915928-15-9

***52. Materials Sciences in Space with Application to Space Processing (1977)**
Leo Steg
General Electric Co.
ISBN 0-915928-16-7

---

*Out of print.

***53. Experimental Diagnostics in Gas Phase Combustion Systems (1977)**
Ben T. Zinn, Editor
Georgia Institute of Technology
Craig T. Bowman, Associate Editor
Stanford University
Daniel L. Hartley, Associate Editor
Sandia Laboratories
Edward W. Price, Associate Editor
Georgia Institute of Technology
James G. Skifstad, Associate Editor
Purdue University
ISBN 0-915928-18-3

***54. Satellite Communication: Future Systems (1977)**
David Jarett
TRW Inc.
ISBN 0-915928-18-3

***55. Satellite Communications: Advanced Technologies (1977)**
David Jarett
TRW Inc.
ISBN 0-915928-19-1

***56. Thermophysics of Spacecraft and Outer Planet Entry Probes (1977)**
Allie M. Smith
ARO Inc.
ISBN 0-915928-20-5

***57. Space-Based Manufacturing from Nonterrestrial Materials (1977)**
Gerald K. O'Neill, Editor
Brian O'Leary, Assistant Editor
Princeton University
ISBN 0-915928-21-3

***58. Turbulent Combustion (1978)**
Lawrence A. Kennedy
State University of New York at Buffalo
ISBN 0-915928-22-1

***59. Aerodynamic Heating and Thermal Protection Systems (1978)**
Leroy S. Fletcher
University of Virginia
ISBN 0-915928-23-X

***60. Heat Transfer and Thermal Control Systems (1978)**
Leroy S. Fletcher
University of Virginia
ISBN 0-915928-24-8

***61. Radiation Energy Conversion in Space (1978)**
Kenneth W. Billman
NASA Ames Research Center
ISBN 0-915928-26-4

***62. Alternative Hydrocarbon Fuels: Combustion and Chemical Kinetics (1978)**
Craig T. Bowman
Stanford University
Jorgen Birkeland
Department of Energy
ISBN 0-915928-25-6

***63. Experimental Diagnostics in Combustion of Solids (1978)**
Thomas L. Boggs
Naval Weapons Center
Ben T. Zinn
Georgia Institute of Technology
ISBN 0-915928-28-0

***64. Outer Planet Entry Heating and Thermal Protection (1979)**
Raymond Viskanta
Purdue University
ISBN 0-915928-29-9

***65. Thermophysics and Thermal Control (1979)**
Raymond Viskanta
Purdue University
ISBN 0-915928-30-2

***66. Interior Ballistics of Guns (1979)**
Herman Krier
University of Illinois at Urbana–Champaign
Martin Summerfield
New York University
ISBN 0-915928-32-9

***67. Remote Sensing of Earth from Space: Role of "Smart Sensors" (1979)**
Roger A. Breckenridge
NASA Langley Research Center
ISBN 0-915928-33-7

***68. Injection and Mixing in Turbulent Flow (1980)**
Joseph A. Schetz
Virginia Polytechnic Institute and State University
ISBN 0-915928-35-3

*Out of print.

***69. Entry Heating and Thermal Protection (1980)**
Walter B. Olstad
NASA Headquarters
ISBN 0-915928-38-8

***70. Heat Transfer, Thermal Control, and Heat Pipes (1980)**
Walter B. Olstad
NASA Headquarters
ISBN 0-915928-39-6

***71. Space Systems and Their Interactions with Earth's Space Environment (1980)**
Henry B. Garrett
Charles P. Pike
Hanscom Air Force Base
ISBN 0-915928-41-8

***72. Viscous Flow Drag Reduction (1980)**
Gary R. Hough
Vought Advanced Technology Center
ISBN 0-915928-44-2

***73. Combustion Experiments in a Zero-Gravity Laboratory (1981)**
Thomas H. Cochran
NASA Lewis Research Center
ISBN 0-915928-48-5

***74. Rarefied Gas Dynamics, Parts I and II (two volumes) (1981)**
Sam S. Fisher
University of Virginia
ISBN 0-915928-51-5

***75. Gasdynamics of Detonations and Explosions (1981)**
J. R. Bowen
University of Wisconsin at Madison
N. Manson
Universite de Poitiers
A. K. Oppenheim
University of California at Berkeley
R. I. Soloukhin
Institute of Heat and Mass Transfer, BSSR Academy of Sciences
ISBN 0-915928-46-9

***76. Combustion in Reactive Systems (1981)**
J. R. Bowen
University of Wisconsin at Madison
N. Manson
Universite de Poitiers
A. K. Oppenheim
University of California at Berkeley
R. I. Soloukhin
Institute of Heat and Mass Transfer, BSSR Academy of Sciences
ISBN 0-915928-47-7

***77. Aerothermodynamics and Planetary Entry (1981)**
A. L. Crosbie
University of Missouri-Rolla
ISBN 0-915928-52-3

***78. Heat Transfer and Thermal Control (1981)**
A. L. Crosbie
University of Missouri-Rolla
ISBN 0-915928-53-1

***79. Electric Propulsion and Its Applications to Space Missions (1981)**
Robert C. Finke
NASA Lewis Research Center
ISBN 0-915928-55-8

***80. Aero-Optical Phenomena (1982)**
Keith G. Gilbert
Leonard J. Otten
Air Force Weapons Laboratory
ISBN 0-915928-60-4

***81. Transonic Aerodynamics (1982)**
David Nixon
Nielsen Engineering & Research, Inc.
ISBN 0-915928-65-5

***82. Thermophysics of Atmospheric Entry (1982)**
T. E. Horton
University of Mississippi
ISBN 0-915928-66-3

***83. Spacecraft Radiative Transfer and Temperature Control (1982)**
T. E. Horton
University of Mississippi
ISBN 0-915928-67-1

***84. Liquid-Metal Flows and Magneto-hydrodynamics (1983)**
H. Branover
Ben-Gurion University of the Negev
P. S. Lykoudis
Purdue University
A. Yakhot
Ben-Gurion University of the Negev
ISBN 0-915928-70-1

*Out of print.

***85. Entry Vehicle Heating and Thermal Protection Systems: Space Shuttle, Solar Starprobe, Jupiter Galileo Probe (1983)**
Paul E. Bauer
McDonnell Douglas Astronautics Co.
Howard E. Collicott
The Boeing Co.
ISBN 0-915928-74-4

***86. Spacecraft Thermal Control, Design, and Operation (1983)**
Howard E. Collicott
The Boeing Co.
Paul E. Bauer
McDonnell Douglas Astronautics Co.
ISBN 0-915928-75-2

***87. Shock Waves, Explosions, and Detonations (1983)**
J. R. Bowen
University of Washington
N. Manson
Universite de Poitiers
A. K. Oppenheim
University of California at Berkeley
R. I. Soloukhin
Institute of Heat and Mass Transfer, BSSR Academy of Sciences
ISBN 0-915928-76-0

***88. Flames, Lasers, and Reactive Systems (1983)**
J. R. Bowen
University of Washington
N. Manson
Universite de Poitiers
A. K. Oppenheim
University of California at Berkeley
R. I. Soloukhin
Institute of Heat and Mass Transfer, BSSR Academy of Sciences
ISBN 0-915928-77-9

***89. Orbit-Raising and Maneuvering Propulsion: Research Status and Needs (1984)**
Leonard H. Caveny
Air Force Office of Scientific Research
ISBN 0-915928-82-5

***90. Fundamentals of Solid-Propellant Combustion (1984)**
Kenneth K. Kuo
Pennsylvania State University
Martin Summerfield
Princeton Combustion Research Laboratories, Inc.
ISBN 0-915928-84-1

91. Spacecraft Contamination: Sources and Prevention (1984)
J. A. Roux
University of Mississippi
T. D. McCay
NASA Marshall Space Flight Center
ISBN 0-915928-85-X

92. Combustion Diagnostics by Nonintrusive Methods (1984)
T. D. McCay
NASA Marshall Space Flight Center
J. A. Roux
University of Mississippi
ISBN 0-915928-86-8

93. The INTELSAT Global Satellite System (1984)
Joel Alper
COMSAT Corp.
Joseph Pelton
INTELSAT
ISBN 0-915928-90-6

94. Dynamics of Shock Waves, Explosions, and Detonations (1984)
J. R. Bowen
University of Washington
N. Manson
Universite de Poitiers
A. K. Oppenheim
University of California at Berkeley
R. I. Soloukhin
Institute of Heat and Mass Transfer, BSSR Academy of Sciences
ISBN 0-915928-91-4

95. Dynamics of Flames and Reactive Systems (1984)
J. R. Bowen
University of Washington
N. Manson
Universite de Poitiers
A. K. Oppenheim
University of California at Berkeley
R. I. Soloukhin
Institute of Heat and Mass Transfer, BSSR Academy of Sciences
ISBN 0-915928-92-2

96. Thermal Design of Aeroassisted Orbital Transfer Vehicles (1985)
H. F. Nelson
University of Missouri-Rolla
ISBN 0-915928-94-9

97. Monitoring Earth's Ocean, Land, and Atmosphere from Space—Sensors, Systems, and Applications (1985)
Abraham Schnapf
Aerospace Systems Engineering
ISBN 0-915928-98-1

---

*Out of print.

98. Thrust and Drag: Its Prediction and Verification (1985)
Eugene E. Covert
Massachusetts Institute of Technology
C. R. James
Vought Corp.
William F. Kimzey
Sverdrup Technology AEDC Group
George K. Richey
U.S. Air Force
Eugene C. Rooney
U.S. Navy Department of Defense
ISBN 0-930403-00-2

99. Space Stations and Space Platforms—Concepts, Design, Infrastructure, and Uses (1985)
Ivan Bekey
Daniel Herman
NASA Headquarters
ISBN 0-930403-01-0

100. Single- and Multi-Phase Flows in an Electromagnetic Field: Energy, Metallurgical, and Solar Applications (1985)
Herman Branover
Ben-Gurion University of the Negev
Paul S. Lykoudis
Purdue University
Michael Mond
Ben-Gurion University of the Negev
ISBN 0-930403-04-5

101. MHD Energy Conversion: Physiotechnical Problems (1986)
V. A. Kirillin
A. E. Sheyndlin
Soviet Academy of Sciences
ISBN 0-930403-05-3

102. Numerical Methods for Engine-Airframe Integration (1986)
S. N. B. Murthy
Purdue University
Gerald C. Paynter
Boeing Airplane Co.
ISBN 0-930403-09-6

103. Thermophysical Aspects of Re-Entry Flows (1986)
James N. Moss
NASA Langley Research Center
Carl D. Scott
NASA Johnson Space Center
ISBN 0-930430-10-X

***104. Tactical Missile Aerodynamics (1986)**
M. J. Hemsch
PRC Kentron, Inc.
J. N. Nielson
NASA Ames Research Center
ISBN 0-930403-13-4

105. Dynamics of Reactive Systems Part I: Flames and Configurations; Part II: Modeling and Heterogeneous Combustion (1986)
J. R. Bowen
University of Washington
J.-C. Leyer
Universite de Poitiers
R. I. Soloukhin
Institute of Heat and Mass Transfer, BSSR Academy of Sciences
ISBN 0-930403-14-2

106. Dynamics of Explosions (1986)
J. R. Bowen
University of Washington
J.-C. Leyer
Universite de Poitiers
R. I. Soloukhin
Institute of Heat and Mass Transfer, BSSR Academy of Sciences
ISBN 0-930403-15-0

***107. Spacecraft Dielectric Material Properties and Spacecraft Charging (1986)**
A. R. Frederickson
U.S. Air Force Rome Air Development Center
D. B. Cotts
SRI International
J. A. Wall
U.S. Air Force Rome Air Development Center
F. L. Bouquet
Jet Propulsion Laboratory, California Institute of Technology
ISBN 0-930403-17-7

***108. Opportunities for Academic Research in a Low-Gravity Environment (1986)**
George A. Hazelrigg
National Science Foundation
Joseph M. Reynolds
Louisiana State University
ISBN 0-930403-18-5

109. Gun Propulsion Technology (1988)
Ludwig Stiefel
U.S. Army Armament Research, Development and Engineering Center
ISBN 0-930403-20-7

---

*Out of print.

110. Commercial Opportunities in Space (1988)
F. Shahrokhi
K. E. Harwell
University of Tennessee Space Institute
C. C. Chao
National Cheng Kung University
ISBN 0-930403-39-8

111. Liquid-Metal Flows: Magnetohydrodynamics and Application (1988)
Herman Branover
Michael Mond
Yeshajahu Unger
Ben-Gurion University of the Negev
ISBN 0-930403-43-6

112. Current Trends in Turbulence Research (1988)
Herman Branover
Micheal Mond
Yeshajahu Unger
Ben-Gurion University of the Negev
ISBN 0-930403-44-4

113. Dynamics of Reactive Systems Part I: Flames; Part II: Heterogeneous Combustion and Applications (1988)
A. L. Kuhl
R&D Associates
J. R. Bowen
University of Washington
J.-C. Leyer
Universite de Poitiers
A. Borisov
USSR Academy of Sciences
ISBN 0-930403-46-0

114. Dynamics of Explosions (1988)
A. L. Kuhl
R & D Associates
J. R. Bowen
University of Washington
J.-C. Leyer
Universite de Poitiers
A. Borisov
USSR Academy of Sciences
ISBN 0-930403-47-9

115. Machine Intelligence and Autonomy for Aerospace (1988)
E. Heer
Heer Associates, Inc.
H. Lum
NASA Ames Research Center
ISBN 0-930403-48-7

116. Rarefied Gas Dynamics: Space Related Studies (1989)
E. P. Muntz
University of Southern California
D. P. Weaver
U.S. Air Force Astronautics Laboratory (AFSC)
D. H. Campbell
University of Dayton Research Institute
ISBN 0-930403-53-3

117. Rarefied Gas Dynamics: Physical Phenomena (1989)
E. P. Muntz
University of Southern California
D. P. Weaver
U.S. Air Force Astronautics Laboratory (AFSC)
D. H. Campbell
University of Dayton Research Institute
ISBN 0-930403-54-1

118. Rarefied Gas Dynamics: Theoretical and Computational Techniques (1989)
E. P. Muntz
University of Southern California
D. P. Weaver
U.S. Air Force Astronautics Laboratory (AFSC)
D. H. Campbell
University of Dayton Research Institute
ISBN 0-930403-55-X

119. Test and Evaluation of the Tactical Missile (1989)
Emil J. Eichblatt Jr.
Pacific Missile Test Center
ISBN 0-930403-56-8

120. Unsteady Transonic Aerodynamics (1989)
David Nixon
Nielsen Engineering & Research, Inc.
ISBN 0-930403-52-5

121. Orbital Debris from Upper-Stage Breakup (1989)
Joseph P. Loftus Jr.
NASA Johnson Space Center
ISBN 0-930403-58-4

122. Thermal-Hydraulics for Space Power, Propulsion and Thermal Management System Design (1990)
William J. Krotiuk
General Electric Co.
ISBN 0-930403-64-9

---

*Out of print.

123. Viscous Drag Reduction in Boundary Layers (1990)
Dennis M. Bushnell
Jerry N. Hefner
NASA Langley Research Center
ISBN 0-930403-66-5

***124. Tactical and Strategic Missile Guidance (1990)**
Paul Zarchan
Charles Stark Draper Laboratory, Inc.
ISBN 0-930403-68-1

125. Applied Computational Aerodynamics (1990)
P. A. Henne
Douglas Aircraft Company
ISBN 0-930403-69-X

126. Space Commercialization: Launch Vehicles and Programs (1990)
F. Shahrokhi
University of Tennessee Space Institute
J. S. Greenberg
Princeton Synergetics Inc.
T. Al-Saud
Ministry of Defense and Aviation Kingdom of Saudi Arabia
ISBN 0-930403-75-4

127. Space Commercialization: Platforms and Processing (1990)
F. Shahrokhi
University of Tennessee Space Institute
G. Hazelrigg
National Science Foundation
R. Bayuzick
Vanderbilt University
ISBN 0-930403-76-2

128. Space Commercialization: Satellite Technology (1990)
F. Shahrokhi
University of Tennessee Space Institute
N. Jasentuliyana
United Nations
N. Tarabzouni
King Abulaziz City for Science and Technology
ISBN 0-930403-77-0

***129. Mechanics and Control of Large Flexible Structures (1990)**
John L. Junkins
Texas A&M University
ISBN 0-930403-73-8

130. Low-Gravity Fluid Dynamics and Transport Phenomena (1990)
Jean N. Koster
Robert L. Sani
University of Colorado at Boulder
ISBN 0-930403-74-6

131. Dynamics of Deflagrations and Reactive Systems: Flames (1991)
A. L. Kuhl
Lawrence Livermore National Laboratory
J.-C. Leyer
Universite de Poitiers
A. A. Borisov
USSR Academy of Sciences
W. A. Sirignano
University of California
ISBN 0-930403-95-9

132. Dynamics of Deflagrations and Reactive Systems: Heterogeneous Combustion (1991)
A. L. Kuhl
Lawrence Livermore National Laboratory
J.-C. Leyer
Universite de Poitiers
A. A. Borisov
USSR Academy of Sciences
W. A. Sirignano
University of California
ISBN 0-930403-96-7

133. Dynamics of Detonations and Explosions: Detonations (1991)
A. L. Kuhl
Lawrence Livermore National Laboratory
J.-C. Leyer
Universite de Poitiers
A. A. Borisov
USSR Academy of Sciences
W. A. Sirignano
University of California
ISBN 0-930403-97-5

134. Dynamics of Detonations and Explosions: Explosion Phenomena (1991)
A. L. Kuhl
Lawrence Livermore National Laboratory
J.-C. Leyer
Universite de Poitiers
A. A. Borisov
USSR Academy of Sciences
W. A. Sirignano
University of California
ISBN 0-930403-98-3

---

*Out of print.

135. Numerical Approaches to Combustion Modeling (1991)
Elaine S. Oran
Jay P. Boris
Naval Research Laboratory
ISBN 1-56347-004-7

136. Aerospace Software Engineering (1991)
Christine Anderson
U.S. Air Force Wright Laboratory
Merlin Dorfman
Lockheed Missiles & Space Company, Inc.
ISBN 1-56347-005-0

137. High-Speed Flight Propulsion Systems (1991)
S. N. B. Murthy
Purdue University
E. T. Curran
Wright Laboratory
ISBN 1-56347-011-X

138. Propagation of Intensive Laser Radiation in Clouds (1992)
O. A. Volkovitsky
Yu. S. Sedenov
L. P. Semenov
Institute of Experimental Meteorology
ISBN 1-56347-020-9

139. Gun Muzzle Blast and Flash (1992)
Günter Klingenberg
Fraunhofer-Institut für Kurzzeitdynamik, Ernst-Mach-Institut
Joseph M. Heimerl
U.S. Army Ballistic Research Laboratory
ISBN 1-56347-012-8

140. Thermal Structures and Materials for High-Speed Flight (1992)
Earl. A. Thornton
University of Virginia
ISBN 1-56347-017-9

141. Tactical Missile Aerodynamics: General Topics (1992)
Michael J. Hemsch
Lockheed Engineering & Sciences Company
ISBN 1-56347-015-2

142. Tactical Missile Aerodynamics: Prediction Methodology (1992)
Michael R. Mendenhall
Nielsen Engineering & Research, Inc.
ISBN 1-56347-016-0

143. Nonsteady Burning and Combustion Stability of Solid Propellants (1992)
Luigi De Luca
Politecnico di Milano
Edward W. Price
Georgia Institute of Technology
Martin Summerfield
Princeton Combustion Research Laboratories, Inc.
ISBN 1-56347-014-4

144. Space Economics (1992)
Joel S. Greenberg
Princeton Synergetics, Inc.
Henry R. Hertzfeld
HRH Associates
ISBN 1-56347-042-X

145. Mars: Past, Present, and Future (1992)
E. Brian Pritchard
NASA Langley Research Center
ISBN 1-56347-043-8

146. Computational Nonlinear Mechanics in Aerospace Engineering (1992)
Satya N. Atluri
Georgia Institute of Technology
ISBN 1-56347-044-6

147. Modern Engineering for Design of Liquid-Propellant Rocket Engines (1992)
Dieter K. Huzel
David H. Huang
Rocketdyne Division of Rockwell International
ISBN 1-56347-013-6

148. Metallurgical Technologies, Energy Conversion, and Magnetohydrodynamic Flows (1993)
Herman Branover
Yeshajahu Unger
Ben-Gurion University of the Negev
ISBN 1-56347-019-5

149. Advances in Turbulence Studies (1993)
Herman Branover
Yeshajahu Unger
Ben-Gurion University of the Negev
ISBN 1-56347-018-7

150. Structural Optimization: Status and Promise (1993)
Manohar P. Kamat
Georgia Institute of Technology
ISBN 1-56347-056-X

---
*Out of print.

151. Dynamics of Gaseous Combustion (1993)
A. L. Kuhl
Lawrence Livermore National Laboratory
J.-C. Leyer
Universite de Poitiers
A. A. Borisov
USSR Academy of Sciences
W. A. Sirignano
University of California
ISBN 1-56347-060-8

152. Dynamics of Heterogeneous Gaseous Combustion and Reacting Systems (1993)
A. L. Kuhl
Lawrence Livermore National Laboratory
J.-C. Leyer
Universite de Poitiers
A. A. Borisov
USSR Academy of Sciences
W. A. Sirignano
University of California
ISBN 1-56347-058-6

153. Dynamic Aspects of Detonations (1993)
A. L. Kuhl
Lawrence Livermore National Laboratory
J.-C. Leyer
Universite de Poitiers
A. A. Borisov
USSR Academy of Sciences
W. A. Sirignano
University of California
ISBN 1-56347-057-8

154. Dynamic Aspects of Explosion Phenomena (1993)
A. L. Kuhl
Lawrence Livermore National Laboratory
J.-C. Leyer
Universite de Poitiers
A. A. Borisov
USSR Academy of Sciences
W. A. Sirignano
University of California
ISBN 1-56347-059-4

155. Tactical Missile Warheads (1993)
Joseph Carleone
Aerojet General Corporation
ISBN 1-56347-067-5

156. Toward a Science of Command, Control, and Communications (1993)
Carl R. Jones
Naval Postgraduate School
ISBN 1-56347-068-3

***157. Tactical and Strategic Missile Guidance Second Edition (1994)**
Paul Zarchan
Charles Stark Draper Laboratory, Inc.
ISBN 1-56347-077-2

158. Rarefied Gas Dynamics: Experimental Techniques and Physical Systems (1994)
Bernie D. Shizgal
University of British Columbia
David P. Weaver
Phillips Laboratory
ISBN 1-56347-079-9

159. Rarefied Gas Dynamics: Theory and Simulations (1994)
Bernie D. Shizgal
University of British Columbia
David P. Weaver
Phillips Laboratory
ISBN 1-56347-080-2

160. Rarefied Gas Dynamics: Space Sciences and Engineering (1994)
Bernie D. Shizgal
University of British Columbia
David P. Weaver
Phillips Laboratory
ISBN 1-56347-081-0

161. Teleoperation and Robotics in Space (1994)
Steven B. Skaar
University of Notre Dame
Carl F. Ruoff
Jet Propulsion Laboratory, California Institute of Technology
ISBN 1-56347-095-0

162. Progress in Turbulence Research (1994)
Herman Branover
Yeshajahu Unger
Ben-Gurion University of the Negev
ISBN 1-56347-099-3

163. Global Positioning System: Theory and Applications, Volume I (1996)
Bradford W. Parkinson
Stanford University
James J. Spilker Jr.
Stanford Telecom
Penina Axelrad, Associate Editor
University of Colorado
Per Enge, Associate Editor
Stanford University
ISBN 1-56347-107-8

164. Global Positioning System: Theory and Applications, Volume II (1996)
Bradford W. Parkinson
Stanford University
James J. Spilker Jr.
Stanford Telecom
Penina Axelrad, Associate Editor
University of Colorado
Per Enge, Associate Editor
Stanford University
ISBN 1-56347-106-X

*Out of print.

165. Developments in High-Speed Vehicle Propulsion Systems (1996)
S. N. B. Murthy
Purdue University
E. T. Curran
Wright Laboratory
ISBN 1-56347-176-0

166. Recent Advances in Spray Combustion: Spray Atomization and Drop Burning Phenomena, Volume I (1996)
Kenneth K. Kuo
Pennsylvania State University
ISBN 1-56347-175-2

167. Fusion Energy in Space Propulsion (1995)
Terry Kammash
University of Michigan
ISBN 1-56347-184-1

168. Aerospace Thermal Structures and Materials for a New Era (1995)
Earl A. Thornton
University of Virginia
ISBN 1-56347-182-5

169. Liquid Rocket Engine Combustion Instability (1995)
Vigor Yang
William E. Anderson
Pennsylvania State University
ISBN 1-56347-183-3

170. Tactical Missile Propulsion (1996)
G. E. Jensen
United Technologies Corporation
David W. Netzer
Naval Postgraduate School
ISBN 1-56347-118-3

171. Recent Advances in Spray Combustion: Spray Combustion Measurements and Model Simulation, Volume II (1996)
Kenneth K. Kuo
Pennsylvania State University
ISBN 1-56347-181-7

172. Future Aeronautical and Space Systems (1997)
Ahmed K. Noor
NASA Langley Research Center
Samuel L. Venneri
NASA Headquarters
ISBN 1-56347-188-4

173. Advances in Combustion Science: In Honor of Ya. B. Zel'dovich (1997)
William A. Sirignano
University of California
Alexander G. Merzhanov
Russian Academy of Sciences
Luigi De Luca
Politecnico di Milano
ISBN 1-56347-178-7

174. Fundamentals of High Accuracy Inertial Navigation (1997)
Averil B. Chatfield
ISBN 1-56347-243-0

175. Liquid Propellant Gun Technology (1997)
Günter Klingenberg
Fraunhofer-Institut für Kurzzeitdynamik, Ernst-Mach-Institut
John D. Knapton
Walter F. Morrison
Gloria P. Wren
U.S. Army Research Laboratory
ISBN 1-56347-196-5

176. Tactical and Strategic Missile Guidance Third Edition (1998)
Paul Zarchan
Charles Stark Draper Laboratory, Inc.
ISBN 1-56347-279-1

177. Orbital and Celestial Mechanics (1998)
John P. Vinti
Gim J. Der, Editor
TRW
Nino L. Bonavito, Editor
NASA Goddard Space Flight Center
ISBN 1-56347-256-2

178. Some Engineering Applications in Random Vibrations and Random Structures (1998)
Giora Maymon
RAFAEL
ISBN 1-56347-258-9

179. Conventional Warhead Systems Physics and Engineering Design (1998)
Richard M. Lloyd
Raytheon Systems Company
ISBN 1-56347-255-4

180. Advances in Missile Guidance Theory (1998)
Joseph Z. Ben-Asher
Isaac Yaesh
Israel Military Industries—Advanced Systems Division
ISBN 1-56347-275-9

181. Satellite Thermal Control for Systems Engineers (1998)
Robert D. Karam
ISBN 1-56347-276-7

---
*Out of print.

182. Progress in Fluid Flow Research: Turbulence and Applied MHD (1998)
Yeshajahu Unger
Herman Branover
Ben-Gurion University of the Negev
ISBN 1-56347-284-8

183. Aviation Weather Surveillance Systems (1999)
Pravas R. Mahapatra
Indian Institute of Science
ISBN 1-56347-340-2

184. Flight Control Systems (2000)
Rodger W. Pratt
Loughborough University
ISBN 1-56347-404-2

185. Solid Propellant Chemistry, Combustion, and Motor Interior Ballistics (2000)
Vigor Yang
Pennsylvania State University
Thomas B. Brill
University of Delaware
Wu-Zhen Ren
China Ordnance Society
ISBN 1-56347-442-5

186. Approximate Methods for Weapons Aerodynamics (2000)
Frank G. Moore
ISBN 1-56347-399-2

187. Micropropulsion for Small Spacecraft (2000)
Michael M. Micci
Pennsylvania State University
Andrew D. Ketsdever
Air Force Research Laboratory, Edwards Air Force Base
ISBN 1-56347-448-4

188. Structures Technology for Future Aerospace Systems (2000)
Ahmed K. Noor
NASA Langley Research Center
ISBN 1-56347-384-4

189. Scramjet Propulsion (2000)
E. T. Curran
Department of the Air Force
S. N. B. Murthy
Purdue University
ISBN 1-56347-322-4

---

*Out of print.